Tight Gas Reservoirs

Tight Gas Reservoirs

Tight Gas Reservoirs

Society of Petroleum Engineers
Richardson, Texas, USA

Printed in the United States of America.

Disclaimer

This book was prepared by members of the Society of Petroleum Engineers and their well-qualified colleagues from material published in the recognized technical literature and from their own individual experience and expertise. While the material presented is believed to be based on sound technical knowledge, neither the Society of Petroleum Engineers nor any of the authors or editors herein provide a warranty either expressed or implied in its application. Correspondingly, the discussion of materials, methods, or techniques that may be covered by letters patents implies no freedom to use such materials, methods, or techniques without permission through appropriate licensing. Nothing described within this book should be construed to lessen the need to apply sound engineering judgment nor to carefully apply accepted engineering practices in the design, implementation, or application of the techniques described herein.

ISBN 978-1-61399-818-2

10 9 8 7 6 5 4 3 2 1

Society of Petroleum Engineers
222 Palisades Creek Drive
Richardson, TX 75080-2040 USA

http://store.spe.org
service@spe.org
1.972.952.9393

Content Summary

Tight Gas Reservoirs

(Available in print or digital format. Parts 1, 2, and 3 are contained in this book. Case Histories of Tight Gas Reservoir Development is available only as a digital supplement. The content summary listed below is for reference purposes only.)

***

Tight Gas Reservoirs

Case Histories of Tight Gas Reservoir Development

Edited by Stephen A. Holditch and John Yilin Wang

(Available as an eBOOK SUPPLEMENTAL to *Tight Gas Reservoirs*. Download access will be granted at the time *Tight Gas Reservoirs* is purchased. The chapters listed below are available only in the supplemental digital download and not in this book. May or may not be available separately.)

23 articles from SPE literature covering different geological formations in the US and around the world. New summaries are included that explain the significance of the research.

Table of Contents

Part 1 — Unconventional Gas Resources

Part 2 — Tight Gas Sands Engineering

Part 3 — Hydraulic Fracturing in Vertical Wells

Foreword

The development of tight gas reservoirs over the previous half-century has profoundly affected and expanded the petroleum industry. Moreover, our improved understanding of tight gas reservoirs—from finding, characterizing, testing, modeling, and developing them to producing their resources economically—can be felt not only throughout our industry but also throughout our economy and, indeed, our daily routines.

Abundant, reliable, and inexpensive natural gas has truly transformed many aspects of our modern lifestyles. Within the last decade, for example, the world has made great strides in switching from coal-fired to gas-fired electricity generation, with a resulting reduction of US carbon-dioxide emissions of approximately 14% since 2005 (2017 BP Statistical Review of World Energy). Our expanded knowledge of natural-gas development and production has further advanced the goal of achieving energy independence, transforming the US from a gas importer into the third-largest liquid natural gas exporter in the world. It is hard to overstate the efficacy of our understanding and exploitation of tight gas reservoirs.

The four parts contained in this book methodically and comprehensively unfold the technical elements of developing tight gas reservoirs. They have been written with an industry-wide audience in mind; to help the student understand fundamental concepts; to provide comprehensive reference material for the experienced engineer; for the practitioner in the field looking for case studies and analogues; and for those readers curious about mathematical detail and theory, surely laying the foundation for many future academic investigations and doctoral theses.

This book is comprehensive enough to apply equally to those readers interested in tight oil reservoirs—common fundamentals, many similar concepts, just larger molecules.

The book's organization supports its methodological approach.

Part 1 introduces tight gas resources, including definitions and beginning concepts. Thorough analyses of tight gas resource types (shale, conventional tight gas, and coalbed methane) and their geographical distribution and reserves are provided. This part describes shale gas plays in North America in detail.

Part 2 begins where the study of all reservoirs begins, with detailed characterization. Chapters within this part discuss geological considerations over various scales, in addition to detailed concepts in well testing and modeling to determine necessary formation properties.

Part 3 details all aspects of designing, planning, modeling, and executing hydraulic fracturing treatments and provides details on fracture initiation, geometry, and propagation.

Case Histories of Tight Gas Reservoir Development (digital only) contains 23 chapters combining field and laboratory experiments in addition to revised and updated SPE manuscripts covering subjects from proppant transport to economic gas-reservoir development.

In summary, this book is the culmination of a lifetime of individual and collaborative work on all aspects of tight gas reservoirs. The material comes from a multitude of sources, including many Gas Research Institute field trials, worldwide consulting reports written over several decades, Schlumberger research and development activities, literally hundreds of graduate-student theses and dissertations and, of course, numerous SPE peer-reviewed articles and conference papers.

I am not capable of personally endorsing the entire content, but you can trust that many of our dedicated industry colleagues have reviewed the material and, in many cases, tested and verified it in tight gas reservoirs around the world. If you are looking for a reason to buy this book, perhaps the best is that it represents our industry's cumulative knowledge of tight gas reservoirs combined with decades of personal experiences in the implementation of the content.

Few people in our field have successfully combined deep intellect and old-fashioned know-how with leadership and vision. Perhaps even fewer have combined success and recognition with humility. Add a touch of stubbornness and a sense of humor, and you have our lead author. I cannot think of anyone with more credibility and respect than Dr. Stephen Holditch to assemble this compendium. When the leadership at SPE asked me to write the foreword for this book, I felt both honored and pleased to do so. I knew Steve for nearly 40 years, in many roles, from professor, friend, neighbor, and Schlumberger colleague to fellow golfer, hunter, and storyteller. Very simply, he was a great man and, like many, I miss him greatly. He has done more for our industry and our society and has impacted more lives, young and old, than anyone I can think of.

Now I am going to break a publishing rule or two and include in this foreword both an acknowledgment and a dedication, as the author sadly cannot. The final few drafts of this collective work, and the published version you now hold, could not have been completed, posthumously as it was, without the herculean efforts of the SPE publishing staff—Jane Eden, editorial services manager, and Rebekah Stacha, assistant director of technical publications, in particular—and also Dr. Ding Zhu, professor at Texas A&M University, who went to great effort to ensure the content was accurate and organized in the very readable manner Steve would have wanted and that you now see. Also, special thanks to David Grant, cover designer; Melinda Mahaffey Icden, editor; Lauren Miller, editor; Shashana Pearson-Hormillosa, editor; Dennis Scharnberg, technical editorial specialist; and Ingrid Scroggins, editor.

Steve Holditch was often a man of few words and rarely sentimental. If I knew him as well as I would like to think, though, I know he would surely want to dedicate this book—his magnum opus—to the loves of his life: his wife, Ann, and their daughters, Katie and Abbie. While reading, please keep in mind one of Steve's favorite maxims: "I retain the right to get smarter."

Jeff Spath
Professor and Holder of the Stephen A. Holditch '69 Department Head Chair
Harold Vance Department of Petroleum Engineering
Texas A&M University
March 27, 2020

Introduction

A Brief History of S. A. Holditch & Associates Inc.

In 1976, Stephen A. Holditch joined the faculty at Texas A&M University, followed by W. John Lee in 1977. That same year, S. A. Holditch & Associates Inc. was formed as a petroleum engineering consulting company. The consulting company initially focused on reservoir and production engineering aspects of tight gas reservoirs, to include well completions and hydraulic fracturing treatment design. One unique aspect of S. A. Holditch & Associates Inc. at the time was its approach, designing the optimal fracturing treatment in the office and then going to the field to make sure the optimal treatment was actually pumped in the ground.

S. A. Holditch & Associates Inc. continued to grow, and in 1983 it became the lead technical contractor for the Gas Research Institute (GRI) in its Tight Gas Sands (TGS) research focus area. The grand objective of the GRI tight gas research was to learn how to make measurements before, during, and after a fracturing treatment so the 3D size and shape of a hydraulic fracture could be calculated in real time. The GRI TGS research project lasted approximately 10 years, involved multiple contractors, and essentially achieved its goal by developing the first versions of the technology the industry commonly applies today.

The technology used today involves detailed data collection before and during a fracturing treatment, real-time analyses of the fracturing treatment data, 3D computations of fracture shape and dimensions, microseismic analyses to evaluate fracturing treatments, and complete data integration to better understand the fracturing treatment process. These technologies and systems are being applied worldwide by virtually all service companies to improve performance from unconventional reservoirs.

In addition to the TGS project, S. A. Holditch & Associates Inc. also became involved with the GRI Gas Shales project area and the GRI Coalbed Methane project area. The engineers and geoscientists at S. A. Holditch & Associates Inc. were fortunate to do a lot of cutting-edge work and were encouraged to publish the work in appropriate technical meetings. As a result, the family of professionals who worked at one time or another at S. A. Holditch & Associates Inc. has published several thousand technical papers pertaining to unconventional gas reservoirs. The group also has worked on several SPE textbook, monograph, handbook, and reprint series volumes. A bibliography of the body of work generated by employees of S. A. Holditch & Associates Inc. is found in *Case Histories of Tight Gas Reservoir Development*, available as a digital document from SPE.

By 1997, S. A. Holditch & Associates Inc. had opened five offices and had more than 50 employees. The business was more and more global, and it was decided that to continue to grow, the company needed to change its business model. At the same time, Schlumberger decided that it needed to focus on the reservoirs. Instead of deciding what to recommend to its clients on a well-by-well basis, Schlumberger needed to first understand what technology is required to maximize reservoir performance, then to determine how to drill, evaluate, complete, and stimulate individual wells to optimize the reservoir performance. As it turned out, the business goals of the two organizations were compatible, and Schlumberger purchased S. A. Holditch & Associates Inc.

Purpose of this Book

The purpose of this book is to document the technology allowing the industry to produce tight gas reservoirs. We highlight the contributions of the many employees at S. A. Holditch & Associates Inc. but do not limit our discussions because we want to include other important contributions to the technology. All of the information in this book can be found in the SPE literature, but it is convenient to pull it together in one place. We could not include everything, but we include enough references that the reader can find what is needed in either this book or the references.

Part 1 - Unconventional Gas Resources

Chapter 1

Introduction to Unconventional Gas Reservoirs

Stephen A. Holditch

Unconventional gas is the term commonly used to refer to low-permeability reservoirs that produce mainly dry natural gas (Holditch 2006). Many of the low-permeability reservoirs developed in the past were sandstone, but significant quantities of gas have also been produced from low-permeability carbonates, shales, and coal seams. The technology required to analyze and develop unconventional gas reservoirs has been developing for more than 50 years, since work began on hydraulic fracturing in the 1950s. In the 1960s, some unconventional gas was produced in places such as South Texas, the San Juan Basin, and the Devonian Shale in the Appalachian Basin.

In general, a vertical well drilled and completed in an unconventional gas reservoir must be successfully stimulated to produce at commercial gas flow rates and produce commercial gas volumes. A large hydraulic fracturing treatment normally is required to economically produce gas. In some unconventional gas reservoirs, horizontal and/or multilateral wells must be drilled, but many of these wells also need to be stimulated.

To optimize the development of an unconventional gas reservoir, a team of geoscientists and engineers must optimize the number of wells drilled, in addition to the drilling and completion procedures for each well. More data and more engineering manpower are often required to understand and develop unconventional gas reservoirs than are required for higher-permeability, conventional reservoirs. On an individual-well basis, a well in an unconventional gas reservoir will produce less gas over a longer period of time than that expected from a well completed in a higher-permeability, conventional reservoir. As such, many more wells must be drilled in unconventional gas reservoirs or smaller well spacing must be used to recover a large percentage of the original gas-in-place when compared with a conventional reservoir.

1.1 Definition of Unconventional Gas

In the 1970s, the US government decided that unconventional gas reservoirs were tight gas reservoirs, shale gas reservoirs, and coalbed methane (CBM) gas reservoirs. The Federal Energy Regulatory Commission (FERC) decided that a tight gas reservoir is one in which the expected value of permeability to gas flow would be less than 0.1 md. This was a political definition used to determine which wells would receive federal and/or state tax credits for producing gas from tight reservoirs. There was considerable discussion between producers and the FERC on the meaning of "expected value of permeability" and how to determine the correct method to average permeability. As will be explained in detail, the permeability in a natural gas reservoir is distributed log-normally. As such, the median or geometric mean value of permeability is the correct measure of the central tendency and should be used to determine the expected value of permeability.

Regardless of the political definition of unconventional or tight gas, the actual definition of an unconventional gas reservoir is a function of many physical and economic factors (Holditch 2006). The physical factors are related by Darcy's law, as shown by

$$q = \frac{kh\left(\bar{p} - p_{wf}\right)}{141.2\,B\mu\left[\ln\left(\dfrac{r_e}{r_w}\right) - 0.75 + s\right]} \quad \text{(1.1)}$$

Eq. 1.1 clearly shows that the flow rate q is a function of permeability k, net-pay thickness h, reservoir pressure $\bar{p}$, flowing pressure p_{wf}, fluid properties $B\mu$, drainage radius r_e, wellbore radius r_w, and skin factor s. Thus, to choose a single value of permeability to define tight gas is not wise. In deep, high-pressure, thick reservoirs, excellent completions can be achieved when the formation permeability to gas is in the microdarcy range (0.001 md). In shallow, low-pressure, thin reservoirs, permeabilities of several millidarcies might be required to produce the gas at economical flow rates, even after a successful fracturing treatment.

An unconventional gas reservoir is one in which "the reservoir cannot be produced at economic flow rates nor recover economic volumes of natural gas unless the well is stimulated by a large hydraulic fracture treatment, a horizontal wellbore, or by using multilateral wellbores" (Holditch 2006).

So, what is a typical unconventional gas reservoir? The answer is that there are no typical unconventional gas reservoirs. An unconventional gas reservoir can be a sandstone, shale, carbonate, or coal seam; deep or shallow; high or low pressure; high or low temperature; blanket or lenticular; homogeneous or naturally fractured; and single layered or multilayered.

The optimal drilling, completion, and stimulation methods for each well are a function of the reservoir characteristics and the economic situation. Some unconventional gas reservoirs are in South Texas, while others are in South America or the Middle East. The cost to drill, complete, and stimulate any particular well, in addition to the gas price and the gas market, affect how unconventional gas reservoirs are developed.

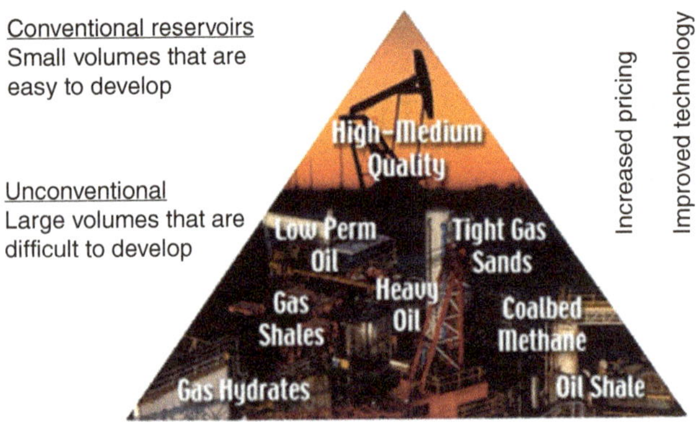

Fig. 1.1—The resource triangle for oil and gas.

1.2 The Resource Triangle

The resource-triangle concept was used by Masters (1979) and Gray (1977) to find a large gas field and build a company in the 1970s. In this concept, all natural resources are distributed log-normally in nature. If you are prospecting for gold, silver, iron, zinc, oil, natural gas, or any resource, you will find that the best or highest-grade deposits are small in size and, once found, easy to extract. The hard part is finding these pure veins of gold or high-permeability gas fields. After you find the high-grade deposit, producing the resource is rather easy and straightforward. **Fig. 1.1** illustrates the principle of the resource triangle.

As you go deeper into the gas resource triangle, the reservoirs are of lower grade, which usually means the reservoir permeability is decreasing. These low-permeability reservoirs, however, are much larger in size than the higher-quality reservoirs. The common theme is that low-quality deposits of natural gas require improved technology and adequate gas prices before they can be developed and produced economically. However, the size of the deposits can be extremely large, when compared to conventional or high-quality reservoirs. The concept of the resource triangle applies to every hydrocarbon-producing basin in the world. You should be able to estimate the volumes of oil and gas trapped in low-quality reservoirs in a specific basin by knowing the volumes of oil and gas that exist in the higher-quality reservoirs and comparing the basin of interest to an analogous basin in North America, where substantial data sets are available concerning the size of the unconventional resource.

1.3 Geographical Information on Unconventional Gas

Unconventional resources, defined as those that have low permeability and require advanced drilling or stimulation technologies to be produced at commercial flow rates, have been an important component of the domestic natural gas supply base in the US for many years. From almost nonexistent production levels in the early 1970s, unconventional resources, particularly tight sands, today provide nearly 30% of the domestic gas supply in the US. Outside the US, with a few exceptions, unconventional gas resources have largely been overlooked and were understudied until approximately 2010. Unconventional gas resources represent a potential long-term global resource of natural gas and have not been appraised in any systematic way.

Unconventional gas resources—including tight sands, CBM, and gas shales—constitute some of the largest components of the remaining natural gas resources in the US. Research and development into the geologic controls of and production technologies for these resources during the past several decades have enabled operators in the US to begin to unlock the vast potential of these challenging resources. Because of their long-lived reserves and stabilizing influence on reserves portfolios, these resources are particularly attractive to natural gas producers.

1.4 Global Gas Resources

From a global perspective, tight gas sands resources are vast but largely undefined. Few systematic evaluations have been carried out on global emerging resources, and the magnitude and distribution of worldwide gas resources in gas shales, tight sands, and CBM formations have yet to be understood. Worldwide estimates, however, are enormous, with an early estimate of gas-in-place of approximately 32,000 Tcf by Rogner (1996). [See **Table 1.1,** Kawata and Fujita (2001).] The probability of this gas resource being in place is supported by information regarding and experience with similar resources in North America. The information in Table 1.1 was compiled in the 1980s and provides a conservative estimate of the volume of gas in unconventional reservoirs worldwide.

Dong et al. (2012), using methodology explained in Dong (2012) and following similar geographical divisions as Rogner (Kawata and Fujita 2001), produced the values shown in **Table 1.2.** Dong (2012) presented research of the published literature for every basin in North America and systematically recorded resource estimates for tight gas, coal seams, and gas shales. She then compared the many estimates and determined the most-likely value of the quantity of gas contained in each type of unconventional reservoir in each basin. Then, she developed correlations among unconventional-resource estimates and production from conventional reservoirs. These correlations were consistent with the theory of log-normal distribution, as explained by the resource triangle. The gas-in-place in coal seams was correlated with the volume of coal known to exist in each basin. The basins included in that research are shown in **Fig. 1.2.**

The OGIP values in Table 1.2 represent original gas-in-place (OGIP) estimates with a 50% probability. These numbers can be compared with the resource estimates in Table 1.1, compiled more than 20 years earlier and with much fewer data for evaluation. The best comparison is that for North America (NAM); Kawata and Fujita (2001) provided 8,228 Tcf as an estimate of the gas-in-place, while Dong et al. (2012) came up with 18,318 Tcf. Globally, Kawata and Fujita (2001) provided 32,559 Tcf of gas-in-place, while Dong et al. (2012) estimated that value to be 125,742 Tcf.

After determining the estimates of gas-in-place, Dong et al. (2012) used reservoir simulation to estimate the volume of gas that could be classified as a technically recoverable resource (TRR). TRR means (1) we know where the gas is located and (2) we have the technology to produce the gas. By "technology," we are referring to the fact that we know how to drill, complete, and stimulate tight gas sands, CBM reservoirs, and shale gas reservoirs.

Region	CBM OGIP (Tcf)	Shale OGIP (Tcf)	Tight Gas OGIP (Tcf)	Total (Tcf)
Australia and Asia (AAO)	1,724	6,151	1,802	9,677
North America (NAM)	3,017	3,840	1,371	8,228
Commonwealth of Independent States (CIS)	3,957	627	901	5,485
Latin America (LAM)	39	2,116	1,293	3,448
Middle East (MET)	0	2,547	823	3,370
Europe (EUP)	275	548	431	1,254
Africa (AFR)	39	274	784	1,097
World	9,051	16,103	7,405	**32,559**

* New emergence of UG plays around the world since 1997
* Uncertainty of Rogner's assessment was not quantified

Table 1.1—Distribution of worldwide unconventional gas (UG) resources (Kawata and Fujita 2001) using the work of Rogner (1996).

	OGIP (P50) (Tcf)				TRR (P50) (Tcf)			
Region	CBM	Tight Gas	Shale Gas	Total	CBM	Tight Gas	Shale Gas	Total
Australia and Asia (AAO)	1,348	6,253	2,690	10,291	483	3,783	676	4,942
North America (NAM)	1,629	10,784	5,905	18,318	584	6,525	1,505	8,614
Commonwealth of Independent States (CIS)	859	28,604	15,880	45,343	308	17,307	3,924	21,539
Latin America (LAM)	13	3,366	3,742	7,122	5	2,037	964	3,006
Middle East (MET)	9	15,447	15,416	30,872	3	9,346	4,000	13,349
Europe (EUP)	176	3,525	2,194	5,895	63	2,133	561	2,757
Africa (AFR)	18	4,000	3,882	7,901	7	2,420	1,007	3,434
World	4,052	71,981	49,709	**125,742**	1,453	43,551	12,637	**57,641**

Table 1.2—Distribution of worldwide unconventional gas resources and TRRs (Dong et al. 2012).

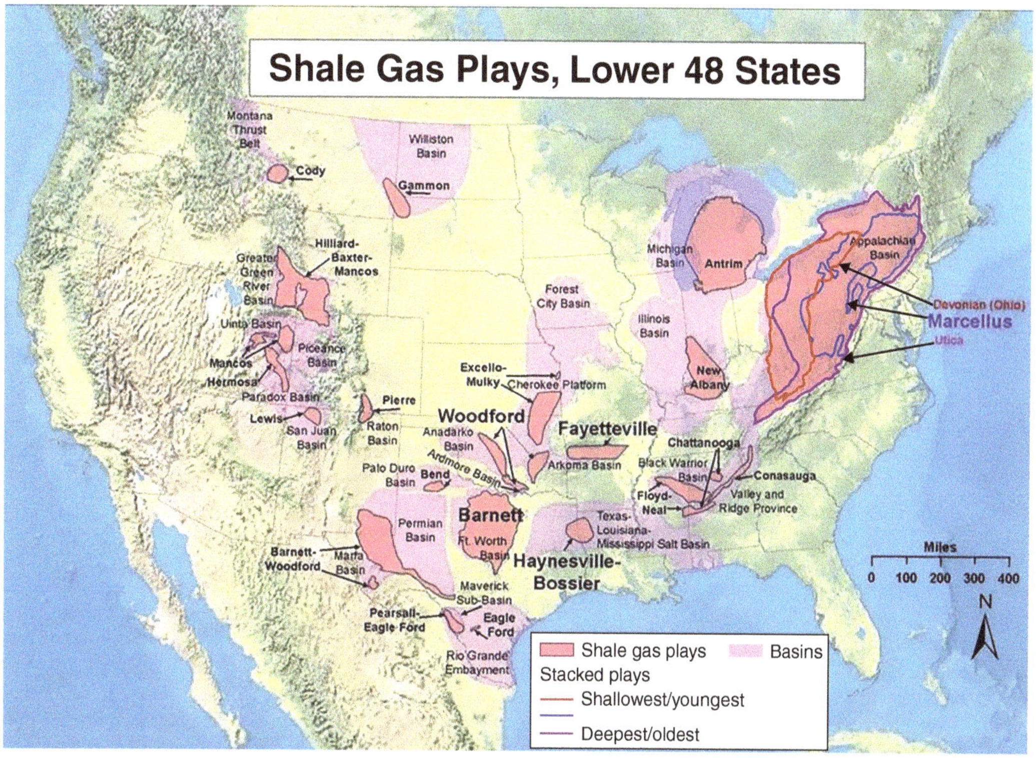

Fig. 1.2—Major unconventional gas basins (US Energy Information Administration 2010).

We are not claiming that all unconventional gas plays can be economically developed or that they will ever be developed. To develop a reservoir that is classified as TRR, one needs to have access to the land, drilling equipment, drilling personnel, completion equipment, completion personnel, hydraulic fracturing equipment, hydraulic fracturing personnel, a market for the natural gas, a gas pipeline, a reasonable gas price, the rule of law to make and enforce contracts, and proper infrastructure—such as roads and housing—just to name some of the considerations.

However, by classifying the gas as TRR—technically recoverable—it is acknowledged that the gas exists and can be drilled and produced if the mineral owners (usually the government) want it to be produced. We know how to do it. The question is: Does the ruling authority want us to produce the gas? Although not discussed in detail, Dong (2012) found that approximately 25 to 50% of the TRR is economically recoverable in basins that have the necessary infrastructure.

The volumes of gas produced from unconventional resources in the US are projected to only increase in importance over the next few decades. The US Energy Information Administration (EIA) forecast shown in **Fig. 1.3** illustrates that shale gas, tight gas, and CBM will comprise 75% of the gas produced in the US by the year 2035. In addition, notice that in 2010 the consumption of dry gas in the US was 23.8 Tcf/yr. For North America, the consumption was approximately 32 Tcf/yr.

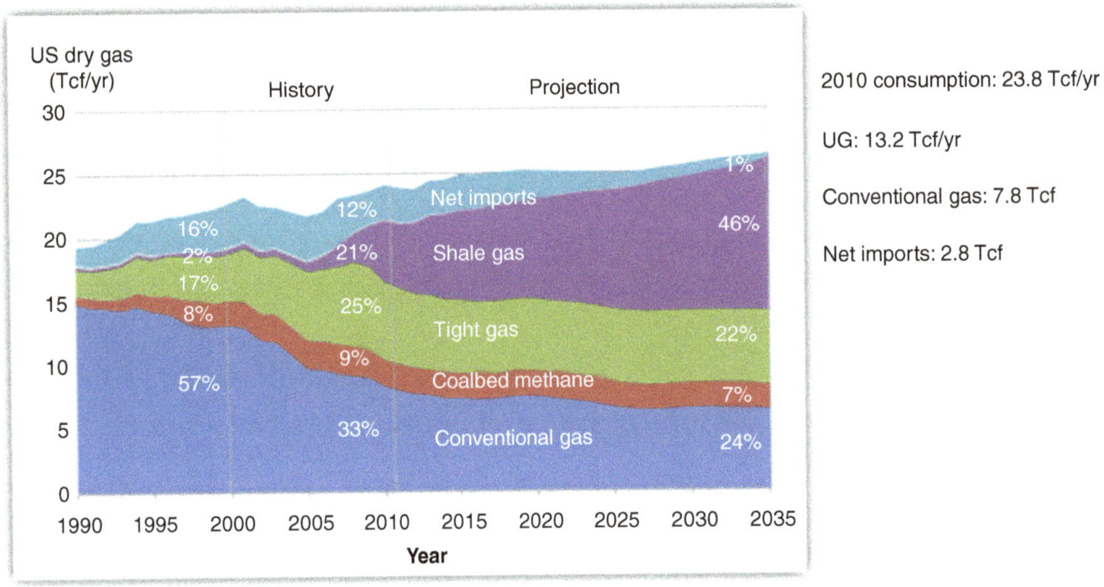

Fig. 1.3—EIA forecast of US dry gas production (US Energy Information Administration 2011).

Using the TRR values in Table 1.2 and the demand for natural gas in North America, one can compute that there are several centuries of natural gas supply to meet demand, as shown in **Table 1.3.**

Area	OGIP (Tcf)	TRR (Tcf)	Usage (Tcf)	Supply (years)
North America	18,318	8,614	32	269
World	125,742	57,641	118	488

Table 1.3—Centuries of natural gas available.

1.5 Coal Seams

CBM is one of the best examples of how technology can have an impact on the understanding and eventual development of a natural gas resource. Gas has been known to exist in coal seams since the coal mining industry began, but significant gas production has been realized only since 1989 **(Fig. 1.4).**

CBM was drilled through and observed for many years yet never produced and sold as a resource. New technology and focused CBM research ultimately solved the resource-complexity riddle and unlocked its production potential. CBM now provides more than 1.6 Tcf of gas production per year in the US and is under development worldwide, including in Canada, Australia, India, and China.

The factors controlling CBM production behavior are similar to those for conventional gas resources in many respects; however, they differ considerably in important areas. One prominent difference is in the understanding of the resource, especially with regard to values of gas-in-place. Natural gas in coal seams adsorbs to the coal surface, allowing for significantly more gas to be stored than in conventional rocks in shallow, low-pressure formations. To release the adsorbed gas for production, we have to substantially reduce the pressure in the reservoir. Adsorbed gas volumes are not important for conventional gas resources but are very important for CBM reservoirs. Significant research was required in the 1990s to fully understand how to produce the adsorbed gas in coal seams and to develop the technology required to explore and produce CBM reservoirs.

A major difference between CBM reservoirs and sandstone gas reservoirs is that many of the coal seams are initially saturated with water. Thus, large volumes of water must be pumped out of the coal seams to reduce the pressure so that desorption will occur before seeing any significant gas production. The technology developed in the 1990s for understanding and dewatering coal seams paved the way for significant CBM development in several US geologic basins.

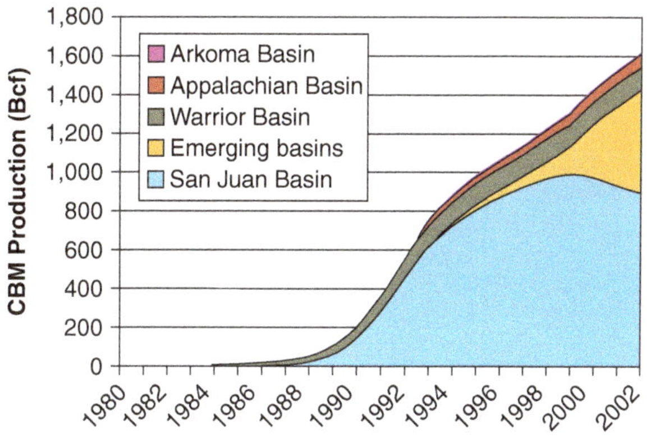

Fig. 1.4—Gas production from coal seams in the US.

1.6 Shale Gas

Shale formations act as both a source of gas and its reservoir. Natural gas is stored in shale in three forms: free gas in rock pores, free gas in natural fractures, and adsorbed gas on organic matter and mineral surfaces. These different storage mechanisms affect the speed and efficiency of gas production.

A global energy study (Rogner 1996) estimated that abundant shale gas resources are distributed mostly in North America, Latin America, and Asia Pacific (Table 1.1). Estimates suggest that the resource ranges from 1,483 to 1,859 Tcf in the US and 500 to 600 Tcf in Canada. In other regions of the world, this resource has been studied to a limited extent. Dong et al. (2012) raised those estimates, as shown in Table 1.2.

Commercial shale gas production first occurred in the US and was distributed in the Appalachian Basin, Michigan Basin, Illinois Basin, Fort Worth Basin, and San Juan Basin. Production increased rapidly in the 1990s and 2000s. In 2004, shale gas production in the US reached approximately 700 Bcf/yr, a huge increase from 350 Bcf/yr in 2000. Since the late 1990s, shale gas has been produced mainly in the Barnett, Haynesville, Fayetteville, Marcellus, and Eagle Ford; in a number of shales in the Rocky Mountain basins; and in western Canada. While technological innovations have increased per well with gas recovery efficiency, most of the increases in reserves have come from the increase in well density.

1.7 Nomenclature

$B\mu$ = fluid properties, cp
h = net-pay thickness, ft
k = permeability, md
$\bar{p}$ = reservoir pressure, psi
p_{wf} = flowing pressure, psi
q = flow rate, Mcf/D
r_e = drainage radius, ft
r_w = wellbore radius, ft
s = skin factor, dimensionless

1.8 References

Dong, Z. 2012. *A New Global Unconventional Natural Gas Resource Assessment.* Doctoral dissertation, Texas A&M University, College Station, Texas (October 2012). https://oaktrust.library.tamu.edu/handle/1969.1/ETD-TAMU-2012-08-11875.

Dong, Z., Holditch, S. A., McVay, D. A. et al. 2012. Global Unconventional Gas Resource Assessment. *SPE Econ & Mgmt* **4** (4): 222–234. SPE-148365-PA. https://doi.org/10.2118/148365-PA.

Gray, J. K. 1977. Future Gas Reservoir Potential Western Canadian Sedimentary Basin. Paper presented at the 3rd National Technical Conference, Canadian Gas Association.

Holditch, S. A. 2006. Tight Gas Sands. *J Pet Technol* **58** (6): 86–93. SPE-103356-JPT. https://doi.org/10.2118/103356-JPT.

Kawata, Y. and Fujita, K. 2001. Some Predictions of Possible Unconventional Hydrocarbons Availability Until 2100. Paper presented at the SPE Asia Pacific Oil and Gas Conference and Exhibition, Jakarta, Indonesia, 17–19 April. SPE-68755-MS. https://doi.org/10.2118/68755-MS.

Masters, J. A. 1979. Deep Basin Gas Trap, Western Canada. *AAPG Bull.* **63** (2): 152–181. https://doi.org/10.1306/C1EA55CB-16C9-11D7-8645000102C1865D.

Rogner, H.-H. 1996. An Assessment of World Hydrocarbon Resources. Working paper, WP-96–56, IIASA, Laxenburg, Austria. http://pure.iiasa.ac.at/id/eprint/4970/1/WP-96-056.pdf.

US Energy Information Administration. 2010. Shale Gas Plays, Lower 48 States (updated 10 March 2010), https://www.eia.gov/maps/maps.htm.

US Energy Information Administration. 2011. Annual Energy Outlook 2011, with Projections to 2035. Annual report, DOE/EIA-0383(2011), US Department of Energy, Washington, DC (April 2011).

Chapter 2

Tight Gas Reservoirs

Stephen A. Holditch

Tight gas is the term commonly used to refer to low-permeability reservoirs that produce mainly dry natural gas. Tight gas reservoirs have one thing in common—a vertical well drilled and completed in a tight gas reservoir must be successfully stimulated to produce at commercial gas flow rates and produce commercial gas volumes. Normally, a large hydraulic fracturing treatment is required to produce gas economically. In some naturally fractured tight gas reservoirs, horizontal wells and/or multilateral wells can provide the stimulation required for commerciality.

To optimize the development of a tight gas reservoir, the geoscientists and engineers must optimize the number of wells drilled, as well as the drilling and completion procedures for each well. Often, more data and more engineering manpower are required to understand and develop tight gas reservoirs than are required for higher-permeability, conventional reservoirs. On an individual-well basis, a well in a tight gas reservoir will produce less gas over a longer period of time than one expects from a well completed in a higher-permeability, conventional reservoir. As such, many more wells must be drilled (or smaller well spacing must be attained) in a tight gas reservoir to recover a large percentage of the original gas-in-place (OGIP) when compared to a conventional reservoir.

In the 1970s, the US government decided that a tight gas reservoir is one in which the expected value of permeability to gas flow would be less than 0.1 md. This was a political definition used to determine which wells would receive federal and/or state tax credits for producing gas from tight reservoirs. Actually, the definition of a tight gas reservoir is a function of many factors, each relating to Darcy's law, as in Eq. 2.1:

$$q = \frac{kh\left(\bar{p} - p_{wf}\right)}{141.2\,B\mu\left[\ln\left(\frac{r_e}{r_w}\right) - 0.75 + s\right]} . \qquad (2.1)$$

The main problem with tight gas reservoirs is that they do not produce at economic flow rates unless they are stimulated—normally by a large hydraulic fracturing treatment. Eq. 2.1 illustrates the main factors controlling flow rate and clearly shows that the flow rate q is a function of permeability k; net-pay thickness h; reservoir pressure $\bar{p}$; flowing pressure p_{wf}; fluid properties $B\mu$; drainage area r_e; wellbore radius r_w; and skin factor s. Thus, to choose a single value of permeability to define "tight gas" is not wise. In deep, high-pressure, thick reservoirs, excellent completions can be achieved when the formation permeability to gas is in the microdarcy range (0.001 md). In shallow, low-pressure, thin reservoirs, permeabilities of several millidarcies might be required to produce the gas at economic flow rates, even after a successful fracturing treatment.

The best way to define tight gas is that the reservoir cannot be produced at economic flow rates nor recover economic volumes of natural gas unless a special technique is used to stimulate production. Specifically, large hydraulic fracturing treatments, a horizontal wellbore, or multilateral wellbores must be used to stimulate flow rates and increase the recovery efficiency in the reservoir.

The optimal drilling, completion, and stimulation methods for each well are a function of the reservoir characteristics and the economic situation. Some tight gas reservoirs are in South Texas, while others are in the deserts of Egypt. The costs to drill, complete, and stimulate the wells, in addition to the gas price and the gas market, affect how tight gas reservoirs are developed. As with all engineering problems, the technology used is a function of the economic conditions surrounding the project.

2.1 Tight Gas in the US

Since the 1950s, the oil and gas industry has completed and fracture treated low-permeability wells in the US. However, it was the natural-gas price increase in the 1970s that spurred significant activity in low-permeability gas reservoirs. Since the 1970s, sustained increases in natural-gas prices—along with advances in evaluation, completion, and stimulation technology—have led to substantial development of low-quality gas reservoirs. **Fig. 2.1** shows the locations of the major tight gas basins in the US.

Fig. 2.2 illustrates the tight gas resource base estimates from the Gas Technology Institute (GTI) (GTI E&P Services 2001). The gas produced through the year 2000, from tight gas reservoirs, is estimated to be 58 Tcf. Proven reserves in tight gas reservoirs are 34 Tcf. Thus, the produced gas plus proven reserves adds up to 92 Tcf. GTI estimates the volume of technically recoverable gas from

Fig. 2.1—Major US tight gas sands basins (after GTI E&P Services 2001).

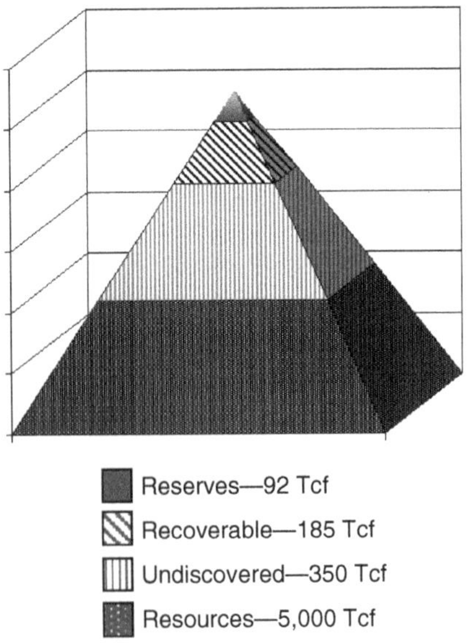

Fig. 2.2—Resource triangle for tight gas in the US (GTI E&P Services 2001).

known US tight gas accumulations at 185 Tcf. The term "technically recoverable" means the gas is known to exist and the technology is available to drill, complete, stimulate, and produce this gas; however, the gas cannot be booked as reserves until the wells are drilled and the reservoirs are developed. The next category in Fig. 2.2 is "undiscovered," which represents the GTI estimate of gas that is likely to be discovered in known tight gas basins. Finally, the largest category is "resources," which represents the gas-in-place in US tight gas basins. Substantial improvements in technology or changes in the gas market are required before the gas in the resources category can be produced economically.

2.2 Geologic Considerations

The analysis of any reservoir, including a tight gas reservoir, should always begin with a thorough understanding of the geologic characteristics of the formation. The important geologic parameters for a trend or basin are the structural and tectonic regime, the regional thermal gradients, and the regional pressure gradients. Knowing the stratigraphy in a basin is very important and can affect the drilling, evaluation, completion, and stimulation activities. Important geologic parameters that should be studied for each stratigraphic unit are the depositional system, genetic facies, textural maturity, mineralogy, diagenetic processes, cements, reservoir dimensions, and presence of natural fractures.

According to Fisher and McGowen (1967), a depositional system is a group of lithogenetic facies linked by depositional environment and associated processes. Each lithogenetic facies has certain attributes—including porosity, permeability, and special relations to other facies—that affect the migration and distribution of hydrocarbons. The nine principal clastic depositional systems reviewed by Fisher and Brown (1972) can be classified into three major groups, as illustrated in **Table 2.1.** Most tight gas sandstones that are being developed and produced in the US are located in barrier strand plains, deltaic systems, or fluvial systems (GTI E&P Services 2001). A few plays are found in shelf and fan-deltaic systems. Knowing the depositional system is important because it will affect the reservoir morphology and both the lateral and vertical continuity one expects in a reservoir. Details concerning clastic depositional systems can be found in Berg (1986) and Galloway and Hobday (1983). Refer also to Ch 5: Reservoir Characterization in Part 2 of this book.

Continental Environments
Folian systems
Lacustrine systems
Fluvial systems
Terrigenous fan (alluvial-fan and fan-deltaic systems)
Shoreline (Marginal Marine) Environments
Deltaic systems
Barrier-strand-plain systems
Lagoon, bay, estuarine, and tidal-flat systems
Marine Environments
Continental and intracratonic shelf systems
Continental and intracratonic slope and basinal systems

Table 2.1—Classification of clastic depositional systems by environment.

2.2.1 Diagenesis. When most sands are deposited, the pores and pore throats are well-connected, resulting in high permeability. As explained by Berg (1986), sands are composed of mineral particles called grains, which usually consist of quartz, feldspars, and rock fragments. The finer particles between the grains are called matrix. The original porosity and permeability of a sandstone are determined by characteristics such as mineral composition, pore type, grain size, and texture. After deposition and burial, the grains and matrix are commonly altered by the physical effects of compaction and chemical changes. These changes are broadly referred to as diagenesis. **Table 2.2** describes common diagenetic changes, as explained by Berg (1986).

Process	Result	Depth	Effect
Compaction	Mechanical rearrangement of grains	Shallow	Reduces porosity from 40% to less than 30%
Cementation	Intergranular calcite, silica overgrowths in quartz	Shallow to moderate	Consolidation (lithification) and porosity reduction
Clay precipitation	Kaolinite, chlorite, and illite as grain coatings	Shallow to moderate	Slight decrease in porosity
Clay dehydration	Loss of interlayered water in expandable clays (smectite/illite)	Moderate	Slight increase in porosity
Dissolution	Leaching of calcite cement and feldspar grains	Moderate	Significant increase in porosity
Cementation	Intergranular ferroan calcite, zeolites	Moderate to deep	Significant decrease in porosity
Pressure solution	Styolitization by silica dissolution	Moderate to deep	Local decrease in porosity (along laminae)
Microfracturing	Breaking of grains	Deep	Slight increase in porosity

Table 2.2—Common diagenetic changes in sandstones (after Berg 1986).

In addition to the grains and the matrix, cement is normally introduced into the rock during diagenesis. Cement is precipitated between the grains and holds the rock together. Nearly every low-permeability gas reservoir has been severely altered by diagenesis. Clay and quartz overgrowths are commonly found to be filling some of the original porosity and reducing the size of the pore throats. Normally, quartz overgrowths cause the most-severe permeability reduction in tight sands. Pore-filling clays reduce the permeability more than pore-lining clays. Detailed geologic analyses are required to understand the effects of diagenesis on the formation and determine how diagenesis information can be used to optimize the completion and stimulation fluids.

2.2.2 Reservoir Continuity. One of the most-difficult parameters to evaluate in tight gas reservoirs is the drainage-area size and shape of a typical well. In tight reservoirs, months or years of production are normally required before the pressure transients are affected by reservoir boundaries or well-to-well interference. As such, the engineer often has to estimate the drainage-area size and shape for a typical well to estimate reserves. Knowledge of the depositional system and the effects of diagenesis on the rock is needed to estimate the drainage-area size and shape for a specific well.

In blanket tight gas reservoirs, the average drainage area of a well largely depends on the number of wells drilled, the size of the fracturing treatments pumped on the wells, and the time frame being considered. In lenticular or compartmentalized tight gas reservoirs, the average drainage area is likely a function of the average sand-lens size or compartment size and might not be a strong function of the size of the fracturing treatment.

A main factor controlling the continuity of the reservoir is the depositional system. Generally, reservoir drainage per well is small in continental deposits and larger in marine deposits. Fluvial systems tend to be more lenticular. Barrier-strand-plain systems tend to be more blanket and continuous. If one looks at the tight gas plays that have been more-successfully developed—such as the Vicksburg in South Texas, the Cotton Valley Taylor in East Texas, the Mesaverde in the San Juan Basin, and the Frontier in the Green River Basin, just to name a few—all these sandstones are marine deposits. Marine deposits tend to be more blanket and continuous. Most of the more-successful tight gas plays are those in which the formation is a thick, continuous marine deposit.

There are other formations—such as the Travis Peak in East Texas, the Abo in the Permian Basin, and the Mesaverde in parts of the Rocky Mountains—that are fluvial systems and tend to be very lenticular. The Wilcox Lobo in South Texas is highly compartmentalized because of faulting. In lenticular or compartmentalized reservoirs, the drainage area is controlled by the geology and must be estimated by the geologist or engineer.

The best way to determine the depositional system is to cut and analyze cores. Coring in the shales, mudstones, and nonreservoir rock above and below the main pay interval is recommended. A geologist can tell much more about the depositional system by studying the entire stratigraphic sequence. The core descriptions can be correlated with openhole logging data to determine the logging signature for various depositional environments. When these correlations have been made, logs from additional wells can be analyzed to generate maps of the depositional patterns in a specific area. These maps can be useful in developing field optimization plans.

Tectonic activity during deposition can affect reservoir continuity and morphology. In addition, regional tectonics affect the horizontal stresses in all rock layers. The horizontal stresses, in turn, affect faulting, rock strength, drilling parameters, hydraulic fracture propagation, natural fracturing, and borehole stability. The main concerns for tight gas reservoirs are the effects of regional tectonics on hydraulic fracture propagation and natural fracturing in the formation.

Natural fractures affect both the overall level of permeability in a reservoir and the degree of permeability anisotropy in the reservoir. If a reservoir is naturally fractured, it is possible that a horizontal well or multilateral wellbores will be more effective in producing gas than a vertical well with a hydraulic fracture. If a fracturing treatment is performed in a reservoir containing an abundance of natural fractures, problems with multiple hydraulic fractures, near-wellbore tortuosity problems, and excessive fluid leakoff can occur during the fracturing treatment. The engineer and geologist should work together to understand the current and past tectonic activity in a basin. Knowledge of the tectonic history is important in designing the field optimization plan and developing drilling and completion procedures. A good way to begin is to study the fault systems in a basin. Hydraulic fractures tend to parallel normal faults and run perpendicular to reverse faults. The engineer should use data from openhole caliper logs and from injection tests from prior hydraulic fracturing treatments to better understand the total in-situ stresses and the tectonic stress component in a given area. By combining engineering data with geologic data, a team of geologists and engineers can develop an understanding of the regional tectonics in an area. This understanding is important to the analysis and development of any tight gas reservoir.

2.3 Reservoir Considerations

Normally, a tight gas reservoir can be described as a layered system. In a clastic depositional system, the layers are composed of sandstone, siltstone, mudstone, and shale. In carbonate systems, layers of limestone, dolomite, shale, and, perhaps, halite or anhydrite occur. To optimize the development of a tight gas reservoir, a team of geoscientists, petrophysicists, and engineers must fully characterize all the layers of rock above, within, and below the pay zones in the reservoir. Data concerning gross pay thickness, net-pay thickness, permeability, porosity, water saturation, pressure, in-situ stress, and Young's modulus for all layers are required to use 3D reservoir and fracture propagation models to evaluate the formation, design the fracturing treatment, and forecast production rates and ultimate recovery.

The speed at which pressure transients move through porous media is a function of the formation permeability, fluid viscosity, and fluid compressibility, as well as other variables. In a high-permeability gas reservoir (say, 100 md), a pressure transient will reach the reservoir boundary in a matter of hours or days. Well-to-well interference in high-permeability blanket gas reservoirs is quite common. However, in a gas reservoir with a permeability of 0.1 md, the pressure transients move 1,000 times slower than the transients in a 100-md reservoir. As such, it might take years of production before well-to-well interference, or a boundary, can be recognized by studying pressure-transient or production data.

In high-permeability gas reservoirs, the semisteady-state form of Darcy's law works well. Methods such as the McGuire and Sikora graph and Prats equations can be used to design and analyze hydraulic fractures in medium- to high-permeability gas reservoirs (McGuire and Sikora 1960; Prats 1961). Short (24- to 72-hour) pressure buildups, analyzed using a Horner graph, can provide accurate estimates of formation properties in medium- to high-permeability gas reservoirs (Horner 1951).

However, in tight gas reservoirs, semisteady-state analysis methods cannot be used alone to analyze short-term (days, weeks, or months) data. The best methods for analyzing transient production or pressure data are type curves, analytical models, or finite-difference models (Agarwal et al. 1979; Cinco L. et al. 1978). Transient-flow analyses can be used to estimate values of formation permeability, skin, fracture half-length, and fracture conductivity, and a minimum value of drainage area.

2.4 Drilling and Completion Considerations

The most-important part of drilling a well in a tight gas reservoir is to drill a gauge hole. A gauge hole is required to obtain an adequate suite of openhole logs and to obtain an adequate primary cement job. In low-porosity shaly reservoirs, the analyses of gamma ray (GR), spontaneous potential (SP), porosity, and resistivity logs to determine accurate estimates of shale content, porosity, and water saturation can be difficult. If the borehole is washed out ("out of gauge"), the log readings will be affected, and it will be even more difficult to differentiate the pay from the nonpay portions of the formation. If the borehole is washed out, obtaining a primary cement seal is difficult, which could affect zonal isolation and cause the well to have to be cement squeezed before running tests or pumping stimulation treatments.

Formation damage and drilling speed should be secondary concerns. Some wells are drilled underbalanced to increase the bit-penetration rate or to minimize mud-filtrate invasion. However, if the wellbore is severely washed out because the well was drilled underbalanced, it is probable that a large amount of money will be wasted as a result of inaccurate logs and an inadequate primary cement job. It is best to drill a tight gas well near balanced to minimize borehole washouts and mud-filtrate invasion.

The completion strategy and stimulation strategy required for a tight gas reservoir very much depend on the number of layers of net gas pay and the overall economic assessment of the reservoir. In almost every case, a well in a tight gas reservoir is not economic to produce unless the optimal fracturing treatment is both designed and pumped into the formation. The well can be perfectly drilled, cased, and perforated, but it will be uneconomic until the optimal fracturing treatment is pumped. As such, the entire well prognosis should focus on drilling and completing the well so that it can be successfully fracture treated. The hole sizes, casing sizes, tubing sizes, wellhead, flowlines, and perforation scheme should be designed to accommodate the fracturing treatment.

2.5 Formation Evaluation

To properly complete, fracture treat, and produce a tight gas reservoir, each layer of the pay zone, and the formations above and below the pay zone, must be thoroughly evaluated. The most-important properties to evaluate are pay-zone thickness, porosity, water saturation, permeability, pressure, in-situ stress, and Young's modulus. The raw data used to estimate values for these important parameters come from logs, cores, well tests, drilling records, and production from offset wells.

Because tight gas reservoirs are normally also low-porosity reservoirs, detailed log analyses become critical to understanding the reservoir. For example, if an error of 2 porosity units (p.u.) occurs when the porosity is 30%, it is normally not critical. The difference between 28 and 30% porosity will not lead to much error in net gas pay, water saturation, or gas-in-place. However, the same 2-p.u. error applied to a reservoir in which the porosity is 8% is a much-more-significant problem. The difference between 6 and 8% porosity can cause significant errors in estimates of net gas pay, water saturation, and gas-in-place. As such, careful preprocessing of log data and detailed petrophysical analyses of all openhole logging data are very important in the analyses of tight gas reservoirs.

2.5.1 Log Analyses. The logs provide the most-economical and -complete source of data for evaluating layered, complex, low-porosity tight gas reservoirs. The recommended logging suite for a tight gas reservoir consists of the SP, GR, density (FDC), neutron (CNL), sonic (SON), and dual (or array) induction logs (DILs). All openhole logging data should be preprocessed before the data are used in any detailed computations. The steps required for preprocessing logs are (1) digitize all log data, (2) depth shift the data as required, (3) perform all environmental corrections, and (4) normalize data so that all logs from different wells are reading the same in the same zones. A good example would be the thick marine shales because one expects the log readings to be consistent from well to well (Aly et al. 1997; Howard and Hunt 1986).

After the data have been preprocessed and stored in a digital database, a series of statistical analyses must be conducted to quantify certain evaluation parameters. These statistical analyses consist of a Pickett plot to determine estimates of water resistivity (R_w), cementation factor (m), and saturation exponent (n); shale histograms to find the shale endpoints on all logs; sand and/or limestone histograms to determine the clean zone endpoints on all logs; linear regressions between each porosity log and any core data to establish correlation constants; and linear regressions among the porosity logs to develop correlations that can be used to correct for bad hole effects on one or more of the logs. The steps required to preprocess the logs, develop the correlation parameters, and analyze logs in shaly, low-porosity formations are described in a series of articles published in 1996 and 1997 (Hunt 1996a, 1996b, 1997; Hunt et al. 1996a, 1996b, 1996c; Hunt and McCain 1997; Hunt and Pursell 1996, 1997a, 1997b, 1997c; Pursell and Hunt 1997).

2.5.1.1 Computing Porosity. To correctly compute porosity in tight, shaly (clay-rich) reservoirs, one of the first values to compute is the volume of clay in the rock. The clay volume is normally computed using either the SP or the GR log readings. The following equations are commonly used to compute the clay volume in a formation:

$$V_{SH} = \frac{SP - SP_{CL}}{SP_{SH} - SP_{CL}}, \quad (2.2)$$

$$I_{RA} = \frac{GR - GR_{CL}}{GR_{SH} - GR_{CL}}, \quad (2.3)$$

$$V_{SH} = 0.33\left[2^{(2I_{RA})} - 1.0\right]_{\text{old rocks}}, \quad (2.4)$$

$$V_{SH} = 0.083\left[2^{(3.7I_{RA})} - 1.0\right]_{\text{young rocks}}. \quad (2.5)$$

The SP provides reasonable estimates of V_{SH} if the formation water and the mud filtrate do not have the same salinities. The GR log provides reasonable estimates of V_{SH} as long as all the radioactive materials in the formation are part of the clays and not part of the sandstone, such as potassium feldspar.

When the values of V_{SH} are known as a function of depth, the petrophysicists can compute values of clay-corrected porosity from the FDC, CNL, and SON logs with Eq. 2.6, Eq. 2.7, or Eq. 2.8:

$$\phi_{DC} = \frac{(\rho_b - \rho_{ma}) - V_{SH}(\rho_{SH} - \rho_{ma})}{(\rho_f - \rho_{ma})}, \quad (2.6)$$

$$\phi_{SC} = \frac{(\Delta t - \Delta t_{ma}) - V_{SH}(\Delta t_{SH} - \Delta t_{ma})}{(\Delta t_f - \Delta t_{ma})}, \quad (2.7)$$

$$\phi_{NC} = \phi_N - V_{SH}\,\phi_{SH}. \quad (2.8)$$

If the petrophysicist has only an FDC, SON, or CNL log, the clay-corrected estimates of porosity from Eq. 2.6, Eq. 2.7, or Eq. 2.8 should be used to determine the porosity. However, if two or all three logs are available, crossplots should be used to determine the best estimate of porosity (Hunt and Pursell 1997b).

2.5.1.2 Computing Water Saturation. There have been numerous water-saturation equations published in the petroleum engineering and petrophysical literature. Worthington (1985) presents a complete review of all the commonly used water-saturation equations. For tight gas sandstones, the best method to compute the value of water saturation is normally the dual-water model (Clavier et al. 1984). Eq. 2.9 and **Fig. 2.3** illustrate the dual-water model.

$$C_t = C_w'' \frac{\phi_t^m}{A} S_{wt}^n \quad (2.9)$$

and

$$C_w'' = \frac{S_{wb}}{S_{wt}} C_{wb} + \left(1 - \frac{S_{wb}}{S_{wt}}\right) C_{wf}. \quad (2.10)$$

It is possible to use a clay-corrected Archie equation, the Simandoux equation, the Waxman-Smits equation, or any number of other equations as described by Worthington (1985); however, for many situations, the dual-water model provides accurate estimates of water saturation.

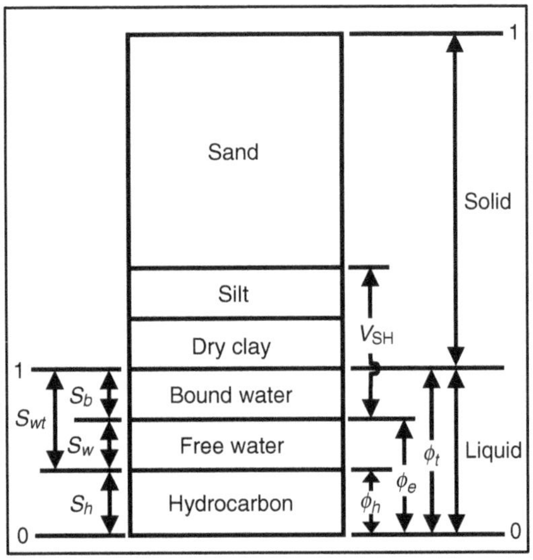

Fig. 2.3—Partitioning of a shaly sand in the dual-water model.

In the Archie equation, all the electrical conductivity in the formation is assumed to be transmitted through the water in the pore space. The rock is assumed to be an insulator and does not conduct current. However, in clay-rich formations, the clays conduct an electric current. The Simandoux and Waxman-Smits equations provide for a conductive rock but assume the water associated with the pore space and the water associated with the clays have the same properties. In the dual-water model, there is free water and bound water. The free water is in the pores, and the bound water is associated with the clays. More-accurate estimates of water saturation can be achieved by taking into account the current conducted by the clays using the dual-water model.

When the formation permeability in a gas reservoir is between 0.01 and 10 md, mud-filtrate invasion from freshwater mud into a formation with saline interstitial water can substantially alter the resistivity profile near the wellbore during the time period before the openhole logs are normally run (Semmelbeck and Holditch 1988). In such cases, DILs, or array induction logs, should be run and used to make corrections to determine the true resistivity, R_t, of the formation. The log readings change with time because of mud-filtrate invasion.

Most tight gas reservoirs are tight because they are highly cemented and have low porosity. The low porosity and cementation cause many tight gas reservoirs to become hard and abrasive, which could prevent the use of logging-while-drilling (LWD) equipment. In addition, the flow rates and ultimate recovery from individual wells are low, and the operator must control drilling, completion, and operating costs to improve the profitability of each well. For these reasons, LWD is not often used when drilling tight gas reservoirs. Most of the logging data comes from openhole logs run after the well reaches total depth. See more discussion on logging practices in Ch 5: Reservoir Characterization in Part 2 of this book.

2.5.2 Core Analyses. Obtaining and analyzing cores are crucial to the proper understanding of any layered, complex reservoir system. Obtaining the data needed to understand the fluid flow properties, the mechanical properties, and the depositional environment of a specific reservoir requires that cores be cut, handled correctly, and tested in the laboratory using modern and sophisticated laboratory methods. Of primary importance is to measure the rock properties under restored reservoir conditions. The effect of net overburden (NOB) pressure must be reproduced in the laboratory to obtain the most-accurate quantitative information from the cores.

To provide all the data needed to characterize the reservoir and depositional system, a core should be cut in the pay interval and in the layers of rock above and below the pay interval. Core from the shales and mudstones above and below the pay interval helps the geologist determine the environment of deposition. Knowing more about the deposition allows the reservoir engineer to better estimate the morphology and size of the gas-bearing reservoir layers. Also, mechanical-property tests can be run on the shales to determine estimates of Poisson's ratio and Young's modulus. Additional tests can be run to measure the shale density and the sonic travel time in the shale to assist in the analyses of the FDC and SON log data.

After cutting the core in the field, it is important to handle the core properly. The core should be pumped out of the barrel, not hammered out. After the core is laid out on the pipe racks, it should be wiped with rags to remove the mud (do not wash with water), then described as quickly as possible. Bedding features, natural fractures, and lithology should be described foot by foot. Permanent markers should be used to label the depth of the core and clearly mark the up direction on the core. As quickly as is feasible, the core should be wrapped in heat-shrinking plastic and then sealed in paraffin for the trip to the core-analysis laboratory. Precautions should be taken to minimize alteration of the core properties while retrieving and describing the core in the field.

When it arrives in the laboratory, the core is unwrapped and slabbed, and plugs are cut for testing. Normally, a core plug should be cut every foot in the core, trying to properly sample all the rock—not just the cleaner pay zones. Routine core analyses can be run on these core plugs. When the routine core analyses are completed, additional core plugs are cut for special core analyses. Sometimes, samples of whole core are used for testing. Both the routine and the special core analyses are required to calibrate the openhole logging data and to prepare the data sets required to design the optimal completion. The core plugs must also be treated with care. For example, if a core plug from a shaly sand is placed in a standard oven, it is likely that the clays in the pores will be altered as they dry out. A more-accurate core analysis usually is achieved if the core plugs are dried in a humidity-controlled oven in which the free water is evaporated but the bound clay water is not affected.

2.5.2.1 ***Routine Core Analyses.*** Routine core analyses should be run on core plugs cut every foot along the core. Routine core analyses should consist of measurements of grain density, porosity and permeability to air (both unstressed and stressed), and cation-exchange capacity, and fluid-saturation analysis. In addition, each core plug should be described in detail to understand the lithology and grain size, and to note any natural fractures and other details that could be of importance to the geologist, petrophysicist, or engineer.

The porosity is used to determine values of gas-in-place and to develop correlations with permeability. The grain density should be used to determine how to correlate the density log values and to validate any calculation of lithology from log data. The cation-exchange capacity can be used to determine how much electric current can be transmitted by the rock rather than the fluid in the pore space. The cation-exchange capacity must be measured in the laboratory, using samples of rock, and is a function of the amount and type of clay in the rock. Saturation analysis measures the amount of water, oil, and gas in the core plugs in the laboratory. Saturation analysis can be misleading in rocks that are cored with water-based mud because of mud-filtrate invasion during the coring process and problems that occur with core retrieval and handling before running the laboratory tests. However, the values of water saturation from the core analysis of cores cut with an oil-based mud can be used to calibrate the log data and to estimate values of gas-in-place in the reservoir.

The measurements of porosity and permeability are a function of the net stress applied to the rock when the measurements are taken. For low-porosity rock, it is very important to take measurements at different values of net stress to fully understand how the reservoir will behave as the gas is produced and the reservoir pressure declines. The data in **Fig. 2.4** illustrate how the values of porosity changed in Travis Peak sandstone cores when the cores were tested both at low net stress and at simulated NOB pressure. Notice that the measurements of porosity are 1 to 2 p.u. lower when measured under NOB pressure than when measured under minimal stress.

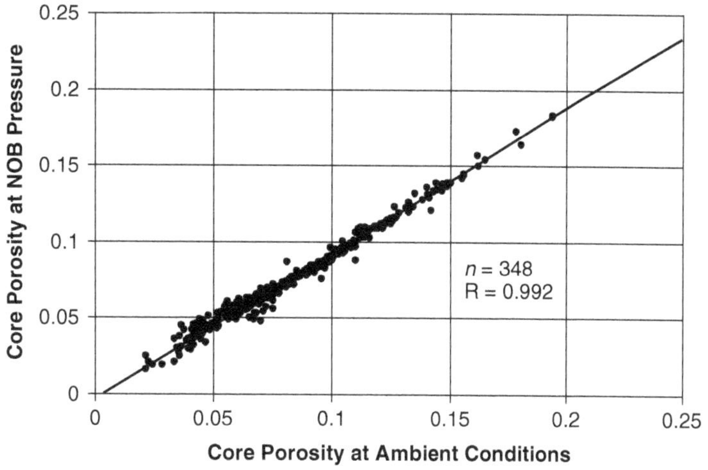

Fig. 2.4—Best-fit relation of core porosity at NOB pressure to porosity at ambient conditions, four wells. *n* = number of samples.

Fig. 2.5 illustrates the effect of NOB pressure on the measurement of air permeability on Travis Peak cores from a well in East Texas. For high-permeability (10 to 100 md) core plugs, the permeability under the original overburden pressure is slightly less than the value of unstressed permeability for that same core plug. However, as the permeability of the core plugs decreases, the effect of NOB pressure on the core plug increases substantially. For the core plugs with values of unstressed permeability of approximately 0.01 md, the values of permeability under NOB pressure were approximately an order of magnitude lower, or 0.001 md. The lower-permeability rocks are the most stress sensitive because the lower-permeability core samples have smaller pore-throat diameters than the higher-permeability rocks. As overburden stress increases, the diameter of the pore throat decreases. Because the permeability of a rock is roughly proportional to the square of the diameter of the pore throat, the permeability reduction in low-permeability rocks is much more dramatic than in high-permeability rocks. In other words, if you make a big pore throat slightly smaller by adding stress to the rock, the permeability is not reduced by much. If you make a very small pore throat even smaller by adding stress to the rock, the permeability is reduced substantially. The reduction is typically an order of magnitude or more, as illustrated in Fig. 2.5.

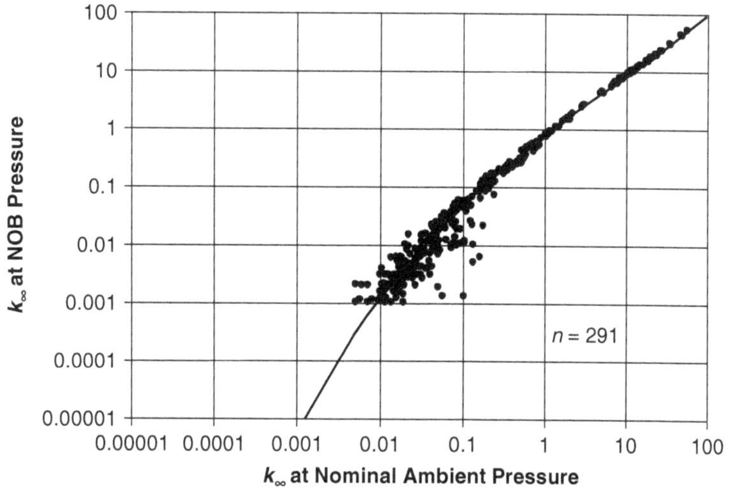

Fig. 2.5—Gas permeability at NOB pressure vs. gas permeability at ambient pressure (both Klinkenberg-corrected) for Howell No. 5 and SFE No. 2, invalid permeability data excluded. *n* = number of samples.

After the values of porosity and permeability under NOB conditions are measured in the laboratory, the values can be correlated. **Fig. 2.6** illustrates a typical correlation of permeability at NOB pressure vs. porosity at NOB pressure for a tight gas reservoir. The data in Fig. 2.6 came from the Travis Peak four-well data set and can be used to estimate values of permeability from estimated values of porosity. For example, when the values of porosity have been determined from openhole log data, the correlation in Fig. 2.6 can be used to estimate permeability for the same rock type in the Travis Peak Formation. However, it should be remembered that these estimates

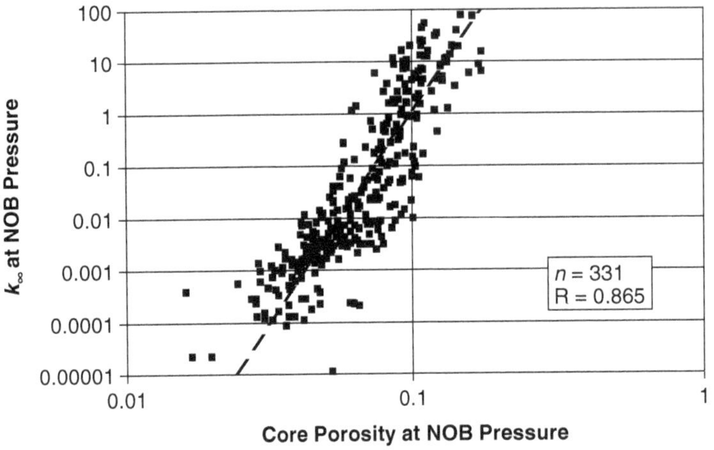

Fig. 2.6—Relation of Klinkenberg-corrected gas permeability measured at NOB pressure to porosity at NOB pressure, fluvial/deltaic channel sands only, invalid data excluded. *n* = number of samples.

are from routine core analyses, which means the core has been tested dry (with no water). If similar measurements are made at connate water saturation, the permeability in the core is further reduced, perhaps by a factor of 2, or even an order of magnitude in some cases. As such, in tight gas reservoirs, it is often found that in-situ permeabilities to gas are 10 to 100 times lower than gas permeabilities measured at ambient conditions on dry core plugs cut from whole core (Jones and Owens 1980; Thomas and Ward 1972). If cores come from a percussion sidewall device, the core plugs are typically altered, and the values of permeability under unstressed conditions can be even more optimistic.

2.5.2.2 Special Core Analyses. To fully understand the properties of tight gas formations, special core analyses must be run on selected core plugs to measure values of gas permeability vs. water saturation, resistivity index, formation factor, capillary pressure, acoustic velocity, and rock mechanical properties (Soeder and Randolph 1987). The values of resistivity index and formation factor are used to better analyze the porosity and resistivity logs. The acoustic velocity can be used to better estimate porosity and to determine how to estimate the mechanical properties of the rock from log data. The mechanical properties are measured and correlated to log measurements and lithology. The capillary-pressure measurements and the gas permeability vs. water saturation relative-permeability measurements are required to properly simulate fluid flow in the reservoir and to design hydraulic fracturing treatments.

It is important to choose the correct core samples for conducting the special core analyses. Special core analyses tests are expensive and require weeks or months of special laboratory measurements. As such, the core samples must be chosen with care to provide the optimal data for designing the well completion and the well stimulation treatment, and for forecasting future gas recovery. A good way to achieve this is to gather a team of geologists, engineers, and petrophysicists in a room; lay out the core on a table; have the routine core analysis and log analysis available; determine how many rock types or lithology types contained in the core are important to the completion and stimulation process; and pick three to six locations for each rock type or lithology in which core plugs are cut for testing.

In the SFE No. 3 Well that was part of the Gas Research Institute (GRI) Tight Gas Sands Project, special core analyses were run and described in detail (Gas Research Institute 1991). **Fig. 2.7** shows how the log analysis, the routine core analysis, and the special core analysis can be combined to develop a detailed description of a layered tight gas reservoir.

2.5.3 Mud-Filtrate Invasion. Many tight gas formations require the use of drilling mud mixed with fresh water in order to drill. Commonly, the formation water is more saline than the water in the drilling mud. When the drill bit penetrates a permeable formation, filtrate from the drilling mud invades the formation (Ferguson and Klotz 1954). The factors that affect mud-filtrate invasion are mudcake properties, reservoir pressure, mud weight, formation permeability, formation porosity, relative permeability, and capillary pressure. In addition to these factors, those that affect the resistivity profile around the well are the formation-water salinity, the mud-filtrate salinity, and the initial water saturation in the formation.

In low-permeability gas reservoirs, mud-filtrate invasion during drilling can affect the results from both drillstem tests and openhole logs (Holditch et al. 1983; Tobola and Holditch 1991). The mud filtrate invades the permeable zones, and the mud-filtrate invasion profile changes with time. Therefore, the values recorded by logging tools are a function of when those logs are run. In addition, values such as mud weight, mud-filtrate salinity, and mud-circulation rate can change hourly or daily. Therefore, it is important to measure the mud properties daily and keep accurate records during drilling operations.

The fact that mud-filtrate invasion in low-porosity rocks does affect openhole logs can be used to the advantage of the log analyst. Semmelbeck and Holditch (1988) explained how mud-filtrate invasion in low-permeability formations affects the deep induction (R_{ild}) and the medium induction (R_{ilm}) logs differently as a function of time. Thus, if one has multiple logging runs, one can evaluate how the ratio of R_{ild}/R_{ilm} varies and can correlate that ratio with formation permeability. **Fig. 2.8** shows simulated data that describe how the ratio of R_{ild}/R_{ilm} for one set of reservoir and drilling-mud parameters varies over time as a function of reservoir permeability. Notice that the resistivity ratio changes with time as the mud filtrate continues to invade the formation. It is clear that the mud-filtrate invasion affects the different resistivity logs more in high-permeability formations than in low-permeability formations. As such, evidence of

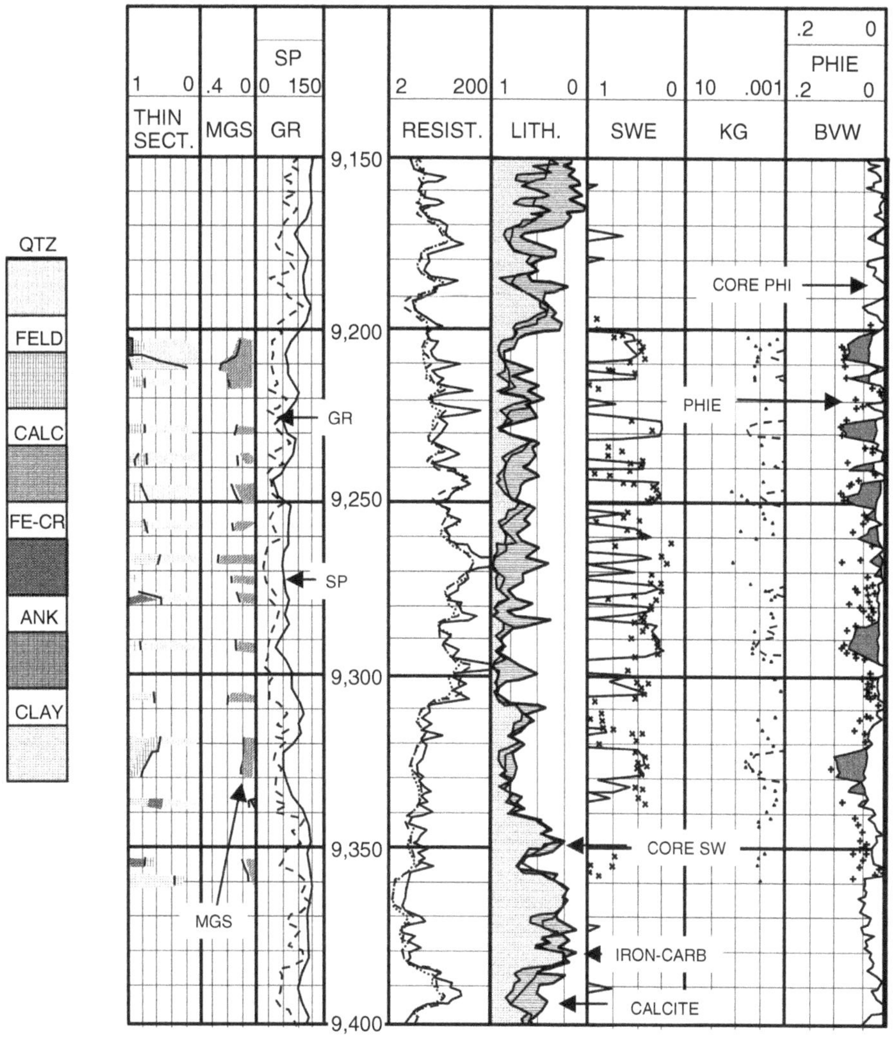

Fig. 2.7—Presentation of the log results over the completed interval in SFE No. 3, displayed with core analysis.

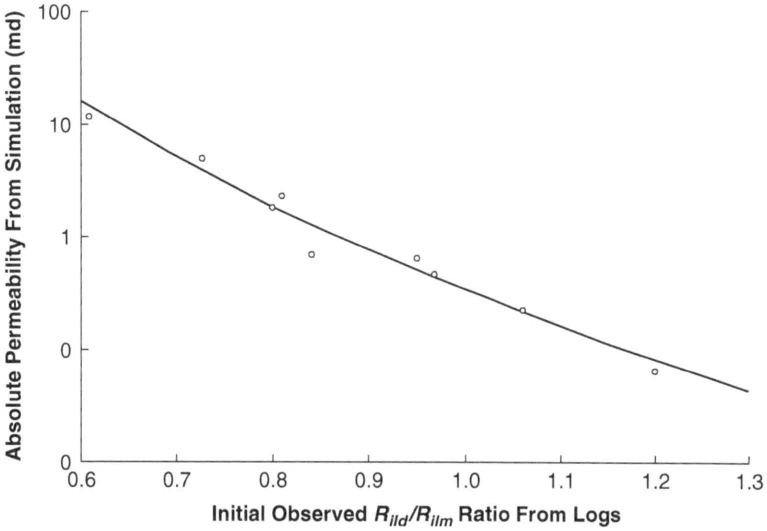

Fig. 2.8—Relation between history-matched absolute permeability and initial observed R_{ild}/R_{ilm} ratio.

mud-filtrate invasion from log analyses can be used to estimate values of formation permeability (Semmelbeck and Holditch 1988; Tobola and Holditch 1991; Yao and Holditch 1996).

The SFE No. 3, a GRI research well in East Texas, was logged four times while drilling to measure the effects of mud-filtrate invasion on the readings from openhole logs (Gas Research Institute 1991). **Fig. 2.9** presents some of the data for a portion of the hole in SFE No. 3. Because the resistivity measurements are changing with every logging run, it is clear that mud-filtrate invasion affects the openhole resistivity logging readings in the permeable zones. However, in the shales, where minimal invasion occurred, the effects of invasion are minimal because the resistivity readings do not vary between logging runs. As a rule of thumb, if the analyst sees evidence of mud-filtrate invasion on the resistivity logs and/or mudcake across a gas-bearing zone, that zone should have enough permeability to produce gas at measurable flow rates.

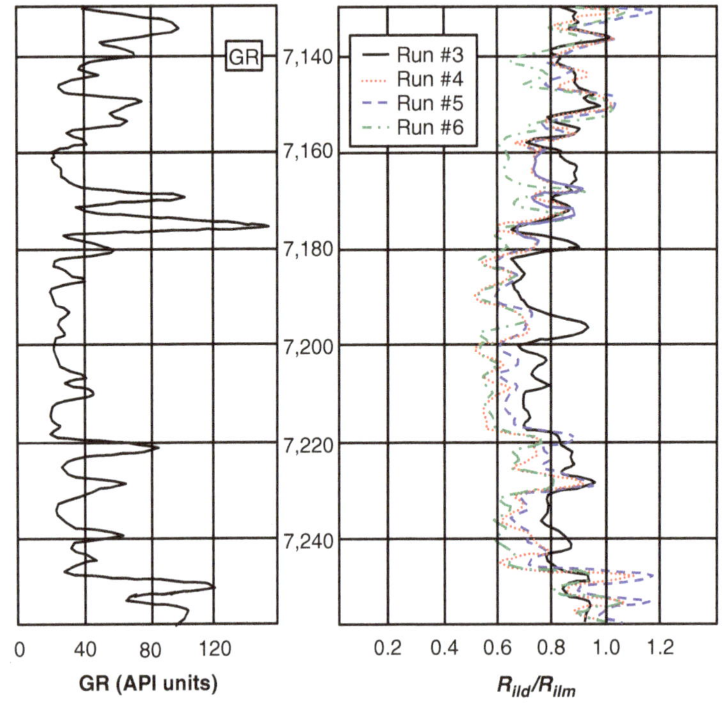

Fig. 2.9—SFE No. 3 logs showing effects of mud-filtrate invasion.

Mud-filtrate invasion also affects the sonic velocities, the bulk densities, and the hydrogen content of the portion of the rock near the wellbore that is invaded. As such, mud-filtrate invasion also affects the SON, FDC, and CNL log readings. As mud-filtrate invasion proceeds, the properties and the readings from the SON, FDC, and CNL logs will change with time (Gas Research Institute 1991).

2.6 Formation Mechanical Properties

Many tight gas reservoirs are thick, layered systems that must be hydraulically fractured to produce at commercial gas flow rates. To optimize the completion, it is necessary to understand the mechanical properties of all the layers above, within, and below the gas pay intervals. Basic rock properties such as in-situ stress, Young's modulus, and Poisson's ratio are needed to design a fracturing treatment. The in-situ stress of each rock layer affects how much pressure is required to create and propagate a fracture within the layer. The values of Young's modulus relate to the stiffness of the rock and help determine the width of the hydraulic fracture. The values of Poisson's ratio relate to the lateral deformation of the rock when stressed. Poisson's ratio is a parameter required in several fracture-design formulas. The definition of these mechanical properties, the importance of these parameters, and how to determine values for each property are discussed in Part 3 of this book.

The most-important mechanical property is in-situ stress, often called the minimum compressive stress, or the fracture-closure pressure. When the pressure inside the fracture is greater than the in-situ stress, the fracture is open. When the pressure inside the fracture is less than the in-situ stress, the fracture is closed. We can determine values of in-situ stress using logs, cores, or injection tests. To optimize the completion, it is very important to know the values of in-situ stress in every rock layer.

2.7 Estimating Permeability

In addition to knowing the values of in-situ stress, it is also extremely important to know the values of formation permeability in every rock layer. The values of permeability control everything from gas flow rate to fracturing-fluid leakoff. It is impossible to optimize the location of the perforations, the length of the hydraulic fracture, the conductivity of the hydraulic fracture, and the well spacing if one does not know the values of formation permeability in every rock layer. In addition, one must know the formation permeability to forecast gas reserves and to analyze post-fracturing pressure-buildup (PBU) tests. To determine the values of formation permeability, one can use data from logs, cores, production tests, and prefracturing PBU tests, or injection falloff tests.

2.7.1 Log-Derived Permeability. The most data that are available vs. depth come from openhole logs. If the logs are analyzed correctly, it is often possible to generate estimates of formation permeability vs. depth using the logging data. However, the correlations used must be calibrated with core, production data, or pressure transient data to ensure the values are representative of the permeability of the particular formation. The following equations have been used in the industry over the years to correlate logs with permeability (Ahmed et al. 1991; Berg 1970; Coates and Dumanoir 1974; Schlumberger 1989; Timur 1968; Yao and Holditch 1996).

Kozeny (1927) and Carman (1937):

$$k = \frac{\phi^3}{5A_g^2 (1-\phi)^2}. \quad (2.11)$$

Berg (1970):

$$k = 0.0053\,\phi^3\,D_g^2. \quad (2.12)$$

Timur (1968):

$$k = 0.136\,\frac{\phi^{4.4}}{S_{wir}^2}. \quad (2.13)$$

Coates and Dumanoir (1974):

$$\sqrt{k} = 100\,\phi_e^2\,\frac{(1-S_{wir})}{S_{wir}}. \quad (2.14)$$

Coates and Denoo (1981):

$$S_{wir} = 1 - \left[(1-S_w)^{2.1} / k_{rh}\right]^{0.5}. \quad (2.15)$$

To use Eqs. 2.11 through 2.15, the values of porosity, water saturation, and irreducible water saturation are obtained from logging data. The various authors suggested ways of determining the values of surface area, grain diameter, and relative permeability. Eqs. 2.13 and 2.14 are the most widely used correlations.

Yao and Holditch (1993) presented another method for correlating formation permeability with log data, as shown by

$$k = U\left[\frac{\phi^{e_1}(1-I_{GR})^{e_2} R_{ild}^{e_3}}{(R_{ild}/R_{sfl})^{e_4}}\right], \quad (2.16)$$

where e_1 ranges at 5.87 to 6.89, e_2 ranges at 0.2 to 0.3, e_3 is 1.18 to 2.54, e_4 is 1.08 to 1.65, and U is a correlation factor.

Eq. 2.17 is derived using Eq. 2.16 to correlate log and core data from SFE No. 3 (Gas Research Institute 1991). The correlation is presented in **Fig. 2.10.**

$$k = 6{,}900\,\frac{\phi^{6.23}(1-I_{GR})^{0.2} R_{ild}^{2.13}}{(R_{ild}/R_{sfl})^{1.65}}. \quad (2.17)$$

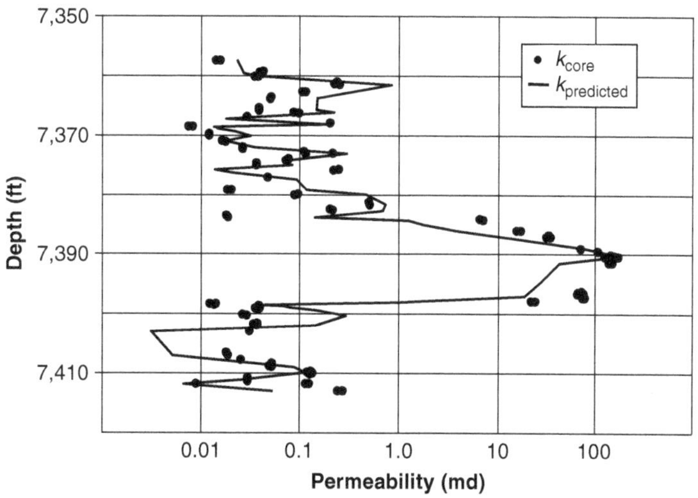

Fig. 2.10—Correlation of log-derived permeability with the core permeability for SFE No. 3.

Notice that when the correlation has been developed, only log data from the GR logs, DILs, and porosity logs are used to estimate permeability.

In summary, obtaining permeability from logging data is very useful because it provides the engineer with estimates of permeability vs. depth. However, to be accurate, the engineer must correlate the logging data with permeabilities measured from core or computed from production or PBU data.

2.7.2 Production Data Analyses. Usually, production data are available for technical calculations. Production data can be measured from a well after it is perforated and before it is fracture treated. Also, production data could be available from other nearby wells producing from similar intervals. Using a computer model based on Darcy's law, one can estimate values of formation permeability. The model can be a simple single-layer, single-phase, single-well analytical model, or it can be a complicated multiwell, multilayered, multiphase finite-difference model (Holditch et al. 1984). For a typical tight gas reservoir, the simple analytical model is usually adequate.

Fetkovich (1980) published type curves that are commonly used to analyze production data, as illustrated in **Fig. 2.11.** One can either manually analyze production data using the type curves, or one can use an analytical model. **Fig. 2.12** illustrates how one set of production data was analyzed to determine estimates of formation permeability, skin factor, and drainage area using an analytical model.

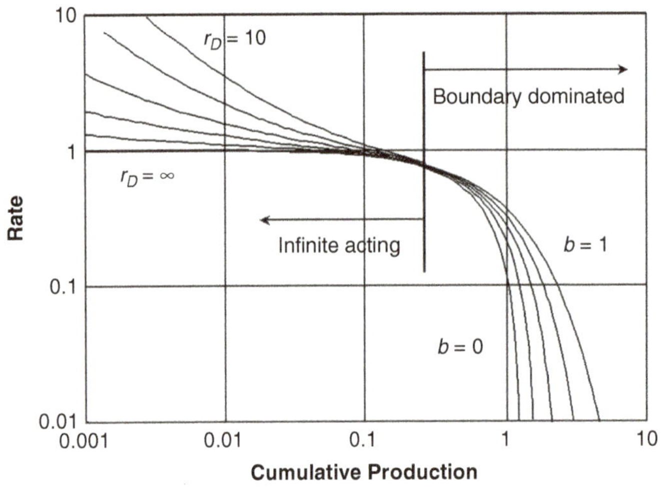

Fig. 2.11—Rate/cumulative-production type curve.

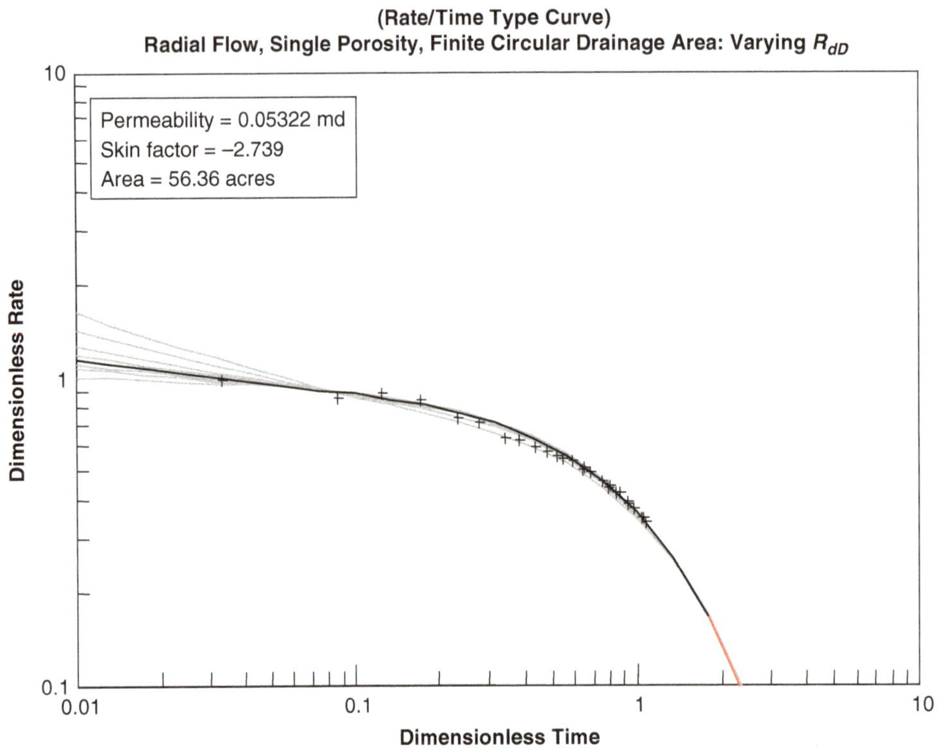

Fig. 2.12—History matching of production data using an analytical model.

Several papers have been written to better explain how to analyze production data using models (Ansah et al. 2000; Fetkovich et al. 1986; Watson et al. 1990).

In tight gas reservoirs, especially when analyzing prefracturing production data, it is often difficult to flow the well to produce at rates high enough to measure. In addition, because the well has to be fractured to be economical, prefracturing flow tests are often not even run, or, if they are run, the flow period is very short. As such, the main goal is to measure flow rates and pressures and to analyze those data to determine an estimate of formation permeability and, perhaps, the skin factor. Seldom do we have enough data to estimate the drainage area, as shown in Fig. 2.12.

2.7.3 PBU Testing. Perhaps the most-accurate method to determine the value of formation permeability is to run a prefracturing PBU test. The literature is voluminous on pressure transient testing (Ansah et al. 2000; Fetkovich et al. 1986; Watson et al. 1990), so that material is not repeated here. Instead, this chapter discusses several issues concerning PBU testing that are important when testing low-permeability gas reservoirs.

A PBU test works well when the formation is well-connected to the wellbore, the flow rate is large enough for accurate measurement, and there are no liquid-loading effects in the wellbore. The well must be produced long enough so that the radius of investigation of the test is meaningful. Eq. 2.18 is used to estimate the radius of investigation of any transient in the reservoir, for radial flow.

$$r_i = \left(\frac{kt}{948\, \phi\, \mu\, c_t} \right)^{\frac{1}{2}} \quad \text{(2.18)}$$

Thus, to determine the length of the production test to sample a reasonable portion of the reservoir, followed by the PBU test, one can select a desired radius and then determine the duration of the test using the "best guess" for the value of permeability. Obviously, the permeability is unknown before running the test. **Table 2.3** illustrates typical test times required on the basis of the desired radius and the best guess at formation permeability.

As seen in Table 2.3, substantial flow times, followed by equal PBU times, are required to sample a large portion of the reservoir in low-permeability gas reservoirs. In most cases, the engineer analyzing the reservoir wants the production and PBU test run as long as possible. On the other hand, because the well is more than likely producing at uneconomic flow rates and a fracturing treatment is required to improve productivity, the operations personnel want to minimize the duration of these tests to minimize costs and get the well producing to sales as soon as possible.

In addition to running the test long enough, the PBU tests in tight gas reservoirs should be analyzed using modern concepts such as pseudopressure, pseudotime, effective pseudotime, producing pseudotime, adjusted pressure, and adjusted time (Lee et al. 1984). Using these concepts helps increase accuracy when large pressure drawdowns exist in the reservoir and changing wellbore-storage constants complicate the analyses of the PBU data.

Desired Radius (ft)	Formation Permeability (md)	Required Test Times (hours)	Area Tested (acres)
100	0.001	142.2	0.7
100	0.1	1.4	0.7
100	1	0.1	0.7
500	0.001	3,555.0	18.0
500	0.1	35.6	18.0
500	1	3.6	18.0
100	0.01	14.2	0.7
500	0.01	355.5	18.0
750	0.01	799.9	40.6
1,000	0.01	1,422.0	72.1

Assuming that

porosity = 0.1

viscosity = 0.015 cp

compressibility = 0.00001 psi^{-1}

Table 2.3—Required test times in tight gas reservoirs.

2.7.4 One Point Testing. In many cases, there are no long-term production data, and operational or cost-related problems prevent one from running a long-term PBU test to quantify the formation permeability. However, it is very important to obtain a rough estimate of formation permeability before designing the fracturing treatment. Sometimes, the well can be perforated and produced for several hours or days before designing and pumping the fracturing treatment. If the production and flowing data are accurately measured, the one point method can be used to estimate the value of formation permeability (Lee et al. 1984).

In this method, the semisteady-state gas-flow equation and the radius-of-investigation equation are solved simultaneously for both permeability and the radius of investigation. The semisteady-state gas-flow equation is

$$\frac{kh}{1{,}422\, q_g\, T}\left(\psi_i - \psi_{wf}\right) = \ln\left(\frac{r_d}{r_w}\right) - 0.75 + s', \quad \text{(2.19)}$$

where

$$r_d = \left(\frac{kt}{376\phi\, \mu\, c_t} \right)^{\frac{1}{2}} \quad \text{(2.20)}$$

and

$$s' = s + D\, q_g. \quad \text{(2.21)}$$

Four steps are used to solve Eqs. 2.19 through 2.21.

1. Assume values of s and D on the basis of the well-completion history, then compute a value for s' with the measured flow rate using Eq. 2.21.

2. Estimate a value for the permeability. An assumption of 1 md for a tight gas reservoir is usually a reasonable guess.
3. Using the values of s' and k, solve Eq. 2.20 for r_d.
4. The value of r_d can be used in Eq. 2.19 to compute a new estimate of permeability.

One can iterate until the values of r_d and k converge. A weakness in this method is that one has to estimate the value of skin factor; therefore, the procedure should be repeated by assuming different values of skin, s. One can generate a range of permeabilities for a range of assumed values of skin factor.

2.8 Statistical Correlations

Tight gas reservoirs generate many difficult problems for geologists, engineers, and managers. Cumulative gas recovery (thus, income) per well is limited because of low gas flow rates and low recovery efficiencies when compared to most high-permeability wells. To make a marginal well into a commercial well, the engineer must increase the recovery efficiency by use of optimal completion techniques and decrease the costs required to drill, complete, stimulate, and operate a tight gas well.

To minimize the costs of drilling and completion, many managers want to reduce the amount of money spent to log wells and totally eliminate money spent on extras such as well testing. However, in these low-permeability layered systems, the engineers and geologists often need more data than are required to analyze high-permeability reservoirs. To balance the need for more data with the need to minimize costs, the logical solution is to spend money gathering accurate data on a few wells, then to use correlations developed from the data to evaluate the wells to be drilled and completed thereafter. When acceptable correlations have been developed for specific reservoirs in specific geographic areas, the correlations can be applied to all wells in the area. By use of these "calibrated" correlations, accurate data sets can be developed for new wells at a minimal cost.

For example, if one spends USD 100,000 to cut cores, analyze the cores, and generate core/log correlations, and these correlations can be used to create and conduct an infill drilling plan for 100 wells, the cost per well to generate accurate data sets is only USD 1,000. Normally, the most-critical data items are formation permeability and in-situ stress. If accurate correlations, in which logs can be used to estimate permeability and in-situ stress, can be developed, the well-completion and stimulation plans can be optimized.

2.8.1 Correlating Core and Log Data. All cores from tight gas reservoirs must be properly handled and tested to obtain the accurate data required for developing useful correlations between log and core data. Information concerning how to cut and test core-plug samples and information concerning how to develop correlations for determining permeability from logs have been previously discussed. One can use Eqs. 2.11 through 2.17 to generate accurate correlations between log data and formation permeability derived from core or well tests. These correlations can then be used to determine values of formation permeability using log data from wells that have not been cored or well tested.

To generate valid correlations in most layered tight gas reservoirs, the core and log data normally must be subdivided by lithology, rock type, or flow units before finalizing the correlations. If one tries to correlate the core and log data for the entire reservoir, the correlation coefficient is usually not very high. For example, consider the data set in **Fig. 2.13,** which contains 1,078 data points from a large Travis Peak data set. The correlation coefficient between core permeability and porosity at NOB pressure is only 0.692. However, if the cores containing visible, natural, or coring-induced fractures are removed and only cores from clean, fluvial-deltaic channel sands are correlated, the correlation coefficient between permeability and porosity increases to 0.865, as illustrated in Fig. 2.6.

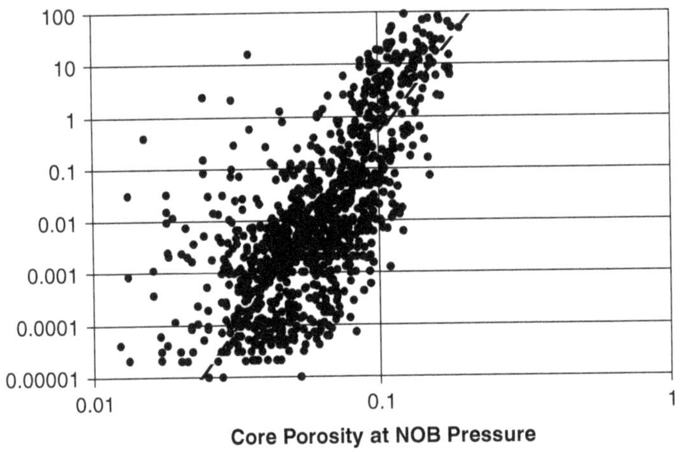

Fig. 2.13—Relation of Klinkenberg-corrected gas permeability measured at NOB pressure to porosity at NOB pressure, all data.

Some have used flow units to segregate core and log data to develop better correlations (Al-Ajmi and Holditch 2000; Amaefule et al. 1993). Amaefule et al. (1993) used the same Travis Peak data set as that illustrated in Fig. 2.13 and analyzed the data using a flow-unit concept (Amaefule et al. 1993; Gas Research Institute 1991). Amaefule et al. (1993) defined a rock quality index and a flow zone indicator (FZI). Using these two parameter groups, they developed a scheme to correlate formation permeability with effective porosity as a function of the FZI.

2.8.2 OGIP Distribution. As suggested by the resource triangle, the distribution of any natural resource is skewed in nature. For natural gas, the distribution is log-normal. As the value of reservoir permeability decreases, the value of OGIP increases exponentially. There is obviously a difference between OGIP and reserves. The OGIP represents all the gas in the rocks that comprise the reservoir layers. Reserves represent the amount of gas that can be produced economically. The value of reserves is a function of gas prices, costs, and the level of technology used to develop the resource.

Often, the amount of OGIP is computed by use of porosity, water saturation, and shale volume cutoffs. In high-permeability reservoirs, using such cutoffs might be appropriate, especially if the reservoir produces water above a certain water-saturation cutoff and the OGIP estimates are not very sensitive to the cutoff values chosen. However, in most tight gas reservoirs, only dry gas and small volumes of water that condense in the wellbore are produced. Very seldom are large volumes of water produced in tight gas reservoirs.

A good rule of thumb for selecting cutoffs to determine net-pay to determine gas-in-place for tight gas reservoirs is to use 3% gas porosity. The first step is to compute the value of porosity, after making clay corrections using Eqs. 2.6 through 2.8. The porosity can then be used to compute the water saturation, normally using the dual-water model. One can compute gas porosity and include all zones with gas-porosity values of 3% or greater in the net-pay count. In the Tight Gas Sands research project sponsored by the GRI, special core analyses on numerous core samples indicated that gas could flow at 3% gas saturation in typical tight gas cores (Gas Research Institute 1991).

2.8.3 Permeability Distribution. Permeability within a gas reservoir, field, or basin is distributed log-normally. To illustrate this concept, four data sets obtained from public records are presented for discussion. The data in **Fig. 2.14** are from the Travis Peak Formation in East Texas, the Cotton Valley Formation in East Texas, the Wilcox Lobo Formation in South Texas, and the Cleveland Formation in Northwest Texas. These reservoirs are in different basins but, remarkably, have very similar log-normal permeability distributions. More information concerning the permeability distribution for these four data sets is presented in **Table 2.4.** The median permeability for all four formations ranges from 0.028 to 0.085 md, while the arithmetic mean values of permeability range from 0.179 to 7.378 md.

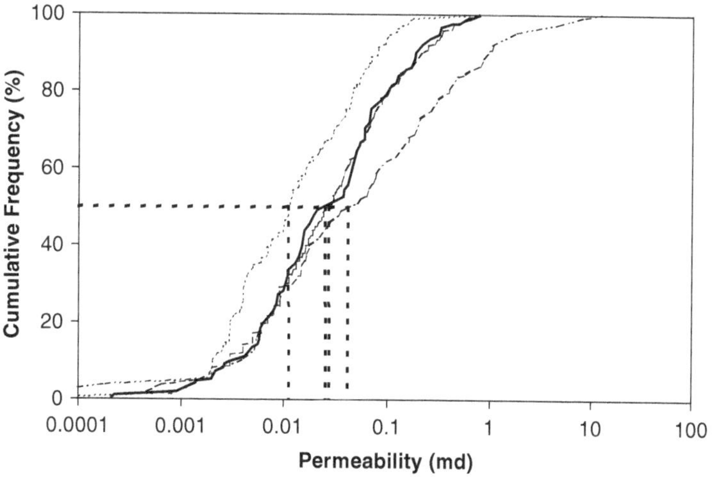

Fig. 2.14—Comparison of permeability distribution from four tight gas formations in Texas.

		Permeabilities			
Formation	Number of Wells	Arithmetic Mean (md)	Median (md)	Minimum (md × 10^5)	Maximum (md)
Cleveland	391	0.179	0.028	3.0	19.6
Cotton Valley	395	7.378	0.045	8.0	358.0
Lobo	112	0.235	0.056	26.0	7.9
Travis Peak	191	1.035	0.085	7.9	27.9

Table 2.4—Arithmetic means and medians for four field cases.

When the permeability is distributed log-normally, the median value rather than the arithmetic mean should be used to determine the average value of permeability for the formation (Holditch and Spivey 1993; Holditch et al. 1991; Rollins et al. 1992). Statistical distributions of porosity, permeability, and net-pay can be used to determine the expected gas recovery from a tight gas reservoir (Holditch and Spivey 1993; Holditch et al. 1991; Rollins et al. 1992). The statistical distribution of permeability, porosity, and net gas pay for the Travis Peak Formation from one study is illustrated in **Figs. 2.15, 2.16, and 2.17,** respectively. In Fig. 2.15, the permeability distribution is clearly log-normal. There is a positive correlation between porosity and permeability, as illustrated in Fig. 2.16. As porosity increases, the permeability increases. There is a negative correlation between net gas pay and permeability, as illustrated in Fig. 2.17. As the value of permeability increases, the net thickness of the layer decreases. This correlation leads to a log-normal distribution of OGIP. The layers of rock that are the most permeable are thin compared to the

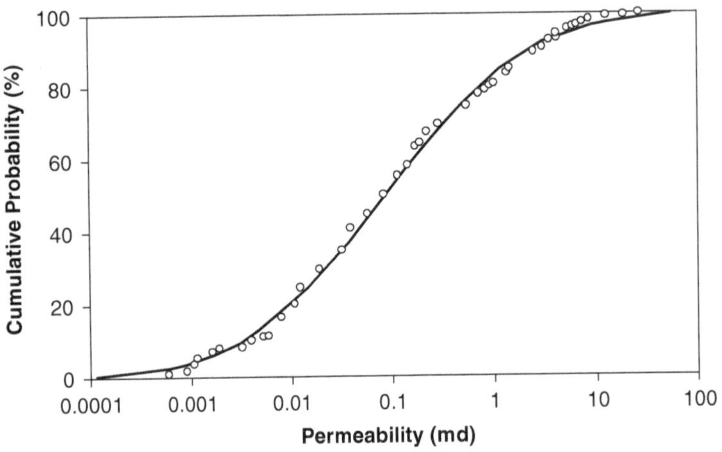

Fig. 2.15—Permeability distribution in the Travis Peak from Bureau of Economic Geology study.

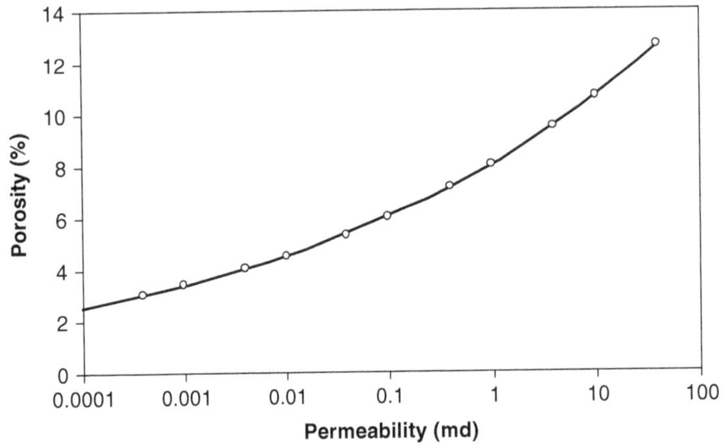

Fig. 2.16—Porosity distribution in the Travis Peak from Bureau of Economic Geology study.

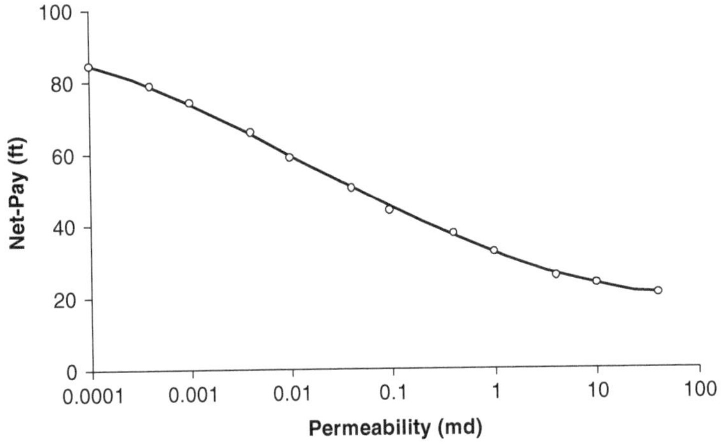

Fig. 2.17—Net-pay distribution in the Travis Peak from Bureau of Economic Geology study.

layers of rock that have low permeability. If one uses the data in Figs. 2.15 through 2.17 in a reservoir simulator, along with other appropriate data for the Travis Peak Formation **(Tables 2.5 and 2.6)**, one can determine the gas recovery for the average well in the Travis Peak Formation (Holditch and Spivey 1993; Holditch et al. 1991; Rollins et al. 1992). The results from the calculations are shown in **Table 2.7.**

First, one must recognize that the gas recovery from a well completed in a tight gas reservoir is a function of the average drainage area and the hydraulic fracture length, as well as of the values of permeability, porosity, net gas pay, reservoir pressure, and other

Average depth	8,000 ft
Reservoir pressure	4,000 psi
Reservoir temperature	225°F
Initial gas flow rate	5,000 Mcf/D
Minimum bottomhole flowing pressure	1,000 psi
Maximum well life	40 years
Economic limit	10 Mcf/D

Table 2.5—Travis Peak Reservoir fixed data.

Gas Permeability (md)	Number of Wells	Gas Porosity (%)	Net-Pay (ft)
40	11	12.65	20
10	25	10.65	23
4	58	9.50	27
1	107	7.99	32
0.4	159	7.13	37
0.1	188	6.00	44
0.04	178	5.35	50
0.01	134	4.50	59
0.004	80	4.02	66
0.001	39	3.38	74
0.0004	15	3.02	79
0.0001	5	2.54	84
0.00004	1	2.25	90

Table 2.6—Travis Peak variable data for the simulation.

Well Spacing (acres)	Fracture Length (ft)	Actual Average Recovery (Bcf/well)	Recovery for Median *k* (Bcf/well)	Recovery for Arithmetic Mean *k* (Bcf/well)
160	0	1.97	2.37	2.92
640	0	5.16	4.32	11.02
160	500	2.63	2.97	2.97
640	500	7.95	9.33	11.70

Table 2.7—Computation of average gas recovery.

reservoir parameters. In Table 2.7, the column labeled "Actual Average Recovery" was computed for each case using 52 simulation runs, representing 1,000 wells drilled for each well spacing, and fracture length using the data in Tables 2.5 and 2.6. The results show the average recovery varies from 1.97 Bcf for an unfractured well that is draining 160 acres to 7.95 Bcf for a well containing a 500-ft-long fracture half-length draining 640 acres. The column labeled "Recovery for Median *k*" illustrates the values of gas recovery that one computes using the median values of all parameters, including permeability, porosity, and net gas pay. The column labeled "Recovery for Arithmetic Mean *k*" illustrates the gas recovery one computes using the arithmetic mean values for permeability, porosity, and net gas pay. The data clearly show that the median values best represent the "average" values when the parameters are log-normally distributed.

2.8.4 Reserves Distribution. Even though the permeability distribution and the OGIP distribution are log-normally distributed, the distribution of reserves might or might not be log-normally distributed because of the changing recovery efficiency vs. permeability and the number of wells drilled in each permeability range. Reserves represent the volume of gas that can be produced economically using existing technology. Reserves are a function of the permeability, net gas pay, porosity, drainage area, initial reservoir pressure, flowing bottomhole pressure, gas prices, operating costs, effective fracture half-length, effective fracture conductivity, and other economic factors such as taxation rates and overhead charges. The data in **Figs. 2.18, 2.19, and 2.20** illustrate how the abandonment pressure and recovery efficiency vary as functions of permeability, net gas pay, and fracture half-length for a specific

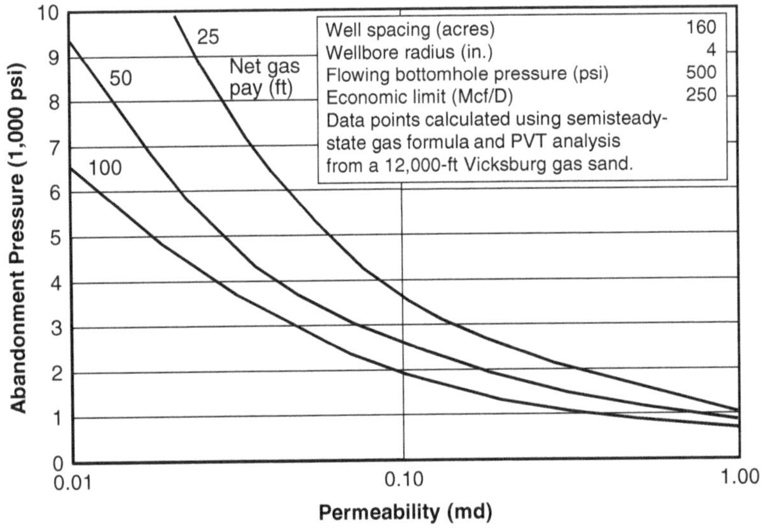

Fig. 2.18—Illustration of abandonment pressure for Vicksburg. PVT is pressure/volume/temperature.

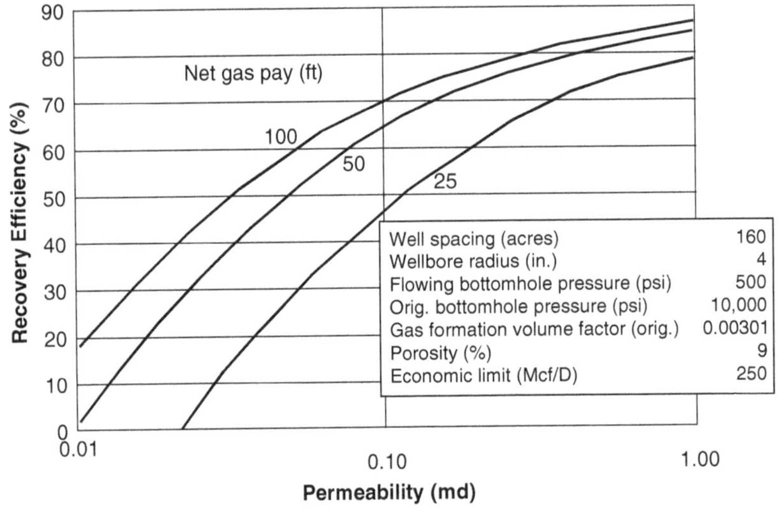

Fig. 2.19—Recovery efficiency for Vicksburg.

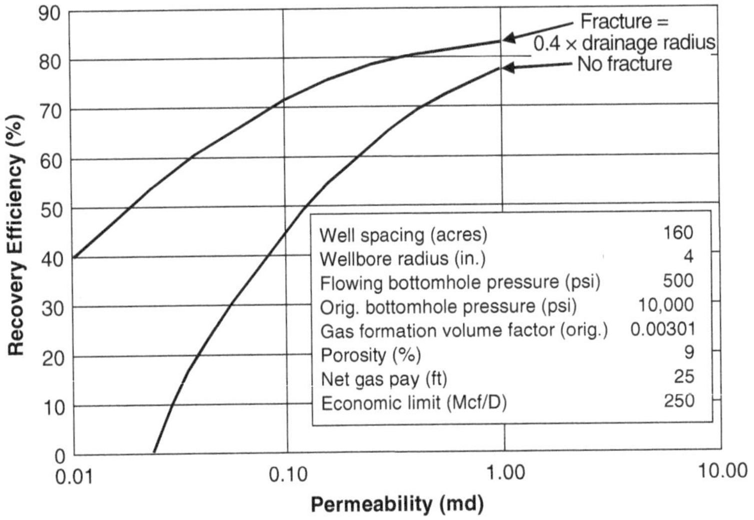

Fig. 2.20—Influence of hydraulic fracture on recovery efficiency for Vicksburg.

set of Vicksburg data (Holditch 1974). At the time the graphs were generated, an economic limit of 250 Mcf/D was being used in the Vicksburg because of low gas prices. If these same cases were computed with a lower value of economic limit, the abandonment pressure would decrease, and the recovery efficiency would increase. These examples illustrate how one should use reservoir engineering to evaluate the effects of drainage area, hydraulic fracture properties, and economic parameters to determine values of recovery efficiency and, thus, the distribution of reserves.

Fig. 2.18 shows that as the permeability and the net gas pay increase, the abandonment pressure in the reservoir, when the economic limit is reached, decreases. The data in Fig. 2.19 illustrate the recovery efficiencies for the same cases as shown in Fig. 2.18. For thick, high-permeability reservoirs, the recovery efficiency can be 80% or more of the OGIP. However, as the value of permeability decreases to less than 0.1 md, the recovery efficiency decreases substantially. For the case in which the net gas pay was only 25 ft and the permeability was between 0.02 and 0.1 md, the recovery efficiency varied from 0 to 45% of the OGIP. The data in Figs. 2.18 and 2.19 are for semisteady-state radial flow. Fig. 2.20 illustrates the effect of a hydraulic fracture on the recovery efficiency for the 25 ft of net-gas-pay case. It is clear that a hydraulic fracture that extends out to 40% of the drainage area substantially increases the recovery of gas in a tight gas reservoir.

2.8.5 In-Situ Stress Correlations. It is important to generate correlations between logs, cores, and measured values of in-situ stress. The values of in-situ stress are very important to the engineer planning the well completion and stimulation treatment. The engineer can usually correlate values of in-situ stress measured from pump-in tests with data measured using logs and cores. A common equation used to correlate lithology (using Poisson's ratio) with the in-situ stress is given in Eq. 2.22.

$$\sigma_{\min} \approx \frac{\nu}{1-\nu}\left(\sigma_1 - \alpha\sigma_p\right) + \alpha\sigma_p + \sigma_{\text{ext}}, \quad (2.22)$$

where $\sigma_{\min}$ is the minimum horizontal stress (in-situ stress), ν is Poisson's ratio, σ_1 is overburden stress, α is Biot's constant, σ_p is reservoir fluid pressure (pore pressure), and σ_{ext} is tectonic stress.

To use Eq. 2.22, one must determine the values of Poisson's ratio using log data. Poisson's ratio can be correlated with SON log data or estimated using the lithology of a formation layer. **Table 2.8** illustrates typical ranges for Poisson's ratio as a function of the lithology. Thus, it is possible to estimate values of Poisson's ratio from correlations with log data, then use those estimates to compute estimates of in-situ stress.

Rock Type	Range of Poisson's Ratio
Gas-bearing sandstone	0.1–0.25
Liquid-filled sandstone	0.25–0.3
Wet siltstone	0.2–0.3
Limestone	0.30–0.35
Shale	0.28–0.43

Table 2.8—Typical ranges of Poisson's ratio for various types of rocks.

Another correlation that usually works for tight gas sands is one between the GR log and values of in-situ stress. Gongora (1995) used data from the Travis Peak Formation collected during the GRI Tight Gas Sands research program. Data from two wells, SFE No. 1 and SFE No. 2, are shown in **Figs. 2.21 and 2.22,** respectively. SFE No. 1 was an infill well drilled into the Travis Peak Formation. There were several zones that were partially depleted and many other zones that were at original pressure. Thus, in Fig. 2.21, the zones were correlated using both the GR-log reading and the reservoir pressure. SFE No. 2 was drilled in a location where little drainage had occurred. As such, the correlation between in-situ stress and the GR log was accomplished using a single correlation, as illustrated in Fig. 2.22.

The correlations included in this chapter were generated using log, core, and well-test data for the Travis Peak Formation; hence, one cannot use these correlations for other formations in other basins around the world. These correlations are included in this chapter to illustrate how values of permeability and in-situ stress can be correlated with log and core data. The methods explained in this chapter can be used to generate other correlations in other formations in other basins.

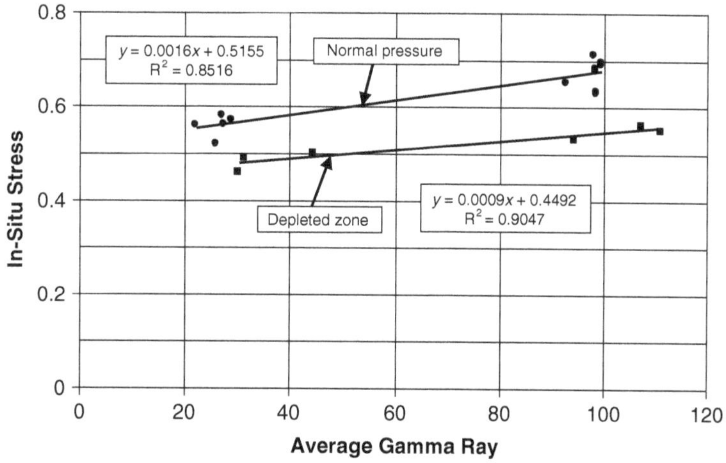

Fig. 2.21—Correlation of in-situ stress value with average GR reading for two different reservoir pressures for SFE No. 1.

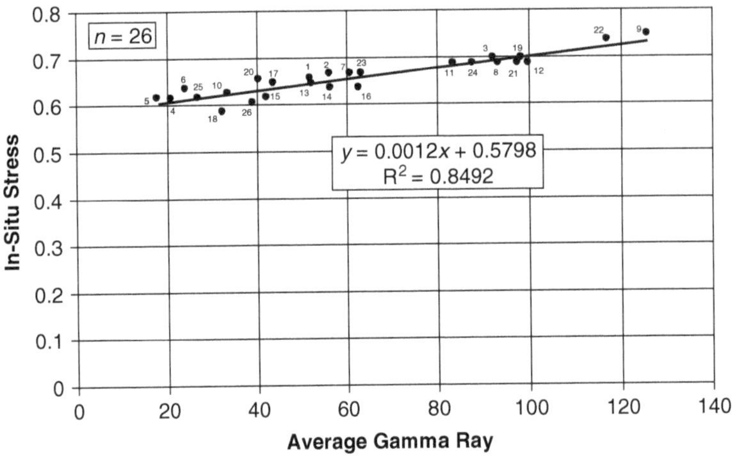

Fig. 2.22—Correlation of in-situ stress value with average GR reading for SFE No. 2. *n* = number of samples.

When specific correlations have been developed and verified, they can be used to evaluate layered tight gas reservoirs to make basic decisions, such as if casing should be set. After the casing is set, the correlations can be used to generate the data required to design the completion and the stimulation treatment for the reservoir layers that are determined to be commercially viable.

2.9 Developing Databases

To evaluate a layered tight gas reservoir and design the well completion, the operator must use both a reservoir model and a hydraulic fracture propagation model. The data required to run both models are similar and can be divided into two groups (Holditch and Rahim 1994). One group consists of data that can be "controlled." The second group reflects data that must be measured or estimated but cannot be controlled.

The primary data that can be controlled by the engineer are the well-completion details and the fracturing-treatment details, such as fluid volume and injection rate. The data that must be measured or estimated by the design engineer are formation depth, formation permeability, in-situ stresses in the pay zone, in-situ stresses in the surrounding layers, formation modulus, reservoir pressure, formation porosity, formation compressibility, and the thickness of the reservoir. There are actually three thicknesses that are important to the design engineer: the gross thickness of the reservoir, the net thickness of the gas-producing interval, and the permeable thickness that accepts fluid loss during the hydraulic fracturing treatment.

2.9.1 Data for Reservoir Simulation Models. The data required to run a reservoir model depend on the type of model one chooses to use. The engineer can use (listed in order of simplest to most complex) semisteady-state flow equations, material-balance methods, single-layer analytical solutions, multilayered analytical solutions, or numerical reservoir simulation models. As one might expect, the amount and complexity of the data required to use these models increase as the complexity of the model increases.

Interestingly, most of the data required to run a reservoir simulation model are also required to run a 3D hydraulic fracture propagation model. **Table 2.9** lists the data required to run both the reservoir model and the fracturing treatment design model. Because a typical tight gas reservoir is a layered formation, it is necessary to determine values of reservoir properties—such as permeability, net-pay, porosity, and water saturation—on a layer-by-layer basis. Many problems can be solved using a single-layer model; however, in other cases, better completions are achieved by developing multilayer models of the reservoir.

2.9.2 Hydraulic Fracture Propagation Models. The fracture propagation model requires information on the rock mechanical properties, such as in-situ stress, modulus, and Poisson's ratio. We also need data on the fracturing-fluid properties and the propping-agent properties.

The most-critical data for the design of a fracturing treatment are (roughly in order of importance) the in-situ stress profile, formation permeability, fluid-loss characteristics, total fluid volume pumped, propping-agent type and amount, pad volume size, fracturing-fluid viscosity, injection rate, and formation modulus. The design engineer should focus his or her time on the most-important parameters. In hydraulic fracturing treatment design, by far the two most-important parameters are the in-situ stress profile and the permeability profile of the zone to be stimulated, in addition to the layers of rock above and below the target zone that affect fracture height growth.

In new fields or reservoirs, most operating companies are normally willing to spend money to run logs, cut cores, and run well tests to determine important factors such as the in-situ stress and the permeability of the reservoir layers. With such data, along with fracturing treatment records and production records, accurate data sets for a given reservoir in a given field can normally be compiled. These data sets can be used on subsequent wells to optimize the fracturing treatment designs. It is normally not practical to cut cores and run well tests on every well. Thus, the data obtained from cores and well tests from a few wells must be correlated to log parameters so the logs on subsequent wells can be used to compile accurate data.

2.9.3 Vertical Profiles. To use either a multilayered reservoir model or a pseudo-3D (P3D) hydraulic fracture propagation model, the data must be entered by reservoir layer. **Fig. 2.23** illustrates the profiles of important input data required by either the reservoir model or the P3D model. For the situation in Fig. 2.23, the well is completed and the fracturing treatment is started in the sandstone reservoir.

	Units	Model	Sources
Formation permeability	md	Both	Cores, well tests, production data, correlations
Formation porosity	%	Both	Cores, logs
Reservoir pressure	psi	Both	Well tests, well files, regional data
Formation compressibility	psi	Both	Cores, logs, correlations
Formation depth	ft	Both	Logs, drilling records
Formation temperature	°F	Both	Logs, well tests, correlations
Water saturation	%	Both	Logs, cores
Net-pay thickness	ft	Both	Logs, cores
Gross pay thickness	ft	Both	Logs, cores, drilling records
Formation lithology		Both	Cores, drilling records, logs, geologic records
Wellbore completion		Both	Well files, completion prognosis
Reservoir fluid properties		Reservoir	Fluid samples, correlations
Relative permeabilities		Reservoir	Core samples, correlations
Formation modulus	psi	Fracture	Cores, logs, correlations
Poisson's ratio		Fracture	Cores, logs, correlations
In-situ stress	psi	Fracture	Well tests, logs, correlations
Fracture toughness	psi$\sqrt{\text{in.}}$	Fracture	Cores, correlations
Fracturing fluids		Fracture	Service company information
Fracturing proppants		Fracture	Service company information

Table 2.9—Sources of data for both the reservoir model and the fracture model.

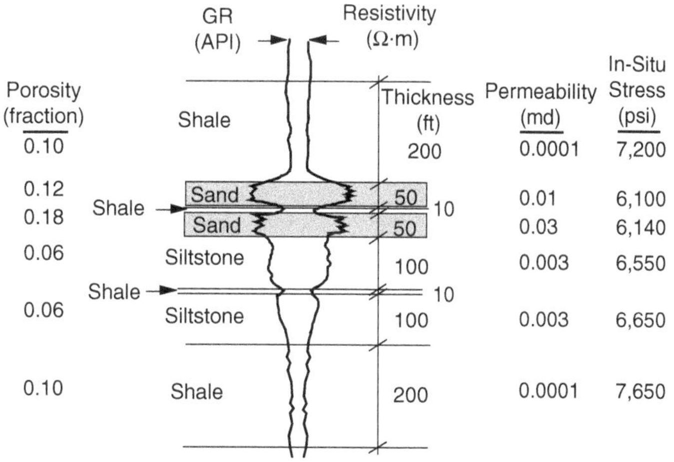

Fig. 2.23—Sources of data for reservoir and fracture models.

The fracture typically grows up and down until a barrier is reached to prevent vertical fracture growth. In many cases, thick marine shales, which tend to have in-situ stresses that are higher than the sandstones, are barriers to vertical fracture growth. In other cases, coal seams prevent fractures from growing vertically. Many coal seams are highly cleated, and when the fracturing fluid enters the coal seam, vertical fracture growth remains contained.

The data used to design a fracturing treatment can be obtained from a number of sources, such as drilling records, completion records, well files, openhole logs, cores and core analyses, well tests, production data, geologic records, and other public records, such as publications. In addition, service companies provide data on their fluids, additives, and propping agents.

2.9.4 Economic Models. To design the optimal well completion and fracturing treatment design in a tight gas reservoir, detailed economic calculations must be conducted. The first decision is usually to determine if there is enough net gas pay, porosity, and permeability to justify setting casing after the well reaches total depth. After casing is set, engineering and economic calculations are required to determine the optimal completion method and the optimal fracturing treatment design. The data required to run an economic model are given in **Table 2.10.** Essentially, one must determine the net cash flow for a variety of completion scenarios. The net cash flow can then be evaluated using a number of ways to determine the optimal completion design. The data required to run an economic model are specific to each situation. Gas prices, operating costs, royalty payments, taxes, and many other factors can vary widely among leases

	Units	Sources
Gas flow rate vs. time for various completion scenarios	Mcf vs. years	Reservoir model is used to generate flow rates
Cost of the fracturing treatment vs. propped fracture length with various fracture conductivities	USD vs. ft for various md-ft	Fracture propagation models and service company cost estimates
Completion costs for various completion scenarios	USD	Generated by the completion engineer
Gas price vs. time	USD vs. years	Normally dictated by corporate policy
Operating costs	USD	Computed for specific field location
Royalty	% of revenue	Value is determined by lease or contract with mineral owner
Taxes	% of revenue	Value is determined by laws and/or contracts with governments
Discount rates	%	Normally dictated by corporate policy
Evaluation parameters, such as NPV, PO, ROI, DCF	USD, years, or %	Normally dictated by corporate policy

Table 2.10—Sources of data for an economic model. NPV = net present value, PO = payoff time, ROI = return on investment, and DCF = discounted cash flow.

in the same field and, especially, in different geologic basins in different countries or even continents. Hence, it is extremely important to gather the appropriate economic data and perform detailed economic calculations to design the optimal well completion.

2.10 Well Construction

Well construction is a term used to incorporate the activities required to drill, complete, and stimulate a well as it goes from spud to production. Well construction is a very broad topic, and a discussion of every aspect in detail is outside the scope of this chapter. Instead, we will concentrate on the aspects of well construction that are unique to tight gas reservoirs. Many of the items discussed next are included in Ch 10: Drilling and Completion Design for Fracturing in Part 2 of this book.

2.10.1 Drilling and Completion. A tight gas reservoir is one that does not produce at commercial gas flow rates or recover commercial volumes of natural gas unless a hydraulic fracturing treatment is properly designed and pumped. As such, the entire drilling and completion procedures should focus on making sure the optimal fracturing treatment can be designed and pumped in the field.

When drilling a tight gas well, the most-important aspect of the drilling operation is to drill a gauge hole. Many times this means the well should be drilled at a balanced, or slightly overbalanced, mud weight. In other cases, air drilling, or underbalanced drilling, works best, as long as the hole remains in gauge. If a gauge hole is drilled, we can run openhole logs and obtain valid data, which are required to properly evaluate the formation and to design the completion. If the hole is washed out and rugose, the logs are difficult or impossible to accurately evaluate, and the net gas pay is difficult to identify. Also, if the borehole is in gauge, the chances of obtaining a satisfactory primary cement job on the production casing increase when compared to trying to cement casing in a washed-out borehole. Obtaining a good primary cement job is extremely important when completing a well in a multilayered reservoir that must be fracture treated.

Some drilling personnel want to drill underbalanced in tight gas reservoirs because the penetration rate is faster, formation invasion of mud filtrate is minimized, and there is little chance of a gas kick because of the low-permeability nature of the formations. However, underbalanced drilling is acceptable only if a gauge hole can be maintained. Speed to reach total depth is not important if the borehole is washed out and the engineer cannot properly evaluate the reservoir layers or obtain an adequate primary cement job. Also, formation damage is not an important consideration in tight gas reservoirs. It does not matter if the near-wellbore formation has been damaged during drilling. In every case, we still back up multiple pump trucks and pump rather large fracturing treatments. The hydraulic fracture breaks through any near-wellbore damage.

Investment required
Number of reservoir layers encountered by the borehole
Desired producing rates
Reserves in the various reservoir layers
Rock and fluid properties in the reservoir layers
Rock properties in the potential barriers to vertical fracture growth
Stimulation requirements
Future workover requirements
Artificial-lift requirements
Economic constraints

Table 2.11—Data required to determine completion strategy.

2.10.1.1 Completion Strategy. To complete a tight gas well successfully, the engineer should consider the items included in **Table 2.11.** The ideal completion is the one that produces the most gas for the lowest cost—considering both the initial completion cost and the subsequent operating costs. This definition implies that a prudent engineer will attempt to provide a functional completion for many years to come at the lowest possible cost to the operator.

Of concern in the design of the completion is always the number of producing zones that are separated in the reservoir by vertical flow barrier layers. To determine whether different producing intervals should actually be treated as a single reservoir, one must first determine if these various intervals are all connected by a single hydraulic fracture. If a particular zone is separated from another pay zone by a thin silt or shale layer with little in-situ stress contrast among the layers, one can use a model to determine if all the zones can be connected by a single hydraulic fracture. If a single fracturing treatment is used to stimulate multiple layers, and no reservoir damage occurs by commingling the different zones, the well should be completed as if all the layers are actually a single reservoir. Normally, in dry gas reservoirs, no reservoir damage occurs by commingling zones. In fact, it is likely that more gas will be recovered by producing all the layers commingled because the abandonment pressure is lower at any given economic limit when the zones are commingled vs. producing the zones one at a time.

If two or more productive intervals are separated by a thick, clean shale (of, say, 50 ft or more) and this shale has enough in-situ stress contrast to be a barrier to vertical fracture growth, the design engineer might need to design the completion and stimulation treatments to consider the fact that multiple hydraulic fractures will be created. In such cases, fracturing treatment diverting techniques must be used to properly stimulate all producing intervals. More information concerning completion design in multilayered reservoirs is available in the technical literature (Gidley et al. 1989; Rahim and Holditch 1995).

2.10.1.2 Tubular Concerns. The two main concerns with tubular design are pumping the optimal fracturing treatment and liquid loading as the gas flow rate declines. These two concerns must be balanced to achieve the optimal well completion. As previously stated, a tight gas well is uneconomic to drill, complete, and produce unless a successful fracturing treatment is designed and pumped. In general, fracturing treatments are more successful when pumped at higher injection rates. When pumping a treatment at a high injection rate, we normally like to use large tubulars.

After the treatment is pumped and the well is put on production, the gas flow rate begins to decline. All wells, even dry gas wells, produce liquids in the form of condensate or water. Regardless of how little liquid is produced, the well eventually loads up with liquids as the flow rate declines. Liquid loading is a function of gas velocity. Therefore, to minimize liquid-loading problems, we must use small tubing.

Thus, the challenge is that we need large tubulars to pump the fracturing treatment and small tubulars to minimize liquid loading. The solutions to this challenge can be as varied as the number of fields in which we work. Many considerations and the computational techniques needed to solve these problems are presented in Gidley et al. (1989). In some cases, when the reservoir pressure is at or less than normal pressure, we can fracture treat the formation down casing, then run small tubing after the treatment to produce the well. If the reservoir is geopressured, we might have to fracture treat the well down tubing at injection rates that are less than optimal.

The topic of how to design casing and tubing and how to design the optimal tubular configuration in a tight gas well is too large to deal with completely in this chapter. The completion engineer should, however, try to design the fracturing treatment and the completion before spudding the well. If, during the design, the engineer determines a certain size casing or a certain size tubing is required to implement an optimal design, the completion engineer should provide that feedback to the drilling engineer. The drilling engineer can then design the bit program and casing program to accommodate the needs of the completion engineer. Once the hole is drilled and the production casing is set and cemented, it is too late to redesign the completion if you discover you needed larger casing to implement the optimal completion.

In the same manner, the fracturing treatment should be designed before spudding the well so a reasonable estimate of fracturing treatment pressures, from the bottom of the hole to the surface, can be made as a function of the casing size, the injection rate, and the fracturing-fluid friction and density properties. It is very important to know the maximum injection pressure during the fracturing treatment for a variety of completion scenarios. The drilling engineer can use that information to select the correct size, weight, and grade of casing. A fracturing treatment is usually not successful if the injection rate or fluid viscosity is compromised when the casing cannot withstand the desired injection pressure. Again, working the problem before spudding and designing the casing correctly can prevent problems and allow the service company to actually pump the optimal fracturing treatment.

2.10.1.3 Perforating Concerns. Perhaps the least-understood part of well completions and hydraulic fracturing revolves around how to perforate a well. Again, there is no simple solution, and the best perforating scheme varies depending on the specific reservoir situation. Two factors seem to be very important. First, the number of layers and the number of fracturing treatment stages affect how we perforate the well. Second, the in-situ stress anisotropy and the presence or lack of natural fractures have a strong bearing on how we perforate the well.

A problem associated with hydraulic fracturing treatments has been identified in the petroleum literature as "near-wellbore tortuosity" (Cleary et al. 1993). This occurs when multiple hydraulic fractures are created near the wellbore. These multiple hydraulic fractures are usually caused by the presence of natural fractures, or the fact that too many perforations are shot in multiple directions over a long perforated interval. When multiple fractures occur near the wellbore, each fracture is narrower than a single fracture, and problems occur when trying to pump the propping agent down the narrow fractures. In many cases, a near-wellbore screenout occurs when near-wellbore tortuosity problems occur.

There are several ways to minimize near-wellbore tortuosity problems. The best solution might be to minimize the length of the perforated interval and to orient the perforations 180° in the direction in which the fracture propagates (which is perpendicular to the minimum principal horizontal stress for a vertical fracture). More information concerning stresses and stress orientations is found in Gidley et al. (1989).

Again, the main concern when perforating a tight gas well is to perforate in such a way that the optimal fracturing treatment(s) can be successfully pumped. The completion engineer must be concerned with choosing the correct zones and perforating those zones to accommodate any diversion techniques that will be used in multistage fracturing treatments.

In the perforating literature, there are many papers discussing how many holes are needed per foot of casing so that the productivity index is not reduced because of too few holes. In a tight gas well that is fracture treated, the number of holes per foot of casing is really not much of a consideration. More importantly, the number of holes with respect to the fracturing treatment injection rate should control the perforation operation. A good rule of thumb is that the number of holes should be such that the injection rate per hole is between 0.25 and 0.5 bbl/min per perforation. For example, if you plan to pump the fracturing treatment at 20 bbl/min, you should consider putting between 40 and 80 holes in the pipe in the zone where you want the fracture to begin. In general, the more compact the perforated interval the better, and perforations oriented 180° in the direction of maximum horizontal stress provide the best situation for hydraulic fracturing. The worst situation is to shoot 4 or 6 shots/ft over a long interval. When too many holes are shot over too long an interval, the engineer loses control of where the fracture begins, and the chances of creating multiple fractures at the wellbore increase substantially.

2.10.2 Hydraulic Fracturing Treatment Considerations. As stated many times in this chapter, a tight gas reservoir is one that must be successfully fracture treated to produce economic volumes of gas at economic flow rates. In this section, we will discuss a few basic considerations for fracturing treatment design and application. More information can be found in Part 3 of this book.

2.10.2.1 Candidate Selection. The success or failure of a hydraulic fracturing treatment often depends on the quality of the candidate well selected for the treatment. Choosing an excellent candidate for stimulation often ensures success, while choosing a poor candidate normally results in economic failure. To select the best candidate for stimulation, the design engineer must consider many variables. The most-critical parameters for hydraulic fracturing are formation permeability, the in-situ stress distribution, reservoir fluid viscosity, skin factor, reservoir pressure, reservoir depth, and the condition of the wellbore. The skin factor refers to whether the reservoir is already stimulated or, perhaps, damaged. If the skin factor is positive, the reservoir is damaged and will likely be an excellent candidate for stimulation.

The best candidate wells for hydraulic fracturing treatments in a tight gas reservoir have a substantial volume of OGIP and good barriers to vertical fracture growth above and below the net-pay intervals. Such reservoirs have a thick pay zone, medium to high pressure, in-situ stress barriers to minimize vertical height growth, and substantial areal extent.

Tight gas reservoirs that are not good candidates for hydraulic fracturing are those with small volumes of gas-in-place because of thin reservoirs, low reservoir pressure, or small areal extent. Also, reservoirs that do not have enough clean shale above or below the pay interval to suppress vertical fracture growth are considered to be poor candidates. Reservoirs with extremely low permeability might not produce enough hydrocarbons to pay all the drilling and completion costs, even if successfully stimulated; thus, such reservoirs would not be good candidates for stimulation.

2.10.2.2 Fracturing Treatment Optimization. The goal of every design engineer is to design the optimal fracturing treatment for each well. Holditch et al. (1978) discuss the optimization of both the propped fracture length and the drainage area (well spacing) for low-permeability gas reservoirs. **Fig. 2.24** illustrates the method used to optimize the size of a fracturing treatment (Britt 1985; Veatch 1983). Fig. 2.24 clearly shows the following:

- As the propped length of a fracture increases, the cumulative production increases, and the revenue from hydrocarbon sales increases.
- As the fracture length increases, the incremental benefit (USD revenue per foot of additional propped fracture length) decreases.
- As the treatment volume increases, the propped fracture length increases.
- As the fracture length increases, the incremental cost of each foot of fracture (USD cost per foot of additional propped fracture length) increases.
- When the incremental cost of the treatment is compared to the incremental benefit of increasing the treatment volume, an optimal propped fracture length can be found for every situation.

Additional economic calculations can be made to determine the optimal fracturing treatment design. However, in all cases, the design engineer must consider the effect of the fracture upon flow rates and recovery, the cost of the treatment, and the investment guidelines of the company that owns and operates the well.

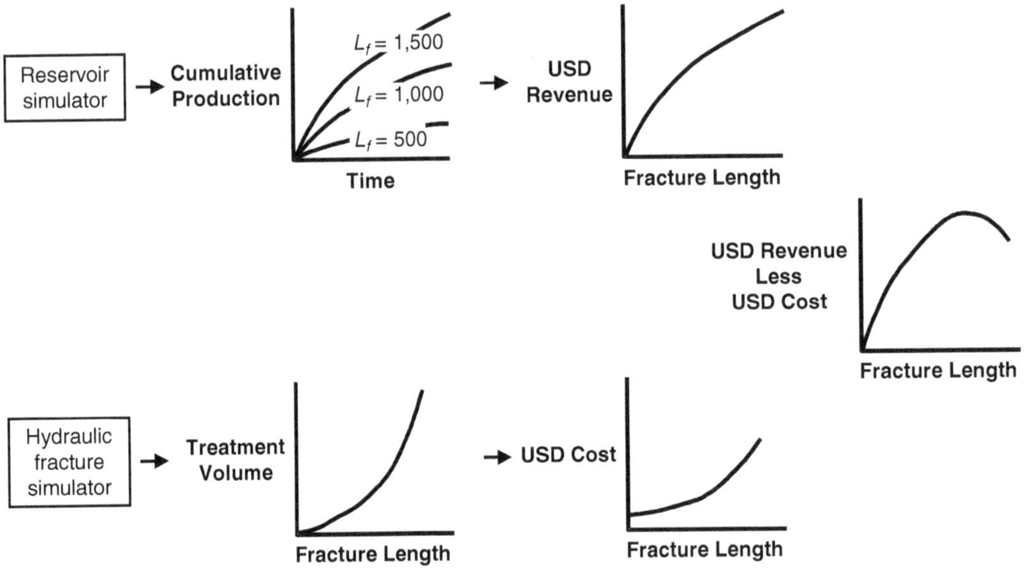

Fig. 2.24—Process that leads to an optimized fracturing treatment.

2.10.2.3 Fracturing Treatment Design Considerations. The most-important data for designing a fracturing treatment are the in-situ stress profile, formation permeability, fluid-loss characteristics, total fluid volume pumped, propping-agent type and amount, pad volume, fracturing-fluid viscosity, injection rate, and formation modulus. It is very important to quantify the in-situ stress profile and the permeability profile of the zone to be stimulated, in addition to the layers of rock above and below the target zone that influence fracture height growth.

There is a structured method that should be followed by the engineer to design, optimize, execute, evaluate, and reoptimize the fracturing treatments in any reservoir (Gidley et al. 1989). The first step is always the construction of a complete and accurate data set. Table 2.9 lists the sources for the data required to run fracture propagation and reservoir models. Notice that the design engineer must be capable of analyzing logs, cores, production data, and well-test data, and reviewing well files, to obtain all the information needed to design and evaluate the well to be fracture treated.

Design Procedures. To design the optimal treatment, the engineer must determine the effect of fracture length and fracture conductivity upon the productivity and the ultimate recovery from the well. As in all engineering problems, sensitivity runs must be made to evaluate uncertainties, such as estimates of formation permeability and drainage area. The production data obtained from the reservoir model should be used in an economic model to determine the optimal fracture length and conductivity. Then, a fracturing treatment must be designed using a fracture propagation model to achieve the desired length and conductivity at minimum cost. The most-important concept is to design a fracture using all data and appropriate models that results in the optimal economic benefit to the operator of the well, as shown in Fig. 2.24. A hydraulic fracture propagation model should be run to determine what needs to be mixed and pumped into the well to achieve the optimal values of propped fracture length and fracture conductivity. The base data set should be used to make a base-case run. Then, the engineer determines which variables are the most uncertain. Many times, the values of in-situ stress, Young's modulus, permeability, and the fluid-loss coefficient, for example, are not known with certainty and have to be estimated. The design engineer should acknowledge these uncertainties and make sensitivity runs with the fracture propagation model to determine the effect of these uncertainties on the design process. As databases are developed, the number and magnitude of the uncertainties will diminish.

In practice, the design engineer should simulate the fracturing treatment of the well many times on his or her computer. Making these sensitivity runs leads to a better design and also educates the design engineer on how certain variables affect both the created- and the propped-fracture dimensions.

2.10.2.4 Fracturing-Fluid Selection. A critical decision by the design engineer is the selection of the fracturing fluid for the treatment. Economides and Nolte (2000) developed a flow chart that can be used to select the category of fracturing fluid required to stimulate a gas well on the basis of factors such as reservoir temperature, reservoir pressure, the expected value of fracture half-length, and a determination of whether the reservoir is water sensitive. That chart is included in Ch 14: Fracture Initiation, Geometry, and Propagation in Part 3.

2.10.2.5 Propping-Agent Selection. Economides and Nolte (2000) also produced a flow chart for selecting propping agents, which is included in Ch 17: Types of Fracture Propping Agents in Part 3. The selection of the propping agent is based on the maximum effective stress that is applied to the propping agent during the life of the well. The maximum effective stress depends on the minimum value of flowing bottomhole pressure that one expects during the life of the well. If the maximum effective stress is less than 6,000 psi, sand is usually recommended as the propping agent. If the maximum effective stress is between 6,000 and 12,000 psi, one should use either resin-coated sand or intermediate-strength proppant, depending on the temperature of the reservoir. For cases in which the maximum effective stress is greater than 12,000 psi, high-strength bauxite should be used as the propping agent.

Of course, any rule of thumb should only be used as a guide because there will be exceptions. For example, even if the maximum effective stress is less than 6,000 psi, the design engineer could choose to use resin-coated sand or other additives to "lock" the proppant in place when proppant flowback becomes an issue. Also, in high-flow-rate gas wells, intermediate-strength proppants might be needed as a result of inertial flow. For fracturing treatments in countries that do not mine sand for fracturing, the largest cost for the proppant is often the shipping charges. Thus, if one has to import the propping agent, one might choose to use intermediate-strength proppants, even for relatively shallow wells, because the cost differential between the intermediate-strength proppants and sand is not a significant factor.

When the optimal fracture half-length has been determined and the fracturing fluid and fracture propping agent have been selected, the design engineer needs to use a P3D model to determine the details of the design, such as the optimal injection rate, the optimal pad volume, the need for fluid-loss additives, and the proper location for the perforations. After designing the optimal treatment, the design engineer must compute the costs of the proposed treatment to be certain the costs are not much different than the costs assumed during the treatment-optimization process. If the treatment costs are substantially different, the entire optimization loop (Fig. 2.24) should be retraced using the correct cost data.

2.10.2.6 Fracturing Treatment Execution in the Field. After the optimal fracturing treatment has been designed, it must be pumped into the well successfully. A successful field operation requires planning, coordination, and the cooperation of all parties. Treatment supervision and the use of quality-control measures improve the successful application of hydraulic fracturing. Safety is always the primary concern in the field, and safety begins with a thorough understanding by all parties of their duties. A safety meeting is always held to review the treatment procedure, establish a chain of command, make sure everyone knows his or her job responsibilities for the day, and establish a plan for emergencies.

The safety meeting should also be used to discuss the well-completion details and the maximum allowable injection rate and pressures, as well as the maximum pressures to be held as backup in the annulus. All casing, tubing, wellheads, valves, and weak links—such as liner tops—should be thoroughly tested before beginning the fracturing treatment. Mechanical failures during a treatment can be costly and dangerous. All mechanical problems should be discovered during testing and repaired before pumping the fracturing treatment.

Before pumping the treatment, the engineer in charge should conduct a detailed inventory of all the equipment and materials on location. The inventory should be compared to the design and the prognosis. After the treatment is concluded, the engineer should conduct another inventory of all the materials left on location. In most cases, the difference between the two inventories can be used to verify what was mixed and pumped into the wellbore and the hydrocarbon-bearing formation.

In addition to conducting an inventory, samples of the base fracturing fluid (usually water) should be taken and analyzed. Typically, a water analysis is conducted on the base fluid to determine the minerals present and the type of bacteria in the water. The data from the water analysis can be used to select the additives required to mix the viscous fracturing fluid required to create a wide fracture and to transport the propping agent into the fracture. In addition to testing the water, samples of the additives used during a treatment, and of the fracturing fluid after all additives have been added, should be taken during the job and saved in case future analyses are required.

2.11 Post-Fracturing Reservoir Evaluation Methods

Analyzing post-fracturing production and pressure data requires a thorough understanding of the flow patterns in the reservoir (Gidley et al. 1989). The technique applied to analyze the data must be compatible with the flow regime that occurs when the data are collected. For a well containing a finite-conductivity hydraulic fracture, the flow regimes that occur consist of bilinear flow, linear flow, transitional flow, and pseudoradial flow. These flow regimes can be defined in terms of dimensionless time. The times that encompass bilinear flow, linear flow, and transitional flow can be termed "transient flow." The pseudoradial-flow data can be analyzed using semisteady-state methods. In most tight gas reservoirs containing a finite-conductivity hydraulic fracture, the flow rate and pressure data measured during well tests fall somewhere in the transient-flow category. Seldom can semisteady-state analysis techniques, such as the Horner analyses of PBU data, be used to successfully analyze well-test data in tight gas reservoirs containing a hydraulic fracture. As such, transient-flow analysis methods should be used to analyze such data. If long-term (years) production data are available, semisteady-state methods can be used successfully to analyze the production and pressure data.

2.11.1 Transient-Flow Considerations. In the 1950s and 1960s, several papers were published containing semisteady-state analysis methods to analyze wells containing hydraulic fractures (Horner 1951; McGuire and Sikora 1960; Prats 1961). During the time those papers were published, most wells that were fracture treated were moderate- to high-permeability wells that had been damaged during drilling or production. The fracturing treatments were designed to be short and only break through the near-wellbore damaged zone. For such situations, the wells would reach semisteady-state flow in a matter of days or weeks, and the semisteady-state analysis methods of Horner, Prats, or McGuire and Sikora could be used successfully to analyze the production and pressure data. However, in low-permeability gas reservoirs containing long hydraulic fractures, months or years of production must occur before the well approaches pseudoradial flow.

The flow regimes of a vertical well containing a finite-conductivity vertical fracture can be defined using the dimensionless time equation:

$$t_D = \frac{0.000264\, kt}{\phi\, \mu\, c_t\, L_f^2}. \qquad (2.23)$$

Lee and Holditch (1981) showed that linear flow occurs between dimensionless times of 0.0225 and 0.1156. Pseudoradial flow of a well containing a finite-conductivity hydraulic fracture does not begin until a dimensionless time of 2 to 5, depending on the value of dimensionless fracture conductivity. Before reaching linear flow, the flow is often characterized as bilinear flow. Between the end of linear flow and the beginning of pseudoradial flow, the regime is often called transitional flow. The data in **Table 2.12** illustrate the actual times required to reach linear flow and pseudoradial flow for typical reservoir situations.

Formation Permeability (md)	Fracture Half-Length (ft)	Start of Linear Flow (days)	End of Linear Flow (days)	Start of Pseudoradial Flow (days)
0.001	100	341	1,752	4,545
	500	8,523	43,788	113,636
	1,000	34,091	175,152	454,545
0.01	100	34	175	455
	500	852	4,379	11,364
	1,000	3,409	17,515	45,455
0.1	100	3	18	45
	500	85	438	1,136
	1,000	341	1,752	4,545
1	50	0.1	0.4	1.1
	100	0.3	1.8	4.5
	250	2.1	10.9	28.4

Assuming that

porosity = 0.1

viscosity = 0.02 cp

compressibility = 0.0002 psi^{-1}

Table 2.12—Times required to reach linear and pseudoradial flow for a well containing a hydraulic fracture.

Notice that for gas reservoirs with permeabilities of 1.0 md containing short hydraulic fractures, linear- and pseudoradial-flow techniques can be used to analyze data during the first month of production. However, as the permeability decreases to less than 0.1 md, long hydraulic fractures are required to produce the well at commercial flow rates, and years of data are required to use linear- and pseudoradial-flow analysis procedures.

So, how do we analyze early-time data from low-permeability gas reservoirs containing finite-conductivity fractures? The answer is we must use analytical or numerical solutions of Darcy's law to properly analyze data in the transient-flow period, which are all the pressure and flow-rate data before reaching pseudoradial flow. Many analytical transient-flow solutions for hydraulically fractured wells have been derived and published. In fact, there are too many to list in the references, but the first and most-important analytical solutions were published by Russell and Truitt (1964), Gringarten et al. (1974), Cinco L. et al. (1978), and Agarwal et al. (1979).

In addition to the analytical solutions, Lee and Holditch (1981) showed that finite-difference modeling could be used to analyze data from tight gas reservoirs containing a finite-conductivity hydraulic fracture. Actually, the ideal solution is to first use the analytical models to analyze the data to determine first-order estimates of formation permeability, fracture half-length, and fracture conductivity and then enter those values into a realistic finite-difference model. The finite-difference model can be used to determine the final estimates of the formation and fracture properties, taking into account effects such as non-Darcy flow, fracture closure, and formation compaction. The key is to use transient-flow models to analyze transient-flow data. If one tries to analyze data in linear flow with a pseudoradial-flow model (such as the Horner graph), one gets incorrect estimates of formation and fracture properties.

2.11.2 Types of Models. As briefly discussed, there are several types of models that can be used to analyze flow and pressure data from a tight gas well containing a finite-conductivity hydraulic fracture.

2.11.2.1 Semisteady-State Models. Early models by Horner, Prats, and McGuire and Sikora, or simply the semisteady-state flow equation, can be used successfully to analyze data from wells that actually reach pseudoradial flow (Horner 1951; McGuire and Sikora 1960; Prats 1961). In the 1950s and 1960s, very few wells with permeabilities less than 1.0 md were completed because of the low gas prices at the time. As illustrated in Table 2.12, the time required to reach pseudoradial flow in reservoirs of 1.0 md or greater was short enough so that semisteady-state models could be used successfully to analyze the data.

2.11.2.2 Semianalytical Models. Analytical and semianalytical models published in the 1960s and 1970s allowed the engineer to analyze both production and pressure data during the transient-flow period (Agarwal et al. 1979; Cinco L. et al. 1978; Gringarten et al. 1974; Russell and Truitt 1964). Originally, these semianalytical solutions were presented in the form of type curves. To analyze the field data, it had to be plotted on a log-log graph made of transparent paper, and then the field data would be placed on top of the type curve and shifted horizontally and vertically until the shape of the field data could be matched with the shape of one of the type curves. At the time, even though the solution method was time consuming, type curves revolutionized pressure-transient analyses by allowing engineers to analyze transient-flow data from wells containing a finite-conductivity hydraulic fracture. A complete discussion on how to use type curves to analyze data from wells containing a hydraulic fracture is found in Gidley et al. (1989).

Even though type curves (semianalytical solutions) have proved to be very useful, they do have their limitations. For example, most semianalytical solutions were derived using the following assumptions:

- Single-phase, Darcy flow.
- Single-layer, horizontal, homogeneous, isotropic reservoir.
- The reservoir is under pressure-depletion drive and no waterdrive or water influx is included.
- The reservoir permeability is constant with time.
- The hydraulic fracture conductivity is constant with both time and distance.

These assumptions do fit many reservoir situations; hence, the engineer analyzing the data must realize these limitations. In most cases, the semianalytical solutions can be used to analyze transient-flow data and derive first-order estimates of formation permeability, fracture half-length, and fracture conductivity.

2.11.2.3 Finite-Difference Models. Finite-difference, or numerical, models can also be used to analyze transient-flow data from a well containing a finite-conductivity hydraulic fracture (Lee and Holditch 1981). Finite-difference models can be used to overcome many of the limitations that accompany the semianalytical solutions. For example, if a finite-difference reservoir model is used, we can model multiphase flow, water influx, multiple layers, and anisotropic reservoir properties (Rahim et al. 1998). In addition, the finite-difference reservoir model can be used to simulate non-Darcy flow and fracture closure because of the crushing and embedment of the propping agent (Holditch and Morse 1976). Other factors such as water blocking—caused by the fracturing fluid—and the effects of fracturing fluid on the cleanup behavior of the reservoir after the fracturing treatment can also be investigated (Holditch 1979; Voneiff et al. 1996).

Finite-difference reservoir models are more difficult to use and more time consuming than the semianalytical models, especially if multiphase flow and multilayered reservoirs are going to be simulated. However, when these properties are important, the extra time and effort required to use a finite-difference reservoir model result in a better understanding of the reservoir and better estimates of the reservoir and fracture properties. The best solution technique is normally to use the semianalytical reservoir models to obtain a first-order estimate of the formation and fracture properties and then use those values as input in the finite-difference reservoir model. The finite-difference model can then be used to determine the final answers.

2.11.3 Analyzing Post-Fracturing Production Data. In many cases, after a well is fracture treated (especially in the early days of fracturing), the well is produced to a pit until the fracturing fluid is cleaned up sufficiently to turn the well to sales. After the well quits producing proppant and fracturing fluid, a test separator is installed to measure the gas flow rate. The gas flow rate and flowing pressures are subsequently analyzed using transient-flow models to estimate values of the reservoir and fracture properties. If possible, the well is shut in to run a PBU test.

In the 1980s, we found we could better analyze the hydraulic fracture and the reservoir if all the pressure and flow-rate data were measured from the time the well is opened for cleanup. The GRI sponsored a multiyear research project to learn how to evaluate fracturing treatments in tight gas reservoirs. The research was focused on four staged field experiments (SFEs) in which four wells named SFE Nos. 1, 2, 3, and 4 were drilled and tested extensively (Holditch et al. 1988; Robinson et al. 1992; Robinson et al. 1991; Saunders et al. 1992). During the GRI Tight Gas Sands research project, it was learned that we need to measure all flow rates (gas, water, and condensate) and the flowing pressures from the time the well is opened for cleanup. Especially in SFE No. 1, it was found that the only acceptable match of the pressure transient data could be obtained by modeling multiphase flow and fracturing-fluid cleanup in a multilayer reservoir (Holditch et al. 1988). Such a complicated analysis can only be performed using a finite-difference reservoir simulator.

If a finite-difference reservoir simulator is used to analyze field data, or just to conduct parametric studies, it is clear that the production and pressure transient data, if measured accurately, can lead to a much better characterization of the reservoir and the hydraulic fracture (Gidley et al. 1989; Holditch et al. 1988; Robinson et al. 1992; Robinson et al. 1991; Saunders et al. 1992). The very early flow-rate data are mostly affected by the fracture conductivity, often called bilinear flow. Later, during linear flow, the flow-rate data are most affected by the fracture half-length. If pseudoradial flow is reached, the flow-rate data are most affected by the formation permeability. As such, if the early-time flow-rate and pressure data, during the first few days and weeks, are not measured accurately, it is possible that one might not truly understand the properties of the hydraulic fracture.

Post-fracturing production data can be analyzed with decline curves, type curves, semianalytical reservoir models, or finite-difference reservoir models. When analyzing the production data, the analyst tries to determine the values of formation permeability, fracture half-length, and fracture conductivity. Experience has shown it is best to measure the formation permeability using pre-fracturing well testing (Gidley et al. 1989; Lee and Holditch 1981). Then, when analyzing the post-fracturing data, one is only trying to determine the properties of the hydraulic fracture.

Experience has also shown that the analysis of post-fracturing production data only provides estimates of the length of the fracture that has cleaned up (Montgomery and Berthelot 1990; Sherman and Holditch 1991; Voneiff et al. 1996). After the fracturing treatment has been pumped, the hydraulic fracture is filled with fluid. The fracturing fluid flows into the wellbore, imbibes into the formation, or stays in the fracture (Holditch 1979). If the fracturing fluid stays in the fracture, that part of the fracture does not allow gas to flow into the fracture from the formation. Consequently, only the part of the fracture near the wellbore that has been cleaned up is seen by the analyses of production or pressure transient data. In essence, there are three fracture lengths: the created fracture length, the propped fracture length, and the effective fracture length. The propped fracture length is the part of the created fracture that contains proppant at the end of the treatment. The effective fracture length is that part of the propped fracture length that has cleaned up enough to allow gas flow. When we analyze production and pressure transient data, we obtain estimates of only the effective fracture length.

2.11.4 Post-Fracturing Pressure Transient Analyses. Lee and Holditch (1981) presented discussions concerning how to analyze post-fracturing pressure transient data (Gidley et al. 1989; Lee and Holditch 1981). In most cases, the analysis of post-fracturing PBU data, along with the analysis of the post-fracturing production data, leads to accurate descriptions of the formation and the hydraulic fracture only if the correct portion of the data is analyzed with the correct model. For example, the analyst must be sure to use only the linear-flow data when analyzing the transient test using a linear-flow model. It might be possible, when one is analyzing data from moderate-permeability reservoirs containing short fractures, that linear-flow or pseudoradial-flow methods can be used to correctly analyze the PBU data. However, in most cases involving tight gas reservoirs containing long hydraulic fractures, transient-flow methods—such as type curves, semianalytical reservoir models, or finite-difference reservoir models—must be used to correctly analyze the post-fracturing PBU data. To account for factors such as multiphase flow, multilayered reservoirs, non-Darcy flow, and fracture-closure effects, the analyst should use a finite-difference reservoir model that is capable of modeling a finite-conductivity fracture, plus the other features previously listed.

2.12 Estimating Reserves in Tight Gas Reservoirs

The resource triangle, seen in Fig. 1.1 in Ch 1: Introduction to Unconventional Gas Reservoirs in Part 1, describes the distribution of OGIP in a typical basin (Masters 1979). At the top of the triangle are the high-permeability reservoirs. These reservoirs are small, and, once discovered, as much as 80 to 90% of the OGIP can be produced using conventional drilling and completion methods. As we go deeper into the resource triangle, the permeability decreases, but the size of the resource increases. Higher gas prices and better technology are required to produce significant volumes of gas from these tight gas reservoirs.

The recovery efficiency is computed by dividing the cumulative gas produced by the OGIP volume. In a tight gas reservoir, the recovery efficiency varies from less than 10% to more than 50% of the OGIP. The recovery efficiency is a function of permeability, net gas pay thickness, drainage area, effective fracture half-length, economic limit, and well life.

2.12.1 Reserves-Evaluation Methods. The most-common methods used by reservoir engineers to determine reserves are volumetric, material balance, decline curves, and reservoir models. **Table 2.13** presents information concerning how these methods are used to evaluate high- and low-permeability gas reservoirs.

Method	Conventional Gas Reservoir	Tight Gas Reservoir
Volumetric	Accurate in blanket reservoirs	Used only when no wells have been drilled
Material balance	Accurate in depletion-drive reservoirs	Should never be used
Decline curves	Exponential decline usually accurate	Must use hyperbolic decline
Reservoir models	Used to simulate the field	Used to simulate individual wells

Table 2.13—Reserves-estimation methods.

2.12.1.1 Volumetric Method. Volumetric methods can be used to estimate reserves from high-permeability, blanket, and depletion-drive gas reservoirs. In such reservoirs, the drainage area and gas recovery efficiency are usually known with reasonable certainty; thus, the volumetric method can provide relatively accurate estimates of OGIP and reserves.

In tight gas reservoirs, the volumetric method might provide reasonable estimates of OGIP; however, estimates of gas reserves are not as reliable because it is very difficult to estimate both the drainage area of a given well and the recovery efficiency. Because the drainage area and recovery efficiency are so difficult to estimate in tight gas reservoirs, the volumetric method of estimating reserves should be used only before drilling any wells and only as a last resort. When drilling and production data become available, production-data analyses should be used to estimate reserves.

2.12.1.2 Material-Balance Method. The material-balance method should be used only in high-permeability gas reservoirs when accurate gas-production and reservoir-pressure data are available. In high-permeability gas reservoirs, the wells can be shut in for hours or days, and accurate estimates of the average reservoir pressure can be measured or computed using Horner graphs. If the high-permeability reservoir is connected to a strong aquifer or the reservoir rock is very compressible, material-balance methods can still be used but are less accurate because of the complexity of the problem and the difficulty in developing an accurate data set.

In tight gas reservoirs, material-balance methods should never be used because it is impossible to obtain accurate data to describe how the reservoir pressure declines as gas is produced. In a tight gas reservoir, a well (or the entire reservoir) must be shut in for months, or years, before enough pressure data are collected to accurately estimate the average reservoir pressure. As such, virtually all shut-in pressure measurements in tight gas reservoirs underestimate the value of average reservoir pressure. If the data are used, the estimate of OGIP and ultimate gas recovery will be too low.

2.12.1.3 Decline-Curve Method. In most gas reservoirs, the decline-curve analysis method can be used to estimate reserves. For high-permeability reservoirs, the decline-curve method works even with limited production data using the exponential equation, which is written as

$$q = q_o \, e^{-at}, \quad \text{(2.24)}$$

where a is the (constant) instantaneous decline factor, q is the flow rate at time t, and q_o is the initial flow rate.

When Eq. 2.24 applies, a graph of gas flow rate vs. the logarithm of time is a straight line. The straight line can be extrapolated to an economic limit or a fixed well life to determine the ultimate gas recovery. **Fig. 2.25** illustrates typical data that can be described using an exponential decline.

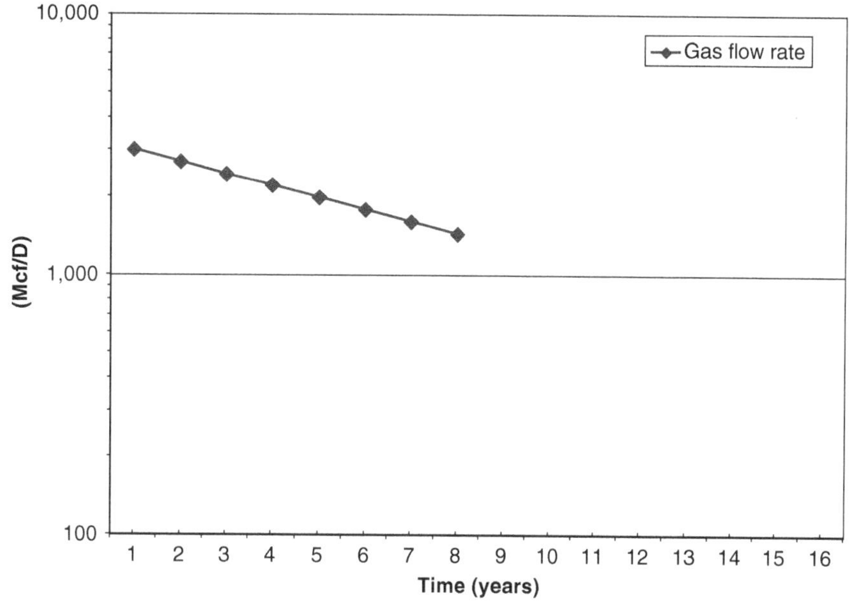

Fig. 2.25—Exponential decline.

For tight gas reservoirs, especially layered reservoirs that have been stimulated with a large hydraulic fracture, decline-curve analysis methods can be used, but a hyperbolic equation must be used to curve fit the data and to extrapolate the data to an economic limit. The hyperbolic decline equation is

$$q = \frac{q_o}{\left(1 + \frac{a_o}{2} t\right)^2}, \quad \text{(2.25)}$$

where a_o is the initial instantaneous decline factor. The decline factor a decreases with time, as given by

$$a = \frac{a_o}{1 + \frac{a_o}{2} t}. \quad \text{(2.26)}$$

Near the end of the life of the well, the decline becomes exponential again. Usually, if the decline rate decreases to less than 6 to 8%, the user sets the decline rate constant (at 6 to 8%) for the remaining life of the well. **Fig. 2.26** illustrates a typical exponential decline for a tight gas well. This is a Cotton Valley well in East Texas that was originally completed and fracture treated in the early 1980s, in the lower Cotton Valley zone called the Taylor Sand. In the early 1990s, the well was completed and fracture treated in the upper Cotton Valley. The gap in gas-production data in the early 1980s was because of the gas market and a curtailment of production.

Even when using the hyperbolic equation to analyze production from tight gas reservoirs, one must carefully analyze all the data. For example, many wells begin producing at high gas flow rates, along with high flowing tubing pressure. During the first few weeks and months, both the gas flow rate and the flowing tubing pressure decline. If the analyst only analyzes the gas flow-rate data, the extrapolation into the future is optimistic. Whenever the flowing tubing pressure reaches the pipeline pressure, and the flowing tubing

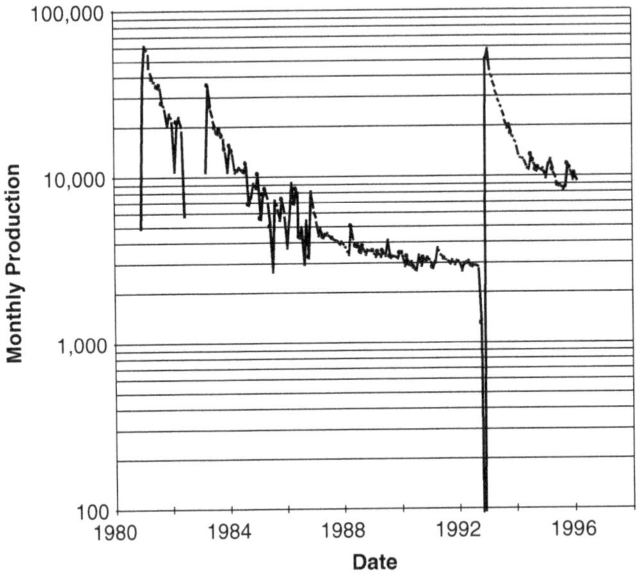

Fig. 2.26—Typical decline curve for a tight gas well.

pressure stops declining, the gas flow-rate decline rate increases. Thus, when both the gas flow rate and the flowing tubing pressure are declining, the analyst needs to compute values of *q*/*p*, or flow rate divided by pressure drop, and use the decline-curve model to match both the decline in flow rate and the decline in flowing tubing pressure.

2.12.1.4 Reservoir-Modeling Method. The most-accurate method of estimating gas reserves in tight gas reservoirs is to use a reservoir model, such as a semianalytical model or a numerical reservoir model, to history match production data from the well. The model should be capable of simulating layered reservoirs, a finite-conductivity hydraulic fracture, and a changing flowing tubing pressure. In some cases, the analyst might also need to simulate non-Darcy flow, formation compaction, fracture closure, and/or fracturing-fluid cleanup effects.

Normally, a reasonable approach to estimating reserves is to use decline curves to review and quality check the data, semianalytical models to history match existing data and estimate reserves, and finite-difference models to analyze the data, especially if factors such as non-Darcy flow, fracture closure, and fracturing-fluid cleanup need to be included in the analyses. **Fig. 2.27** illustrates how the saturation profile around a hydraulic fracture can be simulated to better understand fracturing-fluid cleanup and its effect on gas production vs. time.

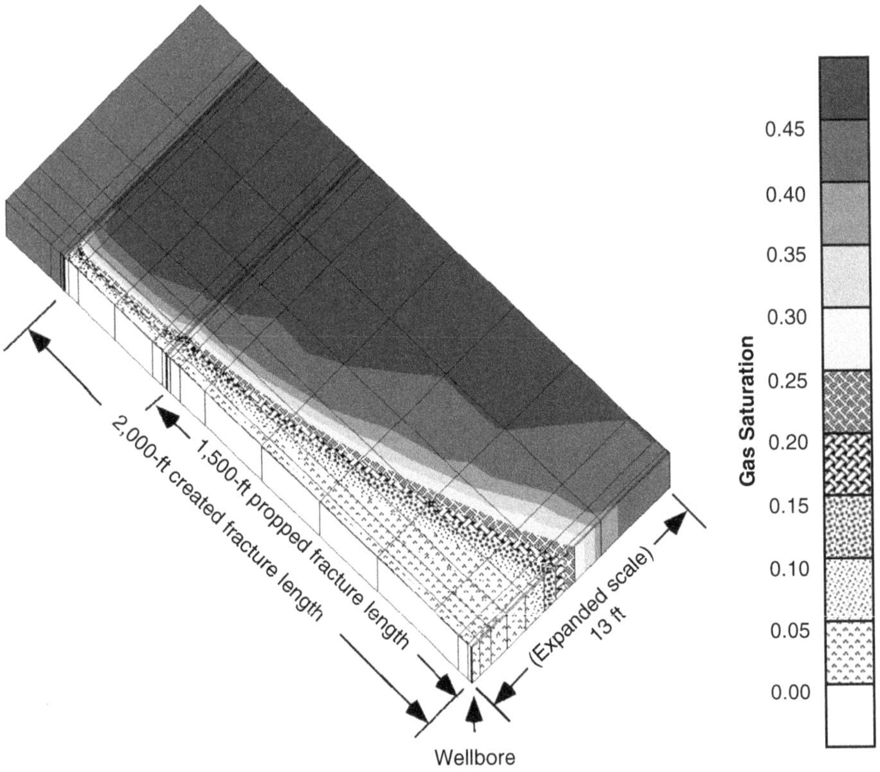

Fig. 2.27—Gas-saturation profile around a hydraulic fracture immediately after a fracturing treatment.

2.13 Nomenclature

a = instantaneous decline factor
A = surface area, L^2
b = decline curve exponent
B = formation volume factor, rcf/scf
c = compressibility, 1/psi
C = conductivity, millimho/m
D = diameter (for grain size), μm
D = constant for computing s'
e = exponent
h = net-pay thickness, ft
I = index
k = permeability, md
L = fracture half-length, ft
m = cementation factor, dimensionless
n = saturation exponent, dimensionless
p = pressure, psi
$\bar{p}$ = average reservoir pressure, psi
q = flow rate, Mcf/D
r = radius, ft
R = resistivity, Ω·m
s = skin, dimensionless
s' = effective skin factor, dimensionless
S = saturation, fraction
t = time, hours or days
T = temperature, °F
U = correlation factor
V = volume, fraction
α = Biot's constant, dimensionless
Δt = travel time, μsec/ft
μ = gas viscosity, cp
ν = Poisson's ratio, dimensionless
ρ = density, g/cm³
σ = in-situ stress, psi
ϕ = porosity, fraction
ψ = pseudopressure, psi

Subscripts

b = bulk
CL = clean
d = drainage
D = dimensionless
DC = density corrected for shale
e = at the extremity of the reservoir
ext = externally applied
f = fluid
f = fracture
g = grain
g = gas (for flow rate)
h = thickness
i = initial (in Eq. 2.19)
i = investigation (for radius)
ild = induction log deep
ilm = induction log medium
ma = matrix
min = minimum
N = neutron log
NC = neutron corrected for shale
o = original (for flow rate)
p = pore pressure
rh = relative to hydrocarbon flow
RA = radioactive
sfl = spherically focused log
SC = sonic corrected
SH = shale
t = true (for conductivity)
t = total (for compressibility)
w = wellbore (for radius)
w = water (for saturation)
wb = bound water (for conductivity and water saturation)
wf = well flowing
wf = free water (for conductivity)
wir = irreducible water
wt = total water

Superscripts

a = (constant) instantaneous decline factor
e = exponent
m = cementation factor
n = saturation exponent
t = time

2.14 References

Agarwal, R. G., Carter, R. D., and Pollock, C. B. 1979. Evaluation and Performance Prediction of Low-Permeability Gas Wells Stimulated by Massive Hydraulic Fracturing. *J Pet Technol* **31** (3): 362–372. SPE-6838-PA. https://doi.org/10.2118/6838-PA.

Ahmed, U., Crary, S. F., and Coates, G. R. 1991. Permeability Estimation: The Various Sources and Their Interrelationships. *J Pet Technol* **43** (5): 578–587. SPE-19604-PA. https://doi.org/10.2118/19604-PA.

Al-Ajmi, F. A. and Holditch, S. A. 2000. Permeability Estimation Using Hydraulic Flow Units in a Central Arabia Reservoir. Paper presented at the SPE Annual Technical Conference and Exhibition, Dallas, Texas, USA, 1–4 October. SPE-63254-MS. https://doi.org/10.2118/63254-MS.

Aly, A. M., Hunt, E. R., Pursell, D. A. et al. 1997. Application of Multi-Well Normalization of Open Hole Logs in Integrated Reservoir Studies. Paper presented at the SPE Western Regional Meeting, Long Beach, California, USA, 25–27 June. SPE-38263-MS. https://doi.org/10.2118/38263-MS.

Amaefule, J. O., Altunbay, M., Tiab, D. et al. 1993. Enhanced Reservoir Description: Using Core and Log Data to Identify Hydraulic (Flow) Units and Predict Permeability in Uncored Intervals/Wells. Paper presented at the SPE Annual Technical Conference and Exhibition, Houston, Texas, USA, 3–6 October. SPE-26436-MS. https://doi.org/10.2118/26436-MS.

Ansah, J., Knowles, R. S., and Blasingame, T. A. 2000. A Semi-Analytic (*p*/*z*) Rate-Time Relation for the Analysis and Prediction of Gas Well Performance. *SPE Res Eval & Eng* **3** (6): 525–533. SPE-66280-PA. https://doi.org/10.2118/66280-PA.

Berg, R. R. 1970. Method for Determining Permeability From Reservoir Rock Properties. *GCAGS Transactions* **20**: 303–317.

Berg, R. R. 1986. *Reservoir Sandstones*. Englewood Cliffs, New Jersey: Prentice-Hall, Inc.

Britt, L. K. 1985. Optimized Oilwell Fracturing of Moderate-Permeability Reservoirs. Paper presented at the SPE Annual Technical Conference and Exhibition, Las Vegas, Nevada, USA, 22–26 September. SPE-14371-MS. https://doi.org/10.2118/14371-MS.

Carman, P. C. 1937. Fluid Flow Through Granular Beds. *Trans. Inst. Chem. Eng.* **15**: 150–167. https://doi.org/10.1016/S0263-8762(97)80003-2.

Cinco L., H., Samaniego V., F., and Dominguez A., N. 1978. Transient Pressure Behavior for a Well With a Finite-Conductivity Vertical Fracture. *SPE J.* **18** (4): 253–264. SPE-6014-PA. https://doi.org/10.2118/6014-PA.

Clavier, C., Coates, G., and Dumanoir, J. 1984. Theoretical and Experimental Bases for the Dual-Water Model for Interpretation of Shaly Sands. *J Pet Technol* **24** (2): 153–168. SPE-6859-PA. https://doi.org/10.2118/6859-PA.

Cleary, M. P., Johnson, D. E., Kogsbøll, H.-H. et al. 1993. Field Implementation of Proppant Slugs to Avoid Premature Screen-Out of Hydraulic Fractures With Adequate Proppant Concentration. Paper presented at the Low Permeability Reservoirs Symposium, Denver, Colorado, USA, 26–28 April. SPE-25892-MS. https://doi.org/10.2118/25892-MS.

Coates, G. R. and Denoo, S. A. 1981. Mechanical Properties Program Using Borehole Analysis and Mohr's Circle. Paper presented at the SPWLA 22nd Annual Logging Symposium, Mexico City, Mexico, 23–26 June. SPWLA-1981-DD.

Coates, G. R. and Dumanoir, J. L. 1974. A New Approach to Improved Log-Derived Permeability. *The Log Analyst*.

Economides, M. J. and Nolte, K. G. ed. 2000. *Reservoir Stimulation,* third edition. Chichester, England: John Wiley & Sons, Ltd.

Ferguson, C. K. and Klotz, J. A. 1954. Filtration From Mud During Drilling. *J Pet Technol* **6** (2): 30–43. SPE-289-G. https://doi.org/10.2118/289-G.

Fetkovich, M. D., Guerrero, E. T., Fetkovich, M. J. et al. 1986. Oil and Gas Relative Permeabilities Determined From Rate-Time Performance Data. Paper presented at the SPE Annual Technical Conference and Exhibition, New Orleans, Louisiana, USA, 5–8 October. SPE-15431-MS. https://doi.org/10.2118/15431-MS.

Fetkovich, M. J. 1980. Decline Curve Analysis Using Type Curves. *J Pet Technol* **32** (6): 1065–1077. SPE-4629-PA. https://doi.org/10.2118/4629-PA.

Fisher, W. L. and Brown, L. F., Jr. 1972. Clastic Depositional Systems—A Genetic Approach to Facies Analysis. Austin, Texas: Bureau of Economic Geology, The University of Texas at Austin.

Fisher, W. L. and McGowen, J. H. 1967. Depositional Systems in the Wilcox Group of Texas and Their Relationship to Occurrence of Oil and Gas. *GCAGS Transactions* **17**: 105–125.

Galloway, W. E. and Hobday, D. K. 1983. *Terrigenous Clastic Depositional Systems*. New York City: Springer-Verlag. https://doi.org/10.1007/978-1-4684-0170-7.

Gas Research Institute. 1991. Staged Field Experiment No. 3: Application of Advanced Technologies in Tight Gas Sandstones—Travis Peak and Cotton Valley Formation, Waskom Field, Harrison County, Texas Reservoirs. Report, GRI-91/0048, CER Corporation, S. A. Holditch & Associates, Inc, and Gas Research Institute, Chicago.

Gidley, J. L., Holditch, S. A., Nierode, D. E. et al. ed. 1989. *Recent Advances in Hydraulic Fracturing*, Vol. 12. Richardson, Texas: Monograph Series, Society of Petroleum Engineers.

Gongora, C. 1995. *Correlations to Determine In-Situ Stress From Open-Hole Logging Data in Sandstone Reservoirs*. MS thesis, Texas A&M University, College Station, Texas.

Gringarten, A. C., Ramey, H. J., Jr., and Raghavan, R. 1974. Unsteady-State Pressure Distribution Created by a Well With a Single Infinite-Conductivity Vertical Fracture. *SPE J.* **14** (4): 347–360. SPE-4051-PA. https://doi.org/10.2118/4051-PA.

GTI E&P Services. 2001. Tight Gas Resource Map of the United States. Quicksilver Resources, Houston.

Holditch, S. A. 1974. Economic Production of Tight Gas Reservoirs Looks Better. *Oil & Gas Journal*: 99.

Holditch, S. A. 1979. Factors Affecting Water Blocking and Gas Flow From Hydraulically Fractured Gas Wells. *J Pet Technol* **31** (12): 1515–1524. SPE-7561-PA. https://doi.org/10.2118/7561-PA.

Holditch, S. A., Gatens, J. M., III, McVay, D. A. et al. 1984. An Automated Method of Matching Production Performance Using Dimensionless Solutions. Paper presented at the SPE Unconventional Gas Recovery Symposium, Pittsburgh, Pennsylvania, USA, 13–15 May. SPE-12846-MS. https://doi.org/10.2118/12846-MS.

Holditch, S. A., Jennings, J. W., Neuse, S. H. et al. 1978. The Optimization of Well Spacing and Fracture Length in Low Permeability Gas Reservoirs. Paper presented at the SPE Annual Fall Technical Conference and Exhibition, Houston, Texas, USA, 1–3 October. SPE-7496-MS. https://doi.org/10.2118/7496-MS.

Holditch, S. A., Lee, W. J., Lancaster, D. E. et al. 1983. Effect of Mud Filtrate Invasion on Apparent Productivity in Drillstem Tests in Low-Permeability Gas Formations. *J Pet Technol* **35** (2): 299–305. SPE-9842-PA. https://doi.org/10.2118/9842-PA.

Holditch, S. A., Lin, Z.-S., and Spivey, J. P. 1991. Estimating the Recovery From an Average Well in a Tight Gas Formation. Paper presented at the SPE Gas Technology Symposium, Houston, Texas, USA, 22–24 January. SPE-21500-MS. https://doi.org/10.2118/21500-MS.

Holditch, S. A. and Morse, R. A. 1976. The Effects of Non-Darcy Flow on the Behavior of Hydraulically Fractured Gas Wells. *J Pet Technol* **28** (10): 1169–1170. SPE-5586-PA. https://doi.org/10.2118/5586-PA.

Holditch, S. A. and Rahim, Z. 1994. Developing Data Sets for 3D Fracture Propagation Models. *SPE Prod & Fac* **9** (4): 257–261. SPE-26155-PA. https://doi.org/10.2118/26155-PA.

Holditch, S. A., Robinson, B. M., Whitehead, W. S. et al. 1988. The GRI Staged Field Experiment. *SPE Form Eval* **3** (3): 519–533. SPE-16429-PA. https://doi.org/10.2118/16429-PA.

Holditch, S. A. and Spivey, J. P. 1993. Estimate Recovery From Tight Gas Formation Wells. *Pet. Eng. Intl.*: 20.

Horner, D. R. 1951. Pressure Build-Up in Wells. *Proc.*, Third World Petroleum Congress, The Hague, Netherlands, 28 May–6 June, Sec. II, 503.

Howard, W. E. and Hunt, E. R. 1986. Travis Peak: An Integrated Approach to Formation Evaluation. Paper presented at the SPE Unconventional Gas Technology Symposium, Louisville, Kentucky, USA, 18–21 May. SPE-15208-MS. https://doi.org/10.2118/15208-MS.

Hunt, E. 1996a. Fundamentals of Log Analysis. I: Overview of Log Analysis. *World Oil* **217** (6): 81–82.

Hunt, E. 1996b. Fundamentals of Log Analysis. II: Data Gathering. *World Oil* **217** (7): 81–84.

Hunt, E. 1997. Fundamentals of Log Analysis. IX: Water Saturation Model Selection, Calculation Methods/Examples. *World Oil* **218** (9).

Hunt, E., Aly, A., and Pursell, D. 1996a. Fundamentals of Log Analysis. IV. *World Oil* **217** (10): 101–102.

Hunt, E., Aly, A., and Pursell, D. 1996b. Fundamentals of Log Analysis. V: Normalizing Logs with Crossplots. *World Oil* **217** (11): 85–88.

Hunt, E. and McCain, W. D., Jr. 1997. Fundamentals of Log Analysis. X: Determining Rock Mechanical Property Values from Log Analysis. *World Oil* **218** (10).

Hunt, E. and Pursell, D. 1996. Fundamentals of Log Analysis. III: Database Editing: Building a Stable Foundation. *World Oil* **217** (9): 81–84.

Hunt, E. and Pursell, D. 1997a. Fundamentals of Log Analysis. VII: Determining Shaliness from Logs. *World Oil* **218** (3): 55–58.

Hunt, E. and Pursell, D. 1997b. Fundamentals of Log Analysis. VIII: Determining Porosity from Density, Neutron and Acoustic Logs. *World Oil* **218** (7): 173–176.

Hunt, E. R. and Pursell, D. A. 1997c. Fundamentals of Log Analysis. XI: Determining Permeability. *World Oil* **218** (11): 127–130.

Hunt, E., Pursell, D., and Aly, A. 1996c. Fundamentals of Log Analysis. VI: Determining Basic Log Analysis Parameters: a, m, n and Rw. *World Oil* **217** (12): 59–60.

Jones, F. O. and Owens, W. W. 1980. A Laboratory Study of Low-Permeability Gas Sands. *J Pet Technol* **32** (9): 1631–1640. SPE-7551-PA. https://doi.org/10.2118/7551-PA.

Kozeny, J. 1927. Uber kapillare Leitung des Wassers im Boden. *Sitzungsber. Akad. Wiss.* **136**: 271–306.

Lee, W. J. and Holditch, S. A. 1981. Fracture Evaluation With Pressure Transient Testing in Low-Permeability Gas Reservoirs. *J Pet Technol* **33** (9): 1776–1792. SPE-9975-PA. https://doi.org/10.2118/9975-PA.

Lee, W. J., Kuo, T. B., Holditch, S. A. et al. 1984. Estimating Formation Permeability From Single-Point Flow Data. Paper presented at the SPE Unconventional Gas Recovery Symposium, Pittsburgh, Pennsylvania, USA, 13–15 May. SPE-12847-MS. https://doi.org/10.2118/12847-MS.

Masters, J. A. 1979. Deep Basin Gas Trap, Western Canada. *AAPG Bulletin* **63** (2): 152–181. https://doi.org/10.1306/C1EA55CB-16C9-11D7-8645000102C1865D.

McGuire, W. J. and Sikora, V. J. 1960. The Effect of Vertical Fractures on Well Productivity. *J Pet Technol* **12** (10): 72–74. SPE-1618-G. https://doi.org/10.2118/1618-G.

Montgomery, K. T. and Berthelot, J. M. 1990. Effects of Fracture Fluid Invasion on Cleanup Behavior and Pressure Buildup Analysis. Paper presented at the SPE Annual Technical Conference and Exhibition, New Orleans, Louisiana, USA, 23–26 September. SPE-20643-MS. https://doi.org/10.2118/20643-MS.

Prats, M. 1961. Effect of Vertical Fractures on Reservoir Behavior-Incompressible Fluid Case. *SPE J.* **1** (2): 105–118. SPE-1575-G. https://doi.org/10.2118/1575-G.

Pursell, D. A. and Hunt, E. R. 1997. Fundamentals of Log Analysis. XII: Integrating Core Data with Log Analysis Results. *World Oil* **218** (12): 185–188.

Rahim, Z. and Holditch, S. A. 1995. The Effects of Mechanical Properties and Selection of Completion Interval Upon the Created and Propped Fracture Dimensions in Layered Reservoirs. *Journal of Petroleum Science and Engineering* **13** (1): 29–45. https://doi.org/10.1016/0920-4105(94)00061-8.

Rahim, Z., Holditch, S. A., Zuber, M. D. et al. 1998. Evaluation of Fracture Treatments Using a Layered-Reservoir Description: Field Examples. *SPE Prod & Fac* **13** (1): 21–28. SPE-26187-PA. https://doi.org/10.2118/26187-PA.

Robinson, B. M., Holditch, S. A., and Peterson, R. E. 1991. The Gas Research Institute's Second Staged Field Experiment: A Study of Hydraulic Fracturing. Paper presented at the SPE Gas Technology Symposium, Houston, Texas, USA, 22–24 January. SPE-21495-MS. https://doi.org/10.2118/21495-MS.

Robinson, B. M., Holditch, S. A., Whitehead, W. S. et al. 1992. Hydraulic Fracturing Research in East Texas: Third GRI Staged Field Experiment. *J Pet Technol* **44** (1): 78–87. SPE-22878-PA. https://doi.org/10.2118/22878-PA.

Rollins, J. B., Holditch, S. A., and Lee, W. J. 1992. Characterizing Average Permeability in Oil and Gas Formations. *SPE Form Eval* **7** (1): 99–105. SPE-19793-PA. https://doi.org/10.2118/19793-PA.

Russell, D. G. and Truitt, N. E. 1964. Transient Pressure Behavior in Vertically Fractured Reservoirs. *J Pet Technol* **16** (10): 1159–1170. SPE-967-PA. https://doi.org/10.2118/967-PA.

Saunders, B. F., Robinson, B. M., Holditch, S. A. et al. 1992. Hydraulic Fracturing Research in the Frontier Formation Through the Gas Research Institute's Fourth Staged Field Experiment. Paper presented at the SPE Annual Technical Conference and Exhibition, Washington, DC, USA, 4–7 October. SPE-24854-MS. https://doi.org/10.2118/24854-MS.

Schlumberger. *Schlumberger Log Interpretation Principles/Applications*, 10. 1989. Sugar Land, Texas: Schlumberger Educational Services.

Semmelbeck, M. E. and Holditch, S. A. 1988. The Effects of Mud-Filtrate Invasion on the Interpretation of Induction Logs. *SPE Form Eval* **3** (2): 386–392. SPE-14491-PA. https://doi.org/10.2118/14491-PA.

Sherman, J. B. and Holditch, S. A. 1991. Effect of Injected Fracture Fluids and Operating Procedures on Ultimate Gas Recovery. Paper presented at the SPE Gas Technology Symposium, Houston, Texas, USA, 22–24 January. SPE-21496-MS. https://doi.org/10.2118/21496-MS.

Soeder, D. J. and Randolph, P. L. 1987. Porosity, Permeability, and Pore Structure of the Tight Mesaverde Sandstone, Piceance Basin, Colorado. *SPE Form Eval* **2** (2): 129–136. SPE-13134-PA. https://doi.org/10.2118/13134-PA.

Thomas, R. D. and Ward, D. C. 1972. Effect of Overburden Pressure and Water Saturation on Gas Permeability of Tight Sandstone Cores. *J Pet Technol* **24** (2): 120–124. SPE-3634-PA. https://doi.org/10.2118/3634-PA.

Timur, A. 1968. An Investigation of Permeability, Porosity, and Residual Water Saturation Relationships. Paper presented at the SPWLA 9th Annual Logging Symposium, New Orleans, Louisiana, USA, 23–26 June. SPWLA-1968-J.

Tobola, D. P. and Holditch, S. A. 1991. Determination of Reservoir Permeability From Repeated Induction Logging. *SPE Form Eval* **6** (1): 20–26. SPE-19606-PA. https://doi.org/10.2118/19606-PA.

Veatch, R. W., Jr. 1983. Overview of Current Hydraulic Fracturing Design and Treatment Technology—Part I. *J Pet Technol* **35** (4): 677–687. SPE-10039-PA. https://doi.org/10.2118/10039-PA.

Voneiff, G. W., Robinson, B. M., and Holditch, S. A. 1996. The Effects of Unbroken Fracture Fluid on Gaswell Performance. *SPE Prod & Fac* **11** (4): 223–229. SPE-26664-PA. https://doi.org/10.2118/26664-PA.

Watson, A. T., Lane, H. S., and Gatens, J. M., III. 1990. History Matching With Cumulative Production Data. *J Pet Technol* **42** (1): 96–100. SPE-17063-PA. https://doi.org/10.2118/17063-PA.

Worthington, P. F. 1985. The Evolution of Shaly-Sand Concepts in Reservoir Evaluation. *The Log Analyst.*

Yao, C. Y. and Holditch, S. A. 1993. Estimating Permeability Profiles Using Core and Log Data. Paper presented at the SPE Eastern Regional Meeting, Pittsburgh, Pennsylvania, USA, 2–4 November. SPE-26921-MS. https://doi.org/10.2118/26921-MS.

Yao, C. Y. and Holditch, S. A. 1996. Reservoir Permeability Estimation From Time-Lapse Log Data. *SPE Form Eval* **11** (2): 69–74. SPE-25513-PA. https://doi.org/10.2118/25513-PA.

SI Metric Conversion Factors

acre	× 4.046 873	E + 03 = m^2
°API	141.5/(131.5 + °API) = g/cm^3	
bbl	× 1.589 873	E – 01 = m^3
cp	× 1.0*	E – 03 = Pa·s
ft	× 3.048*	E – 01 = m
°F	(°F – 32)/1.8	= °C
in.	× 2.54*	E + 00 = cm
psi	× 6.894 757	E + 00 = kPa

*Conversion factor is exact.

Chapter 3

Coalbed Methane*

Sunil Ramaswamy and Stephen A. Holditch

3.1 Introduction

To meet an increasing demand for energy, the oil and gas industry is turning toward unconventional oil and gas reservoirs. Unconventional reservoirs are the oil and gas reservoirs that cannot be produced at an economic rate or cannot produce economic volumes of oil and gas without assistance from massive stimulation treatments, special recovery processes, or advanced technologies. Unconventional reservoirs include tight gas reservoirs, coalbed methane (CBM) reservoirs, gas shales, oil shales, tar sands, heavy oil, and gas hydrates (Holditch 2006).

All natural resources—such as gold, zinc, oil, and gas—are distributed log-normally in nature. John Masters introduced the concept for oil and gas resources in the form of a resource triangle (**Fig. 3.1**) (Holditch 2006).

High-quality resources that are less abundant but easy to produce occur at the top of the triangle, whereas the unconventional resources that are more abundant but difficult and expensive to produce occur at the base of the triangle.

With growing demand for energy and the depletion of conventional energy supplies, the emphasis is shifting toward the lower part of the triangle, and unconventional gas resources are assuming greater importance worldwide. CBM resources occur in the lower portion of this triangle.

CBM is methane produced from coalbeds. Most commonly, a coalbed gas system is a self-sourcing reservoir. Sometimes it generates more gas than it can hold and thus can be a source for nearby traps. The gas generated by the thermal maturation of the coal is stored in the coal matrix as adsorbed gas. The hydraulic pressure in the coal keeps the gas adsorbed. Thus, the coal matrix acts as the primary reservoir rock, with secondary gas stored in cleats as free gas or as solution gas in water.

Worldwide energy demand is predicted to increase from 400 quadrillion Btu/yr in 2004 to 600 quadrillion Btu/yr by 2020 (**Fig. 3.2**). By the year 2020, approximately 47.5% of the energy demand is expected to be satisfied by gas resources. Of this 47.5%, approximately 20% is expected to be fulfilled by CBM.

At the time of this writing, CBM was one of the major unconventional resources fulfilling the demands of the US. In 2006, CBM contributed approximately 9.7% of the total dry gas reserves of the US (US Energy Information Administration 2007a).

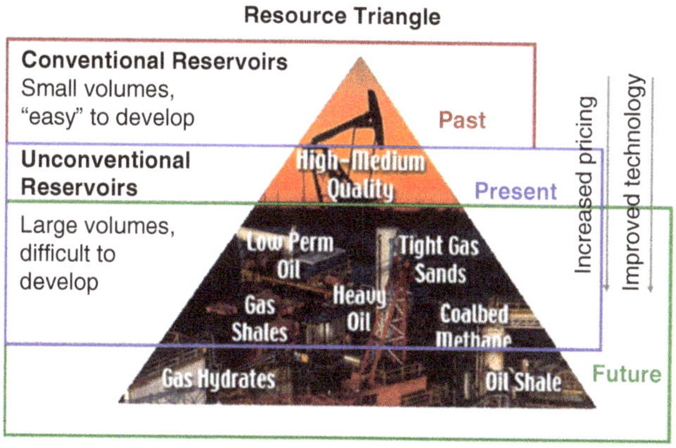

Fig. 3.1—Natural-gas resource triangle (Holditch 2006).

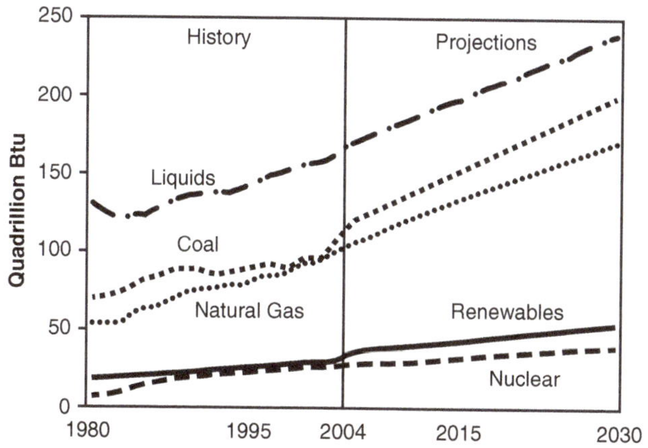

Fig. 3.2—World energy demand.

*This chapter is an edited version of an MS degree thesis in petroleum engineering by Sunil Ramaswamy at Texas A&M University in 2007.

Significant coal reserves underlie approximately 13% of the US land area. Of the coal regions **(Fig. 3.3)**, several produce CBM, and exploration is active in others. The US is the world leader in coalbed gas exploration, booked reserves, and production. At the time of this writing, 12 US basins had commercial coalbed gas production or exploration. The major producing areas are the San Juan, Powder River, Black Warrior, Raton, Central Appalachian, and Uinta basins (Fig. 3.3). Other US areas with significant exploration or production are the Cherokee, Arkoma, Illinois, Hanna, Gulf Coast, and Greater Green River basins. Internationally, commercial coalbed gas is produced in Canada and the Bowen Basin of Queensland, Australia. At the time of this writing, exploration, test wells, or pilot projects were ongoing in several countries, including Russia, the United Kingdom, China, and India.

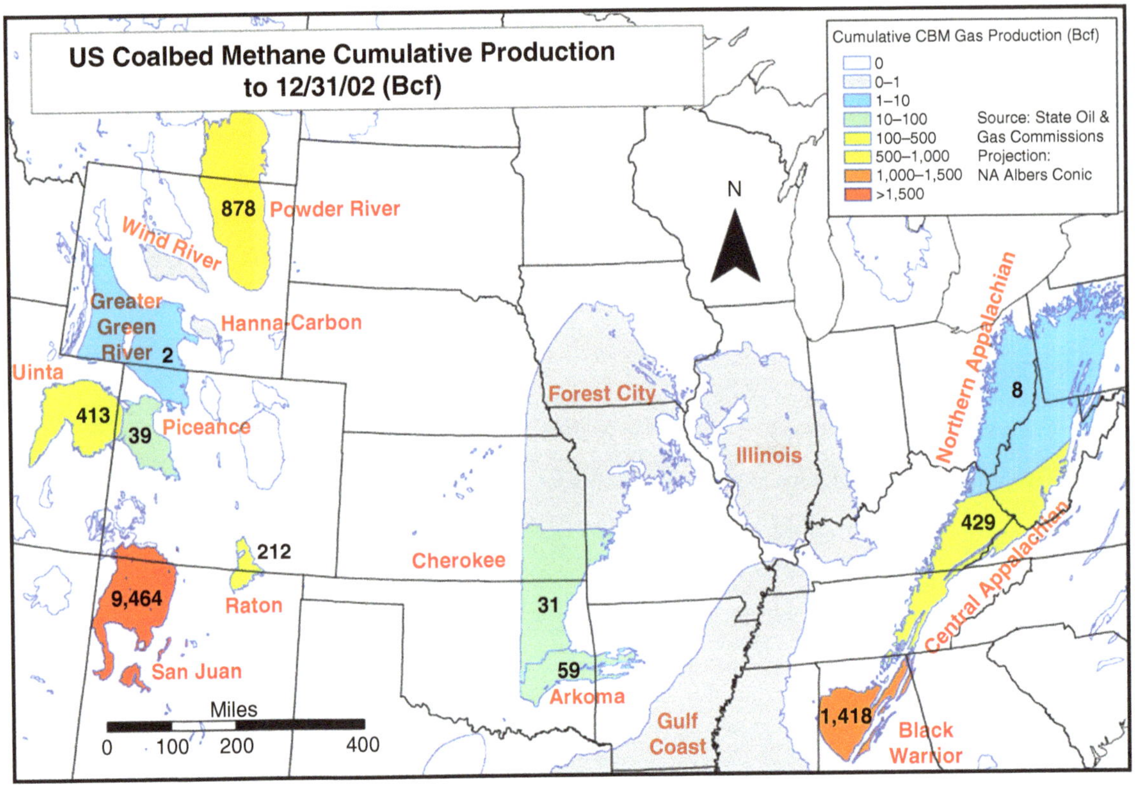

Fig. 3.3—US basins with active CBM wells as of 2002 (US Energy Information Administration 2007a).

The methods used for CBM production vary across basins and from one basin to another, depending on the local geology and reservoir properties **(Fig. 3.4)** (US Energy Information Administration 2007b). To select optimal engineering applications to maximize well performance, it is crucial to determine the influence of these geologic parameters on the success of specific drilling, completion, or stimulation practices.

Depending on the geologic setting, CBM wells can be vertical or horizontal, and the selection of completion and stimulation methods will further depend on the number of coalbeds to be produced, the depth of occurrence, permeability, the compressive strength of coal, and other factors **(Figs. 3.5 and 3.6).** This chapter aims to clarify how various reservoir properties influence the selection of specific drilling, completion, and stimulation applications.

3.2 Evolution of CBM Engineering Practices

3.2.1 Completion Methods. CBM has been produced for many years in the US, and engineering practices have evolved over time. CBM was produced successfully in Oklahoma in 1926, then from Appalachian Basin coals in the mid-1940s (Palmer et al. 1993a; Patchen et al. 1991). The first commercial CBM well in the San Juan Basin was drilled in 1953. In the late 1970s, CBM wells were drilled in the Black Warrior Basin (Palmer et al. 1993a). The US industry expanded rapidly in the 1980s, to take advantage of the Section 29 tax credit.

Most of these CBM wells were vertical wells, and gravel packs were used for completions. Commonly, coal fines plugged the gravel packs, resulting in reduced production. This led to the use of casedhole completions with the hydraulic fracturing stimulation of coalbeds by the late 1970s in the Black Warrior and San Juan basins. Today, openhole completions are seldom used for coalbed wells (Palmer et al. 1993a).

However, some modified versions of openhole completions, such as the openhole cavity completion and the topset-and-underream method, are still used. The openhole cavity completion method was developed for the San Juan Basin "Fairway" coals in 1985, by Meridian Petroleum Company (Palmer et al. 1993b). This is one of the most-successful methods for producing coalbed gas, but it has proved to work only in the specific geologic conditions that occur in the San Juan Basin Fairway. The topset-and-underream method of coalbed completion was developed in the 1990s for producing gas from the shallow coals of the Powder River Basin. In this method, wells are drilled to the top of the coal, and casing is set. Then, the well is drilled through the coal and underreamed. Wells are then stimulated by pumping a small quantity of water (approximately 160 bbl) to remove the damage caused to the coal by drilling.

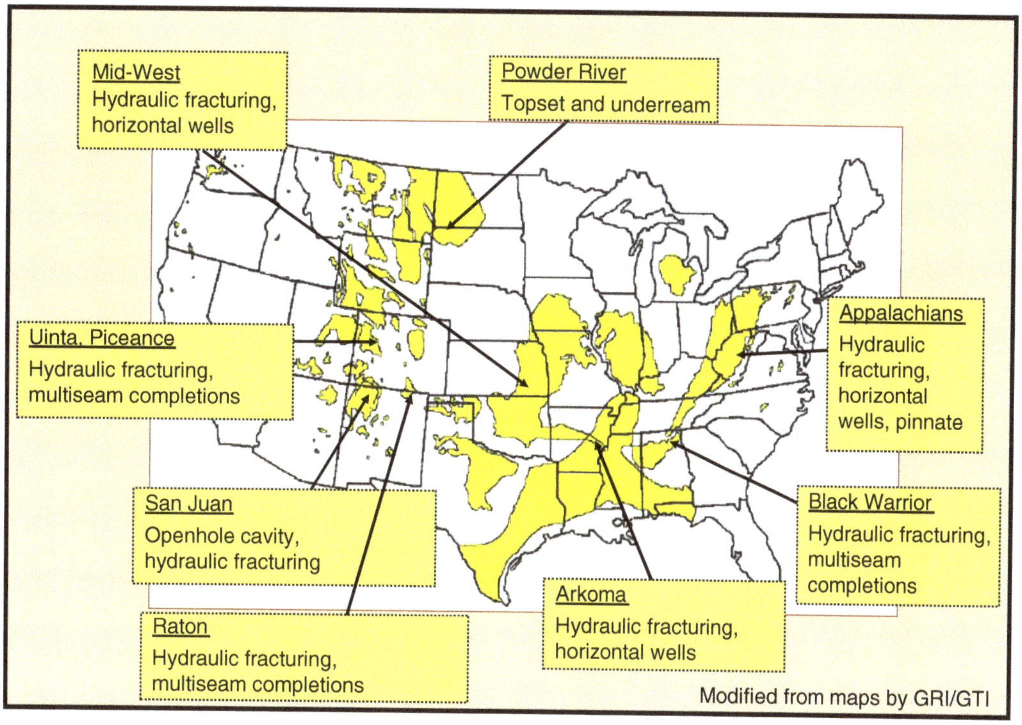

Fig. 3.4—CBM basins and completion and stimulation methods used in the US (after US Energy Information Administration 2007b).

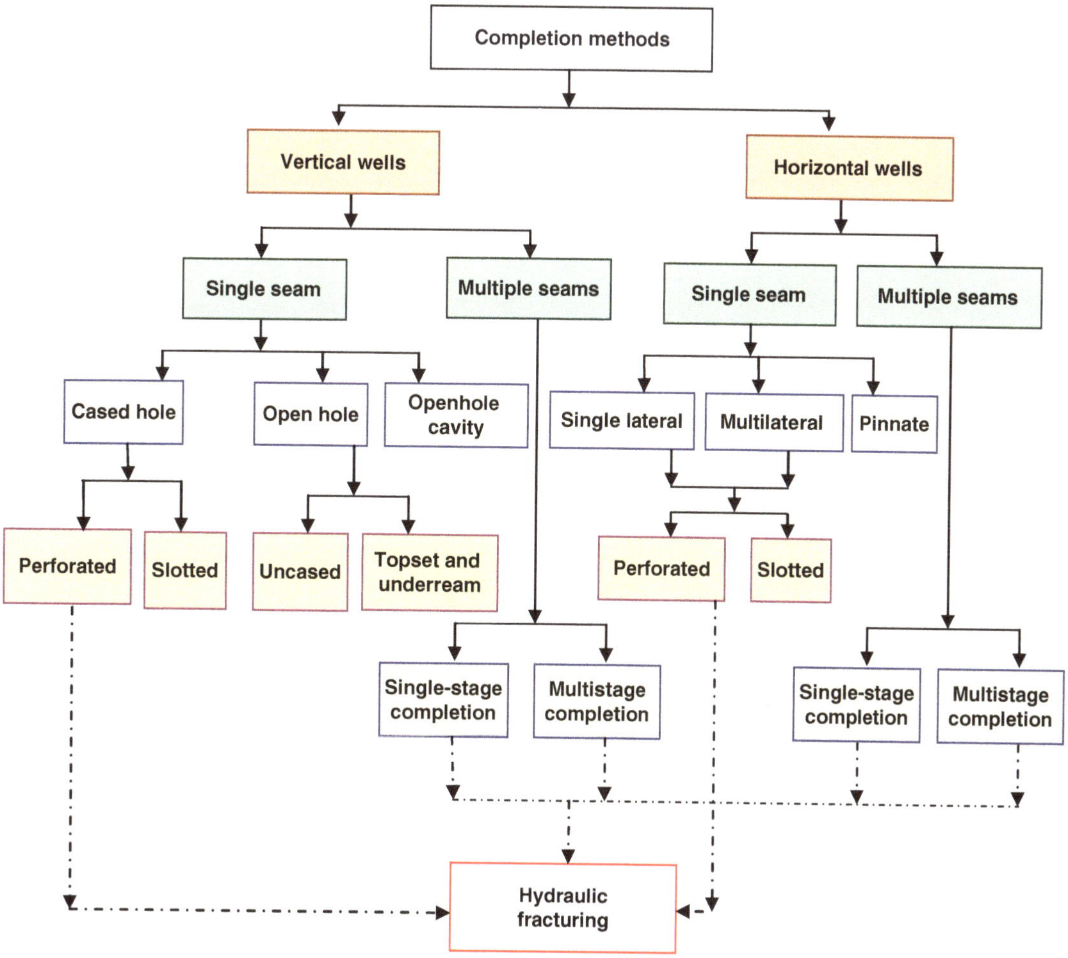

Fig. 3.5—Drilling and completion methods for CBM reservoirs.

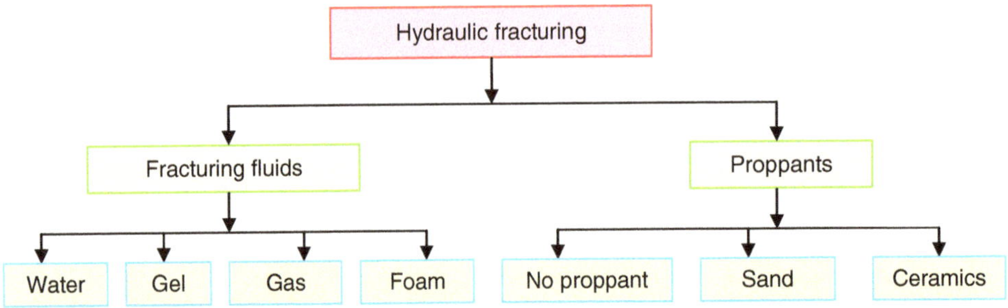

Fig. 3.6—Hydraulic fracturing stimulation fluids and proppants used for CBM reservoirs.

Casedhole completions are the most commonly used completion method for CBM wells. Most cased wells are stimulated using hydraulic fracturing techniques. However, the hydraulic fracturing designs vary from basin to basin and, sometimes, even from place to place within one basin.

Horizontal coalbed wells have long been successfully drilled inside mines for degasifying the coals before a mining operation. In the late 1980s, horizontal CBM wells drilled from the surface were tried in the Black Warrior Basin, but they were considered uneconomic. However, with advances in drilling technology in the early 2000s, horizontal coalbed wells have become more common. These horizontal CBM wells are drilled in thin coal seams to enable the wellbore to contact the maximum possible reservoir area. Multilateral wells have also been successfully drilled in the Arkoma and Appalachian basins.

Several types of hydraulic fracturing methods have been used to stimulate CBM wells (Fig. 3.6). These stimulation methods, and the types of fracturing fluids and proppants, have also evolved over time. The hydraulic fracturing of coalbeds was tried first in the San Juan and Black Warrior basins in the late 1970s. The initial fracturing stimulation treatments in the Black Warrior Basin used slickwater with proppant. Later, linear-gel fluids with proppant were used during fracturing treatments (US Environmental Protection Agency 2004). However, the increase in production observed with the use of linear gels was insignificant, owing to the damage caused to the formation by the gels. Because the gel fracturing fluids were not very successful, operators returned to slickwater, but it was used without proppant. However, even this method was not found to be very successful. Because of further improvements in technology and the development of crosslinked fluids, better gel breakers, and cleaning agents, crosslinked fluids are accepted as the most-suitable fluids (US Environmental Protection Agency 2004).

Drilling, completion, and stimulation methods in CBM reservoirs vary with the different geological parameters. Seam thickness affects the decision on whether to drill a vertical well or a horizontal well. The depth of occurrence and formation permeability further affect engineering decisions, such as whether to complete the well as open hole with underreaming, as an openhole cavity, or as a casedhole completion. If one selects a casedhole completion, further choices must be made concerning the type and volume of the hydraulic fracturing fluid and proppant that will be used. Similarly, in horizontal wells, coalbed permeability and the number of coal seams to be completed affect the decision on whether to drill a single-lateral or multilateral well.

Because more CBM fields are developed in diverse geologic settings, we face difficult decisions concerning the optimal drilling, completion, and stimulation methods. Moreover, the development of new technology further complicates the selection process. On the basis of the geology of the CBM reservoir, one must select the best engineering practices to maximize gas recovery and profits. The best practices identified can be applied to both new and existing fields. By evaluating the geologic setting of producing areas, we can reassess, and possibly increase, reserves on the basis of the best technology applications.

A review of CBM literature was conducted to determine the important CBM reservoir properties that influence CBM engineering practices. In conducting this study, only North American CBM basins were considered because the industry started in North America and this area has been the site of most advancements in CBM technology **(Fig. 3.7).** The different drilling, completion, and stimulation practices used in CBM reservoirs were analyzed. The best engineering practices for the North American CBM basins, and the geological parameters that contribute to the success of these practices, were identified. On the basis of the literature review, a questionnaire was prepared and circulated among industry experts to determine the different geological conditions that affect the selection of specific drilling, completion, and stimulation methods in coalbeds.

On the basis of the industry response and the literature review, a decision chart has been developed to help engineers select the appropriate drilling, completion, and stimulation methods for developing CBM reservoirs in various geologic settings.

3.3 Overview of Coalbed Gas Systems

Owing to differences in reservoir quality, coalbed production varies across individual basins; commonly, the fairways or sweet spots comprise less than 10% of the area of producing basins. An economic CBM project requires the convergence of several geologic factors, as well as acceptable gas prices and operational and environmental conditions. CBM reservoir properties are determined by a number of factors, including the coal properties, depositional setting, and geological processes that occur over time. An understanding of coalbed gas systems helps clarify the complexity and variability of coalbed reservoirs.

A petroleum system is defined as a natural system that encompasses active source rock related to oil and gas, and it includes all the geologic elements and processes that are essential for a hydrocarbon accumulation to exist (Magoon and Dow 1994). The most-important elements of a petroleum system are the hydrocarbon source rock, reservoir, seal rocks, and geological process that occurs over time. Many CBM petroleum systems differ from conventional petroleum systems in a number of ways. Most commonly, a coalbed gas system is a self-sourcing reservoir. Gas generated by the thermal maturation of the coal is stored in the coal matrix as adsorbed gas. The hydraulic pressure in the coal cleats (fractures) assists in keeping the gas adsorbed. Thus, the coal matrix acts as the primary reservoir rock, with secondary gas storage in cleats as free gas or as solution gas in water (Magoon and Dow 1994).

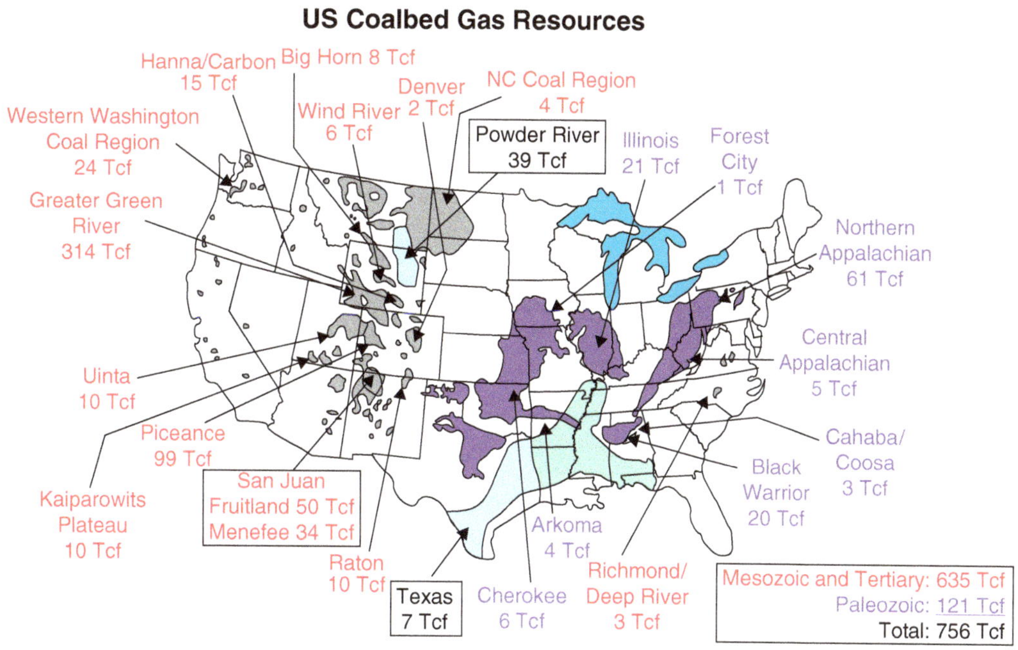

(a)

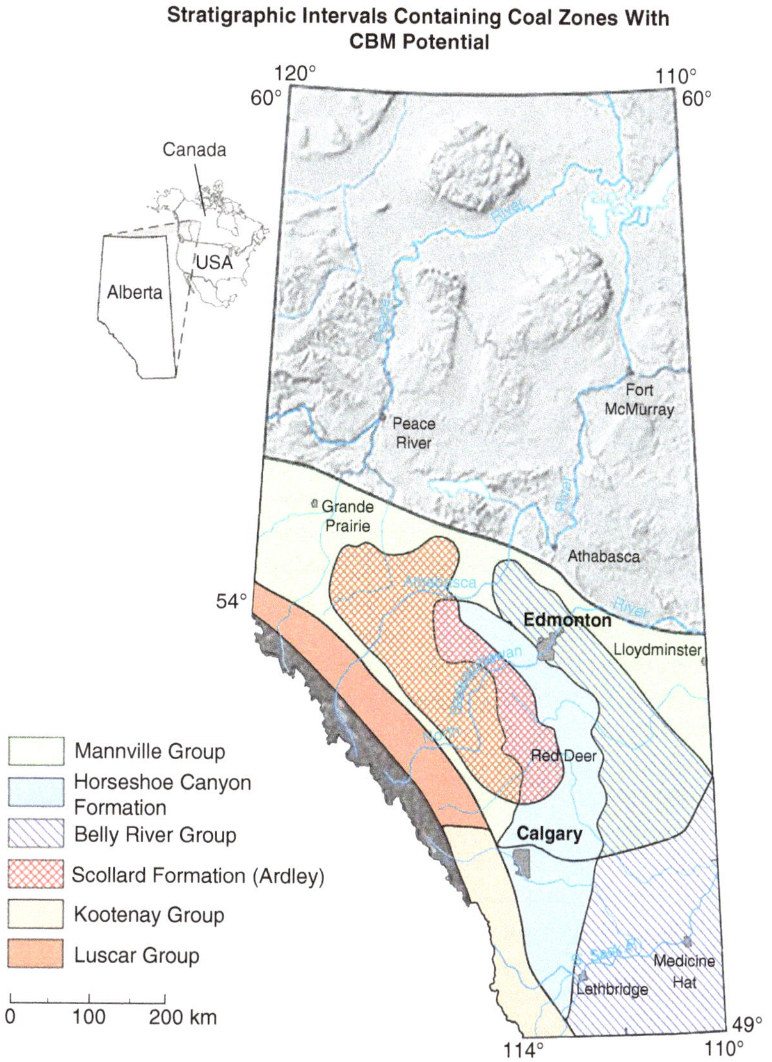

(b)

Fig. 3.7—Basins with CBM resources in North America: (a) US basins; (b) Horseshoe Canyon CBM play in the Western Canada Sedimentary Basin (after Beaton 2003).

Coalbed gas is classified on the basis of origin as primary biogenic gas, secondary biogenic gas, or early thermogenic, thermogenic, migrated thermogenic, or mixed gas (Scott et al. 1994). Primary biogenic gas is generated in peat at a relatively low temperature and shallow burial depth. Most primary biogenic gas is lost during burial and compaction. Early thermogenic gas is generated by the thermal maturation of the coal, generally at a vitrinite reflectance < 0.78%. Thermogenic gas is generated by the further burial and thermal maturation of coal, at a vitrinite reflectance > 0.78%. Thermogenic gas is the source of most gas in thermally mature coals. Secondary biogenic gas is generated by the activity of methanogenic microbes present in meteoric water moving through the coal-cleat system. These microbes are introduced to the coal after the formation of the coal. Migrated thermogenic gas is transported to a location in the coal from other places in the coalbed by hydrologic flow. Alternatively, it can be transported to the coal from other source rock, such as shales or other coals. Mixed gas is a mixture of gas from two or more thermogenic or biogenic sources (Scott et al. 1994).

The majority of coalbed gas is adsorbed on the surface of organic matter in pores of the coal matrix. However, some coalbed gas is stored in the cleats as solution gas in water, or as free gas in the absence of water. Seals in coalbed gas systems maintain formation pressure, and formation pressure holds gas in an adsorbed state preventing gas desorption and escape. Although conventional traps can be present in coalbed gas systems, their presence is unnecessary because gravity separation of gas and water is not required. Thus, coalbed gas can be produced from structurally low sites, such as synclines.

The structural complexity of coal basins can affect CBM project economics. In small basins that are highly faulted, for example, reservoir properties might vary markedly from one fault compartment to the next. In some cases, it might be difficult to develop projects with sufficient numbers of wells to support the required infrastructure.

3.4 CBM Reservoir Properties

Among the CBM reservoir properties that play important roles in determining engineering best practices are the (1) depth of coal occurrence, (2) thickness of individual coal seams and net coal thickness, (3) number of coal seams and their vertical distribution, (4) lateral extent of the coal, (5) thermal maturity, (6) structural dip, and (7) adjacent formations (e.g., aquifer sandstones, fracture barriers).

The number of effective coal seams and their vertical distribution affect the type of completion that will be used. The completion could be single zone or multizone. The areal extent of the coal also plays an important role in selecting well locations and in deciding whether to drill a vertical or a horizontal well. If the dip of the coal is greater than 15°, keeping a horizontal wellbore inside the coal seam is very difficult, and drilling a horizontal well could be uneconomical (Palmer 2007).

The distance to fracture barriers or aquifers above or below the coalbeds influences the selection of fluids when hydraulic fracturing stimulation is used (Holditch et al. 1990). Values of reservoir fluid compressibility and formation compressibility are also important when selecting the type of hydraulic fracturing stimulation.

3.4.1 Depth of Occurrence. Depth of coal occurrence is important to the selection of completion and stimulation methods. With an increase in depth of coal occurrence, overburden stress, formation pressure, and the thermal maturity of coal increase, and gas content might also increase. When designing a fracturing treatment, depth determines the drilling cost, and it is an important factor in determining the surface injection pressure and the bottomhole pressure (Holditch 1990).

3.4.2 Gas Content. Thermal maturity, moisture and ash contents, and the maceral composition of coal directly affect the ability of coal to adsorb gas. The gas content of coal is governed by the adsorption capacity and formation pressure. **Fig. 3.8** shows the gas generation during the coalification process from peat to anthracite. The amount of gas retained depends on the reservoir pressure and coal properties.

3.4.3 Coal Rank. Coal rank, or thermal maturity, can be described on the basis of the percentage of carbon or moisture in the coal, vitrinite reflectance, or other measures **(Table 3.1).** The amount of gas that might be stored in coal is directly dependent on coal rank. Low-volatile bituminous coals might be better suited for CBM production than high-volatile bituminous coals because low-volatile bituminous coals have the potential to adsorb greater amounts of gas and are more cleated than high-volatile bituminous coals. Although the gas content of semi-anthracite and anthracite coals can be very high, there are no economical coalbed gas projects in these coals, owing to low permeability and very slow rates of gas desorption (Ryan and Mannhardt 2007).

3.4.4 Reservoir Pressure. Reservoir pressure affects the gas storage capacity, the amount of depressurization required to initiate gas desorption (critical desorption pressure), effective in-situ stress, the leakoff coefficient, and well productivity. Where coal cleats are water saturated, it is necessary to dewater the coalbed to enable desorption and gas production **(Fig. 3.9)** (Holditch 1990).

When depressurization progresses to the critical desorption pressure, gas desorbs from the coal matrix adjacent to the cleat and moves by Darcy flow to the wellbore (Fig. 3.9). The desorption of coalbed gas from the coal matrix adjacent to the cleat creates a concentration gradient, and gas within the matrix diffuses to the cleat. Over time, water production declines and gas production increases. Thus, the fluid flow in CBM formations is controlled by two flow mechanisms: Darcy flow and diffusion (Olsen et al. 2003).

Any pressure gradient in the reservoir caused by low permeability or poor reservoir access causes a reduction in the amount of gas released (Palmer et al. 1993a). Reservoir pressure is an important factor when selecting the completion and stimulation method because it affects the selection of the fracturing fluid that will be used.

3.4.5 Reservoir Fluid Saturation. In most coals, the cleat is water saturated. Coalbed water is important because it (1) might contribute microbes that generate biogenic gas, (2) might cause artesian overpressure, (3) reduces the relative permeability to gas in the coal cleats, (4) must be removed to enable coalbed gas desorption, and (5) must be disposed of, which adds to operation costs. The quantity of water to be pumped is one of the most important factors in determining the economics of a coalbed gas well.

3.4.6 In-Situ Stress. Coal is highly compressible, and the in-situ stress acting on coal affects reservoir permeability and production. Generally, permeability decreases with depth owing to increased in-situ stress, and most coalbed gas production is from coals at less

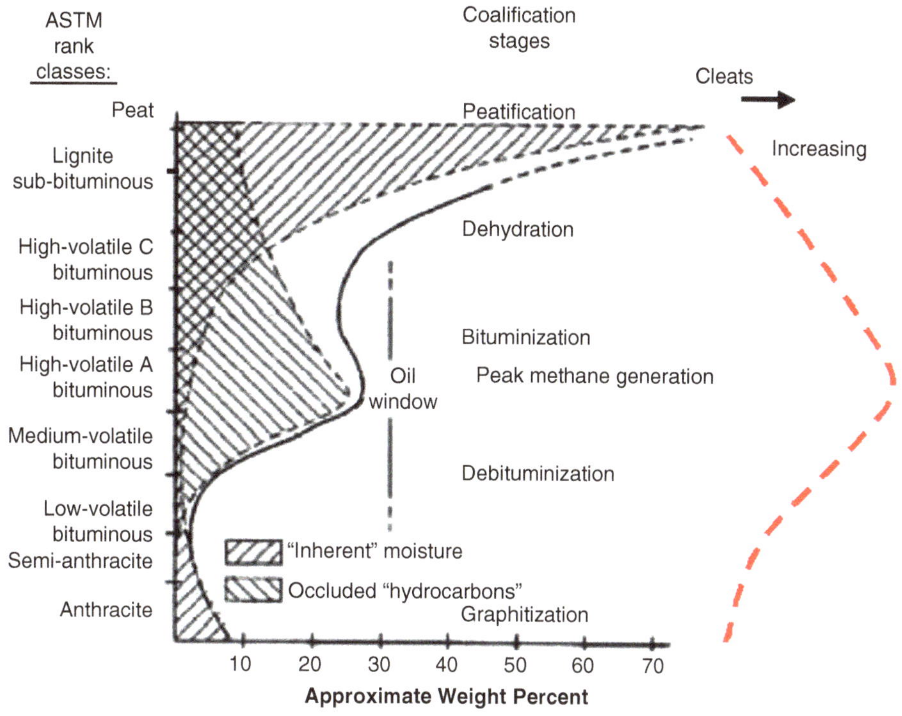

Fig. 3.8—Coalification, cleats, and hydrocarbon generation (Olsen et al. 2003).

Coal Rank	% Carbon	Special Energy (MJ/kg)	Vitrinite Reflectance (max%)
Anthracite	95	35.2	Up to 7.0
Semi-anthracite	92	36	2.83
Low-volatile bituminous coal	91	36.4	1.97
Medium-volatile bituminous coal	90	36	1.58
High-volatile bituminous coal	86	35.6	1.03
Sub-bituminous coal	80	33.5	0.63
Brown coal	71	23	0.42

Table 3.1—Carbon percentage, heating value, and vitrinite reflectance based on coal rank (Ramage 1997).

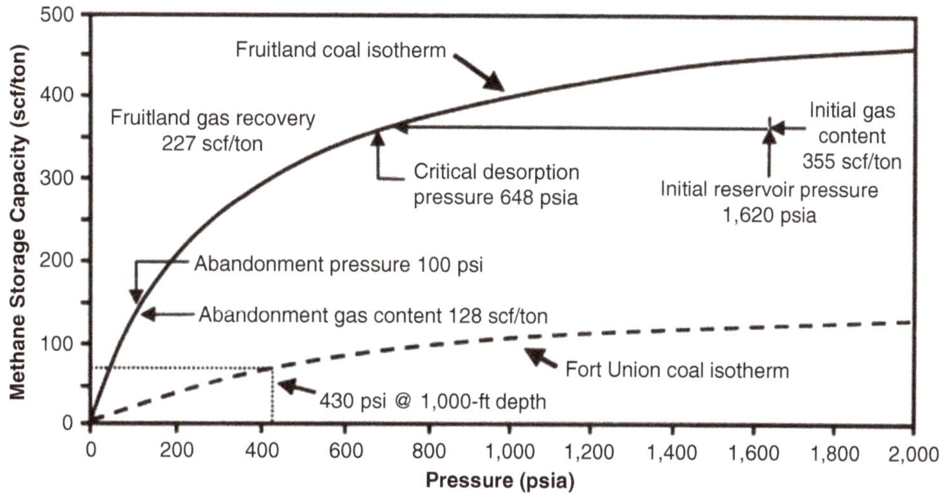

Fig. 3.9—Effect of pressure on methane storage for San Juan Basin Fruitland coal and Powder River Basin Fort Union coals (Pratt et al. 1992).

than 4,000 ft depth. Knowledge of the in-situ stress is used in calculating the surface injection pressure and the bottomhole treatment pressure when designing fracture stimulation treatments (Holditch et al. 1990).

In-situ stress orientation might also impact CBM production. The orientation of the horizontal stress relative to cleat orientation can affect coalbed permeability. Moreover, in-situ stress orientation determines the orientation of induced fractures.

3.4.7 Permeability. Coal has very low matrix permeability (< 1 md). Fluid and pressure transmission in CBM reservoirs is dependent on the coal cleats (Remner et al. 1986). Thus, cleat properties affect the type of completion used. Cleats in the coal seam are thought to form as a result of coal dehydration, local and regional stresses, and the unloading of overburden. Two orthogonal sets of cleats develop perpendicular to bedding in coals **(Fig. 3.10).** Face cleats are the dominant fracture set, and they are more continuous and laterally extensive; face cleats form parallel to maximum compressive stress and perpendicular to fold axes (Olsen et al. 2003). Butt cleats are secondary and terminate against face cleats. Butt cleats are strain-release fractures that form parallel to fold axes.

Cleat spacing is related to coal rank, bed thickness, maceral composition, and ash (inorganic) content. Coals with well-developed cleat sets are brittle. In general, cleats are more closely spaced with increasing coal rank. One study suggests that the average cleat-spacing values for three coal ranks are sub-bituminous (2 to 15 cm), high-volatile bituminous (0.3 to 2 cm), and medium- to low-volatile bituminous (< 1 cm) (Cardott 2001). Cleat spacing is tighter in thin coals, vitrinite-rich coals, and low-ash coals. Where biogenic gas is present, cleats serve multiple purposes: conveying microbes to the coal/water interface, sweeping microbial gas along the groundwater flow paths, and during production transporting water and gas to the wellbore. The fact that the face cleat is more continuous than the subordinate butt cleat leads to permeability anisotropy and elliptical reservoir-drainage patterns (Cardott 2001).

Permeability influences the type and design of the stimulation treatment. Coal-seam permeability is used to calculate the leakoff coefficient, size of the pad volume, and injection rate of a fracturing treatment (Holditch et al. 1990).

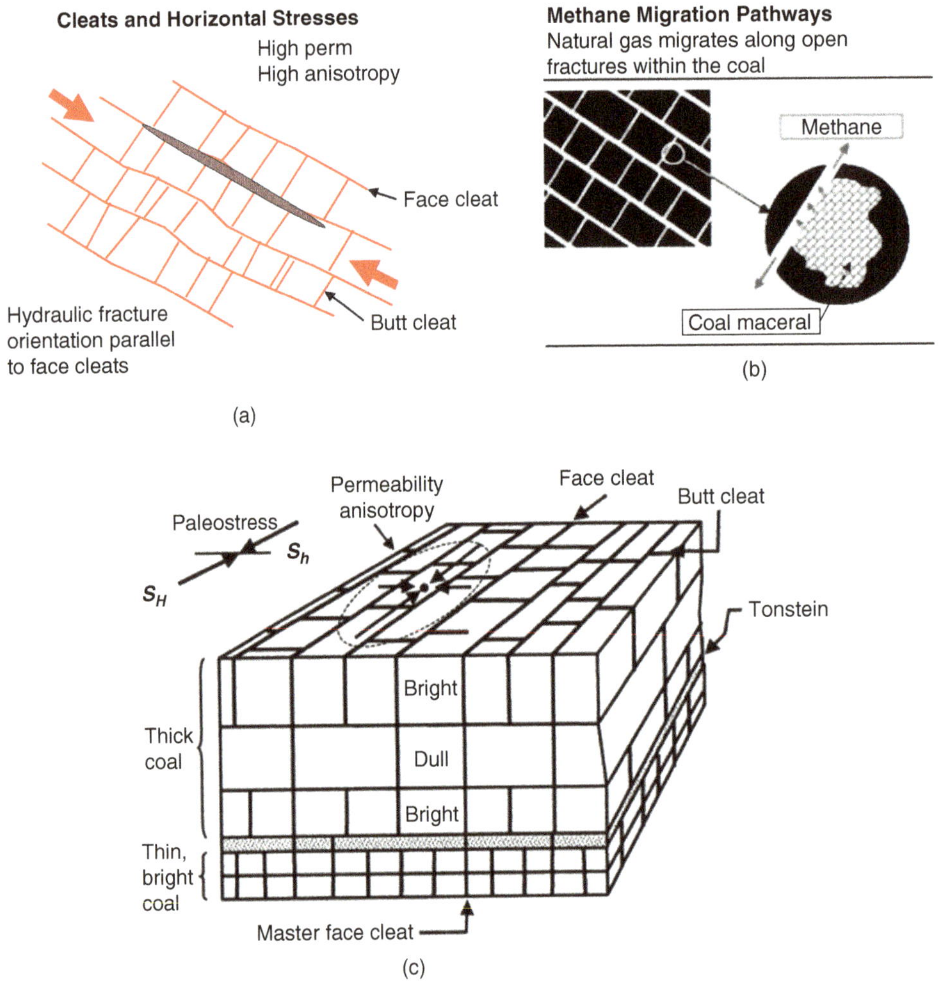

Fig. 3.10—Cleats in coal seam: (a) shows the butt cleats and the face cleats (Olsen et al. 2003); (b) shows the methane migration pathways through coal; (c) shows the permeability anisotropy in coal. S_H and S_h are maximum and minimum horizontal stress, respectively.

3.4.8 Coalbed Thickness. Coal-seam thickness affects the quantity of coalbed gas. Also, thickness affects the decision on whether to drill a vertical or a horizontal well. If the coal seam is thick, drilling a horizontal well is not considered to be the best economic choice because the wellbore might not access all parts of the reservoir (Holditch et al. 1990).

3.4.9 Reservoir Temperature. Reservoir temperature is important in selecting the stimulation design to use for hydraulic fracturing. The fracturing fluid used is directly dependent on the reservoir temperature because it affects the fluid stability.

3.4.10 Coal Porosity. Typically, coal seams have a macroporosity of 1 to 2%, resulting from the presence of cleats. The value of coal-seam porosity is not critical to the selection of completion and stimulation type (Hollub and Schafer 1992).

3.5 CBM Production Practices

CBM wells are drilled, completed, and stimulated in a manner similar to that used for conventional reservoirs. However, engineering practices differ somewhat because of the differences in the reservoir properties between conventional and coalbed reservoirs, and because of differences in coalbed properties from one case to the next. Identifying and understanding the geological and reservoir parameters of coal are necessary for the optimal design of the drilling, completion, stimulation, and production operations. The appropriate completion technique depends upon the specific reservoir characteristics, and each technique requires a different drilling procedure.

After completion, coal reservoirs typically undergo dewatering to reduce reservoir pressure and enable gas to desorb. Therefore, the wellbore configuration and completion technique must be designed to accommodate water and gas production needs. The types of drilling, completion, and stimulation methods that are currently used for producing CBM gas are discussed next.

3.5.1 Drilling Methods. The primary concerns in selecting the appropriate coalbed drilling method are formation damage, lost circulation because of high permeability, overpressure, gas/water flow, and wellbore stability. To address these problems, the following factors and data are considered when designing the drilling program:

- Formation depth, pressure, and production
- Type of coal and noncoal formations
- Well logs
- Drilling fluid specifications
- Casing program
- Drilling problems encountered
- Stimulation and the completion method that will be used

Any depleted zones, or other conditions that can cause a loss of circulation, must be determined. Also, other potential problem zones, such as sloughing shale zones and freshwater aquifers, must be identified.

The selection of the casing setting depth is an important factor in determining the casing string and drilling equipment. Some factors that affect drilling depth are the fracture gradient of the coal seams and adjacent formations, regulatory requirements, and drilling problems.

To determine the hole size for drilling, the following factors are considered:

- Expected production rates of water and gas
- Type of artificial-lift method that will be used
- Casing and tubing size
- Completion method that will be used
- Stimulation method that will be used
- Type of drilling fluid that will be used
- Expected future workover and recompletion requirements

3.5.1.1 Vertical Drilling. Most CBM wells are vertical. The commonly used methods for drilling vertical CBM wells are rotary percussion drilling and conventional rotary drilling. The formation hardness determines the type of drilling method that will be used. For softer formations, the rotary method is used, whereas for harder formations, rotary percussion drilling is used for a faster rate of penetration. The most commonly used drilling fluids in coal are air/mist, aerated mud, and formation water. The selection of fluid is dependent on the coal-seam reservoir properties.

To prevent formation damage while drilling, the coal is drilled underbalanced. This prevents the drilling fluid, chemical additives, and drilling solids from being injected into and plugging the cleat system of the coal. In the case of overpressured reservoirs, a slightly overbalanced water-based drilling fluid is used to maintain well control (Hollub and Schafer 1992).

3.5.1.2 Horizontal Drilling. Horizontal drilling is used to increase the footage of the production zone contacted by the borehole. Horizontal drilling increases the production rate and ultimate reserves recovered. The drilling equipment used for most horizontal wells includes a drilling bit, positive-displacement motor, logging while drilling, measurement while drilling, nonmagnetic drill collars, lateral "push" drillpipe, heavyweight drillpipe or drill collars used for weight, and drillpipe from surface (Lightfoot 2007).

Types of horizontal drilling are

- Long-radius hole (LRH)
- Medium-radius hole (MRH)
- Short-radius hole (SRH)

Horizontal wells have a kickoff point (KOP); a directionally drilled curve section to an inclination within 70 to 110°, depending on the dip of the coal; and a lateral section. The lateral section is drilled while changing the true vertical depth (TVD) of the well and the wellbore direction by adjusting the inclination and azimuth, respectively. Several types of CBM horizontal wells can be drilled **(Table 3.2).**

Horizontal Class	Horizontal Class Identifier	Horizontal Build Rate (°/100 ft)	Hole Radius (ft)	Wellbore Size Diameter (in.)
Long radius	LRH2	2°/100 ft	2,865	8½ 6½ 4¾
(up to 6°/100 ft)	LRH4	4°/100 ft	1,432	
	LRH6	6°/100 ft	955	
Medium radius	MRH8	8°/100 ft	716	
(7°/100 ft to	MRH12	12°/100 ft	477	
40°/100 ft)	MRH16	16°/100 ft	358	
	MRH20	20°/100 ft	286	6½ 4¾
	MRH25	25°/100 ft	229	
	MRH30	30°/100 ft	143	
	MRH35	35°/100 ft	164	4¾
	MRH40	40°/100 ft	143	
Short radius	SRH45	45°/100 ft	127	
(40°/100 ft to	SRH50	50°/100 ft	115	
60°/100 ft)	SRH55	55°/100 ft	104	
	SRH60	60°/100 ft	95	

Table 3.2—Classification of horizontal wells and well specifications.

MRH profiles are generally the preferred design, but for smaller hole sizes, SRH is selected with the appropriate drilling tools. MRH designs cover the widest range of build rates (6°/100 ft to 40°/100 ft) and can be drilled using the most-common drilling tool sizes (Lightfoot 2007).

LRH design is not suitable for CBM and many other unconventional horizontal drilling applications because the KOP above the desired lateral TVD is greater than 950 ft, as is the distance from the surface location to the start of the lateral section in the desired reservoir zone **(Fig. 3.11).** This excessive distance impacts the well's ability to produce and limits the lateral footage able to be drilled because of additional geological zones exposed in the curve. In addition, the extra distance on the build portion of the well is much longer. This increases the section of high contact forces on the drilling assembly (Lightfoot 2007).

Ultra-SRH wells have curve build rates greater than 60°/100 ft (radius less than 95 ft) and are not used for CBM wells because of the limited lateral section achievable. Ultra-SRH profiles are complex and expensive to drill, requiring specialized equipment (Lightfoot 2007).

3.5.2 Completion Methods. Before selecting a completion method for a CBM well, nine factors should be considered (Holditch et al. 1990). These are: investment required, number of coal seams encountered by the borehole, expected production rate, reserves in the various coal intervals, coal-seam permeability and gas content, type of stimulation treatment expected, wellbore-stability problems, future workover requirements, and artificial-lift requirements.

Well-completion design should be coordinated with the stimulation strategy. The need for and the type of pumping equipment must be considered. Besides efficiently removing liquids from the borehole, pump selection should recognize the effects of coal fines and fracturing sand that can migrate back to the wellbore. The estimation of gas-production rates expected after stimulation is also important. In most coalbed reservoirs, early flow rates are small, but they increase with time, when gas desorbs from the coal. The tubing in the well must be designed to maximize the lifting of liquids early in the life of the well, to help dewater the coal.

To help decide which zones to complete, a reserves analysis should be performed on each potential interval to determine the commercial value of the well. The effects of various sizes of stimulation treatments, the type of artificial lift, and the size of the tubulars have to be determined on the basis of reserves and the expected commercial value.

Other factors, such as surface injection pressures for the different wellbore configurations and the volumes of fluids required for stimulation, must be estimated. The pumping of fluids affects stresses in the tubular goods, and the changes in stress caused by the stimulation treatment must be computed to design the tubing and the casing.

The choice of completion type is made on the basis of these factors. After the completion type is chosen, the number of completions within the wellbore is determined, and the final tubing and casing configurations are designed. The different completion methods used in CBM are described next.

3.5.2.1 Casedhole Completion. The casedhole completion is the most commonly used coalbed completion type. It is used somewhere in most producing basins, and it is the most-common completion in medium- to low-permeability coalbeds. This completion is successful because it gives the best control over coal integrity and the stimulation of individual seams. Casedhole completions replaced openhole completions to solve coal-sloughing problems and to allow fracture stimulation treatments (Palmer et al. 1993a).

In most CBM casedhole completions, the casing is perforated and the coal is hydraulically stimulated. Thus, hydraulic fracturing design is an integral part of the casedhole completion design. The various aspects of hydraulic fracturing completion are discussed in Section 3.5.3.

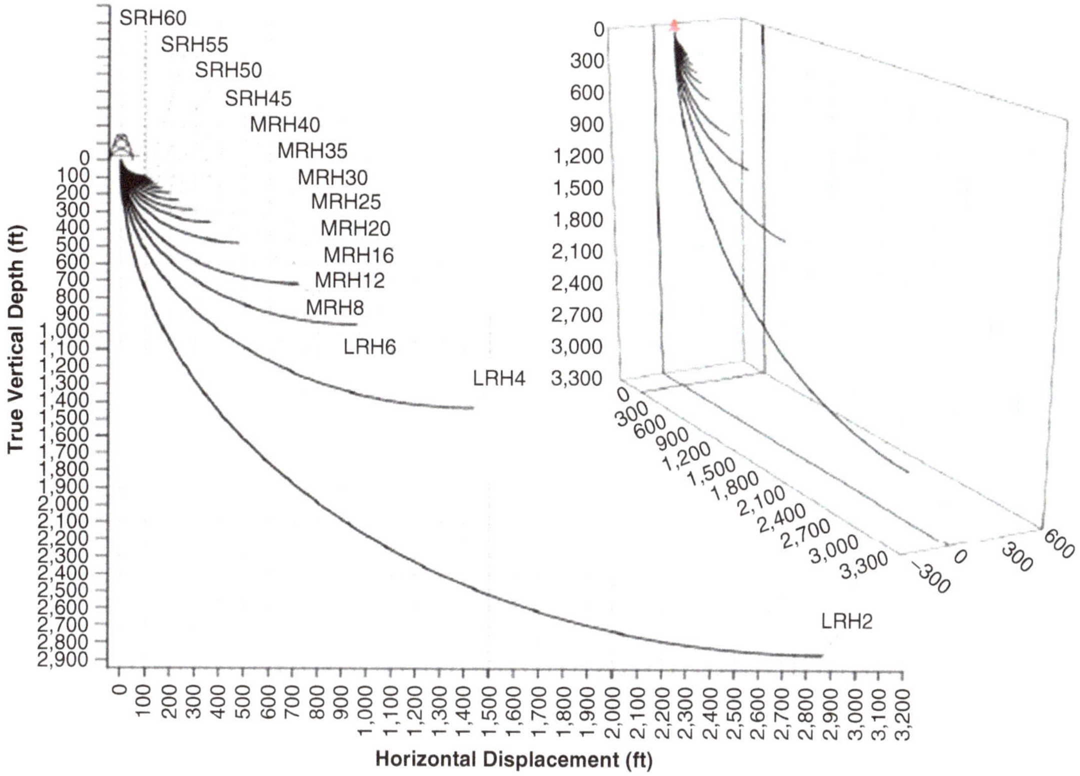

Fig. 3.11—Horizontal well profiles (Lightfoot 2007).

The casedhole completion is suitable for nearly all types of coal seams, with the exception of high-permeability coal seams. The most-important factors in selecting a casedhole completion are the economics involved and the type of stimulation treatment. Depending on the number of seams to be produced, the casedhole completion could be a single-seam or multiseam completion.

*3.5.2.2 **Multiseam Completions.*** Some disadvantages associated with the single-seam completion are (US Department of Energy 2003)

- It could cause thin coal seams to be ignored and, thus, cause large areas in the basin to remain uneconomic.
- It requires a much larger number of CBM wells, with increased capital costs and land disturbance, to produce the same quantity of gas that can be produced from fewer wells using multiseam completions.

Multiseam completions are used in the Appalachian, Black Warrior, Raton, and Uinta basins. The wells can be stimulated in a single fracturing treatment or several treatments, depending on the distance between the seams.

Multiseam technology for completing numerous coals was developed in the Black Warrior Basin. This technology improved the economics of CBM recovery, and also increased the estimated ultimate recovery of wells, because even the thin coal seams were accessed (US Department of Energy 2003).

*3.5.2.3 **Openhole Cavity Completion.*** CBM wells completed by the openhole cavity technique in the Fairway of the Fruitland Formation in the San Juan Basin have gas production rates nearly 10 times greater than those from wells completed by the fracturing stimulation of vertical wells in the same area. However, the openhole cavity completion method works only under favorable reservoir, geologic, and geomechanical conditions (Palmer and Cameron 2003).

The openhole cavity technique involves setting the casing to only the top of the coal formation with the drilling rig. Then, the coal is drilled out using a special completion rig (**Fig. 3.12**). To enhance the flowback and to encourage coal sloughing in the wellbore, compressed air is injected into the reservoir. Then, the well is allowed to rapidly flow back water, gas, and coal. This results in the formation of a cavity in the coal. The generated coal fines can be removed by circulating the drill bit to the total depth from time to time. This process is repeated until the cavity is stable. After stability is attained, the well is left open hole, or it is completed using a perforated, uncemented liner (Logan et al. 1993).

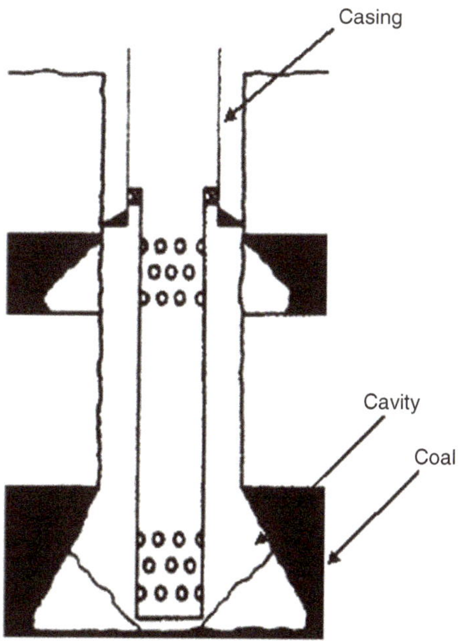

Fig. 3.12—Cross section of cavity completion method (Holditch 1990).

The cavitation process affects the wellbore in the following ways:

- It removes the drilling damage and increases the connectivity of the reservoir to the wellbore.
- It removes stress damage resulting from stress concentration around the wellbore.
- It enlarges the physical wellbore radius.
- It enhances the permeability in a zone beyond the cavity surface by up to five times the actual cavity radius in very weak coals (Palmer et al. 1993b).

This process has been successful only in the San Juan Basin Fairway region in the US and in a limited region of the Bowen Basin in Australia. The specific geologic conditions of the Fairway region that make this method successful are highly compressible coal, high permeability, formation overpressure, and high-gas-content coal.

3.5.2.4 ***Topset and Underream.*** A modified version of the openhole completion is the topset-and-underream method, which is used to produce coalbed gas from shallow coal seams in the Powder River Basin **(Fig. 3.13).** In this method, wells are drilled to the top of the coal and casing is set. Then, the well is drilled through the coal and underreamed to enlarge the borehole to enhance production and to remove permeability damage caused by drilling (Palmer and Cameron 2003).

Wells are then stimulated by pumping a small quantity of water (approximately 160 bbl) to remove the damage caused to the formation by drilling (Prime Western 2007). No proppant is used because the permeability of the reservoir is already very high.

Topset-and-underream wells in the Powder River Basin are successful because

- The permeability of the coalbed is very high.
- Coalbeds are thick and continuous.
- Coals are shallow, and, as a result, the cost of the drilling involved is low.
- Completions are very simple.
- The stimulation treatment used is simple and inexpensive.

Some disadvantages of this method are that the gas decline rate is very high when the reservoir permeability falls because fines can cause plugging because there is no proppant to trap them. The total cost for completing topset-and-underream wells in 2003 ranged from USD 65,000 to 135,000, making these wells very economical and easy to drill (Palmer et al. 1993a).

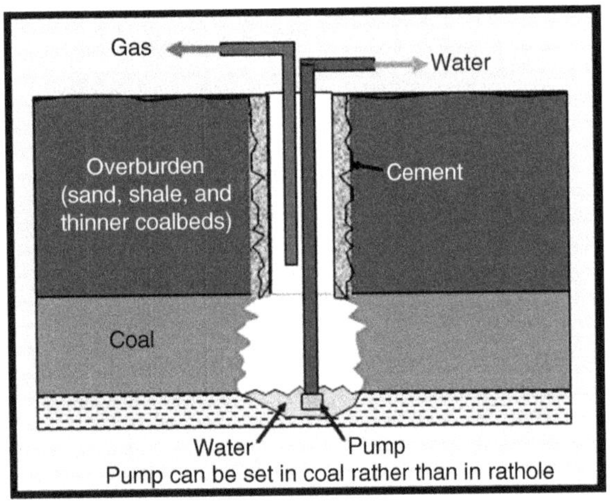

Fig. 3.13—Cross section of topset-and-underream well.

3.5.2.5 ***Horizontal Wells.*** Horizontal wells are drilled to maximize borehole contact with the reservoir. Fracture wings in vertical CBM wells have average half-lengths of less than 200 ft. Such short half-lengths are associated with the creation of complex fracture geometries, such as multistranded, jointed, and T-shaped fractures (Logan et al. 1987).

Increasing the footage of the production zone increases production and ultimate reserves recovered. Horizontal wells contact the main fracture system of the coal because they are drilled perpendicularly to the face cleats. This significantly increases production and ultimate gas recovery because of the large drainage area connected with the lateral. In conventional reservoirs, horizontal wells are most commonly used in formations that are somewhat flat, with thicknesses ranging from less than 10 ft to tens or hundreds of feet. However, in CBM, horizontal wells are drilled in seams ranging from 3 ft to approximately 20 ft thick. In thicker coal seams, horizontal wells are not effective because the wellbore is not able to contact the complete reservoir. To increase the connectivity to the reservoir, the well must be hydraulically fractured or more laterals must be drilled. Hydraulic fracturing has not been successful in horizontal CBM wells because the costs have not been justified by the limited increase in production. Drilling multilateral wells increases the drilling cost and the chances of the wells collapsing during drilling, but production is very high (Palmer and Cameron 2003).

The advantages of horizontal wells over vertical fracture-stimulated wells are that they

- Can be drilled to a length of 8,000 ft, whereas the effective CBM fracture lengths are usually less than 200 ft tip to tip
- Can be oriented in the direction of maximum horizontal stress to intersect face cleats, to provide maximum wellbore stability
- Are better in reservoirs with high-permeability anisotropy
- Can be better controlled to stay in seam (to avoid wet zones) than induced fractures can
- Could provide accelerated cash flow and a small footprint
- Can be expanded to various combinations (multilateral or pinnate designs and multiple fracturing options)

Some disadvantages of horizontal wells are that they are costly when there are many seams that require the drilling of multiple horizontals, and the chances of horizontals collapsing during drilling and production are high. A liner is highly recommended to prevent borehole collapse. In most cases, a preperforated liner is used.

3.5.2.6 ***Multilateral Horizontal Wells.*** Multilateral horizontal wells are drilled in cases in which the ratio of horizontal- to vertical-well gas production rates is less than unity. In these cases, the total contact area for a vertical well is more than that for a single horizontal well. For cases in which a number of thin coal seams are to be accessed, multiple lateral wells will provide greater production than a vertical well.

3.5.2.7 Pinnate Wells. Pinnate-pattern multilateral wells have proved very successful in producing coalbed gas from low-permeability coals **(Fig. 3.14).** Pinnate wells can have a 20-fold increase in production rate compared to fracture-stimulated vertical wells (CDX Gas 2007).

The pinnate well pattern was developed by CDX Gas to produce CBM from low-permeability coals. This method is extensively used in the Arkoma Basin. Some advantages of pinnate wells are that

- Wells can drain up to 2,000 acres from a single drill pad.
- Gas is produced immediately.
- Peak gas production is reached quickly, unlike in vertical wells in a CBM reservoir.
- Wells can drain a reservoir in 2 to 4 years.
- Gas recovery is high (80 to 90%).
- High gas flow rates (1 to 5 MMcf/D) can be achieved (von Schoenfeldt et al. 2004).

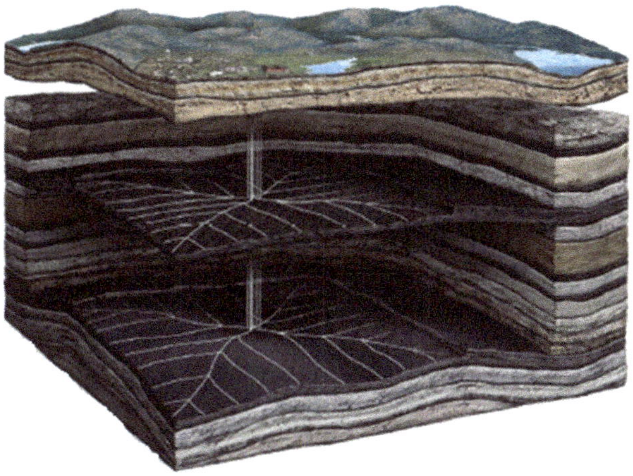

Fig. 3.14—Pinnate-pattern drilling (CDX Gas 2007).

Because many cases of lateral collapse have occurred, these wells are not suitable in high-permeability coals.

3.5.3 Fracturing Stimulation. Hydraulic fracturing is the most commonly used stimulation method in the CBM industry. The stimulation design depends on the reservoir properties. Stimulation treatments are used in casedhole wells to

- Bypass near-wellbore formation damage
- Stimulate production and accelerate dewatering by creating a high-conductivity path in the reservoir
- Distribute the pressure drawdown and thus reduce coal-fines production
- Connect the wellbore effectively to the natural-fracture system of the coal reservoir

Various fracturing techniques, fluid types, and procedures have been developed for coals (Holditch 1990).

Coal-seam fracturing can be compared to the hydraulic fracturing of a naturally fractured carbonate reservoir. In such a reservoir, the matrix permeability is low, and nearly all the productive capacity of the formation is controlled by the natural-fracture system (Holditch et al. 1990). To properly stimulate a naturally fractured carbonate reservoir, one must connect the natural-fracture system to the wellbore. In a coal-seam reservoir, the same goal must be achieved. We must connect the cleat system to the wellbore (Holditch 1990).

The complex stratigraphy of many coal seams complicates completion and stimulation procedures. In some areas, coal seams are relatively thick, uniform layers that are bounded by formations that are barriers to fracture growth. In other cases, however, coal seams occur in thin, multiple layers with essentially no barriers to vertical fracture growth between the seams.

The fracturing procedures and fluids used to stimulate CBM wells differ from operator to operator in a single basin because of local characteristics of geology and because of perceived advantages of cost, effectiveness, production characteristics, or other factors. Moreover, CBM projects in different basins could share common rock types and characteristics, but the fracturing stimulation treatments and fracture behavior might differ significantly.

Aspects of induced fractures, such as fracture dimensions (height, length, and width), are affected by the different fracturing approaches taken by the operator. Generally, the larger the volume of fracturing fluids injected, the greater the potential fracture dimensions. Fluid injection rates and viscosity also affect fracture dimensions. The interconnected cleat system might cause the volume of fluid leakoff to be large during a fracturing treatment. In a permeable coal seam, high injection rates, large pad volumes, and solid fluid-loss additives are needed to pump a fracturing treatment successfully.

If a hydraulically induced fracture has a relatively constant height (resulting from a geologic layer acting as a barrier to fracture propagation) and if the fracture is forced to grow and increase in volume (through an increased volume of fracturing fluid), the fracture will mainly grow in length. Increasing fluid viscosity can increase the injection pressure, resulting in greater fracture width and, thus, often shorter fractures. The type of stimulation treatment selected is a function of the depth, thickness, and permeability of a coal seam (Holditch et al. 1990). The different scenarios and various aspects of the hydraulic fracturing design process are further discussed in Part 3 of this book.

Fracturing-fluid selection is an important part of hydraulic fracturing design. Fracturing-fluid selection is based on site-specific characteristics, including formation geology, field production characteristics, and economics. The fracturing fluid should be able to carry high proppant concentrations, and it should not damage the formation. Hydraulic fracturing operations vary widely with the types of fracturing fluids used, the volumes of fluid required, and the pump rates at which they are injected. We can classify hydraulic fracturing fluids used for CBM wells as

- Fresh water and potassium chloride (KCl) water
- Gelled fluids
- Foam
- Acids
- Gas
- Hybrids (any combination of two or more of the aforementioned fluids)

3.5.3.1 Water Fracturing. Either groundwater pumped directly from the formation or treated water is used for fracturing CBM wells. In some CBM well stimulations, proppants are not needed to prop the fractures, so simple water or slightly thickened water can be a cost-effective substitute for an expensive polymer or foam-based fracturing fluid with proppant. Fresh water has a lower viscosity than gelled water, and, thus, fresh water has a lower proppant-transport capacity (US Environmental Protection Agency 2004). Hydraulic fracturing performance is not exceptional with fresh water, but, in some cases, the production rates achieved are adequate and the lower costs make the well economical to produce. The proppant-carrying capacity of water ranges from 1 to 2 lbm/gal. The fracture conductivity attained with water is good but not better than that with gelled fluids. The biggest advantage of using water fracturing is that it is inexpensive (Holditch et al. 1990).

3.5.3.2 Gelled Fluids. Two types of gelled fluids commonly used in CBM wells are linear gels and crosslinked gels. Gelled fracturing fluids are used because they have a much better proppant-carrying capacity than water and can thus attain better fracture conductivity. The most-common problem faced with gelled fracturing fluids is the polymer residue left behind. This residue causes permeability damage to the coal. To solve this problem, newer and better cleaning agents are being used with gelled fluids. These cleaning agents suppress the fines movement and plugging of the proppant packs that result from coal-fines production during CBM production. The disadvantage of gelled fluid is that it is expensive to use. The proppant-carrying capacity of gelled fluid ranges from 5 to 12 lbm/gal. Linear gel can carry up to 8 lbm/gal, whereas crosslinked gel can carry as much as 12 lbm/gal (Holditch et al. 1990).

3.5.3.3 Linear Gels. The most commonly used gelling agents in fracturing fluids are guar gum, guar derivatives such as hydroxypropyl guar and carboxymethyl hydroxypropyl guar, or cellulose derivatives such as carboxymethyl guar or hydroxyethylcellulose. Gelling agents are biodegradable in nature (US Environmental Protection Agency 2004).

3.5.3.4 Crosslinked Gels. Crosslinking agents can be added to linear gels to form high-viscosity fracturing fluids called crosslinked gels. Crosslinked gels provide higher proppant-transport performance than linear gels. Crosslinking reduces the need for fluid thickener and extends the viscous life of the fluid indefinitely. The fracturing fluid remains viscous until a breaking agent is introduced to break the crosslinker and, eventually, the polymer. Crosslinked gels are normally metal-ion crosslinked guar. Metal ions such as chromium, aluminum, titanium, and others are used to achieve crosslinking (US Environmental Protection Agency 2004).

3.5.3.5 Foam. Foam fracturing technology uses foam bubbles to transport and place proppant in fractures. The most widely used foam fracturing fluids use nitrogen (N_2) or carbon dioxide (CO_2) as their base gas. Incorporating inert gases with foaming agents and water reduces the amount of fracturing liquid required. Foamed gels use fracturing fluids with higher proppant concentrations to achieve highly effective fracturing. The gas bubbles in the foam fill voids that would otherwise be filled by fracturing fluid. The high concentrations of proppant allow for an approximate 75% reduction in the overall amount of fluid that would be necessary using a conventional linear or crosslinked gel. Foaming agents can be used in conjunction with gelled fluids to achieve an extremely effective fracturing fluid. They are more generally used in cases in which there is low water content in the coal cleats and low reservoir pressure gradients (less than 0.2 psi/ft) (US Environmental Protection Agency 2004). Gelled foams have a proppant-carrying capacity of up to 8 lbm/gal. Some of the advantages of using foam as a fracturing fluid are that it causes less permeability damage because less fluid is involved and it has better cleanup than gelled fracturing fluids. Foam fracturing is expensive, and, thus, the use has to be justified economically.

3.5.3.6 Acid Fracturing. Acids are used in carbonate formations that overlie or are interbedded within coals to dissolve the rock and create a conduit through which formation water and CBM can travel (US Environmental Protection Agency 2004). The stimulation fluid is hydrochloric acid, or a combination of hydrochloric and acetic or formic acids. For acid fracturing to be successful, thousands of gallons of acid must be pumped far into the formation to etch the face of the fracture. Some of the cellulose derivatives used as gelling agents in water and water/methanol fluids can be used in acidic fluids to increase treatment distance. Acids can also be used as a component of breaker fluids, and they can be used to clean up perforations of the cement surrounding the well casing before fracturing-fluid injection (US Environmental Protection Agency 2004).

3.5.3.7 Gas Fracturing. When coal comes in contact with water, it can swell, closing the cleats and causing the formation to lose permeability. Therefore, CO_2 or N_2 can be used as the fracturing fluid in CBM reservoirs that have water-sensitive coal. Coiled tubing is commonly used to pump these fluids.

All these variations of hydraulic fracturing fluids are used in North American coal basins **(Table 3.3).** The reader is referred to Ramaswamy (2007) for details regarding specific basins and more information on how the industry has drilled, completed, and stimulated coal seams in the following basins:

- Black Warrior Basin
- Central Appalachian Basin
- Northern Appalachian Basin
- Arkoma Basin
- Cherokee Basin
- Forest City Basin
- Powder River Basin
- San Juan Basin
- Uinta and Piceance basins
- Raton Basin
- Western Canada Sedimentary Basin

Basin	Formation	Thickness (ft) of Coal (all formations)	Depth of Coal (ft)	Fracturing Fluids Used
San Juan	Menefee, Fruitland	20–40	550–4,000	Slickwater, crosslinked gels, N_2 and CO_2 foam
Black Warrior	Pottsville (Mary Lee/Pratt/Black Creek)	2–20	350–2,500	Water, linear gel, crosslinked gel
Piceance	Mesaverde	5.5–12.1	5,000–6,000	Water, linear gel
Uinta	Mancos,	10–50	150–750	Crosslinked gel
	Mesaverde		1,000–7,000	
Powder River	Wasatch,	15–100	0–1,000	Water
	Fort Union		450–6,500	
Central Appalachian	Pocahontas		1,000–2,000	Foam, water
North Appalachian	Pottsville, Allegheny, Monongahela	< 1–8	0–2,000	Water, N_2 foam
Arkoma	Hartshorne, McAlester	2–12	600–2,300	Linear gel, N_2 foam
Cherokee	Weir-Pittsburgh, Riverton	2–5	600–1,200	Linear gel, N_2 foam
Raton	Vermejo, Raton	5–35	3,000–4,000	Linear gel, water
Alberta	Horseshoe Canyon	1.5–9.8	490–2,780	N_2 and CO_2 gas

Table 3.3—Fracturing fluids used in CBM operations in North American basins (Palmer et al. 1993a).

3.6 Summary of CBM

- Drilling, completion, and stimulation methods used in CBM reservoirs differ from basin to basin, and areally within basins, owing to variations in geologic setting and coal-seam properties.
- The key geologic parameters that affect the selection and success of drilling, completion, and stimulation practices in CBM reservoirs are coal depth, thickness, areal extent, dip, permeability, rank, gas content, formation pressure, water saturation, and compressive strength, as well as the vertical distribution of coalbeds and distance to fracture barriers and aquifers.
- Topset-and-underream openhole wells exist primarily in the Powder River Basin. Key reservoir parameters and the associated range of values that influence the selection and the success of this method are
  - Depth < 1,800 ft
  - Permeability > 100 md
  - Thickness > 30 ft
- The openhole cavity completion method is successful primarily in the Fairway region of the San Juan Basin. Key reservoir parameters and the corresponding range of values that influence the selection and the success of this method are
  - Compressive strength < 1,000 psi
  - Permeability > 10 md
  - Coal rank: high- to medium-volatile bituminous
- Horizontal wells are successful in the Arkoma, Appalachian, and San Juan basins. Key reservoir parameters and the corresponding range of values that influence the selection and the success of this method are
  - Thickness of 3 to 20 ft
  - Areal extent of the coal ≥ 1,500 ft
  - Dip of the coal seam < 15°
  - Depth of 500 to 4,000 ft
- Multilateral wells are successful in the Arkoma and Appalachian basins. The key reservoir parameters that influence the selection of this method are similar to those of horizontal wells. A key factor is that multilateral wells are successful when permeability is less than 1 md.
- Multilateral horizontal wells drilled in a pinnate pattern have the highest recovery efficiency of any type of CBM completion method, with a more than 85% recovery.
- The casedhole completion with hydraulic fracturing stimulation is the most commonly used completion and stimulation method and is applicable to all coalbed reservoirs with permeabilities less than 100 md.
- The following are geologic parameters and the recommended stimulation method:
  - For coalbeds with permeabilities greater than 100 md, water fracturing without proppant.
  - For dry coals, gas fracturing is the best stimulation method.
  - For coalbeds with low water saturation and low reservoir pressure, foam fracturing.
  - For coalbed reservoirs with permeabilities of 10 to 100 md, crosslinked-gel fracturing with proppant.
  - For coalbeds with permeabilities less than 1 md, water fracturing with proppant.
- For CBM wells with more than one coalbed, multistage fracturing is the best stimulation method if the seams are separated by a distance of more than 40 ft.

3.7 References

Beaton, A. 2003. Production Potential of Coalbed Methane Resources in Alberta. EUB/AGS Earth Sciences Report 2003-03, Alberta Energy and Utilities Board/Alberta Geological Survey (EUB/AGS), Edmonton, Alberta, Canada (September 2003).

Cardott, B. J. 2001. Introduction to Coal as Gas Source Rock and Reservoir. Oklahoma Coalbed-Methane Workshop 2001 presented 10 October 2001, Carl Albert State College, Poteau, Oklahoma, USA. Oklahoma Geological Survey Open-File Report 2-2001, 1–27.

CDX Gas. 2007.
Holditch, S. A. 1990. Completion Methods in Coal Seam Reservoirs. Paper presented at the SPE Annual Technical Conference and Exhibition, New Orleans, Louisiana, USA, 23–26 September.
Holditch, S. A. 2006. Tight Gas Sands. *J Pet Technol* **58** (6): 86–93. SPE-103356-JPT. https://doi.org/10.2118/103356-JPT.
Holditch, S. A., Ely, J. W., Carter, R. H. et al. 1990. Coal Seam Stimulation Manual. Topical report, contract No. 5087-214-1469, Gas Research Institute, Chicago, Illinois (April 1990).
Hollub, V. A. and Schafer, P. S. 1992. Drilling and Casing the Wellbore. In *A Guide to Coalbed Methane Operations*, V. A. Hollub and P. S. Schafer, Chap. 2, 1–41. Gas Research Institute.
Lightfoot, J. D. 2007. Drilling Sideways in Coal Seams: Development of New "Fit for Purpose" Technology to Optimize CBM Horizontal Drilling. Paper presented at the International Coalbed Methane Symposium, Tuscaloosa, Alabama, USA, 22–25 May.
Logan, D. L., Mavor, M. J., and Khodaverdian, M. 1993. Optimization and Evaluation of Openhole Cavity Completion Techniques for Coal Gas Wells. Paper presented at the Low Permeability Reservoirs Symposium, Denver, Colorado, USA, 26–28 April. SPE-25859-MS. https://doi.org/10.2118/25859-MS.
Logan, T. L., Schwoebel, J. J., and Horner, D. M. 1987. Application of Horizontal Drainhole Drilling Technology for Coalbed Methane Recovery. Paper presented at the SPE/DOE Joint Symposium on Low Permeability Reservoirs, Denver, Colorado, USA, 18–19 May. SPE-16409-MS. https://doi.org/10.2118/16409-MS.
Magoon, L. B. and Dow, W. G. 1994. *The Petroleum System—From Source to Trap*, Vol. 60, 3–24. AAPG Memoir, American Association of Petroleum Geologists.
Olsen, T. N., Brenize, G., and Frenzel, T. 2003. Improvement Processes for Coalbed Natural Gas Completion and Stimulation. Paper presented at the SPE Annual Technical Conference and Exhibition, Denver, Colorado, USA, 5–8 October. SPE-84122-MS. https://doi.org/10.2118/84122-MS.
Palmer, I. D. 2007. Coalbed Methane Well Completions and Production: Field Data, Best Practices, New Aspects. Presented as a pre-symposium short course, International Coalbed Methane Symposium, Tuscaloosa, Alabama, USA, 22 May.
Palmer, I. D. and Cameron, J. 2003. Coalbed Methane Completions: Update on Recent Practices. Paper presented at the Coalbed Methane Symposium, Calgary, Alberta, Canada, 7–9 April.
Palmer, I. D., Lambert, S. W., and Spitler, J. L. 1993a. Coalbed Methane Well Completions and Stimulations. In *Hydrocarbons from Coal*, Vol. 38, ed. B. E. Law and D. D. Rice, Chap. 14, 303–339. AAPG Studies in Geology, American Association of Petroleum Geologists.
Palmer, I. D., Mavor, M. J., Seidle, J. P. et al. 1993b. Openhole Cavity Completions in Coalbed Methane Wells in the San Juan Basin. *J Pet Technol* **45** (11): 1072–1080. SPE-24906-PA. https://doi.org/10.2118/24906-PA.
Patchen, D. G., Schwietering, J. F., Avary, K. L. et al. 1991. *Coalbed Gas Production Big Run and Pine Grove Fields, Wetzel County, West Virginia*. Charleston, West Virginia: West Virginia Geological and Economic Survey.
Pratt, T. J., Close, J. C., and Mavor, M. J. 1992. Summary of the Cooperative Research Well FC Federal #12 Operated by Mesa Operating Limited Partnership. Topical report GRI-92/0015, Gas Research Institute, Chicago, Illinois (February 1992).
Prime Western. 2007.
Ramage. 1997. Coal Fact Sheet. Australian Institute of Energy.
Ramaswamy, S. 2007. *Selection of Best Drilling, Completion and Stimulation Methods for Coalbed Methane Reservoirs*. Master's thesis, Texas A&M University, College Station, Texas (December 2007).
Remner, D. J., Ertekin, T., Sung, W. et al. 1986. A Parametric Study of the Effects of Coal Seam Properties on Gas Drainage Efficiency. *SPE Res Eng* **1** (6): 633–646. SPE-13366-PA. https://doi.org/10.2118/13366-PA.
Ryan, B. and Mannhardt, K. 2007. Diffusivity and Diffusion Coefficients of Anthracite. Paper presented at the International Coalbed Methane Symposium, Tuscaloosa, Alabama, USA, 22–25 May.
Scott, A. R., Kaiser, W. R., and Ayers, W. B., Jr. 1994. Thermogenic and Secondary Biogenic Gases, San Juan Basin, Colorado and New Mexico—Implications for Coalbed Gas Producibility. *AAPG Bulletin* **78** (8): 1186–1209. https://doi.org/10.1306/A25FEAA9-171B-11D7-8645000102C1865D.
US Department of Energy. 2003. Multi-Seam Well Completion Technology: Implications for Powder River Basin Coalbed Methane Production. DOE/NETL-2003/1193, US Department of Energy and National Energy Technology Laboratory, Washington, DC (September 2003).
US Energy Information Administration. 2007a. US Coalbed Methane: Past, Present, and Future. Panel 2 of 2. Washington, DC: US Department of Energy. www.eia.doe.gov/oil_gas/rpd/cbmusa2.pdf.
US Energy Information Administration. 2007b. https://www.eia.gov/.
US Environmental Protection Agency. 2004. Evaluation of Impacts to Underground Sources of Drinking Water by Hydraulic Fracturing of Coalbed Methane Reservoirs. Final report, contract no. EPA 816-R-04-003. US Environmental Protection Agency, Washington, DC (June 2004).
von Schoenfeldt, H., Zupanik, J., Wight, D. R. et al. 2004. Unconventional Drilling Methods for Unconventional Reservoirs in the US and Overseas. Paper presented at the International Coalbed Methane Symposium, Tuscaloosa, Alabama, USA, 3–7 May.

Chapter 4

Shale Gas*

Archna Agrawal, Yao Tian, and Stephen A. Holditch

4.1 Introduction

Although gas shales have become an important strategic play for many companies, there is still uncertainty in determining the optimal drilling and completion techniques for wells drilled in various shale reservoirs. The industry seems to optimize development in each different play by trial and error. Instead, the industry needs to develop predictive models to help determine the optimal drilling, completion, and stimulation options as functions of the shale reservoir properties and economic conditions. In short, the industry needs to determine best practices for various shale scenarios. Another reason that the compilation of knowledge and data transfer are so imperative is that our workforce is aging, and many of the personnel working in the oil and gas industry will reach retirement age relatively soon; hence, without proper documentation, much of the knowledge needed to develop unconventional gas reservoirs will retire with those employees.

As companies continue to increase their activity in shale gas reservoirs, it will be important to decide the best way to drill and complete these wells because technology and understanding are constantly evolving. Thus, there can be a steep learning curve for new companies or new shale plays, which can affect profitability.

The greatest uncertainties and the largest cost in most shale plays are the drilling, completion, and stimulation methods that should be used in any development program.

In gas shale reservoirs, as in most unconventional reservoirs, the characteristics can vary substantially, both vertically and areally. However, despite these variations, shale reservoirs in different basins can be evaluated and compared, and best practices should be somewhat transferable. If engineers and geologists look for differences and similarities in reservoir and geologic properties, it should be possible to use best practices to design the drilling and completion operations to optimize gas recovery and economics. Most reservoir engineers working in gas shale development do not have the time to conduct a thorough study of industrywide activity to help determine what works in other areas and what does not. However, if such information is made available, shale gas development could potentially escalate and help the US meet its energy demand, as well as help the development of international gas shale fields.

Using information in the oil and gas literature on existing gas shale reservoirs, we can develop flow charts/diagrams to allow the application of best practices in new or even existing gas shale plays. This study will help engineers and geologists better understand which drilling and completion practices are successful in plays similar to theirs, as well as determine which completion technique is best in a given economic environment.

Gas shale reservoirs are a very important part of current industry activity. Of the many facets being scrutinized, the most important is how to economically, effectively, and safely drill and complete these wells while maximizing the value of each. There are several factors to explore to determine how to optimally complete and produce from any specific gas shale formation. Depositional environment, total organic content (TOC), gas content, shale mineralogy, shale thickness, and reservoir pressure are the main properties that affect how the shale will produce and the technology that works in each play. After these factors have been determined and coupled with the price environment, it is easier to decide which completion technique is most suitable, whether it be vertical or horizontal drilling. After this decision has been made, the completion and stimulation procedures become crucial to well success.

4.2 Gas Shale Properties

4.2.1 Overview of Gas Shale Systems. As a result of the successful production of large volumes of gas from the Barnett Shale, and then the Haynesville Shale, the petroleum industry switched gears from looking at shales as a source rock to analyzing them as possible gas reservoirs. Most believe that gas shale systems are vastly different from other unconventional plays because of the nature of producing from source rocks, making it difficult to determine optimal drilling, completion, and stimulation methods. Because these source rocks are also reservoir rocks, the environment during deposition must have been anoxic, meaning that organic-rich material could settle with minimal oxygen contact, enabling the material to later generate hydrocarbons. As shale is buried, it

*This chapter is an edited version of an MS degree thesis in petroleum engineering by Archna Agrawal at Texas A&M University in 2009. It also includes excerpts from a PhD degree dissertation by Yao Tian at Texas A&M University in 2014.

undergoes two main processes to generate gas. Biogenic gas can be formed through the action of anaerobic micro-organisms, or thermal gas can be formed through the thermal breakdown of kerogen. With the use of vitrinite reflectance (VR) and core analysis, the origin of the gas can be determined as either biogenic or thermal. Gas shale rocks that are organic rich are usually dark in color (brown/black) with high TOC (can be higher than 10%) and high gamma ray (GR) signatures (greater than 140 API units). The porosity and permeability of these organic shales will be a function of compaction during burial history.

While under production, most shales will produce gas that is stored in one of two places: free gas in the pores and natural fractures, or gas that has been adsorbed to the organic material in the shale. Free gas is what is found in the pores and natural fractures of most formations. Adsorbed gas is that attached to the surface of the organic matter and is released only as the pressure in the reservoir declines. Laboratory tests must be run using cores to determine how much gas will be desorbed from the surface of the shale as the reservoir pressure declines. After this is known, the volume of desorbed-gas production can be easily modeled and predicted.

The main geologic parameters that must be known to determine the quality of a gas shale reservoir are depositional environment, TOC, average gas content, shale mineralogy, thickness, and reservoir pressure. The depositional environment is important for determining a commercially viable shale gas reservoir because this affects how the hydrocarbons are formed, establishes if hydrocarbons are even present, and determines what type of hydrocarbons might exist in the shale. TOC is another factor to evaluate because it is indicative of the quantity of organic material available for the formation of hydrocarbons. It can be directly proportional to the yield of gas and enables the evaluation of organic-matter transformation.

The average gas content is important because it is indicative of what is in place and can be used to forecast what is recoverable. In general, shales with high values of gas content also have higher values of gas permeability. Obviously, a more porous and permeable shale will contain more gas and will enable production at higher gas flow rates. Shale mineralogy is vital to the success of any gas shale play. If there is high clay content, the shale will be more difficult to fracture treat, and it will be more difficult to keep a fracture propped open over time. If a shale contains more quartz, it will be more brittle, and it will be easier to fracture treat and keep a hydraulic fracture propped open.

Thickness is another important parameter for commercial gas shales. It is difficult to produce from shales less than 50 ft thick because there might not be enough gas-in-place to justify the costs to drill and complete the well. Most thinner shales tend to be uneconomic. Likewise, if a shale is too thick, it can be difficult to determine the best layers from which to produce, and the effectiveness of horizontal drilling is reduced unless large fracturing treatments are pumped or multiple horizontal holes are drilled. Reservoir pressure also plays a key role in determining the gas-in-place and gas recovery of the shale. Gas shales with high pressure gradients might never have been compacted, and it is likely that the porosity and permeability will be better than in low-pressure gas shales, all other things being equal.

These geologic parameters will be discussed for five emerging gas shale basins: the Michigan Basin (Antrim Shale), Fort Worth Basin (Barnett Shale), North Louisiana Salt Basin (Haynesville Shale), Appalachian Basin (Marcellus Shale), and Arkoma Basin (Woodford Shale).

4.2.2 Antrim Shale. The Antrim Shale is a gas shale in which production occurs mainly along a belt that crosses the northern part of the Michigan Basin **(Fig. 4.1),** which has been producing since the early 1940s (Goodman and Maness 2008). Despite the early discovery, the Antrim Shale did not become active (economic) until the 1980s. Goodman and Maness (2008) indicate that as of 2008, the Antrim Shale Formation has produced more than 2.6 Tcf of gas from nearly 10,000 wells. On the basis of a US shale gas study (Goodman and Maness 2008), the estimated gas-in-place ranges from 35 to 76 Tcf. The typical depth where this formation is encountered in the basin is between 500 and 2,000 ft and extends areally approximately 30,000 sq miles (Goodman and Maness 2008).

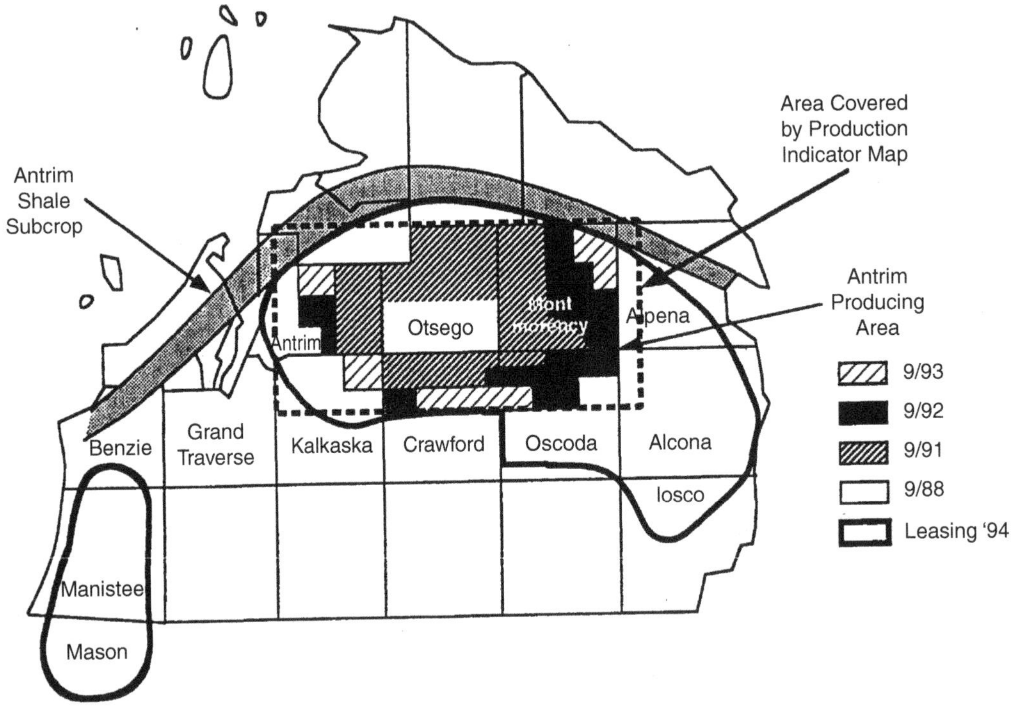

Fig. 4.1—Antrim Shale, located in the Michigan Basin (Zuber et al. 1994).

Through the use of VR and core analysis, it was concluded that the Antrim Shale produces biogenic gas, as opposed to thermogenic gas (Goodman and Maness 2008). From the Antrim Shale outcrop, it is black in color, organic rich, brittle, and radioactive, and contains bitumen (Kuuskraa et al. 1992). Because of radioactivity in the shale, it is easily spotted on a GR log **(Fig. 4.2).**

Understanding the potential recovery of the Antrim Shale is difficult because of three main factors (Kuuskraa et al. 1992). First, this shale produces a substantial amount of water along with the gas production, similar to many coalbed methane reservoirs. The presence of this water complicates the system because it behaves like a two-phase system. The second reason is that there is such a large amount of adsorbed gas and the system of storage, release, and diffusion of gas in shales is poorly understood. Third, the Antrim is a naturally fractured formation, which makes it difficult to determine where the highly fractured areas exist and where the "sweet spots" are (Matthews 1993).

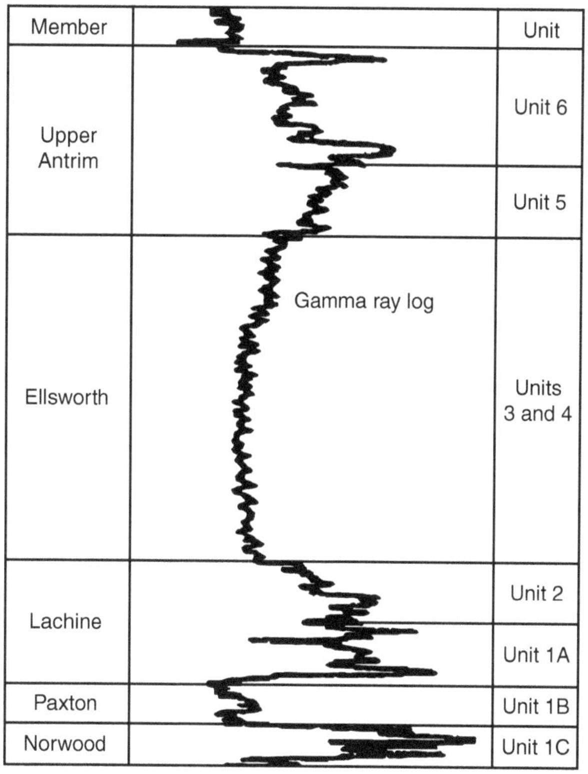

Fig. 4.2—GR log of Antrim Shale (after Kuuskraa et al. 1992).

4.2.2.1 Depositional Environment. The stratigraphy in the Michigan Basin, the Antrim Shale, was deposited in the Late Devonian and Early Mississippian (Matthews 1993). This shale was generated as part of a large Devonian–Mississippian "Black-Shale Sea," which deposited this organic-rich formation from the Transcontinental Arch, found in the west, to the Appalachians, found in the east. When the Antrim Shale began depositing in the Michigan Basin, it was deposited as a shale/sand sequence. In modern times, it has been observed that, moving from the center of the basin to the margins, the shale rises but then truncates under a cover of glacial drift on land or under the lake-sediment deposits of the Great Lakes. Because there is an absence of the Devonian–Mississippian shales across the basin margins, it is conclusive that extensive erosion took place throughout much of the continental interior during the Late Devonian. Also, evidence of the Paleozoic rocks that were once present in the Michigan Basin is found in the elevated organic maturity of the rocks now present in the basin (Matthews 1993).

During its deposition and burial history, four distinct layers, as seen in Fig. 4.2, were formed, which are now considered the Antrim Shale: from top to bottom, the Upper Antrim, Lachine, Paxton, and Norwood (Kuuskraa et al. 1992). The Upper Antrim is made up of several different unit formations. Traditionally, the Upper Antrim is picked at the top of the first black shale beneath the Bedford Formation. The Ellsworth Shale occurs between the Upper Antrim section and the Lower Antrim section that compose the Antrim Shale. The Ellsworth is gray to greenish gray, distinguishing itself from the Antrim members. The Lachine is marked by an increase in the GR log. The top of the Paxton can be identified by a sharp decrease in the GR log, and the top of the Norwood is distinguished by a sharp increase in the GR log. These deflection differences in the GR make it easy to determine where the different layers are in a new well and allow one to correlate between wells. Fig. 4.2 also depicts these attributes for the different stratigraphic members of the Antrim Shale.

4.2.2.2 TOC. The organic content in the Antrim Shale can be as high as 20%. Matthews (1993) indicates that most of this organic material is algal, meaning the hydrocarbons-in-place come from mostly plant material, rather than animal. Plant materials encountered in this shale include tree branches, Callixylon newberryi, and Tasmanite spores. Some fish scales and other small-animal fossil impressions are also present in the shale (Matthews 1993).

4.2.2.3 Gas Content. The total average porosity found throughout the Antrim Shale is approximately 9%, but it can range from 3 to 10%. Of that 9%, the average gas porosity is from 2 to 6%. This porosity accounts for only the free gas that is available in the pores and not the adsorbed gas. The Antrim Shale is shallow compared with other producing shale gas reservoirs, so it is important to know the amount of adsorbed gas because it will probably be liberated when the well begins production and the reservoir pressure decreases. Approximately 60 to 70% of the gas in this gas shale system is adsorbed (Kuuskraa et al. 1992). Gas content is an important factor when determining cubic feet of gas content per ton of rock. Phasis Consulting (2008) states that the gas content ranges from 40 to 100 scf/ton in the Antrim Shale, depending on where the well is located in the basin. Because there is such a large range in the gas content, the estimated ultimate recovery (EUR) varies significantly. Through core analysis, pressure transient testing, and well testing, it has been indicated that the EUR could range from as low as 20% to as high as 60%. If the recovery factor actually reaches 60%, it can be deduced that operators will produce the adsorbed gas that is released.

4.2.2.4 Shale Mineralogy. The mineralogy of the Antrim Shale, as stated by Matthews (1993), is rather uniform and contains both quartz and clay. The Antrim Shale has unusually high quartz content compared with many other shales, making it ideal as a reservoir rock. It can contain up to 50 to 60% quartz in its mineralogy, which is almost twice the amount found in other source-rock shales in the US. The next-most-abundant mineral in this shale is illite, which makes up approximately 20 to 35% of the shale. Remaining minerals found in various quantities are kaolinite, chlorite, and pyrite. Calcite and dolomite can be present in the form of limestone nodules and lenses in the lower half of the shale, as thick as 5 ft. Bitumen is also found in this shale at approximately 0.2 to 0.8%.

4.2.2.5 Thickness and Reservoir Pressure. The shale thickness increases to the northwest and to the east. An isopach map of the entire Antrim Shale found in the Michigan Basin is shown in **Fig. 4.3.**

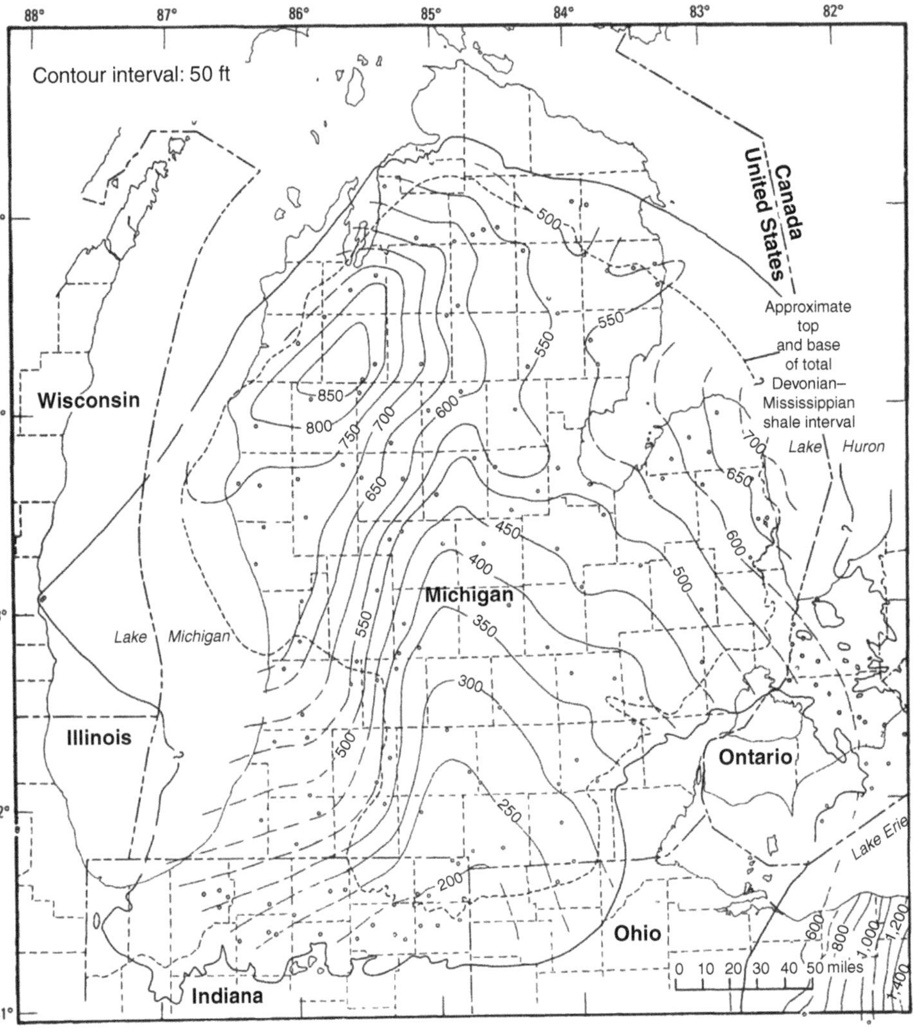

Fig. 4.3—Antrim Shale isopach map (Matthews 1993).

The Antrim Shale is found at depths from 500 to 2,000 ft, depending on the location. The greater depths (> 1,500 ft) are located at the center of the basin, and it gets more shallow toward the margins. The gross thickness of the Antrim Shale ranges from 200 to 850 ft. Of the gross thickness, 70 to 120 ft is considered to be net-pay, yielding a high net-to-gross ratio (Phasis Consulting 2008).

The average reservoir pressure in the Antrim Shale (Hayden and Pursell 2005) is approximately 400 psi at 1,150 ft, yielding a pressure gradient between 0.35 and 0.38 psi/ft. This is lower than normal hydrostatic pressure, but because of low pipeline pressures, the industry still produces the Antrim Shale economically. It should be noted that because the pressure is now lower than hydrostatic pressure, some wells must be producing adsorbed gas in addition to free gas in the pore volume. Most of the wells are on artificial lift to remove water from the wellbore.

4.2.3 Barnett Shale. The Barnett Shale was the most-active gas shale in the US in the early 2000s and is located in the Fort Worth Basin. Givens and Zhao (2008) reveal not only the history of the Barnett Shale but also provide some geologic information regarding the Fort Worth Basin. The Fort Worth Basin covers approximately 15,000 sq miles of North Texas **(Fig. 4.4),** of which the Barnett Shale covers approximately 5,000 sq miles and has depths ranging from 6,500 to 8,500 ft.

The first Barnett well was drilled in 1981, in Wise County, but not much additional drilling activity occurred in the 1980s (Brackett 2006). Airhart (2007) states that the reason the Barnett drilling program took off in the late 1990s was the higher prices for natural gas and improvements in drilling technology. Based on published data (Hayden and Pursell 2005), the production and rig-count increase is shown in **Fig. 4.5.** As of July 2008, the Barnett had produced approximately 4.5 Tcf of gas (Brackett 2006). The estimated gas-in-place is between 25 and 30 Tcf (NKNT 2008). The Barnett Shale is also present in other basins, such as the Permian; however, an overwhelming majority of the wells have been drilled in the Fort Worth Basin. Under the same geologic conditions, the same completion techniques found to be the most economical in the Barnett Shale can be used in the other basins where the Barnett Shale exists. However, the areal extent is large, so there is no guarantee that the Barnett Shale in Oklahoma will be similar to the Barnett Shale in the Fort Worth Basin or the Barnett Shale in West Texas. It is important to look at all the shale properties before determining an analogy.

Although the Fort Worth Basin is very large, so far most of the wells have been drilled in and around the Newark East Field **(Fig. 4.6).** The Newark East Field is bounded by the Muenster Arch to the north and east and the Ouachita Structural Front to the east and south. The Barnett Shale outcrops at the Llano Uplift, located in Central Texas. There are small faults throughout the field, as shown in Fig 4.6. At the time of this writing, the Newark East Field was the largest shale gas field in the world.

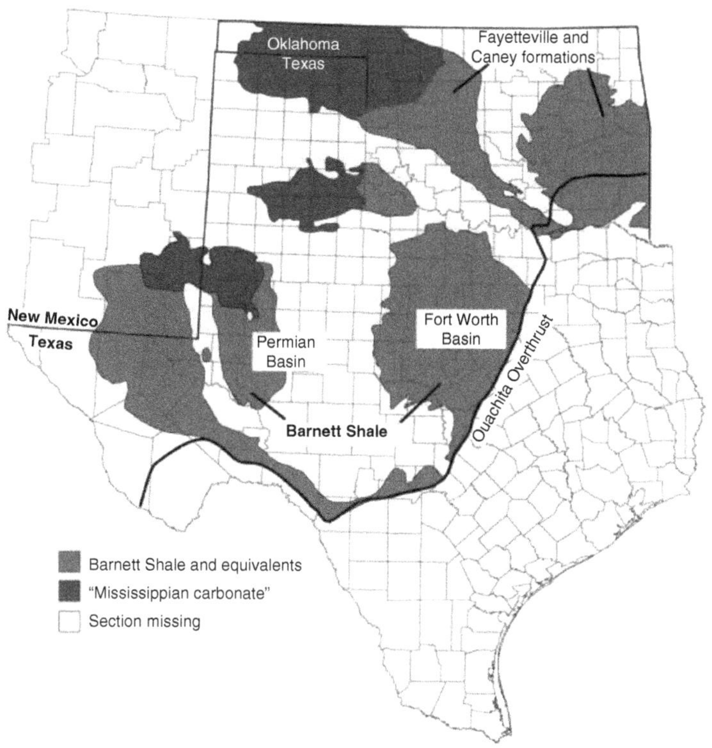

Fig. 4.4—Barnett Shale, located in the Fort Worth Basin.

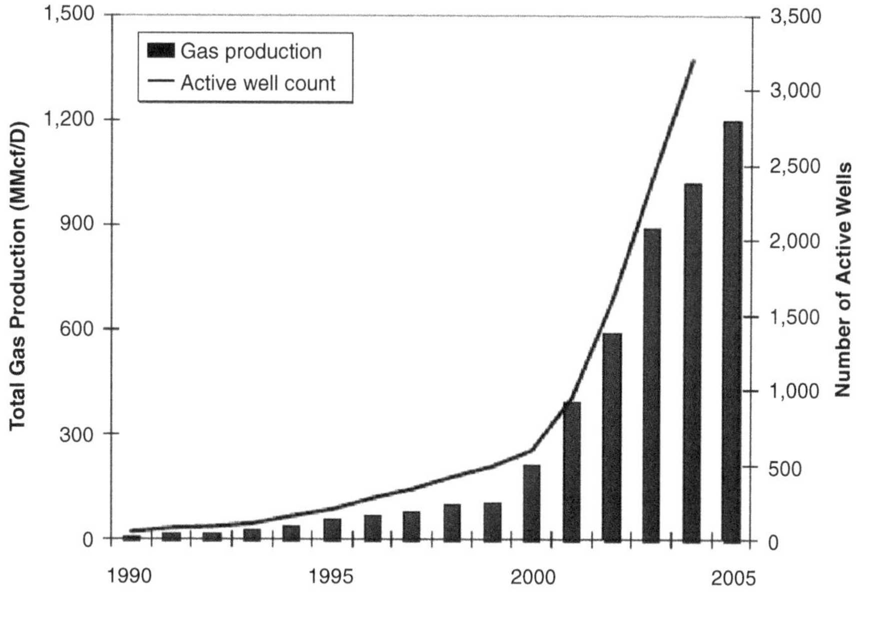

Fig. 4.5—Yearly production curve for the Barnett Shale (after Hayden and Pursell 2005).

Through the use of VR and core analysis, it has been concluded (Jarvie 2004) that the Barnett Shale is capable of producing either gas or oil, depending on where the well is drilled. This chapter will discuss only the gas component, which is the primary hydrocarbon produced from the Barnett Shale in the Fort Worth Basin. Hayden and Pursell (2005) state that the Barnett produces thermogenic gas, as opposed to the biogenic gas created in the Antrim Shale. The Barnett Shale has original organic richness and a hydrocarbon-generation capability, primary and secondary cracking of kerogen, the retention of gas by adsorption, and porosity caused by the decomposition of organic matter. The Barnett Shale is black, organic rich, siliceous, and very hard (Jarvie et al. 2007). A typical Barnett Shale well log is provided in **Fig. 4.7.**

Givens and Zhao (2008) discuss the difficulty in truly understanding the potential gas recovery from the Barnett Shale, resulting from three factors. First, the shale comprises a variety of minerals and multiple layers with varying lithology, making it difficult to correlate reservoir parameters regionally. Second, to understand the distribution of hydrocarbons in this basin, it is also important to understand the faulting and fracturing that occur throughout the area. The faults and fractures provide the necessary channels for gas flow and migration. Because the Barnett Shale has an intricate and multifaceted fracture system, it is often challenging to determine fracture length and how well the wellbore is connected to the reservoir. The third factor is the regional faulting and underlying Ellenburger karsting.

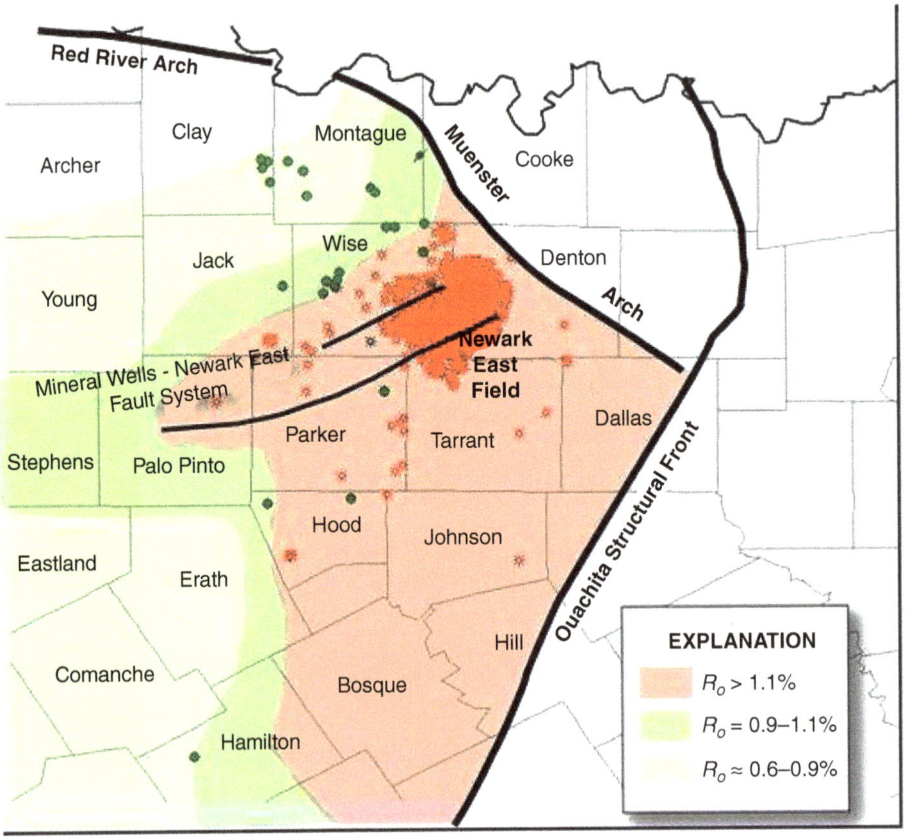

Fig. 4.6—Newark East Field location in the basin and the thermal maturity of the Fort Worth Basin (after Pollastro et al. 2007, *AAPG Bulletin*, AAPG©2007, reprinted by permission of the AAPG whose permission is required for further use). R_o = vitrinite reflectance.

4.2.3.1 Depositional Environment. Givens and Zhao (2008) and Lancaster et al. (1992) discuss the depositional environment of the Barnett Shale, which consists of sedimentary rocks that were deposited during the Late Mississippian. During this time, a marine transgression took place, originated by the closing of the Iapetus Ocean Basin. The Oklahoma Aulacogen, on which the Barnett Shale is deposited, subsided as a result of the Middle or Late Mississippian collision of the North American plate with the South American plate. As the Pennsylvanian came to an end, the Ouachita Thrust Belt began to infringe upon the Barnett Shale sediments and created the foreland basin along the front of the thrust **(Fig. 4.8).**

In modern times, the Barnett thickens north and east and is the thickest just south of the Muenster Arch. This increased thickness is caused by the interstratifications of the shales, limy shales, and the lime beds that were unevenly deposited over the basin. The Barnett Shale rests directly on top of the Ordovician Viola Limestone, or Ellenburger Limestone. Before the Barnett Shale sediments were deposited, the Viola began eroding and pinched out, exposing the Ellenburger Formation **(Fig. 4.9).** A major unconformity is present between the Ellenburger/Viola and the Barnett on top of the Pre-Cambrian crystalline Llano Uplift (Gasparrini et al. 2014).

During deposition and the burial process, two distinct layers—the Upper and Lower Barnett—formed, and these are considered the Barnett Shale **(Fig. 4.10).** These two layers are quite distinct in the northeastern part of the basin, where the Forestburg Limestone separates the shale into the two zones. The Upper Barnett is uniform in thickness, at approximately 60 to 70 ft throughout the entire northeastern basin. The Lower Barnett ranges in thickness, from less than 50 ft near the Bend Arch in the western part of the basin to greater than 600 ft in the northeast portion near the Muenster Arch. In the remaining area of the basin, there is no identifiable difference between the Upper and Lower Barnett Shale, as discussed by Loucks and Ruppel (2007).

4.2.3.2 TOC. The organic material present in the Barnett Shale can be as high as 12%, but, on average, it is approximately 4.5%. Loucks and Ruppel (2007) investigated the mineral makeup of the Barnett Shale. The thinner grains contain accumulations of silty biomicrite, fossiliferous silty shale, and fossiliferous intraclast conglomerates containing abundant phosphate clasts. The coarser material found contains skeletal components derived from an assorted group of marine invertebrates that includes brachiopods, sponges, pelecypods, gastropods, cephalopods, conodonts, and echinoderms. Other species found in the Barnett are crinoids, burrows, benthic foraminifera, filibranch mollusks, bryozoans, ophiuroids, agglutinated foraminifera, and radiolarians. This shows that plant and animal matter form the Barnett Shale, and, depending on which part of the basin is being studied, the organic matter changes (Hickey and Henk 2007).

4.2.3.3 Gas Content. The total average porosity of the Barnett Shale ranges from 3 to 8%, with an average of 5%. Of this 5%, the average gas porosity is approximately 2 to 6%. This porosity accounts for only the free gas that is available in the pores, not the adsorbed gas. Nearly 70 to 85% of the gas in the Barnett Shale is found in the free state (Hayden and Pursell 2005). Approximately 15 to 30% of the gas in this gas shale system is adsorbed, which is not much considering the low recovery factors here. Because less gas is adsorbed, the gas content for the Barnett is higher than that of many other shale plays. It is not uncommon to find a gas content from 300 to 350 scf/ton in the Barnett Shale, which can yield high EURs. From core analysis, pressure transient testing, and well testing, the industry believes the EUR is 10 to 20% (Hayden and Pursell 2005).

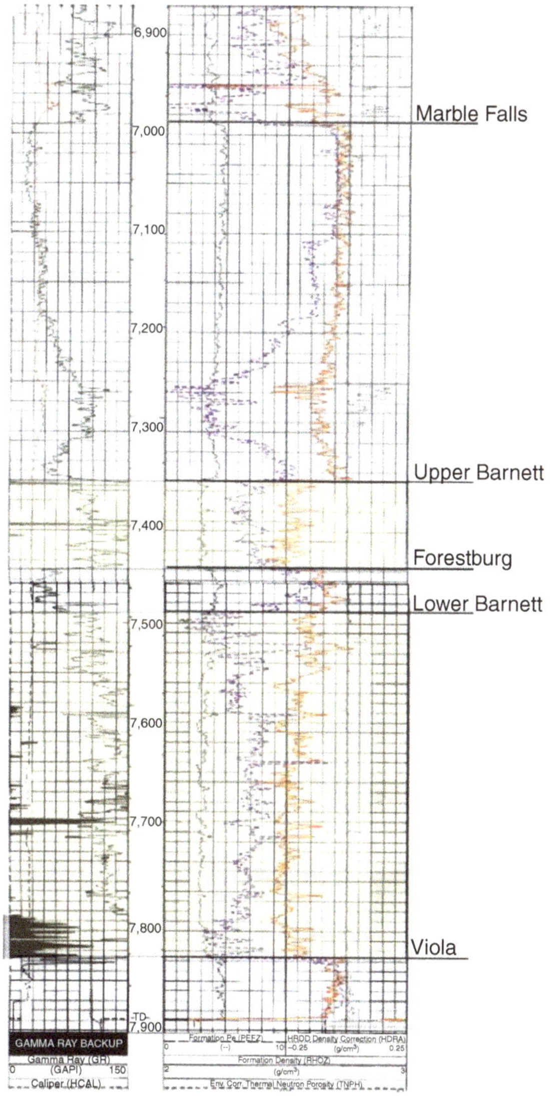

Fig. 4.7—Type log of the Barnett Shale (Givens and Zhao 2008).

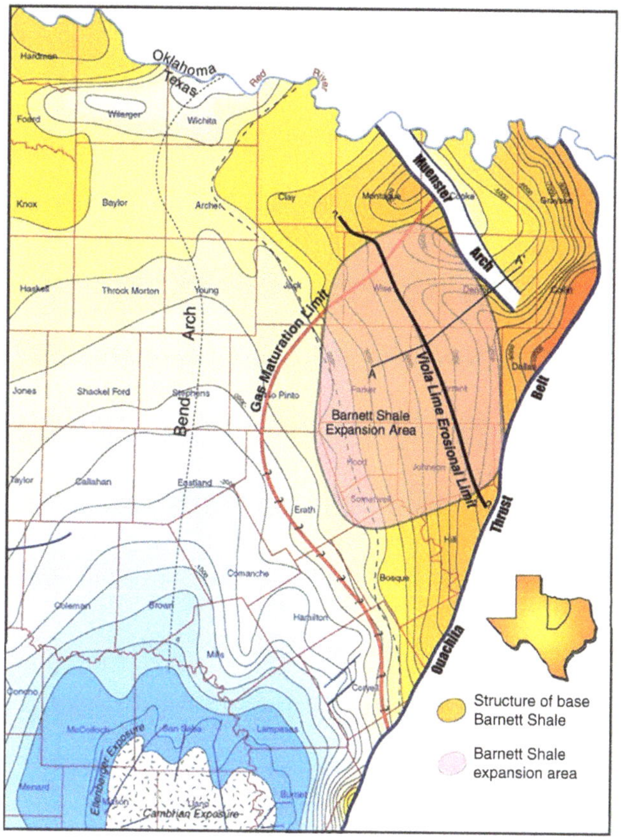

Fig. 4.8—Extent and boundaries of the Fort Worth Basin (Givens and Zhao 2008).

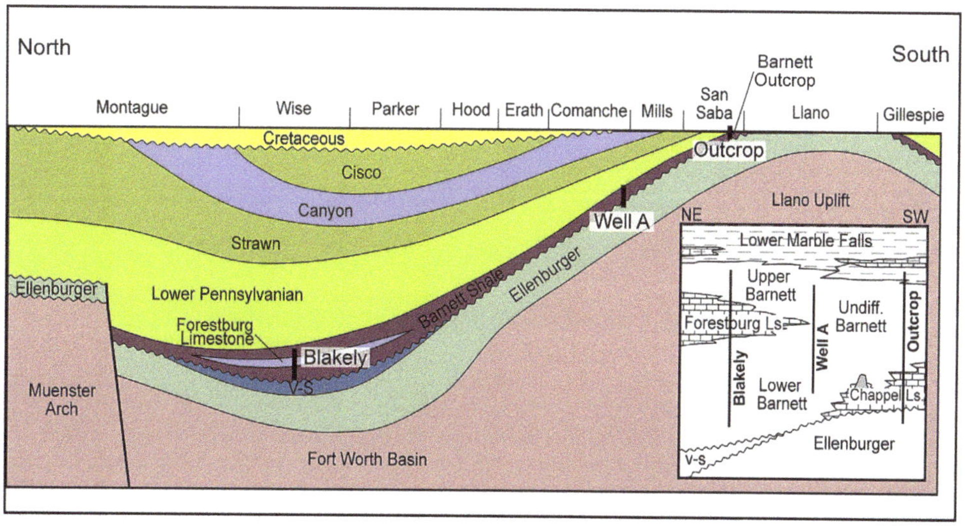

Fig. 4.9—North/south cross section of the Fort Worth Basin (Gasparrini et al. 2014). V-S = Viola-Simpson.

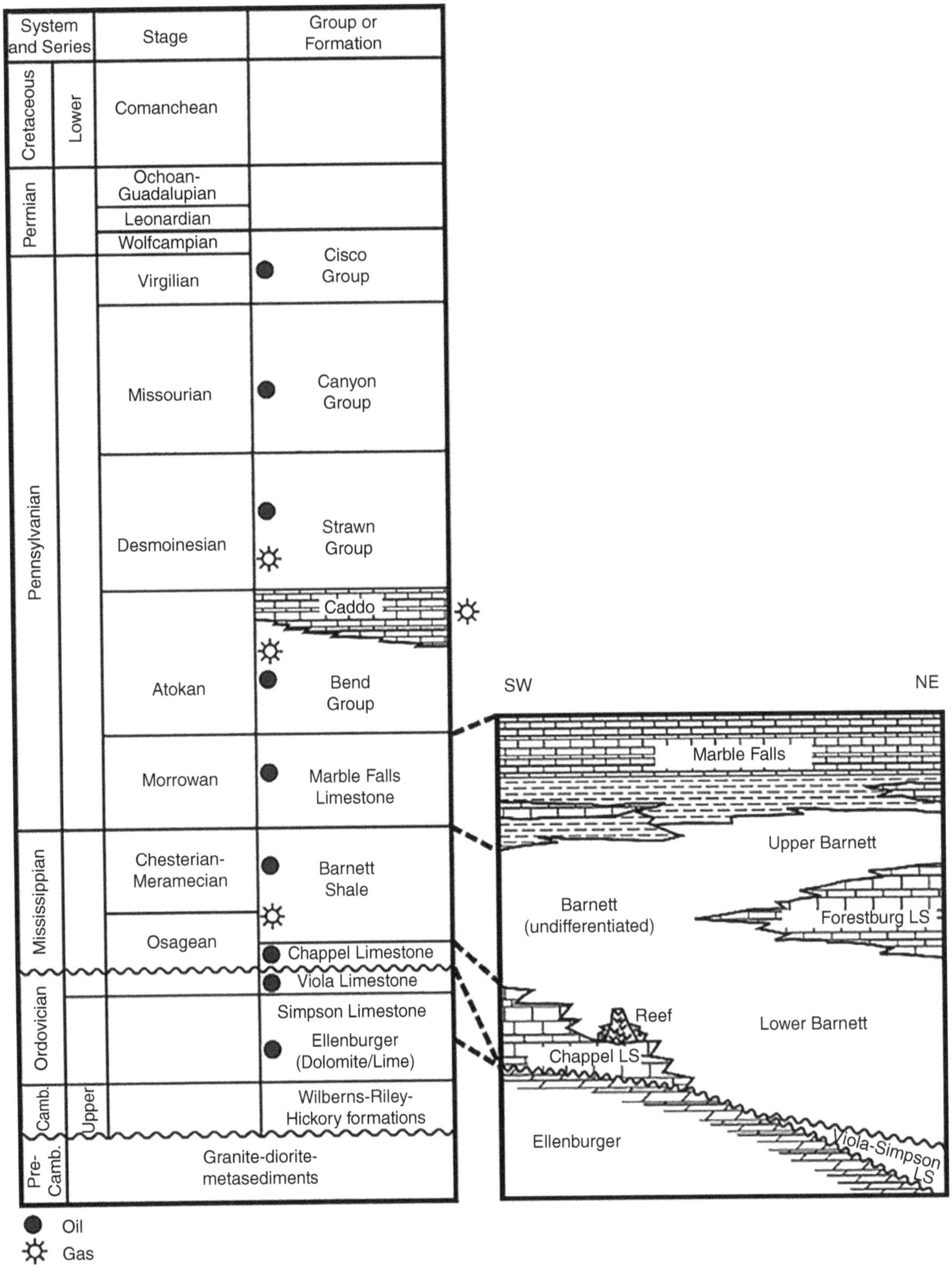

Fig. 4.10—Stratigraphic column of the Fort Worth Basin.

4.2.3.4 Shale Mineralogy. Hickey and Henk (2007) report that although the Barnett is sometimes called a homogeneous black shale, it contains an assortment of organic-rich lithofacies. Some of the most-common lithofacies are of siliceous, calcareous, and phosphatic composition. The Barnett Shale has a high quartz content that makes up approximately 40 to 45% of the formation. The next-most-abundant mineral type is clay, which comprises 20 to 40% of the formation. In addition, pyrite content can be as high as 5%. Remaining minerals, found in various quantities, are feldspar, calcite, dolomite, siderite, and ankerite.

4.2.3.5 Thickness and Reservoir Pressure. The Barnett Shale thickness increases toward the Muenster Arch, which is located to the east/northeast of the Fort Worth Basin (**Fig. 4.11**).

Barnett Shale gas production occurs at depths of 6,500 to 8,500 ft, depending on the location in the basin. The gross thickness of the Barnett Shale varies from 200 to more than 1,000 ft, and the net thickness is approximately 100 to 600 ft (Hayden and Pursell 2005).

The original reservoir pressure in the Barnett Shale Basin is 3,000 to 4,000 psi, yielding a pressure gradient between approximately 0.42 and 0.46 psi/ft (Hayden and Pursell 2005). This variation in pressure is caused by depth. At shallower depths, the pressure is lower, and it is imperative to ensure that there are no surface-pressure issues (e.g., pipeline pressures, pipeline-capacity constraints), to prevent long-term production delays.

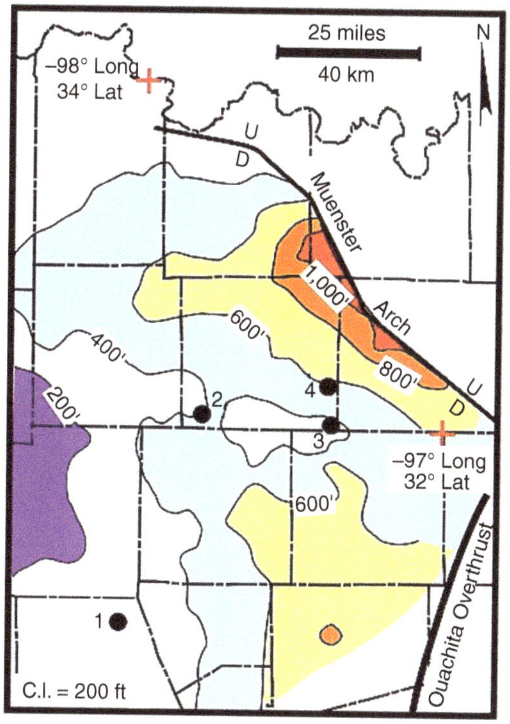

Fig. 4.11—Barnett Shale isopach map (Loucks and Ruppel 2007).

4.2.4 Haynesville Shale. The Haynesville Shale is a gas shale present in many different basins in the southeastern US (**Fig. 4.12**), but most of the recent drilling success has been in the North Louisiana Salt Basin (Mancini et al. 2006). This shale is also referred to as the Shreveport Shale because most of the activity is in or around the Shreveport, Louisiana, area.

The first Haynesville well was drilled in 1905, but the play did not receive much attention until EnCana Oil & Gas drilled in January 2006 (Durham 2008). Samson Contour drilled the next Haynesville well in June 2007. After the success of the vertical Haynesville well, the first horizontal well was permitted and drilled by Chesapeake Energy in August 2007, which ignited interest in the oil industry and many land acquisitions and trades. The production from the Haynesville Shale, shown in **Fig. 4.13,** exponentially increased in a matter of months as a result of horizontal drilling.

At the time of this writing, the Haynesville Shale was producing 25 MMcf/D from the North Louisiana Salt Basin and approximately 40 MMcf/D from the entire play, which includes Texas and Louisiana areas (Goddard et al. 2008). Because this is a new play, there is still some uncertainty as to how much gas is in place, but current estimates suggest it ranges from 250 to 320 Tcf. The typical depth at which this formation is encountered is between 9,000 and 16,500 ft, and it extends areally approximately 9,000 sq miles. A common phenomenon seen is that the Haynesville Shale gets deeper as the field dips toward the Gulf of Mexico (Goddard et al. 2008).

Through the use of gas samples collected, it was concluded that the Haynesville Shale produces thermogenic gas, like the Barnett does in East Texas. The Haynesville Shale is gray black to blackish green in color and contains muddy fine-grained sandstone, laminated fine-grained sandstone, sandy mudstone, and silty mudstone. These characteristics can be seen in the type log for the

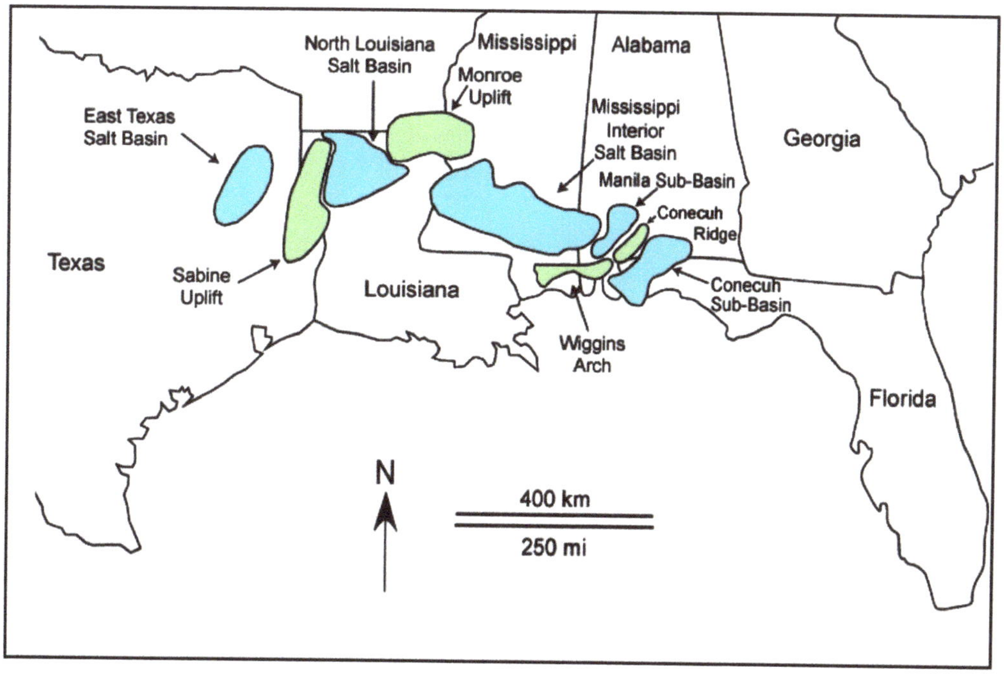

Fig. 4.12—Haynesville Shale, located across the Southeast US (Mancini et al. 2006).

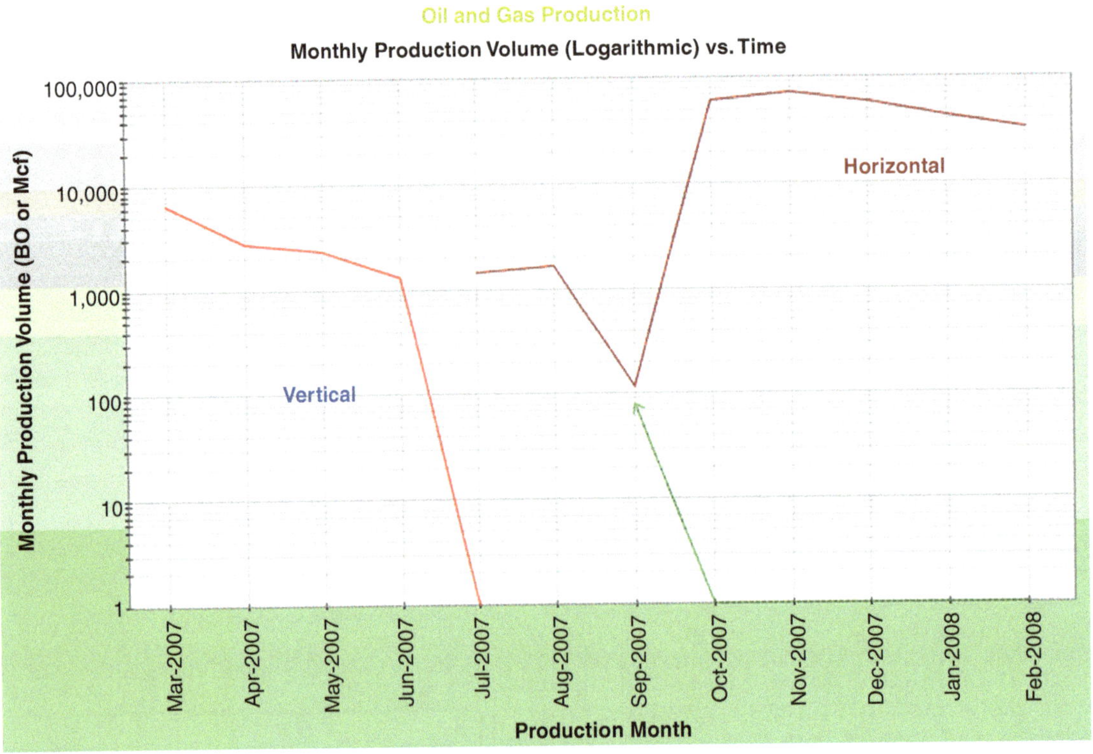

Fig. 4.13—Haynesville Shale production.

Haynesville Shale **(Fig. 4.14).** Fig 4.14 illustrates that, typically, the shale will have a high GR count of approximately 135 API units (Goddard et al. 2008).

Because of its size, the Haynesville is a fascinating play. As shown previously (Fig. 4.12), this play spans Louisiana, Alabama, Mississippi, and Texas. Many shale plays seem to be isolated to one basin, or at the very least to one state, where considered productive, but the Haynesville is found to be productive throughout the southeastern portion of the US. This discussion will focus on the Haynesville in the North Louisiana Salt Basin, where most of the publicized activity lies. However, it is worth noting that Texas and Mississippi are both undergoing large development because of the blanket deposits of this shale. Like most thick, black shales, the Haynesville was deposited in a deepwater marine environment, so it will cover a rather large areal extent. The potential gas recovery of the Haynesville is still unknown, but with the help of new technology, most believe it has enormous potential. The difficulties involved in producing the Haynesville include a lack of public data on production and geology; areal extent because it is a new play area; and depth because this formation can be found at depths ranging from 9,000 to 16,500 ft. However, the Haynesville in many areas is overpressured and permeable, so it can be produced even at such depths. There will be limitations in well design, data collection, and completions resulting from the costs to develop these depths; however, with recent advances in drilling and completion technologies, operators are successfully developing the Haynesville at depths of more than 12,000 ft.

4.2.4.1 Depositional Environment. Magner and Wren (2008) indicate the Haynesville Shale was deposited during the Upper Jurassic (Kimmeridgian), which took place approximately 150 million years ago. During this deposition, the majority of the sediments found as part of its makeup flowed in from the north and northeast rivers that were present at the time. In fact, traces of the three rivers are still found through drilling and seismic activity. Also, several geologic structures have been identified that show a lagoon could have been present at the time of the Haynesville deposition and that carbonate shoals surrounded the deposition of the organic matter into the basin. Most of the deposition of the fine-grained-sand to silt sequence took place across the marine slope and the basin environment over an extended period of time, which is the reason the Haynesville Shale is thick and blanketed over such a large area. The Bossier Shale sits on top of the Haynesville Shale, and there is still some debate as to whether it is considered part of the Haynesville Shale. The Bossier Shale has roots similar to those of the Haynesville Shale, as far as deposition is concerned, and, in fact, the Haynesville Shale is sometimes referred to as the Lower Bossier Shale. Above the Bossier and Haynesville shales lies a thick sand formation called the Cotton Valley Group, which is continuous as well, providing an effective seal that prevents any gas leaks.

Fig 4.15 depicts the depositional environment for the Haynesville Shale. This is just one interpretation of how the depositional environment appeared during the Jurassic in this area. As industry knowledge expands through seismic interpretation, drilling, and other processes, this interpretation will more than likely evolve.

4.2.4.2 TOC. The organic material present in the Haynesville is approximately 2.83%, but it can range from 0.5 to 4% (Goddard et al. 2008). It is uncommon to find the lower end of the spectrum, but existing reports show that it could be that low. Most of the organic material in the Haynesville is terrestrial plant debris, Type III kerogen. Large fossil fragments such as textularid-type foraminifera tend to collect together and can occupy 2 to 10% of the rock volume. Filaments of organic material have also been found in samples gathered throughout the basin (Magner and Wren 2008).

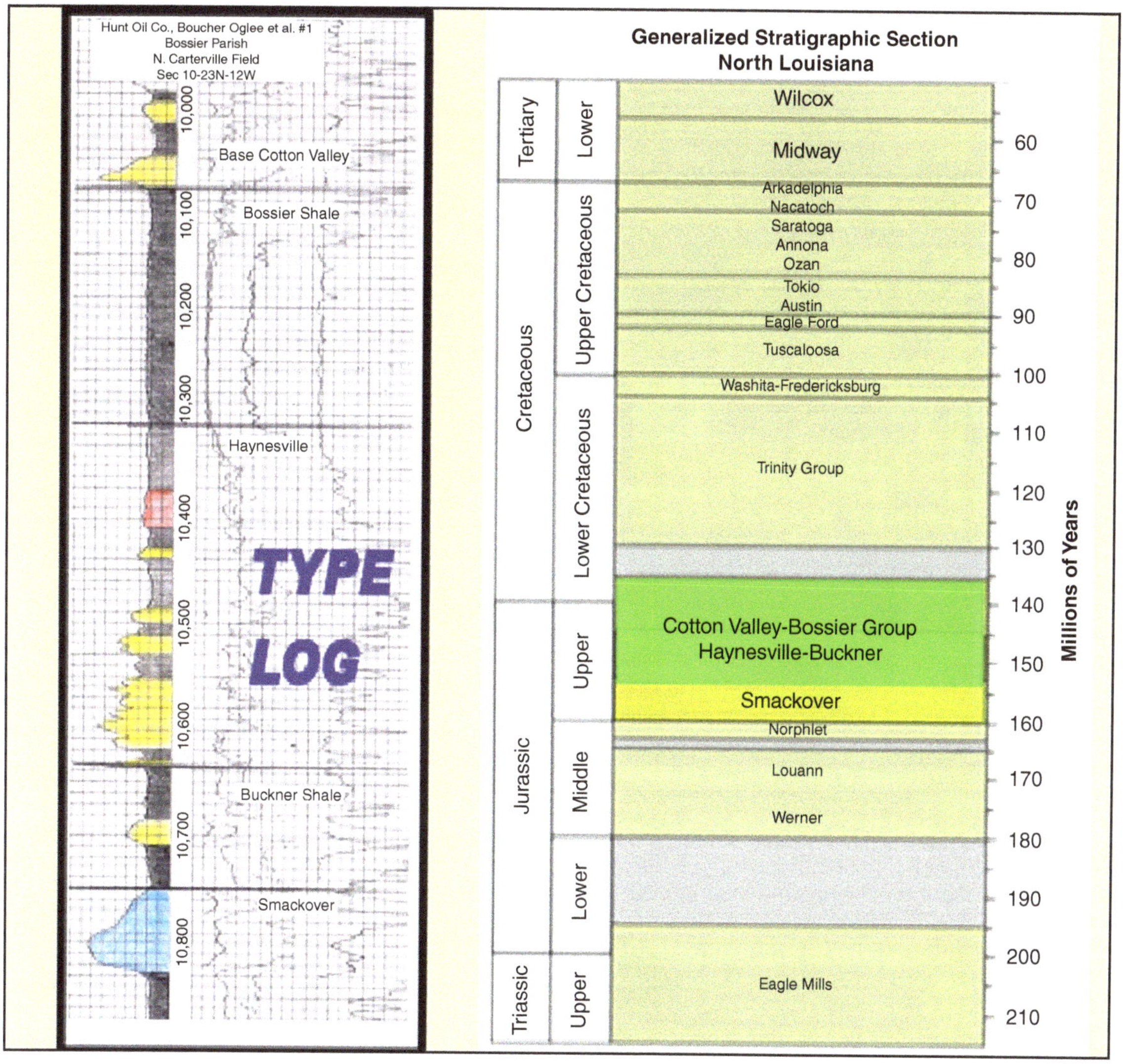

Fig. 4.14—Type log of the Haynesville Shale (Goddard et al. 2008).

4.2.4.3 Gas Content. Total porosity throughout the Haynesville Shale ranges from 6 to 15%. In many of the producing areas, the porosity averages 12%. Of this 12%, the average gas porosity is approximately 5 to 11% (Magner and Wren 2008). This porosity accounts for only the free gas available in the pores and natural fractures and does not include adsorbed gas. The adsorbed gas in this play ranges from 20 to 30% of the total gas-in-place. Because of high pressures and the depth of the Haynesville Shale, there is only a small possibility that the adsorbed gas will be tapped into, through the use of current technology. The Haynesville has similar gas content to the Barnett, which made this play very successful during the past few decades. The gas content ranges from 350 to 475 scf/ton throughout the North Louisiana Salt Basin. Because the Haynesville Shale is deep, free gas is the only gas that produces with present-day technology. Therefore, recovery factors are based only on the volumes of free-gas production and not adsorbed-gas production. Given that the Haynesville play is emerging and there is no established long-term production from these wells, there are theories regarding EURs based on knowledge compiled from other plays. Many gas shale plays have a 1 to 0.7 ratio between initial production (IP) and EUR. For example, if a well has an IP of 2 MMcf/D, the rule of thumb is the EUR will be approximately 1.4 Bcf. Therefore, the industry is looking at an average EUR for each wellbore to be 2 to 8 Bcf in horizontal completions and 0.5 Bcf in vertical completions. In most Haynesville wells, the initial decline rate is high, then decreases to the range of 50 to 80% (Goddard et al. 2008). Through core analysis and well testing, the industry believes the EUR for the Haynesville Shale in the North Louisiana Salt Basin is 34 Tcf, which is a 13.6% recovery factor. With the development of new technology, the recovery factor could increase to 3 to 30% in the future.

4.2.4.4 Shale Mineralogy. Goddard et al. (2008) reported that the mineralogy of the Haynesville Shale contains both quartz and mudstone, which is mostly clay. The Haynesville has a good percentage of quartz content, approximately 28 to 33%. The amount of quartz varies throughout the Haynesville, contributing to areas in the play known as "sweet spots." The next-most-abundant material is clay minerals, which range from 25 to 33% of the mineralogy. Remaining minerals, found in various quantities, are calcite, dolomite, siderite, feldspar, and pyrite.

4.2.4.5 Thickness and Reservoir Pressure. The thickness of the Haynesville Shale is relatively uniform throughout the North Louisiana Salt Basin. At the time of this writing, studies were under way to ensure the proper characterization of the Haynesville Shale (Hutchinson 2009). **Fig. 4.16** shows an isopach map of the Haynesville Shale trend and highlights key players throughout not only the North Louisiana Salt Basin but also other areas where the Haynesville is gaining momentum.

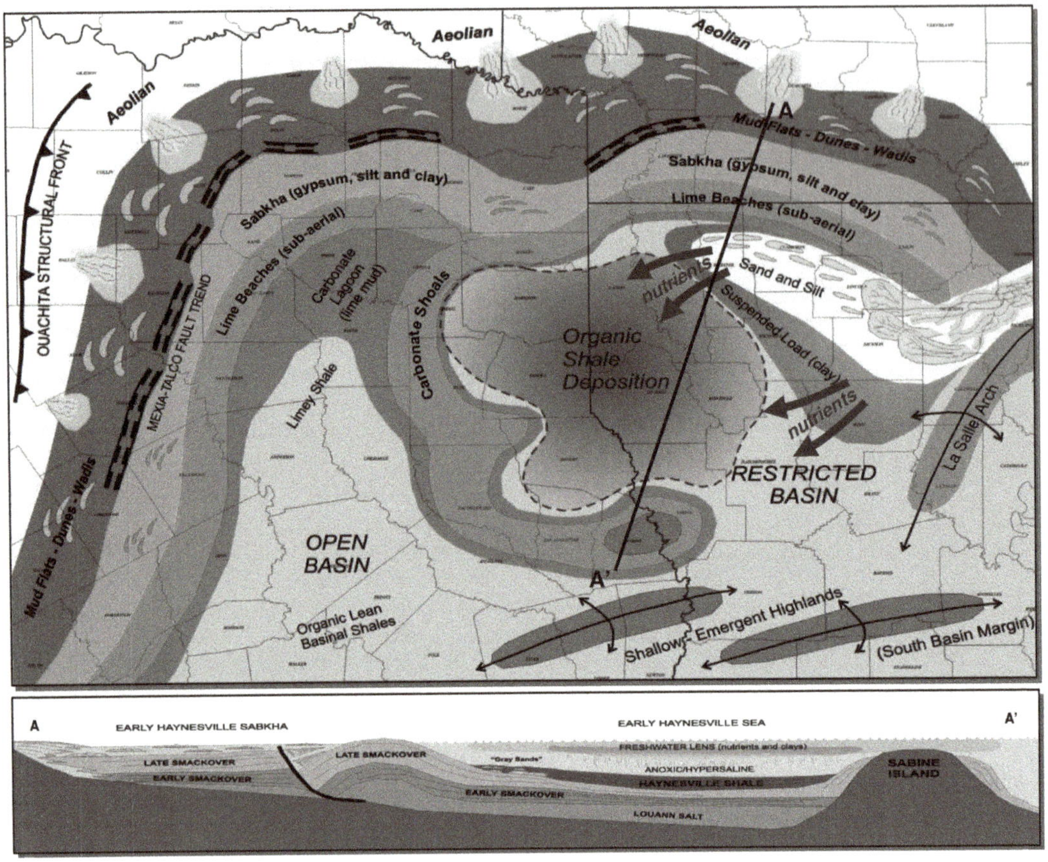

Fig. 4.15—Haynesville Shale depositional environment (after Chesapeake Energy 2009).

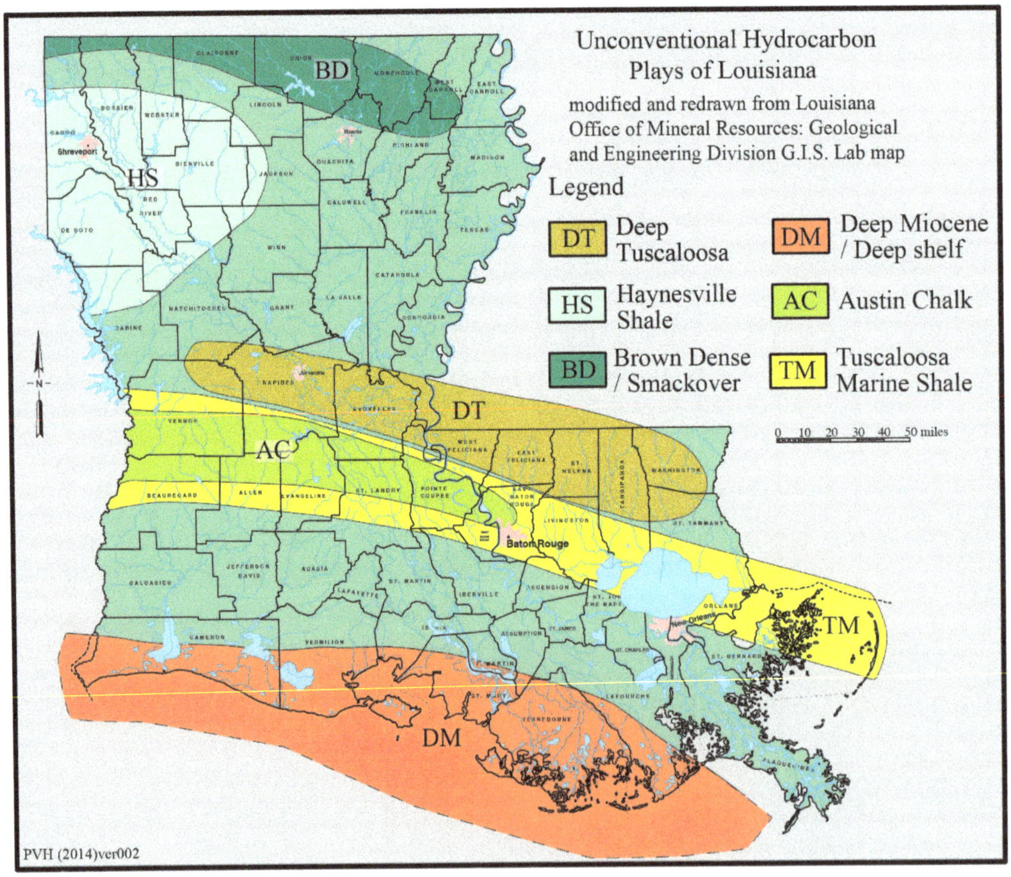

Fig. 4.16—Haynesville Shale isopach: trend and activity map (Cristellaria 2014).

The Haynesville Shale occurs at depths from 9,500 to 16,500 ft, depending on the location in the North Louisiana Salt Basin. The gross thickness of the Haynesville Shale is, on average, approximately 270 ft. Of this 270 ft, 100 to 200 ft is considered net-pay, showing a high net-to-gross ratio (Hutchinson 2009; Magner and Wren 2008).

Average Haynesville reservoir pressure in the North Louisiana Salt Basin is approximately 8,000 to 10,000 psi, yielding a pressure gradient between 0.7 and 0.92 psi/ft (Goddard et al. 2008; Chesapeake Energy 2009). These pressure gradients are 1.5 to 2.0 times the normal gradient. As a result of the depth, the burial in this basin, and the overpressured shale, this area produces dry gas.

4.2.5 Marcellus Shale. The Marcellus Shale is a gas shale found throughout the Allegheny Plateau region of the northern Appalachian Basin **(Fig. 4.17).** This shale runs across the southern portion and Finger Lakes region of New York, northern and western Pennsylvania, eastern Ohio, and western Maryland, and throughout most of West Virginia (Bertola 2008). The Marcellus Shale was first completed in the 1980s with foam fracturing but really kicked off in 2000. The estimated gas-in-place is approximately 1,300 to 1,900 Tcf, with approximately 350 to 500 Tcf recoverable, which could be a low estimate because of activity in this area. The typical depth in which this formation is encountered is between 1,500 and 8,500 ft—of which the producing depth is usually 4,000 to 8,500 ft (Chesapeake Energy 2009)—and it extends areally approximately 54,000 sq miles and covers more than 15 to 20 counties (Brackett 2006).

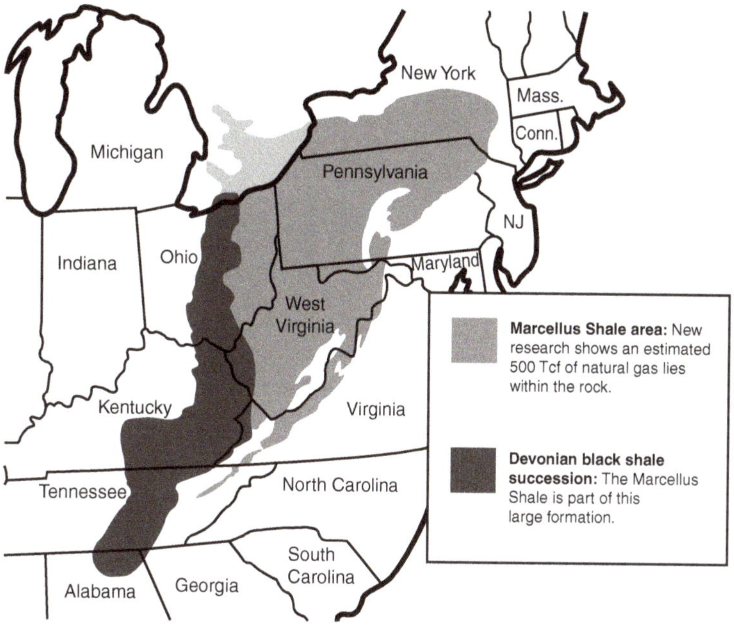

Fig. 4.17—Marcellus Shale, located in the Appalachian Basin (after Bertola 2008).

Through the use of core analysis, it was discovered that the Marcellus Shale produces thermogenic gas, similar to the Barnett Shale. If seen in an outcrop, the shale is a black, carbonaceous shale interbedded with well-developed limestone beds (King 2009). It also contains bitumen but is too old to contain any bituminous coal.

The Marcellus Shale is highly fractured, and fractures can be seen throughout the outcrop of the formation, as shown in **Fig. 4.18.** Because of radioactivity in the shale, the Marcellus makes a strong positive deflection on the GR curve, allowing it to be easily spotted on a GR log **(Fig. 4.19).** Engelder (2009) states this deflection is even more prominent if the shale lies directly on top of carbonate rocks. The GR log is vital to the development of this shale because of the different depths, thickness, and extent of the Marcellus Shale throughout the Appalachian Basin. The Marcellus Shale sits under a disconformity, and if not for this strong deflection, it could be mistaken for another formation. Also, because of the varying thicknesses throughout the basin, this GR signature helps determine gross thickness as one moves from state to state.

Fig. 4.18—Marcellus Shale outcrop.

Developing and comprehending the potential of the Marcellus is difficult because of the fractured characteristic of the shale. This reservoir is very brittle and falls apart easily, as seen in outcrops. Another potential hurdle to overcome is the lack of oilfield infrastructure in place in the Northeast US. This area is growing its infrastructure—such as building roadways to sites, pipelines, and plants—but it is a slow process. With more studies, coupled with industry knowledge, the Marcellus Shale can have a vital impact on the financial health of the companies vested in the play, as well as the oil and gas economy as a whole.

4.2.5.1 Depositional Environment. As mentioned, the Marcellus Shale is a radioactive, carbonaceous black shale that was deposited during the Middle Devonian, in the Paleozoic **(Fig. 4.20).** Deposition of the Marcellus Shale took place when the Acadian Mountains jutted upward, and the black and gray shales of the Hamilton Group sediments began collecting in the sea, as a result of the mountains eroding (Martin 2009). More sediment fragments were carried into the Appalachian Basin by ancient rivers that meandered to the ocean. While deposition was taking place, the Appalachian Basin was hundreds of miles away (Barrett 2009; Mazzullo 1973).

This shale was formed in a relatively deep, sediment-starved, and oxygen-deficient trough that formed comparable to the mountain chain. This deposition produced transgressive black shale because it was deposited in deepening conditions when the basin floor dropped, resulting from the rising of the mountain (Patrick 2004). The dark facies of this shale materialized from flysch, which was a mud

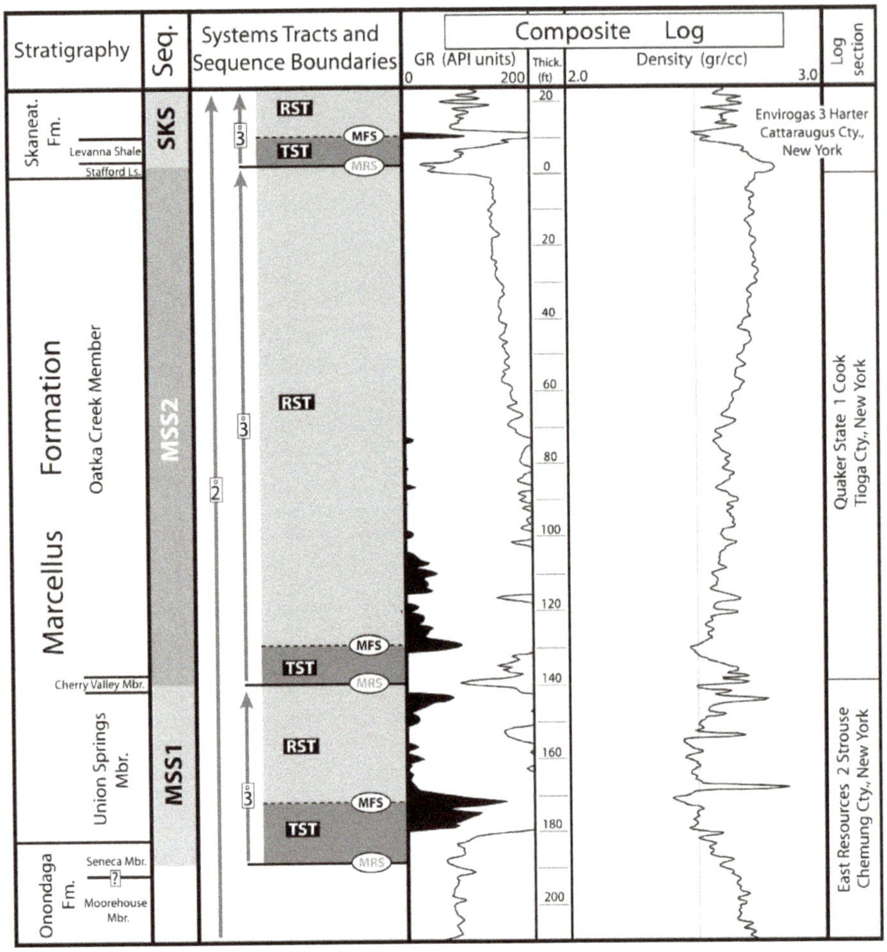

Fig. 4.19—GR log of the Marcellus Shale (Lash and Engelder 2011).

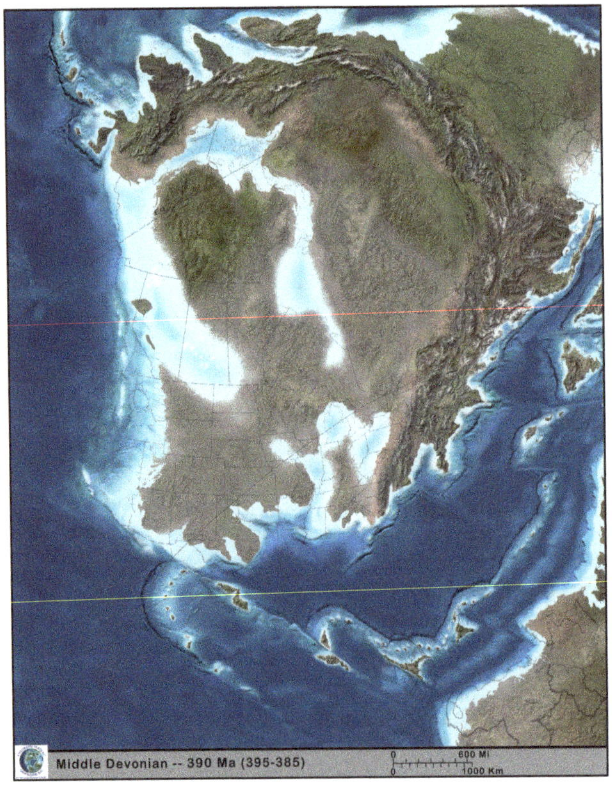

Fig. 4.20—Depositional environment of the Marcellus Shale (Deep Time Maps 2013).

deposited in the deep water. Studies show the Marcellus Shale has two composite depositional sequences (Arthur and Sageman 2005). One is the general coarsening-upward cycle and continues to the base of the overlying formations, and the other is the short-term oscillations in the basin depth, which are depicted by the interbedding of lighter shales and limestones in the shale.

Grasso (1968) states the Marcellus Shale is a member of the Hamilton Group, which consists of black and dark gray shales deposited during the Middle Devonian. Other shales that make up the Hamilton Group include the Skaneateles, Ludlowville, and Moscow shales. All three of these formations have characteristics similar to the Marcellus Shale and are limestone beds found within the main shale reservoir. The Marcellus Shale consists of three distinct nonshale layers imbedded within that appear in different regions of the Appalachian Basin: the Cherry Valley Limestone, Purcell Limestone, and Tioga Ash Bed **(Fig. 4.21).**

<table>
<tr><td colspan="2" rowspan="4">Oatka Creek Shale</td><td rowspan="3">Cardiff dark gray shale</td><td>Pecksport Shale</td><td>(Mt. Marion Fm.)</td></tr>
<tr><td colspan="2">Solsville Shale and Sandstone</td></tr>
<tr><td>Bridgewater Shale</td><td>Otsego</td></tr>
<tr><td colspan="2">Chittenango black shale</td><td>Berne</td></tr>
<tr><td colspan="4">Cherry Valley Limestone</td><td rowspan="2">Stony Hollow Shale and Limestone</td></tr>
<tr><td rowspan="2">(Seneca Mb.)
(Onondaga Fm.)</td><td colspan="3">Union Springs Shale and Limestone</td></tr>
<tr><td colspan="2">(Onondaga Fm.)</td><td colspan="2">Union Springs Shale and Limestone</td></tr>
</table>

Fig. 4.21—Members of the Marcellus Shale.

The Cherry Valley Limestone is found throughout New York, Pennsylvania, and West Virginia (de Witt et al. 1993). This limestone merges with the Union Springs and the Oatka Creek, which are two other members found within the Marcellus Shale, under Lake Erie. The Cherry Valley is composed of skeletal limestones with shaly intervals between its lower massive limestone layer and the upper limestone layer. In central Pennsylvania, Maryland, and West Virginia, the Purcell Limestone is found within the Marcellus Shale. Here, the shale comprises gray silty sand and mudrock, with abundant limestone and barite nodules dispersed throughout. The Purcell Limestone is approximately 50 to 100 ft of interbedded calcitic shale and limestone. The Purcell is equivalent to the Cherry Valley Limestone found in New York. Last, the Tioga Ash Bed is found at the base of the Marcellus Shale in much of the northern part of the Appalachian Basin. There was much speculation that the Tioga Ash Bed was one ash fall, but over time and through research, it appears now to encompass three definite ash falls (de Witt et al. 1993). These beds are used for correlation purposes throughout the basin to locate not only the Marcellus but other formations as well. Other layers found within the Marcellus Shale, which do not make up much of the shale, are the Bakoven Shale, Cardiff Shale, Chittenango Shale, and Solsville Sandstone (de Witt et al. 1993).

4.2.5.2 TOC. The organic material present in the Marcellus Shale ranges from 0.3 to 11% throughout the basin (Martin 2009). Most of the organic materials that created the shale are land plants, such as trunks of conifer trees, that were present during deposition. MacFarlane (1875) indicated that the Marcellus contains the oldest-known diverse set of thin-shelled mollusks, and he also found an external-shelled marine animal similar to a squid, called goniatites. Other species found in the Marcellus include brachiopods, called Spinocyrtia; crinoids; sea lilies, which are plant-like animals; limonite; gastropods; and bivalves (Schneider 1894; Burns 1991; Pabian and Strimple 1976).

4.2.5.3 Gas Content. The total average porosity found throughout the Marcellus Shale is approximately 9 to 13%, with an average of 11%. Of this 11%, the average gas porosity is approximately 3 to 6%. Approximately 20 to 40% of the gas in this shale system is adsorbed gas. The Marcellus Shale has a wide range of gas content that differs throughout the different states, and even within a particular state. The gas content ranges from 275 to 325 scf/ton (Phasis Consulting 2008). The different values result from adsorbed gas, thermal maturity, and depth. Studies and industry proven trends show the EUR is approximately 2 to 4.6 Bcf per well. Using this recovery for a Marcellus well led to the development of vertical wells on 40-acre spacing and horizontal wells on 80-acre spacing (DeWitt 2008).

4.2.5.4 Shale Mineralogy. Hosterman (1993) examined the mineralogy and determined the Marcellus Shale is heterogeneous throughout the basin, and it covers most of the Northeast US. Therefore, the composition is based on what is most commonly found throughout this area, but it is important to remember that these percentages can vary from outcrop to outcrop. The Marcellus Shale bedrock contains approximately 18 to 24% quartz. This quartz percentage is lower than that of other plays discussed in this chapter, a result of the sheer amount of calcite present in the shale, 25%. The other major mineral present, other than clay, is pyrite, which makes up 5% of the mineralogy. The clay minerals make up 40 to 60% of the mineralogy, one of the largest percentages in what is considered a "commercial" gas shale play. The minerals that make up the bulk of this bedrock are chlorite (15%), illite (70%), and illite-smectite. Some samples that have been captured show traces of illite-chlorite. Uranium is also present, making this shale radioactive and, hence, easier to identify in well logs.

4.2.5.5 Thickness and Reservoir Pressure. Because of the extensive magnitude of the Marcellus Shale, the thickness of the formation fluctuates throughout the basin (Brackett 2006). An isopach map of the entire Marcellus Shale found in the Appalachian Basin is shown in **Fig. 4.22.**

The Marcellus Shale is found as thin as 40 ft near Canada to as thick as 900 ft in New Jersey. In West Virginia, the shale is approximately 200 ft thick. North of West Virginia, in eastern Pennsylvania the shale is 800 ft thick and thins to the west to approximately 50 ft along the Ohio River. This thinning is caused by the decrease in grain size in the clastic deposits, because the

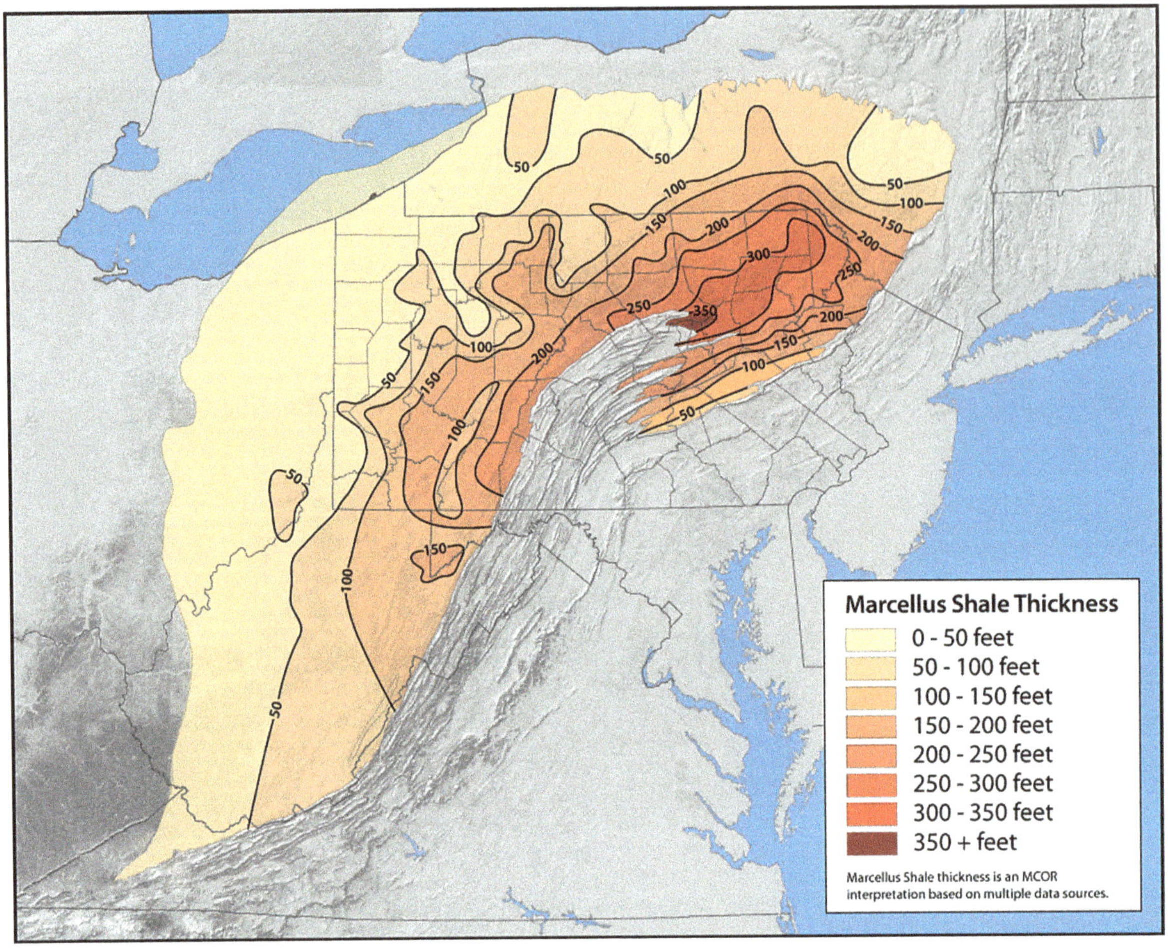

Fig. 4.22—Marcellus Shale isopach map (Penn State Marcellus Center for Outreach and Research 2017).

sediments entered on the eastern flank of the basin during deposition (de Witt et al. 1993). The Marcellus Shale pinches out at the Cincinnati Arch.

The average reservoir pressure found in the Appalachian Basin of the Marcellus varies in depth from approximately 1,600 to 5,000 psi, yielding a pressure gradient between 0.42 and 0.70 psi/ft. The Marcellus Shale pressure gradient ranges from less than the hydrostatic gradient to 1.5 times the hydrostatic; therefore, it is imperative to understand pressures and depths to ensure successful drilling and completions (Chesapeake Energy 2009).

4.2.6 Woodford Shale. The Woodford Shale was deposited during the Late Devonian/Early Mississippian and is now found throughout Oklahoma and Arkansas (Comer 2008). Comer (2008) notes that several major basins where the Woodford is found are the Arkoma Basin, Ardmore Basin, and Anadarko Basin **(Fig. 4.23).** This section will focus primarily on the Arkoma Basin.

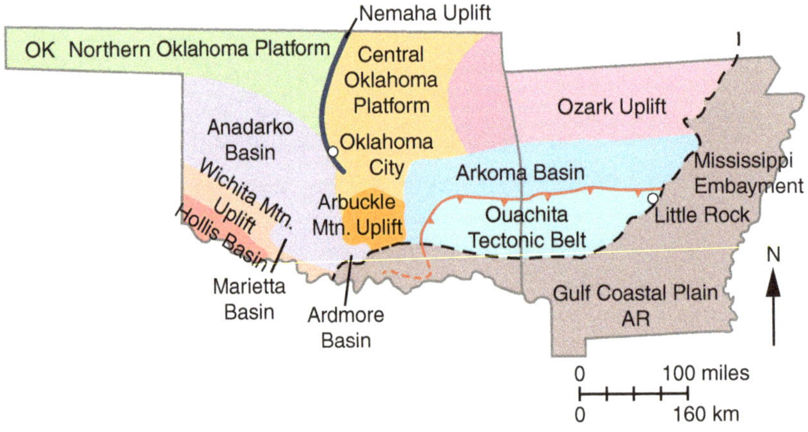

Fig. 4.23—Oklahoma basins (after Comer 2008).

Many of the operators producing the Woodford from the Ardmore Basin have reported oil/condensate production with minimal gas associated with liquids. Only approximately 10 to 15 wells have been drilled in this basin, likely because it is liquid production instead of gas. Most operators shy away from liquids because they seem to lock up the system quickly. Regardless, the Ardmore Basin still receives much attention (Ardmore Basin Woodford Gas Plan Takes Off 2008). On the other hand, the Anadarko Basin produces mostly gas, but there are limitations because of depth. The majority of the Woodford Shale production comes from depths between 7,000 and 18,000 ft. Marathon stated in a January 2009 press release that their first successful Woodford horizontal well drilled in a new area of the Anadarko Basin yielded a normalized IP of approximately 5.2 MMcf/D (Gulfoilandgas.com 2009). Even with the success in the Anadarko Basin, only 50 to 60 wells have been drilled here. In contrast to the Ardmore Basin and the Anadarko Basin, the Arkoma Basin contained more than 150 wells producing from the Woodford Shale in 2008. As mentioned, the Woodford Shale is located in the Arkoma Basin, which covers the southeastern section of Oklahoma and western portion of Arkansas (Fig. 4.23). Wickstrom (2008) discussed in depth his observations and conclusions regarding the Woodford Shale at the 2008 American Association of Petroleum Geologists conference held in San Antonio, Texas. The Woodford Shale was first completed in 1939, but this play did not gain momentum until 2003, as a result of new completion techniques and an increase in gas prices. Before the boom in 2003, only approximately 24 wells were producing from the Woodford Shale, throughout Oklahoma. During this time, IP for these wells ranged from 0 to 200 Mcf/D. After 2003, production rates climbed almost exponentially, as seen in **Fig. 4.24.** By early 2008, there were more than 750 Woodford wells in production, most of which were drilled in Hughes, Coal, Pittsburg, and McIntosh counties, located on the western side of the basin. The major players in the Woodford were Newfield Exploration, Devon, Continental, and Chesapeake Energy. The estimated gas-in-place for the Woodford at the time of this writing was approximately 45 to 60 Tcf. In the Arkoma Basin, the Woodford is typically encountered at depths between 6,000 and 11,000 ft, and it extends areally nearly 1,500 sq miles (Coffey 2007).

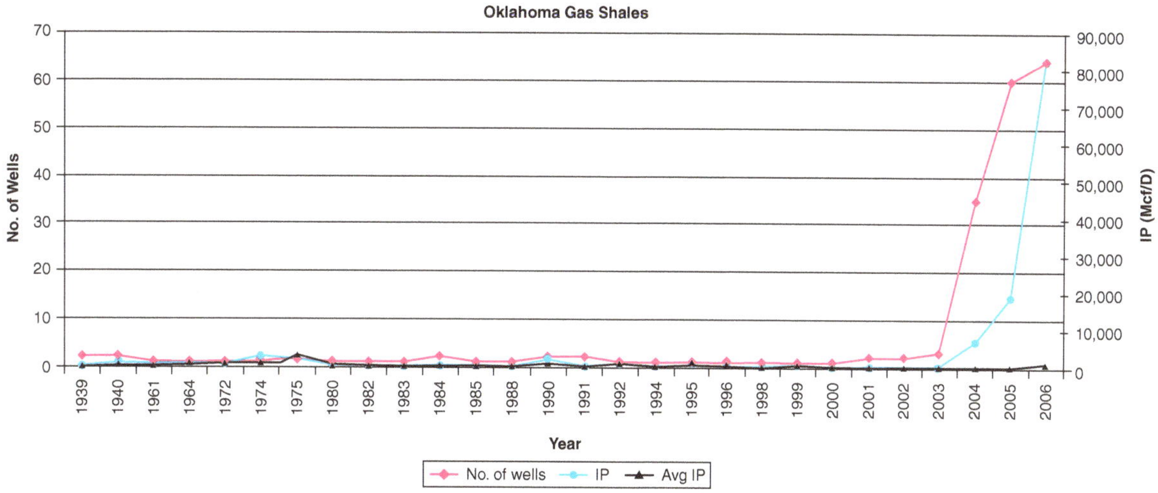

Fig. 4.24—Number of wells and IP rates for the Woodford Shale in Oklahoma (after Wickstrom 2008).

Through the use of core analysis coupled with VR, it is known that the Woodford Shale produces thermogenic methane and biogenic gas (in the cherty portion of the Woodford), similar to the Antrim Shale (Andrews 2007). The Woodford is a brittle gray to black shale found throughout the basin. Cardott (2007) indicates, in the outcrop of the Woodford Shale shown in **Fig. 4.25,** the most distinguishing feature is its natural fractures. Three sets of natural fractures are observed in Fig. 4.25. The first set is approximately parallel to the dip direction of the bedding plane. The second set is parallel to the dip of the bedding, and the third set is oblique to the strike-and-fold axis of the formation.

The Woodford Shale is easily recognized on well logs because of the level of radioactivity, as well as the placement of the formation between two carbonates **(Fig. 4.26).** The Woodford is characterized by relatively low sonic, resistivity, and neutron-induced radiation signatures (Cardott 2007; Comer 2008). The Woodford overlies a regional unconformity, which can be observed on well logs. The Woodford Shale is broken into three units: upper, middle, and lower. The lower unit, which sits on top of the unconformity, has the lowest radioactivity and thus contains more carbonate, silt, and sand. The middle unit has the highest radioactivity; it also contains the most organic content and pyrite. The upper unit's radioactivity falls between that of the middle and that of the lower unit. This unit consists of black shale with few spores and mostly parallel laminae.

Developing the Woodford has its challenges because of several factors (Cemen et al. 2007). First, this reservoir is naturally highly fractured because the rock is very brittle. There is little knowledge regarding how the fracture system contributes to the overall performance of the play or the number of fractures one might encounter when drilling. The second vital challenge that operators face is the vast spread of drilling depths in which the Woodford is encountered (Coffey 2007). Shallow gas plays (approximately 6,000 ft) are much easier to develop than deep gas plays (approximately 11,000 ft). Operators are overcoming these challenges with the development of new drilling and completion technologies that not only unlock ways to develop this reservoir but look to do it economically.

Fig. 4.25—Woodford Shale outcrop (Cardott 2007).

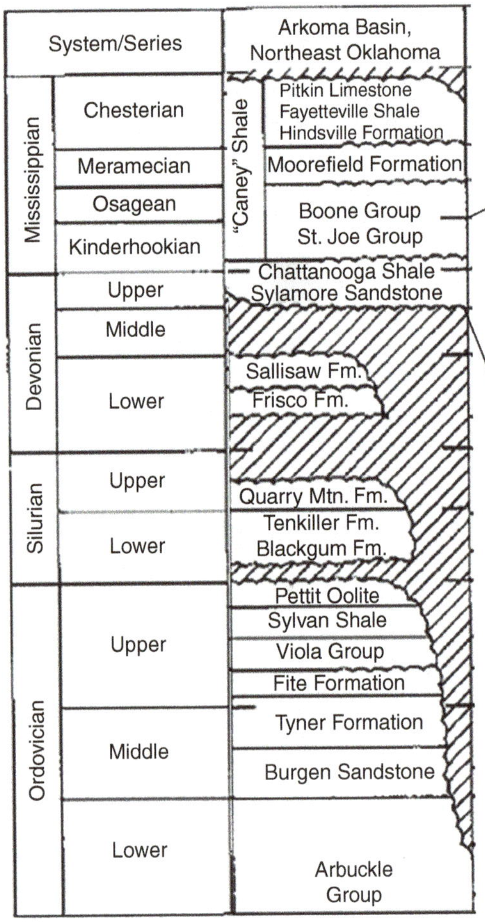

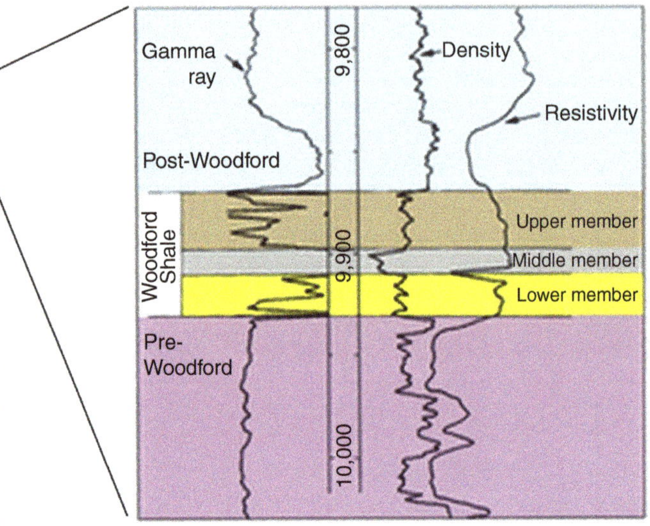

Fig. 4.26—(a) Stratigraphic column of the Woodford Shale (Cardott 2007) paired with (b) the GR log for the target section (Hester et al. 1990).

4.2.6.1 Depositional Environment. As mentioned previously, the Woodford Shale was deposited during the Late Devonian/Early Mississippian (Comer 2008). During this time, the southern midcontinent lay along the western margin of North America, in what was considered warm and dry tropics. This theory is confirmed by the hypersalinity found within the cracks of the present-day Woodford. Development of the Woodford Shale began in the epeiric seas when the sea level began to rise and drowned the marine embayment that was present at the time. The sea level continued to rise, covering the subareal, eroded, and dissected terrain that consisted of Ordovician to Middle Devonian carbonate rocks.

Comer (2008) also discussed the three members that make up the Woodford Shale: the upper, middle, and lower units. The upper unit is chertier than the other two formations, but it is not as radioactive as the middle member. On the other hand, the middle unit is more widespread than the other two regionally, but it is also the one that contains the most organic content. The lower unit is known for the abundance of other lithofacies, such as carbonates.

4.2.6.2 TOC. The organic material present in the Woodford Shale can be as high as 20%. This TOC is higher than even that of the Barnett Shale, which was the frontier play into the modern-day gas shale arena. The average TOC for the Woodford in the Arkoma Basin ranges from 3 to 10% (Cardott 2007). Most of the organic material found in the Woodford Shale is marine because it was once deposited on an embayment. The fossils discovered in the Woodford Shale are conodonts, ammonoids, fish debris, shark debris, and radiolarians. One of the most fascinating findings in the Woodford is that the formation is the oldest rock in Oklahoma containing wood from the progymnosperm Archaeopteris. This wood was transported to the depositional area by rivers present at that time (Andrews 2007).

4.2.6.3 Gas Content. The total average porosity found throughout the Woodford Shale can range from 7.5 to 14%, with an average of 10%. Of this 10%, the average gas-filled porosity ranges from 3 to 6.5% (Andrews 2007). This range is comparable to that seen in the Barnett and Marcellus shales, is slightly higher than that of the Antrim Shale, and is lower than that observed in the Haynesville Shale. The Woodford Shale contains approximately 20 to 40% adsorbed gas. The gas content in the Woodford Shale ranges from 60 to 115 scf/ton, resulting from the depth and thermal maturity found in the Arkoma Basin (Phasis Consulting 2008). This has led to much debate as to what the correct well spacing should be here. Langford (2009) discusses that while this play was developed, most companies drilled on the basis of 80-acre spacing; however, by 2008, there was an indication that 40-acre spacing is possible because minimal depletion occurred when companies drilled infill wells. The average recovery from these wells ranges from less than 1 to 6 Bcf, depending on whether an operator drills vertically or horizontally.

4.2.6.4 Shale Mineralogy. Cardott (2007) explained how the mineralogy of the Woodford Shale is very similar to that of the other shales previously discussed. The most-abundant mineral found in the Woodford is quartz, making up 55 to 87% of the formation.

The second most abundant is illite, found in quantities ranging from 8 to 34%. Less-abundant minerals are kaolin, which is present at approximately 3 to 7%, and plagioclase feldspar, found up to 7%. Other minerals present are dolomite (0 to 3%), apatite (0 to 1%), and pyrite (0 to 1%).

4.2.6.5 Thickness and Reservoir Pressure. The Woodford Shale ranges in thickness from 50 to 220 ft throughout the Arkoma Basin, with an average net-pay of 100 to 150 ft. An isopach map of the Woodford Shale found in the Arkoma Basin is shown in **Fig. 4.27** (Comer 1992).

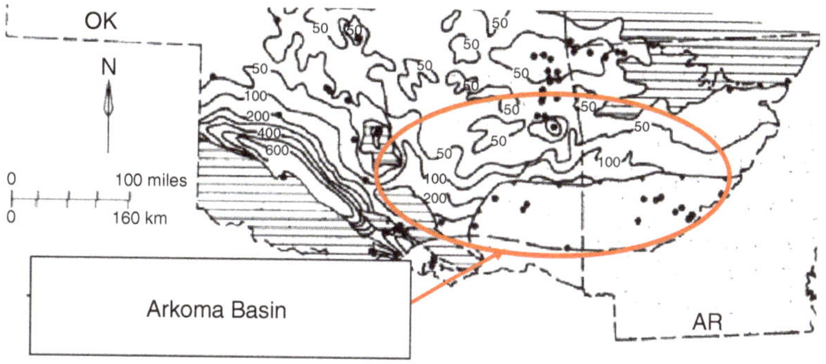

Fig. 4.27—Woodford Shale isopach map (after Comer 1992).

The thinnest portion of the Woodford Shale is located in the center of the Arkoma Basin. As one moves to the outskirts of the basin, the Woodford begins to thicken. There is also a regional thickening to the southwest of the basin. The Woodford Shale juts up against the Ouachita Thrust Belt, which is seen on the south side of the Arkoma Basin (Comer 1992).

Average reservoir pressure found in the Arkoma Basin varies in depth from approximately 3,100 to 5,800 psi, yielding a pressure gradient between 0.48 and 0.55 psi/ft (Langford 2009). Like the Haynesville Shale, and other gas shale plays, the Woodford is considered an overpressured reservoir because the pressure gradient is higher than the hydrostatic gradient. To ensure successful drilling and completions, it is vital to understand pressures and the depths at which the Woodford will be encountered.

4.2.7 Summary of Gas Shale Geological Parameters. Many of the geological parameters for the five shale plays analyzed are similar. Ranges for the geological parameters previously discussed for all five plays are displayed in **Table 4.1** for comparison purposes.

The information in Table 4.1 can help an operator compare the qualities of a new, untested gas shale basin with the qualities of these five plays and determine if an investment of time and money is worth pursuing. While there is no guaranteed success or failure either way, this information can offer some confidence in decision making. With the remarkable amount of technological advances taking place in the oil and gas industry, plays that fall short of what is considered economic right now will most likely be unlocked in the future.

Property	Antrim Shale	Barnett Shale	Haynesville Shale	Marcellus Shale	Woodford Shale
Basin	Michigan	Fort Worth	North Louisiana Salt	Appalachian	Arkoma
Fairway depth (ft)	500 to 2,000	6,500 to 8,500	9,000 to 16,500	4,000 to 8,500	6,000 to 11,000
Thickness (ft)	200 to 850 (net: 70 to 120)	200 to 1,000 (net: 100 to 600)	270 (net: 100 to 200)	40 to 900 (varies between states)	50 to 220 (net-pay: 100 to 150)
TOC (%)	Up to 20	Up to 12 (average: 4.5)	0.5 to 4	0.3 to 11	Up to 20 (average: 3 to 10)
Total porosity (%)	3 to 10	3 to 8	6 to 15	9 to 13	7.5 to 14
Gas-filled porosity (%)	2 to 6	2 to 6	5 to 11	3 to 6	3 to 6.5
Gas content (scf/ton)	40 to 100	300 to 350	350 to 475	275 to 325	60 to 115
Quartz content (%)	50 to 60	40 to 45	28 to 33	18 to 24	55 to 87
Clay content (%)	20 to 35	20 to 40	25 to 33	40 to 60	8 to 34
Reservoir pressure (psi)	400 to 1,150	3,000 to 4,000	8,000 to 10,000	1,600 to 5,000	3,100 to 5,800
Reservoir pressure gradient (psi/ft)	0.33 to 0.38	0.42 to 0.46	0.7 to 0.92	0.42 to 0.7	0.48 to 0.55
Adsorbed gas (%)	60 to 70	15 to 30	20 to 30	20 to 40	20 to 40
Well spacing (acres as of 1/1/09)	40 to 160	80 to 160	40 to 160	40 to 80	40 to 80
Gas-in-place (Tcf as of 1/1/09)	35 to 76	25 to 30	250 to 320	1,300 to 1,900	45 to 60

Table 4.1—Summary of key properties from five shale gas basins.

4.3 Overview of Completion Techniques

For gas shale plays, as with most plays, operators must experience a learning curve before drilling and well-completion operations can be optimized. Many failures have occurred, and uneconomic wells have been drilled and completed in every gas shale play, but as technology improves and we gain more experience, the success ratio and quality of well completions continue to improve. Failures can include anything from losing a wellbore to ending up with a subeconomical well. The objective of the literature review is to determine the best practices for drilling, completing, and stimulating wells in all the important gas shale plays. Then, using the geologic and reservoir characteristics of each gas shale play, guidelines are established, in the form of flow charts, that help operators obtain a first estimate regarding how to drill, complete, and stimulate the next well in a gas shale play. By use of these guidelines, it will be easy to transition these technologies to application in other basins/plays not discussed herein. These other basins/plays can exist in North America or anywhere in the world such gas shales exist.

4.3.1 Vertical Completions. In essentially every shale gas play, the first wells drilled, completed, and tested were vertical wells. In the Haynesville, most vertical wells were drilled to this formation for data-acquisition purposes but never produced long-term. In the future, we expect nearly all gas shale plays will be developed using horizontal, or even multilateral, wells.

4.3.1.1 Antrim Shale. The Antrim Shale was one of the first shale gas productions in the US, coming online in 1926 (Blakey 2009). Before the Antrim, many Devonian Shale wells were drilled in the Appalachian Basin. One of the main differences between the Antrim and other shales discussed in this chapter is the amount of water produced. On average, the Antrim Shale produces approximately 110 BWPD, along with gas. Goodman and Maness (2008) discuss the history of completions for the Antrim Shale, as well as the fracture stimulation methods that were used in the past and those used at the time of this writing. The first wells in the Antrim Shale were drilled to the Lachine member and completed open hole. These wells were not drilled deeper into the Traverse because it was believed this zone produced water. As more data were collected, it was determined that the water was actually in the shale fractures and that drilling the Traverse zone was acceptable and added reserves. Most wells are fracture stimulated with a multistage treatment that has an average of two to five stages. The most-common stimulation design used by operators is an N_2 foam that carries approximately 25,000 to 50,000 lbm of 20/40 sand into the fracture (Goodman and Maness 2008). With this completion design, the well cost for the Antrim ranges from USD 200,000 to 700,000 for drilling and completions (Phasis Consulting 2008). In **Fig. 4.28,** Drake (2007) illustrates the average performance of the vertical wells producing from the Antrim. Fig. 4.28 shows the typical production rate for one well. Not shown is the water production discussed above, which is approximately 110 B/D.

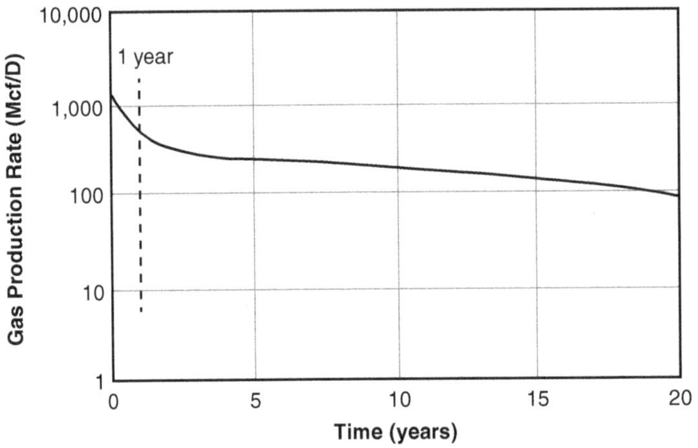

Fig. 4.28—Antrim Shale vertical-well performance (after Drake 2007).

4.3.1.2 Barnett Shale. Martineau (2003) described the evolution of the Barnett Shale, in which only 100 vertical wells were completed during the 1980s. During this time frame, the vertical wells were fracture treated with approximately 200,000 gal of crosslinked-gel fluid and approximately 300,000 lbm of sand, which was usually 20/40 mesh. With this fracturing treatment design, the IP for most of these wells averaged 600 to 700 Mcf/D. Near the end of the decade, the size of the average fracturing treatment increased to 1 million gal of crosslinked-gel fluid and 1 million lbm of sand, resulting in an IP approximately twice that achieved with the smaller fracturing treatments. In the 1990s, more than 2,000 wells were drilled in the Barnett Shale. This increase largely resulted from changes made to fracturing stimulation designs. Many operators switched from crosslinked gel to slickwater fracturing fluid and reduced the proppant amount to 100,000 lbm with no change in well deliverability, but they achieved a cost reduction of 30% (Martineau 2003).

As drilling activity picked up in the Barnett, many operators focused their attention on improving productivity by improving the fracturing treatments. Operators started fracturing with N_2 and eventually moved to fracturing with 123,000 gal of CO_2 foam and 188,000 lbm of Ottawa sand. The most-successful stimulation treatment, widely used since the early 2000s, was 31,000 gal of slickwater and approximately 95,000 lbm of 20/40 Ottawa sand (Lancaster et al. 1992). With this new completion design, the average well cost for the Barnett was approximately USD 1 million (Hayden and Pursell 2005). The average performance of the vertical wells from 1999 to 2003 is depicted in **Fig. 4.29.**

Fig. 4.29 shows that although there is little change in the decline profile for wells drilled in the Barnett in five consecutive years, there is a change in IP. The year 2000 had the best IP and decline profile among all the other years shown. The reason for the lower IP after 2000 is a noticeable depletion of the reservoir and smaller fracturing treatments. Although it seems to be less prolific from the graph, the cost savings for drilling this way actually enable more drilling (tighter well spacing) and higher field recovery.

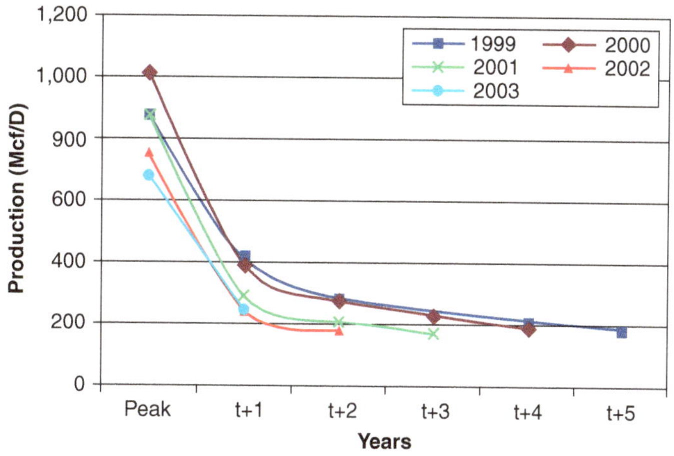

Fig. 4.29—Barnett Shale vertical-well performance.

4.3.1.3 Marcellus Shale. When the Marcellus Shale was discovered to be productive, all the wellbores that penetrated this shale were vertical. There are three schools of thought concerning the effectiveness of completing and stimulating this formation: straight N_2, N_2 foam stimulation, or a slickwater fracturing treatment (Sumi 2008). Sumi (2008) addresses when and where these different methods were used. In 1997, the first slickwater fracturing treatment was tested. This fracturing stimulation treatment used approximately 800,000 gal of water in addition to 250,000 lbm of 20/40-mesh sand. Water fracturing treatments are most commonly used in deeper, high-pressure shales, while N_2 foam treatments are pumped in shallower and lower-reservoir-pressure shales.

Many operators still choose between slickwater fracturing techniques and N_2 foam to treat the Marcellus wellbores. Operators are trending toward using slickwater fracturing treatments because of the cost reduction, but if a well is drilled in an area that has lower bottomhole pressure than expected, companies move to an N_2 foam (Myers 2008). Most wells will be completed with a three- to four-stage treatment to maximize coverage of the reservoir in contact with the wellbore. With the use of an N_2 foam fracturing treatment, the average EUR is approximately 0.4 to 1 Bcf per well, which costs approximately USD 800,000 to 1 million to drill and complete (Sumi 2008). Miller (2008) depicts the average performance of the vertical wells, shown in **Fig. 4.30.**

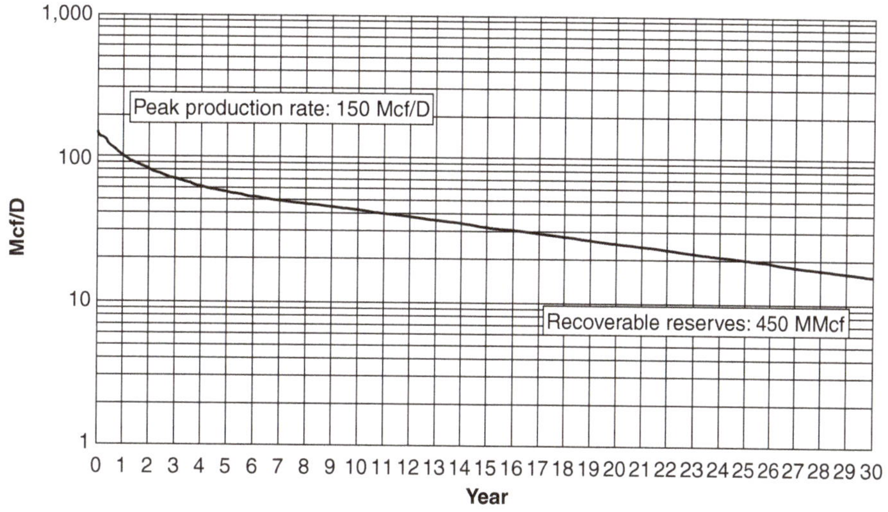

Fig. 4.30—Marcellus Shale vertical-well performance (after Miller 2008).

4.3.1.4 Woodford Shale. Vertical Woodford Shale wells have been completed over the years, but it is believed these wells have been commingled with other reservoirs, such as the Hunton Limestone. In the late 1990s, Newfield Exploration began to drill Woodford vertical wells. From the late 1990s to 2006, approximately 100 wells were drilled, with 60% of them drilled by Newfield Exploration (Shale Gas 2006). Most of these wells were completed with cemented casing and then perforated with three to four clusters throughout the interval. Some companies fracture treated all the perforations together; however, most wells were stimulated with three- to four-stage fracturing treatments. The most-common stimulation design used by operators drilling vertical Woodford wells has been slickwater treatments, the lowest-cost method to complete. Usually, approximately 100,000 gal of slickwater would be used to transport 40,000 to 50,000 lbm of 20/40-, 40/70-, or 100-mesh sand per stage.

In general, the average EUR has been approximately 1.35 Bcf per well. The cost of the vertical wells was approximately USD 2 million to drill and complete. In **Fig. 4.31,** Andrews (2007) illustrates the average performance of the vertical wells for the Woodford Shale.

4.3.2 Horizontal Completions. Since 2000, many operators have introduced new horizontal drilling and completion designs to improve the productivity of gas shale wells. After reviewing the costs to drill and complete vertical wells in plays such as the Barnett and Woodford shales, and factoring in the EURs from most wells, the industry turned to the use of horizontal drilling technology to

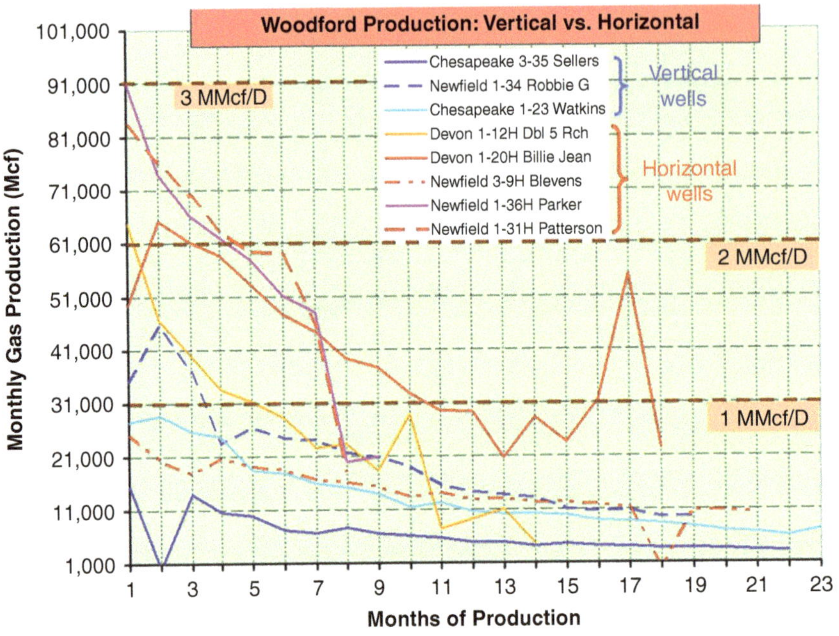

Fig. 4.31—Woodford Shale production performance: horizontal vs. vertical (Andrews 2007).

improve gas flow rates and gas recovery. Horizontal drilling has clearly become one of the greatest improvements in gas shale basins. The biggest challenge has been determining the best way to complete these horizontal wells. Continued studies help the industry understand drilling and placement techniques, such as where to land the lateral in the formation and what size casing to use. As a rule of thumb, most operators will land the lateral in the middle of the formation, or relatively close to the middle of the formation.

For example, the Woodford, Barnett, and most of the Marcellus wellbores are landed in the middle of their particular formations, while the Haynesville is landed closer to the bottom, or approximately 55 to 60% from the top (DeWitt 2008; Wickstrom 2008). There is not much difference in placement techniques, but most operators believe fracture growth occurs throughout the entire interval if the horizontal hole is in the middle. With the use of microseismic technology, many operators have confirmed that height growth during the fracturing treatment is sufficient to connect the entire shale to the horizontal well, provided that the shale is not too thick. The optimal result achieved by the use of microseismic technology is to have seismic activity follow the path of the perforations the fracturing treatment is trying to stimulate without too much overlap with the prior treatment and the next treatment. Therefore, the entire lateral has some seismic activity without too much overlapping. It is imperative to understand that if there is a significant change in the rock mechanical strengths in the reservoir, landing the lateral in the middle of the section might not be the optimal choice.

4.3.2.1 Antrim Shale. The Antrim Shale is one of the formations that has been developed primarily using vertical wells rather than horizontal wells. The Antrim Shale vertical wells cost approximately USD 350,000 to drill and complete because the formation is shallow. Usually, horizontal wells cost approximately twice as much to drill as vertical wells. In addition, the reservoir pressure in the Antrim Shale is low (400 psi), potentially causing issues with wellbore stability. Finally, the Antrim Shale has high water production, making the wells difficult to operate in the long term. Many of these wells would not be able to lift the water effectively; thus, the laterals would water out early in the life of the well.

4.3.2.2 Barnett Shale. In 2002, development of the Barnett Shale with horizontal wells increased substantially. Many operators steered away from vertical wells and shifted their focus to drilling horizontal wells. In the early stages of horizontal drilling, operators looked at both cemented and uncemented laterals. Uncemented laterals were soon phased out because shorter laterals were required, thus leading to fewer fracturing stages; also, stimulations in the uncemented lateral were difficult to design and understand. As such, most wells are now completed with cemented laterals to maintain better control of where the hydraulic fractures are created. As horizontal drilling became increasingly common, many operators rather quickly shortened the learning curve on fracture stimulation as a result of information shares, conferences, and relationships with service companies.

Schein (2008) states that the general design for a Barnett horizontal well is to drill a 3,000- to 4,000-ft lateral, then place a fracturing treatment approximately every 500 ft down the lateral. As such, it is common to pump 6 to 10 fracturing stages in any one wellbore. For the total well, it would not be uncommon to pump 1 million gal of slickwater, carrying 300,000 lbm of sand, in each stage. The sand is normally 20/40, 40/70, or 100 mesh. Generally, a 4½- or 5½-in. casing is run in hole in the lateral portion of the well, so the average pump rate is 40 to 120 bbl/min. Also, many operators are now adding friction reducer to keep the rate higher, as well as biocide (to prevent corrosion-causing bacteria) and scale inhibitors (to prevent the buildup of calcium carbonate) to avert issues that will reduce short- and long-term productivity. In many treatments, some gel is added to the slickwater to increase viscosity so all the sand can be pumped away (Lancaster et al. 1992). There have been some attempts to use simu-fracturing treatments on the Barnett Shale wells. In simu-fracturing treatments, at least two laterals are drilled next to each other and are fracture stimulated simultaneously. **Fig. 4.32** shows the typical performance of a Barnett horizontal well that has been fracture treated with four stages and the optimal fracturing design.

4.3.2.3 Haynesville Shale. Most Haynesville Shale wells are horizontal completions rather than vertical. Vertical pilot holes are drilled to gather data, by means of cores and logs, and then plugged back to kick off a horizontal leg. Most Haynesville laterals are between 4,000 and 4,600 ft in length, which enables more fracturing stages in each wellbore. The fracturing stimulation is designed with perforation clusters approximately 300 to 350 ft apart, with each perforation cluster fracture treated separately. Most of the Haynesville horizontals have approximately 10 to 12 fracturing stages per well. Typically, these wells are fracturing stimulated down cased, cemented liners with either 4½- or 5½-in. casing to enable a high-enough treating pressure to pump the fracturing job away without fighting too much friction in the process. Each stage is treated with approximately 8,000 to 12,000 bbl of slickwater, which carries approximately 300,000 lbm of proppant into the formation. However, in practice, much of the so-called slickwater actually contains gel to increase the fluid viscosity to open the fractures and enable the propping agent to be pumped into the fracture. The proppant pumped into the reservoir is a combination of Hydro Prop, ceramic, or resin-coated sand. Most operators generally have used 20/40 or 40/70 sand.

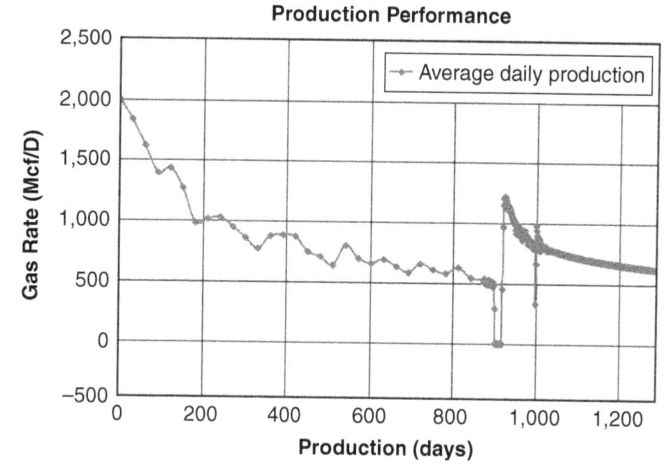

Fig. 4.32—Barnett Shale horizontal-well performance.

At the time of this writing, typical horizontal wells in the Haynesville Shale had thus far yielded an IP from 5 to 20 MMcf/D, but they had steep decline rates (Hutchinson 2009). The industry is predicting these wells will recover 4.5 to 8.5 Bcf per well. The cost to drill and complete them is approximately USD 6.9 million to 11 million (Chesapeake Energy 2009). Many operators are moving to closer perforation clusters, such as 60 to 100 ft, so they can contact more of the reservoir when they fracture stimulate these wellbores, which could be a mistake because these might cause narrower fractures instead of broader ones. Kapchinske and Sharp (2008) portray the typical performance of a Haynesville horizontal well in **Fig. 4.33.**

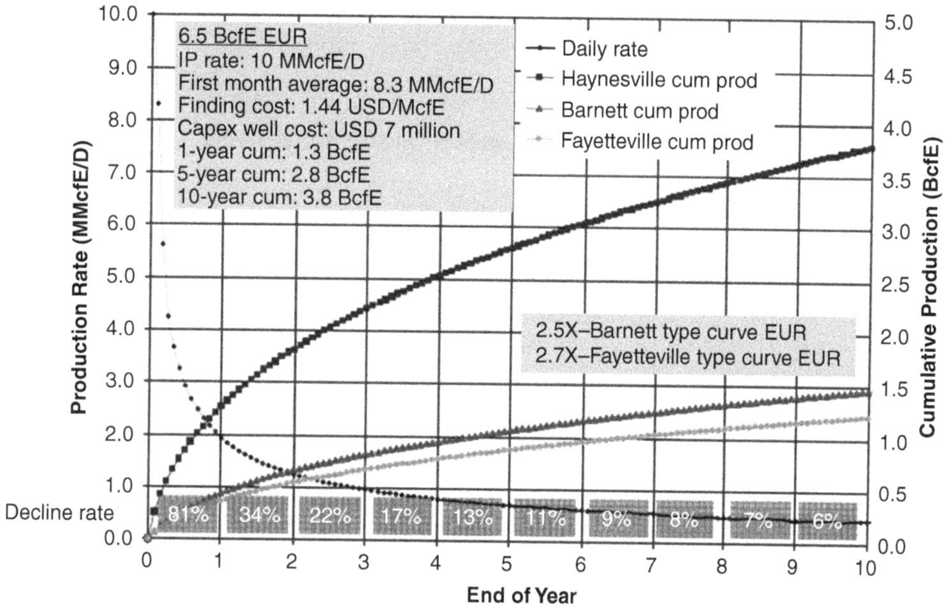

Fig. 4.33—Haynesville Shale horizontal-well performance (after Kapchinske and Sharp 2008). Capex = capital expenditure. cfE = cubic feet equivalent.

4.3.2.4 Marcellus Shale. The Marcellus Shale play now uses horizontal drilling to develop the reservoir because this technique can be used to intersect the vertical fracture network. Many companies have realized they could recover more reserves by connecting to more fractures by using horizontal wellbores because the wellbore is now perpendicular to the fracture system. In the 1980s, many operators drilled horizontal wells and were technically successful, but they did not produce commercial quantities of gas. The main reason for the economic failure was the high cost of fracturing stimulation and the limitations of fracturing technology during that time (Miller 2008). As a result of the improvements to fracturing technology over the 10 years previous to this writing, horizontal drilling can now be used successfully in the Marcellus Shale. Even with horizontal drilling as the preference, many facets of it are being tested in the Marcellus Shale. The biggest issue is whether to drill these wells open hole or to case and cement them. Most operators choose to case and cement because of issues with mechanically separating the fracturing stages. Even with the decision to use cemented-hole completions, problems with fracture initiation still arise, resulting from stress anisotropy (Gottschling 2007). Arthur et al. (2008) state that operators are experimenting with the phasing of the perforations in an attempt to reduce the fracturing-treatment breakdown pressure. They use hydrochloric acid to open the perforations before stimulating, and they reperforate if there is difficulty pumping into the formation. The laterals for Marcellus Shale wells typically range from 2,000 to 4,500 ft in length, depending on the number of fracturing stages to be completed (DeWitt 2008). As previously discussed, there are three schools of thought regarding the fracturing fluid used in these wells—straight N_2, N_2 foam, or slickwater. Gottschling (2007) discusses how most horizontal wellbores are stimulated using large foam fracturing treatments in which approximately 5 to 10 MMcf of N_2 is pumped, transporting from 728,000 to more than 1.8 million lbm of sand into a well that will be completed with 4 to 10 stages. The sand typically used in this type of stimulation treatment is 100 mesh or a 20/40 mesh.

The Marcellus Shale wells have an IP from 1 to 4.3 MMcf/D. Most companies believe a horizontal well in the Marcellus Shale will eventually recover 1 to 4.5 Bcf per well. The costs to drill and complete a Marcellus Shale well are generally USD 3 to 4 million (Sumi 2008). The Marcellus Shale has nearly a 1:1 ratio between IP and recoverable reserves; if a well's IP is 1 MMcf/D, it is likely to recover approximately 1 Bcf of gas. As the industry moves to optimizing these fracturing methods, some companies are looking at using CO_2 to help assist in the flowback after the fracturing treatment is completed. CO_2 is water soluble, which enables a higher water-recovery percentage, thus helping recover fluid in low-bottomhole-pressure wells. DeWitt (2008) shows a typical Marcellus horizontal-well performance in **Fig. 4.34.**

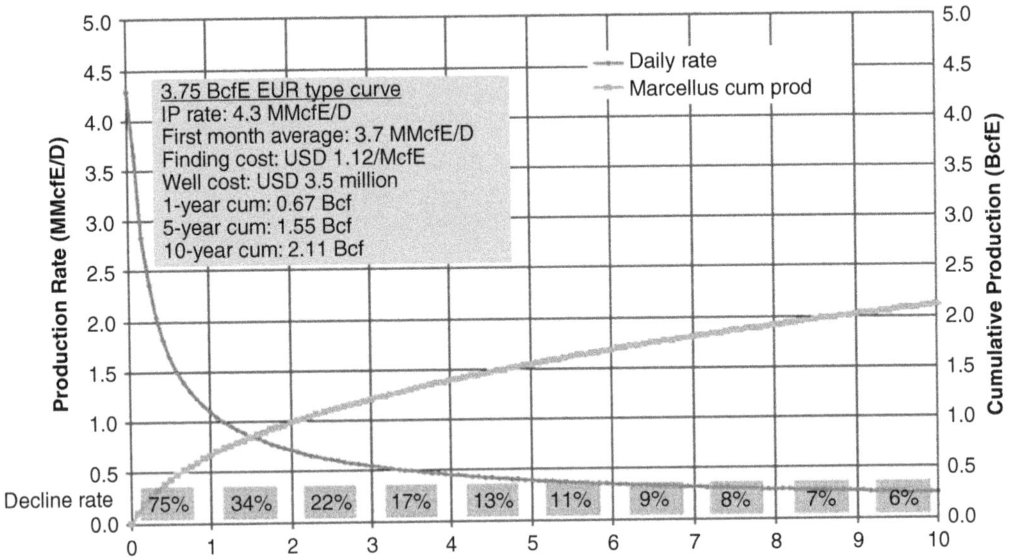

Fig. 4.34—Marcellus Shale horizontal-well performance (after DeWitt 2008). cfE = cubic feet equivalent.

4.3.2.5 Woodford Shale. In 2004, the Woodford Shale horizontal program began but did not gain much momentum until 2006, when more than 130 Woodford horizontal wells were completed (Wickstrom 2008). As with other shale plays, there is a learning curve, leading to better completions in the Woodford (Coffey 2007). It is now common practice to find cased, cemented laterals in this play ranging in length from 2,500 to 4,000 ft. The difference between these two lengths is not only cost but also EURs, because of the number of fractures involved. The 2,500-ft lateral lengths tend to enable only five stages, which yield an average of approximately 2.5 Bcf per well that is recovered over the life of that well. In 4,000-ft laterals, operators can execute eight to nine stages, which can yield recoveries of 6 Bcf on average. From this information, many deduce that it is beneficial to drill longer laterals. However, price is still an issue because shorter laterals cost approximately USD 4.75 million to drill and complete, while longer laterals cost approximately USD 7.75 million (Wickstrom 2008).

Wickstrom (2008) states that a common practice in the Arkoma Woodford is to perforate 300-ft sections at a time, then acidize the perforations with 15% hydrochloric acid before fracturing the formation. Each stage pumps approximately 18,000 bbl of slickwater and 200,000 lbm of sand (100 mesh, 40/70 Ottawa, or 20/40 Ottawa). Generally, a 5½-in. casing is run in hole in the lateral portion of the well, so the average pump rate ranges from 60 to 95 bbl/min and the average treating pressure is approximately 5,600 to 5,800 psi. The typical performance of a 4,000-ft lateral Woodford horizontal well is seen in Fig. 4.31. Similar to the Barnett Shale, many operators are moving to simu-fracturing treatments in the Woodford Shale, and some operators have performed zipper fracture stimulations. In a zipper fracture, two laterals are next to each other and the fracturing stages mirror each other. Starting from the toe of each lateral, one fracturing stage is performed on the first well, and then the same stage is performed on the second well. Then, the operator completes the next stage on the first well, then stimulates the second stage on the second well. The operator continues going back and forth between the two wells until both wells are stimulated from toe to heel. Some studies have indicated that this method helps stimulate more of the reservoir, but there are not yet enough data to indicate whether this is true or false.

More details regarding the economics of producing shale reservoirs (as of 2009) can be found in Agrawal (2009).

4.4 Eagle Ford Shale

The following section has been taken from a 2014 PhD dissertation in petroleum engineering by Yao Tian at Texas A&M University. More information is available in Tian (2014). This text is included herein to provide additional information generated since the publication of Agrawal (2009).

4.4.1 Eagle Ford Shale Geological Features. The Eagle Ford Shale extends from the Mexican border northeastward to the Texas/Louisiana border. By 2014, exploration and production were focused on the Rio Grande Embayment of South Texas, where more than a dozen operators have Eagle Ford leases and/or vertical and horizontal production wells. The Eagle Ford Shale wells produce oil, condensate, and dry gas, depending on the geological setting.

From its outcrop, the Eagle Ford extends basinward (southeastward) to the Edwards and Sligo Reef trends, where its depth exceeds 14,000 ft **(Fig. 4.35)** (Hentz and Ruppel 2010). There are several main structural features in the Gulf Province, which extends across the Texas and Louisiana borders (Ewing and Lopez 1991; Condon and Dyman 2006). Just northwest of the province boundary is the elevated Llano Uplift **(Fig. 4.36).** The San Marcos Arch is a subsurface extension of the Llano Uplift that extends southeastward toward the Gulf of Mexico.

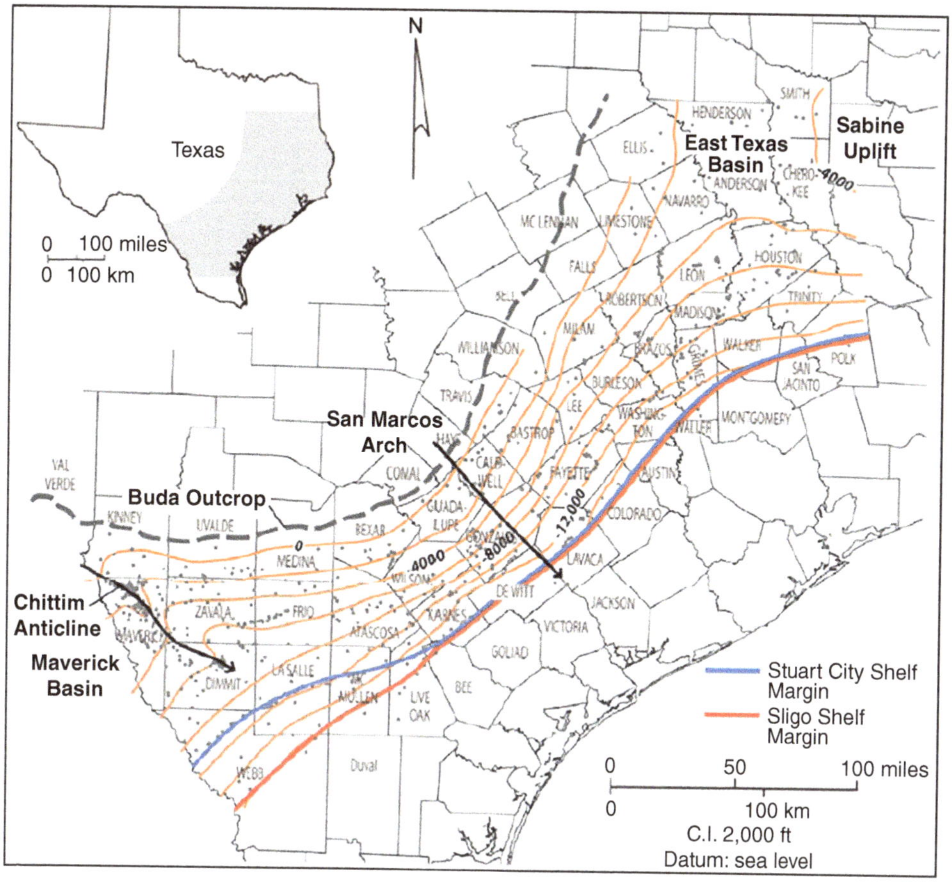

Fig. 4.35—Structural top of the Eagle Ford Shale (Hentz and Ruppel 2010).

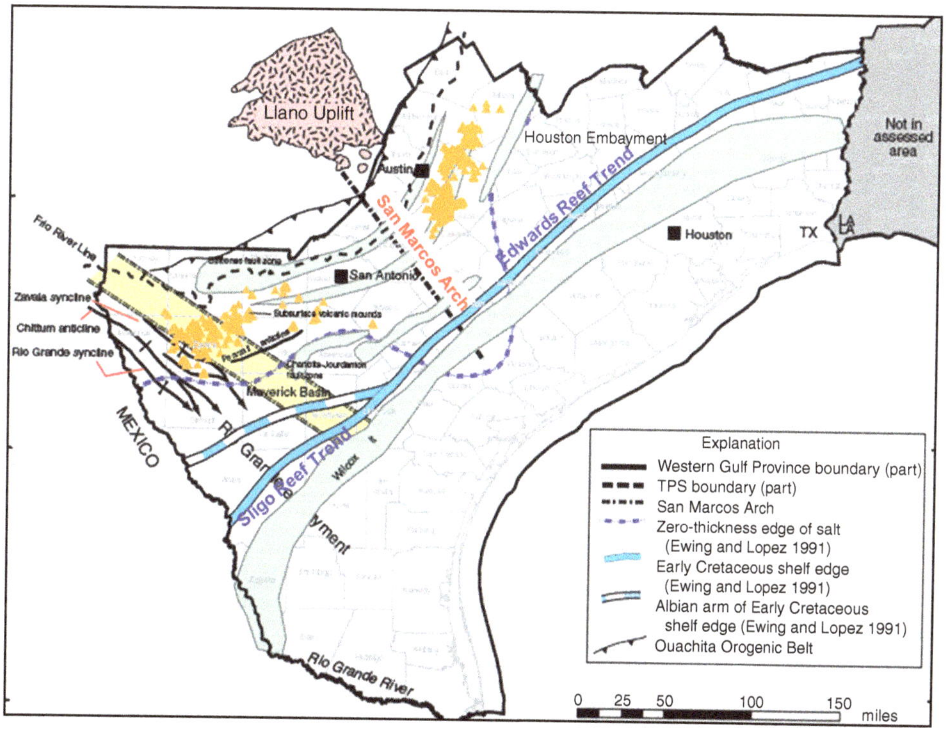

Fig. 4.36—Map of the western part of the Western Gulf Province showing the main structural features (after Condon and Dyman 2006). TPS = total petroleum system.

The Rio Grande Embayment lies southwest of the San Marcos Arch (Fig. 4.36) (Ewing and Lopez 1991; Condon and Dyman 2006). The Edwards and Sligo Reef trends are two prominent factors that influence the structure in the Rio Grande Embayment (Fig. 4.36). Most of the folds in the western part of the Maverick Basin are of the Late Cretaceous to Tertiary and are the result of Laramide compression (Ewing and Lopez 1991; Condon and Dyman 2006). The main folds in the western part of the basin are the Rio Grande and Zavala synclines, separated by the southeastward-plunging Chittum Anticline (Fig. 4.36) (Condon and Dyman 2006).

The Eagle Ford Group is Late Cretaceous **(Fig. 4.37)** (Hentz and Ruppel 2010, after Childs et al. 1988). In the Rio Grande Embayment, the Eagle Ford Shale overlies the Buda Formation and is overlain by the Austin Chalk. In the Houston Embayment and San Marcos Arch region, the Eagle Ford Shale can be divided into the Turonian Eagle Ford and the Cenomanian Eagle Ford. The Cenomanian Eagle Ford unconformably overlies the Buda Limestone and is disconformably overlain by the Austin Chalk. In the East Texas Basin, only the Turonian Eagle Ford is present, and it is overlain by the Austin Chalk (Fig. 4.37).

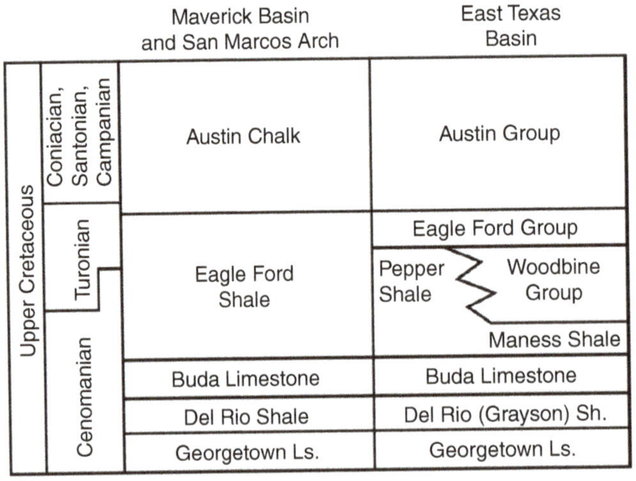

Fig. 4.37—Lithostratigraphy of the Eagle Ford Shale in Texas (Hentz and Ruppel 2010, after Childs et al. 1988).

From the well-log responses, the Eagle Ford Shale can be divided into the Upper and Lower Eagle Ford Shale. The Lower Eagle Ford Shale has a high GR response, which indicates high shale content and a low-energy deposition environment (Liro et al. 1994; Dawson 2000; Condon and Dyman 2006). The Upper Eagle Ford Shale shows low GR response, indicating a relatively high carbonate composition and low shale content.

The total Eagle Ford Shale thins from > 600 ft in the Maverick County depocenter to < 50 ft to the northeast. The Lower Eagle Ford is present throughout the study area **(Fig. 4.38)** (Hentz and Ruppel 2010). The thickness of the Lower Eagle Ford Shale is > 200 ft

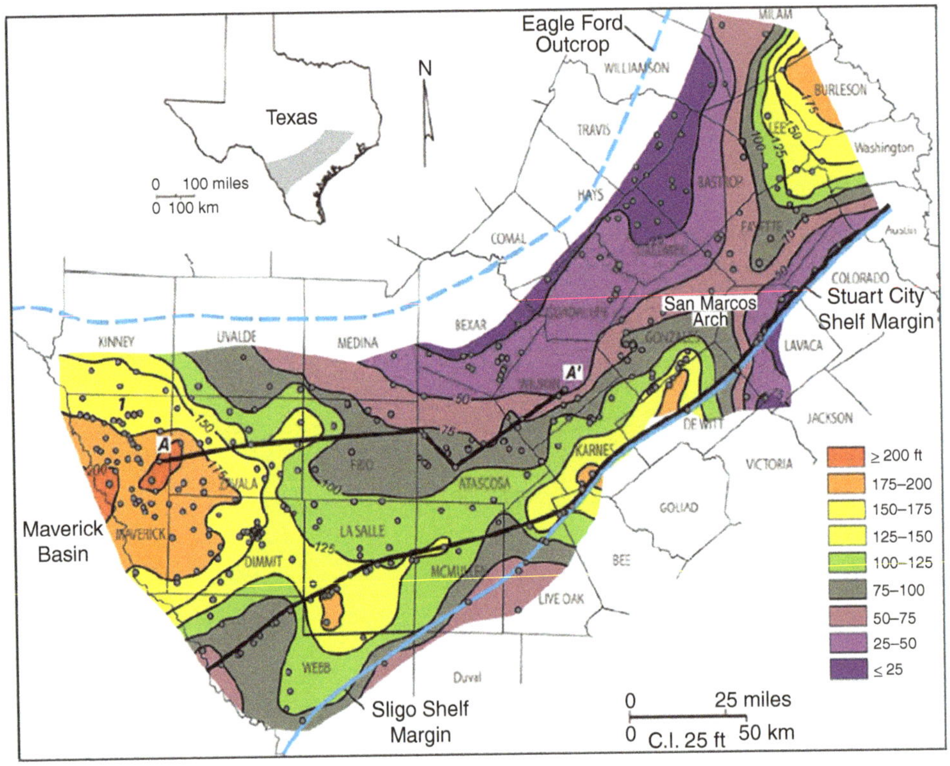

Fig. 4.38—Isopach map of the Lower Eagle Ford Shale (Hentz and Ruppel 2010).

in the western Maverick County depocenter (Fig. 4.38) (Hentz and Ruppel 2010). It thins northward to < 50 ft at the updip limit of well control in Bexar County (Fig. 4.38) (Hentz and Ruppel 2010). The Upper Eagle Ford Shale is present in only the southwest region, and its thickness is > 425 ft in the Maverick County depocenter. From Frio County, the Upper Eagle Ford Shale thins northeastward and pinches out in Atascosa County **(Fig. 4.39)** (Hentz and Ruppel 2010).

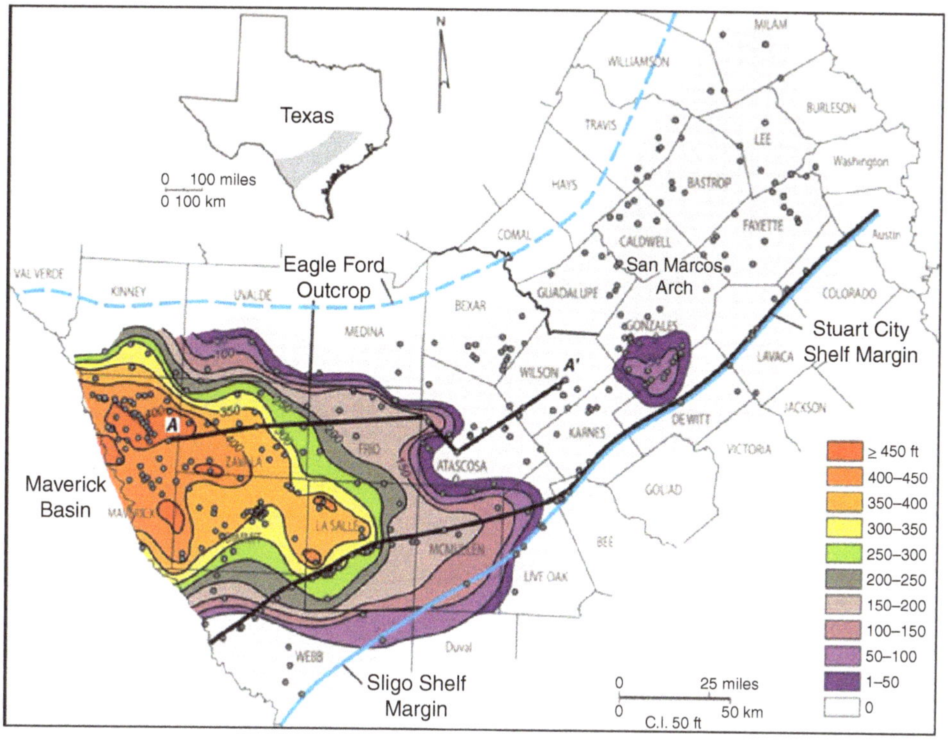

Fig. 4.39—Isopach map of the Upper Eagle Ford Shale (Hentz and Ruppel 2010).

4.4.2 Reservoir Properties. The Eagle Ford Shale varies significantly in reservoir properties, including thickness, mineral composition, and hydrocarbon type (Hentz and Ruppel 2010; Mullen 2010). Thermal maturity increases basinward, and the formation passes through the oil and condensate windows; along the Edwards and Sligo Reef trends, it is in the dry-gas window (Lewan 2002).

Previous outcrop studies indicate three major lithologic units in the formation, including (1) organic-rich, pyritic, and fossiliferous marine shales and bituminous claystone (in the lower part) that were deposited during a transgressive episode (Condon and Dyman 2006); (2) a condensed section of pyritic, phosphatic, and bentonitic shale beds (in the middle part) (Condon and Dyman 2006); and (3) shales, limestones, and carbonaceous siltstones (in the upper part) that were laid down during a regressive highstand (Condon and Dyman 2006; Dawson 2000).

TOC is typically 4 to 5% (Condon and Dyman 2006); kerogen composition is mixed Type 2 and Type 3 (Liro et al. 1994). Outcrop samples near Austin, Texas, were reported to have an average TOC of 5.15 weight percent in the Lower Eagle Ford Shale (Dawson 2000). The Eagle Ford Shale is thermally immature at outcrop.

Eagle Ford porosity ranges from 5 to 10%; the core-derived permeability from samples from two wells was 1 to 2 md (Petrohawk 2009; TXCO Resources 2009; Mullen 2010). In the production region, the Eagle Ford Shale is predominantly calcite rich (40 to 68%). Clay content averages 15%, quartz-plus-feldspar content is approximately 15%, and TOC averages 4% (Mullen et al. 2010; Mullen 2010; Quirein et al. 2013). In the downdip region, Eagle Ford Formation temperatures range from 280 to 315°F, and the formation pressure gradient is 0.65 to 0.75 psi/ft (Petrohawk 2009; TXCO Resources 2009).

4.4.3 Production. The Eagle Ford Shale produces oil, gas condensate, and dry gas **(Fig. 4.40).** Fracturing stages range from 10 to 20 for both gas-rich and liquid-rich wells **(Table 4.2)** (Fan et al. 2011). The most-productive gas wells are located in La Salle County, whereas the most-productive oil wells are located in Karnes and Gonzales counties (Fan et al. 2011).

There are also similarities in lateral horizontal length and clusters per stage between gas-rich and liquid-rich wells (Table 4.2). However, compared to liquid-rich wells, stimulation of gas-rich wells requires more total fluid and less proppant. The average pump rate of gas-rich wells is as high as 70 bbl/min, whereas it is only 50 bbl/min for liquid-rich wells (Fan et al. 2011).

Stimulation design has evolved with the maturation of the Eagle Ford Shale play development. In 2009, 96% of the stimulation operations by Petrohawk were slickwater, which used linear gels or other friction reducers in the water treatment (Rushing and Sullivan 2003), pumped at 12 stages per well (Rhein et al. 2011). The completion strategy evolved into a hybrid fracturing treatment with 16 stimulation stages, which captures the advantages of both conventional gel and water-fracturing treatments. A hybrid treatment provides only a modest improvement in production but a significant reduction of water consumption. In 2010, channel-fracturing technology was introduced in the Eagle Ford Shale, which significantly improved the completion efficiency. Channel fracturing is the hydraulic fracturing technique that applies intermittent pumping of proppant-laden and proppant-free gelled fluid at a high frequency, generating a heterogeneous placement of proppant and open channels throughout the proppant pack (Gillard et al. 2010; Rhein et al. 2011). Channel fracturing improved the cumulative production and reduced the required amount of fracturing fluid and proppant (Rhein et al. 2011).

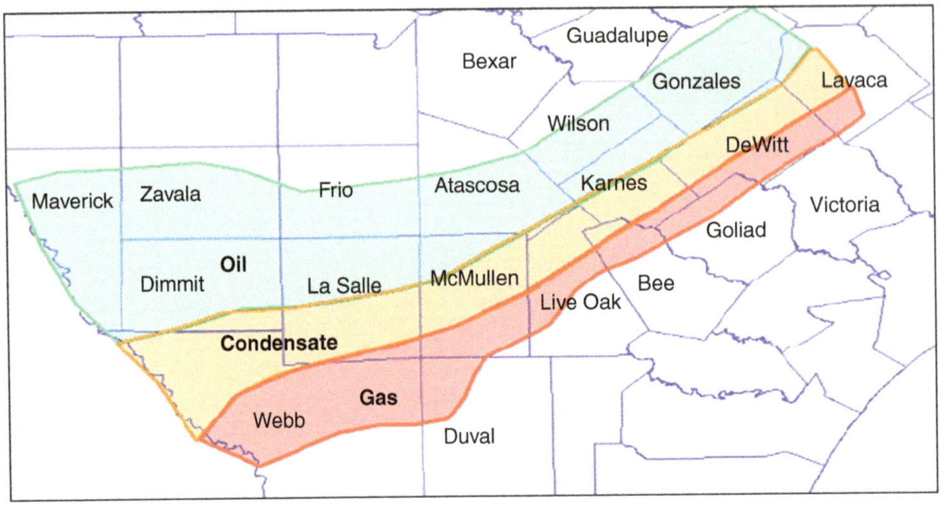

Fig. 4.40—Maturation window of Eagle Ford Shale hydrocarbon (after Fan et al. 2011).

	Eagle Ford Gas-Rich Area	Eagle Ford Liquid-Rich Area
Number of stages	10–20	10–20
Lateral length (ft)	Approximately 4,500	Approximately 4,500
Number of clusters per stage	4–8	4–8
Stage length (ft)	200–400	200–400
Distance between clusters (ft)	30–80	30–80
Number of clusters	50–120	50–120
Fluid total (1,000 gal)	6,112	4,032
Fluid per stage (1,000 gal)	509	252
Fluid per cluster (1,000 gal)	127	50
Fluid per ft (1,000 gal)	1.4	0.8
Proppant total (1,000 lbm)	3,432	5,120
Proppant per stage (1,000 lbm)	246	320
Proppant per cluster (1,000 lbm)	61	64
Proppant per ft (1,000 lbm)	0.7	1
Max proppant concentration (PPA)	1.5	4
Average pump rate (bbl/min)	70	51
Average pump rate per cluster (bbl/min)	17.5	10
Pump rate per perforation (bbl/min)	1.5–2.5	1.28

Table 4.2—Typical completion for horizontal Eagle Ford wells in the gas- and liquid-rich areas (after Fan et al. 2011). PPA = lbm of proppant added per gal.

4.5 Summary

From the information presented in this chapter, the following conclusions have been reached:

- A systematic evaluation of completion techniques in gas shale is important to the energy industry because gas shale plays will be an important global resource for the 21st century. With our conventional reservoirs dwindling, it is important to find new technologies and techniques to economically extract oil and gas from unconventional reservoirs. The need for this has sparked interest in the factors that make up a commercially viable gas shale play.
- Through the geologic analysis of the Antrim Shale, Barnett Shale, Haynesville Shale, Marcellus Shale, and Woodford Shale, this chapter looks at the similarities as well as the differences in the key parameters. From the literature review, the key geologic parameters include depositional environment, depth, TOC, gas content, clay content, quartz content, shale mineralogy, adsorbed-gas percentage, pressure gradient, and thickness.
- Table 4.1 summarizes the geological parameters found for the five shales evaluated by Agrawal (2009). Although there were some stark differences (e.g., the depth between the Haynesville and the Antrim shales), the shale plays were very similar in terms of other geologic properties.

- Also, the geologic parameters that differ did change completion designs, showing that even if the parameters of other plays not discussed fall out of the range shown, the wells are not necessarily uneconomic.
- Completion techniques are discussed because the geologic parameters are not all the same. With the learning curve the industry has experienced over the years, certain aspects were not tested because of economic viability, such as horizontal Antrim Shale wells.
- Completions are key to the success of many of the gas shale plays, and with new developments in this arena and more-efficient costs, gas shale plays will take off even faster.
- Because the energy industry is volatile, economic sensitivity was performed on the price, capital, and operating cost to see what effect it would have on these five plays. Of all the observations made, it seems that horizontal drilling in almost every environment tested is economic, except in the Woodford Shale, which had only one scenario that was not economic.
- With the development of newer technology, these plays will become more competitive with other conventional plays in differing economic environments, but this should not stop operators from testing different shale plays not only in the US but throughout the world because this could one day become one of the primary resources for gas production.

4.6 References

Agrawal, A. 2009. *A Technical and Economic Study of Completion Techniques in Five Emerging US Gas Shale Plays*. Master of Science thesis, Texas A&M University, College Station, Texas (December 2009).

Airhart, M. 2007. Barnett Boom Ignites Hunt for Unconventional Gas Resources. Jackson School of Geosciences news (15 January 2007), http://www.jsg.utexas.edu/news/2007/01/barnett-boom-ignites-hunt-for-unconventional-gas-resources/.

Andrews, R. 2007. Production and Outcrop/Reservoir Characteristics of the Woodford Shale in South-Central and Southeast Oklahoma. *Search and Discovery* article no. 10137 (posted 4 October 2007). Adapted from a paper presented at the Tulsa Geological Society Conference, Tulsa, Oklahoma, USA, 11 September.

Ardmore Basin Woodford Gas Plan Takes Off. 2008. *Oil & Gas Journal* **106** (20).

Arthur, J. D., Bohm, B., and Layne, M. 2008. Hydraulic Fracturing Considerations for Natural Gas Wells of the Marcellus Shale. Paper presented at the Ground Water Protection Council 2008 Annual Forum, Cincinnati, Ohio, USA, 21–24 September.

Arthur, M. A. and Sageman, B. B. 2005. Sea-Level Control on Source-Rock Development: Perspectives from the Holocene Black Sea, the Mid-Cretaceous Western Interior Basin of North America, and the Late Devonian Appalachian Basin. In *The Deposition of Organic-Carbon-Rich Sediments: Models, Mechanisms, and Consequences*, Vol. 82, ed. N. B. Harris, 35–59. SEPM Society of Sedimentary Geology. https://doi.org/10.2110/pec.05.82.0035.

Barrett, R. 2009. The Depositional Setting of the Marcellus Black Shale. Independent Oil and Gas Association of West Virginia.

Bertola, D. 2008. Researchers: Shale Holds Vast Supply of Natural Gas. *Buffalo Business First* (11 February 2008), https://www.bizjournals.com/buffalo/stories/2008/02/11/story2.html.

Blakey, R. 2009. Paleography. https://www2.nau.edu/rcb7/.

Brackett, W. 2006. Developing the Marcellus Shale: What We've Experienced in the Barnett Shale. *Powell Barnett Shale Newsletter*.

Burns, J. 1991. *Fossil Collecting in the Mid-Atlantic States With Localities, Collecting Tips, and Illustrations of More Than 450 Fossil Specimens*, 96. Baltimore: Johns Hopkins University Press.

Cardott, B. J. 2007. Overview of Woodford Gas-Shale Play in Oklahoma. PowerPoint presentation given at the Woodford Gas Shale Conference, Oklahoma City, Oklahoma, USA, 23 May.

Cemen, I., Ataman, O., Puckette, J. et al. 2007. Natural Fractures in Woodford Shale, Arbuckle Mountains, Oklahoma. Paper no. 131–6 presented at the 2007 GSA Denver Annual Meeting, Denver, Colorado, USA, 28–31 October.

Chesapeake Energy. 2009. 2009 Institutional Investor and Analyst Meeting. Presentation for the 2009 Investor and Analyst Meeting, New York City, New York, USA, 14 October.

Childs, O. E., Steele, G., and Salvador, A. 1988. Correlation of Stratigraphic Units in North America (COSUNA) Project, Gulf Coast Region. Tulsa, Oklahoma: American Association of Petroleum Geologists correlation chart series.

Coffey, B. 2007. Gas Resource Potential of the Woodford Shale, Arkoma Basin, Oklahoma. Article no. 90071 presented at the AAPG Rocky Mountain Meeting, Snowbird, Utah, USA.

Comer, J. B. 1992. Organic Geochemistry and Paleogeography of Upper Devonian Formations in Oklahoma and Northwestern Arkansas. In *Source Rocks in the Southern Midcontinent, 1990 Symposium*, ed. K. S. Johnson and B. J. Cardott, 70–93. Norman, Oklahoma: Oklahoma Geological Survey, Circular 93.

Comer, J. B. 2008. Reservoir Characteristics and Production Potential of the Woodford Shale. Pete'sPlace on Global Warming/Climate Change, 22 August 2008. http://petesplace-peter.blogspot.com/2008/08/woodford-shale-major-new-unconventional.html.

Condon, S. M. and Dyman, T. S. 2006. 2003 Geologic Assessment of Undiscovered Conventional Oil and Gas Resources in the Upper Cretaceous Navarro and Taylor Groups, Western Gulf Province, Texas. In *Petroleum Systems and Geologic Assessment of Undiscovered Oil and Gas, Navarro and Taylor Groups, Western Gulf Province, Texas*, US Geological Survey Western Gulf Province Assessment Team, Chap. 2. Reston, Virginia: US Geological Survey Digital Data Series DDS–69–H.

Cristellaria. 2014. Unconventional Hydrocarbon Plays of Louisiana - Haynesville Shale, Tuscaloosa Marine Shale, Brown Dense (Smackover), https://commons.wikimedia.org/wiki/File:LAHydrocarbonPlaysVer001.png (accessed 13 January 2020).

Dawson, W. C. 2000. Shale Microfacies: Eagle Ford Group (Cenomanian-Turonian) North-Central Texas Outcrops and Subsurface Equivalents. *Gulf Coast Association of Geological Societies Transactions* **50**: 607–621.

Deep Time Maps. 2013. North American Key Time-Slices Series: Middle Devonian – 390 Ma (395–385) map. https://deeptimemaps.com.

DeWitt, H. 2008. Marcellus Shale Overview. Presented at the Chesapeake Energy 2008 Investor and Analyst Meeting, Oklahoma City, Oklahoma, USA, 15–16 October.

de Witt, W., Jr., Roen, J. B., and Wallace, L. G. 1993. Stratigraphy of Devonian Black Shales and Associated Rocks in the Appalachian Basin. In *Petroleum Geology of the Devonian and Mississippian Black Shale of Eastern North America*, Bulletin 1909, ed. J. B. Roen and R. C. Kepferle, Chap. B, 1–57. Denver: US Geological Survey. https://doi.org/10.3133/b1909.

Drake, S. 2007. Unconventional Gas Plays. Presented at the SPEE Convention, 6 December.

Durham, L. S. 2008. Complex Targets Now in Sight: Prices, Technology Make Shales Hot. *AAPG Explorer* **29** (7): 10.

Engelder, T. 2009. The Marcellus Black Shale Formation.

Ewing, T. E. and Lopez, R. F. 1991. Plate 2, Principal Structural Features, scale 1: 2,400,000. In *The Gulf of Mexico Basin*, Volume J, ed. A. Salvador. Boulder, Colorado: Geological Society of America. https://doi.org/10.1130/DNAG-GNA-J.

Fan, L., Martin, R., Thompson, J. et al. 2011. An Integrated Approach for Understanding Oil and Gas Reserves Potential in Eagle Ford Shale Formation. Paper presented at the Canadian Unconventional Resources Conference, Calgary, Alberta, Canada, 15–17 November. SPE-148751-MS. https://doi.org/10.2118/148751-MS.

Gasparrini, M., Sassi, W., and Gale, J. F. W. 2014. Natural Sealed Fractures in Mudrocks: A Case Study Tied to Burial History From the Barnett Shale, Fort Worth Basin, Texas, USA. *Marine and Petroleum Geology* **55**: 122–141.

Gillard, M., Medvedev, O., Peña, A. et al. 2010. A New Approach to Generating Fracture Conductivity. Paper presented at the SPE Annual Technical Conference and Exhibition, Florence, Italy, 19–22 September. SPE-135034-MS. https://doi.org/10.2118/135034-MS.

Givens, N. and Zhao, H. 2008. The Barnett Shale: Not So Simple After All.

Goddard, D. A., Mancini, E. A., Talukar, S. C. et al. 2008. Bossier-Haynesville Shale: North Louisiana Salt Basin. Presentation, Center for Energy Studies, Louisiana State University, Baton Rouge, Louisiana.

Goodman, W. R. and Maness, T. R. 2008. Michigan's Antrim Gas Shale Play—A Two-Decade Template for Successful Devonian Gas Shale Development. *Search and Discovery* article no. 10158 (posted 25 September 2008). Adapted from an oral presentation given at the AAPG Annual Convention, San Antonio, Texas, USA, 20–23 April.

Gottschling, J. 2007. Appalachian Basin Black Shale Exploitation: Past, Present, and Future. Presented at the Independent Oil & Gas Association of Pennsylvania Annual Meeting, 16–17 May.

Grasso, T. X. 1968. A New Coral Bed in the Hamilton Group (Middle Devonian) of Central New York. *Journal of Paleontology* **42** (1): 84–87.

Gulfoilandgas.com. Marathon Announces Natural Gas Discovery in New Area of Oklahoma's Woodford Shale Resource Play, https://www.gulfoilandgas.com/webpro1/main/mainnews.asp?id=7012 (accessed 16 January 2020).

Hayden, J. and Pursell, D. 2005. The Barnett Shale: Visitor's Guide to the Hottest Gas Play in the US. Pickering Energy Partners Inc., Houston, Texas.

Hentz, T. F. and Ruppel, S. C. 2010. Regional Lithostratigraphy of the Eagle Ford Shale: Maverick Basin to East Texas Basin. *Gulf Coast Association of Geological Societies Transactions* **60**: 325–337.

Hester, T. C., Schmoker, J. W., and Sahl, H. L. 1990. Log-Derived Regional Source-Rock Characteristics of the Woodford Shale, Anadarko Basin, Oklahoma. US Geological Survey Bulletin 1866-D.

Hickey, J. J. and Henk, B. 2007. Lithofacies Summary of the Mississippian Barnett Shale, Mitchell 2 T. P. Sims Well, Wise County, Texas. *AAPG Bulletin* **91** (4): 437–443. https://doi.org/10.1306/12040606053.

Hosterman, J. W. 1993. Illite Crystallinity As an Indicator of the Thermal Maturity of Devonian Black Shales in the Appalachian Basin. In *Petroleum Geology of the Devonian and Mississippian Black Shale of Eastern North America*, Bulletin 1909, ed. J. B. Roen and R. C. Kepferle, Chap. G, 1–9. Denver: US Geological Survey. https://doi.org/10.3133/b1909.

Hutchinson, R. 2009. Haynesville Play: Keeping an Eye on the Haynesville Shale.

Jarvie, D. 2004. Evaluation of Hydrocarbon Generation and Storage in the Barnett Shale, Ft. Worth Basin, Texas. Presentation, Humble Geochemical Services.

Jarvie, D. M., Hill, R. J., Ruble, T. E. et al. 2007. Unconventional Shale-Gas Systems: The Mississippian Barnett Shale of North-Central Texas as One Model for Thermogenic Shale-Gas Assessment. *AAPG Bulletin* **91** (4): 475–499. https://doi.org/10.1306/12190606068.

Kapchinske, J. and Sharp, J. 2008. Haynesville Shale Overview. Presented at the Chesapeake Energy 2008 Investor and Analyst Meeting, Oklahoma City, Oklahoma, 15–16 October.

King, H. M. 2009. Marcellus Shale – Appalachian Basin Natural Gas Play. Geology.com, https://geology.com/articles/marcellus-shale.shtml.

Kuuskraa, V. A., Wicks, D. E., and Thurber, J. L. 1992. Geologic and Reservoir Mechanisms Controlling Gas Recovery From the Antrim Shale. Paper presented at the SPE Annual Technical Conference and Exhibition, Washington, DC, 4–7 October. SPE-24883-MS. https://doi.org/10.2118/24883-MS.

Lancaster, D. E., McKetta, S. F., Hill, R. E. et al. 1992. Reservoir Evaluation, Completion Techniques, and Recent Results From Barnett Shale Development in the Fort Worth Basin. Paper presented at the SPE Annual Technical Conference and Exhibition, Washington, DC, 4–7 October. SPE-24884-MS. https://doi.org/10.2118/24884-MS.

Langford, S. 2009. The Woodford Shale. Presentation for the Newfield Exploration Company Investor Conference 2008.

Lash, G. G. and Engelder, T. 2011. Thickness Trends and Sequence Stratigraphy of the Middle Devonian Marcellus Formation, Appalachian Basin: Implications for Acadian Foreland Basin Evolution. *AAPG Bulletin* **95** (1): 61–103. https://doi.org/10.1306/06301009150.

Lewan, M. D. 2002. New Insights on Timing of Oil and Gas Generation in the Central Gulf Coast Interior Zone Based on Hydrous-Pyrolysis Kinetic Parameters. *Gulf Coast Association of Geological Societies Transactions* **52**: 607–620.

Liro, L. M., Dawson, W. C., Katz, B. J. et al. 1994. Sequence Stratigraphic Elements and Geochemical Variability Within a "Condensed Section": Eagle Ford Group, East-Central Texas. *Gulf Coast Association of Geological Societies Transactions* **44**: 393–402.

Loucks, R. G. and Ruppel, S. C. 2007. Mississippian Barnet Shale: Lithofacies and Depositional Setting of a Deep-Water Shale-Gas Succession in the Fort Worth Basin, Texas. *AAPG Bulletin* **91** (4): 579–601.

MacFarlane, J. 1875. *The Coal-Regions of America: Their Topography, Geology, and Development*, third edition, 612. New York: D. Appleton and Company.

Magner, J. and Wren, J. 2008. Welcome to Haynesville, Population 60 Tcf. *Energy Investment Research*, 20 June.

Mancini, E. A., Goddard, D. A., Aharon, P. et al. 2006. Resource Assessment of the In-Place and Potentially Recoverable Deep Natural Gas Resource of the Onshore Interior Salt Basins, North Central and Northeastern Gulf of Mexico. Final technical report, Project No. DE-FC26-03NT41875, US Department of Energy (November 2006).

Martin, J. P. 2009. The Middle Devonian Hamilton Group Shales in the Northern Appalachian Basin: Production and Potential. New York State Energy Research and Development Authority.

Martineau, D. 2003. The Expansion of the Barnett Shale Play in the Fort Worth Basin.

Matthews, R. D. 1993. Review and Revision of the Devonian-Mississippian Stratigraphy in the Michigan Basin. In *Petroleum Geology of the Devonian and Mississippian Black Shale of Eastern North America*, Bulletin 1909, ed. J. B. Roen and R. C. Kepferle, Chap. D, 1–85. Denver: US Geological Survey. https://doi.org/10.3133/b1909.

Mazzullo, S. J. 1973. Deltaic Depositional Environments in the Hamilton Group (Middle Devonian, Southeastern New York State). *Journal of Sedimentary Research* **43** (4): 1061–1071. https://doi.org/10.1306/74D728F1-2B21-11D7-8648000102C1865D.

Miller, M. J. 2008. Marcellus Shale. Presented to Fireside Pumpers, Bradford, Pennsylvania, USA, 12 March.

Mullen, J. 2010. Petrophysical Characterization of the Eagle Ford Shale in South Texas. Paper presented at the Canadian Unconventional Resources and International Petroleum Conference, Calgary, Alberta, Canada, 19–21 October. SPE-138145-MS. https://doi.org/10.2118/138145-MS.

Mullen, J., Lowry, J. C., and Nwabuoku, K. C. 2010. Lessons Learned Developing the Eagle Ford Shale. Paper presented at the Tight Gas Completions Conference, San Antonio, Texas, USA, 2–3 November. SPE-138446-MS. https://doi.org/10.2118/138446-MS.

Myers, R. R. 2008. Stimulation and Production Analysis of Underpressured (Marcellus) Shale. Paper presented at the SPE Shale Gas Production Conference, Fort Worth, Texas, USA, 16–18 November. SPE-119901-MS. https://doi.org/10.2118/119901-MS.

NKNT. 2008. Geology of the Barnett Shale.

Pabian, R. K. and Strimple, H. L. 1976. Middle Devonian Crinoids from Western Maryland. *Journal of Paleontology* **50** (4): 759–762.

Patrick, K. 2004. *Pennsylvania Caves & Other Rocky Roadside Wonders*, 18. Mechanicsburg, Pennsylvania: Stackpole Books.

Penn State Marcellus Center for Outreach and Research. 2017. Marcellus Shale Thickness map, http://www.marcellus.psu.edu/resources/images/marcellus-thickness.gif (accessed 14 January 2020).

Petrohawk. 2009. Leaders in Industry Luncheon.

Phasis Consulting. 2008. US Shale Gas Brief. Phasis Consulting, Calgary, Alberta, Canada (September 2008).

Pollastro, R. M., Jarvie, D. M., Hill, R. J. et al. 2007. Geologic Framework of the Mississippian Barnett Shale, Barnett-Paleozoic Total Petroleum System, Bend Arch–Fort Worth Basin, Texas. *AAPG Bulletin* **91** (4): 405–436. https://doi.org/10.1306/10300606008.

Quirein, J., Praznik, G., Galford, J. et al. 2013. A Workflow to Evaluate Mineralogy, Porosity, TOC, and Hydrocarbon Volume in the Eagle Ford Shale. Paper presented at the SPE Unconventional Resources Conference and Exhibition-Asia Pacific, Brisbane, Australia, 11–13 November. SPE-167012-MS. https://doi.org/10.2118/167012-MS.

Rhein, T., Loayza, M., Kirkham, B. et al. 2011. Channel Fracturing in Horizontal Wellbores: The New Edge of Stimulation Techniques in the Eagle Ford Formation. Paper presented at the SPE Annual Technical Conference and Exhibition, Denver, Colorado, USA, 30 October–2 November. SPE-145403-MS. https://doi.org/10.2118/145403-MS.

Rushing, J. A. and Sullivan, R. B. 2003. Evaluation of a Hybrid Water-Frac Stimulation Technology in the Bossier Tight Gas Sand Play. Paper presented at the SPE Annual Technical Conference and Exhibition, Denver, Colorado, USA, 5–8 October. SPE-84394-MS. https://doi.org/10.2118/84394-MS.

Schein, G. 2008. Hydraulic Fracturing in Gas Shale Plays – Are They All the Same? Presented at the Barnett Shale Breakfast Symposium, 29 February.

Schneider, P. F. 1894. *Notes on the Geology of Onondaga County, NY: Embracing a Short Description of the Various Eras, Periods and Groups, Together with Statements as to Their (1) Outcrop, (2) Thickness, (3) Economic Importance, (4) Fossils, (5) Favorable Localities for Study*, 37. Syracuse, New York: Hall & McChesney, Printers. https://doi.org/10.5962/bhl.title.66893.

Shale Gas. 2006. Supplement to *Oil and Gas Investor*.

Sumi, L. 2008. Shale Gas: Focus on the Marcellus Shale. Report, Oil & Gas Accountability Project/Earthworks, Durango, Colorado.

Tian, Y. 2014. *Occurrence of Multiple Fluid Phases Across a Basin, in the Same Shale Gas Formation – Eagle Ford Shale Example*. PhD dissertation, Texas A&M University, College Station, Texas (May 2014).

TXCO Resources. 2009. The Emerging Resource Company, TXCO Resources (accessed 20 May 2009).

Wickstrom, C. W. 2008. Woodford Shale Gas in Oklahoma. *Search and Discovery* article no. 110068 (posted 20 August 2008). Prepared for oral presentation at the 2008 AAPG Annual Convention, San Antonio, Texas, USA, 20–23 April.

Zuber, M. D., Voneiff, G. W., Frantz, J. H., Jr. et al. 1994. Characterization of Michigan Antrim Shale Reservoirs Based on Analysis of Field-Level Data. Paper presented at the SPE Regional Meeting, Charleston, West Virginia, USA, 8–10 November. SPE-29169-MS. https://doi.org/10.2118/29169-MS.

Part 2 - Tight Gas Sands Engineering

Chapter 5

Reservoir Characterization

Stephen A. Holditch

5.1 Geologic Considerations

The study of all reservoirs, including tight gas reservoirs, begins with a thorough understanding of the geologic characteristics of the reservoir. In this section, we will concentrate on the geology of sandstone bodies. However, tight gas reservoirs can also be found in carbonates, coal seams, and shales.

The technical team of engineers involved in drilling, completing, and stimulating a tight gas reservoir should work closely with geologists and geophysicists to understand the geologic setting of the reservoir under development. The geologic considerations are important to determine many topics that will be covered in this book, such as which layers to target, where to drill the wells, and what well spacing is required for optimal development planning. In addition, the layers of rock above and below the pay zone must be characterized, because they will affect the vertical fracture height growth and the design of the optimal fracture treatment.

Important geologic parameters for a trend or basin are

- Structural and tectonic regime
- Regional thermal gradient
- Regional pressure gradient

Important geologic parameters for each stratigraphic unit are

- Depositional system and genetic facies
- Textural maturity
- Mineralogy
- Diagenetic processes and cements
- Reservoir dimensions
- Pressure and temperature range
- Natural fractures

The basic sedimentary framework of a basin can be understood by using lithogenetic facies as stratigraphic units. Each facies is a 3D body of rock whose environmental origin is inferred by studies of the rock characteristics, such as

- External and internal geometry
- Sedimentary structure
- Lithology
- Organic content
- Stratigraphic relations
- Associated sedimentary facies

5.1.1 Depositional Environments. According to Fisher and McGowen (1967), a depositional system is "a group of lithogenetic facies linked by depositional environment and associated processes." Each lithogenetic facies has certain attributes, including porosity, permeability, and spatial relations to other facies, that affect the migration and distribution of hydrocarbons (Galloway et al. 1982).

The nine principal clastic depositional systems reviewed by Fisher and Brown (1972) can be classified into three major groups established by Selley (1978). The classification is listed in **Table 5.1.**

Continental Environments
Eolian systems
Lacustrine systems
Fluvial systems
Terrigenous fan (alluvial fan and fan-deltaic systems)
Shoreline (Marginal Marine) Environments
Delta systems
Barrier-strand plain systems
Lagoon, bay, estuarine, and tidal-flat systems
Marine Environments
Continental and intracratonic shelf systems
Continental and intracratonic slope and basinal systems

Table 5.1—The nine principal clastic depositional systems (Selley 1978).

Fig. 5.1 illustrates the different clastic depositional systems. It is important to determine the depositional environment because the system can be one of the primary drivers in determining the drainage area in a tight gas sands reservoir. For example, if the

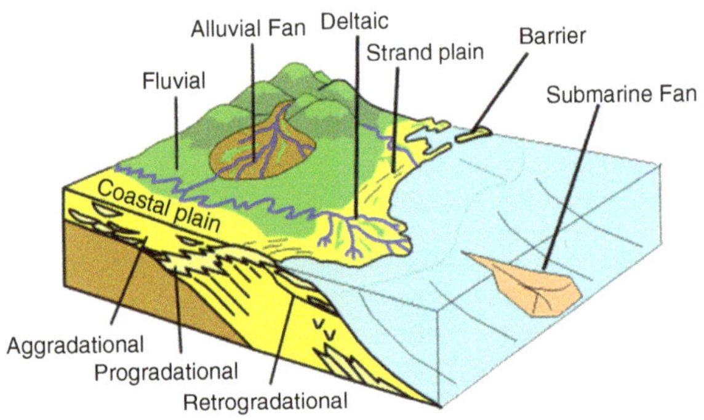

Fig. 5.1—Clastic depositional systems. Adapted from Hentz et al. (1997).

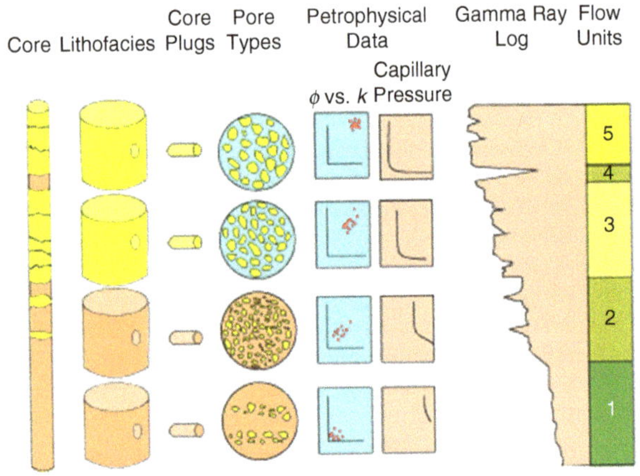

Fig. 5.2—Example of geologic flow units.

Environment	Agent of Transportation Deposition	Sediments
Alluvial	Rivers	Sand, gravel, mud
Lake	Lake currents, waves	Sand, mud
Desert	Wind	Sand, dust
Glacial	Ice	Sand, gravel, mud
Delta	River + waves, tides	Sand, mud
Beach	Waves, tides	Sand, gravel
Shallow shelf	Waves, tides	Sand, mud
Deep sea	Ocean currents, setting	Sand, mud

Table 5.2—Clastic sedimentary environment transportation systems.

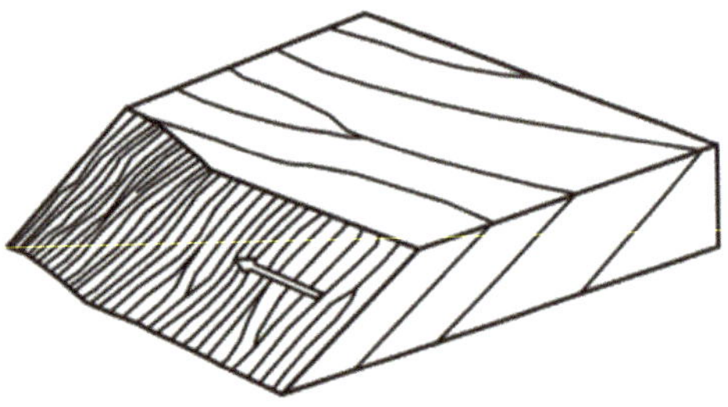

Fig. 5.3—Eolian foresets with transposed ripples directed at right angles to the foreset dip. Adapted from Walker and Harms (1972).

depositional system is a strand plain or a barrier island, the formation tends to be more of a blanket reservoir. In blanket reservoirs, the drainage area can be influenced by the number of wells drilled and the size of the fracture treatment, so it is usually best to choose the best pay zone and pump a large fracture treatment in that zone. If there are multiple zones, consider completing the deepest zone first.

If the depositional system is fluvial, then the sand bodies are expected to be lenticular. In lenticular reservoirs, the drainage area is more controlled by the size of the sand lenses. In many cases, if multiple lenses are penetrated with the vertical wellbore, then a multistage fracture treatment can be designed and pumped to complete several lenses with the wellbore.

Alluvial fans, deltaic fans, and submarine fans are similar in terms of sand body, size, and shape, but can have differences that need to be considered. Some deltaic and submarine fans can be quite large, with sand bodies that can provide large drainage volumes.

In summary, the completion and reservoir engineers must work with the geoscientists to understand the size and shape of the sand bodies that will be developed.

Within most sand bodies, specific geologic flow units can be defined. These flow units are defined by the lithofacies, the pore types, the data on porosity vs. permeability, and the capillary pressure. Cores can usually be taken to define the flow units, then the flow units can be correlated to logs, such as the gamma ray (GR), spontaneous potential (SP), and resistivity logs. Porosity logs, such as the sonic log (SON), neutron log (CNL), and density log (FDL), can also be used in the correlation.

Petrophysical analyses of core samples are used to identify reservoir flow units and nonflow units, as shown in **Fig. 5.2.** The results are used to calibrate well logs, then those well logs can be used to map flow units throughout a field. Core data from each flow unit can be measured in the laboratory, then the results can be used to select samples for special core analyses, such as mechanical properties and relative permeability. When the geologist and engineer have examples of what the routine and special core analyses should be for each flow unit, the flow units can be identified from the log data, and the data can be mapped to get a better idea of the size and shape of each flow unit. The properties of each flow unit can be estimated for use in reservoir calculations and the design of fracture treatments.

Clastic sedimentary rocks are composed of rock fragments that were eroded from existing rocks; transported by gravity, ice, wind, or water; and laid down as sediments. Sediments can be derived from igneous, metamorphic, or sedimentary rocks. **Table 5.2** includes information on these sedimentary environments.

In Table 5.2, we find all the depositional environments—alluvial, desert, deltaic, beach, and shallow shelf sediments can result in producible hydrocarbon reservoirs. In fact, many of the tight gas sand reservoirs in the world are in clastic sedimentary formations.

The best source rocks are shale, silty rocks, or mudstones that are rich in organic materials. These source rocks are found in deepwater deposits, large-lake deposits, or deltaic environments.

Some tight gas sand formations are eolian. Eolian systems (shown in **Fig. 5.3**) are windblown sand dune deposits, usually formed in hot, arid regions in land-locked continental interiors. Eolian sands produce in the US western interior and the Permian Rotliegendes of the North Sea. Elsewhere, eolian sandstone is usually not productive because it lacks source beds and seals. In the Rotliegendes, the formation permeability is heterogeneous. Thus, horizontal drilling has revived the development of several fields that were previously drilled with vertical wells and hydraulically fractured.

Fig. 5.4 shows crossbedding in the Navaho Sandstone, located in Zion National Park, Utah, USA. Dune sediments deposited by wind action (eolian) are excellent reservoirs (the best of the clastics). Generally, they are composed of well-sorted, texturally mature sands. There are a number of different classic dune types (e.g., barchans, seif dunes). The changing wind directions during deposition of sand dunes can be seen in the layers, and the different directions of crossbedding (primary sedimentary structures) reflect the movement of the dunes during their deposition, just as dunes march across deserts today. Water-deposited sediments might have similar but smaller crossbedding.

Note the redness of the sediments below the sandstone. Generally, if a sediment is red, it is because of oxidization and is terrestrial. If it is green or grey, it is marine and is a result of reducing conditions.

There are some eolian deposits that can be called tight gas sands, but in general these sands are normally of higher porosity and permeability than the tight gas sands from other depositional systems, as shown in **Fig. 5.5.** If an eolian tight gas sand is encountered, the drainage area might be small—as boundaries to fluid slow, they can be deposited as the dunes shift with time.

Tight gas formations can also be associated with lakes. Terrigenous lacustrine systems (lakes) can produce oil and gas, coal, and uranium. Coal seams are widely developed in lacustrine and associated systems. Shales and coals in lacustrine systems can often be source beds for the low-permeability sandstones, as shown in **Fig. 5.6.** Lacustrine systems are often constituent parts of other depositional systems, such as fluvial or deltaic systems. Most large lakes are located in depressions caused by tectonics (rifts, continental collisions, pull-apart basins).

Depending on where the well is drilled, one could encounter continuous sands in the middle of the lake, continuous mudstones in the middle of the lake, or fluvial deltaic sediments at the edge of the lake where sediments are being fed into the lake. There could be a sizable difference in the expected drainage area depending on the environment of deposition.

Fluvial depositional systems consist of river-associated deposits inland from a delta or shoreline. Genetic facies associated with fluvial systems are channel-fill, channel-margin, and flood-basin deposits, as shown in **Fig. 5.7.** Channel-fill deposits contain most of the sediment retained in the system. There can be both low- and high-sinuosity channels.

In fluvial systems, the sizes of the sand bodies are usually limited. If the sand layer is thick and in a low-sinuosity channel, the sand body will usually be larger than if the sand layer is thin and in a high-sinuosity channel. There are studies published in the geologic literature in which outcrops have been analyzed to determine estimates and shapes of the various fluvial sand bodies as a function of the thickness of the sand. The reservoir and completion engineers should work with a geologist who is familiar with such work and can help the team estimate the reservoir drainage volume using the logging data.

Fig. 5.4—Outcrop picture of eolian bed dune structure and layering.

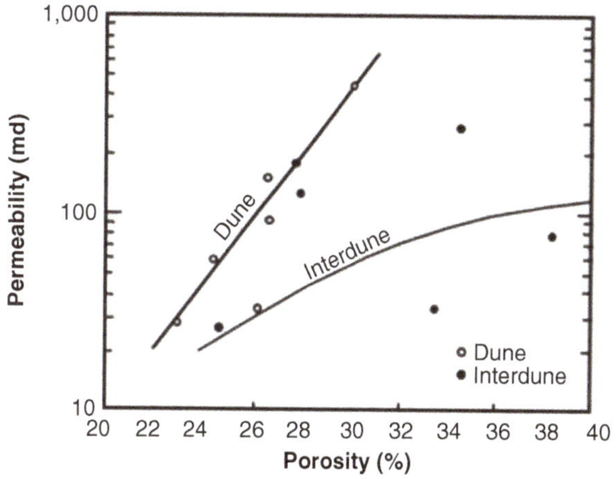

Fig. 5.5—Graph showing typical porosity vs. permeability in eolian deposits. Notice the different behavior between dune and interdune deposits. Adapted from Lupe and Ahlbrandt (1979).

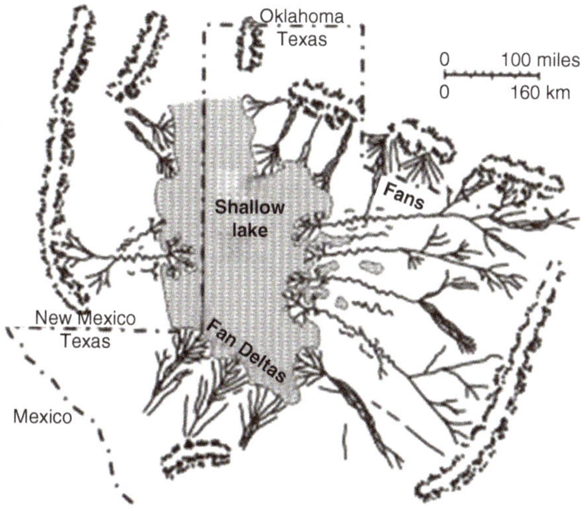

Fig. 5.6—Lacustrine system with rivers and deltas feeding sediment into the lake. Adapted from McGowen et al. (1976).

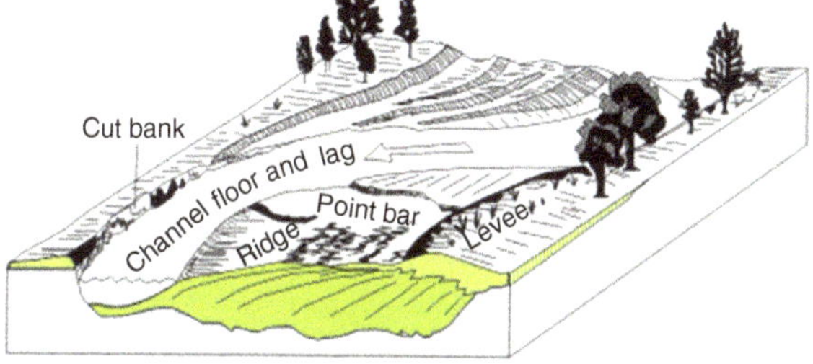

Fig. 5.7—Generalized depositional model and vertical sequences of a meander bolt sand body produced by a high-sinuosity channel.

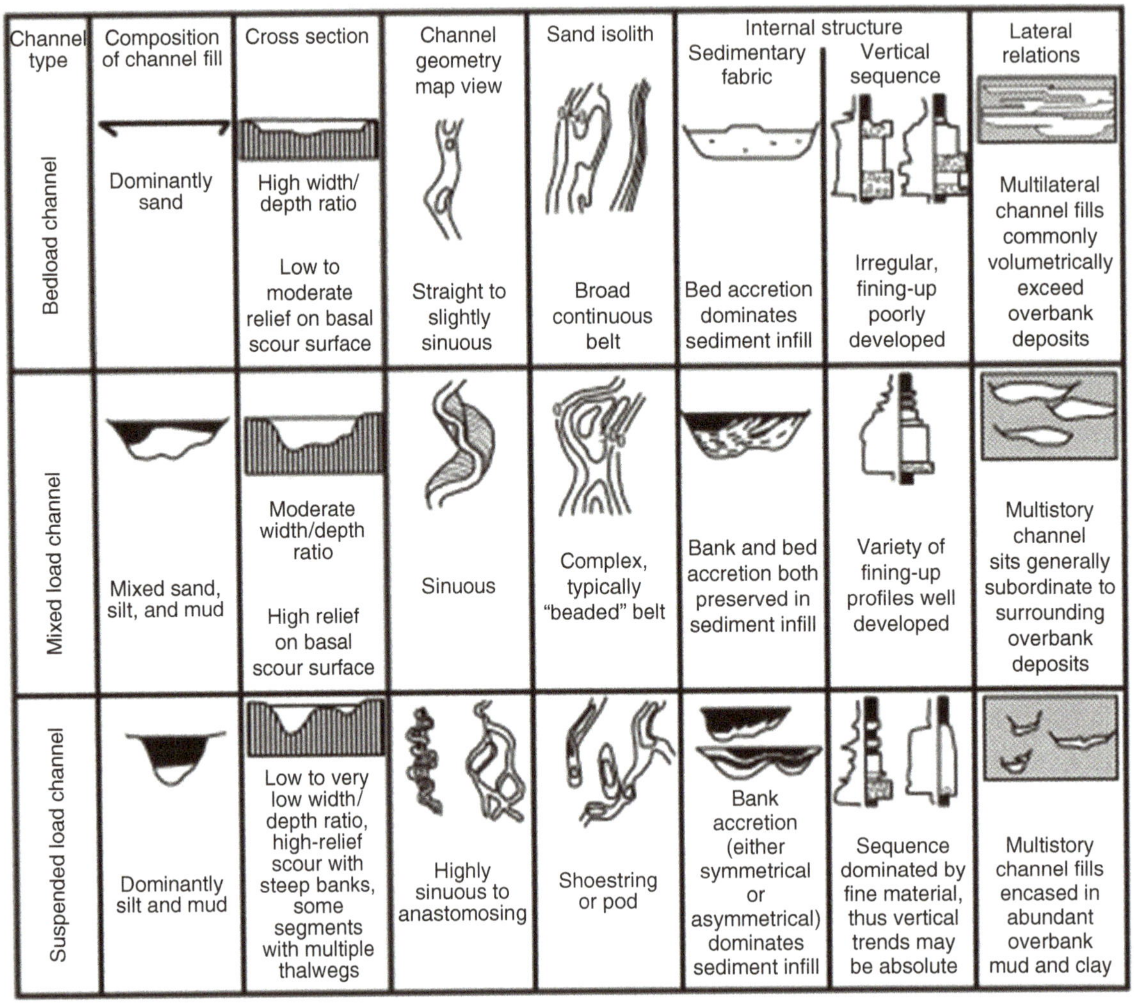

Fig. 5.8—The geomorphic and sedimentary characteristics of various fluvial systems. Adapted from Galloway (1977).

Fig. 5.8 shows the geomorphic and sedimentary characteristics of various fluvial systems.

Alluvial fans are accumulations of predominantly coarse-grained clastics extending from a mountain front or escarpment across adjacent lowland. Alluvial fans are most numerous in tectonically active areas, such as the Rocky Mountains in North America.

The onshore fan shown in **Fig. 5.9** starts at the upper center of the photo, where the stream leaves the confined channel in the mountains for the broad lowlands.

Shoreline environmental depositional systems include deltas (**Fig. 5.10**). A delta forms where a river transporting significant quantities of sediment enters a marine basin near a shoreline. The depositional architecture of a deltaic system is characteristically progradational. Sediments are derived from definable point sources. Deltaic systems develop around the periphery of large basins. Famous deltaic systems include those associated with the Mississippi River, the Orinoco River, and the Nile River. **Fig. 5.11** shows an example of what a deltaic deposit might resemble.

Fig. 5.9—An alluvial fan in Cooper Canyon, Death Valley, California, USA.

A delta can also be described as a mound of sediments deposited where a river enters a still-standing body of water, either a lake or ocean, and loses the velocity required to transport the sediments. The shape of the delta and the facies characteristics depend on the interplay of fluvial current, sediment load, wave action, and tidal strength. Fig. 5.11 shows an example of an elongate or bird-foot deltaic system, where fluvial processes are stronger than either tidal or wave processes.

Deltaic systems are important in the production of coal, oil, and gas. Many of the large oil and gas fields around the world are found in deltaic deposits. Many deltaic reservoirs are compartmentalized because of their many facies.

Sometimes a formation can be a mixture of different depositional systems, and the engineering and geologic team must sort out the geology to help form a reservoir development plan. One example of this mix of depositional

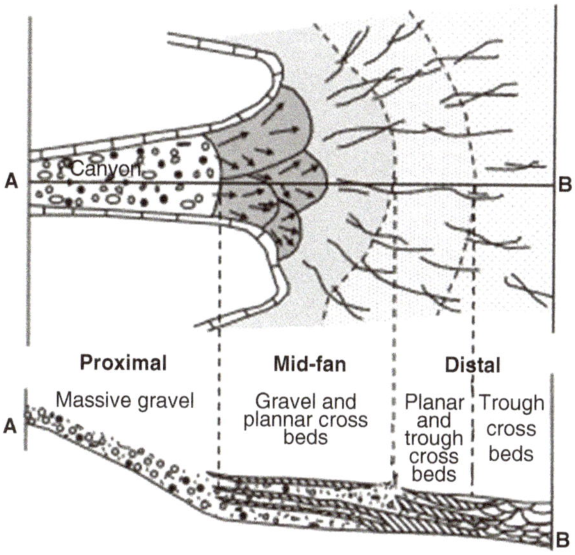

Fig. 5.10—Idealized cross section and plan view through the wet alluvial-fan system of the Cambrian Van Horn Sandstone. Adapted from McGowen and Groat (1971).

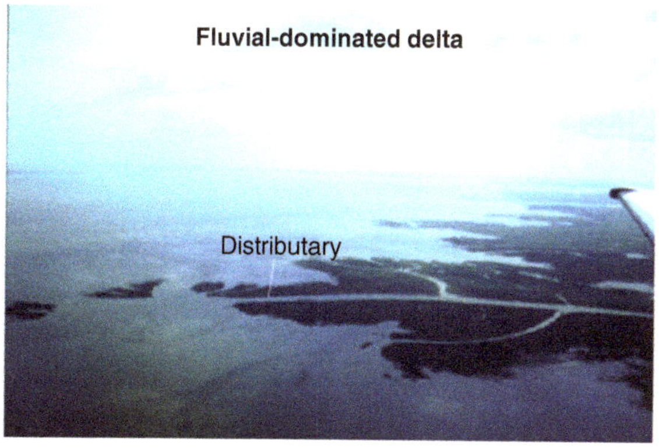

Fig. 5.11—Athabasca Delta in Canada.

systems is the Travis Peak Formation in East Texas and Louisiana, USA. **Fig. 5.12** shows a cross section of the Travis Peak Formation, where part of the formation was deposited as a deltaic fringe, but most of the sand was deposited in a fluvial system. Thus, depending on the location and depth, the depositional environment must be determined to estimate the potential drainage area for the wells.

Shoreline deposits, excluding deltas, comprise the narrow, high-energy transitional environment that extends from the wave base (approximately 10 m of water) to the seaward edge of the alluvial coastal plain. Barrier-island and strand-plain systems can be continuous blanket sands that provide excellent reservoir rock, especially when capped with marine shale or other seals.

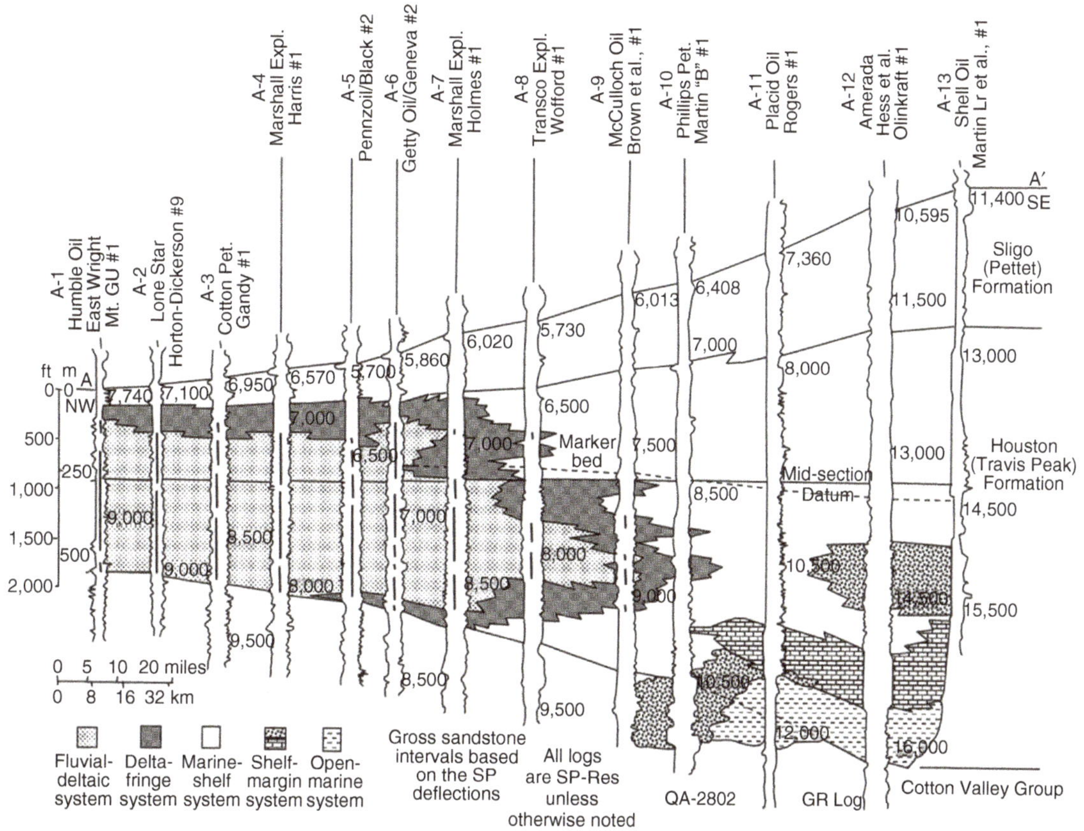

Fig. 5.12—Cross section of the Travis Peak Formation, showing how the depositional system can vary with location and depth.

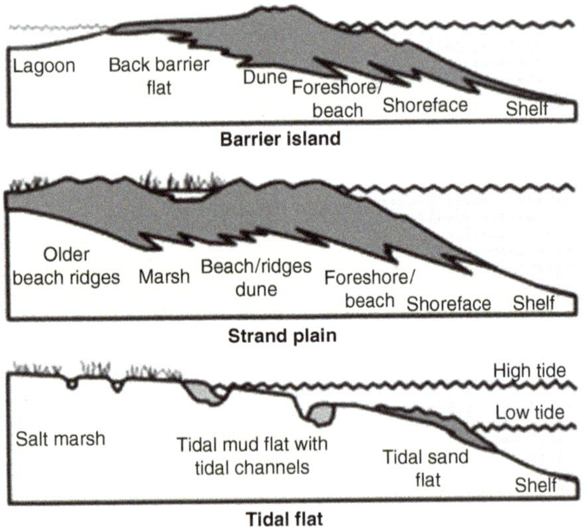

Fig. 5.13—Cross section of various shoreline systems.

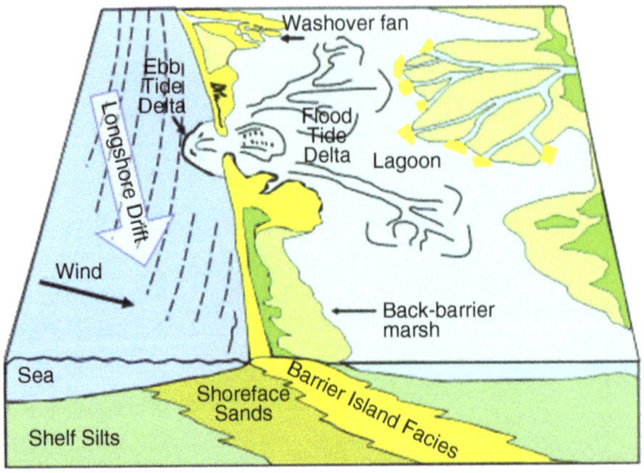

Fig. 5.14—Various shoreline deposits. After Tavener-Smith (1982).

Barrier/lagoon and strand-plain deposits form between the barrier systems and the continental systems, as shown in **Fig. 5.13.** These barrier and strand-plain systems do not normally produce reservoir-quality rock. However, they can be important source beds and/or seals.

A barrier island is a beach ridge formed along a wave-dominated shoreline and separated from the mainland by a lagoon. The main core of barrier-island depositional systems is elongated parallel to the shoreline, but associated facies, such as a tidal inlet, might be oriented perpendicularly to the shoreline, as shown in **Fig. 5.14.** A strand plain is a series of welded beach and dune ridges (lacking intervening lagoonal deposits) deposited along a wave-dominated shoreline.

Because the sea level has risen and fallen over geological time, the sand deposits along the shoreline have tended to be transgressive when the sea level is rising and regressive when the sea level is falling. These moving shorelines cause the various sand, silt, and shale layers to move with time, but the movement can be understood and mapped to better understand the reservoir system. **Fig. 5.15** shows how facies can change with geologic time and location.

Marine deposits provide both high-quality reservoir rock and source beds that are normally organic-rich shales **(Fig. 5.16).** The sandstones are normally continuous in the inner-shelf area. In the outer-shelf and shelf-break areas, turbidite deposits provide reservoir morphology similar to that of deltas. One way to determine the difference between a deltaic deposit and a turbidite deposit is to study the lifeform fossils that are found in the rocks. There is a difference between the plants and animals found in deltas and those found in deep water. Also, in many deltas, there will be coal seams that will not be found in turbidites.

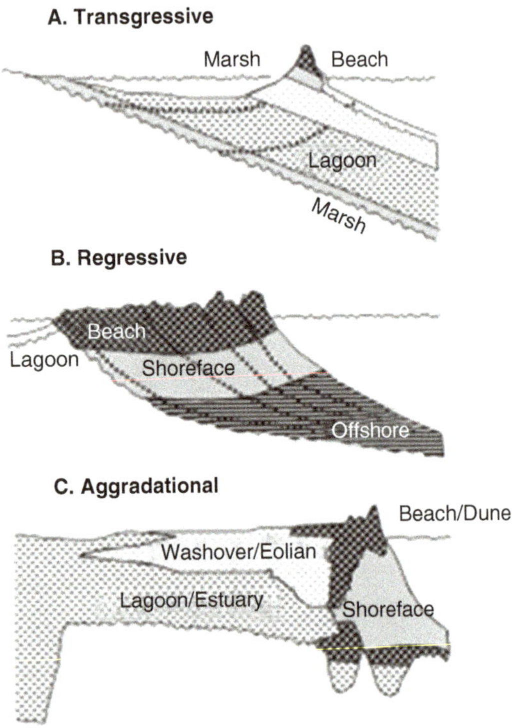

Fig. 5.15—Relationship of various barriers and sand bodies. Adapted from Kraft and John (1979), Bernard et al. (1970), Fisk (1959), and Morton and McGowen (1980).

5.1.2 Diagenesis. Berg (1986), a book-length study of reservoir sandstones, discusses diagenesis in detail. We have summarized some of that work in this section of the book. Sandstones are composed of mineral particles called grains, which consist of quartz, feldspar, and rock fragments. The finer particles between the grains is called matrix. After deposition and burial, the grains and matrix are commonly altered by the physical effects of compaction and by chemical changes.

These changes are broadly referred to as diagenesis. Early chemical changes are called anthogenesis. **Table 5.3** shows diagenetic changes in sandstones.

In addition to the grains and matrix, a third major component is the cement introduced into the rock during diagenesis. Cement is normally precipitated between the grains. The original porosity and permeability of sandstones are determined by depositional characteristics, such as mineral composition, pore type, grain size, and texture. However, porosity and permeability will be altered over time by diagenetic processes. Tight gas sandstones have had the porosity occluded by diagenesis, which results in reduced permeability; thus, the rock is tight.

As a rule, low-permeability sandstones have been severely diagenetically altered. Clays and quartz overgrowths are commonly found to be controlling the pores and pore-throat geometry. Normally, quartz overgrowths cause the most permeability reduction. Pore-filling clays reduce the permeability more than pore-lining clays.

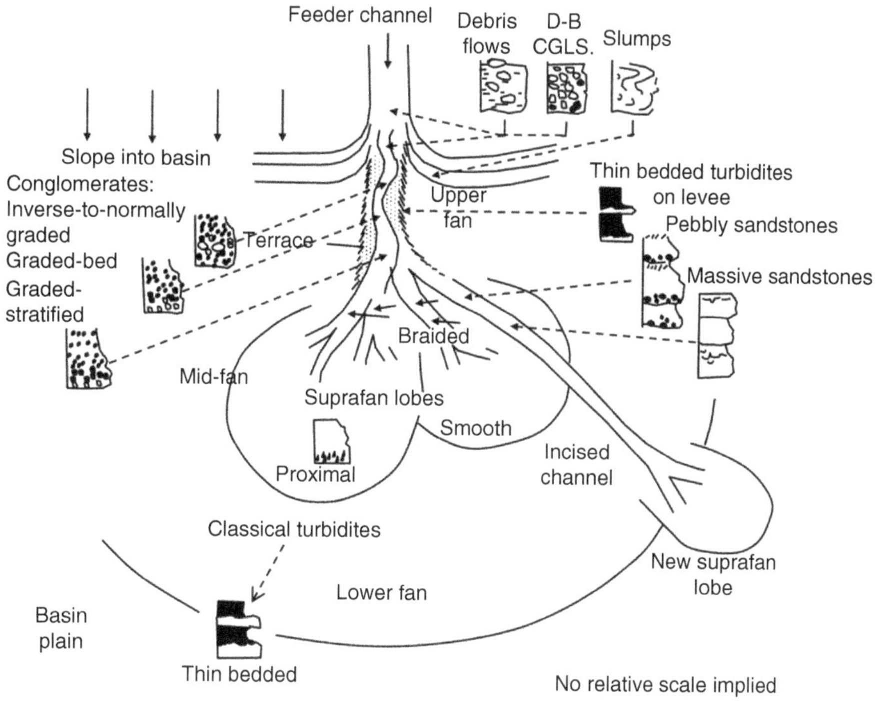

Fig. 5.16—Drawing of a turbidite system showing depositional model of sand-rich, unconfined submarine fan, relating morphologic features, deposition environments, and resultant facies. Adapted from Walker (1978).

Process	Result	Depth	Effect
Compaction	Mechanical rearrangement of grains	Shallow	Reduction in porosity from 40% to less than 30%
Cementation	Intergranular calcite, silica overgrowths in quartz	Shallow to moderate	Consolidation (lithification and porosity reduction)
Clay precipitation	Kaolinite, chlorite, and illite as grain coating	Shallow to moderate	Slight decrease in porosity
Clay dehydration	Loss of interlayered water in expandable clays (smectite/illite)	Moderate	Slight increase in porosity
Dissolution	Leaching of calcite cement and feldspar grains	Moderate	Slight increase in porosity
Cementation	Intergranular ferroan calcite, zeolites	Moderate to deep	Significant decrease in porosity
Pressure solution	Stylolitization by silica dissolution	Moderate to deep	Local decrease in porosity (along laminae)
Microfracturing	Breaking of grains	Deep	Slight increase in porosity

Table 5.3—Diagenetic changes in sandstones.

Clays can be authigenic or detrital. Detrital clays are deposited along with the sand grains during deposition. Authigenic clays are precipitated over geologic time as water moves through the rock.

Figs. 5.17 and 5.18 illustrate examples of pore-lining and pore-filling clay. Normally, pore-filling clay is more damaging to the permeability than pore-lining clay. In general, the more clay in the pore spaces, the more the permeability is reduced. But the permeability is also affected by the location and type of clay.

Illite in reservoirs typically occurs as submicrometer-diameter fibers. Illite is important in petroleum geology because illite fibers can block pore throats, thereby significantly reducing porosity. Fibrous illite might not be recognized in samples from cores that have been air dried. The drying process could cause the delicate fibers to collapse against grain surfaces. Thus, permeability measurements taken with illite-bearing core samples might be in error.

Authigenic chlorite in sandstone reservoirs is typically iron-rich, as shown in **Fig. 5.19.** This iron-rich chlorite can react with acid used during reservoir stimulation.

Pore throats in sandstone might be lined with a variety of cement minerals that affect petrophysical properties

Scanning Electron Micrograph
Tordillo Sandstone, Neuquen Basin, Argentina

Fig. 5.17—Example of pore-lining clay in sandstone.

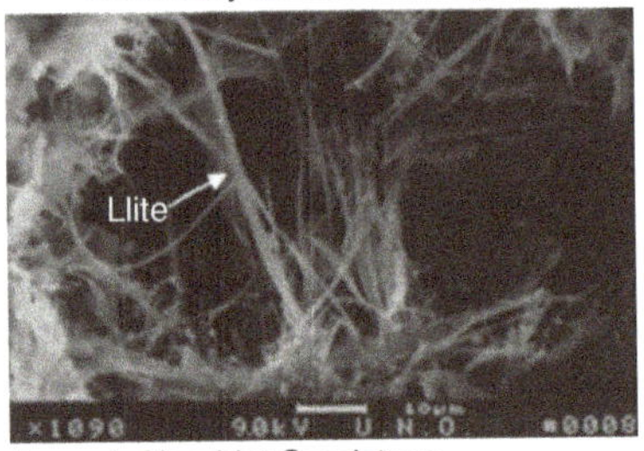

Fig. 5.18—Example of pore-filling clay (fibrous authigenic illite). (Photo by R. L. Kugler.)

Fig. 5.19—Example of authigenic chlorite. (Photo by R. L. Kugler.)

This reaction results in the formation of gelatinous blobs that have large diameters relative to pore-throat sizes. Thus, gelatinous blobs that block pore throats might reduce permeability.

Kaolinite might fill pores or replace detrital grains, such as feldspar, as shown in **Fig. 5.20.** Kaolinite is relatively inert and does not react with fluids that could be introduced into the reservoir. However, kaolinite platelets have large diameters relative to pore throats. As a result, under certain conditions, kaolinite could migrate and block pore throats, thus reducing permeability. Kaolinite also contains water in micropores between platelets and stacks of platelets. The presence of this water in kaolinite-bearing sandstone must be acknowledged during well-log analysis to properly interpret water saturation.

Many geologists and engineers are aware of clay content and the fact that clay reduces the permeability of a tight gas sandstone. However, some clays are worse than other clays, depending on where the clay is deposited and if the clay tends to swell when exposed to fresh water. It has been our experience that clays in deep, hot reservoirs are not particularly prone to freshwater swelling but can still reduce the permeability of the rock if the clays are loosened and allowed to migrate and block pore throats.

However, what many geologists and engineers need to know is that a more severe problem for reducing permeability is the presence of quartz overgrowths, as shown in **Fig. 5.21.** Over geologic time, water migrates through the sandstone, dissolving silica in one formation and precipitating the silica as quartz overgrowths that severely reduce porosity and permeability. The quartz overgrowths result in slot pores that are extremely sensitive to increased stress on the rock as the reservoir pressure is reduced during production, as shown in **Fig. 5.22.**

The detrimental effects of clay content on the permeability of a sandstone are uncertain, but, if the sandstone is filled with quartz overgrowths, the permeability will be much lower at a given porosity, and the permeability will be highly stress-sensitive.

In summary, a core must be cut, and detailed petrological analyses are required, to understand diagenesis. Often, the presence of hydrocarbons in the rocks will alter the process of diagenesis.

5.1.3 Estimating Drainage Area. One of the most difficult values to obtain is reservoir size or drainage area on a well-by-well basis in a tight formation. Because it takes months or years before a pressure transient is affected by reservoir boundaries or well-to-well interference in tight reservoirs, the engineer often must estimate the drainage area to estimate reserves.

In blanket tight gas reservoirs, the average drainage area of a well is largely a function of the number of wells drilled and the size of the fracturing treatment pumped. In lenticular or compartmentalized tight gas reservoirs, the average drainage area is a function of the lens size or compartment size and is not a strong function of the size of the hydraulic fracturing treatment.

Fig. 5.20—Example of authigenic kaolinite. (Photo by R. L. Kugler.)

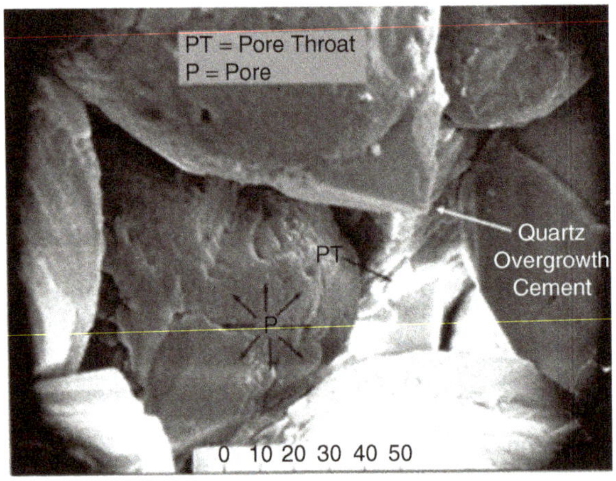

Fig. 5.21—Example of a quartz overgrowth in cementation (Jorden and Campbell 1984).

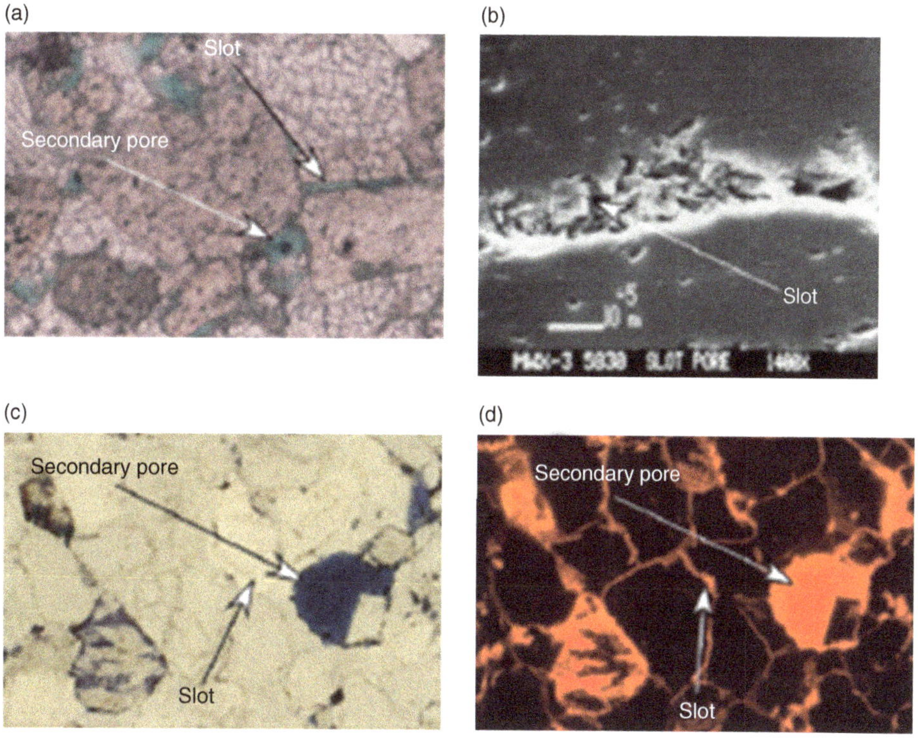

Fig. 5.22—Example of slot pores caused by quartz overgrowths. Adapted from Shanley et al. (2004).

The main factor controlling the continuity of the reservoir is the depositional environment. Generally, reservoir drainage area per well is small in continental deposits. Fluvial systems also typically result in lenticular or highly compartmentalized reservoirs. Compartmentalized reservoirs are also part of some deltaic systems and deepwater turbidite systems.

Blanket sands are normally associated with barrier islands, strand-plain systems, marine longshore bars, or shallow shelf systems. Thus, knowledge of the depositional system is important when conducting an integrated reservoir study. The best data for studying the depositional system are core and log data.

Cutting core in the shales, mudstones, and nonreservoir rock is recommended. Core descriptions of the nonreservoir rock could tell the geologist more about the depositional system than core descriptions only from the sandstone. Thick, continuous, homogeneous layers of rock are needed to normalize log data. Generally, thick, clean marine shales can be used to normalize logs. In many fluvial systems, it is difficult to correlate any single layer of rock across an entire field, but it might be possible to correlate shales across a large areal extent.

In summary, both geologic and engineering data are required to estimate reservoir continuity. Estimates of the drainage area are normally based on geology early in the life of a well and on production and pressure data later in the life of a well.

5.1.4 Regional Tectonics. Knowledge of basinwide tectonic activity is important for the analysis and development of tight gas reservoirs. To study the tectonics, we must study the faulting systems, estimate the in-situ stress regime, and study the prior depositional and burial history of a basin. Galloway (1977) and Galloway et al. (1982) describe how basins fall into eight categories that are based on tectonic and depositional style. These eight categories are

- Simple intracratonic basins
- Composite intracratonic basins
- Rift basins
- Extracontinental downwarps
- Passive continental margin depressions
- Subduction basins
- Median basins
- Later Tertiary deltaic basins

Plate tectonics provides a systematic approach to analyzing sedimentary basins and hydrocarbon occurrence. The Earth's crust is mobile and can be analyzed using maps showing the age of oceanic crust. The youngest oceanic crustal rocks occur along mid-ocean ridges, where the seafloor is spreading. **Fig 5.23** shows an example of the age of the ocean floor. Plate tectonics affects mountain building, causes earthquakes, changes how sedimentary basins are formed, and creates crustal deformation and petroleum traps. As the Earth's tectonic plates move, they can cause stresses in the Earth that in turn cause faulting and natural-fracture systems in some cases. We can study regional faulting and natural fractures to learn about the tectonic stresses there and how those stresses have changed over geologic time.

Fig. 5.24 shows that the types of basins and the dominant fault types in the basins vary with the type and orientation of stress. Sedimentary basin fill and structural traps are related to the tectonic setting. Structural features can be used to infer causative stress types and orientations.

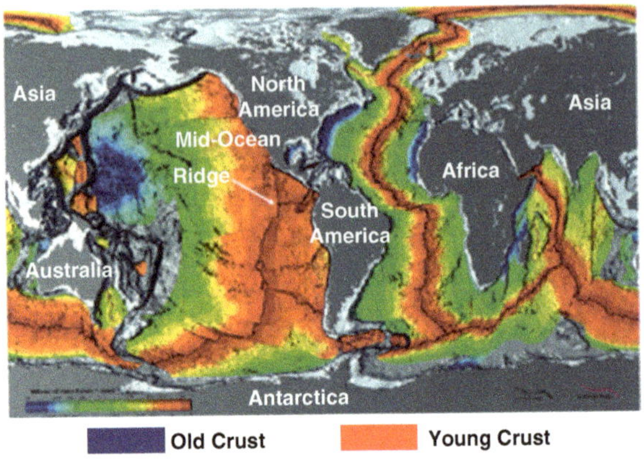

Fig. 5.23—Example of the age of the ocean floor.

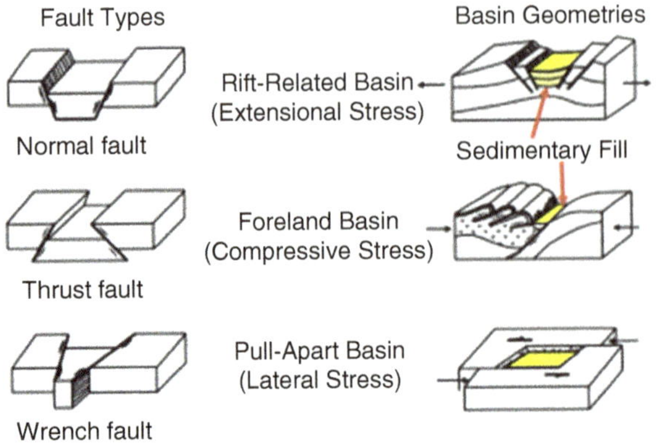

Fig. 5.24—Examples of stress fields in sedimentary basins and how the stress affects faulting.

In addition to affecting the depositional basin, regional tectonics affect the in-situ stresses in all rock layers. In-situ stresses affect

- Faulting
- Rock strength
- Drilling parameters
- Hydraulic fracturing
- Natural fracturing
- Borehole stability

The main effects of regional tectonics and in-situ stress are the effects of stress on the design and pumping of hydraulic fracturing treatments. The tectonics also affects the abundance of natural fracturing in the rocks that in turn affects the permeability and therefore the design of the hydraulic fracturing treatments. Natural fractures will also affect the overall permeability, the permeability anisotropy, the well completion methodology, and the ultimate recovery.

The geologists and engineers should gather data to determine the minimum-principal-stress orientation. Studying the fault maps in a basin can help determine stress directions. The minimum principal stress is not necessarily a vertical plane—the principal stress direction might be slightly inclined from vertical. Horizontal fractures are possible for fracture gradients near to or higher than 1.0 psi/ft or the average density of the Earth from the surface to the formation of interest. An FDL can be used to determine the overburden. Horizontal stress will change as the reservoir is depleted.

In an extension zone, the minimum-principal-stress component is usually directed perpendicularly to the orientation of the normal fault. Hydraulic fractures will be vertical and will propagate in a plane (parallel) along the existing fault pattern, as shown in **Fig. 5.25.**

The principal-stress components are of the following order: $s_{h,\min} < s_{h,\max} < s_v$, where $s_{h,\min}$ = minimum horizontal stress, $s_{h,\max}$ = maximum horizontal stress, and s_v = vertical stress.

In a compression zone, the minimum-principal-stress component is usually directed parallel to the orientation of the reverse fault. Hydraulic fractures therefore propagate perpendicularly to the existing fault pattern, as shown in **Fig. 5.26.** If $s_{h,\max} > s_{h,\min} > s_v$, then the fracture will be horizontal.

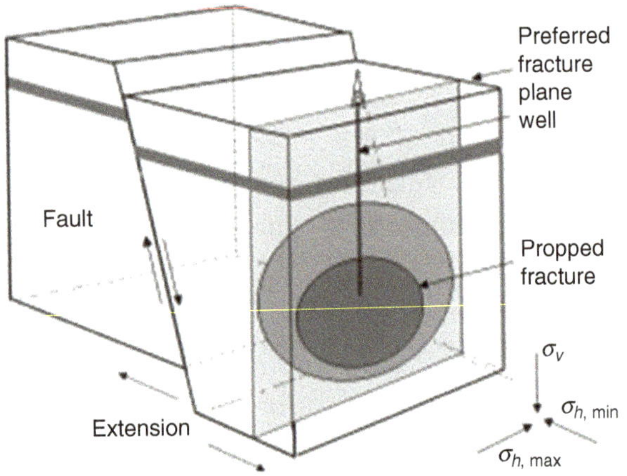

Fig. 5.25—Fracture growth along normal faults.

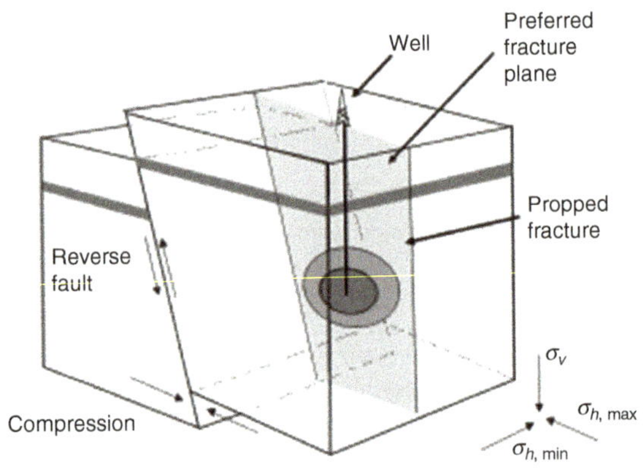

Fig. 5.26—Fractures grow perpendicularly to reverse faults.

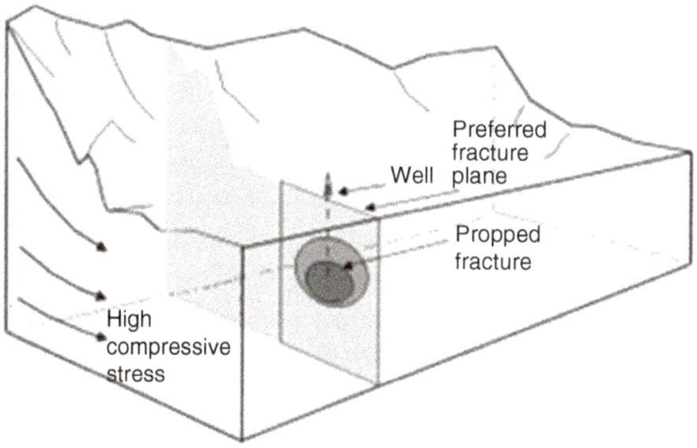

Fig. 5.27—Fracture propagation near a mountain range.

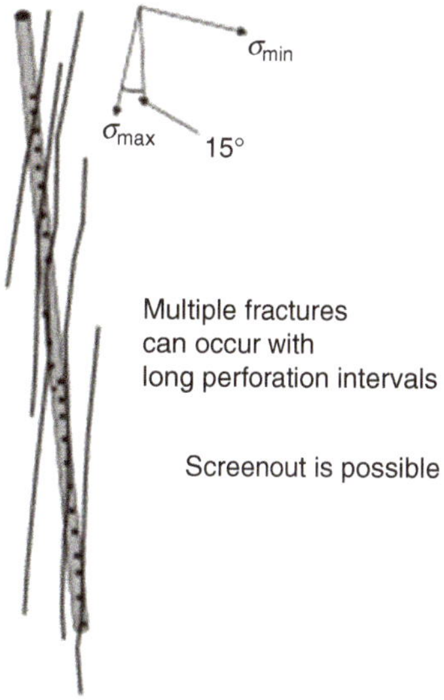

Fig. 5.28—Preferred fracture plane not aligned with wellbore.

The vertical and horizontal stress components that grow toward mountains will increase because of the increased overburden load from the mountains. The vertical stress component might become somewhat higher than the average normal overburden gradient of 1 psi/ft. The principal stress component in the horizontal direction along the mountain range will be the minimum principal stress, as shown in **Fig. 5.27.** Therefore, fractures tend to grow in the direction running toward the mountains.

In practical fracture growth, the exact vertical plane will not often be the case. The fracture will have some inclination with respect to the vertical plane because the principal stress components are not perfectly vertically aligned with gravity.

When the well is not perfectly aligned with the preferred fracture plane, the growth of multiple hydraulic fractures could be enhanced, and near-wellbore fracture tortuosity might occur, as shown in **Fig. 5.28.** If a thick sand is going to be fracture treated, the use of perforation schemes that restrict the injection point so multiple fractures are not created should be considered. If too many fractures are being propagated simultaneously, it is possible to have injection pressures higher than if a single fracture was being propagated, and the chances of a screenout could increase.

Horizontal wells are not common in tight gas sands because in many cases the abrasive nature of the rock on the drill bit makes drilling the horizontal well difficult. In addition, in many tight gas sands, the reservoir is compartmentalized, and it is difficult to know where the bit is located in the reservoir. Even though horizontal wells are not common in tight gas sands, it is important to understand how in-situ stresses affect the preferred direction of the horizontal well.

The preferred fracture plane refers to the plane in which the fracture will propagate in the "far field" (several wellbore radii away from the well). The orientation of this plane is determined by the minimum-principal-stress orientation and is independent of the orientation of the wellbore.

In **Fig. 5.29,** it was assumed that the minimum principal stress was horizontal, and therefore that the fracture orientation was vertical. The figure shows the fracture orientation for two horizontal wells: one well in the direction of the minimum horizontal stress and one well in the direction of the maximum horizontal stress. In the first case, fractures grow perpendicularly to the well (transverse fractures), and in the latter case, fractures grow parallel to the well (longitudinal fractures).

Near-wellbore tortuosity could become a problem for a well that is not aligned with the maximum principal stress.

In general, the fracture orientation is controlled by the regional stress field, but reservoir depletion can alter the field over time. Regional stress orientation is usually consistent over large areas in tectonically relaxed areas under original pressure conditions. In tectonically and geologically complex areas, it is possible to have variable stress orientation, such as when a well is located near a thrust fault. Also, production will result in pressure depletion, which will alter local stress orientation with time. As shown in **Fig. 5.30,** the orientation of hydraulic fractures for infill wells could be different from the orientation of hydraulic fractures in the original wells drilled in a field.

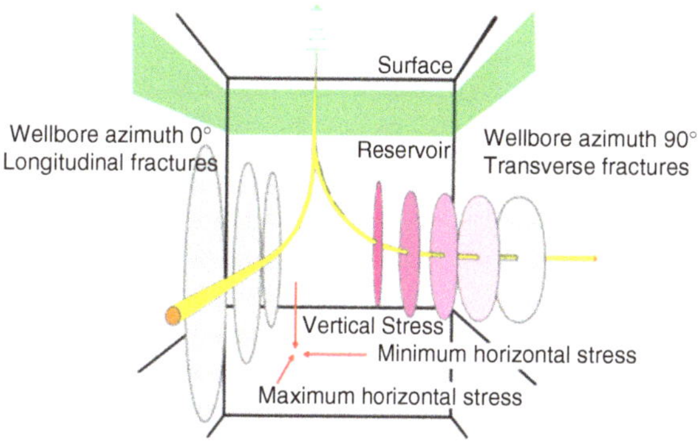

Fig. 5.29—Fracture orientation in horizontal wells.

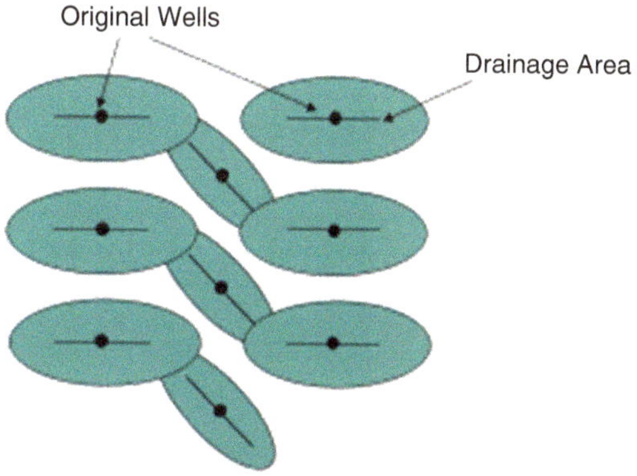

Fig. 5.30—Infill wells might have different fracture orientations.

5.1.5 Summary. Understanding the depositional system, geologic structure, and tectonic history is important for proper development of tight gas reservoirs. The geologic information should be used to develop

- Optimal drilling patterns
- Well completion designs
- Well stimulation designs
- Forecasting of production and economics

5.2 Logging and Log Analyses

5.2.1 Basic Log Analysis. In this section, we discuss the special considerations that must be addressed to analyze log data in tight gas sands properly. Please note that this portion of the book is not intended to be a comprehensive discussion of log analyses. That discussion is beyond the scope of this book, but the reader should better understand how to analyze logs to obtain the values required to characterize a tight gas reservoir.

After reviewing the geology, the engineer should perform a thorough log analysis to determine the reservoir fluid flow properties of porosity and water saturation. Wellbore, shale, hydrocarbon, and lithology corrections must be made to calculate the values of total and effective porosity. Shaly-sand water saturation must also be used to calculate water saturation.

For a tight gas reservoir, a hydraulic fracturing treatment must be successfully designed and pumped to produce gas at a commercially viable rate and recover commercial volumes of natural gas. Accurate measurement or estimation of mechanical rock properties, such as in-situ stress, Poisson's ratio, and Young's modulus, are required to design the fracturing treatment. Mechanical properties are also required in the evaluation of wellbore stability.

It is possible to calculate the values for fluid flow properties and mechanical properties from a comprehensive openhole logging suite. A standard logging suite is shown in **Fig. 5.31.** This log, commonly called a triple combo, includes

- Correlation curves: GR and SP
- Resistivity curves: shallow, medium, and deep
- Porosity logs: bulk FDL, CNL, and acoustic

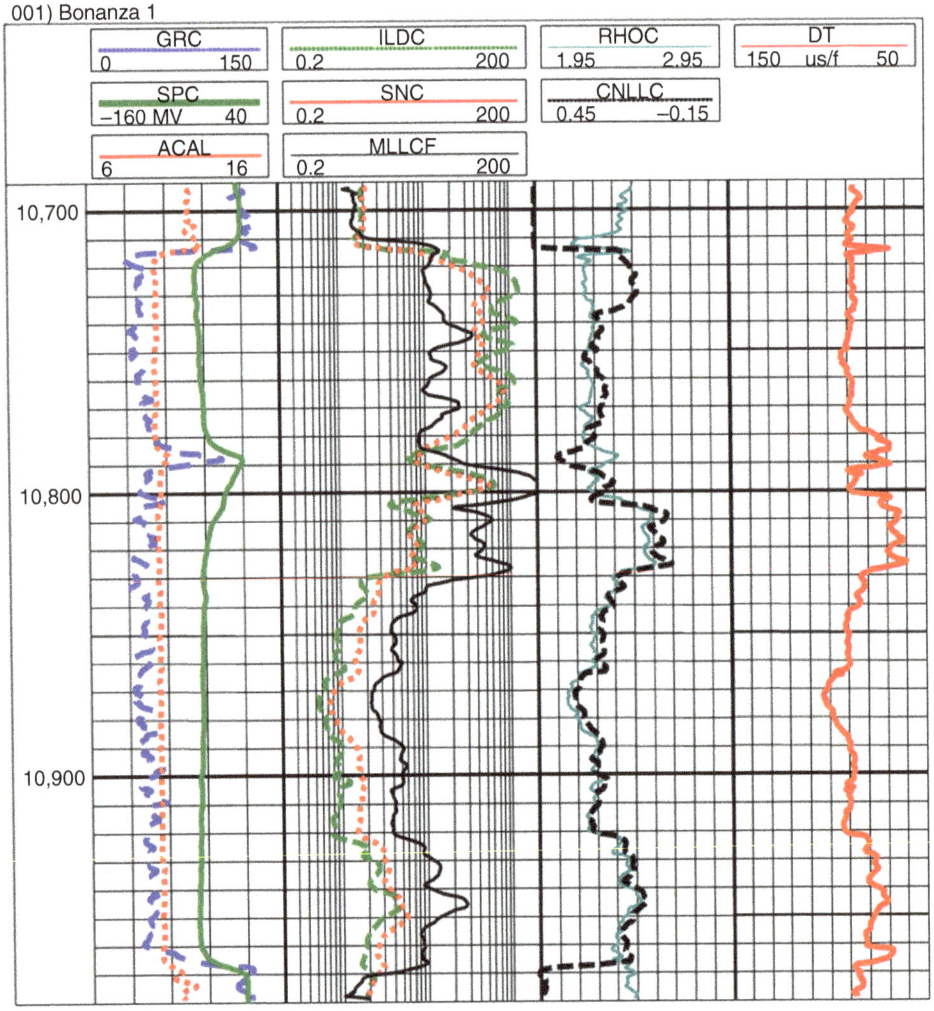

Fig. 5.31—Typical triple-combo log.

5.2.1.1 Layered Formations. Tight gas reservoirs are typically layered, have low porosity, and are naturally fractured deposits of sandstone, siltstone, and shale. There are wide variations in the layer properties, and some of the layers will be naturally fractured. To properly complete, fracture treat, and produce a tight gas well, the properties of each layer must be evaluated for the inclusion of the entire pay interval and the layers above and below the pay interval that affect fracture height growth.

The properties of each layer that we most need to know, at a minimum, are permeability, porosity, water saturation, pressure, gross and net thickness, in-situ stress, and Young's modulus for each layer of rock above, within, and below the productive interval.

The sources of information we can use to estimate values for all of these parameters are openhole and casedhole well logs, cores, well tests, drilling records, and geologic information from seismic data or outcrops.

In addition to the values of porosity, water saturation, and mechanical properties discussed above, we also must determine values for gross and net-pay, formation permeability, and reservoir pressure.

Gross and net-pay can be estimated from logs or drilling data. Gross thickness is important in defining in-situ stress profiles. Net thickness is used to calculate values for oil- and gas-in-place. Net thickness is also important when calculating the leakoff profile for designing a hydraulic fracturing treatment. Net-pay is a function of the porosity and permeability that will support oil and gas flow at measurable flow rates. The caliper log, showing mudcake buildup, can be useful in estimating the net-pay.

Formation permeability is used to calculate oil- and gas-flow rates, fracturing fluid leakoff rates, optimal fracture length, and optimal fracture conductivity and is used to estimate fracture half-length from post-fracture production and well test data.

Reservoir pressure is used to calculate oil- and gas-in-place, in-situ stress, oil- and gas-flow rates, and fracturing fluid loss into the formation. The reservoir pressure also affects fracturing fluid cleanup, the choice of fracturing fluid, and the choice of propping agent. These effects will be discussed in Part 3 of this book, which describes how hydraulic fracturing treatments must be designed.

5.2.1.2 Required Logging Suite. Basic log analysis refers to the standard treatment of logs and provides accurate results in high-permeability reservoirs that do not contain high volumes of clay or shale. We will explain how basic log analyses should be performed as a foundation for the next steps, which are shaly-sand analyses and corrections for mud-filtrate invasion.

Openhole well logs provide the most economical source of information for evaluating layered, low-porosity, tight gas reservoirs. Typically, we must run a full suite of logs to properly characterize the various rock layers. The logs must be calibrated with core, well test, and production data.

Fig. 5.31 shows the "triple-combo logging suite." In new areas, the triple combo should be run on every well. However, after the geology has been well-defined, one can use fewer openhole logs to plan the completion, as a way to reduce costs. But, in new areas or areas where the geology varies measurably in the lateral or vertical direction, it is not a good idea to try to save a little money on logging and risk not having the data needed to optimize the completion.

Logs that should be run on every well, with few exceptions, are the SP log, GR log, FDL, CNL, dual-induction log (DIL), and SON. The newer versions of some of these logs should be understood. For example, there are many new logs that might also provide valuable data, such as the array induction log (AIT), array sonic log (AST), electromagnetic propagation log (EPT), nuclear magnetic resonance log (NMR), and formation imager log (FMI). Data from these logs can be used to improve the evaluation of each formation layer, but it can be expensive to acquire these data routinely.

One interesting fact about logging and log analyses is that the logging tools do not actually measure what we want to know. We must take the measurements and correct them for borehole and environmental effects, then calculate what we are really trying to measure. For example, the SP log measures the naturally occurring electrical current flowing from the formation into the borehole because of salinity differences in the fluids. We use this information to pick formation boundaries, estimate formation-water salinity, and estimate the volume of shale in the reservoir.

GR logs (shown in **Fig. 5.32**) measure the natural radioactivity emitted by a formation and can be used to pick formation boundaries and estimate the volume of shale in the reservoir. The FDL actually measures the electron density in a formation, while the SON measures the acoustic travel. The CNL is mainly affected by the hydrogen content. These three logs are then used to calculate formation porosity, after making several assumptions.

Standard GR logs measure the total natural radioactivity of a formation. Special GR logs, called spectral gamma rays, can determine the amount of gamma rays from potassium, uranium, and thorium. Although these logs are not commonly run, they are often useful for the geologist when trying to correlate various shales. The standard GR log can be used to calculate the fraction of the reservoir that is shale by using a straight-line calculation.

Because the GR log measures natural radioactivity of a formation and is mainly affected by potassium, uranium, and thorium (elements normally found in shale), it is an important indicator of the amount of shale present in the tight gas sand. On occasion, a radioactive element, such as potassium feldspar, can be found in fairly clean sandstone, and just using the GR log to calculate shale content can be misleading. In such cases, core data are needed to recognize the presence of radioactive elements in otherwise clean sandstone.

However, in general, clean sandstone, limestone, and dolomites should exhibit low natural radioactivity, while shale formations exhibit high natural radioactivity. GR logs can be run in both openhole and casedhole environments. Although the overall reading of the GR log is affected by the casing, the character of the log is not, so that the casedhole GR log is used to correlate with the openhole GR log.

The logging suite can be tailored to the type of well that is being drilled. The well types discussed in this chapter are

- Exploration wells: These are the initial wells drilled in a new area, and proper formation evaluation dictates that a full suite of logs is required to properly characterize the formation.
- Development wells: After the exploration phase has proved to be successful and before full-scale production, it is common to drill development wells to better understand the formation to be produced.
- Production wells: When the formation is understood and enough development wells have been drilled to proceed with full-scale drilling, a decision to use a minimum logging suite to reduce costs can be determined.

For exploration wells, you might want to obtain as much information as possible so that some or all of the following logs or testing tools can be run: GR log, SP log, CNL, FDL, DIL, AST (compressional and shear slowness), NMR, borehole imaging, formation tester tool, and side-wall coring tool.

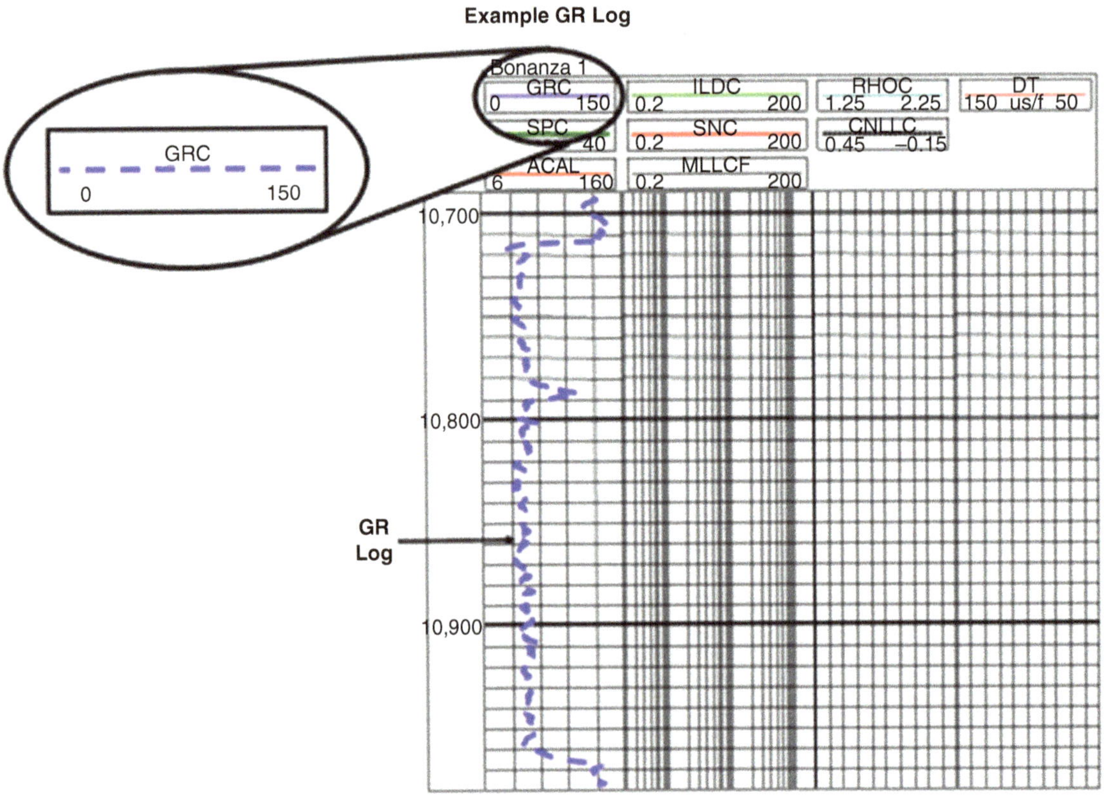

Fig. 5.32—Typical GR log presented in API units.

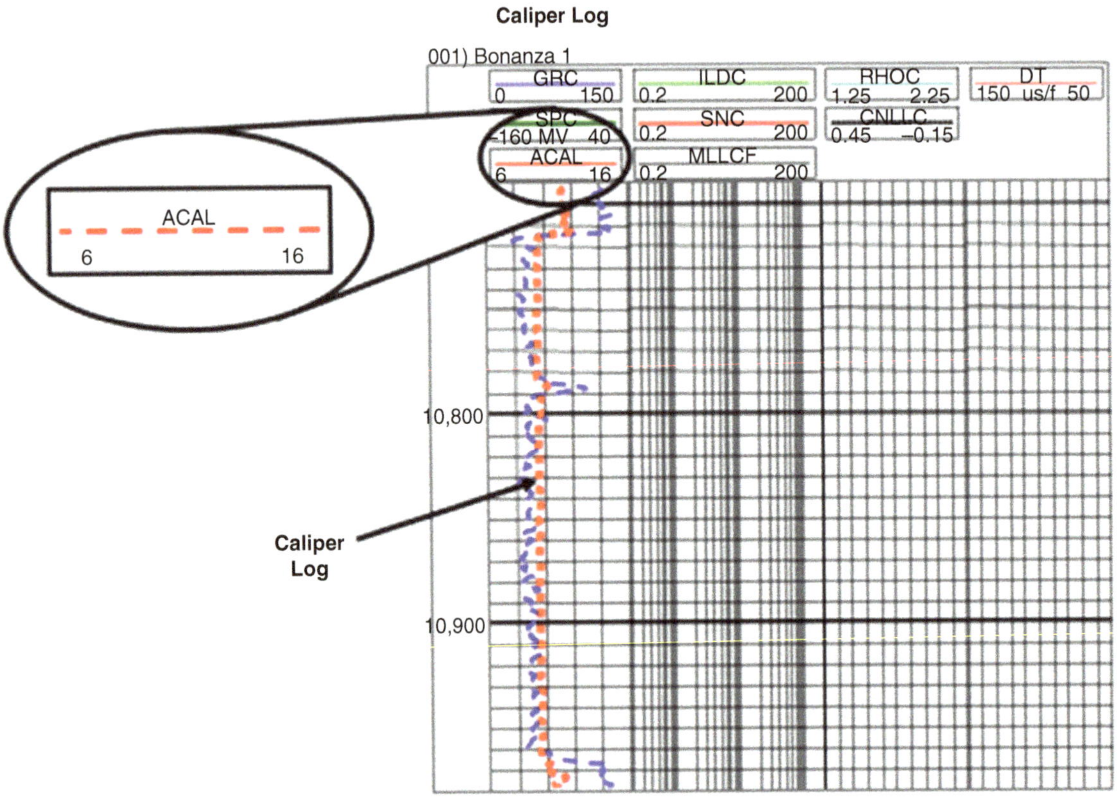

Fig. 5.33—Example of a caliper log with a scale of 6 to 16 in.

For development wells, it might be possible to reduce the costs of the well by reducing the openhole logging suite. The following logs (essentially the triple combo) should be considered for development wells: GR, SP, sonic compressional and shear slowness, DIL or CNL, and FDL.

For production wells, it is likely that the logging suite will need to be minimized to reduce costs and improve the profit margin. To get the necessary information from a minimal logging suite, correlations between logs and cores will have to be developed; then, those correlations can be used along with the following logs: GR, SP, DIL, and sonic compressional and shear slowness logs.

5.2.1.3 Steps Required To Analyze Logs in Tight Gas Reservoirs. If older logs are available only as paper copies, a substantial amount of work is required before any meaningful reservoir properties can be calculated. For these older paper logs, the steps required to preprocess the log data are digitizing all data, depth shifting, environmental corrections, and normalization. When all data have been preprocessed and are stored in a digital database, a series of statistical analyses should be conducted to quantify certain parameters.

These statistical analyses include: Pickett plotting, to determine water resistivity (R_w), cementation factor (m), and saturation exponent (n); shale histograms, to determine shale endpoints; sand or limestone histograms, to determine clean zone endpoints; environmental corrections; linear regression of SON, FDL, and CNL data with core porosity data when available; and linear regression of SON vs. FDL and CNL vs. FDL readings.

For tight gas reservoirs, more than 80% of the work required to conduct proper log analyses, especially with older logs, involves preprocessing of the data before any actual analyses can be performed. Hunt et al. (1981, 1985) published a series of articles in *World Oil* that clearly show the steps required to process logs correctly so that the true values of porosity, shale content, and water saturation can be calculated in low-porosity formations. The reader is encouraged to find and read these articles for further details. We have summarized the work of Hunt et al. in this chapter.

The first step in log analysis is to study the log to understand the headings, to include the type of logs included and the scales of the logs. The next step, in our opinion, is to study the caliper log. The caliper log is important for several reasons. One is that, if you can find mudcake from the caliper log, it is an indicator of permeability. In tight gas sands, if there is enough permeability to allow a mudcake to build, then there is enough permeability to produce gas. Second, if the hole is severely washed out, then the accuracy of the logs in the washed-out section comes into question. This problem is especially crucial for pad tools such as the FDL and any micrologs or imaging logs using pads. **Fig. 5.33** shows a caliper log.

5.2.1.4 Porosity Logs. As discussed previously, the three main logs for determining porosity are the bulk FDL, compensated CNL, and SON (acoustic log). None of these logs actually measures porosity. The electron density affects the FDL, the volume of hydrogen in the rock affects the CNL, and the SON measures the speed of sound in the rock. The details of how these measurements are used to estimate porosity are presented below. More-detailed explanations can be found in a typical log analysis textbook. **Figs. 5.34, 5.35, and 5.36** are examples of how these logs are typically presented.

The FDL has a radioactive source that emits gamma rays at a certain energy level. The particles collide with electrons in the rock, and the population of gamma rays remaining at a certain time can be correlated with the electron density.

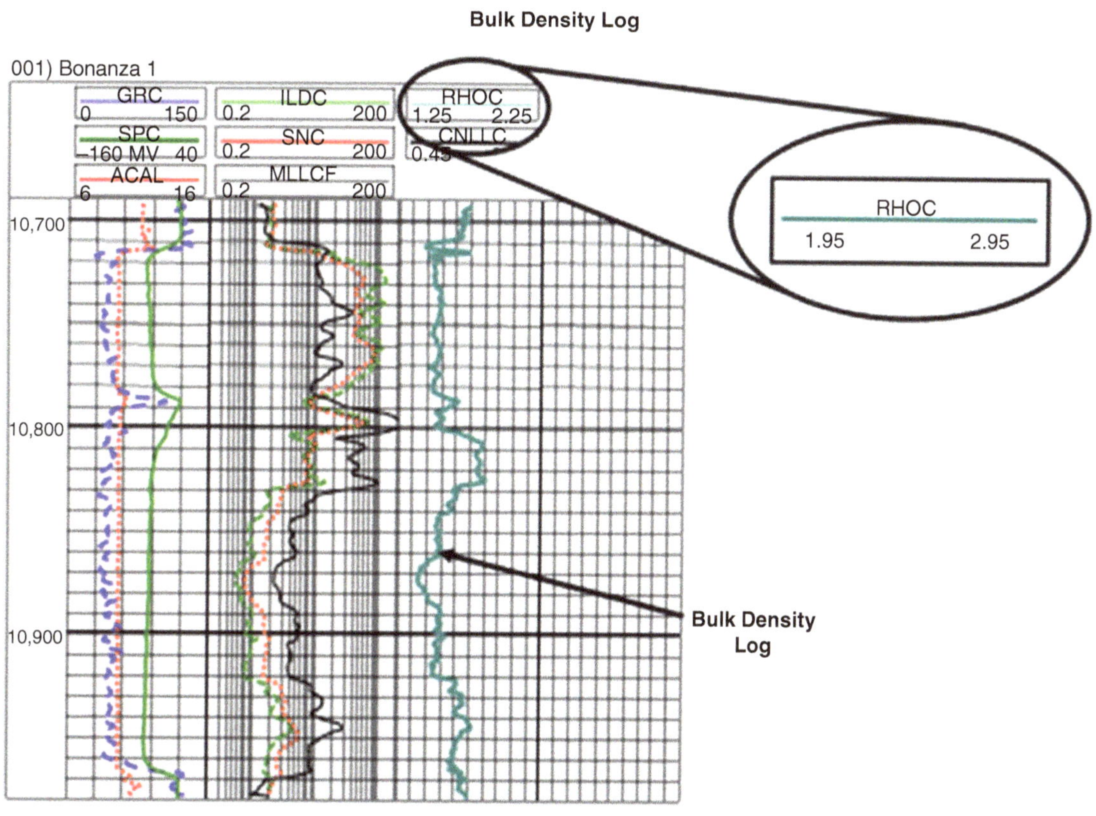

Fig. 5.34—Example of a bulk FDL presentation.

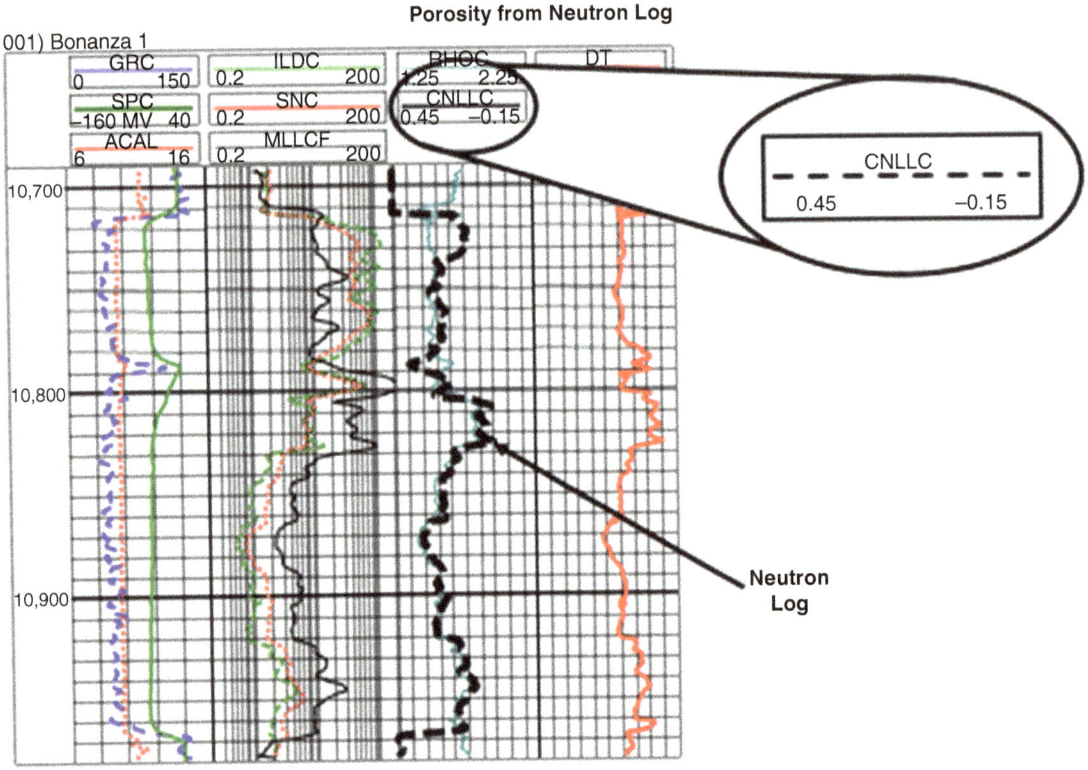

Fig. 5.35—Example of a CNL presentation with a scale of apparent limestone porosity.

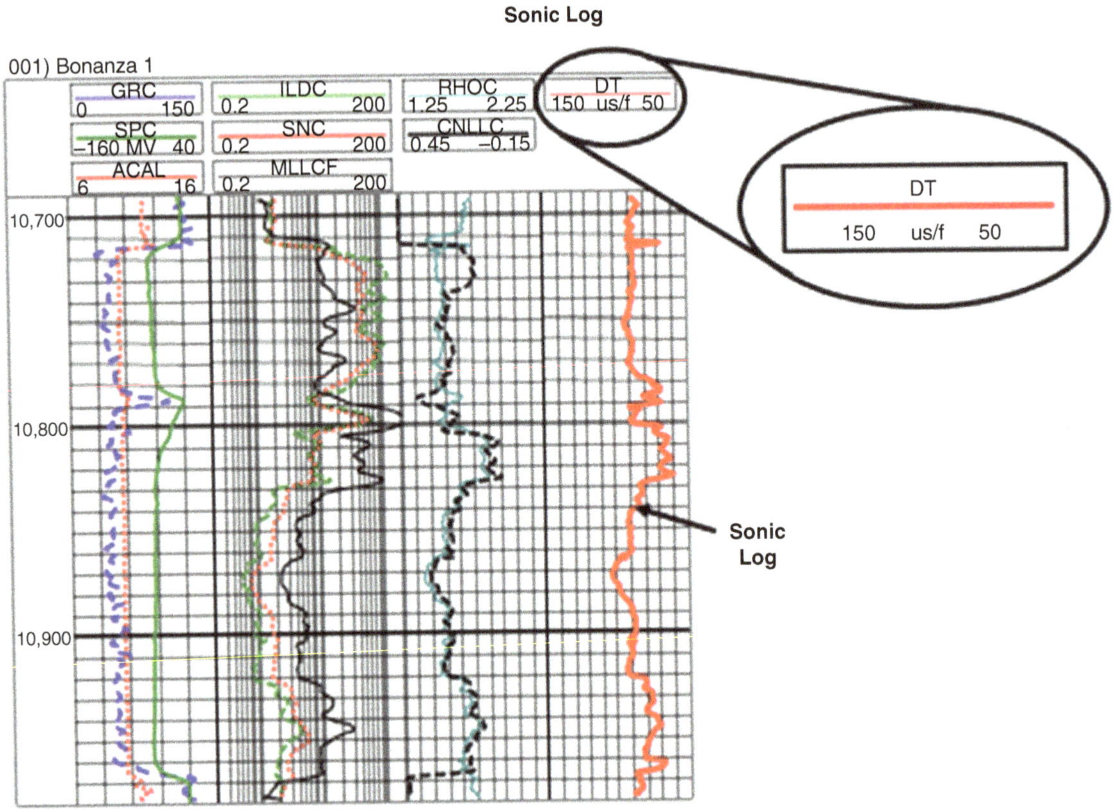

Fig. 5.36—Example of an SON that measures the compressional velocity.

The electron density is then correlated with the formation bulk density. Then, with the use of the lithology, the porosity can be calculated, assuming the formation is filled with water. The next step is to make corrections to the porosity if the formation is only partially filled with water and contains gas or oil.

The scale of most FDLs is in g/cm^3. A scale that is in terms of porosity can be calculated by making an assumption of lithology and assuming the pores are water filled.

A CNL tool emits bursts of high-energy neutrons into the formation, where the neutrons collide with the nuclei of the formation's atoms. Neutrons lose energy (velocity) with each collision. If a neutron hits a hydrogen atom head-on, it will be adsorbed, and a gamma ray will be emitted. The number of neutrons left or the number of gamma rays emitted over time is a direct measure of the amount of hydrogen in the formation. Again, if the lithology can be estimated, the porosity can be calculated. The scale for the neutron log is apparent limestone units. If the formation is a sandstone, then a lithology correction must be made.

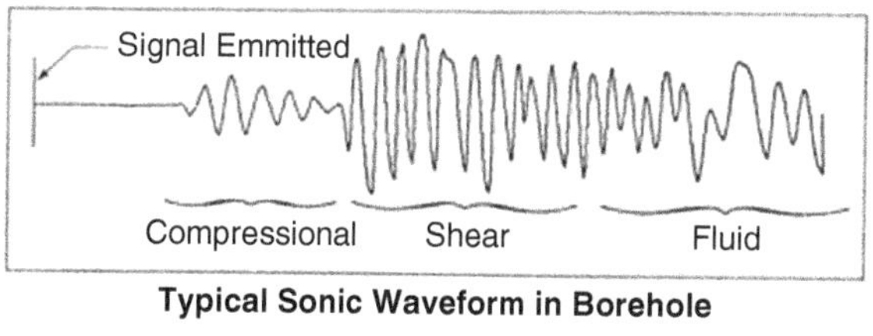

Fig. 5.37—Typical sound wave emitted from a sonic tool.

The SON tool has a sonic emitter (source) and a receiver a few feet from the source. The number of source and receiver sondes and the distance between the sondes vary depending on the tool and the service company. The basic principle is to measure the velocity of the compressional and shear waves as the sound passes through the rock near the borehole. Again, by making an assumption concerning the lithology and the fluid content, the estimated porosity can be calculated. **Fig. 5.37** shows the nature of an individual sound wave.

The sonic tool measures the travel time between the sondes and records the measurement in microseconds per foot. This is the reciprocal of the velocity of the sound wave. Formations with high porosity will be slow, and formations with low porosity will be fast. Tight gas sandstones with low porosity will result in faster sonic wave velocities.

5.2.1.5 ***Calculating Porosity.*** The following equations are used to calculate the porosity from the logging data. Before these calculations are performed, the analyst should make the necessary corrections for environmental issues, such as mud weight and mud salinity. In addition, corrections for borehole size and other issues as discussed by Whitehead et al. (1986, 1987) are recommended. After these basic calculations are completed, shale corrections and hydrocarbon corrections must be made. These corrections are discussed in Section 5.3 of this book and in the articles by Whitehead et al. (1986, 1987).

5.2.1.6 ***Density Porosity.***

$$\phi_D = \frac{\rho_b - \rho_{ma}}{\rho_f - \rho_{ma}}, \quad (5.1)$$

where ϕ_D = density porosity, ρ_b = bulk density from the log, ρ_{ma} = assumed matrix density of the rock, and ρ_f = assumed fluid density in the pores of the rock.

Hydrocarbon corrections are made by assuming the expected fluid density when there is a mixture of water and hydrocarbons in the pore space.

5.2.1.7 ***Sonic Porosity.***

$$\phi_S = \frac{\Delta t - \Delta t_{ma}}{\Delta t_f - \Delta t_{ma}}, \quad (5.2)$$

where ϕ_s = sonic porosity, Δt = log travel time in microseconds per foot, Δt_f = travel time of sound in the fluid in the pore space, and Δt_{ma} = travel time in the rock.

The SON was first introduced for oilfield use in the 1950s, but the log only recorded the compressional wave velocity (v_c). Later, the full-waveform acoustic log was developed to allow the determination of both compressional velocity and shear velocity (v_s). With both shear and compressional velocity, we can calculate mechanical properties, including the elastic modulus. Even-newer logging tools can determine the velocity anisotropy in the formation.

5.2.1.8 ***Calculation of Water Saturation.*** When the value of porosity is estimated, the analyst must calculate the volume of hydrocarbon (oil or gas) that is in the pore space. The computation actually centers on estimating the water saturation and then inferring that the remainder of the pore space is filled with gas. The key measurement to allow for the estimation of water saturation is the measurement of resistivity.

The resistivity of a formation is an important property for understanding the rock and calculating water saturation and thus the hydrocarbon saturation. The resistivity that is measured depends on the volume of rock, the composition of the rock, the volume of pore space, and the composition of the fluid in the pore space. The resistivity of the water in the pore space depends on the ions in the water, the concentration of the ions, and the temperature.

Fig. 5.38 shows a resistivity log. The resistivity log actually measures the conductivity of an electric current in the formation. In some logs, an electric current is generated, and the conductivity of the formation to that electric current is measured between two points. The distance between the points of measurement affects the radius of investigation of the particular measurement. Electric logs work only in water-based muds.

For oil-based muds, the industry invented induction logs, which use a magnetic field to induce a current in the formation. Like the resistivity logs, the spacing between the signal source and the receivers can vary to generate resistivity measurements with a variety of depths of investigation. The resistivity measurements are given in Ω·m.

For clean sandstone, we assume the formation is composed of three parts: matrix (the rock) and water and hydrocarbon in the pore space, as shown in **Fig. 5.39.**

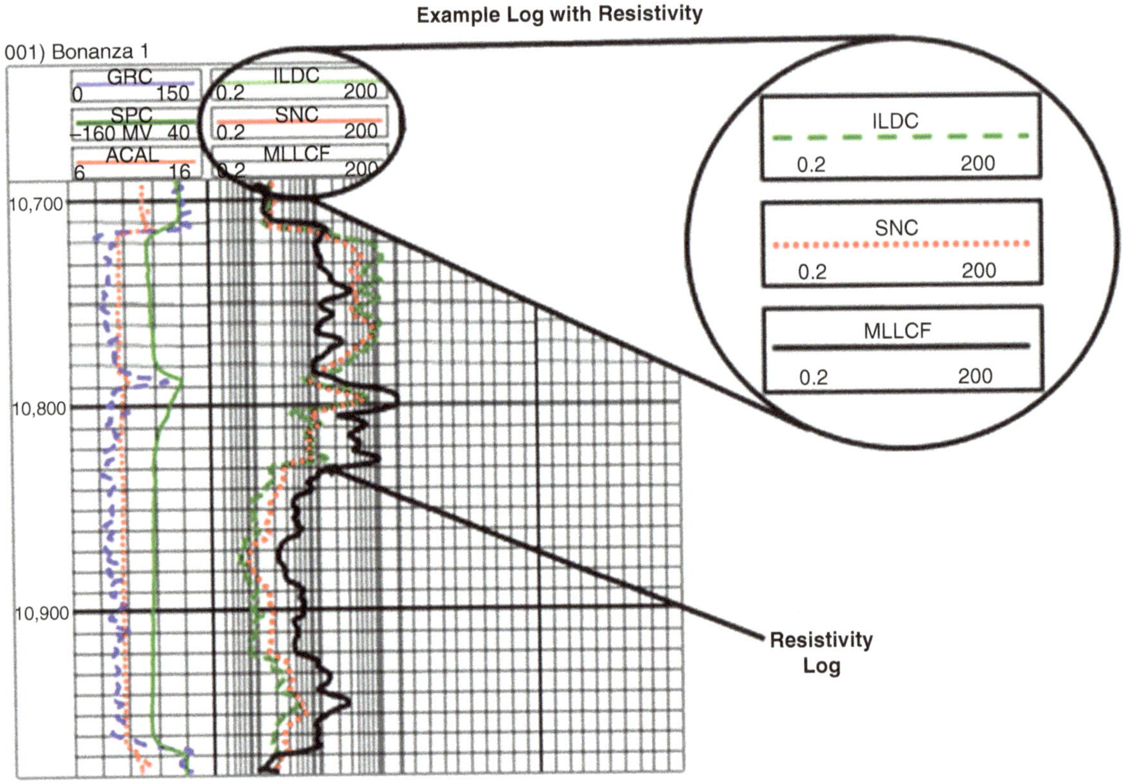

Fig. 5.38—Example of a resistivity log.

To calculate water saturation, it should be understood that the most resistive part of the formation is the matrix (i.e., the rock itself). In most cases, the rock (such as clean sandstone) is an insulator, and no current is carried by it. The exception, which will be addressed in the Shaly-Sand Analyses section of this chapter, is whether there is shale or clay in the rock—if so, the shale or clay can transmit some of the current. However, for clean sandstone, the current generated to measure the resistivity is assumed to flow through only the pore space containing water of varying salinity and hydrocarbon content. Thus, the more conductive a formation appears to be, the higher its porosity will be and the more water saturation it will have. More-resistive rocks have low porosity and/or low water saturation.

To calculate water saturation, we use the Archie empirical equation:

$$S_w^n = \frac{R_o}{R_t} = \frac{F \times R_w}{R_t} = \frac{a \times R_w}{\phi^m R_t}, \qquad (5.3)$$

where S_w = water saturation, fraction; R_t = true resistivity from log readings, Ω·m; R_o = resistivity of the formation if 100% saturated with formation water, Ω·m; F = formation factor, dimensionless; R_w = formation water resistivity, Ω·m; ϕ = porosity, dimensionless; a = empirical constant, dimensionless; m = cementation factor, dimensionless; and n = saturation exponent, dimensionless.

For clean sands, the empirical constant (a), cementation factor (m), and saturation exponent (n) are often assumed to be equal to a value of 2 if no other data are available. However, for tight gas sands, it is best to measure a, m, and n in the laboratory using formation samples, as shown in **Fig. 5.40.** If $R_t = R_o$, then the formation is 100% saturated with formation water. However, if $R_t > R_o$, then the formation contains oil or gas. R_o is the value of the resistivity of the formation if it is 100% filled with formation water whose resistivity is R_w.

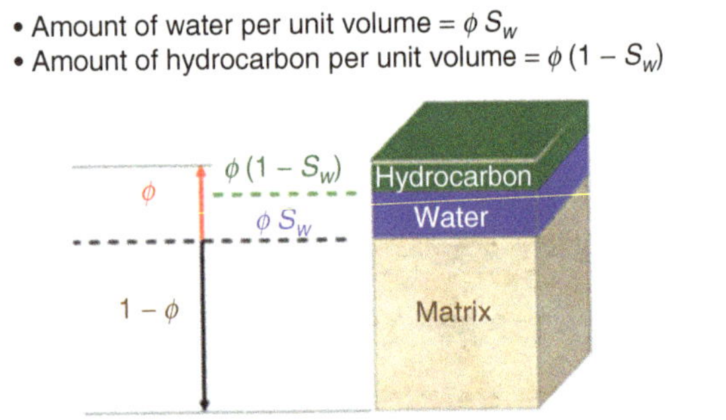

Fig. 5.39—Composition of a clean formation (no shale).

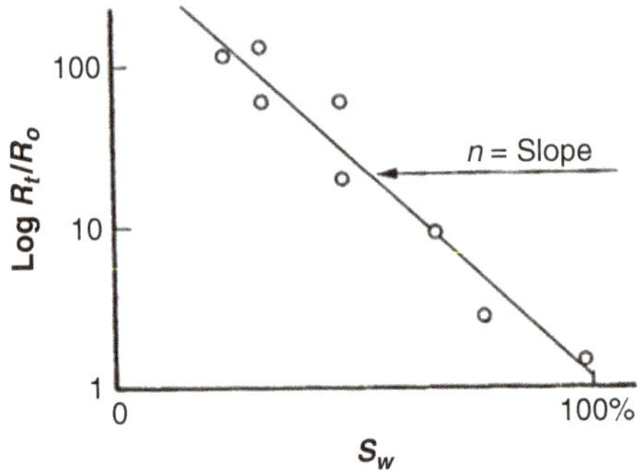

Fig. 5.40—Graph to determine the value of *n* from laboratory measurements.

The Archie equation is the basic equation used by petrophysicists to determine whether a formation has water, hydrocarbons, or both in the pore space (**Fig. 5.41**). As you can see, the water saturation is a function of several variables: the resistivity of the water in the pore space, the total composite formation resistivity, porosity, and several other variables that are a function of the rock.

R_o is the resistivity of a formation 100% saturated with water. If we think about a clean core sample, there is only one major component that conducts electricity, and that is the water in the pore space. The magnitude of the resistivity is a function of the amount of fluid (porosity and water saturation), the resistivity of the saturating fluid, and the interconnectedness of the pore space. The graph in **Fig. 5.42** shows that there is a relationship between resistivity index (I_R) and water saturation for different rock types. The slope of the line in Fig. 5.42 is a function of the pore structure (or tortuosity), which is called the saturation exponent, *m*. The saturation exponent is a measure of the degree of interconnectivity of the pore space.

As we stated earlier, the resistivity of a formation is affected by the amount of water, the resistivity of that water, and the interconnectivity of the pore space. Both porosity and water saturation control the amount of water in the pore space. When porosity is constant, the resistivity relationships are solely affected by water saturations and water salinity. Fig. 5.42 clearly shows that the ratio of R_t/R_o is a function of water saturation where the saturation exponent is a measure of the interconnectivity of the fluids in the pore space, which is related to capillary pressure and relative permeability. The saturation exponent *n* is related to the cementation factor *m*, as discussed previously.

The value of formation water resistivity, R_w, can be determined directly by measuring the resistivity from a produced-water sample or indirectly by calculating it from openhole-well-log data. The SP log can be used along with a Pickett plot and/or using the apparent water resistivity (R_{wa}) technique.

Eq. 5.4 is used to calculate R_w using the R_{wa} technique.

$$R_{wa} = \phi^m R_t. \quad (5.4)$$

If we assume S_w = 100% and a = 1.0, then $R_{wa} = \phi^m R_t$.

In clean water zones, $R_{wa} = R_w$; in hydrocarbon-bearing zones, $R_{wa} >> R_w$.

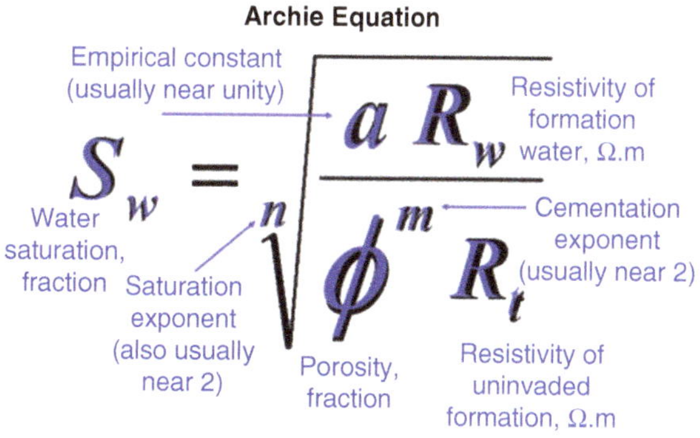

Fig. 5.41—Explanation of the Archie equation.

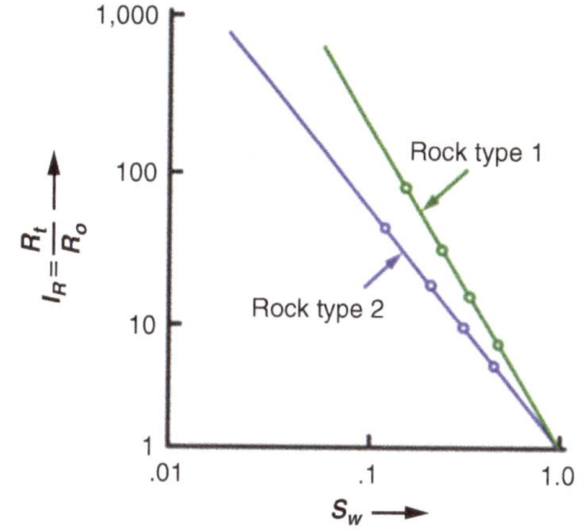

• When ϕ is constant

Note: When S_w = 100%, R_t must equal R_o.

Fig. 5.42—Graph showing how the slope (*n*) will vary with the rock type.

This method is similar to the Pickett plot, except the cementation factor, *m*, is assumed. This technique also assumes that the well has penetrated 100%-water sand in a clean interval (i.e., no shale). The assumption the log analyst also makes is that the formation water in the water zone has the same salinity as (or similar salinity to) the formation water in the adjacent hydrocarbon-bearing zones.

As an example of how to determine R_w from a water zone, consider the information in the example log in **Fig. 5.43.** A quick look at the log suggests that a hydrocarbon/water contact is located at approximately 10,830 ft. Above that depth, the formation has oil or gas, and below that depth, the formation appears to transition into a water zone. If it is assumed that the formation below 10,850 ft is a water zone, then the R_{wa} method should be used to calculate the water resistivity, R_w.

The Archie equation and the data from a well log, such as that shown in Fig. 5.36, can be used to construct a Pickett plot that can be used to determine the water resistivity, R_w.

Here, we discuss how to use a Pickett plot (Fig. 5.43) to evaluate R_w from the Archie equation,

$$S_w = \left(\frac{aR_w}{\phi^m R_t} \right)^{1/n} \quad (5.5)$$

Using mathematical operation and rearranging, we have

$$\log(R_t) = -m\log(\phi) + \log(aR_w) - n\log(S_w). \quad (5.6)$$

For water sands,

$$\underbrace{\log(R_t)}_{y} = \underbrace{-m}_{\text{slope}}\underbrace{\log(\phi)}_{x} + \underbrace{\log(aR_w)}_{\text{intercept}}. \quad (5.7)$$

When S_w = 100%, the term "$n \log (S_w)$" = 0.

A Pickett plot is a graphical representation of the Archie equation, as shown in **Fig. 5.44.** Log analysts typically use Pickett plots to determine both the formation resistivity and the cementation factor. The assumption that is made when using a Pickett plot to determine formation water resistivity is that the well has penetrated a zone that is clean or has no shale and is 100% saturated with water.

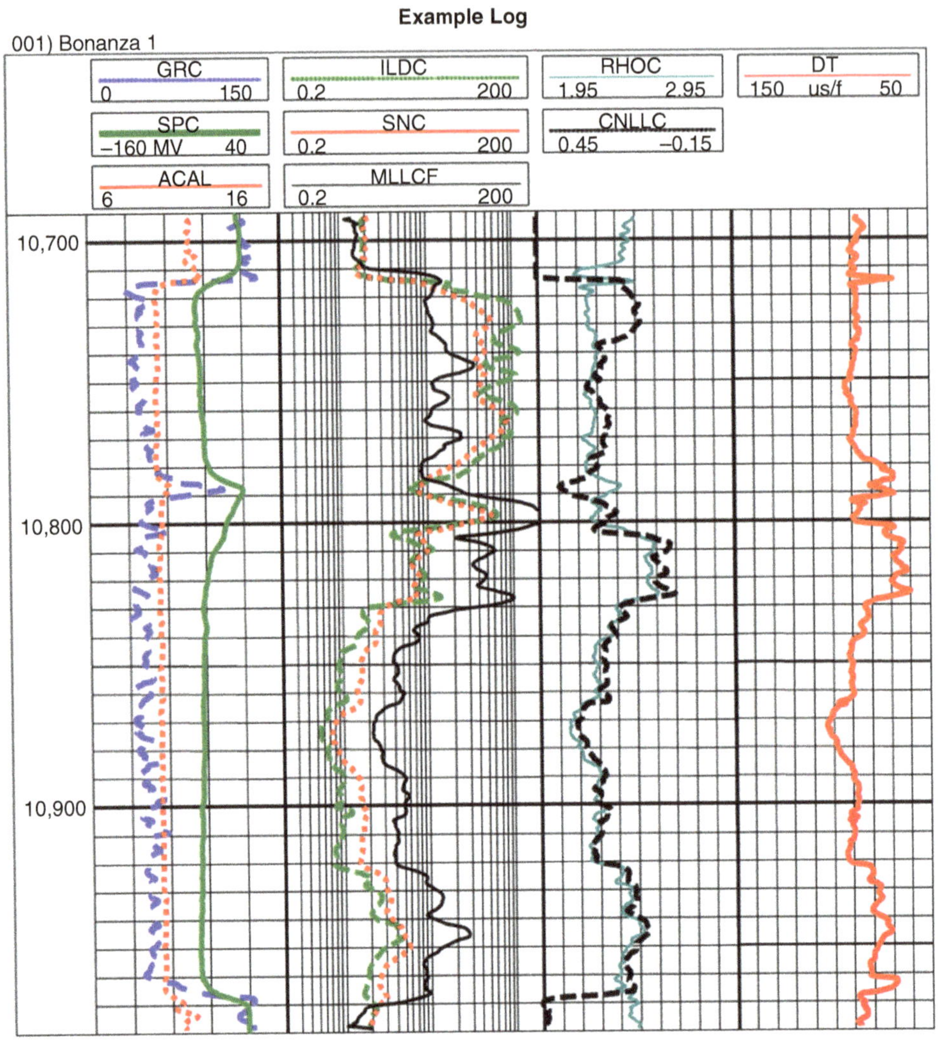

Fig. 5.43—Typical log showing a hydrocarbon/water contact.

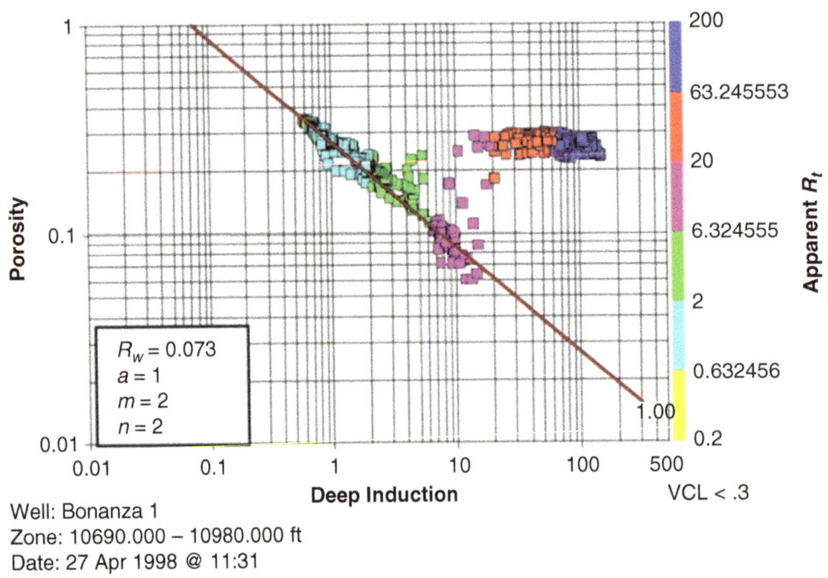

Fig. 5.44—Example of a Pickett plot used to determine values of R_w and m in a water sand.

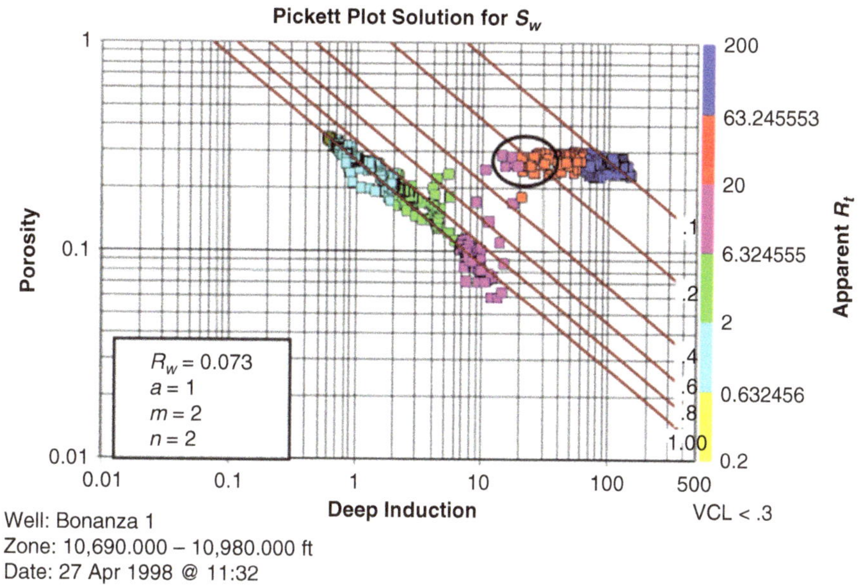

Fig. 5.45—Pickett plot used to determine values of R_w, m, and S_w.

The other assumption is that the formation water resistivity in the water sand is the same as the formation water resistivity in the adjacent hydrocarbon-bearing zones. This assumption is usually accurate.

The Pickett plot can also be used to estimate water saturation graphically for data points that do not fall on the 100%-water-saturation line, as shown in **Fig. 5.45.** The red lines with the scale at the right of the graph represent the water saturation for the data points shown.

5.2.1.9 Formation Compressibility. An often-overlooked rock characteristic with major impact is that of stress sensitivity and formation compressibility. Although stress sensitivity cannot be completely mitigated, early awareness of its likelihood and potential consequences allows for better field-development planning and economic evaluation. Phenomena that can result from stress sensitivity include subsidence (e.g., in the Ekofisk Oil Field), casing collapse (e.g., in the Vicksburg Oil Field), fracture healing through proppant embedment in the fracture face, and loss of deliverability. If pore-volume compressibility is underestimated, methods such as material balance, reservoir limits testing, or rate transient analysis might severely overestimate original fluid in place (by a factor of 10 in one worst scenario known to the author). Software default values for pore-volume compressibility (often 3×10^{-6} or 4×10^{-6} 1/psi) tend to be on the extreme low end of the range of likely values for even conventional reservoir rock. These default values should be treated with great skepticism in unconventional reservoirs.

The following criteria should be used to screen for possible stress-sensitive formations. Stress sensitivity is likely when any of the following criteria are met:

1. Porosity is less than 12% or higher than 25%.
2. Rock is friable or unconsolidated.
3. Net overburden (NOB) pressure is less than 3,000 psi (a rough rule-of-thumb equivalent is when the initial pressure in psia is within 3,000 psi of the formation depth in feet).
4. Initial pressure gradient is higher than 0.465 psi/ft.

The first two criteria are based on work by Newman (1973) and incorporating earlier work by Hall (1953) and Van der Knaap (1959). The third criterion is based on observations by Fatt and Davis (1952), Fatt (1958), and Jones (1988). The fourth criterion is based on the fact that, in overpressured reservoirs, the fluid supports at least part of the weight of the overburden, coupled with the assumption that the rock matrix has not previously been subjected to the portion of the overburden stress that is supported by the fluid.

These screening criteria are based on studies of both conventional and unconventional reservoir rocks. On the basis of the first and fourth criteria, most unconventional reservoirs would be considered candidates for high stress sensitivity. The first and fourth criteria can also be used to estimate the degree of stress sensitivity in unconventional reservoirs, with lower porosities or higher initial pressure gradients indicating the likelihood of higher stress sensitivity.

5.2.1.10 Summary. The most economical way to evaluate a thick, layered formation is to run openhole well logs and correlate the logs with cores and well tests. For unconventional gas wells, the log readings must be corrected for shale, hydrocarbons, wellbore effects, and other factors before the values from logs can be correlated with cores and well tests. Shale corrections are discussed in Section 5.2.2 of this chapter. Openhole logs create a continuous record of how certain fluids flow and mechanical properties vary with depth. The logs can be used to identify the contrast of rock properties between layers. In-situ data from logs can be correlated with core data to calibrate the logs that will be discussed later in this chapter.

The log analyst should use the available log analysis methods to determine acceptable values of a, m, n, and R_w as the porosity and water saturation by zone are calculated. The log analyst should estimate rock properties not only for the pay zone but also for the zones above and below the pay zone that will affect the vertical growth of a hydraulic fracture.

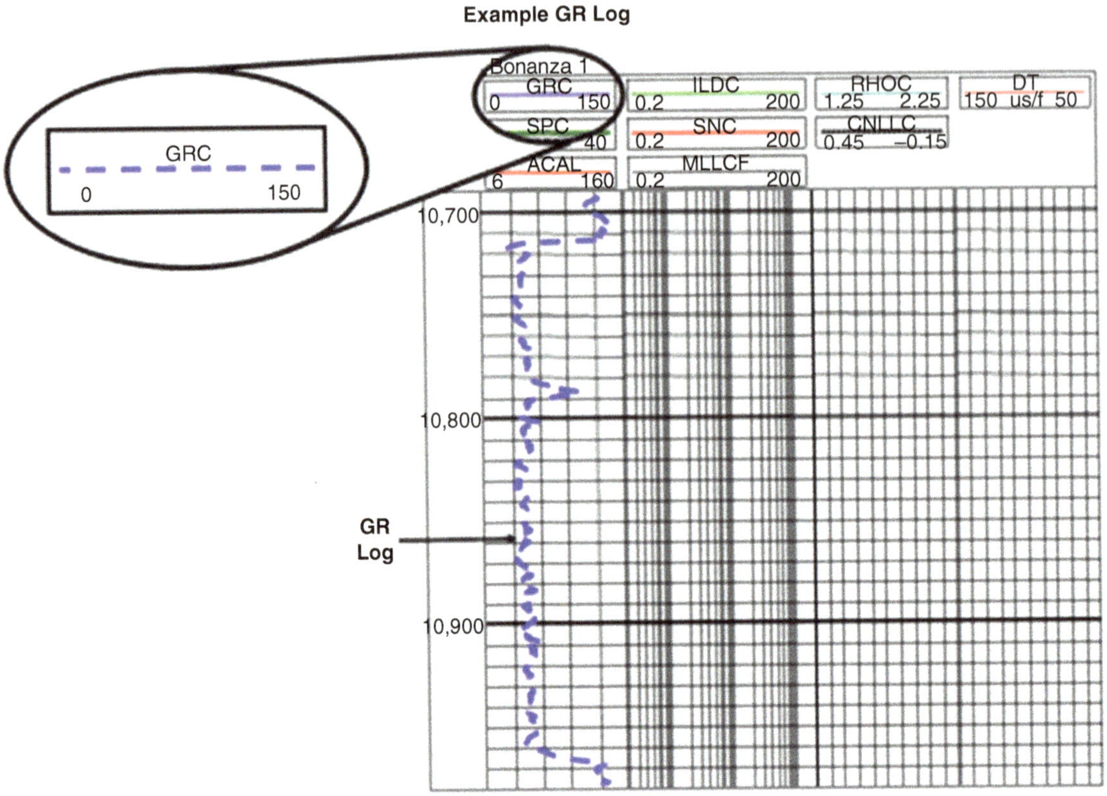

Fig. 5.46—Example of a GR log. Please note that the scale is in API units.

5.2.2 Shaly-Sand Analyses.

5.2.2.1 Logs to Estimate Shale Volume. One of the most important components of analyzing well logs in tight gas reservoirs is to properly determine the shale volume and lithology of the various rock layers. Tight gas reservoirs are tight because the original pore space is filled with clays and/or quartz overgrowths. Equations are available to estimate shale volume (V_{SH}) from a variety of logs. The two logs most often used are the SP and GR logs.

The GR log measures the natural radioactivity of the formation. In the formation rock, the radioactivity is mostly affected by the presence of the potassium, uranium, and thorium that is found in the shale layers. High levels of natural radioactivity indicate the presence of shale in the rock.

GR logs are calibrated using API standard logging boreholes, and the results are presented in API units, as shown in **Fig. 5.46.** Zero on the scale is indicative of no natural radioactivity. A high value, such as 150 API units, is indicative of a formation that is 100% shale and contains radioactive elements. Caution should be used, however, because sometimes radioactive elements, such as potassium feldspar, can be found in clean sandstones, so core is always useful.

GR logs are used to help delineate bed boundaries and to determine the volume of clay or shale in the formation. There are many ways to use the GR log that will be discussed later in this section.

As mentioned earlier, the log analyst should always look at what kind of log is being presented and the scale at the top of the log in the header. Many times, a caliper log (shown in **Fig. 5.47**) is presented in the same track as the GR log. The caliper log should be examined for mudcake and washouts. The size of the borehole drilled could also affect the environmental corrections to the raw log data.

The SP log is the other log that is often used to estimate the boundaries between layers, the water resistivity, and the volume of shale in the formation. The SP log measures the electrical potential in the formation caused by the salinity difference between the drilling mud and the formation water. The SP log is also a general indicator of permeability, because some permeability is required to allow current flow through the formation.

Fig. 5.48 shows a typical presentation for an SP log. The units are in millivolts (mV). The absolute value of the SP log reading is not used to calculate any variable. Instead, the analyst must choose a layer of rock that is considered a "clean shale" and a layer of rock that is considered a "clean sand." Then, the difference between the two layers of rock, in millivolts, is used to interpolate the volume of shale in the layer of interest. It is not always easy to find a clean sand in some formations, so the analyst uses the cleanest sand or estimates what a clean-sand reading might be using the GR log to help with the determination.

To calculate the volume of shale in a particular zone, use the linear interpolation of the values read from the SP log. The endpoints—the reading for a clean shale and the reading for a clean sand—must be determined. Then, the SP log reading in the zone of interest can be used to estimate the shale volume in that zone of interest.

$$V_{SH} = \frac{SP - SP_{CL}}{SP_{SH} - SP_{CL}}, \quad (5.8)$$

where V_{SH} = volume of shale (clay) in the formation, fraction; SP = deflection of the SP log from the clean formation line, dimensionless; SP_{CL} = the reading of the SP log in a clean sand, dimensionless; and SP_{SH} = the reading of the SP log in a clean shale, dimensionless.

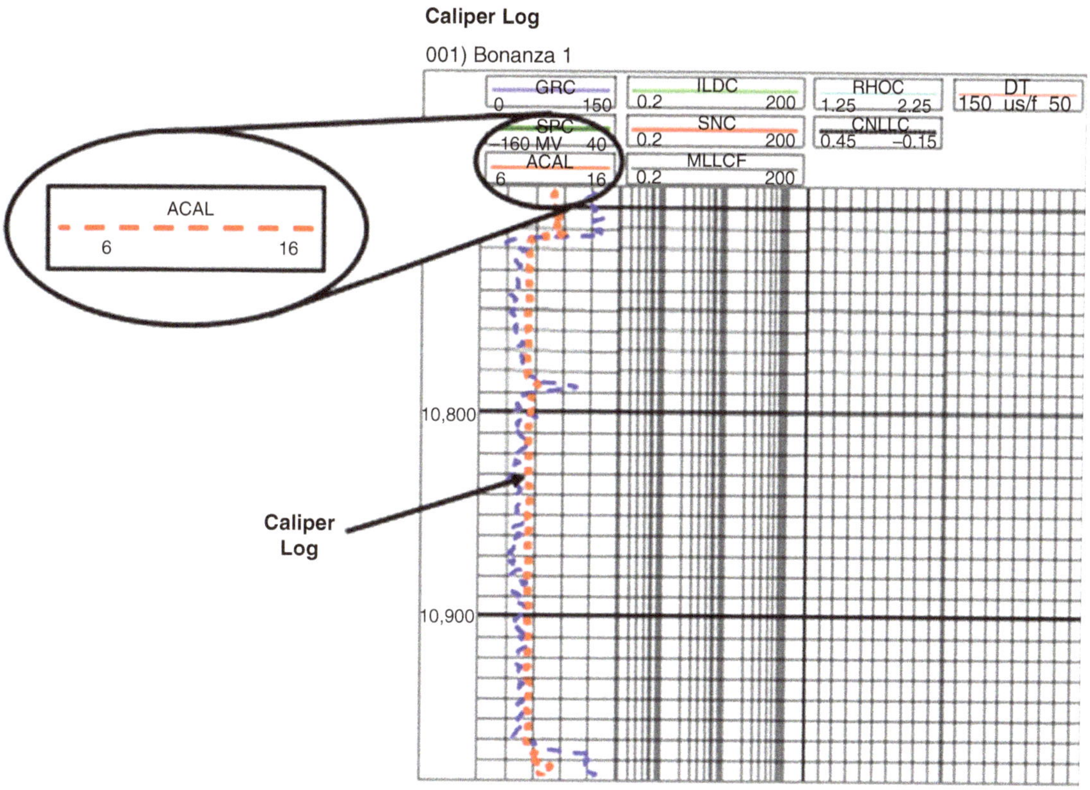

Fig. 5.47—Example of a GR log and caliper log.

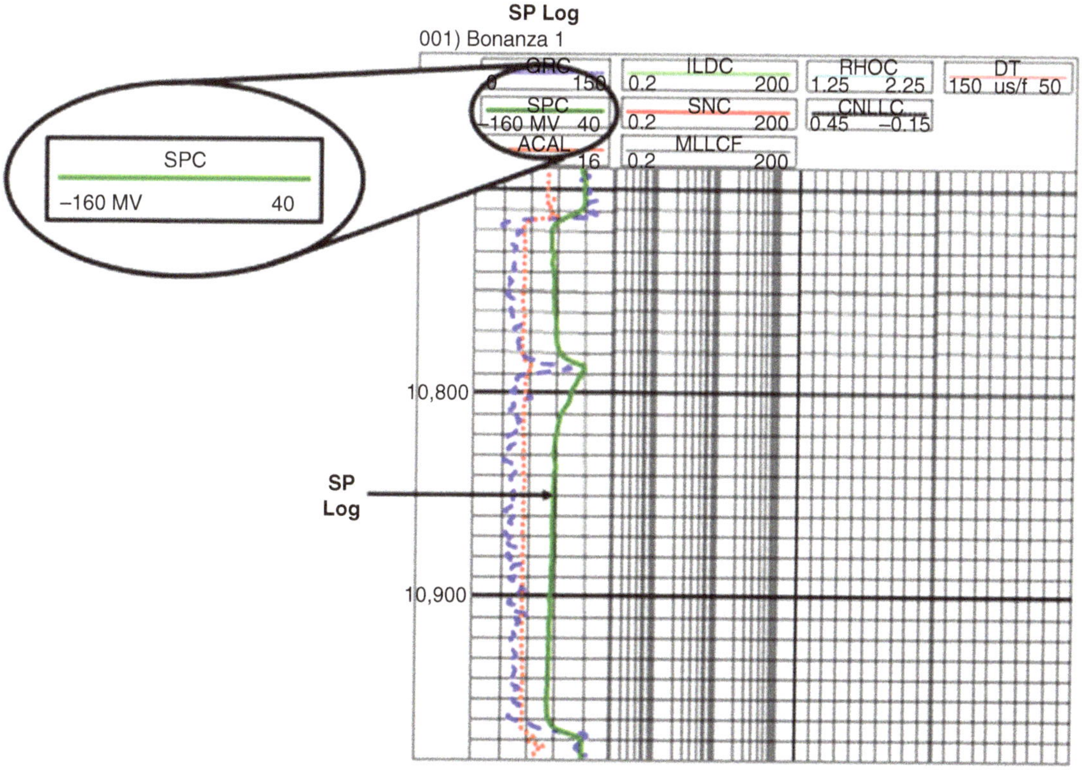

Fig. 5.48—Example of an SP log with units in millivolts (mV).

The procedure for estimating the volume of shale using the readings from a GR log requires an additional step. The process is to calculate the GR index (I_{RA}) first and then use that value to calculate the volume of shale, V_{SH}.

$$I_{RA} = \frac{GR - GR_{CL}}{GR_{SH} - GR_{CL}} \quad (5.9)$$

I_{RA} = the GR index; GR = the reading of the GR log in the zone of interest, API units; GR_{CL}= the reading of the GR log in a clean sand, API units; and GR_{SH} = the reading of the GR log in a clean shale, API units.

The entire computational procedure is based on empirical correlations that were developed using core data and log data. Eqs. 5.10 and 5.11 are called Larionov correlations, after the author of the paper where the correlations were first published (Asquith and Krygowski 2004):

$$V_{SH} = 0.33[2^{(2I_{RA})} - 1.0], \text{ for older rocks} \quad (5.10)$$

$$V_{SH} = 0.083[2^{(3.7I_{RA})} - 1.0], \text{ for younger rocks,} \quad (5.11)$$

where V_{SH} = volume of shale (clay) in the zone of interest, fraction.

The analyst first calculates I_{RA} using the GR readings in a clean shale, a clean sandstone, and the formation of interest. As with the selection of the endpoints for the SP log, it is not always possible to define a clean sandstone to estimate the GR reading in a clean sandstone, GR_{CL}. However, once I_{RA} is determined for a particular layer of rock, V_{SH} can be calculated using one of the Larionov correlations (Eqs. 5.10 and 5.11). The choice of which equation to use should be based on the age of the formations. For a given amount of shale, the younger formations will be more radioactive than older formations, where more time has allowed the radioactivity to subside.

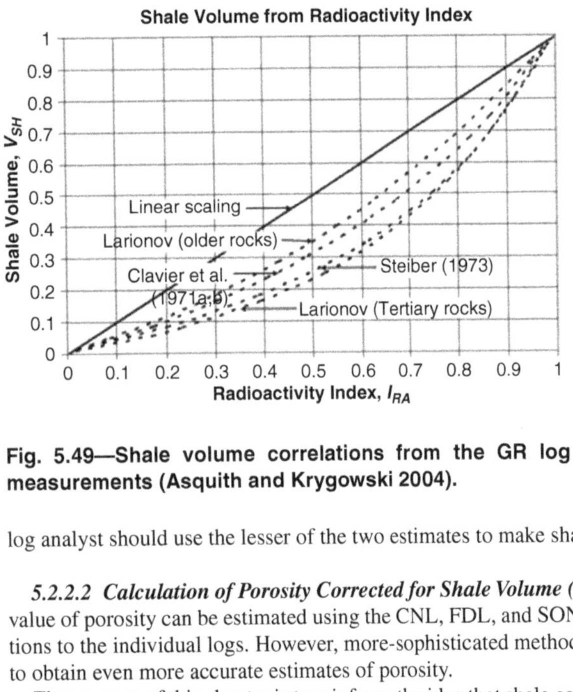

Fig. 5.49—Shale volume correlations from the GR log measurements (Asquith and Krygowski 2004).

Fig. 5.49 shows the various empirical correlations relating the radioactive index to the shale volume in sandstone. All of these correlations were developed using core and laboratory measurements.

As more is learned about calculating the volume of shale in a tight gas sandstone using logging data, it will become evident that every method is an approximation. In fact, one paper was published that presented several methods to calculate V_{SH} from log data.

Determining the correct value of V_{SH} is what the analyst should be trying to accomplish. There are at least seven ways to calculate V_{SH} from logs as documented in the literature. However, the values from the SP log or the GR log usually provide reasonable estimates, and these calculations are simple to use when there are SP and GR logs on most wells. V_{SH} calculated from the logs should be compared (when possible) to V_{SH} measured using cores.

In the literature, several authors have published results in which different methods to calculate V_{SH} have been compared to V_{SH} measured with core. In these papers, the consensus is that the analyst should calculate V_{SH} from logs in as many ways as feasible, then choose the smallest value of V_{SH} as the most realistic. So, if V_{SH} is calculated from the SP log and the GR log, then the log analyst should use the lesser of the two estimates to make shale corrections to the porosity.

5.2.2.2 Calculation of Porosity Corrected for Shale Volume (V_{SH}). When the value of V_{SH} has been calculated, the shale-corrected value of porosity can be estimated using the CNL, FDL, and SON. In this chapter, we will only discuss how to make the shale corrections to the individual logs. However, more-sophisticated methods should be used when using crossplots of the various porosity logs to obtain even more accurate estimates of porosity.

The purpose of this chapter is to reinforce the idea that shale corrections are required in tight gas sandstone log analyses. But a much more sophisticated analysis should be performed before investing in multimillion-dollar development plans for tight gas reservoirs.

When the values of V_{SH} have been calculated for specific layers of rock, shale-corrected values of neutron porosity, density porosity, and sonic porosity can be calculated. These calculations can be made on a layer basis. However, with computers and digitized data, the calculation can be made on a foot-by-foot basis.

Eq. 5.12 is used to calculate the CNL porosity corrected for V_{SH}.

$$\phi_{NC} = \phi_{N_{LOG}} - V_{SH}\phi_{NSH}, \quad (5.12)$$

where ϕ_{NC} = porosity from the CNL corrected for shale content, dimensionless; $\phi_{N_{LOG}}$ = porosity from the CNL, dimensionless; and ϕ_{NSH} = porosity of a clean shale, dimensionless.

Eq. 5.13 is used to calculate the FDL porosity corrected for V_{SH}.

$$\phi_{DC} = \left[\frac{(\rho_b - \rho_{MA}) - V_{SH}(\rho_{SH} - \rho_{MA})}{(\rho_{fl} - \rho_{MA})}\right] \quad (5.13)$$

where ϕ_{DC} = porosity calculated using an FDL, dimensionless; ρ_b = bulk FDL reading for the zone of interest, g/cm^3; ρ_{MA} = matrix density, g/cm^3; ρ_{SH} = density of a clean shale, g/cm^3; and ρ_{fl} = density of the fluid in the pore space, g/cm^3.

Eq. 5.14 can be used to calculate the SON porosity corrected for V_{SH}.

$$\phi_S = \left[\frac{\left(\Delta t_{LOG} - \Delta t_{MA}\right) - V_{SH}\left(\Delta t_{SH} - \Delta t_{MA}\right)}{\left(\Delta t_{fl} - \Delta t_{MA}\right)} \right], \quad (5.14)$$

where ϕ_S = porosity calculated from sonic log, dimensionless; Δt_{LOG} = sonic travel time in the zone of interest, μs/ft; Δt_{MA} = sonic travel time in the matrix, μs/ft; Δt_{SH} = sonic travel time in a clean shale, μs/ft; and Δt_{fl} = sonic travel time in the fluid that is in the pore space, μs/ft.

Common values used in the porosity equations are ρ_{ma} = 2.65 g/cm³ for sandstone, 2.71 g/cm³ for limestone, and 2.87 g/cm³ for dolomite; and Δt_{ma} = 55.55 μs/ft for sandstone, 47.62 μs/ft for limestone, and 43.48 μs/ft for dolomite.

It is always wise to compare/correlate log-calculated values of any parameter to/with values calculated or measured from cores. For V_{SH}, the method is to calculate V_{SH} from logs using the GR and SP logs and then measure V_{SH} from thin-sections in core from the same interval. When reasonable correlations are obtained, it boosts confidence in the formation-evaluation process. **Fig. 5.50** shows a comparison of lithology from thin-section analysis and log analysis. **Fig. 5.51** shows a comparison data set. The dotted line confirms the estimated shale volume.

5.2.2.3 Hydrocarbon Corrections. To calculate accurate values of water saturation in shaly tight gas sandstone reservoirs, after the shale-corrected values of porosity have been calculated, corrections must be made for the fact that the pore space is filled with both gas and water. It should be noted that, in the shale-corrected porosity equations (Eqs. 5.12 through 5.14), there are values that must be included for the correct value of fluid density, fluid travel time, and fluid effects in the CNL. The original equations assume the pore space is filled with water of a specific density. However, when the pore space contains a mixture of water and gas, the values needed in the porosity equation to account for the gas saturation must be estimated.

5.2.2.4 Water-Saturation Calculations. After determining the shale-corrected and hydrocarbon-corrected values of ϕ_{DC}, ϕ_{SC}, and ϕ_{NC}, use density-neutron or sonic-neutron crossplots to estimate values of actual porosity and lithology.

The final steps involve the computation of the values of water saturation. We can use the Archie equation, the Simandoux equation, the Waxman-Smits model, or the dual-water model. If the analysis is conducted using a computer, it is simple to calculate the value of water saturation using all four models to compare the results. The analyst should realize that all log analyses are empirical and are approximations, so using multiple methods to calculate porosity and water saturation can only help to understand the reservoir properties.

There are several concepts that must be understood when using the different water-saturation models. Shaly-sand water-saturation equations might use different definitions of porosity. The correct definition of porosity should be used in the equation selected to calculate water saturation. Those porosity definitions are

- ϕ_e = the effective porosity containing only oil, gas, and free water.
- ϕ_z = total porosity containing oil, gas, free water, and disseminated clay.
- ϕ_t = the volume of oil, gas, and bound water for the disseminated shale.

The Archie equation is for a clean zone where 100% of the current flows through the pore fluids. We can use the Archie equation after making shale and hydrocarbon corrections to determine the effective porosity, ϕ_e.

The Simandoux equation was the first to recognize that part of the conductivity of the formation can be conducted through the shale contained within the sandstone. The Simandoux model includes a term for evaluating the effects of the volume and conductivity of the shale. For this equation, use ϕ_e.

The Waxman-Smits equation uses an effective water conductivity based on the pore-water conductivity, the cation exchange capacity of the clays, and the volume of water and clay in the pore space. For the Waxman-Smits equation, use ϕ_z.

The dual-water model is a more general form of the Waxman-Smits model. In the dual-water model, the salinity of the free water can be different from the salinity of the clay-bound water. For the dual-water model, use ϕ_t.

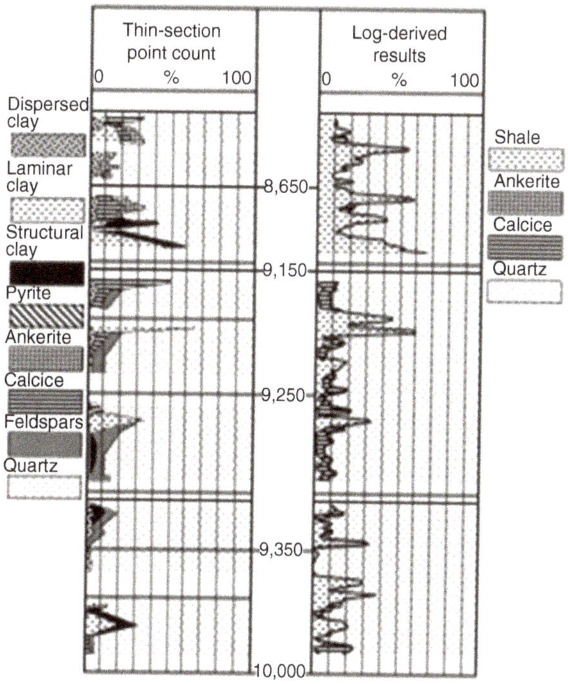

Fig. 5.50—Shale volume correlations between log calculations and core measurements.

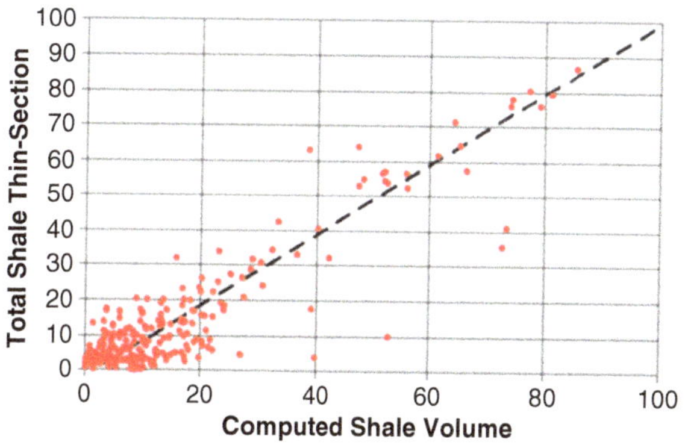

Fig. 5.51—Correlation of shale volume calculated from core to thin-section values.

Depending on the water-saturation model that is chosen, the selection of porosity (ϕ) must match with the saturation equation. The various equations are given below.

Eq. 5.15 is the Archie equation:

$$C_t = C_w \frac{\phi_e^m}{A} S_w^n, \qquad (5.15)$$

where C_t = total conductivity of the formation, mho/m; C_w = conductivity of the formation water, mho/m; A = empirical constant, dimensionless; m = cementation factor, dimensionless; and ϕ_e = the effective porosity containing only oil, gas, and free water.

Eq. 5.16 is the Simandoux equation:

$$C_t = \frac{C_w \phi^m S_w^n}{\left(1 - V_{SH}\right)^{m-1}} + V_{SH} C_{SH}, \qquad (5.16)$$

where C_{SH} = the conductivity of the shale in the zone of interest, mho/m.

Eqs. 5.17 and 5.18 are the Waxman-Smits equations:

$$C_t = C'_w \frac{\phi_z^m}{A} S_w^n \qquad (5.17)$$

$$C'_w = C_w + \frac{BQ_V}{S_w}, \qquad (5.18)$$

where ϕ_z = total porosity containing oil, gas, free water, and disseminated clay, dimensionless; B = empirical factor for the conductance of the clay per exchangeable cation, mmho-cm^2/cm^3; and Q_v = cation-exchange capacity per unity pore volume, meq/cm^3.

Eqs. 5.19 and 5.20 are the dual-water equations:

$$C_t = C''_w \frac{\phi_t^m}{A} S_{wt}^n, \qquad (5.19)$$

$$C''_w = \frac{S_{wb}}{S_{wt}} C_{wb} + \left(1 - \frac{S_{wb}}{S_{wt}}\right) C_{wf}, \qquad (5.20)$$

where ϕ_t = volume of oil, gas, and bound water for the disseminated shale; S_{wb} = saturation of the clay-bound water, dimensionless; S_{wt} = saturation of the free water, dimensionless; C_{wb} = conductivity of the clay bound water, mho/m; and C_{wf} = conductivity of the free water, mho/m.

The dual-water model is the most sophisticated, and many believe it is the most accurate method for determining the water saturation in a shaly gas sand. It is not that difficult to program, and virtually all logging companies and log analysis software programs you can purchase have an option for the dual-water model. **Fig. 5.52** shows the various parts of the formation that are considered in the dual-water model. The shale volume is composed of silt, dry clay, and bound water. The pore volume is composed of hydrocarbon porosity, effective porosity, and total porosity. Only the fluid in the effective porosity can flow. However, the bound water does affect the porosity log and resistivity log readings.

5.2.2.5 Corrections for Invasion. Mud-filtrate invasion will be discussed in detail in Section 5.2.3 of this chapter. However, it should be mentioned here that mud-filtrate invasion can substantially affect the resistivity log readings, especially when freshwater mud filtrate invades a low-porosity gas interval that also contains salt water. Deep filtrate invasion affects the log readings as a function of time because mud-filtrate invasion will occur during drilling and is a dynamic process, as shown in **Fig. 5.53.** Mud-filtrate invasion also affects the porosity logs. One benefit of mud-filtrate invasion is that it is an indicator of permeability—if mud-filtrate invasion is identified, it is a sure sign that the formation is permeable.

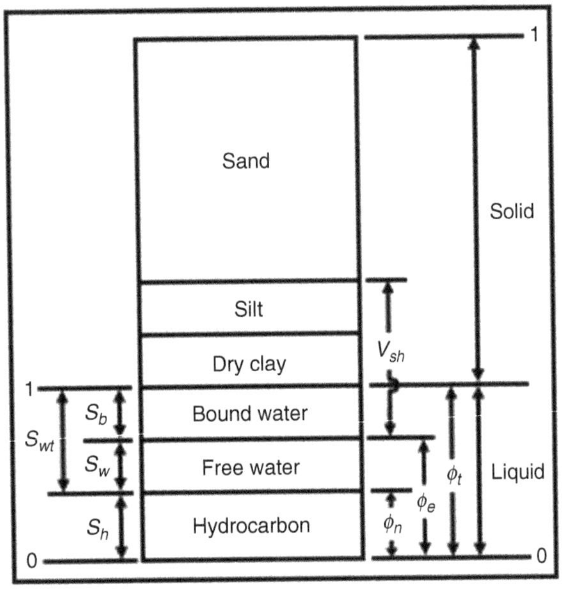

Fig. 5.52—Concept behind the dual-water model.

5.2.2.6 Summary. A basic log analysis will not provide the values needed to understand a shaly tight gas reservoir. Wellbore environmental, shale, hydrocarbon, and lithology corrections must be made to the log data to correctly calculate porosity in unconventional reservoirs. When the correct values of porosity have been calculated, the shaly-sand water-saturation equations should be used to determine the water saturation. Most analysts consider the dual-water model to be the best approach, but all the models discussed in this chapter can be used. In fact, we recommend that the analyst calculate shale content, porosity, and water saturation using all available methods and then compare the results. None of the methods is perfect. They are all just estimates, and comparing the estimates can be useful.

5.2.3 Mud-Filtrate Invasion.

5.2.3.1 Introduction. In low-permeability reservoirs, mud-filtrate invasion during drilling can affect the results from both logs and drillstem tests (DSTs). When a well is drilled with freshwater mud and penetrates a formation containing salt water

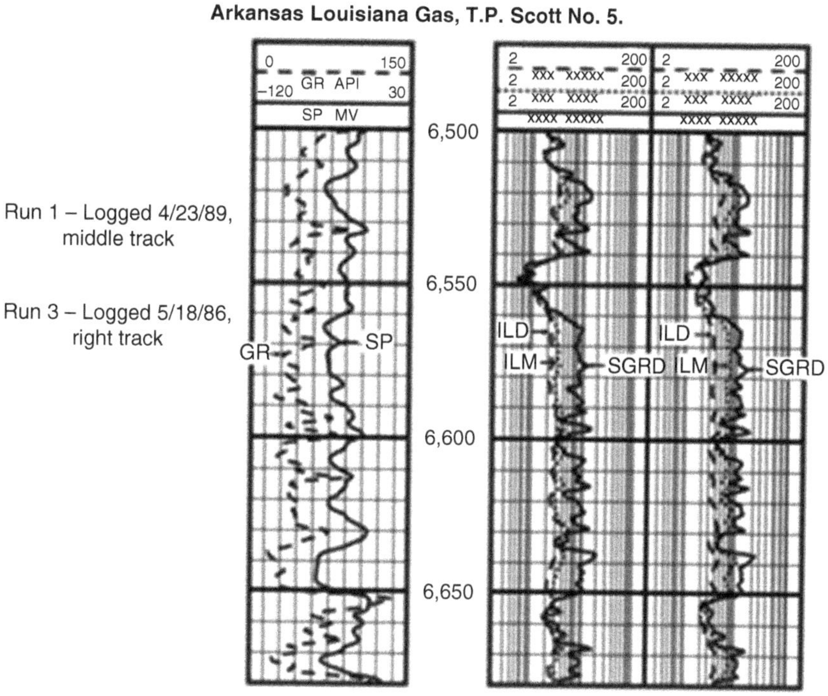

Fig. 5.53—Effects of mud-filtrate invasion on the induction log.

and gas, the freshwater mud filtrate will invade permeable zones, as shown in **Fig. 5.54.** The mud filtrate will invade continuously, but the mudcake seldom shuts off invasion completely. In the research work conducted for the Gas Research Institute (GRI), we have found that the salinity of the mud filtrate can vary on a daily basis. Thus, to properly evaluate resistivity logs, the value of the mud-filtrate resistivity, R_{mf}, should be measured daily.

Mud-filtrate invasion occurs in all permeable reservoirs. In high- to moderate-permeability reservoirs, invasion occurs, but once the mudcake seals the mud filtrate imbibes quickly away from the wellbore. As such, the effects of mud-filtrate invasion might not be noticed on many logs. However, in low-permeability reservoirs, mud-filtrate invasion during drilling can affect the results from both logs and DSTs. Invasion will occur, but because of the lower permeability, the filtrate will not imbibe into the formation for days or weeks and thus the mud filtrate is near the wellbore when the logs are run.

In the staged-field-experiment (SFE) Well 3, sponsored by the GRI, we were looking at the effects of mud-filtrate invasion using time-lapse logging. During the SFE for this well and other GRI wells, we ran logging suites over the same openhole intervals several times over several weeks. As part of that work, we measured the drilling-mud properties every day to give us the information we needed to interpret the logs. In a normal well, the logging company will measure the resistivity of the mud and mud filtrate immediately before logging the well. What we found was that it might be a good idea to measure these values more often. **Fig. 5.55** shows the change of resistivity as a function of time for this well.

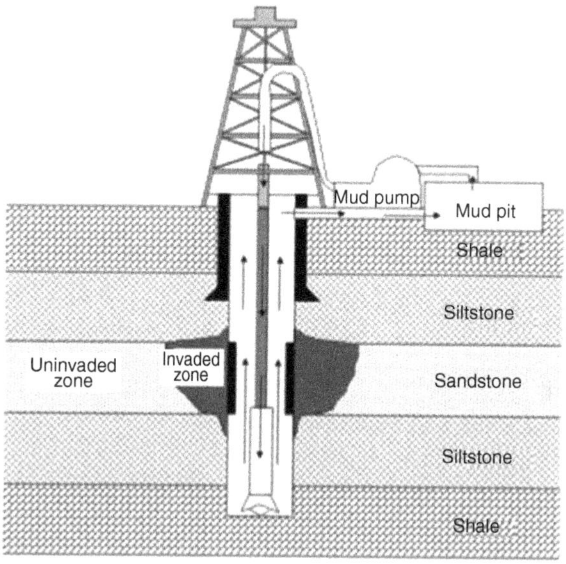

Fig. 5.54—Mud-filtrate invasion in permeable sandstone.

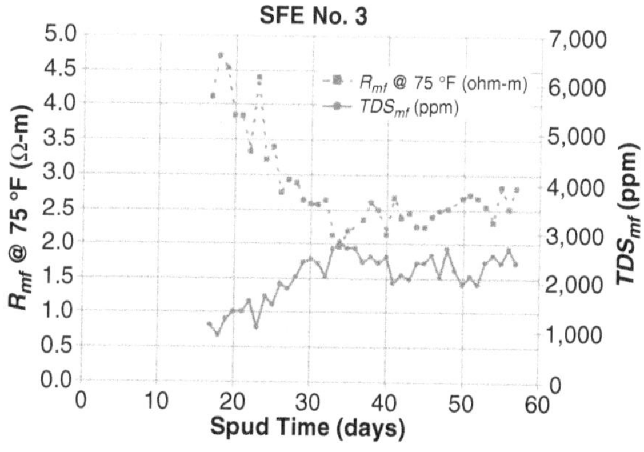

Fig. 5.55—Measurement of mud-filtrate resistivity vs. time for SFE Well 3.

After realizing that mud-filtrate invasion affects logs and DSTs, a series of research projects was sponsored by the GRI to investigate mud-filtrate invasion in more detail. These projects led to multiple PhD degree dissertations and SPE papers. In this section, we summarize some of this work, but all of the work can be found in the references to this chapter.

The factors that affect filtrate invasion are mudcake properties, reservoir pressure, mud weight, formation permeability, formation porosity, near gas/water permeability, and capillary pressure. The factors that affect the resistivity profile near the well are formation-water salinity, mud-filtrate salinity, and initial water saturation.

5.2.3.2 Effect of Mud-Filtrate Invasion on Prefracture Well Tests. One critical issue in evaluating tight gas reservoirs is to run well tests to characterize the formation before pumping a large hydraulic fracturing treatment. It is much easier to evaluate the well after a fracturing treatment if the formation has been properly characterized before the treatment. When freshwater mud invades a gas-bearing zone that contains salt water, the resistivity profiles surrounding the wellbore will be substantially altered. But the invasion also affects the permeability to gas, which can alter the behavior of gas flow during well tests.

Before fracture treating a tight gas well, several prefracture well tests should be run to properly evaluate the reservoir. Typical tests that can be run are a DST, a pressure-buildup test, an in-situ stress test, a step-down test, and a mini-fracture test.

5.2.3.3 Invasion Affects Drillstem Tests (DSTs). We have documented the effects of mud-filtrate invasion on the results of a DST in a tight gas sand. Canadian Hunter drilled dozens of wells in the tight gas sands in the Deep Basin in the 1970s and 1980s. The logs and cores showed many zones that should be gas productive, but the DSTs would barely flow gas. We studied the problem to find out what might be affecting the DST results.

Using reservoir simulation and data from Canadian Hunter, we found that mud-filtrate invasion will suppress initial gas-flow rates and result in calculated values of permeability that are too low. We found that, when initial gas-flow rates are extremely low, a closed-chamber DST should be run. Even with a closed-chamber test with extremely low flow rates, the permeability-thickness product (kh) and skin (s) can be calculated with confidence if the flow rates and pressures are accurately measured.

To model the effects of mud-filtrate invasion on the results from a DST in a tight gas reservoir, we used a finite-difference model to simulate the invasion, imbibition, and saturation profiles near the wellbore that develop during drilling. With the saturation profile, we could then determine the relative permeability to gas and water in the invaded zone and simulate how the DST would behave because of invasion.

Fig. 5.56 shows what the invasion profile looks like for a 12-hour imbibition period and a 7-day imbibition period, as well as what the profiles look like after producing the well for 90 days. The calculations were made for the following data, which simulate the conditions in the Canadian Hunter reservoir:

- Reservoir pressure, 2,275 psi
- Water saturation, 45%
- Total porosity, 7%
- Reservoir permeability, 0.1 md

Fig. 5.57 illustrates what the gas-flow rates would be for the following three cases:

- No filtrate invasion
- Twelve hours of invasion
- Seven days of invasion before initiating flow

For the case simulated, it is clear that mud-filtrate invasion can substantially reduce the gas-flow rates at early times, especially during DSTs.

Table 5.4 shows the results of 10 DSTs in the Falher Formation in Canada. In all cases, the DST flow rates were low. The operator set pipe on all of these wells because of a favorable log analysis. After stimulation, all 10 wells produced at commercial gas-flow rates. The conclusion is that all data should be carefully analyzed and the effects of mud-filtrate invasion on logs, cores, and DSTs should be determined when evaluating tight gas reservoirs. In general, if the DST produced any measurable gas, it would probably produce at commercial flow rates after a large fracture treatment. Also, if the logs indicate the formation is porous, permeable, and gas-bearing,

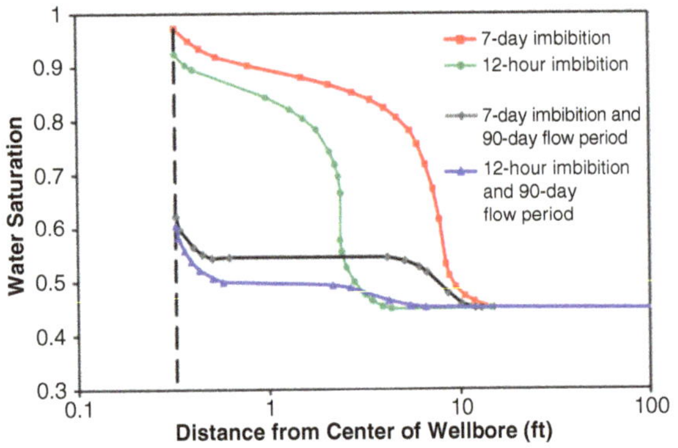

Fig. 5.56—Calculated saturation profiles at various times around the wellbore.

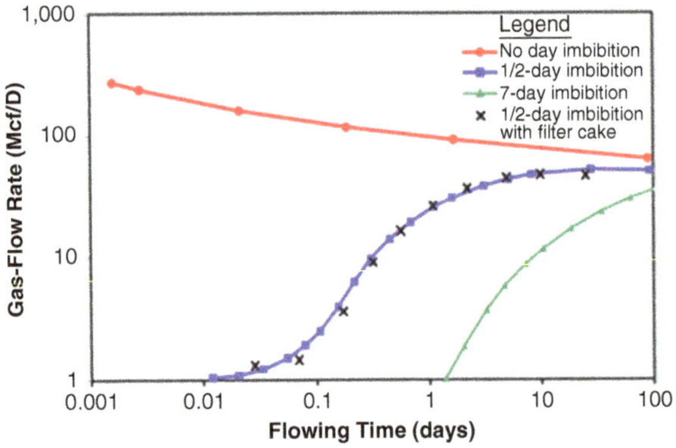

Fig. 5.57—Calculated gas-flow rates over time for no mud-filtrate invasion and substantial mud-filtrate invasion.

Mud-Filtrate Invasion Affects DSTs

Well	DST Flow Rate (Mcf/D)	Interval (m)	Post Completion Rate, (Mcf/D)
Well 1	3.5	2,525 to 2,542	784
Well 2	0.5	2,335 to 2,353	1,100
Well 3	9.25	2,842 to 2,854	675
Well 4	10.0	2,325 to 2,354	1,075
Well 5	25.0	2,435 to 2,450	1,680
Well 6	11.5	2,500 to 2,526	1,437
Well 7	10.0	2,209 to 2,219	2,125
Well 8	100.0	1,843 to 1,865	4, 040
Well 9	TSTM*	1,820 to 1,832	1,100
Well 10	6.0	2,835 to 2,863	593

Table 5.4—Results from 10 wells in the Falher Formation showing DST gas-flow rates and gas-flow rates after a fracturing treatment for the same zones.

DST results should not be considered if those results are caused by mud-filtrate invasion. The mud-filtrate invasion should be easy to identify from the logging data.

On the basis of our analyses and experience, a DST is normally not recommended if the reservoir is known beforehand to be 1.0 md or less. Instead, the decision to complete the well should be made using log and/or core data; then, reservoir properties can be calculated by running casedhole pressure-buildup tests.

*5.2.3.4 **Effects of Mud-Filtrate Invasion on Openhole Logs.*** After we performed the work on DSTs for Canadian Hunter, we concluded that, if mud-filtrate invasion can be so severe in tight sands, it must be affecting the well logs. We began a series of research projects at Texas A&M University, sponsored by the GRI, to look at the effects of mud-filtrate invasion on openhole well logs. We discovered that, if the formation permeability is very low (<< 0.001 md), not much filtrate invades the formation. If the permeability is large (several microdarcies), a large amount of filtrate invades the formation, but it imbibes away from the borehole quickly and does not affect the logs much. But if the permeability is between 0.001 and 1.0 md, then there is substantial invasion, it does not imbibe quickly, and it will affect the readings from the openhole logs.

To illustrate how the mud-filtrate invasion profile affects the resistivity logging readings, we used a finite difference model to simulate a typical invasion profile. Then, knowing the saturation distribution, we could calculate the salinity profiles and how they change with time. The saturation and salinity profiles can be used along with radial geometric functions published by the logging service companies to predict what the resistivity logs would read as a function of time. For these calculations, we used the data in **Table 5.5.**

Mud filtrate in the reservoir is simulated using a finite-difference model to investigate the effect of mud filtrate on the log analysis as function of time, and **Figs. 5.58 through 5.63** show the results.

The salinity of the mud filtrate is less than the salinity of the formation. Therefore, mud filtrate reduces the salinity of formation water in the invaded zone with time. As the saturation and salinity profiles change with time and distance around the wellbore, the formation resistivity profile will also change.

Formation Flow Parameters	
Original reservoir pressure, psia	2,275
Original water saturation, fraction	0.475
Formation temperature, °F	100
Formation-water salinity, ppm	100,000
Mud-filtrate salinity, ppm	2,000
Porosity, fraction	0.070
Permeability, md	0.100
Wellbore radius, ft	0.300
Wellbore pressure, psia	2,950
Permeability of the Mud Filter Cake, md	
For first 24 hours	0.001
For after 24 hours	0.00001
Formation Electrical Properties	
a, constant in Archie equation	1.00
m, cementation exponent	2.00
n, saturation exponent	2.00
Formation-water resistivity, Ω·m	0.0707
Mud-filtrate resistivity, Ω·m	2.5800
True formation resistivity, Ω·m	64.000

Table 5.5—Base-case input data for simulating the effects of mud-filtrate invasion on the readings from a resistivity log.

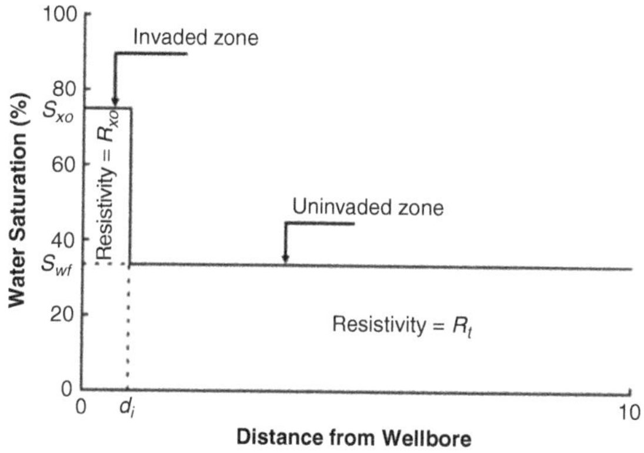

Fig. 5.58—Ideal invasion profile used by the logging service companies to construct their resistivity log correction (tornado) charts.

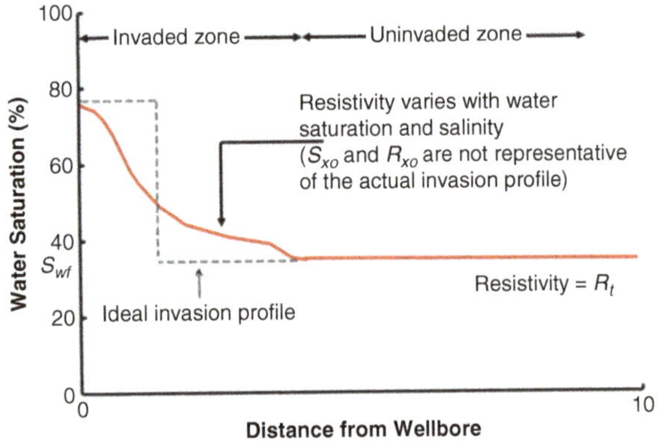

Fig. 5.59—A more typical invasion profile in a low-permeability gas reservoir.

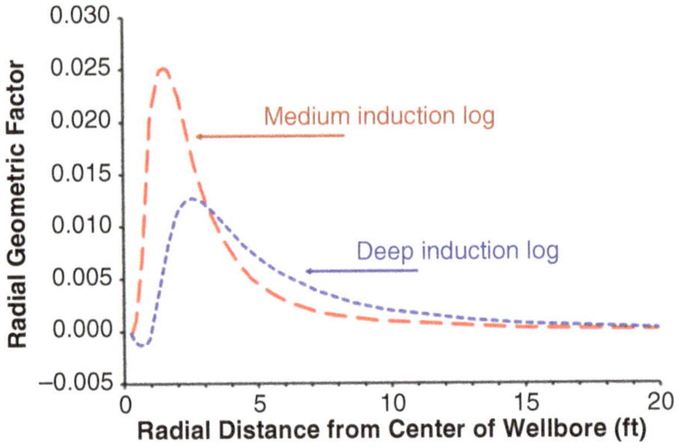

Fig. 5.60—Radial geometric factors for deep and medium induction logs.

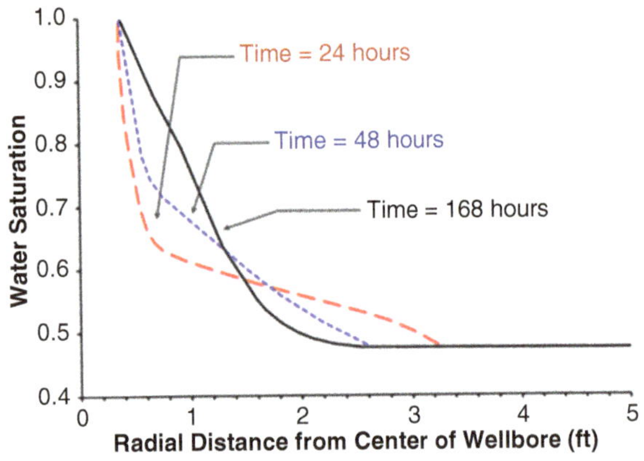

Fig. 5.61—Mud filtration is a dynamic process, and water saturation increases near the wellbore with time.

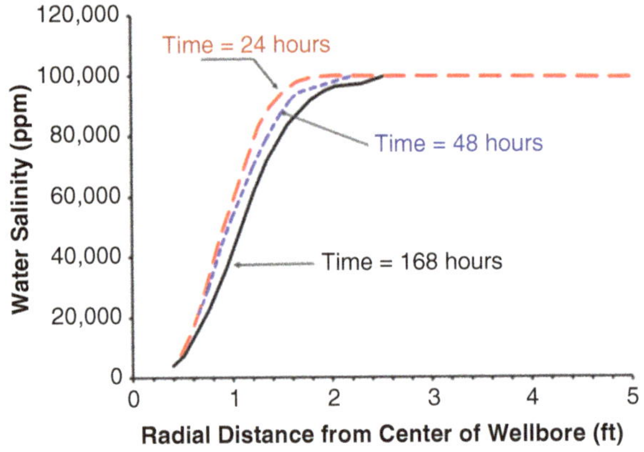

Fig. 5.62—Water-salinity profile during mud-filtrate invasion. The changes occur with time.

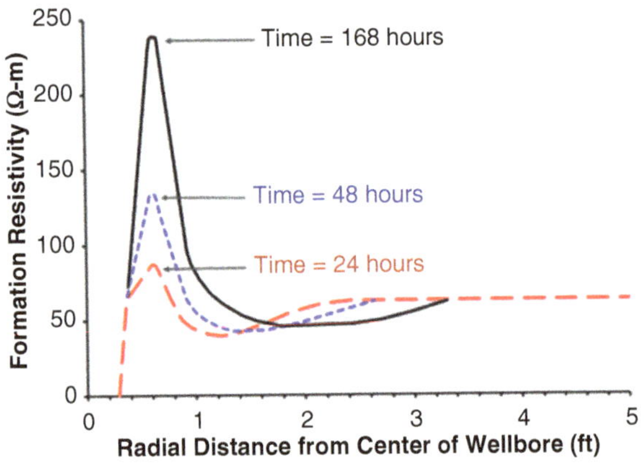

Fig. 5.63—Resistivity profiles during mud-filtrate invasion. This change occurs with time and distance from the borehole.

It must be emphasized that this is a dynamic process, so the log readings obtained depend on when the logging tool is run.

Fig. 5.63 shows how the resistivity profile changes with time and distance. If the zone is logged a short time after the bit penetrates the formation, the resistivity values will differ from those recorded if the logging tool is run later after more invasion occurs. Because the low-salinity, high-resistivity mud filtrate is displacing the high-salinity, low-resistivity formation water (in this example), a low-resistivity water bank moves out into the reservoir radially from the wellbore. We often see this effect on the triple-combo log and the AIT.

Formation resistivity readings are a function of three parameters: tool geometrical reading, saturation of water, and salinity of water. For this simulated case, although the water saturation increases with time, the formation resistivity increases with time because the salinity of the water decreases with time, and the water decreases in salinity because of the invasion of fresh water that displaces the salt water A bank or region of lower resistivity is created in front of the mud-filtrate invasion, where the salinity of the water is greater than the original salinity of the formation water. Therefore, the resistivity is less than the original resistivity of the formation.

To determine which of these resistivity profiles in Fig. 5.63 affects the reading from the resistivity logs, we can calculate what the medium and deep induction logs would read as a function of time for the case simulated in this work.

The resistivity profiles in Fig. 5.63 and the radial geometric factors in Fig. 5.60 can be used to calculate what the deep and medium induction logs should read as a function of time. Those computations are shown in **Fig. 5.64.** Notice there is a crossover between 48 and 72 hours where the medium induction log is more resistive than the deep induction log. Again, these readings from the logging tool are being affected by the saturation and salinity changes that are occurring radially from the wellbore as a function of time. We found that this crossover point can be correlated with formation permeability. If the value of the resistivity deep-induction log (RDIL) is divided by the value of the resistivity medium-induction log (RMIL), a unique-shaped curve, which will be a function of formation permeability when all other factors are held constant, will be obtained. **Fig. 5.65** shows this relationship.

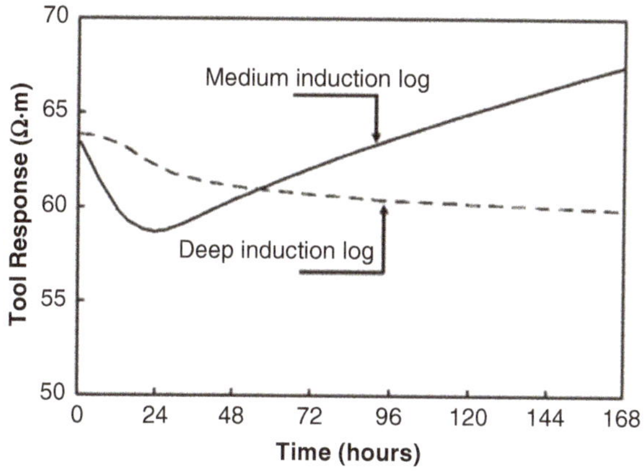

Fig. 5.64—Induction log tool response with time and distance from the borehole.

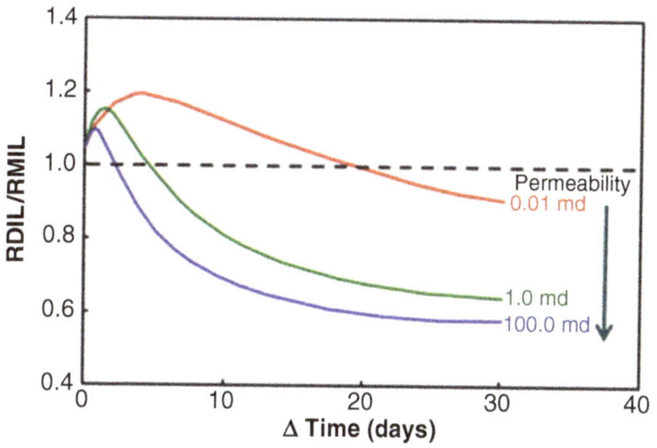

Fig. 5.65—The ratio of deep to medium resistivity changes with time and can be an indicator of the formation permeability.

The ratio of RDIL/RMIL turns out to be an indicator of the reservoir permeability if all other reservoir properties are known. As shown in Fig. 5.65, the time for the resistivity ratio to reach a value of unity can be correlated to the higher value of the formation permeability.

We have found that the time required to reach the crossover from RDIL/RMIL > 1.0 to < 1.0 can be correlated to formation permeability. In the Travis Peak Formation, we have also developed a correlation between the RDIL/RMIL ratio from actual logs to the absolute permeability of individual sands (see **Fig. 5.66**).

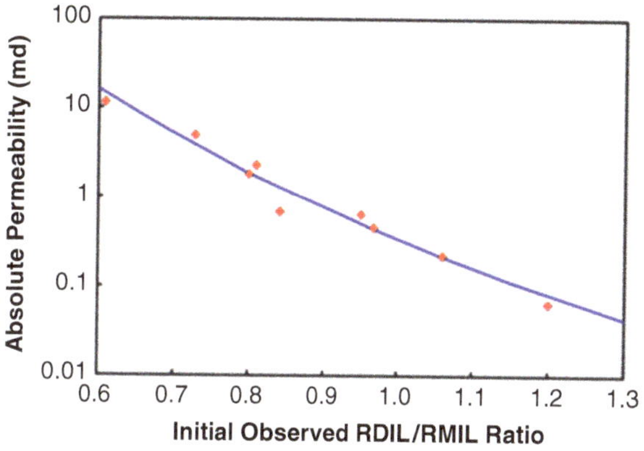

Fig. 5.66—Correlation of permeability to the initial ratio of RDIL/RMIL for the Travis Peak Formation in East Texas. The curve in the figure history matches the result.

In the Travis Peak Formation, we have also developed a correlation between the RDIL/RMIL ratio from actual logs to the absolute permeability of individual sands. Knowing the resistivity profiles and radial geometric factors, deep and medium induction logs can be calculated. By dividing the value of RDIL by RMIL, a unique-shaped curve is obtained that is a function of formation permeability when all other factors are held constant.

To clearly illustrate that mud-filtrate invasion does affect resistivity logs in real wells and not just in our finite-difference model, data are presented from four logging runs in SFE Well 3 (see **Table 5.6** and **Fig. 5.67**).

The log data in Fig. 5.67 are from four logging runs in SFE Well 3 that were run during September, October, and November 1988. The first three logs were run in three consecutive weeks. The last log was run two weeks after the fifth run.

The log and field data illustrate that mud-filtrate invasion does affect resistivity logs with time. Running several logs at different times can help determine permeable formations and estimate the reservoir properties. In the areas where the resistivity changes with time, the formation is permeable, and the height of the permeable formation can be determined. Also, simulation runs can be performed to determine the formation properties.

Of course, most operators will not run multiple logging trips for the sake of science. The GRI wells were drilled to gather data to learn how to characterize the shape and size of a hydraulic fracture in real time, so they were science wells. However, in new areas, where the formations are not well-characterized, an operator might consider time-lapse logging as another tool to characterize the reservoir.

Mud-filtrate invasion also affects other logs that rely on the fluid properties in the pore space. For example, the CNL is really measuring the hydrogen content in the formation. If filtrate, with high hydrogen density, displaces natural gas, with low hydrogen density, then the readings from the CNL will be affected by mud-filtrate invasion. Similar issues will affect the FDL and SON, to a certain degree. **Fig. 5.68** shows that filtrate invasion in gas reservoirs will also affect porosity logs, such as the CNL.

Date	29-Sep 1988	05-Oct 1988	14-Oct 1988	02-Nov 1988
Run No.	Three	Four	Five	Six
Depth – Driller (ft)	7,351.0	7,868.0	9,017.0	9,700.0
Depth – Logger (ft)	7,354.0	7,868.0	9,018.0	9,690.0
Bottom Logged Interval (ft)	7,352.0	7,866.0	9,016.0	9,688.0
Top Logged Interval (ft)	6,310.0	6,300.0	6,310.0	4,600.0
R_m (Ω·m)	1.15 @ 86°F	2.4 @ 74°F	1.82 @ 56°F	1.9 @ 75°F
R_{mf} (Ω·m)	0.86 @ 89°F	1.3 @ 74°F	1.49 @ 56°F	1.4 @ 75°F
R_{mc} (Ω·m)	1.44 @ °F	3.0 @ 74°F	2.16 @ 56°F	2.3 @ 75°F

Table 5.6—Comparison of deep induction (four logging runs) measurement while drilling resistivity in the SFE Well 3.

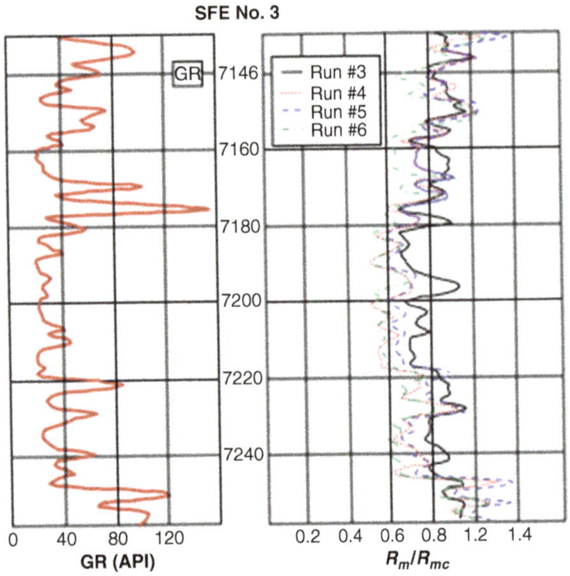

Fig. 5.67—Results from four logging runs in SFE Well 3.

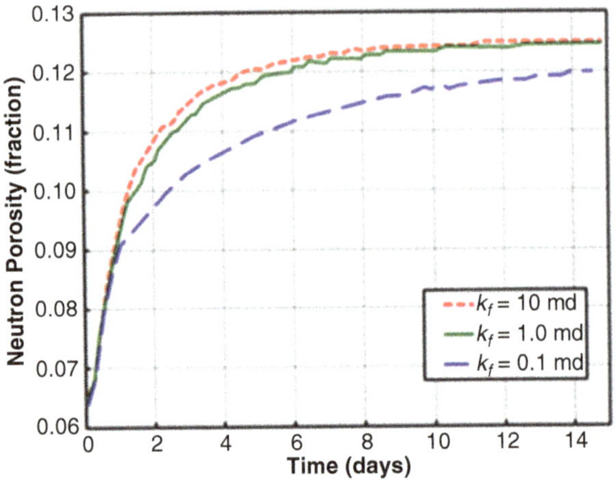

Fig. 5.68—Effect of mud-filtrate invasion on the neutron porosity log.

The effect of mud-filtrate invasion on resistivity logs as a function of time is more significant when compared to the effect of filtrate invasion on CNLs. Therefore, the use of some type of resistivity log is recommended for determining reservoir properties.

5.2.3.5 Summary. Mud-filtrate invasion affects the logs and DSTs most when the formation permeability is between 0.001 and 1.0 md. The following helps to explain that there is an optimal permeability range for seeing the effects of mud-filtrate invasion on the logs.

- For k << 0.001 md, not much filtrate imbibes.
- For k >> 1.0 md, filtrate does invade the formation but imbibes into the formation before the log can be run.
- If k is between 0.001 and 1.0 md, the filtrate stalls near the wellbore and affects the logs and DSTs.

Mud-filtrate invasion is a dynamic process and changes the resistivity of the formation, measured by logging, with time. The ratio of medium and deep resistivity readings can be used to determine the permeability of the reservoir if other reservoir parameters are known. In tight gas formations, decisions should be made using log and core data, not using DSTs.

5.3 Coring and Core Analysis

5.3.1 Coring Techniques.

5.3.1.1 Need for Core. Obtaining and analyzing core are crucial to the proper understanding of any unconventional gas reservoir system. To understand these low-permeability reservoirs, modern core analysis methods are required. Of primary importance is the measurement of rock properties under restored-state conditions: simulated reservoir pressure and in-situ stress.

Before the development of tight gas reservoirs, when conventional, high-permeability reservoirs were the only reservoirs that could be developed commercially, the geologist and engineer would core the "pay interval" and throw any shale or silt core into the mud pit. However, because we are now developing these low-quality reservoirs, it is important to cut the core, preserve the core in the field, transport the core properly, and run modern core analyses to obtain the information required to optimize development.

In general, core should be cut above, within, and below the producing portion of the gas reservoir. Core should be cut to obtain samples of all important lithologies, including clean sandstones, clean shales, limestones, siltstones, mudstones, and shaly sands. Information on all the rock types will be useful for designing the completion and stimulation treatments.

Core should be handled properly in the field and preserved using heat-shrinking wrap or other methods before being shipped quickly to the core analysis laboratory, where it will be described and the routine core analysis processes to gather data for designing the completion will begin.

For example, during the design phase of a hydraulic fracturing treatment, a large volume of data and information is required to characterize the formation layers above, within, and below the pay zone. Obtaining and analyzing core is crucial to the proper understanding of any complex reservoir system, such as is often found in tight gas reservoirs.

Most measurements should be made under restored-state conditions to simulate the condition of the rock in situ. The effect of NOB pressure must be reproduced in the laboratory to obtain the most-accurate quantitative information from the cores.

Core testing is an important tool that provides information necessary for proper reservoir description. **Table 5.7** shows the information, obtained from cores and other sources, that is needed to design a well completion.

In addition to the design of the well completion, core data are useful for the design of a fracture treatment. The core data reduce uncertainty in reservoir evaluation and give the engineer a better description of the lithology. Also, knowing the true values of permeability, porosity, and water saturation is crucial to computing hydrocarbons-in-place, where the pay zone exits, and inputs into the fracture treatment design. Additionally, values of rock mechanical properties, such as Poisson's ratio and Young's modulus, are crucial to the design of a fracture treatment.

Sources of Data

Data	Units	Sources
Formation permeability	md	Cores, well tests, correlations, production data
Formation porosity	%	Cores, logs
Formation modulus	psi	Cores, logs, correlations
Formation compressibility	psi	Cores, logs, correlations
Poisson's ratio		Cores, logs, correlations
Fracture toughness	psi-$\sqrt{\text{in}}$.	Cores, correlations
Water saturation	%	Logs, cores
Net-pay thickness		Logs, cores
Gross-pay thickness	ft	Logs, cores, drilling records
Formation lithology		Cores, drilling records, logs, geologists

Table 5.7—Logs and cores can be used to obtain useful data.

Core data provide only a limited sampling in relation to reservoir size, so core data must be correlated with log and production data. Core data represent a few cubic inches of rock, log data provide information on a few cubic feet of rock, and production data provide information on a much larger volume of the reservoir rock.

Pressure-drawdown and -buildup testing will be more representative of true reservoir permeability on a large scale. Pressure transient testing is the preferred method for determining total permeability values in most reservoirs. However, permeability can be estimated from logs and measured from core. As will be discussed later in this book, the preferred method is to develop correlations of reservoir properties that can be measured or estimated using core, logs, and reservoir testing.

*5.3.1.2 **Types of Core.*** The engineer and geologist must determine where to core and how much core is needed for a particular well or play. If the well is an early delineation well, then the whole core should be considered to learn as much as is feasible about the formation that is to be developed. As discussed earlier, the core can be the best source of data to determine the environment of deposition, which helps delineate the type and size of the reservoir under development. The engineer also must obtain data on what layers might be barriers to fracture growth.

Conventional coring will allow a larger-diameter core to be obtained, but it will be the most expensive. However, the data from the core can be used to plan future development wells, so the actual cost of obtaining the whole core should be allocated, to some degree, to the cost of future wells that will benefit from the information. **Fig. 5.69** shows the standard core barrel.

Wireline coring will allow core over long intervals to be obtained without having to trip the drill string because the core barrel is retrieved by wireline. However, the diameter of the wireline core might not be as large as could be obtained using a conventional core barrel. **Fig. 5.70** shows a wireline coring system. **Table 5.8** shows the specifications for various types of inner barrels that can be used while coring.

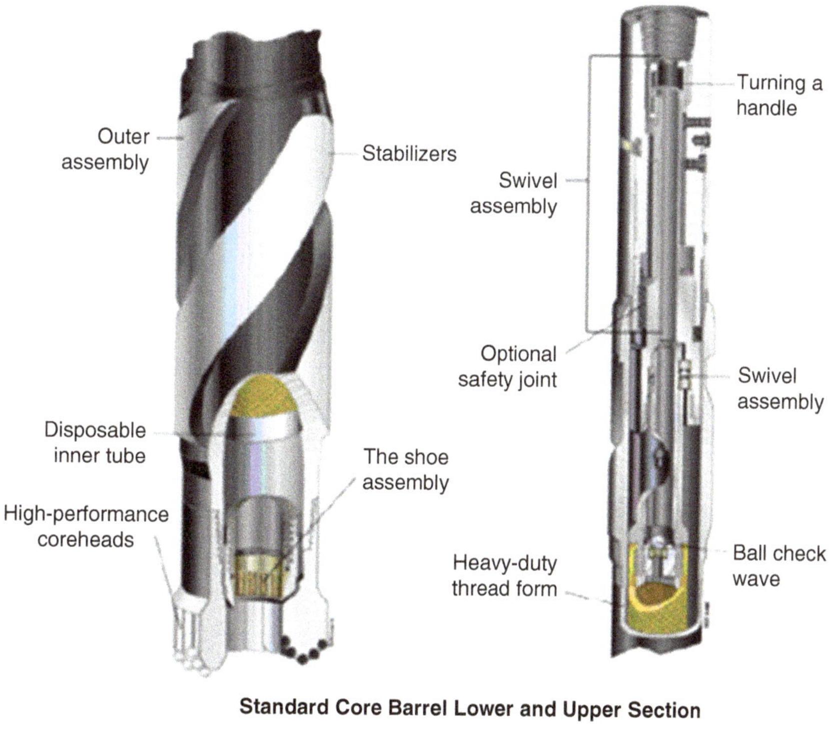

Fig. 5.69—A standard core barrel (Courtesy of Halliburton).

Inner barrel	Core Length	Special Features
Mild steel	30–120 ft	Ready-made core preservation system. High-temperature applications
Mild steel	1.5 ft	Designed for short-radius coring
High-strength steel	120–400 ft+	Stronger barrel, includes additional inner and outer core barrel stabilization
Fiberglass	30–90 ft	Ready-made core preservation system. Used for consolidated and unconsolidated formations. Maximum operating temperatures: normal resin 250°F, high-temperature resin 350°F
Aluminum	30–90 ft	Ready-made core preservation system. High-temperature applications, maximum 350°F
Steel with plastic liner	30 ft	Ready-made core preservation system. Maximum temperature of 180°F. Reduces core diameter by ½ in.
Steel with a fiberglass liner	30 ft	Ready-made core preservation system. Maximum temperature of 250°F. Reduces core diameter by ½ in.
Steel with an aluminum liner	30 ft	Ready-made core preservation system. Maximum temperature of 350°F. Reduces core diameter by ½ in.

Table 5.8—Properties of inner core barrels.

Sidewall coring **(Fig. 5.71)** can be less expensive if all goes as planned, but you will obtain only small core plugs. You can retrieve sidewall cores using either a percussion tool (which can cause some fracturing of the core plug) or a tool that retrieves core plugs with a drill. The drilled core plugs usually preserve the porosity and permeability better than the percussion core plugs. However, either type of core plug can give the engineer and the geologist useful information on lithology.

As mentioned above, there is also a tool that will drill sidewall cores using an openhole wireline logging unit. Sidewall coring with a drilling tool **(Fig 5.72)** will obtain small-diameter cores, but these cores can be used to accurately determine lithology and can be a cost-effective means for determining Young's modulus. It is even possible to perform oriented sidewall coring.

The percussion sidewall coring system **(Fig. 5.73)** can acquire up to 116 formation samples in one logging run. Precise depth control for log analysis and formation characterization is possible, and these plugs are also useful for determining lithology. However, the plugs are not so useful for determining permeability, because the percussion tends to compact or fracture the rock. For high-permeability formations, the percussion could compact the rock and reduce the permeability. For low-permeability formations, the percussion tends to fracture the rock, resulting in higher permeability. However, if the engineer knows this, corrections can be made for such effects.

5.3.1.3 Core Preservation. Preserving the core in the field is the first step to obtaining useful information about the formation. Preservation of a core sample correctly in the field affects the quality of data obtained in the laboratory. Core-handling and preservation procedures should be designed to maintain both the fluid in the core sample and the distribution of that fluid. This requires minimizing the time the core is exposed to surface conditions before sealing the core for transport.

In past years, the core would be laid out on the catwalk at the rig and described by the geologist before boxing the core and sending it to the laboratory. This is no longer the preferred procedure. The core should be sealed quickly, packaged for shipping, and properly marked to be sure the core can be correlated with depth and the openhole logs.

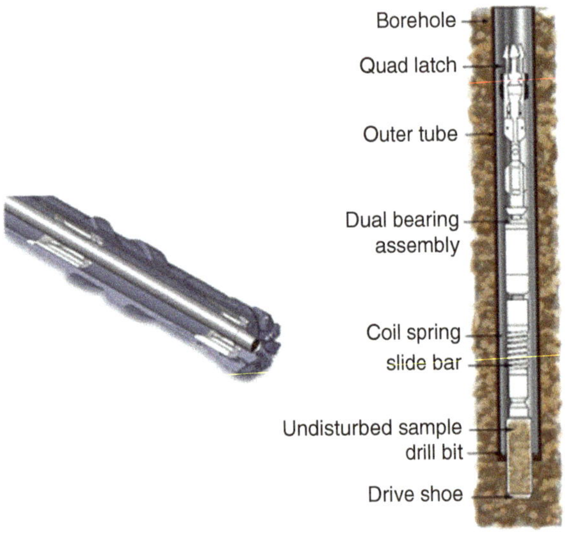

Fig. 5.70—A wireline coring system.

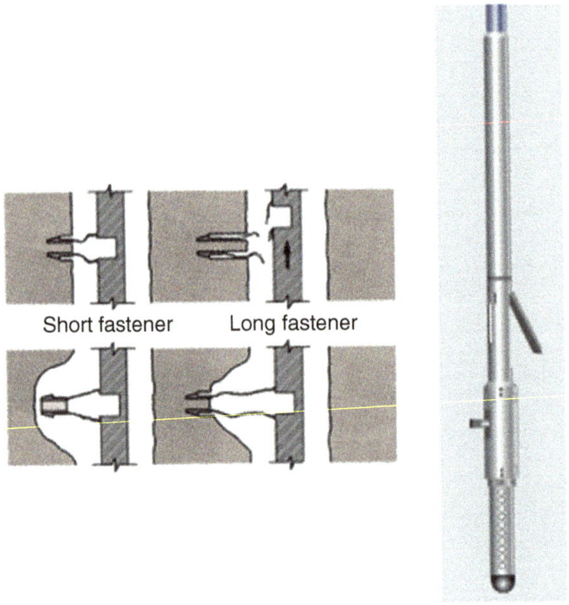

Fig. 5.71—A sidewall coring system.

Fig. 5.72—A sidewall coring system that drills the core plugs (Courtesy of Halliburton).

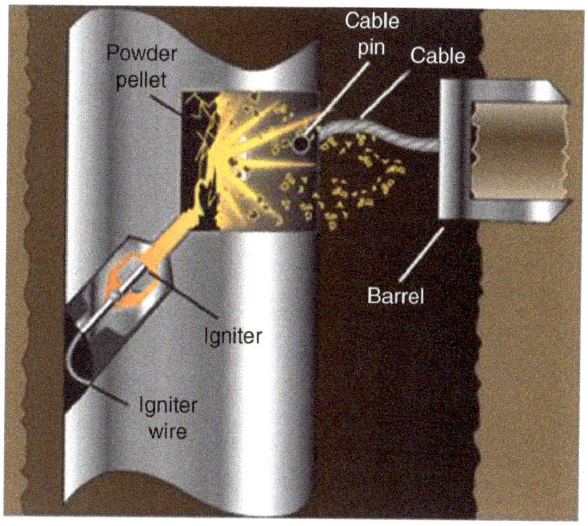

Fig. 5.73—A percussion sidewall coring system (Courtesy of Halliburton).

5.3.2 Routine Core Analyses. Obtaining and analyzing core is crucial to properly understanding any complex reservoir system. To properly understand low-permeability reservoirs, modern core analysis methods are required. Of primary importance is to measure rock properties under restored-state conditions.

In general, core should be cut above, within, and below the producing portion of the gas reservoir. The core in the potential pay zone will provide data on gas-in-place, deliverability calculations, and rock mechanical properties. Core in the layers above and below the pay zone are required to model hydraulic fracture growth and determine the optimal fracture treatment design.

Using logs from other wells in the general area, the geologist should identify the various rock types that are expected to be in or near the pay zone, then core should be cut to obtain samples of all important lithologies, including clean sandstones, clean shales, limestones, siltstones, mudstones, and shaly sands. The core analyses of these various lithologic intervals can eventually be correlated with log, well-test, and production data to establish correlations of important properties by rock type.

Cores should be handled properly in the field and preserved using heat-shrinking wrap or other methods before being shipped to the core laboratory. Once at the laboratory, routine core analyses should be run on every foot of core and should include the following, at a minimum:

- Porosity
- Grain density
- Unstressed permeability to air
- Stressed permeability to air
- Cation-exchange capacity
- Saturation analysis

As mentioned, the routine core analyses should be run on every foot of core and done as quickly as possible to provide data for the completion engineer to design the completion and fracture treatment and for the reservoir engineer to estimate flow rates and the ultimate recovery expected from the well. Then, the core should be analyzed to determine the different rock types and/or flow units that were sampled by the core. This can be done by a geologist describing the core and using information from openhole logs that show the properties of the various rock layers.

Once the different rock types (e.g., shale, silt, sand) are identified, then five to ten core plugs should be cut in every rock type and/or flow unit to be sent off for special core analyses. The special core analyses take longer than the routine core analyses to obtain reliable results, but special core analyses are required to correlate in-situ rock properties with log and production data and to better determine the rock properties in the reservoir vs. the properties at the surface.

Special core analyses should be run on four to five core plugs for each lithology identified as a separate rock type. The tests that need to be run include:

- Permeability vs. NOB pressure
- Permeability vs. water saturation (relative permeability)
- Resistivity index
- Formation factor vs. NOB pressure
- Capillary pressure
- Porosity vs. NOB pressure
- Acoustic velocity
- Mechanical and elastic properties

More discussion on the importance of special core analyses will be presented in Section 5.3.3. Here, we discuss measurements that are considered part of the routine core analyses for tight gas sandstones.

5.3.2.1 Porosity. The measurement of porosity in a tight gas reservoir is not straightforward. A rock is considered tight when the permeability is low, usually in the microdarcy range. The reason the permeability is so low is that the porosity is very low, usually less than 10%. As such, standard laboratory equipment and procedures might have trouble obtaining correct values of porosity, especially if the porosity is less than approximately 5%. Consequently, the laboratory technician should be diligent and make sure the equipment is working, the equipment is properly calibrated, and procedures are followed precisely.

Using a 1.5-in.-diameter core will provide more-accurate estimates of porosity than using smaller-diameter cores. The result should be compared with other core properties, such as grain density and permeability. Eventually, the effect of NOB must be reproduced in the laboratory to obtain the most accurate information from the cores, including the porosity estimate.

The effect of NOB pressure must be reproduced in the laboratory to obtain the most accurate information from the cores. The data in **Fig. 5.74** show values of porosity measured at NOB pressure vs. values measured at ambient conditions for a Travis Peak data set. In all cases, the values at NOB pressure will be lower than those measured in the unstressed condition.

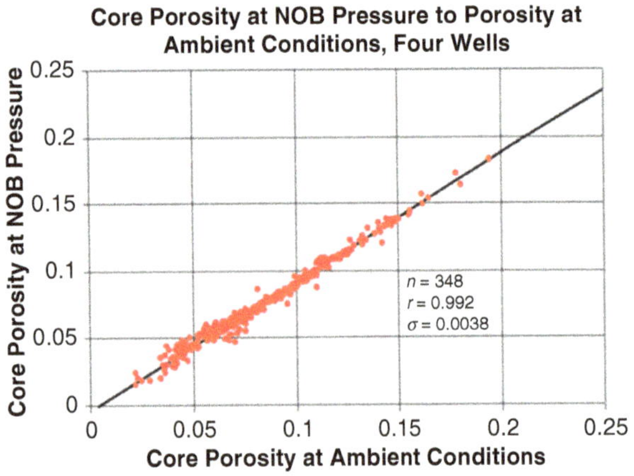

Fig. 5.74—The effect of NOB pressure on porosity measurements for four wells.

5.3.2.2 Unstressed Permeability. Just like porosity, the measurement of permeability in a tight gas core is difficult in the laboratory—perhaps even more difficult than measuring porosity. One aspect of proper laboratory procedures is how the core is handled before any measurements. To measure the true permeability of the rock in a gas reservoir, it must have zero water saturation and be free of any oil or condensate. It should be filled with gas only—air or nitrogen. To obtain zero free-water saturation, the core must be dried. The drying procedure in the laboratory can influence the accuracy of measured permeability. Some laboratories use humidity-controlled ovens to remove the free water and leave the clay-bound water.

If the core is not put under expected in-situ overburden stress, the laboratory measurements of permeability will be larger than expected in the reservoir under overburden stress. One method that will be discussed in Section 5.3.3 is the use of pulse testing across the core to better characterize the permeability. Properties of the rock matrix and fractures can be simultaneously measured in a naturally fractured, low-permeability core using pulse testing.

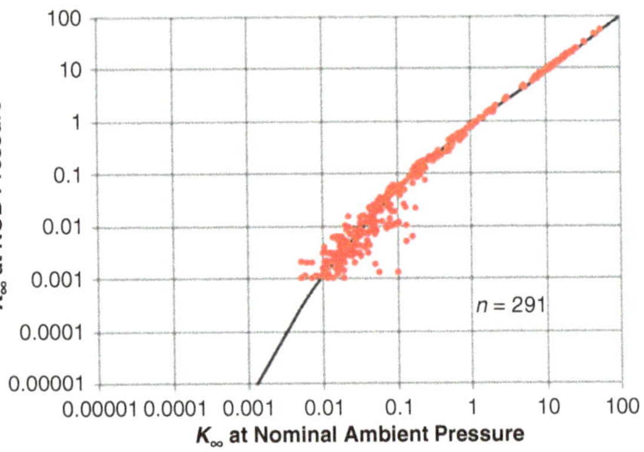

Fig. 5.75—Effect of NOB pressure on permeability in a set of Travis Peak core plugs for Howall Well 5 and SFE Well 2 (invalid data excluded).

5.3.2.3 Stressed Permeability. Permeability is largely controlled by the size of the pores and the pore throats. The distribution of the pores and pore throats, the geometric shape of the pores, and the tortuosity of the pore system will affect the permeability. Because the pore throats are small and permeability is a strong function of the size of the pore throats, the effect of the overburden on the permeability is significant. This is especially true if the pore throats contain quartz overgrowths and slot pores, as discussed in Section 5.1.

In addition, as a reservoir is produced, there is an increase in net effective overburden stress, which tends to compact the rock even more. Some rock types might lose 10% of original permeability, while others could lose 90% or more as the reservoir is depleted. The amount of permeability reduction is controlled by the amount of reduction in the pore distribution. However, with core, this effect can be measured in the laboratory.

The values of permeability measured at NOB pressure are substantially lower than values measured at ambient conditions, for the same Travis Peak data set as shown in **Fig. 5.75.** As mentioned above, the degrees of stress sensitivity will be a function of the lithology and pore throat size distributions. Rocks with slot pores will be more stress sensitive than rocks with more round pore throats. Slot pores are created by quartz overgrowths during diagenesis, as discussed in Section 5.1.

Fig. 5.75 shows the effect of NOB pressure on the measurement of permeability in dry cores.

In Section 5.4.2 of this chapter, we will discuss the methods for correlating various data sets to determine formation permeability. For the data set used to generate Figs. 5.68 and 5.69, we have also developed a correlation between porosity and permeability from these Travis Peak core plugs. This correlation is shown in **Fig. 5.76.**

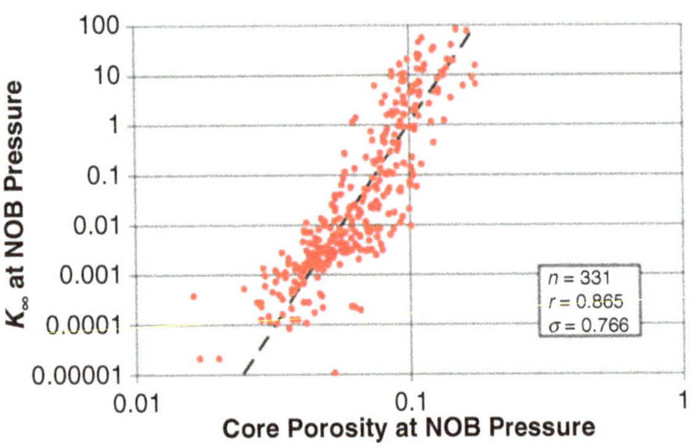

Fig. 5.76—Correlation of porosity and permeability at NOB pressure using core from the Travis Peak Formation for fluvial/deltaic channel sand (invalid data excluded).

Even though the correlation in Fig. 5.76 is a reasonable correlation, the correlation can be improved by further subdividing the data by flow units. The discussion of flow units is presented in a later part of this book.

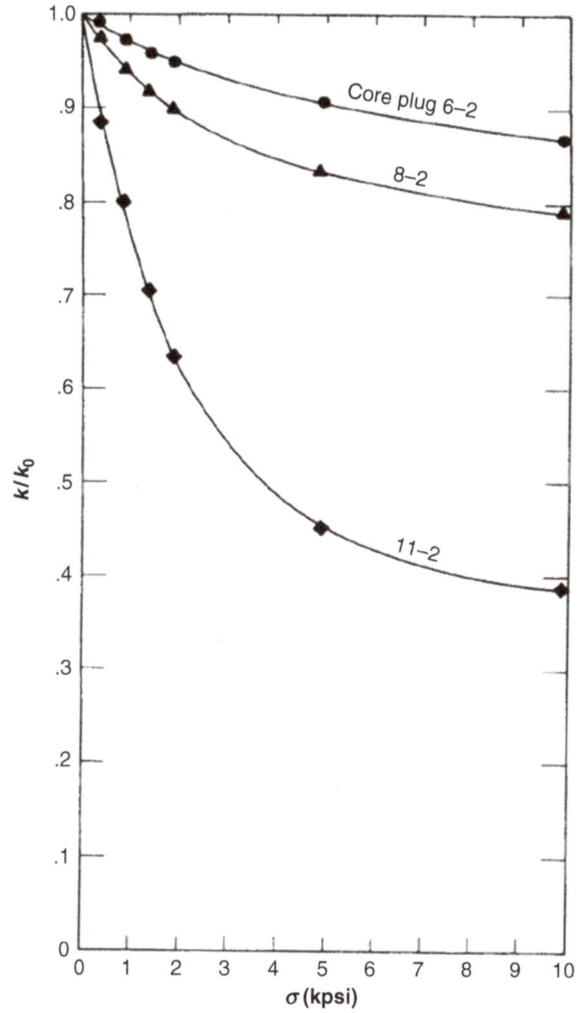

Fig. 5.77—Permeability vs. net stress for three typical core plugs in a tight gas sand reservoir.

Compound	Formula	Grain Density (g/cm^3)
Quartz	SiO_2	2.65
Calcite	$CaCO_3$	2.71
Dolomite	$CaMg(CO_3)_2$	2.84
Sylvite	KCl	1.98
Halite	NaCl	2.16
Gypsum	$CaSO_4\ 2H_2O$	2.32
Anhydrite	$CaSO_4$	2.96
Pyrite	FeS_2	5.01
Barite	$BaSO_4$	4.48
Siderite	$FeCO_3$	3.96

Table 5.9—The grain density of minerals that are typically found in a tight gas sand.

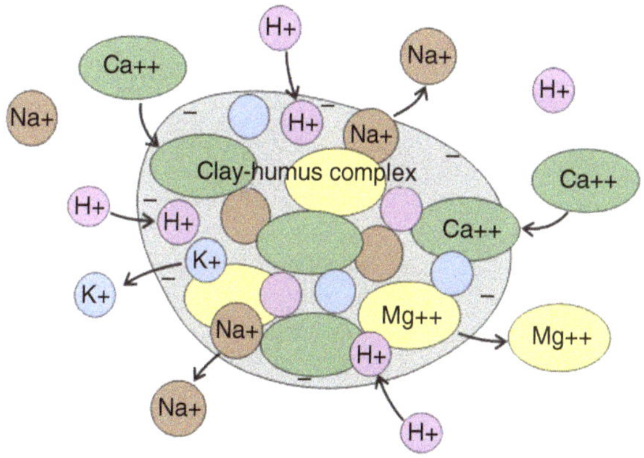

Fig. 5.78—Schematic showing the ions that could be part of a typical clay particle.

*5.3.2.4 **Permeability vs. Net Stress.*** For any rock, the porosity and permeability will decrease as the stress on the rock increases. In high-permeability cores, the reduction in permeability is not significant. However, as previously discussed, in low-permeability cores, with small pores and pore throats, the permeability reduction because of increasing stress can be substantial. **Fig. 5.77** illustrates the results or measurements on three core plugs. The data have been normalized to present the ratio of k at a given stress divided by k_0, which is the permeability at zero stress.

Core Plug 6-2 retains approximately 80–90% of its original permeability, while Core Plug 8-2 retains approximately 80% of its original permeability at 10,000-psi closure stress. For practical purposes, this amount of permeability reduction would be negligible when it comes to predicting flow rates vs. time and ultimate recovery. However, Core Plug 11-2 has only 40% of its original permeability at 10,000-psi closure stress. Such a large reduction in permeability must be included when performing reservoir simulation and predicting flow rates and recovery. Some reservoir rocks are even more stress sensitive. Rocks with slot pores caused by quartz overgrowths can lose more than 90% of their original permeability when compressed.

Measuring permeability as a function of closure stress using core from a tight gas reservoir is highly recommended to obtain the data required to evaluate the formation. These measurements are normally made using nitrogen in a dry core. When these same cores have been restored to original water saturation, the permeability to gas will be even lower.

Grain density is an important rock property because it is used to interpret the FDL and calculate the overburden stress in a given geographical area. To measure grain density, the cores should be cleaned, extracted, and dried to remove all hydrocarbons, saline fluid, salt crystals, and drilling- and coring-fluid solids.

Table 5.9 presents a list of the minerals that a tight gas reservoir often comprises, along with the grain density of each mineral. We can run logs to determine the minerals present in the formation, but core is preferred to determine the exact mineral composition

*5.3.2.5 **Cation-Exchange Capacity (CEC).*** The capacity of clays to allow for exchange of cations between the clay surfaces **(Fig. 5.78)** and surrounding water is another measurement that should be made using core from a tight gas reservoir. These cation-exchange-capacity (CEC) measurements are necessary for shaly formations. In most formations, all the electrical current generated and measured by resistivity logging tools flows through the pore space in the rock. The pore space is filled with gas, water, and clay. The water can be a mixture of formation water and mud filtrate. The clay can be in the pore space or part of the rock matrix. If the clay is in the pore space, it will affect the resistivity measurements and must be accounted for when computing water saturation from log data.

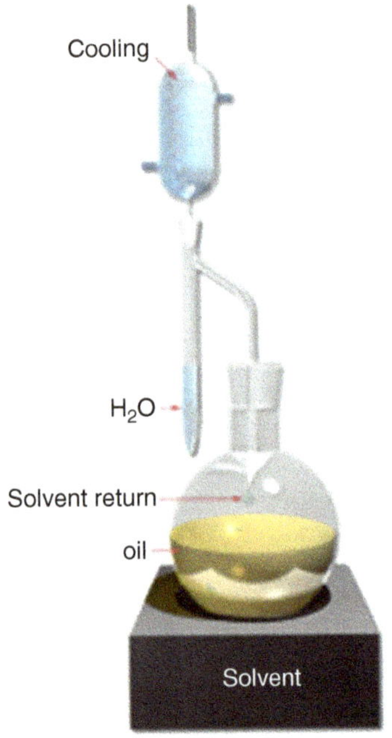

Fig. 5.79—A Dean-Stark apparatus.

In Section 5.2.2 of this chapter, we discussed the distribution of clays in tight gas sand and the use of the Waxman-Smits equation to calculate water saturation in shaly sand. One of the parameters in the Waxman-Smits equation is the CEC. The CEC is a function of type and amount of clay. All core in shaly sands should be described by the petrophysicists, and simple laboratory tests can be run to measure CEC.

5.3.2.6 Saturation Analysis. As we will discuss in more detail in Section 5.4 of this chapter, the results from log analyses must be confirmed and correlated with the results from core analyses. Not only porosity and permeability but also estimates of water saturation must be made using cores. When formation core is taken in a well drilled with water-based mud, the saturation estimates can be skewed because of mud-filtrate invasion. To minimize this problem, drill and core the well with oil-based mud. Alternatively, a low-level radioactive material can be added in precise quantities to the water-based mud and then laboratory tests can be run to determine how much of the water in the core is mud filtrate.

The two most common methods of measuring saturation in cores are the distillation/extraction method (Dean-Stark method) and the retort method (summation-of-fluids method). The Dean-Stark method **(Fig. 5.79)** will provide precise quantitative water saturations. The water collected can be analyzed to estimate the volume of formation water and/or mud filtrate that is in the core if a tracer was used in the mud. The retort method (summation-of-fluids method) is acceptable if the saturations are used mainly to locate fluid contacts.

5.3.2.7 Correlation of Log and Core Data. The correlation of log and core data will be discussed in detail in Section 5.4 of this chapter. To obtain results from openhole-log analyses that are trusted, the log data and the log calculations should be correlated and calibrated with core data. However, before the core data can be trusted, you must ensure that the core was properly cut and marked in the field, preserved and shipped correctly, and properly measured in the laboratory. On occasion, the core data might not be as reliable as is desired. However, if handled properly, the core data can be considered the "ground truth" for correlating log data.

Core analysis must also be consistent with the production data. The results of both the routine and special core analyses should be correlated with all well-test, log, and seismic data. **Fig. 5.80** illustrates how the log and core data can be compared and correlated.

5.3.2.8 Summary. Obtaining and analyzing core is a crucial step to understanding layered, complex, tight gas reservoir. Cores should be handled properly in the field and in the laboratory. Once the important values are known for each rock layer, estimates of gas-in-place for each rock layer using the values of porosity (ϕ), net thickness (h), and water saturation (S_w) can be calculated. The permeability-thickness distribution (kh) can be estimated from logs and calibrated with pressure buildup tests. With estimates of gas-in-place and permeability by layer, the completion, stimulation, forecast flow rates, and recovery can be designed. However, the routine core analysis is just the beginning in evaluating a formation, especially if it is a formation that has not been widely developed. The second step is to submit core and core plugs for special core analyses.

5.3.3 Special Core Analyses. Obtaining and analyzing core are crucial to the proper understanding of any complex tight gas reservoir system—to properly understand low-permeability reservoirs, modern core analysis methods are required. Of primary importance is the measurement of rock properties under restored-state conditions. We discussed what is expected for routine core analyses in the immediately preceding subsections of this chapter. Routine core analyses should be run on every foot of core and should include the following: porosity, grain density, unstressed permeability to air, stressed permeability to air, CEC, and saturation analysis.

The following properties are typical of what is measured during special core analysis. Permeability under in-situ stress is required both to interpret well tests and to predict the future flow rates and recovery of natural gas vs. time. The reduction of permeability must be included in any reservoir simulation of a tight gas reservoir. It is also important to measure relative permeability to gas as a function of water saturation. These data are also crucial to any reservoir simulation study, as well as measurements of capillary pressure as a function of water saturation. Capillary pressure can be a factor in mud-filtrate invasion, in fractureing-fluid-filtrate invasion, and in how fluid affects formation damage near the wellbore and the fracture face.

Other important parameters that can be measured in the laboratory that affect how tight gas reservoirs are developed are natural-fracture orientation and formation mechanical properties. These parameters will be crucial in designing the completion and understanding the reservoir. The mechanical properties will help us understand the in-situ stress in the formation.

From the routine core analysis results, the core should be analyzed to determine the different rock types and flow units that make up the interval that was cored. Then, 5 to 10 core plugs should be cut in every rock type and/or flow unit to be sent off for special core analyses. Most of the tests should be measured as a function of NOB pressure.

The special core analysis for each lithology should comprise some or all the following tests: permeability vs. NOB pressure, permeability vs. water saturation, resistivity index, formation factor vs. NOB pressure, capillary pressure, porosity vs. NOB pressure, acoustic velocity, and mechanical and elastic properties.

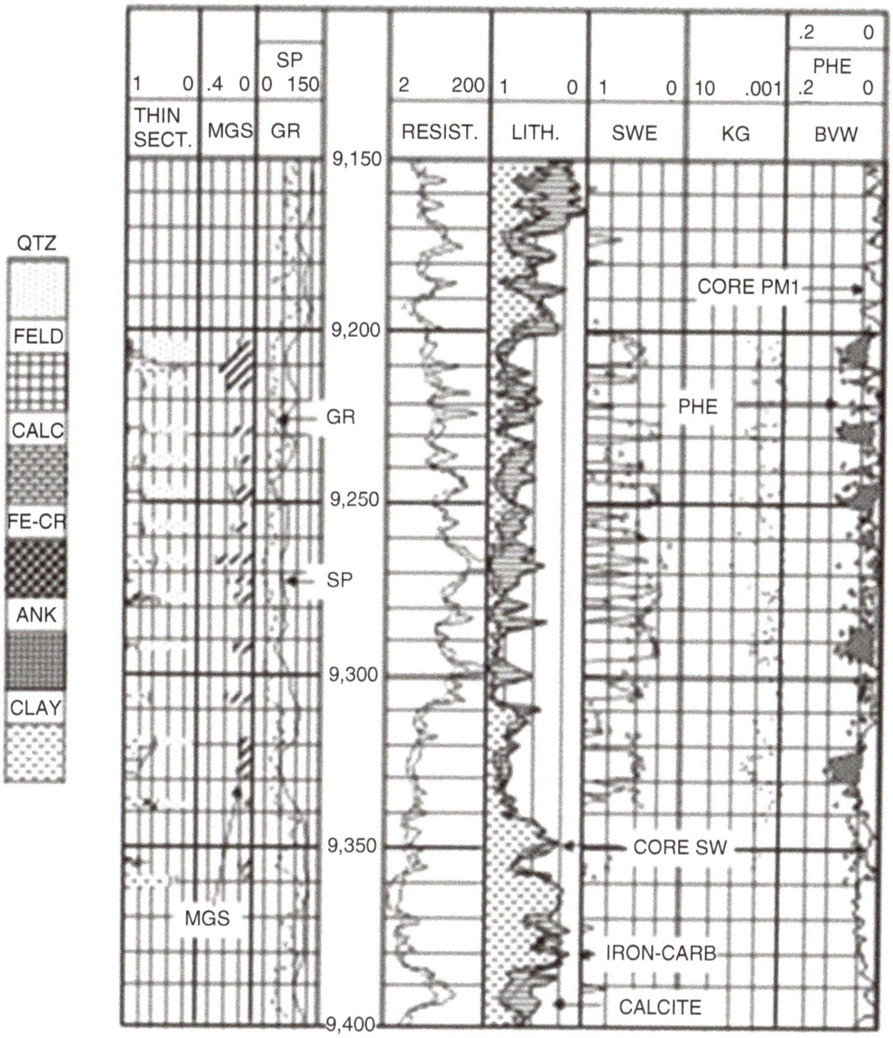

Fig. 5.80—Example of how log and core data can be compared (SFE #3 well).

5.3.3.1 Computed-Tomography Scanning and Image Processing. A recently developed method for permeability and porosity estimation in core samples involves mapping the pore structure using a computed-tomography (CT) scan of the whole core (**Fig. 5.81**). In this method, the core samples are put under CT scanning devices, and their pore volumetric distributions are digitized. The 3D processed distribution of the pore sizes is then used to estimate porosity and permeability. Correlations can also be used to estimate other properties, such as relative permeability, capillary pressure, and electrical properties.

Even when this is not the primary use of CT scanning, we have found that, when testing cores involving fluid invasion, the CT images can be useful for understanding the laboratory results. Also, CT scans can be useful in describing natural fractures and the distribution of natural fractures.

2 D X-Ray Attenuation Picture of a Core Sample

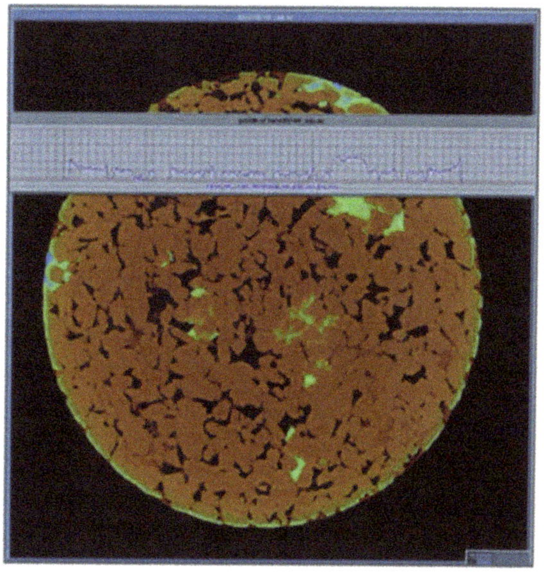

Fig. 5.81—CT scan showing the distribution of the pores in the rock (Knackstedt et al. 2004).

5.3.3.2 Stress-Dependent Porosity and Permeability. Values of the permeability and porosity in tight sandstones are highly dependent on the NOB stress magnitude and how the stress changes during production as the formation pressure is reduced. Factors such as lithology, pore-throat-size distributions, and natural fractures affect the initial values of in-situ permeability (k) and porosity (ϕ). The decrease in pressure in the reservoir as production occurs causes an increase in net effective reservoir stresses.

Stress-dependent k (**Fig. 5.82**) and ϕ studies can be implemented in reservoir-evaluation, reservoir-productivity, and

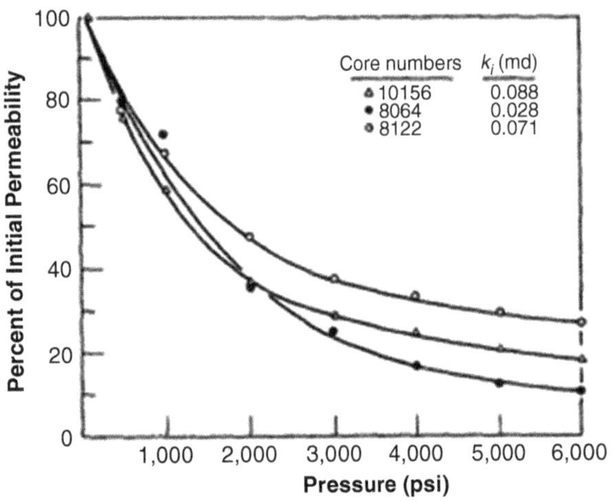

Fig. 5.82—Example of stress-dependent permeability in three cores from Project Gasbuggy.

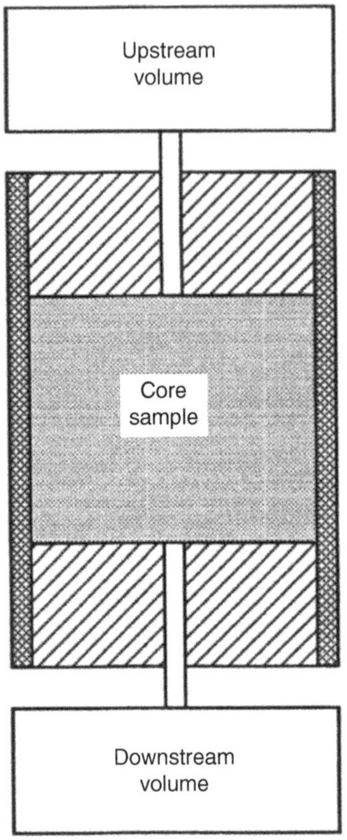

Fig. 5.83—Schematic for a pressure-pulse test.

reservoir-management studies. In fact, if stress-dependent k, in particular, is not modeled correctly, gas-flow rates and ultimate gas recovery can be substantially overpredicted.

*5.3.3.3 **Pulse Testing in Tight Gas Cores.*** Conventional core analysis methods for determining permeability and stress-sensitive permeability are not practical for rocks with permeabilities on the order of 0.1 md or lower. In fact, many tight gas formations have been produced economically with permeabilities ranging from 0.001 to 0.05 md.

Conventional core testing for permeability relies on developing a steady-state flow of fluid through the core and then measuring the rate and pressure drop to calculate permeability. Such an approach would not work with microdarcy rock because the time to reach steady state is prohibitive and the flow rates would be too small to measure.

The concept of pulse testing in a core is simple. The first paper on pulse testing was by Brace et al. (1968). They developed the method to measure the permeability of granite under high pressure. Yamada and Jones (1980) conducted an analytical review of the

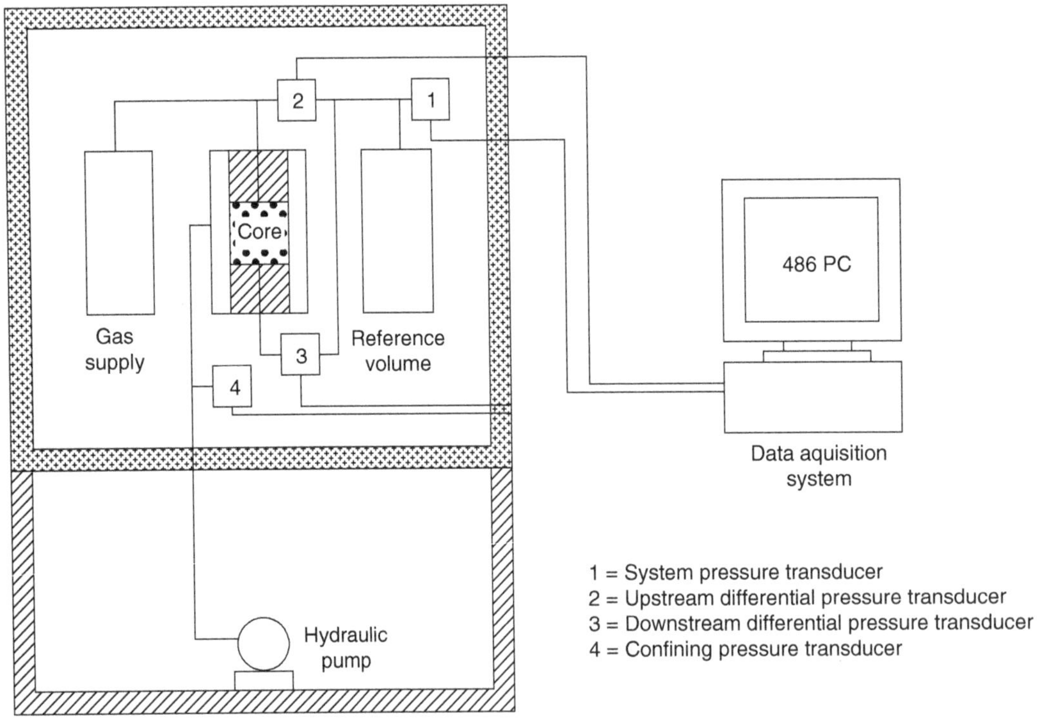

Fig. 5.84—Schematic of the pulse-testing laboratory equipment.

solution developed by Brace et al. They concluded that the upstream volume should be 10 times greater than the pore volume and the downstream volume should be larger than the upstream volume.

Fig. 5.83 is a schematic of the pulse-testing method. A core is isolated between two volumes, as shown in Fig. 5.83, and the entire apparatus is allowed to reach pressure equilibrium before the test. At equilibrium, the pressures in the upstream volume, in the core, and in the downstream volume are equal. The valve between the upstream volume and the core is closed to start a test; then, the pressure in the upstream volume is increased to a specified value. A pressure pulse is created by quickly opening the valve between the upstream volume and the core. The pressure then decays as the fluid (nitrogen) enters the core and flows through the core. With the knowledge of the volumes of the apparatus and the dimensions of the core, a transient equation can be used to calculate the permeability of the core. This method is like a pressure transient test in a wellbore.

A more detailed schematic of the laboratory apparatus needed to conduct a pulse-test permeability measurement is shown in **Fig. 5.84.** The fluid normally used is nitrogen, and the core is normally cleaned and dried before the testing. The permeability can be measured under overburden stress by applying pressure on the outside of the core using a hydraulic pump. The upstream chamber, the core, and the downstream chamber must be in a temperature-controlled oven. The flow through these cores is so small that a person's body temperature can affect the pressure transducers if the temperature of the equipment is not kept constant.

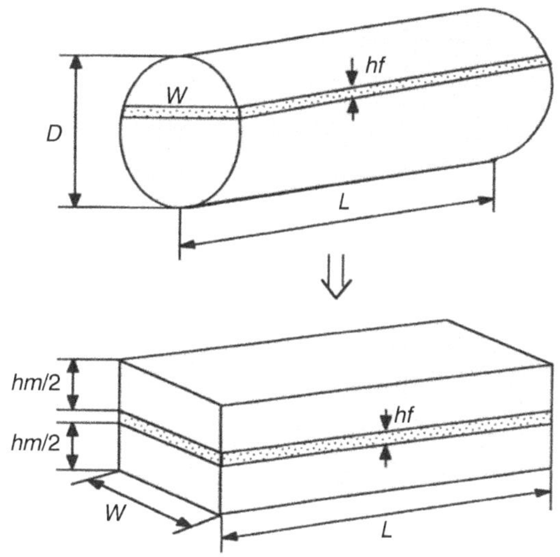

Fig. 5.85—Drawing of the core sample used in the modified pulse test by Ning et al. (1993).

In 1992, a PhD degree dissertation by Ning (1992) took the pulse-test method to a more-accurate and -useful level. Ning developed a test in which he cut the core to create a path between the upstream and downstream chambers. He then developed a method such that the volumes of both the upstream and downstream chambers were on the same order of magnitude as the expected pore volume in the core. When these changes in the laboratory equipment had been made and the pulse-testing methods were applied, it became possible to calculate not only the permeability of the core, but also the porosity of the core. The test also allowed for the computation of the fracture porosity and permeability of a fractured core.

Ning (1992) developed analytical solutions to the problem and documented those in his dissertation. Using the analytical solutions, he wrote a computer program to analyze the pulse test. Again, all the equipment had to be in a controlled-temperature oven so that the test was not affected by the temperature fluctuations in the laboratory. **Fig. 5.85** shows a core with a man-made fracture.

Ning et al. (1993) details the modified pulse test and the analytical solutions developed for it. **Fig. 5.86** shows the graphical solution of these equations.

Fig. 5.87 shows the sensitivity of the analytical solutions to the porosity of the matrix. Notice the time scale on the graphs. The time required for steady-state tests in the laboratory would be prohibitive for a core with a permeability of 10^{-6} md.

Finally, **Fig. 5.88** shows the sensitivity of the pulse test to the matrix permeability. It is clear from Fig. 5.88 that, if the test is run properly, the matrix porosity and permeability can be estimated. More useful information, including the analytical solutions and results from numerous core plugs, can be found in the dissertation by Ning (1992) and subsequent publications. In his dissertation, Ning also shows how the pulse technique can be used to measure relative permeability to gas.

5.3.3.4 Relative Permeability. The measurements of permeability either at ambient conditions or at NOB conditions are normally conducted on cores that have been cleaned and dried. However, the effective permeability to gas flow in the reservoir is also affected by water saturation. As such, we need to measure the relative permeability to both gas and water in cores, if possible.

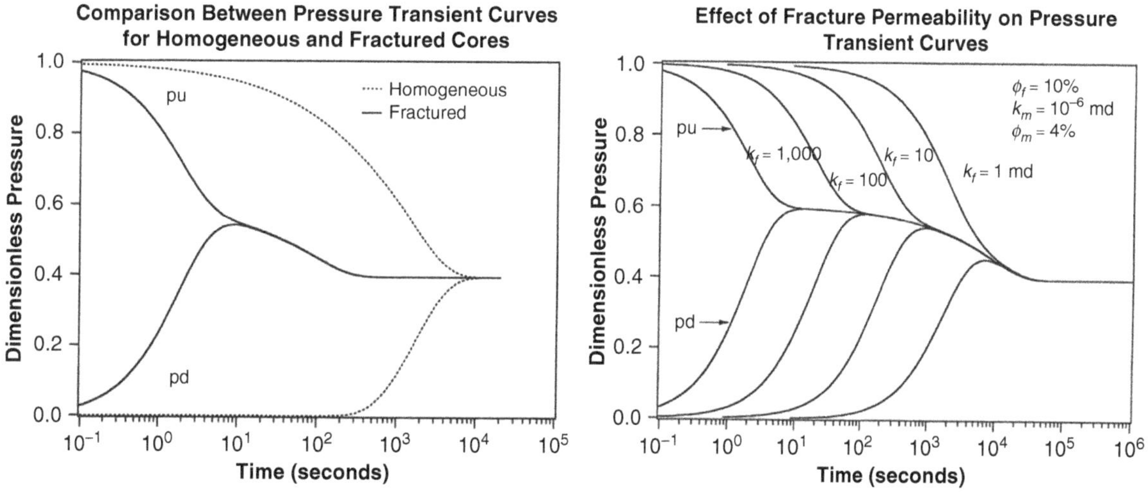

Fig. 5.86—Graphical presentation of the analytical solutions showing sensitivity to fracture permeability.

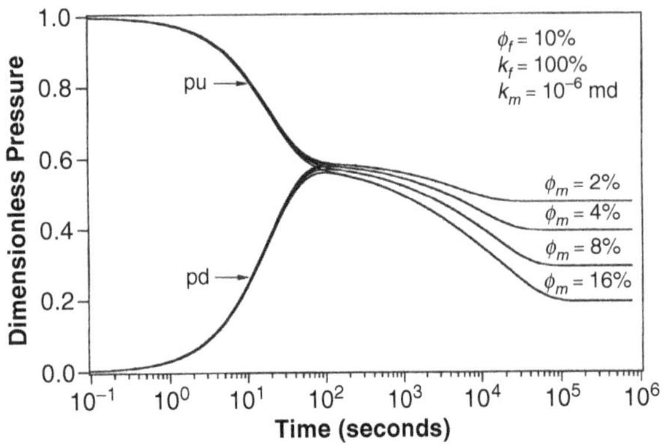

Fig. 5.87—Effect of matrix porosity on the pressure transient curves.

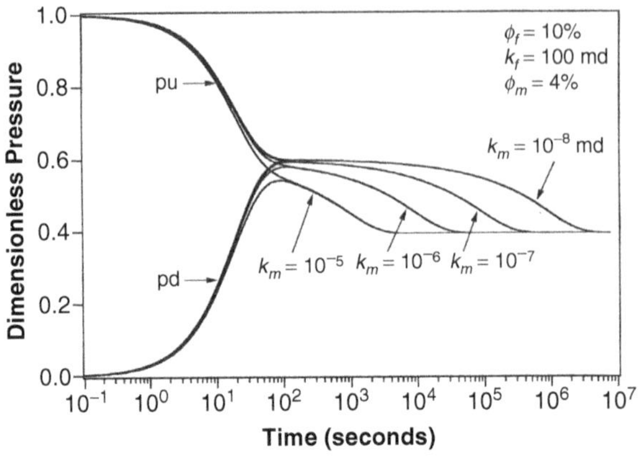

Fig. 5.88—Effect of matrix permeability on the pressure transient curves.

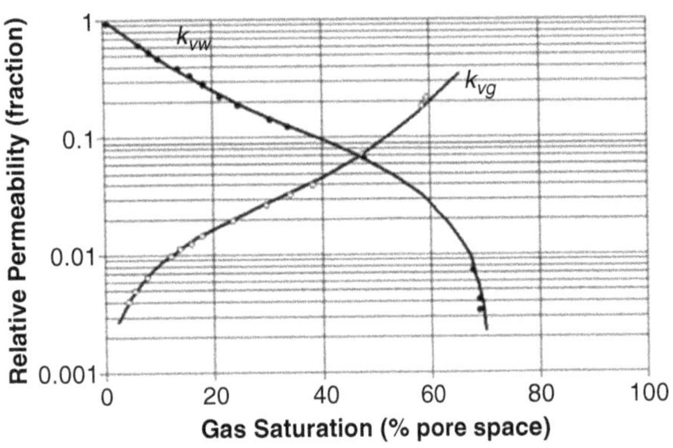

Fig. 5.89—Relative permeability for an 11.8-md core sample in the Travis Peak Formation.

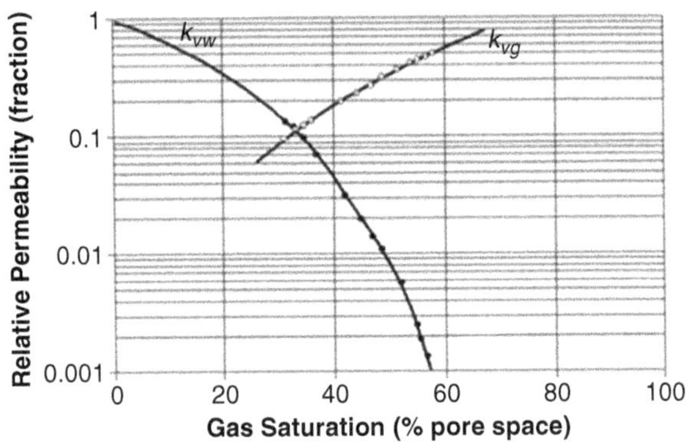

Fig. 5.90—Relative permeability for a 0.028-md core sample in the Travis Peak Formation.

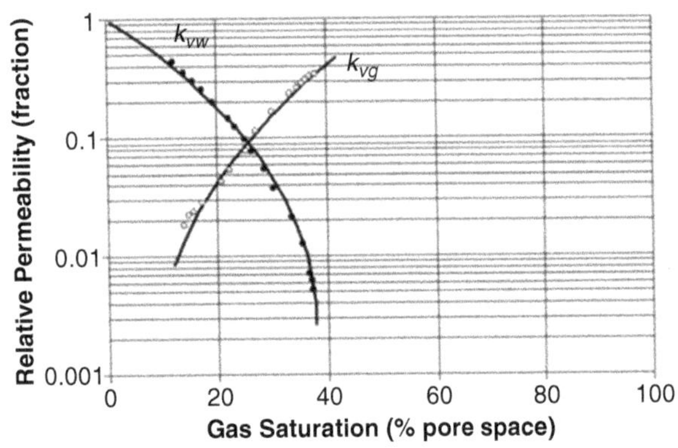

Fig. 5.91—Relative permeability for a 0.0022-md core sample in the Travis Peak Formation.

Figs. 5.89 through 5.91 illustrate data for Travis Peak core samples with permeabilities of 11.8, 0.028, and 0.0022 md, respectively. The data illustrate that low values for residual gas saturation and high values for water-saturation cutoff should be used to determine net-pay in these tight rocks.

These data were measured in the laboratory using transient testing methods. The results clearly indicate that, for formations with permeabilities below 0.1 md, gas can flow in the rock even when the water saturation is 50–80%. Therefore, caution should be taken when using a low water-saturation cutoff to determine net-pay. The cutoff, if used at all, should be determined for each formation. There are many cases where substantial gas flow has occurred in formations whose water saturation was calculated to be 60% or more.

5.3.3.5 Wettability and Capillary Pressure. Wettability is a rock characteristic that determines which fluid phase is in contact with the rock. There is a wetting phase and a nonwetting phase. The wetting phase has the tendency to spread on the solid surface and to displace the nonwetting phase. Rocks are often categorized as water-wet, oil-wet, or intermediate-wet. Most tight gas reservoirs are water-wet.

Wettability affects capillary pressure, relative permeability, electrical properties, irreducible water saturation, and residual oil saturation. Surfactants are sometimes used to alter wettability, especially when injecting fracturing fluids into a formation. If the proper surfactant is used, the well will clean up better after the fracture treatment.

Capillary pressure is the difference in pressure inside the rock between the wetting phase and the nonwetting phase. In tight gas reservoirs, capillary pressure is

$$P_c = P_g - P_w, \quad (5.21)$$

where P_c = capillary pressure, P_g = pressure in the gas phase, and P_w = pressure in the water phase.

In a conventional gas reservoir, the expectation is to find a gas/water contact. Capillary pressure data have been used for decades to determine the height of the hydrocarbon column and the level of the gas/water or oil/water contact (**Fig. 5.92**).

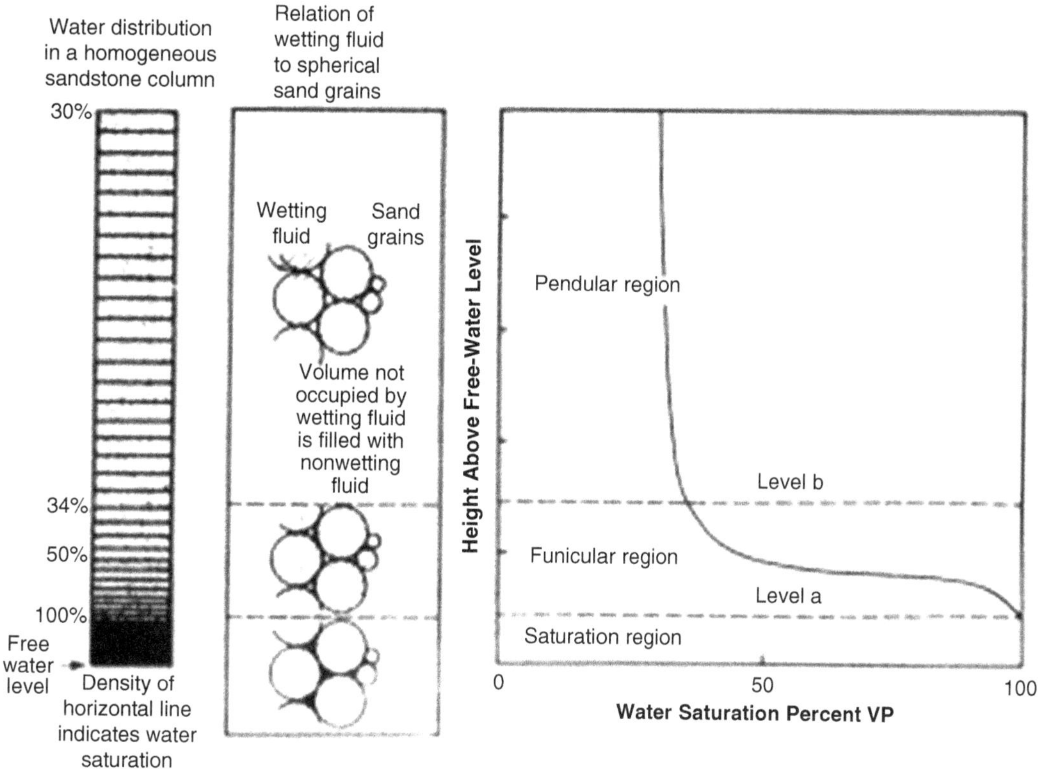

Fig. 5.92—How a capillary pressure curve can be used to estimate fluid distribution in a homogeneous reservoir.

There are various methods used to measure capillary pressure in core samples. One such method is the semipermeable disk, but this method works only on high-permeablility core samples.

Another method, mercury injection, works for low-permeability core samples but permanently damages the core. Mercury must be disposed of properly because mercury contamination is dangerous to human health. The best method to measure capillary pressure in low-permeability core samples is to use a centrifuge. This method is faster, but the fluid distribution along the core is not uniform.

The value of capillary pressure at any given value of water saturation will depend on the saturation history of the test **(Fig. 5.93).** If the core is saturated 100% with water and that water is displaced with nitrogen gas, then there will be a capillary pressure curve for the drainage case. We consider the drainage curve that is most appropriate for the reservoir. Virtually all reservoirs were deposited in an ocean, river, or lake, and the rock was filled with water initially. Over geologic time, the rock was buried, and oil or gas migrated into the formation. This is what we simulate with the drainage curve.

However, when we pump a fracture treatment, fracturing-fluid filtrate invades the formation, causing the water saturation to increase. The same is true of a waterflood in a formation. When the water saturation increases, this is what we simulate with the imbibition capillary pressure curve. So, if a reservoir simulator is used to model a fracturing treatment, fluid cleanup, and long-term gas production, then both the drainage and imbibition capillary pressure curves in the model must be included. **Fig. 5.94** shows the relationship of the two capillary pressure curves. Notice that, as water saturation increases in the imbibition test, the capillary pressure reduces quickly.

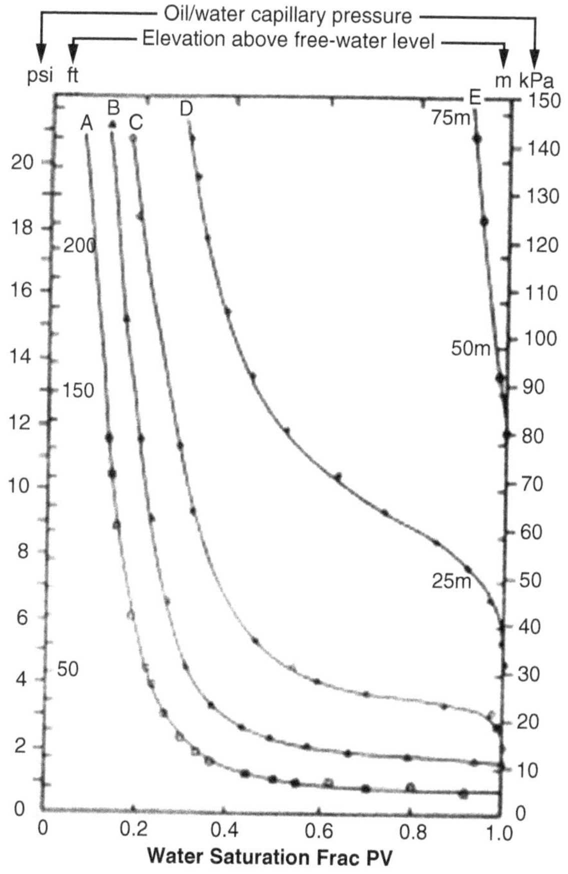

Fig. 5.93—Example of capillary pressure curves for different values of permeability.

5.3.3.6 Fracture Orientation. The orientation of natural fractures and the expected orientation of the induced hydraulic fractures are extremely important for planning drilling,

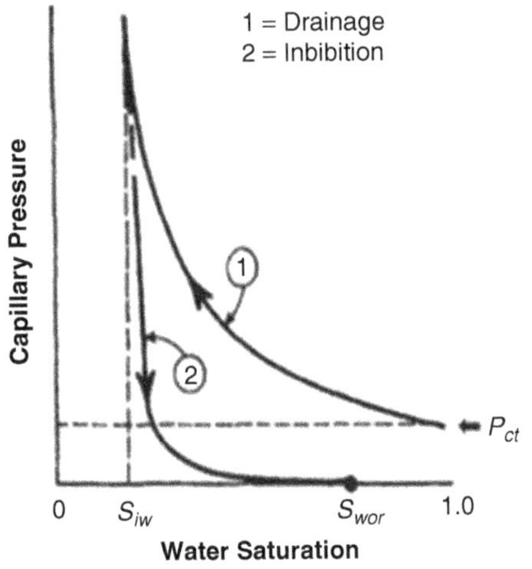

Fig. 5.94—Example of capillary pressure curves for drainage and imbibition.

completion, and stimulation procedures. Core can be useful in determining the fracture orientation. There are several methods to predict fracture orientation from a core sample. The core can be visually inspected for natural or coring induced fractures to obtain some information. One special method for determining the orientation of the natural fractures is to cut oriented core.

It is common to find two or more sets of natural fractures in different orientations. The reason for multiple sets of natural fractures is that the stress and strain on the formation have changed over time and have been a function of the loading rate and time. Typical tight gas sandstone formations are hundreds of millions of years old and have undergone subsidence and other tectonic events. As such, the core should be analyzed by the geologists to help define the sets of natural fractures so the information can be used in developing the reservoir.

Microcracks in the core can be induced as the core relaxes from the in-situ stress conditions to ambient conditions and as it is pulled in the core barrel from where it was cut to the surface. A laboratory measurement method called strain relaxation can be used to determine the magnitude and direction of minimum and maximum stress, but oriented core must be used to know the orientation of the core sample in the reservoir.

Other tests that can be used to help define natural-fracture orientation are the differential strain analysis, differential wave velocity analysis, residual-strain over coring, and thermal-expansion anisotropy.

*5.3.3.7 **Formation Mechanical Properties.*** In developing plans for drilling, completing, and fracturing a tight gas sandstone, the engineers need information on the mechanical properties of the pay zone containing the gas reservoir and the layers of rock above and below the pay zone that could affect fracture height growth. The primary mechanical properties we must characterize are Poisson's ratio, Young's modulus, and in-situ stress. Of course, there are other rock properties, such as fracture toughness, tensile stress, and compressive stress, that are also important.

Poisson's ratio (**Fig. 5.95**) is required to calculate in-situ stress from logging data. Young's modulus is a measure of the stiffness of the formation. Compressive strength can also be important in deep geopressured reservoirs where formation collapse can be a problem when the pressure drawdown on the formation exceeds the rock strength.

Poisson's ratio is a measure of lateral strain to longitudinal strain for a given stress in the direction of the longitudinal direction. In a formation, the longitudinal stress is effectively the overburden stress. Unless a fault occurs, the formation at any depth cannot move laterally, so the lateral strain that should occur will result in a lateral stress. For a high value of Poisson's ratio (such as that found in a shale), the lateral stress will be higher than for a sandstone or stiffer rock with a low value of Poisson's ratio. As such, we expect to see higher values of in-situ stress in formations with high values of Poisson's ratio and lower values of in-situ stress in formations with lower values of Poisson's ratio.

*5.3.3.8 **Young's Modulus.*** Young's modulus is a measure of the stiffness of the rock layer. The stiffness of the formation will affect the fracture-width calculations. To create a hydraulic fracture, use a fluid to create a crack and propagate the crack to essentially push the earth apart. For a given set of conditions, a stiff formation will have hydraulic fracture widths that are less than those that would occur for formations that are less stiff.

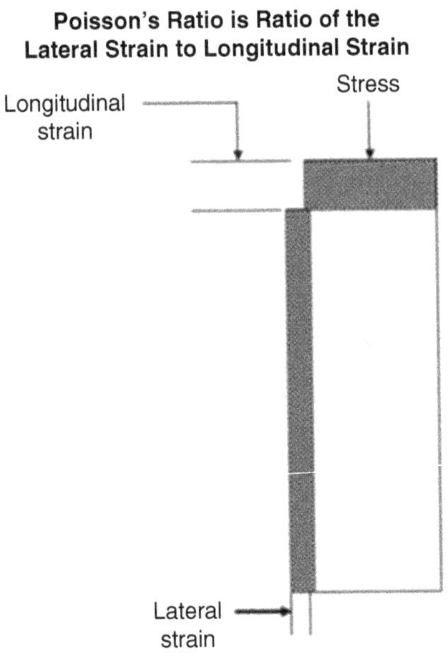

Fig. 5.95—Explanation of Poisson's ratio.

Young's modulus can be measured with cores or estimated from logs. Young's modulus can be correlated with acoustic travel time from the openhole logs. However, the SON measurements are defined as dynamic measurements because they are measured using high-frequency sound waves that are not affected much by fractures. The sonic measurements are mainly affected by the travel time in the matrix portion of the formation.

Stress/strain tests using cores also provide an estimate of Young's modulus. These measurements are described as static measurements because the test involves loading the core slowly (low frequency). In fact, the slow loading of the core more closely resembles the loading during a hydraulic fracturing treatment. The static values of Young's modulus are what the fracture design engineer needs for running the 3D fracture design model.

It is important to understand that we can measure both static and dynamic measurements simultaneously in the laboratory using cores. Then, once we establish the correlation between static and dynamic Young's modulus and the Poisson's ratio using core data, we can use those correlations to interpret SON data on wells that do not have core.

Fig. 5.96 shows how Young's modulus is measured in the laboratory. A rock sample is subjected to a confining stress. The stress is then increased in one direction, and the strain (deformation) that results from the stress is measured with strain gauges. The data are then graphed and analyzed to determine Young's modulus, as shown in **Fig. 5.97.**

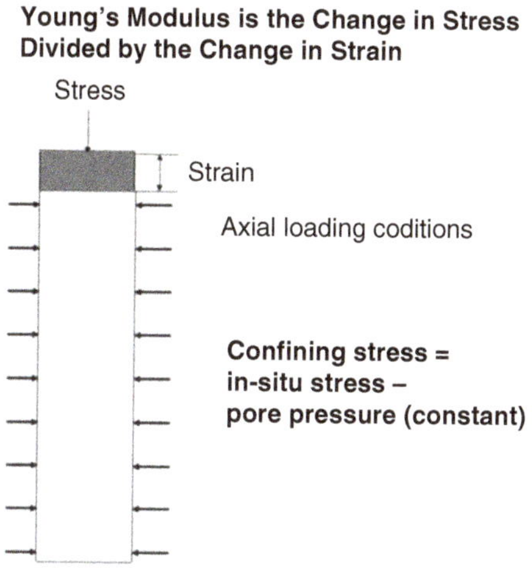

Fig. 5.96—How to measure Young's modulus.

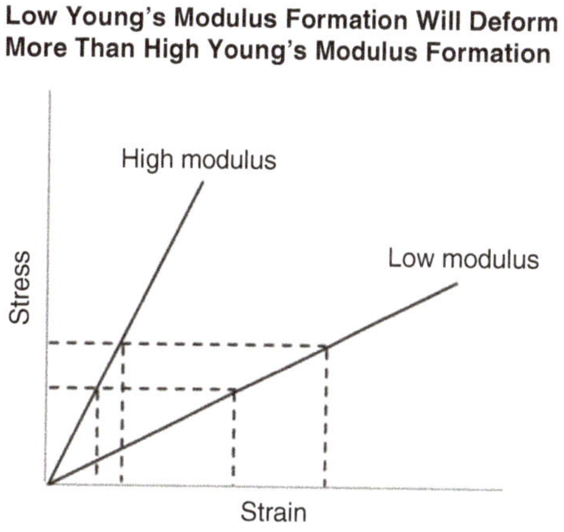

Fig. 5.97—Graph showing the stress/stain relationship for high- and low-modulus rocks.

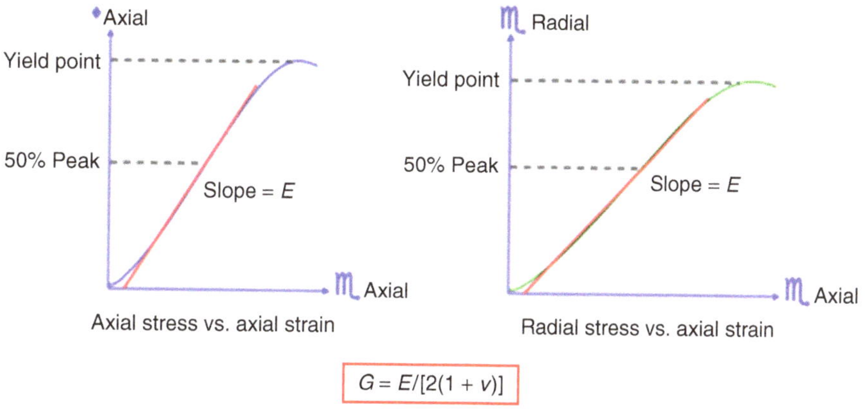

Fig. 5.98—Graphs showing the stress/stain relationship for actual whole core samples.

Young's modulus is calculated as the slope of the stress/strain curve. High-modulus rocks are stiffer (they deform less with the same amount of stress) when compared to low-modulus rocks. Stress/strain curves for rocks often are not linear and sometimes cannot be described by a single Young's modulus.

Fig. 5.98 shows two graphs that are more typical than the idealized graph in Fig. 5.97. The actual stress/strain relationship might not be linear at low stress values, and at some point the stress will cause the material to fail and the strain will actually decrease. This is called the yield point of the material being tested.

A graph published by Halliburton in 1971 **(Fig. 5.99)** interestingly presents a correlation of Young's modulus to compressive travel time that can be obtained from an SON.

In some of the work for the GRI tight gas research project, we compared data from logs and cores to the correlation in Fig. 5.99 and found that it was quite accurate.

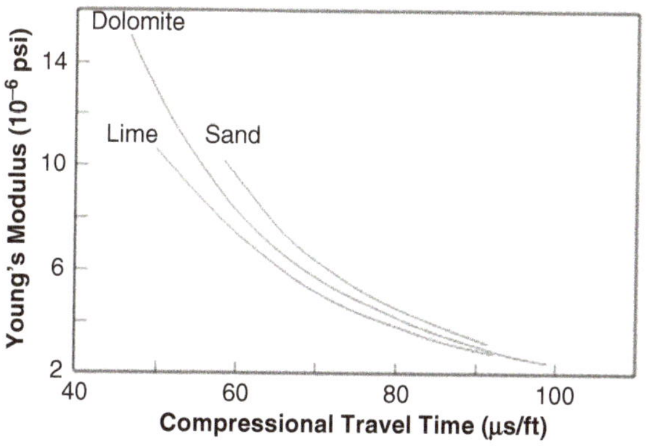

Fig. 5.99—Correlation of Young's modulus to compressional travel time.

5.3.3.9 Mechanical Properties from Logs. As discussed above, we often use SON data to estimate rock mechanical properties. Poisson's ratio can be estimated using values of Δt_s and Δt_c from the sonic waveform to calculate the velocity ratio, R_v. We use a modern sonic tool to measure the waveform, as shown in **Fig. 5.100.** We can use dipole or quadra-pole tools to measure directional sonic velocities if we want to obtain mechanical properties in different directions downhole.

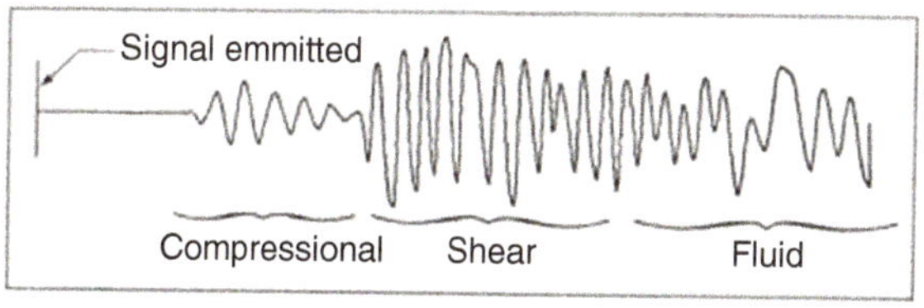

Fig. 5.100—Full waveform showing the compressional, shear, and fluid arrival times.

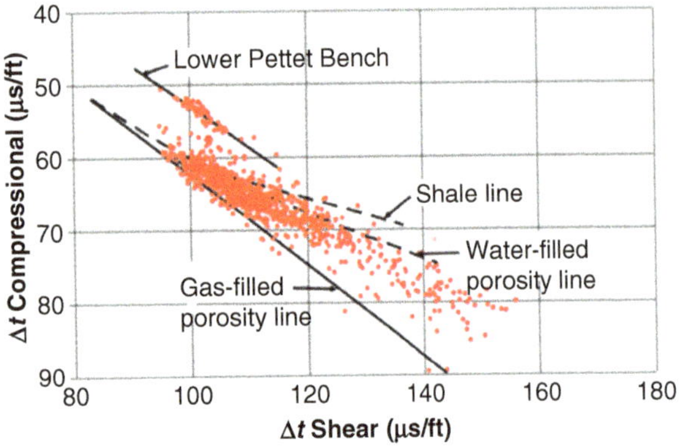

Fig. 5.101—Correlation of compressional travel time to shear travel time for cores in the Travis Peak Formation.

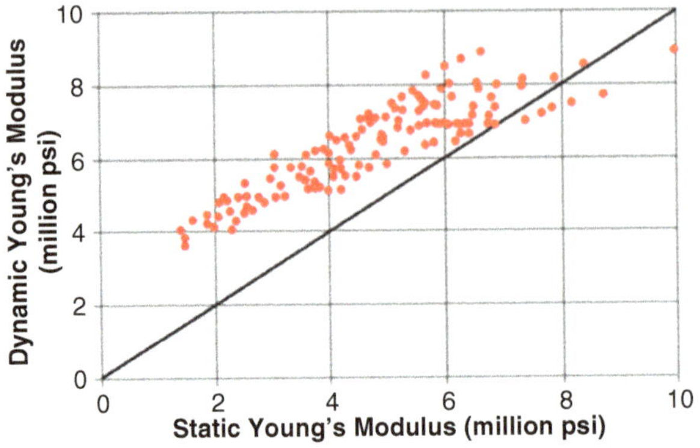

Fig. 5.102—Correlation of dynamic vs. static measurements of Young's modulus for cores in the Travis Peak Formation. Static value was measured while axial load decreased.

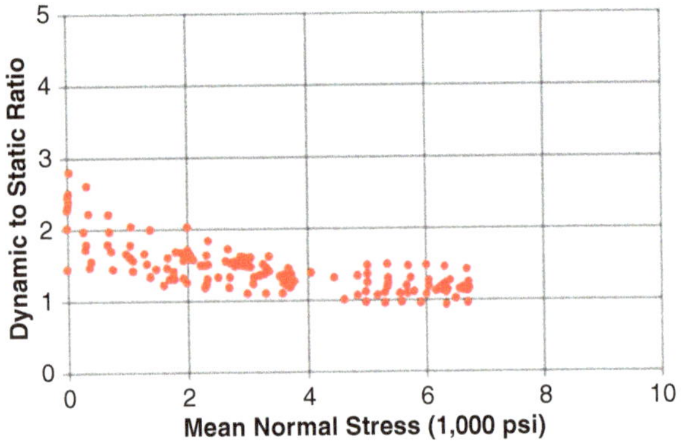

Fig. 5.103—Correlation of dynamic to static ratio of Young's modulus vs. normal stress.

The advantage of computing mechanical properties from logs is that it creates continuous mechanical-property profiles by depth, the contrast between layers can be easily identified, and obtaining in-situ data from logs costs less than obtaining the same data from cores.

The disadvantages of computing mechanical properties from logs is that log measurements are taken at high frequencies and provide dynamic estimates of Young's modulus and Poisson's ratio. The rock deformation or hydraulic fracture propagation process is relatively slow, and the static measurements from core are more useful. However, the dynamic moduli values can be calibrated using static core analyses to improve the usefulness of log-calculated values. The dynamic values of Young's and shear moduli are calculated using sonic travel times. The equation for shear modulus is

$$G = 1.34 \times 10^{10}\, \rho_b / \Delta t_s^2, \quad (5.22)$$

and the equation for Young's modulus is

$$E = 2G\,(1 + \nu), \quad (5.23)$$

where ϕ_b = bulk density, g/cm³, and ν = Poisson's ratio.

Fig. 5.101 shows data from SONs run in the Travis Peak Formation, where we have measured both the compressional- and shear-wave velocities. These values are measured at high frequency using a full waveform SON.

In fracture propagation models, we need static values of modulus and other rock properties. **Fig. 5.102** shows the relationship between static and dynamic values of Young's modulus that were measured using cores from the Travis Peak Formation.

Fig. 5.103 shows how the ratio of dynamic modulus to static modulus changes as a function of mean normal stress on the core. These data show that for low-modulus (less-stiff) rocks the factor required to correct the dynamic modulus to the static modulus needed for fracture models can be as much as 2 at low stress. However, as the rock stresses, the correction factor approaches a value of unity.

It is not the purpose of this chapter to document everything we know about measuring or calculating mechanical properties from logs and cores. The reader should conduct more research on these rock properties that are examined in many papers and books. The purpose of this chapter is to stress the importance of making special core analyses measurements to better understand the formation.

Below are several correlations of Young's modulus that have been published:

Linear correlation (Eissa and Kazi 1988),

$$E_l = 0.74E_d - 0.82. \quad (5.24)$$

Log linear correlation (Eissa and Kazi 1988),

$$\log(E_l) = 0.02 + 0.71\log(\rho_b E_d). \quad (5.25)$$

Modified log linear correlation (Barree et al. 2009),

$$\log(E_l) = \log(\rho_b E_d) - 0.55. \quad (5.26)$$

Van-Heenden's power-law model (Barree et al. 2009),

$$E_l = 0.1832E_d^{1.5795}. \qquad (5.27)$$

A more complicated model (Nikitin et al. 2009),

$$E_l = 80.36\log^2(E_d) - 195.86\log(E_d) + 132.69. \qquad (5.28)$$

5.3.3.10 In-Situ Stress. In-situ stress is the local stress in the formation. The in-situ stress is often characterized as three principal stresses: two horizontal stresses (σ_2, σ_3) and one vertical stress (σ_1). **Fig. 5.104** presents the terms of in-situ stress.

All rocks in the subsurface are subjected to stress because of overburden weight and compressive (mountain-building) stresses. Because of changes in rock properties (e.g., Poisson's ratio, porosity) and rock fabric (e.g., natural fractures, cementation) in both the areal and vertical directions, in-situ stress is normally anisotropic and nonhomogeneous. In Fig. 5.104, σ_3 is the maximum horizontal stress and σ_2 is the minimum horizontal stress.

In a vertical well containing a vertical hydraulic fracture, the most important parameter controlling fracture growth is the in-situ stress distribution in the layers of rock above, within, and below the productive interval. We have found that we can correlate in-situ stress with lithology and reservoir pressure. We assume that we can also correlate Poisson's ratio with lithology.

$$\sigma_2 \approx \frac{v}{1-v}(\sigma_1 - P_P) + P_P + \sigma_{ext}, \qquad (5.29)$$

where v = Poisson's ratio, dimensionless; σ_1 = vertical overburden stress, psi; P_p = reservoir pressure, psi; and σ_{ext} = external stress, psi.

To properly design a fracture treatment, the values of in-situ stress in each rock layer must be known. We can estimate in-situ stresses from SON data using the following equations:

$$R_v = \frac{V_c}{V_s} = \frac{\Delta t_s}{\Delta t_c} \qquad (5.30)$$

and

$$v = \frac{0.5R_v^2 - 1}{R_v^2 - 1}, \qquad (5.31)$$

where R_v = ratio of compressional wave and shear wave velocity, dimensionless; V_c = compressional velocity, ft/sec; V_s = shear velocity, ft/sec; Δt_s = shear-wave travel time, µs/ft; Δt_c = compressional-wave travel time, µs/ft; and v = Poisson's ratio, dimensionless.

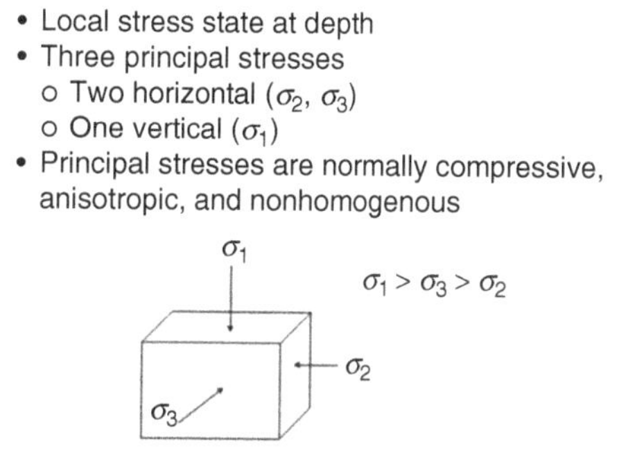

Fig. 5.104—Three principal in-situ stresses.

Fig. 5.105 presents data measured by Amoco in the various tight gas sand plays where they compared in-situ stress values calculated from core to data measured from injection tests in the field. There is a correlation that allows the engineer to take log-calculated values and use them in engineering calculations.

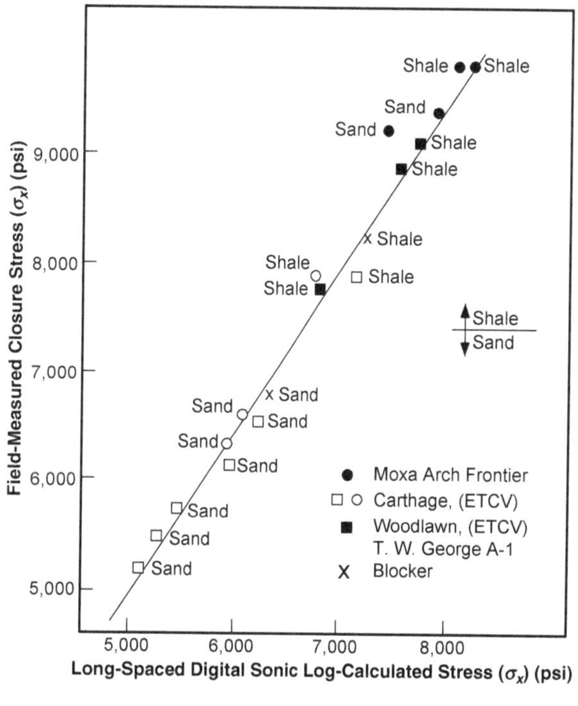

Fig. 5.105—Correlation of in-situ stress measured in the field pump-in test to stress calculated from long-space sonic logs (Veatch and Moschovidis 1986).

5.3.3.11 Summary. Engineers and geologists need to use logs and cores to calculate the mechanical rock properties such as Poisson's ratio, Young's modulus, and in-situ stress. The engineer needs to understand the difference between static modulus and dynamic modulus and how the values are used in engineering calculations.

The geologist must use core to define the lithology; then, the properties measured from cores and logs should be correlated as a function of lithology. Core testing is an important tool that provides information necessary for proper reservoir description. Reliable core data aid engineers in the design of an optimal hydraulic fracturing treatment for successful hydrocarbon recovery.

5.4 Correlations

5.4.1 In-Situ Stress. All rocks in the subsurface are subjected to stress because of overburden weight and compressive stresses, which are caused by plate movements or other geologic conditions that add stress to the rock layers, such as fault movement. Because of changes in rock properties, such as Poisson's ratio and porosity, and rock fabric issues, such as natural fractures and cementation, in the areal and vertical directions, in-situ stress is normally anisotropic and nonhomogeneous.

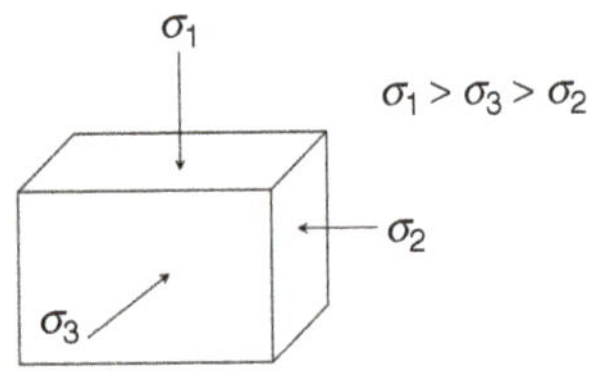

Fig. 5.106—Drawing depicting the three principal compressive stresses.

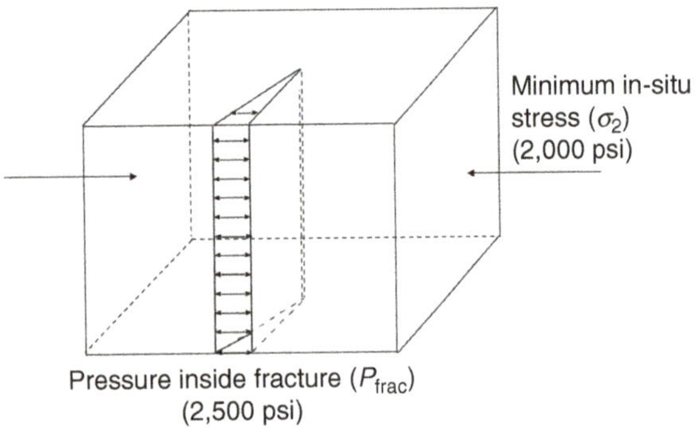

Fig. 5.107—Illustration of net pressure.

There are three principal stresses: two horizontal (σ_2 and σ_3) and one vertical (σ_1), as shown in **Fig. 5.106.**

Fig. 5.107 conceptually illustrates the term net pressure. It is important to realize that, by definition, net pressure is the magnitude of pressure that physically pushes the formation apart once the fracture has been created. Net pressure is an important variable in computing fracture width and creating graphs of net pressure vs. time that allows the fracture design engineer to interpret field data from pumping operations of all types, to include formation injection tests, mini-fracture tests, and the main fracturing treatment of a zone.

As illustrated in Fig. 5.107, the pressure inside the fracture is 2,500 psi. This is the pressure created from injecting fluid into the fracture. The fracture is held open against a minimum in-situ stress (σ_2) of 2,000 psi. The net pressure is the pressure inside the fracture minus the minimum stress, σ_2. Thus, for the values in Fig. 5.107, the net pressure is 500 psi.

5.4.1.1 Calculating In-Situ Stress from Log and Core Measurements. In-situ stress, specifically the minimum horizontal stress, in each rock layer above, within, and below the main pay zone is perhaps the most important parameter controlling fracture growth. The in-situ stress can be determined from injection tests or calculated from logs; then, in-situ stress can be correlated with lithology and reservoir pressure.

Eq. 5.29 is the equation most used to calculate minimum σ_2. As illustrated, Poisson's ratio, calculated in Eq. 5.31, can often be correlated to the lithology of a rock layer. The vertical overburden stress in Eq. 5.29 can be calculated from an FDL run from total depth (TD) to the surface. Because the pore pressure will decline as the reservoir is depleted, the in-situ stress in the pay zone will also decline with production. The external stress is caused by extra horizontal forces, such as thrust faults or nearby mountain ranges. The value of the external stress must be determined empirically by comparing actual values of stress measured during injection tests to the theoretical value using poroelastic theory, as shown in Eq. 5.29.

As mentioned several times so far in this book, the values of in-situ stress in each rock layer must be known to properly design a fracture treatment. We can estimate the value of minimum stress, σ_2, from SON data using Eqs. 5.29 through 5.31.

The values of compressional and shear velocities are obtained from a full waveform SON as described in Section 5.3.3 of this book. **Fig. 5.108** shows the relationship between the compressional and shear velocities measured with Travis Peak core plugs. Notice that this graph shows clear trends for gas-filled sand, shale, and shaly sand in between these two endpoints. The Lower Pettet is a carbonate.

5.4.1.2 Correlating Stress to Lithology Using Poisson's Ratio. In-situ stress can be correlated to the value of Poisson's ratio as calculated using Eq. 5.31 and data from a full waveform SON. We found in our work for the GRI that Poisson's ratio can be correlated to lithology in tight gas reservoirs, as the data in Fig. 5.108 suggest. Typical ranges of Poisson's ratio for various types of rocks are given in **Table 5.10.**

As will be illustrated in the following examples, the value of in-situ stress increases as the value of Poisson's ratio also increases. High values of Poisson's ratio mean that there will be more horizontal stresses because the rock strain would be higher if unconfined. The rock, however, is not unconfined, which results in a higher horizontal stress.

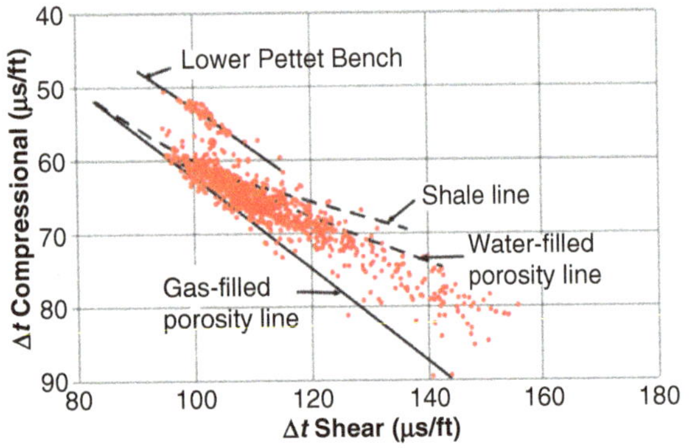

Fig. 5.108—Graph showing the relationship of compressional slowness vs. shear slowness for a set of Travis Peak core plugs.

Rock Type	Range of Poisson's Ratio
Gas-bearing sandstone	0.1 – 0.25
Liquid-filled sandstone	0.25 – 0.3
Wet siltstone	0.2 – 0.3
Limestone	0.30 – 0.35
Shale	0.28 – 0.43

Table 5.10—Typical ranges of Poisson's ratio for several rock types.

Lithology	Fluid Content	Poisson's Ratio	In-Situ Stress Gradient (psi/ft)
Sandstone	Gas	0.17	0.67
Sandstone	Oil or water	0.25	0.75
Siltstone	Water	0.30	0.81
Limestone	Water	0.35	0.87
Shale	Water	0.4	0.95

Table 5.11—Calculated values of in-situ stress gradient using values of Poisson's ratio for several rock types.

Source	Stress Correlation Equation
Anderson et al. (1973)	$\sigma_x = \frac{\nu}{1-\nu}\left(\sigma_v - \alpha p_p\right) + \alpha p_p.$
Amadei et al. (1988)	$\sigma_{H\min} - Xp_p = \frac{E'\nu}{E'(1-\nu)}\left[\sigma_v - X(1-\zeta)p_p\right].$
Holditch and Rahim (1993)	$\sigma_x = \frac{\nu}{1-\nu}\left(\sigma_v - sp\right) + sp + \sigma_{\text{ext}}.$
Whitehead et al. (1987)	$\sigma_c = \left(1-p_f'\right)\left(\frac{\nu}{1-\nu}\right) + p_f'.$
Whitehead et al. (1986)	$\sigma_c = 0.80342\left(\frac{\nu}{1-\nu}\right) + 0.47399.$

Table 5.12—Several correlations for in-situ stress that have been published for various conditions.

Following are some example calculations to illustrate how lithology affects horizontal stress.

Assume that $\sigma_1 = 1.05$ psi/ft, $\rho_p = 0.45$ psi/ft, and $\sigma_{\text{ext}} = 0.1$ psi/ft.

To calculate the values of in-situ stress gradient as a function of lithology and fluid content, use Table 5.10 as a guide along with Eq. 5.29 and the average values of Poisson's ratio.

The in-situ stress data in **Table 5.11** were calculated using Eq. 5.31. However, there have been numerous authors who have also presented correlations of stress under different conditions. **Table 5.12** shows some of those correlations.

5.4.1.3 Measuring In-Situ Stress in the Field. Amoco was very active in developing tight gas reservoirs in many areas, including fields in East Texas and some in the Rocky Mountain Basins. As many organizations did in the early days of developing tight gas reservoirs, in-situ stress was measured in both openhole and casedhole situations using some sort of injection test. Sometimes these tests were called in-situ stress tests. But in-situ stress could also be measured using a step-down test or a mini-fracture test.

All of these tests and more are described in detail in Part 3 Ch 15: Data Acquisition and Core Measurements. However, we found that it is important to conduct in-situ stress tests correctly, with limited fluid volumes, limited wellbore-storage effect, and using downhole pressure gauges. If the tests are conducted properly, then the calculated in-situ stresses from logs can be compared to the "measured" in-situ stresses from in-situ stress tests (**Fig. 5.109**). In tectonically relaxed areas, we find the two values correlate well. In tectonically active areas, the measured value of in-situ stress is often larger than the calculated value, and the difference is attributed to the tectonics and this is how the value of σ_{ext} can be determined.

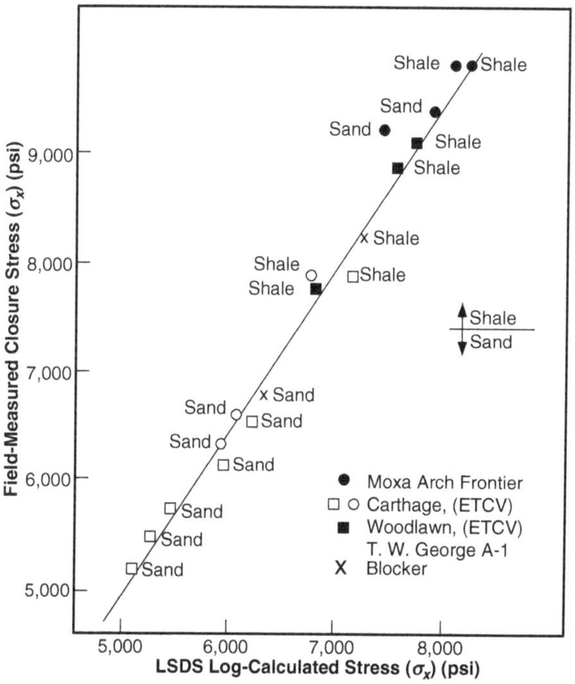

Fig. 5.109—Comparison of in-situ stresses calculated from long-spaced SONs vs. data measured by pump-in tests (Veatch and Moschovidis 1986).

As mentioned, the in-situ stress gradient will change in a reservoir as the pore pressure decreases and gas is produced from the formation. We clearly saw this behavior in 1972–73 when working for Shell Oil Company in the McAllen Ranch Field. Salz, with Shell, published SPE 6870 with field data that clearly show the relationship between in-situ stress and reservoir pressure in the tectonically relaxed Gulf Coast Basin (Salz 1977). Even though many companies have recognized this in-situ stress behavior as a function of reservoir pressure, the Salz paper is important because it was the first paper to document the situation.

Data are shown in the Salz paper for initial fracture treatments and wells that were re-treated with fracturing. The data can be used to verify the poroelastic in-situ stress equation (Eq. 5.29) as mentioned above.

Porosity	13–20%
Permeability	0.1 – 1 md
Young's Modulus	1–54 × 10^6 psi
Poisson's Ratio	0.2 – 0.25
Compressive Strength	4,000 – 15,000 psi
Quartz	40%
Feldspar	20%
Volcanics	20%
Carbonate/Quartz Cement	20%

Table 5.13—Range of properties for the Vicksburg Formation in South Texas.

Table 5.13 shows the range of properties for the Vicksburg Formation.

It is clear from **Table 5.14** that the fracture gradient (in-situ stress gradient) is significantly affected by the pore pressure. This behavior brings up many possibilities. One possibility is that the engineer could choose to pump a small fracture treatment initially, produce the zone, and reduce the in-situ stress. Then, the zone could be refractured, which should allow for better containment of the vertical fracture height growth, resulting in a longer propped fracture.

However, it is not that simple. For example, there are multiple risks in drilling and completing a deep, high-pressure gas well. One major risk is the mechanical risk to the casing, tubing, packers, and liner tops. We have found that, to minimize mechanical risk, it is best to fracture treat a well upon initial completion rather than 1 or 2 years later. There can be unforeseen problems, such as corrosion or corroded inhibitors, that might cause formation damage if these products are in the casing and swept into the new fracture.

Also, when fracture treating a partially depleted zone, the fracturing fluid filtrate loss into the formation can be substantial, leading to low fluid efficiency, formation damage, slow cleanup, and even excessive screenouts if the pad volume is not calculated correctly. For all of these reasons, it is recommended to fracture treat a well upon initial completion to minimize mechanical risks.

However, there are many documented cases of successful refracture treatments, so the problems just discussed should not be used to rule out refracturing of any particular interval. The engineer should be aware that mechanical risks are higher, and that fluid leakoff could be an issue.

5.4.1.4 Rock-Property Profiles with Depth. So far, we have discussed how to make calculations for a single zone or even a portion of that zone. To apply rock-property calculations to the design of a completion and fracture treatment, it is necessary to generate a rock-property profile by depth. Of course, these profiles by necessity must be calculated using log data. We have included considerable discussion in this chapter on how to interpret logs and correlate the log data to the core and well data.

The ultimate objective of the prefracturing formation evaluation in a layered, complex, tight gas reservoir is to estimate values of formation pressure (P), permeability (k), porosity (ϕ), water saturation (S_w), pay thickness (h), in-situ stress (σ_2), and Young's modulus (E) for each rock layer.

Fig. 5.110 shows vertical profiles of many of these reservoir properties that were measured during the GRI tight gas sands research projects in East Texas and Wyoming.

Fig. 5.110 shows the perforations in the Cotton Valley Taylor Sandstone along with the lithology, the estimated permeability, and the in-situ stress on a 2-ft basis. The dashed lines are the calculated values of in-situ stress, and the heavy black, blocky line is the average of the log calculations into layers that could be entered into the fracture design model and the reservoir model. There are three black squares that show the measured in-situ stress obtained from injection falloff testing. This log was one of the first attempts in the GRI project to develop profiles that are necessary for designing a fracturing treatment using a 3D fracture propagation model.

The data in **Fig. 5.111** are for the same vertical interval as the data in Fig. 5.110. However, Fig. 5.111 shows calculated values of Poisson's ratio and Young's modulus from the logs. The dashed lines were the log-derived results, and the solid, blockier lines were the analyst's interpretation of the data into layers that could be used in a fracture design model.

After SFE Well 3, the GRI project turned to the Rocky Mountain Basins to see how the technology developed in the Cotton Valley and Travis Peak Formations could be applied in another formation in another basin. We used the same methods of logging, coring, and testing to generate property profiles in the experimental wells. **Fig. 5.112** shows the lithology, porosity, permeability, and in-situ

Well	Reservoir	Depth (ft)	Pore Pressure Gradient (psi/ft)	Fracture Gradient (psi/ft)
1	Q0–Q1	10,959–11,052	0.856	0.946
			0.396	0.656
2	S	11,437–11,615	0.904	0.968
			0.279	0.629
3	S	11,556–12,090	0.942	0.972
			0.493	0.766
4	S	11,954–12,124	0.914	0.958
			0.229	0.721
5	T	12,142–12,223	0.911	0.979
			0.249	0.651
6	T	11,824–12,026	0.878	0.951
			0.194	0.623

Table 5.14—Measured fracture gradients from six wells that were fractured upon initial completion and then re-treated with fracturing after significant production had occurred.

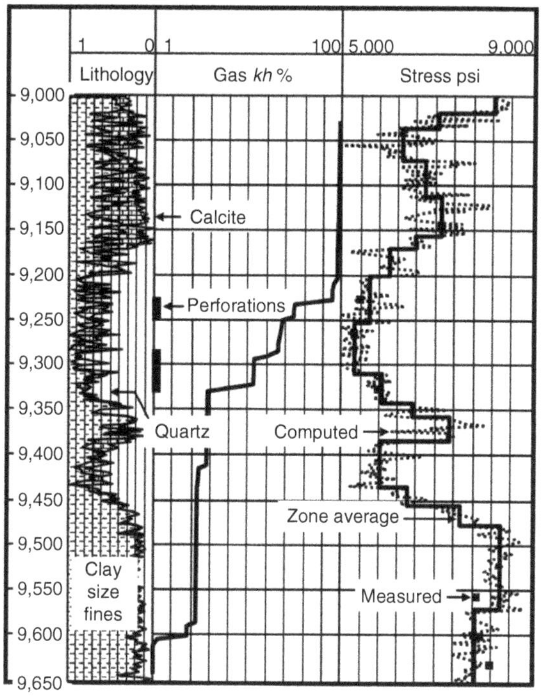

Fig. 5.110—Permeability-thickness and stress profile of completion interval in SFE Well 3.

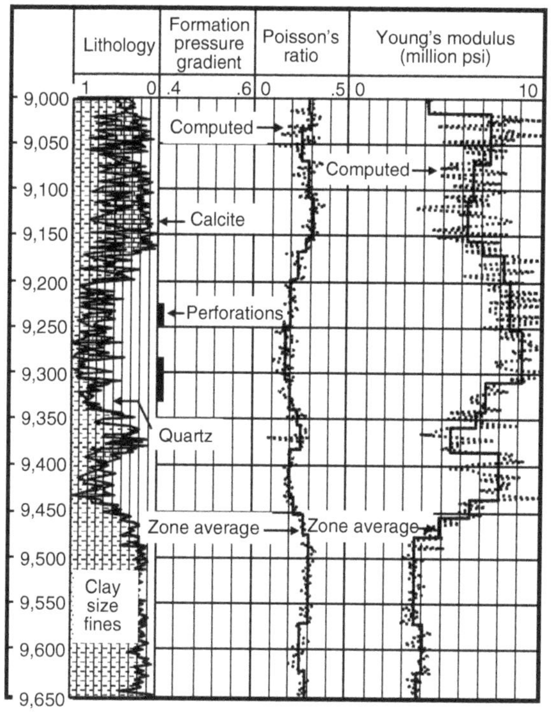

Fig. 5.111—Calculated values for Poisson's ratio and Young's modulus for the completion interval in SFE Well 3.

stress profile for the Anderson Canyon 3-17 well. More data on all the GRI co-operative wells and SFE wells have been included in the published literature and in GRI reports available from GRI.

It is of interest to identify the pay zone from approximately 9,080 ft to 9,150 ft on the basis of the permeability log trace (*kh*). So, when the operator perforates the pay zone, it can be seen from the stress log that the shale immediately above the pay zone has a very high stress value that will likely be a barrier to fracture growth. The stress increases gradually down from the pay zone but should be high enough at approximately 9,200 ft to blunt vertical fracture growth. However, the engineer does not have to guess, because these layer profile data can be entered into a 3D fracture propagation model, which will tell the engineer what the fracture dimensions should be. This was the purpose of the GRI tight gas sand project—to develop data to use in a 3D fracture model to calculate fracture dimensions in real time.

The log profile of the Wyoming U NCT-1 Well 1 depicted in **Fig. 5.113** clearly shows the pay zone with the porosity and permeability profiles and the likely barriers to fracture height growth. Again, these data were used as input for a 3D fracture model to design the fracturing treatment.

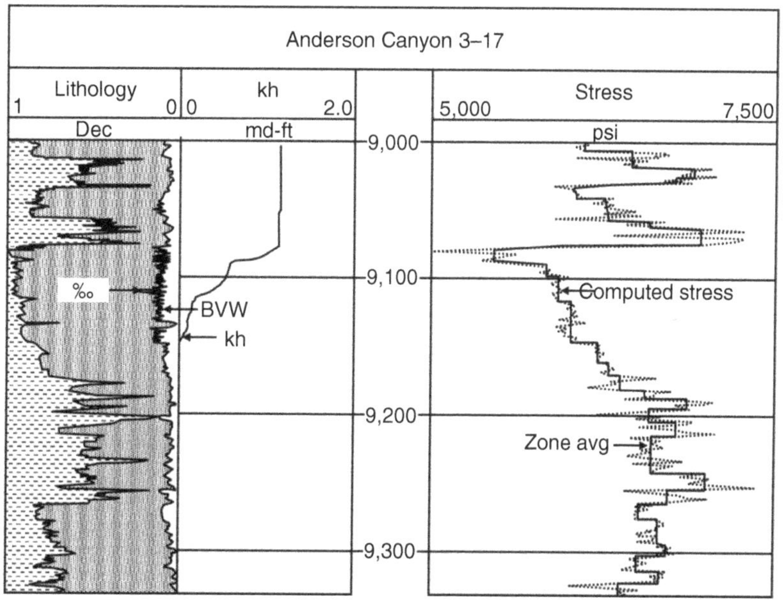

Fig. 5.112—Log analysis, Anderson Canyon Well 3-17 in Wyoming, USA. BVM = bulk volume of water.

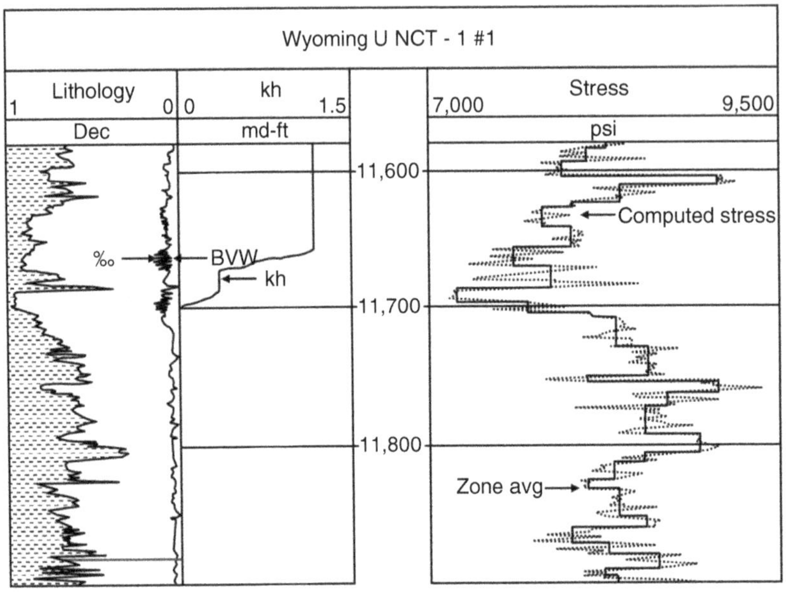

Fig. 5.113—Log analysis, Wyoming U NCT-1 Well 1.

The log of the South Hogsback Well 13-8A shows a profile different from that of the other examples. In **Fig. 5.114,** the same data are presented: lithology, porosity, permeability, and in-situ stress. In this well, there were seven casedhole stress tests that were used to correlate the log-calculated values on stress. In the South Hogsback well, the in-situ stress increases in the downward direction but does not going up. As such, substantial height growth above the pay zone would be expected.

5.4.1.5 Example of In-Situ Stress as a Function of Reservoir Pressure Using Travis Peak Data. A graduate student at Texas A&M University (Gongora 1995) evaluated log and in-situ stress test data from the GRI tight gas sands project in East Texas. Gongora took data from the Travis Peak Formation and developed correlations between lithology and in-situ stress using the GR log. The following data are from Gongora (1995).

Using the results of the 20 in-situ stress tests [16 casedhole stress tests (CHSTs) and 4 openhole stress tests (OHSTs)], a simple plot of in-situ stress vs. the values from the openhole GR log was developed to evaluate the tendency of the relationship. **Fig. 5.115** shows a correlation between the instantaneous GR reading and the in-situ stress value in the Travis Peak Formation for SFE Well 1. It can be established from this plot that there is a definitive correlation between these variables. However, the correlation coefficient in Fig. 5.115 is only 0.1237. It is important to establish that the GR value was directly taken from the log (instantaneous GR) at the point where the well was perforated for the in-situ stress test. GR readings (instantaneous value), closure pressures, and in-situ stress values for each zone are given in **Table 5.15.**

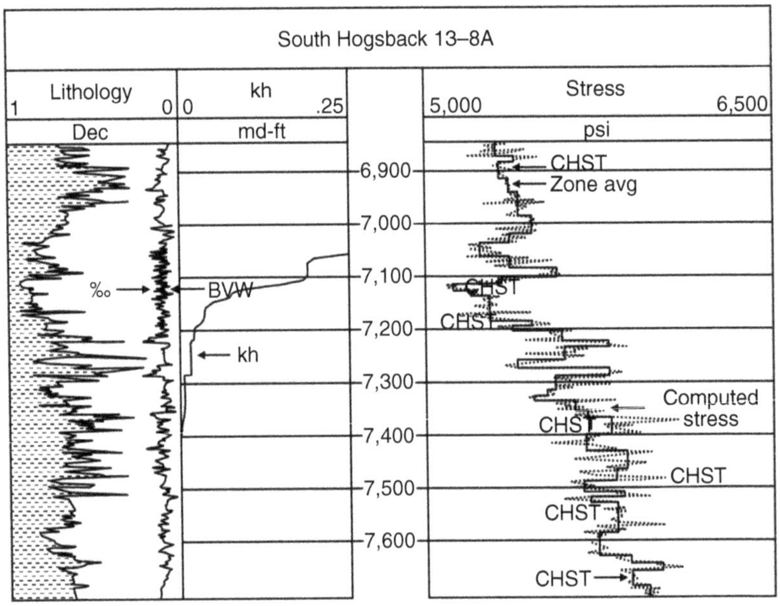

Fig. 5.114—Log analysis, South Hogsback Well 13-8A.

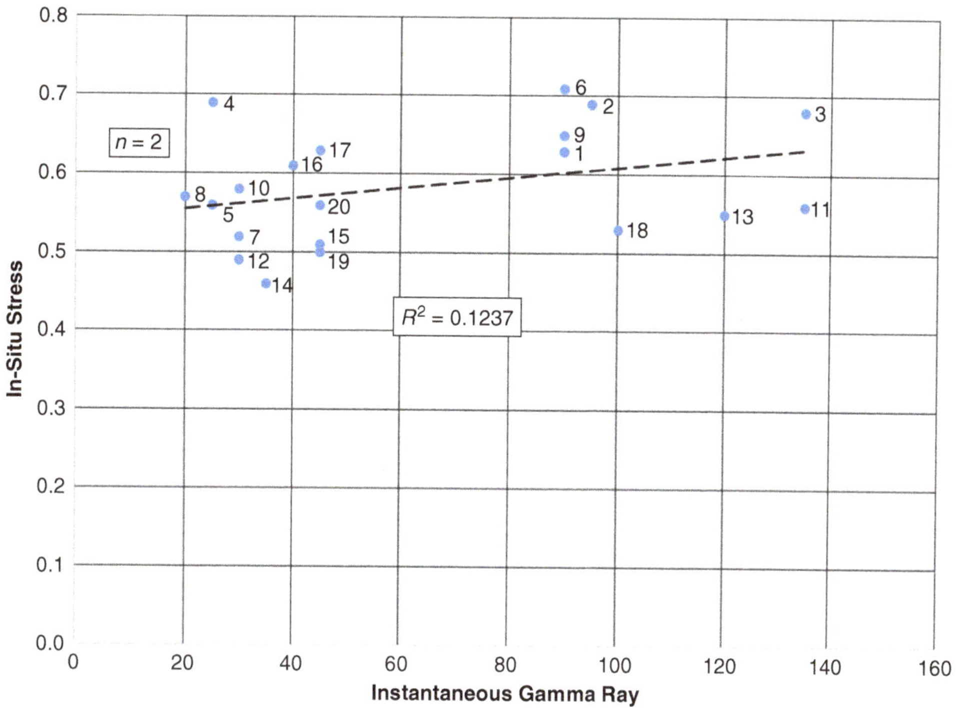

Fig. 5.115—Correlation of in-situ stress value with instantaneous GR reading for SFE Well 1.

Test No.	Test Type	Original Interval (ft)	Lithology	Instantaneous GR Reading	Closure Pressure (psia)	In-Situ Stress (psi/ft)
1	CHST 1	7,784–7,786	Shale	90	5,450±80	0.63
2	CHST 2	7,124–7,726	Shale	95	5,220±115	0.69
3	CHST 3	7,575–7,577	Shale	135	5,510+70,–140	0.68
4	CHST 4	7,560–7,662	Sandstone	25	5,170±180	0.69
5	CHST 5	7,503–7,505	Sandstone	25	3,980±20	0.56
6	CHST 6	7,487–7,489	Shale	90	5,230±50	0.71
7	CHST 7	7,465–7,467	Sandstone	30	3,920±20	0.52
8	CHST 8	7,307–7,309	Sandstone	20	4,350±100	0.57
9	CHST 9	7,279–7,281	Siltstone	90	5,175±40	0.65
10	CHST 10	7,209–7,211	Sandstone	30	4,280±100	0.58
11	CHST 11	6,243–6,245	Shale	135	6,980+60,–90	0.56
12	CHST 12	6,193–6,195	Sandstone	30	3,110±100	0.49
13	CHST 13	6,163–6,165	Shale	120	3,650±60	0.55
14	CHST 14	6,116–6,118	Sandstone	35	2,970+70,–130	0.46
15	CHST 15	5,991–5,993	Sandstone	45	2,910±10	0.51
16	CHST 16	5,847–5,849	Limestone	40	3,600+25,–90	0.61
17	OHST 1	5,846–5,852	Limestone	45	3,850±60	0.63
18	OHST 2	6,144–6,150	Siltstone	100	3,540±80	0.53
19	OHST 3	6,194–6,200	Sandstone	45	3,230±15	0.50
20	OHST 4	7,550–7,560	Sandstone	45	7,990±20	0.56

Table 5.15—Instantaneous GR reading and in-situ stress test for SFE Well 1.

A better correlation can be obtained by reprocessing the GR values. The GR readings in **Table 5.16** are a 10-ft average value of log readings for each of the tested levels. Averaging the log values is necessary to better fit the thickness of the zone of investigation of the microfracture stress test measurements. The information in Table 5.16 also shows the new interval for the average GR reading. **Fig. 5.116** depicts a better relationship after the GR log readings have been adjusted; however, there are still some outliers that should be evaluated. The correlation coefficient of averaging the GR reading in Fig. 5.116 (R^2=0.2173) is better than that in Fig. 5.115, but it is not good enough to conclude a solid relationship between in-situ stress and GR log value.

Test No.	Test Type	Original Interval (ft)	Lithology	Instantaneous GR Reading	Average GR Reading	Closure Pressure (psia)	In-Situ Stress (psi/ft)	New Interval (ft)[1]
1	CHST 1	7,784–7,786	Shale	90	98	5,450±80	0.63	7,781–7,789
2	CHST 2	7,124–7,726	Shale	95	99	5,220±115	0.69	7,723–7,727
3	CHST 3	7,575–7,577	Shale	135	98	5,510+70,–140	0.68	7,573–7,580
4	CHST 4	7,560–7,662	Sandstone	25	46	5,170±180	0.69	7,500–7,568
5	CHST 5	7,503–7,505	Sandstone	25	27	3,980±20	0.56	7,500–7,505
6	CHST 6	7,487–7,489	Shale	90	98	5,230±50	0.71	7,484–7,490
7	CHST 7	7,465–7,467	Sandstone	30	26	3,920±20	0.52	7,465–7,467
8	CHST 8	7,307–7,309	Sandstone	20	29	4,350±100	0.57	7,307–7,309
9	CHST 9	7,279–7,281	Siltstone	90	92	5,175±40	0.65	7,277–7,283
10	CHST 10	7,209–7,211	Sandstone	30	27	4,280±100	0.58	7,209–7,211
11	CHST 11	6,243–6,245	Shale	135	107	6,980+60,–90	0.56	6,232–6,251
12	CHST 12	6,193–6,195	Sandstone	30	31	3,110±100	0.49	6,193–6,195
13	CHST 13	6,163–6,165	Shale	120	111	3,650±60	0.55	6162–6171
14	CHST 14	6,116–6,118	Sandstone	35	30	2,970+70,–130	0.46	6,116–6,119
15	CHST 15	5,991–5,993	Sandstone	45	47	2,910±10	0.51	5,991–5,993
16	CHST 16	5,847–5,849	Limestone	40	40	3,600+25,–90	0.61	5,847–5,849
17	OHST 1	5,846–5,852	Limestone	45	41	3,850±60	0.63	5,846–5,852
18	OHST 2	6,144–6,150	Siltstone	100	94	3,540±80	0.53	6,144–6,150
19	OHST 3	6,194–6,200	Sandstone	45	44	3,230±15	0.50	6,194–6200
20	OHST 4	7,550–7,560	Sandstone	45	22	7,990±20	0.56	7,550–7,560

[1] Interval at which average GR reading is determined.

Table 5.16—Average GR reading and in-situ stress test for SFE Well 1.

At this time, a discussion of the in-situ stress tests is warranted. Fig. 5.116 is a graph of in-situ stress vs. the average GR reading calculated as a 10-ft average value of log readings for each of the tested values. The reasons for the scatter are probably twofold: first, several tests resulted in somewhat questionable values for the in-situ stress, and, second, the actual reservoir pressure in several sands was substantially lower than normal because of pressure partial depletion from offsetting wells producing from the same interval.

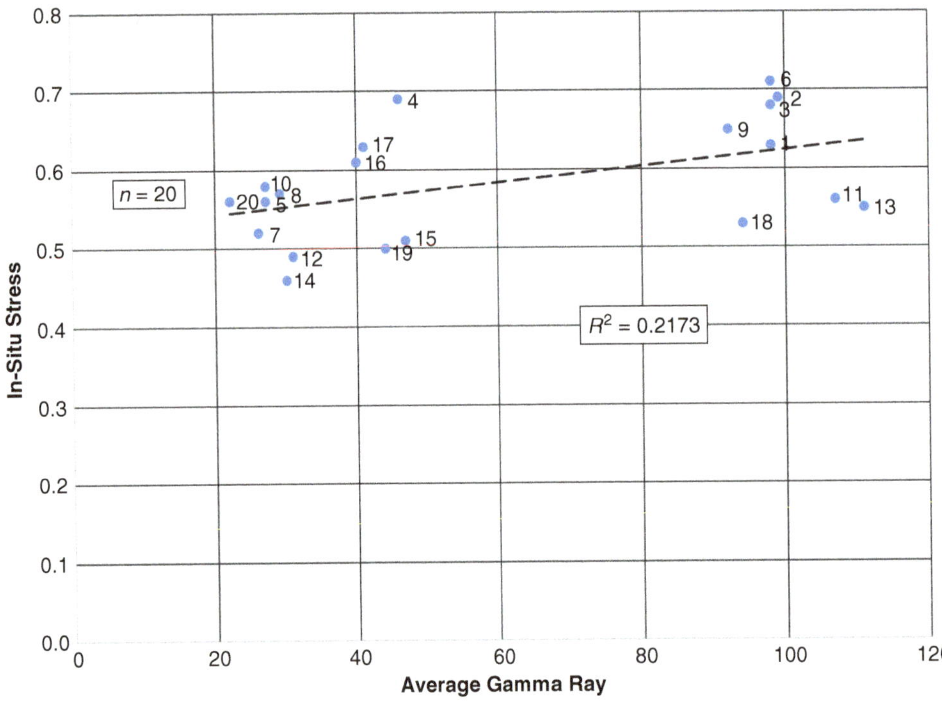

Fig. 5.116—Correlation of in-situ stress value with average GR reading for SFE Well 1.

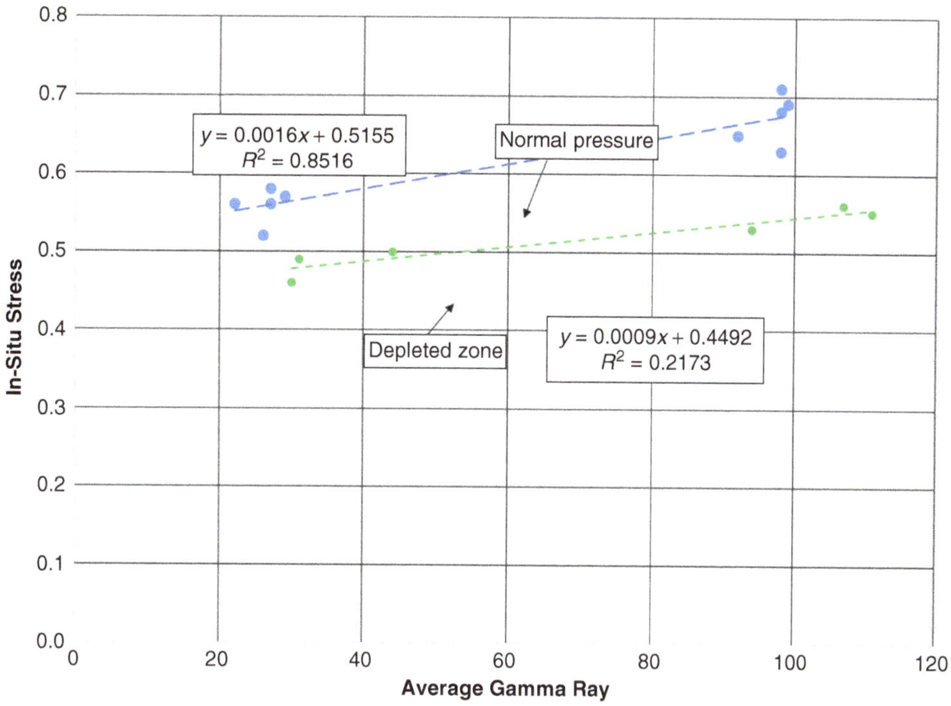

Fig. 5.117—Correlation of in-situ stress value with average GR reading for two different reservoir pressures for SFE Well 1.

Clearly, the determination of in-situ stress is affected by pore pressure. Therefore, by concluding that points 11, 12, 13, 14, 18, and 19 were located in sands that had been pressure depleted and eliminating the questionable tests CHST 4 and 15, which are examples of multiple fracture closures in a single test, a better correlation between the in-situ stress value and GR reading can be obtained. These correlations exclude not only the questionable tests, but also the tests that were conducted in limestone formations (Tests 16 and 17). **Fig. 5.117** shows two different correlation coefficients, one of which is for the data points from normally pressured zones. The other correlation is for the partially depleted zones.

5.4.1.6 Empirical Approach for SFE Well 2. **Fig. 5.118** is the result of plotting 29 results of the in-situ stress tests (26 CHST and 3 OHST) vs. instantaneous GR log readings from another well, SFE Well 2. Once again, it can be calculated from Fig. 5.118 that there

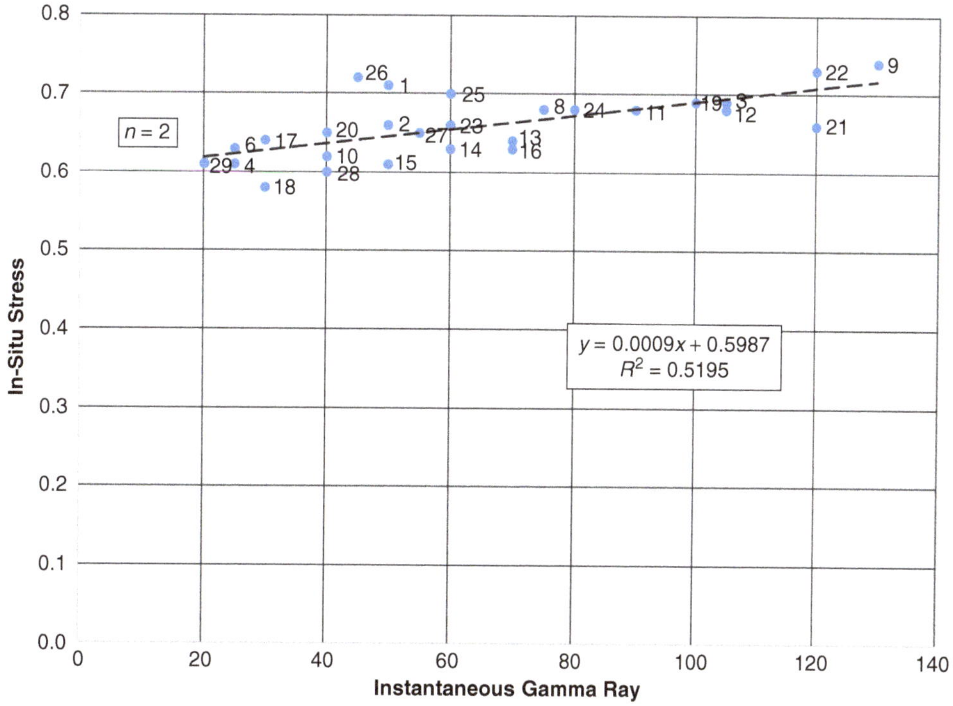

Fig. 5.118—Correlation of in-situ stress value with instantaneous GR reading for SFE Well 2.

Test No.	Test Type	Interval (ft)	Lithology	Instantaneous GR Reading	Closure Pressure (psia)	In-Situ Stress (psi/ft)
1	CHST 1	10,102–10,104	Limestone	50	7,230±30	0.71
2	CHST 2	10,025–10,027	Shaly Sand	50	6,700+140,–60	0.66
3	CHST 3	9,985–9,987	Shale	105	6,860±40	0.69
4	CHST 4	9,944–9,946	Sandstone	25	6,270±30	0.61
5	CHST 5	9,887–9,889	Sandstone	20	6,245±75	0.61
6	CHST 6	9,823–9,825	Sandstone	25	6,355±105	0.63
7	CHST 7	9,754–9,756	Siltstone	60	6,400±200	0.66
8	CHST 8	9,676–9,678	Shale	75	6,600±500	0.68
9	CHST 9	9,590–9,592	Shale	130	7,100±500	0.74
10	CHST 10	9,508–9,510	Sandstone	40	5,900±100	0.62
11	CHST 11	9,425–9,427	Shaly Sand	90	6,200+50,–200	0.68
12	CHST 12	9,065–9,067	Shale	105	6,235+100,–200	0.68
13	CHST 13	8,962–8,964	Shaly Sand	70	5,750±20	0.64
14	CHST 14	8,844–8,846	Shaly Sand	60	5,730±130	0.63
15	CHST 15	8,702–8,704	Sandstone	50	5,270±60	0.61
16	CHST 16	8,516–8,518	Sandy Shale	70	5,505±95	0.63
17	CHST 17	8,416–8,418	Sandstone	30	5,320+130,–60	0.64
18	CHST 18	8,329–8,331	Sandstone	30	4,910±10	0.58
19	CHST 19	8,304–8,306	Shale	100	5,775±15	0.69
20	CHST 20	8,265–8,267	Sandstone	40	5,410±40	0.65
21	CHST 21	8,233–8,235	Shale	120	5,745+45,–95	0.66
22	CHST 22	8,161–8,163	Shale	120	6,080±40	0.73
23	CHST 23	8,111–8,113	Sandstone	60	5,785±65	0.66
24	CHST 24	8,054–8,056	Shaly Sand	80	5,510+95,–30	0.68
25	CHST 25	7,997–7,999	Sand/Lime	60	5,715±70	0.70
26	CHST 26	7,973–7,975	Limestone	45	5,790±50	0.72
27	OHST 1	8,259–8,266	Sandstone	55	5,360±165	0.65
28	OHST 2	9,505–9,510	Sandstone	40	5,740+50,–90	0.60
29	OHST 3	9,814–9,822	Sandstone	20	6,100±50	0.61

Table 5.17—Instantaneous GR reading and in-situ stress test for SFE Well 2.

is a relationship between the in-situ stress and the GR log. However, the correlation coefficient in Fig. 5.118 is only 0.5195. The value of the instantaneous GR reading was directly taken from the log. **Table 5.17** shows the lithology, instantaneous GR reading, closure pressure, and in-situ stress value for each interval.

The correlation in Fig. 5.118 was improved by using the same procedure as in SFE Well 1. **Table 5.18** shows the 10-ft average GR reading and the new analyzed intervals. **Fig. 5.119** represents a better correlation after the GR reading has been averaged. However, there are some outliers that must be explained and will be discussed later. The correlation coefficient in Fig. 5.119 is 0.5787.

As mentioned, three tests were conducted in limestone formations (Tests 1, 25, and 26); these tests are the outlying points in Fig. 5.119. Because of the nature of the limestone formation and the irregular results that are obtained when using these values, the points have been excluded in **Fig. 5.120.** Without the limestone values, the correlation between the in-situ stress value and GR reading increases considerably from 0.5787 to 0.8492. This correlation indicates a direct relationship between the two variables. As shown in Fig. 5.120, the relationship developed is valid for the sandstone/siltstone/shale intervals within the Travis Peak Formation. Very little difference is seen between the openhole and casedhole tests. All zones in this well are at normal pressures or near-normal pressures.

After arriving at the empirical relationships shown in Figs. 5.117 and 5.120, the results of the individual in-situ stress tests are compared with the log-derived in-situ stress measurements on the basis of lithology type to see if the same relationship holds true for all lithologies (excluding limestone). **Fig. 5.121** is a crossplot validation that presents values of the log-derived in-situ stress vs. the measured in-situ stress tests for every interval that is considered in Figs. 5.117 and 5.120.

5.4.1.7 Summary of In-Situ Stress Correlations. Correlations calculated from GR log data and empirically corrected with information from actual in-situ stress tests provide the most useful source of in-situ stress determination for routine application in the Travis Peak Formation. **Table 5.19** summarizes parameter correlations for SFE Well 1 and SFE Well 2.

Table 5.20 shows the in-situ stress and GR classification for both the normal-pressure zone and the depleted zone for both SFE Well 1 and SFE Well 2.

Test No.	Test Type	Interval (ft)	Lithology	Instantaneous GR Reading	Average GR Reading	Closure Pressure (psia)	In-Situ Stress (psi/ft)	New Interval (ft)[1]
1	CHST 1	10,102–10,104	Limestone	50	49	7,230±30	0.71	10,102–10,104
2	CHST 2	10,025–10,027	Shaly Sand	50	56	6,700+140,–60	0.66	10,025–10,027
3	CHST 3	9,985–9,987	Shale	105	92	6,860±40	0.69	9,984–9,987
4	CHST 4	9,944–9,946	Sandstone	25	20	6,270±30	0.61	9,944–9,946
5	CHST 5	9,887–9,889	Sandstone	20	17	6,245±75	0.61	9,887–9,889
6	CHST 6	9,823–9,825	Sandstone	25	23	6,355±105	0.63	9,821–9,828
7	CHST 7	9,754–9,756	Siltstone	60	61	6,400±200	0.66	9,754–9,756
8	CHST 8	9,676–9,678	Shale	75	93	6,600±500	0.68	9,675–9,678
9	CHST 9	9,590–9,592	Shale	130	126	7,100±500	0.74	9,590–9,592
10	CHST 10	9,508–9,510	Sandstone	40	33	5,900±100	0.62	9,508–9,511
11	CHST 11	9,425–9,427	Shaly Sand	90	83	6,200+50,–200	0.68	9,424–9,427
12	CHST 12	9,065–9,067	Shale	105	100	6,235+100,–200	0.68	9,065–9,067
13	CHST 13	8,962–8,964	Shaly Sand	70	52	5,750±20	0.64	8,960–8,966
14	CHST 14	8,844–8,846	Shaly Sand	60	56	5,730±130	0.63	8,842–8,847
15	CHST 15	8,702–8,704	Sandstone	50	42	5,270±60	0.61	8,702–8,710
16	CHST 16	8,516–8,518	Sandy Shale	70	63	5,505±95	0.63	8,505–8,518
17	CHST 17	8,416–8,418	Sandstone	30	43	5,320+130,–60	0.64	8,408–8,418
18	CHST 18	8,329–8,331	Sandstone	30	32	4,910±10	0.58	8,329–8,331
19	CHST 19	8,304–8,306	Shale	100	98	5,775±15	0.69	8,304–8,306
20	CHST 20	8,265–8,267	Sandstone	40	40	5,410±40	0.65	8,265–8,267
21	CHST 21	8,233–8,235	Shale	120	98	5,745+45,–95	0.66	8,230–8,240
22	CHST 22	8,161–8,163	Shale	120	117	6,080±40	0.73	8,161–8,163
23	CHST 23	8,111–8,113	Sandstone	60	63	5,785±65	0.66	8,109–8,113
24	CHST 24	8,054–8,056	Shaly Sand	80	88	5,510+95,–30	0.68	8,054–8,057
25	CHST 25	7,997–7,999	Sand/Lime	60	52	5,715±70	0.70	7,996–8,000
26	CHST 26	7,973–7,975	Limestone	45	47	5,790±50	0.72	7,973–7,975
27	OHST 1	8,259–8,266	Sandstone	55	51	5,360±165	0.65	8,259–8,266
28	OHST 2	9,505–9,510	Sandstone	40	39	5,740+50,–90	0.60	9,505–9,510
29	OHST 3	9,814–9,822	Sandstone	20	27	6,100±50	0.61	9,814–9,822

[1] Interval at which average GR reading is determined.

Table 5.18—Average GR reading and in-situ stress test for SFE Well 1.

One complication in the evaluation of stress models is related to problems of scale. In-situ stress data, as typically acquired in the petroleum industry, represent the stress averaged throughout vertical distances that extend over distinct lithological variations. Consequently, it is questionable whether the interval of cores tested is representative of the lithology that has been stress tested.

Heterogeneity is also typical in rocks tested within a few feet. Rock properties might change significantly even when lithology is apparently the same. Furthermore, it was found that the variability affects the in-situ stress distribution as well as the GR reading.

When evaluating these correlation models, the accuracy of the in-situ stress values and GR log readings must be studied in detail. In-situ stress measurements are uncertain when the closure pressure is not well-defined or when multiple closures are observed. A GR log reading can be variable because of the heterogeneity of the formation. Variability of several pounds per square inch is not uncommon for stress interpretations. However, accurate in-situ stress measurements and average GR log readings in reservoir sandstones are difficult to achieve owing to their nature.

Despite the uncertainty associated with the model parameter selected, this study shows that the prediction of these correlations is in reasonable quantitative agreement with the measured in-situ stress values in the Travis Peak Formation. These correlations predict higher stress in the shales than in the sandstones. Note that these correlations describe the contribution of overburden, pore pressure, and any other external force to the minimum effective in-situ stress.

Some of the sandstones penetrated by SFE Well 1 have been significantly depleted owing to production from adjacent wells. Therefore, for the formation in SFE Well 1, two correlations of in-situ stress have been indicated: The first was obtained at normal pore pressure, and the second was obtained from depleted layers. The correlation model for SFE Well 2 provides an accurate in-situ stress prediction throughout the normal-pressure zone.

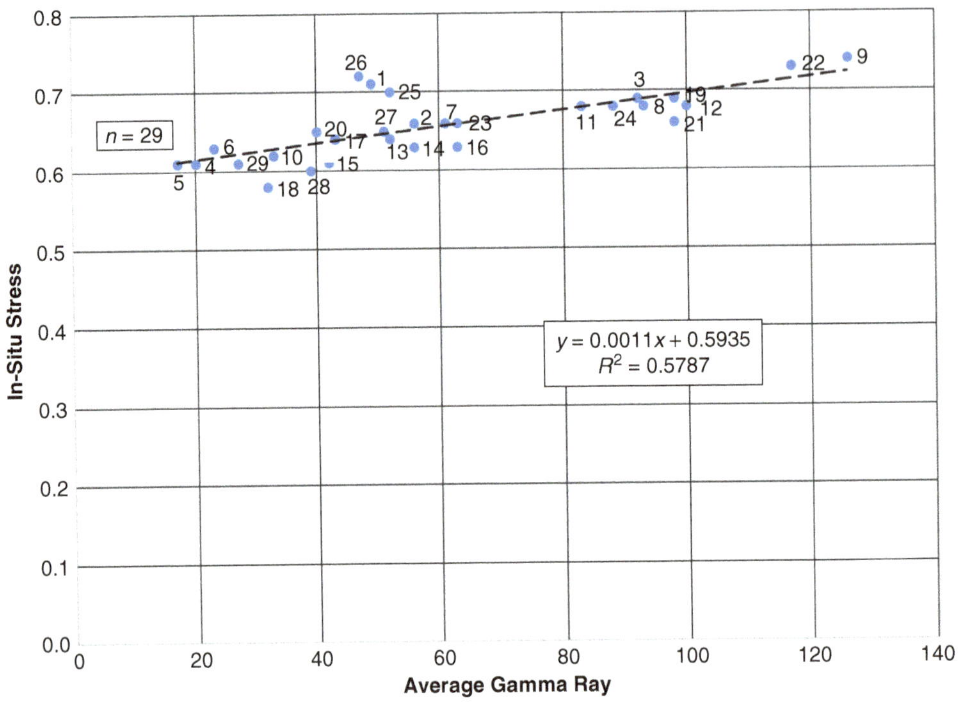

Fig. 5.119—Correlation of in-situ stress value with average GR reading for SFE Well 1.

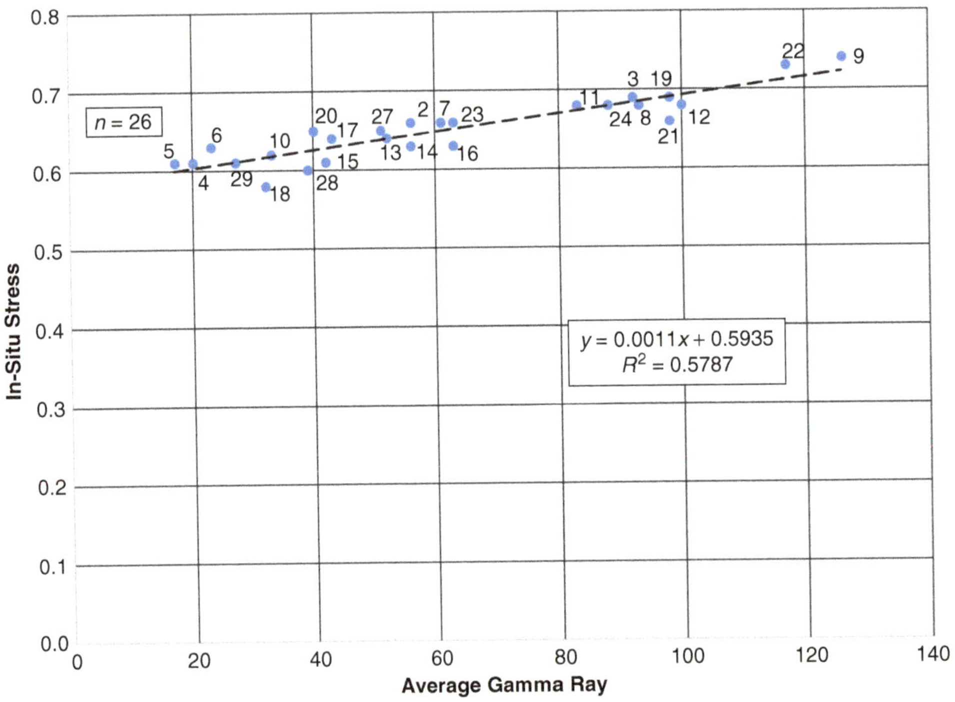

Fig. 5.120—Correlation of in-situ stress value with average GR reading for SFE Well 1.

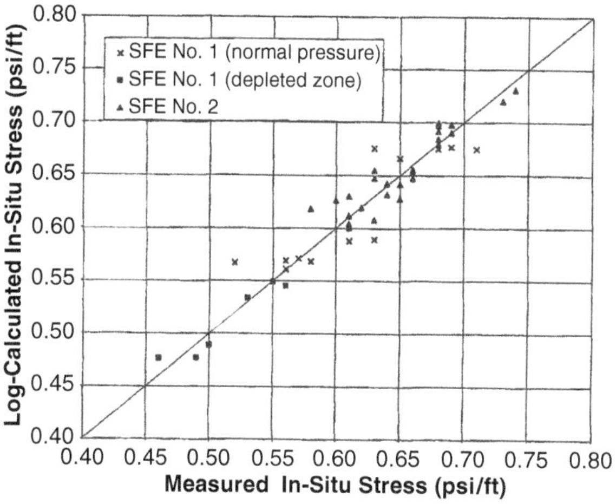

Fig. 5.121—Correlation of calculated in-situ stress and measured in-situ stress for SFE Well 1 and SFE Well 2.

5.4.1.8 Summary. We must use logs and cores to calculate the mechanical rock properties, such as Poisson's ratio, Young's modulus, and in-situ stress. The design engineer must understand the difference between static and dynamic moduli and how to calculate and use the static modulus in the fracture design model. The design engineer also must work with petrophysicists to calculate rock properties from logs and cores and to generate vertical profiles of the rock properties to use in the model. All rock properties should be correlated to rock type (i.e., lithology).

5.4.2 Permeability Distribution. Permeability can be measured with cores or with well tests, such as DSTs, pressure-buildup tests, or production tests. Permeability can also be calculated using the interpretation of openhole well log data. These calculated estimates of permeability are derived from estimates of porosity, water saturation, grain size, and other properties that can be measured with core or estimated from logs. There are many permeability correlations included in the literature as shown in **Table 5.21.**

In Table 5.21, k = permeability, ϕ =porosity, A_g = grain area, D_g = grain diameter, S_{wir} = irreducible water saturation, S_w = water saturation, and k_{rh} = hydrocarbon relative permeability.

Well	In-Situ Stress Correlation	R^2	Formation
SFE Well 1—Normal Pressure	$\sigma = 0.0016\ GR + 0.5155$	0.8516	Travis Peak
SFE Well 1—Depleted Pressure	$\sigma = 0.0009\ GR + 0.4492$	0.9047	Travis Peak
SFE Well 2—Normal Pressure	$\sigma = 0.0012\ GR + 0.5798$	0.8492	Travis Peak

Table 5.19—Summary of in-situ stress correlations.

	Normal Pressure SFE Well 1 and SFE Well 2		Depleted Zone SFE Well 1	
Lithology	σ (psi/ft)	GR (API Units)	σ (psi/ft)	GR (API Units)
Sandstone	0.52–0.66	20–60	0.46–0.51	26–47
Shale	0.63–0.74	90–130	0.55–0.56	107–111
Siltstone	0.65–0.66	60–92	0.53	94
Limestone	0.70–0.71	40–60	N/A	N/A

Table 5.20—In-situ stress and GR classification.

Authors (year)	Permeability Correlations
Kozeny (1927) Carmen (1937)	$k = \phi^3 / \left[5A_g^2(1-\phi)^2\right]$
Berg (1970)	$k = 0.0053\ \phi^3 D_g^2$
Timur (1968)	$k = 0.136 \dfrac{\phi^{4.4}}{S_{wir}^2}$
Coates and Dumanoir (1974)	$\sqrt{k} = 100\phi_e^2 \dfrac{(1-S_{wir})}{S_{wir}}$
Coates and Denoo (1981)	$S_{wir} = 1-\left[(1-S_w)^{2.1} / k_{rh}\right]^{0.5}$

Table 5.21—Classic permeability correlations.

Notice that these correlations are based on values of porosity, grain size, and water saturation. All of these correlations were developed essentially for clean, high- to moderate-permeability formations and do not apply directly to tight gas reservoirs. However, equations similar in form to those in Table 5.21 can be used to develop correlations by determining new values of the coefficients, such as the 0.0053 in Berg's equation. Correlations for shaly, tight gas sands are presented in Section 5.4.3 of this chapter. In the following discussion, we will focus on how permeability is distributed in a tight gas sand formation.

To illustrate the distributions of permeability in tight gas reservoirs, information from three tight gas formations was evaluated using data submitted to the Texas Railroad Commission (RRC) to provide proof that these formations met the definition of a tight gas reservoir. If designated as a tight gas reservoir, the operator of the gas wells could either negotiate a higher price for the gas or federal income tax credits for the gas produced from these formations. These issues were part of the Natural Gas Policy Act of 1978.

Here, we present data from the Cotton Valley Formation in East Texas, the Cleveland Formation in North Texas, and the Wilcox Lobo Formation in South Texas. The data that we used for the permeability distributions came from the following hearings:

- Cotton Valley Formation, RRC Docket No. 20-75, 144
- Cleveland Formation, RRC Docket No. 10-77, 222
- Wilcox Lobo Formation, RRC Docket No. 20-75, 144

Figs. 5.122 through 5.128 illustrate the permeability distributions for the three formations. These figures are classic examples of skewed permeability distributions that are essentially distributed in a log-normal pattern.

The definition of a tight gas reservoir by the Federal Energy Regulatory Commission (FERC) was that the expected permeability of the next well drilled would be 0.1 md or less. Some of the FERC examiners wanted to use the arithmetic average permeability, but, according to the permeability distribution, the correct value for expected permeability would be the median value. As you can see in Fig. 5.122, the median for all three formations was 0.011–0.027 md. If the analyst had to use the arithmetic average, then the

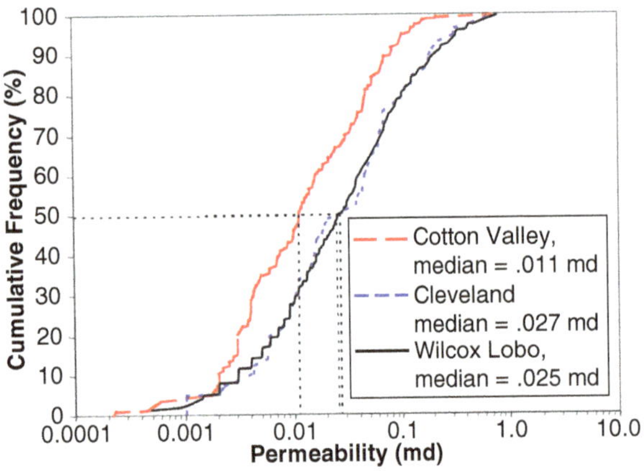

Fig. 5.122—Permeability distribution in three tight gas sand reservoirs.

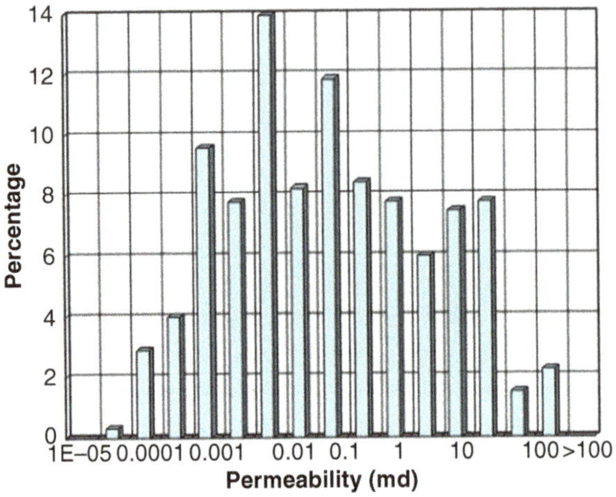

Fig. 5.123—Permeability distribution in the Cotton Valley Formation.

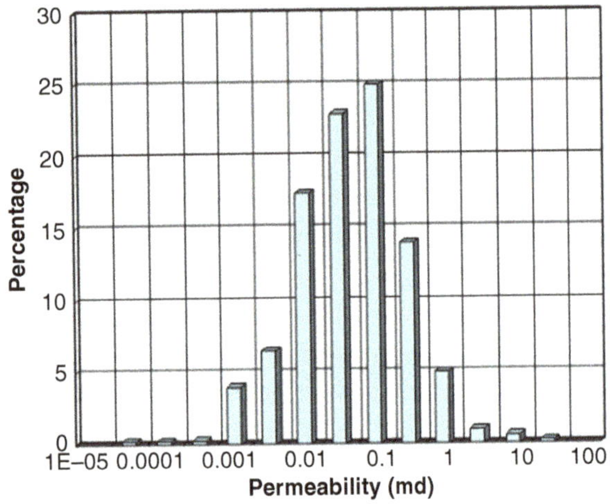

Fig. 5.124—Permeability distribution in the Cleveland Formation.

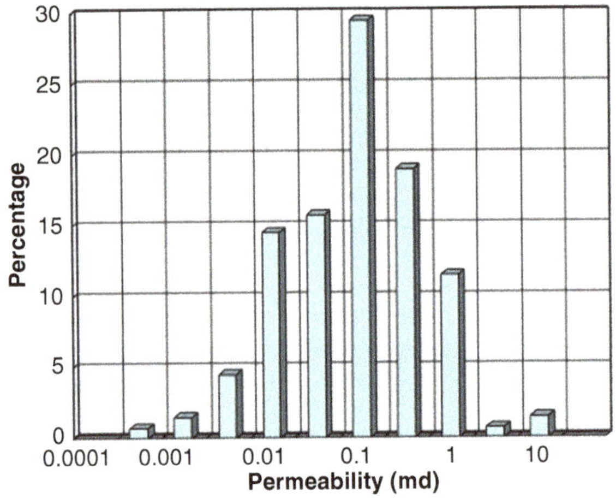

Fig. 5.125—Permeability distribution in the Wilcox Lobo Formation.

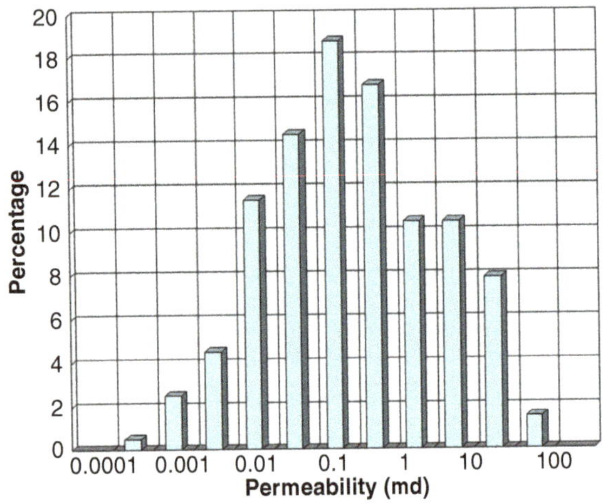

Fig. 5.126—Permeability distribution in Travis Peak Formation.

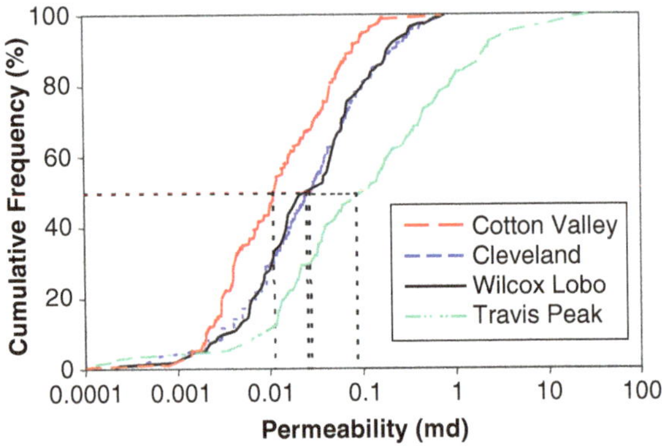

Fig. 5.127—Permeability distribution of four tight gas sand reservoirs.

Formation	Number of Wells	Arithmetic Mean (md)	Permeabilities Median (md)	Minimum (md × 10^5)	Maximum (md)
Cleveland	391	0.179	0.028	3.0	19.6
Cotton Valley	395	7.378	0.045	8.0	358.0
Wilcox Lobo	112	0.235	0.056	26.0	7.9
Travis Peak	191	1.035	0.085	7.9	27.9

Table 5.22—Arithmetic means and medians for the four field cases shown in Fig. 5.127.

few high-permeability data points would increase this expected number to an unreasonably high estimate. The FERC, however, did not use logic and made companies eliminate the high-permeability wells from the tax credits. Politics overruled science.

In Figs. 5.123 through 5.125, we have shown the distribution of permeability using a log scale for permeability. Although the permeability data we analyzed are not exactly in a log-normal distribution, the information in Figs. 5.123 through 5.125 indicate the distribution is approximately log-normal. It is believed that, if all the information on the reservoir were available (and it never will be), the data would be exactly log-normal.

As clearly discussed in this book, the GRI tight gas sands research program collected substantial amounts of data in the Travis Peak Formation. We analyzed the Travis Peak data to also determine what the average production should be from a Travis Peak well. We were trying to determine what the next well drilled would be expected to produce. Fig. 5.126 illustrates the permeability distribution in the Travis Peak Formation, which is also log-normal.

We next added the Travis Peak distribution to the other three formations shown in Fig. 5.122. The median permeability in the Travis Peak Formation was 0.085 md, so it also meets the FERC definition of a tight gas reservoir.

As briefly discussed above, there was a disagreement between the FERC staff and the Texas RRC along with some of the analysts on how to average permeability. Even though the answer is clear—use the median in a log-normal distribution—significant work went into illustrating why the median should be used.

The question we addressed was "What is the definition of average?" The nature of the distribution will affect how the average of the data determines which value of permeability should be used to represent the average well.

Table 5.22 illustrates these measures of the central tendency for the four data sets included in Fig. 5.127. Notice that the median values of permeability—the most likely value of permeability to be encountered by the next well drilled—are all less than the 0.1-md threshold for being a tight gas reservoir. However, because there were a few wells that had values of permeability significantly greater than the median, these few wells caused the arithmetic mean to far exceed the threshold. If the arithmetic mean were used as the average, the gas-flow rates and recoveries would be substantially overpredicted. To understand this better, we used reservoir simulation to perform an analysis of which value of permeability should be used to estimate ultimate gas recovery. We performed this work using the Travis Peak data.

To estimate the recovery of an average well in the Travis Peak Formation, we must be aware that the values of net-pay and porosity will also be a function of permeability, as illustrated in the following data. The data in **Figs. 5.128 and 5.129** and in **Table 5.23** were determined using information from the GRI tight gas sands research project.

Table 5.24 presents the results of this simulation study. The Actual Average Recovery column shows the cumulative recovery of all wells divided by the total number of wells. The median value of permeability was k = 0.085 md, and the mean value of permeability was k = 0.907 md. Using these values of permeability, runs were made for fracture half-lengths of 0 and 500 ft and two different values of well spacing to determine average gas recovery. It is clear that the median value of permeability results in gas production estimates that are closer to the actual permeability.

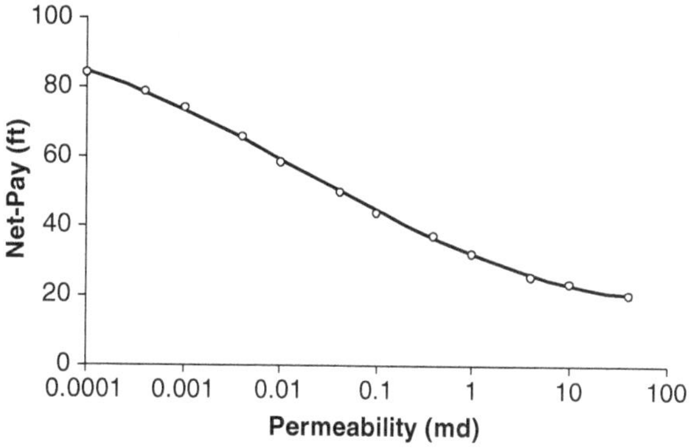

Fig. 5.128—Net-pay distribution in the Travis Peak Formation.

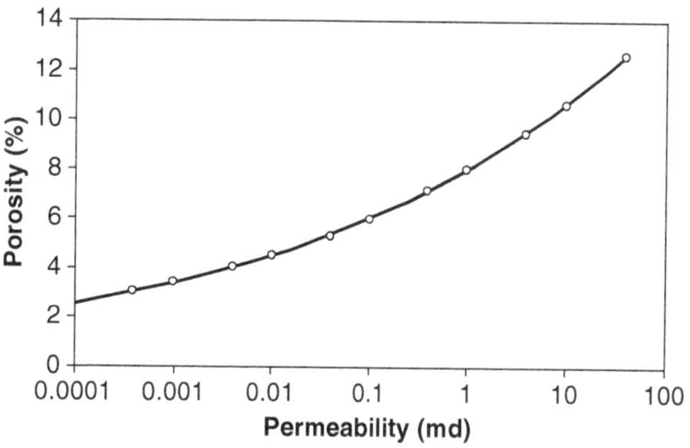

Fig. 5.129—Porosity distribution in the Travis Peak Formation.

Gas Permeability (md)	Number of Wells	Gas Porosity (%)	Net-Pay (ft)
40	11	12.65	20
10	25	10.65	23
4	58	9.50	27
1	107	7.99	32
0.4	159	7.13	37
0.1	188	6.00	44
0.04	178	5.35	50
0.01	134	4.50	59
0.004	80	4.02	66
0.001	39	3.38	74
0.0004	15	3.02	79
0.0001	5	2.54	84
0.00004	1	2.25	90

Table 5.23—Distributions used to perform reservoir simulation results to estimate average gas recovery of a well drilled in the Travis Peak Formation.

Well Spacing (acres)	Fracture Length	Actual Average Recovery (Bcf/Well)	Recovery for Median (Bcf/well)	Recovery for Arithmetic Average (Bcf/well)
160	0	1.97	2.37	2.92
640	0	5.16	4.32	11.02
160	500	2.63	2.97	2.97
640	500	7.95	9.33	11.70

Table 5.24—Results from a simulation study comparing permeability estimates to actual permeability.

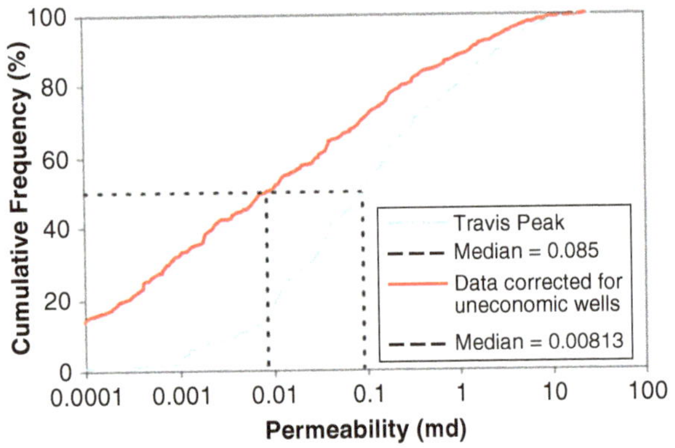

Fig. 5.130—Permeability distribution in the Travis Peak Formation, corrected to include uneconomic wells.

However, there is still a problem with these recovery values. The average values of permeability and the data in Table 5.24 are from wells that were completed and produced. These data are indicative of the best parts of the Travis Peak Formation.

Typically, data from producing wells are used to evaluate the overall economics of a basin or field. It is important to remember that the data are biased toward the better wells. Better estimates should include all wells drilled, including the uneconomic wells. The information in the bulleted list below about the Travis Peak Formation illustrates the importance of using all available data on a reservoir.

We need to reanalyze the Travis Peak data set by correcting for dry holes. Many of the early Travis Peak wells drilled in the 1930s, 1940s, and 1950s produced oil and gas from the very porous and permeable sections within the Travis Peak Formation. However, most of the permeable formations associated with known structures have been developed, and current exploration focuses on the deeper, lower-porosity, lower-permeability portion of the formation.

In **Fig. 5.130,** we can see from the permeability distribution that the median permeability in the Travis Peak Formation is less than 0.1 md. However, there are many wells that contain rock layers with permeabilities greater than 1 md. There are also many layers with permeabilities in the microdarcy range. We can obtain a clearer picture of the average permeability and average gas recovery by analyzing all wells drilled into the Travis Peak Formation, including dry holes. We also understand that some of the dry holes were just uneconomic wells at the time they were drilled.

The following are some statistics we derived from public data bases.

- The Travis Peak Formation in East Texas has been under development for more than 50 years.
- According to public information, the following statistics represent activity through 1 July 1990:
  - 12,308 wells have been drilled to or through the Travis Peak Formation.
  - 3,731 of the 12,308 wells were tested (30%).
  - 2,932 of the 12,308 wells were completed.
  - 4,030 of the 12,308 wells were uneconomic wells (33%).
  - 467 of the 4,030 wells were fracture treated (12%).
  - 4,785 of the 12,308 wells reached TD.
  - 2,500 of the 4,785 wells were tested (52%).
  - 2,087 of the 4,785 wells were completed.
  - 2,698 of the 4,785 wells were not completed (56%).
  - 1,465 of the 2,500 wells are still producing (59%).
  - 1,465 of the 4,785 wells that reached TD are still producing (31%).

We took this information and calculated the following Travis Peak ratios:

- (4,030/12,308) or 33% of all wells drilled to or through the Travis Peak Formation were uneconomic wells.
- (467/4,030) or 12% of the uneconomic wells had a test in the Travis Peak Formation.
- (3,731/12,308) or 30% of all wells drilled were tested in the Travis Peak Formation.
- (2,500/4,785) or 52% of all wells reaching TD in the Travis Peak Formation actually tested the Travis Peak Formation
- (2,698/4,785) or 56% of the wells that reached TD in the Travis Peak Formation were not completed in the Travis Peak Formation.
- (1,465/4,785) or 31% of the wells reaching TD in the Travis Peak Formation are still producing.
- (1,465/2,500) or 59% of all wells reaching TD in the Travis Peak Formation and that were tested in the Travis Peak Formation are still producing.

We can then use these ratios to correct the permeability distribution in the Travis Peak Formation for the uneconomic wells. Notice that the median value of permeability decreased from 0.085 to 0.00813 md.

5.4.2.1 Summary. We summarize with the following statements:

- Statistical correlations can be used with complete data sets from a few select wells to estimate critical reservoir properties for remaining wells.
- The cost of obtaining the data sets can be distributed to the other wells so that gathering the data is economical.

5.4.3 Correlation of Permeability.

5.4.3.1 Defining the Problem. Optimal development of the resource is important in any oil and gas play. Tight gas reservoirs are challenging to develop economically mainly because of low gas recovery per well and the high cost of completions. Cumulative gas recovery (thus income) per well is limited because of low permeability, low recovery efficiencies, and smaller drainage areas when compared to high-permeability gas wells. As such, the data required must be developed to optimize the reservoir at the lowest feasible cost.

When planning the development and considering options, the engineer must consider the costs required to acquire the data for planning. To drill and complete a well at low cost, most managers want to minimize the money spent to log wells and eliminate extra expenses such as coring and well testing. However, in these layered unconventional gas reservoirs, we often need more data than is normally required to analyze a typical high-permeability reservoir.

Well	Permeability Correlations	R^2
SFE No. 1	$k = 2{,}800\,\dfrac{\phi^{5.87}(1-I_{GR})^{0.2} R_{ild}^{\;2.54}}{(R_{ild}-R_{sfl})^{1.08}}$	0.79
SFE No. 3	$k = 6{,}900\,\dfrac{\phi^{6.23}(1-I_{GR})^{0.2} R_{ild}^{\;2.13}}{(R_{ild}-R_{sfl})^{1.65}}$	0.81
LCU No. 48	$k = 20{,}160\,\dfrac{\phi^{6.89}(1-I_{GR})^{0.3} R_{ild}^{\;1.18}}{(R_{ild}-R_{sfl})^{1.51}}$	0.70

Table 5.25—Correlations of core permeability to log data for three wells in the GRI tight gas sand project.

The solution to effective formation characterization is to pay for and obtain the data required to fully characterize the layered, tight gas reservoir from a few wells, then develop correlations with logging data to use on the remainder of the wells drilled. Once acceptable correlations are developed for specific reservoirs in specific geographic areas and these correlations are applied to other wells, we can develop accurate data sets for analysis at minimum cost.

For example, if USD 100,000 is spent to cut and analyze core and generate core/log correlations and these correlations are then used in a 10- to 100-well infill-drilling project, the cost per well to obtain accurate data sets is only USD 1,000–10,000. Normally, the two most critical data items that must be characterized for every layer above, below, and within the pay zone are permeability and in-situ stress. We discussed in-situ stress correlations in Section 5.4.1. Here, we will focus on permeability correlations.

We need to develop the relationship between permeability and porosity as a function of lithology or rock type, or flow units, and rock mechanical properties. There have been many correlations published since 1927; however, most were developed using data from high-permeability formations and might not apply to low-permeability reservoirs.

Even though these correlations were developed essentially for clean, high- to moderate-permeability formations and do not apply directly to tight gas reservoirs, we have developed other correlations that do work in low-permeability, shaly formations. Other equations similar in form to those in Table 5.21 can be used to develop correlations by determining new values of the coefficients, such as the 0.0053 in Berg's equation.

Correlations by Yao (1992) in his PhD dissertation used data from the Travis Peak and Wilcox Formations; again, the data were generated during the GRI tight gas research project **(Table 5.25).** Yao and Holditch (1993) used the following relationship to correlate data from low-permeability sandstones. The logic is that permeability should be a function of porosity, shale volume, water saturation, and the degree of mud-filtrate invasion as reflected by the ratio of deep to shallow resistivity measurements.

$$k = U \times f(\phi, V_{SH}, \frac{R_D}{R_S}, S_w). \quad (5.32)$$

The final correlation, developed by Yao mainly using raw log data from Travis Peak and Wilcox cores, was as follows:

$$k = U\left(\frac{\phi^{e^1}(1-I_{GR})^{e^2} R_{ild}^{\;e^3}}{(R_{ild}/R_{sfl})^{e^4}}\right), \quad (5.33)$$

where U = unit conversion factor that varies for different reservoirs and drilling mud systems, e^1 = 5.87–6.89, e^2 = 0.2–0.3, e^3 = 1.18–2.54, and e^4=1.08–1.65.

Figs. 5.131 through 5.133 show the results of the correlation effort. The solid line was calculated from logs and was empirically fit to the core-measured values of permeability. The details of Yao's work can be found in his dissertation and subsequent SPE papers.

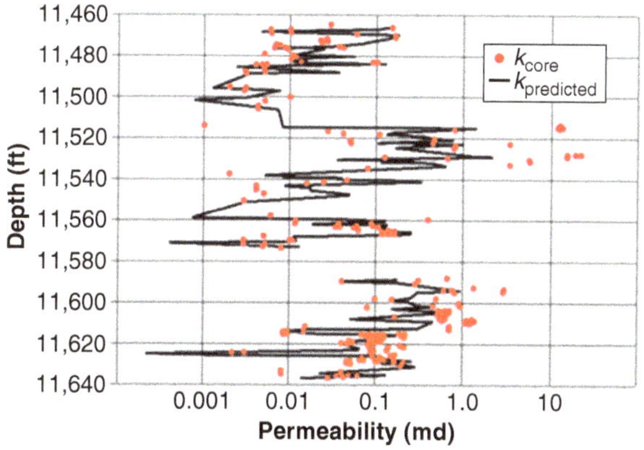

Fig. 5.131—Log-derived permeability vs. core permeability for LCU Well 48.

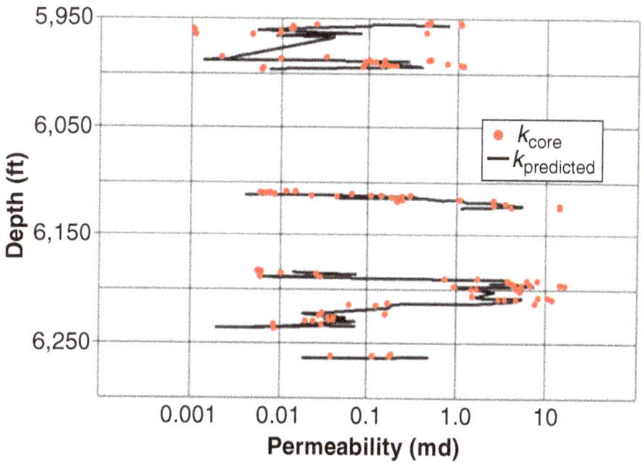

Fig. 5.132—Log-derived permeability vs. core permeability for SFE Well 1.

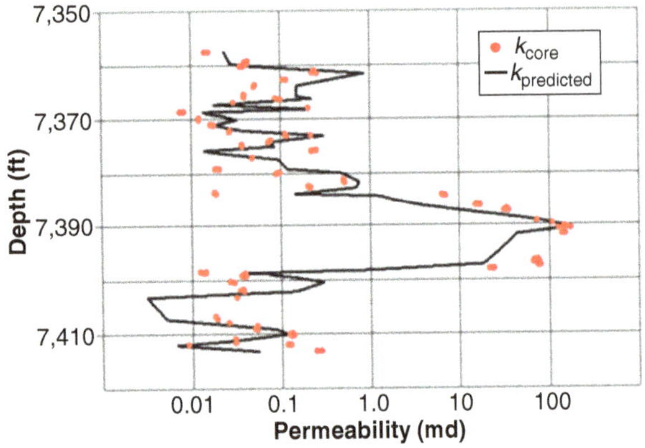

Fig. 5.133—Log-derived permeability vs. core permeability for SFE Well 3.

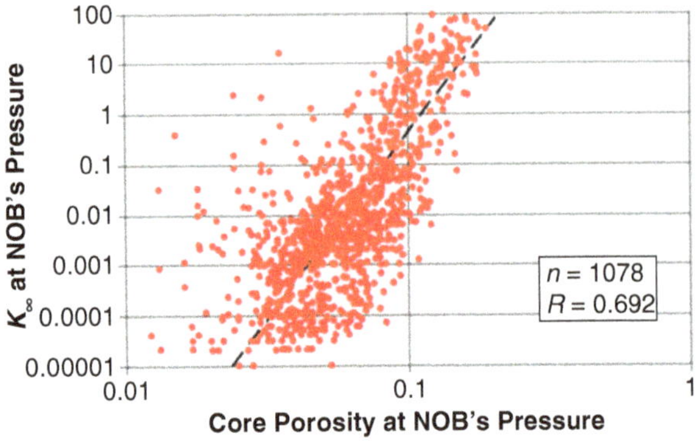

Fig. 5.134—Correlation using all 1,078 Travis Peak core plugs, where $R^2 = 0.692$.

The conclusion from the work by Yao is that it is possible to develop basic correlations between log data and core permeability for a given formation in a given field. The parameters controlling permeability are porosity, shale content, and water saturation. The presence or absence of mud-filtrate invasion can also be used in the permeability correlations. Subsequent work by Semmelbeck and Diyashev (1996) and others clearly shows that there can be a direct link between the mud-filtrate-invasion profile and permeability.

In addition to developing correlations of permeability using log data, we also must refine how we generate correlations between permeability and porosity data from cores as a function of lithology, which some authors call rock type. We have developed the correlations of permeability and porosity data using the Travis Peak data set to illustrate the process.

In layered reservoirs, the core and log data must often be subdivided by lithology or rock type to properly correlate the data. **Fig. 5.134** is a graph of k_{NOB} vs. ϕ_{NOB} using the complete Travis Peak core data set with 1,078 data points. The correlation coefficient is 0.692, which implies that the permeability can be 69% explained by the value of porosity alone.

However, many of the cores used to generate the correlation were fractured. Some fractures were natural, while others were coring-induced. We had a geologist analyze each core plug, select the plugs that were not fractured, and eliminate the cores that contained fractures.

Fig. 5.135 shows the same correlation with all fractured cores removed. Notice that the number of data points is 615, and the correlation coefficient has increased to 0.784.

The data in Fig. 5.135 still represent many different depositional systems. To further improve the correlation, we will look only at data from clean, fluvial-deltaic channel sands. Again, we had a geologist inspect each core plug and pick the core from the channel sands for the correlation. The correlation coefficient is now 0.865.

As suggested in **Figs. 5.134 through 5.136,** permeability vs. porosity correlations can be improved by dividing the formation rocks into flow units. Amaefule et al. (1993) used the same Travis Peak data to show how flow units should correlate data.

The authors start with the generalized form of the Kozeny-Carmen relationship (Carmen 1937), where F_s = shape factor, two for a circular cylinder, and S_{gv} = surface area per unit grain volume:

$$k = \frac{\phi_e^{\,3}}{1-\phi_e^{\,2}}\left(\frac{1}{F_s\tau^2 S_{gv}^{\,2}}\right) \qquad (5.34)$$

The term $F_s\tau^2$ is referred to as the Kozeny constant, and it usually varies from 5 to 100. The Kozeny equation is difficult to use because the constant varies by rock type, and its value is often not known.

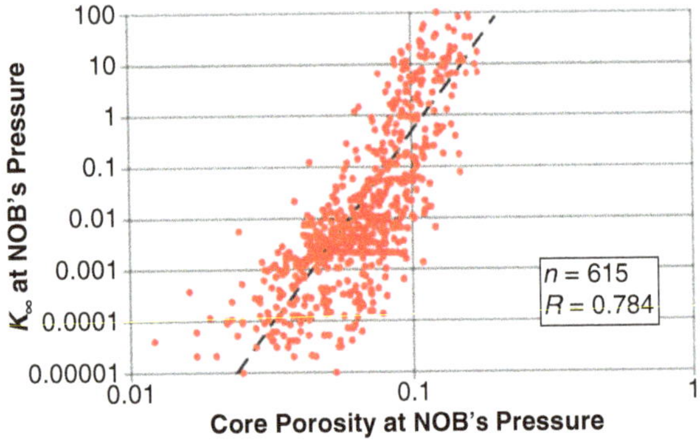

Fig. 5.135—Correlation using 615 Travis Peak core plugs that were not fractured, where $R^2 = 0.784$. Invalid data are excluded.

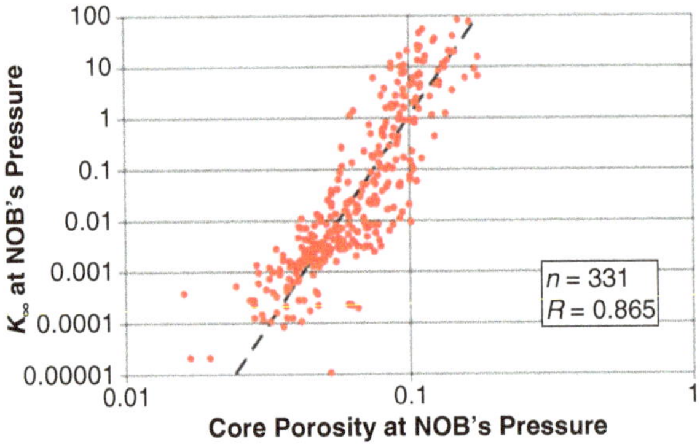

Fig. 5.136—Correlation using 331 Travis Peak core plugs that were not fractured and from fluvial-deltaic channel sands, where $R^2 = 0.865$. Invalid data are excluded.

We can rewrite the Kozeny-Carmen equation as

$$\sqrt{\frac{k}{\phi_e}}=\left(\frac{\phi_e}{1-\phi_e}\right)\left(\frac{1}{\sqrt{F_s\tau S_{gv}}}\right). \quad (5.35)$$

We define the reservoir-quality index (RQI) (mm) as

$$\mathrm{RQI\,(mm)}=0.0314\sqrt{\frac{k}{\phi_e}}. \quad (5.36)$$

Also, the pore-volume to grain-volume ratio is

$$\phi_2=\frac{\phi_e}{1-\phi_e}. \quad (5.37)$$

Then, the flow-zone indicator (FZI) is

$$FZI=\frac{1}{\sqrt{F_s\tau S_{gv}}}=\frac{RQI}{\phi_2}. \quad (5.38)$$

When previous equations are combined, take the log of both sides and use Eq. 5.39:

$$\log(RQI)=\log(\phi_2)+\log(FZI). \quad (5.39)$$

On a log-log graph of RQI vs. ϕ_z, all samples with similar FZI values will lie on a straight line with unit slope. Samples with different FZI values will lie on other parallel lines, and the value of FZI can be determined at the value of $\phi_z = 1$ on the graph. Samples with the same value of FZI are assumed to have similar pore-throat attributes and thereby constitute a hydraulic unit.

If k(md), FZI (mm), and ϕ(fraction), then

$$k=1{,}014(FZI)^2\left[\frac{\phi_e^{\,3}}{(1-\phi_e)^2}\right]. \quad (5.40)$$

The following example shows the results of using the Amaefule et al. (1993) method on the GRI Travis Peak core data. **Fig. 5.137** shows the values of core data k_{NOB} vs. ϕ_{NOB} for SFE Well 1, which was Howell Well 5. The two solid lines are conventional statistical fits for the two cores, while the dashed lines are labeled using expected relations on the basis of clay morphology, as described by Amaefule et al. (1993).

The correlation of RQI vs. ϕ_2 for the SFE Well 1 data set is shown in **Fig. 5.138.** As many as six flow units have been defined by Amaefule et al. (1993).

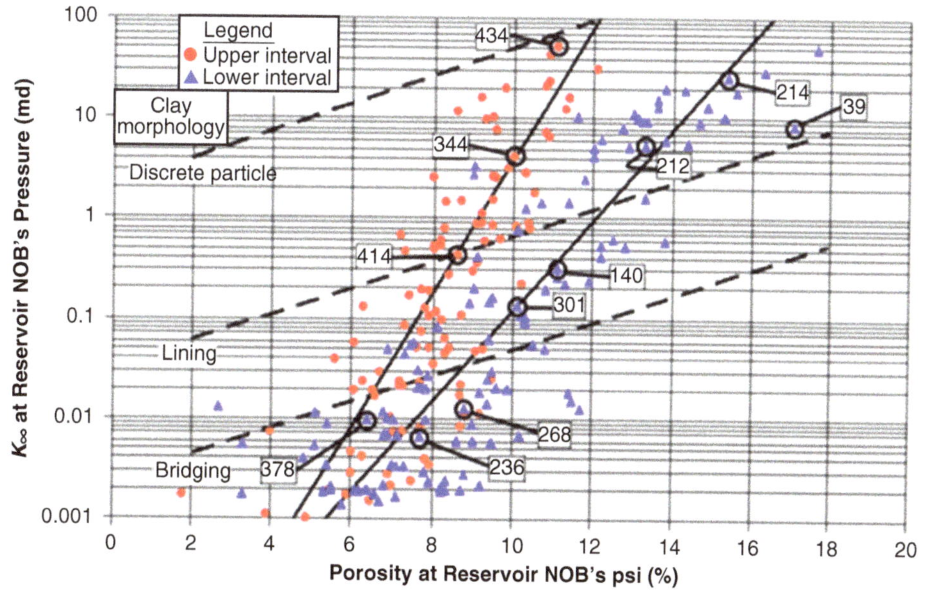

Fig. 5.137—Permeability at NOB pressure vs. porosity at NOB pressure for two sets of cores in the Travis Peak Formation.

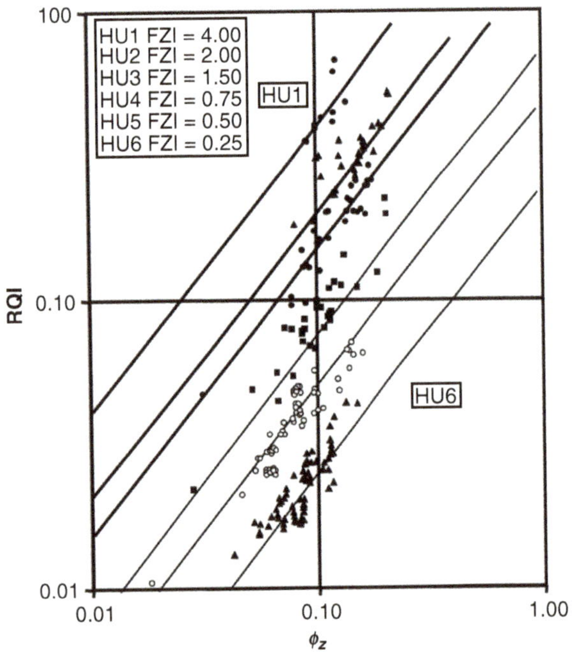

Fig. 5.138—Correlation between RQI and porosity for core data taken from SFE Well 1.

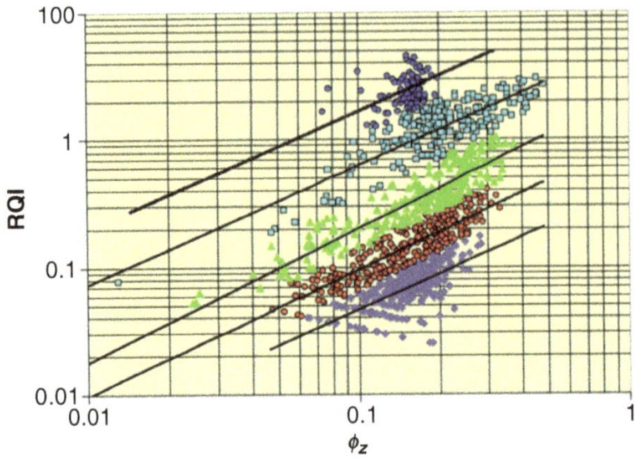

Fig. 5.139—Clustering the core data into the optimal number of HFUs.

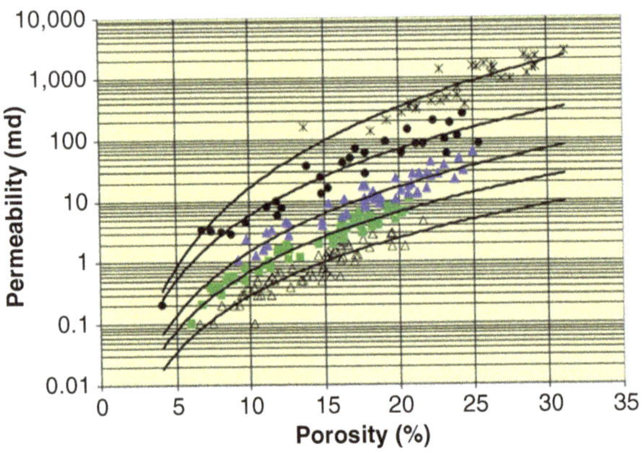

Fig. 5.140—Porosity vs. permeability relationship derived using the HFU clustering technique.

Permeability estimation using the hydraulic-flow-units (HFU) technique was later improved by developing an unbiased clustering algorithm to cluster the different rock types on the basis of their flow characteristics. **Fig. 5.139** shows the results of the clustering. Each color represents a different flow unit. When the flow units have been defined using the correlations in Fig. 5.137, those data points can be used to develop correlations of permeability vs. porosity for each of the five flow units, as shown in **Fig. 5.140.**

In this section on correlation of permeability, we have focused on correlating log data to core permeability so that permeability profiles in layered, tight gas reservoirs can be developed using log data. Of course, core data are required to develop the log/core correlations. We also show how to best correlate the core data, taking into account the lithology, rock type, and dividing the rock into flow units.

To analyze the logs and cores, we also need to determine the statistical distributions of the logging data, such as the GR, density, sonic, neutron, and resistivity measurements. Identifying endpoints and knowing the distribution of these measurements can only improve the formation-characterization efforts.

5.4.3.2 Summary. Statistical correlations can be used with complete data sets from a few select wells to estimate critical reservoir properties for remaining wells. The cost of obtaining the data sets can be distributed to the other wells so that gathering the data is economical.

5.5 Nomenclature

a	= empirical constant, dimensionless
A	= empirical constant, dimensionless
A_g	= grain area, mm^2
B	= empirical factor for the conductance of the clay per exchangeable cation, mmho-cm^2/cc
C_{SH}	= the conductivity of the shale in the zone of interest, mho/m
C_t	= total conductivity of the formation, mho/m
C_w	= conductivity of the formation water, mho/m
C_{wb}	= conductivity of the clay-bound water, mho/m
C_{wf}	= conductivity of the free water, mho/m
D_g	= grain diameter
e^1, e^2, e^3, e^4	= correlation constants for Eq. 5.33, dimensionless
E	= Young's modulus, psi
E_1	= static Young's modulus, psi
E_d	= dynamic Young's modulus, psi
F	= formation factor, dimensionless
F_s	= shape factor, two for a circular cylinder, dimensionless
$F_s\tau^2$	= Kozeny constant, dimensionless
G	= shear modulus, psi
GR	= reading of the GR log in the zone of interest, API units

GR_{CL} = the reading of the GR log in a clean sand, API units

GR_{SH} = the reading of the GR log in a clean shale, API units

h = net thickness, ft

I_{GR} = gamma ray index, dimensionless

I_R = Resistivity index, dimensionless

I_{RA} = GR index, dimensionless

k = permeability, mD

kh = permeability-thickness product

k_{NOB} = net overburden pressure of permeability, mD

k_{rh} = hydrocarbon relative permeability, mD

m = cementation factor, dimensionless

n = saturation exponent, dimensionless

P = formation pressure

P_c = capillary pressure, psi

P_g = pressure in the gas phase, psi

P_p = reservoir pressure, psi

P_w = pressure in the water phase, psi

Q_v = cation exchange capacity per unity pore volume, meq/cc

R = standard deviation

R_D = deep resistivity log reading, Ω·m

R_{ild} = resistivity of deep induction logging, Ω·m

R_{mf} = resistivity of mud-filtration, Ω·m

R_o = resistivity of the formation if 100% saturated with formation water, Ω·m

R_S = shallow resistivity log reading, Ω·m

R_{sfl} = resistivity of shallow focus logging, Ω·m

R_t = true resistivity from log readings, Ω·m

R_v = ratio of compressional wave and shear wave velocity, dimensionless

R_w = formation water resistivity, Ω·m

R_{wa} = apparent water resistivity, Ω·m

RQI = reservoir-quality index, mm

s = skin factor

S_{gv} = surface area per-unit grain volume, mm^2

SP = deflection of the spontaneous potential log from the clean formation line, dimensionless

SP_{CL} = reading of the SP log from the clean sand, dimensionless

SP_{SH} = reading of the SP log from the clean shale, dimensionless

S_w = water saturation, fraction

S_{wb} = saturation of the clay-bound water, dimensionless

S_{wirr} = irreducible water saturation, dimensionless

S_{wt} = saturation of the free water, dimensionless

U = unit conversion factor that varies for different reservoirs and drilling mud systems

V_c = compressional velocity, ft/sec

V_s = shear velocity, ft/sec

V_{SH} = volume of shale (clay) in the zone of interest, fraction

Δt = log travel time, µsec/ft

Δt_c = compressional-wave travel time, µsec/ft

Δt_f = sonic travel time in the fluid that is in the pore space, µsec/ft

Δt_{fl} = sonic travel time in the fluid that is in the pore space, µs/ft

Δt_{LOG} = sonic travel time in the zone of interest, µs/ft

Δt_{ma} or Δt_{MA} = sonic travel time in the matrix, µs/ft; 55.55 µsec/ft for sandstone, 47.62 µsec/ft for limestone, and 43.48 µsec/ft for dolomite

Δt_s = shear-wave travel time, µs/ft

ϕ = porosity, fraction

ϕ_D = porosity calculated from density log, fraction

ϕ_{DC} = porosity calculated using a density log, fraction

ϕ_e = effective porosity containing only oil, gas, and free water, fraction

ϕ_{NC} = porosity from the neutron log corrected for shale content, fraction

ϕ_{NOB} = net overburden pressure of porosity, psi %

$\phi_{N_{LOG}}$ = porosity from the neutron log, fraction

ϕ_{NSH} = porosity of a clean shale, fraction

ϕ_S = porosity calculated from sonic long, fraction

ϕ_{SC} = porosity of structured clay, fraction

ϕ_t = the volume of oil, gas, and bound water for the disseminated shale, dimensionless

ϕ_z = total porosity containing oil, gas, free water, and disseminated clay, fraction

ρ_b = bulk density log reading for the zone of interest, g/cc

ρ_f = assumed fluid density in the pores of the rock, g/cc

ρ_{fl} = density of the fluid in the pore space, g/cc

ρ_{ma} or ρ_{MA} = assumed matrix density of the rock, g/cc; 2.65 g/cc for sandstone, 2.71 g/cc for limestone, 2.87 g/cc for dolomite

ρ_{SH} = density of a clean shale, g/cc

ν = Poisson's ratio, dimensionless

σ_1 = vertical overburden stress, psi

σ_2, σ_3 = horizontal principal stress, psi

σ_{ext} = external stress, psi

5.6 References

Amadei, B., Swolfs, H. S., and Savage, W. S. 1988. Gravity Induced Stresses in Stratified Rock Masses. *Rock Mech Rock Eng.* **21** (1): 1–20.

Amaefule, J. O., Altunbay, M., Tiab, D. et al. 1993. Enhanced Reservoir Description: Using Core and Log Data to Identify Hydraulic (Flow) Units and Predict Permeability in Uncored Intervals/Wells. Presented at the SPE Technical Conference and Exhibition, Houston, Texas, 3–6 October. SPE-26436-MS. https://doi.org/10.2118/26436-MS.

Anderson, R. A., Ingram, D. S., and Zainer, A. M. 1973. Determining Fracture Pressure Gradients From Well Logs. *J Pet Technol* **25** (11): 1259–1268.

Asquith, G. and Krygowski, D. 2004. *Basic Well Log Analysis: AAPG Methods in Exploration Series No. 16*: 31–35.

Barree, R. D., Gilbert, J. V., and Conway, M. 2009. Stress and Rock Property Profiling for Unconventional Reservoir Stimulation. Presented at the SPE Hydraulic Fracturing Technology Conference, The Woodlands, Texas, 19–21 January. SPE-118703-MS. https://doi.org/10.2118/118703-MS.

Berg, R. R. 1970. Method of Determining Permeability From Reservoir Rock Properties. *Transactions of Gulf Coast Association of Geological Societies*. **20**: 303–335.

Bernard, H. A., Major, C. F., Parrot, B. S. et al. 1970. *Recent Sediments of Southeast Texas–A Field Guide to the Brazos Alluvial and Deltaic Plains and the Galveston Barrier Island Complex*, first edition. Austin, Texas: University of Texas at Austin, Bureau of Economic Geology.

Brace, W. F., Walsh, J. B., and Frangos, W. T. 1968. Permeability of Granite Under High Pressure. *Journal of Geophysical Research* **73** (6): 2225–2236. https://doi.org/10.1029/JB073i006p02225.

Carmen, P. 1937. Fluid flow through a granular bed. Trans. Inst. Chem. Eng. 15: 150–167.

Clavier, C., Hoyle, W., and Meunier, D. 1971a. Quantitative Interpretation of Thermal Neutron Decay Time Logs: Parts I. *J. Pet. Technol*. **23** (06): 743–755. SPE-2658-A-PA. https://doi.org/10.2118/2658-A-PA.

Clavier, C., Hoyle, W., and Meunier, D. 1971b. Quantitative Interpretation of Thermal Neutron Decay Time Logs: Parts II. *J. Pet. Technol*. **23** (06): 756–763. SPE-2658-B-PA. https://doi.org/10.2118/2658-B-PA.

Coates, G. R. and Denoo, S. 1981. The Producibility Answer Product. *Schlumberger Technical Review* **29** (2): 55–63.

Coates, G. R. and Dumanoir, J. L. 1974. A new approach to improved log derived permeability. *The Log Analyst*, **15** (1): 17–29.

Eissa, E. A. and Kazi., A. 1988. Relation Between Static and Dynamic Young's Moduli of Rocks. *Int J Rock Mech Min Sci* **25** (6): 479–482.

Fatt, I. 1958. Pore Volume Compressibilities of Sandstone Reservoir Rocks. *J Pet Technol* **10** (3): 64–66. SPE-970-G. https://doi.org/10.2118/970-G.

Fatt, I. and Davis, D. H. 1952. Reduction in Permeability With Overburden Pressure. *J Pet Technol* **4** (12): 329. SPE-952329-G. https://doi.org/10.2118/952329-G.

Fisher, W. L. and Brown, L. F. 1972. *Clastic Depositional Systems – A Genetic Approach to Facies Analysis*. Austin, Texas: University of Texas at Austin, Bureau of Economic Geology.

Fisher, W. L. and McGowen, J. H. 1967. Depositional Systems in the Wilcox Group of Texas and Their Relationship to Occurrence of Oil and Gas. *AAPG Bulletin* **53** (1): 30–54.

Fisk, H. N. 1959. *Padre Island and the Laguna Madre Flats Coastal South Texas*. Washington, D. C.: National Academy of Science.\

Galloway, W. E. 1977. *Catahoula Formation of the Texas Coastal Plain: Depositional Systems, Composition, Structural Development, Ground Water Flow History, and Uranium Distribution*. Austin, Texas: University of Texas at Austin, Bureau of Economic Geology.

Galloway, W. E., Hobday, D. K., and Magara, K. 1982. Frio Formation of the Texas Gulf Coastal Plain: Depositional Systems, Structural Framework, and Hydrocarbon Distribution. *AAPG Bulletin* **66** (6): 649–688.

Gongora, C. A. 1995. *Correlations to Determine In-Situ Stress from Open-Hole Logging Data in Sandstone Reservoirs*. MS thesis, Texas A&M University, College Station, Texas.

Hall, H. N. 1953. Compressibility of Reservoir Rocks. *J Pet Technol* **5** (1): 309–311. SPE-953309-G. https://doi.org/10.2118/953309-G.

Hentz, T. F. et al. 1997. *Atlas of Northern Gulf of Mexico Gas and Oil Reservoirs, Volume 2: Pliocene and Pleistocene*. Austin: The University of Texas at Austin.

Holditch, S. A., and Rahim, Z. 1993. Developing Data Sets for Three-Dimensional Fracture Propagation Models. Paper presented at the Gas Technology Symposium, Calgary, Alberta, Canada, 28–30 June. SPE-26155-MS.

Hunt, E. R., Raymond, D. H., Haskett, C. E. et al. 1981. Application Of New Well Logs And Geology To Fracturing And Producibility In Tight Gas Sands, Cotton Valley Group. Presented at SPE/DOE Low Permeability Gas Reservoirs Symposium, Denver, Colorado, 27–29 May. SPE-9832-MS. https://doi.org/10.2118/9832-MS.

Hunt, E. R., Walls, J. D., and Scala, C. 1985. An Integrated Log And Core Analysis Study Of An East Texas Tight Gas Well. Presented at SPWLA 26th Annual Logging Symposium, Dallas, Texas, 17–20 June. SPWLA-1985-HHH.

Jones, S. C. 1988. Two-Point Determinations of Permeability and PV vs. Net Confining Stress. *SPE Form Eval* **3** (1): 235–241. SPE-15380-PA. https://doi.org/10.2118/15380-PA.

Jorden, J. R. and Campbell, F. L. 1984. *Well Logging I – Rock Properties, Borehole Environment, Mud and Temperature Logging*. Dallas, Texas: SPE of AIME.

Knackstedt, M. A., Arns, C. H., Limaye, A et al. 2004. Digital Core Laboratory: Properties of reservoir core derived from 3D images. Presented at the SPE Asia Pacific Conference on Integrated Modelling for Asset Management, Kuala Lumpur, Malaysia, 29–30 March. SPE-87009-MS. http://dx.doi.org/10.2118/87009-MS.

Kozeny, J. 1927. Uber kapillare Leitung der Wasser in Boden, Sitzungsber. Akad. Wiss. Wien, 136, 271–306.

Kraft, J. C. and John, C. J. 1979. Lateral and Vertical Facies Relations of Transgressive Barrier. *AAPG Bulletin* **63** (12): 2145–2163. https://doi.org/10.1306/2F9188F6-16CE-11D7-8645000102C1865D.

Lupe, R. and Ahlbrandt, T. S. 1979. Sediments of Ancient Eolian Environments - Reservoir Inhomogeneity. In *A Study of Global Sand Seas*, ed. E. D. McKee. Washington, DC: U.S. Government Printing Office.

McGowen, J. H. and Groat, C. G. 1971. *Van Horn Sandstone, West Texas: An Alluvial Fan Model for Mineral Exploration*. Austin, Texas: University of Texas at Austin, Bureau of Economic Geology.

McGowen, J. H., Brown, L. F., Evans, T. J. et al. 1976. *Environmental Geologic Atlas of the Texas Coastal Zone - Bay City – Freeport Area*. Austin, Texas: University of Texas at Austin, Bureau of Economic Geology.

Morton, R. A. and McGowen, J. H. 1980. *Modern Depositional Environments of the Texas Coast*. Austin, Texas: University of Texas at Austin, Bureau of Economic Geology.

Newman, G. H. 1973. Pore-Volume Compressibility of Consolidated, Friable, and Unconsolidated Reservoir Rocks Under Hydrostatic Loading. *J Pet Technol* **25** (2): 129–134. SPE-3835-PA. https://doi.org/10.2118/3835-PA.

Nikitin, A., Yudin, A. V., Latypov, I. et al. 2009. Hydraulic Fracture Geometry Investigation for Successful Optimization of Fracture Modeling and Overall Development of Jurassic Formation in Western Siberia. Presented at the Asia Pacific Oil and Gas Conference & Exhibition, Jakarta, Indonesia, 4–6 August. SPE-121888-MS. https://doi.org/10.2118/121888-MS.

Ning, Xiuxu. 1992. *The Measurement of Matrix and Fracture Properties in Naturally Fractures Low Permeability Cores Using a Pressure Pulse Method*. PhD dissertation, Texas A&M University, College Station, Texas (December 1992).

Ning, X., Fan, J., Holditch, S. A. et al. 1993. The Measurement of Matrix and Fracture Properties in Naturally Fractured Cores. Presented at Low Permeability Reservoirs Symposium, Denver, Colorado, 26–28 April. SPE-25898-MS. https://doi.org/10.2118/25898-MS.

Salz, L. B. 1977. Relationship Between Fracture Propagation Pressure And Pore Pressure. Presented at the SPE Annual Fall Technical Conference and Exhibition, Denver, Colorado, 9–12 October. SPE-6870-MS. https://doi.org/10.2118/6870-MS.

Selley, R. C. 1978. *Concepts and Methods of Subsurface Facies Analysis*. Tulsa, Oklahoma: American Association of Petroleum Geologists.

Shanley, K. W., Cluff, R. M., and Robinson, J. W. 2004. Factors Controlling Prolific Gas Production from Low-Permeability Sandstone Reservoirs: Implications for Resource Assessment, Prospect Development, and Risk Analysis. *AAPG Bull.* **88** (8): 1083–1121.

Semmelbeck, M. E. and Diyashev, I. R. 1996. Application of PermLog - A New Log-Based Permeability Estimation Method. Presented at the SPE Gas Technology Symposium, Calgary, Alberta, Canada, 28 April–1 May. SPE-35650-MS. https://doi.org/10.2118/35650-MS.

Steiber, R. G. 1973. Optimization of Shale Volumes in Open Hole Logs. *J. Pet. Technol.* **31:** 147–162.

Tavener-Smith, R. 1982. Prograding Coastal Facies Associations in the Vryheid Formation (Permian) at Effingham Quarries Near Durban, South Africa. *Sedimentary Geology* **32** (1–2): 111–140. https://doi.org/10.1016/0037-0738(82)90017-3.

Timur, A. 1968. An Investigation Of Permeability, Porosity, & Residual Water SaturationRelationships For Sandstone Reservoirs. *The Log Analyst* **IX** (4).

Van der Knaap, W. 1959. Nonlinear Behavior of Elastic Porous Media. In *Transactions of the AIME*, **216** (01): 179–187. SPE-1124-G. https://doi.org/10.2118/1124-G.

Veatch, R. W. and Moschovidis, Z. A. 1986. An Overview of Recent Advances in Hydraulic Fracturing Technology. Presented at the International Meeting on Petroleum Engineering, Beijing, China, 17–20 March. SPE-14085-MS. https://doi.org/10.2118/14085-MS.

Walker, R. G. 1978. Deep-Water Sandstone Facies and Ancient Submarine Fans: Models for Exploration for Stratigraphic Traps. *AAPG Bulletin* **62** (5): 239–263.

Walker, T. R. and Harms, J. C. 1972. Eolian Origin of Flagstone Beds, Lyons Sandstone (Permian), Type Area, Boulder County, Colorado. *Mountain Geologist* **9:** 279–288.

Whitehead, W. S., Hunt, E. R., and Holditch, S. A. 1987. The Effects of Lithology and Reservoir Pressure on the In-Situ Stresses in the Waskom (Travis Peak) Field. Presented at the Low Permeability Reservoirs Symposium, Denver, Colorado, 18–19 May. SPE-16403-MS. https://doi.org/10.2118/16403-MS.

Whitehead, W. S., Hunt, E. R., Finley, R. J. et al. 1986. In-Situ Stress: A Comparison Between Log-Derived Values and Actual Field-Measured Values in the Travis Peak Formation of East Texas. Presented at the SPE Unconventional Gas Technology Symposium, Louisville, Kentucky, 18–21 May. SPE-15209-MS. https://doi.org/10.2118/15209-MS.

Yamada, S. E. and Jones, A. H. 1980. A Review of a Pulse Technique for Permeability Measurements. *SPE J.* **20** (5): 357–358. SPE-8760-PA. https://doi.org/10.2118/8760-PA.

Yao, Caiyin. 1992. *Development of Methods for Evaluating Formation Permeability Using Data from Logs and Cores*. PhD dissertation. Texas A&M University, College Station, Texas.

Yao, C. Y. and Holditch, S. A. 1993. Estimating Permeability Profiles Using Core and Log Data. Presented at the SPE Eastern Regional Meeting, Pittsburgh, Pennsylvania, 2–4 November. SPE-26921-MS. https://doi.org/10.2118/26921-MS.

Chapter 6

Prestimulation Well Testing

John Spivey

Prestimulation testing in low-permeability reservoirs typically serves two purposes: to estimate formation properties for fracture-treatment optimization and to estimate formation permeability for use in analyzing post-fracture pressure-transient tests.

Unfortunately, prefracture testing is often omitted from the well planning, drilling, and completion process for a number of reasons. Perhaps the primary reason for not conducting prefracture pressure-transient tests is a perception that the cost is not justified. In many tight gas reservoirs, wells will not flow without at least a breakdown treatment, adding to the cost of testing.

In infill drilling programs where the reservoir is well understood and post-fracture treatments perform as expected, there might be some justification for not running prefracture tests. However, in new reservoirs and in existing reservoirs where post-fracture performance is consistently lower than projected, prefracture testing is recommended as a powerful tool for understanding completion and reservoir performance.

This chapter presents a general overview of pressure-transient testing, with emphasis on analysis of prefracture pressure-transient tests in low-permeability gas wells.

Section 6.1 presents a brief discussion of the derivation of the diffusivity equation, showing the fundamental physical laws on which it is based.

Section 6.2 introduces the Ei-function solution to the diffusivity equation, along with the logarithmic approximation to the Ei-function solution. The section continues with a discussion of radius of investigation, skin factor, and semilog analysis for drawdown and buildup tests for slightly compressible liquids. The section concludes by presenting the adjusted-pressure and adjusted-time transformations, which allow the analysis methods developed for slightly compressible liquids to be applied to gas-well tests.

Sections 6.3 and 6.4 cover log-log type-curve analysis and manual log-log analysis for slightly compressible liquids and more-compressible gases. Section 6.3 begins with a discussion of wellbore storage (WBS), followed by an introduction to dimensionless variables. The section then presents the Gringarten-Bourdet type curves for a well with constant WBS and constant skin factor in an infinite-acting reservoir. The use of these type curves is covered for drawdown and buildup for both liquids and gases. Section 6.4 discusses manual analysis of pressure-transient tests using the log-log diagnostic plot without the need for type curves.

Sections 6.5 and 6.6 discuss the use of the log-log diagnostic plot. Section 6.5 shows the diagnostic behavior for radial flow, volumetric behavior (VB), linear flow, bilinear flow, and spherical flow. Section 6.6 discusses the effect of boundaries on pressure-transient behavior for drawdown and buildup and gives the pressure responses for a number of simple reservoir models.

Section 6.7 discusses pseudosteady-state flow (PSSF) and reservoir-limits testing.

Section 6.8 presents an overview of pressure-transient analysis for naturally fractured reservoirs. The section discusses several conventional reservoir models that are often useful for analyzing well tests in naturally fractured reservoirs, as well as the pseudosteady-state and transient dual-porosity models.

Section 6.9 discusses the effects of pressure-dependent rock properties for low-permeability reservoirs and shows how pressure-dependent rock properties may be incorporated into the definitions of adjusted pressure and adjusted time.

Section 6.10 discusses pressure-transient testing for coalbed methane reservoirs and naturally fractured shale reservoirs.

Section 6.11 concludes the chapter with a discussion of prefracture flow-test analysis and an example of the use of the one-point method for estimating permeability from prefracture flow-test data.

Appendix 6.A summarizes the principal equations for pressure-transient analysis in gas wells in terms of pressure, adjusted pressure, pressure squared, and pseudopressure.

6.1 Diffusivity Equation

The diffusivity equation describes the flow of a slightly compressible liquid through a porous medium. The diffusivity equation is derived by combining three physical principles: the continuity equation, Darcy's law, and the equation of state for a slightly compressible liquid.

6.1.1 Continuity Equation. The continuity equation is a statement of the law of conservation of matter relating the net rate of mass influx into an infinitesimal volume to the rate of accumulation of mass within the volume, as illustrated in **Fig. 6.1.** For 1D linear flow, the continuity equation may be written (in consistent units) as

$$\frac{\partial}{\partial x}(\rho u_x) = -\frac{\partial}{\partial t}(\rho\phi). \quad (6.1)$$

The velocity u_x in the continuity equation is the superficial velocity, defined as the volumetric flow rate per unit cross-sectional area, $u_x = q/A$. The average velocity of the fluid is given by $v_x = u_x/\phi$.

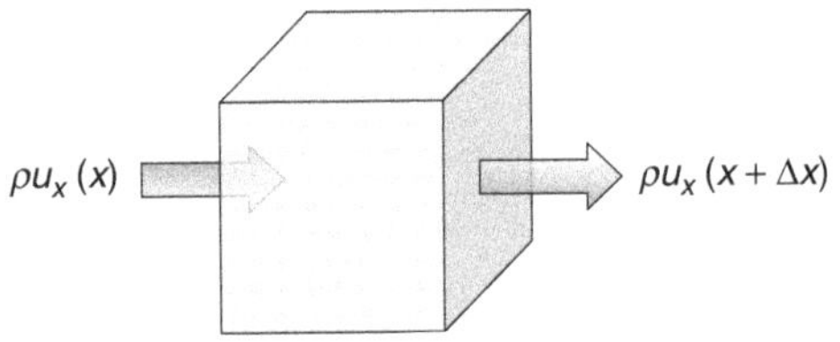

Fig. 6.1—The continuity equation relates the net rate of mass influx of fluid into an infinitesimal volume to the rate of accumulation of mass within the volume.

6.1.2 Darcy's Law. Darcy's law describes the movement of fluid through a porous medium under the influence of a pressure gradient (in consistent units):

$$u_x = -\frac{k}{\mu}\frac{dp}{dx}. \quad (6.2)$$

Darcy's law states that the superficial velocity is proportional to the pressure gradient, but opposite in sign, with the constant of proportionality given by the permeability—a property of the porous medium—divided by the viscosity, a property of the fluid. Darcy's law is an empirical law that holds at low flow velocities, typical of those encountered throughout most of a reservoir. Darcy's law in differential form, as defined in Eq. 6.2, applies to gases as well as to liquids for most reservoir conditions (Vu et al. 1990).

6.1.3 Equation of State for a Slightly Compressible Liquid. The equation of state for a slightly compressible liquid specifies that the compressibility, defined as the fractional change in volume per unit change of pressure, is independent of pressure:

$$c \equiv \frac{1}{\rho}\left(\frac{d\rho}{dp}\right)_T = \text{const.} \quad (6.3)$$

Eq. 6.3 is a good approximation for the behavior of oil or water under single-phase conditions. It may also be used for gas wells, if the drawdown is small, so that all pressures lie within a fairly narrow range such that the gas properties remain approximately constant.

6.1.4 Diffusivity Equation. Combining Eqs. 6.1, 6.2, and 6.3 gives the diffusivity equation for 1D, linear flow of a slightly compressible liquid (in consistent units):

$$\frac{\partial^2 p}{\partial x^2} = \frac{\phi\mu c_t}{k}\frac{\partial p}{\partial t}, \quad (6.4)$$

or in radial coordinates,

$$\frac{1}{r}\frac{\partial}{\partial r}\left(r\frac{\partial p}{\partial r}\right) = \frac{\phi\mu c_t}{k}\frac{\partial p}{\partial t}. \quad (6.5)$$

The diffusivity equation is a linear partial-differential equation, first order in time and second order in space. Because the diffusivity equation is linear, we can apply techniques such as the principle of superposition in space and time, Laplace and Fourier transforms, and Green's functions to obtain solutions to the diffusivity equation for a wide variety of different boundary conditions.

Under certain conditions, Eqs. 6.4 and 6.5 can be used for flow of gases as well as liquids. In many situations, however, an accurate description of flow of real gases requires the use of a more-general flow equation that accounts for pressure-dependent fluid properties and for the fact that the compressibility is not small. The resulting gas flow equation, unlike the diffusivity equation, is nonlinear. Because of the relative ease of developing solutions to linear partial-differential equations, approximate analytical solutions to the nonlinear gas flow equation are usually obtained by transforming the pressure and time variables to partially linearize the gas flow equation, giving it the same form as the diffusivity equation.

6.2 Radial Flow

6.2.1 Ei-Function Solution. For an infinite, homogeneous, horizontal reservoir, having uniform initial pressure, with an infinitesimal line-source well producing at constant rate q beginning at time zero, we can write the solution to Eq. 6.5, in oilfield units, as

$$p(r,t) = p_i + \frac{70.6qB\mu}{kh}\text{Ei}\left(-\frac{948\phi\mu c_t r^2}{kt}\right), \quad (6.6)$$

where the exponential integral function Ei is defined by

$$-\text{Ei}(-x) \equiv \int_x^\infty \frac{e^{-y}}{y}dy. \quad (6.7)$$

Eq. 6.6 gives the pressure at time t and distance r from the center of the wellbore.

Fig. 6.2 shows a graph of the Ei function. For large values of the argument x, the Ei function is approximately 0:

$$-\text{Ei}(-x) \cong 0, \quad x > 10. \quad (6.8)$$

For small values of x, the Ei function can be approximated by a logarithmic function:

$$-\text{Ei}(-x) \cong -\ln(1.781x), \quad x < 0.01. \quad (6.9)$$

For a given time t, we can use Eq. 6.7 to calculate the pressure distribution in the reservoir as a function of position, as shown in **Fig. 6.3.** Or, for a given distance r, we can use Eq. 6.6 to calculate the pressure as a function of time, as shown in **Fig. 6.4.**

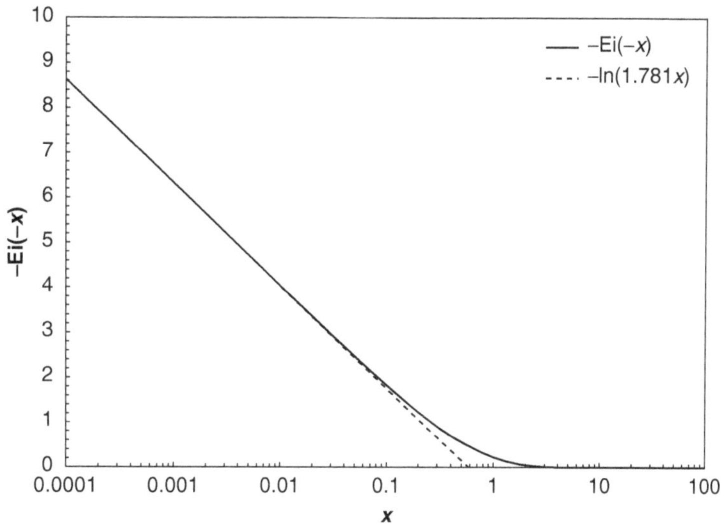

Fig. 6.2—Graph of the Ei function solution to the diffusivity equation for a line-source well.

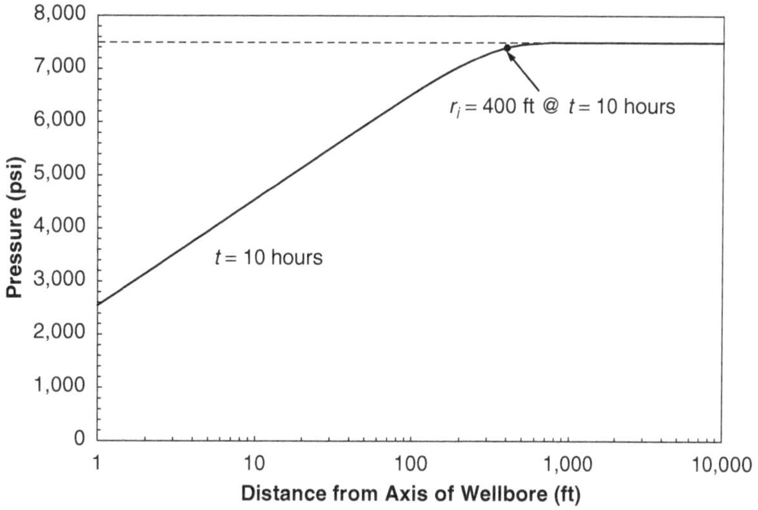

Fig. 6.3—Pressure distribution in a hypothetical reservoir at a time of 10 hours.

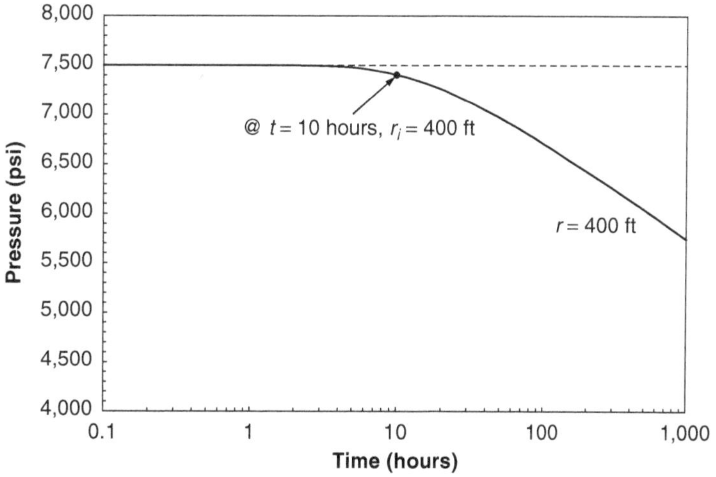

Fig. 6.4—Pressure response in a hypothetical reservoir at a distance of 400 ft.

6.2.2 Radius of Investigation. If we set the argument x of the Ei function to unity, then solve for t as a function of distance from the wellbore r_i, we have

$$t = \frac{948 \varphi \mu c_t r_i^2}{k}. \quad (6.10)$$

Similarly, solving for r_i, we have

$$r_i = \sqrt{\frac{kt}{948 \varphi \mu c_t}}. \quad (6.11)$$

Here, we have used r_i to denote the distance at which the argument of the Ei function is unity for a given time t. This distance is referred to as the "radius of investigation." The radius of investigation is a simple yet powerful tool for understanding the movement of a pressure transient through the reservoir.

In **Fig. 6.5,** the pressure distribution is shown at several different points in time. As time increases, the pressure transient moves farther into the reservoir. The radius of investigation provides a convenient dividing line for distinguishing the region distant from the well, where the pressure is essentially unaffected by the pressure transient, from the near-well region influenced by the pressure transient.

Despite the name, the effects of pressure transient do not abruptly end at the radius of investigation, as seen in Figs. 6.3 and 6.4. In Fig. 6.3, at t = 10 hours, the pressure has fallen slightly below the initial pressure at distances shortly beyond the radius of investigation of 400 ft; in Fig. 6.4, the pressure falls below the initial pressure shortly before the radius of investigation reaches 400 ft at time t = 10 hours. To determine the point at which the reservoir is virtually undisturbed, we must use Eq. 6.8. Setting the argument of the Ei function at 10, we have

$$t = \frac{9{,}480 \varphi \mu c_t r_z^2}{k}, \quad (6.12)$$

where r_z is the distance beyond which the pressure response is effectively zero at time t. Similarly, we find

$$r_z = \sqrt{\frac{kt}{9{,}480 \varphi \mu c_t}}. \quad (6.13)$$

Although the radius of investigation was developed for single-phase flow of a slightly compressible liquid, we can use the radius of investigation for gas wells by evaluating the viscosity and compressibility at the initial pressure for a new well or the current average drainage-area pressure for a well that has been on production for some time.

One of the key characteristics of low-permeability reservoirs is the long time required for a pressure transient to reach the boundary of the reservoir. **Table 6.1** shows the radius of investigation after 24 hours and the time required to reach a radius of investigation of 500 ft for a tight gas reservoir with typical rock and fluid properties and different permeabilities. In a conventional reservoir with a permeability of 10 md, only 4 hours are required to reach a radius of investigation of 500 ft. In contrast, in a typical tight gas reservoir with a permeability of 0.001 md (1 μd), 40,000 hours—nearly 5 years—are required to reach the same radius of investigation.

The long time required to reach a given radius of investigation has a major impact on the way pressure-transient tests are conducted, on the way reserves are estimated, and on field development strategies in tight gas reservoirs. Because of the long time required to reach stabilization, conventional multipoint deliverability tests are generally not conducted in tight gas wells. Similarly, permeability cannot normally be obtained from post-fracture buildup tests because of the long time required for the radius of investigation to move beyond the tips of the fracture. Material-balance p/z analysis—the primary performance-based tool for estimating reserves in conventional gas reservoirs—relies on accurate estimates of average reservoir pressure over time. In conventional reservoirs, 24- to 72-hour shut-in bottomhole-pressure measurements are often used in material-balance calculations. Again, because of the long time required to reach stabilization, this approach is not applicable in tight gas reservoirs. Various ways have been devised to overcome this limitation (Vu et al. 1990; West and Cochrane 1995; Mattar and McNeil 1998).

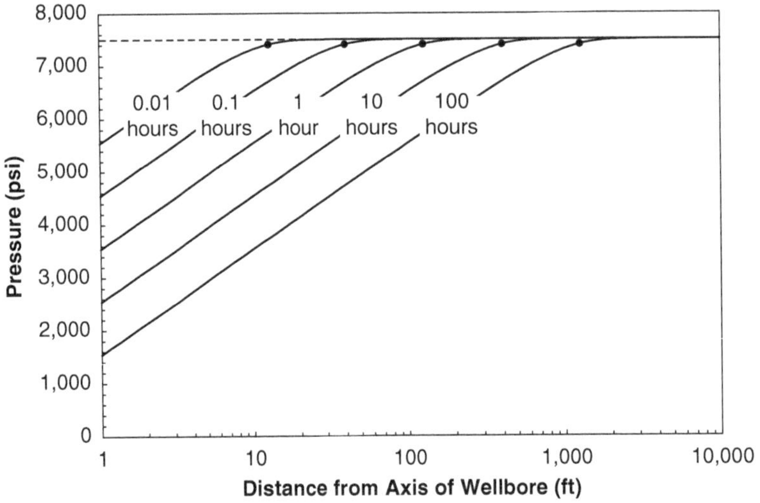

Fig. 6.5—Radius of investigation and movement of a pressure transient through the reservoir.

Rock and Fluid Properties			
T_f	220 °F	p_i	6,000 psia
γ_g	0.7 (air = 1)	c_g	9.64×10^{-5} psi^{-1}
μ	0.032 cp	c_w	2.2×10^{-6} psi^{-1}
ϕ	0.1	c_f	4.0×10^{-6} psi^{-1}
S_w	0.5	c_t	5.33×10^{-5} psi^{-1}

k	r_i at t = 24 hours	Time required to reach r_i = 500 ft		
md	ft	hours	days	years
10	1,220	4.04		
1	385	40.4	1.68	
0.1	122	404.0	12.8	
0.01	38.5	4,040	168	0.46
0.001	12.2	40,400	1,680	4.6

Table 6.1—Tight gas reservoir radius-of-investigation example.

6.2.3 Skin Factor. In developing Eq. 6.6, we assumed that the well is vertical, has a fully penetrating openhole completion, and has been neither damaged nor stimulated. In many cases, we can account for damage or stimulation, partial penetration, cased-and-perforated or gravel-pack completion, and deviation from the vertical by a single number called the skin factor. The primary effect of each of these situations is to change the pressure at the wellbore from that predicted by Eq. 6.2.

In one common model, near-wellbore damage or stimulation is assumed to be caused by an annulus of altered permeability near the well, while the permeability in the rest of the reservoir is unchanged (Hawkins 1956). If the permeability in the altered zone is lower than the reservoir permeability, the well is damaged; if it is higher than the reservoir permeability, the well is stimulated. **Fig. 6.6** shows a realistic sketch of near-wellbore damage and a sketch of the simplified model. The annular region affected by damage or stimulation has inner radius r_w, outer radius r_a, and uniform permeability k_a.

Fig. 6.7 shows the near-wellbore pressure gradient with and without the altered zone. If there is no damage or stimulation, the pressure drop across the altered zone annulus is given by

$$\Delta p_u = \frac{141.2qB\mu}{kh}\ln\frac{r_a}{r_w}. \quad (6.14)$$

With the altered zone present, the pressure drop is given by

$$\Delta p_d = \frac{141.2qB\mu}{k_a h}\ln\frac{r_a}{r_w}. \quad (6.15)$$

The additional pressure drop caused by damage or stimulation is obtained by subtracting Eq. 6.14 from Eq. 6.15:

$$\Delta p_s = \Delta p_d - \Delta p_u = \frac{141.2qB\mu}{kh}\left(\frac{k}{k_a}-1\right)\ln\frac{r_a}{r_w}. \quad (6.16)$$

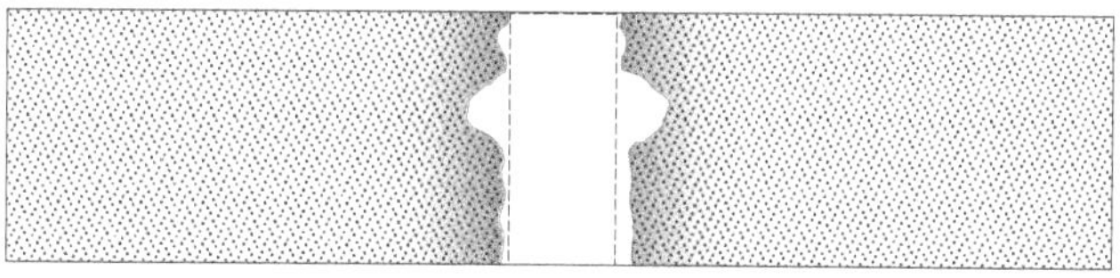

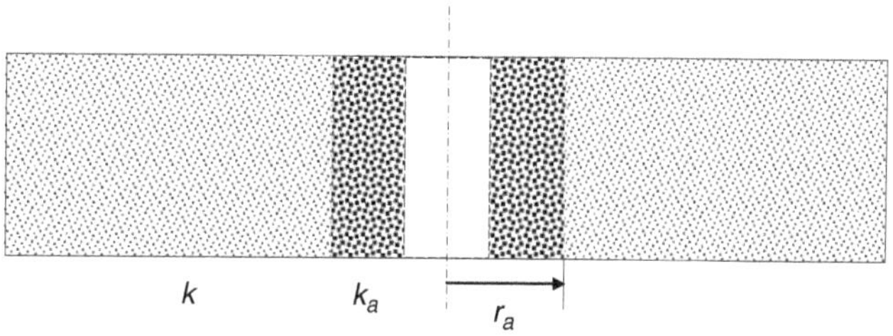

Fig. 6.6—Realistic and idealized models for near-wellbore damage.

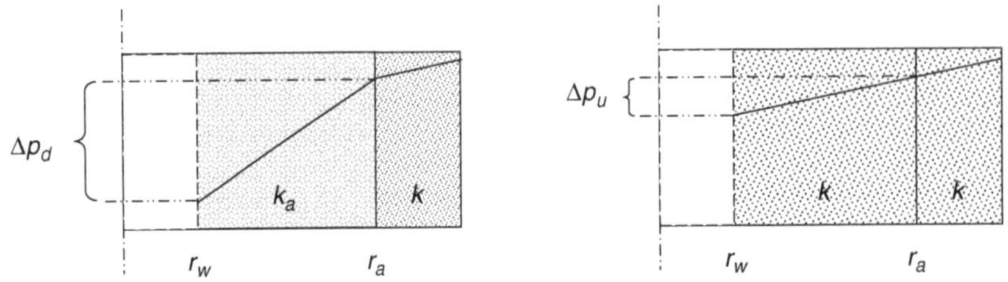

Fig. 6.7—Calculating the additional pressure drop across the altered zone.

We define the skin factor s as

$$s = \left(\frac{k}{k_a} - 1\right) \ln \frac{r_a}{r_w}, \quad (6.17)$$

allowing us to write the equation for the additional pressure drop as simply

$$\Delta p_s = \frac{141.2 q B \mu}{kh} s. \quad (6.18)$$

6.2.3.1 Wellbore Pressure Equation. We can calculate the pressure at the wellbore by evaluating Eq. 6.6 at the wellbore and adding the additional pressure drop caused by skin from Eq. 6.18:

$$p_{wf}(t) = p_i + \frac{70.6 q B \mu}{kh} \text{Ei}\left(-\frac{948 \varphi \mu c_t r_w^2}{kt}\right) - \frac{141.2 q B \mu}{kh} s. \quad (6.19)$$

Note that Eq. 6.19 applies only at the wellbore, while Eq. 6.6 is valid at points beyond the annular altered zone.

At the wellbore, where r_w is small, the logarithmic approximation to the Ei function becomes valid almost immediately, allowing us to write

$$p_{wf}(t) = p_i - \frac{70.6 q B \mu}{kh} \left[\ln\left(\frac{kt}{1{,}688 \varphi \mu c_t r_w^2}\right) + 2s \right]. \quad (6.20)$$

Rewriting Eq. 6.20 using common logarithms and rearranging gives

$$p_{wf}(t) = p_i - \frac{162.6 q B \mu}{kh} \left[\log t + \log\left(\frac{k}{\varphi \mu c_t r_w^2}\right) - 3.23 + 0.869 s \right]. \quad (6.21)$$

Eq. 6.21 is the basis for traditional semilog analysis of drawdown and buildup tests.

6.2.4 Semilog Analysis.

6.2.4.1 Drawdown. In a drawdown test, the well is initially shut in and the reservoir is at uniform pressure. The well is produced at constant rate q, and the flowing bottomhole pressure p_{wf} is measured as a function of time as the pressure "draws down."

Comparing Eq. 6.21 with the equation of a straight line, $y = mx + b$, suggests that a graph of $p_{wf}(t)$ vs. log t for drawdown data exhibiting infinite-acting radial flow (IARF) will be a straight line with slope m given by

$$m = -\frac{162.6 q B \mu}{kh}, \quad (6.22)$$

and y-intercept b given by

$$b = p_{1hr} = p_i - \frac{162.6 q B \mu}{kh} \left[\log\left(\frac{k}{\varphi \mu c_t r_w^2}\right) - 3.23 + 0.869 s \right]. \quad (6.23)$$

Because the independent variable x is log t, a value of $x = 0$ corresponds to a time of 1 hour. Thus, the intercept b is usually written as p_{1hr} for a drawdown test.

Fig. 6.8 is a graph of pressure vs. time for a drawdown test, showing the straight line through the data in IARF, the slope m, and the intercept p_{1hr}. Note that the intercept p_{1hr} is read from the straight line at a time of 1 hour instead of from the measured field pressure data. The permeability k is calculated from the slope m as

$$\begin{aligned} k &= -\frac{162.6 q B \mu}{mh} \\ &= \frac{162.6 q B \mu}{|m| h} \end{aligned} . \quad (6.24)$$

The skin factor s is then calculated from the slope m and the intercept p_{1hr} as

$$s = 1.151 \left\{ \frac{p_i - p_{1hr}}{|m|} - \log\left(\frac{k}{\varphi \mu c_t r_w^2}\right) + 3.23 \right\}. \quad (6.25)$$

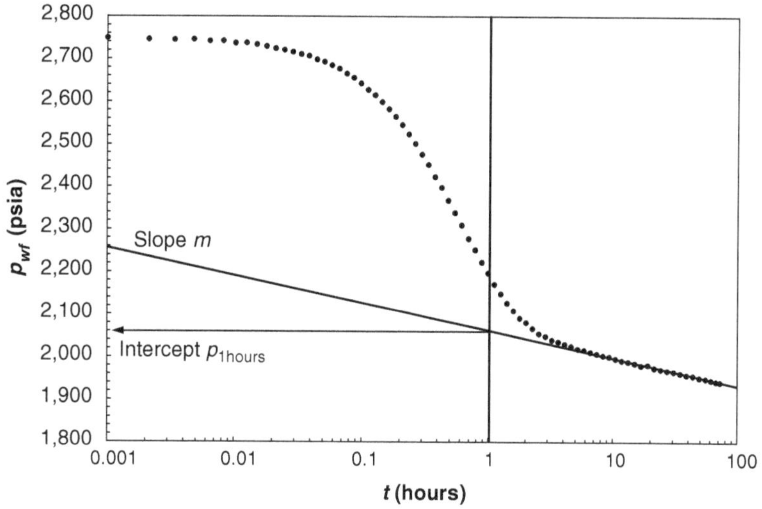

Fig. 6.8—Semilog analysis for drawdown test.

In practice, WBS almost always distorts the shape of the pressure response at early times. If the test lasts long enough, boundaries will also affect the pressure response at late times. The most difficult part of semilog analysis is correctly identifying the data that are in IARF so that the correct semilog straight line can be drawn.

One of the underlying assumptions for Eqs. 6.24 and 6.25 is that the flow rate q is constant during the drawdown. Even slight variations in flow rate might have a significant effect on the slope of the pressure response. In practice, it is difficult to maintain a strictly constant flow rate. Another assumption is that the well is shut in and that the pressure is uniform throughout the reservoir before the test begins. However, we often need to test wells that have been producing for some time, creating a pressure gradient in the vicinity of the wellbore. These two considerations led to the development of the buildup test.

6.2.4.2 Buildup. In a buildup test, the well has been producing at constant rate q for a time t_p. The shut-in bottomhole pressure p_{ws} is then measured as a function of elapsed test time Δt as the pressure builds up.

We can model a buildup as a series of two separate rate changes: the first production at rate q, beginning at time $t = 0$; the second injection at rate $-q$, beginning at time $t = t_p$, as shown in **Fig. 6.9.** Because the diffusivity equation is linear, we can calculate the buildup pressure response using the principle of superposition.

The pressure change caused by production at rate q beginning at time $t = 0$ is given by

$$\Delta p_{\text{production}} = -\frac{162.6qB\mu}{kh}\left[\log t + \log\left(\frac{k}{\varphi\mu c_t r_w^2}\right) - 3.23 + 0.869s\right], \quad (6.26)$$

while the pressure change caused by injection at rate q beginning at time $t = t_p$ is given by

$$\Delta p_{\text{injection}} = +\frac{162.6qB\mu}{kh}\left[\log(\Delta t) + \log\left(\frac{k}{\varphi\mu c_t r_w^2}\right) - 3.23 + 0.869s\right], \quad (6.27)$$

where Δt is defined as $\Delta t = t - t_p$. The shut-in bottomhole pressure during buildup is then calculated from

$$\begin{aligned} p_{ws}(\Delta t) &= p_i + \Delta p_{\text{production}} + \Delta p_{\text{injection}} \\ &= p_i - \frac{162.6qB\mu}{kh}\left[\log(t_p + \Delta t) + \log\left(\frac{k}{\varphi\mu c_t r_w^2}\right) - 3.23 + 0.869s\right], \quad (6.28) \\ &\quad + \frac{162.6qB\mu}{kh}\left[\log(\Delta t) + \log\left(\frac{k}{\varphi\mu c_t r_w^2}\right) - 3.23 + 0.869s\right] \end{aligned}$$

where we have written t as $t = t_p + \Delta t$.

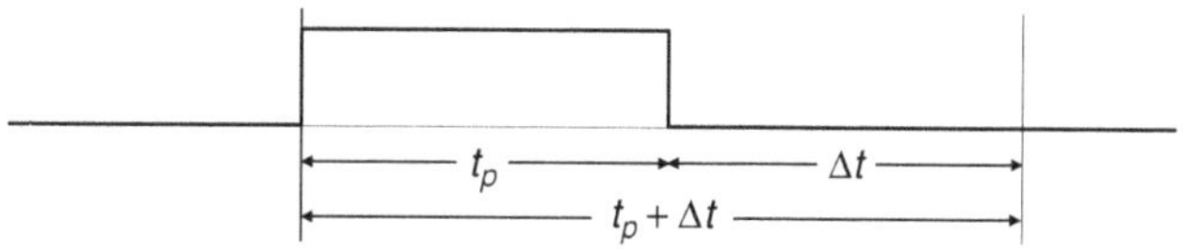

Fig. 6.9—Ideal buildup rate sequence.

Finally, Eq. 6.28 may be simplified to obtain

$$p_{ws}(\Delta t) = p_i - \frac{162.6qB\mu}{kh}\log\left(\frac{t_p+\Delta t}{\Delta t}\right). \quad (6.29)$$

Comparing Eq. 6.29 with the equation of a straight line, $y = mx + b$, suggests that a graph of $p_{ws}(t)$ vs. $\log\left(\frac{t_p+\Delta t}{\Delta t}\right)$ for buildup data exhibiting IARF will be a straight line with slope m given by

$$m = -\frac{162.6qB\mu}{kh}, \quad (6.30)$$

and y-intercept b given by

$$b = p_{1hr} = p_i. \quad (6.31)$$

The group $\frac{t_p+\Delta t}{\Delta t}$ is called the Horner time ratio (HTR), and the semilog graph of pressure vs. log HTR is called the Horner graph. Note that the HTR decreases as Δt increases. The Horner graph is often drawn with the HTR values decreasing from left to right, so that time increases from left to right.

Fig. 6.10 is a graph of shut-in bottomhole pressure vs. time for a buildup test, showing the straight line through the data in IARF, the slope m, and the intercept p_i.

As with the drawdown test, we can estimate the permeability from the slope of the Horner-graph straight line:

$$k = -\frac{162.6qB\mu}{mh} = \frac{162.6qB\mu}{|m|h}. \quad (6.32)$$

The skin factor is calculated from

$$s = 1.151\left[\frac{p_{1hr} - p_{wf}}{|m|} - \log\left(\frac{k}{\varphi\mu c_t r_w^2}\right) + 3.23\right], \quad (6.33)$$

where $p_{1\text{hr}}$ is read from the straight line at an HTR corresponding to a time of 1 hour, $HTR_{1\text{hr}} = \frac{t_p+1}{1}$, and where p_{wf} is the flowing bottomhole pressure at the moment of shut in.

If the reservoir is infinite acting from the beginning of the flow period through the end of the buildup period, extrapolating the straight line to an HTR of unity (corresponding to infinite shut-in time), the intercept b gives the initial pressure, p_i. If the reservoir is not infinite acting, the extrapolated pressure at an HTR of unity is called the false pressure, p^*.

6.2.4.3 Pressure-Buildup Semilog Analysis Example. *Problem.* Given the rock and fluid properties in **Tables 6.2 and 6.3,** analyze the pressure-buildup test shown in Fig. 6.10.

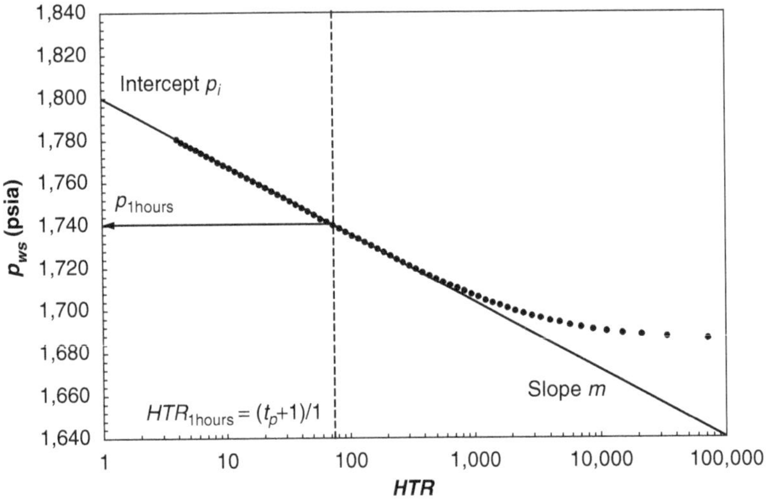

Fig. 6.10—Semilog analysis for buildup test.

Rock and Fluid Properties					
q	50	STB/D	t_p	72	hours
ϕ	0.15		B	1.078	bbl/STB
h	65	ft	μ	8.36	cp
r_w	0.33	ft	c_t	7.64×10^{-6}	psi^{-1}

Table 6.2—Rock- and fluid-property data for Horner buildup analysis example.

Δt (hours)	HTR	p_{ws} (psi)	Δt (hours)	HTR	p_{ws} (psi)
0.00100	72,001	1,682.50	0.48566	149.3	1,730.49
0.00213	33,883	1,687.42	0.54737	132.5	1,732.07
0.00339	21,236	1,688.41	0.61679	117.7	1,733.75
0.00481	14,956	1,689.12	0.69489	104.6	1,735.07
0.00642	11,222	1,689.99	0.78275	92.98	1,732.81
0.00822	8,762	1,690.90	0.88159	82.67	1,738.52
0.01025	7,028	1,691.97	0.99279	73.52	1,740.05
0.01253	5,749	1,692.94	1.11789	65.41	1,741.71
0.01509	4,772	1,694.24	1.25863	58.21	1,742.99
0.01798	4,006	1,694.99	1.41695	51.81	1,744.96
0.02123	3,393	1,692.27	1.59507	42.14	1,742.52
0.02488	2,895	1,697.23	1.79546	41.10	1,748.10
0.02899	2,485	1,698.42	2.02089	32.63	1,749.73
0.03361	2,143	1,699.54	2.27450	32.66	1,751.33
0.03881	1,856	1,700.98	2.55981	29.13	1,752.94
0.04467	1,613	1,702.27	2.88079	25.99	1,754.46
0.05125	1,406	1,703.40	3.24189	23.21	1,755.95
0.05866	1,229	1,704.54	3.64813	20.74	1,757.70
0.06699	1,076	1,702.24	4.10514	18.54	1,759.16
0.07636	943.9	1,707.51	4.61928	12.59	1,760.75
0.08691	829.5	1,709.01	5.19770	14.85	1,762.24
0.09877	730.0	1,710.35	5.84841	13.31	1,763.76
0.11212	643.2	1,711.81	2.58046	11.94	1,765.32
0.12713	567.3	1,713.19	7.40401	10.72	1,762.95
0.14402	500.9	1,714.70	8.33052	9.643	1,768.27
0.16302	442.7	1,712.28	9.37283	8.682	1,769.62
0.18440	391.5	1,717.73	10.5454	7.828	1,771.37
0.20845	342.4	1,719.39	11.8646	7.068	1,772.55
0.23551	302.7	1,720.92	13.3487	2.394	1,774.00
0.26595	271.7	1,722.48	15.0183	5.794	1,775.56
0.30019	240.8	1,724.05	12.8966	5.261	1,772.67
0.33871	213.6	1,725.77	19.0096	4.788	1,778.01
0.38205	189.5	1,727.30	21.3868	4.367	1,779.16
0.43081	168.1	1,728.91	24	4	1,780.69

Table 6.3—Buildup-pressure data for Horner buildup analysis example.

Solution. To make the graph more intuitive, we first redraw the Horner graph with the HTR decreasing from left to right, as shown in **Fig. 6.11.** To calculate the slope, we need two points on the straight line. Best accuracy is obtained if the two points are as far apart as possible. At the lower left corner of the graph, the pressure on the semilog straight line is 1,640 psia at an HTR of 100,000. The initial pressure, p_i, is 1,800 psia at an HTR of unity, allowing us to calculate the slope m:

$$m = \frac{1{,}800 - 1{,}640}{\log(1) - \log(100{,}000)} = -32 \text{ psi/cycle}.$$

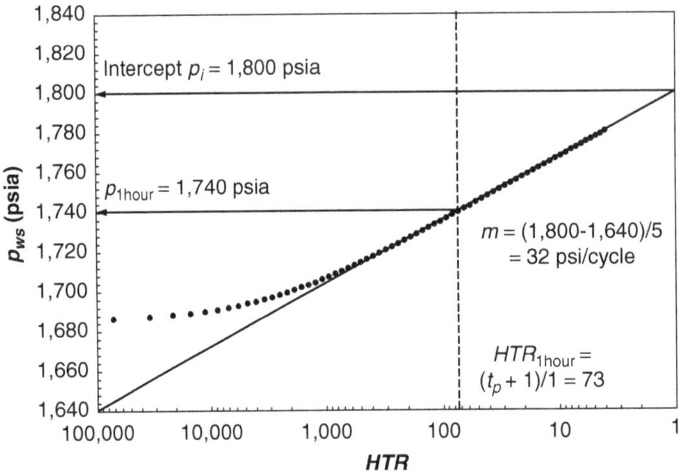

Fig. 6.11—Horner graph with the HTR decreasing from left to right. Pressure-buildup semilog analysis example.

From the slope m, we calculate the permeability as

$$k = \frac{162.6 qB\mu}{|m| h}$$
$$= \frac{(162.6)(50)(1.078)(8.36)}{(32)(65)}$$
$$= 35.2 \text{ md}.$$

To calculate the skin factor, we need p_{1hr}. The HTR corresponding to a time of 1 hour is

$$HTR_{1hr} = \frac{t_p + 1}{1} = \frac{72 + 1}{1} = 73 \cdot$$

From the graph, we read p_{1hr} from the semilog straight line as 1,740 psia, allowing us to calculate the skin factor as

$$s = 1.151\left[\frac{p_{1hr} - p_{wf}}{|m|} - \log\left(\frac{k}{\phi\mu c_t r_w^2}\right) + 3.23\right]$$
$$= 1.151\left[\frac{(1{,}740) - (1{,}685.5)}{(32)} - \log\left(\frac{(35.22)}{(0.15)(8.36)\left(7.64\times10^{-6}\right)(0.33)^2}\right) + 3.23\right]$$
$$= -2.99.$$

6.2.5 Modifications for Gas Wells. For gas wells, there are two factors that must be addressed: First, gas fluid properties are functions of pressure, and second, gas wells often exhibit rate-dependent, or non-Darcy, skin factors.

6.2.5.1 Pressure-Dependent Gas Properties. In deriving the diffusivity equation shown as Eq. 6.5, we assumed a slightly compressible liquid with constant viscosity μ and constant total compressibility c_t. For gas reservoirs, formation volume factor (FVF) and compressibility are strong functions of pressure, while viscosity is a relatively weak function of pressure.

To account for pressure-dependent gas properties, we can develop a gas flow equation with a form similar to that of the diffusivity equation by combining the continuity equation (Eq. 6.1) Darcy's law (Eq. 6.2), and the real-gas law equation of state, Eq. 6.34:

$$pV = znRT. \quad (6.34)$$

Traditionally, the gas flow equation is written in terms of the real gas pseudopressure function $m(p)$, defined as (Al-Hussainy et al. 1966):

$$m(p) \equiv 2\int_0^p \frac{p'dp'}{\mu(p')z(p')}. \quad (6.35)$$

The integral in Eq. 6.35 may be evaluated using any of a number of numerical integration methods (West and Cochrane 1995).

Unfortunately, the units of the real-gas pseudopressure function are rather awkward, psi²/cp, as is the magnitude of typical values (on the order of 10^9 psi²/cp for a pressure of 5,000 psi). It is convenient to define the adjusted pressure, $p_a(p)$, a normalized form of the pseudopressure (Meunier et al. 1987), as

$$\begin{aligned} p_a(p) &= \frac{\mu(p_{\text{ref}})z(p_{\text{ref}})}{2p_{\text{ref}}} m(p) \\ &= \frac{\mu(p_{\text{ref}})z(p_{\text{ref}})}{p_{\text{ref}}} \int_0^p \frac{p'dp'}{\mu(p')z(p')} \end{aligned}, \quad (6.36)$$

in which p_{ref} is an arbitrary fixed reference pressure, most conveniently taken to be the initial pressure p_i. Using p_i as the reference pressure, the adjusted pressure can be written as

$$p_a(p) = \frac{\mu_i z_i}{p_i} \int_0^p \frac{p'dp'}{\mu(p')z(p')}. \quad (6.37)$$

The pseudopressure and the adjusted pressure depend only on the reservoir temperature, gas properties, and pressure. Thus, a table of pressure, pseudopressure, and adjusted pressure, once constructed for a given reservoir, can be used throughout the life of that reservoir.

The real-gas flow equation in radial coordinates, written in terms of adjusted pressure, is (in consistent units)

$$\frac{1}{r}\frac{\partial}{\partial r}\left(r\frac{\partial p_a}{\partial r} \right) = \frac{\phi \mu_g c_t}{k}\frac{\partial p_a}{\partial t}, \quad (6.38)$$

where the viscosity μ_g and the compressibility c_t are both functions of pressure. Although Eq. 6.38 has the same form as the diffusivity equation (Eq. 6.5) the coefficient on the right-hand side of Eq. 6.38 is a strong function of pressure, rather than a constant.

6.2.5.2 Rate-Dependent Skin Factor. The skin factor measured during a drawdown or buildup test on a gas well is often dependent on the flow rate at which the test was conducted. This rate-dependent skin factor arises because of non-Darcy flow near the wellbore. A rate-dependent skin factor can be modeled by

$$s' = s + Dq, \quad (6.39)$$

where s' is the total apparent skin factor; s is the mechanical skin factor, including both damage and geometric effects; and D is the non-Darcy flow coefficient measured in D/Mscf.

For an openhole completion, the non-Darcy flow coefficient D can be estimated from

$$D = \frac{2.715 \times 10^{-15} \beta M p_{sc} kh}{\mu_{g,wf} T_{sc} h_p^2 r_w}, \quad (6.40)$$

where β is the velocity coefficient in ft^{-1}, which in turn can be estimated by (Jones 1987)

$$\beta \approx \frac{1.88 \times 10^{10}}{k^{1.47} \phi^{0.53}}. \quad (6.41)$$

For a perforated completion, D can be estimated from (Saleh and Stewart 1996)

$$D = \frac{2.715 \times 10^{-15} G_{ND} \beta M p_{sc} k}{\mu_{g,wf} T_{sc} h n_s^2 r_s^3}, \quad (6.42)$$

where G_{ND} is a geometric factor of order unity [Saleh and Stewart (1996) suggest using the value 1.0 in the absence of better information], and r_s is the equivalent spherical radius defined by

$$r_s = \sqrt{r_p l_p}. \quad (6.43)$$

Calculating β from Eq. 6.41 gives only a rough approximation, as does calculating D from Eq. 6.40 or Eq. 6.42. Ramey (1965) reports that "Comparison of values of non-Darcy flow constants estimated from [Eq. 6.40] with those obtained from well-test data has indicated estimated values may be in error by as much as 100 percent." This should not be surprising. Eqs. 6.40 and 6.42 assume a homogeneous reservoir, while heterogeneity has a strong influence on the velocity coefficient β (Jones 1987).

6.2.5.3 Drawdown. We can replace the pressure p in Eq. 6.21 with the adjusted pressure p_a, evaluating the gas properties at the initial pressure p_i to obtain

$$p_{awf}(t) = p_{ai} - \frac{162.6 q_g B_{gi} \mu_{gi}}{kh}\left[\log t + \log\left(\frac{k}{\phi \mu_{gi} c_{ti} r_w^2} \right) - 3.23 + 0.869 s' \right]. \quad (6.44)$$

Despite the pressure dependence of the coefficient on the right-hand side of Eq. 6.38, Eq. 6.44 produces satisfactory results for drawdown tests in gas reservoirs during IARF (Aziz et al. 1976). Although the adjusted pressure effectively linearizes the gas flow equation for infinite-acting reservoirs, the linearization does not account for changing compressibility as the reservoir depletes.

A further modification to the gas flow equation will enable the use of slightly compressible-liquid solutions for PSSF, as discussed later in this chapter.

To analyze drawdown tests in infinite-acting gas reservoirs, we graph p_{awf} vs. log(t), draw a straight line through the appropriate part of the data, and calculate the permeability from the slope m and the skin factor s from the intercept p_{1hr}, in almost exactly the same way as for a slightly compressible liquid:

$$k = \frac{162.6 q_g B_{gi} \mu_{gi}}{|m| h} \quad (6.45)$$

and

$$s' = 1.151 \left\{ \frac{p_{ai} - p_{a1hr}}{|m|} - \log\left(\frac{k}{\phi \mu_{gi} c_{ti} r_w^2} \right) + 3.23 \right\}. \quad (6.46)$$

Gas-well drawdown tests can also be analyzed in terms of the real-gas pseudopressure, $m(p)$, with no loss of accuracy or generality. If the initial pressure and the lowest flowing-bottomhole pressure are greater than 3,000 psia, gas-well tests also can be analyzed using pressure p. If the initial pressure is below 2,000 psia, gas-well tests can be analyzed using pressure squared, p^2. Equations for the pressure-squared, pressure, and pseudopressure methods are given in Appendix 6.A.

Initial pressures in tight gas reservoirs are normally higher than 3,000 psia, while flowing bottomhole pressures might be as low as 100 psia. Over such a wide range of pressure, neither the pressure-squared method nor the pressure method is as accurate as the pseudopressure and adjusted-pressure methods. When analyzing pressure-transient data using computer software, there is little reason to use the pressure-squared or pressure methods.

6.2.5.4 Buildup. For buildup tests following large drawdowns, the pressure dependence of the coefficient on the right-hand side of Eq. 6.38 can no longer be ignored. In this case, the flow equation can be written in terms of adjusted pressure $p_a(p)$ and adjusted time Δt_a, defined as (Agarwal 1979),

$$\Delta t_a \equiv \mu_{gi} c_{ti} \int_{t'=0}^{\Delta t} \frac{dt'}{\mu_g c_t}. \quad (6.47)$$

The fluid properties in the denominator of Eq. 6.47 are evaluated at the shut-in wellbore pressure $p_{ws}(t)$. The adjusted time must be calculated from measured test pressure data using a numerical method, such as the trapezoidal method, to evaluate the integral in Eq. 6.47.

With the adjusted pressure, the buildup equation for IARF in a gas reservoir becomes

$$p_{aws}(\Delta t_a) = p_{ai} - \frac{162.6 q_g B_{gi} \mu_{gi}}{kh} \log\left(\frac{t_p + \Delta t_a}{\Delta t_a} \right). \quad (6.48)$$

Note that the actual producing time, t_p, is used in the adjusted HTR, $HTR_a \equiv \frac{t_p + \Delta t_a}{\Delta t_a}$. This is consistent with the use of adjusted pressure along with actual time in analyzing drawdown tests.

The recommended procedure for analyzing gas-well buildup tests is as follows:

1. Estimate the initial pressure p_i to use as a reference pressure in calculating the adjusted pressure and adjusted time.
2. Calculate the adjusted shut-in pressure for each measured pressure data point.
3. Calculate the adjusted time for each measured pressure data point, using the trapezoidal method to evaluate the adjusted time integral.
4. Calculate HTR_a for each data point.
5. Graph p_{aws} vs. HTR_a on a semilog scale.
6. Identify the data in IARF and fit a straight line through the data.
7. Calculate the slope m of the straight line.
8. Estimate the permeability k from the slope m:

$$k = \frac{162.6 q_g B_{gi} \mu_{gi}}{|m| h}.$$

9. Read p_{a1hr} from the semilog straight line at an HTR_a, corresponding to an adjusted time of 1 hour.
10. 10. Estimate the total skin factor $s' = s + Dq$ from

$$s' = 1.151 \left[\frac{p_{a1hr} - p_{awf}}{|m|} - \log\left(\frac{k}{\phi \mu_{gi} c_{ti} r_w^2} \right) + 3.23 \right].$$

11. Extrapolate the semilog straight line to an HTR_a of unity to estimate the initial adjusted pressure, p_{ai}.
12. Calculate the initial pressure p_i from the initial adjusted pressure p_{ai} by reverse interpolation on the table of p_a vs. p.

6.2.5.5 Gas-Well Buildup Test Semilog-Analysis Example. *Problem.* Given the rock and fluid properties in **Tables 6.4 and 6.5,** analyze the gas-well pressure-buildup test shown in **Fig. 6.12.**

Rock and Fluid Properties					
q	5,000	Mscf/D	t_p	168	hours
p_i	7,500	psia	γ_g	0.7	(air = 1.0)
T_f	220°F				
ϕ	15%		B_{gi}	0.5495	bbl/Mscf
h	125	ft	μ_{gi}	0.0346	cp
r_w	0.25	ft	c_{gi}	2.48×10^{-5}	psi^{-1}
S_w	35%		c_w	3.60×10^{-6}	psi^{-1}
c_f	4.00×10^{-6}	psi^{-1}	c_{ti}	4.74×10^{-5}	psi^{-1}
p_{wf}	5,607	psia	p_{awf}	4,173.9	psi

Table 6.4—Rock- and fluid-property data for gas-well pressure-buildup analysis example.

Δt (hours)	Δt_a (hours)	Adjusted HTR	p_{ws} (psi)	p_{aws} (psi)
0.00100	0.00076	221,915	5,611.5	4,178.5
0.00213	0.00161	104,388	5,615.9	4,183.0
0.00339	0.00257	65,391	5,621.9	4,189.1
0.00481	0.00365	46,028	5,622.8	4,194.1
0.00642	0.00487	34,518	5,634.2	4,201.6
0.00822	0.00624	26,931	5,641.8	4,209.4
0.01025	0.00778	21,587	5,649.5	4,217.3
0.01253	0.00952	17,642	5,660.1	4,228.0
0.01509	0.01149	14,629	5,669.7	4,237.8
0.01798	0.01370	12,267	5,681.5	4,249.9
0.02123	0.01619	10,379	5,694.6	4,263.2
0.02488	0.01900	8,844	5,710.4	4,279.3
0.02899	0.02217	7,578	5,722.4	4,295.6
0.03361	0.02575	6,526	5,744.5	4,314.0
0.03881	0.02979	5,641	5,764.3	4,334.2
0.04467	0.03435	4,892	5,787.2	4,357.5
0.05125	0.03950	4,254	5,812.3	4,383.1
0.05866	0.04532	3,708	5,839.9	4,411.2
0.06699	0.05190	3,238	5,869.3	4,441.1
0.07636	0.05934	2,832	5,903.2	4,475.6
0.08691	0.06776	2,480	5,940.3	4,513.4
0.09877	0.07730	2,174	5,981.1	4,554.9
0.1121	0.0881	1,908	6,024.1	4,598.7
0.1271	0.1003	1,675	6,072.6	4,648.0
0.1440	0.1142	1,472	6,124.1	4,700.4
0.1630	0.1300	1,294	6,179.4	4,752.7
0.1844	0.1479	1,137	6,239.6	4,817.8
0.2085	0.1682	1,000	6,301.2	4,880.4
0.2355	0.1913	879.4	6,367.5	4,947.7
0.2659	0.2175	773.5	6,432.4	5,017.7
0.3002	0.2473	680.4	6,507.0	5,089.3
0.3387	0.2812	598.5	6,579.4	5,162.6
0.3821	0.3197	522.4	6,652.4	5,232.5
0.4308	0.3636	463.1	6,724.9	5,309.9
0.4857	0.4133	407.5	6,793.6	5,379.2
0.5474	0.4699	358.5	6,860.0	5,442.2

Table 6.5—Pressure-buildup data and plotting functions for gas-well pressure-buildup analysis example.

Δt (hours)	Δt_a (hours)	Adjusted HTR	p_{ws} (psi)	p_{aws} (psi)
0.6168	0.5340	315.6	6,921.6	5,508.4
0.6949	0.6068	277.9	6,978.6	5,565.8
0.7827	0.6893	244.7	7,028.0	5,615.6
0.8816	0.7826	215.7	7,073.2	5,661.1
0.9928	0.8883	190.1	7,111.3	5,699.5
1.118	1.008	167.7	7,142.7	5,731.0
1.259	1.142	148.1	7,169.5	5,758.0
1.417	1.295	130.8	7,189.7	5,778.3
1.595	1.466	115.6	7,208.3	5,792.9
1.795	1.659	102.2	7,222.5	5,811.2
2.021	1.877	90.49	7,234.7	5,823.4
2.275	2.123	80.14	7,245.5	5,834.3
2.560	2.399	71.02	7,254.7	5,843.5
2.881	2.711	62.97	7,263.7	5,852.5
3.242	3.062	55.87	7,271.6	5,860.5
3.648	3.457	49.60	7,280.2	5,869.1
4.105	3.902	44.06	7,288.3	5,877.2
4.619	4.403	39.16	7,295.4	5,884.3
5.198	4.967	34.82	7,302.4	5,891.3
5.848	5.602	30.99	7,308.7	5,897.7
2.580	2.318	27.59	7,312.2	5,905.2
7.404	7.123	24.59	7,324.4	5,913.4
8.331	8.030	21.92	7,330.1	5,919.1
9.373	9.051	19.56	7,337.7	5,922.8
10.55	10.20	17.47	7,342.6	5,931.6
11.86	11.50	15.62	7,349.9	5,939.0
13.35	12.95	13.97	7,355.7	5,944.8
15.02	14.59	12.51	7,362.5	5,951.6
12.90	12.44	11.22	7,369.4	5,958.5
19.01	18.52	10.07	7,373.9	5,963.0
21.39	20.86	9.053	7,380.4	5,969.5
24.06	23.50	8.149	7,382.2	5,975.4
27.07	22.47	7.348	7,392.2	5,981.3
30.45	29.81	2.636	7,397.3	5,982.5
34.26	33.57	2.005	7,402.7	5,991.9
38.55	37.80	5.444	7,408.4	5,997.6
43.37	42.57	4.946	7,413.8	6,003.0
48.79	47.94	4.505	7,419.3	6,008.5
54.89	53.98	4.112	7,424.1	6,013.3
61.75	60.78	3.764	7,429.0	6,018.2
69.47	68.43	3.455	7,431.9	6,021.1
78.15	77.05	3.180	7,437.4	6,022.6
87.92	82.75	2.937	7,441.9	6,031.1
98.91	97.66	2.720	7,445.5	6,034.7
111.3	110.0	2.528	7,449.3	6,038.5
125.2	123.8	2.357	7,453.0	6,042.2
140.8	139.3	2.206	7,452.9	6,042.2
158.4	152.9	2.071	7,461.9	6,051.1
168	162.4	2.010	7,461.9	6,051.1

Table 6.5—(*Continued*).

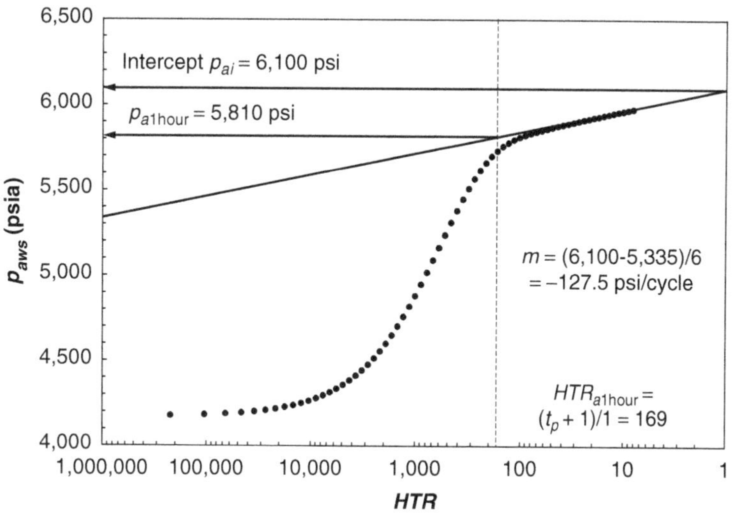

Fig. 6.12—Gas-well pressure buildup semilog analysis example.

Solution. Because Steps 1 through 5 have been completed, begin the example with Step 6:

6. The data from an adjusted HTR of 50 through the end of the test appear to fall on a straight line, so we draw a straight line through this portion of the data.
7. Next, calculate the slope of the straight line. Reading two points on the straight line as (1,000,000; 5,335) and (1; 6,100) gives

$$m = \frac{6,100 - 5,335}{\log(1) - \log(1,000,000)} = -127.5 \text{ psi/cycle}.$$

8. Next, estimate the permeability from

$$k = \frac{162.6 q B_{gi} \mu_{gi}}{|m| h}$$
$$= \frac{(162.6)(5,000)(0.5495)(0.0346)}{(127.5)(125)}$$
$$= 0.970 \text{ md}.$$

9. Read p_{a1hr} from the semilog straight line at an adjusted HTR, HTR_a = 169, corresponding to an adjusted time of 1 hour, as 5,810 psi.
10. Estimate the total skin factor s' from p_{a1hr} as

$$s' = 1.151\left[\frac{p_{a1hr} - p_{awf}}{|m|} - \log\left(\frac{k}{\phi \mu_{gi} c_{ti} r_w^2}\right) + 3.23\right]$$
$$= 1.151\left[\frac{(5,810) - (4,173.9)}{(127.5)} - \log\left(\frac{(0.97)}{(0.15)(0.0346)(4.74 \times 10^{-5})(0.25)^2}\right) + 3.23\right]$$
$$= 9.5.$$

11. Extrapolate the semilog straight line to an adjusted HTR of unity to estimate the initial adjusted pressure, p_{ai}, as 6,100 psi.
12. Reverse interpolate from the table of p_a vs. p to obtain p_i = 7,500 psi. If the estimated initial pressure had been significantly different from the original estimate, we would have repeated Steps 2 through 12 to refine the analysis.

6.3 Log-Log Type Curve Analysis

One of the major drawbacks to straight-line analysis methods, such as semilog analysis for radial flow, is that the flow regime of interest must be fully developed for straight-line analysis methods to be valid. Often, especially in low-permeability reservoirs, this condition requires longer test times than are practical. One of the more useful methods for overcoming this difficulty is the use of log-log type curves. Type curves can account for multiple flow regimes and the intervening transitions.

One of the most useful type curves for analyzing prefracture tests was introduced by Gringarten et al. (1979) and was later enhanced by Bourdet et al. (1983). This type curve describes the pressure response for a well in an infinite-acting reservoir, with constant WBS coefficient and constant skin factor.

Before introducing the Gringarten-Bourdet type curve, we will take a slight detour to discuss WBS and dimensionless variables.

6.3.1 Wellbore Storage (WBS). The Ei-function solution assumes that the sandface flow rate rises instantly from zero to the final constant rate q at time $t = 0$ at the beginning of a drawdown test. In using superposition to calculate the buildup-pressure response using the Ei-function solution, we assume that the sandface flow rate drops instantly from q to zero at time $\Delta t = 0$ at the beginning of the buildup test. Because of the finite volume of the wellbore between the sandface and the valve used to control the flow rate **(Fig. 6.13),** the sandface rate does not change immediately upon opening or closing the valve.

Instead, the first fluid produced upon opening the valve comes from the wellbore. As the pressure in the wellbore decreases because of withdrawal of fluid, the sandface flow rate rises, rapidly at first, asymptotically approaching a constant value equal to the surface flow rate q, as shown in **Fig. 6.14.** Similarly, when the well is shut in at the surface following a period of production at constant rate, fluid continues to flow into the wellbore, asymptotically approaching zero, as shown in **Fig. 6.15.** During buildup, this phenomenon is often referred to as afterflow.

In drawdown and buildup, WBS distorts the shape of the pressure response. The effects of WBS can be seen in Figs. 6.8 and 6.10 through 6.12. In Figs. 6.8 and 6.12, the well is damaged, causing a characteristic "lazy S" shape. In Figs. 6.10 and 6.11, the lazy S shape is absent. Instead, the pressure response approaches the semilog straight line from above, typical of a buildup test on a well that has been stimulated.

We can write a material-balance equation for the relationship between the surface and sandface flow rates for a wellbore filled with a single-phase fluid as

$$24c_{wb}V_{wb}\frac{dp_w}{dt} = -\left[q(t) - q_{sf}(t)\right]B_{wb}, \quad (6.49)$$

where c_{wb} is the compressibility of the wellbore fluid in psi^{-1}, B_{wb} is the FVF of the fluid in the wellbore, V_{wb} is the wellbore volume in bbl, p_w is the wellbore pressure, and q_{sf} is the sandface flow rate. By introducing a WBS coefficient C, defined as

$$C \equiv c_{wb}V_{wb}, \quad (6.50)$$

we can rewrite Eq. 6.49 as

$$24C\frac{dp_w}{dt} = -\left[q(t) - q_{sf}(t)\right]B_{wb}. \quad (6.51)$$

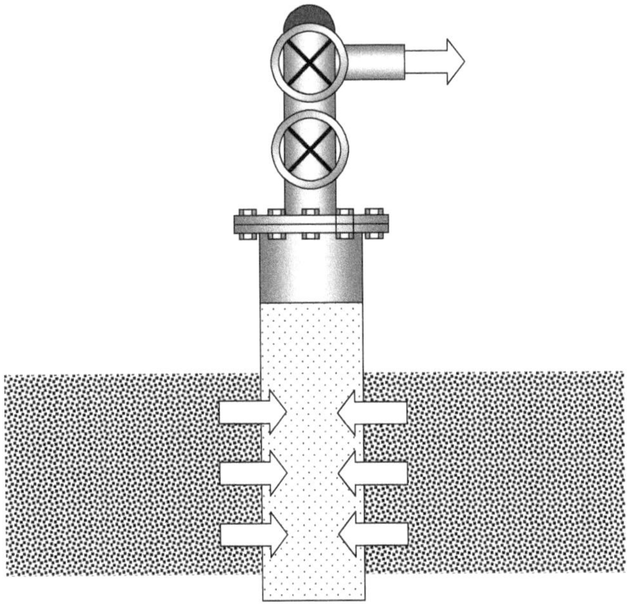

Fig. 6.13—WBS, fluid-filled wellbore.

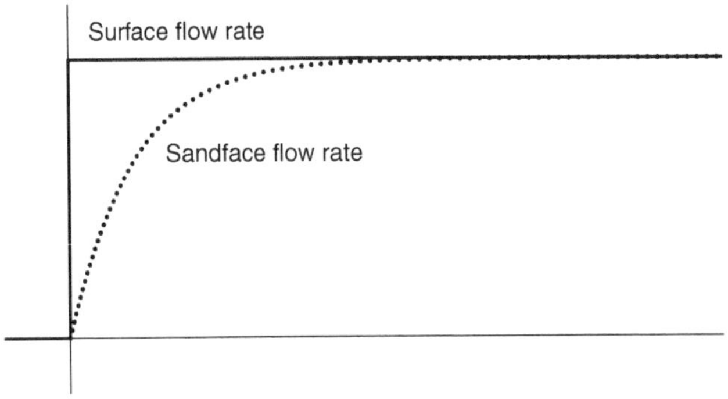

Fig. 6.14—Effect of WBS on sandface flow rate during production at constant surface rate.

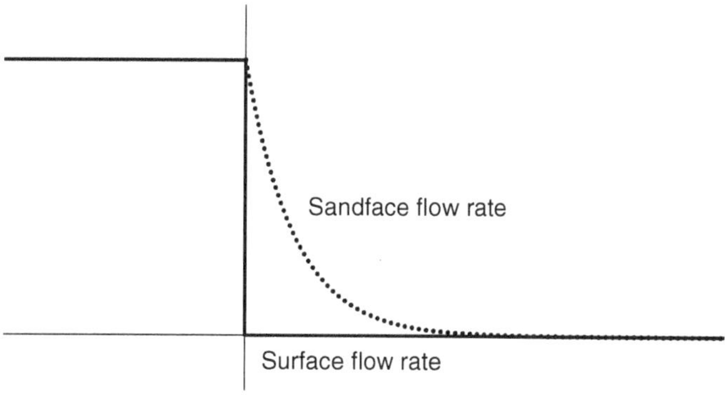

Fig. 6.15—Effect of WBS on sandface flow rate during buildup following constant rate production.

Eq. 6.51 is general, applying to variable flow-rate production, as well as drawdown and buildup.

Just before the beginning of a drawdown, the sandface flow rate and surface flow rate are both zero. Upon opening the valve, the surface flow rate rises instantly to a constant value q. The rate of pressure change immediately after the valve is opened, when the sandface flow rate is approximately zero, is then given by

$$24C\frac{dp_w}{dt} \approx -(q-0)B_{wb} \approx -qB_{wb}. \quad (6.52)$$

The flowing wellbore pressure p_{wf} during this WBS-dominated period is then given by

$$p_i - p_{wf} \approx \frac{qB_{wb}}{24C}t. \quad (6.53)$$

Similarly, just before shutting in for a buildup, the sandface and surface flow rates are both q. Immediately after shutting in the well, before the sandface flow rate has begun to drop, the rate of pressure change in the wellbore is given by

$$24C\frac{dp_w}{dt} \approx -(0-q_{sf})B_{wb} = qB_{wb}, \quad (6.54)$$

while the shut-in wellbore pressure p_{ws} is given by

$$p_{ws} - p_{wf} \approx \frac{qB_{wb}}{24C}\Delta t, \quad (6.55)$$

where p_{wf} is the flowing bottomhole pressure at the moment of shut in.

The analytical solution for the pressure response for production at constant rate q from a well with constant WBS and skin in an infinite-acting reservoir is rather complex, with no simple closed form. However, the solution can be presented graphically as a set of drawdown type curves.

6.3.2 Dimensionless Variables. Central to the use of type curves is the concept of dimensionless variables. Dimensionless variables allow general solutions to the diffusivity equation to be defined independently from specific reservoir properties. By finding a general solution, or type curve, that has the same shape on a log-log graph as the field data to be analyzed, and by shifting the field data axes to overlay the type curve, we can analyze pressure-transient test data in more detail than can be achieved with using only semilog methods.

The Ei-function solution to the diffusivity equation for production at constant rate is

$$p_{wf}(t) = p_i + \frac{70.6qB\mu}{kh}\mathrm{Ei}\left(-\frac{948\varphi\mu c_t r^2}{kt}\right). \quad (6.56)$$

We can rearrange Eq. 6.56 as

$$\frac{0.00708kh}{qB\mu}\left[p_i - p_{wf}(t)\right] = -\frac{1}{2}\mathrm{Ei}\left(-\frac{1}{4\left(\dfrac{0.0002637kt}{\phi\mu c_t r_w^2}\right)}\left(\frac{r}{r_w}\right)^2\right), \quad (6.57)$$

then define the dimensionless variables, dimensionless radius r_D, $r_D \equiv \frac{r}{r_w}$, dimensionless time t_D, $t_D \equiv \frac{0.0002637k}{\phi\mu c_t r_w^2}t$, and dimensionless pressure p_D, $p_D \equiv \frac{0.00708kh}{qB\mu}(p_i - p)$.

We can rewrite the Ei-function solution given in Eq. 6.57 in general form as

$$p_D = -\frac{1}{2}\text{Ei}\left(-\frac{r_D^2}{4t_D}\right). \quad (6.58)$$

A graph of p_D as a function of $r_D^2/4t_D$, on a log-log scale, is one of the simplest type curves and is often used in analyzing interference tests.

Many different type curves have been developed to account for different inner boundary conditions (i.e., constant-rate production, constant-pressure production, constant or variable WBS), different well models (i.e., vertical wells, hydraulically fractured wells, horizontal wells), different reservoir media (i.e., homogeneous, dual porosity, dual permeability), and different outer boundary conditions (e.g., infinite acting, linear no-flow boundary, well in a rectangle).

6.3.3 Gringarten-Bourdet Type Curves. One of the most useful type curves for analyzing prefracture tests is the Gringarten-Bourdet type curve, which describes the pressure response for a vertical well with constant WBS and constant skin in an infinite-acting reservoir. The Gringarten-Bourdet type curve is a combination of the Gringarten type curve (Gringarten et al. 1979), shown in **Fig. 6.16,** and the Bourdet pressure-derivative type curve (Bourdet et al. 1983), shown in **Fig. 6.17.**

The Gringarten type curve is a graph of dimensionless wellbore pressure, p_{wD}, vs. t_D/C_D, for different values of the dimensionless group C_De^{2s}. The dimensionless pressure is defined as $p_{wD} \equiv \frac{0.00708kh}{qB\mu}(p_i - p_{wf})$, the dimensionless time as $t_D \equiv \frac{0.0002637k}{\phi\mu c_t r_w^2}t$, and the dimensionless WBS coefficient as $C_D = \frac{0.894}{\phi c_t h r_w^2}C$.

On the Gringarten type curve, stimulated wells will have values of C_De^{2s} less than 100. Unstimulated wells will have values of C_De^{2s} greater than 100 and 10,000, while damaged wells will have values of C_De^{2s} above 10,000. The curve for a damaged well follows a unit-slope line, then breaks fairly sharply to the right, becoming almost horizontal. The curve for a stimulated well leaves the unit-slope line very early and bends slowly to the right. Because the shapes of the curves for values of C_De^{2s} greater than 1,000 are so similar, it is often difficult to obtain a unique match of field data using the Gringarten type curve alone.

The introduction of the logarithmic pressure derivative along with the Bourdet derivative type curve (Fig. 6.17) revolutionized the practice of well test interpretation by allowing the unambiguous determination of the WBS-dominated period, identified by an initial unit-slope line, and the IARF period, identified by a horizontal derivative, as well as several other flow regimes. As seen in Fig. 6.17, the Bourdet derivative type curve also clearly differentiates the various curves having values of C_De^{2s} greater than 1,000.

The logarithmic derivative, defined as

$$t\frac{dp}{dt} = \frac{dp}{d\ln t} \quad (6.59)$$

has the same units as pressure. This characteristic of the derivative allows the dimensionless pressure, p_D, and the dimensionless pressure derivative, $t_D(dp_D/dt_D)$, to be graphed together on a single set of axes, as shown in **Fig. 6.18.**

The derivative curves are particularly useful for identifying wells with stimulation or damage simply by inspecting the graph. The derivative curve for a highly stimulated well approaches the horizontal IARF line from below, without ever exhibiting a peak in the derivative. On the other hand, the derivative curve for a damaged well continues along the WBS unit-slope line for some time, then reaches a peak in the derivative before falling a log cycle or more to the horizontal IARF line. The derivative curve for a well with

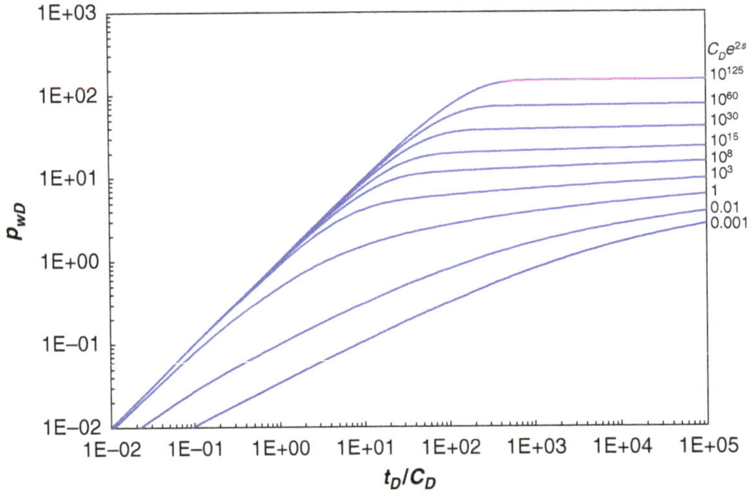

Fig. 6.16—Gringarten type curve for a vertical well with constant WBS and constant skin factor in an infinite-acting reservoir.

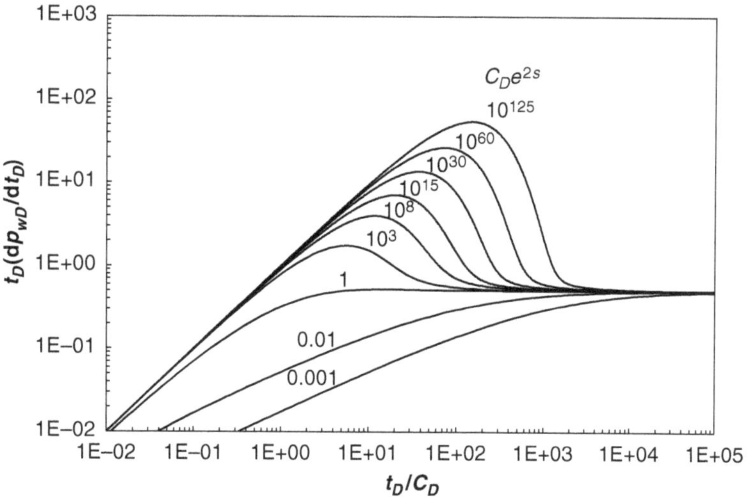

Fig. 6.17—Bourdet pressure-derivative type curve for a vertical well with constant WBS and skin factor in an infinite-acting reservoir.

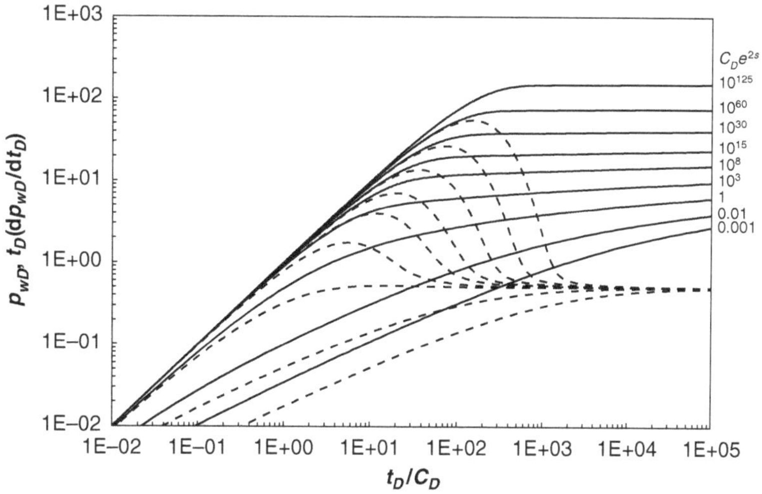

Fig. 6.18—Gringarten-Bourdet type curve for a vertical well with constant WBS and skin factor in an infinite-acting reservoir.

neither damage nor stimulation also exhibits a peak in the derivative before falling to the horizontal IARF line, but the maximum is approximately ½ log cycle higher than the horizontal line. The separation between the pressure-change and pressure-derivative curves also provides a useful clue to the degree of damage or stimulation. For a stimulated well, the separation will be less than a log cycle at the time WBS ends. For a damaged well, the separation between the pressure change and pressure derivative curves could be 1½ log cycles or more.

6.3.4 Drawdown-Test Analysis Using the Gringarten-Bourdet Type Curves. Analyzing constant-rate drawdown-test data using the Gringarten-Bourdet type curve is fairly straightforward. The recommended procedure is as follows:

1. Graph the pressure change defined as $\Delta p \equiv p_i - p_{wf}$ and the logarithmic derivative defined by Eq. 6.59 vs. the test time t on translucent log-log graph paper having the same size log cycles as the type curve. This graph will be referred to as the field data graph in the rest of this section.
2. Identify the WBS-dominated period, indicated by the unit-slope portion of the derivative, and the IARF period, indicated by the horizontal portion of the derivative.
3. Place the field data graph on top of the type curve graph, as shown in **Fig. 6.19.**
4. Keeping the axes of the field data graph and the type curve parallel, move the field data graph vertically until the horizontal part of the field data derivative overlies the horizontal part of the type curve derivative, as shown in **Fig. 6.20.**
5. Still keeping the axes of the two graphs parallel, move the field data graph horizontally until the unit slope portion of the field data pressure change and pressure derivative overlies the type curve pressure and pressure-derivative curves, as shown in **Fig. 6.21.**

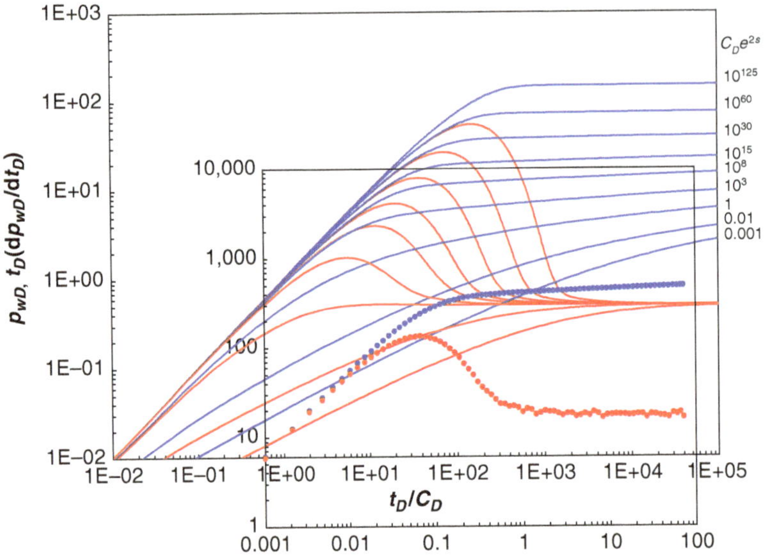

Fig. 6.19—Type curve match procedure, Step 3.

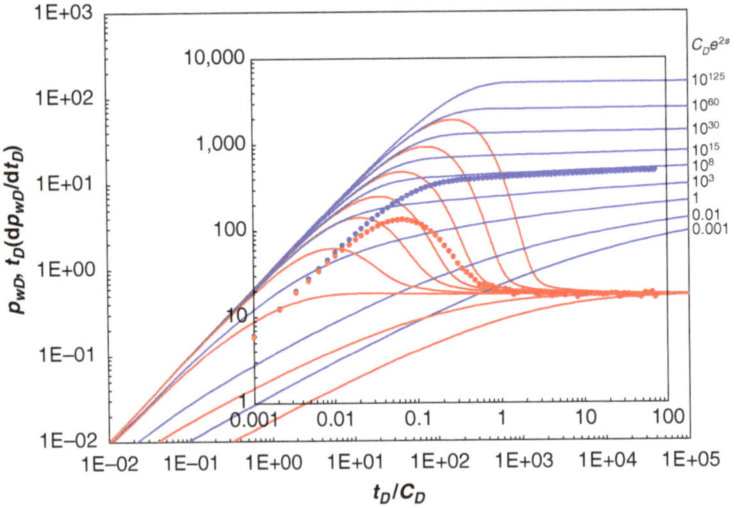

Fig. 6.20—Type curve matching procedure, Step 4.

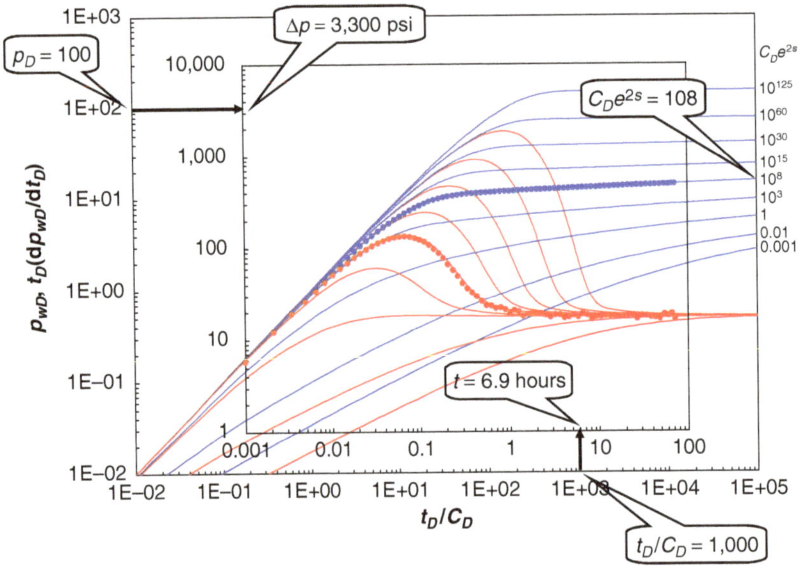

Fig. 6.21—Type curve match procedure, Steps 5–8.

6. On the type curve graph, find the pressure and pressure-derivative curve pair that best match the observed field pressure-change and pressure-derivative data, as shown in Fig. 6.21. It might be necessary to interpolate between curves to obtain a good fit. Record the value of C_De^{2s} corresponding to the chosen type curve pair.
7. Without changing the alignment of the field data and type curve graphs, select a convenient point on the x-axis of the type curve graph, t_D/C_D, then read the coordinates of that point on the x-axis of the field data graph, t, as shown in Fig. 6.21. Because the logarithm of the product of two numbers is the sum of the logarithms of the numbers, the time match-point ratio $\left(\dfrac{t}{t_D/C_D}\right)_{MP}$ will be independent of the specific values of t_D/C_D and t obtained.
8. Similarly, select a convenient point on the y-axis of the type curve graph, p_D, and read the corresponding coordinates from the field data graph, Δp, as shown in Fig. 6.21. Again, the pressure match-point ratio $\left(\dfrac{p_D}{\Delta p}\right)_{MP}$ will be independent of the specific values of p_D and Δp selected.
9. Calculate the permeability from the pressure match-point ratio:

$$k = \frac{qB\mu}{0.00708h}\left(\frac{p_D}{\Delta p}\right)_{MP}. \qquad (6.60)$$

10. Calculate the dimensionless WBS coefficient C_D from the time match-point ratio:

$$C_D = \frac{0.0002637k}{\varphi\mu c_t r_w^2}\left(\frac{t}{t_D/C_D}\right)_{MP}. \qquad (6.61)$$

11. Calculate the WBS coefficient from

$$C = \frac{\phi c_t h r_w^2}{0.894}C_D. \qquad (6.62)$$

12. Finally, calculate the skin factor from the value of C_De^{2s} obtained in Step 6:

$$s = 0.5\ln\left(\frac{C_De^{2s}}{C_{aD}}\right). \qquad (6.63)$$

6.3.5 Buildup-Test Analysis Using Drawdown Type Curves. Type curves such as the Gringarten-Bourdet type curve are referred to as drawdown type curves because they are constructed specifically for production at constant rate beginning at time zero. We can analyze buildup tests using drawdown type curves if one of the following two conditions is met:

- If the buildup test is much shorter than the preceding flow period ($\Delta t \le 0.1t_p$), we can use the shut-in time Δt as the time plotting variable and the pressure change $\Delta p = p_{ws} - p_{wf}$ as the pressure plotting variable.
- If the reservoir is infinite acting from the beginning of the flow period through the end of the shut-in period, we can use the radial-flow equivalent time (Agarwal 1980), defined as $\Delta t_e = \dfrac{t_p\Delta t}{t_p + \Delta t}$, as the time plotting variable and the pressure change $\Delta p = p_{ws} - p_{wf}$ as the pressure plotting variable.

In practice, the radial-flow equivalent time is often used, even for cases in which the flow period is much longer than the shut-in period, or in which the reservoir is not infinite acting. In the latter case, Bourdet recommends calculating the derivative with respect to the equivalent time [i.e., $\Delta t_e(\mathrm{d}p/\mathrm{d}\Delta t_e)$], but graphing the derivative vs. the shut-in time, Δt (Bourdet et al. 1989). The pressure change is graphed vs. the equivalent time. Although not mathematically rigorous, in many cases this approach gives the buildup pressure derivative a shape similar to that of the drawdown derivative for the same reservoir model. The resulting graph can be used as an aid to flow-regime and model identification, but should not be used for estimating parameters such as distance to boundaries.

6.3.6 Modifications for Gas Wells. For log-log analysis of gas-well drawdown tests, we use the adjusted pressure change $\Delta p_a \equiv p_{ai} - p_{awf}$ as the pressure plotting variable and the actual time t as the time plotting variable. For buildup tests on gas wells, we use the adjusted pressure change $\Delta p_a \equiv p_{aws} - p_{awf}$ as the pressure plotting variable and the adjusted equivalent time $\Delta t_{ae} \equiv \dfrac{t_p\Delta t_a}{t_p + \Delta t_a}$ as the time plotting variable. Note the use of the actual producing time t_p.

The use of adjusted pressure and adjusted time is especially important in type curve analysis of low-permeability gas wells, which often have large drawdowns. The WBS coefficient for a wellbore filled with a single-phase fluid is given by

$$C = c_{wb}V_{wb}. \qquad (6.64)$$

The wellbore pressure might change from less than 500 psia to more than 5,000 psia during a buildup test in a gas well following a large drawdown. The gas compressibility will thus roughly decrease by an order of magnitude, causing a corresponding decrease in the WBS coefficient C. **Fig. 6.22** shows a log-log graph of a synthetic gas-well buildup test following a large drawdown. The decreasing WBS coefficient causes the data to deviate upward from the unit-slope line representing constant WBS.

However, if we graph the adjusted pressure change vs. the adjusted equivalent time, as shown in **Fig. 6.23,** the data fall on a unit-slope line, characterized by the adjusted WBS coefficient (Lee and Holditch 1982),

$$C_a = c_{gi}V_{wb}, \qquad (6.65)$$

where we have replaced the compressibility of the wellbore fluid c_{wb} with the gas compressibility c_{gi}, evaluated at the initial pressure p_i used as the reference pressure to calculate the adjusted pressure.

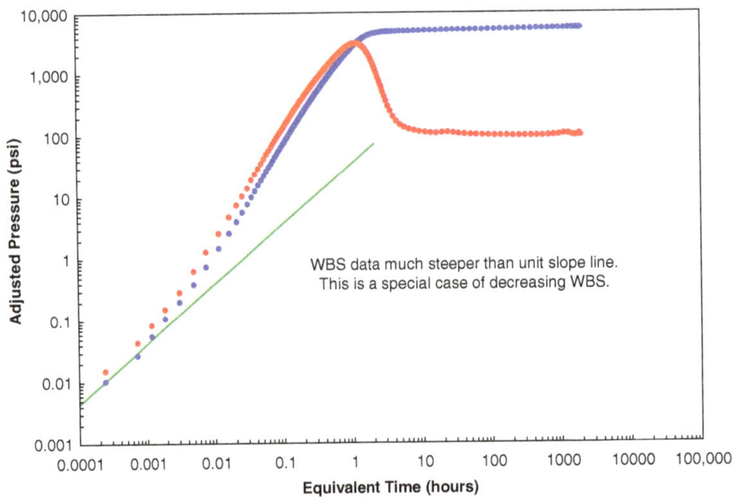

Fig. 6.22—Buildup test on low-permeability gas well following large drawdown, graphed as adjusted pressure change vs. equivalent time. The changing WBS coefficient causes the data to deviate upward from the unit-slope line.

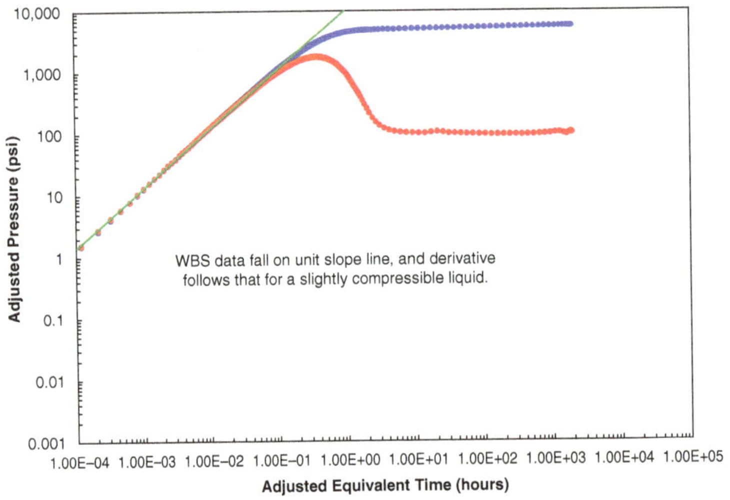

Fig. 6.23—Same buildup test as in Fig. 6.22, graphed as adjusted pressure change vs. adjusted equivalent time. With the adjusted equivalent time, the data follow the unit-slope line, just as in the liquid case.

Thus, adjusted pressure and adjusted equivalent time enables the analysis of gas-well buildup-test data using drawdown type curves for slightly compressible liquids. With these modifications, the procedure for analyzing a buildup test on a gas well is similar to that for an oil well:

1. Construct the field data graph by plotting the adjusted pressure change, defined as $\Delta p_a \equiv p_{ai} - p_{awf}$, and its logarithmic derivative with respect to the adjusted equivalent time Δt_{ae} vs. the adjusted equivalent time on translucent log-log graph paper with the same size log cycles as the type curve.
2. Identify the WBS-dominated period.
3. Place the field data graph on top of the type curve graph.
4. Move the field data graph vertically to align the horizontal part of the field data derivative with the horizontal part of the type curve derivative.
5. Move the field data graph horizontally to align the unit slope portions of the field data graph and type curve graph.
6. Find the pressure-change and pressure-derivative curve pair that best matches the field data. Record the value of $C_D e^{2s}$ corresponding to the chosen type curve pair.
7. Select a convenient point on the x-axis of the type curve graph, t_D/C_D, then read the coordinates of that point on the x-axis of the field data graph, Δt_{ae}.
8. Similarly, select a convenient point on the y-axis of the type curve graph, p_D, and read the corresponding coordinates from the field data graph, Δp_a.

9. Calculate the permeability from the pressure match-point ratio:

$$k = \frac{q_g B_{gi} \mu_{gi}}{0.00708h} \left(\frac{p_D}{\Delta p_a} \right)_{MP}.$$

10. Calculate the dimensionless adjusted WBS coefficient C_{aD} from the time match-point ratio:

$$C_{aD} = \frac{0.0002637k}{\varphi \mu_{gi} c_{ti} r_w^2} \left(\frac{\Delta t_{ae}}{t_D / C_D} \right)_{MP}.$$

11. Calculate the adjusted WBS coefficient from

$$C_a = \frac{\phi c_{ti} h r_w^2}{0.894} C_{aD}.$$

12. Finally, calculate the skin factor from the value of $C_D e^{2s}$ obtained in Step 6:

$$s' = 0.5 \ln \left(\frac{C_D e^{2s}}{C_{aD}} \right).$$

If there is a high non-Darcy skin factor, the decreasing sandface flow rate during the buildup will cause a decreasing skin factor during the buildup test (Spivey et al. 2002). This decreasing skin factor will cause the gas-well pressure-derivative data to briefly rise above the liquid curve, then fall very steeply toward the horizontal IARF portion of the curve, as shown in **Fig. 6.24.** This phenomenon can be seen in a field example discussed in Section 6.6.2.

If both unit-slope WBS and IARF regimes are present in a buildup test, type curve matching can still be performed, ignoring the transition period between WBS and IARF. If the IARF does not appear during the test, the data can still be analyzed using an analytical or numerical simulator incorporating non-Darcy skin factor.

6.4 Manual Parameter Estimation Using the Log-Log Diagnostic Plot

Although the log-log graph of pressure change and pressure derivative vs. time is often associated with type curve analysis, it can be used to estimate permeability, skin factor, and WBS coefficient without the use of type curves. This technique allows the analyst to quickly double check the results of a type curve analysis. In other cases, the original pressure data cannot be available, while a log-log graph of the data is. A quick analysis can be performed using only the log-log graph.

Figs. 6.25 and 6.26 illustrate the procedure for estimating permeability and skin factor for a vertical well in an infinite-acting reservoir from the log-log graph. The procedure is as follows:

1. Identify the horizontal part of the derivative corresponding to IARF.
2. Draw a horizontal line through the IARF data and read the value of the derivative during IARF, $(t\Delta p')_R$.
3. Pick a convenient time, Δt_R, during IARF and draw a vertical line to intersect the pressure-change curve.
4. Draw a horizontal line from the point of intersection of the vertical line to the y-axis and read the pressure change, Δp_R.
5. Calculate the permeability from

$$k = \frac{70.6 q B \mu}{h (t\Delta p')_R}. \qquad (6.66)$$

6. Calculate the skin factor from

$$s = \frac{1}{2} \left[\frac{\Delta p_R}{(t\Delta p')_R} - \ln \left(\frac{k \Delta t_R}{1688 \varphi \mu c_t r_w^2} \right) \right]. \qquad (6.67)$$

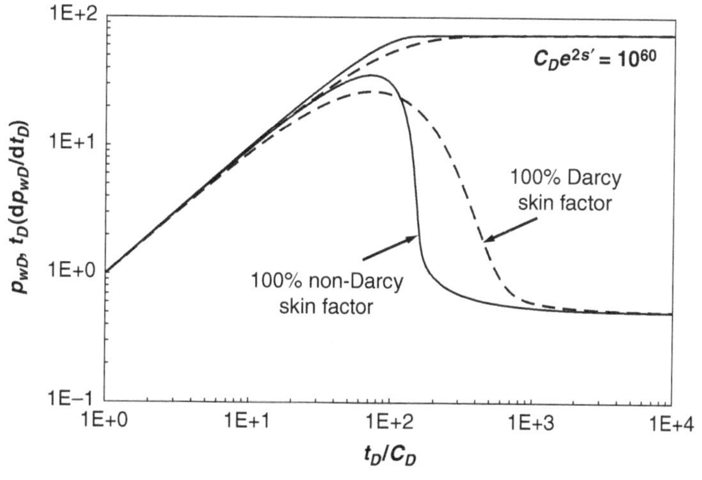

Fig. 6.24—Effect of non-Darcy skin factor on pressure response during buildup.

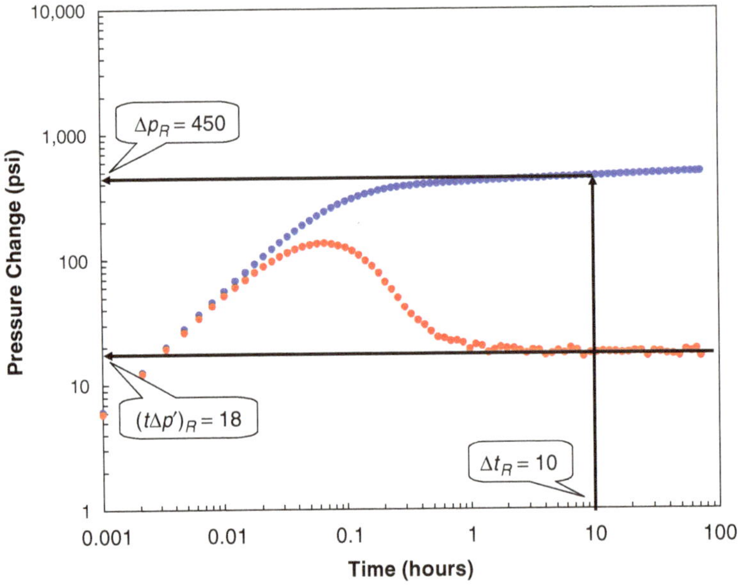

Fig. 6.25—Estimating permeability and skin factor from the log-log diagnostic plot.

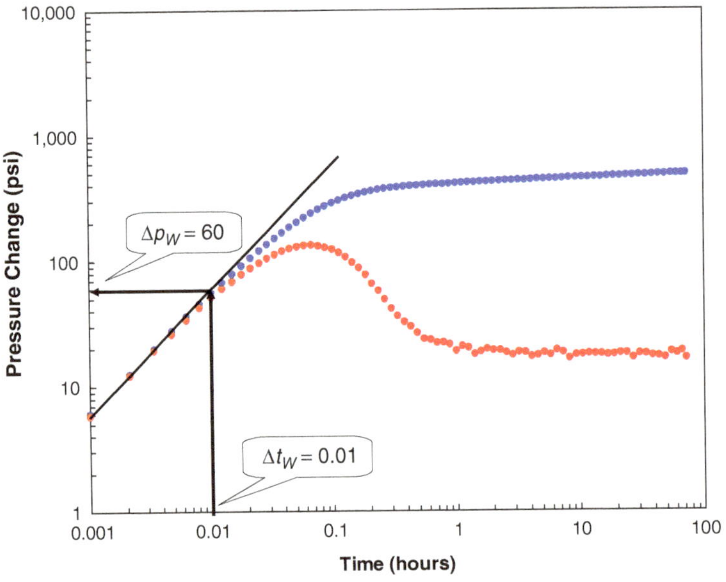

Fig. 6.26—Estimating WBS coefficient from the log-log diagnostic plot.

7. Draw a unit-slope line through data corresponding to WBS.
8. Read the coordinates of any point on the unit-slope line (Δt_W, Δp_W).
9. Calculate the WBS coefficient from

$$C = \frac{qB\Delta t_W}{24\Delta p_W}. \qquad (6.68)$$

6.4.1 Manual Log-Log Analysis Example.

6.4.1.1 Problem. Given the rock and fluid properties in **Table 6.6 and 6.7,** analyze the drawdown test shown in Figs. 6.25 and 6.22.

6.4.1.2 Solution.

1. The IARF period begins at hour 3 and lasts through the end of the test.
2. From the derivative during IARF, read $(t\Delta p')_R = 18$ psi.
3. Select a convenient value of time during IARF, $\Delta t_R = 10$ hr.
4. From the pressure-change curve, read $\Delta p_R = 450$ psi at $\Delta t_R = 10$ hr.

Rock and Fluid Properties

q	125	STB/D	B	1.152	bbl/STB
ϕ	0.22		μ	2.12	cp
h	50	ft	c_o	2.94×10^{-6}	psi^{-1}
S_w	0		c_f	4×10^{-6}	psi^{-1}
T_f	120	deg F	c_t	1.094×10^{-5}	psi^{-1}
r_w	0.25	ft	p_i	2,750	psi

Table 6.6—Rock- and fluid-property data for manual log-log analysis example.

t (hours)	p_{wf} (psia)	Δp_{wf} (psi)	$t(dp/dt)$ (psi)	t (hours)	p_{wf} (psia)	Δp_{wf} (psi)	$t(dp/dt)$ (psi)
0.00100	2,743.96	2.04	5.82	0.8816	2,341.74	408.3	21.09
0.00213	2,737.52	12.48	12.24	0.9928	2,339.53	410.5	18.78
0.00339	2,730.16	19.84	19.20	1.118	2,337.28	412.7	20.48
0.00481	2,722.32	27.68	22.01	1.259	2,334.67	415.3	19.96
0.00642	2,713.83	32.17	33.67	1.417	2,332.55	417.4	17.69
0.00822	2,704.46	45.54	42.10	1.595	2,330.48	419.5	18.58
0.01025	2,694.23	55.77	51.01	1.795	2,328.15	421.8	19.36
0.01253	2,683.03	62.97	59.53	2.021	2,325.90	424.1	18.99
0.01509	2,671.26	78.74	68.33	2.275	2,323.66	422.3	18.95
0.01798	2,658.36	91.64	78.00	2.560	2,321.42	428.6	18.24
0.02123	2,644.71	105.3	82.97	2.881	2,319.35	430.7	17.32
0.02488	2,630.14	119.9	95.97	3.242	2,317.33	432.7	18.29
0.02899	2,614.84	135.2	104.50	3.648	2,315.03	435.0	18.26
0.03361	2,598.72	151.3	112.20	4.105	2,313.02	437.0	12.82
0.03881	2,582.14	167.9	118.90	4.619	2,311.06	438.9	17.24
0.04467	2,564.96	185.0	125.10	5.198	2,308.95	441.1	18.10
0.05125	2,547.42	202.6	129.00	5.848	2,302.79	443.2	17.30
0.05866	2,529.89	220.1	131.70	2.580	2,304.87	445.1	18.94
0.06699	2,512.19	237.8	133.00	7.404	2,302.32	447.7	18.17
0.07636	2,494.89	255.1	131.00	8.331	2,300.59	449.4	12.36
0.08691	2,478.16	271.8	127.50	9.373	2,298.46	451.5	17.39
0.09877	2,462.14	287.9	122.70	10.55	2,292.49	453.5	17.56
0.1121	2,442.97	303.0	115.80	11.86	2,294.32	455.7	17.86
0.1271	2,432.97	317.0	107.10	13.35	2,292.28	457.7	17.61
0.1440	2,420.21	329.8	92.45	15.02	2,290.17	459.8	17.69
0.1630	2,409.04	341.0	82.94	12.90	2,288.11	461.9	17.44
0.1844	2,398.75	351.3	72.81	19.01	2,282.06	463.9	17.48
0.2085	2,390.20	359.8	65.33	21.39	2,283.99	462.0	17.91
0.2355	2,382.79	367.2	52.92	24.06	2,281.84	468.2	17.95
0.2659	2,372.35	373.7	48.81	27.07	2,279.76	470.2	12.51
0.3002	2,370.96	379.0	40.93	30.45	2,277.95	472.1	17.81
0.3387	2,362.46	383.5	35.72	34.26	2,275.56	474.4	17.71
0.3821	2,362.35	387.7	32.11	38.55	2,273.78	472.2	17.05
0.4308	2,358.74	391.3	29.35	43.37	2,271.54	478.5	17.03
0.4857	2,355.31	394.7	22.07	48.79	2,269.77	480.2	12.48
0.5474	2,352.50	397.5	23.35	54.79	2,267.69	482.3	18.42
0.6168	2,349.73	400.3	23.13	60.79	2,265.73	484.3	17.85
0.6949	2,342.98	403.0	21.79	62.79	2,264.46	485.5	18.66
0.7827	2,344.54	405.5	22.01	72	2,262.99	487.0	12.23

Table 6.7—Drawdown pressure data and plotting functions for manual log-log analysis example.

5. Calculate the permeability as

$$k = \frac{70.6 qB\mu}{h\left(t\Delta p'\right)_R}$$

$$= \frac{(70.6)(125)(1.152)(2.12)}{(50)(18)}$$

$$= 23.9 \text{ md}.$$

6. Calculate the skin factor as

$$s=\frac{1}{2}\left[\frac{\Delta p_R}{(t\Delta p')_R}-\ln\left(\frac{k\Delta t_R}{1688\phi\mu c_t r_w^2}\right)\right]$$

$$=\frac{1}{2}\left[\frac{450}{18}-\ln\left(\frac{(23.9)(10)}{(1688)(0.22)(2.12)(1.094\times10^{-5})(0.25)^2}\right)\right]$$

$$=6.0.$$

7. Draw a unit-slope line through the data corresponding to WBS between $t = 0.001$ and $t = 0.01$ hour.
8. Read $(\Delta t_W, \Delta p_W)$ as (0.01, 60).
9. Calculate the WBS coefficient from

$$C=\frac{qB\Delta t_W}{24\Delta p_W}$$

$$=\frac{(125)(1.152)(0.01)}{(24)(60)}$$

$$=0.001\ \text{bbl/psi}.$$

6.4.2 Modifications for Gas Wells. Manual log-log analysis for gas wells relies on the same transformations as log-log type curve analysis (i.e., use of adjusted pressure and real time for drawdown tests and adjusted pressure and adjusted equivalent time for buildup tests). The procedure is essentially the same, with the following equations for gas-well buildup tests:

1. Calculate the permeability from

$$k=\frac{70.6q_gB_{gi}\mu_{gi}}{h(t\Delta p')_{aR}}. \quad (6.69)$$

2. Calculate the total skin factor s' from

$$s'=\frac{1}{2}\left[\frac{\Delta p_{aR}}{(t\Delta p')_{aR}}-\ln\left(\frac{k\Delta t_{aR}}{1688\varphi\mu_{gi}c_{ti}r_w^2}\right)\right]. \quad (6.70)$$

3. Calculate the adjusted WBS coefficient from

$$C_a=\frac{q_gB_{gi}\Delta t_{aW}}{24\Delta p_{aW}}. \quad (6.71)$$

6.5 Flow-Regime Identification Using the Log-Log Diagnostic Plot

One of the most important steps in using semilog analysis to analyze pressure-transient data is to identify the correct portion of the data (i.e., data in IARF) for the analysis. The log-log diagnostic plot is the best tool currently available for identifying IARF, as well as other flow regimes such as WBS, linear flow, bilinear flow, PSSF, and spherical flow. Each of these flow regimes exhibits a characteristic behavior on the diagnostic plot.

The log-log diagnostic plot is a graph of the pressure change and logarithmic derivative of pressure vs. time on a log-log scale. On the log-log diagnostic plot, the logarithmic derivative follows a straight line having a characteristic slope during each flow regime. In many cases, the pressure change will also follow on a straight line having the same slope. This behavior will, in general, be absent if the flow regime of interest is preceded by another flow regime or if there is some restriction to flow between the wellbore and the portion of the reservoir causing the flow regime in question. Thus, the logarithmic derivative should be used as the primary tool for flow-regime identification, with the pressure-change curve serving as a consistency check.

6.5.1 Radial Flow. As we have seen, during IARF, the logarithmic derivative is a horizontal line. **Fig. 6.27** illustrates IARF for a vertical well with constant WBS and skin factor; **Fig. 6.28** shows pseudoradial flow (PRF) for a well with a vertical hydraulic fracture.

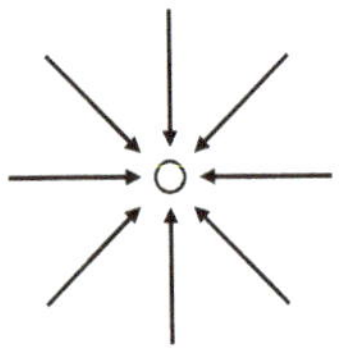

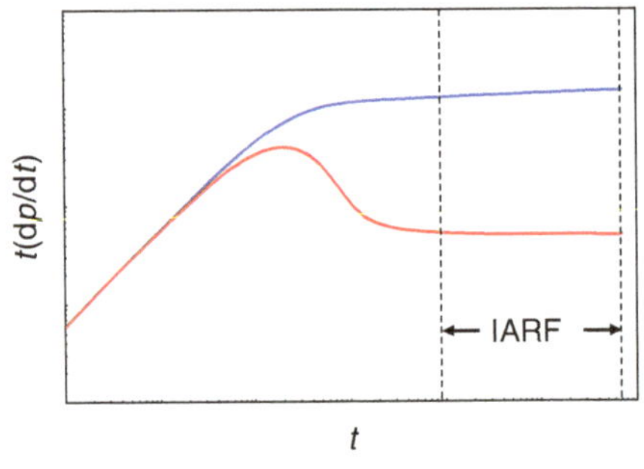

Fig. 6.27—Diagnostic plot, IARF.

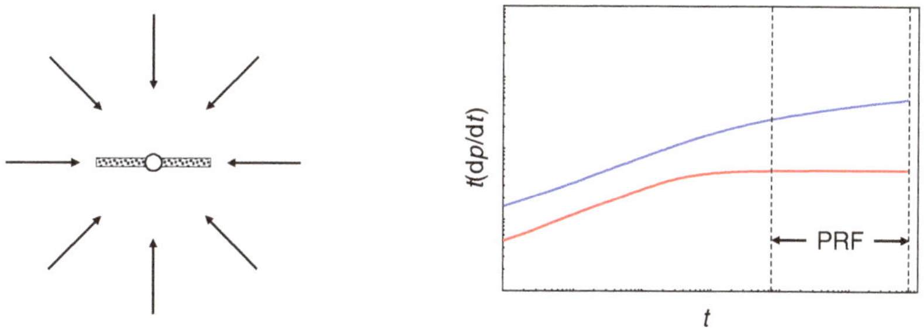

Fig. 6.28—Diagnostic plot, hydraulically fractured well exhibiting PRF.

6.5.2 Volumetric Behavior. Volumetric behavior (VB) includes WBS and PSSF, as well as other flow regimes. The logarithmic derivative during VB is a unit-slope line. When VB is the first flow regime to appear during WBS, the pressure-change curve will also be a unit-slope line, overlying the pressure-derivative curve, as shown in **Fig. 6.29.**

During other flow periods exhibiting VB, the pressure-change curve will asymptotically approach a unit-slope line from above, as shown in **Figs. 6.30 and 6.31.**

In Fig. 6.30, PSSF during a constant-rate drawdown is a special case of VB. For most of the reservoir configurations shown in this section, a short buildup test following a long flow period will exhibit the same behavior on a diagnostic plot as a drawdown test. For a closed reservoir, a buildup will not exhibit PSSF; instead, once the pressure transient has reached all reservoir boundaries, the pressure will stabilize, approaching the average reservoir pressure, while the pressure derivative decreases exponentially.

In contrast, a well in a radial composite reservoir where the outer region has much lower mobility than the inner region, as shown in Fig. 6.31, might exhibit behavior almost identical to that of PSSF for buildup as well as drawdown.

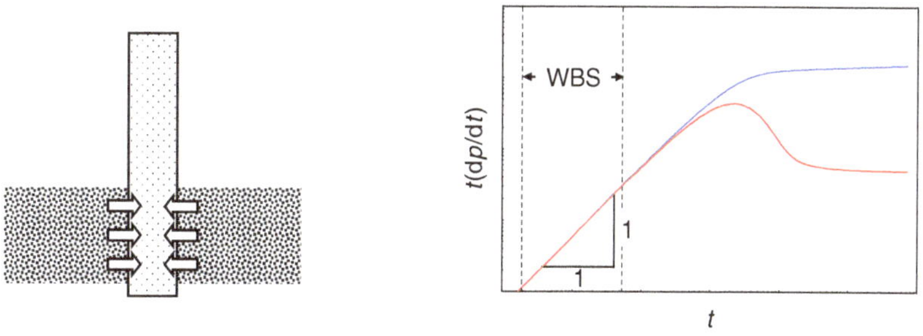

Fig. 6.29—Diagnostic plot, vertical well exhibiting constant WBS.

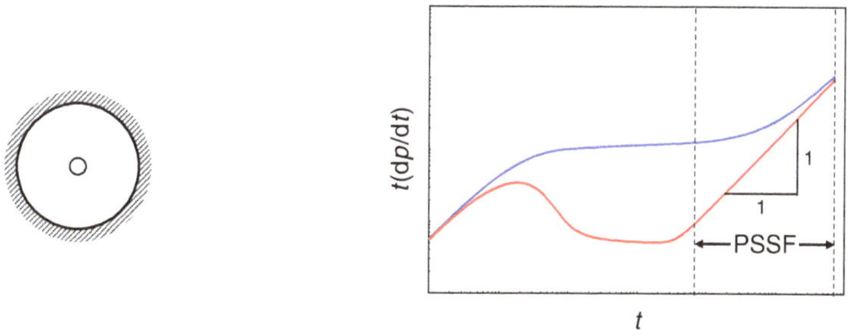

Fig. 6.30—Diagnostic plot, vertical well in a closed circular reservoir, exhibiting PSSF during a drawdown test.

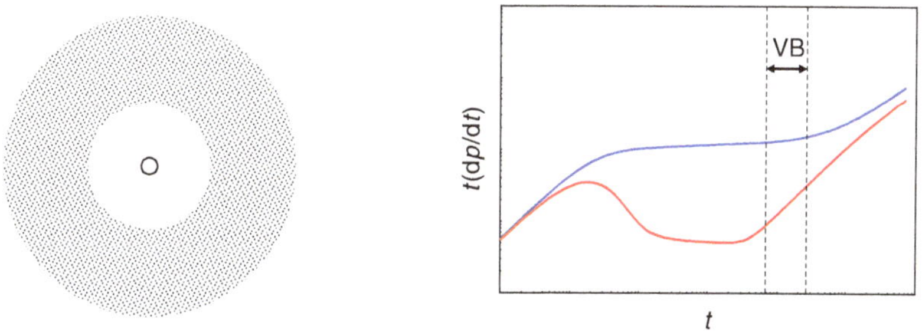

Fig. 6.31—Diagnostic plot, vertical well in an infinite radial composite reservoir, exhibiting VB.

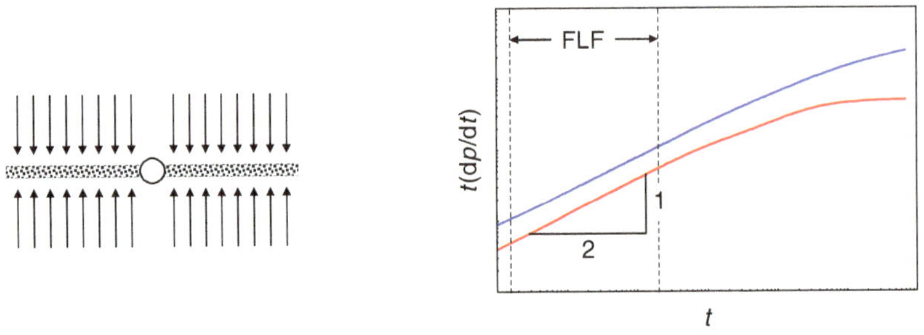

Fig. 6.32—Diagnostic-plot well with high-conductivity vertical hydraulic fracture exhibiting formation linear flow (FLF).

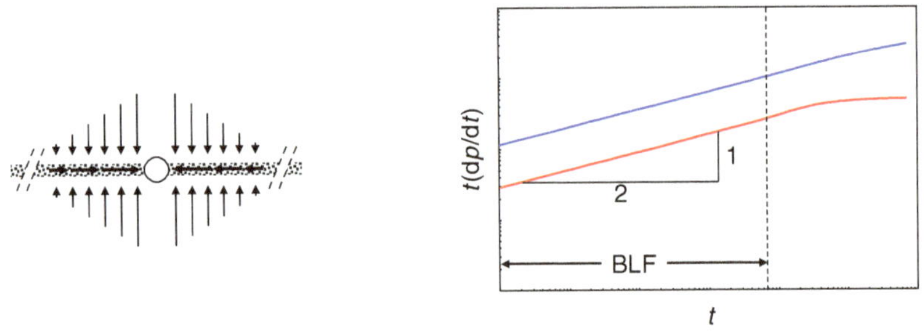

Fig. 6.33—Diagnostic plot, well with finite-conductivity vertical hydraulic fracture, exhibiting bilinear flow (BLF).

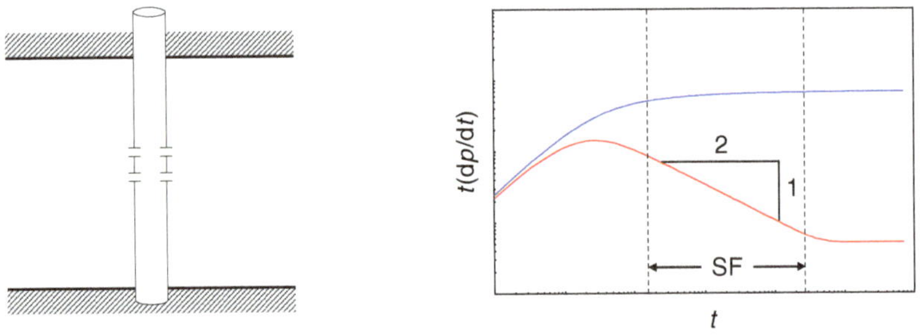

Fig. 6.34—Diagnostic plot, vertical well with limited-entry completion, exhibiting spherical flow.

During buildup, this behavior can be referred to as repressurization, or recharge, because fluid influx from the lower-mobility outer region repressurizes, or recharges, the higher-mobility inner zone. Because of the much higher viscosity of water compared to that of gas, a waterdrive gas reservoir exhibits behavior similar to that shown in Fig. 6.31 (Chen et al. 1996).

6.5.3 Linear Flow. Linear flow is one of the most common flow regimes in tight gas reservoirs because of its association with high-conductivity hydraulic fractures. During linear flow, the logarithmic derivative has a slope of one-half, as shown in **Fig. 6.32.** If linear flow is the first flow regime to appear, and there is no restriction to flow, the pressure-change curve will also exhibit a slope of one-half, as in Fig. 6.32. The pressure-change curve will be displaced upward from the derivative curve by a factor of two. If other flow regimes precede the linear-flow regime, as in a well in a channel, or if there is a restriction to flow, as in a choked fracture, the pressure-change curve will asymptotically approach a straight line with a slope of one-half.

6.5.4 Bilinear Flow. Bilinear flow occurs in finite-conductivity fractures as the pressure transient simultaneously moves in two directions—down the fracture and out into the formation. During bilinear flow, the logarithmic derivative curve will exhibit a slope of one-quarter, as shown in **Fig. 6.33.** If the fracture is undamaged, the pressure-change curve will also exhibit a slope of one-quarter, but shifted upward from the derivative curve by a factor of four.

6.5.5 Spherical Flow. Spherical flow occurs when the pressure transient is free to move in three dimensions, as in a vertical well with a limited-entry completion, as in **Fig. 6.34,** or in a formation-test tool. During spherical flow, the logarithmic derivative has a slope of minus one-half, while the pressure-change curve approaches a constant value, as shown in Fig. 6.34.

6.6 Boundary Effects

Because of their low permeability, it often requires months or years for the effects of reservoir boundaries to affect the pressure response at the wellbore for tight gas reservoirs. However, because permeability is log-normally distributed, a well might encounter permeability high enough for boundary effects become to important, even within the duration of a pressure-transient test. Because such a well is likely to be among the most prolific wells in a field, it is important for the tight gas pressure-transient analyst to understand the effects of reservoir boundaries.

All physical reservoir-boundary phenomena can be placed in one of three broad categories: closed boundaries, boundaries characterized by a reduction in flow capacity, and boundaries characterized by an increase in flow capacity.

A change in flow capacity might be caused by a change in reservoir geometry or by a change in mobility. A change in reservoir geometry results from the pressure transient encountering one or more no-flow boundaries or a change in reservoir thickness. A change in mobility, defined as k/μ, can be caused by a change in absolute or relative permeability, a change in reservoir thickness, or a change in fluid phase with an associated change in viscosity.

Although a closed reservoir can be considered as a special case of a decrease in flow capacity, the buildup-pressure response for a closed reservoir is quite different from that of a reservoir with a less-extreme decrease in flow capacity.

6.6.1 Closed Boundary. Fig. 6.35 shows the log-log diagnostic plot for a well in a reservoir with a closed circular boundary. The buildup derivative is calculated with respect to the shut-in time, and the buildup pressure-change and derivative curves are graphed vs. the shut-in time.

A constant-rate drawdown test exhibits a unit-slope line in the derivative of the diagnostic plot, indicating the onset of PSSF, after the pressure transient has reached all boundaries of the reservoir. On the diagnostic plot, the pressure curve asymptotically approaches the unit-slope line for the pressure derivative. A Cartesian graph of pressure vs. time exhibits a straight line during PSSF.

For a buildup test on the same reservoir, the derivative curve (defined in terms of the shut-in time Δt) decreases exponentially while the pressure curve approaches a horizontal line, corresponding to pressure stabilization throughout the reservoir. The shape of the buildup curve is independent of the duration of the flow period, provided that the flow period is long enough to reach PSSF.

Fig. 6.36 shows a field example of a prefracture pressure-buildup test for a gas well in a closed reservoir. The well produced approximately 2 MMscf/D during the 3-day flow period; after which, the well was shut in for a 7-day buildup test.

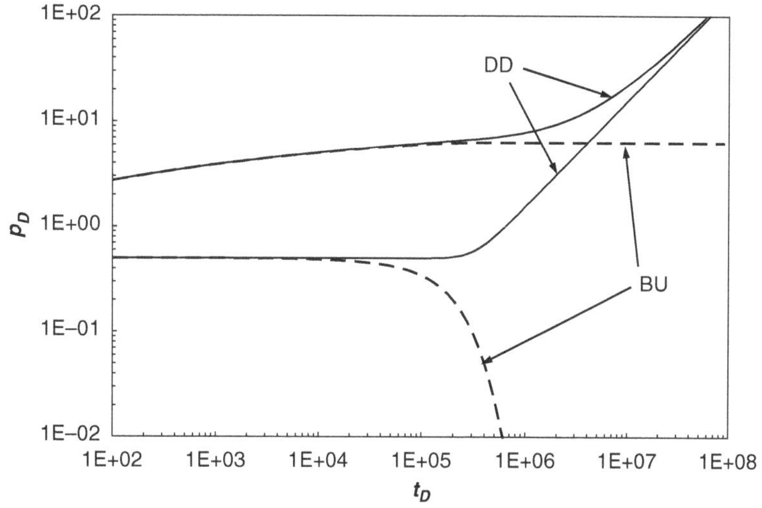

Fig. 6.35—Drawdown (DD)- and buildup (BU)-pressure responses for a closed circular reservoir. The buildup derivative is calculated with respect to the shut-in time, and the buildup-pressure and derivative curves are graphed vs. the shut-in time.

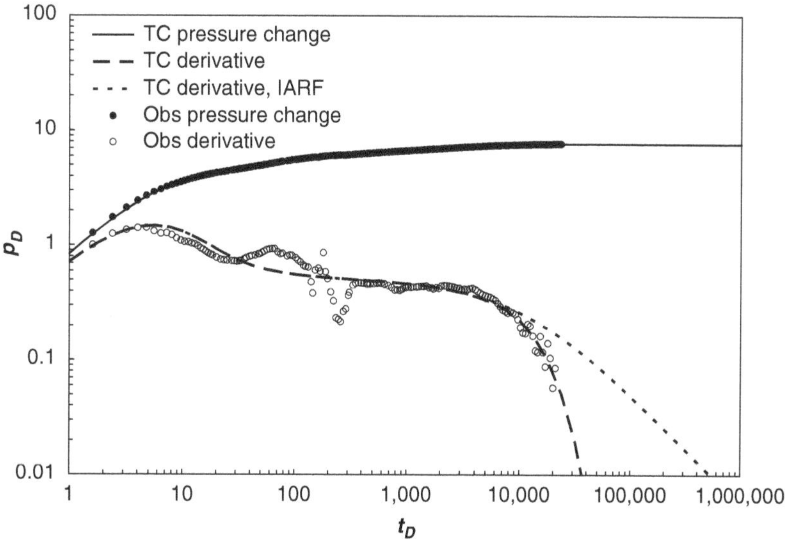

Fig. 6.36—Buildup test on a gas well in a closed reservoir, analyzed using buildup type curves (TC). The derivative decreases exponentially as the pressure stabilizes within the drainage area.

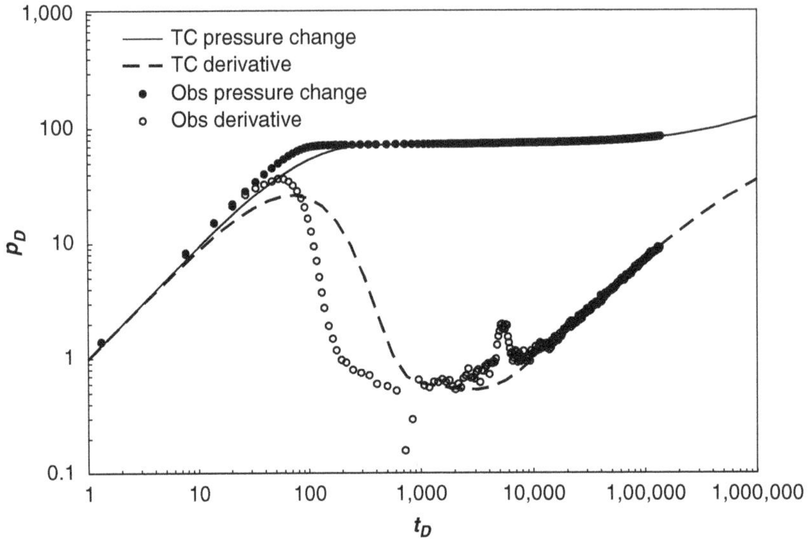

Fig. 6.37—Buildup test on a well in a gas reservoir with an edgewater drive. The pressure derivative follows a unit-slope line as water influx from the aquifer repressurizes the gas reservoir.

Even though this well was completed in a tight formation, the permeability was high enough that the pressure-derivative response clearly showed closed reservoir behavior during the buildup. Post-fracture well performance confirmed that the well had a limited drainage area.

6.6.2 Decrease in Flow Capacity. Any reduction in flow capacity will cause the pressure derivative to deviate upward from the horizontal line corresponding to IARF.

In a strong waterdrive, the contrast in viscosity between gas and water will cause a significant decrease in mobility as the pressure transient moves from the gas phase to the water phase (Chen et al. 1996). If the pressure transient has reached all other boundaries of the reservoir and the flow period preceding the buildup is sufficiently long, the decrease in mobility will cause the pressure derivative to exhibit a unit-slope line.

While a unit-slope line on the derivative in a constant-rate flow test suggests the presence of a closed reservoir, a unit-slope line in a buildup test can only occur if the reservoir is not closed. There must be some source of additional energy to recharge or repressurize the reservoir. The additional energy might be supplied by water influx from a strong waterdrive or by gas influx from a lower-permeability zone or from another reservoir compartment.

Fig. 6.37 shows a field example of a pressure-buildup test for a gas well in a reservoir with an edgewater drive. After producing for 32 months, the well was shut in for a 2-day buildup. The test showed a high skin factor, indicating a significant degree of rate-dependent skin. The pressure derivative exhibited a unit-slope line during the latter part of the test, indicating some form of pressure support. An earlier buildup test had also shown a unit-slope derivative. However, the unit-slope line from the first buildup test corresponded to a larger reservoir volume than that from the second buildup, suggesting that water influx was the source of the pressure support. Water breakthrough was observed during subsequent production, confirming the presence of a strong waterdrive.

6.6.3 Increase in Flow Capacity. Any increase in flow capacity will cause the pressure derivative to deviate downward from the IARF horizontal line. This behavior is often attributed to the presence of one or more constant-pressure boundaries.

Although mathematically convenient, true constant-pressure boundaries do not exist in nature. However, a large increase in flow capacity might appear to be a constant-pressure boundary. An increase in flow capacity might arise from a change in facies from a less-permeable to a more-permeable rock, or by the presence of a highly conductive fault near the wellbore. It should be emphasized that it is the responsibility of the well test analyst to identify the cause of the increase in flow capacity before accepting an analysis using a constant-pressure boundary as valid.

In particular, as discussed in the previous section, an aquifer in a gas reservoir will appear as a reduction in mobility and should never be modeled as a constant-pressure boundary. In the following sections, we will discuss the drawdown-pressure response for the following boundary models: linear no-flow boundary, linear constant-pressure boundary, closed circular boundary, circular constant-pressure boundary, infinite channel, intersecting no-flow boundaries, rectangular reservoir, infinite radial composite reservoir, and infinite linear composite reservoir.

6.6.4 Linear No-Flow Boundary. One of the simplest bounded reservoir models is that of a well in a reservoir with a single, linear, no-flow boundary, sometimes referred to as a well in a semi-infinite reservoir, as shown in **Fig. 6.38.** For the pressure response from a single linear no-flow boundary to be observed in the field, the well must be much closer to one boundary than to any of the remaining reservoir boundaries.

The single, linear, no-flow boundary model provides a good approximation of the behavior of a real reservoir in which the well is much closer to one of the reservoir boundaries than to any of the remaining boundaries. A no-flow boundary might be structurally caused by a sealing fault or stratigraphically caused by a lateral facies change from reservoir to nonreservoir rock.

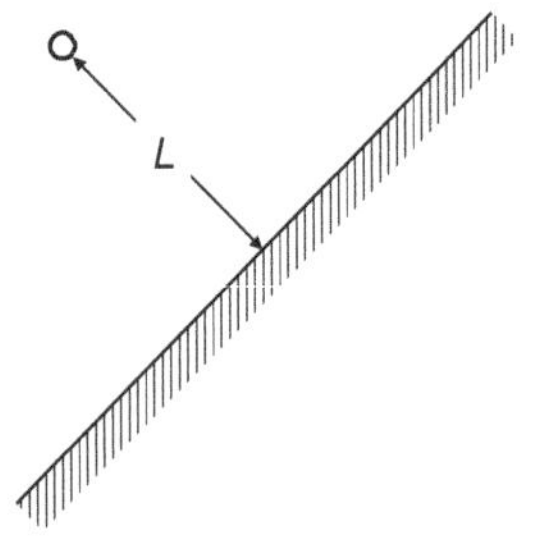

Fig. 6.38—Well near a single, linear, no-flow boundary.

The pressure response for constant-rate production from a line-source well in such a reservoir is readily obtained using superposition in space as

$$p_{wf}(t) = p_i + \frac{70.6qB\mu}{kh}\left[\mathrm{Ei}\left(-\frac{948\phi\mu c_t r_w^2}{kt}\right) + 2s\right] + \frac{70.6qB\mu}{kh}\mathrm{Ei}\left(-\frac{948\phi\mu c_t (2L)^2}{kt}\right) \quad (6.72)$$

The first Ei-function term on the right-hand side describes the pressure response at the sandface for a well in an infinite-acting reservoir; while the second Ei-function term describes the pressure response from an image well located a distance $2L$ from the actual well, producing at the same rate and beginning at the same time as the actual well.

It is often convenient to write the pressure response in terms of the logarithmic approximation to the Ei function, as we did for the infinite-acting reservoir. However, while the logarithmic approximation for the first Ei function in Eq. 6.72 becomes very quickly valid, that for the second Ei function does not. Replacing the first Ei function in Eq. 6.72 with its logarithmic approximation, we obtain

$$p_{wf}(t) = p_i - \frac{162.6qB\mu}{kh}\left[\log\left(\frac{kt}{\phi\mu c_t r_w^2}\right) - 3.23 + 0.869s\right] + \frac{70.6qB\mu}{kh}\mathrm{Ei}\left(-\frac{948\phi\mu c_t (2L)^2}{kt}\right) \quad (6.73)$$

Fig. 6.39 shows the pressure response predicted by Eq. 6.73 for a well in a reservoir with a single, linear, no-flow boundary. Before the boundary begins to affect the pressure response, the pressure response is identical to that of an infinite-acting reservoir. As the boundary begins to affect the pressure response, the derivative begins to rise, eventually reaching a value of unity—twice that of the derivative for the infinite-acting case. Thus, we expect a semilog graph of p_{wf} vs. $\log_{10}(t)$ to exhibit a doubling of slope when a single, linear, no-flow boundary is present.

At early times, when the argument of the Ei function in Eq. 6.72 is greater than 10, the Ei function is approximately zero. The pressure response is the same as that for a well in an infinite-acting reservoir, exhibiting IARF, as shown in Fig. 6.39.

At late times, when the argument of the Ei function is less than 0.01, the Ei function can be replaced by its logarithmic approximation, giving

$$p_{wf}(t) = p_i - \frac{162.6qB\mu}{kh}\left[\log\left(\frac{kt}{\phi\mu c_t r_w^2}\right) - 3.23 + 0.869s\right] - \frac{162.6qB\mu}{kh}\left[\log\left(\frac{kt}{\phi\mu c_t (2L)^2}\right) - 3.23\right], \quad (6.74)$$

which may be simplified to obtain

$$p_{wf}(t) = p_i - \frac{325.2qB\mu}{kh}\left\{\log\left(\frac{kt}{\phi\mu c_t r_w^2}\right) - 3.23 + 0.434\left[s - \ln\left(\frac{2L}{r_w}\right)\right]\right\}. \quad (6.75)$$

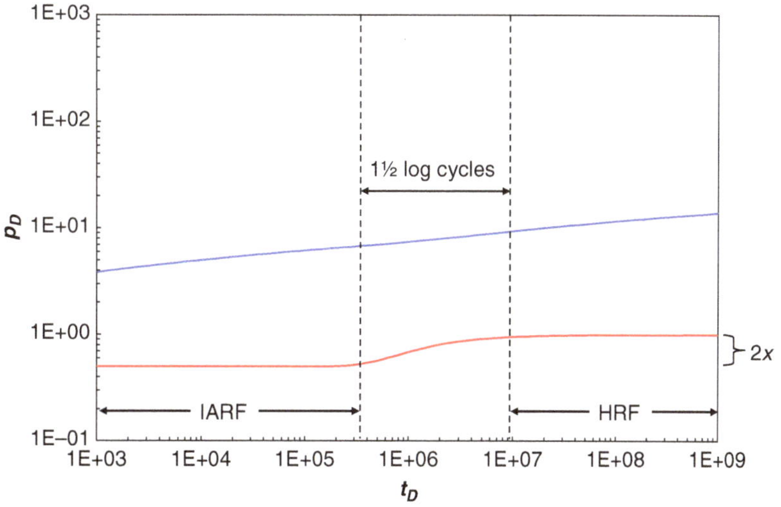

Fig. 6.39—Pressure response for a well near a single, linear, no-flow boundary.

Eq. 6.75 describes hemi-radial flow (HRF) to a well in a semi-infinite reservoir. As with IARF, the logarithmic derivative of the pressure response described by Eq. 6.75 is a constant, but with twice the magnitude of that for IARF:

$$t\frac{dp}{dt}=\frac{141.2qB\mu}{kh}. \quad (6.76)$$

Thus, the derivative in Fig. 6.39 shows two horizontal segments corresponding to two radial-flow regimes: IARF and HRF. There is a long transition, approximately 1½ log cycles in duration, between the two flow regimes.

Different criteria have been used to define flow-regime limits in the literature. Because we usually identify the beginning and end of flow regimes using the derivative on the log-log diagnostic plot, the most logical criterion would seem to be one based on deviation of the derivative from the value predicted by a particular flow regime. In our experience, the smallest deviation that is readily apparent to the unaided eye on a log-log scale is approximately 5%. Using a 5% deviation of the derivative from the value predicted for the flow regime as the criterion for the flow-regime limits, the IARF period ends at

$$t_{eIARF}=\frac{1{,}266\phi\mu c_t L^2}{k}, \quad (6.77)$$

and the HRF period begins at

$$t_{bHRF}=\frac{36{,}000\phi\mu c_t L_e^{\,2}}{k}. \quad (6.78)$$

To see fully developed HRF described by Eq. 6.75, the second-closest boundary must be a distance of at least $10L$ away from the well, a situation that is seldom encountered in practice.

6.6.5 Linear Constant-Pressure Boundary. Another simple boundary model is that of a well in a reservoir with a single, linear, constant-pressure boundary, as shown in **Fig. 6.40.**

While the conditions necessary for a no-flow boundary are often closely approximated in real reservoirs, the conditions necessary to create a constant-pressure boundary occur much less frequently.

The constant-pressure boundary model provides a good approximation to the behavior of a real reservoir in which there is a large increase in mobility along a linear boundary near the well. In oil reservoirs, a linear constant-pressure boundary can sometimes be used to model the pressure response for a well located in the oil leg downdip from a gas/oil contact. In an oil or gas reservoir, a linear constant-pressure boundary might also be used to model a well near a single, linear, conductive fault (Abbaszadeh and Cinco-Ley 1995), as might occur when there is extensive fracturing or brecciation along the fault. This model should never be used without external evidence to justify a boundary with a large increase in mobility.

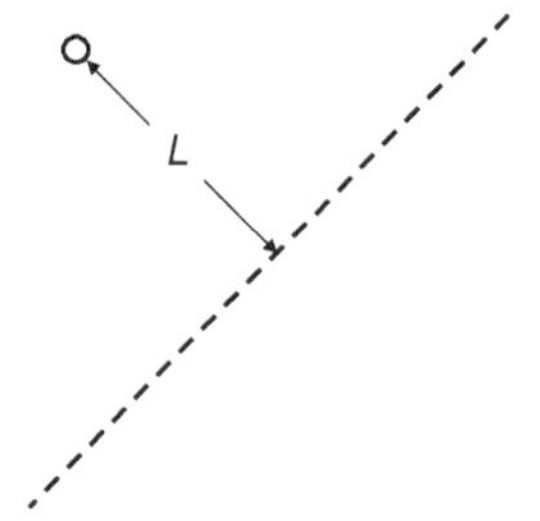

Fig. 6.40—Well near a single, linear, constant-pressure boundary.

There is a common misconception that a strong waterdrive will always act like a constant-pressure boundary. In fact, unless the mobility in the aquifer is much lower than that in the reservoir, the aquifer will not act like a constant-pressure boundary. In particular, for a gas reservoir, where the gas viscosity is typically 0.01–0.035 cp compared with the typical water viscosity of 1 cp, the mobility in the aquifer is much lower than that in the reservoir (Chen et al. 1996).

Superposition in space can be used to calculate the pressure response for a well near a single, linear, constant-pressure boundary as

$$p_{wf}(t)=p_i-\frac{162.6qB\mu}{kh}\left[\log\left(\frac{kt}{\phi\mu c_t r_w^{\,2}}\right)-3.23+0.869s\right] -\frac{70.6qB\mu}{kh}\mathrm{Ei}\left(-\frac{948\phi\mu c_t(2L)^2}{kt}\right). \quad (6.79)$$

Fig. 6.41 shows the pressure response predicted by Eq. 6.72 for a well in a reservoir with a single linear, constant-pressure boundary. Before the boundary begins to affect the pressure response, the pressure response is identical to that of an infinite-acting reservoir. The time to the end of IARF can be calculated using the same equation as for a single linear no-flow boundary (Eq. 6.77).

After the boundary begins to affect the pressure response, the pressure approaches a constant value given by

$$p_{wf}=p_i-\frac{141.2qB\mu}{kh}\ln\left(\frac{2L}{r_w}\right), \quad (6.80)$$

while the pressure derivative asymptotically approaches a straight line with a slope of –1, labeled "radial stabilization" in Fig. 6.41.

6.6.6 Closed Circular Reservoir. The simplest closed-reservoir model is that for a well centered in a closed, circular drainage area, as shown in **Fig. 6.42.**

Fig. 6.43 shows the drawdown-pressure response for a well centered in a closed circular reservoir. The first flow regime to appear is IARF, before the boundary affects the pressure response. As the boundary begins to affect the pressure response, the derivative deviates upward, approaching a unit-slope line after a brief transition period lasting approximately one quarter of a log cycle.

Geologically, there is little motivation for a well centered in a closed, circular reservoir. However, a well located near the center of any reasonably symmetric drainage area will exhibit a pressure response very similar to that of the closed, circular reservoir model. For example, the pressure response for a well in the center of a square is almost indistinguishable from that of a closed, circular

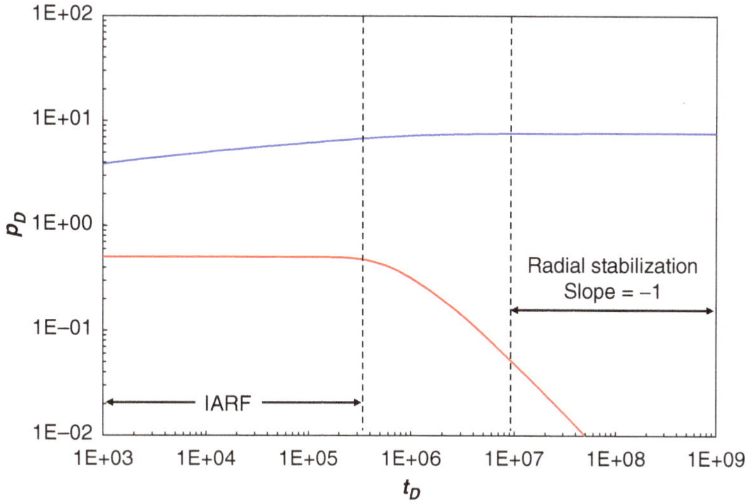

Fig. 6.41—Pressure response for a well near a single, linear, constant-pressure boundary.

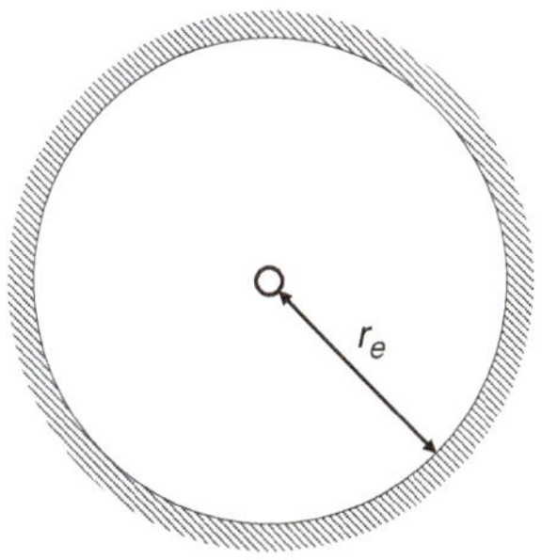

Fig. 6.42—Well centered in a closed, circular reservoir.

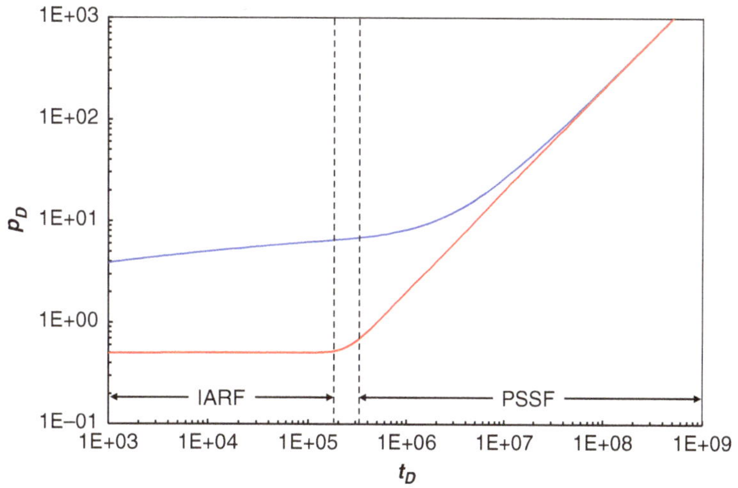

Fig. 6.43—Drawdown-pressure response for a well centered in a closed, circular reservoir.

reservoir having the same drainage area, differing by less than 3% in the derivative response and less than 0.3% in the pressure response.

The pressure response for the closed, circular reservoir model has the shortest transition between IARF and PSSF of any 2D reservoir model. This characteristic makes the closed circular reservoir model important in designing and interpreting reservoir tests on exploration wells, where establishing a minimum reservoir size is the primary test objective. Rate-transient solutions for the well centered in a closed, circular reservoir also play a prominent role in production-data analysis—a topic considered in greater depth in Part 2 Ch 8: Rate Transient Analysis.

Traditionally, a criterion of 1% deviation in the pressure response has been used to define the end of IARF and the start of PSSF. Using the traditional criterion (Earlougher 1977), IARF ends at a time given by a dimensionless time based on drainage area as

$$t_{AD} = 0.10, \quad \text{(6.81)}$$

or in field units,

$$t_{e\text{IARF}} = \frac{1{,}191\phi\mu c_t r_e^2}{k}, \quad \text{(6.82)}$$

while PSSF begins at

$$t_{AD} = 0.06, \quad \text{(6.83)}$$

or in field units,

$$t_{b\text{PSSF}} = \frac{715\phi\mu c_t r_e^2}{k}. \quad \text{(6.84)}$$

Using the criterion of 1% deviation in the pressure response, the IARF and PSSF periods overlap by almost one quarter of a log cycle.

Using the criterion of 5% deviation in the derivative, the time at which IARF ends is given in dimensionless form as

$$t_{eIARFD} = 0.1905 r_{eD}^{\ 2}, \quad (6.85)$$

or in field units,

$$t_{eIARF} = \frac{722 \phi \mu c_t r_e^{\ 2}}{k}, \quad (6.86)$$

which is slightly earlier than the time to reach a radius of investigation equal to the drainage radius r_e. The time to reach PSSF is given by

$$t_{ePSSFD} = 0.324 r_{eD}^{\ 2}, \quad (6.87)$$

or in field units,

$$t_{ePSSF} = \frac{1{,}229 \phi \mu c_t r_e^{\ 2}}{k}. \quad (6.88)$$

Using the criterion of 5% deviation in the derivative, the transition from IARF to PSSF for a closed, circular reservoir lasts slightly less than one-quarter of a log cycle.

6.6.7 Circular Constant-Pressure Boundary. A well in a circular reservoir with a constant-pressure outer boundary, as shown in **Fig. 6.44,** is another mathematically convenient model. In spite of its mathematical convenience, there are very few situations in which this model is appropriate.

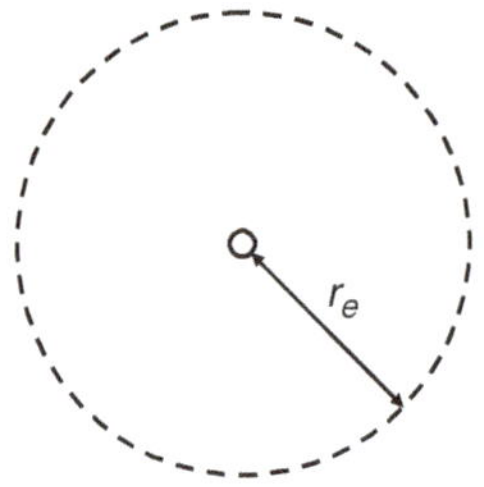

Fig. 6.44—Well centered in a reservoir with a circular, constant-pressure boundary.

One possible application would be in testing injection or production wells in a pattern waterflood, such as five-spot, regular or inverted seven-spot, or regular or inverted nine-spot patterns. For an isolated production well, it is difficult to conceive of a geologically plausible scenario that would produce a pressure response that might be approximated by this model. This is particularly true for gas reservoirs, where a downdip aquifer will act like a decrease in mobility, rather than the large increase in mobility required for the behavior of a radial-composite reservoir model to approximate that of a circular, constant-pressure boundary.

This model should never be used for a gas reservoir without good geological evidence for the presence of a large increase in mobility in all directions surrounding the well. This is especially important when analyzing buildup tests, because the buildup-pressure response for a well in the center of a closed, circular drainage area is almost identical to the drawdown-pressure response for a well in a reservoir with a circular, constant-pressure boundary.

Fig. 6.45 shows the pressure response for a well in the center of a reservoir with a circular, constant-pressure boundary. Initially, the pressure response exhibits IARF lasting until a time given by

$$t_{eIARF} = \frac{745 \phi \mu c_t r_e^{\ 2}}{k}. \quad (6.89)$$

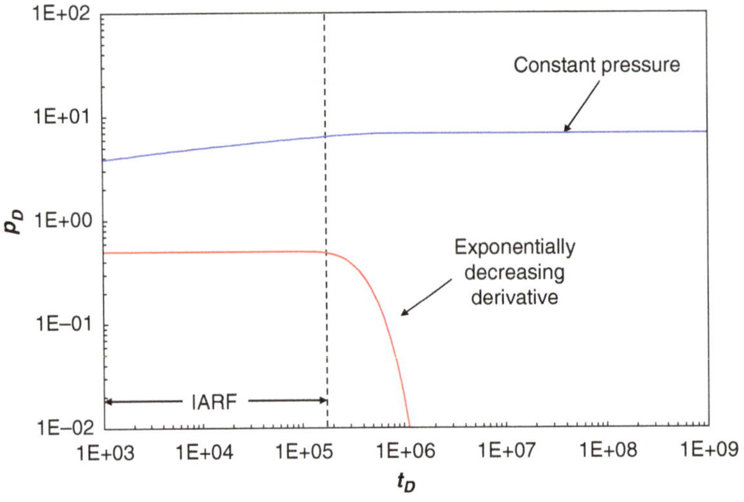

Fig. 6.45—Pressure response for a well centered in a reservoir with a circular, constant-pressure boundary.

After the boundary effects become apparent, the derivative decreases approximately exponentially with time, and the pressure stabilizes to a value given by

$$p_{wf} = p_i - \frac{141.2qB\mu}{kh}\ln\left(\frac{r_e}{r_w}\right). \qquad (6.90)$$

6.6.8 Infinite Channel. Many reservoir bodies are long and narrow and may be modeled by a well between two parallel, linear, no-flow boundaries, often called an infinite-channel reservoir model, as shown in **Fig. 6.46.**

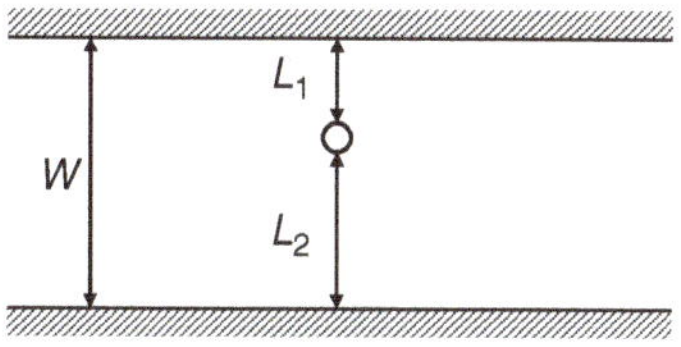

Fig. 6.46—Well in an infinite channel reservoir.

Many different geologic processes can create narrow, elongated reservoirs that could be modeled by the well in a channel model. Examples include stratigraphic traps, such as fluvial channel deposits—from which the model derives its name—point-bar deposits, and offshore sandbars, and structural features such as closely spaced parallel sealing faults. Levorsen (1967) gives two especially interesting examples. The South Ceres Pool in Oklahoma produces from the Red Fork Sand, which appears to be an offshore sandbar 10.5 miles long with an average thickness of 25 ft and an average width of 1,000 ft (Levorsen 1967). The Scipio-Albion Field in Michigan produces from a dolomitized zone in the Trenton Limestone; dolomitization seems to be controlled by fracturing and minor faulting over a deeper fault. The trend averages 3,500 ft wide and is more than 25 miles long.

Fig. 6.47 shows the pressure response for a well centered in an infinite-channel reservoir, while **Fig. 6.48** shows the pressure response for wells at several different positions in an infinite-channel reservoir.

For a well centered in a channel, as shown in Fig 6.47, IARF lasts until the sides of the channel are felt. Channel linear flow (CLF) then begins following a very brief transition period. IARF ends at a time given by

$$t_{eIARF} = \frac{256\phi\mu c_t w^2}{k} = \frac{1{,}024\phi\mu c_t L^2}{k}, \qquad (6.91)$$

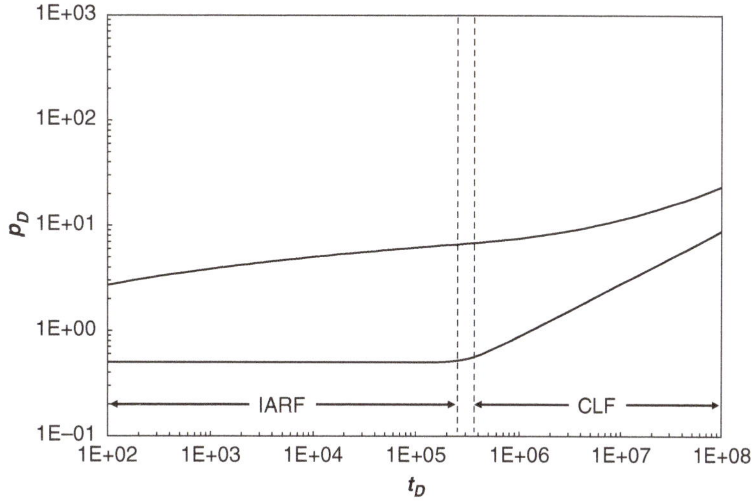

Fig. 6.47—Pressure response for a well centered in an infinite-channel reservoir.

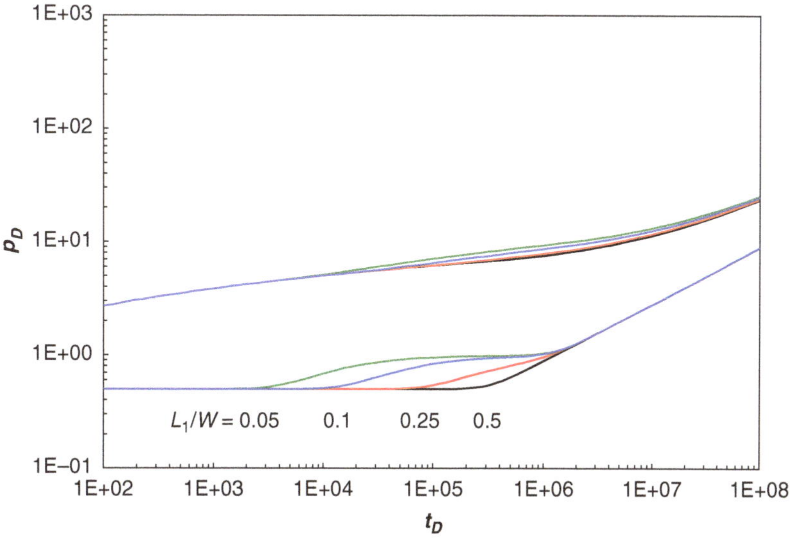

Fig. 6.48—Pressure response for a well at an arbitrary position in an infinite-channel reservoir.

where w is the width of the channel, and L is the distance from the well to either side of the channel. CLF begins at a time given by

$$t_{b\mathrm{CLF}} = \frac{350\phi\mu c_t w^2}{k} = \frac{1{,}400\phi\mu c_t L^2}{k}. \quad (6.92)$$

As seen in Fig. 6.47, the transition from IARF to CLF is very short when the well is centered in the channel.

For an off-center well in a channel reservoir, as shown in Fig. 6.48, IARF lasts until the nearer side of the channel is seen. If the well is much closer to one boundary than to the other, $L_1/W \le 0.1$, HRF will then appear, following a relatively long (1½-log-cycle) transition. If $0.1 < L_1/W < 0.5$, HRF will not occur; instead, the derivative during the transition from IARF to CLF will deviate from that for IARF, once the closer boundary begins to affect the pressure response. CLF will develop after the farther boundary begins to affect the pressure response. The time to end of IARF can be estimated by

$$t_{e\mathrm{IARF}} = \min\left\{ \begin{array}{c} \dfrac{256\phi\mu c_t w^2}{k} \\ \dfrac{1{,}266\phi\mu c_t L_1^{\,2}}{k} \end{array} \right., \quad (6.93)$$

where L_1 is the distance to the closer boundary. The approximate time to the beginning of CLF can be estimated from

$$t_{b\mathrm{CLF}} = \frac{1{,}400\phi\mu c_t L_2^{\,2}}{k}, \quad (6.94)$$

where L_2 is the distance to the farther boundary.

6.6.9 Intersecting No-Flow Boundaries. Two no-flow boundaries may intersect at an angle θ, as in **Fig. 6.49,** creating a so-called wedge-reservoir model. This model is especially useful in areas with extensive faulting, where sealing faults provide the no-flow boundaries.

Fig. 6.50 shows the pressure response for a well centered between two no-flow boundaries intersecting at an angle of 60°. This model shows IARF until the boundaries are encountered—a very long transition period (more than two log cycles), followed by fractional radial flow (FRF). The derivative during FRF will be larger than the derivative during IARF by a factor of 360/θ.

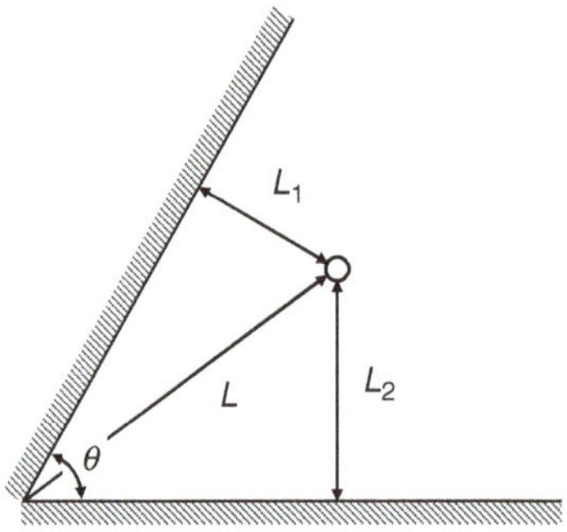

Fig. 6.49—Well between intersecting no-flow boundaries.

The smaller the angle between the boundaries, the longer the transition region, as shown in **Fig. 6.51,** which depicts the pressure response for boundaries intersecting at 10°. For an angle this small, the transition from IARF to FRF is very long, exhibiting a period during which the derivative has a slope that is almost exactly one-half, which is characteristic of linear flow. This flow period might be called pseudolinear flow (PLF). During PLF, the pressure transient is affected by both boundaries of the wedge, but the transient has not reached the apex of the wedge. Thus, the pressure response during PLF is the same as that for a well in a channel with sides that are slightly out of parallel. This is encouraging, because even nonideal channels exhibit a pressure response that can be analyzed using techniques developed for parallel no-flow boundaries.

In **Fig. 6.52,** the distance L_1 from the well to the closer boundary is one-tenth the distance L from the well to the apex of the wedge. Not surprisingly, the pressure response shows IARF, followed by a transition into HRF, followed by a longer transition into FRF.

6.6.10 Rectangular Reservoir. One of the most-frequently used bounded reservoir models is that of a well at an arbitrary location in a rectangular reservoir, as seen in **Fig. 6.53.** Because of its flexibility in matching many different pressure responses, this model is often used, even when there is no external information to justify its use.

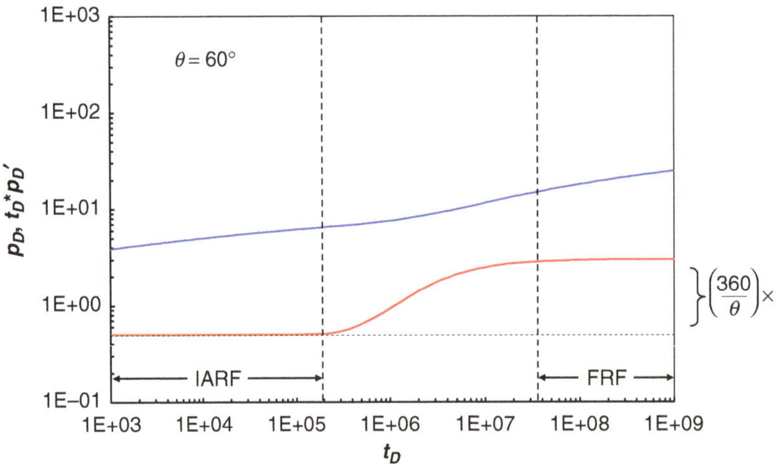

Fig. 6.50—Pressure response for a well centered between intersecting no-flow boundaries.

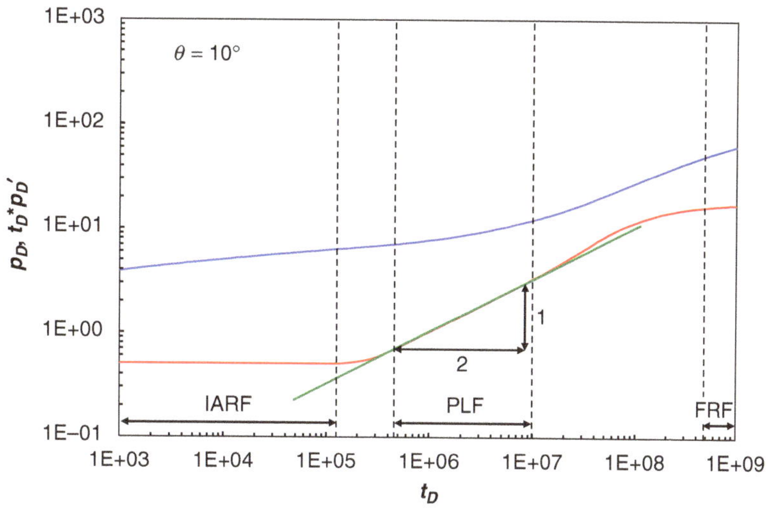

Fig. 6.51—Pressure response for a well between no-flow boundaries intersecting at a small angle.

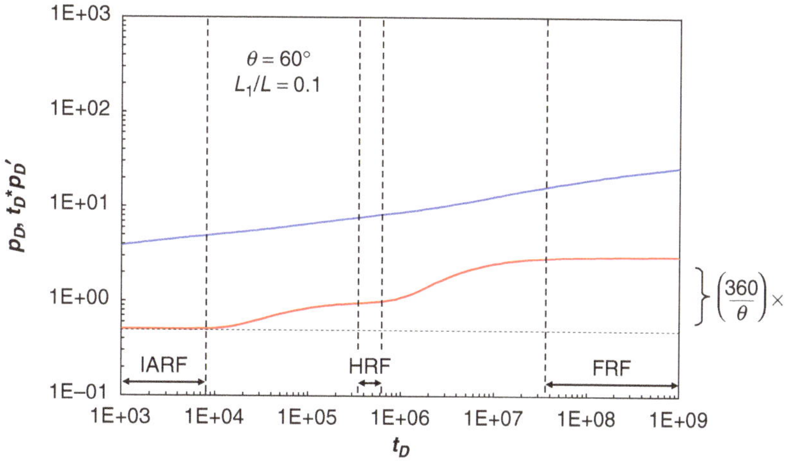

Fig. 6.52—Pressure response for a well between intersecting no-flow boundaries, much closer to one boundary than the other.

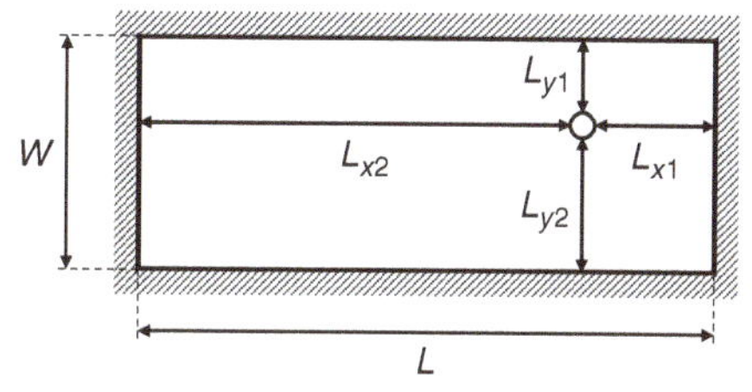

Fig. 6.53—Well in a closed, rectangular reservoir.

Geologically, the rectangular reservoir model is applicable to wells in fault blocks delineated by sealing faults: in fluvial channel, point bar, and offshore bar deposits, where the ends of the reservoir affect the test response; and in development wells drilled in a regular pattern.

Figs. 6.54 through 6.58 show a few of the wide range of pressure responses possible with the rectangular reservoir model with no-flow boundaries.

The drawdown-pressure response of a well in the center of a square, as shown in Fig. 6.54, is virtually identical to that of a well in a closed, circular drainage area, as shown in Fig. 6.43, which has the same drainage area, first exhibiting IARF and then followed by PSSF after a brief transition period.

Fig. 6.55 shows the drawdown-pressure response for a well centered in a 10×1 rectangle. The pressure response shows IARF until the pressure transient reaches the nearer pair of sides, then there is a short transition into CLF, which lasts until the pressure transient reaches the farther pair of sides and there is another short transition into PSSF.

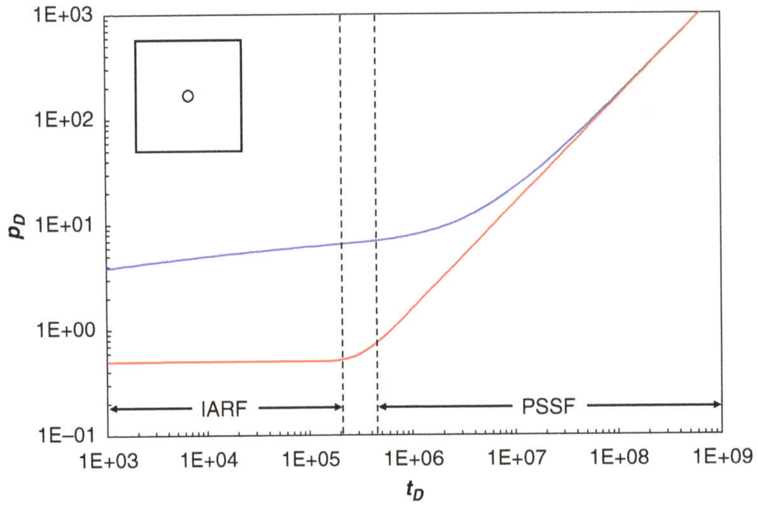

Fig. 6.54—Drawdown-pressure response for a well centered in a square reservoir.

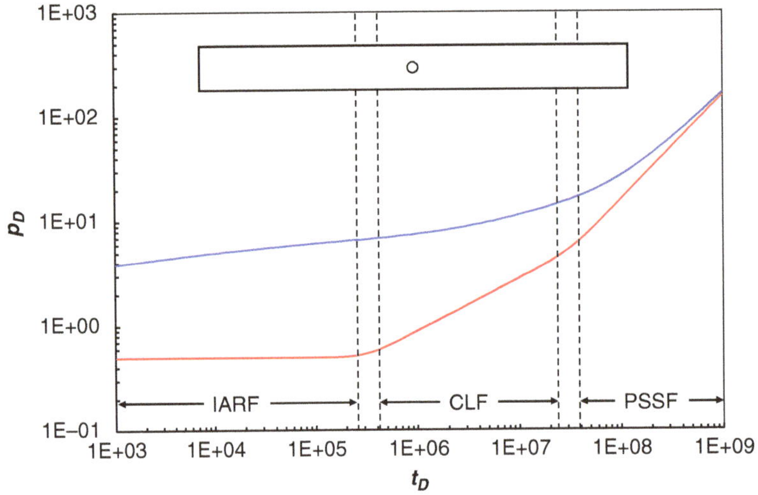

Fig. 6.55—Drawdown-pressure response for a well centered in a 10 X 1 rectangular reservoir.

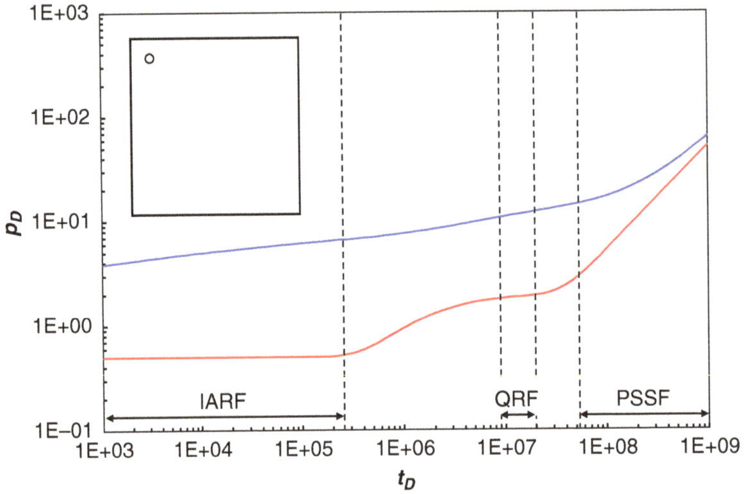

Fig. 6.56—Drawdown-pressure response for a well in one corner of a square reservoir.

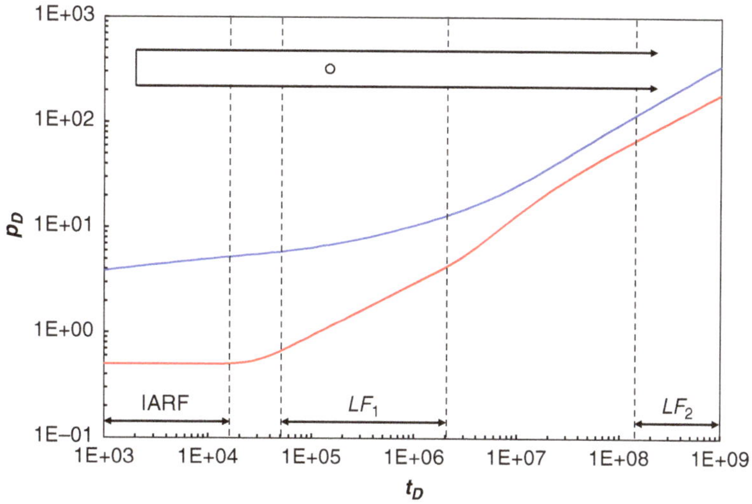

Fig. 6.57—Pressure response for a well in a semi-infinite channel.

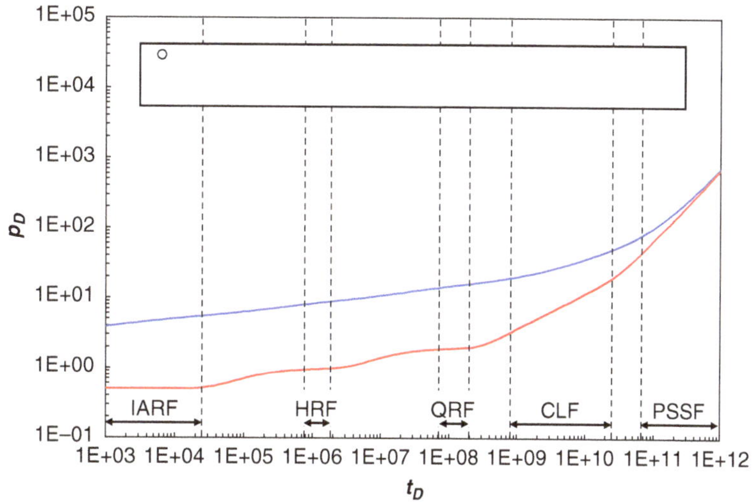

Fig. 6.58—Drawdown-pressure response for a well located so that each boundary is 10 times farther away than the next-closest boundary.

Fig. 6.56 shows the drawdown-pressure response for a well in a corner of a square reservoir. The IARF regime lasts until the pressure transient reaches the nearer pair of intersecting boundaries. There is a long transition into FRF or quarter radial flow (QRF) within the 90° angle formed by the adjacent sides. After the pressure transient reaches the opposite sides of the reservoir, there is a brief transition period followed by PSSF.

Fig. 6.57 shows the pressure response for a well in a long, narrow reservoir closed at one end, sometimes called a semi-infinite channel or a U-shaped reservoir. This reservoir model exhibits two linear-flow regimes. During the first linear-flow period, the transient moves in both directions along the channel (LF_1) before the pressure response reaches the end of the channel. The first linear-flow period is followed by a long transition into a second linear-flow period, LF_2, during which the transient moves in one direction toward the open end of the channel. The derivative follows a half-slope line for both linear-flow periods; the line for the second linear-flow period is shifted upward from that of the first flow period by a factor of two.

Fig. 6.58 shows the drawdown-pressure response for a well off-center in a long, narrow reservoir. This example is specifically constructed to show the flexibility of the rectangular reservoir model and is not intended to represent a pressure response expected to be observed in practice. The reservoir geometry and the well placement are chosen so that each boundary is 10 times farther from the well than the previous boundary; thus, the effects of each boundary on the pressure response can be separately identified. This model shows IARF, followed by HRF after the pressure transient reaches the closest boundary. After the transient reaches the second boundary, QRF appears. When the transient reaches the third boundary, the flow pattern becomes linear as the transient moves toward the remaining boundary. Finally, PSSF begins when the transient reaches the farthest boundary.

6.6.11 Infinite Radial-Composite Reservoir. Two models of importance for low-permeability gas reservoirs are the radial-composite and linear-composite models. We have already introduced the infinite radial-composite model as a model for damage or stimulation skin in Figs. 6.6 and 6.7. In this section, we consider the infinite radial-composite model shown in **Fig. 6.59** as a model for bounded reservoir behavior. In the next section, we will look at the infinite linear-composite model.

Both radial- and linear-composite models are characterized by a distance to the boundary, r_a; for the radial-composite model, a mobility ratio M is defined as

$$M \equiv \frac{M_1}{M_2} = \frac{(k/\mu)_1}{(k/\mu)_2}. \quad (6.95)$$

where the subscripts 1 and 2 refer to the inner and outer regions, respectively, a storativity ratio, S, is defined as

$$S \equiv \frac{S_1}{S_2} = \frac{(\phi c_t)_1}{(\phi c_t)_2}. \quad (6.96)$$

and a diffusivity ratio, D, is defined as

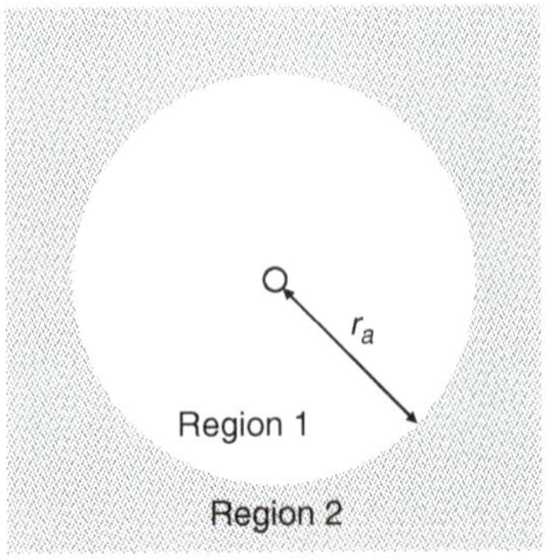

Fig. 6.59—Well in an infinite radial-composite reservoir.

$$D \equiv \frac{D_1}{D_2} = \frac{(k/\phi\mu c_t)_1}{(k/\phi\mu c_t)_2}. \quad (6.97)$$

Only two of the three ratios M, S, and D may be chosen independently. Usually, the independent parameters are chosen as M and D; however, we find it more convenient to use M and S as the independent parameters, because M and S contain no terms in common. Changing only permeability affects both the mobility ratio M and the diffusivity ratio D, without affecting the storativity ratio S; changing only porosity affects both the storativity ratio S and the diffusivity ratio D, but not the mobility ratio M. If M and S are known, the diffusivity ratio can be calculated as

$$D = \frac{M}{S}. \quad (6.98)$$

Fig. 6.60 shows the pressure response for a well in a radial-composite reservoir having uniform storativity ratio, S, and varying mobility ratio, M. This model shows two radial-flow periods, $IARF_1$ and $IARF_2$, representing IARF in the inner and outer zones respectively. If the mobility of the outer zone is higher than that of the inner zone, M is less than unity; if the outer zone mobility is lower than that in the inner zone, M is greater than unity. The value of the derivative during the second IARF period will be M times that of the first radial-flow period.

As the mobility of the outer zone becomes much larger than that of the inner zone, the mobility ratio M approaches zero, and the radial-composite model becomes equivalent to that of a reservoir with a circular, constant-pressure boundary. As the mobility of the outer zone decreases to zero, M approaches infinity, and the radial-composite model becomes equivalent to the closed, circular reservoir model.

Fig. 6.61 shows the pressure response for an infinite radial-composite reservoir with fixed mobility ratio, M, of 30 and storativity ratio, S, varying from 0.01 to 100. For small values of S, the transition from the first radial-flow period to the second radial-flow period shows an approximately one-half-slope line; for large values of S, the transition shows an approximately unit-slope line.

The radial-composite model is often misused because many different pressure responses can be matched by treating the parameters M and S (or M and D) simply as fitting parameters, without regard to their physical significance. In reality, although mathematically the parameters M and S contain no physical variables in common, M and S are inevitably correlated because both depend on the rock and fluid properties of the system in question.

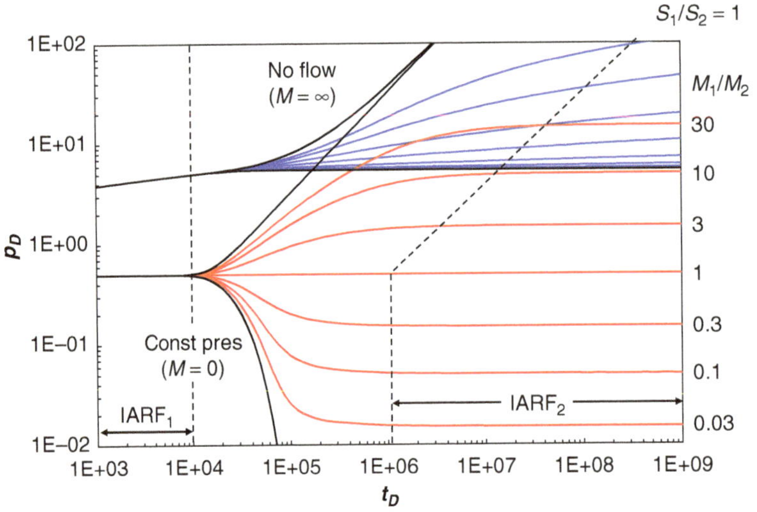

Fig. 6.60—Pressure response for a well in infinite radial-composite reservoir; fixed storativity ratio, S = 1, varying mobility ratio, M.

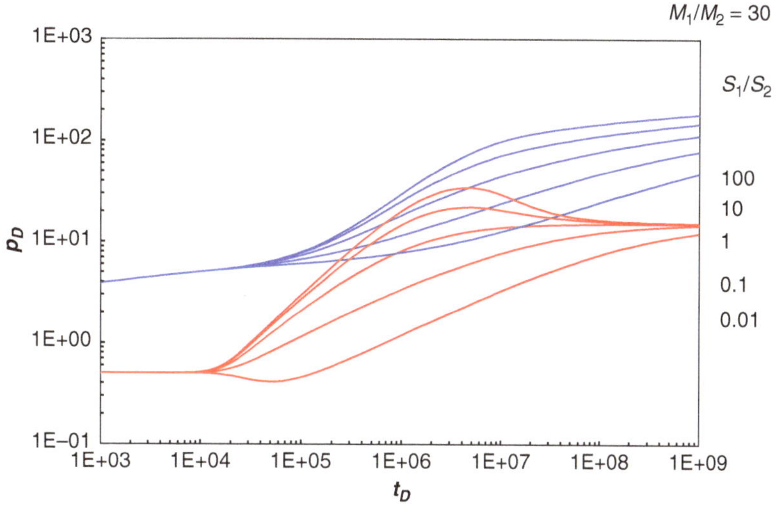

Fig. 6.61—Pressure response for a well in an infinite radial-composite reservoir; fixed mobility ratio, *M* = 30, varying storativity, *S*.

For example, it is difficult to envision a plausible reservoir setting that would cause a mobility ratio of 30 combined with storativity ratio of 0.01, corresponding to the bottommost pair of pressure and derivative curves in Fig. 6.61. For a system with a single-phase fluid, a mobility ratio of 30 might be obtained with a high-permeability inner region embedded in a low-permeability outer region, but it is unlikely that the porosity-compressibility product of the inner region in such a situation would be lower than that of the outer region, or even much lower by a factor of 100. For a gas reservoir with a downdip aquifer and uniform rock properties, the gas-filled inner region might have a mobility 30 times that of the water-filled outer region, but the low compressibility of water relative to that of the gas would likely cause the storativity ratio S to be between 10 and 100, corresponding to the topmost pairs of pressure and derivative curves in Fig. 6.61.

When a composite reservoir model is chosen for a well test interpretation, the engineer has the responsibility of finding a physical explanation for the particular combination of M and S (and r_a) obtained from the analysis, preferably with corroborating evidence from external data.

6.6.12 Infinite Linear-Composite Reservoir. The infinite linear-composite reservoir model shown in **Fig. 6.62** has many similarities to the infinite radial-composite reservoir model. Linear- and radial-composite models are described by mobility, storativity, and diffusivity ratios (Eqs. 6.95 through 6.97). Both models exhibit two radial-flow regimes, as shown in **Figs. 6.60 and 6.63.** Both models have no-flow and constant-pressure boundaries as limiting cases.

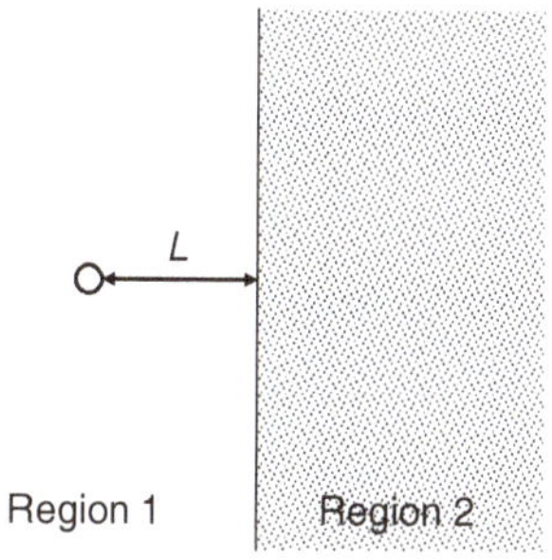

Fig. 6.62—Well in an infinite linear-composite reservoir.

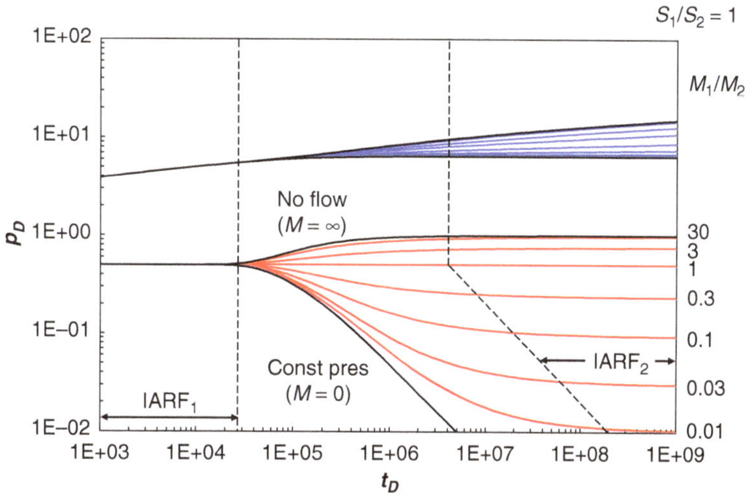

Fig. 6.63—Pressure response for a well in an infinite linear-composite reservoir; fixed storativity ratio *S* = 1, varying mobility ratio *M*.

6.7 Pseudosteady-State Flow (PSSF)

Fig. 6.64 shows the pressure profile in a hypothetical closed, circular reservoir producing at constant rate, during the IARF and PSSF periods. During the IARF period, the profiles are shown at times forming a geometric series (equally spaced on a log scale), while during the PSSF period, the profiles are shown at times forming an arithmetic series (equally spaced on a linear scale).

During IARF, a graph of pressure vs. time will form a straight line on a semilog scale, as shown in **Fig. 6.65.** During PSSF, a graph of pressure vs. time will form a straight line on a Cartesian scale, as shown in **Fig. 6.66.**

The equation for PSSF for a closed, circular reservoir can be written as

$$p_{wf}(t) = p_i - \frac{0.0744 qBt}{r_e^2 h\phi c_t} - \frac{141.2 qB\mu}{kh}\left[\ln\left(\frac{r_e}{r_w}\right) - \frac{3}{4} + s\right]. \quad (6.99)$$

For a closed reservoir of arbitrary geometry, we write

$$p_{wf}(t) = p_i - \frac{0.234 qBt}{Ah\phi c_t} - \frac{141.2 qB\mu}{kh}\left[\frac{1}{2}\ln\left(\frac{10.06A}{C_A r_w^2}\right) - \frac{3}{4} + s\right], \quad (6.100)$$

where C_A is a shape factor that depends on the shape of the reservoir and the position of the well in the reservoir. Eq. 6.100 is the basis for reservoir-limits testing, often used in exploration wells to estimate or to place a lower bound on hydrocarbons in place.

6.7.1 Drainage Area Shape Factor. Shape factors for several simple drainage area geometries are given in **Table 6.8.** Other, more-comprehensive shape factor tables are found in Earlougher (1977), Lee and Holditch (1982), Fetkovich and Vienot (1985), and Lee et al. (2003).

Noting that the average reservoir pressure for a closed, circular reservoir can be written as

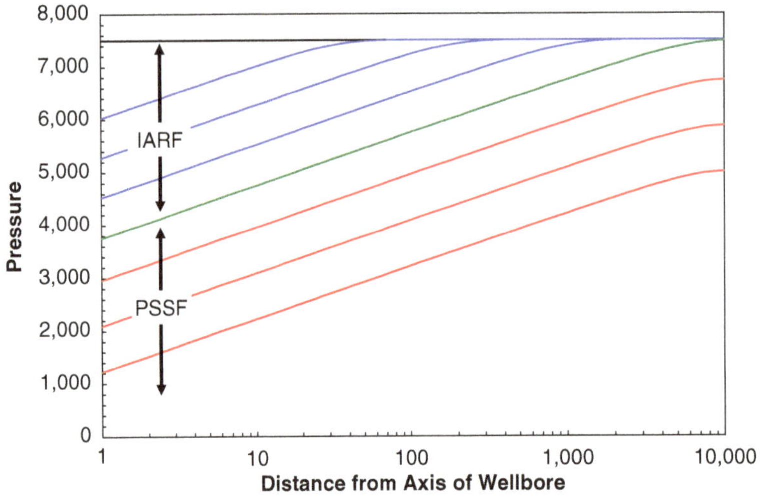

Fig. 6.64—Reservoir pressure profiles for IARF and PSSF.

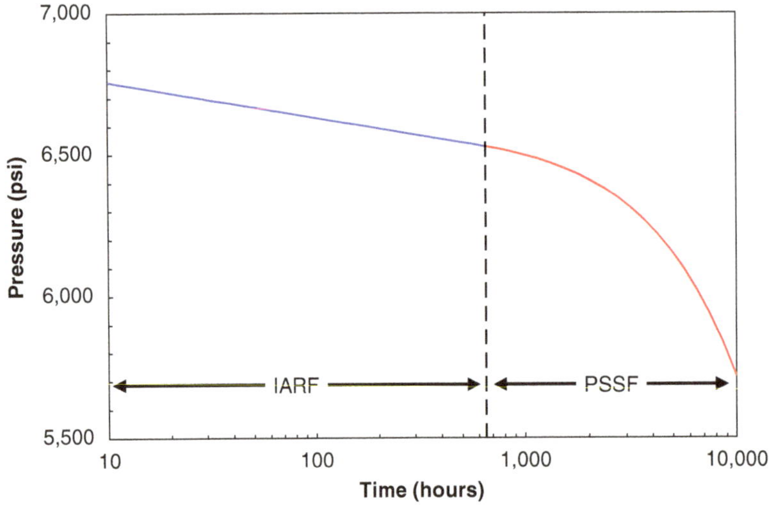

Fig. 6.65—A graph of pressure vs. time forms a straight line on a semilog scale during IARF.

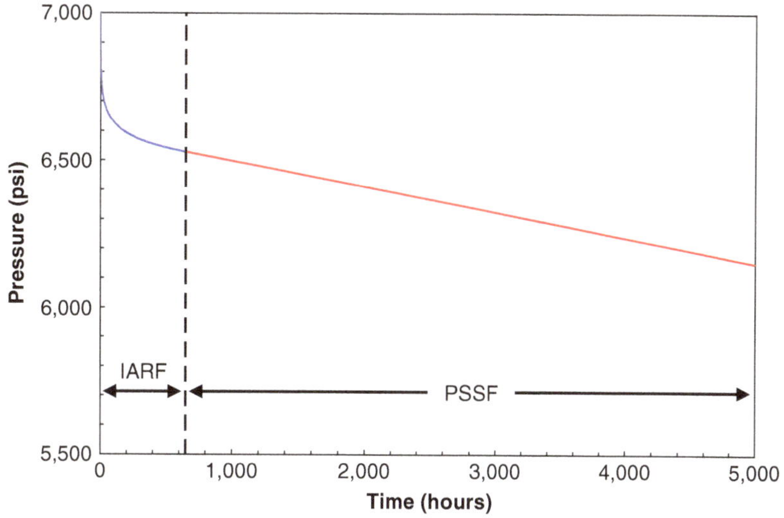

Fig. 6.66—A graph of pressure vs. time forms a straight line on a Cartesian scale during PSSF.

Reservoir Shape, Well Location	Shape Factor C_A	Reservoir Shape, Well Location	Shape Factor C_A
Circle 	31.62	2×1 Rectangle 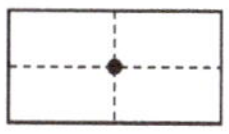	21.8369
Square 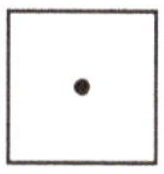	30.8828	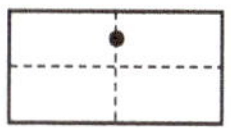	10.8374
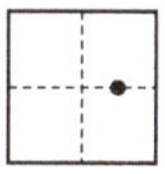	12.9851	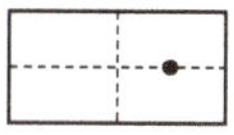	4.5141
	4.5132	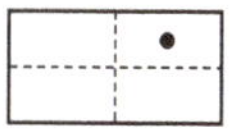	2.0769

Table 6.8—Shape factors for simple drainage-area geometries.

$$\bar{p}(t) = p_i - \frac{0.0744 qBt}{r_e^2 h\phi c_t}, \quad (6.101)$$

we can rewrite Eq. 6.99 in terms of the average pressure, thereby eliminating the explicit time dependence, as

$$\bar{p}(t) - p_{wf}(t) = \frac{141.2 qB\mu}{kh}\left[\ln\left(\frac{r_e}{r_w}\right) - \frac{3}{4} + s\right]. \quad (6.102)$$

6.7.2 Modifications for Gas Wells. For gas wells, we can write Eq. 6.102 in terms of adjusted pressure as

$$\bar{p}_a(t) - p_{awf}(t) = \frac{141.2 q_g B_{gi}\mu_{gi}}{kh}\left[\ln\left(\frac{r_e}{r_w}\right) - \frac{3}{4} + s\right]. \quad (6.103)$$

Eq. 6.103 is valid for gas wells producing from closed, circular reservoirs at any point during the life of the well. However, we cannot simply write Eq. 6.99 in terms of time and adjusted pressure. For a gas well, the analog to Eq. 6.101,

$$\bar{p}_a(t) = p_{ai} - \frac{0.0744 q_g B_{gi} t}{r_e^2 h\phi c_{ti}}, \quad (6.104)$$

becomes increasingly inaccurate as reservoir pressure declines because of the strong pressure dependence of the FVF, B, and the total compressibility, c_t. For a reasonably symmetric drainage-area shape and well location, such as a well in the center of a circle or a square, Fraim and Wattenbarger (1987) defined a reservoir-adjusted time as

$$t_a = \mu_{gi} c_{ti} \int_0^t \frac{dt}{\mu_g(\bar{p}) c_t(\bar{p})}, \quad (6.105)$$

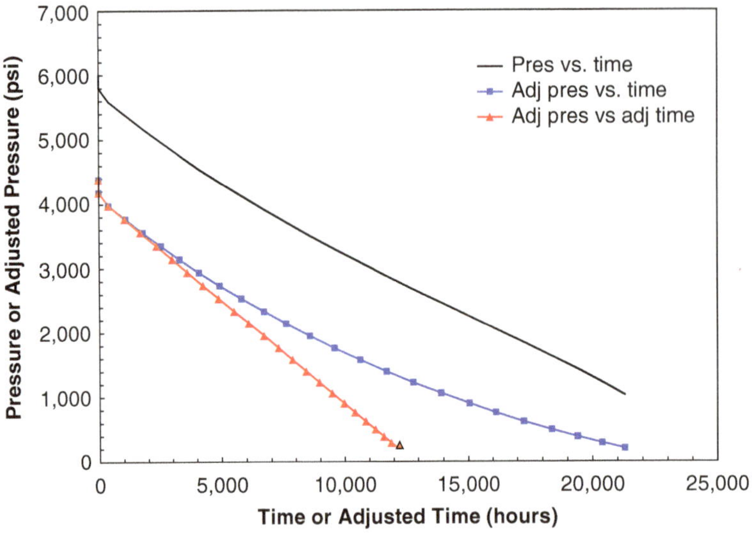

Fig. 6.67—Using adjusted pressure and adjusted time based on average reservoir pressure effectively linearizes the gas flow equation for PSSF.

where the gas properties inside the integral in Eq. 6.105 are evaluated at the average reservoir pressure as a function of time. With adjusted time defined as in Eq. 6.105, we may write Eq. 6.99 in terms of adjusted pressure, p_a, and adjusted time, t_a, as

$$p_{awf}(t) = p_{ai} - \frac{0.0744 q_g B_g t_a}{r_e^2 h \phi c_t} - \frac{141.2 q_g B_g \mu_g}{kh}\left[\ln\left(\frac{r_e}{r_w}\right) - \frac{3}{4} + s\right]. \quad (6.106)$$

Fig. 6.67 illustrates the problem with using pressure vs. time or adjusted pressure vs. time to analyze data in PSSF for a gas well. Neither pressure nor adjusted pressure give a straight line when graphed vs. the actual producing time. However, graphing the adjusted pressure vs. the adjusted time defined by Eq. 6.106 does give a straight line during PSSF.

Just as the adjusted time defined by Eq. 6.47 enabled the use of type curves for slightly compressible liquids to analyze gas-well buildup data, we will see in Part 2 Ch 8: Rate Transient Analysis that Eq. 6.105 enables the use of decline type curves for slightly compressible liquids to analyze gas-well production data.

6.8 Naturally Fractured Reservoirs

The terms "naturally fractured reservoir" and "dual-porosity reservoir" are often used synonymously; however, this usage is misleading. Many naturally fractured reservoirs do not exhibit dual-porosity behavior; instead, they act like homogeneous reservoirs. On the other hand, many unfractured layered reservoirs exhibit so-called dual-porosity behavior.

"Naturally fractured reservoir" refers to any reservoir in which naturally existing fractures provide a significant fraction of the total flow capacity and/or storage capacity.

"Dual porosity" refers to a specific class of mathematical models of heterogeneous behavior; a dual-porosity reservoir is a reservoir that exhibits behavior characteristic of a dual-porosity model. Dual porosity derives from the fact that the model includes two porosity systems, usually referred to as the matrix and the fractures. The matrix-porosity system contains most of the fluid in place, but has extremely low permeability. The fracture-porosity system has extremely limited storage capacity relative to the matrix, but provides a flow path for fluid to flow from the matrix to the wellbore.

In this section, we first consider a number of general-purpose reservoir models that have been found useful in analyzing pressure-transient data for naturally fractured reservoirs. We then discuss the two primary dual-porosity models, along with variations on those models.

6.8.1 Conventional Reservoir Models. Cinco-Ley (1996) discusses a number of reservoir models that might have application in naturally fractured reservoirs: the homogeneous reservoir model, the anisotropic reservoir model, the composite reservoir model, the single-fracture model, and the dual-porosity model.

6.8.1.1 Homogeneous Reservoir Model. In the homogeneous reservoir model shown in **Fig. 6.68,** the reservoir medium acts like a homogeneous, single-porosity system. This might occur in reservoirs in which fracturing is intensive and the matrix blocks are small enough, or the matrix permeability high enough, that the matrix and fracture porosity systems act as one. Occasionally, fractured basement rock is found to contain oil or gas; in these reservoirs, the fractures contribute 100% of both flow and storage capacity, and a homogeneous reservoir model is appropriate.

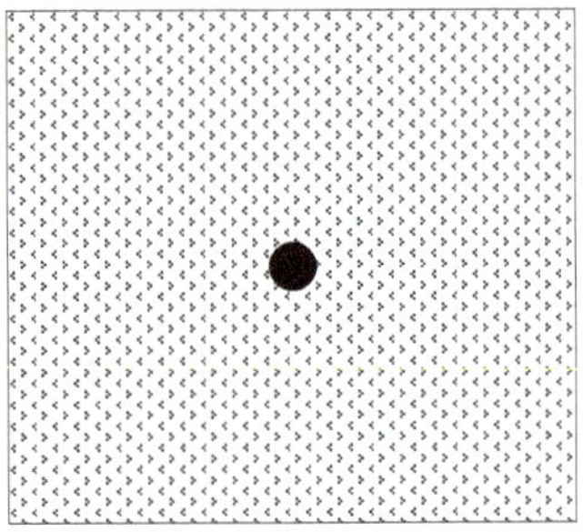

Fig. 6.68—A naturally fractured reservoir with high matrix permeability or small matrix blocks might act like a homogeneous reservoir.

6.8.1.2 Anisotropic Reservoir Model. In some reservoirs, because of the regional tectonic stresses, natural fracturing forms a single dominant set of parallel fracture planes, as shown in **Fig. 6.69a.** If the fractures are spaced close enough or the matrix permeability is high enough, the reservoir might act as a single-porosity system, but with an anisotropic permeability, as shown in **Fig. 6.69b.**

A pressure-transient test on a well in an anisotropic reservoir will look exactly like a pressure-transient test on a well in an isotropic reservoir. Any homogeneous anisotropic system can be converted to an equivalent isotropic system that will have exactly the same pressure response.

6.8.1.3 Radial-Composite Model. **Fig. 6.70a** shows a well in an area of intensive fracturing in an otherwise-relatively unfractured reservoir. A radial-composite model in which the mobility ratio M_1/M_2 (and possibly the storativity ratio S_1/S_2) is much greater than unity, as shown in **Fig. 6.70b,** is likely to be appropriate for this situation.

6.8.1.4 Well Near a Single Natural Fracture or Conductive Fault. **Figs. 6.71a and b** show a well near a single natural fracture or conductive fault. When the effects of the fault first become apparent, the pressure response follows that of a well near a linear constant-pressure boundary, as shown in **Fig. 6.71c.** Eventually, the pressure response will be similar to that of a hydraulically fractured well with a choked fracture skin, as shown in **Fig. 6.71d.**

Cinco-Ley et al. (1976) calculated the pressure response for a well near a single infinite-conductivity natural fracture of length L in an otherwise infinite-acting reservoir, as shown in Fig. 6.71a. In the Cinco-Ley et al. model, the fracture can be at any orientation with respect to the well.

The largest pressure response occurs when the well is close to the center of the fracture. L_{fD} is the dimensionless fracture half-length, defined as

$$L_{fD} = \frac{L_f}{L}, \qquad (6.107)$$

where L is the distance from the well to the fracture and L_f is the half-length of the natural fracture. This reservoir model can exhibit as many as four flow regimes, as shown in **Fig. 6.72:** IARF; radial stabilization, with a derivative slope of –1; linear flow, with a derivative slope of ½; and a second late radial-flow period.

During radial stabilization, the fracture acts like a constant-pressure boundary, exhibiting a derivative slope of –1. Radial stabilization occurs only for small values of L_D (less than 0.01). If radial stabilization appears, the distance to the boundary can be estimated.

The linear-flow period corresponds to formation linear flow to a high-conductivity hydraulic fracture. The linear-flow period will occur only for large values of L_{fD}. During the linear-flow period, the pressure response is that of a well with an infinite-conductivity hydraulic fracture (see Part 2 Ch 7: Post-Fracture Treatment Well Testing), with a choked-fracture skin factor given by

$$s_f = \ln\left(\frac{2L}{r_w}\right), \qquad (6.108)$$

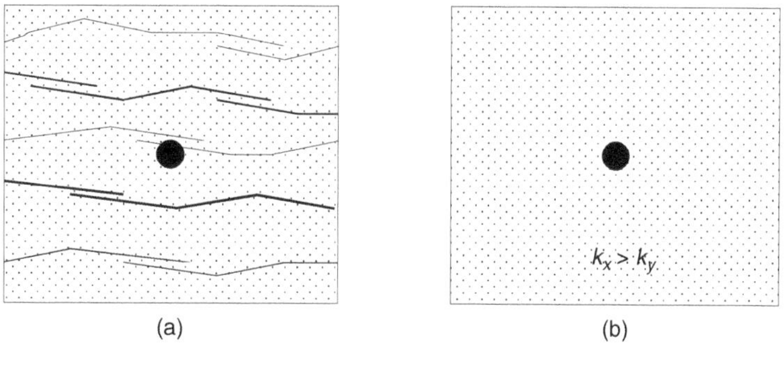

Fig. 6.69—A naturally fractured reservoir with one dominant set of fracture planes can often be modeled as a homogeneous reservoir with permeability anisotropy.

Fig. 6.70—A naturally fractured reservoir with extensive localized fracturing can often behave like a radial- or linear-composite reservoir model.

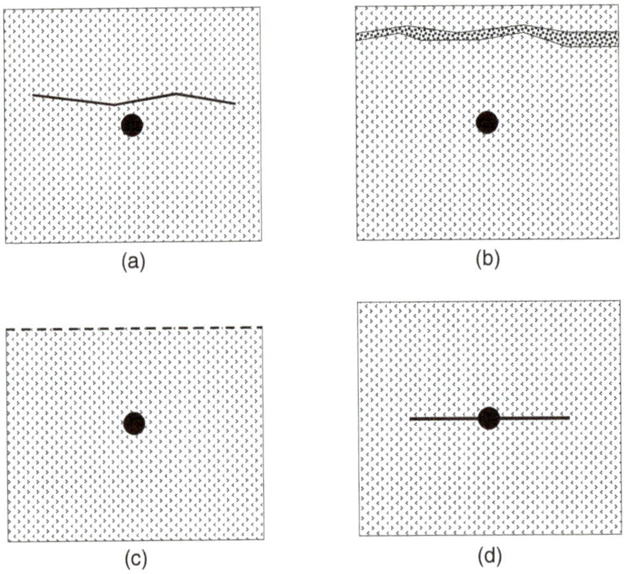

Fig. 6.71—A well near a single natural fracture (a, b) acts first like a well near a linear constant-pressure boundary (c), then like a hydraulically fractured well with choked fracture skin (d).

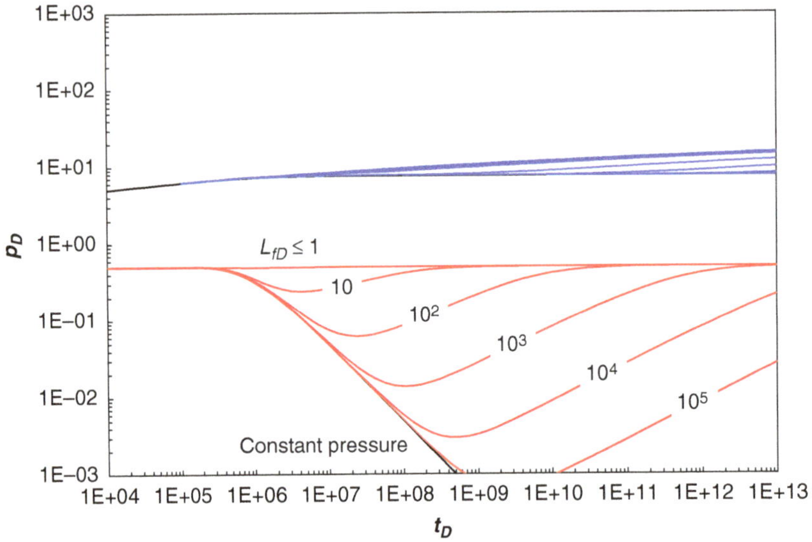

Fig. 6.72—Pressure response for a well near a single, high-conductivity natural fracture in an otherwise infinite-acting reservoir.

where L is the distance from the well to the fracture and r_w is the wellbore radius. If either of the two radial-flow periods appears along with the linear-flow period, the fracture length and choked-fracture skin factor can be estimated. The distance to the fracture can be estimated from the choked-fracture skin factor using Eq. 6.108.

During the second radial-flow period, a semilog graph of the pressure response will have the same slope as that during the first radial-flow period, but with a negative geometric skin factor. The total skin factor is the combined effect of the positive choked fracture skin factor and the negative skin factor for a well with an undamaged fracture. The stimulation skin factor for a well with an undamaged, infinite-conductivity hydraulic fracture is

$$s = -\ln\left(\frac{L_f}{2r_w}\right). \qquad (6.109)$$

Combining Eqs. 6.108 and 109, we find the total skin factor for a well near a high-conductivity natural fracture is given by

$$s_{tot} = \ln\left(\frac{4L}{L_f}\right). \qquad (6.110)$$

Eq. 6.110 is valid only for L_{fD} greater than 10. For L_{fD} less than 10, the flow pattern is distorted, and the total skin factor is lower than predicted by Eq. 6.110. The total skin factor for $L_{fD} = 1$ is zero for all practical purposes, as shown in Fig. 6.72.

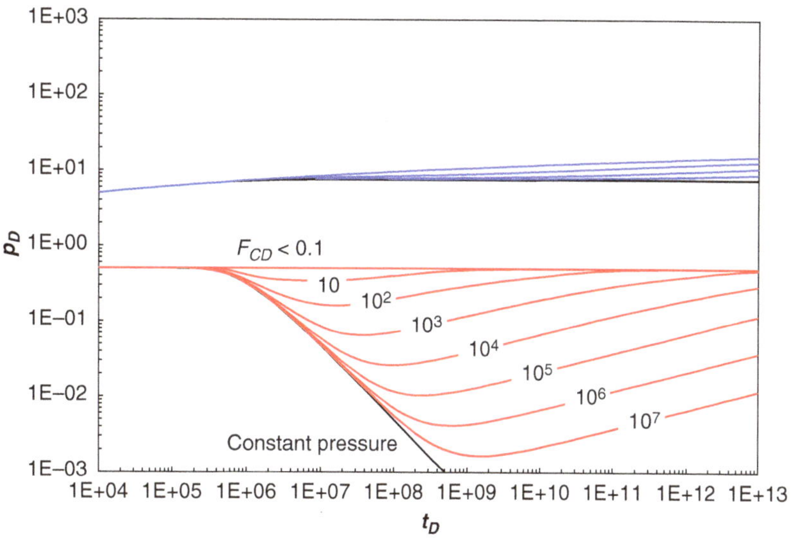

Fig. 6.73—Pressure response for a well near a finite-conductivity natural fracture in an infinite-acting reservoir.

Abbaszadeh and Cinco-Ley (1995) studied the pressure response for a well in an infinite-acting reservoir with a single, finite-conductivity fault of infinite length. The pressure derivative for this model shows four flow regimes: IARF, with horizontal derivative; radial stabilization, with a derivative slope of –1, like a well near a linear constant-pressure boundary; bilinear flow, with a derivative slope of ¼; and a second IARF period. When the reservoir properties are the same on both sides of the fault, the derivative will have the same constant value for both radial-flow periods.

The pressure responses shown in Fig. 6.72, for a fracture of finite length and infinite conductivity, and **Fig. 6.73,** for a fracture of infinite length and finite conductivity, are very similar.

6.8.1.5 Dual-Porosity Models. A number of different dual-porosity models have been presented in the literature. Most models have assumed two distinct pore systems: a fracture system, with high permeability, but relatively little (or no) storage capacity; and a matrix system, with low permeability but high storage capacity. The models differ in the treatment of interporosity flow.

Warren and Root (1963) noted that dual-porosity behavior must be characterized by two parameters in addition to those required for single-porosity systems. The first of these parameters is the storativity ratio, ω, which is defined as

$$\omega \equiv \frac{(\phi c_t)_f}{(\phi c_t)_f + (\phi c_t)_m}. \qquad (6.111)$$

The second parameter is the interporosity flow coefficient, λ, defined as

$$\lambda \equiv \frac{\alpha k_m r_w^2}{k_f}, \qquad (6.112)$$

where α is defined as

$$\alpha \equiv \frac{4n(n+2)}{L_m^2}, \qquad (6.113)$$

where n is the number of sets of orthogonal fractures (1, 2, or 3) and L_m is the characteristic fracture spacing.

6.8.2 Warren and Root Pseudosteady-State Dual-Porosity Model. Warren and Root proposed a model of a naturally fractured reservoir as shown in **Fig. 6.74.** The pressure within the matrix and within the fracture systems within a representative elementary volume is assumed to be uniform, but fluid flow between the two systems occurs under pseudosteady-state conditions. In the Warren and Root model, the shape of the individual matrix blocks has no effect on the pressure response, as shown in **Figs. 6.75 and 6.76.** The pressure transient response for the Warren and Root model exhibits two parallel semilog straight lines: the first corresponding to transient flow in the fracture system alone, before the matrix begins to respond; the second to transient flow in the total system, where the matrix and fractures act together as a homogeneous system.

6.8.3 Kazemi-de Swaan Transient Dual-Porosity Model. Kazemi (1969) and de Swaan (1976) presented models similar to that of Warren and Root, but ones in which transient fluid flow was allowed within the matrix, as shown schematically in **Fig. 6.77.** De Swaan provided solutions for the pressure-transient response for reservoirs in which the matrix blocks were either slabs or spheres. **Figs. 6.78 and 6.79** show the pressure-transient response for the de Swaan solution for the case of slab matrix blocks. The Kazemi-de Swaan model exhibits three semilog straight lines. The first and third semilog straight lines are parallel and correspond to fracture and total system behavior, respectively. The second semilog straight line appears during the transition from fracture behavior to total system behavior. This transition straight line has one-half the slope of the first and third semilog straight lines.

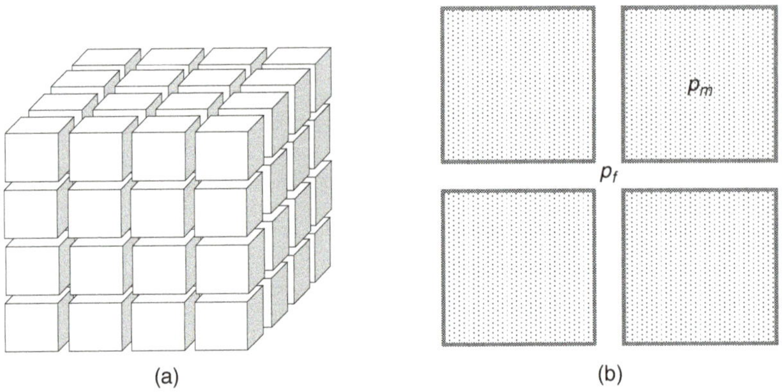

Fig. 6.74—Warren and Root pseudosteady-state dual-porosity model.

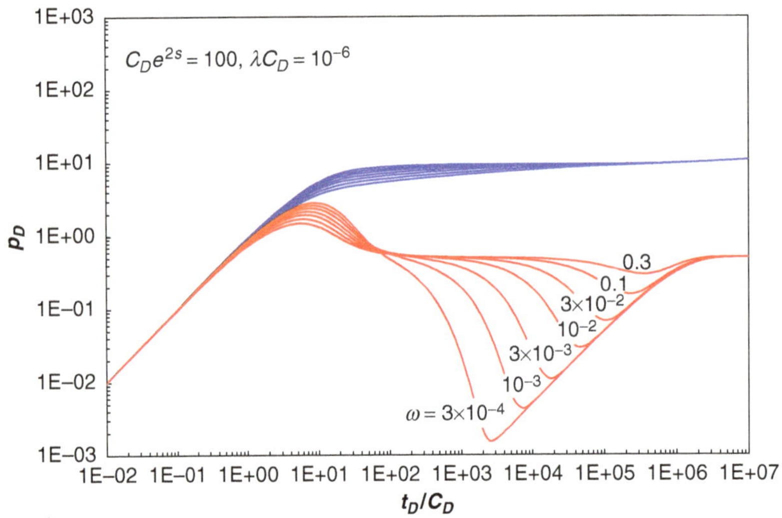

Fig. 6.75—Pressure response for pseudosteady-state dual-porosity model, varying ω.

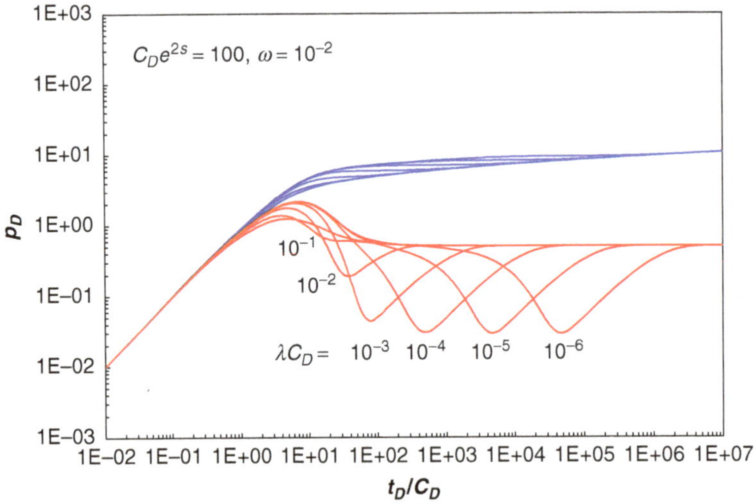

Fig. 6.76—Pressure response for pseudosteady-state dual-porosity model, varying λC_D.

In contrast to the pseudosteady-state model, the pressure response for the transient dual-porosity model depends on the shape of the matrix blocks. Solutions have been presented for slab matrix blocks with one set of fracture planes (Gringarten 1984; Abdassah and Ershagi 1986; Stewart and Ascharsobbi 1988; Kulatilake et al. 1993), column or matchstick (Cinco-Ley et al. 1985) matrix blocks (two mutually orthogonal sets of fracture planes), and spherical (Abdassah and Ershagi 1986) or cube (Cinco-Ley et al. 1985)

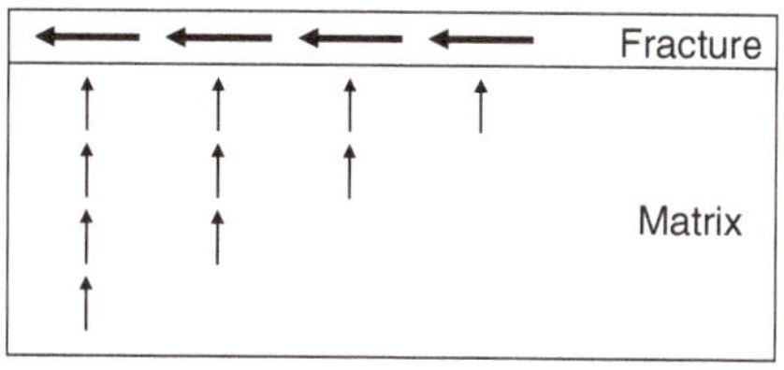

Fig. 6.77—Kazemi-de Swaan transient dual-porosity model.

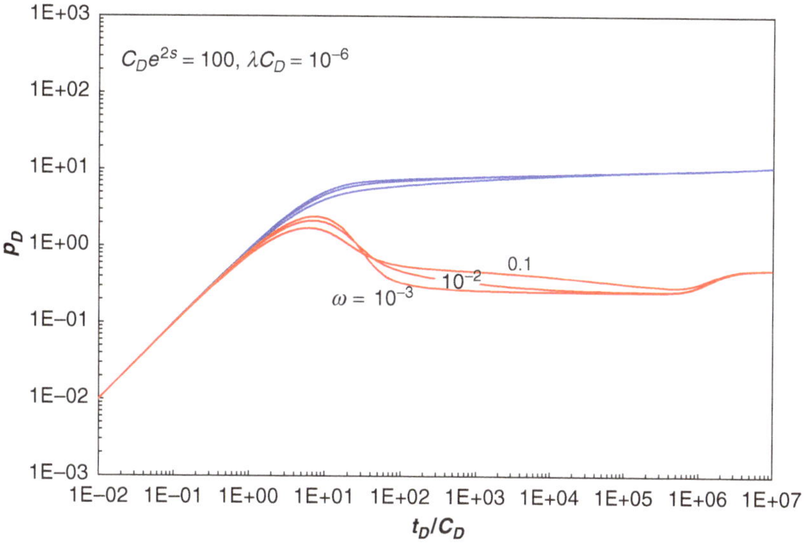

Fig. 6.78—Pressure response for transient dual-porosity model; varying ω.

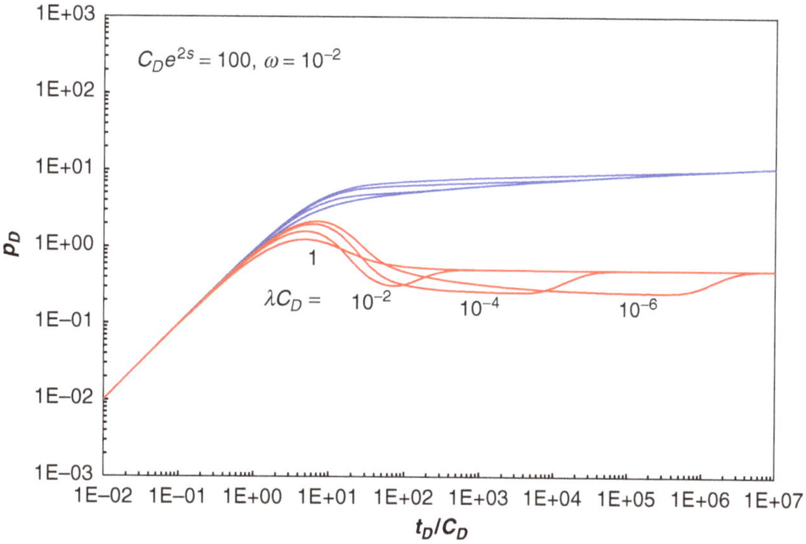

Fig. 6.79—Pressure response for transient dual-porosity model; varying λC_D.

matrix blocks (three mutually orthogonal sets of fracture planes). **Fig. 6.80** shows the pressure derivative for pseudosteady-state dual-porosity models and for transient dual-porosity models with slab, column, or cube matrix blocks. There is negligible difference between the pressure responses for columns and cubes and little difference between pressure responses for slabs, columns, and cubes.

6.8.4 Transient Dual-Porosity Model With Interporosity Skin. One of the puzzling factors about the pseudosteady-state dual-porosity model is that, while the transient dual-porosity model would appear to be more realistic physically, behavior of naturally

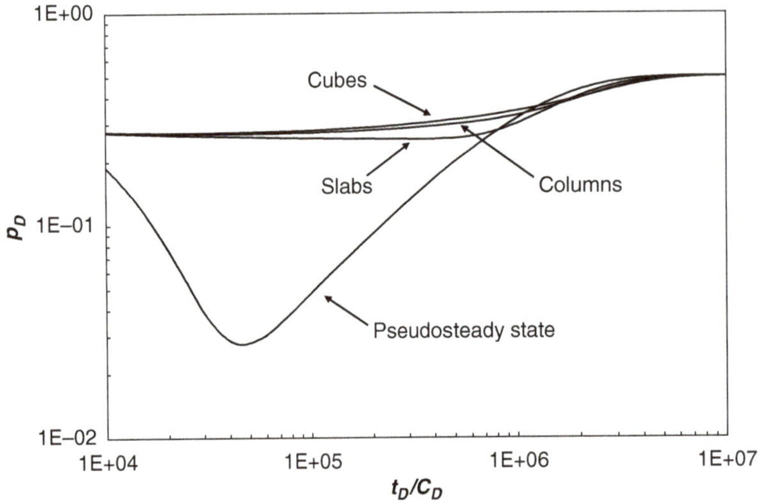

Fig. 6.80—Pressure response for transient dual-porosity model for different block shapes.

fractured reservoirs often follows the pseudosteady-state model instead (Stewart and Ascharsobbi 1988). One possible explanation for this phenomenon is the existence of a pseudosteady-state skin between the matrix and the fracture system.

Moench (1984) generalized the transient interporosity flow model to include the effects of an interporosity skin. This model reduces to that of Kazemi and de Swaan as the matrix skin approaches zero. As the matrix skin increases so that by comparison the pressure drop within the matrix becomes negligible, the pressure-transient behavior of the Moench model approaches that of the Warren and Root pseudosteady-state model.

Stewart and Ascharsobbi (1988) defined the interporosity skin s_{ma} as

$$s_{ma} \equiv \frac{2k_{mi}h_s}{k_s h_m}, \quad (6.114)$$

where k_{mi} is the intrinsic matrix permeability, k_s is the permeability of the skin, h_m is the matrix block width, and h_s is the thickness of the skin. **Figs. 6.81 and 6.82** show the pressure response for slab matrix blocks with interporosity skin s_{ma} ranging from 0 to 10. In Fig. 6.81, the interporosity flow coefficient λ is held constant. Stewart and Ascharsobbi (1988) showed that an effective interporosity flow coefficient, λ_{eff}, can be defined as

$$\lambda_{\text{eff}} \equiv \frac{\lambda}{1+(n+2)s_{ma}}, \quad (6.115)$$

where n is the number of mutually orthogonal sets of fracture planes. In Fig. 6.82, s_{ma} is varied so that λ_{eff} remains constant. As the interporosity skin s_{ma} increases from 0 to 10, the pressure-transient response changes from that of transient dual porosity to pseudosteady-state dual porosity.

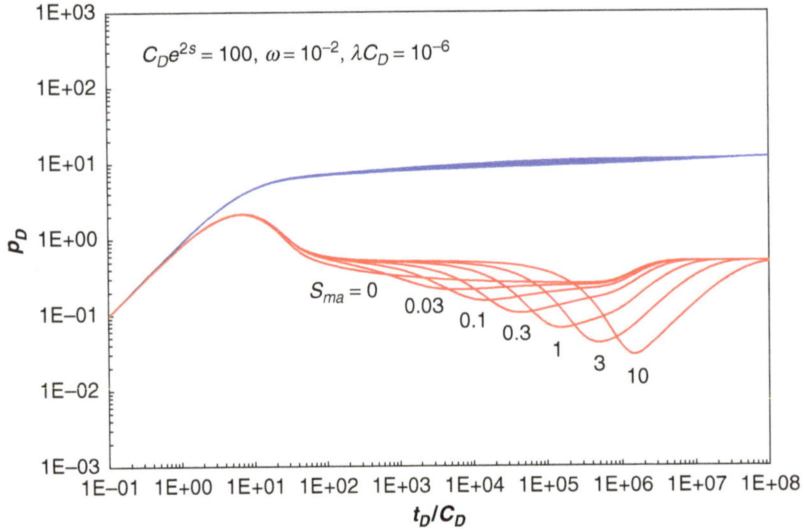

Fig. 6.81—Pressure response for transient dual-porosity model with interporosity skin; varying S_{ma}.

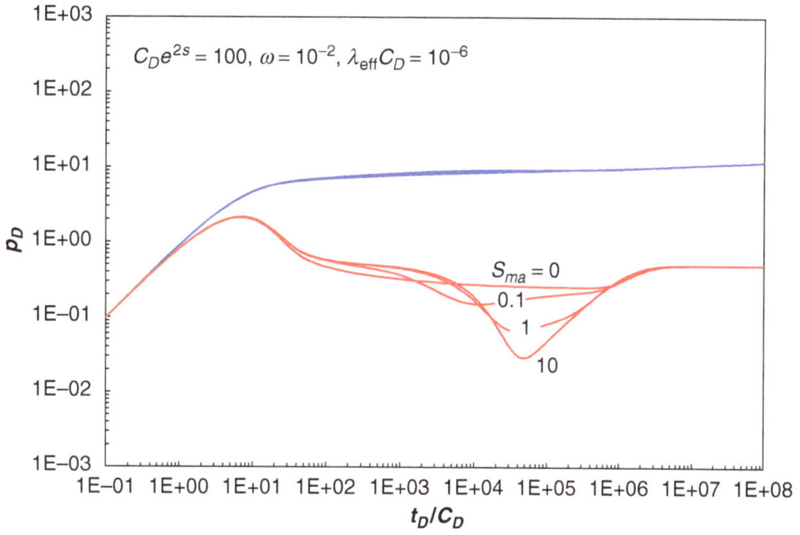

Fig. 6.82—Pressure response for transient dual-porosity model with interporosity skin; varying S_{ma} with constant λ_{eff}.

6.8.5 Dual-Porosity Model With Variable Block Size. A major objection to the Warren and Root model is the assumption of identical matrix blocks separated by identical fractures. A brief look at any outcrop showing natural fractures reveals this assumption to be somewhat unrealistic.

Baecher et al. (1977) investigated the distribution of joint length and spacing at two field locations: one sandstone/siltstone/shale sequence, and one metamorphic sequence with igneous intrusions. They concluded that joint trace length from both sites could be described with a log-normal distribution, while the joint spacing was best described with an exponential distribution. Kulatilake et al. (1993) studied a joint system in a granite mass, intersected by a ventilation drift in the Stripa mine, Sweden. They examined a number of different joint parameters and considered several different distributions. They concluded that joint length follows a gamma distribution, while joint spacing follows either a gamma distribution or an exponential distribution.

A number of authors have considered ways to relax the assumption of uniform block size; however, most of these studies have not attempted to model a realistic distribution of block sizes.

Abdassah and Ershagi (1986) presented a triple-porosity model with two distinct sets of matrix blocks having different properties. The two sets of matrix blocks interact only with the fracture. Transient interporosity flow is assumed between either set of matrix blocks and the fracture. Cinco-Ley et al. (1985) developed analytical solutions for the pressure-transient response for a reservoir with an arbitrary discrete distribution of block sizes. Their model is general in that interporosity flow can be pseudosteady state, transient, or transient with an interporosity skin. They found that the shape of the distribution had a significant effect on the shape of the pressure response. In particular, if there is a broad distribution of block sizes, the classic V-shaped minimum in the derivative predicted by the Warren and Root model will not occur. Belani and Yazdi (1988) studied the effects of uniform and bimodal distributions of block size on the pressure transient response.

Spivey and Lee (2000) considered the effects of exponential fracture spacing on block size and interporosity flow coefficient for one, two, and three sets of exponentially spaced, infinite fracture planes. They assumed PSSF between the fracture and the matrix. They considered cases in which λ scales as $1/h_m$, corresponding to interporosity skin thickness being independent of matrix block size; and as $1/h_m^{\,2}$, corresponding to interporosity skin thickness proportional to matrix block size. **Fig. 6.83** shows the pressure response for slab matrix blocks and exponential fracture spacing.

6.8.6 Fractal Reservoir Model. Chang and Yortsos (1990) presented a formulation for computing the pressure-transient response of a naturally fractured reservoir with a fractal fracture network embedded in a Euclidean matrix. They considered the pressure response for two cases: from the fracture network only, and from the fracture network with contribution from the matrix. Fluid flow from the matrix to the fracture system was assumed to occur under pseudosteady-state conditions, as in the Warren and Root model. Beier (1990) and Acuna et al. (1995) presented field data from naturally fractured reservoirs consistent with a fractal model.

6.9 Pressure-Dependent Rock Properties

As has been known for many years, both porosity and permeability of reservoir rock decrease as the net confining stress increases (Botset and Reed 1935; Fatt and Davis 1952; Hall 1953; Fatt 1958; Vairogs et al. 1971; Thomas and Ward 1972; Jones 1975; Jones and Owens 1980; Jones 1988).

Thomas and Ward (1972) studied the effects of both overburden pressure and water saturation on low-permeability sandstones. They found that most of the permeability reduction took place at low overburden pressure less than 3,000 psi; fractured samples experienced significantly greater permeability reduction than unfractured samples; and overburden pressure has little effect on relative permeability.

In the laboratory, stressed porosity and permeability data are normally measured under hydrostatic conditions. A number of methods have been used to convert data expressed in terms of hydrostatic stress to reservoir stress conditions.

Fatt and Davis (1952) state: "The assumption which is usually made is that the rocks in the earth are under a uniform pressure equal to the weight per unit area of the vertical overburden column from the rock in question to the surface of the earth less the pressure of the liquid in the pores of the rock."

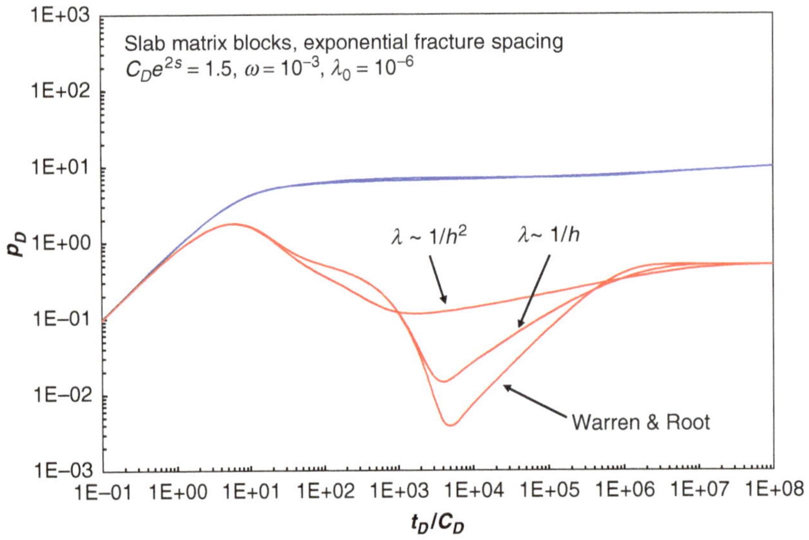

Fig. 6.83—Pressure response for pseudosteady-state dual-porosity model with variable block size and slab blocks with exponential fracture spacing.

Fatt (1958) used the net overburden pressure (NOB) defined as

$$p_{\mathrm{NOB}} = p_e - 0.85p. \quad (6.116)$$

Vairogs et al. (1971) used the effective stress, defined (in circular cylindrical coordinates) as

$$\sigma'''_{r,\theta,z} = \sigma_{r,\theta,z} - (1-\phi)p. \quad (6.117)$$

Teeuw (1971) proposed relating uniaxial compaction to hydrostatic compaction by means of

$$\varepsilon_Z = \frac{1}{3}\left(\frac{1+v}{1+v}\right)e, \quad (6.118)$$

where ε_Z is the uniaxial compaction, e is the hydrostatic compression, and v is Poisson's ratio. Jones (1975) defined the net stress as the difference between hydrostatic confining pressure and the pore pressure. Ostensen (1986) defined the net confining stress for a fractured rock as

$$\sigma_n = \sigma - \alpha p, \quad (6.119)$$

where σ is the rock stress perpendicular to the crack face, σ_n is the net confining stress, p is the gas pressure, and α is a coefficient of order unity.

6.9.1 Porosity and Permeability. Jones (1988) proposed a four-parameter empirical equation to correlate stress-dependent permeability and pore volume:

$$k = k_0 \frac{\exp\left(a_k\left[\exp\left(-\frac{\sigma}{\sigma^*}\right)-1\right]\right)}{1+C\sigma}, \quad (6.120)$$

where σ is the net stress, and k_0, a_k, C, and σ^* are empirical fitting parameters. Jones showed that two of the parameters, C and σ^*, could be fixed at 3×10^{-6} psi^{-1} and 3,000 psi, respectively, with little effect on permeabilities for net stresses up to 10,000 psi. Eq. 6.120 can be rearranged as

$$\ln\left(k(1+C\sigma)\right) = \ln k_0 - a_k\left[1-\exp\left(-\frac{\sigma}{\sigma^*}\right)\right]. \quad (6.121)$$

For fixed C and σ^*, Eq. 6.121 provides a graphical technique for estimating k_0 and a_k from a graph of $\ln\left(k(1+C\sigma)\right)$ on the y-axis vs. $1-\exp\left(-\frac{\sigma}{\sigma^*}\right)$ on the x-axis. The negative of the slope of the graph gives a_k, and the intercept gives $\ln(k_0)$. Jones (1988) proposed an expression of the same form as Eq. 6.121 for correlating pore volume:

$$V_p = V_{p0} \frac{\exp\left(a_V\left[\exp\left(-\frac{\sigma}{\sigma^*}\right)-1\right]\right)}{1+C\sigma}, \quad (6.122)$$

where C and σ^* have the same values as in Eq. 6.105, and a_V and V_{p0} are fitting parameters. Taking the derivative of Eq. 6.122 with respect to stress gives the pore-volume compressibility as

$$c_{PV} = \frac{a_V \exp\left(-\frac{\sigma}{\sigma^*}\right)}{\sigma^*} + \frac{C}{1+C\sigma}. \quad (6.123)$$

Jones (1988) also proposed that the porosity be correlated as

$$\frac{\phi}{1-\phi} = \frac{\phi_0}{1-\phi_0} \frac{\exp\left(a_V\left[\exp\left(-\frac{\sigma}{\sigma^*}\right)-1\right]\right)}{1+C\sigma}, \quad (6.124)$$

where ϕ_0 is a fitting parameter.

6.9.2 Modifications to Adjusted Pressure. The definitions of pseudopressure and adjusted pressure can be modified to account for pressure-dependent rock properties (Al-Hussainy et al. 1966). Allowing permeability to be a function of pore pressure, we write the adjusted pressure as

$$p_a \equiv \frac{\mu_i z_i}{k_i p_i} \int_0^p \frac{kp dp}{\mu z}. \quad (6.125)$$

6.9.3 Modifications to Adjusted Time. Similarly, the adjusted time can be modified to account for pressure-dependent rock properties:

$$t_a \equiv \frac{\phi_i \mu_i c_{ti}}{k_i} \int_0^t \frac{k dt}{\phi \mu c_t}. \quad (6.126)$$

Eq. 6.126 may be used for buildup-test interpretation, in which case the pressure-dependent properties in the integral are evaluated at the wellbore pressure. Eq. 6.126 may also be used for production from a closed, circular reservoir, in which case the pressure-dependent properties are evaluated at the average drainage-area pressure.

For pressure-dependent rock properties, there are some subtleties that should be noted when evaluating Eq. 6.122. First, when the porosity is a function of pressure, the water saturation will not be constant, even for an immobile water phase. Assuming that the water is immobile, the water saturation is also a function of pressure:

$$S_w = \frac{\phi_i}{\phi} \frac{B_w}{B_{w_i}} S_{wi}, \quad (6.127)$$

where the subscript i denotes properties at initial pressure. The gas saturation is also pressure dependent:

$$S_g = 1 - S_w. \quad (6.128)$$

The total compressibility c_t in Eq. 6.126 should be evaluated using the pressure-dependent water and gas saturations as defined by Eqs. 6.127 and 6.128:

$$c_t = c_f + S_w c_w + S_g c_g. \quad (6.129)$$

Second, the virtually universal assumption that formation thickness is constant implies that the bulk volume is also independent of pressure. Thus, to preserve material balance, the porosity in Eq. 6.126 must be defined relative to the bulk volume at initial pressure V_{bi} rather than the pressure-dependent bulk volume at current pressure V_b. This, in turn, implies that stress-dependent porosity should be correlated by

$$\phi' = \phi_0' \frac{\exp\left(a_V\left[\exp\left(-\frac{\sigma}{\sigma^*}\right)-1\right]\right)}{1+C\sigma}, \quad (6.130)$$

where ϕ' is the porosity defined relative to the bulk volume at initial pressure:

$$\phi' \equiv \frac{V_p}{V_{bi}}, \quad (6.131)$$

and where $\phi_0' = V_{p0}/V_{bi}$, and V_{p0}, a_V, C, σ, and σ^* are used as in Eq. 6.122.

Third, the fact that the water saturation is a function of pressure implies that the relative permeability to gas is also an indirect function of pressure:

$$k_{rg} = k_{rg}\left(S_w(p)\right). \quad (6.132)$$

Thus, the permeability in Eqs. 6.125 and 6.126 should be calculated as

$$k_g(p) = k_{rg}\left(S_w(p)\right)k(p), \quad (6.133)$$

where the absolute permeability $k(p)$ is obtained from Eq. 6.120, and the relative permeability k_{rg} is obtained from Eq. 6.132.

Finally, if the net overburden pressure, external stress, and pore pressure are related by an equation of the form of Eq. 6.119, the formation compressibility c_f in Eq. 6.126 must be calculated as

$$c_f = \frac{1}{\alpha} c_{\mathrm{PV}}, \qquad (6.134)$$

where c_{PV} is the pore-volume compressibility obtained from Eq. 6.123, and α is a coefficient of order unity, as used in Eq. 6.119.

6.10 Coalbed Methane and Naturally Fractured Shale

In coals and in naturally fractured shales, much of the gas is found adsorbed on the surface of the matrix rather than as free gas within a conventional porosity system. This adsorbed gas content can be described by the Langmuir isotherm, which relates the adsorbed gas volume, V, to the pressure of the gas phase:

$$V = \frac{V_L p}{p_L + p}. \qquad (6.135)$$

Coals vary widely in their adsorbed gas content. Sorption isotherms for coal can have a Langmuir volume, V_L, of 200–800 scf/ton and a Langmuir pressure, p_L, of 100–500 psi. The matrix density for coal ranges from 1.2 to 1.45 g/cm^3.

As with coalbed methane reservoirs, much of the gas content in naturally fractured shales is present as adsorbed gas. Sorption isotherms for shale can have a Langmuir volume of 60–200 scf/ton and a Langmuir pressure of 400–600 psi. The matrix density for shale is approximately 2.45 g/cm^3.

Naturally fractured shales are characterized by a network of natural fractures, which provides virtually all the flow capacity, but only a small fraction of the storage capacity, of the reservoir rock. Most of the gas content in a naturally fractured shale is found either as free gas in a conventional porosity system within the matrix or as gas adsorbed on surfaces within the shale. As the reservoir is produced, gas desorbs from the shale into the matrix porosity system. It then flows through the matrix to the natural fracture system, and then through the fracture system to the wellbore, as in a conventional dual-porosity system. Gatens et al. (1985) developed a transient dual-porosity model to history match production data from Eastern Devonian Shale wells. Lee and Gatens (1985) and and Gatens et al. (1989) used the model to analyze shale well production.

Bumb and McKee (1988) showed that the total compressibility for coals and naturally fractured shales must be modified to include a term accounting for desorption of gas from the matrix:

$$c_t = c_t + S_w c_w + (1 - S_w) c_g + \frac{p_{sc} V_L p_L T z}{\phi p T_{sc} z_{sc} (p_L + p)^2}, \qquad (6.136)$$

where V_L is expressed in scf/rcf.

Spivey and Semmelbeck (1995) showed that Eq. 6.136 should be used in the definition of adjusted time to account for sorption effects in dry coals and naturally fractured shales during buildup or during production from a closed reservoir.

6.11 Flow Tests in Low-Permeability Gas Reservoirs

6.11.1 Deliverability Testing. Traditional methods of deliverability testing, such as deliverability tests, isochronal tests, and modified isochronal tests, are of little use in low-permeability reservoirs. Deliverability tests were developed for predicting production from high-permeability reservoirs in which PSSF is reached in a matter of hours or days. For low-permeability reservoirs, estimating flow rates with a pseudosteady-state relationship such as the Rawlins-Schellhardt deliverability equation or the Hoeupert deliverability equation will underestimate flow rates.

6.11.2 Analyzing Production Data Using Pressure-Transient Methods. Analysis of production data using methods developed for pressure-transient analysis of variable-rate/variable-pressure production is an extremely useful tool for reservoir characterization of low-permeability reservoirs. Often, permeability, skin factor or fracture half-length, and drainage area/original gas-in-place can be estimated using such methods. Discussion of these methods is deferred to Part 2 Ch 8: Rate Transient Analysis of this book.

6.11.3 One-Point Method. The one-point method (Lee et al. 1984) is a simple way of estimating permeability from single-point-flow test data in which the reservoir is still infinite acting.

6.11.3.1 Permeability From Data in PSSF. Start with the equation for PSSF from a closed, circular reservoir given by

$$\bar{p} - p_{wf} = \frac{141.2 q_g \bar{\bar{B}}_g \bar{\bar{\mu}}_g}{kh} \left[\ln\left(\frac{r_e}{r_w}\right) - \frac{3}{4} + s' \right], \qquad (6.137)$$

where $\bar{p}$ is the average drainage-area pressure, and $\bar{\bar{B}}_g$ and $\bar{\bar{\mu}}_g$ are evaluated at $\bar{\bar{p}} = 0.5(\bar{p} + p_{wf})$.

Eq. 6.137 can be rearranged to give the permeability k as

$$k = \frac{141.2 q_g \bar{\bar{B}}_g \bar{\bar{\mu}}_g}{h(\bar{p} - p_{wf})} \left[\ln\left(\frac{r_e}{r_w}\right) - \frac{3}{4} + s' \right]. \qquad (6.138)$$

For single-point flow test data in which the reservoir has reached PSSF, Eq. 6.138 can be used to estimate the permeability, provided the drainage radius r_e and the total skin factor s' can be estimated.

6.11.3.2 Permeability From Data in IARF. For IARF, the pressure response can be written as

$$p_i - p_{wf} = \frac{70.6 q_g B_{gi} \mu_{gi}}{kh}\left[\ln\left(\frac{kt}{1{,}688 \phi \mu_{gi} c_{ti} r_w^2}\right) + s'\right]. \quad (6.139)$$

Unlike Eq. 6.137, Eq. 6.139 cannot be solved for permeability k in closed form. We can, however, develop a fixed-point iteration method to evaluate Eq. 6.139, as follows.

First, we rearrange Eq. 6.139 as

$$p_i - p_{wf} = \frac{141.2 q_g B_{gi} \mu_{gi}}{kh}\left[\ln\left(\sqrt{\frac{kt}{377 \phi \mu_{gi} c_{ti}}}\frac{1}{r_w}\right) - \frac{3}{4} + s'\right]. \quad (6.140)$$

If we define the transient drainage radius r_d as

$$r_d \equiv \sqrt{\frac{kt}{377 \phi \mu_{gi} c_{ti}}}, \quad (6.141)$$

we can rewrite Eq. 6.140 as

$$\bar{p} - p_{wf} = \frac{141.2 q_g B_{gi} \mu_{gi}}{kh}\left[\ln\left(\frac{r_d}{r_w}\right) - \frac{3}{4} + s'\right]. \quad (6.142)$$

Eq. 6.142 may be solved for k to give

$$k = \frac{141.2 q_g B_{gi} \mu_{gi}}{h(p_i - p_{wf})}\left[\ln\left(\frac{r_d}{r_w}\right) - \frac{3}{4} + s'\right]. \quad (6.143)$$

Eqs. 6.141 and 6.143 provide a simple fixed-point iteration method for estimating permeability that converges extremely fast, often within two or three iterations. The procedure is as follows:

1. Calculate the Horner pseudoproducing time from $t_p = \frac{24 G_p}{q_g}$.
2. Assume a starting estimate for permeability. A value of 0.1 md works well if no better estimate is available.
3. Calculate a trial value of the transient drainage radius r_d from Eq. 6.141. On the first iteration, use the estimate from Step 2. During subsequent iterations, use the most recent estimate from Step 4.
4. Using the value of r_d calculated in Step 3, calculate a new estimate of permeability k from Eq. 6.143.
5. Repeat Steps 3 and 4 until the value of k converges.

6.11.3.3 One-Point Method Example. Problem. Given the data in **Table 6.9,** estimate the permeability using the one-point method.

Solution. First, we calculate the producing time using the Horner pseudoproducing time approximation:

ϕ = 0.199	h = 40 ft	r_w = 0.25 ft
B_{gi} = 0.546 bbl/Mscf	μ_{gi} = 0.0356 cp	c_{ti} = 2.98 × 10^{-5} psi^{-1}
p_i = 9,665 psia	p_{wf} = 9189 psia	s' = 0
G_p = 5,850 Mscf	q_g = 1,950 Mscf/D	

Table 6.9—Data for one-point method example.

$$t_p = \frac{24 G_p}{q_g} = \frac{(24)(5{,}850)}{1{,}950} = 72 \text{ hr}.$$

Iteration 1. Starting with a permeability estimate of 0.1 md, we calculate the transient drainage as

$$r_d = \sqrt{\frac{kt}{377 \phi \mu_{gi} c_{ti}}}$$
$$= \sqrt{\frac{(0.1)(72)}{(377)(0.199)(0.0356)\left(2.98\times10^{-5}\right)}}$$
$$= 300.8 \text{ ft}.$$

Next, we calculate a new estimate of permeability as

$$k = \frac{141.2 q_g B_{gi} \mu_{gi}}{h(p_i - p_{wf})}\left[\ln\left(\frac{r_d}{r_w}\right) - \frac{3}{4} + s'\right]$$
$$= \frac{(141.2)(1{,}950)(0.546)(0.0356)}{(40)(9{,}665 - 9{,}189)}\left[\ln\left(\frac{300.8}{0.25}\right) - \frac{3}{4} + 0.0\right]$$
$$= 1.78 \text{ md}.$$

Iteration 2. The new estimate of permeability is much greater than the previous estimate, so we continue iterating. The new transient drainage radius is

$$r_d = \sqrt{\frac{kt}{377\phi\mu_{gi}c_{ti}}}$$
$$= \sqrt{\frac{(1.78)(72)}{(377)(0.199)(0.0356)\left(2.98\times10^{-5}\right)}}$$
$$= 1{,}270 \text{ ft.}$$

Next, we calculate an updated permeability estimate as

$$k = \frac{141.2q_g B_{gi}\mu_{gi}}{h\left(p_i - p_{wf}\right)}\left[\ln\left(\frac{r_d}{r_w}\right) - \frac{3}{4} + s'\right]$$
$$= \frac{(141.2)(1{,}950)(0.546)(0.0356)}{(40)(9{,}665-9{,}189)}\left[\ln\left(\frac{1{,}270}{0.25}\right) - \frac{3}{4} + 0.0\right]$$
$$= 2.19 \text{ md.}$$

Iteration 3. The new estimate of permeability is approximately 20% larger than the previous estimate, so we will do one more iteration. The new transient drainage radius is

$$r_d = \sqrt{\frac{kt}{377\phi\mu_{gi}c_{ti}}}$$
$$= \sqrt{\frac{(2.19)(72)}{(377)(0.199)(0.0356)\left(2.98\times10^{-5}\right)}}$$
$$= 1{,}407 \text{ ft.}$$

The new permeability estimate is

$$k = \frac{141.2q_g B_{gi}\mu_{gi}}{h\left(p_i - p_{wf}\right)}\left[\ln\left(\frac{r_d}{r_w}\right) - \frac{3}{4} + s'\right]$$
$$= \frac{(141.2)(1{,}950)(0.546)(0.0356)}{(40)(9{,}665-9{,}189)}\left[\ln\left(\frac{1{,}407}{0.25}\right) - \frac{3}{4} + 0.0\right]$$
$$= 2.22 \text{ md.}$$

The permeability from the third iteration is only 1.5% higher than that from the second iteration, so we accept the estimate as correct. For comparison, the permeability was estimated from an extended buildup following the flow period as 2.12 md, which is close to the value estimated from the one-point method.

6.11.3.4 Limitations of the One-Point Method. Because the one-point method assumes IARF, it will not give accurate results before the end of WBS or after reservoir boundaries are encountered.

In developing the theory for the one-point method, the skin factor s' was assumed to be known. If the skin factor s' is not known, the permeability calculated from the one-point method will be inaccurate. However, even for a rough estimate of s', the one-point method will normally give an order-of-magnitude estimate for permeability.

For hydraulically fractured wells, the test duration must be long enough to reach PRF. If the one-point method is used on data in which the test duration is not long enough to reach PRF, calculated permeabilities could be negative, even if the correct equivalent skin factor is known.

6.12 Summary

Prefracture pressure-transient testing provides one of the best ways of estimating formation permeability for use in stimulation-treatment design and post-fracture pressure-transient analysis. This chapter can be summarized by the following points:

- Pressure-transient tests in gas wells can be analyzed using the semilog straight-line method, the log-log type curve method, or the manual log-log method.
- Semilog analysis of gas-well pressure-buildup tests can be used to estimate permeability, skin factor, and initial reservoir pressure.
- Log-log type curve analysis of gas-well pressure-buildup test data can provide estimates of permeability, skin factor, and adjusted WBS coefficient.
- Manual log-log analysis can provide estimates of permeability, skin factor, and adjusted WBS coefficient from the log-log graph alone.
- Adjusted pressure and actual time should be used to analyze gas-well drawdown tests, especially for flow periods with large drawdowns.

- Adjusted pressure and adjusted time should be used to analyze gas-well buildup tests, especially following flow periods with large drawdowns.
- The log-log diagnostic plot is the primary tool for flow-regime identification, allowing radial flow, linear flow, bilinear flow, spherical flow, and VB to be identified from the characteristic slope of the pressure derivative.
- Boundary effects include closed boundaries, boundaries characterized by a reduction in flow capacity, and boundaries characterized by an increase in flow capacity. In this context, the term "flow capacity" includes both mobility and flow geometry.
- A strong waterdrive acts like a large reduction in mobility. A unit-slope derivative at late times might be an indication of the presence of a waterdrive.
- A constant-pressure boundary model should never be used for a gas well without a physical explanation for the cause of a large increase in mobility away from the well.
- Adjusted pressure can be used in conjunction with a reservoir-adjusted time to linearize data in PSSF for wells in reasonably symmetric gas-drainage areas.
- Many unconventional gas reservoirs are naturally fractured. Reservoir models that find application in naturally fractured reservoirs include homogeneous, homogeneous with permeability anisotropy, well near a single conductive fault or fracture, radial composite, linear composite, pseudosteady-state dual-porosity, and transient dual-porosity reservoir models.
- Low-permeability reservoir rock is especially sensitive to net confining stress, with significant reduction of both porosity and permeability. The definitions of adjusted pressure and adjusted time can be modified to account for pressure-dependent rock properties.
- Much of the gas present in coalbed methane and naturally fractured shale reservoirs is adsorbed on the matrix. Gas content as a function of pressure can be described with a Langmuir isotherm. Desorption of the gas from the matrix must be included as an additional term in the total compressibility equation.
- Conventional multilayer pressure-transient testing finds little application for wells in unconventional gas reservoirs.
- The one-point method can be used to estimate formation permeability from prestimulation flow data when prefracture buildup data are not available.

6.13 Nomenclature

A = cross-sectional area open to flow, cm^2
B = FVF, bbl/STB
c = compressibility, psi^{-1}
C = WBS coefficient, bbl/psi
C_a = adjusted WBS coefficient, bbl/psi
C_A = drainage area shape factor, dimensionless
D = non-Darcy flow coefficient, D/Mscf
G_{ND} = geometric factor, dimensionless
G_p = cumulative gas production, Mscf
h = net-pay thickness, ft
h_p = height of perforated interval, ft
k = permeability, md
k_r = relative permeability, dimensionless
l_p = length of perforation tunnel, ft
L = distance to boundary, ft
L_f = fracture half-length, ft
L_m = dual porosity model characteristic fracture spacing, ft
m = slope of semilog straight line, psi/cycle
$m(p)$ = real-gas pseudopressure, psi^2/cp
M = molecular weight, lbm/lbm-mol
n = transient dual-porosity model, number of sets of mutually orthogonal fracture planes (1, 2, or 3)
n_s = shot density, shots/ft
p = pressure, psia [atm]
p_{1hr} = pressure read from semilog straight line at a time of 1 hour
p_a = adjusted pressure, psi
p_L = Langmuir pressure, psia
p_i = initial pressure, psia
p_w = wellbore pressure, psia
q = volumetric flow rate, STB/D [cm^3/s]
q_g = gas production rate, Mscf/D
r = radial spatial coordinate, ft [cm]
r_d = transient drainage radius; $r_d \equiv \sqrt{\dfrac{kt}{377\phi\mu_{gi}c_{ti}}}$, ft
r_e = drainage radius, ft
r_i = radius of investigation, ft
r_p = radius of perforation tunnel, ft
r_s = equivalent spherical radius of perforation tunnel, ft
r_w = wellbore radius, ft
R = universal gas constant, 10.732 psi·ft^3/lbm-mol·°R
s = skin factor, dimensionless
s' = total skin, including damage, geometric, and non-Darcy flow components
s_f = fracture damage skin factor, dimensionless
s_{ma} = interporosity skin factor, dimensionless
S_g = gas saturation, fraction
S_w = water saturation, fraction
t = time variable, hours (field units); seconds (consistent units)
t_p = producing time, hours
T = temperature in real gas law equation of state, °R
T_f = formation temperature, °F [°R]
T_{sc} = standard temperature, °F (or °R in equations)
u = superficial velocity, cm/s (consistent units)
v = average fluid velocity, cm/s
V = volume, ft^3

V_b = bulk volume, ft^3

V_L = Langmuire volume, scf/ton

V_p = pore volume, ft^3

w = width, ft

x = Cartesian spatial coordinate, cm (consistent units)

z = real-gas deviation factor, dimensionless

β = velocity coefficient, ft^{-1}

Δp_d = pressure drop across altered zone

Δp_s = additional pressure drop resulting from the skin factor

Δp_u = pressure drop across altered zone if the permeability were unchanged

Δt_a = adjusted time, hours

Δt_e = radial-flow equivalent time, hours

θ = angle between intersecting sealing faults, degrees

λ = dual porosity model interporosity flow coefficient, dimensionless

μ = viscosity, cp

ρ = density, lbm/ft^3 (field units); g/cm^3 (consistent units)

ϕ = porosity, fraction

ω = dual-porosity model storativity ratio, dimensionless (not to be confused with radial composite model storativity ratio S!)

Subscripts

a = characteristic of altered zone

b = beginning (of flow regime)

D = dimensionless quantity

e = end (of flow regime)

f = fracture system

g = gas

i = initial conditions

m = matrix system

MP = match point

ref = reference pressure

R = point on pressure-change or derivative curve during infinite-acting radial flow

sc = standard conditions

sf = sandface

t = total

w = water

wb = wellbore

wf = flowing wellbore conditions

ws = shut-in wellbore conditions

W = point on unit-slope line

6.14 References

Abbaszadeh, M. and Cinco-Ley, H. 1995. Pressure-Transient Behavior in a Reservoir With a Finite-Conductivity Fault. *SPE Form Eval* **10** (1): 26–32. SPE-24704-PA. https://doi.org/10.2118/24704-PA.

Abdassah, D. and Ershagi, I. 1982. Triple-Porosity Systems for Representing Naturally Fractured Reservoirs. *SPE Form Eval* **1** (2): 113–127. SPE-13409-PA. https://doi.org/10.2118/13409-PA.

Acuna, J. A., Ershagi, I., and Yortsos, Y. C. 1995. Practical Application of Fractal Pressure-Transient Analysis in Naturally Fractured Reservoirs. *SPE Form Eval* **10** (3): 173–179. SPE- 24705-PA. https://doi.org/10.2118/24705-PA.

Agarwal, R. G. 1979. Real Gas Pseudo-Time—A New Function for Pressure Buildup Analysis of MFH Gas Wells. Presented at the 54th Annual Fall Technical Conference and Exhibition of the SPE of AIME, Las Vegas, Nevada, 23–26 September. SPE- 8279-MS. https://doi.org/10.2118/8279-MS.

Agarwal, R. G. 1980. A New Method to Account for Producing Time Effects When Drawdown Type Curves Are Used to Analyze Pressure Buildup and Other Test Data. Presented at the 55th Annual Fall Technical Conference and Exhibition of SPE of AIME, Dallas, Texas. 21–24 September. SPE-9289-MS. https://doi.org/10.2118/9289-MS.

Al-Hussainy, R., Ramey, H. J. Jr., and Crawford, P. B. 1962. The Flow of Real Gases Through Porous Media. *J Pet Technol* **18** (4): 624–632. SPE-1243-A-PA. https://doi.org/10.2118/1243-A-PA.

Aziz, K., Mattar, L., Ko, S. et al. 1972. Use of Pressure, Pressure-Squared or Pseudo-Pressure in the Analysis of Transient Pressure Drawdown Data from Gas Wells. *J Can Pet Technol* **15** (2): 58–65. PETSOC-76-02-02. https://doi.org/10.2118/76-02-02.

Baecher, G. B., Lanney, N. A., and Einstein, H. H. 1977. Statistical Description of Rock Properties and Sampling. Presented at the 18th U.S. Symposium on Rock Mechanics (USRMS), Golden, Colorado, 22–24 June. ARMA-77-0400.

Beier, R. A. 1990. Pressure Transient Field Data Showing Fractal Reservoir Structure. Presented at the CIM/SPE International Technical Meeting, Calgary, Alberta, 10–13 June. PETSOC-90-04. https://doi.org/10.2118/90-04.

Belani, A. K. and Yazdi, Y. J. 1988. Estimation of Matrix Block Size Distribution in Naturally Fractured Reservoirs. Presented at the SPE Annual Technical Conference and Exhibition, Houston, Texas, 2–5 October. SPE-18171-MS. https://doi.org/10.2118/18171-MS.

Botset, H. G. and Reed, D. W. 1935. Experiment on Compressibility of Sand. *AAPG Bulletin* **19:** 1053–60.

Bourdet, D., Whittle, T. M., Douglas, A. A. et al. 1983. A New Set of Type Curves Simplifies Well Test Analysis. *World Oil* **(May).**

Bourdet, D., Ayoub, J. A., and Pirard, Y. M. 1989. Use of Pressure Derivative in Well-Test Interpretation. *SPE Form Eval* **4** (2): 293–302. SPE-12777-PA. https://doi.org/10.2118/12777-PA.

Bumb, A. C. and McKee, C. R. 1988. Gas-Well Testing in the Presence of Desorption for Coalbed Methane and Devonian Shale. *SPE Form Eval* **3** (1): 179–185. SPE-15227-PA. https://doi.org/10.2118/15227-PA.

Chang, J. and Yortsos, Y. C. 1990. Pressure-Transient Analysis of Fractal Reservoirs. *SPE Form Eval* **5** (1): 31–38. SPE-18170-PA. https://doi.org/10.2118/18170-PA.

Chen, C. C., Chu, W. C., and Sadighi, S. 1992. Pressure-Transient Testing of Gas Reservoirs With Edge-Waterdrive. *SPE Form Eval* **11** (4): 251–252. SPE-28381-PA. https://doi.org/10.2118/28381-PA.

Cinco-Ley, H. 1992. Well-Test Analysis for Naturally Fractured Reservoirs. *J Pet Technol* **48** (1): 51–54. SPE-31162-JPT. https://doi.org/10.2118/31162-JPT.

Cinco-Ley, H., Samaniego V. F., and Dominguez A. N. 1972. Unsteady-State Flow Behavior for a Well Near a Natural Fracture. Presented at the Annual Fall Technical Conference and Exhibition, New Orleans, Louisiana, 3–6 October. SPE-6019-MS. https://doi.org/10.2118/6019-MS.

Cinco-Ley, H., Samaniego V. F., and Kucuk, F. 1985. The Pressure Transient Behavior for Naturally Fractured Reservoirs With Multiple Block Size. Presented at the SPE Annual Technical Conference and Exhibition, Las Vegas, Nevada, 22–25 September. SPE-14168-MS. https://doi.org/10.2118/14168-MS.

de Swaan, O. A. 1972. Analytic Solutions for Determining Naturally Fractured Reservoir Properties by Well Testing. *SPE J.* **16** (3): 117–122. SPE-5346-PA. https://doi.org/10.2118/5346-PA.

Earlougher, R. C., Jr. 1977. *Advances in Well Test Analysis*. Chap. 5, 203. Dallas, Texas: Monograph Series, SPE.

Fatt, I. 1958. Pore Volume Compressibilities of Sandstone Reservoir Rocks. *J Pet Technol* **10** (3). SPE-970-G. https://doi.org/10.2118/970-G.

Fatt, I. and Davis, D. H. 1952. Reduction in Permeability with Overburden Pressure. *J Pet Technol* **4** (12). SPE-952329-G. https://doi.org/10.2118/952329-G.

Fetkovich, M. J. and Vienot, M. E. 1985. Shape Factor, C_A, Expressed as Skin, s_{CA}. *J Pet Technol* **37** (2): 321–322. SPE-13304-PA. https://doi.org/10.2118/13304-PA.

Fraim, M. L. and Wattenbarger, R. A. 1987. Gas Reservoir Decline-Curve Analysis Using Type Curves With Real Gas Pseudopressure and Normalized Time. *SPE Form Eval* **2** (4): 671–682. SPE-14238-PA. https://doi.org/10.2118/14238-PA.

Gatens, J. M., III, Lee, W. J., and Rahim, Z. 1985. Application of an Analytical Model to History Match Devonian Shales Production Data. Presented at the SPE Eastern Regional Meeting, Morgantown, West Virginia, 6–8 November. SPE-14509-MS. https://doi.org/10.2118/14509-MS.

Gatens, J. M., Lee, W. J., Lane, H. S. et al. 1989. Analysis of Eastern Devonian Gas Shales Production Data. *J Pet Technol* **41** (5): 519–525. SPE-17059-PA. https://doi.org/10.2118/17059-PA.

Gringarten, A. C. 1984. Interpretation of Tests in Fissured and Multilayered Reservoirs With Double-Porosity Behavior: Theory and Practice. *J Pet Technol* **36** (4): 549–564. SPE-10044-PA. https://doi.org/10.2118/10044-PA.

Gringarten, A. C., Bourdet, D. P., Landel, P. A. et al. 1979. A Comparison Between Different Skin and Wellbore Storage Type-Curves for Early-Time Transient Analysis. Presented at the 54th Annual Fall Technical Conference and Exhibition of SPE of AIME, Las Vegas, Nevada, 23–25 September. SPE-8205-MS. https://doi.org/10.2118/8205-MS.

Hall, H. N. 1953. Compressibility of Reservoir Rocks. *J Pet Technol* **5** (1). SPE-953309-G. https://doi.org/10.2118/953309-G.

Hawkins, M. F., Jr. 1952. A Note on the Skin Effect. *J Pet Technol* **8** (12): 65–62. SPE-732-G. https://doi.org/10.2118/732-G.

Jones, F. O. 1975. A Laboratory Study of the Effects of Confining Pressure on Fracture Flow and Storage Capacity in Carbonate Rocks. *J Pet Technol* **27** (1): 21–27. SPE-4569-PA. https://doi.org/10.2118/4569-PA.

Jones, F. O. and Owens, W. W. 1980. A Laboratory Study of Low-Permeability Gas Sands. *J Pet Technol* **32** (9): 1631–1640. SPE-7551-PA. https://doi.org/10.2118/7551-PA.

Jones, S. C. 1987. Using the Inertial Coefficient, β, To Characterize Heterogeneity in Reservoir Rock. Presented at the 62nd Annual Technical Conference and Exhibition of SPE, Dallas, Texas, 27–30 September. SPE-16949-MS. https://doi.org/10.2118/16949-MS.

Jones, S. C. 1988. Two-Point Determinations of Permeability and PV vs. Net Confining Stress. *SPE Form Eval* **3** (1): 235–241. SPE-15380-PA. https://doi.org/10.2118/15380-PA.

Kazemi, H. 1969. Pressure Transient Analysis of Naturally Fractured Reservoirs with Uniform Fracture Distribution. *SPE J.* **9** (4): 451–462. SPE-2156-A. https://doi.org/10.2118/2156-A.

Kulatilake, P. H. S. W., Wathugala, D. N., and Stephansson, O. 1993. Joint Network Modelling with a Validation Exercise in Stripa Mine, Sweden. *International Journal of Rock Mechanics and Mining Sciences & Geomechanics Abstracts* **30** (5): 503–522. https://doi.org/10.1016/0148-9062(93)92217-E.

Lee, W. J. and Holditch, S. A. 1982. Application of Pseudotime to Buildup Test Analysis of Low-Permeability Gas Wells With Long-Duration Wellbore Storage Distortion. *J Pet Technol* **34** (12): 2877–2887. SPE-9888-PA. https://doi.org/10.2118/9888-PA.

Lee, W. J. and Gatens, J. M., III, 1985. Analysis of Eastern Devonian Gas Shales Production Data. Presented at the SPE Eastern Regional Meeting, Morgantown, West Virginia, 6–8 November. SPE-14506-MS. https://doi.org/10.2118/14506-MS.

Lee, W. J., Kuo, T. B., Holditch, S. A. et al. 1984. Estimating Formation Permeability From Single-Point Flow Data. Presented at the SPE/DOE/GRI Unconventional Gas Recovery Symposium, Pittsburgh, Pennsylvania, 13–15 May. SPE-12847-MS. https://doi.org/10.2118/12847-MS.

Lee, W. J., Rollins, J. B., and Spivey, J. P. 2003. *Pressure Transient Testing*, Vol. 9. SPE Textbook Series.

Leftkovits, H. C., Hazebroek, P., Allen, E. E. et al. 1961. A Study of the Behavior of Bounded Reservoirs Composed of Stratified Layers. *SPE J.* **1** (1): 43–58. SPE-1329-G. https://doi.org/10.2118/1329-G.

Levorsen, A. I. 1967. *Geology of Petroleum*, second edition. San Francisco, California: W.H. Freeman and Co.

Mattar, L. and McNeil, R. 1998. The 'Flowing' Gas Material Balance. *J Can Pet Technol* **37** (2): 52–55. PETSOC-98-02-02. https://doi.org/10.2118/98-02-02.

Meunier, D. F., Kabir, C. S., and Wittmann, M. J. 1987. Gas Well Test Analysis: Use of Normalized Pseudovariables. *SPE Form Eval* **2** (4): 629–632. SPE-13082-PA. https://doi.org/10.2118/13082-PA.

Moench, A. F. 1984. Double-Porosity Models for a Fissured Groundwater Reservoir With Fracture Skin. *Water Resources Research* **20** (7) 831–832. https://doi.org/10.1029/WR020i007p00831.

Ostensen, R. W. 1982. The Effect of Stress-Dependent Permeability on Gas Production and Well Testing. *SPE Form Eval* **1** (3): 227–235. SPE-11220-PA. https://doi.org/10.2118/11220-PA.

Ramey, H. J. Jr. 1965. Non-Darcy Flow and Wellbore Storage Effects in Pressure Build-Up and Drawdown of Gas Wells. *J Pet Technol* **17** (2): 223–233. SPE-1058-PA. https://doi.org/10.2118/1058-PA.

Russell, D. G. and Prats, M. 1962. The Practical Aspects of Interlayer Crossflow. *J Pet Technol* **14** (6): 589–594. SPE-369-PA. https://doi.org/10.2118/369-PA.

Saleh, A. M. and Stewart, G. 1992. New Approach Towards Understanding of Near Well Bore Behaviour of Perforated Completions. Presented at the SPE European Petroleum Conference, Milan, Italy, 22–24 October. SPE-36866-MS. https://doi.org/10.2118/36866-MS.

Spivey, J. P. and Lee, W. J. 2000. Pressure Transient Response for a Naturally Fractured Reservoir With a Distribution of Block Sizes. Presented at the SPE Rocky Mountain Regional/Low Permeability Reservoirs Symposium, Denver, Colorado, 12–15 March. SPE-60294-MS. https://doi.org/10.2118/60294-MS.

Spivey, J. P. and Semmelbeck, M. E. 1995. Forecasting Long-Term Production of Dewatered Coal Seams and Fractured Shales. Presented at the SPE Rocky Mountain Regional/Low-Permeability Reservoirs Symposium, Denver, Colorado, 20–22 March. SPE-29580-MS. https://doi.org/10.2118/29580-MS.

Spivey, J. P., Brown, K. G., Sawyer, W. K. et al. 2002. Estimating Non-Darcy Flow Coefficient From Buildup-Test Data With Wellbore Storage. Presented at the SPE Annual Technical Conference and Exhibition, San Antonio, Texas, 29 September–2 October. SPE-77484-MS. https://doi.org/10.2118/77484-MS.

Stewart, G. and Ascharsobbi, F. 1988. Well Test Interpretation for Naturally Fractured Reservoirs. Presented at the SPE Annual Technical Conference and Exhibition, Houston, Texas, 2–5 October. SPE-18173-MS. https://doi.org/10.2118/18173-MS.

Teeuw, D. 1971. Prediction of Formation Compaction from Laboratory Data. *SPE J.* **11** (3): 263–271. SPE-2973-PA. https://doi.org/10.2118/2973-PA.

Thomas, R. D. and Ward, D. C. 1972. Effect of Overburden Pressure and Water on Gas Permeability of Tight Sandstone Cores. *J Pet Technol* **24** (2): 120–24. SPE-3634-PA. https://doi.org/10.2118/3634-PA.

Vairogs, J., Hearn, C. L., Dareing, D. W. et al. 1971. Effect of Rock Stress on Gas Production from Low-Permeability Rocks. *J Pet Technol* **23** (9): 1161–1167. SPE-3001-PA. https://doi.org/10.2118/3001-PA.

Vu, T., Chen, H. Y., and Poston, S. W. 1990. Using Short-Term Pressure Buildup Tests to Calculate Gas Reservoirs. Presented at the Permian Basin Oil and Gas Recovery Conference, Midland, Texas, 8–9 March. SPE-20111-MS. https://doi.org/10.2118/20111-MS.

Warren, J. E. and Root, P. J. 1963. The Behavior of Naturally Fractured Reservoirs. *SPE J.* **3** (3): 245–255. SPE-426-PA. https://doi.org/10.2118/426-PA.

West, S. L. and Cochrane, P. J. R. 1995. Reserves Determination Using Type-Curve Matching and EMB Methods in the Medicine Hat Shallow Gas Field. *SPE Res Eng* **10** (2): 82–87. SPE-28609-PA. https://doi.org/10.2118/28609-PA.

Appendix 6.A

6.A.1 Summary of Equations in Terms of Pressure, Adjusted Pressure, Pressure Squared, and Pseudopressure

The following tables present the more important equations for analyzing drawdown and buildup tests in gas wells using pressure and time, adjusted pressure and time (or adjusted time), pressure squared and time, and pseudopressure and time.

The first column in each table gives the equation number in the text for the slightly compressible liquid case, while the second column gives the corresponding equation in terms of the appropriate time and pressure plotting functions.

Equations for type curve analysis using the pressure/time, pressure-squared/time, and pseudopressure/time formulations are included for completeness. However, field data with large drawdown will not follow the characteristic type curve shapes, making type curve analysis using these formulations less useful than analysis using an adjusted-pressure/adjusted-time formulation.

Equation	Pressure Formulation
6.18	$\Delta p_s = \dfrac{141.2qB\mu}{kh}s$
6.21	$p_{wf}(t) = p_i - \dfrac{162.6qB\mu}{kh}\left[\log t + \log\left(\dfrac{k}{\phi\mu c_t r_w^2}\right) - 3.23 + 0.869s\right]$
6.24	$k = \dfrac{162.6qB\mu}{\lvert m\rvert h}$
6.25	$s = 1.151\left[\dfrac{p_i - p_{1hr}}{\lvert m\rvert} - \log\left(\dfrac{k}{\phi\mu c_t r_w^2}\right) + 3.23\right]$
6.29	$p_{ws}(\Delta t) = p_i - \dfrac{162.6qB\mu}{kh}\log\left(\dfrac{t_p + \Delta t}{\Delta t}\right)$
6.32	$k = \dfrac{162.6qB\mu}{\lvert m\rvert h}$
6.33	$s = 1.151\left[\dfrac{p_{1hr} - p_{wf}}{\lvert m\rvert} - \log\left(\dfrac{k}{\phi\mu c_t r_w^2}\right) + 3.23\right]$
6.60	$k = \dfrac{qB\mu}{0.00708h}\left(\dfrac{p_D}{\Delta p}\right)_{MP}$
6.61	$C_D = \dfrac{0.0002637k}{\phi\mu c_t r_w^2}\left(\dfrac{t}{t_D/C_D}\right)_{MP}$
6.62	$C = \dfrac{\phi c_t h r_w^2}{0.894}C_D$

Table 6.A-1—Analysis equations—pressure/time formulation.

Equation	Pressure Formulation
6.63	$s = 0.5\ln\left(\frac{C_D e^{2s}}{C_D}\right)$
6.66	$k = \frac{70.6qB\mu}{h(t\Delta p')_r}$
6.67	$s = \frac{1}{2}\left[\frac{\Delta p_r}{(t\Delta p')_r} - \ln\left(\frac{k\Delta t_r}{1688\phi\mu c_t r_w^2}\right)\right]$
6.68	$C = \frac{qB\Delta t_w}{24\Delta p_w}$

Table 6.A-1—(*Continued*).

Equation	Adjusted-Pressure/Adjusted-Time Formulation
6.18	$\Delta p_{as} = \frac{141.2q_g B_{gi}\mu_{gi}}{kh}s$
6.21	$p_{awf}(t) = p_{ai} - \frac{162.6q_g B_{gi}\mu_{gi}}{kh}\left[\log t + \log\left(\frac{k}{\phi\mu_{gi}c_{ti}r_w^2}\right) - 3.23 + 0.869s'\right]$
6.24	$k = \frac{162.6q_g B_{gi}\mu_{gi}}{\lvert m\rvert h}$
6.25	$s' = 1.151\left[\frac{p_{ai} - p_{a1hr}}{\lvert m\rvert} - \log\left(\frac{k}{\phi\mu_{gi}c_{ti}r_w^2}\right) + 3.23\right]$
6.29	$p_{aws}(\Delta t_a) = p_{ai} - \frac{162.6q_g B_{gi}\mu_{gi}}{kh}\log\left(\frac{t_p + \Delta t_a}{\Delta t_a}\right)$
6.32	$k = \frac{162.6q_g B_{gi}\mu_{gi}}{\lvert m\rvert h}$
6.33	$s' = 1.151\left[\frac{p_{a1hr} - p_{awf}}{\lvert m\rvert} - \log\left(\frac{k}{\phi\mu_{gi}c_{ti}r_w^2}\right) + 3.23\right]$
6.60	$k = \frac{q_g B_{gi}\mu_{gi}}{0.00708h}\left(\frac{p_D}{\Delta p_a}\right)_{MP}$
6.61	$C_{aD} = \frac{0.0002637k}{\phi\mu_{gi}c_{ti}r_w^2}\left(\frac{\Delta t_{ae}}{t_D/C_D}\right)_{MP}$
6.62	$C_a = \frac{\phi c_{ti}hr_w^2}{0.894}C_{aD}$
6.63	$s' = 0.5\ln\left(\frac{C_D e^{2s}}{C_{aD}}\right)$
6.66	$k = \frac{70.6q_g B_{gi}\mu_{gi}}{h(t\Delta p')_{ar}}$
6.67	$s' = \frac{1}{2}\left[\frac{\Delta p_{ar}}{(t\Delta p')_{ar}} - \ln\left(\frac{k\Delta t_{ar}}{1,688\phi\mu_{gi}c_{ti}r_w^2}\right)\right]$
6.68	$C_a = \frac{q_g B_{gi}\Delta t_{aw}}{24\Delta p_{aw}}$

Table 6.A-2—Analysis equations—adjusted-pressure/adjusted-time formulation.

Equation	Pressure-Squared/Time Formulation
6.18	$\Delta p^2{}_s = \dfrac{1,422 q_g T \bar{z} \bar{\mu}_g}{kh} s'$
6.21	$p_{wf}{}^2(t) = p_i{}^2 - \dfrac{1,637 q_g T \bar{z} \bar{\mu}_g}{kh}\left[\log t + \log\left(\dfrac{k}{\phi \bar{\mu}_g \bar{c}_t r_w{}^2}\right) - 3.23 + 0.869 s'\right]$
6.24	$k = \dfrac{1,637 q_q T \bar{z} \bar{\mu}_g}{\lvert m \rvert h}$
6.25	$s' = 1.151\left[\dfrac{p_i{}^2 - p^2{}_{1\text{hr}}}{\lvert m \rvert} - \log\left(\dfrac{k}{\phi \bar{\mu}_g \bar{c}_t r_w{}^2}\right) + 3.23\right]$
6.29	$p_{ws}{}^2(\Delta t) = p_i{}^2 - \dfrac{1,637 q_g T \bar{z} \bar{\mu}_g}{kh} \log\left(\dfrac{t_p + \Delta t}{\Delta t}\right)$
6.32	$k = \dfrac{1,637 q_g T \bar{z} \bar{\mu}_g}{\lvert m \rvert h}$
6.33	$s' = 1.151\left[\dfrac{p^2{}_{1\text{hr}} - p_{wf}{}^2}{\lvert m \rvert} - \log\left(\dfrac{k}{\phi \bar{\mu}_g \bar{c}_t r_w{}^2}\right) + 3.23\right]$
6.60	$k = 1,422 \dfrac{q_g T \bar{z} \bar{\mu}_g}{h}\left(\dfrac{p_D}{\Delta p^2}\right)_{\text{MP}}$
6.61	$C_D = \dfrac{0.0002637 k}{\phi \bar{\mu}_g \bar{c}_t r_w{}^2}\left(\dfrac{t}{t_D / C_D}\right)_{\text{MP}}$
6.62	$C = \dfrac{\phi \bar{c}_t h r_w{}^2}{0.894} C_D$
6.63	$s' = 0.5 \ln\left(\dfrac{C_D e^{2s}}{C_D}\right)$
6.66	$k = \dfrac{711 q_g T \bar{z} \bar{\mu}_g}{h\left(t \Delta p^2\right)_r}$
6.67	$s' = \dfrac{1}{2}\left[\dfrac{\Delta p^2{}_r}{\left(t \Delta p^2\right)_r} - \ln\left(\dfrac{k \Delta t_r}{1,688 \phi \bar{\mu}_g \bar{c}_t r_w{}^2}\right)\right]$
6.68	$C = \dfrac{0.419 q_g T \bar{z} \Delta t_w}{\Delta p^2{}_w}$

Table 6.A-3—Analysis equations—pressure-squared/time formulation.

Equation	Pseudopressure/Time Formulation
6.18	$\Delta m(p)_s = \frac{1,422 q_g T}{kh} s'$
6.21	$m(p_{wf}(t)) = m(p_i) - \frac{1,637 q_g T}{kh}\left[\log t + \log\left(\frac{k}{\phi \bar{\mu}_g \bar{c}_t r_w^2}\right) - 3.23 + 0.869 s'\right]$
6.24	$k = \frac{1,637 q_g T}{\lvert m \rvert h}$
6.25	$s' = 1.151\left[\frac{m(p_i) - m(p)_{1\text{hr}}}{\lvert m \rvert} - \log\left(\frac{k}{\phi \bar{\mu}_g \bar{c}_t r_w^2}\right) + 3.23\right]$
6.29	$m(p_{ws}(\Delta t)) = m(p_i) - \frac{1,637 q_g T}{kh}\log\left(\frac{t_p + \Delta t}{\Delta t}\right)$
6.32	$k = \frac{1,637 q_g T}{\lvert m \rvert h}$
6.33	$s' = 1.151\left[\frac{m(p)_{1\text{hr}} - m(p_{wf})}{\lvert m \rvert} - \log\left(\frac{k}{\phi \bar{\mu}_g \bar{c}_t r_w^2}\right) + 3.23\right]$
6.60	$k = \frac{1,422 q_g T}{h}\left(\frac{p_D}{\Delta m(p)}\right)_{\text{MP}}$
6.61	$C_D = \frac{0.0002637 k}{\phi \bar{\mu}_g \bar{c}_t r_w^2}\left(\frac{t}{t_D / C_D}\right)_{\text{MP}}$
6.62	$C = \frac{\phi \bar{c}_t h r_w^2}{0.894} C_D$
6.63	$s' = 0.5 \ln\left(\frac{C_D e^{2s}}{C_D}\right)$
6.66	$k = \frac{711 q_g T}{h[t \Delta m'(p)]_r}$
6.67	$s' = \frac{1}{2}\left\{\frac{\Delta m(p)_r}{[tm'(p)]_r} - \ln\left(\frac{k \Delta t_r}{1,688 \phi \bar{\mu}_g \bar{c}_t r_w^2}\right)\right\}$
6.68	$C = \frac{0.419 q_g T \Delta t_w}{\bar{\mu}_g \Delta m(p_w)}$

Table 6.A-4—Analysis equations—pseudopressure/time formulation.

Chapter 7

Post-Fracture Treatment Well Testing

John Spivey

Many factors affect the production response for a hydraulically fractured well in a low-permeability reservoir:

- Reservoir quality (net pay, in-situ permeability, permeability anisotropy, reservoir pressure)
- Fracture dimensions and geometry (length, width, height, conductivity, near-wellbore tortuosity)
- Fracture damage during the treatment (proppant flowback, overdisplacement of proppant, proppant settling, unbroken gel, reaction of fracture fluid with formation water or clays)
- Post-treatment issues (proppant embedment, proppant crushing)

Because of the number of factors affecting the production response and the inherent heterogeneity in low-permeability reservoirs, post-treatment fracture evaluation is a difficult and complex process. Integrated analysis of data from a wide variety of different sources is required for a unique interpretation. Fracture treating pressure analysis, microseismic analysis, pre- and post-fracture pressure transient analysis, and production data analysis all provide data vital to the evaluation process. While fracture treating pressure analysis and microseismic analysis provide information about the created dimensions of the fracture, only post-fracture pressure transient analysis and production data analysis provide performance-based data that reflect the effective propped dimensions of the fracture.

The relative success of fracture treatments from two wells in different parts of a field cannot be determined simply by comparing production rates of the wells. Even for offset wells, relative success can be evaluated only in a statistical sense using methods such as moving-domain analysis, which is discussed in Section 11.3 of Part 2 Ch 11: Development Economics.

As has been recognized for some time, the best wells in a field are often the best stimulation or restimulation candidates (Veatch et al. 2017). However, convincing company management to fracture or refracture the best well in a field might be difficult (Chu 1996). Post-fracture transient testing is a key component in the decision-making process.

This chapter provides an overview of post-fracture pressure transient testing. The closely related topic of post-fracture rate transient analysis, the analysis of production data to estimate reservoir properties, will be covered in Part 2 Chapter 8: Rate Transient Analysis.

Section 7.1 discusses three of the most common models for hydraulically fractured wells: the uniform-flux fracture, the infinite-conductivity fracture, and the finite-conductivity fracture. Of these models, the uniform-flux model is of primarily mathematical interest, while the infinite-conductivity fracture model could be considered as a finite-conductivity fracture model in the limit of high conductivity.

Section 7.2 outlines the flow regimes that can occur in a well with a finite-conductivity hydraulic fracture. The section then discusses the fracture linear flow, bilinear flow, formation linear flow, elliptical flow, and pseudoradial flow (PRF) periods.

Section 7.3 presents straight-line methods for analyzing data in formation linear flow, bilinear flow, PRF, and elliptical flow, illustrating each method with an example problem.

Section 7.4 discusses the use of log-log type curve analysis and then gives a worked example using a finite-conductivity-fracture type curve.

Section 7.5 shows how to estimate fracture half-length and fracture damage skin factor from manual log-log analysis of data in formation linear flow and how to estimate fracture conductivity and skin factor from data in bilinear flow.

Section 7.6 shows how the observed start and end of the bilinear and linear flow regimes can be used to estimate or to place bounds on the formation permeability, fracture half-length, and fracture conductivity.

Sections 7.7 through 7.10 briefly discuss several topics of practical interest in post-fracture well testing. Section 7.7 defines an area of investigation for a hydraulically fractured well. Section 7.8 discusses two models of fracture damage: choked fracture damage and fracture face damage. Section 7.9 discusses the effects of non-Darcy flow in the fracture on post-fracture tests. Finally, Section 7.10 illustrates the effect of permeability anisotropy on apparent fracture half-length.

7.1 Hydraulically Fractured Well Models

Fig. 7.1 shows the conceptual model of a finite-conductivity hydraulic fracture (Cinco-L. et al. 1978), which is used as the basis for most analytical solutions in use today. The fracture has half-length (L_f) and width (w_f). The permeability of the proppant pack is k_f. Often, the fracture width (w_f) and permeability (k_f) are combined into a single parameter, the fracture conductivity (wk_f).

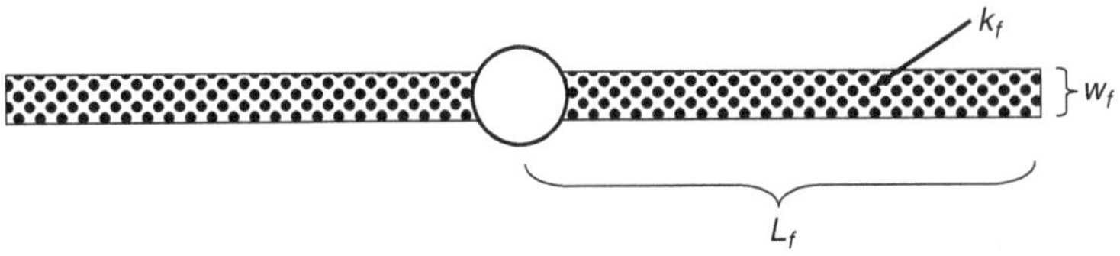

Fig. 7.1—Conceptual model of hydraulic fracture.

Two other models are also mentioned often enough in the literature that they merit some discussion: the uniform-flux fracture model and the infinite-conductivity fracture model.

In the uniform-flux fracture model, the same amount of fluid enters each unit length of the fracture wings per unit time, hence the name "uniform flux." The uniform-flux fracture model is particularly convenient, mathematically speaking; solutions for the uniform-flux fracture can be easily obtained using methods such as the Green's-function method and the Newman-product method.

In the infinite-conductivity fracture (Gringarten et al. 1974), there is no pressure drop along the fracture. The early-time solution behavior for the infinite-conductivity fracture is the same as that for the uniform-flux fracture; as the pressure transient moves away from the fracture into the formation, the flux distribution changes, with the flux highest at the tips of the fracture and lowest at the wellbore.

7.1.1 Dimensionless Parameters. A number of dimensionless parameters are useful in describing pressure responses for hydraulically fractured wells.

For fractured wells, it is convenient to define the dimensionless time in terms of the fracture length, rather than the wellbore radius:

$$t_{LfD} = \frac{0.0002637k}{\phi\mu c_t L_f^2} t. \quad (7.1)$$

The dimensionless pressure for hydraulically fractured wells is defined the same way as for vertical wells:

$$p_D = \frac{0.00708kh}{qB\mu} \Delta p. \quad (7.2)$$

The dimensionless fracture conductivity has been defined two different ways and has been given a number of different symbols. Some authors define the dimensionless fracture conductivity as

$$C_{fD} \equiv \frac{wk_f}{kL_f}, \quad (7.3)$$

while others add a factor π in the denominator:

$$C_r \equiv \frac{wk_f}{\pi kL_f} = \frac{C_{fD}}{\pi}. \quad (7.4)$$

Both Eq. 7.3 and Eq. 7.4 are called the "dimensionless fracture conductivity" in the literature. Further, historically, there has been no consistent symbol for either definition. Other symbols for the term defined by Eq. 7.3 include F_{cD}, $(k_f b_f)_D$, and $(k_{fD} b_{fD})$. The reader is cautioned to verify how the dimensionless fracture conductivity is defined in any context in which it is used, regardless of the symbol used.

The dimensionless fracture diffusivity (η_{fD}) is defined as

$$\eta_{fD} \equiv \frac{(k/\phi c_t)_f}{(k/\phi c_t)} \quad (7.5)$$

and occurs in describing the fracture linear flow regime.

7.2 Flow Regimes in Hydraulically Fractured Wells

A large number of different flow regimes can occur in wells with finite-conductivity hydraulic fractures (Cinco-Ley and Samaniego-V. 1981a; Hale and Evers 1981; Rahim and Lee 1989; Poe et al. 1999; Manrique and Poe 2007). These include

- Wellbore storage
- Fracture linear flow **(Fig. 7.2a)**
- Bilinear flow **(Fig. 7.2b)**
- Formation linear flow **(Fig. 7.2c)**
- Formation pseudolinear flow
- Elliptical flow **(Fig. 7.2d)**
- PRF **(Fig. 7.2e)**
- Pseudosteady-state flow

Of the five flow regimes shown in Fig. 7.2, fracture linear flow is rarely, if ever, observed in field data, and thus plays virtually no role in well test interpretation. The remaining flow regimes shown in Fig. 7.2 have specific analysis methods that can be used to estimate fracture or reservoir properties. Unfortunately, it is impossible to determine all the fracture and reservoir properties of interest using data from only a single flow regime. Only by combining data from multiple flow regimes can a complete fracture and reservoir characterization be performed.

Fig. 7.3 shows the pressure response for a well with a vertical hydraulic fracture and no wellbore storage, with the bilinear, formation linear, and PRF regimes identified, while **Fig. 7.4** shows the corresponding pressure derivative response.

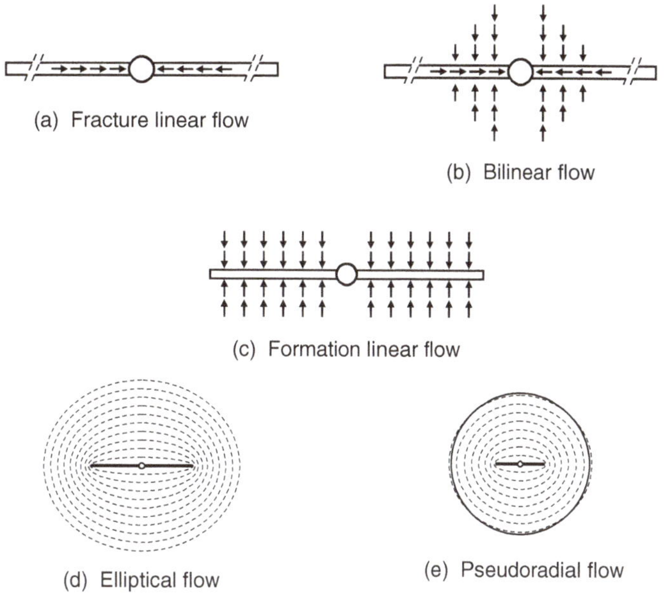

Fig. 7.2—Flow regimes in hydraulically fractured wells.

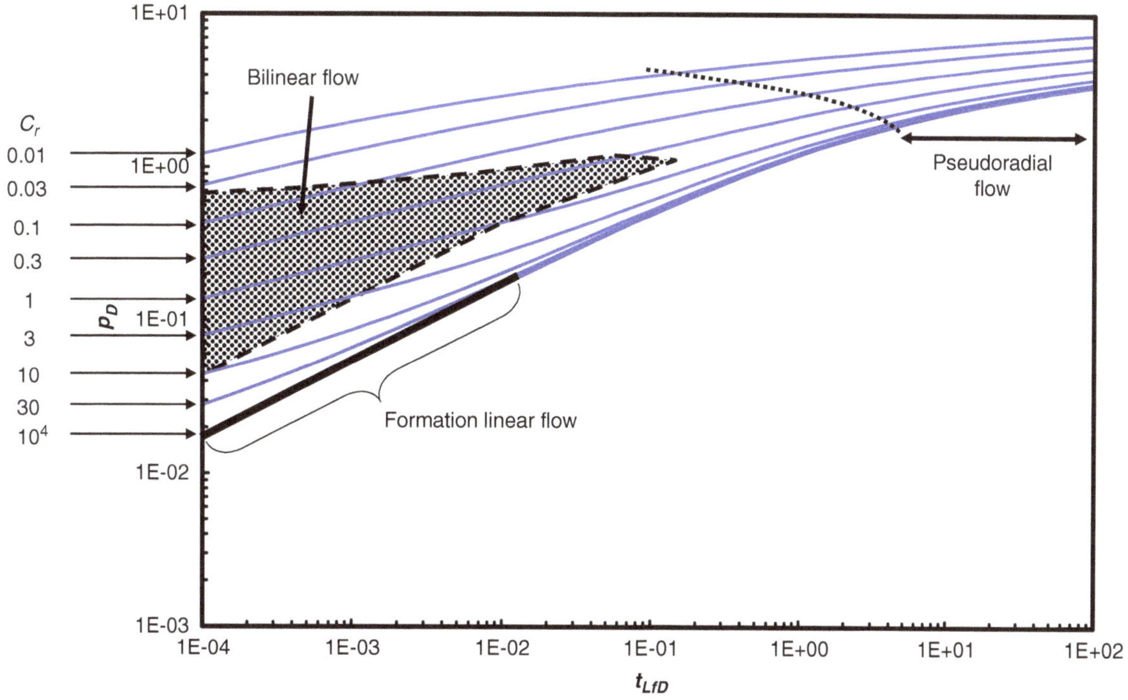

Fig. 7.3—Pressure response for a finite-conductivity vertical fracture in an infinite-acting reservoir.

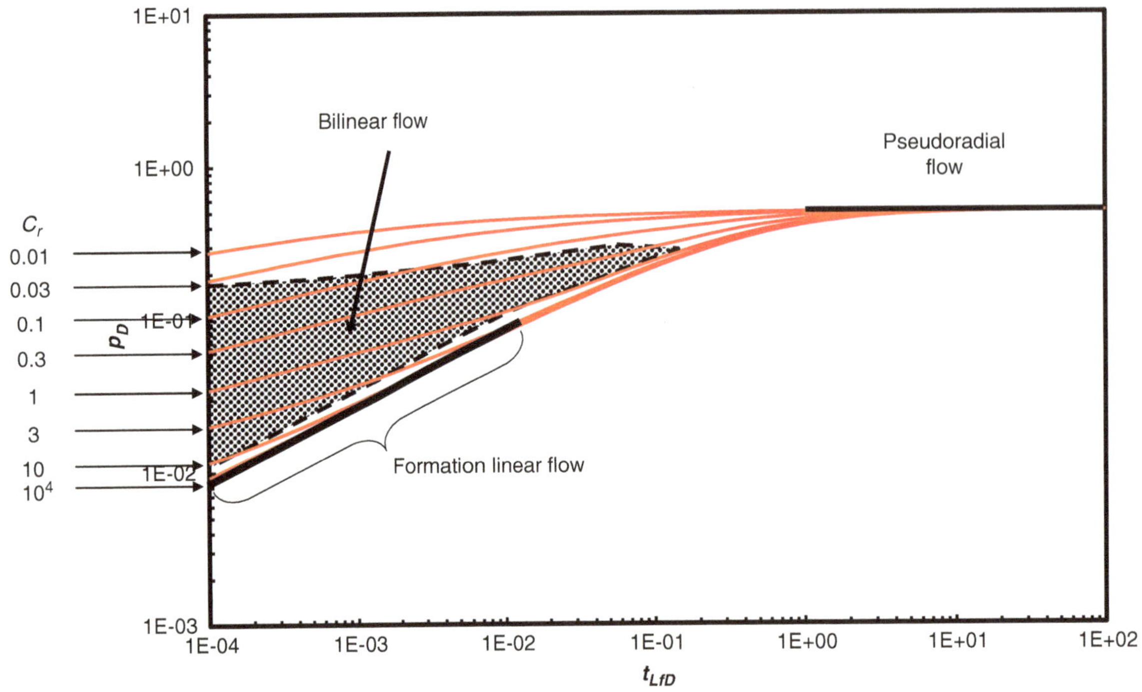

Fig. 7.4—Pressure derivative for a finite-conductivity vertical fracture in an infinite-acting reservoir.

7.2.1 Fracture Linear Flow. During fracture linear flow, shown in **Fig. 7.5,** the pressure transient moves down the length of the fracture because of fluid expansion within the fracture, with no fluid flowing into the fracture from the formation. Fracture linear flow ends at a dimensionless time given by Cinco-Ley and Samaniego-V. (1981a):

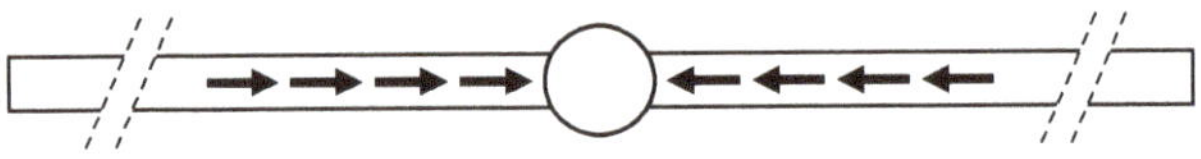

Fig. 7.5—Fracture linear flow.

$$t_{LfD} \approx \frac{0.01 C_{fD}^{2}}{\eta_{fD}^{2}}, \quad (7.6)$$

where C_{fD} is the dimensionless fracture conductivity, defined by Eq. 7.3, and η_{fD} is the dimensionless fracture diffusivity, defined by Eq. 7.5. The fracture linear flow regime ends so quickly it will virtually always be obscured by wellbore storage and therefore plays no significant role in practical well test interpretation.

7.2.2 Bilinear Flow. Bilinear flow occurs in a finite-conductivity fracture when the pressure transient simultaneously moves down the fracture and out into the formation, as shown in **Fig. 7.6.** Bilinear flow ends when the pressure transient reaches the tips of the fracture. Because the transient has not reached the tips of the fracture during bilinear flow, the fracture half-length, L_f, cannot be estimated from

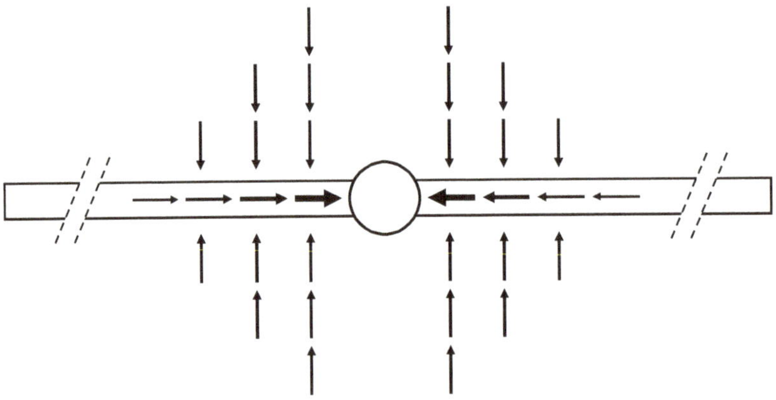

Fig. 7.6—Bilinear flow in finite-conductivity fracture.

the data in the bilinear flow regime. However, it might be possible to obtain a lower bound for L_f from the time at which bilinear flow ends. Bilinear flow is the first flow regime likely to be encountered following the end of wellbore storage.

According to Cinco-Ley and Samaniego-V. (1981a), bilinear flow ends at a time given by

$$t_{eblfD} \approx \left[\frac{4.55}{\sqrt{C_{fD}}} - 2.5\right]^{-4}, \; C_{fD} \leq 1.6 \quad (7.7a)$$

$$t_{eblfD} \approx \frac{0.0205}{\left(C_{fD} - 1.5\right)^{1.53}}, \; 1.6 \leq C_{fD} \leq 3 \quad (7.7b)$$

$$t_{eblfD} \approx \frac{0.1}{C_{fD}^{\;2}}, \; C_{fD} \geq 3. \quad (7.7c)$$

Bilinear flow exhibits a derivative with a slope of ¼ on the log-log diagnostic plot, as shown in Fig. 7.4. If the fracture skin factor s_f is zero, the pressure response will also exhibit a ¼ slope on the diagnostic plot, but will be shifted upward from the derivative curve by a factor of 4, as in Fig. 7.3. If s_f is not zero, the pressure change curve will approach the ¼ slope line asymptotically from above.

The pressure response during bilinear flow is given by

$$p_{wf} = p_i - \frac{44.1qB\mu}{h\sqrt{wk_f}}\left(\frac{t}{\mu\phi c_t k}\right)^{1/4} - \frac{141.2qB\mu}{kh}s_f \quad (7.8)$$

for a drawdown test, and by

$$p_{ws} = p_{wf} + \frac{44.1qB\mu}{h\sqrt{wk_f}}\left(\frac{\Delta t_{eB}}{\mu\phi c_t k}\right)^{1/4} + \frac{141.2qB\mu}{kh}s_f \quad (7.9)$$

for a buildup test. The term Δt_{eB} is the bilinear equivalent time, defined as

$$\Delta t_{eB} = \left[t_p^{1/4} + \Delta t^{1/4} - \left(t_p + \Delta t\right)^{1/4}\right]^4. \quad (7.10)$$

Inspection of Eqs. 7.8 and 7.9 suggests that, if we know the formation permeability, we can use data in bilinear flow to estimate the fracture conductivity wk_f from a graph of p_{wf} vs. t or p_{ws} vs. Δt_{eB}.

7.2.3 Formation Linear Flow. Formation linear flow occurs in a high-conductivity fracture when the pressure drop along the fracture can be neglected, as shown in **Fig. 7.7.** The pressure transient moves away from the fracture face, out into the reservoir, but has not moved beyond the tips of the fracture. Because the pressure drop is negligible during formation linear flow, it is impossible to get a unique estimate of fracture conductivity from data in formation linear flow. It might be possible to obtain a lower bound on the fracture conductivity from the time at which formation linear flow begins.

For a high-conductivity fracture ($C_{fD} \geq 100$), formation linear flow begins at a time given by Lee et al. (2003):

$$t_{L_f D,bflf} \approx \frac{0.8}{C_{fD}^{\;2}}. \quad (7.11)$$

According to Gringarten et al. (1975), formation linear flow ends at a time given by

$$t_{L_f D,eflf} \approx 0.016. \quad (7.12)$$

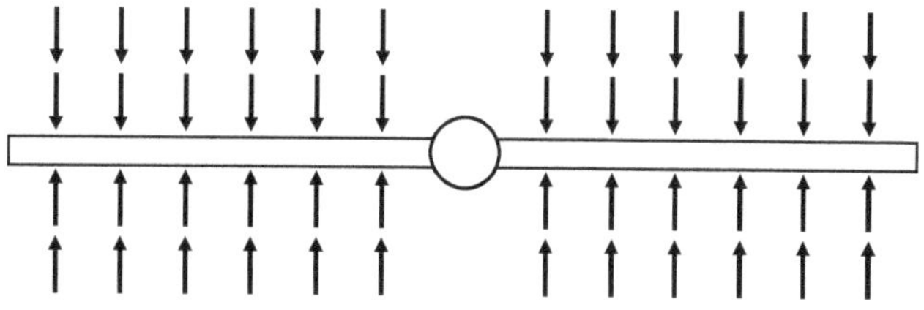

Fig. 7.7—Formation linear flow in high-conductivity fracture.

Formation linear flow exhibits a derivative with a slope of ½ on the log-log diagnostic plot, as shown in Fig. 7.4. If the fracture skin factor s_f is zero, the pressure response will also exhibit a ½ slope on the diagnostic plot, but shifted upward from the derivative curve by a factor of 2, as in Fig. 7.3. If s_f is not zero, the pressure change curve will approach the ½ slope line asymptotically from above.

The flowing wellbore pressure during formation linear flow is given by

$$p_{wf} = p_i - 4.064\frac{qB}{hL_f}\sqrt{\frac{\mu t}{k\phi c_t}} - \frac{141.2qB\mu}{kh}s_f \quad (7.13)$$

for a drawdown test. For a buildup test, the shut-in wellbore pressure is given by

$$p_{ws} = p_{wf} + 4.064\frac{qB}{hL_f}\sqrt{\frac{\mu \Delta t_{eL}}{k\phi c_t}} + \frac{141.2qB\mu}{kh}s_f, \quad (7.14)$$

where Δt_{eL} is the linear equivalent time, defined as

$$\Delta t_{eL} = \left(\sqrt{t_p} + \sqrt{\Delta t} - \sqrt{t_p + \Delta t}\right)^2. \quad (7.15)$$

Inspection of Eqs. 7.13 and 7.14 suggests that we can estimate the fracture half-length, L_f, from data in formation linear flow from a graph of p_{wf} vs. t or p_{ws} vs. Δt_{eL}, provided that we know the formation permeability, k.

7.2.4 Elliptical Flow. Elliptical flow begins when the pressure transient has moved beyond the tips of the fracture and the flow pattern departs from a linear geometry. For a high-conductivity fracture, the contours of the pressure transient are elliptical, as shown in **Fig. 7.8.**

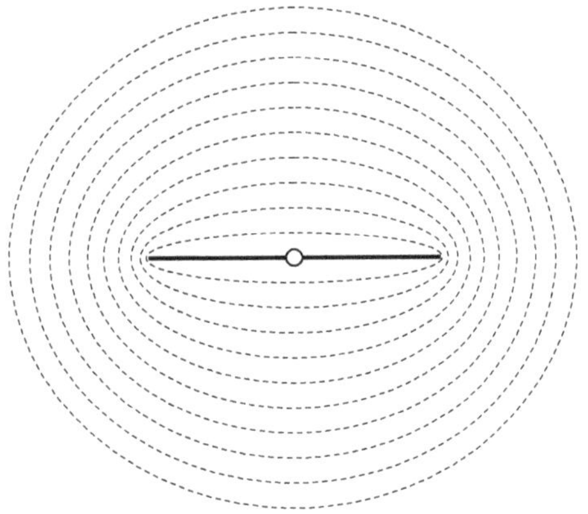

Fig. 7.8—Elliptical flow.

Although the elliptical flow regime has been regarded as a transitional period by most authors, Hale and Evers (1981) presented an approximate solution for constant-rate production from a well with an infinite-conductivity vertical fracture in an infinite reservoir:

$$P_D = \ln\left(\frac{A_t + B_t}{L_f}\right), \quad (7.16)$$

where B_t is defined as

$$B_t = 0.02878\left(\frac{kt}{\phi\mu c_t}\right)^{1/2}, \quad (7.17)$$

and A_t is defined as

$$A_t = \sqrt{B_t^2 + L_f^2}. \quad (7.18)$$

Cheng et al. (2007) later modified the Hale and Evers solution to improve the accuracy during elliptical flow and PRF by changing the coefficient for the term B_t:

$$B_t = 0.02634\left(\frac{kt}{\phi\mu c_t}\right)^{1/2}. \quad (7.19)$$

The author of this chapter prefers to define B_t as

$$B_t = 0.02434\left(\frac{kt}{\phi\mu c_t}\right)^{1/2}, \quad (7.20)$$

so the pressure response defined by Eq. 7.16 will be continuous with that for PRF. This allows the elliptical flow analysis procedure to include data from both elliptical flow and PRF regimes.

Unlike the other flow regimes we will consider (bilinear flow, formation linear flow, and PRF), an iterative procedure must be used to analyze data in elliptical flow. If sufficient data are available, both formation permeability and fracture half-length can be estimated from data in elliptical flow for an infinite-conductivity fracture. For a finite-conductivity fracture, the permeability and the equivalent infinite-conductivity fracture half-length can be estimated. The equivalent infinite-conductivity fracture half-length is the length of an infinite-conductivity fracture that will give the same pressure drop as the given finite-conductivity fracture during PRF and can be calculated from Meyer and Jacot (2005):

$$L_{f\text{eq}} = \frac{2L_f}{2+\frac{1}{C_r}}. \qquad (7.21)$$

The author defines the beginning of elliptical flow as the time at which Eq. 7.16 can be used to estimate formation permeability with a relative error, δ. With this definition, the time to the start of elliptical flow can be calculated from

$$t_{L_f D,bef} = \frac{0.0703+\frac{0.1257}{C_r^{\,0.7632}}}{\delta^{0.8811}}. \qquad (7.22)$$

7.2.5 Pseudoradial Flow (PRF). As the pressure transient continues to move away from the fracture, the pressure contours become less elongated. Eventually, the contours can be treated as circular, as seen in **Fig. 7.9.** This is the PRF regime. Because the elliptical flow pressure response given by Eq. 7.16 is also valid during the PRF regime, PRF can be regarded as a limiting case of elliptical flow.

The pressure response during PRF is given by

$$p_{wf}(t) = p_i - \frac{70.6qB\mu}{kh}\left[\ln\left(\frac{kt}{1{,}688\phi\mu c_t r_w^{\,2}}\right)+s+s_f\right], \qquad (7.23)$$

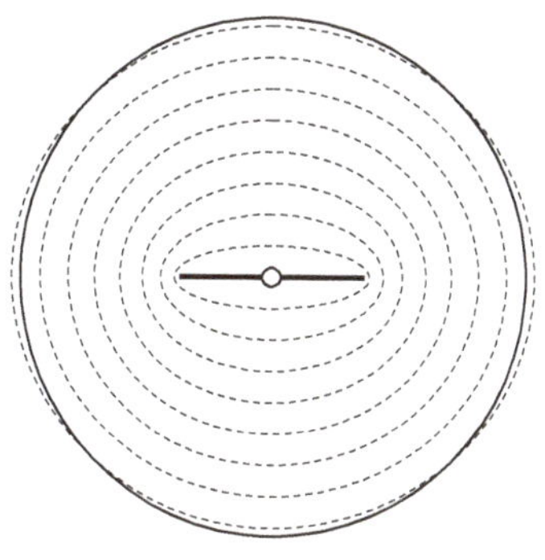

Fig. 7.9—PRF.

where the skin factor, s, describes the effect of stimulation from the fracture treatment, while s_f accounts for the additional skin factor from damage to the fracture. The skin factor, s, can be estimated from

$$s = -\ln\left(\left(\frac{r_{wa}}{L_f}\right)\frac{L_f}{r_w}\right), \qquad (7.24)$$

where the term r_{wa}/L_f can be estimated graphically from **Fig. 7.10** or calculated from Meyer and Jacot (2005):

$$\frac{r_{wa}}{L_f} = \frac{1}{2+\frac{1}{C_r}}. \qquad (7.25)$$

PRF begins at a time dependent on the fracture conductivity; the time at which PRF begins can be estimated from

$$t_{L_f D,bprf} \approx \frac{0.2}{\delta^{1.055}\left(1+\frac{0.256}{C_r}\right)}, \qquad (7.26)$$

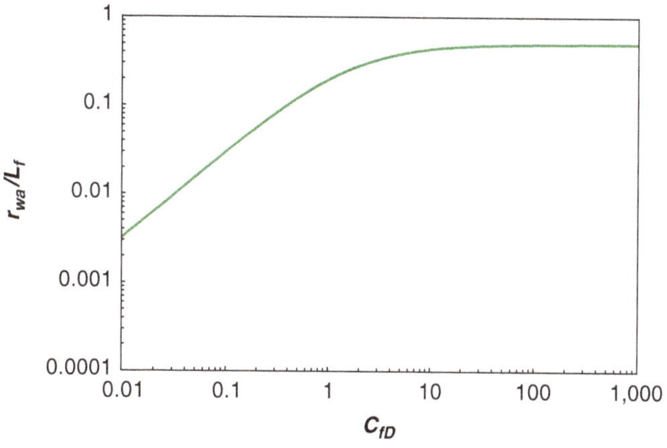

Fig. 7.10—Apparent wellbore radius for a finite-conductivity fracture.

valid for $C_r \geq 0.1$. Eq. 7.26 gives the dimensionless time at which the logarithmic derivative of the pressure response is within a fraction, δ, of the value corresponding to infinite-acting radial flow (IARF). For example, for a dimensionless conductivity C_r of 100 and a maximum error δ of 5%, we have $t_{L_f D,bprf} \approx \frac{0.2}{\delta^{1.055}\left(1+\frac{0.256}{C_r}\right)} = \frac{0.2}{(0.05)^{1.055}\left(1+\frac{0.256}{100}\right)} = 4.7.$

PRF ends when the pressure transient reaches the nearest reservoir boundary, causing the pressure response to deviate from that of an infinite-acting reservoir.

7.3 Straight-Line Analysis Methods

Straight-line methods are used to analyze data in specific flow regimes. None of the flow regimes can provide all the desired information about the fracture. If formation permeability is known, bilinear flow provides an estimate of fracture conductivity, while formation linear flow allows fracture half-length to be estimated. Formation permeability and total skin factor can be estimated from data in PRF, and formation permeability and effective fracture half-length can be determined from data in elliptical flow.

7.3.1 Formation Linear Flow. Analysis of data in formation linear flow is based on Eq. 7.13 for drawdown tests and Eq. 7.14 for buildup tests. If the formation permeability is known, the fracture half-length, L_f, can be estimated, as can a lower bound on fracture conductivity. If formation permeability is not known, data in formation linear flow can be used to estimate the product, $\sqrt{k}L_f$. A rough upper bound on permeability and lower bound on fracture half-length can also be estimated.

The procedure for analyzing data in formation linear flow is as follows:

1. Identify data in linear flow from the log-log diagnostic plot.
2. Construct a square-root-of-time graph. For a drawdown test, graph the flowing bottomhole pressure, p_{wf}, vs. the square root of the test time, t. For a buildup test, graph the shut-in bottomhole pressure, p_{ws}, vs. the square root of the linear equivalent time, Δt_{eL}, defined as

$$\Delta t_{eL} \equiv \left(\sqrt{t_p} + \sqrt{\Delta t} - \sqrt{t_p + \Delta t}\right)^2. \quad (7.27)$$

3. Draw a straight line through the data in formation linear flow. Find the slope, m_L, and the intercept, b_L, of the straight line.
4. If formation permeability is known, calculate the fracture half-length from the slope, m_L:

$$L_f = 4.064 \frac{qB}{|m_L| h} \sqrt{\frac{\mu}{k \phi c_t}}. \quad (7.28)$$

If formation permeability is not known, calculate the product $\sqrt{k}L_f$:

$$\sqrt{k} L_f = 4.064 \frac{qB}{|m_L| h} \sqrt{\frac{\mu}{\phi c_t}}. \quad (7.29)$$

5. If the formation permeability is known, calculate the fracture skin factor, s_f, from the intercept, b_L. For a drawdown test, calculate s_f from

$$s_f = \frac{0.00708 kh}{qB\mu}\left(p_i - b_L\right). \quad (7.30)$$

For a buildup test, calculate s_f from

$$s_f = \frac{0.00708 kh}{qB\mu}\left(b_L - p_{wf}\right). \quad (7.31)$$

6. If formation permeability is known, calculate the time to the end of formation linear flow from the dimensionless time to the end of formation flow, Eq. 7.12, as a consistency check:

$$t_{eflf} = \frac{\phi \mu c_t L_f^2}{0.0002637 k} t_{LfD,eflf} = 60.7 \frac{\phi \mu c_t L_f^2}{k}. \quad (7.32)$$

7. If formation permeability is not known, calculate a rough estimate of the lower bound on fracture half-length/upper bound on permeability from the observed time to the end of linear flow (or from the duration of the test if the well is still in linear flow at the end of the test) and the product $\sqrt{k}L_f$ from Step 4:

$$k_{\max} = \sqrt{60.7 \frac{\phi \mu c_t}{t_{eflf}}}\left(\sqrt{k} L_f\right). \quad (7.33)$$

$$L_{f,\min} = \frac{\left(\sqrt{k} L_f\right)}{\sqrt{k_{\max}}}. \quad (7.34)$$

8. Calculate a rough estimate of the lower bound on the fracture conductivity from the time to the beginning of formation linear flow, Eq. 7.11:

$$t_{L_f D, bflf} = \frac{0.0002637 k}{\phi \mu c_t L_f^2} t_{bflf}. \quad (7.35)$$

$$C_{fD,\min} = \sqrt{\frac{0.8}{t_{L_f D, bflf}}}. \quad (7.36)$$

$$\left(wk_f\right)_{\min} = kL_f C_{fD,\min}. \quad (7.37)$$

7.3.2 Formation Linear Flow Analysis Example.

7.3.2.1 Example Problem 7.1. Given the rock and fluid properties provided in **Table 7.1,** analyze the gas-well pressure-buildup-test data, given in **Table 7.2,** in formation linear flow to estimate the product $\sqrt{k}L_f$. Calculate a rough estimate of the upper bound on permeability and the lower bound on fracture half-length.

Rock and Fluid Properties

q_g	2,500	Mscf/D	t_p	4,320	hours
ϕ	0.085		h	75	ft
S_w	45	%	c_f	4×10^{-6}	psi^{-1}
c_w	3.6×10^{-6}	psi^{-1}			
T_f	225	°F	r_w	0.25	ft
γ_g	0.67	(air = 1)	B_{gi}	0.586	bbl/Mscf
μ_{gi}	0.0309	cp	c_{ti}	5.13×10^{-5}	psi^{-1}
p_i	6,700	psia	p_{awf}	2,306.6	psi

Table 7.1—Rock- and fluid-property data for linear flow analysis example.

Δt (hours)	p_{ws} (psia)	Δt_a (hours)	p_{aws} (psi)	Δt (hours)	p_{ws} (psia)	Δt_a (hours)	p_{aws} (psi)
0.0010	3,898.80	0.0006	2,310.7	0.4857	4,115.96	0.3059	2,517.6
0.0021	3,902.84	0.0013	2,314.6	0.5474	4,118.36	0.3451	2,519.9
0.0034	3,907.36	0.0020	2,318.8	0.6168	4,121.08	0.3892	2,522.5
0.0048	3,913.18	0.0029	2,324.3	0.6949	4,123.89	0.4388	2,525.2
0.0064	3,918.24	0.0039	2,329.1	0.7827	4,127.07	0.4947	2,528.2
0.0082	3,924.39	0.0050	2,335.0	0.8816	4,129.97	0.5576	2,531.0
0.0102	3,931.02	0.0062	2,341.2	0.9928	4,133.56	0.6285	2,534.5
0.0125	3,938.59	0.0076	2,348.4	1.1180	4,137.83	0.7082	2,538.5
0.0151	3,946.04	0.0092	2,355.5	1.2590	4,141.39	0.7980	2,542.0
0.0180	3,953.57	0.0109	2,362.6	1.417	4,145.08	0.8991	2,545.5
0.0212	3,962.54	0.0129	2,371.1	1.595	4,149.85	1.013	2,550.1
0.0249	3,971.32	0.0152	2,379.5	1.795	4,153.43	1.141	2,553.5
0.0290	3,981.18	0.0177	2,388.9	2.021	4,158.81	1.286	2,558.7
0.0336	3,990.02	0.0205	2,397.3	2.275	4,162.59	1.448	2,562.3
0.0388	4,000.40	0.0237	2,407.1	2.560	4,167.15	1.631	2,566.7
0.0447	4,010.26	0.0274	2,416.5	2.881	4,173.51	1.838	2,572.8
0.0512	4,020.4	0.0314	2,426.2	3.242	4,178.67	2.070	2,577.8
0.0587	4,030.14	0.0360	2,435.5	3.648	4,185.07	2.332	2,584.0
0.0670	4,040.24	0.0412	2,445.1	4.105	4,192.96	2.626	2,591.6
0.0764	4,049.25	0.0471	2,453.7	4.619	4,199.77	2.959	2,598.1
0.0869	4,056.67	0.0537	2,460.8	5.198	4,205.61	3.333	2,603.7
0.0988	4,065.20	0.0611	2,469.0	5.848	4,214.36	3.755	2,612.2
0.1121	4,071.48	0.0695	2,475.0	6.580	4,220.72	4.230	2,618.3
0.1271	4,078.58	0.0790	2,481.8	7.404	4,230.92	4.766	2,628.1
0.1440	4,084.14	0.0896	2,487.1	8.331	4,239.05	5.370	2,636.0
0.1630	4,088.67	0.1016	2,491.4	9.373	4,250.03	6.051	2,646.6
0.1844	4,092.52	0.1151	2,495.1	10.55	4,260.12	6.818	2,656.3
0.2085	4,097.93	0.1303	2,500.3	11.86	4,271.02	7.684	2,666.9
0.2355	4,100.13	0.1474	2,502.4	13.35	4,282.41	8.660	2,677.9
0.2659	4,102.35	0.1666	2,504.5	15.02	4,295.26	9.761	2,690.3
0.3002	4,105.29	0.1883	2,507.3	16.90	4,307.86	11.00	2,702.5
0.3387	4,107.99	0.2127	2,509.9	19.01	4,321.41	12.41	2,715.6
0.3821	4,110.88	0.2402	2,512.7	21.39	4,335.88	13.99	2,729.6
0.4308	4,113.71	0.2711	2,515.4	24.06	4,350.37	15.77	2,743.7

(Continued)

Δt (hours)	p_{ws} (psia)	Δt_a (hours)	p_{aws} (psi)	Δt (hours)	p_{ws} (psia)	Δt_a (hours)	p_{aws} (psi)
27.07	4,367.59	17.79	2,760.4	78.15	4,555.07	52.82	2,943.1
30.45	4,384.57	20.06	2,776.9	87.92	4,580.43	59.66	2,967.9
34.26	4,402.49	22.63	2,794.3	98.91	4,608.13	67.39	2,995.1
38.55	4,421.38	25.53	2,812.7	111.28	4,635.94	76.14	3,022.4
43.37	4,441.80	28.81	2,832.6	125.19	4,665.93	86.04	3,051.8
48.79	4,462.30	32.51	2,852.5	140.84	4,695.11	97.24	3,080.5
54.89	4,484.52	36.70	2,874.2	158.45	4,727.91	109.93	3,112.7
61.75	4,506.03	41.43	2,895.2	168	4,743.08	116.84	3,127.7
69.47	4,530.67	46.77	2,919.3				

Table 7.2—Buildup data for linear flow analysis example.

7.3.2.2 Solution. Because this is a gas-well buildup test, we use adjusted pressure and adjusted time for the analysis. Following the recommended procedure, we proceed as follows.

1. Identify data in linear flow from the log-log diagnostic plot. The diagnostic plot is shown in **Fig. 7.11.** From the derivative, we identify 0.35 and 16 hours as the start and end of formation linear flow, respectively.

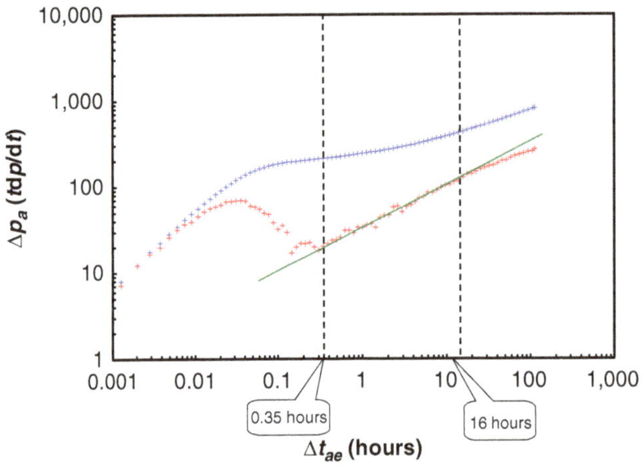

Fig. 7.11—Diagnostic plot, formation linear flow analysis example.

2. Construct the square-root-of-time graph. **Fig. 7.12** shows the graph of adjusted shut-in bottomhole pressure, p_{aws}, vs. the square root of the linear equivalent adjusted time, Δt_{aeL}. The indicated formation linear flow period is transferred from the diagnostic plot.

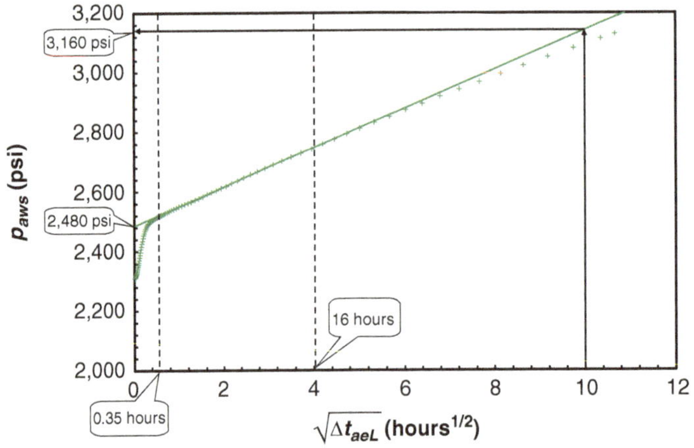

Fig. 7.12—Square-root-of-time plot, formation linear flow analysis example.

3. Draw a straight line through the data in formation linear flow. Read the intercept, b_L, as 2,480 psi. Calculate the slope, m_L, as

$$m_L = \frac{3{,}160 - 2{,}480}{10 - 0} = 68\,\text{psi/hr}^{1/2}.$$

   Note the importance of identifying the linear flow period from the log-log diagnostic plot rather than relying on the square-root-of-time plot.
4. Calculate L_f (if k is known) or $\sqrt{k}L_f$ (if k is unknown). We do not know k, so we calculate $\sqrt{k}L_f$ as

$$\begin{aligned}\sqrt{k}L_f &= 4.064\frac{q_g B_{gi}}{|m_L|h}\sqrt{\frac{\mu_{gi}}{\phi c_{ti}}}\\ &= 4.064\frac{(2{,}500)(0.586)}{(68)(75)}\sqrt{\frac{(0.0309)}{(0.085)(5.13\times10^{-5})}}\\ &= 98.3\ \text{md}^{1/2}\text{–ft.}\end{aligned}$$

5. If the formation permeability is known, calculate the fracture skin factor, s_f, from the intercept, b_L. We do not know the permeability, so we neglect this step.
6. If formation permeability is known, calculate the time to the end of formation linear flow as a consistency check. Again, we do not know k, so we also neglect this step.
7. If formation permeability is not known, calculate a rough estimate of the lower bound on fracture half-length/upper bound on permeability from the observed time to the end of linear flow and the product $\sqrt{k}L_f$ from Step 4:

$$\begin{aligned}k_{\max} &= \sqrt{60.7\frac{\phi\mu_{gi}c_{ti}}{t_{elf}}}\left(\sqrt{k}L_f\right)\\ &= \sqrt{(60.7)\frac{(0.085)(0.0309)(5.13\times10^{-5})}{(16)}}(98.3)\\ &= 0.07\ \text{md}.\end{aligned}$$

   We then calculate the lower bound on the fracture length as

$$\begin{aligned}L_{f,\min} &= \frac{\left(\sqrt{k}L_f\right)}{\sqrt{k_{\max}}}\\ &= \frac{98.3}{\sqrt{0.07}}\\ &= 370\ \text{ft}.\end{aligned}$$

   For this synthetic example, the actual permeability was 0.05 md, and the actual fracture half-length was 450 ft.
8. Calculate a rough estimate of the lower bound on the fracture conductivity from the time to the beginning of formation linear flow. We neglect this final step because we do not know the formation permeability.

7.3.3 Bilinear Flow. If reservoir permeability is known, data in bilinear flow can be used to estimate fracture conductivity, wk_f. If permeability is not known, the product $wk_f\sqrt{k}$ can be estimated. The procedure for analyzing data in bilinear flow is as follows:

1. Identify data in bilinear flow from the log-log diagnostic plot.
2. Construct a fourth-root-of-time graph. For a drawdown test, graph the flowing bottomhole pressure, p_{wf}, vs. the fourth root of the test time, t. For a buildup test, graph the shut-in bottomhole pressure, p_{ws}, vs. the fourth root of the bilinear equivalent time, Δt_{eB}, defined as

$$\Delta t_{eB} \equiv \left(\sqrt[4]{t_p} + \sqrt[4]{\Delta t} - \sqrt[4]{t_p + \Delta t}\right)^4. \quad (7.38)$$

3. Draw a straight line through the data in bilinear flow. Find the slope, m_B, and the intercept, b_B.
4. If the formation permeability is known, calculate the fracture conductivity, wk_f, from the slope, m_B:

$$wk_f = \left(\frac{44.1qB\mu}{h|m_B|}\right)^2\sqrt{\frac{1}{\phi\mu c_t k}}. \quad (7.39)$$

   If the formation permeability is not known, calculate the product $wk_f\sqrt{k}$:

$$wk_f\sqrt{k} = \left(\frac{44.1qB\mu}{h|m_B|}\right)^2\sqrt{\frac{1}{\phi\mu c_t}}. \quad (7.40)$$

5. If the formation permeability is known, calculate the fracture skin factor, s_f, from the intercept, b_B. For a drawdown test, calculate s_f from

$$s_f = \frac{0.00708kh}{qB\mu}\left(p_i - b_B\right). \quad (7.41)$$

For a buildup test, calculate s_f from

$$s_f = \frac{0.00708kh}{qB\mu}\left(b_B - p_{wf}\right). \quad (7.42)$$

6. If the formation permeability is known, calculate a rough estimate of the fracture half-length from the time at which bilinear flow ends, t_{eblf}. If bilinear flow lasts through the end of the test, using the test duration in place of t_{eblf} will give a lower bound on the fracture half-length.

$$L_f \approx 0.2266\left(\frac{kt_{eblf}}{\phi\mu c_t}\right)^{1/4}\left(\frac{wk_f}{k}\right)^{1/2}. \quad (7.43)$$

7.3.4 Bilinear Flow Analysis Example.

7.3.4.1 Example Problem 7.2. Given the rock and fluid properties in **Table 7.3,** analyze the gas-well pressure-buildup-test data in **Table 7.4** to estimate the fracture conductivity using bilinear flow analysis. Calculate a rough estimate of the lower bound on fracture half-length.

Rock and Fluid Properties

q_g	3,500	Mscf/D	t_p	4,320	hours
ϕ	0.075		h	135	ft
S_w	40	%	c_f	4×10^{-6}	psi^{-1}
c_w	3.6×10^{-6}	psi^{-1}	k	0.03	md
T_f	220	°F	r_w	0.25	ft
γ_g	0.725	(air=1)	B_{gi}	0.545	bbl/Mscf
μ_{gi}	0.0364	cp	c_{ti}	4.08×10^{-5}	psi^{-1}
p_i	7,700	psia	p_{awf}	3,707.3	psi

Table 7.3—Rock- and fluid-property data for bilinear flow analysis example.

Δt (hours)	p_{ws} (psia)	Δt_a (hours)	p_{aws} (psi)	Δt (hours)	p_{ws} (psia)	Δt_a (hours)	p_{aws} (psi)
0.0010	5,038.2	0.0007	3,713.4	0.0764	5,252.3	0.0520	3,934.4
0.0021	5,044.2	0.0014	3,719.6	0.0869	5,268.2	0.0593	3,950.8
0.0034	5,048.4	0.0023	3,723.9	0.0988	5,281.3	0.0676	3,964.3
0.0048	5,056.9	0.0032	3,732.7	0.1121	5,296.5	0.0769	3,980.0
0.0064	5,064.4	0.0043	3,740.4	0.1271	5,308.8	0.0874	3,992.7
0.0082	5,072.1	0.0055	3,748.3	0.1440	5,325.3	0.0992	4,009.7
0.0102	5,079.5	0.0068	3,756.0	0.1630	5,336.2	0.1126	4,021.0
0.0125	5,090.2	0.0084	3,767.1	0.1844	5,347.5	0.1277	4,032.6
0.0151	5,099.5	0.0101	3,776.6	0.2085	5,356.7	0.1446	4,042.1
0.0180	5,111.1	0.0120	3,788.6	0.2355	5,367.0	0.1638	4,052.8
0.0212	5,121.3	0.0142	3,799.2	0.2659	5,378.1	0.1853	4,064.2
0.0249	5,136.0	0.0167	3,814.4	0.3002	5,388.7	0.2096	4,075.1
0.0290	5,147.5	0.0195	3,826.1	0.3387	5,398.2	0.2371	4,085.0
0.0336	5,161.3	0.0226	3,840.4	0.3821	5,405.1	0.2679	4,092.1
0.0388	5,177.5	0.0262	3,857.1	0.4308	5,415.3	0.3027	4,102.6
0.0447	5,191.6	0.0302	3,871.8	0.4857	5,422.8	0.3419	4,110.3
0.0512	5,205.1	0.0347	3,885.6	0.5474	5,430.5	0.3861	4,118.3
0.0587	5,221.2	0.0398	3,902.3	0.6168	5,439.6	0.4358	4,127.7
0.0670	5,237.0	0.0455	3,918.5	0.6949	5,448.2	0.4919	4,136.6

(Continued)

Δt (hours)	p_{ws} (psia)	Δt_a (hours)	p_{aws} (psi)	Δt (hours)	p_{ws} (psia)	Δt_a (hours)	p_{aws} (psi)
0.7827	5,458.7	0.5551	4,147.3	13.35	5,728.4	9.894	4,425.4
0.8816	5,465.0	0.6263	4,153.9	15.02	5,742.0	11.16	4,439.3
0.9928	5,475.3	0.7065	4,164.5	16.90	5,759.9	12.58	4,457.8
1.118	5,483.3	0.7969	4,172.7	19.01	5,773.7	14.19	4,472.0
1.259	5,491.2	0.8988	4,180.9	21.39	5,791.9	16.00	4,490.7
1.417	5,497.9	1.014	4,187.8	24.06	5,808.1	18.05	4,507.4
1.595	5,510.4	1.143	4,200.7	27.07	5,827.2	20.35	4,527.1
1.795	5,519.5	1.289	4,210.1	30.45	5,843.8	22.96	4,544.1
2.021	5,529.5	1.453	4,220.4	34.26	5,864.4	25.90	4,565.4
2.275	5,539.0	1.638	4,230.2	38.55	5,883.3	29.21	4,584.7
2.560	5,551.8	1.847	4,243.4	43.37	5,902.4	32.96	4,604.4
2.881	5,562.0	2.082	4,253.9	48.79	5,921.9	37.18	4,624.5
3.242	5,572.1	2.347	4,264.3	54.89	5,945.8	41.95	4,649.0
3.648	5,582.4	2.646	4,274.9	61.75	5,966.0	47.34	4,669.8
4.105	5,593.7	2.983	4,286.6	69.47	5,990.2	53.42	4,694.7
4.619	5,606.4	3.362	4,299.6	78.15	6,011.2	60.29	4,716.3
5.198	5,620.3	3.790	4,314.0	87.92	6,035.1	68.04	4,740.7
5.848	5,629.3	4.273	4,323.2	98.91	6,058.8	76.80	4,765.1
6.580	5,642.1	4.817	4,336.4	111.28	6,082.1	86.69	4,789.0
7.404	5,655.8	5.431	4,350.6	125.19	6,110.2	97.86	4,817.8
8.331	5,670.7	6.122	4,366.0	140.84	6,137.0	110.48	4,845.4
9.373	5,683.9	6.903	4,379.5	158.45	6,160.8	124.74	4,869.8
10.55	5,698.6	7.783	4,394.7	168.00	6,176.2	132.50	4,885.6
11.86	5,711.5	8.775	4,408.0				

Table 7.4—Buildup data for bilinear flow analysis example.

7.3.4.2 ***Solution.*** Because this is a gas-well buildup test, we use adjusted pressure and adjusted time for the analysis. Following the recommended procedure, we proceed:

1. Identify data in bilinear flow from the log-log diagnostic plot. The diagnostic plot is shown in **Fig. 7.13.** From the derivative, we identify 5 hours as the start of bilinear flow. We will assume bilinear flow ends at the end of the test at 130 hours.

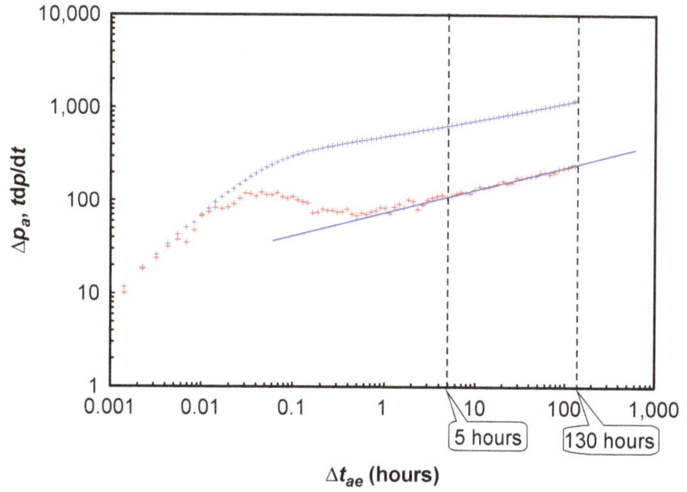

Fig. 7.13—Diagnostic plot, bilinear flow analysis example.

2. Construct a fourth-root-of-time graph. **Fig. 7.14** shows the fourth-root-of-time plot, constructed using the bilinear equivalent adjusted time. The start and end of bilinear flow have been transferred from the log-log diagnostic plot, as shown in Fig. 7.13.

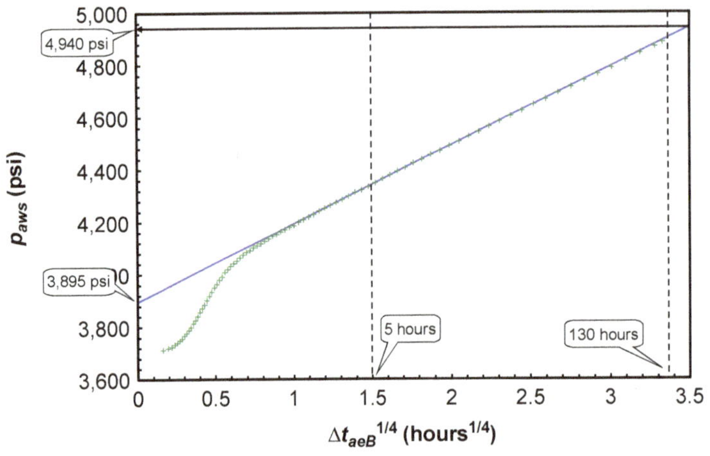

Fig. 7.14—Fourth-root-of-time plot, bilinear flow analysis example.

3. Draw a straight line through the data in bilinear flow. Read the intercept b_B as 3,895 psi. Calculate the slope, m_B, as

$$m_B = \frac{4{,}940 - 3{,}895}{3.5 - 0} = 299\,\text{psi/hr}^{1/4}.$$

Note the importance of identifying the bilinear flow period from the log-log diagnostic plot rather than relying on the square-root-of-time plot.

4. Calculate wk_f (if k is known) or $wk_f\sqrt{k}$ (is k is unknown) from the slope, m_B. We know k, so we calculate the fracture conductivity, wk_f, from the slope:

$$wk_f = \left(\frac{44.1 q_g B_{gi}\mu_{gi}}{h|m_B|}\right)^2 \sqrt{\frac{1}{\phi\mu_{gi}c_{ti}k}}$$

$$= \left(\frac{(44.1)(3{,}500)(0.545)(0.0364)}{(135)(299)}\right)^2 \sqrt{\frac{1}{(0.075)(0.0364)(4.08\times10^{-5})(0.03)}}$$

$$= 99.5\,\text{md-ft}.$$

5. If the formation permeability is known, calculate the fracture skin factor, s_f, from the intercept, b_B. For a buildup test, calculate s_f from

$$s_f = \frac{0.00708kh}{q_g B_{gi}\mu_{gi}}\left(b_B - p_{awf}\right)$$

$$= \frac{(0.00708)(0.03)(135)}{(3{,}500)(0.545)(0.0364)}(3{,}895 - 3{,}707.3)$$

$$= 0.078.$$

6. If the formation permeability is known, calculate a rough estimate of the fracture half-length from the time at which bilinear flow ends, t_{eblf}. If bilinear flow lasts through the end of the test, using the test duration in place of t_{eblf} will give a lower bound on the fracture half-length:

$$L_f \approx 0.2266\left(\frac{kt_{eblf}}{\phi\mu_{gi}c_{ti}}\right)^{1/4}\left(\frac{wk_f}{k}\right)^{1/2}$$

$$= 0.2266\left(\frac{(0.03)(130)}{(0.075)(0.364)(4.08\times10^{-5})}\right)^{1/4}\left(\frac{99.5}{0.03}\right)^{1/2}$$

$$= 1{,}000\,\text{ft}.$$

The calculated lower bound of 1,000 ft is somewhat lower than the actual length of 1,250 ft used to generate the data for this synthetic example.

7.3.5 Pseudoradial Flow (PRF). Eq. 7.23 is the basis for analysis of pressure transient data exhibiting PRF. Analysis of data in PRF gives estimates of formation permeability and total skin factor, s_t, which includes the effect of stimulation from the fracture and damage. The total skin factor can be used to calculate an equivalent infinite-conductivity fracture half-length under the assumptions that the fracture is undamaged and that fracture conductivity is infinite. Data in PRF alone cannot be used to separate the effects of fracture half-length, fracture conductivity, and fracture damage.

The following procedure can be used for analyzing data in PRF:

1. Identify data in PRF from the log-log diagnostic plot.
2. Construct a semilog graph. For a drawdown test, graph the flowing bottomhole pressure, p_{wf}, vs. the test time, t, on a semilog graph with time on the logarithmic axis. For a buildup test, graph the shut-in bottomhole pressure, p_{ws}, vs. the equivalent time, Δt_e, defined as

$$\Delta t_e = \frac{t_p \Delta t}{t_p + \Delta t}. \quad (7.44)$$

3. Draw a straight line through the data in PRF. Find the slope m and the intercept $p_{1\mathrm{hr}}$ of the straight line. The intercept is read from the straight line at a time or equivalent time of 1 hour, corresponding to the vertical axis where log(1) = 0.
4. Calculate the formation permeability from the slope m as

$$k = \frac{162.6 qB\mu}{|m|h}. \quad (7.45)$$

5. Calculate the total skin factor from the intercept, $p_{1\mathrm{hr}}$. For a drawdown test, calculate the total skin factor, s_t, from

$$s_t = 1.151\left[\frac{p_i - p_{1hr}}{|m|} - \log\left(\frac{k}{\phi\mu c r_w^2}\right) + 3.23\right]. \quad (7.46)$$

For a buildup test, calculate the total skin factor, s_t, as

$$s_t = 1.151\left[\frac{p_{1hr} - p_{wf}}{|m|} - \log\left(\frac{k}{\phi\mu c r_w^2}\right) + 3.23\right]. \quad (7.47)$$

6. Calculate the equivalent infinite-conductivity fracture half-length from the total skin factor, s_t, as

$$L_{f\mathrm{eq}} = 2 r_w e^{-s_t}. \quad (7.48)$$

7. If the fracture conductivity, wk_f, is known, the fracture half-length, L_f, can be estimated, assuming the fracture damage skin factor, s_f, is zero. First, calculate the apparent wellbore radius, r_{wa}:

$$r_{wa} = r_w e^{-s}. \quad (7.49)$$

Estimate the fracture half-length, L_f, from

$$L_f = \frac{2 r_{wa}}{\left(1 - \dfrac{\pi k r_{wa}}{w k_f}\right)}. \quad (7.50)$$

Note that Eq. 7.50 becomes ill-behaved as $\pi k r_{wa}$ approaches wk_f and is meaningless for $\pi k r_{wa} \geq wk_f$.

7.3.6 PRF Analysis Example.

7.3.6.1 Example Problem 7.3. Given the rock and fluid properties in **Table 7.5,** analyze the gas-well pressure-buildup-test data in **Table 7.6** to estimate formation permeability and total skin factor using PRF analysis. Calculate the equivalent infinite-conductivity-fracture half-length. Calculate the actual fracture half-length from the given fracture conductivity.

Rock and Fluid Properties

q_g	8,325	Mscf/D	t_p	168	hours
ϕ	18	%	h	36	ft
S_w	32	%	c_f	4×10^{-6}	psi^{-1}
c_w	3.6×10^{-6}	psi^{-1}	wk_f	4,000	md-ft
T_f	120	°F	r_w	0.25	ft
γ_g	0.723	(air=1)	B_{gi}	0.978	bbl/Mscf
μ_{gi}	0.0195	cp	c_{ti}	2.91×10^{-4}	psi^{-1}
p_i	2,350	psia	p_{awf}	1,161	psi

Table 7.5—Rock- and fluid-property data for PRF analysis example.

Δt (hours)	p_{ws} (psia)	Δt_a (hours)	p_{aws} (psi)	Δt (hours)	p_{ws} (psia)	Δt_a (hours)	p_{aws} (psi)
0.0010	2,109.1	0.0009	1,164.9	0.4857	2,173.0	0.4510	1,226.6
0.0021	2,112.0	0.0019	1,167.7	0.5474	2,175.8	0.5087	1,229.3
0.0034	2,114.3	0.0031	1,169.9	0.6168	2,178.7	0.5737	1,232.1
0.0048	2,116.1	0.0044	1,171.7	0.6949	2,181.6	0.6469	1,234.9
0.0064	2,117.9	0.0059	1,173.4	0.7827	2,184.7	0.7294	1,238.0
0.0082	2,119.5	0.0075	1,174.9	0.8816	2,187.7	0.8222	1,241.0
0.0102	2,120.8	0.0094	1,176.2	0.9928	2,191.0	0.9269	1,244.2
0.0125	2,122.1	0.0115	1,177.4	1.118	2,194.1	1.045	1,247.2
0.0151	2,123.8	0.0138	1,179.0	1.259	2,197.4	1.177	1,250.4
0.0180	2,124.8	0.0165	1,180.0	1.417	2,200.9	1.327	1,253.8
0.0212	2,126.3	0.0194	1,181.5	1.595	2,204.2	1.495	1,257.0
0.0249	2,128.0	0.0228	1,183.1	1.795	2,207.4	1.685	1,260.1
0.0290	2,129.2	0.0266	1,184.3	2.021	2,210.9	1.899	1,263.6
0.0336	2,130.6	0.0308	1,185.6	2.275	2,214.6	2.139	1,267.2
0.0388	2,132.3	0.0356	1,187.2	2.560	2,217.9	2.411	1,270.4
0.0447	2,133.6	0.0410	1,188.5	2.881	2,221.4	2.716	1,273.9
0.0512	2,135.3	0.0471	1,190.1	3.242	2,225.0	3.060	1,277.4
0.0587	2,136.7	0.0539	1,191.5	3.648	2,228.5	3.448	1,280.8
0.0670	2,138.5	0.0616	1,193.2	4.105	2,231.9	3.885	1,284.1
0.0764	2,140.1	0.0702	1,194.8	4.619	2,235.7	4.377	1,287.9
0.0869	2,141.9	0.0800	1,196.4	5.198	2,239.1	4.931	1,291.2
0.0988	2,143.6	0.0909	1,198.1	5.848	2,242.5	5.555	1,294.5
0.1121	2,145.4	0.1033	1,199.9	6.580	2,246.2	6.258	1,298.2
0.1271	2,147.3	0.1172	1,201.7	7.404	2,249.9	7.051	1,301.8
0.1440	2,149.3	0.1328	1,203.7	8.331	2,253.4	7.943	1,305.3
0.1630	2,151.2	0.1504	1,205.5	9.373	2,257.0	8.949	1,308.8
0.1844	2,153.2	0.1703	1,207.4	10.55	2,260.2	10.08	1,312.0
0.2085	2,155.5	0.1926	1,209.7	11.86	2,263.8	11.36	1,315.5
0.2355	2,157.8	0.2177	1,211.9	13.35	2,267.4	12.79	1,319.1
0.2659	2,159.9	0.2460	1,213.9	15.02	2,270.8	14.41	1,322.5
0.3002	2,162.4	0.2779	1,216.3	16.90	2,274.1	16.24	1,325.8
0.3387	2,164.9	0.3138	1,218.8	19.01	2,277.3	18.29	1,328.9
0.3821	2,167.5	0.3542	1,221.3	21.39	2,280.7	20.61	1,332.3
0.4308	2,170.2	0.3997	1,223.9	24.00	2,283.9	23.15	1,335.5

Table 7.6—Buildup data for PRF analysis example.

7.3.6.2 Solution. Because this is a gas-well buildup test, we use adjusted pressure and adjusted time for the analysis. Following the recommended procedure, we proceed:

1. Identify data in PRF from the log-log diagnostic plot. The diagnostic plot is shown in **Fig. 7.15.** From the horizontal part of the derivative, we see that the PRF period begins at approximately 2.3 hours and lasts through the end of the test at 21 hours.

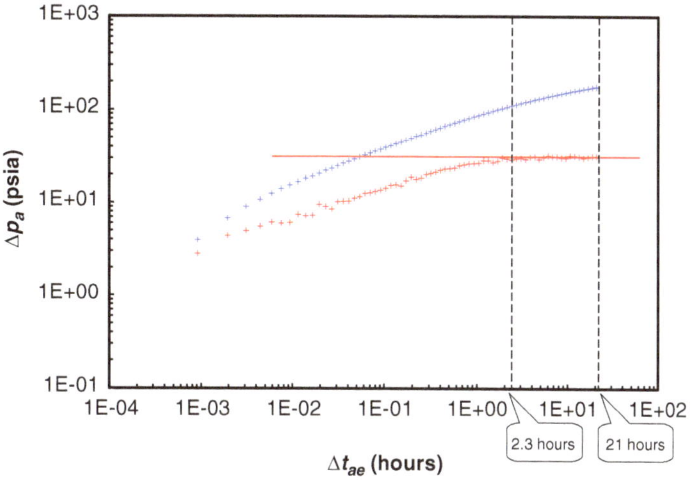

Fig. 7.15—Diagnostic plot, PRF analysis example.

2. Construct a semilog graph. **Fig. 7.16** shows the semilog graph of shut-in bottomhole adjusted pressure, p_{aws}, vs. the equivalent adjusted time, Δt_{ae}.

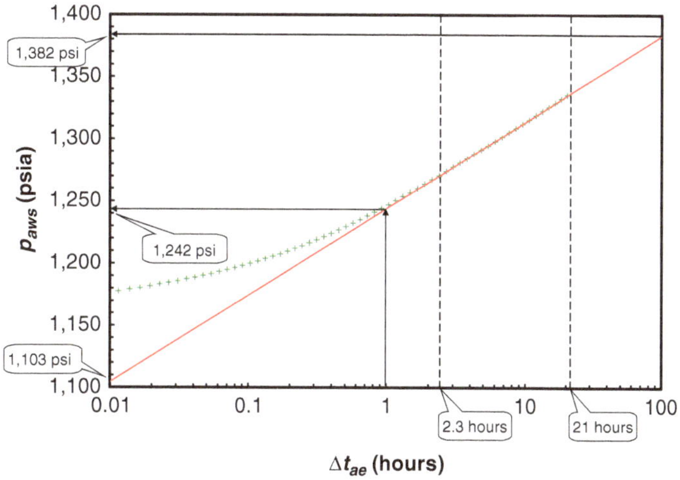

Fig. 7.16—Semilog plot, PRF analysis example.

3. Draw a straight line through the data in PRF. Find the slope, m, and the intercept, p_{a1hr}, of the straight line. We read p_{a1hr} as 1,242 psi. The slope, m, is given by

$$m = \frac{1,382 - 1,103}{\log(100) - \log(0.01)} = \frac{279}{4} = 69.75 \text{ psi/log cycle.}$$

4. Calculate the formation permeability from the slope, m:

$$\begin{aligned} k &= \frac{162.6 q_g B_{gi} \mu_{gi}}{|m| h} \\ &= \frac{(162.6)(8,325)(0.978)(0.0195)}{(69.75)(36)} \\ &= 10.3 \text{ md.} \end{aligned}$$

5. Calculate the total skin factor from the intercept, p_{a1hr}. Because this is a buildup test, we calculate

$$s_t = 1.151\left[\frac{p_{a1hr} - p_{awf}}{|m|} - \log\left(\frac{k}{\phi\mu_{gi}c_{ti}r_w^2}\right) + 3.23\right]$$

$$= 1.151\left[\frac{1{,}242 - 1{,}161}{69.75} - \log\left(\frac{10.3}{(0.18)(0.0195)(2.91\times10^{-4})(0.25)^2}\right) + 3.23\right]$$

$$= -4.39.$$

6. Calculate the equivalent infinite-conductivity fracture half-length from the total skin factor, s_t, as

$$L_{feq} = 2r_w e^{-s_t} = (2)(0.25)e^{-(-4.39)} = 40.3 \text{ ft.}$$

7. If the fracture conductivity, wk_f, is known, the fracture half-length, L_f, can be estimated, assuming the fracture damage skin factor s_f is zero. First, calculate the apparent wellbore radius, r_{wa}:

$$r_{wa} = r_w e^{-s} = 0.25e^{-(-4.39)} = 20.2 \text{ ft.}$$

Estimate the fracture half-length, L_f, from

$$L_f = \frac{2r_{wa}}{\left(1 - \frac{\pi k r_{wa}}{wk_f}\right)}$$

$$= \frac{(2)(20.2)}{\left(1 - \frac{\pi(10.3)(20.2)}{(4{,}000)}\right)}$$

$$= 48.3 \text{ ft.}$$

7.3.7 Elliptical Flow. Historically, elliptical flow was analyzed primarily using type curves or history matching. In a recent paper, Cheng et al. (2007) present an iterative method for analyzing data in elliptical flow to estimate formation permeability and effective fracture half-length. As with PRF, the effects of fracture half-length, fracture conductivity, and fracture skin factor cannot be separated using only data in elliptical flow. If bilinear flow data are also available, the fracture half-length can be estimated.

In this section, we present the following modification of the Cheng et al. (2007) method:

1. Assume initial estimates for permeability, k, and effective fracture half-length, L_{feq}.
2. Calculate A_t and B_t using Eqs. 7.18 and 7.20. For a drawdown test, use Eq. 7.20:

$$B_t = 0.02434\left(\frac{kt}{\phi\mu c_t}\right)^{1/2}.$$

For a buildup test, use Eq. 7.51:

$$B_t = 0.02434\left(\frac{k\Delta t_e}{\phi\mu c_t}\right)^{1/2} \quad (7.51)$$

Then, use Eq. 7.18:

$$A_t = \sqrt{B_t^2 + L_{feq}^2}.$$

3. Graph $p_i - p_{wf}$ vs. $\ln(A_t + B_t)$ for a drawdown test, and $p_{ws} - p_{wf}$ vs. $\ln(A_t + B_t)$ for a buildup test.
4. Draw a straight line through the data in elliptical flow. Find the slope, m_E, and the x-intercept, b_{xE}.
5. Calculate the formation permeability k from the slope, m_E, as

$$k = \frac{141.2qB\mu}{m_E h}. \quad (7.52)$$

Calculate the effective fracture half-length, L_{feq}, from the x-intercept, b_{xE}, as

$$L_{feq} = \exp(-b_{xE}). \quad (7.53)$$

6. Repeat Steps 2 through 5, updating the estimates of k and L_{feq} at each iteration until the values converge.
7. If the bilinear flow regime occurs during the test, analyze the data in bilinear flow to obtain wk_f using the bilinear flow procedure given in Section 7.3.3.
8. If the fracture conductivity wk_f is known (independently or from bilinear flow analysis), the fracture half-length L_f can be estimated, assuming the fracture damage skin factor s_f is zero. First, calculate the apparent wellbore radius, r_{wa}:

$$r_{wa} = \frac{L_{feq}}{2}. \quad (7.54)$$

Estimate the fracture half-length, L_f, from

$$L_f = \frac{2r_{wa}}{\left(1 - \dfrac{\pi k r_{wa}}{wk_f}\right)}. \quad (7.55)$$

Estimate the dimensionless fracture conductivity from

$$C_r = \frac{wk_f}{\pi k L_f}. \quad (7.56)$$

9. Calculate the dimensionless time at which elliptical flow begins on the basis of the earliest time at which the elliptical flow method gives the desired level of accuracy, from

$$t_{L_f D,\,bef} = \frac{0.0703 + \dfrac{0.1257}{C_r^{\,0.7632}}}{\delta^{0.8811}}, \quad (7.57)$$

where δ is the desired maximum fractional error. Eq. 7.56 is valid for $C_r \geq 2$.

Calculate the dimensionless time at the beginning of the data analyzed as elliptical flow from Eq. 7.1:

$$t_{L_f D} = \frac{0.0002637 kt}{\phi \mu c_t L_f^{\,2}}.$$

Then compare the value from Eq. 7.1 with the start of elliptical flow estimated from Eq. 7.56. If necessary, update the range of data used in the analysis and repeat Steps 2 through 9.

This last step is particularly important because the elliptical flow period has no characteristic signature that can be identified on the diagnostic plot.

Any of the three definitions for B_t (Eqs. 7.17, 7.19, and 7.20) can be used in Step 2. Interestingly, the choice of B_t has no effect on the resulting estimate of permeability, k, from Eq. 7.52, but it does affect the estimate of effective fracture length, L_{feq}, from Eq. 7.53. Of the definitions of B_t, Eq. 7.20 gives the best results for fracture length if Eq. 7.57 is used to define the start of elliptical flow. Eq. 7.20 has the additional advantage that data in PRF can be included in the elliptical flow analysis. This is potentially very useful in situations in which the data appear to reach PRF during the test, but the PRF period lasts less than the one-third to one-half log cycle required to confirm the presence of a particular flow regime.

7.3.8 Elliptical Flow Analysis Example.

7.3.8.1 Example Problem 7.4. Analyze the data exhibiting elliptical flow in **Table 7.6** to estimate permeability, effective fracture half-length, and actual fracture half-length.

7.3.8.2 Solution. We use the data during the two-third log cycle before PRF begins, from $\Delta t_{ae} = 0.4$ to $\Delta t_{ae} = 2.3$. These data are given in **Table 7.7,** along with the intermediate results from the first iteration. We could include data in PRF in the elliptical flow analysis, if desired. For this example, we chose not to include such data to illustrate the use of elliptical flow analysis when PRF data are not available.

Δt_a (hours)	p_{aws} (psi)	Δt_{ae} (hours)	Δp_{aws} (psi)	B_t (ft)	A_t (ft)	X
0.4510	1,226.6	0.4498	65.6	16.15	101.30	4.766
0.5087	1,229.3	0.5071	68.3	17.15	101.46	4.776
0.5737	1,232.1	0.5717	71.1	18.21	101.64	4.786
0.6469	1,234.9	0.6444	73.9	19.33	101.85	4.797
0.7294	1,238.0	0.7262	77	20.52	102.08	4.809
0.8222	1,241.0	0.8182	80	21.78	102.35	4.821
0.9269	1,244.2	0.9218	83.2	23.12	102.64	4.834
1.045	1,247.2	1.038	86.2	24.54	102.97	4.848

(Continued)

Δt_a (hours)	p_{aws} (psi)	Δt_{ae} (hours)	Δp_{aws} (psi)	B_t (ft)	A_t (ft)	X
1.177	1,250.4	1.169	89.4	26.04	103.33	4.863
1.327	1,253.8	1.317	92.8	27.64	103.75	4.878
1.495	1,257.0	1.482	96	29.32	104.21	4.894
1.685	1,260.1	1.668	99.1	31.10	104.73	4.911
1.899	1,263.6	1.877	102.6	33.00	105.30	4.929
2.139	1,267.2	2.113	106.2	35.01	105.95	4.948

Table 7.7—Elliptical flow analysis example calculations, first iteration.

1. Assume initial estimates for permeability, k, and effective fracture half-length, L_{feq}. We will start with k = 1 md and L_{feq} = 100 ft.
2. Calculate A_t and B_t using Eqs. 7.17 and 7.18. For the first point in elliptical flow, we have

$$B_t = 0.02434\left(\frac{k\Delta t_{ae}}{\phi\mu_{gi}c_{ti}}\right)^{1/2}$$

$$= 0.02434\left(\frac{(1)(0.4498)}{(0.18)(0.0195)(2.91\times10^{-4})}\right)^{1/2}$$

$$= 16.15 \text{ ft.}$$

$$A_t = \sqrt{B_t^2 + L_{feq}^2}$$

$$= \sqrt{(16.15)^2 + (100)^2}$$

$$= 101.30 \text{ ft.}$$

3. Graph $p_{aws} - p_{awf}$ vs. $\ln(A_t + B_t)$. The graph for the first iteration is shown in **Fig. 7.17.**

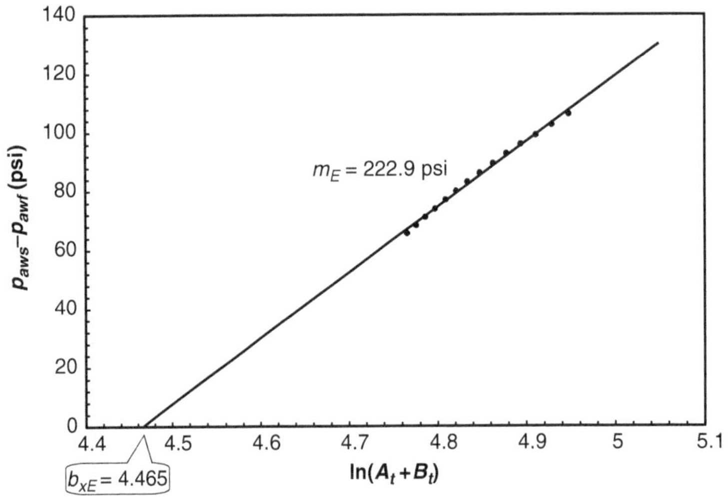

Fig. 7.17—Elliptical flow example, first iteration.

4. Drawing a straight line through the data in elliptical flow, we find the slope, m_E, to be 222.9 psi and the x-intercept, b_{xE}, to be 4.465.

5. Calculate the formation permeability, k, from the slope, m_E, as

$$k = \frac{141.2qB_{gi}\mu_{gi}}{m_E h}$$
$$= \frac{(141.2)(8,325)(0.978)(0.0195)}{(222.9)(36)}$$
$$= 2.793.$$

Calculate the effective fracture half-length, L_{feq}, from the x-intercept, b_{xE}, as

$$L_{feq} = \exp(-b_{xE})$$
$$= \exp(-4.465)$$
$$= 87.0 \text{ ft.}$$

6. Repeat Steps 2 through 5, updating the estimates of k and L_{feq} at each iteration until the values converge. After ten iterations, we converge to k = 10.52 md and L_{feq} = 39.1 ft. The elliptical flow graph for the final iteration is shown in **Fig. 7.18.**

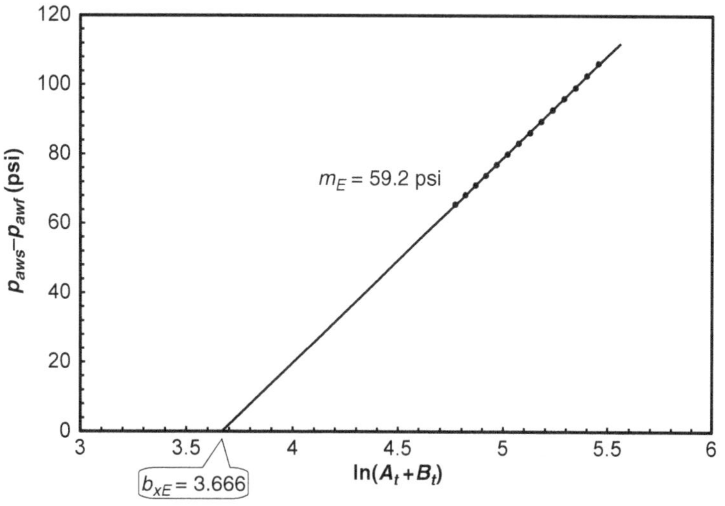

Fig. 7.18—Elliptical flow example, final iteration.

7. If the bilinear flow regime occurs during the test, analyze the data in bilinear flow to obtain wk_f using the bilinear flow procedure given earlier in this section. We neglect this step, because we are given wk_f = 4,000 md-ft.
8. If the fracture conductivity, wk_f, is known (independently or from bilinear flow analysis), the fracture half-length, L_f, can be estimated, assuming that the fracture damage skin factor, s_f, is zero. First, calculate the apparent wellbore radius, r_{wa}:

$$r_{wa} = \frac{L_{feq}}{2}$$
$$= \frac{49.1}{2}$$
$$= 19.55 \text{ ft.}$$

Estimate the fracture half-length, L_f, from

$$L_f = \frac{2r_{wa}}{\left(1 - \dfrac{\pi k r_{wa}}{wk_f}\right)}$$
$$= \frac{(2)(19.55)}{\left(1 - \dfrac{\pi(10.52)(19.55)}{4,000}\right)}$$
$$= 46.6 \text{ ft.}$$

Estimate the dimensionless fracture conductivity from

$$C_r = \frac{wk_f}{\pi k L_f} = \frac{4{,}000}{\pi(10.52)(57.1)} = 2.6.$$

9. Calculate the dimensionless time at which elliptical flow begins on the basis of the earliest time at which the elliptical flow method gives the desired level of accuracy. For a maximum error of 10%, we have

$$t_{L_f D,bef} = \frac{0.0703 + \dfrac{0.1257}{C_r^{\,0.7632}}}{\delta^{0.8811}} = \frac{0.0703 + \dfrac{0.1257}{(2.6)^{0.7632}}}{(0.10)^{0.8811}} = 0.996.$$

Calculate the dimensionless time at the beginning and end of the elliptical flow period. We analyzed data from $\Delta t_{ae} = 0.4498$ to 2.113 hours. At the start of the data interpreted as elliptical flow, we have

$$t_{L_f D} = \frac{0.0002637kt}{\phi \mu c_t L_f^{\,2}} = \frac{(0.0002637)(10.52)(0.4498)}{(0.18)(0.0195)(2.91\times10^{-4})(46.6)^2} = 0.563.$$

Because we used data that was earlier than calculated time to give the desired 10% accuracy, we repeat Steps 1 through 9 using only data between $\Delta t_{ae} = 0.4498\dfrac{0.996}{0.563} = 0.796$ and 2.113 hours. The elliptical flow graph for the final iteration is shown in **Fig. 7.19.** All the data used for the first round of iterations are shown Fig. 7.19, but only the data using the revised time range were used for the straight-line fit. After iterating with the updated data range, the final results are $k = 10.3$ md, $L_{feq} = 40.0$ ft, and $L_f = 47.7$ ft. In this example, the results using the revised data range for analysis are quite close to those obtained using the original data range. The final results from elliptical flow analysis are almost identical to those from analysis of PRF: $k = 10.3$ md, $L_{feq} = 40.3$ ft, and $L_f = 48.3$ ft. The agreement is especially good considering that the data sets used in the two methods do not overlap.

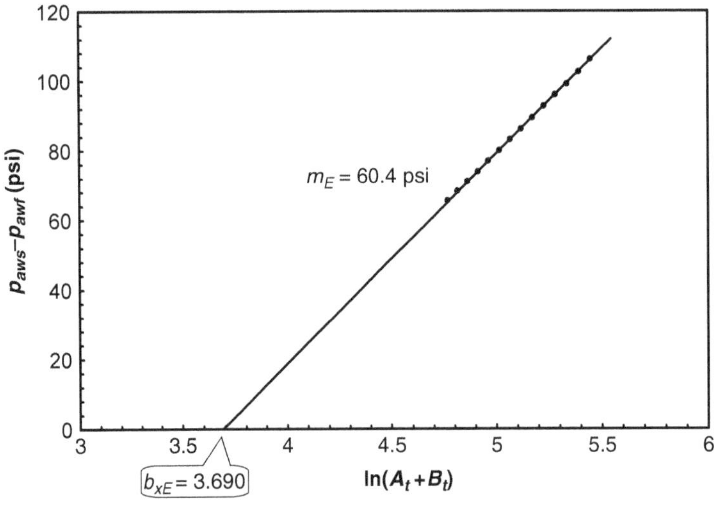

Fig. 7.19—Elliptical flow example, final iteration with revised data range.

7.4 Log-Log Type Curve Analysis

A number of different log-log type curves have been presented for hydraulically fractured gas wells, taking into account various phenomena such as finite fracture conductivity (Cinco-Ley and Samaniego-V. 1981a), wellbore storage (Ramey and Gringarten 1975), reservoir size (Gringarten et al. 1974), and fracture damage skin factor (Cinco-Ley and Samaniego-V. 1981b) and combinations of these (Wong et al. 1986). Because the analysis procedure is basically the same for any of these type curves, we will illustrate the method using the Cinco-Ley and Samaniego-V. (1981a) type curves for wells with finite-conductivity vertical fractures, supplemented with derivative curves as in Wong et al. (1986).

The basic type-curve matching procedure for most fractured-well type curves is as follows, assuming formation permeability is not known:

1. Graph field data pressure change and pressure derivative plotting functions vs. the field time plotting function on a log-log scale. For constant-rate drawdown tests, the elapsed test time is used. For buildup tests following long drawdown periods, the equivalent time (for oil wells) or adjusted equivalent time (for gas wells) should be used if the flow period is less than 100 times the duration of the subsequent shut-in.

   If PRF has been reached during the flow period before shut-in or if the flow regime is unknown, the radial equivalent time can be used. In the latter case, when the analysis is complete and the flow regimes can be identified, it might be necessary to repeat the analysis using a better choice of equivalent time.

   If data are in bilinear flow at the end of the flow period and the shut-in period, the bilinear equivalent time defined by Eq. 7.38 is recommended. Similarly, if data are in linear flow at the end of the flow period and the shut-in period, the linear equivalent time, Eq. 7.27, should be used.

   An elliptical equivalent time has even been defined (Liao and Lee 1993). However its calculation requires both permeability and fracture length be known, thus requiring an iterative procedure for type-curve matching. Its chief advantage is that it reduces to linear equivalent time for linear flow and to radial equivalent time for PRF and is thus applicable to any test in which the flow and buildup periods end in any combination of linear flow, elliptical flow, and PRF.
2. Identify any flow regimes exhibited by the test data.
3. Place the field data on top of the type curve.
4. If PRF is present, move the field data graph vertically to align the horizontal part of the field data derivative with the horizontal part of the derivative type curve corresponding to PRF.
5. Move the field data graph horizontally to align the field data pressure and derivative curves with the pressure and pressure derivative type curves that best fit the data.

   If PRF is not present, move the field data both vertically and horizontally to obtain the best match between the field data pressure and derivative curves and the pressure and pressure derivative type curves. In this case, there will, in general, be no unique match.
6. Record the matching stem parameter corresponding to the pressure/pressure-derivative type curve pair best matching the field data. For finite-conductivity-fracture type curves, the matching stem parameter is the dimensionless fracture conductivity, C_{fD} or C_r.
7. Calculate the formation permeability from the pressure match-point ratio:

$$k = \frac{141.2qB\mu}{h}\left(\frac{p_D}{\Delta p}\right)_{MP} \quad (7.58)$$

8. Calculate the fracture half-length, L_f, from the time match-point ratio:

$$L_f = \sqrt{\frac{0.0002637k}{\phi\mu c_t}\left(\frac{\Delta t}{t_{L_fD}}\right)_{MP}} \quad (7.59)$$

9. Calculate the physical parameter corresponding to the matching stem parameter. For finite-conductivity-fracture type curves, calculate the fracture conductivity as

$$wk_f = kL_fC_{fD} = \pi kL_fC_r. \quad (7.60)$$

Note the requirement that PRF be present to obtain a unique match if formation permeability is unknown. If permeability is known, a unique match can be obtained by precalculating the pressure match-point ratio to fix the vertical position of the field data relative to the type curve. In this case, Steps 4 through 7 are modified as follows.

4. If permeability is known, calculate the pressure match point ratio as

$$\left(\frac{\Delta p}{p_D}\right)_{MP} = 141.2\frac{qB\mu}{kh}. \quad (7.61)$$

5. Move the field data vertically to align the value corresponding to the calculated pressure match-point ratio on the field data axis with $p_D = 1$ on the type curve axis.
6. Move the field data graph horizontally to align the field data pressure and derivative curves with the pressure and pressure derivative type curves that best fit the data.
7. Record the matching stem parameter corresponding to the pressure/pressure-derivative type curve pair best matching the field data.

7.4.1 Type Curve Analysis Example.

7.4.1.1 Example Problem 7.5. Given the rock- and fluid-property data in **Table 7.8,** analyze the pressure-buildup data in **Table 7.9** to estimate fracture half-length L_f, dimensionless fracture conductivity C_r, and fracture conductivity wk_f.

7.4.1.2 Solution. Because this is a gas-well buildup test, we use adjusted pressure and adjusted equivalent time for the analysis. Because the flow period is much longer than the buildup, we can use radial equivalent adjusted time with little error.

Rock and Fluid Properties					
q_g	2,500	Mscf/D	t_p	4,320	hours
ϕ	8.5	%	h	75	ft
S_w	45	%	c_f	$4{\times}10^{-6}$	psi^{-1}
c_w	$3.6{\times}10^{-6}$	psi^{-1}	k	0.035	md
T_f	225	°F	r_w	0.25	ft
γ_g	0.67	(air=1)	p_i	6,700	psia
B_{gi}	0.586	bbl/Mscf	μ_{gi}	0.0309	cp
c_{ti}	$5.133{\times}10^{-5}$	psi^{-1}			
p_{wf}	2,199	psia	p_{awf}	860	psi

Table 7.8—Rock- and fluid-property data for type curve analysis example.

Δt (hours)	p_{ws} (psia)	Δt_a (hours)	p_{aws} (psi)	Δt (hours)	p_{ws} (psia)	Δt_a (hours)	p_{aws} (psi)
0.0010	2,256.1	0.0004	901.7	0.3002	2,434.3	0.1182	1,035.1
0.0021	2,268.5	0.0008	910.8	0.3387	2,441.3	0.1337	1,040.4
0.0034	2,276.8	0.0013	916.8	0.3821	2,448.9	0.1511	1,046.2
0.0048	2,283.8	0.0018	922.0	0.4308	2,456.8	0.1707	1,052.3
0.0064	2,290.2	0.0024	926.7	0.4857	2,463.9	0.1928	1,057.8
0.0082	2,295.8	0.0031	930.8	0.5474	2,472.1	0.2178	1,064.2
0.0102	2,301.3	0.0039	934.9	0.6168	2,480.3	0.246	1,070.5
0.0125	2,306.8	0.0048	939.0	0.6949	2,488.5	0.2778	1,076.9
0.0151	2,311.6	0.0057	942.6	0.7827	2,496.9	0.3136	1,083.4
0.0180	2,316.7	0.0068	946.3	0.8816	2,505.8	0.3541	1,090.3
0.0212	2,321.7	0.0081	950.0	0.9928	2,514.8	0.3997	1,097.3
0.0249	2,326.4	0.0095	953.5	1.118	2,523.8	0.4512	1,104.4
0.0290	2,331.4	0.0111	957.3	1.259	2,533.8	0.5093	1,112.2
0.0336	2,336.3	0.0129	960.9	1.417	2,543.4	0.5749	1,119.8
0.0388	2,340.9	0.0149	964.4	1.595	2,553.7	0.6489	1,127.8
0.0447	2,346.3	0.0172	968.4	1.795	2,564.5	0.7325	1,136.4
0.0512	2,351.3	0.0197	972.2	2.021	2,575.1	0.8268	1,144.8
0.0587	2,356.4	0.0226	976.0	2.275	2,586.2	0.9333	1,153.6
0.0670	2,361.7	0.0258	979.9	2.560	2,597.9	1.054	1,162.9
0.0764	2,367.4	0.0295	984.2	2.881	2,609.6	1.189	1,172.2
0.0869	2,372.5	0.0336	988.1	3.242	2,621.8	1.343	1,182.0
0.0988	2,378.5	0.0383	992.6	3.648	2,634.8	1.516	1,192.3
0.1121	2,384.1	0.0435	996.8	4.105	2,648.3	1.712	1,203.2
0.1271	2,389.7	0.0494	1,001.0	4.619	2,661.3	1.933	1,213.8
0.1440	2,395.9	0.0561	1,005.8	5.198	2,675.6	2.183	1,225.2
0.1630	2,401.9	0.0636	1,010.3	5.848	2,690.3	2.465	1,237.2
0.1844	2,407.8	0.0721	1,014.8	6.580	2,706.3	2.784	1,250.2
0.2085	2,414.5	0.0816	1,019.9	7.404	2,721.9	3.145	1,262.9
0.2355	2,420.9	0.0924	1,024.8	8.331	2,738.6	3.553	1,276.6
0.2659	2,427.4	0.1045	1,029.8	9.373	2,755.6	4.014	1,290.6

(Continued)

Δt (hours)	p_{ws} (psia)	Δt_a (hours)	p_{aws} (psi)	Δt (hours)	p_{ws} (psia)	Δt_a (hours)	p_{aws} (psi)
10.55	2,773.8	4.536	1,305.5	34.26	3,009.6	15.55	1,504.2
11.86	2,792.7	5.127	1,321.2	38.55	3,040.5	17.61	1,530.8
13.35	2,812.2	5.795	1,337.3	43.37	3,072.3	19.95	1,558.2
15.02	2,833.4	6.551	1,355.0	48.79	3,105.7	22.60	1,587.3
16.90	2,855.1	7.407	1,373.1	54.89	3,140.9	25.62	1,618.1
19.01	2,877.8	8.377	1,392.1	61.75	3,177.4	29.04	1,650.2
21.39	2,901.3	9.476	1,411.9	69.47	3,216.5	32.94	1,684.6
24.06	2,926.4	10.72	1,433.1	78.15	3,256.6	37.37	1,720.2
27.07	2,953.1	12.13	1,455.8	87.92	3,298.3	42.41	1,757.4
30.45	2,980.3	13.73	1,479.0	96.00	3,331.1	46.62	1,786.8

Table 7.9—Buildup data for type curve analysis example.

1. Graph field data pressure change and pressure derivative vs. the test time on a log-log scale.
2. Identify any flow regimes exhibited by the test data.
3. Place the field data on top of the type curve.
4. Because permeability is known, calculate the pressure match-point ratio as

$$\left(\frac{\Delta p}{p_D}\right)_{MP} = 141.2\frac{qB\mu}{kh}$$

$$= 141.2\frac{(2{,}500)(0.586)(0.0309)}{(0.035)(75)} \quad \text{(7.62)}$$

$$= 2{,}435 \text{ psi}$$

5. Move the field data vertically to align the value corresponding to the calculated pressure-match point-ratio 2,435 psi on the field data axis with $p_D = 1$ on the type curve axis, as shown in **Fig. 7.20**.

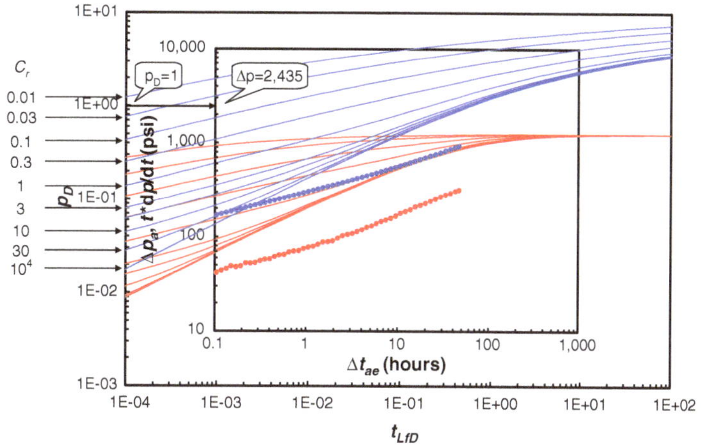

Fig. 7.20—Type curve match example, Step 5.

6. Move the field data graph horizontally to align the field data pressure and derivative curves with the pressure and pressure derivative type curves that best fit the data, as shown in **Fig. 7.21**.
7. Record the matching stem parameter corresponding to the pressure/pressure-derivative type curve pair best matching the field data. From Fig. 7.21, we read $C_r = 3$.

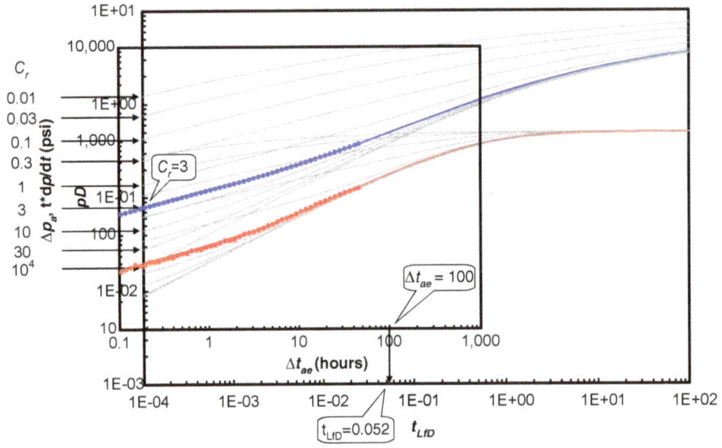

Fig. 7.21—Type curve match example, Steps 6 and 7.

8. Calculate the fracture half-length L_f from the time match-point ratio:

$$L_f = \sqrt{\frac{0.0002637k}{\phi\mu_{gi}c_{ti}}\left(\frac{\Delta t_{ae}}{t_{L_f D}}\right)_{MP}}$$

$$= \sqrt{\frac{(0.0002637)(0.035)}{(0.085)(0.0309)(5.133\times10^{-5})}\left(\frac{100}{0.052}\right)}$$

$$= 363 \text{ ft.}$$

9. Calculate the physical parameter corresponding to the matching stem parameter; in this case, calculate the fracture conductivity wk_f from the dimensionless fracture conductivity C_r:

$$wk_f = \pi k L_f C_r$$

$$= \pi(0.035)(363)(3)$$

$$= 120 \text{ md-ft.}$$

7.5 Parameter Estimation Using the Log-Log Diagnostic Plot

In this section, we show how to use the log-log diagnostic plot to estimate fracture length and fracture damage skin from data in linear flow and to estimate fracture conductivity and fracture damage skin from data in bilinear flow. Data in PRF can be analyzed the same way as data in radial flow, as shown in Part 2 Ch 6: Prestimulation Well Testing.

7.5.1 Linear Flow Manual Log-Log Analysis. The log-log diagnostic plot can be used to estimate fracture length and fracture damage skin factor from data in linear flow. **Fig. 7.22** illustrates the procedure.

1. Graph the pressure and pressure derivative plotting functions vs. the time plotting function on a log-log scale.
2. Identify the linear flow period.

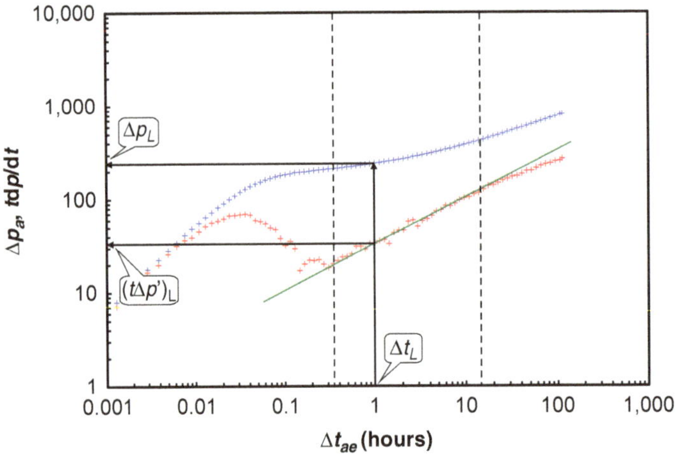

Fig. 7.22—Linear flow manual log-log analysis.

3. Select a time Δt_L during the linear flow period. The selected time may be chosen for convenience, but should be near the start of the linear flow period for highest accuracy in estimating the fracture damage skin.
4. At the selected time Δt_L, read the pressure, Δp_L, and pressure derivative, $(t\Delta p')_L$, plotting functions.
5. Calculate the linear flow slope m_L from the time Δt_L and pressure derivative $(t\Delta p')_L$ as

$$m_L = \frac{2}{\sqrt{t_L}}\left(t\frac{dp}{dt}\right)_L . \quad (7.63)$$

This is the same slope, m_L, that would be obtained from a linear flow graph of pressure vs. the square root of time.

6. Calculate the product $\sqrt{k}L_f$ from the slope, m_L:

$$\sqrt{k}L_f = 4.064\frac{qB}{hm_L}\sqrt{\frac{\mu}{\phi c_t}}. \quad (7.64)$$

7. If permeability is known, calculate the fracture half-length from

$$L_f = \frac{\sqrt{k}L_f}{\sqrt{k}}. \quad (7.65)$$

8. Calculate the additional pressure drop caused by fracture damage skin as

$$\Delta p_{s_f} = \left[\Delta p_L - 2\left(t\frac{dp}{dt}\right)_L\right]. \quad (7.66)$$

This pressure drop includes the pressure drop caused by finite fracture conductivity and the pressure drop from fracture damage proper. However, because linear flow appears only for high-conductivity fractures, the pressure drop caused by finite fracture conductivity will be small.

9. Calculate the fracture damage skin from

$$s_f = \frac{0.00708kh}{qB\mu}\Delta p_{s_f}.$$

If the dimensionless fracture conductivity is known or can be estimated, the component of the skin caused by finite fracture conductivity can be estimated from

$$s_{fcf} = -\ln\left(\frac{2}{2+C_r}\right). \quad (7.67)$$

7.5.2 Linear Flow Manual Log-Log Analysis Example.

7.5.2.1 Example Problem 7.6. Find the fracture half-length and skin factor for the buildup data set given in Tables 7.1 and 7.2.

7.5.2.2 Solution.

1. Graph the pressure and pressure derivative plotting functions vs. the time plotting function on a log-log scale. Because this is a gas-well buildup test, we use the adjusted pressure and the adjusted radial equivalent time, as shown in **Fig. 7.23.**

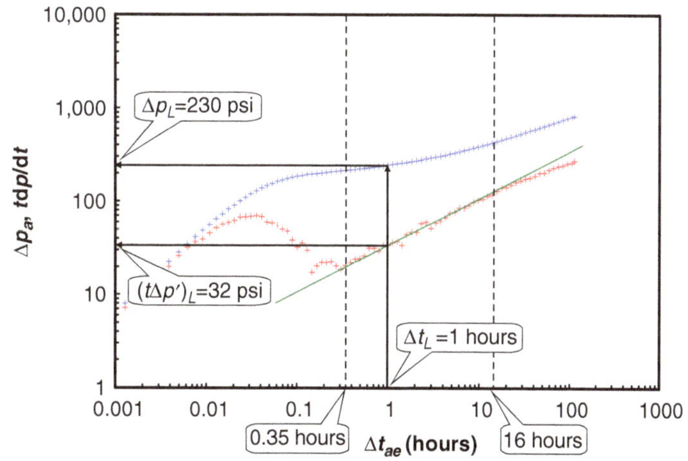

Fig. 7.23—Linear flow manual log-log analysis example.

2. Identify the linear flow period. We have previously identified the linear flow period as 0.35 to 16 hours.
3. Select a time, Δt_L, during the linear flow period. For this example, we chose $\Delta t_L = 1$ hour, but a time closer to the start of the linear flow period might give higher accuracy for the fracture damage skin pressure drop.
4. At the selected time, Δt_L, read the pressure, Δp_L, and pressure derivative, $(t\Delta p')_L$, plotting functions. From Fig. 7.23, we read $\Delta p_L = 230$ psi and $(t\Delta p')_L = 32$ psi at $\Delta t_L = 1$ hour.
5. Calculate the linear flow slope m_L from the time Δt_L and pressure derivative $(t\Delta p')_L$ as

$$m_L = \frac{2}{\sqrt{t_L}}\left(t\frac{dp}{dt}\right)_L = \frac{2}{\sqrt{1}}(32) = 64 \text{ psi/hr}^{1/2}.$$

6. Calculate the product $\sqrt{k}L_f$ from the slope, m_L:

$$\sqrt{k}L_f = 4.064\frac{qB_{gi}}{hm_L}\sqrt{\frac{\mu_{gi}}{\phi c_{ti}}}$$
$$= (4.064)\frac{(2{,}500)(0.586)}{(75)(64)}\sqrt{\frac{0.0309}{(0.085)(5.13\times10^{-5})}}$$
$$= 104.4 \sqrt{\text{md}}\text{ ft}.$$

7. If permeability is known, calculate the fracture half-length from

$$L_f = \frac{\sqrt{k}L_f}{\sqrt{k}} = \frac{104.4}{\sqrt{0.05}} = 467 \text{ ft}.$$

8. Calculate the additional pressure drop caused by fracture damage skin as

$$\Delta p_{s_f} = \left[\Delta p_L - 2\left(t\frac{dp}{dt}\right)_L\right] = 230 - (2)(32) = 166 \text{ psi}.$$

This pressure drop includes both the pressure drop caused by finite fracture conductivity and the pressure drop from fracture damage proper. However, because linear flow appears only for high-conductivity fractures, the pressure drop caused by finite fracture conductivity will be small.

9. Calculate the fracture damage skin from

$$s_f = \frac{0.00708kh}{q_g B_{gi}\mu_{gi}}\Delta p_{s_f}$$
$$= \frac{(0.00708)(0.05)(75)}{(2{,}500)(0.586)(0.0309)}(166)$$
$$= 0.097.$$

7.5.3 Bilinear Flow Manual Log-Log Analysis. Bilinear flow manual log-log analysis is very similar to linear flow manual log-log analysis. Data in bilinear flow can be used to estimate the fracture conductivity and fracture damage skin factor from the log-log diagnostic plot. **Fig. 7.24** illustrates the procedure.

1. Graph the pressure and pressure derivative plotting functions vs. the time plotting function on a log-log scale.
2. Identify the linear flow period.

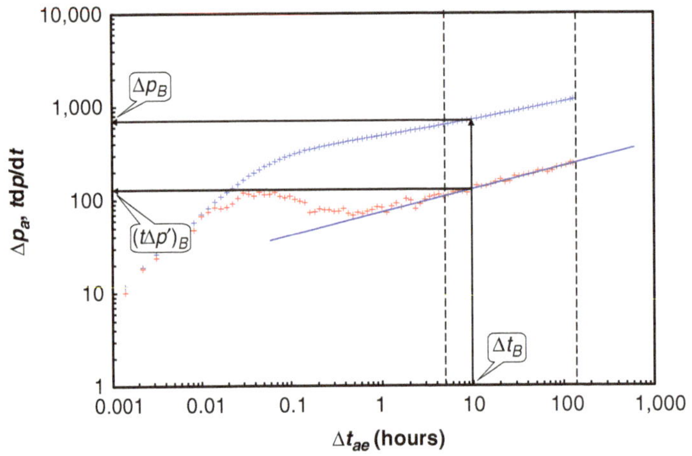

Fig. 7.24—Bilinear flow manual log-log analysis.

3. Select a time, Δt_B, during the bilinear flow period. The selected time may be chosen for convenience but should be near the start of the bilinear flow period for highest accuracy in estimating the fracture damage skin.
4. At the selected time Δt_B, read the pressure Δp_B and pressure derivative $(t\Delta p')_B$ plotting functions.
5. Calculate the bilinear flow slope m_B from the time Δt_B and pressure derivative $(t\Delta p')_B$ as

$$m_B = \frac{4}{t_B^{0.25}}\left(t\frac{dp}{dt}\right)_B . \quad (7.68)$$

This is the same slope, m_B, that would be obtained from a bilinear flow graph of pressure vs. the fourth root of time.

6. Calculate the product $wk_f\sqrt{k}$ from the slope, m_B:

$$wk_f\sqrt{k} = \left(\frac{44.1qB\mu}{h|m_B|}\right)^2\sqrt{\frac{1}{\phi\mu c_t}} . \quad (7.69)$$

7. If permeability is known, calculate the fracture conductivity from

$$wk_f = \frac{wk_f\sqrt{k}}{\sqrt{k}} . \quad (7.70)$$

8. Calculate the additional pressure drop caused by fracture damage skin as

$$\Delta p_{s_f} = \left[\Delta p_B - 4\left(t\frac{dp}{dt}\right)_B\right] . \quad (7.71)$$

Unlike the case with manual linear flow analysis, this pressure drop represents only the pressure drop from fracture damage proper.

9. Calculate the fracture damage skin from

$$s_f = \frac{0.00708kh}{qB\mu}\Delta p_{s_f} .$$

7.5.4 Bilinear Flow Manual Log-Log Analysis Example.

7.5.4.1 Example Problem 7.7. Find the fracture half-length and skin factor for the buildup data set given in Tables 7.3 and 7.4.

7.5.4.2 Solution.

1. Graph the pressure and pressure derivative plotting functions vs. the time plotting function on a log-log scale. Because this is a gas-well buildup test, we use the adjusted pressure and the adjusted radial equivalent time, as shown in **Fig. 7.25.**

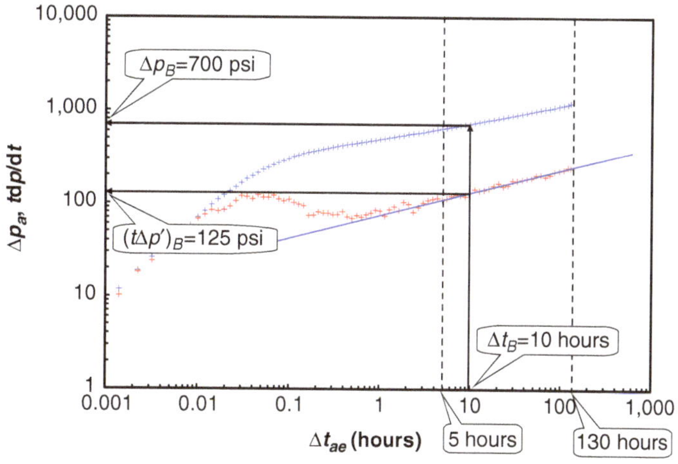

Fig. 7.25—Bilinear flow manual log-log analysis example.

2. Identify the bilinear flow period. We have previously identified the bilinear flow period as 5 hours to the end of the test at 130 hours.
3. Select a time Δt_B during the linear flow period. For this example, we chose $\Delta t_B = 10$ hours, but a time closer to the start of the bilinear flow period might give higher accuracy for the fracture damage skin pressure drop.

4. At the selected time, Δt_B, read the pressure, Δp_B, and pressure derivative, $(t\Delta p')_B$, plotting functions. From Fig. 7.25, we read $\Delta p_B = 700$ psi and $(t\Delta p')_B = 125$ psi at $\Delta t_B = 10$ hours.
5. Calculate the linear flow slope m_L from the time Δt_B and pressure derivative $(t\Delta p')_B$ as

$$m_B = \frac{4}{\Delta t_B^{1/4}}\left(t\frac{dp}{dt}\right)_B = \frac{4}{10^{1/4}}(125) = 281 \text{ psi/hr}^{1/4}.$$

6. Calculate the product $wk_f\sqrt{k}$ from the slope, m_B:

$$wk_f\sqrt{k} = \left(\frac{44.1 q_g B_{gi}\mu_{gi}}{h|m_B|}\right)^2 \sqrt{\frac{1}{\phi\mu_{gi}c_{ti}}}$$

$$= \left(\frac{(44.1)(3{,}500)(0.545)(0.0364)}{(135)(281)}\right)^2 \sqrt{\frac{1}{(0.075)(0.0364)(4.08\times10^{-5})}}$$

$$= 19.5.$$

7. If permeability is known, calculate the fracture conductivity from

$$wk_f = \frac{wk_f\sqrt{k}}{\sqrt{k}} = \frac{19.5}{\sqrt{0.03}} = 113 \text{ md-ft}.$$

8. Calculate the additional pressure drop caused by fracture damage skin as

$$\Delta p_{s_f} = \left[\Delta p_B - 4\left(t\frac{dp}{dt}\right)_B\right] = 700 - (4)(125) = 200 \text{ psi}.$$

9. Calculate the fracture damage skin from

$$s_f = \frac{0.00708kh}{q_g B_{gi}\mu_{gi}}\Delta p_{s_f}$$

$$= \frac{(0.00708)(0.03)(135)}{(3{,}500)(0.545)(0.0364)}(200)$$

$$= 0.083.$$

7.6 Estimating Fracture Properties from Flow Regime Limits

In many cases, the formation permeability is unknown. Although it is impossible to obtain a unique estimate of formation permeability from data exhibiting only bilinear or linear flow, some information about the ranges of permeability, fracture half-length, and fracture conductivity can be obtained from the times at which the bilinear and linear flow regimes start and end.

Poe et al. (1999) showed that it is theoretically possible to obtain unique estimates of formation permeability, fracture half-length, and fracture conductivity from the slope of the formation linear flow plot and the times at which formation linear flow starts and ends. However, their method makes two implicit assumptions:

- The start of formation linear flow occurs at the end of the transition period from bilinear flow to formation linear flow.
- The end of formation linear flow occurs as a transition into elliptical or PRF.

If either of these assumptions is violated, the Poe et al. (1999) method will not in general give the correct permeability, fracture half-length, and fracture conductivity. Wellbore storage often masks the transition from bilinear flow to formation linear flow, causing linear flow to begin later than it would have otherwise. The test might not be long enough for the end of formation linear flow to be observed.

Poe et al. (1999) recommends defining the start and end of specific flow regimes using the criterion of a 5% deviation of the true solution from the straight-line approximation for the flow regime. Using the 5% deviation criterion, Poe et al. (1999) developed a log-log piecewise linear approximation for the time for the start and end of formation linear flow:

$$t_{L_f D, bflf} = \frac{a}{C_{fD}^{\ b}}, \quad (7.72)$$

and

$$t_{L_f D, eflf} = \frac{c}{C_{fD}^{\ d}}, \quad (7.73)$$

where the coefficients for Eqs. 7.72 and 7.73 are given in **Table 7.10.**

$C_{fD\,min}$	$C_{fD\,max}$	*a*	*b*	*c*	*d*
0.03	0.3	0.523612	0.748563	0.723758	0.689208
0.3	1	0.43	0.882160	0.613	0.827161
1	3	0.43	1.200472	0.613345	1.074699
3	30	0.613871	1.522896	0.613345	1.074699
30	100	1.423943	1.770279	0.0414235	0.282308
100	1,000	3.423759	1.959880	0.0112884	0

Table 7.10—Coefficients for start and end of formation linear flow regime.

The following procedure can be used to estimate the formation permeability, fracture half-length, and fracture conductivity if the start and end of formation linear flow can be identified and the assumptions discussed above are met. If wellbore storage obscures the beginning of formation linear flow, the calculated fracture conductivity will be a lower bound on the true value. If the test ends before the end of formation linear flow, the formation permeability obtained will be an upper bound on the true value, while the calculated fracture half-length and calculated fracture conductivity will be lower bounds on the true values.

1. Use the straight-line analysis method discussed in Section 7.3 to obtain the slope, m_L, and the product $\sqrt{k}L_f$.
2. From the formation linear flow square-root-of-time graph, identify the time at which formation linear flow begins, t_{bflf}, and the time at which formation linear flow ends, t_{eflf}.
3. Calculate the ratio $R = \dfrac{t_{eflf}}{t_{bflf}}$. Find the row in **Table 7.11** such that $R_{min} \le R \le R_{max}$.

$C_{fD\,min}$	$C_{fD\,max}$	R_{min}	R_{max}	*e*	*f*
0.03	0.3	1.122518	1.334241	1.382239	0.02935
0.3	1	1.334241	1.426385	1.425581	0.055
1	3	1.426385	1.634827	1.426385	0.12577
3	30	1.634827	4.588478	0.999143	0.4482
30	100	4.588478	27.39754	0.029091	1.48797
100	1,000	27.39754	2,497.995	0.003297	1.95988

Table 7.11—Coefficients for duration of formation linear flow regime.

4. Estimate the dimensionless fracture conductivity, or its lower bound, C_{fD}, from

$$C_{fD} = \left(\frac{R}{e}\right)^{\frac{1}{f}}, \qquad (7.74)$$

where the coefficients *e* and *f* are obtained from Table 7.11.

5. Using Eq. 7.73 and the appropriate values for *c* and *d* from Table 7.10, calculate the dimensionless time to the end of formation linear flow, $t_{L_fD,eflf}$:

$$t_{L_fD,eflf} = \frac{c}{C_{fD}^{\;d}}.$$

6. Estimate the formation permeability, or its upper bound, from

$$k = \frac{141.2qB\mu}{h}\left(\frac{\sqrt{\pi t_{L_fD,eflf}}}{m_L\sqrt{t_{eflf}}}\right). \qquad (7.75)$$

7. Estimate the fracture half-length, or its lower bound, from

$$L_f = \sqrt{\frac{0.0002637k}{\phi\mu c_t}\left(\frac{t_{eflf}}{t_{L_fD,eflf}}\right)} \qquad (7.76)$$

8. Estimate the fracture conductivity, or its lower bound, from

$$wk_f = kL_fC_{fD}. \qquad (7.77)$$

7.6.1 Linear Flow Manual Log-Log Analysis Example.

7.6.1.1 Example Problem 7.8. Estimate the formation permeability, fracture half-length, and fracture conductivity from the time to start and end of linear flow for the buildup data set given in Tables 7.1 and 7.2.

7.6.1.2 Solution.

1. Using the straight-line analysis method for formation linear flow, we earlier found the slope, m_L, to be 68 psi/hr$^{1/2}$ and the product $\sqrt{k}L_f$ to be 98.3 md$^{1/2}$-ft.
2. From the formation linear flow square-root-of-time graph, identify the time at which formation linear flow begins, t_{bflf}, and the time at which formation linear flow ends, t_{eflf}. For this example, we will use the start and end of formation linear flow as determined from the log-log diagnostic plot: t_{bflf} = 0.35 hours and t_{eflf} = 16 hours.
3. Calculating the ratio $R = \frac{t_{eflf}}{t_{bflf}}$, we have $R = \frac{t_{eflf}}{t_{bflf}} = \frac{16}{0.35} = 45.7$, from which we find $100 \le C_{fD} \le 1{,}000$.
4. Estimate the dimensionless fracture conductivity, or its lower bound, C_{fD}, from

$$C_{fD} = \left(\frac{R}{e}\right)^{\frac{1}{f}} = \left(\frac{45.7}{0.003297}\right)^{\frac{1}{1.95988}} = 129.8.$$

This is a lower bound on fracture conductivity, because wellbore storage obscured the true beginning of formation linear flow.

5. Using Eq. 7.73 and the appropriate values for c and d from Table 7.10, we calculate the dimensionless time to the end of formation linear flow, $t_{L_fD,\,eflf}$, as

$$t_{L_fD,\,eflf} = \frac{c}{C_{fD}{}^{d}} = \frac{0.0112884}{(129.8)^0} = 0.0112884.$$

6. We estimate the formation permeability, or its upper bound, as

$$\begin{aligned} k &= \frac{141.2qB_{gi}\mu_{gi}}{h}\left(\frac{\sqrt{\pi t_{L_fD,eflf}}}{m_L\sqrt{t_{eflf}}}\right) \\ &= \frac{(141.2)(2{,}500)(0.586)(0.0309)}{(75)}\left(\frac{\sqrt{\pi(0.0112884)}}{68\sqrt{16}}\right) \\ &= 0.059 \text{ md.} \end{aligned}$$

7. We estimate the fracture half-length, or its lower bound, as

$$\begin{aligned} L_f &= \sqrt{\frac{0.0002637k}{\phi\mu_{gi}c_{ti}}\left(\frac{t_{eflf}}{t_{L_fD,\,eflf}}\right)} \\ &= \sqrt{\frac{(0.0002637)(0.059)}{(0.085)(0.0309)(5.13\times10^{-5})}\left(\frac{16}{0.0112884}\right)} \\ &= 405 \text{ ft}. \end{aligned}$$

8. We estimate the fracture conductivity, or its lower bound, as $wk_f = kL_fC_{fD} = (0.059)(405)(129.8) = 3{,}099$ md-ft. For comparison, the values used to generate this synthetic example are k = 0.05 md, L_f = 450 ft, wk_f = 5,000 ft, and C_{fD} = 222.

   Considering that the true start of formation linear flow was obscured by wellbore storage and that the end of formation linear flow was picked from the derivative on the log-log diagnostic plot instead of the square-root-of-time plot, the estimates obtained from the start and end of formation linear flow are in good agreement with the true values.

7.7 Area of Investigation

The radius of investigation was defined as a convenient way of estimating the region of the reservoir affecting the pressure response for a well exhibiting IARF. For a hydraulically fractured well, the pressure transient contours are elliptical rather than circular. In this case, an elliptical area of investigation can be defined as follows (Hale and Evers 1981). For linear flow, pseudosteady state begins at a dimensionless time, t_{bD}, of $1/\pi$:

$$\frac{0.0002637kt}{\phi\mu c_t b_i^2} = \frac{1}{\pi}. \quad (7.78)$$

Solving for b_i, we find

$$b_i = 0.02878\left(\frac{kt}{\phi\mu c_t}\right)^{1/2}. \tag{7.79}$$

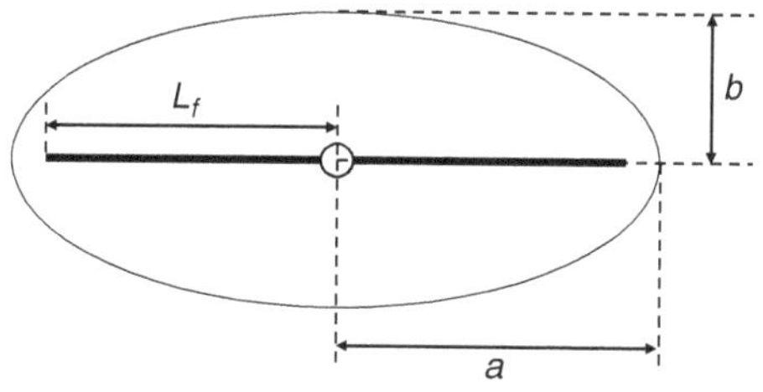

Fig. 7.26—Area of investigation for a hydraulically fractured well.

If we assume that Eq. 7.79 describes the depth of investigation along the semiminor axis of the ellipse shown in **Fig. 7.26,** we can calculate the semimajor axis, a_i, as

$$a_i = \sqrt{L_f^{\,2} + b_i^{\,2}}. \tag{7.80}$$

The area of investigation is then given by the area of the ellipse:

$$A_i = \pi a_i b_i. \tag{7.81}$$

7.8 Fracture Damage

Cinco-Ley and Samaniego-V. (1981b) proposed two models for fracture damage: a choked fracture damage skin factor model and a fracture face damage skin factor model. In the choked fracture skin model, the portion of the fracture nearest the wellbore is damaged; while in the fracture face damage skin factor model, the formation permeability adjacent to the fracture face is damaged.

7.8.1 Choked Fracture Damage Skin Factor Model. In the choked fracture damage skin factor model, the fracture conductivity near the wellbore is $(wk_f)_d$ over a distance L_d away from the wellbore, as shown in **Fig. 7.27.**

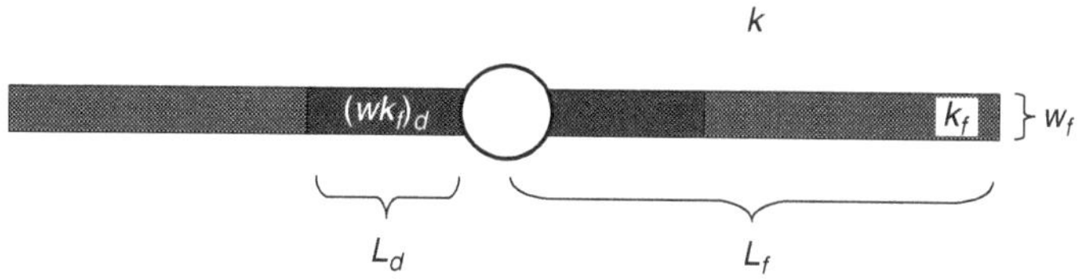

Fig. 7.27—Choked fracture damage skin factor model.

Different mechanisms that might cause choked fracture damage include proppant crushing or embedment because of excessive stress, overflushing of the proppant pushing the proppant pack away from the well, and producing back proppant leaving the near-wellbore portion of the fracture unpropped.

The skin factor caused by choked fracture damage can be calculated from

$$s_f = \frac{\pi k L_d}{(wk_f)_d}. \tag{7.82}$$

The total skin factor representing the combination of the hydraulic fracture stimulation skin and the choked fracture skin is obtained simply as the algebraic sum of the skin factor for an undamaged fracture of the same length, s, from Eq. 7.24, and the choked fracture skin factor, s_f, calculated from Eq. 7.82:

$$s_{tot} = s + s_f. \tag{7.83}$$

Fig. 7.28 shows the pressure response for an infinite-conductivity fracture with choked fracture damage skin factor and no wellbore storage.

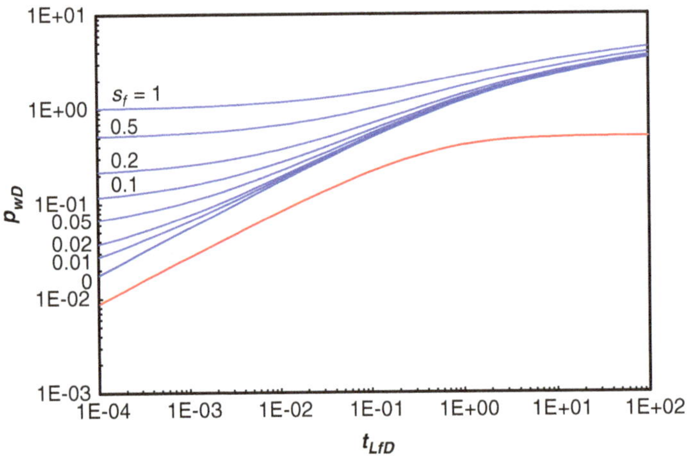

Fig. 7.28—Choked fracture skin factor type curve, infinite-conductivity fracture, no wellbore storage.

7.8.2 Fracture Face Damage Skin Factor Model. In the fracture face damage skin factor model, the formation permeability is reduced to k_d with a distance w_d around the face of the fracture, perhaps because of a reaction of fracturing fluid with the formation water, as shown in **Fig. 7.29.**

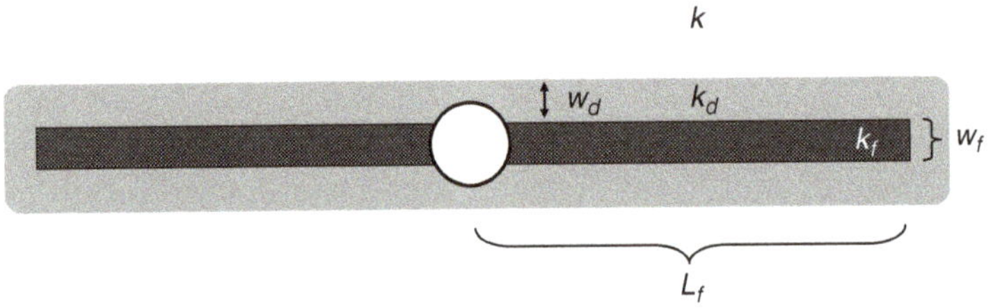

Fig. 7.29—Fracture face damage skin factor model.

The fracture face damage skin factor can be calculated from

$$s_f = \frac{\pi w_d}{2L_f}\left(\frac{k}{k_d} - 1\right). \quad \text{(7.84)}$$

The effect of fracture face damage skin factor on the pressure response for an infinite-conductivity fracture is similar to that of a choked fracture skin. It is unlikely that the two cases could be distinguished from field data.

7.9 Non-Darcy Effects

It has long been recognized that non-Darcy flow in the fracture might have a significant effect on the performance of a hydraulically fractured well (Wattenbarger and Ramey 1969; Holditch and Morse 1976). Guppy et al. (1982a) showed that, under constant-rate-drawdown conditions, non-Darcy flow in the fracture will cause the apparent fracture conductivity, as determined from data in bilinear flow, to be lower than it would be if there were only Darcy flow. They defined a dimensionless flow rate constant, $(q_{DND})_f$, in field units as

$$\left(q_{DND}\right)_f = \frac{1.027336\times10^{-15} k_f \rho\beta q}{w_f h\mu} \quad \text{(7.85)}$$

for oil wells, and

$$\left(q_{DND}\right)_f = \frac{4.64\times10^{-16} k_f \beta M q_g}{w_f h\mu_{gi}} \quad \text{(7.86)}$$

for gas wells. They proposed a correlation for the ratio of the apparent fracture conductivity to the true conductivity as a function of $(q_{DND})_f$:

$$\left(\frac{C_{fD,\,\text{true}}}{C_{fD,\,\text{apparent}}}\right) \approx 1 + 0.31\left(q_{DND}\right)_f, \quad (7.87)$$

valid for the following ranges:

- $C_{fD,\text{apparent}} \geq 2$ and $C_{fD,\text{true}} \leq 10$
- $1 \leq (q_{DND})_f \leq 20$ and $10 \leq C_{fD,\text{true}} \leq 100$
- $(q_{DND})_f \leq 10$ and $100 \leq C_{fD,\text{true}} \leq 500$

In a later paper, Guppy et al. (1982b) found a similar effect for buildup tests following production at constant flowing bottomhole pressure. However, a different correlation was required for the buildup case:

$$\left(\frac{C_{fD,\,\text{true}}}{C_{fD,\,\text{apparent}}}\right) \approx 1 + 0.55\left(q_{DND}\right)_f, \quad (7.88)$$

with a maximum error of 5% over the following ranges:

- $C_{fD,\text{true}} < 10$ and $(q_{DND})_f \leq 2$
- $C_{fD,\text{true}} \geq 10$ and $1 \leq (q_{DND})_f \leq 10$

Guppy et al. (1982b) proposed a method of obtaining the true dimensionless fracture conductivity if the apparent dimensionless fracture conductivity can be estimated from tests at two different rates. First, calculate the ratio of the apparent dimensionless conductivities at the two rates:

$$a \equiv \frac{C_{fD,\,\text{apparent},\,1}}{C_{fD,\,\text{apparent},\,2}}. \quad (7.89)$$

The true dimensionless conductivity can then be calculated from either of the following two equivalent expressions:

$$C_{fD} = C_{fD,\,\text{app},\,1}\frac{q_2 - q_1}{q_2 - aq_1} \quad (7.90)$$

and

$$C_{fD} = aC_{fD,\,\text{app},\,2}\frac{q_2 - q_1}{q_2 - aq_1}. \quad (7.91)$$

An alternative procedure, which can be used for two or more tests, is to graph $1/C_{fD,\,\text{apparent}}$ vs. q_g. The y-intercept of a straight line through the data points will be $1/C_{fD,\,\text{true}}$. This procedure can be used with either flow or buildup tests, but should not be used with a mixture of flow and buildup tests.

With regard to the effect of non-Darcy flow on estimated fracture half-length, the literature reports conflicting results. Holditch and Morse (1976) show that calculating the fracture half-length using an analysis method based on PRF will give a much smaller fracture length than would be obtained if there were no non-Darcy flow. However, for the test durations considered, the fracture half-length is underestimated even for the Darcy case. Guppy et al. (1982a, 1982b) do not address the issue of the effect of non-Darcy flow on apparent fracture half-length. Umnuayponwiwat et al. (2000) conclude that conventional methods that ignore non-Darcy flow would yield apparent fracture half-lengths that are lower than the true lengths. However, Gil et al. (2001) implicitly assume that the only effect of non-Darcy flow on the pressure response during formation linear flow is to introduce a constant-rate-dependent offset. Because the slope of the linear flow plot would be unchanged, linear flow analysis would yield the true fracture length even in the presence of non-Darcy flow.

7.10 Effect of Permeability Anisotropy on Estimated Fracture Half-Length

Many reservoirs have anisotropic permeability, often caused by the presence of one or more sets of pre-existing natural fractures. The pressure transient response for a vertical or hydraulically fractured well in an anisotropic reservoir cannot be distinguished from that for a well in an isotropic reservoir. If a well test on a well with a hydraulic fracture in an anisotropic reservoir is analyzed using methods assuming an isotropic reservoir model, the resulting fracture length and conductivity will not be the actual fracture length and conductivity (Gatens et al. 1991; Spivey and Lee 1999). Instead, the apparent fracture length and conductivity will be those of the equivalent isotropic system. If the hydraulic fracture parallels the existing natural fractures, both the apparent fracture length and the apparent fracture conductivity obtained using an isotropic model will be smaller than the true values.

If the x-axis is oriented parallel to the maximum principal permeability and the fracture lies at an angle θ to the x-axis, the apparent fracture half-length is given by

$$L_f' = L_f\sqrt{\frac{\bar{k}}{k_x}\cos^2\theta + \frac{\bar{k}}{k_y}\sin^2\theta}, \quad (7.92)$$

where $\bar{k}$ is the geometric mean permeability in the horizontal plane,

$$\bar{k} = \sqrt{k_x k_y}\,. \qquad (7.93)$$

For the common case where the hydraulic fracture is parallel to the existing natural fractures, the apparent fracture length is given simply as

$$L_f' = L_f\sqrt{\frac{\bar{k}}{k_x}} = L_f\left(\frac{k_y}{k_x}\right)^{1/4}. \qquad (7.94)$$

The apparent fracture conductivity, $w'k_f'$, is obtained from

$$w'k_f' = wk_f\sqrt{\frac{\bar{k}}{k_x}\cos^2\theta + \frac{\bar{k}}{k_y}\sin^2\theta}\,. \qquad (7.95)$$

Table 7.12 illustrates the effect of permeability anisotropy; even a moderate degree of anisotropy of 16:1 reduces the apparent length by a factor of two.

k_y/k_x	L_f'/L_f
4	0.707
10	0.562
16	0.5
81	0.333
100	0.3162
256	0.25

Table 7.12—Effect of permeability anisotropy on apparent fracture half-length.

7.11 Summary

The following points summarize this chapter:

- In the finite-conductivity fracture model, the fracture is characterized by its half-length, L_f, and conductivity, wk_f.
- A large number of flow regimes have been identified for fractured wells. Five of the flow regimes are particularly important: fracture linear flow, bilinear flow, formation linear flow, elliptical flow, and PRF.
- Fracture linear flow is rarely, if ever, observed because of its short duration.
- Bilinear flow occurs in finite-conductivity fractures.
- Data in bilinear flow exhibit a quarter-slope line on the derivative plot.
- If formation permeability is known, analysis of data in bilinear flow gives the best estimates of fracture conductivity, wk_f.
- During bilinear flow, the pressure transient has not reached the tips of the fracture.
- Formation linear flow occurs in high-conductivity fractures.
- Data in formation linear flow exhibit a half-slope line on the derivative plot.
- If formation permeability is known, analysis of data in formation linear flow gives the best estimates of fracture half-length.
- Analysis of data in elliptical flow or PRF gives estimates of the formation permeability and the composite skin factor, including the effects of the fracture, fracture conductivity, and fracture damage.
- The Cinco-Ley and Samaniego-V. (1981a) log-log type curves can be used to estimate formation permeability, fracture half-length, and fracture conductivity.
- Data in linear flow can be analyzed using the log-log diagnostic plot to estimate fracture half-length and fracture damage skin factor.
- Data in bilinear flow can be analyzed using the log-log diagnostic plot to estimate fracture conductivity and fracture damage skin factor.
- The observed time to the beginning and end of the formation linear flow regime can be used to place an upper bound on formation permeability, a lower bound on fracture half-length, and a lower bound on fracture conductivity.
- An elliptical area of investigation can be used to evaluate the depth to which a pressure transient has moved.
- Fracture damage models include fracture face damage and choked fracture damage.
- Non-Darcy flow in the fracture reduces the apparent fracture conductivity for both drawdown and buildup.
- Analysis of data from a fractured well in an anisotropic reservoir using an isotropic reservoir model might give an apparent fracture half-length that is much shorter than the true length.

7.12 Nomenclature

a_i = semimajor axis of ellipse in area of investigation equation, ft

a = ratio of apparent dimensionless fracture conductivities

A_t = semimajor axis of ellipse in elliptical flow equation, ft

A_i = area of investigation ellipse, ft^2

b_i = semiminor axis of ellipse in area of investigation equation, ft

B_t = semiminor axis of ellipse in elliptical flow equation, ft

B = formation volume factor, bbl/STB

b_B = y-intercept of bilinear flow graph, psi

b_{xE} = x-intercept of elliptical flow graph, dimensionless

B_g = gas formation volume factor, bbl/Mscf

b_L = y-intercept of linear flow graph, psi

c_f = formation compressibility, psi^{-1}

C_{fD} = dimensionless fracture conductivity (without π), $C_{fD} \equiv \dfrac{wk_f}{kL_f}$

C_r = dimensionless fracture conductivity (with π), $C_r \equiv \dfrac{wk_f}{\pi kL_f}$

c_t = total compressibility, psi^{-1}

k = formation permeability, md

k_f = fracture permeability, md

h = net-pay thickness, ft

L_f = fracture half-length, ft

L_{feq} = equivalent infinite-conductivity fracture half-length, ft

m = slope of PRF graph, psi/log cycle

m_B = slope of bilinear flow graph, psi/hr$^{1/4}$

m_E = slope of elliptical flow graph, psi

m_L = slope of linear flow graph, psi/hr$^{1/2}$

p_{1hr} = pressure reading from semilog straight line at a time of 1 hour

p_D = dimensionless pressure, $p_D \equiv \dfrac{0.00708kh}{qB\mu}\Delta p$

q = volumetric oil-flow rate, STB/D

q_g = volumetric gas-flow rate, Mscf/D

r_w = wellbore radius, ft

r_{wa} = apparent wellbore radius, ft

s = skin factor

s_f = fracture damage skin factor

t = time, hours

Δt_{eB} = radial equivalent time, $\Delta t_e = \dfrac{t_p \Delta t}{t_p + \Delta t}$, hours

Δt_{eB} = bilinear equivalent time, $\Delta t_{eB} = \left[t_p^{1/4} + \Delta t^{1/4} - \left(t_p + \Delta t\right)^{1/4}\right]^4$, hours

Δt_{eL} = linear equivalent time, $\Delta t_{eL} = \left(\sqrt{t_p} + \sqrt{\Delta t} - \sqrt{t_p + \Delta t}\right)^2$, hours

t_{eblf} = time at end of bilinear flow, hours

t_{eflf} = time at end of formation linear flow, hours

T_f = formation temperature, °F

t_{LfD} = dimensionless time based on fracture half-length, $t_{L_fD} \equiv \dfrac{0.0002637k}{\phi\mu c_t L_f^2} t$

$t_{LfD,bef}$ = dimensionless time at beginning of elliptical flow

$t_{LfD,bflf}$ = dimensionless time at beginning of formation linear flow

$t_{LfD,eblf}$ = dimensionless time at end of bilinear flow

$t_{LfD,eflf}$ = dimensionless time at end of formation linear flow

$t_{LfD,bprf}$ = dimensionless time at beginning of PRF

t_p = producing time, hours

w_f = fracture width, ft

wk_f = fracture conductivity, md-ft

β = velocity coefficient, ft^{-1}

γ_g = reservoir gas specific gravity (air = 1)

δ = maximum fractional error, dimensionless

η_{fD} = dimensionless fracture diffusivity, $\eta_{fD} \equiv \dfrac{\left(k/\phi c_t\right)_f}{\left(k/\phi c_t\right)}$

θ = angle between fracture and direction of maximum principal permeability

ϕ = porosity, fraction

μ = viscosity, cp

ρ = density, lbm/ft^3 (field units); g/cm^3 (consistent units)

Subscripts

a = adjusted variable

b = beginning of flow regime

B = bilinear flow

blf = bilinear flow

d = damage

D = dimensionless

e = end of flow regime

f = fracture property

g	=	gas	*ND*	=	non-Darcy
i	=	evaluated at initial pressure	*w*	=	water
L	=	linear flow	*wf*	=	flowing wellbore conditions
MP	=	match point	*ws*	=	shut-in wellbore conditions
max	=	maximum value or upper bound	*x,y,z*	=	spatial coordinates
min	=	minimum value or lower bound			

7.13 References

Cheng, Y., Lee, W. J., and McVay, D. A. 2007. A New Approach for Reliable Estimation of Hydraulic Fracture Properties in Tight Gas Wells. Presented at the SPE Hydraulic Fracturing Technology Conference, College Station, Texas, 29–31 January. SPE-105767-MS. https://doi.org/10.2118/105767-MS.

Chu, W. C. 1996. Application of Pressure Transient Testing for Stimulation Decisions. Presented at the SPE Annual Technical Conference and Exhibition, Denver, Colorado, 6–9 October. SPE-36529-MS. https://doi.org/10.2118/36529-MS.

Cinco-Ley, H. and Samaniego-V., F. 1981a. Transient Pressure Analysis for Fractured Wells. *J Pet Technol* **33** (9): 1749–1766. SPE-7490-PA. https://doi.org/10.2118/7490-PA.

Cinco-Ley, H. and Samaniego-V., F. 1981b. Transient Pressure Analysis: Finite Conductivity Fracture Case Versus Damaged Fracture Case. Presented at the SPE Annual Technical Conference and Exhibition, San Antonio, Texas, 5–7 October. SPE-10179-MS. https://doi.org/10.2118/10179-MS.

Cinco-L., H., Samaniego-V., F., and Dominguez-A., N. 1978. Transient Pressure Behavior for a Well With a Finite-Conductivity Vertical Fracture. *SPE J.* **04** (18): 253–264. SPE-6014-PA. https://doi.org/10.2118/6014-PA.

Gatens, J. M., Lee, W. J., Hopkins, C. E. et al. 1991. The Effect of Permeability Anisotropy on the Evaluation and Design of Hydraulic Fracture Treatments and Well Performance. Presented at the SPE Gas Technology Symposium, Houston, Texas, 23–25 January. SPE-21501-MS. https://doi.org/10.2118/21501-MS.

Gil, J. A., Ozkan, E., and Raghavan, R. 2001. Fractured-Well-Test Design and Analysis in the Presence of Non-Darcy Flow. Presented at the SPE Annual Technical Conference and Exhibition, New Orleans, Louisiana. 30 September–3 October. SPE-71573-MS. https://doi.org/10.2118/71573-MS.

Gringarten, A. C., Ramey, H. J., Jr., and Raghavan, R. 1974. Unsteady-State Pressure Distributions Created by a Well With a Single Infinite-Conductivity Vertical Fracture. *SPE J.* **14** (4): 347–360. SPE-4051-PA. https://doi.org/10.2118/4051-PA.

Gringarten, A. C., Ramey, H. J., Jr., and Raghavan, R. 1975. Applied Pressure Analysis for Fractured Wells. *J Pet Technol* **27**(7): 887–892. SPE-5496-PA. https://doi.org/10.2118/5496-PA.

Guppy, K. H., Cinco-Ley, H., Ramey, H. J., Jr. et al. 1982a. Non-Darcy Flow in Wells With Finite-Conductivity Vertical Fractures. *SPE J.* **22** (5): 681–698. SPE-8281-PA. https://doi.org/10.2118/8281-PA.

Guppy, K. H., Cinco-Ley, H., and Ramey, H. J., Jr. 1982b. Pressure Buildup Analysis of Fractured Wells Producing at High Flow Rates. *J Pet Technol* **34** (11): 2656–2666. SPE-10178-PA. https://doi:10.2118/10178-PA.

Hale, B. W. and Evers, J. F. 1981. Elliptical Flow Equations for Vertically Fractured Gas Wells. *J Pet Technol* **33** (12): 2489–2497. SPE-8943-PA. https://doi.org/10.2118/8943-PA.

Holditch, S. A. and Morse, R. A. 1976. The Effects of Non-Darcy Flow on the Behavior of Hydraulically Fractured Gas Wells. *J Pet Technol* **28** (10): 1169–1179. SPE-5586-PA. https://doi.org/10.2118/5586-PA.

Lee, W. J., Rollins, J. B., and Spivey, J. P. 2003. *Pressure Transient Testing*, Vol. 9. Richardson, Texas: Society of Petroleum Engineers.

Liao, Y. and Lee, W. J. 1993. Equivalent Drawdown Time for Pressure Buildup Analysis for Hydraulically Fractured Wells. Presented at the SPE Gas Technology Symposium, Calgary, Alberta, 28–30 June. SPE-26179-MS. https://doi.org/10.2118/26179-MS.

Manrique, J. F. and Poe, B. D., Jr. 2007. Evaluation and Optimization of Low Conductivity Fractures. Presented at the SPE Hydraulic Fracturing Technology Conference, College Station, Texas, 29–31 January. SPE-106317-MS. https://doi.org/10.2118/106317-MS.

Meyer, B. R. and Jacot, R. H. 2005. Pseudosteady-State Analysis of Finite Conductivity Vertical Fractures. Presented at the SPE Annual Technical Conference and Exhibition, Dallas, Texas, 9–12 October. SPE-95941-MS. https://doi.org/10.2118/95941-MS.

Poe, B. D., Jr., Conger, J. G., Farkas, R. et al. 1999. Advanced Fractured Well Diagnostics for Production Data. Presented at the SPE Annual Technical Conference and Exhibition, Houston, Texas, 3–6 October. SPE-56750-MS. https://doi.org/10.2118/56750-MS.

Rahim, Z. and Lee, W. J. 1989. A New Technique for Analyzing Buildup Tests From Hydraulically Fractured Wells. *SPE Form Eval* **4** (4): 535–540. SPE-15929-PA. https://doi.org/10.2118/15929-PA.

Ramey, H. J., Jr., and Gringarten, A. C. 1975. Effect of High Volume Vertical Fractures on Geothermal Steam Well Behavior. Presented at the Proceedings 2nd United Nations Symposium on the Use and Development of Geothermal Energy, San Francisco, California, 20–29 May.

Spivey, J. P. and Lee, W. J. 1999. Estimating the Pressure-Transient Response for a Horizontal or a Hydraulically Fractured Well at an Arbitrary Orientation in an Anisotropic Reservoir. *SPE Res Eval & Eng* **2** (5): 462–469. SPE-58119-PA. https://doi.org/10.2118/58119-PA.

Umnuayponwiwat, S., Ozkan, E., Pearson, C. M. et al. 2000. Effect of Non-Darcy Flow on the Interpretation of Transient Pressure Responses of Hydraulically Fractured Wells. Presented at the SPE Annual Technical Conference and Exhibition, Dallas, Texas, 1–4 October. SPE-63176-MS. https://doi.org/10.2118/63176-MS.

Veatch, R. W., King, G. E., and Holditch, S. A. 2017. Introduction. In *Essentials of Hydraulic Fracturing: Vertical and Horizontal Wellbores,* ed. S. Hill, Chap. 1, 24. Tulsa, Oklahoma: PennWell Corporation.

Wattenbarger, R. A. and Ramey, H. J., Jr. 1969. Well Test Interpretation of Vertically Fractured Gas Wells. *J Pet Technol* **21** (05): 625–632. SPE-2155-PA. https://doi.org/10.2118/2155-PA.

Wong, D. W., Harrington, A. G., and Cinco-Ley, H. 1986. Application of the Pressure Derivative Function in the Pressure Transient Testing of Fractured Wells. *SPE Form Eval* **1** (5): 470–480. SPE-13056-PA. https://doi.org/10.2118/13056-PA.

Appendix 7.A

7.A.1 Summary of Equations in Terms of Pressure, Adjusted Pressure, Pressure Squared, and Pseudopressure

The following tables present the more important equations for analyzing post-fracture drawdown and buildup tests in gas wells using pressure and time, adjusted pressure and time or adjusted time, pressure squared and time, and pseudopressure and time. In several cases, similar equations with the same numerical coefficients have been omitted.

The first column in each table gives the equation number in the text for the slightly compressible liquid case, while the second column gives the corresponding equation in terms of the appropriate time and pressure plotting functions.

If using software for analysis, there is no technical reason to use anything other than the adjusted-pressure/adjusted-time formulation or the pseudopressure/pseudotime formulation.

Eq.	Pressure Formulation
7.1	$t_{LfD} = \dfrac{0.0002637k}{\phi\mu c_t L_f^{\,2}}t$
7.2	$p_D = \dfrac{0.00708kh}{qB\mu}\Delta p$
7.8	$p_{wf} = p_i - \dfrac{44.1qB\mu}{h\sqrt{wk_f}}\left(\dfrac{t}{\mu\phi c_t k}\right)^{1/4} - \dfrac{141.2qB\mu}{kh}s_f$
7.13	$p_{wf} = p_i - 4.064\dfrac{qB}{hL_f}\sqrt{\dfrac{\mu t}{k\phi c_t}} - \dfrac{141.2qB\mu}{kh}s_f$
7.20	$B = 0.02434\left(\dfrac{k\Delta t_e}{\phi\mu c_t}\right)^{1/2}$
7.23	$p_{wf}(t) = p_i - \dfrac{70.6qB\mu}{kh}\left[\ln\left(\dfrac{kt}{1{,}688\phi\mu c_t r_w^{\,2}}\right) + s + s_f\right]$
7.28	$L_f = 4.064\dfrac{qB}{\lvert m_L\rvert h}\sqrt{\dfrac{\mu}{k\phi c_t}}$
7.30	$s_f = \dfrac{0.00708kh}{qB\mu}\left(b_L - p_{wf}\right)$
7.39	$wk_f = \left(\dfrac{44.1qB\mu}{h\lvert m_B\rvert}\right)^2\sqrt{\dfrac{1}{\phi\mu c_t k}}$
7.43	$L_f \approx 0.2266\left(\dfrac{kt_{eblf}}{\phi\mu c_t}\right)^{1/4}\left(\dfrac{wk_f}{k}\right)^{1/2}$
7.45	$k = \dfrac{162.6qB\mu}{\lvert m\rvert h}$
7.47	$s_t = 1.151\left[\dfrac{p_{1hr} - p_{wf}}{\lvert m\rvert} - \log\left(\dfrac{k}{\phi\mu c r_w^{\,2}}\right) + 3.23\right]$
7.52	$k = \dfrac{141.2qB\mu}{m_E h}$

Table 7.A-1—Analysis equations—pressure/time formulation.

Eq.	Adjusted-Pressure/Adjusted-Time Formulation
7.1	$t_{LfD} = \dfrac{0.0002637k}{\phi\mu_{gi}c_{ti}L_f^{\,2}}\Delta t_a$
7.2	$p_D = \dfrac{0.00708kh}{q_g B_{gi}\mu_{gi}}\Delta p_a$
7.8	$p_{awf} = p_{ai} - \dfrac{44.1q_g B_{gi}\mu_{gi}}{h\sqrt{wk_f}}\left(\dfrac{t}{\mu_{gi}\phi c_{ti}k}\right)^{1/4} - \dfrac{141.2q_g B_{gi}\mu_{gi}}{kh}s_f$
7.13	$p_{awf} = p_{ai} - 4.064\dfrac{q_g B_{gi}}{hL_f}\sqrt{\dfrac{\mu_{gi}t}{k\phi c_{ti}}} - \dfrac{141.2q_g B_{gi}\mu_{gi}}{kh}s_f$
7.20	$B = 0.02434\left(\dfrac{k\Delta t_{ae}}{\phi\mu_{gi}c_{ti}}\right)^{1/2}$
7.23	$p_{awf}(t) = p_{ai} - \dfrac{70.6q_g B_{gi}\mu_{gi}}{kh}\left[\ln\left(\dfrac{kt}{1{,}688\phi\mu_{gi}c_{ti}r_w^{\,2}}\right) + s + s_f\right]$
7.28	$L_f = 4.064\dfrac{q_g B_{gi}}{\lvert m_L\rvert h}\sqrt{\dfrac{\mu_{gi}}{k\phi c_{ti}}}$
7.30	$s_f = \dfrac{0.00708kh}{q_g B_{gi}\mu_{gi}}\left(b_L - p_{awf}\right)$
7.39	$wk_f = \left(\dfrac{44.1qB\mu}{h\lvert m_B\rvert}\right)^2\sqrt{\dfrac{1}{\phi\mu c_t k}}$
7.43	$L_f \approx 0.2266\left(\dfrac{kt_{aeblf}}{\phi\mu_{gi}c_{ti}}\right)^{1/4}\left(\dfrac{wk_f}{k}\right)^{1/2}$
7.45	$k = \dfrac{162.6q_g B_{gi}\mu_{gi}}{\lvert m\rvert h}$
7.47	$s_t = 1.151\left[\dfrac{p_{a1hr} - p_{awf}}{\lvert m\rvert} - \log\left(\dfrac{k}{\phi\mu_{gi}c_{ti}r_w^{\,2}}\right) + 3.23\right]$
7.52	$k = \dfrac{141.2q_g B_{gi}\mu_{gi}}{m_E h}$

Table 7.A-2—Analysis equations—adjusted-pressure/adjusted-time formulation.

Eq.	Pressure-Squared/Time Formulation
7.1	$t_{LfD} = \dfrac{0.0002637k}{\phi \bar{\mu}_g \bar{c}_t L_f^{\,2}} \Delta t$
7.2	$p_D = \dfrac{kh}{1422 q_g T \bar{z} \bar{\mu}_g} \Delta p^2$
7.8	$p_{wf}^{\,2} = p_i^{\,2} - \dfrac{444 q_g T \bar{z} \bar{\mu}_g}{h\sqrt{wk_f}} \left(\dfrac{t}{\bar{\mu}_g \phi \bar{c}_t k} \right)^{1/4} - \dfrac{1422 q_g T \bar{z} \bar{\mu}_g}{kh} s_f$
7.13	$p_{wf}^{\,2} = p_i^{\,2} - 40.93 \dfrac{q_g T \bar{z}}{hL_f} \sqrt{\dfrac{\bar{\mu}_g t}{k \phi \bar{c}_t}} - \dfrac{1422 q_g T \bar{z} \bar{\mu}_g}{kh} s_f$
7.20	$B_t = 0.02434 \left(\dfrac{k \Delta t_e}{\phi \bar{\mu}_g \bar{c}_t} \right)^{1/2}$
7.23	$p_{wf}^{\,2}(t) = p_i^{\,2} - \dfrac{711 q_g T \bar{z} \bar{\mu}_g}{kh} \left[\ln\left(\dfrac{kt}{1{,}688 \phi \bar{\mu}_g \bar{c}_t r_w^{\,2}} \right) + s + s_f \right]$
7.28	$L_f = 40.93 \dfrac{q_g T \bar{z}}{\lvert m_L \rvert h} \sqrt{\dfrac{\bar{\mu}_g}{k \phi \bar{c}_t}}$
7.30	$s_f = \dfrac{kh}{1422 q_g T \bar{z} \bar{\mu}_g} \left(b_L - p_{wf}^{\,2} \right)$
7.39	$wk_f = \left(\dfrac{444 q T \bar{z} \bar{\mu}_g}{h \lvert m_B \rvert} \right)^2 \sqrt{\dfrac{1}{\phi \bar{\mu}_g \bar{c}_t k}}$
7.43	$L_f \approx 0.2266 \left(\dfrac{k t_{eblf}}{\phi \bar{\mu}_g \bar{c}_t} \right)^{1/4} \left(\dfrac{wk_f}{k} \right)^{1/2}$
7.45	$k = \dfrac{1637 q_g T \bar{z} \bar{\mu}_g}{\lvert m \rvert h}$
7.47	$s_t = 1.151 \left[\dfrac{p_{1hr}^{\,2} - p_{wf}^{\,2}}{\lvert m \rvert} - \log\left(\dfrac{k}{\phi \bar{\mu}_g \bar{c}_t r_w^{\,2}} \right) + 3.23 \right]$
7.52	$k = \dfrac{1422 q_g T \bar{z} \bar{\mu}_g}{m_E h}$

Table 7.A-3—Analysis equations—pressure-squared/time formulation.

Eq.	Pseudopressure/Time Formulation
7.1	$t_{LfD} = \dfrac{0.0002637k}{\phi\bar{\mu}_g\bar{c}_t L_f^{\,2}}\Delta t$
7.2	$p_D = \dfrac{kh}{1422 q_g T}\Delta m(p)$
7.8	$m(p_{wf}) = m(p_i) - \dfrac{444 q_g T}{h\sqrt{wk_f}}\left(\dfrac{t}{\bar{\mu}_g\phi\bar{c}_t k}\right)^{1/4} - \dfrac{1422 q_g T}{kh}s_f$
7.13	$m(p_{wf}) = m(p_i) - 40.93\dfrac{q_g T}{hL_f}\sqrt{\dfrac{t}{k\phi\bar{\mu}_g\bar{c}_t}} - \dfrac{1422 q_g T}{kh}s_f$
7.20	$B = 0.02434\left(\dfrac{k\Delta t_e}{\phi\bar{\mu}_g\bar{c}_t}\right)^{1/2}$
7.23	$m(p_{wf}(t)) = m(p_i) - \dfrac{711 q_g T}{kh}\left[\ln\left(\dfrac{kt}{1{,}688\phi\bar{\mu}_g\bar{c}_t r_w^{\,2}}\right) + s + s_f\right]$
7.28	$L_f = 40.93\dfrac{q_g}{\lvert m_L\rvert h}\sqrt{\dfrac{1}{k\phi\bar{\mu}_g\bar{c}_t}}$
7.30	$s_f = \dfrac{kh}{1422 q_g T}\left(b_L - p_{wf}^{\,2}\right)$
7.39	$wk_f = \left(\dfrac{444 qT}{h\lvert m_B\rvert}\right)^2\sqrt{\dfrac{1}{\phi\bar{\mu}_g\bar{c}_t k}}$
7.43	$L_f \approx 0.2266\left(\dfrac{kt_{eblf}}{\phi\bar{\mu}_g\bar{c}_t}\right)^{1/4}\left(\dfrac{wk_f}{k}\right)^{1/2}$
7.45	$k = \dfrac{1637 q_g T}{\lvert m\rvert h}$
7.47	$s_t = 1.151\left[\dfrac{m(p_{1hr}) - m(p_{wf})}{\lvert m\rvert} - \log\left(\dfrac{k}{\phi\bar{\mu}_g\bar{c}_t r_w^{\,2}}\right) + 3.23\right]$
7.52	$k = \dfrac{1422 q_g T}{m_E h}$

Table 7.A-4—Analysis equations—pseudopressure/time formulation.

Chapter 8

Rate Transient Analysis

John Spivey

8.1 Introduction

In conventional gas reservoirs, pressure transient analysis is arguably the best tool for estimating permeability, quantifying formation damage, evaluating stimulation treatment effectiveness, and tracking average reservoir pressure. Unfortunately, in unconventional gas reservoirs, the time required to reach boundary-dominated flow, or even to see infinite-acting radial flow (IARF) in stimulated wells, makes routine use of pressure transient testing economically unjustifiable.

Gas material balance is one of the most accurate analytical tools for estimating initial gas-in-place for conventional gas reservoirs. However, without the ability to obtain accurate estimates of average reservoir pressure over time through pressure transient analysis, traditional material-balance methods cannot be applied to unconventional gas reservoirs.

Conventional decline-curve analysis (DCA) using the hyperbolic decline-curve equations (Arps 1945) remains the most common tool for production forecasting and reserves estimation for both unconventional and conventional oil and gas reservoirs. Although DCA was developed empirically, it has been placed on a solid theoretical foundation for depletion-drive reservoirs producing in boundary-dominated flow under constant-backpressure operating conditions for a wide range of reservoir drive mechanisms. Traditionally, hyperbolic exponents have been limited to the range of $0 \leq b < 1$, because production cannot be sustained indefinitely for $b \geq 1$. However, unconventional reservoirs often exhibit b values greater than unity during transient flow, which might last years. Thus, the use of DCA, like pressure transient and material-balance analysis, becomes problematic for unconventional reservoirs.

Modern rate transient analysis (RTA) combines elements of pressure transient analysis, material-balance analysis, and DCA to enable accurate formation evaluation, estimation of original gas-in-place, production forecasting, and reserves estimation at relatively low cost. As a result, RTA has almost completely replaced the use of pressure transient analysis and material-balance analysis for tight gas and other unconventional gas reservoirs. While RTA has not replaced DCA, nor is it likely to do so, it is often used in conjunction with DCA for production forecasting.

Section 8.2 gives a quick review of the theory and mathematics of DCA, followed by a discussion of the influence of reservoir drive mechanism on the hyperbolic exponent. The section closes with a consideration of the question "Can b be greater than unity?"

Section 8.3 introduces the rate vs. time type curve for a well producing at constant flowing bottomhole pressure from a closed circular reservoir and gives a procedure for estimating permeability, skin factor, and drainage area using the type curve.

The birth of modern RTA methods should be dated to the introduction of the Fetkovich type curves at the SPE Annual Fall Meeting in 1973, although the paper was not published until 1980 (Fetkovich 1980). Section 8.4 begins with a discussion of the development of the Fetkovich type curves by combining the theoretical rate-transient solution for constant-pressure production with the empirical Arps hyperbolic decline-curve equations, then continues with a brief discussion of how the accuracy of the method compares with results from pressure transient analysis from Fetkovich's original paper. The section then presents a procedure for using the Fetkovich type curve, along with a worked-out field example of a West Virginia gas well. The section concludes with a discussion of the use of the Arps decline-curve parameters q_i and D_i to back into an estimate of drainage area and gas-in-place, with the West Virginia gas well again featured as an example.

The usefulness of the Fetkovich type curves, along with the phenomenal success of the logarithmic derivative in pressure transient testing, prompted many authors to attempt to improve on the Fetkovich type curve, with varied degrees of success. Section 8.5 provides a brief discussion of one of the most successful of these attempts, the McCray time-averaged rate, and integral derivative type curves. Again, a worked-out example featuring the West Virginia gas well is presented. The section concludes with a brief discussion of cumulative-production vs. time type curves.

While the theory underpinning modern RTA is based on slightly compressible liquids, perhaps the dominant application is in low-permeability gas and other unconventional gas reservoirs, in which pressure-dependent fluid properties, and often even rock properties, must be included in the analysis. Section 8.6 begins by introducing adjusted pressure and adjusted time, integral transforms of the pressure and time plotting variables that allow the Fetkovich type curves to be used for gas wells by accounting for the effects of pressure-dependent gas properties on both flow and storage considerations. The section then shows how the adjusted pressure and adjusted time transforms can be modified to incorporate pressure-dependent rock properties as well. The section then discusses the impact of stress-dependent rock properties, giving an overview of the Jones (1988) two-point model for stress-dependent porosity and permeability. The section continues with a discussion of how the standard gas material balance equations can be modified to include pressure-dependent porosity and water formation volume factor, both factors that might become very important for overpressured, low-porosity, high-water-saturation tight gas reservoirs. The section concludes with the presentation of an iterative procedure for using the Fetkovich type curves to analyze gas-well production data using the adjusted pressure and adjusted time transforms to account for pressure-dependent fluid and rock properties.

Although both the original and the modified Fetkovich type curves proved to be successful for wells producing at essentially constant pressure, their application to real wells experiencing fluctuations in both pressure and rate was problematic. Two innovations, rate-normalization and material-balance time, enabled variable-rate/variable-pressure production data to be analyzed as equivalent constant-rate cases, as discussed in Section 8.7. The section then gives procedures for using rate-normalization and material-balance time, with modifications for pressure-dependent fluid and rock properties, for both infinite-acting and boundary-dominated flow, illustrating each procedure with a worked-out example for a synthetic, variable-rate/variable-pressure production data set. The section concludes by applying rate-normalization and material-balance time to analyze boundary-dominated flow in the West Virginia gas-well production data.

Section 8.8 extends the application of rate-normalized pressure change and material-balance time to the use of pressure vs. time type curves for a well producing at constant rate from a closed circular reservoir, again illustrating their use with the West Virginia gas-well data set.

Section 8.9 covers the Agarwal-Gardner family of type curves based on solutions for production at constant pressure, but drawn to emphasize the correspondence to DCA.

Section 8.10 discusses the extended material-balance method for analyzing boundary-dominated flow and illustrates yet another analysis method of the West Virginia gas-well data set.

Section 8.11 summarizes the results obtained by the various methods for the West Virginia gas-well example.

Section 8.12 covers practical considerations and recommendations.

8.2 Conventional Decline-Curve Analysis

Forecasting future performance and estimating reserves have been primary concerns of the reservoir engineer since the inception of the petroleum industry. One of the simplest tools for forecasting performance and estimating reserves is DCA. Empirical equations are used to fit the historical production trend of a well or field. Extrapolation of the trend into the future gives the desired production forecast; extrapolation to an economic limit or abandonment rate gives the desired reserves estimate.

8.2.1 Arps Decline-Curve Equations. The equations most commonly used are often referred to as the Arps decline-curve equations for hyperbolic decline, characterized by a parameter b called the decline-curve exponent (Arps 1945). Two special cases of hyperbolic decline are commonly used: exponential decline, where b is zero, and harmonic decline, where b is unity.

The decline rate, D, is defined as the fractional decrease in flow rate per unit time, or

$$D=-\frac{1}{q}\frac{\mathrm{d}q}{\mathrm{d}t}=-\frac{\mathrm{d}\ln q}{\mathrm{d}t}. \qquad (8.1)$$

In general, the decline rate will be a function of time.

8.2.1.1 Exponential Decline. If the decline rate is constant, Eq. 8.1 can be integrated to obtain the equation for exponential decline (also called constant percentage decline):

$$q(t)=q_i e^{-D_i t}, \qquad (8.2)$$

where D_i is the initial decline rate. For exponential decline, because the decline rate is constant, the decline rate at any time t is equal to D_i.

Integrating Eq. 8.2 with respect to time gives the cumulative production at time t for a well or field experiencing exponential decline as

$$N_p(t)=\frac{q_i-q(t)}{D_i}. \qquad (8.3)$$

If the economic limit rate, q_a, is known, the cumulative production at abandonment, N_{pa}, can be calculated as

$$N_{pa}=\frac{q_i-q_a}{D_i}. \qquad (8.4)$$

8.2.1.2 Hyperbolic Decline. In many cases, forecasts produced using Eqs. 8.2 and 8.3 are too conservative. Arps (1945) defined the equation for hyperbolic decline as

$$q(t)=\frac{q_i}{(1+bD_i t)^{1/b}}. \qquad (8.5)$$

Integrating Eq. 8.5 with respect to time gives the cumulative production at time t for a well experiencing hyperbolic decline as

$$N_p(t)=\frac{q_i^b}{D_i(1-b)}\left[q_i^{1-b}-q(t)^{1-b}\right]. \qquad (8.6)$$

The cumulative production at abandonment for hyperbolic decline is

$$N_{pa}=\frac{q_i^b}{D_i(1-b)}\left(q_i^{1-b}-q_a^{1-b}\right). \qquad (8.7)$$

	Exponential (b = 0)	Hyperbolic	Harmonic (b = 1)
$q(t)$	$q(t)=q_ie^{-D_it}$	$q(t)=\frac{q_i}{(1+bD_it)^{1/b}}$	$q(t)=\frac{q_i}{1+D_it}$
$N_p(t)$	$N_p(t)=\frac{q_i-q(t)}{D_i}$	$N_p(t)=\frac{q_i^b}{D_i(1-b)}(q_i^{1-b}-q^{1-b})$	$N_p(t)=\frac{q_i}{D_i}\ln\frac{q_i}{q}$
r	$r=\frac{q_i}{q_a}$	$r=\frac{q_i}{q_a}$	$r=\frac{q_i}{q_a}$
t_a	$t_a=\frac{\ln r}{D_i}$	$t_a=\frac{r^b-1}{bD_i}$	$t_a=\frac{r-1}{D_i}$
N_{pa}	$N_{pa}=\frac{q_i-q_a}{D_i}$	$N_{pa}=\frac{q_i^b}{D_i(1-b)}(q_i^{1-b}-q_a^{1-b})$	$N_{pa}=\frac{q_i}{D_i}\ln\frac{q_i}{q_a}$
$D(t)$	$D(t) = D_i$	$D(t)=\frac{D_i}{1+bD_it}$	$D_{(t)}=\frac{D_i}{1+D_it}$
D_{ei}	$D_{ei}=1-e^{-D_i}$	$D_{ei}=1-\frac{1}{(1+bD_i)^{1/b}}$	$D_{ei}=\frac{D_i}{1+D_i}$
D_i	$D = -\ln(1-D_e)$	$D_i=\frac{1}{b}\left[\frac{1}{(1-D_{ei})^b}-1\right]$	$D_i=\frac{D_{ei}}{1-D_{ei}}$

Table 8.1—Summary of Arps decline-curve equations.

8.2.1.3 Harmonic Decline. The special case where $b = 1$ is called harmonic decline. For harmonic decline, the rate can be calculated from

$$q(t)=\frac{q_i}{1+D_it}, \qquad (8.8)$$

while the cumulative production N_p is calculated from

$$N_p(t)=\frac{q_i}{D_i}\ln\frac{q_i}{q}. \qquad (8.9)$$

The cumulative production to an abandonment rate of q_a is given by

$$N_{pa}=\frac{q_i}{D_i}\ln\frac{q_i}{q_a}. \qquad (8.10)$$

Table 8.1 summarizes the equations for exponential, hyperbolic, and harmonic decline. Equations are included for rate, q; cumulative production, N_p; cumulative production at abandonment, N_{pa}; time to abandonment, t_a; and decline rate, D, as functions of time, and equations for converting initial effective decline rate, D_{ei}, to initial decline rate, D_i, and vice versa. The time to abandonment is the time at which the production rate has declined to the abandonment rate, q_a.

The effective decline rate D_e is defined as the fractional change in production rate during one unit of time:

$$D_e=\frac{1}{q(t)}\left[\frac{q(t)-q(t+1)}{(t+1)-t}\right]=\frac{q(t)-q(t+1)}{q(t)}. \qquad (8.11)$$

The initial effective decline rate is thus

$$D_{ei}=\frac{q_i-q(1)}{q_i}. \qquad (8.12)$$

8.2.1.4 Exponential Decline—Semilog Rate vs. Time Straight Line Analysis. For exponential decline, a semilog graph of rate vs. time (with rate on the logarithmic axis) will show a straight line. The intercept of the straight line gives the initial rate, q_i. The initial decline rate can be calculated from

$$D_i=-2.303m_1, \qquad (8.13)$$

where m_1 is the slope of the straight line on the semilog graph of rate vs. time in log cycles per unit time.

8.2.1.5 Exponential Decline—Cartesian Rate–Cumulative Straight Line Analysis. A graph of flow rate vs. cumulative production on a Cartesian graph will also give a straight line. The initial rate q_i can be read directly from the intercept. The initial decline rate can be obtained from

$$D_i = -m_2, \quad (8.14)$$

where m_2 is the slope of the straight line on the Cartesian graph of rate vs. cumulative production in units of inverse time.

8.2.1.6 Harmonic Decline—Semilog Rate–Cumulative Straight Line Analysis. For harmonic decline, a graph of flow rate vs. cumulative production will give a straight line with slope m and intercept b on a semilog graph, where the flow rate is graphed on the logarithmic axis. The initial rate q_i is read directly from the intercept b, while the initial decline rate is obtained from

$$D_i = -2.303 q_i m_3, \quad (8.15)$$

where m_3 is the slope of the straight line on the semilog graph of rate vs. cumulative production in log cycles per unit volume of fluid produced.

8.2.2 Hyperbolic Exponent and Drive Mechanism. The hyperbolic exponent is difficult to determine from field production data, especially during the early life of a field or well when major investment decisions must be made, and production forecasts and estimates of ultimate recovery are particularly important. A number of studies (Mead 1956; Matthews and Lefkovits 1956; Lefkovits and Matthews 1958; Fetkovich 1980; Maley 1985; Fetkovich et al. 1990; Fetkovich et al. 1996; Spivey et al. 2001) have shown that the hyperbolic exponent is largely controlled by the reservoir drive mechanism. Thus, if the drive mechanism can be identified, the hyperbolic exponent can be preselected when the production history is limited. **Table 8.2** shows typical values for the hyperbolic exponent for a variety of drive mechanisms for both oil and gas.

Cutler (1924) published data for a large number of hyperbolic field decline curves. Using Cutler's data, Arps (1945) found that b ranged from 0.0 to 0.7, with 80% of the fields having b less than 0.4. Although none of the decline curves in Cutler's data set exhibited b greater than 0.7, Arps noted that, in his own experience, harmonic decline (b = 1) did occur occasionally.

On the basis of empirical evidence, Mead (1956) suggested ranges for the hyperbolic exponent for oil reservoirs producing under solution-gas drive, gas-cap drive, gravity drainage, gas-pressure maintenance, and water-pressure maintenance. Matthews and Lefkovits (1956) and Lefkovits and Matthews (1958) showed that an oil reservoir producing under gravity drainage should have a b value of 0.5.

Fetkovich (1980) showed that single-layer undersaturated oil reservoirs producing at constant flowing bottomhole pressure would exhibit exponential decline. Fetkovich also showed that, for oil reservoirs producing by solution-gas drive, b should be between 0.333 and 0.667. Fetkovich et al. (1996) recommended using a value of b = 0.3 for a solution-gas-drive reservoir unless b can clearly be determined from field data.

Fetkovich (1980) showed that b should be between 0 and 0.5 for single-layer gas reservoirs producing under depletion drive at constant flowing bottomhole pressure. For Darcy flow, b is zero in the limit of small drawdowns and 0.5 in the large-drawdown limit. Fetkovich suggested a b value of 0.4 for gas wells producing under typical operating conditions.

Oil	
Drive Mechanism	b
Single-phase, volumetric depletion (Fetkovich 1980)	0
Gravity drainage (Matthews and Lefkovits 1956)	0.5
Solution-gas drive (Fetkovich 1980)	0.333–0.667
Two-layer, single-phase oil, volumetric depletion (Fetkovich et al. 1990)	0.0–0.7
Gas	
Drive Mechanism	b
Single-phase, volumetric depletion, high drawdown, backpressure exponent n = 1 (100% Darcy flow) (Fetkovich 1980)	0.5
Single-phase, volumetric depletion, high drawdown, backpressure exponent n = 0.5 (100% non-Darcy flow) (Fetkovich 1980)	0
Dry-gas reservoir with constant backpressure equal to 10% of initial reservoir pressure (Fetkovich et al. 1990)	0.4
Two-layer, single-phase gas, volumetric depletion, n = 1 (Fetkovich et al. 1990)	0.4–0.8
Coalbed methane, $p >> p_L$ (Spivey et al. 2001)	1.5
Coalbed methane, $p = p_L$ (Spivey et al. 2001)	1.0
Coalbed methane, $p << p_L$ (Spivey et al. 2001)	0.5
Low-permeability reservoir with long, high-conductivity hydraulic fracture during formation linear flow (Maley 1985)	2
Low-permeability reservoir with long, finite-conductivity hydraulic fracture during bilinear flow (Spivey et al. 2001)	4
Naturally fractured shale, matrix linear flow (Spivey et al. 2001)	2

Table 8.2—Influence of drive mechanism on decline-curve hyperbolic exponent.

In general, multilayer reservoirs with contrasting layer properties exhibit higher hyperbolic exponents than do single-layer reservoirs with the same drive mechanism. Fetkovich et al. (1990) showed that a two-layer gas reservoir, in which each layer would have a hyperbolic exponent of 0.4 if produced independently, would exhibit a composite b value between 0.4 and 0.8 if produced commingled. Similarly, Fetkovich et al. (1996) reported that a two-layer oil reservoir, in which each layer would decline exponentially if produced separately, would exhibit a composite b value between 0 and 0.7 if production were commingled. A reservoir in which the layers depleted at the same rate would act like a single-layer reservoir, while a reservoir in which one layer depleted significantly faster than the other would give a much higher b-value than either layer alone.

Maley (1985) showed that a well with a massive hydraulic fracture in a low-permeability reservoir would exhibit a hyperbolic exponent of 2 during transient linear flow.

Spivey et al. (2001) showed that a coalbed methane reservoir producing with a backpressure exponent of unity (Darcy flow) would exhibit a b-value of 1.5 at average reservoir pressures much greater than the Langmuir pressure, declining to a b-value of 0.5 at reservoir pressure much lower than the Langmuir pressure.

Spivey et al. (2001) also showed that a transient dual porosity reservoir will exhibit a hyperbolic exponent of 2 if the pressure transient in the fracture system reaches the reservoir boundaries before pseudosteady-state flow is established in the matrix. They found that this phenomenon would occur for reservoirs in which $\omega \leq 0.01$ and $\lambda R_D^2 \leq 0.1$, where ω and λ are the storativity ratio and interporosity flow coefficient, respectively, and R_D is the dimensionless drainage radius, defined as

$$R_D = \frac{r_e}{r_w}. \qquad (8.16)$$

Many of the phenomena that cause high values of the hyperbolic exponent, such as gas wells with large drawdowns, multilayer reservoirs, massive hydraulic fractures, desorption, and natural fracturing, are usually found individually or in combination in unconventional gas reservoirs. Thus, unconventional gas reservoirs will almost always exhibit high b-values, particularly during early production.

8.2.3 Can *b* Be Greater Than Unity? One question that has been debated extensively is whether the hyperbolic exponent b can be greater than unity. Using the Arps decline-curve equations, it is easy to show that

$$\lim_{t\to\infty} N_p(t) = \begin{cases} \dfrac{q_i}{D_i(1-b)}, & b<1 \\ \infty, & b \geq 1 \end{cases}; \qquad (8.17)$$

that is if $b \geq 1$, the cumulative production is infinite for infinite producing time.*

Obviously, infinite cumulative production is not a physically realistic result, even assuming infinite producing time. Nevertheless, considering the theoretical work discussed in Section 8.2.1, b-values larger than unity are theoretically justified and are often observed in practice. How can this apparent paradox be resolved?

Significantly, in every situation listed in Table 8.2 for which b is greater than unity, the requisite conditions cannot be sustained indefinitely. For a well with a long hydraulic fracture, linear flow or bilinear flow will end when the pressure transient reaches the reservoir boundaries and boundary-dominated flow is established. The hyperbolic exponent will then decrease to a value less than unity. For a coalbed methane reservoir, the hyperbolic exponent will decline to a value less than unity after the reservoir pressure has dropped below the Langmuir pressure. For a transient dual-porosity reservoir, the hyperbolic exponent will fall below unity after pseudosteady-state flow has been established in the matrix.

In each of these cases, the duration of the flow period in which b is greater than unity cannot be determined solely from production data during that flow period. Thus, reserves must not be estimated by extrapolating the curve for $b \geq 1$ to an abandonment rate. It is common practice to switch to a lower value of b at some point during a forecast. External data can be used to determine when to switch from $b \geq 1$ to $b < 1$.

A rigorous method to determine the time to switch to $b < 1$ would be to estimate reserves volumetrically, then to constrain the decline-curve extrapolation. A volumetric reserves estimate would require external data, such as the drainage area or well spacing, for a well with a long hydraulic fracture; the drainage area and storativity ratio ω for a naturally fractured shale; and the drainage area and Langmuir pressure for a coalbed methane reservoir.

A more common method to extrapolate decline curves is to switch to $b < 1$ after a specified time or after a specified minimum decline rate has been reached. When this method is used, the reserves estimate obtained by extrapolating the final hyperbolic segment to an abandonment rate is an analogy-based reserves estimate rather than a performance-based reserves estimate.

8.3 Rate Type Curves for a Well Produced at Constant Flowing Bottomhole Pressure

Fig. 8.1 shows the dimensionless rate for a well producing at constant flowing bottomhole pressure from a closed circular reservoir. The dimensionless time t_D is defined as

$$t_D = \frac{0.00633k}{\phi \mu c_t r_{wa}^2} t, \qquad (8.18)$$

where time t is in days and r_{wa} is the apparent wellbore radius, defined as

$$r_{wa} = r_w e^{-s}. \qquad (8.19)$$

*It is interesting to note that the case of harmonic decline, with $b = 1$, exhibits the same behavior that is often cited in objections to the use of $b > 1$.

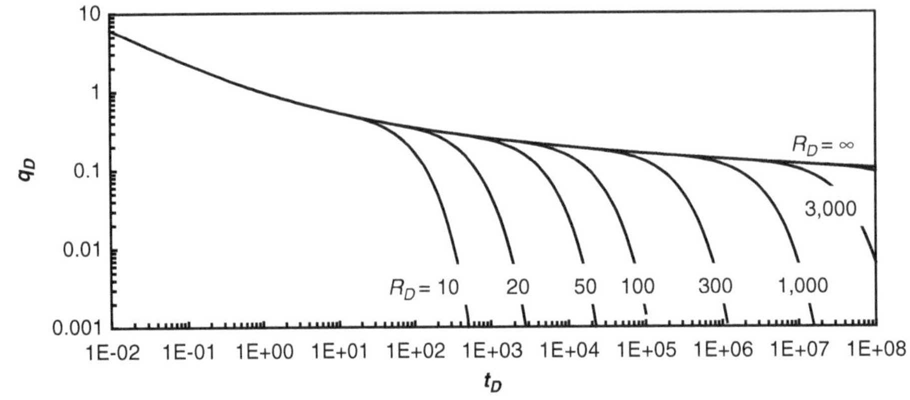

Fig. 8.1—Rate transient type curve for constant-pressure production from a closed circular reservoir.

The dimensionless flow rate q_D is defined as

$$q_D = \frac{141.2\mu B}{kh(p_i - p_{wf})} q. \quad (8.20)$$

The dimensionless drainage radius R_D is defined as

$$R_D = \frac{r_e}{r_{wa}} = \frac{r_e}{r_w e^{-s}}. \quad (8.21)$$

Fig. 8.1 can be used to analyze rate transient data for wells produced at constant flowing bottomhole pressure. The following procedure is recommended:

1. Construct the field-data graph by plotting the production rate q vs. time t on a log-log scale.
2. Identify the transient flow period from the concave upward portion of the field-data graph. Identify the boundary-dominated flow period from the concave downward portion of the field-data graph.
3. Place the field-data graph on top of the rate vs. time type curve.
4. Move the field-data graph vertically and horizontally as necessary to match the field data during the transient flow period to the $R_D = \infty$ stem, keeping the field-data and type curve axes parallel.
5. Read the time and rate match points (t, t_D) and (q, q_D).
6. Calculate the permeability from the rate match point as

$$k = \frac{141.2B\mu}{h(p_i - p_{wf})} \left(\frac{q}{q_D} \right)_{MP}. \quad (8.22)$$

7. Calculate the apparent wellbore radius r_{wa} from the time match point as

$$r_{wa} = \sqrt{\frac{0.00633k}{\phi\mu c_t} \left(\frac{t}{t_D} \right)_{MP}}. \quad (8.23)$$

8. Calculate the skin factor s from r_{wa} as

$$s = -\ln\left(\frac{r_{wa}}{r_w} \right). \quad (8.24)$$

9. If boundary-dominated flow is present, identify the late-time R_D stem that best fits the field data, interpolating if necessary.
10. Calculate the drainage radius from

$$r_e = R_D r_{wa}. \quad (8.25)$$

11. Calculate the drainage area from

$$A = \pi\left(r_e^2 - r_{wa}^2\right). \quad (8.26)$$

Although useful, the constant-pressure-production type curve in Fig. 8.1 is difficult to use in practice because the R_D stems differ only in the slope of the curve at the dimensionless time at which a given stem deviates from infinite-acting behavior. After the onset of boundary-dominated flow, all stems exhibit exponential decline behavior. This observation motivated the development of the Fetkovich type curve, as discussed in the next section.

8.4 Fetkovich Type Curve

8.4.1 Development of Fetkovich Type Curve. Fetkovich (1980) noted that the Arps equations could be written in dimensionless form by defining a decline-curve dimensionless time, t_{dD}; dimensionless rate, q_{dD}; and dimensionless cumulative production, Q_{dD}, as

$$t_{dD} = D_i t, \quad (8.27)$$

$$q_{dD} = \frac{q}{q_i}, \quad (8.28)$$

and

$$Q_{dD} = \frac{D_i}{q_i} N_p, \quad (8.29)$$

respectively. Substituting Eqs. 8.27 and 8.28 into Eqs. 8.2, 8.5, and 8.8, we have

$$q_{dD} = e^{-t_{dD}}, \quad (8.30)$$

for exponential decline,

$$q_{dD} = \frac{1}{(1 + bt_{dD})^{1/b}} \quad (8.31)$$

for hyperbolic decline, and

$$q_{dD} = \frac{1}{1 + t_{dD}} \quad (8.32)$$

for harmonic decline.

Fig. 8.2 shows Eqs. 8.2, 8.5, and 8.8 plotted as dimensionless rate, q_{dD}, vs. dimensionless time, t_{dD}, for various values of the hyperbolic exponent b. Fig. 8.2 can be used to analyze rate transient production data by type-curve matching in much the same way pressure transient test data are analyzed. The time-match-point ratio gives the initial decline rate, D_i, as

$$D_i = \left(\frac{t_{dD}}{t} \right)_{MP}; \quad (8.33)$$

the rate-match-point ratio gives the initial rate, q_i, as

$$q_i = \left(\frac{q}{q_{dD}} \right)_{MP}; \quad (8.34)$$

and the matching stem gives the hyperbolic exponent, b.

8.4.2 Combining the Arps Type Curves with the Constant-Pressure Solution. Fetkovich (1980) also noted that the dimensionless rate for a well producing at constant flowing bottomhole pressure from a closed circular reservoir declines exponentially after boundary-dominated flow has been reached, as shown in Fig. 8.1.

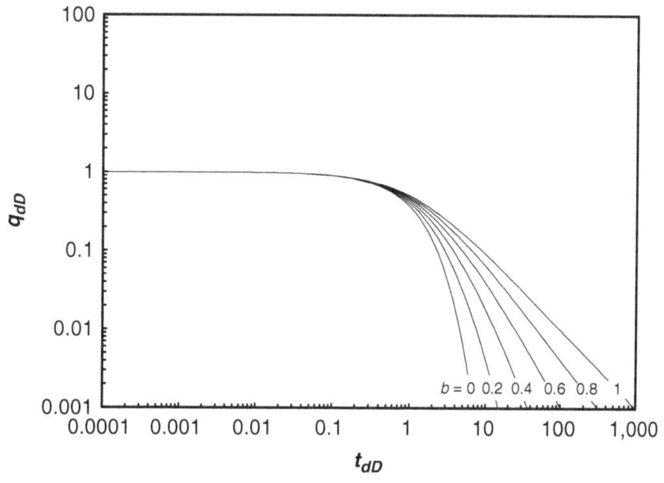

Fig. 8.2—Decline type curves based on the dimensionless Arps equations.

Fetkovich related the decline-curve dimensionless time t_{dD} and rate q_{dD} to t_D and q_D as

$$t_{dD}=\frac{t_D}{\frac{1}{2}\left(R_D^2-1\right)\left[\ln(R_D)-\frac{1}{2}\right]} \quad (8.35)$$

and

$$q_{dD}=\left[\ln\left(R_D\right)-\tfrac{1}{2}\right]q_D. \quad (8.36)$$

When the analytical solutions in Fig. 8.1 are plotted as q_{dD} vs. t_{dD}, as defined by Eqs. 8.35 and 8.36, the late-time boundary-dominated portions of each curve fall on a single exponential decline curve, as shown in **Fig. 8.3.**

The Fetkovich decline type curve in **Fig. 8.4** shows the analytical solution for constant-pressure production (Fig. 8.3) plotted on the same scale as the Arps decline curves (Fig 8.2).

The Fetkovich type curve displays two flow regimes:

- IARF, often called transient flow
- Boundary-dominated flow

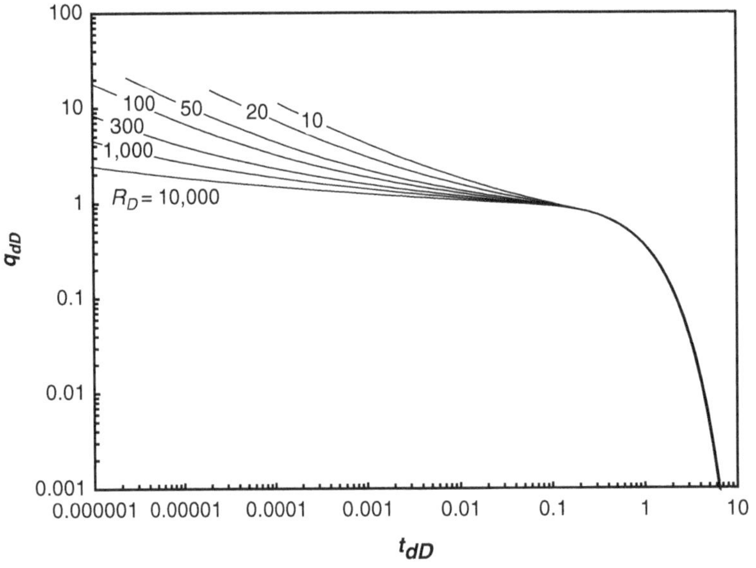

Fig. 8.3—Rate transient solution for constant-pressure production from a closed circular reservoir plotted as decline dimensionless rate q_{dD} vs. decline dimensionless time t_{dD}.

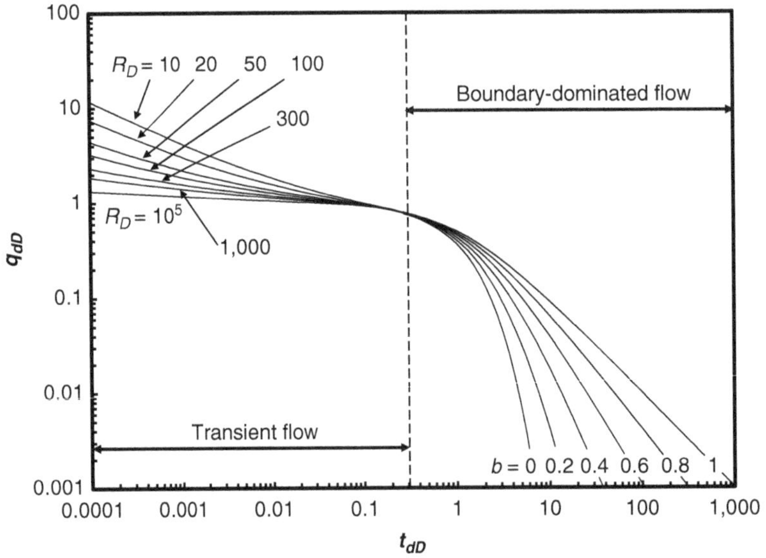

Fig. 8.4—Fetkovich decline type curve.

The transient flow period lasts until a dimensionless time t_{dD} of 0.3, at which time the boundary-dominated flow period begins. During transient flow, the rate is concave upward; during boundary-dominated flow, the rate is concave downward. The Fetkovich type curve has two sets of type-curve stems, corresponding to the transient and boundary-dominated flow periods.

The terms boundary-dominated flow, pseudosteady-state flow, and semisteady-state flow are often used interchangeably. The author of this chapter prefers to reserve the use of pseudosteady-state flow or semisteady-state flow to refer to constant-rate production. Pseudosteady-state flow has been defined as a state in which a period of constant-rate production has lasted long enough that the rate of pressure change is the same at every point in the reservoir. This state allows the reservoir to be treated as a tank at uniform average pressure. Boundary-dominated flow includes pseudosteady-state flow as a special case but can also be applied to constant-pressure or variable-rate/variable-pressure production.

The type-curve stems during transient flow correspond to different values of the dimensionless drainage radius R_D. Wells with large negative skin factors will have small values of R_D, while wells with damage will have large values of R_D. At first, it might seem counterintuitive that the dimensionless drainage radius describes the type-curve stems for IARF. However, the transient stems for different values of R_D have different slopes only because they reach boundary-dominated flow at different times, as shown in Fig. 8.1. The type-curve stems during boundary-dominated flow correspond to different values of the Arps hyperbolic exponent b.

The Fetkovich type curves are much more sensitive to negative skin factors than to positive skin factors. In Fig. 8.4, the separation between the curves for $R_D = 1{,}000$ and $R_D = 10^5$ is approximately the same as that between the curves for $R_D = 10$ and $R_D = 20$. A change in R_D from 10 (corresponding to a highly stimulated well) to 20 represents an increase in skin factor of 0.7, while a change in R_D from 1,000 (corresponding to a well with little or no stimulation) to 10^5 represents an increase in skin factor of 4.6. The lack of sensitivity to differences in positive skin factors makes it difficult to separate the effects of permeability and skin factor for damaged wells using the Fetkovich type curves, or other rate-transient methods.

8.4.3 Accuracy of Fetkovich Type-Curve Analysis. Fetkovich (1980) compared the results obtained from type-curve matching with the permeability and skin factor for pressure-buildup tests for 22 wells. For 10 of the wells, both permeability and skin factor were estimated from the type-curve match. For the remaining 12 wells, the skin factor from the buildup analysis was used to calculate an apparent wellbore radius, r_{wa}. The apparent wellbore radius was then used to calculate R_D using an assumed value of 1,490 ft for r_e.

Fig. 8.5 shows a comparison of the permeability derived from production type-curve analysis to that from buildup analysis. Although the agreement is not perfect, the results are quite good, taking into consideration the difference in radius of investigation between the relatively short duration buildup-test data and the much longer duration production data.

Fig. 8.6 shows a comparison of the skin factors estimated from decline type-curve matching and pressure-buildup analysis for those wells for which the type curve was used to estimate skin factor. All of the wells had negative skin factors derived from buildup test analysis. The values for skin factor estimated from Fetkovich decline type-curve analysis were systematically lower (more negative) than those obtained from pressure-buildup analysis.

8.4.4 Procedure for Using the Fetkovich Type Curves. The following procedure is recommended for analyzing production data from a well producing at constant bottomhole pressure using the Fetkovich type curves.

1. Construct the field-data graph by plotting the production rate q vs. time t on a log-log scale.
2. Identify the transient flow period from the concave upward portion of the field-data graph. Identify the boundary-dominated flow period from the concave downward portion of the field-data graph. If all data are in transient flow, instead of using the Fetkovich type curve, the data should be matched to the $R_D = \infty$ stem of the rate vs. time type curve shown in Fig. 8.1.
3. Place the field-data graph on top of the Fetkovich type curve, aligning the transition from transient flow to boundary-dominated flow on the field-data graph with $t_{dD} = 0.3$ on the type curve.

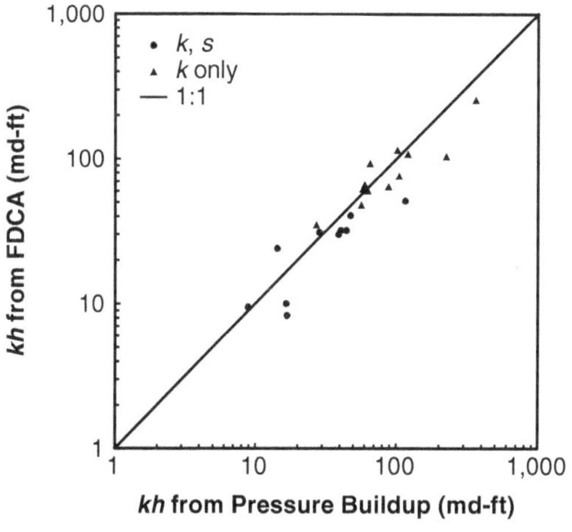

Fig. 8.5—Comparison of permeability calculated from Fetkovich type-curve match to permeability from pressure-buildup test. FDCA = Fetkovich decline-curve analysis and s = skin factor.

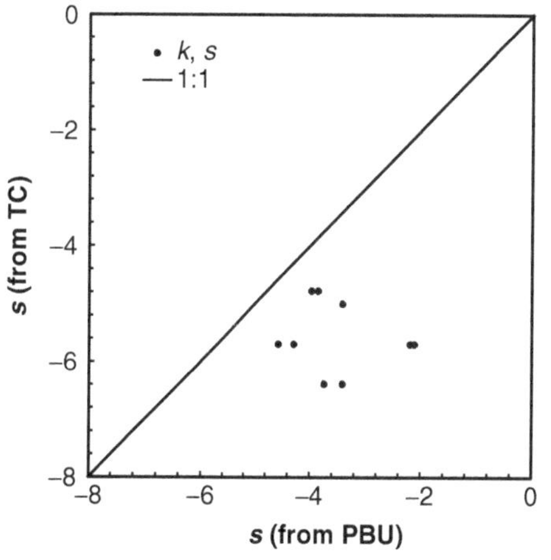

Fig. 8.6—Comparison of skin factor from Fetkovich type-curve match to skin factor from pressure-buildup test and well spacing. TC = type curve; s = skin factor.

4. Move the field-data graph vertically and horizontally as necessary to obtain the best match of the field data to the early- and late-time stems, keeping the field-data and type-curve axes parallel. If the drive mechanism is known, the field data should be matched to the late-time stem corresponding to the appropriate value of the Arps hyperbolic exponent b from Table 8.2, if possible.
5. Read the time and rate match points, (t, t_D) and (q, q_D).
6. Calculate the drainage area from the time and rate match points as

$$A = -\frac{5.6145B}{\phi h c_t (p_i - p_{wf})}\left(\frac{t}{t_{dD}}\right)_{MP}\left(\frac{q}{q_{dD}}\right)_{MP}. \quad (8.37)$$

7. Estimate the initial oil-in-place from

$$N_i = -\frac{1 - S_w}{c_t (p_i - p_{wf})}\left(\frac{t}{t_{dD}}\right)_{MP}\left(\frac{q}{q_{dD}}\right)_{MP}. \quad (8.38)$$

Note that Eqs. 8.37 and 8.38 can be used to estimate the drainage area and original oil-in-place even if the transient data do not give a unique value of R_D. Comparison of Eqs. 8.37 and 8.38 shows that an accurate estimation of fluid-in-place is possible without knowing the net pay or porosity, provided that water saturation, compressibility, and pressure data are available, while accurate estimation of drainage area requires both net pay and porosity.

8. If the dimensionless drainage radius R_D can be identified, calculate the permeability from the rate match point as

$$k = \left[\ln(R_D) - \tfrac{1}{2}\right] - \frac{141.2\mu B}{h(p_i - p_{wf})}\left(\frac{q}{q_{dD}}\right)_{MP}. \quad (8.39)$$

9. Calculate the apparent wellbore radius r_{wa} from the time match point as

$$r_{wa} = \sqrt{\frac{\dfrac{0.00633k}{\phi \mu c_t}\left(\dfrac{t}{t_{dD}}\right)_{MP}}{\tfrac{1}{2}\left(R_D^2 - 1\right)\left[\ln(R_D) - \tfrac{1}{2}\right]}}, \quad (8.40)$$

or more simply from the drainage area A as

$$r_{wa} = \sqrt{\frac{A}{\pi\left(R_D^2 - 1\right)}}. \quad (8.41)$$

10. Calculate the skin factor s from r_{wa} as

$$s = -\ln\left(\frac{r_{wa}}{r_w}\right). \quad (8.42)$$

11. If the Arps hyperbolic exponent b was not determined from the known drive mechanism in Step 4, read the value of b from the late-time stem.
12. If the R_D stem cannot be identified uniquely, the productivity index can still be estimated as

$$J = \frac{1}{p_{ai} - p_{awf}}\left(\frac{q}{q_{dD}}\right)_{MP}. \quad (8.43)$$

It should be noted that the value estimated from Eq. 8.43 will be lower than that calculated from k and s by a factor $[\ln(R_D) - 1/2]/[\ln(R_D) - 3/4]$. Without knowing R_D, the productivity index cannot be estimated as accurately.

The fluid properties in Eqs. 8.37 through 8.40 should be evaluated at the initial reservoir pressure. The reservoir pressure is essentially constant until after boundary-dominated flow has been reached. Thus, the time at which boundary-dominated flow begins (and therefore the time match point) is controlled by fluid properties at the original reservoir pressure. This allows use of the Fetkovich type curve for both gas reservoirs and undersaturated solution gas–drive oil reservoirs.

8.4.5 Field Example Problem 8.1—West Virginia Gas Well, Fetkovich Type-Curve Match. Use Fetkovich type-curve matching to estimate drainage area, initial gas-in-place, permeability, and skin factor given the rock, fluid, and production data provided in **Tables 8.3 and 8.4.** Use liquid equations with gas properties at initial pressure, and use adjusted pressure to linearize the steady-state gas-flow equation.

8.4.5.1 Solution. We follow the procedure for using the Fetkovich type curves given in the previous section. **Fig. 8.7** shows the completed match. The Arps hyperbolic exponent b was chosen to be 0.4 (Fetkovich et al. 1990).

Estimate the drainage area from Eq. 8.37 as

$$A = \frac{5.6145B_{gi}}{\phi h c_{ti}(p_{ai} - p_{awf})}\left(\frac{t}{t_{dD}}\right)_{MP}\left(\frac{q}{q_{dD}}\right)_{MP}$$

$$= \frac{(5.6145)(0.7535)}{(0.032)(100)(1.51\times10^{-4})(2384.4 - 244.0)}(775)(934)$$

$$= 2{,}957{,}000 \text{ ft}^2$$

$$= 67.9 \text{ acres.}$$

Rock and Fluid Properties					
p_i	3,814	psia	T_f	156	°F
f	0.032		S_w	0.3	
h	100	ft	r_w	0.365	ft
c_f	4.0×10^{-6}	psi^{-1}	c_w	3.5×10^{-6}	psi^{-1}
γ_g	0.57	(air=1)	B_{gi}	0.7535	bbl/Mscf
c_{gi}	2.09×10^{-4}	psi^{-1}	μ_{gi}	0.0215	cp
c_{ti}	1.51×10^{-4}	psi^{-1}	p_{ai}	2,384.4	psi
p_{wf}	1,100.0	psia	p_{awf}	244.0	psi

Table 8.3—Rock- and fluid-property data for Fetkovich type-curve match example, West Virginia gas well.

Production Data					
Time (days)	Cumulative Production (Mscf)	Time (days)	Cumulative Production (Mscf)	Time (days)	Cumulative Production (Mscf)
18	20,590	283	239,830	558	398,430
46	48,620	314	259,930	582	410,140
72	75,510	342	277,770	616	426,750
99	101,940	373	296,940	636	435,750
127	125,940	404	314,880	668	450,510
158	151,680	432	331,330	699	463,950
188	173,830	461	347,920	731	476,870
219	196,350	497	367,130	761	488,970
250	217,800	527	382,430	771	492,910

Table 8.4—Production data for Fetkovich type-curve match example, West Virginia gas well.

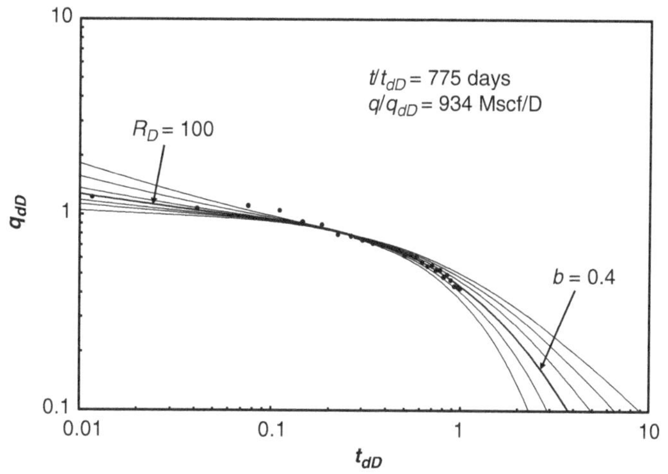

Fig. 8.7—Field example of West Virginia gas well, Fetkovich type-curve match.

Use Eq. 8.38 to estimate the initial gas-in-place as

$$\begin{aligned}G_i &= \frac{1-S_w}{c_{ti}\left(p_{ai}-p_{awf}\right)}\left(\frac{t}{t_{dD}}\right)_{MP}\left(\frac{q}{q_{dD}}\right)_{MP}\\ &= \frac{1-0.3}{\left(1.51\times10^{-4}\right)\left(2384.4-244.0\right)}(775)(934)\\ &= 1.566\times10^{6}\,\text{Mscf}\\ &= 1.566\ \text{Bcf}.\end{aligned}$$

Use Eq. 8.39 to estimate the permeability from the rate match point and the early-time R_D stem as

$$k = \left[\ln(R_D) - \tfrac{1}{2}\right] \frac{141.2\mu_{gi}B_{gi}}{h(p_{ai} - p_{awf})}\left(\frac{q}{q_{dD}}\right)_{MP}$$
$$= \left[\ln(100) - \tfrac{1}{2}\right] \frac{(141.2)(0.0215)(0.7535)}{(100)(2,384.4 - 244.0)}(934)$$
$$= 0.041 \text{ md.}$$

Estimate the apparent wellbore radius from the drainage area with Eq. 8.41:

$$r_{wa} = \sqrt{\frac{A}{\pi\left(R_D^2 - 1\right)}}$$
$$= \sqrt{\frac{2,956,000}{\pi\left((100)^2 - 1\right)}}$$
$$= 9.7 \text{ ft.}$$

Using Eq. 8.42, estimate the skin factor s from the apparent wellbore radius:

$$S = -\ln\left(\frac{r_{wa}}{r_w}\right)$$
$$= -\ln\left(\frac{9.7}{0.365}\right)$$
$$= -3.28.$$

Estimate the productivity index J from Eq. 8.43:

$$J = \frac{1}{p_{ai} - p_{awf}}\left(\frac{q}{q_{dD}}\right)_{MP}$$
$$= \frac{1}{2,384.4 - 244}(934)$$
$$= 0.436 \text{ Mscf/D/psi.}$$

8.4.6 Connection Between Arps Equations and Fetkovich Decline-Curve Analysis. Not surprisingly, the Arps equations and Fetkovich type-curve matching are closely related. Eqs. 8.33 and 8.34 give the Arps parameters D_i and q_i directly from the match-point ratios for a Fetkovich type-curve match. The Arps hyperbolic exponent is given by the Fetkovich matching stem parameter, b, for boundary-dominated flow.

Conversely, if q_i and D_i have been determined from DCA, the Fetkovich type-curve match-point ratios are obtained by rearranging Eqs. 8.33 and 8.34:

$$\left(\frac{t}{t_{dD}}\right)_{MP} = \frac{1}{D_i} \qquad (8.44)$$

and

$$\left(\frac{q}{q_{dD}}\right)_{MP} = q_i. \qquad (8.45)$$

The drainage area and initial oil-in-place can then be estimated from the match-point ratios:

$$A = \frac{5.6145B}{\phi h c_t(p_i - p_{wf})}\frac{q_i}{D_i} \qquad (8.46)$$

and

$$N_i = \frac{1 - S_w}{c_t(p_i - p_{wf})}\frac{q_i}{D_i}. \qquad (8.47)$$

If skin factor is known, the permeability can be estimated as follows. First, calculate the apparent wellbore radius, r_{wa}:

$$r_{wa} = r_w e^{-s}. \qquad (8.48)$$

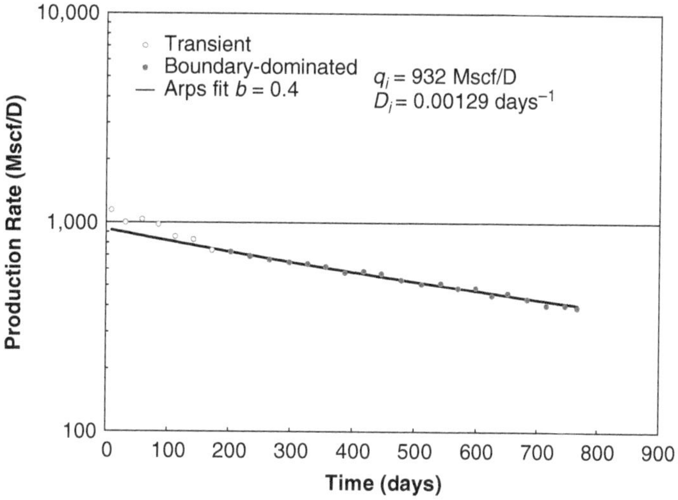

Fig. 8.8—Field example of West Virginia gas well, Arps decline-curve fit.

Next, calculate the dimensionless drainage radius R_D from

$$R_D = \sqrt{\frac{A}{\pi r_{wa}^2} + 1}. \quad (8.49)$$

Finally, calculate the permeability k from

$$k = \left[\ln(R_D) - \tfrac{1}{2}\right] \frac{141.2 \mu B}{h(p_i - p_{wf})} q_i. \quad (8.50)$$

As an alternative, if the permeability is known, calculate the dimensionless drainage radius from

$$R_D = \exp\left(\frac{kh(p_i - p_{wf})}{141.2 q_i \mu B} + \frac{1}{2} \right), \quad (8.51)$$

the apparent wellbore radius r_{wa} from Eq. 8.41, and the skin factor s from Eq. 8.42.

If neither skin factor nor permeability is known, calculate the productivity index J from

$$J = \frac{q_i}{p_i - p_{wf}}. \quad (8.52)$$

As with the Fetkovich analysis, the value of J estimated from Eq. 8.52 will be lower than that estimated directly from k and s.

8.4.7 Example Problem 8.2. Repeat the analysis of the West Virginia gas-well example using the Arps decline curve rather than the Fetkovich type curve.

***8.4.7.1 Solution*.** **Fig. 8.8** shows an Arps fit of the data in boundary-dominated flow, with the Arps exponent b set to 0.4. The Arps equations implicitly assume boundary-dominated flow. Thus, the transient period was ignored when performing the decline-curve fit.

The Fetkovich match-point ratios are obtained from Eqs. 8.44 and 8.45:

$$\left(\frac{t}{t_{dD}}\right)_{MP} = \frac{1}{D_i}$$
$$= \frac{1}{0.00129}$$
$$= 775 \text{ days}$$

and

$$\left(\frac{q}{q_{dD}}\right)_{MP} = q_i$$
$$= 932 \text{ Mscf/D}.$$

From this point, the solution for the drainage area and original gas-in-place follows as in the Fetkovich type-curve example, giving almost identical results, with A = 67.7 acres, G_i = 1.562 Bcf, and J = 0.435 Mscf/D/psi. Analysis using the Arps decline match alone cannot separate the effects of permeability and skin factor.

8.5 Modifications of the Fetkovich Type Curve

A number of modifications of the Fetkovich type curve have been proposed. One of the drawbacks of the Fetkovich type curves is the limited ability to discriminate between different values of the matching stem parameter R_D.

8.5.1 McCray Time-Averaged Rate Type Curves. McCray (1990) proposed adding two additional curves to the Fetkovich type curve: the time-averaged rate and the integral derivative curves. Both curves have the dimensions of rate vs. time and can therefore be plotted on the same scale as the Fetkovich type curve.

The time-averaged rate curve is defined as the cumulative production divided by the time, Q_{dD}/t_{dD}. **Fig. 8.9** shows the Fetkovich type curve with the addition of McCray's time-averaged rate curve.

The dimensionless cumulative production Q_D is defined as

$$Q_D = \frac{(141.2)(0.00633)B}{\phi h c_t r_{wa}^2 \left(p_i - p_{wf} \right)} N_p, \quad (8.53)$$

and the dimensionless decline cumulative production Q_{dD} is defined as

$$Q_{dD} = \frac{Q_D}{\frac{1}{2}\left(R_D^2 - 1\right)}. \quad (8.54)$$

8.5.2 McCray Integral Derivative Type Curves. The time-averaged rate curve smooths the data and thus does not help discriminate between values of R_D. McCray proposed the integral derivative curve to provide greater separation between the various R_{dD} curves.

The integral derivative is defined as the negative of the logarithmic derivative of the time-averaged rate:

$$\begin{aligned} q_{dDi} &= -\frac{d}{d\ln t_{dD}} \frac{Q_{dD}}{t_{dD}} \\ &= -t_{dD} \frac{d}{dt_{dD}} \frac{Q_{dD}}{t_{dD}} \\ &= -t_{dD} \left(\frac{q_{dD}}{t_{dD}} - \frac{Q_{dD}}{t_{dD}^2} \right) \\ &= \frac{Q_{dD}}{t_{dD}} - q_{dD}. \end{aligned} \quad (8.55)$$

Fig. 8.10 shows the Fetkovich type curve with McCray's time-averaged rate and integral derivative type curves.

When high-quality data are available, the integral derivative curve provides more separation between stems for different values of R_D. Unfortunately, for a variety of reasons, production data are usually of much lower quality than pressure-transient data. Although the integral derivative does not require taking a numerical derivative, as its name suggests, the difference operation magnifies relative errors in the underlying field data in much the same way as would numerical differentiation.

The McCray type curves are used almost exactly the same way as the Fetkovich type curves, with the added step of calculating and plotting the time-averaged rate and integral derivative curves on the field-data graph.

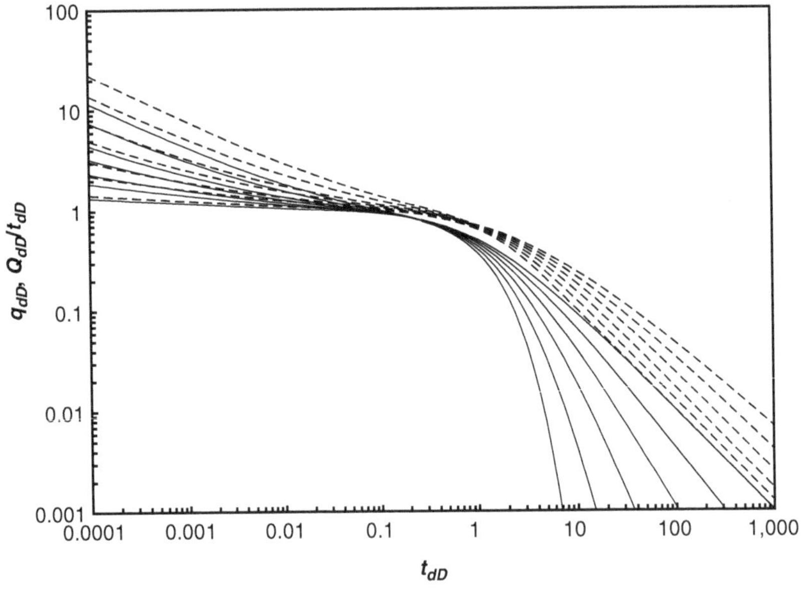

Fig. 8.9—Fetkovich-McCray rate and time-averaged rate type curves.

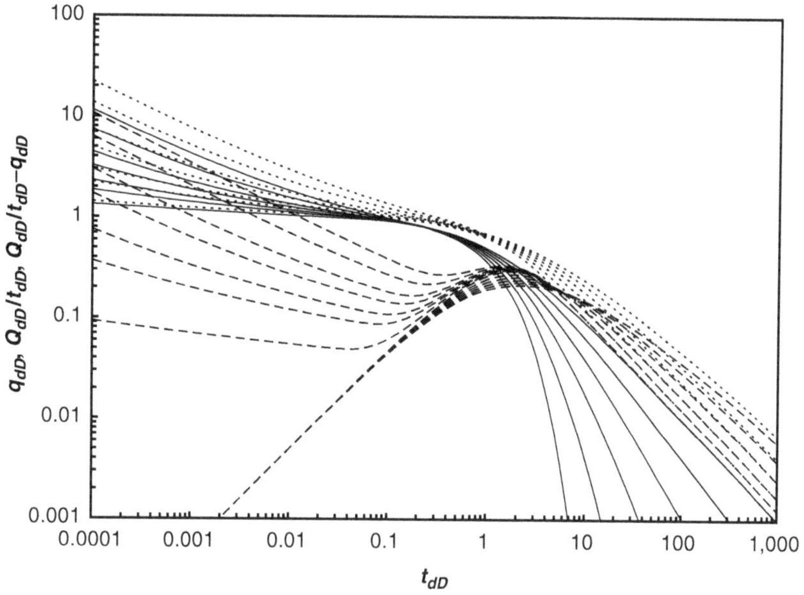

Fig. 8.10—Fetkovich-McCray rate, time-averaged rate, and integral-derivative type curves.

8.5.3 Field Example Problem 8.3—West Virginia Gas Well, McCray Type-Curve Match. Match the field data given in Tables 8.3 and 8.4 with the McCray type curve.

8.5.3.1 Solution. The procedure for using the McCray type curve is the same as that for the Fetkovich type curve, with the additional constraint that the match includes consideration of the time-averaged rate and integral derivative curves as well as the rate curve. **Fig. 8.11** shows the completed match.

For this match, the best fit of the data was found on the $R_D = 50$ transient stem and the $b = 2$ depletion stem, with match-point ratios $t/t_{dD} = 860$ days and $q/q_{dD} = 911$ Mscf/D. Repeating the calculations as for the Fetkovich example, we obtain $A = 73.5$ acres, $G_i = 1.694$ Bcf, $k = 0.033$ md, $r_{wa} = 20.2$ ft, and $s = -4.01$.

There are several things to note in this match. First, the first point on the time-averaged-rate curve overlaps the first point on the rate curve. This is an artifact of the way the field-data plotting functions have been calculated for this example. Second, the calculated value of the integral derivative for the first production data point is zero, thus it does not appear on the log-log graph. Again, this is an artifact of the way the field-data plotting functions have been calculated. Third, the integral derivative points for the second through fourth rows of the production data fall well below the $R_D = 50$ stem selected for the match. This might be caused by the fact that the backpressure during the first few weeks or months of production is often decreasing rather than strictly constant, as assumed in development of the type curves. Unfortunately, the early-time data are critical for accurately separating the effects of permeability and skin factor. Finally, no attempt was made to honor the integral derivative curve for the Arps b stems. These stems are based strictly on the Arps hyperbolic decline equations and do not include the flush production observed during the transient flow period for an effectively stimulated well.

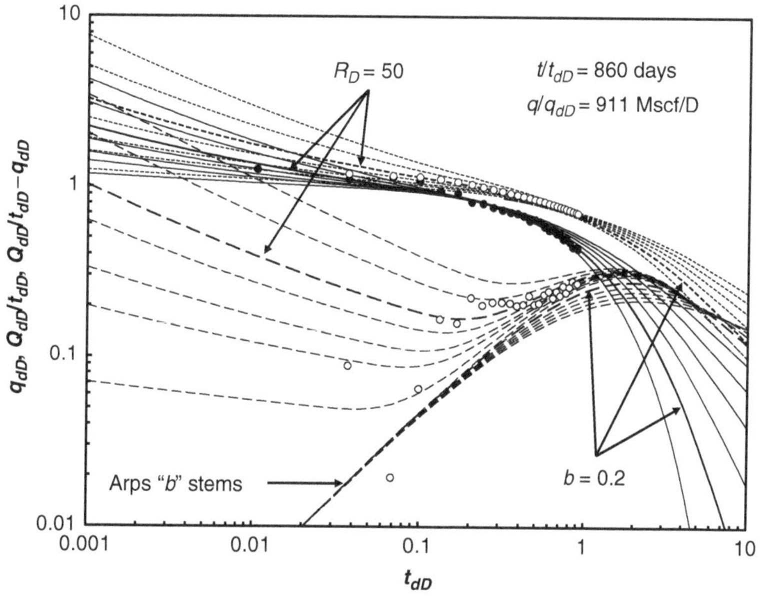

Fig. 8.11—Field example of West Virginia gas well, match using McCray rate, time-averaged rate, and integral derivative type curves.

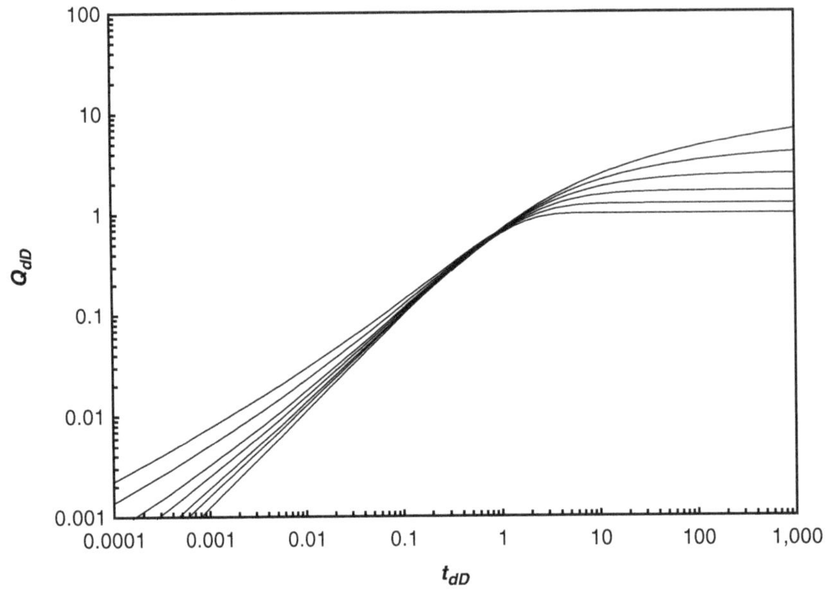

Fig. 8.12—Cumulative-production type curve.

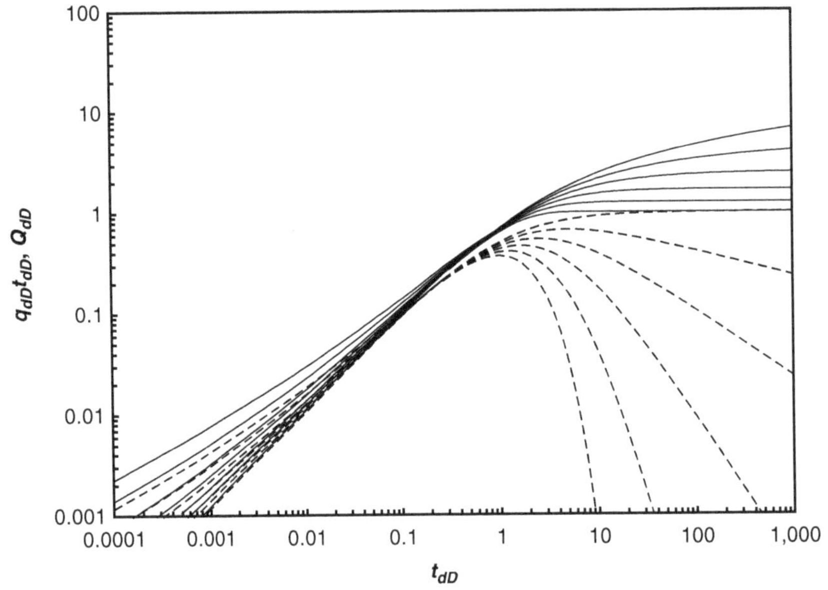

Fig. 8.13—Cumulative-production and cumulative-derivative type curves.

8.5.4 Cumulative-Production Type Curves. Other modifications of the Fetkovich type curves are the cumulative-production (Fraim and Lee 1987; Fraim 1988) type curves, shown in **Fig. 8.12,** and the cumulative-derivative (Spivey et al. 1992) type curves, shown in **Fig. 8.13.**

The cumulative-production type curve is a graph of the cumulative production Q_{dD} vs. the dimensionless time t_{dD}. The cumulative-derivative type curve is the logarithmic derivative of the cumulative production:

$$\frac{dQ_{dD}}{d\ln t_{dD}} = t_{dD}\frac{dQ_{dD}}{dt_{dD}} = t_{dD}q_{dD}. \quad (8.56)$$

The procedure for using the cumulative-production type curves is closely similar to that of the Fetkovich type curves:

1. Construct the field-data graph by plotting the cumulative production N_p vs. time t on a log-log scale.
2. Identify the transient flow period from the concave upward portion, and the boundary-dominated flow period from the concave downward portion, of the field-data graph. It might be more difficult to identify the transient and boundary-dominated flow periods from cumulative-production data. If the boundary-dominated flow period is absent, the cumulative-production type curve should not be used. Instead, the rate vs. time data should be matched to the $R_D = \infty$ stem of the rate vs. time type curve shown in Fig. 8.1.
3. Place the field-data graph on top of the cumulative-production type curve, aligning the transition from transient flow to boundary-dominated flow on the field-data graph with $t_{dD} = 0.3$ on the type curve.

4. Move the field-data graph vertically and horizontally as necessary to obtain the best match of the field data to the early- and late-time stems, keeping the field-data and type-curve axes parallel.
5. Read the time (t, t_{dD}) and cumulative-production match points (N_p, Q_{dD}).
6. Calculate the drainage area from the cumulative-production match point as

$$A = \frac{5.1645B}{\phi h c_t (p_i - p_{wf})}\left(\frac{N_p}{Q_{dD}}\right)_{MP} \quad \text{(8.57)}$$

7. Estimate the initial oil-in-place from

$$N_i = \frac{1 - S_w}{c_t (p_i - p_{wf})}\left(\frac{N_p}{Q_{dD}}\right)_{MP}. \quad \text{(8.58)}$$

8. If the dimensionless drainage radius R_D can be identified, calculate the permeability from the time and cumulative-production match points as

$$k = \left[\ln(R_D) - \tfrac{1}{2}\right]\frac{141.2\mu B}{h(p_i - p_{wf})}\left(\frac{N_p}{Q_{dD}}\right)_{MP}\left(\frac{t_{dD}}{t}\right)_{MP}. \quad \text{(8.59)}$$

9. Calculate the apparent wellbore radius r_{wa} from the time match point as

$$r_{wa} = \sqrt{\frac{\dfrac{0.00633k}{\phi \mu c_t}\left(\dfrac{t}{t_{dD}}\right)_{MP}}{\tfrac{1}{2}\left(R_D^2 - 1\right)\left[\ln(R_D) - \tfrac{1}{2}\right]}}, \quad \text{(8.60)}$$

or more simply from the drainage area A as

$$r_{wa} = \sqrt{\frac{A}{\pi\left(R_D^2 - 1\right)}}. \quad \text{(8.61)}$$

10. Calculate the skin factor s from r_{wa} as

$$s = -\ln\left(\frac{r_{wa}}{r_w}\right). \quad \text{(8.62)}$$

11. Read the Arps hyperbolic exponent b from the late-time stem.

8.6 Modifications for Pressure-Dependent Fluid and Rock Properties

Dimensionless rate vs. time type curves based on analytical solutions to the diffusivity equation, such as the Fetkovich type curve, make two assumptions that limit their direct application to low-permeability gas reservoirs:

- A slightly compressible liquid
- Constant flowing wellbore pressure

In this section, we will discuss an approximate linearization of the gas-flow equation that accounts for pressure-dependent rock and gas properties for production from a closed reservoir. Variable-pressure/variable-flow-rate production will be discussed in a later section.

The adjusted pressure is defined by

$$p_a(p) = \frac{\mu_i Z_i}{P_i}\int_0^p \frac{p' dp'}{\mu(p')z(p')}, \quad \text{(8.63)}$$

which provides an exact linearization of the gas-flow equation for steady-state flow. Fraim and Wattenbarger (1987) defined a reservoir adjusted time as

$$t_a = \mu_{gi} c_{ti}\int_0^t \frac{dt'}{\mu_g(\bar{p})c_t(\bar{p})}, \quad \text{(8.64)}$$

where the gas properties inside the integral in Eq. 8.64 are evaluated at the average reservoir pressure as a function of time. Use of adjusted pressure (Eq. 8.63) with adjusted time (Eq. 8.64) linearizes the gas-flow equation for constant-rate production during pseudosteady-state flow from a reservoir with a reasonably symmetric drainage area shape and well location, such as a well in the center of a circle or a square. Not surprisingly, the use of adjusted pressure and adjusted time also linearizes the gas-flow equation for constant-pressure production during boundary-dominated flow.

8.6.1 Modified Adjusted Pressure and Adjusted Time. Pressure-dependent rock properties can also be taken into account by modifying the definitions of adjusted pressure and adjusted time:

$$p_a = \frac{\mu_i z_i}{k_i p_i}\int_0^p \frac{k(p')p'(dp')}{\mu(p')z(p')}. \quad \text{(8.65)}$$

and

$$t_a \equiv \frac{\phi_i \mu_i c_{ti}}{k_i} \int_0^t \frac{k(\bar{p})\mathrm{d}t'}{\phi(\bar{p})\mu(\bar{p})c_t(\bar{p})}, \quad (8.66)$$

respectively. The rock and fluid properties in Eq. 8.66 must be evaluated at the average reservoir pressure as a function of time, $\bar{p}(t')$. The porosity in Eq. 8.66 is defined as the pore volume, V_p, at pressure p, divided by the bulk volume at initial reservoir pressure, V_{bi}:

$$\phi(p) \equiv \frac{V_p(p)}{V_{bi}}. \quad (8.67)$$

Defining the porosity with Eq. 8.67 allows us to ignore changes in bulk volume with pressure and treat the formation thickness as a constant.

The use of adjusted pressure and adjusted time effectively linearizes the gas-flow equation for both pseudosteady-state flow at constant rate and boundary-dominated flow at constant wellbore pressure. However, the integrands of Eqs. 8.64 and 8.66 depend on the average reservoir pressure, which is in general unknown as a function of time. As a result, use of adjusted pressure and adjusted time requires an iterative approach.

First, an initial gas-in-place is assumed. The average pressure is then calculated using material balance, the plotting functions are calculated, and the estimate of gas-in-place is updated from the type-curve match. The process is repeated until the initial-gas-in-place estimate converges.

8.6.2 Stress-Dependent Rock Properties. Eqs. 8.65 and 8.66 account for pressure-dependent rock and fluid properties. While fluid-property correlations are widely used for calculating gas properties as functions of pressure, the pressure dependence of rock properties is often glossed over. Pressure-dependent porosity is usually treated by assuming constant compressibility, while permeability is assumed constant. Many unconventional gas reservoir rocks are highly sensitive to stress; therefore, stress-dependent rock properties should be accounted for when performing production-data analysis on such reservoirs.

A number of different stress-dependent porosity and permeability relationships have been proposed. Unlike fluid-property correlations, stress-dependent rock-property models should be tuned to laboratory measurements if such data are available.

The Jones (1988) model gives realistic stress-dependent porosity and permeability and requires only a limited amount of laboratory data. Jones proposed two four-parameter models for porosity and permeability:

$$\phi(p) = \phi_o \frac{\exp\left(-A_V\left[1-\exp(-\sigma/\sigma^*)\right]\right)}{1+C\sigma} \quad (8.68)$$

and

$$k(p) = k_o \frac{\exp\left(-A_k\left[1-\exp(-\sigma/\sigma^*)\right]\right)}{1+C\sigma}, \quad (8.69)$$

where ϕ_0, A_V, k_0, A_k, σ^*, and C are adjustable parameters, and where σ is the net confining stress. For convenience, the net confining stress is often estimated as

$$\sigma - \rho_{\text{ith}} d - p, \quad (8.70)$$

where ρ_{ith} is the lithostatic gradient in psi/ft, d is the true vertical depth in ft at reservoir datum, and p is the pore pressure referenced to reservoir datum.

If sufficient laboratory data are available, all four model parameters can be adjusted to fit the laboratory data. When only limited laboratory data are available, or to reduce the amount of laboratory measurements required, Jones suggested fixing σ^* at 3,000 psi and C at 3.0×10^{-6} 1/psi.

Eq. 8.68 can be differentiated to obtain an equation for the pore-volume compressibility:

$$c_f(p) = \frac{A_V \exp(-\sigma/\sigma^*)}{\sigma^*} + \frac{C}{1+C\sigma}. \quad (8.71)$$

The parameters ϕ_0 and k_0 in the Jones model are the (extrapolated) porosity and permeability at zero net confining stress, respectively. The parameters A_v and A_k are shape parameters that describe the curvature of the porosity and permeability with pressure. Larger values indicate higher stress sensitivity. Values for A_v could be as high as 0.3 in stress-sensitive formations, while those for A_k might be 3.0 or higher.

Figs. 8.14 and 8.15 show the effect of A_V and A_k on the amount of porosity and permeability reduction with pressure. For both porosity and permeability, the reduction is steepest at net confining stresses below 3,000 psi.

Because of this, for a given value of A_V or A_k, the initial pressure gradient for overpressured reservoirs places the reservoir depletion path further to the left on the net-confining-stress axis. **Figs. 8.16 and 8.17** show the impact of highly overpressured vs. only slightly overpressured initial conditions. For both cases, the depth was assumed to be 10,000 ft, and the lithostatic gradient to be 1.0 psi/ft. The porosity or permeability is graphed as a fraction of that at initial reservoir pressure on the vertical axis, while the confining stress is expressed as the pressure change from initial reservoir pressure.

A quick estimate of the relative importance of pore-volume, connate water, and gas compressibility can be made by calculating the fraction of the total reservoir energy attributable to each mechanism:

$$C_t = c_f + S_w c_w + (1 - S_w)c_g, \quad (8.72)$$

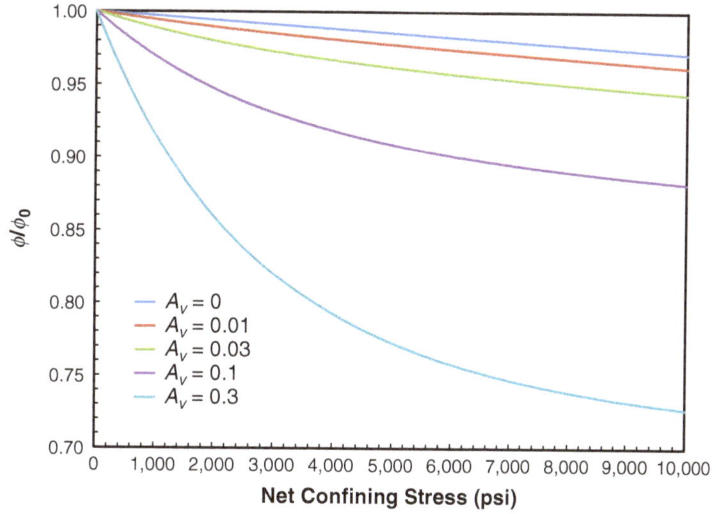

Fig. 8.14—Jones stress-sensitive porosity model, effect of A_v.

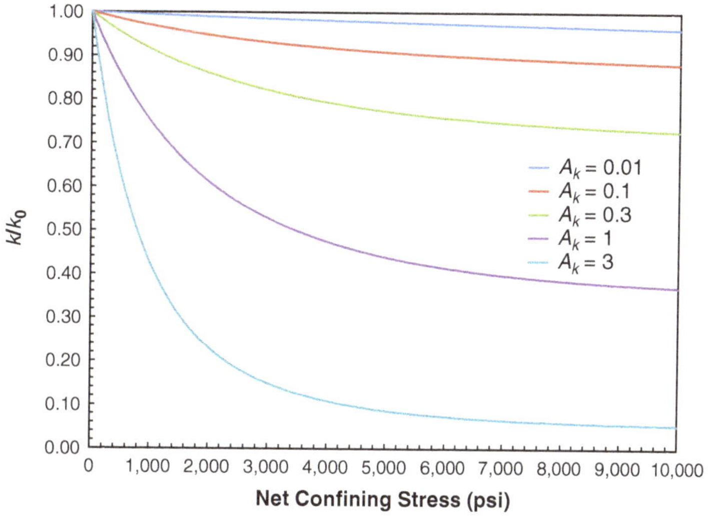

Fig. 8.15—Jones stress-sensitive permeability model, effect of A_k.

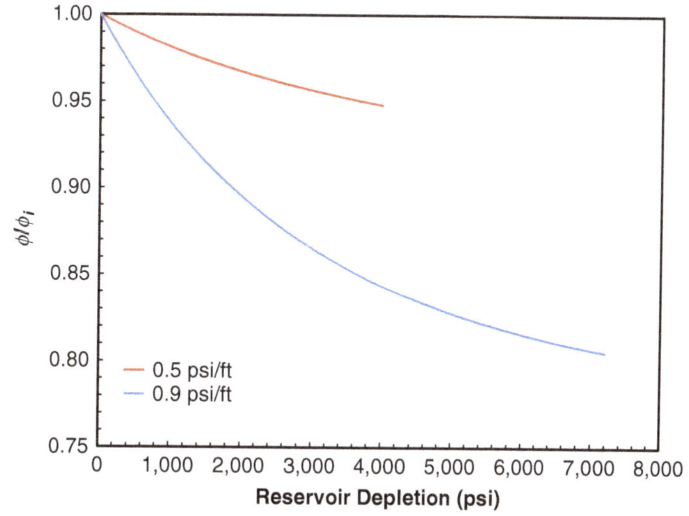

Fig. 8.16—Jones stress-sensitive porosity model, effect of initial pore-pressure gradient, $A_v = 0.3$.

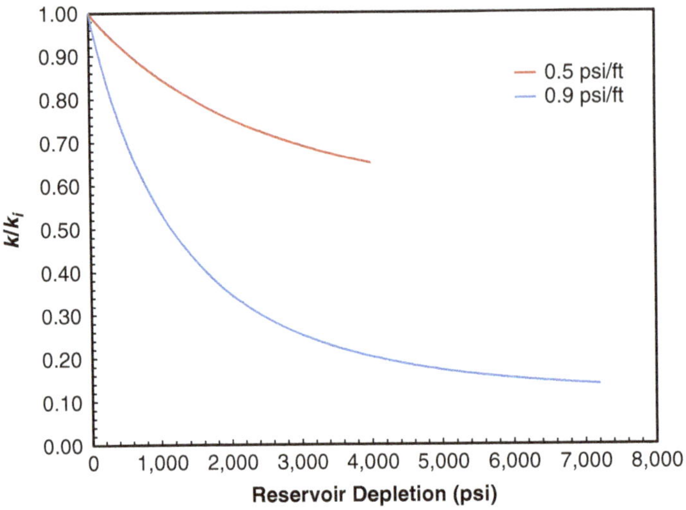

Fig. 8.17—Jones stress-sensitive permeability model, effect of initial pore-pressure gradient, A_k = 3.0.

$$E_f = \frac{c_f}{c_t} = \frac{c_f}{c_f + S_w c_w + (1 - S_w) c_g}, \tag{8.73}$$

$$E_w = \frac{S_w c_w}{c_t}, \tag{8.74}$$

and

$$E_g = \frac{(1 - S_w) c_g}{c_t}. \tag{8.75}$$

Keep in mind that c_f, c_w, c_g, and even S_w are functions of pressure, thus making E_f, E_w, and E_g also functions of pressure. The gas and water compressibilities will always be lowest at initial reservoir pressure, while the pore-volume compressibility will always be highest. Thus, evaluating Eq. 8.73 at initial pressure will almost always give the maximum fraction of reservoir energy attributable to pore-volume compressibility (i.e., the worst-case scenario).

A set of criteria to screen for likely stress-sensitive formations is suggested in Section 5.3 of Part 2 Ch 5: Reservoir Characterization. Summarized, these criteria are as follows:

- Porosity is less than 12% or higher than 25%.
- Rock is friable or unconsolidated.
- Net overburden pressure is less than 3,000 psi (a rough rule-of-thumb equivalent is when the initial pressure, in psia, is within 3,000 psi of the formation depth in feet).
- Initial pressure gradient is higher than 0.465 psi/ft.

Although not suggested as a screening criterion, the high water saturations often present in low-permeability gas reservoirs accentuate the effects of high pore-volume compressibility.

8.6.3 Material Balance. It is traditional to incorporate stress-dependent porosity in the material-balance equation by defining a cumulative or secant compressibility (Fetkovich et al. 1998; Spivey et al. 2007). However, it seems easier and less confusing (not to mention less prone to error) to write the material-balance equation in terms of the stress-dependent porosity itself. The mass balance can be written simply as

$$\frac{43{,}560 A \phi(\bar{p}) h \left[1 - S_w(\bar{p})\right]}{B_g(\bar{p})} = \frac{43{,}560 A \phi_i h (1 - S_{wi})}{B_{gi}} - G_p, \tag{8.76}$$

where A is reservoir area in acres, B_g is in ft³/Mscf, and G_p is in Mscf. Using the definition of B_g and accounting for the change in water saturation caused by pore-volume compaction and connate water expansion, Eq. 8.76 can be rearranged to obtain

$$\frac{\left[\frac{\phi(\bar{p})}{\phi_i} - S_{wi}\frac{B_w(\bar{p})}{B_{wi}}\right]}{1 - S_{wi}} \frac{\bar{p}}{\bar{z}} = \frac{p_i}{z_i} - \frac{1{,}000 T_f p_{sc}}{43{,}560 A \phi_i h (1 - S_{wi}) T_{sc}} G_p, \tag{8.77}$$

or simply

$$\left(\frac{\bar{p}}{\bar{z}}\right)^* = \frac{p_i}{z_i} - m G_p, \tag{8.78}$$

where $\left(\frac{\bar{p}}{z}\right)^*$ is defined as

$$\left(\frac{\bar{p}}{\bar{z}}\right)^* \equiv \frac{\left[\frac{\phi(\bar{p})}{\phi_i} - S_{wi}\frac{B_w(\bar{p})}{B_{wi}}\right]}{1-S_{wi}}\frac{\bar{p}}{\bar{z}}. \quad (8.79)$$

A modified p/z graph may be constructed by graphing $\left(\frac{\bar{p}}{\bar{z}}\right)^*$ vs. G_p, then finding the best-fitting straight line through the data. The x-intercept gives the initial gas-in-place, just as in the standard p/z material-balance plot.

In conventional reservoirs, the average drainage-area pressure can be estimated from buildup tests of limited duration, thereby allowing estimation of initial gas-in-place from either the standard or modified p/z material-balance plots. In unconventional reservoirs, the length of time required to estimate average drainage-area pressure from buildup tests renders their use cost-prohibitive. The material-balance equation remains valid, but an iterative procedure must be used to estimate initial gas-in-place and average drainage-area pressure. In addition, an accurate estimate of initial reservoir pressure is required.

For finding average reservoir pressure as a function of cumulative production, a first estimate of the initial gas-in-place, G_i, must be assumed. The slope m of the straight line on the modified p/z plot is calculated from

$$m = \frac{p_i/z_i}{G_i}, \quad (8.80)$$

and then Eq. 8.78 is solved numerically for the pressure $\bar{p}$ as a function of the cumulative production G_p.

8.6.4 Type-Curve Analysis Procedure. The following procedure is recommended for analyzing production data from gas wells producing at constant flowing bottomhole pressure.

1. Estimate the initial pressure p_i to use as a reference pressure in calculating the adjusted pressure and adjusted time.
2. Build a table of p_a vs. p by integrating Eq. 8.65 numerically.
3. Make an initial estimate of the drainage area and gas-in-place.
4. Integrate Eq. 8.66 numerically to find the adjusted time, t_a, using Eq. 8.78 to solve for $\bar{p}$ as a function of G_p.
5. Use the adjusted time, t_a, and the actual flow rate, q, as the time and rate plotting variables, respectively.
6. Perform the type-curve match. For a single-layer gas reservoir, the late-time data should fall on the $b = 0$ stem.
7. Calculate the drainage area from the time and rate match points as

$$A = \frac{5.6145B_{gi}}{\phi h c_{ti}\left(p_{ai} - p_{awf}\right)}\left(\frac{t_a}{t_{dD}}\right)_{MP}\left(\frac{q}{q_{dD}}\right)_{MP}. \quad (8.81)$$

8. Estimate the initial gas-in-place from

$$G_i = \frac{1-S_w}{c_{ti}(p_{ai} - p_{awf})}\left(\frac{t_a}{t_{dD}}\right)_{MP}\left(\frac{q}{q_{dD}}\right)_{MP}. \quad (8.82)$$

9. If the dimensionless drainage radius R_D can be identified, calculate the permeability from the rate match point as

$$k = \left(\ln(R_D) - \tfrac{1}{2}\right)\frac{141.2\mu_{gi}B_{gi}}{h(p_{ai} - p_{awf})}\left(\frac{q}{q_{dD}}\right)_{MP}. \quad (8.83)$$

10. Calculate the apparent wellbore radius r_{wa} from the drainage area A as

$$r_{wa} = \sqrt{\frac{A}{\pi(R_D^2-1)}}. \quad (8.84)$$

11. Calculate the skin factor s from r_{wa} as

$$S = -\ln\left(\frac{r_{wa}}{r_w}\right). \quad (8.85)$$

12. Repeat Steps 4 through 11 until the drainage area estimate converges.

8.7 Modifications for Variable-Rate/Variable-Pressure Production

Analytical solutions to the diffusivity equation are usually developed assuming that production occurs at either constant-rate or constant-pressure conditions. The assumption of production at constant rate is usually made for development of pressure transient type curves. On the other hand, production at constant flowing bottomhole pressure is usually assumed for rate or cumulative production decline type curves. The assumption of constant flowing bottomhole pressure is a reasonable approximation of the way in which many, if not most, unconventional gas wells are produced.

Pressure transient tests are conducted with reservoir characterization as a primary objective. Best practices thus dictate careful control of rate throughout a test to ensure that the collected data can be interpreted successfully. On the other hand, RTA typically relies on data collected during normal production operations, with reservoir characterization being at best simply an afterthought. Thus, RTA methods must be capable of handling variable-rate/variable-pressure production typical of routine production operations.

During the past 7 decades, many different ways have been developed to adapt pressure transient analysis methods to analyze variable-rate histories. For example, superposition time functions ranging from the Horner time ratio (Horner 1951) to the Agarwal multirate radial equivalent time function (Agarwal 1980) have been proposed to allow (piecewise constant) variable-rate histories to be analyzed using methods developed for constant-rate production. For RTA, one of the most useful methods combines two concepts, rate-normalization and material-balance time, to allow production with smoothing changing rates and pressures to be modeled as an equivalent constant-rate production history.

8.7.1 Rate Normalization. Rate normalization, sometimes called Gladfelter deconvolution (Gladfelter et al. 1955), can be used to analyze variable-rate/variable-pressure data for infinite-acting reservoirs when the rate is changing slowly and smoothly. Jones (1957) reported using the time derivative of the rate-normalized pressure change to estimate fluid-in-place for both oil and gas reservoirs. Winestock and Colpitts (1965) showed that the pressure response for a varying flow rate in an infinite-acting reservoir can be modeled as

$$\frac{p_i - p_{wf}(t)}{q(t)} = \frac{162.6B\mu}{kh}\left[\log t + \log\left(\frac{k}{\phi\mu c_t r_w^2}\right) - 3.23 + 0.869s\right], \quad (8.86)$$

provided that the flow rate is changing slowly and smoothly.

For a gas well, using adjusted pressure, Eq. 8.86 becomes

$$\frac{p_{ai} - p_{awf}(t)}{q_g(t)} = \frac{162.6B_{gi}\mu_{gi}}{kh}\left[\log t + \log\left(\frac{k}{\phi\mu_{gi}c_{ti}r_w^2}\right) - 3.23 + 0.869s\right]. \quad (8.87)$$

8.7.2 Material-Balance Time. Blasingame and Lee (1986, 1988) introduced a time plotting function, now usually referred to as the material-balance time, to model variable-rate/variable-pressure production after boundary-dominated flow has been achieved. For a slightly compressible liquid, the material-balance time is defined as

$$t_{mb} = \frac{24N_p(t)}{q(t)}. \quad (8.88)$$

The Horner pseudoproducing time (Horner 1951) for modeling a variable-rate history before a buildup test as an equivalent constant-rate history is an early use of the material-balance-time function. For a gas well, the adjusted cumulative production must be used instead of the cumulative production. The material-balance adjusted time is given by

$$t_{amb} = \frac{24G_{pa}(t)}{q(t)}, \quad (8.89)$$

where G_{pa} is the adjusted cumulative production defined as

$$\begin{aligned} G_{pa}(t) &= \int q\,\mathrm{d}t_a \\ &= \mu_{gi}c_{ti}\int_0^t \frac{q(t')\mathrm{d}t'}{\mu_g(\overline{p})c_t(\overline{p})}, \\ &= \mu_{gi}c_{ti}\int_0^{G_p} \frac{\mathrm{d}G_p'}{\mu_g(\overline{p})c_t(\overline{p})} \end{aligned} \quad (8.90)$$

where the gas viscosity and total compressibility are evaluated at the average reservoir pressure as a function of time, $\overline{p}$. For an infinite-acting reservoir, the average reservoir pressure is the initial pressure, making the adjusted cumulative production numerically equal to the cumulative production and the material-balance adjusted time numerically equal to the material-balance time.

For formations exhibiting significant stress sensitivity, Eq. 8.90 must be modified to include porosity:

$$G_{pa}(t) = \phi_i\mu_{gi}c_{ti}\int_0^{G_p} \frac{\mathrm{d}G_p'}{\phi(\overline{p})\mu_g(\overline{p})c_t(\overline{p})}. \quad (8.91)$$

Although the material-balance-time function was developed for analyzing data in the boundary-dominated period, it also works well for data in the transient period.

It should always be kept in mind that rate-normalization/material-balance-time plotting functions allow use of analysis methods developed for constant-rate production rather than constant-pressure production. Blasingame et al. (1991) summarized and discussed use of both constant-rate analog time (i.e., material-balance time) and constant-pressure analog time functions. They noted that, for boundary-dominated flow, a constant-pressure analog time function could be defined in terms of the constant-rate analog time and that the resulting function also seemed to work for transient flow.

8.7.3 Using Rate-Normalization/Material-Balance Time—Infinite-Acting Radial Flow (IARF). The following procedure is recommended for analyzing gas-well drawdown test data or gas-well production data exhibiting IARF when the flow rate is changing slowly and smoothly:

1. Calculate the rate-normalized adjusted pressure change as

$$\Delta p_a/q = \left[p_{ai} - p_{awf}(t)\right]/q_g(t). \quad (8.92)$$

2. Calculate the material-balance time as

$$t_{mb} = \frac{24G_p(t)}{q(t)}. \quad (8.93)$$

3. Graph the rate-normalized adjusted pressure change, $\Delta p_a/q$, vs. the material-balance time, t_{mb}, on a semilog scale.
4. Identify data in IARF.
5. Draw (or fit) a straight line through the selected data, and find the slope m' and intercept $(\Delta p_a/q)_{1\text{hr}}$. Note that the intercept $(\Delta p_a/q)_{1\text{hr}}$ must read from the straight line or its extrapolation, not from the measured field pressure data.
6. Calculate the permeability k from the slope m' from

$$k = \frac{1{,}626 B_{gi}\mu_{gi}}{m'h}. \quad (8.94)$$

Note that the slope m' must always be positive.

7. Calculate the skin factor s from the slope m' and the intercept $(\Delta p_a/q)_{1\text{hr}}$ as

$$s = 1.151\left[\frac{(\Delta p_a/q)_{1\text{hr}}}{m'} - \log\left(\frac{k}{\phi\mu_{gi}c_{ti}r_w^2}\right) + 3.23\right]. \quad (8.95)$$

Note that the intercept $(\Delta p_a/q)_{1\text{hr}}$ might be negative for a highly stimulated well.

8.7.4 Example Problem 8.4.

8.7.4.1 Rate-Normalization/Material-Balance-Time Example for IARF. Analyze 45 days of daily production data. Given the rock and fluid properties in **Table 8.5,** analyze the gas-well daily production data given in **Table 8.6** and shown in **Fig. 8.18.**

Rock and Fluid Properties

p_i	5,000	psia	T_f	150	°F
ϕ	0.085		S_w	0.5	
h	100	ft	r_w	0.25	ft
c_f	2.50×10^{-6}	psi^{-1}	c_w	2.07×10^{-6}	psi^{-1}
γ_g	0.62	(air=1)	B_{gi}	0.6076	bbl/Mscf
c_{gi}	1.215×10^{-4}	psi^{-1}	μ_{gi}	0.0279	cp
c_{ti}	6.429×10^{-5}	psi^{-1}	p_{ai}	3,596.6	psi

Table 8.5—Rock- and fluid-property data for gas-well production data analysis examples, IARF and boundary-dominated flow.

Daily Production Data

t (days)	t (hours)	p_{wf} (psia)	q (Mscf/D)	G_p (Mscf)	p_{awf} (psi)	t_{mb} (hours)	$\Delta p_{awf}/q$ (psi/Mscf/D)
0	0	5,000.0	0	0	3,596.6	0.0	0.0000
1	24	4,742.0	551	349	3,339.0	15.2	0.4676
2	48	4,500.7	799	1,026	3,098.4	30.8	0.6236
3	72	4,274.9	979	1,913	2,874.0	46.9	0.7381
4	96	4,063.7	1,122	2,960	2,665.0	63.3	0.8303
5	120	3,866.1	1,238	4,137	2,470.7	80.2	0.9095
6	144	3,681.3	1,335	5,420	2,290.4	97.4	0.9785
7	168	3,508.4	1,415	6,791	2,123.4	115.2	1.0412
8	192	3,346.6	1,482	8,237	1,969.0	133.4	1.0983
9	216	3,195.2	1,538	9,743	1,826.3	152.0	1.1511
10	240	3,053.7	1,585	11,302	1,694.7	171.1	1.2000
11	264	2,921.2	1,623	12,903	1,573.4	190.8	1.2466
12	288	2,797.3	1,655	14,539	1,461.7	210.8	1.2900
13	312	2,681.4	1,680	16,203	1,359.4	231.5	1.3317
14	336	2,573.0	1,700	17,890	1,265.4	252.6	1.3713
15	360	2,471.5	1,715	19,595	1,179.0	274.2	1.4097
16	384	2,376.6	1,727	21,314	1,099.9	296.2	1.4457

(Continued)

Daily Production Data

t (days)	t (hours)	p_{wf} (psia)	q (Mscf/D)	G_p (Mscf)	p_{awf} (psi)	t_{mb} (hours)	$\Delta p_{awf}/q$ (psi/Mscf/D)
17	408	2,287.8	1,735	23,043	1,027.3	318.8	1.4809
18	432	2,204.8	1,740	24,779	960.8	341.8	1.5148
19	456	2,127.1	1,743	26,518	900.3	365.1	1.5469
20	480	2,054.4	1,744	28,260	844.8	388.9	1.5779
21	504	1,986.4	1,743	30,002	793.9	413.1	1.6080
22	528	1,922.8	1,740	31,742	747.6	437.8	1.6374
23	552	1,863.3	1,737	33,479	705.3	462.6	1.6646
24	576	1,807.6	1,732	35,212	666.2	487.9	1.6919
25	600	1,755.5	1,726	36,940	630.9	513.7	1.7183
26	624	1,706.8	1,720	38,661	598.1	539.5	1.7433
27	648	1,661.2	1,713	40,377	568.5	565.7	1.7677
28	672	1,618.6	1,706	42,085	541.1	592.1	1.7910
29	696	1,578.7	1,698	43,786	516.0	618.9	1.8143
30	720	1,541.3	1,690	45,480	493.0	645.9	1.8365
31	744	1,506.4	1,682	47,165	471.6	673.0	1.8579
32	768	1,473.8	1,674	48,842	452.4	700.2	1.8783
33	792	1,443.2	1,665	50,511	434.6	728.1	1.8991
34	816	1,414.6	1,657	52,172	418.0	755.7	1.9183
35	840	1,387.9	1,649	53,824	402.9	783.4	1.9368
36	864	1362.9	1,640	55,468	389.2	811.7	1.9558
37	888	1,339.5	1,632	57,104	376.5	839.8	1.9731
38	912	1,317.6	1,624	58,731	364.5	867.9	1.9902
39	936	1,297.1	1,616	60,351	353.4	896.3	2.0069
40	960	1,277.9	1,607	61,962	343.6	925.4	2.0243
41	984	1,260.0	1,599	63,565	334.4	954.1	2.0402
42	1,008	1,243.2	1,592	65,160	325.8	982.3	2.0545
43	1,032	1,227.6	1,584	66,748	317.8	1,011.3	2.0700
44	1,056	1,212.9	1,576	68,328	310.3	1,040.5	2.0852
45	1,080	1,199.1	1,569	69,900	303.3	1,069.2	2.0990

Table 8.6—Production data for gas-well production-data-analysis example, IARF.

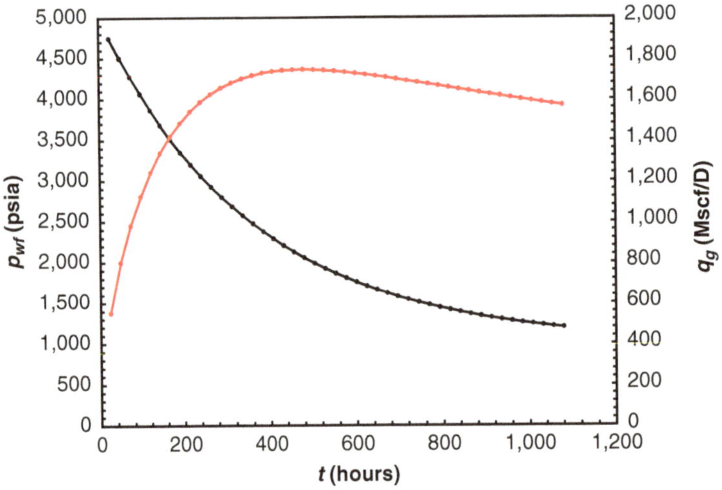

Fig. 8.18—Daily production data for example of using rate normalization with material-balance time, IARF.

These are synthetic data, generated using an analytical model, so the true reservoir properties are known. The data were generated using a decreasing flowing bottomhole pressure, approaching a constant value of 1,000 psia. The tabulated flow rate is the instantaneous rate at the end of each day of production, rather than an average flow rate for the period.

8.7.4.2 Solution. We will use the rate-normalized adjusted pressure change for the pressure plotting variable and the material-balance time for the time plotting variable. The adjusted pressure has been calculated and tabulated. For this example, we show calculations for only the first row of data.

1. Calculate the rate-normalized adjusted pressure change using Eq. 8.92. For the first row of data, we have

$$\begin{aligned}\Delta p_a/q &= \left[p_{ai} - p_{awf}(t)\right]/q_g(t)\\ &= (3{,}596.6 - 3{,}339.0)/551\\ &= 0.4676\ \text{psi/Mscf/D}.\end{aligned}$$

2. Calculate the material-balance time from Eq. 8.93. For the first row,

$$\begin{aligned}t_{mb} &= \frac{24G_p(t)}{q(t)}\\ &= \frac{(24)(349)}{551}\\ &= 15.2\ \text{hours}.\end{aligned}$$

3. Graph the rate-normalized adjusted pressure change, $\Delta p_a/q$, vs. the material-balance time, t_{mb}, on a semilog scale, as shown in **Fig. 8.19.**
4. Identify data in IARF. The data from t = 25 days to t = 45 days appear to fall on a straight line.
5. Draw (or fit) a straight line through the selected data. We find the best-fit straight line has slope m' = 1.198 psi/Mscf/D/cycle and intercept $(\Delta p_a/q)_{1\text{hr}}$ = −1.530 psi/Mscf/D.
6. Calculate the permeability k from the slope m' from Eq. 8.94 as

$$\begin{aligned}k &= \frac{162.6B_{gi}\mu_{gi}}{m'h}\\ &= \frac{(162.6)(0.6059)(0.0269)}{(1.198)(100)}\\ &= 0.0221\ \text{md}.\end{aligned}$$

7. Calculate the skin factor s from the slope m' and the intercept $(\Delta p_a/q)_{1\text{hr}}$ from Eq. 8.95 as

$$\begin{aligned}s &= 1.151\left[\frac{(\Delta p_a/q)_{1\text{hr}}}{m'} - \log\left(\frac{k}{\phi\mu_{gi}c_{ti}r_w^2}\right) + 3.23\right]\\ &= 1.151\left[\frac{-1.530}{1.198} - \log\left(\frac{0.0221}{(0.085)(0.0269)(6.429\times10^{-5})(0.25)^2}\right) + 3.23\right]\\ &= -5.10.\end{aligned}$$

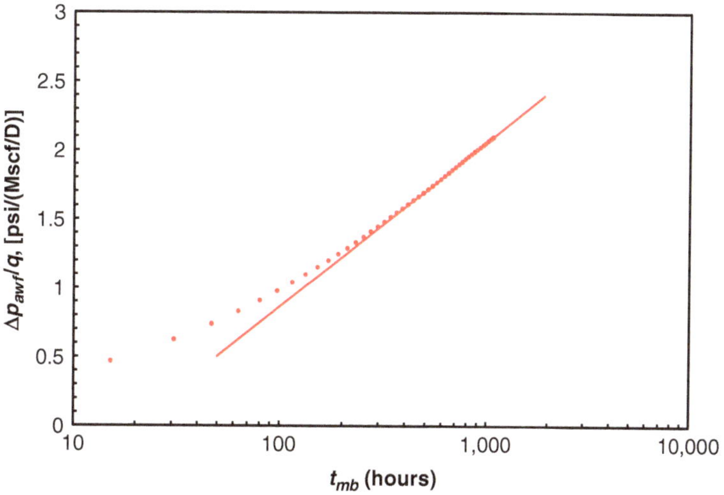

Fig. 8.19—IARF, semilog analysis using rate normalization with material-balance time.

8. Finally, we estimate the apparent infinite-conductivity fracture length from Eq. 8.48 as:

$$L_{f,\text{eq}} = 2r_w \exp(-s)$$
$$= (2)(0.25)\exp(-(-5.10)).$$
$$= 82 \text{ ft}.$$

These results agree very well with the permeability of 0.023 md and fracture half-length of 80 ft used to generate this synthetic data set.

8.7.5 Using Rate Normalization and Material-Balance Time—Boundary-Dominated Flow. Use of rate-normalization and material-balance time for gas wells exhibiting boundary-dominated flow requires a trial-and-error method because of the dependence of the adjusted cumulative production, Eq. 8.90, on the average reservoir pressure as a function of time, which is almost always unknown.

The following procedure may be used to analyze variable-rate/variable-pressure gas well production data during boundary-dominated flow:

1. Calculate the rate-normalized pressure change, $\Delta p_a/q$, as

$$\Delta p_a/q = \left[p_{ai} - p_{awf}(t)\right]/q_g(t). \quad \text{(8.96)}$$

2. Make an initial guess of the original gas-in-place, G_i. The guess should be larger than the most recent cumulative production. A more rigorous lower bound can be obtained by rearranging the material-balance equation, recognizing that the average reservoir pressure must be higher than the flowing bottomhole pressure:

$$G_i > \frac{G_p}{1 - \dfrac{p_{wf}}{z_{wf}} \Big/ \dfrac{p_i}{z_i}}. \quad \text{(8.97)}$$

3. Using the gas material-balance equation, calculate the average reservoir pressure as a function of time from the cumulative gas produced:

$$\frac{\bar{p}}{\bar{z}} = \left(1 - \frac{G_p}{G_i}\right)\frac{p_i}{z_i}. \quad \text{(8.98)}$$

For stress-sensitive formations, Eq. 8.98 should be modified as

$$\left(\frac{\bar{p}}{\bar{z}}\right)^* = \left(1 - \frac{G_p}{G_i}\right)\frac{p_i}{z_i}, \quad \text{(8.99)}$$

where $\left(\frac{\bar{p}}{\bar{z}}\right)^*$ is defined in Eq. 8.79.

4. Calculate the adjusted cumulative production by integrating Eq. 8.90 numerically using the trapezoidal rule:

$$G_{pa,k} = \mu_{gi} c_{ti} \sum_{j=1}^{k} \frac{1}{2}\left(\frac{1}{\mu_{g,j-1} c_{t,j-1}} + \frac{1}{\mu_{g,j} c_{t,j}}\right)\left(G_{p,j} - G_{p,j-1}\right). \quad \text{(8.100)}$$

For stress-sensitive formations, Eq. 8.100 should be modified to incorporate stress-dependent porosity as

$$G_{pa,k} = \phi_i \mu_{gi} c_{ti} \sum_{j=1}^{k} \frac{1}{2}\left(\frac{1}{\phi_{j-1}\mu_{g,j-1} c_{t,j-1}} + \frac{1}{\phi_j \mu_{g,j} c_{t,j}}\right)\left(G_{p,j} - G_{p,j-1}\right). \quad \text{(8.101)}$$

5. Calculate the material-balance adjusted time, t_{amb}, from

$$t_{amb} = \frac{24 G_{pa}(t)}{q(t)}. \quad \text{(8.102)}$$

6. Graph the rate-normalized adjusted pressure change, $\Delta p_a/q$, vs. the material-balance adjusted time, t_{amb}, on Cartesian coordinates.
7. Draw a straight line through the data in boundary-dominated flow.
8. Find the slope m'_{pss} and the intercept b_{pss} of the straight line.
9. Calculate the original gas-in-place, G_i, from the slope, m'_{pss}, as

$$G_i = \frac{(1 - S_w)}{24\left|m'_{pass}\right| c_{ti}}. \quad \text{(8.103)}$$

10. Using the updated estimate of G_i from Eq. 8.103, repeat Steps 3 through 9 until the original-gas-in-place estimate converges.

11. Calculate the reservoir pore volume, V_p, from the slope, m'_{pss}, as

$$V_p = \frac{0.234 B_{gi}}{|m'_{pass}| c_{ti}}. \quad (8.104)$$

12. If the formation net pay and porosity are known, calculate the drainage area from the pore volume:

$$A = \frac{V_p}{\phi_i h}. \quad (8.105)$$

13. If desired, calculate the productivity index from the intercept, b_{pss}:

$$J = \frac{1}{b_{pss}}. \quad (8.106)$$

8.7.6 Example Problem 8.5.

8.7.6.1 Rate-Normalization/Material-Balance-Time Example for Boundary-Dominated Flow. Analyze 1 year of daily/weekly production data. Given the rock and fluid properties in Table 8.5, analyze the gas-well production data provided in **Tables 8.6 and 8.7** and shown in **Fig. 8.20.**

Additional Daily/Weekly Production Data

t (days)	t (hours)	p_{wf} (psia)	q (Mscf/D)	G_p (Mscf)	p_{awf} (psi)	t_{mb} (hours)	$\Delta p_{awf}/q$ (psi/Mscf/D)
46	1,104	1,186.3	1,561	71,465	297.2	1,098.8	2.1137
47	1,128	1,174.3	1,554	73,022	291.5	1,127.8	2.1269
48	1,152	1,163.0	1,547	74,572	286.1	1,156.9	2.1400
49	1,176	1,152.5	1,540	76,115	281.1	1,186.2	2.1529
56	1,344	1,095.7	1,493	86,725	254.3	1,394.1	2.2387
63	1,512	1,060.0	1,451	97,028	238.6	1,604.9	2.3143
70	1,680	1,037.6	1,413	107,052	228.8	1,818.3	2.3835
77	1,848	1,023.6	1,379	116,822	222.6	2,033.2	2.4467
84	2,016	1,014.8	1,346	126,357	218.8	2,253.0	2.5095
91	2,184	1,009.3	1,315	135,673	216.4	2,476.2	2.5705
98	2,352	1,005.8	1,286	144,775	214.8	2,701.9	2.6297
105	2,520	1,003.6	1,258	153,677	213.9	2,931.8	2.6890
112	2,688	1,002.3	1,232	162,392	213.3	3,163.5	2.7462
119	2,856	1,001.4	1,207	170,929	212.9	3,398.8	2.8034
126	3,024	1,000.9	1,183	179,296	212.7	3,637.5	2.8605
133	3,192	1,000.6	1,160	187,498	212.6	3,879.3	2.9173
140	3,360	1,000.4	1,138	195,542	212.5	4,123.9	2.9738
147	3,528	1,000.2	1,117	203,434	212.4	4,371.0	3.0298
154	3,696	1,000.1	1,096	211,178	212.3	4,624.3	3.0879
161	3,864	1,000.1	1,076	218,778	212.3	4,879.8	3.1453
168	4,032	1,000.1	1,056	226,240	212.3	5,141.8	3.2049
175	4,200	1,000.0	1,037	233,567	212.3	5,405.6	3.2636
182	4,368	1,000.0	1,019	240,764	212.3	5,670.6	3.3212
189	4,536	1,000.0	1,001	247,833	212.3	5,942.0	3.3809
196	4,704	1,000.0	984	254,779	212.3	6,214.1	3.4394
203	4,872	1,000.0	967	261,605	212.3	6,492.8	3.4998
210	5,040	1,000.0	950	268,315	212.3	6,778.5	3.5625
217	5,208	1,000.0	934	274,911	212.3	7,064.1	3.6235
224	5,376	1,000.0	919	281,396	212.3	7,348.8	3.6826
231	5,544	1,000.0	904	287,774	212.3	7,640.0	3.7437
238	5,712	1,000.0	889	294,046	212.3	7,938.2	3.8069
245	5,880	1,000.0	874	300,217	212.3	8,243.9	3.8722

(Continued)

Additional Daily/Weekly Production Data

t (days)	t (hours)	p_{wf} (psia)	q (Mscf/D)	G_p (Mscf)	p_{awf} (psi)	t_{mb} (hours)	$\Delta p_{awf}/q$ (psi/Mscf/D)
252	6,048	1,000.0	860	306,288	212.3	8,547.6	3.9353
259	6,216	1,000.0	847	312,262	212.3	8,848.0	3.9957
266	6,384	1,000.0	833	318,141	212.3	9,166.1	4.0628
273	6,552	1,000.0	820	323,927	212.3	9,480.8	4.1272
280	6,720	1,000.0	807	329,624	212.3	9,802.9	4.1937
287	6,888	1,000.0	795	335,232	212.3	10,120.2	4.2570
294	7,056	1,000.0	783	340,754	212.3	10,444.6	4.3223
301	7,224	1,000.0	771	346,192	212.3	10,776.4	4.3895
308	7,392	1,000.0	759	351,548	212.3	11,116.1	4.4589
315	7,560	1,000.0	748	356,824	212.3	11,448.9	4.5245
322	7,728	1,000.0	737	362,022	212.3	11,789.0	4.5920
329	7,896	1,000.0	726	367,143	212.3	12,137.0	4.6616
336	8,064	1,000.0	716	372,189	212.3	12,475.6	4.7267
343	8,232	1,000.0	705	377,162	212.3	12,839.6	4.8005
350	8,400	1,000.0	695	382,063	212.3	13,193.5	4.8695
357	8,568	1,000.0	685	386,894	212.3	13,555.4	4.9406
364	8,736	1,000.0	676	391,657	212.3	13,905.0	5.0064

Table 8.7—Production data for gas-well production-data-analysis example, boundary-dominated flow.

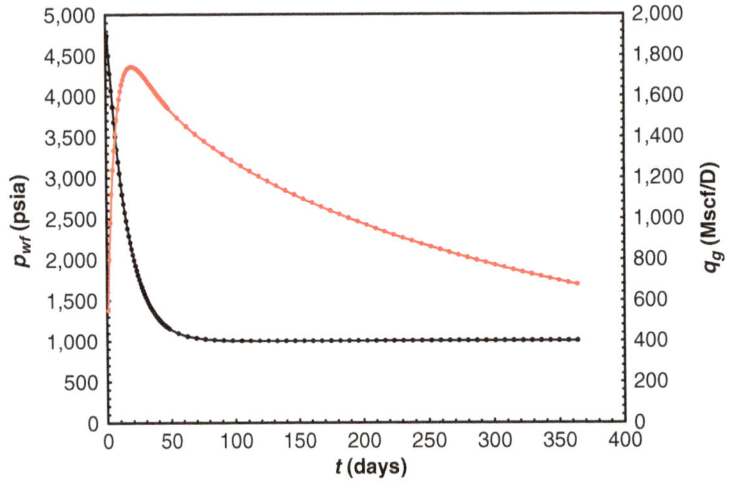

Fig. 8.20—Production data for example of using rate normalization with material-balance time, boundary-dominated flow.

8.7.6.2 Solution. **Fig. 8.21** shows a Cartesian graph of the rate-normalized adjusted pressure change vs. the material-balance adjusted time for the first and last iterations of the procedure. Because we have no information about the size of the reservoir, we assume an initial gas-in-place of 1,000 Bcf. Fitting a straight line through the data from 168 days to 364 days for the first iteration gives an original gas-in-place, G_i, of 1.576 Bcf. The final iteration gives an original gas-in-place of 1.055 Bcf, approximately 3.3% smaller than the true value of 1.088 Bcf. Incorporating stress-dependent porosity in the material-balance and adjusted cumulative production equations (Eqs. 8.78 and 8.91, respectively) gives an initial gas-in-place of 1.081 Bcf, only 0.7% smaller than the true value. Given the low pore-volume compressibility for this example, 2.50×10^{-6} psi^{-1}, the improvement in accuracy reiterates the importance of incorporating stress dependence for cases involving high pore-volume compressibility.

8.7.7 Example Problem 8.6.

8.7.7.1 Rate-Normalization/Material-Balance-Time Example for Boundary-Dominated Flow. Analyze monthly production data. Given the rock and fluid properties in Table 8.3, analyze the gas-well production data provided in Table 8.4 using rate normalization and material-balance adjusted time.

8.7.7.2 Solution. An initial guess of 10 Bcf gas-in-place was assumed, giving a first-iteration result of 1.786 Bcf. The final iteration gave a gas-in-place of 1.431 Bcf and a productivity index of 0.462 Mscf/D/psi, as shown in **Fig. 8.22.**

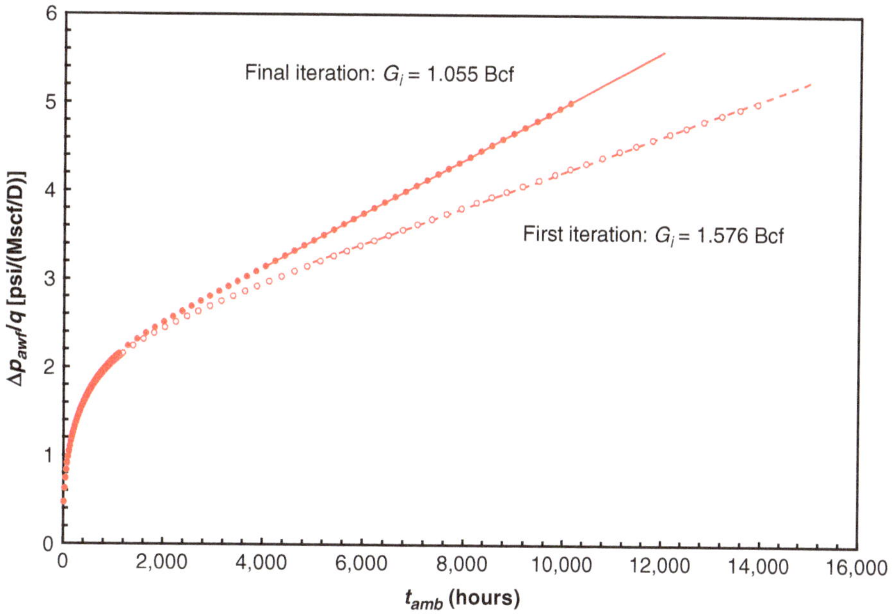

Fig. 8.21—Reservoir limits analysis, rate normalization with material-balance time.

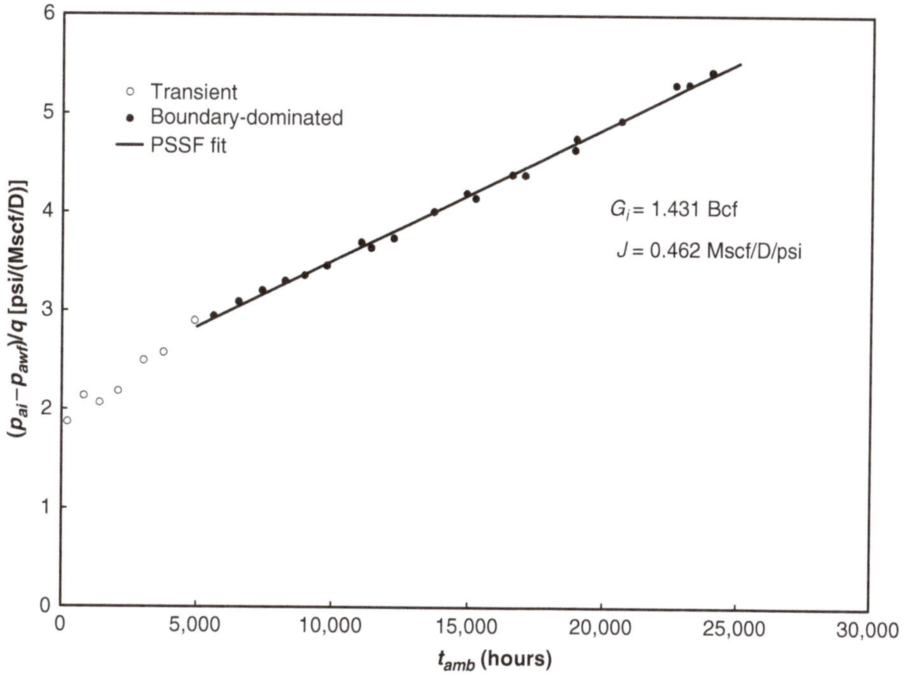

Fig. 8.22—Reservoir limits analysis field example, rate normalization with material-balance time. PSSF = pseudosteady state flow.

The drainage area can be calculated from the initial gas-in-place by rearranging the familiar volumetric equation:

$$A = \frac{G_i B_{gi}}{7{,}758 \phi_i h (1 - S_{wi})}$$

$$= \frac{(1.431)(0.7535)}{(7{,}758)(0.032)(100)(1-0.3)}$$

$$= 62.1 \text{ acres.}$$

8.7.8 Discussion. In spite of its simplicity, rate normalization is one of the most effective methods for analyzing variable-rate/variable-pressure production. However, there are a couple of characteristics that should be kept in mind when using rate-normalization/material-balance-time methods.

First, both pressure and rate should be changing slowly and smoothly. Following an abrupt change in pressure, as in **Fig. 8.23,** there will be a transient period during which the rate-normalization/material-balance-time method will not follow the constant-rate solution, as shown in **Fig. 8.24.** At first glance, the data in Fig. 8.24 appear to exhibit a period of wellbore storage following the pressure change at 180 days. However, this phenomenon is not wellbore storage. Instead, it is simply an artifact of the rate-normalization/material-balance-time graphing technique.

Second, the material-balance time is not a monotonic function of time. When the rate is low (e.g., following an increase in backpressure), the material-balance time might be very large—as the rate increases, the material-balance time decreases, as shown in **Fig. 8.25.**

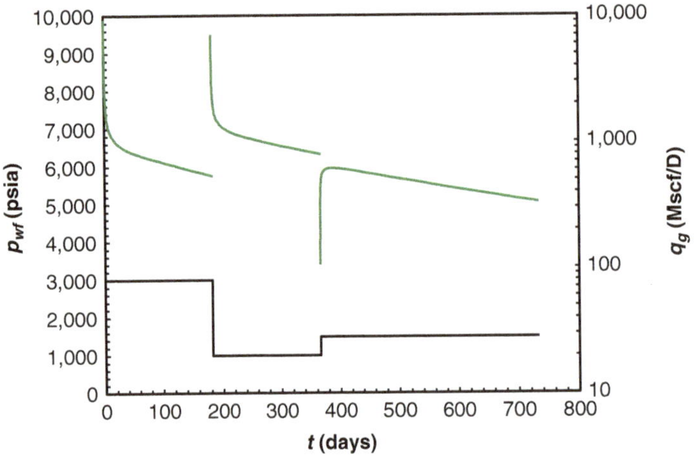

Fig. 8.23—Pressure and rate history, rate-normalization/material-balance-time pressure-step-change example.

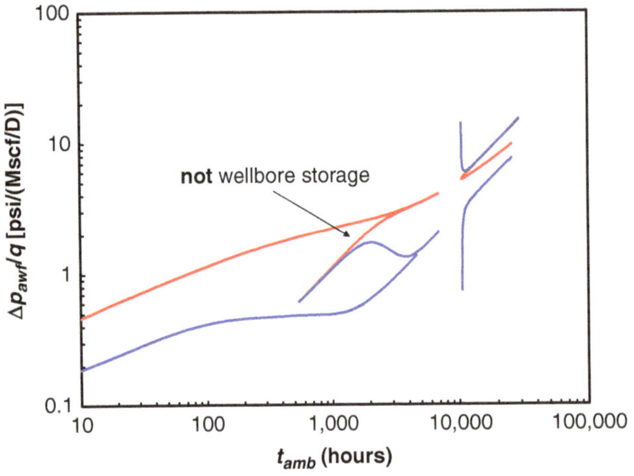

Fig. 8.24—Rate normalization with material-balance time applied to the rate/pressure history in Fig. 8.23.

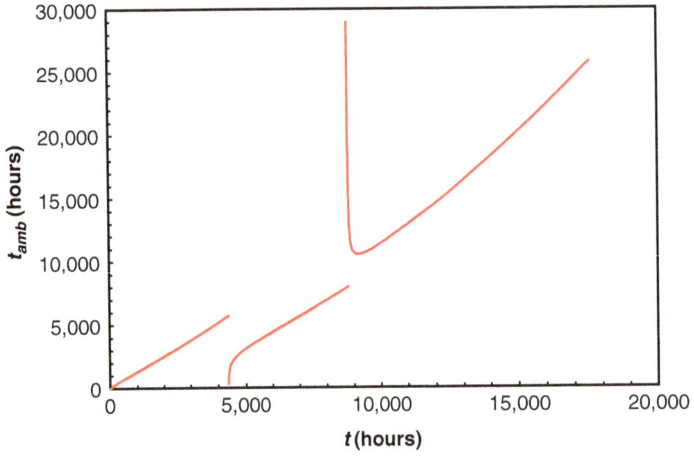

Fig. 8.25—Material-balance time is not a monotonic function of time.

8.8 Analyzing Production Data With Pressure Transient Type Curves

Although we have presented the rate-normalization/material-balance-time technique in the context of semilog analysis of data in IARF, the method is also applicable to pressure and pressure-derivative type-curve analysis methods as well as analysis of data exhibiting linear or bilinear flow. **Fig. 8.26** shows pressure and pressure-derivative type curves for a closed circular reservoir producing at constant rate. The dimensionless pressure, dimensionless time, and dimensionless drainage radius are defined by

$$p_D = \frac{kh}{141.2qB\mu}\left(p_i - p_{wf}\right), \quad (8.107)$$

$$t_D = \frac{0.00633kt}{\phi\mu c_t r_{wa}^{\ 2}}, \quad (8.108)$$

and

$$R_D = \frac{r_e}{r_{wa}}, \quad (8.109)$$

respectively, where time t in Eq. 8.108 is in days for convenience for analyzing production data, and r_{wa} is the apparent wellbore radius defined by

$$r_{wa} = r_w e^{-s}. \quad (8.110)$$

The use of rate normalization along with material-balance time allows the use of pressure transient type curves, such as those shown in **Figs. 8.26 and 8.27,** for analysis of production data. For Fig. 8.26, the permeability is calculated from the pressure-match-point ratio, the apparent wellbore radius and skin factor from the time-match-point ratio, and the drainage radius from the matching stem parameter R_D.

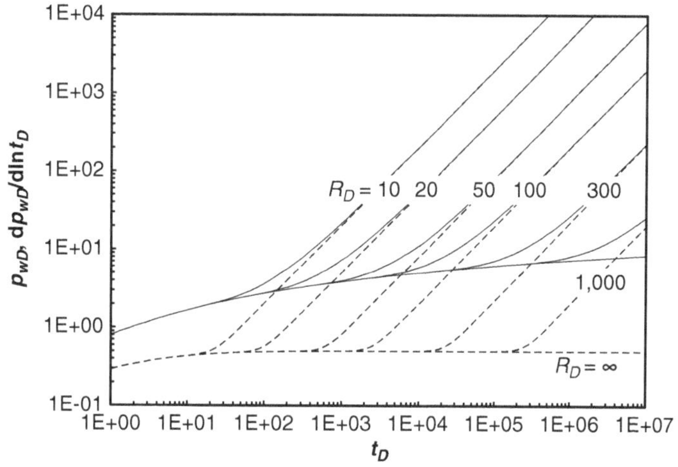

Fig. 8.26—Pressure and pressure-derivative type curves, closed circular reservoir, dimensionless time based on apparent wellbore radius.

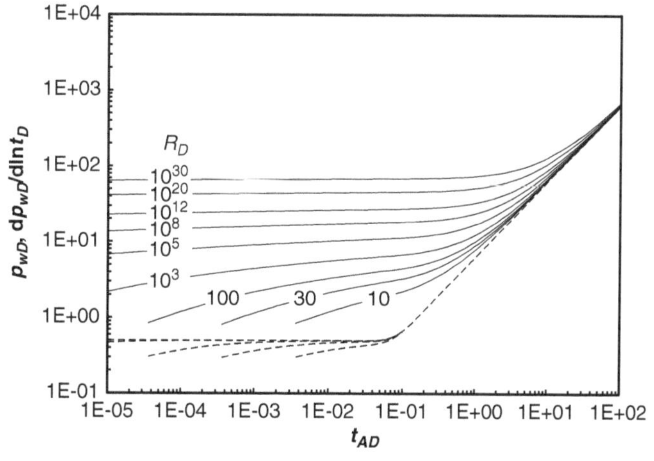

Fig. 8.27—Pressure and pressure-derivative type curves, closed circular reservoir, dimensionless time based on drainage area.

$$t_{AD}=\frac{0.00633kt}{\phi\mu c_t A}, \quad (8.111)$$

where A is the drainage area in ft^2,

$$A=\pi r_e^2. \quad (8.112)$$

For production data analysis, it is convenient to graph the pressure and pressure-derivative vs. dimensionless time based on drainage area, t_{AD}, as shown in Fig. 8.27. In this figure, the permeability is calculated from the pressure-match-point ratio, the drainage area from the time-match-point ratio, and the apparent wellbore radius and skin factor from the matching stem parameter R_D.

The following procedure can be used to analyze variable-rate/variable-pressure production for a gas well using the constant-rate-production pressure-change type curves shown in Figs. 8.26 and 8.27:

1. Calculate the rate-normalized pressure change, $\Delta p_a/q$, from Eq. 8.96.
2. Estimate the initial gas-in-place, G_i.
3. Calculate the average drainage-area pressure as a function of time from the cumulative gas produced from Eq. 8.98 or Eq. 8.99.
4. Calculate adjusted cumulative production using the trapezoidal rule, using Eq. 8.100 or Eq. 8.101.
5. Calculate the material-balance adjusted time, t_{amb}, from Eq. 8.102.
6. Calculate the logarithmic derivative of the rate-normalized adjusted pressure change with respect to the material-balance adjusted time. For data sets without significant noise or abrupt changes in pressure, the Bourdet method (Bourdet et al. 1989) using adjacent points can be used:

$$\left(\frac{dy}{dx}\right)_j=\frac{\left(x_{j+1}-x_j\right)\dfrac{y_j-y_{j-1}}{x_j-x_{j-1}}+\left(x_j-x_{j-1}\right)\dfrac{y_{j+1}-y_j}{x_{j+1}-x_j}}{x_{j+1}-x_{j-1}}, \quad (8.113)$$

where x is the natural logarithm of the material-balance adjusted time,

$$x=\ln t_{amb}, \quad (8.114)$$

and y is the rate-normalized adjusted pressure change,

$$y=\frac{p_{ai}-p_{awf}}{q}. \quad (8.115)$$

Unfortunately, production data are usually much noisier and lower in frequency than pressure transient data, resulting in lower-quality derivative data.

7. Graph the rate-normalized adjusted pressure change and its logarithmic derivative vs. the material-balance adjusted time on a log-log scale.
8. If the data exhibit boundary-dominated flow, indicated by the presence of a unit-slope line in the derivative at late time, use the type curve in Fig. 8.27. Otherwise, if only transient flow is present, indicated by the presence of a horizontal derivative at early time, use the type curve in Fig. 8.26.

For the p_D vs. t_D type curves in Fig. 8.26,

9. Move the data vertically to align the transient part of the field-data derivative with the horizontal part of the derivative type curve.
10. Move the data horizontally to align the rate-normalized adjusted-pressure-change curve with the pressure type curve.
11. If data in boundary-dominated flow are present, find the R_D stem that best fits the unit-slope field-data derivative at late time, interpolating if necessary. If data are not present, select the smallest R_D stem consistent with the data.
12. Calculate the permeability from the pressure-match-point ratio:

$$k=\frac{141.2B_{gi}\mu_{gi}}{h}\left[\frac{p_D}{\left(p_{ai}-p_{awf}\right)/q}\right]_{MP}. \quad (8.116)$$

13. Calculate the apparent wellbore radius from the time-match-point ratio:

$$r_{wa}=\sqrt{\frac{0.00633k}{\phi_i\mu_{gi}c_{ti}}\left(\frac{t_{amb}}{t_D}\right)_{MP}}. \quad (8.117)$$

14. Calculate the drainage radius from the R_D stem, then calculate the drainage area:

$$r_e=R_D r_{wa}, \quad (8.118)$$

and

$$A=\pi r_e^2. \quad (8.119)$$

For the p_D vs. t_{AD} type curves in Fig. 8.27,

15. Move the data horizontally to align the boundary-dominated, unit-slope portion of the field-data derivative with the unit-slope portion of the derivative type curve.
16. Move the field data diagonally, keeping the unit-slope part of the field-data derivative aligned with the derivative type curve, to find the best fit of the field-data pressure change and field-data derivative with pressure and pressure-derivative type curves, respectively.
17. Find the R_D stem that best fits the field-data rate-normalized adjusted pressure change.
18. Calculate the permeability from the pressure-match-point ratio using Eq. 8.116.

19. Calculate the drainage area from the time-match-point ratio:

$$A = \frac{0.00633k}{\phi_i \mu_{gi} c_{ti}} \left(\frac{t_{amb}}{t_{AD}} \right)_{MP}. \quad (8.120)$$

20. Calculate the drainage radius from the area:

$$r_e = \sqrt{\frac{A}{\pi}}. \quad (8.121)$$

21. Calculate the apparent wellbore radius from the R_D stem:

$$r_{wa} = \frac{r_e}{R_D}. \quad (8.122)$$

For either type curve,

22. Calculate the skin factor from the apparent wellbore radius:

$$s = -\ln\left(\frac{r_{wa}}{r_w} \right). \quad (8.123)$$

23. Calculate the initial gas-in-place, G_i, from the drainage area:

$$G_i = \frac{7,758 A \phi_i h (1 - S_{wi})}{B_{gi}}. \quad (8.124)$$

If the initial gas-in-place G_i calculated from Eq. 8.124 is significantly different from the initial guess in Step 2, repeat Steps 3 through 23 until the assumed and calculated values are consistent.

8.8.1 Field Example Problem 8.7—West Virginia Gas Well, Pressure-Transient Type-Curve Analysis. Analyze Monthly production data. Given the rock and fluid properties in Table 8.3, match the gas-well production data provided in Table 8.4 using rate normalization and material-balance adjusted time with the pressure-change type curves, shown in Figs. 8.26 and 8.27.

8.8.1.1 Solution. Because we have already estimated the initial gas-in-place using boundary-dominated flow, we use that value, 1.431 Bcf, as our starting estimate.

Fig. 8.28 shows the resulting match, with a pressure-match-point ratio of 0.865 psi/Mscf/D, a time-match-point ratio of 1,686 days, and a dimensionless drainage radius of 30.

1. Calculate the permeability from the pressure-match-point ratio using Eq. 8.116:

$$k = \frac{141.2 B_{gi} \mu_{gi}}{h} \left[\frac{p_D}{(p_{ai} - p_{awf})/q} \right]_{MP}$$

$$= \frac{(141.2)(0.7535)(0.0279)}{(100)} \left[\frac{1}{0.865} \right]$$

$$= 0.0264 \text{ md}.$$

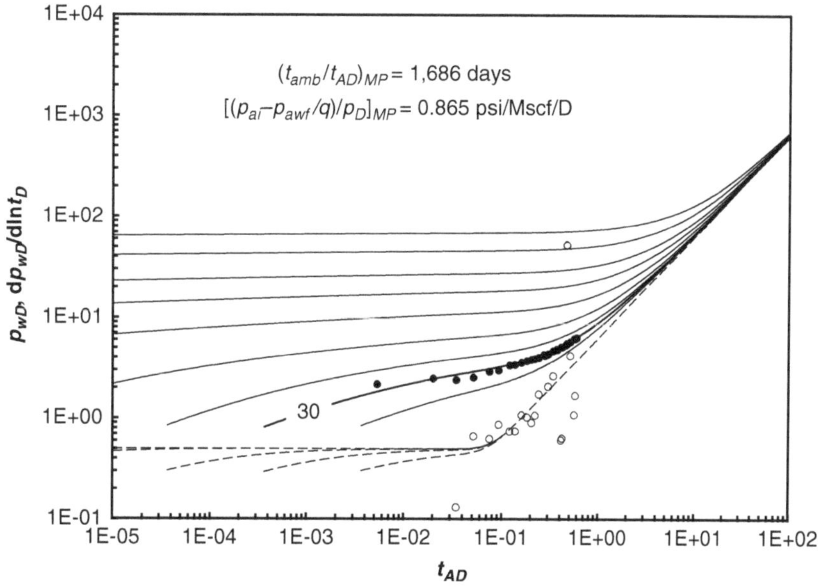

Fig. 8.28—Field example, analysis using rate-normalized adjusted pressure change and material-balance adjusted time, pressure-change type curves.

2. Calculate the drainage area from the time match point ratio using Eq. 8.120:

$$A = \frac{0.00633k}{\phi_i \mu_{gi} c_{ti}} \left(\frac{t_{amb}}{t_D} \right)_{MP}$$
$$= \frac{(0.00633)(0.0264)}{(0.032)(0.0215)(1.51\times10^{-5})}(1,686)$$
$$= 2,708,000 \text{ ft}^2$$
$$= 62.2 \text{ acres.}$$

3. Calculate the drainage radius from the area:

$$r_e = \sqrt{\frac{A}{\pi}}$$
$$= \sqrt{\frac{2,708,000}{\pi}}$$
$$= 928 \text{ ft.}$$

4. Calculate the apparent wellbore radius from the drainage radius and the R_D stem:

$$r_{wa} = \frac{r_e}{R_D}$$
$$= \frac{928}{30}$$
$$= 30.9 \text{ ft.}$$

5. Calculate the skin factor from the apparent wellbore radius using Eq. 8.123:

$$s = -\ln\left(\frac{r_{wa}}{r_w}\right)$$
$$= -\ln\left(\frac{30.9}{0.365}\right) \quad \text{...(8.125)}$$
$$= -4.44$$

6. Calculate G_i using the volumetric equation for initial gas-in-place, Eq. 8.124:

$$G_i = \frac{7,758 A \phi_i h (1 - S_{wi})}{B_{gi}}$$
$$= \frac{(7,758)(62.2)(0.032)(100)(1-0.3)}{0.7535} \quad \text{...(8.126)}$$
$$= 1,434,000 \text{ Mscf}$$
$$= 1.434 \text{ Bcf}$$

The calculated G_i is in excellent agreement with the assumed G_i of 1.431 Bcf, so there is no need for an additional iteration.

8.8.2 Extension to Other Flow Regimes and Models. Although the rate-normalization/material-balance-time approach has been presented here within the context of its use for data in transient flow/IARF and boundary-dominated/pseudosteady-state flow and for the vertical well in a closed circular reservoir, its use is much more broadly applicable. Because it effectively transforms a variable-rate/variable-pressure production history to an equivalent constant-rate production history, it can be used with flow regimes and models, such as those presented in Part 2 Ch 7: Post-Fracture Treatment Well Testing, for post-fracture pressure transient analysis, often with little or no modification.

8.9 Agarwal-Gardner Type Curves

One of the most popular sets of production decline type curves is the Agarwal-Gardner family of type curves (Agarwal et al. 1999). The Agarwal-Gardner type curves can may be divided into three categories:

- Rate vs. time type curves
- Rate vs. cumulative-production type curves
- Cumulative-production vs. time type curves

Agarwal et al. (1999) presented type curves for radial flow and both infinite-conductivity and finite-conductivity fractures. In this section, we will consider only the radial flow type curves (the type curves for hydraulically fractured wells are similar to those for radial flow and are used in much the same way).

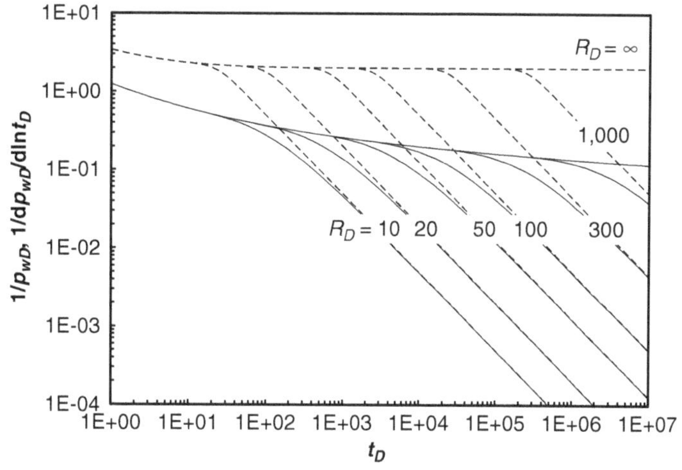

Fig. 8.29—Agarwal-Gardner rate vs. time decline type curves, $1/p_{wD}$ vs. t_D.

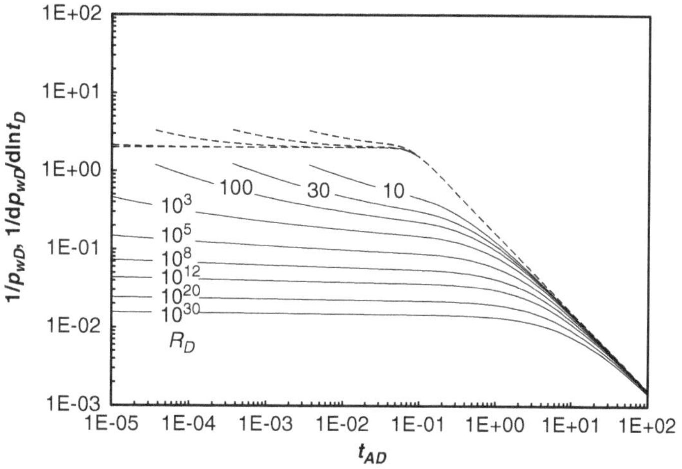

Fig. 8.30—Agarwal-Gardner rate vs. time decline type curves, $1/p_{wD}$ vs. t_{AD}.

8.9.1 Agarwal-Gardner Rate vs. Time Type Curves. Fig. 8.29 is a graph of $1/p_{wD}$ and $1/(dp_{wD}/d\ln t)$ vs. t_D for production at constant rate from a closed circular reservoir. **Fig. 8.30** is a graph of $1/p_{wD}$ and $1/(dp_{wD}/d\ln t)$ vs. t_{AD}. Fig. 8.29 is simply Fig. 8.26 turned upside down. Similarly, Fig. 8.30 is just Fig. 8.27 plotted upside down. Effectively, the Agarwal-Gardner presentation turns the unit-slope lines during pseudosteady-state flow in Figs. 8.26 and 8.27 into negative-unit-slope lines in Figs. 8.29 and 8.30.

The Fetkovich rate vs. time type curves also exhibit a negative-unit-slope line during boundary-dominated flow for harmonic decline. Thus, Figs. 8.29 and 8.30 provide a more natural decline presentation for analyzing declining production data, while Figs. 8.26 and 8.27 emphasize the connection to pressure transient analysis. The choice of whether to use Figs. 8.26 and 8.27 or Figs. 8.29 and 8.30 is strictly a matter of personal preference; there is no technical difference between the two methods. Critically, the constant-rate analog time plotting function, more commonly known as material-balance or material-balance adjusted time, must be used for the Agarwal-Gardner type curves (Figs. 8.29 and 8.30) as well as for the dimensionless pressure type curves (Figs. 8.26 and 8.27).

8.9.2 Agarwal-Gardner Pressure-Derivative vs. Time Type Curves. Fig. 8.31 is a graph of the derivative dp_{wD}/dt_D for constant-rate production from a closed circular reservoir, plotted vs. dimensionless time. Jones (1957) presented several field examples including estimation of initial oil- or gas-in-place using a log-log plot of the derivative of the rate-normalized pressure change vs. time, basically the same as Fig. 8.31.

8.9.3 Agarwal-Gardner Rate vs. Cumulative Production Type Curve. Fig. 8.32 is a Cartesian graph of the reciprocal of the dimensionless pressure, $1/p_{wD}$, vs. the dimensionless cumulative production, Q_{DA}, defined as

$$Q_{DA} = \frac{t_D}{p_{wD}}. \quad (8.127)$$

On the Agarwal-Gardner Cartesian rate vs. cumulative-production type curve, data in pseudosteady-state flow falls on a straight line with the x-intercept given by the anchor point value,

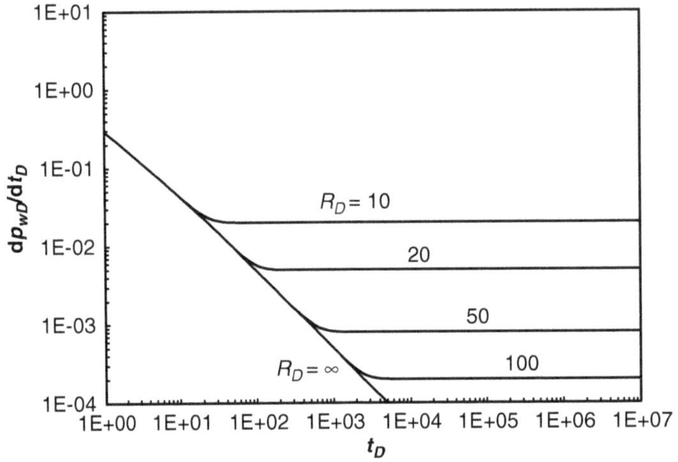

Fig. 8.31—Agarwal-Gardner type curve, dp_{wD}/dt_D vs. t_D.

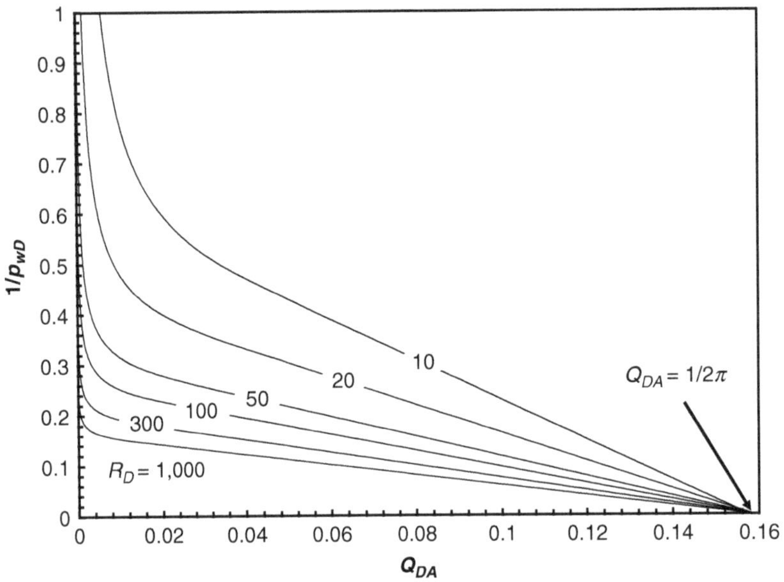

Fig. 8.32—Agarwal-Gardner rate vs. cumulative-production type curve, $1/p_{wD}$ vs. Q_{DA}.

$$Q_{DA} = \frac{1}{2\pi}\left(1 - \frac{1}{R_D^2}\right) \approx \frac{1}{2\pi}, \qquad (8.128)$$

and the y-intercept given by the dimensionless productivity index,

$$\frac{1}{p_{wD}} = \frac{1}{\ln(R_D) - 3/4}. \qquad (8.129)$$

The Agarwal-Gardner rate vs. cumulative-production type curve is particularly useful when used in conjunction with the pressure vs. time type curve shown in Fig. 8.27 or the Agarwal-Gardner rate vs. time decline type curve, shown in Fig. 8.30. After matching the field data using either the pressure vs. time type curve (Fig. 8.27) or the rate vs. time type curve (Fig. 8.30), the field data can be plotted in dimensionless form as $1/p_{wD}$ vs. Q_{DA} where

$$Q_{DA} = \frac{t_D}{p_{wD}}. \qquad (8.130)$$

As Agarwal et al. (1999) observed, overestimating gas-in-place will cause field data crossplotted on Fig. 8.32 to undershoot the anchor point, while underestimating gas-in-place will cause the field data to overshoot the anchor point. Thus, the anchor point serves as a sensitive tool to verify that the pressure or rate type-curve match honors the correct initial gas-in-place. This is especially important when the data are too noisy for the derivative to clearly establish a unit-slope line during boundary-dominated flow.

In addition, the R_D value can be estimated with greater sensitivity from Fig. 8.32 than from the pressure vs. time and rate vs. time type curves (Figs. 8.27 and 8.30). While this does not establish a unique value for R_D (and thus skin factor), it does ensure that the permeability (calculated from the pressure vs. time or rate vs. time type-curve vertical match-point ratio) and the skin factor (calculated from R_D) combine to give a productivity index consistent with observed field performance and estimated initial gas-in-place.

Fig. 8.33 shows the field data from the field example pressure type-curve match in Fig. 8.28, crossplotted on the Agarwal-Gardner rate vs. cumulative-production type curve. The boundary-dominated data fall on a straight line corresponding to an R_D value of 26, rather than a value of 30 as chosen from Fig. 8.28. Updating the apparent wellbore radius and skin factor estimates for the revised value of $R_D = 26$, we find $r_{wa} = 35.7$ ft and $s = -4.58$.

To reiterate that this match is not unique, **Figs. 8.34 and 8.35** show the pressure vs. time and rate vs. cumulative-production type curves for an alternative match. The alternative match was obtained by moving the field data upward and to the right, parallel to the unit-slope pseudosteady-state line, corresponding to dividing the pressure- and rate-match-point ratios by a factor of 1.5. As a result of moving the data parallel to the unit-slope line, the calculated gas-in-place and drainage area remain unchanged: 1.434 Bcf and 62.2 acres, respectively. On the other hand, the alternative match gives a permeability and skin factor of 0.0396 md and –3.34, respectively. Note that the permeability from the alternative match is, not coincidentally, 1.5 times the value (0.0264 md) obtained from the original match.

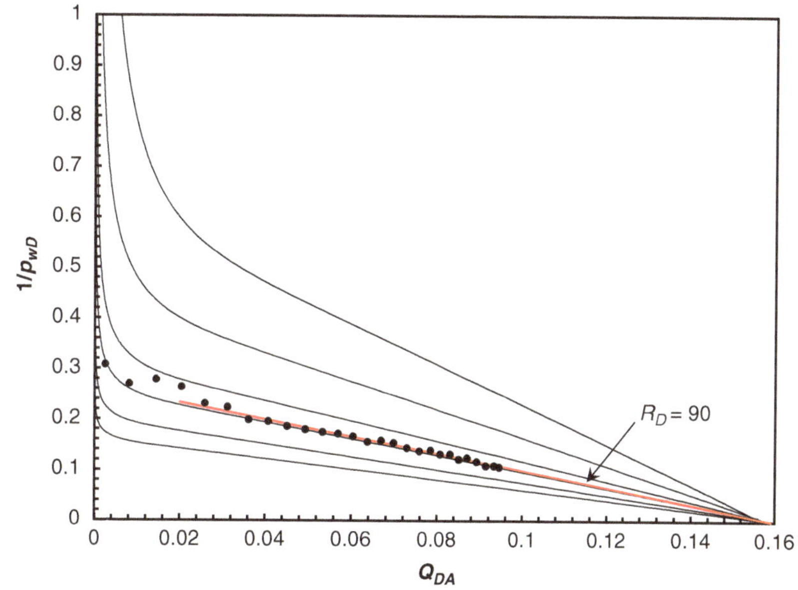

Fig. 8.33—West Virginia gas-well example, Agarwal-Gardner rate vs. cumulative-production type curve.

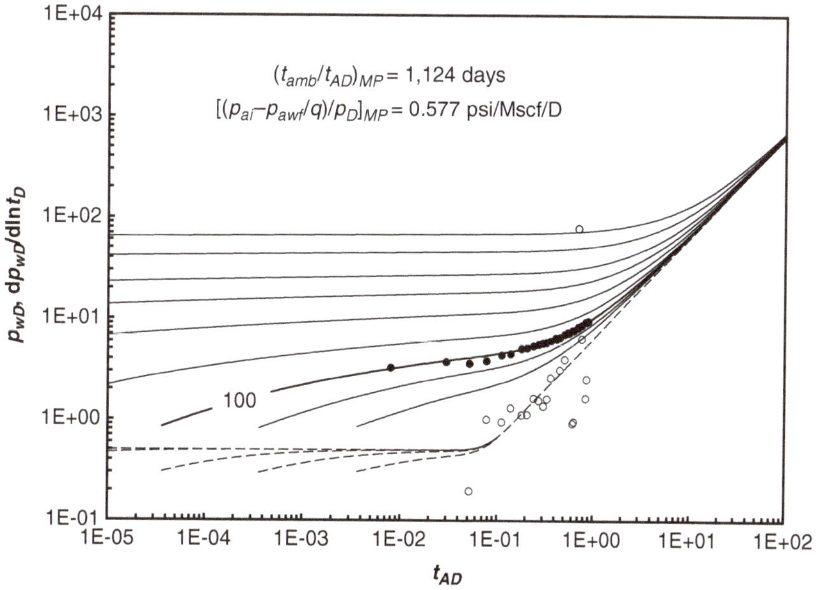

Fig. 8.34—Field example, alternative match using rate-normalized adjusted pressure change and material-balance adjusted time, pressure change type curves.

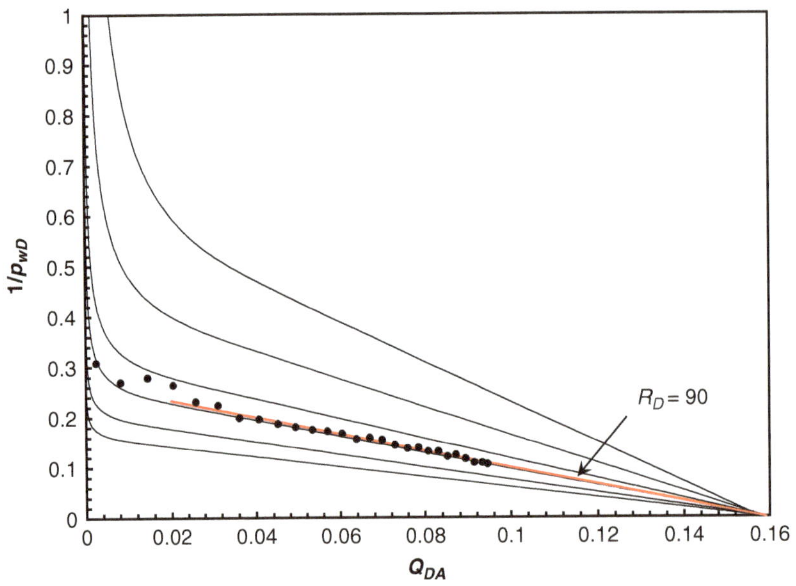

Fig. 8.35—Field example, alternative match using Agarwal-Gardner rate vs. cumulative-production type curve.

8.10 Extended Material Balance

The extended material-balance method was proposed by West and Cochrane (1995) for low-pressure, moderate-permeability gas wells in the Medicine Hat Field. The method assumes boundary-dominated flow, depletion drive, negligible non-Darcy skin factor, and constant deliverability. Although originally developed for low-pressure, moderate-permeability gas wells, its use is applicable to any depletion drive gas reservoir producing under boundary-dominated flow.

As with other methods presented in this chapter, a preliminary estimate of initial gas-in-place must be made. The average pressure is then calculated as a function of time, and the single-phase productivity index is calculated as

$$J(t) = \frac{q(t)}{\bar{p}_a(t) - p_{awf}(t)}. \quad \text{(8.131)}$$

The calculated productivity index is then graphed vs. time on Cartesian coordinates. If the gas-in-place estimate is accurate, the productivity index will be approximately horizontal. However, if the gas-in-place estimate is too low, the calculated productivity index will trend upward with time. Conversely, if the gas-in-place estimate is too high, the calculated productivity index will trend downward with time.

When a reliable baseline has been established, the extended material-balance plot can be used for reservoir surveillance. Deviation downward from the horizontal line would suggest either loss of productivity or interference from offset wells. On the other hand, upward deviation might occur following a successful stimulation treatment or extended shut-in of one or more offset wells.

Extended material-balance can be used to estimate original gas-in-place with the following procedure:

1. Make an initial guess of the original gas-in-place, G_i, that satisfies the lower bound given by Eq. 8.97:

$$G_i > \frac{G_p}{1 - \frac{p_{wf}}{z_{wf}} \Big/ \frac{p_i}{z_i}}.$$

2. Using the gas material-balance equation, Eq. 8.98, calculate the average reservoir pressure as a function of time from the cumulative gas produced:

$$\frac{\bar{p}}{\bar{z}} = \left(1 - \frac{G_p}{G_i}\right)\frac{p_i}{z_i}.$$

For stress-sensitive formations, use Eq. 8.99:

$$\left(\frac{\bar{p}}{\bar{z}}\right)^* = \left(1 - \frac{G_p}{G_i}\right)\frac{p_i}{z_i}.$$

3. Find the average drainage area pressure, $\bar{p}$ and the average adjusted pressure, $\bar{p}_a$.

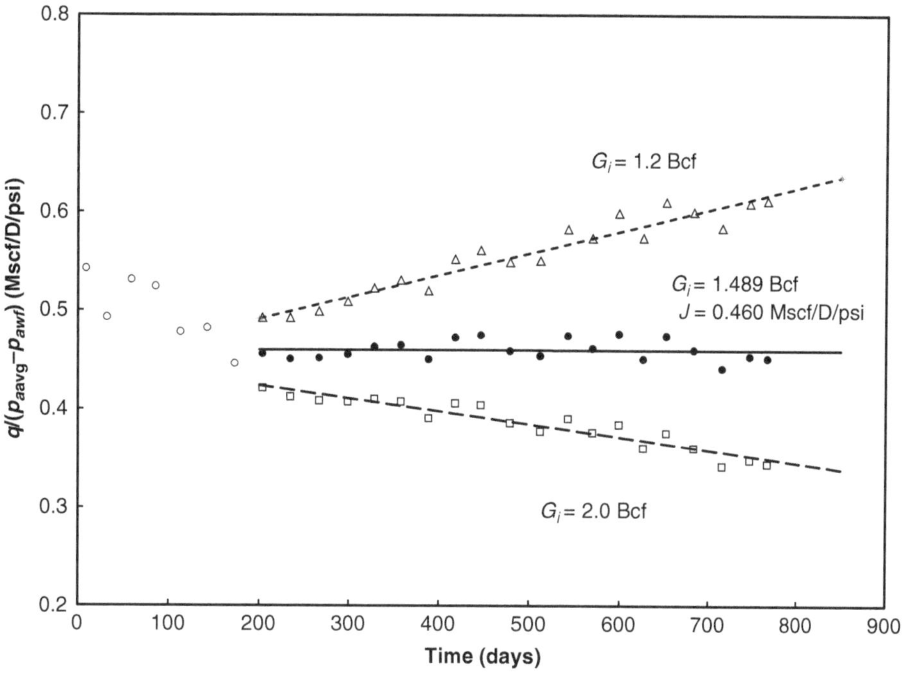

Fig. 8.36—West Virginia gas well, extended material-balance analysis.

4. Calculate and graph the single-phase productivity index as a function of time using Eq. 8.131:

$$J(t) = \frac{q(t)}{\bar{p}_a(t) - p_{awf}(t)}.$$

5. Identify boundary-dominated flow on the productivity-index vs. time graph.
6. If a line drawn through the data in boundary-dominated flow is horizontal, the calculated productivity index is stable with time, and the estimate of initial gas-in-place is correct. If the line trends downward, the estimate of initial gas-in-place must be too high; if the line trends upward, the estimate must be too low.
7. The productivity index J can be read directly from the y-intercept of the horizontal line.

8.10.1 Field Example Problem 8.8—West Virginia Gas Well, Extended Material-Balance Analysis. Use extended material balance to estimate initial gas-in-place for the field data given in Tables 8.3 and 8.4.

8.10.1.1 Solution. We used an initial estimate of 2 Bcf, then adjusted the estimate until a trend line through the data in boundary-dominated flow (200 days through the end of production) gave a horizontal line. The final match, giving an initial gas-in-place of 1.489 Bcf and a productivity index J of 0.460 Mscf/D/psi, is shown in **Fig. 8.36,** along with curves showing the trend for estimates of 2.0 and 1.2 Bcf.

Rearranging the volumetric equation for gas-in-place to solve for drainage area, we have

$$\begin{aligned} A &= \frac{G_i B_{gi}}{7{,}758\phi h(1 - S_{wi})} \\ &= \frac{(1.489)(0.7535)}{(7{,}758)(0.032)(100)(1 - 0.3)} \\ &= 64.6 \text{ acres.} \end{aligned}$$

8.11 Comparison of Results from West Virginia Gas-Well Example

We have analyzed the data given in Tables 8.3 and 8.4 using a variety of different methods. Given that all the methods discussed in this chapter rely on the same basic physical principle, the results should be similar for all methods. For comparison, we have also history matched the data using a single-phase analytical simulator for analyzing production data, using superposition of constant-pressure solutions to model variable-rate/variable-pressure production.

Table 8.8 summarizes the results of the various methods. For those methods that gave estimates of permeability and skin factor, the productivity index in the last column was calculated from

$$J = \frac{kh}{141.2 B_{gi}\mu_{gi}\left[\ln\left(\frac{r_e}{r_w}\right) - \frac{3}{4} + s\right]}. \quad \text{(8.132)}$$

The last two rows of Table 8.8 give the results from automatic history matching with and without consideration of pore-volume and connate water compressibility. Using Eqs. 8.72 through 8.75 to estimate the relative importance of pore-volume, water, and gas compressibilities gives

$$\begin{aligned} c_t &= c_f + S_w c_w + (1 - S_w)c_g \\ &= 4.0 + (0.3)(3.5) + (1 - 0.3)(209) \\ &= 151\ \mu\text{sip}. \end{aligned}$$

$$\begin{aligned} E_f &= \frac{c_f}{c_t} \\ &= \frac{4.0}{151} \\ &= 2.6\%. \end{aligned}$$

$$\begin{aligned} E_w &= \frac{S_w c_w}{c_t} \\ &= \frac{(0.3)(3.5)}{151} \\ &= 0.7\%. \end{aligned}$$

$$\begin{aligned} E_g &= \frac{(1-0.3)(209)}{151} \\ &= 96.7\%. \end{aligned}$$

For comparison, the difference in initial gas-in-place estimated using automatic history matching with PROMAT® with and without water and pore-volume compressibility was approximately 2%, compared to an estimated contribution of 3.3% from Eqs. 8.73 and 8.74 combined. In this case, the 2% error caused by ignoring water and pore-volume compressibility is well within the range of uncertainty of the field data and the methods used to estimate gas-in-place.

8.12 Practical Considerations

8.12.1 Data Requirements and Recommendations. Data requirements for RTA are much the same as for pressure transient interpretation. Spivey and Lee (2013) discuss data requirements for well-test interpretation in some detail. Broadly speaking, the data required for RTA can be categorized as petrophysical data, reservoir data, fluid-property data, well data, and production data.

Petrophysical data comprise wireline logs and core data. Wireline logs must be interpreted to obtain net pay, average porosity, and average water saturation over the producing interval. The analytical models used in RTA assume that net pay is taken along a direction perpendicular to the reservoir bedding planes. In many cases, net pay can be taken in the vertical direction, corresponding to true vertical depth.

Core data are particularly useful for reservoirs exhibiting significant stress sensitivity. Although conditions under which stress sensitivity is likely can be easily identified, quantification of the degree of stress sensitivity is dependent on special core analysis laboratory measurements.

Reservoir data of particular relevance to RTA in low-permeability gas reservoirs include reservoir temperature and initial reservoir pressure, both of which are essential for estimating gas and water properties from correlations.

Fluid-property data include gas formation volume factor, real-gas deviation factor, gas compressibility, gas viscosity, water formation volume factor, and water compressibility. These data can be obtained for most low-permeability gas reservoirs from gas- and water-property correlations that predict fluid properties as functions of pressure and temperature (McCain et al. 2011). Data required to use gas-property correlations include gas gravity and nonhydrocarbon-impurities content, along with condensate yield and API gravity.

Method	G_i (Bcf)	A (acres)	R_D	k (md)	s	J (Mscf/D/psi)
Fetkovich TC, Fig. 8.7	1.566	67.9	100	0.041	–3.28	0.465
Arps, Fig. 8.8	1.562	67.7				0.435
Fetkovich-McCray TC, Fig. 8.11	1.694	73.5	50	0.033	–4.01	0.456
Reservoir limits analysis,Fig. 8.22	1.431	62.1				0.462
Pressure/time TC, Fig. 8.28	1.434	62.2	30	0.0264	–4.44	0.435
AG Rate/Cum. TC, Fig. 8.33			26		–4.58	0.460
Pressure/time TC alternative match, Fig. 8.34	1.434	62.2	100	0.0396	–3.24	0.449
AG Rate/Cum. TC alternative match, Fig. 8.35			90		–3.34	0.461
Extended material balance, Fig. 8.36	1.489	64.6				0.460
Single-phase analytical	1.401	61.7	59.2	0.0375	–3.68	0.481
Single-phase analytical simulator, $c_w = c_f = 0$	1.426	62.8	65.1	0.0378	–3.67	0.482

Table 8.8—Summary of results, West Virginia gas well example. AG = Agarwal-Gardner, TC = type curve.

The gas gravity should be that of the reservoir gas, not the separator gas. For wet gas and gas/condensate reservoirs at pressures above the dewpoint, the reservoir gas gravity can be calculated from the separator gas gravity, the condensate gravity, and the initial gas/condensate ratio (Craft et al. 1991). Water-property correlations typically require only salinity as input data.

Production data include days on production and flowing wellhead pressure as well as produced volumes of gas, condensate, and water vs. time. Ideally, daily production is preferred for data in transient flow to provide the resolution to better separate the effects of permeability and skin factor or fracture length and fracture conductivity, while monthly production data are usually adequate for estimating gas-in-place and drainage area from boundary-dominated flow.

For wet-gas reservoirs and gas/condensate reservoirs above the dewpoint pressure, the gas and condensate volumes should be recombined to obtain the total wellstream production removed from the reservoir (Craft et al. 1991). The water production should be checked against gas water-vapor-content correlations to verify that produced water is in the vapor phase in the reservoir rather than free water. In principle, produced water existing in the vapor phase should be recombined along with gas and condensate, but this is often simply ignored.

Surface pressures should be converted to bottomhole pressure, taking into account gas, condensate, and water production, and accounting for any changes in production-string configuration over time. Some older methods for converting surface pressures to bottomhole pressure, including the average temperature and pressure and the Cullendar and Smith methods, were developed for hand calculations. Despite having significant drawbacks (Young 1967), many software implementations still retain these methods. Use of up-to-date implementations that use an adaptive step-size algorithm to integrate the fundamental flow equations is strongly recommended. This is imperative for situations involving possible downward flow (i.e., injection wells or commingled reservoirs that might experience differential depletion).

Well data should include a wellbore schematic and stimulation, workover, and recompletion histories. These histories should be reviewed to ensure that their impact on the production history is accounted for in the RTA. Location and production histories of offset wells should also be reviewed.

8.12.2 Nonideal Situations. There are a number of nonideal situations commonly encountered in RTA, including fracture cleanup, pressure fluctuations or changes, shut-in periods, wellbore loading, and commingled production.

Fracture cleanup distorts both the rate and pressure histories following a fracturing stimulation treatment. Unfortunately, this part of the production profile is the most important for separating permeability and stimulation effects (for tight gas reservoirs on the high end of the low-permeability range) as well as for separating the effects of fracture conductivity and fracture half-length.

Pressure fluctuations can arise for a number of reasons. Controlled flowback is often used to prevent damage to the fracture or formation in stress-sensitive formations. As with fracture cleanup, this tends to distort the data necessary for separating permeability and skin factor or fracture conductivity and fracture half-length.

Wells brought online early in field development might experience rising backpressure from line pack in the gathering system as production from newer wells comes on line. If constant pressure is assumed during RTA, the decline in production rate arising from the increasing backpressure can cause severe underestimation of gas-in-place and drainage area.

Line pressure can fluctuate seasonally in response to changes in demand. Normally, these changes can be handled well by rate-normalization/material-balance-time methods, provided that pressure data are available at the same intervals as production data.

Abrupt pressure changes, in contrast to smooth pressure fluctuations, distort the rate-normalization/material-balance-time-transformed data, causing it to deviate from the predicted constant-rate pressure response. These distortions dissipate as boundary-dominated flow is re-established following a pressure change.

Shut-in periods allow pressure to build up near the wellbore. Once the well is brought back on production, it experiences a spurt of transient, or flush, production, very similar to the effect caused by an abrupt decrease in backpressure. As with the distortion caused by an abrupt pressure change, transient production following shut-in will dissipate as boundary-dominated flow is re-established.

Commingled production is not typically amenable to analysis using the methods presented in this chapter, which assume single-layer reservoirs. One method for analyzing wells with commingled production uses an analytical simulator to history match total well surface production and data from multiple production logs (Spivey 2006). With good data, it is possible to estimate individual-layer properties including permeability, equivalent infinite-conductivity fracture half-length, and drainage area.

Wellbore loading occurs when reservoir pressure falls so low that a well cannot continue to produce at a rate high enough to remove all produced liquids from the wellbore, as discussed below.

8.12.3 Wellbore Loading and Critical Rate to Lift Liquids. A common issue with low-permeability gas wells is their ability to lift liquids, specifically condensate and water. For gas wells producing limited amounts of liquid, Turner et al. (1969) found that the critical rate to lift liquids for wells exhibiting mist flow was determined primarily by the interfacial tension between the gas and the liquid and the difference in density between the gas and the liquid. They found that the mist-flow model was valid for gas/liquid ratios up to 130 MMscf/bbl. Coleman et al. (1991) verified the Turner et al. (1969) model, looking primarily at wells with flowing tubing pressures below 500 psi. Coleman et al. (1991) found that wells intentionally driven into load up by increasing the flowing tubing pressure exhibited rapid exponential decline after onset of loading. Because this behavior is not representative of bounded reservoir behavior, any rate data that fall below the critical rate to lift liquids should be omitted from RTA.

Turner et al. (1969) and Coleman et al. (1991) provided nomographs for estimating the critical rate to lift either water or condensate. Because most gas wells produce some condensed water, a quick estimate of the critical rate should assume water production. A quick estimate of the critical rate to lift water can be obtained from the following equation (Spivey and Lee 2013):

$$q_{g,\,\mathrm{crit}} = 7D_{\mathrm{tbg}}^2\sqrt{p_{whp}}, \quad \text{(8.133)}$$

where D_{tbg} is the internal diameter of the tubing in inches and p_{whp} is the flowing tubing pressure at the wellhead. Onset of liquid loading is also suggested by an abrupt decrease in, and erratic behavior of, the gas rate, accompanied by absence of any produced water.

8.12.4 Example Problem 8.9. Estimate the critical rate to lift liquids for a gas well producing through 2 ⅜-in., 4.6-lbm/ft tubing, with a flowing tubing pressure of 900 psia.

8.12.4.1 Solution. From any of a number of references, the internal diameter of 2 ⅜-in., 4.6-lbm/ft tubing is found to be 1.995 in. Plugging this number into Eq. 8.133, we have:

$$\begin{aligned} q_{g,\,crit} &= 7D_{tbg}^2\sqrt{p_{whp}} \\ &= (7)(1.995)^2\sqrt{900} \\ &\approx (7)(2)^2(30) \\ &= 840 \text{ Mscf/D.} \end{aligned}$$

For comparison, estimating the critical rate to lift water using the nomograph in Coleman et al. (1991) gives a rate of 900 Mscf/D.

8.12.5 Interpretation Issues. A major issue in RTA, as in pressure transient interpretation, is that of nonuniqueness. This issue is amplified for RTA by the higher degree of noise present in production data, which makes the logarithmic derivative less useful than in pressure transient interpretation. In pressure transient interpretation, nonuniqueness often refers to the fact that multiple models often match the data equally well. In rate transient interpretation, we find that multiple interpretations are often possible with a single model (i.e., that of a well in a closed circular reservoir).

Obtaining a unique result using the methods discussed in this chapter requires data in transient flow of frequency and quality adequate enough to allow the effects of skin factor and permeability to be separated, and data in boundary-dominated flow of quality and duration sufficient to establish gas-in-place and drainage area.

If boundary-dominated flow is present, but transient flow data are limited (as was the case in the West Virginia gas-well example), an accurate estimate of gas-in-place and drainage area might be possible, but it will not be possible to obtain a unique answer for permeability and skin factor. However, as noted in Section 8.11, the productivity index was determined quite accurately, although the estimates of permeability and skin factor individually were not.

If transient flow is present but boundary-dominated flow is not, a unique answer for gas-in-place, and hence drainage area, cannot be obtained, but it is usually possible to find a lower bound on the gas-in-place (e.g., by fitting the data as far as possible to the right on the Fetkovich type curve).

For hydraulically fractured wells, it might be difficult to separate the effects of fracture half-length and fracture conductivity. For convenience, when fracture conductivity cannot be estimated uniquely, it might be convenient to express results in terms of a total skin factor or an equivalent infinite-conductivity-fracture half-length. For a finite-conductivity hydraulic fracture, the apparent wellbore radius might be estimated from (Meyer and Jacot 2005)

$$r_{wa} = \frac{L_f}{2+\dfrac{1}{C_r}}, \quad \text{(8.134)}$$

and the equivalent skin factor can then be estimated from

$$s = -\ln\left(\frac{r_{wa}}{r_w}\right). \quad \text{(8.135)}$$

The equivalent infinite-conductivity-fracture half-length for a finite-conductivity fracture can be estimated from

$$L_{f,\inf} = \frac{2}{2+\dfrac{1}{C_r}}L_{f,\,\mathrm{fin}} \quad \text{(8.136)}$$

8.12.6 Recommendations. Although we have presented a wide variety of different methods for RTA in this chapter, the choice of which method or methods to use is really a matter of personal choice and the software available to the engineer. Anyone working regularly with RTA, either as an analyst or a consumer of the results so obtained, should have a basic familiarity with all of these methods.

The author of this chapter has a slight preference for the constant-rate pressure type curve methods because of his background in pressure transient analysis and because of the wide availability of constant-rate solutions for different inner and outer boundary conditions. Extended material balance is a recent addition to his toolbox, but quickly became such because of its simplicity of application for well performance monitoring.

8.13 Nomenclature

A = drainage area, ft^2 or acres, depending on context

A_k = permeability coefficient in Jones model, dimensionless

A_V = porosity coefficient in Jones model, dimensionless

b = hyperbolic exponent, dimensionless

B = formation volume factor, bbl/STB or bbl/Mscf

B_g = gas formation volume factor, bbl/Mscf or ft^3/Mscf, depending on context

c_f = pore-volume compressibility, 1/psi

c_g = gas compressibility, 1/psi

c_t = total compressibility, 1/psi

c_w = water compressibility, 1/psi

C = compressibility parameter in Jones model (suggested default value 3.0×10^{-6} 1/psi), 1/psi

C_r = dimensionless fracture conductivity, defined as

$$C_r = \frac{w_f k_f}{\pi k L_f}$$

d = true vertical depth, ft

D = instantaneous decline rate, 1/days

D_e = effective decline rate, 1/days

D_{ei} = initial effective decline rate, 1/days

D_i = initial decline rate, 1/days

D_{tbg} = internal diameter of tubing, in.

E_f = fraction of reservoir energy attributable to pore-volume compressibility

E_g = fraction of reservoir energy attributable to gas compressibility

E_w = fraction of reservoir energy attributable to water compressibility

G_i = initial gas-in-place, Mscf

h = net-pay thickness, ft

J = productivity index, STB/D/psi or Mscf/D/psi

k = permeability, md

k_0 = permeability at zero net confining stress, Jones model, md

k_f = fracture permeability, md

L_f = fracture half-length, ft

m = slope of semilog straight line, psi/log cycle

m' = slope of semilog straight line for rate-normalized pressure change, psi/Mscf/D/log cycle

N_i = initial oil-in-place, STB

N_p = cumulative production, STB or Mscf

N_{pa} = cumulative production to abandonment, STB or Mscf

p = pressure, psi

p_a = adjusted pressure, psi

p_i = initial reservoir pressure, psi

p_{wf} = flowing bottomhole or sandface pressure, psi

p_{whp} = wellhead tubing pressure (absolute), psi

q = production rate, STB/D or Mscf/D

q_a = production rate at abandonment, STB/D or Mscf/D

q_D = dimensionless production rate

q_{dD} = decline dimensionless production rate

$q_{g,\text{crit}}$ = critical gas rate to lift liquids in mist flow, Mscf/D

q_i = initial production rate, STB/D or Mscf/D

Q_{dD} = decline dimensionless cumulative production

r_e = drainage radius, ft

r_w = wellbore radius, ft

r_{wa} = apparent wellbore radius, ft

R_D = dimensionless drainage radius

s = water saturation, fraction

S_w = skin factor, dimensionless

t = time, days (hours in Section 8.7), depending on context

t_a = adjusted time, days (hours in Section 8.7), depending on context

t_D = dimensionless time

t_{dD} = decline dimensionless time

t_{mb} = material-balance time, days (hours in Section 8.7), depending on context

T_f = formation temperature, °F (or °R in equations)

T_{sc} = standard temperature, °F (or °R in equations)

V_b = bulk volume, ft^3

V_p = pore volume, ft^3

w_f = fracture width, ft

x = graph abscissa, units irrelevant

y = graph ordinate, units irrelevant

z = real-gas deviation factor, dimensionless

ϕ = porosity, fraction

ϕ_0 = porosity at zero net confining stress, Jones model, fraction

μ = viscosity, cp

ρ = density, lbm/ft^3

ρ_{lith} = lithostatic gradient, psi/ft

σ = net confining stress, psi

σ^* = stress parameter in Jones model (suggested default value 3,000 psi), psi

Subscripts

a = adjusted

A = area

D = dimensionless

f = fracture

fin = finite conductivity

g = gas

i = initial

inf = infinite conductivity

MP = match point

pss = pseudosteady state

wf = flowing bottomhole conditions

8.14 References

Agarwal, R. G. 1980. A New Method To Account For Producing Time Effects When Drawdown Type Curves Are Used To Analyze Pressure Buildup And Other Test Data. Presented at the SPE Annual Technical Conference and Exhibition, Dallas, Texas, 21–24 September. SPE-9289-MS. https://doi.org/10.2118/9289-MS.

Agarwal, R. G., Gardner, D. C., Kleinsteiber, S. W. et al. 1999. Analyzing Well Production Data Using Combined-Type-Curve and Decline-Curve Analysis Concepts. *SPE Res Eval & Eng* **2** (5): 478–486. SPE-57916-PA. https://doi.org/10.2118/57916-PA.

Arps, J. J. 1945. Analysis of Decline Curves. *Trans. AIME* **160** (1): 228–247. SPE-945228-G. https://doi.org/10.2118/945228-G.

Blasingame, T. A. and Lee, W. J. 1986. Variable-Rate Reservoir Limits Testing. Presented at the Permian Basin Oil and Gas Recovery Conference, Midland, Texas, 13–15 March. SPE-15028-MS. https://doi.org/10.2118/15028-MS.

Blasingame, T. A. and Lee, W. J. 1988. The Variable-Rate Reservoir Limits Testing of Gas Wells. Presented at the SPE Gas Technology Symposium, Dallas, Texas, 13–15 June. SPE-17708-MS. https://doi.org/10.2118/17708-MS.

Blasingame, T. A., McCray, T. L., and Lee, W. J. 1991. Decline Curve Analysis for Variable-Pressure Drop/Variable Flowrate Systems. Presented at the SPE Gas Technology Symposium, Houston, Texas, 22–24 January. SPE-21513-MS. https://doi.org/10.2118/21513-MS.

Bourdet, D., Ayoub, J. A., and Pirard, Y. M. 1989. Use of Pressure Derivative in Well Test Interpretation. *SPE Form Eval* **4** (2): 293–302. SPE-12777-PA. https://doi.org/10.2118/12777-PA.

Coleman, S. B., Clay, H. B., McCurdy, D. G. et al. 1991. A New Look at Predicting Gas-Well Load-Up. *J Pet Technol* **43** (3): 329–333. SPE-20280-PA. https://doi.org/10.2118/20280-PA.

Craft, B. C., Hawkins, M., and Terry, R. E. 1991. *Applied Petroleum Reservoir Engineering*, second edition. Englewood Cliffs, New Jersey: Prentice Hall.

Cutler, W. W., Jr. 1924. Estimation of Underground Oil Reserves by Oil-Well Production Curves. 228, Department of the Interior, Bureau of Mines, Washington, D.C. (August 1924).

Fetkovich, M. J. 1980. Decline Curve Analysis Using Type Curves. *J Pet Technol* **32** (6): 1065–1077. SPE-4629-PA. https://doi.org/10.2118/4629-PA.

Fetkovich, M. J., Bradley, M. D., Works, A. M. et al. 1990. Depletion Performance of Layered Reservoirs Without Crossflow. *SPE Form Eval* **5** (3): 310–318. SPE-18266-PA. https://doi.org/10.2118/18266-PA.

Fetkovich, M. J., Fetkovich, E. J., and Fetkovich, M. D. 1996. Useful Concepts for Decline Curve Forecasting, Reserve Estimation, and Analysis. *SPE Res Eng* **11** (1): 13–22. SPE-28628-PA. https://doi.org/10.2118/28628-PA.

Fetkovich, M. J., Reese, D. E., and Whitson, C. H. 1998. Application of a General Material Balance for High-Pressure Gas Reservoirs (includes associated paper 51360). SPE J. 3 (1): 3–13. SPE-22921-PA. https://doi.org/10.2118/22921-PA.

Fraim, M. L., and Lee, W. J. 1987. Determination Of Formation Properties From Long-Term Gas Well Production Affected By Non-Darcy Flow. Presented at the SPE Annual Technical Conference and Exhibition, Dallas, Texas, 27–30 September. SPE-16934-MS. https://doi.org/10.2118/16934-MS.

Fraim, M. L. and Wattenbarger, R. A. 1987. Gas Reservoir Decline-Curve Analysis Using Type Curves With Real Gas Pseudopressure and Normalized Time. *SPE Form Eval* **2** (4): 671–682. SPE-14238-PA. https://doi.org/10.2118/14238-PA.

Fraim, M. L. 1988. *Multi-phase Decline Curve Analysis with Normalized Rate and Time*. MS thesis, Texas A&M University, College Station, Texas (August 1988).

Gladfelter, R. E., Tracy, G. W., and Wilsey, L. E. 1955. Selecting Wells Which Will Respond to Production-stimulation Treatment. Presented at the Drilling and Production Practice, New York, New York, 1 January. API-55-117.

Horner, D. R. 1951. Pressure Build-Up in Wells. Presented at the 3rd World Petroleum Congress, The Hague, The Netherlands, 28 May–6 June. WPC-4135.

Jones, P. 1957. Drawdown Exploration Reservoir Limit, Well and Formation Evaluation. Presented at the Permian Basin Oil Recovery Conference, Midland, Texas, 18–19 April. SPE-824-G. https://doi.org/10.2118/824-G.

Jones, S. C. 1988. Two-Point Determinations of Permeability and PV vs. Net Confining Stress. *SPE Form Eval* **3** (1): 235–241. SPE-15380-PA. https://doi.org/10.2118/15380-PA.

Lefkovits, H. C. and Matthews, C. S. 1958. Application of Decline Curves to Gravity-Drainage Reservoirs in the Stripper Stage. *Trans. AIME* **213**: 275–280. SPE-969-G.

Maley, S. 1985. The Use of Conventional Decline Curve Analysis in Tight Gas Well Applications. Presented at the SPE/DOE Low Permeability Gas Reservoir Symposium, Denver, Colorado, 19–22 May. SPE-13898-MS. https://doi.org/10.2118/13898-MS.

Matthews, C. S. and Lefkovits, H. C. 1956. Gravity Drainage Performance of Depletion-Type Reservoirs in the Stripper Stage. *Trans. AIME* **207**: 265–274. SPE-665-G-P.

McCain, W. D., Jr., Spivey, J. P., and Lenn, C. P. 2011. *Petroleum Reservoir Fluid Property Correlations*. Tulsa, Oklahoma: PennWell Corporation.

McCray, T. L. 1990. *Reservoir Analysis Using Production Decline Data and Adjusted Time*. MS thesis, Texas A&M University, College Station, Texas (December 1990).

Mead, H. N. 1956. Modifications to Decline Curve Analysis. *Trans. AIME* **207**: 11–16. SPE-464-G.

Meyer, B. R. and Jacot, R. H. 2005. Pseudosteady-State Analysis of Finite Conductivity Vertical Fractures. Presented at the SPE Annual Technical Conference and Exhibition, Dallas, Texas, 9–12 October. SPE-95941-MS. https://doi.org/10.2118/95941-MS.

Spivey, J. P. 2006. Estimating Layer Properties for Wells in Multilayer Low-Permeability Gas Reservoirs by Automatic History Matching Production and Production Log Data. Presented at the SPE Gas Technology Symposium, Calgary, Alberta, Canada, 15–17 May. SPE-100509-MS. https://doi.org/10.2118/100509-MS.

Spivey, J. P. Frantz, J. H., Jr., Williamson, J. R. et al. 2001. Applications of the Transient Hyperbolic Exponent. Presented at the SPE Rocky Mountain Petroleum Technology Conference, Keystone, Colorado, 21–23 May. SPE-71038-MS. https://doi.org/10.2118/71038-MS.

Spivey, J. P., Gatens, J. M., Semmelbeck, M. E. et al. 1992. Integral Type Curves for Advanced Decline Curve Analysis. Presented at the SPE Mid-Continent Gas Symposium, Amarillo, Texas, 13–14 April. SPE-24301-MS. https://doi.org/10.2118/24301-MS.

Spivey, J. P. and Lee, W. J. 2013. *Applied Well Test Interpretation*. Richardson, Texas: Society of Petroleum Engineers.

Spivey, J. P., Valko, P. P., and McCain, W. D. 2007. Applications of the Coefficient of Isothermal Compressibility to Various Reservoir Situations With New Correlations for Each Situation. Presented at the SPE Annual Technical Conference and Exhibition, Dallas, Texas, 9–12 October. SPE-96415-MS. https://doi.org/10.2118/96415-MS.

Turner, R. G., Hubbard, M. G., and Dukler, A. E. 1969. Analysis and Prediction of Minimum Flow Rate for the Continuous Removal of Liquids From Gas Wells. *J Pet Technol* **21** (11): 1475–1482. SPE-2198-PA. https://doi.org/10.2118/2198-PA.

West, S. L. and Cochrane, P. J. R. 1995. Reserves Determination Using Type-Curve Matching and EMB Methods in the Medicine Hat Shallow Gas Field. *SPE Res Eng* **10** (2): 82–87. SPE-28609-PA. https://doi.org/10.2118/28609-PA.

Winestock, A. G. and Colpitts, G. P. 1965. Advances in Estimating Gas Well Deliverability. *J Can Pet Technol* **4** (3): 111–119. PETSOC-65-03-01. https://doi.org/10.2118/65-03-01.

Young, K. L. 1967. Effect of Assumptions Used to Calculate Bottom-Hole Pressures in Gas Wells. *J Pet Technol* **19** (4). SPE-1626-PA. https://doi.org/10.2118/1626-PA.

Chapter 9

Well and Reservoir Numerical Modeling

John Yilin Wang and Stephen A. Holditch

On the basis of our experience in evaluating tight gas wells and reservoirs, we found that it is best to approach a complex engineering problem by first simplifying the problem and then solving the problem using a simple model, such as decline curve analysis, type curves, and/or analytical solutions. These models usually provide an approximate solution and give reasonably accurate answers for the values of permeability, fracture half-length, original gas-in-place, and other important parameters. We call this the 80% answer. To obtain more-accurate estimates of parameters such as permeability, skin, hydraulic fracture length, and fracture conductivity and to predict well performance more accurately, we then use finite-difference modeling, especially when we must model layered reservoirs.

It requires more time and effort to conduct numerical simulation studies of tight gas wells, but it is the best way to model features such as non-Darcy flow, changing fracture conductivity, changing fluid properties, multilayered reservoirs, and other complexities. This chapter presents a proven methodology of modeling tight gas wells and reservoirs using finite-difference reservoir simulation technologies in the following order. The methodology and technologies could be applied directly, or with tuning, to solve science and practical engineering problems in tight oil and gas reservoirs, shale oil and gas reservoirs, and other types of reservoirs. The overview for each section is below.

- Section 9.1 provides a high-level overview of the fundamentals of finite-difference simulation.
- Section 9.2 provides a high-level overview of modeling hydraulically fractured well performance in tight gas formations.
- Section 9.3 discusses how to model fracturing-fluid cleanup.
- Section 9.4 discusses how to model non-Darcy flow and changing fracture conductivity.
- Section 9.5 discusses how to evaluate reservoir and fracture properties by history matching pressure transient data and/or production data using reservoir simulation technologies, especially in reservoirs with multiple layers, multiphase flow, or other complex geological and engineering attributes. Section 9.5 is different from and complements the 80% answer, which was presented in Part 2 Ch 6: Prestimulation Well Testing, Part 2 Ch 7: Post-Fracture Treatment Well Testing, and Part 2 Ch 8: Rate Transient Analysis.
- Section 9.6 discusses how to model multilayer reservoirs.
- Section 9.7 discusses how to model whole-field tight gas reservoirs with larger numbers of wells.
- Section 9.8 provides the chapter summary.

9.1 Fundamentals of Finite-Difference Simulation

Fluid flow in tight gas reservoirs is governed by the diffusivity equation, which are derived using continuity equations, Darcy's law, and equations of state. The diffusivity equation becomes highly nonlinear when we add other physics, including non-Darcy flow through the hydraulic fracture, multiphase flow, capillary pressure, gravity, change of permeability with pressure, change of fluid properties with pressure, temporal and spatial differences of the parameters, and change of boundary conditions. Different numerical methods can be used to solve the partial-differential equation, but the finite-difference method is commonly used in reservoir simulation. The equation must be discretized, linearized, and then solved using direct or iterative linear-equation-solving techniques. Papers or books with details on finite-difference methods can be used as a reference. In general, the simulator chosen for use could be black-oil or compositional, 2D or 3D, and one to three phases, depending on the well conditions and geology. For most tight gas reservoirs, a single-phase model is all we need to conduct the analyses of the pressure and production data. If fracturing-fluid injection and cleanup are modeled, then a two-phase gas/water model is required.

9.2 Modeling Hydraulically Fractured Well Performance

To model hydraulically fractured well performance, we need to incorporate additional physics in the reservoir model. Depending on the complexity of the problem, consideration is normally given for fracturing-fluid cleanup, non-Darcy flow inside the fractures, fracture-conductivity changes, multilayer effect, and well interference if there are natural fractures or if the hydraulic fracture from one well intersects an offset well. Best practices in simulation techniques, such as selection of gridding, initiation, timesteps, convergence criteria, and the analysis of results, must also be considered. It takes extra care and attention to simulate tight gas well performance.

This section describes how to simulate and analyze the performance of hydraulically fractured wells using an improved numerical reservoir simulator, which will be discussed in Sections 9.3 through 9.7. We choose to focus on the simulation technologies and practices that have been developed and applied at Texas A&M University and S. A. Holditch & Associates Inc. to solve real-world problems. Each subsequent section first discusses the physics of pertinent phenomena, then presents mathematical equations to characterize the physics, then shows how to integrate the physics into a reservoir simulator, and finally explains how to analyze the results. Important input variables for simulation models are pressure distribution, saturation distribution, fluid properties, hydraulic-fracture properties, and reservoir properties at time zero (the end of a fracturing treatment). The outputs are pressure distribution, saturation distributions, and gas/water-flow rates over a period of time of interest. We will explain how to gain insights on well and reservoir performance by analyzing the simulation results in a logical, consistent, and systematic way.

9.3 Modeling Fracturing-Fluid Invasion and Cleanup

A successful fracturing treatment depends not only on the treatment design and field implementation, but also on flowback and cleanup operations after a treatment. During a hydraulic fracturing treatment, a large volume of fracturing fluid is pumped into the formation. The fracturing fluid is normally a high-viscosity, water-based polymer gel transporting a propping agent. When fluid is pumped down the wellbore and into the perforations, the pressure in the reservoir at the perforations increases, and fractures are initiated in the formation when fluid pressure is greater than the least principal stress plus rock strength. As pumping continues, the fracture propagates up, down, and outward, and a portion of the fracturing fluid leaks off along the fracture face and into the formation. The open fracture is supported by the injected fluid and continues to grow until pumping is stopped. The fracture length, width, and height propagated during the fracturing treatment are referred to as the created fracture dimensions.

After the hydraulic fracturing treatment, the well is often shut in to allow the fracture to close and trap the proppant before flowback of the fracturing fluid begins. The high temperature of the surrounding formation heats the fracturing fluid, causing the gelling agents to break. In many cases, chemical breakers are also used to reduce the viscosity. The remaining lower-viscosity fluid continues to imbibe into the formation. Pressure in the fracture decreases and the fracture closes as leakoff continues, but propping agents keep the fracture open and maintain fracture conductivity. When the fracture has closed, flow is reversed to flow back the fracturing fluid, clean up the fracture, and produce fracturing fluid and gas. As flow to the surface continues, the pressure in the fracture decreases further. To fully understand fracturing-fluid cleanup and well performance, there needs to be an understanding of fracture damages and the related simulation technologies.

9.3.1 Damage in and around Fractures. Van Poollen (1957) performed laboratory tests to determine whether fracturing fluid damages the productivity of hydraulically fractured wells. He addressed the effect of fracturing-fluid invasion on the formation permeability in the invaded zone. The different fracturing fluids and their additives were found to reduce the permeability around the fracture. The permeability reduction extended into the formation at a depth of approximately 1 in. in Van Poollen's tests. The damage to the formation immediately surrounding the fracture had a minor effect on the productivity of a hydraulically fractured well. Cooke (1973) used laboratory tests to determine how certain phenomena affect fracture conductivity and, subsequently, the well productivity. He examined the effect of reservoir environment on fracture conductivity under stress. Cooke measured the fracture permeability as a function of closure stress. The presence of a high-temperature brine resulted in a decrease in fracture conductivity when stress was applied to the proppant. The reduction in fracture conductivity was essentially irreversible.

Holditch (1979) investigated various factors affecting water blocking in fractured gas wells and found, using finite-difference modeling, that reservoir properties such as capillary pressure, change of capillary pressure, and relative permeability are extremely important in determining the cleanup behavior. Holditch found that, if the reservoir-rock permeability near the hydraulic fracture is damaged by fracturing-fluid invasion, a large pressure drawdown will be needed to overcome the capillary end effect, or the water is trapped and a complete block of gas flow will occur. If reservoir permeability is not damaged, no serious water block will occur if the pressure drawdown is much greater than capillary pressure or capillary pressures are large enough to imbibe the fracture water further into the formation. Otherwise, the gas production can be severely curtailed.

Lolon et al. (2003, 2004) studied the effect of fracture conductivity using water-based fracturing fluids and drew conclusions that higher fracture conductivity would result in faster fracturing-fluid cleanup, longer effective fracture length, and greater cumulative gas production. The fracture cleans up faster with higher gas-flow rates, and effective length is affected more by fracture conductivity than by formation permeability, fracture closure effects, and reservoir water mobility. Lolon et al. looked only at Newtonian fluids. Most fracturing fluids are actually non-Newtonian and are affected by yield stress. The effect of yield stress on fracturing-fluid cleanup has recently been investigated by several groups. Balhoff and Miller (2005) derived an analytical model for yield stress to predict fracturing-fluid cleanup by incorporating fracture geometry (length and width) and fracture and reservoir properties (pressure, permeability, porosity, and proppant diameter). Yi (2004) derived a model for non-Newtonian fracturing fluid and used parametric analysis to study the effects of rheological parameters and injection rate. He concluded that, if a pressure gradient is not high enough to overcome the yield stress, the polymer will not flow, and high flow rate increases cleanup. May et al. (1997) studied the effect of yield stress on fracturing-fluid cleanup using a numerical simulation model. For the case they studied, May et al. concluded that increasing drawdown did not affect the amount of cleanup, but increasing the fracture conductivity did increase the effective fracture length. A study by Cooke (1975) showed that the polymer residue can be trapped in the fracture. Other work suggests that polymer residue might leak off into the formation and be trapped in the invaded zone around the fracture. Siddiqui et al. (2004) reported that some gel residue is present in the rock and cannot be displaced by production—it degrades and is removed slowly. This residue occupies the pore spaces of the rock and reduces the rock permeability. They concluded that the structure and mineralogy of the formation rock play a significant role in determining the amount of gel residue in the formation. A certain portion of the gel residue (as high as 25%) remains in the core plug and cannot be recovered by the cleanup process.

Generally, fracture damages include

- Damages inside the fracture, which can be caused by proppant crushing, proppant embedding, fracture-face damage, or fracture plugging with chemicals and polymers.
- Damages inside the reservoir, which can be caused by excessive fluid leakoff, clay swelling, relative permeability hysteresis, and capillary effects.

Of course, other factors that should be considered include pressure drawdown, fracture geometry, non-Darcy flow effect, fracture conductivity, heterogeneity, formation temperature, fracturing-fluid viscosity, viscous fingering, gel residue, breaker, and operational procedures. All of these referenced studies have led to a better understanding of what causes fracture damage. Fracturing fluids and proppants have been improved in recent years to enhance the cleanup process. However, many fracturing treatments, particularly in shallow, low-pressure, tight gas reservoirs, seem to fall short of expectations and produce gas at a lower flow rate than expected.

There are actually three values of fracture length that must be evaluated in any design. First, we must determine the created fracture length. This is the value of the crack length in the rock and can be estimated with fracture-propagation models. Second, we need to estimate the propped fracture length. The propped length is the distance in the fracture from the wellbore that contains propping agents. In many cases, the propped length will be 70 to 80% or more of the created length. However, what really counts is the effective fracture length. The effective fracture length is the part of the propped length that has a high-enough proppant concentration to allow the fracturing fluid to clean up so natural gas can flow in the fracture. Unfortunately, the effective fracture length is often 10 to 50% of the propped fracture length.

In many situations, the created length was probably achieved, and perhaps even the propped length was achieved. However, because of insufficient proppant concentration, insufficient proppant transport, the use of the wrong propping agent, or a fracturing fluid that does not break to a low-viscosity fluid, the effective fracture length does not provide optimal production results.

Insufficient fracturing-fluid cleanup is the primary cause of poor results when the optimal effective fracture length is not achieved. Some conclusions from literature are summarized as follows:

- Fracture damage of 90% or higher of the reservoir formation permeability will decrease the gas production rate (Gdanski et al. 2005).
- High fracture conductivity will help fracturing-fluid cleanup after the fracturing treatment (Lolon et al. 2003; May et al. 1997; Tannich 1975).
- Capillary pressure will affect fracturing-fluid cleanup. The higher the capillary pressure, the smaller the amount of fluid will be recovered (Holditch and Morse 1976). To recover more of the fracturing fluids, we must use surfactants and increase the pressure gradient in the fracture.
- A large pressure drawdown helps clean up the fracturing fluid more quickly and produce more gas as long as the fracture propping agent is not crushed. However, not all of the fracturing fluid can be recovered during the cleanup process (Lolon et al. 2003).
- It is necessary to take relative permeability hysteresis into account in the mathematical analysis of the cleanup process and well productivity performance (Holditch 1979).

9.3.2 Simulation Methodology. Reservoir and hydraulic fracture simulation methodology was documented by Wang et al. (2010) after reviewing fracturing treatment data and previous research conducted at Texas A&M University.

*9.3.2.1 **Fractured Well Model.*** Wang et al. (2010) used a 2D, three-phase black-oil simulator to model fracturing-fluid cleanup and long-term gas production in tight gas reservoirs. They assumed that the fracture extends an equal distance on two sides of the wellbore and fully penetrates the formation. Fracture length and width and fracturing-fluid volumes were obtained by running fracture design models. The width of the fracture is assumed to be constant from the wellbore to the tip, while the porosity of the fracture is 30%. The reservoir is assumed to be homogeneous and isotropic. Because of the symmetry of the well configuration, they simulated one-quarter of the drainage area from a fractured well, as shown in **Fig. 9.1.**

Grid dimensions in the fracture direction are varied so that the grids are very small near the wellbore, increase toward the middle of the propped fracture half-length, decrease toward the tip of the propped fracture, and finally increase again to the drainage boundary. This gridding technique was used by and Wang et al. (2010) and Voneiff et al. (1996) for research work and field studies. The wellbore is in cell (1, 1, 1) and is operated at constant bottomhole pressure (BHP). Gridding with a hydraulic fracture is shown in Fig. 9.1.

*9.3.2.2 **Reservoir, Fracture, and Fluid Properties.*** Reservoir and fracture properties used in this study are summarized in **Table 9.1.** To model the gas/water/fracturing-fluid multiphase flow, relative permeability curves from tight gas cores in East Texas, as shown in **Fig. 9.2,** were used.

*9.3.2.3 **Initiation of Simulation.*** The fracturing fluid was injected into the reservoir to create a filtrate-invasion zone near the fracture. The fracturing fluid leaked off into the reservoir formation, creating a highly water-saturated zone near the fracture, and built a polymer cake inside the fracture. Because of the tight formation, the polymer stays inside the fracture and displaces all gas and water out of the propped fracture. A successful fracturing treatment requires that fracturing fluids degrade rapidly after the treatment to prevent plugging the proppant pack with highly concentrated gel. However, because of the small pore-throat sizes in tight gas reservoirs, virtually all the polymers pumped in the treatments end up in the fracture. The polymer inside the fracture pack will create a filter cake, as shown in **Figs. 9.3 and 9.4.**

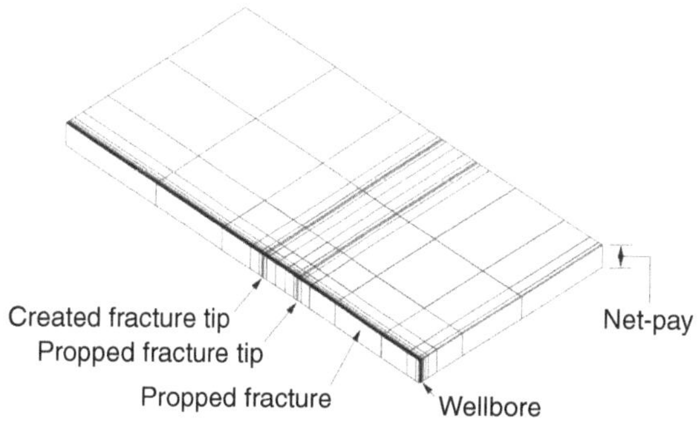

Fig. 9.1—Rectangular gridding used for simulation of fracturing-fluid cleanup (Voneiff et al.1996).

Base Data Set for Parametric Simulation Study

Parameter	Value
Drainage area (acres)	160
Reservoir thickness (ft)	100
Formation permeability (md)	0.001, 0.01, 0.1
Formation porosity (%)	10
Formation depth (ft)	4,000, 8,000, 12,000
Reservoir pressure (psi)	1,860, 3,720, 5,580
Reservoir temperature (°F)	130, 190, 250
Formation water saturation	0.4
Water compressibility (psi^{-1})	3.00E-06
Gas specific gravity	0.6
Fracture half-length (ft)	264, 528, 924
Dimensionless fracture conductivity	0.1, 1, 10, 100
Fracturing fluid (1,000 gal)	105, 235, 420
BHP	10% of reservoir pressure
Fracture fluid (gal)	80,000

Table 9.1—Reservoir and fracture parameters.

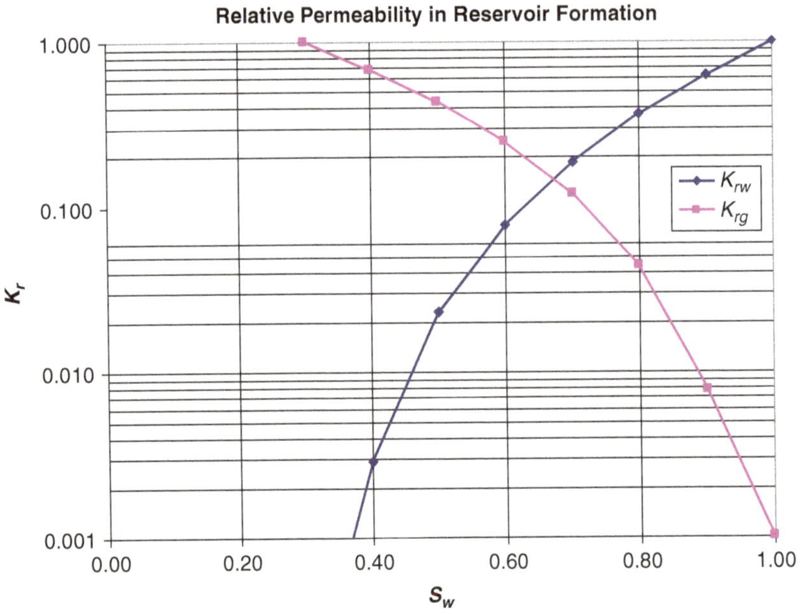

Fig. 9.2—Relative permeability curve in the reservoir formation from East Texas (Holditch 2005).

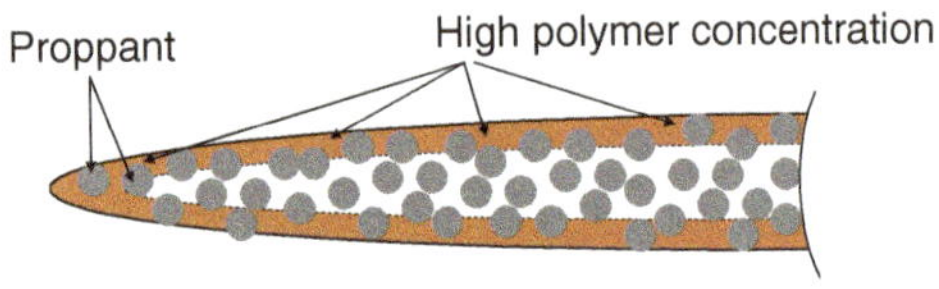

Fig. 9.3—Top view of a half-fracture with proppant and polymer inside.

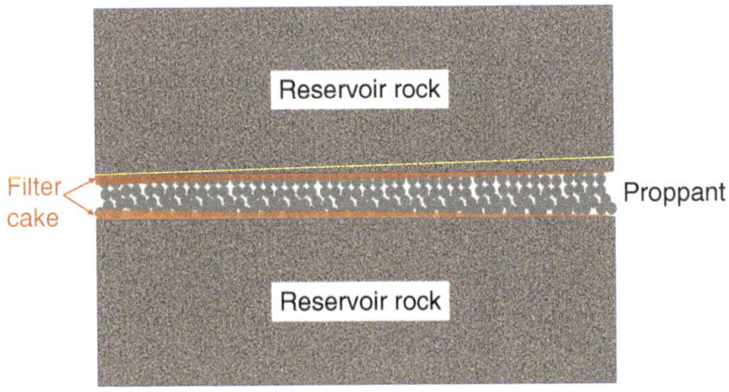

Fig. 9.4—Schematic of polymer filter cake inside a hydraulic fracture at the end of a treatment.

When the initiations were complete, Wang et al. (2010) conducted a parametric study of eight cases shown in **Table 9.2** at a total of 576 simulations following the steps below.

9.3.2.4 Steps in Modeling Fracture-Fluid Cleanup.

1. Simulation of single-phase flow
2. Simulation of gas/water two-phase flow
3. Simulation of proppant-crushing effect
4. Simulation of polymer filter cake
5. Simulation of gel damage

Case	
1	Single phase
2	Gas/water two phase
3	Proppant crushing
4	Filter cake
5	Yield stress
5a	3 Pa
5b	10 Pa
5c	20 Pa
5d	100 Pa

Table 9.2—Different simulation cases.

9.3.2.5 Simulation Results and Analyses. The single-phase-flow solution for a vertical well containing a vertical fracture has been well-known for 50 years. A single-phase simulation was made as the base case for each run. As we develop the scenarios associated with fracturing-fluid cleanup, we can compare each case to the base case (single-phase flow) to evaluate its effect on both cleanup and gas recovery. We present our results in graphs of log (gas-flow rate) vs. log (time) to evaluate fracturing-fluid cleanup and cumulative gas produced, which show the early-time transient flow and late pseudosteady-state flow.

Fig. 9.5 shows the gas production rate vs. time after a fracturing treatment for a reservoir where

- $L_f = 528$ ft
- $p_r = 3{,}720$ psi
- $k = 0.1$ md
- $S_{wi} = 0.4$
- $h = 100$ ft

The assumption for single-phase flow is that a hydraulic fracture has been created, but without any fracturing-fluid damage, so the gas production rate reaches a maximum value when the well is put on production. As expected, with higher fracture conductivity, the gas production rate is higher. If the fracture conductivity is greater than 10, a higher conductivity does not substantially affect the gas-flow rate in the idealized case. This means that, from a production point of view, a proppant that can provide dimensionless fracture conductivity of 10 or greater is sufficient for optimized gas production. A better, stronger proppant may be a waste of money and investment once a C_r of 10 is obtained.

Fig. 9.6 shows the gas production rate vs. time after a fracturing treatment, where

- $L_f = 528$ ft
- $p_r = 3{,}720$ psi
- $k = 0.1$ md
- $S_{wi} = 0.4$
- $h = 100$ ft with added effect of multiphase flow

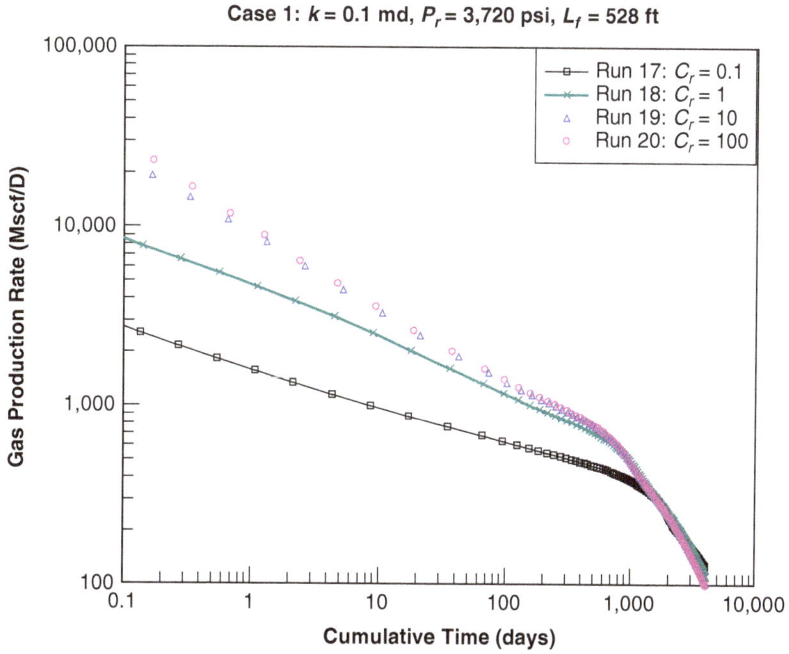

Fig. 9.5—Gas production rate at different fracture conductivity where L_f = 528 ft, p_r = 3,720 psi, k = 0.1 md, S_{wi} = 0.4, and h = 100 ft.

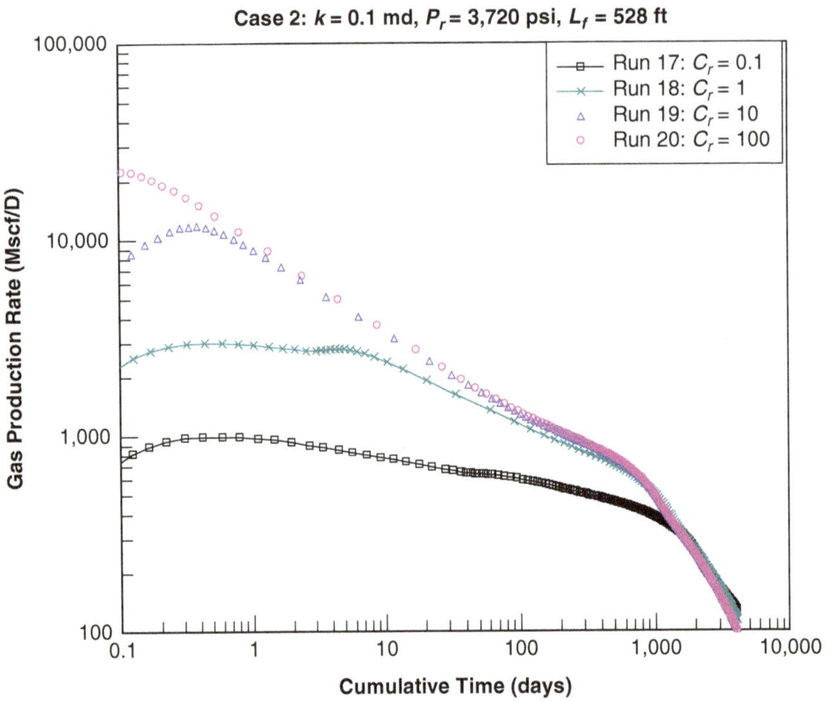

Fig. 9.6—Gas production rate at different fracture conductivity where L_f = 528 ft, p_r = 3,720 psi, k = 0.1 md, S_{wi} = 0.4, and h = 100 ft with added effect of multiphase flow.

We can see that the gas production rate increases for a period of time before it reaches a maximum value, which is the cleanup time. The fracture cleans up faster if the fracture conductivity is higher. For fracture conductivity of 0.1, the gas production rate is much lower during the first 3 years of production. However, if the fracture conductivity is greater than 10, a higher conductivity does not make much further difference on the gas-flow rate. Cumulative gas production will be higher if the fracture conductivity is higher. However, the cumulative gas production does not differ much when the fracture conductivity is greater than 10. If the fracture conductivity can be increased from 0.1 to 10, the cumulative gas production will increase by 36% in 10 years, which is approximately 1.3 Bcf for one well. Thus, strong proppants and high fracture conductivity are critical to the success of fracturing treatments and long-term gas production. Thus, from a production point of view, proppants that can provide dimensionless fracture conductivity of 10 or greater are good enough for gas production when only two-phase flow is considered and the fracturing-fluid-filtrate viscosity is the same as that of the formation water.

Just like for single-phase flow in a hydraulically fractured well, there have been many papers published showing the effects of two-phase flow on gas recovery in a hydraulically fractured well. As such, Fig. 9.6 is not new, but needs to be created to expand the base-case scenario. Later, when we evaluate effects of other factors upon gas-flow rate, we will be able to compare those results to the single-phase and two-phase simulation results. The effect of fracture conductivity on the early-time transient flow and long-term cumulative gas production can be seen in Fig. 9.6. The assumption for two-phase flow is that a hydraulic fracture has been created, and the fracturing fluid that leaks off into the reservoir formation has the same properties as the formation water. So, after a fracturing treatment, the region near the fracture is saturated by the water phase. The water phase temporarily blocks the gas from flowing into the fracture and the wellbore. The well takes time to clean up the fracturing fluid before gas can flow at a maximum rate.

It is known that the value of fracture conductivity will decrease as the closure pressure on the proppant increases. When the proppant crushes, the analysis of cleanup becomes more complicated. **Fig. 9.7** shows the gas production rate vs. time for different values of fracture conductivity after a fracturing treatment with the added effect of proppant crushing on the cases in Fig. 9.6. The early gas-flow rate for a C_r of 10 is near 4,000 Mcf/D and peaks at near 8,000 Mcf/D after 1 day. For the infinite fracture conductivity of C_r = 100, the gas-flow rate starts out at 20,000 Mcf/D. After a few days, productions at C_r = 10 and at C_r = 100 is the same, which indicates that fracture conductivity with C_r = 10 is sufficient for optimized gas production before other factors are considered. For a fracture conductivity of 0.1, the gas production rate is much lower during the first 3 years of production. If the fracture conductivity can be increased from 0.1 to 10, the cumulative gas production will be increased by 50% in 10 years, which is approximately 2.0 Bcf for one well. Cumulative gas production does not differ much when the fracture conductivity is greater than 10. So, strong proppants and high fracture conductivity are critical to the success of fracturing treatment and long-term gas production.

The preceding agrees with observations made by others. For example, loss of fracturing fluid to the formation adjacent to the fracture face creates a zone of immobile fracturing liquid and reduces the flow of gas through the formation face because of the reduction of gas relative permeability (Holditch and Morse 1976; Ahmed et al. 1979). The formation damage near the fracture also results in longer gas-breakthrough times and slower fracturing-fluid cleanup in hydraulically fractured gas wells (Montgomery and Berthelot 1990).

In **Fig. 9.8,** the effect of filter cake does reduce the gas production rate during early time. After 30 days, there is not much difference between the curves with and without filter cake. And it also shows that, if the gel has a yield stress of 20 Pa or greater, the gas-flow rate never peaks and remains much lower compared to Cases 1 through 4. Gas production rate becomes lower and cleanup time is longer when more factors are considered from Cases 1 through 5. For Cases 1 through 4, we can see that the gas production rate increases for a certain period of time before it reaches a maximum value, which is the cleanup time.

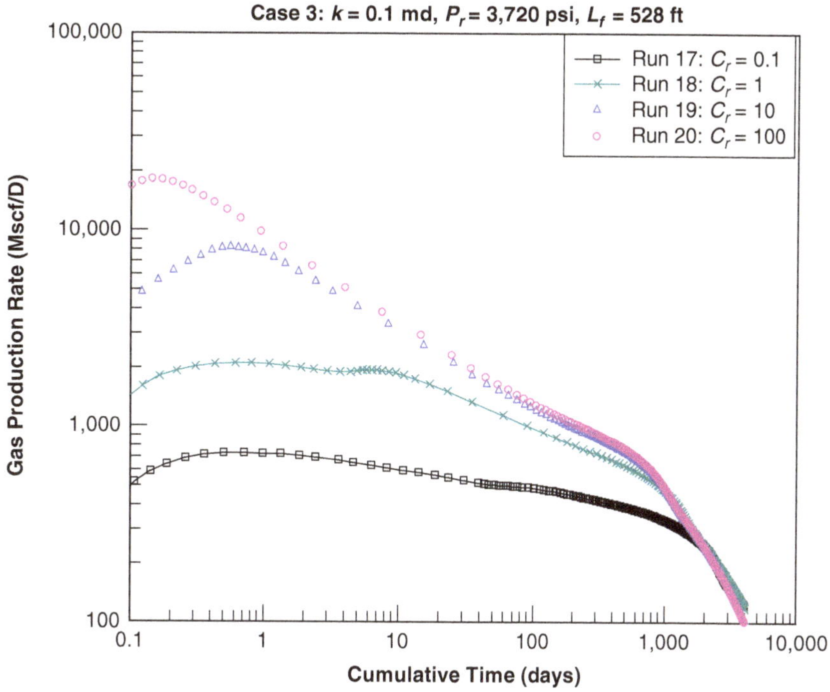

Fig. 9.7—Gas production rate for different fracture conductivities where L_f = 528 ft, p_r = 3,720 psi, k = 0.1 md, S_{wi} = 0.4, and h = 100 ft with added effect of proppant crushing.

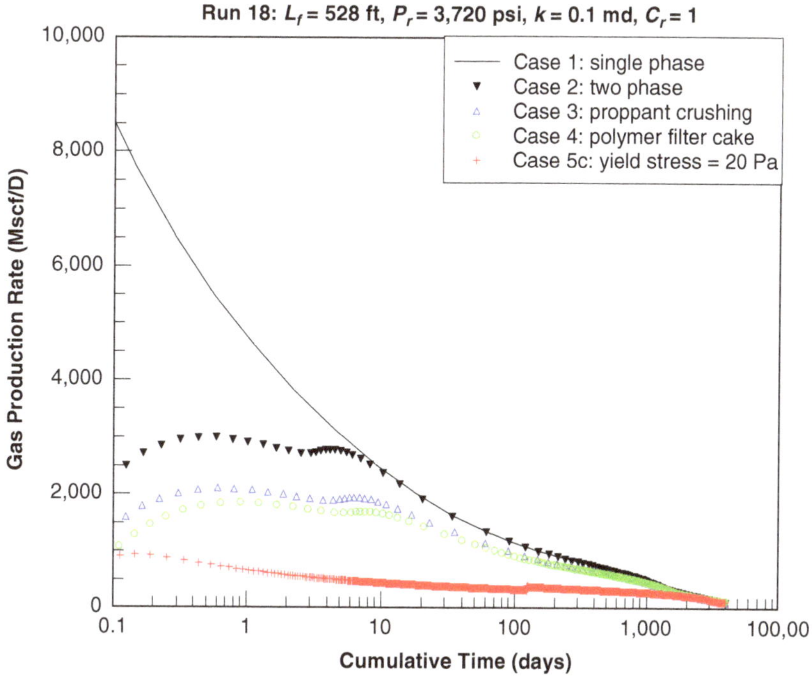

Fig. 9.8—Gas production rate for different cases where L_f = 528 ft, p_r = 3,720 psi, k = 0.1 md, C_r =1, S_{wi} = 0.4, and h = 100 ft with added effect of gel.

For Case 2 (two-phase flow) it takes approximately 10 days to clean up, and for Case 3 (proppant crushing) it takes approximately 1,000 days. So, when more factors are considered, the cleanup process takes longer, and results are more realistic. This should help both researchers and engineers understand how the cleanup process has been affected. In 10 years, proppant crushing reduces the cumulative gas production to 93% of the ideal case. The addition of filter cake does not make much difference, but the inclusion of gel yield stress of 20 Pa in the fracture will reduce cumulative gas production to 63% compared to the ideal single-phase case. For Case 1 of single-phase flow, there is no formation water and no fracturing fluid, so water production is zero. For Cases 2 through 5, the produced water can come from either formation water or fracturing-fluid filtrate, which is considered to have the same properties as formation water. We can see that most water is produced during early time; after approximately 10 days, the water production levels off at a very low rate. For the case with yield stress of 20 Pa, the water production rate is even lower.

In summary, if the damage mechanisms and simulation technologies and techniques discussed here are understood, simulation of how a hydraulically fractured well cleans up and its impact on well performance can be performed; then, practical problems associated with tight gas development can be solved.

9.4 Modeling Non-Darcy Flow and Fracture-Conductivity Reduction

Laminar flow of fluid through porous media can be described using Darcy's law:

$$\frac{\Delta p}{\Delta L} = \mu v / k. \quad (9.1)$$

Eq. 9.1 indicates that, if the resistance (μ/k) remains constant, the pressure gradient ($\Delta p/\Delta L$) is proportional to the velocity of the fluid (v). However, when the velocity is increased such that the flow is not laminar, the pressure drop will increase more than the proportional increase in velocity.

Forchheimer (1901) proposed a generalized equation that takes into account the additional pressure drop resulting from high flow velocities:

$$\frac{\Delta p}{\Delta L} = \frac{\mu}{k} v + \beta \rho v^2. \quad (9.2)$$

In hydraulically fractured gas wells, where large gas-flow rates create high velocities, the pressure drop in the hydraulic fracture is dominated by non-Darcy flow. The resulting pressure distribution in the fracture affects the pressure distribution in the entire reservoir. Cooke (1973) concluded that "it is very important to consider turbulent flow in designing gas well fracturing treatment." He presented several example calculations that indicated that the actual pressure drop in a packed hydraulic fracture can be many times higher than the value predicted by Darcy's law. Wattenbarger and Ramey (1969) considered the effects of turbulent flow in the reservoir on the behavior of fractured wells. They assumed the fracture was infinitely conductive and used an effective wellbore radius in their solution using the method proposed by Prats (1961). One of the conclusions in Wattenbarger and Ramey (1969) is of particular interest: "it is likely that turbulence within the fracture itself is sometimes more important than turbulence within the formation."

9.4.1 Non-Darcy Flow Coefficient. Katz et al. (1959) were the first to develop a correlation for the non-Darcy coefficient that can be correlated with porosity and permeability of the porous medium. They used sandstone, limestone, and dolomite cores in their laboratory experiments. They found that there was no difference among the turbulence factors for the various rock types.

Tek et al. (1962) proposed a correlation after analyzing the Katz et al. data:

$$\beta = \frac{5.5 \times 10^9}{k^{5/4} \phi^{3/4}}. \quad (9.3)$$

Cooke (1973) investigated non-Darcy flow in packed, hydraulically induced fractures. He noted that β factors for fractures packed with multiple layers of sand had not been reported in the petroleum literature and suggested the following equation for calculating β factors:

$$\beta = \mathrm{b} / (1{,}000\ k)^a. \quad (9.4)$$

The values for a and b can be found in **Table 9.3.**

Geertsma (1974) concluded that the deviation from Darcy's law resulted from the convective acceleration and deceleration of the fluid particles. He defined a new correlation for the non-Darcy coefficient that relates the porosity and permeability of the formation for dry gas flow in porous media:

$$\beta = \frac{0.005}{k^{0.5} \phi^{5.5}}. \quad (9.5)$$

Geertsma (1974) also calculated the non-Darcy coefficient for a formation containing an immobile liquid saturation and proposed the following correlation:

$$\beta = \frac{0.005}{k^{0.5} \phi^{5.5}} \left[\frac{1}{(1 - s_w)^{5.5} k_{re}^{\ 0.5}} \right]. \quad (9.6)$$

9.4.2 Numerical Model. A single-phase, 2D, finite-difference reservoir simulator was used by Holditch and Morse (1976) to evaluate non-Darcy flow in a tight gas reservoir. The reservoir was assumed to be horizontal, homogeneous, isotropic, and initially filled with gas at a constant pressure (p_i). The drainage pattern is in the form of a square with closed external boundaries, as shown in **Fig. 9.9.**

The fracture extends on two opposing sides of the wellbore and is propped over the entire vertical extent of the formation. Only one-fourth of the drainage pattern is modeled. A grid pattern consisting of 300 cells is used to provide fine breakup near the wellbore and along the fracture. Such gridding is required to improve the accuracy of the model during the early transient flow periods. For the pressure-buildup simulations, the timesteps ranged from a minimum of 0.0001 days to a maximum of 1.0 day. A direct solution technique—alternating diagonal matrix inversion—was used in the model. This method provides the best means for solving the equation that describes a fractured reservoir.

Mesh size	*a*	*b*
8/12	1.24	3.32
10/20	1.34	2.63
20/40	1.54	2.65
40/60	1.60	1.10

Table 9.3—Constants *a* and *b* for Cooke's correlation (Cooke 1973).

9.4.3 Non-Darcy Factor. To obtain the correct pressure in the reservoir under non-Darcy flow conditions, a correction term called the non-Darcy factor was included in the model. It is defined as

$$F_{ND}=\frac{k_T}{k_D}=\frac{1}{1+5.655\times10^{-9}\dfrac{\beta\rho v k_D}{\mu}}. \quad (9.7)$$

The non-Darcy factor is a measure of the degree of non-Darcy flow through the porous media. If the β factor or velocity (v) approaches zero, then the non-Darcy factor approaches unity, and the pressure drop can be calculated using Darcy's law. If the velocity is high, then the value of the non-Darcy factor will be smaller than unity. As the degree of non-Darcy flow increases, the value of the non-Darcy factor decreases. For example, if the non-Darcy factor is equal to 0.1, the pressure gradient will be 10 times the value calculated using Darcy's law.

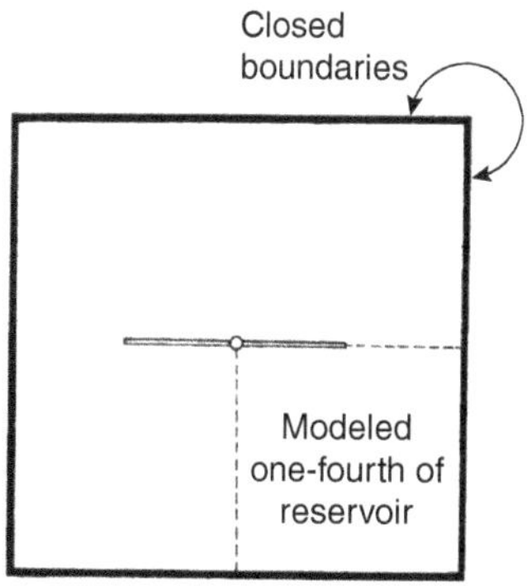

Fig. 9.9—View of fractured reservoir (Holditch and Morse 1976).

The numerical model was based on Darcy's law. Therefore, the non-Darcy factor was used to modify the equations so that the effects of non-Darcy flow could be included in the solution. To apply Eq. 9.7, the density (ρ), velocity (v), permeability (k_D), and viscosity (μ) were calculated in the model for each gridblock on the basis of the data from the previous timestep. Eq. 9.7 was solved for each gridblock, and the non-Darcy factor was then used to modify the flow coefficients between the gridblocks. This procedure allowed the model to calculate the correct pressures and rates under non-Darcy flow conditions.

9.4.4 Non-Darcy Flow in the Fracture. The initial objective was to determine the reservoir conditions in which non-Darcy flow can occur in a fracture. Using the data for a deep-well case in **Table 9.4,** 48 simulations were made that covered the range of data in **Table 9.5.**

Results of three simulations are presented in **Fig. 9.10.** It is illustrated that the non-Darcy factors in the fracture are a function of production rates for three different reservoir conditions. These figures are typical of the overall results and indicate that substantial non-Darcy flow can exist in the fracture over the entire life of a well. In relatively low conductivities, large pressure gradients are obtained only near the wellbore; therefore, the flow pattern in the reservoir does not deviate substantially from radial flow. A large percentage of the gas being produced enters the fracture close to the wellbore. The effect of the gas entering the fracture can be seen in the decrease of the non-Darcy factors near the wellbore.

9.4.5 Effect of Non-Darcy Flow on the Gas-Well Productivity Index. The productivity index for a gas well can be defined by the following equation:

$$J_g=q\mu z/\left(\bar{p}_R^2-p_{wf}^2\right). \quad (9.8)$$

Eq. 9.8 is similar to the productivity-index expression for an oil well. The difference is that the pressure-dependent terms of viscosity and gas compressibility are included and the pressures are squared in the equation.

Fig. 9.11 illustrates how non-Darcy flow affects the productivity index (J_g) of a fractured well, which was constructed from 48 simulations with original fracture conductivities ranging from 500 to 5,000 md-ft. When non-Darcy flow was not considered, the gas-well productivity index exhibited a decline of approximately 50% over the life of the well. Part of this decline occurs as the well goes from transient flow to stabilized flow. Most of the productivity-index decline can be attributed to decreasing fracture conductivity because of increasing closure stress. From the curves in Fig. 9.11, the ratio of J_g for non-Darcy flow over J_g for Darcy flow can be calculated. It is found that the ratios are identical for a given proppant type, relative conductivity, and stage of production, regardless of the absolute values of fracture and formation permeabilities. For relatively low-conductivity fractures, non-Darcy flow reduces the ratio to half; for high relative conductivities, the fracture efficiency is high and the increased pressure drop resulting from non-Darcy flow does not significantly affect the gas-well productivity index.

9.4.6 Effects of Non-Darcy Flow on Pressure-Buildup Analysis. While studying the relationship between non-Darcy flow in the fracture and the well productivity index, it is observed that the pressure distribution in the entire reservoir was affected when

	Deep-Well Case	Shallow-Well Case
Depth (ft)	13,000	4,000
Reservoir pressure (psi)	7,500	1,860
Bottomhole treating pressure (psi)	11,000	2,800
Bottomhole temperature (°F)	300	125

Table 9.4—Reservoir data for reservoirs of different depths.

Formation permeability (md)	0.001–1.5
Initial fracture conductivity (md-ft)	500–5,000
Fracture proppant size (mesh)	20/40 and 40/60
Fracture length (ft)	132 and 660

Table 9.5—Input data that were varied.

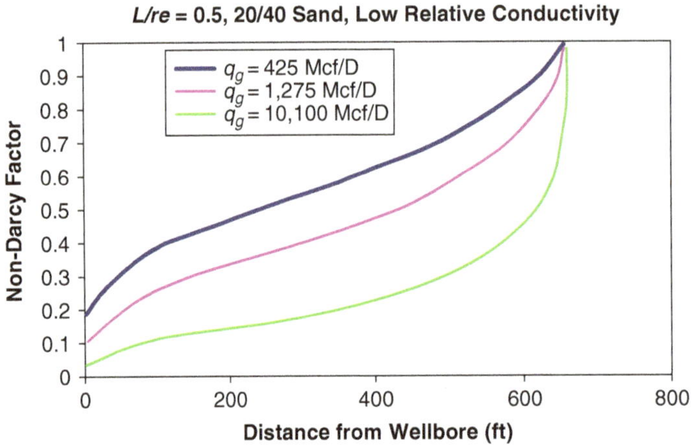

Fig. 9.10—Non-Darcy factors in fracture for 20/40 mesh sand and low relative conductivity (Holditch and Morse 1976).

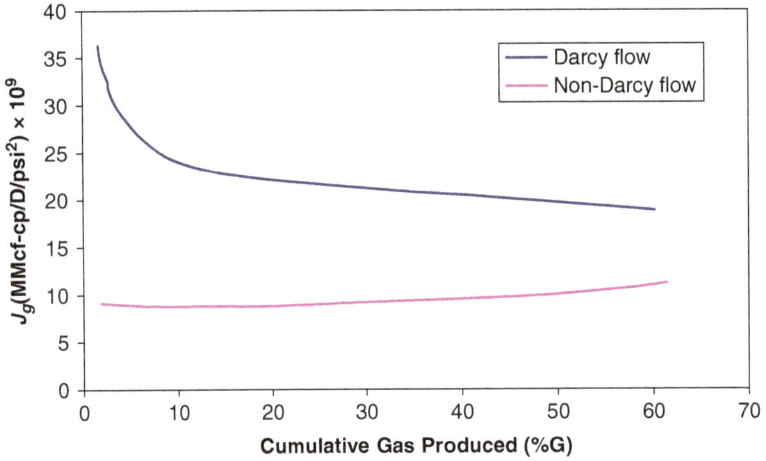

Fig. 9.11—Effect of non-Darcy flow and closure stress on the gas-well productivity index for low relative conductivity. (Holditch and Morse 1976). %G = percentage of gas-in-place.

non-Darcy flow was included in the simulations. Because the shape of the pressure-buildup curve is influenced by the pressure distribution, a calculation was made to determine the effects of non-Darcy flow in the fracture on the analysis of the pressure-buildup test.

Fig. 9.12 shows buildup curves for a deep well. It illustrates the effects of non-Darcy flow in a fractured reservoir of 0.6-md permeability. It is apparent from Fig. 9.12 that non-Darcy flow in the fracture substantially changes the shape of the pressure-buildup curve, causing the flowing BHP at the time of shut-in to be reduced by 2,079 psi.

9.4.7 Fracture-Conductivity Reduction. Reduction in fracture permeability with increasing closure stress should be considered when modeling hydraulically fractured gas wells. When a fractured well begins to produce, the pressure in the fracture decreases, closure pressure increases, and the permeability in the fracture decreases. Data on fracture permeability as a function of closure pressure were given by Cooke (1973), as shown in **Fig. 9.13.**

The effective stress on the proppant is equal to the in-situ stress minus the flowing BHP. The value of the in-situ stress is very important for fracturing treatment design and evaluation. The in-situ stress depends on the overburden pressure, pore pressure, rock properties, and tectonic forces. In-situ stress can vary in different local geological settings and can change in response to reservoir depletion. One equation used to calculate in-situ stress is as follows:

$$\sigma_{\min} = \frac{\nu}{1-\nu}(\sigma_z - p) + p + \sigma_{\text{ext}}, \quad (9.9)$$

where ν = Poisson's ratio, σ_z = overburden pressure, p = pore pressure, and σ_{ext} = any externally applied stress.

We have already discussed the effect of proppant crushing on short-term cleanup and long-term gas production. The effect of proppant crushing could be added to existing simulation models; for example, you could consider different cases: no proppant crushing, slight crushing, and high crushing, as shown in **Fig. 9.14.** The results from finite-difference reservoir models are made more realistic by adding the proppant-crushing effect in the fracture. The results are shown in Fig. 9.7.

In summary, hydraulic fracture conductivity reduction because of non-Darcy flow and increasing stress should be considered in analyzing tight gas wells. Our discussions on the physics, mechanisms, and example results help in understanding the importance of non-Darcy flow and stress-dependent conductivity and provide a solid starting point on how to integrate these factors into reservoir simulation models.

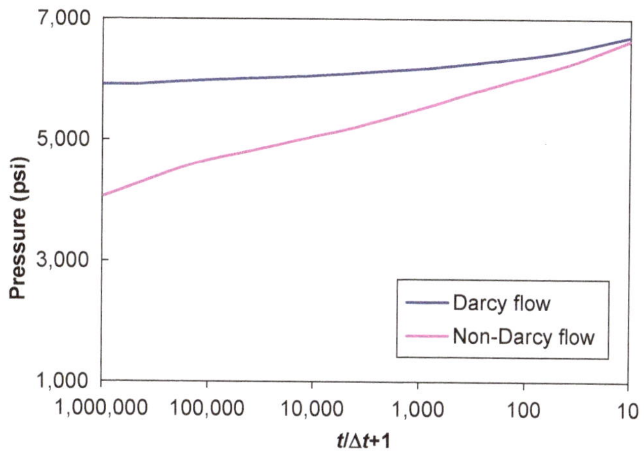

Fig. 9.12—Effect of non-Darcy flow in the fracture on the shape of a buildup curve for low relative conductivity (Holditch and Morse 1976).

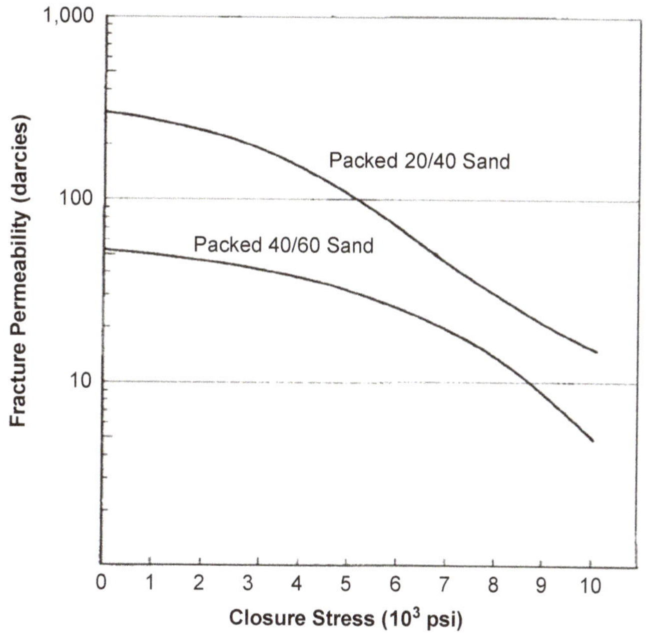

Fig. 9.13—Effect of stress on the permeability of 20/40 and 40/60 mesh sand at 250°F (Cooke 1973).

9.5 Analyzing Post-Fracturing Flow and Buildup Data

This section discusses how to evaluate reservoir and fracture properties by history matching transient pressure data and/or production data using reservoir simulation technologies, especially in reservoirs with multiple layers, multiphase flow, or other complex geological and engineering attributes that make it difficult or impossible to obtain analytical solutions. Thus, this section is different from and complements the 80% answer, which was discussed in detail in Part 2 Ch 6: Prestimulation Well Testing, Part 2 Ch 7: Post-Fracture Treatment Well Testing, and Part 2 Ch 8: Rate Transient Analysis.

Evaluation of a well after a hydraulic fracturing treatment can be significantly influenced by the different methods used to compute reservoir characteristics, such as formation permeability, fracture length, and fracture conductivity. Discussions on basic techniques and their strengths and weaknesses were provided by Holditch et al. (1983) and Lee and Holditch (1979, 1981). Results from combining post-fracturing performance data with post-fracturing buildup data were reported by Bostic et al. (1980). Similar techniques have been described by

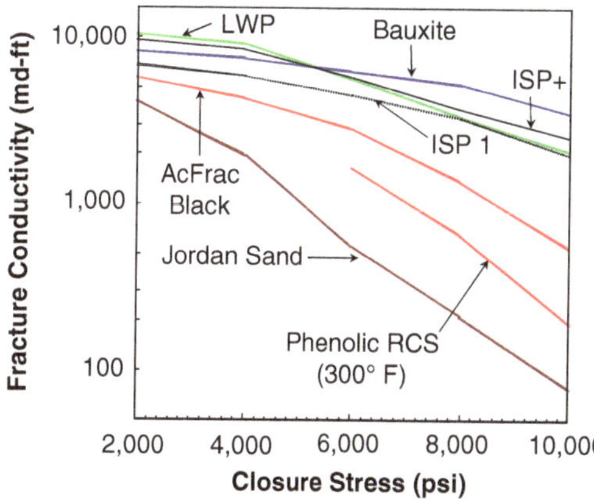

Fig. 9.14—Different proppant conductivity vs. closure stress (Carbo Ceramics 2007).

other researchers; for example, combining conventional pressure transient test analysis with material-balance decline type curves (MBDTCs) for gas wells with finite-conductivity vertical fractures by Rushing et al. (2005). Fracturing-fluid interactions (Ahmed et al. 1979), fluid invasion (Montgomery and Berthelot 1990), and non-Darcy flow (Alvarez et al. 2002) in hydraulically fractured wells in tight gas reservoirs have also been investigated. Field cases of fracture evaluation with pressure transient testing have been documented (Holditch and Lee 1979; Holditch et al. 1983; Hopkins et al. 1991; Lee et al. 1984).

Holditch et al. (1987a) and Rahim and Holditch (1993) presented prefracture and post-fracture formation evaluation to obtain the 3D shape of a hydraulic fracture. They described the importance of collecting and analyzing data before, during, and after a hydraulic-fracturing treatment so that the 3D properties of both the formation and the fracture can be determined. Similarly, it has been mentioned that 2D models can be used only if the reservoir can be described in 5 to 10 layers, and a single fracture will propagate through all pay zones (Rahim and Holditch 1993).

Furthermore, the pressure transient responses in multilayer reservoirs have been studied in detail. Sullivan et al. (1987) presented estimations of induced fracture properties and reservoir geometry in multilayered gas reservoirs in East Texas. Holditch et al. (1987) thoroughly discussed problems associated with complex Travis Peak Reservoirs in East Texas. Similarly, Frantz et al. (1992) summarized a multilayer reservoir description for the (Gas Research Institute) comprehensive study of the Devonian Shales. Gao et al. (1994) stated that an equivalent single-layer reservoir model is not as accurate when describing a well completed in a layered reservoir. They concluded that the single-layer model will be optimistic. Jatmiko et al. (1994) investigated the use of a nonlinear programming approach to solve the nonlinearity problems arising in stratified commingled reservoirs. Lolon et al. (2008) presented approximate (semianalytical) solutions for the wellbore pressure and fractional flow-rate responses of commingled layered reservoirs.

Long hydraulic fractures are usually required to optimize recovery from low-permeability gas reservoirs. Because these fractures can be quite expensive to create and there has always been a significant amount of art associated with fracturing treatment design, it is frequently helpful to use pressure transient tests to estimate effective fracture properties. In this way, optimized fracturing treatments could be achieved for a specific tight gas formation. Holditch et al. (1983) provided a brief but comprehensive summary of four basic techniques used to analyze post-fracture pressure transient tests to be understood for the different solutions. The four basic techniques are

- Semilog [pseudoradial flow (PRF)] analysis
- Square-root-of-time (linear flow) analysis
- Type-curve analysis
- Reservoir simulator history matching

9.5.1 Pseudoradial flow (PRF). A definition of PRF is that sufficient time has elapsed in a buildup or drawdown test so that BHP varies linearly with the logarithm of flow time (drawdown) or the Horner time group $(t_p + \Delta t)/\Delta t$ (buildup), as expected for radial flow in an unfractured reservoir. In an infinite-acting (unbounded) reservoir, the analysis technique is based on the use of skin factor s, which can be calculated as provided by Lee and Holditch (1981):

$$s = 1.151\left[\frac{p_{1\text{hr}} - p_{wf}}{m} - \log\left(\frac{k}{\phi\mu_a c_{ta} r_w^2}\right) + 3.23\right], \quad \text{(9.10)}$$

$$\log(L_f) = \frac{1}{2}\left[\log\left(\frac{k}{\phi\mu_a c_{ta}}\right) + \frac{p_{1\text{hr}} - p_{wf}}{m} - 2.63\right], \quad \text{(9.11)}$$

with an infinite-conductivity vertical fracture,

$$L_f = 2r_w e^{-s}. \quad \text{(9.12)}$$

9.5.2 Linear Flow. For a buildup test, with linear flow continuing to time

$$t_p + \Delta t,\ p_i - p_{ws} = m'\sqrt{t_p + \Delta t} - \sqrt{\Delta t}, \quad \text{(5.13)}$$

where $t_p >> \Delta t$, $p_{ws} - p_{wf} = m'\sqrt{\Delta t}$, and where m' is the slope of the straight line when BHP vs. the square root time is plotted, which relates fracture half-length and formation permeability (Lee and Holditch 1981),

$$L_f\sqrt{k} = 4.064\frac{q_g B_{ga}}{h(m')}\left(\frac{\mu_a}{\phi c_{ta}}\right)^{1/2}. \quad \text{(5.14)}$$

9.5.3 Superposition. Superposition can be used to apply known pressure solutions for a single-well constant-rate system to multiwell and/or multirate systems. The general form of the superposition equation describing the pressure history of an undamaged well in an infinite system producing a slightly compressible fluid is after (Bostic et al. 1980):

$$p(t_o) + p(t_n) = \Delta p(t_n) = \frac{141.2\mu}{kh}\sum_{j=1}^{n}\left\{\left(q_j B_j - q_{j-1}B_{j-1}\right)\left[p_D(t_n - t_{n-1})_D\right]\right\}, \quad \text{(9.15)}$$

where $q(0) = 0$, $t_0 = 0$, and $q(t_0) = p_i$.

For a fractured gas well, it can be given as

$$m[p(t_o)]-m[p(t_n)]=m[\Delta p(t_n)]=\frac{141.2T}{kh}\sum_{j=1}^{n}\{(q_j-q_{j-1})[p_D(t_n-t_{n-1})_{Dxf}]\}, \quad (9.16)$$

where $m(p)$ is the real-gas pseudopressure.

The transformation function that is equivalent to variable pressure history with a constant rate is given by

$$P_{FCN}=\frac{\{m[p(t_o)]-m[p_c(t)]\}}{q_c}, \quad (9.17)$$

where q_c = constant rate and p_c = flowing BHP that would exist if the rate q_c were imposed on the well.

The strengths and weaknesses of four techniques are discussed by Lee and Holditch (1979, 1981). They commented on the theoretical basis of the test analysis methods as

- Test analysis methods based on achieving PRF (Russel and Truitt method) have limited applicability because of the long times required to achieve the proper straight line and because they assume infinite fracture conductivity.
- The Millheim-Cichowicz (M-C) method, based upon the assumption of linear flow into the fracture, is limited because it requires an independent estimate of formation permeability and assumes a constant flux into the fracture.
- Type curves that include finite fracture conductivity (Bostic et al. 1980) frequently produce nonunique well-test data interpretations. This problem is diminished, however, if an independent permeability estimate is available, as demonstrated in a field case study in **Fig. 9.15.**
- The modified M-C method appears to be useful because it can provide estimates of fracture length and fracture conductivity using early-time pressure-buildup data. This method also requires an independent permeability estimate.
- Simulator history matching is the most generally applicable analysis method for determining the formation and fracture properties. Lee and Holditch (1981) conducted a comprehensive comparison study of fracture lengths obtained using different methods as shown in **Table 9.6.**

The data in Table 9.6 clearly show that history matching using a numerical simulator or the modified M-C method resulted in hydraulic fracture lengths that were closer to the design length than could be achieved with the Russel and Truitt method. In general, it has been learned and documented over the years that the propped fracture length will be underestimated unless a finite-difference model is used to history match the production and/or pressure-buildup data when the well has been put on production.

The dimensionless fracture flow capacity can be defined as (Bostic et al. 1980)

$$F_{CD}=\frac{k_f w_f}{kL_f}. \quad (9.18)$$

A well with a hydraulic fracture performs as an infinite-flow-capacity fracture only for high F_{CD} values. Massive hydraulic fracturing (MHF) treatments are usually associated with long finite-flow-capacity fractures. So, any technique used to analyze MHF-stimulated wells should be applicable to finite-flow-capacity fractures.

As described by a number of authors (Lee and Holditch 1979; Holditch and Lee 1979; Bostic et al. 1980; Rushing et al. 2005), the conventional techniques for analyzing production and/or pressure-buildup data from a well containing a hydraulic fracture require that the formation permeability be known a priori. But wells completed in tight gas sands require very long shut-in times to reach PRF, and operators are reluctant to shut in a well for extended periods. However, if we have an independent estimate of reservoir permeability, then shorter-duration pressure-buildup tests become practical (Rushing et al. 2005).

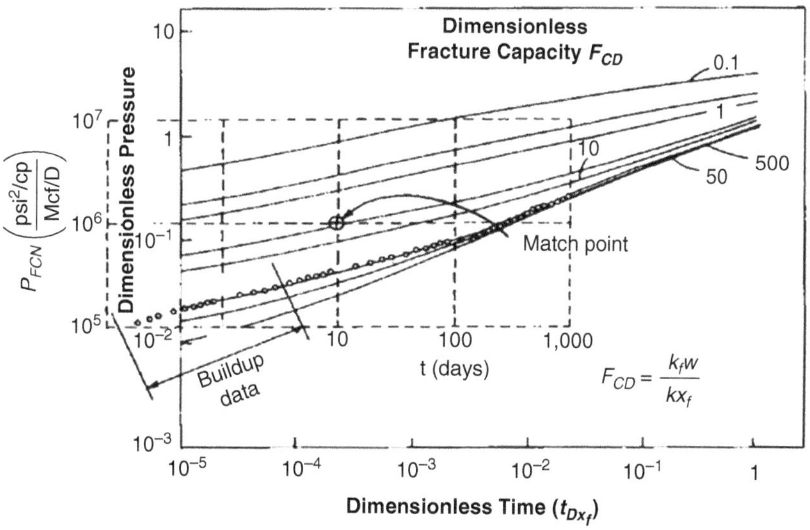

Fig. 9.15—Type curve match (Bostic et al. 1980).

Well	Design Length (ft)	L_{fRT}/L_{fD}* (%)	L_{fHM}/L_{fD}* (%)	L_{fMC}/L_{fD}* (%)
1	400	9.5	12.5	19.0
2	700	10.7	57.1	61.0
3	640	40.2	23.4	26.0
4	900	20.3	61.1	79.0
5	1,500	9.6	13.3	16.3
6	500		80.0	
7	1,500	9.1	66.7	63.2
8	1,000	2.5	50.0	40.4
9	1,500	5.0	100.0	125.0
10	700	1.0	71.4	88.0
11	1,000	5.0	60.0	85.0
12	800	12.5	87.5	103.0
13	800	2.0	50.0	65.0
Average 1**		10.6	56.4	64.2
Average 2***		7.6	68.4	78.8

* L_{fD} = design fracture length
L_{fRT} = Russell and Truitt method
L_{fHM} = history match
L_{fMC} = modified M-C method
** includes all wells
*** wells 1, 3 and 5 eliminated from average

Table 9.6—Comparison of fracture length (Lee and Holditch 1981).

Furthermore, as demonstrated in Fig. 9.12 (Holditch and Morse 1976), high gas-flow rates will create high velocities in a fracture, and the pressure drop in a hydraulic fracture is dominated by non-Darcy flow. The resulting pressure distribution in the fracture affects the pressure distribution in the entire reservoir and BHP responses. Moreover, the non-Darcy coefficient relates to rock type, fluid type, size of propping agents, temperature, permeability of the formation, liquid saturation, and so on.

9.5.4 Post-Fracture Flow Rate Transient Analysis. Decline type curve analysis of production data is one common method for estimating reservoir permeability without shutting in the well. Fetkovich (1980) was the first to incorporate transient flow models with production-data decline curve analysis. He developed the standard decline type curves by combining an analytical model for transient, radial flow at constant BHP with the Arps (1945) empirical exponential and hyperbolic decline models. Palacio and Blasingame (1993) developed the MBDTC methodology that also overcomes the constant-BHP constraint using rate (or pressure) normalization and a material-balance time (i.e., cumulative production divided by instantaneous rate) (Rushing et al. 2005).

MBDTCs (Rushing et al. 2005) were developed using a desuperposition technique that mathematically combines transient and boundary-dominated flow models, as shown in **Fig. 9.16.** The transient model is that for a well with a finite-conductivity, vertical fracture producing at a constant rate, while the boundary-dominated model is for a well centered in a finite, volumetric circular reservoir. The type curves are correlated using dimensionless fracture conductivity (F_{CD}) and dimensionless reservoir radius (r_{eD}).

There are a number of field case studies of fracture evaluation using pressure transient analysis (Holditch and Lee 1979; Holditch et al. 1983; Hopkins et al. 1991; Lee et al. 1984). In addition, these studies have profound observations related to field cases and identification of potential well issues. Poor well performance can be caused by

- An extremely low formation permeability of 0.001 md or less
- Insufficient fracture length, normally caused by fracturing out of zone or from poor sand transport
- Insufficient fracture conductivity normally caused by crushing or embedment of the propping agent
- Insufficient gas-in-place

When analyzing pressure-buildup tests, several performance parameters are normally considered as fixed for a specific problem. These fixed parameters are

- Producing time before shut-in
- Volume of gas produced before shut-in
- Producing rates during the production period
- Flowing BHP at the time of shut-in

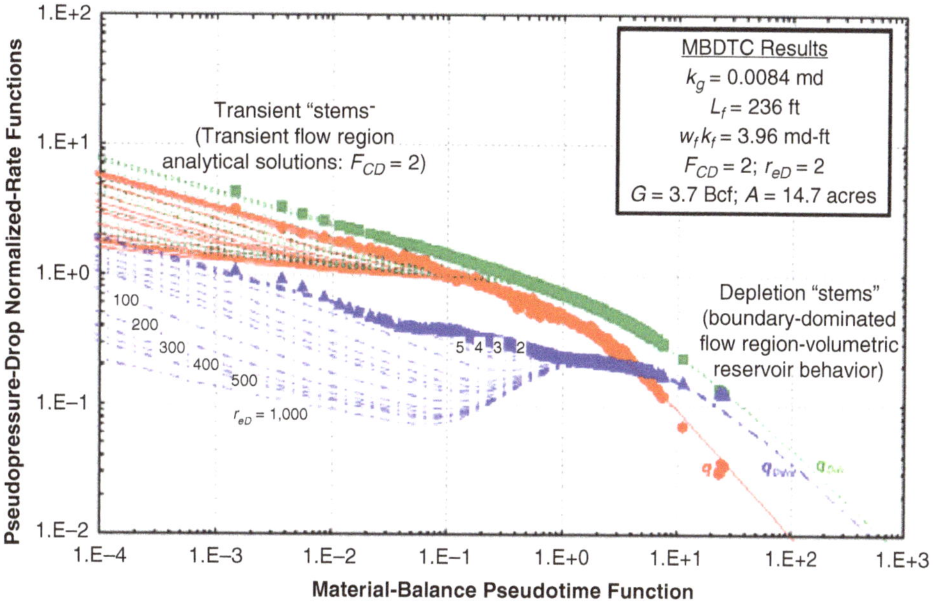

Fig. 9.16—MBDTCs of post-fracturing gas production in a well completed in Lower Cotton Valley sands (Rushing et al. 2005).

For a fixed set of producing conditions and known values of porosity, original reservoir pressure, and net-pay, the unique values of formation permeability, fracture length, and fracture conductivity can be obtained from long-term (14+ days) pressure-buildup data, as shown in **Fig. 9.17** (Holditch and Lee 1979).

Holditch et al. (1983) proposed an iterative procedure to solve for fracture length, formation permeability, and fracture conductivity using a semilogarithmic graph, a square-root-of-time graph of test data, and two correlation curves developed from the analytical solutions for finite-conductivity fractures presented by Cinco-Ley and Samaniego-V. (1981). The analysis of a buildup test generated using a computer reservoir simulator has been shown in **Fig. 9.18,** with a production time of 30 days and with 27 days of pressure buildup test. After the iteration, the converged result for reservoir permeability is 0.121 md against true formation permeability of 0.1 md.

9.5.5 Effect of Multiple Layers on Pressure-Buildup Analysis. Hopkins et al. (1991) evaluated post-stimulation well performance in the Devonian Shales and made the following conclusions.

- Large intervals in a single wellbore probably cannot be stimulated effectively by a single-stage treatment.
- A combination of layering effects and partial stimulation of the perforated interval accounted for the less-than-expected effective fracture half-length determined from post-fracturing well test analysis.

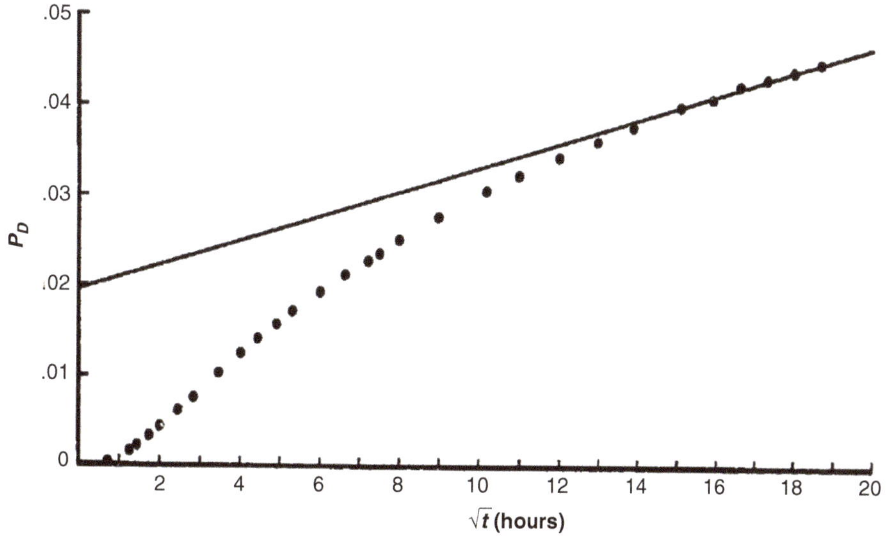

Fig. 9.17—Pressure-buildup data for long fracture, modified M-C graph using a permeability of 0.001 md (Holditch and Lee 1979).

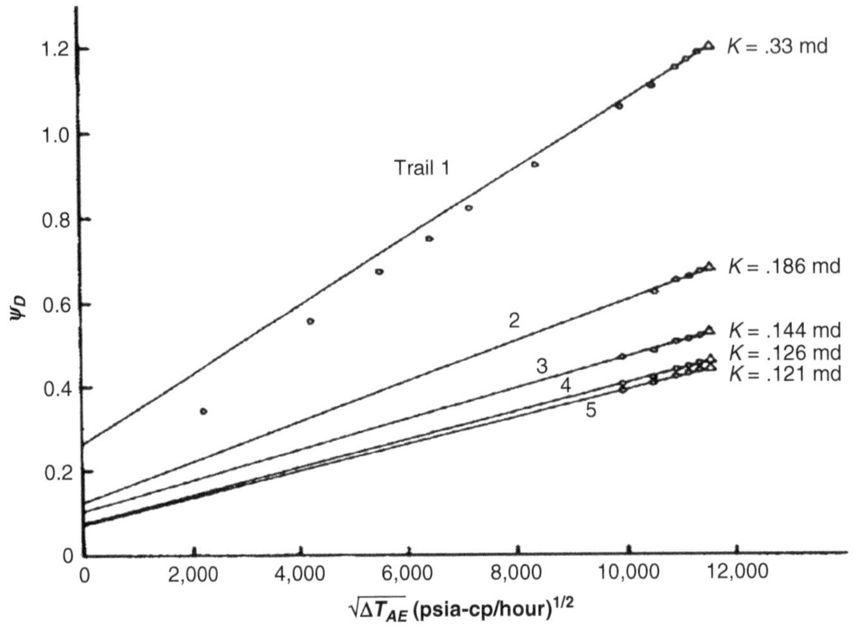

Fig. 9.18—Pressure-buildup analysis using iterative method (Holditch et al. 1983).

If these effects are ignored, post-fracturing performance projections result in significant overestimation of well performance. Lee et al. (1984) analyzed gas-well buildup tests with changing wellbore storage (WBS) in tight gas and showed that the WBS coefficient can cause serious difficulties in recognizing when, if ever, the semilog straight line with slope related to formation permeability appears.

A number of authors (Gao 1985; Sullivan et al. 1987; Holditch et al. 1987b; Frantz et al. 1992; Gao et al. 1994; Jatmiko et al. 1994; Lolon et al. 2008) reported studies of pressure transient response in multilayer reservoirs. In the layered, low-permeability reservoirs, 3D modeling, which will be discussed in the next section, is normally required to history match the production data and pressure-buildup data measured after a well has been fracture stimulated. To use a 3D model successfully, the permeability, porosity, net-pay, and fracture properties of all the various layers of rock must be known with accuracy (Holditch et al. 1987b). Holditch et al. (1987b) demonstrated the importance of using 3D hydraulic fractures for post-fracturing formation evaluation, which was also observed by Rahim and Holditch (1993). **Fig. 9.19** shows the production data after the fracturing treatment and before shut-in for pressure buildup. **Fig. 9.20** is a Horner plot of post-fracturing BHP buildup data in terms of pseudopressure and pseudotime. Conventional analysis of the pressure data did not provide reliable results. Holditch et al. (1987b) concluded that a 3D reservoir simulator is needed to analyze the post-fracturing performances, because of lateral and vertical heterogeneity.

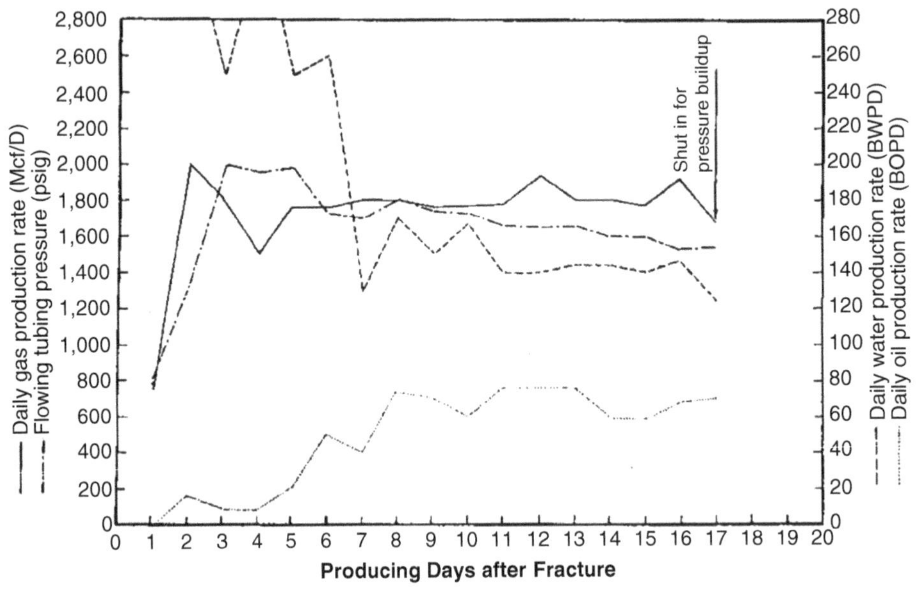

Fig. 9.19—Production test after fracturing treatment in Smith County (Holditch et al. 1987b).

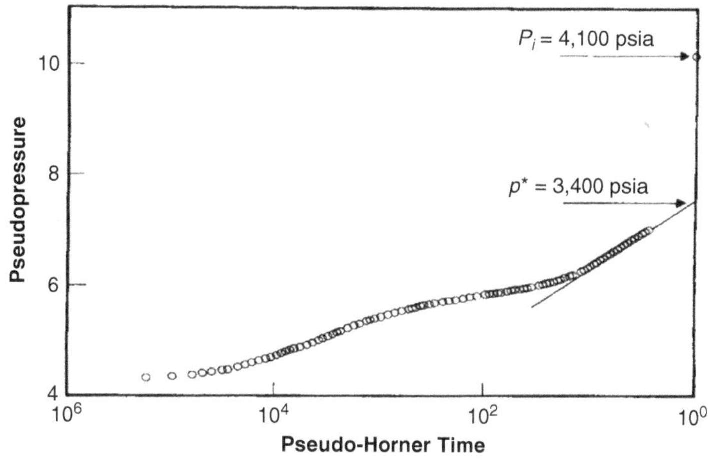

Fig. 9.20—Horner plot of post-fracturing pressure-buildup data in Smith County (Holditch et al. 1987b).

9.6 Modeling Multilayer Reservoirs

Many tight gas reservoirs are geologically complex and layered because of the depositional environments. Therefore, the petrophysical, geomechanical, reservoir, and fluid properties vary dramatically vertically and laterally, which cannot be modeled using 2D models.

Holditch et al. (1987b) thoroughly discussed problems associated with complex Travis Peak Reservoirs in East Texas, USA. The Travis Peak Formation in East Texas and north Louisiana is a complex sand dispersal system that is typically 1,500 to 2,000 ft thick and contains both lenticular and blanket sandstones. The majority of sands are of low permeability and are gas bearing. Holditch et al. (1987b) used complex reservoir models to history match the production and pressure-buildup data that had been measured both before and after the hydraulic fracturing treatment. Three-dimensional hydraulic fracture propagation models were also used to history match the bottomhole treating pressure and injection-rate data measured during the fracturing treatment. Multilayer reservoir modeling has been used by a number of authors in the past (Gao 1985; Sullivan et al. 1987; Holditch et al. 1987b; Frantz et al. 1992; Gao et al. 1994; Jatmiko et al. 1994; Lolon et al. 2008).

The steps in building 3D reservoir models are as follows:

1. Understand the regional geology to obtain ideas of the environment of the deposition and the sources and migration of fluids.
2. When the regional geologic studies have been completed, localized geologic studies should be performed. Geologists must generate maps of structure and isopach for each layer/facies, to be exported into reservoir models.
3. Oil- and gas-bearing formations should be broken into a number of layers (5 to 20 layers) and evaluated to obtain representative petrophysical, geomechanical, and fluid properties using core, log, and field tests.
4. Run 3D hydraulic fracture modeling to obtain hydraulic fracture geometries and properties, including length, height, width, and its conductivity in three dimensions.
5. Integrate and build a dynamic reservoir model.
6. Validate the model with available pressure and/or production data. If the model could not be validated, go back to Steps 2 and 3 to improve the geological analysis and formation evaluation.
7. Use the model to analyze post-fracturing flow and/or pressure-buildup data and to forecast well performance.

Following Steps 1 through 3, a multilayer 3D reservoir model can be built, such as the one shown in **Fig. 9.21.** A rule of thumb is to have 5 to 20 layers for the reservoir model, but a decision must be made on the basis of experience, regional geology, number of pay zones/layers, number of fracturing treatment stages or number of hydraulic fractures, fluid distribution vertically, and areal extent of all rock layers, including sandstone, siltstone, and shale intervals. It is important to understand the size, shape, and areal extent of the permeable producing intervals if the expectation is to use a 3D reservoir model to analyze production and pressure transient data to calculate created, propped, and effective fracture dimensions. When a static model has been built, a 3D hydraulic fracture propagation model is required to generate created and propped hydraulic fracture dimensions and properties for each layer with accuracy, as shown in **Fig. 9.22,** which also requires deliberate calibration and validations with extensive of knowledge and skill sets. Now, all these geologic, engineering, hydraulic fracture, and production/pressure data can be integrated into the static reservoir model to build a dynamic reservoir model, as shown in **Fig. 9.23.** A modern software package enables generation of richer and more-colorful reservoir models, but the key concept is precisely demonstrated in Fig. 9.23. We need to have accurate geologic interpretations, formation evaluation, and estimation of created hydraulic fracture dimensions for all layers. If conducted correctly, the model will now be able to validate with pressure and/or production data, as in Step 6, and predict well performances, as shown in **Fig. 9.24.** If any discrepancy exists, Steps 1 through 5 can be performed again; all the evaluations, data integrations, and constructions of models can be thoroughly revisited; and the analysis can be continually revised until a good match is achieved. This is an iterative process, but it allows for a better understanding of the geology, reservoir, hydraulic fracturing, and well performances to improve completions and operations of existing and subsequent wells.

Using the validated model, well performances can be forecast, as shown in Fig. 9.15, reserves can be assessed, pressure transient data can be analyzed, and bypassed reserves and ways for improved well productivity can be identified.

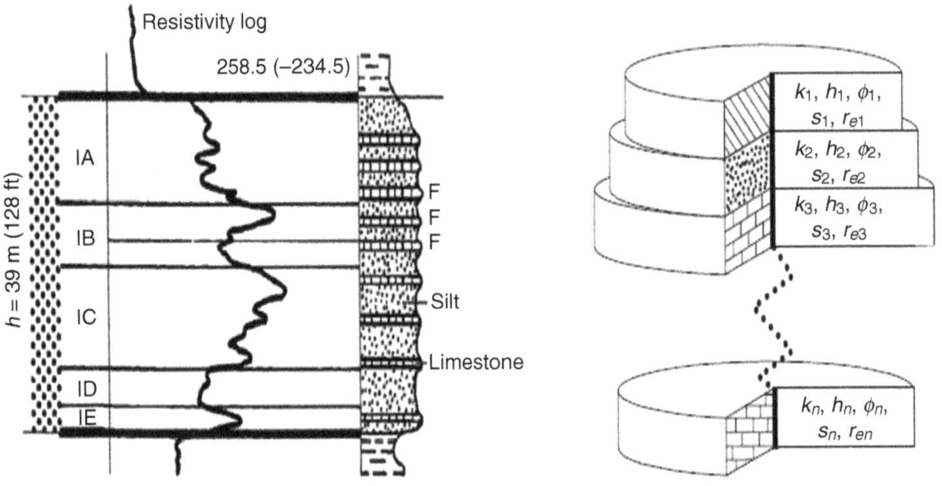

Fig. 9.21—Multilayer reservoir model (left: schematic of an actual multilayer reservoir; right: idealized multilayer reservoir model) (Lolon et al. 2008).

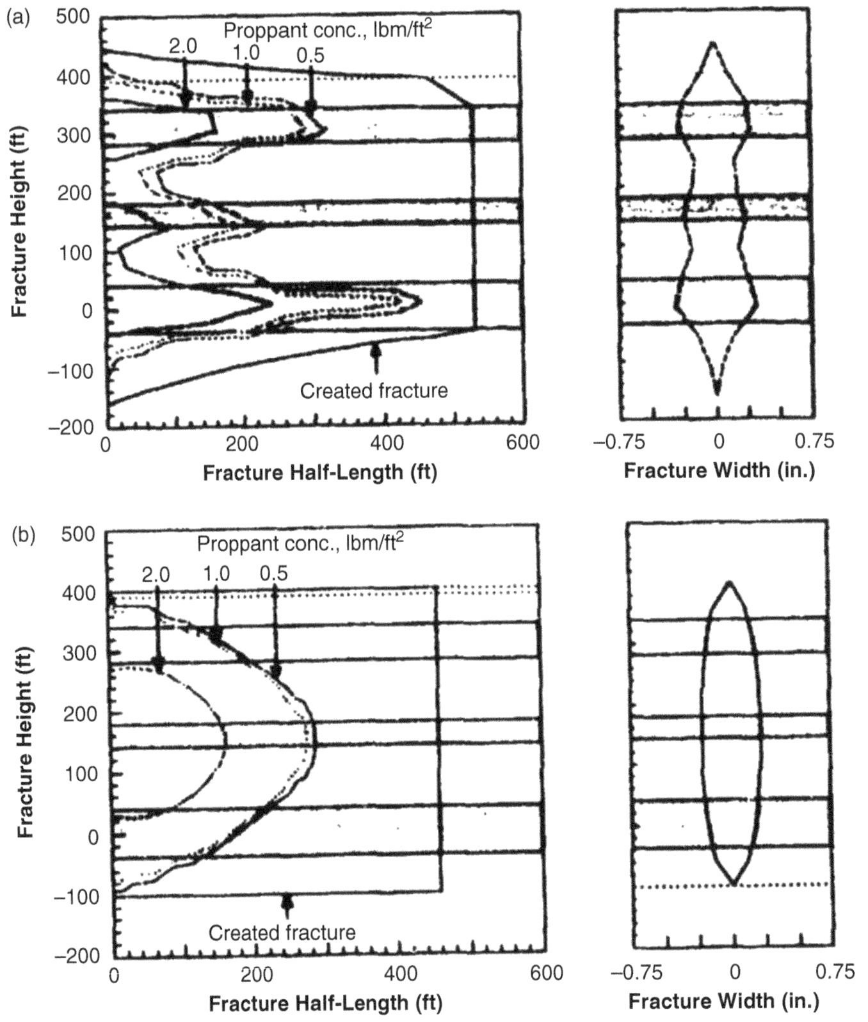

Fig. 9.22—Comparison of fracture profile between 3D and simplified 2D model: (a) computation using 3D model, (b) computation using 2D Perkins-Kern-Nordgren model (Rahim and Holditch 1993).

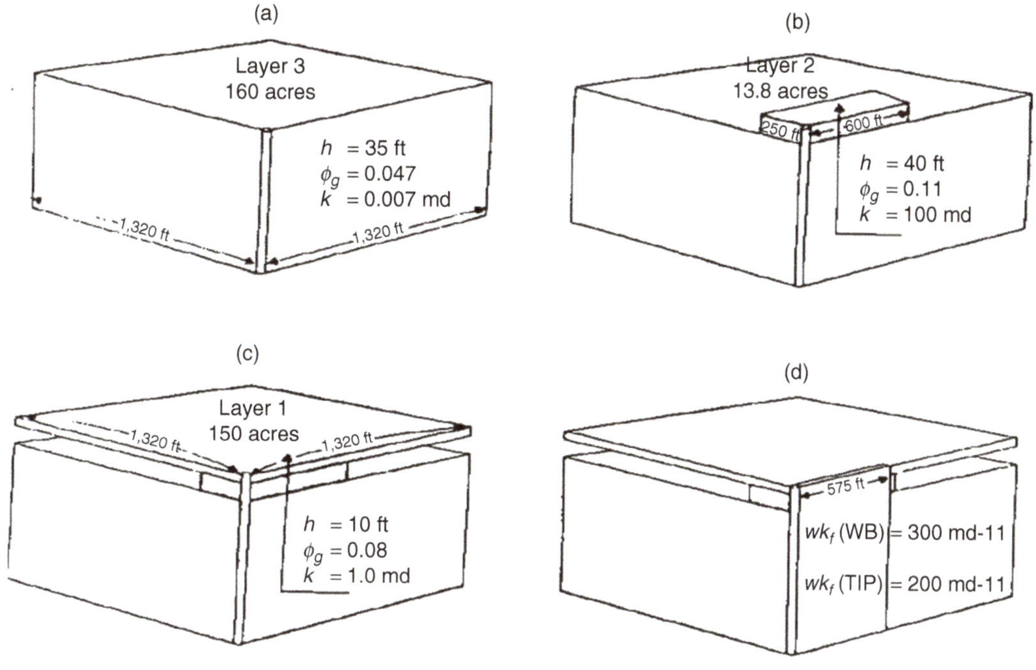

Fig. 9.23—Three-dimensional reservoir model (Holditch et al. 1987b).

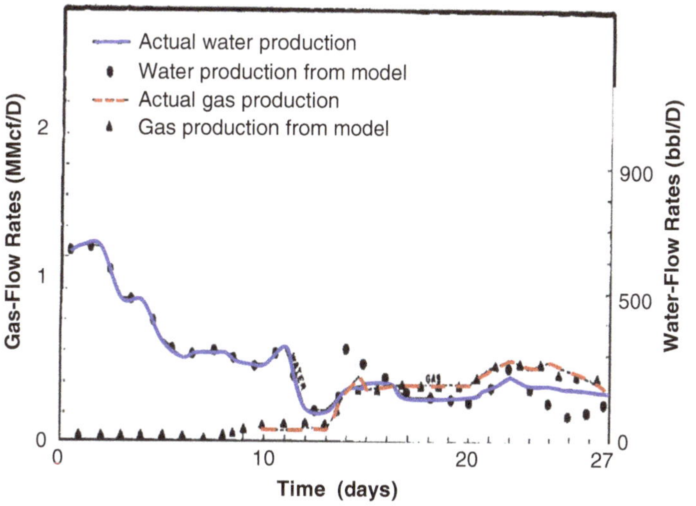

Fig. 9.24—Actual and forecast post-fracturing production (Holditch et al. 1987b).

Frantz et al. (1992) used a multilayer reservoir model in a low-permeability hydraulically fractured shale reservoir. They showed that a multilayer reservoir description provides better estimates of post-fracturing performance compared to a more conventional, single-layer reservoir description. Predictions with the single-layer models are 106% higher than those with the two-layer model. Gao et al. (1994) presented an analytical simulator capable of modeling the performance of commingled reservoirs. The reservoir model can have different initial pressures in different layers and allows each layer to be homogeneous or dual-porosity, to be hydraulically fractured, to have radial flow (with skin), or to be finite or infinite in extent. Jatmiko et al. (1994) used a deconvolution method to handle the multivariable rate, which enables the determination of the layered reservoir properties. To calculate the effective oil permeability, their work proposes a polynomial curve fitting to represent the whole pressure response of the drawdown test. Lolon et al. (2008) presented semianalytical solutions using the Laplace transform for wellbore pressure and fractional-flow-rate responses of commingled layered reservoirs without interlayer crossflow.

In summary, layered and geologically complex low-permeability gas reservoirs, together with complicated hydraulic fracturing propagation, make it necessary to build 3D reservoir models. Modeling a multilayer reservoir in three dimensions requires additional efforts and expertise in geologic analysis, sophisticated formation evaluation techniques, and hydraulic fracture reservoir modeling. However, through it all, insights are gained on how to optimize completions, stimulations, and operations to increase gas production and reserves.

9.7 Fieldwide Modeling of Tight Gas Reservoirs

Integrated fieldwide reservoir modeling has been useful in reservoir management and development planning of conventional oil and gas fields. The modeling of fields is not applied widely in development of tight gas reservoirs for two primary reasons:

- Drainage area of a tight gas well is smaller compared to conventional oil and gas wells, transient flow period will be years, and well spacing is dictated by the effective fracture length. This means that tight gas wells probably will not interfere like wells in conventional reservoirs. As such, modeling of single wells is sufficient to develop optimized development strategies.
- Computational challenges are a result of the need for a large number of grids surrounding hydraulic fractures, additional complicated physics and flow mechanisms inside and near hydraulic fractures as discussed in previous sections, and a large number of wells in whole-field simulations. However, there are certainly merits in conducting a fieldwide simulation study in highly heterogeneous tight gas reservoirs to meet certain business needs, such as infill drilling.

It seems that the only solution to computational challenges is to minimize the number of grids surrounding hydraulic fractures, because we should not cut short physics or the number of wells. This is probably the primary difference between fieldwide simulation of tight gas reservoirs and that of conventional reservoirs. Abacioglu et al. (2013) reviewed different methods used to simulate hydraulic fractures in large reservoir models and concluded that fine gridding in the fracture and near-wellbore region provides an accurate solution but is not practical for fieldwide simulation studies. They summarized six different methods available to approximate hydraulic fractures:

1. Use of a large wellbore radius or negative skin if the hydraulic fracture length is smaller than the wellbore grid; this method is simple to implement but results in radial flow
2. Increased permeability of coarse grids representing the fractures
3. Increased transmissibility for coarse grids representing the fractures
4. Treating a fracture as an infinite-conductivity line, but this method needs calibrations with fine-grid models
5. Increased well index and transmissibility by a multiplier, but this method cannot be applied in heterogeneous reservoirs
6. Local grid refinements (LGRs)

The authors then chose to use Method 6 for its accuracy and developed an LGR technique to allow for and automate the grid refinement of only coarse grids that have fractures and wells, as shown in **Fig. 9.25.** Gridding in Fig. 9.25 is not different from that in Fig. 9.1 if we just look at the drainage area, but it does improve computational efficiency because the fine grids are not needed throughout the whole reservoir. Abacioglu et al. (2013) conducted a comparison study of the different methods and proved that their LGR technique is accurate, especially in modeling early transient production. They conducted field case studies of a Cotton Valley Field in East Texas and generated pressure maps as shown in **Fig. 9.26.**

Gallego (2011) worked with a multidisciplinary group of experts at BP and developed frameworks and a decision tree to provide step-by-step guidelines for a selection of modeling frameworks to meet reservoir management objectives starting from the number of wells in the model. Frameworks 1 and 2 suggest single-well modeling, Frameworks 3 and 4 suggest multiwell models with up to 50 wells, and Framework 5 suggests a fieldwide model. The author then applied Framework 5 to build a fieldwide reservoir model to simulate the Moxa Arch development area in southwestern Wyoming, which is 12 miles wide and 39 miles long, with 1,500 active wells. Given the number of wells, the author used hybrid methods for hydraulic fracture representations: Method 1 (r_w and s) was used in old wells, Method 2 (k) was used in newer wells, and Method 6 (LGR) was used in infill wells. The author concluded that the estimated ultimate recovery for infill wells was predicted, drilling schedules were prioritized, and infill drilling and well spacing were determined on the basis of saturation maps, as shown in **Fig. 9.27.**

Abacioglu et al. (2009) chose to use Method 5 (tuning well index) by perforating multiple grids up to the length of a hydraulic fracture in the (I, J) plane in a commercial simulator. They found that the well index should vary with fracture conductivity, fracture half-length, formation permeability, and grid size. A fieldwide simulation study was conducted on a field of 54 sections with 52 wells

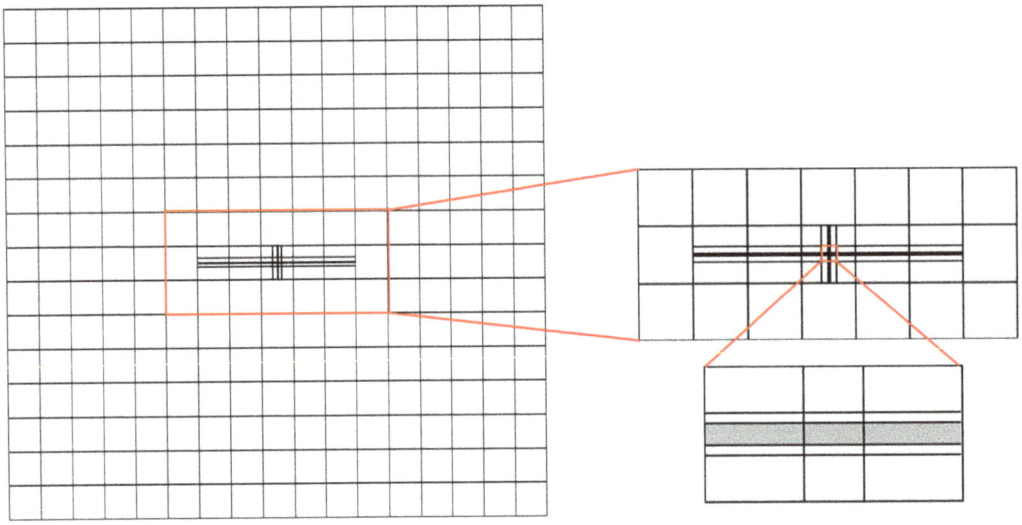

Fig. 9.25—Example LGR for a single vertical well in the fieldwide reservoir model (Abacioglu et al. 2013).

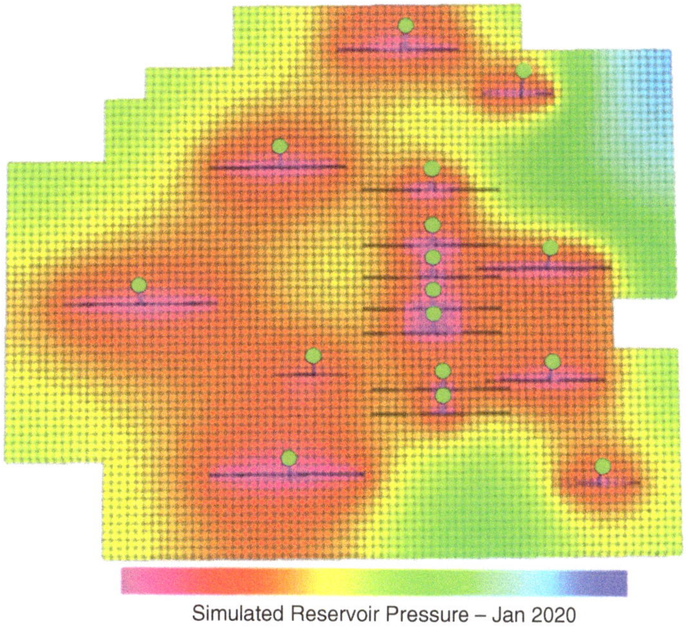

Fig. 9.26—Simulated pressure map of the reservoir (Abacioglu et al. 2013).

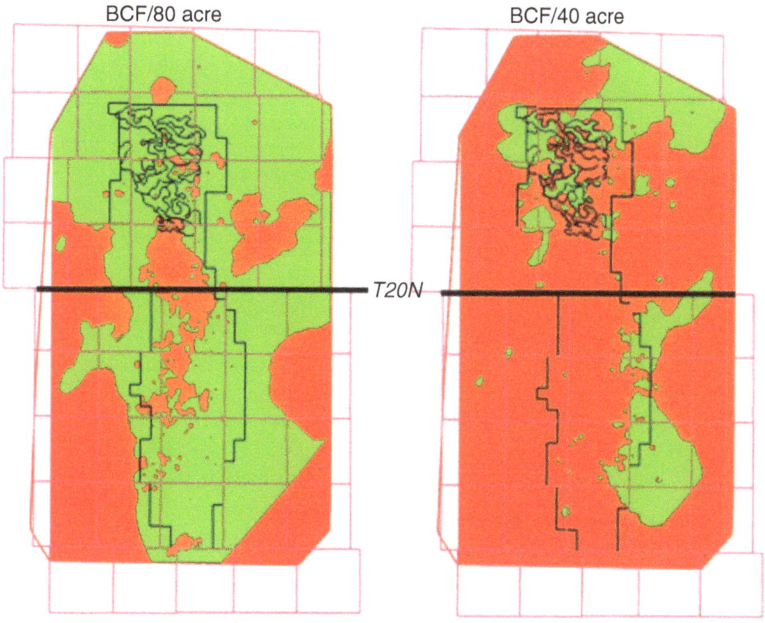

Fig. 9.27—Recoverable-gas maps for one of the model realizations, where green areas have gas volumes higher than the threshold (Gallego 2011).

and between 3 and 5 years of production history. The authors concluded that extra care must be taken to construct and check the models and crossflow must be implemented in the coarse-grid fieldwide model to simulate observed shut-in pressures. King et al. (2013) also reported that Method 6 using LGR is able to capture transient flow rates, while other methods could not. These authors developed spatially adaptive upscaling techniques to preserve the vertical discontinuities of the sands and also the heterogeneity within sands. Combining LGR with upscaling, they built a high-resolution reservoir model based on a high-resolution 3D geologic model, as shown in **Fig. 9.28.** Simulation was performed using an unstructured commercial simulator with accuracy and computational efficiency.

In summary, fieldwide modeling of tight gas reservoirs could be conducted, but extra care and effort must be taken because of complicated physics and mechanics inside and surrounding hydraulic fractures. Consideration should be given to the limitations and advantages of the six methods for hydraulic fracture representations, but it seems clear that a localized LGR should be used if possible. If a reservoir is highly heterogeneous vertically and laterally, upscaling techniques and geostatistical methods should be used in building reservoir models. Other than this, the steps for modeling multilayer reservoirs could be used in conducting fieldwide simulation studies with the purpose of meeting certain business needs and increasing gas recovery.

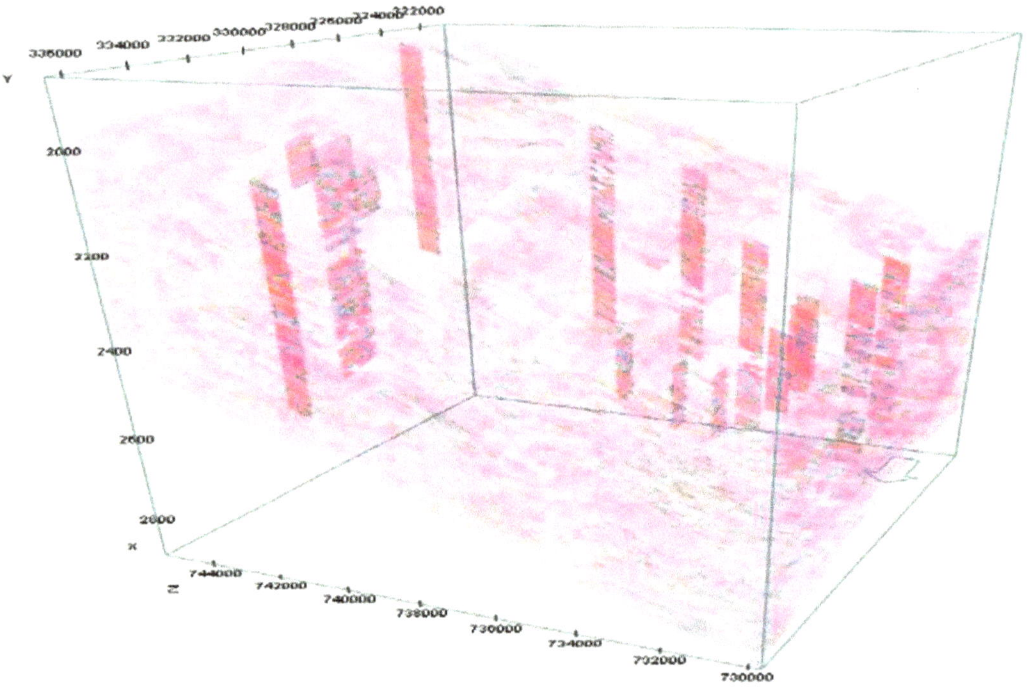

Fig. 9.28—Hydraulic fracture representations in 11 wells in an upscaled reservoir model (King et al. 2013).

9.8 Summary

In this chapter, we shared a proven methodology and reservoir simulation technologies for modeling tight gas wells and reservoirs using a finite-difference reservoir simulator with a focus on six topics of interest for practicing engineers and researchers:

- Fracturing-fluid cleanup
- Non-Darcy flow
- Changing fracture conductivity
- Integration of post-fracturing flow and pressure data
- Multilayered formations
- Fieldwide simulation

In each section, we started with the physics and mechanisms, followed by mathematical equations, then implementation into reservoir simulators, and then example results and analyses. This is a solid starting point for conducting numerical simulation of tight gas wells and reservoirs. With insights gained from modeling and parametric studies, the drilling, completion, stimulation, and production of tight gas wells can be optimized.

9.9 Nomenclature

B = formation volume factor, RB/STB

B_{ga} = gas formation volume factor evaluated at p_a, bbl/Mscf

c_{ta} = total compressibility evaluated at p_a, psi^{-1}

h = net formation thickness, ft

k = formation permeability, md

k_f = fracture permeability, md

L_f = fracture half-length, ft

m = slope of p_{ws} vs. $\log\left[\left(t_p+\Delta t\right)/\Delta t\right]$, psi/cycle

m' = slope of p_{wf} vs. $\sqrt{\Delta t}$ plot or p_{ws} vs. $\sqrt{t_p+\Delta t}-\sqrt{\Delta t}$ plot, psi/ $\sqrt{\text{hours}}$

p_{1hr} = shut-in BHP at 1-hour shut-in time on straight line in buildup-test plot, psi

P_{FCN} = pressure function (psi^2/cp)/(Mcf/D)

p_i = initial reservoir pressure, psi

p_{wf} = flowing BHP, psi

p_{ws} = shut-in BHP, psi

q_g = gas-flow rate, Mscf/D

r_w = wellbore radius, ft

s = skin factor, dimensionless

t_p = producing time, hours

w = fracture width, ft

L_f = half-fracture length, ft

ϕ = total porosity, fraction

Δt = shut-in time, hours

μ_a = viscosity evaluated at p_a, cp

9.10 References

Abacioglu, Y., Sebastian, H. M., and Bonney, K. E. 2013. Case History of Using Integrated Reservoir Modeling Workflow for Tight Gas Reservoirs. Presented at the SPE Unconventional Gas Conference and Exhibition, Muscat, Oman, 28–30 January. SPE-163989-MS. https://doi.org/10.2118/163989-MS.

Abacioglu, Y. Sebastian, H. M., and Oluwa, J. B. 2009. Advancing Reservoir Simulation Capabilities for Tight Gas Reservoirs. Presented at the SPE Rocky Mountain Petroleum Technology Conference, Denver, Colorado, 14–16 April. SPE-122793-MS. https://doi.org/10.2118/122793-MS.

Ahmed, U., Abou-Sayed, A. S., and Jones, A. H. 1979. Experimental Evaluation of Fracturing Fluid Interaction with Tight Reservoir Rocks and Propped Fractures. Presented at the Symposium on Low Permeability Gas Reservoirs. Denver, Colorado, 20–22 May. SPE-7922-MS. https://doi.org/10.2118/7922-MS.

Alvarez, C. H., Holditch, S. A., and McVay, D. A. 2002. Effects of Non-Darcy Flow on Pressure Transient Analysis of Hydraulically Fractured Gas Wells. Presented at the SPE Annual Technical Conference and Exhibition, San Antonio, Texas, 29 September–2 October. SPE-77468-MS. https://doi.org/10.2118/77468-MS.

Arps, J. J. 1945. Analysis of Decline Curves. *Transactions of the AIME* **160** (1): 228–247. SPE-945228-G. https://doi.org/10.2118/945228-G.

Balhoff, M. T. and Miller, M. J. 2005. An Analytical Model for Cleanup of Yield-Stress Fluids in Hydraulic Fractures. *SPE J.* **10** (1): 5–12. SPE-77596-PA. https://doi.org/10.2118/77596-PA.

Bostic, J. N., Agarwal, R. G., and Carter, R. D. 1980. Combined Analysis of Post-fracturing Performance and Pressure Buildup Data for Evaluating an MHF Gas Well. *J Pet Technol* **32** (10): 1711–1719. SPE-8280-PA. https://doi.org/10.2118/8280-PA.

CarboCeramics.com. 2007. Technical sheet on conductivity data, https://carboceramics.com/news-and-resources/resources/technical-data-sheets/CARBOLITE (accessed 20 October 2007).

Cinco-Ley, H. and Samaniego-V., F. 1981. Transient Pressure Analysis for Fractured Wells. *J Pet Technol* **33** (9): 1749–1766. SPE-7490-PA. https://doi.org/10.2118/7490-PA.

Cooke, C. E., Jr. 1973. Conductivity of Fracture Proppants in Multiple Layers. *J Pet Technol* **25** (9): 1101–1107. SPE-4117-PA. https://doi.org/10.2118/4117-PA.

Cooke, C. E., Jr. 1975. Effect of Fracturing Fluids on Fracture Conductivity. *J Pet Technol* **27** (10): 1273–1282. SPE-5114-PA. https://doi.org/10.2118/5114-PA.

Forchheimer, P. 1901. Wasserbewegung durch boden. *Zeitschrift des Vereines deutscher Ingenieure* **45** (50): 1781–1788.

Fetkovich, M. J. 1980. Decline Curve Analysis Using Type Curves. *J Pet Technol* **32** (6): 1065–1077. SPE-4629-PA. https://doi.org/10.2118/4629-PA.

Frantz, J. H., Gatens, J. M. III, and Hopkins, C. W. 1992. Using a Multilayer Reservoir Model to Describe a Hydraulically Fractured, Low-Permeability Shale Reservoir. Presented at the SPE Annual Technical Conference and Exhibition, Washington, D.C., 4–7 October. SPE-24885-MS. https://doi.org/10.2118/24885-MS.

Gallego, F. 2011. Tight Gas Modeling Frameworks for Improved Reservoir Management and its Application to the Moxa Development Area. Presented at the SPE Middle East Unconventional Gas Conference and Exhibition, Muscat, Oman 31 January–2 February. SPE-142822-MS. https://doi.org/10.2118/142822-MS.

Gao, C. 1985. Determination of Individual Layer Properties by Layer-by-Layer Well Tests in Multilayer Reservoirs with Crossflow. SPE-15321-MS.

Gao, C., Jochen, J. E., and Lee, W. J. 1994. Interpretation of Field Pressure and Production Data with a New Analytical Layered Reservoir Simulator. Presented at the Permian Basin Oil and Gas Recovery Conference, Midland, Texas, 16–18 March. SPE-27667-MS. https://doi.org/10.2118/27667-MS.

Gdanski, R. D., Weaver, J. D., Slabaugh, B. F. et al. 2005. Fracture Face Damage – It Matters. Presented at SPE European Formation Damage Conference, Sheveningnen, The Netherlands, 25–27 May. SPE-94649-MS. https://doi.org/10.2118/94649-MS.

Geertsma, J. 1974. Estimating the Coefficient of Inertial Resistance in Fluid Flow Through Porous Media. *SPE J.* **14** (5): 445–450. SPE-4706-PA. https://doi.org/10.2118/4706-PA.

Holditch, S. A. 1979. Factors Affecting Water Blocking and Gas Flow From Hydraulically Fractured Gas Wells. *J Pet Technol* **31** (12): 1515–1524. SPE-7561-PA. https://doi.org/10.2118/7561-PA.

Holditch, S. A. 2005. Well Stimulation Lecture Notes. Texas A &M University, College Station, Texas, USA.

Holditch, S. A. and Lee, W. J. 1979. Fracture Evaluation with Pressure Transient Tests in Low-permeability Gas Reservoirs. Part II: Field Examples. Presented at the Symposium on Low Permeability Gas Reservoirs, Denver, Colorado, 20–22 May. SPE-7930-MS. https://doi.org/10.2118/7930-MS.

Holditch, S. A. and Morse, R. A. 1976. The Effects of Non-Darcy Flow on the Behavior of Hydraulically Fractured Gas Wells (Includes Associated Paper 6417). *J Pet Technol* **28** (10): 1169–1179. SPE-5586-PA. https://doi.org/10.2118/5586-PA.

Holditch, S. A., Lee, W. J., and Gist, S. R. 1983. An Improved Technique for Estimating Permeability, Fracture Length, and Fracture Conductivity From Pressure-Buildup Tests in Low-Permeability Gas Wells. *J Pet Technol* **35** (5): 981–990. SPE-9885-PA. https://doi.org/10.2118/9885-PA.

Holditch, S. A., Robinson, B. M., and Whitehead, W. S. 1987a. Prefracture and Postfracture Formation Evaluation Necessary To Characterize the Three-Dimensional Shape of a Hydraulic Fracture. *SPE Form Eval* **2** (4): 523–534. SPE-14086-PA. https://doi.org/10.2118/14086-PA.

Holditch, S. A., Robinson, B. M., and Whitehead, W. S. 1987b. The Analysis of Complex Travis Peak Reservoirs in East Texas. Presented at the Low Permeability Reservoirs Symposium, Denver, Colorado, 18–19 May. SPE-16427-MS. https://doi.org/10.2118/16427-MS.

Hopkins, C. W., Lancaster, D. E. and Foster, J. M. 1991. The Use of Injection/Falloff Tests and Pressure Buildup Tests to Evaluate Fracture Geometry and Post-Stimulation Well Performance in the Devonian Shales. Presented at the SPE Eastern Regional Meeting, Lexington, Kentucky, 22–25 October. SPE-23433-MS. https://doi.org/10.2118/23433-MS.

Jatmiko, W., Daltaban, T. S., and Archer, J. S. 1994. Determination of Relative Permeability From Well Testing in Multi-Layer Reservoirs. Presented at the SPE Asia Pacific Oil and Gas Conference, Melbourne, Australia, 7–10 November. SPE-28752-MS. https://doi.org/10.2118/28752-MS.

Katz, D. L., Cornell, D., Kobayashi, R. et al. 1959. *Handbook of Natural Gas Engineering*, first edition. New York, New York: McGraw Hill.

King, M. J. Abacioglu, Y., Ballin, P. R. et al. 2013. Multiscale Grid Design and Upscaling for a Full Field Tight Gas Reservoir Model on a Next Generation Simulator. Presented at the SPE Unconventional Gas Conference and Exhibition, Muscat, Oman, 28–30 January. SPE-163993-MS. https://doi.org/10.2118/163993-MS.

Lee, W. J. and Holditch, S. A. 1979. Fracture Evaluation with Pressure Transient Testing in Low-Permeability Gas Reservoirs. Part I: Theoretical Background. Presented at the Symposium on Low Permeability Gas Reservoirs, Denver, Colorado, 20–22 May. SPE-7929-MS. https://doi.org/10.2118/7929-MS.

Lee, W. J. and Holditch, S. A. 1981. Fracture Evaluation with Pressure Transient Testing in Low-Permeability Gas Reservoirs. *J Pet Technol* **33** (9): 1776–1792. SPE-9975-PA. https://doi.org/10.2118/9975-PA.

Lee, W. J., Retna, T., Holditch, S. A. et al. 1984. Analyzing Gas Well Buildup Tests with Changing Wellbore Storage in Tight Gas. Presented at the SPE Unconventional Gas Recovery Symposium. Pittsburgh, Pennsylvania, 13–15 May. SPE-12845-MS. https://doi.org/10.2118/12845-MS.

Lolon, E. P., Archer, R. A., Ilk, D. et al. 2008. New Semi-Analytical Solutions for Multilayer Reservoirs. Presented at the CIPC/SPE Gas Technology Symposium 2008 Joint Conference, Calgary, Alberta, Canada, 16–19 June. SPE-114946-MS. https://doi.org/10.2118/114946-MS.

Lolon, E. P., Chipperfield, S. T., McVay, D. A. et al. 2004. The Significance of Non-Darcy and Multiphase Flow Effects in High-Rate, Frac-Pack Gas Completions. Presented at the SPE Annual Technical Conference and Exhibition, Houston, Texas, 26–29 September. SPE-90530-MS. https://doi.org/10.2118/90530-MS.

Lolon, E. P., McVay, D. A., and Schubarth, S. K. 2003. Effect of Fracture Conductivity on Effective Fracture Length. Presented at the SPE Annual Technical Conference and Exhibition, Denver, Colorado, 5–8 October. SPE-84311-MS. https://doi.org/10.2118/84311-MS.

May, E. A., Britt, L. K. and Nolte, K. G. 1997. The Effect of Yield Stress on Fracture Fluid Cleanup. Presented at the SPE Annual Technical Conference and Exhibition, San Antonio, Texas, 5–8 October. SPE-38619-MS. https://doi.org/10.2118/38619-MS.

Montgomery, K. T. and Berthelot, J. M. 1990. Effects of Fracture Fluid Invasion on Cleanup Behavior and Pressure Buildup Analysis. Presented at the SPE Annual Technical Conference and Exhibition, New Orleans, Louisiana, 23–26 September. SPE-20643-MS. https://doi.org/10.2118/20643-MS.

Palacio, J. C. and Blasingame, T. A. 1993. Decline-Curve Analysis With Type Curves - Analysis of Gas Well Production Data. Presented at the Low Permeability Reservoirs Symposium, Denver, Colorado, 26–28 April. SPE-25909-MS. https://doi.org/10.2118/25909-MS.

Prats, M. 1961. Effect of Vertical Fractures on Reservoir Behavior-Incompressible Fluid Case. *SPE J.* **1** (2): 105–118. SPE-1575-G. https://doi.org/10.2118/1575-G.

Rahim, Z. and Holditch, S. A. 1993. Using a Three-Dimensional Concept in a Two-Dimensional Model to Predict Accurate Hydraulic Fracture Dimensions. Presented at the SPE Eastern Regional Meeting, Pittsburgh, Pennsylvania, 2–4 November. SPE-26926-MS. https://doi.org/10.2118/26926-MS.

Rushing, J. A., Sullivan, R. B., and Blasingame, T. A. 2005. Post-Fracture Performance Diagnostics for Gas Wells with Finite-Conductivity Vertical Fractures. Presented at the SPE Eastern Regional Meeting. Morgantown, West Virginia, 14–16 September. SPE-97972-MS. https://doi.org/10.2118/97972-MS.

Siddiqui, M. A., Nasr-El-Din, H. A., Al-Anazi, M. S. et al. 2004. Formation Damage in Gas Sandstone Formations by High-temperature Borate Gels Due to Long Term Shut-in Periods. Presented at the SPE/DPE Symposium on Improved Oil Recovery, Tulsa, Oklahoma, 17–21 April. SPE-89476-MS. https://doi.org/10.2118/89476-MS.

StimLab Consortia. 2006. Fracture conductivity data, https://corelab.com/stimlab/proppant-consort (accessed 20 November 2006).

Sullivan, R. B., Lee, W. J., and Holditch, S. A. 1987. Pressure Transient Response in Multilayer Gas Reservoirs Containing Hydraulic Fractures. Presented at the Low Permeability Reservoirs Symposium, Denver, Colorado, 18–19 May. SPE-16399-MS. https://doi.org/10.2118/16399-MS.

Tannich, J. D. 1975. Liquid Removal from Hydraulically Fractured Gas Wells. *J Pet Technol* **27** (11): 1309–1317. SPE-5113-PA. https://doi.org/10.2118/5113-PA.

Tek, M. R., Coats, K. H., and Katz, D. L. 1962. The Effect of Turbulence on Flow of Natural Gas Through Porous Reservoirs. *J Pet Technol* **14** (7): 799–806. SPE-147-PA. https://doi.org/10.2118/147-PA.

Van Poollen, H. K. 1957. Do Fracture Fluids Damage Productivity? *Oil Gas J.* (19 April 1957).

Voneiff, G. W., Robinson, B. M., and Holditch, S. A. 1996. The Effects of Unbroken Fracture Fluid on Gas Well Performance. *SPE Prod & Fac* **11** (4): 223–229. SPE-26664-PA. https://doi.org/10.2118/26664-PA.

Wang, J. Y., Holditch, S. A., and McVay, D. A. 2010. Modeling of Fracture Fluid Cleanup in Tight Gas Wells. *SPE J.* **15** (3): 783–793. SPE-119624-PA. https://doi.org/10.2118/119624-PA.

Wattenbarger, R. A. and Ramey, H. J., Jr. 1969. Well Test Interpretation in Vertically Fractured Gas Wells. *J Pet Technol* **21** (5): 625–632. SPE-2155-PA. https://doi.org/10.2118/2155-PA.

Yi, X. 2004. Model for Displacement of Herschel-Bulkley Non-Newtonian Fluid by Newtonian Fluid in Porous Media and Its Application in Fracturing Fluid Cleanup. Presented at the SPE International Symposium and Exhibition on Formation Damage Control, Lafayette, Louisiana, 18–20 February. SPE-86491-MS. https://doi.org/10.2118/86491-MS.

Chapter 10

Drilling and Completion Design for Fracturing

Stephen A. Holditch

The material comprising Chapter 6 is partially taken from Chapter 12 in SPE Monograph 12 (Gidley et al. 1989), titled *Recent Advances in Hydraulic Fracturing*. SPE Monograph 12 was published in 1989, so the content is not all that recent. However, the information in Chapter 12 is still relevant to drilling vertical wells in tight gas reservoirs and pumping fracturing treatments that are also vertical. If we were to discuss how to drill and complete horizontal wells, that information would be somewhat different, and it is not included in this chapter. You can find such information in other books.

10.1 Drilling Aspects

Drilling personnel are sometimes judged by how fast they can drill a well to a specified depth. However, when one is drilling a well that will be completed and fracture treated in a low-permeability reservoir, the speed of the rig personnel is of secondary importance. Of primary importance is drilling a gauge borehole. If the borehole is in gauge, it substantially improves the chances of obtaining both an accurate suite of openhole geophysical logs and a satisfactory primary cementing job. When an accurate set of logs is obtained so that the potential producing zones can be easily identified, and when the primary cementing operation successfully isolates the potential producing zones, most completion problems will have been eliminated.

To design a casing program for the drilling operation, an engineer needs to estimate several completion parameters before spudding the well. In consultation with a geologist and a reservoir engineer, the engineer designing the fracturing treatment should estimate

- Values for depth to the pay zone
- Initial reservoir pressure
- Fracture gradient
- Created fracture height
- Young's modulus of the pay zone
- Optimal fracturing-fluid viscosity

With estimates of these parameters, an engineer can calculate values of surface injection pressure during a fracturing treatment for a variety of tubular configurations. From these calculations, a preliminary completion strategy can be chosen, and, from this strategy, the casing string and bit program can be designed. It is imperative to design a casing string large enough and strong enough to allow the proper size of fracturing treatment to be performed at the optimal injection rate. Because virtually all low-permeability wells must be fracture treated to provide economic flow rates and ultimate recoveries, the entire drilling program must be centered on the fracture treatment. If a few dollars are saved by running small casing or casing that is not strong enough to withstand fracturing pressures and the ability to fracture treat the well successfully is restricted, then the overall profit from the well might be greatly reduced or completely eliminated.

Perhaps the most important contribution to formation evaluation from drilling operations is a gauge borehole. An accurate log evaluation is a necessity in deciding how to complete a low-permeability reservoir. Because the logs are affected by hole size and mud-filtrate invasion and because the hole size is often a function of the mud system and bit hydraulics, the drilling-mud system should be monitored closely. In most low-permeability reservoirs, formation damage near the wellbore caused by mud-filtrate invasion is not of major importance. Shallow damage around the wellbore can easily be removed or bypassed with stimulation treatment. Therefore, the mud system and bit hydraulics should be designed to minimize hole washout problems and to maximize the chances of obtaining a gauge borehole and thus an accurate log suite.

Another important aspect of successful stimulation is the ability to pump the fracture fluid into specified intervals behind the casing. This can be accomplished only if the primary cement job isolates the multiple intervals. It is well-understood by the industry that, to maximize the chances of obtaining an acceptable primary cementing job, the borehole must be in gauge. If the borehole is in gauge and the casing can be both rotated and reciprocated during the cementing operation, the primary cementing operation will usually be successful.

In summary, it is always of concern that a well be drilled for the least feasible cost while safety regulations are observed. The term *least cost* is usually synonymous with *fastest drilling*, but, when all costs including evaluation and completion costs are considered, fast drilling times might not necessarily mean that the total well costs are minimized. Of primary importance to the success of a completion are

- An accurate log suite
- An effective primary cementing job

Both of these factors can be accomplished with a high degree of certainty if the drilling personnel drill a gauge borehole while minimizing mud-filtrate invasion into the potential producing intervals.

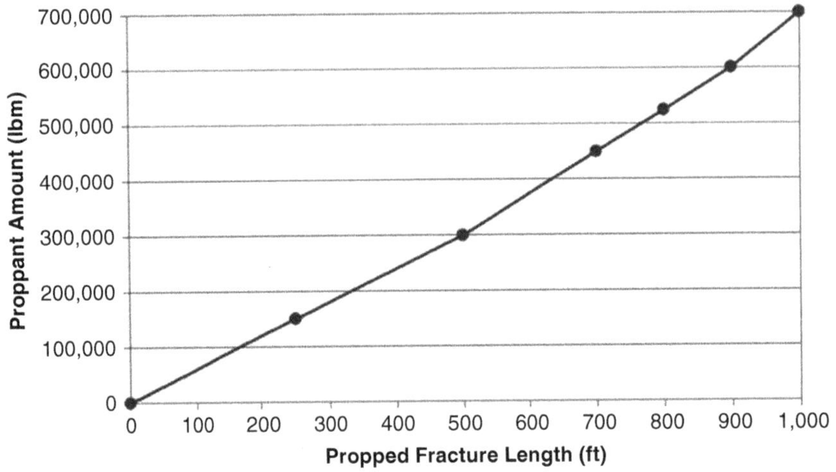

Fig. 10.1—Example of the proppant amount required for different propped lengths.

10.2 Completion Aspects

Tight gas reservoirs have been defined as formations whose expected permeability is 0.1 md or less. We recognize that permeability is distributed log-normally and that there will be some parts of the reservoir with high permeability (greater than 0.1 md) and some with low permeability (less than 0.1 md), but the median value of permeability will be 0.1 md or less. The geologic environment (deposition and diagenesis) are the main factors that affect reservoir quality. These factors, along with natural-fracture systems, affect the permeability, both vertically and horizontally, within the reservoir. Because these formations have low permeability by definition, we have to fracture treat the formations to achieve commercial flow rates and recoveries.

If Darcy's law were the only consideration, many blanket tight gas reservoirs could be developed using horizontal wells. However, because many tight gas reservoirs are sandstones, the abrasive nature of the rock rapidly wears out drill bits and prevents development through use of horizontal wells. In lenticular reservoirs or other reservoirs that are compartmentalized by faults or depositional features, it is also not wise to drill horizontal wells. As such, most tight-gas-sand reservoirs are developed using vertical wells.

In tight gas reservoirs, long propped hydraulic fractures are required to efficiently drain the reservoir and produce enough gas to make the development profitable. By *long propped fracture*, we mean the propped fracture length is greater than 50% of the drainage radius.

In lenticular reservoirs, the drainage radius is defined by the size of the lens. In reservoirs such as the Mesa Verde in the Rocky Mountain basins, the completion will commonly involve perforating multiple sand lenses and pumping small fracturing treatments in each lens using a diversion technique. The size of the treatment is dictated by the perceived size of the lens.

In blanket sands, it is necessary to pump large volumes of fracturing fluid and proppant. The size of the treatment is dictated by the perceived drainage area, which is dictated by the well spacing. In most blanket sands, there are no or few horizontal permeability barriers, although some of the blanket sands can be compartmentalized by faults.

In summary, the fracturing treatment design engineer should work closely with the geoscientists to estimate the drainage area of the tight gas sand to be fracture treated. When the drainage size and shape have been estimated, a treatment should be designed to penetrate the optimal distance in the reservoir. This should typically be a length of 50% of the drainage area or more. Because large fracturing treatments are required, the cost of the fracturing treatments can be 10 to 50% of total well costs.

Hydraulic fracturing designs have advanced significantly since the industry began using 3D fracture design models to design treatments. These models, along with better data collection, have led to a better understanding and modeling of the physical processes that occur as the fracture is created and grows in the formation. Improved diagnostic methods and better models have led to more-accurate estimates of fracture dimensions. **Fig. 10.1** shows how a model can be used to estimate how much proppant is required to develop a specific propped fracture length for a given set of input data. This graph was the result of a case study for wells in northern Mexico by Robinson et al. (1998).

Similar graphs can be developed for the required fluid volume and can then be combined with information about equipment and chemical needs to formulate a reasonable cost estimate for developing certain propped lengths in specific situations. As discussed in Part 3: Hydraulic Fracturing in Vertical Wells of this book series, the engineer needs to optimize the fracturing treatment along with drainage area and well spacing.

10.3 Casing and Wellbore Configuration

Fracturing treatment conditions should be considered in the casing design. Many times, the fracturing treatment will be pumped down the casing or a live annulus will be used to monitor the bottomhole pressure (BHP) during the fracturing treatment. In those situations, the casing must be designed to withstand the full fracturing BHP. In other situations, the fracturing treatment will be pumped down tubing, and, to minimize the burst pressure on the top joint of tubing, backup pressure will be held on the casing or tubing annulus. In virtually every situation, pressure will be exerted on the casing at the surface; therefore, the burst pressure of the casing string should be designed to withstand the maximum pressure expected during the fracturing treatment.

10.4 Cementing

To maximize the chances of success, each well should be completed so that the hydraulic fracture can be initiated in the desired reservoir. The fracture might eventually propagate out of zone; however, the design engineer should minimize the chances of out-of-zone migration at or near the wellbore by obtaining a primary cement seal around the production casing or liner.

To determine whether the primary cementing operation was successful, a cement-bond log can be run to estimate the quality of the cement bond near the zone of interest. Many papers are available to explain the use and interpretation of cement-evaluation logs

(Brown et al. 1970; Cooke 1979; Froelich et al. 1982; Grosmangin et al. 1961; Pickett 1966). Special care should be taken to ensure that the log data are not misinterpreted to conclude that a poor primary cement bond exists when the problem is actually a microannulus or poor jog data (Fitzgerald 1985).

10.5 Perforating

Perforations are required to connect the wellbore with the formation so a fracturing treatment can be initiated in the proper location. The main concerns involving perforations are safety and obtaining the correct number of perforations in the proper formations. The safety aspect refers mainly to well control. In highly geopressured areas or highly depleted formations, perforation might best be accomplished using a through-tubing gun with the wellhead installed. By using a proper size of lubricator and perforating the well underbalanced, one should maximize the chances of obtaining penetration through the casing and the cement and into the formation. A well would normally be perforated through tubing in an underbalanced condition when the fracturing treatment is going to be pumped down the tubing.

In a normally pressured reservoir that will be fracture treated down the casing, it is usually better to perforate the formation overbalanced with a casing gun. Clean fluid should be circulated into the wellbore before perforating to minimize the amount of debris left in the perforation tunnel. In the following, we present a general discussion on perforating. However, a more detailed discussion can be found in Part 3: Hydraulic Fracturing in Vertical Wells of this book series.

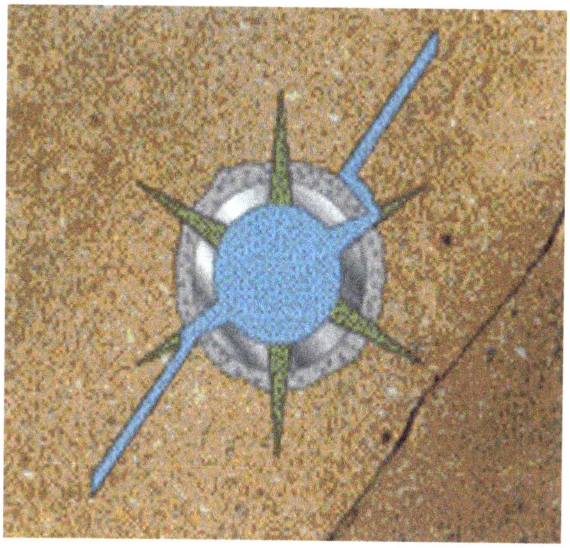

Fig. 10.2—Using 60° perforation phasing to increase connectivity with the fracture.

To state it more formally: "The objective of perforating is to provide a conductive flow path from the formation, through the cement sheath and casing, and into the completion tubulars. In a fracture-stimulated reservoir, the perforations connect the fracture with the wellbore." (Van Gijtenbeek and Pongratz 2004). Perforations provide communication among the fracture, the wellbore, and the reservoir, as shown in **Fig. 10.2.** Proper perforating design for hydraulic fracturing will allow the designed treatment to be pumped with minimal near-wellbore pressure drop.

In conventional reservoirs, the industry was concerned with perforating for production, trying to connect the best portions of the reservoir to the wellbore. Or, in some cases, keeping the perforations away from any gas/oil or water/oil contact. However, in a tight gas reservoir, perforating for production might not be the best way to ensure success in a fracturing treatment.

We often have to compute the net pressure in the fracture to understand how the hydraulic fracture is propagating in the formation. It is best to have a downhole gauge, but, when we have to work with surface pressure, we must calculate the frictional pressure drop in the tubing or casing, the hydrostatic head of the fluid, and the near-wellbore pressure drop. The near-wellbore pressure drop is caused by perforation friction and near-wellbore tortuosity.

A schematic of a typical vertical well containing a vertical hydraulic fracture is shown in **Fig. 10.3.** We can measure the surface pressure, calculate the perforation friction to determine the number of holes that are open, and estimate whether there is any pressure drop resulting from tortuosity.

The details of these calculations are given in Part 3: Hydraulic Fracturing in Vertical Wells of this book series.

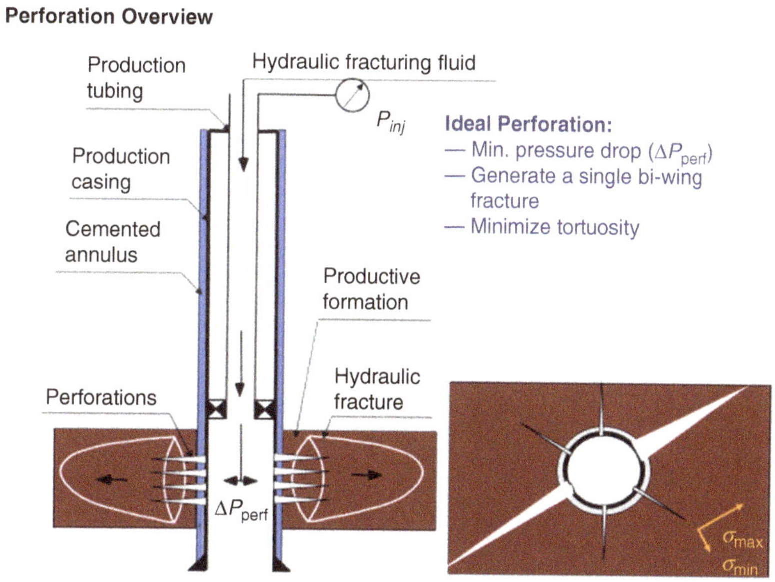

Fig. 10.3—Schematic of the fluid flow in a vertical well containing a vertical fracture.

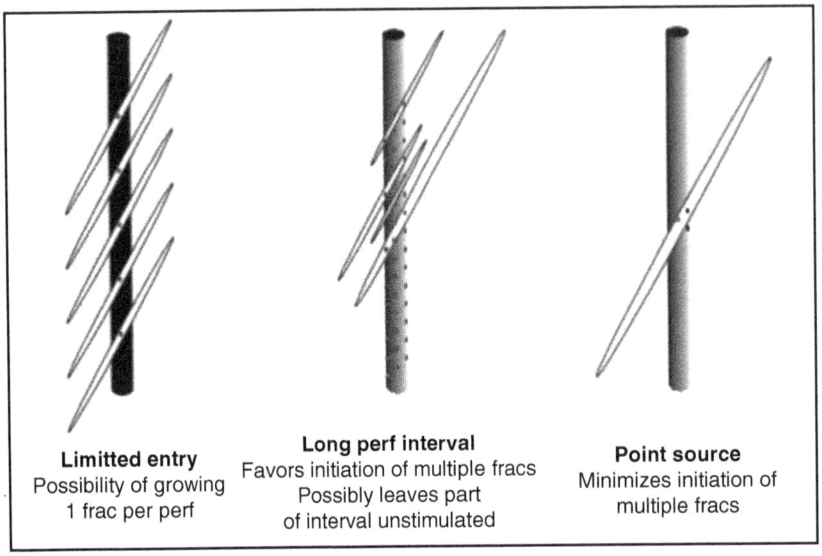

Fig. 10.4—Possible conditions when perforating a vertical wellbore (Lestz et al. 1999).

Fig. 10.3 is a typical well completion when fracture treating a single layer with a single-stage fracturing treatment. The most important perforation parameters determining success of a hydraulic fracturing treatment are

- Perforation interval
- Perforation shot density
- Perforation phasing
- Type of charge

The perforated interval length can be a very important factor in the success of a treatment. It is undeniable that no wellbore is straight, no wellbore is perfectly vertical, most wellbores spiral down as they are drilled, and most fractures are not perfectly vertical. Because of these conditions, it is probable that multiple vertical fractures will be created if the perforated interval is too long, as shown in **Fig. 10.4.**

There are several other issues to consider when choosing how to perforate the well so you can fracture treat the zone containing the natural gas. If you have a single zone, you will want to run a 3D model to determine how the hydraulic fracture will grow under several different perforating scenarios. In many cases, a point-source perforating scheme will provide the best opportunity to create a single fracture that can penetrate deeply into the formations, as shown in Fig. 10.4.

To stimulate several sand layers separated by shale layers that provide barriers to vertical fracture growth, the industry has used the limited-entry perforation technique, as shown in Fig. 10.4. Limited entry means you control the number of holes in each zone to pump fracture fluid into each layer. The example in Fig. 10.4 contemplates five fractures in five intervals.

If you are trying to fracture a single zone, then a point-source scheme is usually better than perforating the entire interval, especially if the interval is several hundred feet thick. The location of the perforations should be selected to maximize the growth of the hydraulic fracture into the formation. If you create multiple fractures, the average width of each fracture will be decreased, and the penetration of the fractures will also be decreased. The fractures push against each other, causing the width to be small compared to propagating a single fracture, and this could cause screenout problems when pumping larger-mesh proppants.

A simple chart, **Fig. 10.5,** can be used as a rough guide for how to perforate for fracturing. However, as stated previously, the design engineer should use a 3D reservoir description of the rock layers along with a 3D fracture-propagation model to determine where to perforate the well to achieve maximum fracture dimensions within the pay zone. It is simple to use the model on a trial-and-error basis to see what the model tells you about where to perforate. If you have a reasonable reservoir description and run the 3D fracture model correctly, you can solve the problem of where to perforate. Fortunately, the answer often is to use a point source in the center of the most productive layer.

Fig. 10.5 is a 2D chart that offers advice on how to select the perforation interval. We found that pay-zone thickness and number of separate fractures are the most important parameters for the interval selection. In Fig. 10.5, we have several options:

- Perforate the entire interval
- Perforate the most-porous zone
- Use point-source perforation or limited-entry techniques

If there is only one productive layer, we should consider whether the upper and lower boundaries are good barriers for fracture growth. If so, we can perforate either an entire interval if it is short or one of the most-porous zones. However, if the upper or lower boundaries are not likely to confine fracture growth, using the point-source approach should be considered.

In multilayer formations, we should consider the strength of the barrier between the layers. If the barrier is weak, only one layer should be perforated because the fracture will overcome the weak barrier and will cover all intervals. However, if the barrier is strong, each interval should be perforated so separate fractures will be created in each productive layer.

Number of Separate Fractures per Stage → / Pay-Zone Thickness ↓	1	2	3+
Thin	Entire interval	Entire interval in each layer	Limitted entry
Moderate	Most-porous zone	Point source in each layer	Limitted entry in zones with most gas-in-place
Thick	Most-porous zone	Point source in each layer	Limitted entry in zones with most gas-in-place

Fig. 10.5—Example of how to select the perforation interval.

The goal is to achieve good communication between the wellbore and the hydraulic fracture through the perforations. Good communication means excessive near-wellbore pressure drop is not created as a result of a tortuous path from inside the wellbore to inside the propagating fracture. To minimize the issues with tortuosity, the orientation of the perforations should be within 30° of the preferred fracture plane. For this discussion, only 60, 180, and 0° phasing are considered, as shown in **Fig. 10.6.** Phasing of 60° guarantees that some of the perforations will be within the necessary angle. If properly oriented, 180° phasing also satisfies this requirement. Even though 0° phasing might be significantly deviated from the preferred fracture plane, it should be used if perforation is carried out by a small-diameter through-tubing gun.

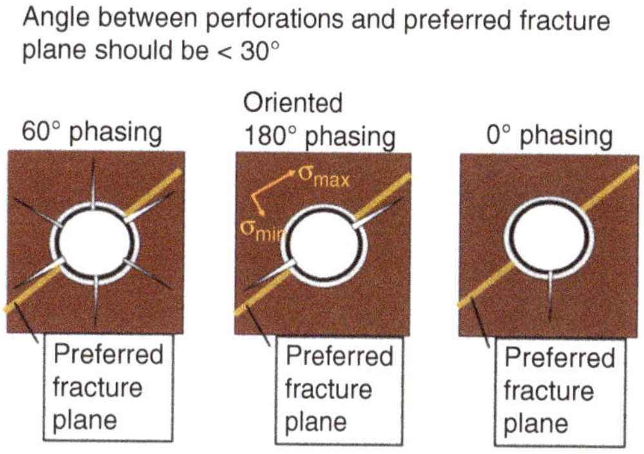

Fig. 10.6—Orientation of the perforations.

No well is perfectly vertical. As such, a perforation gun will always lie on the lower side of the casing. But perforation efficiency is a function of clearance (distance between the gun and the casing). In this case, the diameter of the gun is much smaller than that of the casing, so perforations toward the farther wall of the casing will not be efficient. Moreover, a small-diameter gun cannot be centralized in the hole, so 0° perforation is probably the best solution in this case. If casing is perforated through tubing with a small gun, then a 0° phasing and a magnetically decentralized gun should be used so the shots are facing the casing.

More information on perforating and an example calculation of perforation friction can be found in Part 3: Hydraulic Fracturing in Vertical Wells of this book series.

10.6 Completion Strategy

To complete a well successfully, the engineer should consider

- Investment required
- Number of pay intervals encountered by the borehole
- Desired producing rates
- Reserves in the various intervals
- Reservoir rock and fluid properties
- Stimulation requirements
- Sand-control requirements
- Future workover requirements
- Artificial-lift considerations
- Future secondary-recovery needs

The ideal completion is one that can be obtained for the lowest cost—considering both the initial completion cost and the subsequent operating costs—that meets or most nearly meets the demands placed on the well for most of its life. Although this definition is somewhat academic, it correctly implies that a prudent engineer will attempt to provide a functional completion for many years to come at the lowest possible cost to the operator.

Always of concern in the design of a completion is the number of producing zones that are separated vertically in the reservoir. To determine whether different producing intervals should be treated as a single reservoir, one must first determine whether these intervals will all be connected by a single hydraulic fracture. If a particular zone is separated from a second zone by a thin shale or silt layer with little stress contrast among the various intervals, then it can usually be assumed that a single hydraulic fracture will connect both intervals. In such a situation and when these different reservoir layers can be commingled with no damage to either layer, the well should be completed as if all the different layers were a single reservoir.

If two productive intervals are separated by a shale member 50 to 100 ft (15 to 300 m) thick, which will clearly be a barrier to fracture growth, then the decision to commingle these intervals must be based on the potential for reservoir damage and the governing rules from the government regulatory agency in charge of such matters. In essence, it is necessary for an engineer to analyze the logs and core to determine the number of distinctly different producing horizons that have been contacted by the borehole. Then, depending on the net hydrocarbon pay, hydrocarbon content, presence or lack of barriers to fracture growth among the various intervals, and government regulatory rules, a decision must be made concerning the number of pay zones to be completed and produced in the well.

10.7 Treatment Diversion

The efficient, simultaneous treatment of multiple sets of perforations spread over a long vertical section has been a problem in well stimulation for many years. Numerous treatment diversion methods have been used to treat multiple zones, with various degrees of effectiveness. Harrison (1972) presented histories of example applications for most diverting agents. He explained many of the techniques used since 1936 and included information on diverting both acid and fracturing treatments. The following sections briefly discuss most of the currently applied diverting techniques.

10.7.1 Limited Entry. One method for diverting a fracturing treatment is the limited-entry technique. This diverting method was first reported by Lagrone and Rasmussen (1963), who reported that the limited-entry technique was used in 363 wells in Texas and New Mexico. In the limited-entry technique, the number of perforations in a well is limited, and sufficient injection rate is provided for the restricted flow capacity of the perforations to divert the treatment to a greater portion of the perforated interval.

10.7.2 Multistage Treatment with Ball Sealers. The limited-entry technique has been used successfully in many situations to divert fracturing treatments; however, some intervals are best treated with a pseudolimited-entry technique in which the treatment stages are diverted with ball sealers. This technique requires a prescribed number of perforations to be placed in two or more intervals. A multistage fracturing treatment is then pumped, and a prescribed number of ball sealers are dropped between each stage. Such a technique has been described by Webster et al. (1965) and by Stipp and Williford (1968). These authors demonstrated that, with an effective cement bond surrounding the pipe and proper design of the perforating and treatment diversion system, it is possible to stimulate multiple sets of perforations effectively.

10.7.3 Packer-and-Bridge Plug. The most reliable technique for diverting fracturing treatments into multiple intervals is the packer-and-bridgeplug method. A typical treatment would involve perforating a lower zone, performing a fracturing treatment of that zone, setting a bridge plug above that particular interval, and then perforating and fracture treating the next zone. This process can be repeated several times before the well is cleaned out and all the zones previously fracture stimulated are commingled. In certain situations, the well can be completed as a dual completion. The main advantage of the packerand-bridge-plug method is that it is the most positive way to divert a fracturing treatment. When the method is performed correctly, each zone should be optimally stimulated, and income from the well should be maximized.

The disadvantages with this method are that a workover rig is required to move the packer-and-bridge-plug hardware in the well, the method is time-consuming, and mechanical problems can occur with the retrievable bridge plugs normally used to isolate intervals.

10.7.4 Baffles. Baffles installed in the casing string have been used for many years to divert stimulation treatments. **Fig. 10.7** shows a two-stage fracturing procedure that uses a casing baffle and a fracturing ball. The baffle is placed between the two intervals to be fracture stimulated. The lower zone can be fractured, then a ball can be dropped to sit in the baffle so the upper zone can be perforated and fracture treated. Flowing the well will remove the fracturing ball; then, the baffle can be drilled up with a normal rock bit. This technique has been widely used in shallow areas in Texas and the Appalachian Basin.

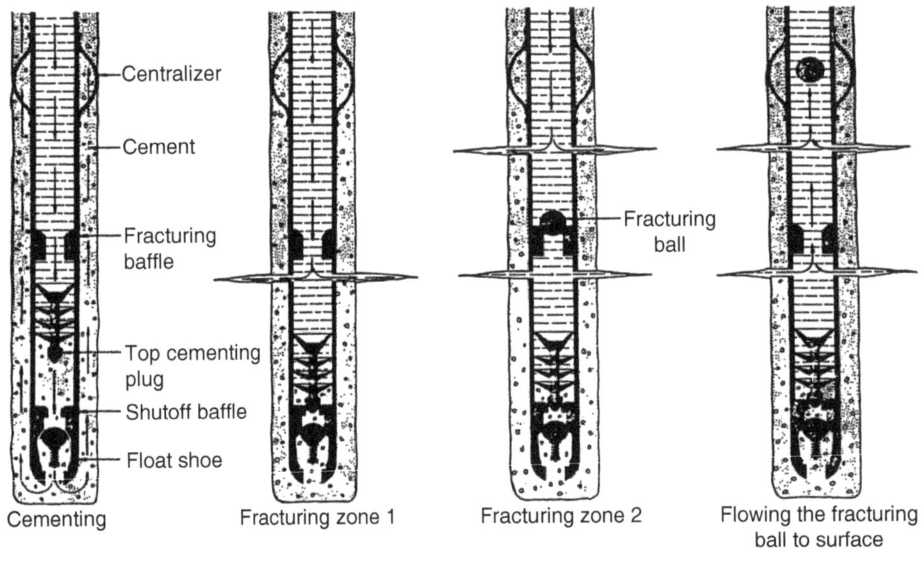

Fig. 10.7—Two-step fracturing procedure.

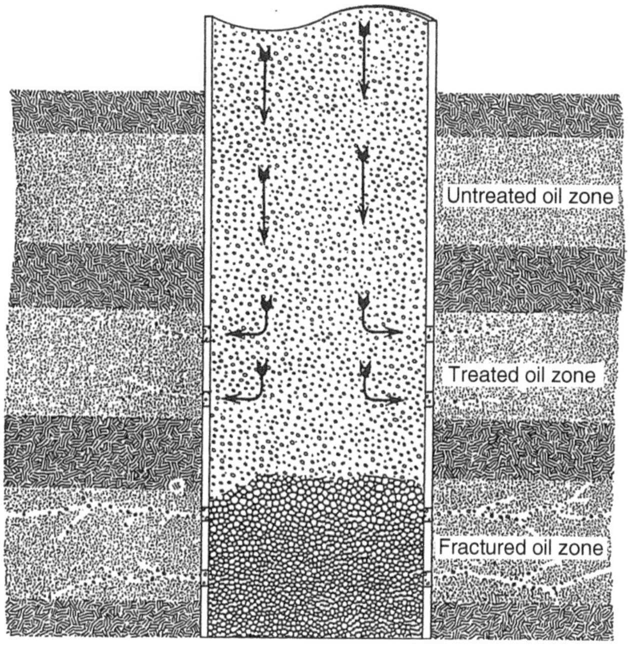

Fig. 10.8—Pine Island technique.

10.7.5 Pine Island Technique. Another widely applied treatment diversion method is the Pine Island fracturing technique. **Fig. 10.8** shows how this method is applied. A typical well might have three zones to be stimulated. The lower interval can be perforated and fracture treated, then a sand plug is placed over the first zone. After the sand plug has been pressure tested, the second interval can be perforated and fracture treated. The second zone would also be covered with a sand plug, and the third zone would be perforated and stimulated. After the third fracturing treatment, the wellbore can be cleaned out with coiled tubing or a conventional workover rig.

More information on treatment diversion can be found in Part 3: Hydraulic Fracturing in Vertical Wells of this book series.

10.8 References

Brown, H. D., Grijalva, V. E., and Raymer, L. L. 1970. New Developments in Sonic Wave Train Display and Analysis in Cased Holes. Presented at the SPWLA 11th Annual Logging Symposium, Los Angeles, California, 3–6 May. SPWLA-1970-F.

Cooke, C. E., Jr. 1979. Radial Differential Temperature (RDT) Logging – A New Tool for Detecting and Treating Flow. *J Pet Technol* **31** (6): 676–682. SPE-7558-PA. https://doi.org/10.2118/7558-PA.

Fitzgerald, D. D. 1985. Guidelines for 90% Accuracy in Zone-Isolation Decisions. *J Pet Technol* **37** (11): 2013–2022. SPE-12141-PA. https://doi.org/10.2118/12141-PA.

Froelich, B., Dumont, A., Pittman, D. et al. 1982. Cement Evaluation Tool: A New Approach to Cement Evaluation. *J Pet Technol* **34** (8): 1835–1841. SPE-10207-PA. https://doi.org/10.2118/10207-PA.

Gidley, J. L., Holditch, S. A., Nierode, R. W. et al. 1989. *Recent Advances in Hydraulic Fracturing*. SPE Monograph Series, Vol. 12. Richardson, Texas: SPE.

Grosmangin, M., Kokesh, F. P., and Majani, P. 1961, February 1. A Sonic Method for Analyzing the Quality of Cementation of Borehole Casings. *J Pet Technol* **13** (2): 165–171. SPE-1512-G-PA. https://doi.org/10.2118/1512-G-PA.

Harrison, N. W. 1972. Diverting Agents-History and Application. *J Pet Technol* **24** (5): 593–598. SPE-3653-PA. https://doi.org/10.2118/3653-PA.

Lagrone, K. W. and Rasmussen, J. W. 1963. A New Development in Completion Methods-The Limited Entry Technique. *J Pet Technol* **15** (7): 695–702. SPE-530-PA. https://doi.org/10.2118/530-PA.

Lestz, R. S., Clarke, J. N., Plattner, D. et al. 1999. Perforating for Stimulation: An Engineered Solution. Presented at the SPE Annual Technical Conference and Exhibition, Houston, Texas, 3–6 October. SPE-56471-MS. https://doi.org/10.2118/56471-MS.

Pickett, G. R. 1966. Prediction of Interzone Fluid Communication Behind Casing by Use of Cement Bond Log. Presented at the SWPLA 7th Annual Logging Symposium, Tulsa, Oklahoma, 9–11 May. SPWLA-1966-J.

Robinson, B. M., Rascon, A. R., Hernandez, C. P. et al. 1998. Evaluating Fracture Stimulation Treatments in Northern Mexico by Integrated Reservoir and Fracturing treatment Analysis. Presented at the SPE Annual Technical Conference and Exhibition, New Orleans, Louisiana, 27–30 September. SPE-49079-MS. https://doi.org/10.2118/49079-MS.

Stipp, L. C. and Williford, R. A. 1968. Pseudolimited Entry: A Sand Fracturing Technique for Simultaneous Treatment of Multiple Pays. *J Pet Technol* **20** (5): 457–462. SPE-1903-PA. https://doi.org/10.2118/1903-PA.

Van Gijtenbeek, K. A. W. and Pongratz, R. 2004. Perforating and Hydraulic Proppant Fracturing in Western Siberia, Russia. Presented at the SPE Annual Technical Conference and Exhibition, Houston, Texas, 26–29 September. SPE-90238-MS. https://doi.org/10.2118/90238-MS.

Webster, K. R., Goins, W. C., Jr., and Berry, S. C. 1965. A Continuous Multistage Tracing Technique. *J Pet Technol* **17** (6). SPE-977-PA. https://doi.org/10.2118/977-PA.

Chapter 11

Development Economics

John Yilin Wang and Stephen A. Holditch

The first step in developing a tight gas reservoir play is to evaluate the volume of gas-in-place and potential gas recovery, as discussed in Part 1: Unconventional Gas Resources of this book series. The next phase of work involves a thorough engineering and geologic study to determine the reservoir properties of the tight gas sand reservoir, which is covered here in Part 2: Tight Gas Sands Engineering. When the reservoir properties are understood to the extent allowed by the available data, the next step is to conduct a detailed and comprehensive economic analysis to justify the proposed development. In addition, planning must be optimized before capital can be approved and allocated for a tight gas development project. Because tight gas projects (like many projects) are usually marginal economic projects and these projects must compete with other projects within a company for funding, it is critical that one clearly understands the income, expenses, and potential profitability of the development.

The steps to carry out an economic evaluation of a tight gas development project are as follows:

1. Perform an in-depth evaluation of resources and accurate reserves estimates (Sections 11.1 through 11.6).
2. Estimate cost, including finding and development (F&D) costs and operating costs (Section 11.7).
3. Generate field-development strategies and economics (Section 11.8).
4. Understand the benefits of applying new technologies (Section 11.9).

This chapter provides a high-level overview of reserves estimation and economic evaluation, with particular emphasis on what one must understand in successful tight gas developments. This chapter was written to help our readers—managers, engineers, geologists, investors, financial institutions, regulators, land and minerals owners, or interested public—understand how to push the boundaries in improving tight gas economics and recovery. A reader with basic knowledge of reserves or economics might find this useful.

11.1 Estimating Reserves in Unconventional Reservoirs

Tight gas development requires an understanding of the volume of the resource and the properties of the reservoirs, including regional geology, petrophysics, reservoir fluids, drilling and completions (D&C) methods and costs, stimulation, and production mechanisms. With solid geological and engineering foundations, one can forecast future production and estimate reserves for different development plans, which will be incorporated in an economic analysis and optimization study. In this section, we summarize six commonly used methods for reserves estimates: analogy, volumetric methods, material balance, decline curves, type curves, and reservoir modeling.

11.1.1 Analog. Analog can provide a reliable reserves estimate if a truly analogous well or reservoir is identified. One must pay attention to two aspects when choosing an analogy: geological attributes and engineering parameters. The geological attributes should be similar, including depositional environment, trapping mechanism, depths, lithology, pore type and sizes, permeability, formation pressures, formation temperatures, and reservoir-fluid type, just to name a few. Engineering parameters should be checked, including D&C methods, stimulation treatment types and sizes, and operational conditions. As we gain deeper understanding of the geology, stimulation methods, and production mechanisms, we will improve our use of analogy to obtain reasonable estimates of oil and gas reserves.

11.1.2 Volumetric Methods. These have been used to estimate resources and reserves in conventional gas reservoirs with success when drainage areas and recovery factors can be estimated with reasonable certainty. In tight gas reservoirs, volumetric methods might provide reasonable estimates of original gas-in-place (OGIP), but should be implemented with care because it can be difficult to estimate the drainage area of a reservoir, especially when the depositional environment is fluvial, which can result in lenticular formations with small areal extent or layers of sand that pinch out at some unknown distance from the wellbore, or in basin-centered reservoirs because the permeability distribution is difficult to estimate. The permeability of a tight gas sand controls the effective drainage area over a given time frame, such as 20 to 30 years. In addition, because of layered effects and high capillary pressures in tight gas sands, initial gas saturation and net-pay are difficult to estimate. As such, acceptable geological and petrophysical analyses are needed to improve the accuracy of the volumetric estimates and minimize the uncertainties. In tight gas reservoirs, the more difficult problem is to estimate the fraction of gas that can be recovered. Drainage area and recovery factor in the drainage area depend on the permeability of the reservoir and the effective hydraulic fracture length and conductivity, which can vary by location and from well to well. We will discuss recovery factor further in a later section. Even with these associated uncertainties, analysis using volumetric methods is essential and provides a credibility check on other reserves estimates using other methods, as noted by Lee and Sidle (2010).

11.1.3 Material-Balance Methods. These are widely used in conventional gas reservoirs, especially on a total-reservoir basis, when accurate gas-production and reservoir-pressure data are available. In these conventional medium- to high-permeability gas reservoirs, wells can be shut in for hours or several days, and accurate estimates of the average reservoir pressure can be measured or computed. Material-balance methods assume that drainage area/volume does not change and that flow is boundary dominated. If the reservoir is connected to a strong aquifer, modified material-balance methods should be used to account for added physics and to develop an accurate production and p/z data set.

In tight gas reservoirs, the two assumptions of material-balance methods are not satisfied because transient flow is extended—often for a period of months to years because of the low permeability. There are three reasons why material-balance methods do not work in tight gas reservoirs:

- The well drainage area changes with time.
- It takes years to reach boundary-dominated flow.
- The average reservoir pressure is almost impossible to estimate.

Cox et al. (2002) agreed with these three issues and said that it is impossible to obtain accurate data to describe how the reservoir pressure declines as gas is produced, which is a basic requirement for using the material-balance method. A well (or the entire reservoir) must be shut in for months or years before enough pressure data are collected to accurately estimate the average reservoir pressure. As such, virtually all shut-in pressure measurements in tight gas reservoirs underestimate the value of average reservoir pressure. If the data are used, OGIP and gas reserves will also be underestimated.

Mattar and McNeil (1998) designed flowing material-balance methods to establish the presence of boundary-dominated flow and to estimate the OGIP. When the flow becomes dominated by the boundaries, pressure at every point in the reservoir declines at the same rate. This method provides an equation that gives the relationship between the well flowing pressure and average reservoir pressure. As with any rigorous approach, the outcome is affected by data quality. Ilk et al. (2008) suggested that this approach will always provide some diagnostic value, even in the presence of significant or extreme data bias (e.g., liquid loading in a gas well). Mattar et al. (2006) improved the flowing material-balance method and proposed the dynamic material-balance method. Their method is not limited to constant flow rate, which was a disadvantage of the flowing material-balance method. But neither method can be used during transient flow because the formulation of the material-balance time and pseudotime are precise only during boundary-dominated flow. Also, the dynamic material-balance method underpredicts OGIP for the case of pressure-dependent permeability or continuously changing skin.

Gas material balance in conventional reservoirs shows a linear relationship between p/z (where p is pressure and z is compressibility) and cumulative production, but tight gas reservoirs exhibit a nonlinear trend. This trend was recognized by Stewart (1970) and Fetkovich et al. (1987) and was discarded as impractical, as shown in **Fig. 11.1.** Engler (2000) showed that this nonlinearity is a function of two parameters: the pressure measurement technique and the reservoir characteristics. He described that, even though the magnitude of the pressure measurements is doubtful, the trend can be applied to obtain a reasonable estimate of gas-in-place. He identified three types of tight gas reservoir behavior and developed relationships with the desired responses by using key intersection points and slopes from a tight gas material-balance plot.

Sullivan et al. (1988) developed a method to estimate OGIP in a tight gas reservoir using transient pressure data. Their method is similar to an isochronal flow test. Pressure data recorded at specific shut-in intervals are collected during the production life of the reservoir. A p/z plot is constructed using these pressure data and the corresponding production data. The p/z curve will plot as a straight line with a slope similar to the nonstabilized p/z curve through a point on the stabilized p/z curve. This new line is extrapolated to the point of intersection on the x-axis (G_p), as shown in **Fig. 11.2.**

Aprilia et al. (2006) presented an application of Bayes' theorem to combine volumetric and material-balance methods and quantify uncertainty in OGIP. In the Bayesian approach, an attempt is made to use all available information to reduce the amount of uncertainty present in a decision-making problem. The formal mechanism used to combine the new information with the previously available information is known as Bayes' theorem.

In summary, reserves estimates based on material-balance methods should not be used for reporting and development planning purposes but do provide reasonable checks on other methods.

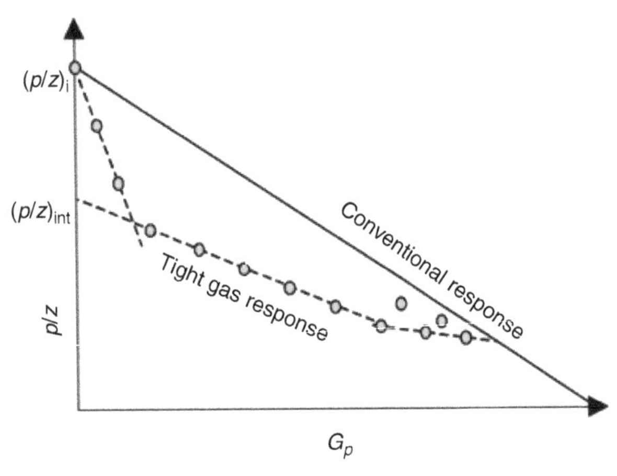

Fig. 11.1—*p/z* and cumulative production (*Gp*) in conventional gas reservoirs vs. tight gas reservoirs. After Engler (2000).

11.1.4 Decline-Curve Methods. The predominant method used by the industry to determine the reserves in tight gas wells is that of decline curves. The industry professionals performing reserves work for companies, banks, and the US Securities and Exchange Commission (SEC) typically use a hyperbolic equation that is fit to the actual production data to determine the constants in the equation. Then, this empirically fit curve is used to predict future production. Typically, the hyperbolic equation is switched to a constant-rate decline of 3–6% to model the last years of production. We recognize that there is nothing theoretical about this method of computing estimated ultimate recovery (EUR) or remaining reserves. It is purely a curve-fitting exercise to mathematically model the actual production data to predict future flow rates. However, there are numerous papers on decline-curve analyses, from the simple Arps paper to the much more complicated methods developed to be used with spreadsheets. The following text describes what is in the literature. Normally, using decline-curve equations to curve fit the actual production data results in the best estimate of near-term production rates.

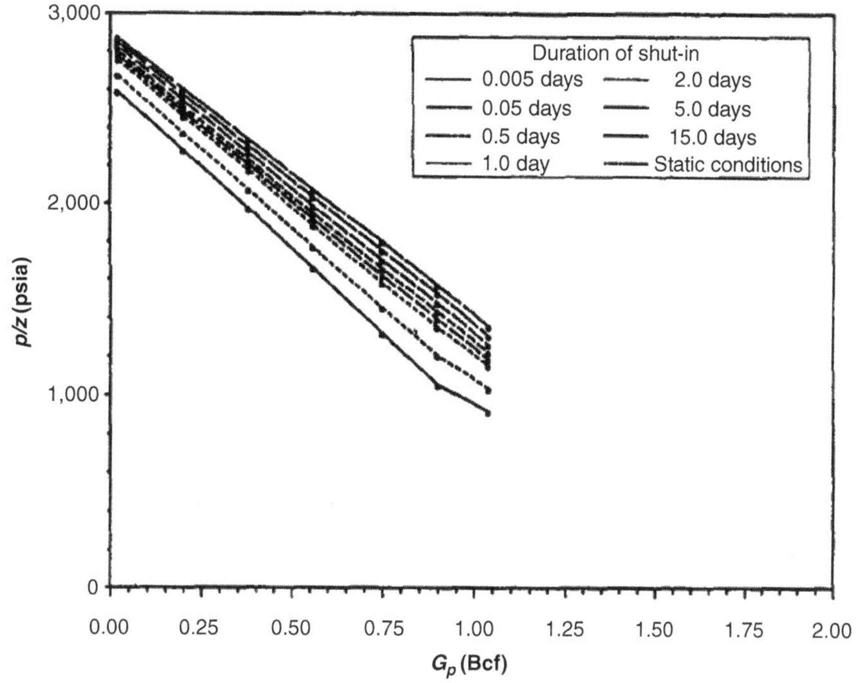

Fig. 11.2—Isochronal *p/z* curves for a tight gas formation with permeability of 0.3 md (Sullivan et al. 1988). G_p = cumulative production.

Decline-curve analysis, as shown in Eq. 11.1, was developed by Arps in 1945 and is still one of the most widely used techniques to predict future production performance and estimate reserves using available production data from wells completed in tight gas reservoirs (Arps 1945). Basic decline-curve analysis assumes a single-layer reservoir, with the well producing at constant bottomhole pressure (BHP) and with stabilized flow conditions (Fetkovich et al. 1996). In addition, the use of Arps' equations implies that there are no changes in completion and operating conditions. It is well-documented that the decline exponent b should range between 0.0 and 1.0 when these assumptions are satisfied (Arps 1945; Fetkovich et al. 1996). When $b = 0$, it is called exponential decline, and, when $b = 1$, it is called harmonic decline, as shown in **Fig. 11.3.**

$$q(t) = \frac{q_i}{\left(1 + bD_i t\right)^{1/b}}, \quad (11.1)$$

where q = gas flow rate at time t, q_i = initial gas flow rate, D_i = initial decline rate, and b = decline exponent.

However, tight gas wells require successful hydraulic fracturing treatments to produce economically. As explained in the section on material-balance methods, long time periods—often years—are required to reach pseudosteady-state flow (more precisely, boundary-dominated flow, because the term "pseudosteady state" implies constant-rate production). The production data available for decline-curve analysis are, therefore, often not stabilized. Therefore, analysis of decline behavior in tight gas wells presents unique technical challenges (Cox et al. 2002; Neal and Mian 1989). It is often very difficult, if not impossible, to estimate reserves accurately in a timely and consistent fashion using traditional Arps' decline-curve analysis.

Maley (1985) found that the value of exponent b does exceed 1.0 during linear flow in tight gas reservoirs. Using this value to forecast flow rates could result in unreasonably long well lives and reserves that are too large. In practice, some engineers simply use the b value obtained from the matching of production data, while others force the b value to be 1.0. Still others use the hyperbolic decline and the matched value of b, but, when the decline rate reaches a predetermined limit, they switch to exponential decline for the remainder of the forecast. Even though the procedure of using a hyperbolic equation and then switching to an exponential decline late in life has no physical basis, it seems to work well and is accepted as perhaps the best method for computing future production in tight gas reservoirs.

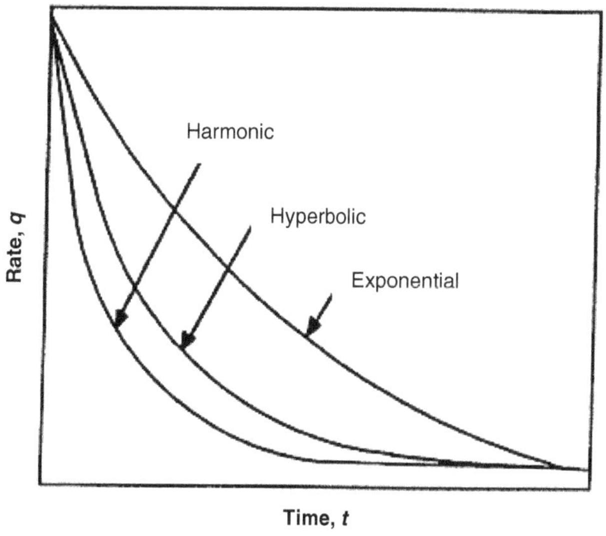

Fig. 11.3—Characteristics of typical decline curves.

Cox et al. (2002) showed that precise reserves estimation of tight gas reservoirs can be achieved only after a well has reached boundary-dominated flow. Four field-data analyses and two simulation cases demonstrated that decline-curve analysis can provide reasonable reserves estimation only after the well has reached boundary-dominated flow. Otherwise, Cox et al. say that this method is inaccurate. Cox et al. (2002) concluded that before boundary-dominated flow, integrating decline-curve, material-balance, and pressure-transient analysis techniques into a comprehensive method can be used to forecast reserves.

Cox et al. (2003) extended their earlier work to examine common reserves analysis techniques for noncommunicating multilayered tight gas reservoirs. They found that decline-curve analysis is inaccurate for these reservoirs and that hyperbolic decline forecasts with b exponents greater than 1.0 should not be performed. Maley (1985) and Robertson (1988) suggested constraining a hyperbolic or superbolic decline curve to an exponential decline curve at a certain time or at a specific decline rate. Other authors suggest that traditional decline-curve analysis be avoided altogether during transient flow and instead recommend using modern decline analysis methods (Fetkovich 1980; Agarwal et al. 1998; Mattar et al. 2006) because they are applicable to both transient and boundary-dominated flow. However, these methods are complicated and dependent on a complete (pressure/rate) data set, and the data for these analysis methods are not always available in practice. Cheng et al. (2007) showed that, with existing decline-curve methods, analysis of stabilized-flow data is sufficient to produce accurate production forecasts and reserves estimates, but analysis of transient flow production data can result in large errors. Cheng et al. minimized these errors by determining the b value for stabilized (boundary-dominated) flow a priori, history matching multiple periods of late production data in a backward fashion to determine a set of decline parameters (q_i and D_i), and then extrapolating these parameters to q_i and D_i corresponding to the end of history for use in projecting future production. Valkó (2009) developed an empirical decline-curve analysis that was completely different from Arps' methods and was based on analysis of a large number of tight gas and shale gas wells in Texas. Ilk et al. (2008) proposed similar models to better fit data in both transient and boundary-dominated flow.

11.1.4.1 New Schools of Thought and Approaches. Although Arps' decline equations were developed empirically, the parameter b in the decline equations was proved to be related to fluid properties and production conditions (Chen and Teufel 2002; Fetkovich et al. 1996). Hyperbolic decline behavior commonly observed in tight gas wells relates to both reservoir and hydraulic fracture properties (Rushing et al. 2007a). Spivey et al. (2001) introduced a method of calculating the hyperbolic decline exponent instantaneously on a rate time production curve. They showed that the slope of the instantaneous decline rate, shown in Eq. 11.2, can be used to calculate the instantaneous decline exponent, as shown in Eq. 11.3:

$$D = -\frac{1}{q}\frac{\mathrm{d}q}{\mathrm{d}t}, \quad \text{(11.2)}$$

$$b = \frac{\mathrm{d}}{\mathrm{d}t}\left(\frac{1}{D}\right). \quad \text{(11.3)}$$

Kupchenko et al. (2008) showed that traditional decline-curve analysis can provide acceptable forecasts for tight gas reservoirs provided that the behavior of the Arps equation under transient and boundary-dominated flow is well-understood. They summarized the behavior of the decline exponent during transient flow for bilinear, linear, and pseudoradial flow and introduced a method to forecast production throughout the life of a tight gas well using the Arps equation in its original form by tuning b. The method uses simple plots to limit superbolic decline exponents on the basis of drainage area, permeability, and hydraulic fracture size. Rushing et al. (2007b) developed an appraisal workflow process (shown in **Fig. 11.4**) that has two fundamental stages: quantification and validation.

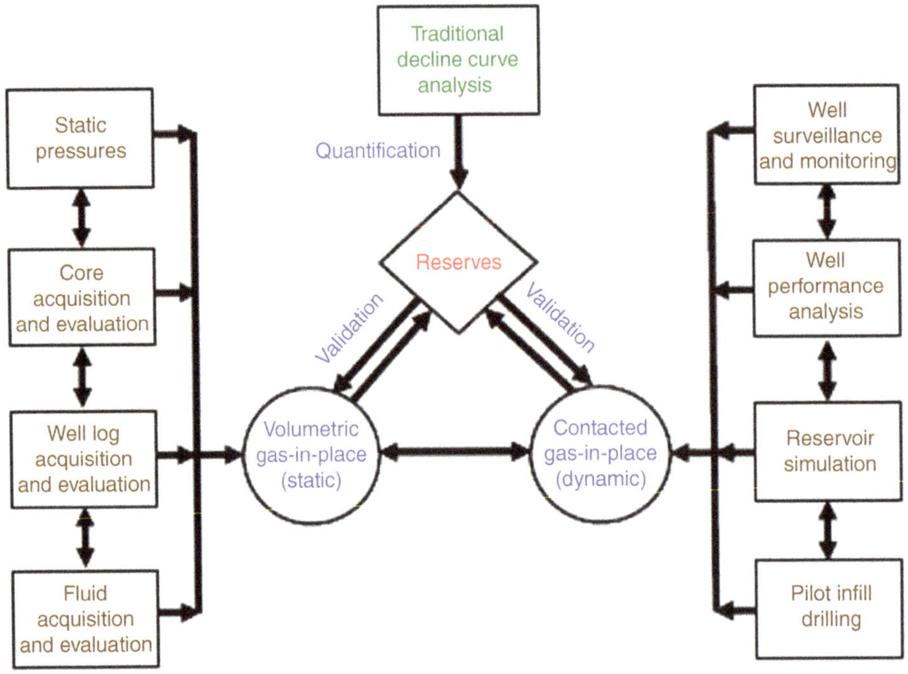

Fig. 11.4—Workflow for evaluating reserves in tight gas sands by Rushing et al. (2007b).

Reserves from producing wells are initially quantified using a traditional decline-curve methodology. Depending on field maturity, data quality, and data quantity, the Arps decline-curve parameters (e.g., initial decline rate D_i, decline exponent b) can be either determined from existing producers in the field or estimated from suitable field analogies. The next step in the workflow process is a reserves validation stage in which reserve volumes are substantiated by comparing to both volumetric gas-in-place and contacted gas-in-place volumes. Rushing et al. (2007b) defined volumetric gas-in-place as a static volume representing the total resource-in-place and the contacted gas-in-place as a dynamic volume representing the fraction of volumetric gas-in-place in pressure communication with producing wells at any point in time. Aniemena (2014) further developed this physics-based b approach for tight gas and shale gas application, where b was correlated to reservoir properties, fluid properties, stimulation treatments, flow regimes during both transient and boundary-dominated flows, and time. This approach was able to history match production data during the transient-flow period, which yields a function to predict b and then gas rates over time.

When using the decline-curve equations to analyze production from tight gas reservoirs, one must carefully analyze all the data. For example, many wells begin producing at high gas-flow rates along with high flowing tubing pressure. During the first few weeks and months, both the gas-flow rate and the flowing tubing pressure decline. If the analyst analyzes only the gas-flow-rate data, the extrapolation into the future is optimistic. Whenever the flowing tubing pressure reaches the pipeline pressure and the flowing tubing pressure is no longer declining, the gas-flow-rate decline rate increases. Thus, when both the gas-flow rate and the flowing tubing pressure are declining, the analyst must compute values of $q/\Delta p$ or flow rate divided by pressure drop and use the decline-curve model to match the decline in both flow rate and flowing tubing pressure. Another complication in analysis of decline data in tight gas wells is that, in most cases, production is commingled from multilayered formations that are hydraulically fractured with multiple stages. Because of variations in formation permeability and fracture half-length, different flow regimes might coexist in different layers. Lower-permeability zones might be in transient flow, while higher-permeability zones have established stabilized and boundary-dominated flow. The profile of production contribution by individual layers changes constantly with time. Given these complications, decline-curve analysis of multilayer tight gas wells is especially difficult, particularly with regard to estimating long-term production and reserves.

The most common misapplications of traditional decline-curve analysis in low-permeability gas reservoirs are (Rushing et al. 2007b):

- Application of the Arps decline curves during transient and/or transitional flow periods. The traditional Arps equations are strictly valid when applied to boundary-dominated data. There is simply no basis for predicting long-term performance from an extrapolation of the curve fit through data during transient or transitional flow periods. Until all drainage-area boundaries have influenced the well's decline characteristics, predictions of long-term behavior are probably both nonunique and incorrect.
- Failure to modify reserves estimates when operational conditions change. The Arps models assume implicitly that the well is produced at a constant-flowing-BHP condition, so any operational changes that alter this flowing condition could affect the extrapolated decline.
- Application of decline-curve models with changing reservoir and/or hydraulic fracture properties. The Arps models also assume implicitly that reservoir and fracture properties will remain constant. Stress-dependent reservoir properties are a common characteristic of tight gas sands, especially those that are initially abnormally overpressured. It is also not uncommon to have changes in fracture properties (i.e., effective fracture conductivity and/or length) over time, as discussed in Part 2 Ch 9: Well and Reservoir Numerical Modeling. Common causes include two-phase flow, ineffective cleanup of fracturing fluids, and loss of effective fracture conductivity because of proppant crushing or embedment.

11.1.5 Type Curves. If used correctly, type curves can be used to forecast production, estimate reserves, and evaluate reservoir and fracture properties. Either an analogy-based or a model-based type curve can be useful when appropriate analogies are identified and model calibrations are performed. Type curves are easy to apply in practice, but often matching the real data to the theoretical type curves is nonunique and depends on the experience of the analyst.

11.1.6 Modeling. The use of a reservoir model, including analytical models, semianalytical models, and numerical reservoir models, to history match production data from a well or a whole field is the most accurate method of estimating gas reserves in tight gas reservoirs, especially in complex geological or engineered tight reservoir systems. The model should be capable of simulating layered reservoirs, a finite-conductivity hydraulic fracture, and a changing flowing BHP. In many cases, the analyst might also need to simulate non-Darcy flow, fracture closure, and/or fracturing fluid-cleanup effects. Normally, a reasonable approach to estimating reserves is to use decline curves to review and quality check the data, semianalytical models to history match existing data and estimate reserves, and finite-difference models to perform the final analyses of the data, especially if factors such as non-Darcy flow, fracture closure, and fracturing fluid cleanup need to be included in the analysis (Holditch et al. 1987; Holditch and Rahim 1994; Holditch and Tschirhart 2005; Holditch 2006, 2009).

The primary advantage of a reservoir simulator over analytical or semianalytical production-data-analysis models is the ability to incorporate complex geological attributes and heterogeneities, such as reservoir anisotropy, layered systems, and faulted reservoirs. It also includes changes in flowing pressures, multiphase flow in wellbores (Rushing et al. 2007b), and the Klinkenberg effect. The reservoir simulation model and fracture properties should be updated when historical data (core, well-log, test, and production data) become available, thus improving reserves estimates and performance forecasts (Hatzignatiou and McKoy 2000).

Jordan et al. (2006) proposed that a simpler reservoir model (a single well completed in the center of a circular reservoir) can be used to represent a far more complex reservoir system and still provide some representative reservoir characterization and accurate production forecasting. The Jordan model approach reduces a complex and heterogeneous reservoir to a simplified reservoir, and, if OGIP is honored correctly, then generates accurate production forecasts as precisely as possible. Rushing and Newsham (2001) modified the petrophysical integration process model, originally developed by Gunter et al. (1997a, 1997b) to describe and characterize tight gas reservoirs. As shown in **Fig. 11.5,** the model has four separate analysis stages, and each stage describes different reservoir scales ranging from large-scale geologic elements and heterogeneities to small-scale properties at the pore level. The model uses and combines data from multiple and apparently independent data sources, which link together with a petrophysical model. Holmes et al. (2007) improved the traditional one-time simulation study with a continuous process potentially spanning the life of the tight gas reservoir.

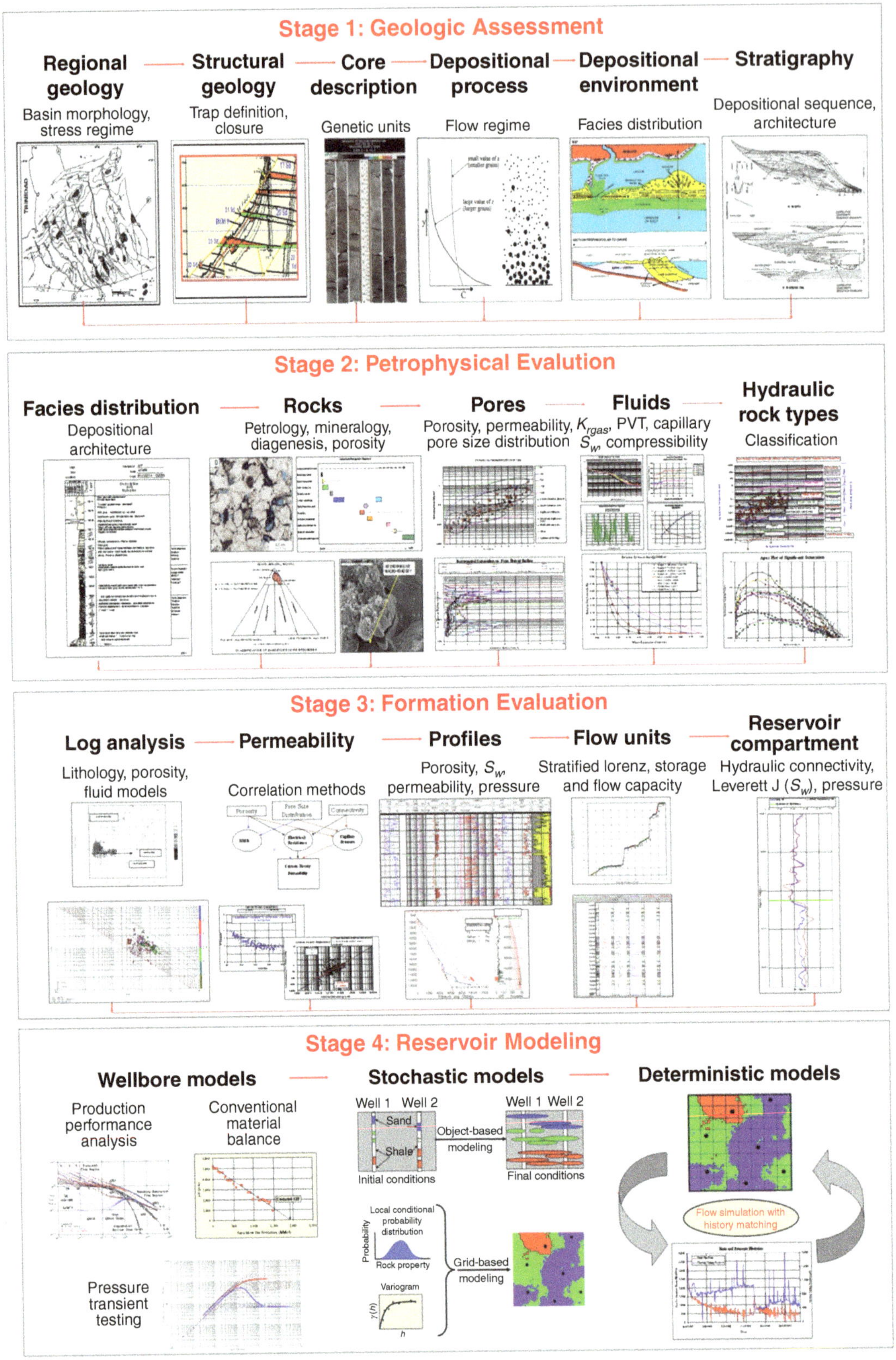

Fig. 11.5—Modified petrophysical integration model. Stage 1 is a geological assessment. Stage 2, the petrophysical-evaluation stage, describes the rock and fluid systems on a pore scale. Stage 3, the formation-evaluation stage, integrates Stages 1 and 2 and describes the reservoir through upscaling techniques. Stage 4 calibrates the geological and petrophysical models from Stages 1, 2, and 3 and constructs various 2D and 3D wellbore and reservoir models (Rushing and Newsham 2001). PVT = pressure/volume/temperature; K_{rgas} = gas relative permeability; S_w = water saturation.

In this process, reservoir models are generated and run 24 hours a day, seven days a week, allowing many more runs than previously possible and yielding a more thorough exploration of possible reservoir descriptions. This is combined with real-time acquisition of production and pressure data, which is automatically integrated into simulation runs to allow the process to run continuously with little human interaction.

So why are reservoir simulation models not used more often to estimate reserves? The answer is that they are difficult to use and take too much engineering and geologic effort, which results in higher costs. However, more importantly, if an engineer needs to generate reserves for hundreds or thousands of wells for a company's annual report, it is impossible to use even semianalytical reservoir models to do the work. So, the hyperbolic curve fitting equation is the method the industry uses because it gives reasonable estimates of near-term future flow rates and can be performed quickly for hundreds of wells.

Either finite-difference or semianalytical reservoir simulation can be used to analyze a sample of typical wells. The samples can be chosen from groups of wells with different geologic attributes and D&C methods. The computer reservoir models can then be used to model these sample wells to help fine tune the decline-curve methods one must use to obtain acceptable reserves estimates.

In summary, it is very important that production forecasts and reserves be estimated using the appropriate and most accurate methods before developing economic evaluations. To do so, we must keep the following factors in mind:

- The geological, stimulation, and production mechanisms must be understood in depth. We will then know how to identify appropriate analogous wells and reservoirs.
- Even with associated uncertainties, volumetric methods are essential and provide a credibility check on other reserves estimates using other methods, as noted by Lee and Sidle (2010).
- Material-balance methods are not recommended for reporting and development planning purposes but do provide reasonable checks on other methods.
- Decline-curve methods should be used if production data are available: traditional Arps decline, advanced decline curves, or recent decline curves. One should first understand the flow regimes and production mechanisms of a well or reservoir, then try to understand the physical meaning of coefficients (such as D_i and b) in the decline equations. Most engineers use the hyperbolic equation for tight gas wells containing a hydraulic fracture.
- If used correctly, type curves can be used to forecast production, estimate reserves, and evaluate reservoir and fracture properties. However, matching is often nonunique and depends on the experience of the analyst.
- The most accurate method of estimating gas reserves in tight gas reservoirs is use of a reservoir model, including analytical models, semianalytical models, and numerical reservoir models, to history match production data from a well or a whole field and then to forecast future production. Besides reserves from status quo, one could evaluate increase in production and reserves if production operations such as recompletion, artificial lift, acidizing, refracturing, and chemical treatments were to improve performance. One could then evaluate the cost/benefit results of different scenarios. We have applied this method in many situations and found it to be useful in leading to a better understanding of the well, the fracture properties, and the reservoir.

11.2 US SEC Rules on Reserves in Unconventional Resources

On 29 December 2008, the SEC adopted modernized rules for reporting oil and gas reserves in filings with the SEC. The SEC posted the full text of these rules and commentary on its website. The new rules went into effect for documents filed with the SEC for fiscal years ending on or after 1 January 2010.

These new rules are the culmination of an effort by the SEC to review the oil-and-gas reporting requirements announced publicly in May 2007, when the SEC published a notice that it intended to appoint an Academic Engineering Fellow to examine reserves-reporting requirements—an appointment for which John Lee was chosen. Following an effort that began with Lee's arrival at the SEC in October 2007, the SEC issued a "Concept Release" in December 2007 in which it asked for public comments on the need for revisions of the oil-and-gas reporting requirements and suggestions for what the revisions, if any were made, should include. Eighty comments from the public were received through February 2008. The SEC staff studied these comments and, in June 2008, issued proposed rules that would modernize oil-and-gas reserves definitions and reporting requirements. Sixty-eight comments from the public were received through September 2008, and analysis of these comments led to the final rules adopted in December 2008.

SEC staff then issued major guidance in October 2009 and guidance on a single issue in May 2013. In addition, Morales and Lee (2014) reviewed the rules, guidance, and comment letters that the SEC has made public for filing covering the 2009 to 2012 fiscal years, as well as public presentations by SEC staff from 2009 to 2014. They then discussed the issues subject to most comments by the SEC and provided guidelines to improve the reporting consistency and comparability to help minimize SEC comments.

The new SEC rules have had a significant impact on reserves in unconventional resources, including

- Use of annual average prices, determined as an unweighted numerical average of first-day-of-month prices for each month during the filer's fiscal year, instead of a single-day, year-end price. This helps diminish reserves write-down resulting from decreased prices at the end of the year.
- Recognition of nontraditional resources, such as bitumen from oil sands and synthetic oil from oil shales and coal.
- Optional disclosure of probable and possible reserves.
- Replacement of the "certainty" criterion for proved undeveloped reserves from immediate offsets with a "reasonably certain" criterion. This indicates that more proved, undeveloped drilling locations may be recognized by filers, especially in a continuous petroleum reservoir.
- Modern, reliable technologies that have been demonstrated in practice to provide a repeatable and consistent basis being allowed to establish proved reserves instead of only flow tests or sustained production in the previous 1978 rules. The SEC allows reliable technologies to include proprietary technology without disclosing the details to the public—only, on request, to SEC staff.

In summary, one must follow the modernized 2008 SEC rules for reporting oil and gas reserves and economic evaluations, especially for tight gas sands.

11.3 Use of Moving Domain

Operating companies collect detailed information on their wells and production histories. Production data are normally published and available to the public. For an analysis of the data, it is helpful to have production data on a well basis. However, in some states, the production data may be published on a lease basis because the data are measured at tank batteries and not at the well. Production data by well can be analyzed to better understand the reservoir and make prudent operating decisions. Ideally, the well data collected would include daily gas, water, and condensate volumes; tubing pressures; production duration; choke size; and artificial-lift information. As discussed in Section 11.1 on reserves, there are different methods available for analyzing production data, such as conventional decline-curve analysis (Arps), advanced decline-curve analysis (Fetkovich), automatic history matching, and finite-difference numerical reservoir simulation. It is often difficult to quantify infill drilling and recompletion potential in producing gas fields because of the large variability in rock quality, well spacing, and well completion practices, as well as the large number of wells involved. Given the marginal nature of many of these fields, it is often prohibitively expensive to conduct conventional reservoir characterization and simulation studies to determine infill potential. There is a need for rapid, cost-efficient technology to evaluate infill potential in gas reservoirs, particularly tight gas reservoirs.

McCain et al. (1993) used a statistical, moving-domain method of production-data analysis to determine infill potential in East Texas Cotton Valley in 1993. Voneiff and Cipolla (1996) applied a similar method to analyze well-location and production data for rapid assessment of infill and recompletion potential in the Ozona (Canyon) gas sands. Later, Kyte and Meehan (1996) successfully used the moving-domain technology to quantify infill potential in the Austin Chalk. On the basis of the work above, Hudson et al. (2001) applied the method to the Mesaverde, Morrow, and Cotton Valley formations. Guan et al. (2002) assessed the accuracy of the moving-window technique for selecting infill candidate wells in low-permeability gas reservoirs by analyzing synthetically generated production data. The primary advantages of the moving-window technique are that it is easy to apply and it relies only on publicly available well-location and production data. Gao and McVay (2004) presented a fast method, which is an extension of the moving-domain method described by Voneiff and Cipolla (1996), and analyzed the accuracy of the methodology. The fast method was demonstrated to be more accurate than moving-domain statistical methods in synthetic cases, with comparable analysis times and costs.

11.3.1 What is Moving-Domain Analysis? A domain is defined as a group of wells grouped around a center single well. The wells in the group are determined by drawing a circle around the center well and including all the wells in the circle as part of the domain. In a moving-domain analysis, the production data for the well in the center is compared with the production data for all the wells in the domain. It is implied that, by comparing the well in the center with the other wells that are close by, the effect of changing geology is minimized. Then, another well is chosen as the center well, and the new wells inside its domain are determined. The well in the center of the new domain is compared to its neighbors. The method tries to determine if the center well is better or worse than the other wells in the domain. By moving the domain so that every well is compared with its neighbors, one can find areas in a field that are better or worse than the average wells and find out where one might drill an infill well, for example. One can also find a well or two that need to be restimulated or that require a workover.

The moving-domain method is a technique to rapidly process large production and completion databases to help determine depletion, infill potential, well-quality distribution, technology benefits, and restimulation potential. The moving domain is a set of empirically derived approximations, comparisons, and statistical tests that attempt to mimic what a reservoir engineer does when faced with a single infill location evaluation. This technique looks at surrounding well performance, compares new wells with old wells for signs of depletion, calculates effective well density, and, once linked to a scattering of conventional estimates of drainage area, provides estimates of undrained acreage and infill reserves. The process restricts its basic comparisons to small groups of wells to minimize the effects of areal changes in rock properties. The analysis is repeated, well by well, until the entire area of interest has been evaluated. Typically, results are presented in map form with a mix of dot, bubble, contour, and color-based presentations to highlight various data and conclusions. Results range from purely statistical in nature, such as evidence of depletion, to precisely deterministic in nature, such as effective well density. The infill potential computed by this technique does have a statistical component, which is significantly reduced by linking this technique to detailed conventional reservoir engineering (Voneiff and Cipolla 1996).

11.3.2 How to Apply the Moving-Domain Method. The moving-domain technique can be applied in two phases. The first phase is treated as a scoping study that can be completed in a matter of weeks to provide preliminary infill estimates, identify areas of depletion, and determine areas that will likely require further analysis with detailed conventional engineering. The second phase involves a much greater degree of detailed engineering and includes linking the moving-domain technique to detailed engineering. With the detailed engineering, the drainage areas, undrained acreage, recovery per acre, and infill reserves for a scattering of wells across the field can be calculated to improve infill estimates (Voneiff and Cipolla 1996).

11.3.3 Production Indicators. Ideally, the engineer would have an EUR for each well to perform the study. However, because of the large number of wells, the goal of achieving preliminary conclusions rapidly, and the variability in EUR that typically arises when estimating recoveries for 1,000 or more wells, the method computes and uses production indicators. The goals of the production indicators are threefold (Voneiff and Cipolla 1996):

- They provide a good estimate of long-term performance.
- They can be calculated on virtually all the wells, both new and old, with a minimum of subjectivity.
- They are programable.

In the moving-domain method, "best-year" (BY) production is used as a proxy for production rate in the pseudosteady-state-flow model. BY is simply the best 12 consecutive months of production divided by 12. **Fig. 11.6** shows how one can calculate these production indicators. Another production indicator is virgin best year (VBY), which is the BY of the well at virgin conditions. Depletion effects are removed by computing the BY of a local area at a time before depletion using a 2D regression of BY vs. well start date. These indicators are plotted against the date of first production, and a statistically fit (least-squares) straight line is shown through each data set **(Fig. 11.7)**. With this plot, the question, "Are the new wells performing differently from the older wells?" can be converted

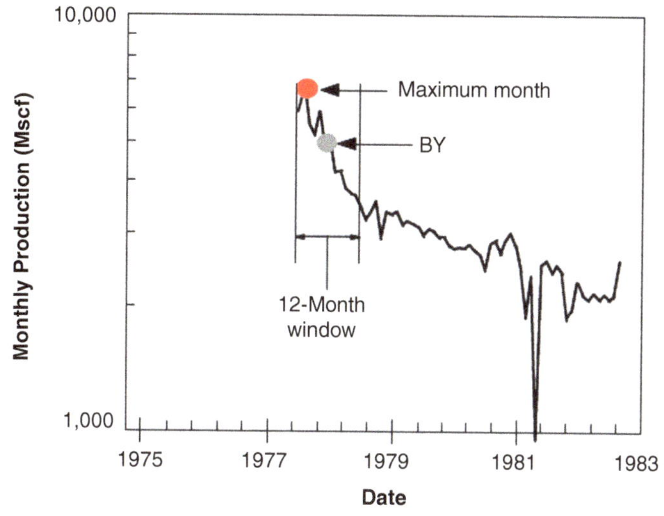

Fig. 11.6—Illustration of production indicators (Voneiff and Cipolla 1996).

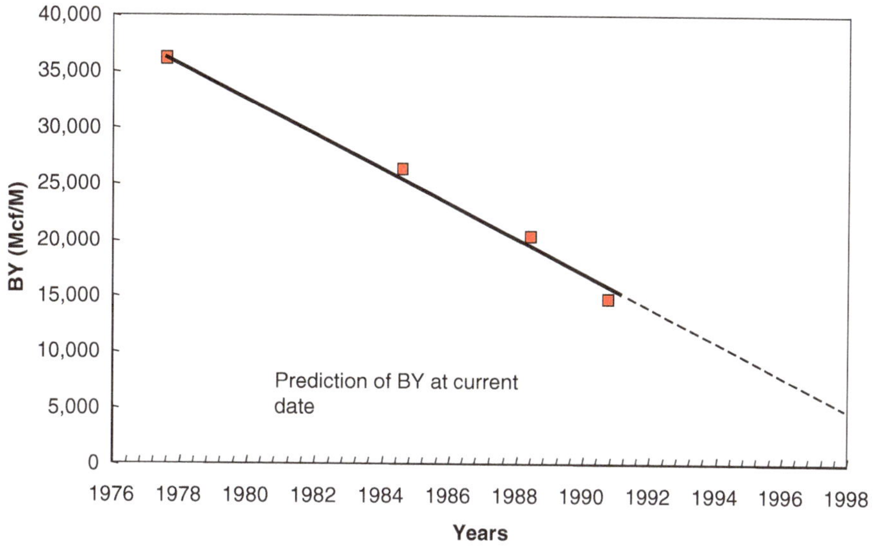

Fig. 11.7—Production indicator of BY (Guan and Du 2004).

into the statistical question, "Is the slope of the indicator plot line significantly different from zero?". Slopes of zero mean that the new wells are performing identically to the older wells, indicating no interference or depletion at infill well locations. Negative slopes indicate some interference between wells. The hypothesis was answered with the student's t-test. If there appears to be a positive or negative slope, the student's t-test determines if the slope is significant, considering the amount of scatter in the data (Voneiff and Cipolla 1996).

11.3.3.1 ***Evidence of Depletion.*** One use of the production indicators is to identify the area experiencing depletion. One can construct a production-indicator graph by plotting the BY production vs. the date of first production. One can identify areas of probable depletion by comparing the production from new wells to that from older wells. If the new wells are poorer producers than the older wells, one can conclude that depletion has probably occurred at the new well locations (Voneiff and Cipolla 1996).

11.3.3.2 ***Well Density and Undrained Acreage.*** The next step in the scoping study is to calculate the effect of well density. When one knows the effective well density for a well, one can easily calculate the undrained acreage surrounding that well given the drainage area. One can compute the undrained acreage as the effective well density minus the drainage area (Voneiff and Cipolla 1996).

11.3.3.3 ***Infill Potential.*** The infill potential is calculated as the undrained acreage times the recovery per acre. One can determine recovery per acre as part of the drainage-area analysis. The recovery per acre is adjusted downward in those areas where one finds depletion to reflect the reduced amount of gas per acre available to a new infill well. One can make this adjustment by comparing the recovery per acre of each round of infill wells, and, if it is decreasing, one can continue reduction into the proposed round of infill wells (Voneiff and Cipolla 1996).

It has been demonstrated that the moving-domain technique is more accurate when the permeability is smaller (Guan et al. 2002), the reservoir is less heterogeneous, the search area is smaller than 3,000 acres, a larger number of wells are in the domain, and the timing of application is closer to infill drilling (Guan and Du 2004).

11.3.4 Application in the Carthage (Cotton Valley) Field.

11.3.4.1 Problems. Evaluation of the need for infill drilling in complex, low-permeability gas reservoirs requires the application of advanced technology. Use of typical engineering tools such as conventional decline curve analysis, conventional pressure transient analysis, and single-layer reservoir descriptions often creates inaccurate results and misleading conclusions. Despite the accuracy of advanced analysis techniques, these techniques are typically not performed on a well-by-well basis in fieldwide studies because the cost cannot be justified. As a result, accurate conclusions, valid for all areas of a field, are not obtained (McCain et al. 1993).

11.3.4.2 Solutions. The moving-domain technique evaluates the performance of each well with surrounding wells to determine if infill wells are performing similarly to or differently from older offsets. The statistical analysis is performed by constructing and evaluating graphs of well performance plotted against the date of first production. To minimize bias, McCain et al. (1993) characterized well performance by using a performance indicator that did not rely on input that could be subjective (such as decline-curve extrapolations) and was available for all wells in the field. Because this technique was new, they performed the evaluation with three performance indicators to determine if one particular indicator was best. The three productivity indicators selected were

- Maximum monthly production rate
- Average monthly production rate for the most-productive 12-month period
- Monthly production rate (5-month average) at the time a cumulative production of 250 MMscf was reached

Fig. 11.8 shows "indicator plots" for each of the three indicators for nine wells in a 1,280-acre area. None of the three lines in Fig. 11.8 were found to have a slope different from zero; thus, in this area of the field, infill wells are finding virgin, not-depleted (McCain et al. 1993).

11.3.4.3 Results. The entire field was divided into equal-sized study units of approximately 3,300 acres. Each study unit contained between 10 and 20 wells. The analysis showed that the Carthage Field has two areas where infill wells are inferior to older wells, suggesting interference and depletion. These two areas comprise nearly one-third of the field as shown in **Fig. 11.9.** Note the correlation between these areas and the locations of the sands with more continuity, the Taylor and the A-sand. In the remaining two-thirds of the field, the productive behavior of wells indicates that there is no interference between wells, possibly because of separate reservoir compartments as suggested by the geological description and by the pressure measurements (McCain et al. 1993).

11.3.5 Application in the Ozona (Canyon) Gas Sands.

11.3.5.1 Problems. Quantifying the infill potential in the Ozona (Canyon) Field is difficult for several reasons: the layered and low-permeability nature of the reservoir, the hydraulic fracturing of wells, the variation in well spacing, and the local and regional variation in rock quality. These complication factors often cloud the extent of depletion and benefits of improved completion and operation techniques and make the quantification of infill potential rather difficult. Because of the complexities and the fact that the study area contained more than 100 wells, the time and cost requirements of a conventional reservoir engineering study were unacceptable. Instead, the authors applied the moving-domain method to the large tight gas field (Voneiff and Cipolla 1996).

11.3.5.2 Solutions. In the Ozona (Canyon) Field, two production indicators are required. One indicator is the BY, and the other is the rate of decline during the BY, which was termed the decline ratio. The best year indicators in the Ozona Field range from less than 100 Mscf/M to more than 50,000 Mscf/M. The average for the entire field is 8,500 Mscf/M, representing a well that averages approximately 280 Mcf/D during its best 12 consecutive months of production. Voneiff and Cipolla (1996) made a plot of the BY indicator vs. the 10-year cumulative production, as shown in **Fig. 11.10.** They found 962 wells in the Ozona (Canyon) Field with 10 years of production. The decline ratio indicator further enhances our abilities to predict long-term performance from only 1 year of production from a well. Because both of these indicators can be computed automatically, they had a set of values that could be rapidly obtained and used as a guide to EUR in the study (Voneiff and Cipolla 1996).

Fig. 11.11 shows the production indicator plot by plotting the best year of production vs. the date of first production for a 15-well area. In the figure, the significance test gives a 98% confidence level that new wells are worse than old wells. A combination of the dots from the production indicator plot using the best year and the production indicator plot using the decline ratio shows the evidence of depletion (Voneiff and Cipolla 1996).

11.3.5.3 Results. The result of the field study culminated in a map locating infill potential. The authors identified 636 infill locations, where each infill location has a minimum reserves expectation of 200 MMcf of gas (Voneiff and Cipolla 1996).

11.3.6 Conclusions.

- The moving-domain analysis of production data was developed to evaluate wells when conventional decline curve analysis is not appropriate (McCain et al. 1993).
- The use of production indicators in place of of EUR provides an acceptable alternative for gauging (Voneiff and Cipolla 1996).
- The production indicator plot was able to rapidly identify areas of depletion before drainage areas had been calculated through conventional reservoir engineering techniques (Voneiff and Cipolla 1996).
- Accurate quantification of infill potential cannot be made using only the moving-domain scoping study but requires some level of detailed conventional engineering to quantify drainage areas. Without the input of drainage areas, an infill estimate from the moving-domain analysis should be considered an upside estimated to infill potential (Voneiff and Cipolla 1996).

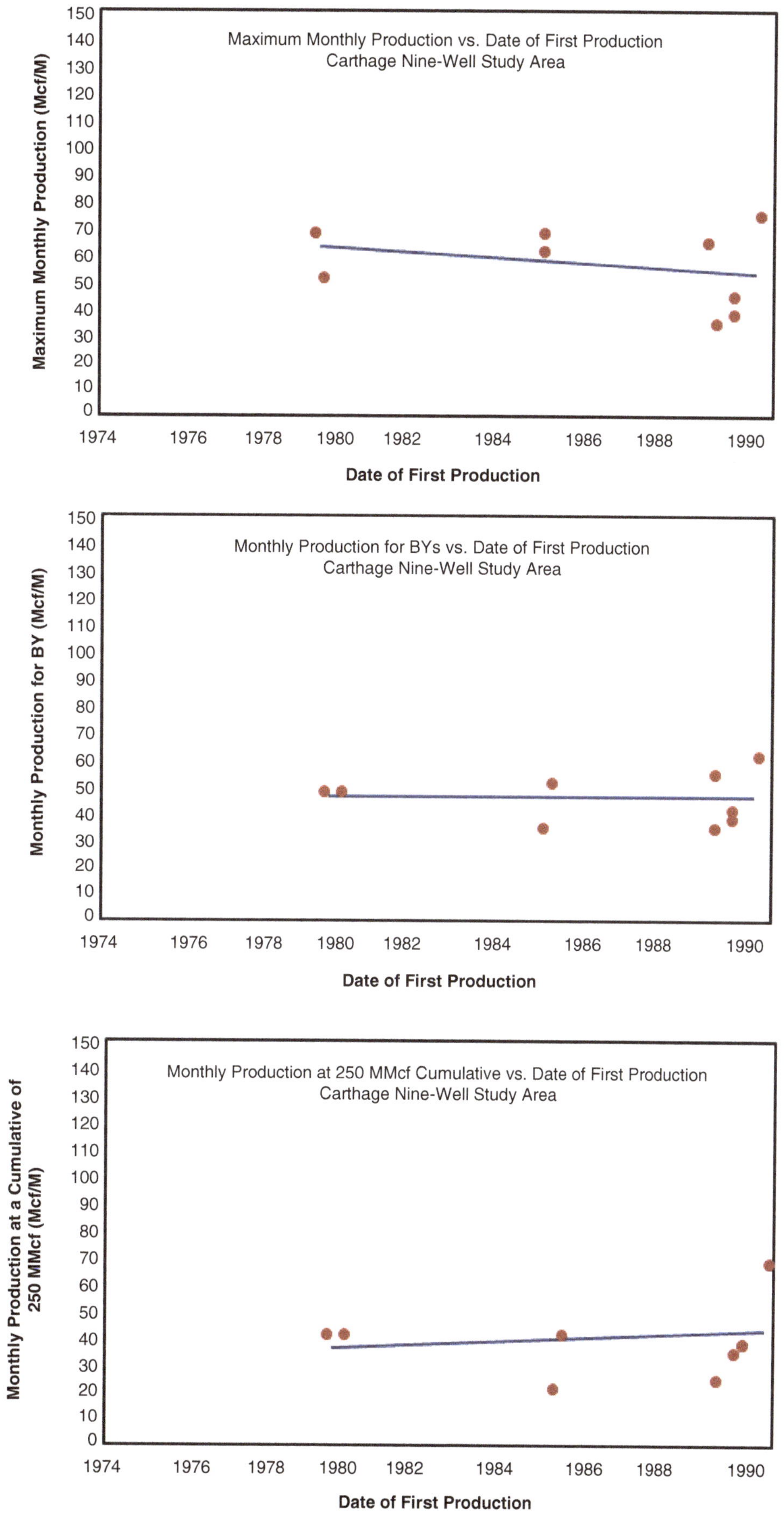

Fig. 11.8—Indicator plots for a nine-well, 640-acre study area in the Carthage (Cotton Valley) Field (McCain et al. 1993).

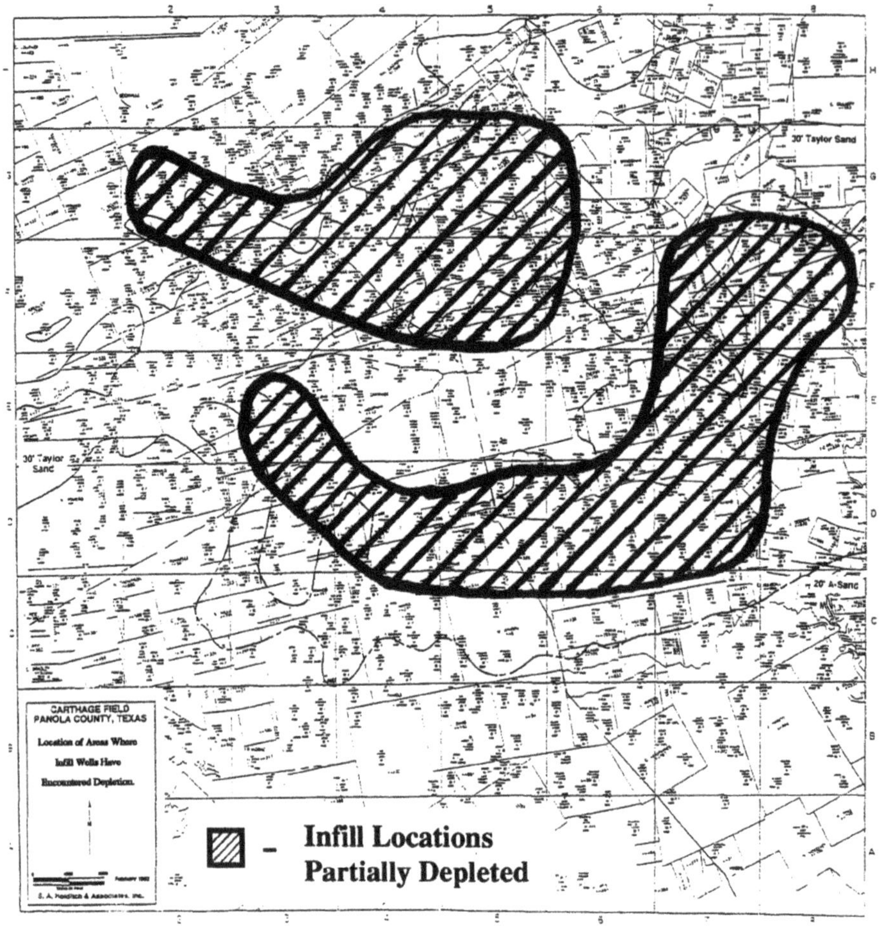

Fig. 11.9—Base map of the Carthage (Cotton Valley) Field showing areas where infill wells indicate partial depletion. This map is centered on the field and contains all but a few fringe wells (McCain et al. 1993).

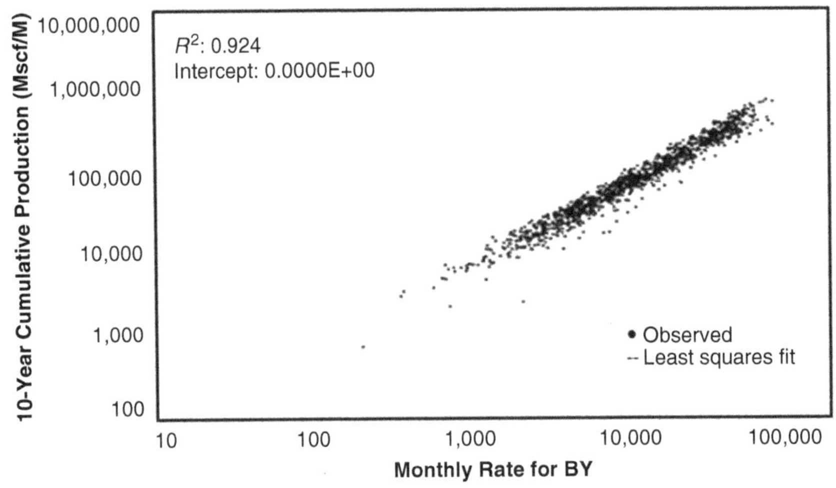

Fig. 11.10—BY vs. 10-year cumulative correlation (Voneiff and Cipolla 1996).

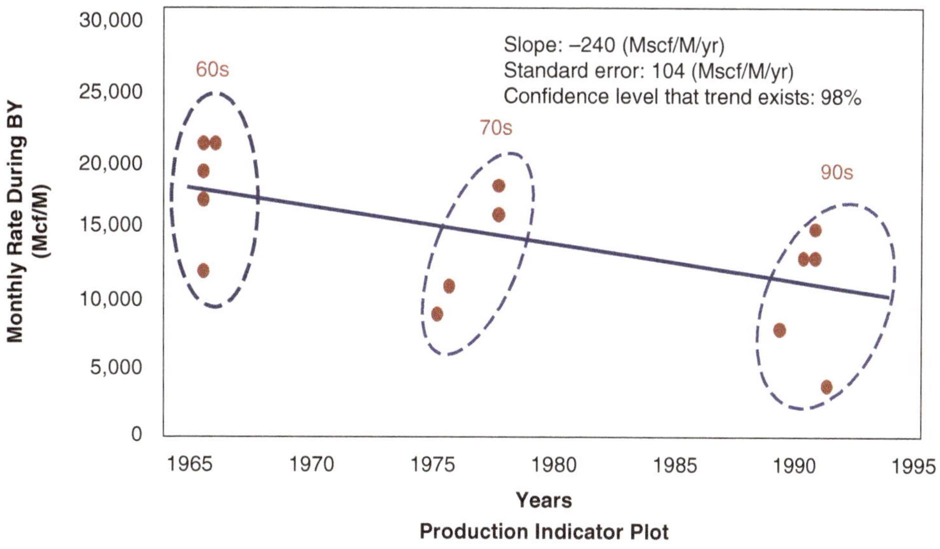

Fig. 11.11—Production indicator plot (Voneiff and Cipolla 1996).

11.4 Average Recovery

In tight gas reservoirs, one must understand the distribution of reservoir properties and then determine the proper method of averaging these reservoir properties to be able to predict the performance of additional wells accurately. Knowledge of how reservoir properties are distributed is critical, especially knowledge of the most important properties, including formation permeability and porosity, net-pay, and areal size of the reservoir. In most tight gas reservoirs, the values of permeability will vary vertically from layer to layer. To obtain an average permeability for a single vertical well, it is best to use the thickness-weighted averaging method. The same approach usually applies to other petrophysical properties. When average permeability and other reservoir properties have been obtained for all wells in a tight gas formation, the next task is to look at the areal distributions and determine an average for the entire formation. Rollins et al. (1992) said that distribution of permeability in a formation can be generally characterized as unimodal, rightward skewed, and similar to a log-normal distribution. They concluded that a median value is a better estimate of average formation permeability than other methods.

Holditch et al. (1991a) discussed how to obtain average reservoir properties and how to estimate ultimate gas recovery from an average well in a tight gas formation. They used the permeability distribution of a data set containing information from 191 wells in the Travis Peak Formation, as shown in **Fig. 11.12.** They correlated porosity to permeability, then correlated net-pay to permeability, as shown in **Figs. 11.13 and 11.14.**

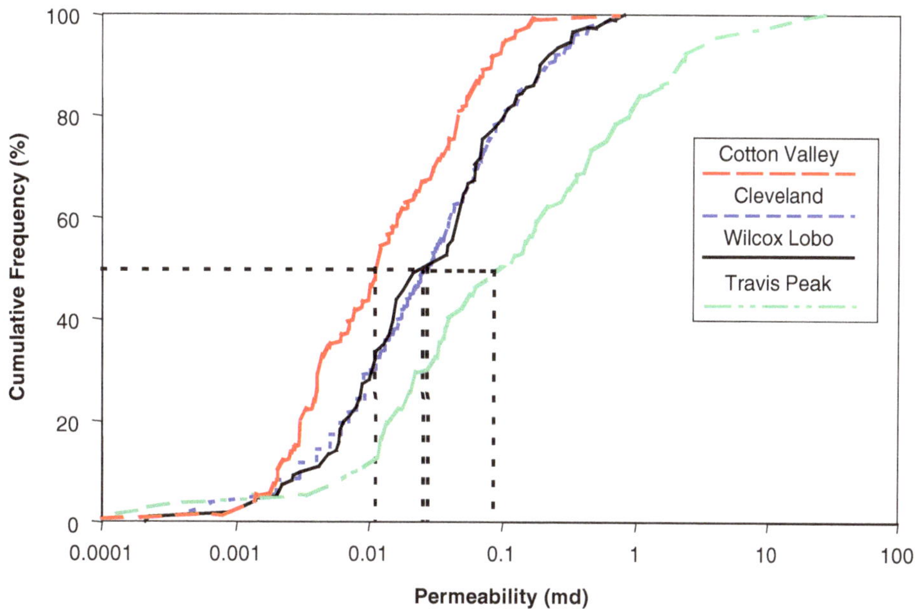

Fig. 11.12—Areal permeability distribution.

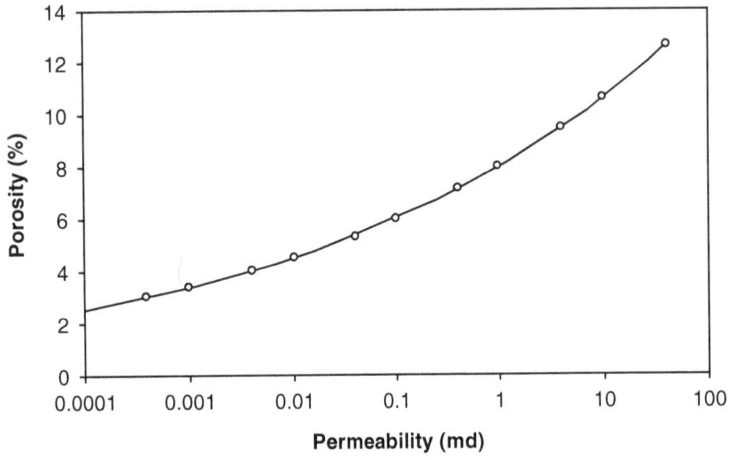

Fig. 11.13—Correlation of porosity with permeability.

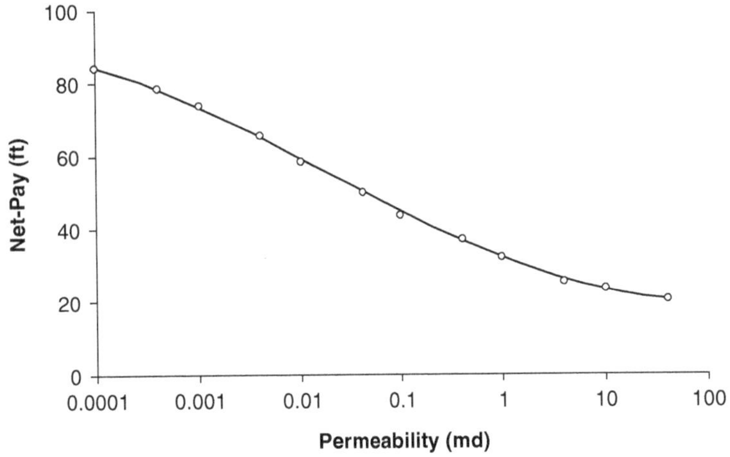

Fig. 11.14—Correlation of net-pay with permeability.

With this information, Holditch et al. (1991a) conducted a reservoir simulation study to generate ultimate gas recovery from 1,000 wells by changing permeability, porosity with respect to permeability, net-pay with respect to permeability, drainage area, hydraulic fracture length, and effects of uneconomic wells. On the basis of the systematic study, they concluded that

- Areal distribution of permeability in a tight gas reservoir can be described as unimodal, rightward skewed, and approximately log-normal.
- A median value of permeability should be used to predict gas recovery from an average well in a tight gas formation.
- Uneconomic wells should be included in the permeability-distribution evaluation.
- The average recovery for all wells will be the average recovery based on the median permeability of economic wells times the fraction of wells that are productive.
- In the Travis Peak Formation, an average well should be fracture treated and produce approximately 1.84 Bcf in a 160-acre spacing.

11.5 Infill Drilling Potential

To fully develop a tight gas field and/or to improve economics, one must drill infill wells. It is not uncommon to have hundreds or thousands of infill candidates to choose from in a given field. The key is to identify the locations that will produce the most gas per well.

Quantifying infill potential requires a comprehensive assessment of the geology and production history of the wells. One must understand the nature of the layered, low-permeability formation and the history of how the wells were drilled, completed, and stimulated. The technical team also needs to review the well spacing and estimate current drainage volumes for the existing wells.

When the geological, engineering, and production data have been obtained, one can choose to conduct a reservoir simulation study of the field and then identify infill locations by analyzing saturation maps, pressure maps, and permeability maps, which were discussed in detail in Part 2 Ch 9: Well and Reservoir Numerical Modeling. After infill locations are identified, one can make predictive simulations to quantify reserves from infill wells and rank these wells to help schedule drilling and development programs. Finite-difference reservoir modeling is probably the most accurate method to identify infill locations, but it can be performed only if the data are good enough to justify the cost of a fieldwide simulation study. Many times, the data are not good enough, the time allowed for such a study is limited, or the engineering organization does not have the expertise to conduct a reservoir simulation study. Under these conditions, other methods must be used to determine infill locations.

If a field does not have many wells, one can apply reservoir-engineering analysis methods, including volumetric methods, type curves, analytical models, and/or numerical models, to evaluate each well and to obtain the effective drainage area. One can then compare these drainage areas to well spacing on a map to identify areas with infill potential. Additionally, one can map EUR to overlap with infill locations to identify and rank areas with highest to lowest infill potential.

If a field does have hundreds or thousands of wells, the time and cost requirements of a conventional reservoir engineering study might be too high to justify a detailed reservoir engineering study. In these situations, the moving-domain technique, initiated and developed at S. A. Holditch and Associates, Inc., could be used to identify infill locations and might require only a couple of weeks. Details of the moving-domain technique were discussed in Section 11.3. The key is to find a reliable production indicator, for example the BY, as a proxy for EUR. By doing this, one can save much time and effort in estimating EUR for all wells because the production indicator is readily available. However, the production indicator must be checked and validated on a fraction of wells with an in-depth reservoir engineering study to make sure it does reflect physics and reservoir performance. When the production indicator has been obtained and validated, one can investigate

- Evidence of depletion by comparing new wells with old wells and by plotting production indicators vs. time. However, one must consider heterogeneity of the reservoir and stimulation effectiveness in analyzing depletion.
- Drainage area vs. well spacing. It is important to evaluate wells that have sufficient petrophysical and production data and are already in boundary-dominated flow, to obtain drainage areas. One can then estimate EUR and drainage area using production

indicators and validated decline-curve models. **Fig. 11.15** shows 1,384 drainage areas, where 325 drainage areas were obtained based on type-curve analysis and 1,059 were estimated using calibrated decline-curve analysis (Voneiff and Cipolla 1996). It shows that 84% of the drainage areas are less than 80 acres per well. There are many infill locations because well spacings were 80 to 310 acres per well. **Fig. 11.16** is a bubble map of the drainage areas of all wells in the field. It is obvious that Area 1A has few undrained areas and Area 2B probably has well interferences. **Fig. 11.17** shows 636 infill locations among 1,246 wells. One can see that sparsely drilled areas, such as 4A and 4B, have a large number of infill locations.

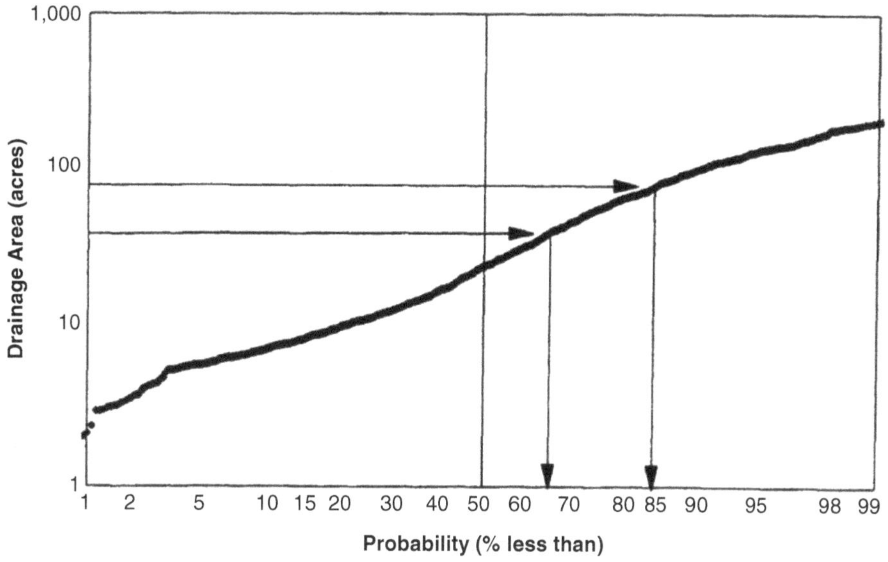

Fig. 11.15—Probability of drainage area in Ozona (Voneiff and Cipolla 1996).

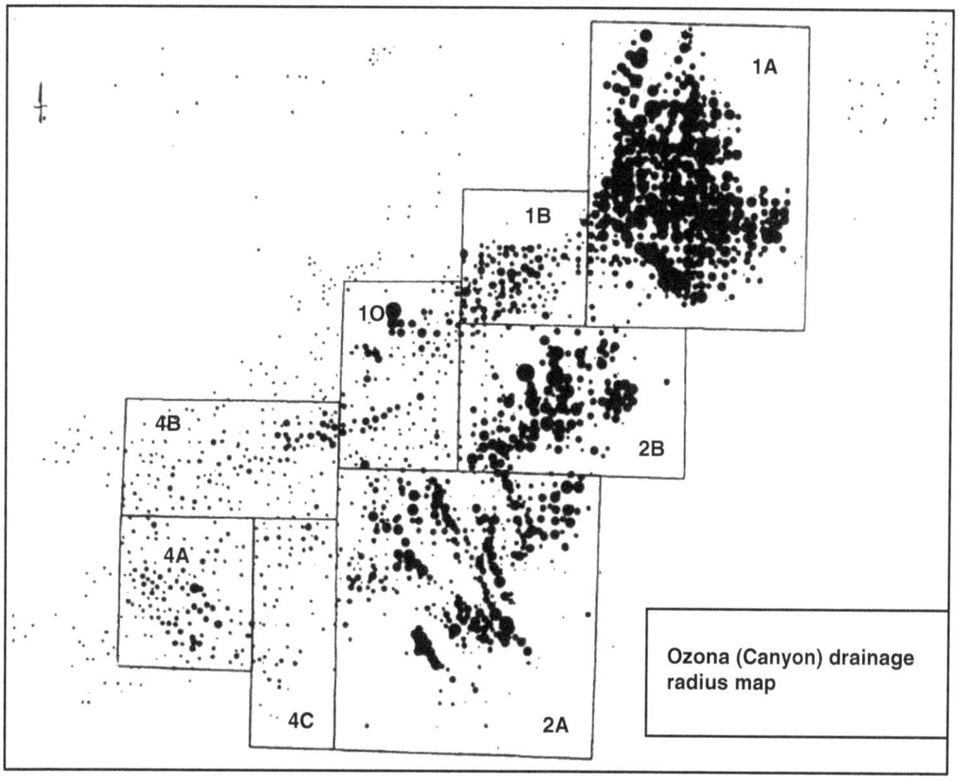

Fig. 11.16—Ozona (Canyon) drainage radius map (Voneiff and Cipolla 1996) developed at S. A. Holditch & Associates Inc.

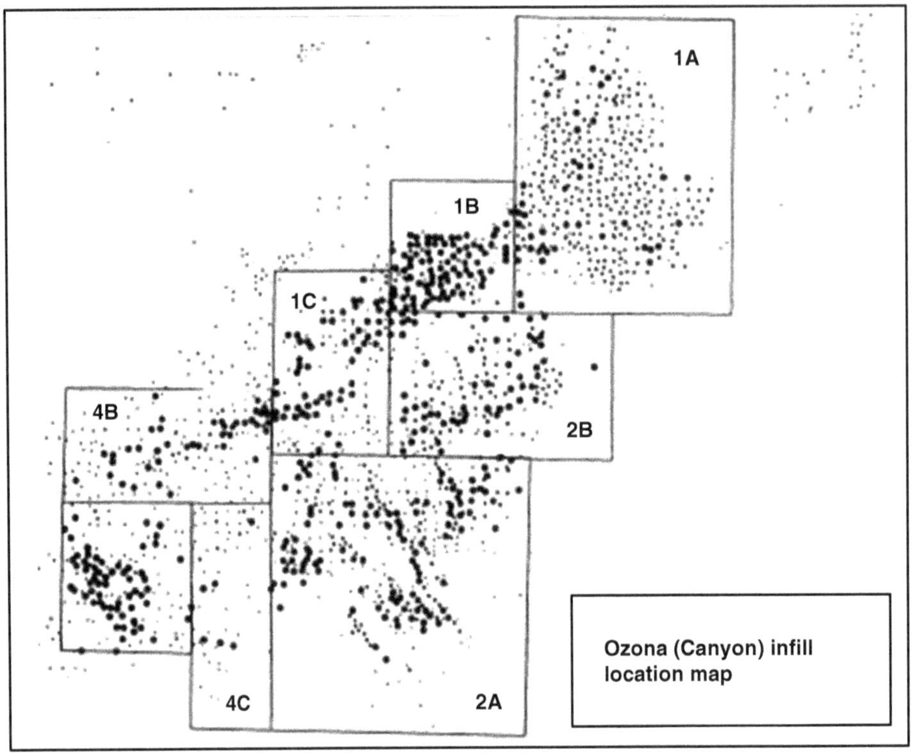

Fig. 11.17—Ozona(Canyon) infill location map (Voneiff and Cipolla 1996) developed at S. A. Holditch & Associates, Inc.

11.6 Well Completion Information

To improve development economics and to increase reserves and recovery, we must work on and optimize well completions and recompletions. Completions refers to all operations starting from when the bit reaches the total depth (drilling engineers in charge) to when the well is put on production (production engineers in charge). Because most tight gas formations are layered, it is important to complete each layer in an optimized way to maximize gas flow into the wellbore. This is more important in thick formations, such as the Mesaverde Formation. This section discusses well completions and recompletions from the perspective of reservoir performance for economic analysis.

In new infill wells, it is important that one not overdesign the well completion, which increases costs and might even result in uneconomic wells. Overdesign includes using higher-than-necessary casing grade, completing layers with a limited amount of gas-in-place, designing a fracture treatment that is larger than needed, and selecting large-volume surface equipment to handle short-term high flow rates instead of long-term flow rates. By the same token, it is as important that one not underdesign well completions to save a penny and then lose a dollar. We have seen wells where casing and tubulars are not strong enough or large enough to accept certain artificial-lift systems or stimulation treatments and gas-bearing layers are not being stimulated effectively or at all. The most common mistakes are not pumping enough propping agent and not designing the treatment to properly transport the proppant deep into the hydraulic fracture.

In existing wells, one can and should apply the best petroleum-engineering technologies to evaluate and understand well performance to identify development potentials and increase economic returns with minimal recompletion costs. Past experience shows that one might be able to increase production substantially by doing one or a combination of the following:

- Recompleting bypassed layers with sufficient gas-in-place
- Installing effective artificial-lift systems to remove liquid loading
- Removing near-wellbore and/or fracture damages by chemical or solvent injections
- Refracturing existing wells

The key to completing new wells or recompleting existing wells successfully is to apply the formation evaluation technologies discussed in Part 2 Ch 5: Reservoir Characterization, Part 2 Ch 6: Prestimulation Well Testing, Part 2 Ch 7: Post-Fracture Treatment Well Testing, Part 2 Ch 8: Rate Transient Analysis, and Part 2 Ch 9: Well and Reservoir Numerical Modeling, and the reserves estimation techniques in Sections 11.1 through 11.4 of this chapter. The moving-domain technique can be used to identify wells with potential and evaluate the effectiveness of different completions. We understand that advances have been made in the industry, but what we have discussed here provides a framework that can be built upon.

11.7 Estimating Costs to Develop Unconventional Reservoirs

We discussed how to estimate benefits—total value (reserves) and gross revenues (production)—of tight gas development in Sections 11.1 through 11.6. Costs to develop a tight gas field include many items and details, which change dynamically and depend upon geographical areas. We do not plan to discuss all the details of cost estimates, but will discuss important aspects of costs in two categories:

- F&D costs
- Operating costs

We will also discuss what must be considered to minimize these costs without sacrificing revenue. We will also discuss the associated risks that must be considered. F&D costs are often referred to as capital expenditure (CAPEX), which must be paid for by the operating company and its working-interests partners and investors (at a total of 100% working interests) to get a well drilled, completed, and put on production. Operating costs are often referred to as operational expenses (OPEX), which will be paid for from cash flow after the well is producing. Of course, the key is to have a positive cash flow.

11.7.1 F&D Costs. Before we explain these costs in more detail, one should keep in mind that the objective of a prudent operator is to control costs and complete the well in an optimal manner so we can achieve the lowest possible F&D costs. F&D costs are reported as USD/Mcf in tight gas sands, where the cost of drilling and completing the well is compared to the EUR of the well. A prudent operator will maintain low F&D costs while also considering any associated risks with a specific development project. Low F&D costs mean the gas was found and developed at a cost that, when compared to the gas price, will allow the operator to make enough profit to justify the development. It is important to control the costs and, at the same time, to drill, complete, and fracture treat the well using the best technology to maximize the ultimate recovery. Investors look for companies who can maximize reserves and keep the F&D costs to a minimum.

Some of the items one must consider in computing the costs involved with a project are

- Land acquisition costs. In oil and gas mineral leases, the owners are compensated at market prices and are given a royalty interest in the production. It is important that the operator determine the market conditions in an area and obtain leases that are in line with the ability of the wells to produce. In active areas, some operators tend to overpay for some leases. There is always a tradeoff between the lease bonus and the royalty rate. Some mineral owners want more upfront cash, while others would rather have a higher royalty rate. As with most issues in the oil business, these details can be negotiated and put in the lease.
- Well pad and roads. Pad drilling is often used when drilling multiple horizontal wells in shale reservoirs. In tight gas sand reservoirs, most wells are drilled vertically with one well per pad. There are exceptions, such as the way the Mesaverde Reservoir is developed in several Rocky Mountain Basins. A major independent producer has been pad drilling in developing its tight gas fields in Rocky Mountain areas for decades. Pad drilling decreases the cost of building additional well pads, roads, and gathering lines. At the same time, pad drilling minimizes environmental impacts and decreases D&C and processing costs because of centralized operations. However, in most tight gas sand reservoirs, the fields are developed with one well per pad, and the costs of the pads and roads are included in the costs associated with the F&D costs.
- D&C and stimulation cost. Drilling, completions, and stimulation should be designed to optimize production and reserves. Many times, with an accurate formation evaluation, one can increase gas reserves and production by a large percentage with just a small added cost. It is not smart to reduce costs if the reduction results in lower production rates and recoveries.
- Costs of processing equipment and gathering lines. Although it is not a major cost overall, it is still important to design processing equipment properly to increase its effectiveness and efficiency. Even without pad drilling, centralized processing, gathering, and compression can decrease operational costs, lower the economic limit, and increase the well or field life.
- Risks. Risks exist, and most people tend to be overconfident. In the development phase, risks associated with costs are usually lower than the risks that occurred during the exploration phase. However, there is still a risk of drilling high-volume wells vs. low-volume wells or even uneconomic wells because of changing geology or operational issues that result in lower flow rates. A common question is, "How many wells should we drill, and where should we drill these wells to achieve the best economics in a given time?"

Hudson et al. (2001) quantified F&D costs and the associated risks of infill drilling in the Cotton Valley Formation in East Texas, the Morrow Formation in the Permian Basin, and the Mesaverde Formation in the San Juan Basin using two different methods: average prediction and decision tree. In the Cotton Valley Field study, they first identified infill locations with information from 1,505 existing wells using the moving-domain method, then divided the field into nine subareas, as shown in **Fig. 11.18.** Afterward, they calculated average EUR for each subarea using average prediction methods. Using a representative set of existing wells, they calculated cash flow and discounted cash flow (DCF) for each area using a typical production decline for Cotton Valley, which matched EUR for each subarea. They then tested their calculations with a subset of 90 wells that were not included in the original correlation. **Table 11.1** shows the costs and assumptions. **Table 11.2** shows the economics for each subarea ranked on the basis of risks and cash flow. The total F&D costs and risk ratio for the 83 infill wells are USD 0.47/Mcf and 82%, respectively. Average EUR per well is 2.1 Bcf per well. The authors also applied the decision-tree method and achieved results similar to average prediction methods.

Hudson et al. (2001) then tried to answer the question of how many wells should be included in an infill drilling program. **Fig. 11.19** shows the distribution of predicted EUR for a one-well drilling program in Subareas 3 and 8. They concluded that, although Subarea 3 has a higher possibility of higher EUR wells than subarea 8, it also has a higher risk of having poorer wells. In other words, Subarea 8 is less risky. **Fig. 11.20** is noteworthy because it helped operators decide the number of wells in a drilling program with respect to associated risks. If just one infill well is drilled in Subarea 8, it has an 80% chance of being an economic well; if a drilling program of five infill wells is implemented, the wells (on average) have a 96% probability of being economic wells.

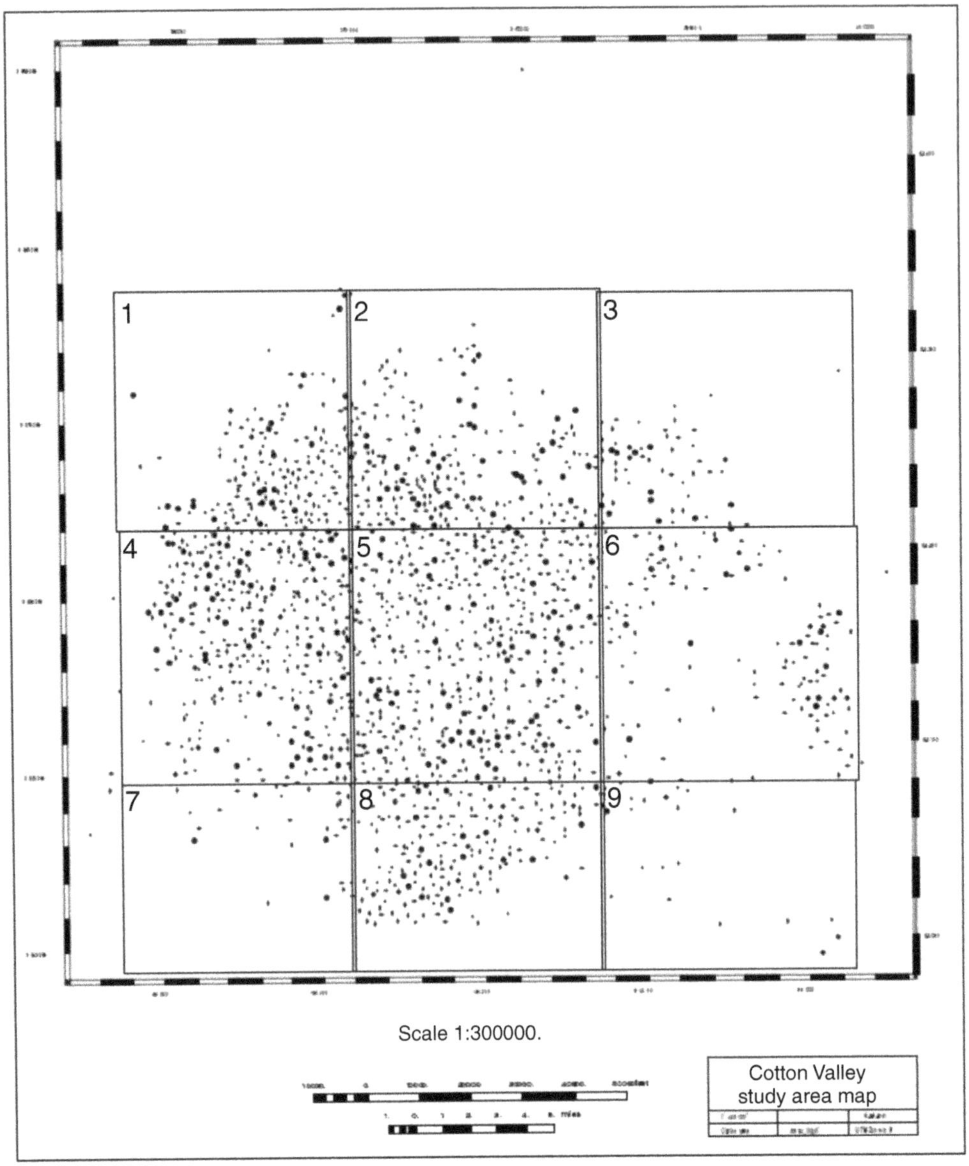

Fig. 11.18—A base map showing nine subareas created for the Cotton Valley study (Hudson et al. 2001).

	Cotton Valley	Mesaverade	Morrow
D&C cost (million USD)	1	0.6	1.2
Gas price (USD/Mcf)			
Lease operating expense (USD/well/month)	1,000	800	1,200
Net revenue interest (%)	75	75	75

Table 11.1—Economic assumptions (Hudson et al. 2001).

In summary, F&D costs in units of USD/Mcf and the associated risks should be minimized by conducting a geological, engineering, and economic study for a tight gas development project. The F&D costs can be minimized by controlling costs and increasing the reserves. The engineer should try to accomplish both.

11.7.2 Operating Costs. When estimating operating costs, one must consider

- Cost of electricity and or fuel gas. One can try to optimize equipment and its operations, including pumps, heaters, dehydrators, and compressors, to minimize costs. If fuel gas is used, in some cases royalty interest holders might need to be compensated for this, depending on the lease documents.
- Cost of water disposal. Water disposal can be quite expensive in some geographical areas, but advances in water treatment and reuse have improved the situation. However, when possible, one must complete a well in ways that minimize unnecessary water production, especially from layers above or below pay zones. Water disposal costs are not always an issue in tight gas

Infill potential by subarea						Case-study wells drilled after June 1998			Predicted totals for all case-study wells			Theoretical drilling program—only drill in areas where average predicted DCF is positive				
Area	Actual number of wells per area	Potential number of infill candidates	Average predicted DCF (USD 1,000)	Expected value DCF per area (USD 1,000)	Probability of first well being economic	Actual number of wells drilled	Assumed drilling cost (USD 1,000)	Number of wells with negative NPV	EUR for actual wells drilled (MMcf)	Actual total CF (USD 1,000)	Actual total DCF (USD 1,000)	Theoretical number of wells drilled	Number of wells with negative DCF for theoretical drilling campaign	Projected EUR for theoretical drilling campaign (MMcf)	Projected CF for theoretical drilling campaign (USD 1,000)	Projected DCF for theoretical drilling campaign (USD 1,000)
Areas where average predicted DCF is positive																
1	152	18	1,857	1,870	99.24%	6	6,000	1	19,431	35,129	13,261	6	1	19,431	35,129	13,261
2	214	54	1,405	1,328	94.23%	28	28,000	1	73,055	125,840	44,337	28	1	73,055	125,840	44,337
4	352	33	1,190	1,169	93.18%	26	26,000	4	52,465	83,625	25,862	26	4	52,465	83,625	25,862
5	458	50	748	752	77.44%	14	14,000	6	17,759	22,352	3,479	14	6	17,759	22,352	3,479
3	55	29	502	460	72.62%	5	5,000	1	9,996	15,881	4,881	5	1	9,996	15,881	4,881
8	196	20	448	398	79.66%	3	3,000	1	3,833	4,850	773	3	1	3,833	4,850	773
7	29	11	225	208	64.22%	1	1,000	1	506	−51	−511	1	1	506	−51	−511
9	17	8	146	130	60.64%	0	0	0	0	0	0	0	0	0	0	0
Areas where average predicted DCF is negative																
6	121	54	(38)	(22)	47.50%	7	7,000	5	7,387	7,950	253	0	0	0	0	0
Total						90	90,000	20	184,433	295,575	92,335	83	15	177,046	287,625	92,082

Table 11.2—Economic summary by study area for the Cotton Valley Formation (Hudson et al. 2001). NPV = net present value.

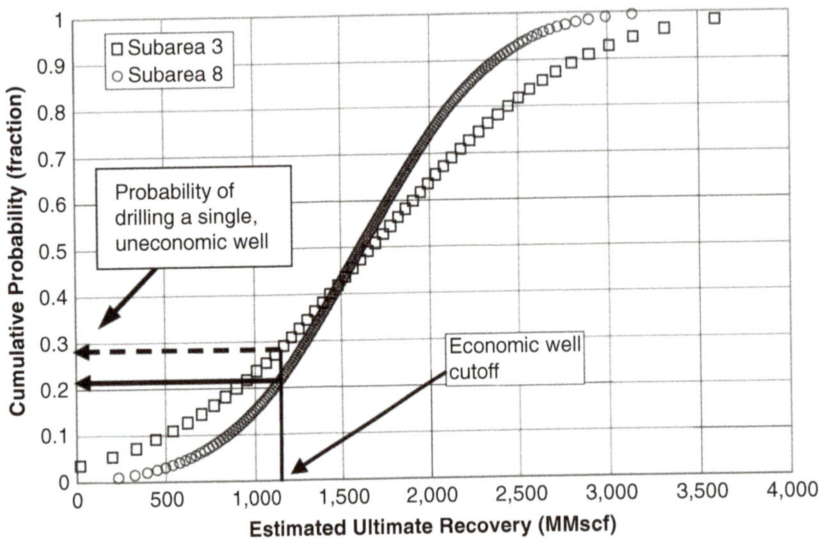

Fig. 11.19—Cumulative probability of predicted infill EURs for a one-well program in Subareas 3 and 8 (Hudson et al. 2001).

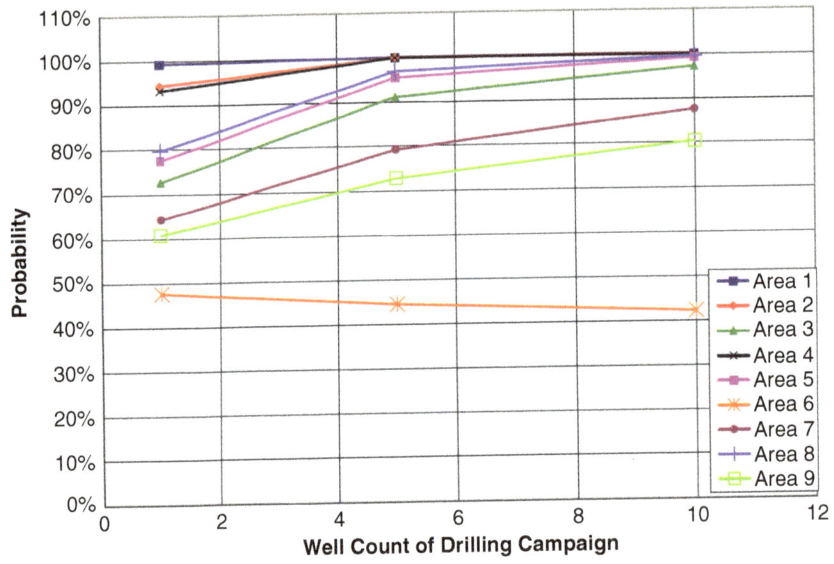

Fig. 11.20—Probability that an average well within a drilling program of one, five, and ten wells will be economic (Hudson et al. 2001).

sands because many of the wells produce very little water. However, if a well does produce substantial volumes of water, water disposal costs can affect well economics, especially when gas prices are depressed and water-disposal costs are high.

- Well services. Over time, wells need to be worked over, so well services are always necessary and there should be a budget for workovers. One should also budget for the maintenance of roads, well pads, and equipment.
- Personnel. Oilfield automation can decrease this cost and help optimize production, but a human will always need to check on the wells, so these costs must be estimated.
- Engineering. This is not a fixed cost, but money should be budgeted for engineering studies to optimize production and to design workovers as needed.
- Administration (well management fee). This cost depends on the operating agreement and structure of the operating company. It might be a fee per well or a fraction of revenue.
- Insurance. Insurance companies provide a variety of insurance products to help operating companies cover the costs of operational accidents if they occur.

As an illustration of what is typically a part of operating costs, a Travis Peak lease of seven wells has an OPEX breakdown of 9% for electricity, 7% for water disposal, 43% for well services, 12% for pumpers, 10% for insurance, and 19% for administration fees in East Texas as of April 2018.

In summary, the operating costs depend on the operating accompany, fields, and geographical areas. With experience, companies tend to find ways to minimize OPEX.

11.8 Field Development Strategies and Economics

Although geologic exploration risks in developing tight gas are lower those that in conventional oil and gas development, the risks of drilling uneconomic wells is high in many reservoirs because of the heterogeneity and complexity of geological systems in tight gas formations. To drill just one well and achieve economic goals successfully, one probably needs some good luck in addition to good geological and engineering work. However, to develop a tight gas field successfully, one must understand that many wells will have to be drilled in order to find the "sweet spots," which are the more-permeable portions of the tight gas sands. To succeed, the best strategy is to drill a number of wells to balance out the risks by following the rule of 80/20: 20% of wells will produce 80% of the reserves. We discussed the impact of numbers of wells on risks in the Finding and Development Costs subsection and how to obtain a minimum number of wells to achieve a low and acceptable risk. Another successful business strategy is to apply technologies, such as recompletions, workovers, and seismic imaging, to improve economics and reserves.

It is important to compute project economics to optimize development planning and to help rank projects using economic yardsticks including earnings before interest, taxes, depreciation, and amortization; net present value; rate of return; internal rate of return; and payback time. One could consult a book or a paper on the details of how to compute these yardsticks, which is straightforward once reserves, production, and costs are estimated accurately. Taxes might include the impact fee, severance tax, state franchise or net income tax, and federal income tax.

In summary, project economics is a primary way to evaluate costs and benefits of a project, optimize development, and rank projects. To develop a tight gas reservoir, one must drill enough wells to identify the 20% of the wells that will pay 80% of the F&D costs plus operating costs.

11.9 Benefits of Applying New Technology

New technologies are developed to help increase gas production (benefits) or to decrease costs, or both, in tight gas development. Technologies can also help quantify and reduce the risks associated with a development project, such as environmental or safety risks, which in turn will eventually increase the profitability of a project. However, there is an associated cost for deployment of any new technology; we must therefore evaluate the cost/benefit ratio of any new technology.

To obtain optimal results from hydraulic fracturing, several new fracturing technologies have been applied to prevent inappropriate stimulation designs and fracturing-fluid problems. Despite industry awareness of these technologies, they are not applied routinely because of their added expenses. In addition, the benefits are difficult to quantify and often are not clearly understood. Voneiff and Holditch (1994) presented an economic assessment of fracture-treatment quality control, in-situ stress profiling, and 3D fracture-propagation models. The following text explains the work and what Voneiff and Holditch computed as part of the Gas Research Institute's tight gas sands research portfolio.

Fluids in a hydraulic fracturing treatment are complex mixtures of chemicals that must meet or exceed the specifications designed by the engineer in the office. One can know the magnitude of the actual fluid problems that occur during a treatment only by being in the field and making fracturing fluid-property measurements. Fracture-treatment quality control can be applied to prevent and/or correct most problems. The fluids used to stimulate tight gas reservoirs are complex systems that require intense quality control for correct mixing and pumping correctly. Fracturing fluids are very sensitive to chemical concentrations and mixing procedures. **Table 11.3** shows the costs of performing each level of quality control. When the design engineer does not know the in-situ stress contrast, accurately predicting the length and height of the created fracture becomes difficult. Without knowledge of the created fracture geometry, the optimal treatment volume cannot be computed consistently, resulting in stimulation treatments that can be inappropriate for a given reservoir. To compute an in-situ stress profile economically, one must use data primarily from logs and data from formation tests. **Table 11.4** shows the cost of applying in-situ stress profiling. A typical operator applies technology at either the none or standard level. Most of the time in 2D modeling, when engineers use in-situ stress and permeability thickness profiles to design the well completion, they tend to underestimate fracture height and thus overestimate fracture length in the 2D fracture design model. However, using all three profiles in a 3D fracture-propagation model, an engineer can use 3D fracture and reservoir models to optimize the completion of a given well. **Table 11.5** shows an example of the benefit calculations for applying the technologies in the Travis Peak Formation. **Fig. 11.21** summarizes the benefit/cost ratios of applying the technologies in six different formations. Voneiff and Holditch (1994) concluded that these technologies increase gas reserves by 10 to 20% for a 5 to 10% increase in the well cost. The application of these technologies is profitable when the formation gas permeability is between 0.001 and 0.3 md (Voneiff and Holditch 1994).

Hoover et al. (2009) reported that a new water-based mud system for depleted tight gas reservoirs had been successfully introduced to provide increased drilling performance as compared to conventional water-based mud systems. This technology is applicable to any depleted tight gas reservoir where losses or stuck pipe occurs. Bobatchev and Holditch (2008) developed a software package that can be used to help engineers make decisions while completing and stimulating tight gas reservoirs. The modules include perforation selection and proppant selection. On the basis of input of well or reservoir parameters as input, these subroutines provide unambiguous recommendations for perforation and proppant selection.

In summary, the costs and benefits of applying new technologies can be evaluated using the discussed methodologies. The key is to understand the technology and then to estimate the oil and gas production before and after deploying a new technology by use of reservoir engineering techniques.

Level	Frequency of Significant Problems (%)	Per-Treatment Cost USD)	Description
None	50	0	Do not test fluids or inventory materials.
Standard	32	1,000	Inventory materials and check fluid rheology using rotational viscometer and heated cup.
Intense	5	10,000	Inventory materials; test gel before and during treatment at bottomhole temperature and share rate for length of pump time.

Table 11.3—Definition and cost of fracture-treatment quality control.

Level	95% of Stress Contrast Errors are Between (psi)	Per-Well Cost (USD)	Description
None	± 600	0	Use nonacoustic openhole logs to estimate fracture height.
Standard	± 300	10,000	Use acoustic logs to compute and define in-situ stress profile.
Intense	± 100	15,000	Calibrate acoustic log stresses with stresses measured from pump-in/falloff tests performed on one out of every six wells.

Table 11.4—Definition and cost of in-situ stress profiling.

Outcome of Fracture Design and Implementation Process	Discounted Revenue (USD)	Technology Cost (USD)
Range of outcomes	2,651,000	NA
Optimal	1,800,000	NA
Worst conditions		
Statistical evaluation of outcomes	2,475,000	25,000
Average with intense technology	2,034,000	—
Average without technology	441,000	25,000
Technology benefit and cost		
NA = not applicable		

Table 11.5—Example benefit calculation for Travis Peak Formation when the formation permeability is 0.04 md and technology is applied at an intense level.

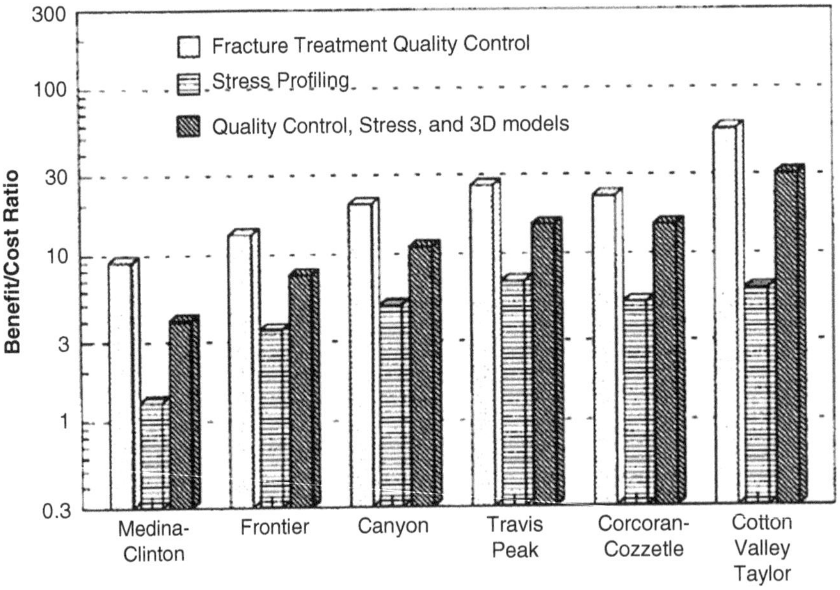

Fig. 11.21—Benefit/cost ratio of new fracture technology for six formations.

11.10 Summary

We first provided a high-level overview of the practical methods to estimate reserves and forecast production, 2009 SEC regulatory rules on reserves, and infill drilling and completions for improved economics and the associated risks. We then discussed cost estimates in the categories of F&D costs and operating costs and what one needs to pay attention to. We finally discussed development strategies and economics and their importance in successful tight gas development. This gives a solid foundation and perspective in computing and applying development economics.

11.11 References

Agarwal, R. G., Gardner, D. C., Kleinsteiber, S. W. et al. 1998. Analyzing Well Production Data Using Combined Type Curve and Decline Curve Analysis Concepts. Presented at the SPE Annual Technical Conference and Exhibition, New Orleans, Louisiana, 27–30 September. SPE-49222-MS. https://doi.org/10.2118/49222-MS.

Aniemena, C. 2014. *New Predictive Models for Reserve Assessment of Shale Gas Wells*. Ph.D. Dissertation, The Pennsylvania State University, State College, Pennsylvania (May 2014).

Aprilia, A. W., Li, Z., McVay, D. A. et al. 2006. Quantifying Uncertainty in Original-Gas-in-Place Estimates With Bayesian Integration of Volumetric and Material Balance Analyses. Presented at the SPE Gas Technology Symposium, Calgary, Alberta, Canada, 15–17 May. SPE-100575-MS. https://doi.org/10.2118/100575-MS.

Arps, J. J. 1945. Analysis of Decline Curves. *Trans. AIME* **160** (1): 228–247. SPE-945228-G. https://doi.org/10.2118/945228-G.

Bogatchev, K. Y. and Holditch, S. A. 2008. Developing Tight Gas Sand Adviser for Completion and Stimulation in Tight Gas Sand Reservoirs Worldwide. Presented at the CIPC/SPE Gas Technology Symposium 2008 Joint Conference. Calgary, Alberta, Canada, 16–19 June.

Chen, H. Y. and Teufel, L. W. 2002. Estimating Gas Decline-Exponent Before Decline-Curve Analysis. Presented at the SPE Gas Technology Symposium, Calgary, Alberta, Canada, 30 April–2 May. SPE-75693-MS. https://doi.org/10.2118/75693-MS.

Cheng, Y., Lee, W. J., and McVay, D. A. 2007. Improving Reserves Estimates From Decline Curve Analysis of Tight and Multilayer Gas Wells. Presented at the Hydrocarbon Economics and Evaluation Symposium, Dallas, Texas, 1–3 April. SPE-108176-MS. https://doi.org/10.2118/108176-MS.

Cox, S. A., Gilbert, J. V., Sutton, R. P. et al. 2002. Reserve Analysis for Tight Gas. Presented at the SPE Eastern Regional Meeting, Lexington, Kentucky, 23–26 October. SPE-78695-MS. https://doi.org/10.2118/78695-MS.

Cox, S. A., Stoltz, R. P., Wilson, A. S. et al. 2003. Reserve Analysis for Multilayered Tight Gas Reservoirs. Presented at the SPE Eastern Regional Meeting, Pittsburgh, Pennsylvania, 6–10 September. SPE-84814-MS. https://doi.org/10.2118/84814-MS.

Engler, T. W. 2000. A New Approach to Gas Material Balance in Tight Gas Reservoirs. Presented at the SPE Annual Technical Conference and Exhibition, Dallas, Texas, 1–4 October. SPE-62883-MS. https://doi.org/10.2118/62883-MS.

Fetkovich, M. J. 1980. Decline Curve Analysis Using Type Curves. *J Pet Technol* **32** (6): 1065–1077. SPE-4629-PA. https://doi.org/10.2118/4629-PA.

Fetkovich, M. J., Fetkovich, E. J., and Fetkovich, M. D. 1996. Useful Concepts for Decline Curve Forecasting, Reserve Estimation, and Analysis. *SPE Res Eng* **11** (1): 13–22. SPE-28628-PA. https://doi.org/10.2118/28628-PA.

Fetkovich, M. J., Vienot, M. E., Bradley, M. D. et al. 1987. Decline Curve Analysis Using Type Curves: Case Histories. *SPE Form Eval* **2** (4): 637–656. SPE-13169-PA. https://doi.org/10.2118/13169-PA.

Gao, H. and McVay, D. A. 2004. Gas Infill Well Selection Using Rapid Inversion Methods. Presented at the SPE Annual Technical Conference and Exhibition, Houston, Texas, 26–29 September. SPE-90545-MS. https://doi.org/10.2118/90545-MS.

Guan, L. and Du, Y. 2004. Fast Method Finds Infill Drilling Potentials in Mature-Tight Reservoirs. Presented at the SPE International Petroleum Conference in Mexico, Puebla Pue., Mexico, 7–9 November. SPE-91755-MS. https://doi.org/10.2118/91755-MS.

Guan, L., McVay, D. A., Jensen, J. L. et al. 2002. Evaluation of a Statistical Infill Candidate Selection Technique. Presented at the SPE Gas Technology Symposium, Calgary, Alberta, Canada, 30 April–2 May. SPE-75718-MS. https://doi.org/10.2118/75718-MS.

Gunter, G. W., Finneran, J. M., Hartmann, D. J. et al. 1997a. Early Determination of Reservoir Flow Units Using an Integrated Petrophysical Method. Presented at the SPE Annual Technical Conference and Exhibition, San Antonio, Texas, 5–8 October. SPE-38679-MS. https://doi.org/10.2118/38679-MS.

Gunter, G. W., Pinch, J. J., Finneran, J. M. et al. 1997b. Overview of an Integrated Process Model to Develop Petrophysical Based Reservoir Descriptions. SPE Annual Technical Conference and Exhibition, San Antonio, Texas, 5–8 October. SPE-38748-MS. https://doi.org/10.2118/38748-MS.

Hatzignatiou, D. G. and McKoy, M. L. 2000. Probabilistic Evaluation of Horizontal Wells in Stochastic Naturally Fractured Gas Reservoirs. Presented at the SPE/CIM International Conference on Horizontal Well Technology, Calgary, Alberta, Canada, 6–8 November. SPE-65459-MS. https://doi.org/10.2118/65459-MS.

Holditch, S. A. 2006. Tight Gas Sands. *J Pet Technol* **58** (6): 86–93. SPE-103356-JPT. https://doi.org/10.2118/103356-JPT.

Holditch, S. A. 2009. SS: Unlocking the Unconventional Oil and Gas Reservoirs: Stimulation of Tight Gas Sands. Presented at the Offshore Technology Conference, Houston, Texas, 4–7 May. OTC-20267-MS. https://doi.org/10.4043/20267-MS.

Holditch, S. A. and Rahim, Z. 1994. Developing Data Sets for 3D Fracture Propagation Models. *SPE Prod & Fac* **9** (4): 257–261. SPE-26155-PA. https://doi.org/10.2118/26155-PA.

Holditch, S. A. and Tschirhart, N. 2005. Optimal Stimulation Treatments in Tight Gas Sands. Presented at the SPE Annual Technical Conference and Exhibition, Dallas, Texas, 9–12 October. SPE-96104-MS. https://doi.org/10.2118/96104-MS.

Holditch, S. A., Lin, Z.-S., and Spivey, J. P. 1991a. Estimating the Recovery From an Average Well in a Tight Gas Formation. Presented at the SPE Gas Technology Symposium, Houston, Texas, 22–24 January. SPE-21500-MS. https://doi.org/10.2118/21500-MS.

Holditch, S. A., Robinson, B. M., and Whitehead, W. S. 1987. The Analysis of Complex Travis Peak Reservoirs in East Texas. Presented at the SPE/DOE Joint Symposium on Low Permeability Reservoirs, Denver, Colorado, 18–19 May. SPE-16427-MS. https://doi.org/10.2118/16427-MS.

Holditch, S. A., Whitehead, W. S., and Rahim, Z. 1991b. Using In-Situ Stress and Permeability Thickness Profiles To Design the Well Completion. Presented at the SPE Gas Technology Symposium, Houston, Texas, 22–24 January. SPE-21493-MS. https://doi.org/10.2118/21493-MS.

Holmes, J., McVay, D. A., and Senel, O. 2007. A System for Continuous Reservoir Simulation Model Updating and Forecasting. Presented at the Digital Energy Conference and Exhibition, Houston, Texas, 11–12 April. SPE-107566-MS. https://doi.org/10.2118/107566-MS.

Hoover, E. L., Trenery, J. B., Mullen, G. A. et al. (2009). New Water-Based Fluid Designed for Depleted Tight Gas Sands Eliminates NPT. Presented at the SPE/IADC Drilling Conference and Exhibition, Amsterdam, The Netherlands, 17–19 March. SPE-119363-MS. https://doi.org/10.2118/119363-MS.

Hudson, J. W., Jochen, J. E., and Spivey, J. P. 2001. Practical Methods to High-Grade Infill Opportunities Applied to the Mesaverde, Morrow, and Cotton Valley Formations. Presented at the SPE Hydrocarbon Economics and Evaluation Symposium, Dallas, Texas, 2–3 April. SPE-68598-MS. https://doi.org/10.2118/68598-MS.

Ilk, D., Perego, A. D., Rushing, J. A. et al. 2008. Integrating Multiple Production Analysis Techniques To Assess Tight Gas Sand Reserves: Defining a New Paradigm for Industry Best Practices. Presented at the CIPC/SPE Gas Technology Symposium 2008 Joint Conference, Calgary, Alberta, Canada, 16–19 June. SPE-114947-MS. https://doi.org/10.2118/114947-MS.

Jordan, C. L., Fenniak, M., and Smith, C. 2006. An Efficient Model for Gas-Rate Forecasting in Complex Reservoirs. Presented at the SPE Gas Technology Symposium, Calgary, Alberta, Canada, 15–17 May. SPE-99352-MS. https://doi.org/10.2118/99352-MS.

Kupchenko, C. L., Gault, B. W., and Mattar, L. 2008. Tight Gas Production Performance Using Decline Curves. Presented at the CIPC/SPE Gas Technology Symposium 2008 Joint Conference, Calgary, Alberta, Canada, 16–19 June. SPE-114991-MS. https://doi.org/10.2118/114991-MS.

Kyte, D. G. and Meehan, D. N. 1996. Horizontal Spacing, Depletion, and Infill Potential in the Austin Chalk. Presented at the SPE Annual Technical Conference and Exhibition, Denver, Colorado, 6–9 October. SPE-36721-MS. https://doi.org/10.2118/36721-MS.

Lee, W. J. and Sidle, R. E. 2010. Gas-Reserves Estimation in Resource Plays. Presented at the SPE Unconventional Gas Conference, Pittsburgh, Pennsylvania, 23–25 February. SPE-130102-MS. https://doi.org/10.2118/130102-MS.

Maley, S. 1985. The Use of Conventional Decline Curve Analysis in Tight Gas Well Applications. Presented at the SPE/DOE Low Permeability Gas Reservoirs Symposium, Denver, Colorado, 19–22 March. SPE-13898-MS. https://doi.org/10.2118/13898-MS.

Mattar, L. and McNeil, R. 1998. The "Flowing" Gas Material Balance. *J Can Pet Technol* **37** (2): 52–55. PETSOC-98-02-06. https://doi.org/10.2118/98-02-06.

Mattar, L., Anderson, D., and Stotts, G. 2006. Dynamic Material Balance-Oil-or Gas-in-Place Without Shut-Ins. *J Can Pet Technol* **45** (11): 7–10. PETSOC-06-11-TN. https://doi.org/10.2118/06-11-TN.

McCain, W. D., Voneiff, G. W., Hunt, E. R. et al. 1993. A Tight Gas Field Study: Carthage (Cotton Valley) Field. Presented at the SPE Gas Technology Symposium, Calgary, Alberta, Canada, 28–30 June. SPE-26141-MS. https://doi.org/10.2118/26141-MS.

Morales, E. C. and Lee, J. 2014. Interpretation of Recent SEC Reserves Reporting Guidance. Presented at the SPE Annual Technical Conference and Exhibition, Amsterdam, The Netherlands, 27–29 October. SPE-170616-MS. https://doi.org/10.2118/170616-MS.

Neal, D. B. and Mian, M. A. 1989. Early-Time Tight Gas Production Forecasting Technique Improves Reserves and Reservoir Description. *SPE Form Eval* **4** (1): 25–32. SPE-15432-PA. https://doi.org/10.2118/15432-PA.

Robertson, S. 1988. Generalized Hyperbolic Equation. SPE-18731-MS.

Rollins, J. B., Holditch, S. A., and Lee, W. J. 1992. Characterizing, Average Permeability in Oil and Gas Formations (includes associated papers 25286 and 25293). *SPE Form Eval* **7** (1): 99–105. SPE-19793-PA. https://doi.org/10.2118/19793-PA.

Rushing, J. A., Perego, A. D., Sullivan, R. et al. 2007a. Estimating Reserves in Tight Gas Sands at HP/HT Reservoir Conditions: Use and Misuse of an Arps Decline Curve Methodology. Presented at the SPE Annual Technical Conference and Exhibition, Anaheim, California, 11–14 November. SPE-109625-MS. https://doi.org/10.2118/109625-MS.

Rushing, J. A. and Newsham, K. E. 2001. An Integrated Work-Flow Model to Characterize Unconventional Gas Resources: Part II - Formation Evaluation and Reservoir Modeling. Presented at the SPE Annual Technical Conference and Exhibition, New Orleans, Louisiana, 30 September–3 October. SPE-71352-MS. https://doi.org/10.2118/71352-MS.

Rushing, J. A., Newsham, K. E., Perego, A. D. et al. 2007b. Beyond Decline Curves: Life-Cycle Reserves Appraisal Using an Integrated Work-Flow Process for Tight Gas Sands. Presented at the SPE Annual Technical Conference and Exhibition, Anaheim, California, 11–14 November. SPE-109836-MS. https://doi.org/10.2118/109836-MS.

Spivey, J. P., Frantz, J. H., Jr., Williamson, J. R. et al. 2001. Applications of the Transient Hyperbolic Exponent. Presented at the SPE Rocky Mountain Petroleum Technology Conference, Keystone, Colorado, 21–23 May. SPE-71038-MS. https://doi.org/10.2118/71038-MS.

Stewart, P. R. 1970. Low-Permeability Gas Well Performance At Constant Pressure. *J Pet Technol* **22** (9): 1149–1156. SPE-2604-PA. https://doi.org/10.2118/2604-PA.

Sullivan, S. A., Poston, S. W., and Piper, L. D. 1988. Using Short-Term Pressure Buildup Tests for Reserves Estimation in Tight Gas Reservoirs. Presented at the SPE Gas Technology Symposium, Dallas, Texas, 13–15 June. SPE-17707-MS. https://doi.org/10.2118/17707-MS.

Valkó, P. P. (2009, January 1). Assigning Value to Stimulation in the Barnett Shale: A Simultaneous Analysis of 7000 Plus Production Histories and Well Completion Records. Presented at the SPE Hydraulic Fracturing Technology Conference, The Woodlands, Texas, 19–21 January. SPE-119369-MS. https://doi.org/10.2118/119369-MS.

Voneiff, G. W. and Cipolla, C. 1996. A New Approach to Large-Scale Infill Evaluations Applied to the OZONA (Canyon) Gas Sands. Presented at the Permian Basin Oil and Gas Recovery Conference, Midland, Texas, 27–29 March. SPE-35203-MS. https://doi.org/10.2118/35203-MS.

Voneiff, G. W. and Holditch, S. A. 1994. Economic Assessment of Applying Advances in Fracturing Technology. *J Pet Technol* **46** (1): 51–57. SPE-24888-PA. https://doi.org/10.2118/24888-PA.

Part 3 - Hydraulic Fracturing in Vertical Wells

Chapter 12

Introduction to Hydraulic Fracturing

Stephen A. Holditch

Hydraulic fracturing is the process of pumping a fluid into a wellbore at an injection rate too high for the formation to accept in a radial flow pattern. As resistance to flow in the formation increases, the pressure in the wellbore increases to a value that exceeds the breakdown pressure of the formation adjacent and open to the wellbore. Once the formation breaks down, a crack or fracture is formed, and the injected fluid begins moving down the fracture. In most formations, a single, vertical fracture is created that propagates in two directions from the wellbore. These fracture wings are 180° apart and normally are assumed to be identical in shape and size at any point in time. In naturally fractured or cleated formations, such as gas shales or coal seams, multiple fractures can be created and propagated during a hydraulic fracturing treatment.

Fluid that does not contain any propping agent, often called pad, is injected to create a fracture that grows up, out, and down and creates a fracture wide enough to accept a propping agent. The purpose of the propping agent is to prop open the fracture. Once the pumping operation ceases, the pressure in the fracture decreases, and the fracture closes. In deep reservoirs, we use manmade ceramic beads to prop open the fracture. In shallow reservoirs, sand is normally used as the propping agent.

When wells are drilled and completed, the formation is often damaged by the drilling fluid, by the cementing operation, or by the completion fluids. The producing fluids can also damage existing wells. Damage is more detrimental to the productivity of high-permeability oil or gas wells; however, if the damage can be removed by a stimulation method, such as hydraulic fracturing, the well productivity and oil and gas recovery can be substantially improved.

In many cases, wells are completed in low-permeability reservoirs. In those cases, flow must be stimulated deep in the reservoir to improve well performance. Hydraulic fracturing works in low-permeability reservoirs by changing the flow pattern in the reservoir. As illustrated in **Fig. 12.1,** creating a long, high-conductivity fracture into the matrix can cause the flow in the reservoir to change from radial to linear, and, after significant production, to elliptical. This fundamental change in the reservoir flow pattern can substantially improve the productivity index from the reservoir.

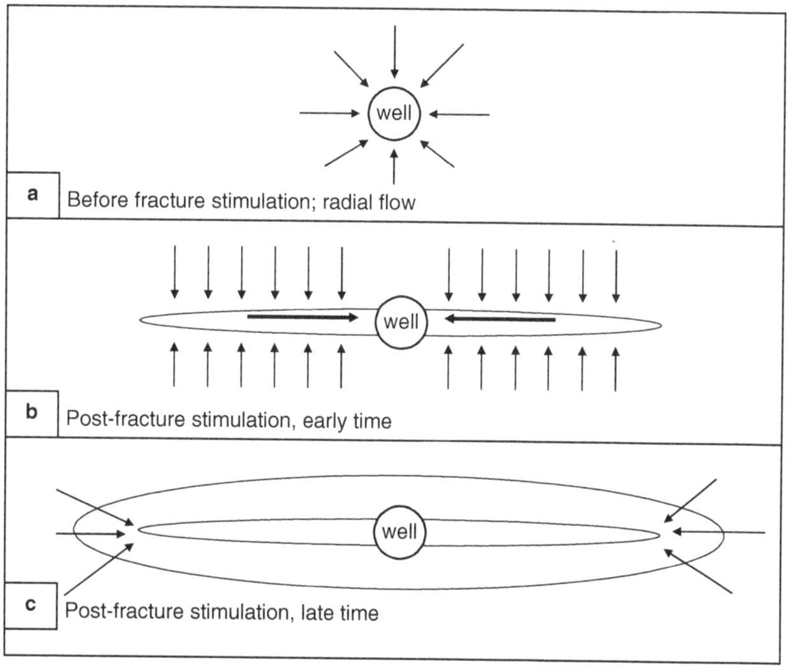

Fig. 12.1—Flow around the fracture changes from radial to linear to elliptical.

Chapter 13

Selection of Candidates

Stephen A. Holditch

13.1 Fundamental Requirements

Although the selection of the best wells to stimulate is difficult and time consuming, selecting the right candidate is important and can significantly increase the likelihood of success (Shelley 1999). The selection of candidates for well stimulation is a complex process that requires well-data gathering, data analysis, screening, and ranking of potential stimulation candidates (Hopkins and Hill 1994). Despite the complexities, the importance of choosing a proper stimulation candidate cannot be overemphasized. Designing a perfect treatment on a well that does not meet the criteria for stimulation wastes time and resources. Because various wells will be selected for different treatment methods, candidate selection also should be tailored to treatment type: A well selected for matrix acidizing might not be fit for proppant or acid fracturing, for example. In addition to minimizing failure rates, candidate selection is necessary in budget restrictions and mutually exclusive situations (Nnanna and Ajienka 2005).

During the Gas Research Institute (GRI) study of restimulation of tight gas sands, Reeves et al. (1999) identified a need to develop cost-effective, efficient, and reliable methodologies to detect stimulation-preferred candidates because they hold the greatest potential for profitable rewards in the industry. The GRI team postulated the 85/15 rule: 85% of the restimulation potential lies in 15% of the wells in the field, which means that being able to identify the 15% high-potential wells is crucial to project success. McLeod (1989) showed that the greatest cost savings in matrix acidizing resulted from better engineering evaluation of the acidizing candidate. McLeod's evaluation improved success rates from 54% in an example field to between 70 and 95%.

Candidate selection requires a sound understanding of reservoir and geological information and a detailed well history (Nnanna and Ajienka 2005). The best strategy is to study and understand the interactions between the various factors and parameters that flag a potential stimulation candidate. For example, using underperformance alone to select a candidate completed in a poor reservoir can lead to a worse outcome than randomly picking candidates (Moore and Ramakrishnan 2006); nothing we can do will improve performance from a poor reservoir. Candidate selection is all about identifying and ranking both underperforming and superior wells that have the potential for performing better when effectively stimulated.

Our goal in most stimulation treatments is to increase productivity and increase well recovery through the removal of wellbore damage, creating fractures in low-permeability sands, connecting natural fractures, and increasing the effective drainage area. To identify stimulation candidates that would benefit from these remedial activities, we need to know the parameters and factors that caused the adverse effects in the first place. The next step is to weigh the effects of the parameters individually or in combination with other parameters or factors. The combined effects of these weighted variables are subsequently used to screen, rank, and select the best candidates.

The key variables that affect well productivity are permeability, porosity, viscosity, skin factor, reservoir pressure, net-pay thickness, reservoir depth, water saturation, drainage area, and wellbore condition (Xiong 1992). Other factors include the producing gas/oil ratio (GOR), water/oil ratio (WOR), sand production, and limited connectivity to natural fractures (Moore and Ramakrishnan 2006; Jin and Seah 2006).

For restimulation candidates, the key parameters include ineffective initial completion, unstimulated horizons, insufficient fracture length and conductivity, inappropriate fracture orientation, incompatible stimulation fluids, premature screenout, proppant crushing, perforation density, type and amount of fracturing fluid, limited-entry completion, number of zones and layers involved in each fracture job, and the job details (Reeves et al. 1999; Moore and Ramakrishnan 2006).

13.2 Stimulation Selection Criteria

Reservoir permeability has a huge effect on well productivity. This effect is even more pronounced when we consider the ratio of permeability to viscosity. A moderate-to-high k/μ ratio indicates a good stimulation candidate. A well with a very high or very low ratio might not need, or benefit from, stimulation (Xiong and Holditch 1995, 2006a).

Skin damage, average reservoir pressure, and net-pay thickness are also important selection criteria. High skin damage caused by high reservoir pressure and thick net-pay all point to a high-potential candidate. Because stimulation can remove the skin, a damaged well or an underperforming well completed in a thick reservoir with high reservoir pressure would be a good stimulation candidate; however, a well with a skin of less than –4 should not be stimulated (Xiong et al. 1993; Xiong and Holditch 2006b).

The interaction between the porosity and pay thickness (ϕh) indicates the storage capacity of the reservoir. As with the k/μ ratio, a well with a very low ϕh product or low cumulative production/ϕh ratio might not need stimulation (Fleming 1992; Moore and Ramakrishnan 2006).

Screening criteria such as pay thickness, drainage area, porosity, and water saturation have a combined effect on reserves and hydrocarbon in place. Lack of wellbore connectivity in a naturally fractured reservoir with substantial hydrocarbon in place is a good indication of potential stimulation rewards. These reservoirs can be fractured to increase the effective well drainage area by connecting to natural fractures and isolated pools of hydrocarbon accumulation.

The mechanical condition of the wellbore can cause problems during a stimulation treatment. Potential candidates should be reviewed for wellbore restriction such as nipple joints, buckled tubing, crossovers, and sand buildup. Other pertinent data such as the cement quality, deviation survey, well schematics, sliding sleeves, lift method, tubing and casing size, age of the completion, corrosion problems, packer information, and records of historical well jobs also should be reviewed. A good candidate should have mechanical integrity and should be free from well restriction from the surface to the target stimulation depth.

In locations where gas and water handling are problematic, wells with high GOR and high WOR might not make good candidates. For restimulation treatments, Reeves et al. (1999) recommended production data analysis, data analytics, and field-specific expertise for selecting candidates and short-duration diagnostic tests for identifying damage and refracturing potential.

The screening-criteria data are derived from production-data analysis, pressure surveys, well files, historical completion data, stimulation data, petrophysical analysis, core analysis, well logs, and industry correlations. Special tests, such as annulus injection tests, can be conducted to establish communication between a high-conductivity fracture or natural fracture and the wellbore (Hopkins and Hill 1994). Production type curves can be used to estimate permeability, skin, and drainage in single-layered reservoirs, although the result might not be unique. Artificial neural networks also can be used to estimate parameters such as porosity, permeability, thickness, reservoir pressure, stimulation effectiveness, and drainage area. **Table 13.1** gives a summary of some of the parameters and their sources.

Parameter	Source
Permeability, k (md)	Flow/buildup tests, cores, correlations
Porosity, ϕ (fractional)	Logs, cores
Viscosity, μ (cp)	Pressure/volume/temperature studies, well files, correlations
Skin factor, s (dimensionless)	Flow/buildup tests, well files
Reservoir pressure, pe (psi)	Pressure survey, offset wells
Net thickness, h (ft)	Logs, cores
Reservoir depth, D (ft)	Logs, well schematics
Drainage area, A (ft^2)	Logs, cores, offset wells
Wellbore condition	Well tests
GOR (scf/STB)	Production-test well files
WOR (STB/STB)	Production-test well files
Wellbore connectivity	Annulus injection tests

Table 13.1—Summary of parameters and their source.

13.3 Productivity Improvement Factor (PIF)

When hydraulic fracturing began in the 1950s, the two most-pressing problems involved how to predict fracture dimensions for a given fluid volume and how to forecast oil- and gas-flow rates as a function of the fracture dimensions. During this time, most wells being drilled were oil wells. These oil wells were completed in moderate- to high-permeability formations, and the main reason for stimulation was to remove near-wellbore damage.

Darcy's law is used to compute fluid flow from oil and gas wells. When a well is opened to flow, the early flow rates are transient and, with sufficient production, the well will settle into semisteady-state flow for bounded reservoirs and, in some cases, steady-state flow. For a well that has been fracture treated, the flow regimes go from bilinear flow to linear flow to elliptical flow, and finally to pseudoradial flow or even elliptical flow. To analyze data properly or even to forecast flow rates, we need to be certain to use the equations corresponding to the flow regimes.

In high-permeability reservoirs, semisteady-state flow can be reached in days or even hours. In low-permeability reservoirs, especially those that have been fracture treated, reaching pseudoradial flow could take weeks or months or years. We can use several different methods to analyze reservoirs and predict flow rates, and we can use those reservoir engineering calculations to help design and evaluate fracture treatments.

Eq. 13.1 is a version of Darcy's law for steady-state flow in a radial system. The equation relates the pressure drop between the extremity of the reservoir and the wellbore to the flow rate, the formation permeability-thickness product, and the drainage area. Also included are reservoir-fluid properties, the wellbore radius, and the skin factor.

$$P_e - p_{wf} = 141.2\frac{qB\mu}{kh}\left(\ln\frac{r_e}{r_w}+S\right), \quad \text{(13.1)}$$

where B is formation volume factor, h is thickness, k is permeability, p_e is effective pressure, p_{wf} is wellbore flowing pressure, q is flow rate, r_e is effective radius, r_w is wellbore radius, S is skin, and μ is fluid viscosity.

If we rearrange the Eq. 13.1 as shown next, we can compute a factor J, which is the productivity index of the well. The units of productivity index for an oil well are barrels of oil per day (B/D) for each pound per square inch (psi) of pressure drop between the extremity of the reservoir and the wellbore.

$$J_o = \frac{q_o}{\left(p_e - p_{wf}\right)} = \frac{q_o}{\left(p_e - p_{wf} + \Delta p_{\text{skin}}\right)} = \frac{7.08\times10^{-3}kh}{\mu B_o\left(\ln\frac{r_e}{r_w}+S\right)}, \quad \text{(13.2)}$$

where J_o *is* productivity index, q_o is original flow rate, Δp_{skin} is pressure change caused by skin, and B_o is the oil formation volume factor.

Darcy's law for steady-state flow can be used for high-permeability reservoirs. Because almost all reservoirs being drilled and stimulated in the 1950s and 1960s were high-permeability reservoirs, several techniques to design fracture treatments and analyze the resulting flow rates were based on either the semisteady-state or the steady-state flow equations.

A fracture treatment will increase the productivity index, J, of a well. If the productivity of a well is already a high value, then any stimulation treatment that increases the productivity index should substantially increase production. However, if the productivity index of a well is very low, even a large increase in that index might still result in minimal production increase. After all, ten times zero is still zero.

In the early days of hydraulic fracturing, three groups published papers on the PIF—sometimes referred to as the productivity-index increase ratio—but all made different assumptions. The three groups were McGuire and Sikora (1960) from Atlantic Richfield Company, Prats (1961) from Shell Oil Company, and Tinsley et al. (1969) from Halliburton Company. The method of McGuire and Sikora and that of Prats are detailed in the subsections that follow.

13.3.1 McGuire and Sikora Method. The most-used method is the one by McGuire and Sikora (1960), who used an electrical analog computer to generate a graphical solution. The pair assumed a square drainage area, and the results are for semisteady-state flow. The McGuire-Sikora graph is shown in **Fig. 13.1.** As seen in the figure, they compared the productivity index of a stimulated well, J_s, to the productivity index of the well under radial flow, J_o. The graph shows the importance of fracture length and fracture conductivity on the productivity-index increase ratio.

When the McGuire-Sikora graph was first published, it was the only logical method for engineers to design a fracture treatment and analyze the production data from a well that had been fracture treated. The graph shows the stimulation ratio or PIF vs. the conductivity ratio as a function of the fracture half-length and the drainage radius.

To design a fracture treatment, the McGuire-Sikora graph can be used to determine the effects of fracture length and fracture conductivity on the resulting flow rate from a well. The objective of a fracture treatment is to increase the productivity index and, thus, the flow rate of oil and gas from the well. By putting a few numbers to the graph, we can see that for low-permeability reservoirs, the conductivity ratio will be typically large—for example, 10^5 or more. For these low-permeability cases, creating longer fractures is more important than creating higher fracture conductivity.

For high-permeability reservoirs and low fracture conductivity, we find that we are working on the low end of the conductivity-ratio scale—down to a value of 10^4 or less, for example. For high-permeability reservoirs, creating greater fracture conductivity should be more important than creating a long fracture.

In addition to design, the McGuire-Sikora graph can be used to analyze well production data. To do so, we must produce the well before the hydraulic fracturing treatment to measure the productivity index under radial flow conditions. Then, we can analyze the well performance after the treatment to determine the productivity index. All the determinations of the productivity index, both before and after the treatment, should be at semisteady-state flow. Once we know the value of J/J_o, then we can use the McGuire-Sikora graph to determine the most likely combinations of fracture half-length and fracture conductivity that will produce the observed results.

13.3.2 Prats Method. The second-most-used method was developed by Prats (1961) at Shell Oil Company. Prats published a number of papers on hydraulically fractured wells that are classic papers in the field. Prats developed an analytical solution for steady-state flow assuming a circular drainage area. The equation below includes an assumption of infinite fracture conductivity, although Prats provided a method to look at results if the fracture conductivity was finite.

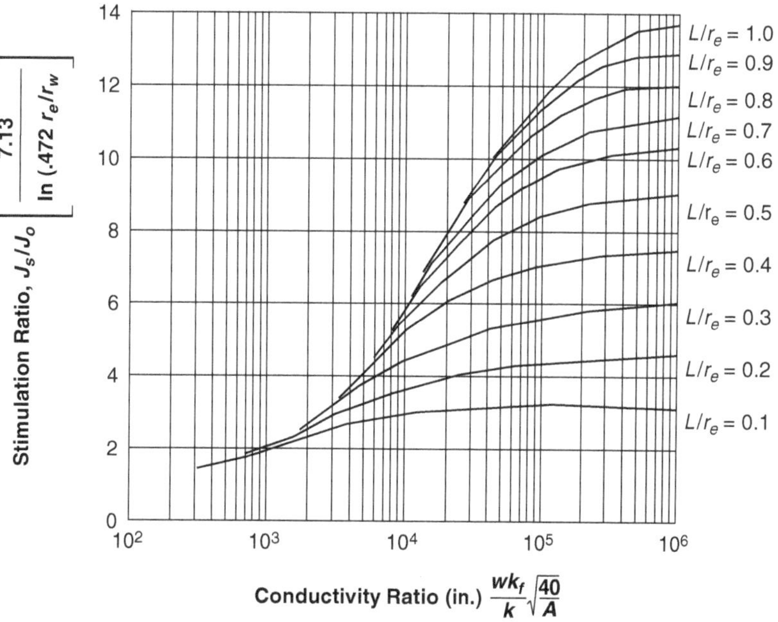

Fig. 13.1—McGuire-Sikora graph for computing the productivity-index increase ratio shows that low-permeability reservoirs call for longer fractures, while high permeabilities need increased conductivities.

$$\frac{q_f}{q} = \frac{\ln \frac{r_e}{r_w}}{\ln \frac{r_e}{r_w'}}, \qquad (13.3)$$

where q_f *is* flow through the fracture and $r_w' = 0.5L_f$, in which L_f is fracture half-length.

With the Prats equation, we can compare the flow-rate ratio of a fractured well to the radial flow case at steady-state conditions. Because Eq. 13.3 assumes that the pressure drops before and after fracturing are the same, then the ratio of q_f/q is also the PIF for the case being evaluated.

Notice that the effective wellbore radius is equal to 50% of the fracture half-length. What that means is that if (for example) we have an infinite-conductivity fracture that has a half-length of 500 ft (1,000 ft tip to tip), then the well will flow oil or gas as if the wellbore radius were 250 ft (500-ft diameter) when it reaches steady-state flow.

Thus, to stimulate a well, we can choose to fracture treat the well and create a high-conductivity fracture that stretches 1,000 ft (tip to tip), or we can drill the well using a drill bit that is 500 ft in diameter. Which one would you choose?

As mentioned, the Prats equation (Eq. 13.3) is for an infinite-conductivity fracture. However, Prats' papers show the effect of finite conductivity on the effective wellbore radius, r_w', and show the effects of damage around the fracture on the productivity index. Prats' work was remarkable at the time and has led to a long-lasting understanding of how wells containing hydraulic fractures behave.

13.4 Summary

Within this chapter, we have reviewed the literature on stimulation candidate selection and outlined the various stimulation selection criteria. We presented a summary of the methods used in the industry to screen, rank, and select stimulation candidates on the basis of the criteria and outlined field cases in which the different methodologies were applied.

On the basis of the reviews/summary, we can conclude the following:

- Most screening and selection methodologies are field/reservoir/well specific and must be customized for site-specific application.
- We can build a framework for proceeding with candidate selection for different fields using the existing methodologies, as described by Xiong (1992) and Xiong et al. (1993).
- There is no universal methodology for selecting stimulation candidates, and no selection criteria can be applied to every situation.
- Poor candidate selection usually will result in unsuccessful stimulation projects.

13.5 Nomenclature

A	=	drainage area, acres	m	=	viscosity, cp
B	=	formation volume factor, res bbl/STB	p_e	=	effective pressure, psi
B_o	=	oil formation volume factor, res bbl/STB	p_{wf}	=	wellbore flowing pressure, psi
D	=	reservoir depth, ft	q	=	flow rate, B/D
ϕ	=	porosity, fractional	q_f	=	flow through the fracture, B/D
h	=	formation thickness, ft	q_o	=	original flow rate, B/D
J	=	productivity index, (B/D)/psi	r_e	=	effective radius, ft
J_o	=	productivity index of the well under radial flow, (B/D)/psi	r_w	=	wellbore radius, ft
J_s	=	productivity index of a stimulated well, (B/D)/psi	S	=	skin factor, dimensionless
k	=	permeability, md	w	=	fracture width, ft
k_f	=	fracture permeability, md	Δp_{skin}	=	pressure change caused by skin, psi
L	=	fracture length, ft	μ	=	fluid viscosity, cp

13.6 References

Fleming, M. E. 1992. Successful Refracturing in the North Westbrook Unit. Paper presented at the SPE Permian Basin Oil and Gas Recovery Conference, Midland, Texas, USA, 18–20 March. SPE-24011-MS. https://doi.org/10.2118/24011-MS.

Hopkins, C. W. and Hill, D. G. 1994. Screening Restimulation Candidates in the Antrim Shale. Paper presented at the SPE Eastern Regional Meeting, 8–10 November, Charleston, West Virginia, USA. SPE-29172-MS. https://doi.org/10.2118/29172-MS.

Jin, L. and Seah, M. 2006. Stimulation in Asia Pacific Region—Challenges and Opportunities. Paper presented at the SPE Annual Technical Conference and Exhibition, San Antonio, Texas, 24–27 September. SPE-103275-MS. https://doi.org/10.2118/103275-MS.

McGuire, W. J. and Sikora, V. J. 1960. The Effect of Vertical Fractures on Well Productivity. *J Pet Technol* **12** (10): 72–74. SPE-1618-G. https://doi.org/10.2118/1618-G.

McLeod, H. O. 1989. Significant Factors for Successful Matrix Acidizing. Paper presented at the SPE Centennial Symposium at New Mexico Tech, Socorro, New Mexico, 16–19 October. SPE-20155-MS. https://doi.org/10.2118/20155-MS.

Moore, L. P. and Ramakrishnan, H. 2006. Restimulation: Candidate Selection Methodologies and Treatment Optimization. Paper presented at the SPE Annual Technical Conference and Exhibition, 24–27 September, San Antonio, Texas. SPE-102681-MS. https://doi.org/102681-MS.

Nnanna E. J. and Ajienka J. A. 2005. Critical Success Factors for Well Stimulation. Paper presented at the Nigeria Annual International Conference and Exhibition, 1–3 August, Abuja, Nigeria. SPE-98823-MS. https://doi.org/10.2118/98823-MS.

Prats, M. 1961. Effect of Vertical Fracture on Reservoir Behavior—Incompressible Fluid Case. *J Pet Technol* **1** (02): 105–118. SPE-1575-G. https://doi.org/10.2118/1575-G.

Reeves, S. R., Hill D. G., Hopkins C. W. et al. 1999. Restimulation Technology for Tight Gas Sand Wells. Paper presented at the SPE Annual Technical Conference and Exhibition, Houston, Texas, 3–6 October 1999. SPE-56482-MS. https://doi.org/10.2118/56482-MS.

Shelley R. F. 1999. Artificial Neural Networks Identify Restimulation Candidates in the Red Oak Field. Paper presented at the SPE Mid-Continent Operations Symposium, Oklahoma City, Oklahoma, USA, 28–31 March. SPE-52190-MS. https://doi.org/10.2118/52190-MS.

Tinsley, J. M., Williams Jr., J. R., Tiner, R. L. et al. 1969. Vertical Fracture Height-Its Effect on Steady-State Production Increase. *J of Pet Technol* **21** (05): 633–638. SPE-1900-PA. https://doi.org/10.2118/1900-PA.

Xiong, H. 1992. *STIMEX – An Expert System Approach to Well Stimulation Design*. PhD dissertation, Texas A&M University, Lubbock, Texas.

Xiong, H. and Holditch, S. A. 1995. An Investigation Into the Application of Fuzzy Logic to Well Stimulation Treatment Design. *SPE Computer Applications* **7** (01): 18–24. SPE-27672-PA. https://doi.org/10.2118/27672-PA.

Xiong, H. and Holditch. S. A. 2006a. Will the Blossom of Unconventional Natural Gas Development in North America be Repeated in China? Paper presented at SPE International Oil and Gas Conference and Exhibition in China, 5–7 December, Beijing. SPE-103775-MS. https://doi.org/10.2118/103775-MS.

Xiong, H. and Holditch, S. A. 2006b. Using A Fuzzy Expert System to Choose Target Wells and Target Formations for Stimulation, Lecture Materials, Texas A&M University Course PETE 602, College Station, Texas.

Xiong, H., Holditch, S. A., and Rahim, Z. 1993. Using a Fuzzy Expert System To Choose Target Wells and Target Formations for Stimulation. Paper presented at the Conference on Artificial Intelligence in Petroleum Exploration and Production, Dallas, Texas, 19–21 May.

Chapter 14

Fracture Initiation, Geometry, and Propagation

Stephen A. Holditch

This chapter presents the fundamental concepts concerning in-situ stress, near-wellbore fracture geometry, fracture containment issues, fracture growth analyses, and the effect of fractures on the productivity index of a well. In addition, we describe the various 2D and 3D hydraulic fracture propagation models in use in the oil and gas industry.

One of the most important variables to characterize when designing hydraulic fracturing treatments is the in-situ stress in each layer of rock above, within, and below the pay zone, especially if those layers materially affect fracture propagation. The techniques used to estimate in-situ stress can be placed into three categories:

- Logging techniques: These techniques are based on semiempirical relationships between rock properties and stresses at or near the wellbore, or the inference of stress through a mechanistic model that uses openhole logging measurements.
- Injection tests: In these tests, the in-situ stresses are inferred from the pressure responses with the opening—and sometimes the closing—of the fracture. The most common techniques to measure in-situ stress are step-rate tests and minifrac tests. Small-volume hydraulic fracturing tests still give the most-reliable and -accurate measurement of in-situ stresses, and they should be used in conjunction with the other methods for more accuracy.
- Core tests: Core techniques to measure in-situ stress are based on performing a series of strain tests such as the anelastic-strain-recovery test (ASRT) on cores recovered from the borehole. The interpretation process is based on assumptions that cannot easily be verified. So, the results of these methods should always be checked with a more-reliable stress-determination technique such as minifracturing.

14.1 In-Situ Stresses

The first step in determining the in-situ stresses is to estimate the orientation of the stresses. In a normal situation with no active tectonic force, the minimum principal far-field stress is horizontal, and the maximum principal stress is vertical. However, if the area is tectonically active, the maximum principal stress can be horizontal. Generally, stress orientation can be determined from field observations, general geologic information, or the world stress map.

Several techniques have been developed to measure in-situ stresses. The most common methods are borehole breakouts, hydraulic fracturing techniques, and the study of focal mechanisms of induced seismicity, overcoring, and core relaxation tests.

14.1.1 Vertical Stress. The vertical or overburden stress, S_v, at a depth of z is defined as the pressure exerted by the weight of the overlying formation and expressed as

$$S_v = \int_0^z \rho_b(z) g dz, \quad \text{(14.1)}$$

where g is acceleration resulting from gravity; S_v is vertical (overburden) stress; z is depth; and ρ_b is bulk density. Bulk density, ρ_b, can be obtained from bulk-density or sonic logs.

14.1.2 Minimum Horizontal Stress. The most reliable calculation of minimum horizontal stress can be obtained by hydraulic fracturing tests (Hillis 2001). In such tests, increasing the fluid pressure within an isolated part of the wellbore causes a tensile fracture. If we assume that the fluid is keeping the fracture open against the least principal stress, the fluid pressure at which the hydraulic fracture closes is a reasonable estimation of the minimum horizontal stress, S_h. Because hydraulic fracturing tests are not widely undertaken during exploration drilling, leakoff tests are used for estimating the minimum horizontal stress. These tests are routinely performed to determine the maximum mud weight that can be used without generating fractures. The results of leakoff tests are not as reliable as those from hydraulic fracture tests, because leakoff is controlled by the disturbed stress field at the borehole wall. It is accepted that the lower bound of leakoff pressure gives a good and reasonable estimation of minimum horizontal stress.

In the absence of tectonic stresses, we commonly use a linear-elastic relationship in which the horizontal stress, S_h, increases with depth as a fraction of the vertical stress:

$$S_h = \left(\frac{v}{1-v}\right)(S_v - \alpha p_p) + \alpha p_p, \quad (14.2)$$

where p_p is pore pressure, S_h is horizontal stress, S_v is vertical (overburden) stress, α is Biot's poroelastic coefficient, and v is Poisson's ratio.

Eq. 14.2 is an isotropic stress equation and assumes uniaxial strain, $S_h = S_H$. The equation only considers the effects of overburden stress, pore pressure, and rock properties; it does not include the effect of horizontal strains, but other forms can be derived with $S_h \neq S_H$.

14.1.3 Maximum Horizontal Stress. The maximum horizontal stress, S_H, cannot be determined directly. The best way to estimate its direction and quantity is by observing image logs, which give information about the occurrence and orientations of tensile borehole-wall fractures and the orientations and widths of breakouts. Then, by applying quantitative techniques of wellbore-stress analysis, we can constrain the magnitude of S_H (Moos 2001). In addition, the four-arm caliper (dipmeter) can be used to determine the orientation of borehole breakouts. The disadvantage of using this device is that it provides little information about the detailed shape of the borehole and it is not easy to differentiate between key seats and washouts.

An accurate and proper approach may use observed or calculated values for S_v, p_p, S_h, unconfined strength of rock, width and position of wellbore failure, and the recorded mud weight used in the well at the depth of the observed failure.

Modeling in deviated wellbores first requires determining the azimuth of maximum horizontal stress consistent with the location of observed borehole breakouts. Afterward, this azimuth will be used to verify and then constrain the magnitude of maximum horizontal stress using wellbore-breakout and drilling-induced-tensile-fracture observations. Drilling-induced tensile fractures develop only for certain combinations of horizontal stresses and can provide very accurate boundaries on maximum-horizontal-stress magnitudes (Khaksar et al. 2004).

14.1.4 Stress-Measurement Techniques. Many different techniques have been developed to measure in-situ stresses. They can be divided into core-based methods, borehole-based techniques, near-wellbore techniques, and log-analysis techniques. The list of techniques for determining in-situ stress is quite long (Bloch et al. 1997):

- ASRT
- Axial point-load test
- Borehole-breakout analysis
- Borehole deformation
- Borehole imaging
- Circumferential-velocity anisotropy
- Differential-strain-curve analysis
- Direct observation of overcored, openhole microfractures
- Directional gamma ray logging
- Drilling-induced fractures in the core
- Earth-tilt surveys
- Leakoff-inversion technique
- Microfracture pressure analysis
- Microseismic logging
- Overcoring of archived core
- Petrographic examination of microcracks
- World stress mapping

Small-volume hydraulic fracturing tests still give the most-reliable and -accurate measurement of in-situ stresses, although they can be used in conjunction with the other methods for added completeness (Ahmed et al. 1989).

14.2 Near-Wellbore Fracture Geometry

Near-wellbore fracture geometry refers to how the fracture is connected to the well. Near-wellbore fracture geometry can affect the pressure drop of fluid flowing from the reservoir into the fracture and then into the well. The near-wellbore geometry can also affect propping-agent injection and pumping pressures during the fracture treatment. Under ideal conditions, a single fracture propagates in two directions from the wellbore. Once initiated, the fracture will grow up, out, and down until it hits a barrier to fracture growth. Normally, barriers to vertical fracture growth consist of thick shale zones. When the reservoir layer can be described as homogeneous with respect to its mechanical properties and when the reservoir dip angle is low, the fracture will equally propagate as two wings from the wellbore. If the reservoir is steeply dipping, then the fracture wing that grows up the dip might do so at a faster rate than the fracture wing that grows down the dip.

The industry has developed equations to compute fracture growth for a number of idealized conditions. Some of the equations are for 2D fracture growth, which usually means the fracture height is fixed; we compute the fracture width and length as a function of how much fluid is pumped into the fracture at specified injection rates and fluid properties. In practice, however, most fractures do not have a constant height with respect to length, nor can they be described in simple geometric shapes. We must understand the layers of rock that affect the fracture growth patterns, as shown in **Fig. 14.1.**

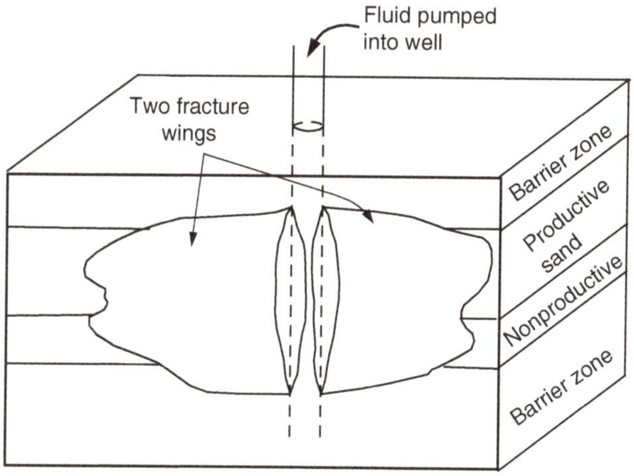

Fig. 14.1—In ideal conditions, the two wings of a vertical fracture equally propagate toward barrier zones.

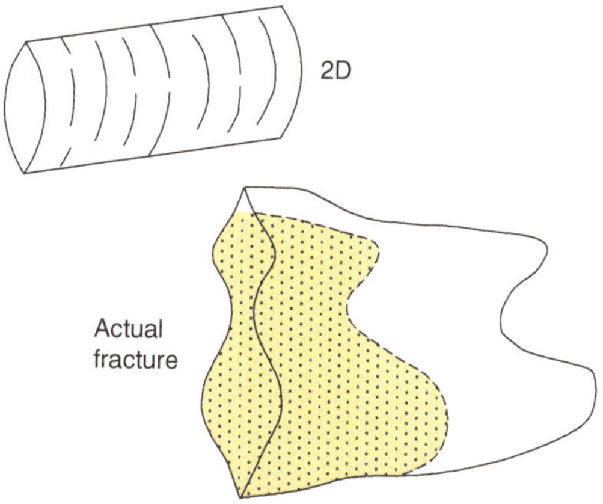

Fig. 14.2—Actual fracture shape is likely to be far different from idealized 2D theory.

In many situations, it is still feasible to use 2D models to design fracture treatments; however, the industry uses 3D models more often for the design of most fracture treatments. As shown in **Fig. 14.2,** for a typical situation, the fracture height is not constant, and the fracture width, length, and height will vary down the fracture, mainly affected by the mechanical properties of the formation layers.

The connection of the fracture to the wellbore is affected by the angle of the wellbore, the angle of the induced hydraulic fractures, and the length of the perforated interval. **Fig. 14.3** shows that in some cases multiple fractures will be created during a single-stage treatment. No well is truly vertical, nor is it perfectly straight; wellbores will spiral down, and virtually every section of a wellbore deviates from vertical. Therefore, if we use limited-entry perforating, or if we perforate a long interval (100 ft or more), multiple fractures likely will be created. The possibility that multiple fractures will be created substantially increases if the formation contains an abundance of natural fractures. One method the industry uses to limit the number of fractures created during a single-stage treatment is a point-source perforating technique. In vertical wells, we try to limit the perforating interval to 10 ft or so. In horizontal wells, a single perforation cluster should be limited to 2 to 4 ft.

Because no well is perfectly straight, no well is perfectly vertical, and no hydraulic fracture is perfectly vertical, there is no reason to believe that a hydraulic fracture should propagate in the same plane as the wellbore. As shown in **Fig. 14.4,** in many cases the plane of the propagating fracture can be substantially different from the trajectory of the wellbore, even for reasonably straight and vertical wells. Creating and propagating multiple fractures near the wellbore will cause many problems; a single, wide, vertical fracture is better because it will penetrate more deeply into the formation and the proppant will get into the fracture more easily. With multiple fractures, the effective fracture length will be less and the width of the individual fractures will be small, often preventing propping agents from entering the fracture. In many cases, having multiple fractures leads to high pumping pressures during the treatment and to screenouts.

In vertical wells, the connectivity of the multiple vertical fractures with the wellbore can cause many problems. **Fig. 14.5** shows that the problem is even more pronounced in horizontal wells. Multiple fractures not only can be created in horizontal wells, but the

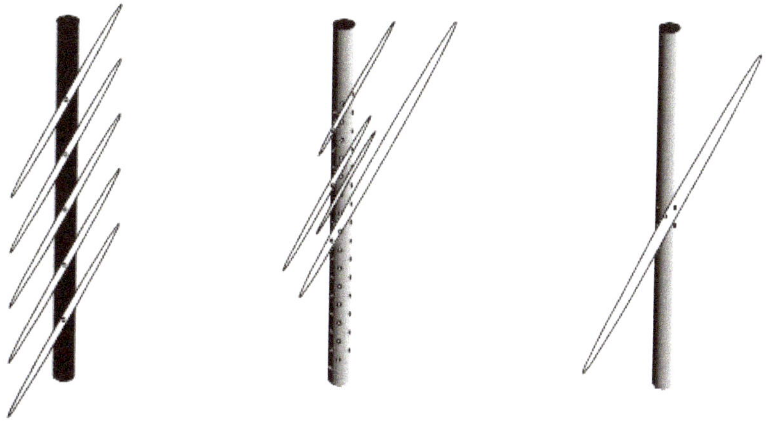

Fig. 14.3—Long perforation intervals can lead to multiple fractures.

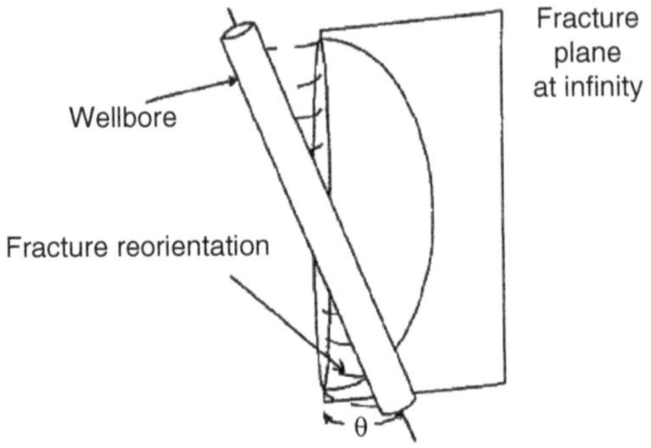

Fig. 14.4—Because the fracture plane and the wellbore might not be parallel, the hydraulic fracture might not propagate in the same plane as the wellbore.

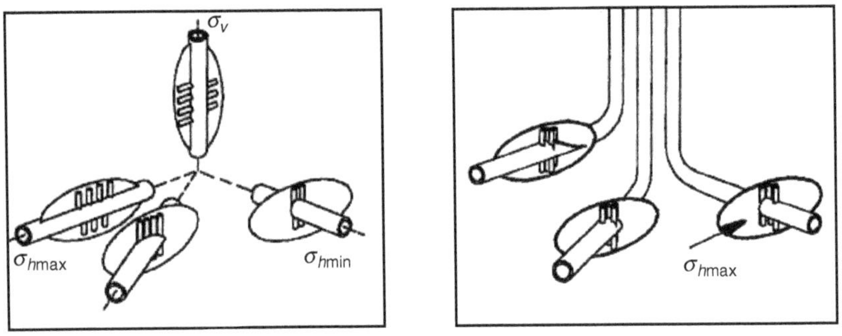

Fig. 14.5—The connection between wellbores and fractures in horizontal wells will be more complex than in vertical wells.

problems with connectivity between the created fractures and the wellbore are even more pronounced. If the propped fracture is not well-connected to the wellbore, the productivity-index increase expected from the completion might never be achieved. Because of these connectivity problems, point-source perforating often is used when pumping fracturing treatments to transport large volumes of propping agent.

In addition to problems associated with the creation of multiple fractures, serious issues are raised when the plane of the fracture changes direction near the wellbore. The hydraulic fracture will be created near the wellbore on the basis of stresses in the formation at the wellbore. Many times stresses at the wellbore are affected by the wellbore itself, the perforations, and the fluids pumped into the well. Thus, the hydraulic fracture is often created and leaves the wellbore in such a manner that it changes direction at some point away from the wellbore. As we know, the hydraulic fracture eventually will propagate perpendicular to the least principal far-field stress, which we call the minimum in-situ stress. If the fracture leaves the wellbore in another direction, then gets approximately 10 wellbore diameters from the wellbore, the fracture will begin to rotate so that further propagation will be in the direction perpendicular to the minimum in-situ stress. Because of this rotation, the fracture width will be restricted in size as it turns.

A restriction in fracture width close to the well can occur in the presence of natural fractures that cause fracture reorientation. The near-wellbore fracture geometry might also be affected by multiple hydraulic fractures induced by perforations, natural fractures that compete for fracture width, or fracture growth along the annulus between the cement and the stiffer casing.

Fig. 14.6 illustrates one situation in which multiple fractures can be created from a single wellbore, showing that the fracture might need to reorient and commingle to connect the main hydraulic fracture with the wellbore. Sometimes, these complicated flow patterns cause issues the industry has named "near-wellbore tortuosity," which implies that the fracturing fluid being injected into the fracture from the wellbore must follow a tortuous, or twisted, path.

Wellbore tortuosity can be the cause of high near-wellbore friction and high injection pressure. Tortuosity can also be the cause of premature screenouts, if the proppant bridges either in the narrow multiple fractures or in the part of the fracture that bends.

The near-wellbore pressure drop might prevent the identification of true net pressures in the fracture. We require the true net pressure as a function of time if we choose to use a fracture-propagation model to analyze minifracture treatments and the main fracturing treatment. As shown in **Fig. 14.7,** the pressure profile down a fracture at any point in time will be a function of the pressure drop at the fracture tip, the viscous pressure drops down the fracture, and the near-wellbore pressure drop. The near-wellbore pressure drop will be affected by the perforation and the near-wellbore tortuosity issues previously described.

What we need to know is the true net fracturing pressure, which is the pressure in the fracture proportional to the width and height of the fracture at the point at which the net pressure is measured. We attempt to measure the net pressure near the wellbore. To do so, we must be able to quantify the near-wellbore pressure drop.

The near-wellbore pressure drop is affected by perforation friction and near-wellbore tortuosity. In essence, the total near-wellbore pressure drop is the summation of the perforation-friction pressure drop and the pressure drop resulting from tortuosity. Both components are functions of the injection rate but are affected quite differently by changes in the injection rate.

A field test and a computational technique have been developed to differentiate between the two causes of near-wellbore pressure drop. The field test is called a step-down test. We begin the test with an injection rate large enough to create and propagate a fracture.

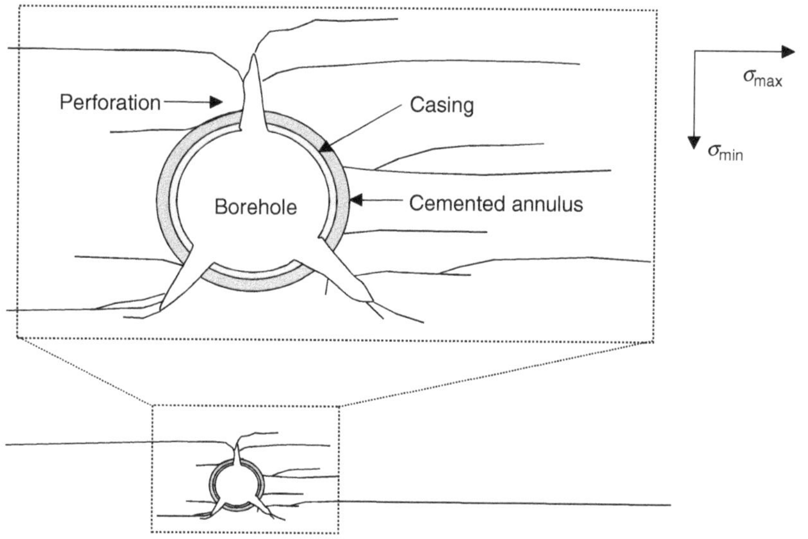

Fig. 14.6—Multiple fractures near the wellbore might need to commingle to connect the main fracture with the wellbore (from Wright et al. 1999).

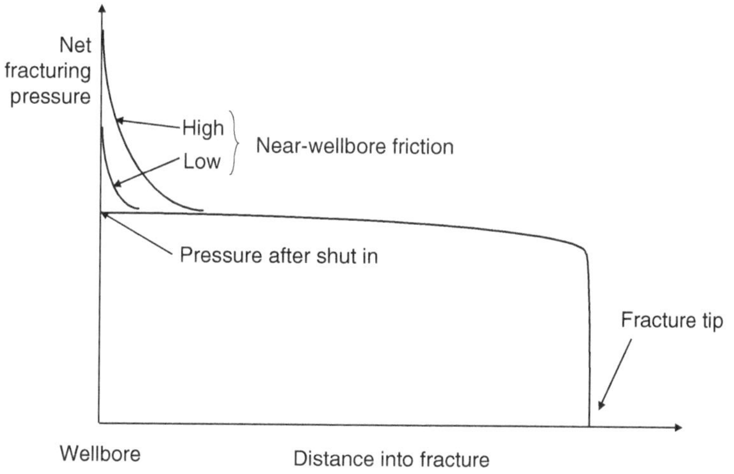

Fig. 14.7—Pressure profile down a fracture is affected by perforation friction, tortuosity, and injection rate.

We then progressively reduce the injection rate in steps and measure the change in injection pressure. If we pump long enough to create a reasonably long fracture, then step the rates down fast enough, the change in pressure measured during the test can be attributed to a change in frictional pressure drop. If we measure the downhole pressures or can properly account for wellbore friction, then the change in friction has to be caused by the near-wellbore friction.

The perforation friction component can be computed using Eq. 14.3. Notice that the perforation friction increases as a function of the injection rate squared.

$$\Delta p_{\text{perf}} = 0.2369\rho\left(\frac{q}{C_d N_{\text{perf}} D_{\text{perf}}^2}\right)^2, \qquad (14.3)$$

where C_d is the discharge coefficient; D_{perf} is the perforation diameter, in.; N_{perf} is the number of perforations; Δp_{perf} is the perforation pressure drop, lbf/in²; q is the injection rate, bbl/min; and ρ is the density, lbm/gal.

The other component, the near-wellbore-tortuosity component, is affected by the width of the fractures near the wellbore. Fluid flowing through a narrow, pressure-dependent opening in the near-wellbore region will react much differently compared with fluid flow through a perforation. The pressure drop is proportional to the injection rate raised to the ½ power. Therefore, the total near-wellbore pressure drop can be broken into its two components and compared graphically to determine the cause of the near-wellbore pressure drop, as shown in **Fig. 14.8.**

If the near-wellbore pressure drop is believed to be too large, and we can determine the cause of the pressure drop, then we can engineer a solution. If the perforation friction is too large, we can either add perforations or pump a hole-opening ballout treatment. If multiple fractures are causing near-wellbore tortuosity, we can use smaller perforated intervals. If we are already propagating multiple fractures, we can use proppant slugs to erode some of the fractures and to try to plug off multiple fractures. When narrow fracture width is the problem, we can try higher injection rates and higher fluid viscosity to obtain a larger fracture width in the near-wellbore region.

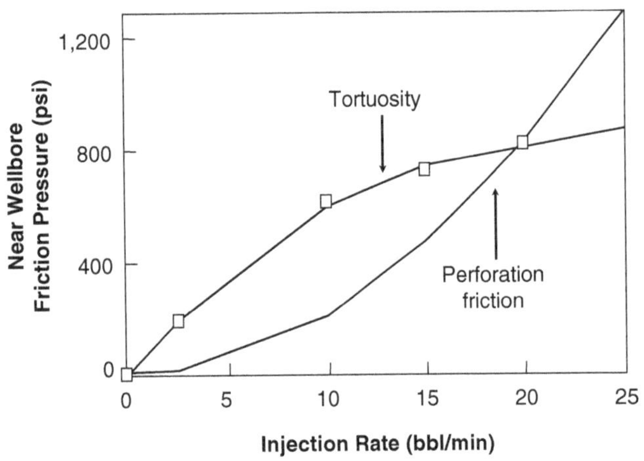

Fig. 14.8—Near-wellbore pressure drop depends the effects of tortuosity and perforation friction.

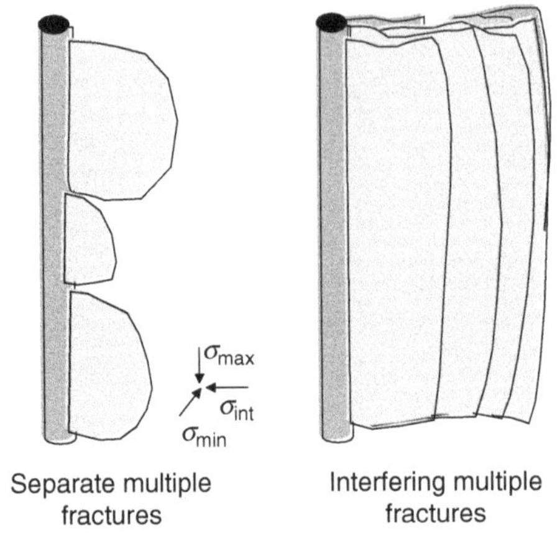

Fig. 14.9—Near-wellbore multiple fractures might interfere with each other, thereby reducing fluid-flow rates.

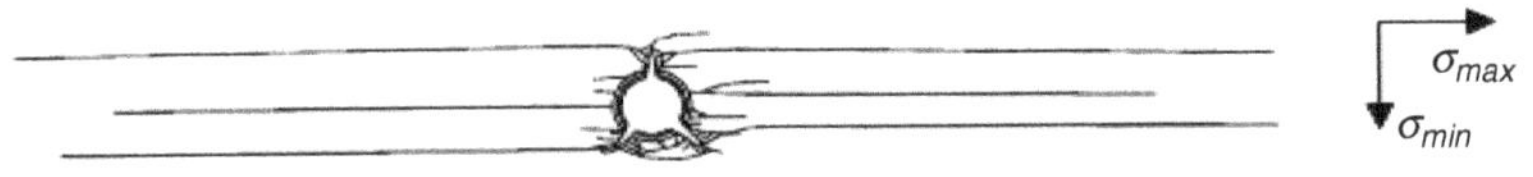

Fig. 14.10–Multiple fractures that propagate away from the wellbore reduce the ability to propagate deeply into the reservoir.

Sometimes creating multiple fractures near the wellbore can be advantageous, and sometimes it can be the opposite. For example, as shown in **Fig. 14.9,** sometimes multiple fractures are useful to fracture treat multiple layers in either a single-stage treatment or in a multiple-stage treatment. In such a case, the fractures do not interfere with each other, either during the fracture propagation stages or during production. However, if the multiple fractures do interfere with each other, then the results of the treatment will be less than optimal. First, when multiple competing fractures are propagated, the created fracture length will be reduced compared to a single fracture. Also, it is very difficult to pump propping agents into multiple, narrow fractures. Even worse, once the gas near the fractures has been produced, the fracture set really is behaving as a single, short fracture in which oil- and gas-flow rates will be lower than those expected from a longer fracture.

As shown in **Fig. 14.10,** multiple fractures can propagate out into the reservoir away from the wellbore. When this occurs, each of the fractures will affect the others, and the fractures will not propagate deeply into the reservoir, as is usually required.

If multiple fractures are created, as shown in **Fig. 14.11,** then as each fracture pushes the Earth apart, the strain created by one fracture will affect the strain or fracture opening of all the fractures, as shown in **Fig. 14.12** (Jeffrey et al. 1987). The result is that each individual fracture will be less wide than if only a single fracture had been created.

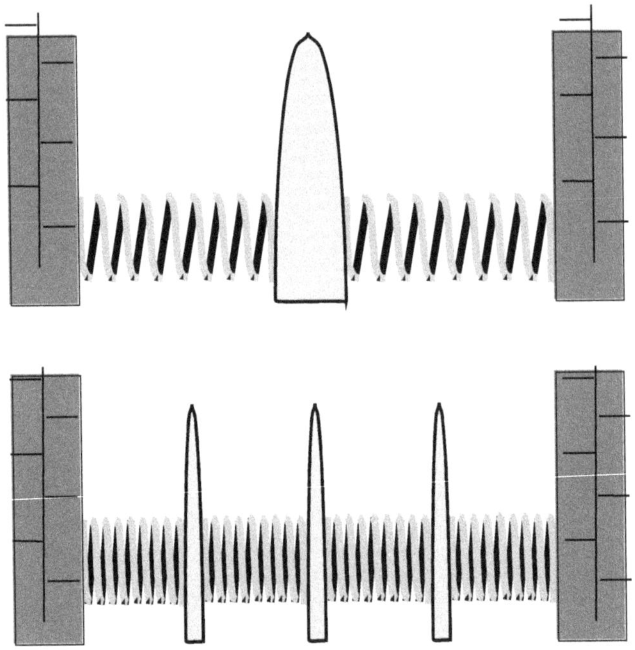

Fig. 14.11—Multiple fractures compete for fracture width, reducing the width of each individual fracture (from Jeffrey et al. 1987).

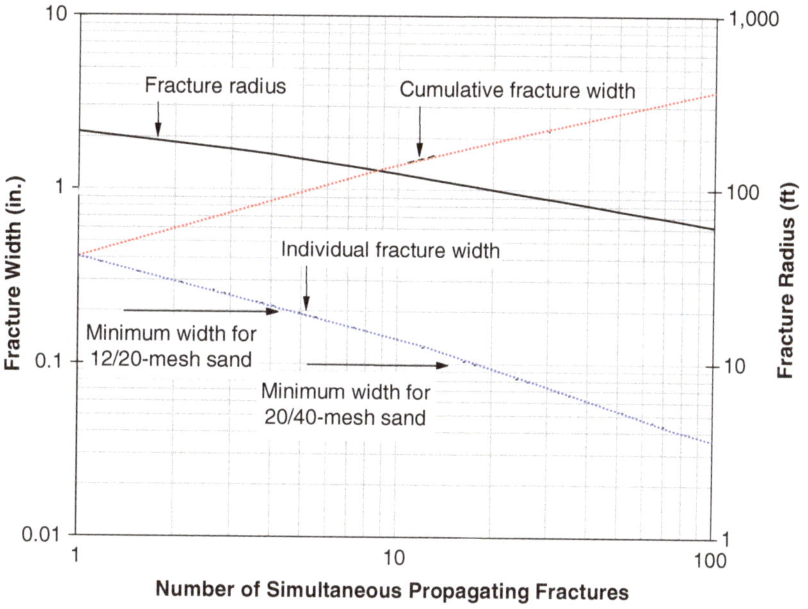

Fig. 14.12—Multiple fractures reduce individual-fracture width, leading to lower fluid efficiency.

As the number of simultaneously propagating fractures increases, we see reduced individual-fracture width, increased screenout potential, and increased fracturing-fluid leakoff, which leads to lower fracturing-fluid efficiency. In addition, we have reduced effective fracture length, which affects oil- and gas-flow rates. The resulting higher net treating pressures cause more fracture-height growth. On a positive note, the multiple fractures, if propped, will result in higher fracture conductivity close to the well.

In summary, the design engineer must be cognizant of the issues surrounding near-wellbore fracture geometry. If the fracture is not satisfactorily connected to the wellbore, problems are likely during the fracture treatment and the production of oil and gas.

Some issues the design engineer should consider are summarized as follows:

- The likelihood that multiple hydraulic fractures will be created and propagated increases in naturally fractured reservoirs.
- If long perforated intervals exist, the likelihood that multiple hydraulic fractures will be created and propagated increases.
- In horizontal wells, multiple hydraulic fractures normally will occur during a treatment.
- Sometimes, smaller perforated intervals should be used to reduce the number of multiple-fracture-initiation points.
- Breaking down an interval with a high-viscosity fluid rather than a low-viscosity fluid might lead to the propagation of fewer fractures near the wellbore.
- High viscosities and increased injection rates might help link up multiple fractures away from the wellbore.
- Dropping proppant slugs early in the treatment might plug off multiple fractures near the wellbore.
- Near-wellbore fracture geometry might not affect far-field fracture geometry.

14.3 Fracture Containment Considerations

Fracture containment encompasses the stress and rock properties that control the vertical height of a hydraulic fracture; it is a very important concept in the design of a fracturing treatment. Fracture height is considered one of the critical factors that can define the success or failure of a hydraulic fracturing treatment. Fracture length essentially is inversely proportional to fracture height, which explains the importance of having good fracture-height information in treatment designs. Prediction of hydraulic fracture patterns is essential in hydraulic fracture modeling. Failure in fracture containment during treatment often results in poor production and recovery, which influence economic profit in a negative way.

14.3.1 Effect of In-Situ Stresses on Hydraulic Fracture Containment. Because fracture containment plays a significant role in hydraulic fracturing treatment, it is important to control the fracture growth and, consequently, know the factors that affect it.

Warpinski et al. (1982b) explained that containment of massive hydraulic fracturing in tight gas sands can refer to confinement of the fracture to specified intervals comprising both gas-bearing sandstone lenses and surrounding shale zones or to the more-common concept of confinement within a single reservoir zone. In either case, the containment of hydraulic fractures requires designers to define the parameters and conditions that will limit the height of the fracture and control its direction of propagation to obtain the necessary fracture length. With sufficient understanding of these conditions, operators can design later treatment parameters to help control containment, at least in those formations in which the likelihood of containing the fractures is good. If the fractures cannot be contained, the operators might still be able to determine zones that are most likely to provide economic production.

Several studies have been conducted on the effects of formation Young's modulus, in-situ stress, fracture toughness, and layer interfaces on hydraulic-fracture-height containment in layered formations. These studies agree that in-situ stress—the local stress state in a given rock mass at depth—is the first parameter controlling fracture-height growth; a stress contrast between the pay zone and a bounding layer is the most important factor controlling fracture height.

Warpinski et al. (1982b) demonstrated that the work of Simonson et al. (1978) showed the increase in fracturing pressure required to continue propagating a fracture into a layer of greater in-situ stress than the fracture zone provides a barrier to propagation that can be more or less effective, depending on the difference in stress in the two regions. They also showed that the hydrostatic gradient of the fracturing fluid can influence the upward or downward migrations of the fractures. Drawing down the reservoir can create an effective stress differential between the porous reservoir rock and the impermeable barrier, thereby reducing pore pressure and lateral confining stress and encouraging the fracture to preferentially propagate into the lower-stress reservoir rock.

Warpinski et al. (1982a) present results of several hydraulic fracturing experiments to show a clearer picture of the factors that influence containment. These results show that a material-property difference between the reservoir rock and the bounding formation is not sufficient to contain the fracture. These authors noted that the minimum in-situ stress not only dictates the orientation of the fracture, but also can provide barriers to fracture propagation.

Yew and Chiou (1983) showed that the containment effect of in-situ rock stresses depends upon four factors: distribution of in-situ stresses within a region near the fracture that ranges from three to four times the fracture height; ratio of the fracture height to the thickness of the stress zone adjacent to the fracture; relative magnitude of in-situ stresses between the reservoir zone and the adjacent zones; and the relative distance between the fracture tip and the boundary of stress discontinuity.

Even where the in-situ stress has the same ratio as the layer material difference, the in-situ stress difference has a greater effect on the movement of a fracture. Yew and Chiou (1983) showed that the stress-intensity factors calculated at the fracture tips are greater from the in-situ stress ratio than from the material difference. Using Tennessee and nugget sandstone and ash-full tuff for comparison with mineback experiments, Warpinski et al. (1982a) showed that a 300- to 400-psi (2- to 3-MPa) contrast in in-situ stress across adjacent rock strata is sufficient to restrict the growth of a hydraulic fracture. Although pore-pressure effects appear to be very important, mechanical properties of the rock did not significantly affect stress levels. Warpinski et al. (1982a) noted that a biomaterial interface under the loading conditions likely to be encountered in the field also can result in stress sufficient to contain fractures.

14.3.2 Effect of Young's Modulus Contrast on Hydraulic-Fracture-Height Containment. Although in-situ stress is clearly the dominant factor in controlling fracture-height growth, a contrast in the Young's modulus can have significant implications on fracture growth. Young's modulus (E) essentially is an index of the rock's resistance to external force, defined as the ratio of the applied uniaxial stress to the resulting strain.

$$E = \frac{\sigma}{\varepsilon}, \qquad (14.4)$$

where E is Young's modulus; σ is applied uniaxial stress; and ε is strain.

The formation modulus and in-situ stress contrasts together can affect fracture height. Gu and Siebrits (2006) considered two situations for modulus contrast with respect to the position of the fracture relative to a material interface; one was a fracture approaching the interface, and the other was a fracture spanning the interface **(Fig. 14.13).**

Gu and Siebrits (2006) found that the stress-intensity factor decreases—theoretically, to zero—as a fracture approaches a high-modulus layer from a low-modulus layer. The high-modulus layer can begin to fracture before the fracture in the low modulus layer even reaches the interface **(Fig. 14.14).**

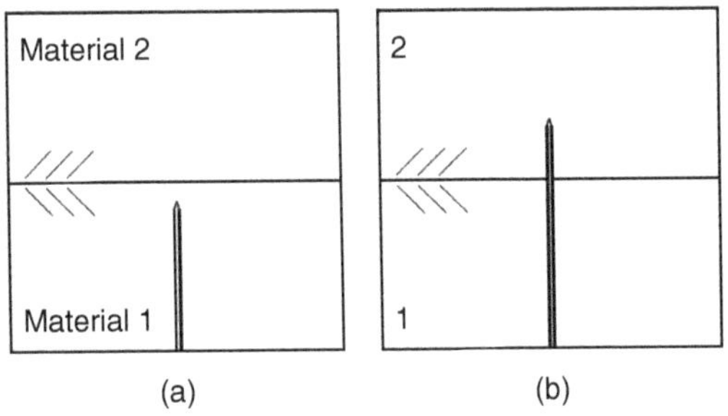

Fig. 14.13—As the fracture approaches the interface between a low-modulus layer and a high-modulus layer (a), the stress-intensity factor decreases, so that the fracture can span the interface (b).

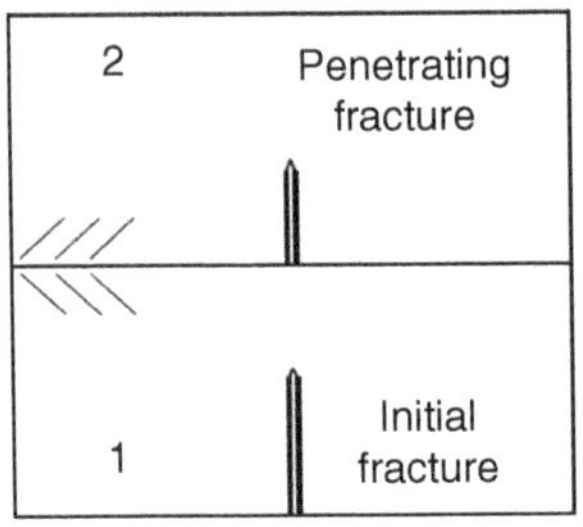

Fig. 14.14—The penetrating fracture can start before the initial fracture reaches the interface.

Experimental and mineback evidence shows that a hydraulic fracture can cross the interface and extend to the high-modulus layer.

When a fracture approaches a low-modulus layer from a high-modulus layer, the stress-intensity factor increases to infinity. Therefore, the fracture grows toward the lower-modulus layer. However, once the fracture tip reaches the lower-modulus layer, the stress-intensity factor decreases and reduces further fracture growth into the layer.

Gu and Siebrits (2006) also studied the effect of fracture-toughness contrast on height containment. In the case of sedimentary reservoir rocks, fracture toughness does not depend on rock type; sandstones have higher formation modulus than shales, but they can have higher or lower fracture toughness. Gu and Siebrits (2006) showed that higher fracture toughness in the outer layer of a three-layer model limited the height growth, whereas lower fracture toughness in the outer layer enhanced it. They also found that when a fracture approaches a material interface, many other possibilities exist, including delamination and penetration in various forms **(Fig. 14.15).**

Hydraulic fractures behave differently than ideal hydraulic fractures under static and constant fluid pressure, because fluid pressure directly influences fracture deformation and height growth, and at the same time fluid pressure itself is strongly affected by fracture geometry.

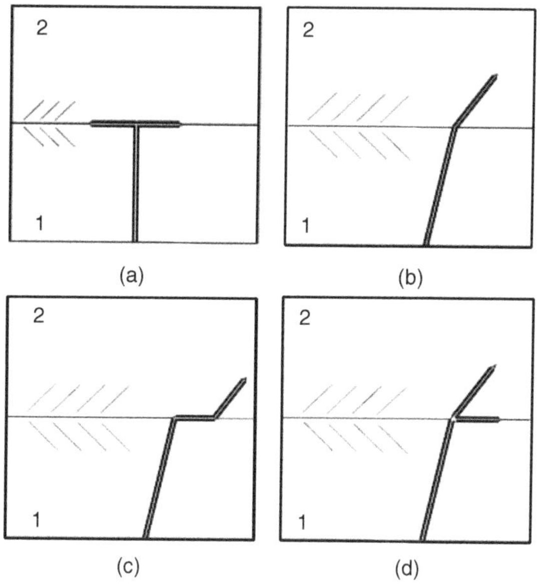

Fig. 14.15—Depending on the interface, fracture propagation can lead to (a) delamination, (b) kinked penetration, (c) jog formation, or (d) bifurcation.

14.3.3 Control of Fracture Height. Prediction and control of fracture height are among the main problems in hydraulic fracturing design and execution. Success in field operations requires us to understand the effects of barriers to vertical fracture propagation. Yudin et al. (2007) identified three techniques to control fracture-height growth:

- Restricting net-pressure increase by using fluids with lower viscosity and density to minimize pressure applied to barriers. Yudin et al. (2007) suggest low-viscosity fluids, such as viscoelastic fracturing fluid; low-polymer-concentration linear gels, including fibers; or compressible fluids.
- Placing an artificial barrier for fracture propagation.
- Depleting the reservoir pressure to increase the stress difference.

Any combination of these methods can improve the success ratio, although depleting the reservoir pressure by producing the well before the fracture stimulation is unlikely when oil prices are high or when production forecasts are to be met.

Limiting the injection rate can limit fracture extension only where high fluid leakoff is encountered. Restricting fluid viscosity throughout the treatment and on treatments in which extended pump times are required can result in decreased width, excessive fracturing-fluid loss to formation, unstable fluids, and, finally, premature screenout in high-temperature wells.

14.3.4 Summary. Understanding factors that affect hydraulic fracture containment is important when determining effects on fracture growth. Experiments have shown that the minimum in-situ stress is the predominant influence on hydraulic fracture containment. Fractures might still occur in places where stress gradients or discontinuities (e.g., faults) are sufficiently large. Field measurement of in-situ stress helps improve the success of hydraulic fracturing in oil and gas wells.

14.4 Fracture Growth Analysis

Hydraulically induced fractures in all cases will propagate on the basis of the lithology of the porous media in which they are propagating and the state of stress surrounding the fracture. Near the wellbore, stress in the formation can be altered by factors in the hole, by the perforations, and by the presence of natural fractures. However, away from the wellbore, the fracture will propagate in a manner dictated by the far-field stresses.

14.4.1 Fracture Orientation. Whether a fracture will propagate vertically or horizontally is determined by the orientation and magnitude of the three principal stresses. The fracture will always propagate perpendicular to the minimum principal stress. Vertical fractures are the most common, but not all fractures follow a vertical plane. The principal-stress direction might also be inclined from the vertical.

Horizontal fractures are possible in shallow formations at depths of 3,000 ft or less. Horizontal fractures have been documented at greater depths when a well is drilled and completed near a thrust fault.

The azimuth of a vertical fracture is controlled by the direction of the minimum horizontal principal stress, which we often call the in-situ stress. A hydraulic fracture will always propagate perpendicular to the minimum in-situ stress. We can determine the direction of the in-situ stress in many ways, which will be covered in Part 3 Ch 15: Data Acquisition and Core Measurements.

One of the first things a fracture-design engineer should do is sit down with the geologic staff, and ask them to explain the tectonics in the region surrounding the well to be fracture treated. It is always a very good idea to look at the faulting in the area. If we assume that the current stresses are like the stresses that created the faulting, we can then estimate the probable fracture orientation. As shown in **Fig. 14.16,** we expect a hydraulic fracture near a normal fault to propagate parallel to the fault.

We can tell if the fault is a normal fault, because when we drill through the fault plane, a section will be missing. For wells in normally faulted areas, the tectonic-force component essentially is zero. For such cases, the in-situ stress can be accurately calculated using the poroelastic stress/strain equations.

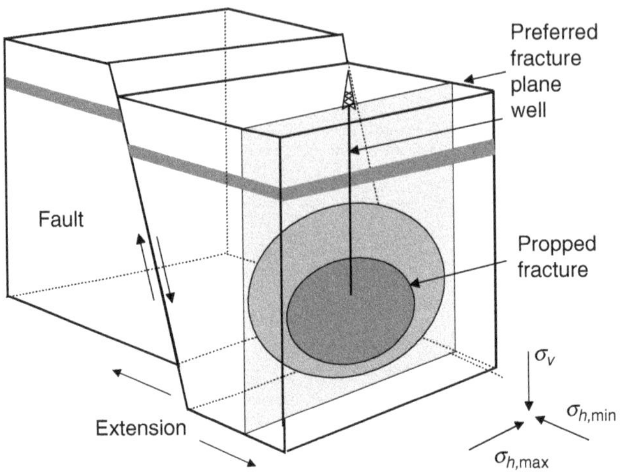

Fig. 14.16—Fracture orientation typically is parallel to a normal fault.

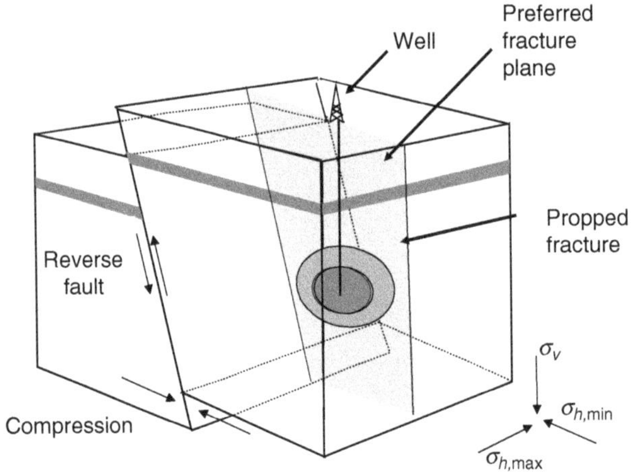

Fig. 14.17—Fracture orientation usually is perpendicular to a reverse or thrust fault.

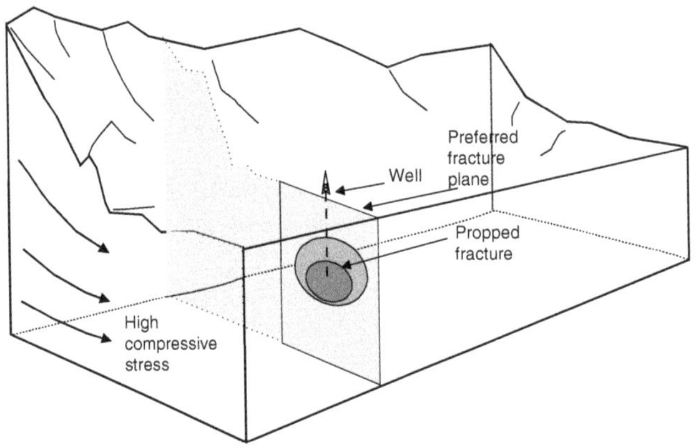

Fig. 14.18—Fracture orientation can be strongly affected by mountain ranges.

For a thrust fault, a vertical fracture will propagate perpendicular to the fault plane, as shown in **Fig. 14.17.** Very high horizontal stresses must be present to thrust one part of a formation on top of itself. We can tell when we drill through a thrust fault, because we will drill the same interval twice. In severely thrusted areas, the same zone might be drilled through more than twice. In thrusted areas, the in-situ stress gradient will be higher than computed using just the poroelastic uniaxial strain equations because of a substantial tectonic-stress component.

In wells very close to active thrust faults, it is possible to create horizontal fractures. Shell Corporation has used tiltmeter data from several wells in California to record horizontal fracture propagation in wells deeper than 3,000 ft.

Other factors can also affect the orientation of hydraulic fractures. **Fig. 14.18** illustrates a case in which larger-than-normal horizontal stresses in a certain direction might occur in a formation that is in a well drilled near the base of a mountain range.

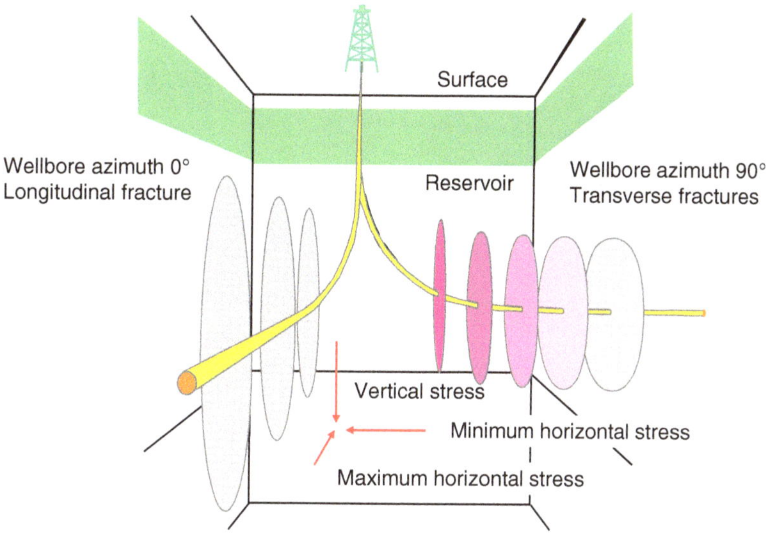

Fig. 14.19—Multiple vertical fractures can lead to an almost infinite combination of orientations in a horizontal well.

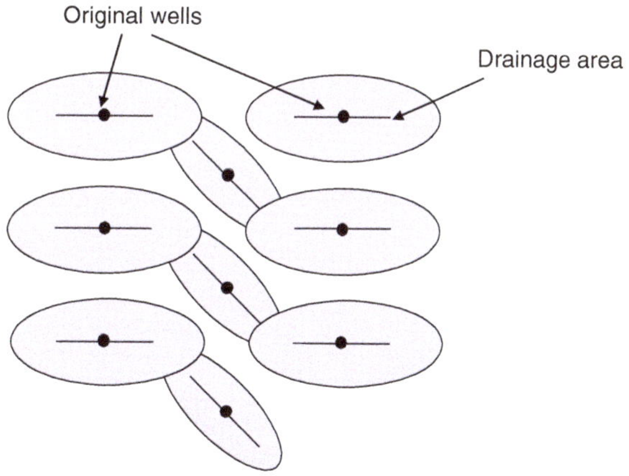

Fig. 14.20—Partial pressure depletion can lead to fracture reorientation.

In such cases, the stresses will control not only the orientation, but also the magnitude, of the in-situ stress. Other factors affecting fracture growth patterns are controlled by the interface between the wellbore and the individual fractures created during a treatment.

In horizontal wells, the fracture growth analysis is very complicated, as shown in **Fig. 14.19.** In horizontal wells, almost an infinite combination of results is possible. The simplest situation is when the hydraulic fractures propagate in a longitudinal direction along the wellbore. However, this orientation might not always be possible.

Fracture growth analysis becomes more complicated when the wellbore direction is different from the fracture direction. The number of vertical fractures might be controllable if a cemented casing string is run, and the number of perforations is limited. In openhole or uncemented-liner completions, we have no control on the number of vertical fractures initiated and propagated. In horizontal wells, the critical portion of the fracture is the part that connects the propped fracture to the borehole. If for some reason that connection is lost, the effects of the fracture treatment will be greatly diminished.

In general, regional stresses are usually consistent over large areas in tectonically relaxed basins. Therefore, if we can determine the orientation of the regional stresses in one location, we can usually transfer that information to nearby wells, reservoirs, and fields. In tectonically active areas or geologically complex areas, the local stress orientation might not be so consistent. For example, in tectonically thrusted areas or in areas near mountain ranges, the orientation of the minimum in-situ stress can change from location to location.

Oil and gas production causing pressure depletion can also alter local stress orientation with time. In some areas, the orientation of fractures in wells that are partially depleted has been measured to be slightly different from the orientation of the original fractures in those same wells or formations. Such a situation is shown schematically in **Fig. 14.20.**

14.4.2 Fracture-Height Growth. When we begin pumping fluid into a fracture, the fracture will grow up, out, and down until it hits a barrier to fracture growth. The reservoirs we stimulate are always layered systems. A fracture will propagate through a layer unless the in-situ stresses, the modulus of elasticity, or other factors such as the permeability of the layer prevent propagation. Of all the factors, the in-situ stress of the layer plays the major role in causing a certain layer to become a barrier to vertical fracture growth.

The in-situ stress will be mainly controlled by how much of the vertical or overburden stress is transposed into horizontal stress. One factor in this conversion is the value of Poisson's ratio for each layer. As the value of Poisson's ratio increases, more of the vertical stress is transposed into horizontal stress. As it turns out, in general, a plastic shale will have the largest value of Poisson's ratio, which will translate into a higher value of horizontal in-situ stress, which simply means a good, clean shale will be a barrier to vertical fracture growth. For a shale layer to be a barrier, it needs to be approximately 50-ft thick; a thin shale, say 10 ft, is not thick enough to be a fracture barrier. A propagating fracture will cut through a thin shale before it knows it is there.

To understand vertical-fracture growth analysis, we need to understand the concept of net pressure and how to use the concept to analyze field data. Net pressure is the magnitude of pressure inside the fracture with respect to the minimum in-situ stress and is directly related to the width and height of the fracture.

$$p_{net} = p_{frac} - \sigma_{min}, \quad (14.5)$$

where p_{net} is net pressure, psi; p_{frac} is pressure inside fracture, psi; σ_{min} is minimum in-situ stress, psi.

Net pressure is the pressure inside the fracture minus the minimum in-situ stress. For the example shown in **Fig. 14.21,** the net pressure (2,500 – 2,000) is 500 psi. The net pressure will vary as a function of the fracture length. At any point down the fracture, the vertical distribution of net pressure will be affected only by gravity and the width and height of the fracture then can be related to the net pressure at that point in the fracture.

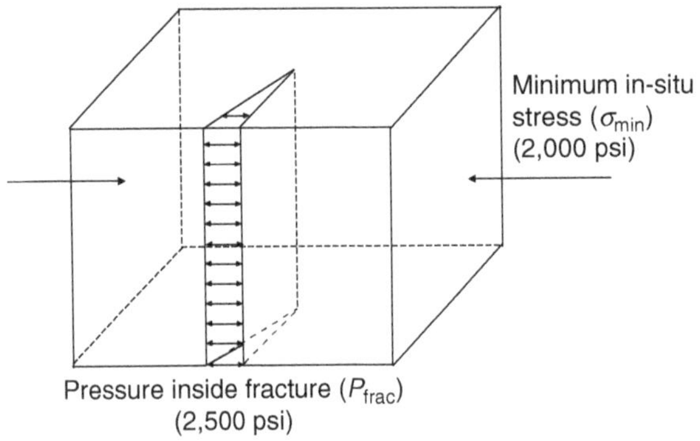

Fig. 14.21—In-situ stress and pressure interact in the fracture.

We use the net pressure near the wellbore in our fracture-propagation models to history match the fracture-treatment data. To determine the correct value of net pressure in the fracture near the wellbore, we must be certain we can correctly compute the near-wellbore pressure drop.

Nolte and Smith (1981) developed a method to use the change in net pressure near the wellbore as a function of time to determine how a fracture is growing in terms of length, width, and height. For the method to provide accurate results, the net pressure must be properly computed. The most important issue is that the near-wellbore pressure drop must be properly accounted for in the analyses. In Fig. 14.7, the pressure in the fracture that affects the height and width distribution is the true net pressure down the fracture after subtracting the near-wellbore pressure drop. This illustrates why a step-down test should be run on every well before treatment to determine the near-wellbore pressure drop. Then at the end of the treatment, the shutdown pressures also should be analyzed to determine the near-wellbore pressure drop at the end of the treatment because the near-wellbore pressure drop can change during the treatment.

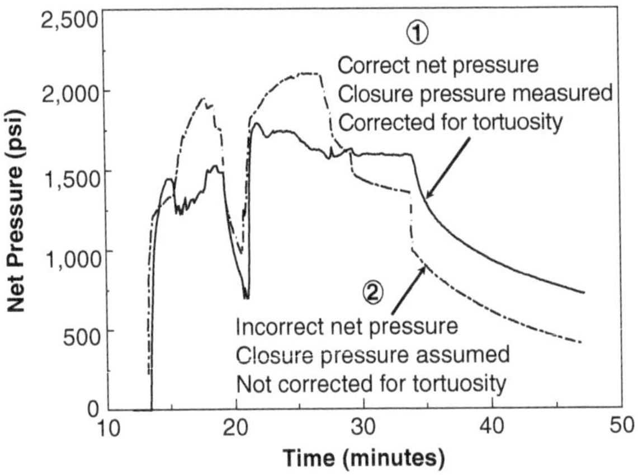

Fig. 14.22—Without the correction for tortuosity, estimates of fracture dimensions based on net pressure and time can be far from accurate.

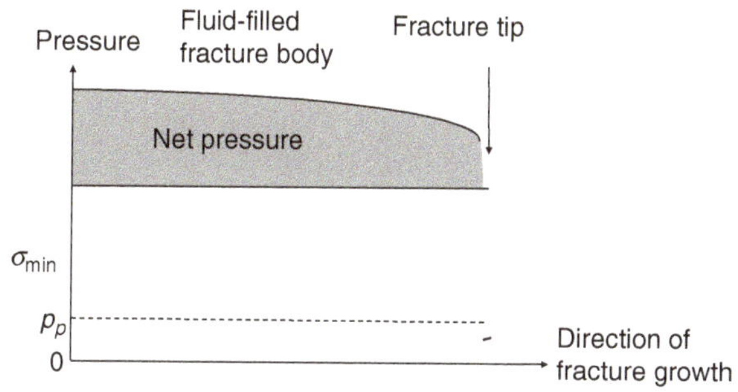

Fig. 14.23—The net-pressure profile should be analyzed with a fracture model.

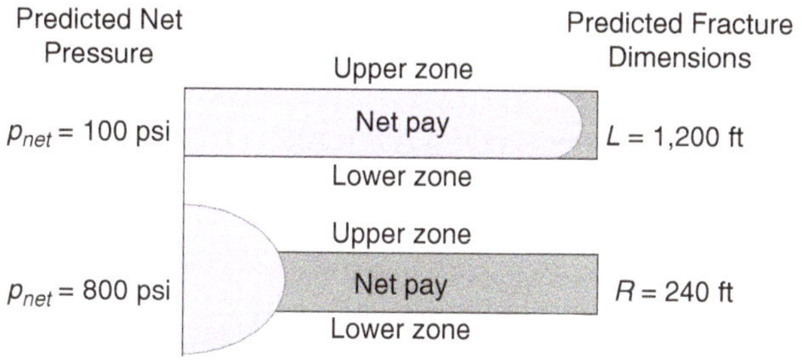

Fig. 14.24—Excessive net pressure might cause the fracture to grow out of the pay zone, thereby decreasing its length.

The graph in **Fig. 14.22** comes from Wright et al. (1995) and illustrates the problems that can occur if the in-situ stress and the near-wellbore pressure drop are not accurately known. As shown in Fig. 14.21, we must know the value of the in-situ stress to correctly compute the net pressure. Fig. 14.7 illustrates why we must also know the value of the near-wellbore pressure drop to compute accurate values of net pressure. If we make a mistake in either estimation, we could end up with a graph of net pressure vs. time that is not correct. If we try to match field-measured (i.e., observed) pressures that are in error with a fracture model, the resulting analyses will be incorrect as well.

In **Fig. 14.23,** we illustrate the true net pressure from the tip of the fracture to the wellbore. As the fracture continues to grow, the magnitude of the net pressure and the distribution of the net pressure will change. By modeling the change in net pressure vs. time, we obtain reasonable estimates of fracture dimensions vs. time. **Fig. 14.24** shows the ideal estimate of the net pressure down a fracture. A fracture-propagation model should be run to history match this net pressure. When a reasonable history match of net pressure is obtained, then one will have a reasonable estimate of created-fracture dimensions.

To illustrate how net pressure can be used to help determine fracture properties, consider the simple example in Fig. 14.24. Suppose the difference between the in-situ stresses in the upper and lower zones (the potential barriers) and the net-pay zone is 500 psi. If a fracture treatment is pumped and a value of net pressure of 100 psi is recorded, then the fracture will stay in zone and a relatively long fracture will be created. If the value of net pressure during the treatment exceeds 500 psi, then the fracture will grow up, out, and down, and the resulting length will be much less than the case in which the fracture is contained.

Thus, to analyze fracture-growth patterns, we must know the in-situ stresses in all the rock layers above, within, and below the pay zone. We also must be able to characterize the value of net pressure near the wellbore vs. time as the treatment is being pumped. It is best to measure pressures downhole, so the friction and gravity effects of the fluid flowing in the casing or tubing do not have to be accounted for in the analyses.

Rarely, however, do we have the luxury of measuring downhole pressures during a treatment. To compensate, we must be able to accurately estimate the friction and gravity forces during a treatment. Such computations are simple when pumping the pad. The computations become much more complicated when we pump the fluid and proppant slurry, especially at high proppant concentrations.

Nolte and Smith (1981) and Nolte (1986) explained a theory that we can use to interpret net pressures measured during injection. They noticed that reservoir engineers could measure flow rates, flowing pressures, and fluid properties, then use those data to estimate reservoir properties such as permeability, skin, and drainage areas. They used the Perkins and Kern equations to allow an engineer to estimate how a fracture is growing by measuring the injection rate, injection pressure, and fluid properties during a fracture treatment.

Nolte and Smith (1981) based their analyses on a graph of the log of net pressure vs. the log of Δt. They noted that in the Perkins and Kern equations, for a fixed fracture height, the value of net pressure will steadily increase with time, as shown in the following equation, if the fracture is growing unrestricted in length and width:

$$p_{net} \,\alpha\, t^{e}, \qquad (14.6)$$

where p_{net} is the net pressure, psia; t is time; $1/8 < e < 1/5$; e is 1/5 for small leakoff; and e is 1/8 for high leakoff.

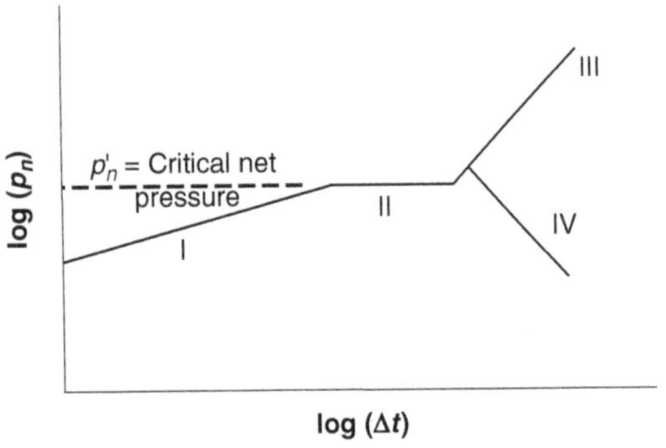

Fig. 14.25—Nolte-Smith net-pressure graph (Nolte and Smith 1981) will yield a slope between 1/8 and 1/5.

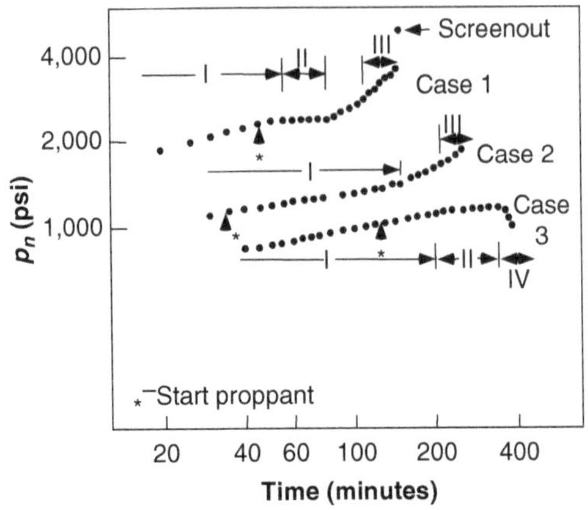

Fig. 14.26—Field data illustrate that the Nolte-Smith net-pressure graph can identify four slopes that define restrictions in fracture growth (Gidley et al. 1989).

What this equation means is that if we graph up the log of net pressure vs. the log of time, the graph should be a straight line with a slope of between 1/8 and 1/5. Such a graph is illustrated in **Fig. 14.25.**

Mode I in Fig. 14.25 represents the portion of the graph predicted from Eq. 14.6. When Nolte and Smith from Amoco Corporation went into the field to measure downhole pressures during a fracture treatment to confirm the analyses concept, they were successful, but they also met a few surprises that led to the identification of Modes II, III, and IV. **Fig. 14.26** is one of the field-data sets they published to illustrate the use of net pressure in the analyses of fracture-growth patterns.

Figs. 14.25 and 14.26 illustrate four distinct slopes that can be identified in a Nolte-Smith net-pressure graph. These modes have been identified as follows:

- Mode I: Confined height, unrestricted length, and width growth
- Mode II: Stable height growth or increased fluid loss
- Mode III: Growth restriction in length, but possible width growth
- Mode IV: Rapid height growth or near-wellbore change in pressure drop

These four modes can be explained using one of the Perkins and Kern equations (Gidley et al. 1989):

$$p_{wb} = \sigma_{\min} + \frac{\alpha}{h}\left[\frac{G^3 q \mu L}{(1-\nu)}\right]^{1/4} + \Delta p_{\text{perf}}, \qquad (14.7)$$

where G is the modulus, psi; h is the height, ft; L is the length, ft; Δp_{perf} is the pressure change in the perforations, psia; p_{wb} is the wellbore pressure, psia; q is flow rate, bbl/min; σ = Biot constant, dimensionless; μ is viscosity, cp; ν is Poisson's ratio, dimensionless; and $\sigma_{\min}$ is minimum stress, psia.

The Nolte-Smith analysis assumes that only the value of the wellbore pressure changes as a function of length. Thus, it is implicitly assumed that everything else remains constant, which in Eq. 14.7 would be the in-situ stress, the fracture height, the modulus, the injection rate, the injection-fluid viscosity, Poisson's ratio, and the near-wellbore pressure drop. Obviously, several of these variables can change with time during the fracture treatment. The engineer supervising and analyzing the fracture treatment must be aware of any such changes and factor them into the analysis of the data.

14.5 Hydraulic Fracturing Models

14.5.1 Two-Dimensional Hydraulic Fracturing Modeling. An appropriate fracture-propagation model is crucial to design and optimize a fracture treatment that can effectively increase oil- and gas-flow rates and, ultimately, hydrocarbon recovery. Two-dimensional fracturing models were developed in the early 1960s (Gidley et al. 1989) as an attempt to predict fracture dimensions such as width, w, and length, L, given certain rock properties, formation-fluid properties, the injection rate, and the time of hydraulic fracturing treatment.

Currently, we use not only 2D fracture-propagation models, but also fully 3D models and pseudo-3D models. However, the scope of this subsection is to describe 2D fracturing models, predict fracture geometry by simultaneously using 2D and 3D models, use 2D models in formations exhibiting stress-dependent mechanical properties, and use existing 2D industry practices.

The most common 2D models used to design treatments are the Geerstma-deKlerk (GdK) and Perkins-Kern-Nordgren (PKN) models. The GdK model uses a constant-height fracture and assumes a horizontal plane strain condition **(Fig. 14.27).** The cross-sectional area perpendicular to the direction of fracture propagation is rectangular and has a constant width (vertically unbounded geometry). The pressure profile decreases with time. This model is applicable for fractures with ratios of length to half-height less than unity (L/H<1) (Meyer 1986).

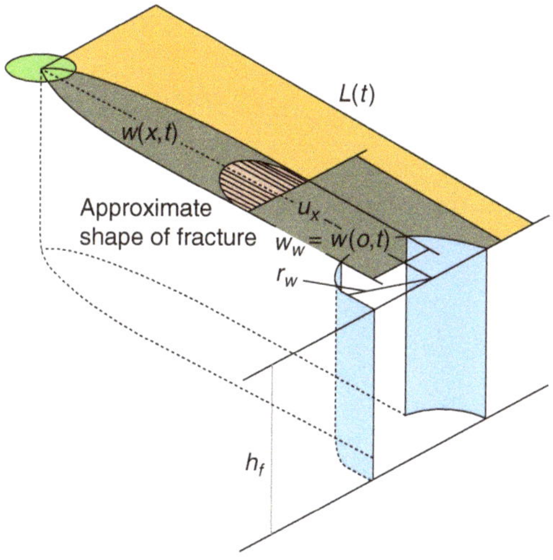

Fig. 14.27—GdK models use a constant-height fracture in a vertically unbounded geometry.

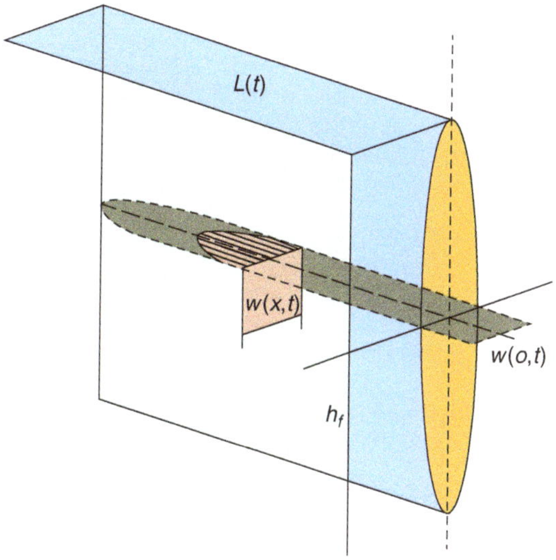

Fig. 14.28—PKN fracture geometry assumes elliptical fractures in both the horizontal and the vertical directions.

The PKN model assumes a constant-height fracture with a vertical elliptical shape and a horizontal elliptical length **(Fig. 14.28)**. The pressure profile increases with time. This model is most applicable for fractures with length to half-height ratios greater than unity (*L/H*>1) (Meyer 1986).

Meyer (1986) presented the equations for mass conservation, continuity, width-opening pressure, and momentum, taking into consideration the constitutive relationships for the fracture propagation and proppant transport. The major fracture- and fluid-mechanics phenomena include asymmetrical confining stress and moduli contrast, adjacent-layer stress gradients, rock fluids with leakoff and spurt loss, proppant transport with settling and closure, and production simulation. Meyer's parametric studies identified the effects parameters have on fracture characteristics:

- Fracture efficiency at the end of pumping for large and moderate leakoff decreased with increasing confining stress. The major reason for this is that with leakoff only to the pay zone, the leakoff area increases with increasing confining stress because of the increased fracture length. Thus, GdK modeling's efficiency is higher where fracture lengths are shorter.
- Fracture height decreased with increasing confining stress. The fracture approaches a penny shape for low confining stresses and a constant height for large confining stresses, with a wellbore height equal to the pay-zone height.
- The GdK pressure decreased and the PKN pressure increased with time. The wellbore pressure at the end of pumping for increasing confining stress increased more significantly for the PKN than for the GdK model.
- The GdK wellbore width increased with increasing confining stress for no fluid loss. The PKN width initially increased and then decreased with increasing confining stress for large fluid loss, and the width increased for increasing values of confining stress for low leakoff.
- The PKN model is more applicable at larger length-to-height ratios, whereas GdK is more accurate for predicting fracture characteristics at low confining stresses.
- In proppant scheduling, the pad volume required in both models was greater for large leakoff coefficients at a given confining stress. However, GdK underestimated the pad volume for large confining stresses with their larger efficiency, greater width, and shorter length. Meanwhile, PKN required more pad volume because it had lower efficiency and narrower fracture width near the tip.
- Propped-fracture length increased, and conductivity decreased with increasing confining stresses. The GdK model showed shorter propped length and greater conductivity because the length/height (*L/H*) ratio increases, whereas PKN modeling showed lower conductivity and longer half-length. Thus, less slurry volume was required to achieve a certain length by using PKN.

14.5.1.1 Industry Practices for 2D Fracturing Modeling. In 1990, the main fracture-design technology was 2D modeling. By 1995, it seemed that 3D modeling had become the preferred design tool. A hydraulic fracturing survey conducted in 1995 (Carter et al. 1996) found a slight decrease—from 58 to 37%—in the use of 2D modeling in fracture design. However, 3D modeling was used in 49% of all fracture treatments, and the use of 3D models increased from 1990 (15%) to 1995 (38%).

More than 10 years after this survey was performed, 2D modeling has become a secondary tool because fracture geometry is assumed known and height is constant. However, 2D modeling is a very reliable fracture-geometry simulation tool when strong barriers that can ensure fracture containment are found.

14.5.1.2 Summary of 2D Hydraulic Fracturing Modeling. Under certain circumstances, when strong barriers are likely to inhibit fracture growth, or for short fractures in thick intervals, 2D fracture-propagation models can be used to design a successful hydraulic fracturing treatment. Engineers must be aware that a GdK model is more suitable for fracture-geometry modeling in some cases than a PKN model. The GdK model is appropriate for low confining stresses, whereas PKN is for high confining stresses.

To use 2D modeling and ensure consistent results, the design engineers must be certain of all input data and should have knowledge of the reservoir and any stimulation treatments performed in the past on offset or similar wells. This will allow them to predict with a high degree of confidence which type of geometry likely will be obtained.

14.5.2 3D Fracture-Propagation Models. Hydraulic fracturing modeling is one tool that engineers use to design and analyze hydraulic fracturing treatments. For simplicity and convenience, the industry initially used 2D models developed from simplifying assumptions to solve this complicated, 3D process in a 2D format. However, with advances in the industry, well-completion techniques, and computational power of computers, we can now develop and solve complicated 3D problems that better describe the fracturing process (Rahim and Holditch 1993; Gidley et al. 1989).

The hydraulic fracturing process is complicated, because it includes different mechanisms: fluid flow and leakoff, rock deformation, fracture propagation, and proppant transportation (Gidley et al. 1989; Yamamoto et al. 1999). The best way to minimize assumptions and simulate this process is by developing a set of equations that describe each process in 3D domains at different timesteps (Gidley et al. 1989).

The advantage of using 3D models rather than 2D models is to better simulate hydraulic fractures and to capture fracture geometry more accurately (Valkó and Economides 1995). The 3D models allow the fracture geometry to vary in three dimensions, taking into consideration the variation in properties of different layers of the formation (Economides and Nolte 2000; Holditch and Rahim 1993). The 3D model can predict fracture-geometry elements including the distribution of fracture length, height, and width.

*14.5.2.1 **Governing Equations in 3D Modeling.*** The governing equations used in 3D modeling are as follows (Gidley et al. 1989; Yamamoto et al. 1999; Valkó and Economides 1995):

- An elasticity equation that relates the pressure distribution in the fracture face with the crack opening
- A fluid-flow equation that relates the flow to the pressure gradient in the fluid
- A fracture-initiation criterion to relate the intensity of the stress state ahead of the fracture to the critical intensity required to create a tensile fracture in the rock

Because the GdK and PKN models assume a fixed height independent of fracture length, 2D models are successful only at predicting fracture dimensions if strong barriers above and below the pay zone are present and can prevent excessive vertical growth, and if a realistic fracture-height input value is used.

In the past, these models mostly had been run when reservoirs did not include many layers because 3D computing was very time consuming, regardless of constant-height input data, which was considered a drawback in the simulation. However, Rahim and Holditch (1993) found that 2D models can be run if reservoirs are described in five to ten layers and a single fracture will propagate through all the pay zones. Some benefits provided by the 2D model are ease and speed of computation.

The methodology consists of determining the number of layers that should be included in the solution; obtaining in-situ stress, fracture toughness, and the thickness of each layer; and determining net pressure from either previous fracture treatments or output from 3D models. To compute fracture height, Rahim and Holditch used mathematical equilibrium equations:

$$K_{it} = \frac{1}{\sqrt{\pi\alpha}} \int_{-\alpha}^{\alpha} p(y) \left(\frac{\alpha + y}{\alpha - y} \right) dy^{1/2}, \qquad (14.8)$$

$$K_{ib} = \frac{1}{\sqrt{\pi\alpha}} \int_{-\alpha}^{\alpha} p(y) \left(\frac{\alpha + y}{\alpha - y} \right) dy^{1/2}. \qquad (14.9)$$

To estimate a value of fracture height, they used other mathematical equations (Gidley et al. 1989) when the values of in-situ stress, fracture toughness, layer thickness, and net pressure could be determined. Once fracture height was calculated for the given value or range of values of net pressure obtained by 3D modeling, it was used as input data in the GdK and PKN models.

*14.5.2.2 **Preparing Input Data for 3D Models.*** The quality of the output of any fracture model is a function of the quality of input data. In 3D models, the formation is divided into several layers. Then, both mechanical and flow properties are entered for each layer. Holditch and Rahim (1994) developed a systematic way to prepare data sets for 3D models. They defined critical parameters and developed correlations to obtain the required data when needed. The most-critical parameters needed for stimulation-treatment design using 3D models include formation permeability, Young's modulus, average pressure, minimum in-situ stresses, gross- and net-pay thickness, reservoir temperature, and total fluid-loss coefficient. Rahim and Holditch (1993) assigned these parameters for each rock layer. **Table 14.1** illustrates each parameter, how it will affect the model output, and how it is obtained.

The procedure to prepare the model requires defining the formation layers by using logs, cores, and well-testing data. After identifying these layers, the parameter should be assigned for each layer before running the model.

*14.5.2.3 **Types of 3D Models and Solution Methodology.*** *Full 3D Models.* 3D fracture-propagation models use 3D elastic equations and 2D fluid-flow equations that include flow in horizontal and vertical directions (Gidley et al. 1989; Economides and Nolte 2000). These nonplanar models have no assumption about the fracture orientation. Fractures can initiate in one direction because of well-completion-related factors, such as perforation direction, before turning to another orientation, perpendicular to minimum horizontal stress (Economides and Nolte 2000). This type of model uses no assumptions, but applies a coupled solution for the fracture and fluid flow to simultaneously solve the equations.

A numerical method is used to solve for the grids developed on the fracture face (Gidley et al. 1989; Economides and Nolte 2000), as shown in **Fig. 14.29.** Other numerical methods have been suggested by different people to solve for this complicated problem (Valkó and Economides 1995).

Pseudo-3D Models. The pseudo-3D models assume there is no fluid flow in the vertical direction in the fracture, so a 1D fluid-flow equation is solved along the fracture height. However, the height is predicted from the net pressure using the fracture-mechanics criteria

Parameter	Data Source and Method	Note
Formation permeability	By correlating permeability with porosity, lithology, and shale content (developed for a specific area in a formation)	This is applicable for sandstone reservoirs. Permeability affects hydrocarbon flow from each layer and leakoff of fracture fluid.
Average pressure	New field: determine pressure gradient and average layer depth for pressure calculation. Old field: repeat formation tester, pressure buildup, material-balance estimate	This will affect the reserves assigned for each layer, fracturing-fluid selection and cleanup; in-situ stress.
In-situ stress	By historical data and correlating in-situ stress data with lithology, Poisson's ratio, and reservoir pressure. Also, by minifracture tests	Very important for 3D fracture-propagation model.
Gross and net thickness	By log interpretation	Gross thickness to determine fracture height and perforation location. Net thickness to determine productive intervals.
Young's modulus	Obtained from logs and cores with information about lithology and porosity.	Important for determining fracture width. Dynamic Young's modulus determined by the logs should be converted to static values using correlations between dynamic and staic laboratory data.
Poisson's ratio	Computed from sonic logs	Important for in-situ stress distribution in layered reservoir.
Reservoir temperature	Obtained from logs	Used to calculate hydrocarbon in place and effect on fracture fluids.
Fluid-loss coefficient	Obtained from historical data from previous treatments and minifrac tests.	Used to predict fracture geometry.

Table 14.1—Critical parameters used in 3D models to design and optimize hydraulic fracturing treatments.

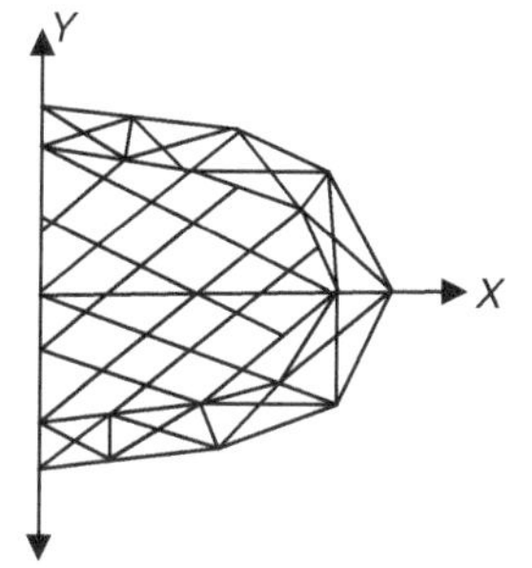

Fig. 14.29—This 3D fracture model is divided into elements to solve for grids developed on the fracture face.zbe analyzed with a fracture model.

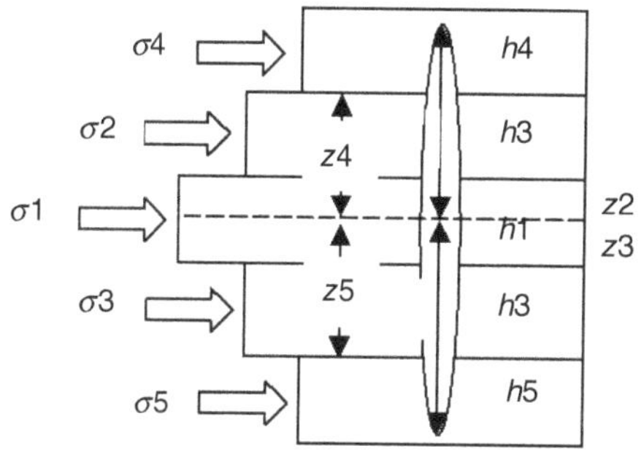

Fig. 14.30—3D methods solve for hydraulic fracture propagation in five-layer multistress media through separate 1D models of fluid flow.

of stress intensity at the top and bottom of the crack. **Fig. 14.30** shows a formation divided into five layers with different properties and in-situ stresses.

The methodology to solve the problem using the pseudo-3D model is to first solve for z_2 by iterating a value using the equation

$$\frac{\sqrt{\pi a}}{2(K_{ica}-K_{icm})}=\sum_{i=2,z}^{m} s_i\sqrt{a^2-z_i^a}-\sum_{i=2,z}^{m} s_i\sqrt{a^2-z_i^2}. \quad (14.10)$$

S_o is solved using z_2 and Eq. 14.10,

$$S_o=\frac{2}{\pi}\left[\frac{1}{2}\sqrt{\frac{m}{a}}(K_{ica}+K_{icm})-\sum_{i=2,z}^{m} s_i\sin^{-1}\left(\frac{Z_i}{a}\right)-\sum_{j=3.2}^{m} s_j\sin^{-1}\left(\frac{Z_i}{a}\right)\right]. \quad (14.11)$$

And then using Eq. 14.12 to calculate net pressure,

$$S_o=2p-\sigma_4-\sigma_5. \quad (14.12)$$

The upper and the lower heights are calculated using the following:

$$h_u = a - Z_2 - h_1/2, \quad (14.13)$$

$$h_d = 2a - h_u. \quad (14.14)$$

After the fracture height has been determined, it can be used in a 2D (PKN or GdK) model to compute other fracture-geometry elements (width and length).

14.5.3 Coupling the Hydraulic Fracture Model with Geomechanical and Reservoir Models. The concept of coupling the hydraulic fracture model with a reservoir and geomechanical model was introduced to handle unconventional fracturing applications, such as water fractures or fracturing caused by waterflooding. Coupling the hydraulic fracture with reservoir and geomechanical models focuses on the effect of the fracture on the fluid flow and geomechanics in the reservoir by introducing pressure/stress dynamic-transmissibility multipliers and treating them as properties of the matrix. This approach allows modeling of fracture propagation, dynamic multiphase fracture conductivity, cleanup, and pre- and post-fracturing well performance in a changing stress, pressure, and temperature environment in one simulator (Ji et al. 2006).

14.5.4 Post-Treatment Analysis and Fracture Model Calibration. To complete the circle, post-treatment analysis should be conducted to compare the predicted fracture geometry with the actual treatment-pressure data. A multi-isotope tracer survey can be conducted to estimate the fracture height near the wellbore and compare it with model results. If the height from the tracer survey is different from what is predicted by the model, the in-situ stresses should be adjusted for different layers and lithologies to match field results. Permeability estimates from pressure-buildup tests, core data, and minifracture analysis should be used to further validate the model. The total leakoff coefficient during the fracture operation should be monitored, compared, and adjusted for the model to better predict the fracture geometry (Larkin et al. 2002).

14.6 Summary

- Developing a database for each formation layer with mechanical, flowing, and lithological properties will improve the 3D models.
- Logs, cores, pressure-buildup and minifracture testing, and geological data should be correlated to estimate some of the rock properties.
- Simplifying assumptions used to develop 2D models will be less representative in situations in which deviated and horizontal wells are to be fractured.
- The new approach of coupling hydraulic fracture modeling with reservoir and geomechanical simulators will enable better understanding of the well performance before and after stimulation.
- Post-treatment analysis should be conducted to calibrate the 3D model and correct for some of the estimated rock properties during data input for different layers.

14.7 Nomenclature

a	=	fracture half-height, ft
C	=	discharge coefficient, dimensionless
D_{perf}	=	perforation diameter, in.
e	=	slope in Nolte-Smith plot
E	=	Young's modulus, psi
g	=	acceleration factor
G	=	modulus, psi
h	=	height, ft
h_f	=	fracture height, ft
h_d	=	fracture lower height, ft
h_u	=	fracture upper height, ft
i	=	number of a layer above a crack center
j	=	number of a layer below a crack center
K_{ib}	=	stress-intensity factor at the bottom of crack, psi
K_{ica}	=	critical stress-intensity factor where a is fracture half-height upward, psi
K_{icm}	=	critical stress-intensity factor where m is fracture half-height downward, psi
K_{it}	=	stress-intensity factor at the top of crack, psi
L	=	length, ft
m	=	layer where top tip resides
o	=	refers to crack center
n	=	layer where bottom tip resides
N_{perf}	=	number of perforations
$p(y)$	=	vertical distribution of net pressure, psi
p_n	=	net pressure, psi
p_p	=	pore pressure, psi
p_{frac}	=	pressure inside fracture, psi
p_{net}	=	net pressure, psi
p_{perf}	=	perforation pressure, lbf/ft^2
p_{wb}	=	wellbore pressure, psi
q	=	flow rate, bbl/min
r_w	=	wellbore radius, ft
R	=	fracture radius, ft
s	=	stress, psi
s_i	=	$\sigma_i - \sigma_i - 2$ for $i > 2$
s_j	=	stress in layer j, psi
S_h	=	minimum horizontal stress, psi
S_H	=	maximum horizontal stress, psi
S_v	=	vertical (overburden) stress, psi
u_x	=	fluid flow velocity inside a fracture, ft/min

v = Poisson's ratio, dimensionless
w = width, ft
w_w = maximum fracture width at wellbore, ft
x = x-axis in the fracture-length direction, ft
y = y-axis in the fracture-width direction, ft
z = depth, ft
Z = distance from the center of crack, ft
z_i or Z_i = distance of layer from middle of fracture, ft
α = Biot constant; fracture half-height in feet in Eqs. 14.8 and 14.9
ε = strain, dimensionless
Δp_{perf} = pressure change in the perforations, psi
Δt = time, mins
ρ = density, lbm/gal
ρ_b = bulk density, lbm/gal
μ = viscosity, cp
σ = applied uniaxial stress, psi
σ_{int} = intermediate stress, psi
σ_{max} = maximum in-situ stress, psi
σ_{min} = minimum in-situ stress, psi
σ_{Hmax} or $\sigma_{h,max}$ = maximum horizontal stress, psi
σ_{Hmin} or $\sigma_{h,min}$ = minimum horizontal stress, psi
σ_v = vertical stress, psi

14.8 References

Ahmed, U., Markley, M. E., and Crary, S. F. 1989. Enhanced In-situ Stress Profiling with Microfracture, Core, and Sonic-Logging Data. Paper presented at the SPE Rocky Mountain Regional/Low-Permeability Reservoirs Symposium, Denver, Colorado, 6–8 March. SPE-19004-MS.

Bloch, M., Siqueira, C. A. M., Ferreira, F. H. et al. 1997. Techniques for Determining In-situ Stress Direction and Magnitudes. Paper presented at the SPE Latin American and Caribbean Petroleum Engineering Conference, Rio de Janeiro, Brazil, 30 August–3 September. SPE-39075-MS. https://doi.org/10.2118/39075-MS.

Carter, R. H., Holditch, S. A., and Wolhart, S. L. 1996. Results of a 1995 Hydraulic Fracturing Survey and a Comparison of 1995 and 1990 Industry Practices. Paper presented at the SPE Annual Technical Conference and Exhibition, Denver, Colorado, 6–9 October. SPE-36483-MS. https://doi.org/10.2118/36483-MS.

Economides, M. and Nolte, K. G. 2000. *Reservoir Stimulation*. Third edition. Chichester, West Sussex, UK: John Wiley & Sons, Ltd.

Gidley, J. L., Holditch, S. A., Nierode, R. W. et al. 1989. *Recent Advances in Hydraulic Fracturing*. SPE Monograph Series, Vol. 12. Richardson, Texas: SPE.

Gu, H. and Siebrits, E. 2006. Effect of Formation Modulus Contrast on Hydraulic Fracture Height Containment. Paper presented at the International Oil & Gas Conference and Exhibition in China, Beijing, China, 5–7 December. SPE-103822-MS. https://doi.org/10.2118/103822-MS.

Hillis, R. R. 2001. Coupled Changes In Pore Pressure And Stress In Oil Fields And Sedimentary Basins. *Petroleum Geoscience* 7 (4): 419–425.

Holditch, S. A. and Rahim, Z. 1994. Developing Data Sets for 3D Fracture Propagation Models. *SPE Prod & Fac* **9** (04): 257–261. SPE-26155-PA. https://doi.org/10.2118/26155-PA.

Holditch, S. A. and Rahim, Z. 1993. Developing Data Sets for 3D Fracture Propagation Models. Paper presented at the Gas Technology Symposium, Calgary, Ontario, Canada, 28–30 June.

Jeffrey, R. G., Vandamme, L., Hinkel, J. J. et al. 1987. Propagation of Fractures in Coal: Equivalent Modulus and Parallel Fracture Effects. *Proc.*, Coalbed Methane Symposium, Tuscaloosa, Alabama, Paper 8726, 113–118.

Ji, L., Settari, A. and Sullivan, R. B. 2006. A New Approach to Hydraulic Fracturing Modeling - Fully Coupled with Geomechanical and Reservoir Simulation. Paper presented at the SPE Europec/EAGE Annual Conference and Exhibition, Vienna, Austria, 12–15 June. SPE-99428-MS. https://doi.org/10.2118/99428-MS.

Khaksar, A., Warrington, A. H., Magee, M. E. et al. 2004. Coupled Pore Pressure and Wellbore Breakout Analysis in the Complex Papua New Guinea Fold Belt Region. Paper presented at the SPE Asia Pacific Oil and Gas Conference and Exhibition, Perth, Australia, 18–20 October. SPE-88607-MS. https://doi.org/10.2118/88607-MS.

Larkin, S. D., Brown, E. K., Bazan, L. W. et al. 2002. Stimulation Design and Post Fracture Production Analysis: A Tight Gas Sand Case History. Paper presented at the SPE International Petroleum Conference and Exhibition in Mexico, Villahermosa, Mexico, 10–12 February. SPE-74361-MS. https://doi.org/10.2118/74361-MS.

Meyer, B. R. 1986. Design Formulae for 2-D and 3-D Vertical Hydraulic Fractures: Model Comparison and Parametric Studies. Paper presented at the SPE Unconventional Gas Technology Symposium, Louisville, Kentucky, 18–21 May. SPE-15240-MS. https://doi.org/10.2118/15240-MS.

Moos, D. 2001. Wellbore Stability in Deep Water—Handling Geomechanical Uncertainty. Paper presented at the AADE National Drilling Technical Conference, Houston, Texas, 17–19 February. http://www.aade.org/app/download/7238823193/AADE+43.pdf.

Nolte, K. G. 1986. A General Analysis of Fracturing Pressure Decline With Application to Three Models. *SPE Form Eval* **1** (06): 571–583. https://doi.org/10.2118/12941-PA.

Nolte, K. G. and Smith, M. B. 1981. Interpretation of Fracturing Pressures. *J Pet Tech* **33** (09): 1,767–1,775. SPE-8297-PA. https://doi.org/10.2118/8297-PA.

Rahim, Z. and Holditch, S. A. 1993. Using a Three-Dimensional Concept in a Two-Dimensional Model To Predict Accurate Hydraulic Fracture Dimensions. Paper presented at the SPE Eastern Regional Meeting, Pittsburgh, Pennsylvania, 2–4 November. SPE-26926-MS. https://doi.org/10.2118/26926-MS.

Simonson, E. R., Abou-Sayed, A. S., and Clifton, R. J. 1978. Containment of Massive Hydraulic Fractures. *SPE J.* **18** (01): 27–32. SPE-6089-PA. https://doi.org/10.2118/6089-PA.

Valko, P. and Economides, M. J. 1995. *Hydraulic Fracture Mechanics*. First edition. Chichester, West Sussex, UK: John Wiley & Sons.

Warpinski, N. R., Clark, J. A., Schmidt, R. A. et al. 1982a. Laboratory Investigation on the -Effect of In-Situ Stresses on Hydraulic Fracture Containment. *SPE J.* **22** (03): 333–340. SPE-9834-PA. https://doi.org/10.2118/9834-PA.

Warpinski, N. R., Schmidt, R. A., and Northrop, D. A. 1982b. In-Situ Stresses: The Predominant Influence on Hydraulic Fracture Containment. *J Pet Technol* **34** (03): 653–664. SPE-8932-PA. https://doi.org/10.2118/8932-PA.

Wright, C. A., Conant, R. A., Golich, G. M. et al. 1995. Hydraulic Fracture Orientation and Production/Injection Induced Reservoir Stress Changes in Diatomite Waterfloods. Paper presented at the SPE Western Regional Meeting, Bakersfield, California, 8–10 March. SPE-29625-MS. https://doi.org/10.2118/29625-MS.

Wright, C. A., Weijers, L., Davis, E. J. et al. 1999. Understanding Hydraulic Fracture Growth: Tricky but Not Hopeless. Paper presented at the SPE Annual Technical Conference and Exhibition, Houston, Texas, 3–6 October. SPE-56724-MS. https://doi.org/10.2118/56724-MS.

Yamamoto, K., Shimamoto, T., and Maezumi, S. 1999. Development of a True 3D Hydraulic Fracturing Simulator. Paper presented at the SPE Asia Pacific Oil and Gas Conference and Exhibition, Jakarta, Indonesia, 20–22 April. SPE-54265-MS. https://doi.org/10.2118/54265-MS.

Yew, C. H. and Chiou, Y. J. 1983. The Effects Of In-Situ Stresses And Layer Properties On The Containment Of A Hydraulic Fracture. Unsolicited paper. SPE-12332-MS.

Yudin, A. Y., Butula K. K., and Novikov Y. V. 2007. A Novel Approach to Fracturing Height Control Enlarges the Candidate Pool on the Ryabchik Formation of West Siberia's Mature Oilfields. Paper presented at the European Formation Damage Conference, Scheveningen, The Netherlands, 30 May–1 June. SPE-107604-MS. https://doi.org/10.2118/107604-MS.

Chapter 15

Data Acquisition and Core Measurements

Stephen A. Holditch

15.1 Log Data

15.1.1 Breakout Analysis. Breakouts are zones of failure of the borehole wall that symmetrically form at the azimuth of the least principal horizontal stress **(Fig. 15.1).** Breakouts are the best indicators of maximum horizontal stress when enough data are available. The breakouts frequently are extended in the direction of the borehole axis and can be described by orientation in the borehole (θ_b), opening angle ($2\phi_b$), and radial depth (r_b) (Lund and Zoback 1999).

Two hypotheses are used to explain the formation of breakouts. The first hypothesis suggests that if the difference in magnitude of two horizontal principal stresses is high enough, compressive shear failure will cause breakouts to form in the borehole wall. The opening angle can be empirically related to the far-field in-situ stresses:

$$\sigma_{\theta\theta}\left(\theta_b\right) = S_H + S_h - 2\left(S_H - S_h\right)\text{Cos}2\theta_b - \Delta p = C_{\text{eff}}, \quad (15.1)$$

where C_{eff} is effective rock strength, S_h is minimum horizontal stress, S_H is maximum horizontal stress, Δp is the difference between the borehole-fluid pressure and the pore pressure in the host rock, $\sigma_{\theta\theta}$ is principal horizontal stresses, and θ_b is orientation in the borehole.

The second hypothesis assumes that the formation of the breakouts involves tensile splitting. In vertical wells, breakouts reliably can be used to indicate the direction of the least horizontal principal stress. However, deviated wells are more complicated, and a nonlinear inversion scheme is needed to analyze the stress state. The geometry of breakouts is affected by the rock strength, lithology, pre-existing fractures, drilling intensity, and hole diameter.

Different logging tools are available to analyze borehole breakouts. Four-arm-caliper or dipmeter logs, formation microscanner, and formation microimager (FMI) are some of the most common tools used to compute breakouts in borehole walls. The last two methods measure the microresistivity profiles in the borehole wall and give much higher resolution. The best coverage of the borehole wall is through a borehole televiewer instrument.

15.2 Fracture Mapping

Fracture mapping is a relatively new science still under development by the industry. In 1983, the Gas Research Institute started a project to learn how to image and control hydraulic fractures in real time in tight gas sands. At the same time, the US Department of Energy and several industrial companies were also doing research in that area. Fracture mapping techniques can be divided into two major categories:

- Near-wellbore methods that use instruments in the borehole to try to image the fracture orientation and fracture height near the wellbore
- Far-field methods that try to look at the orientation, fracture-height growth, fracture-length growth, and evidence of anisotropic growth patterns

In addition, the fracture mapping techniques can be divided into direct and indirect methods. For example, microseismic imaging is a direct method. The analysis of production or pressure-buildup data to determine the propped-fracture dimensions would be an indirect method. The capabilities and limitations of each fracture-diagnostic technique are shown in **Table 15.1** (from Cipolla and Wright 2000a).

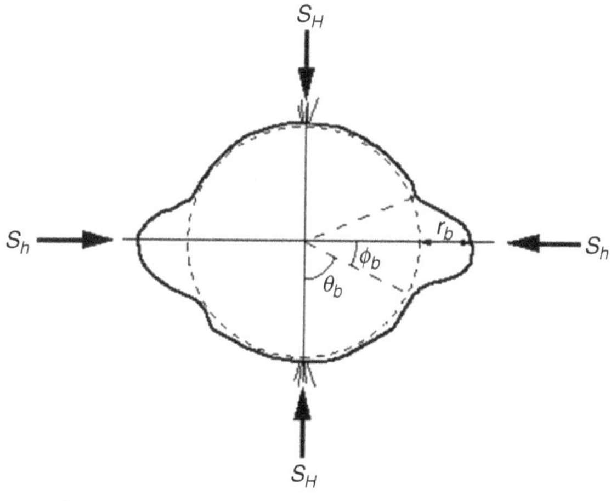

Fig. 15.1—A breakout in the wall of a vertical borehole indicates zones of failure (from Lund and Zoback 1999).

Type	Limitations and Applications for Direction Measurement Techniques	Results from Measurement
	Far-Field Techniques	
1. Surface-tiltmeter fracture mapping	–Cannot precisely resolve individual-fracture dimension –Mapping resolution decreases with depth –Application for shallow reservoirs	Fracture azimuth and dip
2. Downhole-tiltmeter fracture mapping	–Resolution of fracture height and length decreases as offset-well distance increases –Cannot provide proppant distance and productive fracture-geometry information	Fracture height, length, and width
	Near-Field Techniques	
1. Impression packers	–Application for only openhole well completion –Only the lower edge of the created fracture is obtained if the fracture is not perfectly aligned –Not sensitive to formation fluid	Fracture azimuth
2. Borehole televiewer	–Application for only openhole completion –Logging-tool decentralization causes incomplete fracture-orientation data –Not sensitive to formation fluids	Fracture orientation
3. Downhole video	–Clean and clear fluid in wellbore required to obtain high-quality image –Mostly used in casedhole well completion	Fracture images, zone, and perforation contribution
4. Wellbore ellipticity measurements	–Caliper log is usually run in openhole wells for fracture evaluation purpose –Formation washouts skew interpretation	Minimum-stress direction and fracture azimuth
5. Radioactive-tracer technology	–Depth of investigation only 1 to 2 ft away from fractured well –Only lower edge of created fracture obtained if the fracture is not perfectly aligned –Mostly used in casedhole well completion	Pad and proppant volume in each treated interval
6. Temperature logging	–Temperature-log results can be deviated by thermal conductivity of different formations –Only lower edge of created fracture will be obtained if the fracture is not perfectly aligned –Required multiple passes soon after treatment	Fracture height
7. Production logging	–Only provide production contribution from each perforated zone in casedhole wells	Fracture height in openhole completion
	Indirect Measurement Techniques	
1. Fracture-pressure analysis	–Results depend on reservoir description and model assumptions –Calibration with direct observation required	Fracture length, height, and conductivity
	Predicted-/Oriented-Core Techniques	
1. Differential-strain analysis	–Presumes no major pre-existing in-situ microcracks present	Fracture azimuth
2. Residual-strain overcoring	–Averaging data obtained from 6 to12 coring pieces improves accuracy and reliability	Fracture azimuth
3. Thermal-expansion anisotropic	–Averaging data obtained from 6 to12 coring pieces improves the accuracy and reliability	Maximum in-situ stress plane

Table 15.1—Capabilities and limitations of each fracture-diagnostic technique (from Cipolla and Wright 2000a).

15.2.1 Direct Measurement, Near-Wellbore Techniques. Direct, near-wellbore fracture mapping techniques are used in the treatment well after the hydraulic fracturing operation. Only the fracture information within approximately 1 to 2 ft of the wellbore is obtained (Cipolla and Wright 2000b). If the fracture is not perfectly aligned, the techniques will obtain only the information of the lower edge of the created fracture. However, these techniques are widely applied for fluid-/proppant-entry identification and production contribution from each zone in commingled completions (Cipolla and Wright 2000b).

15.2.1.1 Temperature Logging. Temperature logs commonly are used to determine the fracture height near the borehole. The cooling effect of the fracturing fluid during the fracture treatment creates cooling anomalies in the temperature log running after the fracturing. The pattern and magnitude of the fluid displacement will affect the temperature behavior. The cooling section on the log indicates the top and bottom of the fracture near the borehole. However, in a deviated well, the cooling region is always less than the fracture height.

To prevent misinterpretation, shutting in the well for a longer period of time before temperature logging after the treatment will minimize this effect (Davis et al. 1997).

15.2.1.2 Impression Packers. An inflatable production packer with a deformable rubber and flexible steel element is run on tubing and set across the desired or treated formation in the open hole. The rubber equipped in the toolstring takes an impression of the wellbore features and created fractures. The wireline-conveyed gyroscope is set inside the orienting profile above the packer to determine the fracture azimuth. This technique is normally run in openhole well completions. The advantage of this direct-measurement technique is its ease of interpretation without sensitivity to formation fluids. This method provides reliable fracture azimuth data (Gidley et al. 1989).

15.2.1.3 Borehole Televiewer. The sonic borehole-logging tool generates sonic pulses and then produces a direct record from the reflection of the sonic pulses from the borehole wall in the open hole. This logging tool is conveyed by electric-line cable across the openhole section, in which the magnetic compass is used to orient the tool. The lack of reflections of sonic pulses points to the potential existence of a fracture (Gidley et al. 1989). The limitation of this particular technique is logging-tool decentralization. If the tool is taken off-center by well deviation or configuration, it will result in blind areas. However, this kind of sonic logging tool is not sensitive to formation fluid. Although highly viscous fluid or gas-cut fluids present in the well can cause the dwindling of the signal or the making of false signals, reliable fracture orientation can be obtained.

15.2.1.4 Downhole Video. Downhole videos are mostly used in casedhole well completions. The fracture, perforation tunnel, and contributing zones can be examined from the pictures taken from the sandface. However, clean, clear fluid is required for this particular operation. Injecting the visibly clean fluid into the well by bullheading or using a coiled-tubing unit to displace the wellbore fluid might be necessary to obtain high-quality images from downhole.

15.2.1.5 Wellbore Ellipticity Measurements. The caliper log is usually run in the openhole well before the hydraulic fracturing operation to investigate the wellbore ellipticity and formation breakouts. The difference between the horizontal stresses creates a wellbore cross-sectional shape that is elliptical when drilling a well. The ellipse major axis represents the minimum-stress direction or the in-situ stress field perpendicular to the fracture growth.

15.2.1.6 Radioactive-Tracer Technology. A radioactive tracer is pumped with the fracturing fluid and proppant during the treatment. After that, several gamma ray log and tracer-ejector tools are run into the well adjacent to the treated zone. When tracer pulses are injected, the time discrepancy between the injection and detection can be recorded. The injecting flow rate into the formation can be estimated because the volume between injector and detector is known. The relative gamma ray intensities at various detectors can be used to determine proportional fluid entry into the fractured zone (Abernathy et al. 1994). As a result, the pad and proppant volume in each treated interval can be estimated. This technique can be implemented in casedhole well completions.

15.2.1.7 Production Logging. The production log consists of the spinner and multiple sensors with nuclear density, capacitance, and gamma ray that can be used to identify the amount and type of fluid being produced from each interval. This technique can be used to estimate the fracture height in an openhole well. For a casedhole-completion well, it can evaluate only the production contribution from the perforated zone being fractured (Cipolla and Wright 2000a).

15.2.2 Indirect Measurement Techniques. Indirect measurement techniques can provide fracture dimensions, fracture length, and fracture conductivity (Cipolla and Wright 2000a). The fracture information can be estimated from the pressure response during the treatment or during a well test after the well is put on production. The results from indirect measurement techniques are necessary to properly calibrate the direct measurements. However, the calibrated results from the indirect measurements are tremendously useful for fracture optimization.

The bottomhole pressure during the fracture treatment, the subsequent pressure decline until flowback for cleanup, and an estimate of fracture-closure pressure are required for fracture-pressure analysis. The best way to measure the bottomhole pressure is to position the pressure sensor below the treated zones or perforation intervals. The closure pressure that can be obtained from the bottomhole-pressure data during the treatment is the minimum pressure needed to open the existing natural fracture and the fracture created from drilling or perforation. This pressure equals the minimum principal stress of the formation, which is perpendicular to the fracture plane (Nolte 1986).

Fracture length and fracture conductivity can be calculated from the pre- and post-fracturing pressure-buildup tests (Cipolla and Wright 2000a). Problems such as poor fracture-cleanup efficiency, damage from the fracturing fluid, or poor proppant displacement can be identified from the pressure-buildup test. The causes of poor fracturing performance are very crucial for future fracturing treatment.

15.2.3 Tiltmeter Surveys. Direct, far-field fracture-diagnostic techniques will generate far-field fracture growth and fracture geometry to determine the total coverage of fracture growth. These techniques do not provide detailed fracture geometry, fracture perimeter, effective propped-fracture length, or conductivity. Fracture mapping from surface and downhole tiltmeters is widely used in many oil fields.

The tiltmeter techniques determine the deformation of the rock surrounding the fracture by measuring the gradient of the displacement above and around the fracture **(Fig. 15.2)** (Cipolla and Wright 2000a).

The fracture-induced deformation magnitudes detected by the surface tiltmeter are small and cannot precisely resolve the individual-fracture dimensions because surface tiltmeters normally are installed very far away from the created fracture. As a result, the surface tiltmeter is limited to relatively shallow reservoirs (Wright et al. 1998).

Downhole tiltmeters are wireline-conveyed tools that are run in one or more offset wells. The downhole tiltmeter can more accurately resolve the fracture dimension than the surface tiltmeter. Downhole meters can be positioned much closer to the created facture

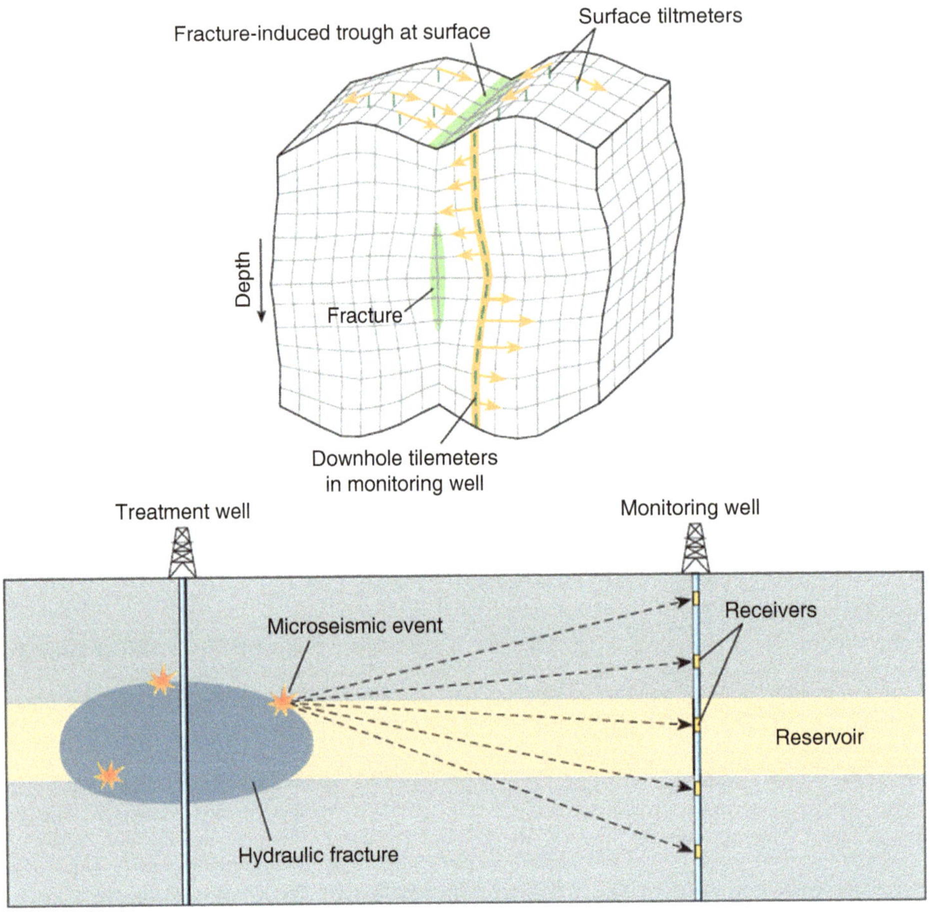

Fig. 15.2—Tiltmeters for fracture mapping should be located within 1.5 times the fracture length from the well being tested (from Wright et al. 1998).

than the surface tiltmeters; however, the resolution of the fracture height and length decreases as the offset-well distance increases (Cipolla and Wright 2000b). The offset well should be located no more than 1.5 times the fracture length away in the direction of fracture growth (Wright et al. 1998).

Surface tiltmeters are mostly appropriate for determining the fracture azimuth and dip, while downhole tiltmeters are more applicable in diagnosing the fracture height, length, and width (Wright et al. 1998).

15.2.4 Fracture Mapping with Microseisms. Engineers have found that the increase in pore pressure and formation stress near the fracture causes microseismic events that can be measured by carefully designed microphones placed in the wellbore of the treatment well or in a nearby wellbore (Warpinski 2009). The science used to describe these microseisms uses wave mechanics to create a relationship between the properties of the waves and the locations of the microseismic events. However, knowing how to measure a microseism is only the first step in accurately predicting a fracture's shape.

For several decades, engineers have been improving geophones and other devices to accurately measure the microseismic events, which are essentially miniearthquakes caused by the propagation of the hydraulic fracture. Researchers have created several different methods for locating the origins of the microseisms, including the placing of microphones in the stimulated wellbore or in remote monitoring wells. Both methods are currently being used; however, multiple case studies have suggested that remote monitoring methods might not be as reliable as placing the detection equipment in the stimulated well.

15.2.4.1 Microseismic-Event Detection. The ability to detect and accurately locate the origin of a microseismic event is the foundation of fracture mapping research. Researchers have found that microseisms emit elastic waves that are composed of compressional waves (*P*-waves) and shear waves (*S*-waves) (Warpinski 2009). Each of these events is between 100 and 500 milliseconds in duration and, in a hard sandstone formation, gives off between 5 and 20 W of acoustic power (Lacy and Smith 1989).

Once a microseismic event has been detected, the properties of its associated *P*- and *S*-waves must be analyzed to calculate the distance from the geophone to the event and the direction from which the wave arrived (Thorne and Morris 1988). Researchers have identified the following relationship between the distance D to a microseismic event and the arrival times of the *S*-wave, t_s, and *P*-wave, t_p:

$$D_1 = F_v \times (t_s - t_p), \qquad (15.2)$$

where D_1 is distance to a microseismic event, F_v is velocity factor, t_s is arrival time of the *S*-wave, and t_p is arrival time of the *P*-wave.

The velocity factor, F_v, is unique to every formation; it is correlated to the velocity of the *S*-wave, v_s, and the velocity of the *P*-wave, v_p, by the following relationship:

$$F_v = \left(v_s \times v_p\right) / \left(v_p - v_s\right), \quad (15.3)$$

where F_v is velocity factor, v_s is velocity of the *S*-wave, and v_p is velocity of the *P*-wave.

The F_v term cannot accurately be estimated from core samples, so it is often empirically determined in each individual reservoir (Thorne and Morris 1988). Engineers commonly estimate the value of F_v by detonating a perforation shot in a nearby well of known distance *D* from the observation well. The difference observed between the arrival times of the *S*- and *P*-waves is used with Eq. 15.2 to establish the value of F_v (Thorne and Morris 1988).

Determining the vector that links the origin of the microseismic event with the location of the geophone requires analyzing horizontal components of the *P*-wave in both the horizontal and vertical planes (Thorne and Morris 1988). The horizontal components of the *P*-wave in the horizontal plane enable engineers to determine the 2D projection of the vector, parallel to the surface of the Earth. This value is between 0 and 180º, so the processing computer must guess whether the waves are coming from the front or back of the geophone. The horizontal components of the *P*-wave in the vertical plane provide insight into the vertical azimuth of the *P*-wave. When both of these vectors are combined, a 3D vector is obtained; this vector supplies the axis of propagation (from either the front or the back of the geophone).

15.2.4.2 Design Considerations. The ability to detect small microseismic events by means of a wireline tool has posed several unique challenges for designers. The multiwell experiment (MWX) was conducted in 1985 and to provide fracture information to engineers working in the lenticular tight gas resource (Thorne and Morris 1988). The MWX identified several key considerations to aid engineers in developing a successful microseismic monitoring device.

The initial test phase of the MWX project indicated a dominating frequency of 200 Hz, suggesting a resonance frequency in the device (Thorne and Morris 1988). Examining the tool's stabilizing arm (used to wedge the tool against the edge of the casing) showed that an arm length of 17.25 in. should have a resonance of 195 Hz—almost exactly what was being observed. Most microseismic events are observed in the 15- to 600-Hz range, so designers modified the arm to be 6.75 in. with a resonance frequency of 2,000 Hz (Thorne and Morris 1988). The designers also found that the gain of the geophones needed to properly match the input of the frequency modulator. To achieve this, they implemented both 100- and 112-dB geophones to increase the listening range by a factor of four (Thorne and Morris 1988).

The wireline tool needs to have geophones on three different axes so that the horizontal component of the *P*-wave can be calculated in the horizontal plane and the vertical plane (Thorne and Morris 1988). In addition to having geophones on three different axes, we must know the overall orientation of the wireline tool and whether it is perfectly vertical or is leaning in one direction. To accomplish this, an inclinometer is incorporated into the wireline tool so that engineers know the orientation of the tool downhole (Brady et al. 1994).

Once the vertical angle is known, engineers must calculate the horizontal orientation of the tool. They can determine this by one of two ways. The first method is to lower the wireline tool into the hole, set the stabilizing bar, and calibrate the orientation of the tool from a perforation shot in a nearby test well. This perforation shot is often the same one used to calculate the formation velocity of the producing zone (Thorne and Morris 1988). The second way is to incorporate a gyroscope into the tool so that the orientation of the tool compared to the control set at the surface is always known. This is the less accurate of the two methods but is common in offshore microseismic analysis in which access to remote wells is limited and costly (Mahrer et al. 2007).

15.2.4.3 Operational Considerations. Engineers implementing the geophone wireline tool are faced with several operational considerations. The main decision is whether to place the geophones in the stimulated well or in a nearby remote well. Another common operational consideration involves deploying either a single wireline geophone or multiple geophones at varying heights. Multiple case studies have been conducted to test the effectiveness of both remote-well monitoring and stimulated-well monitoring. Mahrer et al. (2007) conducted an onshore test to determine whether remote-well monitoring provided better results than stimulation-well monitoring. The remote well was offset by only 700 ft and had an array of 15 three-axis geophones deployed in it. The stimulated well had a string of three three-axis geophones deployed in it. During pumping, intense background noise obscured most of the microseismic events, but afterward more than 400 events were detected by the geophones in the stimulated well. No events were detected from the remote well.

Because nothing indicated equipment failure in the case study, the reliability of the remote monitoring system can be questioned. However, a later study on microseismic fracture monitoring in the Barnett Shale suggests that the use of a remote monitoring well can be highly effective (Le Calvez et al. 2007). This study used remote monitoring to analyze the effectiveness of multiple-stage fracture treatments in horizontal wells. The study successfully monitored microseismic events up to 2,000 ft away from the remote monitoring well and enabled engineers to modify fracturing treatments to prevent fracture communication with an aquifer (Le Calvez et al. 2007). Cases such as the Barnett Shale, which were highly successful, promote the use of remote-well monitoring; however, other tests suggest that the reliability of the remote monitoring technique is not as high as desired.

A test performed on the Williams Fork Formation in the Piceance Basin in 2002 showed how microseismic fracture mapping from a remote test well can provide valuable insight into a fractured formation (Wolhart et al. 2005a, 2005b). This test involved two different fields (Rulison and Grand Valley), each of which had two different fractures mapped, for a total of four mapped fractures. **Figs. 15.3 and 15.4** show the layout of the fractured wells in both fields in correlation to each field's observational well. Notice that in both fields, only one observation well was used to monitor two different fractures. Microseismic fracture mapping from the observation well provided a significant amount of data from both fields and demonstrated how multiple pay zones can be monitored effectively for fracture effectiveness (Wolhart et al. 2005a). **Figs. 15.5 and 15.6** display how microseismic fracture mapping was used to evaluate the height of the fracture in each of the stimulated zones (three zones in each of the four wells). The tests conducted on the Williams Fork Formation illustrate how remote monitoring of fractures can be as effective as stimulation-well monitoring while being more cost-effective (two fractures monitored from the same remote well).

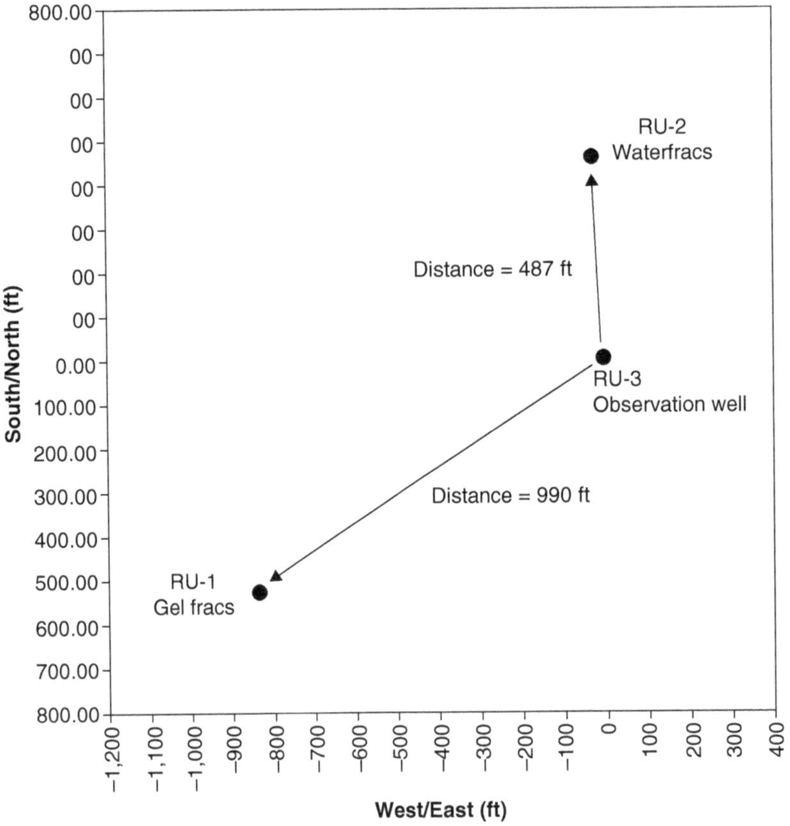

Fig. 15.3—Well location and distances for the Rulison (RU) mapping project (based on Wolhart et al. 2005a).

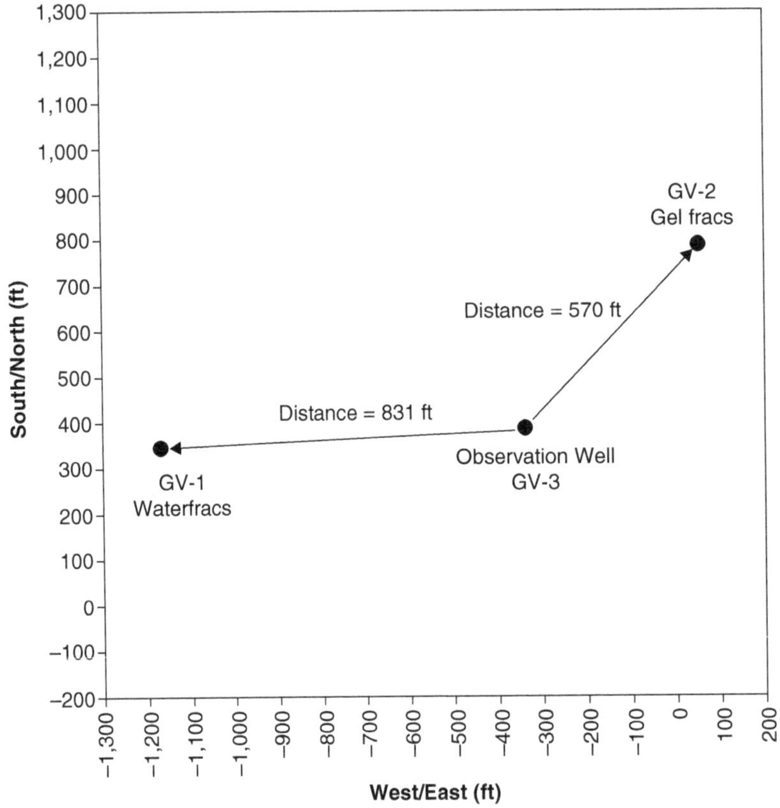

Fig. 15.4—Well location and distances for Grand Valley (GV) mapping project (based on Wolhart et al. 2005a).

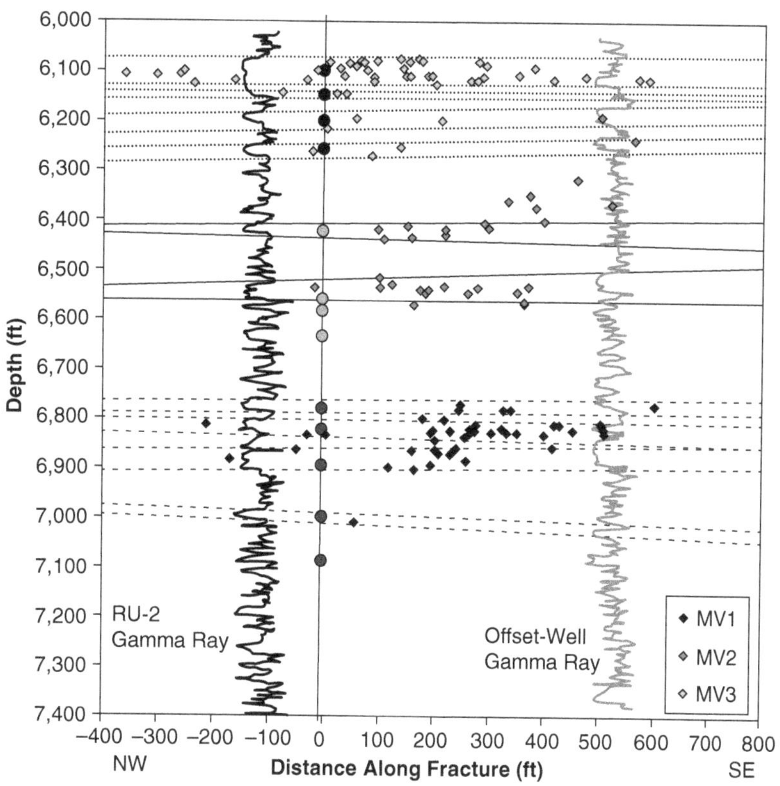

Fig. 15.5—View of microseismic mapping results for Rulison Well No. 1 (based on Wolhart et al. 2005a).

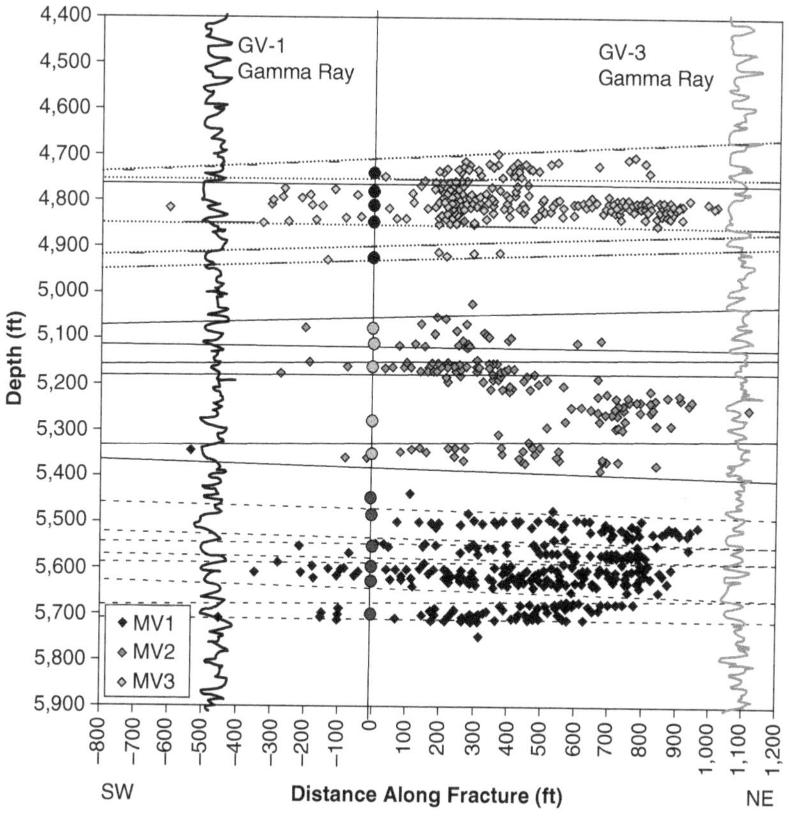

Fig. 15.6—Side view of microseismic mapping results for Grand Valley Well No. 1 (based on Wolhart et al. 2005a).

In some formations, the *P*-waves are not strong enough for the direction of propagation to be calculated accurately (Phillips et al. 1998). These formations often need multiple geophones in the observation well. The additional data obtained from the extra geophones allows engineers to monitor the reflected phases of the *P*-wave off the formation boundaries (Phillips et al. 1998). The engineers then use this information, along with known depths of the formation boundaries, to estimate the origin of the microseismic events. In other cases in which the *P*-wave is sufficiently strong, additional geophones can increase the overall accuracy of the data.

The use of a remote monitoring well or multiple geophones should be evaluated on a case-by-case basis. However, for a general case, monitoring the fracture from the stimulated well with several geophones would be a good starting point. The use of multiple geophones, as opposed to only one, should not increase the cost by a very large margin, and the success rate of monitoring from the stimulated well is higher than that from a remote observation well.

15.2.4.4 Data Collection. Several data collection techniques should be considered before commencement of the hydraulic fracturing. During a hydraulic fracturing operation, the noises from pumping the fluid drown out the majority of the microseismic events and tend to falsely trigger data recording (Thorne and Morris 1988). For this reason, engineers typically do not start listening for microseisms until after the fracturing job is completed. If an engineer desires to detect microseismic events while the fracturing is taking place, the triggering levels in the data recording software must be increased. This way, only large microseismic events that overpower the pumping noises are detected and no false events are recorded.

The engineer needs to make sure that the data collection software is set to sample at a high enough rate that the frequency of the events is accurately recorded. A typical sampling rate is in the neighborhood of 5 kHz, roughly 10 times the highest frequency that can be expected from a microseism (Lacy and Smith 1989; Phillips et al. 1998).

15.2.4.5 Summary. The following is a summation of fracture mapping with microseisms:

1. Understanding how to diagnose *P*- and *S*-waves is essential in determining the location of the microseismic events.
2. An effectively designed downhole listening tool will consider resonance-frequency issues and tool orientation.
3. The number of geophones used and the decision to use a remote observation well should be determined on a case-by-case basis.
4. Knowing when to expect noise and when to expect useful data is essential in analyzing microseismic events.

15.3 Microfracture Testing

Knowledge of in-situ stress direction and stress magnitude is a valuable tool for successful drilling, completion, and stimulation of oil and gas wells. These factors are of even greater importance for horizontal or highly deviated wells. Problems such as wellbore collapse, breakout, and breakdown during drilling and completion can be minimized by knowing in-situ stress orientation and magnitude to optimize wellbore direction and/or mud weight.

Recent improvements in rock mechanics have led to sophisticated modeling of sedimentary rocks. These models can solve problems associated with borehole stability, hydraulic fracturing, fracture confinement, sand production, and subsidence. Proper use of these models requires knowledge of minimum in-situ stress, σ_{min}.

A microfracture test is considered the best method to directly measure in-situ stress. Because measurement is made in situ with actual fracture opening occurring, it is more representative of real conditions than measurements through coring or logging.

Two methods most widely used to determine in-situ stress in open holes and cased holes are the pump-in/flowback (PI/FB) and pump-in/shut-in (PI/SI) tests (Miller et al. 1994). While PI/FB tests usually are performed as an integral part of determining in-situ stress, the preferred method is the PI/SI test. This typically consists of pumping a small volume of fluid (e.g., <100 gal of KCl water) at injection rates of 5 to 20 gal/min to contain the created fracture to the zone of interest. After initial formation breakdown is observed, injection is continued until the bottomhole pressure stabilizes. A wireline pressure gauge with a downhole shutoff tool is then lowered into a shutoff nipple to acquire the pressure-decline data.

To determine the in-situ stress from the PI/SI test, the pressure decline after fluid injection is assumed to be caused by fluid leakoff into the formation. This should represent linear flow as long as the fracture is open. Therefore, a plot of shut-in pressure vs. the square root of shut-in time should be a straight line during this period. At the point of fracture closure, a distinct slope change should occur; the pressure at this point is inferred to be the shut-in stress or closure pressure. Unfortunately, many times the change in slope is not distinct, making interpretation difficult. For this reason, a series of stresses (typically four or five) is required in a single interval to establish repeatability in the measured stress value.

The orientation of a fracture is a significant factor in production optimization and field development strategy. Determining fracture orientation requires effective techniques, to achieve the well economics. Numerous advanced analytical and diagnostic measurements have been developed and improved to characterize the fracture geometry and improve hydraulic fracturing design.

The crucial reservoir rock properties and accurate reservoir description affect the fracture model results. Calibration of the fracture models by use of the available fracture mapping data will provide a realistic fracture model that will enhance the hydraulic fracturing performance and well production.

The fracture determination and prediction techniques can be categorized by type of measurement: direct measurements, indirect measurements, and predictive oriented-core techniques.

The direct measurements use oriented surface or downhole tools to analyze the fracture at the wellbore, while the indirect measurements determine the fracture geometry from the formation signals. For the predictive oriented-core techniques, the oriented core samples are used to measure particular rock-mechanics properties, such as strain relaxation and wave-velocity anisotropy. Each technique has different capabilities and limitations; however, the maximum principal horizontal stress and the in-situ stress concept are primary perceptions for fracture characterization. The large variation in stress generates more-complicated fracture modeling that can significantly impact the optimal well spacing and location.

15.3.1 Performing a Microfracture Test. A microfracture test makes use of injection along a very short section of the wellbore (4 to 15 ft) to induce failure in rocks. The injection rates necessary to induce fracturing pressures in the tested formations are low (3 to 25 gal/min).

Fracturing pressures are reached a minute or two after injection begins. After fracture propagation occurs, injection continues at the same rate for several minutes, and then pumping is stopped and pressure-falloff-test methods are used to find the fracture-closure pressure.

The minimum principal in-situ stress usually is determined from analysis of the pressure decline after shut-in. Because the fracture-closure pressure or fracture-reopening pressure is equal to an applied stress that would nominally open an existing fracture, and thereby overcome the least principal in-situ stress, these two pressures generally are defined as equal to the minimum principal in-situ stress. This is not the case for wellbores in which natural fractures near the wellbore are opened during the microfracture test, if those fractures are not perpendicular to the minimum principal stress, nor for the case in which wellbore deviation causes stress concentration that forces the induced fractures to be at some angle with the natural plane of fractures in the near-wellbore vicinity.

Microfracture tests are usually performed with three injection/shut-in cycles. Analysis of the pressure response from these tests can yield formation-breakdown pressure, fracture-reopening pressure, and fracture-closure pressure values. The reopening and closure data can then be used to determine minimum principal stress.

Fracture azimuth (maximum-horizontal-stress direction) can be determined from openhole microfracture testing. When the microfracture test is performed at the bottom of an isolated openhole wellbore section, a portion of the resulting fracture can be recovered by coring. If core-orientation data are available, the created-fracture direction can be correlated with the core-orientation data, yielding the fracture azimuth. Image logs also can be used for determining fracture azimuth.

15.3.2 Field Procedures. Microfracture tests can be performed on cased holes and on open holes. The procedures differ, and each has its own advantages and disadvantages. Special equipment also is required for obtaining and ensuring adequate-quality data.

*15.3.2.1 **Openhole Microfracture.*** Openhole microfracturing is often performed alongside an extensive oriented-core study. An unfractured core may be used in an inelastic strain-relaxation test that can identify the fracture azimuth. Another method of obtaining fracture azimuth is to create a microfracture at the bottom of a drilled wellbore and to retrieve part of the induced fracture within an oriented core. The created-fracture angle is then correlated to the core-orientation data, yielding fracture azimuth. A borehole televiewer or downhole video camera also can be used to obtain the azimuth.

Theoretically, the openhole microfracture is preferred because there is no interference from perforations and cement. In openhole testing, however, fluid loss at the wellbore sometimes complicates determination of minimum principal in-situ stress.

The tested zone should be mostly homogeneous, so it is good practice to isolate the center of the zone of interest. The smallest possible zone should be isolated, usually 3 to 6 ft.

A small segment is required for better resolution of σ_{min} and helps in ensuring that only one zone is fractured. If the interest is to observe the out-of-zone fracture growth, then the interface may be isolated for fracturing.

The isolation might be accomplished with an impermeable plug, such as an openhole-testing packer. One packer may be used if the microfracture is to be created at the bottom of the wellbore. **Fig. 15.7** shows a microfracture-setup diagram.

A surface readout, downhole pressure/temperature gauge, or a memory gauge can be used to obtain downhole pressure data. Locating the gauges downhole is recommended because of the problems associated with excessive pressure drop in the test string.

Because of the small volumes injected during a microfracture test, all surface lines and pumps should be fluid tight. The lines should be tested to the maximum anticipated pressure.

After testing the surface lines and the work string, the packer seat should be tested by pressuring up the work string at a nonfracturing rate.

The initial injection can now begin. A constant injection rate is established. Using acid or cyclically pressurizing the formation is recommended if the formation does not break down after injection.

After breakdown **(Fig. 15.8)**, a constant injection rate is maintained for a few minutes to create a small penny-shaped fracture. The fracture likely will propagate around the packer, and hydraulic communication might be established with the annulus. The well is then shut in, and the instantaneous shut-in pressure (ISIP) is noted. The well is shut in until fracture closure is observed.

Subsequent tests are performed on the same zone by reopening the fracture at constant rates. The small fracture is created, and the ISIP and minimum horizontal stress are noted as before. A minimum of two such tests should be performed to verify results **(Fig. 15.9)**.

*15.3.2.2 **Casedhole Microfracture.*** Casedhole microfracturing is performed in a fashion similar to that of the openhole procedure. It is mechanically simpler because proper setting and fastening of the casedhole packers is more certain.

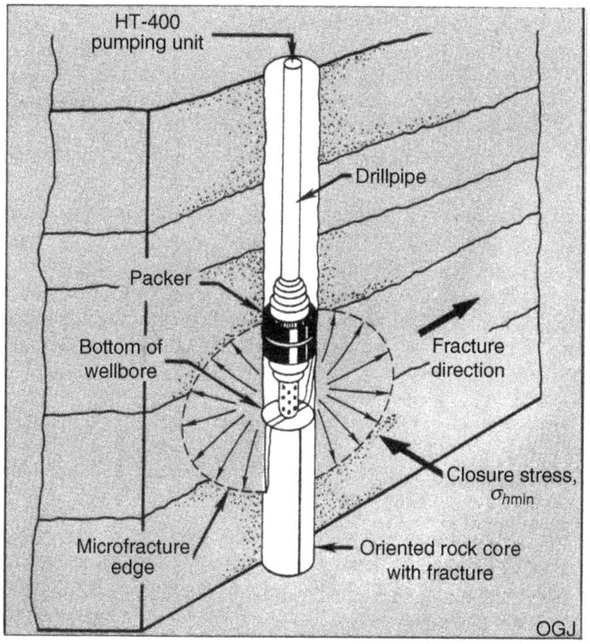

Fig. 15.7—Microfracture-setup diagram. (Note: HT is high temperature.)

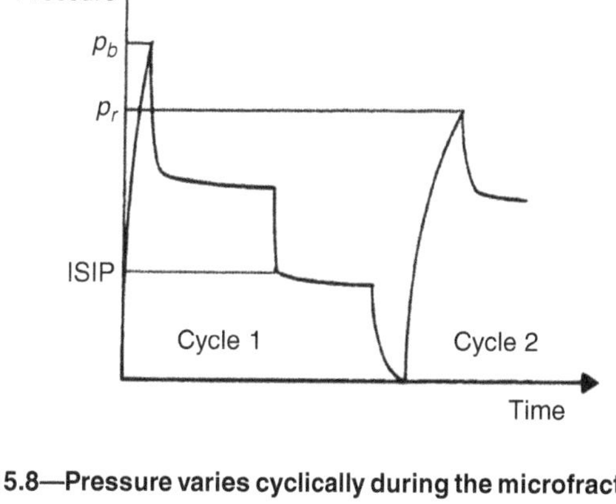

Fig. 15.8—Pressure varies cyclically during the microfracture test.

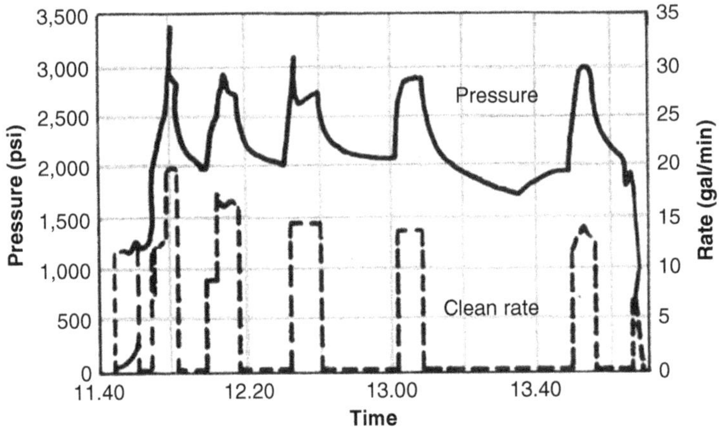

Fig. 15.9—Multiple stress tests during reopening of the fracture reveal effects of injection program.

A 1- or 2-ft interval of perforations should be phased at 60, 120, or 90°, closely spaced (4 shots/ft) and of the largest allowable charge.

The casedhole microfracture test has some advantages. Tests can be performed on weak or incompetent formations that would otherwise be prone to instability during openhole tests. Also, this method is faster compared with openhole testing because packers are mechanically fastened to the casing, and because of the ease of maneuvering the downhole tools, five or six different zones can be tested per day.

The casedhole microfracture is performed following the same injection procedure detailed in the openhole section. The main disadvantage of the casedhole microfracture is that the fracture azimuth cannot be determined. Results also can be affected by casing, cement, perforations, and wellbore inclination.

Special equipment has been developed to enable better execution of the microfracture tests. Special injection manifolds are used to obtain the low injection rates required for these tests. Conventional pumps can be modified with plunger-assembly kits for small-rate delivery and with enough horsepower.

*15.3.2.3 **Microfracture Test Fluids.*** The objective of a microfracture test is to propagate a small fracture away from the wellbore and to measure the pressure at which it closes during the pressure-decline period. So, the fluid used must be able to fracture the formation. In openhole tests, the mud system is usually suitable and convenient; however, compatibility between the fracturing fluid and the drilling fluid is usually a concern. If introducing mud into the test string or formation is undesirable, viscosified brine is a viable alternative.

If drilling mud is used, cuttings should be removed and fluids should be circulated to establish a more-uniform density. These tests should be used with a clean, nondisturbing fluid with stable and predictable rheological properties. Selection of a fluid depends on the same formation properties considered in conventional fracturing.

In casedhole applications, usually KCl water (brine) is used, but the density should be chosen carefully so that it is large enough to control downhole wellbore pressure.

*15.3.2.4 **Microfracture Interpretation.*** The theoretical pressure/time curve for microfracturing allows for identifying the breakdown pressure at which the crack begins to propagate (p_b), the ISIP immediately following when the pumping stops, and eventually when a reopening pressure (p_r) occurs, if more than one cycle is performed. ISIP is sometimes referred to in literature as the least principal stress. This use of ISIP is acceptable in cases in which fluid loss is very high and the fracture closes quickly.

A straightforward method is used to determine ISIP. The logarithm of the difference between the last pumping pressure and subsequent pressures vs. the logarithm of time is plotted. The pressure data before ISIP that characterize the wellbore-storage effect are on this plot with a line of unit slope. This plot should yield a line with a one-half slope, which usually characterizes a linear or infinite-conductivity-fracture flow from a fracture treatment. The end of linear flow, which is the end of the one-half slope on the log-log plot, usually defines closure pressure. If the fracture is with finite conductivity, we should observe a one-quarter slope instead of the one-half slope that characterizes bilinear flow.

The interpretation of pressure records is not always straightforward, especially if the rock permeability allows high fluid flow into the formation, disturbing the pattern of the pressure/time curve. Even though several methods have been developed for improving the pressure-decline analysis, and the correct ISIP determination, the stress computation might become rather subjective.

*15.3.2.5 **Limitations.*** As mentioned earlier, a microfracture test is considered the best method to directly measure in-situ stress; however, this technique has certain limitations. Sometimes it might not be practical to break down and measure in-situ stress in adjacent nonproducing layers. In cased holes, this technique depends on a good bond between cement and casing and the formation. Microfracture-measured in-situ stress in a well without an adequate cement bond might represent an adjacent layer that was in hydraulic communication with the perforated interval. Use of this technique for intervals of varying stresses can also lead to inconclusive results. When injected volumes are increased to the point at which the fracture migrates into adjacent layers of varying stress, the measured in-situ stress value might not represent the originally perforated layer. Other techniques, such as core testing and well logging, can be used as alternatives.

*15.3.2.6 **Use of In-Situ Stress Profiling with Microfracture, Core, and Sonic-Logging Data.*** One of the biggest difficulties in establishing a correlation between microfracture-, log-, and core-derived stress data results from irregularities in the depth reference. Microfracture stresses might not reflect the same horizon to which the log or core measurement refers. This problem also arises when deciding where to perform a microfracture test. Thus, efforts should be directed to carefully cross correlate between the various measurements.

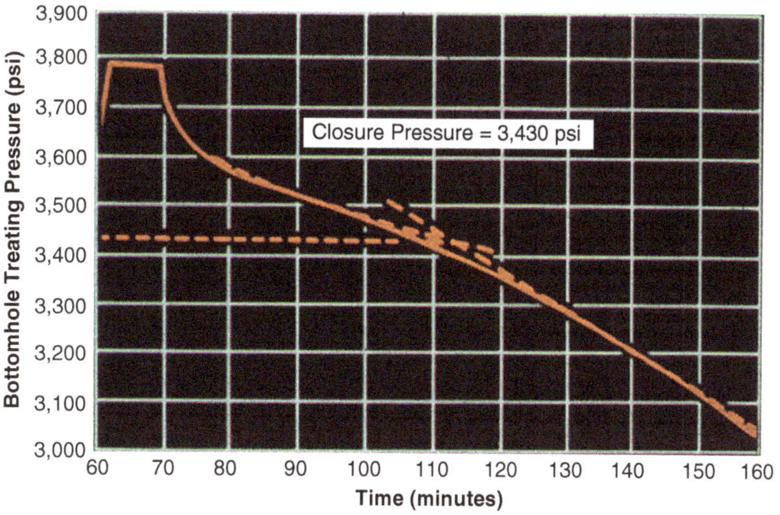

Fig. 15.10—Pump-in/flowback test indicate the estimation of fracture-closure pressure.

In a microfracture test, the injected volume of fluid typically is limited to approximately 10 to 20 gal per stage. In such a case, the created fracture usually is confined to within several feet of the perforated interval. A significant majority of operators, however, performed the test with tens to hundreds of gallons of injected fluid. Under such circumstances, the created fracture is likely to migrate into adjoining layers or zones. If these layers or zones are at a stress level different from that of the intended zone, then the stress estimation can be significantly influenced. Therefore, extra care should be taken in the design and evaluation of microfracture tests from layers and zones with varying stress levels.

During the design phase, the microfracture-test zone selection should carefully consider how pressure varies, from the sonic logs and/or available core data. Experience has shown that microfracture test zone selection should be such that the pressure variation within the interval results in less than a 100-psi variation in in-situ stress. Otherwise, the resulting pressure-vs.-time record following test shut-in might not clearly demonstrate the ISIP.

15.4 Minifracture Testing

A minifracture is a technique that initiates a hydraulically induced fracture at the borehole wall, propagates it through the rock mass, and monitors its closure. Then, the measured pressure responses are analyzed. The fracture propagation is perpendicular to the direction of minimum horizontal stress acting at the borehole wall. Furthermore, the initial in-situ stresses will affect the fracture when it has extended away from the borehole by a length of approximately four times the borehole radius.

To determine which locations are promising to conduct the stress test, gamma ray logs, cores, and FMI logs are used. These logs make it possible to detect the zones at which the borehole wall is heterogeneous, naturally fractured, or out of gauge. These locations are not suitable for performing the minifracture test, and they can cause packer failure, abnormal data, and lost seal.

15.4.1 General Aspects.

15.4.1.1 Pump-In/Flowback Test. This test primarily was designed to determine the fracture-closure pressure or minimum principal in-situ stress of a formation. The test involves pumping a low-viscosity fluid into the formation at a sufficient rate to initiate a fracture. Following the pump in, the fluid is removed from the fracture by flowing back at a steady rate. During flowback, the pressure follows a gradual decline at first and then follows an accelerated decline. This pressure is plotted against time **(Fig. 15.10)**, and the point at which the slope of the curve changes is the fracture-closure pressure. From this, we can interpret the net wellbore pressure, which is the difference between bottomhole pressure and closure pressure (Plahn et al. 1997).

15.4.1.2 Step Rate. A step-rate test traditionally has been used to obtain an estimate of the fracture-extension pressure in high-permeability formations. A typical step-rate test starts by pumping brine or slickwater into the formation at a rate low enough to ensure that fracture initiation does not take place. The rate is maintained for a short period of time until the pressure stabilizes, and then the rate periodically increases. The rate is held steady at each new rate for a short period of time, and the pressure is continuously monitored at each rate **(Fig. 15.11)**. The break point of the step-rate test in the curve in the figure is the extension pressure, which is larger than the closure pressure because of the loss of pressure resulting from friction in the fracture.

A step-down rate test, on the other hand, is used to identify near-wellbore restrictions. As the name suggests, fluid is pumped at a higher rate for a certain period of time until the wellhead pressure stabilizes, and then the rate is dropped. Until the rate goes to zero, the pressure is monitored to determine whether fluid entry into the reservoir is good or restricted.

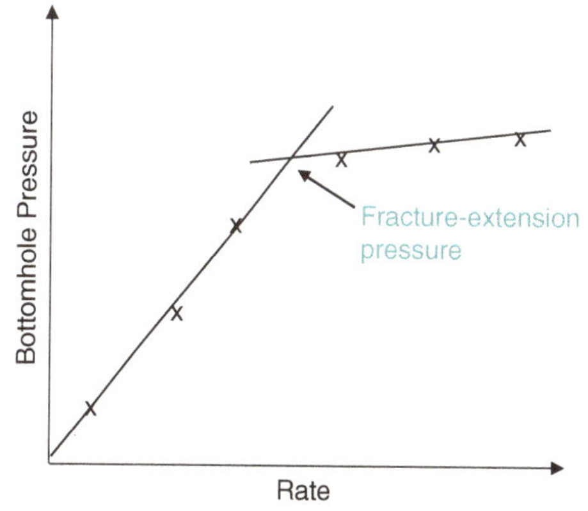

Fig. 15.11—Fracture-extension pressure appears where the extension pressure causes a break in the curve.

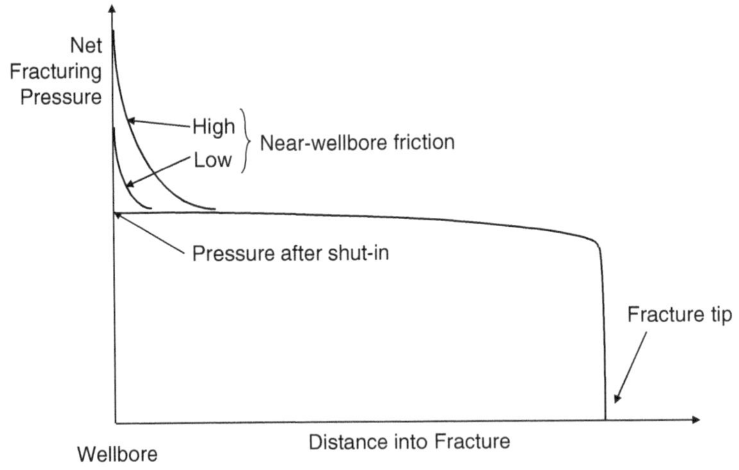

Fig. 15.12—Typical pressure profile down a vertical fracture.

15.4.1.3 Pump-In/ Shut-In Test. The shut-in test is used to determine fluid-loss parameters that are later used in the fracture-model simulator. Fluid efficiency, leakoff coefficient, and net pressure can be determined from the pressure-decline curve obtained from this test. Nolte (1986) analysis, which uses a plot of pressure against time (or square root of time), is the most commonly used analysis to find fluid parameters.

One of the key results of a shut-in test is the fluid efficiency (FE), which determines the percentage of fluid volume in the fracture at the end of pumping. The FE greatly impacts the volume of the fracture-treatment pad to be pumped. A minifracture is very helpful, especially in low-permeability reservoirs in which light gel loads can be pumped; with certain correlations, the behavior of high gel loads can be determined. In some low-permeability reservoirs, a minifracture has reduced the leakoff coefficient by approximately 35%. The reduction could be because of an increase in pore pressure in the leakoff zone resulting from the minifracture and the low permeability of the rock, but because minifractures often are pumped less than 24 hours before the actual fracture treatment, they could lead to a decrease in actual pad volume (Tinker et al. 1997).

15.4.1.4 Step-Down Test. A step-down test is run to better determine the near-wellbore pressure drop, which will be a combination of perforation-friction- and tortuosity-induced pressure drop.

Fig. 15.12 shows the typical net-pressure profile down the fracture. Near the wellbore, increasing pressure can be caused by pressure drop through the perforations or pressure drop resulting from the tortuosity between the wellbore and the main fracture. To use the Nolte-Smith graph of net pressure vs. time to interpret the vertical growth of a hydraulic fracture, we must be able to calculate the near-wellbore pressure drop.

The near-wellbore pressure drop could be a result of a restriction in fracture width close to the well or a change in fracture direction as it moves toward the preferred fracture plane. Also, sharp fracture reorientation caused by natural fractures can occur, as well as multiple hydraulic fractures induced by perforations or natural fractures that compete for fracture width, causing an extra pressure drop. Regardless of the cause, we must be able to compute the near-wellbore pressure drop to compute the actual net pressure in the fracture that is causing fracture width.

Fig. 15.13 is an illustration of what issues could be causing the near-wellbore pressure drop.

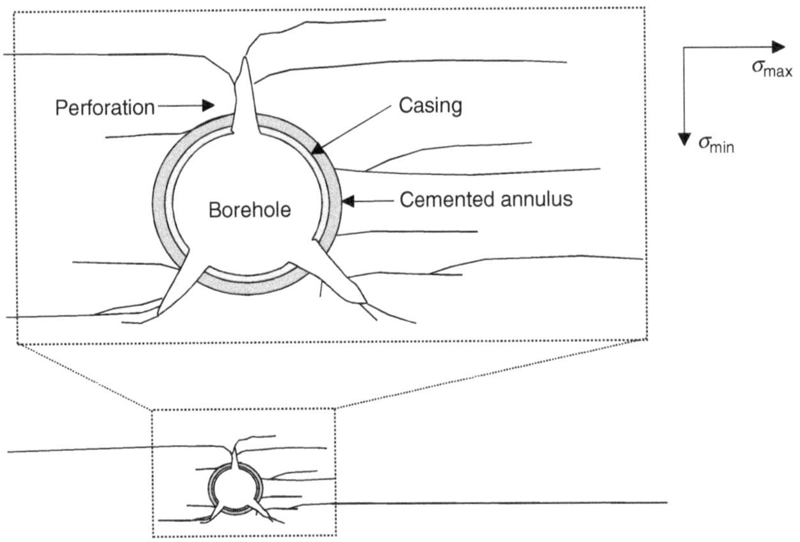

Fig. 15.13—Illustration of why we expect near-wellbore pressure drop because of tortuosity.

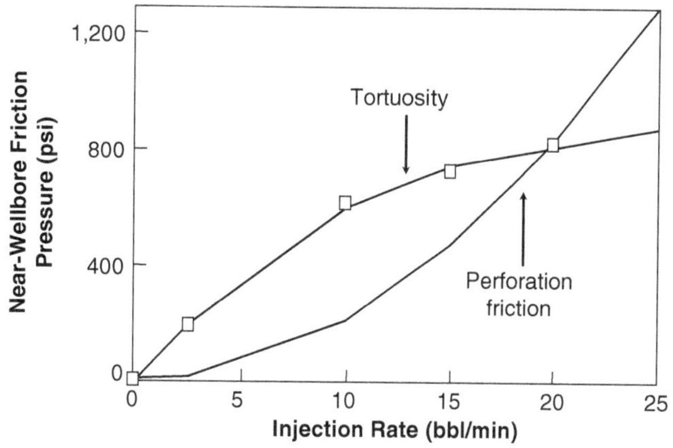

Fig. 15.14—Near-wellbore friction pressure vs. injection rate for different situations.

Near-wellbore pressure drop is an all-encompassing term, regardless of whether it is caused by perforation friction or by tortuosity. The issues with high near-wellbore pressure drop could be that it causes premature screenouts because of proppant bridging in narrow fractures, and it prevents or complicates the identification of true net pressures in the fracture. However, with a step-down test, we can determine which mechanism is causing the near-wellbore pressure drop, and then decide on a course of action, if needed. One key to identifying the cause of the pressure drop is that perforation friction varies with the square of the injection rate, as shown in Eq. 15.4:

$$p_{\text{perf}} = 0.2369\rho\left(\frac{Q}{CN_{\text{perf}}D_{\text{perf}}^2}\right)^2, \quad (15.4)$$

where ρ is density (lbm/gal), Q is flow rate (bbl/min), N_{perf} is number of perforations, D_{perf} is perforation diameter (in.), and C is discharge coefficient.

The second key is that near-wellbore pressure drop caused by tortuosity varies with $Q^{1/2}$. Fluid flows through a narrow pressure-dependent opening (slot) in the near-wellbore region that will change differently than observed through a fixed orifice, such as a perforation. The pressure drop will be proportional to Q^n, with values of n between 1/4 and 1, but the typical value expected is $n = ½$. **Fig. 15.14** shows how the pressure drop varies depending on the cause of the pressure drop.

A step-down test is run before the main fracture treatment, or even before the minifracture treatment, usually on the same day as the fracture treatment. The step-down test is run with slickwater and at fairly low injection rates, compared with the main fracture treatment. First, the injection begins at a rate that breaks down the formation and creates a propagating fracture. Once the injection rate and the pressure stabilize, the injection rate is reduced in steps and the pressure drop caused by the reduction in the injection rate is recorded. The time for each step depends on how long it takes for the pressure to level off or stabilize. A series of steps, usually four, are carried out to develop a graph of bottomhole pressure vs. injection rate. The shape of the graph enables the engineer to interpret the near-wellbore pressure drop and the contribution of the perforations and/or the tortuous paths near the wellbore. Typical examples of a step-down test are shown in **Figs. 15.15 and 15.16.**

If the near-wellbore pressure drop is too large, the engineer has a few options to reduce the pressure drop. If the excess pressure drop is a result of perforation friction, one can reperforate the zone to be fracture treated. If the high pressure is caused by tortuosity, some researchers in the literature have reported that using flow-rate changes and proppant slugs can help by eroding the small fractures and/or plugging off multiple fractures. Other remedies for tortuosity include higher injection rates and higher viscosities to obtain larger fracture widths in the near-wellbore region. Tortuosity can also be avoided by limiting the footage of perforations. Long perforated intervals can increase the chances of creating multiple fractures near the wellbore.

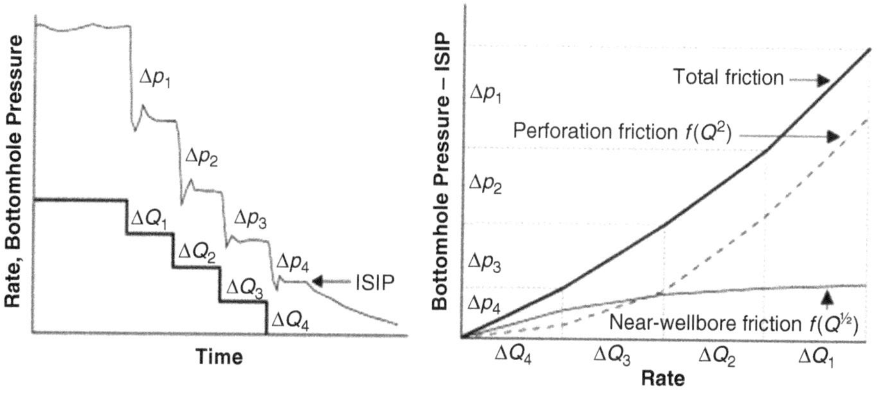

Fig. 15.15—Typical data for a step-down test in which the friction pressure is dominated by the perforations (Wright et al. 1998).

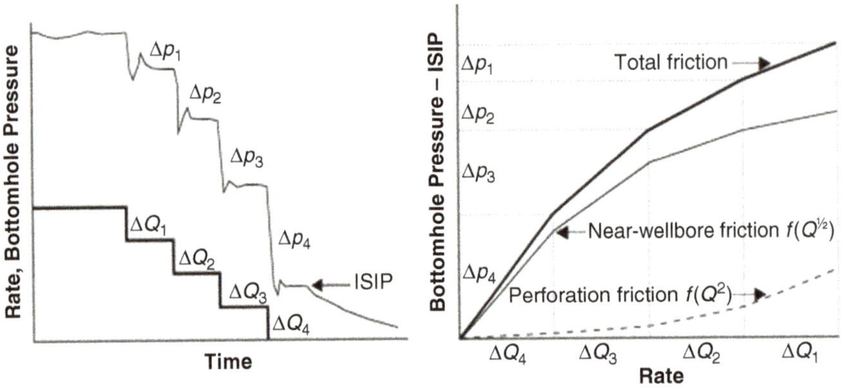

Fig. 15.16—Typical data for a step-down test in which the friction pressure is dominated by tortuosity.

15.4.2 Operational Procedures. A typical PI/SI test requires several injection cycles in each test, which usually consists of six steps, as follows and as depicted in **Fig. 15.17** (Fourmaintraux et al. 2005):

1. Inflate packers.
2. Perform a sequence of short, low-pressure increases to investigate the leak off of a specific interval.
3. Pressurize the interval at constant flow rate until tensile failure occurs at the borehole wall and a tensile fracture begins to propagate.
4. Keep the flow rate constant to extend the fracture in a steady mode for 1 to 5 minutes.
5. Shut down the injection, which causes the pressure to fall off and the fracture to close. In shaly formations, because of the low fluid filtration, pressure takes a very long time to fall off. Sometimes a flowback/pressure-rebound test can be included in some cycles to check whether a hydraulic fracture has been created.
6. After obtaining required data, deflate packers and perform the test on the next interval.

A minifracture test is the creation of a small fracture in the desired formation without the use of any propping agent. This test is performed from a few hours to several days in advance of the main fracture treatment. The objectives of a minifracture test are to gain knowledge of fluid loss, in-situ stress, and the fracture geometry, which is critical for use in a 3D fracture simulator. Key components obtained from a minifracture test are leakoff coefficient, fracture-extension pressure, fluid efficiency, and closure pressure/net pressure (Soliman et al. 1990).

A minifracture test can be carried out in different forms, including a PI/SI test, a PI/SI test, and a step-up/step-down rate test. The minifracture test involves a relatively short pump-in near the planned treatment rate followed by a shut-in period. These pump-ins might be carried out at variable rates, and they typically use the same fluid planned for the main-fracture-treatment job. Nolte (1986) developed an analysis of the pressure decline from these tests to identify the various attributes such as FE and fracture geometry. While Nolte assumed pressure-independent leakoff, which is not always valid, Castillo (1987) developed the *G*-function, which provides a better analysis of the closure mechanism, assuming the leakoff is pressure dependent.

15.4.3 *G*-Function Analysis. The *G*-function is a tool that shows the closure and leakoff mechanism. It was developed by Castillo (1987) because Nolte (1986) assumed that the leakoff coefficient is not pressure dependent, which might hold true for cases in which leakoff is largely controlled by an incompressible filter cake. In cases in which leakoff is controlled by reservoir permeability or by

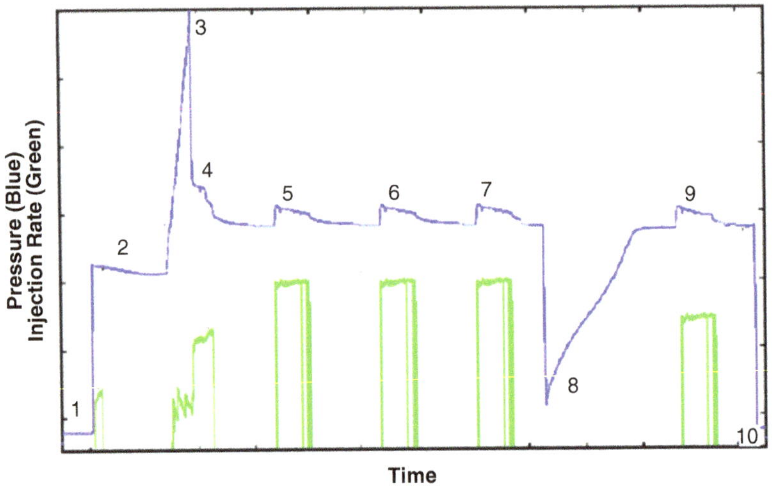

Fig. 15.17—Minifracture tests in shale follow six basic steps: (1) inflate packers; (2) leakoff; (3) breakdown; (4) fracture propagation and fall off; (5 through 7) cycle through fracture reopening/propagation/falloff cycles, (8) rebound, (9) repeat cycle through fracture reopening/propagation/falloff cycles; and (10) deflate packers (Fourmaintraux et al. 2005).

filtrate viscosity, leakoff is more pressure dependent, and the Castillo *G*-function analysis gives a better idea of the closure mechanism. The key plots used for this analysis (**Fig. 15.18**) are the ratio of dimensionless pressure to a dimensionless difference function vs. the *G*-function (Castillo 1987):

$$G(t_{D,}0)=\frac{4}{\pi}\left[g(t_D)-g(0)\right], \quad (15.5)$$

where $G(t_D, 0)$ is dimensionless difference function, $g(t_D)$ is dimensionless-loss volume function, and g is total fluid efficiency of the upper and lower bounds.

$$g = g_u(FE) + g_l(l\text{-}FE), \quad (15.6)$$

where FE is fluid efficiency.

The upper bound, g_u, and lower bound, g_l, are

$$g_u(X) = 4/3[(1+X)^{1.5} - X^{1.5}] \quad (15.7)$$

and

$$g_l(X) = (1+X)\sin^{-1}(1+X)^{-0.5} + X^{0.5}). \quad (15.8)$$

The plot illustrates how to estimate closure pressure/net pressure from shut-in pressure data when plotted against the square root of time. The point at which the decline curve deviates from the straight line gives the estimate of closure pressure.

Castillo's *G*-function analysis shows the characteristic closure and leakoff mechanism of the fracture. Often a plot of *G*d*P*/d*G* vs. *G* and d*p*d*G* vs. *G* is used to identify these characteristics.

15.4.3.1 Normal Leakoff. Normal leakoff occurs when the fracture area is constant during shut-in, and leakoff is through a homogeneous rock matrix. In **Fig. 15.19,** leakoff is indicated by a constant derivative of the d*P*/d*G* plot and when the *G*d*P*/d*G* plot lies on a straight line through the origin.

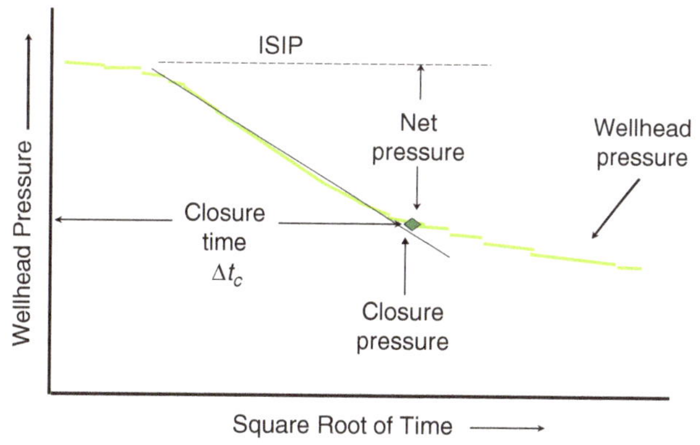

Fig. 15.18—Closure pressure occurs when the decline curve deviates from a straight line in Nolte-Smith analysis.

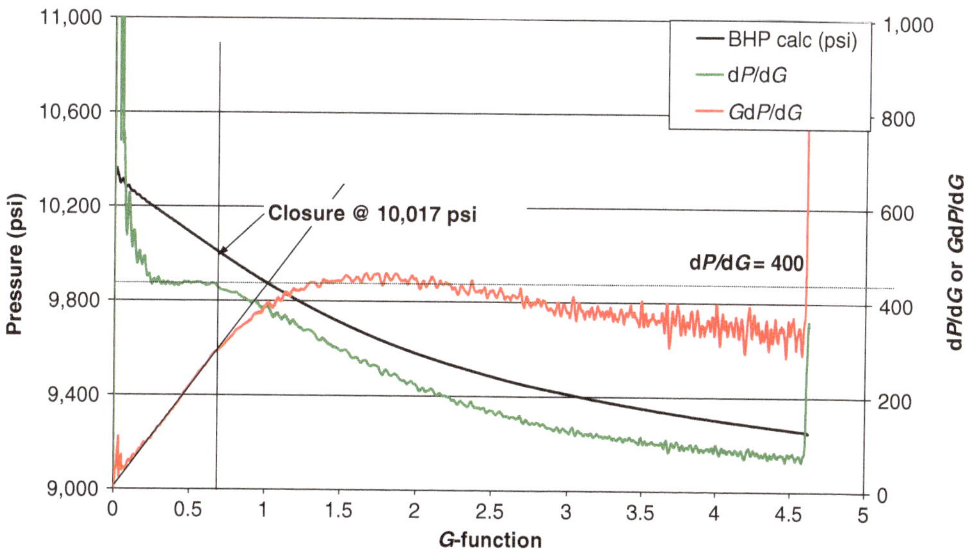

Fig. 15.19—Normal leakoff occurs where the *G*d*P*/d*G* plots as a straight line through the origin and the derivative of the d*P*/d*G* is constant. BHP = bottomhole pressure.

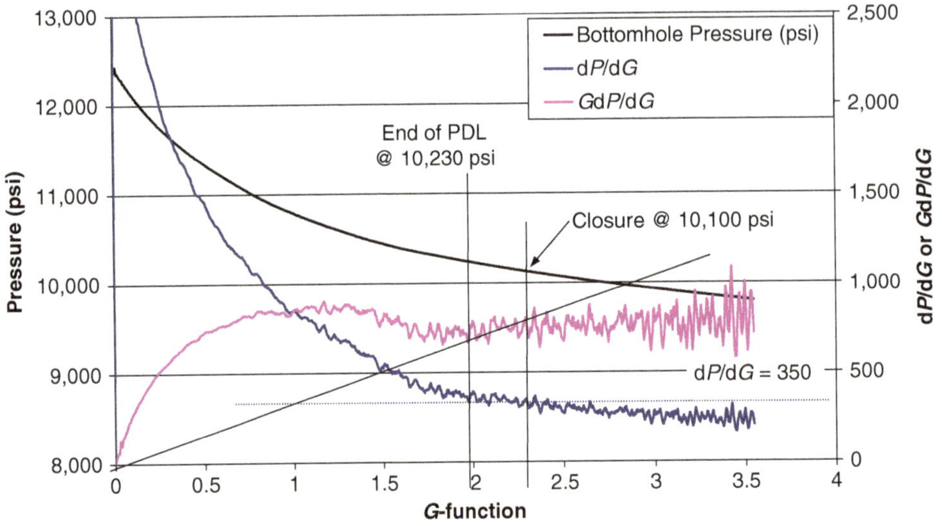

Fig. 15.20—Pressure-dependent leakoff (PDL) is indicated by a hump in the GdP/dG plot that lies above the extrapolated straight line that runs through the normal leakoff data.

15.4.3.2 Pressure-Dependent Leakoff. Pressure-dependent leakoff occurs when leakoff is controlled by filtrate viscosity or by the reservoir. In **Fig. 15.20,** it is indicated by a characteristic hump in the *G*d*P*/d*G* plot that lies above the extrapolated straight line through the normal leakoff data.

15.4.3.3 Fracture-Tip Extension. Fracture-tip extension occurs when the width of the created fracture continues to grow after the injection is stopped. In **Fig. 15.21,** it is indicated when the *G*d*P*/d*G* data lie along a straight line that extrapolates above the origin.

15.4.3.4 Fracture-Height Recession. After the minifracture treatment has been pumped, we typically see a fracture-height recession. In **Fig. 15.22,** this characteristic is indicated when the *G*d*P*/d*G* data fall below the extrapolated line and the pressure curve follows a downward-concave path leading to an increasing d*P*/d*G* curve.

15.4.3.5 Summary. The following is a summation of *G*-function analysis:

1. Minifracture tests can involve PI/SI, flowback, or step-rate tests.
2. Pump-in/flowback tests allow fracture-closure pressure to be observed faster than simply shutting in the well.
3. Accurately estimating the leakoff coefficient is critical to the fracture-treatment pad volume. Minifractures can also alter (reduce) the leakoff coefficient in some low-permeability reservoirs.
4. Nolte analysis assumes pressure-independent leakoff, while Castillo *G*-function analysis assumes pressure-dependent leakoff.
5. The *G*-function analysis is a dimensionless function that helps identify leakoff and closure mechanisms of the fracture.

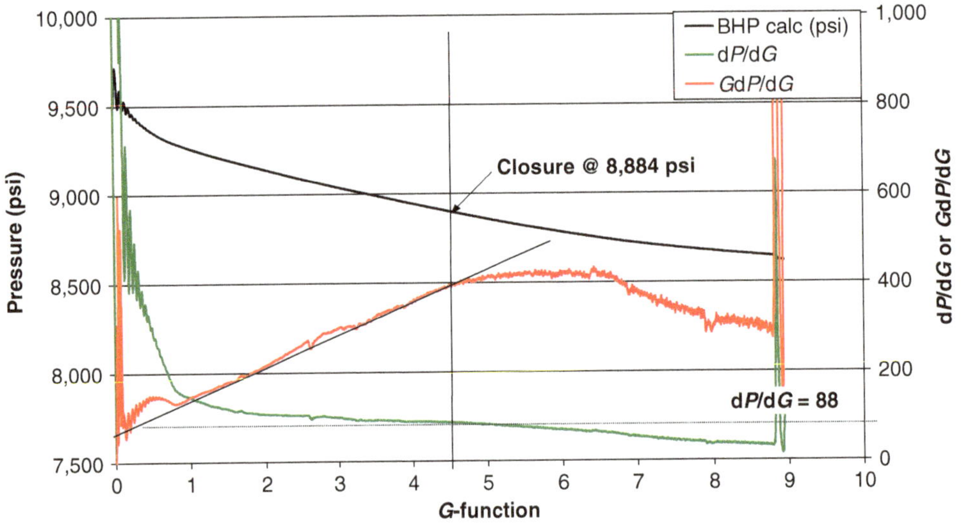

Fig. 15.21—Fracture-height extension is indicated by the slope that extrapolates above the origin. BHP = bottomhole pressure.

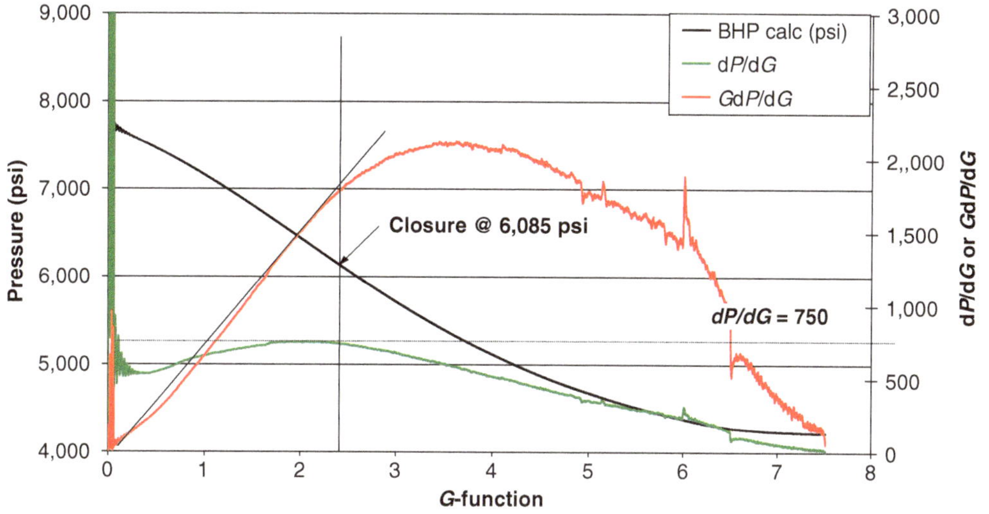

Fig. 15.22—Fracture-height recession is indicated by the downward-concave path of the pressure data and the GdP/dG data below the extrapolation. BHP = bottomhole pressure.

15.5 Core Testing

During the design phase of a hydraulic fracturing treatment, much information is required to characterize complex heterogeneous formations. An adequate description of the various layers of the well is needed to design the optimal treatment. Among the methods currently used for reservoir description, core analysis has yielded insights into complex reservoirs, furnishing data useful for estimations of reservoir in-situ conditions.

Core testing is used to help evaluate the amount of oil- and gas-in-place in the reservoir, to determine effective values of permeability, and to obtain correlations between the log readings and the core. Furthermore, core analysis provides a varied menu of laboratory data for reservoir description, including measured values, visual observations, and photographs that help describe the heterogeneity, degree of consolidation, and other formation characteristics to aid in understanding the reservoir anatomy.

Technology has played a key role in the development of new equipment and better practices to obtain core samples that would most likely represent bottomhole conditions. Costs and the need for accuracy come together in determining the proper selection of conventional or special techniques for the specific scenario in which the well stimulation might be taking place.

Several authors have defined the meaning of core analysis. Anderson (1975) defines it as a process of obtaining representative samples of the productive formation to conduct a variety of laboratory testing that provides information on porosity, permeability, phase saturations, lithology, and texture of petroleum reservoirs. Fleckenstein and Eustes (2003) complement this concept by adding the need of recovering the formation material in as unaltered and undamaged a state as possible.

Various techniques are used to collect core samples, depending on the detail of the reservoir description and the availability of tools to perform the operations. Geological data, as well as completion data (Keelan 1985), can be obtained using core analysis through different methods that include conventional coring, pressure coring, sidewall coring, oriented coring, and recovery of cuttings generated from the drilling operation.

Oriented cores can be used to predict fracture orientation. A core barrel is usually used to collect core sample during the drilling operation. An orientation tool is attached above the core barrel to determine relative orientation of the core. To obtain accurate and reliable fracture information, at least six coring pieces from one standard-deviation interval are essential. The mapped fracture usually occurs within one standard deviation.

The principle of the oriented-core techniques is that stress and strain of the rock are a function of the loading rate and time (Davis 1987). Strains stored in the rock will not be immediately released when the core is cut, but will relax over time. The anisotropy within the rock keeps increasing with time. Microcracks in the rock are induced after the rock is released from the in-situ stress conditions.

Generally, the core sample is available 4 to 6 hours after it is cut. The strain-response measurement can last approximately 6 to 8 hours, or until the responses have ended. The measurement should be performed at room temperature and at isothermal conditions. The dehydration of the core normally is assumed to have little impact on in-situ stress results.

Because the strain is time dependent, the measurement of the relaxation will provide the magnitude and direction of minimum and maximum strain. The minimum and maximum stresses can then be predicted. Several techniques have been proposed to diagnose the fracture orientation.

15.5.1 Core Tests Useful in Fracture Design.

15.5.1.1 Anelastic Strain Recovery Test (ASRT). The principal-in-situ-stress direction and magnitudes can be determined by applying this method. This method is based on strain-relaxation measurements in oriented-core samples, from which the principal strain can be derived. According to early researchers, the principal-stress orientation can be directly computed by assuming the same orientation of the principal strains, with the higher strain orientation being associated to the orientation of the higher principal in-situ stress (Bloch et al. 2005).

Computing the original principal-stress state assumes that the rock is homogeneous, isotropic, and linearly viscoelastic. In vertical wells, one of the principal directions is vertical, and principal strains can be determined by monitoring the core relaxation in only four directions (one vertical and three horizontal). However, in inclined wells, the strain tensor must be fully determined, and 12 gauges must be set up in six different directions around the rock sample, as shown in **Fig. 15.23** (Bloch et al. 1997).

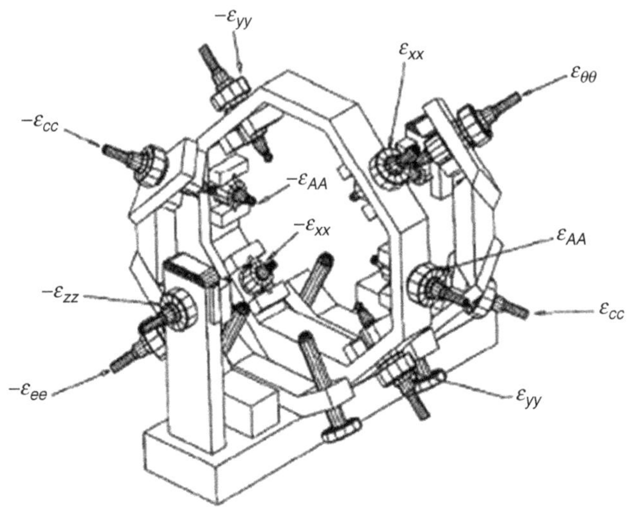

Fig. 15.23—Sample support and ASRT measurement assembly (Bloch et al. 1997).

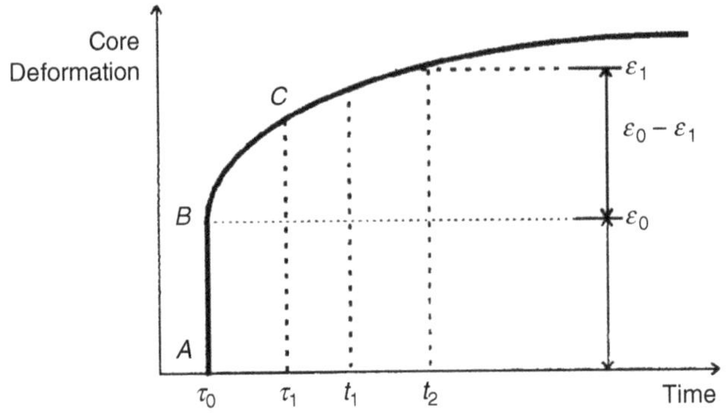

Fig. 15.24—Ideal behavior of rock strain vs. time (from Bloch et al. 1997).

ASRT gives a very reliable principal-in-situ-strain direction, because it does not depend on the rock strain absolute values. Normally, two samples from close depths are tested at the same time to accomplish the experimental procedure effects.

Strain relaxation decays exponentially with time **(Fig. 15.24);** it is vital to run the ASRT immediately after the core is available at the drilling rig and the rock is still undergoing measurable strain.

In this graph, *AB* is the instantaneous elastic deformation when the core is cut (τ_0); *BC* is the relaxation during recovering during the period of τ_0 to τ_1; and from t_1 to t_2 is the ASRT acquisition period. The initial elastic recovery and some of the anelastic-strain recovery are lost while the core is still underground (from τ_0 to τ_1 in Fig. 15.24), and the test is able to measure only a small range of deformation from t_1 to t_2.

To reduce the time between coring and testing, we commonly test the bottom part of the whole core, which is the last to be cored before retrieving the core from underground. Vibration, temperature, and humidity changes must be carefully controlled during the test because they can easily affect the results. For this purpose, the sample is placed inside an isolated chamber and the strain is computer monitored for 24 to 36 hours, or until the strain becomes stable.

If we assume that the vertical stress is equal to the overburden stress, horizontal stress magnitudes can be obtained by (Bloch et al. 1997):

$$\sigma'_{ij} = \frac{1}{V_o}\left[\frac{(1-2v)\Delta\varepsilon_{ij} + v\delta_{ij}\Delta\varepsilon_{kk}}{D(1+v)(1-2v)} - V_m(1-\alpha)\delta_{ij}p_p\right], \quad (15.9)$$

where *D* is compliance coefficient (inverse of Young's modulus), p_p is pore pressure (from field data), α is Biot's coefficient, δ_{ij} is Kronecker Delta, $\Delta\varepsilon_{ij}$ is strain variation, and *v* is Poisson's ratio (from uniaxial compression test).

Also,

$$V_0 = D_n\left[e^{-\frac{(t_1-\tau_0)}{n}} - e^{-\frac{(t_2-\tau_0)}{n}}\right], \quad (15.10)$$

$$V_m = -D_n\left\{e^{-\frac{(t_1-\tau_0)}{n}} - e^{-\frac{(t_2-\tau_0)}{n}} + \frac{n}{\Delta\tau}\left[e^{-\frac{(t_2-\tau_1)}{n}} - e^{-\frac{(t_2-\tau_0)}{n}}\right] - \frac{n}{\Delta\tau}\left[e^{-\frac{(t_1-\tau_1)}{n}} - e^{-\frac{(t_1-\tau_0)}{n}}\right]\right\}, \quad (15.11)$$

where *n* and D_n are creep coefficients from vertical stress data and $\Delta\tau = \tau_1 - \tau_0$ (from Fig. 15.24).

Some difficulties related to this method are retrieving good samples or handling the rock sample in an appropriate way so that the fluid content in the sample is not lost. This can be done through use of wax instead of plastic wrapping to prevent moisture changes. One important assumption that makes it possible to relate the relaxation phenomena to the rock-stress history is that the relaxation process is totally caused by cracks opening under the stress relief when cutting the rock and isolating it from the rock mass. However, temperature variations and pore-pressure diffusion are unavoidable while the core is still traveling from the cutting place in the wellbore to the surface, and it could take up to 10 hours. To overcome this problem, pore-pressure diffusion and thermal effects are often assumed to be isotropic, and the overall rock relaxation alone can be attributed to the crack opening, and orientations can be calculated, even while having rock shrinkage.

15.5.1.2 ***Differential-Strain Analysis.*** When the core sample is reloaded in the pressure vessel and restressed with the hydrostatic confinement pressure under laboratory conditions, pressure vs. strain is recorded. The differential strains in each direction over time are observed. The strain relaxation induces microcracks that are normally oriented by the in-situ stress.

This technique assumes that the formation has no major, pre-existing in-situ microcracks (Smith et al. 1986). Because of rock expansion from the downhole confinement, the difference of the strain and number of microcracks of various directions indicate the magnitude and direction of the in-situ stress. The direction of the greater number of the cracks is perpendicular to the direction of maximum principal stress (Gidley et al. 1989).

15.5.1.3 ***Differential Wave Velocity Analysis.*** On the basis of the microcracking concept as a differential-strain analysis technique, differential wave velocity analysis is used to analyze the in-situ stress direction by use of the compressional acoustic-wave velocity. The velocity of a *P*-wave in a rock is inversely proportional to the number of microcracks; the *P*-wave velocity will decrease if oriented microcracks exist. The degree of velocity variation is also proportional to the density of the microcracking in a specific direction (Smith et al. 1986).

Once the oriented core is submitted under laboratory condition, *P*-waves are transmitted in different directions through the rock sample. The travel time of each direction is recorded. The direction of the highest travel time will be perpendicular to the majority of the microcracks, which will be parallel to the maximum-stress direction and to the fracture azimuth (Gidley et al. 1989). The results from both differential-strain analysis and differential wave velocity are consistent on the fracture azimuth prediction (Smith et al. 1986).

15.5.1.4 ***Residual-Strain Overcoring.*** The fracture azimuth can be obtained from residual-strain overcoring techniques. Three strain gauges are mounted at 45° to each other in the horizontal plane of each core. The vertical strain is also monitored by a gauge mounted parallel to the core axis. The data of displacement on each gauge are recorded and converted to strain. The magnitude and orientation of the horizontal and vertical strain are estimated. The measurement can then determine the minimum-horizontal-stress plane.

15.5.1.5 ***Thermal Expansion Anisotropy.*** Under the concept that the majority of microcracks will be perpendicular to the maximum-stress direction, the oriented core samples will be monitored in the expansion direction after the rock reaches the cooling condition, which is approximately 40°F. The rock expansion rate of each direction will then be measured when the rock sample has warmed to the room temperature (Gidley et al. 1989). The direction of the maximum expansion indicates the maximum in-situ stress.

15.5.2 Importance of Coring for Hydraulic Fracturing Treatment Design. The main goal of core analysis is to reduce uncertainty in reservoir evaluation by providing data representative of the reservoir at in-situ conditions, if possible. This process has to be integrated with field and production data to minimize reservoir uncertainties that cannot be addressed with other data sources such as well logging, well testing, or seisms.

In the design of a hydraulic fracturing treatment, it is important to have an adequate description of the layers encountered in a well. These layers usually consist of sandstones, siltstones, limestones, and shales. The properties of each of these rock types can be critical to the hydraulic fracturing treatment design. Therefore, when a well in a low-permeability area is cored, some of the shales, siltstones, and mudstones found above and below the potential productive intervals also should be cored (Gidley et al. 1989). This methodology is applied to estimate the behavior of critical fracture geometrical parameters, such as height, width, and growth. These geometrical parameters interact and vary depending on the physical properties encountered in the layers (which might act as seals/barriers) surrounding the pay zone. In many cases, coring these potential barriers to fracture growth is more important than coring the productive interval.

Cores not only provide information for fracture designs, but also are used to evaluate the amount of oil- and gas-in-place in the reservoir, determine effective values of permeability, and obtain correlations between the log readings and the core material. New cost-effective techniques have been developed on the basis of the economics of the petroleum industry combined with the need to evaluate thin-bed and unconventional reservoirs.

15.5.3 Special Considerations for Fracture Treatment Design. Relying on core data when limited sampling in relation to reservoir size is available can result in an inferior, less-cost-effective hydraulic fracturing design that can result in a serious underestimation of production potential (Robinson et al. 1991). In these cases, pressure-buildup testing can be more representative of true reservoir permeability and other properties measured from cores.

Because of analytical difficulties, core analysis for reservoir properties is not often conducted in shales. However, Luffel and Guidry (1990) developed a methodology of crushing the core samples to accurately measure porosity and fluid content.

Permeability (specifically in tight gas shales) is a key factor in stimulation design and production prediction. To properly evaluate a gas shale reservoir, the total matrix permeability can be obtained from core analysis on whole core or on multiple rotary sidewall cores. However, pressure-transient testing is the preferred method for determining total permeability values in gas shale reservoirs (Frantz et al. 2005).

15.5.3.1 ***Conventional Coring.*** A conventional core analysis usually is performed to determine values of porosity, permeability, and water saturation at (essentially) atmospheric conditions. The measurements are made at room temperature under moderate pressure (± 100 psi). This coring technique has proved to be quite useful in conventional reservoirs; however, in tight gas reservoirs, conventional coring is not very useful.

Thomas and Ward (1972) demonstrated that in low-permeability reservoirs, the permeability can decrease by one or two orders of magnitude when the core is tested under restored-state conditions. Furthermore, the results obtained from conventional core analysis showed that the values of permeability in low-permeability reservoirs easily can be too optimistic, typically overstated by a factor of 50 to 100.

Using conventional coring methods requires tripping the drillstring and drilling bottomhole assembly (BHA) and changing over to the coring assembly with a coring BHA. This represents an expensive operation when considering the daily rig costs. Fleckenstein and Eustes (2003) show a comparison between wireline coring and conventional coring, in which the latter increases total costs by a factor of 220%. Other problems, such as core jam and identification of core point, can be evidenced (Halliburton 2007).

15.5.3.2 Special Core Analysis. Several authors (for example, Davies and Holditch 1998; Holditch et al. 1991; and Tobola and Holditch 1991) have confirmed that formation permeability substantially decreases in low-permeability reservoirs when the confining stress increases. The industry has recognized that special core-analysis procedures are needed to properly describe low-permeability reservoirs and to obtain the data necessary for evaluating the formation and designing fracture treatments. Techniques are being used in the industry to measure virtually any reservoir property under restored-state (in-situ) conditions.

Special core analysis can be used to determine values for permeability; porosity; capillary pressure; relative permeabilities to oil, gas, and water; saturation exponent; and cementation factor under simulated reservoir conditions. Other parameters, such as compressional travel time, shear travel time, and formation density, can also be measured in the laboratory. Each of these values can be extremely important to the overall formation evaluation. Shear and compressional travel time and bulk density can be used to compute mechanical properties and to correlate these properties to values measured from openhole geophysical logs and in-situ stress tests. Obviously, these types of tests will have a direct impact on the hydraulic fracturing treatment design.

In addition to measuring formation properties under restored-state conditions, it is also useful at times to determine the possible interaction of fracturing fluids and fracturing proppants with the formation (Ahmed et al. 1979). Such tests are difficult to run under restored-state conditions, but can provide useful information concerning possible adverse reactions between the fluids used during a stimulation treatment and materials in the core. In some cases, the formation can be unusually soft or unusually hard, and the effects of proppant embedment or proppant crushing should be studied in the laboratory. Therefore, in addition to the special core analysis performed to evaluate the fluid flow properties and mechanical properties of the formation, special tests also should be run in special instances to confirm the choice of fracturing fluid and proppant.

15.5.3.3 Pressure Coring. Treinen et al. (1994) defined pressure coring as the only method for recovering cores that have not undergone saturation changes caused by depressurization as they are lifted to the surface. Pressure coring is recognized as a reliable technique to evaluate residual oil saturation and vertical conformance for miscible gas enhanced-oil-recovery (EOR) processes and has proved to be the best method for coalbed methane applications.

Depending on the goals of the application, pressure coring or in-situ data-gathering services can obtain information on reservoir fluids, formation saturations, and mechanical properties of target rocks. If required, downhole pressure can be maintained in the barrel, while fluids are retained until they can be captured for analysis.

Additional studies have used oil saturations from pressure-cored wells to help validate correlations suitable for different types of reservoirs used for predicting oil saturation from well log data in which accurate total hydrocarbon composition and saturation data are needed.

15.5.3.4 Wireline Coring. Wireline coring is a hybrid between a completely wireline-operated system and conventional coring. It provides significant rig time savings by recovering cores without tripping the whole drillstring. It is ideal for long sections and continuous coring. The system also provides an instant indication of jamming, and its use has proved to be successful in coalbed methane applications (Halliburton 2007; Warren et al. 1996), because the system quickly gets the core to the surface to minimize gas loss for best measurement.

15.5.3.5 Sidewall Coring. Cores are obtained during the logging of a well when a tool is lowered on wireline to a zone of interest that has been identified from preliminary analysis of the logs obtained before sidewall coring. A small-diameter core plug is obtained from the side of the borehole, as seen in **Fig. 15.25,** either through a core-plug sidewall-coring assembly or by a plug shot into the sidewall. Sidewall cores have a minimal volume for analysis.

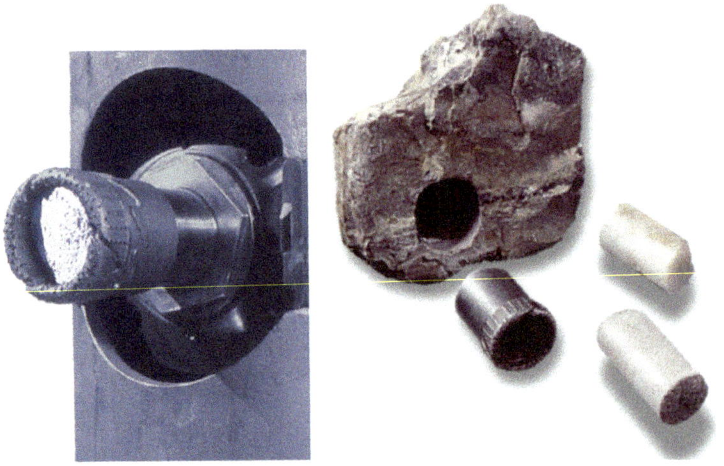

Fig. 15.25—Sidewall coring tool and core samples (courtesy of Halliburton).

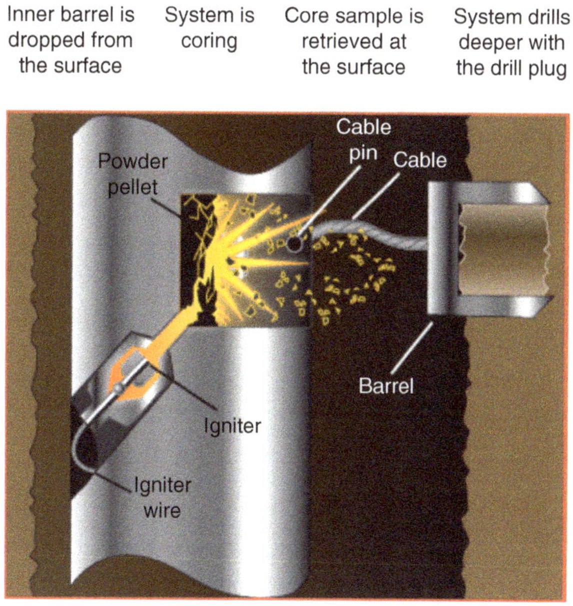

Fig. 15.26—Components of a percussion sidewall coring tool.

Sidewall coring can be used to recover cores in hard rock formations that are inaccessible with percussion tools **(Fig. 15.26),** but can also be used in soft rocks. They include a gamma ray tool positioning mechanism that provides an accurate core-point location. The samples can then be extracted undistorted, with a consistent cylindrical geometry, which allows for a wide range of petrophysical testing and analyses.

Sidewall coring has many applications (Halliburton 2007). Britt et al. (2004) demonstrated that sidewall coring represented a cost-effective means for determining Young's modulus—an important parameter to the fracturing process that has a direct relationship with the net treating pressure, fracture geometry, and fracture width.

Another application of sidewall coring is oriented sidewall coring. Laubach and Doherty (1999) described oriented sidewall coring as a successful procedure to evaluate large natural fractures by use of microfracture indicators in gas wells. Successful orientation of sidewall cores involves measuring core azimuth, determining polarity of the core, and determining which side of the core is the top. This last step is considered the most challenging task, because a mistake could result in a measured fracture strike of as much as 90°.

15.5.3.6 Percussion Sidewall Coring. Percussion sidewall coring is a versatile and reliable mechanism that allows acquisition of up to 116 formation samples in a single run. It is compatible with virtually any tool, and it can be positioned anywhere within a tool string to enable sampling with precise depth control for improved log analysis and formation characterization.

The outer diameter body of the tool houses propelling-explosive material and hollow core barrels, which are lowered to a predetermined depth and electrically fired into the sidewall one at a time from bottom charge to top charge. Each barrel, with a ⅞- or 1-in. core plug intact, is then extracted by moving the body uphole so the attached cable pulls the barrel from the wall.

15.5.3.7 Oriented Coring. To determine the direction of certain reservoir characteristics, such as natural fractures or stress patterns, oriented coring techniques often are useful. Special coring equipment is used to obtain an oriented core. The core barrel can be oriented with respect to the magnetic north pole by use of standard borehole surveying tools. A scribe line is etched onto the face of the core, and periodic surveys can be run to confirm that the core orientation is properly maintained.

Knowledge of the core orientation after the core is retrieved at the surface can be useful when planning the location of development wells in blanket reservoirs. When fracture orientation is known, the wells can be drilled to obtain adequate drainage in a reservoir; however, if the well spacing and locations, with respect to the fracture azimuth, are not well-planned, the resulting drainage pattern might be incomplete and will not sufficiently drain a particular reservoir (Gidley et al. 1989).

The use of oriented cores to determine the stress state of the Earth and the expected azimuth of both natural and induced fractures has been well-documented in the industry (Lacy 1997). In particular, the use of anelastic strain-recovery measurements from oriented cores to determine the direction of in-situ stresses has proved quite useful to the industry. Anelastic strain-recovery-measurement methods enable even measuring of the strain in three directions. As a result, we can estimate the magnitude of the in-situ stresses in three directions, and at the same time, help estimate the direction of fracture propagation.

15.5.3.8 Summary. Core testing is an important tool that provides information necessary for proper reservoir description and determination of rock mechanical properties. Reliable core data aid engineers in the design of an optimal hydraulic fracturing treatment for successful hydrocarbon recovery. However, different methods are available in the market and the applicability of each is subject to the conditions for which a core should be taken for analysis.

The selection of the appropriate method for the coring system should be evaluated using a technical/economic approach. Some methods (such as conventional coring) increase tripping time compared with other methods (such as wireline coring). Nevertheless, the accuracy and coring situation will play key roles during the selection process.

Ample technology is available for the industry; whether or not it is properly used is up to the engineers in charge of the study. Furthermore, special attention should be placed on handling the cores at the surface, because if they are if mishandled, errors could yield inaccurate results.

15.6 Nomenclature

C	=	discharge coefficient
C_{eff}	=	effective rock strength
D_1	=	distance to a microseismic event, ft
D	=	compliance coefficient (inverse of Young's modulus)
D_{perf}	=	perforation diameter, in.
D_n	=	creep coefficients from vertical-stress data
e	=	natural exponent
F_v	=	velocity factor, ft/sec
$g(t_D)$	=	dimensionless-loss volume function
g_u	=	fluid efficiency of upper bound
g_l	=	fluid efficiency of lower bound
G	=	G-function
$G(t_D, 0)$	=	dimensionless difference function
n	=	creep coefficients from vertical stress data
N_{perf}	=	number of perforations
p	=	bottomhole pressure, psi
p_b	=	breakdown pressure at which a crack begins to propagate, psi
p_p	=	pore pressure (from field data), psi
p_{perf}	=	frictional pressure loss through perforations, psi
p_r	=	reopening pressure, psi
q or Q	=	flow rate, bbl/min
r_b	=	radial depth of borehole breakout, ft
s	=	in-situ stress, psi
S_h	=	minimum horizontal stress, psi
S_H	=	maximum horizontal stress, psi
t	=	time of anelastic strain test, minutes
t_c	=	arrival time of the P-wave, seconds
t_p	=	arrival time of the P-wave, seconds
t_s	=	arrival time of the S-wave, seconds
v	=	Poisson's ratio
v_p	=	velocity of the P-wave, ft/sec
v_s	=	velocity of the S-wave, ft/sec
V_o	=	a variable calculated using Eq. 15.10
V_m	=	a variable calculated using Eq. 15.10
α	=	Biot's coefficient
δ_{ij}	=	the Kronecker Delta
ε	=	strain
ε_{AA}	=	strain in AA direction
ε_{cc}	=	strain in cc direction
ε_{ee}	=	strain in ee direction
ε_{ll}	=	strain in ll direction
ε_{xx}	=	strain in xx direction
ε_{yy}	=	strain in yy direction
$\varepsilon_{\theta\theta}$	=	strain in $\theta\theta$ direction
$\Delta\varepsilon_{ij}$	=	strain variation in ij direction
$\Delta\varepsilon_{kk}$	=	strain variation in kk direction
Δp	=	difference between borehole-fluid pressure and pore pressure in the host rock
ΔQ	=	change in injection rates during step-rate testing, bbl/min
Δt_c	=	time elapsed from fracture closure, minutes
$\Delta\tau$	=	time elapsed from when core is cut, minutes
ρ	=	density, lbm/gal
υ	=	Poisson's ratio (from uniaxial compression test)
σ_{ij}	=	horizontal stress, psi
$\sigma_{\max}$	=	maximum in-situ stress, psi
$\sigma_{\min}$	=	minimum in-situ stress, psi
$\sigma_{h\min}$	=	closure stress, psi
$\sigma_{\theta\theta}$	=	principal horizontal stresses, psi
θ_b	=	orientation of borehole, degrees
ϕ_b	=	opening angle of borehole breakout, degrees
τ	=	time of elastic deformation and relaxation, minutes

15.7 References

Abernathy, S. E., Woods, S. E. and Taylor, J. L. III. 1994. Radioactive Tracers in Oil and Gas Production: Practical Considerations in the 1990's. Presented at the SPE Health, Safety and Environment in Oil and Gas Exploration and Production Conference, Jakarta, Indonesia, 25–27 January. SPE-27236-MS. https://doi.org/10.2118/27236-MS.

Ahmed, U., Abou-Sayed, A. S., and Jones, A. H. 1979. Experimental Evaluation Of Fracturing Fluid Interaction With Tight Reservoir Rocks And Propped Fractures. Paper presented at the Symposium on Low-Permeability Gas Reservoirs, Denver, Colorado, 20–22 May. SPE-7922-MS. https://doi.org/10.2118/7922-MS.

Anderson, G. 1975. *Coring and Core Analysis Handbook*, Tulsa, Oklahoma: Petroleum Publishing Company.

Bloch, M., Freitas, S. M. S, Souza, M. S. et al. 2005. In-Situ Stress Determination Based Upon Borehole Imaging and Rock-Sample Analyses: A Comparison Among Anelastic Strain Recovery (ASR), Acoustic Velocity, and Acoustic Tomography. Presented at the SPE Latin American and Caribbean Petroleum Engineering Conference, Rio de Janeiro, Brazil, 20–23 June. SPE-94919-MS. https://doi.org/10.2118/94919-MS.

Bloch, M., Siqueira, C. A. M., Ferreira, F. H. et al. 1997. Techniques for Determining In-situ Stress Direction and Magnitudes. Presented at the SPE Latin American and Caribbean Petroleum Engineering Conference, Rio de Janeiro, Brazil, 30 August–3 September. SPE-39075-MS. https://doi.org/10.2118/39075-MS.

Brady, J. L., Withers, R. J., Fairbanks, T. D. et al. 1994. Microseismic Monitoring of Hydraulic Fractures in Prudhoe Bay. Presented at the SPE Annual Technical Conference and Exhibition, New Orleans, LA, 25–28 September. SPE-28553-MS. https://doi.org/10.2118/28553-MS.

Britt, L. K., Smith, M. B., Haddad, Z. et al. 2004. Rotary Sidewall Cores-A Cost Effective Means of Determining Young's Modulus. Presented at the SPE Annual Technical Conference and Exhibition, Houston, Texas. 26–29 September. SPE-90861-MS. https://doi.org/10.2118/90861-MS.

Castillo, J. L. 1987. Modified Fracture Pressure Decline Analysis Including Pressure Dependent Leakoff. Presented at the SPE Low Permeability Reservoir Symposium, Denver, 18–19 May. SPE-16417-MS. https://doi.org/10.2118/16417-MS.

Cipolla, C. L. and Wright, C. A. 2000a. Diagnostic Techniques to Understand Hydraulic Fracturing: What? Why? and How? Presented at the SPE/CERI Gas Technology Symposium, Calgary, Alberta, Canada, 3–5 April. SPE-59735-MS. https://doi.org/10.2118/59735-MS.

Cipolla, C. L. and Wright, C. A. 2000b. State-of-the-Art in Hydraulic Fracture Diagnostics. Paper presented at the SPE Asia Pacific Oil and Gas Conference and Exhibition, Brisbane, Australia, 16–18 October. SPE-64434-MS. https://doi.org/10.2118/64434-MS.

Davies, J. P. and Holditch, S. A. 1998. Stress Dependent Permeability in Low Permeability Gas Reservoirs: Travis Peak Formation, East Texas. Paper presented at the SPE Rocky Mountain Regional/Low-Permeability Reservoirs Symposium, Denver, Colorado, 5–8 April. SPE-39917-MS. https://doi.org/10.2118/39917-MS.

Davis, R. J. 1987. Use of Log-Derived Parameters For Completion Design Enhancement In the Medina Formation. Paper presented at the SPE Eastern Regional Meeting, Pittsburgh, Pennsylvania, USA, 21–22 October. SPE-17035-MS. https://doi.org/10.2118/17035-MS.

Davis, E. R., Ding Zhu, and Hill, A. D. 1997. Interpretation of Fracture Height From Temperature Logs – The Effect of Wellbore/Fracture Separation. *SPE Form Eval* **12** (02): 119–124. SPE-29588-PA. https://doi.org/10.2118/29588-PA.

Fleckenstein, W. W. and Eustes, A. W. 2003. Novel Wireline Coring System. Presented at the SPE Annual Technical Conference and Exhibition, Denver, 5–8 October. SPE-84358-MS. https://doi.org/10.2118/84358-MS.

Fourmaintraux, D. M., Bois, A.-P., le Goff, C. et al. 2005. Detailed In-Situ Stress Measurements for Qualifying the Safety of Underground Gas Storage Overburden. Presented at the 2005 SPE Europe/EAGE Annual Conference, Madrid, Spain, 13–16 June. SPE-94253-MS. https://doi.org/10.2118/94253-MS.

Frantz, J., Sawyer, W., MacDonald, R. J. et al. 2005. Evaluating Barnett Shale Production Performance—Using an Integrated Approach. Presented the SPE Annual Technical Conference and Exhibition, Dallas, 9–12 October. SPE-96917-MS. https://doi.org/10.2118/96917-MS.

Gidley, J. L., Holditch, S. A., Nierode, R. W. et al. 1989. *Recent Advances in Hydraulic Fracturing*. SPE Monograph Series, Vol. 12. Richardson, Texas: SPE.

Halliburton. 2007. Rotary Coring. *Halliburton.com*. https://www.halliburton.com/en-US/ps/solutions/formation-evaluation/rotary-coring.html (accessed 10/10/2017).

Holditch, S. A., Lin, Z-S., and Spivey, J. P. 1991. Estimating the Recovery From an Average Well in a Tight Gas Formation. Paper presented at the SPE Gas Technology Symposium, Houston, Texas, 22–24 January. SPE-21500-MS. https://doi.org/10.2118/21500-MS.

Keelan, D. 1985. Coring – Why it's Done. *World Oil*. **200** (4): 83–86, 88, and 90.

Lacy, L. L. 1997. Dynamic Rock Mechanics Testing for Optimized Fracture Designs. Paper presented at the SPE Annual Technical Conference and Exhibition, San Antonio, Texas, 5–8 October. SPE-38716-MS. https://doi.org/10.2118/38716-MS.

Lacy, L. and Smith, M. 1989. Fracture Azimuth and Geometry Determination. In *Recent Advances in Hydraulic Fracturing*, SPE Monograph Series, Vol. 12, pp. 344–346. Richardson, Texas: SPE.

Laubach, S. and Doherty, E. 1999. Oriented Sidewall Cores for Natural Fracture Evaluation. Presented at the SPE Annual Technical Conference and Exhibition, Houston, 3–6 October. SPE-56801-MS. https://doi.org/10.2118/SPE-56801-MS.

Le Calvez, J. H., Craven, M., Klem, R. C. et al. 2007. Real-Time Microseismic Monitoring of Hydraulic Fracture Treatment: A Tool To Improve Completion and Reservoir Management. Presented at the SPE Hydraulic Fracturing Technology Conference, College Station, Texas, 29–31 January. SPE-106159-MS. https://doi.org/10.2118/106159-MS.

Luffel, D. and Guidry, F. 1990. New Core Analysis Methods for Measuring Reservoir Rock Properties of Devonian Shale. Paper presented at the Annual Technical Conference and Exhibition, New Orleans, Louisiana, 23–26 September. SPE-20571-MS.

Lund, B. and Zoback, M. D. 1999. Orientation and Magnitude of *In Situ* Stress to 6.5 km Depth in the Baltic Shield. *International Journal of Rock Mechanics and Mining Sciences*, **36** (2): 169–190. https://doi.org/10.1016/S0148-9062(98)00183-1.

Mahrer, K. D., Zinno, R. J., Bailey, J. R. et al. 2007. Simultaneous Recording of Hydraulic-Fracture-Induced Microseisms in the Treatment Well and in a Remote Well. Presented at the SPE Hydraulic Fracturing Technology Conference, College Station, Texas, 29–31 January. SPE-106025-MS. https://doi.org/10.2118/106025-MS.

Miller II, W. K., Peterson, R. E., Stevens, J. E. et al. 1994. In-Situ Stress Profiling and Prediction of Hydraulic Fracture Azimuth for the West Texas Canyon Sands Formation. *SPE Prod & Fac* **9** (03): 204–210. SPE-21848-PA. https://doi.org/10.2118/21848-PA.

Nolte, K. G. 1986. Determination of Proppant and Fluid Schedules From Fracturing-Pressure Decline. *SPE Prod. Eng.* **1** (04): 255–265. SPE-13278-PA. https://doi.org/10.2118/13278-PA.

Phillips, W. S., Rutledge, J. T., Fairbanks, T. D. et al. 1998. Reservoir Fracture Mapping Using Microearthquakes: Austin Chalk, Giddings Field, TX and 76 Field, Clinton Co., KY. *SPE Res Eval & Eng* **1** (02): 114–121. SPE-36651-PA. https://doi.org/10.2118/36651-PA.

Plahn, S. V., Nolte, K. G., and Miska, S. 1997. A Quantitative Investigation of the Fracture Pump-In/Flowback Test. *SPE Prod & Fac* **12** (01): 20–27. SPE-30504-PA. https://doi.org/10.2118/30504-PA.

Robinson, J., Chu, W.-C., Woods, M. et al. 1991. Pressure Buildup Testing an Aid in Well Stimulation: Case History. Presented at the SPE Annual Technical Conference and Exhibition, Dallas, 6–9 October. SPE-22750-MS. https://doi.org/10.2118/22750-MS.

Smith, M. B., Ren, N. K., Sorrells, G. G. et al. 1986. A Comprehensive Fracture Diagnostics Experiment: Part 2-Comparison of Fracture Azimuth Measuring Procedures. *SPE Prod Eng* **1** (06): 423–431. SPE-13894-PA. https://doi.org/10.2118/13894-PA.

Soliman, M. Y., Kuhlman, R. D., and Poulsen, D. K. 1990. Minifrac Analysis For Heterogeneous Reservoirs. Paper presented at the CIM/SPE International Technical Meeting, Calgary, Alberta, Canada, 10–13 June. SPE-21554-MS. https://doi.org/10.2118/21554-MS.

Thomas, R. D. and Ward, D. C. 1972. Effect of Overburden Pressure and Water Saturation on Gas Permeability of Tight Sandstone Cores. *J Pet Tech* **24** (02): 120–124. SPE-3634-PA. https://doi.org/10.2118/3634-PA.

Thorne, B. J. and Morris, H. E. 1988. An Assessment of Borehole Seismic Fracture Diagnostics. Paper presented at the SPE Annual Technical Conference and Exhibition, Houston, 2–5 October. SPE-18193-MS. https://doi.org/10.2118/18193-MS.

Tinker, S. J., Baycroft, P. D., Ellis, R. C. et al. 1997. Mini-Frac Tests and Bottomhole Treating Pressure Analysis Improve Design and Execution of Fracture Stimulations. Paper presented at the SPE Production Operations Symposium, Oklahoma City, Oklahoma, 9–11 March. SPE-37431-MS. https://doi.org/10.2118/37431-MS.

Tobola, D. P. and Holditch, S. A. 1991. Determination, of Reservoir Permeability From Repeated Induction Logging. *SPE Formation Evaluation,* **6** (01): 20–26. SPE-19606-PA. https://doi.org/10.2118/19606-PA.

Treinen, R., Bone, R., Vogel, H. et al. 1994. Hydrocarbon Composition and Saturation from Pressure Core Analysis. Presented at the SPE/DOE Improved Oil Recovery Symposium, Tulsa, Oklahoma, 17–20 April. SPE-27802-MS. https://doi.org/10.2118/27802-MS.

Warpinski, N. 2009. Microseismic Monitoring: Inside and Out. *J Pet Tech.* **61** (11): 80–85. SPE-118537-JPT. https://doi.org/10.2118/118537-JPT.

Warren, T., Powers, J., Bode, D. et al. 1996. Development of a Commercial Wireline Retrievable Coring System. Paper presented at the SPE Annual Technical Conference and Exhibition, Denver, Colorado, 6–9 October. SPE-52993-MS.

Wolhart, S., Davis, E., Roadarmel, W. et al. 2005a. Reservoir Deformation Monitoring to Enhance Reservoir Characterization And Management. Paper Presented at the SEG Annual Meeting, Houston, Texas, 6–11 November. SEG-2005-2512.

Wolhart, S. L., Odegard, C. E., Warpinski, N. R. et al. 2005b. Microseismic Fracture Mapping Optimizes Development of Low-Permeability Sands of the Williams Fork Formation in the Piceance Basin. Presented at the SPE Annual Technical Conference and Exhibition, Dallas, 9–12 October. SPE-95637-MS. https://doi.org/10.2118/95637-MS.

Wright, C. A., Davis, E. J., Weijers, L. et al. 1998. Downhole Tiltmeter Fracture Mapping: A New Tool for Directly Measuring Hydraulic Fracture Dimensions. Presented at the SPE Annual Technical Conference and Exhibition, New Orleans, Louisiana, 27–30 September. SPE-49193-MS. https://doi.org/10.2118/49193-MS.

Chapter 16

Fracturing Fluids and Additives: Properties and Selection

Stephen A. Holditch

16.1 Introduction

Stimulation using hydraulic fracturing methods has been applied in the petroleum industry since the 1950s to improve the productivity and reserves from wells completed in low-permeability reservoirs or wells in higher-permeability reservoirs that were damaged during drilling, completion, or production. Fracture stimulation can be a complex field operation. Fracturing fluid is the medium that creates the fracture and transports proppants into the fracture to create a highly permeable flow path for the reservoir fluid to enter and then move to the wellbore. In addition, the fracturing fluid should not cause severe damage to either the reservoir or the proppant pack. Different types of fracturing fluid are available in the industry. We present general characteristics of fracturing fluid and discuss the pros and cons of different fluid types.

Fracturing fluid is a critical component of a hydraulic fracturing treatment. Selection of the fracturing fluid, job design, and well-turnaround procedure all determine the productivity of a well after stimulation by hydraulic fracturing. A fracturing fluid should provide sufficient viscosity to suspend proppant and transport it into the fracture and should break into a low-viscosity fluid after the job is completed. This will facilitate fracture cleanup by allowing rapid flowback of fluid to the surface.

Fracturing fluids are classified according to their physical/chemical state:

- Fluids that comprise solvents containing uncrosslinked polymers are referred to as solutions. Often they are called linear or base gels, even though this is not correct in the colloidal sense.
- When the polymer molecules are crosslinked, the colloidal system can be categorized as a gel.
- Fracturing fluids consisting of gaseous or liquid internal phases dispersed in a liquid external phase are referred to as foams and emulsions, respectively.
- A fracturing fluid having proppants in itself is referred to as proppant-laden fluid or slurry.

16.2 Properties and Rheology of Fracturing Fluids

Fracturing fluid serves as a medium to create a fracture and to transport and place proppant into the fracture. To achieve this goal, fracturing fluid should exhibit several characteristics that might contradict each other. McGowen et al. (1993) lists the following as primary characteristics of fracturing fluid:

- Able to suspend and transport proppants
- Low fluid loss
- Easily recovered after the stimulation operation
- Minimum damage to the formation and fracture conductivity
- Compatible with formation and formation fluid
- Maintains rheological properties throughout the operation
- Low friction pressure
- Environmentally friendly
- Safe and simple preparation
- Prepared and operated at reasonable cost

The ideal fracturing fluid should have low viscosity for liquid handling at the surface, while having enough viscosity to transport proppants all along the surface facility through the tubulars. Relatively high-viscosity fluid is required to transport and suspend proppants through the operation. The viscosity of fluid is very important at the tortuous region near the wellbore where high risk of premature screenout exists (Rae and Di Lullo 1996). A viscous fluid creates wider fractures, reduces fluid loss into the formation, and evenly distributes proppants into the fracture. However, high viscosity might increase formation damage, friction pressure, and cost because it requires higher pumping pressures.

Fluid loss affects the volume of fluid and the injection rate of fracturing fluid required to achieve a specified fracture volume. McGowen et al. (1993) considered fluid loss a critical factor in the success of fracturing operations in formations with permeability higher than 10 md. Additives to fracturing fluids control fluid loss. Moreover, high fluid loss translates to higher cost and formation damage.

To reduce formation damage, the fracturing fluid is flowed back to surface. Sherman and Holditch (1991) and Rae and Di Lullo (1996) studied formation damage during fracturing treatment. Fracturing fluid damages the formation by altering relative permeability and leaving residue in the formation. To reduce the damage, the time that fluid remains in the formation should be minimized and fluid should be flowed back to clean the formation. Siddiqui et al. (2004) showed that most of damage from fluid residue occurs in a few hours. After that time, formation damage does not change over time during shut-in. Also, fracturing fluid causes damage because of the relative permeability and hysteresis in the invaded zone.

Fracturing fluid should be environmentally friendly and safe, to reduce the effect on the environment and to reduce the cost of handling fluid before and after the operation. In addition, fracturing fluid should be prepared in simple steps to reduce the complexity of operation in the field, which in return increases the success rate. Straightforward procedures can be performed easier and faster, eliminating the risk of unscheduled problems and saving time.

In the past, the focus has been on the rheological properties of fracturing fluid, and damage optimization has been the secondary consideration. Brannon et al. (2006) refers to the immediate effect of rheological properties of fracturing fluid on field operations and the equipment limitations as the main reasons for prioritizing rheological properties of a fracturing fluid. Post-fracture analyses indicate that the formation damage that occurs during a fracturing job can be significant. Recent changes in the oil industry, such as high oil prices and marginal reservoirs, require more reservoir consideration during the optimization process.

Rae and Di Lullo (1996) and Brannon et al. (2006) studied the effect of guar residue on formation damage. Natural guar leaves a significant amount of residue in the formation and damages the formation and proppant pack. Modified polymers derived from guar are used to reduce formation damage. Although new polymers have fewer residues by weight, the volume ratio is almost the same for all of them.

Laboratory tests conducted by Brannon et al. (2006) indicated that crosslinked guar-based fluid might severely damage the proppant pack and reduce the conductivity by more than 80%. To mitigate this problem, breakers evolved in the industry. Recent advances have resulted in fracturing fluids with lower concentrations of polymers, which reduce viscoelastic-surfactant gelled fluids.

Fig. 16.1 illustrates the amount of residues in guar by volume and weight percent. From weight percent, prospective polymers derived from guar contain fewer residues by more than 50%. Volume-wise, natural guar exhibits the same or less residue. This has caused confusion in the industry. Natural guar is less expensive than derived products; so, natural guar is used to reduce costs wherever possible.

Properties of fracturing fluid depend on operational and bottomhole conditions. Several authors (Harris 1989; Harris and Reidenbach 1987; Rae and Di Lullo 1996) have discussed how the rheological properties of fracturing fluid change with shear history, temperature, pH, concentration of polymer, bottomhole condition, and proppant concentration. Most fracturing fluids exhibit shear-dependent properties. Behavior change resulting from shear rate might be reversible, such as with borate-crosslinked fluid systems, or irreversible, such as with organometallic systems.

Fig. 16.2 illustrates the effect of shear history on apparent viscosity of an organometallic system at different temperatures. This figure shows that as fluid experiences high shear rates, its effective viscosity at specific conditions drops with time.

Rheological properties of some fracturing fluids depend on pH. Most fracturing fluids require a certain range of pH for optimal performance. In addition, pH should be adjusted with respect to additives; therefore, additives and gelling polymers should be pH compatible. **Fig. 16.3** illustrates the pH range for different fracturing fluid systems.

Temperature reduces viscosity, increases pH, activates delayed crosslinkers, and complicates the rheology of fracturing fluid. Stability of additives and fluid components at different temperatures should be monitored, especially while dealing with high-temperature formations.

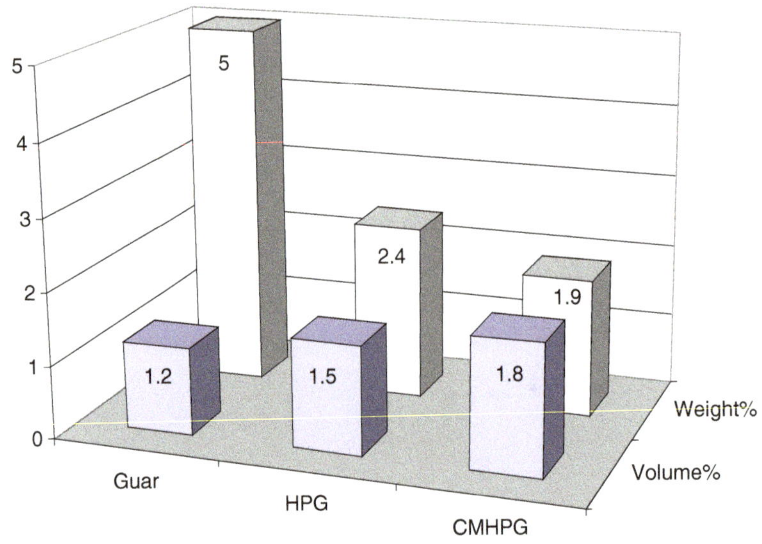

Fig. 16.1—Residue of guar and guar derivatives yield opposite results in weight percent and volume percent (modified). HPG = hydroxypropyl guar; CMHPG = carboxymethyl HPG.

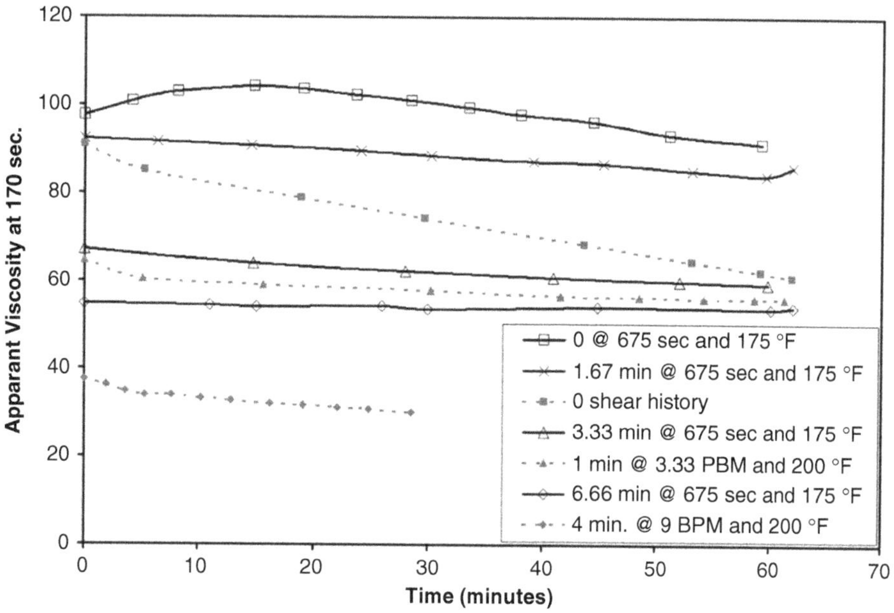

Fig. 16.2—Effective viscosity of an HPG/titanium acetylacetonate crosslinked fluid drops with time (Economides and Nolte 1989).

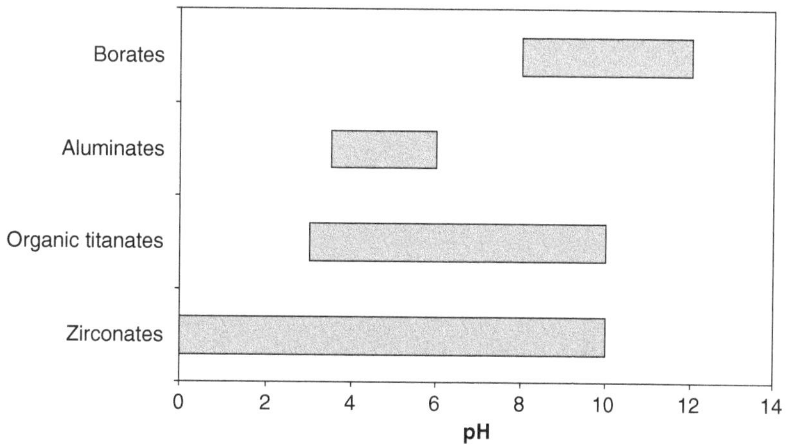

Fig. 16.3—Range of pH for different crosslinking systems (Rae and Di Lullo 1996).

Some additives can degrade with temperature, destabilize the fluid, and damage the formation.

Most of the time, properties of fracturing fluids are measured without the addition of proppant. Studies have shown the effect of the proppant load on the properties of fracturing fluid. **Fig. 16.4** illustrates the effect of the proppant on the rheology of the fluid. As the amount of proppant increases, the viscosity increases; therefore, more horsepower is required to pump the fluid into the fracture. This extra pressure should be considered during planning.

Currently, five different liquid-based fracturing fluids are available. These systems are water-, oil-, emulsion-, foam-, and alcohol-based systems. Water-based systems are less expensive and widely used. Oil-based and emulsion fluids are used in water-sensitive formations to mitigate formation damage and in low-temperature areas. Foam systems are used to fracture low-pressure formations. Alcohol-based fluid systems are used when formation damage occurs in the formations because of a water block. Different types of fracturing fluid are prepared using base liquids. **Table 16.1** is a summary of different gelled-fluid systems with their benefits and drawbacks, as proposed by Rae and Di Lullo (1996) and Holditch et al. (1993).

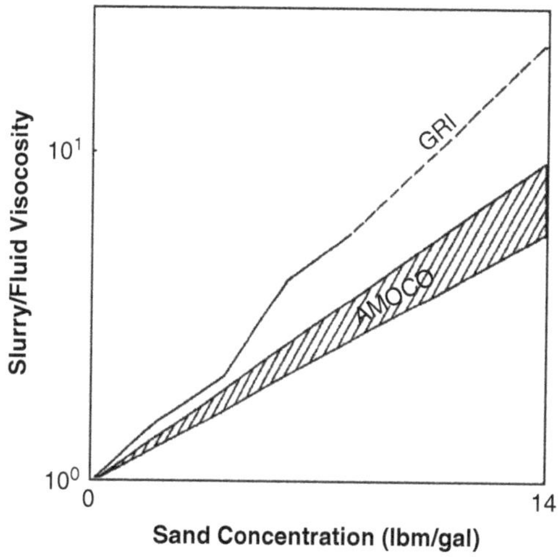

Fig. 16.4—Increasing sand concentration increases viscosity of the slurry (Holditch et al. 1991).

Gelling System	Benefits	Drawbacks
Slickwater	Inexpensive.	Poor proppant transport.
	Low damage to the formation.	Requires large volumes of fluid at a high rate.
Linear gel	Inexpensive.	Damages fracture permeability.
	Easy to prepare on the field.	Poor proppant transport.
	Great fluid-loss properties in low-permeability formations.	Thick filter cake in low-permeability formations.
Crosslinked gel	Stable up to 300°F.	Performance depends on pH.
	Good proppant transport.	
	Stable rheology.	
Organometallic crosslinked gel	Stable up to 300°F.	Hard to remove filter cake.
	Stable at large range of pH.	Rheology depends on shear rate history.
Foam	More-stable rheological behavior at high temperatures.	Foam rheology is a function of quality and behavior of liquid phase.
		High-friction pressure.
		Expensive.

Table 16.1—Benefits and drawbacks of different fracturing fluid systems.

Primary considerations to select fracturing fluid are permeability/viscosity, fracture half length, reservoir pressure gradient, fracture height, and reservoir temperature. On the basis of these considerations and characteristics of different fluid systems, guidelines and expert systems are developed to select appropriate fluid systems for fracturing operations. Holditch et al. (1993) developed **Table 16.2** as a guideline on which to build their expert system.

Fluid Base	Fluid Type	Main Application
Water	Linear fluid	Short fracture in low temperature.
	Crosslinked fluids	Long fractures in high temperatures.
Oil	Linear fluid	Short fracture in water-sensitive formation.
	Crosslinked fluids	Long fracture in water-sensitive formation.
	Emulsion	Long fractures with a low-fluid-loss system.
Foam	Acid-based	Low-pressure, water-sensitive formation.
	Water-based	Low-pressure formation.
	Alcohol-based	Low-pressure formation with water-blocking problems.
Alcohol	Linear system	Removing/preventing water blocks.

Table 16.2—Guidelines for selecting base fluid/type of fracturing fluid (Holditch et al. 1993).

16.2.1 Summary.

- Fracturing fluid is one parameter that can be controlled by the design engineer and tailored for the fracturing operation. Success of the fracturing job depends on the rheology of the fracturing fluid.
- The desired properties of fracturing fluid will be different for every formation. The effects of several factors, such as temperature and desired fracture length, that influence the properties of fracturing fluid should be determined to optimize the operation.
- The addition of proppant changes the properties of the fracturing fluid. Therefore, properties of the fluid/proppant mixture should be considered when designing the treatment.
- Currently, the industry is searching for fluids that are nondamaging and inexpensive.

16.3 Types of Fracturing Fluids

Hydraulic fracturing has been boosting hydrocarbon production for more than 70 years. Fluids are pumped into wells at pressures high enough to make deep cracks in the formation. Once the fracture has been created, it is held open with a sand or ceramic proppant that has traveled in the fluid.

Depending on the makeup of the rock formation, fracturing fluid can come in several different forms. The base liquid is most commonly water, but could also be acid or alcohol. It can contain a variety of additives to strengthen the proppant or to help break down the fluid once the fracture is complete. The fracture must be cost effective, must optimize production, and be physically stable over a long time.

16.3.1 Water-Based Fracturing Fluids. In the 1950s, oil-based fluids were the most-common fracturing fluids, because operators were concerned with inserting water into oil-based formations. This has changed after research and new technologies made water-based fluids more useful to operators. If water can perform as the base of a fracturing fluid, it becomes the most-desirable option for several reasons.

First, water is much less costly than other fluids, such as oil or acid, which in turn makes it more readily available to operators. Second, it is a much-safer substance for service companies to mix with because it is incombustible. Third, water has a heavier hydrostatic head than other options. Finally, it is much easier for service companies to manipulate viscosified water (Ely 1990).

16.3.1.1 Linear Fluids. For the water-based fluids to transport the proppant while avoiding fluid loss, operators had to be able to viscosify the water. Starch was tested in this role, but it was quickly discontinued, because it could not withstand much temperature variation. A popular polymer product is guar gum, first used in the 1960s. The original guar polymer would uncoil during hydration, which thickened the liquid (Ely 1990). This same guar product is still used in some form today.

There has been a progression from guar gum to hydroxypropyl guar (HPG). HPG became very popular as a water-based fluid viscosifier, because it had a much greater temperature-stability range, while being more viscous. It was initially produced to reduce the residue that occurred in the guar gum during hydration. As shown in **Table 16.3,** Guar products have 8 to 13% residue, while HPG has 1 to 4%, which is an excellent improvement (Ely 1990). This leaves less residue and increases flow into the fracture. While in many cases HPG would be preferred over guar gum, if the formation has high permeability and temperature stability is not an important element, then guar gum is preferable because the increased residue decreases fluid loss from the formation.

	Gelling Agents	% Residue
High residue	Guar	8 to 13
Moderate residue	Refined guar	5 to 6
Low residue	HPG and CMHPG	1 to 4
No residue	HEC, CMC, CMHEC, and Polyacrylamides	0

Table 16.3—Residue of gelling agents.

Hydroxyethylcellulose (HEC), carboxymethylcellulose (CMC), and carboxymethylhydroxyethylcellulose (CMHEC) are considered synthetics, whereas guar gum is considered a natural polymer. Unlike HPG and guar gum, these products leave no residue (Ely 1990). Because they must be created, they are also more expensive to use. Typically, HEC is used in gravel-pack situations in which no leftover residue is desired. The high cost and sensitivity to salts has prohibited common use of CMC today, although it was used briefly in the 1970s. CMHEC was developed to improve CMC and is essentially the same product with the salt sensitivity removed. CMHEC is still relatively uncommon, apart from occasional crosslinked-acid systems in low-temperature environments. Another high-cost product is xanthan gum, which has been used as a thickener for hydrochloric acid (Ely 1990).

Polyacrylamides are also linear viscosifiers, but are more often used to reduce friction of water or acids, even at very low concentrations. They can be cost prohibitive, but generally they are stable at most temperatures.

16.3.1.2 Crosslinked Fluids. Crosslinked fluids are an advance from linear fluids in which viscosity can be increased using the same concentration of product. Essentially, the molecular weight of a product is increased by joining together polymer molecules and metal crosslinkers.

Early crosslinked fracturing fluids had field test problems of poor pumpability, because the crosslinking reaction happened too quickly. Another problem was the base gel being too viscous and not completely degrading during the treatment. This made it possible for the fracturing fluid to then transport the proppant back out of the fracture (Ely 1990).

As development continued, HPG and CMHEC had titanium and aluminum chelates used as the crosslinkers. By the 1980s, zirconium chelates became more common because they increased temperature stability and cost effectiveness (Ely 1990).

Although the crosslinked products were an improvement over their base counterparts, issues still arose. If the product was crosslinked at the surface, or the reaction occurred too early, viscosity was lost because of the shear pressures placed on the fluid (Armstrong et al. 1995). The replacement, and the more commonly used product, is the delayed crosslink system. Research has shown benefits, including improved temperature stability and increased viscosity, when the crosslink reaction is delayed. In practice, the delay also has reduced the pumping friction (Ely 1990).

Crosslinked fluids have the following advantages over linear fluids (Ely 1990):

- Can achieve a much higher viscosity in the fracture.
- More efficiently handles fluid loss.
- Holds better proppant-transport capabilities.
- Holds better temperature stability.
- Is more cost effective per pound of polymer.

Overall, linear fracturing fluids are relatively easy and safe to use. Although they are generally less costly than crosslinked fluids, they might have less temperature stability. In general, these fluids will be ideal for near-wellbore, high-permeability situations, whereas crosslinked fluids might be more useful for deep-penetrating fractures.

16.3.1.3 Micellar Fracturing Fluids. As an alternative to using high-residue polymers, micellar fluids are polymer-free and provide a simpler fluid-mixing approach. A quaternary ammonium salt builds viscosity and viscoelacity by the entanglement of wormlike micelles in a brine solution (Samuel et al. 1997). Micelles have a structure similar to that of a polymer and can be broken down with a change of structure, thus eliminating the need for breaker additives. The reduced number of necessary additives saves equipment and is less complex for operators to perform.

16.3.1.4 Hybrid Fracturing Fluids. Recent developments have occurred using hybrid stimulations, which are a water-based-fluid prepad followed by a crosslink-gelled fluid. Hybrid treatments have shown the ability to create longer fractures, making them useful for low-permeability situations. Coronado (2007) simulated a 12-ft fracture located at the Red Fork Zone in Roger Mills County, Oklahoma, USA, to test the differences between linear, crosslinked, and hybrid treatments. The first simulation, **Fig. 16.5,** showed that the crosslinked system created a theoretical fracture 699 ft long and 115 ft high, with an average width of 0.0855 in., based on a 0.5 lbm/ft^2 cutoff. A hybrid simulation performed using the same simulation settings, but with a water-based prepad, created a fracture of 783 ft **(Fig. 16.6),** which was 84 ft longer than the fracture of the crosslinked system.

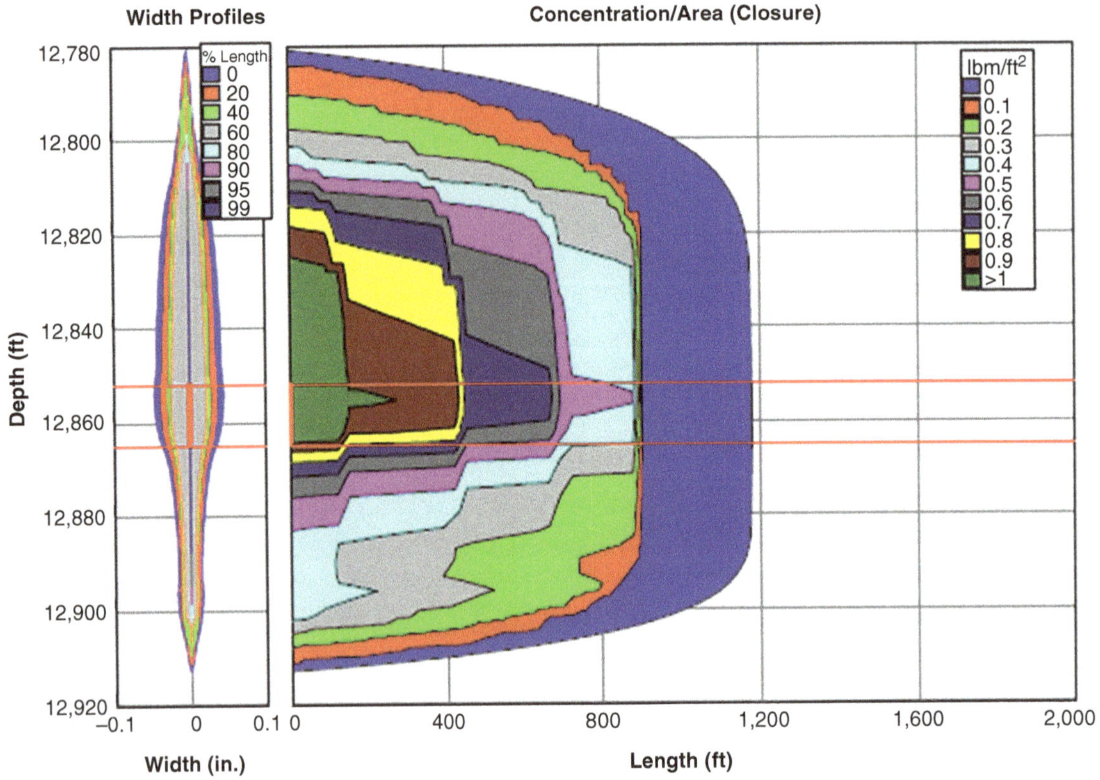

Fig. 16.5—A crosslinked-treatment simulation of the Red Fork Zone (from Coronado 2007).

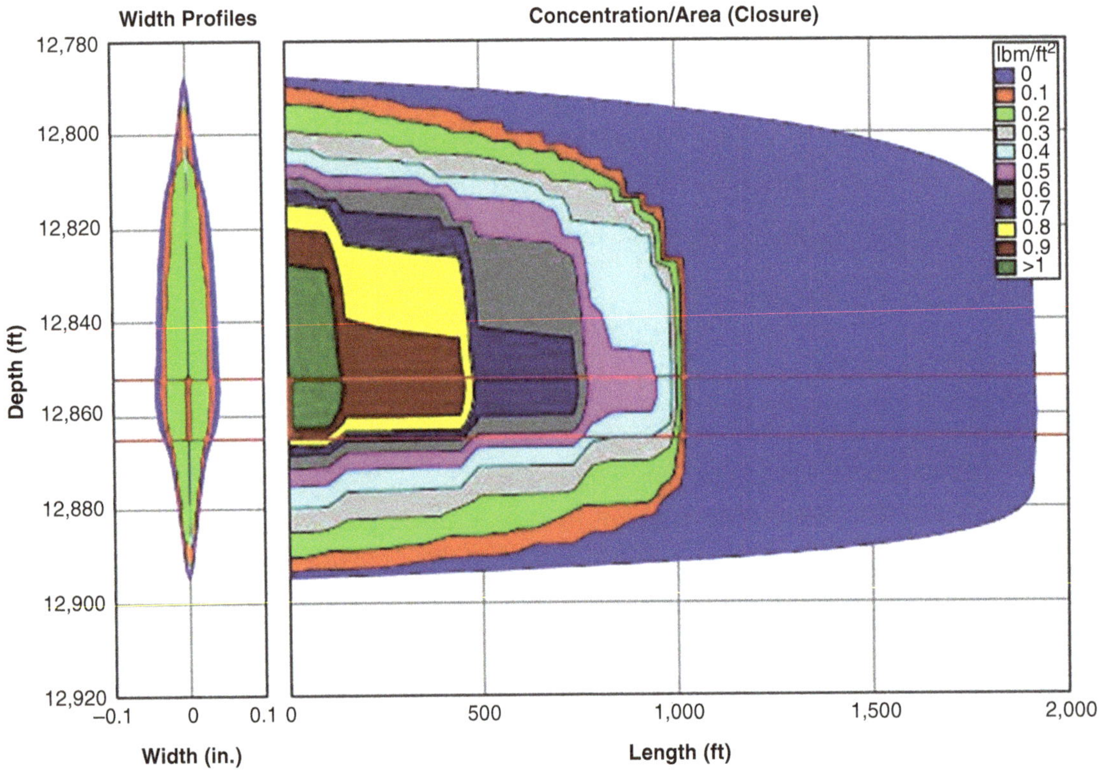

Fig. 16.6—A hybrid-fracturing-treatment simulation of the Red Fork Zone (from Coronado 2007).

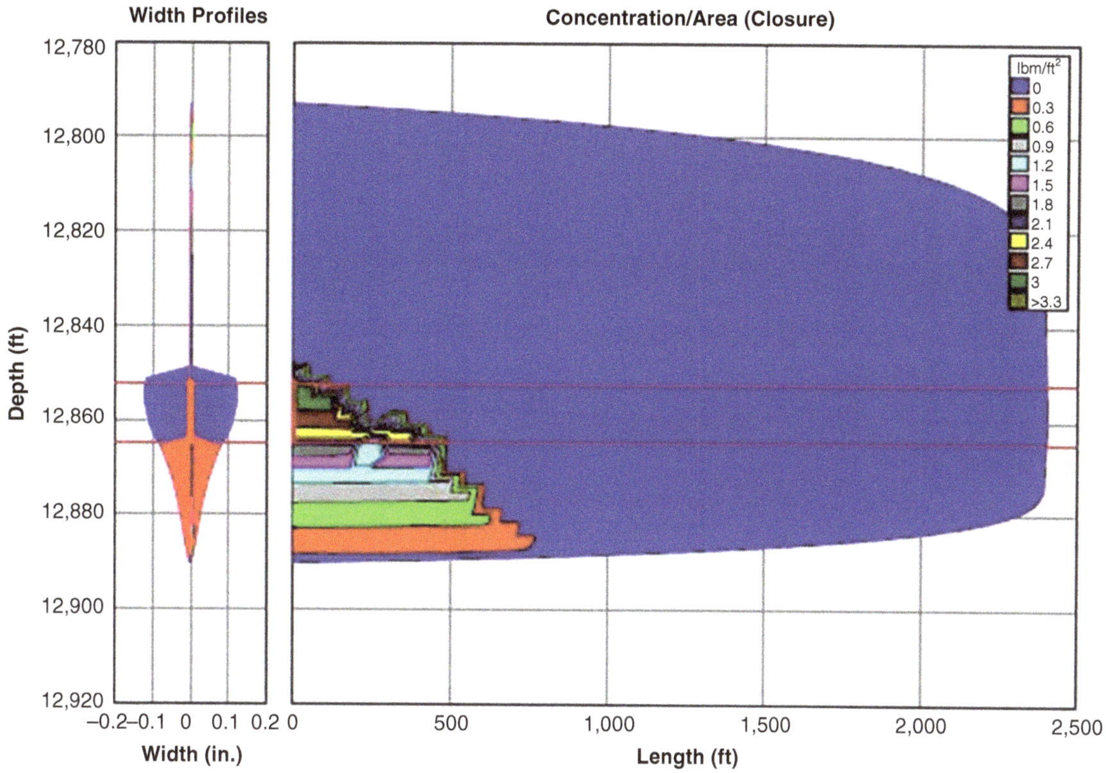

Fig. 16.7—A water-based fracturing-treatment simulation of the Red Fork Zone (from Coronado 2007).

An advantage highlighted by the simulation was a smaller fracture height using the hybrid treatment, which reduced wasted fluids through the fracture. Coronado (2007) estimated material-cost savings of up to 38% because of the different materials used in the fluid, in particular the smaller amount of base polymer. Smaller polymer concentrations also quickened the fluid break and prevented more proppant from flowing back out of the fracture.

As previously discussed, a water-based linear fluid is not as useful if a long fracture is desired, because these fluids struggle to carry the proppant into the fracture. The water-based treatments have high settling velocities, which leads to a bottom-heavy proppant distribution throughout the fracture, as shown in **Fig. 16.7** (Coronado 2007). In high-pressure-formation systems, this treatment becomes inefficient, because more fluid is needed at higher injection rates. These problems led operators to search for the hybrid-fluid alternative. Operators using the hybrid treatment saw material-cost savings up to 58% more than those of equivalent water-based treatments.

16.3.2 Oil-Based Fracturing Fluids. Today, the reaction product of aluminum phosphate ester, with a base of sodium aluminate, is commonly used for oil-based fracturing (Burnham et al. 1980). Similar to the water-based fluids, the oil-based fluids have been developed to improve viscosity as gelled crude oils and to increase the range of temperature stability. In particular, the aluminum phosphate ester can successfully operate in wells with a bottomhole temperature of more than 260°F (Ely 1990). As previously discussed, oil-based fracturing systems present a much greater fire hazard, but can be useful if the formation is water sensitive. Mixing and preparing the oil-based solution also can be much more difficult (Ely 1990). In general, pumping friction and pumping pressures are higher than in crosslinked water-based fluids, because of the smaller hydrostatic head of the oil.

16.3.3 Emulsion Fracturing Fluids. Although a common product used for many years, emulsion fluids have had problems with high friction pressure. Once water-external-emulsion fluids were created, which require one-third of the equivalent water-based gels, they became much more useful to the industry, because they are cost-effective and fracture the formation well. Water-external-emulsion fluids are easy to mix, use, and break; the surfactant enters the formation once the emulsion is complete. Water-external emulsions contain the water as a continuous phase, while oil is discontinuous. Oil-external emulsions contain the oil as a continuous phase, while water is discontinuous. Because it retains properties of the continuous product, water-external models have low-friction pressures, but surprisingly, have pumping pressures higher than a typical crosslinked water-based product (Ely 1990). Water-external emulsions are cost-effective treatment options when temperature stability is not as important.

16.3.4 Foam-Based Fluids. As a development from emulsion fluids, foam fracturing fluids are gas-in-liquid emulsions. The gas bubbles form the desired viscous, proppant-transporting medium. Ideally, between 70 and 80% of the fracturing fluid should be gas (Ely 1990). A surfactant can be added as a stabilizer, which coats the bubbles and increases the half-life of the foam. It can increase the half-life by nearly 10 times.

Foam fracturing fluids offer obvious advantages: 65 to 80% less liquid is required for preparation and pumping into the wellbore, and the gas in the fluid acts as an energizer, thereby improving the fracturing fluid recovery. These advantages are offset by the difficulty associated with correctly initiating a foam-based treatment. Depending on the gas used, a much higher pumping pressure might be necessary. More importantly, the two-phase mixture must be very precise to obtain the expected results. A foam-based treatment should be used in shallow wells with a low pressure that might need an energized fluid, or in a water-sensitive formation (Ely 1990).

16.3.5 Energized Fracturing Fluids and CO_2 Fluids. Energized fluids include gases leaving the liquid and are useful in wellbores with low bottomhole pressure. The energy from this reaction allows for a quicker removal of the fracturing fluid, while also helping prevent fluid loss.

Carbon dioxide (CO_2) and N_2 are gases commonly used over the past 40 years (Ely 1990). Their differences are important when choosing which to use. CO_2 is soluble in water, and it might not easily dissipate into the formation. This is useful when removing the treatment fluid, because it creates a gas drive that does not otherwise happen when using N_2, which is an immiscible inert gas. N_2 mostly will dissipate into the formation during treatment.

Tulissi and May (2002) showed that using a CO_2 foam system instead of a conventional CO_2 fluid system can improve long-term production. In addition, increasing the percentage of CO_2 in the fluid does not necessarily increase the proppant flowback or increase long-term production. Overall, energized fluids reduce the volume of fluid needed, while helping in fluid removal once the treatment is complete. Energized fluids could cost more than other treatment options and might be more complex to complete.

16.3.6 Summary. When deciding which fracturing fluids to use, the operator must carefully evaluate the needs and expected response of the target formation. The operator should also know the experience and technical expertise of the fracturing team, because some treatments can be very challenging to mix to perform an optimal treatment. New hybrid-fluid developments have led to many innovations, including cost-effective use of materials, and should be more commonly used in the future. The advantages of each type of fluid must be balanced with a treatment-cost analysis to determine an optimal fracturing fluid for the treatment.

16.4 Fracturing Fluid Additives

Fracturing fluids are mixtures of the base fluid and, in many cases, a substantial number of other chemicals. Each chemical should have a specific role in tailoring the fluid for a specific well or reservoir. For water-based fluids, the selection of fluid additives can come from a long list, as shown in **Table 16.4.**

Polymers	Crosslinkers
Breakers	Buffers
Foamers	Temperature stabilizers
Biocides	Surfactants
Clay stabilizers	Friction reducers
Diverting agents	Fluid-loss additives

Table 16.4—Common additives in fracturing fluids.

As a general rule, we should use the minimum number of additives required for each situation. As more and more additives are used during a treatment, the potential for unwanted consequences increases. It is important always to check for compatibility of all additives by mixing in the laboratory ahead of time the exact fluid using the exact chemicals to be used on the actual treatment and to look for unwanted or unplanned chemical reactions. Many additives are temperature and/or pH sensitive, so it is very important to check out all additives before using them during an actual fracturing treatment.

For example, when the industry first started using resin-coated sand, there were many problems with the crosslinking systems and clumping of the proppant when it was added to the blender. These unwanted chemical reactions caused many jobs to fail. The industry had to learn which surfactants were required to keep the fracturing fluid compatible with the resins.

16.4.1 Viscosity. Adding polymers usually controls the viscosity of a fracturing fluid. In some new fluids, surfactants and salt are used to form micelles. We also can increase viscosity by forming emulsions. In all but the water-fracture treatments being used in gas shale reservoirs, some form of additive is used to increase viscosity. We need viscosity to create fracture width, control leakoff, and transport the propping agent.

16.4.2 Polymers. In most fracture treatments, polymers are used to create viscosity in water. Polymers are long-chain molecules that interfere with each other in the fluid, causing resistance to movement that increases the viscosity. Viscosity can be increased by increasing the concentration of the polymer. Typically, concentrations of 20 to 50 lbm of polymer per 1,000 gal of water are used to create a fracturing fluid. The most inexpensive polymer, and one of the best, is guar. Guar is a plant product, and most guar comes from India or Pakistan. Guar contains residue that can act as a fluid-loss additive by building a filter cake on the fracture wall. Guar cannot be in reservoirs in which the temperature is above 250°F.

As the temperature increases, we switch to refined guar polymers such as HPG and CMHPG. These polymers maintain viscosity better than guar at high temperature and also have less residue. Another polymer often used in high-permeability reservoirs for fracture treatments is HEC.

16.4.3 Crosslinkers. Guar, HPG, and CMHPG can be crosslinked with borate, titanate, or zirconate. Different types of polymers and crosslinking agents can be used with success. It is important to work with service companies to select a base fluid and a crosslinking system compatible with reservoir rock and reservoir fluid and that will be stable at the formation temperature. Many crosslinkers are pH sensitive. Other crosslinkers are temperature activated. To provide for maximum viscosity in the fracture, it is best to delay the crosslinking reaction for as long a time period as possible, so the majority of the crosslinking reaction occurs near the bottom of the wellbore and near the mouth of the fracture. Sometimes, dual crosslinking systems are used. A borate can be used to provide some crosslinking in the wellbore, and either titanate or zirconate can be used to cause additional crosslinking in the fracture. Again, the service company engineer and the operating company engineer should discuss basic fluid and crosslinking options in detail before designing the treatment.

16.4.4 Bactericide. All water contains bacteria, no matter the source. In water-based fluids that will be viscosified with polymers, we always add biocide to the water before any other additive is used. Many times, we first put the biocide in the storage tanks and then add the water, so we are able to achieve maximum mixing of the two fluids. Biocides control growth of both aerobic and anaerobic bacteria. To know which biocide to use, we run tests on the mix water ahead of time. It is very important to identify the specific bacteria in the mix water and to use the correct biocide. If that does not happen, then the bacteria will attack the organic polymers, destroying the bonds and reducing the viscosity. The bacteria will also generate enzymes that will consume the polymers.

16.4.5 Breakers. For water-based, crosslinked-polymer fluids, it is very important to ensure that the fluid breaks after the fracture treatment, so it properly cleans up. Perhaps the single-largest problem in hydraulic fracturing is that unbroken fracturing fluid can plug the fracture and not let the fracture transport oil and gas along the fracture to the wellbore. For guar and refined-guar polymers, these polymers will not break down on their own at temperatures below 250°F, so breakers should always be used in a treatment in which the reservoir temperature is below 250°F. Of course, when large treatments are being pumped, problems can occur if too much breaker is added too quickly to the fluid. It is a very difficult engineering problem. We must have sufficient viscosity during the treatment, but we want the gel to break after the treatment.

For formations of 270°F or more, the polymers eventually will break down without using breaker. Many times, the problem there is trying to stabilize the fluid, so it does not break down too soon.

A large number of breakers can be used, and the choice is very specific to the fluid, temperature, and formation. The type of breaker used will be a function of pH and temperature, as shown in **Table 16.5.**

	pH Range	Temperature Range
Conventional enzyme	3–7.5	70–130°F
High-temperature enzyme	3–7.5	100–250°F
pH-tolerant enzyme	3–14	100–250°F
Oxidizer	3–15	130–260°F*
Catalyst oxidizer	3–16	70–120°F
Encapsulated oxidizer	3–17	180–250°F
Delay oxidizer	3–18	100–300°F
Weak acid	3–19	200°F±**

*very rapid breakdown and sensitive above 180°F
**not applicable in carbonate reservoirs

Table 16.5—Conditions for breakers.

The three basic types of breakers are enzyme breakers, oxidizing breakers, and acidic breakers. Clearly, we must select a breaker for the specific conditions of the treatment. Table 16.5 is, by no means, inclusive and might be slightly out of date. It is important for the design engineer with the operating company to work closely with the design engineer at the service company to choose the correct type and amount of breaker for a specific treatment. In fact, it is often necessary to run laboratory pilot tests of the fluids and additives, including breakers, to ensure that the mixture provides the desired results.

As previously mentioned, breaker loading and type is a function of pH, temperature, and polymer loading. To prevent the breaker from acting too fast, encapsulated breakers can be used to delay the release of the breaker into the fluid. In many low-temperature reservoirs, it is difficult to obtain the optimal gel break time. Without good engineering, the gel either will break too fast or too slow. On-location pilot testing is very important.

Oxidative breakers (ammonium and sodium persulfate) cause a chemical reaction with metal ions used to crosslink the gel. Enzyme breakers (sugar) split the crosslink bonds. Enzymes can react with multiple crosslink sites, causing the gel to break.

Craig and Holditch (1992) shows the effect of an ammonium persulfate (AP) breaker on the apparent viscosity of an uncrosslinked, 40 lbm/1,000 gal HPG gel at 200°F **(Fig. 16.8).** If no breaker is added, the apparent viscosity will decrease from approximately 300 cp to 150 cp in 3 hours. As very small amounts of AP breaker are used, the apparent viscosity decreases much more rapidly. Notice that if only 0.5 lbm/1,000 gal of AP breaker is added, the apparent viscosity decreases to near 10 cp in 30 minutes.

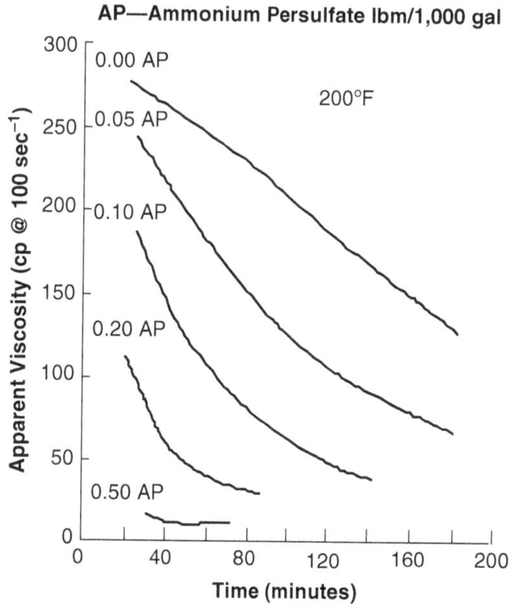

Fig. 16.8—Effect of ammonium persulfate breaker on a 40 lbm/1,000 gal HPG polymer gel (Craig and Holditch 1992).

Craig and Holditch (1992) pointed out that to completely break down the polymers so they can be cleaned up properly, the fluid viscosity should be decreased to a value of less than 3 cp. Using this criterion, they ran a series of tests to determine how much AP breaker was needed to completely break down various gels under various conditions. In one of the cases **(Fig. 16.9),** notice that when the gel is crosslinked, it takes as much as 1 to 2 lbm of AP breaker per 1,000 gal of water to completely break the gel.

Herein lies one of the most important problems in designing and pumping a hydraulic fracturing treatment: If we add enough breaker to really break down the polymers so they can clean up, we might break the fluid too rapidly and cause a screenout because of high leakoff or proppant bridging.

16.4.6 pH. The pH of a fluid can be important, and we use additives to control it. Many of the crosslinking additives will be affected by pH; most fluids will break down more quickly when pH is lower, but the temperature stability of many fluids will be higher at higher pH values. All design engineers need to be familiar with pH requirements for the specific fluids being pumped in a specific well.

16.4.7 Surfactants. Surfactants are added to control reactions among other additives. Surfactants can also be used to affect the reactions between the fracturing fluid and the reservoir rock or reservoir fluid. When a water-based fluid is being pumped into an oil reservoir, we use surfactants to prevent emulsions from forming. In low-permeability reservoirs, we use surfactants to water-wet the formation to improve fracture-fluid cleanup. In the past, many different types of surfactants have been used. Today, we primarily use nonionic surfactants, because they have less adsorption in the reservoir and are most compatible with clay-control additives.

16.4.8 Fluid-Loss Control. Fluid-loss control additives are used to control leakoff of the fracturing-fluid filtrate into the matrix or natural fractures. In high- or moderate-permeability reservoirs, fluid-loss control additives can be extremely important to the success of the treatment.

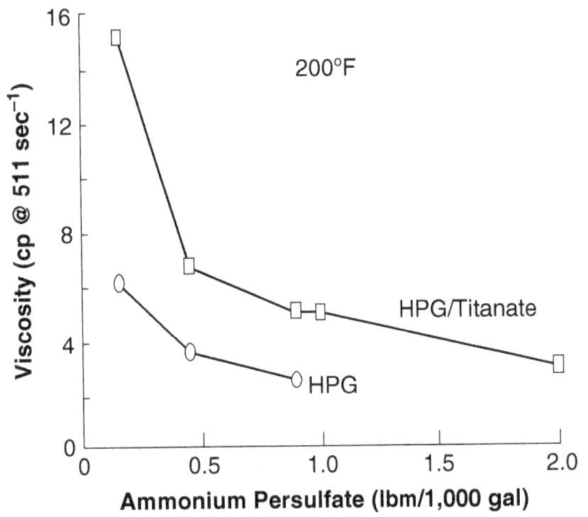

Fig. 16.9—Amount of ammonium persulfate breaker needed to completely break the gel (Craig and Holditch 1992).

In low-permeability reservoirs, little or no fluid-loss additive might be required. We can run a minifracture test to determine the fluid efficiency of different types of fluids pumped at different injection rates. If the fluid efficiency is too low, we should consider using some sort of fluid-loss additive.

Several types of fluid-loss additives are available. One category falls under the title of bridging materials, such as silica flour or 100-mesh sand. Silica flour is sand that has the consistency of a facial cosmetic powder. Silica flour works well in fairly permeable formations. We use 100-mesh sand in naturally fractured reservoirs, because it will invade many natural fractures and will bridge in those fractures, forming a base to let other fluid-loss additives, such as plastering agents, work.

Plastering materials are soft materials, such as resins and starch. As the name implies, these fluid-loss additives act by plastering themselves against the fracture wall and minimizing leakoff. The polymers in the gel can also form a filter cake that will substantially reduce leakoff in many cases.

Another mechanism for leakoff control is use of multiphase fluids, such as hydrocarbon emulsions and foams. For these fluids, the small bubbles of hydrocarbon or gas tend to plug pores and reduce leakoff. **Tables 16.6 and 16.7** summarize the additives and fracturing fluids commonly used in the field.

16.5 Fracturing Fluid Rheology

16.5.1 Definition of Rheology. Rheology is the study of deformation and flow—that is, the relationship between stresses imposed on a material (such as a fracturing fluid) and its resulting motion or, inversely, the stresses generated when a material moves under an imposed deformation. For a simple Newtonian fluid, the viscosity is the only material function required to calculate pressure drop and flow rate. For viscoelastic-polymer solutions, gels, and foams, more-complex descriptions of the material properties usually are required.

Type of Additives	Function Performed	Typical Products
Biocide	Kill bacteria	Gluteridehyde carbonates
Breaker	Reduce fluid viscosity	Acid, oxidizer, enzyme breaker
Buffer	Control pH value	Sodium bicarbonate, fumaric acid
Clay stabilizer	Prevent clay swelling	KCl, NH, Cl, KCl substitute
Diverting agent	Divert flow	Ball sealer, rock salt, flake boric acid
Fluid-loss agent	Improve fluid efficiency	Diesel, particulate, fine sand
Fraction reducer	Reduce fraction	Anionic copolymer
Iron controller	Keep iron in solution	Acetic and citric acid, ethylene diamine tetra-acetic acid, nitrilo triacetic acid
Surfactant	Lower-surface tension	Fluorocarbon, nonionic
Gel stabilizer	Reduce thermal degradation	methanol sodium thiosulfate

Table 16.6—Summary of chemical additives.

Base Fluid	Fluid Type	Main Composition	Application
Water	Linear fluids	Gelled water, GUAR<HPG, HEC, CMHPG	Short fracture, low temperature
	Crosslinked fluids	Crosslinked + GUAR, HPG, CMHPG, CMHEC	Long fracture, high temperature
Foam	Water-based foam	Water and foamer + N_2 or CO_2	Low-pressure formation
	Acid-based foam	Acid and foamer + N_2	Low-pressure, water-sensitive formations
	Alcohol-based foam	Methanol and foamer + N_2	Low-pressure formation with water-blockage problems
Oil	Linear fluids	Oil, gelled oil	Water-sensitive formations, short fractures
	Crosslinked fluids	Phosphate ester gels	Water-sensitive formations, long fractures
	Water-external emulsions	Water + oil + emulsifier	Good for fluid-loss control

Table 16.7—Fracturing fluids summary.

In addition to viscous stresses, these systems can demonstrate elastically generated stresses that act perpendicular (normal) to the direction of flow.

One goal of rheology is to mathematically represent the material properties responsible for these complex-flow phenomena, so that data collected in simple laboratory instruments can be generalized and used to predict performance under process conditions in more-complex geometries. The mathematical equation describing the relationship between stress and deformation is called a constitutive equation, or rheological equation of state.

Additionally, rheological relationships between gel-crosslink density and dynamic moduli can be used to interpret rheological data to aid in the characterization of gels. Rheological measurements are often the most-sensitive ways to characterize materials or to compare one class of materials with another.

16.5.2 Newtonian and Non-Newtonian Fluid Models. Flow in a hydraulic fracture, which must certainly be irregular, is often modeled as flow between smooth parallel plates, and flow in a wellbore is modeled as flow in a tube.

The simplest rheological model is the Newtonian model that has one constant parameter, a shear-rate-independent viscosity. Examples of Newtonian fluids include fresh water, salt water, acid, kerosene, diesel, produced oil, alcohol, and gases.

However, many fluids used in the oil and gas industry do not exhibit Newtonian behavior; that is, they cannot be characterized by the simple relationship between shear stress and shear rate. Those fluids that cannot be classified as Newtonian fluids are called non-Newtonian fluids. Fracturing fluids, drilling fluids, cementing slurries, emulsions, and polymeric solutions are a few examples of non-Newtonian fluids. Four general categories of non-Newtonian fluids are pseudoplastic, dilatant, Bingham plastic, and Herschel-Bulkley. Dilatant fluids are rare in the oil industry and, therefore, of limited interest. Most of the fluids used today for hydraulic fracturing treatments can be adequately described by pseudoplastic non-Newtonian fluid behavior. **Fig. 16.10** summarizes different kinds of fluids.

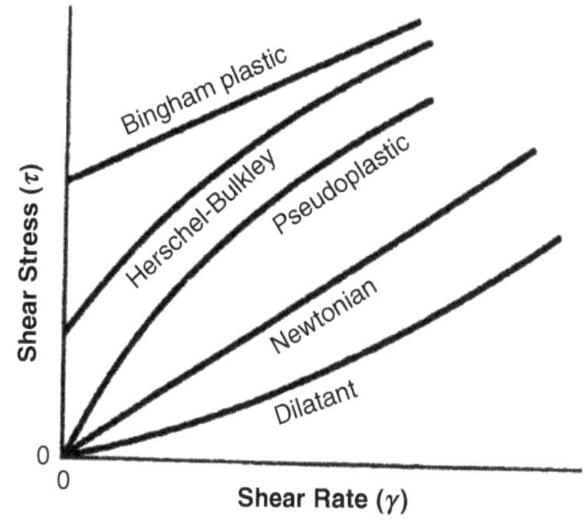

Fig. 16.10—Typical shear-stress/shear-rate relationships for fluids.

Various useful constitutive equations or models are presented in the literature to describe the behavior of pseudoplastic and dilatant fluids. Among these, the simplest and, hence, the most popular is the two-parameter power-law model. The three-parameter Ellis model describes fluid behavior for a wider range of shear rate than the power-law model. However, choosing a fluid model with more parameters and implementing it in the description of the flow geometry makes the mathematical analysis more cumbersome. Other models available in the literature have more than three parameters, but in most instances, they have very limited usefulness. Surprisingly, the literature offers more than 30 models or constitutive equations. The first choice, however, is still the power-law model.

Another model is the Bingham plastic model, which describes the behavior of fluids having a yield stress. The most common examples of Bingham plastic fluid are drilling mud, cement slurries, and foam fracturing fluids. The Herschel-Bulkley model also describes the behavior of fluids having a yield stress. It satisfies the condition of a yield point, similar to Bingham plastic, and also ties in the nonlinear-viscosity feature of the power-law-model fluid, describing the behavior of many yield pseudoplastic fluids. However, it is not possible to derive an explicit relationship between the shear rate and the volumetric flow rate in a pipe or an annulus.

16.6 Measurement of Fracturing Fluid Rheology

For steady-shear flow, the relationship between stresses and velocity gradients can be specified to define material functions (also called viscometric functions). The shear stress is related to the shear rate by the viscosity, and the normal stresses are related to the shear rate squared by the primary and secondary normal-stress coefficients. The material functions can be obtained by a number of experimental geometries. The choice of geometry is usually governed by such considerations as the amount of material available, the shear-rate range to be studied, and available equipment.

The following is a discussion of various instruments used to measure rheological properties of fracturing fluids and their relative advantages and disadvantages.

16.6.1 Rotational Viscometers. Rotational viscometers can be used with several different geometries: couette (concentric cylinders), cone and plate, or parallel plate. The advantage of the narrow-gap couette and the cone-and-plate geometries is that the shear rate throughout the fluid is uniform. The viscosity can be directly calculated from the measured torque and known geometry. With the parallel-plate geometry, the viscosity also can be determined, but the calculation involves differentiation of the torque vs. angular-velocity data. In most instances, rotational instruments are limited to lower shear rates for solutions, because inertia causes secondary flows and the sample can be thrown out of the gap. However, Kramer et al. (1984) showed how to obtain data to 50,000 seconds using very small gaps between parallel plates.

The fracturing-fluid-flow behavior, or rheology, is studied most often with rotational viscometers with a concentric bob-and-sleeve. This couette-type configuration allows simple calculations and easy control of the shear rate and shear stress on the test fluid. Highly viscoelastic fluids must be carefully evaluated. Inaccurate results might be caused by slippage, by the Weissenberg effect, and by nonhomogeneous shear over the entire fluid volume, resulting in static plugs.

Rotational viscometers are widely used to provide rheological characterizations of fracturing fluids, because they are easy to operate, require small fluid samples, and are relatively inexpensive. Rotational viscometers appear to be ideal instruments for evaluating Newtonian fluids, viscous polymer solutions, and emulsions. However, they do not appear to be as reliable for evaluating highly viscoelastic gels (crosslinked polymer solutions) and foams. Elastic instabilities, such as spindle climbing and slippage, limit the effectiveness of these viscometers with fluids such as borate-crosslinked gels. Because these viscometers typically are limited to pressures of approximately 1,000 psi (6,898 kPa) and less, they cannot effectively be used with foams at elevated temperatures.

Numerous examples can be cited showing the use of rotational viscometers to characterize the rheology of fracturing fluids. Researchers have combined the use of two rotational viscometers, a capillary viscometer, and laser Doppler anemometry to characterize the rheology of a 0.48 wt% HPG solution. Others have used transparent viscometer cups for flow-field visualization during their investigations of crosslinked fluids.

Gdanski et al. (1991) used a viscometer to determine the effects of crosslinking shear rate and gap width on the rheology of a delayed-crosslink HPG fluid. Flow-through viscometers have been used at the wellsite as part of a quality-control program. The rheometrics pressure rheometer (RPR) has been used to perform both steady- and oscillatory-shear measurements with crosslinked fluids. The RPR has been used to collect rheological data with crosslinked HPG solutions containing CO_2. Using shear-history simulation, it is possible to obtain reproducible viscosity measurements from rotational viscometers (RPR) and a reciprocating capillary device.

16.6.2 Pipe Viscometer. While rotational viscometers previously described are limited to laminar-flow evaluations, pipe viscometers can be used to describe both laminar- and turbulent-flow behavior. Time-dependent flow behavior is easily evaluated using a pipe viscometer by measuring pressure loss over a flow-rate range that includes both laminar- and turbulent-flow regimes. Time- and energy-dependent crosslinked fracturing fluid evaluations require simulation of the field-mixing process and the shear-history experience in the wellbore and fracture. Centrifugal and positive-displacement pumps used in pipe-viscometer operations can be used to simulate the continuous crosslinker addition that occurs in the field-mixing process.

To simulate shear history, the pipe viscometer is placed in a closed-recirculation mode. In this mode of operation, the fluid initially is in turbulent flow to simulate wellbore traverse, then it is placed in a simulated fracture flow by reducing the recirculation rate. While the fluid is in the closed-recirculation-flow mode, pressure loss is recorded as a function of time to characterize both wellbore- and fracture-flow behavior. The principal disadvantage of using the closed-recirculation-flow mode for shear-history simulation is that the fluid makes repeated passes through a pump in which the shear-history contribution is poorly defined. Overall, principal disadvantages of pipe-viscometer usage are large fluid-volume requirements, equipment size, and operational complexity.

Several investigators have designed, assembled, and operated pipe viscometers for various fluid-characterization tasks:

- Evaluating the thermal stability of various crosslinked fluids at temperatures up to 245°F
- Characterizing various rapid- and delayed-crosslink fracturing fluids
- Examining shear and proppant effects on crosslinked fluid viscosity
- Conducting tests for crosslinked and uncrosslinked fluids
- Simulating field crosslinker additions, turbulent flow through the wellbore and perforations, and laminar flow in the fracture
- Characterizing the rheology of N_2 and CO_2 foams
- Determining the rheological properties of N_2 foams under simulated downhole conditions
- Evaluating mixed-gas (N_2 and CO_2) fracturing foams

16.6.3 Capillary Viscometers. Capillary viscometers generally are most useful for obtaining high-shear-rate data. Generally, larger volumes of fluid are required for capillary viscometers than for rotational viscometers. Data from a capillary viscometer can cover the range of shear rates that characterize process conditions, but cannot be used to determine other material functions that might be useful in polymer characterization. Capillary viscometers also have the advantage of being able to make measurements at elevated pressures and temperatures.

16.6.4 Oscillatory Measurements. Studying viscoelastic-fluid behavior, such as that exhibited by crosslinked gels, requires dynamic oscillatory measurements, which can be generated in couette, cone-and-plate, or parallel-plate geometries. For linear viscoelastic measurements, the linear regime must have been achieved by performing tests at several strain amplitudes. In the linear regime, the calculated regime and the calculated moduli are independent of strain amplitude. For polymer solutions, linear behavior is the rule, but for some dispersions and gels, it is the exception. These systems can display nonlinear behavior at strains as low as 0.1%, where the modulus of a polyacrylamide gel, which shows linear viscoelastic behavior, is contrasted with the behavior of an HPG gel, which shows structural breakdown under even small strains. In the linear regime, the moduli can be determined with the parallel-plate geometry without having to resort to differentiation, as is the case of steady-shear flow.

16.6.5 Proppant-Laden-Fluids Rheology. Measuring fluids containing proppant is difficult in both rotational and pipe viscometers. In rotational viscometers, solid particles can cause torque readings to fluctuate. As the proppant settles during a test and fills up the annular space between the bob and sleeve, torque significantly increases. This technique has been used to monitor relative proppant-suspension characteristics, but it is not useful in determining slurry rheologies. The risk of settled proppant forming a solid mass prevents the use of resin-coated proppants (RCPs). The pipes used in laboratory-scale pipe viscometers usually have diameters that are too small relative to the proppant grain diameter.

In addition, the closed-loop model with multiple passes through a pump could result in problems if solid particles are present. Because a fracturing fluid should be intimately mixed with proppant, it is important to know how the proppant and the fluid interact and how these interactions affect the slurry rheology. The interaction is both physical and chemical, especially when RCPs are used. RCPs have been reported to interfere with the fluid pH, titanate crosslinkers, and persulfate breaker, all of which greatly influence the fluid's rheological behavior. Therefore, compatibility testing between RCPs and crosslinked fracturing fluids is a critical part of the fluid design for a specific application.

Using a new helical-screw (HS) bob enables accurate measuring of the rheology of proppant-laden fluids. The HS bob produces crucial information about the interaction of proppants with fluids used during hydraulic fracturing operations. The HS bob was successfully modeled as a straight bob, enabling accurate calculation of rheological parameters. The technique gives engineers good insight into the way a hydraulic fracturing fluid transports and supports proppant (especially RCPs) into a created fracture. Because the HS bob keeps the proppant mixed with the fluid during tests, it prevents the formation of a consolidated proppant plug at the bottom of the test cup, as can occur when the proppant is resin coated.

The HS bob viscometer is useful for studying real-time interaction between RCPs and hydraulic fracturing fluids. The screw configuration can help suspend the proppant while providing enough surface area for a measurable torque reading. **Fig. 16.11** shows the HS bob.

16.6.6 Kinetic Rheological Modeling. A comprehensive rheological model of a complex fracturing gel prepared with a guar derivative and dual crosslinkers uniquely integrates the nonlinear kinetics of crosslinker chemistry with the in-situ changes in fluid rheology. The model accurately predicts the effects of shear rate, shear history, temperature, temperature/time history, pH, gel loading, crosslinker concentration, and breaker concentration on the apparent-viscosity function. The model predicts the early noncrosslinked, pseudoplastic behavior of the linear gel fluid, and then accurately predicts the rapid increase in apparent viscosity caused by the thermal- and shear-induced kinetic-crosslink reaction. The model also accurately predicts thermal thinning at higher temperatures and the exponential decline in apparent viscosity that occurs when the breaker chemistry begins to dominate.

Morgan et al. (1989) used a pseudo-first-order polymerization-reaction model to predict thermal-history effects of protein denaturation on the apparent viscosity of extruded molten proteins. They used dimensionless functions to combine the effects of thermal thinning, polymer concentration, moisture content, thermal history, and shear history. Dolan et al. (1989) used a similar approach to model the effects of starch gelatinization on the apparent viscosity function and developed the following generalized kinetic-viscosity model by expanding the approach of Morgan et al. (1989) to crosslinked polymer gels.

$$\eta = \eta_{ref} \gamma T \psi \Phi Z, \quad (16.1)$$

where η is apparent viscosity, η_{ref} is reference viscosity module, γ is shear-thinning module, T is thermal-thinning module, ψ is thermal-history module, Φ is shear-history module, and Z is breaker module.

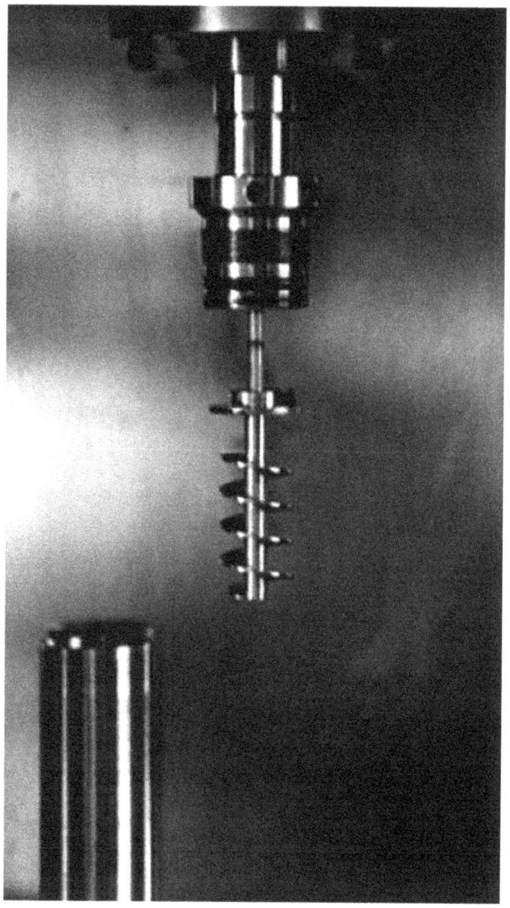

Fig. 16.11—HS bob mounted on the viscometer bob shaft.

16.6.7 Summary. In recent years, many different fracturing fluids have been developed. Some advancements have been made in characterization methods and testing procedures for these fluids. Significantly increased efforts need to be devoted to achieve a thorough understanding of their true behavior under field conditions. The fundamental knowledge of fluid behavior is key to better characterization of today's fracturing fluids. A part of the problem, however, is that no standard instrument can be used to simulate field conditions and better evaluate these fluids.

Several methods and procedures have been made available in recent years to predict friction pressures of fracturing fluids. Some of these are new and others are improvements over existing methods, but, overall, some progress has definitely been made.

16.7 Fluid-Leakoff Analysis

One primary design criterion is that the volume of fluid we pump into a fracture minus the volume of fluid that leaks off into the formation has to equal the fracture volume. At any point during the fracture treatment, the ratio of the volume of fluid still in the fracture divided by the total volume of fluid that has been pumped is considered the fluid efficiency. We can estimate the fluid efficiency using a fracture-propagation model. We can also estimate fluid efficiency in an injection/leakoff test or a minifracture test. After the pumps have been shut down and the fluid continues to leakoff, the fracture will close and virtually 90% of the fluid pumped into the fracture will leak off into the formation. The rate of leakoff mainly will be a function of the permeability of the formation exposed to the fracture. The rate will also be influenced by the viscosity of the fluid filtrate and fluid-loss additives used during the treatment.

Fracturing-fluid efficiency is an interesting parameter. Fracturing fluid efficiency mainly depends on reservoir permeability, fracturing fluid rheology, fracture pressure, reservoir pressure, and leakoff additives. If the fracturing-fluid efficiency at the end of pumping is 80 to 90%, then we might consider this a good number, because we created a lot of fracture volume during the treatment. However, in actuality, it might not be so good, because it probably means we are fracture treating a very-low-permeability reservoir and the flow rates of oil and gas will not be very high.

If the fracturing-fluid efficiency at the end of pumping is 10 to 20%, we might consider this a poor result, because we did not create much fracture volume. However, in practice, this might mean we have a very permeable reservoir, and we could expect the flow rates of oil and gas to be high.

In many reservoir situations, we like to see fluid efficiencies of 40 to 60% at the end of pumping. In this range, we should have a reasonable fracture volume, and we probably have reasonable formation permeability. In practice, however, it is difficult to have much effect on fracturing-fluid efficiency. The primary factor controlling efficiency is the formation permeability. We can reduce leakoff by adding fluid-loss agents. We can also reduce leakoff by pumping the treatment at the highest possible injection rate to shorten the time allowed for leaking. **Fig. 16.12** illustrates the concept of fracturing-fluid efficiency.

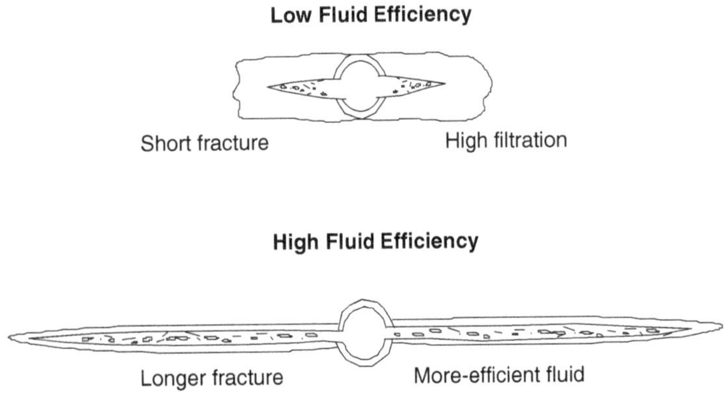

Fig. 16.12—Shorter fractures typically have lower fluid efficiency than longer fractures, often because of their higher level of filtration.

16.8 Fluid-Loss Equations

The first paper to explain the concept of leakoff as a fracture treatment design concept was written by Howard and Fast (1957). In fact, the Appendix of that paper described the leakoff process that was developed by Carter (1957). The following fluid-loss equations were derived by Carter.

$$C_v = 0.0469\left(\frac{k_i \Delta p \phi}{\mu_a}\right)^{1/2}, \quad (16.2)$$

$$C_c = 0.0374\Delta p\left(\frac{k_i C_r \phi}{\mu_r}\right)^{1/2}, \quad (16.3)$$

$$C_w = \frac{0.0164m}{A}, \quad (16.4)$$

where A is area of the filter-cake medium; C_c is fluid-loss coefficient controlled by the reservoir fluid viscosity; C_r is reservoir compressibility, psi^{-1}; C_v is fluid-loss coefficient controlled by fluid-filtrate viscosity, C_w is wall-building fluid-loss coefficient, k_i is formation permeability with respect to the fluid filtrate; k_r is formation permeability with respect to the reservoir fluid; m is slope of graph from laboratory measurements; Δp is pressure in the fracture minus pressure in the reservoir, psi; ϕ is formation porosity, fraction; μ_a is viscosity of the fluid filtrate, cp; and μ_r is reservoir fluid viscosity, cp.

Howard and Fast (1957) explained that leakoff can be controlled by any one of three factors. If the fracturing-fluid filtrate is very viscous, then the fluid will not leak off very much into the reservoir. On the other hand, if the reservoir fluid is viscous, it will be difficult for the fracturing-fluid filtrate to push and compress the reservoir fluid away from the fracture. And in the case in which the formation is permeable and neither fluid is particularly viscous, then we need to add a fluid-loss additive to build a filter (wall) cake along the fracture walls to reduce leakoff.

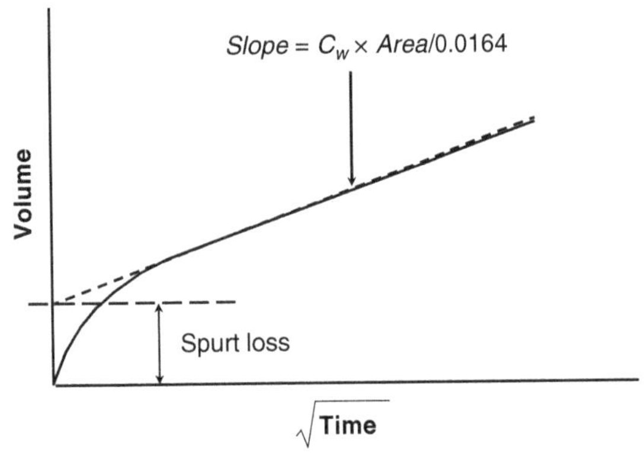

Fig. 16.13—Graph of fluid-leakoff volume during a typical fluid-loss test in the laboratory.

The method for measuring filter-cake leakoff effects is to run a fluid-loss test in the laboratory. We use equipment similar to what is used in testing drilling mud. A sample is put into a cell and a pressure drop is placed across a piece of filter paper or a slab of core. The amount of fluid that leaks out as a function of time is graphed, as shown in **Fig. 16.13.** Because the leakoff is controlled by linear flow, then a graph of leakoff volume as a function of the square root of time will be linear. We can use the equation of the line to predict leakoff in the fracture as a function of the amount of time any segment of the fracture is exposed to the fluid.

To be useful, the laboratory tests run to determine the spurt loss and fluid-loss coefficient should be run using fluids similar to those that will be pumped during the treatment, with cores from the formation if possible, at the reservoir temperature, and at the correct pressure differential. Standard tests are run with a 1,000-psi pressure differential across the filter medium. However, in practice, the pressure drop between the fracture and the formation can be several thousand psi, especially in tectonically active areas.

In low-permeability reservoirs, the fluid loss during a treatment will usually be small and the addition of fluid-loss additives might not be needed. The exception is the case in which the reservoir contains abundant natural fractures. In those cases, 100-mesh sand and a plastering fluid-loss agent might be needed to reduce leakoff down the natural fractures, enough to be able to pump the treatment without screening out. In high-permeability reservoirs, fluid-loss additives should always be used to improve fluid efficiency.

To illustrate how effectively fluid-loss additives can work, we have included **Fig. 16.14** from work published by (Gidley et al. 1989). The graph shows the wall-building fluid-loss coefficient, C_w, for various temperatures, as functions of both polymer-loading and fluid-loss additives. Notice that the fluid-loss coefficient rapidly increases as the temperature increases. Using more gelling agent or polymer will allow the fluid to be used at a higher temperature. When a fluid-loss additive is used, the fluid-loss coefficient is substantially decreased, especially at the higher temperatures.

Information in Fig. 16.14 is just one example of data that can be collected in the laboratory. The service companies have spent countless hours in the laboratory over the past 40 years making such measurements for a variety of fluids, additives, gels, core samples, and temperatures. It is important to work with the service companies to determine the best value of C_w to use in any particular design.

Because the leakoff process is a sequential-flow process, the leakoff coefficients can be averaged in a series-average method to determine the value of C_t, which is the total fluid-loss coefficient. Thus, using Eqs. 16.2 through 16.4 to compute C_v and C_c and then using data from a laboratory test to estimate C_w, one can compute an estimate of C_t using either of the equations that follow. The first equation is a simple series-average equation; Gidley et al. (1989) proposed the second equation:

$$C_t = \frac{1}{\frac{1}{C_v} + \frac{1}{C_c} + \frac{1}{C_w}}, \quad \text{(16.5)}$$

$$C_t = \frac{1C_c C_v C_w}{C_v C_w + \left[C_w^2 C_r^2 + 4C_c^2\left(C_v^2 + C_w^2\right)\right]^{1/2}}. \quad \text{(16.6)}$$

The design engineer will enter data to compute C_v and C_c; C_w is a typical value based on laboratory tests. In the design program, data will be entered to describe the layers of rock exposed to the fracturing fluid during the treatment. Some of the layers will be permeable and accept fluid loss; some will not be permeable and will not affect the leakoff volume. So, in a computer-designed model, the value of C_t should be applied only to the percentage of the fracture face exposed to permeable rock.

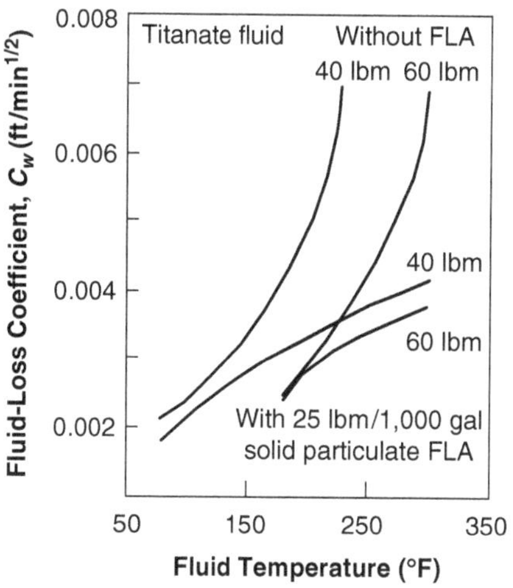

Fig. 16.14—Example of the data that can be collected in the laboratory with and without fluid-loss additives (FLA).

16.8.1 Net-to-Gross Fracture-Height Estimates for Leakoff Volume. The design engineer should always review the well log for the formation to be fracture treated. It is important to estimate the gross fracture height from logs, as well as let the fracture-design model predict the height growth on the basis of in-situ stresses and fracturing pressures. The design engineer should also estimate the amount of the gross height permeable enough to accept leakoff. We call this permeable thickness the net thickness. Logs can then be used to estimate the net-to-gross ratio for purposes of computing the correct estimate of leakoff volume.

16.8.2 Field Measurement of Leakoff Volume. One of the most important aspects of designing and pumping the optimal fracture treatment is to be certain we use the correct value of fluid-loss coefficient in the fracture-propagation model. Often, we have to estimate many of the variables as we are designing the treatment. Fortunately, the industry has developed a good method for computing fluid-loss efficiency from injection data. We can use the method to verify the fluid-loss assumptions used in the design of the treatment. If there is a large discrepancy, we can redesign the treatment before it is pumped to maximize the chance of success. Many different methods to compute fluid efficiency have been published. Nolte (1986) introduced the first method.

The data in **Fig. 16.15** are typical of what would be measured during a fracture treatment. Similar data can also be measured during a minifracture treatment. There are three major areas of interest as a function of time. Area one is when the fracture is being propagated. During this time, we see the full fracture height, and we are growing the fracture. At the end of the pumping period, the pressure begins to decline, and we measure the change in pressure vs. time to determine the time at which the fracture closes. These data yield estimates of near-wellbore pressure, net pressure, and fracturing-fluid efficiency.

Nolte (1986) published a method to compare the closure time to the pumping time, to estimate fracturing fluid efficiency (Gidley et al. 1989).

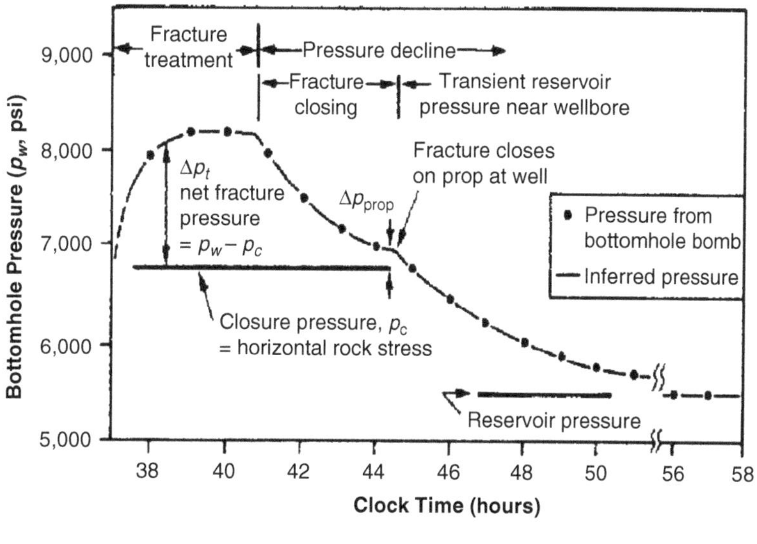

Fig. 16.15—Typical data measured during a fracture treatment.

16.9 Nomenclature

A	=	area of the filter-cake medium, ft^2
C_c	=	fluid-loss coefficient controlled by the reservoir fluid viscosity, ft/$\sqrt{\text{min}}$
C_r	=	reservoir compressibility, psi^{-1}
C_t	=	total fluid-loss coefficient, ft/$\sqrt{\text{min}}$
C_v	=	fluid-loss coefficient controlled by fluid-filtrate viscosity, ft/$\sqrt{\text{min}}$
C_w	=	wall-building fluid-loss coefficient, ft/$\sqrt{\text{min}}$
k_i	=	formation permeability with respect to the fluid filtrate, md
k_r	=	formation permeability with respect to the reservoir fluid, md
m	=	slope of graph from laboratory measurements
p_c	=	closure pressure, psi
p_w	=	bottomhole pressure, psi
T	=	absolute temperature, K
Z	=	breaker effect, dimensionless
Δp	=	pressure in the fracture minus pressure in the reservoir, psi
Δp_{prop}	=	net fracturing pressure when fracture closes on proppant pack
Δp_f	=	net fracturing pressure
γ	=	shear rate, seconds^{-1}
ψ	=	integral temperature/time history function, dimensionless
Φ	=	integral strain history function, dimensionless
ϕ	=	formation porosity, fraction
η	=	apparent viscosity, Pa·s
η_{ref}	=	reference viscosity module, Pa·s
μ_a	=	viscosity of the fluid filtrate, cp
μ_r	=	reservoir-fluid viscosity, cp

16.10 References

Armstrong, K., Card, R., Navarrete, R. et al. 1995. Advanced Fracturing Fluids Improve Well Economics, *Oilfield Review* **07** (03): 34–51.

Brannon, H. D., Wood, W. D., and Wheeler, R. S. 2006. Large Scale Laboratory Investigation of the Effects of Proppant and Fracturing Fluid Properties on Transport. Paper presented at the SPE International Symposium and Exhibition on Formation Damage Control, Lafayette, Louisiana, USA, 15–17 February. SPE-98005-MS. https://doi.org/10.2118/98005-MS.

Burnham, J. W., Harris, L. E., and McDaniel, B. W. 1980. Developments in Hydrocarbon Fluids for High-Temperature Fracturing. *J Pet Technol* **32** (02): 217–220. SPE-7564-PA. https://doi.org/10.2118/7564-PA.

Carter, E. D. 1957. Appendix to Optimum Fluid Characteristics for Fracture Extension, In *Drilling and Production Practices*, ed. G. C. Howard and C. R. Fast, **24:** 261–270. New York City: American Petroleum Institute.

Coronado, J. A. 2007. Success of Hybrid Fracs in the Basin. Paper presented at the Production and Operations Symposium, Oklahoma City, Oklahoma, USA, 31 March–3 April. SPE-106758-MS. https://doi.org/10.2118/106758-MS.

Craig, D. and Holditch, S. A. 1992. The Degradation of Hydroxypropyl Guar Fracturing Fluids by Enzyme, Oxidative, and Catalyzed Oxidative Breakers. Topical Report. Southwest Petroleum Short Course, Lubbock, Texas.

Dolan, K. D., Steffe, J. F., and Morgan, R. G. 1989. Back Extrusion and Simulation of Viscosity Development During Starch Gelatinization. *Journal of Food Process Engineering* 11: 79–101. https://doi.org/10.1111/j.1745-4530.1989.tb00023.x.

Economides, M. and Nolte, K. G. 1989. *Reservoir Stimulation.* Second edition. Chichester, West Sussex, UK: John Wiley & Sons, Ltd.

Ely, J. W. 1990. Fracturing Fluids and Additives. In *Recent Advances in Hydraulic Fracturing*, ed. J. Gidley, Chap. 7, 131–146. Richardson, Texas: Society of Petroleum Engineers.

Gdanski, R. D., Harris, P. C., Lord, D. L. et al. 1991. Analysis of Fann Model 50 for Fracturing Fluid Rheology. Paper presented at the SPE Production Operations Symposium, Oklahoma City, Oklahoma, 7–9 April. SPE-21647-MS. https://doi.org/10.2118/21647-MS.

Gidley, J. L., Holditch, S. A., Nierode, R. W. et al. 1989. *Recent Advances in Hydraulic Fracturing*. SPE Monograph Series, Vol. 12. Richardson, Texas: SPE.

Harris, P. C. 1989. Effects of Texture on Rheology of Foam Fracturing Fluids. *SPE Prod Eng* **4** (03): 249–257. https://doi.org/10.2118/14257-PA.

Harris, P. C. and Reidenbach, V. G. 1987. High-Temperature Rheological Study of Foam Fracturing Fluids. *J Pet Technol* **39** (05): 613–619. https://doi.org/10.2118/13177-PA.

Holditch, S. A., Robinson, B. M., Ely, J. W. et al. 1991. The Effect of Viscous Fluid Properties on Excess Friction Pressures Measured During Hydraulic Fracture Treatments. *SPE Prod Eng* **6** (01): 9–14. SPE-18208-PA. https://doi.org/10.2118/18208-PA.

Holditch, S. A., Xiong, H., Rueda, J. et al. 1993. Using an Expert System to Select the Optimal Fracturing Fluid and Treatment Volume. Paper presented at the SPE Gas Technology Symposium, Calgary, Alberta, Canada, 28–30 June. SPE-26188-MS. https://doi.org/10.2118/26188-MS.

Howard, G. C. and Fast, C. R. 1957. Optimum Fluid Characteristics for Fracture Extension. Paper presented at the American Petroleum Institute Drilling and Production Practice, New York, New York, 1 January. API-57-261.

Kramer, J., Prud'homme, R. K., and Chu, A. 1984. Rheological Characterization of Fracturing Fluids. Paper presented at the American Petroleum Institute Annual Meeting Papers, Division of Production, New Orleans, Louisiana, 1–4 April. API-84-G001.

McGowen, J. M., Vitthal, S., Parker, M. A. et al. 1993. Fluid Selection for Fracturing High-Permeability Formations. Paper presented at the SPE Annual Technical Conference and Exhibition, Houston, Texas, 3–6 October. SPE-26559-MS. https://doi.org/10.2118/26559-MS.

Morgan, R. G., Steffe, J. F., and Ofoli, R. Y. 1989. A Generalized Viscosity Model for Extrusion of Protein Doughs. *Journal of Food Process Engineering*, 11: 55–78. https://doi.org/10.1111/j.1745-4530.1989.tb00022.x.

Nolte, K. G. 1986. Determination of Proppant and Fluid Schedules From Fracturing-Pressure Decline. *SPE Prod. Eng.* **1** (04): 255–265. SPE-13278-PA. https://doi.org/10.2118/13278-PA.

Rae, P. and Di Lullo, G. 1996. Fracturing Fluids and Breaker Systems - A Review of the State-of-the-Art. Paper presented at the SPE Eastern Regional Meeting, Columbus. Ohio, 23–25 October. SPE-37359-MS. https://doi.org/10.2118/37359-MS.

Samuel, M. M., Card, R. J., Nelson, E. B. et al. 1997. Polymer-Free Fluid for Fracturing Applications. Paper presented at the SPE Annual Technical Conference and Exhibition, San Antonio, Texas, 5–8 October. SPE-59478-MS.

Sherman, J. B. and Holditch, S. A. 1991. Effect of Injected Fracture Fluids and Operating Procedures on Ultimate Gas Recovery. Paper presented at the SPE Gas Technology Symposium, Houston, Texas, 22–24 January. SPE-21496-MS. https://doi.org/10.2118/21496-MS.

Siddiqui, M. A., Nasr-El-Din, H. A., Al-Anazi, M. S. et al. 2004. Formation Damage in Gas Sandstone Formations by High-Temperature Borate Gels Due to Long Term Shut-in Periods. Paper presented at the SPE/DOE Symposium on Improved Oil Recovery, Tulsa, Oklahoma, 17–21 April. SPE-89476-MS. https://doi.org/10.2118/89476-MS.

Tulissi, M. G. and May, R. E. 2002. A Comparison of Results of Three Different CO2 Energized Frac Fluids: A Case History. Paper presented at the SPE Gas Technology Symposium, Calgary, Alberta, Canada, 30 April–2 May. SPE-75681- MS. https://doi.org/10.2118/75681-MS.

Chapter 17

Types of Fracture Propping Agents

Stephen A. Holditch

17.1 Introduction

The function of a propping agent is to hold the fracture open after the fluid injection is stopped and the fracturing fluid has been removed or has leaked off into the reservoir. Proppants are sized particles that hold fractures open after a hydraulic fracturing treatment. In addition to naturally occurring sand grains, man-made or specially engineered proppants, such as resin-coated sand or high-strength ceramic materials such as sintered bauxite, also can be used. Proppant materials are carefully sorted for size and sphericity to provide an efficient conduit for production of fluid from the reservoir to the wellbore.

A well proppant-packed fracture should have a much higher permeability than the reservoir rock, having linear flow and negligible pressure drop through the fracture to the wellbore, giving rise to increased well productivity. The reservoir can now flow from the extremities of the fracture to the wellbore through the highly conductive proppant pack. The effectiveness of the propping agent has great bearing on the success of a fracture treatment.

Sand has been used as a propping agent since 1947. Sand has been the most widely used propping agent because it is inexpensive, readily available, and chemically inert. Sand used for fracturing is mined from shallow depths. Ceramic proppants were developed to meet the need for proppants that provide high fracture-flow capacity in multilayer placement and can withstand high closure pressure in formations.

Sintered bauxite—a ceramic—is a much stronger proppant used in high close-stress applications. It has a higher density than sand and, therefore, it is not as easy to transport in the fracturing fluids. Intermediate-density proppants (IDPs) have lower densities than sintered bauxite. IDPs are therefore less-expensive alternatives to bauxite for intermediate wells. In resin-coated proppants (RCPs), resin is generally applied to proppant to increase its load-carrying capacity and also to prevent proppant flowback.

According to Gidley et al. (1989), commercial proppants used in hydraulic operations include sand, sintered bauxite, IDPs, and RCPs. Sand was the earliest proppant used and is by far the most popular proppant used in wells with low closure stress. Sands are available in different mesh sizes and have specified American Petroleum Institute (API) qualities.

Over the past decade, several proppants have been developed and tested in hydraulic fracturing operations. These proppants either are different from or are modifications of previously developed proppants. These include non-API sand, surface-modification-agent (SMA) proppants, and ultralightweight (ULW) proppants. **Fig. 17.1** shows the evolution of proppant types.

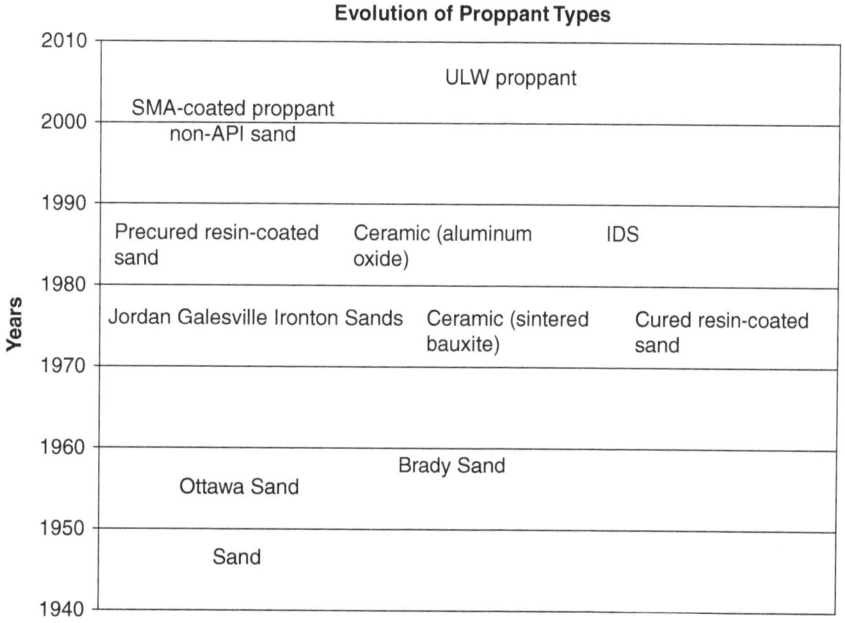

Fig. 17.1—Evolution of proppants.

17.2 Commonly Used Proppants in Industry

Propping agents can be classified as naturally occurring (e.g., sand varieties), commercial (e.g., ceramics), or a combination of naturally occurring and commercial (e.g., resin-coated sands). Sand is the most commonly used propping agent in North America. However, outside of North America, bauxite and ceramic proppants are most commonly used during a fracturing treatment. For areas without an adequate source of API-quality sand, the cost to ship propping agents to where they are needed generally equals or exceeds the cost to mine or manufacture the propping agent on-site.

17.2.1 Sand. Sand used for fracturing is mined from shallow deposits, which include ancient river beds, aeolian dunes, and other depositional environments. Several sand varieties are used in the US and sometimes are exported for treatments in which sand is the most optimized and cost-effective propping agent. These sand types include premium sand, such as Ottawa or Jordan, and lower-strength sands, such as Brady. The names reflect the location where the sand is mined. API has set the standards for fracture sand.

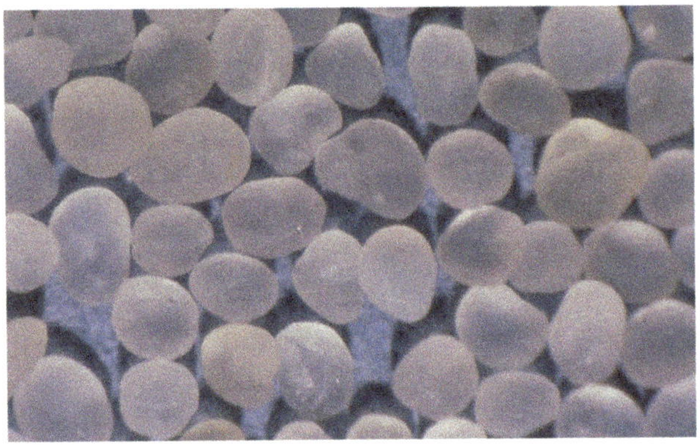

Fig. 17.2—Ottawa Sands are smooth, uniform, and generally white.

Fig. 17.3—Brady Sands are coarse, faceted, and generally dark.

Fig. 17.4—Intermediate-density proppant has a lower density and specific gravity than its bauxite source.

17.2.1.1 Premium Sand. Ottawa or Jordan Sand is the general name for fracturing sand mined from deposits found near Ottawa, Illinois. These sands are considered by many to be the highest-quality fracturing sand in North America.

Ottawa Sands are characterized by high purity, whiteness (although some variations in color does occur), high roundness and sphericity, and lack of dust **(Fig 17.2).** Ottawa Sands are monocrystalline (single crystal phase), which results in individual grain strength exceeding nearly all other sands.

Ottawa Sands are mined primarily from the St. Peter Formation near Ottawa, Illinois, and the Jordan Sand is mined from the Jordan Formation in south central Minnesota. These sands were formed from quartz freed by the erosion of granite in the Canadian Shield. The eroded sand grains accumulated on the beaches of an ancient Cambrian Sea, where they were repeatedly washed, sifted, sorted, and resorted by tides and winds over millions of years before they were covered and protected from erosion and contamination. This repeated washing and sorting yielded deposits particularly suited for use as fracturing sand.

17.2.1.2 Brady Sands. Brady Sand is mined from the Hickory Formation that outcrops near Brady, Texas. Brady Sand is slightly darker in color than Ottawa Sands; hence, "brown sand" is a name often used when referring to Brady Sand. These Sands are characterized by their impurities, color, and angularity **(Fig. 17.3).** Brady Sands are polycrystalline; that is, each whole grain is composed of quartz crystals bonded together into one unit, leaving cleavage planes in the whole grain. The deposits contain impurities such as feldspars and clays, but the sands are washed thoroughly during processing to create a high-purity product.

17.2.1.3 Other Sands Used. Results of an initial series of tests on various-sized subsamples of Manigotagan Silica Sand (mined in Winnipeg, Canada) have exceeded all of the minimum standards for fracture sand used by the oil and gas industry. The Manigotagan Sand could become a viable alternative to other fracturing sands, particularly with increasing demand for sand proppants.

Silica Sands mined in Australia and the United Kingdom are also being used as fracturing sands in stimulation treatments. The Badger Sand from Poland is very similar to Ottawa Sand and has been used frequently.

In general, fracturing sands are used at net closure stresses below 6,000 psig. The proppant must adhere to strong rules and requirements, as stipulated by API.

17.2.2 Intermediate-Strength Proppant. Intermediate-strength (density) proppant (ISP) was developed to produce a strong propping agent not as dense as sintered bauxite. It is manufactured using a less-dense form of aluminum oxide called mullite. The proppant is brownish, with lower bulk density and specific gravity than sintered bauxite **(Fig. 17.4).** The proppant particles have good resistance to corrosion and particle shape (good roundness and sphericity). ISPs are very useful in intermediate well depth.

17.2.3 Sintered Bauxite. Sintered bauxite is an inert, high-strength ceramic proppant **(Fig. 17.5)** made from aluminum ore. Sintered bauxite is particularly useful in high-closure-stress wells in which high crush resistance is necessary to enable good fracture conductivity. This is a result of the proppant's ability to be only partially crushed under high stress (good elastic properties). A drawback to this type of proppant is the high density that reduces its ability to be transported long distances.

Fig.17.5—Sintered bauxite resists stress but can be transported over only short distances into the formation.

17.3 Resin-Coated Proppants

To increase the strength of a proppant and to keep the proppant in place, the propping agent can be coated or encapsulated with resin. The coating on the surface of the proppant grains controls movement of the grains and dispersion of particulate fragments or particulate fines, should the grain inside the coating crush under increased stress on the grain. Controlling the migration of particulate fines enhances formation conductivity by preventing free particulate fines from plugging interstitial flow passages therein. Proppants that normally disintegrate upon crushing benefit most from coating. Coating or encapsulating a proppant generally improves its overall downhole stability.

RCPs improve well-stimulation results by preventing flowback, or fracture evacuation, near the wellbore. In addition, the deformable coating protects the RCPs from crushing and helps resist embedment in softer formations. The tough coating is chemically inert in crude oils, brines, and most acids **(Fig. 17.6).** The resin-coated materials work best under actual downhole conditions, because the closure stress and temperature help lock the materials in the fracture for the life of the well.

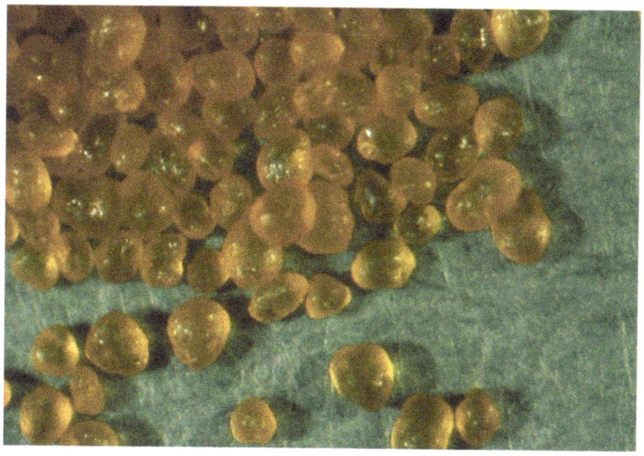

Fig. 17.6—Resin-coated sands resist stress and are easier to control.

17.3.1 Resin-Coated Sands.

17.3.1.1 Curable Resin-Coated Sands. This type of proppant uses the application in which the curable coating bonds the sand grains together after they are placed in the fracture. This application is very useful in preventing flowback. Curable resin coating is not limited to sand; other proppants such as sintered bauxite and IDPs can be used.

17.3.1.2 Precured Resin-Coated Sands. These are similar to cured resin-coated sand; however, the resin coating is cured by heating the coating during the manufacturing process. The resin coating is helpful in preventing fines migration of the crushed fines during manufacture. It is very resistive to destruction from hot formation brines and crude oils up to 150°F.

17.3.2 Surface-Modification-Agent-Coated Proppants. SMA proppants are modified proppants developed to address the need for sustained conductivity. Rapid loss of fracture conductivity after hydraulic fracturing has been a frequent occurrence. This loss has been attributed to several factors, such as generation of porosity-filling materials through geochemical-type reactions between the proppant and the formation, generation of fines from proppant crushing, and migration of formation fines (Weaver et al. 2006).

SMA proppants improve conductivity resulting from hydraulic fracturing treatments that use water-based fluids. Directly applied onto the dry proppant at the wellsite before it is added to the fracturing fluid, SMAs chemically alter proppant surfaces to dramatically increase the intergranular surface friction of individual proppant grains **(Fig. 17.7).** Proppant grains coated with the SMA achieve better proppant-pack porosity and permeability at reservoir closure pressures as great as 4,000 psi.

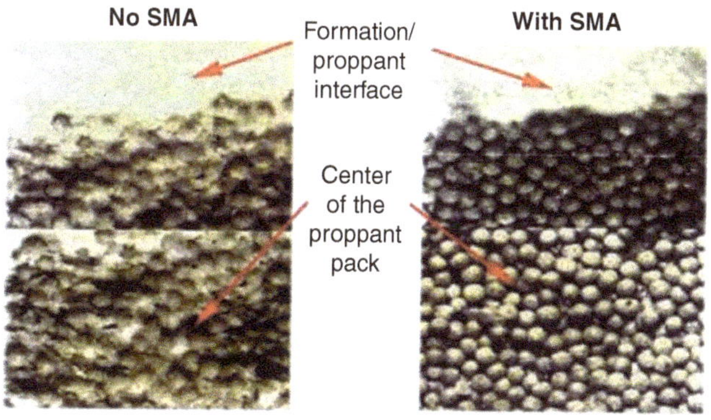

Fig. 17.7—SMA-coated proppants increase intergranular surface friction, porosity, and permeability within the fracture.

Fig. 17.8—ULW proppants have low specific gravity but high potential for excellent performance.

Variable	API Sand	Resin-Coated Sand	Ceramic High-Strength Bauxite
	USD320/m³	$1,200/ m³	$3,000/ m³
Average cost	$200/t	$750/t	$1,500/t
	$.09/lbm	$.34/lbm	$.75/lbm
Shape	cubical	irregular	round
Conductivity	limited	improved	best
Closure Stress, psi	5,000	8,000	15,000
Bulk Density, g/cm³	1.6	1.6	2.0
Particle Density, g/cm³	2.65	2.65	3.56
Temp. Range, °F	60–175	70–375	175–450

Table 17.1—Comparison of different properties for key proppants used in stimulation treatment.

In addition, proppant grains coated with SMA stick together better in the pack and exhibit little or no movement in response to high flow rates and extreme cyclic loading. This improves the stability of both the proppant pack and the pack/formation interface, preventing formation material from entering the proppant bed and better sustaining effective fracture width and fracture conductivity in high-stress reservoirs. SMAs also can be used with RCPs (Nguyen et al. 1998).

17.3.3 ULW Proppant. ULW proppant is a new product, boasting a specific gravity of < 2.0 **(Fig. 17.8).** The potential of this material in fracturing applications is truly staggering, and it is believed that many of the industry standard procedures and processes potentially are made obsolete by the application of ULW proppant.

Lightweight proppants have much lower densities than conventional proppants, while still possessing sufficient particle strength. They, therefore, have much lower settling velocity and better suspension capability. This allows for increased propped length and greater conductivity.

Wood et al. (2003) tested walnut and ceramic ULW proppants. These are manufactured through a two-step process of impregnating the material with epoxy resin and then coating with phenol or other resins. Walnut ULW proppant has a bulk density of 0.85 g/cm3 and could withstand up to 6,000-psi closure pressure at 175°F. Ceramic-based ULW proppants, on the other hand, have a bulk density of 1.25 to 2.0 g/cm3 and can withstand a closure stress greater than 8,000 psi and a bottomhole temperature of 275°F. ULW proppants have been used the Barnett Shale, Appalachian tight gas sandstones, and the Big Sandy Field in Kentucky.

17.3.4 Summary. Table 17.1 summarizes different properties for key proppants used in stimulation treatment. These include cost, shape, fracture conductivity, closure stress, bulk density, particle density, and working temperature range. This table gives useful information that could be used in selecting a proppant for a fracturing treatment.

17.4 Proppant Flowback Considerations

Proppant flowback is a critical issue surrounding many hydraulic fracturing treatments in today's hydrocarbon recovery efforts. The common occurrence of this problem has led to a variety of attempted treatments and innovations, all of which seek to remediate or mitigate the backflow of proppants out of formation fractures and into wellbores and production equipment, where they frequently cause a multitude of issues from erosion to loss of conductivity. In nearly any situation, the displacement of proppants resulting from flowback is both costly and damaging from a production standpoint. This section describes many of the reasons for proppant flowback, the problems inherent in this phenomenon, and the methods currently used to mitigate the flowback of proppants placed during or after fracture treatments.

Proppant flowback, first reported in the 1960s (Di Lullo and Rae 2001), is the formation's release of granules through production or cleanup activity (Aidagulov et al. 2007). The phenomenon is frequently seen in the field and can negate positive effects entailed in expensive fracturing jobs. Back production of proppant from fractures can be a continuous operational nuisance throughout the life of a well (Romero and Feraud 1996). For this reason, over the past 45+ years, proppant flowback has been studied in relation to production loss.

17.4.1 Proppant Flowback Problems. The loss of proppant materials from hydraulically created fractures is an area of concern for several key reasons. Among these is the consistent erosion and damage inherently caused by a fluid with solid additives being produced within a wellbore (Aidagulov et al. 2007). The abrasion that takes place while proppants are released from the fracture is known for its ability to damage surface equipment and pumps (Vincent et al. 2004). Rapid flowback after a fracturing treatment can necessitate a lengthy cleanup process, which is costly because of lost production and process considerations.

Perhaps the most concerning problem exhibited by the flowback phenomenon is the loss in fracture conductivity as the proppants begin to exit the fracture during production, leaving less proppant to support the fracture and/or resulting in crushed proppant grains (Aidagulov et al. 2007). The lack of proper support for the fracture reduces the capability for the process to result in increased production, and it has been found that proppant flowback becomes a monetary concern more often than not from a well-performance standpoint. Production is not only hurt by lack of fracture conductivity as proppant material is lost, but it is also affected by the clogging of perforations by the fines generated from proppant crushing and subsequent flowback.

If proppant flows back and is produced at the surface, it can cause severe damage to downhole rod pumps, electrical submersible pumps, chokes, pipelines, and storage facilities (Ingram et al. 2007). These occurrences all tend to raise the cost of maintenance and usually result in loss of production from the well being shut in.

*17.4.1.1 **Failure Quantification.*** Several factors have been shown to affect the stability of the proppant pack when fracture treatments take place. These considerations include, but are not limited to, the original fracture width, closure stress of the formation, pressure drawdown, and fluid properties of the proppant mixture and formation (Aidagulov et al. 2007). Proppant grains form arches across the fracture, which can be seen in **Fig. 17.9.**

With proppant flowback, the most critical concern is the stability of this arch system, of which the breakdown caused by high flow rate or weak support from neighboring particles results in a failed or crippled fracturing treatment (Aidagulov et al. 2007). The three main forces incumbent upon the outermost compressive arch are the fracture-produced force holding the arch in place, the frictional forces between grains, and the viscous drag force produced by fluid flowing out of the fracture face. The latter can become destructive and cause an inward migration of the arch, leaving the outtermost proppant grains unsupported. The greatest area of material failure takes place in the fracture mouth, seen on the right side near the arch in Fig. 17.9. Being able to determine the rate and duration at which the proppant will be produced from a fracture is critical in choosing the proper completion method.

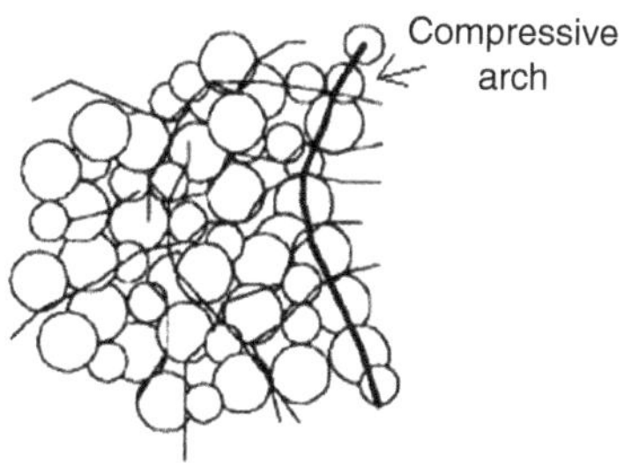

Fig. 17.9—Proppant grains in a fracture compressive arch that resists breakdown as fluids flow through it (Romero and Feraud 1996).

17.4.2 Treatments. Several solutions have been proposed to resolve the issues stemming from proppant flowback. Many recent advances have led to some highly effective and economical solutions.

*17.4.2.1 **RCPs.*** Introduced in the early 1980s, RCPs have been implemented in the oil field with the highest rate of success of all treatments and were one of the very first introduced solutions to the proppant flowback issues (Di Lullo and Rae 2001; Ingram et al. 2007). This treatment involves coating propping sands with phenolics, which are curable through temperature modification only once and which are pumped into fractures and subsequently hardened to increase proppant-pack stability. This is achieved by placing the RCPs at low temperature, increasing the temperature of the fracturing fluid, and allowing the resin on each particle to bridge between sand particles. As it turns out, this activity reduces the porosity in the fracture, but this negative effect is offset by the lack of damage caused by proppant flowback.

More-successful RCP treatments have been achieved through use of large tail-in stages, because this activity prevents the otherwise high flowback rate when smaller tail-in stages have been implemented (Di Lullo and Rae 2001). Additionally, deformable isometric particles (DIPs) have been included at approximately 5 to 30 wt% of proppant. One important property of DIPs includes deformability when exposed to pressure or temperature extremes, which allows them to redistribute loads in and around the proppant pack. This enables avoidance of fracture-width loss from embedment and precludes the generation of fines because of the lowered point-to-point contact forces.

RCPs have some drawbacks in that they do not completely remediate proppant flowback in many cases. This occurs because the bonded grains of RCP do not always have the strength required to withstand drag forces in the formation (Ingram et al. 2007). Naturally, propping agents that have been modified create a greater expense for the operator, which is why entire treatments are not performed using RCPs. Di Lullo and Rae (2001) estimated that 215,000 tons of RCP material was used to support hydraulic fractures through the year 2000.

*17.4.2.2 **SMAs.*** SMAs are liquid additives applied to hydraulic fracturing fluid as needed during pumping (Terracina et al. 2001). These agents coat the proppant grains with a thin, tacky substance that enables greater vertical distribution of the propping agent within the fracture and increases the porosity of the proppant pack. Much like RCPs, SMAs increase cohesion between proppant grains without the need for pressure- or temperature-based curing.

The SMA material itself is composed of water and an oil-insoluble polymer that creates a very stable molecule with nearly no solubility in standard well fluids (Terracina et al. 2001). Proppant packs treated with SMAs exhibit porosity and permeability increases of up to 30%. Though SMA-coated proppant is still susceptible to flowback, studies have indicated that proppant is far less likely to be produced when treated, and it is resistant to production surges. **Fig. 17.10** shows the differences in produced proppant vs. fluid rate.

The main benefits of SMAs include reduced fines production in high-rate gas/condensate wells and lower-pressure oil wells (Terracina et al. 2001). Additionally, this treatment allows for more-aggressive and, hence, faster well-cleanup efforts, if coupled with a coiled-tubing approach.

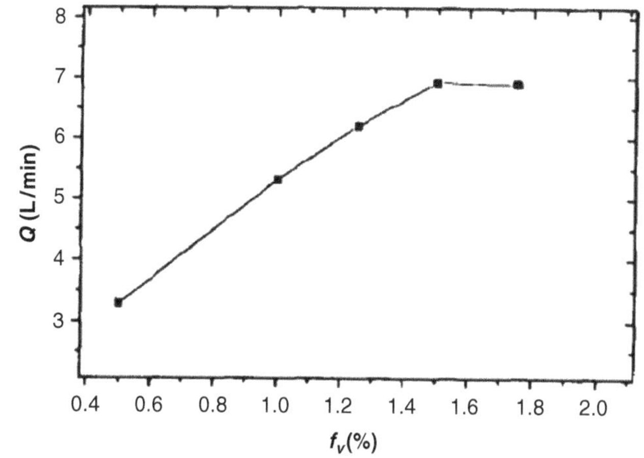

Fig. 17.10—Proppant production rate increases with produced-fluid rate (Romero and Feraud 1996).

*17.4.2.3 **Fiber Reinforcement.*** The act of fiber reinforcement initially was an effort to increase the stability of soils through the use of tensile-resisting fibers (Romero and Feraud 1996). We commonly find this in civil engineering, which is where the idea originated. For hydraulic fracturing treatments, supplementing the fracturing fluid and proppant with short fiber reinforcements enhances compatibility within the fracture geometry. Fibers have been found to improve the resilience of the outer compressive arch for larger fracture widths. This allows the fracture to be wider without producing any proppant up to a critical-failure flow rate.

In terms of makeup, fibers usually entail chopped fibers or thermoplastic strips, both of which have similar effects on the pack. Each of these solid additives creates a mutual contact among proppant grains, which in turn increases the frictional capability each contacted grain has with the pack (Ingram et al. 2007). As a result, proppant-pack failure occurs at a higher critical production rate because of the added resistance to breakdown based on drag effects. These fibers can be seen in **Fig. 17.11.**

Increasing the length of fibers beyond a critical limit will fail to incrementally increase pack stability. Volumetric concentration of these fibers in the fracturing fluid also fails to increase performance after approximately a 2% limit is reached (Romero and Feraud 1996). This effect can be seen in **Table 17.2,** which shows an experimental result for a 12-mm fiber with a 20/40-mesh sand. The addition of these fibers to the proppant material enables a fracture width greater than five grain diameters without proppant flowback problems, which increased the conductivity of the fracture and increased production.

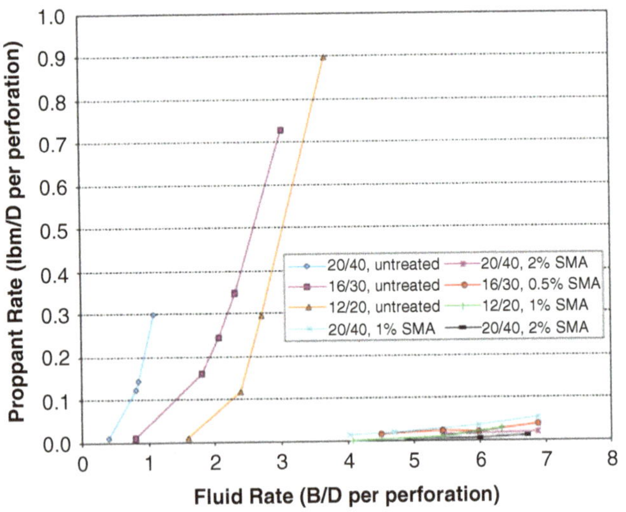

Fig. 17.11—Increasing quantity of fibers in fracturing fluid has little effect on production after a critical level (Terracina et al. 2001).

Closure Stress (psi)	100% Sand	Sand + DIP 10% by Weight	Sand + DIP 25% by Weight
	Fine % by Weight		
5,000	4.9	4.8	3.5
6,000	21.8	13.6	8.5
7,000	25.9	15.2	11.5
8,000	32.9	22.6	13.8
10,000	40.4	28.5	15.5

Table 17.2—Reduction in produced fines when different percentages of DIPs are added to the proppant pack (Di Lullo and Rae 2001).

17.5 Nomenclature

f_v = volumetric concentration of fibers in proppant, %

Q = critical flow rate, l/min

17.6 References

Aidagulov, G. R., Thiercelin, M. J., Nikolaevskiy, V. N. et al. 2007. Prediction of Long-Term Proppant Flowback in Weak Rocks. Paper presented at the SPE Hydraulic Fracturing Technology Conference, College Station, Texas, 29–31 January. SPE-106264-MS. https://doi.org/10.2118/106264-MS.

Di Lullo, G. and Rae, P. 2001. New Technique Prevents Proppant Flowback and Improves Fracture Conductivities. Paper presented at the SPE Asia Pacific Oil and Gas Conference and Exhibition, Jakarta, Indonesia, 17–19 April. SPE-86856-MS. https://doi.org/10.2118/68656-MS.

Gidley, J. L., Holditch, S. A., Nierode, R. W. et al. 1989. *Recent Advances in Hydraulic Fracturing*. SPE Monograph Series, Vol. 12. Richardson, Texas: SPE.

Ingram, S. R., Nguyen, P. D., and Hood, A. L. 2007. Remediation of Production Loss Due to Proppant Flowback Through Coiled-Tubing Intervention. Paper presented at the SPE Production and Operations Symposium, Oklahoma City, Oklahoma, 31 March–3 April. SPE-106532-MS. https://doi.org/10.2118/106532-MS.

Nguyen, P. D., Weaver, J. D., Dewprashad, B. T. et al. 1998. Enhancing Fracture Conductivity Through Surface Modification of Proppant. Paper presented at the SPE Formation Damage Control Conference, Layfayette, Lousiana, USA, 18–19 February. SPE-39428-MS. https://doi.org/10.2118/39428-MS.

Romero, J. and Feraud, J. P. 1996. Stability of Proppant Pack Reinforced With Fiber for Proppant Flowback Control. Paper presented at the SPE Formation Damage Control Symposium, Lafayette, Louisiana, USA, 14–15 February. SPE-31093-MS. https://doi.org/10.2118/31093-MS.

Terracina J., Parker, M., Pongratz et al. 2001. How to Reduce Proppant Flowback in High-Rate Wells. Paper presented at the SPE Middle East Oil Show, Manama, Bahrain, 17–20 March. SPE-68202-MS. https://doi.org/10.2118/68202-MS.

Vincent, M. C., Miller, H. B., Milton-Tayler, D. et al. 2004. Erosion by Proppant: A Comparison of the Erosivity of Sand and Ceramic Proppants During Slurry Injection and Flowback of Proppant. Paper presented at the SPE Annual Technical Conference and Exhibition, Houston, Texas USA, 26–29 September. SPE-90604-MS. https://doi.org/10.2118/90604-MS.

Weaver, J. D., van Batenburg, D. W., and Nguyen, P. D. 2006. Sustaining Conductivity. Paper presented at the SPE International Symposium and Exhibition on Formation Damage Control, Lafayette, Louisiana, USA, 15–17 February. SPE-98236-MS. https://doi.org/10.2118/98236-MS.

Wood, D. W., Brannon, H. D., Rickards, A. R. et al. 2003. Ultra-Lightweight Proppant Development Yields Exciting New Opportunities in Hydraulic Fracturing Design. Paper presented at the SPE Annual Technical Conference and Exhibition, Denver, Colorado, 5–8 October. SPE-84309-MS. https://doi.org/10.2118/84309-MS.

Chapter 18

Hydraulic Fracturing Treatment Design and Guidelines

Stephen A. Holditch

18.1 Introduction to Treatment Design

Hydraulic fracturing of wells is a common practice used in the industry to improve well productivity either by changing the flow pattern in low-permeability formations or by overcoming drilling and completion damage near the wellbore. Several variables must be considered in a fracture-optimization design; however, effective fracture length and conductivity are two important variables that determine the effectiveness of a fracture treatment.

Optimal fracture conductivity is desirable, because it represents an overall resistance to fluid flow in the proppant pack. Economic design of a fracture treatment is required to select the fracture lengths and conductivities in which the income and costs combine to maximize economic returns (Gidley et al. 1989).

Optimization can range from simple approaches to those requiring the use of highly sophisticated economic, fracturing, and reservoir performance simulators. The approach depends on the reservoir, the fracturing requirement, and the economic parameters associated with a given field. No single procedure is considered universally applicable to investigating field-development optimization, other than comparing the results that evolve from a number of scenarios and selecting the set that yields the most attractive economic return (Mohan et al. 2006).

Holditch et al. (1978) studied three examples that represented high-, low-, and medium-permeability gas reservoirs. They concluded that the most important parameters in determining the optimal fracture length are the formation permeability and the initial gas-in-place. For high-permeability reservoirs, they found that short fractures and large well spacing provide the optimal profit, whereas, for tight gas reservoirs, long hydraulic fractures and small well spacing are required to optimize profit from the reservoir.

Mohan et al. (2006) presented analytical expressions that can be used as a simple means of obtaining optimal fracture lengths for gas or gas/condensate wells with or without Darcy effects in the fracture. They discovered that optimal fracture length mainly depends on the following:

- Fracture conductivity
- Formation permeability
- Size of the condensate bank around the fracture
- Proppant volume per unit pay-zone thickness
- Reservoir thickness

Because the fracture length increases at a fixed volume of proppant per unit of thickness, the value of the well productivity was found to go through a maximum at a particular fracture length.

Indriati et al. (2002) proposed a model to predict the performance of hydraulically fractured gas/condensate reservoirs and to calculate optimal fracture length. Their work showed that condensation accumulation causes the optimal fracture length to be larger than the zero-fracture-face skin optimum. For every flowing bottomhole pressure, they found an optimal fracture geometry that maximizes the dimensionless productivity index.

Aggour and Economides (1998) conducted economic optimization studies combining high-permeability fracture modeling with expected well performance and net present value (NPV) calculations. They found that in high-permebility reservoirs, fracture conductivity is more critical than fracture length. They also found that larger drainage areas require larger designed fractures to achieve maximum economic returns.

Richardson (2000) proposed a methodology that optimizes fracture stimulation design for any reservoir type. The optimization method uses a constant dimensionless fracture conductivity, F_{CD}, to estimate fracture conductivity and length. Richardson found that a constant value of 30 for F_{CD} can be used in the design and optimization.

18.2 Well Conditions and Parameters

Most of the time spent in the design of a fracture treatment involves the analyses of the formation and the expected response to a fracture treatment. The four basic steps to designing a fracture-treatment success are summarized next.

1. Select the candidate. We must determine the reason a well should be stimulated. Evaluating the formation, reservoir, and wells to select the best candidates for well stimulation is important. Often a prefracture production or pressure-buildup test is required to select the best simulation candidates from a reservoir point of view.
2. Design the optimal fracture. To design the optimal fracture, we must evaluate drilling data, openhole logs, production data, core data, and any pressure-transient tests available to ensure that we have the best description of the reservoir layers, as well as all layers above and below the pay zones. We use these data in 3D reservoir models and 3D fracture models to design the optimal fracture treatment.
3. Pump the optimal treatment. Once the optimal treatment is designed, it must be pumped into the well as close to the design as possible. To do this, we must have on-location personnel who can plan and conduct quality-control procedures to ensure that the correct materials are on hand and that those materials are mixed and pumped into the hole as designed.
4. Analyze the treatment data during and after the fracture treatment. After the treatment, the well has to be produced back properly to allow the fracturing fluid to clean up without overstressing the propping agent and causing it to be crushed or to embed. The flowback sequence after the treatment is crucial to success. The data measured during the treatment and during the flowback period should be analyzed with 3D fracture models and 3D reservoir models to estimate the created and propped fracture dimensions. The results of these analyses should be beneficial when the next fracture treatment is being designed.

18.2.1 Condition of the Well. Implicit in the selection of a candidate and in the pumping operation is that the wellbore conditions are such that the treatment can be pumped without any mechanical issues. The well completion aspects of designing and pumping a fracture treatment are discussed in detail in Chapter 12 of Gidley et al. (1989). The first thing to consider is how the well is drilled. It is important to drill the well with a mud system that causes minimal damage and results in a gauge borehole. A gauge borehole is required to obtain an accurate suite of openhole logs and to obtain an adequate primary-cementing operation.

The casing and tubing configuration must be designed with the fracture treatment in mind. There must be adequate borehole size to allow treatments to be pumped using high injection rates with low injection pressures. Liner tops and sliding sleeves should not be exposed to high injection pressures, if at all possible.

To control the initiation of the fracture as it leaves the wellbore, the well must have a good primary-cementing job. In a new well, it is advisable to run a cement-bond log after the cement has set, but before any completion operations. If there are questions about the quality of the bonding, then a cement-squeeze operation must be considered. In older wells, if there are questions concerning the cement bond, it might be necessary to pull the tubing and packer and run a cement-bond log before deciding exactly how to fracture treat the well.

18.2.2 Burst, Collapse, and Tension. In new wells, it is important to determine how the well will be treated, and then to be certain there will be no harm done to the casing, tubing, or packer in the well. Maximum burst pressure is at the top of the well. The casing must be designed so it will not burst if the maximum surface-injection pressure is reached during the treatment. Or, to put it another way, if the treatment is being pumped down the casing, then the maximum allowable surface-injection pressure should not exceed the burst of the casing (after correcting for a safety factor).

Maximum collapse will occur downhole. When pumping a treatment down tubing, we must hold backup pressure on the annulus, and we should be certain we do not collapse any of the pipe downhole.

When tubing is in the hole—and especially if it is anchored to a packer—we must perform buckling computations to ensure we do not exceed the tension force on the top joint of the tubing or the maximum tension force, as applied to the packer. A complete discussion of these helical buckling equations, and an example computation, are given in Chapter 12 of Gidley et al. (1989). In addition, virtually every service company has a computer program to check out wellbore conditions to ensure no harm is done to the wellbore during the treatment.

18.2.3 Wellhead Conditions. In many cases, the wellhead will put an upper limit on the allowable treating pressure. The working pressure of a wellhead must be high enough to allow for the fracture treatment of the formation at optimal injection rates. As such, it is common either to remove the wellhead and replace it with a fracture value or to use a wellhead isolation tool to bypass the wellhead. In either case, the result is that we can increase the maximum allowable surface-injection pressure without harming the wellhead.

18.2.4 Older Wells. In older wells that have been producing for several years, we must be especially careful to not damage the casing, tubing, or packers during a treatment. The mechanical risk of successfully pumping a treatment substantially increases in older wells, especially those that might be producing corrosive fluids. In older wells, it is often advisable to work over the well before the fracture treatment. The best approach is to pull the well, run in a new packer, and test the tubing to ensure that the packer, tubing, and wellhead can withstand the maximum allowable surface-injection pressure. If there is a possibility of a mechanical failure, it is best that it happens with a workover rig on location before the treatment, so it can easily be fixed.

A mechanical failure of the casing, tubing, packer, or wellhead during a treatment can become a safety and economic disaster.

18.3 Reservoir and Rock Parameters

Initial reservoir data for fracturing treatment design generally is acquired with well logs, cores, well tests, and drilling records and includes the following (Gidley et al. 1989):

- Permeability
- Porosity

- Pressure
- Temperature
- In-situ stress level
- In-situ stress orientations
- Fluid properties
- Rock properties
- Formation thickness

18.3.1 Reservoir Permeability. Reservoir permeability is one of the most important parameters to quantify, because it is a controlling variable in the performance of the stimulated reservoir. Wireline measurements have made great improvements in recent years; however, the volume they are able to sample is limited, and it might not be representative of the reservoir as a whole. Ideally, some sort of flow-to-surface test, such as a drillstem test, would be performed. The minifracture test performed before pumping a treatment also can be analyzed for permeability information, but it is useful to have these data earlier in the design process.

18.3.2 Reservoir Thickness, Porosity, Pressure, and Water Saturation. Reservoir thickness, porosity, pressure, and water saturation are critical in the economic analysis of fracturing performance. Because the fracturing program is optimized for economic return, the original-hydrocarbons-in-place and the reservoir pressure can be limiting factors in the amount of investment put into stimulation. Reservoir thickness also plays a role in the volume of treatment needed.

18.3.3 In-situ Stress. An understanding of the stress levels and orientation is needed to predict the growth and character of the fracture. Reservoir fractures often do not follow predicted paths because of variable, such as natural fractures, but it is still necessary to estimate the result of injecting fluid at various pressures. Fluid injected will follow the path of least resistance, which corresponds to the planes of maximum stress. The rock can most easily be compacted in the direction of minimum stress, so quantification of that value is necessary. If possible, the minimum stress needs to be identified, as well as the levels of maximum stress.

18.3.4 Reservoir Temperature. The reservoir temperature is a factor in fracturing-fluid design. When a polymer gel is used, it is desirable to adjust the fluid properties to maintain a low viscosity while pumping through the completion, and then to crosslink to obtain a higher viscosity as the fluid enters the formation.

18.3.5 Additional Parameters. In addition to the importance of the data for the reservoir, the thickness and stress levels of adjacent beds are important, because the vertical growth of the fracture must be determined. While not a factor in calculating the productivity increase, information regarding the adjacent beds defines the volume of the fluid pumped. Any water/oil contacts also must be identified because this might be the limiting consideration in fracture growth. It is possible to limit fracture growth through methods such as tip screenout by reducing the pad-stage volume and concentrating the proppant into making a wider fracture to prevent growth into a water zone. This has been done successfully with water contacts within 15 m of the lower perforations (Engels and England 2001).

If data are available from previous fracturing programs in the reservoir, they can be used to verify the simulated response to treatment. In addition, some values normally obtained in the prejob minifracture, such as the fracture gradient, fracture-extension pressure, leakoff coefficient, and fluid efficiency, can be used in developing the design. Data such as frictional effects from previous jobs must be used with care, because any changes in the completion and perforation orientation can dramatically change those values.

Additional data requirements include anything that restricts flow rate or reduces working pressure. Any operational restrictions, such as pump capacity, pump horsepower, or space limitations, also should be recognized at this point. If the service company performing the treatment is not designing the fracturing program, at minimum they must be involved in ensuring that logistical concerns are addressed, and not overlooked or ignored. Proper identification of constraints in the fracturing program limits the combinations of treatment parameters, reduces the computing time, and avoids problems when pumping the design (Wright et al. 1999).

18.4 Determination of Optimal Fracture Length and Conductivity

The design engineer should strive to optimize the fracture treatment to maximize one or more economic parameters. Some organizations try to maximize the NPV, while others look at parameters such as pay out, profit-to-investment ratio, or return on investment. NPV has been extensively used in petroleum literature as a tool for optimizing field operations.

A hydraulic fracturing simulator is usually required to compute treatment volumes, types of materials, and pumping schedules necessary to achieve various fracture lengths and conductivities (Gidley et al. 1989). With these data, a relationship between fracture length and conductivity can be generated, as depicted in **Fig. 18.1.** To further investigate the net profit, the income and costs can be combined to relate net revenue to fracture length. As the treatment becomes larger, the performance of the well improves up to the point at which additional revenue becomes marginal and is overwhelmed by increasing costs (Veatch 1983). An NPV analysis determines the fracture treatment corresponding to the optimal economic returns and is used in optimization schemes.

Fig. 18.2 shows a graph of NPV vs. fracture length for multiple fracture conductivities. This graph usually is used in optimal fracture-stimulation designs (Richardson 2000).

18.4.1 Optimization Theory. A major factor in optimization involves achieving an appropriate balance between the fracture characteristics and the formation properties that govern reservoir performance (Gidley et al. 1989). High-permeability reservoirs require high fracture conductivities, but usually do not need deeply penetrating fractures. Low-permeability formations require deeply penetrating fractures, and some can tolerate lower fracture conductivities.

Fracture-optimization procedures usually begin with an estimate of the anticipated production results for a given completion at various fracture lengths and at various fracture conductivities. The goal of any optimization problem is based primarily on determining the combination of control variables that would maximize a prestated objective function, with possible constraints on the range that the variables can assume.

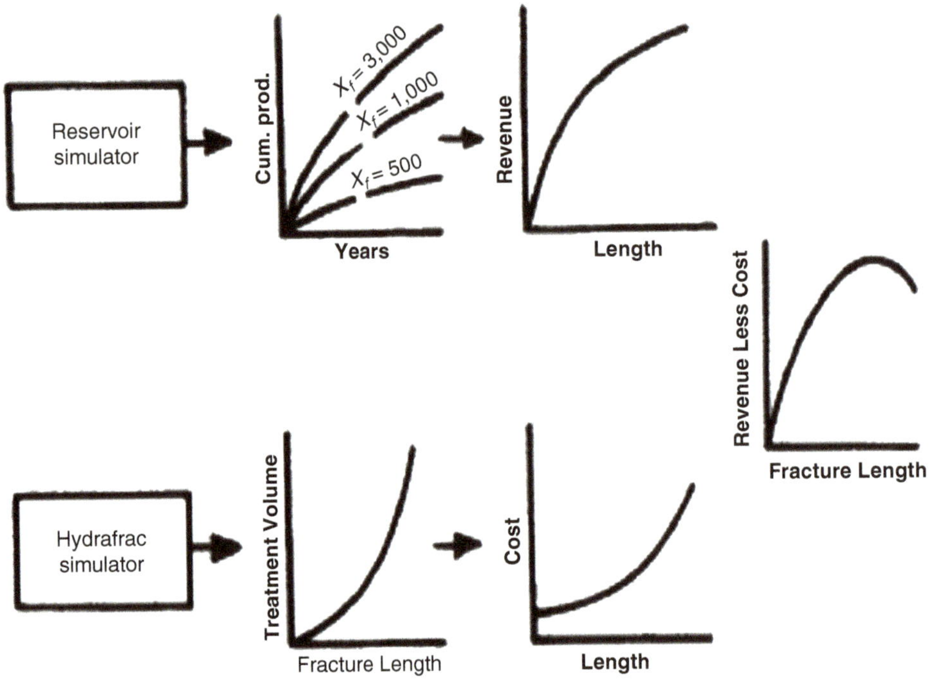

Fig. 18.1—Fracture-stimulation design: the total concept for optimization (from Veatch 1983).

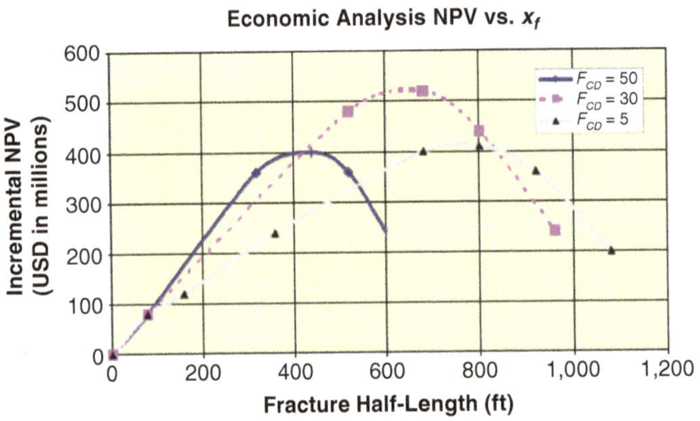

Fig. 18.2—Generalized graph of fracture half-length vs. incremental NPV at various dimensionless fracture conductivities (from Richardson 2000).

The combination of control variables usually includes finding the combination of parameters that optimize a given quantity, possibly subject to some restrictions on the allowed parameter ranges.

The quantity to be optimized (in this case, the fracture length and conductivity) is termed the objective function; the parameters that can be changed in the quest for the optimum are called control or decision variables; the restrictions on allowed parameters are known as constraints. Any combination of decision variables for which all of the constraints are satisfied is known as a feasible solution. An algorithm known as an exhaustive search, in which all known feasible solutions are evaluated and the optimal solution is selected, is usually used in the industry (Aggour 2001). This optimization procedure is time consuming and tedious, because it requires many sensitivity runs.

For the optimization of fracture length and conductivity discussed in this section, dimensionless fracture conductivity is used to optimize the values of fracture length and conductivity. By making dimensionless fracture conductivity, F_{CD}, a constant, the number of sensitivity runs is reduced and will provide an initial starting point to begin the fracture-length sensitivity analysis (Valko 2007).

18.4.2 Making F_{CD} a Constant. Several authors have suggested that the ability of the reservoir to deliver hydrocarbons and fracture flow capacity occurs at a dimensionless conductivity value between 1.3 (Hunt and Soliman 1994) and 1.6 (Cinco-Ley and Samaniego-V 1981), as shown in **Fig. 18.3.** The range represents the best compromise between the capacity of the fracture to conduct fluids and the ability of the reservoir to deliver hydrocarbons to the fracture.

The optimal dimensionless conductivity theoretically is constant for any given reservoir, and physical optimizations have resulted in dimensionless conductivity values near unity. In addition, many authors have shown that the maximization of fracture conductivity

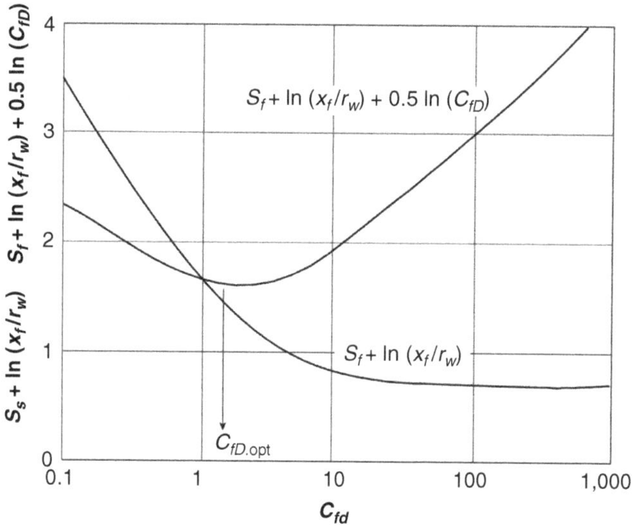

Fig. 18.3—Graph of dimensionless fracture conductivity vs. the productivity index showing the optimal dimensionless fracture conductivity of 1.6 (Cinco-Ley and Samaniego-V 1981).

is the key to increased well performance for those reservoirs with relatively high permeability (Aggour 2001; Cinco-Ley and Samaniego-V 1981). If F_{CD} can be fixed, then only one optimal fracture conductivity exists for a given fracture length.

If

$$F_{CD} = k_f w/k\, x_f = \text{Constant}, \quad (18.1)$$

then

$$k_f w = \text{Constant}\,(k\, x_f). \quad (18.2)$$

From Eqs. 18.1 and 18.2, it is obvious that the number of sensitivity runs is significantly reduced, because neither F_{CD} nor the fracture conductivity needs to be varied.

18.4.2.1 Optimal F_{CD}. Although it has been proposed that the optimal dimensionless fracture conductivity is in the range between 1.3 and 1.6, a value of F_{CD} outside this range might be required for several reasons. A value of dimensionless fracture conductivity greater than 1.3 is essential for adequate productivity in tight reservoirs.

Soliman et al. (2004) and Montgomery and Berthelot (1990) showed that the optimal dimensionless fracture conductivity to clean up the formation near a created fracture might be much higher than that necessary to produce the reservoir. This is shown to be especially true in tight gas reservoirs and reservoirs with low mobile-water saturations.

Soliman and Boonen (1996) also showed that in low-F_{CD} fractures, gas breakthrough occurs early and near the wellbore. As a result, the effective fracture length and the ultimate productivity of the well are substantially reduced. Soliman et al. (1999) and Montgomery and Berthelot (1990) recommended a dimensionless fracture conductivity of ≥ 50 to ensure that invaded fracturing fluid is effectively cleaned up and that severe and early gas breakthrough does not occur. Robinson et al. (2000) later showed that a value of F_{CD} greater than 30 is unnecessary, if proper flowback procedures are followed after fracture stimulation.

As these authors have shown, the optimal F_{CD} for hydraulic fracturing stimulation is much greater than 1.6. A constant F_{CD} value of 30 and a slow and controlled flowback following stimulation is recommended for the fracture-optimization procedure described in this subsection.

18.4.3 Estimating Effective Fracture Conductivity. The fracture design-optimization procedure recommended by Richardson (2000) sets F_{CD} at a constant value and solves for the required fracture conductivity at various fracture lengths. Because F_{CD} is constant, only one optimal fracture conductivity exists for a given fracture length. Although realistic values of fracture conductivity can be obtained from this method, the effects of closure pressure, temperature, embedment, gel damage, non-Darcy turbulent flow, and non-Darcy multiphase flow must be considered (Richardson 2000).

Penny (1987) found that time, temperature, and closure stress are major contributors to long-term fracture conductivity decline. A combination of these variables can reduce long-term conductivity by a factor of 10 compared with static, short-term, room-temperature conditions. Penny (1987) also found that embedment of a proppant agent, which depends on rock and proppant type, proppant hardness, and closure stress, can dramatically reduce the effective fracture conductivity, particularly at high closure stress (> 8,000 psi).

It is well-known that gel damage reduces effective conductivity by as little as 10% to as much as 90%. Most of this impairment occurs at the formation/proppant interface, where the residual filter cake and fines act to diminish the pack width. Penny (1987) found that the impact of gel damage depends on the gel concentration in the fracture, the inertia force available for cleanup, and the presence of multiphase flow. In addition, non-Darcy flow can reduce well productivity by as much as 50%.

A realistic estimation of effective fracture conductivity is therefore critical to the overall process and can be achieved only when the effects of closure pressure, temperature, embedment, gel damage, non-Darcy turbulent flow, and non-Darcy multiphase flow have been considered.

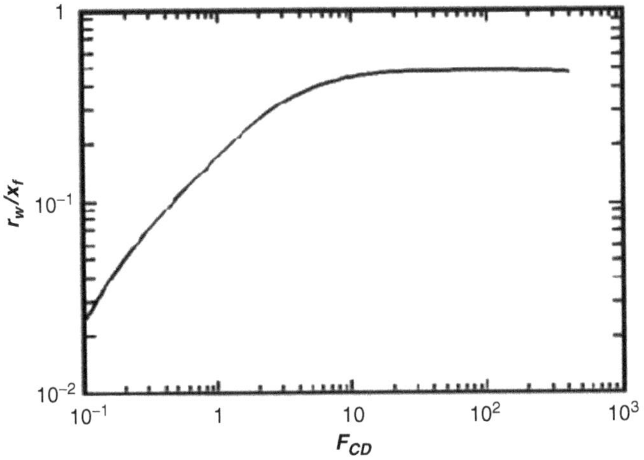

Fig. 18.4—A relationship between effective wellbore radius, fracture half-length, and dimensionless fracture conductivity (from Cinco-Ley and Samaniego-V 1981).

18.4.4 Estimating Fracture Length. Once F_{CD} is set at a constant value, a minimum required fracture length can be determined easily. The minimum required half-length can be calculated using the Cinco-Ley and Samaniego-V (1981) relationship between effective wellbore radius, fracture half-length, and F_{CD}, as well as the productivity improvement factor (PIF). From **Fig. 18.4,** it is clear that at $F_{CD} = 30$, the ratio of the effective wellbore radius to the fracture half-length (r_w'/x_f) is 0.45; therefore,

$$r_w' = 0.45 \times x_f. \quad (18.3)$$

The PIF for hydraulically fractured vertical wells to unstimulated vertical wells is given by

$$PIF = \frac{J_{\text{frac}}}{J_v} = \frac{\ln(r_e / r_w)}{\ln(r_e / r_w')}. \quad (18.4)$$

Substituting Eq. 18.4 into Eq. 18.3, we obtain an equation to calculate the fracture half-length as

$$x_f = \frac{r_e}{0.45 \times \left\{ e^{\left[\ln(r_e/r_w)/PIF\right]} \right\}}. \quad (18.5)$$

With an estimate of the minimum PIF to produce an economic stimulation, the corresponding minimum fracture half-length can be determined easily.

Richardson (2000) proposed an eight-step procedure for the design-optimization method. The procedure is summarized as follows:

1. Set the dimensionless fracture conductivity, F_{CD}, equal to a constant value of 30.
2. Estimate the minimum PIF necessary to achieve an economic stimulation, and then calculate the corresponding minimum required fracture half-length.
3. Calculate three additional fracture half-lengths at progressively higher PIFs.
4. Calculate the optimal fracture conductivity for each of the four fracture half-lengths determined in Step 2 using constant $F_{CD} = 30$.
5. Calculate the required proppant concentration to achieve each of the four fracture conductivities determined in Step 4, taking into account closure pressure, temperature, embedment, gel damage, non-Darcy turbulent flow, and non-Darcy multiphase flow.
6. Use the best tools and information available to design stimulations to achieve each of the four fracture half-lengths and corresponding fracture conductivities determined in Steps 2 through 4.
7. Estimate costs and run economics for each of the four designs. Plot NPV vs. fracture half-length and determine the optimal fracture half-length from this graph.
8. Use the optimal fracture half-length determined in Step 7 to calculate the required fracture conductivity.

In summary, the fracture length and conductivity can be optimized from the preceding eight-step procedure. The procedure assumes that for any given fracture length there exists only one optimal fracture conductivity, which can be calculated by assuming a constant F_{CD} value of 30. From the required fracture conductivity, the corresponding proppant concentration can be calculated and, finally, the optimal stimulation can be designed.

18.5 Perforating for Fracturing

"The objective of perforating is to provide a conductive flow path from the formation, through the cement sheath and casing, and into the completion tubulars. In a fracture-stimulated reservoir, the perforations connect the fracture with the wellbore," van Gijtenbeek and Pongratz (2004).

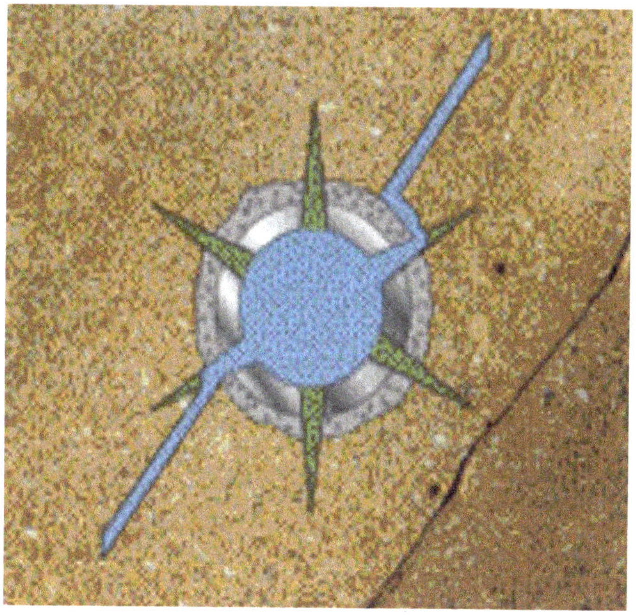

Fig. 18.5—Perforations among fracture, wellbore, and reservoir (van Gijtenbeek and Pongratz 2004).

Perforations provide the communication among the fracture, wellbore, and reservoir, as shown in **Fig. 18.5.** Proper perforating design for hydraulic fracturing not only will increase the production, but also will save money. In recent years, successful cases in Western Siberia (Russia), North America, and China, which represent the highest concentration of hydraulic fracturing activity, show the importance and economic effect of perforating.

In this section, we review perforating for hydraulic fracturing. A new philosophical concept will first be presented, and compared with the traditional concept of perforation; the key parameters and other considerations are listed following the new concept. Before concluding, we present case histories that successfully applied perforating for fracturing.

18.5.1 A New Philosophical Concept. In recent years, a new concept called "perforating for stimulation" has been gradually accepted instead of "perforating for production." As we know, perforating for production places the key point at the production, but it sometimes has resulted in no incremental production, because we found that not all perforated layers were being stimulated. The new concept of perforating for stimulation concentrates on the best perforation method for stimulation fracturing parameters that will increase production.

We should consider four important parameters when we design the perforating for hydraulic fracturing: perforated-interval length, type of charge [big hole (BH) vs. deep penetration (DP)] perforation phasing, and shot density.

18.5.2 Perforated-Interval Length. Determining how to choose the perforated zone is the first step of perforating design. In a field case study several years ago, an attempt was made to perforate each foot of pay zone, but the result was that not all layers were stimulated. Hence, the limited-entry perforation technique became more and more popular. In **Fig. 18.6,** the limited perforations might add excess perforation-friction pressure to break down a higher-stress zone, thereby making the fracture treatment extend to all zones. The proper perforated-interval length, regardless of the wellbore deviation, is needed. If a large amount of coarse proppants is placed, the length will be reduced.

18.5.3 Type of Charge. BH is a shaped charge that gives priority to the entrance-hole diameter over the depth of penetration, while DP gives priority to penetration depth instead of the entrance-hole diameter. BH charges or DP charges aim at providing the minimum perforation entry-hole diameter (EHD) for propped hydraulic fracturing. EHD is at least six times the proppant size. A BH charge creates larger entry holes than a DP charge. In general, the EHD generated by BH is 0.5 in. to 1.2 in., nearly twice the diameter created by DP. But, the BH charge is not recommended for hydraulic fracturing in hard rock formations because it might greatly increase the tortuosity of fractures. **Table 18.1** shows examples of EHD using BH or DP charges under differing conditions.

18.5.4 Perforation Phasing. Experiments show that the tortuosity (shown in **Fig. 18.7**) of preferred fractures will greatly be reduced by oriented perforation, and the best angle is less than 30°, which means a perforation gun should use a phasing of 60° or less. For hard-rock hydraulic fracturing, the perforation phasing will differ, depending on whether the well has no microannulus or an open microannulus and whether the well is vertical, horizontal, or deviated.

18.5.5 Shot Density. Shot density is an important parameter for proper perforating design, because it determines the number of holes that can accept fluids per unit distance. It is affected by the EHD, injection rate, and perforation friction pressure. Behrmann and Nolte (1999) provide clear guidelines on how to calculate the minimum shot density for fracture stimulation. **Table 18.2** summarizes these numbers as reference.

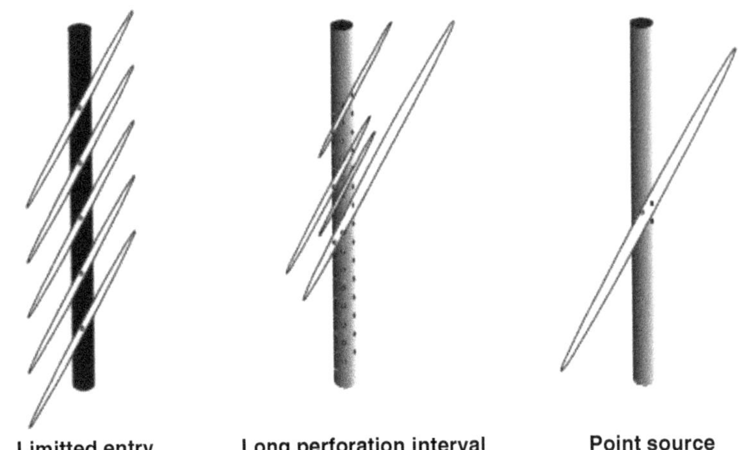

Fig. 18.6—Different perforating strategies (Lestz et al. 1999).

Phasing	EH (2½ BH)	EH (3⅜ DP)
API data	0.61 in.	0.40 in.
0°	0.58 in.	0.38 in.
60 and 300°	0.58 in.	0.37 in.
120 and 240°	0.49 in.	0.30 in.
180°	0.36 in.	0.27 in.

Table 18.1—Casing hole diameters (Behrmann and Nolte 1999). API = American Petroleum Institute.

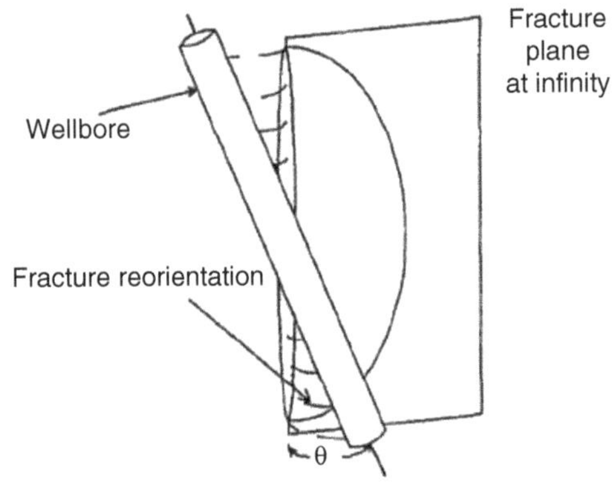

Fig. 18.7—Near-wellbore fracture tortuosity (from Behrmann and Nolte 1999).

Gun/Phase	Shots per Foot (spf) Using API EH spf	Actual ΔP Using API EH psi	Minimum spf, ΔP = 25-psi spf
2½ BH 180°	0.86	70.5	1.5
2½ BH 60°	2.57	70.5	4.5
3⅜ DP 180°	2.18	58.5	3.4
3⅜ DP 60°	6.53	58.5	10

Table 18.2—Minimum gun shot density (Behrmann and Nolte 1999).

In addition to the four parameters discussed, there are still other considerations for perforating for fracturing. For example, in the case of the Reef Ridge Diatomite Formation in the Midway-Sunset Field, Kern County, California, only 5 of 13 studied wells used the limited-entry technique, while 5 of them used a point-source technique (shown in Fig. 18.6). And the properties of rock also affect the results of perforating for fracturing.

Perforating strategies include analyzing the reservoir using all data available. The perforating strategies can affect the fracture geometry. We find that, in Fig. 18.6, a limited-entry perforation provides even fractures per perforation, while the long perforating interval might leave part of the interval unstimulated. The point source per interval is sufficient for total zone coverage.

Actually, the perforating-interval length, perforating phasing, and the type of charge (BH or DP) are related to the properties of rocks, especially the in-situ stress. In vertical wells, we conclude that the perforations should be placed in the direction of maximum horizontal stress (**Fig. 18.8**); while to eliminate the near-wellbore tortuosity in horizontal wells, angular and clustered perforations should be considered (**Fig. 18.9**).

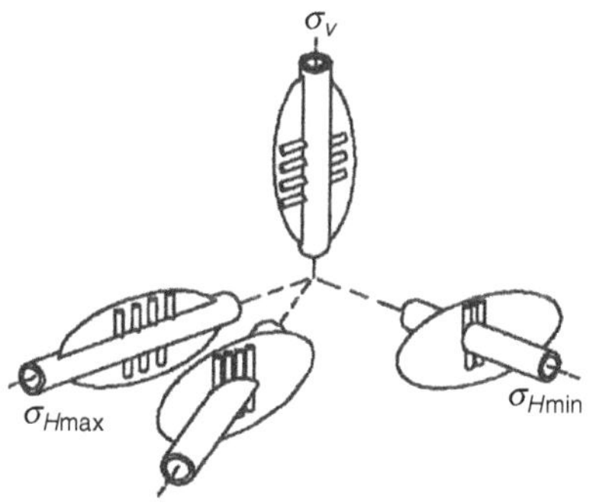

Fig. 18.8—Fracture orientation for vertical and horizontal wellbores with oblique perforations (Abass et al. 1994).

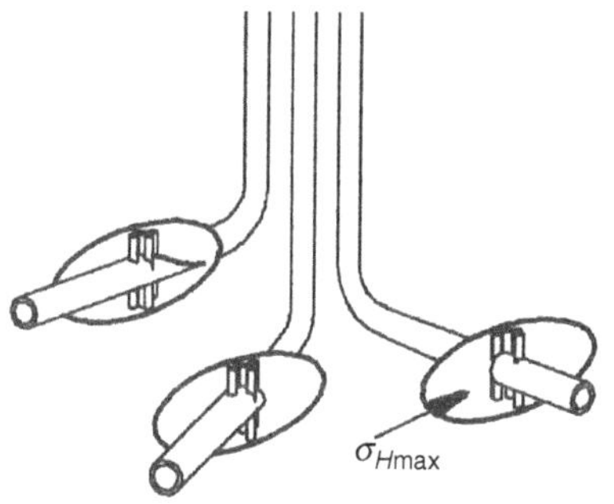

Fig. 18.9—Fracture orientation for horizontal wellbores with clustered perforations (Abass et al. 1994).

18.6 Selecting a Fracturing Fluid

The fracturing fluid selected not only has to keep the majority of proppants in suspension, but also must consider the leakoff coefficient, closure time, and degradation with time and temperature. It is simple to increase the viscosity of a fracturing fluid to carry more proppant, but the possibility of increasing the fracture-height growth with a more-viscous fluid and increasing frictional losses must be considered. Generally, the gel loading selected is the lowest one that provides the minimum apparent viscosity (Gidley et al. 1989). There are many different kinds of fracturing fluid available, including foam, gelled oil, and gelled water, that might meet the engineering requirements; in that case, one must select the most economical fracturing fluid.

The ability of the fluid to clean up, as well as any interaction with the reservoir fluid or the proppant (Soliman 1986), should be included as controllable variables in the design. Important aspects in fracture cleanup are rapid breakup of the chemical bond in the polymer, minimal residue in the proppant pack, and low-enough fluid density for the liquids to flow back to surface (Engels and England 2001). The pressure/volume/temperature properties of the reservoir fluid also must be considered when pumping cool fluids directly from the surface and reducing the reservoir temperature. Solids might precipitate and negatively affect permeability by plugging pore throats. It is possible in cases like this to heat the fracturing fluids at the surface before injection or to limit the volume of the treatment.

To select the best fluid for a given reservoir and well, one must consider the following aspects:

- The fracturing fluid must be compatible with reservoir fluid.
- The fracturing fluid must be compatible with the formation material.
- The fluid must be capable of suspending and transporting proppants.
- The fluid must be viscous enough to create fracture width and height.
- The fluid should have low friction loss in the tubulars.
- The fluid should break to a low viscosity after the treatment so it will clean up.

Compatibility with reservoir rock and fluid is critical. The fracturing fluid must not plug pore channels and cause swelling of clays. Viscosity is needed to suspend and transport proppants and to develop the fracture width needed to create and to prop long fractures. The fracturing fluid also should be moderately efficient, with the greatest percentage of the fluid remaining in the formation.

Another important property is cleanup. After the treatment is finished, the fluid must rapidly break down to a very low-viscosity system that is easily produced back or can imbibe into the reservoir, away from the fracture.

Table 18.3 shows the different kinds of fluids available.

Fluid Base	Fluid Type	Typical Composition	Mainly Used for
Water-based	Linear fluids	Water gelled with HPG HEC or CMHPG	Short fractures and/or low temperatures
	Crosslinked fluids	Crosslinker + HPG, Guar or CMHPG	Long fractures and/or high-temperature formations
Oil-based	Linear fluids	Oil, gelled oil	Short fractures in water-sensitive formations
	Crosslinked fluids	Crosslinker + oil	Long fractures in water-sensitive formations
	Water external emulsion	Emulsified + oil + water	Creating long fractures using a low fluid-loss system
Foam-based	Acid-based foams	Acid + foamer + N_2	Low pressure, water-sensitive formations
	Water-based foams	Water + foamer + CO_2 or N_2	Low-pressure formations
	Alcohol-based foams	Methanol + foamer + N_2	Low-pressure formations with water-blocking problems
Alcohol-based	Linear system	Gelled water + alcohol	Removal of water blocks
	Crosslinked system	Crosslinked system + alcohol	

Table 18.3—Fracturing-fluid categories.

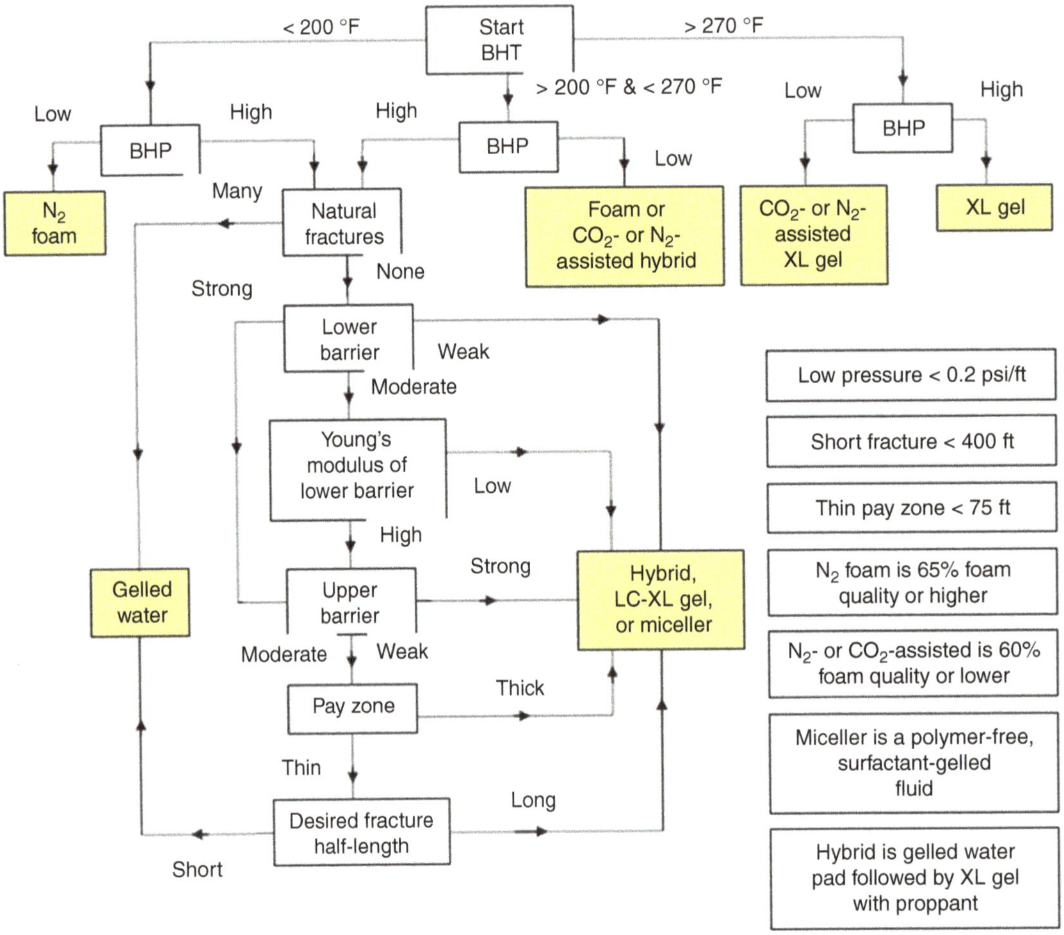

Fig. 18.10—Malpani-Holditch flow chart for selecting tight-gas-sand fracturing fluid (Malpani 2007). BHP = bottomhole pressure; BHT = bottomhole temperature.

18.6.1 Gas-Well Fracturing-Fluid Selection. Malpani (2007) published a flow chart for selecting the appropriate fracturing fluid for a particular set of conditions. The flow chart, shown in **Fig. 18.10,** includes eight key parameters to guide engineers to the appropriate fluid. The eight key parameters include the following:

- Bottomhole temperature
- Bottomhole pressure
- Presence of natural fractures
- Type of lower and upper barriers
- Modulus of the formation
- Height of the pay
- Desired fracture half-length

Each fluid used in the chart in Fig. 18.10 is described next:

- Crosslinked gel—Fluid created by use of water gelled with polymers able to be crosslinked; large volumes of propping agents at high concentrations (8 to 10 lbm/gal) are commonly pumped.
- Low-concentration crosslinked gel—Fluid created by use of water gelled with low-concentration polymers (20 to 25 lbm/1,000 gal of fluid) able to be crosslinked.
- Gelled water—Fluid created by use of a low-viscosity fracturing fluid composed of water gelled with linear polymers, clay stabilizers, surfactants, and friction reducers.
- Hybrid fracturing-treatment fluid—Gelled gelled water is pumped as pad followed by a crosslinked gel with proppant.
- Miceller fluid—Fluid created by adding an electrolyte, such as quaternary ammonium salt, to water along with a special surfactant that creates long, worm-like micelles. The micelles create viscosity in a way similar to the way that long-chain polymers create viscosity in gel fluids. In this case, hydrocarbons (oil or gas) are the breaker system, so when hydrocarbons are produced, they break the micelles and help clean up.
- Foam fluids—Nitrogen foam is 65% foam quality or higher. At lower temperatures—200°F or lower—and for low reservoir pressures, the industry typically uses foam fluids. Foam fluids will break and clean up when the bottomhole pressure is reduced during flowback.

18.7 Selecting a Propping Agent

Once the reservoir permeability is known and the optimal fracture length has been computed, then the desired conductivity of the fracture can be determined (Fig.13.1). If we assume that the optimal F_{CD} is 1.6, then in a low-permeability reservoir of 0.2 md, with a design fracture half-length of 500 ft, a fracture conductivity of 160 md-ft would be optimal for gas production. However, this low value of fracture conductivity might not allow for efficient cleanup of a partially broken fracturing fluid. In tight formations such as this, a higher value of dimensionless conductivity is required (Soliman and Hunt 1985) to be certain the fracturing fluid can clean up. Other considerations in selecting a propping agent include the degradation of conductivity with time and limitations imposed by flow profiles.

Many authors (Richardson 2000; Barree et al. 2003; Vincent et al. 1999) have stated the primary variables in fracture conductivity:

- Multiphase-flow effects that reduce hydrocarbon permeability
- High closure stress that causes
  - Proppant crushing
  - Proppant embedment
- Partially broken fracturing fluids that plug the fracture because of high yield points
- Non-Darcy flow in the fracture that causes extra pressure drop
- Gel damage that reduces fracture permeability
- Filter-cake deposition that reduces fracture width
- Fines migration from the reservoir or from crushed proppant grains

To compensate for these effects, Richardson (2000) proposed that a dimensionless conductivity (F_{CD}) of 30 should be used. In higher-permeability formations, however, this might prove to be cost prohibitive or impossible, even when using methods such as tip screenout to increase fracture width. Lopez-Hernandez et al. (2004) presented a method to account for non-Darcy flow and to calculate required conductivity. This model usually results in recommending shorter and wider fractures as drawdown increases, but can prevent overdesigning to achieve a high dimensionless conductivity that might not be necessary.

Although materials selection and fracture profiles are designed around a desired conductivity, when specifying the conductivity, the probable materials must be kept in mind. This is similar to questioning whether the chicken or the egg came first. Some assumptions can be made about the ability of the gel to clean up, the sanding potential of the reservoir, and the fluid flow that should aid an experienced designer in specifying conductivity.

Table 18.4 lists five types of propping agents, their composition, and a few comments on each.

The main aspects of proppant selection for design are

- Proppant permeability
- Proppant transportability
- Cost-effective

The three main proppants available are sand, intermediate density, and high strength. It is possible, in certain cases, to go from a conductivity-limited condition to a length-limited condition by changing from sand to high-strength proppant. The fracturing fluids must have a moderate viscosity and treating rate so that there is no difficulty in transporting the proppants.

The factors that affect the fracture-flow capacity developed with proppants are

- Closure stress
- Proppant particle size
- Proppant concentration
- Proppant strength
- Proppant grain shape

In the presence of high in-situ stresses, the proppant might crush, thereby reducing particle size, increasing surface area, and reducing the permeability of the propped fracture. Larger particles, especially when sand is used as the propping agent, provide greater conductivity, but large sand particles crush more easily. Also, proppant size is important in the design of fracturing treatments, because

Proppant Type	Typical Composition	Highlights
Brady-type sand	Polycrystalline quartz	The deposit yields acceptable fracturing sand in the 20/40-mesh size ranges and larger.
Ottawa-type sand	Pure quartz sand	It is monocrystalline with a sand size of 20/40 and small. It is well-rounded.
Sintered bauxite	High alumina bauxite shaped into pellets using water and binder	Inert, high-strength, high-density ceramic proppant.
Intermediate density	Corundum and mullite	Medium strength and moderately inert with a lower density and more economical.
Resin coated	Ottawa-type sand coated with resin	High resistance to brittle failure. Moderate strength and moderate toughness.

Table 18.4—Proppant type categories.

a minimum fracture width is needed to allow the proppant to enter the fracture. Fracture conductivity increases with increasing concentration of proppant in the fracture. The shape of the proppant is important, because the surface stresses are quite uniform, and a well-rounded spherical particle is capable of carrying higher loads without crushing.

18.8 Determination of Fluid Properties

To design the optimal fracture treatment, it is necessary to pick the best fluid for the reservoir conditions. To select the correct fluid, the design engineer must consider the following fracturing fluid requirements, and, at minimum, the fracturing fluid should be

- Compatible with the formation rock
- Compatible with the formation fluid
- Viscous enough to generate a pressure drop in the fracture that will produce enough fracture width to allow propping agents to enter and move into the fracture
- Capable of suspending propping agents and transporting them deep into the fracture
- Capable of maintaining viscosity during the pumping portion of the treatment, but also able to break to a lower-viscosity fluid so it can either imbibe into the formation or flow back into the wellbore
- Easy to mix in the field
- Easy to control from a chemistry viewpoint
- Cost effective

The first decision the design engineer must make is to determine the base fluid required for the formation. One can choose from water-based, foam-based, or oil-based fluid. Most treatments are pumped with water-based fluids. In low-pressure reservoirs, foam-based fluids might be used to aid in fracturing fluid cleanup. In some oil reservoirs, oil-based fluid might be used. However, water-based fluids can also be successfully used to stimulate oil reservoirs. In fact, one of the best ways to get more oil out of the ground is to design a waterflood. As such, one should not worry about pumping water into an oil reservoir, as long as the proper fluid compatibility tests are run. **Table 18.5** shows the fracturing-fluid selection summary.

Once the base fluid has been chosen, the next step is to determine the additives required. The best way to select the fracturing fluids and additives required for a specific situation is to work closely with the service company expert on a case-by-case basis. It is important that both parties agree on which additives are required, the concentration of the additive, and why a particular additive is needed for this particular treatment. **Table 18.6** summarizes the chemical additives.

Base Fluid	Fluid Type	Main Composition	Application
Water	Linear fluids	Gelled water, guar < HPG, HEC, CMHPG	Short fracture, low temperature
	Crosslinked fluids	Crosslinked + guar, HPG, CMHPG, CMHEC	Long fracture, high temperature
Foam	Water-based foam	Water and foamer + N_2 or CO_2	Low-pressure formation
	Acid-based foam	Acid and foamer + N_2	Low-pressure, water-sensitive formations
	Alcohol-based foam	Methanol and foamer + N_2	Low-pressure formation with water-blockage problems
Oil	Linear fluids	Oil, gelled oil	Water-sensitive formations, short fractures
	Crosslinked fluids	Phosphate-ester gels	Water-sensitive formations, long fractures
	Water-external emulsions	Water + oil + emulsifier	Good for fluid-loss control

Table 18.5—Fracturing fluids summary.

Biocide	Kill bacteria	Gluteridehyde carbonates
Breaker	Reduce fluid viscosity	Acid, oxidizer, enzyme breaker
Buffer	Control pH value	Sodium bicarbonate, fumaric acid
Clay stabilizer	Prevent clay swelling	KCl, NH, Cl, KCl substitute
Diverting agent	Divert flow	Ball sealer, rock salt, flake boric acid
Fluid-loss agent	Improve fluid efficiency	Diesel, particulate, fine sand
Fraction reducer	Reduce fraction	Anionic copolymer
Iron controller	Keep iron in solution	Acetic and citric acid, ethylene diamine tetra-acetic acid, nitrilo triacetic acid
Surfactant	Lower surface tension	Fluorocarbon, nonionic
Gel stabilizer	Reduce thermal degradation	methanol, sodium thiosulfate

Table 18.6—Summary of chemical additives.

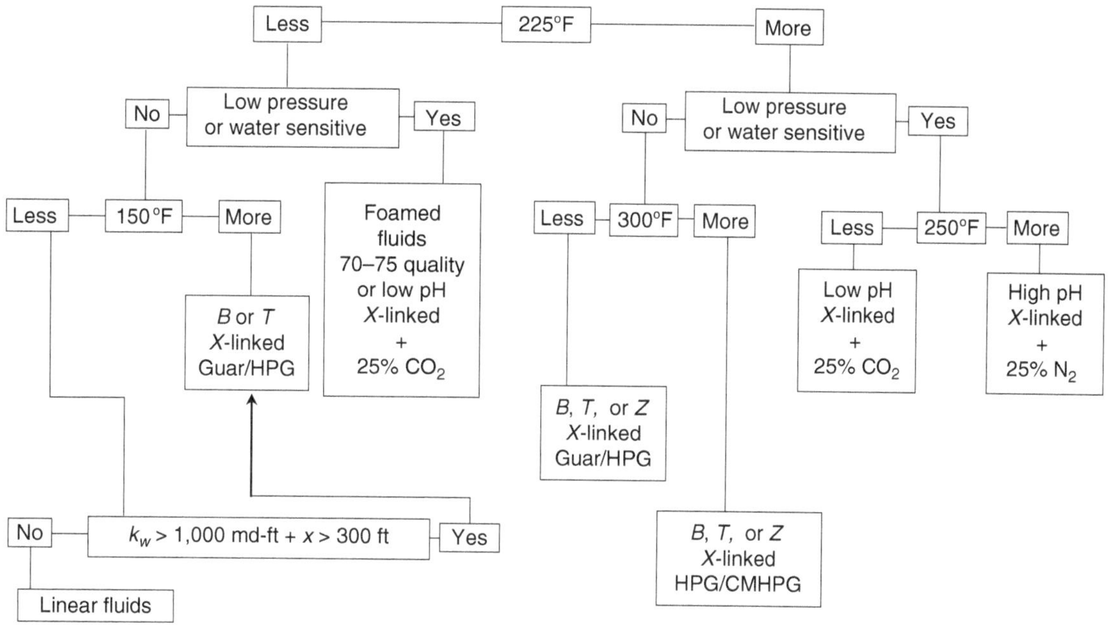

Fig. 18.11—Selecting a fracturing fluid.

We present two flow charts that can be used to help select when to use water-based fluids and foam fluids for a fracturing treatment. The first flow chart (Fig. 18.10) has been available for 10 years or more and, essentially, considers only crosslinked fluids and foams. Notice that the decision on which base fluid to use is mainly determined on the basis of the bottomhole temperature and an ill-defined term to decide whether the reservoir you are working with is water sensitive. In truth, most reservoirs really are not water sensitive, but some are water sensitive.

The section-fracture fluid-selection flow chart was recently developed at Texas A&M University and includes choices for water-fracture treatments and hybrid treatments (**Fig. 18.11**). In this second flow chart, the decision of which base fluid to select is based on the following factors:

- Bottomhole temperature
- Reservoir pressure
- Influence of natural fractures
- Status of the barriers above and below the pay zone
- Pay-zone thickness
- Expected fracture half-length

It is recommended that the process of selecting the fluid for a particular treatment begin with the second flow chart.

18.9 Fracturing Fluid Additives

Table 18.7 summarizes the function of most common additives for fracturing fluid.

Type of Additive	Function Performed	Typical Products
Biocide	Prevents viscosity loss to bacterial degradation; protects formation from anaerobic bacteria growth.	Gluteraldehyde Carbamates
Breaker	Enables viscous fracturing fluids to be controllably degraded to thin fluid that can be produced back out of fractures; enhances proppant distribution.	Acid breakers Oxidizer breakers Enzyme breakers
Buffer	Controls the pH for specific crosslinkers and crosslink times; speeds up or slows down the hydration of certain polymers.	Sodium bicarbonate Fumaric acids
Clay stabilizer	Prevents fines migration; prevents clay swelling; prevents clay/sand matrix from disaggregating.	KCl NH_4
Diverting agent	Diverts flow of the fracturing fluid to a different perforated interval by plugging off either the perforation or some part of the formation.	Ball sealers Rock salt Oil-soluble resin

Table 18.7—Function of most-common additives for fracturing fluid. (*Continued*)

Type of Additive	Function Performed	Typical Products
Fluid-loss control	Improves the fracturing-fluid efficiency.	Diesel
		Silica flour
		Resin/polymer mixtures
Friction reducer	Reduces the friction resistance caused by fluid flowing in the tubing or casing.	Polyacrylamide
		Citric acid
Iron control	Keeps earth iron ion products in solution; prevents formation damage from iron ion products.	Ethylenediaminetetraacetic acid (EDTA)
Surfactant	Assists in fluid cleanup by lowering surface tension; minimizes emulsion problems; enhances maintenance of relative permeability.	Fluorosurfactants
Gel stabilizer	Protects the fracturing fluid from degradation at high bottomhole temperatures by removing free oxygen from the system.	Sodium thiosulfate

Table 18.7—Function of most-common additives for fracturing fluid.

18.10 Determining the In-Situ Stress Profile

In-situ stress plays a critical role in the design and implementation of a hydraulic fracture. Hydraulic fracture simulators need the design engineer to supply the in-situ stress profile of the well, including the stress at the horizon where the hydraulic fracture is being placed as well as formations above and below the interval of interest. The stress contrast between different formations is a key ingredient in the prediction and actual geometry of a hydraulic fracture.

18.11 Fracture Treatment Design Using a Fracture Model

Hydraulic fracture fluids usually are composed of a mixture of water, chemicals, and proppants such as sand or bauxite. The proppants are used to keep the fracture open after the end of the injection and to procure a high permeability in the fracture. Therefore, the proppant agent must

- Be of certain size to generate a sufficient permeability in the fracture
- Be strong enough to resist in-situ stresses
- Be placed in the fracture to achieve the desired propped height, length, and width

The first two criteria are fulfilled through the choice of proppant agent. To fulfill the last requirement, an adequate proppant pumping schedule must be computed.

Two concepts govern the flow of the proppant inside the fracture:

- For the proppant agent to freely move in the fluid, its horizontal velocity must be large enough to allow the proppant to move along the fracture and to not settle on the bottom of the fracture. The vertical settling velocity depends on the densities of the fluid ρ_f and the proppant ρ_{sd}, the viscosity of the fluid, the proppant concentration, and the diameter of the proppant particle.

$$v = \frac{(2n+1)d}{108n}\left[\frac{\left(\rho_{sd}-\rho_f\right)^d}{72K_a}\right]^{\frac{1}{n}} \qquad (18.6)$$

- If the fracture width $w(t)$ at the proppant front is too small, then the proppant will bridge in the fracture and a near-wellbore screenout may occur.

If we consider that the geometric dimensions of the fractures are modeled,

$$w_o(t)' = A_1 t^{i1} \qquad (18.7)$$

and

$$L(t)' = A_2 t^{i2} \qquad (18.8)$$

and that we are injecting with a constant rate,

$$V_i = q_i t. \qquad (18.9)$$

To ensure that the proppants are able to penetrate the fracture, the width of the fracture must be three times bigger than the proppant agent diameter when it reaches the fracture entrance. For these reasons, we usually start by injecting the fracturing fluid without the proppant (pad). We can compute a lower limit for the pad volume:

$$V_{\text{pad}} = q_i \left(\frac{\alpha t}{A_2} \right)^{\frac{1}{i1}} \quad (18.10)$$

A few other reasons to inject a pad volume are

- As the slurry liquid leaks off through the walls of the facture, the concentration of the proppant increases while approaching the fracture tip, and it might became immobile inside the fracture.
- The proppant agent might reach the fracture tip and be trapped. The sand accumulates and the fluid moves forward. The fracture will not be filled with the proppant and then will close. This also creates a sudden and significant restriction to fluid flow, which causes a rapid rise in pump pressure.
- The pad can be used to test the integrity of the surface connections and can cool down the formation to minimize degradation of the fluid viscosity.

The optimal proppant schedule allows distribution of the proppants over the desired length, and the proppant bank reaches the fracture tip when L = expected length **(Figs. 18.12 and 18.13).**

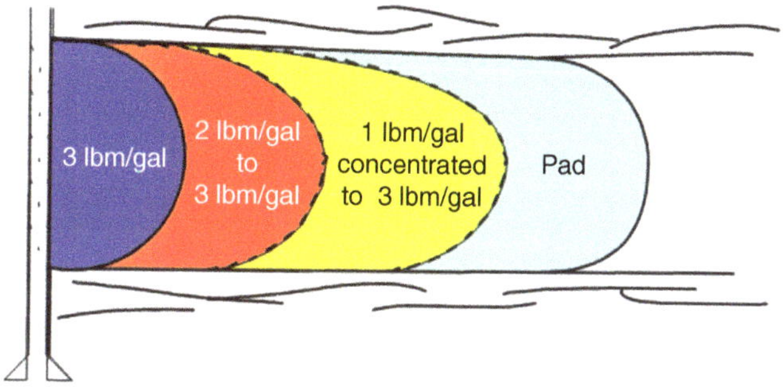

Fig. 18.12—Stage at intermediate time.

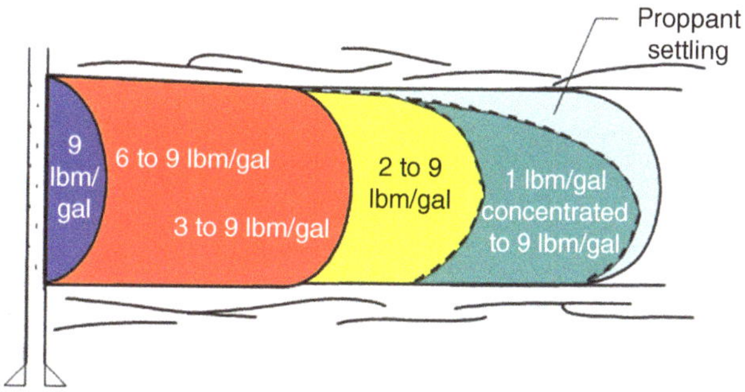

Fig. 18.13—Stage at end of pumping.

After the pad stage, we pump slurry stages—fracturing fluid mixed with proppant—in which the proppant concentration and size can be increased during the pumping operations to maximize the volume of proppant put into the fracture. The next subsection explains how to optimize the pad volume and the proppant schedule to obtain the desired proppant concentration.

18.11.1 Step-by-Step Computation for Optimal Fluid and Proppant Schedule: Volume-Injected Simplified Computation. First, we must compute the fracture geometry. In this case, we can use one of the analytical models to compute the fracture width and length. The Perkins-Kern-Nordgren model provides the following value of fracture width at the fracture entrance:

$$w_{w,0} = 9.15^{\frac{1}{2n+2}} \times 3.98^{\frac{n}{2n+2}} \left(\frac{1+2.14n}{n} \right)^{\frac{n}{2n+2}} K^{\frac{1}{2n+2}} \left(\frac{q_i^n H^{1-n} L}{E'} \right)^{\frac{1}{2n+2}} \quad (18.11)$$

We compute the average width of the fracture:

$$\overline{w}_e = 0.628 w_{w,0}. \quad (18.12)$$

We apply a material-balance equation, taking into account the leakoff fluid:

$$\left(\frac{q_i}{HL}\right)t' - (2\kappa C_L)\sqrt{t} - (\overline{w} + 2S_p) = 0. \qquad (18.13)$$

We compute the injection time t_i and, therefore, the injected volume:

$$V_i = q_i t_i. \qquad (18.14)$$

18.11.1.1 Pad Volume Computation. As previously explained, the optimal pad volume ensures the entrance of the proppants into the fracture, as well as their movement, until they have reached the aimed-for length. This means that the slurry reaches the fracture tip when the injection stops. Nolte (1988) showed that this optimal pad (free \of proppant) volume corresponds to

$$V'_{pad} = (1 - E_f)^2 V_i, \qquad (18.15)$$

where E_f is the fluid efficiency, defined as V_f/V_i, and can be computed using the fracture-design model or by closure-time computation.

From Fracture Model. Having the fracture length and width, and assuming that the fracture height is constant and equal to H,

$$E_f = \frac{HL\overline{W_e}}{V_i}. \qquad (18.16)$$

From Closure Time. The closure time is the time from the end of the injection to closure of the fracture and the settlement of the proppants. The closure time depends on the fluid efficiency **(Fig. 18.14).** If the leakoff is small, then the closure time will be long. Nolte (1986) developed methods and experiments to link the closure time to the fluid efficiency.

Therefore, by carrying out a minifracture using a fracturing fluid free from proppants, and by measuring the pressure decline, we can compute the closure time and estimate the fluid efficiency.

18.11.1.2 Conclusion on Pad Volume Computation. The previous subsection showed the computation of the pad volume, which allows an optimal settlement of the proppants, and therefore, the injection time of the pad:

$$t_{fp} = (1 - E_f)^2 t_i. \qquad (18.17)$$

The injection of proppant starts at $t = t_{fp}$. The concentration will start gradually and will increase during the treatment. The next subsection shows how to compute the proppant concentration schedule.

18.11.1.3 Proppant Concentration Schedule. The proppant's concentration should increase gradually. Nolte (1988) suggested a power-law function to compute the proppant concentration **(Fig. 18.15).**

Below, we give a short presentation of Nolte's computations: W is the quantity of proppants injected during the treatment, c_f is the desired proppant concentration, and $\overline{c}$ is the average concentration.

$$\overline{c} = \frac{W}{V_i}. \qquad (18.18)$$

and

$$c_f = \frac{W}{V'_f} = \frac{W}{E_f V_i}. \qquad (18.19)$$

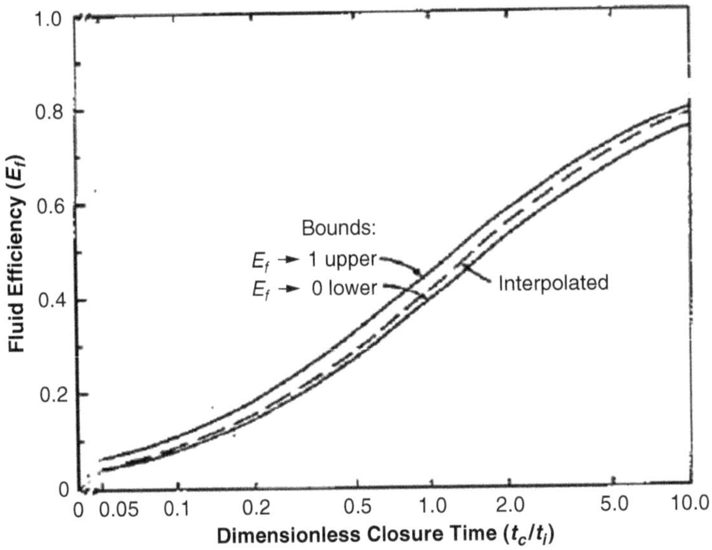

Fig. 18.14—Relationship between the fluid efficiency and the closure time.

We define the dimensionless variable c_D as c/c_f, and, therefore,

$$\overline{C_D} = \frac{\overline{c}}{c_f} = E_f. \qquad (18.20)$$

Dimensionless time is

$$\zeta = \frac{t - t_{fp}}{t_i - t_{fp}}. \qquad (18.21)$$

A power-law function that satisfies the previous conditions is

$$c_D(\zeta) = \zeta^{1-E_f}. \qquad (18.22)$$

Therefore,

$$c(t > t_{fp}) = c_f \left(\frac{t - t_{fp}}{t_i - t_{fp}} \right)^{1-E_f}. \qquad (18.23)$$

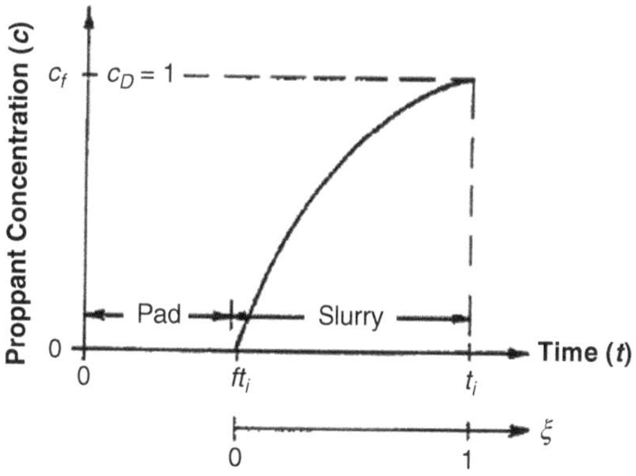

Fig. 18.15—The proppant concentration profile (from Nolte 1988).

18.11.2 Computing Pump Schedule Using Simulators. For more-complex fracture geometry, the previous computations are oversimplified. Therefore, fracture simulators compute fracture geometry and proppant distribution on the basis of characteristics of the fluid and the proppant agent and on the basis of the pump schedule. By assuming a given pump schedule, the simulator will compute the fracture length and conductivity. The design process of a pump schedule is based on experience and trial-and-error. If the results do not correspond to the desired goals, we can modify the pump schedule until we obtain a satisfactory result.

18.11.3 Presentation of Pumping Schedule Generator. Pumping schedules can be generated by computer. These computer-generated models take into account a more-complex geometry. The fracture length and the proppant distribution are entered as an input, and the program solves an inverse problem to obtain the pump schedule. An example of such a generator is given in Gu and Desroches (2003). The schedule is generated by iteration.

The first iteration starts with two stages: a free-pad volume and a slurry stage of the given concentration **(Fig. 18.16a).** The slurry stage is then divided into substages with the same concentration. A simulator executes this schedule (Fig. 18.16b). After the end of every substage, the concentration is examined. If it exceeds the maximum concentration, the concentration of the substage in the schedule is adjusted (Fig. 18.16c) to achieve the correct concentration (Fig. 18.16d). A substage might be depleted and become a pad stage. This iteration repeats until it reaches the optimal concentration in the fracture.

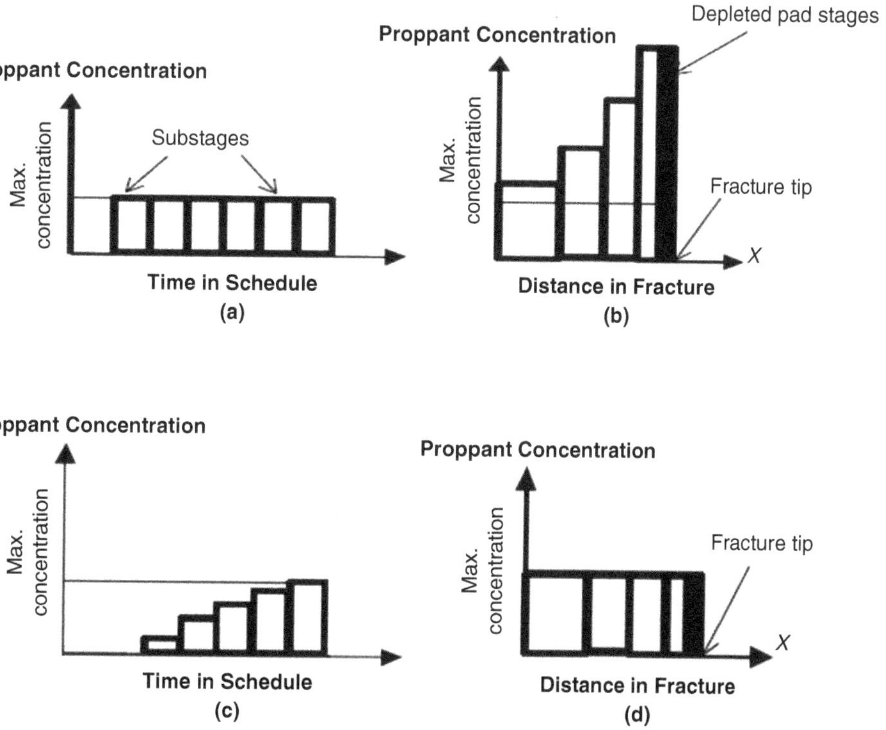

Fig. 18.16—Iteration steps in the automatic pump generator: (a) initial proppant schedule; (b) result in the fracture; (c) final proppant schedule; and (d) final concentration in the fracture.

18.11.4 Design Modification. As more data become available through monitoring reservoir responses, or before a job by incorporating minifracture data, the design can be modified to account for differences between simulated and realized results. Limitations in changing a design immediately before pumping are imposed by the physical presence of fracturing equipment and material on location. However, it is common practice to pump a minifracture job with a step-down test once the equipment is in place. This can accurately determine the near-wellbore friction, closure pressure, net pressure, and fluid efficiency (Lizak et al. 2006). These updated parameters can be used on-site to adjust the fluid and proppant in each stage to obtain the desired fracture geometry. After the fracture treatment, post-job analysis such as net pressure matching, measurements of fracture growth, and production data can be used to fine tune the simulation to improve future fracture designs.

Even as the simulations and models of fracturing are improving, post-job observations have shown a high complexity and variability in hydraulic fracturing (Wright et al. 1999). Fracturing is not an exact science because the reservoir itself is a wildcard. For the design to improve, it is important to analyze the results. Fortunately, things do not have to go exactly as designed to be a good stimulation job.

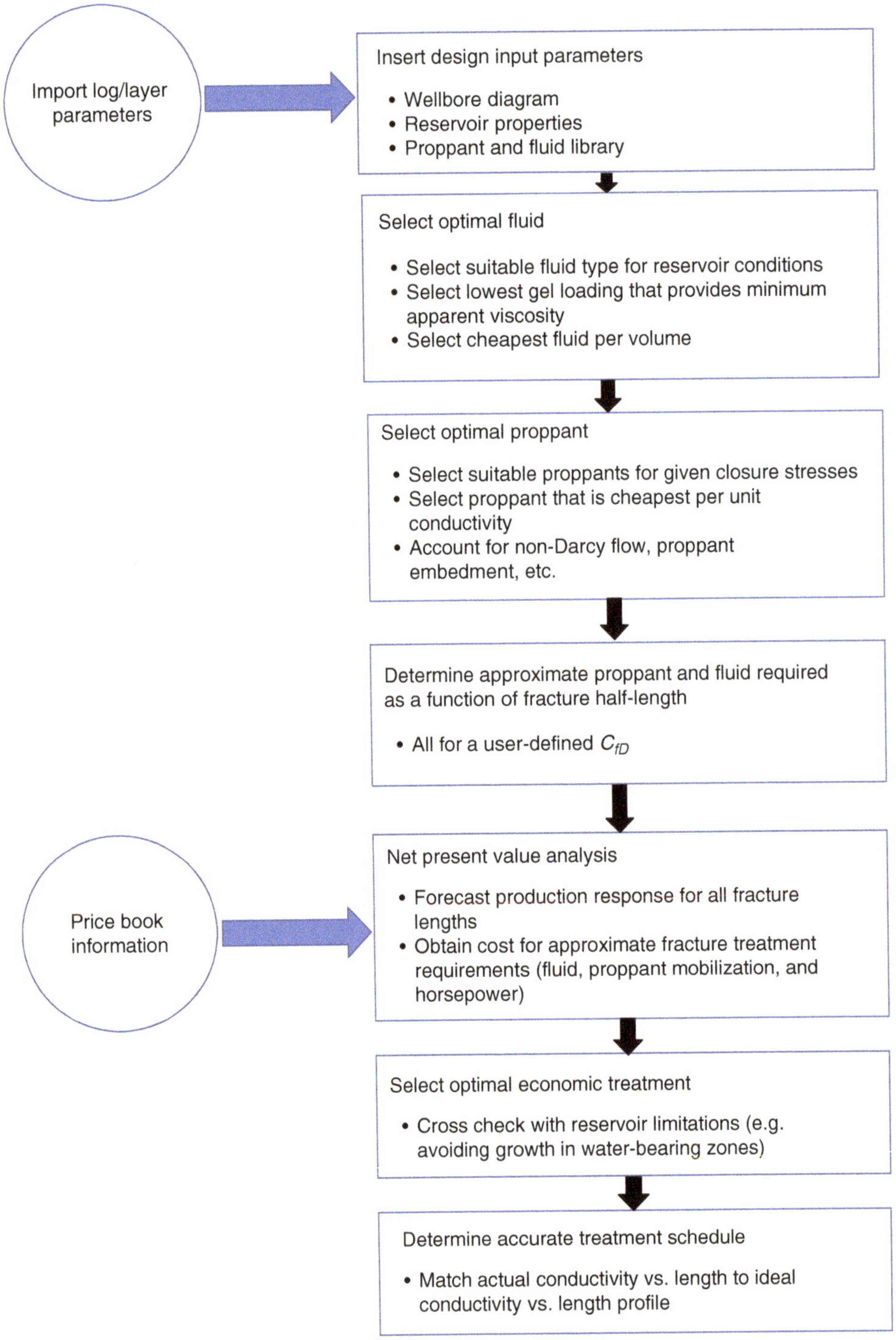

Fig. 18.17—Flow of fracture design (Soliman et al. 2004).

The result might pay off a little sooner or later than expected, but a successful end result is an increase in production that was worth the effort and investment.

18.12 Summary

Hydraulic fracturing treatment design integrates five major steps:

1. Identifying the physical constraints
2. Selecting a treatment that falls within the constraints
3. Predicting fracture geometry for the treatment
4. Forecasting the production increase
5. Quantifying the profit in terms of an economic indicator such as net present value

As presented by Soliman et al. (2004), the flow of the fracture design can be shown as **Fig. 18.17.** While computers are invaluable in processing complex simulations and aiding in the overall design, there still exists the need for a knowledgeable designer. There are so many variables and unique considerations from well to well that the best solution for a treatment program is still the human touch. Not only engineering, but also logistical requirements, must be met. Regardless of the completeness and possible brilliance of the design, it must be implemented at the wellsite to become a success.

18.13 Nomenclature

A_1	=	constant
A_2	=	constant
c	=	proppant concentration, lbm/gal
$\overline{c}$	=	average proppant concentration, lbm/gal
c_f	=	desired proppant concentration, lbm/gal
c_D	=	dimensionless proppant concentration
C_{fD}	=	fracture conductivity, dimensionless
C_L	=	leakoff coefficient, ft/min$^{0.5}$
C_r	=	dimensionless fracture conductivity
d	=	diameter of proppant, in.
E'	=	plane-strain modulus, psi
E_f	=	fracture efficiency, dimensionless
F_{CD}	=	dimensionless fracture conductivity
h_f or H	=	fracture height, ft
J_{frac}	=	productivity index of fractured well
J_v	=	productivity of unfractured vertical well
k	=	permeability of formation, md
k_f	=	permeability on fracture pack, md
K_a	=	power-law rheological properties of fracture fluid
L	=	fracture length, ft
n	=	power-law rheological properties of fracture fluid
q_i	=	injection rate, bbl/min
r_e	=	reservoir drainage radius, ft
r_w	=	wellbore radius, ft
r'_w	=	equivalent wellbore radius, ft
s_s or s_f	=	skin factor, dimensionless
S_p	=	spurt loss per unit area, ft
t	=	time, min
t_c	=	closure time, min
t_{fp}	=	injection time for pad fluid, min
t_i	=	total injection time, min
v	=	proppant settling velocity, ft/min
V_f	=	fracture volume, ft^3
V_i	=	injection volume, ft^3
V_{pad}	=	pad volume, ft^3
w	=	fracture width, in.
$\overline{w}$ or $\overline{w}_e$	=	average fracture width, in.
w_o or $w_{w,o}$	=	fracture width at the wellbore, in.
W	=	quantity of proppants injected during the treatment, lbm
x_f	=	fracture half-length, ft
α	=	constant
κ	=	constant
ρ_f	=	density of fluid, lbm/ft^3
ρ_{sd}	=	density of proppant, lbm/ft^3
σ_v	=	vertical stress, psi
$\sigma_{H\max}$	=	maximum horizontal stress, psi
$\sigma_{H\min}$	=	minimum horizontal stress, psi
ζ	=	dimensionless time

Subscripts

$i1$	=	exponent
$i2$	=	exponent

18.14 References

Abass, H. H., Brumley, J. L., and Venditto, J. J. 1994. Oriented Perforations—A Rock Mechanics View. Paper presented at the SPE Annual Technical Conference and Exhibition, New Orleans, Louisiana, 25–28 September. SPE-28555-MS. https://doi.org/10.2118/28555-MS.

Aggour, T. 2001. Optimization Strategies for Hydraulic Fractures in High Permeability Reservoirs. Paper presented at the SPE Middle East Oil Show, Manama, Bahrain, 17–20 March. SPE-68131-MS. https://doi.org/10.2118/68131-MS.

Aggour, T. M. and Economides, M. J. 1998. Optimization of the Performance of High-Permeability Fractured Wells. Paper presented at the SPE Formation Damage Control Conference, Lafayette, Louisiana, 18–19 February. SPE-39474-MS. https://doi.org/10.2118/39474-MS.

Barree, R. D., Cox, S. A., Barree, V. L. et al. 2003. Realistic Assessment of Proppant Pack Conductivity for Material Selection. Paper presented at the SPE Annual Technical Conference and Exhibition, Denver, Colorado, 5–8 October. SPE-84306-MS. https://doi.org/10.2118/84306-MS.

Behrmann, L. A. and Nolte, K. G. 1999. Perforating Requirements for Fracture Stimulations. SPE Drilling & Completion **14**(04): 228–234. SPE-59480-PA. https://doi.org/10.2118/59480-PA.

Cinco-Ley, H. and Samaniego-V, F. 1981. Transient Pressure Analysis for Fractured Wells. *J Pet Tech* **33**(09): 1,749–1,766. https://doi.org/10.2118/7490-PA.

Engels, J. N. and England, K. W. 2001. Tip Screenout Designs and Alternative Height Growth Containment Techniques. Paper presented at the SPE Production and Operations Symposium, Oklahoma City, Oklahoma, 24–27 March. SPE-67230-MS. https://doi.org/10.2118/67230-MS.

Gu, H. and Desroches, J. 2003. New Pump Schedule Generator for Hydraulic Fracturing Treatment Design. Paper presented at the SPE Latin American and Caribbean Petroleum Engineering Conference, Port-of-Spain, Trinidad and Tobago, 27–30 April. SPE-81152-MS. https://doi.org/10.2118/81152-MS.

Gidley, J. L., Holditch, S. A., Nierode, R. W. et al. 1989. *Recent Advances in Hydraulic Fracturing*. SPE Monograph Series, Vol. 12. Richardson, Texas: SPE.

Holditch, S. A., Jennings, J. W., Neuse, S. H. et al. 1978. The Optimization of Well Spacing and Fracture Length in Low Permeability Gas Reservoirs. Paper presented at the SPE Annual Fall Technical Conference and Exhibition, Houston, Texas, 1–3 October. SPE-7496-MS. https://doi.org/10.2118/7496-MS.

Hunt, J. L. and Soliman, M. Y. 1994. Reservoir Engineering Aspects of Fracturing High Permeability Formations. Paper presented at the SPE Asia Pacific Oil and Gas Conference, Melbourne, Australia, 7–10 November. SPE-28803-MS. https://doi.org/10.2118/28803-MS.

Indriati, S., Wang, X., and Economides, M. J. 2002. Adjustment of Hydraulic Fracture Design in Gas-Condensate Wells. Paper presented at the International Symposium and Exhibition on Formation Damage Control, Lafayette, Louisiana, 20–21 February. SPE-73751-MS. https://doi.org/10.2118/73751-MS.

Lestz, R. S., Clarke J. N., Plattner D. et al. 1999. Perforating for Stimulation: An Engineered Solution. Paper presented at the 1999 SPE Annual Technical Conference and Exhibition, Houston, Texas, 3–6 October. SPE-76812-MS.

Lizak, K. F., Bartko, K. M., Self, J. F. et al. 2006. New Analysis of Step-Rate Injection Tests for Improved Fracture Stimulation Design. Paper presented at the SPE International Symposium and Exhibition on Formation Damage Control, Lafayette, Louisiana, USA, 15–17 February. SPE-98098-MS. https://doi.org/10.2118/98098-MS.

Lopez-Hernandez, H. D., Valko, P. P., and Pham, T. T. 2004. Optimum Fracture Treatment Design Minimizes the Impact of Non-Darcy Flow Effects. Paper presented at the SPE Annual Technical Conference and Exhibition, Houston, Texas, 26–29 September. SPE-90195-MS. https://doi.org/10.2118/90195-MS.

Malpani, R. V. 2007. *Selection of Fracture Fluid for Stimulating Tight Gas Reservoirs*. MS thesis, Texas A & M University, College Station, Texas (December 2006).

Mohan, J., Sharma, M. M., and Pope, G. A. 2006. Optimization of Fracture Length in Gas/Condensate Reservoirs. Paper presented at the SPE Gas Technology Symposium, Calgary, Alberta, Canada. 15–17 May. SPE-100543-MS. https://doi.org/10.2118/100543-MS.

Montgomery, K. T. and Berthelot, J. M. 1990. Effects of Fracture Fluid Invasion on Cleanup Behavior and Pressure Buildup Analysis. Paper presented at the SPE Annual Technical Conference and Exhibition, New Orleans, Louisiana, 23–26 September. SPE-20643-MS. https://doi.org/10.2118/20643-MS.

Nolte, K. G. 1986. A General Analysis of Fracturing Pressure Decline With Application to Three Models. *SPE Formation Evaluation* **1** (06): 571–583. SPE-12941-PA. https://doi.org/10.2118/12941-PA.

Nolte, K. G. 1988. Application of Fracture Design Based on Pressure Analysis. *SPE Prod Eng* **3** (01): 31–42. SPE-13393-PA. https://doi.org/10.2118/13393-PA.

Penny, G. S. 1987. An Evaluation of the Effects of Environmental Conditions and Fracturing Fluids Upon the Long-Term Conductivity of Proppants. Paper presented at the SPE Annual Technical Conference and Exhibition, Dallas, Texas, 27–30 September. SPE-16900-MS. https://doi.org/10.2118/16900-MS.

Richardson, M. 2000. A New and Practical Method for Fracture Design and Optimization. Paper presented at the SPE/CERI Gas Technology Symposium, Calgary, Alberta, Canada. 3–5 April. SPE-59736-MS. https://doi.org/10.2118/59736-MS.

Robinson, B. M., Diyashev, I. R., Rahim, Z. et al. 2000. Fracture Fluid Damage to High-Permeability Gas Reservoirs: It Can Make a Difference. Paper presented at the SPE Annual Technical Conference and Exhibition, Dallas, Texas, 1–4 October. SPE-63106-MS. https://doi.org/10.2118/63106-MS.

Soliman, M. 1986. Fracture Conductivity Distribution Studied. *Oil & Gas J.* **84** (06).

Soliman, M. Y. and Boonen, P. 1996. Review of Fractured Horizontal Wells Technology. Paper presented at the Abu Dhabi International Petroleum Exhibition and Conference, Abu Dhabi, United Arab Emirates, 13–18 October. SPE-36289-MS. https://doi.org/10.2118/36289-MS.

Soliman, M. Y., Byrd, A., Walters, H. et al. 2004. Reservoir-Based Fracture Optimization Approach. Paper presented at the SPE International Thermal Operations and Heavy Oil Symposium and Western Regional Meeting, Bakersfield, California, 16–18 March. SPE-86991-MS. https://doi.org/10.2118/86991-MS.

Soliman, M. Y. and Hunt, J. L. 1985. Effect of Fracturing Fluid and Its Cleanup on Well Performance. Paper presented at the SPE Eastern Regional Meeting, Morgantown, West Virginia, 6–8 November. SPE-14514-MS. https://doi.org/10.2118/14514-MS.

Soliman, M. Y., Hunt, J. L., and Azari, M. 1999. Fracturing Horizontal Wells in Gas Reservoirs. *SPE Prod & Fac* **14** (04): 277–283. SPE-59096-PA. https://doi.org/10.2118/59096-PA.

Valko, P. 2007. Production Enhancement. Lecture Notes, PETE 416. Texas A&M University.

van Gijtenbeek, K. A. W. and Pongratz, R. 2004. Perforating and Hydraulic Proppant Fracturing in Western Siberia, Russia. Paper presented at the SPE Annual Technical Conference and Exhibition, Houston, Texas, 26–29 September. SPE-90238-MS. https://doi.org/10.2118/90238-MS.

Veatch, R. W. 1983. Overview of Current Hydraulic Fracturing Design and Treatment Technology—Part 1. *J Pet Tech*, **35** (04): 677–687. SPE-10039-PA. https://doi.org/10.2118/10039-PA.

Vincent, M. C., Pearson, C. M., and Kullman, J. 1999. Non-Darcy and Multiphase Flow in Propped Fractures: Case Studies Illustrate the Dramatic Effect on Well Productivity. Paper presented at the SPE Western Regional Meeting, Anchorage, Alaska. 26–27 May. SPE-54630-MS. https://doi.org/10.2118/54630-MS.

Wright, C. A., Weijers, L., Davis, E. J. et al. 1999. Understanding Hydraulic Fracture Growth: Tricky but Not Hopeless. Paper presented at the SPE Annual Technical Conference and Exhibition, Houston, Texas, 3–6 October. SPE-56724-MS. https://doi.org/10.2118/56724-MS.

Chapter 19

Planning and Executing a Hydraulic Fracture Treatment

Stephen A. Holditch

19.1 Pretreatment Quality Control Considerations

Hydraulic fracturing is a tool oil companies use to enhance production from existing and new wellbores. Although fracturing is not used on all wells, a vast majority of them are in need of this type of stimulation. It is used in situations ranging from overcoming near-wellbore damage to increasing productivity in tight microdarcy reservoirs. The theory behind a conventional (hydraulic) stimulation method is to pump fracturing fluid blended with proppant, which is usually sand, down the wellbore at a high-enough rate to overcome the breakdown pressure of the formation, while staying within the bounds of the casing strength and uphole equipment. When the pressure in the wellbore exceeds the formation pressure, the formation breaks, allowing the proppant-filled fluid to travel into the reservoir through a large fracture in which fluids from the reservoir can travel more readily into the wellbore and back up to surface.

Although this process might seem simple, in reality, it is becoming more complicated to perform as the industry starts to step away from the conventional and easily obtainable reservoirs. One way to deal with some of these less-conventional, low-pressure reservoirs or water-sensitive zones is to use foamed or energized stimulation in which a gas, commonly nitrogen or carbon dioxide, is used in conjunction with the fracturing fluids.

For a fracture stimulation to be as effective as possible, many checkpoints must be met before, during, and after the fracturing job; the most crucial checks should be made during the pretreatment phase of the job because this is controllable, unlike the characteristics of the formation. Therefore, for the job to go smoothly, it is imperative to have on-site quality control (QC) before the treatment to ensure all necessary ingredients are uncontaminated and in the appropriate location for the job to go off correctly. To ensure that quality is not compromised, it is important to have one person—usually a person from the operating company—in charge of QC on the wellsite.

As Hall and Larkin (1989) observed, QC must be put in place to ensure maximum productivity from the stimulated well, to obtain quality services and materials from service companies, to eliminate substandard performance from service companies, to improve the quality of future jobs, and to keep learning about fracturing. On-site pretreatment QC includes preparing the location and checking fracturing equipment; fluids; foam considerations, if applicable; proppants; and monitoring equipment.

19.1.1 Fracture Equipment. As fracture tanks, which are used to segregate various fluid types, additives, and proppants, are brought in and placed, the QC engineer should check the cleanliness of the tanks and that their sizes are accurate. This is the starting point to having a successful fracturing job when deciding where to spot equipment, as previously discussed.

Fracture tanks come in many different sizes and shapes. Depending on the location, size, and scale of the job, fracture tank dimensions can play a major role in the design. Tanks can be rectangular or cylindrical, vertical or horizontal, and can hold different volumes. When talking to a service company, the QC engineer should make sure to check how many tanks are needed and the availability of these tanks in that area. Most service companies have this information, but it is essential for the QC engineer to question this to make sure there are no delays when stimulating the formation(s).

Cleanliness is a major issue that cannot be overlooked by the QC engineer. This is crucial because dirty tanks can lead to dirty fracturing fluids, lowering the permeability in the proppant being pumped and causing gunk to fill pore spaces, thereby defeating the purpose of stimulating the well.

To start the QC procedure, the engineer first can visually check the tanks to ensure there are no fluids or sands left from previous jobs that could contaminate the fluids being stored. Also, the tanks should be lined to help safeguard them from possible iron contamination in the water that could lead to problems with gel hydration and/or the crosslinking mechanisms of fracturing fluids. When there is a split or holes in the tank lining, the spaces beneath the lining promote bacterial growth or introduce an incompatibility component to the fluids that are in place when attempting to blend them with other additives. Therefore, checking the fracture tank can save money by stopping potential negative impacts resulting from contaminated fluids. Although there are steps in place to check for bacterial growth after fracturing fluid is placed in the tanks, such as rheology measurements, many QC engineers add a small amount of bactericide to help kill any bacteria that might be present in the tanks.

Along with the tanks, valves should also be checked to ensure they are not leaking. Leaky valves can cause too much water to enter the system, which can dilute the gel system needed in the formation. All valves should be checked to ensure they are in good condition before loading fluids and additives into the tanks.

19.1.2 Fracturing Fluids. Fracturing fluids are among the main components pumped down the well to help carry proppants to the fracture opening and into the reservoir; therefore, uncontaminated and carefully chosen fracturing fluids must be pumped to ensure conductivity of the fracture and effective flow rates into the wellbore.

As previously stated, fracturing fluids must be free of contaminants because the gel systems are so sensitive to pH, temperature, and foreign substances. In addition to iron, other contaminants include rust, oil, drilling mud, residue, gravel, old gel, bacteria, phosphates, and reducing agents. Temperature and pH affect the polymer system when it comes to dispersion of the fluids. In colder temperatures and alkaline pH levels, dispersion is severely reduced.

When checking for the quality of the fracturing fluids, several things must be reviewed. It is very important to ensure that all tests on the water source are correctly executed. Typically, water sources include lakes and reservoirs near the location. The engineer should visually check the water in each tank for oil slicks, suspended particles, and debris.

Before adding any chemicals or pumping the job, each tank should be strapped and analyzed to ensure the correct specific gravity, pH, and temperature are present. After that, it is essential to run small-scale pilot tests on the fluids with a gelling agent to ensure proper viscosity and crosslink time. Checking for base-gel viscosity is done by allowing a small sample to flow through a funnel and recording the time from when the sample enters the funnel to when it exits. These samples should be taken from the top and bottom of each fracture tank to ensure that the outcomes are uniform for each tank. The acceptable ranges for viscosity and crosslink time are a function of mixing procedure and downhole temperature.

When fracturing fluids have been analyzed for quality, they are then sent to the blender to get mixed with the proppant before being pumped down to the formation. As the fracturing fluid travels through the blender, the QC engineer can visually observe the fluid for formation of unhydrated polymer lumps during the gelling operations to ensure nothing contaminated flows into the formation below.

19.1.3 Proppants. The delivery methods of proppants differ depending on the size of the fracture treatment. In small jobs, dump trucks are placed next to the blender to gravity feed the proppant into the blender at a manually controlled rate. In large jobs, companies use storage bins to store their proppant. These storage bins are usually divided into compartments and are hydraulically controlled. The gates to a storage bin open and drop the proppant onto a conveyor belt that transports it to the blender, where it then mixes with the fluids. The proppant is used to keep the formation propped open so that reservoir fluids can travel down the higher-permeability sand and into the wellbore; thus, high-quality sand is necessary for the stimulation.

The QC engineer must observe the condition of the proppant as it is brought to location. The top hatches should then be inspected to ensure the cleanliness of the proppant. Also, when looking at the bins, the engineer should ensure there are no remaining contaminants from prior jobs, such as cement or other types of sand. If these contaminants are present, they could result in ineffective performance of the fractured well after the stimulation is completed.

When inspecting the proppant, the engineer should confirm the cleanliness of the proppant itself, as well as that it is the correct type and size. Before pumping the job, a proppant sieve analysis and physical parameters test should be conducted on location to guarantee that it meets the expected standards of that proppant type, such as proppant grade and the amount of fines present. Each bin should be individually tested to confirm uniformity among all bins.

19.1.4 Monitoring Equipment. For any job to be successful, the monitoring equipment must be accurate and functional. Therefore, before pumping a job, the QC engineer should test all pumps, flowmeters, density meters, pressure transducers, and thermometers, if applicable. The QC engineer should recalibrate all instruments to fit all the standards requested.

19.2 Planning and Scheduling of Hydraulic Fracturing Treatments

A QC program for hydraulic fracturing starts with the primary-cementing program. The elasticity of the cement must withstand high pressures and temperature changes during the fracturing treatment. Well information such as in-situ stresses, reservoir description, and rock properties is necessary to start the fracture design. In addition, evaluations of fracture propping agents on location, water quality, fracture equipment, and personnel performance are relevant during the design stage and execution of the fracturing treatment.

The on-site QC program requires an inventory of all additives before, during, and after the fracturing treatment to verify the quality of the fracturing fluid system. Health, safety, and environmental issues are very important during any stage of the fracture treatment. Previous experience in the area or in similar wells enables reducing or eliminating most quality risks.

A QC program requires checking water quality, gelled fracturing fluids, fracture propping agents, breakers, and other additives used in the fracture treatment. The materials that primarily affect the fracture treatment are the viscous fracturing fluids and the propping agents. Most of the fracture treatments use a water-based polymer; therefore, we will focus on QC of water-based polymers.

19.2.1 QC Procedures. Careful planning and documentation of all steps and risks in the fracturing treatment process can eliminate human error. Before starting a fracture treatment, it is necessary that all on-site QC criteria be satisfied. Screenout near the wellbore mainly is caused by tortuosity effects, insufficient fracture width, rock properties, and pump failure. A poor perforation design increases the chances of an early screenout.

19.3 Fracturing Fluid Preparation

High-quality fracturing fluids require that the water supplied for the treatment be free of contaminants that affect the proper properties of the system. Sources of contaminants could be residues from previous treatments in the tanks, water transports, or water source. According to Hall and Larkin (1989), common contaminants are rust, oil, drilling mud, gravel, old gel, bacteria, and phosphates. Bacteria and pH of the water can ruin the polymers, causing poor viscous properties of the fracturing fluid. Phosphates reduce the crosslinking effect of hydraulic fluids.

The QC procedure must be conducted during the gel preparation to check for the formation of fisheyes, or unhydrated polymer lumps; if gel is added to water too rapidly, it can lead to improper pH control and blender equipment malfunction. Hall and Larkin (1989) recommend that base-gel viscosity should be measured for fluid samples taken at the top and bottom of each fracturing tank to determine possible problems, such as gel stratification, air trapping, or improper gel loading. The ideal fracturing gel should contain no fisheyes.

Most polymer gelling agents disperse and hydrate when the water has a pH level of 6 to 8. Water with a pH higher than 8 might result in a poorly hydrated gel; water with a pH lower than 6 could hydrate too quickly and create gel balls, lumping, or fisheyes.

After checking the viscosity properties of the fracturing fluid, it is necessary to check the crosslinking ability of the base gel. The crosslink time can range from a few seconds for a borate-crosslinking system to several minutes for a delayed organometallic-crosslinking system. The delayed crosslinking time should be equal to the pumping time from the wellhead to the perforations. The crosslinking tests are at bottomhole-temperature conditions. The downhole temperature conditions can decrease during the fracturing treatment, so the concentration or crosslinker type can change during the fracture treatment. Bicarbonate concentrations greater than 300 ppm can delay crosslinking for some fluids. HCl acid could adjust the fluid pH to solve this problem. Davidson et al. (1994) concludes that overcrosslinking and undercrosslinking of the fracturing fluid systems can result in a lower-viscosity fluid that affects its fracture-propping-agent transport properties.

Fracturing water containing excessive amounts of calcium and magnesium can prevent proper gelling, crosslinking, temperature maintenance, and shear stability. NaOH can adjust fluid pH. Mix water needs to be free of iron, less than 10 ppm, to promote gel degradation. Viscometer tests can include pressure and temperature parameters to verify the rheology of the fracturing fluids before pumping them. In addition, a series of tests can check the effect of crosslinker concentration on the fluid viscosity.

Buffers typically control gel hydration, crosslinking, and fluid stability. The use of encapsulated breakers has become more common in high-temperature formations to reduce fluid viscosity. Screenout could occur if fluid characteristics are not right or are even slightly off.

19.4 Executing the Hydraulic Fracturing Treatment

19.4.1 Job Responsibilities. The most important part of a fracture treatment is ensuring that it is pumped as designed, without causing injury or harm to any individuals, while protecting the integrity of the wellbore. Of course, designing the optimal fracture treatment is important. The design phase is difficult, time consuming, and complicated. We must select the best completion method; the best injection method; the correct fluids, additives, and propping agents; and the best pumping schedule. But designing the optimal treatment will not ensure success until that treatment is successfully pumped into the formation.

Two groups are involved in any well operation: the operator or owner of the well and the service company that will perform the well operation. It is very important to establish the two persons who will be at the top of the chain of command for each group. The ultimate authority will be the operating company representative in charge. This person should be intimately involved in the design and planning of the treatment. All major decisions before, during, and after the treatment should be approved by the operating company supervisor.

The service company also should have a single person in charge. Many times this person is called the fracture treatment operator. This person can advise the operating company supervisor, but should not overrule the operating company supervisor unless a safety issue dictates such action. Normally, the two supervisors work closely together in harmony for the betterment of the operation.

19.4.2 Rig Up of Equipment. Once the location has been made ready for the fracture treatment, the service company will begin bringing in the equipment needed for the treatment. Generally, a location diagram will be made so that everyone knows where to place the pits, tanks, proppant storage containers, blenders, pumps, and other pieces of equipment or chemicals.

It is very important to place the high-pressure equipment and injection lines so that everyone has access to the blenders, the low-pressure lines, and the proppant-delivery systems during the treatment. No one should have to get near the high-pressure equipment during the treatment. The location should have the proper instrumentation to be able to monitor and record everything that occurs on location during the treatment.

19.4.3 Safety Meeting. Safety is important. In most cases, a safety meeting should be held every day to inform all on-location personnel of the activities of the day and what is expected of each of them. Basic safety issues should be discussed. On the day of the fracture treatment, after the equipment has been rigged up and tested, a safety meeting should be held to ensure that everyone knows what to expect during the fracture treatment. The first order of business should be to establish a chain of command. All persons on location should know who is in charge from the operating company and from the service company. In addition, the service company fracture operator should specify to each person with the service company what that person's job assignments are for the day.

The service company should have a clear line of command for the blender operators; the pump operators; and the persons in charge of the fracturing fluid tanks, additives, and propping agents. Everyone should have a role that is clearly defined and understood.

Once the roles are clearly understood, the service company supervisor should go through the treatment procedures with everyone on location. The size of the treatment should be reviewed, as well as the proposed injection rates and maximum allowable surface injection pressures. The crew should understand the order of scheduling for the fluids, additives, and propping agents. The capacity of the wellbore should be discussed, and the operator should be assured that the treatment is not overflushed.

A plan should be discussed in case an emergency occurs, such as a ruptured injection line or wellhead, or a fire in one of the engines. Fire extinguishers should be on location, and everyone should be informed of their location. A meeting place should be established at which all persons can gather, in case the fracture location must be abandoned.

19.4.4 Treatment Schedule. As previously mentioned, the treatment schedule should be discussed in detail before starting the treatment. In many cases, the original treatment schedule will be altered as more data become available or more individuals review the schedule and make changes. It is not unusual to have several versions of a completion and stimulation procedure in circulation before the treatment, so the service company supervisor and the operating company supervisor should verify with the design engineers to be certain everyone is using the correct and final version of the procedures.

Fig. 19.1—A typical well site for fracture treatments in the Appalachia Basin (Courtesy of John Wang).

Once the two supervisors agree on the correct procedure, then they should let everyone know the exact pumping schedule planned for the treatment. The crew should know the fluid schedule, the proppant schedule, the additives required, and the expected injection rates and pressures. Everyone should also know the maximum allowable injection pressure and the displacement volume.

If any discrepancies or questions arise, they should all be answered or cleared up before beginning the pumping operation. For example, once the schedules are known, the crew should be able to verify that enough fluid, proppant, and chemicals are on location to correctly pump the treatment. If anyone expresses any doubts, those doubts should be cleared up at once.

19.4.5 Material Inventory. As part of their job responsibilities, the service company and the operating company should conduct inventories of all fluids, chemicals, and propping agents before and after the fracture treatment. If these inventories are correctly conducted, then the crew should be able to tell exactly what has been pumped into the fracture. In addition, it is a good idea to conduct inventories during the treatment. For example, when the treatment is approximately 50% complete, then half of the fluid should have been emptied from the tanks. This sounds trivial, but in some cases, the flowmeters might not be properly calibrated and the crew might have pumped more or less than the amount of fluid they should have pumped at some point in the treatment. The same logic applies to proppant and chemicals. The crew should perform periodic inventories to verify the readings from densitometers and chemical flowmeters.

19.4.6 Logistics and Site Layout. Well site preparation is one of the most important aspects of pretreatment QC that needs to be closely looked at. The main idea is to ensure that the location is easily accessible and is large enough to hold all the fracturing equipment needed for the job. For example, if the location currently has stock tanks, they can greatly hinder how much equipment can fit on location. Other restrictions on the location can be mud pits, separators, glycol units, and rod pumps.

To guarantee that the location is ready for the fracturing equipment and materials, the service company person in charge of QC can draw a site plan of where all the equipment should be spotted on the wellsite. **Fig. 19.1** shows the elements of such a site plan. Spotting equipment is very important during the actual treatment to ensure that all materials are accessible when they are needed. The equipment shown in Fig. 19.1, including the pumps, blender, fracture tanks, proppant-handling tanks, and monitoring equipment, is needed for nearly all stimulation treatments. The blender is used to prepare the correct proportion of liquid and additives specified in the design.

The placement of equipment also is key to how the job is pumped during the treatment. Usually, tanks are placed so that the one that contains the highest-viscosity gel is pumped first; therefore, it is the closest to the wellhead and blender. As the job continues, the fracturing fluid that has successively less viscosity is pumped, because there is a higher probability of screenout and unwanted fracture-height growth in the zone of interest.

19.5 Fracturing Treatment Diagnostics During the Treatment

Fracture growth depends on the rock properties and in-situ stresses in the formation. Fracture mechanics control fracture-height growth, propagation of the fracture, near-wellbore fracture turning, tortuosity, fracture azimuth, and fracture closure behavior. Although many measurements are available to help to define the fracture behavior, there is seldom a case in which the fracture growth patterns are precisely known.

The following data must be measured and recorded during the treatment, as a function of time:

- Pumping rate
- Surface pressure
- Bottomhole pressure
- Net pressure
- Fracture propping-agent concentrations
- pH and additives rates

Fracture tank levels require strapping and recording as often as possible during the treatment. Fluid samples, including pad, first, middle, and final fracturing-fluid stages, as well as fracture propping samples, must be collected during the treatment and tested for crosslink times. Deviations in expected fluid properties, such as crosslinking time or viscosity, require immediate response. Viscometer tests should include temperature and pressure, as well as rheology. van Gijtenbeek and Pongratz (2004) state that extremely cold conditions could affect crosslink times because of the addition of cold fracture-propping agents to a relative warm fluid.

19.5.1 Assess and Monitor the Fracture. Net-pressure fracture analysis is the most common way to judge the fracture treatment and its geometry. It is possible to make use of net-pressure fracture analysis during or after the fracture treatment. Unfortunately, even with downhole treatment-pressure data, it is not possible to accurately determine the geometry of the hydraulic fractures. However, the real-time fracturing pumping pressure monitoring and analysis could help minimize screenout and optimize a treatment.

Another indirect way to estimate the geometry of the hydraulic fractures is using post-treatment production data. Bennett et al. (2005) suggest that the post-treatment production-data method determines the well productivity, and it reflects the portion of the hydraulic fracture that is open, cleaned up, and contributing to production. Using post-treatment production data requires several months of well production. Well-testing methods (buildup and drawdown) can define the effective production geometry created hydraulically.

Near-wellbore methods such as radioactive tracers, temperature logs, and production logs can be used to investigate the presence of hydraulic fractures. Near-wellbore methods determine the presence of hydraulic fractures and estimate fracture height, but they cannot measure what is occurring away from the borehole.

Radioactive-isotope tagging during injection enables distinguishing of fluid and fracture-propping-agent positions for multiple-stage stimulation treatments. Temperature logs identify near-wellbore regions cooled by injection of fracturing fluids and, therefore, an estimation of the fracture height. Production logs that measure fluid flow, fluid density, and temperature determine perforation intervals opened to production.

Surface and downhole tiltmeters and microseismic monitoring can characterize hydraulic fracture behavior and geometry away from the wellbore. Tiltmeters measure hydraulically induced deformations. By placing the devices in an array of shallow boreholes (20 to 40 ft deep) that measure deformations induced by fracture creations, surface tiltmeters can estimate the azimuth, dip, depth, and width of the hydraulic fractures. Downhole tiltmeters are located in nearby monitoring wells at a depth similar to that of the created fracture. Downhole tiltmeters tend to be more accurate than surface tiltmeters because the sensors are closer to the propagating fracture. According to Barree et al. (2002), downhole tiltmeters can estimate the azimuth, height, length, and width of the fracture. Tiltmeter results depend on the spatial relationship between the tiltmeters (surface or downhole) and the treatment well.

Mapping with surface tiltmeters has limitations in characterizing hydraulic fractures deeper than 10,000 ft. Bennett et al. (2005) suggest that downhole tiltmeters normally lose their effectiveness when the distance from the hydraulic fracture to the tiltmeter exceeds three times the length of the created fracture. **Fig. 19.2** shows the capabilities and limitations of indirect and direct fracture-diagnosis techniques.

Microseismic events, or small earthquakes, happen because the normal stresses can reduce along pre-existing planes of weakness until shear slippage occurs. According to Bennett et al. (2005), shear movements emit both compressional and shear waves detected by geophones. The shearing accompanies the fracture-tip zone; locating the source of these waves in space and time enables constructing a map of the created fracture by plotting the location of the acoustic emissions over time while fracturing. According to Bennett et al. (2005), acoustic emissions might also occur away from the fracture tip, where stress changes cause shear slippage in natural fractures.

Microseismic mapping requires an accurate velocity model from which to calculate event locations, a low-noise environment, highly sensitive geophones to record microseismic events, and knowledge of the exact location and orientation of the receivers. **Fig. 19.3** shows a real-time imaging of microseismic mapping in which the fracture is close to intercepting a thrust fault.

The interwell geology also plays an important role. Vertical seismic profile (VSP) surveys help relate the time domain to the depth domain and, therefore, help calibrate the velocity model. The versatile seismic imager (VSI) tool used to obtain the VSP data also records the microseismic events, ensuring consistency in data acquisition, processing, and interpretation.

Reservoir-fluid type can also affect microseismic activity. Fluid factors can reduce stress and pore-pressure changes in the formation that occur during fracturing. A gas-filled reservoir can produce a narrower band of microseismic events that more clearly defines the geometry of the fracture (Warpinski et al. 2001).

Development oil fields have many sources of noise that could have a negative impact on the microseismic technique, such as electrical noise, nearby drilling activity, hydraulic fracturing treatments, or fluid flowing from perforations in the monitoring well. Filters can eliminate most of the noise during data processing.

Tiltmeters and microseismic techniques enable the monitoring of the hydraulic fracture during the treatment, and it is possible to make changes to the design to prevent early screenout or to divert the treatment fluid away from perforations that can take the majority of the treatment.

One of the major limitations in microseismic monitoring methods is finding a candidate treatment well with nearby monitoring well or wells in which the VSI tool (less than 2,000 ft) can be installed. The well must have good cement and be acoustically quiet during fracturing operations. Microsiesmic events used with seismic inversion can determine stress changes, then the stress changes can constrain geomechanical models and better model the fracture propagation in the reservoir.

Techniques ■ Can determine ▨ May determine □ Cannot determine

Group	Fracture-Diagnostic Method	Main Limitations	Ability To Estimate							
			Length	Height	Asymmetry	Width	Azimuth	Dip	Volume	Conductivity
Far field, during fracturing	Surface tiltmeter mapping	• Cannot resolve individual and complex fracture dimensions. • Mapping resolution decreases with depth (fracture azimuth ±3° at 3,000-ft depth and ±10° at 10,000-ft depth.	May	May	May	Cannot	Can	Can	Can	Cannot
Far field, during fracturing	Downwhole tiltmeter mapping	• Resolution in fracture length and height decreases as monitoring-well distance increases. • Limited by the availability of potential monitoring wells. • No information about proppant distribution and effective fracture geometry.	Can	Can	Can	May	May	Cannot	Can	Cannot
Far field, during fracturing	Microseismic mapping	• Limited by the availability of potential monitoring wells. • Dependent on velocity-model correctness. • No information about proppant distribution and effective fracture geometry.	Can	Can	Can	Cannot	Can	May	Cannot	Cannot
Near wellbore, after fracturing	Radioactive tracers	• Measurement in near-wellbore volume. • Provides only a lower limit for fracture height if fracture and well path are not aligned.	Cannot	May	Cannot	May	May	Cannot	Cannot	Cannot
Near wellbore, after fracturing	Temperature logging	• Thermal conductivity of different formations can vary, skewing temperature log results. • Post-treatment log requires multiple passes within 24 hours after the treatment. • Provides only a lower limit for fracture height if fracture and well path are not aligned.	Cannot	May	Cannot	Cannot	Cannot	Cannot	Cannot	Cannot
Near wellbore, after fracturing	Production logging	• Provides only information about zones or perforations contributing to production in casedhole applications.	Cannot	May	Cannot	Cannot	Cannot	Cannot	Cannot	Cannot
Near wellbore, after fracturing	Borehole image logging	• Run only in open hole. • Provides fracture orientation only near the wellbore.	Cannot	May	Cannot	May	May	May	Cannot	Cannot
	Downhole video	• Run mostly in cased holes and provides information only about zones and perforations contributing to production. • Might have openhole applications.	Cannot	May	Cannot	Cannot	Cannot	Cannot	Cannot	Cannot
Model based	Net-pressure fracture analysis	• Results depend on model assumptions and reservoir description. • Requires calibration with direct observations.	May	May	Cannot	May	Cannot	Cannot	May	May
Model based	Well testing	• Results dependent on model assumptions. • Requires accurate permeability and reservoir pressure estimates.	May	Cannot	Cannot	May	Cannot	Cannot	Cannot	May
Model based	Production analysis	• Results dependent on model assumptions. • Requires accurate permeability and reservoir pressure estimates.	May	Cannot	Cannot	May	Cannot	Cannot	Cannot	May

Fig. 19.2—Capabilities and limitations of fracture diagnostics (from Bennett et al. 2005, after Cipolla and Wright 2000).

19.6 Post-Treatment Reporting and Evaluation

A fracture treatment report should contain the following information:

- Graphical and digital data of treatment parameters
- Listing of treatment parameters
- Prefracture treatment QC form
- Inventory form
- During-fracture treatment QC form
- Fracture-propping-agents inventory form
- Mass-balance/cumulative additive schedule
- Chart of viscosity, pH, and chemical-additive rates

Collected samples during the treatment should be stored for a possible post-treatment inspection. Collecting post-fracture production data and comparing them to the original design can help us better understand the fracture treatment.

19.6.1 Summary.

- QC before, during, and after the hydraulic fracturing treatment leads to a successful placement of fracturing treatments.
- Rigorous QC and quality-assurance procedures must be in place, including fluid testing, for actual reservoir conditions to check the performance of the fracturing fluid downhole.
- Strict inventory practices of the products and fluid tests before, during, and after the fracturing treatment permit better control of the operation.

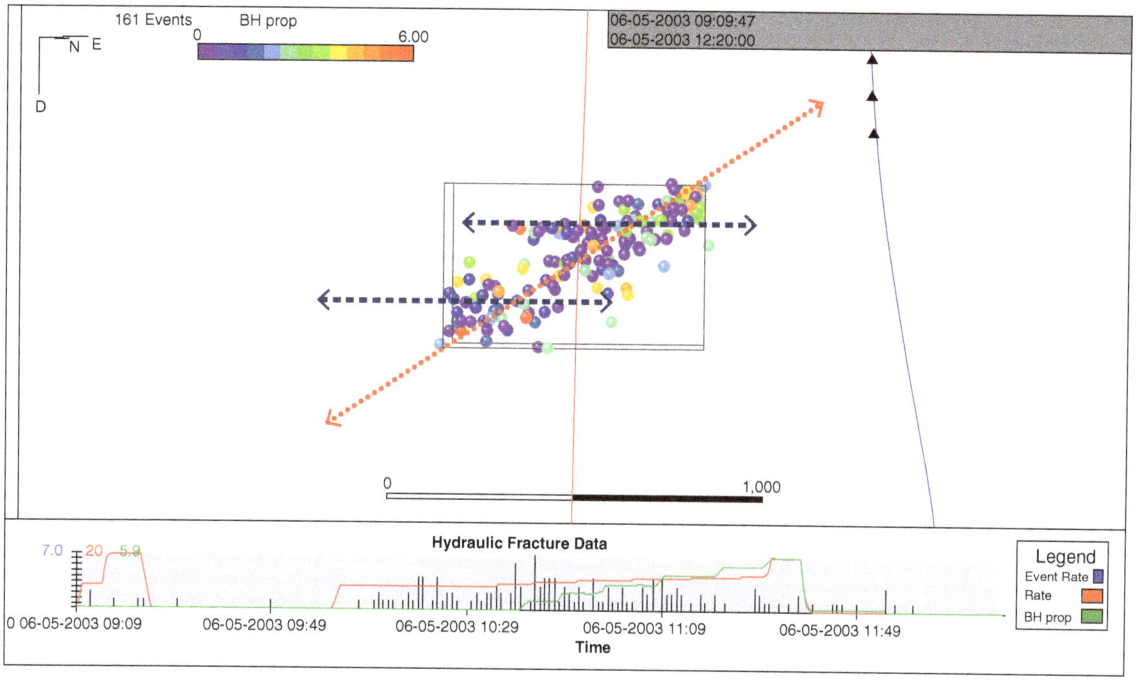

Fig. 19.3—Real-time imaging of microseismic mapping (based on Bennett et al. 2005).

- Fracture mechanics play an important role in hydraulic fracturing, and it is important not to save effort in acquiring and assembling all data required for a fracture design.
- There are more chances of fracture-treatment success when all processes of a hydraulic fracture are controlled.
- Microseismic imaging, perhaps, may offer one of the best resolutions to image fracture complexity, and allow assessing the possibility of altering the fracturing design to control the fracture in real time or to optimize future fracture treatment.

19.7 Flowing Back the Treatment and Forced Closure

The goal of a good production engineer is to maximize well and, ultimately, fieldwide productivity in a cost-effective manner (Economides et al. 1994). This is true whether the producing asset is a vertical, a horizontal, or a fractured well. For fractured wells, a fracture treatment can account for a large portion of total well cost (Ely et al. 1990). This assertion serves to emphasize the necessity of an effective fracturing job.

In the ideal case, once the proppant is placed in the fracture, the fracture should close with the proppant still in suspension. The fracturing fluid then breaks and the well is opened for cleanup.

However, one of the challenges expected when flowing back the treatment is proppant flowback. Proppant flowback results in consequences detrimental to fracture conductivity and, hence, well productivity. It also affects well integrity. These consequences are well-established in the literature and include poor productivity, increased overhead because of damage to surface facilities, and costly well intervention or recompletion operations (Robinson et al. 1988).

Practices to minimize the occurrence of proppant flowback have been applied successfully in the field. The solutions fall under two broad headings: those that deal with pre-job, on-the-job, and post-job design/engineering; and those that adjust the physical properties of the proppant and fracturing fluid to minimize the incidence of proppant flowback. Service companies tend to propose the latter, while contributors from operating companies and academia tend to favor the former.

There is debate regarding the mechanism of proppant flowback and the best way to combat it. The safest conclusion is that, now, there are no general solutions. If a given field has the same rock properties over a large area, then a specific solution can be generalized for the field.

Other challenges associated with flowing back the treatment include excessive fracturing-fluid viscosities after closure resulting from insufficient breaker loading and safety/environmental concerns regarding the disposal of contaminated fracturing fluids.

The subsections that follow review the causes and remediation of proppant flowback during fluid cleanup. They draw special attention to the concepts of forced closure and reverse screenouts and discuss other factors such as excessively high flowback rates that tend to aggravate flowback problems. In addition, they include a description of products from service companies that increase proppant-pack resistance to drag forces. Last, they emphasize the importance of the bottomhole flowing pressure as a parameter for determining choke size and, hence, flowback time.

19.7.1 Flowing Back the Treatment—An Idealized View. The objective of a fracturing treatment is to maximize both the near-wellbore fracture conductivity and the maximum attainable conductive length in communication with the perforations. Therefore, an ideal flowback design should result in a propped fracture that will meet that standard. No matter how good flowback procedures are, if the proppant scheduling is inadequate or the fracturing fluid is unstable, achieving the idealized result is difficult.

19.7.2 Causes of Proppant Flowback. There is no general theory yet regarding how proppant flowback occurs. Parker et al. (1999) suggested that a combination of interactive, hydrodynamic, and mechanical forces tend to determine whether a proppant pack will be stable when subjected to fluid flow. Interactive forces tend to keep the proppant in the fracture, while hydrodynamic and mechanical forces tend to extrude the proppants. Typically, increased fluid viscosity and drawdown worsen the effect of the hydrodynamic force, while increased closure stress worsens the effect of the mechanical force. Closure stress has also been reported to increase proppant-pack stability. Above a critical value, closure pressure becomes a destabilizing factor; below this critical value, it acts as a stabilizing factor. Terracina et al. (2001) developed a material that increases the interactive forces between the proppant grains. Whether or not a proppant pack fails depends on the dominant force present in the fracture system.

Neither Parker et al. (1999) nor Terracina et al. (2001) mention the fracture width as a major factor; however, Canon et al. (2003) reported that fractures wider than six particle grains could be unstable. Whether or not a proppant pack is stable depends on the resultant effect of all factors that affect proppant-pack stability.

19.7.3 Minimizing Proppant Flowback. No universal technique exists for preventing proppant flowback. Because the concept itself is not fully understood, the available solutions are mostly applicable to specific reservoir settings. The concepts currently used to minimize the occurrence of proppant flowback include forced closure, reverse screenout, surface modification agents, resins, or deformable proppants.

Strictly speaking, forced closure might not prevent proppant flowback. It does prevent the proppant from settling out of zone in the fracture. However, because the forced closure procedure ensures fracture closure, it apparently aids proppant stability by confining the proppant pack.

19.7.4 Forced Closure and Reverse Screenouts. Forced closure is a technique used in low-permeability reservoirs when minifracture tests or pressure-falloff data indicate that the fracture closure will not occur quickly. If fracture closure times are long, proppant settling and fracturing fluid degradation become likely. The time allowed for fracture closure depends on the formation characteristics, the settling tendencies of the proppant used, and the suspending properties of the fracturing fluid. The technique of forced closure reduces wellhead pressure and, ultimately, fluid pressure in the fracture. Forced closure has been reported to be very successful at minimizing the occurrence of proppant flowback in low-permeability reservoirs with strong stress contrasts.

In their appraisal of the forced-closure technique, Barree and Mukherjee (1995) explained that forced closure is a necessary result of proppant movement or settling at the end of the job, before the fracture closes. They suggested six factors that enable proppant movement in the fracture at the end of pumping:

- The expansion of wellbore fluids caused by temperature and compressibility. The effect of thermal expansion can be neglected for foamed and nonfoamed fracture fluids; however, the effect of fluid compressibility is appreciable for foamed fluids. This force tends to extrude the proppant as the fluids expand, especially in low-permeability reservoirs.
- The leakoff gradient from the tip of the fracture to the wellbore. Because fluids take the path of least resistance, the fluid in the fracture tends to move toward the area of highest leakoff in the fracture. As the fracturing fluid moves, it carries the proppant away from the perforations, thereby reducing the proppant concentration at the near-wellbore region.
- The instability of the fracture fluid. This accelerates the tendency of the proppant to settle out of zone.
- The pressure fluctuations transmitted downhole as the high-pressure pumps are shut down at the end of the job. High velocities result, leading to proppant redistribution along the fracture.
- The continual fracture extension in low-permeability reservoirs after the pumps are shut down. This tends to cause the proppant-laden fracturing fluid to move away from the perforations.
- The crossflow from high-closure-stress layers to lower-closure-stress layers. This tends to flush the lower-closure-stress layer.

The main drawback to using the forced-closure technique is that there is no set of guidelines for determining where it should or should not be applicable. However, if the fracturing treatment was designed to screen out at the wellbore, then forced closure would no longer be a necessity. This concept is called the "reverse screenout." For this procedure, the treatment is designed so that the last stage before the flush has the highest proppant loading. The assumption is that minimizing the pad volume and maximizing the proppant loading at the tail end of the job lead to tightly packed fractures that are less prone to flowback. But, that assumption is debatable, because if the fracture is tightly packed, the closure stress might act in a way that aids the extrusion of the proppant.

19.7.5 Resin-Coated Proppants, Resins, Fibers, Deformable Proppants, and Surface-Modification Agents. Largely, service companies have responded to the challenge of proppant flowback by developing products that either increase the bonding between individual proppant grains or use deformable proppants to increase the contact frictional force between grains. Sometimes, this leads to increased proppant-pack stability at the expense of fracture conductivity. The use of resins and resin-coated proppant has been limited, because they are perceived to reduce fracture conductivity, and also because the bonded grains do not always have enough strength to ensure proppant-pack stability.

Some success has been reported for deformable proppants. They are used with normal proppants in percentages of 10 to 25% by weight of normal proppant (Stephenson et al. 1999; Brannon et al. 2003). The challenge with these solutions is that their ability to effectively function under high closure stresses and high temperature has not been conclusively proved.

The successful use of surface-modification agents has been reported in literature (Terracina et al. 2001; Nguyen et al. 2006). These agents are soluble in water and in oil. They do not harden under reservoir conditions, but they help minimize proppant flowback by making the proppants tacky. However, they can be rendered ineffective if the critical fluid flow rate for the system has been exceeded.

The main drawback of these approaches is that service companies drive them. They tend to focus on increasing the resistance of the proppant pack to drag forces. The forces that tend to cause the proppant pack to fail are not hydrodynamic. However, regardless of the drawbacks of these approaches, they still provide options to the practicing engineer seeking solutions to challenges related to proppant flowback.

19.7.6 Relationship Between Choke Diameter and Tubinghead Pressure. For given rock and fluid conditions, it is important to identify the optimal bottomhole flowing pressure and then determine the choke size that will fit the particular bottomhole flowing pressure. An increase in the hydrodynamic forces tends to increase the tendency of the proppant pack to fail. An implication of that is that during actual treatment flowback, it is best to impose the minimum drag force on the proppant pack. This usually is achieved by minimizing the flowback rate, which also increases the capacity of the pack to withstand the closure pressure exerted by the formation.

Willberg et al. (1998) reported significant gains in conductivity and polymer recovery if the flowback rates are aggressive, but we must approach the flowback challenge from a desire to optimize flowback time and prevent the proppant instability that often results from an overtly aggressive flowback schedule.

Eq. 19.1 is the expression for pseudosteady-state flow of single-phase liquid into a fracture system that is small compared to the reservoir drainage area. Eq. 19.2, the expression relating choke pressure drop (Δp) in psi and choke diameter (d) in in., is also for single-phase (liquid) flow of fluid.

$$p_{avg} - p_{wf} = \frac{141.2 q B_o \mu / kh}{\ln(0.472 r_e / r_w) + S}, \qquad (19.1)$$

$$q = 22{,}800 C \left(d^2\right)\left(\Delta p / \rho_f\right)^{10.5}. \qquad (19.2)$$

The procedure described next can be generalized for any given set of conditions through use of the appropriate equations or models. The procedure is as follows:

1. Determine the optimal bottomhole flowing pressure on the basis of the values of closure stress, reservoir pressure, proppant strength, and the desired productivity.
2. Use the appropriate inflow performance relationship (IPR) to determine the expected rate under these conditions.
3. Input the value for q into the equation relating choke-pressure drop and choke diameter.
4. Determine the optimal choke size from Eq. 19.2. The choke pressure drop is the extra backpressure applied to the reservoir for any given choke size.

The system described was assumed to have a skin of –5 to model the effect of the fracture. The input parameters are seen in **Table 19.1. Fig. 19.4** depicts the relationship between choke pressure drop and choke diameter at varying values of bottomhole flowing pressure.

Figs. 19.4 and 19.5 can be generated for any fracture system, whether producing oil, fracturing fluid, or gas. The equations used to generate these plots should be as representative of the situation as possible. A tool such as the IPR curve enables the engineer to make informed decisions about the optimal flowback rate for a given system.

19.7.7 Foamed Fracturing Fluids. Depending on the reservoir pressure, flowing back the treatment might be easy or difficult. This is because, in some scenarios, low reservoir pressure can prevent flowback returns. In these scenarios, foamed fracturing fluids are used. The foam is created by mixing nitrogen or carbon dioxide with the fracturing fluid containing foaming agent at high shear rates. The gas component of the foamed fluid can be as high as 80 to 90%. These fracturing fluids have great impact on the hydrodynamic forces inside the fracture. Controlling flowback in this scenario is even more important because of the high gas velocities expected near the wellbore because of expansion.

p_{avg}	6,000	psi	Average reservoir pressure
p_{wf}	varied	psi	Flowing bottomhole pressure
k	100	md	Reservoir permeability
h	10	ft	Reservoir thickness
q	calculated	bbl/D	Flow rate
B_o	1.2	res bbl/STB	Oil-formation volume factor
μ	2	cp	Oil viscosity
r_e	4,000	ft	Drainage radius
r_w	0.2916667	ft	Wellbore radius
S	–5	dimensionless	Skin factor
Δ_p	varied	psi	Choke pressure drop
ρ_f	49.92	lbm/ft3	Density of fluid
C	1	dimensionless	Flow coefficient of the choke
d	calculated	in.	Choke diameter

Table 19.1—Input data.

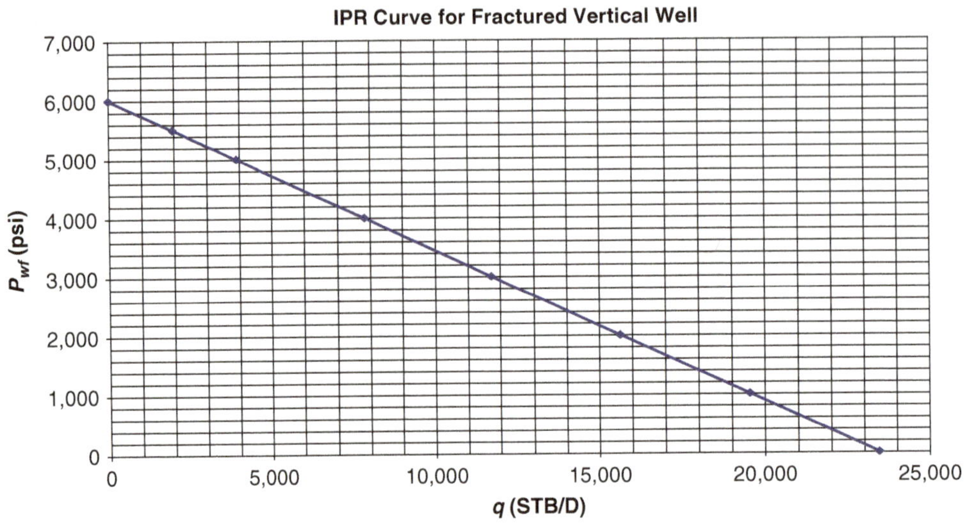

Fig.19.4—Steady-state IPR for fractured vertical wells.

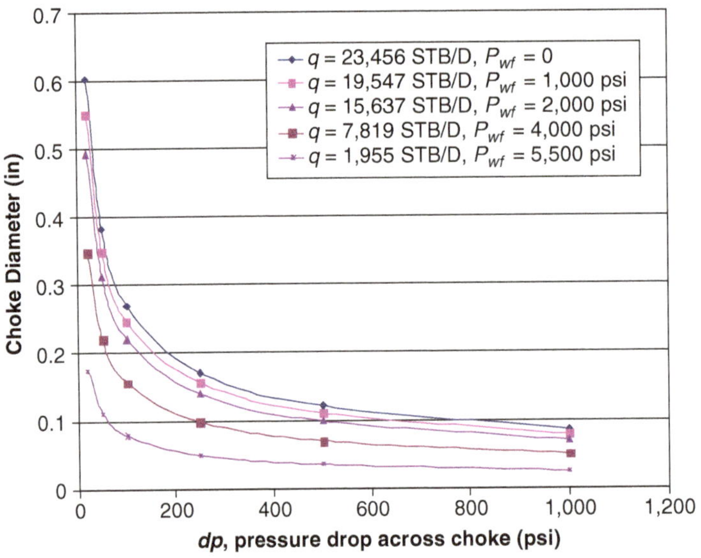

Fig. 19.5—Plot of choke diameter (in.) vs. choke pressure drop (psi).

19.7.8 Summary.

- Designing an optimal flowback schedule is important, because it has implications for well productivity and integrity.
- The mechanism and remediation of proppant flowback is not fully understood, and it is a subject of continuing research.
- Determining the optimal bottomhole flowing pressure and choke size during flowback is important, because it can help optimize flowback time and well productivity.
- Choosing the appropriate method to prevent proppant flowback must be done on a case-by-case basis.

19.8 Fracturing Treatment Diversion Methods

When stimulating a horizontal well, one of the long-standing challenges is how to efficiently treat multiple intervals of perforations simultaneously. Strategies to attack this issue date as far back as 1932 when hydrochloric acid was used to stimulate oil wells, to the present where packers and bridge plugs are used to divert fracture treatments.

19.8.1 Limited-Entry Technique. The concept of limiting the number of perforations, conceived in 1961, is still one of the most-efficient techniques in diverting fracture treatments. Limiting the number of perforations in a well and increasing the injection rate diverts the treatment to a greater portion of the perforated interval.

Perforation friction plays a key role in this method. Friction is used to break down each perforation zone to be treated by increasing the pressure inside the wellbore until it is greater than the threshold pressure. This is achieved by fixing the diameter of the casing to a size that drives an increase in perforation pressure with increasing perforation rates. **Fig. 19.6** shows data obtained from an example in which the same fluid is pumped through a perforation with common coefficients but a different perforation diameters.

Fig. 19.6 shows that the perforation friction of a formation exponentially increases with increasing injection rates for a perforation with a smaller diameter; the increase is slower for a perforation with a larger diameter.

Erosion is a common disadvantage of the application of this technique; however, perforation friction can be maximized by increasing the injection rate to offset the losses from erosion.

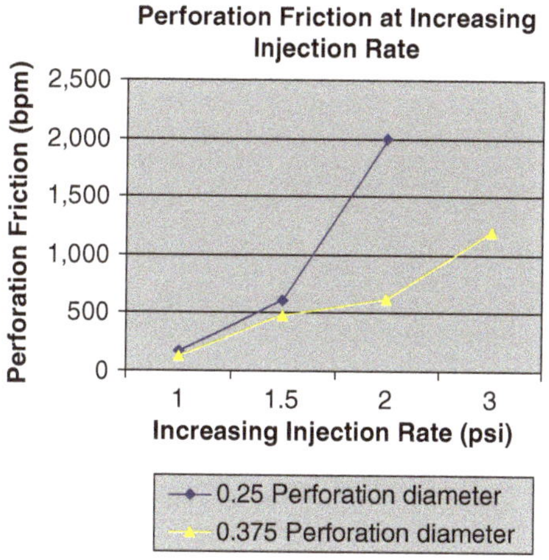

Fig. 19.6—Perforation friction increases more rapidly in smaller perforation diameters.

19.8.2 Multistage Treatment with Ball Sealers. Multistage treatment with ball sealers is a modification of the limited-entry method, in which a set number of perforations is placed in two or more intervals, and then a multistage fracturing treatment is pumped with a fixed number of ball sealers set between each stage. The ball sealers are used to divert fracturing fluid to the desired zone for treatment. Recently, lightweight balls—less dense than the fluid carrying them—have been used to ensure that the balls float back up to the perforation, in the event that they bypass the formation.

A disadvantage to using the less-dense ball sealers is that they tend to return to surface as the well is being produced, so additional equipment, costing more money, is needed to accommodate their undesirable return.

There is high uncertainty in determining the number of ball sealers needed in the case of a well with thick and thin zones and a varying number of perforations; furthermore, it is difficult to determine which zone will first be treated. Therefore, this technique gets complicated with an increasing number of zones.

Because of these issues, this method is most effective when diverting a short, highly conductive fracture to stimulate a fairly high-permeability formation with facture barriers between each formation.

19.8.3 Packer and Bridge Plug. According to Gidley et al. (1989), using a packer and bridge plug is the most-effective method of diverting a fracture treatment into multiple intervals. In the process, a lower zone is perforated and treated, and then a bridge plug is set above the interval. The next zone is perforated and treated with another bridge plug set above the interval, and the procedure is repeated to achieve the desired number of intervals. After all zones are treated, the well is cleaned out and the stimulated fracture zones are commingled.

Compared to other diversion methods, this technique is time consuming, relatively expensive to implement, and presents a mechanical hazard because it requires a workover rig to move the packer and bridge plugs into the well.

19.8.4 Pine Island Technique. The Pine Island technique is used in vertical wells. In this technique, a lower interval is perforated and treated, after which a sand plug is applied over the perforated zone and then pressure treated. The next interval above the first is perforated and treated. Another sand plug is applied over this zone and pressure tested, and the process is repeated until the desired number of perforated zones is achieved. The wellbore can be cleaned out, either after all perforations are completed and treated, or as each perforation zone is treated.

Application of this technique presents a few problems; often, it is difficult to get the sand plug precisely where we want it to be located, fluid might leak through the plug, or the sand plug might permeate through the perforation to voided areas behind the pipe. A mixture of sand is used to minimize the permeability of the sand through the formation.

There are many good publications on hydraulic fracturing, which are included in the lists of references found at the end of each chapter of this book.

19.9 Nomenclature

B_o = formation volume factor of oil, res bbl/STB

C = flow coefficient of choke, dimensionless

d = choke diameter, in.

h = formation thickness, ft

k = reservoir permeability, md

p_{avg} = average reservoir pressure, psi

p_{wf} = bottomhole flowing pressure, psi

q = flow rate, STB/D

r_e = drainage radius, ft

r_w = wellbore radius, ft

S = skin factor

Δ_p = pressure drop across the perforation, psi

ρ_f = fluid density, lbm/ft^3

μ = oil viscosity, cp

19.10 References

Barree, R. D., Fisher, M. K., and Woodroof, R. A. 2002. A Practical Guide to Hydraulic Fracture Diagnostic Technologies. Paper presented at the SPE Annual Technical Conference and Exhibition, San Antonio, Texas, 29 September–2 October. SPE-77442-MS. https://doi.org/10.2118/77442-MS.

Barree, R. D. and Mukherjee, H. 1995. Engineering Criteria for Fracture Flowback Procedures. Paper presented at the Low Permeability Reservoirs Symposium, Denver, Colorado, 19–22 March. SPE-29600-MS. https://doi.org/10.2118/29600-MS.

Bennett, L., Le Calvez, J., Sarver, D. R. et al. 2005. The Source for Hydraulic Fracture Characterization. *Oilfield Review* **17** (04): 42–57.

Brannon, H. D., Rickards, A. R., and Stephenson, C. J. 2003. Exceptional Proppant FlowbackControl for the Most Extreme Well Environments: The Shape of Things to Come. Paper presented at the SPE Asia Pacific Oil and Gas Conference and Exhibition, Jakarta, Indonesia, 9–11 September. SPE-80509-MS. https://doi.org/10.2118/80509-MS.

Canon, J. M., Romero, D. J., Pham, T. T. et al. 2003. Avoiding Proppant Flowback in Tight-Gas Completions with Improved Fracture Design. Paper presented at the SPE Annual Technical Conference and Exhibition, Denver, Colorado, 5–8 October. SPE-84310-MS. https://doi.org/10.2118/84310-MS.

Cipolla, C. L. and Wright, C. A. 2000. Diagnostic Techniques to Understand Hydraulic Fracturing: What? Why? and How? Paper presented at the SPE/CERI Gas Technology Symposium, Calgary, Alberta, Canada, 3–5 April. SPE-59735-MS. https://doi.org/10.2118/59735-MS.

Davidson, B. M., Saunders, B. F., and Holditch, S. A. 1994. Field Measurement of Hydraulic Fracture Fluids: Case Study. Paper presented at the Permian Basin Oil and Gas Recovery Conference, Midland, Texas, 16–18 March. SPE-27694-MS. https://doi.org/10.2118/27694-MS.

Economides, M. J., Hill, A. D., and Ehlig-Economides, C. 1994. Gas Well Testing and Matrix Simulation—Chemistry of Acid Rock Reactions. In *Petroleum Production Systems*, Chaps. 16 and 17. Des Moines, Iowa, United States: Prentice-Hall/Neodata.

Ely, J. W., Arnold, W. T. III, and Holditch, S. A. 1990. New Techniques and Quality Control Find Success in Enhancing Productivity and Minimizing Proppant Flowback. Paper presented at the SPE Annual Technical Conference and Exhibition, New Orleans, Louisiana, 23–26 September. SPE-20708-MS. https://doi.org/10.2118/20708-MS.

Fracfocus. 208. Site Setup. http://fracfocus.org/hydraulic-fracturing-how-it-works/site-setup. (accessed 16 July 2019).

Gidley, J. L., Holditch, S. A., Nierode, R. W. et al. 1989. *Recent Advances in Hydraulic Fracturing*. SPE Monograph Series, Vol. 12. Richardson, Texas: SPE.

Hall, B. E. and Larkin, S. D. 1989. On-Site Quality Control of Fracture Treatments. *J Pet Tech* **41** (05): 526–532. SPE-15014-PA. https://doi.org/10.2118/15014-PA.

Nguyen, P. D., Stegent, N. A., and Ingram, S. R. 2006. Remediation of Production Loss Due to Proppant Flowback in Existing Wellbores. Paper presented at the SPE Annual Technical Conference and Exhibition, San Antonio, Texas, United States, 24–27 September. SPE-102629-MS. https://doi.org/10.2118/102629-MS.

Parker, M., Weaver, J., and Van Batenburg, D. 1999. Understanding Proppant Flowback. Paper presented at the SPE Annual Technical Conference and Exhibition, Houston, Texas, 3–6 October. SPE-56726-MS. https://doi.org/10.2118/56726-MS.

Robinson, B. M., Holditch, S. A., and Whitehead, W. S. 1988. Minimizing Damage to a Propped Fracture by Controlled Flowback Procedures. *J Pet Tech* 40 (06): 753–759. SPE-15250-PA. https://doi.org/10.2118/15250-PA.

Stephenson, C. J., Rickards, A. R., and Brannon, H. D. 1999. Increased Resistance to Proppant Flowback by Adding Deformable Particles to Proppant Packs Tested in the Laboratory. Paper presented at the SPE Annual Technical Conference and Exhibition, Houston, Texas, 3–6 October. SPE-56593-MS. https://doi.org/10.2118/56593-MS.

Terracina J., Parker, M., Pongratz et al. 2001. How to Reduce Proppant Flowback in High-Rate Wells. Paper presented at the SPE Middle East Oil Show, Manama, Bahrain, 17–20 March. SPE-68202-MS. https://doi.org/10.2118/68202-MS.

van Gijtenbeek, K. A. W. and Pongratz, R. 2004. Perforating and Hydraulic Proppant Fracturing in Western Siberia, Russia. Paper presented at the SPE Annual Technical Conference and Exhibition, Houston, Texas, 26–29 September. SPE-90238-MS. https://doi.org/10.2118/90238-MS.

Warpinski, N. R., Wolhart, S. L., and Wright, C. A. 2001. Analysis and Prediction of Microseismicity Induced by Hydraulic Fracturing. Paper presented at the SPE Annual Technical Conference and Exhibition, New Orleans, Louisiana, 30 September–3 October. SPE-71649-MS. https://doi.org/10.2118/71649-MS.

Willberg, D. M., Steinsberger, N., Hoover, R. et al. 1998. Optimization of Fracture Cleanup Using Flowback Analysis. Paper presented at the SPE Rocky Mountain Regional/Low-Permeability Reservoirs Symposium, Denver, Colorado, 5–8 April. SPE-39920-MS. https://doi.org/10.2118/39920-MS.

INDEX

D

E

CASE HISTORIES OF TIGHT GAS RESERVOIR DEVELOPMENT

Edited by
Stephen A. Holditch
and John Y. Wang

TEXTBOOK SERIES

SPE International

Society of Petroleum Engineers

Tight Gas Reservoirs
Case Histories of Tight Gas Reservoir Development

Tight Gas Reservoirs
Case Histories of Tight Gas Reservoir Development

Society of Petroleum Engineers
Richardson, Texas, USA

Printed in the United States of America.

Disclaimer

Society of Petroleum Engineers
222 Palisades Creek Drive
Richardson, TX 75080-2040 USA

http://store.spe.org
service@spe.org
1.972.952.9393

Content Summary

Parts 1, 2, and 3 listed below are contained in the *Tight Gas Reservoirs* book, available both in print and digitally, but not in THIS digital supplement. They are listed here for reference purposes only. You will find ONLY Case Histories of Tight Gas Reservoir Development in this document.)

Tight Gas Reservoirs

Part 1 – Unconventional Gas Reservoirs

By Stephen A. Holditch
Contributions by Archna Agrawal, Sunil Ramaswamy, and Yao Tian

Part 2 – Tight Gas Sands Engineering

By Stephen A. Holditch, John Spivey, and John Yilin Wang

Part 3 – Hydraulic Fracturing in Vertical Wells

By Stephen A. Holditch

***

(*The chapters listed below are available only in this supplemental digital eBOOK, available only as a digital download. Download access may be granted at the time *Tight Gas Reservoirs* is purchased. Download access to only the case histories separate from the remainder of the book may or may not be available at a later date.)

Tight Gas Reservoirs

Case Histories of Tight Gas Reservoir Development

Edited by Stephen A. Holditch and John Yilin Wang

23 articles from SPE literature covering different geological formations in the US and around the world. New summaries are included that explain the significance of the research.

Table of Contents

Case Histories of Tight Gas Reservoir Development

Introduction

This reprint volume provides a series of articles of case histories on evaluation, stimulation, and development of tight gas reservoirs. These articles demonstrate how to develop and apply the technologies and practices in geological analysis and formation evaluation, and to include log interpretation, coring and core analysis, special core analysis, formation test, correlation of log and core analyses, in-situ stress, permeability distribution and correlations, and estimating net-pay and gas-in-place. These papers also include examples of prefracture well testing, post-fracture well testing, analytical and numerical modeling, hydraulic fracture design and evaluation, reserve estimates, and economic analysis, which are discussed in Parts 1, 2, and 3 of this book series.

We have selected 23 articles in the topical areas covering different geographies and geological formations in U.S. and worldwide, including

- Travis Peak, East Texas Basin, Texas
- Cotton Valley, East Texas Basin, Texas
- Frontier, Green River Basin, Wyoming
- Mesaverde, Piceance Basin, Colorado
- Williams Fork, Piceance Basin, Colorado
- Variable formations of Permian, Cooper Basin, Australia
- Muddy "J", D-J Basin, Colorado
- Rottliegendes, offshore, The Netherlands
- Vicksburg, Gulf Coast Basin, Texas
- Vicksburg, Burgos Basin, Mexico
- Upper Devonian Elk Sand, Appalachia Basin, Pennsylvania
- Variable formations of Jurassic, Sichuan Basin, China
- Lance, Green River Basin, Wyoming
- Dakota, San Juan Basin, New Mexico
- Cotton Valley Lime (Haynesville), East Texas Basin, Texas
- Spirit River, Deep Basin, Alberta, Canada
- Wilcox (Lobo), Gulf Coast, Texas
- Wilcox, Burgos Basin, Mexico

These case histories in total illustrate how a team of geologists, engineers, field personnel, and management work together to successfully evaluate and develop tight gas sand reservoirs. It is also made clear that successful hydraulic fracturing is critical for economic gas production and increased reserves. It is important to carry out a comprehensive data collection and evaluation program before, during, and after a fracture treatment, and by doing so, one will learn how to improve and optimize completions and well stimulation. The methodology, and most of the technology, presented in this book and demonstrated in the case histories often can be extended and applied in developing other unconventional reservoirs, including coalbed methane, shale oil and gas, tight oil and gas, and possibly geothermal and gas hydrate.

Our initial search found more than 40,000 articles related to tight gas, of which 79 papers were relevant to case histories based on our review of abstracts. We analyzed these 79 articles in detail and selected 23 to be included in this reprint series along with our review of each. A bibliography of 79 papers, reviewed and considered relevant, is also included at the end of this collection.

It is our hope that these selected 23 articles not only help transfer the technologies and knowledge of increased economic recovery of tight gas resources discussed in Parts 1, 2, and 3 of this book series, but also help inspire the next generation of students and professionals in advancing the profession and the industry.

I.1 List of Papers Included

I.1.1 GRI Staged Field Experiments.

1. Holditch, S. A., Robinson, B. W., Whitehead, W.S. et al. 1988. The GRI Staged Field Experiment. *SPE Form Eval* **3** (3): 519–533. SPE-16429-PA. https://doi.org/10.2118/16429-PA.
2. Robinson, B. M., Holditch, S. A., and Peterson, R. E. 1991. The Gas Research Institute's Second Staged Field Experiment: A Study of Hydraulic Fracturing. Paper presented at the SPE Gas Technology Symposium, Houston, Texas, 22–24 January. SPE-21495-MS. https://doi.org/10.2118/21495-MS.
3. Robinson, B. M., Holditch, S. A., Whitehead, W. S. et al. 1992. Hydraulic Fracturing Research in East Texas: Third GRI Staged Field Experiment. *J Pet Technol* **44** (1): 78–87. SPE-22878-PA. https://doi.org/10.2118/22878-PA.

4. Saunders, B. F., Robinson, B. M., Holditch, S. A. et al. 1992. Hydraulic Fracturing Research in the Frontier Formation Through the Gas Research Institute's Fourth Staged Field Experiment. Paper presented at the SPE Annual Technical Conference and Exhibition, Washington, DC, 4–7 October. SPE-24854-MS. https://doi.org/10.2118/24854-MS.

I.1.2 DOE Multisite Experiment.

5. Warpinski, N. R., Branagan, P. T., Settler, A. R. et al. 1987. Fracturing and Testing Case Study of Paludal, Tight, Lenticular Gas Sands. *SPE Form Eval* **2** (04): 535–545. SPE-13876-PA. https://doi.org/10.2118/13876-PA.
6. Warpinski, N. R., Wright, T. B., Uhl, J. E. et al. 1999. Microseismic Monitoring of the B-Sand Hydraulic-Fracture Experiment at the DOE/GRI Multisite Project. *SPE J* **4** (3): 242–250. SPE-57593-PA. https://doi.org/10.2118/57593-PA.

I.1.3 Hydraulic Fracturing.

7. Holditch, S. A. and Ely, J. 1973. Successful Stimulation of Deep Wells Using High Proppant Concentrations. *J Pet Technol* **25** (8): 959–964. SPE-4118-PA. https://doi.org/10.2118/4118-PA.
8. Parrot, D. I. and Long, M. G. 1979. A Case History of Massive Hydraulic Refracturing In The Tight Muddy "J" Formation. Paper presented at the SPE Symposium on Low Permeability Gas Reservoirs, Denver, Colorado, 20–22 May. SPE-7936-MS. https://doi.org/10.2118/7936-MS.
9. Wyman, R. E., Holditch, S. A., and Randolph, P. L. 1980. Analyses of an Elmworth Hydraulic Fracture in Alberta. *J Pet Technol* **32** (9): 1621–1630. SPE-7935-PA. https://doi.org/10.2118/7935-PA.
10. Reeves, S. R., Hill, D. G., Tiner, R. L. et al. 1999. Restimulation of Tight Gas Sand Wells in the Rocky Mountain Region. Paper presented at the SPE Rocky Mountain Regional Meeting, Gillette, Wyoming, USA, 15–18 May. SPE-55627-MS. https://doi.org/10.2118/55627-MS.
11. Kang, Y., Luo, P., Jin, X. et al. 2000. Employing Both Damage Control and Stimulation: A Way to Successful Development for Tight Gas Sandstone Reservoirs. Paper presented at the International Oil and Gas Conference and Exhibition in China, Beijing, China, 7–10 November. SPE-64707-MS. https://doi.org/10.2118/64707-MS.
12. Xiong, H. and Holditch, S. A. 2006. Will the Blossom of Unconventional Natural Gas Development in North America Be Repeated in China? Paper presented at the International Oil and Gas Conference and Exhibition in China, Beijing, China, 5–7 December. SPE-103775-MS. https://doi.org/10.2118/103775-MS.
13. Scott, M. P., Stephens, T., Durant, R. et al. 2013. Investigating Hydraulic Fracturing in Tight Gas Sand and Shale Gas Reservoirs in the Cooper Basin. Paper presented at SPE Unconventional Resources Conference and Exhibition–Asia Pacific, Brisbane, Australia, 11–13 November. SPE- 167073-MS. https://doi.org/10.2118/167073-MS.

I.1.4 Reservoir and Production Analyses.

14. Eberhard, M. J., Mullen, M. J., Seal, C. A. et al. 2000. Integrated Field Study for Production Optimization: Jonah Field–Sublette County, Wyoming. Paper presented at the SPE/CERI Gas Technology Symposium, Calgary, Alberta, Canada, 3–5 April. SPE-59790-MS. https://doi.org/10.2118/59790-MS.
15. Sunde, A., Chen, H. -Y. and Teufel, L. W. 2000. Producing Characteristics and Drainage Volume of Dakota Reservoirs, San Juan Basin, New Mexico. Paper presented at the SPE Rocky Mountain Regional/Low-Permeability Reservoirs Symposium and Exhibition, Denver, Colorado, 12–15 March. SPE-60288-MS. https://doi.org/10.2118/60288-MS.
16. Arevalo-Villagran, J. A., Wattenbarger, R. A., and El-Banbi, A. H. 2000. Production Analysis of Commingled Gas Reservoirs–Case Histories. Paper presented at the SPE International Petroleum Conference and Exhibition in Mexico, Villahermosa, Mexico, 1–3 February. SPE-58985-MS. https://doi.org/10.2118/58985-MS.

I.1.5 General Case Histories.

17. Kozik, H. G. and Holditch, S. A. 1981. A Case History for Massive Hydraulic Fracturing the Cotton Valley Lime Matrix, Fallon and Personville Fields. *J Pet Technol* **33** (2): 229–244. SPE-7911-PA. https://doi.org/10.2118/7911-PA.
18. Hopkins, C. W., Gatens, J. M., Caballero, J. S. et al. 1991. Stimulation Optimization in a Low-Permeability Upper Devonian Sandstone Reservoir: A Case History. Paper presented at the SPE Gas Technology Symposium, Houston, Texas, 23–25 January. SPE-21499-MS.
19. Robinson, B. M., Holditch, S. A., and Lee, W. J. 1986. A Case Study of the Wilcox (Lobo) Trend in Webb and Zapata Counties, TX. *J Pet Technol* **38** (12): 1355–1366. SPE-11600-PA. https://doi.org/10.2118/11600-PA.
20. Baumgartner, W. E., Shlyapobersky, J., Abou Sayed, I. S. et al. 1993. Fracture Stimulation of a Horizontal Well in a Deep, Tight Gas Reservoir: A Case History From Offshore The Netherlands. Paper presented at the Offshore Europe conference, Aberdeen, Scotland, United Kingdom, 7–10 September. SPE-26795-MS. https://doi.org/10.2118/26795-MS.
21. Berumen, S., Cipolla, C., Finkbeiner, T. 2000. Integrated Reservoir Geomechanics Techniques in the Burgos Basin Mexico: An Improved Gas Reservoirs Management. Paper presented at the SPE Asia Pacific Conference on Integrated Modeling for Asset Management, Yokohama, Japan, 25–26 April. SPE-59418-MS. https://doi.org/10.2118/59418-MS.
22. Patino, A. H., Garcia, E. R., Torne, J. P. et al. 2004. Case Histories–Combining Crossed Dipole Sonic Anisotropy and Oriented Perforating to Optimize Hydraulic Fracturing In the Burgos Basin–Reynosa Mexico. Paper presented at the 2004 SPE International Petroleum Conference in Mexico, Puebla, Mexico, 7–9 November. SPE-92014-MS. https://doi.org/10.2118/92014-MS.

I.1.6 Benefits of Technology.

23. Voneiff, G. W. and Holditch, S. A. 1994. Economic Assessment of Applying Advances in Fracturing Technology. J Pet Technol **46** (1): 281–296. SPE-24888-PA. https://doi.org/10.2118/24888-PA.

Chapter 1

The GRI Staged Field Experiment

S. A. Holditch, Texas A&M University; **B. M. Robinson,**
W. S. Whitehead, and **J. W. Ely,** S. A. Holditch & Associates Inc.

SPE 16429 presents a case history of a Travis Peak tight gas well located in East Texas as part of the Gas Research Institute's (GRI's) first staged field experiment (SFE) during 1986 through 1987. SFE No. 1 provided a comprehensive and valuable data set that can be used to calibrate both 3D hydraulic fracture design models and 3D reservoir simulation models. Holditch et al. first discuss criteria of wellsite selection and provide an overview of regional geology, depositional environment, and petrographic description, and then summarize in detail the design and execution of drilling and completion operations. They go on to summarize the prefracture formation evaluation, which includes log analysis, well tests, and minifrac tests. SPE 16429 also contains detailed discussion of the fracture treatment design and execution and of the data collected before, during, and after the fracture treatment. Finally, the authors discuss how they evaluated treatment effectiveness using gamma ray and temperature log pressure-buildup tests, production performance data, and numerical reservoir simulations. Conclusions, regarding the Travis Peak Formation and hydraulic fracturing, in general, include

- Excessive height growth can be a major problem in Travis Peak, in which lack of fracture containment is a direct result of insufficient stress contrast between the net-pay zones and the rock layers above and below.
- To contain a fracture, the excessive pressure within the fracture must be minimized, if possible.

This study discusses that without accurate measurement of slurry viscosity during the fracture treatment it is difficult to calculate the net pressure in the fracture. The net pressure is required to compute the correct fracture dimensions. Several additional specific conclusions pertaining to GRI's SFE No. 1 were

- Reserves are between 3.5 to 4.0 bcf
- Fracture fluid leak off is high (80%)
- Productivity index increase is 3.6 times
- Effective propped fracture length is 150 ft
- Effective fracture conductivity is 2,550 md-ft
- Drainage area is 320 acres
- Evidence of crossflow between layers is observed

1.1 Summary

The GRI's Tight Gas Sands Program was involved with research on low-permeability formations for 4 years in the 1980s. The main focus of the research was to improve the general understanding of producing tight reservoirs, while a specific focus was to advance the technology involving hydraulic-fracture geometry. The unique aspect of this research was that the laboratories were actual gas wells completed in the Travis Peak Formation in East Texas.

To extend the development of technology fully, GRI—now the Gas Technology Institute—planned four staged field experiments (SFEs) from 1986 through 1989. The SFE program provided the opportunity to collect a wealth of data that could not be obtained from normal cooperative research wells. The purpose of this paper is to summarize all the tests and data collected from the first SFE and to present the available results from the analyses of these data.

1.2 Introduction

Research was directed at reducing the cost and improving the recovery of gas produced from low-permeability reservoirs. Because the most viable method of improving recovery in tight formations is hydraulic fracturing, part of the research was aimed at better understanding the processes that control fracture growth. If the petroleum industry can substantially improve its ability to predict and to control the shape of hydraulic fractures accurately, then the chances of achieving optimum propped-fracture lengths are increased. The end result will be an improved economic incentive for developing tight gas reservoirs.

To understand the hydraulic-fracturing process better, it was first necessary to improve the understanding of both the reservoirs being fracture treated and the formations surrounding the main productive interval. To achieve this goal, a comprehensive program was initiated to perform geologic, coring, logging, well testing, fracture-treatment monitoring, and fracture-diagnostic studies on selected cooperative research wells in the Travis Peak Formation of the East Texas Basin.

The technical approach was to use the necessary electronic equipment and computers in the field so that the GRI contractors could actually analyze the fracture-treatment data in real time. Once this system was perfected, we expected to be able to predict fracture shape and extent. We believed if this could be accomplished successfully, we would then attempt to control fracture growth by controlling the fracture-fluid viscosity or the injection rate.

The success of the program hinged on the cooperation of operating companies. Through their participation, GRI contractors were allowed to collect the data necessary for a complete formation and fracture evaluation from actual gas wells referred to as cooperative wells. With these data, we believed that substantial progress could be made toward the development of tight gas through better stimulation technology.

Table 1.1 lists all GRI contractors, service companies, and producing companies that cooperated with GRI in the field data collection. **Table 1.2** summarizes the data collected during the first 3 years of this research program.

GRI Participants	Service Companies	Producers
Applied Geomechanics Inc.	Dowell-Schlumberger	Amoco Production Co.
BEG (U. of Texas)	Halliburton	Arkla Exploration Co.
CER Corp.	Dresser Atlas	Mobil Producing Texas & New Mexico
Inst. of Gas Technology	Totco	American Petrofina
ResTech Inc.	Exlog	Arco Oil & Gas Co.
Resources Engineering Systems	Terra Tek	Ashland Exploration Co.
S. A. Holditch & Associates Inc.	Schlumberger	Clayton W. Williams Jr.
Dresser Petroleum Engineering Services	Sperry Sun	Barrett Energy
Stanford U.	Core Research	Prairie Producing Co.
Teledyne Geotech Inc. Tomex Corp.	The Western Co. Nowsco Services Welex	Marshall Exploration Co.

Table 1.1—Participating organizations.

Type of Operation	1984	1985	1986	Total
Coring				
Number of wells	3	2	4	9
Core recovered, ft	371	466	1,156	1,993
Logging				
Number of wells	4	2	5	11
Number of surveys	58	29	55	142
In-situ stress tests				
Number of wells	4	4	4	12
Number of zones	6	8	21	35
Prefracture well tests				
Number of wells	4	2	4	10
Fracture treatments				
Number of wells	4	6	4	14
Postfracture well tests				
Number of wells	3	3	1	7

Table 1.2—Summary of data collected from all cooperative research wells.

The cooperative research wells were just the first step in the overall plan to characterize fully the formation and hydraulic fracture in three dimensions. The next phase of the plan was to pull all GRI contractors together on strategic wells to use all possible means of improving the understanding of tight gas reservoirs. This was called the SFEs phase.

1.3 Goals of the SFEs

Four SFEs were planned from 1986 through 1989. An SFE allowed the research contractors to collect the same basic data that had been available from most of the cooperative wells. Additional data, however, that had not been collected on cooperative wells because of high cost or high risk to the operator also was obtained during the SFE. The goals of each SFE are listed below.

SFE 1—1986–87:

1. To bring together all tight gas sand contractors to study a common formation and wellbore.
2. To drill, complete, and stimulate a highly instrumented well to collect and analyze one of the most comprehensive data sets ever obtained in the industry.
3. To implement data-collection methods that will be structured to benefit the research contractors working on fracture-diagnostic techniques and fracture-modeling efforts.
4. To present information obtained during the SFE to the industry in an integrated form to facilitate technology transfer. The various analyses were not linked and were not necessarily consistent between any two of the contractors.

SFE 2—1987–88:

1. A 3D fracture design and evaluation model was used to analyze the data in real time and to offer suggestions concerning changes that should be considered to improve the treatment results.

2. Fracture-diagnostic experiments were evaluated to focus future research efforts on the most promising of the techniques. All fracture-diagnostic methods were separate from the fracture-modeling efforts.
3. Field-measuring and -monitoring systems of the GRI Mobile Testing and Control Facility were superior to all other field systems in versatility, accuracy, and reliability.
4. Experiments focused mainly on the determination of fracture-height and fracture-length measurements and/or calculations of these properties.

SFE 3—1988–89:

1. Fracture-modeling systems computed fracture shape with precision.
2. Fracture-diagnostic systems were functional and provided input to the fracture models in real time.

SFE 4—1989–90:

1. To move to a new location and prove that the technology developed could be transferred to a different environment.

1.4 Site Selection for SFE 1

Rigorous selection criteria set up for the first SFE included the presence of direct offset wells, blanket-type sand deposition over a fairly wide area, the presence of discrete individual sand members separated by thick shale formations, and the need for fracture stimulation of the sands. While these criteria might not have represented a typical Travis Peak well, it was decided to seek an ideal laboratory environment for the first SFE well. As a result of this study, the Waskom Travis Peak Field was selected as an appropriate site.

A farmout agreement was obtained with Arkla Exploration Co. to drill the SFE well on the company's Howell Unit, located essentially in the heart of the Waskom Travis Peak Field. The agreement allowed complete operational control over the well for a specific time period.

1.4.1. Geologic Overview. The Travis Peak Formation of the East Texas Basin is a lobate, highly constructive fluvial-deltaic depositional system that prograded to the southeast across east Texas and north Louisiana. On the basis of extensive depositional studies (Finley 1985), the Travis Peak was divided into three large-scale facies. The middle progradational sequence is generally well developed and represents a predominantly sandy, fluvial channel facies associated with minor flood-plain and back-swamp components. This sand-rich fluvial-deltaic system is gradationally overlain and underlain by delta-fringe facies that contain a much higher mudstone content.

The Travis Peak upper delta-fringe sequence is generally well developed and represents enhanced mudstone accumulation in a mixed delta platform and marginal marine setting. It is gradationally overlain by marine-shelf carbonates of the Sligo and equivalent Pettit Formations. The lower delta-fringe sequence overlies the marine Knowles limestone of the Cotton Valley group, but is generally not well developed. Where developed, however, it too is characterized by a higher mud content with more widely separated sand bodies. This sequence exhibits characteristics that indicate a rapid influx of terrigenous clastic sediments that abruptly smothered the Knowles carbonate. Thus, it is plausible to interpret the lower delta-fringe as a predominantly transgressive sequence.

Domes and structural terraces on the west flank of the Sabine uplift influenced the deposition of the Travis Peak. Most of the Travis Peak gas production is in this area and is from thin sandstones of the relatively thick upper sequence. Trapping mechanisms include structural closure and stratigraphic pinchout of sandstones, porosity zones within the sandstones, or both. These occur updip on the flanks of the structures. Most of the low-permeability gas production is clustered around the western flank of the structures and some over the structure crests.

The Bureau of Economic Geology (BEG) conducted regional geologic studies of the Travis Peak over the eastern portion of the East Texas Basin and the North Louisiana Salt Basin. Most of the BEG's efforts centered around the Sabine uplift, where much of the Travis Peak production occurs. More recent research at the time, however, concentrated on increasingly smaller areas for selecting an appropriate drilling site for the first SFE well.

From the selection criteria, the Waskom Travis Peak Field in Harrison County in Texas was selected as the site for the first SFE. The Waskom Field is a broad, asymmetric dome with the top of the Travis Peak at a drill depth of about 5,900 ft (1798 m) as shown in **Fig. 1.1.** The Upper Travis Peak contains many individual sand members isolated by fairly thick mudstones. Many wells in this field, particularly on the western flank of the dome, must be hydraulically fractured to achieve commercial production.

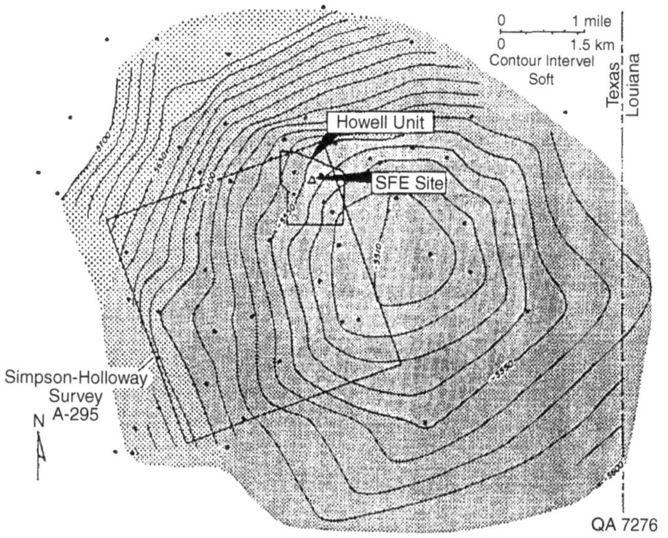

Fig. 1.1—Structure contour map on base of Sligo Formation—Waskom Field (Fracasso 1987).

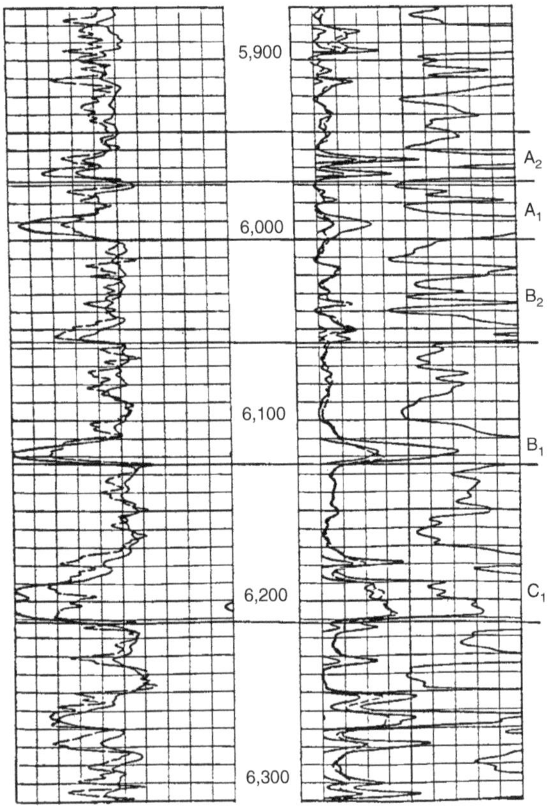

Fig. 1.2—Informal sandstone units of Upper Travis Peak—Howell 5 log.

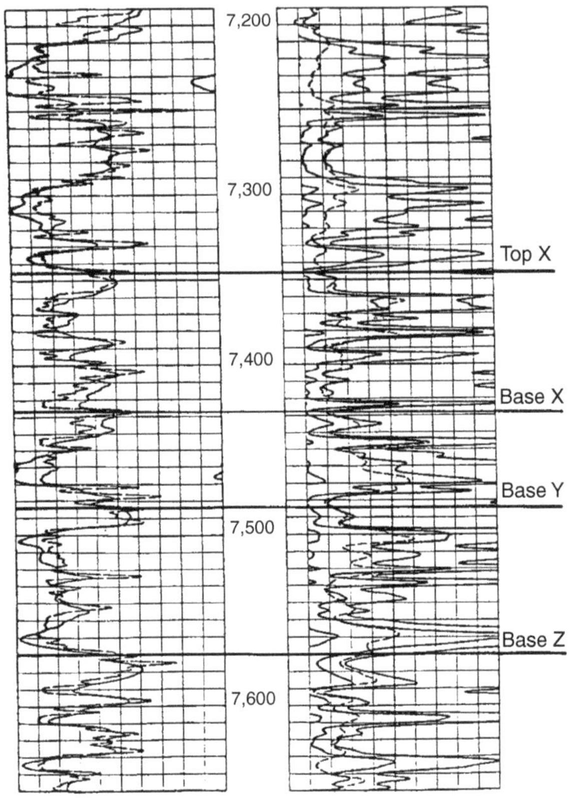

Fig. 1.3—Informal sandstone units of Lower Travis Peak—Howell 5 log.

There are three primary sandstones in the Upper Travis Peak, defined as Sandstone Sequences A, B, and C. These are illustrated in **Fig. 1.2,** which is a log from the first SFE well (Howell 5). Sequence A is about 75 ft (22.9 m) thick and generally contains two well-developed, upward-fining sandstones designated as Sandstones A-1 and A-2. Sandstone Sequence B is about 125 ft (38.1 m) thick and generally contains one very-well-developed sandstone designated as Sandstone B-1. Sandstone Sequence C is about 95 ft (29.0m) thick and consists of two well-developed upward-fining sandstones designated as Sandstones C-1 and C-2. These three sandstone sequences were the primary targets for completion in the first SFE well.

Individual sand units were also identified in the middle, fluvial-deltaic portion of the Travis Peak. These sands are correlatable over most of the Waskom Field, especially in the area surrounding the Howell lease. **Fig. 1.3** is a log from Howell 5 that illustrates the informal sandstone units designated by BEG.

1.4.2 Depositional Environment. Through extensive studies of whole-core and openhole logs, the Upper Travis Peak has been described (Fracasso 1987) as representing deposits of delta platform, meandering distributary channels that have cut over well-drained to poorly drained swampy flood plains. This description is strongly supported by the dominant upward-fining sequences found in the Upper Travis Peak Sandstones in Waskom. There are many variations of the upward-fining characteristics, the complexity of which represents a fluvial rather than tidal channel origin for the sandstones.

One of the most dominant features of the Upper Travis Peak in the Waskom Field is the presence of calcareous nodules that are widespread in red and gray mudstones that cap the upward-fining sequences. These nodules range up to several centimeters in diameter and are composed of dolomite and ankerite. Many of these nodules have grown in situ in the mudstones. Such characteristics are usually associated with flood-plain swamps.

The sand-rich middle fluvial-deltaic division of the Travis Peak is characterized by "blocky" sandstones. The tops and bases of these sands are frequently sharp relative to the thickness of the units, and as a result are often difficult to characterize from logs as either fining upward or coarsening upward. The best-developed blocky sandstones probably formed in sand-rich, braided-stream depositional systems. The low mudstone content of these systems reflects the downdip transport of the finest grain size sediment as the braided streams swept across the delta platform complex during progradation.

1.4.3 Petrographic Description. The sandstones are considered mineralogically very mature and generally are classified as quartz arenites. Most sand grains are fine to very fine, ranging between 0.002 and 0.01 in. (0.063 and 0.25 mm). Most siltstones range between 0.001 and 0.002 in. (0.031 and 0.062 mm), while clay particles are smaller than 0.0002 in. (0.004 mm).

Quartz makes up 94 to 100% of the essential framework constituents, and total feldspar varies from 0 to 4.7%. Authigenic cements constitute between 14.5 and 27.5% of the sandstone volume.

Reservoir bitumen, or dead oil, is common in the upper sections of the Travis Peak. However, little reservoir bitumen is observed in the Lower Travis Peak. This solid precipitate fills pore spaces and coats quartz overgrowths.

Both primary and secondary pores are present, and primary pores are particularly abundant in samples with thick illite cutans. Secondary pores are formed by the dissolution of framework grains and are approximately the same size as detrital grains. The secondary pores commonly contain authigenic clays and fragments of dissolved framework grains, particularly feldspar.

1.5 Drilling Summary

Howell 5 was spudded on 14 August 1986 and reached total depth (TD) on 22 September 1986. A 17½-in. (444.5-mm) hole was drilled from surface to 1,480 ft (451 m) and 13⅜-in. (339.7-mm) casing was set and cemented in place. Below this point, a 12¼-in. (311.2-mm) hole was drilled to a depth of 5,846 ft (1782 m), which corresponded to the base of the Sligo carbonate. A string of 9⅝-in. (244.5-mm) casing was set and cemented at this depth. At this point, an 8¾-in. (222.2-mm) hole was drilled to a TD of 7,900 ft (2408 m). A string of 7-in. (177.8-mm) casing was then run from surface to a depth of 7,895 ft (2406 m) and cemented in place.

During drilling to the top of the Travis Peak Formation, seismic activity associated with drilling the well was measured with geophones set out at the surface. The technique, known as seismic borehole tomography (SBT), was developed to use the drill bit as a downhole seismic source to image formations around the borehole. The main purpose of SBT is to develop a 3D seismic interpretation for a specific well to enhance the overall seismic interpretation in a particular field. The geophones, which are set out at strategic surface locations surrounding the well, monitor the downhole seismic signals created by the drilling bit.

Four openhole stress tests were conducted in different formations during drilling of the SFE well. The tests were conducted in the base of the Sligo carbonate (5,846 to 5,852 ft [1782 to 1784 m]), in a mudstone and a sandstone in the Upper Travis Peak (6,144 to 6,150 ft [1872.7 to 1874.5 m] and 6,194 to 6,200 ft [1888 to 1890 m], respectively), and one test was performed in a Lower Travis Peak sandstone (7,550 to 7,560 ft [2301 to 2304 m]).

After each openhole stress test was conducted within the Travis Peak Formation, an attempt was made to core and recover the hydraulic fracture created during the stress test. Created fractures were successfully recovered in the mudstone and sandstone intervals of the Upper Travis Peak; however, because of severe jamming of the core barrel, only a few small pieces of core were recovered from the sandstone at 7,560 ft (2304 m). Attempts were made to orient the cores at 6,144 to 6,150 ft (1872.7 to 1874.5 m) [mudstone] and at 6,194 to 6,200 ft (1888 to 1890 m) [sandstone]. Core-orientation equipment did not work during coring of the upper mudstone interval; however, the equipment did work properly during coring of the sand at 6,200 ft [1890 m]. As a result, the orientation of the induced fracture was measured to be north 70° east.

More than 500 ft of 4-in. (152.4 m of 101.6-mm) whole core was cut in two main intervals of the Travis Peak Formation: an upper interval from 5,950 to 6,311 ft (1814 to 1924 m) and the lower part of the Travis Peak from 7,395 to 7,562 ft (2254 to 2305 m). A total of 470 ft (143 m) of core was actually recovered, resulting in a recovery efficiency of about 94%.

Attempts were made to orient most of the core using either a continuous multishot technique or paleomagnetic measurements. Orientation was successfully obtained on approximately 25% of the core with the paleomagnetic technique. Because of various equipment and electronic problems, however, only small portions of the cored intervals were successfully oriented with the continuous multishot technique.

Core-strain-relaxation measurements were performed on 18 samples of the Upper Travis Peak Formation ranging in depth from 5,989 to 6,310 ft (1825 to 1923 m). Core orientation, primarily with the paleomagnetic technique, was available on 11 of these samples. From these measurements, the direction of maximum principal stress was determined to be north 77° east.

An extensive openhole logging program was performed on the SFE well. A total of 17 descents was made into the well with various logging tools that required 5 days of logging operations. This logging program is summarized in **Table 1.3.** Note that, from the Micro Electric Scanning Tool™, depositional features and natural (or induced) fractures can be detected over a portion of the borehole. Because this log is oriented, the direction of these fractures can also be determined. In Howell 5, both natural fractures and hydraulically induced fractures from the openhole stress tests were observed. The direction of these fractures was generally east/west, which is in good agreement with the oriented-core measurements.

Dual Laterolog	Electromagnetic Propagation
Dual induction	Natural GR spectroscopy
Digital sonic	Induced GR spectroscopy
Litho-density	Microspherically focused
Compensated neutron	Dipmeter
Acoustic-long spacing	Microelectric scanning
Gamma ray	Borehole televiewer
Spontaneous potential	RFT
Microlog	

1. Twenty-five logging runs.
2. Approximately 50,000 ft of openhole log taken in 7,900-ft hole.
3. Mobil Oil Research Co. and Amoco Oil Research Co. each ran logs including borehole televiewers.
4. Travis Peak logged three times over same interval to evaluate resistivity profile as a result of invasion.

Table 1.3—Logs run on Howell 5.

1.6 Completion Summary

After the SFE well was drilled, an extensive evaluation and testing program was conducted. Additional logs were run in the cased hole, and 16 in-situ stress tests were performed on different intervals within the Travis Peak. Several sands were perforated and production tested before the upper portion of Sand C-l at 6,189 to 6,211 ft (1886 to 1893 m) was selected for the main fracture-stimulation treatment. Two minifracture injection tests and a hydraulic-fracture treatment were performed on Sand C-l.

Additional logging tools run in the cased hole included a gamma ray spectroscopy, digital sonic, and a cement bond/variable density log. A long-spacing sonic tool, provided by Amoco Production Co., was also run in cased hole. The main purpose of running the cased-hole logs was to provide a comparison with similar logs run in the open hole in an attempt to develop a cased-hole evaluation package that could assist operators in the evaluation of older wells.

After the cased-hole logging operations, prefracture seismic calibration data were measured with downhole triaxial geophones. A mobile vibroseis unit, which was spotted at various strategic locations around the wellsite, provided the surface energy source to orient the downhole seismic tool. These data, coupled with similar measurements made during one of the minifracture injection tests, provided the information necessary to determine fracture azimuth.

To describe the vertical-stress profile within the Travis Peak Formation, 16 individual in-situ tests were performed on various types of lithologies such as mudstones, sandstones, siltstones, and the Lower Sligo carbonate. Bottomhole pressures (BHPs) were recorded with a quartz pressure gauge, and multiple injection tests were performed to ensure repeatability. Measured stress values ranged from a low of 0.45 psi/ft (10.18 kPa/m) in an Upper Travis Peak Sandstone to a high of 0.725 psi/ft (16.40 kPa/m) in a Lower Travis Peak Shale. A list of all the intervals tested and a summary of the results are presented in **Table 1.4.**

Depth (ft)	Lithology	In-Situ Stress Gradient (psi/ft)	Comments
5,848	Limestone	0.65	Sligo; also tested in open hole
5,992	Sandstone	0.45	Sand A-1; cored; no flow on RFT
6,117	Sandstone	0.46	Sand B-1; RFT = 558 psi
6,164	Mudstone	0.63	Thin shale, surrounded by silt between Sands B-1 and C-1; also tested in open hole
6,194	Sandstone	0.515	Sand C-1; RFT = 1,210 psi; also tested in open hole
6,244	Shale	0.562	Below Sand C-1; possible lower fracture barrier
7,210	Sandstone	0.605	RFT = 3, 134 psi
7,280	Shale	0.542	Shale above lower zone
7,308	Sandstone	0.57	Possibly wet; no RFT
7,466	Sandstone	0.524	RFT = 2,792 psi; lower completion zone
7,488	Shale	0.714	Possible fracture barrier
7,504	Sandstone	0.565	RFT = 3,037 psi, probably gas productive
7,561	Sandstone	0.65	No flow on RFT; also tested in open hole
7,576	Shale	0.725	Possible lower fracture barrier
7,725	Shale	0.69	Possible lower fracture barrier
7,785	Shale	0.70	Possible lower fracture barrier

Table 1.4—Cased-hole stress tests, Howell 5.

Production tests and pressure-buildup surveys were conducted on several sandstone intervals within the Travis Peak. These included Sand Y from 7,442 to 7,476 ft (2268 to 2279 m), Sand X from 7,359 to 7,424 ft (2243 to 2263 m), and Sand C-l from 6,189 to 6,211 ft (1886 to 1893 m). Each zone was isolated and individually tested.

Two separate minifracture injection tests were performed on Howell 5. The first injection test was conducted specifically to measure seismic activity with downhole geophones. The second test was conducted primarily to obtain data for calibration of a 3D fracture simulator. In addition, the test provided data for other research contractors studying hydraulic-pressure responses that result from the created fracture.

1.7 Prefracture Formation Evaluation

1.7.1 Log Analysis. From the log analysis performed on Howell 5, several intervals were identified as potentially productive. **Table 1.5** summarizes the reservoir properties determined by this analysis. Of these intervals, the top zone (6,100 to 6,125 ft [1859 to 1867 m]) is Sand B-l. A repeat formation tester (RFT) pressure of 558 psi [3.8 MPa] was measured on this sand; thus, it was found to be mostly depleted as a result of offset production. The next zone, 6,170 to 6,211 ft (1881 to 1893 m), was identified as Sand C-l. Although the pressure in this sand was measured to be low (1,200 psi [8.3 MPa]), it was still believed to be productive. None of the other sands in the Upper Travis Peak were considered potential completion intervals. Two Lower Travis Peak Zones were also thought to be gas bearing and potentially productive. These zones were Sands X and Y at 7,350 to 7,426 and 7,430 to 7,476 ft (2240 to 2263 and 2265 to 2279 m), respectively.

1.7.2 Lower Travis Peak Well Tests. The first zone selected for testing was Sand Y. This zone was perforated from 7,442 to 7,447 and from 7,464 to 7,476 ft (2268 to 2270 and from 2275 to 2279 m). A perforation bailout treatment was pumped using treated KCl water. After load fluid was swabbed back for about 1½ to 2 days, the well would not flow at a stabilized rate and was making about 2 bbl/hr (0.3 m^3/h) water with a slight gas cut. Because of the low feed-in rate, a decision was made to perform a mud acid balloff treatment. After the load fluid was swabbed back from this treatment, a flow rate of about 10 Mcf/D (280 m^3/d) was measured from the well. At this point, a BHP-buildup survey was started.

Fig. 1.4 presents the Horner plot of these pressure-buildup data. Note that these data have been plotted in terms of adjusted pressures and adjusted Horner times. These functions, like pseudopressure (Al-Hussainy et al. 1966) and pseudotime (Agarwal 1979), take into account changing gas properties with pressure.

From this graph, it appears that the pressure-buildup data may be very close to reaching the Horner straight line. Assuming 10 ft (3 m) of net-pay and using the straight line illustrated in Fig. 1.4, a permeability of 0.009 md and a skin factor of −1.2 are calculated. If this line is extrapolated to a Horner time of 1, a p^* of 2,182 psi (15.0 MPa) is determined.

A type-curve analysis of these data is illustrated in **Fig. 1.5.** With our best match, a permeability of 0.03 md and a +4.6 skin factor were calculated. These results indicated that the buildup data had not reached the true Horner straight line and were still being affected by wellbore storage.

Interval (ft)	h (ft)	ϕ_t (%)	S_w (%)	$\phi_g h$ (ft)
6,100 to 6,125	16	14.2	45.0	1.25
6,170 to 6,211	20.5	13.1	43.8	1.51
7,350 to 7,426	20	10.6	46.7	1.13
7,430 to 7,476	10	9.3	46.1	0.50

Table 1.5—Results of log analysis.

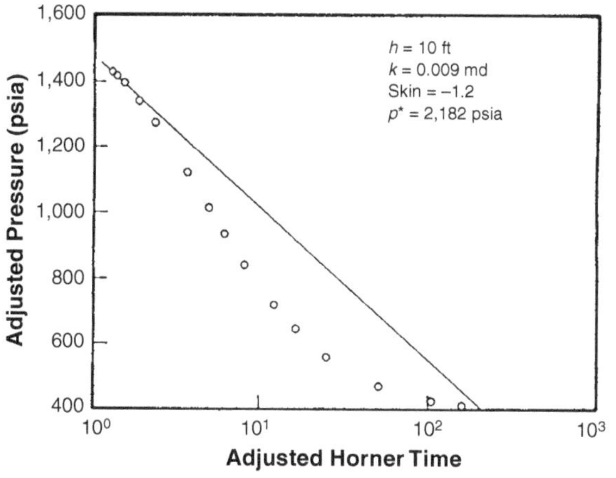

Fig. 1.4—Horner plot of prefracture pressure buildup—Sand Y.

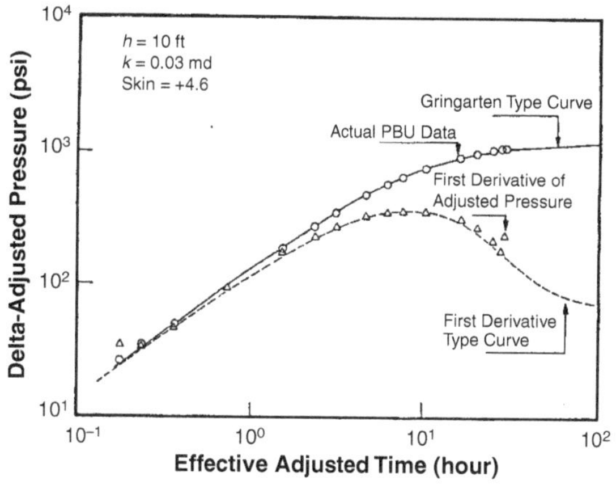

Fig. 1.5—Type-curve plot of prefracture pressure buildup (PBU)—Sand Y.

After Sand Y was tested, a bridge plug was set above the sand and another interval in the Travis Peak, Sand X, was perforated from 7,359 to 7,368, 7,379 to 7,383, and 7,418 to 7,424 ft (2243 to 2246, 2249 to 2250, and 2261 to 2263 m). A perforation bailout treatment was performed with 3,000 gal (11.4 m^3) of treated KCl water. During swabbing of fluid from the wellbore after the treatment, the zone started flowing gas. A 5-day flow test was conducted, during which time production stabilized at 590 Mcf/D (16.7 × 10^3 m^3/d) with 10 B/D (1.6 m^3/d) of condensate. The flowing tubing pressure was reported to be 1,950 psig (13.4 MPa) on an $^8/_{64}$ -in. [3.18-mm] choke and the flowing BHP was measured to be 2,517.7 psia (17.4 MPa). At this point, a 5-day BHP-buildup survey was conducted.

Immediately before shut-in began, a flowmeter and fluid density log were run in the wellbore. These logs indicated that about 78% of the flow was coming from the upper set of perforations (7,359 to 7,368 ft [2243 to 2246 m]), 22% of the flow was coming from the lower perforations (7,418 to 7,424 ft [2261 to 2263 m]), and essentially none of the flow was coming from the middle perforations.

Fig. 1.6 shows a Horner plot of these pressure-buildup data. Assuming 15 ft (4.6 m) of net-pay and using the straight line indicated on this figure, a permeability of 0.56 md was calculated. An apparent skin factor of –3.4 was also determined. Extrapolating the straight-line portion of the curve to a Horner time of 1 yielded a value for p^* of 3,240 psia (22.3 MPa). Even though a skin factor of –3.4 may appear to indicate a highly stimulated formation, previous work with multilayer or dual-porosity reservoirs indicates that skin factors of about –3 are not uncommon in zones that, in fact, do not have a region of enhanced permeability near the wellbore. Current plans are to history match the performance of the well with a 3D reservoir simulator using the production log results and pressure-buildup data to determine the individual layer properties.

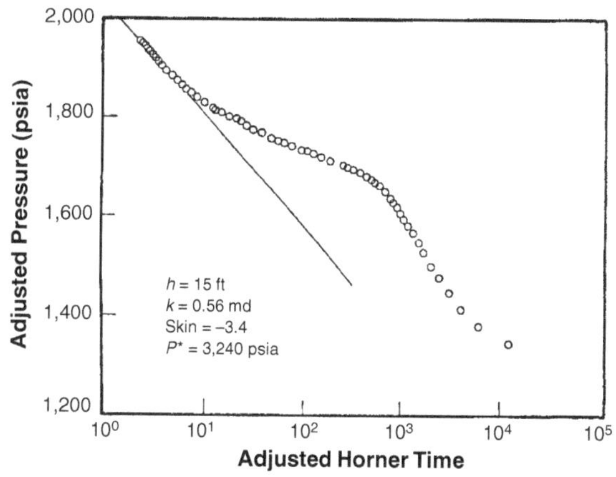

Fig. 1.6—Horner plot of prefracture pressure buildup—Sand X.

1.7.3 Upper Travis Peak Well Tests. Sand X was temporarily abandoned with a packer/plug to test Sand C-l in the Upper Travis Peak. This zone was perforated from 6,189 to 6,211 ft (1886 to 1893 m). A perforation bailout treatment was performed on Sand C-l using treated 2% KCl water, then the well was swabbed in for cleanup and testing. A 5-day flow test was conducted, followed by a 2½-day BHP-buildup survey. Before shut-in for the buildup test, the flow rate was recorded to be 460 Mcf/D (13.0 × 10^3 m^3/d) with 2 BWPD (0.3 m^3/d water) and 1 B/D (0.15 m^3/d) condensate. The flowing tubing pressure was recorded at 225 psig (1.6 MPa) on a $^{24}/_{64}$ -in. (9.5-mm) choke, and the BHP was measured at 401 psia (2.8 MPa).

A Horner plot of the BHP-buildup data is illustrated in **Fig. 1.7.** As can be seen, a reversal occurs in the pressure data, which we believe was caused by multiphase behavior in the tubing. It appears, however, that a Horner straight line might have been achieved before the pressure hump. Assuming a net-pay of 20 ft (6.1 m) and using the straight-line portion of the curve illustrated in Fig. 1.7, the permeability is calculated to be 1.55 md with an apparent skin factor of +1. A value for p^* of 1,236 psi (8.5 MPa) is obtained when these data are extrapolated to a Horner time of 1. Type-curve analyses of these data, **Fig. 1.8,** result in essentially the same values for permeability and skin factor.

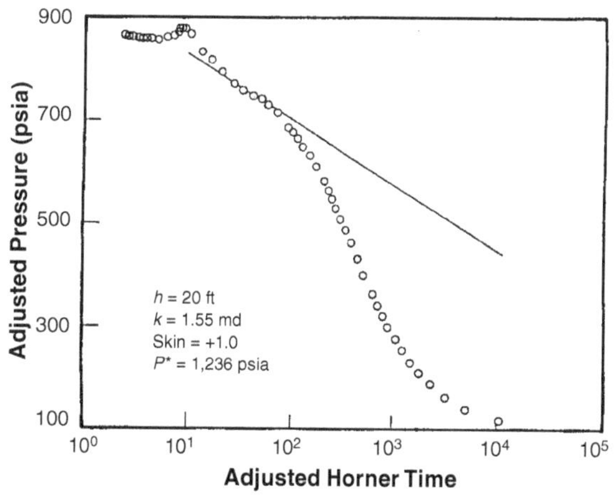

Fig. 1.7—Horner plot of prefracture pressure buildup—Sand C-1.

1.7.4 Minifracture Tests of Sand C-1. After the prefracture production test and buildup were completed, two separate minifracture injection tests were performed on Sand C-l. The first injection test was conducted specifically to measure

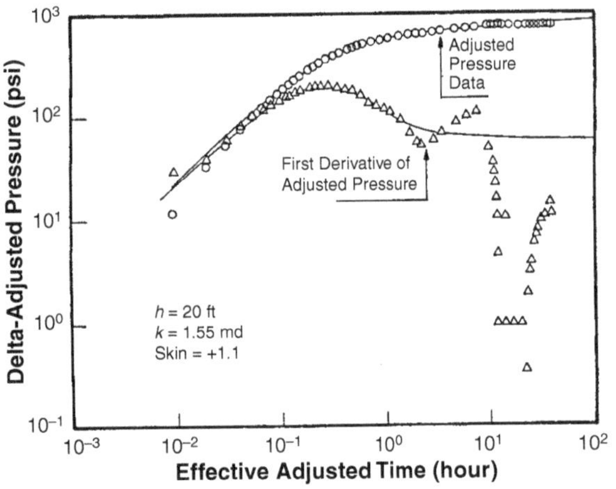

Fig. 1.8—Type-curve plot of prefracture pressure buildup—Sand C-1.

Stage	Fluid	Volume (bbl)	Rate (BPM)
1	30-lbm/1,000-gal HPG* linear gel	50	5
2	30-lbm/1,000-gal HPG linear gel	150	10
3	Shut-in 3 hours		
4	30-lbm/1,000-gal HPG linear gel	50	5
5	30-lbm/1,000-gal HPG linear gel	150	10
6	Shut-in 3 hours		
7	30-lbm/1,000-gal HPG linear gel	400	10
8	Shut-in 3 hours		

*Hydroxypropyl guar.

Table 1.6—Minifracture No. 1 design.

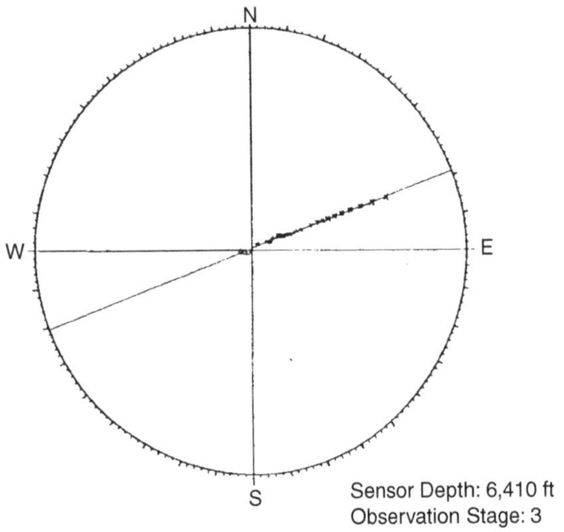

Fig. 1.9—Stereographic (polar) plot of borehole seismic data—Minifracture No. 1.

seismic activity with downhole geophones. The second test was performed primarily to obtain data for calibration of the 3D fracture simulator. In addition, these tests provided data for additional research contractors studying the hydraulic-fracturing process.

During the first minifracture, a total of 800 bbl of 30-lbm/1,000-gal (127 m^3 of 3.6-kg/m^3) linear gel was pumped into the formation. **Table 1.6** presents the design of Minifracture No. 1. Triaxial geophones were anchored in the wellbore during injection. The geophones were moved within the borehole during the shut-in periods to record seismic activity associated with the hydraulic fracture. A preliminary analysis of these seismic data indicated that the fracture azimuth was about north 69° east. These results are illustrated in **Fig. 1.9,** which shows the normalized vector distribution on a stereographic (polar) plot. A slight curvature of these vectors indicates that the fracture is possibly inclined from vertical or, in other words, not parallel with the wellbore.

The second minifracture was performed down a casing/tubing annulus, and BHPs were recorded throughout the treatment with a surface-readout quartz pressure gauge that was run inside the tubing. The design for Minifracture No. 2 is given in **Table 1.7.** This treatment was preceded with a prepad stage that contained materials that were pumped to minimize vertical fracture-height growth. Two separate materials were used and each was tagged with a different radioactive isotope (scandium 46 and iridium 192). The fluid was tagged with a third type of isotope (gold 198). A special radioactive log was run after the minifracture to locate the position of the different materials.

As **Fig. 1.10** shows, it does not appear that there was any separation between the two diverting materials or the fluid, thus indicating that there was no restriction of vertical fracture growth. Note, however, that interpretation of the log is possibly misleading for two reasons. First, there are two strings of casing in the well (5 and 7 in. [127 and 177 mm]) and it was necessary to run a slim-hole log through the tubing. Second, on the basis of preliminary results from downhole geophones in Minifracture No. 1, the fracture could be tilted from vertical or, at least, not parallel with the wellbore; thus, the top and bottom of the fracture (where the materials are

Stage	Fluid	Volume (bbl)	Rate (BPM)
1	Slickwater (bracket fracture)	190	5
2	30-lbm/1,000-gal HPG linear gel	100	5
3	30-lbm/1,000-gal HPG linear gel	90	10
4	30-lbm/1,000-gal HPG linear gel	60	2.5
5	Shut-in 1 hour		
6	30-lbm/1,000-gal HPG crosslinked gel	400	10
7	30-lbm/1,000-gal HPG crosslinked gel	400	20
8	Slickwater	190	20
9	Shut-in 2 hours		

Table 1.7—Minifracture No. 2 design.

supposed to be) may be several feet from the wellbore. These two factors could make it impossible for the log to detect the diverting materials in the fracture.

A total of 250 bbl of 30-lbm/1,000-gal (40 m^3 of 3.6-kg/m^3) linear gel and 800 bbl of 30-lbm/1,000-gal (127 m^3 of 3.6-kg/m^3) crosslinked gel were pumped into the formation at various rates. **Fig. 1.11** illustrates the resulting BHP data recorded during Minifracture No. 2.

Sand C-l was broken down, then the pumps were shut down to measure instantaneous shut-in pressure (ISIP) and closure stress. The second, third, and fourth stages were pumped using a 30-lbm/1,000-gal (3.6-kg/m^3) linear gel at 5, 10, and 2.5 bbl/min (0.8, 1.6, and 0.4 m^3/min), respectively. Because the linear gel was used, the apparent viscosity of the fracture fluid remained essentially constant. Therefore, the changes in pressure illustrated in Fig. 1.11 are caused by changes in fluid flow rate down the fracture. The magnitude of the pressure change should be a function of the fracture aperture (average width and height at the wellbore), while the time required to sense the full pressure change will be a function of fracture length.

Stage 5 was a shut-in period so that data could be collected to estimate the leakoff coefficient. The purpose of pumping Stages 2 through 5 was to gather data using a constant-viscosity fluid that could be used to "calibrate" a fracture-design model. Beginning at Stage 6, a 30-lbm/1,000-gal (3.6-kg/m^3) crosslinked fluid was pumped. Again, the injection rate was varied to provide data to calibrate the fracture-design model. In addition, the second half of Minifracture No. 2 was designed to provide insight into the possibility of vertical-height containment during the main fracture treatment. The last stage, Stage 9, was performed to determine whether the leakoff coefficient changed as a result of the use of crosslinked fluid and silica flour.

These data were analyzed by history matching the excess-pressure curve from the second minifracture treatment. The match is shown in **Fig. 1.12.** The analysis indicated that the created fracture length was 160 ft (49 m) and the fracture height was 275 ft (84 m). From the data, it was also determined that the fluid efficiency was only about 20%.

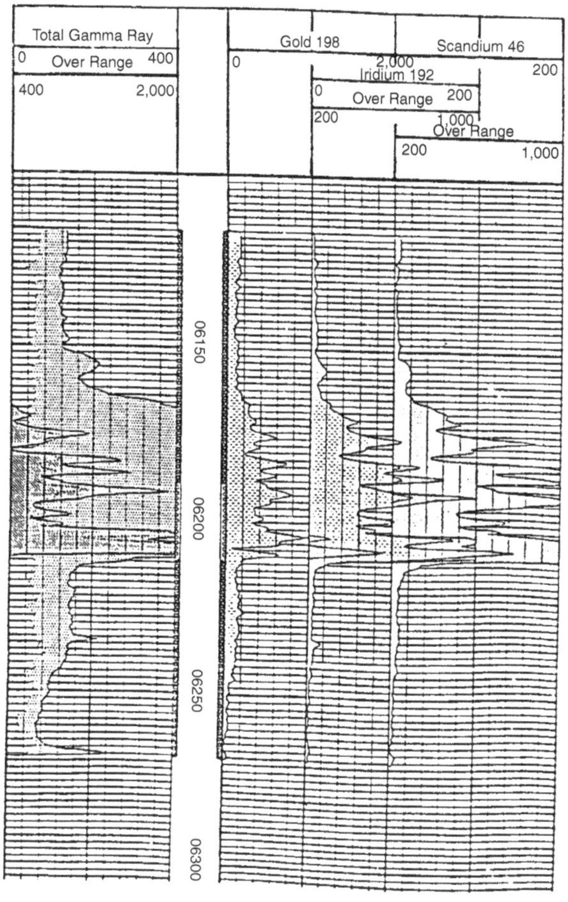

Fig. 1.10—Radioactive isotope log after Minifracture No. 2.

In summary, the data from both minifractures were extremely useful. All research contractors involved with fracture diagnostics were able to obtain data useful to their projects. In addition, an excellent data set was obtained during Minifracture No. 2 that can be and has been used to calibrate 3D fracture-propagation models.

Finally, it was obvious from these minifracture data that vertical fracture-height containment and excess fluid loss could cause problems during the main hydraulic-fracture treatment.

1.8 Summary of Prefracture Formation Evaluation

By the end of 1986, the prefracture formation evaluation had been completed on Well Howell 5. **Table 1.8** shows a summary of the results from the Sand X in the Lower Travis Peak. This zone was not fracture treated because there were no potential fracture barriers to vertical growth above or below this sand. Because one of the primary objectives of SFE 1 was to fracture treat an interval that did contain potential barriers to vertical-height growth, it was decided to plug back and stimulate Sand C-l. In fact, Sand C-l had been the primary interval of interest from the very early planning stages because of the 40-ft (12.2-m) thick mudstone immediately above the zone.

Table 1.9 shows a summary of the results from the prefracture formation evaluation performed on Sand C-1. This interval had obviously been produced in the area by other wells, because the reservoir pressure gradient at the time was only about 0.2 psi/ft [4.5 kPa/m].

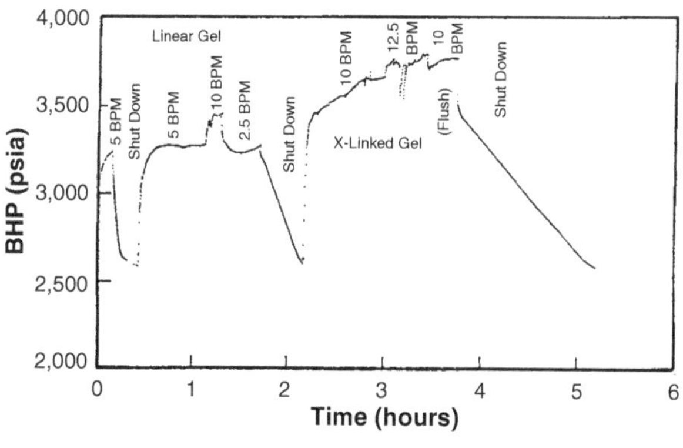

Fig. 1.11—BHPs during Minifracture No. 2.

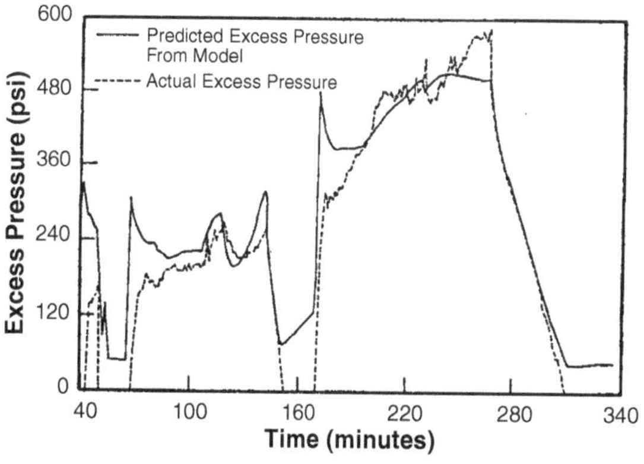

Fig. 1.12—History match of excess pressures from Minifracture No. 2.

Perforations, ft	7,359 to 7,424
Net gas pay, ft	15
Total porosity, %	12.5
Water saturation, %	39.8
Permeability, md	0.56
Skin factor	– 3.4
Initial pressure, psia	3,240
Absolute open flow, Mscf/D	2,150

Table 1.8—Results of Sand X evaluation.

Perforations, ft	9 January 1987
Net gas pay, ft	20
Total porosity, %	12.9
Water saturation, %	45.4
Permeability, md	1.55
Skin factor	+ 1.1
Initial pressure, psia	1,230
Maximum flow rate, Mscf/D	461

Table 1.9—Results of Sand C-1 evaluation.

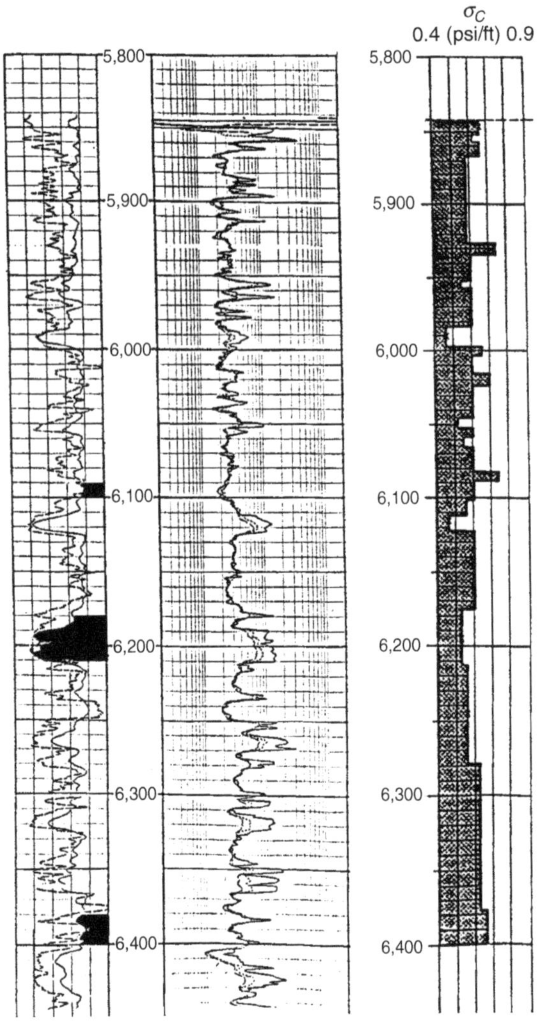

Fig. 1.13—In-situ stress profile for Upper Travis Peak.

Sand C-1, however, did meet all the criteria that had been outlined for SFE 1. It was a blanket zone with mudstones on both sides that could be considered potential barriers to fracture growth.

Fig. 1.13 presents a summary of the stress profile in the Upper Travis Peak Zones near Sand C-l. This figure is a compilation of information obtained from digital sonic log interpretation, openhole stress tests, and cased-hole stress tests. As can be seen by studying the information in Fig. 1.13, one could expect fracture-height containment to be a problem when trying to fracture treat Sand C-l because of a lack of significant stress contrast in the thicker shale sections. Specifically, upward fracture growth appears probable.

Finally, the results from the two minifracture tests clearly indicated that it would be very difficult to pump a large fracture treatment successfully in Sand C-l without successfully controlling the large amount of leakoff that was observed. The fluid efficiency was determined to be only 20%; 80% of the fluid leaked off during pumping. The analysis of Minifracture No. 2, where crosslinked fluid was pumped, indicated that the fracture height grew to 275 ft (84 m) during pumping of only a small volume of fluid at 10 bbl/min (1.6 m³/min). Thus, it was apparent that vertical fracture-height containment would also be a problem.

1.9 Fracture-Treatment Design and Execution

From the prefracture formation evaluation, a fracture treatment was designed that incorporated the following guidelines.

1. Use a large prepad and a 50%-thick pad because of the large leakoff volume expected.
2. Use extra volumes of liquid and solid fluid-loss additives to minimize leakoff.
3. Because of the high formation permeability, pump an intermediate-strength proppant (ISP) at high concentrations to maximize the dimensionless fracture conductivity.

The final fracture-treatment design is presented in **Table 1.10.** This design was entered into a 3D fracture-propagation model. From the fluid-loss data measured during Minifracture No. 2, the model predicted that a screenout would occur before completion of the treatment. The simulation analysis of this case is presented in **Fig. 1.14.**

To investigate the effects of using more fluid-loss additive, a second simulation was performed with the pumping schedule in Table 1.10 but assuming that the fluid-loss coefficient could be reduced by one-half. The result of that scenario is presented in **Fig. 1.15,** which indicates that no screenout would occur.

Fig. 1.16 presents a comparison of the excess pressures predicted by the model for the two possible cases. The first case predicted an early screenout as a result of high leakoff. The second case indicated that if more fluid-loss additive could be used to reduce leakoff substantially, then the treatment could be successfully pumped as designed.

From these results, a decision was made to proceed with the aggressive treatment design, after incorporating additional volumes of fluid-loss additives, to see how much proppant could be pumped into Sand C-l. Because the rationale of the entire SFE project was to gather a data set that could be used to develop and validate 3D fracture-design theory, the aggressive approach was more compatible with the goals of the overall research effort.

The fracture treatment was performed on the SFE well on 9 January 1987. The treatment consisted of a 30-lbm/1,000-gal (3.6-kg/m³) delayed crosslink gel carrying 18/20-mesh intermediate-strength proppant (ISP). The treatment was performed down the casing/tubing annulus so that bottomhole injection pressures could be recorded throughout the treatment.

Stage	Stage Volume (gal)	Proppant Concentration (lbm/gal)	Proppant Type
1	30,000	—	Prepad
2	28,000	—	Gel pad
3	15,000	0.5	100-mesh sand
4	15,000	1.0	100-mesh sand
5	1,000	0.5	18/20 ISP
6	1,000	1.0	18/20 ISP
7	2,000	2.0	18/20 ISP
8	2,000	3.0	18/20 ISP
9	4,000	4.0	18/20 ISP
10	10,000	5.0	18/20 ISP
11	10,000	6.0	18/20 ISP
12	10,000	7.0	18/20 ISP
13	10,000	8.0	18/20 ISP
14	12,000	10.0	18/20 ISP

Total fluid = 150,000 gal. Total proppant (ISP) = 407,500 lbm.

Table 1.10—Proposed fracture treatment for Howell 5 (SFE 1).

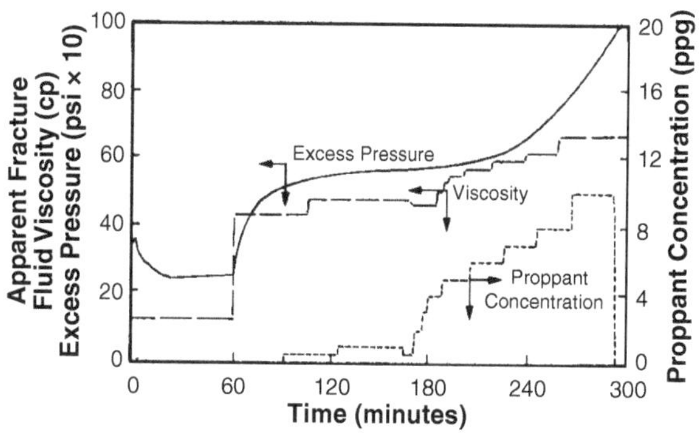

Fig. 1.14—Predicted excess pressure from fracture design—Case 1, high leakoff.

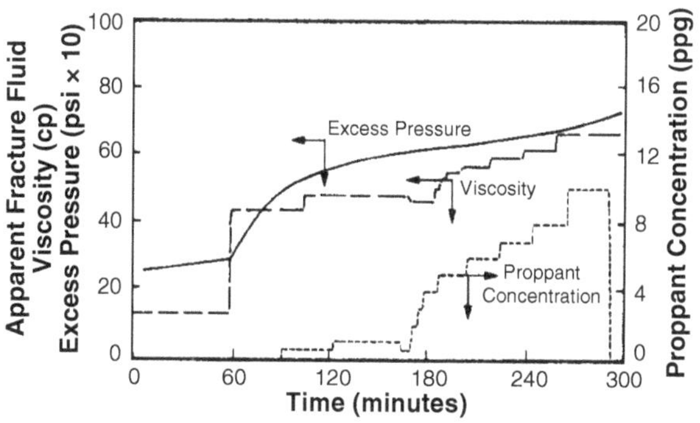

Fig. 1.15—Predicted excess pressure from fracture design—Case 2, low leakoff.

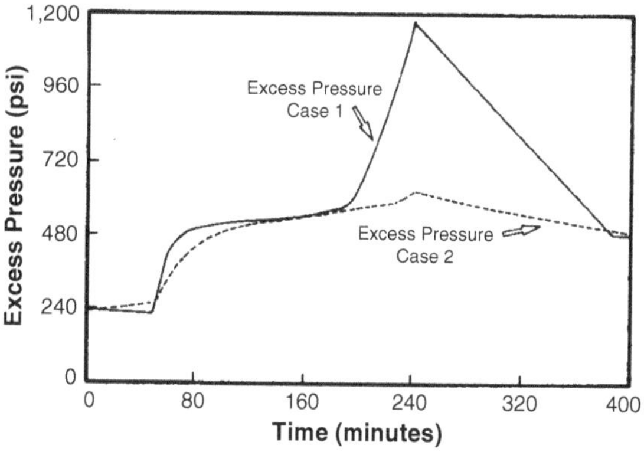

Fig. 1.16—Comparison of predicted excess pressures from Cases 1 and 2.

During the fracture treatment, several fracture-diagnostic and fracture-treatment-monitoring experiments were conducted. All pertinent fracturing data were recorded by the Mobile Testing and Control Facility for use in the 3D fracture simulator. The pump-generated pressure frequencies were recorded for evaluating the fracture volume and, ideally, the fracture dimensions. Hydraulic-fracture impedance testing was also performed to determine the fracture dimensions.

Fig. 1.17 presents the bottomhole treating pressure measured with an electronic pressure gauge suspended in the open-ended tubing. As can be seen, there were some problems during pumping of the slick pad. Some material (which was unknown at the time) began to plug the perforations. The treatment had to be shut down twice to surge the well to clean the debris from the perforations. Even though all of the fracture tanks were cleaned before arriving on location and clean water was added to the tanks, it was later discovered that pieces of rubber material used to line the tanks had evidently broken loose from one of the tanks and were pumped down the well. This is what caused the erratic pressure behavior observed early in the treatment.

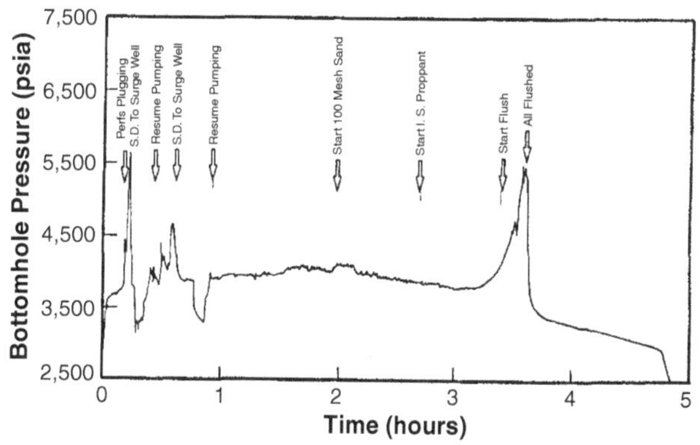

Fig. 1.17—Measured bottomhole treating pressure during main hydraulic fracture.

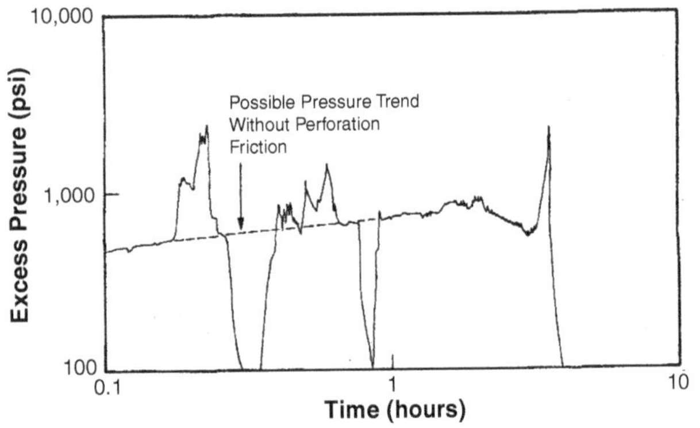

Fig. 1.18—Actual excess pressure during main hydraulic-fracture treatment.

Date	9 January 1987
Service company	Dowell-Schlumberger
Average injection rate, bbl/min	15
Average surface pressure, psig	1,200
Average BHP, psia	3,960
Average proppant concentration, lbm/gal	4.2
Maximum proppant concentration, lbm/gal	5.3
Total prepad volume, gal	30,000
Total fracturing fluid, gal	78,000
Total ISP, lbm	87,500

Table 1.11—Fracture treatment summary.

Fig. 1.18 presents a log-log plot of excess pressure vs. time. From this graph, one can see that the rubber pieces, which plugged the perforations, were certainly annoying but did not really affect the eventual trend of the excess pressure. It could have affected the eventual outcome, however, because the fracture closed during both shutdowns when the well was surged. In addition, it was believed at the time that the heavy concentrations of fluid-loss additives might be causing the problem; thus, the volume of these additives was reduced.

In Fig. 1.18, one can notice that the excess pressure increased steadily until a time of about 2 hours, when the pressure began to decline. This decline in pressure was caused by fracture-height growth, perforation erosion, or both.

After 3 hours and 15 minutes of pumping, the excess pressure began to increase rapidly, indicating that a screenout was imminent. Once it was clear that the treatment was going to screen out, the proppant was cut. The maximum concentration achieved at this point was 5.3 lbm/gal (635 kg/m^3). The well was successfully flushed before a screenout was complete. A summary of what was actually pumped is given in **Table 1.11.**

The final result was accurately predicted by the 3D model using the fluid-loss values obtained from Minifracture No. 2. Because of the early problems such as perforation plugging and shutdowns the chances of pumping the treatment successfully, as shown in Table 1.10, were substantially reduced. The main goal, however, was to pump as much proppant as possible and to monitor and record successfully all data necessary to run a 3D fracture simulator. With these objectives in mind, the stimulation treatment of Sand C-1 in SFE 1 was considered very successful.

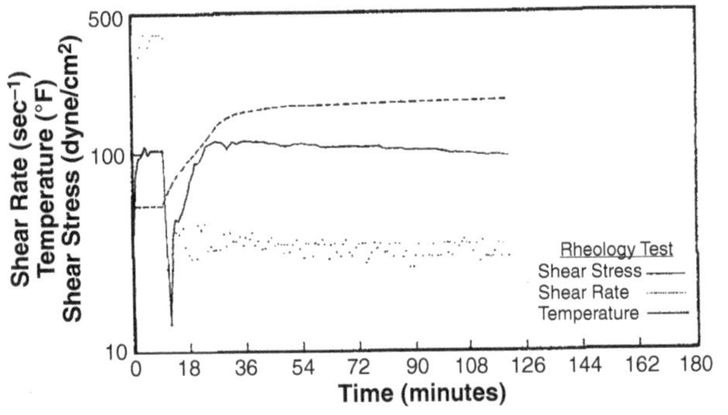

Fig. 1.19—Viscous behavior of fracturing fluid from Minifracture 2 (measured with capillary-tube viscometer).

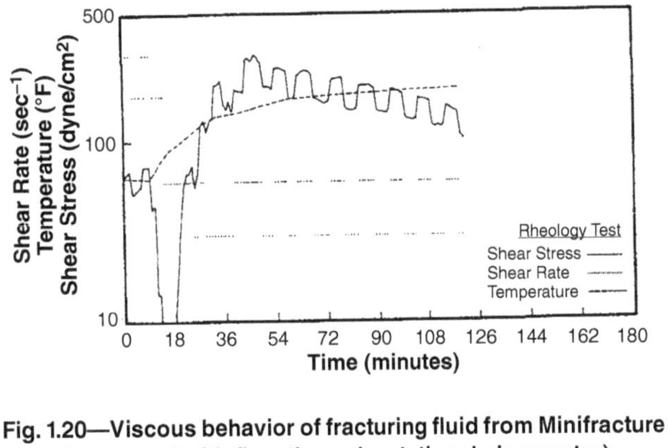

Fig. 1.20—Viscous behavior of fracturing fluid from Minifracture No. 2 (measured with flow-through rotational viscometer).

1.10 Fracture-Fluid Viscosity Measurements

The main purpose of SFE 1 was to generate a data set that can be used in a 3D fracture-propagation model. In previous efforts, research showed that no 3D model can be used properly to history match excess pressures during a fracture treatment if the viscous behavior of the fracture fluid is not known with certainty. In essence, the viscous behavior of the fluid must be known if the fracture shape is to be the main variable used to match excess pressure. If one does not know the actual viscosity of the fluid, then the shape of the excess-pressure curve can be matched with any number of combinations of fracture shape and viscous behavior.

To begin to solve this problem, the GRI Rheology Unit was constructed. This unit contained two main components: the rheology van and the rheology skid. The rheology unit could be taken to the field and used to measure the viscous properties of the actual fracturing fluids under simulated downhole conditions of shear rate and bottom-hole temperature.

Examples of the data measured for SFE 1 are presented in **Figs. 1.19 through 1.21.** The data in Fig. 1.19 were obtained with a capillary-tube viscometer for the gel pumped during Minifracture No. 2. Fig. 1.20 presents similar data measured with a flow-through rotational viscometer. Fig. 1.21 shows data measured using a sample of the gel pumped during the fracture treatment.

The importance of these data is that they are an accurate representation of the viscous behavior that can be expected from the gel mixed for a specific treatment and measured under the shear rate and temperature conditions actually expected during the fracture treatment. These data can be used in a 3D fracture-propagation model to improve the analysis of fracture geometry.

1.11 Postfracture-Treatment Evaluation

After the fracture treatment, gamma ray/temperature logs were run, which indicated that the casing was clear to at least 6,350 ft (1935 m) and the vertical fracture height near the wellbore was at least 190 ft (58 m). **Fig. 1.22** presents the after-fracture production log.

An extended production test, followed by a BHP-buildup survey, was then conducted. The following discussion summarizes the results of the analyses of these data. A more detailed discussion of the postfracture evaluation can be obtained from the 1987 SFE 1 annual report (S. A. Holditch & Associates 1988).

1.11.1 Production Performance. After the fracture treatment, the well was opened to the pit for cleanup. Swabbing was necessary to initiate flow; however, the well began flowing after about 1½ days. It produced for an additional 4 days, then was shut in to be connected to the production and pipeline facilities. SFE 1 produced to sales from 20 January–16 March 1987. At this time, the well was shut in to conduct a BHP-buildup survey.

Fig. 1.23 presents the production history from the well, showing water and gas production and BHP. As can be seen, only water was initially produced; however, gas breakthrough occurred approximately 1½ days after swabbing. Gas flow rate increased to about 500 Mcf/D (14.2×10^3 m^3/d), while water flow rate decreased to about 65 B/D (10.3 m^3/d). After being turned to the permanent facilities, gas production continued to increase and water production continued to decrease. At the end of the flow period, gas production had stabilized at about 1,100 Mcf/D (31.1×10^3 m^3/d), while water production leveled off at 8 to 9 B/D (1.3 to 1.4 m^3/d).

As mentioned, gas production continued to improve throughout the early cleanup period. As a result, the productivity index (PI) also continued to increase. **Table 1.12** presents a summary of the PI calculations for Sand C-1. This table also shows the increase over prefracture conditions. It is interesting to note that because of the low BHP, the well had to produce for about 1½ months before fully cleaning up all the fracturing fluid and achieving its maximum productivity increase of 3.6. Before shut-in for the pressure-buildup survey, Sand C-1 had produced a total of 52,500 Mcf (1490×10^3 m^3) of gas and 1,706 bbl (271 m^3) of water. A small amount of condensate was also produced.

1.11.2 Conventional Pressure-Buildup Analysis. Fig. 1.24 is a Horner plot of the pressure-buildup data. These data were analyzed using the concepts of adjusted pressures and adjusted times previously discussed. **Fig. 1.25** presents an expanded scale of the late-time buildup data. As can be seen, a straight line has been drawn through pressure-buildup data that appear to represent the middle time region (MTR). If this is the true MTR, the permeability is calculated to be 6.3 md, assuming a net-pay of 20.5 ft (6.2 m). A skin factor of –3.6 is also determined.

Fig. 1.26 is a log-log plot of the pressure-buildup data. These data were analyzed using type curves, resulting in a permeability calculation of 8 md with a skin factor of –2.6. The test could not be matched adequately with type curves developed for vertically fractured wells because of the unusual shape during the early portion of the buildup and the late-time region. As a result, it was believed that history matching the well performance using more complex reservoir-modeling techniques would be required to evaluate the reservoir and fracture properties fully.

1.11.3 Reservoir-Simulation Analysis. Many types of reservoir models were used to analyze the well performance. These models included a single-phase, single-layer model; a single-phase, multilayer model; and a two-phase, multilayer model.

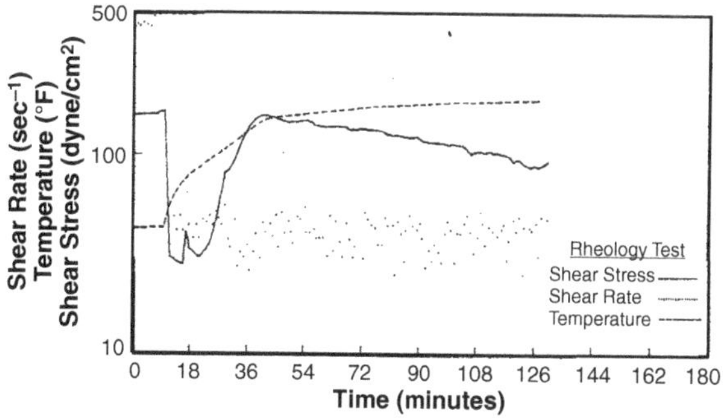

Fig. 1.21—Viscous behavior of fracturing fluid from main fracture treatment.

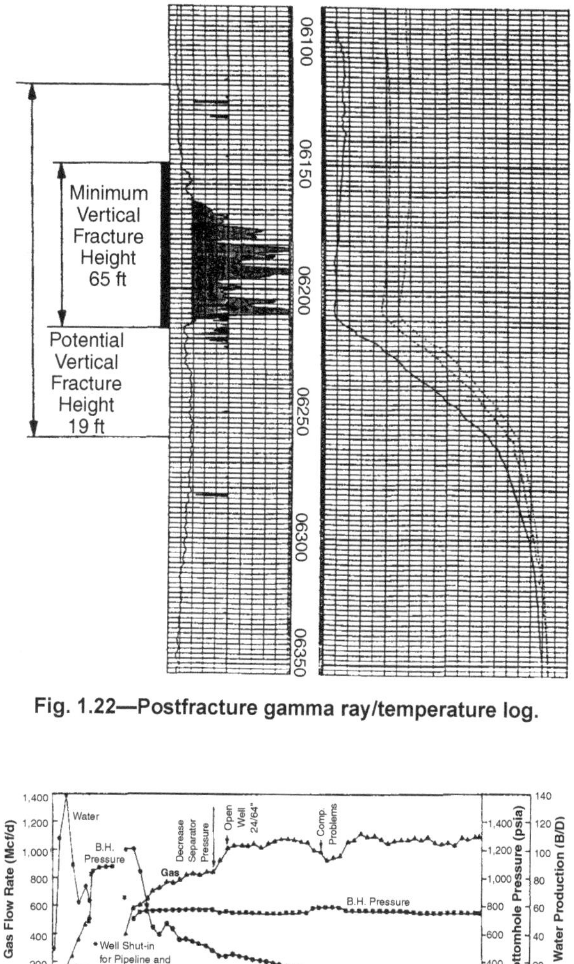

Fig. 1.22—Postfracture gamma ray/temperature log.

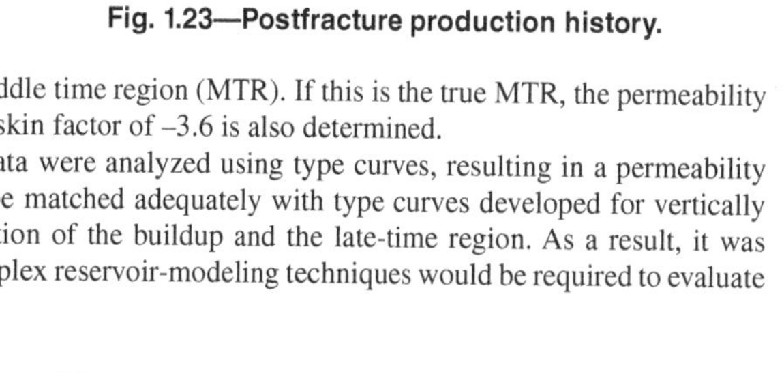

Fig. 1.23—Postfracture production history.

Date	Flow rate (Mscf/D)	Flowing BHP (psia)	Productivity index (scf/psia^2)	PI increase
Dec. 10, 1986	461	401.4	0.36	Prefracture
Jan. 16, 1987	500	687.4	0.52	1.4
Jan. 28,1987	735	762.5	0.85	2.4
Feb. 4, 1987	837	769.6	0.99	2.8
Feb. 11, 1987	1,045	753.7	1.20	3.3
Feb. 28, 1987	1,100	771.8	1.30	3.6

Table 1.12—Comparison of PI before and after fracturing.

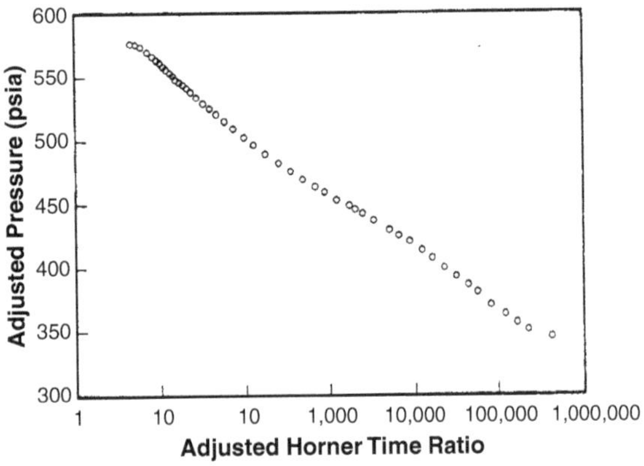

Fig. 1.24—Horner plot of postfracture pressure buildup.

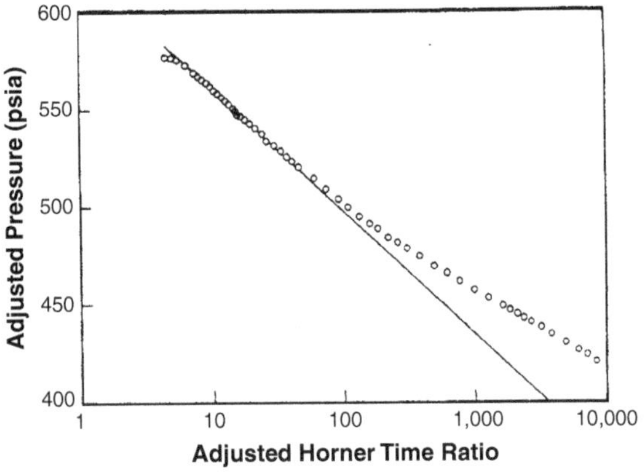

Fig. 1.25—Horner plot of postfracture pressure buildup—expanded scale.

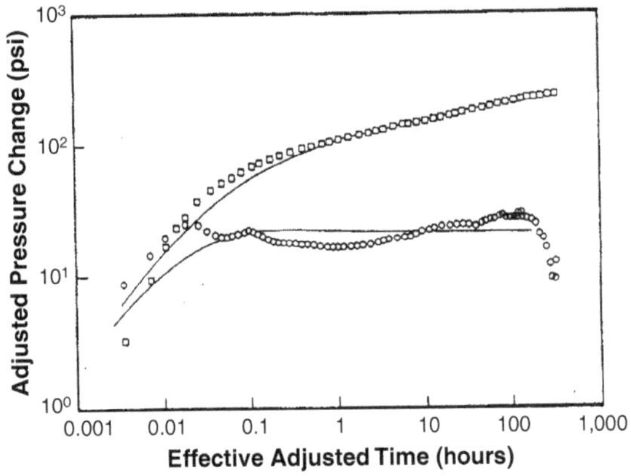

Fig. 1.26—Log-log plot of postfracture pressure buildup.

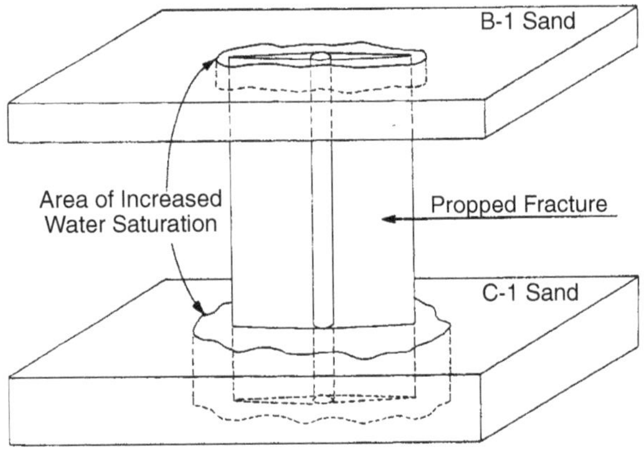

Fig. 1.27—Schematic of reservoir model.

On the basis of all the reservoir-simulation work performed, the two-phase, multilayer model provided the most acceptable results. As a result, only this work will be presented.

The purpose of using a two-phase, multidimensional reservoir simulator was twofold. First, it was believed that the fracture-fluid cleanup rates would need to be analyzed in addition to the gas flow rates and flowing pressures. Previous work (Holditch et al. 1987) indicated that analysis of these data should provide unique results for the reservoir and fracture properties. Secondly, analysis of the fracture-treatment data using fracture-propagation models indicated that the hydraulic fracture grew vertically and penetrated several sandstones, one of which was partially depleted. This sand, known as Sand B-1, is at a depth of about 6,100 ft (1859 m). The reservoir pressure in Sand B-1 was known to be about 550 psi (3.8 MPa). As a result, it was believed that two individual sand layers communicating by a vertical fracture would need to be simulated. **Fig. 1.27** is a schematic of the reservoir model set up to represent the conditions existing in SFE 1. In addition to the two layers of sand intersected by a vertical fracture, an area of increased water saturation was placed around the fracture in each of the formations. This area represented the fracturing fluid that was pumped into the formations.

The depth of invasion of this fluid was found to be an important consideration in history matching postfracture cleanup. For relatively shallow invasion, gas breakthrough occurs fairly rapidly. If deep invasion has taken place, gas breakthrough could take several days. This behavior, coupled with the fact that the gas and water flow rates are strongly controlled by the propped fracture length, provides a means by which to obtain a unique solution for the fracture properties.

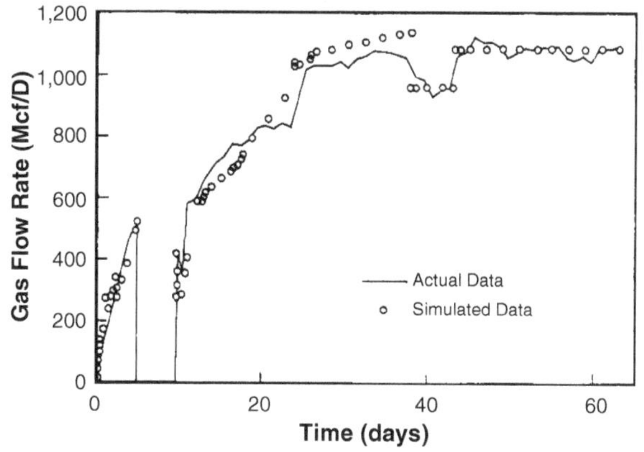

Fig. 1.28—Comparison of actual gas production to simulation results.

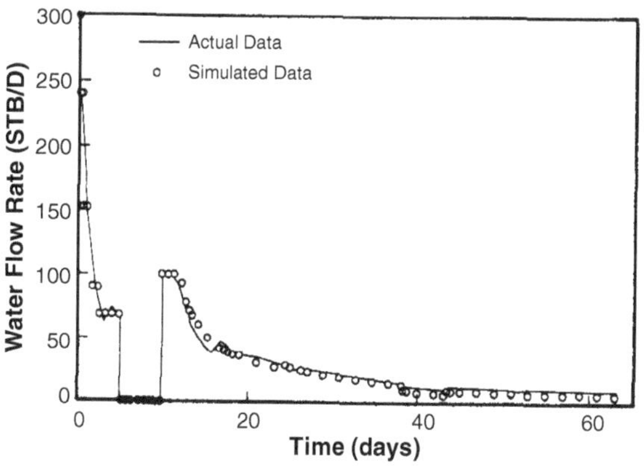

Fig. 1.29—Comparison of actual water production to simulation results.

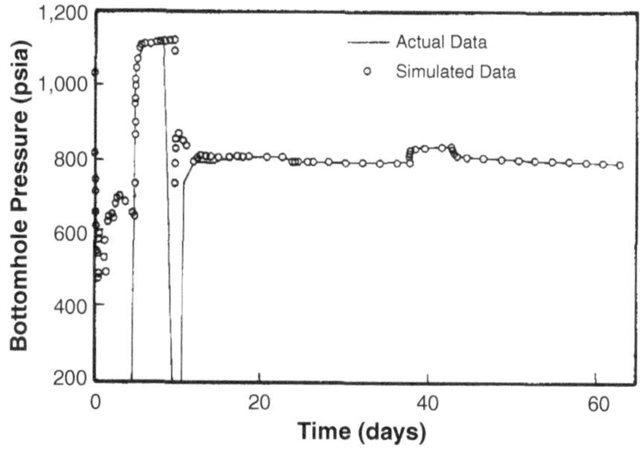

Fig. 1.30—Comparison of actual flowing pressures to simulation results.

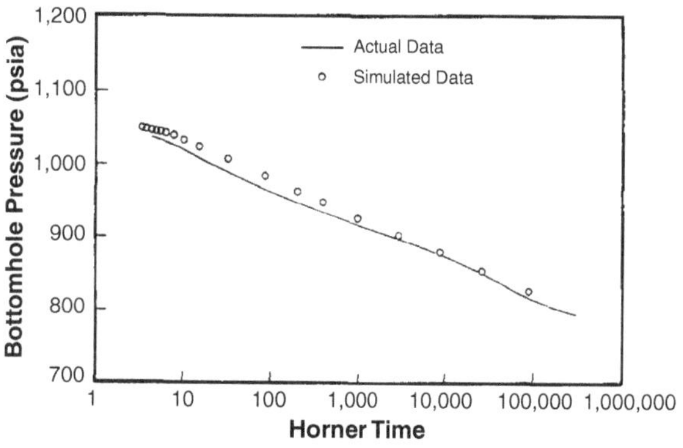

Fig. 1.31—Comparison of actual pressure-buildup test to simulation results.

Figs. 1.28 through 1.30 show the actual gas rates, water rates, and flowing BHP, respectively, during the extended production period. These figures also show the results of our final history match. **Fig. 1.31** shows a Horner plot of the actual pressure-buildup test and, again, the results of our final history match. The reservoir and fracture properties used to achieve these history matches are summarized in **Table 1.13.**

Overall, a very good history match of all the available data was achieved. To obtain the final results, however, it was necessary to vary several other parameters. These were the water viscosity and the amount of crossflow allowed to occur from Sand C-l to Sand B-1. Because the water in the areas of increased saturation represented fracturing fluid and not formation water, it was necessary to simulate higher initial viscosities than normally expected. In addition, these viscosities tended to vary with time as the fracturing fluid degraded. An interrupt feature in the model allowed us to simulate this behavior.

Reservoir Properties	Sand C-1	Sand B-1
Absolute permeability, md	8.0	15.0
Gas permeability, md	5.0	5.0
Net-pay, ft	20	10
Porosity, %	13.1	14.2
Water saturation, %	43.8	45.0
Reservoir pressure, psia	1,160	550
Drainage area, acres	320	320
Gas in place, Bcf	1.4	0.3
Fracture Properties		
Propped length, ft	150	
Fracture conductivity, md-ft	2,550	

Table 1.13—Properties used in reservoir simulation history match.

During the simulation studies, it was found that the amount of crossflow greatly affected the shape of the pressure-buildup curve. This is even more dramatic when analyzing a two-phase system. In general, the overall pressure level of the buildup could be shifted up or down, or the slope of the curve could be changed just by altering the amount of crossflow between the two layers. It was also possible to cause the simulated pressures actually to decrease by allowing more crossflow to occur between the two zones. This type of behavior is also a function of the two-phase-flow characteristics of the formations. As a result, it is necessary to use a two-phase, multidimensional reservoir model to analyze this type of behavior.

Given the final reservoir and fracture properties obtained from the reservoir-simulation work, future performance was projected. It was estimated that SFE 1 would recover 1.14 Bcf (32.2 × 10^6 m^3) of gas over 6.5 years. To obtain this projection, the ultimate flowing BHP was assumed to be 250 psi (1.7 MPa) and an economic limit of 100 Mcf/D (2.8 × 10^3 m^3/d) was used.

1.12 Conclusions

The first GRI SFE provided a very comprehensive and valuable data set that can be used to calibrate both 3D fracture-design models and 3D reservoir-fluid-flow models. The SFE also allowed various research and development organizations the opportunity to measure and to record data before, during, and after a hydraulic-fracture treatment that can be analyzed and integrated into a comprehensive understanding of the hydraulic-fracturing process.

Additional details concerning the data measured during 1986 are available from individual contractors. Specifically, information concerning the coring and logging operations (CER Corp. 1986a, 1986b, 1987a, 1987b) is available, as well as core-analysis results (NL Erco/NL Industries 1986a, 1986b, 1986c, 1986d; Petrophysical Services 1986; Reservoirs Inc. 1986; Terra Tek Research 1986; Litton-Core Lab 1986) and geologic descriptions (Dutton and Finley 1986a, 1986b; Dutton et al. 1986). Detailed log analyses (ResTech 1986a, 1986b, 1986c) are also available, and individual well reports (S. A. Holditch & Associates 1986a, 1987a, 1987b) on each cooperative were issued. Included in these reports are summaries of the coring and logging analyses, prefracture injection tests, well test data, and the fracture-treatment data measured and analyzed on each of the wells.

From our studies of the data collected during this project, several observations and conclusions were made regarding the Travis Peak and hydraulic fracturing in general.

- Excessive vertical height growth during a fracture treatment is considered a major problem in the Travis Peak with respect to achieving long, conductive fractures. Lack of fracture containment is a direct result of insufficient stress contrast between the net-pay zones and the rock layers above and below.
- To contain a fracture within the formation of interest or to minimize the vertical height growth, the excess pressure within the fracture must be minimized. The excess pressure is almost entirely dependent on the friction pressure drop down the fracture, which in turn, is a function of the gel viscosity, the injection rate, and the friction coefficient between the fluid and the fracture wall. Previous work and numerical modeling studies indicate that gel viscosity and the friction coefficient have the most influence over friction pressure drop. Thus, to control vertical fracture-height growth successfully, it will be necessary to quantify the effects of these variables further and to learn how to control these factors during a treatment.
- Without accurate slurry viscosity measurements of the actual fracturing fluids used during a treatment, it will not be possible to calculate and to predict the correct fracture dimensions by analyzing fracture-treatment data. Data currently available are not subjected to the same conditions of shear, temperature, and leakoff that are actually present downhole; thus, these data do not accurately reflect the actual viscosity of the fluids in the fracture.
- From studies performed during 1986, we believe that reproducible slurry viscosity measurements can be made under simulated downhole conditions. A flow loop was built that can simulate the downhole conditions of shear rate, temperature, and fluid leakoff. Plans are to use the flow loop in the field before and during actual fracture treatments to generate viscosity data for 3D fracture-propagation models.

The following additional conclusions pertain more specifically to GRI's first SFE well, Howell 5.

- On the basis of an initial reserves study of the Waskom Field (S. A. Holditch & Associates 1986b), the 320-acre (130-ha) unit surrounding Well Howell 5 should contain about 3.5 to 4.0 Bcf (99×10^6 to 113×106 m^3) of gas in the Upper Travis Peak. Sand C-1, the upper completion interval, contains about one-third of these reserves, but could have greater reserves if a larger area can be drained.
- The hydraulic-fracture orientation was successfully measured by numerous independent techniques. On the basis of these results, the fracture azimuth was determined to be between north 69° east and north 77° east.
- Extremely high leakoff was observed in Sand C-l. A fluid efficiency of only 20% was calculated with the crosslinked gel fluids used during the second minifracture.
- Because of the excessive leakoff and expected lack of fracture containment, it was predicted by a 3D fracture-propagation model that a screenout would occur if a massive hydraulic-fracture treatment was attempted. During the actual treatment, it appeared that a screenout was in fact occurring at about the same time predicted by the model. All the proppant-laden fluid, however, was successfully flushed into the formation.
- Well productivity has been substantially improved as a result of the fracture treatment. The PI is 3.6 times higher than the prefracture conditions.
- Using a two-phase multidimensional reservoir simulator, a history match of the postfracture gas production, water production, flowing BHPs, and pressure-buildup test was performed. The results from this analysis indicate that Sand CM has a permeability to gas of 5.0 md with an average net-pay thickness of 20 ft (6 m). The propped fracture length was determined to be 150 ft (46 m), and the fracture conductivity was estimated to be 2,550 md-ft (777 md-m).
- The history-match analysis indicates that SFE 1 is draining about 320 acres (130 ha); therefore, the ultimate recovery is projected to be 1.14 Bcf (32.3×10^6 m^3) of gas over 6.5 years.
- The analysis also substantiates independent modeling work that indicates that the hydraulic fracture grew vertically and thus penetrated Sand B-l. The well performance and pressure-buildup test were influenced by crossflow between Sands C-l and B-l and a successful history match would not have been obtained if this phenomenon had not been modeled.

1.13 Nomenclature

h = thickness, ft [m]
k = permeability, md
p^* = pressure, psi [kPa]
S_w = water saturation
a_c = closure stress gradient, psi/ft [kPa/m]
ϕ = porosity

Subscripts

g = gas
t = total

1.14 Acknowledgments

We gratefully acknowledge the support of the GRI, who sponsored this work under GRI Contract No. 5083-211-9877. We also appreciate the efforts and dedication of all the organizations listed in Table 1.1 who have contributed to the GRI research program.

1.15 References

Agarwal, R. G. 1979. Real Gas Pseudo-Time - A New Function for Pressure Buildup Analysis of MHF Gas Wells. Presented at the SPE Annual Technical Conference and Exhibition, Las Vegas, Nevada, 23–26 September. SPE-8279-MS. https://doi.org/10.2118/8279-MS.

Al-Hussainy, R., Ramey, H. J. Jr. and Crawford, P. B. 1966. The Flow of Real Gases Through Porous Media. *J Pet Technol* **18** (5): 624–636. SPE-1243-A-PA. https://doi.org/10.2118/1243-A-PA.

CER Corp. 1986a. Summary of Data Acquisition and Field Operations—Arkla Exploration T. P. Scott No. 5, Harrison County, Texas. GRI Contract No. 5082-211-0641, CER Corp., Las Vegas, Nevada (July 1986).

CER Corp. 1986b. Summary of Data Acquisition and Field Operations—Marshall Exploration, Wemer Sawmill No. 5, Panola County, Texas. GRI Contract Nos. 5082-211-0641 and 5083-211-0858, CER Corp., Las Vegas, Nevada (April 1986).

CER Corp. 1987a. Summary of Data Acquisition and Field Operations—Mobil Producing Texas and New Mexico Inc. G. E. Cargill No. 14, Harrison County, Texas. GRI Contract No. 5086-211-1306, CER Corp., Las Vegas (April 1987).

CER Corp. 1987b. Summary of Data Acquisition and Field Operations—S. A. Holditch & Associates Inc., Howell No. 5, Harrison County, Texas. GRI Contract No. 5086-211-1306, CER Corp., Las Vegas, Nevada (April 1987).

Dutton, S. P. and Finley, R. J. 1986a. Preliminary Geologic Description, Arkla Exploration T. P. Scott No. 5. GRI Contract No. 5082-211-0708, Bureau of Economic Geology, The University of Texas at Austin, Austin, Texas (June 1986).

Dutton, S. P. and Finley, R. J. 1986b. Preliminary Geologic Description, Marshall Exploration, Werner Sawmill No. 5. GRI Contract No. 5082-211-0708, Bureau of Economic Geology, The University of Texas at Austin, Austin, Texas (March 1986).

Dutton, S. P., Fracasso, M. A., and Finley, R. J. 1986. Preliminary Geologic Description, Mobil Producing G. E. Cargill No. 14. GRI Contract No. 5082-211-0708, Bureau of Economic Geology, The University of Texas at Austin, Austin, Texas (December 1986).

Finley, R. J. 1985. The Travis Peak (Hosston) Formation: Geologic Framework, Core Studies, and Engineering Field Analysis—Topical Report. GRI Contract No. 5082-211-0708, Bureau of Economic Geology, The University of Texas at Austin, Austin, Texas (January 1985).

Fracasso, M. A. 1987. Staged Field Experiment Site Selection and Geologic Characterization of the Upper Travis Peak Formation, Harrison and Panola Counties, East Texas. Topical Report, GRI Contract No. 5082-211-0708, Bureau of Economic Geology, The University of Texas at Austin, Austin, Texas (February 1987).

Holditch, S. A., Robinson, B. M. and Whitehead, W. S. 1987. The Analysis of Complex Travis Peak Reservoirs in East Texas. Presented at the SPE Symposium on Low Permeability Reservoirs, Denver, 18–19 May. SPE-16427-MS. https://doi.org/10.2118/16427-MS.

Litton-Core Lab. 1986. Core Analysis Report for S. A. Holditch No. 5 Howell, Waskom Field, Harrison County, Texas. Litton-Core Lab, Dallas (November 1986).

NL Erco/NL Industries. 1986a. Combined Special Core Analysis and Petrographic Investigation of the Travis Peak Formation, T. P. Scott No. 5 Well, Waskom Field, Harrison County, Texas. NL Erco/NL Industries Inc., Houston.

NL Erco/NL Industries. 1986b. Routine Core Analysis, Final Report 00161257 for Arkla Exploration T. P. Scott No. 5, Harrison County, Texas. NL Erco/NL Industries Inc., Houston (July 1986).

NL Erco/NL Industries. 1986c. Routine Core Analysis, Revised Report 00161002 for Marshall Exploration Company, Wemer Sawmill No. 5 Well, J. W. Dollahite A-173 Field, Panola County, Texas. NL Erco/NL Industries Inc., Houston (27 March 1986).

NL Erco/NL Industries. 1986d. Tritium Analysis, Final Report 00161344, Arkla Exploration T. P. Scott No. 5 Well, Harrison County, Texas. NL Erco/NL Industries Inc., Houston (22 September 1986).

Petrophysical Services. 1986. Special Analysis Report: Werner Sawmill No. 5, Marshall Exploration Company. Petrophysical Services Inc., Mountain View, California (September 1986).

Reservoirs Inc. 1986. Routine Conventional Core Analysis, Mobil G. E. Cargill No. 14 Well, Harrison County, Texas. Reservoirs Inc., Houston (September 1986).

ResTech Inc. 1986a. Development of Petrophysical Techniques for Evaluation: Tight Gas Sands, Annual Technical Report, March 1985–March 1986. GRI Contract No. 5084-211-1062, ResTech Inc., Houston (April 1986).

ResTech Inc. 1986b. Development of Petrophysical Techniques for Evaluation: Tight Gas Sands, Quarterly Technical Report, April–June 1986. GRI Contract No. 5084-211-1062, ResTech Inc., Houston (June 1986).

ResTech Inc. 1986c. Development of Petrophysical Techniques for Evaluation: Tight Gas Sands, Quarterly Technical Report. July–September 1986. GRI Contract No. 5084-211-1062, ResTech Inc., Houston (September 1986).

S. A. Holditch & Associates Inc. 1986a. Cooperative Well Report: Marshall Exploration Werner Sawmill No. 5, Panola County, Texas. GRI Contract No. 5083-211-0877, S. A. Holditch & Associates Inc., College Station, Texas (November 1986).

S. A. Holditch & Associates Inc. 1986b. Reserve Study for a Portion of the Waskom (Travis Peak) Field, Harrison County, Texas. GRI Contract No. 5083-211-0877, S. A. Holditch & Associates Inc., College Station, Texas (September 1986).

S. A. Holditch & Associates Inc. 1987a. Cooperative Well Report: Arkla Exploration T. P. Scott No. 5, Harrison County, Texas. GRI Contract No. 5083-211-0877, S. A. Holditch & Associates Inc., College Station, Texas (January 1987).

S. A. Holditch & Associates Inc. 1987b. Cooperative Well Report: Mobil Producing Texas and New Mexico, Inc. G. E. Cargill No. 14, Harrison County, Texas. GRI Contract No. 5083-211-0877, S. A. Holditch & Associates Inc., College Station, Texas (January 1987).

S. A. Holditch & Associates Inc. 1988. 1987 Annual Report—Post-Fracture Production and Pressure Transient Analysis of SFE No. 1. GRI Contract No. 5083-211-0877, S. A. Holditch & Associates Inc., College Station, Texas (January 1988).

Terra Tek Research. 1986. Final Report—Special Core Analysis, Arkla's Scott No. 5 Well. Terra Tek Research, Salt Lake City (November 1986).

SI Metric Conversion Factors

acres	×	4.046 873	E–01	=	ha
bbl	×	1.589 873	E–01	=	m^3
cp	×	1.0*	E–03	=	Pa·s
dynes/cm^2	×	1.0*	E–01	=	Pa
ft	×	3.048*	E–01	=	m
ft^3	×	2.831 685	E–02	=	m^3
°F		(°F–32)/1.8	= °C		
gal	×	3.785 412	E–03	=	m^3
in.	×	2.54*	E+00	=	cm
lbm	×	4.535 924	E–01	=	kg
lbm/gal	×	1.198 264	E+02	=	kg/m^3
psi	×	6.894 757	E+00	=	kPa
psi^2	×	4.753 8	E+01	=	kPa^2

*Conversion factor is exact.

Chapter 2

The Gas Research Institute's Second Staged Field Experiment: A Study of Hydraulic Fracturing

B. M. Robinson, S. A. Holditch & Associates Inc.; **S. A. Holditch,** Texas A&M University; and **R. E. Peterson,** CER Corp.

SPE 21495-MS presents a case history of Gas Research Institute's (GRI's) SFE No. 2 well, which was drilled and completed in the Travis Peak Formation in East Texas. The objective was to develop a fracture monitoring and fracture modeling system that could compute the shape and extent of a hydraulic fracture in real time. Robinson et al. perform in-depth and comprehensive analysis of geology, coring, logging, well testing, drilling and completions, and in-situ stress measurements. They then summarize prefracture, real-time fracturing, and post-fracture data collection and analysis in great details and conclude the following:

- Extensive natural fracturing is present in the lower portion of the Travis Peak in SFE No. 2 well
- Multiple hydraulic fractures are created in the Lower Travis Peak when hydraulic pressure is applied
- Integration of acoustic log with actual stress test measurement is first applied to develop a detailed vertical stress profile, which is critical for Optimizing completion and stimulation
- Fracture dimensions for both Lower and Upper Travis Peak are predicted by matching the minifracture and hydraulic fracture treatment data using a 3D fracture-design model
- The production and pressure history are also matched to predict effective fracture dimensions, which are fewer compared with the created- and propped-fracture dimensions.

2.1 Introduction

The GRI—now the Gas Technology Institute—sponsored a field-based research program to better understand the processes that control hydraulic fracture growth. To effectively conduct this research and validate the technologies under development, GRI planned a series of staged field experiment (SFE) wells. These SFE wells, which were drilled specifically for conducting research, were "staged" in the sense that a series of successively higher goals were set for each well. These goals were more completely described in previous publications (Holditch 1986). The first SFE was completed in 1987, the results of which were also published (CER Corp. and S. A. Holditch & Associates Inc. 1988; Holditch et al. 1988).

The first three SFE wells were planned to be drilled through the Travis Peak Formation, while the fourth SFE well was scheduled to be completed in another basin containing significant gas reserves in a tight sand. The purpose of moving SFE No. 4 to another basin was to apply and test any technologies developed in the GRI program in an entirely different geologic environment. Many research technologies were developed and tested in the SFE program, but the primary objective was to develop a fracture monitoring and fracture modeling system that could compute the shape and extent of a hydraulic fracture in real time. To obtain accurate and reliable input data for the fracture models, it was necessary to perform studies on geology, coring, logging, and well testing of tight gas sands. To substantiate and verify the models, additional research was sponsored in the development of instrumentation and analysis techniques for measuring the dimensions of the fracture.

The second SFE, officially known as SFE No. 2, was successfully completed in 1989. An extensive amount of data was collected on this well from two different intervals of the Travis Peak that were tested and fracture stimulated. This paper summarizes all of the data that were collected and the analyses performed on these data. As a result of the work performed on SFE No. 2, much was learned with regard to the completion and stimulation of the Travis Peak Formation. A new strategy with regard to completion and stimulation of wells was developed that used the integration of openhole log and in-situ stress data. These results are presented for the Travis Peak. The hope was that this same approach could be applied on most tight gas sands.

2.2 Objectives of SFE No. 2

As previously mentioned, each of the four SFE wells had a series of specific goals to be accomplished. For SFE No. 2, these goals were

- A 3D-fracture design and evaluation model was developed that could analyze fracture treatment data in real time.
- Fracture diagnostic experiments were evaluated to focus future research efforts on the most promising of the techniques. These fracture diagnostic methods were separate from the fracture modeling efforts.

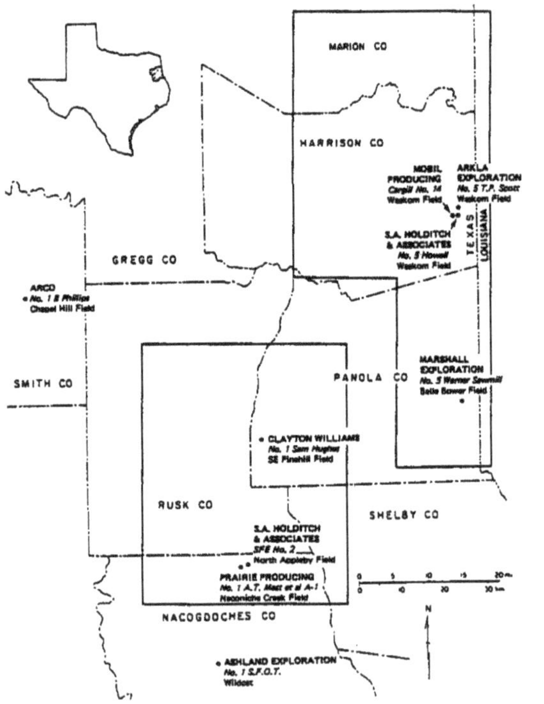

Fig. 2.1—East Texas research areas and location of SFE No. 2.

Organization	Research Area
Applied Geomechanics, Inc. (AGI)	Fracture Diagnostics (Hydraulic Impedance)
Bureau of Economic Geology – UT at Austin (BEG)	Geology
CER Corporation (CER)	Data Acquisition - Coring and Logging
Institute of Gas Technology (IGT)	Special Core Analyses
ICF-Lewin Energy	Economic Studies
Resources Engineering System, Inc. (RES)	Hydraulic Fracture Modeling
ResTech Houston, Inc.	Formation Evaluation
S. A. Holditch & Associates Inc. (SAH)	Reservoir Modeling and Hydraulic Fracturing
Stanford University	Fracture Diagnostics (Hydraulic Resonance)
Teledyne Geotech	Fracture Diagnostics (Borehole Microseismics)

Table 2.1—Participating GRI contractors.

- The GRI Mobil Testing & Control Facility was fully capable of monitoring, measuring, and recording all field data related to the testing and stimulation of tight gas sands.
- Experiments focused mainly on the determination of fracture height and fracture length measurements and/or calculation of these properties.

To achieve these goals, it was necessary to select a SFE site and drill a well in an area of the Travis Peak that would provide the most realistic conditions in which to collect the necessary data and perform the required experiments. As such, SFE No. 2 was targeted for an area that represented conditions that were typical of the Travis Peak. Other specific criteria identified for selecting a site for SFE No. 2 included:

- The location should offset existing Travis Peak production to maximize the probability of encountering a gas productive pay interval.
- Offsetting wells should have been hydraulically fractured to enhance gas production.
- Total depth to the base of the Travis Peak should be less than a threshold depth.
- Several of the producing zones should contain natural fractures so that models of fracture occurrence, as well as the influence of fractures on well completion and stimulation, could be evaluated.
- The diagenetic history of the producing intervals should be typical of the Travis Peak, i.e., extensive quartz overgrowth resulting in extremely low permeabilities.
- The field should be relatively new so that interwell depletion is minimized.

During the program, GRI had participated in several cooperative research wells in east Texas. Producing areas near these wells were evaluated and leases offered by several operators were reviewed. Based on this study and the technical criteria noted above, the North Appleby Field of Nacogdoches County, Texas, was selected as the most favorable site for SFE No. 2. This field was also considered attractive because of previous research conducted in the area. As such, most tight gas sands research contractors were familiar with the field. However, it was also recognized through this experience that the reservoir conditions that existed in the North Appleby Field were probably some of the most severe that could be expected in the Travis Peak. Extremely low formation permeabilities, water production, and difficulties in fracture stimulation were normal problems associated with this area. **Fig. 2.1** is a regional map showing the locations of GRI's cooperative research wells and the site of SFE No. 2. Through the cooperation and assistance of Amoco Production Company, a group of leases were acquired on which to drill SFE No. 2.

Various organizations were contracted by GRI to perform the research on SFE No. 2. These participating organizations are listed in **Table 2.1,** along with their primary area of research. The authors of this paper wish to gratefully acknowledge the efforts of these organizations in conducting the research and providing input into the preparation of this work.

2.3 Geologic Overview

The Bureau of Economic Geology (BEG) performed detailed stratigraphic and petrologic studies on the Travis Peak Formation (Tye 1988) primarily using core and log data from SFE No. 2 and offsetting wells. From these studies, a better understanding of the tight sandstone reservoirs within the Travis Peak Formation was obtained.

Within the paper, references are made to "Upper" and "Lower" Travis Peak intervals. With respect to geology, these intervals represent cored formations from 8,230 to 8,739 ft and 9,480 to 9,942 ft, respectively. During discussions of completion operations, these references pertained to perforated zones within these cored intervals.

2.3.1 Stratigraphy and Depositional Analysis. From regional studies conducted by BEG and other researchers, it has been determined that the Travis Peak Formation represents a 2,000-ft thick sandstone-rich sequence of fluvial, fluvial-deltaic, and associated paralic depositional environments. This Lower Cretaceous wedge of sediment prograded from the north to northwest into the East Texas Basin. The Travis Peak produces primarily from structural traps associated with the Sabine Arch; however, stratigraphic traps do exist. In the North Appleby Field, a combination of stratigraphic and diagenetic traps has formed in multilateral sandstones as a result of mudstone interbeds and extensive quartz cementing.

Much of the Travis Peak section represents an aggradation of alluvial valley deposits. During deposition, fluvial channels evolved from predominantly braided streams near the base to a more mud-rich meandering stream system at the top. Multiple channel belts form a network of broad, stacked sandstones. Six to eight channel belts, each containing 80 to 90% medium- to fine-grained sandstone,

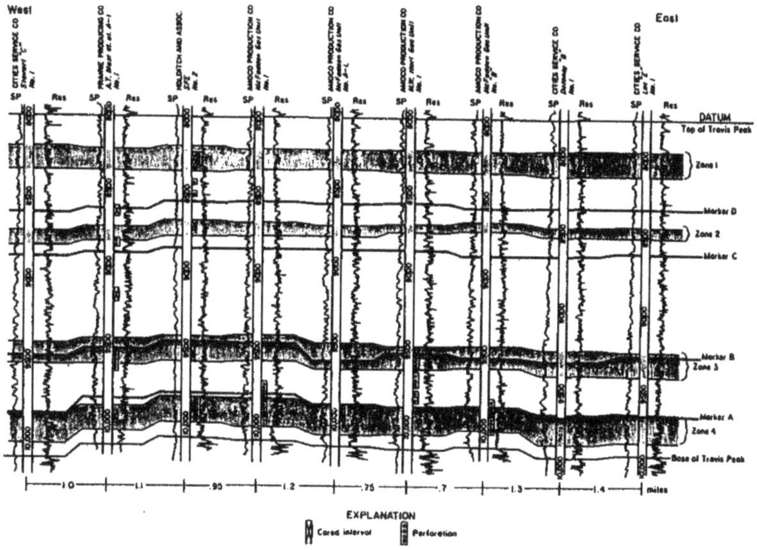

Fig. 2.2—Stratigraphic cross section across the North Appleby Field.

can occupy a 200-ft thick interval. Individual sandstones can cover an area greater than 40 square miles. In a vertical sequence through one channel-belt sandstone, crossbedding grades upward into thinly interbedded sets of planar crossbeds and ripples. Clay clasts conglomerates line the bottom of scoured channels. Interbedded mudstones that are burrowed, rooted, and diagenetically altered accumulated in well drained swamps and lakes adjacent to these channels. Poorly sorted sandstones represent overbank deposition (crevasse, splays, and lacustrine deltas).

Correlation of channel-belt sandstones and interbedded mudstones provides a basis for identifying reservoirs and estimating their extent and geometry. **Fig. 2.2** illustrates a cross section through the North Appleby Field. Even though one productive channel-belt sandstone may cover more than 25,000 acres, maximum drainage areas in the North Appleby Field are generally less than 80 acres. This is primarily because of the extreme heterogeneous nature of the sandstone geometry. Small-scale bed forms, biogenic structures, clay drapes, and scour surfaces create flow barriers and tend to segment the reservoirs. The best quality reservoir sands occur in wide bands or in parallel to depositional dip. Reservoir quality decreases at channel margins and in interchannel areas where siltstones and mudstones accumulated.

The extremely thick sandstone sequences and presence of thin quartz-rich mudstones generally do not provide the lithologic qualities needed for barriers to vertical fracture growth. Created fracture height can easily exceed several hundred feet. The vertical and lateral growth of hydraulic fractures will result in the fracture intersecting many mudstone or channel-belt systems adjacent to the perforated interval. This communication could enhance or detract from production, depending upon the amount of water saturation in the adjacent sandstones. Our research has indicated that many of these channel sandstones exhibit extremely high water saturation.

Analysis of SFE No. 2 core provided detailed petrographic descriptions of the Travis Peak Formation. Thin-section point counts supplied data on the mineral composition for use in calibration of openhole log data. The mineralogic descriptions also indicated zones of potential fluid sensitivity. For example, abundant ankerite cement found locally in some areas of the Travis Peak could form a damaging iron hydroxide gel if it were treated with acid that did not contain an iron sequestering agent. But the most important information developed from these petrographic studies is the extent of cementation by quartz overgrowth. This work by BEG has shown that the deeper part of the Travis Peak is more intensely cemented than the shallower intervals. This greater degree of cementation results in substantial differences in mechanical properties and porosity, but is the primary reason for the much lower permeability observed in the thick, fluvial-deltaic portion of the Travis Peak. Unfortunately, most of the gas is also present in this interval.

2.3.2 Analysis of Natural and Induced Fractures. Extensive natural fractures observed in the core from the Lower Travis Peak of SFE No. 2 provide an opportunity to study fractures in a low-permeability gas reservoir. A total of 169 fractures and fractured zones were observed in SFE No. 2. Of these fractures, 48 were natural and 121 were the result of coring or stress tests. These fractures were studied to determine their influence on reservoir quality and for fracture treatment design and analysis. Induced fractures (from coring or stress tests) were also evaluated as potential indicators of in-situ stress direction.

The presence of natural fractures in any tight sand could influence the design and analysis of the fracture treatment. The potential effects include increased fracture-fluid leakoff or the creation of multiple sets of fractures away from the wellbore that could result in unusual or high treating pressures.

Natural fractures present in the lower completion zone of SFE No. 2 might have affected the hydraulic fracture treatment in this interval—extremely high treating pressures were observed (see the discussion of fracture treatment on Lower Travis Peak completion). These high pressures were probably the result of multiple, poorly interconnected fracture strands that emanated from the wellbore before eventually connecting to a single hydraulic fracture. These types of multiple fracture strands are illustrated in **Fig. 2.3.** This figure represents a sketch of a cored interval that shows hydraulic fractures induced during an openhole stress test in the Lower Travis Peak. Each fracture represented a fresh breakage of the rock. The fractures were created by running an openhole, drillstem-test-type packer and then pressurizing the formation at the bottom of the hole until breakdown occurred. A core barrel was then run in an effort to overcore the fractured interval below the bottom of the hole.

As can be seen, a single vertical fracture was not created. Instead, the rock appears to have "shattered" in various places. It is not clear whether these multiple fractures are the result of mechanical properties of the rock, internal structure, or influenced by the regional stress field. It is clear, however, that if these types of fractures were created during the hydraulic fracture treatment, narrow fluid flow paths would be created near the wellbore potentially resulting in high injection pressures and problems while pumping proppant into the formation.

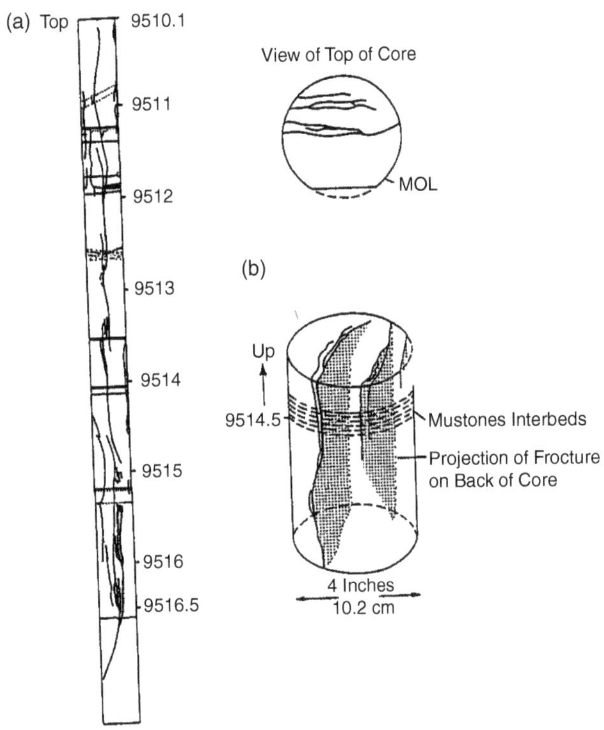

Fig. 2.3—Multiple hydraulic fracture(s) induced by openhole stress test in Lower Travis Peak.

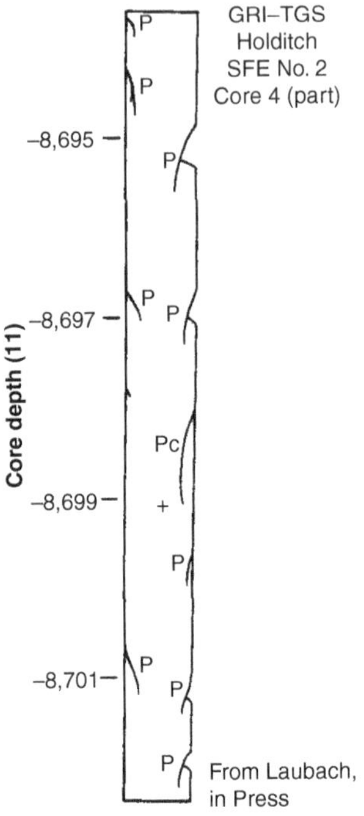

Fig. 2.4—Aligned petal (*p*) and petal-centerline (p_c) fractures from SFE No. 2 core.

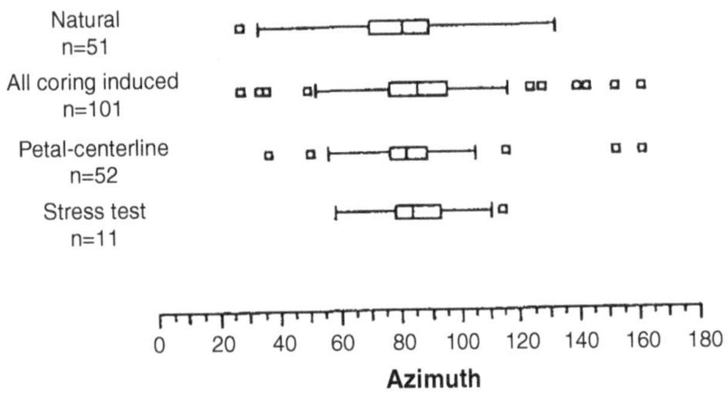

Fig. 2.5—Orientation of fractures observed in SFE No. 2.

Predicting the direction of the hydraulic fracture is a key part of completion strategy. Geologic studies of SFE No. 2 core showed that both natural and coring-induced fractures can be used as an inexpensive and reliable method to predict in-situ stress directions and thus, fracture azimuth. **Fig. 2.4** is a sketch showing aligned petal and petal centerline fractures that were induced as a result of coring operations on SFE No. 2. In general, these coring-induced fractures are recognized by their distinctive curvature into the core in a downward direction. Because much of the Travis Peak core was oriented, the strike of these fractures could be determined. The same is true for the direction of natural extension fractures. **Fig. 2.5** shows the orientation of both the natural and coring-induced fractures. Also indicated on this figure is the orientation of stress-test induced fractures that were measured from openhole logs. As can be seen, excellent agreement is obtained by all of these methods, the results of which indicate the hydraulic fracture direction should be approximately north 80° east.

In general, few natural fractures existed in the upper completion zone of SFE No. 2. Therefore, similar geologic interpretations were not possible. Likewise, no problems were expected during the hydraulic fracture treatment as a result of natural fractures. However, it is obvious that the presence of natural fractures in the lower portion of the Travis Peak could have had a significant effect on the completion and stimulation of this interval. In addition, the natural fractures provided valuable information with regard to hydraulic fracture azimuth.

2.4 Drilling Summary

SFE No. 2 was spudded on 7 July 1987. A 12 1/4-in. hole was drilled from surface to 3,945 ft and 9 5/8-in. casing was set and cemented in place. An 8 3/4-in. hole was then drilled from the 9 5/8-in. casing shoe to a total depth (TD) of 10,169 ft. Drilling operations were completed on 26 August 1987. After an extensive openhole log program (**Table 2.2**), 7-in. casing was then run from surface to TD and cemented in place.

The top of the Travis Peak was encountered at approximately 8,000 ft. While drilling the Travis Peak, a substantial amount of openhole data was collected. A total of 359 ft of 4-in. diameter core was obtained in four main intervals within the Travis Peak: 8,230 to 8,320 ft, 8,678 to 8,739 ft, 9,480 to 9,572 ft, and 9,807 to 9,942 ft. A total of 104 ft of this core was oriented using a paleomagnetic technique. Orientation of core was also verified using a borehole televiewer.

Three openhole stress tests were conducted during drilling operations in SFE No. 2. The objective of these tests was to determine in the openhole environment the average in-situ stress in representative sandstone formations of the Travis Peak.

Phasor Dual Induction (20 kHz)/GR/SP	Aluminum Activation Clay
Digital Sonic/GR/Caliper	Stratigraphical High Resolution Dipmeter
LithoDensity/Compensated Neutron/Electromagnetic Propagation/ MicroSpherically Focused Resistivity/Caliper/GR	Borehole Geometry Tool
Borehole Televiewer (SWS)	Borehole Televiewer (APR)
Digital Waveform Sonic	Long Spaced Acoustical (APR)
Repeat Formation Tester	Full Waveform Sonic/GR (Welex)
Long Spaced Sonic/Stonely Wave Sonic/Q-Sonic (DRL)	Long Spaced Sonic (Welex)

Logging Service Companies: Schlumberger Well Services (SWS) Welex
Research Logging Organizations: Schlumberger Doll Research (SDR), Mobil Dallas Research Lab (DRL), Amoco Production Research (APR)

Table 2.2—Openhole logs run in SFE No. 2.

Test depths were 8,259 to 8,266 ft, 9,505 to 9,510 ft, and 9,814 to 9,822 ft. The core barrel was run immediately after each test in an attempt to overcore the fractured portion of the rock below the bottom of the hole. These cores were oriented such that fracture azimuth could be determined. In addition, borehole televiewer logs also observed these hydraulically induced fractures in the borehole wall. As a result of this work, actual hydraulic fractures were recovered in the core (Fig. 2.3) and the average fracture azimuth was determined to be north 80° east.

While drilling through the Travis Peak Formation, four different logging operations were conducted at various depths. In addition to obtaining the base data needed to describe reservoir properties, the additional logs were run to evaluate any change in formation properties and saturations with time. In addition to the normal services that are commercially available, special experimental tools were also run by research departments of Schlumberger, Mobil, and Amoco Production Company. Standard and special mud-logging services were also performed starting at a depth of 7,200 ft and continuing to TD. Remote sensors and a digital recording unit recorded all pertinent drilling parameters.

2.5 Determining the Vertical Stress Profile

One of the most important factors that controls the vertical growth of a hydraulic fracture is the in-situ stress contrast between the layers of rock above and below the treated zone. Establishing the stress profile in the vertical direction is necessary if one expects to accurately predict and calculate the shape and extent of the hydraulic fracture. If this can be accomplished, then the true optimal fracture length can be designed.

Three openhole and 26 casedhole stress tests were performed to define the in-situ stress in various layers of rock within the Travis Peak interval. Sandstones, siltstones, shale, and limestones were all tested. In general, the casedhole tests were easier to perform and less complicated to analyze than the openhole tests. Therefore, these results were used primarily for determining the in-situ stress values in each of the different layers. The methodology involved in conducting and analyzing these tests has been previously described (Whitehead et al. 1986). The results from the analysis of these data are presented in **Table 2.3.**

Even though 26 different stress tests were conducted, not all of the rock layers could be tested because the Travis Peak is over 2,000 ft thick and is made up of stacked sandstone and mudstone sequences. However, another method for estimating in-situ stress on smaller intervals is to use values of Poisson's ratio calculated from acoustical log data. Using these values of Poisson's ratio, an estimate of the overburden stress, and the reservoir pressure, the in-situ stress for each layer can be calculated on the basis of the properties of an elastic material. The following equation is normally used to compute the values for stress.

$$S_h = \frac{\nu}{1-\nu}\left(S_v - S_p\right) + S_p, \quad \text{.......................(2.1)}$$

where S_h = minimum horizontal stress gradient, S_v = vertical (overburden) stress gradient, S_p = pore pressure gradient, and ν = Poisson's ratio.

There is a potential error associated with calculating in-situ stresses from log data in that only the elastic component can be estimated. Other factors, such as tectonic stresses, thermal history of a formation, and subsidence history of the basin, cannot be included in the computations. Therefore, the log-computed values must be calibrated using actual stress test data. The resulting stress profile would then be more representative of the actual vertical stress distribution. To calibrate the log-computed values, the actual measured stresses are plotted versus $\nu/1-\nu$. A straight line representing the average can then be drawn through the data points. **Fig. 2.6** represents this calibration technique on actual data from SFE No. 2. Note that if the straight line is extrapolated to $\nu/1-\nu = 1$, then S_h reduces to S_v.

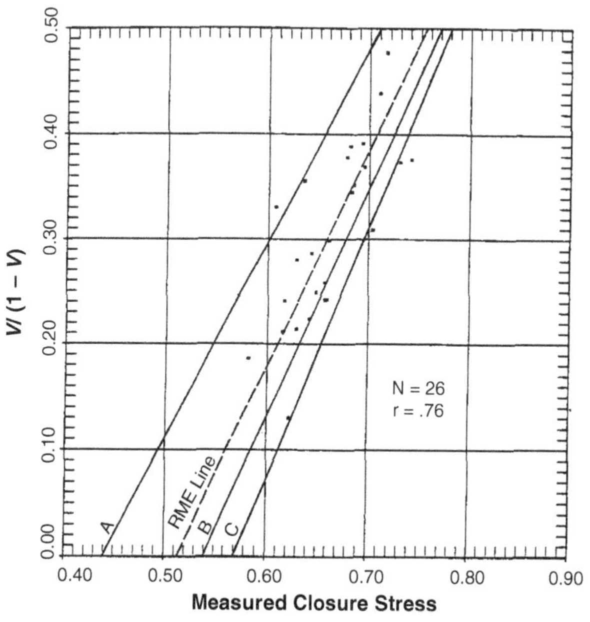

Fig. 2.6—Calibration technique for acoustical-log data using actual measured stresses.

Test	Depth(ft)	Lithology	Fracture Closure Pressure (psi)	In-Situ[1] Stress Gradient (Psi/ft)	Possible Dual Fracture Closure (psi)
CHST 1	10,102-10,104	Limestone	7,160±150	0.709	
CHST 2	10,025-10,027	Shaley Sand	6,580±150	0.656	
CHST 3	9,985-9,987	Shale	6,935+50, −300	0.694	
CHST 4	9,944-9,946	Sandstone	6,100±50	0.613	
CHST 5	9,887-9,889	Sandstone	6,080±100	0.615	
CHST 6	9,823-9,825	Sandstone	6,160±150	0.627	
CHST 7	9,754-9,756	Siltstone	6,400±200	0.656	5,750±50
CHST 8	9,676-9,678	Shale	6,600±500	0.682	
CHST 9	9,590-9,592	Shale	7,100±500	0.740	
CHST 10	9,508-9,510	Sandstone	5,900±100	0.620	
CHST 11	9,425-9,427	Shaley Sand	6,200+50, −200	0.658	
CHST 12	9,065-9,067	Shale	6,200±100	0.684	5,600±100
CHST 13	8,962-8,964	Shaley Sand	5,750±25	0.642	5,600±50
CHST 14[2]	8,844-8,846	Shaley Sand	6,050±150	0.684	5,600±100
CHST 15	8,702-8,704	Sandstone	5,270±50	0.606	
CHST 16	8,516-8,518	Shale	5,400±100	0.634	5,200±100
CHST 17	8,416-8,418	Sandstone	5,390±50, −100	0.640	
CHST 183[3]	8,329-8,331	Sandstone	4,840±50	0.581	
CHST 19[4]	8,304-8,306	Shale	5,750±50	0.692	5,600±50
CHST 20[5]	8,265-8,267	Sandstone	5350±50	0.647	
CHST 21	8,233-8,335	Shale	5,600+50, −100	0.680	5,400±50
CHST 22	8,161-8,163	Shale	5,950±100	0.729	
CHST 23[5]	8,111-8,113	Sandstone	5,350±100	0.660	
CHST 24	8,054-8,056	Shaley Sand	5,450±100	0.677	5,300±50
CHST 25	7,997-7,999	Sand/Lime	5,620±50	0.703	5,500±50
CHST 26	7,973-7,975	Limestone	5,700±50	0.715	5,500±50

1 Gradient calculated on the basis of best estimate.

2 σ_c = 6,050 psi seen only in first 2 stages; best estimate for fracture design $-\sigma_c$ = 5,600±50 psi.

3 Early stage behaved differently than later stages; in later stages, σ_c = 5,250±100 psi; the fracture in later stages may have grown out of zone and tested other intervals.

4 Possible communication between CHST 18, CHST 19, and CHST 20; pressure anomaly seen in all three tests at 4,800±50.

5 σ seems to increase in later stages; could be caused by increase of pore pressure due to leakoff or fracture out of zone into higher stress intervals.

Table 2.3—Results of casedhole stress tests, SFE No. 2.

For this plot, S_v = 0.994 psi/ft, which is a reasonable value for overburden stress. Similarly, if the straight line is extrapolated to a $v/1-v$ value of 0, then S_h is equal to S_p. For these data, S_p = 0.514 psi/ft. By performing these extrapolations, one can achieve empirically derived averages for S_v and S_p. These values should also be calibrated for any external forces present. Assuming these gradients remain fairly constant within the vertical sequence, they can be used in Eq. 2.1 to calculate the horizontal stress for zones that were not stress tested. Because overburden stress is not likely to vary significantly within a well, it is reasonable to use a single value for S_v. However, because of the geologic and production history of an area, the value for S_p might vary considerably from zone to zone. This is known to be the case for several intervals stress tested in SFE No. 2 and is believed to be at least one reason for some of the data scatter in Fig. 2.6.

Different lines can be drawn on such a figure that represent different pore pressure gradients. Three such lines have been drawn for a gradient of 0.44 psi/ft, 0.54 psi/ft, and 0.57 psi/ft. It is known that the data points that fall close to the 0.44 psi/ft line are from an interval of the Travis Peak that was produced in offsetting wells. Therefore, the interval might be partially depleted. (The normal pressure gradient for the Travis Peak is approximately 0.5 psi/ft.) Similarly, several points on the 0.57 psi/ft line are from thick shales that could be slightly overpressured. In this case, the graph was very useful for making corrections to the calculated stress values when the pore pressure is different within various layers.

The SFE No. 2 well was the first well in the GRI Tight Gas Sand Program that used this methodology to develop a vertical stress profile. Profiles were determined for both the Upper and Lower Travis Peak completion intervals and these are illustrated in **Figs. 2.7 and 2.8.** On these figures, the calibrated, log-computed values for stress are shown as the dashed curve. The actual measured values for stress are shown as the black squares. As can be seen, the measured values of stress are in good agreement with the log-computed values using the calibration technique previously described. Once this calibrated curve has been established, then a zone average can

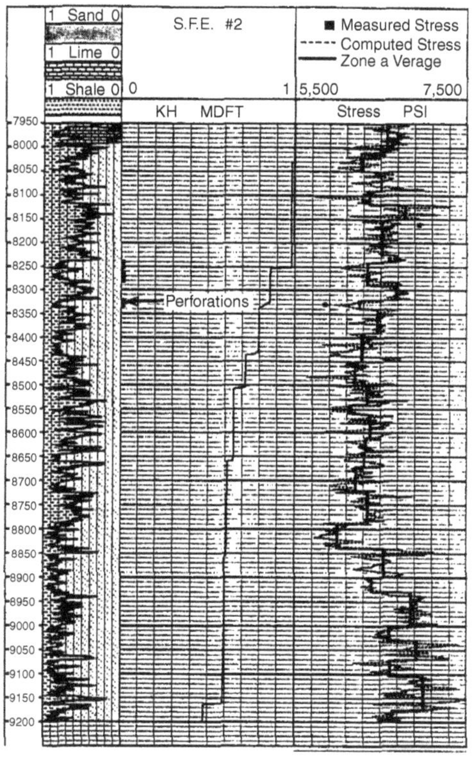

Fig. 2.7—In-situ stress and permeability profile for the Upper Travis Peak.

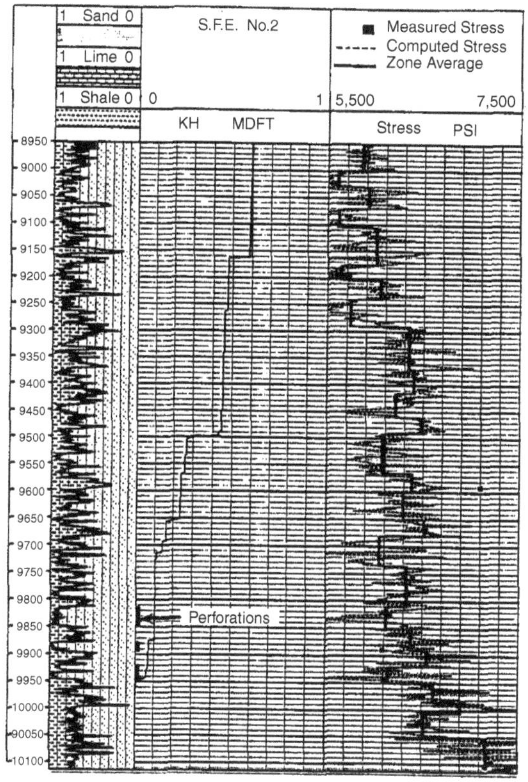

Fig. 2.8—In-situ stress and permeability profile for the Lower Travis Peak.

be established for each interval, as illustrated by the solid black line. This stress profile can now be used in a fully 3D simulator to estimate the shape of a hydraulic fracture created in any interval or intervals.

Through the GRI Tight Gas Sands Program, and especially as a result of work performed on SFE No. 2, the importance of using stress profiles was extended well beyond merely designing a hydraulic fracture treatment. The stress data have been used to analyze minifracture data, as well as data collected during the main hydraulic fracture treatment. The in-situ stress profile was also used to develop a completion strategy for the upper zone in SFE No. 2. As will be discussed later in the paper, we have learned that simply perforating the gas-productive intervals based on conventional log analyses may not allow one to successfully stimulate a formation.

Because fracture growth is controlled principally by the stress field, the perforations should be placed in strategic locations that will maximize the chance of the hydraulic fracture encountering most or all of the net-pay intervals. This is exactly the reason why the permeability distribution and lithologic description are displayed next to the stress profile. In this way, the engineer can see exactly what intervals are likely to be intersected by the hydraulic fracture. As will also be discussed later in this paper, knowledge of the stress profile in this much detail also helped to evaluate the actual hydraulic fracture treatments.

2.6 Completion Summary

Before drilling SFE No. 2, a study had been performed on the completion, stimulation, and production history of wells in the North Appleby Field. Two of these wells were GRI-funded cooperative research wells. On the basis of this study, the following characteristics/problems were identified.

- The average permeability is 0.1 md or less
- The average value for Young's modulus is 7.0 million psi or higher
- No evidence of fracture height containment could be found on any wells that were fracture stimulated
- Screenouts of hydraulic fracture treatments were not uncommon
- Excess water production usually occurs after a stimulation treatment

The best Travis Peak wells in Nacogdoches County would produce at initial rates of approximately 1,000 Mcf/D after a successful stimulation treatment. However, it was not uncommon for these same wells to be producing only 250 to 500 Mcf/D after a few months. In addition, many of these wells would produce 100 to 300 barrels of formation water per day. A large number of wells would not produce at gas rates sufficient to continuously lift the water. As a result, many of these wells would load up with water and the zones were eventually abandoned. The results of this study indicated that many difficult problems existed in the Travis Peak. Reservoir quality was generally quite poor and problems were experienced in fracture stimulation. These problems, however, added to the overall interest in conducting research in this area.

From core data, openhole log analyses, and a general knowledge of the area, it was decided that the first zone to be completed should be in the lower portion of the Travis Peak. As illustrated in Fig. 2.8, the interval from 9,814 to 9,947 was perforated. Openhole data indicated that this Lower zone was, at best, marginally productive. However, core that had been acquired in the interval exhibited a fairly large number of natural fractures. Because one of the primary goals of the project was to study the effect of natural fractures on production, it was decided to test the Lower Travis Peak. Openhole data also indicated another potentially productive interval in the upper part of the Travis Peak. Therefore, it was decided that if a second zone was to be tested, it would include the upper interval from 8,238 to 3,338 ft (Fig. 2.7).

These two zones were chosen strictly upon the basis of openhole data (cores and logs) from which the permeability-thickness (*kh*) and porosity-thickness (ϕh) values were calculated. When selecting the completion intervals and the location of the perforations, the in-situ stress profiles were not considered. As a result of SFE No. 2, however, the importance of using the stress profile to help plan the completion is now recognized. As will be discussed in the next sections of this paper, placing the perforations in the "productive interval" was potentially detrimental to successfully pumping the hydraulic fracture treatments.

2.7 Lower Travis Peak

2.7.1 Well Tests. The Lower Travis Peak interval was perforated from 9,814 to 9,848 ft, 9,878 to 9,896 ft, 9,920 to 9,930 ft, and 9,933 to 9,947 ft. A prefracture production test was conducted during which the zones flowed 66 Mcf/D and 24 BWPD with a flowing tubing pressure of 70 psi. A bottomhole pressure buildup survey immediately followed the production test. An analysis of these was performed; however, the reader is referred to CER Corp. and S. A. Holditch & Associates Inc. (1990) for details of this work. In summary, the in-situ permeability to gas was calculated to be 0.007 md assuming a net gas pay of 42 ft. The skin factor was calculated to be −1.0 and P^* was estimated to be 4,550 psia.

2.7.2 Minifracture Treatments. Two separate minfractures were performed on the lower interval of SFE No. 2. The first minifracture was conducted to provide an opportunity for collecting microseismic data in an effort to estimate fracture height and azimuth. A three-component geophone and gyroscope were run in the well to monitor the seismic activity induced by the fluid injection. The seismic energy was logged at several stations within the wellbore. Teledyne Geotech is the primary contractor who developed this technique and performed the analyses.

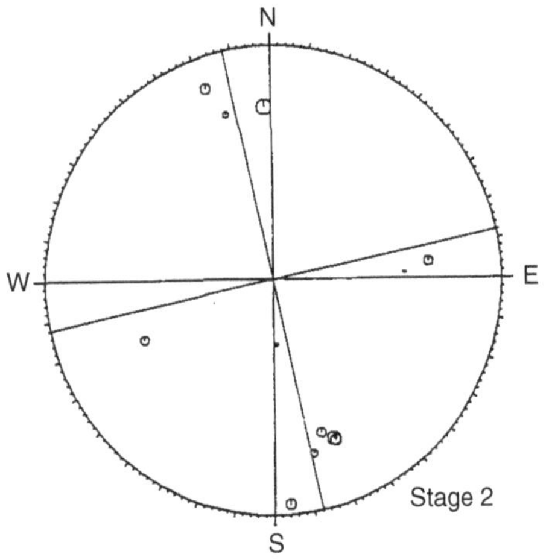

Fig. 2.9—Stereographic (polar) plot of processed seismic data at 10,038 ft.

The orientation of fractures can be determined from both discrete signals and the continuous microseismic radiation (CMR). In SFE No. 2, the CMR data were dominated by tube waves and/or tool-borehole resonances. Therefore, analysis of fracture azimuth used only discrete data signals. A stereographic (polar) plot of processed borehole seismic data is shown in **Fig. 2.9.** As can be seen, there are signals indicated in two principal directions—north 12° west and north 78° east. Although not clearly understood, it is not unusual to detect discrete signals normal to the plane of the hydraulic fracture. However, in these cases, the dominant signals are usually in the direction of the fracture. This was not the case in SFE No. 2, however, because the data showed a significant amount of scatter. It is possible that multiple fractures extending away from the wellbore might have had some effect on the microseismic data.

A new method for locating the top and bottom of a hydraulic fracture was, in part, developed at the SFE No. 2 well. By evaluating the horizontal and vertical components of background motion, Teledyne found that an inversion or discontinuity in the dominant direction correlates with the location of the fracture. Numerical modeling studies performed by Teledyne substantiate this behavior. A simple plot of the horizontal to vertical ratio, as illustrated in **Fig. 2.10,** will then indicate the top and bottom of the fracture.

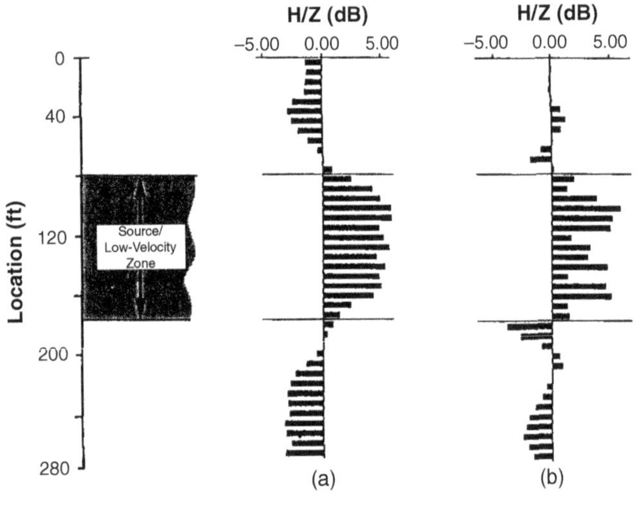

Fig. 2.10—H/Z ratios calculated from simulation of background motion data.

Fig. 2.11 shows the actual data recorded during Minifracture No. 1 on SFE No. 2. The figure shows three different plots representing ratios calculated at different frequencies. A distortion, or reversal, of the H/Z ratio is definitely indicated. The top of the fracture appears to be at a depth of 2,936 meters or 9,635 ft. The bottom of the fracture is not clearly defined, indicating that it might extend below the deepest monitoring depth of 3,060 meters, or 10,040 ft. From this work, the total created fracture height appears to be in excess of 400 ft.

The second minifracture test was conducted primarily to obtain data for calibration of a 3D fracture model, FRACPRO. This model was developed for the GRI Tight Gas Sands program by Resources Engineering Systems (RES). The minifracture was designed to pump various types of fluids at different injection rates. Treated water, linear gel, and crosslinked gel were all used. A total of almost 90,000 gal of fluid was pumped.

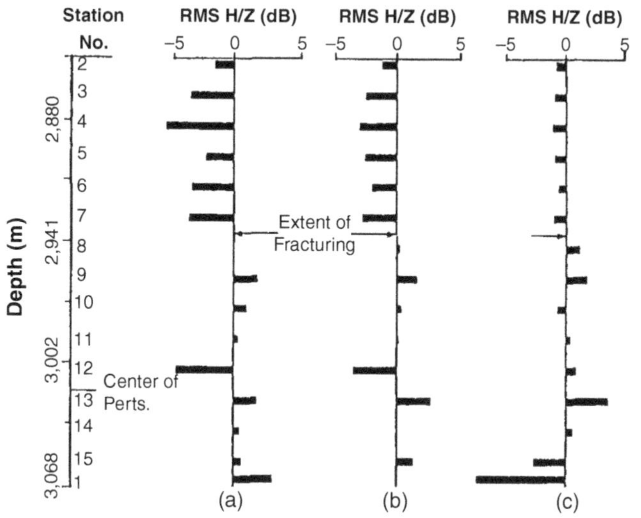

Fig. 2.11—H/Z ratios at different depths for Lower Travis Peak.

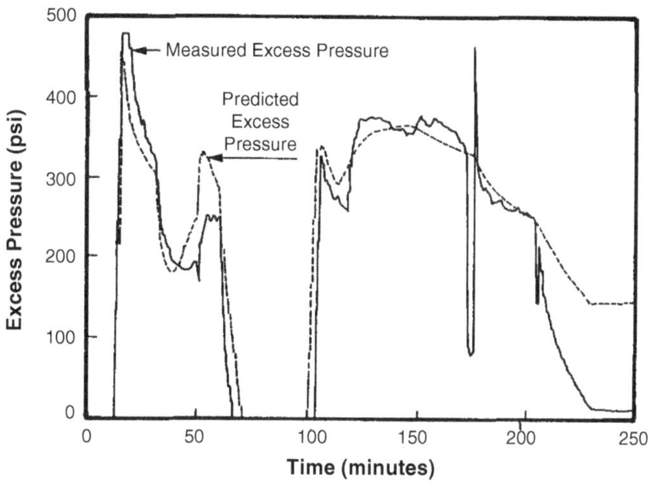

Fig. 2.12—History match of net fracturing pressures during Minifracture No. 2.

An analysis of these minifracture data was performed by history matching the net fracturing pressure measured during the treatment with the 3D fracture model. The final match of the actual net pressures is shown in **Fig. 2.12.** As can be seen, the pressures recorded during the longest injection stage decreased throughout most of the treatment. Uncontained vertical fracture height growth represents this type of pressure behavior. In addition, the bottomhole pressures decline rapidly when injection is shut down. To simulate this pressure decline, it was necessary to add a "thief" zone to the model that had either higher permeability than had been previously indicated by well tests and core analyses or much lower stress than had been measured in the fractured interval.

On the basis of this analysis, the model predicted the fracture half-length to be approximately 300 ft, while the total fracture height was in excess of 500 ft. This estimated fracture height was in excellent agreement with Teledyne Geotech's analysis during the first minifracture. The model results also indicated that additional pressure drop was also present near the wellbore. This pressure drop was not felt to be perforation friction because over 400 holes had been shot and a bailout treatment was successfully performed. As a result, we recognized the potential for multiple fracture strands propagating from the wellbore. This problem could have actually been worsened by the large number of holes in the pipe and the bailout treatment could have caused the formation to break down in multiple directions. This type of completion was common in the area and might offer an explanation for some of the difficulties experienced during fracture stimulation treatments in the North Appleby Field.

2.7.3 Fracture Stimulation Treatment. A fracture treatment of the Lower Travis Peak interval was performed on 2 December 1987. The treatment was pumped down the casing-tubing annulus so that bottomhole treating pressures could be measured. As shown in **Table 2.4,** the treatment was designed to pump 140,000 gal of crosslinked and linear gel with 620,000 lbm of 18/20-mesh intermediate-strength proppant. The gel concentrations were tapered to minimize fluid viscosity in an effort to control fracture height. During the actual treatment, 164,000 gal of gel and only 74,000 lbm of proppant were successfully pumped. The average treating pressure was 4,100 psi at an injection rate of 30 bbl/min.

Stage No.	Fluid Volume (gals)	Fluid Type	Proppant Concentration (lbm/gal)	Proppant Type	Proppant Amount (lbm)
1	40,000	WF-10	0.0	Pre-Pad	0
2	40,000	37.5# XL	0.0	100 Mesh	0
3	40,000	35.0# XL	1.0	Pad	40,000
4	20,000	35.0# XL	0.0	18/20 ISP	0
5	20,000	35.0# XL	Z0	18/20 ISP	40,000
6	20,000	32.5# XL	3.0	18/20 ISP	60,000
7	20,000	30.0# XL	4.0	18/20 ISP	80,000
8	40,000	WF60	5.0	18/20 ISP	200,000
9	40,000	WF60	6.0	18/20 ISP	240,000
10	11,560	WF10	0.0	Flush	0

Table 2.4—Fracture treatment design on Lower Travis Peak first treatment.

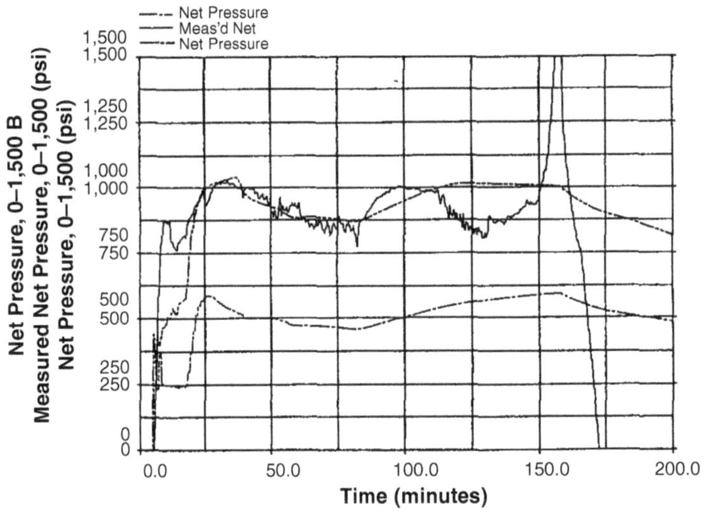

Fig. 2.13—History match of main hydraulic fracture treatment on Lower Travis Peak.

As shown in **Fig. 2.13,** the actual measured net treating pressures were successfully history matched with the 3D model. RES performed this analysis, which resulted in several interesting observations regarding the treatment and the reason for the screenout. Excessive near wellbore pressure drop was again observed during the treatment. This was substantiated by a large increase in bottomhole treating pressure when 100-mesh sand started into the formation. As shown on Fig. 2.13, this occurred at approximately 80 minutes. Pressure decreased when the 100-mesh sand was flushed, but again increased when the intermediate strength proppant was started. This behavior indicates that a near wellbore restriction was present. It is believed that this restriction was a result of multiple narrow fractures extending from the wellbore.

Another interesting observation is that the pressure continuously decreased throughout the treatment except during times when proppant was pumped. This behavior indicates uncontrolled vertical height growth; however, it was also necessary to add a thief zone to simulate the rapid decrease in pressure. The combination of these conditions is believed to be the main reason for the screenout that occurred. The 3D fracture model predicted the created fracture height to be 425 ft. The propped fracture half-length was estimated to be approximately 450 ft. It should be noted, however, that these estimates are probably optimistic because the model included the 100-mesh sand as proppant and does not take into account any severe restrictions in width.

After the first fracture stimulation treatment, a decision was made to attempt to re-establish an injection rate into the Lower Travis Peak. This was successfully accomplished on the following day, indicating that the initial fracture could still be propagating or that a secondary fracture was growing. On the basis of the injection test, the lower zone was refractured on 8 December 1987. This treatment consisted of 171,000 gal of crosslinked gel with 114,000 lbm of 18/20-mesh intermediate-strength proppant. The second treatment used much thicker gel at slightly higher injection rates. The main objective of the second fracture treatment was to determine if a wider fracture could be established with a much thicker fluid, therefore overcoming near wellbore restrictions. Unfortunately, the second treatment also screened out, indicating that even a more viscous fluid was not sufficient to overcome the near-wellbore restrictions.

The final analysis of the second fracture treatment by RES indicated the fracture shape was roughly circular. The created fracture height was in excess of 600 ft and the average propped fracture length was approximately 250 ft. Again, a high-permeability or low-stress thief zone was modeled during the analysis. The combination of high leakoff into this thief zone and the restrictive flow paths

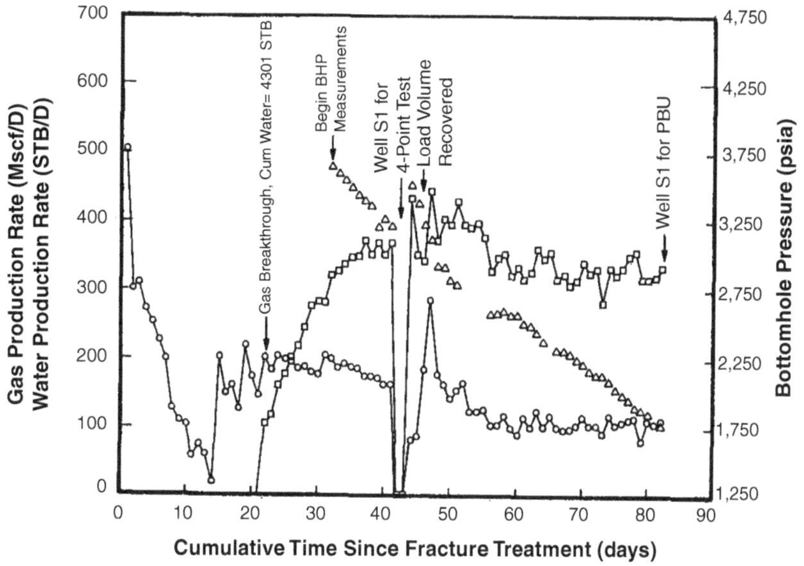

Fig. 2.14—Post-fracture production performance of Lower Travis Peak.

near the wellbore is believed to be the primary cause of the screenout. If any one of these conditions had been the only problem encountered, then the hydraulic fracture treatment may have been successfully pumped. However, the combination of all these factors resulted in reservoir conditions that virtually could not be overcome.

Approximately four months after the two hydraulic fracture treatments, another microseismic survey was conducted by Teledyne Geotech. An analysis of these data using the H/Z method indicated the top of the fracture to be approximately 9,585 ft. This is approximately 50 ft taller than the height predicted during Minifracture No. 1. These results are very encouraging in that it does appear possible to determine fracture height long after a fracture treatment has been pumped.

2.7.4 Post-Fracture Production and Pressure Analysis. Production testing of the Lower Travis Peak began on 8 December 1987 and continued until 1 March 1988. As shown on **Fig. 2.14,** water production was measured from the day the well was open until it was shut in for a pressure buildup test. Even though a test separator was on location during this time, no measurable gas was observed until after 20 days of water production. After gas breakthrough finally occurred, gas production increased over the next 20 days, then gradually declined to a final rate of 330 Mcf/D. Water production stabilized at 110 BWPD. Flowing bottomhole pressures were recorded daily during the production period. These values are also shown on Fig. 2.14 and indicate a continuously declining bottomhole pressure until the well was shut-in for an extended bottomhole pressure buildup survey.

The extended delay in gas breakthrough has been observed on other cooperative research wells in East Texas. This unusual production behavior, coupled with the production of a significant amount of water, indicated that the most reliable method to evaluate these data was to use a 3D, two-phase reservoir simulator. Conventional analysis techniques were performed on these data, and these results are discussed in great detail in CER Corp. and S. A. Holditch & Associates Inc. (1990). Only the final modeling results are summarized here because of space limitations. However, a detailed discussion of the reservoir simulation analysis is presented by Bastian and Sherman (1993).

The most reliable method by which to evaluate reservoir and fracture properties was to use a 3D, two-phase reservoir simulator. The actual production and pressure performance of the well was history matched using reservoir and fracture properties in the model as history matching parameters. Other reservoir input data were supplied by core analyses, openhole log analyses, and input from geological studies. The actual well performance was found to be best predicted using a two-layer model; one layer that is primarily gas saturated, a second that is primarily water saturated. The final estimate of reservoir and fracture properties used in the two-phase, multilayer areal model is shown in **Table 2.5. Figs. 2.15 and 2.16** show the final history match of the actual production data and pressure buildup test on the Lower Travis Peak.

In general, the overall match of all of these data is very good. An exact duplication of the actual production from the well was not obtained with the model, but was not expected. This is because of the complex nature of the Travis Peak. However, a general model using a two-layer system is believed to be a good representation of the formations that produce within the Travis Peak. Using this model, reasonable estimates for propped fracture length were obtained that agree very well with the calculated values from the 3D modeling work performed by RES.

	Layer 1	Layer 2
Absolute Permeability, md	0.008	0.08
Net-Pay, ft	55	41
Total Porosity, %	4.5	4.9
Water Saturation, %	35	60
Gas Saturation, %	65	40
Relative Permeability, fracture	0.1375	0.0238
Gas Porosity, %	2.9	2.0
Gas Permeability, md	0.0014	0.0019
Fracture Length, ft	300	300
Fracture Conductivity, md-ft	6,000	6,000
Area, acres	2.57	160

Table 2.5—Reservoir and fracture properties estimated for SFE No. 2–Lower Zone.

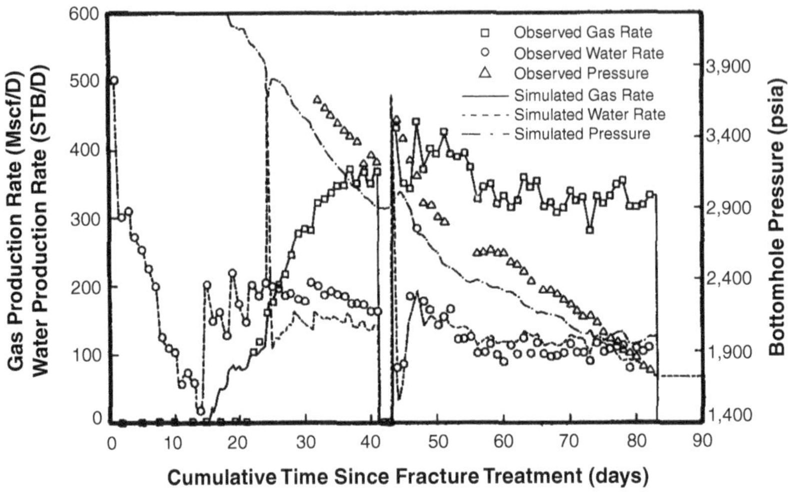

Fig. 2.15—History match of Lower Travis Peak production performance.

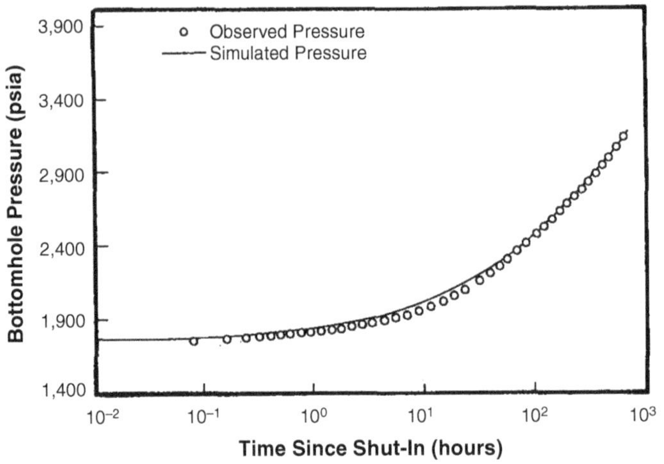

Fig. 2.16—History match of Lower Travis Peak pressure buildup test.

2.8 Upper Travis Peak

Before completing SFE No. 2 in the Upper Travis Peak, a complete analysis of the stimulation treatment on the Lower Travis Peak was performed. On the basis of this work, it was obvious that the fracture height grew vertically upward from the lower perforated interval to the low-stress region at about 9,300 ft (Fig. 2.8). In reviewing the stress and permeability profile of the Upper Travis Peak, it appeared that a similar problem would exist for this interval except that the fracture would grow downward to an area of lower stress. The stress data and minifracture data were supplied to various GRI contractors, as well as Amoco Production, Halliburton Services, and Dowell-Schlumberger. These data were critically reviewed and analyzed by the group, and it was determined that the fracture treatment was likely to screen out for reasons similar to the Lower Travis Peak treatment. A critical issue, however, was the thickness and overall stress magnitude of the shaley interval from 8,130 to 8,230 ft above the perforated intervals (Fig. 2.7). There also appeared to be a thin, high-stress zone at 8,300 ft between the sets of perforations.

On the basis of modeling work performed on the Upper Travis Peak and subsequent discussions, three different scenarios were established. First, the treatment could be pumped in the current perforated interval. This option carried the highest risk of screenout. Second, perforations could be added to the lower-stress region below, say from 8,600 ft down to 8,850 ft. This option minimized the chance of a screenout, but was less attractive for mechanical reasons. The third option included pumping the treatment into the current perforations but also to use a treatment designed by Dowell-Schlumberger to minimize downward fracture growth. The group decided, along with GRI management, that the third option was the best alternative.

2.8.1 Well Tests. The Upper Travis Peak was perforated over the intervals 8,238 to 8,255 ft, 8,262 to 8,284 ft, and 8,321 to 8,337 ft. To eliminate or minimize the possibility of creating multiple fractures during formation breakdown, these perforations were phased 180° apart and oriented in a north 80° east plane. From openhole data, north 80° east was expected to be the direction of fracture propagation. Thus, fluid pressure applied to the formation should initiate a fracture with little or no near-wellbore tortuosity. Fluid flow into the fracture should then be unrestricted (assuming multiple fractures could be prevented).

In the Upper Travis Peak, the minifracture treatments were pumped before the prefracture production test and pressure buildup survey. Afterward, the zone was successfully tested at a gas rate of 35 Mcf/D and a water rate of 32 BWPD; however, it was necessary to install a pumping unit to sustain production. An analysis of the bottomhole pressure buildup test conducted following the production period indicated the permeability to gas was 0.0059 md, assuming 13 ft of net gas pay. A skin factor of −3.9 was also calculated and reservoir pressure was estimated to be 4,200 psi. From these results, the upper zone was obviously marginal; however, we decided to continue with the fracture treatment to verify the fracture model predictions.

2.8.2 Minifracture Treatments. A minifracture treatment was again pumped to allow Teledyne Geotech to collect seismic data for analysis of fracture azimuth and fracture height. Various injection stages were planned with surveys taken during shut-down periods between each injection. During the minifracture, the geophone became stuck and the wireline was pulled out of the rope socket. It was successfully recovered the following day and the minifracture treatment was repeated.

The seismic data recorded during both minifractures was analyzed and the fracture azimuth was determined to be between north 78° east and north 86° east. Even though fracture azimuth was successfully determined, fracture height could not be estimated by the H/Z method because of sporadic cement bond in the liner above the perforated interval.

After these minifracture treatments, the well was produced to clean up as much of the injection fluid as possible. On 12 May, a second minifracture was pumped using the same fluids intended for the main hydraulic fracture treatment. Several different stages were pumped at various injection rates. A total of 43,500 gal of slickwater and 215,500 gal of 60-lbm crosslinked gel were pumped. The treatment, termed a "shadow" fracture, used essentially the same volume of fluid that was planned for the main hydraulic fracture treatment.

RES again analyzed these data with FRACPRO to calibrate the model and determine fracture dimensions. The results of this analysis are illustrated in **Fig. 2.17,** which shows a history match of the net fracturing pressures. To obtain the match, it was necessary to alter the stress profile slightly by increasing the stress contrast by 200 psi within thc perforated interval and in the barriers above and below. During the earlier slickwater stages (from 0 to 125 minutes), the excess pressure increased indicating that the fracture was being contained by the stress barriers. However, once crosslinked gel was injected (after 200 minutes) the pressure decreased throughout the remainder of the injection phase. The decline in pressure indicated that the fracture was growing vertically. However, the pressure decline was not as drastic as it would be for a fracture with uncontained height growth. As such, it appeared that the fracture might still be contained on one side, which was exactly the situation expected based on an evaluation of the stress profile. From the analysis, fracture half-length was calculated to be approximately 800 ft, while the total fracture height was estimated to be less than 400 ft. It should be noted that most of the height growth was projected to be in a downward direction.

2.8.3 Fracture Stimulation Treatment. In preparation for the hydraulic fracture treatment of the Upper Travis Peak, slots were cut in the 5-in. liner and 7-in. casing string over the intervals 8,324 to 8,332 ft, 8,268 to 8,270 ft, and 8,271 to 8,273 ft. These slots were phased 180° apart and oriented in a north 80° east plane. The slots were cut primarily to ensure good communication to the reservoir and minimal perforation friction through the casing strings.

Because of concerns about excessive downward fracture height growth, it was decided to pump a treatment designed by Dowell-Schlumberger. The treatment, known as DivertaFrac, attempts to limit downward growth by allowing proppant to settle to the bottom of the fracture and, thus, impede fluid flow in that direction. The treatment was successfully pumped on 8 September using various stages of thin fluid with 20/40-mesh and 40/70-mesh sand at different injection rates. While injecting the treatment, an increase in the excess pressure was observed indicating that either downward fracture growth was being curtailed or that near-wellbore restrictions were again being experienced. Even though the overall injection pressure level was higher for the DivertaFrac treatment than during the previous minifractures, the net pressure did decline throughout most of the treatment indicating that vertical fracture height growth was probably occurring.

The main hydraulic fracture treatment was pumped on 17 September 1988. The proposed design called for 260,000 gal of fluid with 520,000 lbm of 20/40- and 18/20-mesh intermediate-strength proppant. The proposed injection rate was 50 bbl/min. While pumping the 3 lbm/gal proppant stage, surface injection pressures started to increase. Once the maximum surface pressure was reached, the treatment was shut down. A total of 235,000 gal of fluid and 115,000 lbm of proppant was pumped into the formation. The average treating pressure was 3,950 psi at an average injection rate of 50.5 bbl/min.

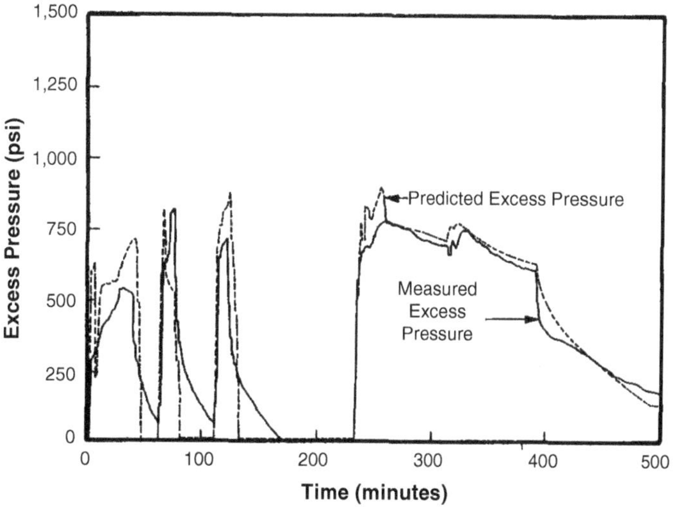

Fig. 2.17—History match of Minifracture No. 2 in the Upper Travis Peak.

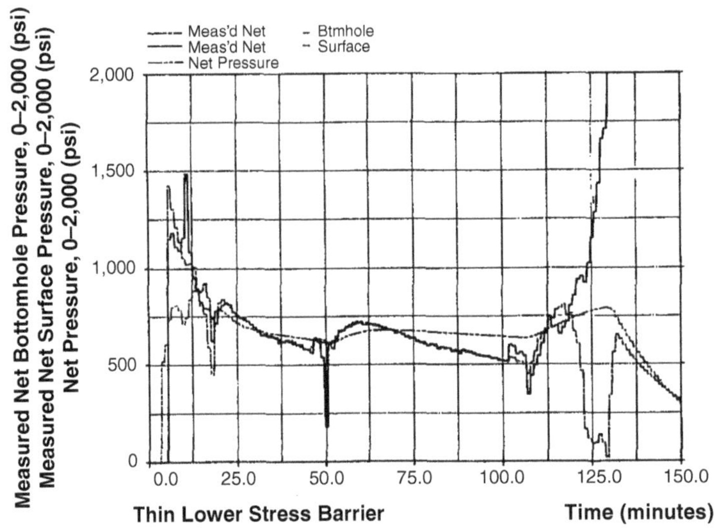

Fig. 2.18—History match of main fracture treatment in the Upper Travis Peak.

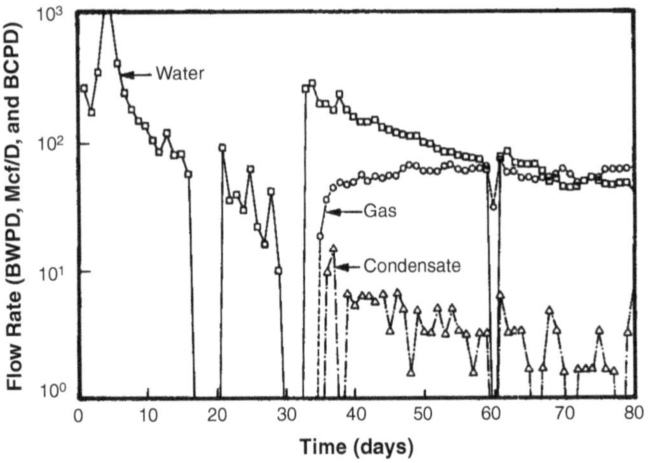

Fig. 2.19—Postfracture production history of the Upper Travis Peak.

The net fracturing pressures were again analyzed to determine the created fracture dimensions and help explain why the treatment screened out. The final results of the history match by RES are shown in **Fig. 2.18.** The best match of the data was obtained with a strong barrier above the perforated interval and a thin high-stress barrier between the perforations. In general, the fracture grew uncontained in the downward direction, but fluid flow was restricted downward by the thin stress barrier. Most likely, excessive downward fracture growth coupled with the restriction near the perforated interval was the cause of the screenout. The amount of downward fracture growth was calculated to be between 300 and 600 ft, depending on the final stress profile used. The propped fracture length was determined to be at least 250 ft and potentially up to 500 ft, again depending on the stress profile used. Because injected fluid was flowing primarily in a downward direction when it entered the fracture, it is possible that most of the proppant was not placed across the gas sands. Such an effect upon proppant distribution could not be analyzed at the time of this project, but research was being conducted in this area and was expected to be a capability in future versions of FRACPRO.

2.8.4 Post-Fracture Production and Pressure Analysis. Immediately after the fracture treatment of the Upper Travis Peak, SFE No. 2 was opened to flow. As illustrated in **Fig. 2.19,** only water was produced for the first 30 days. At this point, the well died after having produced 5,520 bbl. A pumping unit was installed in an attempt to lift the water. Three days after production was resumed, measurable quantities of gas were observed. SFE No. 2 continued to produce for approximately 50 more days at which time the well was shut in for a bottomhole pressure buildup survey. Cumulative production before the buildup was 10,280 bbl of water, 2,580 Mcf of gas, and 170 bbl of condensate. Final flow rates before shut-in were 65 Mcf/D of gas, 42 bbl/D of water, and 3.3 bbl of condensate per day. Bottomhole pressures could not be recorded during the production period because the well was on pump. However, once the well was shut in and the rods and pump pulled, bottomhole pressure gauges were run into the well.

The reader is again referred to CER Corp. and S. A. Holditch & Associates Inc. (1990) and Bastian and Sherman (1993) for a detailed discussion on the analysis of these production and pressure buildup data. A summary of the final results is presented below. As with the Lower Travis Peak, reservoir simulation was used to account for multiphase flow effects, extended gas breakthrough time, and reservoir heterogeneity in the Travis Peak. A two-layer model composed of a gas sand and a water sand was found to best match the actual well performance.

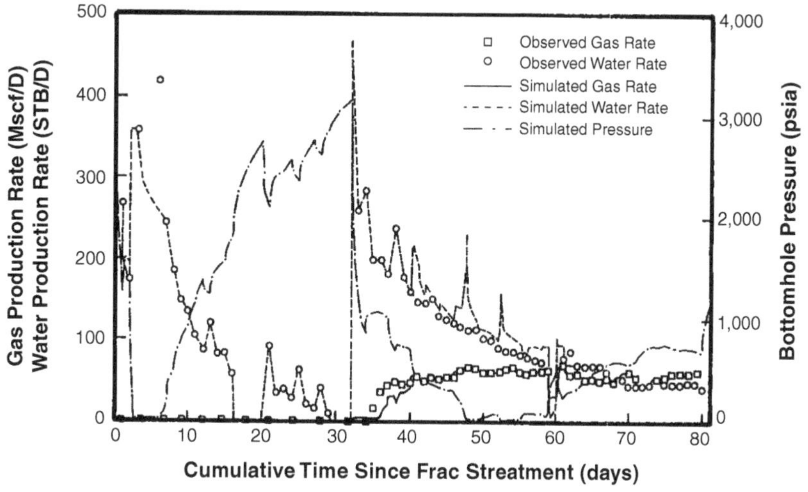

Fig. 2.20—History match of Upper Travis Peak production.

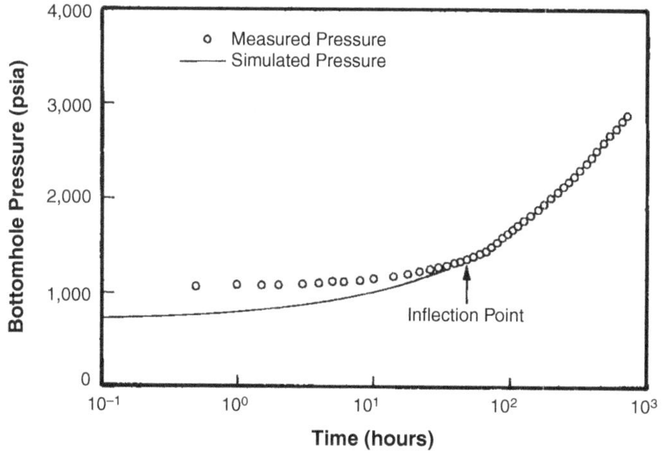

Fig. 2.21—History match of Upper Travis Peak pressure buildup test.

	Layer 1	Layer 2
Absolute Permeability, md	0.0891	0.1434
Net-Pay, ft	10	17.5
Total Porosity, %	5.0	5.8
Water Saturation, %	32.9	80
Gas Saturation, %	67.1	20
Relative Permeability, fracture	0.2565	0.0208
Gas Porosity, %	3.37	1.15
Gas Permeability, md	0.0229	0.0030
Fracture Length, ft	158	158
Fracture Conductivity, md-ft	6,000	6,000
Area, acres	160	160

Table 2.6—Reservoir and fracture properties estimated for SFE No. 2–Upper Zone.

The final history match of production is illustrated in **Fig. 2.20,** and the pressure buildup match is shown in **Fig. 2.21.** The reservoir and fracture properties used to obtain these matches are summarized in **Table 2.6.** As can be seen, a fracture length of 158 ft was determined. Overall, the fracture length was shorter than that predicted by the 3D fracture model; however, as previously discussed, the effect of proppant flow below the net-pay interval could not be considered with the fracture model. In addition, confidence in the reservoir model results for the Upper Travis Peak is somewhat less since actual flowing pressures could not be analyzed. Even so, the results are believed to be reasonable considering the volume of proppant pumped.

2.9 Conclusions

The GRI's second SFE was successfully completed. An extensive amount of data was collected and analyzed by various organizations and researchers for GRI. Detailed results of this work can be obtained from the individual contractors listed in Table 2.1. The following conclusions represent a summary of some of the more important results.

- Extensive quartz overgrowth in the thick fluvial-deltaic section of the Travis Peak results in extremely low permeability and relatively poor lateral continuity of the sands. The quartz cementing is also believed to affect mechanical properties that have a direct impact on hydraulic fracture treatments.
- Because of the thickness of the Travis Peak and similar mechanical and lithologic properties, good barriers to vertical fracture height growth are generally not present.
- Extensive natural fracturing is present in the lower portion of the Travis Peak in SFE No. 2. Geologic studies of the core show that the natural fractures and coring-induced fractures can be used as a reliable method to predict in-situ stress direction and hydraulic fracture azimuth. From this work, the orientation of a hydraulic fracture in SFE No. 2 is estimated to be north 80° east.
- Multiple hydraulic fracture strands were created in the Lower Travis Peak when hydraulic pressure was applied. The presence of these multiple fractures could cause narrow flow paths for injected fracturing fluid, resulting in screenouts during hydraulic fracture treatments.

- By integrating openhole acoustical log data with actual stress test measurements, a method is now available for developing a detailed vertical stress profile. SFE No. 2 was the first well in the GRI Tight Gas Sands program that used this methodology.
- From the work conducted on SFE No. 2, the importance of using the stress profile to help plan the completion is now recognized. In many tight gas wells, the perforations should not be chosen based solely on the analysis of net gas pay. Instead, the perforations should be placed at a location that will allow a successful fracture treatment to be pumped so that most or all of the net gas pay can be stimulated.
- From an analysis of the minifracture and hydraulic fracture treatment data, fracture dimensions for the Lower and Upper Travis Peak intervals were predicted. The analyses, coupled with other formation data from cores and logs, provided reasonable explanations for the screenouts of both treatments. In fact, the screenout for the Upper Travis Peak was actually predicted on the basis of the stress profile and formation description.
- The production and pressure history of the Lower and Upper Travis Peak were successfully matched with a two-layer, two-phase reservoir simulator. This type of model seems to consistently fit the well performance from the Travis Peak.

2.10 Acknowledgments

We gratefully acknowledge the support of the GRI who sponsored this work under GRI Contract No. 5083-211-1734. We also appreciate the efforts and dedication of all of the organizations listed in Table 2.1 whose results were the basis for this paper.

2.11 References

Bastian, P. A. and Sherman, J. B. 1993. Analysis of a Hydraulically Fractured, Low Permeability Gas Reservoir Using Numerical Simulation. *SPE Advanced Technology Series* **1** (2) SPE-21511-PA. https://doi.org/10.2118/21511-PA.

CER Corp. and S. A. Holditch & Associates Inc. 1988. Advancements in Travis Peak Formation Evaluation and Hydraulic Fracture Technology, Staged Field Experiment No. 1, Waskom Field, Harrison County, Texas. Report No. GRI-88/0077 prepared for GRI under Contract No. 5086-211-1306.

CER Corp. and S. A. Holditch & Associates Inc. 1990. Staged Field Experiment No. 2: Application of Advanced Geological, Petrophysical and Engineering Technologies to Evaluate and Improve Gas Recovery From Low Permeability Sandstone Reservoirs. Report Nos. GRI-89/0140 and GRI-90/0093 prepared for GRI under Contract No. 5088-211-1682.

Holditch, S. A. 1986. Staged Field Experiments – An Introduction. *In Focus - Tight Gas Sands* **3** (1).

Holditch, S. A., Robinson, B. M., Whitehead, W. S. et al. 1988. The GRI Staged Field Experiment. *SPE Formation Evaluation* (September 1988): 519–533.

Tye, R. S. 1988. Stratigraphy and Depositional Systems of the Lower Cretaceous Travis Peak Formation, East Texas Basin. Report prepared for GRI under Contract No. 5082-211-0708, Bureau of Economic Geology, The University of Texas at Austin, Austin, Texas.

Whitehead, W. S., Hunt, E. R., Finley, R. J. et al. 1986. In-Situ Stresses: A Comparison Between Log-Derived Values and Actual Field-Measured Values in the Travis Peak Formation of East Texas. Presented at the SPE Unconventional Gas Technology Symposium, Louisville, Kentucky, USA, 18–21 May. SPE-15209-MS. https://doi.org/10.2118/15209-MS.

Chapter 3

Hydraulic Fracturing Research in East Texas: Third GRI Staged Field Experiment

B. M. Robinson, S. A. Holditch & Associates Inc.; **S. A. Holditch,** Texas A&M U.; **W. S. Whitehead,** S. A. Holditch & Associates Inc.; and **R. E. Peterson,** CER Corp.

SPE 22878 presents a case history of Gas Research Institute's (GRI's) SFE No. 3 well, which was drilled and completed in the Cotton Valley Formation in East Texas in late 1988 and 1989. Data collected and analyzed include regional geology, whole cores, openhole logs, in-site stress tests, drilling and completion data, production and pressure-transient tests, fracture stimulation treatments, fracture diagnostic measurements, and post-fracture performance tests. On the basis of comprehensive modeling and matching of prefracture data, fracture treatment data, and post-fracture performance data, the authors concluded that

- Shales above and below the Lower Cotton Valley Taylor Section are good barriers to vertical fracture-height growth
- A long hydraulic fracture (> 1,000 ft) was achieved with good fracture containment
- Application of sophisticated diagnostic technologies and numerical modeling technologies result in more-consistent analysis of the fracture treatment data compared with conventional well test analysis and casedhole logging techniques

3.1 Summary

SPE paper 22878-PA presents results from research conducted on the third GRI—now the Gas Technology Institute—staged field experiment (SFE) well. Research well SFE No. 3 was drilled as part of a field-based research program conducted in East Texas for 7 years. Most of the work before SFE No. 3 involved the Travis Peak Formation; however, the Cotton Valley Sandstone was the primary research target for this well. SFE No. 3 was the last in a series of research wells planned for East Texas. A fourth SFE was conducted in the Frontier Formation of southwestern Wyoming.

3.2 Introduction

A major effort of this program was the SFE project, in which wells were drilled specifically to conduct research in the analysis and stimulation of tight gas sands and to validate technologies under development in other GRI research projects. The first two SFE wells were completed and the results published in various reports (CER Corp. and S. A. Holditch & Associates 1988, 1990) and technical papers (Holditch et al. 1988; Robinson et al. 1991). The fourth SFE project, in the Frontier Formation of southwestern Wyoming, was completed in late 1991. This paper presents the results from the third SFE well.

Data on SFE wells were collected from whole cores, openhole geophysical logs, in-situ stress tests, production and pressure-transient tests, fracture stimulation treatments, fracture-diagnostic measurements, and postfracture performance tests. The main purpose of collecting such a comprehensive set of data was to describe the reservoir and hydraulic fracture fully. The goal was to obtain the most accurate analysis possible to provide a better understanding of the hydraulic fracturing process. The team hoped that this work would lead to better treatment designs and additional gas production from tight gas sands, which they believed could be achieved through the practical application of new technologies and more-rigorous analysis techniques with numerical models. The development of practical and useful tools and methods that benefit the natural gas industry was of primary importance to this research project. SFE No. 3 was drilled and completed in late 1988 and 1989. The Lower Cotton Valley Sand was the primary interval of completion and stimulation in this well. This paper summarizes the work performed and data collected on SFE No. 3 and presents results from the data analyses.

3.3 Objectives of SFE No. 3

These research wells were "staged" in the sense that successively higher goals were set for each SFE well. Before SFE No. 3, almost all the research was conducted in the Travis Peak Formation (immediately above the Cotton Valley). Even though additional work was needed in the Travis Peak, there was also a need to begin testing (in a different formation) some of the new methods and technologies that had been developed during the program. Thus, it was logical to drill through the Travis Peak and complete SFE No. 3 in the Cotton Valley Formation.

Organization	Research Area
Bureau of Economic Geology, University of Texas, Austin	Site selection, geology, core analysis
OER Corp., Las Vegas, Nevada	Data acquisition—coring and logging, project administration and documentation
Resources Engineering Systems	Hydraulic fracture treatment design and analysis
ResTec Houston, Houston	Formation evaluation, in-situ stress profiling
S. A. Holditch & Associates Inc., College Station, Texas	Drilling and completion, reservoir engineering, lead technical contractor
Teledyne Geotech, Garland, Texas	Microseismic analysis for fracture dimensions and azimuth

Table 3.1—Participating GRI contractors.

In terms of specific research goals, two different fracture-diagnostic technologies were to undergo final testing and evaluation. One was a 3D fracture simulator, FRACPRO, developed during the program. This model was developed to the point at which it was expected to rigorously model all pertinent fracturing parameters and provide an accurate calculation of fracture dimensions. The other involved the measurement and processing of microseismic data to determine fracture height and azimuth. The results of the research conducted on SFE No. 3 indicate that these research objectives were met.

Various organizations were contracted to perform the research on SFE No. 3. **Table 3.1** lists these participating organizations, along with their primary area of research.

3.4 Geologic Overview

The Bureau of Economic Geology performed detailed stratigraphic and petrologic studies on the Travis Peak and Cotton Valley Formations primarily from core and log data from SFE No. 3; data from SFE No. 1 and numerous offsetting wells were also available. CER Corp. and S. A. Holditch & Associates (1988, 1990), Holditch et al. (1988), and Robinson et al. (1991) discuss the depositional environments for the Travis Peak in more detail. This paper concentrates primarily on the Cotton Valley.

As **Fig. 3.1** shows, SFE No. 3 is in Harrison County, Texas. The Waskom Field is on the northwest flank of the Sabine Arch, a large basement-cored, low-amplitude anticline that marks the eastern side of the East Texas Basin (see **Fig. 3.2**). In this area, the Cotton Valley Formation is a predominant tight gas reservoir that produces primarily from structural traps associated with the Sabine Arch and numerous other salt structures (Kosters et al. 1989). The Cotton Valley Formation and the overlying Knowles limestone together are about 1,500 ft thick. The Knowles limestone is a thin, transgressive marine deposit that pinches out updip to the northwest. Where the Knowles limestone is absent, the Travis Peak/Cotton Valley contact is considered to be unconformable.

Fig. 3.3 shows a log section from SFE No. 3 in the Lower Cotton Valley and the Bossier Shale. The perforated intervals within the Cotton Valley Taylor Sandstone also are shown.

3.5 Operations Summary

3.5.1 Drilling. Officially permitted as the Mobil Cargill Unit No. 15, SFE No. 3 was drilled from 6 September through 13 November 1988. The top of the Cotton Valley was penetrated at 7,943 ft below kelly bushing. Total depth (TD) of the well was 9,690 ft, which is about 200 ft into the underlying Bossier Shale.

An extensive amount of openhole data was collected during the drilling of SFE No. 3. A total of 390 ft of 4-in. diameter core was planned for this well in five different intervals (CER Corp. 1990). The first two intervals are from the Travis Peak, while the last three intervals are from the Lower Cotton Valley. Scribe marks were cut in an attempt to orient all the core by use of one of three techniques: downhole electronic surveys, paleomagnetic analyses, and borehole imaging. Except in a few intervals where electronic problems and/or poor borehole images were experienced, these operations were successful.

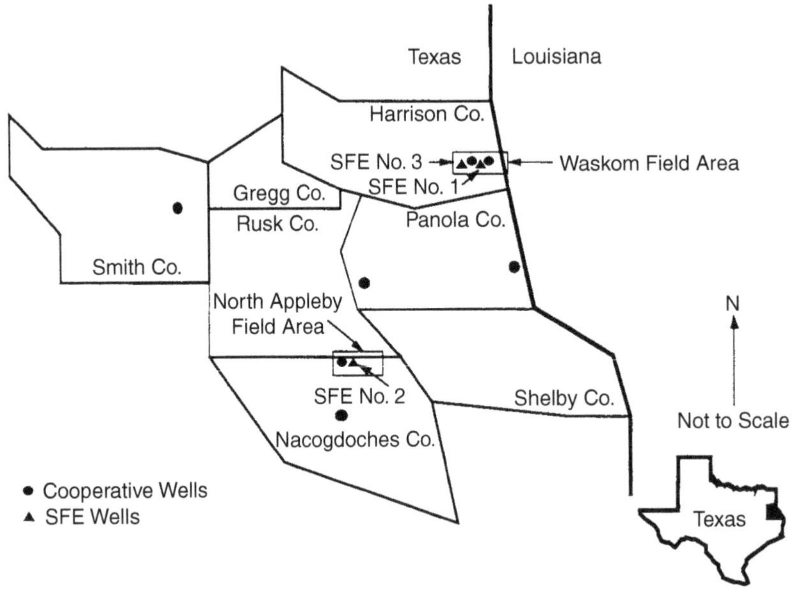

Fig. 3.1—Location of GRI research wells in East Texas.

Six openhole stress tests were conducted during drilling of SFE No. 3 to determine the average in-situ closure stress in both the Travis Peak and Cotton Valley intervals. In addition, an attempt was made to determine the hydraulic fracture orientation created in the openhole by imaging the fracture with borehole televiewer (BHTV) and Formation MicroScanner (FMS) logs.

The first openhole stress test was conducted in the Travis Peak, four were conducted within the Cotton Valley interval, and the last was performed in the Bossier Shale. On one test conducted in the shaley interval above the main Taylor Sand, an attempt was made to overcore the stress-test-induced fracture. Unfortunately, the induced fracture was not recovered in the core. Fractures were observed in the borehole according to BHTV and FMS logs in three of the six tested intervals. The observed azimuth of these hydraulically induced fractures averaged north 86° east. These results are in good agreement with the analysis of natural and coring-induced fracture orientations, indicating that these studies may provide a reliable estimate of hydraulic fracture orientation.

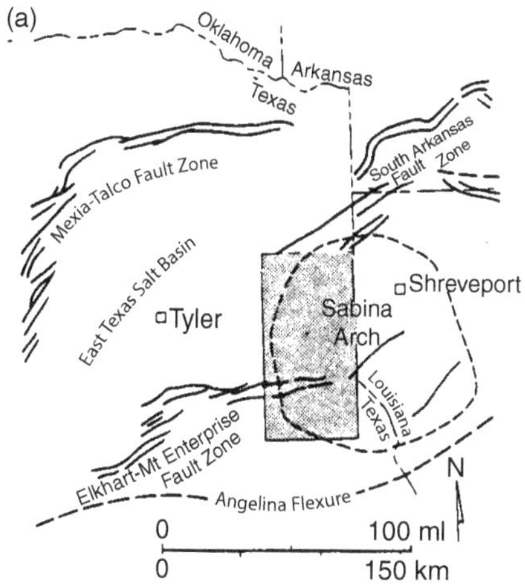

Fig. 3.2—Structural setting of the study area in the northern Gulf of Mexico Basin.

As SFE No. 3 was drilled, openhole logs were run on six different occasions (CER Corp. 1990). Openhole logging data also were obtained by use of measurement-while-drilling systems. Several logging passes were made at different times during drilling to study the effects of mud-filtrate invasion on the readings from logging tools. After reaching TD, a comprehensive set of geophysical logs was run to obtain data for calculating permeability-thickness, kh, porosity-thickness, ϕh, and in-situ stress, σ_x, profiles and formation lithology. Some of the logs were experimental/research tools run by the research departments of Mobil, Amoco, and Schlumberger.

3.5.2 Completion. After casing was set, another logging program, consisting of four different runs, was conducted. The primary purpose was to determine the fracture height following various hydraulic fracturing events (e.g., injection/ballout, minifracture and main fracture treatments). In addition, the casedhole logging provided an indication of the cement-bond quality in potential completion intervals. On one run, borehole gravity/density measurements were made to determine if any formation density anomalies existed. After the casedhole logging program was completed, stress tests were performed in various intervals. Prefracture well testing was then initiated.

SFE No. 3 was perforated in the Cotton Valley Taylor Sand over two intervals: 9,225 to 9,250 and 9,285 to 9,330 ft. Afterward, several prefracture well tests (two flow tests and two pressure-buildup tests) were conducted. The initial flow test occurred after a perforation bailout treatment. The well flowed for 7 days, with an average rate of 50 Mcf/D and 8 BWPD. Although the flowing wellhead pressure was too small to measure, the flowing bottomhole pressure (BHP) at the end of the flow period was measured at 507 psia. After the flow period, SFE No. 3 was shut in for a 7-day pressure-buildup test. BHPs were recorded with an electronic memory gauge run on slickline.

The first minifracture treatment was pumped on 17 February 1989, and was followed by additional casedhole logging. SFE No. 3 then was opened for another flow test. The average flow rate over 6 days was 310 Mcf/D and 8 BWPD on a 24/64-in. choke. The flowing tubing pressure was 150 psi, and the measured flowing BHP was 727 psi. This production period was followed by a 6-day pressure-buildup test.

On 16 March 1989, a second minifracture and the main hydraulic fracture treatment were pumped. Injection was down the casing/tubing annulus so that BHPs could be recorded by a pressure gauge run inside the tubing. **Table 3.2** outlines the treatment design and execution. The treatment consisted of 15 separate stages, the first being Minifracture No. 2. As Table 3.2 shows, the fracture treatment was pumped essentially as designed.

For SFE No. 3, the fracturing service company batch mixed approximately one-half of the fluid before the treatment. Afterward, they used a liquid gel concentrate to mix fluids semicontinuously for the remainder of the treatment.

The GRI rheology van (Ely and Holditch 1988) was on location to measure fracture-fluid properties before and during the treatment. These tests provided a measure of quality control over gel behavior during the treatment.

Production began after the fracture treatment and continued for 89 days, at which time a pressure-buildup test was performed. By 16 June, when the well was shut in for the extended pressure-buildup test, total postfracture production was 34,127 Mcf of gas, 8,992 bbl of water, and 244 bbl of condensate. The final flow rates before shut-in were 320 Mcf/D, 24 BWPD, and 2 B/D condensate.

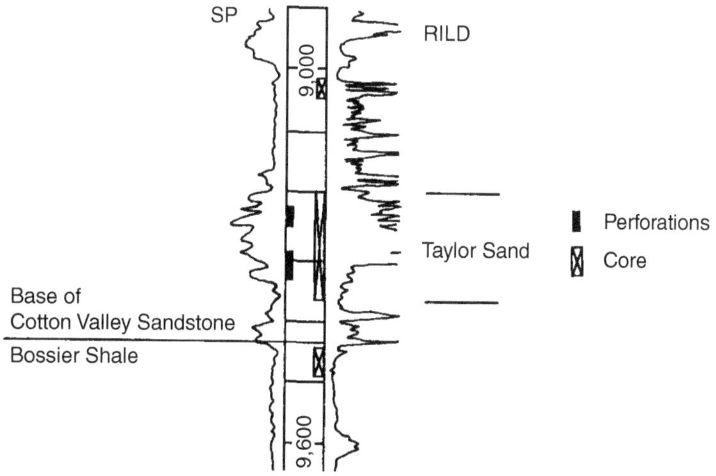

Fig. 3.3—Spontaneous potential (SP)/resistivity (RILD) log of Lower Cotton Valley Sandstone and Upper Bossier Shale.

Stage	Fluid Volume (bbl)		Fluid Type	Proppant Concentration (lbm/gal)	Proppant Volume (lbm)		Proppant Type
	Design	Actual			Design	Actual	
1	1,500	1,571	40 lbm linear	0	0	0	None
2	0	0	Shut-in	0	0	0	None
3	1,000	576	Slickwater	0	0	0	None
4	3,000	2,908	50 lbm crosslinked	0	0	0	None
5	300	293	50 lbm crosslinked	1	12,600	11,110	100-mesh sand
6	400	477	40 lbm crosslinked	0	0	0	None
7	400	337	40 lbm crosslinked	1	16,800	14,150	20/40 Ottawa Sand
8	500	500	40 lbm crosslinked	2	42,000	41,120	20/40 Ottawa Sand
9	500	437	40 lbm crosslinked	3	63,000	54,910	20/40 Ottawa Sand
10	600	623	40 lbm crosslinked	4	100,800	105,330	20/40 Ottawa Sand
11	800	755	40 lbm crosslinked	5	168,000	133,630	20/40 Ottawa Sand
12	900	807	40 lbm crosslinked	6	226,000	227,235	20/40 Ottawa Sand
13	900	934	40 lbm crosslinked	7	264,600	300,100	20/40 Ottawa Sand
14	900	935	40 lbm crosslinked	8	302,400	292,420	20/40 Ottawa Sand
15	300	240	Slickwater	0	0	0	None

Table 3.2—Fracture treatment design and actual treatment volumes for SFE No. 3.

After the pressure-buildup test, SFE No. 3 was loaded with water and a microseismic survey and casedhole logs were run. A commercially available product designed to remove gel filter cake and residue was used in the first 500 bbl of kill fluid while logging was performed. This treatment appeared to help; gas flow rates increased after the logging operations.

3.6 Formation Evaluation

The cores and openhole logs were analyzed to determine such basic reservoir properties as lithology, porosity, and water saturation. However, specialized analyses also were performed (some on a research basis) to evaluate mechanical properties and reservoir flow behavior.

3.6.1 Core Analysis. Routine core analyses were performed on 320 core samples to measure porosity, as-received fluid saturation, permeability, and grain density. These measurements also were made on 21 rotary sidewall cores taken after the well reached TD (6 in the Travis Peak and 15 in the Cotton Valley). Reports from Western Atlas International Core Laboratories (1989a, 1989b) give detailed results of the routine plug and sidewall core analyses.

Special core analyses were performed on 38 core plugs from 18 depth locations (4 in the Travis Peak, 10 in the Cotton Valley, and 4 in the Bossier Shale). The parameters measured by the special tests can be grouped into four categories: dynamic mechanical properties, electrical properties, relative permeability behavior, and capillary pressure properties. Results from these tests have been reported elsewhere (Western Atlas International Core Laboratories 1989c), so details will not be discussed. However, some of the more significant findings are presented.

Research has been conducted on calculating effective gas permeability from openhole logs. Because of the extensive amount of data from numerous cooperative research wells and SFE wells, a methodology was developed that gives reliable results most of the time. The method, which was previously published (ResTech 1988), involves the use of special core analyses and rigorous lithology/environmentally corrected log analyses. Correlations are then developed, from relative permeability data, so that effective gas permeability at reservoir conditions can be predicted. These results are discussed further under Log Analysis.

A considerable amount of discussion and work evolved during years of the project on the proper way to prepare core plugs for testing. The extraction and drying processes can dramatically alter delicate clay structures in some cases, resulting in measured properties that are not representative of reservoir conditions.

Special tests were conducted on cores from 10 depths in the Travis Peak in SFE No. 3 to examine the effects of extraction and air drying on specific reservoir properties. Ten pairs of plugs were collected from the Travis Peak fluvial-channel sandstone at 7,382.5 to 7,390.5 ft. Brine permeability, relative permeability, capillary pressure, and resistivity measurements were made before and after extraction and drying. Scanning electron microscope photos of companion core end trims were used for visual comparison of freeze-dried, air-dried, and toluene-extracted/dried samples.

Upon air drying and extraction, fibrous illites in the Travis Peak samples collapsed and matted against pore walls. However, they partially rebounded when resaturated with brine. The shorter, bladed illite also showed evidence of damage and collapse but generally was better preserved than the fibrous illite.

The only reservoir properties significantly affected were the effective permeability to water, k_w, and capillary pressure, P_c, behavior at intermediate water saturations. Effective gas permeability, k_g, and electrical resistivity as a function of water saturation showed little change.

Fig. 3.4 shows typical gas and water effective permeability behavior for the samples tested. While the k_g curves show little change, the restored k_w curves are significantly greater than the preserved-core k_w curve.

The air/brine capillary pressure, P_c, results supported those obtained from the permeability tests, indicating that pore throats have been enlarged. After extraction and resaturation of samples, about 6 to 7% of the pore spaces have enlarged throat radii that increased from the 0- to 40-µm size group to 4 to 10 µm.

Most of the data previously collected on the Travis Peak involved use of extracted, resaturated cores. In general, the data can be used with confidence except for effective water permeabilities that were measured at intermediate values of S_w. These permeabilities may be too high. Consequently, using these older core data to predict productivity of well completions could, in turn, lead to the prediction of erroneously high water cuts in zones with transition water saturations.

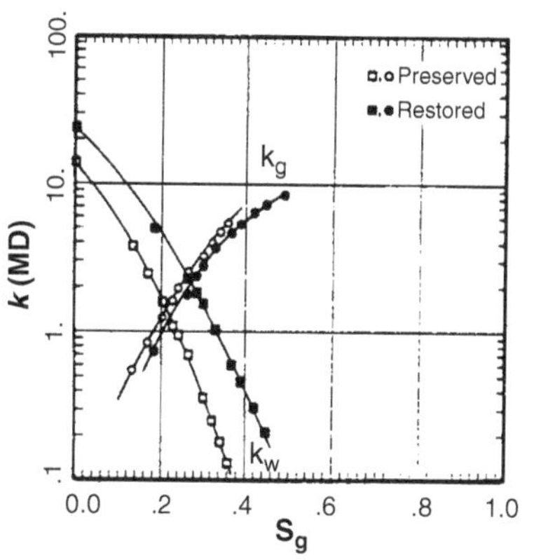

Fig. 3.4—Comparison of effective permeability to gas and water of preserved vs. extracted/restored core.

3.6.2 Log Analysis. The petrophysical analyses for this well used the same methodology previously developed for the Travis Peak Formation (ResTech 1988). The close agreement between log analyses results and core analyses indicates that the methodology developed in the Travis Peak might be successfully applied to the Cotton Valley. Indeed, the fact that the well-test analyses show similar agreement is encouraging.

In SFE No. 3, invalid log readings caused by borehole breakout were compensated for in the data processing. Wellbores drilled through tight gas sands often have stress-induced borehole breakouts. Where these breakouts occur, the density and sonic log data may be invalid because of poor density-tool/pad contact or longer sonic-tool transit times caused by the breakout.

After processing all the openhole log data, reservoir properties for the Taylor Sandstone were estimated. **Table 3.3** summarizes the results. As discussed later, pressure-transient analyses indicated a gas *kh* ranging from 0.38 to 0.88 md-ft. The results agreed very well with the *kh* estimated from openhole log interpretation.

3.7 Vertical Stress Profiling

In-situ stress tests were performed on SFE No. 3 in openhole and casedhole environments. Whitehead et al. (1986) describe the methodology involved in conducting and analyzing these tests. The results were used to calibrate the analyses of acoustical waveforms to develop a vertical-stress profile. The calibration technique was first used on SFE No. 2 and is described in Robinson et al. (1991).

The six casedhole stress tests performed on SFE No. 3 included three tests in the Bossier Shale below the Cotton Valley Taylor Sandstone and three tests in the Taylor Sand. **Table 3.4** summarizes the results. In general, these tests were easier to perform and less complicated to analyze than the openhole stress tests. For these reasons, the confidence level of the casedhole stress tests is much higher than that of the openhole tests. Thus, the casedhole results were used primarily to develop the vertical-stress profile.

To aid the calibration, the elastic equation was modified to include a term for the nonelastic sources of stress (CER Corp. and S. A. Holditch & Associates 1991).

$$\sigma_x = \left(\frac{\nu}{1-\nu}\right)(\sigma_V - g_p) + g_p + F_\sigma . \qquad (3.1)$$

Parameter	Gross Interval	Gas Sand
Thickness, ft	147.5	68.0
Average porosity	0.039	0.060
Average water saturation	0.569	0.413
Permeability-thickness, md-ft	0.34	0.33

Table 3.3—Petrophysical analysis, Cotton Valley (Taylor) Sand.

Test	Depth (ft)	Lithology	Fracture Closure Pressure (psi)	Error Range (psi)	In-Situ Stress Gradient (psi/ft)	Possible Range of Stress Gradient (psi/ft)	$\frac{\nu}{1-\nu}$
CHST 1	9,630 to 9,631	Shale	8,200	±100	0.85	0.84 to 0.86	0.380
CHST 2	9,600 to 9,601	Shale	7,950	±100	0.83	0.82 to 0.94	0.430
CHST 3	9,554 to 9,555	Shale	7,950	±100	0.83	0.82 to 0.84	0.340
CHST 4	9,324 to 9,325	Sandstone	5,800	±250	0.62	0.60 to 0.64	0.205
CHST 5	9,266 to 9,267	Sandstone	5.275	±75	0.57	0.56 to 0.58	0.300
CHST 6	9,227 to 9,228	Sandstone	5,400	±100	0.59	0.58 to 0.60	0.210

Table 3.4—SFE No. 3 results of casedhole stress tests.

The empirical factor F_σ was obtained from a graph of $\Delta\sigma$ vs. neutron porosity for all layers of rock where stress was measured. This factor was used to compensate for changes in stress caused by changes in lithology, porosity, and water saturation. It substantially changed the log-calculated stresses and properly corrected them with the results from casedhole tests. **Fig. 3.5** shows the final resulting stress profile.

A major result of our research effort was the development of a method for choosing perforations in zones to be fracture-stimulated. The conventional method was to perforate solely on the porosity and water saturation calculations. Besides using petrophysical data, the completion engineer should also consider the ϕh, *kh,* and in-situ-stress profiles. When used in conjunction with 3D hydraulic fracture models and 3D reservoir flow models, these profiles enable the engineer to locate the better gas-producing intervals and determine whether they can be connected successfully to the wellbore with a hydraulic fracture.

3.8 Prefracture Analysis

3.8.1 Well Testing. An analysis of the first pressure-buildup test with semilog and type-curve analysis techniques yielded $kh = 0.378$ md-ft, gas permeability = 0.014 md, net-pay thickness = 27 ft, skin = +0.1, and extrapolated reservoir pressure, p_{ext} = 3,400 psi. The value of 27 ft for net-pay was based on ResTech's prefracture analysis log.

After the first minifracture treatment, the well was retested and produced at a much higher gas flow rate (310 Mcf/D) and BHP (727 psia) than before. A second pressure-buildup test was conducted, the analysis of which resulted in the following values: $kh = 0.880$ md-ft, gas permeability = 0.022 md, net-pay thickness = 40 ft, skin = –3.2, and p_{ext} = 3,600 psi.

Because of the higher *kh* observed from the second test, additional net gas sand appeared to be in communication with the wellbore. This is reasonable because all the Taylor Sand was not perforated initially. Most likely, the fracture created during the minifracture opened new pay that had not been exposed previously. This is why a net gas pay of 40 ft was used in the interpretation of this buildup test.

3.8.2 Minifracture Treatments. The main purpose of Minifracture No. 1 was to collect microseismic data associated with the hydraulic fracturing process. From these data, fracture height and azimuth can be estimated (Fix et al. 1988). The treatment consisted of three sets of pump-in/shut-in periods. During each shut-in, microseismic signals were recorded by a three-component geophone and orientation gyroscope on a wireline. After Minifracture No. 1 was completed, acoustic and radiation logging tools were run to gather additional fracture-height area.

The orientation of hydraulic fractures can be determined by measuring the continuous microseismic radiation emitted as a result of the fracturing process. On SFE No. 3, continuous microseismic radiation data were actually measured on four separate occasions: before the injection/ballout treatment (after perforating), before Minifracture No. 1, during the shut-in and flowback periods of

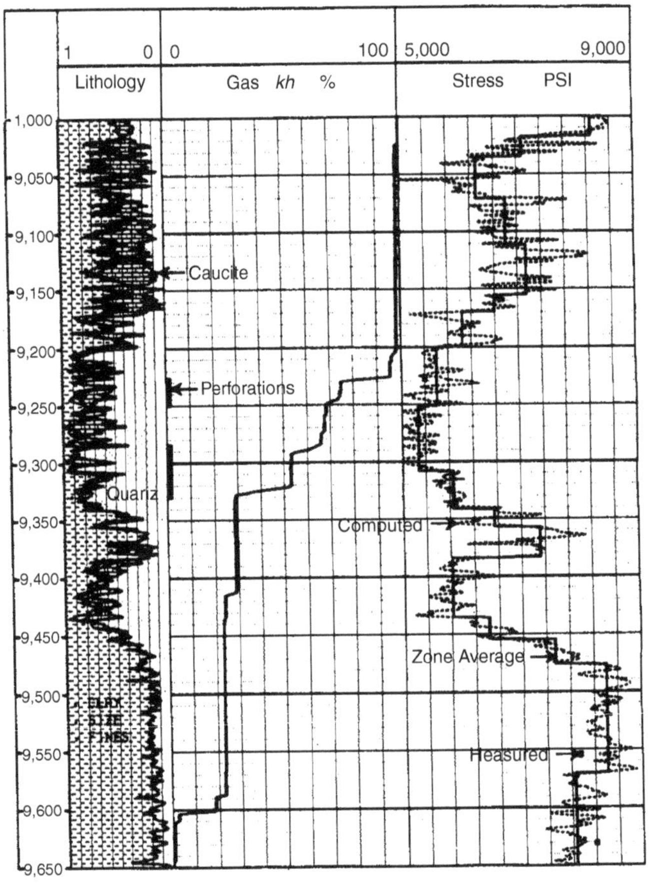

Fig. 3.5—Permeability-thickness profile and stress profile of the completion interval in SFE No. 3.

Minifracture No. 1, and 4 months after the main fracture treatment. The analysis of the survey after the main fracture treatment is discussed later.

After the last shut-in stage of Minifracture No. 1, the well was flowed back at rates of 1 to 5 bbl/min for about 90 minutes. During the flowback, 40 minutes of data were recorded at a depth of 9,590 ft. Some data also were recorded during a 4-minute interruption in flow in the middle of the flowback. These measurements marked the first time that passive microseismic data had been recorded during a flowback after a hydraulic fracture treatment. This procedure appears to provide a class of low-frequency signals that might aid in determining fracture properties.

Fig. 3.6 shows a stereographic (polar) plot of the processed microseismic data. The results for three of the signal sets—those recorded at 9,100 ft during the first shut-in, at 9,000 ft during the last shut-in, and at 9,590 ft during the second shut-in—are consistent with a fracture azimuth of 77° relative to north. Signals recorded at 9,100 ft during the last shut-in, however, are oriented perpendicular to this direction, while signals recorded at 9,000 ft during the first shut-in are oriented 30° from the perpendicular direction. The differences observed for the five sets of low-frequency signals are not understood at this time. Orientations of low-frequency flowback signals also were calculated. Like much of the data from the shut-in periods, the flowback data are oriented 77° relative to north.

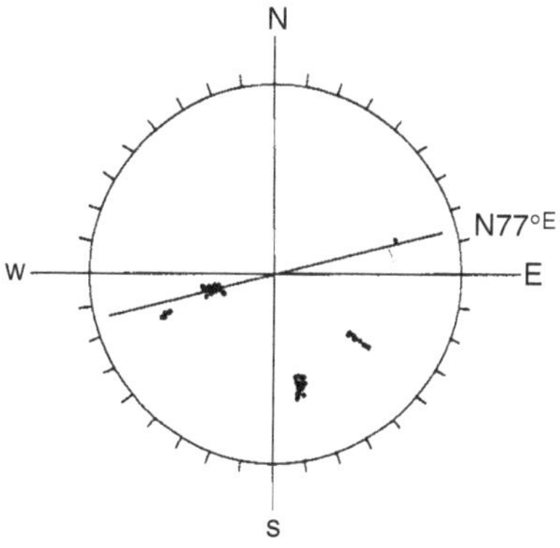

Fig. 3.6—Stereographic (polar) plot of low-frequency seismic signals, Minifracture No. 1.

The fracture height can be determined from geophone data because the background seismic motion changes according to whether the measurement is made within or outside of the treatment-altered region (Mahrer 1993). To evaluate the change, information is required on the average of the background horizontal component of motion, *H,* and the average of the background vertical component of motion, *Z*. Data recorded outside the treatment-affected zone show *Z*>*H*. However, data recorded within the zone show either *Z*<*H* or a marked and consistent reduction in *Z* compared to *H*.

It has been found that a single average value of *H/Z* for each recording depth shows the change and/or inversion of the *H/Z* ratio. The ratio is used because it has the advantage of reducing time-dependent source effects that have been linked to pressure and temperature diffusion of the treatment fluid.

Fig. 3.7 shows preperforation *H/Z* ratios as a function of depth. The most important characteristic in Fig. 3.7 is that no anomalies in the *H/Z* ratios are evident in the data. The *H/Z* ratios are consistently horizontal, showing *H*>*Z* with no abrupt change or inversion. This characteristic is expected because no fracturing had occurred when the data were recorded and is consistent with observations of other pretreatment data.

Fig. 3.8 shows the *H/Z* for the last shut-in. Despite the anomalous *H/Z* value between the perforations, the *H/Z* inversion in Fig. 3.8 gives the top of the fracture at 9,125 ± 25 ft and the bottom of the fracture 9,425 ± 25 ft. The most important aspect of these results is the extensive depth of investigation with this technique; potentially tens or hundreds of feet instead of only a few inches with conventional logging techniques.

Minifracture No. 1 also was analyzed by Resources Engineering Systems (RES) using a 3D fracture model, FRACPRO. A history-match analysis of net pressure (calculated as treating BHP minus closure pressure) was performed to estimate fracture dimensions.

According to this analysis, the maximum growths in upper and lower fracture heights were 133 and 91 ft, respectively, or an interval from 9,144 to 9,368 ft. Overall, the FRACPRO analysis suggested a fractured interval that was smaller than that indicated by microseismic data (224 vs. 300 ft). These results, however, were in much better agreement than any of the other fracture-height logging methods performed.

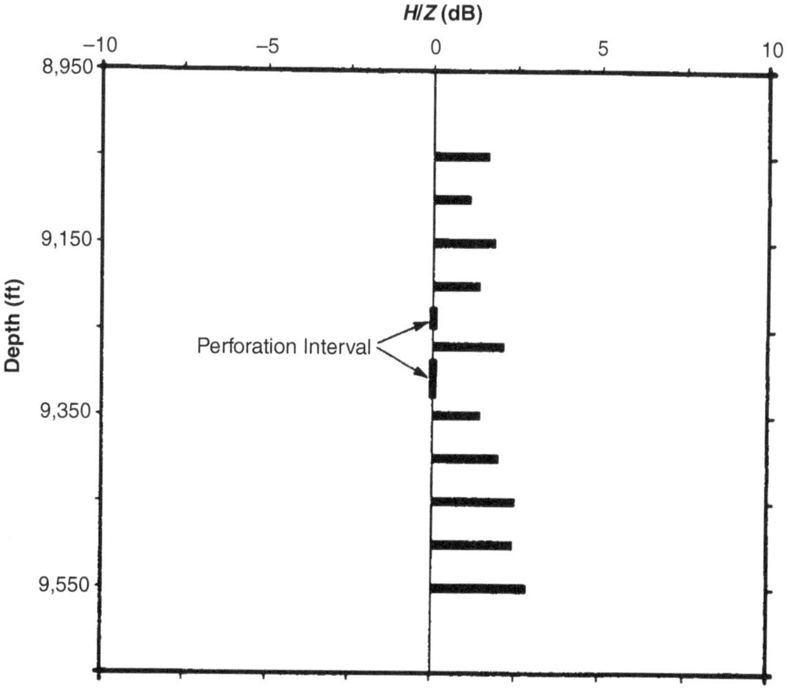

Fig. 3.7—Preperforation *H/Z* ratio.

The objective of the second minifracture treatment was to gather additional data to calibrate the 3D fracture models and to estimate values of fluid-loss coefficient. Minifracture No. 2 was pumped just before the actual fracture treatment on 16 March 1989. Thus, the analysis of these data is discussed with the main fracture treatment.

3.9 Fracture Treatment Analysis

In SFE No. 1 and SFE No. 2, near-circular fractures were created because there was not enough stress contrast between zones within the Travis Peak to limit fracture height successfully. As a consequence, one of the goals of SFE No, 3 was to create a long, elliptical fracture so that fracture-diagnostic tools and techniques could be tested on a different fracture geometry. The Taylor section of the Cotton Valley Sandstone was stimulated because it was believed to be bounded by significant stress barriers.

The fracture treatment was analyzed by history matching the net fracturing pressures with FRACPRO. **Fig. 3.9** illustrates the match. The main fracture treatment actually begins just after 250 minutes; the time before corresponds to Minifracture No. 2 and the subsequent shut-in. For this match, the curve for measured net pressure was corrected for the measured reduction in near-wellbore friction. In addition, the fracture-fluid rheological properties were modified somewhat to account for temperature effects and, more importantly, for the details of leading-edge response (i.e., most of the pressure drop occurs near the fracture perimeter).

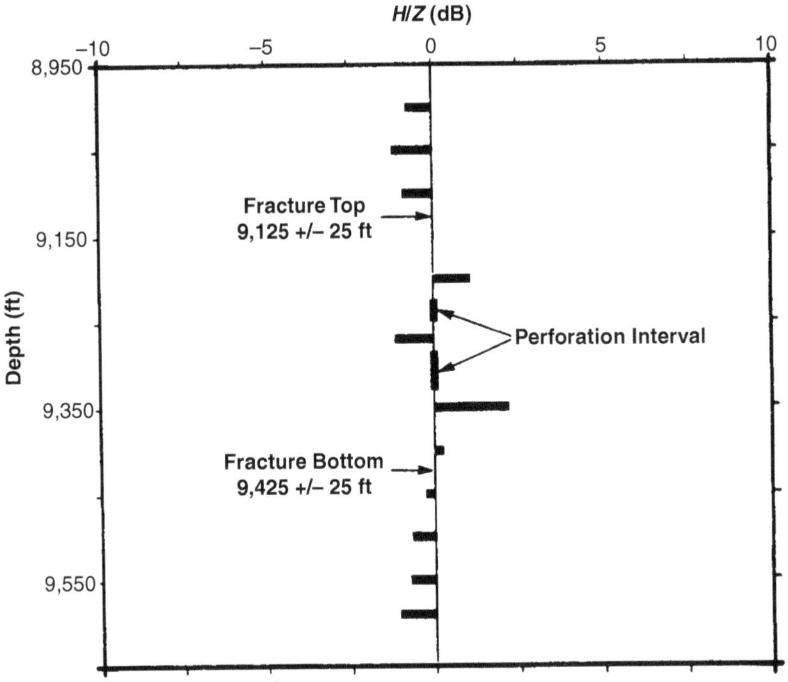

Fig. 3.8—Minifracture *H/Z* vs. depth, last shut-in.

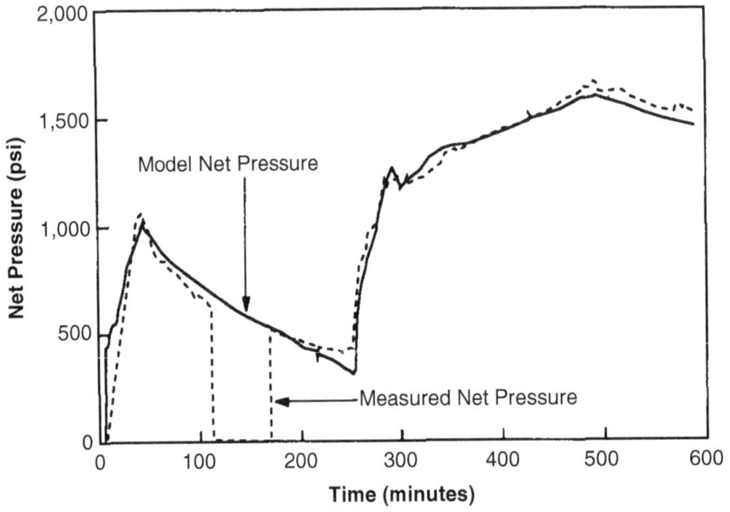

Fig. 3.9—Net pressure match of Minifracture No. 2 and main fracture treatment.

For Minifracture No. 2, the pressure falloff after shut-in was characterized by an initial rapid decline followed by an extended period of slow decline. A simple leakoff coefficient alone (i.e., with constant fracture area or uniform stress) was not sufficient to match this behavior. Either the early or late shut-in portion of the falloff would have to be ignored. The close-in stress contrasts influence the late-time shut-in, a period when the net pressure curve declines very slowly. Thus, when pressure-falloff data are analyzed to determine leakoff coefficients, one should include the effects of stress contrasts and of continued fracture growth/contraction.

As Fig. 3.9 shows, the rapid drop in calculated net pressure at about 300 minutes corresponds with the time when the fracture reached the low-stress zone indicated by the log-derived stress data at 9,380 to 9,455 ft. In the original analysis with the log-derived stress profile, the measured net pressure curve shows a much smaller drop than the modeled pressure. For this reason, it was necessary to increase the stress for this zone in the model. Note that no stress measurements were made in this interval; only log-calculated values were available.

Fracture wing length, ft	1,819
Fracture upper height, ft	166
Fracture lower height, ft	192
Fracture width (maximum), in.	0.72
Fracture aspect ratio	10.2
Propped length, ft	1,441
Propped upper height, ft	130
Propped lower height, ft	154
Proppant concentration, lbm/ft^2	1.88
Propped aspect ratio	10.1

Table 3.5—Fracture dimensions as determined by FRACPRO analysis.

Overall, an excellent match of the fracture treatment was obtained. **Table 3.5** gives the dimensions of the main fracture indicated by fracture modeling. As discussed later, the lower and upper heights compare fairly well with those determined from microseismic data. In addition, the propped length is also in good agreement with postfracture flow-performance analysis.

3.10 Postfracture Analysis

3.10.1 Production Performance and Conventional Well-Test Analysis. Shortly after the fracture treatment, SFE No. 3 was opened for fluid cleanup and flow testing. The well produced for 89 days, at which time the gas flow rate was stable at 320 Mcf/D with 24 BWPD and 2 B/D condensate. The flowing tubing pressure was 1,000 psig on an 18/64-in. choke. At this point, SFE No. 3 was shut in for an extended BHP-buildup survey. After the pressure-buildup survey (and subsequent logging operations), production was resumed. As **Fig. 3.10** shows, however, the gas flow rates were noticeably higher. This increase in production probably can be attributed to the gel-filter-cake-removal treatment added to the kill fluid before logging.

Conventional analysis of the BHP-buildup survey (including both semilog and type-curve analyses) indicated that a short, high-conductivity fracture was achieved. Even though the estimates of permeability-thickness (0.8 md-ft) and permeability (0.02 md) are probably accurate, the estimates of fracture length (200 ft) and fracture conductivity (1,170 md-ft) depict only the portion of the hydraulic fracture that has "cleaned up" and is affecting gas flow at the time of the pressure-buildup test.

3.10.2 Reservoir Modeling. To describe the reservoir and fracture properties more accurately, numerical simulation models must be used to history match the production and pressure-buildup data. Several models were used to simulate flow from the Taylor Sandstone in SFE No. 3. The models can be grouped into three categories: Type 1–1D, radial, and single-phase; Type 2–2D, areal, and single-phase; and Type 3–3D, areal, and two-phase.

Results of the single-phase (Types 1 and 2) history matching were very similar to the conventional analytical results. An excellent match of the observed BHPs during the postfracture pressure-buildup test was obtained, but the propped fracture length (200 ft) was considerably less than the estimate from an analysis of the fracture treatment data (1,400 ft).

The Type 3 modeling of SFE No. 3 postfracture behavior was completed with a 3D, three-phase, finite-difference reservoir simulator. In this case, two-phase (gas/water) flow was sufficient because the amount of condensate produced was negligible.

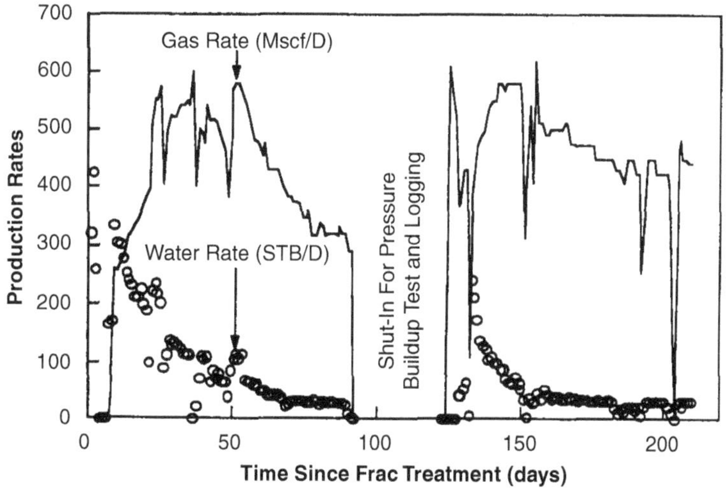

Fig. 3.10—Production history of SFE No. 3.

Reservoir data and rock properties used in the model came from core analyses, log analyses, and the single-phase postfracture pressure-buildup analysis. History matching in this study was divided into three stages: water injection, production pressure buildup, and production after the pressure-buildup test.

The primary reason for modeling water injection was to obtain the proper saturation profile around the fracture. This saturation profile then was used as input for further modeling efforts. In this instance, the key parameter to match was the total volume of water injected (12,247 bbl). When a fracture half-length of 200 ft (based on results from analytical and single-phase model analysis) was used, the formation could not accept the full injected volume in a reasonable timeframe, indicating that this fracture length was unrealistic.

For production modeling, gas flow rates were given generally as input data. The simulator then calculated values of water flow rate and BHP that were used as history-match parameters. The single exception occurred when the well produced only water. For that time period, the water flow rate was entered into the simulator, and gas breakthrough and BHPs were calculated.

Figs. 3.11 and 3.12 show the final matches of water rates and flowing BHPs. Excellent matches were obtained, with the exception of the water rates in the first 3 weeks following the fracture treatment. The discrepancy may be caused by incorrect modeling of the fracture-fluid viscosity. Another possibility is that fracture closure was not modeled.

The properties used to obtain the matches shown in Figs. 3.11 and 3.12 also resulted in an excellent match of the postfracture pressure-buildup tests **(Fig. 3.13).**

Fig. 3.14 shows the reservoir schematic obtained from history matching the production and the pressure-buildup test. Total created fracture length was 1,515 ft, which was similar to the fracture length determined by RES fracture modeling. The propped fracture length was 1,100 ft. The first 600 ft had a conductivity of 1,400 md-ft (k_f = 35,000 md), while the last 500 ft had a conductivity of only 17 md-ft (k_f = 430 md). Thus, the segment of the fracture away from the wellbore was contributing little to production from the well.

Although the higher-conductivity segment of the fracture is much longer than the fracture length determined from the single-phase analysis, it is important to remember that much of the injected water still remained in the fracture and the formation. Inspection of the saturation profile from the final simulation showed that water saturations greater than 80% still existed around most of the fracture. In fact, only the first 200 ft of the propped fracture length had actually cleaned up and exhibited water saturations near the initial water saturation. This explains why the single-phase analysis resulted in a fracture length of only 200 ft when a much longer fracture existed.

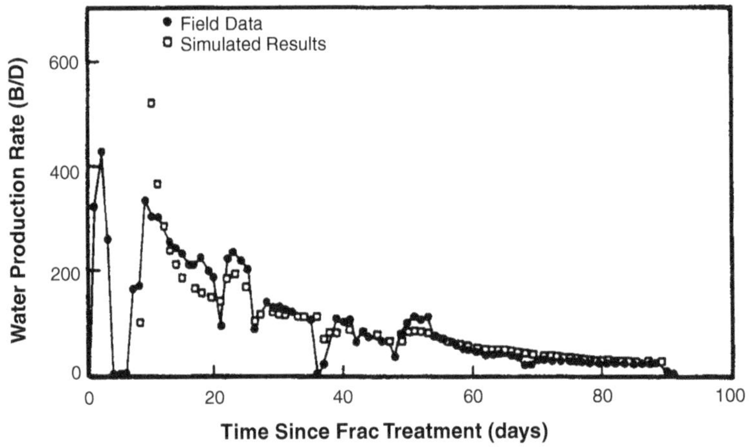

Fig. 3.11—History match of water production before pressure-buildup test, SFE No. 3.

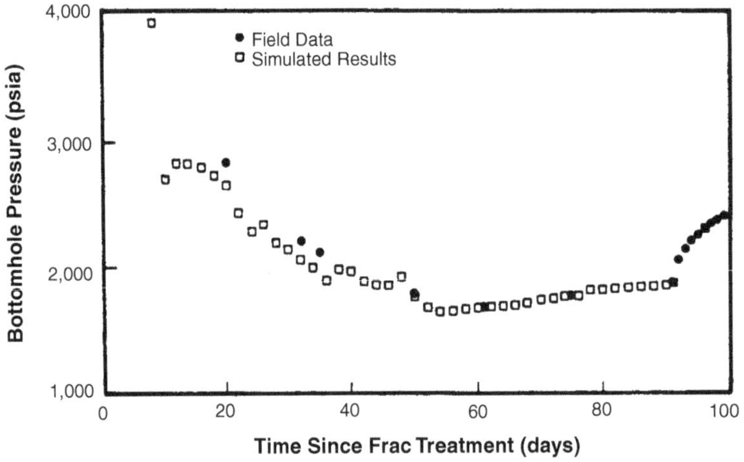

Fig. 3.12—History match of flowing BHP, SFE No. 3.

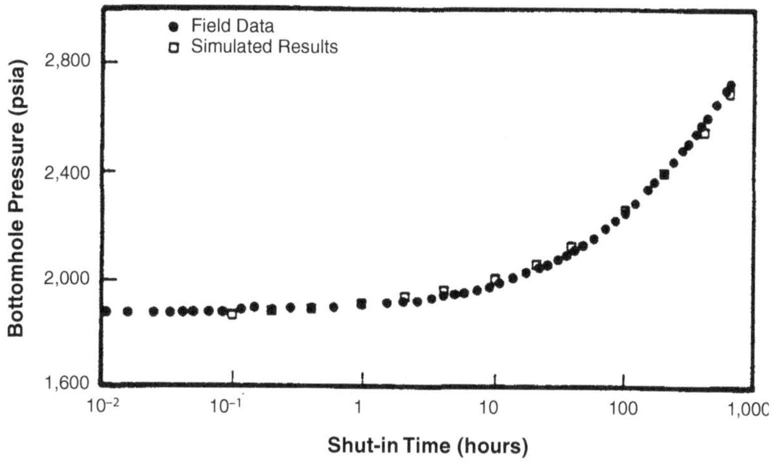

Fig. 3.13—History match of postfracture pressure-buildup tests, SFE No. 3.

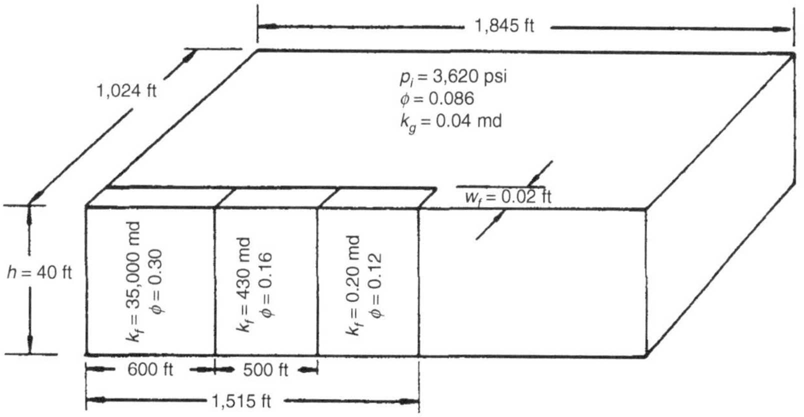

Fig. 3.14—Reservoir model for initial history match, SFE No. 3.

Because of the change in production behavior (after the gel-filter-cake treatment), the 3D, two-phase model was used again to history match the production and pressure data to determine the associated change in fracture and/or reservoir properties. Good matches of the water rate and BHPs were achieved with the higher gas flow rates. These results indicated that the only difference from the original analysis was that the conductivity near the wellbore increased from 1,400 to 1,880 md-ft and the conductivity away from the wellbore increased from 17 to 40 md-ft.

3.10.3 Fracture Height Research. A microseismic survey was conducted on SFE No. 3 about 4 months after the main fracture treatment. The objective was to determine if fracture height and/or azimuth could be determined from microseismic data acquired by reoccupying previously fractured wells. If so, these parameters could be determined at the operator's convenience, provided that the tubing has been removed from the well (e.g., during a workover).

SFE No. 3 was killed with 2% KCl water and the hole was kept full even though a considerable amount of fluid was lost to the fractured interval (about 3,000 bbl). This process in itself was thought to be good for the survey because the fluid repressurization probably tended to re-excite microseismic activity. In addition, because the fracture was propped open, and thus not allowed to equilibrate back to its normal state, microseismic signals should still have been fairly strong. Indeed, excellent signals were recorded during the survey.

Analysis of the microseismic data resulted in two different azimuth predictions: north 40° east and north 61° east. Although the overall directional trend is similar to previous surveys, the wide range of the data is not understood fully at this time. It could be that the seismically disturbed region around a propped fracture (that has been repressurized) is much wider than previously observed from minifractures. Conceivably, more than one fracture could exist (at least, leaving the wellbore) or a wide zone of fractures may have been created. Considering the volume of fluid pumped and the corresponding fracture lengths determined from the production analysis, the idea of multiple fractures is not likely.

The *H/Z* analysis on the post-main-fracture data indicates that the top and bottom of the fracture were at 9,125 ± 25 and 9,375 ± 25 ft, respectively. These height estimates are similar to those obtained from 3D modeling of the fracture treatment. In addition, comparison of the main fracture microseismic analysis with the results from Minifracture No. 1 reveals that the created dimensions of Minifracture No. 1 were only slightly larger than those of the main hydraulic fracture. This makes sense because the fracture height resulting from Minifracture No. 1 should have been the created fluid height, while the dimensions from the main fracture treatment should reflect only the propped height.

As a proof-of-concept experiment, a number of casedhole logs and tracers were run to measure or to infer fracture height near the wellbore. Different logging measurements were conducted: after cementing but before perforating, before Minifracture No. 1, after the last stage of Minifracture No. 1, and after the main fracture treatment. The logs run included the cement-bond log/variable density display (CBL/VDL), the natural gamma ray (GR) log, the cement evaluation tool (CET), and Tracer Scan™. The microseismic tools also were run, as previously described. All these logs were compared to the stress-profile log during the attempt to determine fracture height.

Fig. 3.15 displays estimated top and bottom interpretations after the main fracture treatment. These measurements were made after 4 months of production. Accordingly, we believe that the depths shown in Fig. 3.15 correspond to those for a propped fracture.

The bottom of the fracture interpreted from the conventional logs (including one estimate from the CET log) is at 9,330 ft, the bottom of the perforations. The other possible CET log interpretation agrees with that of the microseismic height log, which is at about 9,375 ft. There is a relatively higher stress value of 7,800 psi at 9,370 ft. The 1,900-psi stress change at this depth suggests that it could be an effective barrier to proppant.

Again, most of the logs predicted the top of the fracture to be much higher than the upper stress barrier, except for the CBL, which predicted it was at the top of the perforations, and the tracer log, which predicted it was at 9,182 ft (within the upper barrier). The microseismic fracture top at 9,125 ft occurs at a high stress peak in the upper barrier.

In general, the conventional casedhole logs (e.g., CBL, VDL, GR, CET, and tracer) were not consistent in estimating fracture height in this well. The microseismic method was more consistent with the known stress profile and was in harmony with the 3D model predictions. Because the microseismic method has an advantage in that it penetrates the formation at dimensions of tens to a few hundred feet, we believe it gives more-realistic estimates of fracture height.

3.11 Conclusions

GRI's third SFE was successfully completed. An extensive amount of data was collected and analyzed by various organizations and researchers. Detailed results of this work can be obtained from contractors listed in Table 3.1. The following conclusions summarize some of the most important results.

- Isolation of the Lower Cotton Valley Taylor section by the marine Bossier Shale below and a carbonaceous shale above provides for good barriers to vertical fracture height growth.
- Natural and coring-induced fractures in the Cotton Valley appear to be oriented in the same direction as hydraulic fractures. Thus, oriented-core analyses of these features can provide an indication of fracture azimuth.
- A long propped fracture (> 1,000 ft) appears to have been achieved in the Lower Cotton Valley Taylor section in SFE No. 3.
- The created fracture height (about 300 ft) appears to be contained primarily within the Taylor interval.
- The technique for estimating fracture height from microseismic data (*H/Z* method) appears to be accurate and capable of measuring the true created fracture height. Seismic signals that are tens or even hundreds of feet from the wellbore can be detected.
- In general, conventional well-test analysis and casedhole logging techniques did not agree with the results obtained by applying sophisticated diagnostic technologies and numerical modeling techniques. The conventional analyses were less consistent and, in some instances, provided results that were unrealistic. Application of new technology in this research program has resulted in a consistent analysis of the fracture treatment data. The dimensions for fracture length and height are approximately the same for the different methods.

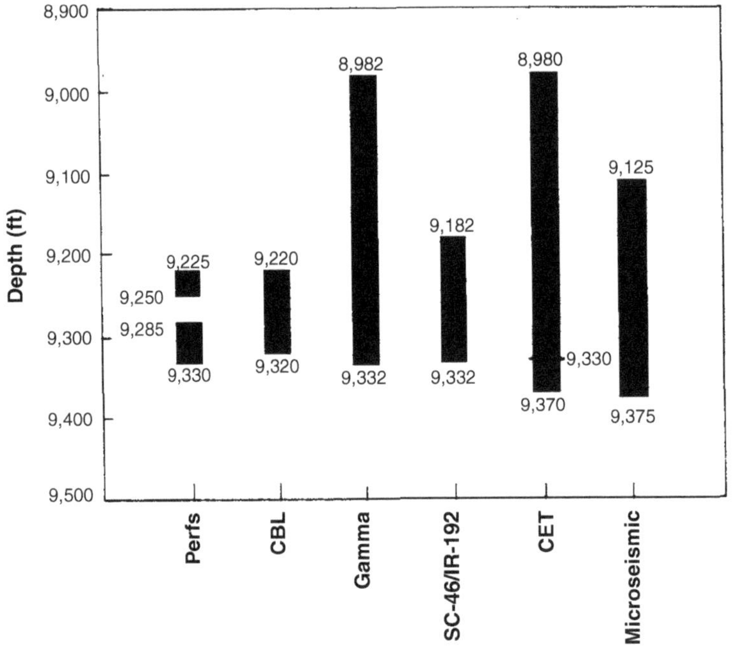

Fig. 3.15—Summary of fracture-height measurements after main treatment.

3.12 Nomenclature

F_a = empirical stress factor, psi/ft
g_p = pore pressure gradient, psi/ft
h = net-pay, ft
H = horizontal component of motion
k = permeability, md
k_f = fracture permeability, md
k_g = gas permeability, md
k_w = water permeability, md
k_∞ = Klinkenberg corrected permeability, md
P_c = capillary pressure, psi
p_{ext} = extrapolated reservoir pressure
p_i = initial reservoir pressure, psi
S_g = gas saturation, fraction
S_w = water saturation, fraction
w_f = fracture width, ft
Z = vertical component of motion
v = Poisson's ratio, dimensionless
σ_v = vertical stress gradient, psi/ft
a_x = horizontal stress gradient, psi/ft
$\Delta\sigma$ = stress difference
ϕ = porosity, fraction

3.13 Acknowledgments

We gratefully acknowledge the support of the GRI, who sponsored this work under GRI Contract No. 5083-211-1734. We also appreciate the efforts and dedication of the organizations listed in Table 3.1 whose results were the basis for this paper. GRI acknowledges Mobil Oil Co., Amoco Production Co., and Schlumberger for the cooperative research efforts.

3.14 References

CER Corp. 1990. Summary of Open-Hole Data Acquisition and Field Operations, Mobil E&P USA, Cargill Unit No. 15, Waskom Field, Harrison County, Texas. Report prepared under GRI Contract No. 5088-211-1682, CER Corp., Las Vegas, Nevada (August 1990).

CER Corp. and S. A. Holditch & Associates Inc. 1988. Advancements in Travis Peak Formation Evaluation and Hydraulic Fracture Technology, Staged Field Experiment No. 1, Waskom Field, Harrison County, Texas. Report No. GRI-88/0077 prepared for GRI under Contract No. 5086-211-1306.

CER Corp. and S. A. Holditch & Associates Inc. 1990. Staged Field Experiment No. 2: Application of Advanced Geological, Petrophysical and Engineering Technologies to Evaluate and Improve Gas Recovery From Low Permeability Sandstone Reservoirs. Report Nos. GRI-89/0140 and GRI-90/0093 prepared under Contract No. 5088-211-1682, Gas Research Institute (Merged into Gas Technology Institute in 2000) Des Plaines, Illinois (1990).

CER Corp. and S. A. Holditch & Associates Inc. 1991. Staged Field Experiment No. 3, Application of Advanced Technologies in Tight Sandstones—Travis Peak and Cotton Valley Formations, Waskom Field, Harrison County, Texas. Report No. GRI-91/0048 prepared under Contract No. 5090-211-1940, Gas Research Institute (Merged into Gas Technology Institute in 2000), Des Plaines, Illinois.

Ely, J. W. and Holditch, S. A. 1988. Field Experience with the GRI Rheology Unit. Presented at the Gas Technology Symposium, Dallas, 13–15 June. SPE-17715-MS. https://doi.org/10.2118/17715-MS.

Fix, J. E. et al. 1988. Development of Microseismic Methods to Determine Hydraulic Dimensions. Annual report prepared under GRI Contract No. 89/0116, Teledyne Geotech, Garland, Texas (November 1988).

Holditch, S. A., Robinson, B. M., Whitehead, W. S. et al. 1988. The GRI Staged Field Experiment. *SPE Form Eval* **3** (3): 519–33. SPE-16429-PA.https://doi.org/10.2118/16429-PA.

Kosters, E. C., Bebout, D. G., Seni, S. J. et al. 1989. *Atlas of Major Texas Gas Reservoirs*. Austin, Texas: Bureau of Economic Geologic Geology, The University of Texas at Austin.

Mahrer, K. D. 1993. Microseismic Logging: A New Hydraulic Fracture Diagnostic Method. SPE Form Eval **8** (1). SPE-21834-PA. https://doi.org/10.2118/21834-PA.

ResTech Inc. 1988. Development of Petrophysical Techniques for Evaluating Tight Gas Sands. Final report prepared under GRI Contract No. 5084-2111062, ResTech Inc., Houston.

Robinson, B. M., Holditch, S. A. and Peterson, R. E. 1991. The Gas Research Institute's Second Staged Field Experiment: A Study of Hydraulic Fracturing. Presented at the SPE Gas Technology Symposium, Houston, 22–24 January. SPE-21495-MS. https://doi.org/10.2118/21495-MS.

Western Atlas International Core Laboratories. 1989a. Core Analysis Report for Mobil Exploration & Production Company, Cargill Unit No. 15, SFE No. 3, Waskom Field, Harrison County, Texas. Report, Western Atlas International. Core Laboratories, Dallas (April 1989).

Western Atlas International Core Laboratories. 1989b. Core Analysis Report for Mobil Exploration & Production Company, Cargill Unit No. 15, SFE No. 3, Waskom Field, Harrison County, Texas. Report, Western Atlas International. Core Laboratories, Dallas (23 June 1989).

Western Atlas International Core Laboratories. 1989c. Special Core Analysis Study on Samples from the Travis Peak and Cotton Valley Formations, Mobil E&P Cargill Unit #15. SFE report, Western Atlas Intl. Core Laboratories, Dallas (20 October 1989).

Whitehead, W. S., Hunt, E. R., Finley, R. J. et al. 1986. In-Situ Stresses: A Comparison Between Log-Derived Values and Actual Field-Measured Values in the Travis Peak Formation of East Texas. Presented at the SPE Unconventional Gas Technology Symposium, Louisville, Kentucky, USA, 18–21 May. SPE-15209-MS. https://doi.org/10.2118/15209-MS.

SI Metric Conversion Factors

bbl × 1.589 873	E–01 = m^3
ft × 3.048*	E–01 = m
ft^2 × 9.290 304*	E–02 = m^2
ft^3 × 2.831 685	E–02 = m^3
gal × 3.785 412	E–03 = m^3
in. × 2.54*	E+00 = cm
lbm × 4.535 924	E–01 = kg
md × 9.869 233	E–04 = μm^2
psi × 6.894 757	E+00 = kPa

*Conversion factor is exact.

Chapter 4

Hydraulic Fracturing Research in the Frontier Formation Through the Gas Research Institute's Fourth Staged Field Experiment

B. F. Saunders and **B. M. Robinson,** S. A. Holditch & Associates Inc.;
S. A. Holditch, Texas A&M University; and **R. E. Peterson,** CER Corp.

SPE 24854-MS presents a case history of Gas Research Institute's (GRI's) SFE No. 4 well, which was drilled and completed in the Frontier Formation in southwestern Wyoming. The objective of the case history is to determine if technology developed in the previous three SFE wells in East Texas can be applied in other geological basins. Data collected include regional and field-scale stratigraphic and petrologic studies, whole cores, openhole logs, in-situ stress measurements, production and pressure-transient tests, fracture treatments, and fracture diagnostic measurements. These data were analyzed using the most-advanced formation evaluation techniques, transient-pressure-analysis techniques, hydraulic fracture modeling and reservoir modeling to fully characterize the formation and hydraulic fracture.

4.1 Introduction

The GRI—now Gas Technology Institute—sponsored a field-based research program through the staged field experiment (SFE) project for several years. SFE wells were drilled specifically for conducting research in the analysis and stimulation of tight sands and for validating technologies under development in other GRI projects.

The majority of this work was spent in the Travis Peak and Cotton Valley Formations in East Texas where the first three SFE wells were drilled. Results of this work were published in GRI reports (CER Corp. and S. A. Holditch & Associates 1988) and Society of Petroleum Engineers papers (Holditch et al. 1988; Robinson et al. 1991, 1992).

This paper summarizes the results from research conducted on the fourth SFE well (**Table 4.1**). A primary objective was to determine if technology developed in previous SFE wells in East Texas could be successfully transferred to other geologic basins.

To fully describe each of the layers of rock on SFE wells, data were collected from whole cores, openhole geophysical logs, in-situ stress measurements, production and pressure transient tests, fracture injection treatments, and fracture diagnostic measurements. These data were then analyzed in detail with advanced formation evaluation techniques, including numerical models, in an effort to obtain the best possible description of the reservoir and hydraulic fracture. On SFE No. 4, there were several new technologies and analysis methods that were evaluated. Specifically, these techniques included in-situ stress profiling, fracture height and azimuth analysis, hydraulic fracture analysis, and post-fracture performance analysis.

Knowledge and techniques developed through GRI research programs were designed to benefit the natural gas industry. A better understanding and more accurate analysis of the hydraulic fracturing process will lead to improved treatment designs and additional gas production from tight sands such as the Frontier. A GRI program goal was to apply rigorous analysis techniques to evaluate hydraulic fracture treatments in any geologic basin. The SFE No. 4 in southwestern Wyoming illustrates that rigorous analysis techniques can be used to analyze and better understand fracture geometries in complex reservoirs that typify tight gas formations.

In-situ stress tests conducted on SFE No. 4 indicated abnormally high stress levels in this wellbore. In-situ measurements had been used in previous SFE wells to calibrate the stress profile derived from openhole logs. However, the stress profile generated assuming purely elastic behavior predicted values that were much lower than the actual in-situ measurements. To successfully replicate the in-situ stress values, it was necessary to incorporate a shaliness term in the elastic equation. Detailed analysis of minifracture data using

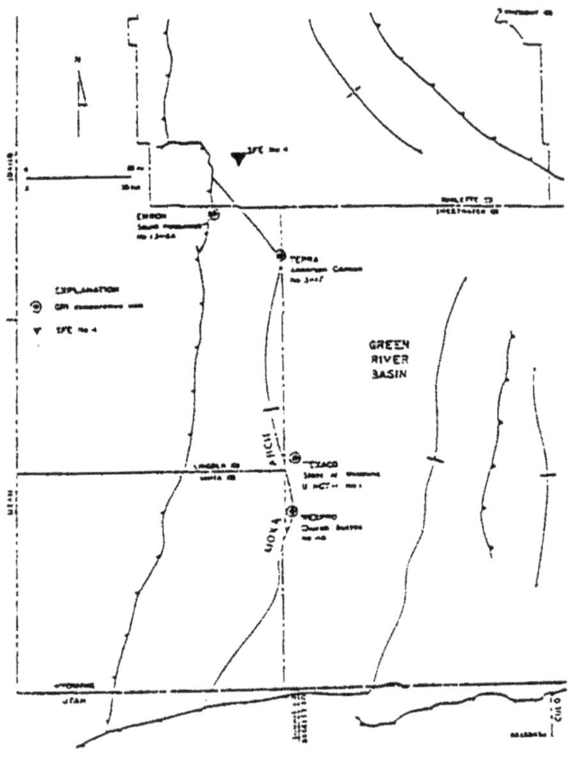

Fig. 4.1—Location map of SFE No. 4 and GRI coop wells in southwestern Wyoming.

3D fracture models indicated a complicated fracture geometry probably exists in conjunction with the abnormally high stress field. Analysis of seismic data and 3D fracture modeling also indicated there is very little fracture height containment. This is contrary to current thinking in the area that fractures are contained in the second Frontier. Because of poor reservoir quality, a proppant-laden fracture treatment was not pumped on SFE No. 4. Thus, postfracture reservoir evaluation was not performed. However, altogether, there was an extensive amount of research performed on SFE No. 4. This paper summarizes that work.

4.2 Geologic Overview

The Bureau of Economic Geology (BEG) performed regional and field-scale stratigraphic and petrologic studies on the Frontier Formation incorporating core and log data from the SFE No. 4 (Dutton et al. 1992). From these studies a better understanding of the physical characteristics of these tight reservoirs was obtained. This information was essential to understanding the reservoir performance and fracture stimulation response seen during testing operations.

As shown in **Fig. 4.1,** SFE No. 4 is located in Sweetwater County, Wyoming. It was drilled within the Chimney Butte field on the northwest flank of the LaBarge Platform on the periphery of established Frontier gas-producing areas. In this area, the Frontier is 6,500 to 8,000 ft deep and dips gently to the northeast. The Thrust Belt, which is a major structural province composed of folded and faulted strata, lies 5 miles west of SFE No. 4, but is no longer tectonically active. More-recent tectonic movements predominantly control current stress directions in the basin.

4.3 Stratigraphic and Depositional Analysis

From regional studies conducted by the BEG and others, it has been found that the Frontier Formation in the LaBarge Platform area contains marine shoreline sandstones and nonmarine, fluvial channel-fill sandstones enclosed in thick, regionally extensive marine shales. The second Frontier member of the Frontier Formation contains most of the prolific Frontier gas reservoirs. The second Frontier sandstone is made up of several sandstone "benches" that are interbedded with shale. The first, fourth, and fifth benches are depositionally similar: laterally discontinuous, fluvial channel-fill sandstones. These fluvial benches also contain bentonite shales, which may contribute to the high in-situ stresses measured during injection testing on SFE No. 4. The second bench has a marine shoreline facies and has widespread continuity. The third bench is also a marine shoreline sandstone that merges with the second bench in the western part of the La Barge Platform.

Analysis of geologic data on SFE No. 4 indicated that second Frontier reservoirs are poorly developed at this wellsite. The first, fourth, and fifth benches include several thin, channel-fill and channel-flank sandstones, but are predominantly interchannel sandy shale. The lateral discontinuity of these formations is illustrated by an offset well 0.5 miles southwest of SFE No. 4 that encountered thick channel-fill sandstones in the fourth and fifth benches. The second bench marine sandstone is 46 ft thick (gross thickness) in SFE No. 4, but net-pay determined from logs amounts to only 10 ft. This is partially because of an atypically study upper member of the second bench.

The marine shoreline sandstones can be further identified as either upper or lower shoreface facies. The lower shoreface facies is characterized by pore-filling, detrital clay matrix. The upper shoreface sandstone is deposited in a relatively clay-free environment and is a more favorable target. As stated previously, Frontier sandstones are poorly developed at SFE No. 4. Petrographic analysis indicates the upper shoreface facies, which once had the highest initial porosity, now has only 1.5 percent primary porosity (thin-section count). Most of the porosity was lost through compaction and cementation during burial. A study of wells in the Frontier indicates that the best permeability in the Frontier along the Moxa Arch occurs where primary porosity is greater than 3 percent. Therefore, SFE No. 4 would be expected to have low in-situ permeability and flow capacity.

Organization	Research Area
Bureau of Economic Geology UT at Austin (BEG)	Site Selection, Geology, Core Analysis
CER Corporation (CER)	Data Acquisition — Coring and Logging Project Administration and Documentation
Resources Engineering Systems (RES)	Hydraulic Fracture Treatment Design and Analysis
ResTech Houston (RTH)	Formation Evaluation, In-Situ Stress Profiling
S. A. Holditch & Associates Inc. (SAH)	Drilling and Completion, Reservoir Engineering Lead Technical Contractor
Teledyne Geotech (TG)	Microseismic Analysis For Fracture Dimensions and Azimuth
University of Texas at Austin/Schlumberger	Core/Log Analysis Research

Table 4.1—Participating GRI contractors.

Array Sonic (STC Processing)/GR	Deep Propagation/GR
Long Spaced Sonic/GR/Caliper	Formation MicroScanner/GR
Compensated Neutron	MicroSFL/GR
Litho Density	Geochemical
Dual Induction/GR/SP	High Resolution Induction
Phasor Induction/GR/SP	Borehole Televiewer/GR
Dual Porosity Neutron/GR/Caliper	XRH High Resolution Deep Resistivity/GR
Vertical Seismic Survey	Full Waveform Sonic/GR
Repeat Formation Tester	Full Wave Shear Sonic/GR
Electromagnetic Propagation	Full Wave Extra Long Spacing Sonic
Micro Log	Thermal Decal
Natural GR/Caliper	Ultra Slow Drawdown Formation Tester/GR

Table 4.2—Summary of final openhole logging operations.

4.4 Analysis of Natural and Induced Fractures

Natural fractures and fractures induced from coring operations were observed in SFE No. 4 core. These fractures were studied to determine their influence on reservoir quality and for fracture treatment design and analysis. Natural fractures include three mineralized extension fractures and 22 gently to moderately dipping striated fractures. The striated fractures are primarily features of compaction and dewatering in mudstones. Extension fracture widths in cores are generally less than 0.5 mm. Fracture heights measured in the core were generally less than 15 cm, but the fractures commonly ended outside of the core so true dimensions cannot be determined.

The three extension fractures were found in sandstone with strikes of north, northeast, and northwest. Outcrop observations indicate that fracture strike dispersion varies considerably in the Frontier. This, in addition to the small number of fractures in the core, makes determination of local fault strike orientation statistically unreliable.

The presence of natural fractures will influence both reservoir analysis and fracture treatment design. The potential effects include increased fracture fluid leakoff during injection and the creation of multiple sets of fractures that could result in abnormal or high treating pressures. This effect would be most prominent if multiple fractures caused inefficient communication between the wellbore and the fracture further away from the wellbore.

Analysis of minifracture injections indicate complex fracture geometries might have been created during the treatments. Natural fractures might have affected the initiation of these hydraulic fractures. Extremely high treating pressures were observed during the minifracture injections. These high pressures are possibly the result of multiple, poorly connected fractures initiated from the wellbore. There are indications from 3D fracture modeling that these fractures never consolidated into a single hydraulic fracture.

4.5 Drilling Operations

SFE No. 4 was spud on 10 May 1990. A 17 1/2 in.-hole was drilled from surface to 976 ft and 13 3/8 inch casing was set. A 12 1/2-in. hole was drilled from the 13 3/8-in. casing shoe to 7,116 ft. Openhole logs were run, whole core was taken, and an openhole in-situ stress was performed before running intermediate 9 5/8-in. casing. An 8 1/2-in. hole was drilled from the 9 5/8-in. casing shoe to a depth of 8,087 ft. Whole core was also recovered over selected intervals. Drilling operations were completed on 23 June 1990. After an extensive openhole log program (**Table 4.2**), 7-in. casing was run from surface to total depth and cemented in place. A wellbore diagram is shown in **Fig. 4.2.**

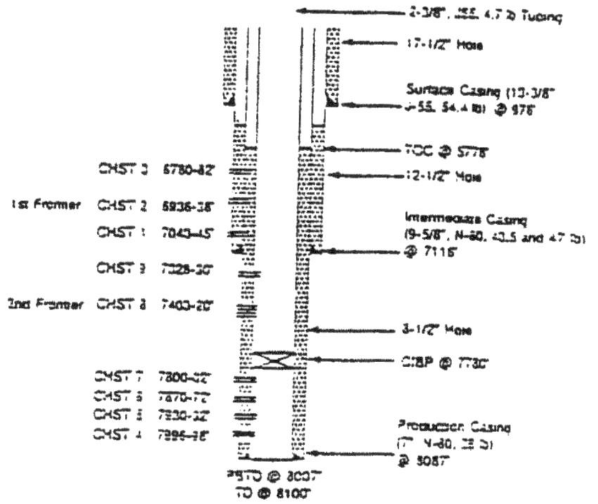

Fig. 4.2—SFE No. 4 wellbore diagram.

The Hilliard Shale above the first Frontier from 6,777 to 6,805 ft was cored with a 4-in. barrel and the top of the first Frontier was encountered at 6,906 ft. Three casedhole stress tests were attempted in the 9 5/8-in. casing at 6,780 to 6,782 ft, 6,936 to 6,938 ft, and 7,043 to 7,045 ft. The 6,936 to 6,938-ft test interval is in the first Frontier, but it could not be broken down at the 5,000 psi maximum surface pressure limitation. The other two intervals were in the shales above and below the first Frontier.

An additional 303 ft of 4-in. core was recovered over five main intervals within the Frontier: 7,226 to 7,240 ft, 7,311 to 7,494 ft, 7,607 to 7,651 ft, 7,753 to 7,783 ft, 7,963 to 8,004 ft. A total of 178.2 ft of this core was oriented. The cored intervals are illustrated in **Fig. 4.3.** Six additional casedhole stress tests (CHSTs) were also performed in the Frontier at 7,329 to 7,330 ft, 7,410 to 7,412 ft, 7,800 to 7,802 ft, 7,870 to 7,872 ft, 7,930 to 7,932 ft, and 7,996 to 7,998 ft. The CHST at the deepest interval was also aborted because it could not be broken down under the 5,000 psi maximum pressure limitation.

As part of the testing program, perforations for two sets of the CHSTs were oriented approximately orthogonal to investigate the relationship between perforation orientation and measured in-situ stress. Perforations at 7,800 to 7,802 ft and 7,930 to 7,932 ft were oriented 9° and 19°, respectively, west of north. At 7,870 to 7,872 ft and 7,996 to 7,998 ft they were oriented 81° and 71° east of north, respectively.

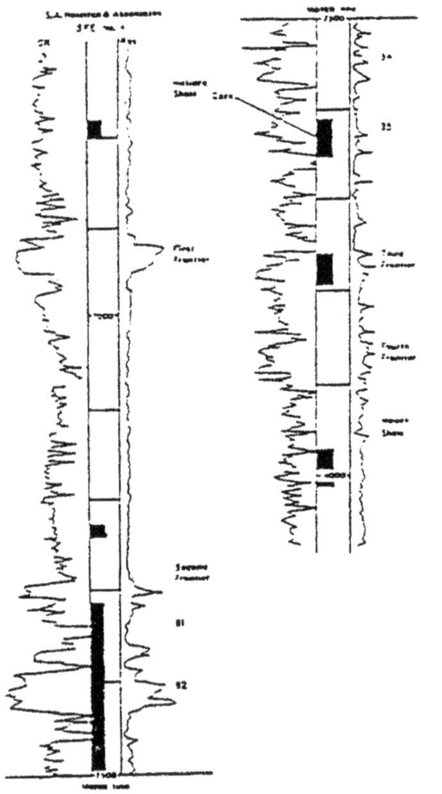

Fig. 4.3—SFE No. 4 GR/induction log showing Frontier Formation and cored intervals.

The perforations within each interval were phased at 180° and had a shot density of 4 per ft. This arrangement resulted in the two sets oriented approximately 90° to each other. Perforations in the other CHST intervals had shots phased at 120 degrees and a shot density of four per ft. Other than the high stresses encountered, there was not any anomalous behavior that could be attributed to orientation.

Three openhole and five casedhole logging operations were carried out in this wellbore at various depths. Data obtained from these logging operations were used to develop permeability-thickness (*kh*) profiles, porosity-thickness (*øh*) profiles, in-situ stress (σ_x) profiles, and formation lithology. In addition to acquiring base data to describe reservoir properties, special experimental tools were run by Halliburton Logging Services, Magnetic Pulse Inc., Schlumberger, and Western Atlas.

4.6 Formation Evaluation

The cores and openhole logs were analyzed to determine basic reservoir properties such as lithology, porosity, and water saturation. In addition, specialized work was also performed to improve log accuracy and evaluate mechanical properties and reservoir flow characteristics. The next two sections summarize these results.

4.7 Core Analysis

Routine core analyses were performed on 316 core samples taken from 323 ft of conventional core to measure porosity, permeability, grain density, and water saturation. Detailed results of these tests were published by Western Atlas International Core Laboratories (1991). In general, almost all of the permeabilities (k_∞) measured at 800 psi confining pressure on samples cut from second bench core were very low, typically less than 0.1 md. At net overburden (NOB) pressure, only two samples measured over 0.1 md. However, measured porosities averaged 10.3%. The NOB gradient was calculated at 0.34 psi/ft.

Special core analysis tests included relative permeability, capillary pressure, and mechanical properties. Thirty samples were selected from 23 depths for mechanical property evaluation and 15 samples from five depths were chosen for relative permeability, capillary pressure, and electrical property testing. The mechanical properties testing can be further categorized as either static or dynamic measurements. Results of these analyses were previously reported; therefore, details will not be discussed here.

GRI has sponsored research for several years with the objective of calculating in-situ gas permeability from openhole logs. The result of this research has been the development of a methodology relating porosity and water saturation to gas permeability. This method, which has been previously published (Western Atlas International Core Laboratories 1989; ResTech Houston 1988), develops correlations to adjust Klinkenberg corrected permeability measured at NOB conditions to in-situ conditions. The first correlation converts Klinkenberg corrected permeability measured at NOB to brine-saturated permeability. The second correlation uses relative permeability data to calculate effective permeability to gas. These correlations are applied to log derived porosity and water saturation values to predict permeability for specific intervals.

4.8 Log Analysis

Drilling SFE No. 4 in a different geologic basin provided an excellent opportunity to evaluate petrophysical methodologies developed for tight gas sands (TGSs) in East Texas. A new petrophysical model (GRI/TGS mineral model) was also developed using log data from the previous SFE wells and then applied to SFE No. 4. Thin-section analysis indicated a large number of different clay size minerals present in the Frontier, which had a dramatic impact on log response. The GRI/TGS mineral model uses 10 elemental relationships to derive the mineral composition of the rock. The accuracy of this model is based on the use of pure mineral endpoints to calculate composite matrix properties such as grain density or cation exchange capacity.

The GRI/TGS mineral model was used to estimate reservoir properties in SFE No. 4. After processing log data to obtain accurate mineralogy, log-derived porosity, permeability, and gas saturation were found to be similar to core measurements. These results are summarized in **Table 4.3.** While there is good agreement between core and log-derived properties, they both give an estimate of *kh* that is approximately a magnitude too large when compared to the prestimulation pressure-transient-analysis results.

Parameter	GRI/TGS Model	Core
Thickness, ft	7.5	7.5
Average porosity		
Total porosity	0.131	0.117
Effective porosity	0.103	0.117
Average water saturation	0.512	0.527
Permeability thickness, md-ft	0.055	0.49

Table 4.3—Summary of reservoir properties in the second bench of the second Frontier.

4.9 Vertical Stress Profiling

In-situ stress tests were performed exclusively in the casedhole on SFE No. 4. The methodology used to perform and analyze the pump-in/falloff/flowback tests is described (Whitehead et al. 1986) and used in previous SFE well analyses; results of these tests were then used to calibrate acoustic data acquired during openhole logging operations to develop a vertical stress profile (CER Corp. and S. A. Holditch & Associates Inc. 1990, 1991).

In-situ stress tests were performed on nine different intervals in various lithologies in SFE No. 4. Fig. 4.2 is a wellbore schematic showing the relative location of these tests. As stated previously, the first three tests were conducted in the 9 5/8-in. casing in the shale above the first Frontier, in the first Frontier Sand, and in the shale below the first Frontier. Data could not be obtained in the first Frontier perforations because they could not be broken down under the maximum surface pressure limitations. The remaining stress tests were conducted in the second Frontier and in the Mowry Shale below the second Frontier. Data could not be obtained from the lowermost shale perforations (7,996 to 7,998 ft) because the formation could not be broken down.

After the data were analyzed, it was apparent that the correlation developed for previous SFE wells (CER Corp. and S. A. Holditch & Associates Inc. 1990, 1991) could not be directly applied to the Frontier Formation; the calculated stress values were much too low. The in-situ measurements in both sands and shale were found to correlate with bulk compressibility according to the following relationship.

$$S_h = \left(\frac{v}{1-v}\right)\left(S_V - S_p\right) + S_p\left(1+\beta\right), \qquad (4.1)$$

where S_h = minimum horizontal stress gradient, psi/ft; S_p = stress gradient produced by pore pressure, psi/ft; v = Poisson's ratio from acoustic logs; and β = unitless shaliness or "softness" factor.

The beta factor is computed by

$$\beta = \frac{\left(C_b - C_{b\,\min}\right)}{\left(C_{b\,\max} - C_{b\,\min}\right)}, \qquad (4.2)$$

where C_b = bulk compressibility computed from sonic density logs, Mpsi^{-1}; $C_{b\max}$ = maximum bulk compressibility expected for a porous sandstone, Mpsi^{-1}; and $C_{b\min}$ = minimum bulk compressibility expected for sandstone with zero porosity, Mpsi^{-1}.

Fig. 4.4 is a graph of the log-derived stress gradient, computed using the bulk compressibility correlation, compared to the stress gradient from stress test analyses. A line has been drawn from the formation pressure gradient of 0.486 psi/ft to the overburden gradient of 1.156 psi/ft. It shows good agreement with the measured in-situ stresses. The resulting stress profile using that relationship is shown in **Fig. 4.5.** The results of this work again emphasize that for accurate results, log data must be calibrated with in-situ stress tests. Re-evaluation of stress profiles was included on all SFE wells using this bulk compressibility relationship.

4.10 Prefracture Testing

After the stress testing was completed in August 1990, the second bench of the second Frontier (7,403 to 7,420 ft) was perforated with 4 shots per ft for production testing, A breakdown treatment with 2% KCl water was attempted, but a stable injection rate could not be established under 5,000 psi surface pressure limitations. At that point, 700 gallons of 7 1/2% HCl acid was spotted at the perforations and displaced into the formation at 3.5 bpm and 4,600 psig. Subsequent injection attempts with 2% KCl water did not indicate improvement in injectivity. Swab testing did not indicate any show of gas.

The tubing string was then changed out to allow higher treating pressures. Formation breakdown was achieved and injection was established down 2 7/8-in. tubing at 5 bpm and 4,600 psig. After a perforation bailout treatment, the well was swabbed for cleanup and eventually produced at a rate of 9 Mscf/D. Flow testing continued for approximately one month at this rate when the well was shut-in for a pressure buildup test. Bottomhole pressure gauges and a downhole shutoff tool were used in the test. The pressure buildup data were analyzed to estimate permeability and skin factor. It should be noted that the low flow rate and flowing pressure preceding the buildup period led to difficulties in determining a unique value for permeability and skin factor.

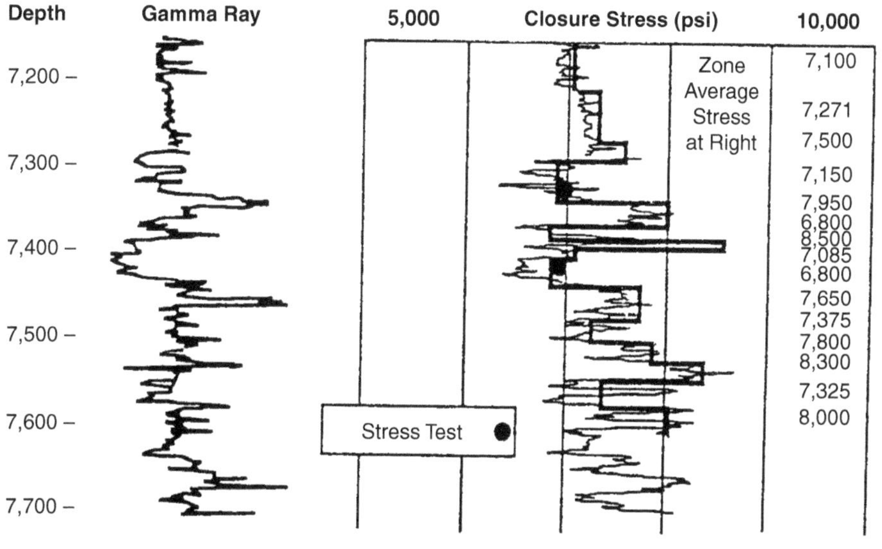

Fig. 4.4—Comparison or measured in-situ stresses to calibrated log values.

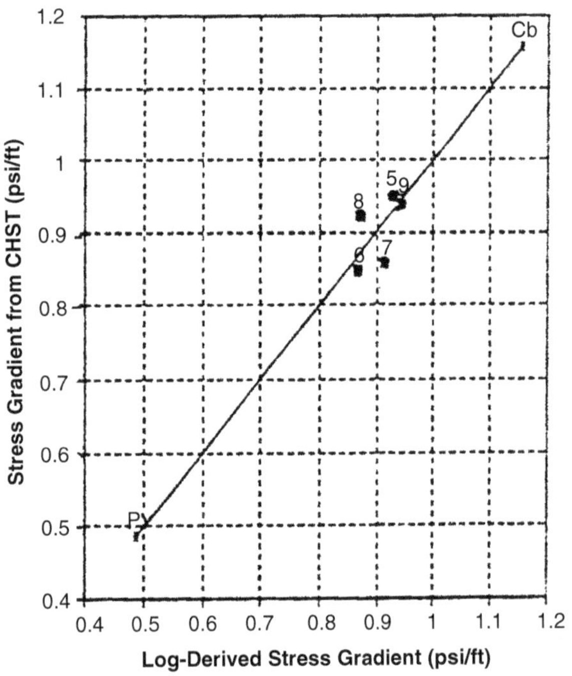

Fig. 4.5—Calibrated vertical stress profile for SFE No. 4.

However, a magnitude and range for these values can be determined using conventional and deconvolution pressure buildup techniques. Results of these analyses are shown below.

Permeability-thickness product:	0.054 md-ft
Gas permeability:	0.007 md
Net-pay thickness:	7.5 ft
Skin factor:	–0.3
Extrapolated reservoir pressure (p^*):	3,628 psia

The net-pay value is based on the GRI/TGS mineral model determination.

4.11 Minifracture Injection Testing

Minifracture injections were performed on each of the SFE wells. These treatments were pumped to gather microseismic data that can be used to determine hydraulic fracture azimuth and created fracture height and obtain calibration data for 3D fracture models. Typically, these data can be obtained from two or three injection treatments. However, on SFE No. 4 injection pressures were unusually high and the initial injections were unsuccessful because of surface pressure limitations. Thus, it was necessary to replace the original wellhead equipment with equipment rated for higher pressures. Once in place, a comprehensive diagnostic testing program was commenced to evaluate the source of the high injection pressures.

The initial phase of the diagnostic testing program was designed to characterize the magnitude and type of pressure behavior through the pumping of multirate treatments with both low and high viscosity fluids. A second phase was designed to further isolate and evaluate the high injection pressure and provide data for 3D-fracture model calibration. Some of the experiments required multiple injections to properly characterize the resultant pressure behavior. Fourteen separate injections were performed after completion of the prefracture testing. **Table 4.5** summarizes all of the treatments that were pumped.

All 14 data sets were evaluated to gain an understanding of the high treating pressures and possible fracture geometry. Four of these injections were selected for in-depth analysis based on three criteria: bottomhole pressures were measured by electronic pressure gauge; treatments consisted of relatively large injected volumes; and crosslinked gel was pumped during some part of the tests (except one, in which only 2% KCl water was pumped). Injection treatments performed on 16 May, 5 June, 5 August, and 10 August were selected for detailed analysis. Results of the testing program generated data that did not readily conform to conventional fracturing theory. Slightly different explanations for this behavior were proposed by the three GRI contractors (NSI Technologies, Resources Engineering Systems, and S. A. Holditch & Associates Inc.) that were responsible for evaluating the data. Other industry experts were also consulted. Analyses of all three of the GRI contractors were plausible and summaries of their results are presented in this paper. Detailed discussions of their analyses are in a comprehensive report on SFE No. 4.

To analyze the minifracture tests, history matching the injection data with numerical models was employed in addition to conventional analytical techniques. The history matching process compares pressures predicted by a numerical model to the actual pressure response seen during the treatment. This is accomplished by varying different physical properties in the model and simulating the injection treatment. Actual injection pressures were much higher than those predicted by the initial modeling efforts. This would indicate either that the input parameters did not represent actual conditions or the models themselves did not accurately describe processes occurring in-situ. Most likely causes of the high injection pressures include low perforation efficiency, near wellbore tortuosity, high stress layers within the second Frontier, and/or multiple fractures propagating simultaneously.

CHST No.	Interval (ft)	Lithology	Closure Pressure (psia)	Average In-Situ Stress (psi/ft)
1	7,043–7,045	Shale	6,590±100	0.94
2	6,936–6,938	Sandstone	*	*
3	6,780–6,782	Shale	6,510±50	0.96
4	7,996–7,998	Shale	*	*
5	7,930–7,932	Shale	7,570±120	0.95
6	7,870–7,872	Shale	6,690±100	0.85
7	7,800–7,802	Shale	6,670±70	0.86
8	7,410–7,412	Sandstone	6,850±50	0.92
9	7,328–7,330	Shaly Sandstone	6,870±100	0.94

*Unable to break down.

Table 4.4—Summary of casedhole stress-test results.

Date	Description of Treatment	Fluid* Type	Injection Rate (BPM)	Treatment Volume (Barrels)
90/11/14	Teledyne minifracture: 1st attempt	2% KCl	8–3	230
90/11/15	Teledyne minifracture: 2nd attempt after reperforating	2% KCl	3	40
90/5/15	Step-rate injection/falloff tests	2% KCl	0.5–10	155
91/5/16	Step-rate injection/minifracture	2% KCl crosslinked gel	5–22	540
91/6/5	Teledyne minifracture–Wellbore filled with 2% KCl water	2% KCl crosslinked gel	10–20	1010
91/8/5	Step-rate injection test (circulate well with crosslinked gel)	crosslinked gel	0.5–20	600
91/8/7	Step-rate injection test	2% KCl	0.5–20	255
91/8/8	Spinner survey during injection	2% KCl	1–11	120
91/8/9 (am)	Injection and flowback with spinner	2% KCl	1–20	115
91/8/9 (pm)	Multiple injection and flowback with spinner	2% KCl	20	130
91/8/10	Multirate minifracture	2% KCl	5–40	1160
91/8/11	Injected 100-mesh sand	2% KCl	10–20	640
91/8/13 (am)	Teledyne minifracture: test aborted–lubricator malfunction	2% KCl crosslinked gel	20	480
91/8/13 (pm)	Teledyne minifracture	2% KCl crosslinked gel	20	405

*Fluid injected into the formation.

Table 4.5—Treatment summary.

Viscous fluid properties were known with relative certainty because proppant was not pumped. The stress profile had been calibrated using casedhole stress tests; thus, it was believed to be fairly accurate. However, it was not possible to measure the stress in all the layers, and there were several thin bentonite beds surrounding the second Frontier in which the stresses could not be described. Because of the softness of these layers, it is possible that the formations behave plastically, thus exhibiting very high stresses. This was probably the case, based on the analysis, and was part of the reason for the high injection pressure.

NSI analyzed the data with a pseudo3D fracture model using values for reservoir properties determined from previous analyses. The stress profile used in their analysis was very similar to the profile shown in Fig. 4.5; however, they did use peak log responses in their model for some stress layers above and within the second Frontier. The stress profile used during NSI's analysis is illustrated by **Fig. 4.6.**

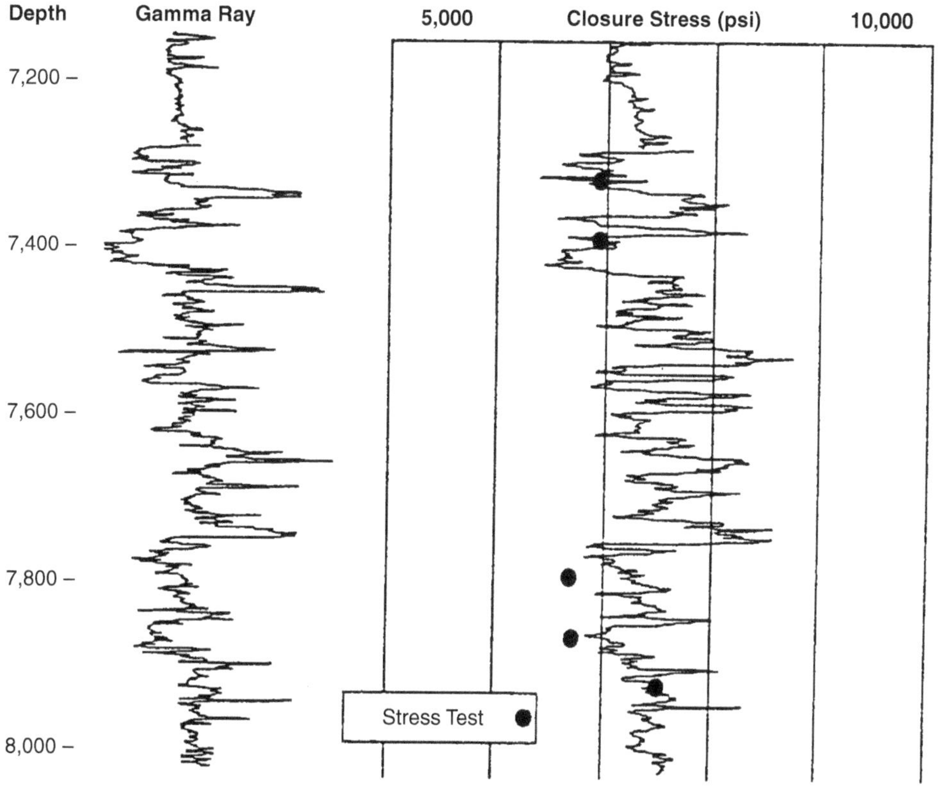

Fig. 4.6—Stress values used by NSI for 3D-fracture modeling.

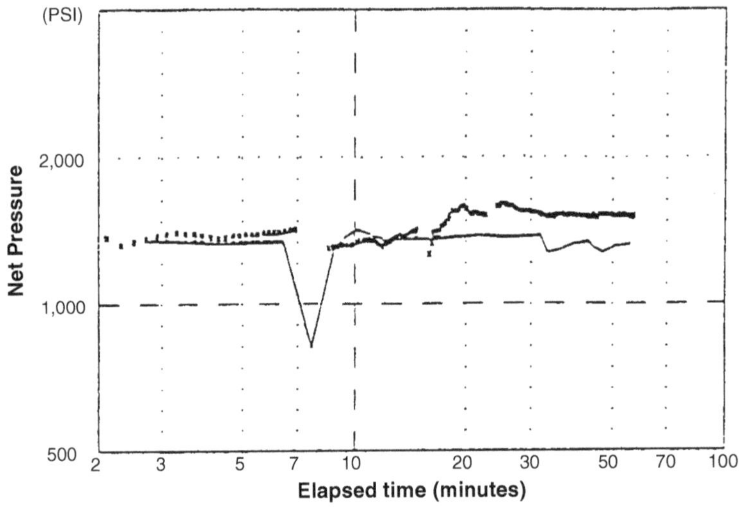

Fig. 4.7—NSI history match of net pressure for injection pumped on 8 June 1991.

Measured bottomhole pressures were corrected to remove the effects of near-wellbore pressure drop. The correction was made by generating correlations for near-wellbore friction and injection rate. This corrected pressure was then used to generate accurate values for history matching true pressure behavior in the fracture. Their final matches simulated the shape of the pressure responses and were at approximately the same level of net pressure as the actual values. **Fig. 4.7** is a graph of the history match for the 5 June injection. NSI's results indicated downward fracture growth was limited to a high stress interval at 7,521 to 7,546 ft, but that upward fracture height growth was essentially unrestrained. Total height was 570 ft with a created length of approximately 300 ft. NSI concluded that matches of the minifracture data obtained with a single planar vertical fracture were close enough to accurately describe fracture growth. Their primary explanation for the high treating pressures were the high stress layers above and below the second Frontier.

Resources Engineering Systems (RES) also emphasized the effects of near-wellbore pressure drop in their analysis. They concluded that this pressure drop appears to be a function of fluid viscosity and injection rate. They pursued the analysis considering two possibilities. One way was to vary the number of fractures that were propagating (simultaneously) to match the observed pressures. The second was to increase stress in the layers above and below the perforated interval, which had the effect of restraining fracture height growth and, consequently, increasing fracturing pressure.

The concept of multiple fractures propagating from a single wellbore is relatively uncommon, but it has become a more widely accepted phenomenon based on mineback experiments that recovered multiple fractures in naturally fractured sandstones. When modeling multiple fractures, it is important to recognize that the actual fracturing process might be much more complicated than the geometry depicted by the model. However, it appears that multiple, tortuous paths were created in SFE No. 4, and that is the primary cause of the high treating pressures.

Fig. 4.8 is RES's match of the pressure data from the injection performed on 5 August using multiple fractures. In comparison, increasing the stress in adjacent layers enough to obtain a match is illustrated by **Fig. 4.9.** The multiple fracture match is superior. Overall, the best matches on all of the injections were obtained using multiple fractures and by allowing near wellbore pressure to vary as a function of injection rate. RES noted instantaneous shut-in pressure (ISIP) data was not significantly affected by injection rate

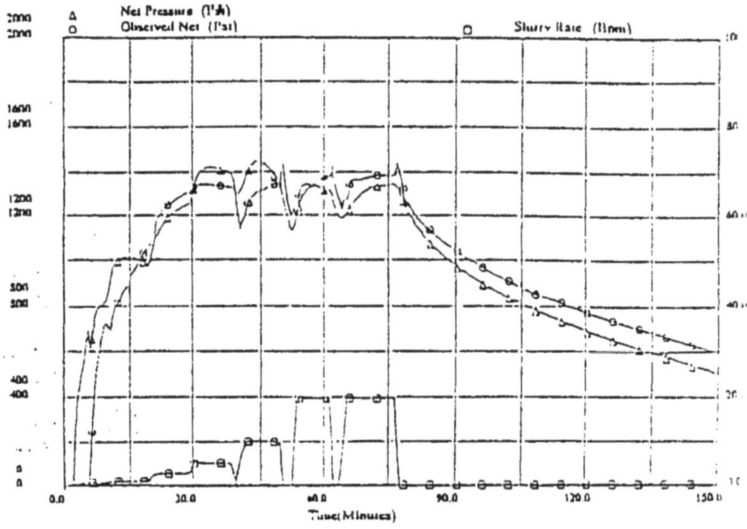

Fig. 4.8—RES history match of sec pressure data using multiple fractures. 8 August 1991 injection.

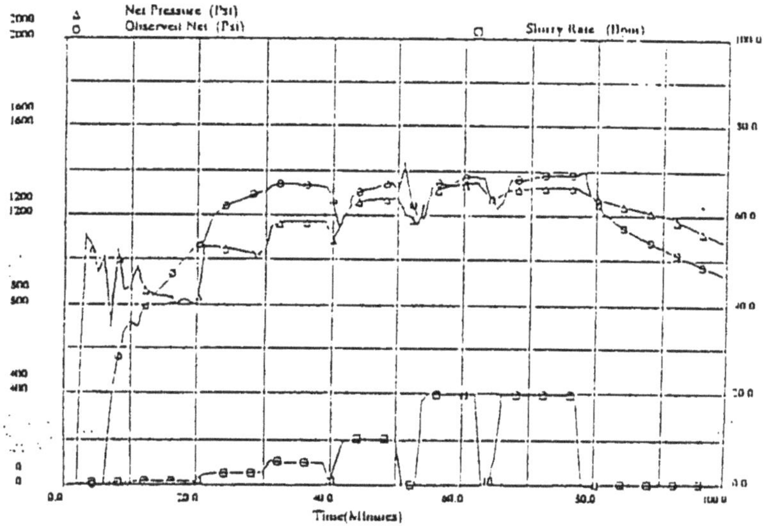

Fig. 4.9—RES history match using high stress layers above and below the second Frontier 8 August 1991 injection.

or volume and that many of their conclusions are based on the ISIP data rather than their history matching of the pressure data.

S. A. Holditch & Associates Inc. (SAH) applied conventional pressure decline analyses and history matching techniques using 3D models. First, a log-log plot of the pressure falloff data was evaluated to determine the presence of specific flow regimes during the pressure decline period after injection. Based on the particular flow regimes indicated, a bilinear and/or linear flow analysis was performed. The 3D history matching was performed using two different fracture model options. One used a conventional fluid rheology and the other used a tip-dominated geometry. **Fig. 4.10** is an idealized depiction of how pressure is distributed along the fracture length for each of the models. Curve 1 illustrates the tip-dominated model and Curve 2 describes the conventional model. The effect of near-wellbore pressure drop is also shown for each. SAH found that certain characteristics of the net-pressure response could be better defined depending on which model was used.

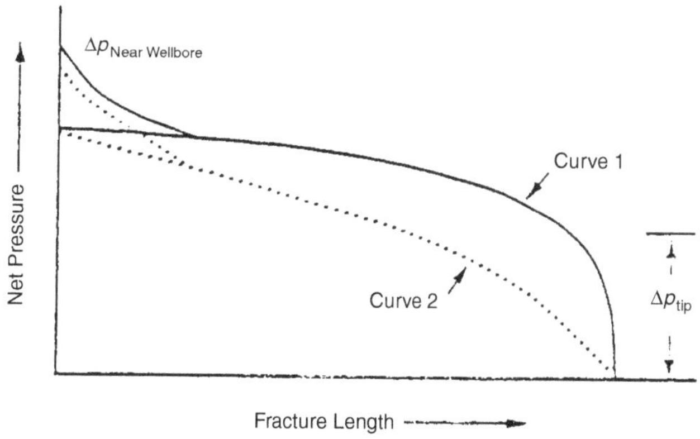

Fig. 4.10—Idealized depiction of pressure distribution along a fracture for two different models.

An example of one of their matches is illustrated by **Fig. 4.11.** The measured bottomhole data was corrected for near-wellbore friction pressure effects before history matching. SAH concluded that multiple propagating fractures were the most plausible explanation

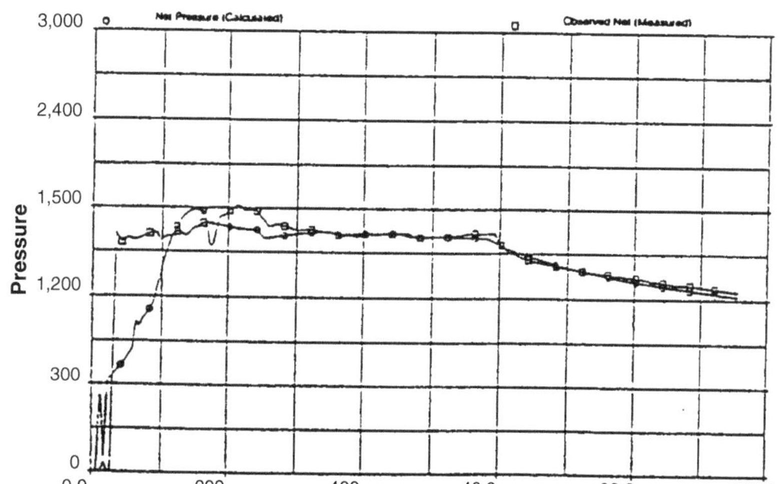

Fig. 4.11—SAH history match of pressure for 5 June 1991 injection.

for the unusual treating pressures. Incorporating high stress layers within the Frontier helped to increase the net pressure somewhat but did not account for all of the extra pressure. The net pressures were matched using two simultaneously propagating fractures while crosslinked gel was being pumped; however, when only 2% KCl water was pumped, many more fractures were modeled (up to 10).

4.12 Conclusions

The GRI SFE No. 4 was generally successful in proving that methods and technology developed on the previous three SFE wells in a different geologic basin can be applied, with some modification, to the Frontier Formation. The Frontier Formation presented significant challenges to the transfer of in-situ stress profiling, fracture height and azimuth analysis, and hydraulic fracture analysis technologies. Poor reservoir quality limited some of our work. For example, a proppant treatment was not pumped and, thus, postfracture reservoir evaluation could not be performed. However, extensive data sets were acquired and analyzed by several organizations and researchers. Detailed results of this work can be obtained from individual contractors listed in Table 4.1. The following conclusions represent a summary of some of the more important results.

- The second Frontier reservoir was poorly developed and very thin in SFE No. 4. Extensive cementation had also occurred, which significantly reduced primary porosity.
- Some natural fractures were encountered in core analysis and might have been the cause of high fracture injection pressures. They could have resulted in a tortuous path between the wellbore and created fracture(s).
- Casedhole stress test results indicated an abnormally high in-situ stress field was present. Calculated stress values were correlated to bulk compressibility to develop a calibrated in-situ stress profile.
- The high injection pressures seen in SFE No. 4 were the result of several complex phenomena. Analysis of the data resulted in the following possible explanations.
    - o The fracturing pressure responses could be matched accurately by modeling multiple propagating fractures and accounting for thin, high stress layers.
    - o Modeling a single planar fracture accounting for thin, high stress layers in the stress profile simulated the shape of the actual pressure response at approximately the same net-pressure level.
    - o The high injection pressures were partially because of poor communication or a tortuous path between the wellbore and fracture(s).
- In general, GRI-sponsored methods and technology developed in previous SFE wells were successfully transferred to another geologic basin.

4.13 Acknowledgments

We gratefully acknowledge the support of the Gas Research Institute who sponsored this work under GRI Contract No. 5091-221-2129. We also appreciate the efforts and dedication of all of the organizations listed in Table 4.1 whose results were the basis for this paper.

4.14 References

CER Corp. and S. A. Holditch & Associates Inc. 1988. Advancements in Travis Peak Formation Evaluation and Hydraulic Fracture Technology, Staged Field Experiment No. 1, Waskom Field, Harrison County, Texas. Report No. GRI-88-0077 prepared for GRI under Contract No. 5086-211-1306.

CER Corp. and S. A. Holditch & Associates Inc. 1990. Staged Field Experiment No. 2: Application of Advanced Geological, Petrophysical and Engineering Technologies to Evaluate and Improve Gas Recovery From Low Permeability Sandstone Reservoirs. Report Nos. GRI-89/0140 and GRI-90/0093 prepared for GRI under Contract No. 5088-211-1682. (April 1990).

CER Corp. and S. A. Holditch & Associates Inc. 1991. Staged Field Experiment No. 3: Application of Advanced Technologies in Tight Gas Sandstones – Travis Peak and Cotton Valley Formations, Waskom Field, Harrison County, Texas. Report No. GRI-91/0048 prepared for GRI under Contract No. 5090-2111940.

Dutton, S. P., Hamlin, H. S., and Laubach, S. E. 1992. Geologist Controls on Reservoir Properties of Low-Permeability Sandstone, Frontier Formation, Moxa Arch, Southwest Wyoming. Contract Report No. GRI-92/0127 prepared for the Gas Research Institute under Contract No. 5082-211-0708.

Holditch, S. A., Robinson, B. M. Whitehead, W. S. et al. 1988. The GRI Staged Field Experiment. *SPE Form Eval* **3** (3): 519–533. SPE-16429-PA. https://doi.org/10.2118/16429-PA.

ResTech Houston 1988. Development of Petrophysical Techniques for Evaluating Tight Gas Sands. Final report prepared for GRI under Contract No. 5084-211-1062.

Robinson, B. M., Holditch, S. A., and Peterson, R. E. 1991. The Gas Research Institute's Second Staged Field Experiment: A Study of Hydraulic Fracturing. Presented at the SPE Gas Technology Symposium, Houston, 23 January. SPE-21495-MS. https://doi.org/10.2118/21495-MS.

Robinson, B. M., Holditch, S. A., Whitehead, W. S. et al. 1992. Hydraulic Fracturing in East Texas: Third GRI Staged Field Experiment. *J Pet Technol* **44** (1). SPE-22878-PA. https://doi.org/10.2118/22878-PA.

Western Atlas International Core Laboratories. 1989. Special Core Analysis Study on Samples from the Travis Peak and Cotton Valley Formations, Mobil E&P Cargill Unit #15 (SFE #3). Report, 20 October.

Western Atlas International Core Laboratories. 1991. Core Analysis Report for Site No. 4–24, Chimney Butte Field, Sublette County, Wyoming.

Whitehead, W. S., Hunt, E. R., Finley, R. J. et al. 1986. In-Situ Stresses: A Comparison Between Log-Derived Values and Actual Field-Measured Values in the Travis Peak Formation of East Texas. Presented at the SPE Unconventional Gas Technology Symposium, Louisville, Kentucky, USA, 18–21 May. SPE-15209-MS. https://doi.org/10.2118/15209-MS.

Chapter 5

Fracturing and Testing Case Study of Paludal, Tight, Lenticular Gas Sands

N. R. Warpinski, SPE, Sandia Natl. Laboratories; **P. T. Branagan,** CER Corp.; **A. R. Sattler,** SPE, Sandia Natl. Laboratories; **J. C. Lorenz,** Sandia Natl. Laboratories; **D. A. Northrop,** SPE, Sandia Natl. Laboratories; **R. L. Mann,** SPE, CER Corp.; **K. H. Frohne,** SPE, U.S. DOE

 Journal paper SPE 13876 was published in *SPE Form Eval* 2 (4): 535–545. Original SPE manuscript received for review 19 May 1985 and accepted for publication 14 March 1986. Revised manuscript received 25 March 1987 and first presented at the SPE/DOE Low Permeability Gas Reservoirs Symposium held in Denver, 19–22 May 1985.

SPE 13876 presents a case history of the formation evaluation, completion, and stimulation of the Paludal Interval (Late Cretaceous Lower Delta Plain deposits) of the Williams Fork Formation at the M-Site in the Piceance Basin in Colorado. Warpinski et al. provided first a detailed description of the regional geology and sedimentology, performed comprehensive coring (> 4,000 ft) and log analysis from two wells, measured in-situ stress throughout the Paludal Interval, conducted prefracture well tests and mini-fracture tests to estimate important parameters (e.g., permeability, fracture behavior, and leakoff coefficient), and discussed fracture design and implementation and treating pressure analysis using the Nolte-Smith method. The authors worked to improve their understanding of what causes a screenout and how to clean out the fracture sand in the wellbore. This study also discusses remedial treatments to break gel, which resulted in precipitation of iron oxides, and after 20 months of shut in, the better-than-expected production performance. Conclusions from the study include the following:

- Effective fracture length of 150 ft
- Created fracture length of approximately 400 ft
- Well completions minimize damage to natural fracture systems
- A back pressure of 1,000 psi is needed to produce and maintain openness of the natural fractures
- High fracture pressure observations and explanations
- Reservoir damage occurs when high fracture pressure forces fluids into natural fractures
- Similar results between geophone, temperature log, and Nolte analyses
- Fracture containment by higher stress layers acting as barriers to vertical fracture growth

5.1 Summary

SPE 13876 is a case study of the stimulation and testing of tight, lenticular sands in the paludal interval of the Mesaverde Group in the Piceance Basin at DOE's Multiwell Experiment (MWX) site in Colorado. Topics discussed include geologic data, stress test results, well testing, laboratory core studies, stimulation and stimulation analyses, and post-fracture operations.

5.2 Introduction

For a number of years, the U.S. government has engaged in research to enhance gas recovery from unconventional reservoirs, such as organically rich, fractured shale and discontinuous, lenticular, tight sandstones. Although large quantities of natural gas are trapped in these formations, the permeabilities are too low to permit economic recovery by conventional technology. In the western U.S., the Greater Green River, Piceance, Wind River, and Uinta Basins have been identified as containing significant amounts of gas in thick sections of lenticular sands. The Natl. Petroleum Council has appraised (Tight Gas Reservoirs 1980) these four basins to hold 136 Tcf [4×10^{12} m^3] of maximum recoverable gas in lenticular reservoirs. This sizeable resource is being investigated by the DOE in the Piceance Basin of western Colorado, where a field laboratory containing three closely spaced wells penetrating the lenticular Mesaverde Formation has been constructed. This facility, near Rifle, Colorado, is the site of the DOE MWX, which has been developed to determine the viability of the lenticular, tight sands as a gas resource.

Massive hydraulic fracturing has demonstrably increased gas production from tight reservoirs. Its performance in lenticular formations is currently unpredictable, however, because of poor definition of reservoir properties and sizes, inadequate understanding of the physics that control fracture propagation and proppant transport, limited ability to measure, describe, or evaluate the created fracture, and uncertainty as to the relationship of stimulation design variables (fluids, proppants, and pumping rates) to the resultant fracture. These difficulties are compounded in the lenticular formations by the uncertainty concerning whether multiple lenses, some remote from the wellbore, can be stimulated by a common treatment. Improved understanding, evaluation, prediction, and possible control of stimulation technology are needed for effective development of tight, lenticular reservoirs.

In this paper, we describe a case study of the testing and fracturing of lenticular sands in the paludal interval (Lorenz 1985) of the Mesaverde Group. Results include (1) geologic studies that delineate the sizes and shapes of the lenses, (2) detailed core and log reservoir/rock property data, (3) stress test data showing the vertical distribution of the horizontal in-situ stress, (4) well tests (Branagan et al. 1985) (drawdown, build up, and interference) to determine in-situ reservoir properties, (5) laboratory data (Sattler et al. 1985) on fracture fluid invasion and damage, and (6) the various analytic and diagnostic (Hart and Engi 1987) techniques used to evaluate the hydraulic-fracture treatment.

5.3 Overview

The objectives of this paludal experiment were to characterize the lenticular sandstones for reservoir quality and size, to determine fracture geometry and behavior in a lenticular environment, and to stimulate lenticular sandstones in this complex, coal-bearing interval successfully. To achieve these objectives, the hydraulic fracture treatment was divided into two phases. The first phase consisted of two minifractures and step-rate/flowback tests, during which several diagnostic techniques were used to map fracture behavior. These tests also provided fracture design information for the main treatment. Associated with this phase were pre- and post-fracture well tests, stress tests, and laboratory fluid-damage tests. The second phase was the full-scale treatment, again with several diagnostic techniques. Associated testing included post-fracture cleanup and well tests, as well as laboratory damage and residue measurements. Detailed core and log analyses were performed to support all tests. The results of these experiments and the conclusions inferred from them will be outlined in this paper.

5.4 MWX Site

The MWX site consists of three closely spaced wells (Wells MWX-1, MWX-2, and MWX-3). We have taken over 4,100 ft [1250 m] of core with more than 1,100 ft [335 m] of oriented, and have performed comprehensive core and log analysis programs, detailed geophysical surveys, and extensive well testing and stress testing programs. The three wells are 115 to 215 ft [35 to 66 m] apart and their depths are 8,350, 8,300, and 7,565 ft [2545, 2530, and 2306 m], respectively. The Mesaverde Group at this location is found at 4,100 to 8,300 ft [1250 to 2530 m].

5.5 Geologic and Sedimentologic Setting

The paludal interval (Lorenz 1985) refers to the Late Cretaceous lower delta plain deposits of the Williams Fork Formation of the Mesaverde Group. It is characterized by muddy and carbonaceous flood-plain deposits, within which narrow distributary channels, channel margin (levee) deposits, and splay (flood) deposits are interbedded. Distributary deposits typically are thick sandstones with few or no shale breaks; channel margin deposits contain thinner sandstone beds with frequent Siltstone and mudstone interfingering; splay deposits are often reservoir-quality clean sandstones near their apex (the channel), but become interbedded with siltstone and mudstone near their extremities; flood-plain deposits consist primarily of coals and mudstones.

Lorenz (1985) statistically estimated the widths (narrow, horizontal channel dimension) of the sand reservoirs on the basis of two procedures. The first procedure (Lorenz et al. 1985) combined the correlation percentages of reservoir sandstones between the three wells with the spacing of the wells to predict an average lens width. With this procedure, lens widths for this delta plain environment are on the order of 100 to 500 ft [30 to 152 m]. The second procedure relies on the relationship between lens thickness (vertical dimension) and maximum lens width as measured in outcrop. A best-fit straight line through an ensemble of outcrop data yielded width = 8.6 thickness (Tight Gas Reservoirs 1980), with a correlation coefficient of 0.62. Major sources of error are the scatter in outcrop data and the uncertainty of penetrating the lens in the location of its maximum width. In the 11 channel lenses encountered in MWX, the predicted widths were 80 to 550 ft [25 to 168 m]. Channel lengths are estimated to be at least one order of magnitude greater than the widths. No method is available to estimate the size of the splay deposits.

Lorenz (1985) also estimates the orientation of the channels on the basis of paleogeography, well-to-well correlations, crossbeds in oriented core, and a high-resolution dipmeter. Paleogeography suggests that the major flow direction was generally east-northeast, and most of the distributary lenses at MWX are in this range.

Fig. 5.1 shows correlated gamma ray logs of the two sandstone reservoirs that were chosen for hydraulic fracture. These reservoirs are labeled Zone 3, which is interpreted as a distributary channel, and Zone 4, which is interpreted as a splay deposit. Zone 3 was perforated from 7,120 to 7,144 ft [2170 to 2177 m] and Zone 4 from 7,076 to 7,100 ft [2157 to 2164 m]. Zone 3 is probably about 350 ft [107 m] wide on the basis of its 28-ft [8.5-m] thickness, and probably is oriented east-northeast because it intersects Wells MWX-1 and MWX-2, but is only marginally evident in Well MWX-3. No size estimates are available for Zone 4 (because it is a splay), but the splay probably originated from a channel to the north-northeast because this zone is thick in Wells MWX-1 and MWX-3 but is marginal in Well MWX-2.

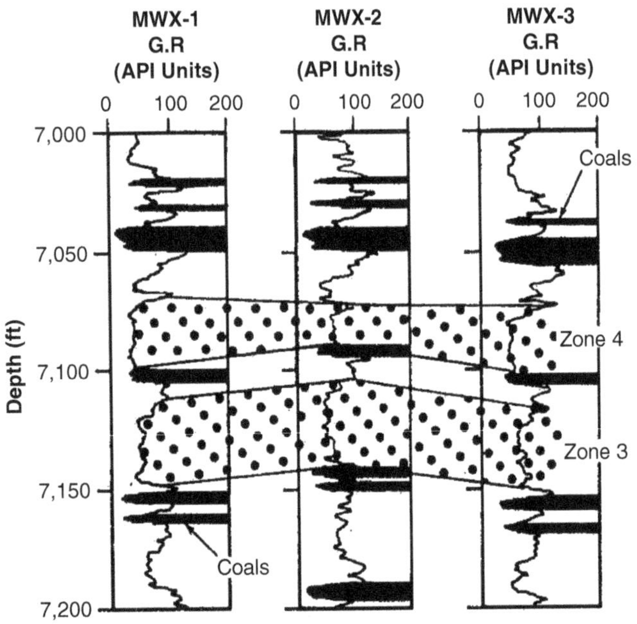

Fig. 5.1—Paludal Zones 3 and 4.

Fig. 5.2 shows a plan view of this representation of Zones 3 and 4. Also shown is the expected hydraulic-fracture azimuth from several stress-orientation measurement techniques (Teufel et al. 1984). The probable intersection of the hydraulic fracture and the sand reservoirs is 200 to 500 ft [61 to 152 m], depending on the true orientation of the channel and the width of the splay.

One other factor further complicates this geometry. A fault with a 12-ft [4-m] throw intersects Well MWX-2 just above Zone 4. The orientation of this fault is not known, but it does not intersect the other wells. It is assumed that it is oriented in the same direction as the natural fractures observed in oriented core, about N70°W.

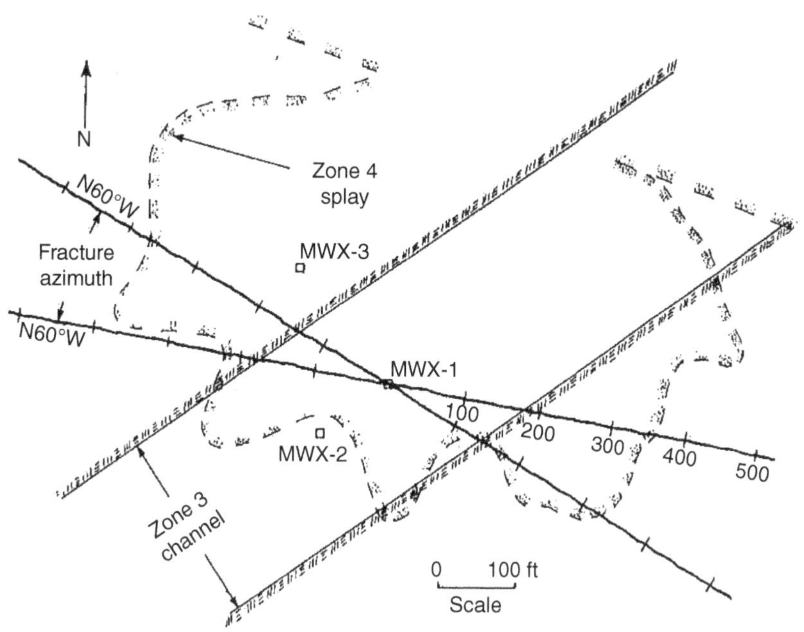

Fig. 5.2—Plan view of sand reservoirs and hydraulic fracture.

The final complication is the presence of coal beds above, below, and in between Zones 3 and 4. The proximity of these coals to the reservoir sands was a cause for concern in designing the fracture treatments.

5.6 Core and Log Analyses

Core was obtained in both zones in Wells MWX-2 and MWX-3. None of it was oriented in this interval. **Fig. 5.3 and Table 5.1** show an example of the results of special core analyses including Boyle's law porosity measurements; dry, Klinkenberg-corrected permeability measurements at confining stress (2,000 psi [14 MPa]); and water saturation, S_w, data. Also shown for correlation are the gamma and porosity logs. In this well, Well MWX-3, a casual look at the logs shows little developed porosity. However, the 10 to 12% porosity measured in core in these zones was the best in the entire Mesaverde section (see Table 5.1). Dry, Klinkenberg-corrected permeabilities of 10 to 20 μd were also the highest matrix permeabilities measured in the section.

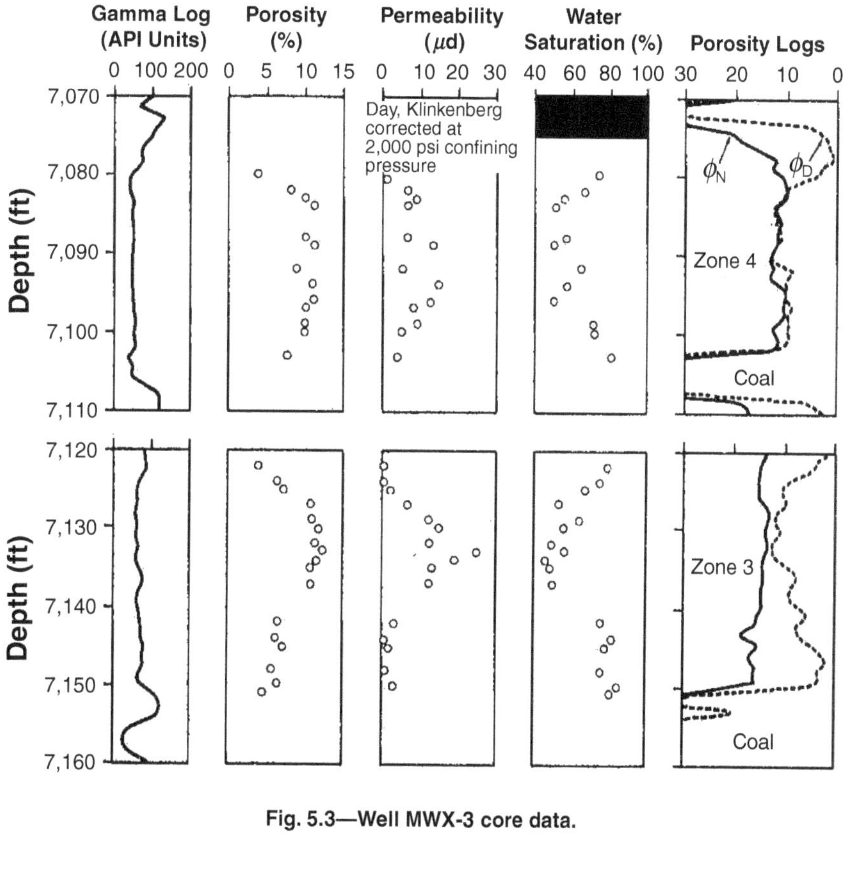

Fig. 5.3—Well MWX-3 core data.

Log Depth (ft)	Porosity (%)	Water Saturation (%)	Klinkenburg Permeability (md) At Effective Confining Stress (psi)		
			1,000	2,000	3,000
7,080	3.8	74	0.0020	0.0006	0.0001
7,082	8.3	66	0.0118	0.0065	0.0046
7,083	10.0	56	0.0203	0.0086	0.0063
7,085	11.1	51	0.0419	0.0062	0.0150
7,088	10.0	57	0.0101	0.0062	0.0051
7,089	11.2	51	0.0161	0.0130	0.0096
7,092	8.9	62	0.0062	0.0051	0.0037
7,094	10.8	57	0.0182	0.0144	0.0083
7,096	10.9	50	0.0165	0.0117	0.0080
7,097	10.0	50	0.0107	0.0080	0.0057
7,099	10.0	73	0.0201	0.0089	0.0060
7,100	9.8	75	0.0104	0,0053	0.0048
7,103	7.6	80	0.0072	0.0038	0.0014
7,122	4.0	86	0.0009	0.0003	0.0002
7,124	6.4	74	0.0017	0.0006	0.0003
7,125	7.1	67	0.0056	0.0021	0.0010
7,127	10.6	53	0.0154	0.0066	0.0063
7,129	10.9	64	0.0133	0.0124	0.0091
7,130	11.6	56	0.0166	0.0152	0.0136
7,132	11.4	48	0.0147	0.0128	0.0088
7,133	12.4	56	0.0323	0.0252	0.0218
7,134	11.5	45	0.0259	0.0189	0.0163
7,135	10.9	45	0.0140	0.0131	0.0115
7,137	10.8	48	0.0139	0.0123	0.0094
7,142	6.5	77	0.0053	0.0029	0.0011
7,144	6.1	80	0.0023	0.0005	0.0003
7,145	7.0	78	0.0031	0.0018	0.0010
7,148	5.6	76	0.0041	0.0011	0.0009
7,150	6.4	84	0.0032	0.0028	0.0010
7,151	4.6	81	0.0021	0.0004	0.0002

Table 5.1—Paludal core data in Well MWX-3.

Relative permeability data from Randolph (1983) show that the gas permeabilities are a factor of 3 to 5 lower at 40% S_w and will be effectively shut off for $S_w > 60\%$. He also found that these paludal rocks are not very stress-sensitive.

Log analyses by Kukal (1984) gave permeability/height, *kh,* values of 5 to 1.0 md-ft [1.5×10^{-4} to 3.0×10^{-4} md·m] for both sands in all wells, with the exception of Zone 4 in Well MWX-2, which was not analyzed for permeability. He calculated water saturations of 52 to 60% for these zones (again with the exception of Zone 4 in Well MWX-2, which was 76%).

The decreased porosity and permeability in the bottom half of Zone 3 indicate that Well MWX-3 has penetrated the channel near the margin. While the top of the zone is a clean sand (as seen in the core), the bottom shows interfingering with siltstone and mudstones and thus poor reservoir quality. These large changes in reservoir quality are typical of these lenticular reservoirs.

Several natural fractures were observed in the core (not oriented), but they were mostly in the fine-grained, thin-bedded materials, and all were calcite-filled. The televiewer log picked up seven possible fractures, but no preferred orientation of these fractures was found.

Rock properties determined from core compression tests under confining pressure resulted in a Young's modulus of about 3.7×10^6 psi [25 GPa] and a Poisson's ratio of 0.22. Poisson's ratio measured in situ with a long-spaced sonic log resulted in values of 0.16 to 0.2. Young's modulus in the bounding mudstones and siltstones was considerably greater than in the sands; values ranged from 4×10^6 to 6×10^6 psi [28 to 41 GPa]. Poisson's ratio in these rocks varied from 0.2 to 0.26. Sonic data in the bounding materials were generally unusable because of the presence of coals.

5.7 Geophysical Surveys

While a three-dimensional (3D) seismic survey, two vertical seismic profiles (one azimuthal), and two cross-well seismic surveys have been performed for this experiment (Searls 1985), the presence of coal in this interval made these data useless for any interpretation of structure and lens morphology. The coals acted as the strongest reflectors in this section and as diffraction gratings to some extent, thus masking all other important features.

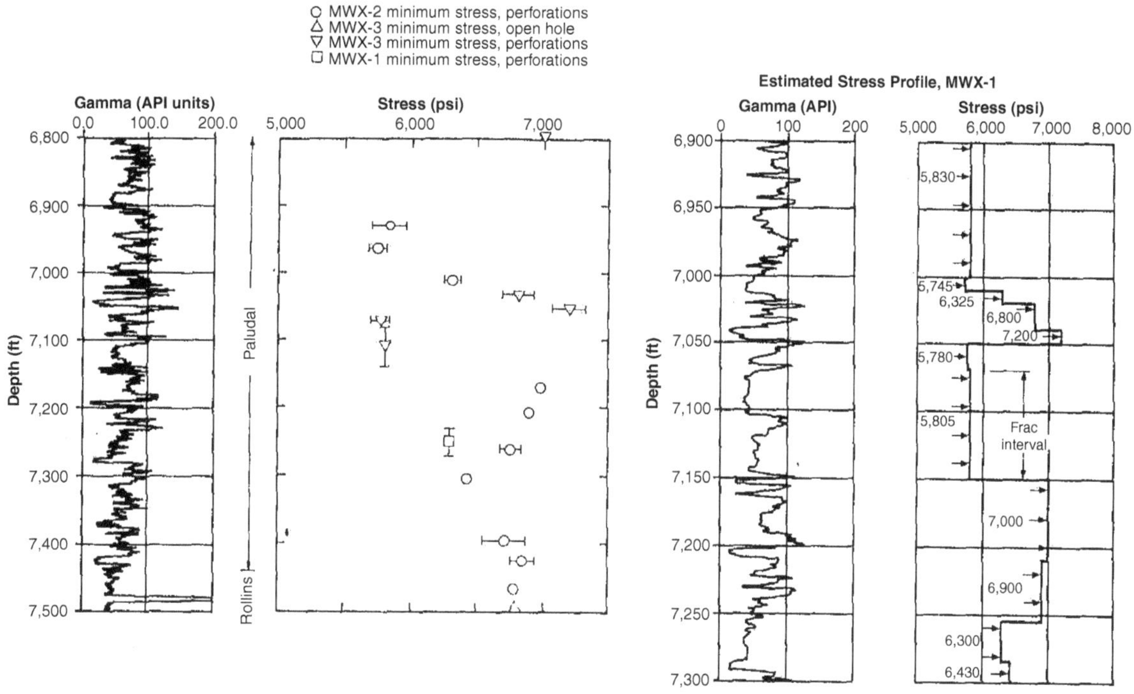

Fig. 5.4—Stress data and estimated stress profile in Well MWX-1.

5.8 Stress Test Data

In-situ stress measurements were performed throughout the paludal section to determine the vertical distribution of the minimum horizontal principal in-situ stress for hydraulic-fracture containment analyses. These measurements used the hydraulic fracturing technique as described in Warpinski et al. (1985). **Fig. 5.4** shows the minimum stress data from all three wells vs. the depth in Well MWX-2, where most of the measurements were made. Additionally, Fig. 5.4 shows our best reconstruction of a stress profile for Well MWX-1, the well that was stimulated. This profile is only an approximation because of the limited number of data points and the good possibility of stress variations in lateral directions.

Fig. 5.4 shows that we were fortunate to be located in low-stress sands with high-stress bounding layers. The upper bounding zone is fairly thin, however, and thus would not be a complete barrier to fracture propagation if the treatment pressures became high. Above this barrier, only low stresses were measured up to 6,800 ft [2073 m]; with so few data points, however, high-stress layers are still quite likely above 7,000 ft [2134 m].

The highest stress measured in the paludal section was in the coal above Zone 4 at 7,050 ft [2149 m] with a 1.02-psi/ft [23-kPa/m] gradient. Sands typically have 0.82- to 0.87-psi/ft [18.5- to 19.7-kPa/m] gradients. Of all the stress measurements performed in these wells, the data in this paludal section are the least accurate and reproducible. We believe it is a result of the complex lithology in this section; core logs show that lithologic changes are found from every few feet to as little as every few inches. Thus, a stress measurement probably "averages" these stresses, and different results should be expected for different volumes. Typical accuracies for these data are only ±50 to 100 psi [±345 to 690 kPa] for most tests.

5.9 Prefracture Well Test Data

Prefracture well testing (Branagan et al. 1985) of commingled Zones 3 and 4 was conducted in Fall 1983. Testing consisted of an initial flow period for cleanup followed by a short buildup test and then an extended interference test. During a 7-day flow period of the interference test, the well produced at 200 to 250 Mcf/D [5665 to 7080 m^3/d] at surface pressures of 800 to 2,800 psi [5.5 to 19.3 MPa], However, no interference was seen in the observation wells at distances of only 112 and 189 ft [34 and 58 m]. The flow period was followed by a 7-day buildup from which an initial *kh* of 0.95 md-ft [0.0003 md·m] was calculated for both zones. This yields an estimated average permeability of 36 μd, and the formation temperature was about 210°F [99°C].

The high permeability and absolute open flow (AOF) compared to the core data show that these zones are naturally fractured. Because there was no evidence of linear flow in the well test data, we are probably dealing with a fairly interconnected natural-fracture system. However, no well-to-well interference was observed with these permeabilities. For a 0.95-md-ft [0.0003-md·m] formation flowing at 250 Mcf/D [7,080 m^3/d], interference should have been observed in 1 or 2 days. The lack of interference may indicate that there is no connection between the natural fractures intersecting individual wells, and the natural fracture system is suggested to be one of subparallel fractures with low-angle intersections (Lorenz et al. 1986). Other factors confusing the results are the fault in Well MWX-2 and the limited thickness of Zone 3 in Well MWX-3 (the channel margin).

5.10 Phase 1 Minifractures

The purpose of the Phase 1 minifractures was two-fold: to attempt to map fracture behavior in a lenticular reservoir by use of two minifractures with different volumes and to obtain design information for the main treatment. The approach was to conduct one small, unpropped minifracture and try to map its important characteristics, and then to conduct a second minifracture with twice the volume. We would like to discern any differences in fracture behavior or geometry that might be attributable to the lens morphology and associated stress and lithologic features. Because we were using no proppant in these tests, we could maximize our information by obtaining careful pressure-decline data after the treatment for a Nolte-type analysis (Nolte 1979). Additionally, we conducted step-rate/flowback and pump-in/flowback tests before conducting the minifractures to obtain additional closure stress data that were averaged over the commingled zones. This series of tests was conducted in Well MWX-1 in December 1983.

5.10.1 Step-Rate, Pump-In, and Flowback Tests. Step-rate/flowback and pump-in/flowback tests with KCl water were conducted to provide a suitably averaged (over the fracture size) initial closure stress for the minifracture analyses and an initial fracture extension pressure for comparison with later data should anomalous behavior occur. The step-rate test resulted in a minimum fracture extension pressure of about 5,950 psi [41 MPa]. The flowback test following the step-rate test resulted in an apparent closure stress of 5,900 psi [40.7 MPa]. The second flowback test yielded an apparent closure stress of 6,100 psi [42 MPa], considerably higher than the first test. This change in closure stress was initially attributed to leakoff of fracture fluid into the near-fracture pore space (back stress), but other interpretations are also possible. The injection pressures were 6,355 psi [43.8 MPa] bottomhole at 8 bbl/min [0.021 m^3/s] for the steprate test and 6,500 psi [44.8 MPa] at 10 bbl/min [0.026 m^3/s] for the pump-in test (~120 bbl [19 m^3] total volume for each test). These values are quite high for pumping KC1 water at these low rates. The closure stress data can be compared with a minimum stress of 5,805 psi [40 MPa] measured with a small breakdown in Well MWX-3.

5.10.2 Minifracture 1. Minifracture 1 was designed to be 200 ft [61 m] long so that the fracture would remain, for the most part, within the sand bodies. A volume of 15,000 gal [57 m^3] of a noncross-linked 30-lbm/1,000-gal [3.6-kg/m^3] WGA-2 gel was chosen on the basis of both pseudo-3D analyses using the measured in-situ stresses (Fig. 5.4) and an estimated maximum fracture height of 120 ft [37 m] (constant-height model) using the same stress data. Smith Energy Services, the stimulation contractor, suggested a prepad of low-pH methanol to reduce formation damage and to aid fluid recovery; 2,100 gal [8 m^3] was used. This was based on a laboratory study of MWX core that showed a near 100% permeability recovery with this prepad as opposed to 40 to 60% damage for short-term data with no prepad. The flow rate was 10 bbl/min [0.026 m^3/s].

During the minifractures (as well as the step-rate/flowback tests), diagnostics consisted of the following:

1. A quartz-crystal-oscillator, bottomhole pressure (BHP) transducer, and a bottomhole temperature gauge at 6,700 ft [2042 m] in Well MWX-1 (the stimulation well) on a wireline in tubing, recording at the surface
2. Borehole seismic geophones (Hart and Engi 1987) in Wells MWX-2 and MWX-3 to map fracture geometry
3. A post-fracture temperature survey

This treatment was conducted as planned, and the important data are shown in **Fig. 5.5.**

Fig. 5.5a shows the treatment pressure vs. time for this treatment. The important aspect of this figure is the high treatment pressure (800 psi [5.5 MPa] above initial closure stress). The Nolte-Smith (Nolte and Smith 1981) log/log plot of pressure behavior (Fig. 5.5b) yields an exponent of 0.28, which is higher than the value of 0.22 expected for a fluid with this rheology ($n' = 0.76$; $k' = 0.00072$ lbf-sec$^{n'}$/ft^2 [0.034N·s$^{n'}$/m^2]). Nevertheless, the plot indicates that the fracture was extending in length without excessive height growth. The temperature log in Fig. 5.5c indicates that total fracture height was on the order of 135 ft [41 m], although the top of the fracture is equivocal. The Nolte pressure-decline analysis (Nolte 1979) is shown in Fig. 5.5d, using a height of 135 ft [41 m], a plane-strain modulus of about 4.5×10^6 psi [31 GPa], a viscosity degradation exponent of 1.0, a leakoff height of 55 ft [16.8 m], a pump time of 43 minutes, and an instantaneous shut-in pressure (ISIP) of 6,670 psi [46 MPa], results in a wing length of 275 ft [84 m] and a leakoff coefficient of 0.0013 ft/$\sqrt{\text{min}}$ [0.0004 m/$\sqrt{\text{min}}$]. These results are longer than the design, but this is probably a result of neglecting the methanol prepad in the design calculations and calculating a lower gross leakoff coefficient than used in the design. Two shut-ins during the treatment were analyzed following Nierode (1985) and resulted in a leakoff coefficient less than 0.001 ft/$\sqrt{\text{min}}$ [0.0003 m/$\sqrt{\text{min}}$]. Results of the borehole seismic geophones will be given in a later section.

5.10.3 Minifracture 2. Minifracture 2 was designed to be twice the size of Minifracture 1 and used 30,000 gal [114 m^3] of a noncross-linked 60-lbm/1,000-gal [7.2-kg/m^3] WGA-2 gel, a more viscous fluid. A 4,500-gal [17-m^3] prepad of low-pH methanol was again used, and the flow rate for this test was also 10 bbl/min [0.027 m^3/s]. The same diagnostics were used to evaluate fracture behavior.

The data from Minifracture 2 are shown in **Fig. 5.6.** Again, the treatment pressures were high for this small-volume treatment, reaching a value of 1,100 psi [7.6 MPa] above the initial closure stress as shown in Fig. 5.6a. The log/log effective-treatment-pressure plot (Fig. 5.6b) shows an initially greater-than-expected exponent of 0.32, as opposed to the value of 0.26 expected for this rheology ($n' = 0.46$; $k' = 0.02$ lbf-sec$^{n'}$/ft^2 [0.96 N·sec$^{n'}$/m^2]). Later in the treatment, the exponent decreases to 0.17, which may indicate some height growth out of zone. The temperature log shows some additional height over Minifracture 1; we estimate it to be 150 ft [46 m]. The Nolte pressure-decline analysis is shown in Fig. 5.6d. The modulus, leakoff height, and degradation are the same as Minifracture 1; the pump time was 88 minutes and the ISIP was 6,890 psi [48 MPa]. The analysis results in an estimated wing length of 420 ft [128 m] and a leakoff coefficient of 0.0007 ft/$\sqrt{\text{min}}$ [0.0002 m/$\sqrt{\text{min}}$]. The analysis of two shut-ins during this treatment gave an estimated leakoff coefficient of less than 0.001 ft$\sqrt{\text{min}}$ [0.0003 m/$\sqrt{\text{min}}$]. These data indicated that fracture-length extension was occurring without excessive height growth even though the treatment pressures were abnormally high. Additionally, the low leakoff coefficient suggested that the nearby coals were not thieving excessive fracture fluid.

5.10.4 Minifracture Borehole Seismic Results. The results of the bore-hole seismic system have been reported by Hart and Engi (1987) and are shown in **Fig. 5.7.** Seismic signals created by the fracturing process were mapped by the geophones in Well MWX-3; the geophones in Well MWX-2 provided no clear data because of noise problems. The locations of origin of the seismic disturbances were mapped in 3D space and then projected on horizontal and vertical planes in Fig. 5.7 to show the fracture azimuth, height, and length. The measured fracture azimuth was N67°W for the one wing of the fracture from which the geophone in Well MWX-3 could accurately resolve locations. The height and length of this one wing were about 140 and 240 ft [43 and 73 m], respectively, for Minifracture 1 and 150 and 370 ft [46 and 113 m], respectively, for Minifracture 2. These numbers are very similar to the Nolte analysis results.

5.11 Post-Minifracture Cleanup and Well Tests

After completion of the minifracture tests, the well was cleaned for a week before cold weather necessitated a winter shutdown. During this week, 60% of the fluid load was recovered. During January, the well was flowed once a week for additional cleanup, and another 10% was recovered. Finally, the well was shut in for 2 months before the post-minifracture well testing operations. In late March, testing began and an additional 22% of the initial load was recovered for a total of 92% (some of this may have been formation water).

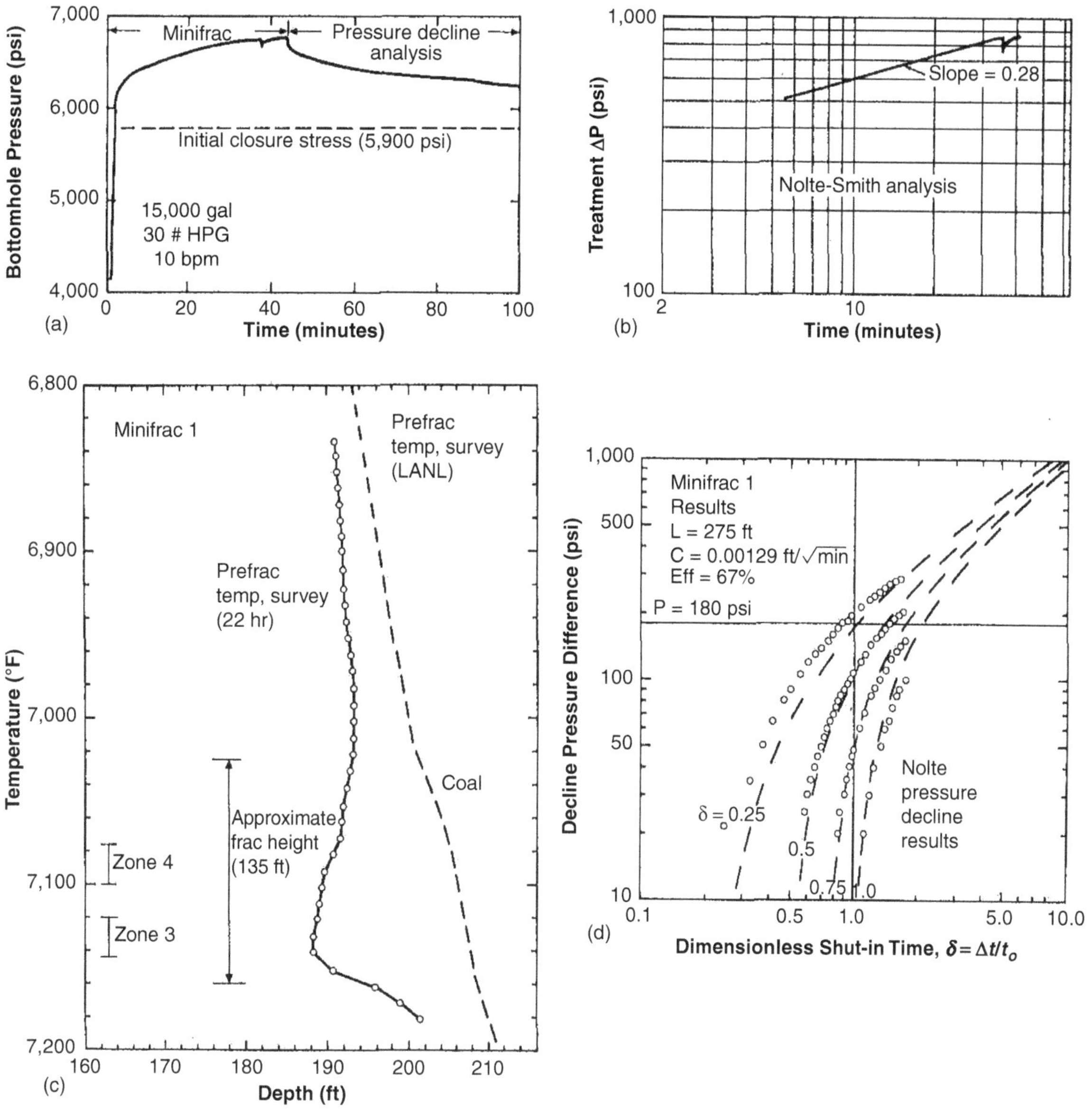

Fig. 5.5—Results of Minifracture 1.

Post-minifracture testing (Branagan et al. 1985) consisted of three consecutive drawdown periods at different rates, a shut-in period, and a final drawdown. The main purpose was to determine the effects of the treatment and the unpropped fracture on reservoir performance. No attempt was made to look for interference. We found that the productive capacity of the reservoir had decreased considerably, at least over the short times of these tests (25 days of testing). The formation capacity was estimated to be 0.64 md-ft [0.0002 md·m], a decrease of about 30%, and the AOF was about 200 Mcf/D [5665 m^3/d], as opposed to 250 Mcf/D [7080 m^3/d] pre-fracture. The un-propped fracture provided a relatively high, linear, conductive flow path with a fracture capacity, $K_f w$, of 75 md-ft [0.022 md·m] and a skin of –3.8. Apparently, the natural fractures had been damaged by the treatment, and test time was not sufficiently long to clean them, if they could be cleaned.

5.12 Hydraulic Fracture Design

The design of the second phase—the main hydraulic-fracture treatment—was influenced by several considerations. First, we wanted to optimize propped fracture length with respect to sand reservoir size. The extent of Zone 3 was probably 200 to 500 ft [61 to 152 m]; the extent of Zone 4 was unknown. We found no reasons for creating a fracture with a propped length greater than 500 ft [152 m]. Second, we were concerned about the high treatment pressures and preferred keeping the viscosities, flow rates, and volumes low to minimize the pressure. Third, we had no information on remote lenses (not intersected by the wells) and we were concerned about the effects of the coals on the treatment and production; this favored a small treatment. Fourth, the fracture diagnostics could only "see" about 400 ft [122 m], so much longer fractures would be difficult to diagnose.

Fracture height for the design models was uncertain because we had broken through the known upper barrier, yet we did not experience excessive out-of-zone growth. This is probably because the upper barrier significantly reduced fracture width there and acted as an efficient flow restriction. Because of these complications, we assumed a constant fracture height of 200 ft [61m] for design purposes (this is based on the minifracture results). A gross leakoff coefficient of 0.00065 ft/$\sqrt{\text{min}}$ [0.0002 m/$\sqrt{\text{min}}$] for a full 200-ft [61-m] height was calculated from the minifracture results. Rock and reservoir properties are the same as before.

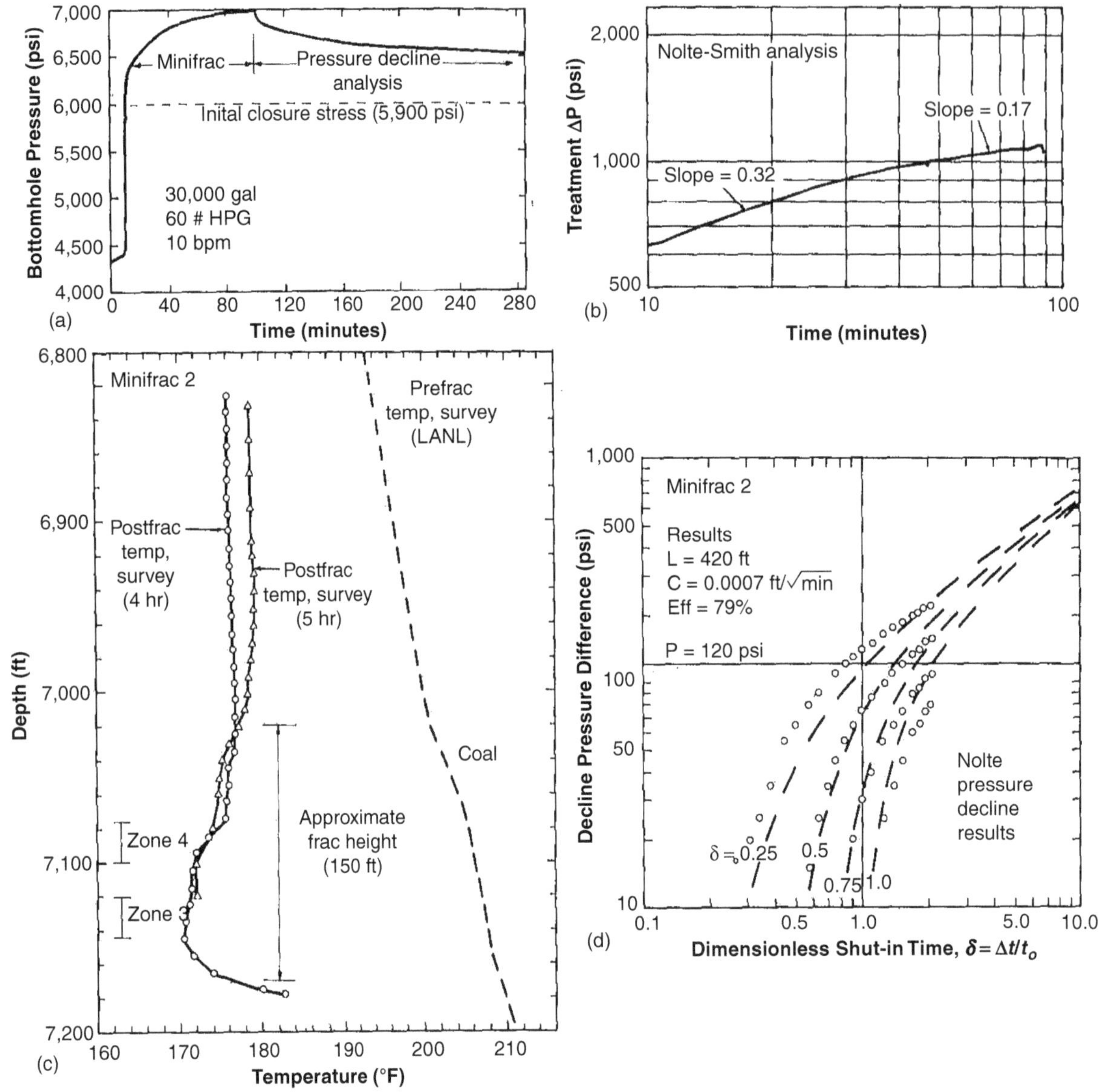

Fig. 5.6—Results of Minifracture 2.

The results of this treatment are shown in **Table 5.2.** The cross-linked water-based fluid system, Apollo, was used and was staged from a concentration of 40 lbm/1,000 gal [4.8 kg/m^3] in the pad to 25 lbm/1,000 gals [3 kg/m^3] in the final proppant-carrying stage (all water is 3% KCl). For analyses requiring rheological data, the 35-lbm/l,000-gal [4.2-kg/m^3] gel at a residence time of 1 hour and temperature of 193°F [89.4°C] was considered average. Under these conditions, $n' = 0.78$ and $k' = 0.0061$ lbf-sec$^{n'}$/ft^2 [0.29 N·sec$^{n'}$/m^2]. Because of the high temperatures, breaker was added only in Stages 7 and 8 in concentrations of 0.25 to 0.5 lbm/1,000 gal [0.03 to 0.06 kg/m^3]; the addition of breaker in the earlier stages would have reduced viscosity sufficiently that sand fall-out would have been expected before the slurry reached the fracture extremities. Because of the success with fluid recovery in the minifractures, a methanol prepad was again used.

Sand concentrations were staged up to 5.5 lbm/gal [660 g/m^3], the maximum that we felt confident of being able to inject at rates of 20 bbl/min [0.053 m^3/s]. This results in an average design concentration in the fracture of about 1 lbm/ft^2 [4.9 kg/m^2]. A 20/40-mesh sand was used for all stages except the final one (Stage 8), in which 12/20 mesh was used for a tail-in. A radioactive sand tag was used in the entire job; iridium 192 was used in the first half and iodine 131 in the last half of the treatment. Sufficient sand concentrations were used so that if the fluid broke before final closure, the resultant sand bank would fill both channel sands. Ammonium thiocyanate in a concentration of about 100 ppm was used as a fluid tag.

We planned to perform three shut-ins during pad pumping to estimate the leakoff coefficient using Nierode's (1985) analysis, a short rate test (15 bbl/min [0.04 m^3/s]) during the pad to see the effect on treatment pressure, and a long post-fracture shut-in to obtain pressure decline data (Nolte 1979). We had the same diagnostic techniques as used in the minifractures, as well as a treatment monitoring vehicle for calculating BHP from surface injection conditions and extended realtime data-monitoring capabilities.

5.13 Phase 2 Hydraulic Fracture Experiment

The Phase 2 hydraulic fracture was conducted in early May 1984. The well configuration consisted of 7-in. [18-cm] casing with a bridge plug at 7,200 ft [2195 m] and $2^7/_8$-in. [7.3-cm] tubing landed open-ended at 6,750 ft [2057 m]. The pressure and temperature transducers were again in the tubing at 6,700 ft [2042 m] while we pumped down the annulus. Surface data included casing treatment pressure, static-tubing pressure, flow rate, sand concentration, base gel viscosity, and surface fluid temperature.

On the treatment day, the borehole geophones were lowered in Wells MWX-2 and MWX-3, oriented, and rechecked. KCl water (3%) was circulated down the annulus to remove gas in the well, and the 7,700 gal [29 m³] of methanol were spotted in the annulus. Finally, the pressure and temperature tools were lowered and stabilized and the treatment begun.

Only one problem occurred during the treatment, and this was a result of problems with the BHP processor. It began to malfunction, and we were forced to shut in for about 40 minutes after pumping 163 bbl [26 m³]. (Only methanol was in the formation at this point.) Once the problem was repaired, the treatment continued as planned. However, this unplanned shut-in probably resulted in a leakoff of much of the methanol in the near-wellbore region and in considerable viscosity degradation of the pad fluid that was sitting in the hot wellbore.

The results from the treatment are shown in **Fig. 5.8.** This includes the actual treatment data in Fig. 5.8a, the Nolte-Smith analysis in Fig. 5.8b, the Nierode analysis in Fig. 5.8c, and the Nolte pressure decline analysis in Fig. 5.8d. The difference in the three BHP measurements is a result of offsets in the three curves so that they could be distinguished. The curve labeled BHP is from the BHP gauge, is corrected to the top of the perforations, and is not offset. Treatment pressures reached about 1,500 psi [10 MPa] above closure stress during the job.

The Nolte-Smith fracture-pressure analysis shows a small slope (0.21) for the first half of the treatment, but then an increasing slope (to 0.57) the final half of the treatment. We were concerned that this was a sign of impending screenout. The three shut-ins during the pad and the final shut-in were used for the Nierode analysis. The initial data showed a low fluid-loss coefficient (the accuracy of those low points is questionable), but the final shut-in indicates a much greater fluid loss. Whether this is real in the sense that much greater fluid loss occurred at late times (possibly into the natural fractures) or whether there is some other explanation is uncertain.

The Nolte analysis provided our most interesting look at fracture behavior. Attempts to fit the pressure-decline data to the type curve were unsuccessful until we realized that something unusual occurred at a dimensionless time of 0.7 (because the pump time was 100 minutes, this is 70 minutes after shut-in). The pre-0.7 data can be fit nicely, while the post-0.7 data flatten significantly. We interpret this as initial fracture closure on the proppant, at least near the wellbore. The only way this could occur so early was if the gel broke, the sand fell to the bottom of the fracture, and any additional incremental leakoff would then result in closure of the bottom of the fracture on the proppant. Using the early data for the analyses, we obtain a wing length of 520 ft [158 m] and a leakoff coefficient of 0.0012 ft $\sqrt{\text{min}}$ [0.0004 m/$\sqrt{\text{min}}$] if the height is 180 ft [55 m], or values of 420 ft [128 m] and 0.0015 ft/$\sqrt{\text{min}}$ [0.0005 m/$\sqrt{\text{min}}$], if the height is 200 ft [61 m]. The fluid efficiency in either case is 74%. Additional parameters for this analysis are a leakoff height of 80 ft [24 m], an ISIP of 7,300 psi [50.3 MPa], and a degradation exponent of 1.0.

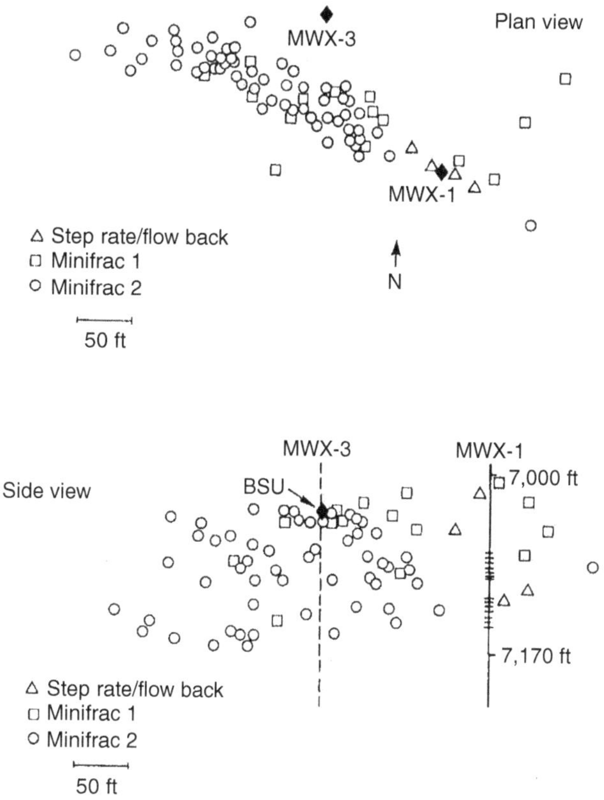

Fig. 5.7—Results of borehole geophones for minifractures.

Stage	Fluid	Amount (gal)	Sand Concentration (lbm/gal)	Sand (lbm)
1	Methanol	7,700	0	0
2	Apollo 40**	18,000	0	0
3	Apollo 35	3,000	1.5	4,500
4	Apollo 35	5,000	2.0	10,000
5	Apollo 35	6,000	3.0	18,000
6	Apollo 35	14,000	4.0	56,000
7	Apollo 25	18,000	5.5	99,000
8	Apollo 25	1,000	5.5	5,500
9	Water flush	8,764	0	0
	Total	81,464		193,000

*Courtesy of The Western Co.

**A 40-lbm/1,000-gal hydroxypropyl guar crosslinked with a titanium salt. Base fluid is a 3% KCl water with ½-gal/1,000-gal bactericide.

Table 5.2—Treatment design schedule*.

The post-fracture surveys were not very helpful in defining the fracture height, as shown in **Fig. 5.9.** We had trouble lowering the temperature tool into the zone because of sand fill, and we have, at best, a gross estimate of fracture top. The gamma survey only showed radioactive sand in the perforated interval and thus gave no indication of fracture height. Furthermore, the borehole geophones were plagued by high noise levels during this treatment, and only a few clear seismic signals were obtained. All that can be said about these data is that fracture azimuth was approximately the same as that of the minifractures, and the few signals that were obtained all fell within a 200-ft [61-m] height window. However, the accuracy of some of these points may have been poor.

5.14 Post-Fracture Cleanup

After completion of the treatment, the well was opened to clean up and produce back as much of the fracture fluid as possible. Recovery in the first few days was more than 50% of the total injected volume, but additional water recovery was much slower. Problems with sand in the wellbore, a stuck packer, and the well head and choke resulted in several shut-in periods and the circulation of water several times.

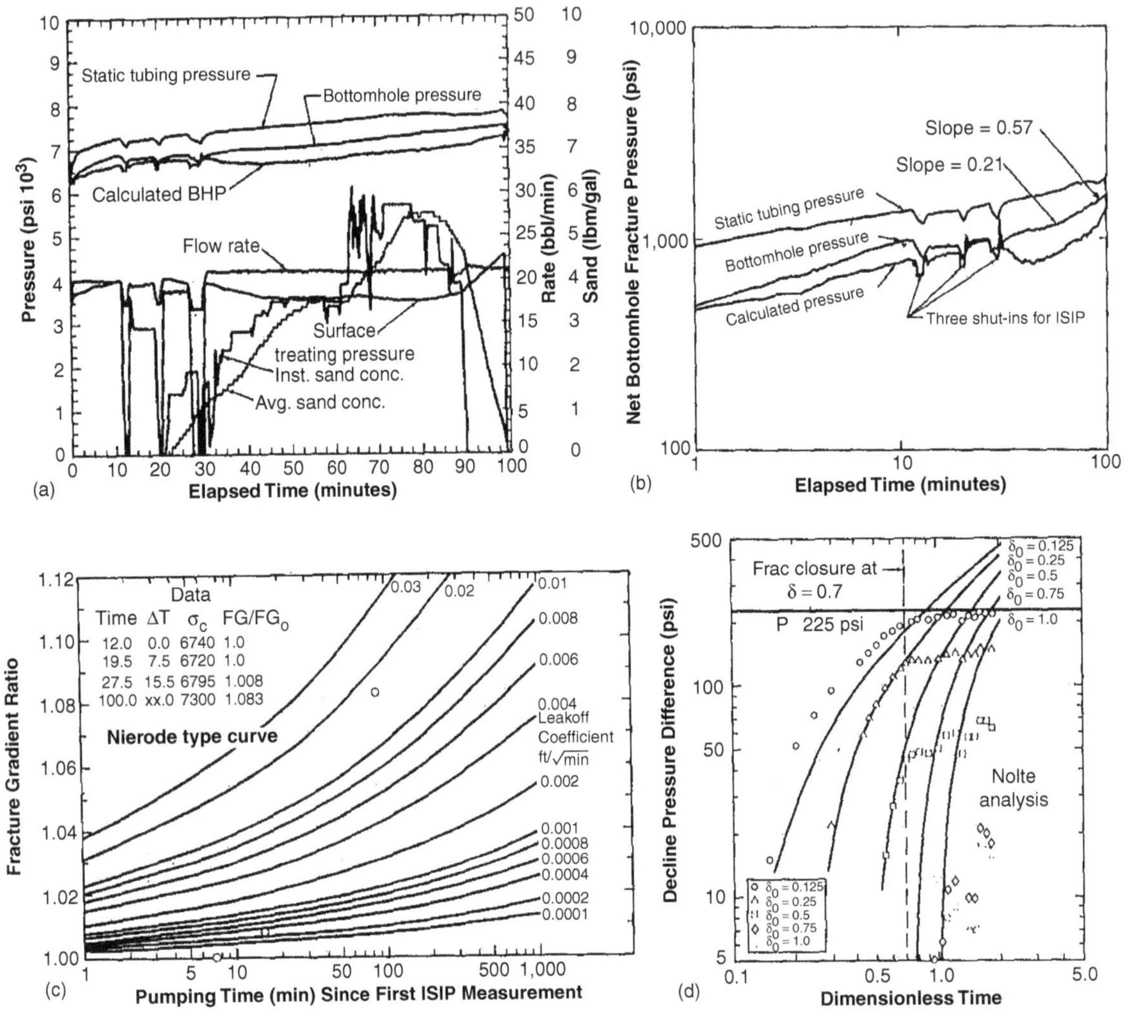

Fig. 5.8—Results of the Phase 2 stimulation.

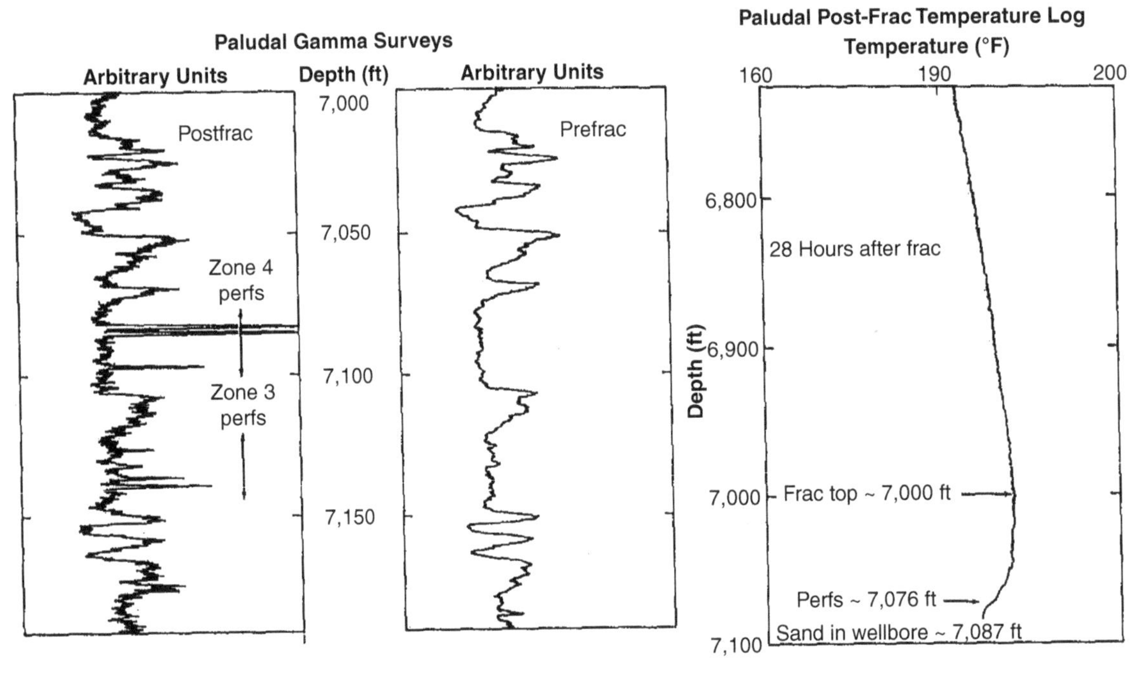

Fig. 5.9—Post-fracture survey results.

These may have hampered the cleanup process somewhat. Over the next 4 weeks, additional water recovery was poor, and the well produced less than 50 Mcf/D [1416 m^3/d]. Furthermore, it showed no signs of improving.

Several buildups over this period seemed to indicate that the fracture was very short (10 ft [3 m]), as if it were clogged or bridged. Continuous returned fluid samples were taken, and analysis of the organics indicated molecular weights on the order of 2×10^6. We were afraid that the minimal amounts of breaker that we used to ensure good proppant transport may have resulted in inadequate breaking of the gel (at least over this period of a few weeks) and possibly a gel plug existed in the near-wellbore region. Because schedule concerns would not allow us to wait indefinitely for the gel to break completely by formation temperature, we proceeded with a remedial breaker treatment in an attempt to break any gel plug and to ensure conductivity in the fracture.

5.14.1 Remedial Treatment. The remedial treatment consisted of 6,500 gal [25 m^3] of 3% KC1 water with 135 lbm/1,000 gal [16 kg/m^3] of ammonium persulfate breaker plus 1,000 gal [3.8 m^3] of 3% hydrogen peroxide. This 7,500-gal [28-m^3] total volume was sufficient to fill the fracture porosity with excess. It was injected at sufficiently low rates (1 to 2 bbl/min [0.002 to 0.005 m^3/s]) to keep the BHP below 6,000 to 6,200 psi [41 to 43 MPa] and thus not open the fracture. This treatment was completed with no problems, but we saw an immediate 1- to 2-psi [7 to 14-kPa] response with a BHP gauge in Well MWX-2. Whether this was an actual interference through connected permeability or a poromechanical response to the pressurized crack is uncertain; both are possible.

The load water from the remedial treatment was recovered within a week, and fluid recovery after this treatment was two to three times better than before. Water samples showed that the molecular weights of the polymer were now less than 200,000, down from about 1,000,000. Unfortunately, while this treatment cleaned up the gel residues, it also resulted in the formation of large amounts of iron oxide precipitates because of the reactive nature of the chemicals. Most of these were from reaction with the tubing and casing; however, core studies after the fact showed a 40% reduction in permeability and more than doubled cleanup times. Permeability reduction in an artificially created fracture in the core was about 90% with cleanup times increasing about two orders of magnitude. We may have exchanged one problem for another. Additionally, ionic analyses of the returned fluids indicated that we were producing increased amounts of formation waters, possibly from the coals.

The well still did not flow readily, and we continued cycling the well to produce as much fluid as possible. Throughout this cleanup phase, the well configuration consisted of open-ended tubing situated just above the perforations. Gas flow rates were too low to give us lift up the tubing. After 1 month of gradual cleanup, we changed the well configuration and inserted a packer with tubing extending below the perforations. The packer gave us better control of BHP (we find that we need to maintain at least 1,000-psi [7-MPa] BHP to keep gas flowing in these formations), and the long tubing tail provided a method to drain the maximum amount of fluid from the fracture and wellbore. We immediately recovered about 50 bbl [8 m^3] of liquid and obtained gas flows in excess of 100 Mcf/D [2832 m^3/d]. At this point, the total recovery of fracture fluid was somewhat less than 70%, while the recovery of all fluids put in the well since the treatment was more than 80%.

5.15 Post-Fracture Well Testing

We continued to flow Well MWX-1 while we prepared the other two wells for another interference test (Branagan et al. 1985). In mid-July 1984, this test was started, and the important data are shown in **Fig. 5.10 and Table 5.3.** Specific reservoir and completion data are given in **Table 5.4.** The testing consisted of three drawdowns with two interspersed buildup pulses and a long, final buildup in Well MWX-1. The maximum sustainable flow rate (AOF) was about 170 Mcf/D [4814 m^3/d], down from the previous flow tests. No clear indication of interference was seen in Well MWX-2 or MWX-3 during these tests. While some pressure disturbances can be seen in Fig. 5.10, none of these are correlatable with flow or shut-in periods in Well MWX-1 on a constant delta-time basis; i.e., if we see a pressure response as a result of one pulse at some delayed time Δt, we should see another pressure response as a result of the next pulse at approximately the same Δt after the pulse. No such behavior is obvious.

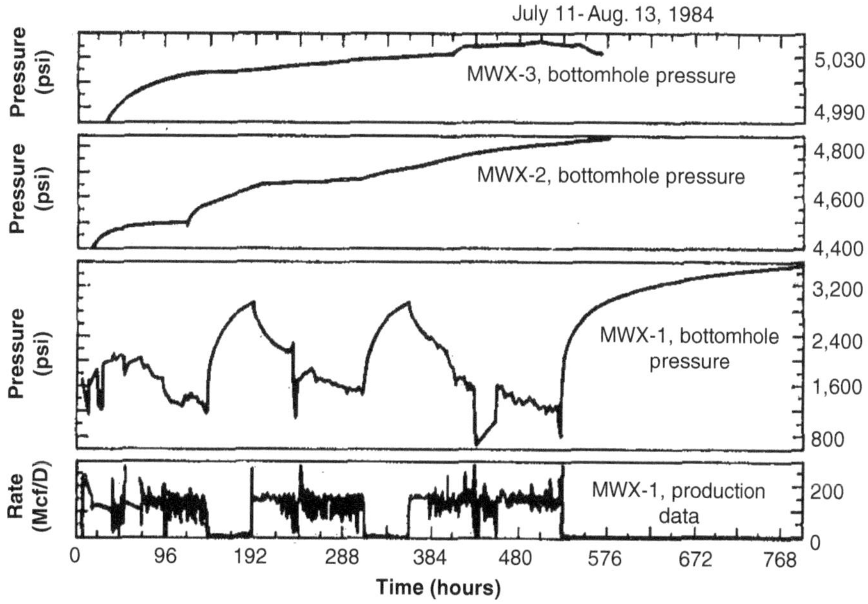

Fig. 5.10—Post-fracture well test data.

	Production		Cumulative
Time (days)	Rate (Mcf/D)	BHP (psi)	Gas (Mcf)
0.5	225	1,307	150
1.0	114	1,514	193
1.5	107	1,971	257
2.0	64	2,071	296
2.5	120	1,958	369
3.0	171	1,932	437
3.5	147	1,726	524
4.0	171	1,504	597
4.5	157	1,319	670
5.0	151	1,400	741
5.5	150	1,327	817
5.87	146	1,222	852
6.02	0	2,004	
6.06	0	2,067	
6.13	0	2,145	
6.28	0	2,281	
6.43	0	2,393	
6.58	0	2,488	
6.73	0	2,566	
7.05	0	2,698	
7.20	0	2,748	
7.36	0	2,792	
7.51	0	2,831	
7.65	0	2,877	
7.80	0	2,901	
7.9	182	2,847	862
8.5	142	2,439	953
9.0	159	2,281	1,047
9.5	120	2,149	1,108
10.0	154	1,702	1,196
10.5	81	1,860	1,233
11.0	126	1,702	1,284
11.5	155	1,658	1,389
12.0	158	1,580	1,458
12.5	152	1,515	1,533
13.0	120	1,521	1,597
13.11	0	2,056	
13.18	0	2,158	
13.29	0	2,265	
13.36	0	2,329	
13.44	0	2,385	
13.51	0	2,435	
13.58	0	2,481	
13.81	0	2,585	
14.25	0	2,766	
14.55	0	2,848	
14.99	0	2,892	
15	145	2,764	1,605
15.5	152	2,469	1,688
16.0	151	2,285	1,755
16.5	171	2,265	1,827
17.0	189	1,738	1,929
17.5	164	1,708	2,004
18.0	173	1,540	2,076
19.0	159	1,556	2,190
20.0	152	1,384	2,346
21.0	162	1,249	2,492
21.93	186	864	2,641
21.96	0	1,019	
21.98	0	1,733	
22.03	0	1,948	
22.08	0	2,077	
22.21	0	2,275	
22.32	0	2,365	
22.54	0	2,510	
22.78	0	2,615	
23.01	0	2,705	
23.24	0	2,778	
23.59	0	2,861	
24.05	0	2,952	
24.49	0	3,022	
24.93	0	3,078	
25.38	0	3,124	
25.80	0	3,164	
26.46	0	3,216	
27.13	0	3,261	
27.79	0	3,299	
28.53	0	3,338	
29.50	0	3,386	
30.70	0	3,438	
31.44	0	3,466	
32.45	0	3,501	

*Water production less than 2 B/D; no condensate.

Table 5.3—(Continued).

Table 5.3—Post-fracture well-test data*.

To interpret these results, both homogeneous and dual-porosity reservoir simulators were used. We assumed that the far-field in-situ permeability was 36 μd because of the natural-fracture system, but the natural-fracture system near the induced hydraulic fracture was plugged and the permeability in this damaged region was equal to the matrix permeability, about 2 μd. We proceeded to match the Well MWX-1 BHP data by varying the hydraulic-fracture length and the thickness of the damaged zone around the fracture. A suitable match (see Branagan et al. 1985) was found with a damage thickness of 9 ft [2.74 m] on both sides of the fracture and an effective fracture wing length of 75 ft [23 m]. This short length could be a result of either a short fracture or a narrow reservoir channel; the two cases are indistinguishable.

Thus, our best match of the well test data shows that we further damaged the existing natural-fracture system, as we also did in the minifractures. Because we believe from the minifracture results that we had a fracture longer than 75 ft [23 m], the channel reservoir size appears to be only about 150 ft [46 m] in total permeable width where the fracture intersects it.

We expect that the reservoir damage would have continued to clean up with much longer production times, but the length of our test time was constrained by other factors. Nevertheless, such long cleanup times will be a problem for any operator.

5.16 Final Production Tests After a Long-Term Shut-In

Testing of these paludal sands was suspended in mid-Aug. 1984, while we tested other zones uphole. In late March 1986 (20 months later), however, an opportunity arose to retest these zones to evaluate any time dependence of the suspected damage. The zone was flow tested for 7 weeks with an initial rate of 420 Mcf/D [11.9 m^3/d], a cumulative produced volume of 15.8 MMscf [447×10^3 std m^3], and an average rate of 320 Mscf/D [9060 std m^3/d]. No liquids were produced initially, but after 5 days, water production started, increased rapidly to about 35 B/D [5.6 m^3/d], and eventually totaled 860 bbl [137 m^3].

These results suggest that the damage after the fracture is reversible and is probably caused by water and gel blockage of the natural fractures. Over the long shut-in, the gel may have degraded further, and imbibition of the water into the matrix rock probably cleared the natural fractures of most water and dehydrated any remaining gel. When production was resumed, gas production through the natural fractures was no longer hindered, and flow rates were much closer to the expected values.

Reservoir Parameters	
Initial pressure, psi	5,400
Temperature, °F	210
Gas specific gravity	0.67
Net porous thickness, ft	26
Average porosity, %	10.2
Average permeability, μd	36
Average water saturation, %	46
Average Core permeability, μd	1
Completion Parameters	
Perforation depths, ft	7,076 to 7,100
	7,120 to 7,144
Casing ID, in.	6,184
Tubing ID, in.	2,441
Packer depth, ft	7,050
Bridge plug depth, ft	7,200
Pressure Buildup Wellbore Volumes, bbl	
Buildup 1 and 2	46
Final buildup with downhole closure	6

Table 5.4—Paludal test parameters.

5.17 Discussion

5.17.1 Lens Morphology. The final well test analysis showed that the intersection of the fracture length and the reservoir was only about 150 ft [46 m]. Because we believe the fracture was much longer than this (the minifractures were about 400 ft [122 m] and the Nolte analysis gave a minimum length of 400 ft [122 m] for the main treatment), the probable reason for this result is a narrow channel width. This is consistent with widths measured in outcrop by Lorenz (1985) and shows the importance of such studies for a complete evaluation of stimulation and well test data, as well as for treatment design and for production and economic forecasts.

5.17.2 Natural Fractures. A comparison of core and well test data leads us to conclude that the major flow paths are natural fractures and that the system must be fairly interconnected to show no sign of linear flow. Any well completion operations must be carefully designed to limit any damage to the natural-fracture system. We also find that these reservoirs will not produce unless a backpressure (in this case about 1,000 psi [7 MPa]) is maintained at the sand face. Branagan et al. (1985) suggest that this may be a stress sensitivity of the natural-fracture system.

5.17.3 Interference. During the prefracture and post-fracture well tests, no certain evidence of interference was found in the pressure data or even in the slopes of the pressure data. While this may have been masked somewhat by the pressure recovery occurring simultaneously in the interference wells (see Fig. 5.10), permeabilities of 36 μd and flow rates of 250 Mcf/D [7180 m^3/d] should have been sufficient to impart an observable superposed pressure response. Factors hampering communication are the thinning of Zone 3 at Well MWX-3, the lack of Zone 4 in Well MWX-2, and the fault in Well MWX-2. Even with these features, good sand connectivity should exist. We can only surmise that this lack of interference is a result of one of two things. Either the fracture networks are isolated and may not intersect all wells, or sufficient shale breaks exist that each well is in communication with an isolated reservoir. If the latter is true, analysis becomes extremely difficult.

5.17.4 Abnormal Treatment Pressures. During all fracturing tests in this zone, we observed very high pressures during the treatments. Several possible reasons exist for such high pressures. The first possibility is complex fracturing, as discussed by Medlin and Fitch (1983) for other wells in the Mesaverde in the Piceance Basin. The second possibility is that the fracture reached the lateral ends of the lens so that treatment pressures would need to increase substantially for excessive fracturing to occur in the higher-stress shales. The short, effective fracture lengths measured in well tests make this credible. The third possibility is backstresses caused by fluid leakoff, but this is typically a much smaller effect. The fourth possibility is the presence of high-stress stringers, possibly between the zones, that would reduce fracture widths and increase pressures somewhat. All these factors may have contributed to the pressure levels. Whatever the cause, high pressures can result in many deleterious effects; the most obvious one is wider, shorter hydraulic fractures.

5.17.5 Reservoir Damage. All the well test data support the premise of damage to the natural fractures as a result of the treatment. The important questions are why this damage occurs and why it is so hard to clean up. We believe that the degree of damage is in large part a result of the high treatment pressures. These high pressures may have either opened the intersecting natural fractures or forced the gel into them under such high differential pressures that production of these gels was difficult under normal reservoir conditions. This indicates that cleanup times will always be excessive when treatments are conducted in this environment with a water-based system. One good alternative is to use a foam fracturing system so the fractures are exposed to less water. Another alternative is not to fracture at all and accept the initial 250 Mcf/D [7080 m^3/d], or possibly to try a treatment, such as tailored pulse loading (Cuderman 1986), where the only intent is to obtain a good connection with the natural-fracture system.

5.17.6 Diagnostics and Analyses. After the minifractures, we were surprised at the similarity of the results of the borehole geophones, the temperature log, and the Nolte analysis. This agreement in independent diagnostic techniques gave us confidence in their accuracy. Additionally, the fracture azimuth determined during the treatment agreed with prefracture predictions obtained from core, log, and surface stress orientation measurements (Teufel et al. 1984). The data from the main treatment, while disappointing, produced consistent trends in fracture behavior.

5.17.7 Containment. We apparently had acceptable containment even though the treatment pressures were much higher than any barrier stress, so any barrier (particularly the top one) should have been penetrated. We attribute this to these high-stress layers acting as restrictions rather than absolute barriers. Because of the high stresses, fracture widths are much smaller in these layers, and thus, pressure drops become very large when we try to force fluid through them at any fast rates. Thus vertical growth occurs quite slowly. In addition, proppant may bridge in these narrow regions.

5.17.8 Coals. We had been concerned about the effect of the coals in the treatment and gas production, but no obvious deleterious developments seemed to occur. One positive effect was the high stress in the coal above Zone 4 (the highest stress in the paludal interval), which probably was a significant factor limiting height growth. No obvious loss of fracture fluid occurred because of the coals, but water production from the coals may have been partly responsible for the hindered cleanup.

5.17.9 Proppant Crushing. We designed the treatment with a sand proppant because we felt that a closure stress of 5,900 psi [40.7 MPa] combined with a minimum wellbore pressure of more than several hundred psi would result in acceptable conductivities for the sand pack, particularly with our formation permeabilities. With the cleanup problems and repeated blowdowns, however, we probably often had effective stresses on the proppant approaching the total closure stress. When combined with the cyclic fatigue loading, these high stresses may have resulted in much reduced conductivity. We suggest that in these situations the proppant should be overdesigned, at least near the wellbore.

5.17.10 Reversibility of the Damage. Our ability to produce this zone at much higher rates after the 20-month shut-in demonstrates that the damage is primarily liquid-induced (gel, water), as opposed to mechanical damage to the natural fracture system. Because the initial production after shut-in was dry, we can also deduce that imbibition of the fracture load water into the matrix pore volume was a positive factor because it cleared the natural fractures.

5.18 Conclusions

We have successfully completed a fracturing and testing experiment in the paludal zone of the Mesaverde. While the experiment was disappointing in terms of gas production, the primary objectives of understanding hydraulic-fracture behavior and reservoir characteristics in lenticular formations have been largely realized. Important considerations—such as lens morphology, reservoir characteristics, gas flow mechanics, fracture containment and lateral extension out of lenses, fracture fluid damage, and many others—have been addressed. We hope that the discussion of these topics and the recommendations that we have made will be helpful to other operators.

5.19 Acknowledgments

We thank Dave and Bill Holcomb of Smith Energy Services, and Larry Harrington, Lacy Lance, and Laddie Harp of The Western Co. for their help and suggestions during these experiments. Special thanks go to the CER field crew, whose hard work made many of these studies successful. We also thank Carolyne Hart for her diagnostic work, Gerry Kukal for his log analyses, and Clarence Raible and Bonnie Gall at the Natl. Inst, of Petroleum Energy Research for many of the laboratory analyses. Special core analyses were provided by Core Laboratories. This work was sponsored by the U.S. DOE's Western Gas Sands Subprogram, which is managed by Morgantown Energy Technology Center under Contract No. DE-AC04-76DP00789.

5.20 References

Branagan, P., Cipolla, C. L., Lee, S. J. et al. 1985. Comprehensive Well Testing and Modeling of Pre- and Post-Fracture Well Performance of the MWX, Lenticular Tight Gas Sands. Presented at the SPE/DOE Low Permeability Gas Reservoirs Symposium, Denver, 19–22 May. SPE-13867-MS. https://doi.org/10.2118/13867-MS.

Cuderman, J. F. 1986. Tailored-Pulse Fracturing in Cased and Perforated Borehole. Presented at the SPE Unconventional Gas Technology Symposium, Louisville, Kentucky, USA, 18–21 May. SPE-15253-MS. https://doi.org/10.2118/15253-MS.

Hart, C. M. and Engi, D. 1987. Fracture Diagnostics Results for the First Multiwell Experiment's Paludal Zone Stimulation. *SPE Form Eval* **2** (03): 320–326. SPE-12852-PA. https://doi.org/10.2118/12852-PA.

Kukal, G. 1984. Multiwell Experiment Well Log Analysis of Paludal Interval. CER Report, Las Vegas, Nevada, USA (October 1984).

Lorenz, J. C. 1985. Predictions of Size and Orientations of Lenticular Reservoirs in the Mesaverde Group, Northwestern Colorado. Presented at the SPE/DOE Symposium on Low Permeability Gas Reservoirs, Denver, 19–22 May. SPE-13851-MS. https://doi.org/10.2118/13851-MS.

Lorenz, J. C., Heinze, D. M., Clark, J. A. et al. 1985. Determination of Widths of Meander-Belt Sandstone Reservoirs from Vertical Downhole Data, Mesaverde Group, Piceance Creek Basin, Colorado. *AAPG Bulletin* **69** (05): 710–721.

Lorenz, J. C., Warpinski, N. R., Branagan, P. T. et al. 1986. Fracture Characteristics and Reservoir Behavior of Stress-Sensitive Fracture Systems in Flat-Lying Lenticular Formations. Paper presented at the SPE Unconventional Gas Technology Symposium, Louisville, Kentucky, USA, 18–21 May. SPE-15244-MS.

Medlin, W. L. and Fitch, J. L. 1983. Abnormal Treatment Pressures in Massive Hydraulic Fracturing Treatments. Presented at the SPE Annual Technical Conference and Exhibition, San Francisco, 5–8 October. SPE-12108-MS.

Nierode, D. E. 1985. Comparison of Hydraulic Fracture Design Methods to Observed Field Results. Presented at the SPE Annual Technical Conference and Exhibition, San Francisco, 5–8 October. SPE-12059-MS.

Nolte, K. G. 1979. Determination of Fracture Parameters from Fracturing Pressure Decline. Presented at the SPE Annual Technical Conference and Exhibition, Las Vegas, Nevada, USA, 23–26 September. SPE-8341-MS. https://doi.org/10.2118/8341-MS.

Nolte, K. G. and Smith, M. B. 1981. Interpretation of Fracturing Pressures. *J Pet Technol* **33** (09): 1,767–1,775. SPE-8297-PA. https://doi.org/10.2118/8297-PA.

Randolph, P. L. 1983. Porosity and Permeability of Mesaverde Sandstone Core from the US DOE Multiwell Experiment, Garfield County, Colorado. Presented at the SPE/DOE Low-Permeability Gas Reservoirs Symposium, Denver, 14–16 March. SPE-11765-MS. https://doi.org/10.2118/SPE-11765-MS.

Sattler, A. R., Raible, C. J., and Gall, B. R. 1985. Integration of Laboratory and Field for Insight on the Multiwell Experiment Paludal Stimulation. Presented at the SPE/DOE Low Permeability Gas Reservoirs Symposium, Denver, 19–22 May. SPE-13891-MS. https://doi.org/10.2118/13891-MS.

Searls, C. A. 1985. The Multiwell Experiment Geophysics Program Final Report. Sandia National Laboratories Report, SAND85-1013, Albuquerque, New Mexico, USA (Sept. 1985).

Teufel, L. W., Hart C. M., Sattler, A. R. et al. 1984. Determination of Hydraulic Fracture Azimuth by Geophysical, Geological, and Oriented-Core Methods at the Multiwell Experiment Site, Rifle, CO. Presented at the SPE Annual Technical Conference and Exhibition, Houston, 16–19 September. SPE-13226-MS. https://doi.org/10.2118/13226-MS.

Tight Gas Reservoirs. 1980. In *Unconventional Gas Sources,* Part I: Resource Assessment, Vol. V. Washington, DC: National Petroleum Council.

Warpinski, N. R., Branagan, P., and Wilmer, R. 1985. In-Situ Stress Measurements at US DOE's Multiwell Experiment Site, Mesaverde Group, Rifle, Colorado. *J Pet Technol* **37** (03): 527–536. SPE-12142-PA. https://doi.org/10.2118/SPE-12142-PA.

SI Metric Conversion Factors

°API	141.5/(131.5 + °API) = g/cm^3
bbl 1.589 873	E–01 = m^3
ft × 3.048*	E–01 = m
ft^3 × 2.831 685	E–02 = m^3
°F (°F-32)/1.8	= °C
gal × 3.785 412	E–03 = m^3
in. × 2.54*	E+00 = cm
lbm × 4.535 924	E–01 = kg
psi × 6.894 757	E+00 = kPa

*Conversion factor is exact.

Chapter 6

Microseismic Monitoring of the B-Sand Hydraulic-Fracture Experiment at the DOE/GRI Multisite Project

N. R. Warpinski, SPE, Sandia Natl. Laboratories; **T. B. Wright***, SPE, Resources Engineering Systems; **J. E. Uhl, B. P. Engler,** and **P. M. Drozda,** Sandia Natl. Laboratories; and **R. E. Peterson,** SPE, and **P. T. Branagan,** SPE, Branagan & Assocs. Inc.

 Journal paper SPE 57593 was published in vol. 4 (3) of *SPE J.,* pages 242–250. Original symposium paper was revised for publication from paper SPE 36450 and presented at the SPE Annual Technical Conference and Exhibition held in Denver, Colorado, 6–9 October 1996.

SPE 57593 presents a case history of microseismic monitoring of six hydraulic fracture injections into the B Sand at the M-Site in the Piceance Basin in Colorado. The test consists of one treatment well and two monitoring wells in which receivers were placed. The six injection tests were step-rate tests, two small-volume pump-in tests, two mini fracture treatments, and a propped fracture treatment using crosslinked gel. Warpinski et al. were able to map fracture geometries in three dimensions for all six injections on the basis of microseismic events and compared with the fracture geometries independently obtained using a lumped 3D fracture model. From the case history, Warpinski et al. draw the following conclusions:

- Microseismic imaging technology may provide reasonable estimate of fracture length, height, and azimuth
- Accurate height measurement requires a multilevel receiver system to account for multilayer formations
- Fracture length changes slightly from small-volume, low-viscosity fluids at low rates to larger-volume, high-viscosity fluids at high rates
- Height grows when the main fracture treatment is pumped
- Gas producers in the Piceance Basin and other western basins with similar high-stress contrast need to work on fracture height containment
- The discrepancies between fracture diagnostics and the fracture modeling suggest the need for more scientific and engineering research and studies to fully understand fracturing, especially in the typical oilfield environment

6.1 Summary

Six hydraulic-fracture injections into a fluvial sandstone at a depth of 4,500 ft were monitored with multilevel, triaxial seismic receivers in two wells, resulting in maps of the growth and final geometry of each fracture based upon microseismic activity. These diagnostic images show that the hydraulic fractures are highly contained for smaller-volume KCl-water injections, but height growth is significant for the larger-volume, higher-rate, higher-viscosity treatments. Fracture lengths for most injections are similar regardless of volume and fluid type.

6.2 Introduction

The imaging of hydraulic fractures at depth has been a long-sought goal of the petroleum industry. Fracturing is an expensive, yet essential, element of production for many gas reservoirs and has significantly improved economics for many oil reservoirs as well. The Gas Research Institute (GRI) and the US Department of Energy have long funded diagnostics programs and have achieved slow but steady progress toward the realization of such technology. Fracture imaging is now attainable due to the improved capabilities of advanced receiver technology, advanced telemetry, and portable high-power computing. It is only characteristics of the reservoir and the well configurations which limit the potential of the technique. This paper describes the results of a series of fracturing experiments in a single reservoir interval that demonstrate the capabilities of this technology and its value to related issues of modeling and fundamental model mechanisms.

---

*Now with Pinnacle Technologies Inc.

6.3 Background

Fracture diagnostics have a long history that includes production history matching, postfrac well testing, radioactive tracers, temperature logs, pressure decline analysis, treatment pressure analysis and modeling, surface tiltmeters, surface electromagnetic techniques, and various seismic techniques. Most of these diagnostics are indirect, in the sense that they measure some parameter associated with the hydraulic fracture and infer fracture characteristics from the parameter or its changes. As with most indirect techniques, the inversion process can introduce large uncertainties.

One of these techniques is different, however, and is capable of producing a highly accurate image of the fracture without the processing difficulties inherent in inversion problems. The microseismic method (Albright and Pearson 1982; Lacy 1987; Dobecki 1983; Thorne and Morris 1988; Sarda and Deflandre 1988; Stewart et al. 1992), one of several seismic technologies, is an indirect technique in that it maps the locations of microseisms (small shear slippages on natural fractures, bedding planes, or other weak features) occurring in the vicinity of the fracture (rather than the fracture itself). However, these microseisms can be accurately located and they do not require any inversion algorithms that suffer from uniqueness and resolution questions. With a proper understanding of the microseismic activity expected for the reservoir under consideration, the relationship of the microseisms to the fracture can generally be clearly established. For example, in reservoirs with highly compressible fluids (e.g., gas reservoirs), the envelope of microseisms approximates the fracture dimensions, with the exception of the width. Thus, the microseismic method can produce relatively accurate images of the fracture length, height, and azimuth.

The microseismic technique is the primary diagnostic method employed to map hydraulic fracture geometry at the M-Site—the focus of this paper. Microseismic imaging technology, conducted in several elaborate field experiments (Albright and Pearson 1982; Vinegar et al. 1992; Truby and Keck 1994; Sleefe et al. 1995; Middlebrook et al. 1993), has evolved to the point where a wireline-run fracture-imaging service is practical. In the process of conducting these field experiments, verification that the imaged fracture geometry closely approximates the actual fracture geometry has been an essential objective to build confidence in this technology. The M-Site experiments have been designed to provide that validation, particularly through the application of downhole inclinometers and intersecting wells used as independent techniques to confirm fracture geometry. In addition to improving fracture diagnostics, a secondary goal of the M-Site experiments has been to improve and validate fracture models and to better understand fracture-growth mechanisms. This goal was accomplished by performing tests in a reservoir with measured properties, fully monitoring the fracture treatments, obtaining time-imaged fracture geometries, comparing the diagnostic results with various models, and using specialized fracture monitoring technology (i.e., an intersecting well) to provide previously unavailable information.

6.4 M-Site

In-depth details of the M-Site layout and fracture treatments are given by Peterson et al. (1996) and are only briefly repeated here. The M-Site Field experiments, located at the previous Multiwell Experiment site in the Piceance Basin of Colorado, are co-funded by the GRI and the US Department of Energy. The reservoirs undergoing testing are fluvial Mesaverde Sand-Shale sequences, so the technologies developed in this difficult environment are easily translatable to many other reservoirs throughout the world. Details of previous work can be found in several papers and reports (Sleefe et al. 1995; Middlebrook et al. 1993; Peterson et al. 1995, 1996).

A schematic of the well, instrument, and Sandstone-reservoir layout is shown in **Fig. 6.1.** The site consists of one treatment well (MWX-2), one monitor well with cemented-in triaxial seismic receivers and biaxial inclinometers, and one cased observation well for wireline-run tools (MWX-3). Not shown in the figure are deviated lateral intersection wells that penetrate through the created hydraulic fractures in each sandstone.

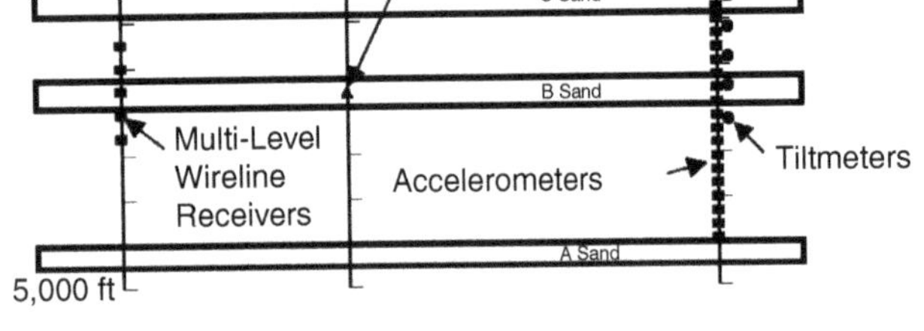

Fig. 6.1—Layout of M-Site wells and instrumentation.

The monitor well provides the instrumentation for validating the seismic results. Thirty triaxial receiver stations, with low- noise, wide-bandwidth accelerometers provide high quality microseismic data that can be accurately located. In the same well six biaxial inclinometers with nanoradian resolution provide information on the mechanical deformation of the formation that is used to validate aspects of the seismic results. The 7-in. cased observation well is used for multilevel, wireline-run, triaxial receiver arrays of the type that would be used in a commercial fracture diagnostic service. This array uses the same accelerometers that are grouted in the monitor well, and the multilevel feature also provides for highly accurate microseismic event location. The monitor well receiver array, with many more levels to apply in location algorithms, generated the reference results used to verify the fracture geometry imaged by the wireline receiver arrays.

Additional information obtained in the treatment well, such as bottomhole pressure, spectral gamma logs of radioactive tracer distributions, and seismic surveys, are used for detailed fracture modeling and additional diagnostic information. Detailed stress, rock property, and reservoir property data are also available for these reservoirs and are used for fracture models, finite element deformation models, and analyses of the mechanical response of the formation to the fracture treatment.

Additionally, crosswell seismic surveys were conducted to determine the compressional (*P*-wave) and shear (*S*-wave) velocity structure at the site. Seismic data were obtained with 5 ft source and receiver spacings in the treatment and monitor well, respectively, and the permanent 30 ft spacing of the cemented receivers in the monitor well. The seismic source was an airgun that provided excellent *P* waves and generally good *S* waves. Both *P* and *S* tomograms have been produced, as well as Poisson's ratio and calculated uniaxial stress tomograms (Branagan et al. 1997; Warpinski et al. 1998). Average velocities used for the microseismic analyses were obtained from the crosswell seismic survey.

The reservoir geometry of the B Sandstone is somewhat complicated by a second sand lobe that lies below the main 30 ft interval of this sandstone. **Fig. 6.2** shows a gamma log of this interval taken from the treatment well. A zone from 4,527 to 4,557 ft was perforated with 2 shots/ft. The stress contrasts around the B Sand, as determined from microfracture stress measurements, are shown in **Fig. 6.3.**

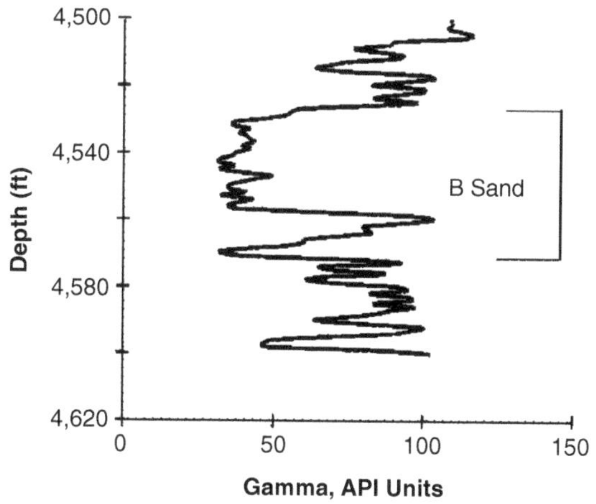

Fig. 6.2—Gamma-ray log showing B Sand.

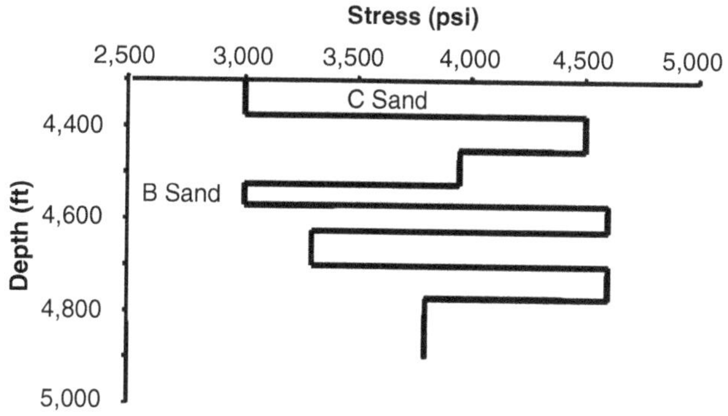

Fig. 6.3—Stress contrasts around B Sandstone.

For modeling purposes, more-detailed calibrated stress logs were also developed by using measured data to correct the log-calculated stresses (with a tectonic stress offset).

During B-Sand testing, six different fracture injections using three different fluids were monitored. Important information about the injections is given in **Table 6.1.**

	Fracture	Volume (bbl)	Fluid	n′	K′ (lb-sec$^{n'}$/ft^2)	Rate (bpm)	Sand (lb)
2B	Step Rate	27	KCl	1	2×10^{-5}	0.5–3	
3B	Pump-in No. 1	100	KCl	1	2×10^{-5}	10	
4B	Pump-in No. 2	210	KCl	1	2×10^{-5}	10	
5B	Minifrac No. 1	400	40# Linear	0.558	6.93×10^{-3}	22	
6B	Minifrac No. 2	400	40# Linear	0.558	6.93×10^{-3}	22	
7B	Propped Frac	670	*X* Link gel	0.55	2.1×10^{-1}	20	77,600

Table 6.1—Treatment data.

6.5 Receiver Orientation

The orientations of cemented-in receivers in the monitor well were determined using three different techniques. During installation of the receivers, a radioactive set screw was inserted in a known location on the receiver and, subsequent to cementing, was azimuthally located using a directional gamma-ray logging tool. This orientation was confirmed using *P*-wave particle motion detected by each receiver from perforations in the treatment well and from a crosswell seismic survey. With the exception of a couple of levels in which one or more receiver axes were clearly nonfunctioning, there was good agreement between the different orientation techniques.

Orientation of the five-level receiver system was accomplished by conducting an airgun orientation scan in the treatment well at 10 ft spacing over the interval in which microseisms were expected. Particle motions of *P*-wave arrivals for each level were analyzed to provide accurate orientation.

6.6 Microseismic Processing

Microseisms were processed in the standard manner, with *P*-wave arrivals, *S*-wave arrivals, and *P*-wave particle motion detected on as many levels from each well as possible. Considerable automatic processing was performed to aid in initial analyses, but all results were quality checked by an analyst. Initial processing started with detection of microseisms using an event-comb algorithm (both spatial and temporal comb) on unfiltered and non-transformed data. The event comb declares events if they

- Have amplitudes and rise times exceeding a threshold factor times the background levels
- Are of sufficiently long duration (at least 1 msec to avoid noise spikes)
- Are detected on at least three levels
- Are first detected within a period of time (starting at the first trigger) not exceeding a specified value (≈ 12 msec)

Some advanced seismic analyses were conducted [using the Nelson and Vidale (1990, 1988) algorithms] for checks, but the initial seismic results using simpler techniques were found to be quite accurate because of the large number of receiver stations.

The initial results assume a single *P*-wave velocity V_p, single *S*-wave velocity V_s, and full azimuthal isotropy. The distance and elevation to a microseism are computed from a joint *P/S* regression of the form

$$F = w_p \sum_n \left[v_p^2 \left(t_{pi} - t_o\right)^2 - \left(r_i - r_0\right)^2 - \left(z_i - z_0\right)^2 \right]^2 + w_s \sum_m \left[v_s^2 \left(t_{si} - t_o\right)^2 - \left(r_i - r_0\right)^2 - \left(z_i - z_0\right)^2 \right]^2, \quad \text{(6.1)}$$

where F is the function to be minimized, t_{pi} is the *P*-wave arrival time, t_{si} is the *S*-wave arrival time, r_i is the distance from the monitor well reference location to the *i*th receiver (usually zero for a vertical well), z_i is the depth of the *i*th receiver, and t_o, r_0, and z_0 are the origin time, radial distance, and depth of the source location, respectively. The parameters w_p and w_s are weighting functions and the summation is

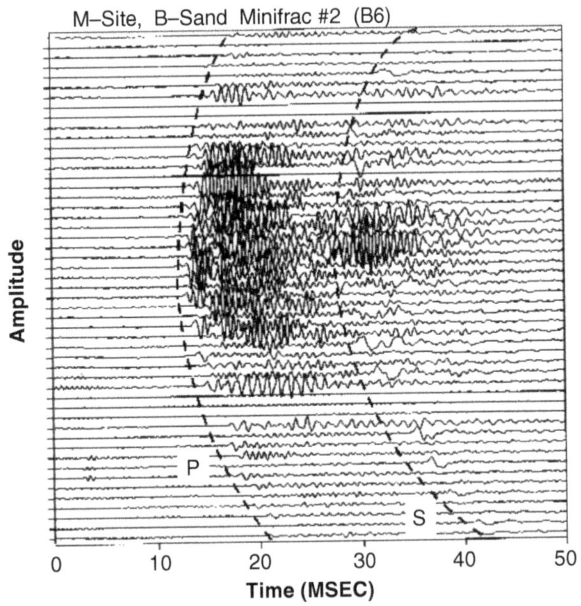

Fig. 6.4—Example microseism.

over n *P*-waves and m *S*-waves given that there are often occasions when one of the phases may not be detectable on a given level. The weighting functions are used if one of the phase arrivals is less reliable than the other, but at the M-Site, both the *P* and the *S* arrivals were clear so weighting values of 1.0 were used for w_s and w_p.

Fig. 6.4 shows an example of microseism detection on the monitor-well array across 17 levels. In this typical example, the *P*-wave arrival is discernible on nearly every level (a few levels have nonfunctioning accelerometers) and *S*-wave arrivals are evident on many of the upper and lower levels. Given such detailed data, accurate locations for most microseisms are routinely calculated. One exception to the height accuracy occurred during the initial injections using KCl water. During these tests only a single receiver was used in MWX-3 and elevation locations from a single receiver are notoriously inaccurate because the velocity structure of the formation results in bent ray paths and misleading vertical locations. To avoid misleading data for fracture height, all single receiver data were scaled back to fit within the height determined by the full array of monitor-well receivers for the same injection. Thus, height on these tests was only determined by the monitor-well instrumentation and the single-level data are only used for length and azimuth information. For the minifracs and propped fracture treatment, the five-level array in MWX-3 was fully capable of providing accurate height data and no such scaling was needed for the later tests.

Microseisms do not necessarily occur on the hydraulic fracture, but can in fact be some distance to the side or even ahead of the fracture (Warpinski 1994). Calculations of the mechanical response of the formation for B-Sand conditions have been performed to estimate the zone of microseismic activity anticipated for this interval. These analyses show that microseismic events at this site can occur (Warpinski 1994)

- As much as 12–15 ft ahead of the tip of the fracture and 15–20 ft normal to the tip of the fracture because of the large stress concentrations at the fracture tip
- Several tens of feet normal to the body of the fracture due to leakoff effects (increase in pore pressure)

The exact distance at which leakoff-induced microseisms might occur is difficult to deduce because of the unknown permeability of the natural fractures in the formation. Nevertheless, these analyses provide guidance in interpreting the microseismic data and show that the microseisms map out an envelope in which the hydraulic fracture is embedded. By examining this envelope, the height, length and azimuth of the fracture can be reasonably estimated, but nothing can be inferred about the width of the fracture, the number of fracture strands, or the width of any complex fracture zone that may or may not exist.

6.7 Fracture Diagnostic Results

Fracture diagnostic results are presented as plan views and side views of the microseismic locations and are presented with the flow rate, net pressure, and pressure match for each treatment. Given the geometry of the arrays, the most accurate position is the elevation of the microseism (because of the large number of receivers in vertical arrays), while the distance from the microseism to the receiver and the azimuth of the microseismic signal have more uncertainty. Accuracy of the elevation position is about ± 5 ft, but results vary from microseism to microseism depending on how many receivers have detected the particular event. Accuracy of the azimuth to the microseism is about ± 5° and, since the distance away is approximately 300 ft, the uncertainty in the azimuthal location is about 25 ft at that 300 ft distance. Uncertainty in the distance from the receiver to the microseism varies considerably depending upon the number of levels on which the 5-wave arrival is accurately detected, but it is on the order of ± 50 ft.

Given the uncertainties in locations and the interpretation of microseisms discussed in the previous section, the microseismic results should be considered to be an envelope that surrounds the fracture. Furthermore, since uncertainties on a few microseisms can become relatively large, the position of any single outlier microseism should always be questioned.

Fig. 6.5 shows the final microseismic images of the 27 bbl step-rate test (maximum rate of 3 bpm), with the side view image having an exaggerated vertical scale. Microseismic results from this test suggest that the fracture

- Is well contained
- Is preferentially located within the upper part of the B Sandstone
- Has a length of 150–200 ft
- Appears to have a slight asymmetry, with greater fracture length on the west wing

The boundaries of the B Sandstone, as shown in the side view of Fig. 6.5, are the top of the clean 30 ft section and the bottom of the thin sand lying beneath the main sand body (see Fig. 6.2).

Fig. 6.6 shows the microseismic images of the first pump-in/ shut-in test, which was conducted with approximately 100 bbl of KCl water pumped at 10 bpm. These images also show a fracture that

- Is well contained
- Is mostly in the thick upper 30 ft of the B Sand
- Has an average length on the order of 300 ft
- Has some asymmetry, with the east wing likely being greater than the west wing of the fracture

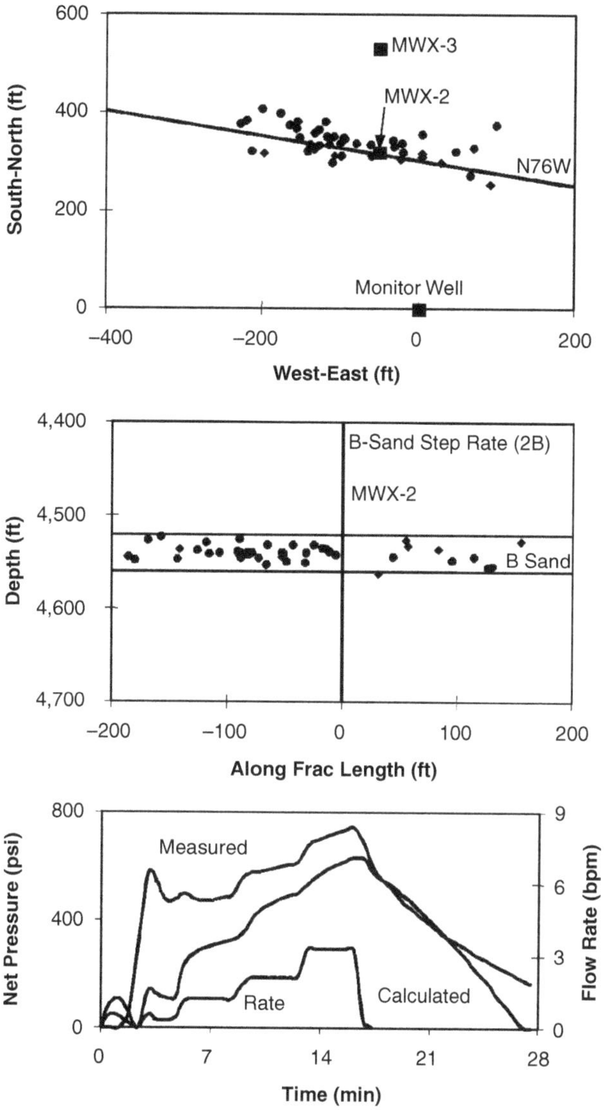

Fig. 6.5—Microseismic images and treatment data for fracture 2B.

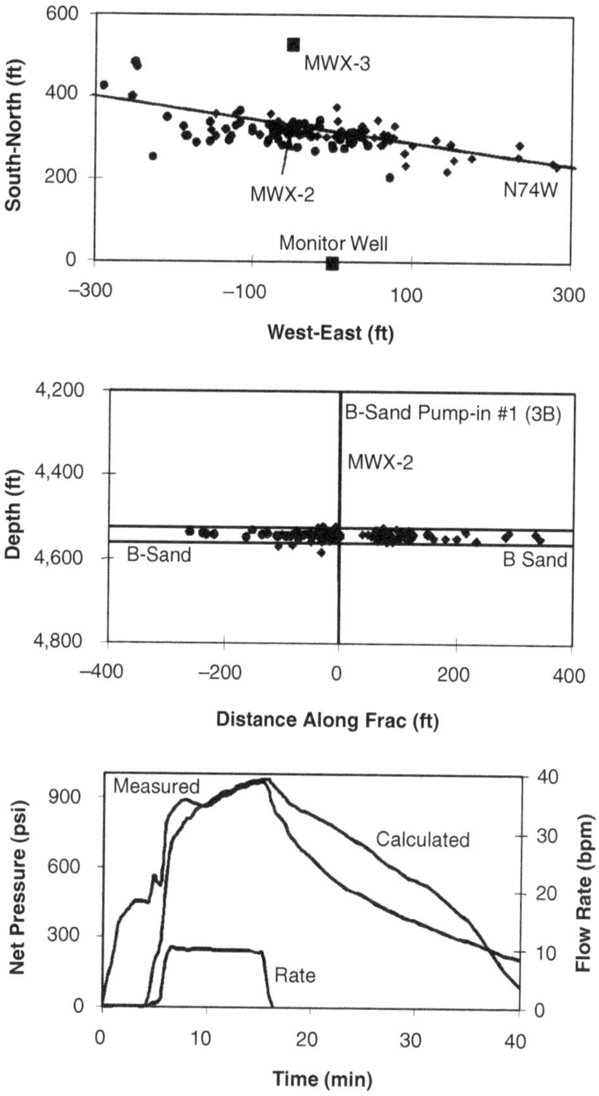

Fig. 6.6—Microseismic images and treatment data for fracture 3B.

The single microseism below the B Sandstone could easily be a point that is in error or a plane of weakness in the formation that is very favorably oriented for slippage to occur with minimal stress loading (microseisms can occur ahead of the fracture tip due to the large stress changes at and ahead of the tip of a fracture). Net pressures were about 900 psi at shut-in compared to stress contrasts of about 1,500 psi, so full containment is realistic and expected.

Fig. 6.7 shows the microseismic images of the second 10 bpm, KCl-water pump-in/shut-in test. The microseismic results from this injection show

- Considerable asymmetry, with the west wing appearing to be less than 300 ft and the east wing greater than 400 ft
- Indications that the volume of this test was sufficient to cause the fracture to at least begin breaking into the surrounding layers

Net pressures during this treatment were about 1,000 psi above closure stress at shut in (and climbing slightly). It is interesting to observe that there are often a few spurious microseismic event locations and they tend to be in locations similar to those for spurious events seen in other injections. Such results could be due to errors, but more likely there is some feature in the formation that is easily disturbed by the stresses induced by the hydraulic fracture. The timing of fracture growth was somewhat unusual because the fracture grew rather symmetrically during most of the treatment, but a surface leak resulted in an 8 minute shutdown during the test.

After this shutdown, the remaining approximately 60 bbl were injected and length growth was observed only on the east wing, resulting in the final asymmetric shape.

The microseismic image maps for the first minifrac (linear gel), injected at 22 bpm, are shown in **Fig. 6.8.** In this injection the net pressure reached 1,300 psi at shut-in (and was relatively flat), close to the confining stress contrast of 1,500 psi. These images show that

- Height growth is clearly evident in the microseismic data
- There is considerably more height growth occurring upward than downward
- This fracture appears to have only slightly greater length dimensions and the same asymmetry as the previous KCl injection, even though both the rate and volumes were doubled

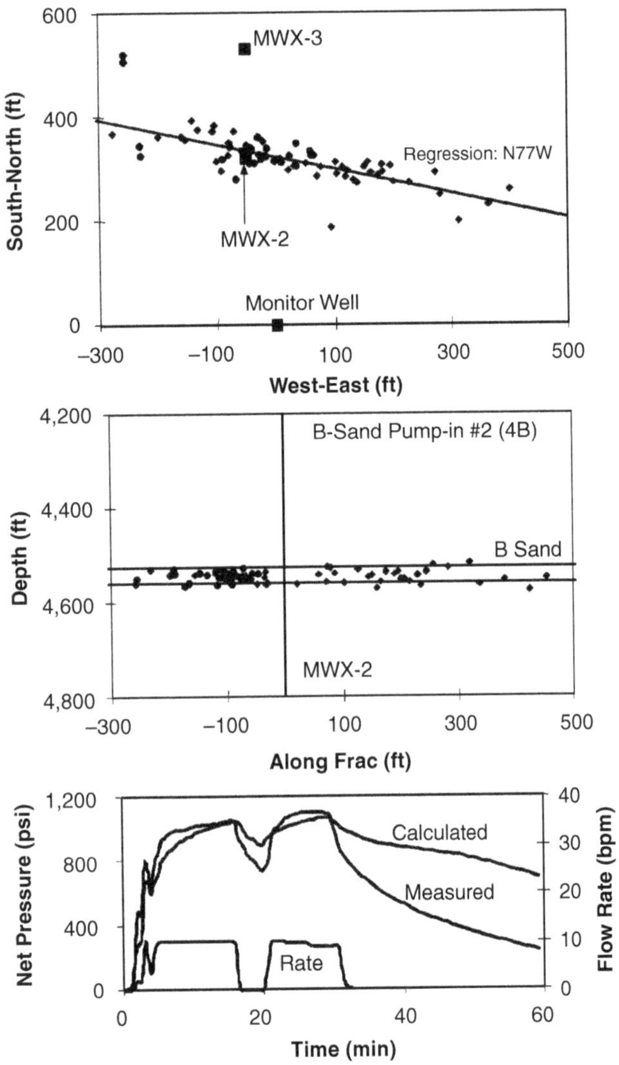

Fig. 6.7—Microseismic images and treatment data for fracture 4B.

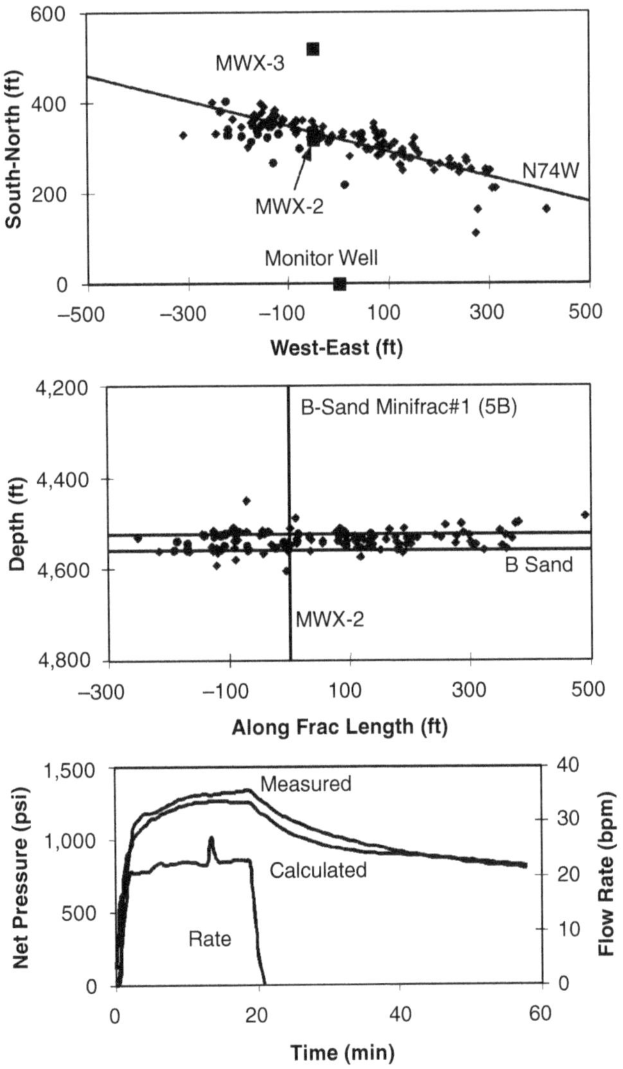

Fig. 6.8—Microseismic images and treatment data for fracture 5B.

Much of the volume was undoubtedly taken up in height growth and wider fracture widths on account of height growth. Indications of height growth appear early in the treatment, so the occurrence of height growth is due to rate or viscosity rather than volume, although the larger volume pumped results in additional height. It should be noted that there is one outlier on the east wing of the fracture that probably should be ignored.

The second minifrac, which was pumped identically to the first minifrac, resulted in very similar pressure and microseismic images, as seen in **Fig. 6.9.** These results indicate that

- There is again evidence of upward fracture growth
- Only limited evidence of any downward growth
- The fracture is asymmetric, with a west wing less than 300 ft and an east wing greater than 400 ft

In this injection, the microseisms are very tightly banded around the fracture with no obvious outliers, whereas in several previous tests there was wide scatter of the microseisms. Reasons for such differences are unknown.

The microseismic images from the propped fracture treatment are shown in **Fig. 6.10.** During this 20 bpm injection, the microseismic results suggest that

- Considerable upward height growth occurred
- Very little downward growth occurred
- The fracture length is relatively symmetric, with approximately 400 ft of length observed on each wing

Although large amounts of upward height growth occurred, all of it happened in the last one third of the treatment. During the first two thirds of the treatment, the fracture height looked very similar to the minifracs. Net fracture pressures reached 2,000 psi during this treatment, but most of the excess over the minifracs (1,300 psi net pressure) occurred after higher concentration sand hit the formation, possibly indicating that some form of proppant bridging may have induced the high pressures. Two outliers suggest that the fracture may have grown up out of zone on the far end of the east wing, but the data are too sparse to confirm this trend. Nevertheless, this treatment is entirely different from the previous injections and shows that fracture treatments with normal fracture fluids can induce height growth in high-stress-contrast environments.

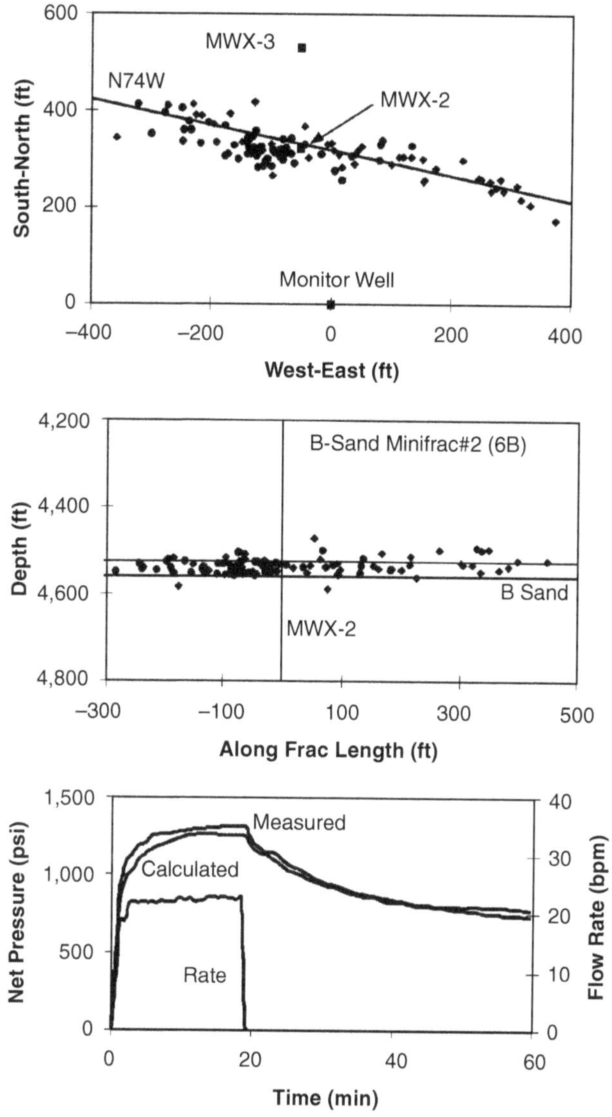

Fig. 6.9—Microseismic images and treatment data for fracture 6B.

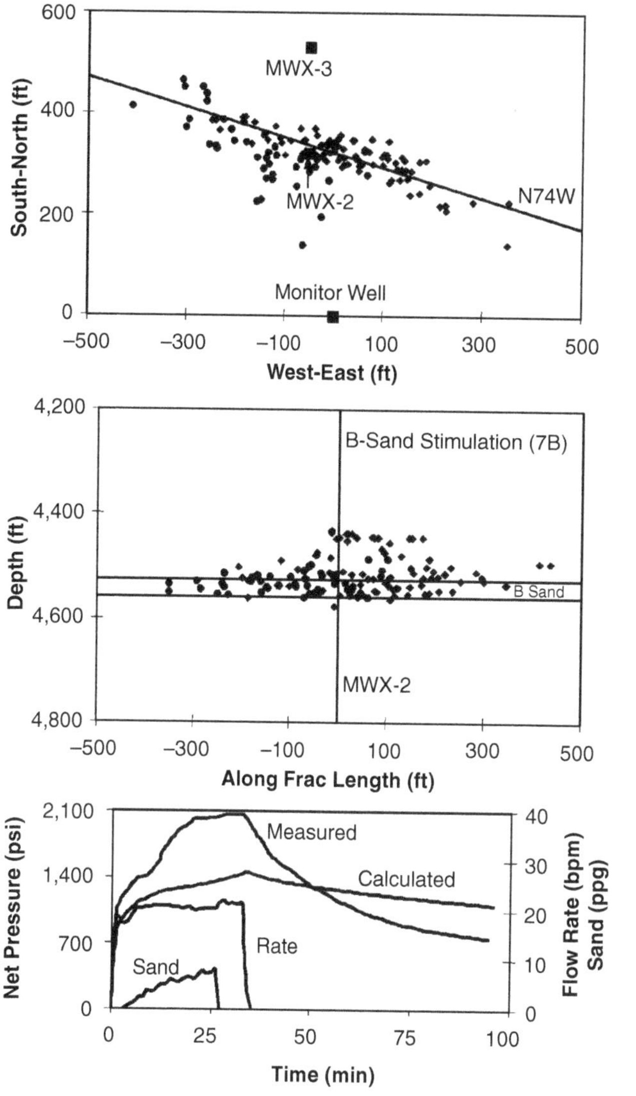

Fig. 6.10—Microseismic images and treatment data for fracture 7B.

6.8 Model Comparison

Comparisons of the microseismic images were made with a lumped fracture model that incorporates increased fracture tip effects (Crockett et al. 1989). The model calculations were made at the site during the treatments, so there were no comparisons with microseismic data at that time to bias the results. In all cases, the net pressure was matched as best as possible, except in the propped fracture treatment. The propped fracture achieved very high net pressures late in the injection (2,000 psi) after higher concentrations of sand were entering the formation, and no attempt was made to match the final pressure (it would have required the addition of an unknown containment mechanism or high levels of near-wellbore restrictions). In all cases the model used the stresses given in Fig. 6.3, the n' and k' values from Table 1, a single fracture, a Young's modulus of 4×10^6 psi in the sand and 3×10^6 psi in the surrounding layers, and a Poisson's ratio of 0.2 in all layers. The pressure match was made on the leakoff coefficient, resulting in a value of 0.00158 ft/$\sqrt{\text{min}}$ in the sandstone and 0.0006 ft/$\sqrt{\text{min}}$ in the shales for all tests.

The results of the diagnostic/model comparison are given in **Table 6.2.** The comparisons are quite interesting in that they show that models do a relatively good job of matching results when the fracture is totally contained within the reservoir zone, but some problems arise when height growth occurs. For the water frac, the model lengths are close to the imaged lengths, but the height is somewhat greater than the image. For the 400 bbl minifracs, the situation is similar to the water injection, except that the model predicts three times more height growth than actually imaged. Length results are relatively close, but the asymmetry masks the match between the model and the diagnostics. For the propped frac, the comparison is not very good and suggests that some features of the formation or the behavior of the fracture in this formation are not well understood. It is particularly difficult to account for the lack of downward growth seen in the imaged data (Fig. 6.10). Stresses in the bounding layers are already modeled as greater than 1.0 psi/ft, so higher stresses in these layers are not reasonable (Warpinski and Teufel 1987). There must be some other containment feature available, or the proppant has played a large role in minimizing downward growth. However, some downward growth was also expected for the minifracs, but was not observed, so the complete answer is probably more complicated. One other possibility is that the microseismic event character is different in the lower shales and any microseisms are not detected, but this explanation requires that the shales above the B Sand be different from the shales below, which is not observed.

	Azimuth (deg)	West Wing Length (ft)	East Wing Length (ft)	Height (ft)	Tiltmeter Height (ft)	Model Length (ft)	Model Height (ft)
2B	N76W	180 ± 20	140 ± 30	40 ± 5	34	154	53
3B	N74W	250 ± 20	340 ± 30	55 ± 5	53	261	91
4B	N77W	260 ± 20	440 ± 30	55 ± 5	53	372	109
5B	N74W	230 ± 25	375 ± 20	80 ± 20	67	425	183
6B	N74W	260 ± 30	420 ± 30	75 ± 15	67	427	185
7B	N74W	350 ± 30	420 ± 30	135 ± 10	135	295	190

Table 6.2—Comparison of microseismic results with tiltmeter and model dimensions.

6.9 Discussion and Conclusions

These results show that microseismic-imaging technology can be used to map out important features of hydraulic fractures in typical oil-field environments. The accuracy of microseismic imaging is sufficient to provide accurate estimates of the length, height and azimuth of a hydraulic fracture. However, accurate height determination requires multilevel receiver systems in order to account for formation layering. Table 6.2 gives the estimated size comparison for the six different injections, as well as a comparison with the height derived from downhole inclinometer measurements (Branagan et al. 1996b).

One comparison of the accuracy and reliability of the measurements that can be made from this set of experiments is the azimuth for each test. The azimuth listed on each plan view is that obtained only from the monitor-well data (only the monitor-well data are used because the instrumentation in MWX-3 changed during the testing from a single receiver during the water injections to a five-level system for later treatments and any comparison would be irrelevant). The average azimuth of the six tests is N75°W with a standard deviation of about 1°. The actual azimuth, based on the location of the fracture in an intersecting well is N72°W (Branagan et al. 1996a). Also, the azimuths using all data (both monitor and MWX-3) differ by as much as 2 to 3° from those given in Table 6.2.

The fracture length variations are interesting, because the length changes only slightly as the test progresses from small volumes of low-viscosity fluids pumped at low rates to larger volumes of high-viscosity fluids pumped at high rates. Of course, some of the volume is taken up by the height growth induced by the higher net pressures in the latter treatments and some additional volume may be lost to higher leakoff at the higher pressures (e.g., possible fissure opening).

The degree of containment during the KCl injections is quite good, but the containment is only effective for small volumes of low viscosity fluid pumped at low rates. Under typical treatment conditions (fluids, volumes, rates), height growth occurred relatively easily and much fluid was lost treating out-of-zone rocks. Since the M-Site location is not likely to be different from other western basins, the high stress contrasts measured here that provided the initial containment may be prevalent throughout these areas. A good strategy for gas producers in these areas would be to work on techniques for fracturing these zones in ways that prolong the containment of the fracture.

Comparisons with other diagnostics (Branagan et al. 1997; Warpinski and Teufel 1987) were also made. Tracers in the treatment well showed no out-of-zone fracturing for the KCl injections and only about 20 ft of growth for the minifracs, which is in good agreement with the microseismic results. However, in the propped fracture the tracers did not show evidence of any more height than seen in the minifracs. Inclinometers (Branagan et al. 1996b) gave direct evidence of the mechanical height of the fracture, yielding values of about 53 ft for the KCl injections and 67 ft for the minifracs. The propped fracture treatment had a geometry that was too complicated to produce a single height value, but the geometry imaged by the microseisms was shown to be able to produce the measured tilts.

Comparisons of a fracture model with the imaged results were quite good for cases where the fracture was contained, but some discrepancies developed for the latter treatments. Of particular significance was the lack of downward growth, as it suggests that there may be other containment features that are not included in current models (i.e., inefficiency as the fracture crosses bedding). Such features can be included in an ad hoc manner after the fact using mechanisms such as a slip condition or a limit on height growth, but they are not usually known beforehand under normal fracturing conditions and are thus difficult to apply. Another possibility is incorrect stress data, but stresses large enough to provide such containment are unlikely.

The discrepancies between the diagnostics and the models, particularly for fracture height, again demonstrate the difficulty in obtaining sufficient information to obtain an accurate design. Even with the vast amount of data available at M-Site, it is difficult to obtain accurate model matches without using effects such as multiple fractures, tortuosity, fracture roughness, and perfect slip. The situation is much more uncertain in a typical oil-field environment where essentially no stress, modulus, or fracture complexity data are typically available.

6.10 Nomenclature

F = distance function to be minimized
n = number of P-wave picks, n
m = number of S-wave picks, n
r_0 = horizontal distance to source, L, ft
t_o = origin time of microseism, L, sec
t_{pi} = P-wave travel time, t, sec
t_{si} = S-wave travel time, t, sec
V_p = P-wave velocity, L/t, ft/sec
V_s = S-wave velocity, L/t, ft/sec
w_p = weighting value for P-wave regression
w_s = weighting value for S-wave regression
z_i = elevation of ith receiver, L, ft
z_0 = elevation of source, L, ft

6.11 Acknowledgments

The authors would like to acknowledge the Gas Research Institute, with funding for this project managed by Steve Wolhart, and the Department of Energy, with funding managed by Karl-H. Frohne, for their continued support of this project. Special thanks are in order for many contributors to this work at Sandia National Labs, Branagan & Associates, Resource Engineering Systems, CER Corporation, Jim Fix & Associates, Bolt Technology Corporation, and many others.

6.12 References

Albright, J. N. and Pearson, C. F. 1982. Acoustic Emissions as a Tool for Hydraulic Fracture Location: Experience at the Fenton Hill Hot Dry Rock Site. *SPE J.* **22** (4). SPE-9509-PA. https://doi.org/10.2118/9509-PA.

Branagan, P. T., Peterson, R. E., Warpinski, N. R., and Wright, T. 1997. Results of Multi-Site Project Experimentation in the B-Sand Interval: Fracture Diagnostics and Hydraulic Fracture Intersection. Gas Research Institute Topical Report, GRI-96/0225 (December 1997).

Branagan, P. T., Peterson, R. E., Warpinski, N. R. et al. 1996a. The Characterization of Remotely Intersected Set of Hydraulic Fractures: Results of Intersection Well No. 1: GRI/ DOE Multi-Site Project. Presented at the SPE Annual Technical Conference and Exhibition, Denver, Colorado, 6–9 October. SPE-36452-MS. https://doi.org/10.2118/36452-MS.

Branagan, P. T., Warpinski, N. R., Engler, B. P. et al. 1996b. Measuring the Hydraulic Fracture-Induced Deformation of Reservoirs and Adjacent Rocks Employing a Deeply Buried Inclinometer Array: GRI/DOE Multi-Site Project. Presented at the SPE Annual Technical Conference and Exhibition, Denver, Colorado, 6–9 October. SPE-36451-MS. https://doi.org/10.2118/36451-MS.

Crockett, A. R., Willis, R. M., and Cleary, M. P. 1989. Improvement of Hydraulic Fracture Predictions by Real-Time History Matching on Observed Pressures. *SPE Prod Eng* **4** (4). SPE-15264-PA. https://doi.org/10.2118/15264-PA.

Dobecki, T. L. 1983. Hydraulic Fracture Orientation Using Passive Borehole Seismics. Presented at the SPE Annual Technical Conference and Exhibit, San Francisco, California, 5–8 October. SPE-12110-MS. https://doi.org/10.2118/12110-MS.

Lacy, L. L. 1987. Comparison of Hydraulic-Fracture Orientation Techniques. *SPE Form Eval* **2** (1). SPE-132254-PA. https://doi.org/10.2118/13225-PA.

Middlebrook, M., Peterson, R., Warpinski, N. R. et al. Multi-Site Project Seismic Verification Experiment and Assessment of Site Suitability. GRI Topical Report, GRI-93/0050 (February 1993).

Nelson, G. D. and Vidale, J. E. 1990. Earthquake Locations by 3D Finite Difference Travel Times. *Bull. Seismol. Soc. Am.* **80** (2): 395–410.

Peterson, R. E., Wolhart, S. L., Frohne, K. H. et al. 1996. Fracture Diagnostics Research at the GRI/ DOE Multi-Site Project: Overview of the Concept and Results. Presented at the SPE Annual Technical Conference and Exhibition, Denver, Colorado, 6–9 October. SPE-36449-MS. https://doi.org/10.2118/36449-MS.

Peterson, R. E., Warpinski, N. R., Wright, T. B. et al. 1995. Results of Multi-Sites Experimentation in the A Sand Interval: Fracture Diagnostics, Fracture Modeling and Crosswell Tomography. GRI Topical Report, GRI-95/0066 (April 1995).

Sarda, J-P. and Deflandre, J-P. 1988. Acoustic Emission Interpretation for Estimating Hydraulic Fracture Extent. Presented at the SPE Gas Technology Symposium, Dallas, Texas, 13–15 June. SPE-17723-MS. https://doi.org/10.2118/17723-MS.

Sleefe, G. E., Warpinski, N. R., and Engler, B. P. 1995. The Use of Broadband Microseisms for Hydraulic-Fracture Mapping. *SPE Form Eval* **10** (4). SPE-26485-PA. https://doi.org/10.2118/26485-PA.

Stewart, L., Cassell, B. R., and Bol, G. M. 1992. Acoustic-Emission Monitoring During Hydraulic Fracturing. *SPE Form Eval* **7** (2). SPE-20970-PA. https://doi.org/10.2118/20970-PA.

Thorne, B. J. and Morris, H. E. 1988. Advances in Borehole Seismic Fracture Diagnostics. *SPE Form Eval* **3** (4). SPE-16405-PA https://doi.org/10.2118/16405-PA.

Truby, L. S and Keck, R. G. 1994. Data Gathering for a Comprehensive Hydraulic Fracturing Diagnostic Project: A Case Study. Presented at the SPE/IADC Drilling Conference, Dallas, Texas, 15–18 February. SPE-27506-MS. https://doi.org/10.2118/27506-MS.

Vidale, J. E. 1988. Finite Difference Calculation of Travel Times. *Bull. Seismol. Soc. Am.* **78** (6): 2062–2076.

Vinegar, H. J. Wills, P. B., DeMartini, D. C. et al. 1992. Active and Passive Seismic Imaging of a Hydraulic Fracture in Diatomite. *J Pet Technol* **44** (1). SPE-22756-PA. https://doi.org/10.2118/22756-PA.

Warpinski, N. R., Peterson, R. E., Branagan, P. T. et al. 1998. In Situ Stress and Moduli: Comparison of Values Derived from Multiple Techniques. Presented at the SPE Annual Technical Conference and Exhibition, New Orleans, Louisiana, 27–30 September. SPE-49190-MS. https://doi.org/10.2118/49190-MS.

Warpinski, N. R. 1994. Interpretation of Hydraulic Fracture Mapping Experiments. Presented at the University of Tulsa Centennial Petroleum Engineering Symposium, Tulsa, Oklahoma, 29–31 August. SPE-27985-MS. https://doi.org/10.2118/27985-MS.

Warpinski, N. R. and Teufel, L. W. 1987. In Situ Stresses in Low-Permeability Nonmarine Rocks. Presented at the SPE-DOE Joint Symposium on Two Permeability Reservoirs, Denver, Colorado, 18–19 May. SPE-16402-MS. https://doi.org/10.2118/16402-MS.

SI Metric Conversion Factors

bbl × 1.589 873	E–01 = m^3
ft × 3.048*	E–01 = m
gal × 3.785 412	E–01 = m^3
lbf × 4.448 222	E+00 = N
lbm × 4.535 924	E–01 = kg
psi × 6.894 757	E+00 = kPa

*Conversion factor is exact.

Appendix 6.A

6.A.1 About the Authors

Norman R. Warpinski is a Distinguished Member of the technical staff at Sandia Natl. Laboratories in Albuquerque, New Mexico, where he has worked since 1977. His current interest is fracture diagnostics, with emphasis on use of state-of-the-art instrumentation to develop real-time and near-real-time systems for imaging hydraulic-fracture geometry and behavior primarily with microseismic technology but also with downhole tiltmeters. In addition, he has worked in the areas of hydraulic fracturing, in situ stress measurements, rock mass behavior, natural fractures, and dynamic-fracturing processes by conducting mineback operations, field and laboratory tests, and simulations. Warpinski holds a PhD degree in mechanical engineering from the U. of Illinois, Urbana. He served on the Editorial Review Committee during 1990-93, as a member of the Forum Series in North America Steering Committee during 1992-93 and 1997-98, and as a member of the Low Permeability Reservoirs/Rocky Mountain Regional Meeting Program Committee during 1990-93 (all SPE positions). Timothy B. Wright, a consultant, currently works primarily for Pinnacle Technologies Inc. in Houston in the area of development and commercialization of software used in the petroleum industry. Wright holds a BS degree in mechanical engineering from Massachusetts Inst. of Technology. James E. Uhl, an instrumentation consultant for JEUhl Consulting, currently consults for GRI, Sandia Natl. Laboratories, and Branagan & Assocs. on a new seismic receiver design. He previously worked for Sandia as a senior member of the technical staff, with responsibilities in electromechanical engineering, including projects involving development of instrumentation for such disciplines as environmental testing, materials science, mining, geothermal energy, and petroleum E&P. He holds an AA degree in electronics from Albuquerque Vocational College and a BS-equivalent degree in electrical engineering through Sandia. Patrick M. Drozda has worked as a principal technologist for Sandia Natl. Laboratories for more than 24 years. His current work is on systems to measure thermal-contact resistance in mechanical fabricated joints for model validation of components in the nuclear-weapons stockpile. He has experience on polymer R&D, nuclear magnetic resonance spectroscopy, explosives and fragmentation experimentation, wellbore-tool development, and nuclear-reactor safety instrumentation and qualification. Drozda holds an AA degree in electrical engineering from North American Technical College. Richard E. Peterson is Vice President and Project Director at Branagan & Assocs. Inc. in Las Vegas, Nevada. Before joining Branagan & Assocs. in 1996, he managed a Western Geophysical shothole crew operating on the Texas Gulf Coast, was a wellsite and exploration geologist with Exxon Corp. in South America, and was a project manager with CER Corp. in Las Vegas. A significant portion of his work at CER and Branagan & Assocs. has been directed at coordinating field research projects with GRI, including the M-Site Hydraulic Fracture Diagnostics Project. Peterson holds a BS degree in geology from Arizona State U. Paul T. Branagan is president and senior scientist at Branagan & Associates Inc. in Las Vegas, Nevada. He initially worked for EG&G Inc. and later transferred to a subsidiary, CER Corp., where he managed reservoir engineering and production assessment projects, including the nuclear stimulation and initial massive hydraulic fracturing projects conducted by the Atomic Energy Commission and US Dept. of Energy. His recent efforts involve GRI-related field studies directed toward development of fracture diagnostic mapping capabilities, including microseismic and downhole tiltmeters. Branagan holds a BS degree in physics and has completed graduate studies in physics. He served as a 1994-95 member of the SPE Forum Series in North America Steering Committee and as a 1992-93 member of the Low Permeability Reservoirs/Rocky Mountain Regional Meeting Program Committee.

Chapter 7

Successful Stimulation of Deep Wells Using High Proppant Concentrations

S. A. Holditch, SPE-AIME, Shell Oil Co.; and **John Ely,** SPE-AIME, Halliburton Services

 Symposium paper SPE 4118 was presented at SPE-AIME Southwest Texas Section Regional Meeting held in Corpus Christi, Texas, 27 April 1973.

SPE 4118 presents a case history of the stimulation of 17 geopressured deep tight gas wells using high proppant concentrations in the Vicksburg Formation in Texas. These wells were some of the first large fracture treatments pumped in South Texas using high-viscosity fracturing fluids to pump large volumes of sand at high proppant concentrations. Holditch and Ely provide a review of the types of stimulation available at that time, the problems in fracturing deep formations, the mechanical issues, and the fracture fluids required. They then summarize and analyze treatment data and production data and compare the productivity in wells stimulated using high-viscosity fluids carrying high proppant concentration with the treatments using low-viscosity fluids carrying low proppant concentrations. Holditch and Ely also compute the normalized productivity index, the reserves based on the decline curve analysis, the reserves based on the material balance method, and the economics. They concluded that high-viscosity fluids carrying high proppant concentration is an effective stimulation of deep, low-permeability gas wells to improve gas flow rates and ultimately result in recovery.

7.1 Introduction

Large portions of the easily recovered hydrocarbons of the world are rapidly being depleted. This is particularly the case in the confines of the western hemisphere where many of the most productive fields are in the latter stages of primary and secondary recovery.

If complete dependence on foreign oil or a large energy gap is to be averted, new supplies of energy must be found soon. One promising source is the oil and gas tightly held in formations that have very low permeability and porosity. If these formations, a large number of which are deep, high-temperature reservoirs, are to be economically produced, an effective means of stimulation must be developed.

7.2 Types of Stimulation

There are basically three different types of stimulation methods available to the industry today. The first, chemical stimulation, has not been useful in tight sandstone reservoirs except in the removal of formation damage (Nierode et al. 1972). In deep, high-temperature and highly soluble formations, it is difficult to use this technique because the acid is rapidly spent near the wellbore. Without deep penetration of live acid into the fracture, effective stimulation cannot be achieved. In some cases the use of large volumes of fluid to cool the formation helps to overcome this deficiency.

The second type of stimulation, which is undergoing limited study today, is the use of both conventional and nuclear explosives (Rogers 1970; Reynolds et al. 1970; Howell and Hille 1970). At this time, two nuclear explosive projects have been conducted. The tests were apparently successful in safely setting off a nuclear device and achieving some stimulation. However, the economic feasibility of nuclear stimulation is still questioned by a large segment of the industry. The use of conventional explosives presents a considerable hazard during injection; detonation in thin layers at large distances from the wellbore is difficult and not always successful.

The third type of stimulation is hydraulic fracturing. Since the inception of fracturing in 1949, many different fluids and proppants have been investigated (Seidel and Stahl 1972; Hassebroek and Waters 1964; Rixe et al. 1963; McGlothlin et al. 1963). In the early 1950s, viscous petroleum base fluids were being used industrywide. From the mid-1950s through the mid-1960s, the trend was to use less viscous fluids pumped at higher rates. Also, the trend has been toward using water-based fluids rather than oil-based fluids for fracturing.

The theory of fracturing developed quite rapidly between the mid-1950s and mid-1960s. In 1957, Howard and Fast (1957) introduced some of the first literature for the calculation of fracture geometry from treatment data. Since that time, a continuous stream of literature, which is much too abundant to reference completely, has been published on fracture dimensions and design criteria. The work that has been published indicates that the ratio of (1) fracture length to the radius drained, and (2) fracture flow capacity to formation permeability are the most critical relationships in the design (van Poollen et al. 1958). The methods most often used for predicting productivity increase are those by McGuire and Sikora (1960), by Prats (1961), and by Tinsley et al. (1969). All of these predict production increase ratios as functions of fracture conductivity and fracture penetration.

A unique viscous fluid has been used in fracturing deep, high-temperature reservoirs. By using large treatments and high concentrations of proppants, reservoirs with low permeability and porosity can be successfully stimulated. Data from widely varying formations are compared, with special emphasis on the Vicksburg…

It is apparent, upon review of the information relating the variables (van Poollen et al. 1958; McGuire and Sikora 1960; Prats 1961; Tinsley et al. 1969) to be considered, that for tighter rocks it is desirable to have a longer fracture with moderate flow capacity. However, obtaining even moderate flow capacity in the deeper formations has always presented problems.

7.3 Problems in Fracturing Deep Formations

In stimulating deep formations, the greatest difficulties lie in creating and maintaining large, highly conductive fractures. Adequate flow capacity is difficult to maintain because the high closure stresses tend to embed or crush the proppant (Huitt et al. 1959; Huitt and McGlothlin 1958). The movement of fines, which tend to plug the fracture, causes additional problems (Coulter and Wells 1972).

This chapter presents the results obtained with a viscous water-based fluid developed for fracturing deep wells (Seidel and Stahl 1972). Before the development of this fluid, gas wells were fractured with fluids having surface viscosities of about 25 cp, which degraded to about 2 cp at formation temperature. These fluids were pumped at rates of 20 bbl/min or greater and carried proppant at a maximum concentration of 1 to 1½ lb/gal. The propped fracture lengths and heights were possibly adequate, but in most cases the propped fracture widths were narrow. In deep reservoirs, a narrow fracture is particularly susceptible to embedment and crushing (Huitt et al. 1959; Huitt and McGlothlin 1958; Coulter and Wells 1972; Darin and Huitt 1960).

7.4 Mechanical Aspects

Extremely large, high-pressure fracturing treatments have been conducted using conventional surface pumping equipment. Often this equipment, even with special high-pressure fluid ends, was not adequate for extended pumping of fluid carrying high concentrations of proppant. Manifolds and ball injectors were not always suitable. However, most of the problems of high-pressure pumping have been solved through the use of pump intensifiers and special surface lines.

In stimulating deep, high-pressure wells, the mode of completion greatly influences the results. Small tubing, low-pressure wellhead equipment, improperly selected packers, and too many perforations can drastically limit the size and effectiveness of the treatment. It is of utmost importance to complete a well so it can be fractured at adequate rates and pressures.

7.5 The Gel

For a fluid to be applicable for stimulating deep, high-temperature wells, it should meet the following criteria:

1. It must have low friction pressure properties.
2. It should be temperature stable up to 450°F.
3. It should be nondamaging; that is, it should leave little or no residue in the formation after breaking.
4. It should be compatible with the various salts used for clay stabilization.
5. Its viscosity in the fracture should be independent of its surface viscosity.
6. It must have good proppant support characteristics for transport into the fracture and for dispersion of the proppant throughout the area of the fracture.
7. It must degrade to the viscosity of the base fluid for ease in cleanup operations.

A fluid was developed that met all the above requirements (Seidel and Stahl 1972). The most rigorous of the requirements—that which calls for downhole viscosity—was met by using a retarded gelling agent that is triggered by bottomhole temperature. By using this retarded gelling agent in conjunction with a batch-mixed gel for surface proppant transport, downhole viscosity can be optimized to meet varying bottomhole temperatures.

Generally, the fluid contains two chemically substituted polymers. One of these is formulated chemically so that it provides the necessary time delay to develop viscosity downhole.

Screenouts have not been a problem with this fluid. Experience has shown that with strict quality control on pH and viscosity and with adequate rates and careful downhole preparation, the chance of bridging at the perforations or of screening out in the fracture is nil. It has been found that inadequate rate—less than 6 bbl/min—or large numbers of perforations have caused most of the screenout problems to date. The actual incidence of screenout is less than 5 percent.

7.6 The Vicksburg Formation

Results from 17 geopressured Vicksburg wells that have been fractured within the past 2½ years provide much of the information presented here. The major cause of difficulties in stimulating the Vicksburg is the softness of the formation. Calculations using the method of Rixe et al. (1963) and laboratory data on Vicksburg cores show that fractures that are propped only one to two grains wide will heal long before the reservoir depletes. The embedment occurs at 50 percent reservoir depletion. In fracturing such a formation, obtaining wide, well propped fractures is a necessity. When fracturing was introduced as a stimulation technique in the Vicksburg Formation, low-viscosity, low-sand-concentration treatments were the trend in the industry. These treatments produced the narrow fractures that were highly susceptible to embedment as the pore pressure was reduced. If sufficient fracture flow capacity is to be obtained, the propped fracture width for the Vicksburg should be several proppant grains wide. With a fracture of this width, the outer layers of proppant still embed, but several layers will remain open until depletion; using the high-viscosity, high-proppant-concentration treatments, fractures with calculated widths of about 0.4 in. are being created. Sand at concentrations of 5 to 6 lb/gal is being pumped for as much as half the treatment. These wide, well propped fractures should remain open even at maximum closure pressures. Of the 17 wells discussed, nine were fractured upon initial completion and the remaining eight wells were produced several years before they were stimulated.

Well	Reservoir Pressure (psi)	Death (ft)	Permeability (md)	Net Pay (ft)	Volume (gal)	Proppant (lb)	Before Fracturing Mcf/D	Before Fracturing psi	After Fracturing Mcf/D	After Fracturing psi
1(1)	5,700	11,700	0.12	127	150,000	225,000	2,690	1,275	6,890	2,200
2(1)	4,290	12,100	0.30	81	52,000	116,400	3,100	1,600	5,000	1,600
*3(1)	5,700	10,700	0.20	100	160,000	376,400	5,178	1,400	11,000	4,000
*4(1)	4,540	10,800	0.10	120	134,000	306,400	2,065	1,250	9,050	2,350
5(1)	6,000	10,500	0.09	86	160,000	375,000	1,418	1,200	4,854	1,700
6(2)	4,200	10,800	0.22	98	159,000	400,000	4,400	1,400	8,015	1,750
7(2)	5,400	11,800	0,21	108	223,000	488,000	5,000	1,375	11,405	3,400
8(2)	5,200	11,800	0.15	124	222,000	500,000	4,500	1,350	10,925	2,700
9(3)	9,800	11,800	0.18	40	146,800	404,000	3,000	4,500	7,000	6,900
*10(3)	10,500	12,250	0.46	42	153,000	397,000	6,000	5,700	7,500	7,600
11(3)	11,500	12,800	0.15	59	138,500	370,000	3,800	4,000	5,700	7,900
*12(3)	10,500	12,250	0.10	44	125,000	322,000	65	1,950	3,500	1,800

*Positive skin.
(1) Fractured in 1970–1971.
(2) Fractured in 1972.
(3) Drilled and fractured in 1972.

Table 7.1—Stabilized production increase, Vicksburg Formation, Field A.

Well	Viscosity	Permeability (md)	Volume (gal)	Proppant (lb)	C-Value × 10^{-9} Before Fracturing	After Fracturing	6 Months After Fracturing	2 Years After Fracturing	Remarks
1	High	0.12	150,000	225,000	2.22	8.47	7.37	7.30	These are the first five wells from Table 7.1, fractured in 1970–71.
2	High	0.30	52,000	116,400	5.11	7.67	9.52	11.60	
3*	High	0.20	160,000	376,400	4.38	35.0	24.0	27.0	
4*	High	0.10	134,000	306,400	2.15	22.0	17.63	18.01	
5	High	0.09	160,000	375,000	0.93	3.53	1.86	1.40	
Average		0.162	131,200	279,920	2.96	15.5	12.08	13.07	
6	High	0.22	159,000	400,000	6.53	19.11	14.0		These are Wells 6, 7, and 8 from Table 7.1, fractured in 1972.
7	High	0.21	223,000	488,000	4.17	9.62	9.45		
8	High	0.15	222,000	500,000	4.11	15.0	16.0		
Average		0.193	201,333	462,666	4.94	14.58	13.15		
9	Low	0.03	315,000	288,000	1.14	9.91	3.64	1.93	These are the 12 wells from Field A that were fractured in 1962–65.
10*	Low	0.05	119,000	88,000	1.50	16.80	5.29	6.05	
11	Low	0.20	118,000	100,000	0.76	0.91	0.78	0.24	
12	Low**	0.10	117,000	100,000	1.13	3.08	2.74	1.62	
13	Low	0.03	94,000	52,000	0.50	2.71	0.85	0.33	
14	Low	0.30	508,000	400,000	2.00	16.01	13.94	9.32	
15	Low	0.20	180,000	150,000	0.78	10.46	5.12	3.94	
16*	Low	0.20	204,000	200,000	0.33	2.63	1.71	0.48	
17*	Low	0.30	210,000	210,000	0.44	11.08	15.81	7.05	
18*	Low	0.25	219,000	300,000	4.96	65.81	19.56	10.99	
19	Low**	0.10	80,000	76,000	1.82	5.31	1.12	1.02	
20*	Low	0.30	200,000	200,000	1.43	24.07	17.67	6.78	
Average		0.171	197,000	180,000	1.40	14.07	7.35	4.13	

*Positive skin.
**Screened out.

Table 7.2—Comparison of low-viscosity treatments with high-viscosity treatments, Field A.

7.7 Comparison With Low-Viscosity Treatments

Tables 7.1 and 7.2 detail most of the fracture treatments that have been performed in one particular South Texas Field. Table 7.1 presents the data on the viscous gel treatments that have been performed. Table 7.2 compares the details of the high- and low-viscosity treatments.

Note that in Table 7.2 the permeabilities and the volumes of fluid pumped for both types of treatment are similar. The main difference is that the high-viscosity treatments carried more than twice the amount of sand per gallon as the low-viscosity treatments. The next four columns of Table 7.2 are the *C*-values for each well, calculated using the following equation (Russell et al. 1966).

$$C\text{-value} = \frac{Q\mu(\tilde{p})Z(\tilde{p})}{(p_s^2 - p_{wf}^2)}$$

where
Q = flow rate, MMcf/D
μ = viscosity, cp
Z = gas deviation factor
p_s = static reservoir pressure, psi
p_{wf} = flowing bottomhole pressure, psi
$\tilde{p} = (p_s + p_{wf})/2$

The majority of the *C*-values on Table 7.2 were calculated using pressures measured on a pressure bomb during annual 90-hour buildups. A few of the *C*-values were calculated using pressures estimated from flowing tubing pressures and plots of *p/Z* vs cumulative gas produced.

It can be seen that the average *C*-values before fracturing for the three categories of wells are quite different. This difference is explained by the number of damaged wells in each category. After the fracture treatments, the average *C*-values are quite close. Since the permeabilities, closure pressures, net pays, and drainage radii of wells in all three categories are similar, it must be concluded that the initial stimulation from either the high-viscosity or the low-viscosity treatments was about equal. But as can be seen, the *C*-values for the high-viscosity treatments have held up, whereas the *C*-values of the low-viscosity treatments show marked decreases with time. As previously explained, the reason for the decrease is embedment of the proppant. The wider fractures are clearly remaining open.

It should be mentioned that the *C*-value equation is applicable only in semisteady-state flow. For the Vicksburg data, semisteady-state flow for an unfractured well is reached in about 15 days. Work published by Morse and Von Gonten (1972) indicates that for reservoirs with very low permeability the productivity of a fractured well may take months to stabilize. Therefore the *C*-values reported on Table 7.2 in the "After Fracturing" column were calculated several months after the jobs had been completed. Using the rates and pressures from Table 7.1 for calculating *C*-values after the job will lead to erroneously high productivity increases. The flow rates are stabilized, but they are not necessarily in semisteady-state flow.

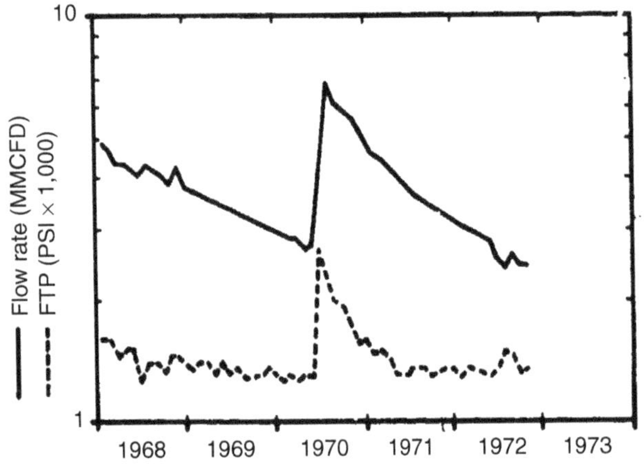

Fig. 7.1—Flow rate performance, Well 1, Table 7.1.

7.8 Effect on Rate and Recovery in the Vicksburg Formation

Tables 7.1 and 7.3 list the stabilized flow rates and tubing pressures for the Vicksburg wells stimulated with the high proppant concentration treatments. **Fig. 7.1** shows the flow rate history of Well 1 in Table 7.1.

All the rates quoted in this chapter are stabilized values taken from plots such as Fig. 7.1.

The after-fracturing rates indicate the gas actually produced to sales. Comparing these rates with the rates before fracturing gives the production increase ratio that must be used to calculate numbers such as payout. The average production increase ratio for Table 7.1 is 3.8 and for Table 7.3 is 3.2.

To determine the actual effect of the fracture treatment on the production of a well it is necessary to consider recovery. **Table 7.4** presents the results of calculations made for Wells 1 through 5 of Table 7.1. Of the wells that were treated using the viscous gel, these five are the only ones with enough production history to validate such calculations.

Well	Reservoir Pressure (psi)	Depth (ft)	Permeability (md)	Net Pay (ft)	Volume (gal)	Proppant (lb)	Before Fracturing Mcf/D	Before Fracturing psi	After Fracturing Mcf/D	After Fracturing psi
1	10,000	12,000	0.82	39	135,000	305,600	3,866	3,600	5,624	6,450
2	10,865	12,200	0.14	54	152,000	365,400	1,000	5,500	3,000	8,800
3*	9,569	11,300	0.15	47	160.000	340,000	1,225	680	9,000	6,800
4*	11,000	12,150	0.23	52	205,000	515,000	2,350	2,200	5,220	8,500
5	10,500	12,300	0.20	95	153,000	267,000	5,000	5,400	10,000	8,400

*Positive skin.

Table 7.3—Stabilized production increase, Vicksburg Formation, Field B.

Well	Before Fracturing (Bcf)	After Fracturing (Bcf)	Increased Recovery (Bcf)
1	3.5	5.7	2.2
2	2.7	3.4	0.7
3	10.0	12.4	2.4
4	2.0	4.5	2.5
5	1.4	1.9	0.5

Table 7.4—Reserves estimates, Wells 1 through 5 of Table 7.1.

								Before Fracturing		After Fracturing	
Well	Formation	Pressure (psi)	Depth (ft)	Permeability (md)	Net Pay (ft)	Volume (gal)	Proppant	Mcf/D	psi	Mcf/D	psi
1	Devonian	4,000	14,600	2.0	120	113,000	113,000	4,600	1,200	10,593	1,200
2	Devonian	4,000	13,520	1.8	112	113,000	60,000	3,913	1,200	7,720	1,200
3	Devonian	4,000	14,670	1.9	120	70,000	48,000	2,199	1,200	6,065	1,200
4	Devonian	4,000	14,040	2.1	95	113,000	42,000	3,800	1,150	7,200	1,150
5*	Wilcox	13,700	14,850	1.0	20	53,000	90,000	200	400	1,500	3,200
6	Wilcox	10,300	12,000	1.0	74	38,000	110,000	2,800	7,300	2,800	8,300
7*	Hosston	3,800	8,000	2.0	9	37,000	50,000	200	50	3,000	1,500
9	Hosston	3,500	8,300	2.0	7	20,000	39,000	200	100	1,600	1,000
8*	Hosston	2,525	5,500	2.0	29	127,000	230,000	4,500	480	10,000	480
10	Hosston	2,525	5,800	2,0	10	127,000	200,000	5,000	500	11,600	500
11	Hosston	3,800	6,200	5.0	75	162,000	376,000	6,900	500	15,000	750

* Positive skin.

Table 7.5—Stabilized production increase.

Two methods were used to calculate reserves. First the production curves, such as Fig. 7.1, were extrapolated to an economic limit. Second, the C-values were used to calculate the abandonment pressures before and after fracturing. Then using the plots of p/Z vs cumulative gas produced, the increase in recovery was determined. The recovery of the five wells considered has been increased by about 8.3 Bcf.

7.9 Results From Other Formations

Listed in **Table 7.5** are results of viscous gel treatments in the Devonian, Wilcox, and Hosston Formations. The four results shown in the Devonian are for wells located near Ft. Stockton, Texas. In this area, the formation is hard, dolomitic limestone with add solubility of 30 percent and higher. In Wells 1 and 2, 15,000 lb of 12–20 glass beads were pumped at 5 lb/gal at the end of the treatment. On the other two wells and for the remainder of Treatments 1 and 2, 20–40 sand alone was used. It should be noted that the large fluid volume pumped on these treatments includes some 60,000 gal of thin prepad fluid plus substantial pad volume.

The Wilcox treatments shown are two examples of treatments on deep, tight sandstone reservoirs in the central and south Texas areas. Again, only 20–40 sand is used. [A comprehensive review of Wilcox stimulation has been given by Seidel and Stahl (1972).] The results from the Hosston Formation came from the northern and central Louisiana areas. Results from Wells 3, 4, and 5 illustrate the use of the process in lower-temperature applications—180°F.

In addition to the aforementioned producing horizons, the Atoka, Morrow, San Andres, Frio, Cotton Valley, Haynesville, Carter and Sanders, Smackover, and Olmos have been successfully stimulated. In some areas, the high proppant concentration treatments have become the standard stimulation method. Results from these treatments have not been included because we do not wish to burden the reader with a large number of results that cannot be objectively compared with those from other forms of stimulation. However, it should suffice to say that of more than 200 treatments performed to date, the success ratio has exceeded 75 percent.

7.10 Economics

There must always be economic justification for stimulating a well. An example of a well that obviously justifies such treatment is one that has adequate gas in place but an unacceptably low flow rate. Several of the wells reported here met those conditions. Consider Well 3 in Table 7.3. A combination of low permeability and high skin caused Well 3 to flow at a rate considered to be uneconomic before stimulation. After the fracture treatment, Well 3 was greatly improved. For this particular case, stimulation made the difference between a dry hole and a producer. The other four wells listed on Table 7.3, although better than Well 3, would probably not have paid out the drilling and completion costs without stimulation. Therefore, for the five wells listed on Table 7.3, the economics was sound, since the fracture treatment converted marginal and unprofitable wells into highly profitable ones.

For a well that produces initially at adequate rates, stimulation in the latter stages of depletion is more difficult to justify. Here, the profitability can be based on acceleration economics. Fracturing can be used to shorten the life of the well, thus recovering the reserves sooner and reducing operating costs. Also, by shortening the life of the well, many mechanical difficulties can be avoided.

When the life of a well is prolonged, there is no assurance that the casing, packer, or tubing will not give way, corrode, or collapse before the reserves are produced.

The stimulation of tight gas reservoirs should be optimized on the basis of the combined effects of accelerating and increasing gas recovery. In this case, it is critical to calculate the potential increase in productivity and recovery for fractures of various lengths. Then with an adequate knowledge of formation flow capacity, drainage radius, economic limit, and the cost of creating fractures of various lengths, one can optimize the size of a stimulation treatment for a given well.

7.11 Conclusions

On the basis of the data presented in this chapter, the following conclusions have been reached:

1. An effective means of stimulating deep, low-permeability gas wells has been developed.
2. The success of the high-viscosity, high-proppant-concentration method is attributed to its ability to create and pack a wide fracture.
3. In low-permeability formations, it is possible to increase recovery by decreasing the abandonment pressure.

7.12 Acknowledgments

We wish to express our appreciation to Shell Oil Co. and Halliburton Services for permission to publish SPE 4118. Also, we thank those who have obtained data and assisted in its preparation.

7.13 References

Coulter, G. R. and Wells, R. D. 1972. The Advantages of High Proppant Concentration in Fracture Stimulation. J *Pet Technol* **24** (06): 643–650. SPE-3298-PA. https://doi.org/0.2118/3298-PA.

Darin, S. R. and Huitt, J. L. 1960. Effect of a Partial Monolayer of Propping Agent on Fracture Flow Capacity. In *Transactions of the Society of Petroleum Engineers,* Vol. 219, 31–37. SPE-1291-G. Richardson, Texas: Society of Petroleum Engineers.

Hassebroek, W. E. and Waters, A. B. 1964. Advancements Through 15 Years of Fracturing. *J Pet Technol* **16** (07): 760–764. SPE-801-PA. https://doi.org/0.2118/801-PA.

Howard, G. C. and Fast, C. R. 1957. Optimum Fluid Characteristics for Fracture Extension. API *Drilling and Production Practice.* API-57-261.

Howell, W. D. and Hille, J. B. 1970. Explosive Detonation Tested in Hydraulically Fractured Gas Wells. J *Pet Technol* **22** (04): 403–408. SPE-2605-PA. https://doi.org/0.2118/2605-PA.

Huitt, J. L. and McGlothlin, B. B. 1958. The Propping of Fractures in Formations Susceptible to Propping-Sand Embedment. API *Drilling and Production Practice.* API-58-115.

Huitt, J. L., McGlothlin, B. B., and McDonald, J. F. 1959. The Propping of Fractures in Formations in Which Propping Sand Crushes. API *Drilling and Production Practice.* API-59-120.

McGlothlin, B. B., Huitt, J. L., and Mackey, J. D. 1963. A High Strength Glass Pellet as a Fracturing Propping Agent. Paper presented at the Spring Meeting the American Petroleum Institute, Amarillo, Texas, USA, 27–29 March. API 851-37-C.

McGuire, W. J. and Sikora, V. J. 1960. The Effect of Vertical Fractures on Well Productivity. J Pet Technol **12** (10): 401–403. SPE-1618-G. https://doi.org/0.2118/1618-G.

Morse, R. A. and Von Gonten, W. D. 1972. Productivity of Vertically Fractured Wells Prior to Stabilized Flow. J *Pet Technol* **24** (07): 807–811. SPE-3631-PA. https://doi.org/0.2118/3631-PA.

Nierode, D. E., Williams, B. R., and Bombardieri, C. C. 1972. Prediction of Stimulation from Acid Fracturing Treatments. Paper presented at meeting of Petroleum Society of CIM, Calgary, Alberta, Canada, 16–19 May. CIM 7228

Prats, M. 1961. Effect of Vertical Fractures on Reservoir Behavior—Incompressible Fluid Case. *SPE J.* **1** (02): 105–118. SPE-1575-G. https://doi.org/0.2118/1575-G.

Reynolds, M. Jr., Bray, B. G., and Mann, R. L. 1970. Project Rulison: A Status Report. Paper presented at SPE Eastern Regional Meeting, Pittsburgh, Pennsylvania, USA, 5–6 November. SPE-3191-MS. https://doi.org/0.2118/3191-MS.

Rixe, F. H., Fast, C. R., and Howard, G. C. 1963. Selection of Propping Agents for Hydraulic Fracturing. Paper presented at the Spring Meeting of the American Petroleum Institute, Casper, Wyoming, USA, 24–26 April. API 875-17-L.

Rogers, L. A. 1970. Determining the Explosion Effects on the Gasbuggy Reservoir from Computer Simulation of the Postshot Gas Production History. Paper presented at Symposium on Engineering with Nuclear Explosives, Las Vegas, Nevada, USA, 14–17 January.

Russell, D. G., Goodrich, J. H., Perry, G. E. et al. 1966. Methods for Predicting Gas Well Performance. *J Pet Technol* **18** (01): 99–108. SPE-1242-PA. https://doi.org/0.2118/1242-PA.

Seidel, W. R. and Stahl, E. J. Jr. 1972. Gas Well Stimulation with a Viscous Water-Base Fracturing Fluid. J *Pet Technol* **24** (11): 1,385–1,390. SPE-3502-PA. https://doi.org/0.2118/3502-PA.

Tinsley, J. M., Williams, J. R. Jr., Tiner, R. L. et al. 1969. Vertical Fracture Height — Its Effect on Steady-State Production Increase. *J Pet Technol* **21** (05): 633–638. SPE-1900-PA. https://doi.org/0.2118/1900-PA.

van Poollen, H. K., Tinsley, J. M., and Saunders, C. D. 1958. Hydraulic Fracturing—Fracture Flow Capacity vs Well Productivity. In *Transactions of the Society of Petroleum Engineers,* Vol. 213, 91–95. SPE-890-G. Richardson, Texas: Society of Petroleum Engineers.

Chapter 8

A Case History of Massive Hydraulic Refracturing in the Tight Muddy "J" Formation

D. I. Parrot and **M. G. Long,** members SPE-AIME, Amoco Production Co.

SPE 7936-MS presents a case history of massive hydraulic refracturing in the tight Muddy "J" Formation in Wattenberg Field in the D-J Basin in Colorado. Parrot and Long summarize the reservoir parameters, discuss the evolution of stimulation to larger-size treatments and its economic benefits, compare well performances using massive fracture treatments with those of small fracture treatments, and select 29 wells for refracturing. On the basis of production performances, the authors concluded that the wells that responded favorably to an initial small fracture treatment can be successfully restimulated with massive hydraulic refracturing. The wells that did not respond to initial stimulation likely are not economic candidates for refracture treatments. Use of advanced pressure-transient techniques and computer modeling to compare individual well fracture geometries and reserves may be a better tool for selecting refracture-treatment candidates than only using performance data.

8.1 Abstract

SPE 7936-MS evaluates the refracturing of wells in the tight (.04 to .0003 md, 40 × 10⁶ to .3 × 10⁻⁶ μm²) Muddy "J" Formation in Wattenberg Field using massive hydraulic fracturing (MHF). Twenty-nine wells that originally received 50,000 gallon (189 m³) gelled water fracture treatments on completion have been refractured with 300,000 gallon (1136 m³) polymer emulsion treatments.

This study shows that wells initially completed successfully with small volume fracture treatments can be successfully restimulated with MHF. Those which did not have successful initial stimulations may not be economic candidates for refracturing.

8.2 Introduction

Initial attempts in Wattenberg Field to achieve gas production in commercial quantities were made with fracturing treatments of up to 50,000 gallons (189 m^3) gelled water and up to 150,000 lb (68,039 Kg) of proppant. Low initial production rates and rapid declines indicated most of the stimulations would result in uneconomical wells. Through laboratory and field research, massive hydraulic fracturing treatments evolved with 300,000 gallons (1136 m^3) polyemulsion fluid and up to 735,000 lb (333,390 Kg) of proppant becoming the predominant treatment. Once this massive treatment was deemed economically feasible for new wells, candidates for restimulation were selected. Though some wells selected for refracturing were commercial producers with 50,000 gallon (189 m^3) initial treatments, it was felt their productivity could be increased substantially with larger treatments. However, most refracturing candidate wells were selected on the basis of unsatisfactory performance. Recently, constant well production rate and pressure type curves have been used to determine percent recovery of gas in place and "effective" fracture length in a refracturing candidate. When compared to offset wells, an assessment of the success of the original stimulation on the candidate well and the potential for refracturing can be made.

8.3 Discussion

8.3.1 Field Description. Wattenberg Field is located 25 miles (40 Km) north of Denver covering 980 square miles (2538 Km^2) in Weld and Adams Counties in the western portion of the D-J Basin (**Fig. 8.1**). Amoco Production Company is the major operator with more than 400 wells. Average gas production per well is 150 MCFD (4247 m^3/d).

The Muddy "J" Formation of Cretaceous age is the gas producing horizon in Wattenberg. This sand is at depths from 7600-8400 feet, 8316–2560 m), averaging 50–100 feet (15–30 m) gross thickness with 10–50 feet (3–15 m) net pay sand thickness. Porosity ranges from 8–12%. Formation, permeability varies from .04 to .0003 md (40×10^{-6} to $.3 \times 10^{-6}$ μm^2). Static bottom hole temperature averages 260°F (400°K) and original bottom hole pressure was 2900 psi (1.99×10^4 kPa). The reservoir is stratigraphically controlled

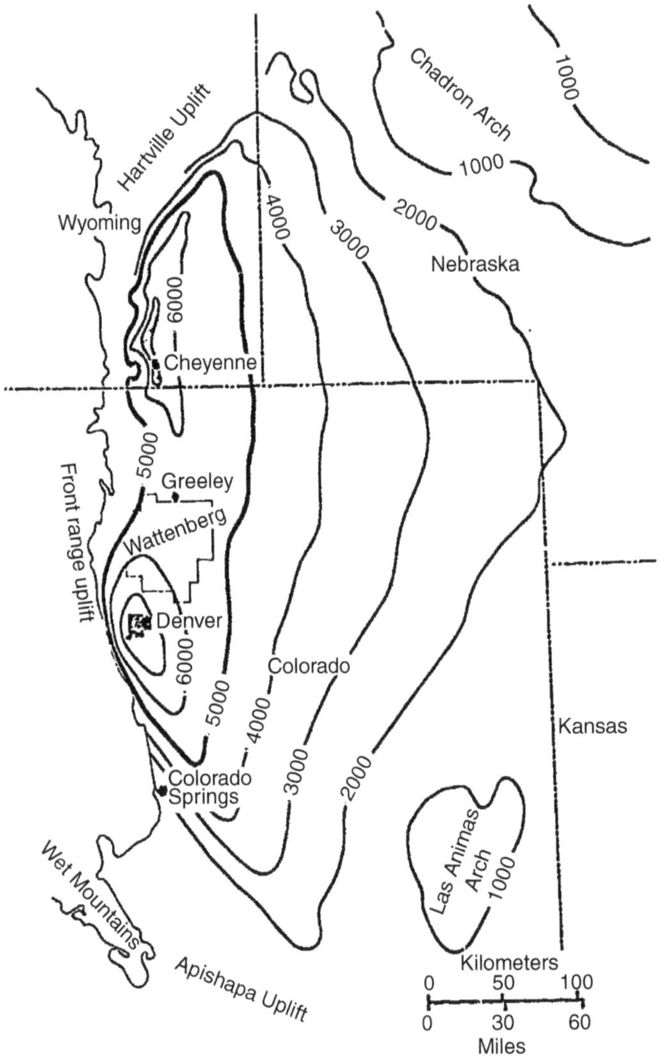

Fig. 8.1—Denver Basin, top pre-Cambrian structure.

by sand pinchout to the southwest and a permeabilty reduction to the northeast (Matuszczak 1973). Unstimulated wells produce from a show of gas to 100 MCFD (2832 m^3/d).

8.3.2 Evolution of Stimulation. During the initial phase of Wattenberg Field development, wells were treated with 30,000 to 50,000 gallons (114–189 m^3) of gelled water down 4½-in. casing with a maximum of 150,000 lb (68,039 Kg) proppant. These treatments created fracture lengths from 100–500 feet (30–152 m) (Fast et al. 1977). Several wells were put on production in late 1971 for long term testing. By early 1973, it was determined that only by creating longer fractures would these wells prove to be economical. A 183,000 gallon (693 m^3) polymer emulsion treatment using 277,000 lb (125,645 Kg) sand evolved as a standard procedure from 1973 to late 1974 which created fracture lengths of 1500–2000 feet (457–610 m) (Fast et al. 1977). Because of up to four times better productivity with this treatment over the smaller treatment, it was concluded that even larger fracture treatments could be economically justified. Designs using up to 623,000 gallons (2358 m^3) of polymer emulsion fluid and 1,041,300 lb (472,326 Kg) sand were used. From 1974 to early 1978, typical stimulations used 300,000 gallons (1136 m^3) and up to 735,000 lb (333,390 Kg) of proppant. Over 100 initial completions were made with a design similar to that shown in **Table 8.1.** Recently, with the advent of efficient cross-linked polymer gels, up to 1.25 million pounds (566,990 Kg) of sand in 570,000 gallons (2158 m^3) of fluid have been pumped in a single stage fracturing treatment during initial completion. As shown by Fast et al. (1977), when treatment size increased, productivity increased and time to payout of drilling and stimulation costs decreased. The upper limit of treatment sizes that can be economically justified in Wattenberg has not been determined.

8.3.3 Refracturing Candidates. Fig. 8.2 is a map of a portion of the Wattenberg Field showing twenty-nine wells that have been refractured. By mid-1974, enough production history was available to definitely establish that some parts of the field were more productive than others. Area A has been determined to be one of these better producing areas and has an average permeability of .005 md (5 × 10^{-6} μm^2). Wells selected for refracturing here were, on initial completion, considered commercial successes (average first month's production greater than 300 MCFD, 8495 m^3/d) even with the 50,000 gallon (189 m^3) initial treatments. However, recently completed offsets fractured with 300,000 gallon (1136 m^3) treatments had up to six times better productivity. This indicated that refracturing these older wells with larger volumes to extend fracture length could yield higher production.

(2/3 "J" Sand Produced Condensate, 1/3 Gelled Water)

1. Clean out to plug back depth.
2. Consider reperfbrating, especially if original perforation density is less than 2 shots/foot (6.5 shots/meter).
3. Set bridge plug above perforations and pressure test casing to 4,500 psi (3.1×10^4 kPa).
4. Set up all equipment and materials and pressure test lines so treatment may begin next morning without delay.
5. Frac down casing at maximum injection rate of 30 bpm (4.8 m³/min) with following design:

# of Stages	Volume gal (m³)	Proppant Concentration lb/gal	(ka/m³)	Mesh Size
a. Pad	29,400 (111)	None		
b. 1	6,300 (24)	2	(240)	100
c. 1	6,300 (24)	3	(360)	100
d. 2	6,300 (24)	4	(479)	100
e. 8	6,300 (24)	4	(479)	20/40
f. 10	8,400 (32)	4	(479)	20/40
g. 2	8,400 (32)	4	(479)	10/20
h. 1	2,016 (7.6)	5	(599)	10/20
i. 1	6,384 (24)	6	(719)	10/20
j. Displace to perforations w/1.5% KCl water.				

Note: Each stage in steps b-g followed by 4,200 gal (16 m³) spacer with no proppant

6. Flow well back immediately
7. When well dies, run tubing and swab to initiate flow.

Material Requirements are as follows:

Emulsion	315,000 gal	(1192 m³)
1/3 water	105,000 gal	(397 m³)
2/3 Muddy 'J' sand produced condensate	210,000 gal	(795 m³)
Sand		
100 mesh	81,900 lb	(37,149 kg)
20/40 mesh	537,600 lb	(243,851 kg)
10/20 mesh	115,584 lb	(52,428 ka)
	735,084 lb	(333,428 kg)

Table 8.1–Typical polymer emulsion refracturing procedure in Wattenberg Field.

Wells in Areas B and C were also initially completed with up to 50,000 gallon (189 m³) treatments, but unlike wells in Area A, these wells exhibited unsatisfactory response to the initial fracturing treatment (average first month's production less than 100 MCFD, 2832 m³/d), probably because of lower permeability in this area. In an attempt to evaluate massive hydraulic fracturing in these areas, eight wells in Area B and two wells in Area C were selected for refracturing with 300,000 gallons (1136 m³) polymer emulsion.

8.3.4 Refracturing Results. Twenty-nine refracturing treatments have been performed to date. A typical procedure for refracturing wells in Wattenberg Field is shown in Table 8.1. The results are summarized in **Fig. 8.3 and Table 8.2.** Of the 19 treatments performed in Area A, six had pre-refracturing production history of at least one year and have post-refracturing data of at least two and a half years. **Fig. 8.4** shows a normalized average performance curve of these six wells. These wells had an average 430 MCFD (12,176 m³/d) increase after refracturing and were still producing an average 55 MCFD (1557 m³/d) above the pre-refrac rate two and a half years after treatment. Fig. 8.4 shows that the average daily gas production for the first full month after refracturing is slightly higher than average daily production for the first full month following initial stimulation (560 vs. 520 MCFD, 15,857 vs. 14,725 m³/d). Incremental reserves added as a result of refracturing average 206 MMCF (5.8×10^6 m³) per well after 2½ years.

Seven more wells in Area A have limited pre- or post-fracturing production data. Four of these wells had pre-refracturing production history of less than one year, but enough data was available (Table 8.2) to judge these wells as having been successfully refractured. The other three were refractured in 1978, and have limited postrefracturing data, but have shown an average stabilized production increase of 240 MCFD (6796 m³/d) after treatment. Five other wells in Area A responded to refracturing; however, payout of the treatments took at least five months longer than other wells in Area A. One well in Area A (A-13) was unsuccessfully refractured in that no response to the restimulation was observed.

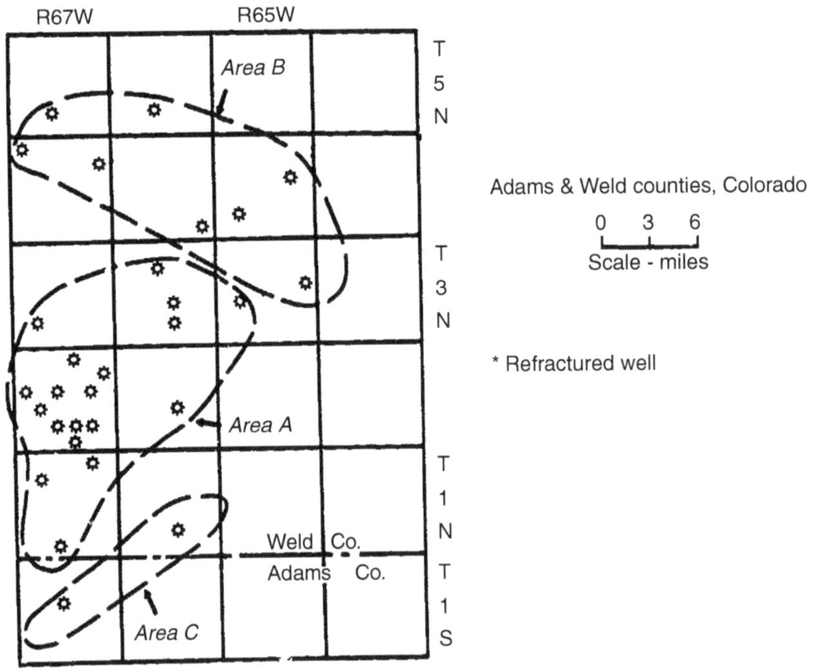

Fig. 8.2—Refracturing areas of Wattenberg Field.

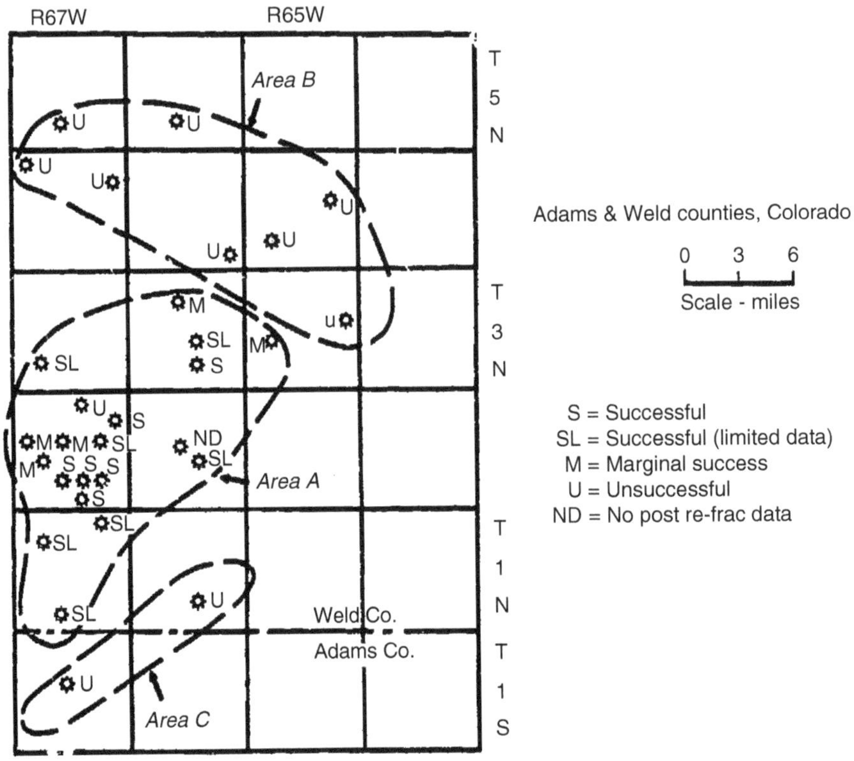

Fig. 8.3—Refracturing results in Wattenberg Field.

Average production history of all the wells refractured in Area A is depicted in **Fig. 8.5.** Fig. 8.5 also shows that producing rates immediately after refracturing are higher than producing rates immediately after the initial stimulation (420 vs. 308 MCFD, 11,893 vs. 8,722 m^3/d). It is obvious from both Fig. 8.4 and 8.5 that refracturing in Area A resulted in significant increased producing rates and reserves.

Table 8.2 and **Fig. 8.6** show production history of wells refractured in Areas B and C. It may be seen from Fig. 8.6 that these wells were very poor gas producers, making less than 100 MCFD (2832 m^3/d) prior to refracturing. Although a response to refracturing was seen, a sustained productivity increase did not occur. From available pressure buildup data in these areas, these wells are known to have much lower permeability (less than .005 md, 5×10^{-6} μm^2) than wells in Area A. It appears that with current fracturing techniques, economical massive hydraulic fracturing treatments may not be possible in these areas of extremely low permeability.

Well No.		Production Prior to Refrac, MCFD	(m^3/d)	Production After Refrac, MCFD	(m^3/d)	No. of Months Since Refrac	Current Production, MCFD	(m^3/d)
A-l		48	(1,359)	360	(10,194)	36	134	(3,794)
A-2	IP	300	(8,495)	1,450	(41.059)	36	330	(9,344)
A-3		91	(2,576)	245	(6,937)	8	315	(8,919)
A-4		135	(3,822)	440	(12,459)	31	172	(4,870)
A-5		45	(1,274)	360	(10,194)	31	122	(3,454)
A-6		50	(1,415)	510	(14,441)	31	132	(3,737)
A-7		48	(1,359)	960	(27,184)	31	139	(3,936)
A-8		24	(679)	116	(3,284)	32	69	(1,953)
A-9		42	(1,189)	170	(4,813)	28	72	(2,038)
A-10		59	(1,670)	125	(3,539)	29	98	(2,775)
A-U		90	(2,548)	300	(8,495)	1	300	(8,495)
A-12	IP	230	(6,512)	560	(15,857)	32	90	(2,548)
A-13		86	(2,435)	90	(3,709)	27	66	(1,869)
A-14		155	(4,389)	640	(18,122)	31	200	(5,663)
A-I5		125	(3,539)	495	(14,016)	36	135	(3,822)
A-16		105	(2,973)	440	(12,459)	31	187	(5,295)
A-17		50	(1,415)	265	(7,503)	5	263	(7,447)
A-18		15	(424)	85	(2.406)	32	23	(651)
A-19		14	(396)	430	(12,176)	34	64	(1,812)
A-20		101	(2,860)	640	(18,122)	1	640	(18,122)
B-l		64	(1,812)	113	(3,199)	34	28	(792)
B-2		86	(2,435)	143	(4,049)	33	45	(1,274)
B-3		66	(1,868)	70	(1.982)	32	29	(821)
B-4	IP	80	(2,265)	123	(3,483)	21	19	(538)
B-5		21	(595)	31	(878)	29	11	(311)
B-6		68	(1,925)	87	(2,463)	33	24	(680)
B-7	IP	470	(13,308)	180	(5,097)	27	58	(1,642)
B-8		8	(226)	128	(3,624)	36	40	(1,132)
C-l		105	(2,973)	160	(4.530)	35	56	(1,585)
C-2		2	(57)	74	(2,095)	33	50	(1,416)

Table 8.2–Summary of refracturing results.

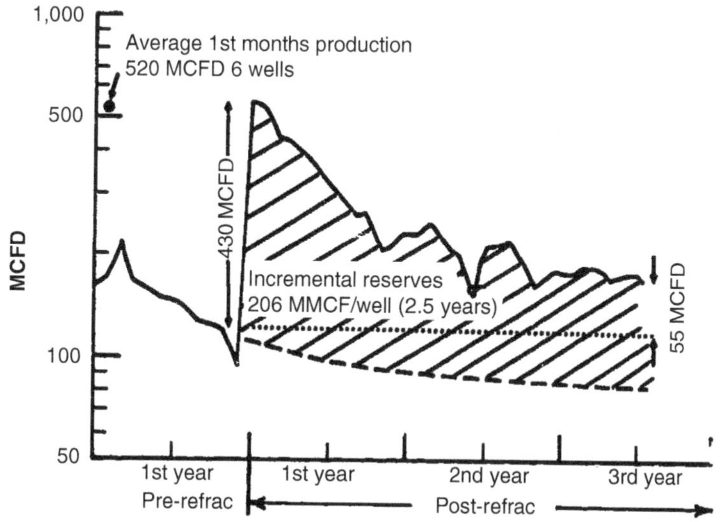

Fig. 8.4—Normalized production curve (six successful refracs, Area A).

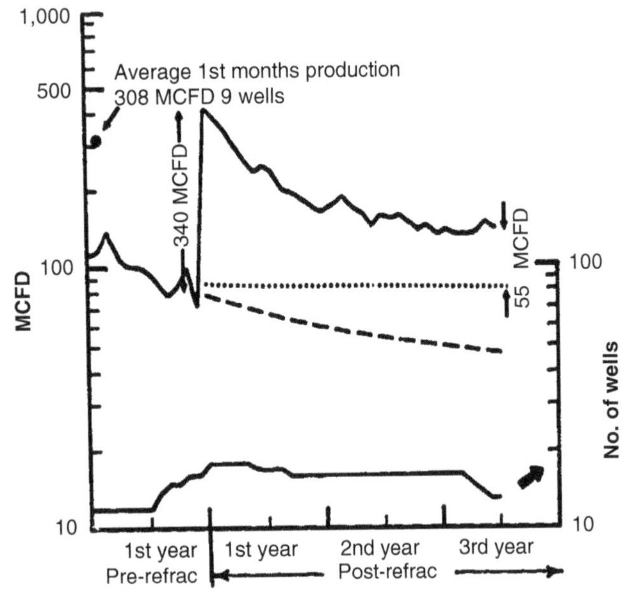

Fig. 8.5—Normalized production curve (nineteen wells, Area A).

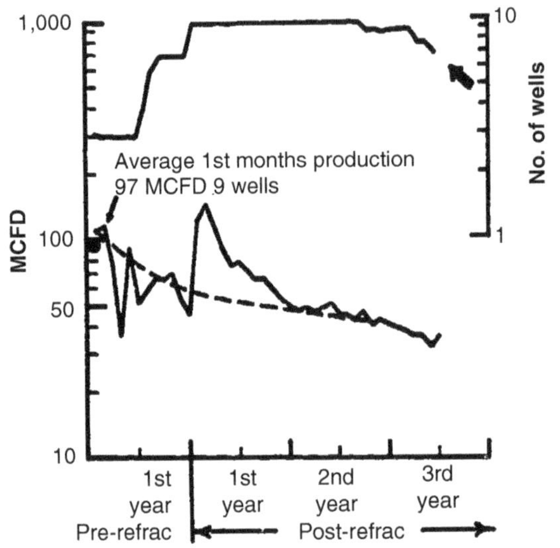

Fig. 8.6—Normalized average production curve 10 wells, Areas B and C.

8.3.5 Load Recovery. An area of major concern in any Wattenberg fracturing treatment, especially when using a hydrocarbon phase, is load fluid recovery. Wells initially completed with 300,000 gallons (1136 m^3) emulsion require a few weeks to several years for load fluid recovery, with those wells in Area A showing faster recovery. Load recovery for refractured wells generally requires even longer, principally because of reduced reservoir pressure due to production.

Primarily because of the value of this slowly recovered condensate, cross-linked gelled water is evolving as the preferred treating fluid in Wattenberg initial completions and refracturing. Cross linked fluids have shown to have equivalent fluid loss characteristics to those exhibited by polymer emulsion fluids and better proppant carrying ability. The higher initial cost (30%) for equivalent volumes of these cross-linked fluids is offset by the fact that the well almost immediately begins production of new condensate. However, similar to the condensate recovery problem experienced with polymer emulsion frac treatments, load water recovery from these cross-linked gelled water treatments is very slow.

The actual effect of long load fluid recovery time on Wattenberg's productivity is unknown. In spite of the fact large quantities of fracturing fluids remain in the formation for some time after initial and refracturing treatments, this has not affected the production of commercial quantities of gas from most Wattenberg wells. It has been suggested (Holditch 1978) that in a tight reservoir such as Wattenberg, a permeability reduction due to possible fracturing fluid invasion does not appreciably affect productivity due to the insignificant pressure drop across the invaded zone compared to the pressure drop from the reservoir to the wellbore. Average reservoir pressure in Wattenberg is 2900 psi (1.99×10^4 kPa) with bottom hole flowing pressure less than 500 psi ($.345 \times 10^4$ kPa).

8.3.6 Improved Techniques for Refracturing Candidate Selection. Recent work (Crowell and Jennings 1978; Agarwal et al. 1979) has allowed development of new pressure transient techniques which along with computer modeling permits reliable estimates of "effective" fracture length and percent recovery of gas in place. This has assisted in refracturing candidate selection. One well in Area A (A-20) was selected for refracturing based on a predicted 25 year recovery of only 1/3 of the gas its offsets would produce and had roughly half the "effective" fracture length of the same offset wells. A larger stimulation of this well appeared warranted in spite of its 182,000 gallon (689 m^3) initial treatment. This well and others scheduled for refracturing in early 1979 will determine the usefulness of these techniques in evaluating wells for restimulation and their potential for success.

8.3.7 Treatment Cost and Payout. Cost of the typical polymer emulsion refrac shown in Table 8.1 averages $90,000 plus the value of condensate used. This includes $70,000 for fracturing treatment costs and $20,000 for tank rental, service unit, fluid hauling, and other miscellaneous costs. The condensate in the treatment is worth approximately $60,000.

Payout for refracturing eighteen of the nineteen wells in Area A ranged from 4 to 41 months using a \$1.37/MCF (\$.05/m^3) gas price, net to the operator after royalty and ad valorem taxes. This does not include credit for produced and plant liquids or BTU adjustments. Since one well in Area A did not respond to refracturing, it of course, will not payout the cost of restimulation. Payout for all wells in Area A averaged 12.1 months which includes expected performance of the three wells refractured in 1978. The value of load condensate is not included in payout as it eventually is recovered.

None of the wells refractured in Areas B and C will realize a payout. As shown in Fig. 8.6, the wells returned to their previous decline shortly after fracturing. The incremental production increase experienced by these wells is insufficient to payout the cost of the job. Further, it is questionable whether these wells will ever recover the condensate used in refracturing.

8.4 Conclusions

This study of performance following refracturing of wells in Wattenberg Field, shows that:

1. Wells initially completed successfully in Area A with small volume fracture treatments can be successfully restimulated by massive hydraulic fracturing (MHF).

2. Wells that responded poorly to small volume fracture treatments upon completion will not respond to refracturing sufficiently to pay out the cost of restimulation. Careful analysis of the reason for poor initial performance must be made. In Wattenberg Areas B and C the extremely low permeability of these areas compared to Area A suggests that a lower permeability limit to successful massive hydraulic fracturing may exist with current fracturing techniques. Further investigation of this lower permeability limit, if it exists, is required.
3. Extremely long load fluid recovery times and the advent of efficient cross-linked water treatments precludes the continued use of a hydrocarbon-based fracturing fluid in Wattenberg Field. In spite of slow load recovery, it does not appear that entrained liquids adversely affect producing rates and recoverable reserves in the better areas of the field.
4. Use of newly developed pressure transient techniques and computer modeling to compare individual well fracture lengths and recoveries may be a better tool for selection of candidates for refracturing than on the basis of poor performance alone.

8.5 References

Agarwal, R. G., Carter, R. D., and Pollock, C. B. 1979. Evaluation and Prediction of Performance of Low-Permeability Gas Wells Stimulated by Massive Hydraulic Fracturing. Paper presented at the Annual Fall Technical Conference and Exhibition, Denver, Colorado, 9–12 October. SPE-6838-MS.

Crowell, R. F. and Jennings, A. R. 1978. A Diagnostic Technique for Restimulation Candidate Selection. Paper presented at the Annual Fall Technical Conference and Exhibition, Houston, Texas, 1–3 October. SPE-7556-MS. https://doi.org/10.2118/7556-MS.

Fast, C. R., Holman, G. B., and Covlin, R. J. 1977. The Application of Massive Hydraulic Fracturing to the Tight Muddy "J" Formation, Wattenberg Field, Colorado. *J Pet Technol* **29** (01): 10–16. SPE-5624-PA. https://doi.org/10.2118/5624-PA.

Holditch, S. A. 1978. Factors Affecting Water Blocking and Gas Flow From Hydraulically Fractured Gas Wells. Paper presented at the Annual Fall Technical Conference and Exhibition, Houston, Texas, 1–3 October. SPE-7561-MS.

Matuszczak, R. A. 1973. Wattenberg Field, Denver Basin, Colorado, The Mountain Geologists, **10** (03): 99–105.

Chapter 9

Analyses of an Elmworth Hydraulic Fracture in Alberta

R. E. Wyman, SPE, Canadian Hunter Exploration Ltd.; **S. A. Holditch,** SPE, Texas A&M University; **P. L. Randolph,** SPE, Inst. of Gas Technology

SPE 7935 presents a case history of a stimulation treatment of a well completed in the Falher interval of the Spirit River Formation of the Lower Cretaceous age within the Deep Basin in Alberta, Canada. Wyman et al. evaluate the formation using openhole logs, core analysis, and testing (drillstem, breakdown, mini-fracture, and well tests). They also design and pump a large foamed fracture treatment and then evaluate the fracture and formation using pressure-buildup analysis. An economic analysis determines the technology and economic conditions needed for commercial development of the low-quality reservoir (k < 0.001 md) in the basin. Through their study, Wyman et al. conclude

- The ratio of net to gross pay is less than 50%
- There is no flow to surface after perforations and swabbing
- The fracture gradient is high—in the range of 0.80 to 0.91 psi/ft, with an average of 0.87 psi/ft
- A foam fracture treatment creates a fracture half-length of 700 ft, with a fracture conductivity of 500 md-ft
- Higher gas prices and larger treatments are needed to economically develop the resources in the Falher Formation

9.1 Introduction

The Elmworth Gas Field is located within the Deep Basin near Grande Prairie, Alberta (**Fig. 9.1**). The Deep Basin covers about 26,000 sq miles and appears to contain an enormous amount of gas in different-quality rocks. Masters (1979) has estimated the Deep Basin could contain more than 400 Tcf of gas, recoverable at various levels of price and technology.

Canadian Hunter visualizes the gas distribution in the Deep Basin as a resource triangle. In such a distribution, the sands and conglomerates of highest porosity and permeability are at the very top of the triangle. As reservoir quality decreases, the resource triangle indicates larger and larger quantities of gas in lower-porosity and lower-permeability rocks. Near the bottom of the triangle are sands approaching 7% porosity and 1 μd in-situ permeability. Within these lower-porosity rocks (7 to 11%) resides a huge ultimate potential. Detailed analyses were made of logs, cores, and tests of about 20 zones from 57 wells in a 60-township area around the Elmworth Field. This study indicated that, with sufficient but reasonable economic incentives, it would be possible to recover more than 200 Tcf from the lower-quality rocks within the Deep Basin (Canadian Hunter Exploration Ltd. 1978).

To obtain more information on the technology and economics required to recover this gas, Canadian Hunter performed a large hydraulic fracture over the Cretaceous Falher Zones A and B in the Canhunter Texcan Elmworth Well 11-12-71-13W6. SPE 7935 presents a description and analysis of the fracture.

9.2 Petrophysical Properties

The Falher Zones A and B were cored completely, and a full suite of openhole logs was run. This suite included the induction-electrical, acoustic with variable density display, compensated neutron-density, and microlog. Selected logs and the core analysis (made at atmospheric conditions) are shown in **Fig. 9.2.** Although the entire cross section of 164 ft is considered gas-bearing, less than one-half of the section is estimated to be net-pay, based on analysis of logs, cores, and pressure buildups. Sixty-six feet of this section have measured core porosities equal to or above 7%; this is shown by the shaded area on the porosity curve in Fig. 9.2. Permeabilities above 0.1 md also are shaded. Sonic log and pressure buildup analysis indicate a net-pay closer to 45 ft. The relation between fractional core porosities (ϕ) measured at 200 psi and sonic travel time (Δt) in μs/ft is described by $\phi = 0.0054\Delta t - 0.2865$.

Permeabilities measured at 200 psi are all below 0.6 md. When the effective overburden is increased to 3,350 psi in the laboratory, the average permeability to gas decreases about threefold, but in-situ permeabilities measured from pressure buildups are closer to 0.001 md, a hundred-fold decrease from dry-core measured permeabilities. This decrease is caused not only by overburden stress but by gas slippage in the fine capillaries (Klinkenberg effect) and by relative permeability effects due to the connate water. **Table 9.1** shows current best estimates of reservoir and hydraulic fracture parameters.

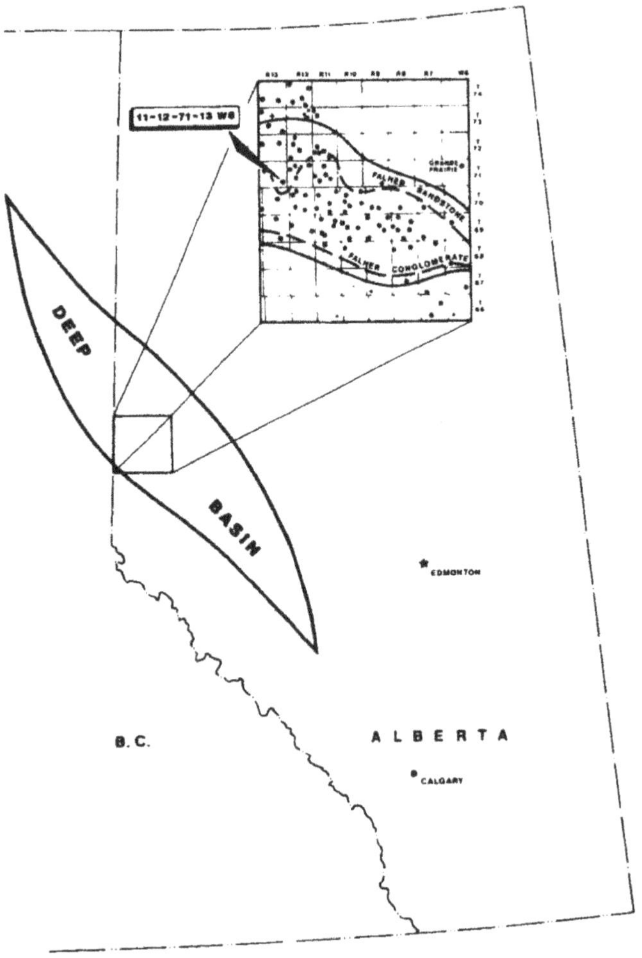

Fig. 9.1—Location of Elmworth Field.

Reservoir pressure, psi	2,200
Reservoir temperature, *F	171
Gross thickness of Falher A and B Sands, ft	164
Gas porosity, %	4.5
Estimated minimum net-pay from pressure buildups, ft	45
Net feet with porosity ≥ 7%, ft	
From whole core analysis	45
From plug analysis	66
Average water saturation of net-pay, %	40
Gas permeability (in situ), md	0.001
Fracture length, ft	700
Fracture conductivity, md-ft	>500
Fracture gradient, psi/ft	0.87
Gas gravity (air = 1.0)	0.68
Original gas formation volume factor, Rcf/scf	0.0068
Original gas viscosity, cp	0.0169
Original gas compressibility, vol/vol/psi	0.000466
Condensate ratio, bbl/MMcf	2
Bottomhole flowing pressure (assumed for simulation), psi	150

Table 9.1—Summary of reservoir and fracture parameters.

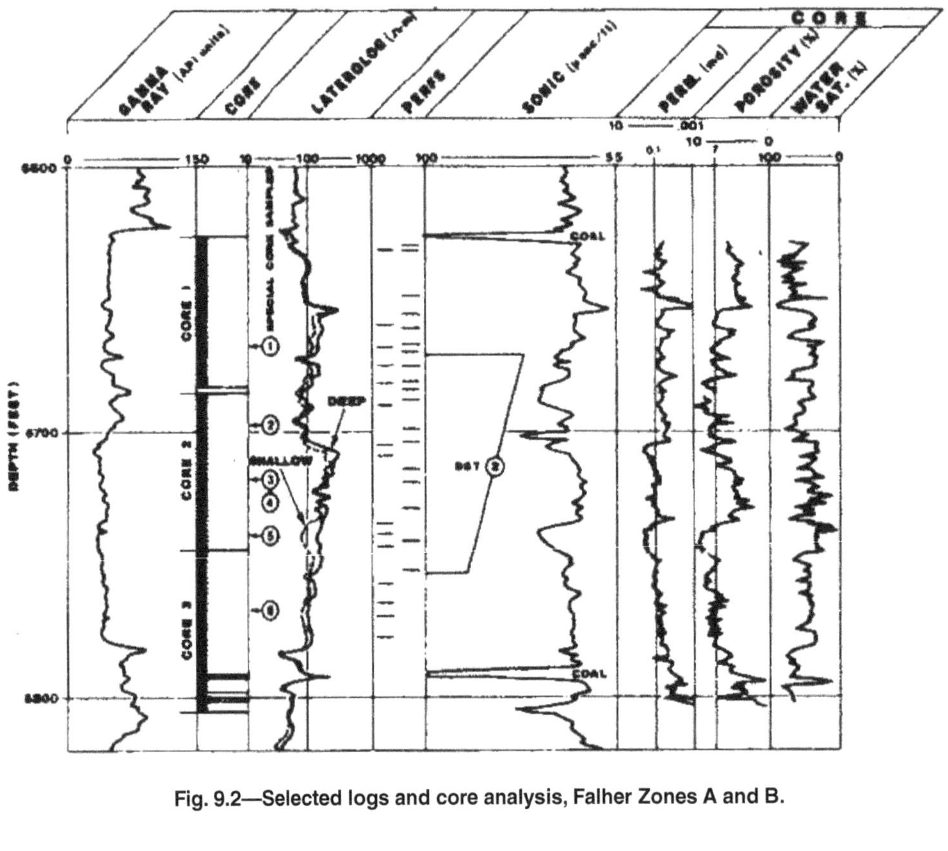

Fig. 9.2—Selected logs and core analysis, Falher Zones A and B.

A tight gas sand with 1-μd permeability was produced, after a large hydraulic foam fracture, at rates close to predicted values. The rates of about 300 Mcf/D (8495 m³/d) after 23 days would not bear the cost of development at 1979 Canadian prices. However, it is estimated that, with higher netbacks and longer fractures, commercial development would be feasible.

9.3 Geologic Description

The Falher Sands are members of the Spirit River Formation of Lower Cretaceous age. They represent part of about 20 potentially productive zones in the Elmworth area. The Falher Reservoirs are upward-coarsening sands, in some areas culminating in chert-pebble conglomerate. These depositional cycles represent high-energy beach or barrier bars within a large, prograding sand delta. The delta was fed by a well-developed distributary channel system that supplied elastics up to pebble size from a source to the southwest. The marine side of this beach/bar system (toward the north) loses the permeable conglomerates but contains a large volume of fine- to coarse-grain low-permeability sand. These sands, originally well-rounded to subrounded, appear to have had substantial porosity at the time of deposition. However, secondary quartz overgrowth and later calcite cementing have reduced the pore space severely. Much of the calcite has been dolomitized. A summary of the mineral content as determined from X-ray analysis and petrographic examination of the thin sections is shown in **Table 9.2.** A typical thin section is shown in **Fig. 9.3.** In this well the upper two members, Falher Zones A and B, contained no conglomerates and provided an excellent opportunity to fracture and test only the tight sands.

Fig. 9.3—Photomicrograph of Falher Sandstone at 6,739 ft. Sample 5–partly crossed Nichols at 150X.

9.4 Sequence of Testing and Stimulation

The Falher Sands first were tested with an openhole drillstem test after coring. After drilling to 8,930 ft, the hole was cased and perforated for limited entry over a gross interval of 146 ft. These holes were broken down individually with a mixture of methanol and water to prepare for subsequent fracturing. After additional perforations, a small fracture was performed to test field procedures and obtain initial *kh* data. This was followed by the main foam fracture. After the main fracture, the well was production-tested for 1 month and then shut in for a pressure buildup over 7½ months. In Nov. 1978, it was tested again for 3 weeks during which time it was reperforated and monitored with production logs. Again it was shut in for a long-term buildup. **Table 9.3** provides the sequence of testing and stimulation activity, which is discussed in detail in the next section.

	Sample											
	1		2		3		4		5		6	
Sonic log depth, ft	6,667.4		6,697.1		6,718.0		6,718.0		6,739.0		6.766.9	
Porosity, alm-fracture %	0.047		0.075		0.035		0.033		0.083		0.079	
Permeability, atm-md	0.007		0.02		0.38		0.02		0.23		0.109	
	X-Ray	Thin Section	X-Ray	Thin Section	X Ray	Thin Section	X-Ray	Thin Section	X-Ray	Thin Section	X-Ray	Thin Section
Quartz	0.75	0.57	0.83	0.62	0.83	0.26	0.82	0.25	0.77	0.52	0.78	0.62
Rock fragments		0.11	0.08	0.14		0.24		0.25		0.05		0.07
Chert				0.11		0.29		0.29		0.13		0.08
Feldspars	0.03	*	0.08	0.03	0.02	*	0.02		0.02	*	0.04	*
Calcite	0.03	0.20**		0.04**	0.03	0.14**	0.03	0.15**				
Dolomite	0.15		0.03		0.05		006		0.17	0.22	0.14	0.15
Pyrite			0.02	*	0.05	*	0.03	*		*		*
Glauconite										*		
Clay	0.04	*	0.04	0.03	0.02	0.05	0.04	0.04	0.04	0.06	0.04	*
Chlorite	(0.4)		(0.79)		(0 34)		(0.72)		(0.20)		(0.64)	
Illite/mica	(0.6)	0.02	(011)	*	(0.66)	*t*	(0.28)	*	(0.80)	*	(0.36)	0.04
Organic material		*				*		*				

Table 9.2—Mineral content determination (Falher A and B Sands).

*Minor amounts.

**Much of the calcite has been dolomitized.

Dates	Events	Remarks
8/24/77	Start coring Falher A and B Sands	
8/24/77	Openhole drillstem test 6,671 to 6,754 ft	Gas to surface too small to measure
8/31/77	Logs run over Falher Zone	
10/12-14/77	Reach total depth, log, and set casing	Cemented 5½-in., 17-lbm N-80 casing at 7,530 ft with 2,445 sacks
11/10/77	Perforate Falher A and B Sands 6,632 to 6,778 ft	15 shots for limited entry
11/11/77	Swab test	No gas to surface; 0 psi after 14 days
11/21/77	Pressure-inject methanol/water into individual perforations	30% methanol/70% water
11/27/77	Fire 17 additional jets 6,630 to 6,753 ft	To supplement smaller than expected holes and some original perforations that did not accept fluid
11/30/77	Test gas to surface	21.4 Mcf/D during 6-½-hour flow
11/30-12/2/77	Buildup with bottomhole packoff	Calculated matrix permeability of 0.0015 md
12/4/77	Mini-fracture	Gelled water and 9,450 lbm of sand
12/6/77 2/15/77	Test of mini-fracture	Tested 21 Mcf/D while cleaning up Dec. 4-14; shut in; flowed again 4 days (Feb. 11-15) with rate of 35 Mcf/D
2/16-28/78	Pressure buildup	Calculated matrix permeability of 0.0015 md
3/1/78	Foam fracture	250,000 lbm of 40/60 sand and 5.7 MMcf of nitrogen
3/1-28/78	First production test	Flowed maximum 933 Mcf/D, decreasing to 283 Mcf/D after 28 days
3/31-11/8/78	First pressure buildup after main fracture	More than 7 months' buildup; maximum bottomhole pressure 2,039 psi; calculated matrix permeability of 0.001 md
11/9-12/2/78	Second production test	Flowed maximum 1,195 Mcf/D, decreasing to 301 Mcf/D after 22 days
12/2/78-5/29/80	Second pressure buildup after main fracture	Built up to 1.574 psi by Jan. 15, 1979, and to 1,954 psi by May 29, 1980; calculated matrix permeability 0.001 md

Table 9.3—Sequence of testing and stimulation.

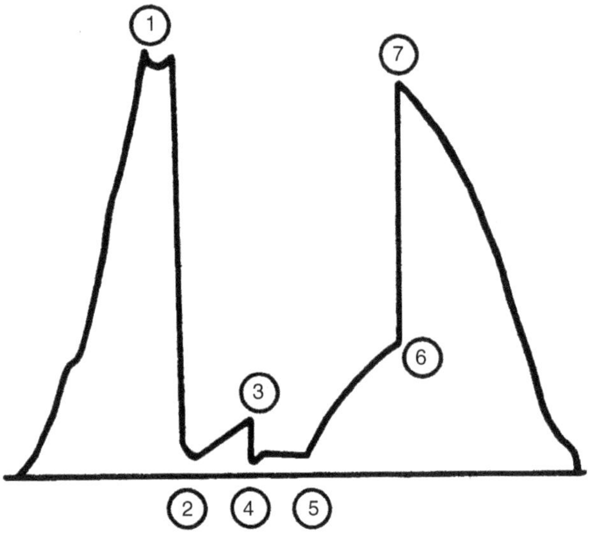

1 IHP 3527 PSIG
2 PF 100 PSIG
3 ISIP 526 PSIG (60 Min.)
4 IFP 139 PSIG
5 FFP 160 PSIG (60 Min.)
6 FSIP 1065 PSIG (90 Min.)
7 FHP 3309 PSIG

Fig. 9.4—Pressure chart for openhole drillstem Test 2 (6,671 to 6,754 ft) on Falher Sandstone.

9.5 Results

9.5.1 Drillstem Test. A drillstem test was run on bottom in the open hole over the 6,671- to 6,754-ft interval. This test was in the middle of the Falher A and B Sand and covered about 50% of the gross interval. There was no gas to the surface during a 15-minute preflow nor during the 60 minutes of main flow time. Ten minutes after the valve was closed, enough gas arrived at the surface to light a 2-ft flare that died after several minutes. The pressure chart shown in **Fig. 9.4** is typical of the very tight sands in the Elmworth area. A final shut-in time of 90 minutes was not enough to allow extrapolation of the pressure buildup **(Fig. 9.5).**

9.5.2 Perforation and Breakdown. The number, size, and depths for perforation were selected using the following criteria.

1. Perforation was performed on each sandstone interval revealed by core analysis to contain natural gas in at least 2% of the gross sandstone volume.
2. Limited-entry treatment, with a pressure drop of roughly 500 psi across the perforations, was chosen in case the different gas-bearing sandstone strata would require different treating pressures for fracture propagation.
3. To the extent practicable, the number of perforations in a continuous sandstone stratum was to be proportional to the gas in place per foot of depth adjacent to the perforations.

The design adopted to meet these criteria involved shooting 15 perforations with a through-tubing magnetically decentralized gun that produces 0.33-in. diameter perforations under API test conditions.

Before perforating, the well was circulated with a mixture of 30% methanol and 70% water and then swabbed to a depth of 6,508 ft. For the anticipated reservoir pressure of 2,220 psi, this provided a pressure drop of about 2,100 psi toward the wellbore at the time of perforation.

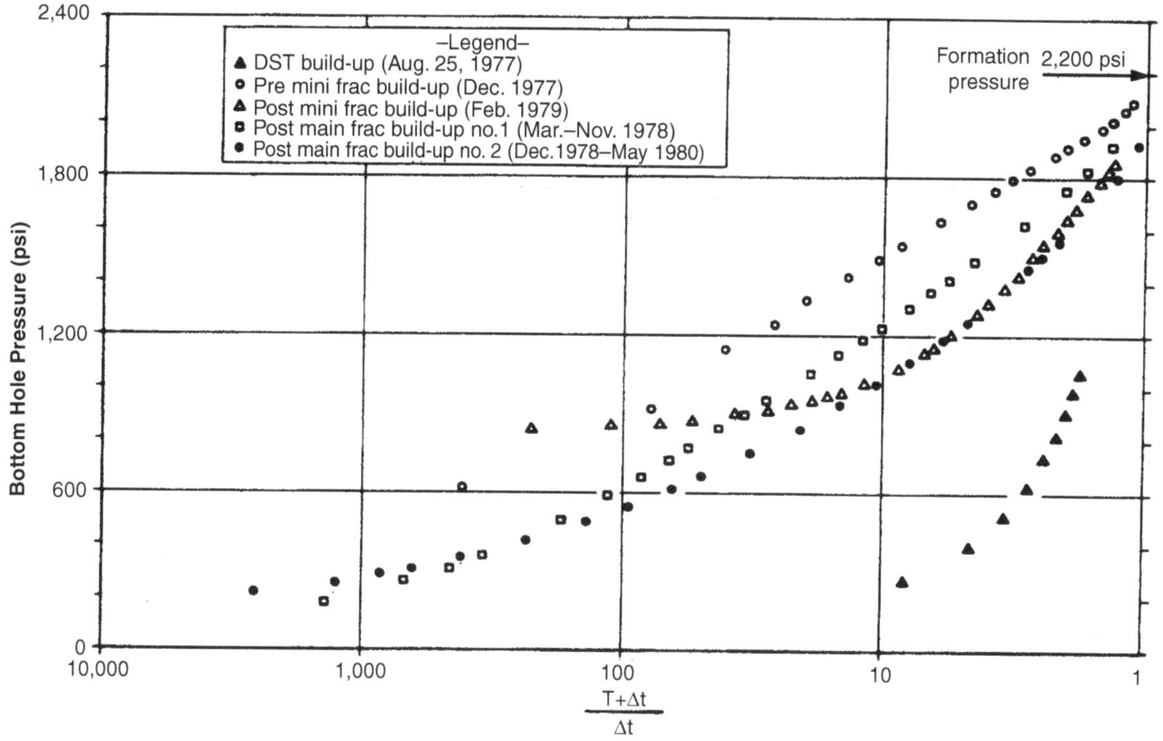

Fig. 9.5—Pressure buildups run before and after main fracture job.

Five successful trips were made. On each trip, up to four perforating charges with the desired spacing were fired simultaneously. After each firing, pressure buildup (if any) was less than 100 psi/D. The total perforated interval was 6,632 to 6,778 ft.

After firing all 15 charges during daylight hours on two successive days, a gamma ray log detected radioactive tracers from all charges at the desired perforation depth. The tubing then was lowered to 6,847 ft, and the well was swabbed to below the deepest perforation (6,778 ft).

After raising the tubing to 6,338 ft, noise logs detected fluid movement in the perforated interval. The liquid level rose to about 6,000 ft during a 24-hour period. The well again was swabbed to below the bottom perforation. Gas production remained too small to measure, and a bottomhole pressure bomb recorded buildup to only 72 psi after 59 hours of shut-in time.

The perforations then were broken down and tested individually by isolating each with packers and pumping a mixture of 30% methanol and 70% water. During this procedure, both tubing and annulus pressures were recorded and flow rate into each perforation was monitored. From this information and the relationships between friction terms and flow rates, calculations were made for the perforation diameter and the pressure required for fracture propagation behind each perforation.

Observations from analysis of this data indicate the following:

1. Although all guns appeared to have fired, the interval opposite one set of four shots did not accept fluid at lithostatic pressure. This may indicate failure of the gun's magnetic decentralization or a zone with a fracture gradient higher than 1.0 psi/ft.
2. Actual perforation diameters were only two-thirds of the 0.33 in. reported from standard API tests with the same gun and casing.
3. The average fracture gradient was calculated to be 0.87. Gradients were deduced from the fracture propagation pressure rather than shut-in pressure. The difference is flowing friction pressure drop in the fracture itself, plus any incremental pressure required to overcome the tensile strength of the rock.
4. Fracture propagation pressures measured at different perforation intervals within the Falher Sand varied by several hundred pounds per square inch. **Fig. 9.6** shows that fracture gradients varied from 0.799 to 0.910. To accept fluid, a bottomhole pressure greater than 6,100 psi inside casing was required for all tested perforations.

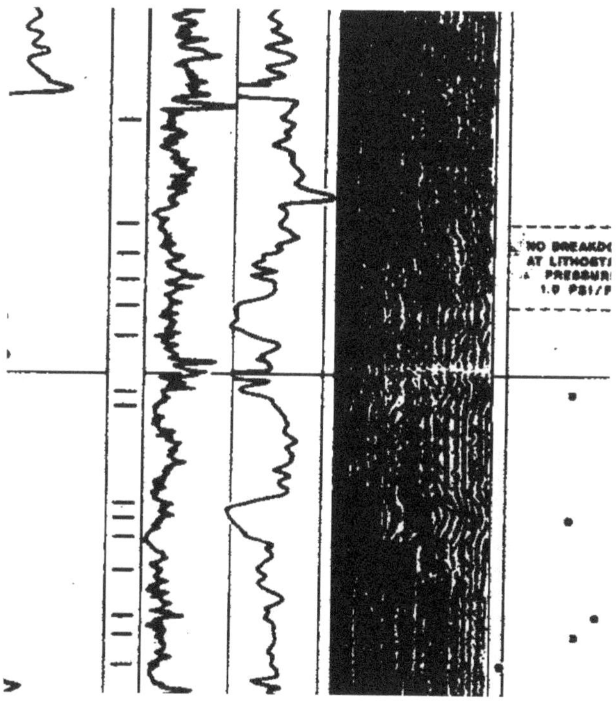

Fig. 9.6—Fracture gradients with acoustic response.

To check these observations, a confirmation test was performed by injecting 89 bbl of 30% methanol/70% water into all open perforations. This test confirmed that only 11 of 15 shots produced perforations that would accept fluid and that the hole diameters were only about two-thirds of the API value. Seventeen additional perforations were requested. The gamma trace log verified that all had fired at the desired depth, but later testing revealed only two or three of the additional perforations accepted fluid at lithostatic pressure (6,700 psi inside casing).

After these perforations were fired, the hole was swabbed to the bottom and natural gas was produced for 6½ hours; a rate of 21.4 Mcf/D was observed. After shut-in with a downhole packoff, bottomhole pressure built up to 2,083 psi in 36 hours (Fig. 9.5). Conventional buildup analysis was used to interpret the data (Matthews and Russell 1967). These data appear to form a straight line from time of shut-in to 7 hours on a Horner plot. The data collected after 7 hours began to deviate from the early straight line. A line drawn from the last data point to the formation pressure of 2,200 psi is believed to approach the correct value. This calculation showed a permeability of about 0.0015 md.

Note the relation of the fracture gradients with the variable density log shown in Fig. 9.6. Anderson and Walker (1972) postulated that formations characterized by a highly attenuated and lower-velocity shear wave indicate a higher degree of "plasticity," which leads to higher fracture gradients. As seen in Fig. 9.6, there is some indication that where the ratio of shear velocity to compressional velocity is lower, the fracture gradients are higher. For instance, note the high-confidence fracture gradient values (0.845 to 0.910 at 6,735 to 6,744 ft) fall within an interval with a distinctly lower shear velocity. In contrast, the high-confidence value (0.799 at 6,778 ft) is in a higher-shear-velocity rock.

9.6 Mini-Fracture

About 1 week after the additional perforations were fired, a small hydraulic fracture operation (mini-fracture) was performed to (1) provide a field test of the equipment and procedures planned for the large foam fracture, (2) determine how many new perforations actually would accept fluid (for final selection of rate and pressure to be used on the large fracture job), and (3) emplace about 10,000 lbm of 40/60 proppant sand near the wellbore (for production and buildup testing to determine *kh*).

The foam mini-fracture was attempted but aborted before any sand was pumped. This was because of problems with lubrication and batteries on the nitrogen pumpers (at surface temperatures of near –40°F). The next day the fracture was performed with gelled water using a prepad of 165 Mcf of nitrogen. This was followed by 145 bbl of gelled water carrying 9,450 lbm of sand.

After 7 days of periodic swabbing, a steady gas production rate of 21.1 Mcf/D was achieved. This reflects little change from the previous rate observed after the methanol/water breakdown. After 2 more weeks of production for cleanup and a shut-in period of 33 days, the well was produced at a rate of about 35Mcf/D for 4 days. It then was shut in for pressure buildup. This pressure buildup (Fig. 9.5) indicated a matrix permeability of about 0.0015 md—the same as the buildup before the mini-fracture.

9.7 Main Fracture Treatment

The large treatment was performed March 1, 1978. During the previous 2 days, all equipment was rigged and all lines were pressure-tested. The equipment on location (**Fig. 9.7**) consisted of eight nitrogen pumpers, six nitrogen tankers, two fracture trucks, two blenders, an acid tanker and a small pumper to inject the foaming agent, a water tanker and transfer pump for the 100 bbl of 30/70 methane/water mixture, three fracture tanks containing gelled water, and two sand silos with 40/60 sand.

Table 9.4 shows the pressures and the schedule of materials pumped, as later deduced from on-line measurement during the fracturing operation. Stages I through III were performed to remove the mini-fracture proppant.

The mini-fracture performed 3 months previously had placed 9,450 lbm of 40/60 sand proppant near the wellbore. To ensure that a sandout would not result from pickup of this proppant by the leading edge of the foam pad during the main fracture, special precautions were taken during the first three stages.

Fig. 9.7—Equipment setup for main foam fracture job.

	Fluid		Rate		Sand	
Time Interval	Type	Volume In Fracture (bbl)	In Fracture (bbl/min)	Liquid (bbl/min)	Concentration In Foam (lbm/gal)	Amount (lbm)
I	N_2	300	7.4	–	–	–
II	H_20 & MEOH	88	4.5	–	–	–
III	N_2	312	15.6	–	–	–
IV	Foam pad	452	11.3	2.3	–	–
V	N_2 Gelled H_20 Sand	156	10.4	2.3	0.48	3,200
VI	N_2 Gelled H_20 Sand	214	11.9	2.3	0.94	8,500
VII	N_2 Gelled H_20 Sand	213	12.5	2.3	1.7	15,000
VIII	N_2 Gelled M_20 Sand	1,134	13.3	2.5	2.4	115,000
IX	N_2	133	3.7	–	–	–
X	N_2 Gelled H_20 Sand	61	10.1	2.6	1.4	3,500
XI	N_2 Gelled H_20 Sand	49	9.8	2.6	2.1	4,400
XII	N_2 Gelled H_20 Sand	737	9.8	2.6	2.9	90,000
XIII	N_2 Flush	46	6.6	–	–	–

Table 9.4—Main fracture.

Stage I. Nitrogen was injected at increasing rates until a stable injection rate was achieved at a bottomhole pressure of about 6,000 psi. Tubing injection rate was decreased to 1,000 scf/min. This small rate, used for the remainder of the operations, was chosen to ensure no liquid leg in the tubing and to permit use of the tubing pressure gauge to monitor bottomhole pressure.

Stage II. Injection was switched from nitrogen to a 30% methanol/70% water mixture. The injection rate of 4.5 bbl/min was lower than desired but was limited by the capacity of the transfer pump between the tanker and blender.

Stage III. The 30% methanol/70% water mixture was displaced through the perforations by injecting nitrogen into the annulus at 20,000 scf/min while maintaining the 1,000-scf/min rate on the tubing. Bottomhole pressure was slightly above 6,000 psi. In the middle of this displacement, a small slug of about 20 bbl of foam was pumped into the annulus to scour any old proppant from the walls. It was hypothesized that the nitrogen following this foam would break it up and allow any entrained proppant to settle to the bottom.

The total materials pumped into the well to displace the mini-fracture proppant consisted of 450 Mcf of nitrogen, 88 bbl of 30% methanol/70% water mixture, and about 5 bbl of gelled water plus foaming agent. Pressures and rates were consistent with mini-fracture results to within previously deduced uncertainties in pressure-gauge calibrations.

At the end of the prepad cleanout procedure, the nitrogen rate was reduced to 13,200 scf/min into the annulus; the rate of 1,000 scf/min into the tubing was continued. The pumping of foaming agent and gelled water commenced at the same time so that flow rate through the perforations remained constant.

When the foam pad reached the perforation, bottomhole pressure jumped from 5,950 to 6,170 psi. This was expected because of the increase in perforation friction due to the density difference between foam and nitrogen gas. At this bottomhole pressure, all perforations were accepting fluid. Further, both rates and pressures were as anticipated from design calculations based on the 13 to 14 open perforations deduced from the mini-fracture analyses.

Stages IV through VIII in Table 9.4 cover the injection of a conservative foam pad and steps to increase sand concentration. Stages IX through XII reflect a field modification that became necessary when centrifugal pump seals failed on the blender. Foam in the fracture was kept in shear by a small nitrogen rate during the 30 minutes required to change blenders. Sand concentration again was increased in stages when foam fracturing was resumed. Stage XIII, the nitrogen flush, was small to ensure that the nitrogen expansion before cleanup would not overdisplace the proppant.

The rates and sand concentrations in Table 9.4 differ substantially from both the plan and on-line information used to control the field operation. Postfracture analyses have revealed the following. Actual nitrogen rates were more than 30% greater than desired. The 6-in. turbine meter on the blender was outside its linear range; tank straps were used to control the gelled water rate of about 2.5 bbl/min when the turbine was reading 3.5 to 4.0 bbl/min. The blender operator controlled the sand feed on the basis of turbine meter reading; as a result, actual average sand concentration in gelled water for Stages VIII and XII was about 12 lbm/gal rather than the planned 8 lbm/gal.

The net result of these metering problems was curtailment of the operation due to lack of nitrogen after about 250,000 lbm of the 330,000 lbm of proppant on location was used. In addition, a total of 5.65 MMcf of nitrogen, 680 bbl of water, 88 bbl of foaming agent, and 88 bbl of 30/70 alcohol/water mixture were used. The average foam quality (percent gaseous phase) in the fracture was about 82, in contrast to the planned 75.

It is recommended that future foam fracturing operations include a control based upon on-line monitoring of both the gelled water/sand and the foam/sand densities using gamma ray absorption techniques.

Despite the several operational problems, the overall objective of the main foam fracture job was accomplished smoothly. It was concluded that even larger jobs in this type of formation could be executed without major problems.

9.8 Testing After Main Fracture

Cleanup with production through the tubing was started 18 minutes after flush. During the first 24 hours, about 1.2×10^6 scf of nitrogen was estimated to have been produced. Temperature and gamma ray tracer logs were run successfully starting 44 hours after flush and after slugs of hydrates were produced as bottomhole pressure declined to 650 psi. Both logs indicated that the fracture was in the desired depth interval.

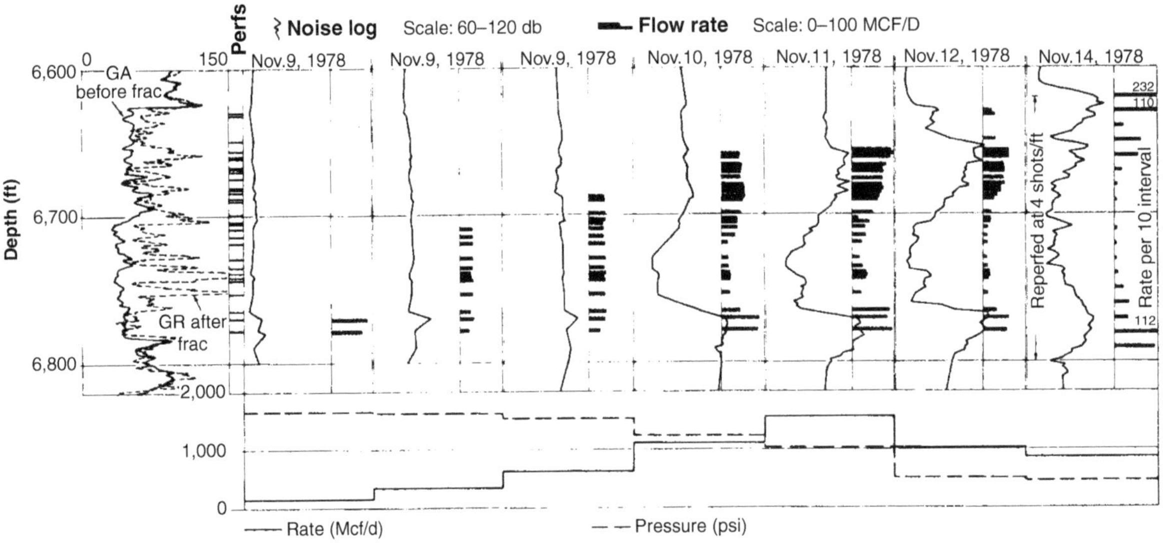

Fig. 9.8—Noise logs and interpreted flow profile for Nov. 1978 test.

Gamma ray response to the radioactively tagged proppant is shown in **Fig. 9.8.** The temperature log revealed cooling in excess of 4°F over the depth interval from 6,623 to 6,788 ft. Within this interval, minimum cooling was in the interval 6,710 to 6,735 ft, where high fracture gradients had been observed in tests of individual perforations (Fig. 9.6). Maximum cooling of about 7°F occurred in the vicinities of 6,660 and 6,775 ft. The cool area near 6,660 ft suggests that the two or three additional perforations that accepted fluid were in the area where breakdown was not achieved after the first set of perforations.

By the fifth day, most of the gas being produced was hydrocarbons, and the percent of nitrogen continued to decrease until it was only about 10% of total gas by the 28th day of production. The calculated production of total gas for the first 28 days in March 1978 is plotted in **Fig. 9.9.** This plot shows reasonable comparison with the expected results from simulation after initial cleanup; when the nitrogen is removed from these rates, the last 2 weeks of production coincide closely with the simulated decline. The cumulative production of hydrocarbon gas for this time was 6.75 MMcf. Also, a total of 3.67 MMcf of nitrogen (or 65% of the total injected) was recovered during this period.

The well then was shut in, and bottomhole pressure readings were monitored over the next 7½ months. This pressure buildup compared with other buildups in this well is shown in Fig. 9.5. By Nov. 8, 1978, the well had built up to a bottomhole pressure of 2,039 psi; it again was opened for testing.

During the first weeks of production, the flow was monitored with 20 noise- and temperature-log runs over the perforated interval. The interpretation from seven of these runs is shown in Fig. 9.8. Note how the initial flow started from the bottom two perforations and gradually worked up each day until it was distributed over the entire perforated interval by the third day of testing. Although only the 1,000-Hz noise curve is shown in Fig. 9.8, other noise curves (200, 600, and 2,000 Hz) and temperature logs were used to make the interpretation of flow rates.

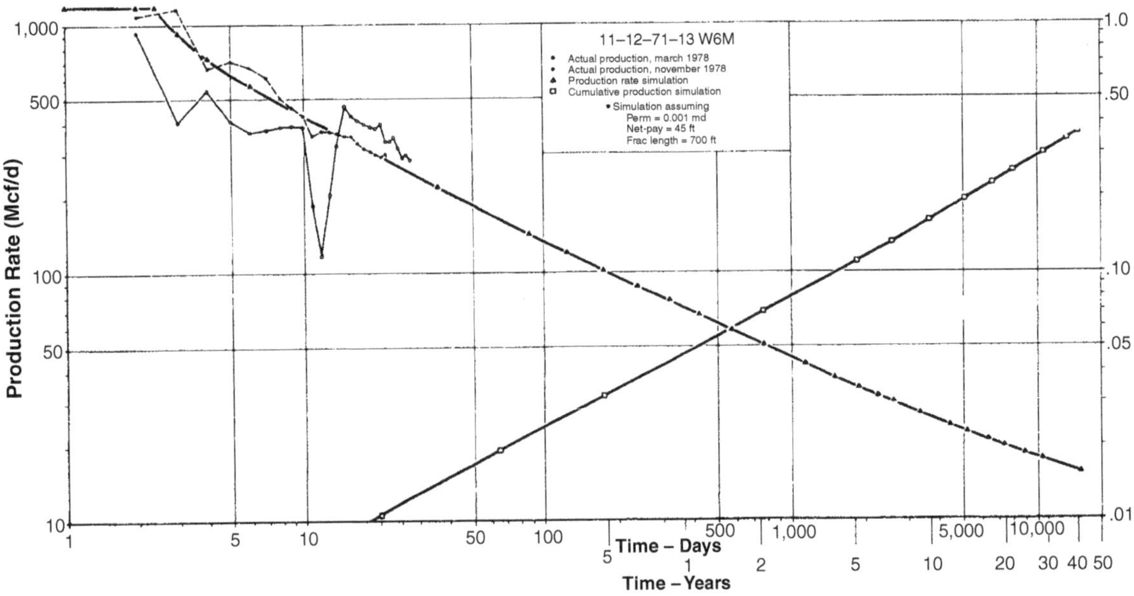

Fig. 9.9—Simulated production rate and cumulative production compared with actual production tests.

After 3 days of production testing, the zone was reperforated at 4 shots/ft between 6,620 and 6,800 ft. This resulted in production of foamy water and emulsion for several days but did not seem to affect the over-all flow rate significantly (Fig. 9.9). Production declined from an initial flow of over 1 MMcf/D to about 300 Mcf/D after 23 days; the cumulative production for this test was 10.94 MMcf. Although the cumulative was greater than from the 27 days of testing in March, the overall decline appeared to be about the same. Undoubtedly, the greater cumulative production was due to reduced cleanup problems during the second production test.

In Dec. 1978 the well again was shut in for a longterm buildup; the Horner plot for the first 18 months of this buildup is shown in Fig. 9.5. A single-phase two-dimensional finite-difference reservoir model was used to obtain the history match of the buildups (Gringarten et al. 1974; Holditch and Lee 1979). The best match on the last two buildups indicated a reservoir with 0.001-md permeability, 45 ft of net-pay, and a fracture length of 700 ft on either side of the wellbore. The permeability is remarkably close to those calculated from the buildups before and after the mini-fracture. The main difference between the last two pressure buildups run after the main fracture was an increase in fracture conductivity from 20 to more than 500 md-ft; this apparently was due to the additional cleanup after reperforation.

Well spacing, acres	640
Well life, years	40
1979 gas price, \$/Mcf	1.55
Operating expense, \$/well/yr	5,000
Drilling/completion cost, \$/well	700,000
Facilities cost, \$/well	50,000
Fracturing cost, \$	
700-ft fracture	250,000
1,000-ft fracture	300,000
1,500-ft fracture	350,000
2,500-ft fracture	450,000
Crown royalty, % of wellhead price	32
Discount factor	0.15
Depreciation calculation	double declining balance (accelerated)
Tangible investment, % of total cost	33

Table 9.5—Economic parameters.

9.9 Economics

Once the reservoir parameters were established and production rates estimated, it was possible to determine the economic conditions necessary to develop and produce this type of reservoir. In addition to the estimated production shown in Fig. 9.9 for a 700-ft fracture, additional schedules were calculated for 1,000-, 1,500- and 2,500-ft fractures using the same reservoir parameters. The cumulative recoveries for these cases are summarized as follows.

Propped Fracture Length (ft)	Cumulative Recovery (Bcf)			
	5 Years	10 Years	20 Years	40 Years
700	0.10	0.15	0.25	0J8
1,000	0.15	0.22	0.32	0.50
1,500	0.22	0.32	0.47	0.70
2,500	0.37	0.53	0.76	1.09

The production schedules were used with the economic parameters shown in **Table 9.5.** Both an unescalated and an escalated case were analyzed. In the unescalated case, the gas price and operating expense were held constant for the entire producing life. The escalated case assumes the gas prices will increase 10%/yr and the operating expense 7%/yr. Both cases assume the well will be completed as a dual, with each completion bearing 50% of the drilling, completion, and facilities cost.

For the escalated case with different fracture length, the percent of present-value profits discounted at 15% for 20 years of production is shown in **Fig. 9.10.** The unescalated case indicates that a wellhead price of \$1.50 to \$3.00 higher would be required to make the same profit.

The economics show that dual completions in tight formations, such as seen in the Falher Sands in Well 11-12-71-13W6, would not pay out at 1979 prices. However, if prices rise to \$3.50/Mcf netback (about \$5.15/Mcf wellhead price at 32% royalty rate) such sands would become economical under the escalated case; the longer the fracture, the more profitable it would be.

Note that while spacing was assumed to be one well per section, only a small portion of the gas in place would be recovered in 40 years. Calculations based on Well 11-12-71-13W6 indicate gas in place in the Falher A and B Sands amounts to 8.2 Bcf/section. If, for instance, one well with a 1,000-ft fracture can recover 6% of the gas in place in 40 years, then four wells per section could recover almost four times that amount. With proper spacing, taking into account the orientation of the fractures, an even higher density of wells could be justified once the economics become satisfactory for one well.

9.10 Conclusions

1. The Falher A and B Sands in the Canhunter Texcan Elmworth Well 11-12-71-13W6 are gasbearing over a 164-ft gross section. Of this, at least 45 ft are considered net-pay with an average gas porosity of 4.5%.
2. Flow rates from an openhole drillstem test in the Falher interval were too small to measure. After setting casing, perforating, and swabbing, there was no flow to surface. After breaking down each perforation, the flow was 21 Mcf/D. After a mini-fracture (9,450 lbm of sand), flow was 35 Mcf/D. Following the main foam fracture and initial cleanup test, production declined from greater than 1 MMcf/D to about 300 Mcf/D in 23 days.
3. The matrix permeability was calculated from four pressure buildups; two were run before the main fracture and two were run afterward. All were remarkably close in predicting in-situ permeability of about 0.001 md.
4. Fracture gradients calculated from breaking down each hole individually varied from 0.799 to 0.910 psi/ft. The average for the zone was 0.87 psi/ft.
5. A large limited-entry foam fracture using 5.65 MMcf of nitrogen injected 250,000 lbm of sand into the formation, creating an estimated fracture length of 700 ft on either side of the borehole with a fracture capacity of more than 500 md-ft.

6. A 700-ft fracture in this zone is expected to produce 0.38 Bcf in 40 years; a 2,500-ft fracture could produce 1.09 Bcf.
7. Such sands would not be economical to develop at present prices. However, economic model runs indicate that dual completions with hydraulic fractures of at least 700 ft would be economical if the netback to the producer were $3.50/Mcf (with the escalated case). Larger fractures yield improved profitability.

9.11 Acknowledgments

Special commendation should be given to the Canadian Hunter Production personnel responsible for the many facets involved in the completion, stimulation, and analysis of this complex job. We also appreciate the permission granted by Texaco Canada Inc. and Canadian Hunter Exploration Ltd. to publish SPE 7935.

9.12 References

Anderson, T. and Walker, T. 1972. Log Derived Rock Properties for Use in Well Stimulation Design. Presented at the Fall Meeting of the Society of Petroleum Engineers of AIME, San Antonio, Texas, 8–11 October. SPE-4095-MS. https://doi.org/10.2118/4095-MS.

Gringarten, A. C., Ramey, H. J. Jr., and Raghavan, R. 1974. Unsteady-State Pressure Distributions Created by a Well With a Single Infinite-Conductivity Vertical Fracture. *SPE J.* **14** (4): 347–360. SPE-4051-PA. https://doi.org/10.2118/4051-PA.

Holditch, S. A. and Lee, W. J. 1979. Fracture Evaluation With Pressure Transient Test in Low-Permeability Gas Reservoirs. Part II: Field Samples. Presented at the Symposium on Low Permeability Gas Reservoirs, Denver, Colorado, 20–22 May. SPE-7930-MS. https://doi.org/10.2118/7930-MS.

Masters, J. A. 1979. Deep Basin Gas Trap, Western Canada. *AAPG Bull.* **63** (2): 152–181.

Matthews, C. S. and Russell, D. G. 1967. *Pressure Buildup and Flow Tests in Wells*, Vol. 1, 13–16. Richardson, Texas: Monograph Series, Society of Petroleum Engineers.

Canadian Hunter Exploration Ltd. 1978. Ultimate Reserves of Gas and Co-Products. Energy Resources Conservation Board Proc. No. 780491, Canadian Hunter Exploration Ltd., Calgary, Alberta, Canada (December 1978).

SI Metric Conversion Factors

acre × 4.046 856	E + 03 = m^2
atm × 1.013 250*	E + 02 = kPa
bbl × 1.589 873	E − 01 = m^3
cp × 1.0*	E − 03 = Pa·s
cu ft × 2.831 685	E − 02 = m^3
°F (°F − 32)/1.8	= °C
ft × 3.048*	E − 01 = m
gal × 3.785 412	E − 03 = m^3
in. × 2.54*	E + 01 = mm
lbm × 4.535 924	E − 01 = kg
psi × 6.894 757	E + 00 = kPa
scf × 2.863 640	E − 02 = std m^3
sq mile × 2.589 988	E + 00 = km^2

*Conversion factor is exact.

Chapter 10

Restimulation of Tight Gas Sand Wells in the Rocky Mountain Region

S. R. Reeves, SPE, Advanced Resources International; **D. G. Hill,** SPE, Gas Research Institute; **R. L. Tiner,** SPE, Ely & Associates; **P. A. Bastian,** SPE, S. A. Holditch & Associates; **M. W. Conway,** SPE, StimLab; and **S. Mohaghegh,** SPE, West Virginia University

SPE 55627-MS presents a case history of restimulation in the Frontier Formation in the Green River Basin in Wyoming. Given the need for restimulation research based on the Gas Research Institute's (GRI's) scoping study on gas-production enhancement by way of restimulation, Reeves et al. conducted a 2-year research project on restimulation to develop methodologies to select candidates, classify wells, and test restimulation techniques. Three candidate-selection methodologies were developed and applied using field data from the Frontier Formation. Those three methods were

- A statistical production-data analysis
- Artificial neural nets
- Production type curve history matching

As a result of the research project, Reeves et al. concluded the following:

- Refracture-treatment well candidates selected from each methodology seem to be different
- Restimulation could enhance production
- No universally applicable methodology is able to select restimulation candidates across different geological basins

10.1 Abstract

In 1996, the GRI performed a scoping study to investigate the potential for natural gas production enhancement via restimulation in the United States. The results indicated that the potential was substantial, particularly in the tight sands of the Rocky Mountain, Mid-Continent and South Texas Regions. However it was also determined that industry's historical experience with restimulation is mixed, and that considerable effort is required in candidate selection, problem diagnosis, and treatment selection/design/implementation for a restimulation program to be successful. As a result GRI initiated a subsequent two-year R&D project with the objectives: 1) to develop efficient, cost-effective, and reliable methodologies to identify wells with restimulation potential, 2) to identify and classify various mechanisms leading to well underperformance, and 3) to develop and test various restimulation techniques tailored to different causes of well underperformance.

The approach adopted for the R&D program is a combination of conceptual methodology development, laboratory studies, and actual field testing of restimulation treatments in tight gas sand reservoirs. In total, four test sites are planned for the project; each site consists of about 300 total wells in a contiguous area and completed in a consistent producing horizon, out of which five are selected for actual restimulation. The first two sites are in the Rocky Mountain Region, the first being in the Big Piney/LaBarge Producing Complex of the northern Moxa Arch in the Green River Basin (Frontier Formation), and the second being the combined Rulison, Parachute and Grand Valley Fields of the eastern Piceance Basin (Williams Fork Formation). At present, restimulation candidates have been selected, verification testing completed, and two restimulations performed at the first site; preliminary candidate selections have been made at the second site.

The results to date from this project show why methodologies to accurately select "high-potential" restimulation candidates have eluded previous investigators. Each of the analytic procedures utilized thus far in the project is providing a different list of candidates, each based upon different criteria and with uncertainty as to the validity (or lack thereof) of each. While the situations being addressed in this project are understandably complex, specifically multilayered, heterogeneous reservoirs that are completed and stimulated in a wide variety of ways, the project experience to date supports that no acceptable methodology currently exists to universally select restimulation candidates across different geologic settings (which is the objective of this project).

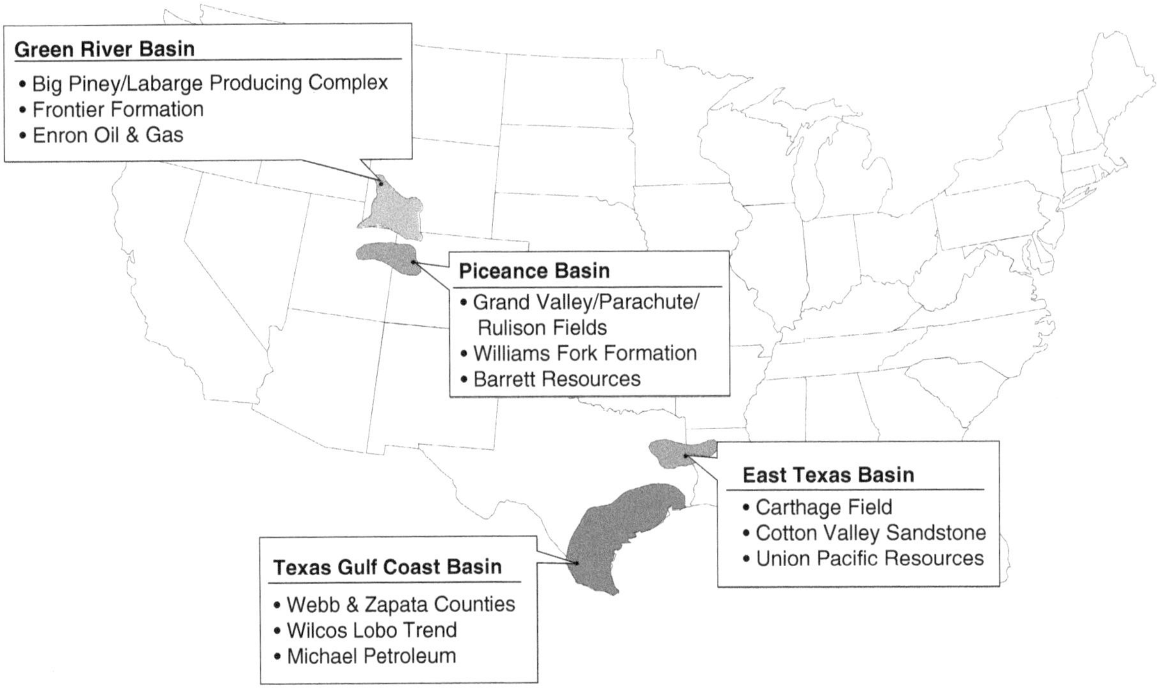

Fig. 10.1—Location of test sites.

What is clear is that restimulation potential does appear to exist, as evidenced from the restimulation treatments performed to-date, but that some degree of site-specific knowledge and methodology customization is required. This is supported by the findings from the individual well reviews and laboratory studies. Clearly, more results from this project are needed to better understand the methodologies being investigated, and how they should be integrated to develop the efficient yet robust methodology sought.

10.2 Introduction

In 1996, GRI performed a scoping study to investigate the potential for natural gas production enhancement via restimulation in the United States (Lower-48 onshore) (Reeves 1996; Hill and Reeves 1998). The results indicated that the potential was substantial (over one Tcf of reserves in five years), particularly in the tight sands of the Rocky Mountain, Mid-Continent and South Texas Regions. Further, when successful, restimulation was found to be an attractive investment, providing incremental reserves at costs of $0.10 to $0.20 per Mcf. However it was also determined that industry's historical experience with restimulation is mixed, and that considerable effort is required in candidate selection, problem diagnosis, and treatment selection/design/implementation for a restimulation program to be successful. Given an overall lack of both specialized (restimulation) technology and "spare" engineering manpower within the industry today to focus on restimulation, GRI initiated a subsequent two-year R&D project in 1998, currently underway, with several objectives. Those objectives are to 1) develop efficient, cost-effective, and reliable methodologies to identify wells with restimulation potential, 2) identify and classify various mechanisms leading to well underperformance, and 3) develop and test various restimulation techniques tailored to different causes of well underperformance. It is believed that with new technologies in these key areas, industry can more effectively capture the incremental reserves that exist via restimulation, and the financial benefits they represent, particularly in today's challenging economic climate.

The approach adopted for the R&D program is a combination of conceptual methodology development, laboratory studies, and actual field testing of restimulation treatments in tight gas sand reservoirs. In total, four test sites are planned for the project; each site consists of about 300 total wells in a contiguous area and completed in a consistent producing horizon, out of which five are selected for actual restimulation. The locations of the selected sites are presented in **Fig. 10.1.** The first two sites are in the Rocky Mountain Region, the first being in the Big Piney/LaBarge Producing Complex of the northern Moxa Arch in the Green River Basin (Frontier Formation), and the second being the combined Rulison, Parachute and Grand Valley Fields of the eastern Piceance Basin (Williams Fork Formation). At present, restimulation candidates have been selected, verification testing completed, and two restimulations performed at the first site; preliminary candidate selections have been made at the second site. The East Texas and South Texas test sites are currently in the earliest stages of data collection and analysis. This paper presents the candidate selection methodology as it currently exists (it is under continuous improvement), and the results-to-date from the two Rocky Mountain test sites.

10.3 Candidate Selection Methodology

Considerable effort in the project has been (and continues to be) focused on the development of the candidate selection methodology itself; it is believed that this is where a technology advancement will provide the greatest industry benefit. In the previously cited GRI study, an important finding was what is referred to as the "85/15" rule. This "rule" states that 85% of the restimulation potential for a field exists in 15% of the wells; the key to any successful restimulation program is therefore being able to correctly and reliably identify that 15% of the total well population that represents the high-potential restimulation candidates. Project economics are highly sensitive to accurate candidate recognition and selection. Poor candidate selection procedures seem to be the reason that many restimulation projects are unsuccessful.

With this understanding, two concepts formed the foundation of the candidate selection methodology. The first concept is that a series of sequential analyses would be used to gradually reduce the total well population to the plus or minus 15% that are actually high-potential candidates. The first analytic level, for example, might be for rapid screening and be low in cost, perhaps eliminating 50% of the wells. Subsequent analytic levels would be incrementally more detailed, labor-intensive and expectedly higher in cost.

The second concept is that to accurately select candidates, well performance must be separated into the reservoir quality and completion efficiency components (i.e., permeability and skin). High-potential candidates, those that would provide the greatest level of incremental production and reserves, are likely to be wells exhibiting both high permeability and high skin. The implication is that the better performing wells in a field may actually be the most attractive restimulation candidates, not those that are underperforming the field or offset well averages. While the practice of restimulating the more productive wells in a field is understandably performed reluctantly, not doing so may be the very reason many restimulation projects do not reach expectations. A candidate selection methodology was therefore prepared with these two concepts in mind, and subsequently field-tested as will be described in the next section. Brief summaries of the methods used for each analytic level are provided below.

10.3.1 Level I Rapid Screening: Statistical Production Data Analysis. In the most simple terms, Level I essentially compares the production performance of each well to its offsets, and identifies underperforming wells that could be restimulation candidates. The process is quite advanced however in that it compares life-cycle production, early-time performance and most-recent performance each separately to provide insights into why a well may be underperforming, and accounts for depletion associated with infill drilling. The drawback of this technique is that it does not identify those "superior" wells that could be improved, and that could be the source of the most favorable restimulation economics. Nevertheless, the selection of underperforming wells, particularly where reservoir properties and completion/stimulation practices are more uniform, is a valid candidate recognition approach (McCain et al. 1993).

10.3.2 Level II Advanced Screening: Artificial Neural Nets and Genetic Algorithms. The second analytic level utilizes artificial neural nets and genetic algorithms to identify potential restimulation candidates. Artificial neural nets are utilized to recognize patterns in how various input parameters (e.g., location, geology, drilling, completion, stimulation and workover data, etc.) impact the well output (i.e., production). This information serves several useful purposes. First, the relative contribution of geologic/reservoir parameters can be separated from drilling, completion, and stimulation parameters. This in effect is the separation of the reservoir and completion components (i.e., permeability and skin). Then, by understanding which of the controllable parameters have the greatest impact on production response (i.e., the "performance drivers"), patterns can be identified to infer "best" (or "worst") practices, and flags raised for wells with restimulation potential (i.e., those where "worst" practices were used). Also, insights can be obtained on the causes of well underperformance and how to treat it. For example, if the use of heavy mud is identified as a "bad" practice, then mud damage may a problem with well performance, and treatments can be designed specifically to remedy that problem. Genetic algorithms are then used to "optimize" the input parameters for any given well (select those "controllable" input parameters that yield the highest predicted production response) and those wells where the greatest discrepancies exist between actual well performance and predicted optimized performance are identified as restimulation candidates. The difficulty with this approach is that frequently the amount, type and quality of data available are insufficient to perform a meaningful and insightful analysis; all technical data used for an analysis must be available for all wells. One piece of missing information for one well requires that either that well or that piece of data (for all wells) be eliminated. Nevertheless, this approach is believed to hold great potential for evaluating restimulation opportunities in tight gas fields (Mohaghegh et al. 1996).

10.3.3 Level III Candidate Identification: Production Type-Curve Matching. The ideal approach to selecting restimulation candidates is to directly understand the relative impact of reservoir and completion properties on the performance of each individual well (i.e., permeability and skin), and select the high-permeability/high-skin wells for restimulation. Unfortunately, obtaining such information can be expensive if approached via pressure transient testing. An alternative approach is to use production type-curves. Such curves have been specifically developed for hydraulically fractured tight sand wells in the western U.S., and can provide estimates of the permeability, skin and drainage area with relatively limited data (Cox et al. 1996). The problems with this approach are severalfold however; first, the models are developed for single-layered reservoirs, and the multi-layered nature of most tight sand plays renders the results suspect. Second, the noise level of the production data normally available, plus the inherent interdependencies of the output parameters, makes the ability to achieve a unique result difficult. Finally, since these methods require values of net pay for each well, some interpretive and potentially labor-intensive petrophysical evaluation may also be required. With an understanding of these limitations, however, such approaches can be used in a relative sense to identify potential restimulation candidates.

10.4 Field Test 1: Green River Basin

10.4.1 Site Description. The first field test site is the Big Piney/LaBarge Producing Complex located in the northern Moxa Arch area of the Green River Basin **(Fig. 10.2).** The cooperative research partner, Enron Oil & Gas, operates close to 300 Frontier Formation (not commingled with production from other horizons) wells in the area. A type-log of the Upper Cretaceous Frontier Formation is shown in **Fig. 10.3;** the productive horizons in the area include the First Frontier, and benches one through five of the Second Frontier. The second (and third) benches of the Second Frontier are the most prolific gas reservoirs, and are marine sandstones. Benches 1, 4 and 5 are primarily fluvial, with the First Frontier being deposited in a distal shoreface environment. Reservoir properties can vary considerably for each horizon, but each reservoir is relatively continuous and can be distinctly correlated throughout the project area. **Table 10.1** provides selected reservoir information for the field.

Well vintages in the study area range from the late 1950s through to the present, representing an interesting blend of completion/stimulation technologies used, stages of infill drilling, and depletion and mechanical (wellbore) issues to consider. The completion intervals for each well are typically based upon log-derived net pays, and hence vary from well to well. Wells can be completed in only one or up to six different pay horizons, and production from multiple pays is commingled.

10.4.2 Candidate Selection and Problem Diagnosis. Being the first field test of the methodology, the three levels of analysis for candidate selection were each performed on all wells at the site. This was to ensure that each analytic level captured all the potential candidates from the subsequent levels (i.e., the preliminary levels did not eliminate candidates that would likely be selected by subsequent, more detailed levels). Once completed, the top 50 candidate wells from each level were compared. That comparison is illustrated in **Fig. 10.4.** What can be observed from this figure is that the coincidence of candidates from one level to the next is inconsistent; further, the top candidates from each level were unique to that level, and are not replicated in the other levels. The implication is that since each level seems to focus on different indicators of restimulation potential, each selects different candidate wells. For example,

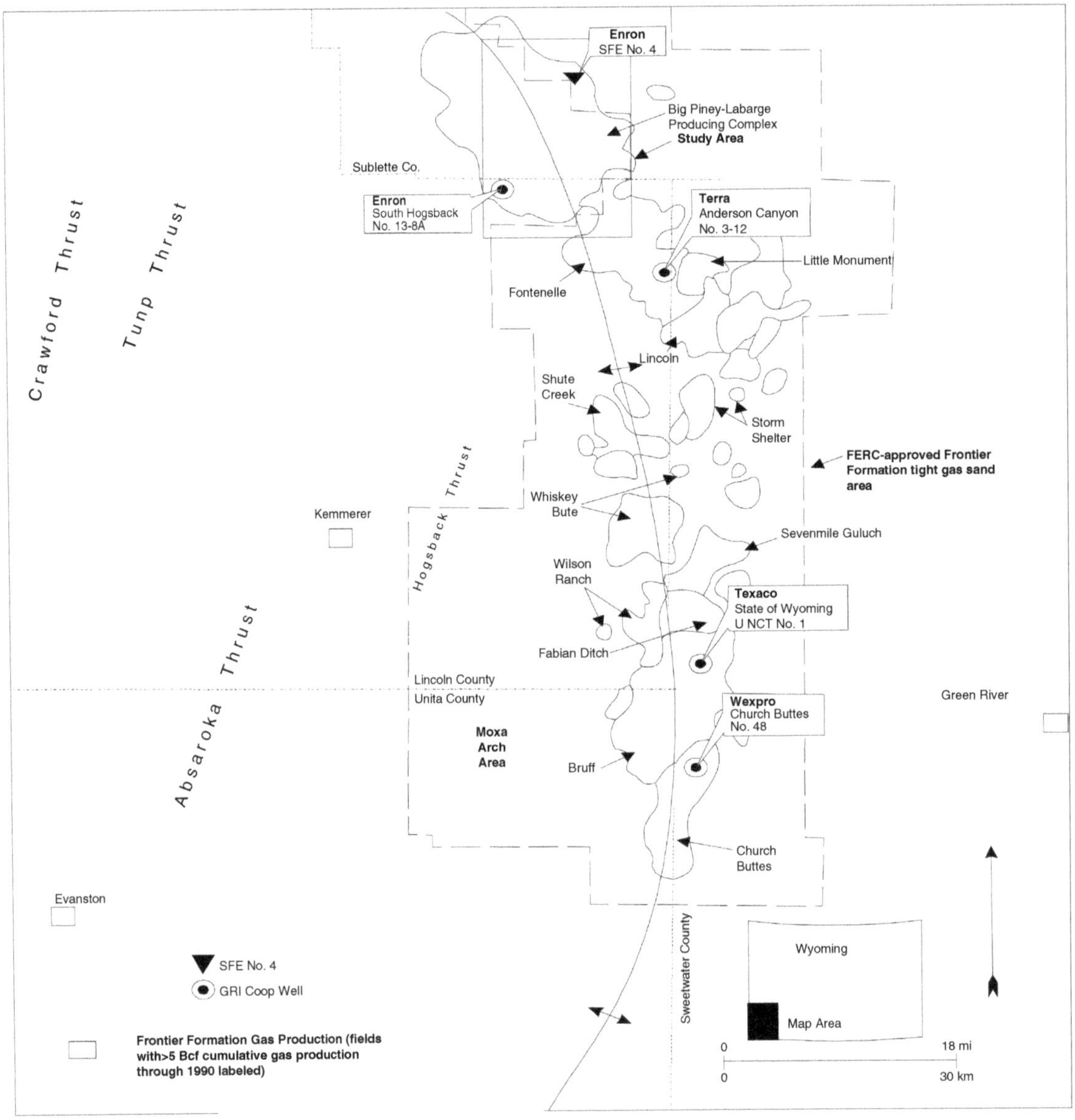

Fig. 10.2—Green River Basin study area.

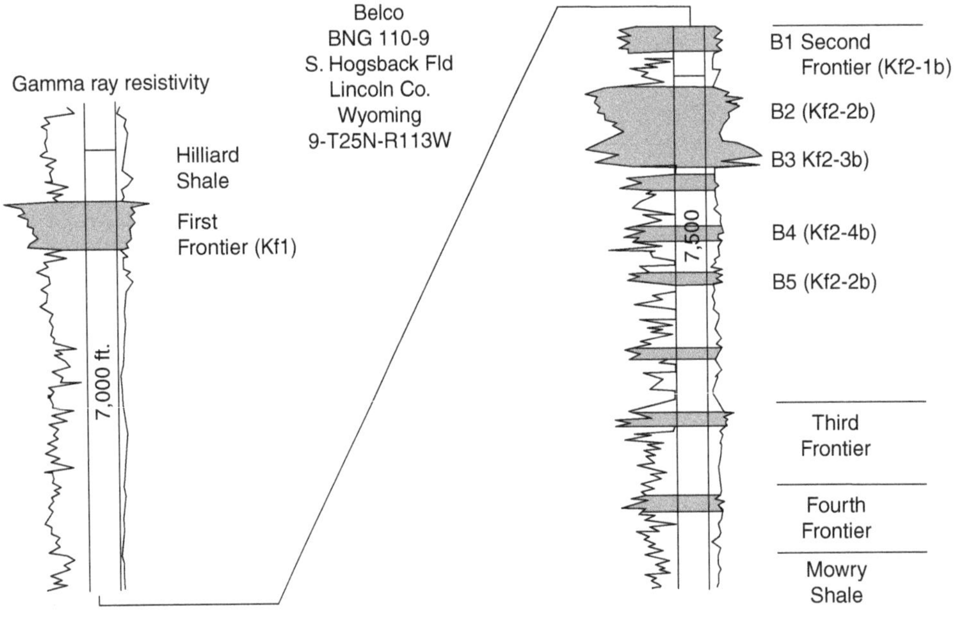

Fig. 10.3—Type log of the Frontier Formation, northern Moxa Arch, Green River Basin.

Basin	Green River	Piceance
Test Site	Big Piney/LaBarge	Rulison, Parachute, Grand Valley
Formation	Frontier	Williams Fork
Average Depth To Top	6,300–7,000 ft	4,800–5,400 ft
Original Pressure	4,200–4,600 psi	3,800–4,200 psi
Thickness (Gross/Net)	100–400 ft/15–90 ft	1,700–2,400 ft/150–300 ft
Permeability	10–150 pd	5–100 pd
Typical Recovery	1–2 Bcf/well	1–2 Bcf/well

Table 10.1—Typical reservoir properties at each test site, pd = microdarcies.

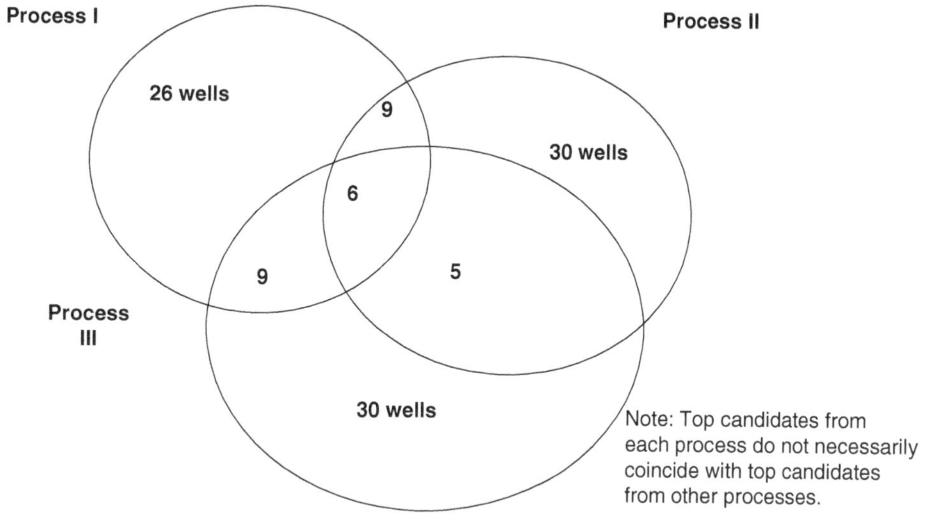

Fig. 10.4—Coincidence of "top 50" candidate selections from each level, Green River Basin.

Level I selects wells that are underperforming their offsets, Level II selects wells where "less-than-best" practices were employed, and Level III selects wells with the greatest estimated upside potential, albeit using non-unique engineering methods.

Without a clear indication of which process was truly providing the best candidates, or if they were each providing high-potential candidates with different characteristics, those candidates that appeared on more than one "top 50" list, plus the top 5 wells that were unique to each list, were selected for further, detailed individual well reviews. Together these wells added up to a short list of about 50 wells for detailed consideration as restimulation candidates.

The candidate wells were then individually evaluated to choose those with the greatest chance of restimulation success. This involved detailed study of all information in the well files as well as interviews with personnel that had direct experience with the well being considered. Information considered included:

- The mechanical integrity and condition of the wellbore, such as pipe size, weight, and grade, perforations or holes in the casing, cement squeeze treatments, primary cement volume and expected top of cement, primary cement results, wellbore deviation, perforation intervals, number of perforations, production packers, production tubing, sliding sleeves, scaling problems, stuck tubing, the existence and location of any fish, and the production method (e.g., plunger lift, intermittent, siphon string, etc.).
- The original completion and stimulation procedures including perforating for limited entry, treating pressures, treating rates, fluid volume, fluid type, proppant volume, proppant type, fluid details (such as polymer type, breakers, clay stabilizers, crosslink type, surfactants, etc.), screen-outs, ISIPs, and anything else noted in the files.
- The flowback procedure, such as shut-in time, use of forced closure, the amount of fluid recovered, and any noted problems such as gel observed for extended periods of time.
- Production history, such as initial production, production decline, production by intervals, water onset, amount of condensate produced, etc.
- Pressures, such as initial pressure, pressure buildups, DSTs, pressures during shut-in periods, etc.
- Any workovers such as new perforations, acid for scale removal, squeeze of perforations, setting of bridge plugs, etc.
- Many of the wells were drilled and completed about 40 years ago. There were no indications of severe pipe deterioration from production fluids and gases, and age therefore was not considered as a factor by itself.

Those wells with mechanical problems, such as squeeze-cement holes above the perforations, stuck tubing or packers, very small primary cement volumes (or those squeeze cemented to increase cement height), and wells completed with low-grade, low-weight casing, were essentially eliminated from further consideration. Another category eliminated were those wells that had major zonal isolation problems due to initial perforating practices. While these wells may indeed be good restimulation candidates, they possessed a high risk of mechanical failure and/or cost. Similarly, deviated wells with large numbers of perforations across a long interval that would require squeeze-cementing and reperforating to minimize multiple fractures (which would increase the risk and cost of the treatments) were also downgraded as candidates.

Reservoir pressure was a major uncertainty for these wells. The initial pressure, cumulative production, latest pressure information and the rate of production since that time were considered in an effort to evaluate present-day pressures in the short list of wells.

	Permeability (md)				Fracture Length (ft)				Drainage Area (acres)			
Well Name	ARI Type[1]	Promat[2]	RPI[3]	Well Test	ARI Type[1]	Promat[2]	RPI[3]	Well Test	ARI Type[1]	Promat[2]	RPI[3]	Well Test
Larson 1-17	<0.01	0.02–0.03	0.02	<0.01	193	1	6	162	42	-	20	-
GRB 45-12	0.02	0.02–0.11	0.10	0.02–0.05	280	1–3	<1	50	258	-	>350	-
NLB 66-04	0.02	-	0.01	-	248	80–104	575[4]	123	23	-	>25	-
SHB 12-04	0.02	-	0.05	-	341	-	540[4]		54	-	-	-
GRB 27-14	0.01	0.01–0.03	0.02	-	200	1	500[4]	-	45	-	-	-
CBU 5-14	<0.01	<0.01	0.01	<0.01	130	63–68	52	50	56	-	60	-
NLB 57-33	0.01	0.02–0.05	0.06	0.06	234	1	1	166	45		100	

Table 10.2—Well test results compared to Level III analytic results, Green River Basin.

[1] Fractured, tight gas sand type-curve
[2] Multipurpose type curve
[3] Reciprocal productivity index method
[4] Linear flow in a channel with a given width

Wells with an expected BHP of 1,500 psi or less were downgraded on the candidate list. Those wells, which were considered to have gel, fluid, and/or scale damage, were rated fairly high (if justified by other considerations) as the cost for remediation would be lower than for wells needing refracturing with additional proppant.

The top 20 candidates following this stage of analysis were submitted to the operator, Enron Oil & Gas, for review. Wells were then further eliminated for various reasons, including uncertainty in the accuracy of the production data, uncertainty on the cause of well underperformance, and production levels being "too high" to risk. In deference to the operators' requirement for "due diligence", these may still be extremely good restimulation candidates. Out of this, 7 wells were chosen as restimulation/remediation test wells. Note that, in general, these wells ranked in the middle to lower end of the "top 50" list from each analysis level.

The original stimulation treatments for these seven wells were then modeled using an industry standard hydraulic-fracturing model. Many early treatments utilized ungelled diesel, while later treatments used various crosslinked fluids, foams, and emulsions. In general, the results indicated the existence of low conductivity fractures, and that proppant settling was probable. This was especially obvious in the Kf1 and Kf2-2b (Fig. 10.3) intervals, where downward fracture growth appears likely.

10.4.3 Pressure Transient Testing. The overall methodology also included a testing component to verify the stimulation effectiveness of the ranked candidates. In addition to applying conventional well testing methods to restimulation candidates, new or novel methods specifically designed to address the problems associated with testing low permeability gas wells were studied. The objective outcome of the new testing method is not a quantitative analysis, but rather a qualitative determination of the current stimulation effectiveness existing in the well. In addition, the testing procedure itself had to be simple, inexpensive and of short duration, to be practical.

For this test site, different testing procedures were designed to handle the various mechanical configurations of the wells, including conventional shut-in tests and a multi-rate test for wells on either plunger lift or intermittent operation. As part of the objective to develop inexpensive testing procedures, the test designs included both bottomhole and surface pressure measurements to determine if surface measurements could be used instead of the more costly and difficult bottomhole pressure measurements.

A total of seven wells were tested; these seven tests were analyzed using conventional methods as well as the newer, novel technique developed to qualitatively determine the stimulation effectiveness of the wells. **Table 10.2** presents a summary of results for the conventional analyses, compared to various production analysis (Level III) methods.

These results show that the interpreted values of permeability are very low, with a wide range of fracture lengths. Unfortunately, the data were often too noisy to apply the newly developed technique, and hence the results were very nonunique. This applied to both the surface and bottomhole data. The conclusion was that it was relatively difficult to assess the degree of stimulation in the wells using the new method, especially with surface-acquired data. A new approach has subsequently been developed for the next site. Using the conventional analysis results, two wells were eliminated from further consideration; the Larson 1-17 was dropped because of the well-test derived long fracture length and low permeability, and the SHB 12-04 was dropped due to a lack of results (plus the long fracture length as assessed by Level III).

10.4.4 Treatment Design, Incremental Production and Economic Forecasts. The key questions to be answered regarding the restimulation treatments were, 1) what intervals to treat and 2) how to treat them. For the five-selected test wells, the following approaches were adopted:

- The GRB 45-12 and GRB 27-14 wells were both originally treated with small proppant volumes and hence larger treatments to create greater length were designed. In the GRB 45-12 well, the original intervals of Kf2-1b, 2b, 3b and 5b each had 8 perforations (perfs), which could all possibly be retreated simultaneously using limited entry. In the GRB 27-14 well, a decision was made to treat only the Kf2-1b and 2b perfs, with a bridge plug set over the 3b perfs. The 3b interval would probably be treated through the 2b perfs as no fracture barrier existed between them. The 4b was nonexistent and the 5b was of small net height and hence not targeted for restimulation.
- The CBU 5-14 well was originally treated over a large interval; it is not believed that the lower benches were effectively stimulated. This well was therefore selected for two treatments since the net pay was widely distributed throughout the Kf2 interval.
- The NLB wells were completed in the Kf1 bench only and with a tubingless completion technique. The NLB 57-33 was deemed to have polymer damage and a fracture clean up treatment was designed for this well. That treatment design for the NLB 57-33 well contains large concentrations of oxidative and enzyme breakers and pH modifiers to break the borate crosslinked and guar gel. The NLB 66-4 was to be refractured to re-establish conductivity as the initial treatment utilized a large pad with proppant settling a high probability.

	GRB 137-4	ELB 55-32
Bench	Kf2-5b	Kf2-2b
Type	Fluvial	Marine
Depth	7262.3 feet	6955.3 feet
Quartz	68%	64%
Plagioclase	19%	12%
Pyrite	Trace	Trace
Calcite	2%	–
Apatite	Trace	–
Total Clay	11%	24%*
I/S Mixed Layer	6%	13%
Illite	3%	5%
Chorite	2%	4%
Kaolinite	–	2%

* Mica was not determined

Table 10.3—X-ray diffraction analysis Frontier Formation.

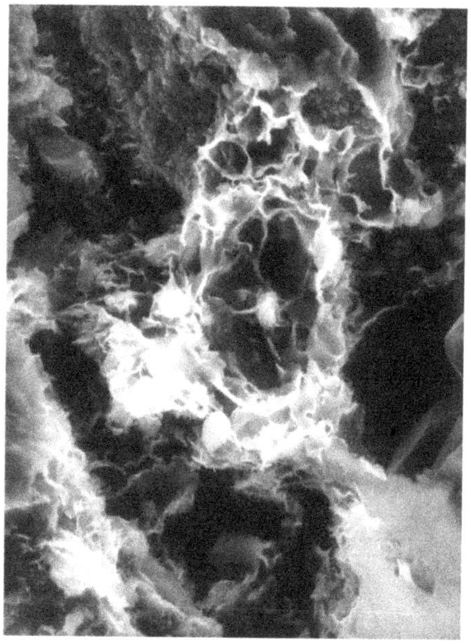

Fig.10.5—SEM photograph of Frontier Formation Core GRB 134-7, Kf2-5b, 1262.7 ft, 800X magnification.

The treatments were designed for ±250 feet of fracture length. The addition of more length appeared uneconomical and could also create water recovery problems due to fluid volumes in the depleted pressure environment. Modeling of the restimulation treatments utilized a reduced stress gradient value for the sandstone intervals, but the original stress gradient (1 psi/ft) in the bounding shale layers. The refracs were therefore much more contained than the original treatments.

The refracturing treatment designs included using a 35lb/Mgal guar gel crosslinked with borate. Nitrogen was not utilized for a flowback assist at the request of the operator; this simplified and reduced treatment cost, and if needed, could still be utilized to unload the well at a later date. Proppant consisted of 20/40 mesh Ottawa Sand that should have sufficient conductivity for the depleted formations.

Laboratory data was also used to aid in the restimulation design process in two ways. The first involved developing general guidelines for conducting remediation treatments that do not involve the use of additional proppant. The second was to address specific issues regarding additives to be used in the stimulation treatments.

X-ray results showed that both the fluvial and marine sequences in the Frontier Formation contain similar clays that line the pores (**Table 10.3**). Both swelling and non-swelling clays are present; a typical example of a pore is shown in **Fig. 10.5.** Fibrous illite is on top of the more general coating of illite/smectite. Some quartz overgrowth is also visible.

The principal stimulation fluid currently in use in this area is a borate crosslinked gel system. Because of the presence of swelling clays, there was concern about the type of clay protection that was required to maximize the compatibility of water-based stimulation fluids with the reservoir. The laboratory data presented in **Fig. 10.6** shows that 2% KCl is required for maximum clay protection.

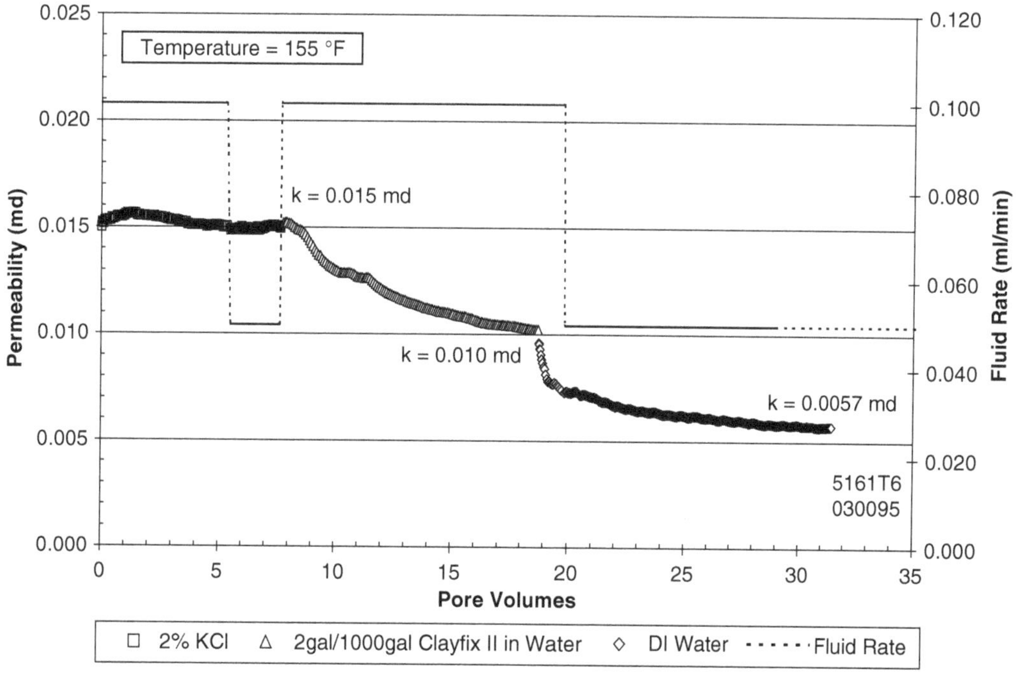

Fig. 10.6—Effect of 2 gal/1,000 gal Clayfix II, KCl substitute on permeability of Frontier Kf2-5b, Well GR B 134-7, depth 7,262.7 ft.

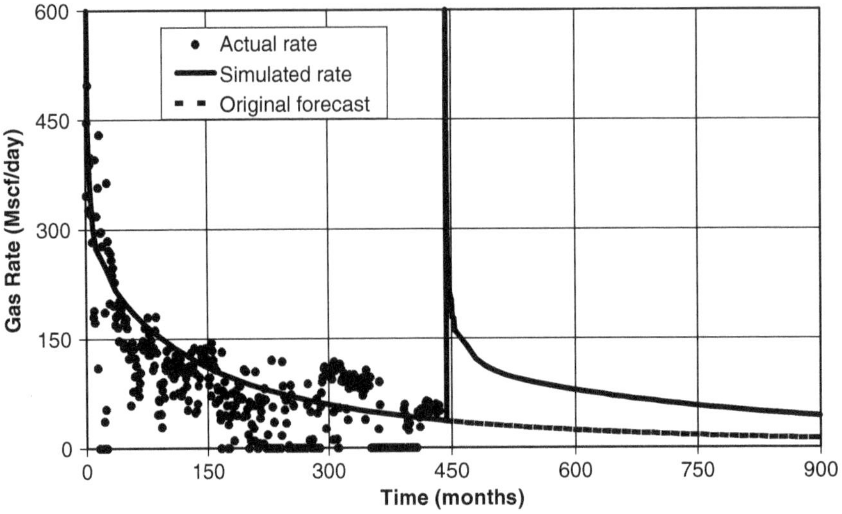

Fig. 10.7—Incremental production forecast, GRB 27-14.

The KCl substitutes provide some improvement over fresh water but clearly suffer leading edge depletion as the limited amount of clay stabilizer is plated out on the clays. For these reasons, the stimulation fluids used in the refracturing operations contained 2% KCl.

Once the treatments were designed, restimulation costs were estimated. Incremental production predictions were also made using type-curve models; an example is shown in **Fig. 10.7.** Restimulation economics were then computed, as provided in **Table 10.4.** For these five wells, 5-year incremental recovery estimates ranged from 0.15 to 0.24 Bcf, with recovery costs ranging from $0.05 to $0.58 per Mcf. Again, note that these wells were not the top picks from any of the analytic levels, but were considered average candidates. The GRB 45-12 was not selected by the methodology at all, but rather was selected by the operator based on their understanding of the geology and completion practices, and represents a "control" case. An incremental production estimate was therefore not available for this well.

10.4.5 Treatment Execution and Results. Restimulation treatments have now been performed on two of the five wells, the GRB 45-12 and the GRB 27-14. These treatments were pumped, with minor exceptions, as originally designed. Both treatments were tagged with radioactive material. The GRB 45-12 was designed to treat four intervals simultaneously, each with 8 perforations. While the rate was limited to 48 bpm in the pad due to wellhead pressure limitations, the pressures were close to expected. The well was flowed back immediately, utilizing the forced closure technique. The flush volume was recovered and the well continued to flow for several hours. Tubing was run and the well cleaned up rather quickly. The poststimulation radioactive tracer log indicated only the two middle sets of perforations took fluid **(Fig. 10.8).** Thus the Kf2-1b and 5b perfs were not treated while the Kf2-2b and 3b perfs were treated. Despite this, the GRB 45-12 treatment is considered successful, with post-restimulation production nearly double the pre-restimulation rate **(Fig. 10.9).**

The GRB 27-14 treatment was pumped through the Kf2-1b and 2b perfs with a bridge plug isolating the Kf2-3b and 5b perfs. Modeling predicted that the Kf2-3b interval would be treated through the Kf2-2b perfs. The poststimulation radioactive tracer log verified this **(Fig. 10.10).** Once again, the well flowed back for several hours using the forced closure technique. However, the well has not cleaned up and continues to produce water without the ability to sustain flow. It is anticipated that, once the water is recovered, improved production will be realized.

		Top "50" List			Restimulation		Estimated 5-Year	Restimulation Cost		
Well No.	Completion	I	II	III	Problem	Treatment Objective	Incremental (Bcf)	$1,000	$/Mcf	Restimulation Status
CBU 5-14	F21,F22, F23,F24, F25	Y	N	Y	Unfrac'd in F24/ F25	Refrac (unstimulated)	0.24	$140	$0.58	Pending
NLB 66-04	F1	Y	Y	Y	Proppant Settling	Refrac (conductivity)	0.23	$70	$0.30	Pending
NLB 57-33	F1	Y	Y	Y	Gel Damage	Gel Cleanup	0.20	$10	$0.05	Pending
GRB 27-14	F21, F22, F23, F25	N	Y	Y	Small Fracs	Refrac (length)	0.15	$87	$0.58	Performed 1/6/99
GRB 45-12	F21, F22, F23, F25	N	N	N	Small Fracs	Refrac (length)	-	$87	-	Performed 1/6/99

Table 10.4—Restimulation economics, Green River Basin.

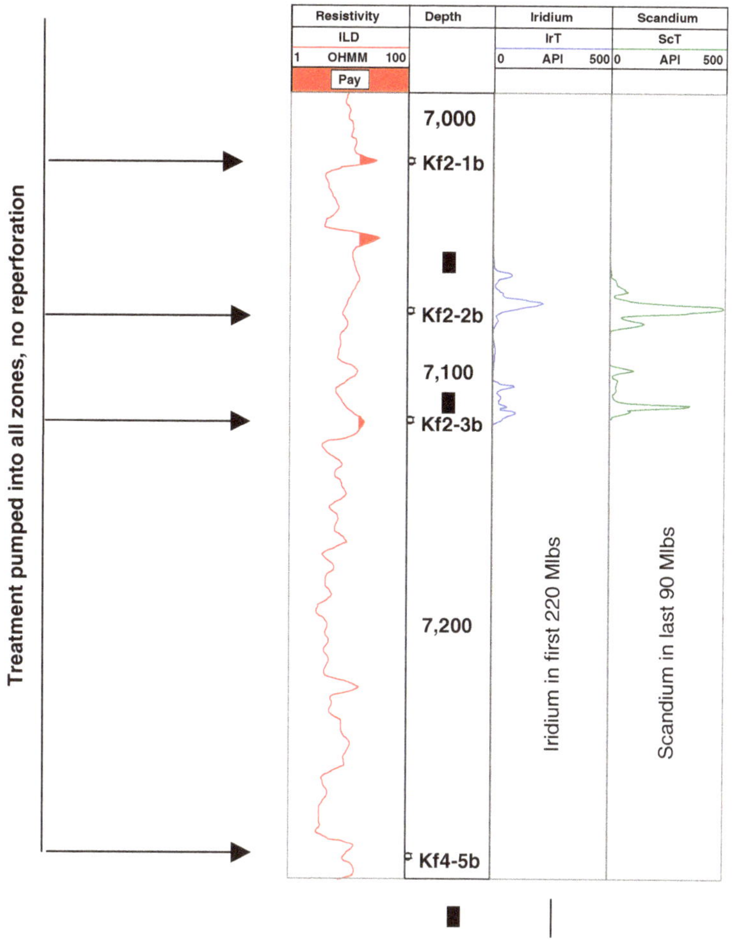

Fig. 10.8—Post-treatment radioactive tracer survey, GRB 45-12.

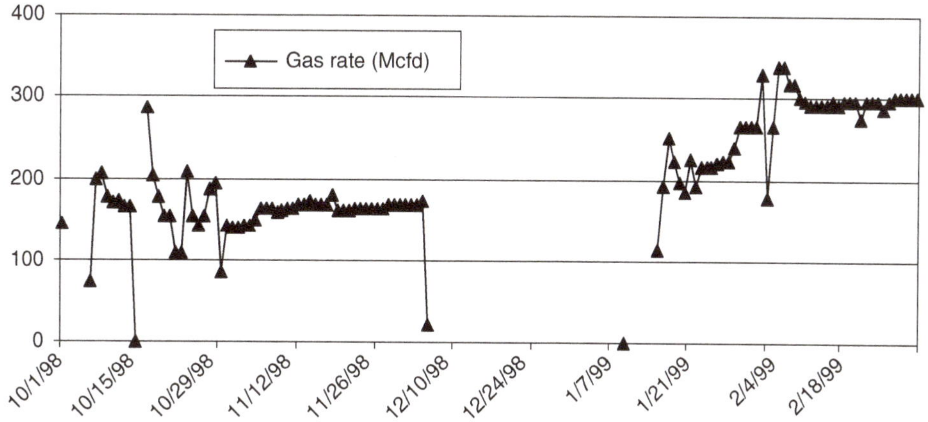

Fig. 10.9—Restimulation results, GRB 45-12.

10.4.6 Next Steps. The remaining tasks to be completed at this site include the restimulation of the final three test wells, and postrestimulation pressure transient testing on all five test wells to evaluate treatment effectiveness.

10.5 Field Test 2: Piceance Basin

10.5.1 Site Description. The second field test site is the combined Rulison, Parachute and Grand Valley Fields located in the eastern portion of the Piceance Basin. The cooperative research partner, Barrett Resources, operates close to 300 Williams Fork Formation wells in the area. A stratigraphic column of the Upper Cretaceous Williams Fork Formation is shown in **Fig. 10.11;** unlike the Frontier Formation in the LaBarge Platform, the productive horizons in the area are discontinuous, massively stacked lenticular sands. This is a basin-centered gas play, with the potential for water production increasing in the upper portions of the gross interval. The lowermost Cameo section is paludal sandstone, which is overlain by coastal sandstones that transgress into fluvial sands.

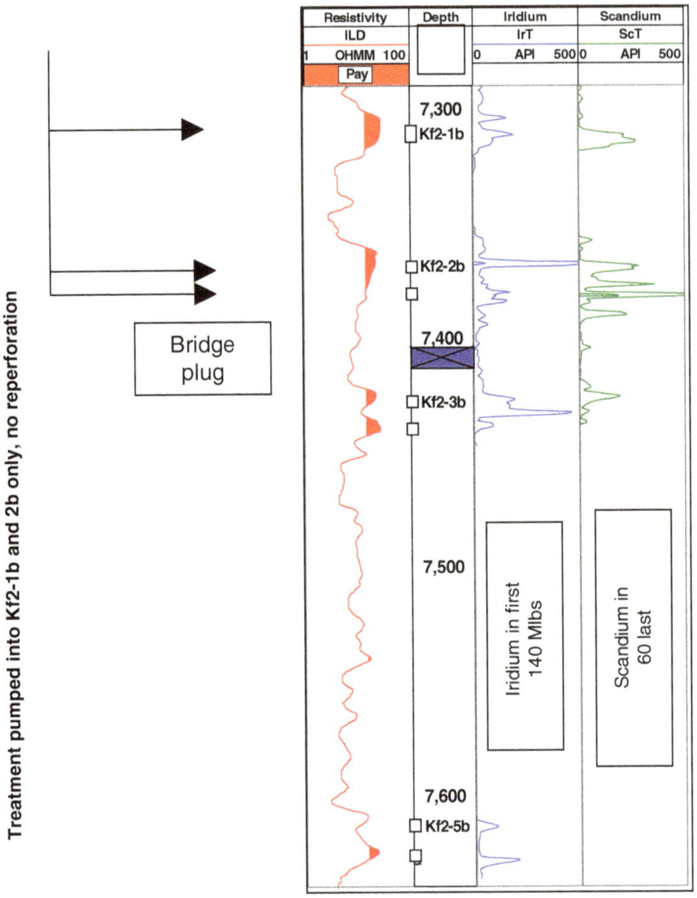

Fig. 10.10—Post-treatment radioactive tracer survey, GRB 27-14.

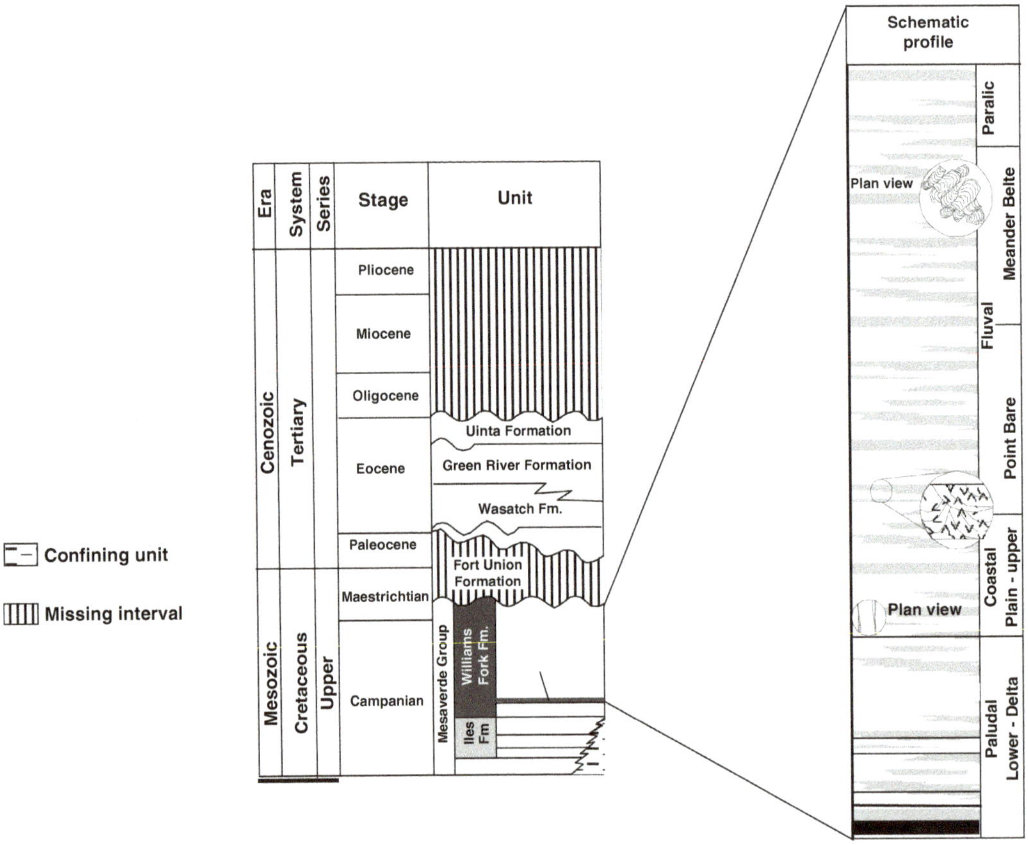

Fig. 10.11—Stratigraphic column, Piceance Basin.

There can be as many as 40 discrete net pay intervals in the several thousand feet of gross interval, but typically they cannot be correlated from one well to the next, even on close spacing. Table 10.1 provides selected reservoir information for the field.

Development of the area began in 1980, but the most intensive drilling has occurred since 1990. Development strategies have evolved, including infill drilling from an original spacing of 160 acres to as little as 20 acres per well recently in some areas of the Rulison and Grand Valley Field. Fracturing and completion strategies have also evolved, including the move to limited entry fracturing and crosslinked gel fluids (up to 4 fracture treatments per well, each targeting several hundred feet of gross interval), and attempting to capture additional pay by more aggressive uphole completions (nearer to the water-wet intervals). Similar to the first site, production from all completed horizons in a well is commingled.

10.5.2 Candidate Selection Results. Because of the uncertainty in candidate selection from the first test site, it was decided to perform each analytic process on all wells at this site also. The purpose was to provide further insights into the uniqueness of selections for each level, and if they could be made more consistent from one level to the next. While refinements and improvements in each analytic level were made, the inconsistencies in selections remained. A similar process for candidate selection, now underway and encompassing both coincidental candidates and the top candidates from each level, is being performed. This is the extent of progress at this time for this site.

10.6 Conclusions

The results to date from this project show why methodologies to accurately select "high-potential" restimulation candidates have eluded previous investigators. Each of the analytic procedures utilized thus far is providing a different list of candidates, each based upon different criteria and with uncertainty as to the validity (or lack thereof) of each. While the situations being addressed in this project are understandably complex, specifically multi-layered, heterogeneous reservoirs that are completed and stimulated in a wide variety of ways, the project experience to date supports that no acceptable methodology currently exists to universally select restimulation candidates across different geologic settings (which is the objective of this project).

What is clear is that restimulation potential does appear to exist, as evidenced from the restimulation treatments performed to-date, but that some degree of site-specific knowledge and methodology customized is required. This is supported by the findings from the individual well reviews and laboratory studies. Clearly, more results from this project are needed to better understand the methodologies being investigated, and how they should be integrated to develop the efficient yet robust methodology sought.

10.7 Acknowledgments

This work is being funded by GRI under the project entitled "Natural Gas Production Enhancement Through Restimulation" (GRI Contract Number 5098-2104090). The authors also wish to thank and acknowledge the valuable contributions of Enron Oil & Gas and Barrett Resources, who provided the test sites, data, experienced personnel and countless insights for the project.

10.8 Nomenclature

°F	=	degrees Farenheit
µd	=	microdarcies
$	=	United States dollars
BHP	=	bottomhole pressure
bpm	=	barrels per minute
DI	=	distilled and ionized
DST	=	drill-stem test
ft	=	feet
gal	=	gallons
lbs	=	pounds
KCl	=	potassium choloride
ISIP	=	instantaneous shut-in pressure
Mcf	=	thousands of cubic feet
Mgal	=	thousands of gallons
md	=	millidarcies
min	=	minutes
ml	=	milliliters
psi	=	pounds per square inch
R&D	=	research and development
SEM	=	scanning electron microscope
Tcf	=	trillions of cubic feet

10.9 References

Cox, D. O., Kuuskraa, V. A., and Hansen, J. T. 1996. Advanced Type Curve Analysis for Low Permeability Gas Reservoirs. Paper presented at the Gas Technology Symposium, Calgary, Alberta, Canada, 28 April–1 May. SPE-35595-MS. https://doi.org/10.2118/35595-MS.

Hill, D. G. and Reeves, S. R. 1998. Restimulation Research to Target Low Cost, Incremental Gas Reserves, *GasTips*, **4** (3), Fall.

McCain, W. D., Voneiff, G. W., Hunt, E. R. et al. 1993. A Tight Gas Field Study: Carthage (Cotton Valley) Field. Paper presented at the Gas Technology Symposium, Calgary, Alberta, Canada, 28–30 June 28-90. SPE-26141-MS. https://doi.org/10.2118/26141-MS.

Mohaghegh, S., Arefi, R., Ameri, S. et al. 1996. Petroleum Reservoir Characterization with the Aid of Artificial Neural Networks. *Journal of Pet. Sci. Eng.* **16** (04): 263–274. https://doi.org/10.1016/S0920-4105(96)00028-9.

Reeves, S. R. 1996. Assessment of Technology Barriers and Potential Benefits of Restimulation R & D for Natural Gas Wells. *Final Report*, Gas Research Institute, 960267 (October 1995–April 1996).

Chapter 11

Employing Both Damage Control and Stimulation: A Way to Successful Development for Tight Gas Sandstone Reservoirs

Kang Yili and **Luo Pingya,** Southwest Petroleum Institute; and **Xu Jin** and **Xu Xinghua,** China National Star Petroleum Corporation

SPE 64707 presents a case history of temporary bridging materials as a drilling fluid additive to build a nonpermeable mudcake in naturally fractured tight gas sands in the western Sichuan Basin. Yili et al. summarize the geological characteristics of tight gas reservoirs, damage mechanisms, and development of temporary bridging materials as a fluid additive. They then discuss field pilot tests and the associated damage control, and reach the following conclusions:

- The naturally fractured tight sands may be damaged easily by stress change
- Fluids and solids from drilling and completions on the basis of geological and core analyses may also cause damage
- Damage control through temporary bridging of natural fractures during drilling seems to work in the tested wells

11.1 Abstract

Tight gas sands exhibit extremely low porosity and permeability, so most people think that it is impossible to produce gas economically without stimulation. During past several decades, great efforts have been paid to improve hydraulic fracturing, but the results are mixed. Formation damages induced by drilling and completion fluids to natural fractures and tight matrix retard artificial fractures to give full conductivity.

Tight gas sands in western Sichuan Basin are characterized with natural fractures developed in heterogeneity, complexity in micro/macro multi-scale structures, and very sensitive to environmental changes. Drilling fluids invasion into tight formation mostly occurred through natural fractures, which leads to solid plugging, water trapping, clay mineral impairments, and chemical adsorption and retention. SPE 64707 presents application cases of temporary plugging fluids for naturally fractured sands in two tight gas fields. As the invasion depth is localized to within 3 cm of the well-bore, perforations can bypass the cake easily and efficiently. Wells with damage control give relatively higher initial production and the production rates after hydraulic fracturing treatments are highest owing to the fact that artificial fractures can connect natural fracture systems and matrix without damages. These wells also show stable and high productivity with longer periods and yield satisfactorily economical results.

11.2 Introduction

The natural gas from tight sands, being with a huge amount of resource, occurs in many petroliferous basins around the world. During the past decades, most of the exploration and production activities of this unconventional resource have been concentrated within North America. For example, the total production of the tight gas reached to 705.3 × 108m^3 in 1994. Nowadays, this kind of resource has been paid much attention in Europe and Asia Pacific areas and tight gas production increases rapidly though the total amount is relatively at low level. The tight gas resource in China is estimated as much as 12 × 10^{12}m^3, which mostly exists in Sichuan, Ordos, Songliao, Tarim, Zhunger, and Chidam Basins. There is an obvious contrast between its huge amount of tight gas resource and production or reserves owing to the fact that the damages during well drilling and completion generally happened, sometimes leading to serious impairments. The situation heavily retards the improvement of exploration and production.

The US and Canada are the pioneers for the successful development of tight gas. At the early and middle stages of tight gas development, the damage control during drilling and completion has not been paid as much attention as that during stimulation. It is widely accepted that the natural productivity is too low to be economically exploited. And it is impossible to successfully produce gas from this type of reservoir without stimulation. Therefore effective stimulation is regarded as the key technology for tight gas development. Under the direction of this opinion, a great deal of effort has focused on the improvements of stimulation. But the application results of stimulation for tight gas formation are mixed, and usually disappointing. During the middle 1960's, the projects for tight gas supported by the US DOE concentrated on the nuclear explosion stimulation. Then it turned to massive hydraulic fracturing (MHF) for some reasons. But most of MHF has not achieved to the expected production, including the multiwell tests (MWX). An explanation about the facts is that the stimulation scale or fracture length is not large enough to connect more volume of reservoirs. Following this idea, stimulation scale increased continuously, and super massive hydraulic fracturing appeared. Application of these techniques show that enlargement of stimulation scale will get half the result with twice the effort unless the damage to natural fracture systems from aqueous based fluids has been prevented (Northrop and Frohne 1990).

Now that the natural fractures are relatively developed in the tight gas sands of the West Sichuan Depression, and that gas wells with high production rates can be met when the conditions of drilling and completion are favorable, the efforts to prevent damage during drilling and completion have been emphasized. In the direction of the temporary shielding principle (Pingya 1997), estimation of fracture parameters, evaluation of the sensitivity and damage mechanisms, and drawing of the damage control program for operation have been conducted. Applications indicate that by damage control during drilling and completion, the wells show not only higher initial productivity, but also highest final production rates after stimulation. SPE 64707 gives two application cases in the studied areas about the drilling/completion/stimulation system, of which temporary shielding for fractures is the key technique.

11.3 Engineering Geological Characteristics of Tight Gas Sandstone Reservoirs

The studied area locates at West Sichuan Depression in Sichuan Basin. The main pay zones are Penglaizhen, Shaximiao, and Qianfoya Formation of the Jurassic. At the early stage of Jurassic, the sedimentary environment was shallow lake to the shore of the lake. Then the environment gradually transfer to alluvial fans-braided streams, to fluvial-delta, and finally to shallow lake to the shore of the lake (Guo 1997). In the section profile, Penglaizhen Formation is dominated with purplish red fine-grained feldspar arkose and lithic fragment arkose, belonging to conventional-nearly tight types. Shaximiao Formation consists of 3–5 layers of green fine to middle grained sandstones with thickness of 1030m, being of typical tight sands. And Qianfoya Formation is characterized with hybrid color conglomerates. The tight gas reservoirs are of complexity in micro/macro multi-scale structures (Kang et al. 1999a).

There exist multiple pay zones differing markedly in thickness and dimension extending in the long gas section. There is a great deal of active gas shows from 300m to more than 5500m at depth.

The gas reservoirs display abnormal high formation pressure and multi-pressure systems. The top depth of the abnormal pressures is about 850m, and the pressure gradient can be as high as 2.05–2.40MPa/100m. At some local regions, abnormal lower pressure reservoirs are present. Generally a sandstone body is a definite pressure unit.

Natural fractures have been developed heterogeneously in different scales with micro-fractures being the major contributors to reservoir permeability. The average apertures of natural fracture range from 5 to 35μm. Well tests and production performances of gas wells show dual-media percolation flow and extensive stress sensitivity.

The residual intergranular pores and dissolution pores are connected with flat micro-fined throats. Micropore content ranges 30–80% and the capillary pressures reach as high as 10–50MPa, which may lead to aqueous trapping damage.

The clay minerals, controlling main sensitivities of reservoirs, include kaolinite, chlorite, illite, interlayered illite/smectite, interlayered chlorite/smectite, and smectite. Grain coating of clay minerals varies from smectite at shallow to interlayered chlorite/smectite and chlorite with increasing depth. Kaolinite occurs as booklet aggregates, and illite and interlayered illite/smectite as pore throat bridging (Kang et al. 1998).

11.4 Damage Mechanisms of Tight Gas Sands

The typical tight gas sands exhibit extreme sensitivity to the changes of surroundings, being liable to be damaged (Kang et al. 1999b). If damages take place, then it is very difficult to remove the impairments induced by operations **(Table 11.1).** Data of core flow tests reflect that the core samples with artificial fractures are of middle-strong to strong sensitivity to fresh water, alkali, and flowing rate. Sensitivity degrees depend on the features of sensitive minerals exposed to the fracture faces and filling in the fractures. For example, quartz and calcite filling can undermine the sensitivity of fractures to working fluids, but the sensitivity of matrix keeps the same with the intact matrix. For tight gas sands, both fracture and matrix show strong to very strong sensitivity to stress variations.

The water wetability of clay minerals and the special pressure evolution of tight gas reservoirs enhance the damage of liquid invasion and capillary suction, so aqueous trapping is one of the major obstacles impeding gas production effectively. The tightness of the rocks is the main factor to control damages from adsorption of high-molecular-weight polymers to rock faces. Natural fractures are very sensitive to the damage of solid invasion. Further matrix of conventional and nearly tight sands may be also sensitive to solid invasion. Therefore, any invasion of both solid and liquid must be controlled in every operation for the reason that when invasion happens in the deep radius around the borehole, it is impossible to flow back easily and clean up thoroughly.

In summary, tight gas sands suffer from an inherent shortage in porosity and permeability, and are liable to damages for its high potential. So damages are vital to gas discovery, accurate evaluation, and economic development. The first action to be tackled is to prevent damages from invasion into natural fractures, as well as to prevent damages induced during stimulation.

11.5 Temporary Shielding Drill-in Fluids and Its Matching Techniques

Considering that natural fractures provide nearly all the permeability, the major objective to be protected ought to be natural fractures, at the same time considering the contribution of matrix porosity (Kang et al. 1999c). The key point of temporary shielding drill-in fluid techniques is to effectively form an inner cake localized within 50mm with no permeability and in very short time. On the consideration of the facts that solids must invade into formation when drilling under positive differential pressure, then following the rule of thumb, the size distribution of solids in the mud can be adjusted by adding some solids to form an inner cake in the formation with no permeability. If the strength of the inner cake reaches as high as 7–9MPa, it can prevent the damage from cement slurry. Theoretically, non-permeable inner cake can not only prevent the invasion of solids and liquid into fractures deeply, but also hinders the effect of aqueous trapping from capillary suction and fluid loss by gravity displacement.

11.5.1 Temporary Shielding Drill-in Fluids for Fractured Reservoirs. The present mud system and technology, which qualified to control shale instability and combat various complexes during drilling the abnormally pressured formations, has been formalized in this area. Because the new drill-in fluids come from modified mud systems, there are two advantages of utilization of the modified mud system. The first is that the drill-in fluids can be well matched with the drilling and mud technology. The other is easy operation with low cost for those additive materials.

Damage type	Origin and Processes	Operation	Prevention and Remedy
Liquid trapping	Increase of water saturation • capillary percolation • imbibition • loss by gravity displacement	Water based mud drilling Well cement Perforation under-balanced drilling workover hydraulic fracturing, acidizing	• avoid to use water based fluids • decrease differential pressure • form cake with no permeable • decrease capillary pressure • inject dry gas or treat thermally • fracturing with liquid N_2,CO_2
(Liquid blocking)	Increase of oil saturation • capillary percolation(positive differential pressure) • condensation	Oil based mud drilling Extremely low flow pressure at well bottom	• avoid to use oil based fluids • control differential pressure • inject dry gas • treat formation thermally • fracturing with liquid N_2,CO_2
Solid invasion	Solid deposits in matrix pore throats Solid deposits in fractures • filling in fractures • inner cake in fracture face	Drilling Well cement Completion (openhole) Hydraulic fracturing Acidizing Workover and well washing	• form cake with no permeable • under-balanced drilling • filtrate fluids strictly • perforate with deep penetration • fracture and acidize formation • treat formation thermally • treat with ultrasonic wave
Clay mineral damages	Foreign fluids—rock reactivity • alkali sensitivity (pH>9) • fresh water sensitivity • salinity sensitivity • acid sensitivity • flow rate sensitivity formation fluids—rock reactivity • flow rate sensitivity	Drilling Well cement Completion Hydraulic fracturing Acidizing Workover and well washing Production	• form cake with no permeable • under-balanced drilling • evaluate potential damage • add clay stabilizers • decrease pH of working fluids • decrease flow rate • avoid pressure fluctuation • produce gas under critical rate
Chemical adsorption	Adsorption and trapping of high molecular polymers • to pore throat surfaces • to fracture faces(walls) • emulsion damage	Drilling, Well cement Completion Hydraulic fracturing Acidizing Workover and well washing	• Evaluate damage in lab • gas based fluids • clean up polymer • degrade polymer microbiologically
Stress sensitivity	Changes in pore throats of matrix Changes in fracture widths (reopening or closure)	Drilling (overbalanced) Drilling (underbalanced) Well cement with long section Perforation completion Production	• form cake with no permeable • control differential pressure • produce with horizontal well • control shale swelling • optimize perforation parameter
Scale	Foreign fluids—formation fluids reactivity inorganic salts $BaSO_4$, $CaSO_4 \cdot 2H_2O$, $CaCO_3$ Formation fluids: • inorganic salts, $CaCO_3$	Drilling Cement Well killing Workover, well washing Production	• evaluate fluids compatibility • control differential pressure • add scale inhibitor • clean scale mechanically or chemically • acidize and hydraulic fracture
Glazing of rock face	Rock powder, water and drilling pipe reaction forming tight layers as porcelain • plugging pore throats • plugging fractures	Drilling with gas Openhole completion Horizontal well drilling	• increase lubricity • increase heat dissipate • use foam as drill-in fluids • perforation • complete in cavity

Table 11.1—Damage mechanisms in the tight gas sands of West Sichuan Depression.

The basic working programs are as follows:

(1) To predict the apertures of natural fractures through integral methods including combination of core description and well tests, lab method by making experiment on cores with artificial fractures under simulated effective stress, and measurement of laser-electrical scanning of fracture faces.
(2) To select a new mud system from all the present mud systems. On the basis of selection and evaluation of additives, by consideration of decreasing cake filtration, inhabitance, damage degree and stability, a mud system of amphoteric polymer has been determined as the candidate drill-in fluid (Jin et al. 1999).
(3) To establish the ratio of fracture width to size of bridging agents through computer simulation. Directed by the temporary shielding principle, a physical model and a mathematical model are formed. Statistical description of fracture apertures provides data for computer simulation. Parameters such as fracture apertures, asperity of fracture surfaces, and size of solid bridging agents were input and the distribution of bridging distances, and ratio of width/solid size could be output.
(4) To suggest a new formulization of drill-in fluids and its realizing conditions. Lab experiments have been conducted to adjust the distribution of whole solid in mud by adding bridging agents, filling agents and soft agents. Then, the effect of temporary shielding was evaluated.
(5) To draw a program for field pilot test and to form the quality control standards.

11.5.2 Matching Technology. At present, techniques matching temporary shielding drill-in fluids are realized in two ways for different geological and engineering environments.

(1) Wells were completed as openhole. And back flow and/or acid washing cleaned up the inner cakes of temporary shielding. This kind of completion method is suited to the wells for which natural fractures are already detected confidently, or predicted to be the areas where natural fractures are well developed. Well X809 at Xinchang gas field is an excellent example.
(2) Wells were cemented and perforated. After optimization of perforation parameters and technical conditions, wells were perforated by SQ-89 gun, which can bypass the inner cakes of temporary shielding efficiently. Next, formations were stimulated through hydraulic fracturing with the characteristics of high pump rate, large volume of proppants and middle ratio of proppant / fluid volume. The treatment can guarantee full connection of artificial fractures with natural fractures. The integral technology is proper to wells in which natural fractures are mostly developed, but also to wells in which natural fractures are relatively less developed. Well CX164 supports this situation.

11.6 Application Cases of the Technology

11.6.1 Shaximiao Gas Reservoirs of Xinchang Field. Well CX164 is the first test well using temporary shielding drill-in fluids for protection of natural fractures in the West Sichuan Basin. The well was cored in Shaximiao Formation. Core plugs were cut with simulated formation water. Permeability of the plugs with inner cakes of temporary shielding was measured at a series of differential pressure. The experimental results show that the strength of the inner cakes reaches 9.0MPa. The thickness of the inner cakes was determined by permeability measurement of plugs when cutting different length. Data in **Table 11.2** indicate that K1 is nearly equal to K2, which illustrates that the thickness of the inner cakes is less than 2.5cm and that perforation hole can penetrate the cakes and recover original permeability of pay zones. These lab tests verify that temporary shielding drill-in fluids for protection can provide intact matrix and natural fracture systems without damage for hydraulic fracturing.

Well CX164 was cemented and perforated at Sand C 2440 to 2448 m of Shaximiao Formation. The initial production rate is 3900 m^3/d. Production logging shows the section of 2440 to 2448 m was poorly perforated, and was re-perforated by SQ-89 gun. The natural open flow capacity was 6000m^3/d. After hydraulic fracturing, the productivity reached to 78415 m^3/d. Comparison of production capacity (**Table 11.3**) reflects that acidizing and fracturing have stimulated many wells, but the increasing of production rate is limited. The density of natural fractures in well CX164 is far less than that in wells such as CX129, CX132, CX133, CX136, and CX160. The natural productivity of CX164 only located at the middle level; after hydraulic fracturing, the productivity reached to the first place and the production rate was more than $3\times10^4m^3/d$ at the beginning. Now the rate keeps at $1.5\times10^4m^3/d$.

After realizing success at well CX164, the technology was propagated to X807, X809, and CX166. All wells for tests have gained expected production rate.

Pay Zones	Samples	First Cutting Length (cm)	Permeability K1 ($10^{-5}\mu m^2$)	Second Cutting Length (cm)	Permeability K2 ($10^{-5}\mu m^2$)
J_2s A	A6	1.80	6.63	0.50	6.65
J_2s A	A9	1.95	5.58	0.45	5.59
J_2s AB	AB1	2.10	1.26	0.52	1.27
J_2s AB	AB12	2.05	2.32	0.51	2.32
J_2s B	B3	2.08	7.76	0.53	7.79
J_2s B	B10	2.06	8.35	0.49	8.32
J_2s C	C2	2.12	5.21	0.52	5.22
J_2s C	C8	2.09	6.38	0.51	6.40

Table 11.2—Determination of the thickness of temporary shielding cakes, Well CX164.

Wells	Pay Zones	Completion	Stimulation Description	Pre-Stimulation Productivity m³/d	Post-Stimulation Productivity m³/d	Skin Factor Before Stimulation	Skin Factor After Stimulation	Remarks
CX129	J_2s B,C	Accident	No stimulation	5.724×10⁴	—	−2.358	—	—
CX132	J_2s B,C	Liner	No stimulation	4.84×10⁴	—	−3.25	—	—
CX133	J_2sB,C	Liner	Acidizing	9021.5	1.41×10⁴	3.757	1.779	Increase
CX151	J_2s	Perfor.	Acidizing	shut in 34h, low P	540	—	—	Increase
CX152-2	J_2s	liner	Acidizing	8022	1.34×10⁴	—	−1.7791	Increase
CX154	J_2s	Liner	Acidizing	no flow	3931	15	—	Increase
CX136	J_2s	Perfor.	Acidizing	low pressure	1.151×10⁴	9.3	−1.6875	Significant increase
CX153-2	J_2s	Liner	Acidizing	3530	8190	8.1	—	Increase
CX158	J_2s	Liner	Acidizing	7244	8383	0	−2.131	Increase
CX160	J_2s A	Perfor.	Frac., proppant	1.3×10⁴	1.1×10⁴	20.789	—	Decrease
CX134-2	J_2s A	Liner	10.2m³	4.01×10⁴	—	9.56	—	—
CX809	J_2s C	Openhole	No stimulation	10.778	—	—	—	—
CX803	J_2sAB, B, C	Liner	Temporary shielding	2486	—	6.4~9.3	—	—
CX805	J_2sA, B, C	Perfor.	—	4000	—	22	—	—
CX808	J_2s A	Perfor.	—	350	1.2×10⁴	—	—	Significant increase
JS2	J_2s	Perfor.	Frac., proppant 30m³	500	1547	—	—	Increase
CX164	J_2s C	Perfor.	Frac., proppant 8m³ temporary shielding, frac. proppant, 25m³	6000	7.8415×10⁴	—	—	Significant increase

Table 11.3—Results of stimulation for Shaximiao Formation at Xinchang Field.

11.6.2 Majing Gas Bearing Structure. MP7 is a test well using temporary shielding drill-in fluids for protection of natural fractures at Majing structure. The pay zones belong to JP3. Compared with the other two wells (MP2, CM601), the gas production of MP7 after fracturing is highest, at 2.16×10⁴m³/d **(Table 11.4).**

11.7 Discussion and Understanding

Whether or not to emphasize the importance of damage control for naturally fractured tight gas sands is under argument. And there are different views on the relationship between damage control during drilling and completion and stimulation. The importance and significance of "employing damage control during both drilling and completion and stimulation" are supported by the successful and failure case wells.

11.7.1 Serious Damage and Deep Invasion Take Place. With this, it is very difficult to reach the prospected results for stimulation, and sometimes stimulation may slow down production rates. The typical example is CX160 of which skin factor is 20.789 (Table 11. 3). The well initial productivity is 1.3×10⁴m³/d; after fracturing it decreased to 1.1×10⁴m³/d. The main reason to explain the phenomenon is that there exists strong damage during drilling and completion. Acidizing can remove damage in some degree, but the treatment scope is limited and acidizing itself can also induce damage. The production rates of wells treated with acidizing are less than 1.41×10⁴ m³/d.

Well CF125 at Fenggu Gas Field is another example to support this opinion. During drilling the lower part of T_3x^4, depth of 3825-3910.5m, gas kick and blow out happened, and killed the well with muds with density of 2.43 g/cm³ and pH value 11.0. The lost volume of mud was estimated as 457.5 m³, and invasion radius is about 30-50m. Treatments such as acidizing, periodical release gas, well washing to accelerate cleaning up have been done, but final outputs were productions of gas at 200m³/d and water at 0.9m³/d.

When drilling section of T_3x^2, (depth of 4388–4351m, formation pressure of 1.61MPa/100m) mud with density of 1.90–2.10g/cm³ was used because of high pressure above the layer. After perforated with YD73 gun, pressures in tube and annulus space were too low to produce gas. Two acidizing treatments were proceeded, and gas production was only 79 m³/d. Torpedo explosion was introduced to stimulate the layer, with no obvious effect. The reason may be that solids and liquids invaded into natural fractures and fine powders created by torpedo explosion plugged fractures, whether natural or artificial. That stimulation radius did not exceed the damage zones is the essential reason.

Well	Pay Zones	Drill-in Mud Density g/cm^3	Final Density of Mud g/cm^3	Productivity of Pre-stimulation $10^4m^3/d$	Productivity of Post-stimulation $10^4m^3/d$	Remarks
MP2	JP_3	1.54	1.65	0.5658	1.4889	Increase
CM601	JP_4	1.17–1.21	1.65	0.4805	1.9347	Increase
CM601	JP_3	1.28–1.53	1.65	0.6255	0.1774	Decrease
MP7	JP_3	1.32–1.56	1.77	0.0700	2.1600	Significant Increase

Table 11.4—Comparison of gas well productivity at Majing Structure.

The invasion of solids and liquids into fractures and matrix leads to fracturing filling of solids and matrix saturated by filtrates. When fracturing, the artificial fracture will orient to the direction that is parallel to the minimum in-situ stress. The direction is the same with the natural fractures. If there is serious damage and deep invasion, the fracturing fluids will transport these solids into deeper parts of reservoir, and plug both natural fractures and artificial fractures. The effective length and width of artificial fractures will be lessened dramatically.

11.7.2 Damage Control Achieved During Drilling and Completion. When this is realized, high production wells can materialize for naturally fractured reservoirs. Two examples are selected to illustrate the view. The first is well CX129 that was completed by accidental openhole. The well located at the top of the structure and density of natural fractures reached to 25–30 per meter. Violent blowout took place when drilling in Shaximiao Formation. The initial production rate was as high as $5.724 \times 10^4m^3/d$. High flow rate of gas removed damage completely, and the skin factor was –2.358. The well has produced gas cumulatively about $6232.4 \times 10^4m^3$ for more than ten years, and nowadays production rate remains nearly $1.6 \times 10^4m^3/d$.

X809 is a well for test of temporary shielding drill-in fluids, and was completed by openhole method. During the third drilling trip, well logging and coring were canceled. When completion was finished, the initial production capacity was $10.778 \times 10^4m^3/d$.

11.7.3 Stimulation Can Reduce Gas Production. Core flow tests show that damage degrees of fracturing fluids range from 29.1–42.3%. Apart from damages of solids and high-molecular-weight polymer absorption, water trapping is an important factor introduced by fracturing fluids, which cut down production from $1.3 \times 10^4m^3/d$ to $1.1 \times 10^4m^3/d$ for well CX160. Although water saturation around fractures will eventually be reduced, the extremely long time for cleaning up fracturing fluids hinders production recovery, no mention to increase gas capacity. J46 and J13 are two wells located at Bajiaochang gas field. These wells were hydraulically fractured by water-base fluids ten years ago. J46 kept $47.87m^3$ residual fracturing fluids after flowing back 8 months, and gas production lowered from $0.5718 \times 10^4m^3/d$ to $0.4268 \times 10^4m^3/d$ (Shaochun et al. 1988). Similarly, J13 kept 268.87 m^3 residual fracturing fluids after flowing back 3 months. Therefore, using gaseous fracturing fluids or high pump rate (quickly pumping in and quickly flowing back) are effective ways to control water trapping (Kang et al. 2000).

11.7.4 Wells of High Production Capacity Can Be Constructed. This can happen when damage control during drilling, completion and stimulation has been truly fulfilled, as well as stimulation technology being reasonable. Wells were cemented and perforated. After optimization of perforation parameters and technical conditions, wells were perforated by deep penetration, high density of perforation holes, which can bypass the inner cakes of temporary shielding efficiently. Formations were stimulated with low damage fracturing fluids and under the condition of high pump rate, large volume of proppants, and middle ratio of proppant/fluid in volume. The treatment can guarantee full connection of artificial fractures with natural fractures. The integral technology is proper to wells for which natural fractures are mostly developed, but also to wells for which natural fractures are relatively less developed. The successes of test wells have demonstrated this.

11.8 Conclusions and Suggestions

1. The tight gas sands suffer from an inherent shortage in porosity and permeability, and are pliable to damages for its high potential. Damages are detrimental to gas discovery, accurate evaluation, and economic development, so effective actions should be tackled to prevent damages from invasion into natural fractures, as well as to prevent damages induced during stimulation.
2. Comprehensive geological and core analysis shows that the tight gas sands in Western Sichuan Basin exhibit abnormal high pressure, multi-pressure system, extremely high capillary pressure in matrix, fracture distribution in heterogeneity, and abundance in clay minerals. These factors lead to damage mainly from liquid phase trapping occurring in matrix, solids invasion into fractures, swelling and dispersion/migration of clay minerals, stress sensitivity for both fractures and matrix, and chemical adsorption and retention on fracture faces.
3. Directed by the principle of temporary plugging, a new drilling mud system has been designed. Laboratory tests of artificially fractured cores indicate that the modified mud can form impermeable inner cake in a few minutes and the strength of the cake is as high as 9MPa. For invasion depth localized to within 3cm of the wellbore, perforation and acidizing can bypass the cake easily and efficiently.
4. The successes of test wells indicate that damage control during drilling and completion is most important. If the damaged zones are localized in minimum radius during drilling and completion, the stimulation would result in satisfactory production for naturally fractured tight gas sands. The facts suggest that employing both techniques for damage control during drilling and completion and stimulation is a viable strategy for tight gas development.

11.9 References

Guo, Z. 1997. Exploration Model of Tight Gas Sands in the Shallow Depth of the Western Sichuan Basin. *Natural Gas Industry* **17** (03): 5–9.

Jin, X., Xinghua, X., and Fangyuan, H. 1999. Drilling and Completion Fluids Suited for Protect the Tight Clasolite Dual Porosity Media Reservoir in West Sichuan. *Natural Gas Industry* **19** (06): 48–50.

Kang, Y., Luo, P. et al. 1998. Clay Minerals and Potential Formation Damage of Tight Gas-Bearing Sandstone in Western Sichuan Basin. *Journal of Southwest Pet Inst* **20** (04): 1–5.

Kang, Y., Luo, P. et al. 1999a. Engineering Geological Characteristics of Tight Gas Sandstone Reservoirs in Western Sichuan Basin and Strategy of Damage Control. *Journal of Southwest Pet Inst* **21** (01): 1–5.

Kang, Y., Luo, P. et al. 1999b. Formation Damage Control Technology for Naturally Fractured Tight Sands in Western Sichuan Basin During Drilling and Completion. Paper presented at the Second International Reservoir Development Conference, Nanchong, China, October 1999.

Kang, Y., Luo, P. et al. 1999c. Strategy for Controlling Formation Damage in the Course of Well Completion in the Gas-Bearing Tight Sandstone in West Sichuan. *Natural Gas Industry* **19** (04): 46–50.

Kang, Y., Luo, P. et al. 2000. Formation Damage Control Technology for Tight Gas Sandstone Reservoirs in Western Sichuan Basin: Developments and Challenges. *Journal of Southwest Pet Inst* **22** (04): 1–4.

Northrop, D. A. and Frohne, K. H. 1990. The Multiwell Experiment-A Field Laboratory in Tight Gas Sandstone Reservoirs. *J Pet Technol* **42** (06): 772–779. SPE-18286-PA. https://doi.org/10.2118/18286-PA.

Pingya, L. 1997. A Temporary Plugging Techniques on Formation Damage Control during Drilling and Completion, *Collected Works of Luo Pingya, China National Engineering Academician*, 68–98. Beijing: China Great Encyclopedia Press.

Shaochun, Z., Xia, X. and Li, Y. 1988. Geological Characteristics of a Low Permeable Sandstone Reservoir and its Stimulation in Xiangxi Formation of Bajiaochang Gas. Paper presented at the International Meeting on Petroleum Engineering, Tianjin, China, 1–4 November. SPE-17831-MS. https://doi.org/10.2118/17831-MS.

Chapter 12

Will the Blossoming of Unconventional Natural Gas Development in North America Be Repeated in China?

Hongjie Xiong, Burlington Resources*, and **Stephen A. Holditch,** Texas A&M U.

SPE 103775-MS provides a high-level overview of tight gas development in the US and discusses the challenges of tight gas development in China. Xiong and Holditch first discuss several key factors contributing to the successful development of tight gas in the US:

- Increased prices for natural gas
- Government incentives for development of tight gas
- New technologies
- A more-efficient services sector

They then go on to summarize key technologies instrumental in the increase of tight gas sand development in North America:

- Better resources assessment
- Improvements in 3D geophysical mapping
- Improved geological modeling
- Advancements in reservoir modeling
- Better drilling technology
- Improved stimulation by hydraulic fracturing
- Improved production technology
- Knowledge/information sharing

Xiong and Holditch discuss what China needs for successful tight gas development, including more drilling rigs and hydraulic fracturing equipment, cost-effective fracture fluids, and trained personnel to run the operations. They argue that China needs an open and innovative services sector—perhaps by allowing US companies to provide consulting and wellsite operations—and more knowledge/information sharing through conferences, training, and consulting. Xiong and Holditch then share the experiences and challenges of tight gas development in China faced by Burlington Resources (now ConocoPhillips) as a case history, which concludes the paper.

12.1 Abstract

There are substantial volumes of unconventional gas resources in China, including tight gas sands and coal bed methane (CBM). Rapid economic growth in China has increased the demand for more sources of clean energy, such as natural gas from unconventional reservoirs.

In the 1980s in North America, the combination of federal tax credits and various technical development programs sponsored by private organizations, public companies and government agencies fostered the development of technology for unconventional natural gas exploration and exploitation. As those technologies in geology, geophysics, drilling, completion, and production have matured, and the price for natural gas has increased, the development of unconventional natural gas has been blossoming for the last decade globally.

Without question, it is certain that the development of unconventional natural gas in China will be blossoming in the coming decades. However, there are significant challenges and hurdles to overcome before that happens.

SPE 103775-MS first analyzes the conditions and environment required to ensure the development of unconventional natural gas sources, including price, the service industry, cost, and the operational environment. We focus on key technologies in geology, geophysics, and engineering, and how they are tied to the success of unconventional natural gas development. The paper also discusses the advantages China offers, and the challenges in China that must be overcome to be sure the blossoming of the unconventional natural gas development actually occurs.

On the basis of the authors' operational experience both in China and North America, we will discuss the impact of different operational environments on unconventional gas development.

---

* Now ConocoPhillips

SPE 103775-MS also discusses

- The impact of new technology on unconventional natural gas development
- The technologies available and unavailable in China for unconventional natural gas development, and how to use and develop those technologies
- The role of foreign operators in developing unconventional natural gas reservoirs in China

and concludes that an open, competitive, and mature service sector is critical to the future blossoming, and that proper government regulations and policies in pricing and taxes, and downstream infrastructure, will impact the development of unconventional natural gas resources.

12.2 Introduction

Unconventional gas resources here refer to natural gas in tight gas sands, shale gas, and CBM (see **Fig. 12.1**). Fig. 12.1 illustrates the principle of unconventional gas resources. Globally, there are large volumes of unconventional gas resources in many gas prone basins. It has been estimated that around 32,560 TCF of gas resources are known to exist worldwide, of which about 5,090 TCF are in China and the Asia-Pacific area; and about 8,230 TCF known to exist in North America (Holditch 2006).

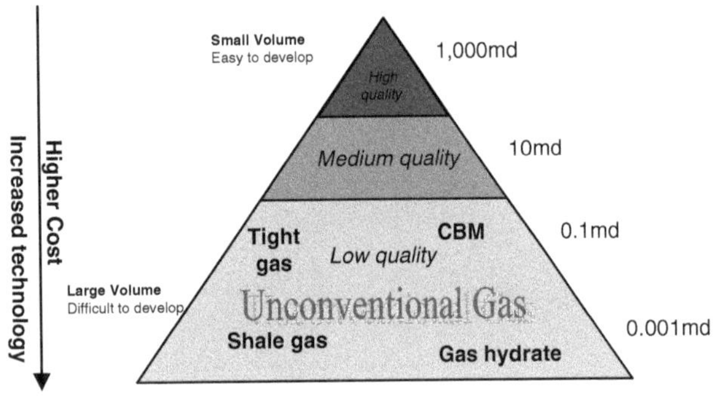

Fig. 12.1—Unconventional natural gas resources (after Holditch 2006).

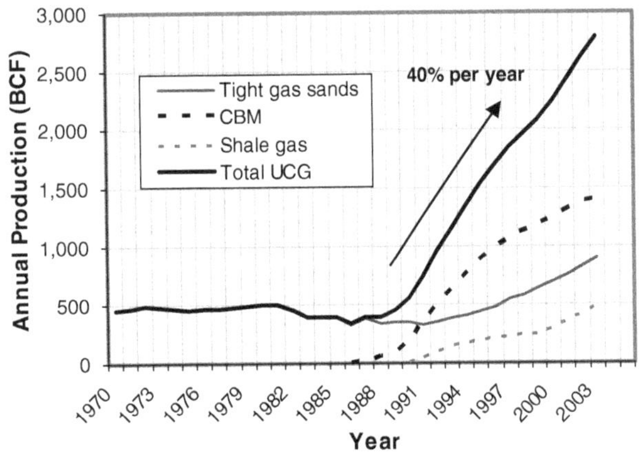

Fig. 12.2—Annual production from selected US unconventional gas plays (after Caldwell 2006).

The development of unconventional gas resources in North American has been booming for last few decades. **Fig. 12.2** presents the annual production rates for selected pays of tight gas sands, shale gas, and CBM in the US (Caldwell 2006). Since 1990, the annual production from unconventional gas resources has increased at 40% per year in the US.

Currently the most active CBM development in the world is in North America, in which there are about 3,000 CBM wells in the US. These wells have led to reserves estimates of 17–35 TCF in the US. In Canada, there are 3,900 CBM wells with reserves of about 700 TCF (Findley and Shirley 2006).

Currently, the demand for more energy is certainly the engine for increasing interest by industry in unconventional gas resource development. Fortunately in the past, tax credit incentive programs, and research sponsored by private organizations, public companies and governmental agencies led to the development of new technology for unconventional natural gas exploration and exploitation. As those technologies in geology, geophysics, drilling, completion, and production were applied by industry, coupled with increases in natural gas price, the development of unconventional natural gas resources in the US has been blossoming for several decades.

SPE 103775-MS discusses the current situation, identifies the key technologies required to explore and produce unconventional gas resources, and then analyzes the current situation and the outlook in China.

12.3 The Reasons Behind the Boom in Development

As shown in **Fig. 12.3**, the "development boom" triangle demonstrates the relationships among three factors behind the boom: Demand, Resource, and Technology (Services). Strong demand for natural gas is the engine for the boom. Better technology allows industry to economically develop unconventional gas resources. To know where to apply the technology, we have to assess the quantity and quality of unconventional resources in gas prone basins. Strong demand puts pressure on natural gas prices. These higher natural gas prices coupled with the technical challenges required to develop these unconventional resources, in turn stimulate new technology innovations. The new technologies enable us improve operation and cost efficiency.

12.3.1 Natural Gas Price. Natural gas prices in North America (shown in **Fig. 12.4**) have been increasing steadily during the last 10 years due to many factors, such as increased demand, higher oil prices, production interruption due to weather, and a general convergence of domestic supply and demand. There is very little, if any, excess capacity. Recent gas prices and reasonable price expectations have been sufficient to stimulate additional activity in natural gas exploration and exploitation, especially from unconventional natural gas resources.

12.3.2 Government Regulations. In 1978, the US Congress passed the Natural Gas Policy Act (NGPA). The Federal Energy Regulatory Commission (FERC) was assigned the task of administering the NGPA. Actions by FERC concerning the natural gas business in the 1980's and early 1990's led to more development in unconventional natural gas resources, such as tight gas, coalbed methane and gas shales.

Federal and state tax credit programs were incentives that definitely led to more development of unconventional natural gas resources. Because of the NGPA, the FERC developed regulations to encourage operators to produce gas from tight reservoirs. The US federal government and some states allowed tax credits or severance tax abatements to companies who developed tight gas, coalbed methane or shale gas. The governmental action played two important roles: it stimulated activity to explore for unconventional gas, and it fostered the development of new technology needed to assess and develop the unconventional gas resources. The result is that a combination of tax incentives, new technology and higher gas prices has led to rapid development of many unconventional reservoirs in the US.

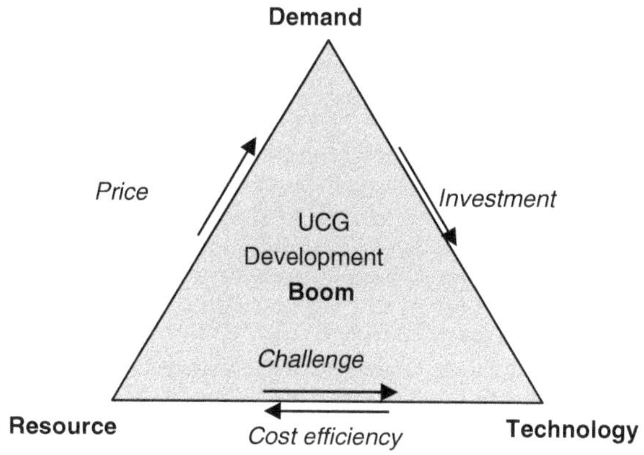

Fig. 12.3—Development boom of unconventional natural gas.

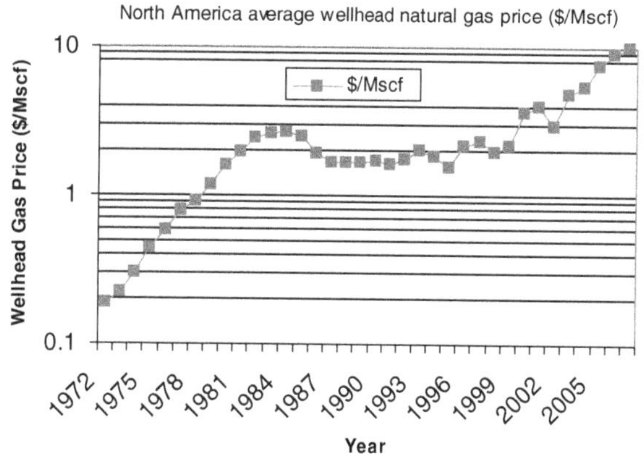

Fig. 12.4—North American average wellhead natural gas history (Source: EIA).

The Gas Research Institute (GRI, now GTI), the US Department of Energy (DOE), and several consortiums have led the way in the development of new technologies related to natural gas development and applications. Since the late 1980's many technical papers and reports have been published in respect to technology development for unconventional gas resources. There have been advancements in resource assessment, drilling and completion methods, and advanced hydraulic fracturing technologies. The new technologies have been used to improve well productivities and cost efficiency. Technology advances have helped to keep development costs low, which is always the key to sustaining the industry during a period of generally low wellhead gas prices in 1980's and 1990's (see Fig. 12.4). The impact of those unconventional gas technologies on production and wellhead gas price is significant. The EIA has estimated **(Fig. 12.5)** that new technology has the potential to increase annual gas production about 3 TCF and lower wellhead gas price by 18% by year 2020 (McCallister 2000).

12.3.3 Efficient Service Sector. One of the biggest challenges we face in our efforts to develop unconventional gas resources is the control of costs. The service industry in North America consists of a combination of large international service companies, small local companies, and a few specialized service companies. The competition usually stimulates efficiency and technology innovations, which ultimately lead to reduced costs for well services. To develop unconventional reservoirs, we must control costs.

An alliance between an operator and a service company is another way to improve operational efficiency. The integration of technologies, experience, and supply chain is often the key to improve performance. A working alliance can be very beneficial in international operations.

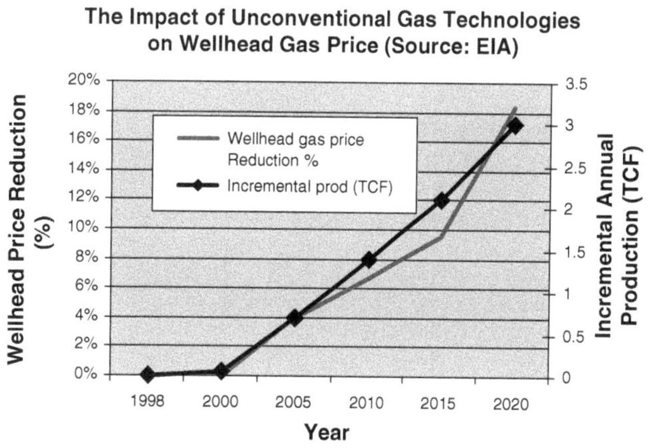

Fig. 12.5—The impact of unconventional gas technologies on wellhead gas price and production.

12.4 Unique Characteristics of Unconventional Gas Resources

Unconventional gas resources include tight gas sands, CBM, and shale gas. Each type of resource has unique characteristics that must be considered in the development plan. One common characteristic, however, is that the stabilized flow rates from all unconventional reservoirs is usually quite low due to low permeability in those reservoirs. The following briefly summarizes those unique characteristics.

12.4.1 Tight Gas Sands. Tight gas sand reservoirs can be either continuous or lenticular sand deposits. Many of the lenticular sands are fluvial channel sands, or low quality sands that are usually deposited in a continental system. Some of the better tight gas sands are more continuous and were deposited as shoreline or marine systems, such as longitudinal bar sands.

Low permeability is the defining feature of tight gas sands (Holditch 2006). Most of the tight gas sands are tight because of diagenesis that has filled the pores with clays or quartz overgrowths. Virtually all wells in low permeability gas reservoirs require massive hydraulic fracturing treatments to achieve commercial production rates. After a large fracture treatment, the gas flow rates can be high, due to formation linear flow, but the wells usually stabilize at a low gas flow rate (see **Fig. 12.6**). Many times, this low gas flow rate causes liquid loading problems.

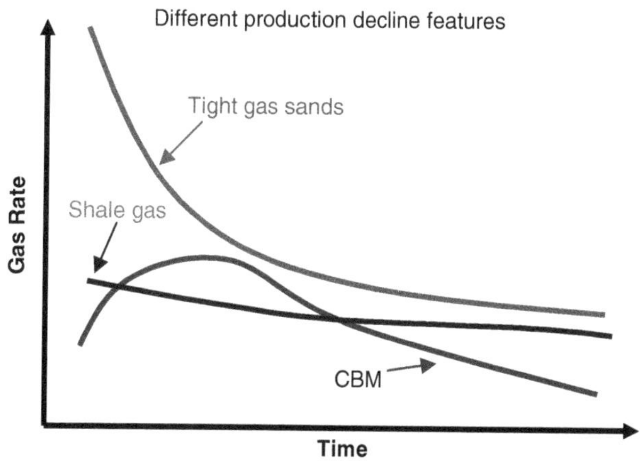

Fig. 12.6—Unique production behaviors of three unconventional gas wells.

12.4.2 Coalbed Methane. In coalbed methane reservoirs, the gas, which is primarily methane, is stored mostly as adsorbed gas on the internal surfaces of the coal. Because of its large internal surface area, coal stores 6 to 7 times more gas than the equivalent rock volume of a conventional gas reservoir at a similar pressure. Gas content generally increases with coal rank, with depth of burial of the coalbed, and with reservoir pressure. Natural fractures (cleats) that often occur in coal seams create the permeable channels that allow the gas to flow. In many cases, the cleats are initially filled with water. There is little free gas in the cleats. The gas in place is mostly the adsorbed gas.

As we begin producing the coal seam reservoirs, by reducing the pressure, gas will begin to desorb, diffuse through the coal to the cleats, then flow through the cleats to the well bore. If the coal cleat system is filled with water, we have to produce the water from the coal cleat system in order to reduce the pressure so gas can desorb and begin flowing. We call this de-watering the coal. To maximize the gas flow rates, we have to minimize the wellbore and the reservoir pressure. In most cases, large volumes of water are produced from coalbed methane wells, especially in the early stages of production. Therefore, gas production rates from CBM wells will start off at very small volumes then the flow rate will increase during the early production period as the reservoir pressure decreases. Eventually, the gas flow rate will reach a maximum before it starts declining (see Fig. 12.6).

Depending on coal type, thickness, porosity, continuity, and gas permeability, a wide variety of completion techniques are used. Some wells are completed with cavity completions. Many wells are drilled vertically and fracture treated. In recent years, horizontal and multi-lateral wells have been successfully used to develop coal seam reservoirs.

12.4.3 Shale Gas. Gas production from shale reservoirs has occurred for most of the 20th Century in the Devonian shale in the Appalachian Basin in the eastern portion of the US. Shale gas reservoirs are both the source rock and the reservoir rock. As with coal seams, some of the gas in shale reservoirs is adsorbed in organic matter in shale and is released by desorbtion when the pressure is reduced by production. Other layers in the shale may be silty or sandy or fractured and contain free gas in pore space or in fractures.

The shale matrix is an ultra low permeability medium. As such, it requires a conductive natural fracture network and hydraulic fracturing to connect the flow networks to the well bore for gas to flow. Many shale gas wells can continue producing gas at a relatively steady rate for long time after the flush production period (see Fig. 12.6).

Horizontal and multi-lateral drilling techniques, along with modern hydraulic fracturing methods, are currently being used to develop many of the new shale gas plays.

12.5 Key Technologies

There are many technology innovations that have occurred during the past 10-20 years that have spurred development of unconventional reservoirs. Better resource assessment strategies can help us find the right places to drill. Other technologies in the geosciences, reservoir engineering, drilling, well completions, stimulation and production have also been very important. The following briefly describes some key technologies that we use to find and develop unconventional gas resources.

12.6 Resource Assessment

Resource assessments for emerging, high risk, high potential unconventional gas plays need to be updated every few years. The reason we need to update our resource assessments for a reservoir, a field, and even a basin is to incorporate recent developments in both technology and new geologic and geophysical information we obtain due to field development activities.

To discover new areas to drill for unconventional resources, we need to perform

- Basin assessments in new, emerging basins
- Play-specific, extended reservoir characterization
- Studies to evaluate tectonics and natural fracture patterns in the plays or basins

We need to continually assess unconventional gas plays. Some unconventional reservoirs, where resource depletion exceeds technology progress, are in decline. It is important to keep such reservoirs producing to eventually take advantage of new technology or higher prices that could someday improve the development economics. Other reservoirs, where technology progress is rapid, such as the remarkable success of horizontal drilling in certain shale gas reservoirs, can be very economic to develop using current technology at current gas prices.

Natural fracture detection and characterization is an extremely important aspect that affects the assessment of shale gas reservoirs. In addition to the abundance of natural fractures, one must know the percentage of total organic carbon within the rock for assessing shale gas reservoirs. With current wireline logging technology, one can quantify the amount of gas in shale, including free gas occupying pores and gas adsorbed onto organic matter in shale. Porosity is the key to quantify free gas amount. Although gas desorbtion and flow mechanisms in shale are still being investigated to improve our understanding of shale gas, we still need to better understand how to estimate shale permeability using log and core data.

For a tight gas reservoir, the size and continuity of the reservoir may be the most difficult parameters that need to be continuously evaluated. Initial data from a new well, including log data, core data, drill cuttings analyses, and gas production, can be used to compute gas-in-place, projected flow rates and estimates of ultimate recovery. We have found that production data should be routinely analyzed to estimate the effective drainage area for each unconventional gas well. The new information from each well, especially data on how drainage areas are evolving with time, needs to be fed back into new resource assessments on a routine basis.

When development of an unconventional gas field begins, the field development plan (FDP) will be constrained with limited data and general uncertainty concerning gas deliverability and ultimate gas recovery. As the development continues and more data and better technologies are available, it will be necessary to re-assess the resource and modify the field development plan. For example, as shown in the left side of **Fig. 12.7,** we may initially develop a tight gas field using only four wells, thinking the four wells will be able to drain the entire field or lease. However, due to sand deposition patterns, or permeability anisotropy, or elliptical flow due to hydraulic fracture patterns, the four wells may only drain a very limited area (as shown in light blue pattern) during a reasonable time period, which could leave significant reservoir undrained. Thus, we may need to drill more wells in a different pattern to effectively recover the gas in the field, as shown in the right hand side of Fig. 12.7.

12.6.1 Geophysics. Geophysical technologies to identify the main pay zones can be very critical to tight gas sand development. Seismic inversion technology is an important tool to identify better sand channels in a fluvial depositional system. Without 3D seismic data, we usually experience a hit-or-miss situation when drilling fluvial systems. One well can be very economic if it is drilled in the middle of good sand channel, while the next well could be marginal if it just misses the fluvial sand body.

A series of techniques in seismic interpretation, visualization, horizon attributes, seismic velocity inversion, and seis-facies are being used to target well locations. Before applying seismic inversion technology to identify main sand channels in the Sichuan Basin in China, we experienced a great variety of hit-or-miss results in the wells we were drilling. However, after applying new 3D seismic technology, we have been able to place wells in the better sand bodies, which has resulted in much better wells. Though it is not certain that every well will be a good well by applying the technology, we have found the new 3D seismic technology definitely helps us find and drill wells to the better channel sands.

Seismic data has also proven to be valuable for detecting and imaging reservoir rock layers; however, there is currently no viable tool for discriminating economic from noneconomic values of gas saturation in gas sands. The sensitivity of electromagnetic methods to water saturation offers the potential to address this issue if electromagnetic data can be successfully combined with higher-resolution seismic data. A new algorithm has been developed for directly imaging reservoir gas saturation from a combination of seismic and electromagnetic geophysical data.

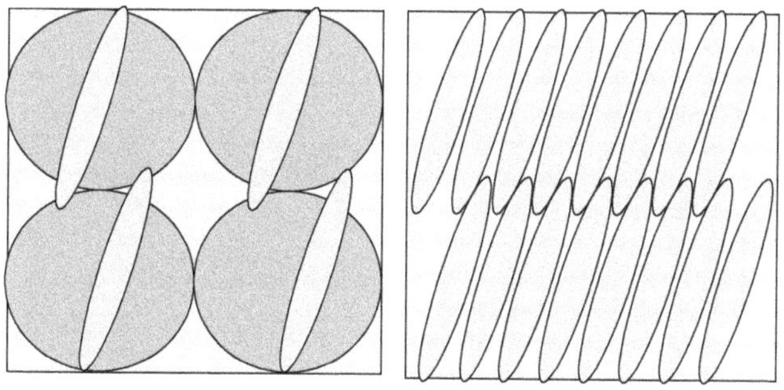

Fig. 12.7—Traditional and extensive development.

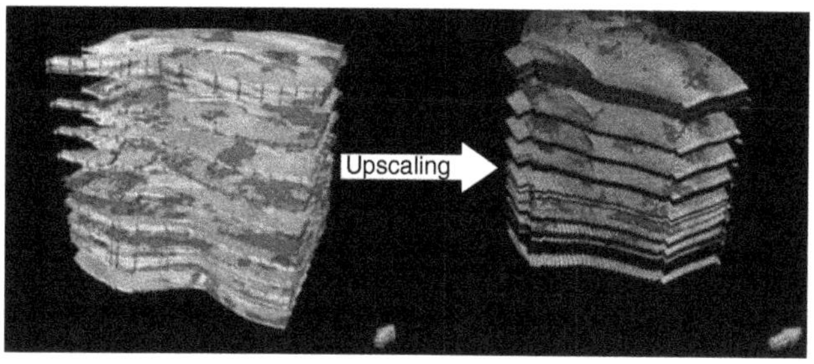

Fig. 12.8—From a geological model to a simulation model.

3D seismic data may greatly help natural fracture reservoir characterization, which requires an accurate delineation of fracture corridors, as well as an estimate of the fracture intensity (Bloch et al. 2003). Analysis of the seismic attribute distributions within each seismic facies allows a preliminary interpretation of these seismic facies in terms of fracture occurrence. The validation of this interpretation with well data delivers a reliable map of fracture intensity.

12.6.2 Geological Modeling. Recently developed fast geological modeling technology can also help us speed up the development of unconventional gas reservoirs. Current geological modeling technology may be summarized as follows:

- All tools from seismic interpretation to simulation are integrated in one application, eliminating import and export problems and promoting collaboration.
- Various and robust geostatistical simulation techniques enable us to study different possibilities of depositions and reservoir property distributions.
- Strong visualization capabilities **(Fig. 12.8)** give us instant quality control of all data in 3D or 4D.
- Models can be updated instantly when new data arrives to make quicker and more reliable decisions, which makes dynamic characterization with time possible and feasible.

The latest emerging geostatistical modeling technologies enable geoscientists and engineers to characterize and model reservoirs with varying degrees of uncertainties, which dramatically speeds-up the modeling process. Geomodeling used to be very time consuming, very expensive, and it requires specialists to run special software tools to build up a geological model. Current technologies make the task much easier and faster. Repeatable work-flows built within a model enable that updating the model becomes more like a few mouse clicks away.

12.6.3 Reservoir Engineering. Most of technologies and techniques in conventional reservoir engineering can be used in unconventional gas resource development. However, there are some exceptions. A good example is using the material balance method to estimate expected ultimate recovery (EUR). One should not usually apply this method to estimate EURs for unconventional gas reservoirs. It is almost impossible to obtain accurate estimates of average reservoir pressure in tight gas reservoirs where gas flows most of the time under transient flow conditions due to low permeability.

12.6.3.1 Perforation Inflow Diagnosis (PID). Permeability is one of key reservoir properties used to determine a reservoir development plan. Since unconventional gas resources have very low values of permeability, it is usually not possible to run a conventional well test to evaluate formation permeability. Many unconventional wells will not flow prior to a fracture treatment, and many operators do not want to shut in wells after they have been fracture treated.

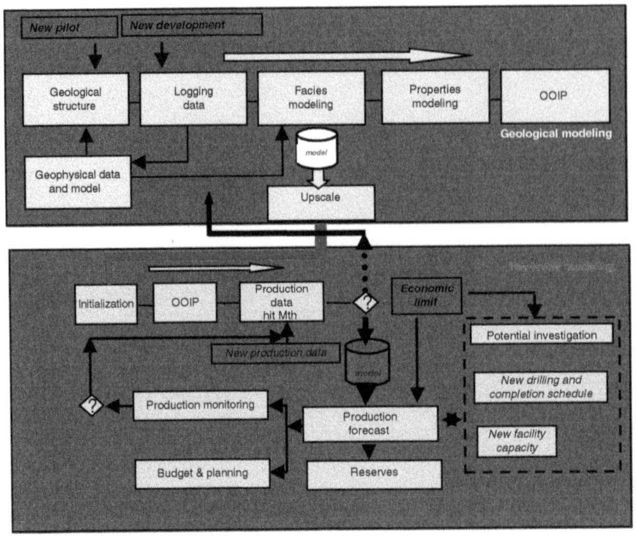

Fig. 12.9—Reservoir management workflow.

Perforation inflow diagnostics, referred to as PID (Hawkes and Hategan 2004) testing, is a modern testing technique designed to deliver, in a cost-effective manner, valuable reservoir information such as: reservoir pressure, formation flow capacity, unstimulated gas inflow rate potential, and near wellbore damage conditions prior to the fracture treatment. So far, in our experience, the method is effective for tight gas sands. PID testing procedure is as follows:

1. The interval is perforated under extreme underbalanced conditions
2. The pressure change response is monitored to evaluate the gas inflow rate potential
3. The transient pressure data are analyzed with the closed chamber drillstem test theory

12.6.3.2 ***Continuous Reservoir Modeling and Management.*** **Fig. 12.9** outlines a basic workflow that one can use to continue updating geological models and reservoir dynamic simulation models. The workflow makes routine reservoir management work easy and quick.

Part of reservoir management includes budgeting, planning, reserve booking, well performance evaluation, candidate selection for in-fill drilling and workovers, and economic and risk assessment. Current modeling technologies make continuous geological and reservoir modeling easy and quick.

12.6.4 Drilling. Directional drilling, to include deviated wells, horizontal wells, and multilateral well, is relatively new technology that makes a real difference in many reservoirs.

Horizontal wells can increase production rates by maximizing reservoir exposure, targeting multiple zones, reducing drawdown, and helping to exploit thin pay zones. Combined with multi-stage hydraulic fracturing treatments, horizontal wells can be a very effective way to develop certain unconventional gas reservoirs, such as shale gas reservoirs. Many wells in Barnett Shale are drilled horizontally and stimulated with large volumes of proppant placed in the single wellbore during multi-stage fracturing treatments.

Based upon a well-established reservoir geological model, well-calibrated 3D seismic interpretation results, and geo-steering and logging-while-drilling technologies, it can be more attractive and feasible to chase sand channels with horizontal wellbores. Multilateral wells are also being utilized to develop CBM and tight gas sands.

Underbalanced drilling (UBD) is a technology we use to reduce formation damage during drilling, improve the rate of penetration of the bit, and reduce drilling and stuck pipe problems in severe lost circulation zones. Drilling induced damage may include fines mobilization, invasion of mud solids, mechanical glazing, phase trapping or chemical reactivity between invading fluids and the formation matrix or in-situ fluids. Open hole completed horizontal wellbores are more sensitive to formation damage caused by the drilling process. Studies (Bennion et al. 1996) illustrate how the productivity of horizontal wells can be reduced dramatically by high near wellbore skins and how this damage effect is attenuated as horizontal to vertical permeability ratio is increased (such as in highly laminated sands). Mud invasion in natural fractures can also be a problem when drilling horizontal wells.

Underbalanced drilling can enhance productivity in tight gas, channel sand reservoirs. As shown in **Fig. 12.10,** drilling filtrate and fines might invade the formation along high perm channels. The resulting formation damage can be permanent and more serious in under-pressured reservoirs.

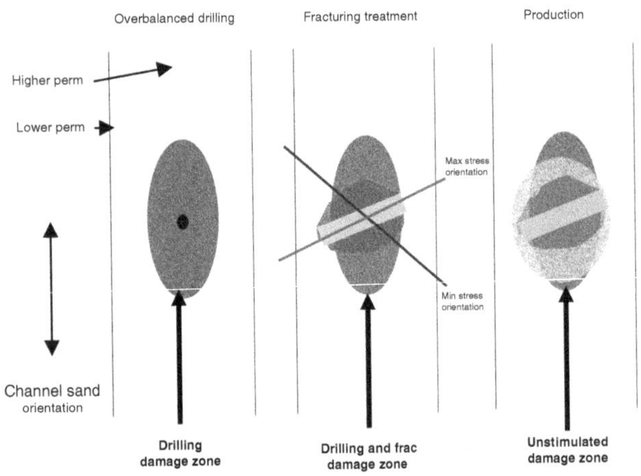

Fig. 12.10—Drilling damage, fracturing, and production.

Also, hydraulic fracture growth orientation may deviate from the damaged high perm direction, which may result in very disappointing production rates and ultimate recovery (EUR).

Underbalanced drilling may also improve well ultimate recovery since the UBD process will not affect absolute and relative permeability (where initial water saturation is higher than connate/critical water saturation). Since the UBD process may help to keep the mud invasion low, it can also benefit reservoir characterization, natural fracture identification, permeability characterization, and productivity evaluation.

12.6.5 Advanced Well Stimulation Technologies. More effective well completion and stimulation technologies are critical to successful development of unconventional gas resources. We need to develop stimulation technologies that also minimize damage to the reservoir from fracture fluids. Successful development of tight gas and shale gas reservoirs is sometimes more affected by the success of the fracturing treatment than by the gas price (McCallister 2000).

Fig. 12.11 illustrates a recommended design workflow for hydraulic fracturing treatments. A preliminary treatment is designed based upon the well bore configuration, well log data, possible fracture initiation intervals (perforation position), and fracture treatment costs. The treatment should be further optimized by re-designing the treatment after a mini-fracture treatment has been pumped and analyzed. The optimum treatment needs to be both designed and pumped in the field as designed to maximize profit. Field quality control must be part of the plan.

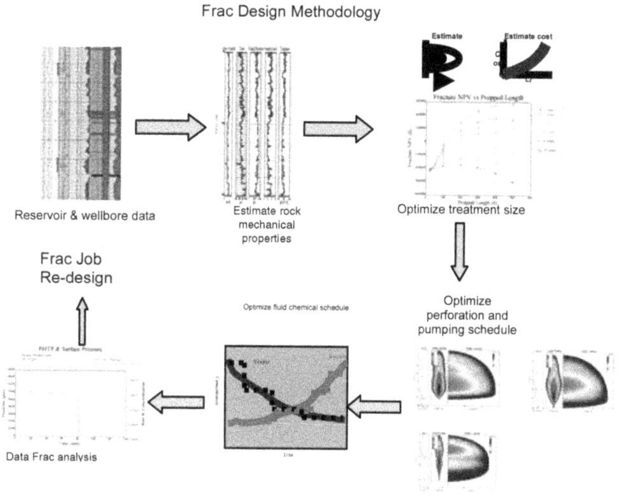

Fig. 12.11—Advanced hydraulic fracturing design approach.

Multiple stage fracture treatments in a single wellbore are being used to develop many layered, tight gas sands and thick, shale gas resources. Recently developed fracture microseismic diagnosis technology can be used to analyze fracture growth patterns, allowing the engineer to optimize well spacing and fracturing treatments (Fisher 2006).

As shown in Fig. 12.7, equipped with knowledge concerning the drainage pattern, and using the latest fracturing technologies, we can improve the development of tight gas sands by placing wells in much smaller spacing, which may result in meaningful improvement in ultimate gas recovery. The Rulison Field study (Kuuskraa and Ammer 2004; Kuuskraa 2006) indicates the gas recovery rate increases from 7% in traditional 80 acre well placing to 48% in an intensive development approach of 20 acre well spacing.

12.6.6 Production Technologies.

12.6.6.1 Compression. Minimizing the flowing bottom hole pressure in a unconventional gas well is extremely important because it helps lift liquids out of the well and helps gas desorb in both coal seams and black shales. A good way to maintain low wellhead pressures is to use a multi-stage compressor to minimize the flowing tubing pressure. To develop an unconventional gas reservoir, compression equipment and related operation expense must be considered in an overall field development plan.

12.6.6.2 Production Water Treatment. In coalbed methane and (sometimes) gas shale wells, the formation has to be dewatered. De-watering and water disposal options that are environmentally acceptable and economically feasible are a concern. Since solids can be present in the fluid, there have been challenges for artificial lift methods in CBM wells. As unconventional gas resource markets have matured, more artificial lift options are available (Findley and Shirley 2006).

In some places, water may be discharged on the surface if it is relatively fresh. Most of the time, however, the produced water is injected into a subsurface formation at a depth where the quality of the injected water is less than that of the host rock.

12.6.6.3 Real-Time Production Optimization. As shown in **Fig. 12.12,** by monitoring daily production data, one can identify underperforming wells and reservoir sections; and then, a series of diagnostic analyses can be performed to determine the reason for the underperformance. Once determined, an optimal remediation solution(s) may be provided to enhance well production and reservoir performance. Real-time production optimization technology is based upon a set of databases containing production data, well completion data, and reservoir properties, and on software tools and other IT technologies, all coordinated by a team of experts.

12.6.7 Knowledge, Experience, and Information Sharing.

12.6.7.1 Technical Meetings. Many technical meetings, organized by SPE and other professional organizations, provide a useful environment to share knowledge and experience.

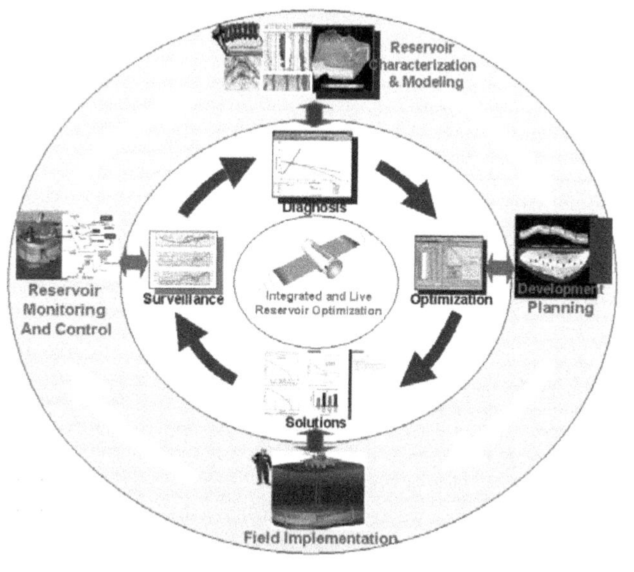

Fig. 12.12—Real-time production optimization.

These professional meetings also provide opportunities for operators and service companies to communicate with each other on new challenges and new research ideas. Local study group meetings are great places to share ideas and/or best practice to improve operations and cost efficiency.

*12.6.7.2 **Basin Excellence.*** Based upon many years of operational practice in different basins in North America, Burlington Resources (now ConocoPhillips) calls such successful experience "Basin Excellence". The main idea behind Basin Excellence is operational and cost efficiency, which includes:

- Critical mass of projects
- Integrated approach to thoroughly understand the subsurface
- Continuous operational process improvement
- Access to infrastructure
- Excellent relationships with host governments, partners, and supers
- Alignment with company competencies

*12.6.7.3 **Rapid Access to the Information.*** In North America, many data related to fields and wells are public ally available, such as well records, drilling and completion, and production data. The significant advances of information technologies make archiving and accessing the information very easy and efficient. Outside of North America, data assimilation can be a challenge.

*12.6.7.4 **Technical Consulting.*** There are numerous consulting companies that oil and gas operators in North America can use, and some of them have years of experience in the development of unconventional gas reservoirs. These consultants have gained experience and knowledge from several decades of working on unconventional gas reservoirs in the North America. Thus, those consultants can also provide independent opinions especially in resource assessment, drilling, completion, and stimulation problems. The challenge is how to use this experience to develop unconventional reservoirs outside of North America.

12.7 Natural Gas Resources and Consumption in China

12.7.1 Abundant Resources. A conservative estimate of proven-probable-possible gas reserves in China is 60–70 TCF (Pietz 2000; Su 2005), with about 35–80 TCF of proven reserve (Pietz 2000; Su 2005). The total resource could be as high as 240–1700 TCF (Pietz 2000; Su 2005) with about 105 TCF of the resources in CBM. There is little information published on shale gas resources in China.

The petroleum industry in China has focused mostly on exploration of oil resources until recently. As natural gas demand increases, undoubtedly, there will be additional gas reserves found soon in mature basins and several unexplored basins spread throughout China (see **Fig. 12.13**).

12.7.2 Strong Demand. Currently, natural gas consumption in China is only about 3% of all energy consumption **(Fig. 12.14).** Natural gas consumption has increased from 0.9 TCF in year 2000 to 1.3 TCF in year 2004 (Su 2005). Due to both strong economic growth and more awareness of environmental concerns, the future demand for natural gas is expected to increase substantially in coming years. The estimated growth in gas consumption ranges from 6.8% (EIA) to 11.8% (PetroChina) per year (Su 2005), which will soon be about 8–10% of total energy consumption.

With the forecast (Su 2005), the Chinese energy sector has started intensifying exploration efforts for unconventional natural gas resources; building nationwide trunk lines such as west-to-east gas pipeline, Shan-Jing Gas Pipeline, and Zhong-Wu Pipeline; constructing gas storage facilities; and encouraging international corporations to participate, including opening more blocks to attract foreign operators to bring more specified technologies and management experience for unconventional gas development.

In addition to abundant resources and strong demand, as shown in Fig. 12.3, successful development of unconventional gas resources also requires an effective service sector to provide qualified and efficient services.

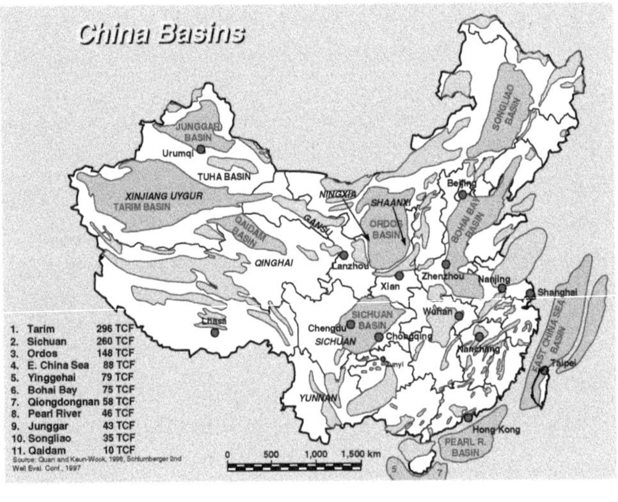

Fig. 12.13—China basins.

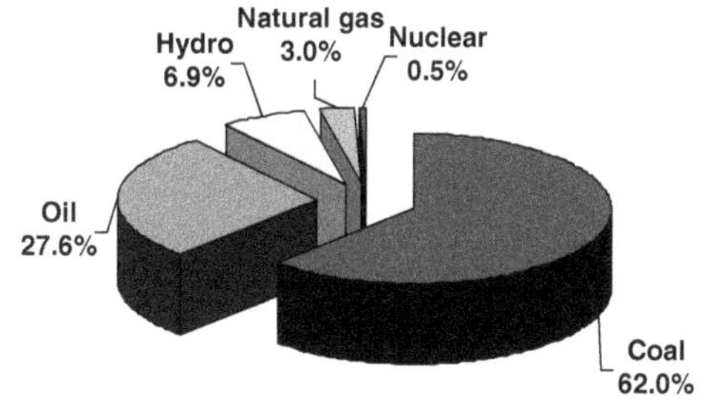

Fig. 12.14—China energy consumption distribution.

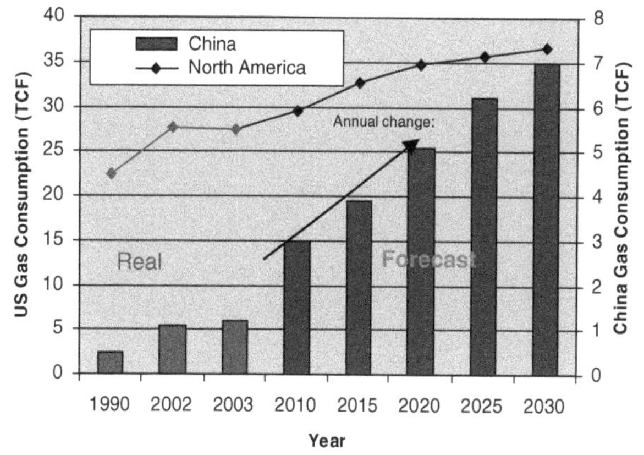

Fig. 12.15—Comparison of gas consumption between US and China (Source: EIA).

There are many service companies in China, each part of one of the national oil companies. These companies have been operating for years and can provide conventional operational services for large scale operations.

Even though the integration between technology applications and reservoirs can always be improved, one can usually find the latest technologies and most of the related tools are available in China. Thus, the blossoming of unconventional gas resource development can be expected in China based upon the triangle relationship among Demand-Resources-Technology, as illustrated in Fig. 12.3.

12.8 The Challenges Ahead

12.8.1 Equipment and Tools. Chinese service companies have been importing equipment for last couple of decades, including fracturing pumping equipment, well logging tools, wellheads, tubulars, compression equipment, pumps, and meters. However, there are still several key tools needed to improve operation efficiency. Here we list a few of very important tools not yet existing in the Chinese market.

12.8.1.1 Precision Continuous Mixers. Hydraulic fracturing is the most critical technology required to economically develop unconventional gas resources. Until now, fracturing fluids have been batch-mixed in China, which can make it difficult to control fluid quality and can often lead to wasted fluid and chemicals. Most of the time, it is almost impossible to pump a multi-stage fracture treatment in a single well site in a period of several days. To improve treatments in China, we need to be using precision continuous mixers (PCMs). A PCM is a pumping and blending system that provides gelled fluid to the blender. The system simultaneously meters and continuously mixes polymer slurry, concentrated KCl solution, and other liquid additives used in fluids.

12.8.1.2 Snubbing Unit. A snubbing unit is a basic tool to control wellhead pressure for underbalanced completion and workover operations. In many cases, snubbing units are needed to complete wells or workover wells without having to kill the well with fluids that can damage the hydraulic fracture.

12.8.1.3 UBD Pressure Control. UBD equipment is not readily available in China. As discussed earlier, UBD can be very useful in low pressure, naturally fractured, unconventional reservoirs.

12.8.1.4 Compressors. Until recently, multi-stage compression in the field has not been applied in China, especially from wellheads to gathering stations. High flowing bottomhole pressure significantly reduces the EUR in unconventional wells. Wellhead compression is very important for those wells, especially in low pressure reservoirs. PCOC (a division of PetroChina operating in the Ordos Bain) has just started testing the wellhead compression concept. Low-cost compression equipment may be one of challenges to be overcome.

12.8.2 Maturity of Technologies.

12.8.2.1 Fracturing Fluids. Many fracturing fluid systems are comprised of complicated combinations of several chemicals that have to be mixed in very small quantities rapidly in the field. All fracturing fluid systems must be compatible with formation conditions, such as rock components, formation fluids, and formation temperature. The fluids also have to be adjustable with treatment sizes, balancing gel stabilization time and gel breaker systems. Many of the fluid systems have been developed in North America using proprietary chemical technology. Such technology is sometimes difficult to import and apply in some areas in the world, like China.

There have been some successes of local, Chinese versions of fracturing fluid systems. However, it has been a challenge to find cost-effective fracturing fluids in China, though a couple of international service companies can provide a few of the fluid systems used in the west, or some chemical additives. It is expected there are additional challenges ahead to provide proper and cost-effective fracturing fluids for large scale operations in China.

12.8.2.2 Operation Efficiency. One may find different completion tools in China now, such as different plugs, perforation guns, and coiled tubing. However, the operational efficiency of well completions and workovers is still a problem. There are also cost-efficiency issues that will need to be resolved.

12.8.3 Integration of Technologies. It is clear that no single technology can be applied everywhere in the world, or even in a single geologic basin. Many times, new technologies have to be developed to fit the needs of the field development plan and the infrastructure and economic limitations. It has been known that an integrated and systematic approach to technology application and field development are necessary to economically and efficiently develop any oil or gas field. It is even more true for unconventional gas resources. The economics of unconventional gas resource development is very sensitive to well placement, drilling and completion cost, and operation efficiency.

To successfully develop unconventional gas resources, we have to form teams. The teams consist of geologists, geophysicists, reservoir, drilling, completion, and production engineers, reservoir management experts, and economists.

In the oil and gas E&P sector in China, one may find most of the latest technologies, tools, techniques, and concepts being applied in North America. Occasionally, one can find some reports on the applications of latest technologies, such as horizontal wells, big fracturing treatments, and underbalanced drilling. However, since those technologies and techniques, like everything else, are not suitable to every single situation, successful applications are difficult to find in the literature. Large-scale application in everyday operation is rare.

12.8.4 Service Sector. In China, an open, independent, competitive, and mature service sector is very critical to the success of the oil and gas E&P industry, and even more important to efficiently and economically develop unconventional gas resources. Though there are a few, small independent private service companies in China, they cannot provide the large scale and efficient operational campaigns that are required to develop unconventional gas reservoirs.

International service companies can provide not only high-tech tools and chemicals, but also their experience from different basins around the world. Almost all international service companies have opened offices in China to provide service in China, and several of them have been providing services for many years to the on-shore oil and gas industry in China. However, most of time, they can only market special tools and chemicals, and/or provide very limited consulting services and well-site operations. International service companies provide essentially no field operations in the drilling and completion areas.

Currently, the majority of local service companies are a part of or are tied to one of the Chinese national oil companies. None of them are independent. One can argue the advantages and disadvantages of each oil company having its own service companies, but the system will probably not change much any time soon. Most Chinese service companies limit their activities to the local market where their oil company operates, which, in return, limits the growth and experience of the service company. There are not a lot of incentives to improve cost efficiency as there is no competition.

Since operational and cost efficiency is the key to developing unconventional gas resources, a fully open, competitive, and independent service industry would speed up the development of such gas resources in China.

12.8.5 Knowledge Sharing and Transfer. Though it is a very competitive market in North America, knowledge sharing often happens. In North America, there are many technical meetings, technical workshops, and local meeting presentations sponsored by professional organizations. Many of the meetings lately focus on the development of unconventional gas resources. Most of those meetings provide a great place to share best practices, lessons learned, real-world challenges, and new ideas for the future.

In China, knowledge sharing, routine technical training, and consulting businesses are still in the developmental stages. It will take time and additional effort to form a consulting and knowledge sharing atmosphere in China like those that exist in the western world.

12.8.6 The Challenges for Foreign Operators to Develop Unconventional Gas in China. China's on-shore unconventional gas resource has been open to foreign operators for many years. A couple of international oil companies are exploring several blocks in the Sichuan Basin (CZ Block and Zitong Block) and the Ordos Basin (Changbei Block, Sulige South Block, and CBM blocks located in the east range of the basin). Also, a new round of bidding activity for 9 blocks in the Tarim Basin is on-going. Some of those blocks are expected to contain tight gas sands.

Burlington Resources (now ConocoPhillips) has the rights to the CZ block in the Sichuan Basin and had been working on the block for about a decade. It has turned out that the biggest challenges were the gas price, operational and cost efficiency.

Natural gas prices are guided by the government, and the price is set by a local partner. Natural gas prices have been low compared with international markets, especially with the North America market (5 – 10 $/MSCF in North America vs. 2 – 3 $/MSCF in China recently). At such low natural gas prices, and low deliverability from tight gas wells, the inability to cover well and production costs discourages foreign exploration activities and makes it difficult for foreign operators to develop unconventional gas resources in China.

Also, the lack of equipment and special tools causes additional difficulty for foreign operators. We need to use the best technology to perform daily drilling, logging, and completion activity. In the last few years, Burlington Resources had to bring a snubbing unit into China to perform necessary well completion activities. The well completion cost increased dramatically due to the limited size operation with that equipment. When we studied the UBD feasibility, we found out that the largest hurdle was to find proper pressure control equipment. We determined that bringing it from North America was too expensive and there was no proper equipment available locally for UBD.

Since the Sichuan Basin is surrounded by mountains, the service market is physically isolated from the other areas in China. Sometimes, our operations have been greatly impacted by the availability of qualified service personnel and equipment. Operational efficiency problems have also negatively impacted cost efficiency.

Accessing data is a formidable task. We often found that gaining access to crucial field data is often either impossible or requires navigating near-impenetrable bureaucracies. Data from seismic and well logs and well test and production reports often are managed by a number of different subsidiaries within the national oil companies that do not routinely communicate with one another. These different bureaucracies can be possessive of their data. Accessing data in areas other than the contracted area is almost impossible.

12.9 Summary and Concluding Remarks

China needs new sources of energy to meet the demand for more energy. China has abundant volumes of gas in unconventional resources; however, we have found it to be a challenge to develop these resources. SPE 103775-MS has explained the main technologies needed to develop unconventional reservoirs, and the status of bringing those technologies to China.

We have discussed

- The impact of new technology on the development of unconventional natural gas reservoirs
- The technologies that are now available and those not yet unavailable in China for unconventional natural gas development
- Some of the problems facing international companies that wish to develop unconventional natural gas reservoirs in China

Economic growth in China demands significant growth in its energy supply. There are abundant volumes of unconventional natural gas resources in China. It is relatively certain that the development of unconventional gas reservoirs in China will soon gain momentum. There will be several tough challenges in the future with access to data, applying new technology, and managing costs. It will be critical to have a more open, competitive, and mature service sector to maximize development of the unconventional gas resources. Also, governmental regulations in pricing, tax, and downstream infrastructure will be required to be favorable to speed up the development of unconventional natural gas resources. It will be necessary to integrate cost effective technologies to improve operational efficiency, just like in every unconventional gas play in the world.

12.10 References

Bennion, D. Brant, Thomas, F. Brent and Bietz, Ronald F. 1996. Formation Damage and Horizontal Wells - A Productivity Killer? Presented at the International Conference on Horizontal Well Technology, Calgary, Alberta, Canada, 18–20 November. SPE-37138-MS. https://doi.org/10.2118/37138-MS.

Bloch, G., El Deeb, M. and Badaam, H. et al. 2003. Seismic Facies Analysis for Fracture Detection: a Powerful Technique. Presented at the Middle East Oil Show, Bahrain 5–8 April. SPE-81526-MS. https://doi.org/10.2118/81526-MS.

Caldwell, R. 2006. Are unconventional resources for real? *E&P*, 68–69.

Findley, M. and Shirley, K. 2006. Options for CBM production. *E&P*, 70–74.

Fisher, K. 2006. Barnett Shale Fracture Faireways Aid E&P. *World Oil*, 83–86.

Hawkes, R. V. and Hategan, F. 2004. Modern Concepts in Perforation Inflow Diagnostic (PID) Testing: A Safe, Green, and Cost-Effective Technique for Evaluating Pre-Frac Reservoir Parameters. Presented the SPE Annual Technical Conference and Exhibition, Houston, Texas, 26–29 September. SPE-90837-MS. https://doi.org/10.2118/90827-MS.

Holditch, S. A. 2006. Tight Gas Sands. *J Pet Technol* **58** (6): 86–94. SPE-103356-JPT. https://doi.org/10.2118/103356-JPT.

Kuuskraa, Vello A. 2006. Unconventional Natural Gas Industry Savoir Or Bridge? Presented at the EIA Energy Outlook and Modeling Conference, Washington, DC, 27 March.

Kuuskraa, V. and Ammer, J. 2004. Tight Gas Sand Development – How to Dramatically Improve Recovery Efficiency. *GasTIPS* **10:** 15–20.

Logan, J. and Chandler, W. 1998. Natural Gas Gains Momentum. US-China Business Council.

McCallister, T. 2000. Impact of unconventional gas technology in the annual energy outlook 2000. Energy Information Administration.

Pietz, D. 2000. China's Natural Gas Sector Generating Great Expectations. *Oil Gas J.* **99** (23).

Su, S. 2005. The Development of China's Natural Gas Market. Presented at the 6th US-China Oil and Gas Industry Forum, New Orleans, Louisiana, June.

Chapter 13

Investigating Hydraulic Fracturing in Tight Gas Sand and Shale Gas Reservoirs in the Cooper Basin

M. P. Scott and **T. Stephens,** Senex Energy; **R. H. Durant** and **J. M. McGowen,** Halliburton; **W. W. Thom,** Pinnacle—A Halliburton Service; **R. A. Woodroof,** ProTechnics Division of Core Laboratories

SPE 167073-MS presents a case study of hydraulic fracturing of three wells in the Cooper Basin of Australia and Queensland. Scott et al. collected data, which include the review of basin structure and offset wells, stress regime, hydraulic fracture treatments, causes of high tortuosity, treatment strategies, and fracture diagnostics. This study also includes results from the prefracture formation evaluation using core analysis, diagnostic injection tests on most intervals, surface tiltmeters to measure fracture azimuth and orientation, tracers to infer fracture height, and chemical tracers to estimate flowback contributions from each stage. A summary of drilling results and analysis of fracture treatment data is presented in detail through use of a 3D hydraulic fracture model, which is compared with results obtained using surface tiltmeters and tracers. The authors of the study conclude the following:

- Fracture treatments encountered high fracture gradients and tortuosity
- It is difficult to pump a slurry with proppant greater than 2 lbm/gallon
- 15% HCl acid and 100-mesh sand was not able to minimize near-wellbore pressure loss
- Lithology change was the main barrier to fracture height growth

13.1 Abstract

The majority of discoveries in the Cooper Basin have been structural traps on regional highs, with significantly less information available from the deep tight sand and shale plays. Recently, exploration and appraisal of the large, prospective gas resources in the deep troughs began. While these plays have been known for some time (Hillis et al. 2001), data has only recently been presented detailing the unconventional targets in the Nappamerri Trough (Pitkin et al. 2012). SPE 167073-MS focuses on unconventional reservoirs within the southern Cooper Basin, specifically, a southern extension of the Nappamerri Trough and the Mettika Embayment, part of the Tenappera Trough.

Senex Energy began to review the potential of the unconventional reservoirs in its Cooper Basin acreage in 2010. SPE 167073-MS focuses on three wells, Skipton 1, Kingston Rule 1, and Hornet 1, which were hydraulically fractured in early 2013. In the well planning stage of the project, it was identified that sufficient data needed to be collected to characterize both the formation and hydraulic fracturing behavior. The requirement for fracturing diagnostics was critical because of the complex nature of hydraulic fracturing within the Cooper Basin. To manage these unknowns, the data that were collected included core testing, diagnostic injection tests on most intervals, surface tiltmeters to measure fracture azimuth and orientation, radioactive (RA) tracers to infer fracture height, and chemical tracers to estimate the flowback contribution from each stage.

The results showed high treating pressures and near-wellbore pressure loss (NWBPL) in many stages of the fracture treatments. Generally, the lower the fracture gradient, the easier the job was to place and the better the reservoir quality. The RA tracers showed good containment in the one well in which they were used, with heights of 20 to 60 ft. The main barriers to height growth appeared to be changes in lithology. The surface tiltmeter results showed an azimuth that was different than expected from borehole breakout data in the basin. This has been seen in other studies in the Cooper Basin. While the exact reason for this requires additional work, possible explanations are discussed within the paper.

13.2 Introduction

The Cooper Basin is a Late Carboniferous to Middle Triassic, non-marine sedimentary basin straddling the border of South Australia and Queensland. The significantly larger, Jurassic to Cretaceous Mesozoic Eromanga Basin overlies the Cooper Basin. (Gravestock et al. 1998) The stratigraphy of the Cooper Basin is shown in **Fig. 13.1.** The Permian section has numerous producing sandstones, which are the Toolachee, Daralingie, Epsilon, Patchawarra, and Tirrawarra, while there are other prospective unconventional formations, such as the Roseneath Shale and Murteree Shale.

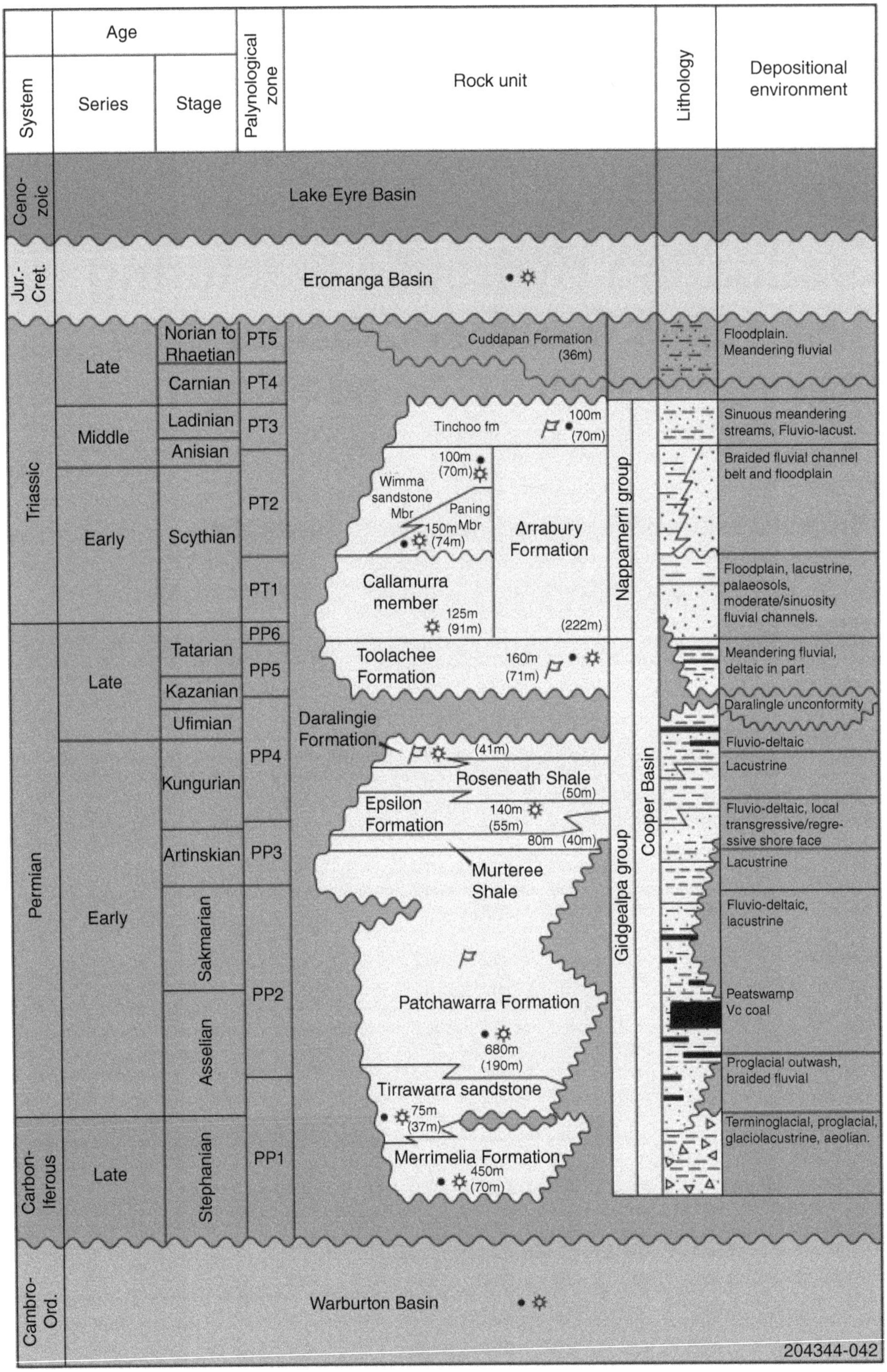

Fig. 13.1—Geological summary of the Cooper Basin (from Menpes 2012).

A map of the Cooper Basin showing the major geological features is presented in **Fig. 13.2.** It is comprised of a series of troughs and ridges that run northeast to southwest. The region is a significant source of Australia's onshore oil and gas, with production starting in the late 1960s. Since that time, a significant amount of geological, reservoir, and production data has been collected. However, to date, the majority of discoveries have been structural traps on regional highs with limited information available from the deep tight gas sand and shale plays.

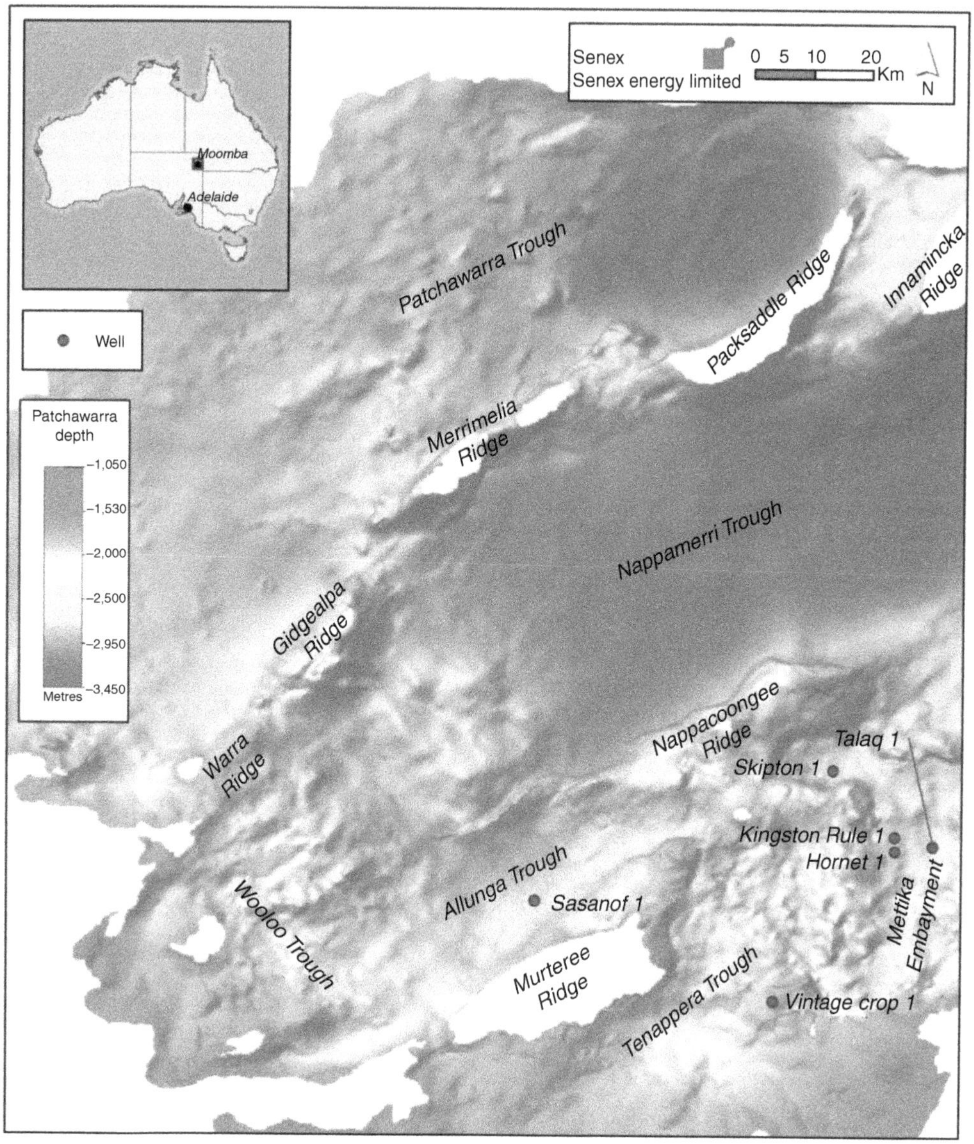

Fig. 13.2—Location of study area and major geological features within the Cooper Basin.

Senex Energy began to review the potential of the unconventional reservoirs in its Cooper Basin acreage in 2010. In May 2011, it began its exploration program with the drilling of Vintage Crop 1. This well was primarily drilled to test the oil potential of the Early to late Cretaceous and mid-Jurassic formations of the Eromanga Basin, but core was also cut from coal in the Toolachee Formation, Roseneath Shale, and Murteree Shale. The exploration program continued in 2012 with the drilling of four more wells. These wells, along with the dates of when they were drilled, the hydraulic fracturing dates, and associated data collected on each well, are listed in **Table 13.1.** To gain an understanding of what was occurring during the hydraulic fracturing, a large program of data collection was implemented. This included surface tiltmeters, RA tracers, and chemical tracers along with rock testing and diagnostic fracture injection tests (DFITs). In addition to the four wells drilled, another well, Hornet 1, which was previously drilled by Victoria Petroleum (later renamed Senex Energy), was fracture stimulated as part of the program. SPE 167073-MS predominantly concentrates on data from three wells in the program—Skipton 1, Kingston Rule 1, and Hornet 1.

Well	Spud Date	Core	Fracture Stimulation	Tiltmeters	RA Tracers	Chemical Tracers
Sasanof 1	January 2012	Yes	May 2012	No	No	No
Talaq 1	April 2012	Yes	N/A	No	No	No
Skipton 1	August 2012	Yes	February 2013	Yes	Yes	Yes
Kingston Rule 1	October 2012	Yes	February 2013	No	No	Yes
Hornet 1	August 2004	No	March 2013	No	No	Yes

Table 13.1—Senex exploration program.

13.3 Cooper Basin Review

There has been extensive development of conventional oil and gas accumulations within the Cooper Basin. Until recently, only limited testing of the productivity of the unconventional gas potential in the deep troughs had been undertaken. In a shift toward unlocking the unconventional resource potential of the Cooper Basin, operating companies have begun testing these plays, predominantly within the Nappamerri Trough (Pitkin et al. 2012; Dunlop 2012). The work in SPE 167073-MS focuses on the southern Cooper Basin, particularly the tight gas and shale gas potential of the Permian sequence outside the central Nappamerri Trough region. These regions are a southern extension of the Nappamerri Trough, referred to here as the Skipton Trough, and the Mettika Embayment, part of the Tenappera Trough.

13.3.1 Lithology and Description. The Cooper Basin is primarily a fluvial-lacustrine sequence consisting of interbedded sandstones, siltstones, shales, and coals. The depositional environments represented by the Cooper Basin fill consist of both braided and meandering fluvial sands, alluvial fans, distributary channels, crevasse splays, and levee deposits (Gravestock et al. 1998). The Early Permian sequence was the primary focus for this work. These units are described below.

13.3.1.1 Roseneath Shale. The Roseneath Shale consists of fine-grained interbedded siltstone, mudstone, and minor sandstone. It is not as extensive throughout the basin as the Murteree Shale because of either non-deposition or erosion. It reaches a maximum thickness of approximately 320 ft and is thickest in the Nappamerri and Tenappera Troughs (Gravestock et al. 1998).

13.3.1.2 Epsilon Formation. The Epsilon Formation is comprised of fine- to medium-grained, quartose sandstone with carbonaceous siltstone, shale, and minor coal interbedded (Gravestock et al. 1998).

13.3.1.3 Murteree Shale. The Murteree Shale is comprised of argillaceous siltstone and fine-grained sandstone. It is relatively constant in thickness, reaching a maximum of approximately 260 ft in the Nappamerri Trough (Gravestock et al. 1998).

13.3.1.4 Patchawarra Formation. The Patchawarra Formation is the thickest Early Permian unit, reaching up to 2230 ft in the Napparmerri Trough. It consists of interbedded sandstone, siltstone, shale, and coal. The unit is interpreted to have been deposited under lacustrine, deltaic, and floodplain conditions, which is reflected in the multistory fluvial channel, point bar, and crevasse splay deposits (Gravestock et al. 1998).

13.3.2 Structural Features and Offset Wells. The wells forming the basis of this investigation are located south of the central Nappamerri Trough, as shown in **Figs. 13.2 and 13.3.** Skipton 1 sits in the deepest portion of a smaller early Permian depocentre, referred to here as the Skipton Trough and shown in Fig. 13.3. This basinal depression was a focus for sediment deposition as fluvial systems drained into the Nappamerri Trough. The thick Permian coals in the Skipton Trough are interpreted to be mature for hydrocarbon expulsion and contributed to the hydrocarbon charge of the surrounding fields. While this area is encircled by producing fields, including Dullingari to the east, Kidman to the south, and Della to the west, only a small portion of the existing well stock has been located off structure, specifically targeting sediments within the trough. The nearest offset wells are Katingawa 1, Gudnuki 2, and the western Dullingari field wells. These wells have indicated the possibility that stratigraphic and basin-centered gas accumulations exist outside of the previously drilled structural closures. Well overviews for the offset wells are provided next.

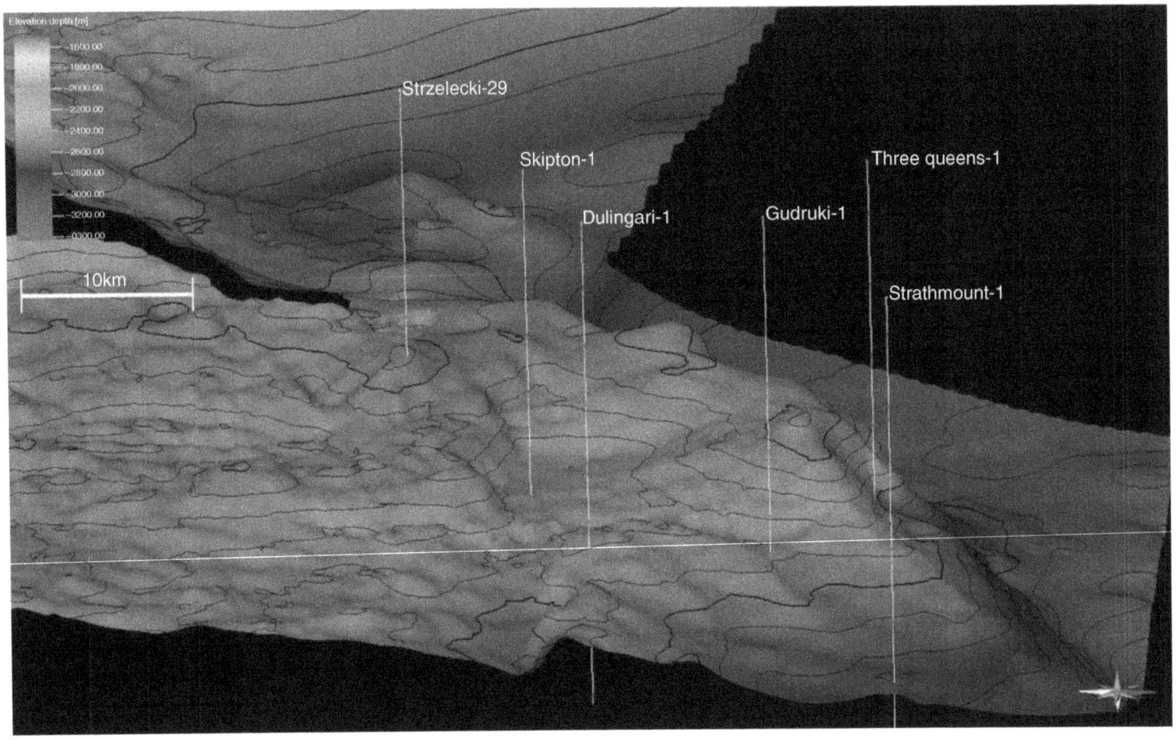

Fig. 13.3—Skipton Trough and surrounding wells.

13.3.2.1 Katingawa 1. Katingawa 1 was drilled by Santos in 1992, targeting a fault-dependent three-way closure at the Patchawarra level. The well was drilled on the downthrown side of a basement offsetting fault, on the northern edge of the Kidman complex. No valid trapping mechanism was interpreted for the younger strata at the time of drilling. Three drillstem tests (DSTs) were conducted, and the results were integrated with petrophysical analysis. The reservoir sandstones were interpreted to be gas saturated but too tight to yield economic rates. The well was plugged and abandoned before casing was set (Bilek 1992).

13.3.2.2 Gudnuki 2. Gudnuki 2 was drilled by Santos in 1998 and is located to the north of the Skipton Trough, on the flanks of the Nappacoongee Ridge. The primary objective of the well was to test the prospectivity of the western extent of the conventional four-way closure tested at Gudnuki 1. Predrill targets were interpreted to be sands developed within the Patchawarra, Epsilon, and Toolachee Formations. One DST conducted in the Patchawarra Formation achieved a stabilized gas flow rate of 0.016 MMscf/D. The well was plugged and abandoned before casing was set, as the DST suggested uneconomic flow rates. The well sustained gas to surface outside of structural closure and proved the existence of reservoir-quality sandstones, with log-derived porosities up to 12% developed in the Patchawarra Formation (Haak 1998).

13.3.2.3 Western Dullingari Field Wells. The basement ridges both surrounding and confining the Skipton Trough (Fig. 13.2) are interpreted to have been present at the time of Early Permian deposition. Well correlations and seismic interpretation of the Patchawarra sequence in the vicinity of the Dullingari high suggest this feature is most likely a post-Permian inversion structure. The wells in the Dullingari Field primarily target the Cretaceous to late-Jurassic oil window, though some wells have penetrated and produced from the Patchawarra, including Dullingari 45, Dullingari 16, and Dullingari 46. Dullingari 45 intersected the Patchawarra Formation and, although poor hole conditions and a limited wireline suite made quantitative analysis difficult, the reservoir interval was interpreted to be unproductive (Nugent 1989). Dullingari 16 was drilled as a gas well and completed as a commingled producer from the Toolachee, Epsilon, and Patchawarra Formations (Courtney and Butler 1982). Dullingari 46 flowed 1.67 MMscf/D from two Patchawarra Sands; however, none of the other sands developed within the sequence were tested. The well was completed as a dual Murta oil and Patchawarra gas producer (Allen 1990).

13.3.2.4 Mettika Embayment. Both Kingston Rule 1 and Hornet 1 were drilled within the Mettika Embayment (Fig. 13.2). In a similar style to the Skipton Trough, the Mettika Embayment is interpreted as being depositional and low-flanked by basement highs present at the time of deposition of the Early Permian units. The Mettika Embayment represents the northern extension of the Tenappera Trough. Hornet 1 was originally sighted by Victoria Petroleum for Jurassic oil potential, but it was deepened to test the stratigraphic trapping potential at the Patchawarra level. A DST over the Upper Patchawarra Sands produced gas to surface at a rate too small to measure, and the well was left cased and suspended. Kingston Rule 1 was drilled by Senex Energy in 2012 to evaluate the shale gas potential and tight gas potential in the Mettika Embayment as part of a corehole program. These wells form an integral part of this study, and the results will subsequently be discussed.

The key offset wells for the Mettika Embayment include Allambi 1, Plantago 1, Kelbrook 1, Boongala 1 and 2, Verbena 1, and the Toolachee Gas Field wells. While these wells were originally located on conventional, structural four-way closures, well results indicate the possible existence of a stratigraphic trapping component to the gas accumulations in the region.

13.3.2.5 Allambi 1. Allambi 1 was drilled by Santos in 1994 on what was thought to be a subtle independent four-way dip closure. The primary objective of the well was to test the Patchawarra Formation, with the Epsilon and Toolachee Formations defined as secondary objectives (Lowman 1995). The well has produced a cumulative volume of 7.35Bcf from the Patchawarra Formation.

13.3.2.6 Kelbrook 1. Kelbrook 1 was drilled by Santos in 1998, targeting a fault-dependent three-way dip closure at the Patchawarra level. While the well intersected good reservoir sand development within the Patchawarra Formation, it was concluded that there was poor evidence for gas-saturated sandstones, and the well was subsequently plugged and abandoned (Ostler 1999).

13.3.2.7 Verbena 1. Verbena 1 was drilled by Santos in 1990, targeting a four-way dip closure connected to the west by means of a spill point to the main Toolachee Field. The primary objective of the well was to test the prospectivity of the Patchawarra Formation. Minor gas shows were observed within the Patchawarra, and a DST over sands developed lower within the sequence flowed gas to surface at a rate too small to measure. The well was plugged and abandoned, with the sands deemed both tight and unproductive (Allen 1991).

13.3.3 Stress Regime. The stress regime within the Cooper Basin has been studied by various authors (Hillis et al. 1998; Hillis et al. 2001; Hillis and Reynolds 2003; Reynolds et al. 2005; Reynolds et al. 2006; Nelson et al. 2007a) based on input from density logs, image logs, injection tests, and wellbore deformation modeling. Although there are variations based on depth, lithology, and localized faulting, the current Cooper Basin stress regime is best characterized as strike-slip ($\sigma_H > \sigma_v > \sigma_h$), with minimum horizontal stresses of 0.6 to 1.0 psi/ft, overburden of 0.95 to 1.05 psi/ft, and maximum horizontal stress of >1.85 psi/ft (Nelson et al. 2007a).

While the direction of absolute plate velocity of the Indo-Australian Plate (IAP) is north-northeast, the maximum horizontal stress orientation, as observed from borehole breakout data, is observed predominantly east-west, as shown in **Fig. 13.4.** The grey shading in this figure indicates a horseshoe-shaped maximum stress rotation across the central eastern portion of Australia that is difficult to explain using plate-scale tectonic forces. This maximum horizontal stress orientation from borehole breakout is further expanded in **Fig. 13.5,** showing the local variations that can occur around the Cooper Basin.

13.3.4 Hydraulic Fracturing. McGowen et al. (2007) present an extensive review of the previous hydraulic fracturing experience in the Cooper Basin, summarizing the initial 650 treatments in the basin covering 1969 to 2006. Since that paper was published, over 600 more treatments have been performed in the conventional fields (i.e., located on stratigraphic traps) plus the expansion of another ~100 treatments in the deeper troughs (targeting basin-centered gas or shale gas), such as those described in SPE 167073-MS and by Pitkin et al. (2012).

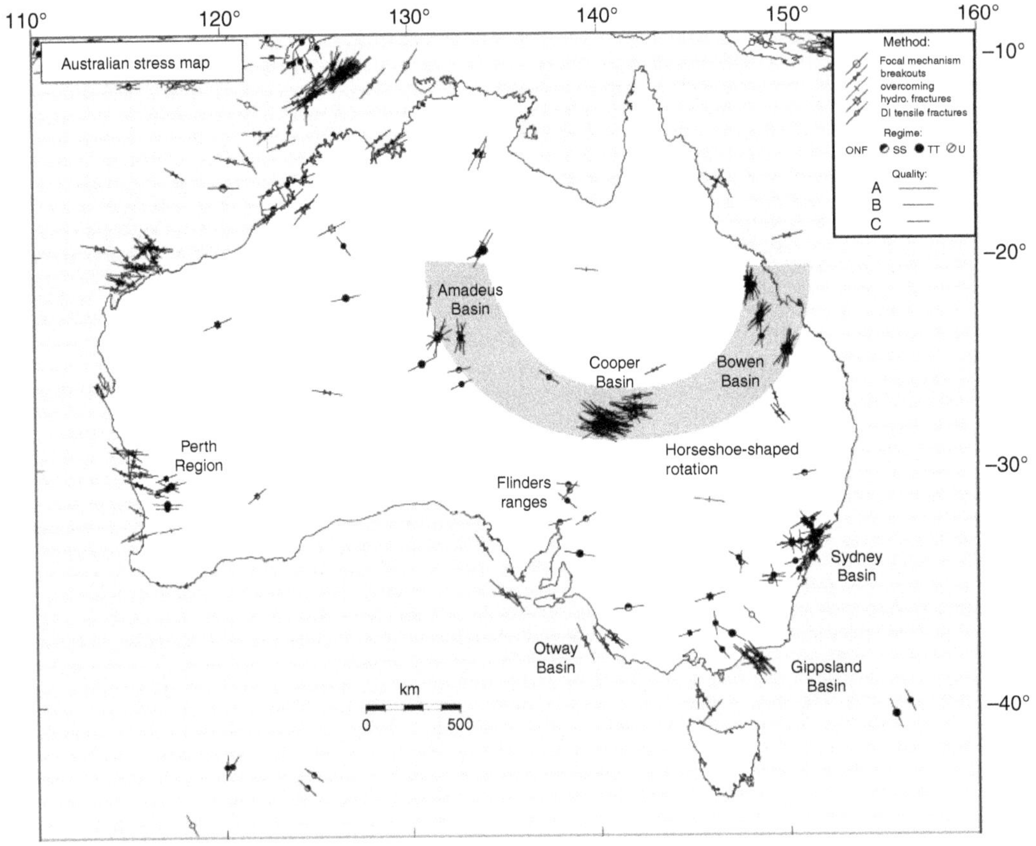

Fig. 13.4—Maximum stress orientations from borehole breakout data in Australia (from Reynolds et al. 2005).

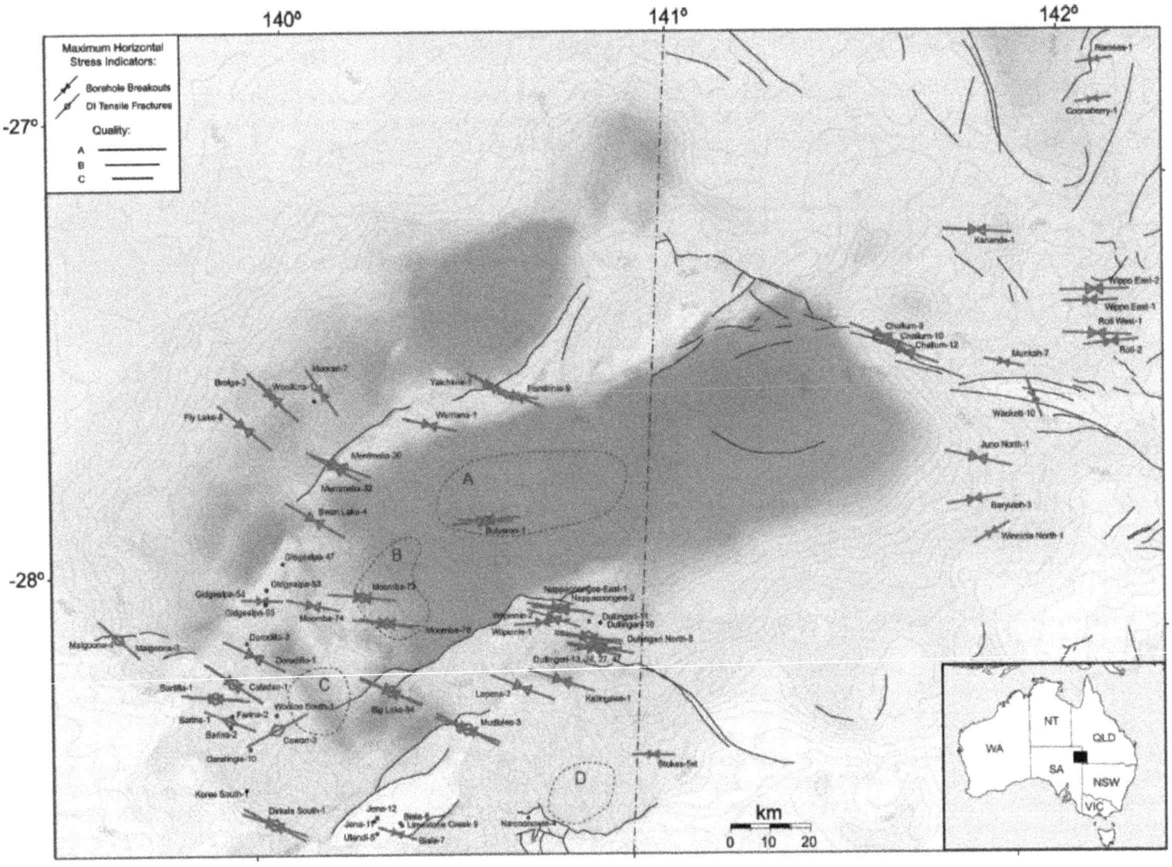

Fig. 13.5—Maximum horizontal stress orientation from borehole breakout data in Cooper Basin (from Reynolds et al. 2005).

The principal issue for successful application of fracturing in this basin is overcoming fracture complexities caused by the stress regime described previously, which is commonly observed by high fracture gradients, high tortuosity, or high pressure-dependent leakoff (PDL). The large maximum horizontal stresses result in elevated fracturing pressures that follow local trends because of faulting, reservoir quality, and depth. Low fracture gradients of approximately 0.8 psi/ft can be obtained in the better-quality shallower formations; however, fracture gradients can range from 0.9 to 1.3 psi/ft as the reservoir quality reduces or formation depth increases. Besides having variable fracturing pressures, it is common to also observe high values of tortuosity in most treatments within the basin. While it would be expected to think that high tortuosity is related to high stresses, no apparent relationship between the fracture gradient and tortuosity was observed by McGowen et al. (2007). Lastly, McGowen et al. (2007) also indicate that PDL was routinely observed in over 65% of treatments. Observing a high fracture gradient, high tortuosity, or high PDL affects treatment designs, as the screenout risk is significantly increased. Some of the main strategies to combat these effects are highlighted below.

13.3.4.1 Causes for High Tortuosity. High tortuosity is most likely the main contributing factor that results in poor treatment placement within the Cooper Basin. The causes for the high tortuosity have been postulated by various authors and are generally described as the following:

- Fracture turning where the fracture remains principally vertical but is forced to counteract the maximum horizontal stress in the NWB region as it reorients after initiating from an unfavorable direction (Chipperfield et al. 2000; Roberts et al. 2000). One cause for this undesired fracture reorientation (not specifically highlighted by these authors) is the expected influence of borehole breakout. Borehole breakout causes the initial wellbore failure (and thus expected fracture initiation) in the minimum horizontal stress orientation and perpendicular to far-field fracture orientation. The fracture would remain principally vertical but would be forced to counteract the maximum horizontal stress in the NWB region. This tortuosity caused by this reorientation would be more severe in areas of severe borehole breakout that are normally observed in hard brittle formations (mainly sandstones) with high deviatoric stress (difference between minimum and maximum horizontal stresses).
- Fracture reorientation near the wellbore caused by generation of vertical fractures at the wellbore and horizontal fractures in the far-field (Chipperfield et al. 2000; Roberts et al. 2000). This can occur if the wellbore orientation is not aligned with the far-field orientation. In vertical wells, this could occur when injection pressures are enough to lift the overburden, most likely in the presence of horizontal natural fractures or laminations.
- Generation of multiple fractures in the NWB region that cause elevated injection pressures. Nelson et al. (2007b) document Cooper Basin cases having high tortuosity, high process zone stress, and high PDL being associated with numerous fabrics dilated at the wellbore wall (from image logs) and high rock tensile strengths (measurements up to 2,176 psi). It was proposed that geological weaknesses inherent in the reservoir and high intact rock tensile strengths were resulting in complex fractures and fractures not opening against the minimum in-situ stress.
- Fracture migration from a layer of high stress at the wellbore into a layer of low stress away from the wellbore (Roberts et al. 2000).

13.3.4.2 Treatment Strategies in the Cooper Basin. There has been a progression in the rationale to tackle the difficult fracturing conditions of the Cooper Basin. There have been numerous papers (Roberts et al. 2000; Johnson et al. 2002; Nelson et al 2007a; and Pitkin et al. 2012) that reviewed these conditions and present recommendations. Some of the main points are summarized below:

- As discussed by Roberts et al. (2000) and Chipperfield et al. (2000), the technique to manage excessive tortuosity was dependent on the fracture gradient. Excessive tortuosity was defined as a value greater than 900 psi for borate-crosslinked gels attempting to place 20/40-mesh proppants at 6 to 8 lbm/gal.
    - For a fracture gradient above 1.05 psi/ft, excessive tortuosity required mitigation, as the condition was deemed extreme. The main method suggested was to produce the interval to promote reservoir depletion, which was observed to reduce the tortuosity and fracture gradient and allow remediation. Below a fracture gradient of 1.05 psi/ft, remediation strategies (such as high-viscosity gel slugs) were observed to be successful in reducing tortuosity to tolerable levels.
    - If the tortuosity was still excessive after remediation attempts, then a tolerance design was recommended that included 100-mesh proppant in the pad, higher-viscosity crosslinked gels, and a more conservative proppant schedule.
    - If the tortuosity was not excessive, then the job proceeded as normal with the designed objective of placing higher concentrations.
- Johnson et al. (2002) review combating high fracture pressures, high tortuosity, and high PDL in high-permeability wells in the Cooper Basin. SPE 167073-MS highlights several methods that were reported to successfully reduce the tortuosity values impacting job placement:
    - Trialing oriented perforations.
    - Initiating the fracture using viscous crosslinked fluid containing 100-mesh sand.
    - Performing the DFIT injections using crosslinked gels and 100-mesh.
    - Flowing back the well after each DFIT to help reduce well pressure before the next injection.
    - Performing the fracture treatment and successively increasing the proppant size (starting with 100-mesh) and concentration until obtaining the desired size and concentration.
- McGowen et al. (2007) state that, around 2002, operations in the basin moved away from reducing placement risk toward improving reservoir evaluation. In general, strategies were to reduce the amount of time and expense on remediating the excessive tortuosity unless the value exceeded 2,000 psi. Treatment designs often incorporated some of the tolerance methods within the standard designs, based on the results of evaluation of initial DFITs.
    - If the fracture gradient was extreme (i.e., above a value of 1.1 psi/ft), a recommended course of action was to move to a low-viscosity slickwater or hybrid-type treatment. High gradients were normally associated with low permeabilities observed from the DFITs, thus making low-viscosity treatments applicable.
    - Incorporate the use of 100-mesh sand within the pad of most treatments to help reduce tortuosity and PDL.
    - Tailor the proppant schedule based on the permeability from the DFIT and the value of the tortuosity observed.

- Pitkin et al. (2012) was the first to specifically address fracturing in the deep Nappamerri Trough. SPE 167073-MS describes multiple methods to handle this difficult area with very high fracturing pressure and high tortuosity.
  - Completions incorporated acid-soluble cement to improve the application of acid spearheads to help reduce treating pressures. Acid was observed to be helping in some but not all of the treatments.
  - Used large slickwater volumes and large 100-mesh proppant stages to reduce treating pressures and tortuosity. This was reported as the most successful method, resulting in significant reductions in most cases.
  - Used smaller-diameter proppants along with linear and crosslinked gels to aid in treatment placement.

13.3.4.3 Fracturing Diagnostics. Only a limited scope of passive diagnostics, such as microseismic and tiltmeter surveys, has been performed in the Cooper Basin. Asanuma et al. (2004) show results from the stimulation of basement granite in the Cooper Basin for geothermal purposes. Over 5.3 million gallons of liquid were injected over a three week period, with a massive seismic cloud detected from an eight-station seismic array consisting of shallow (300 to 1,200 ft) stations and one deeper station (5,900 ft). The seismic cloud showed elongation along a NNE to SSW direction with considerable width and complexity. The results were defined as a sub-horizontal distribution and represented one of the biggest induced fracture networks ever mapped.

Camac et al. (2012) present fracture stimulation results in the Patchawarra Formation using downhole microseismic monitoring in the Greater Tindilpie area of the Cooper Basin. It was shown that the fracture azimuth of N64°W aligned with borehole breakout results in the area. It also showed varying fracture heights but, in general, there was a high degree of height containment in the mid to Upper Patchawarra targets. Asymmetry observed in the microseismic event distribution is linked to fluvial deposition models developed for the area. Pitkin et al. (2012) show the results from two surface tiltmeter surveys in the Napparmerri Trough of the Cooper Basin. The borehole breakout data shows an E-W orientation for maximum horizontal stress, but the far-field tiltmeter data show an induced fracture azimuth in a NW to NNW direction. All three of these papers show significant variation in the induced fracture azimuth across the basin. Two of the papers (Asanuma et al. 2004; Pitkin et al. 2012) show disagreement with expected regional stress and borehole breakout indicators. Variation between fracture azimuth orientations as compared to expected maximum stress orientation from borehole indicators has been previously observed in other basins by Walker et al. (2002) and Wolhart et al. (2000).

13.4 Prefracture Stimulation Planning

In the well planning stage of the project, it was identified that data needed to be collected to characterize both the formation and hydraulic fracturing behavior. The requirement for fracturing diagnostics was critical because of the complex nature of hydraulic fracturing within the Cooper Basin. To manage these unknowns, a large amount of data was collected, including core testing, diagnostic injection tests on most intervals, surface tiltmeters to measure fracture azimuth and orientation, RA tracers to infer fracture height, and chemical tracers to estimate the flowback contribution from each stage.

13.4.1 Petrophysics, Image Logs, and Core Testing. As part of the program, core and dipole sonic as well as acoustic and resistivity image logs were acquired on newly drilled wells. Reservoir pressure estimates were also obtained from a wireline formation tester on Skipton 1, Kingston Rule 1, and Hornet 1. An example log of the Skipton 1 well is provided in **Fig. 13.6.**

Integrating the density log from multiple wells in this work has shown a value of overburden of approximately 1.06 psi/ft at a depth of 9,200 ft. The borehole breakout results from this work were consistent with the regional trend, with the image logs showing a general trend of east-west. Skipton 1 and Kingston Rule 1 both showed significant wellbore breakout, with a maximum horizontal stress direction of N85°W ± 10°.

An extensive suite of core testing was conducted on both the sand and shale. This included rock mechanics, shale rock properties, fluid compatibility studies, and routine and special core analysis (porosity-permeability-grain density, capillary pressure, formation resistivity, and relative permeability).

13.4.2 Well Completion Design. Skipton 1 and Kingston Rule 1 had 5 1/2-in. 26.8 lb/ft P-110 production casing and were stimulated down the casing; a schematic of Skipton 1 is provided in **Fig. 13.7.** While a significant number of wells have used pinpoint fracturing in the basin (Gilbert and Greenstreet 2005; Beatty et al. 2007), a simple plug-and-perforate technique was used in Senex's early exploration phase. One benefit this provided was higher treating rates. Each stage was wireline perforated with 25.0-g shaped charges and isolated from the previous zone using a composite bridge plug. The use of select fire subs allowed the perforating guns and bridge plug to be combined in a single run.

Hornet 1 was an older well drilled in 2004 by Victoria Petroleum and was completed with 7-in. 23 lb/ft N-80 casing. This casing was not sufficiently pressure rated to complete the hydraulic fracturing. A multistage packer and sliding sleeve assembly run on 3 1/2-in. tubing was installed before the job (schematic provided in Fig. 13.7). All stages were perforated before running the completion assembly. The perforating charges used in this well were 22.5 g, with a larger penetration and smaller entry hole size.

13.4.3 Injection Tests. Previous experience in deeper parts of the Cooper Basin has shown high treating pressures and difficulty placing proppant, as detailed by Pitkin et al. (2012). Because of concerns with the difficulty of placing proppant and to help improve understanding of the target formations, injection tests with extended declines and step down tests were performed before each treatment. Throughout the basin, injection tests usually consisted of a large crosslink gelled-water minifrac followed by a slickwater step rate test or a smaller crosslink gelled-water minifrac with a step down test (McGowen et al. 2007). To reduce the time spent monitoring pressure, the injection tests of friction-reduced water were pumped at 5 bbl/min and generally had a volume of less than 500 gal. Once a breakdown occurred and the pressure stabilized, pumping ceased and the well was shut in. Although pressure declines were monitored for multiple hours (often overnight), pseudo-radial flow was not reached on many of the stages.

Before the main fracturing treatment, a second higher-rate, larger (5,000 to 20,000 gal) minifrac was performed. This was performed using friction-reduced water that also contained a 15% hydrochloric (HCl) acid spearhead, concluding with a step down test. The step down test examined the NWBPL and was used to estimate tortuosity and perforation friction. An example pumping chart is shown in **Fig. 13.8a.** Injection and step down tests were performed on every stage. The length of the shut in for the injection test was dependent on operational timing, with times for each stage varying from approximately 30 to 800 min.

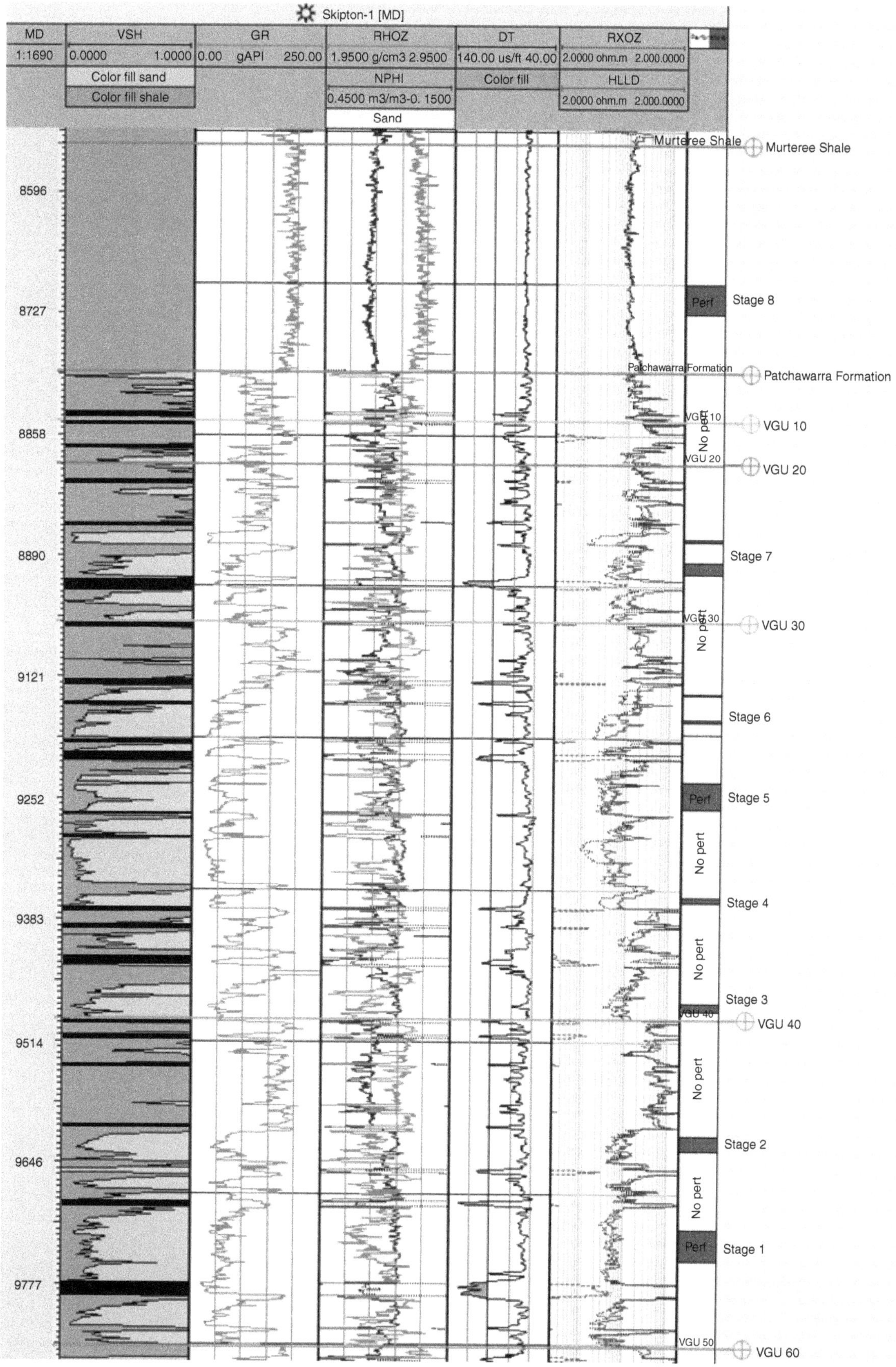

Fig. 13.6—Log of Skipton 1.

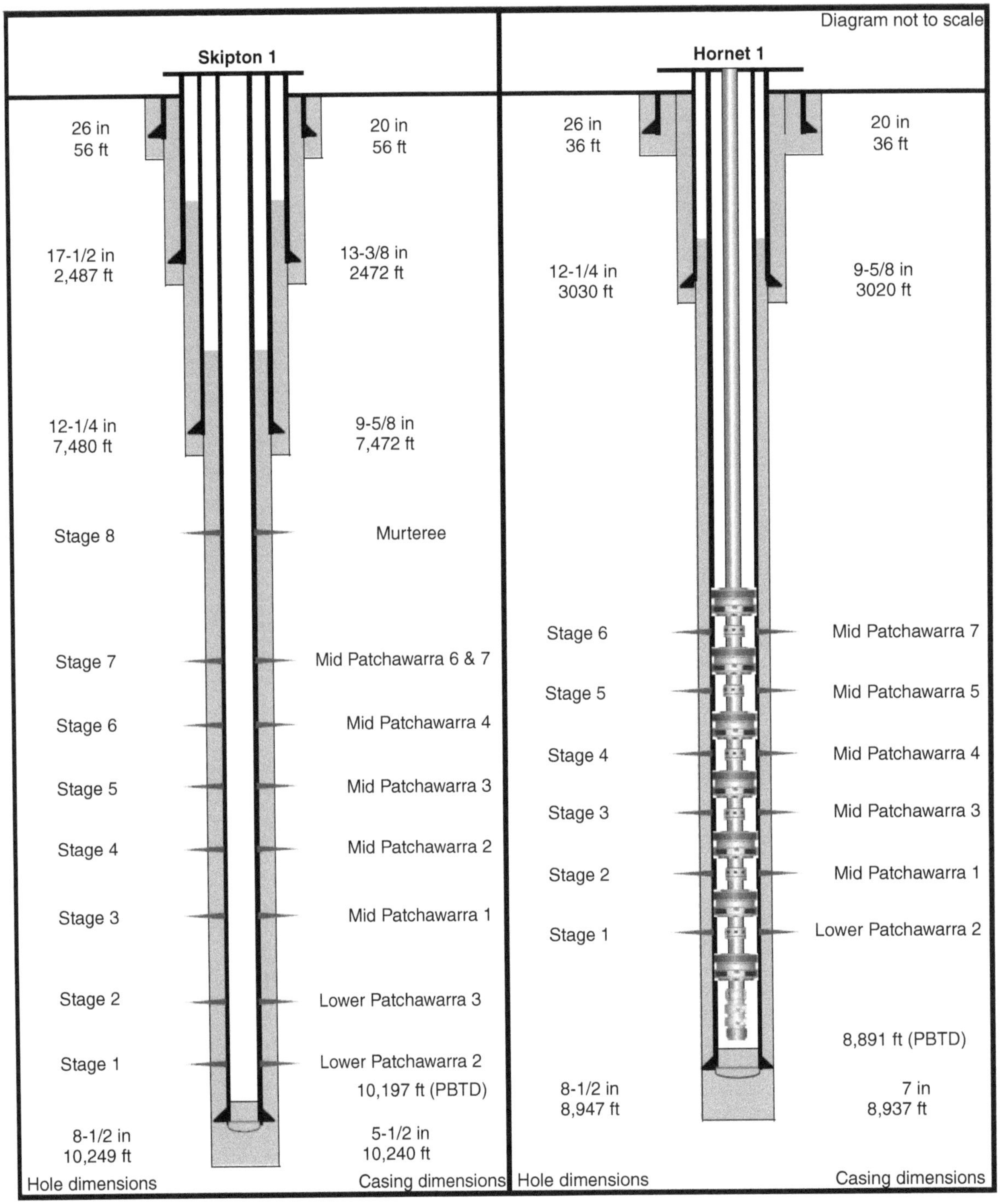

Fig. 13.7—Completion diagrams for Skipton 1 and Hornet 1.

These tests gathered the instantaneous shut-in pressure (ISIP), closure pressure, and leakoff behavior of the fracture. If the shut-in period was sufficient, the reservoir pressure and permeability were estimated from after closure analysis (ACA) (Gu et al. 1993; Nolte et al. 1997; Chipperfield and Britt 2000).

13.4.4 Fracture Stimulation Design. An indicative pumping profile for a shale interval in Skipton 1 is shown in Fig. 13.8b. The shale intervals were stimulated with hybrid jobs consisting of friction-reduced water with alternating sand stages and water sweeps with increasing sand size (100-mesh, then 40/70-mesh intermediate strength proppant [ISP]) as well as increasing concentration (up to 0.75 lbm/gal), followed by a 20-lbm/1,000 gal linear gelled water stage with an increasing concentration of 40/70-mesh ISP.

An indicative pumping profile for a sand interval in Skipton 1 is shown in Fig. 13.8c. The fracture stimulations of the sand intervals typically consisted of a friction-reduced water pad with low concentrations of 100-mesh sand followed by 20-lbm/1,000 gal linear gelled water and 30-lbm/1,000 gal borate-crosslink gelled water with 30/50-mesh ISP and 20/40-mesh ISP. The 100-mesh sand has been used throughout the basin to help reduce NWBPL (Pitkin et al. 2012) and to alleviate PDL issues by eroding the NWB region and plugging off leakoff paths (Johnson et al. 2002).

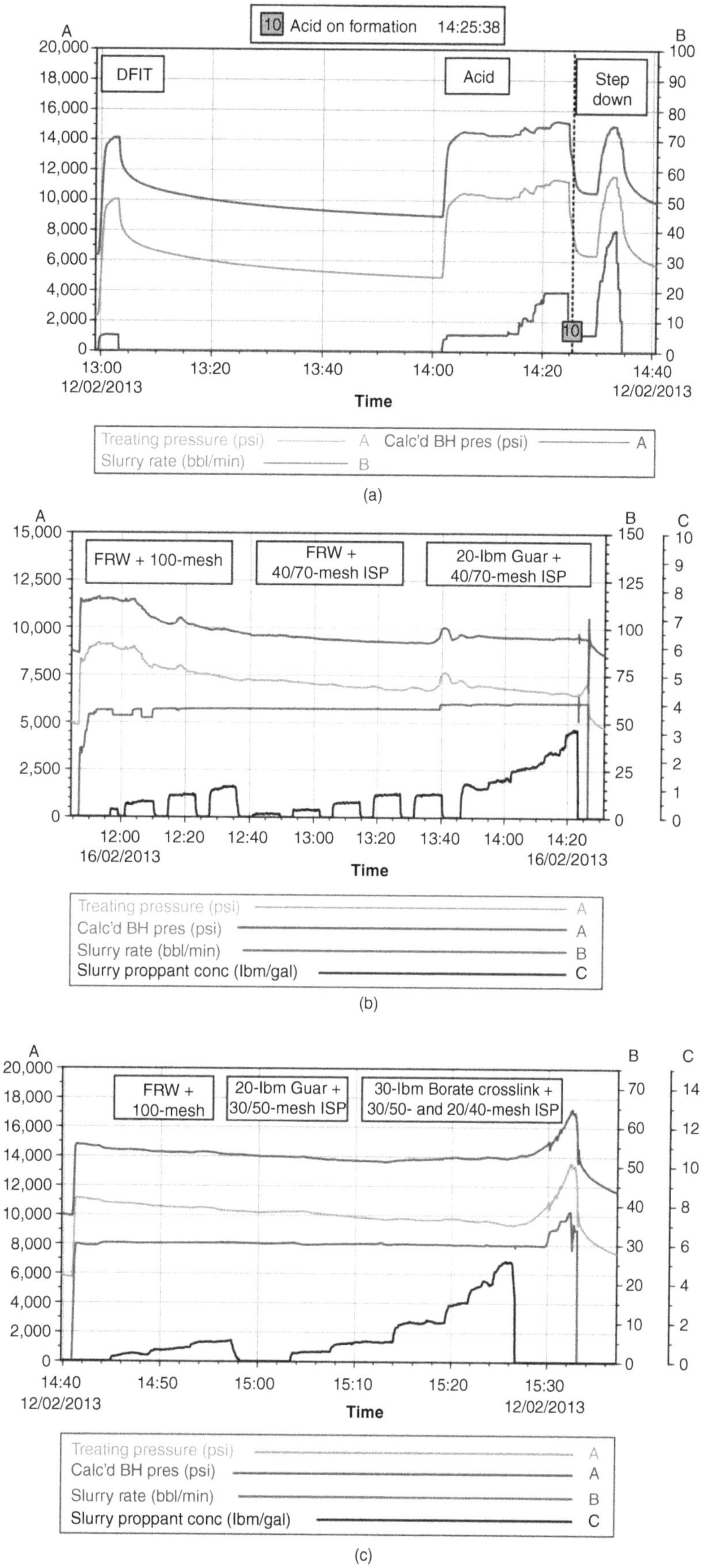

Fig. 13.8—Representative hydraulic fracture treatment: (a) injection test, 15% acid spearhead and step down test, Skipton 1 Stage 3; (b) shale interval, Skipton 1 Stage 8; (c) sand interval, Skipton 1 Stage 3.

13.4.5 Surface Tiltmeter Design. Surface tiltmeters with their nano-radian resolution were used to characterize the hydraulic fracture created during the stimulation treatments on Skipton 1. Surface tiltmeters do not require an offset wellbore to map a fracture, making them an attractive option in remote and low-well-density areas, particularly in Australia, where drilling costs are considerably higher than in North America. A 42-site tiltmeter array was used and setup in a circular pattern around Skipton 1, covering approximately 6 square miles. Each site consisted of a shallow 40-ft borehole cased and cemented with polyvinyl chloride to decouple the sensitive tools from ambient noise sources. The aims of the surface tiltmeter monitoring during this program was to determine fracture far-field azimuth and volumetric components of the created fracture system. Further details on surface tiltmeter theory associated with hydraulic fracture mapping can be found in Siebrits et al. (2000), Wright and Conant (1995), and Wright et al. (1998).

13.4.6 Chemical Tracers. On the three wells, Skipton 1, Kingston Rule 1, and Hornet 1, a different chemical tracer was added during each stage to aid in determining the contribution from each zone during cleanup and initial production. Chemical fracturing tracers are members of a family of organic compounds used to chemically mark individual stage fracturing fluids in a multizone completed well. These compounds, not found in nature, are added to the fracturing fluids at approximately 1 part per million (ppm) throughout all fluid segments of an individual fracture stage, with the exception of prepads and flushes. Upon commencing flowback, individual flowback samples were collected from the treated well according to a pre-established sampling schedule. These flowback samples were analyzed using a mass spectrometer analysis procedure. The analysis procedure automatically identifies and quantifies the flowback concentrations for each chemical tracer pumped in the treatment well or any tracers pumped in fractured offset wells. Using mass balance computations, these recovered chemical tracer concentrations can then be converted into the corresponding fracturing fluid volumes recovered. The computed fracturing fluid volumes recovered can subsequently be divided by the fracturing fluid volumes pumped to determine the flowback efficiencies for each fracturing stage. Chemical tracer technology has been used in the industry for a significant period, and further details can be found in Asadi et al. (2002, 2006, 2008), Leonard et al. (2007), and King and Leonard (2011).

13.4.7 RA Tracers. In Skipton 1, three stable isotopes of RA ceramic bead tracers were used during the treatments. The fracture height, NWB proppant concentrations, and associated propped widths were estimated post-treatment using spectral, gamma ray logging tools. These methods have been used extensively throughout the industry (Anderson et al. 1986; Barree et al. 2002). The spectral gamma ray tool's radius of investigation is approximately 24 to 30 in. from the wellbore. The three isotopes, scandium, iridium, and antimony, were used in different portions of the treatments. Iridium was used on the treated water parts of the treatments while scandium and antimony were alternated as stages progressed up the wellbore on the linear and crosslink gelled-water portions of the treatments.

13.5 Fracture Stimulation Results

Summaries of each stage for the hydraulic fracturing at Skipton 1, Kingston Rule 1, and Hornet 1 are shown in **Tables 13.2 through 13.4.** Each treatment stage was correlated so that chronologically equivalent sand packages were recognized. This was to assist in identifying any trends across the wells in chronologically equivalent sands. The column labeled "Formation" in these tables provides a correlated sand or shale name. The Patchawarra Formation is divided into three packages—the Lower, Mid, and Upper Sand units are numbered sequentially through these packages.

The pumping rates in Hornet 1 were restricted when compared to the other wells because of the well configuration using 3 1/2-in. tubing as the stimulation string. At least one screenout occurred in each well. On Stage 6 in Skipton 1 and Stage 6 in Hornet 1, limited entry design was implemented and elevated amounts of perforation friction were observed. This was thought to contribute to the screenout.

During the main treatments, it was generally difficult to place proppant concentrations above 2 lbm/gal, with 20/40-mesh ISP being harder to place than 30/50-mesh ISP. This problem was most evident on stages with high fracture gradients and large amounts of tortuosity. When the bottomhole pressure (BHP) was observed to be increasing, the rate was generally increased to attempt to aid with proppant placement. If possible, stages were cut short when an impeding screenout was indicated by the net pressure plot to attempt to avoid leaving excess proppant in the wellbore and having to flowback or perform a coiled tubing cleanout.

The main factors that assisted in placing the job were using smaller proppant sizes and higher viscosity fluid. Other possible design changes to improve proppant placement on future treatments include larger pad volumes, placing the majority of the job at lower proppant concentrations, and using 100-mesh sand on stages with PDL.

13.5.1 Injection Test Results. Summaries of the injection tests are shown in **Tables 13.5 through 13.7.** On Hornet 1, there was difficulty on some stages caused by the sleeve and packer assembly determining when the ball had seated, and formation breakdown occurred.

		Perforation Depth, ft						
Stage	Formation	Top	Bottom	Fluid, bbl	Proppant, lbm	Maximum Rate, bbl/min	Screenout	Proppant Placed, %
1	Lower Patch 2	9,717.8	9,750.7	2,921	63,920	45.2	Y	53
2	Lower Patch 3	9,616.1	9,632.5	2,029	75,487	41.8	N	100
3	Mid Patch 1	9,471.8	9,481.6	1,977	68,998	40.2	N	100
4	Mid Patch 2	9,357.0	9,363.5	1,879	52,405	40.6	N	86
5	Mid Patch 3	9,232.3	9,261.8	1,718	75,022	47.6	N	89
6	Mid Patch 4	9,137.1	9,182.1	1,798	67,356	50.0	Y	59
7	Mid Patch 6 and 7	8,973.1	9,007.5	1,951	76,042	48.2	N	93
8	Murteree	8,694.2	8,727.0	9,299	246,486	63.7	N	100

Table 13.2—Skipton 1 fracture stimulation summary.

		Perforation Depth, ft						
Stage	Formation	Top	Bottom	Fluid, bbl	Proppant, lbm	Maximum Rate, bbl/min	Screenout	Proppant Placed, %
1	Mid Patch 1	8,838.6	8,858.3	1,966	82,714	49.6	N	86
2	Mid Patch 3	8,602.4	8,618.8	1,647	59,288	47.1	N	100
3	Mid Patch 4	8,549.9	8,569.6	1,412	42,686	50.7	Y	50
4	Mid Patch 6	8,415.4	8,428.5	2,428	61,559	41.2	N	100
5	Mid Patch 7	8,359.6	8,379.3	1,379	16,863	40.9	Y	17
6	Epsilon	7,926.5	7,942.9	1,865	96,045	41.8	N	117

Table 13.3—Kingston Rule 1 fracture stimulation summary.

		Perforation Depth, ft						
Stage	Formation	Top	Bottom	Fluid, bbl	Proppant, lbm	Maximum Rate, bbl/min	Screenout	Proppant Placed, %
1	Lower Patch 2	8,766.4	8,786.1	1,546	57,237	31.9	N	66
2	Mid Patch 1	8,490.8	8,503.9	1,057	38,913	28.4	N	79
3	Mid Patch 3	8,369.4	8,392.4	1,530	73,086	32.1	N	84
4	Mid Patch 4	8,333.3	8,343.2	1,259	39,619	26.0	N	70
5	Mid Patch 5	8,241.5	8,274.3	1,480	84,754	30.4	N	94
6	Mid Patch 7	8,166.0	8,190.3	894	37,430	19.8	Y	41

Table 13.4—Hornet 1 fracture stimulation summary.

Stage	Formation	Volume, gal	Injection Rate, bbl/min	Monitoring Time, hr	G-Closure Time	BCA KH, md-ft	ACA KH, md-ft
1	Lower Patch 2	443	4.8	14.8	9.70	0.140	
2	Lower Patch 3	368	6.1	0.9	—	0.020	—
3	Mid Patch 1	502	5.2	0.9	—	0.048	—
4	Mid Patch 2	317	5.5	0.9	—	0.033	—
5	Mid Patch 3	460	5.5	0.4	5.18	0.175	—
6	Mid Patch 4	134	5.5	0.3	—	0.050	—
7	Mid Patch 6 and 7	176	4.8	1.2	4.47	0.252	0.504
8	Murteree	502	4.5	11.2	—	0.157	—

Table 13.5—Skipton 1 injection test summary.

Stage	Formation	Volume, gal	Injection Rate, bbl/min	Monitoring Time, hr	G-Closure Time	BCA KH, md-ft	ACA KH, md-ft
1	Mid Patch 1	280	5.0	7.0	7.05	0.192	0.528
2	Mid Patch 3	165	5.2	1.5	13.61	0.042	0.210
3	Mid Patch 4	226	4.9	9.7	13.49	0.140	0.252
4	Mid Patch 6	332	5.0	6.5	5.48	0.176	0.154
5	Mid Patch 7	237	5.1	1.0	—	0.064	—
6	Epsilon	396	5.1	0.8	5.57	0.325	0.925

Table 13.6—Kingston Rule 1 injection test summary.

Stage	Formation	Volume, gal	Injection Rate, bbl/min	Monitoring Time, hr	G-Closure Time	BCA KH, md-ft	ACA KH, md-ft
1	Lower Patch 2	152	5.2	0.5	11.17	0.170	—
2	Mid Patch 1	341	5.1	0.1	3.09	0.216	—
3	Mid Patch 3	705	10.0	1.0	9.33	0.612	—
4	Mid Patch 4	5,167	25.1	3.1	2.42	2.240	8.480
5	Mid Patch 5	7,502	30.2	0.7	4.26	2.886	6.660
6	Mid Patch 7	4,828	15.1	1.2	2.79	0.075	0.650

Table 13.7—Hornet 1 injection test summary.

This resulted in larger volumes being pumped. Permeability was estimated from before closure analysis (BCA) or ACA where possible. Permeability values (KH) are plotted vs. the fracture gradient in **Fig. 13.9.** Low permeabilities generally corresponded with high fracture gradients and high process zone stress. McGowen et al. (2007) previously documented a relationship between the fracture gradient and transmissibility in the Cooper Basin. It is uncertain whether high tectonic stress causes poor reservoir quality or poor reservoir quality results in high tectonic stress. Lower-permeability reservoir rock should have less porosity and/or more cementation, making it more difficult to fracture and resulting in higher fracture gradients.

There was good correlation between the permeability from the core measurements, the log-derived estimates, and from the injection tests. A comparison for three stages in Kingston Rule 1 is shown in **Fig. 13.10.** In this figure, the square symbols represent core plug measurements. The dashed line is an exponential trend fit to the core data, based on porosity. The solid line is a log-derived permeability, again based on porosity. This was calculated before the fracture stimulation treatments and before the core data were obtained.

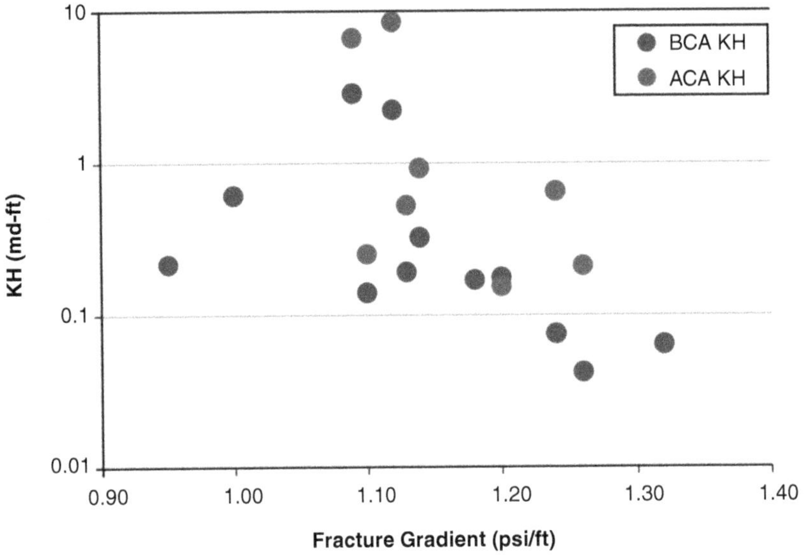

Fig. 13.9—Injection tests KH values (all wells).

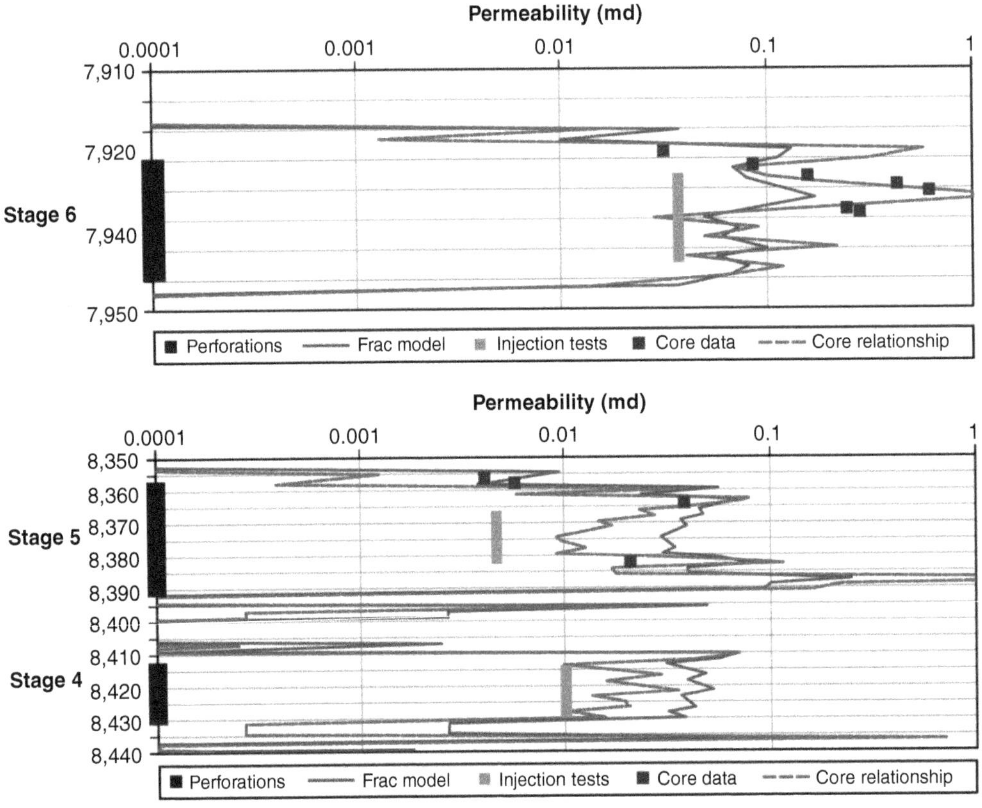

Fig. 13.10—Kingston Rule 1 permeability comparison.

The green bar is the injection test permeability. To obtain this permeability, a fracture height had to be estimated. Based on the height values used, the injection tests provided consistently lower permeability compared to other methods.

13.5.2 NWBPL. Figs. 13.11, 13.12, and 13.13 show the initial fracture gradients, perforation factor, and tortuosity factor for all three wells grouped according to sand package. These results come from step down tests that were conducted after a 15% HCl acid spearhead had been pumped. One interval, Skipton 1 Stage 6, had a second step down test conducted after the friction-reduced water and 100-mesh sand were run. This stage showed almost identical results on the second step down test, indicating that the 100-mesh sand had minimal effect. Overall, stages at Skipton 1 and Kingston Rule 1 had minimal perforation friction. Hornet 1 had much larger perforation friction values and, in general, higher tortuosity. This was attributed to the different perforating charges that were used. While the wells had a different completion, with different amounts of clearance between the center of the wellbore and casing, as

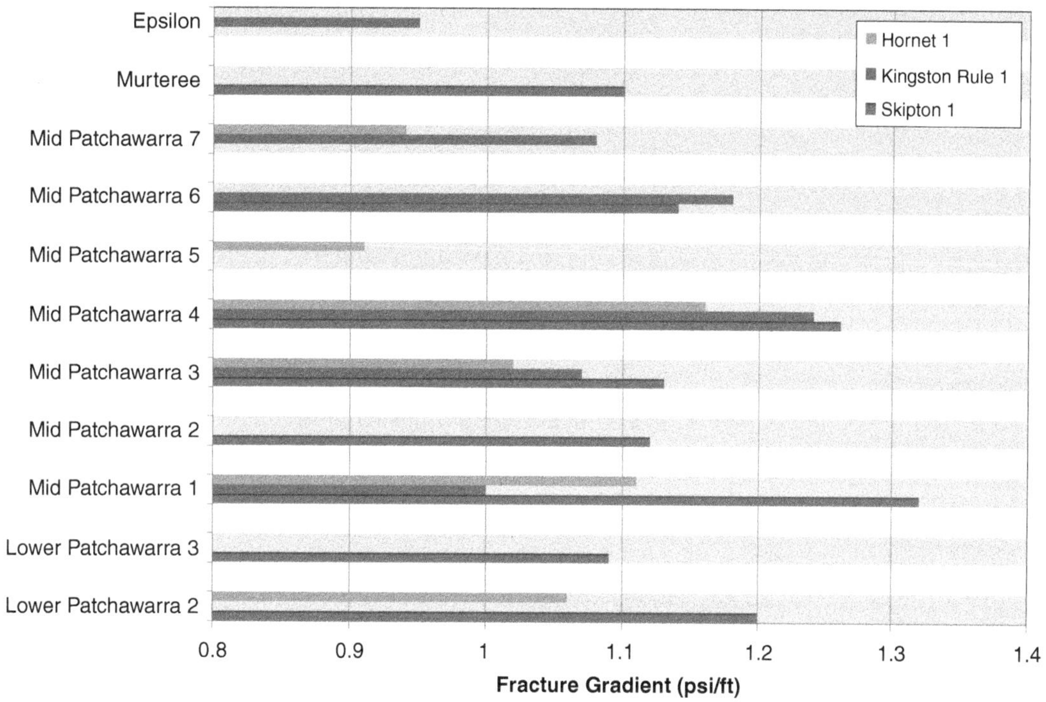

Fig. 13.11—Fracture gradients for all stages.

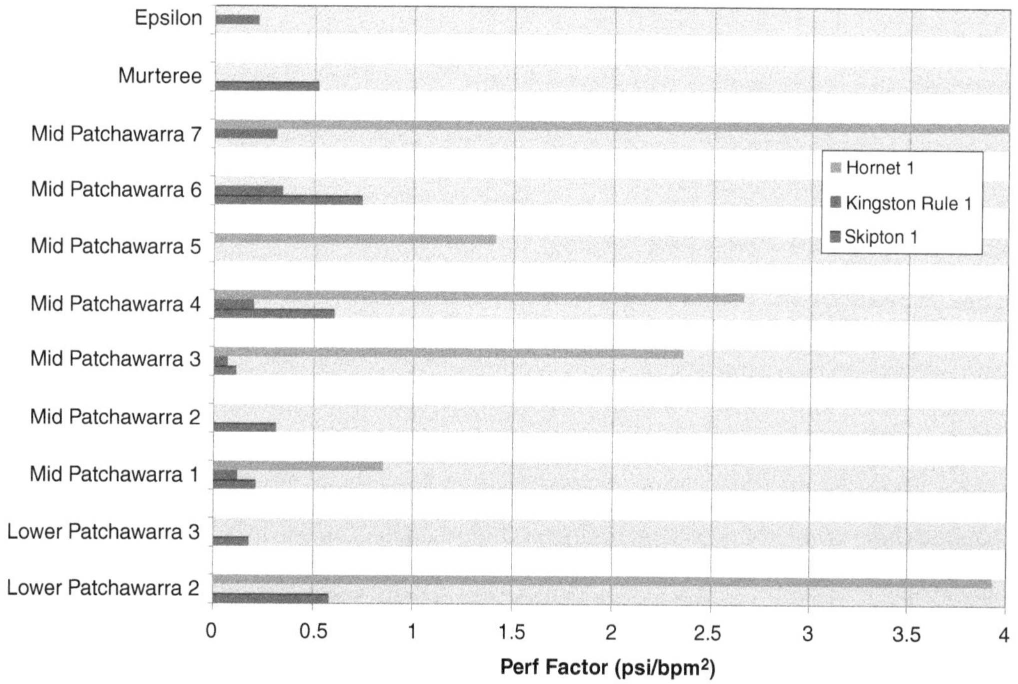

Fig. 13.12—Perforation factor for all stages.

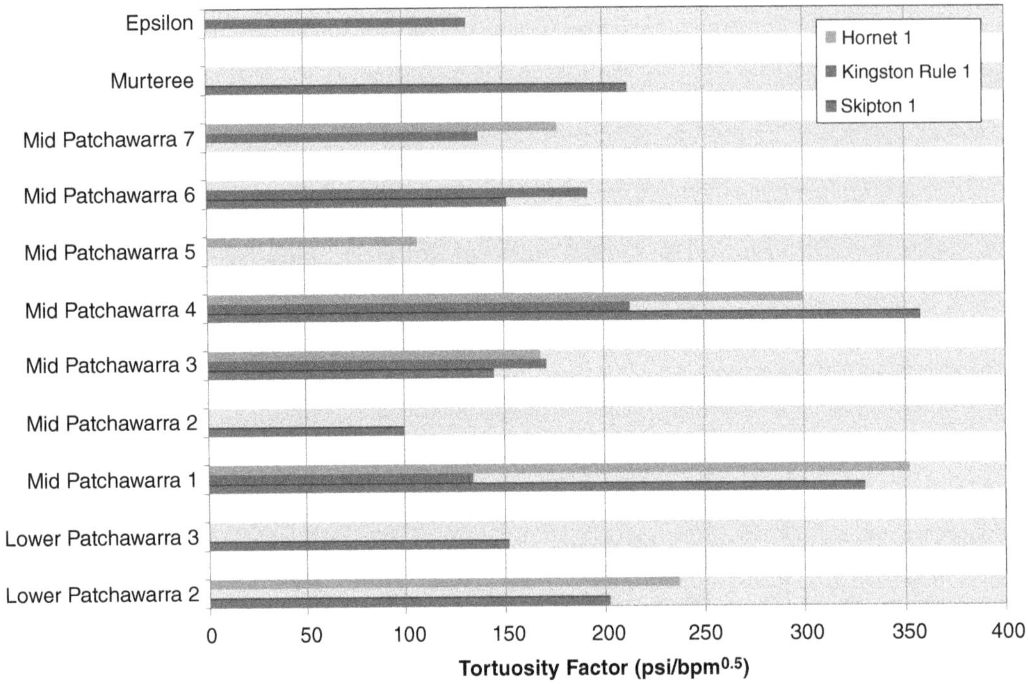

Fig. 13.13—Tortuosity factor for all stages.

well as different thicknesses of cement between the casing and reservoir, modeling work showed that the penetration and entry hole diameter were similar for the same charges in the two different wellbore configurations. This meant the main difference appeared to be a result of the deeper penetrating charges used in Hornet 1 causing more perforation friction.

In this work, a number of zones exhibited extremely high process zone stress. Process zone stress is defined as the excess pressure above the closure stress required to extend the fracture (Ramurthy et al. 2007). It can be estimated by subtracting the closure stress from the ISIP. Some factors thought to influence process zone stress include intact rock tensile strength, fluid-lag region at the fracture tip, leakoff (or pore pressure transient surrounding the fracture), multiple fracture strands at the fracture tip, and non-elastic rock deformation at the fracture tip.

As expected, fracture gradients and tortuosity were high on many of the stages, with high tortuosity generally corresponding with high fracture gradients. The fracture gradient was generally higher in the Mid to Lower Patchawarra Sands on all three wells.

13.5.2.1 15% HCl Acid Spearheads. 15% HCl acid spearheads were pumped before the step down test to attempt to clean the NWB region and reduce the treating pressure. This was successful on some stages but, in general, had little effect based on the pressure responses before, during, and after the spearhead. The acid was more effective on stages with a high fracture gradient, high tortuosity, and exhibiting fracture tip extension during shut-in. This suggests when the acid did have an effect, it was cleaning the NWB region or reducing tip effects that caused high process zone stress. It was also observed that any benefit from pumping acid was achieved in the first half of the spearhead, so the acid stage volume was reduced from 1,500 to 1,000 gal.

13.5.2.2 100-Mesh Sand. Low-concentration (< 1-lbm/gal) 100-mesh sand was run in the friction-reduced water prepad in an attempt to scour the NWB region and control PDL. As the jobs progressed, it became apparent that the 100-mesh sand was not having an effect on the treating pressure and, hence, the mass run on each stage was reduced from approximately 10,000 lbm to approximately 3,300 lbm. Based on the pressure response and the one stage with multiple step down tests, the 100-mesh sand had little effect on the treating pressure on most stages. It appeared to be more effective on stages exhibiting high perforation friction. This suggests the 100-mesh sand was effective at eroding the perforations but did little to scour the formation or plug off competing multiple fractures; however, this might be influenced by the relatively low PDL that was observed.

13.5.3 Hydraulic Fracture Simulator. A planar, three-dimensional (3D), grid-orientated, finite-difference hydraulic fracture modeling software was used to simulate the treatments. A stress profile was calculated from log-derived Poisson's ratio and Young's modulus and calibrated from the core rock testing. The stress log was then calibrated further from the injection tests results. The permeability input was discussed in a previous section. The proppant concentration throughout the interval of Kingston Rule 1 Stage 6, after bottomhole treating pressure (BHTP) history matching, is shown in **Fig. 13.14.** This stage had an isolated flow period of approximately four days with a long-term shut-in period. This allowed the fracture parameters to be estimated from the buildup analysis. The results from the well test were a fracture conductivity of 76 md-ft, an average reservoir permeability of 0.14 md, and a fracture length of 155 ft. The reservoir permeability compared very well to the data presented in Fig. 13.10. A comparison of the well test fracture geometry to the simulator showed that the fracture model predicted more length, with an infinite conductivity fracture length of between 230 to 300 ft. The fracture conductivity was slightly lower at 63 md-ft. Overall, the match was relatively good, and it appeared that the model was providing an acceptable estimate of the fracture geometry.

13.5.4 Surface Tiltmeter Results. The results from the surface tiltmeters are summarized in **Table 13.8.** With the elevated treating pressure gradients as previously discussed, as well as the highly laminated reservoir sections, the presence of significant horizontal

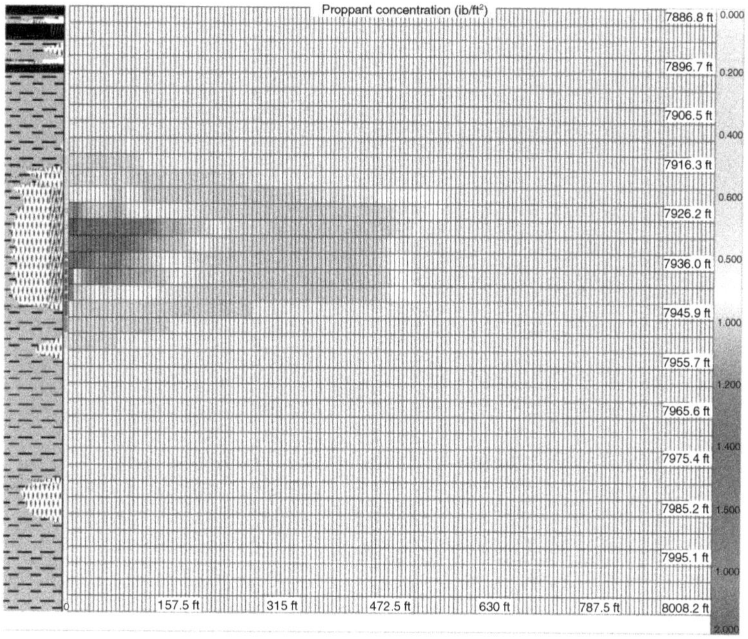

Fig. 13.14—Kingston Rule 1 Stage 6 fracture model proppant concentration.

Stage	Formation	Perforation Depth, ft Top	Bottom	Fracture Azimuth	Fracture Dip	Vertical Component, %	Horizontal Component, %
1	Lower Patch 2	9,717.8	9,750.7	N62°W ± 8°	85° down SW ± 4°	71	29
2	Lower Patch 3	9,616.1	9,632.5	N6°W ± 14°	76° down NE ± 7°	62	38
3	Mid Patch 1	9,471.8	9,481.6	N2°W ± 15°	85° down SW ± 7°	50	50
4	Mid Patch 2	9,357.0	9,363.5	N9°W ± 18°	84° down NE ± 6°	68	32
5	Mid Patch 3	9,232.3	9,261.8	N17°W ± 11°	87° down NE ± 6°	72	28
6	Mid Patch 4	9,137.1	9,182.1	N15°W ± 10°	77° down SW ± 5°	66	34
7	Mid Patch 6 and 7	8,973.1	9,007.5	N3°E ± 15°	86° down NE ± 6°	45	55
8	Murteree	8,694.2	8,727.0	N35°W ± 11°	87° down SW ± 4°	59	41

Table 13.8—Skipton 1 tiltmeters results.

components on several of the stages on Skipton 1 was not a surprise. Through the Patchawarra section, the horizontal volumetric component ranged from 28 to 55%, with an average of 38%. The Murteree showed a higher than average horizontal volumetric component of 41% compared to the Patchawarra section. This observation of elevated horizontal component in the Murteree is consistent with that observed in Pitkin et al. (2012), where the Murteree showed the highest horizontal component in each of the study wells. **Fig. 13.15** depicts a 3D image of the deformation from two stages, 4 and 7. These stages had the second lowest and highest degrees of horizontal component, respectively.

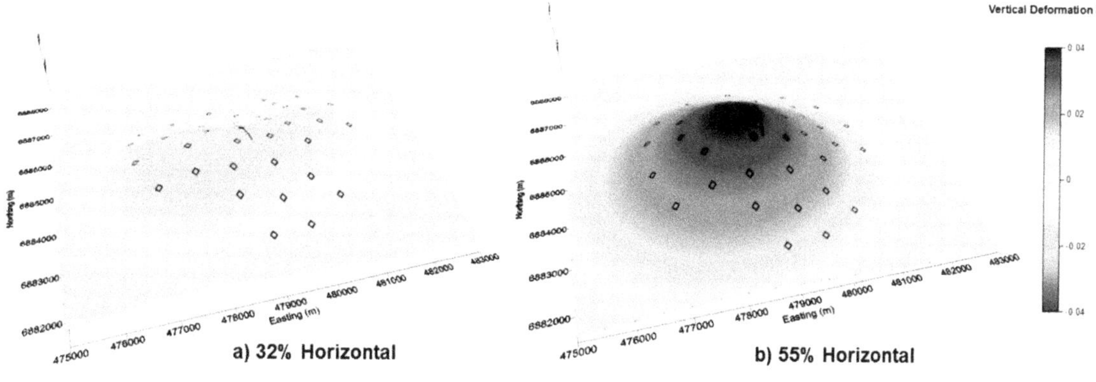

Fig. 13.15—Skipton 1 tiltmeter deformation: (a) Stage 4 and (b) Stage 7.

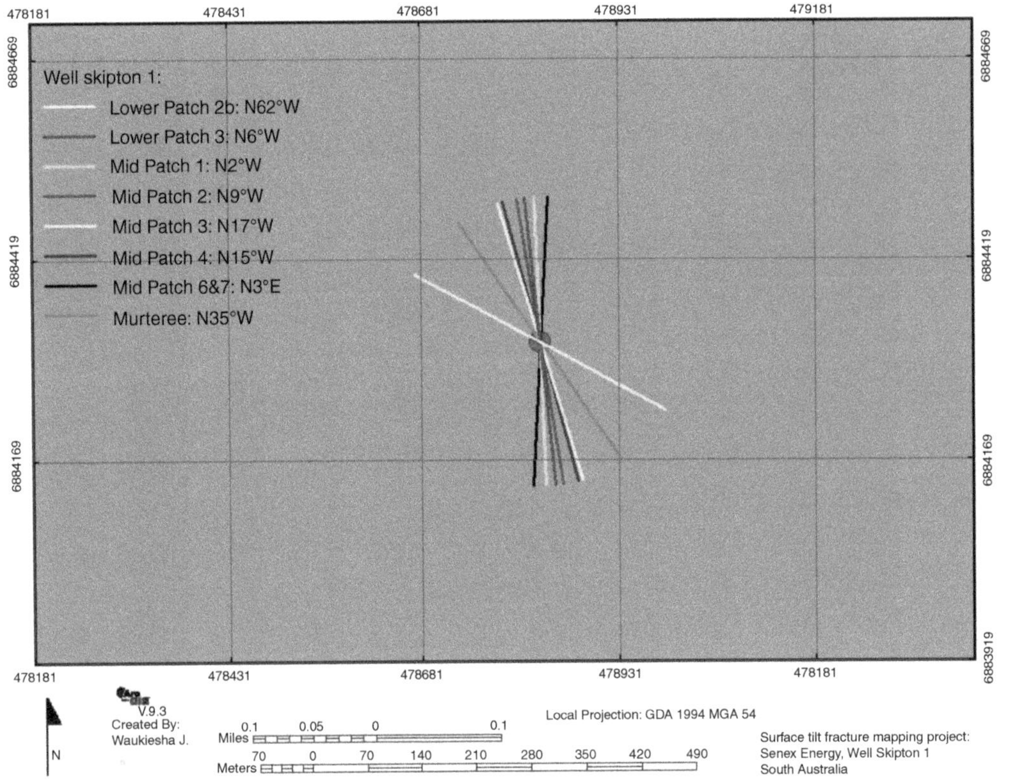

Fig. 13.16—Fracture azimuth in Skipton 1 from the surface tiltmeters.

All of the Patchawarra stages, with the exception of the Lower Patchawarra 2 stage (first stage stimulated) showed a tight azimuth distribution of N17°W to N3°E. The Lower Patchawarra 2 stage on Skipton 1 pressured out during pumping several times and required some flowback between each injection attempt. The fracture azimuth result for this first Lower Patchawarra 2 stage varied between N56°W and N62°W and was considerably different to that of the other six Patchawarra stages treated (but closer to the anticipated fracture orientation from borehole breakout indicators). The Murteree interval presented the only shale interval to be treated on this well, and it displayed an induced fracture azimuth of N35°W. The azimuth for each stage is shown in **Fig. 13.16.**

The dip direction for the vertical component of the fracture was between 77° down to the southwest (Mid Patchawarra 4) and 76° down to the northeast (Lower Patchawarra 3) for all stages. Six of the eight stages returned results within 6° of vertical. These results are within the vertical fracture envelope (60° to 90° dip).

13.5.5 Tracer Results.

13.5.5.1 RA Tracers. The tracer schedule for the stages in Skipton 1 is shown in **Table 13.9.** Unfortunately, the spectral gamma ray tool was unable to be run to total depth (TD) because of an obstruction in the wellbore. This meant only Stages 3 to 8 were logged post-fracture treatment. The fracture heights, as shown in Table 13.9, were fairly contained, with all stages ranging from 20 to 60 ft. **Fig. 13.17** illustrates the results from the RA tracers and includes tracks for lithology and the calibrated stress log. The barriers in the treatment stages appeared to depend little on the stress profile and more on changes in lithology and bulk density of the rock, particularly the coal intervals.

		Perforation Depth, ft					
Stage	Formation	Top	Bottom	Treated Water Isotope	Crosslink Gelled Water Isotope	Single or Multiple Perforations	Estimated Height, ft
1	Lower Patch 2	9,717.8	9,750.7	Iridium	Antimony	Single	NA
2	Lower Patch 3	9,616.1	9,632.5	Iridium	Scandium	Single	NA
3	Mid Patch 1	9,471.8	9,481.6	Iridium	Antimony	Single	39
4	Mid Patch 2	9,357.0	9,363.5	Iridium	Scandium	Single	23
5	Mid Patch 3	9,232.3	9,261.8	Iridium	Antimony	Single	62
6	Mid Patch 4	9,137.1	9,182.1	Iridium	Scandium	Multiple	43
7	Mid Patch 6 and 7	8,973.1	9,007.5	Iridium	Antimony	Multiple	39
8	Murteree	8,694.2	8,727.0	Iridium	Scandium	Single	39

Table 13.9—Skipton 1 RA tracer results.

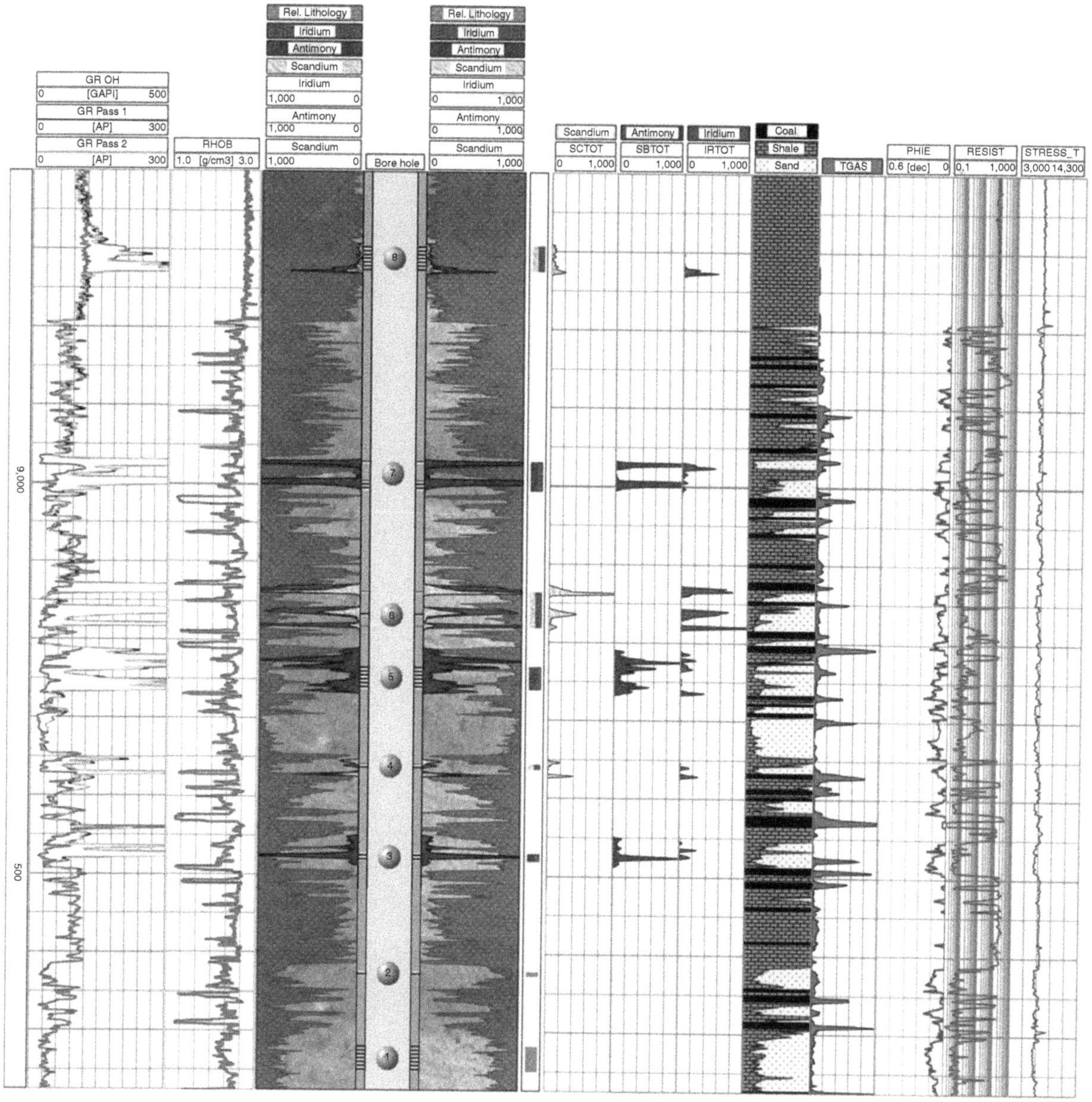

Fig. 13.17—Skipton 1 RA tracer log.

13.5.5.2 Chemical Tracers. The chemical tracers showed that initial flowback was dominated by the more permeable stages. These stages also typically had the better fracture treatments because of the lower BHTPs. Greater than 50% of the fracture stimulation fluid pumped into the wells was produced back post-fracture treatment. This was consistent with the significant amount of chemical tracer returned on Kingston Rule 1 and Hornet 1 (approximately 25% of tracer volume after 10 days of production). However, the exact portion of fracturing fluid produced back after treatment was unclear because of the production of formation fluid. Both Kingston Rule 1 and Hornet 1 had production logs run during flowback. The results from the chemical tracers were predominantly consistent with the zonal contribution from the production logs.

13.6 Discussion

A comparison of the openhole logs, fracture treatment pressure observations, surface tiltmeters, and RA tracers from the eight stages on Skipton 1 showed there was a loose trend of increasing horizontal component with increasing pressure gradient (breakdown pressure, initial ISIP, maximum pressure), as depicted in **Fig. 13.18.** It can also be observed in Fig. 13.18 that pressure and tortuosity strongly trended with each other, meaning that the horizontal component from tiltmeters can in fact be related to tortuosity. It should be noted that no trend with tortuosity was observed in the previous work by Pitkin et al. (2012). In addition, as the horizontal component increased, there was a general trend that the azimuth solution appeared to rotate further toward the north, with the Stages 3 and 7 fracture azimuths closest to due north. The results from the previous work by Pitkin et al. (2012) have been combined with the current work and are shown in **Fig. 13.19.** All three wells show a clear discrepancy between the fracture azimuth observed from the surface tiltmeters and the maximum horizontal stress direction from the borehole breakout in the basin. There is insufficient information in this work to fully describe what is occurring. However, it is currently believed that the fracture direction is related to the high NWBPL and rock strength seen in the Cooper Basin. As previously discussed, some of causes of high tortuosity are fracture turning, fracture reorientation, and multiple fractures.

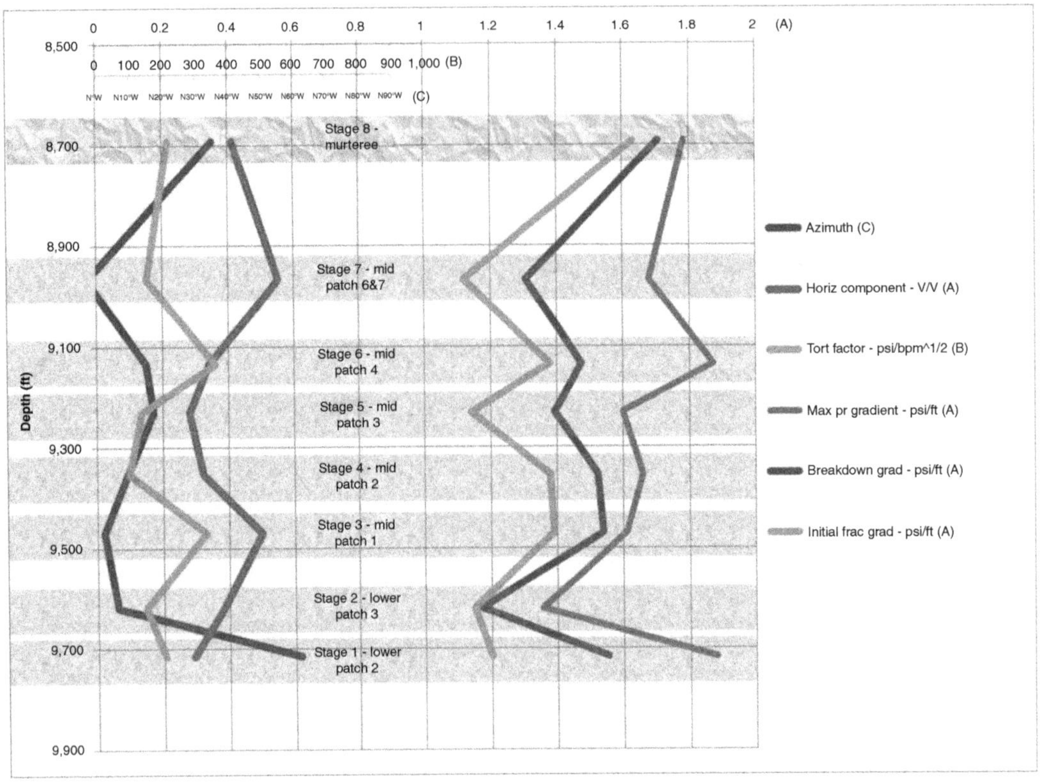

Fig. 13.18—Skipton 1 tiltmeter pressure summary.

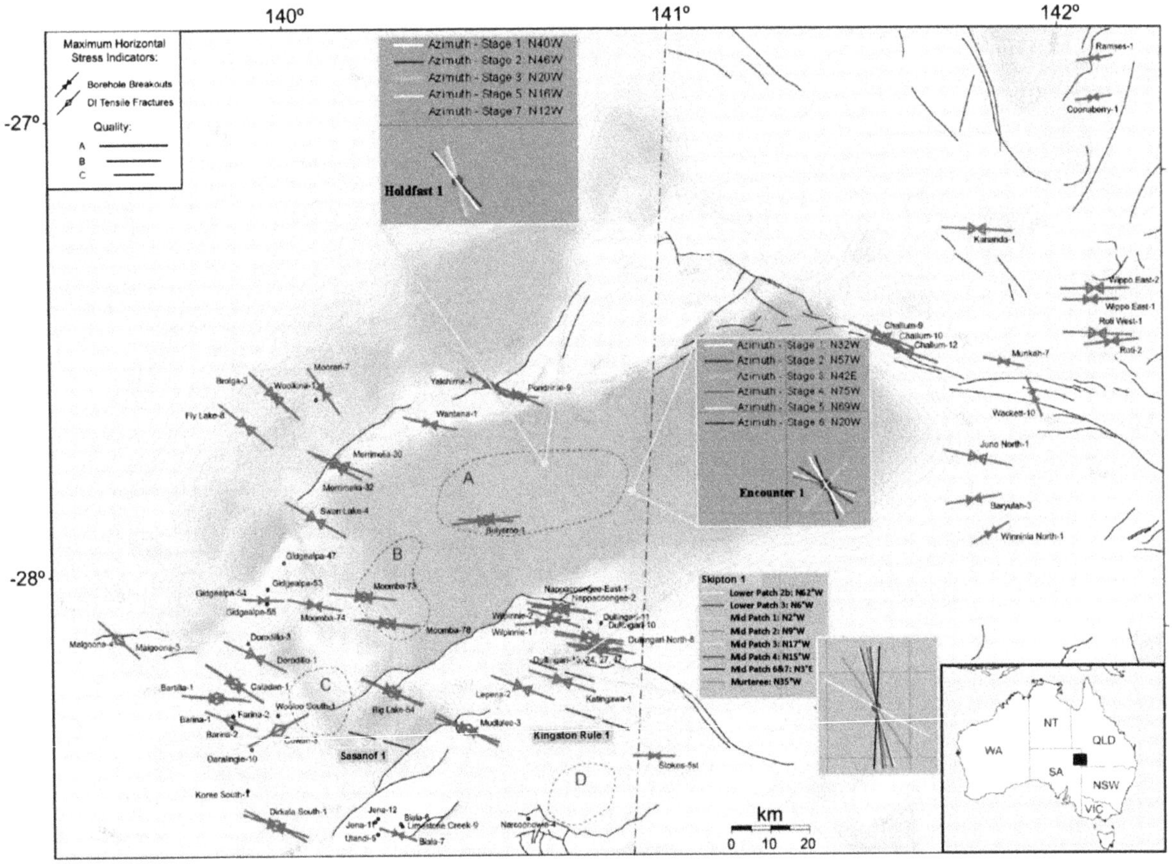

Fig. 13.19—Maximum horizontal stress orientations in the Cooper Basin from borehole breakout data along with surface tiltmeter mapping results (adapted from Reynolds et al. 2005 and Pitkin et al. 2012).

Nelson et al. (2007b) proposed that geological weaknesses inherent in the reservoir and high intact rock tensile strengths were resulting in complex fractures and fractures not opening against the minimum in-situ stress. It would be intuitive that once the fracture moved away from the NWB region into the far field, the fracture would realign itself with the maximum horizontal stress direction. This does not seem to be the case and could mean that the fracture requires a very long distance to realign.

There appears to be good height containment, as indicated from the RA tracer results. For Stage 5 (Mid Patchawarra 3), there appears to be the greatest vertical fracture growth, and this corresponds with the lowest horizontal component, as observed from the surface tiltmeters. The high degree of containment observed with the RA tracers, coupled with the changes in lithology acting as growth boundaries, even without significant changes in the minimum horizontal stress, supports the horizontal component observed with surface tiltmeters. The horizontal component is likely a result of fluid interaction with the rock fabric and delamination of the boundaries between the varying lithology.

13.7 Conclusions

The following conclusions are a result of this work:

- The fracture treatments generally exhibited high fracture gradients and tortuosity, which has previously been observed throughout the Cooper Basin. The main issue with hydraulic fracturing in the basin is overcoming fracture complexity created by the strike-slip stress regime.
- Typically, it was difficult to place proppant concentrations above 2 lbm/gal, with 20/40-mesh ISP being harder to place than 30/50-mesh ISP. This problem was most evident on stages with higher fracture gradients and large amounts of tortuosity. The main factors that assisted in placing the job were using smaller proppant sizes and higher viscosity fluid. Other possible design changes to improve proppant placement on future treatments include larger pad volumes, placing the majority of the job at lower proppant concentrations, and using 100-mesh sand on stages with PDL.
- A 15% HCl acid spearhead and 100-mesh sand were used to help minimize NWBPL, but this appeared to have minimal effect.
- The main barriers for fracture height growth appear to be changes in lithology. The RA tracer results in Skipton 1 showed good containment, with heights ranging from 20 to 60 ft.
- The tiltmeters showed the majority of the fracture treatment in each stage to be vertical, except for the shallowest Patchawarra interval. This stage showed a horizontal component of 55%, with the average horizontal component for all stages being 38%.
- The tiltmeters showed an azimuth that was not consistent with the maximum horizontal stress direction from borehole breakout. While there was insufficient information in this work to provide a definitive conclusion for the mechanism causing this, it is thought that it is related to the high NWBPL and rock strength seen in the Cooper Basin.

13.8 Recommendations

- Further fracture stimulation work should be performed, including microseismic and surface tiltmeters to confirm fracture azimuth within the basin. A fracture treatment in an area of rotated breakout would help in identifying the mechanism orienting the fracture.
- A study of the depositional environment should be conducted to determine if this has an impact on fracture orientation.

13.9 Nomenclature

3D	=	Three dimensional	ISP	=	Intermediate-strength proppant
A CA	=	After closure analysis	NWBPL	=	Near-wellbore pressure loss
BCA	=	Before closure analysis	Patch	=	Patchawarra
BHTP	=	Bottomhole treating pressure	PDL	=	Pressure-dependent leakoff
DFIT	=	Disgnostic fracture injection test	ppm	=	Part per million
DST	=	Drillstem test	RA	=	Radioactive
FRW	=	Friction reduced water	σ_h	=	Minimum horizontal stress
HCl	=	Hydrochloric acid	σ_H	=	Maximum horizontal stress
IAP	=	Indo-Australia Plate	σ_v	=	Vertical stress
ISIP	=	Instantaneous shut-in pressure	TD	=	Total depth

13.10 Acknowledgments

The authors thank Senex Energy and Halliburton for permission to publish the material contained in SPE 167073-MS. They especially thank Cory MacNeill for the geological assistance that went into the work. They also acknowledge Brenton Richards for his proof reading, Mark Warren for preparation of the tracer images, Shaun Moore for his help with the figures, and everyone that assisted with the implementation of the program.

13.11 References

Allen, P. 1990. Dullingari 46 Well Completion Report. Report Number WCR 07276/000, Santos Limited, Australia.
Allen, P. 1991. Verbena 1 Well Completion Report. Report Number WCR 07302/000, Santos Limited, Australia.

Anderson, J. A., Pearson, C. M., Abou-Sayed, A. S. et al. 1986. Determination of Fracture Height by Spectral Gamma Log Analysis. Paper presented at the SPE Annual Technical Conference and Exhibition, New Orleans, Louisiana, USA, 5–8 October. SPE-15439-MS. http://dx.doi.org/10.2118/15439-MS.

Asadi, M., Woodroof, R. W., Malone, W. S. et al. 2002. Monitoring Fracturing Fluid Flowback With Chemical Tracers: A Field Case Study. Paper presented at the SPE Annual Technical Conference and Exhibition, San Antonio, Texas, USA, 29 September–2 October. SPE-77750-MS. http://dx.doi.org/10.2118/77750-MS.

Asadi, M., Woodroof, R. and Himes, R. 2006. Comparative Study of Flowback Analysis Using Polymer Concentrations and Fracturing Fluid Tracer Methods; A Field Study. Paper presented at the SPE International Oil & Gas Conference and Exhibition, Beijing, China, 5–7 December. SPE-101614-MS. http://dx.doi.org/10.2118/101614-MS.

Asadi, M., Woodroof, R. W. and Dumas, J. D. Jr. 2008. Post-Frac Analysis Based on Flowback Results Using Chemical Frac-Tracers. Paper presented at the International Petroleum Technology Conference, Kuala Lumpur, Malaysia, 3–5 December. IPTC11891-MS. http://dx.doi.org/10.2118/11891-MS.

Asanuma, H., Kumano, Y., Izumi, T. et al. 2004. Passive Seismic Monitoring if a Stimulation of HDR Geothermal Reservoir at Cooper Basin, Australia. Paper presented at the SEG Annual Meeting, Denver, Colorado, USA, 10–15 October. SEG-2004-0556.

Barree, R. D., Fisher M. K., and Woodroof, R. A. 2002. A Practical Guide to Hydraulic Fracture Diagnostic Technologies. Paper presented at the SPE Annual Technical Conference and Exhibition, San Antonio, Texas, USA, 29 September–2 October. SPE-77442-MS. http://dx.doi.org/10.2118/77442-MS.

Beatty, K. J., McGowen, J. M., and Gilbert J. V. 2007. Pin-Point Fracturing (PPF) in Challenging Formations. Paper presented at the SPE Hydraulic Fracturing Technology Conference, College Station, Texas, USA, 29–31 January. SPE-106052-MS. http://dx.doi.org/10.2118/106052-MS.

Bilek, M. 1992. Katingawa 1 Well Completion Report. Report Number WCR 07421/000, Santos Limited, Australia.

Camac, B. A., Waldron, A. W., Donley, J. A. et al. 2012. Microseismic and 3D VSP for infill evaluation in Greater Tindilpie, Cooper Basin, Australia. Paper presented at the 46th US Rock Mechanics/Geomechanics Symposium, Chicago, Illinois, USA, 24–27 June. ARMA-12-618.

Chipperfield, S. T. and Britt, L. K. 2000. Application of After-Closure Analysis for Improved Fracture Treatment Optimisation: A Cooper Basin Case Study. Paper presented at the SPE Rocky Mountain Regional/Low Permeability Reservoirs Symposium, Denver, Colorado, USA, 12–15 March. SPE-60316-MS. http://dx.doi.org/10.2118/60316-MS.

Chipperfield, S. T., Roberts, G. A., Miller, W. K. II et al. 2000. Gel Slugs: A Near-Wellbore Pressure-Loss Remediation Technique for Propped Fracturing. Paper presented at the SPE/CERI Gas Technology Symposium, Calgary, Alberta, Canada, 3–5 April. SPE-59777-MS. http://dx.doi.org/10.2118/59777-MS.

Courtney, R. and Butler, G. 1982. Dullingari 16 Well Completion Report. Report Number WCR 04894/000, Santos Limited, Australia.

Dunlop, E. 2012. Santos Unconventional Gas Projects (Past, Present and Future). Presented at the SPE Shale and Unconventional Symposium, Sydney, Australia.

Gilbert, J. and Greenstreet, C. 2005. Applications of Pinpoint Fracturing in the Cooper Basin, Australia. Paper presented at the SPE Annual Technical Conference and Exhibition, Dallas, Texas, USA, 9–12 October. SPE-97004-MS. http://dx.doi.org/10.2118/97004-MS.

Gravestock, D. I., Hibburt, J. E. and Drexel, J. F. eds. 1998. *The Petroleum Geology of South Australia. Volume 4: Cooper Basin.* Adelaide, Australia: Department of Primary Industries and Resources. Report Book 98/9.

Gu, H., Elbel, J. L., Nolte, K. G. et al. 1993. Formation Permeability Determination using Impulse Fracture Injection. Paper presented at the SPE Production Operations Symposium, Oklahoma City, Oklahoma, USA, 21–23 March. SPE-25425-MS. http://dx.doi.org/10.2118/25425-MS.

Haak, K. J. 1998. Gudnuki 2 Well Completion Report. Report Number WCR 07669/000, Santos Limited, Australia.

Hillis, R. R., Meyer, J. J. and Reynolds, S. D. 1998. The Australian Stress Map. *Exploration Geophysics* **29** (3–4): 420–427.

Hillis, R. R., Morton, J. G. G., Warner, D. S. et al. 2001. Deep Basin Gas: A New Exploration Paradigm in the Nappamerri Trough, Cooper Basin, South Australia. Paper presented at the APPEA Conference, Hobart, Australia.

Hillis, R. R. and Reynolds, S. D. 2003. In situ stress field of Australia. *Geological Society of Australia Special Publication* **22:** 43–52. https://doi.org/10.1130/0-8137-2372-8.49.

Johnson, R. L., Aw, K. P., Ball, D. et al. 2002. Completion, Perforating and Hydraulic Fracturing Design Changes Yield Success in an Area of Problematic Frac Placement–The Cooper Basin, Australia. Paper presented at the SPE Asia Pacific Oil & Gas Conference and Exhibition, Melbourne, Australia, 8–10 October. SPE-77906-MS. http://dx.doi.org/10.2118/77906-MS.

King, G. and Leonard, R. 2011. Deciphering Chemical Tracer Results in Multi-Fractured Well Backflow in Shales: A Framework for Optimizing Fracture Design and Application. Paper presented at the SPE Hydraulic Fracturing Technology Conference and Exhibition, The Woodlands, Texas, USA, 24–26 January. SPE-140105-MS. http://dx.doi.org/10.2118/140105-MS.

Leonard, R., Woodroof, R. Bullard, K. et al. 2007. Barnett Shale Completions: A Method for Assessing New Completions Strategies. Paper presented at the SPE Annual Technical Conference and Exhibition, Anaheim, California, USA, 11–14 November. SPE-110809-MS. http://dx.doi.org/10.2118/110809-MS.

Lowman, R. 1995. Allambi 1 Well Completion Report. Report Number WCR 07502/000, Santos Limited, Australia.

McGowen, J. M., Gilbert, J. V., and Samari, E. 2007. Hydraulic Fracturing Down Under. Paper presented at the SPE Hydraulic Fracturing Technology Conference, College Station, Texas, USA, 29–31 January. SPE-106051-MS. http://dx.doi.org/10.2118/106051-MS.

Menpes, S. 2012. Emerging Continuous Gas Plays in the Cooper Basin, South Australia. Paper presented at the APPEA Conference, Adelaide, Australia, 13–16 May.

Nelson, E. J., Chipperfield, S. T., Hillis, R. R. et al. 2007a. Using geological information to optimize fracture stimulation practices in the Cooper Basin, Australia. *Petroleum Geoscience* **13** (1): 3–16. https:// 10.1144/1354-079306-700.

Nelson, E. J., Chipperfield, S. T., Hillis, R. R. et al. 2007b. The relationship between closure pressures from fluid injection tests and the minimum principal stress in strong rocks. *International Journal of Rock Mechanics and Mining Sciences* **44** (5): 787–801. https://doi.org/10.1016/j.ijrmms.2.

Nolte, K. G. 1986. A General Analysis of Fracturing Pressure Decline With Application to Three Models. *SPE Formation Evaluation* **1** (06): 571–583. SPE-12941-PA. https://doi.org/10.2118/12941-PA.

Nolte, K. G., Maniere, J. L. and Owens, K. A. 1997. After-Closure Analysis of Fracture Calibration Tests. Paper presented at the SPE Annual Technical Conference and Exhibition, San Antonio, Texas, USA, 5–8 October. SPE-38676-MS. http://dx.doi.org/10.2118/38676-MS.

Nugent, O. 1989. Dullingari 45 Well Completion Report. Report Number WCR 07228/000, Santos Limited, Australia.

Ostler, S. 1999. Kelbrook 1 Well Completion Report. Report Number WCR 07723/000, Santos Limited, Australia.

Pitkin, M. C., Wadham, T. H., McGowen, J. M. et al. 2012. Taking the First Steps: Stimulating the Nappamerri Trough Resource Play. Paper presented at the SPE Asia Pacific Oil and Gas Conference and Exhibition, Perth, Australia, 22–24 October. SPE-160307-MS. http://dx.doi.org/10.2118/160307-MS.

Ramurthy, M., Lyons, B., Hendrickson, R. B. et al. 2007. Effects of High-Pressure-Dependent Leakoff and High-Process-Zone Stress in Coal Stimulation Treatments. Paper presented at the SPE Rocky Mountain Oil & Gas Technology Symposium, Denver, Colorado, USA, 16–18 April. SPE-107971-MS. http://dx.doi.org/10.2118/107971-MS.

Reynolds, S. D., Mildren, S. D., Hillis, R. R. et al. 2006. Constraining stress magnitudes using petroleum exploration data in the Cooper- Eromanga Basins, Australia. *Tectonophysics* **415** (1–4): 123–140. https://doi.org/10.1016/j.tecto.2005.12.005.

Reynolds, S. D., Mildren, S. D., Hillis, R. R. et al. 2005. Maximum horizontal stress orientations in the Cooper Basin, Australia: implications for plate-scale tectonics and local stress sources. *Geophysical Journal International* **160** (01): 331–343. https://doi.org/10.1111/j.1365-246X.2004.02461.x.

Roberts, G. A., Chipperfield, S. T. and Miller, W. K. II. 2000. The Evolution of a High Near-Wellbore Pressure Loss Treatment Strategy for the Australian Cooper Basin. Paper presented at SPE Annual Technical Conference and Exhibition, Dallas, Texas, USA, 1–4 October. SPE-63029-MS. http://dx.doi.org/10.2118/63029-MS.

Siebrits, E., Elbel, J. L., Hoover, R. S. et al. 2000. Refracture Reorientation Enhances Gas Production in Barnett Shale Tight Gas Wells. Paper presented at the SPE Annual Technical Conference and Exhibition, Dallas, Texas, USA, 1–4 October. SPE-63030-MS. http://dx.doi.org/10.2118/63030-MS.

Walker, T., Kerns, S., Scott, D. et al. 2002. Fracture Simulation Optimization in the Redevelopment of a Mature Waterflood, Elk Hills Field, California. Paper presented at the SPE Western Regional/AAPG Pacific Section Joint Meeting, Anchorage, Alaska, USA, 20–22 May. SPE-76723-MS. http://dx.doi.org/10.2118/76723-MS.

Wolhart, S. L., Berumen, S., Cipolla, C. L. et al. 2000. Use of Hydraulic Fracture Diagnostics to Optimize Fracturing Jobs in the Arcabuz-Culebra Field. Paper presented at the SPE Rocky Mountain Regional/Low Permeability Reservoirs Symposium, Denver, Colorado, USA, 12–15 March. SPE-60314-MS. http://dx.doi.org/10.2118/60314-MS.

Wright, C. A. and Conant, R. A. 1995. Hydraulic Fracture Reorientation in Primary and Secondary Recovery from Low-Permeability Reservoirs. Paper presented at SPE Annual Technical Conference and Exhibition, Dallas, Texas, USA, 22–25 October. SPE-30484-MS. http://dx.doi.org/10.2118/30484-MS.

Wright, C. A., Davis, E. J., Minner, W. A. et al. 1998. Surface Tiltmeter Fracture Mapping Reaches New Depths—10,000 Feet and Beyond? Paper presented at the SPE Rocky Mountain Regional/Low-Permeability Reservoirs Symposium, Denver, Colorado, USA, 5–8 April. SPE-39919-MS. http://dx.doi.org/10.2118/39919-MS.

Chapter 14

Integrated Field Study for Production Optimization: Jonah Field—Sublette County, Wyoming

M. J. Eberhard, SPE, **M. J. Mullen,** SPE, **C. A. Seal,** SPE, Halliburton Energy Services Inc.; **B. P. Ault,** SPE, Ultra Petroleum

SPE 59790 presents a case history of using integrated reservoir studies to optimize completions and production in a multilayer, geopressured Lance Formation in the Green River Basin in Wyoming. Eberhard et al. evaluate well logs from a total of 540 pay zones and production logs from a total 363 pay zones in 21 wells. The case histories they present result in the following conclusions:

- Log-derived petrophysical properties do not correlate with gas production
- Net-pay cutoffs based on the gamma ray and neutron-density porosity is the best indicator of production potentials in the field
- Fracture treatment parameters normalized over net-pay and production log data do not show any correlations
- Completions efficiency is low because of poor isolation practices
- Only 58% of the completed zones are producing at the time of production logging
- The Upper Lance contributed to more than 50% of the gas production

14.1 Abstract

The Lance Formation in the Jonah Field is an overpressured, tight-gas sand that requires hydraulic fracturing for economic production. Because of large gross intervals containing several individual sands, limited entry has been the typical fracturing technique. Wells can have more than 30 individual sands that are completed with multiple fracture treatments; however, production log data indicate that only 58% of the perforated sands contribute to production. Production optimization is dependent on improving the percentage of completed pay contributing to production.

A detailed field study was conducted to determine pay identification and best practices for completions. In the study, 44 wells in the Stud Horse Butte area were analyzed. The study comprised detailed log analysis, fracture treatment data, production data, reservoir analysis, and completion practices. Some of the processes used were log normalization, traditional statistical analysis, and reservoir modeling. Log and treatment data were also analyzed with an artificial neural network (ANN).

14.2 Introduction

Discovered in 1975, Jonah Field is located in the Hoback sub-basin in the far northwest corner of the Green River Foreland Basin (T28 - 29N, R108W) (**Fig. 14.1**). The Hoback Basin lies between the Wyoming Thrust Belt to the west and the Wind River Mountains to the east. The field is 60 miles north of Rock Springs, WY. By mid-1999, more than 100 commercial wells had been drilled on 80-acre spacing. Field boundaries to the north, south, and west have been well defined. The eastern (downdip) edge of the field is still being extended. The known limits of the field exceed 33 square miles.

Jonah Field is bounded on the south and northwest by two wrench faults. Faults within the field boundaries add to the complexity of the reservoirs. Wells within the field encounter overpressured gas at 8,100 to 9,300 ft (0.58 to 0.65 psi/ft gradient), whereas nearby wells drilled across the bounding faults find normal pressure gradients at similar depths.

The Lance Formation is from the Upper Cretaceous age and consists of 2,000 to 3,000 ft of interbedded fluvial sands, mudstones, and coals. A log section of the Lance is shown in **Fig. 14.2.** Individual sandstone units range from 5 ft to more than 50 ft in thickness, and have areal extents ranging from a few acres to 100 acres. Individual sands are geologically heterogeneous reservoirs because of their depositional shapes, but certain stratigraphic intervals consistently have sands developed. Sand-rich intervals are locally called the Upper Lance, Middle Lance, Jonah, Yellow Point, Wardell, and Upper Mesaverde (or Rock Springs). Total net sand in the field ranges from 300 to 600 ft of stacked net-pay. Drilling depths range from 11,000 to 12,500 ft, depending on the number of sand packages an operator believes to be economically feasible to develop. More specific geological descriptions can be found in Warner (1997) and Bowker and Robinson (1997).

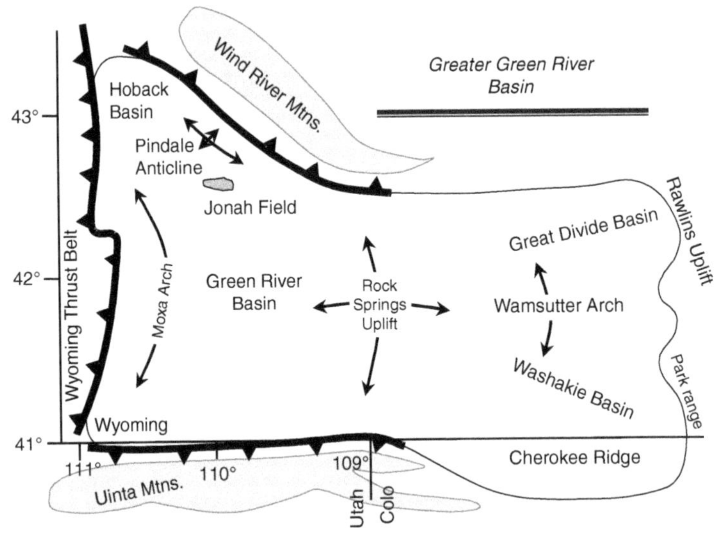

Fig. 14.1—Location and geological setting of Jonah Field.

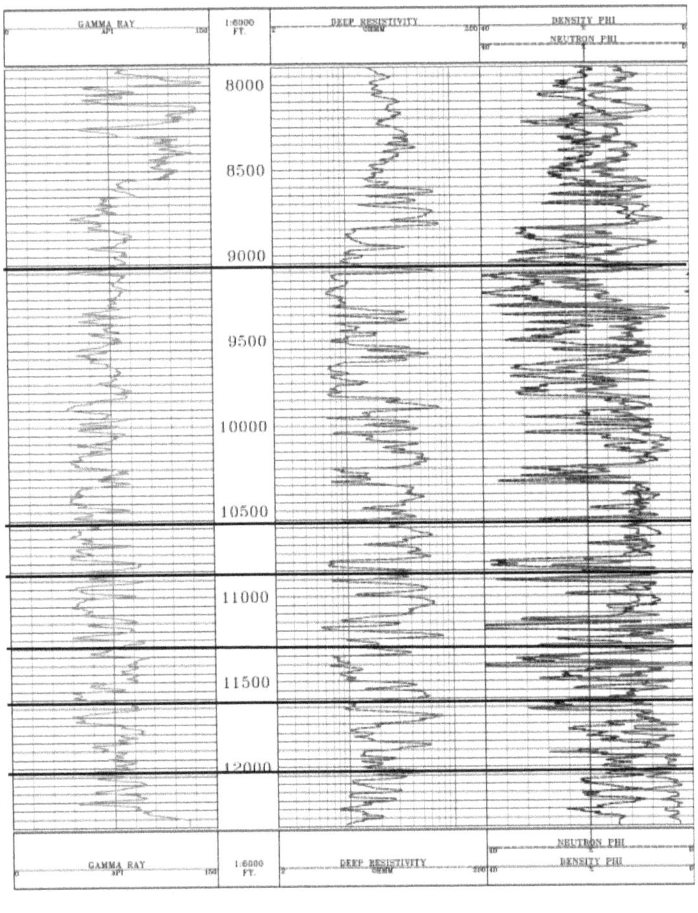

Fig. 14.2—Log of Well 3.

Sand porosity ranges from 5 to 14%, with relative gas permeability ranging from 0.001 to 0.02 md. Water saturation varies from 30 to 60%. Currently, there is no significant water production in the field. The producing condensate yield is between 8 and 10 bbl/MMscf, with an API gravity of 52°. Pressure-volume-temperature (PVT) fluid data are scarce, although it appears that the fluid composition is similar throughout the entire productive section.

At the time of the study, there were five operators in Jonah, each with a different completion strategy; however, because of the low permeability of the Lance Formation in this area, all operators used multiple limited-entry fracture treatments to complete wells. Although all operators used hydraulic fracturing for stimulation, a variety of different treatment types, methods for fracture isolation, and time between treatments (Flack et al. 1997; Integrated Petroleum Technologies, Inc. 1996) were used. Typically, three to six individual sands were completed per fracture treatment (stage), with four to six stages per well.

After a fracture treatment, the well was flowed back for 1 week or longer for cleanup. Pumping a sand plug to cover previous stages was the most common method for isolating additional fracture treatments.

One factor common to all the operators was the 58% completion efficiency. Because stimulation costs typically represent more than 50% of total well costs, this low completion efficiency was a major concern. Before production could be fully optimized, the mechanism(s) causing the low completion efficiency had to be determined. Some preliminary areas of concern were

- Misidentification of productive sands
- Lack of effective stimulation with the limited-entry technique
- Damaging isolation methods
- Improper fracture treatment design

A fully integrated field study was required to evaluate each of these items. The study incorporated detailed log analysis, fracture treatment analysis, production and reservoir analysis, and completion practices. This paper presents the results of this study.

14.3 Analysis Methodology

The first step for any integrated study is to determine what information is available and where these sources are. For this study, information for some wells was provided by the operator of the well. For other wells, only information from public sources was available. Data captured for each well ranged from overall well information to specific information on each log zone. This information was entered into a relational database. Populating the database with accurate information was very time-consuming. This step alone represented approximately 60% of the time spent on this study. Four levels of data were available for these wells:

14.3.1 Zone. Zones were determined by log interpretation based on lithology. There may be 20 to 40 zones throughout the pay interval. Each zone was identified by a number and formation name. Rock properties for each zone were measured or calculated and entered into the database. When production logs were run, a production rate for each zone was added to the database. This allowed reservoir data to be compared on a "per zone" basis.

14.3.2 Stage. A fracture treatment represents a stage, and information for each treatment was entered into the database. Prom the database, zonal information was calculated for each stage, allowing analysis of treatment and reservoir data on a "per stage" basis. Also, if the stage was isolated from other completed stages, flowback and production data for that stage could be analyzed.

14.3.3 Well. Production data, reported per well, were also entered into the database, and overall log and treatment values for the entire well were calculated. Reservoir, treatment, and general data could be compared on a "per well" basis.

14.3.4 Log Normalization. When the logs were compared, the baseline shale readings of bulk density (RhoB), gamma ray (GR), and neutron porosity (PhiN) differed slightly from well to well. For engineers to effectively use one petrophysical model for the field, all the wells were normalized to a "field histogram" of RhoB, GR, and PhiN constructed from the top of overpressuring to the total depth (TD) drilled. Normalization was based around the shale point on the histogram for each well. This normalization process allowed a comparative analysis from well to well.

Next, the normalized log data were calibrated to core data. Core data were available from two wells. Cored intervals captured sand from the Upper Lance, Middle Lance, and the Jonah Sands. Grain density, porosity, permeability, and water saturation were measured in all the sands recovered. Prom the core analysis, engineers decided to use a grain density of 2.68 gr/cc in log calculations.

In this field, density log values overestimate core porosity and the neutron log values underestimate core porosity; therefore, log porosity was calibrated to core porosity using a weighted average of the neutron and density porosity. Eq. 14.1 was used to establish a reasonable match using neutron and density porosity to derive effective porosity from logs.

$$Phi_{xx} = (4\phi_D + 6\phi_N)/10 - 0.01 \quad (14.1)$$

If a correlation existed between the core permeability and porosity, a log-derived porosity could be used to derive a permeability value. Some core permeability measurements were recorded with confining stresses of 0, 800, 2,000, and 4,000 psi. Results from this testing show more than an 80% reduction in air permeability at reservoir conditions. Eq. 14.2 was developed to determine permeability from log-derived porosity based on the relationship between core porosity and permeability.

$$k_{est} = \frac{20.323\left[0.000426e^{22.48Phi_{xx}}\right]}{\left(d \times P_p\right)^{0.5637}} \quad (14.2)$$

Water-saturation (S_w) measurements from the core in tight reservoirs can often be taken as the minimum water saturation in the reservoir. In this study, several standard S_w equations were evaluated. The Simandoux S_w calculation was chosen as the model that consistently yielded favorable results with core results.

Prom the log normalization and calibration procedure, a log model describing porosity and permeability for Jonah Field was developed. Pay cutoffs were established for the log model. These cutoffs were used to determine which sands should be included in net-pay calculations. **Table 14.1** summarizes these cutoffs.

Because a significant amount of production-log data were available, this information was also included in the log analysis. Although production logs are not an exact measure of how each zone will produce over time, they do provide insight into which zones contribute to production. Because production logs were run at different times after a well was completed, a method was needed to normalize production log rates per zone. For this study, the normalization process divided the production log rate for each zone by the total production rate for the well. This gave a percent contribution for each producing zone. The production period used for comparison was then multiplied by this percentage to determine a normalized rate for each zone.

Variable	Value	Unit
Gross Sand	< 65	API units
Litho Type 1	< 50	API units
V_{shale}	< 60	%
Phi_{xx}	> 8	%
Water Saturation	< 70	%

Table 14.1—Net-pay cutoffs for log evaluation.

Subsurface and 3D seismic interpretation had identified numerous faults within the field (Warner 1997). First assumptions in the development in the northern portion of Jonah Field were that these faults were not sealing and overpressured sands would have uniform production characteristics. To determine if compart-mentalization was present, production log surveys, openhole logs, cumulative production, gas/oil ratio (GOR), and flowing bottomhole temperature and pressure were used. If the faults were sealing, as was the case with the two wrench faults, isolated pressure compartmentalization could have affected production.

One of the most critical characteristics in production optimization is the relationship between production and effective fracture half-length. A three-phase, 3D reservoir simulation program was used to determine this relationship by comparing individual well production to treatment size and accounting for reservoir quality. The initial wells chosen for analysis had several months of production history, and were completed with only one or two stages. Choosing these types of wells, in this relatively young field, reduced the number of uncertain reservoir properties such as permeability, drainage area, and effective fracture half-length. In the reservoir simulation model, each stage was represented by one producing layer. For each layer, the total net thickness was determined from logs and multiplied by 65% to account for completion inefficiencies. Porosity, permeability, and water saturation were averaged for input into the model. Once the model was built and initialized, production history matching was usually achieved by varying the effective fracture half-length, gas permeability, and/or drainage area. After an acceptable match was obtained, the effective fracture half-length was compared to treatment size. Gas permeability was compared to the porosity/permeability relationship being developed.

When an adequate amount of data is available, statistical analysis can help researchers understand the impact of different variables on production. A statistical analysis was prepared for each of the three levels of investigation available. Statistical comparisons in this study were based on cumulative production. Cumulative production was chosen over estimated ultimate recovery (EUR) because of the differences in the way each operator calculated EUR. Because completion times for these wells varied significantly, it was necessary to establish a standard time to start cumulative production. For this study, the start of production was considered to be the month of maximum production following the final fracture treatment.

ANNs have been used successfully in other fields to analyze and quantify the effect of reservoir/treatment/operation parameters on production response (Shelley et al. 1998).

ANNs have also been used to predict porosity and permeability from well logs (Olson 1998). Detailed descriptions of ANNs can be found in both references. As with the statistical analysis, ANNs were run for each of the three levels of investigation. For zonal and stage analysis, production log rates were used. For total well comparisons, cumulative production was used.

14.4 Integrated Study Analysis

Forty-five wells in the database had one or more months of production, and 25 wells had more than one year of production data. To determine the minimum production period that would still provide confidence in the analysis, crossplots were created of different cumulative production periods. **Table 14.2** summarizes the comparisons. There were five more wells with 3-month production data than with 6-month cumulative production, and 3-month cumulative production did a reasonable job of predicting 1-year production. Therefore, 3 months was the cumulative production period used for comparison in this analysis.

14.4.1 Log Evaluation Based on Zonal Data. Of the 540 zones that met the pay criteria, 417 were completed. Production logs were run over 363 of these zones in 21 wells, with only 212 zones indicating they were producing. The logs show that only 58% of completed zones were producing at the time the logs were run. **Fig. 14.3** shows a breakdown of percentage of net-pay production for each well; wells are shown in chronological order. Although completion efficiency appeared to improve with time, results were still disappointing.

Production log comparisons of producing to nonproducing zones were evaluated for several properties: net-pay, porosity, resistivity, bulk hydrocarbon volume, etc. In one example of this type of analysis using net-pay, a bar graph was created based on sequential 10-ft net-pay thickness bins. The number of producing and nonproducing zones in each bin (the X-axis label) shows the number of producing and nonproducing zones represented in each bin. As **Fig. 14.4** shows, a zone tends to produce based on increasing thickness. Similar trends were noticed for all the variables analyzed.

One problem with the bar graph is that it does not show if there is a relationship between a zone's net-pay height and production log rate. To evaluate this relationship, a graph of production log rate normalized with 3-month cumulative production vs. net-pay height was created. **Fig. 14.5** shows no clear relationship between net-pay thickness and production log rate on a field wide basis, indicating a significant difference in production from zones identified as pay. To investigate if any log properties would help differentiate pay quality, zones were categorized into eight bins, based on the zone's normalized production log rate, with an additional bin representing all wells with no production log rate. Other than a slight relationship to net-pay height, no log properties appeared to correlate to better production.

Because each zone correlates to a formation in the database, formation to log data were also compared. Again, no log properties appeared to correlate to production; however, this comparison did show definite differences in production response from the different formations as shown in **Table 14.3.** On average, more than half the production comes from the Upper Lance. For most wells that were completed down to the Rock Spring, the Jonah, Middle, and Upper Lance produced more than 70% of the wells' production.

	1-Month	3-Month	6-Month
3-Mo. Cum. (44 wells)	0.957	1	—
6-Mo. Cum. (39 wells)	0.867	0.95	1
1-Yr. Cum. (25 wells)	0.689	0.835	0.958

Table 14.2—R^2 results from cumulative production crossplots.

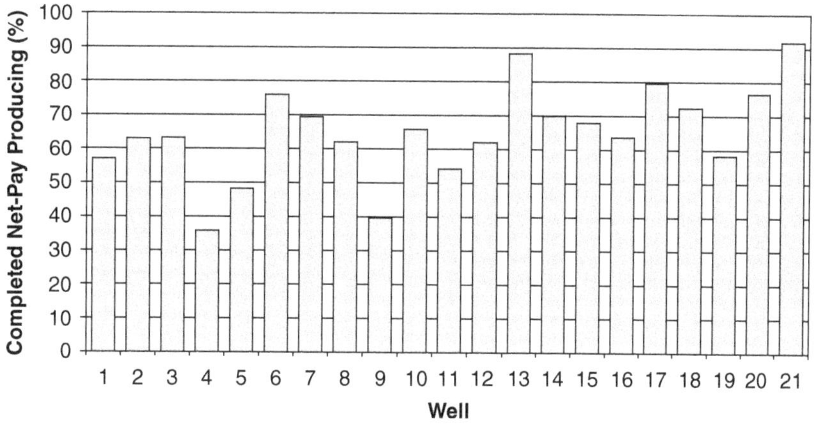

Fig. 14.3—Percent of completed net-pay producing at time the production log was run.

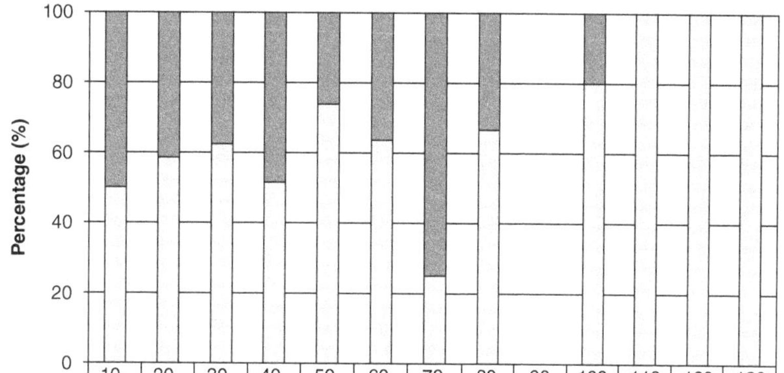

	10	20	30	40	50	60	70	80	90	100	110	120	130
■ Non-Producing	38	55	27	16	6	4	3	1	0	1	0	0	0
□ Producing	38	78	45	17	17	7	1	2	0	4	1	1	1

Zonal Net-Pay Thickness (ft)

Fig. 14.4—Comparison of producing to nonproducing zones based on zone thickness.

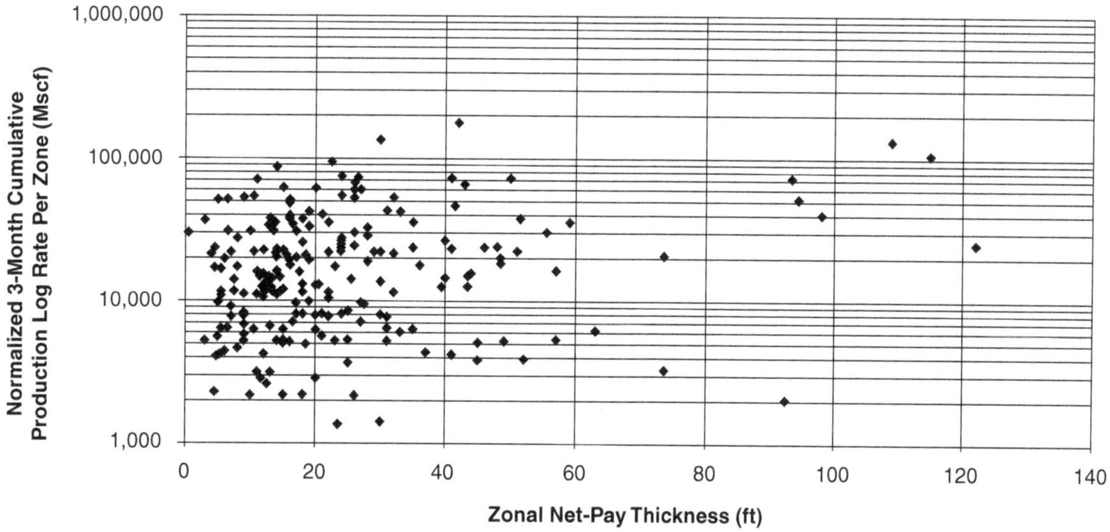

Fig. 14.5—Comparison of a zone's normalized production to the zone's net-pay thickness.

14.4.2 Log Evaluation Based on Stage and Well Data. As with the zonal analysis, there were no clear relationships between production and any of the log properties. The most important observation from the evaluation of production log data on a stage basis was that of 101 stages with production log data, 10 did not show production from any perforations at the time of the production log, yet each stage produced before being isolated to complete additional stages uphole.

Log	Formation					
Data	RS	Wardell	YP	Jonah	ML	UL
Gross Sand (ft)	57.8	92.3	53.8	124	124.9	386.1
Net-Pay (ft)	46.3	64	38.3	95.9	91.3	252.9
PL Rate (Mscf)	42,966	40,332	14,309	72,811	29,170	163,041
PL Rate/Net-Pay (Mscf/ft)	928	630	373	758	319	644
No. of Wells	4	10	13	15	14	16

Table 14.3—Comparison of formation type with log data. PL Rate = normalized production log rate.

The number of entry points per stage varied from 1 to 10, and entry points were shot 3 spf at 120°. For stages with six or more entry points, only 40% of the entry points had a production rate. When five or fewer entry points were used, 63% produced. Of the 61 stages with six or more entry points, there were no cases with production from more than six entry points.

When the total well was examined, there were still no reasonable correlations of production to log properties; however, correlations to gross sand interval and net-pay did improve significantly. **Table 14.4** summarizes correlation coefficients for the log properties evaluated.

14.4.2.1 Compartmentalization. The analysis of the production log data suggests that some compartmentalization may exist between fault blocks. **Fig. 14.6** shows the main fault blocks that were analyzed, overlaid on a map of 3-month cumulative production. Fault blocks were identified by Roman numerals.

If the faults were sealing, noticeable differences should have been observed between the flowing pressure, temperature, and possibly the GOR from the differing fault blocks. The data available are sparse in some of the fault blocks, but in general, all the fault blocks were very similar in their flowing bottomhole pressure, temperature gradient, and GOR. Based on these similarities, no strong evidence of compartmentalization was found.

	GR	SP1	RT	PhiD	PhiAvg	Rwa	SwMS	GSandFt	Net-Pay
3-Mo. Cum.	0.31	−0.34	−0.38	0.39	0.35	−0.39	0.3	0.64	0.69

Table 14.4—Correlation coefficients for log properties to 3-month cumulative production.

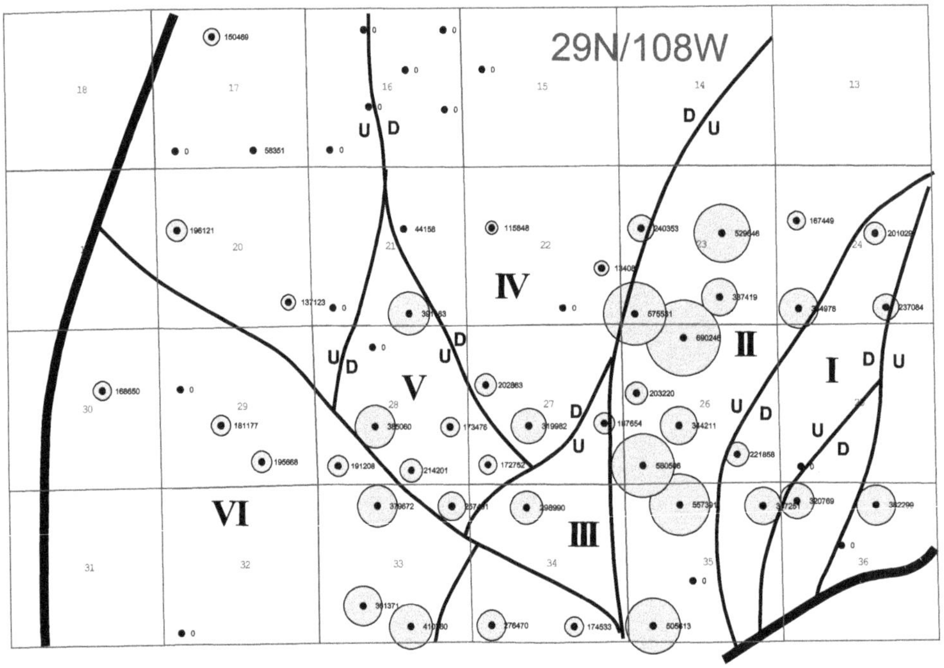

Fig. 14.6—Bubble map of 3-month cumulative production (MCP) overlaid with identified faults. Fault blocks are identified by Roman numerals. U/D signifies upthrown and downthrown areas.

In **Fig. 14.7,** a graph of producing net-pay reflects at least two producing trends of net-pay to production. Wells along the upper trend line are in upthrown fault blocks. Because there was no evidence of compartmentalization, the different producing trends between fault blocks suggested that microfracturing might enhance permeability in the upthrown fault blocks.

14.4.2.2 ANN Analysis of Log Data. An ANN was used to evaluate log data. Several attempts using numerous techniques were made to establish networks to predict normalized production per zone or stage based on log properties. The ANN was not able to make any reasonable prediction of production for either zones or stages. One possible reason for the lack of predictability was the use of normalized production. The ANN was much more successful in predicting production from the log parameters based on the total well. Networks developed based on the total well generally used net-pay thickness as the highest impact variable.

14.4.3 Fracture Treatment Analysis. The database contained 87 stages with fracture treatment information and production log data. From the fracturing treatment information, two levels of data are available: per stage and per well.

14.4.3.1 Stage Analysis. The first step was to look at treatment variables with respect to production. One of the main variables was the interval length completed with a stage. **Fig. 14.8** shows the total interval length completed ranked by each interval's 3-month cumulative production normalized by net-pay thickness. Better production was associated with larger intervals. Because longer intervals typically have more net-pay, a comparison of the interval net-pay ranked by the same normalized production rate was conducted as shown in **Fig. 14.9.** This graph shows an inverse trend of net-pay thickness to normalized production, so that even with less net-pay, longer completion intervals had better production.

Although it was not clear why increased interval length could result in improved interval production, some possible reasons are provided:

- Pumping larger treatments associated with longer intervals may result in placing better fractures in better zones.
- Fewer stages mean fewer cycles of producing/killing each stage before the well is placed on production, thereby reducing the chance of resaturating the reservoir rock, possibly resulting in increased water saturation and decreased relative permeability to gas.
- Improved fluid cleanup means higher gross gas production.

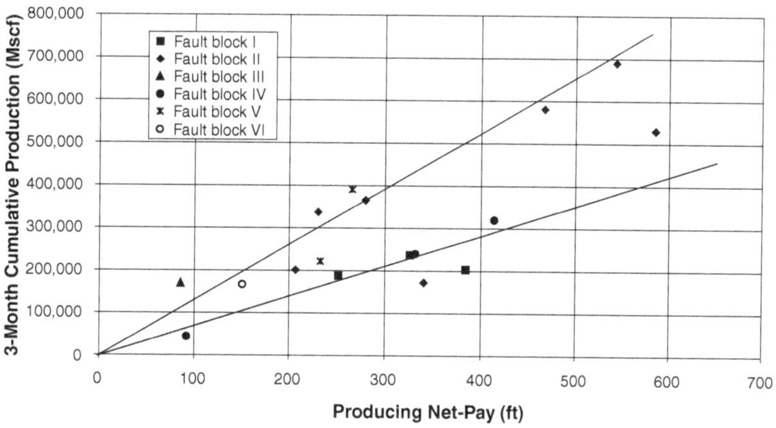

Fig. 14.7—Comparison of producing net-pay to 3-MCP showing producing trends of different fault blocks.

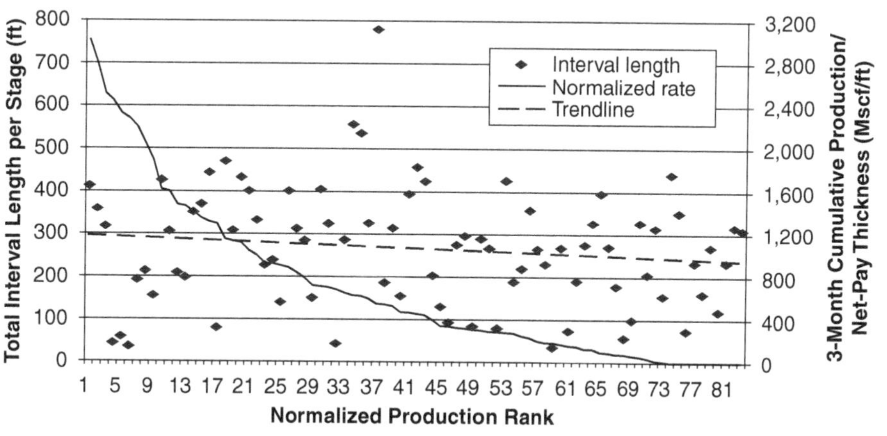

Fig. 14.8—Forced ranking graph of stage interval lengths based on each stage's 3-MCP per net-pay thickness.

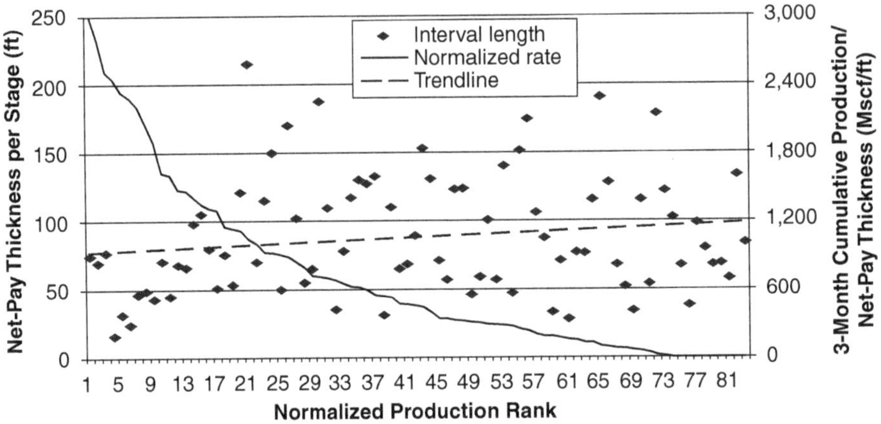

Fig. 14.9—Forced ranking graph of stage net-pay thickness based on each stage's 3-MCP per net-pay thickness.

Similar comparisons were done with treatment variables such as proppant volume, fluid volume, pad percentage, etc. Because of the correlation between net-pay and interval production, all variables were normalized to net-pay for comparison and graphed with a ranking from best to worst. Using this graphical technique, no correlations were evident of treatment volumes and interval production. **Figs. 14.10 and 14.11** show the graphs for proppant and fluid volumes, respectively.

Production rate per stage was broken into five bins of 20,000 Mscf, with a sixth bin representing stages with zero production. **Table 14.5** shows the average value for each variable in each bin. As with the forced rank graphs, there did not appear to be a trend for either proppant or fluid volume based on interval production; however, the increasing trend toward net-pay was again evident. The treatment volumes for the wells in the zero production bin had average job sizes larger than most of the other bins. This raised the question of what effect treatment size has on production.

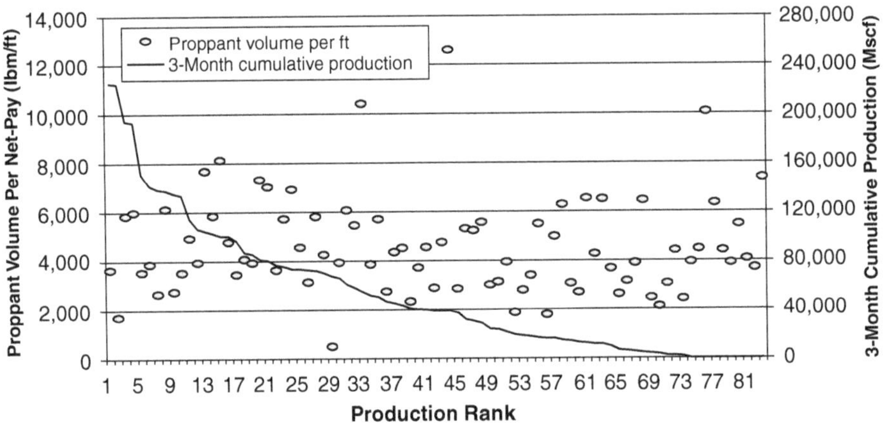

Fig. 14.10—Forced ranking graph of proppant volume pumped per stage based on each stage's 3-MCP.

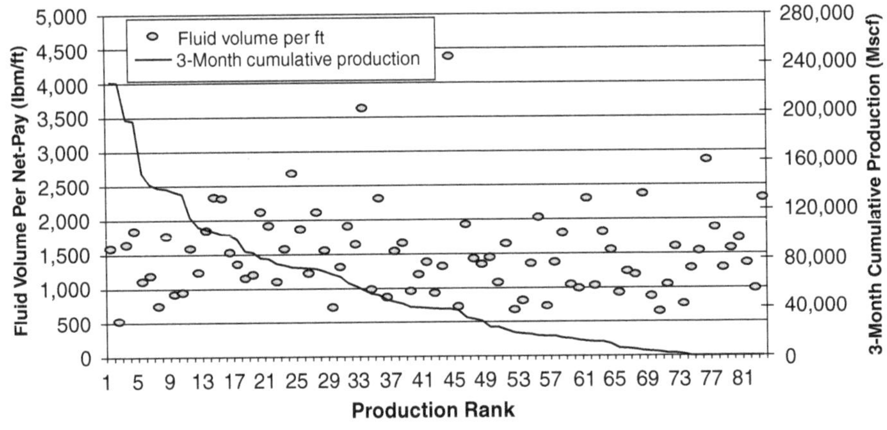

Fig. 14.11—Forced ranking graph of fluid volume pumped per stage based on each stage's 3-MCP.

Data	3-Month Production Rate Per Stage					
	0K	0 to 50K	50 to100K	100 to 150K	150 to 200K	> 200K
Total Fluid	126,576	101,300	131,119	134,976	128,218	116,892
Total Proppant	405,842	313,525	390,463	451,088	433,640	321,100
Gross Pay	105	118	135	142	107	178
Net-Pay	79	83	86	105	73	144
Fluid/Net-Pay	1,683	1,371	1,707	1,464	1,745	1,064
Fluid/Gross Pay	1,313	929	1,199	1,030	1,197	803
Prop/Net-Pay	5,375	4,093	5,042	4,817	5,916	2,686
Prop/Gross Pay	4,182	2,766	3,542	3,422	4,046	2,052
PL Rate	0	21,059	71,240	122,019	193,819	224,895
No. of Stages	10	38	19	12	2	2

Table 14.5—Comparison of treatment volumes per stage.

The effect of discrete variables such as fluid or proppant type, pump rate, additives, etc., was investigated. Two specific variables that were compared were 100-mesh proppant and nitrogen. As **Table 14.6** shows, the use of both of these variables had a negative impact on production. This analysis also showed a positive correlation to using lower gel-loading fracturing fluid.

14.4.3.2 ANN Analysis of Fracture Stage Data. The problem with straight statistical analysis is differentiating between the effect of one variable on another: for example, gel loading and fluid type. This analysis is made easier when an ANN can be trained to predict interval production based on treatment variables. For fracture stage analysis, all treatment data were entered in the ANN. Half the stages were used for training, and the other half were used for testing. No networks were developed with reasonable correlation coefficients, probably because of the use of normalized production based on production log data.

14.4.3.3 Well Analysis. The database contained 31 wells with treatment and production data. Of these, 14 had log interpretation information. The influence of net-pay on production was discussed earlier, and was investigated for the 14 wells used in this section. The overall correlation coefficient between 3-month cumulative production and producing net-pay was 0.78. When the wells were placed into two fault block groups based on the compartmentalization analysis, the correlation coefficients became 0.84 and 0.89 for Groups 1 and 2, respectively. All wells in the upthrown Fault Blocks II, III, and V were placed in Group 1. All other wells were in Group 2.

These groups were used to analyze proppant and fluid volumes. When analyzed in this manner, there was actually a negative correlation to treatment size and 3-month cumulative production, as shown in **Figs. 14.12 and 14.13.** This analysis did not account for proppant type and the different ratios of intermediate/low-strength proppant volumes pumped. However, it does again validate the theory that treatment size does not improve production, and, in fact, may hinder it.

	Yes	No
Nitrogen	43,704 Mscf (52 stages)	70,414 Mscf (31 stages)
100-Mesh	44,579 Mscf (54 stages)	70,632 Mscf (29 stages)

Table 14.6—Comparison of the use of 100-mesh and N_2.

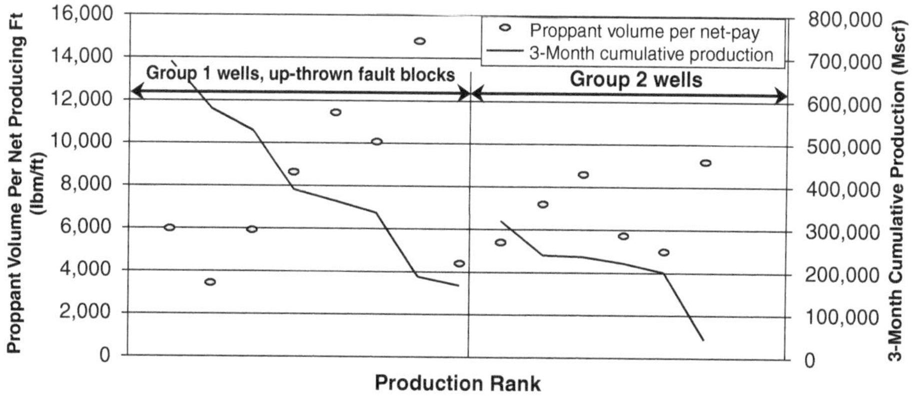

Fig. 14.12—Forced ranking graph of proppant volume pumped per well based on each well's 3-MCP, separated by fault block group.

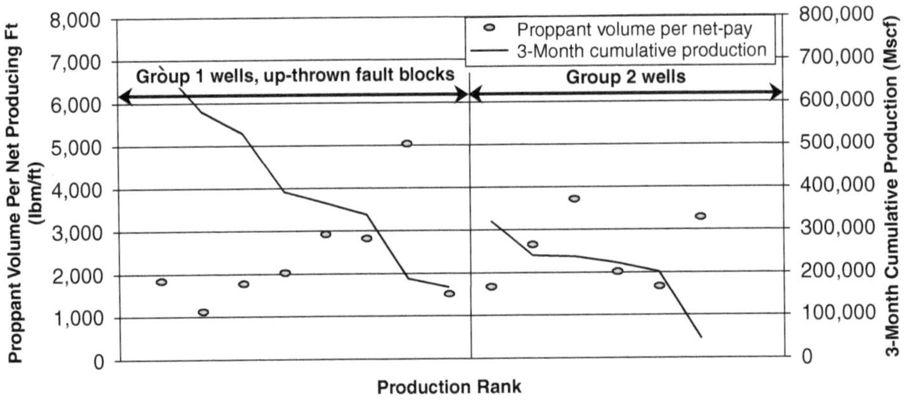

Fig. 14.13—Forced ranking graph of proppant volume pumped per well based on each well's 3-MCP, separated by fault block group.

14.4.3.4 ANN Analysis of Fracture Treatment Per Well Data. This analysis was based on all fracture treatment variables, with the addition of the operator. By including operators, their completion techniques could be indirectly evaluated. Using total well data, the ANNs effectively trained on the data and predicted a result. Correlation coefficients better than 0.90 were achieved for several networks.

For the best network developed, the variables used were operator, number of stages, pad percentage, treatment rate, and proppant volume. These were essentially the same variables used by all ANNs for total well fracture treatment analysis. To evaluate the relative importance of a variable in a network, sensitivity tests were performed. To test sensitivity, the value of a variable is changed and the network is rerun. The new predicted value is then compared to the original predicted value, and the percentage of deviation is calculated. The percentage of deviation is not meant to be an absolute prediction of what the change would mean to production, but rather as a relative indication of which variables could have the biggest impact on production. For variables with actual values, such as fluid volume, the variable is increased or decreased by a percentage, in this case 20%. For yes/no variables, testing was carried out by converting all variables to either yes or no.

Table 14.7 shows the results of the sensitivity analysis for the best network. Pad percentage had the biggest impact on production. Decreasing all pad volumes by 20% resulted in a 30% increase in production. This was the only network where increasing the proppant volume had more than a 9% deviation. Most of the networks also showed that reducing the number of fracture treatment stages increased production.

14.4.4 Log/Fracture Treatment Analysis. When treatment information from 72 stages in 14 wells was combined with log data, the ANNs predicted production more efficiently. In addition to net-pay, the ANNs used both log and treatment variables in all the runs. The most common log property used by the ANNs was density porosity. In all cases, increasing density porosity resulted in better production. Percent pad and fluid volume were the most often used treatment variables. Results were consistent with other ANNs: decreased pad and increased fluid volume resulted in better production. Whenever the ANN chose 100-mesh proppant or nitrogen as a variable, the result was consistent with the statistical analysis. Eliminating both 100-mesh and nitrogen resulted in better production.

14.4.5 Tracer Log Analysis. Tracer log data were available for 23 wells. Treatments in these wells were limited to four to six zones per stage. In most of the treatments, two tracers were run per stage: iridium in the early proppant stage, and either scandium or antimony in the late proppant stage. The scandium and antimony tracers were alternated between stages coming up the wellbore. Results from the surveys indicated that more than 95% of the zones took early and late proppant. Tracers also indicated that fracture height was "contained" within gross sand intervals.

	Predicted	Actual				
Total	9,563,408	9,904,714	Correlation Coefficient = 0.93			
Variable	Operator				No. Stages	
Variance	1	2	Neither		5	10
Total	8,076,967	9,119,581	10,381,602		9,676,123	7,690,146
% Deviation	–0.16	–0.05	0.09		0.01	–0.2
Variable	% Pad		Rate		Proppant Volume	
Variance	80%	120%	80%	120%	80%	120%
Total	12,396,771	7,229,925	9,394,584	8,009,444	7,585,713	11,508,505
% Deviation	30	–24	–2	16	–21	20

Table 14.7—Sensitivity results for treatment data (per well).

	Fault Block	Perm (md)	Net-Pay (ft)	Drainage Area (acre)	Fracture Length (ft)	Proppant Vol/Net-Pay (lb/ft)	3 MCP/ Net-Pay
Well 1	V	0.003	159	15	350	8,800	1,347
Well 2	III	0.005	290	20	375	4,700	1,012
Well 3	II	0.006	585	40	200	4,100	963
Well 4	I	0.003	384	20	150	3,000	523

Table 14.8—Decline curve analysis summary.

14.4.6 Reservoir Analysis. Reservoir analyses from four wells are presented in this paper. Well 1 was originally completed in only the Wardell, Yellow Point, and Jonah Formations, with two limited-entry treatments. A total of 1,400,000 (lbm/ft) of resin-coated sand was placed, with a nitrified borate gel. The well, located in Fault Block V, was produced for more than 18 months from this completion before adding pay uphole. Well 2, located in Fault Block III, was completed from 10,344 to 8,703 ft with four limited-entry fracture treatments. A total of 2,042,000 (lbm/ft) of mixed proppant was placed with a nitrified borate system. Well 3, located in Fault Block II, was completed from 12,230 to 9,310 ft with nine limited-entry fracture treatments. A total of 3,422,000 (lbm/ft) of mixed intermediate-strength proppant was placed with a low gel-loading nitrified borate system. Well 4, located in Fault Block I, was completed from 12,084 to 9,816 ft, with 10 stages limited to two zones per stage. A total of 1,846,000 (lbm/ft) of mixed proppant was placed with a low gel-loading nitrified borate system.

Table 14.8 shows the simulation results for these four wells. The typical drainage area ranges from 15 to 20 acres. The largest drainage area modeled was 40 acres. Based on these matches, fracture lengths should be designed for 20 acres in most areas. This table also shows a contradiction on proppant volumes pumped. Of these four wells, the best performing well on a net-pay basis also had the most proppant pumped per foot of net-pay. Only two stages were pumped in this well; therefore, the bottom stage was isolated only once. Again, the number of times a stage was isolated was a major concern.

The designed fracture length for Wells 3 and 4 was from 500 to 600 ft. The modeled overall effective fracture lengths for these wells were 200 and 150 ft, respectively. All the treatments in these two wells were nitrified limited-entry, making accurate design modeling and post treatment matching very difficult.

A Jonah-type well was developed and entered into the reservoir simulator. The completion efficiency was increased from 65 to 90%. If this efficiency could be achieved, recovered gas could be increased by 800 MMscf (38%) in 3 years. Based on USD 2.50/Mscf, the finding cost for the type well was USD 0.25/Mscf. If the goal was simply to reduce costs with the same expected completion results, the well costs would have to be reduced by 28% to maintain a similar cost benefit.

14.4.7 Microfracturing. Possible microfracturing was investigated, using small volume pump-ins (Barree 1998) in two wells in Section 24. Of 21 stages, pressure-dependent leakoff was observed in only one stage. One of these wells was in Fault Block II, and was the worst producer in that fault block. Two pump-ins were carried out in one well in Section 23. Pressure-dependent leakoff was observed in both stages. Although these data are limited, they help support the case for microfracturing in upthrown fault blocks.

14.5 Summary

No discernable correlation was seen in log properties between zones that showed production at the time the production log was run compared to zones that did not show production. Tracer-log data indicated that 95% of the zones took some amount of proppant. Additionally, entire stages that produced before isolation did not produce at the time of the production log. These facts indicate that the primary cause for the low completion efficiency was the isolation method used.

14.6 Conclusions

1. The low completion efficiency was caused by damaging isolation practices. By increasing the completion efficiency to 90%, production could be improved 38%.
2. Although no compartmentalization was observed, production was dependent on the fault block in which the well was located. Wells in upthrown faults were better producers on a net-foot-of-pay basis.
3. More than one-half of a well's production comes from the Upper Lance. Location should be considered before drilling to the Wardell.
4. Economics based on a 20-acre drainage area should be used for fracture treatment designs, with the possible exception of Fault Block II.
5. Fracture treatment design changes would employ less pad and more sand-laden fluid. Increasing proppant volume does not result in better production.
6. Log-derived net-pay using GR and Phi_{xx} cutoffs is the best indicator of a well's potential.

14.7 Nomenclature

ϕ_d = density porosity, %

ϕ_N = neutron porosity, %

Phi_{xx} = effective porosity from logs, %

D = depth, ft

P_p = pore pressure gradient, psi/ft

k_{est} = log-derived permeability, md

S_w = log-derived water saturation, %

14.8 Acknowledgments

The authors sincerely thank everyone who helped compile the data for this study. Thanks are also extended to Ultra Petroleum and Halliburton Energy Services Inc. for permission to publish this paper.

14.9 References

Barree, R. D. 1998. Applications of Pre-Frac Injection/Falloff Tests in Fissured Reservoirs - Field Examples. Presented at the Rocky Mountain Regional Conference, Denver, Colorado, 5–8 April. SPE-39932-MS. https://doi.org/10.2118/39932-MS.

Bowker, K. A. and Robinson, J. W. 1997. Jonah Field: A Shallow Sweetspot in the Basin-Centered Gas Accumulation of the Northern Green River Basin, Wyoming. Presented at the Rocky Mountain Section Meeting AAPG, Denver, Colorado, 24–27August.

Flack, D. Shaw, J., Martin, J. et al. 1997. Modified Drilling and Completion Procedures Enhance Recovery in South West Wyoming Field. Presented at the Rocky Mountain Regional Conference, Casper, Wyoming, 18–21 May. SPE-38344-MS. https://doi.org/10.2118/38344-MS.

Integrated Petroleum Technologies, Inc. 1996. Advanced Stimulation Technology Deployment Program, Snyder Oil Corporation, Lance Formation, Jonah Field, Sublette County, Wyoming. Topical Report, GRI-96/0371, Golden, Colorado (August 1996).

Olson, T. M. 1998. Porosity and Permeability Prediction in Low- Permeability Gas Reservoirs From Well Logs Using Neural Networks. Presented at the Rocky Mountain Regional Conference, Denver, Colorado, 5–8 April. SPE-39964-MS. https://doi.org/10.2118/39964-MS.

Shelley, R. F., Stephenson, S., Haley, W. et al. 1998. Redfork Completion Analysis with the Aid of Artificial Neural Networks. Presented at the Rocky Mountain Regional Conference, Denver, Colorado, 5–8 April. SPE-39963-MS. https://doi.org/10.2118/39963-MS.

Warner, E. W. 1997. Structural Geology of Jonah Field as Determined from Subsurface Geology and 3-D Seismic. Presented at the Rocky Mountain Section Meeting AAPG, Denver, Colorado, 24–27August.

SI Metric Conversion Factors

acres × 4.047	E – 03 = km^2
bbl × 1.59	E – 01 = m^3
lbm × 4.536	E – 01 = kg
Mscf × 28.32	E + 00 = m^3/d
md × 9.87	E – 04 = mm^2
ft × 3.048*	E – 01 = m
gal × 3.785	E – 03 = m^3

*Conversion factor is exact.

Chapter 15

Producing Characteristics and Drainage Volume of Dakota Reservoirs, San Juan Basin, New Mexico

Arild Sunde, Her-Yuan Chen, and **Lawrence W. Teufel,** New Mexico Institute of Mining and Technology

SPE 60288 presents a case history of evaluating approximately 500 wells in the Dakota Formation in the San Juan Basin in New Mexico. Sunde et al. used type curves to match rate/time data from each well and obtain productivity index (PI), pore volume (PV), initial gas-in-place, drainage area, skin, and permeability, which were then put on maps or graphed vs. time to help identify infill drilling potentials. Through these case histories Sunde et al. were able to determine the following:

- Dakota wells exhibit steep decline during transient flow and then exponential decline during boundary-dominated flow
- The average permeability is on the order of 0.01 md
- The drainage area does not depend on the date of production
- PI does not seem to depend on initial date of production, but is highly heterogeneous.

15.1 Abstract

Approximately 500 wells from the low-permeability, tight-gas Dakota Formation in San Juan Basin, New Mexico, were analyzed to characterize the producing/flow characteristics, to delineate the drainage volume/area, and to infer the infill potential. Production data is from both commercial and public databases. Decline-curve analysis is the primary method. Results indicate that Dakota wells that started producing after the seventies generally have very steep decline during transient flow period. This is due to the negative-skin effect resulted from well stimulation. During boundary-affected flow period, the production decline tends to be exponential-decline-like (low decline exponent). Estimated gas permeability is in the order of 0.01 md. Drainage pore volumes of the wells that started producing during the early eighties are slightly smaller but more uniformly distributed than the wells that started producing during the sixties and seventies. Drainage areas, however, appear to be less dependent on the date of production than the pore volumes. Productivity index is highly variable over both basin and time, suggesting a highly heterogeneous flow/permeability field in the Dakota Reservoirs.

15.2 Introduction

The San Juan Basin is roughly a circular, asymmetrical structure located in northwestern New Mexico and southwestern Colorado; see **Fig. 15.1** (Dutton et al. 1993). Gas production from the San Juan Basin is primarily in reservoirs of the Mesaverde Group and Dakota Sandstone. Only Dakota Reservoirs are considered in this study.

The tight Dakota Reservoirs are generally considered to be naturally fractured. Permeability is in the range of 0.024 to 0.077 md, porosity is 2 to 16%, and water saturation is 40 to 60% (Dutton et al. 1993). Well stimulation (generally hydraulic fracturing) is required for economical production. Approximately 4,000 active Dakota wells are producing from the San Juan Basin, and cumulative production since first delivery production is reported to be 5.4 Tcf by April 1999 (IHS/Dwights LLC 1999). Current well spacing is 160 acres. Haas and Crist (1991) reported that the drainage areas are generally less than 160 acres. Their study also concluded that the Dakota Formation contains the majority of the remaining potential tight gas.

The objectives of this study are: 1) to characterize the producing/flow characteristics of the Dakota wells, 2) to delineate present-day drainage volume and area of Dakota wells, and 3) to investigate the infill potential of Dakota Reservoirs for increased recovery rate and reserves. Both public and commercial production databases are used in this study. Decline-curve analysis is the primary method adopted. Limited reservoir simulations, however, are also conducted to confirm the results obtained from the decline-curve analysis.

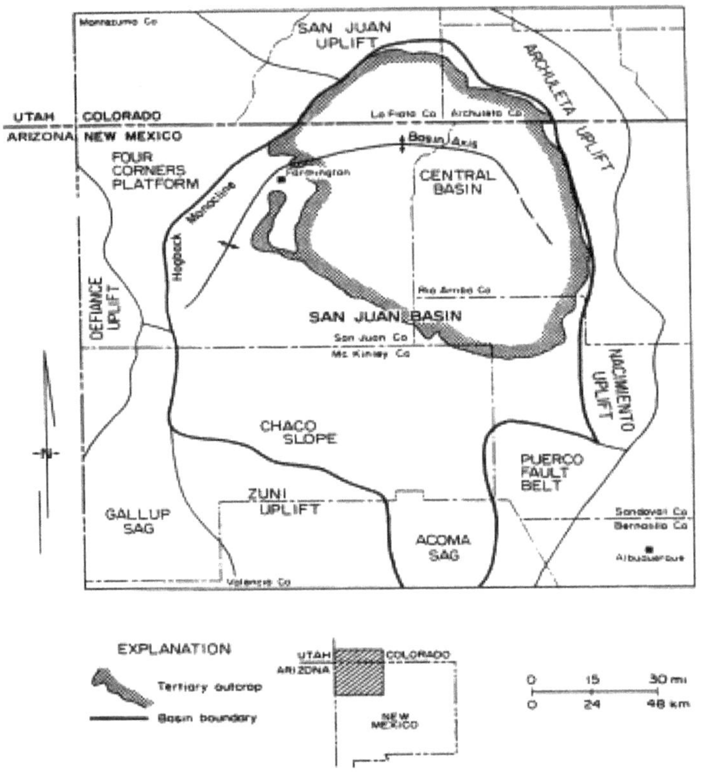

Fig. 15.1—Location map of San Juan Basin, New Mexico (Dutton et al. 1993).

15.3 Dakota Reservoirs: Geology and Study Area

The Dakota Sandstone in the San Juan Basin is the basal sequence created by the southwesterly transgression of the Cretaceous Sea as it entered the western interior. This happened probably during the Late Cretaceous age (Deischl 1973). Dakota Sandstone consists of a complex fluvial and offshore marine sandstone bodies (Hoppe 1978). Most of the Dakota gas is trapped within the upper marine facies, and the underlying facies are generally water saturated. Therefore most of the Dakota gas is trapped in several offshore sandstone lenses surrounded by impermeable marine shale.

Fig. 15.2 shows the study area (shaded area). The area covers approximately 58 townships. Approximately 500 wells distributed over the area were selected for analysis.

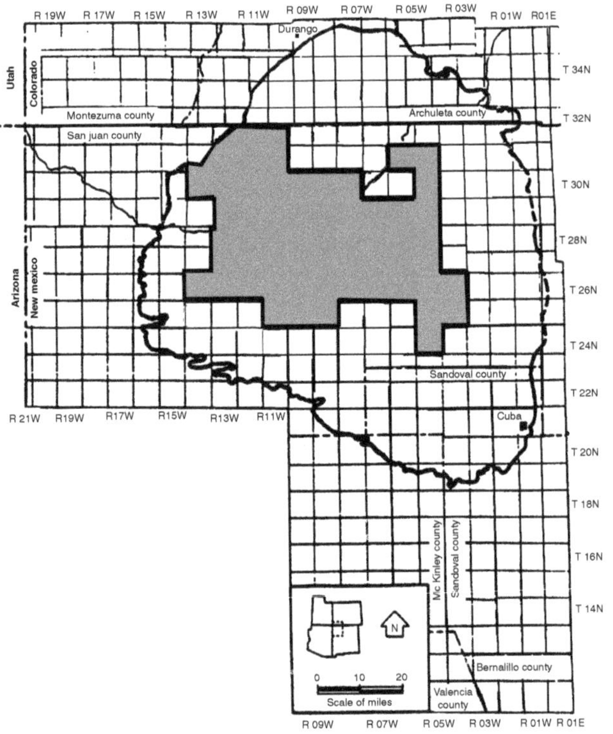

Fig. 15.2—Study area (modified from Haas and Crist 1991).

15.4 Rate/Time Type-Curve Analysis

Dakota wells are generally stimulated with hydraulic fracturing. The transient rate-time behavior thus is expected to be more linear than radial. Both strongly linear and strongly radial flows are observed for the Dakota wells. A set of rate/time type-curves was developed during this study to assist the analysis of the Dakota wells.

15.4.1 Type-Curve. Following the approaches of Fetkovich (1980) and Carter (1985) a set of type-curves covering purely linear flow to fully radial flow is developed for this study as shown in **Fig. 15.3.** Idealized behavior of a fractured well (e.g., half-slope and quarter-slope) generally is not well-defined using monthly production data. The developed type-curve allows us to analyze both linear- and radial-type flows much more easily than the existing type-curves.

The dimensionless rate and time of Fig. 15.3 are defined in a more rigorous manner than the set of Fetkovich (1980). A more detailed description is outlined in Appendix 15.A.

15.4.2 Type-Curve Analysis. A type-curve matching provides four matching parameters, $[q/q_{dD}]_M$ (rate match), $[t/t_D]_M$ (time match), $[r_{eD}]_M = [r_e/r_{wa}]_M$ (match of the ratio of drainage radius to apparent wellbore radius), and b (decline exponent). The first three parameters reflect well-reservoir quality, and are basically performance indicators. With additional information, they may be decomposed to obtain a more detailed reservoir description. The decline exponent is useful in inferring drive mechanism and large-scale reservoir heterogeneity, and in predicting long-term performance and ultimate recovery (Fetkovich et al. 1996).

The two most useful (reliable) properties from a rate/time type-curve matching are the productivity index (PI) and the drainage pore volume (PV). Due of lack of detailed information of fluid properties, pressure-squared approach is adopted instead of the more rigorous pseudopressure approach. Other important properties, which can be extended from PI and PV, include initial gas-in-place (G), drainage area (A), equivalent-circular drainage radius (r_e), skin factor (S), gas flow capacity ($k_g h$), and gas permeability (k_g). Procedures and working equations for our analysis of Dakota wells are presented in Appendix 15.B.

Confidence level of the estimated initial gas-in-place (G) is about the same as that of the PV. Drainage area (A) and the equivalent-circular drainage radius (r_e), however, highly depend on the accuracy of the estimated producing thickness (h). Apparent wellbore radius (r_e) and skin (S) depend on, in addition to the producing thickness (h), the quality of the r_e/r_{wa} matching. Flow capacity ($k_g h$) and gas permeability (k_g) depend on the quality of the r_e/r_{wa} matching. Permeability is further influenced by the producing thickness.

15.5 Producing Characteristics of Dakota Wells

A log-log plot of production history can identify the transient and boundary-affected flow regimes. All the wells analyzed show, more or less, some transient flow. This is to be expected (even of monthly data) because of the low-permeability of Dakota Formation.

15.5.1 Characteristics of Transient Flow. We noted that during transient flow period, Dakota wells exhibit either very steep decline ($r_{eD} \leq 5$) or very gentle decline ($r_{eD} \geq 5{,}000$). The above two types of characteristics are related to the date of first production. The gentle and steep declines correspond to wells producing before and after the seventies, respectively. About half of the wells analyzed belong to the steep decline and the other half to the gentle decline. The steep-decline behavior (low r_{eD} value) is believed to be due to successful well-stimulations (high negative skin). For the case of very gentle decline, the average "flat" flow-rate during the transient period is around 260 Mscf/d. This flow rate translates into an open-flow potential of around 0.26 Mscf/d/psi.

15.5.2 Characteristics of Boundary-Affected Flow. The decline exponents (b) observed during boundary-affected period (either due to well-interference or geological/permeability barriers) range from negative to 0.4. An average value is around 0.2. Exponential decline ($b = 0$), however, appears to be a good approximation.

15.6 Example: A Three-Well Group

A three-well group is presented to show a more detailed analysis of Dakota wells. These three wells are located in the sections 25, 26, and 35 of the township T31N-R14W. **Fig. 15.4** shows the approximate locations and the production histories. The solid lines represent the matched type-curve. Note that these three wells are closely spaced.

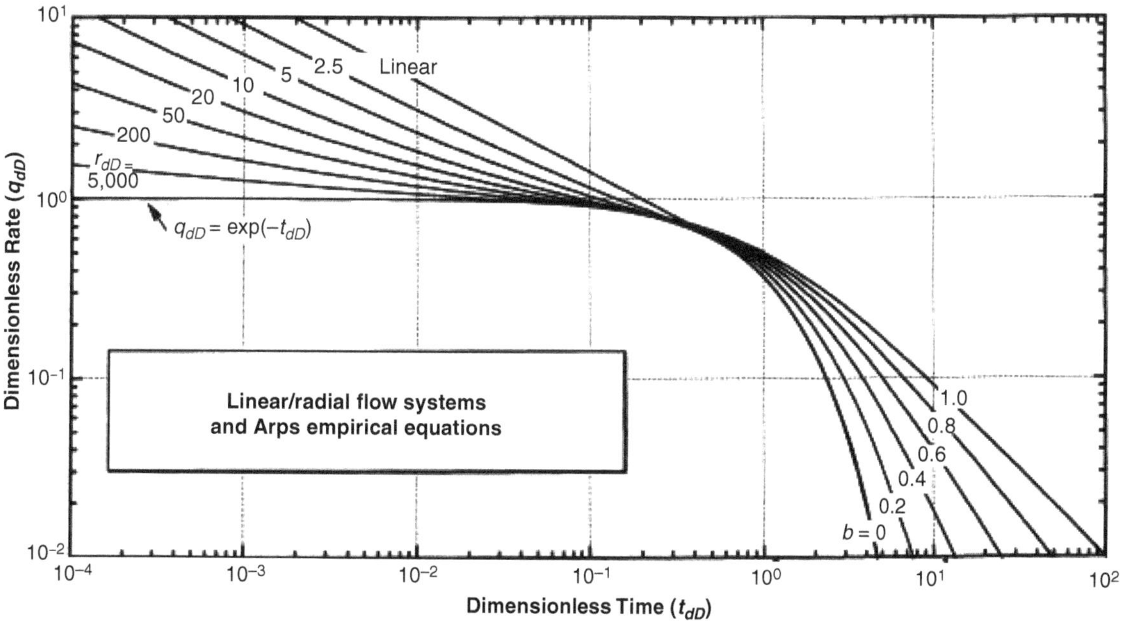

Fig. 15.3—Rate/time decline type curves.

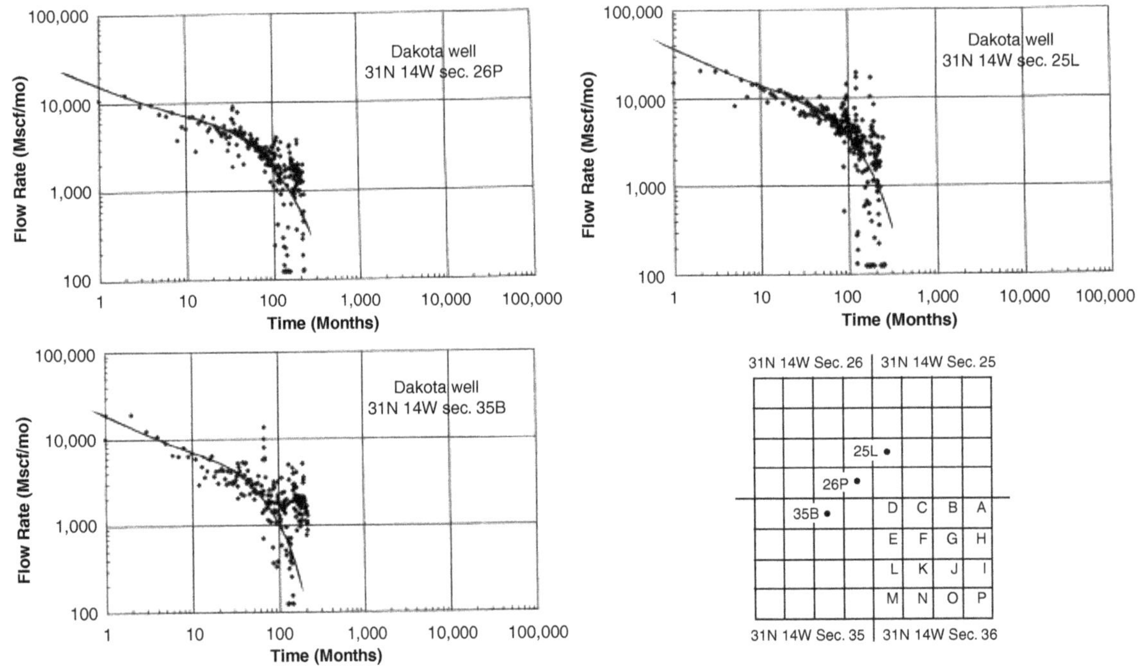

Fig. 15.4—Dakota well example.

15.6.1 Fluid, Well, and Reservoir Information. Average properties assumed in our calculations are listed in **Table 15.1.** Some of the data in Table 15.1 is from Dutton et al. (1993). **Table 15.2** gives the information of first date of production, producing thickness (perforation interval), and cumulative production up to April 1999. Note that Well 25L started producing first (March 1975). Well 26P and 35B were placed on production 4 years later.

15.6.2 Type-Curve Analysis. Table 15.3 summarizes the type-curve analysis of these three wells. Estimated properties are PI, PV, G, A, r_e, k_gh, k_g, and S. Recovery up to April 1999 is also expressed in terms of 1) percentage of initial gas-in-place and 2) per acre. The unrealistic value of the thus-computed recovery is a warning for a questionable type-curve matching.

As shown in Fig. 15.4, all three wells exhibit steep decline during transient period, especially Well 35B. Matched r_{eD} values are 2.5, 5, 2.5 for Well 25L, 26P, and 35B, respectively; see Table 15.3. Significant decline occurs at about 120 months (10 years). It is not clear that this rapid decline is due to well interference or geological (permeability) boundary. The flowrate levels of Well 26P and 35B (newer) are lower than that of Well 25L (older) during transient period. This may be due to poor reservoir properties around the newer Well 26P and 35B or due to partial depletion caused by the older Well 25L.

As shown in Table 15.3, PI and PV of Well 26P and 35B (newer) are inferior to those of Well 25L (older). Partial depletion appears to occur around Well 26P and 35B. Estimated drainage areas for the three wells are smaller than the current well spacing (160 acres). Note that variation (or error) of producing thickness has significant impact on the estimation of drainage area.

15.6.3 Drainage Area Shape. The estimated drainage volume corresponds to a highly uncertain 3D drainage shape/geometry. An assumption of constant thickness reduces a 3D problem to a 2D one (area). Eq. 15.1 relates the estimated drainage area to a simple equivalent circular, elliptical, or rectangular shape,

$$A=\begin{cases}\pi r_e^2 & \text{Circular}\\ \pi ab & \text{Elliptical}\\ wL & \text{Rectangular}\end{cases} \quad \text{(15.1)}$$

Here, a and b are the major and minor semi-axes of an ellipse, respectively, and w and L are the width and length of a rectangle, respectively. In addition, knowledge of orientations is required to completely describe an elliptical-/ rectangular-based pattern.

Eq. 15.2 shows the aspect ratio of an ellipse or a rectangle is related to permeability anisotropy by

$$\frac{a}{b}=\frac{L}{w}=\left(\frac{k_{H\max}}{k_{H\min}}\right)^{1/2}, \quad \text{(15.2)}$$

where $k_{H\max}$ and $k_{H\min}$ are the horizontal maximum and minimum permeabilities, respectively. The permeability estimated from the decline-curve analysis represents the geometric mean of the maximum and minimum permeability, i.e., $k=(k_{H\max}\,k_{H\min})^{1/2}$.

Initial reservoir pressure, p,	2,000 psi
Flowing bottomhole pressure, $\mu_w f$	600 psi
Initial z-factor, z_i	0.82
Initial formation volume factor, B_{gi}	1.414 rb/Mscf
Initial gas viscosity, μ_{gi}	0.0175 cp
Porosity, ϕ	8.5%
Initial total compressibility, c_{ti}	2.5×10^{-4} psi-1
Initial water saturation, S_{wi}	50%
Reservoir temperature, T	210 – 240°F
Wellbore radius, r_w	0.33 ft

Table 15.1—Typical fluid, well, and reservoir properties of Dakota Formation, San Juan Basin, New Mexico.

		Well 25L	Well 26P	Well 35B
Date of first production	Month-year	3-1975	9-1979	9-1979
Producing thickness	ft	102	129	164
Cumulative production (up to April 1999)	MMscf	1,100	578.7	545.6

Table 15.2—Production information of Dakota 31N-14W wells.

		Well 25L	Well 26P	Well 35B
Matching Parameters				
Rate match, q/q_{dD}	Mscf/d	461/1	263/1	263/1
Time match, t/t_{dD}	Day	1976/1	1763/1	1520/1
r_{eD} match	–	2.5	5	2.5
Decline exponent, b	–	0	0.2	0
Estimated Properties				
Productivity index, PI	Mscf/d/psi	0.507	0.289	0.289
Pore volume, PV	MMcf	35.820	17.386	15.719
Initial gas-in-place, G	Bscf	2.256	1.095	0.990
Drainage area, A	Acres	95	36	26
Equivalent drainage radius, r_e	ft	1,147	710	599
Flow capacity, $k_g h$	md-ft	1.013	1.134	0.578
Gas permeability, k_g	md	0.00993	0.00879	0.00352
Skin, S	–	–7.2	–6.1	–6.6
Recovery (up to April 1999)				
% of initial-gas-in-place	%	49	53	55
Per acre	MMscf/acre	11.6	14.5	21.0

Table 15.3—Type curve analysis of Dakota 31N-14W wells.

Most tight-gas reservoirs are heterogeneous and anisotropic (due to natural fractures). An elliptical or rectangular drainage shape thus is more realistic. The following analysis, however, is based on a circular drainage shape for the sake of simplicity and providing a quick picture of well interference.

Fig. 15.5 presents the equivalent circular drainage area of Well 25L, 26P, and 35B (see also Table 15.3). The squares (units) in one section cover 40 acres. Assuming the wells are located in the center of the units, the distance between the wells then is 1,867 ft.

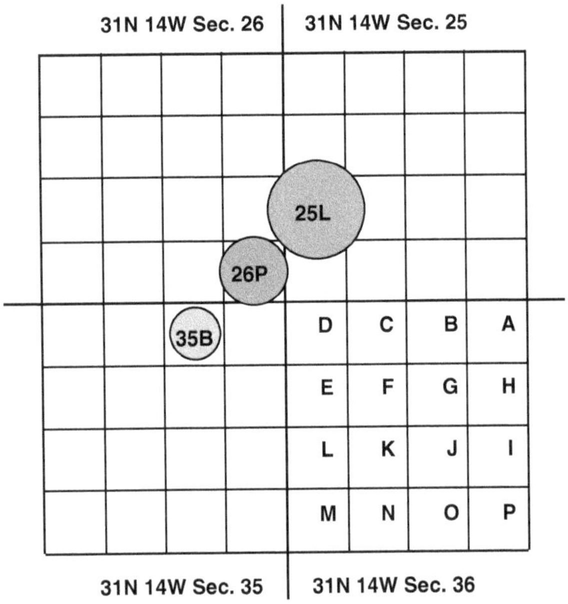

Fig. 15.5—Equivalent circular drainage area of the Dakota well example.

The sum of the drainage radii of Well 25L and 26P is 1,147 + 710 = 1,857 ft, which approximately is the distance between the wells. Thus, the no-flow boundary between Well 25L and 26P could be due to well interference if the drainage area is circular.

The estimated equivalent circular drainage radius of Well 35B is only around 85% of that of Well 26P. These two wells started producing at the same date and have the same PI with similar cumulative production. Producing thickness is the major reason for the difference in the estimated drainage area.

Finally, it should be emphasized again that heterogeneity and anisotropy would distort the idealized circular shape shown in **Fig. 15.6.** For example, a 45° of anisotropy orientation would give an earlier (and also maximum) interference between the three wells shown in Fig. 15.6.

15.7 Drainage Volume, Drainage Area, Initial Gas-In-Place, and Productivity Index of Dakota Wells

Figs. 15.6 through 15.9 present the plots of PV, drainage area (A), initial gas-in-place (G), and PI, respectively, versus first-date-of-production. Only 228 wells are used to construct these figures. Results are preliminary in the sense that we are continuously updating the information and refining the analysis.

Consider Fig. 15.6. The estimated PVs for the wells that started producing during the early eighties appear to be more uniform than those of sixties and seventies. The pattern shown in Fig. 15.6 seems to suggest that the drainage volumes decreases as the date of first production increases. In other words, the PV of the newer wells are smaller those of the older wells. Variability of the newer wells, however, is smaller than the older wells.

Comparing Figs. 15.6 and 15.7 indicates that the overall variability decreases as the PV is converted into the drainage area based on individual well producing-thickness. Fig. 15.7 also indicates that the majority of the wells have a drainage areas below 100 acres.

The pattern of the estimated initial gas-in-place (Fig. 15.8) is similar to that of pore volume (Fig. 15.6). This is to be expected because the same initial water saturation and the initial formation volume factor are used for computations for all wells. Thus, initial gas-in-place is basically proportional to the PV for all wells.

The PI (Fig. 15.9) exhibit higher variability than PV, A, and G. This reflects, considering the lower variability in the drainage area, a high variability in the reservoir permeability (i.e., heterogeneity).

15.8 Conclusions

Approximately 500 wells from the low-permeability, tight-gas Dakota Formation in San Juan Basin, New Mexico, were analyzed by rate/time decline-curve analysis. The objectives are to characterize the producing/flow characteristics, to delineate the drainage volume/area, and to infer the infill potential. Based on current results, the following conclusions are obtained pertinent to the Dakota wells:

1. Dakota wells generally exhibit very steep decline in flow rates during transient period, reflecting a high negative-skin effect due to well stimulation.
2. During boundary-affected period, the decline in flow rates tends to be exponential-decline-like (small value of decline exponent).
3. Average permeability estimated from decline-curve analysis is in the order of 0.01 md.
4. Drainage PV of the wells that started producing during the early eighties are slightly smaller but more uniformly distributed than those of the wells that started producing during the sixties and seventies.
5. Drainage areas appear to be less dependent on the date of production than the drainage PV.
6. PI is highly variable over both basin and time, suggesting a very heterogeneous flow/permeability field of the Dakota Reservoirs.

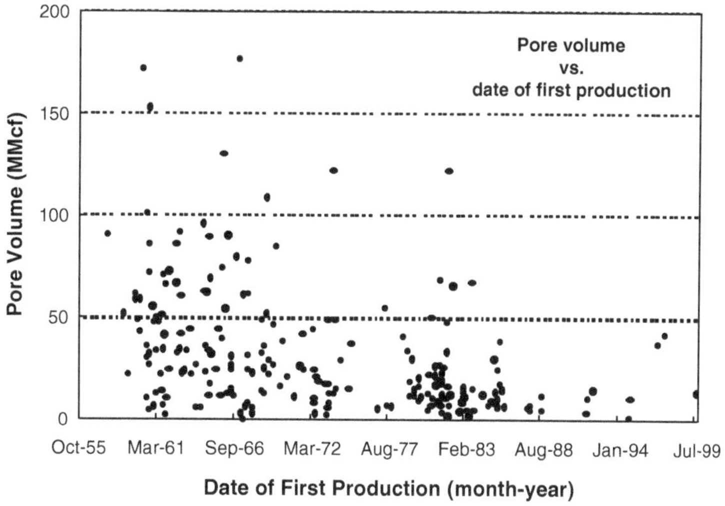

Fig. 15.6—Pore volume vs. date of first production.

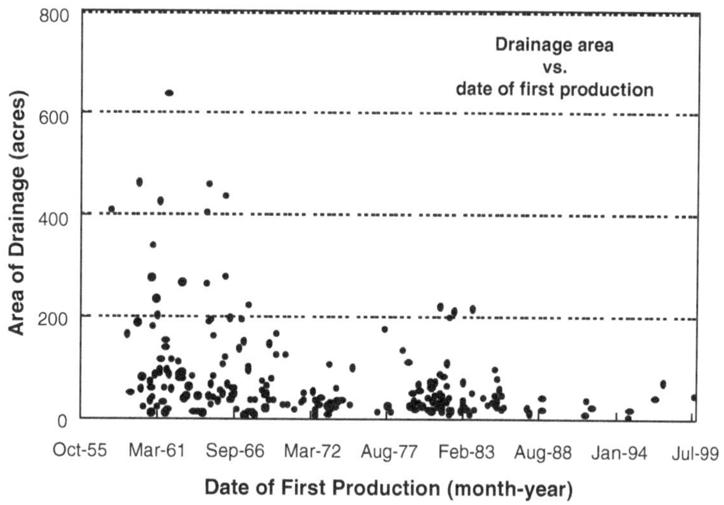

Fig.15.7—Drainage area vs. date of first production.

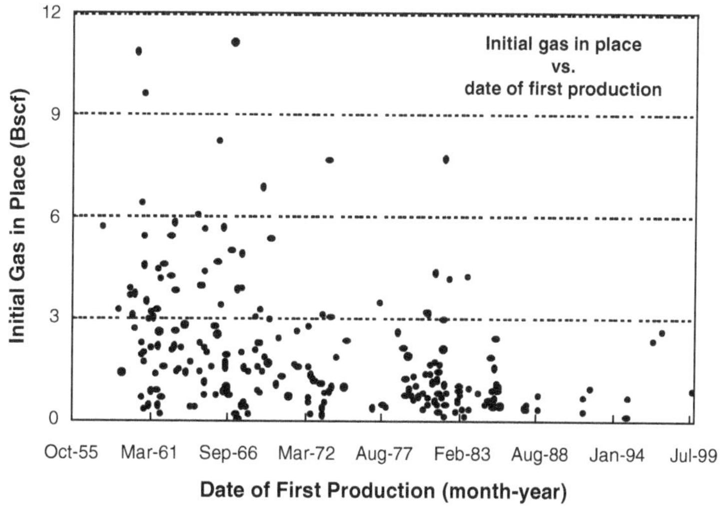

Fig. 15.8—Initial gas in place vs. date of first production.

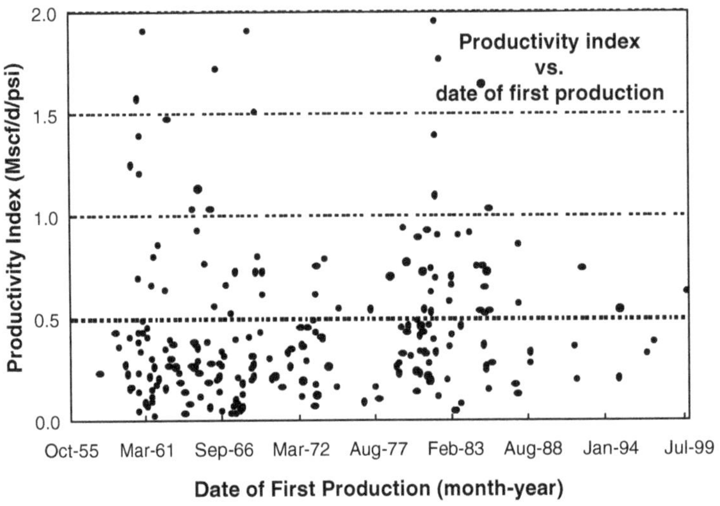

Fig. 15.9—Productivity index vs. date of first production.

15.9 Nomenclature

a, b	=	major and minor semi-axis of an ellipse
b	=	$-d[q/(dq/dt)]/dt$, decline exponent, dimensionless
B	=	FVF, rb/stb (oil) and rb/Mscf (gas)
B_g	=	$(10^3/5.6146)(zT/p)/(T_{sc}/p_{sc})$, gas FVF, rb/Mscf
B_1	=	parameter defined by Eq. A-6
c_t	=	total compressibility, psi^{-1}
G	=	initial gas-in-place, Bscf
h	=	producing thickness, ft
J_0, J_1	=	Bessel function of the first kind, order zero and one
k	=	permeability, md
k_g	=	gas permeability, md
$k_{H\max}$	=	horizontal maximum permeability, md
$k_{H\min}$	=	horizontal minimum permeability, md
L	=	reservoir length, ft
p	=	pressure, psi
Pi	=	initial reservoir pressure, psi
$p_p(p)$	=	dry-gas pseudopressure, psi
p_r	=	reference pressure, psi
$p_w.f$	=	well flowing pressure, psi
PI	=	productivity index, Mscf/d/psi
PV	=	pore volume in MMcf
q	=	flow rate, stb/d (oil) and Mscf/d (gas)
q_{dD}	=	decline-curve dimensionless flow rate
r_e	=	drainage radius, ft
r_{eD}	=	dimensionless drainage radius
r_w	=	wellbore radius, ft
r_{wa}	=	$r_w\exp(-S)$, apparent wellbore radius, ft
S	=	skin factor, dimensionless
t	=	time, day
t_{dD}	=	decline-curve dimensionless time
T	=	temperature, °R
V_p	=	reservoir pore volume, ft^3
w	=	reservoir width, ft
Y_0, Y_1	=	Bessel function of the second kind, order zero and one
z	=	gas compressibility factor, z-factor
α_1	=	smallest root of Eq. A-8
η	=	$(\alpha_1^2 / B_1)(r_{eD}^2 - 1)/2$
μ	=	viscosity, cp
σ	=	fraction of 2π radians open to flow, frac.
f	=	porosity, frac.

Subscripts

g	=	gas
i	=	initial
M	=	match point
sc	=	standard conditions

15.10 Acknowledgments

The authors thank BP-Amoco, Burlington Resources, Conoco, and the Department of Energy (DOE) for their support of this research. They would also like to thank HIS/Dwights for providing the production database, and Dewi Hidayati for the assistance in the type-curve matching and manuscript preparation.

15.11 References

Carter, R. D. 1985. Type Curves for Finite Radial and Linear Gas- Flow Systems: Constant-Terminal-Pressure Case. *SPE J.* **25** (5): 719–728. SPE-12917-PA. https://doi.org/10.2118/12917-PA.

Deischl, D. G. 1973. The Characteristics, History, and Development of the Basin Dakota Gas Field. In *Cretaceous and Teritary Rock of the Southern Colorado Plateau*, ed. J. E. Fassett, 168–173. Durango, Colorado: Four Corners Geological Society.

Dutton, S. P., Clift, S. J., Hamilton, D. S., et al. 1993. Major Low-Permeability-Sandstone Gas Reservoirs in the Continental United States. Report of Investigations No. 211, Bureau of Economic Geology and Gas Research Institute, Austin, Texas.

Fetkovich, M. J. 1980. Decline Curve Analysis Using Type Curves. *SPE JPT* **32** (6): 1065–1077. SPE-4629-PA. http://doi.org/10.2118/4629-PA.

Fetkovich, M. J., Fetkovich, E. J., and Fetkovich, M. D. 1996. Useful Concepts for Decline-Curve Forecasting, Reserve Estimation, and Analysis. *SPE Res Eng* **11** (1): 13–22. SPE-28628-PA. https://doi.org/10.2118/28628-PA.

Haas, M. R. and Crist, T. E. 1991. The Impact of Improved Technology on Potential Tight Gas Reserves in the San Juan Basin: A Case Study. Presented at the SPE Gas Technology Symposium, Houston, Texas, 23–25 January. SPE-21484-MS. https://doi.org/10.2118/21484-MS.

Hoppe, W. F. 1978. Basin Dakota Gas Field. In *Oil and Gas in the Four Corners Area*, ed. J. E. Fasset, 204–206. Durango, Colorado: Four Corners Geological Society.

IHS/Dwights LLC. 1999. Production Data CD-Rom for the Rocky Mountain Area. September1999.

Appendix 15.A

15.A.1 Type Curves

The type curves of Fig. 15.3 cover both linear and radial flow systems. The curves are generated from analytical solutions of a slightly compressible fluid with constant fluid and rock properties. Dimensionless rate/time variables are defined separately for radial and linear flow systems. In field units, they are as follows:

The radial system is defined by Eqs. A-1, A-2, and A-3.

$$q_{dD} = \frac{q(t)}{B_1 \dfrac{\sigma kh\left(p_i - p_{wf}\right)}{141.2\mu B}}, \quad \text{(A-1)}$$

$$t_{dD} = 2\pi\eta B_1 \frac{0.006329\sigma kt}{\mu\phi c_t A}, \quad \text{(A-2)}$$

$$r_{eD} = r_e / r_{wa}. \quad \text{(A-3)}$$

The linear system is defined by Eqs. A-4 and A-5.

$$q_{dD} = \frac{q(t)}{\dfrac{w}{\pi L} \dfrac{kh\left(p_i - p_{wf}\right)}{141.2\mu B}}, \quad \text{(A-4)}$$

$$t_{dD} = \frac{\pi^2 w}{4L} \frac{0.006329 kt}{\mu\phi c_t A}. \quad \text{(A-5)}$$

B_1 (Eqs. A-1 and A-2) and η (Eq. A-2) are defined by (Carter 1985):

$$B_1\left(r_{eD}\right) = \frac{2J_1^2\left(\alpha_1 r_{eD}\right)}{J_0^2\left(\alpha_1\right) - J_1^2\left(\alpha_1 r_{eD}\right)}, \quad \text{(A-6)}$$

and

$$\eta\left(r_{eD}\right) = \frac{\alpha_1^2}{B_1} \frac{r_{eD}^2 - 1}{2}, \quad \text{(A-7)}$$

and Eq. A-8 shows where a is the smallest root of

$$J_0\left(\alpha_n\right) Y_1\left(\alpha_n r_{eD}\right) - Y_0\left(\alpha_n\right) - J_1\left(\alpha_n r_{eD}\right) = 0, \quad \text{(A-8)}$$

$n = 1, 2, ...$, and J_0, J_1, Y_0, and Y_1 are Bessel functions (see Nomenclature). Some selected values of $B_1(r_{eD})$ and $r(r_{eD})$ are given in **Table 15.A-1.** Near-wellbore skin is considered in r_{eD} by the concept of apparent wellbore radius, $r_{wa} = r_w \exp(-S)$.

r_{eD}	B_1	η
1.5	4.416425	1.1819
2.5	1.747429	1.1267
5.0	0.8904994	1.0744
10.0	0.5768081	1.0435
20.0	0.4204741	1.0263
50.0	0.3069877	1.0147
100.0	0.2542351	1.0101
500.0	0.1811789	1.0050
1,000.0	0.1611274	1.0040
5,000.0	0.1281149	1.0025

Table 15.A-1—Values of r_{eD}, B_1, and η.

15.A.1.1 Gas Reservoirs. For gas reservoirs, the pressure is replaced by the pseudopressure defined by Eqs. A-9a and A-9b.

$$p_p(p) = \mu_{gi} B_{gi} \int_{pr}^{p} \frac{1}{\mu_g B_g} \mathrm{d}p', \quad \text{(A-9a)}$$

$$= \frac{\mu_{gi} z_i}{p_i} \int_{pr}^{p} \frac{p}{\mu_g z} \mathrm{d}p'. \quad \text{(A-9b)}$$

The unit of pseudopressure follows that of the actual pressure. In Eq. A-10, for a constant $\mu_g z$ product, say $\mu_g z = (\mu_g z)i$, the pseudopressure reduces to

$$p_p(p) = \frac{p^2 - p_r^2}{2p_i}. \quad \text{(A-10)}$$

Appendix 15.B

15.B.1 Type-Curve Analysis

A type-curve matching provides four matching parameters, $[q/q_{dD}]_M, [t/t_{dD}]_M, [r_{eD}]_M = [r_e/r_{wa}]_M$ and b. In terms of the approach of pressure-squared, PI and PV are estimated by Eqs. B-1 and B-2, respectively.

$$\text{PI} = \frac{2p_i}{p_i^2 - p_{wf}^2}\left[\frac{q}{q_{dD}}\right]_M \quad \text{[Msc/d/psi]}, \quad \text{(B-1)}$$

and

$$\text{PV} = \eta\frac{2(5.6146)B_{gi}p_i}{10^6 c_{ti}\left(p_i^2 - p_{wf}^2\right)}\left[\frac{q}{q_{dD}}\right]_M\left[\frac{q}{q_{dD}}\right]_M\left[\frac{t}{t_{dD}}\right]_M \quad \text{[MMcf]}. \quad \text{(B-2)}$$

Initial gas-in-place (G), drainage area (A), equivalent circular drainage radius (r_e), apparent wellbore radius (r_{wa}), skin factor (S), gas flow capacity (k_gh), and gas permeability (k_g) can be further derived from PI and PV with additional information. They may be computed by Eqs. B-3 through B-9 sequentially,

$$G = \frac{(\text{PV})(1 - S_{wi})}{5.6146B_{gi}}\text{[Bscf]}, \quad \text{(B-3)}$$

$$A = \frac{(\text{PV})\times 10^6}{\phi h}[\text{ft}^2], \quad \text{(B-4)}$$

$$r_e = (A/\pi)^{1/2}\text{[ft]}, \quad \text{(B-5)}$$

$$r_{wa} = \frac{r_e}{[r_e/r_{wa}]_M}\text{[ft]}, \quad \text{(B-6)}$$

$$S = -\ln\frac{r_{wa}}{r_w}, \quad \text{(B-7)}$$

$$k_gh = \frac{141.2\mu_{gi}B_{gi}(\text{PI})}{B_1}\text{[md-ft]}, \quad \text{(B-8)}$$

and

$$k_g = \frac{k_gh}{h}\text{[md]}. \quad \text{(B-9)}$$

Note that constant thickness over the drainage volume is assumed in Eqs. B-4 and B-9. Also, the drainage radius, apparent wellbore radius, and skin calculated from Eqs. B-5 through B-7 are in the sense of an equivalent system.

SI Metric Conversion Factors

bbl × 1.589 873	E-01 = m^3
cp × 1.0*	E- 03 = Pa·s
ft × 3.048*	E-01 = m^2
ft^2 × 9.290 304*	E-02 = m^2
ft^3 × 2.831 685	E-02 = m^3
md × 9.869 233	E-04 = μm
psi × 6.894 757	E+00 = kPa

* Conversion factor is exact.

Chapter 16

Production Analysis of Commingled Gas Reservoirs—Case Histories

Jorge A. Arevalo-Villagran, PEMEX and Texas A&M University; **Robert A. Wattenbarger,** SPE, Texas A&M University; and **Ahmed H. El-Banbi,** SPE, Schlumberger Holditch-Reservoir Technologies

SPE 59885 presents a case history of production data analysis of five commingled gas wells in Mexico. Arevalo-Villagran et al. developed models and a computer program that integrates material-balance equations and boundary-dominated flow equations for each layer in a multilayer reservoir system. The model allows for matching and forecasting production by layer, modeling of recompletions, calculation of original gas-in-place, productivity index, and backflow between layers. It seems to provide a reasonably accurate match with actual field data and reserves assessment in reservoirs with permeability between 0.1 to 20 md. In summary, a commingled gas flow model is developed and demonstrated in solving practical problems.

16.1 Abstract

A method to match and forecast production rates for commingled gas wells has been developed. In this paper, the method previously developed by El-Banbi and Wattenbarger (1996, 1997) was improved and extended to allow modeling of recompletions and different initial pressures in the layers. This new method is based on a layered model that couples the material balance gas equation with the real gas boundary dominated flow equation for each layer of the multilayer-commingled system.

A spreadsheet-based computer program was developed and used to analyze the production history of commingled wells. The program calculates the individual original gas in place, *OGIP,* productivity index, J_g, and production rates for each layer. It also forecasts the production rates of individual layers and the total well. Bottom-hole flowing pressure, p_{wf}, total gas rate history, gas properties, and initial reservoir pressure for each layer are required.

The parameters *OGIP* and J_g for each layer are determined by history-matching the total gas rate using a nonlinear optimization procedure. This new method gives excellent results for cases dominated by stabilized flow, with occasional shut-ins and substantial p_{wf} variations. The method calculates back-flow between layers during shut-in periods. Although this method is based on using stabilized flow in history-calibration data, validation against numerical simulation shows that the method can tolerate many transient flow periods caused by variations in producing conditions. Several wells were used in testing this method. Five Mexican wells are included in this paper.

16.2 Introduction

In commingled gas reservoirs or multi-layer no crossflow gas reservoirs, volumetric methods may not be accurate for evaluating reserves. Similarly, material balance equation (MBE), may not be applied to predict the performance of this type of reservoir because of scarcity of data. In the same way, tight gas reservoir build-up tests may be infeasible as a result of the long time needed for the pressure to stabilize.

Decline curve analysis looks to be a very useful method for performing prediction and evaluating original gas in place, *OGIP,* and reserves. Arps (1945) explained that decline curve analysis began as a rate forecasting technique by extrapolating the production trend. He presented empirical expressions to evaluate rate behavior versus time. Garb et al. (1973) developed techniques to solve the single-layer gas reservoirs problems for production at variable bottomhole flowing pressure, p_{wf}. Later, Fetkovitch (1980) developed a type-curve matching procedure for decline curve analysis. He integrated analytical solutions for oil transient flow with practical decline equations developed by Arps (1945). Rodgers et al. (1983) proposed methods to evaluate single-layer gas reservoirs considering variable p_{wf}. Carter (1985) generated other type curves dealing with the change of gas properties with pressure. Studying gas wells, Fraim (1987) established a normalized pseudo-time function to force the gas reservoir data to match the liquid solution for constant p_{wf}. Fetkovich et al. (1990) analyzed commingled gas reservoirs using type-curves matching. Chen and Poston (1989) tried to develop methods to help in the matching process for one-layer radial gas reservoirs. Aminian et al. (1990) developed other sets of decline curves in multi-gas wells. Blasingame et al. (1991) and Palacio and Blasingame (1993) developed additional pseudo-time approaches that account for variation in p_{wf}. They used type-curve techniques to help in matching one-layer radial reservoirs. Guardia and Hackney (1991), Keating et al. (1994), and West and Cochrane (1995) developed various approximations of coupling the MBE with the gas flow equation for single-layer reservoirs. El-Banbi and Wattenbarger (1996) coupled the stabilized flow equation with the gas MBE to study multi-layer gas reservoirs. Later, El-Banbi and Wattenbarger (1997) included the variation in p_{wf} and non-Darcy flow.

SPE 58985 improves and extends the method used by El-Banbi and Wattenbarger (1996, 1997) to allow modeling of recompletions and different initial pressures in analyzing production data in commingled gas reservoirs.

16.3 The Commingled Gas Flow Model

16.3.1 Commingled Gas Reservoirs. Commingled gas reservoirs are layered reservoirs linked only through the wellbore. These reservoirs do not experience crossflow within reservoir boundaries. Each layer in the commingled system can be defined by its *OGIP* and its gas flow coefficient or productivity index, J_g, if stabilized flow is reached (El-Banbi and Wattenbarger 1996, 1997). If these parameters are different for each layer, the layers' behavior will be also different.

16.3.2 Single Layer Reservoirs. The single-layer model couples the flow equation with the MBE for volumetric gas reservoirs. The flow expression is the analytical solution for flow in gas reservoirs producing against constant p_{wf} (Al-Hussainy et al. 1966). The MBE for volumetric gas reservoir is described by

$$\frac{p}{z}=\left(\frac{p_i}{z_i}\right)\left(1-\frac{G_p}{G}\right). \qquad (16.1)$$

For high-pressure reservoirs we can use the material balance equation suggested by Ramagost and Farshad (1981). The J_g for a gas well (assumed constant with time) defines the relationship between the production rate and the well pseudopressure drawdown, $\left[m(\bar{p})-m(p_{wf})\right]$. Eq. 16.2 is defined as

$$J_g=q_g/\left[m(\bar{p})-m(p_{wf})\right], \qquad (16.2)$$

where $m(p)$ is the real gas pseudo-pressure (Al-Hussainy et al. 1966) defined by

$$m(p)=2\int_{p_o}^{p}\frac{p}{\mu z}\,\mathrm{d}p. \qquad (16.3)$$

Because J_g in Eq. 16.2 is a function of gas rate (non-Darcy flow), q_g, a more practical arrangement is expressed (Lee and Wattenbarger 1996) by

$$m(\bar{p})-m(p_{wf})=aq_g+bq_g^2. \qquad (16.4)$$

Eq. 16.4 is the stabilized flow equation and is a function of a and b flow parameters. The a parameter takes into account Darcy effects and the b parameter is function of non-Darcy flow or inertial effects. The equations that relate these parameters to reservoir and gas properties are shown in Lee and Wattenbarger (1996). We confirmed that non-Darcy parameter, b, can be ignored for most of the field cases we analyzed (permeability between 0.1 and 10 md). Consequently, b parameter in Eq. 16.4 can be approximated to zero and Eq. 16.4 reduces to Eq. 16.2,

$$J_g=\frac{1}{a}=\frac{k_g h}{141.2T\left(\frac{1}{2}\ln\frac{2.458A}{C_A r_w^2}+s\right)}. \qquad (16.5)$$

In addition to Eqs. 16.1 and 16.2, the relation between the gas flow rate and cumulative gas production is given by

$$G_p=\int_0^t q_g(t)\,\mathrm{d}t. \qquad (16.6)$$

The MBE (Eq. 16.1) is used to relate $\bar{p}$ to time, t, and thus relating production rate, q_g to t. J_g (Eq. 16.2) is a function of average reservoir pressure, $\bar{p}$. If the *OGIP* and the flow coefficient (a or J_g) are known for each layer, Eqs. 16.1, 16.2, and 16.6 can be solved simultaneously to find a solution. The present method is based on the same assumption (coupling the J_g equation with the MBE) for commingled gas wells. The method does not require shutting the well in. Alternatively, it estimates the J_g from production rate-time data.

16.3.3 Multi-Layer Reservoirs. The approach of the commingled system is based on calculating the individual layer behavior and adding up the commingled performance. After solving each layer's commingled gas flow model for q_g for every time step, the total q_g of the system can be evaluated by integrating the q_g of each layer at the corresponding time using the following expression.

$$q_{T_g}(t)=\sum_{i=1}^{Nlayer} q_{gi}(t). \qquad (16.7)$$

This expression requires that *OGIP* and J_g (the flow coefficient) be known for each layer in the commingled system. It also requires that the pw be known at each period of time.

16.3.4 Calibration Period of Commingled Reservoirs. The study of the commingled system is performed by guessing values for *OGIP* and J_g (or a parameter) for each layer. Then the q_g for a specified calibration period can be calculated for the commingled well.

A multi-variable nonlinear optimization procedure is applied to minimize the error between the actual q_g data and the calculated q_g for this specified calibration period in the commingled model. The best values of *OGIP* and J_g (or a parameter) for each layer are determined when either of the following objective functions (Eqs. 16.8 or 16.9) is minimized:

$$error_1 = \frac{1}{N}\sum_{j=1}^{N}\left|\frac{Actual\, q_g - Model\, q_g}{Actual\, q_g}\right|, \quad (16.8)$$

$$error_2 = \frac{1}{N}\sqrt{\sum_{j=1}^{N}\left(\frac{Actual\, q_g - Model\, q_g}{Actual\, q_g}\right)}, \quad (16.9)$$

where N is the number of total calibration data points in the history period. These data points do not have to be contiguous. This means that the calibration period may have discontinuities in data allowing for excluding poor quality data points and data points affected by transient flow periods.

16.3.5 Commingled Gas Flow Model Conjectures. The following are assumed in the flow model: (1) gas rate that is coming from each layer of the commingled system is dominantly stabilized flow, (2) p_{wf} is known for each layer, and (3) the effect of non-Darcy flow is ignored.

16.3.6 Commingled Gas Flow Model Essentials. The commingled gas flow model needs initial reservoir pressure for each layer, p_i, reservoir temperature, T, and gas data (gas specific gravity, γ_g, and concentration of H_2S, CO_2, and N_2 in gas composition). The gas properties are calculated into the commingled program from available correlations (Lee and Wattenbarger 1996; Sutton 1985; Dranchuk and Abou-Kassem 1975; Lee et al. 1966).

Total gas production rate and p_{wf} must be input data through the history and calibration periods. When p_{wf} is not known, it can be estimated from surface from readily available correlations (Cullender and Smith 1956). We also need to specify an initial guess for *OGIP* per layer.

16.4 Case Histories

To verify the commingled gas flow model improved in this work, we used several field examples. We display applications of the commingled flow model in matching and forecasting actual field examples. We make a selection of five wells from Mexico, which are producing in distinct gas reservoirs and fields. Four well cases are examples of recompletion in new layers. One well case is an example of a two-layer model. Producing surface rate measurements, q_g, were available on daily basis. Producing p_{wf}, was periodically available and sometimes evaluated from wellhead flowing pressure on monthly basis (Cullender and Smith 1956).

In all examples selected we used part of the history for calibration and the rest of known production history for checking the commingled gas model forecast. Good agreement between the model results and the actual production was obtained in all five cases. Later, the parameters evaluated in the production history match, *OGIP* and J_g, were used to forecast both the total gas rate and gas production from each layer for all wells studied.

16.4.1 Field Example A. This case illustrates the application of the commingled gas model on two different reservoirs opened to exploitation at different times. **Fig. 16.1** shows the schematic completion for Well A. This well had produced dry gas for almost 20 years from the Deposito Reservoir (layer 1). After 20 years a recompletion was made in the Encanto Reservoir (layer 2). **Table 16.1** gives reservoir and fluid information for each layer in Well A. Surface q_g measurements were available on daily basis. Bottom-hole flowing pressures, p_{wf}, were periodically available. Now, the well is producing through the same tubing with gas produced from both formations. The total G_p, including both reservoirs after 30 years, is 10,121 MMscf.

16.4.1.1 Layer 1. The comparison between the model q_g and the actual performance of layer 1 is shown in **Fig. 16.2.** We used 19 years of production and p_{wf} data to calibrate the commingled gas model using the one layer option. During the first year of production the maximum production q_g of 1,589 Mscf/D was reached. Later, the q_g was slightly declining to less than 500 Mscf/D after 19 years of exploitation. The p_{wf} periodically fluctuated due to changes in operation conditions.

The well started producing at p_{wf} of 5,423 psia and the pressure dropped to 1,256 psia in 19 years after producing 3,892 MMscf of gas. The parameters estimated by the model (*OGIP* and J_g) are given in **Table 16.2.**

16.4.1.2 Commingled Production. For the commingled production system (layers 1 and 2) we used 11 years of production and pressure data to calibrate the commingled gas model with the two-layer option. The real G_p and the *OGIP* and average pressure evaluated in layer 1 by the model in the first 19 years were used as data for layer 1 in the two-layer model alternative.

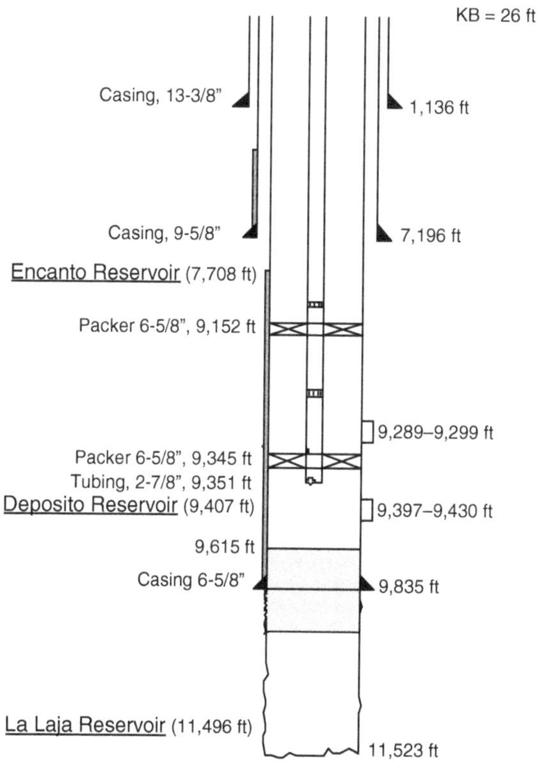

Fig. 16.1—Schematic completion for Well A.

Well	Reservoir	Layer	Opened intervals(s) (ft subsea)	p_i (psia)	T (°F)	ϕ (fraction)	k (md)	γ_g
A	Deposito	1	9,371–9,404	5,463	230	0.07	0.80	0.57
	Encanto	2	9,263–9,273	5,156	230	0.04	0.70	0.57
B	Guzmantla Pelagica	1	7,707–7,740	2,500	194	0.07	0.70	0.77
	S. Felipe Breccia	2	7,232–7,297	2,400	194	0.05	0.10	0.73
C	La Laja	1	9,602–9,737	5,463	194	0.05	0.10	0.57
	La Laja	2	9,514–9,553	2,543	194	0.06	0.20	0.57
	La Laja	3	9,350–9,488	2,543	194	0.70	0.40	0.57
D	Guzmantla Pelagica	1	7,849–7,889	4,000	194	0.07	0.70	0.77
	Guzmantla Pelagica	2	7,570–7,741	2,543	194	0.07	0.70	0.77
	S. Felipe Breccia	3	7,220–7,275	2,630	194	0.05	0.10	0.73
E	Guzmantla	1	7,733–7,874	3,523	210	0.07	0.70	0.74
	Guzmantla	2	7,661–7,727	2,000	210	0.07	0.70	0.74
	S. Felipe Breccia	3	7,494–7,553	1,950	210	0.05	0.10	0.73

Table 16.1—Reservoir and fluid information.

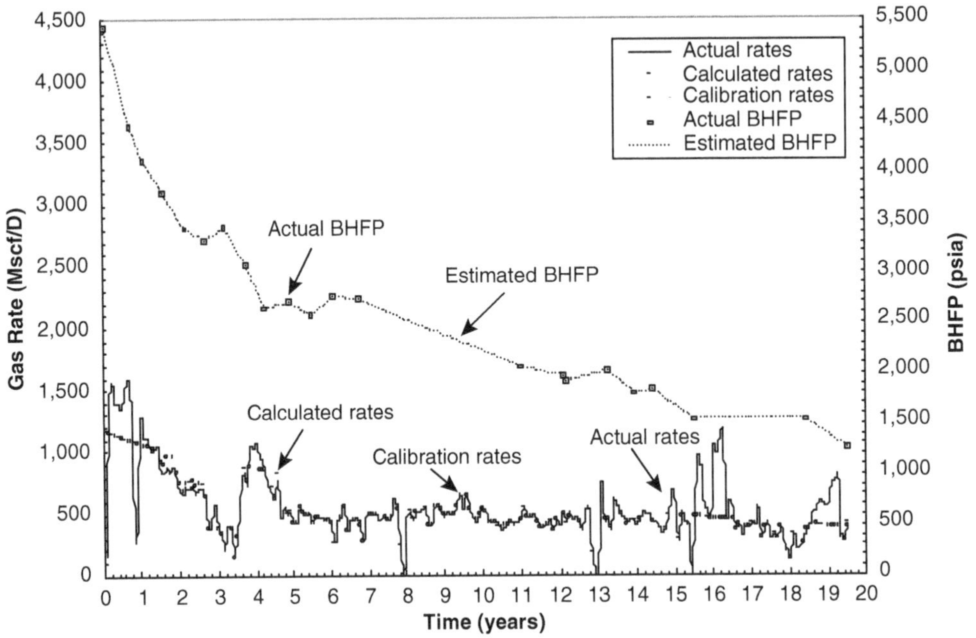

Fig. 16.2—History match and calculated gas rates for Well A. Layer 1 parameters are determined.

Well	Reservoir	Layer	Opened interval(s) (ft subsea)	G_p (MMscf)	*OGIP* (MMscf)	J_g (cp-Mscfd/psia^2)
A	Deposito	1	9,371–9,404	7,856	9,800	8.08E-07
	Mendez	2	9,263–9,273	2,265	4,372	5.39E-07
B	Guzmantla Pelagica	1	7,707–7,740	3,867	31,733	3.13E-05
	S. Felipe Breccia	2	7,232–7,297	484	13,095	3.75E-06
C	La Laja	1	9,602–9,737	4,618	9,889	2.28E-06
	La Laja	2	9,514–9,553	3,072	6,182	4.97E-06
	La Laja	3	9,350–9,488	5,625	13,992	9.46E-06
D	Guzmantla Pelagica	1	7,849–7,889	4,309	11,064	3.17E-06
	Guzmantla Pelagica	2	7,570–7,741	2,293	16,495	1.09E-05
	S. Felipe Breccia	3	7,220–7,275	1,166	9,745	4.39E-06
E	Guzmantla	1	7,733–7,874	12,643	23,061	1.64E-05
	Guzmantla	2	7,661–7,727	316	4,352	2.64E-06
	S. Felipe Breccia	3	7,494–7,553	1,291	5,845	1.34E-05

Table 16.2—Actual G_p and parameters evaluated.

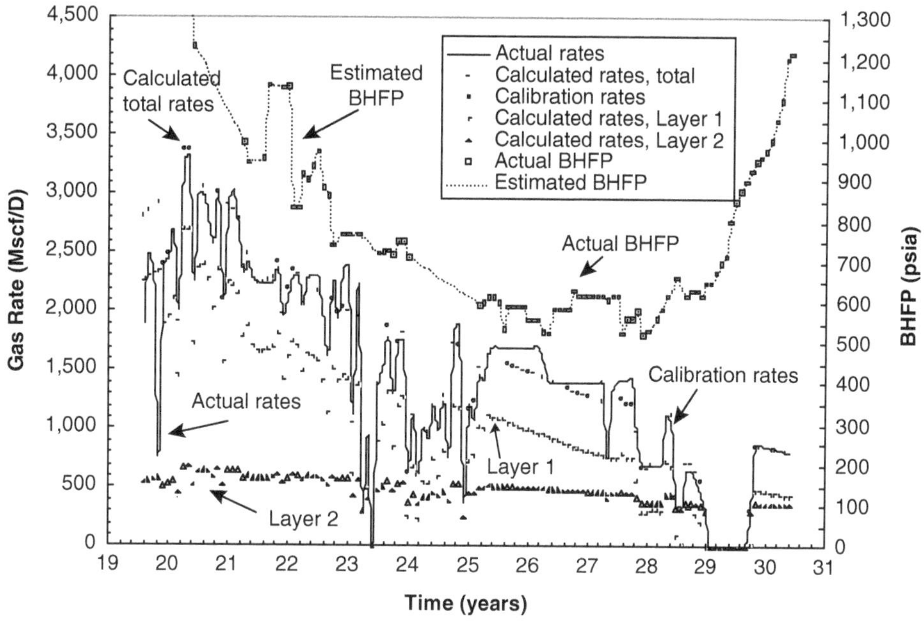

Fig. 16.3—History match and calculated gas rates for Well A. Layer 2 parameters are detemined.

The comparison between the total q_g calculated and the actual behavior of the two-layer system is given in **Fig. 16.3.** We can see that total q_g was continuously declining, and the maximum q_g of 3,377 Mscf/D was reached by the two-layer system during the first 10 months of production (year 20.4). The p_{wf} was changing during the period of production due to several changes in surface conditions. Production q_g for layer 1 is declined from 2,696 Mscf/D in the year 20.4 to less than 500 Mscf/D after eleven years of production. During this period of time the q_g in layer 2 is slightly declining from 680 Mscf/D to less than 400 Mscf/D. Later, we used only p_{wf} data with the two-layer model parameters calculated from the calibration period to forecast q_g.

The parameters evaluated by the two-layer model are shown in Table 16.2. **Fig. 16.4** shows the history match and forecasting q_g for Well A including both layer 1 and the commingled system.

16.4.2 Field Example B. This example shows the application of the model on two reservoirs opened to production at the same time. This well produced a moderate permeability gas reservoir. **Fig. 16.5** displays the schematic completion for Well B. The well has produced wet gas for 2.2 years from both Guzmantla Pelagica and San Felipe Breccia Reservoirs.

Table 16.1 displays reservoir information and fluid information for the opened intervals in both layers. p_{wf} was occasionally available; q_g was available on daily basis.

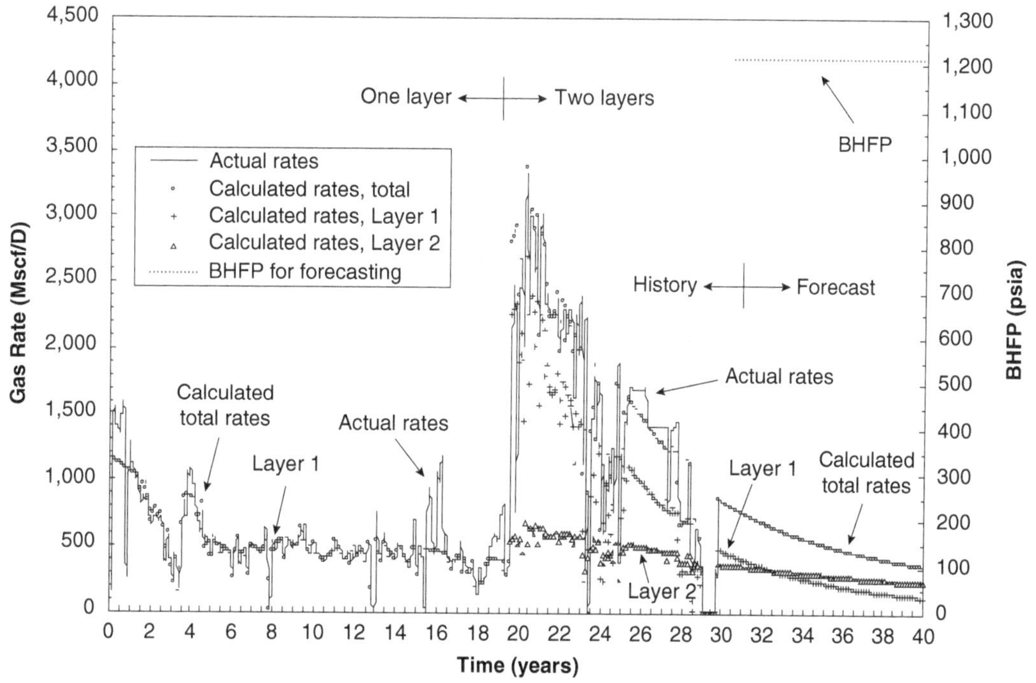

Fig. 16.4—History and forecast gas rates for Well A.

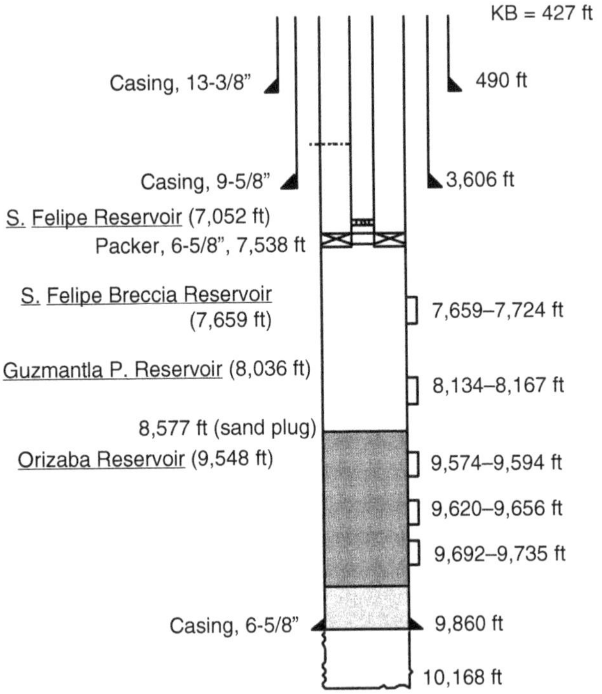

Fig. 16.5—Schematic completion for Well B.

The comparison between the actual q_g performance and the evaluated q_g with the commingled model during the history and forecast periods is shown in **Fig. 16.6.** Production q_g was constant around 7,500 Mscf/D during the first year of exploitation. Later, a change in the operation conditions in the surface gave a variation of production q_g between 5,000 to 4,000 Mscf/D during the following year. As a consequence of this change, the p_{wf} was altered from 1,600 to 1,900 psia after producing 4,351 MMscf of gas.

We used 2.2 years of production and pressure data to calibrate the commingled gas flow model with the two-layer option. We did not use the first 3 months data in the calibration period to decrease transient flow effect. After that, we used the model parameters calculated from the calibration period to estimate the forecast of the production q_g for the following 3 years using a constant p_{wf} of 1,900 psia.

The G_p for each layer as well as the parameters estimated by the commingled gas model (*OGIP* and J_g) are given in Table 16.2. In general, we can see in Fig. 16.6 that q_g from layer 1 represents almost 80% of the total q_g in the well. However, in the forecast period of 3 years and using the p_{wf} established we can see that layer 1 is going to decline continuously while layer 2 tends to maintain its q_g almost constant around 500 Mscf/D.

16.4.3 Field Example C. This example illustrates the purpose of the commingled model on three opened intervals or layers, which are producing in the same reservoir. This reservoir is characterized by its low permeability and intermediate porosity.

Table 16.1 exposes reservoir information and gas specific gravity for this well. The comparison between the actual q_g behavior and the estimated q_g with the gas model during the history and forecast steps is shown in **Fig. 16.7.** The well has produced dry gas for 8 years in La Laja Reservoir in the opened interval called layer 1. After 8 years of exploitation a recompletion job was made in the same reservoir including two upper intervals called layers 2 and 3. The producing p_{wf} measurements were sparse. We evaluated the corresponding p_{wf} from the surface pressure data using Cullender and Smith (1956) method.

Actually this well is exploited through the three opened intervals in the same formation. Total q_g is obtained using single tubing. After 10 years of production the total G_p obtained from the opened intervals is 13,315 MMscf.

16.4.3.1 Layer 1. For the first layer we used the first 8 years of production and p_{wf} data to calibrate the gas model in one layer step. We did not use the first 8 months data in the calibration period to reduce transient flow effect. In Fig. 16.7 we can see that layer 1 reached its maximum production of 2,700 Mscf/D in the first year to decline at 600 Mscf/D after 8 years of exploitation. Table 16.2 discloses the real G_p and the parameters evaluated for layer 1.

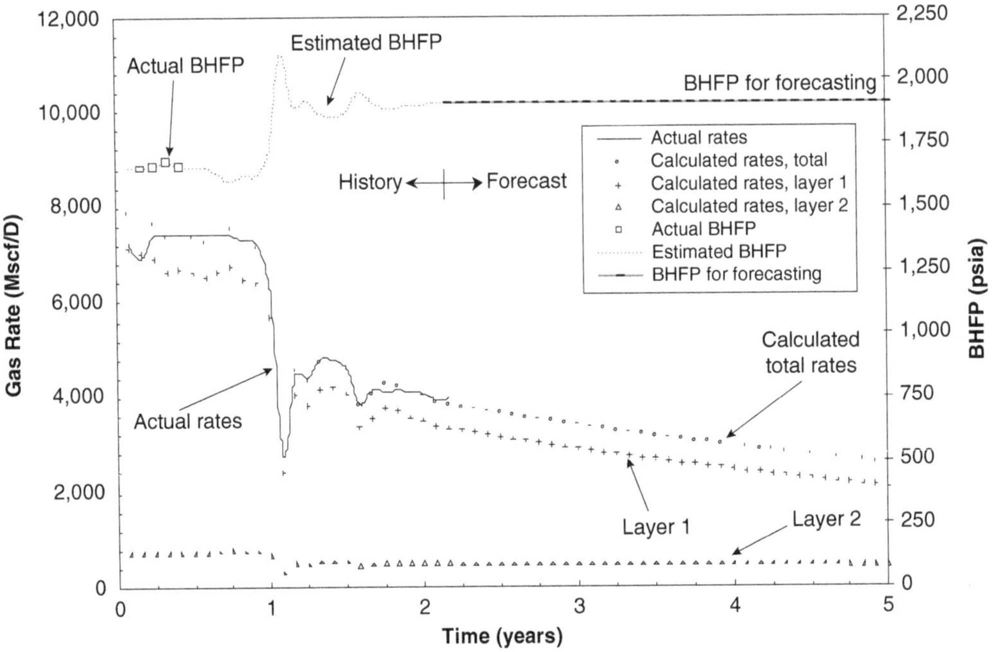

Fig. 16.6—History and forecast gas rates for Well B.

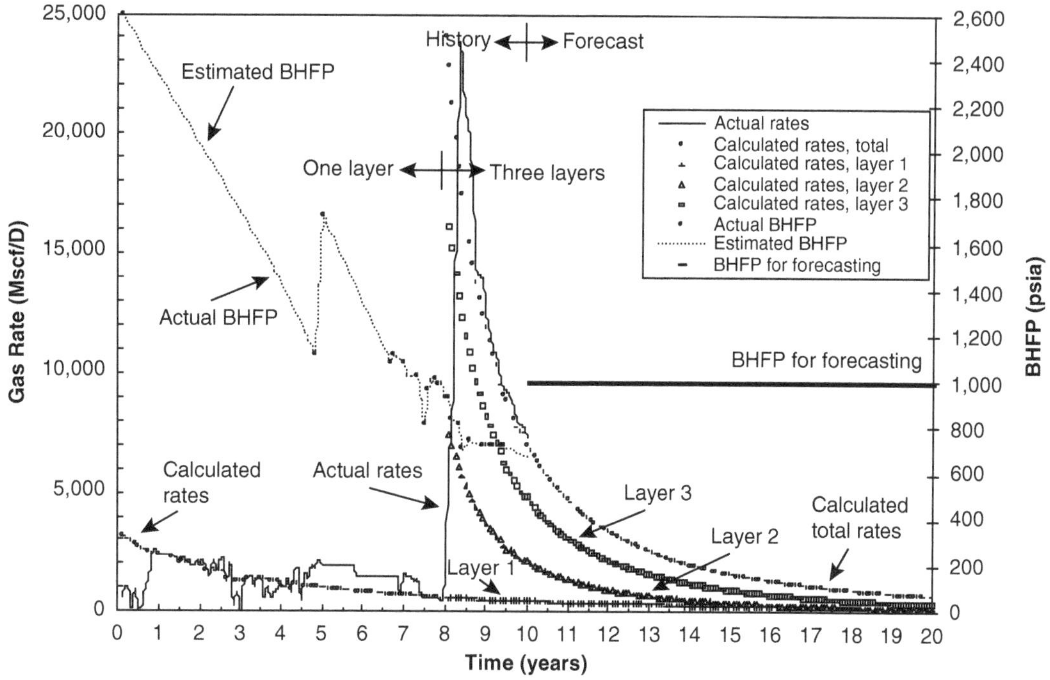

Fig. 16.7—History and forecast gas rates for Well C.

16.4.3.2 Total System. The real G_p and the *OGIP* and pressure calculated by the model in the calibration of the first layer were used as data for layer 1 in the three-layer model option.

For the commingled production system (layers 1, 2 and 3) we used 2 years of production and pressure data to calibrate the model with the three-layer option.

The G_p, *OGIP,* and J_g for each layer calculated from the calibration period are given in Table 16.2. We can see in **Fig. 16.8** that layers 2 and 3 declined from 6,500 and 14,000 Mscf/D in the year 8.3 to less than 4,000 and 8,000 Mscf/D in the year 10, respectively. During this period of time the q_g in layer 1 was slightly declining from 500 to less than 400 Mscf/D. Later, we then used the three-layer model to forecast the q_g in each layer for more than 10 years using the parameters calculated from the calibration period and 1,000 psia of p_{wf} data (Fig. 16.7).

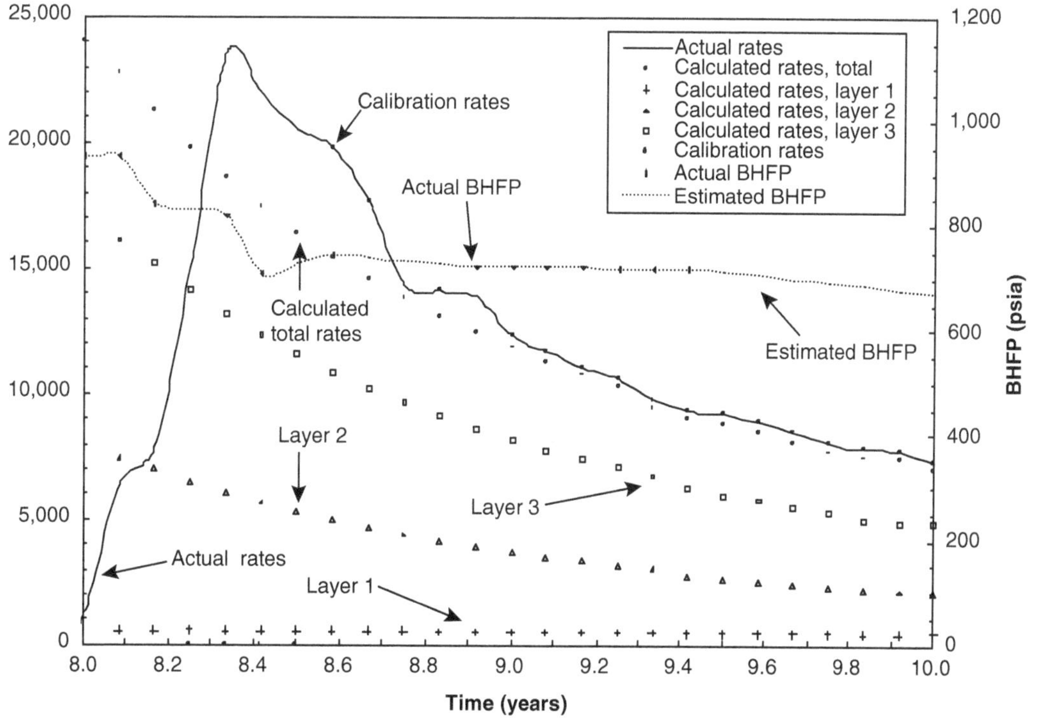

Fig. 16.8—History and forecast gas rates for Well C. Parameter for Layers 2 and 3 are evaluated.

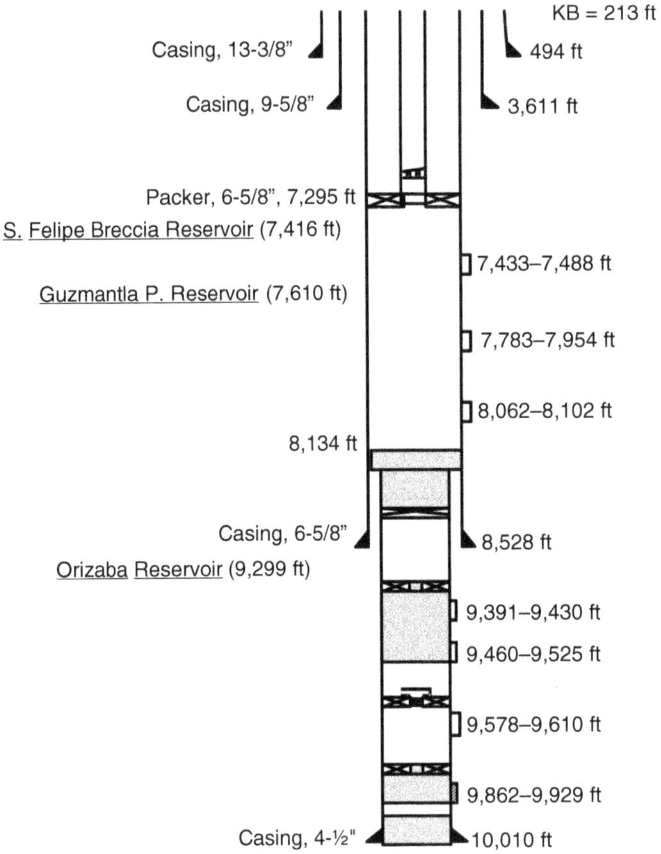

Fig. 16.9—Schematic completion for Well D.

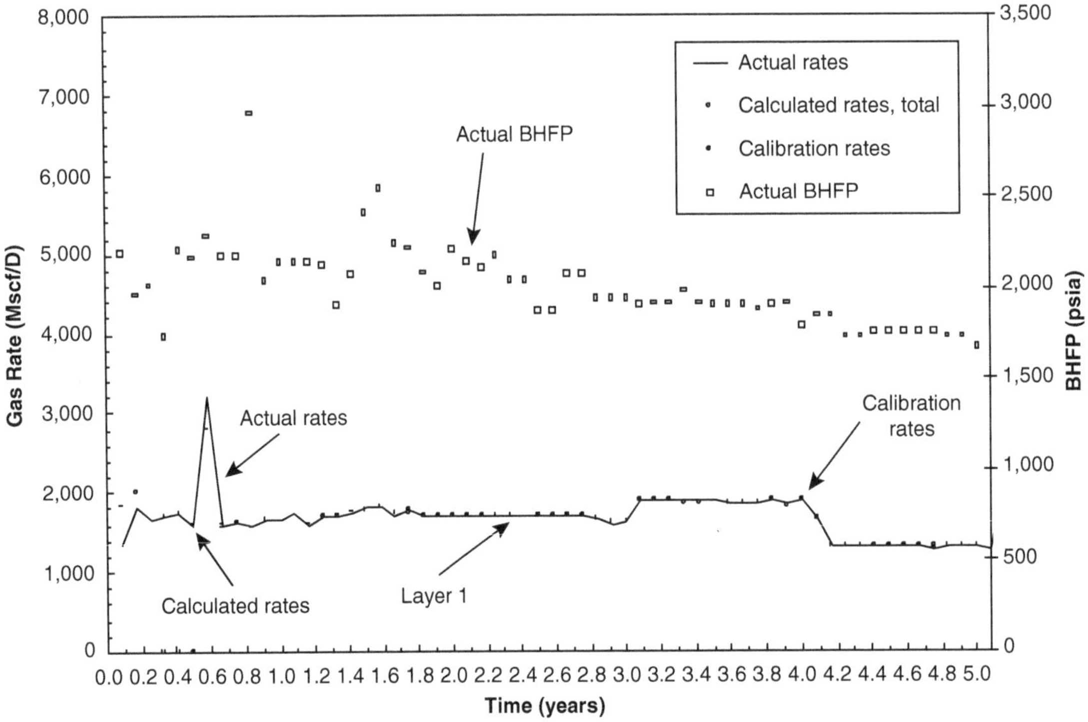

Fig. 16.10—History match and calculated rates for Well D. Layer 1 parameters are determined.

16.4.4 Field Example D. This case gives the value of the commingled gas model on two different reservoirs with several producing intervals opened at different times. This well is a moderate rate well producing a large reservoir. The well was described as being a two-commingled reservoir with no crossflow between Guzmantla Pelagica and San Felipe Breccia Reservoirs.

Fig. 16.9 shows the schematic completion for Well D. The well has produced wet gas for 5.1 years in the Guzmantla Pelagica Reservoir in the lowest opened interval called layer 1. After 5.1 years of production two recompletions jobs were made in the well. One recompletion was in the same reservoir consisting in two more opened intervals. In the commingled model we assumed these intervals to be layer 2 (the upper layer in Guzmantla Pelagica Reservoir). The other recompletion was made in the upper reservoir (San Felipe Breccia) including two opened intervals. These opened intervals were assumed to be layer 3 in the gas model. Table 16.1 gives reservoir and fluid information for each pseudo-layer assumed.

Total production and surface pressure data were available for 8.2 years. Surface rate measurements were available on monthly basis. All p_{wf} measurements were evaluated from wellhead flowing pressure measurements (Cullender and Smith 1956). Actually Well D is producing through the same tubing with gas produced from both reservoirs with the five opened intervals. After 8.2 years of production the cumulative gas production for the well is given as 7,768 MMscf.

16.4.4.1 Single Layer. The comparison between the calculated q_g using the model and the actual performance of the Well D is shown in **Fig. 16.10.** Layer 1 had a cumulative gas production of 2,981 MMscf during the first 5.1 years of production. Gas production was almost constant around 2,000 Mscf/D but the p_{wf} was slightly declining. Unfortunately, the pressure data were not accurate in the first 7 months. However, we did not use this 7 months to calibrate the model due to the behavior of transient flow.

The data we input into our commingled gas program was the monthly production in daily basis and p_{wf} for 5.1 years and the reservoir and fluid data shown in Table 16.1 for layer 1 in Well D. The parameters calculated by the model are given in Table 16.2.

16.4.4.2 Commingled Production. The commingled system is formed for 3 pseudo-layer. Two of them are producing gas from Guzmantla Pelagica Reservoir and the other one is producing from San Felipe Breccia Reservoir. All these layers were opened in the same date with exception of the first layer, which was opened 5.1 years before.

The comparison between the total production calculated by the model and the actual performance of the three-layer system is given in **Fig. 16.11.** For the layered system we used 3 years of q_g and p_{wf} data. The *OGIP*, average pressure, and G_p estimated by the model in the first 5.1 years for opened interval 1 of Guzmantla Pelagica Reservoir were used as data for both layers 1 and 2 in the three-layer model option.

In Fig. 16.11 we can see that the maximum q_g of 7,200 Mscf/D was reached by the three-layer system during the first 9 months of production (year 5.9). Later, total q_g was continuously declining to reach a value of 3,100 Mscf/D after 2.5 years of commingled production. Some data from 5.2 to 6.9 years were not used in the calibration period in order to minimize the effect of transient flow. We can see that Guzmantla Reservoir (layers 1 and 2) is producing almost 75% of total rate, and remaining production, 25%, is given by San Felipe Breccia Reservoir (layer 3). The production q_g for layer 2 declined from 3,800 Mscf/D in the 5.8 year to 1,900 Mscf/D after 2.5 years of exploitation. During this time the q_g from layer 1 was slightly greater than production obtained from layer 3. However, the q_g in both layers 1 and 3 was almost constant around 1,000 Mscf/D. Later, we then used only constant p_{wf} data of 1,620 psia with the three-layer model parameters calculated from calibration period to estimate future production up to 10 years for each layer.

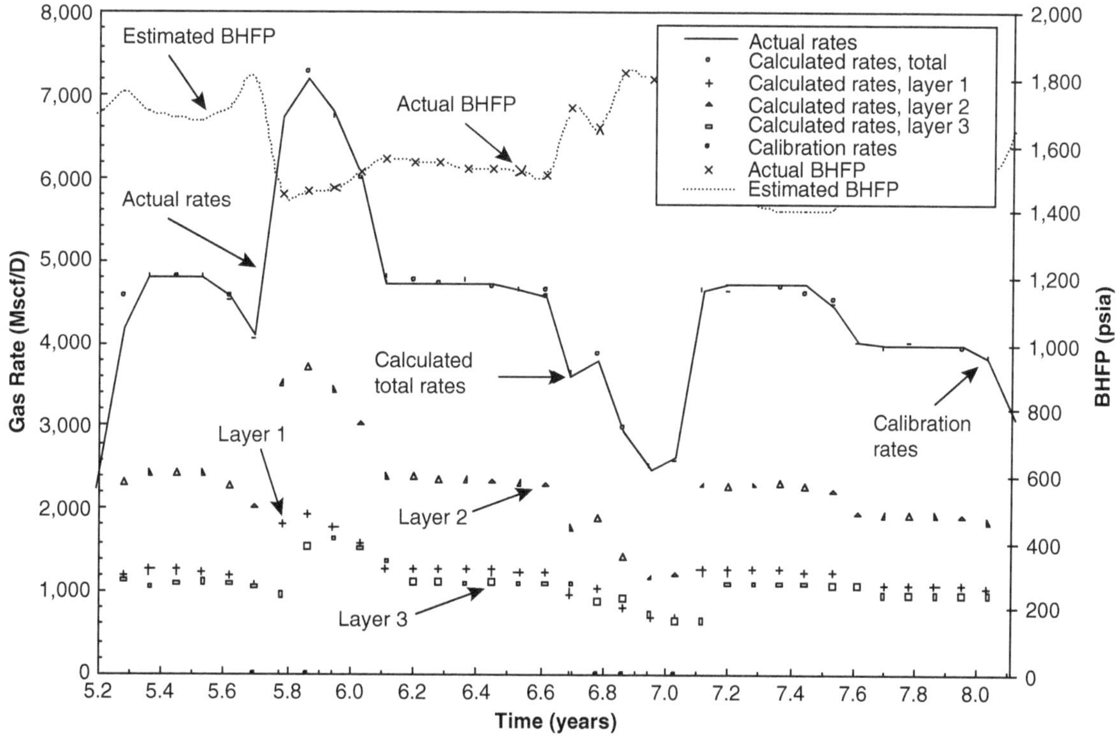

Fig. 16.11—History match and calculated rates for Well D. Parameters for Layers 2 and 3 are determined.

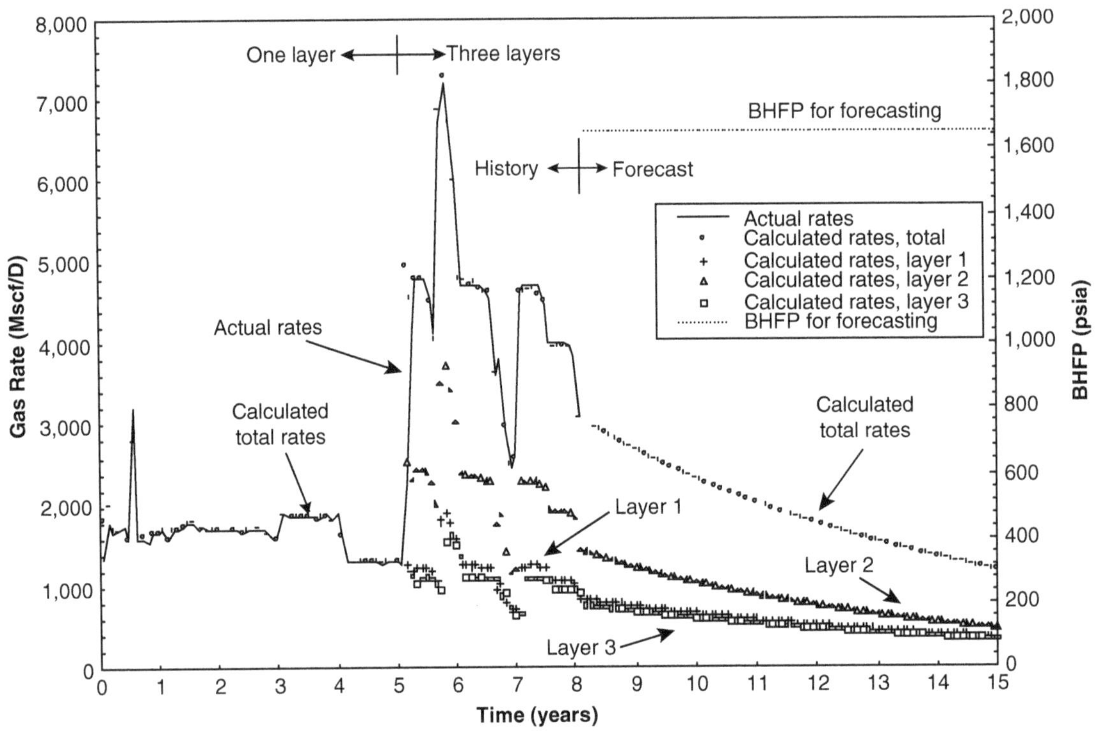

Fig. 16.12—History and forecast gas rates for Well D.

The parameters evaluated by the three-layer system are presented in Table 16.2. **Fig. 16.12** displays the history and forecasting pressure production for Well D including both single layer and commingled production.

16.4.5 Field Example E. This real case illustrates the use of the commingled gas model on two reservoirs. Table 16.1 discloses reservoir and gas specific gravity for each layer.

This well has produced wet gas for 8.3 years in the Guzmantla in the opened interval described in Table 16.1 (**Fig. 16.13**). After 8.3 years of exploitation a recompletion job was made in both Guzmantla Pelagica and San Felipe Breccia Reservoirs where two more intervals were opened respectively.

Surface q_g measures were available on daily basis. Several P_{wf} measurements were not accessible. We evaluated corresponding p_{wf} from surface pressure measurements. In fact, the well is producing through a single tubing with gas produced from three opened intervals in both reservoirs. The total G_p after 10.5 years of exploitation is 14,250 MMscf (Table 16.2).

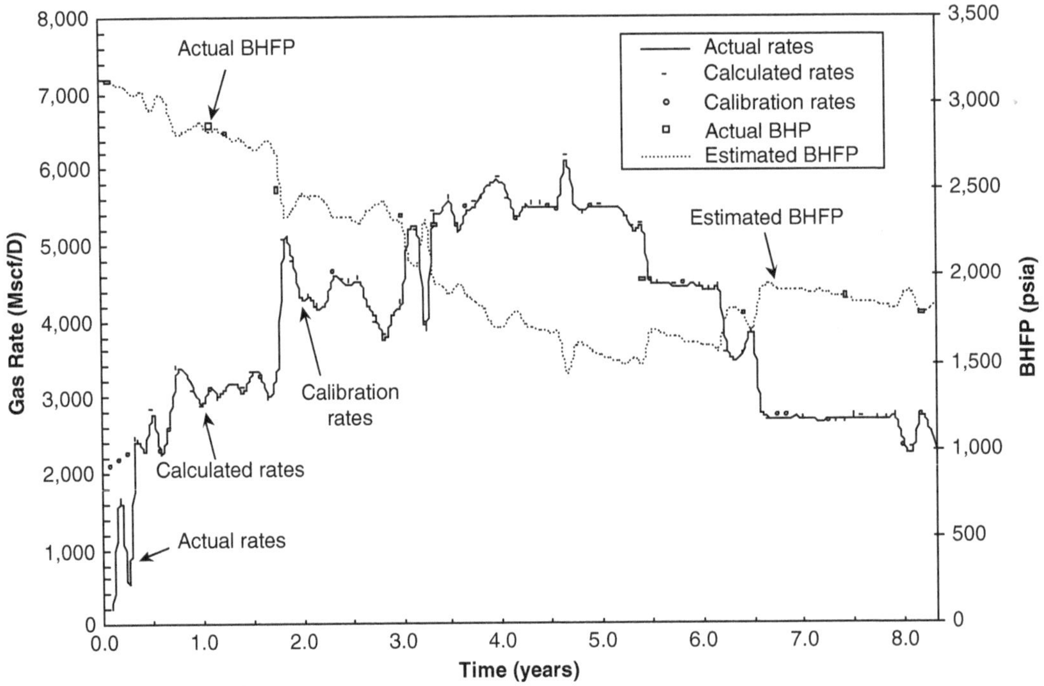

Fig. 16.13—History match and calculated gas rates for Well E. Layer 1 parameters are estimated.

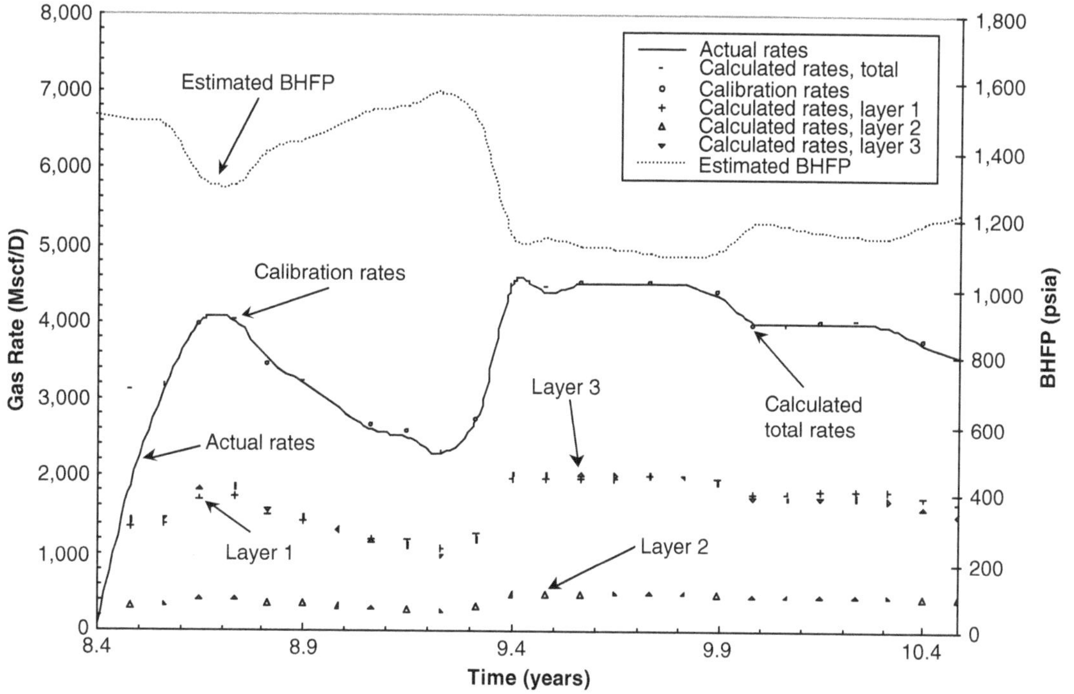

Fig. 16.14—History match and calculated gas rates for Well E. Parameters for Layers 2 and 3 are estimated.

16.4.5.1 Layer 1. The comparison between actual q_g and the calculated with the model is shown in Fig. 16.13. We used 8.3 years of production and p_{wf} data to calibrate the model in one-layer option. We did not use the first 8 months of history data to calibrate the model due the behavior of transient flow. The real G_p and the *OGIP* and J_g parameters estimated for layer 1 are given in Table 16.2. p_{wf} was periodically fluctuated due to changes in operation conditions in the surface.

The well started producing at p_{wf} of 3,200 psia and the p_{wf} dropped to 1,800 psia in 8.3 years. In the same period of time, the q_g was changing to reach the maximum of 6,200 Mscf/D after 4.5 years of exploitation. After 8.3 years of production q_g declined to 2,300 MMscf/D.

16.4.5.2 Commingled Production. For the commingled production (layers 1, 2, and 3) we used 2.2 years of history data to calibrate the model in three-layer option. The real G_p, *OGIP*, and average pressure evaluated in layer 1 by the model in the first 8.3 years were used as data for layers 1 and 2 in the three-layer model.

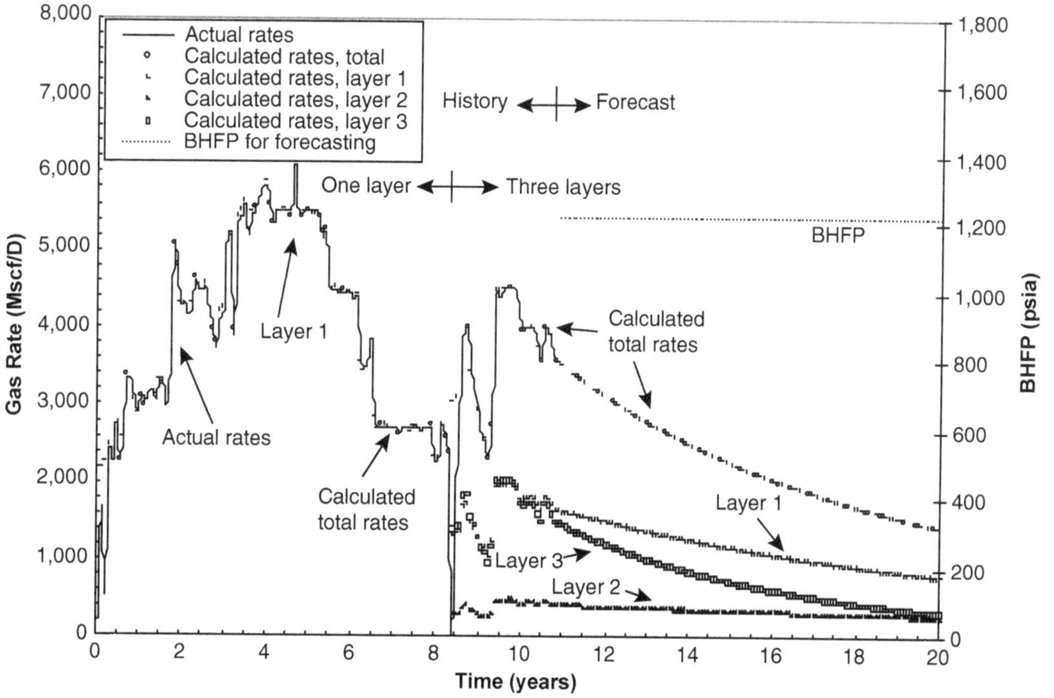

Fig. 16.15—History and forecast gas rates for Well E.

Fig. 16.14 illustrates the comparison between the actual q_g and total q_g calculated with the three-layer system. The maximum production of 4,500 Mscf/D was reached by the three-layer system in the first year of production (year 9.4). The p_{wf} was changing from 1,600 to 1,100 psia through this period of time. Later, we then used only p_{wf} of 1,200 psia as data with the three-layer model parameters calculated from calibration period (Table 16.2) to estimate the forecast of q_g for each layer. **Fig. 16.15** gives the history and forecast q_g for Well E.

16.5 Discussion

Decline curve is the most recognized method tried to effectively analysis multi-layer gas reservoirs from little field data (El-Banbi and Wattenbarger 1996). On the other hand, for commingled gas reservoirs decline curve analysis is absolutely empirical. In the same way, the use of conventional decline curve method would not permit for analyzing production data for gas wells producing at variable p_{wf} with constant q_g production because the q_g is not declining. As a consequence, the commingled gas flow model improved and extended from El-Banbi and Wattenbarger (1997) permits us to analyze, calibrate, and evaluate production data to predict the performance of this kind of wells. Additionally, the method allows modeling of recompletions and different initial pressures in commingled gas reservoirs.

16.5.1 Computer Program. The gas flow model presents an easy technique for modeling commingled gas wells and it is based on established fundamentals, taking into account real gas properties and the variation in p_{wf}.

The improved model was programmed in Visual Basic code for Microsoft Excel. This allows the use of Excel capabilities in nonlinear optimization and plotting. The model was programmed in a sense to permit selection of non-consecutive calibration points. This feature allows eliminating poor quality data points and production data affected by transient flow.

16.5.2 Results. This study shows that the model gives good results if stabilized flow dominates in the commingled system. Consequently, the model is recommended to study low to moderate permeability (between 0.1 and 10 md) reservoirs. In intensely tight reservoirs (less than 0.01 md of permeability) the results calculated by the model may not be reliable because the transient flow might disturb the pressure and production data. However, if the commingled model is applied to analyze transient flow data, it may give conservative estimates of reservoir performance in the history and forecast periods.

To validate and calibrate the commingled gas flow model results we should use well logs and geological information to assist in knowing the layered reservoir and qualitatively judge the relative behavior of each layer (El-Banbi and Wattenbarger 1997). In this work, we confirm that non-Darcy parameter, *b*, could be ignored for most of the field cases we studied. The parameter *b* is expected to be significant for fractured gas wells and for high permeability reservoirs that are produced at very high rates (Holditch and Morse 1976).

The commingled gas flow model calculates the *OGIP*, J_g, reserves, and behavior by individual layer. As a consequence, with this information we can describe the specific layers and evaluate the relative productivity of the layer in the total performance. Additionally, the J_g or *a* parameter can be used to find any parameter in Eq. 16.5 (e.g., reservoir conductivity, $k_g h$, and skin factor, *s*) if other parameters are known. This information can be used to design either stimulation or fracturing treatments.

16.5.3 Limitations. The commingled gas flow model has the following limitations:

1. When the quality of production and pressure data is not precise enough the two-parameter (*OGIP* and J_g) non-linear optimization procedure may converge to incorrect outcomes or local minima. In the same manner, if the commingled system consist of more than 4 layers, the results may not be unique.
2. In gas reservoirs that are exploited predominantly under transient flow effects the results would be not accurate. In this case other techniques can be used to analyze this kind of data. After studying several cases in this work, we verify that a permeability range of 0.1 to 10 md is a suitable range for the application of this model. For gas reservoirs which have permeability less than 0.01 md (extremely tight reservoirs) the transient effects may be so long and they cannot have enough stabilized flow to calibrate our model. On the other hand, for gas reservoirs above 20 md (high permeability) non-Darcy flow may be important and the use of the J_g equation without the inclusion of non-Darcy flow term, *b*, may give inaccurate results.

16.6 Conclusions

The work presented in this paper is based on calibrating or matching production data for stabilized gas flow theory. It considers commingled wells producing against variable p_{wf} and ignores non-Darcy flow. The following conclusions can be drawn from this study.

1. An improved method to estimate *OGIP*, J_g, and reserves for each layer of a commingled gas system was developed. It allows modeling of recompletions and different initial pressures in the layers.
2. The method forecasts the total production of layered gas wells and calculates individual layer contributions. Applications of the method were shown by five Mexican gas wells.
3. Production data from commingled gas wells can be analyzed by commingled gas flow model with the use of a nonlinear optimization procedure.
4. Commingled gas flow models can be used to analyze gas wells if stabilized flow conditions dominate.
5. *OGIP* and J_g for each layer are determined by history matching. The calibration period used for the history-calibration period does not have to be adjacent.
6. History-matching selective intervals of non-consecutive calibration points rather than the entire production history allows identifying and eliminating poor quality data points and production data affected by transient flow effects.
7. The commingled gas flow model can allow many transient flow periods originated by fluctuations in operation conditions.
8. *OGIP* is under-estimated if transient data are included. Total *OGIP* for the commingled system is underestimated if a smaller number of layer models is used in the analysis.

9. The model is easy to use and it can be used as a spreadsheet application.
10. Effect of non-Darcy flow on the analysis using the commingled gas flow model for gas reservoirs between 0.1 and 10 md of formation permeability can be depreciated.
11. Case histories or examples may take place where solutions corresponding to local minima are evaluated by the optimization procedure.

16.7 Nomenclature

A = reservoir drainage area, L^2, ft^2
a = stabilized deliverability coefficient, ($psia^2$–cp)/(Mscf/D)
$BHFP$ = bottom-hole flowing pressure, m/Lt^2, psia
b = stabilized deliverability coefficient, ($psia^2$–cp)/$(Mscf/D)^2$
C_A = shape factor
CO_2 = carbon dioxide, fraction
$error_1$ = normalized error measure given by Eq. 16.8
$error_2$ = normalized error measure given by Eq. 16.9
G = original gas in place, L^3, MMscf
G_p = cumulative gas produced, L^3, MMscf
h = formation thickness, L, ft
H_2S = hydrogen sulfide, fraction
J_g = real gas flow coefficient, L^4t^2/m, cp-Mscfd/$psia^2$
k = permeability, L^2, md
KB = Kelly bushing height, L, ft
$m(\overline{p})$ = real gas pseudo-pressure at average reservoir pressure, m/Lt^3, $psia^2$/cp
$m(p_{wf})$ = pseudo-pressure at bottom hole flowing pressure, m/Lt^3, $psia^2$/cp
$Nlayer$ = number of layers in a commingled reservoir
N = number of gas rate data points used in calibration
N_2 = nitrogen, fraction
$OGIP$ = original gas in place, L^3, MMscf
p = pressure, m/Lt^2, psia
p_i = initial reservoir pressure, m/Lt^2, psia
p_o = base pressure, m/Lt^2, psia
p_{sc} = pressure at standard conditions, m/Lt^2, psia
p_{wf} = bottom hole flowing pressure, m/Lt^2, psia
$\overline{p}$ = material-balance average reservoir pressure, m/Lt^2, psia
q_g = gas production rate, L^3/t, Mscf/D
q_{T_g} = total flow rate in a commingled system, L^3/t, Mscf/D
r_w = wellbore radius, L, ft
s = skin factor
t = time, t, days
T = reservoir temperature, T, °R
T_{sc} = temperature at standard conditions, T, °R
z = real gas compressibility factor
$\varnothing$ = porosity
γ_g = specific gravity of gas
μ_g = gas viscosity, m/Lt, cp

Subscripts

g = gas
i = initial
i = layer index
j = time period index

16.8 Acknowledgments
We thank the management of PEMEX Exploracion-Produccion for funding this study and providing field data. This work was done within the Reservoir Modeling Consortium at Texas A&M University. Special appreciation is due to Angel Meza-Rios and Activo Veracruz of PEMEX and Bryan Maggard and Wael Helmy of Texas A&M University for their invaluable help in this work.

16.9 References

Al-Hussainy, R., Ramey, H. J., and Crawford, P. B.1966. The Flow of Real Gas Through Porous Media, *J Pet Technol* **18** (5): 624–636. SPE-1243-A-PA. https://doi.org/10.2118/1243-A-PA.

Aminian, K., Ameri, S., Srark, J. J., et al. 1990. Gas-Well Production Decline in Multiwell Reservoirs. *J Pet Technol* **42** (12): 1573–1579. SPE-18268-PA. https://doi.org/10.2118/18268-PA.

Arps, J. J. 1945. Analysis of Decline Curves. *Trans. AIME* **160** (01): 228–247. SPE-945228-G. http://dx.doi.org/10.2118/945228-G.

Blasingame, T. A., McCray, T. L., and Lee, W. J. 1991. Decline Curve Analysis for Variable Pressure Drop/Variable Flowrate Systems. Presented at the SPE Gas Technology Symposium, Houston, Texas, 23–24 January. SPE-21513-MS. https://doi.org/10.2118/21513-MS.

Carter, R. D. 1985. Type Curves for Finite Radial and Linear Gas-Flow Systems: Constant-Terminal-Pressure Case. *SPE J.* **25** (5): 1719–728. SPE-12917-PA. https://doi.org/10.2118/12917-PA.

Chen, H. Y. and Poston, S. W. 1989. Application of a Pseudotime Function To Permit Better Decline-Curve Analysis. *SPE Form Eval* **4** (3): 441–448. SPE-17051-PA. https://doi.org/10.2118/17051-PA.

Cullender, M. H. and Smith, R. V. 1956. Practical Solution of Gas-Flow Equations or Wells and Pipelines with Large Temperature Gradients. *Trans, AIME* **207:** 281–287. SPE-696-G.

Dranchuk, P. M. and Abou-Kassem, J. H. 1975. Calculation of Z Factors for Natural Gases Using Equations of State. *J Can. Pet Technol* **14** (3): 34–36. PETSOC-75-03-03. https://doi.org/10.2118/75-03-03.

El-Banbi, Ahmed H. and Wattenbarger, Robert A. 1996. Analysis of Commingled Tight Gas Reservoirs. Presented at the SPE Annual Technical Conference and Exhibition, Denver, Colorado, 6–9 October. SPE-36736-MS. https://doi.org/10.2118/36736-MS.

El-Banbi, Ahmed H. and Wattenbarger, Robert A. 1997. Analysis of Commingled Gas Reservoirs With Variable Bottom-Hole Flowing Pressure and Non-Darcy Flow. Presented at the SPE Annual Technical Conference and Exhibition, San Antonio, Texas, 5–8 October. SPE-38866-MS. https://doi.org/10.2118/38866-MS.

Fetkovich, M. J. 1980. Decline Curve Analysis Using Type Curves. *SPE JPT* **32** (6): 1065–1077. SPE-4629-PA. http://dx.doi.org/10.2118/4629-PA.

Fetkovich, M. J., Bradley, M. D., Works, A. M. et al. 1990. Depletion Performance of Layered Reservoirs Without Crossflow. *SPE Form Eval* **5** (3): 310–318. SPE-18266-PA. https://doi.org/10.2118/18266-PA.

Fraim, M. L. 1987. Gas Reservoir Decline-Curve Analysis Using Type Curves With Real Gas Pseudopressure and Normalized Time. *SPE Form Eval* **2** (4): 671–682. SPE-14238-PA. https://doi.org/10.2118/14238-PA.

Garb, F. A., Rodgers, J. S, and Prasad, R. K. 1973. Find Gas In Place from Shut-In or Flowing Pressures. *Oil & Gas J.* 58–64.

Guardia, M. A. and Hackney, R. M. 1991. A Practical Approach to Original Gas-In-Place Estimation: Development of the South Wilburton Field. Presented at the 66th Annual Fall Technical Conference and Exhibition, Dallas, Texas, 6–9 October. SPE-22925-MS. https://doi.org/10.2118/22925-MS.

Holditch, S. A. and Morse, R. A. 1976. The Effects of Non- Darcy Flow on the Behavior Of Hydraulically Fractured Gas Wells. *J Pet Technol* **28** (10): 1169–1179. SPE-5586-PA. https://doi.org/10.2118/5586-PA.

Keating, J. F., Chen, H. Y., Wattenbarger, R. A. 1994. Original Gas in Place and Decline Curves From Early Stabilized Rate-Time Data. Presented at the SPE Permian Basin Oil and Gas Recovery, Midland, Texas, 16–18 March. SPE-27666-MS. https://doi.org/10.2118/27666-MS.

Lee, A. L., Gonzalez, M. H., and Eakin, B. E. 1966. The Viscosity of Natural Gases." *J Pet Technol* **218** (08): 997–1000. SPE-1340-PA. https://doi.org/10.2118/1340-PA.

Lee, W. J. and Wattenbarger, R. A. 1996. *Gas Reservoir Engineering,* Vol. 5. Richardson, Texas: SPE Textbook Series, Society of Petroleum Engineers.

Palacio, J. C. and Blasingame, T. A. 1993. Decline-Curve Analysis With Type Curves – Analysis of Gas Well Production Data. Presented at the SPE Rocky Mountain Regional/Low Permeability Reservoirs Symposium, Denver, Colorado, 12–14 April. SPE-25909-MS. https://doi.org/10.2118/25909-MS.

Ramagost, B. P. and Farshad, F. F. 1981. P/Z Abnormally Pressured Gas Reservoirs. Presented at the 56th Annual Technical Conference and Exhibition, San Antonio, Texas, 5–7 October. SPE-10125-MS. https://doi.org/10.2118/10125-MS.

Rodgers, J. S., Boykin, R. S., and Coble, L. E. 1983. Nonstatic Pressure History Analyses for Gas Reservoirs. *SPE J.* **23** (2): 209–218. SPE-10356-PA. https://doi.org/10.2118/10356-PA.

Sutton, R. P. 1985. Compressibility Factors for High- Molecular-Weight Reservoir Gases. Presented at the SPE Annual Technical Meeting and Exhibition, Las Vegas, Nevada, 22–25 September. SPE-14265-MS. https://doi.org/10.2118/14265-MS.

West, S. L., and Cochrane, P. J. R. 1995. Reserves Determination Using Type-Curve Matching and EMB Methods in the Medicine Hat Shallow Gas Field. *SPE Res Eng* **10** (2): 82–87. SPE-28609-PA. https://doi.org/10.2118/28609-PA.

Chapter 17

A Case History for Massive Hydraulic Fracturing of the Cotton Valley Lime Matrix, Fallon, and Personville Fields

H. G. Kozik, SPE, Mitchell Energy Corp.; **S. A. Holditch,** SPE, Texas A&M U.

 Journal paper SPE 7911 was published in vol. 33 (02) of *J. Pet. Technol.* Original manuscript received by SPE 27 March 1979 and accepted for publication 21 December 1979. SPE 7911 was first presented at the SPE Symposium on Low-Permeability Gas Reservoirs held in Denver, 20–22 May 1979.

SPE 7911 presents a case history of massive hydraulic fracture treatments in two Cotton Valley Lime (Haynesville) wells located in Limestone County, Texas—on the west flank of the East Texas Basin. Kozik and Holditch provide an overview of the regional geology and summarize the reservoir and fluid properties on the basis of prefracture formation evaluation that included a log analysis, routine core analysis, and special core analysis. They also present the stimulation history and its corresponding production performance of wells in the area. Kozik and Holditch document the fracture design process and evaluated the fracture effectiveness using transient pressure analysis. Last, but not least, fracture length and well spacing were optimized on the basis of parametric studies of fracture length, conductivity, well spacing, and economics. The authors conclude that massive hydraulic fracture techniques have successfully been applied in the Cotton Valley Lime Formation in Central Texas.

17.1 Summary

SPE 7911 summarizes the results of an active stimulation program on the Cotton Valley Lime as evaluated using reservoir production and pressure transient data. Using standard economic parameters and reservoir permeabilities determined by history matching, a detailed study was made to determine the well spacing and fracture length radius necessary for optimum development of the Fallon and North Personville Fields.

17.2 Introduction

Considerable emphasis has been placed in recent years on recovering gas from "tight-gas" basins. One step that shows promise for unlocking tight gas is massive hydraulic fracturing (MHF). A case in point is the evolution of the stimulation in the Cotton Valley Lime at Fallon and North Personville Fields, about 10 miles southeast of Mexia in Limestone County, TX (**Fig. 17.1**).

The Jurassic Cotton Valley group on the west flank of the East Texas Basin generally consists of a succession of about 1,000 ft of terrestrial and marine sands and shales, 800 ft of dark (Bossier) shale, and 300 to 500 ft of limestone at the bottom known as the Cotton Valley Lime or Haynesville. The group covers more than 250,000 sq miles in the East Texas Basin and adjacent parts of Louisiana and southeast Arkansas.

The Cotton Valley Lime in the Fallon and North Personville areas has been a known gas area since its discovery in 1969. However, its poor permeability pay at a depth of about 11,000 ft had produced at rates too low for commercial development under the price schedules and technology of the past decade.

Due to fracture porosity, beginning rates were moderately good (1 to 4 MMcf/D), but the natural fracture system was not sufficient to maintain the high rates of flow; consequently, the rates dropped to 0.200 to 0.500 MMcf/D in less than 2 years.

Lately, advances in stimulation technology and improved economic incentives renewed interest in this area. Conventional acid stimulations and small fracturing jobs brought some improvement. Bigger fracturing jobs brought further improvement, but MHF appears worthwhile.

These are the objectives of this report:

1. Document the evolution of stimulation technology of the Cotton Valley Lime at the Fallon and North Personville Fields.
2. Using a computer reservoir simulator, history match the production and pressure transient data of the Muse No. 1 well before and after MHF and determine the uniqueness of this solution.
3. History match the production of the Muse-Duke No. 1 after a super-MHF treatment.
4. Optimize the development of the Cotton Valley Lime Reservoir with respect to fracture length and well density using reservoir data and economic guidelines.
5. Document the design of a super-MHF job (see Appendix17.A.1).

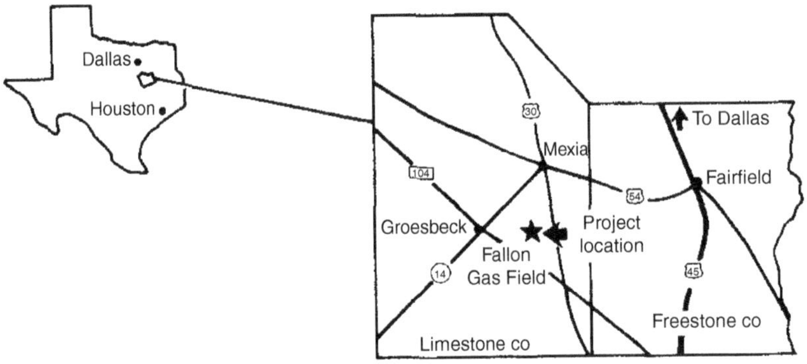

Fig. 17.1—Wellsite location.

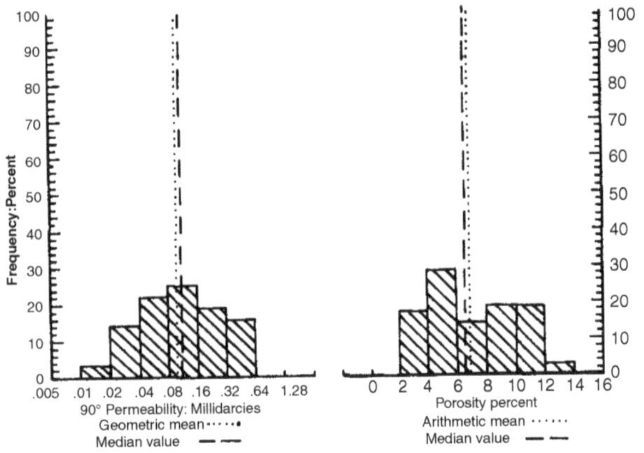

Fig. 17.2—Permeability and porosity histograms for Lawrence No. 1 and Muse-Duke No. 1 (combined).

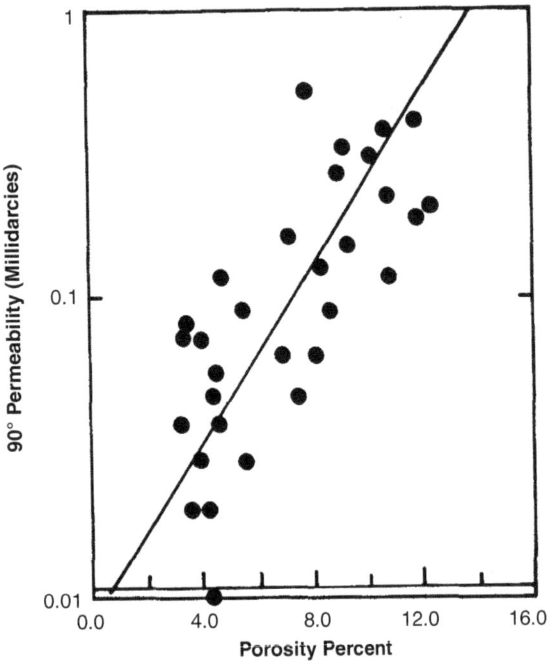

Fig. 17.3—Permeability vs. porosity for Lawrence No. 1 and Muse-Duke No. 1 (combined).

17.3 Reservoir Characteristics

17.3.1 Rock Parameters. The Cotton Valley Lime Formation is generally gray, massive, oolitic to pistolitic, and finely crystalline to micritic. The better porosities appear to be related to the oolitic zones, ranging from 2 to 12% with some local thin zones of 14%. Core analyses indicate that the average matrix permeabilities range from about 0.01 to 0.7 md in the better rock and from 0.003 to 0.7 md when considering the total gross section. An occasional very thin streak of permeability may be as high as 80 to 90 md.

Fig. 17.2 shows histograms of the permeability and porosity of selected cores recovered from the Lawrence No. 1 and Muse-Duke No. 1. From a total of 216 ft of core, 31 ft of the best reservoir rock were selected for a sample statistical analysis. The samples selected were those that were taken from the core 90° to the axis of the maximum measured permeability in the core. The permeability of these samples had a geometric mean of 0.01 md and a harmonic mean of 0.06 md. The average porosity was 6.9%.

17.3.2 Net-Pay. Net-pay was determined as two categories—i.e., the total porosity storage space available and the commercially producible porosity. The total porosity storage space was determined by counting porosity intervals of 2% and greater because some permeability was beginning to develop at 2%. Commercially producible porosity was counted starting at 4% because 98% of the cored rock with 4% or better porosity had permeabilities of 0.01 md or more. The relationship of permeability to porosity is shown in **Fig. 17.3.** Overall, the average producible net-pay is running between 25 to 50 ft per well, using 4% porosity as a cutoff.

17.3.3 Fracture Orientation. Fig. 17.4 shows the distribution of fracture orientation of cores from the Muse-Duke No. 1. The density distribution of the oriented fractures indicated two major fracture strike trends bearing northwest to southeast. Using this information, the fracture strike distribution of the cores that were not oriented could be oriented in relation to true north by fitting the nonoriented section of the core to the sections that were oriented. A comparison of the two density distributions of both oriented and nonoriented fracture strikes shows that the azimuth readings on the nonoriented fractures are equivalent to those readings represented by one of the major strike trends; therefore, they were included in the plot. Also, histograms indicate that the majority of fractures occur in the range of 85 to 90°, with no fracture dip angles less than 50° being noted.

17.3.4 Absolute Permeability. Absolute permeability measurements were made using a transient technique (Brace et al. 1968; Yomada and Jones 1980) which is well suited for permeabilities in the microdarcy range. The results are given in **Table 17.1.** The test conditions of 8,000-psi confining pressure, 6,400-psi pore pressure, and 285°F were commensurate with the in-situ reservoir conditions. Application of simulated in-situ conditions reduces permeability by an order of magnitude.

17.3.5 Effective Permeability. The transient technique also can be used to measure effective permeability to gas at different water saturations by using partially saturated samples and keeping the pulsing pressure lower than the capillary pressure of water in the sample. In the tests performed, the pulsing pressure was always low compared with the expected capillary pressure of more than 10 psi.

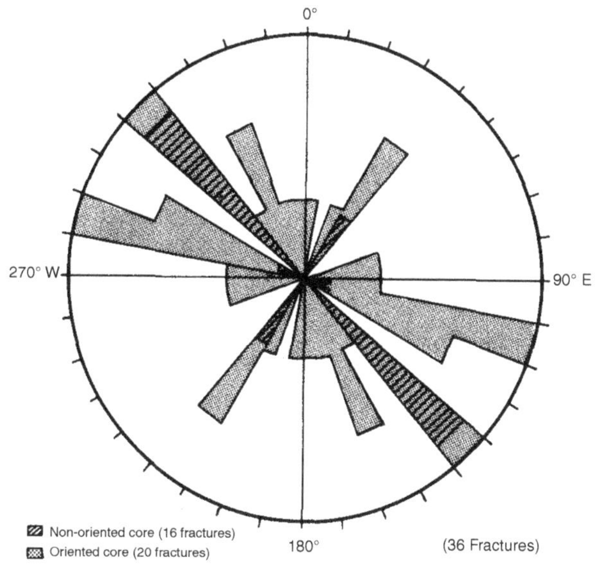

Fig. 17.4—Distribution of fracture orientation, Cotton Valley Lime, Muse-Duke No. 1.

Formation Sample Depth (ft)	Confining Pressure (psi)	Pore Pressure (psi)	Temperature (°F)	Permeability (md)
11,313 A	100	0	75	0.159
	8,800	6,400	285	0.023
11,313 B	100	0	75	0.811
	8,800	6,400	285	0.082
11,3130	100	0	75	0.991
	8,800	6,400	285	0.031

Table 17.1—Absolute permeability measurements, Muse-Duke No. 1.

Table 17.2 lists the effective and relative permeabilities to nitrogen gas for samples from the 11,302- and 11,313-ft depths. The measurements were taken at in-situ conditions of 8,800-psi confining pressure, 6,400-psi pore pressure, and 285°F and at various water saturation levels. **Fig. 17.5** illustrates the variation of relative permeability to gas with increasing water saturation. Both samples show a similar trend. Up to 20% water saturation, the relative permeability to gas virtually is unaffected by the presence of water. Gas permeability then decreases linearly with increasing water saturation, and at about 85% water saturation, the relative permeability to gas approaches zero.

17.3.6 Ultrasonic Tests. Both the compressional (*P*-wave) and shear wave (*S*-wave) velocities were measured on samples subjected to simulated in-situ conditions. Sample axes were normal to the bedding plane, and results are directly comparable, therefore, with the *P*-wave logs. Young's modulus and Poisson's ratio derived from the measured densities and *P*- and *S*-waves assuming isotropic elastic material are compared with data derived from mechanical properties logs in **Fig. 17.6.** The results derived show very good agreement.

Core Sample (ft)	Water Saturation (%)	Effective Permeability k_g (md)	Relative Permeability k_{rg} (md)
11,302	0	0.0022	1.0
	20	0.0021	0.95
	40	0.0013	0.59
	60	0.0008	0.36
	80	<0.0001	–
	100	<0.0001	–
11,313	0	0.0880	1.0
	20	0.0850	0.96
	40	0.0450	0.51
	60	0.0330	0.38
	80	0.0150	0.17
	100	<0.0001	–

Table 17.2—Effective permeability measurements, Muse-Duke No. 1.

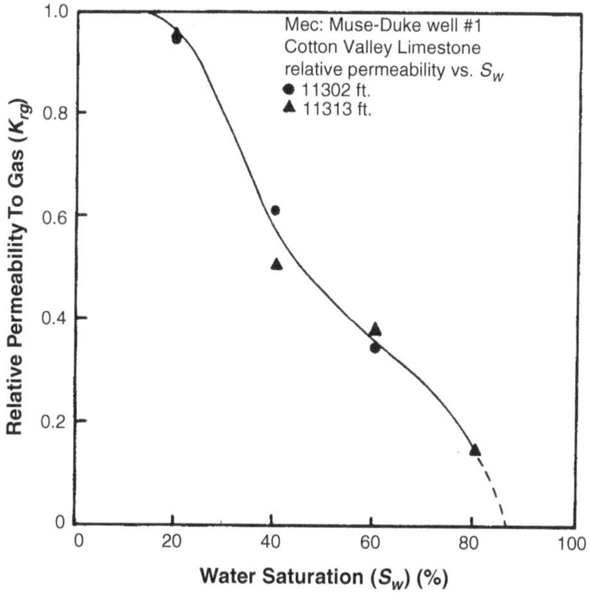

Fig. 17.5—Relative permeability to gas as a function of water saturation, Muse-Duke No. 1.

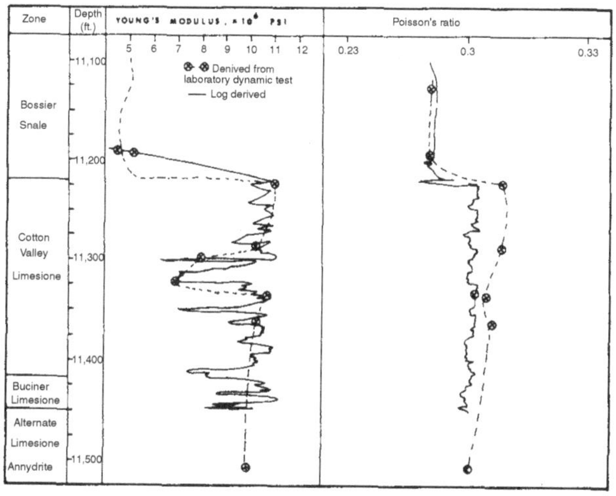

Fig. 17.6—Dynamic property variation with depth, Muse-Duke No. 1.

17.4 Stimulation History

17.4.1 Early Stimulations. The first six wells completed in the Cotton Valley Lime received either a light acid fracture or a limited water/sand fracture (under 50,000 gal). The best well, Burleson No. 1, after a fracture treatment of 48,000 gal of gelled water and 178,000 lbm of sand, began producing 4.4 MMcf/D, which declined to 2.5 MMcf/D in 3 months and to 0.500 MMcf/D in 2 years. The other five wells, after normal acid stimulation treatments, began to produce at about 1 MMcf/D per well, which soon declined to 0.200 to 0.300 MMcf/D. After the initial flush production, the decline rate for Burleson No. 1 was 45% for about the first 2 years and then dropped to 9%. The average of the other five wells declined at a rate of 57% and then dropped to 15%.

17.4.2 Massive Stimulations. The Muse No. 1 was the first well to receive an MHF stimulation treatment. A total of 340,000 gal of gelled water and 450,000 lbm of 20-40 mesh sand was pumped. An additional 48,000 lbm of 100 mesh sand was pumped ahead of the 20-40 sand to help control fluid loss. The initial after-fracture rate was in excess of 4 MMcf/D, and the rate has declined gradually to its present capability in excess of 1.1 MMcf/D after 28 months. After the initial flush production, the annual decline rate was only 29%, and the current decline rate is about 13% per year.

The super-MHF stimulation treatment was performed on the Muse-Duke No. 1. Following an acid stimulation treatment, the well flowed in excess of 4 MMcf/D, declining rapidly to 1.7 MMcf/D. Immediately after the fracture treatment consisting of 2.8 million lbm of sand, the well produced more than 6 MMcf/D. Two months after the fracture treatment, it still had more than 5-MMcf/D deliverability. Six months later it was flowing 4 MMcf/D, and 22 months later it still was flowing 2.1 MMcf/D with 700-psi flowing tubing pressure. The initial decline rate began at 25%, and its present decline rate is 5% per year.

Four other wells have received MHF treatments with sand volumes ranging from 450,000 to 1 million lbm. Initial production indicates performances similar to the Muse No. 1.

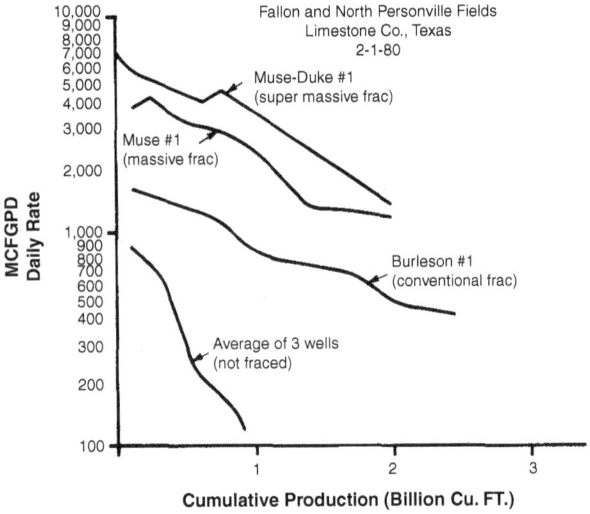

Fig. 17.7—Performance comparisons of various stimulations, Cotton Valley Lime.

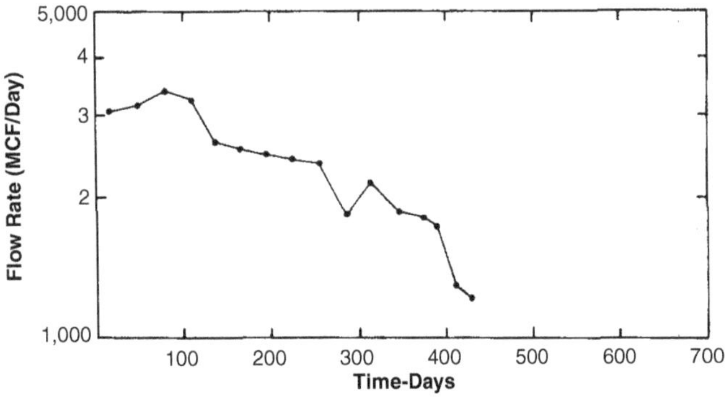

Fig. 17.8—Production decline curve, Muse No. 1.

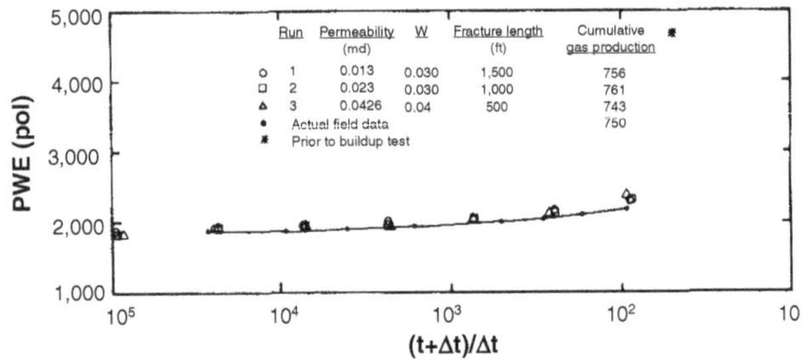

Fig. 17.9—Postfracture BHP buildup, Muse No. 1.

17.4.3 Comparison Plots. **Fig. 17.7** presents a comparison of the daily flow rates vs. cumulative production for the representative wells discussed. The bottom curve illustrates the average well performance for the best three wells that were acidized only. In 11 years, the average cumulative production per well is less than 1 Bcf of gas. The next curve represents the performance of the Burleson No. 1, which was fracture treated conventionally. This well has produced more than 2.4 Bcf of gas in 10 years. The next curve above the Burleson No. 1 presents the production data for the Muse No. 1, which had the first MHF treatment. The Muse No. 1 has produced more than 1.9 Bcf of gas in 2.7 years. The top curve represents the performance of the Muse-Duke No. 1 after its super-MHF treatment. Its cumulative production is more than 1.9 Bcf of gas in just 1.8 years.

17.5 Reservoir Treatment

A computer reservoir simulator (Holditch et al. 1978; Holditch and Laufer 1978) was used to history match the production and pressure transient data of Muse No. 1 and Muse-Duke No. 1. Basic formation permeability, fracture length, and fracture conductivity were varied to match well performance and assess the sensitivity of project profit potential. Also, the study was made to optimize development of the Cotton Valley Lime Reservoir with respect to fracture length and well density.

17.5.1 History Match of Muse No. 1. A history-matching process was performed on the gas production and pressure buildup test data of the Muse No. 1. Inspection of the field data indicated that both the rate and flowing bottomhole pressure had declined with time. A numerical reservoir model was used to simulate the production history with a series of constant flow rates followed by constant-pressure production as the well was pulled down toward line pressure. **Fig. 17.8** illustrates the production history for the Muse No. 1. The line with the solid dots in **Fig. 17.9** represents the buildup data. The model was used to match the data in both these graphs. **Table 17.3** presents the exact production values input into the model.

Time Period (days)	Flow Rate (MMcf/D)	Flowing Bottomhole Pressure (psi)
0 to 108	3,210	Decreasing
108 to 231	2,536	Decreasing
231 to 388	Decreasing	1,080
388 to 398	0	Increasing (buildup)
398 to 430	Decreasing	2,000

Table 17.3—Production values.

In the first history matching of the Muse No. 1, the design fracture half-length of 1,500 ft was used. Several computer runs were necessary before an adequate history match was obtained. First, it was evident that, using the average core derived permeability of 0.09 md, the shape of the buildup curve was not similar to the actual field data. The permeability had to be reduced and the fracture conductivity increased to match the field data. A successful match finally was obtained with a 1,500-ft fracture half-length, a permeability of 0.013 md, and a fracture conductivity of 450 md-ft (Run 1, Fig. 17.9).

In an effort to test the uniqueness of this solution, additional computer runs were made. Using a 1,000-ft fracture half-length and a fracture conductivity of 450 md-ft, an acceptable match was obtained with a permeability of 0.023 md (Run 2, Fig. 17.9).

Also, using a 500-ft fracture half-length and a fracture conductivity of 450 md-ft, an acceptable match was obtained with a permeability of 0.0425 md (Run 3, Fig. 17.9).

Previous experience in history matching has led to the following conclusions:

1. Short-term production data can be history matched adequately using numerous combinations of fracture length and formation permeability.
2. Unique values of permeability and fracture length can be obtained only by history matching both the production data and data from a long-term pressure buildup.

A unique history match was not obtained for the Muse No. 1 because of insufficient pressure buildup data. The buildup data presented in Fig. 17.9 was dominated completely by linear flow into the fracture, and the effects of fracture length had not been felt at the wellbore when the bombs were pulled.

To illustrate this point, consider the buildup data in **Fig. 17.10.** These data were calculated by the numerical model using the three history-match combinations.

History Match	Formation Permeability (md)	Fracture Half-Length (ft)
1	0.013	1,500
2	0.023	1,000
3	0.0425	500

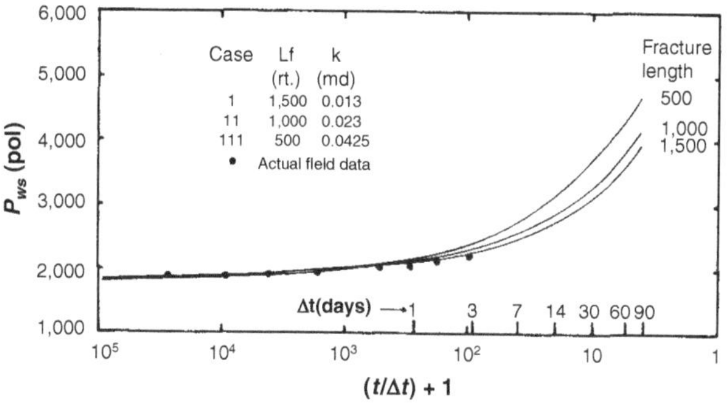

Fig. 17.10—Postfracture BHP buildup test, Muse No. 1.

Notice that for buildup of 1 to 3 days, the shapes of the curves are essentially identical. It requires a buildup of 14 to 30 days before a unique history match can be obtained.

17.5.2 Two-Rate Test. A two-rate test also was run on the Muse No. 1 before the pressure buildup. The two-rate test was included in all of our history-match runs. **Fig. 17.11** compares the pressures from the three history-match runs with the actual field data.

The bottomhole pressures of the simulated cases following the rate change are smooth curves, due to the second flow rate being held precisely constant by the reservoir model. The somewhat erratic behavior of the field data after the rate change is most likely due to the inability to maintain an exactly constant rate. The most important match point of the two-rate test was obtained, that being the pressure at the end of the second flow period.

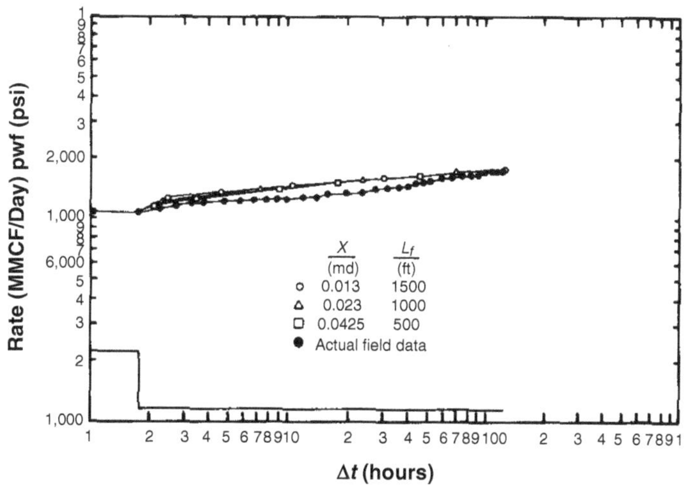

Fig. 17.11—History match of two-rate test, Muse No. 1.

Permeability, md	0.013
Gas porosity	0.036
Original pressure, psi	6,750
Bottomhole temperature, °F	286
Net-pay, ft	50
Fracture conductivity, md-ft	450
Fracture half-length, ft	2,500

Table 17.4—History match parameters.

17.5.3 History Match of Muse-Duke No. 1. The Muse-Duke No. 1 was completed in the Cotton Valley Lime in Sept. 1978. After the well was acidized lightly, the initial production was in excess of 4 MMcf/D at 4,700-psi flowing tubing pressure, but the rate and pressure declined steadily to 3,200 MMcf/D and 3,900 psi within 18 hours. The high initial rates and rapid decline are most likely due to the presence of natural fractures in communication with the wellbore. Muse-Duke No. 1 was tested for approximately 3 weeks at various rates and flowing tubing pressures. During this period, both flowing and shut-in bottomhole pressure surveys were obtained. At the end of the period the flow rate was down to 1,700 MMcf/D at 1,800-psi flowing tubing pressure.

In Nov. 1978, the well received an MHF treatment. The initial production rate after the treatment was in excess of 6 MMcf/D. Six months later, it was flowing 4 MMcf/D, and 22 months later it still was flowing 2.1 MMcf/D at a line pressure of 700 psig.

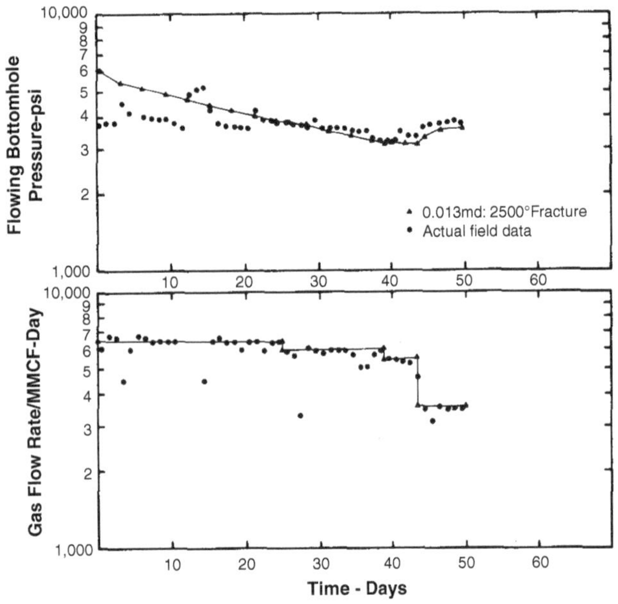

Fig. 17.12—History match of first drawdown, Muse-Duke No. 1.

An attempt was made to history match the Muse-Duke No. 1 after 50 days of production using the parameters developed from the history match of the Muse No. 1 (**Table 17.4**).

Fig. 17.12 shows that an acceptable match was obtained using the parameters in Table 17.4. During the first 20 days, the simulated flowing bottomhole pressure was considerably higher than the estimated flowing bottomhole pressure. This difference was probably due to the large volumes of fracturing water that were produced during the cleanup period. Therefore, the difference between the simulated and actual flowing bottomhole pressures is not as significant as depicted. The simulated case produced 290 MMcf during the 50-day period, while the actual production was 270 MMcf.

The significance of this history match is that by using (1) the reservoir parameters as determined from the Muse No. 1 data and (2) the designed fracture half-length (2,500 ft) for the Muse-Duke No. 1, the actual production history from the Muse-Duke No. 1 was matched successfully. However, as discussed later, this fact does not prove that the designed 2,500 ft fracture half-length was created, because the early-life drawdown can be matched by numerous combinations of fracture-length permeability and conductivity. It does point out, however, that a long fracture was created, even though more data are needed to determine uniquely the fracture parameters. Therefore, the gas production from the Muse-Duke No. 1 was history matched again after 519 days of production. **Table 17.5** presents a summary of the gas production. Also, these data have been presented in **Fig. 17.13** as flow rate and bottomhole pressure vs. cumulative time. Furthermore, a postfracturing pressure buildup was run in Aug. 1979 after 256 days of production. These data are presented in **Table 17.6** and graphed in **Fig. 17.14** as the line connecting the open circles.

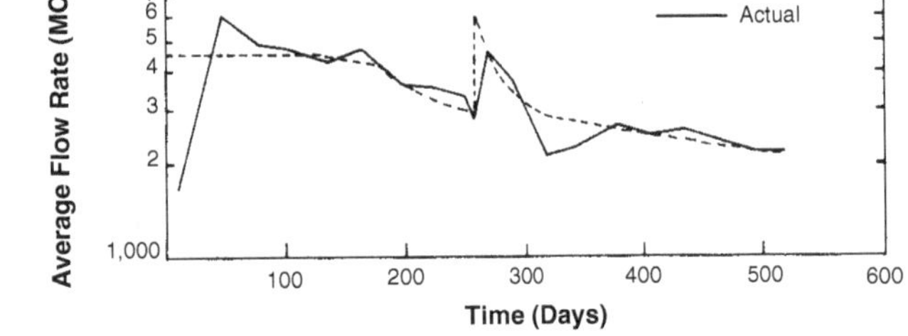

Fig. 17.13—Average flow rate and BHFP vs. time, Muse-Duke No. 1.

The results of the latest history match are illustrated as Cases I, II, and III in Fig. 17.14. These three cases on the graph were generated using these data:

Case	Fracture Length (ft)	Permeability (md)	Fracture Conductivity (md-ft)
I	750	0.04	1,000
II	1,500	0.022	1,000
III	2,500	0.01	2,000

For each of these cases, the correct cumulative production and flowing bottomhole pressure were computed by the model. Notice in Fig. 17.14, however, that the 750-ft fracture is too short, the 2,500-ft fracture is too long, and only the 1,500-ft fracture adequately overlays the field buildup data.

Based on the results of this computer model study, data in **Table 17.7** best describe the Muse-Duke No. 1.

Month	Month Production (Mcf)	Producing Days	Average Rate (Mcf/D)	Cumulative Production (Mcf)	Cumulative Days
Oct. 1978	16,544	10	1,654	16,544	10
Nov.	21,950	10	2,195	38,494	20
Dec.	171,000	28	6,107	209,494	48
Jan. 1979	149,500	31	4,822	358,994	79
Feb.	121,260	26	4,664	480,254	105
March	121,150	29	4,178	601,404	134
April	139,650	30	4,655	741,054	164
May	106,850	30	3,561	847,904	194
June	100,200	29	3,455	948,104	223
July	85,600	26	3,292	1,033,704	249
Aug.	18,700	7	2,671	1,052,404	256
Sept.	55,000	12	4,583	1,107,404	268
Oct.	70,725	19	3,722	1,178,129	287
Nov.	60,425	29	2,083	1,238,554	316
Dec.	96,775	30	3,225	1,335,329	346
Jan. 1980	81,075	31	2,615	1,416,404	377
Feb.	68,975	28	2,463	1,485,379	405
March	69,150	27	2,561	1,554,529	432
April	69,575	30	2,319	1,624,104	462
May	60,650	28	2,166	1,684,754	490
June	63,050	29	2,174	1,747,804	519

Shut in Aug. 8,1979, for pressure buildup.

Cumulative production = 1,052,404 Mcf

Cumulative days = 256 days

Last rate = 2,671 Mcf/D

$$\text{Horner Time} = \frac{1{,}052{,}404\ \text{Mcf}}{2{,}671\ \text{Mcf/D}} = 394\ \text{days}$$

Table 17.5—Muse-Duke No. 1 production.

Type of stimulation: post-MHF

Shut-In Time (hours)	BHP (psia)	Horner Time	Pseudopressure (psia^2/CP)	Delta Pseudopressure
0.00000	1262.0	0.00000	0.11257 E + 09	0.00000
0.25000	1406.0	24577.0	0.13963 E + 09	0.27065 E +08
0.50000	1446.0	12289.0	0.14715 E + 09	0.34583 E + 08
0.75000	1467.0	8193.0	0.15110 E + 09	0.38530 E +08
1.0000	1494.0	6145.0	0.15617 E + 09	0.43604 E + 08
1.5000	1528.0	4097.0	0.16337 E + 09	0.50796 E + 08
2.0000	1556.0	3073.0	0.16943 E + 09	0.56861 E + 08
2.5000	1580.0	2458.0	0.17463 E + 09	0.62059 E +08
3.0000	1599.0	2049.0	0.17874 E + 09	0.66174 E + 08
4.0000	1633.0	1537.0	0.18611 E + 09	0.73537 E + 08
6.0000	1684.0	1025.0	0.19715 E + 09	0.84583 E + 08
8.0000	1727.0	769.0	0.20647 E + 09	0.93896 E + 08
10.000	1761.0	615.40	0.21412 E + 09	0.10155 E + 09
12.000	1791.0	513.00	0.22141 E + 09	0.10885 E + 09
16.000	1840.0	385.00	0.23333 E + 09	0.12076 E + 09
20.000	1886.0	308.20	0.24451 E + 09	0.13194 E + 09

Table 17.6—Data for pressure buildup test analysis.

Type of stimulation: post-MHF				
Shut-In Time (hours)	BHP (psia)	Horner Time	Pseudopressure (psia^2/CP)	Delta Pseudopressure
24.000	1923.0	257.00	0.25351 E + 09	0.14094 E + 09
33.000	1963.0	187.18	0.26323 E + 09	0.15066 E + 09
41.000	2045.0	150.85	0.28429 E + 09	0.17172 E + 09
49.000	2091.0	126.39	0.29663 E + 09	0.18406 E + 09
53.000	2112.0	116.92	0.30226 E + 09	0.18969 E + 09
61.000	2154.0	101.72	0.31352 E + 09	0.20095 E + 09
69.000	2189.0	90.043	0.32290 E + 09	0.21033 E + 09
77.000	2224.0	80.792	0.33229 E + 09	0.21972 E + 09
85.000	2257.0	73.282	0.34130 E + 09	0.22873 E + 09
93.000	2284.0	67.065	0.34918 E + 09	0.23661 E + 09
101.00	2315.0	61.832	0.35822 E + 09	0.24565 E + 09
109.00	2339.0	57.367	0.36522 E + 09	0.25265 E + 09
112.00	2352.0	55.857	0.36901 E + 09	0.25644 E + 09
121.00	2379.0	51.777	0.37689 E + 09	0.26432 E + 09
133.00	2415.0	47.195	0.38739 E + 09	0.27482 E + 09
145.00	2450.0	43.372	0.39760 E + 09	0.28503 E + 09
157.00	2481.0	40.134	0.40664 E + 09	0.29407 E + 09
169.00	2510.0	37.355	0.41532 E + 09	0.30275 E + 09
181.00	2539.0	34.945	0.42440 E + 09	0.31183 E + 09
193.00	2563.0	32.834	0.43192 E + 09	0.31935 E + 09
205.00	2588.0	30.971	0.43975 E + 09	0.32718 E + 09
217.00	2616.0	29.313	0.44852 E + 09	0.33595 E + 09
229.00	2640.0	27.830	0.45604 E + 09	0.34347 E + 09
241.00	2668.0	26.494	0.46481 E + 09	0.35224 E + 09
253.00	2689.0	25.285	0.47139 E + 09	0.35882 E + 09
265.00	2711.0	24.185	0.47828 E + 09	0.36571 E + 09
277.00	2732.0	23.180	0.48486 E + 09	0.37229 E + 09
289.00	2754.0	22.260	0.49182 E + 09	0.37925 E + 09
301.00	2775.0	21.412	0.49879 E + 09	0.38622 E + 09
313.00	2794.0	20.629	0.50510 E + 09	0.39253 E + 09
325.00	2812.0	19.905	0.51108 E + 09	0.39851 E + 09
337.00	2830.0	19.231	0.51705 E + 09	0.40448 E + 09
349.00	2852.0	18.605	0.52436 E + 09	0.41179 E + 09
360.00	2864.0	18.067	0.52834 E + 09	0.41577 E + 09
374.00	2915.0	17.428	0.54527 E + 09	0.43270 E + 09
398.00	2909.0	16.437	0.54328 E + 09	0.43071 E + 09
422.00	2932.0	15.559	0.55092 E + 09	0.43835 E + 09
446.00	2955.0	14.776	0.55855 E + 09	0.44598 E + 09
470.00	2975.0	14.072	0.56519 E + 09	0.45262 E + 09
494.00	3005.0	13.437	0.57523 E + 09	0.46266 E + 09
518.00	3024.0	12.861	0.58185 E + 09	0.46928 E + 09
542.00	3044.0	12.336	0.58881 E + 09	0.47625 E + 09
566.00	3065.0	11.855	0.59613 E + 09	0.48356 E + 09
602.00	3092.0	11.206	0.60553 E + 09	0.49296 E + 09
614.00	3101.0	11.007	0.60867 E + 09	0.49610 E + 09
638.00	3124.0	10.630	0.61668 E + 09	0.50411 E + 09
662.00	3139.0	10.281	0.62190 E + 09	0.50933 E + 09
686.00	3155.0	9.9563	0.62747 E + 09	0.51490 E + 09

Table 17.6—(*Continued*).

Operator	Mitchell Energy
Well name	Muse-Duke No.1
Depth to top of formation, ft	11,220.0
Perforated interval, ft	11,220 to 11,420
Total net-pay, ft	50.0
Perforated net-pay, ft	50.0
Wellbore radius, ft	0.3300
Production time, hours	9456.0
Pressure at shut-in, psi	1262.0
Initial reservoir pressure, psi	6750.0
Buttomhole temperature, °F	286.0
Dry-gas flow rate, Mcf/D	2671.0
Wet-gas flow rate, Mcf/D	2672.0
Total porosity, fraction	0.03600
Water saturation, fraction	0.50000
Gas porosity, fraction	0.01800
Separator gas gravity	0.65000
Gas/condensate ratio, scf/STB	900,000.0
Condensate gravity, °API	55.0
Gas equivalent of condensate, scf/STB	820.0
Wet-gas specific gravity	0.65327
Gas properties at initial pressure and temperature	
Gas viscosity, cp	0.02700
Gas compressibility, psi^{-1}	0.0000892
Z-factor	1.15117
Gas formation volume factor, RB/Mcf	0.64053
Pressure buildup test analysis results	
Horner slope, psia^2/cp-cycle	0.39891 E+ 09
Pressure at 1 hour, psia^2/cp	–0.48940 E+ 09
Formation gas permeability, md	0.16346
Apparent skin factor	–6.7011
Radius of investigation, ft	
End of production period	6134.0
End of Hornor straight line	1652.0
End of buildup test	1652.0
Average reservoir depth, ft	11,320
Original reservoir pressure, psi	6,750
Current reservoir pressure, psi	5,635
Formation gas permeability, md	0.022
Formation gas porosity, %	3.6
Net gas pay, ft	50
Propped fracture half-length, ft	1,500
Average fracture conductivity, md-ft	1,000

Table 17.7—Muse-Duke No. 1.

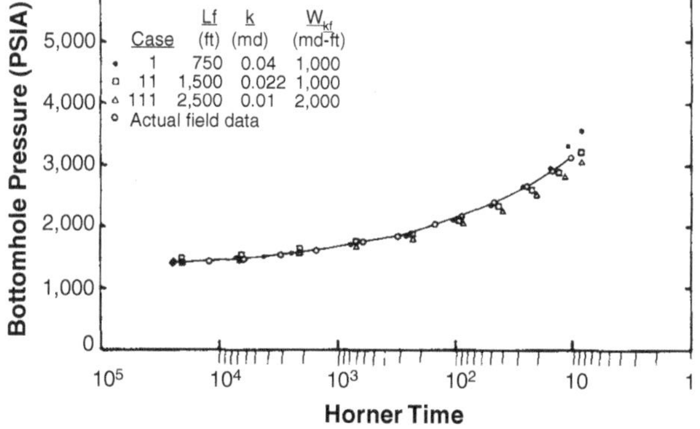

Fig. 17.14—Computer history-match runs, Muse-Duke No. 1.

17.6 Optimization of Fracture Length and Well Spacing

17.6.1 Economic Parameters. The reservoir stimulation history-matching analysis of Muse No. 1 and Muse-Duke No. 1 has defined the formation permeability to gas as being in the range of 0.01 to 0.04 md. Using the computer model FRACOP™ (Daneshy 1978), a series of evaluations was made to determine the optimum fracture length and well spacing for a well of parameters similar to Muse No. 1 and Muse-Duke No. 1. Any evaluation which involves economic analyses is quite sensitive to each company's economic parameters, such as current product price, price escalation, operating costs, initial investment, net working interest, etc., which may not be the same as used here. **Table 17.8** lists the economic parameters used in this study. **Table 17.9** lists the fracturing costs.

Fixed investment, $	1,000,000
Product price, $/Mcf constant	1.52
Operating expenses, $/yr constant	8,400
Net revenue interest, %	79.7
Production tax rate, %	13
Federal income tax rate, %	0.0

Table 17.8—Economic parameters.

Designed Fracture-Length Radius (ft)	Costs ($)
250	75,000
500	100,000
750	150,000
1,000	200,000
1,250	250,000
1,500	300,000
1,750	400,000
2,500	700,000

Table 17.9—Fracturing costs.

17.6.2 Production Functions. The FRACOP model uses input reservoir, well, and fracture-length parameters to generate a production function. This production function then is used to determine net cash flow and net present value for any desired discount rate.

Based on the two permeabilities of 0.01 and 0.04 md, FRACOP runs were made using several fracture half-lengths and for 160-, 320-, and 640-acre drainage areas.

Fig. 17.15 illustrates the effect of permeability on the 30-year ultimate gas recovery (per well) as a function of fracture half-length. As would be expected, ultimate recovery increases as permeability, fracture length, and drainage area (well spacing) increase. Note, however, that for 160-acre spacing, increasing fracture half-length from 250 to more than 1,250 ft does not influence the 30-year ultimate recovery significantly. The effect of the higher permeability becomes more pronounced as the drainage area increases.

Fig. 17.16 illustrates the effect of permeability on the 30-year recovery *per section* as a function of fracture half-length. **Table 17.10** summarizes the results for a 1,000-ft fracture half-length.

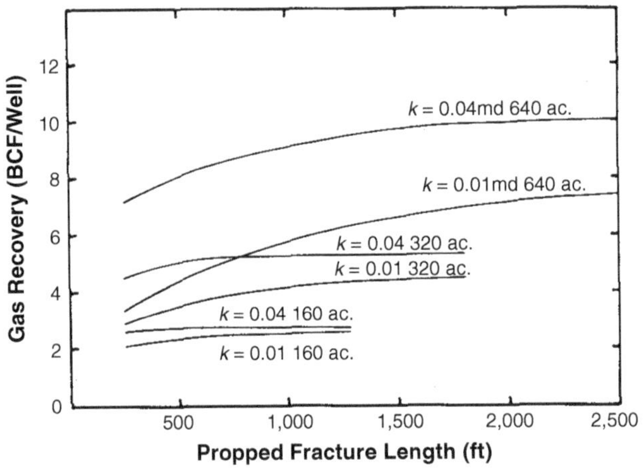

Fig. 17.15—Ultimate gas recovery vs. propped fracture length.

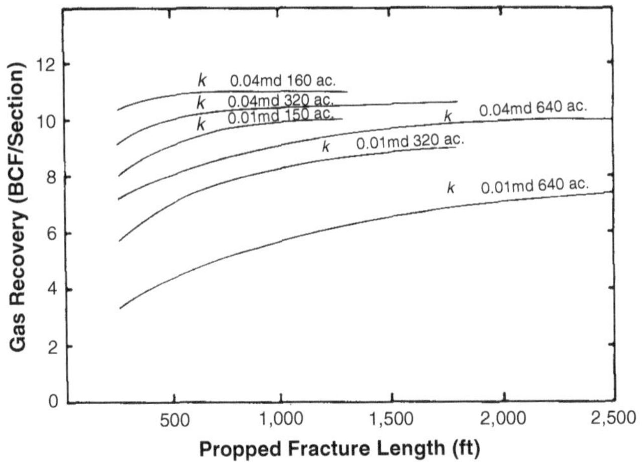

Fig. 17.16—Ultimate gas recovery vs. fracture length.

Well Spacing	Wells per Section	Permeability (md)	Ultimate Recovery (Bcf/section)	Recovery Efficiency (% OGIP)
160	4	0.04	10.9	77.4
160	4	0.01	9.9	70.2
320	2	0.04	10.5	74.5
320	2	0.01	8.3	58.7
640	1	0.04	9.1	65.1
640	1	0.01	5.6	40.2

Table 17.10—Results for 1,000-ft fracture half-length.

Permeability (md)	Discount Factor (%)	Well Spacing (acres)	Optimum Fracture Half-Length (ft)
0.01	0	320	1,400 to 1,600
0.04	0	640	1,500
0.01	10	320	1,500 to 1,700
0.04	10	320	1,400 to 1,600
0.01	15	320	1,700 +
0.04	15	320	1,400 to 1,600

Table 17.11—Results of computer runs.

From a pure recovery basis, it appears that the best well spacing would be four wells per section since that well density results in the maximum recovery efficiency. However, to optimize properly the well spacing and fracture length from an economic viewpoint, the maximum discounted present value profit of each case must be determined. **Figs. 17.17 and 17.18** illustrate the 10% discounted profit per section for permeabilities of 0.01 and 0.04 md, respectively.

Table 17.11 summarizes the results of the FRACOP computer runs. Based on the economic considerations listed earlier, the optimum well spacing and propped fracture half-length are 320 acres and 1,500 ft, respectively.

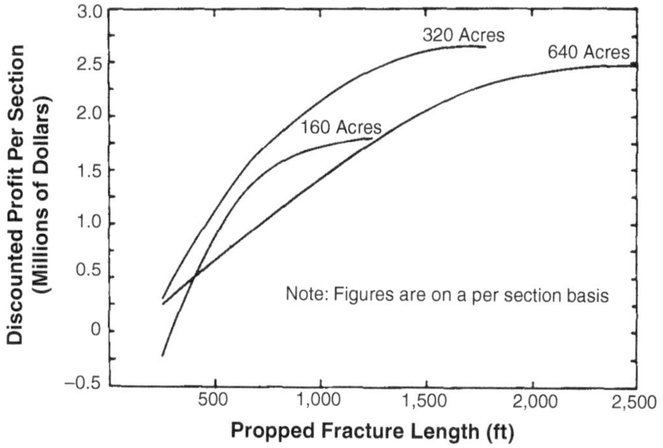

Fig. 17.17—Percent discounted present value profit vs. fracture length (0.01-md permeability, 10% discounted).

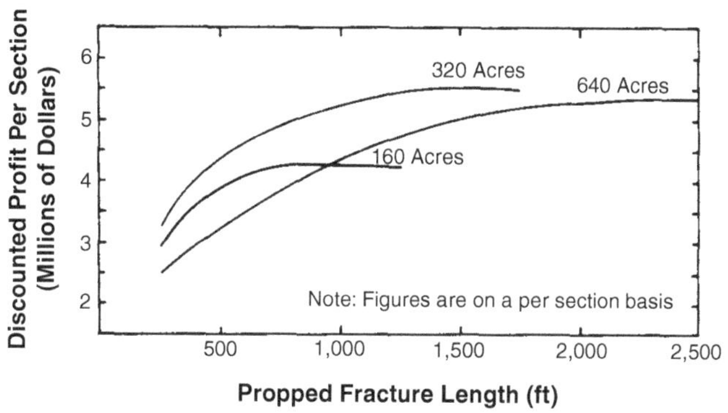

Fig. 17.18—Percent discounted present value profit vs. fracture length (0.04-md permeability, 10% discounted).

17.7 Conclusions

Based on the results of this study, it is obvious (from the data in Fig. 17.7) that MHF techniques have been applied successfully in the Cotton Valley Lime Formation in Limestone County, Texas.

Even though a unique solution could not be obtained for the Muse No. 1 due to the short duration of the pressure buildup test, the history match did result in a reasonable range of reservoir and fracture parameters: the in-situ formation permeability is about 0.01 to 0.025 md and the fracture half-length is about 1,000 to 1,500 ft.

A unique match was found for the Muse-Duke No. 1 which resulted in an estimated in-situ gas permeability of 0.022 md and a fracture half-length of 1,500 ft.

Based on a permeability range of 0.01 to 0.04 md, the optimum well spacing and fracture half-length for this reservoir appears to be 320 acres and 1,500 ft, respectively.

Finally, and very importantly, no unique solution to reservoir behavior in a fractured reservoir can be obtained without long-term pressure buildup data. Prefracture pressure buildup tests should be run to obtain estimates of the true formation capacity (kh). Long-term postfracture pressure buildup tests should be run to obtain estimates of propped fracture half-length and fracture conductivity.

17.8 Nomenclature

k = formation permeability, md

k_{rg} = relative permeability to gas, md

K_{ic} = critical stress intensity factor, psi (MPa$\sqrt{}$mm)

L_f = fracture half-length, ft (m)

p_{wf} = flowing wellbore pressure, psia (MPa)

p_{ws} = shut-in wellbore pressure, psia (MPa)

S_w = water saturation, %

W_f = fracture width, ft (m)

W_{kf} = fracture conductivity, md-ft (md-m)

Δt = shut-in time, hours

17.9 References

Abou-Sayed, A. S., Brechtel, C. E., and Clifton, R. J. 1978. In Situ Stress Determination by Hydro-fracturing – A Fracture Mechanics Approach. *J. Geophys.* **83** (B6): 2851–2862. https://doi.org/10.1029/JB083iB06p02851.

Ahmed, U., Buchholdt, L. M., Abou-Sayed, A. S. et al. 1980. Laboratory and Field Studies in Support of Mitchell Energy Corporation Muse-Duke Well No. 1 Massive Hydraulic Fracturing. Report TR 80-43, Terra Tek, Salt Lake City, Utah (March 1980).

Brace, W. F., Walsh, S. B., and Frangos, W. T. 1968. Permeability of Granite Under High Pressure. *J. Geophys.* **83** (6): 2225–2236. https://doi.org/10.1029/JB073i006p02225.

Clifton, R. J., Simonson, E. R., Jones, A. H. et al. 1976. Determination of the Critical Stress – Intensity Factor K_k In a Circular Ring, *Exp Mech.* **16:** 233–238.

Daneshy, A. A. 1973. On the Design of Vertical Hydraulic Fractures. *J Pet Technol.* **25** (1): 83–93. SPE-3654-PA. https://doi.org/10.2118/3654-PA.

Daneshy, A. A. 1978. Numerical Solution of Sand Transport in Hydraulic Fracturing. *J Pet Technol.* **30** (1): 132–140. SPE-5636-PA. https://doi.org/10.2118/5636-PA.

Holditch, S. A., Jennings, J. W., and Neuse, S. H. et al. 1978. The Optimization of Well Spacing and Fracture Length in Low Permeability Gas Reservoirs. Presented at the SPE Annual Fall Technical Conference and Exhibition, Houston, Texas, 1–4 October. SPE-7496-MS. https://doi.org/10.2118/7496-MS.

Holditch, S. A. and Laufer, W. L. 1978. The Analysis of Fractured Gas Wells Using Reservoir Stimulation. Presented at the SPE Annual Fall Technical Conference and Exhibition, Houston, Texas, 1–3 October. SPE-7473-MS. https://doi.org/10.2118/7473-MS.

Simonson, E. R., Abou-Sayed, A. S., and Clifton, R. J. 1978. Containment of Massive Hydraulic Fractures. *SPE J.* **18** (1): 27–32. SPE-6089-PA. https://doi.org/10.2118/6089-PA.

Yomada, S. E. and Jones, A. H. 1980. A Review of a Pulse Technique for Permeability Measurements. *SPE J.* **20** (5): 357–358. SPE-8760-PA. https://doi.org/10.2118/8760-PA.

Appendix 17.A—Mechanics of a Super-MHF Job

One question always is asked during the evolutionary process of fracturing: How big a job can be performed? The answer: Just about as big a job as you would wish. The technology is available today. The real question: Will it be effective?

The objective of this appendix is to outline some of the major details in designing and executing a super-MHF job for the Muse-Duke No. 1.

17.A.1 Physical Rock Properties. An important factor in the basic design of a fracturing job is a knowledge of the in-situ stresses (Ahmed et al. 1980; Abou-Sayed et al. 1978; Clifton et al. 1976) of the rock—not only of the objective formation but also the bounding formations. The in-situ rock stresses will control the behavior of the induced fracture away from a wellbore.

In an attempt to determine these stresses, the following measurements were made.

17.A.1.1 Triaxial Tests. Triaxial compression tests were performed on core samples with the sample axis both perpendicular and parallel to the original core axis. **Table 17.A-1** lists the results of the measurements as well as the simulated in-situ environment under which the tests were run. The elastic moduli quoted were determined from the initial slopes of the stress/strain curves. As expected, the static moduli were lower than the dynamic moduli for the samples tested (shown in Fig. 17.6), the difference being between 15 and 25%.

Formation	Depth (ft)	Sample Orientation*	Sample Density (g/cm^3)	Confining Pressure (psi)	Pore Pressure (psi)	Young's Modulus (10^6 psi)	Poisson's Ratio	Maximum Compressive Strength (psi)
Bossier	11,195	Perpendicular	2.66	3,200	0	6.85	0.15	23,879
Shale	11,207	Perpendicular	2.67	3,200	0	2.50	0.35	21,491
	11,207	Along	2.67	3,200	0	3.68	0.18	21,405
	11,207	Along	2.68	3,200	0	2.81	0.14	23,440
Cotton	11,239	Along	2.70	3,200	0	5.30	0.16	–
Valley	11,239	Perpendicular	2.69	3,200	0	5.84	0.18	37,023
Lime	11,313	Along	2.51	8,800	5,600	5.15	0.22	29,014
	11,313	Perpendicular	2.52	8,800	5,600	6.67	0.23	23,146
Anhydrite	11,510	Along	2.95	3,200	0	7.69	0.23	24,949
	11,510	Along	2.94	3,200	0	7.94	0.25	26,004
	11,510	Perpendicular	2.93	3,200	0	10.20	0.29	27,000
	11,510	Perpendicular	2.93	3,200	0	11.40	0.23	27,539

*Orientation with respect to the core axis.

Table 17.A-1—Triaxial compression test data, Muse-Duke No. 1.

Zone	Depth (ft)	Confining Pressure (psi)	Burst Pressure (psi)	Temperature (° F)	Critical Stress Intensity Factor K_{ic} (psi $\sqrt{\text{in.}}$)
Bossier Shale	11,121	415	3,360	285	580
	11,195	235	1,960	285	415
	11,207	170	1,400	285	296
Cotton Valley	11,239	390	3,105	285	657
Lime	11,336	290	2,300	285	397
	11,374	315	2,525	285	534
Anhydrite	11,510	315	2,375	285	410

Table 17.A-2—Proportional loading critical stress intensity factor k_{ic}, Muse-Duke No. 1.

Zone	Depth (ft)	Burst Pressure (psi)	Temperature (°F)	Critical Intensity Factor K_{ic} (psi $\sqrt{\text{in.}}$)
Bossier Shale	11,207	1,850	75	568
Cotton Valley Lime	11,313	1,640	75	503
Anhydrite	11,505	2,340	75	718

Table 17.A-3—Direct loading critical stress intensity factor k_{ic}, Muse-Duke No. 1.

17.A.1.2 Fracture Toughness. Fracture toughness measurements were carried out on samples using the technique developed by Clifton et al. (1976). In the technique, a core sample about 3 in. long is used. A small hole is drilled along the axis of the specimen, and diametrically opposite prenotches are placed at the internal wells to specify the fracture initiation points. A bladder is placed in the hole to prevent fluid from entering the sample or the notches, and then pressure is applied in the bladder until the sample bursts. **Tables 17.A-2 and 17.A-3** list the results of proportional and direct loading tests, respectively. The results are shown in **Fig. 17.A-1** illustrating the Cotton Valley Lime Formation to be more susceptible to fracturing in comparison with the bounding layers.

17.A.1.3 Containment Analysis. Final analyses of the laboratory measurement program indicates the possiblity of fracture containment within the Cotton Valley Lime Formation. Variation in elastic moduli between the bounding Bossier Shale at 11,100 ft and the bounding limestone/anhydrite sequence at 11,500 ft to the Cotton Valley Lime is about a factor of two. Fig. 17.6 presents the variation of moduli with depth. Lower moduli are observed between 11,280 and 11,340 ft associated with a higher-porosity gas zone. Based on field experience, a modulus contrast of at least three is required for successful containment. Therefore, upper containment cannot be guaranteed for the Cotton Valley Lime based on moduli data. However, critical stress intensity factor variation with depth (Fig. 17.A-1) does indicate the Cotton Valley Lime to be the more susceptible formation to fracturing in comparison with the bounding layers. Actual postfracturing pressure analyses have indicated that the majority of the induced fractures appear to be staying within the perforated zone.

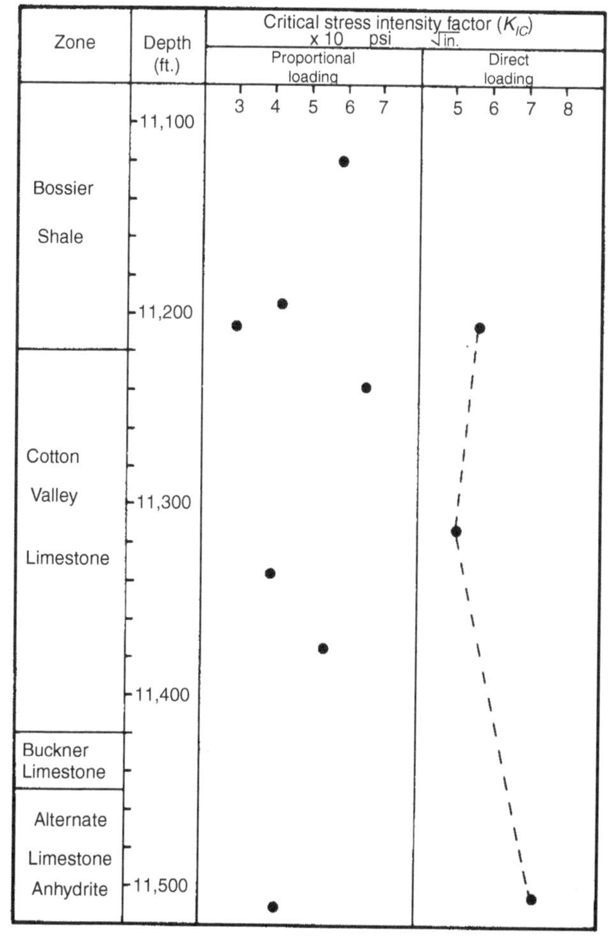

Fig. 17.A-1—Critical stress intensity factor variation with depth, Muse-Duke No. 1.

Appendix 17.B—Fracture Design

17.B.1 Objective. The objective was to create a fracture in the Cotton Valley Lime to drain 640 acres effectively. Also, the fracturing fluid had to be pumped down 2 ⅞-in. tubing under 9,600-psi maximum pressure through 15 perforations from 11,235 to 11,418 ft. Basically, this involves programming a pumping schedule to create a fracture at least 210 ft high with a radius of at least 2,500 ft, keeping it within the section, and packing it properly to keep it open.

The design details were developed by using Halliburton Services' Prop™ computer programs, which integrate Danshey's (Daneshy 1973, 1978) concept of wide fracture development. Danshey's concept is in line with our belief that an induced fracture in the Cotton Valley Lime would have to be wide to extend the fracture any appreciable distance. This also was indicated by analysis of measured mechanical properties of the rock.

Well Completion Data	
Depth of formation, ft	11,200
Depth of packer, ft	11,130
Tubing, ID	2.441
Casing, ID	4.778
Number of perforations	15
Perforation diameter, in.	0.38
Treatment Data	
Type of gel	Versagel™
Gel concentration	1,500 to 1,600
Injection rate, bbl/min	22.0
Treatment fluid, specific gravity	1.020
Behavior index *N*	0.5100
Consistency index *K* (slot), lbf/s-*N*/sq ft	0.037000
Fluid loss coefficient C_W, ft/√min.	0.00260
Spurt volume, gal/sq ft	0.0
Spurt loss coefficient C_{VC}, ft/√min.	0.00029
Spurt time, minutes	0.0
Damage ratio	1.0
Apparent viscosity, cp at 0.995-in. width	869.0
Formation Data	
Young's modulus, psi	5.76 E + 06
Permeability, md	0.09
Porosity, %	8.5
Reservoir fluid compressibility, psi^{-1}	1.66 E–04
Reservoir fluid viscosity, cp	0.02
BHP, psi	8607.0
Reservoir fluid pressure, psi	6350.0
Closure pressure, psi	7100.0
Gross fracture height, ft	210.0
Net fracture height, ft	36.0
Wellbore diameter, in.	5.50
Drainage radius, ft	2640.0
Well spacing, acres	640

Table 17.B-1—Fracturing process input data.

17.B.2 Fracture Containment. All fracture designs, including the one shown for this well, are set up to create fractures in a homogenous rock having constant mechanical properties. Very little consideration is given to the bounding formations above and below the section to be fractured. For short fractures, the effect of the bounding formations on the vertical propagation of an induced fracture may be ignored, but for long fractures, it cannot be ignored. Consideration must be given to the vertical propagation (Clifton et al. 1976; Simonson et al. 1978) of the fracture into the bounding rock. Laboratory measurements indicate that the anhydrite limestone sequence of the Buckner Formation below the Cotton Valley Lime would impede downward vertical migration of the fracture. On the other hand, the top-bounding Bossier Shale is less apt to contain entirely a fracture initiated in the Cotton Valley Lime.

Recent work (Simonson et al. 1978) has demonstrated that the vertical propagation can be constrained by controlling the density of the fracturing fluid, the rates, and the pumping pressure. High-density fluids and low rates will tend to migrate a fracture downward; conversely, a lighter fluid and high rates will tend to migrate a fracture upward. Therefore, to minimize the upward migration of the fracture, the pumping program was modified to initiate the fracture by pumping a weighted prepad (10 lbm/gal salt water) followed by sand-laden gelled water down 2 1/2-in. tubing. This was to be pumped at rates no greater than 24 bbl/min at a maximum of 9,600-psi pumping pressure. Whether these modifications were entirely successful is undecided since production history matches after the fracturing job indicated a propped 1,500- to 1,800-ft fracture radius instead of 2,500 ft that was created in the Cotton Valley Lime Pay.

17.B.3 Well Data. The relevant well completion, formation, and treatment input data used in the program are listed in **Table 17.B-1.** The location of the perforations in the section is shown in **Fig. 17.B-1.**

17.B.4 Program Results. The results of the Prop program are listed in **Tables 17.B-2, 17.B-3, and 17.B-4.** These tables outline the calculated fracture dimensions, the pumping schedule, and the sand depositional profile. The program calculated that it would take 974,000 gal of fluid and 2,965,200 lbm of 20-40 mesh sand to create and prop a 2,708-ft fracture. Also, it would take 5,182 hhp for 18 hours of pumping at 22 bbl/min at 9,610 psi.

Appendix 17.C–Logistics

Bringing in and arranging the equipment for this job was a bit involved but not unwieldy. After the location was enlarged to accommodate the equipment, it was placed as shown in **17.C-1.** All the equipment was trucked in and in place in 10 days.

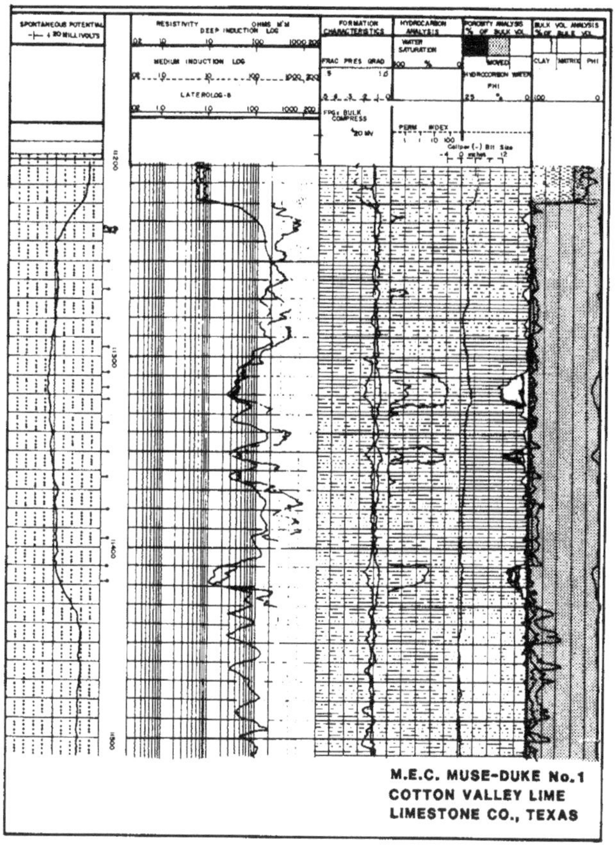

Fig. 17.B-1—Log showing perforations in Cotton Valley Lime, Muse-Duke No. 1.

Design	Volume (gal/1,000 gal) Total	Pad	Created Length (ft)	Width Average (in.)	Proppant Length (ft)	Proppant Height (ft)
1	974.3	200.0	3,522.2	0.995	2,708.4	199.7

Proppant		Relative Production	
Total Sacks	Capacity (ft)	Increase	Efficiency (%)
29,652.0	5.53	4.5	94

Table 17.B-2—Calculated fracture dimensions.

95,000 gal	treated water prepad*
70,000 gal	Versagel 1525 pad
70,000 gal	Versagel 1525,1 lbm/gal Oklahoma No. 1
60,000 gal	Versagel 1525 spacer
15,000 gal	Versagel 1524,1 lbm/gal 20-40 R/A**sand
15,000 gal	Versagel 1524, 2 lbm/gal 20-40 R/A sand
20,000 gal	Versagel 1524, 3 lbm/gal 20-40 R/A sand
90,000 gal	Versagel 1524,4 lbm/gal 20-40 R/A sand
180,000 gal	Versagel 1622, 5 lbm/gal 20-40 R/A sand
200,000 gal	Versagel 1621,5 lbm/gal 20-40 R/A sand
120,000 gal	Versagel 1600,5 lbm/gal 20-40 R/A sand
2,850 gal	treated water displacement*

*The treated water will use 10 lbm/gal brine water as the base fluid.
**R/A is the abbreviation for radioactive.

Table 17.B-3—Pumping schedule.

Appendix 17.D–Execution

Mechanically, the job went well. No breakdown of equipment occurred in the 18 hours of pumping even after pumping 5 lbm of sand per gallon for 10 hours.

As shown by the pressure/rate chart in **Fig. 17.D-1,** the pumping went smoothly without interruption except toward the end. About 4 hours before termination, the pressure dropped slightly and then began to rise gradually while the rates began to drop. We are not sure exactly what caused this behavior. We suspect the fracture penetrated a zone of higher permeability, which increased the fluid loss to the formation, and the fracture began to close thereafter. Even so, 2.8 million lbm of sand (a record amount) was in place before screening out. A comparison of the design job and the actual performance is made in the treatment summary of **Table 17.D-1.**

At the End of Pumping	
Carry distance, ft	2708.4
Maximum bed height, ft	3.9
Average bed height, ft	2.2
Prop deposited, %	4.1

	Deposited Proppant Bed Height (ft)		Suspended Proppant	
Distance From Well (ft)	End of Pumping	Final	Height (ft)	Concentration (lbm/gal)
12.1	3.6	62.1	210.0	5.0
109.3	3.6	61.9	209.3	5.0
206.5	3.8	61.9	208.7	5.0
303.6	3.7	61.8	208.1	5.0
400.8	3.6	61.7	207.4	5.1
498.0	3.4	62.6	206.4	5.1
655.9	3.3	63.0	205.4	5.1
813.7	3.1	63.1	204.4	5.1
971.6	2.9	63.3	203.3	5.1
1,129.5	2.7	63.7	202.3	5.2
1,287.4	2.5	63.1	201.4	5.2
1,433.2	2.3	63.3	200.5	5.2
1,278.9	2.1	64.2	199.6	5.3
1,724.7	1.8	61.7	198.8	5.3
1,882.5	1.5	61.6	197.9	5.3
2,040.4	1.3	51.2	194.5	4.3
2,125.5	1.1	47.5	194.0	4.3
2,210.5	1.0	48.3	193.5	4.3
2,295.5	0.8	49.3	193.0	4.4
2,380.5	0.5	50.3	192.0	4.4
2,465.5	0.4	37.9	187.7	3.3
2,502.0	0.3	34.1	187.5	3.3
2,538.4	0.3	33.7	187.2	3.3
2,574.8	0.2	27.2	108.6	2.2
2,661.3	0.1	25.5	180.3	2.2
2,647.7	0.1	15.7	170.9	1.1
2,684.1	0.0	12.4	170.5	1.1

Equivalent Bed	
Length, ft	2,708
Height, ft	199.7
Bed concentration, lbm/1,000 sq ft	2,741.0

Table 17.B-4—Depositional profiles.

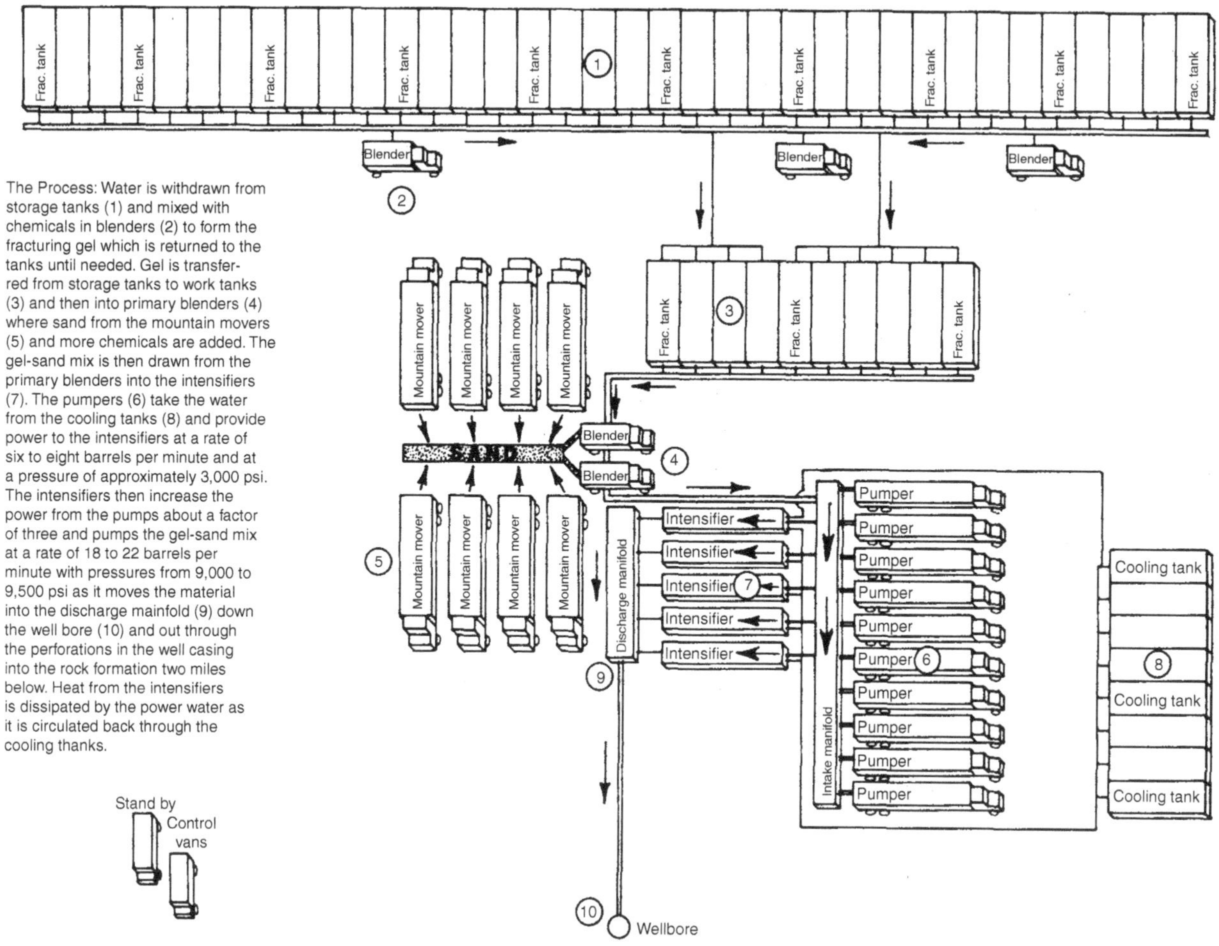

Fig. 17.C-1—Equipment layout for fracturing job, Muse-Duke No. 1.

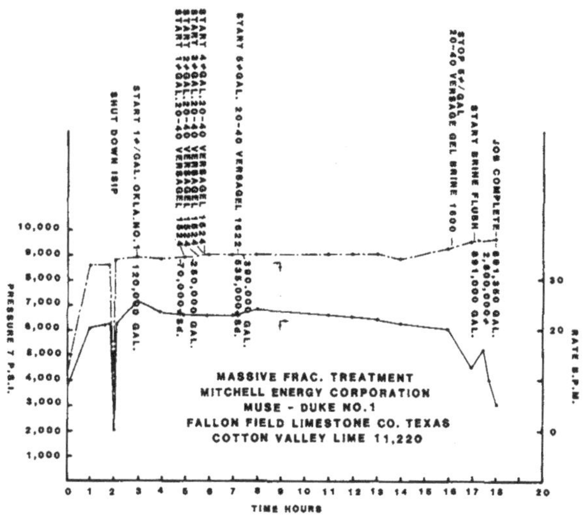

Fig. 17.D-1—Pressure/rate recordings during fracturing job, Muse-Duke No. 1.

	Design	Actual
Hydraulic horsepower, hhp	5,182	5,085
Surface injection pressure, psi	9,610	9,100
Injection rate, bbl/min	22	22.8
Pipe friction, psi	5,785	–
Total Versagel, gal	840,000	793,000
Total 20-40 Ottawa Sand, lbm	2,965,000	2,730,000
Total Oklahoma No. 1, lbm	70,000	70,000
Total water required, bbl	22,500	21,223
Propped fracture length, ft	2,708	1,500
Gel pumping time, hours	16.1	17

Table 17.D-1—Treatment summary.

Appendix 17.E–Summary

The design and execution of a super-MHF job is within the bounds of practicality with today's technology. The gel chemistry and equipment have been developed to the state that almost any of the tight-gas reservoirs can be fractured physically. One question remains: How effectively will the fracture produce the gas?

SI Metric Conversion Factors

acre × 4.046 873	E + 03 = m^2
°API × 141.5/(13.15 +°API)	= g/cm^3
bbl × 1.589 873	E – 0 1 = m^3
cp × 1.0*	E – 03 – Pa·s
cu ft × 2.831 685	E – 02 = m^3
°F (°F-32/1.8)	= °C
ft × 3.048*	E – 01 = m
gal × 3.785 412	E + 00 = dm^3
hhp × 7.460 43	E – 01 = kW
in. × 2.54*	E + 01 = mm
lbf × 4.448 222	E + 00 = N
lbm × 4.535 924	E – 01 = kg
md-ft × 3.008 142	E – 01 = md-m
mile × 1.609 344*	E + 00 = km
psi × 6.894 757	E – 03 = MPa
psi^{-1} × 1.450 377	E – 04 = Pa^{-1}
sq ft × 9.290 304*	E – 02 = m^2
sq mile × 2.589 988	E + 00 = km^2

*Conversion factor is exact.

Chapter 18

Stimulation Optimization in a Low-Permeability Upper Devonian Sandstone Reservoir: A Case History

The Systematic Approach Presented for Optimizing Stimulation Results Is Applicable for Most Low Permeability Gas Formations

Christopher W. Hopkins and **J. Michael Gatens III,** S. A. Holditch & Associates Inc.; and **James S. Caballero** and **Bruce E. Jankura,** Eastern States Exploration Inc.

SPE 21499 presents a case history of stimulation of the Upper Devonian Sandstone (Fifth Elk) in the Appalachia Basin in Centre County of Pennsylvania. Hopkins et al. use a systematic approach to optimize stimulation treatments, including prefracture well tests and production data analysis, fracture treatment design optimization, and post-fracture well tests and production data analysis. The authors evaluate 10 wells using production data analysis methods to obtain estimates of permeability. Hopkins et al. then evaluate fracture treatments of 5 wells using a pseudo-3D fracture design model to obtain propped fracture geometries and identify the existing fracturing problems. On the basis of these findings, 5 wells are stimulated using higher-viscosity fluid and higher-sand concentrations and yield better-than-expected gas production. The authors then conduct a full-blown optimization study of 4 wells and document very detailed analysis and results, including permeability, skin, designed fracture half-length, achieved fracture half-length, and gas flow rates before and after stimulations. On the basis of these studies, Hopkins et al. conclude the following:

- A prefracture estimate of permeability from production data or well test data is essential for optimizing stimulation
- Radial uncontained fracture-height growth model is optimal for fracture design in this particular field
- Higher viscosity and sand concentration up to 6 lbm/gal leads to successful stimulation of wells in this field
- Well test and production data exhibits dual-porosity behaviors

18.1 Summary

SPE 21499 presents the results of a successful stimulation optimization study performed on the Fifth Elk Sand Reservoir in the Council Run Field, Centre Co. and Clinton Co., PA. The systematic approach used to optimize stimulation results included pre-fracture well test and production data analysis, fracture treatment design optimization, and post-fracture treatment evaluation. This study led to a substantial improvement in stimulation results and an engineered approach for fracturing future wells.

18.2 Introduction

Low permeability Devonian gas sands are found throughout the Appalachian Basin. In most areas, these low permeability sands require successful fracture stimulation to achieve optimal economic performance. The principal target formation in the Council Run Field, located in central Pennsylvania near State College, is the Devonian Elk Sand group. **Table 18.1** summarizes the range of basic reservoir properties for the Elk Sands in this field. In general, net pay thicknesses range from 5 to 20 ft with porosities ranging from 0.06 to 0.16. The field is normally pressured with initial reservoir pressures ranging from 1,700 to 2,000 psia. High fracture gradients of 0.9 to 1.2 psi/ft are typical.

Several wells in the field have produced "naturally" at sufficiently high flow rates that they can be produced economically without stimulation. As the field was developed many wells would not produce economic

Completion Depth, ft	4,000 to 4,700
Net Pay Thickness, ft	5 to 20
Porosity, fraction	0.06 to 0.16
Water Saturation, fraction	0.15 to 0.30
Initial Reservoir Pressure, psia	1,700 to 2,000
Reservoir Temperature, 'F	85
Gas Gravity, (air = 1.0)	0.59
Well Spacing, ac	40
Fracture Gradient, psi/ft	0.9 to 1.2

Table 18.1—Average reservoir and well properties in the Council Run Field.

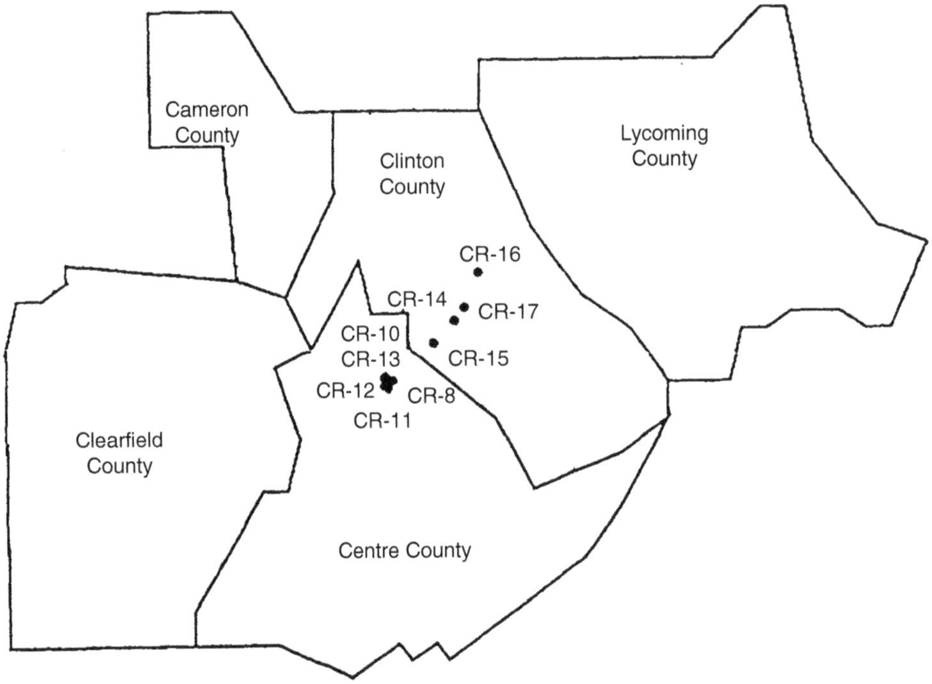

Fig. 18.1—Well location map for Council Run Field study wells in Pennsylvania.

amounts of gas without some stimulation. Fracturing these wells provided economic flow rates, but because these wells were not tested prior to fracturing, the effect of fracturing on a well's performance was unclear. A stimulation optimization study was performed to evaluate the range of formation permeability in the field, to determine when wells should be fracture stimulated to optimize performance, and to develop a method for designing and implementing optimal fracture treatments. **Fig. 18.1** shows the location of the Council Run Field study wells in Clinton Co. and Centre Co., PA. Stimulation results in the Council Run Field were significantly improved as a result of this study, and the systematic fracture design approach used in this case study example can be applied to other low permeability gas formations in the Appalachian Basin and elsewhere.

The fracture optimization study was performed in two stages. Existing production and fracture treatment data were analyzed to determine if different fracturing practices could improve production. Results from this preliminary analysis suggested that production could be increased by making several basic changes to the historical fracture treatment design. Implementing these changes on five test wells resulted in substantial increases in production, but calculated fracture half-lengths were still less than we designed for. We found that our fracture design assumption of contained vertical fracture height growth was not appropriate for this field. We performed a second study to (1) refine our fracture design methodology, and (2) better predict post-fracture well performance using a reservoir description determined from pre-fracture test data and fracture half-lengths estimated with radial fracture design models. Post-fracture performance was close to predicted in this second study. This project led to an "engineered" fracture optimization process for wells in this field. Based on estimates of formation permeability from pre-fracture open-flow data, the optimal treatment for a particular well can be selected and achieved successfully, maximizing economic results.

18.3 Historical Reservoir and Stimulation Treatment Evaluation

Many wells drilled initially in the Council Run Field produced at economic flow rates without stimulation. Based on geologic well log and drilling data, we believe most of these "natural" wells are in a naturally fractured Elk Sand Reservoir. Most of the wells in the field require stimulation to achieve economic well performance, and the operator believed that initial stimulation results could be improved. As part of our stimulation optimization study of this reservoir, we first conducted a preliminary review of existing reservoir and stimulation treatment data to determine the possible reasons for the poorer-than-expected well performance achieved by early stimulation treatment.

18.3.1 Reservoir Evaluation. To evaluate the Council Run Elk Sand Reservoir using existing data, we collected post-fracture initial open-flow test and production data from 10 Council Run Field wells. Using a single-rate flow test analysis technique (Lee et al. 1984) and a production data analysis method (Murtha et al. 1987) similar to that proposed by Fetkovich (1980) and Fetkovich et al. (1987), we analyzed these data in an attempt to determine formation quality (permeability) and stimulation effectiveness (skin factor or fracture half-length). Unfortunately, we found that we could not determine permeability and stimulation effectiveness uniquely from the available data. It is often difficult to quantify both skin factor and permeability uniquely from analysis of post-fracture open-flow and production data. Normally, pre-fracture estimates of permeability are needed for unique estimates of post-fracture stimulation effectiveness. By assuming different potential levels of completion efficiency (radial flow skin factors ranging from +3 to –5), however, we were at least able to estimate a range of permeabilities based on actual performance data for each well.

Results of our preliminary reservoir evaluation indicate the range of permeabilities encountered in these 10 wells is 0.0015 to 1.1 md (**Table 18.2**). We could not determine stimulation effectiveness uniquely from the data available, but we were able to at least qualitatively determine whether individual wells appeared to be stimulated. We found that five of the wells appeared to be stimulated. Two wells exhibited production behavior which was described best with a dual-porosity model, which we attributed to natural fractures or layers.

Reservoir-property-based well performance forecasts indicated that with successful hydraulic fracture stimulation, much greater flow rates could be achieved.

18.3.2 Historical Stimulation Treatment Evaluation. The next step in our historical-data review was to evaluate the historical stimulation practices which had been used, to determine why the treatments were not more effective. The most critical flaws in the original designs were insufficient fracture fluid viscosity and low sand concentrations, which resulted in low fracture conductivity within the pay zone. Also, high treating pressures and insufficient pad volumes resulted in the premature termination of many treatments. These factors resulted in less then optimal post-fracture production. In this section, we discuss the original designs, the flaws as we perceived them, and corrective measures which we recommended to ensure adequate fracture stimulation.

We collected and reviewed stimulation treatment records on five wells as part of our preliminary evaluation. A typical treatment consisted of pumping 39,000 lb of 20/40 mesh sand as proppant, carried by 25,000 gal of 10 lb/1,000 gal linear gel fluid system with a maximum downhole sand concentration of 2 lb/gal. To determine the fracture half-lengths we might expect in the field from pumping such a treatment, we simulated these treatments using a pseudo-three-dimensional fracture design model (Palmer and Luiskutty 1985) to compare the predicted fracture design half-lengths to those estimated from analysis of well performance data.

Well No.	Permeability Range (md)	Qualitative Indication of Stimulation (Y/N)
CR-1	0.0015 to 0.014	yes
CR-2	0.1 to 0.4	yes
CR-3	0.0016 to 0.023	yes
CR-4	0.003 to 0.022	yes
CR-5	0.022 to 0.067	yes
CR-6	0.003 to 0.029	no
CR-7*	0.68 to 1.1	no
CR-8	0.017 to 0.054	no
CR-9	0.036 to 0.12	no
CR-10	0.005 to 0.013	no

*Natural well, completed openhole.

Table 18.2—Preliminary flow test and production data analysis results.

In our preliminary fracture design simulations, we assumed that a created vertical fracture in the Fifth Elk Sand would be reasonably contained. Based on the analysis of an old mechanical properties well log and discussions with service companies and operators familiar with the area, we expected good barriers to vertical fracture height growth existed and that the gross fracture height would not exceed about 150 ft. The fracture conductivity profile for a typical historical stimulation treatment simulation is summarized in **Fig. 18.2.** Fig. 18.2 shows that we predicted a long created fracture half-length, due primarily to the thin fracturing fluid used and the fracture height containment assumption. Because the 10 lb/1,000 gal linear gel is a low-viscosity fluid (approximately 16 cp), however, most of the proppant fell to the bottom of the fracture. Depending upon the extent of the actual fracture height growth in the field, much of the sand could potentially fall out of zone during and after the fracture treatment. This would result in poorer-than-expected performance or, in some severe cases, may be damaging due to fracturing fluid leakoff effects in the pay zone. Without adequate proppant concentrations placed within the pay zone, post-fracture flow rates could actually be less than pre-fracture flow rates.

Using a similar pumping schedule, but with higher viscosity fluids, we optimized the fracture design as a function of fracturing fluid viscosity and determined that a viscosity of at least 50 cp should provide adequate proppant transport in these Fifth Elk Sand wells. We also determined that higher sand concentrations (5 to 6 lb/gal) should be pumped to achieve sufficient fracture conductivity to stimulate the reservoir. Published data provided by stimulation service companies for various fracturing fluids indicated that a 40 to 50 lb/1,000 gal linear gel fluid should provide 50 cp viscosity at the low reservoir temperature (85°F) which exists in the Council Run Field. Foam was investigated as an alternate treating fluid. Due to high expected surface treating pressures and the associated high cost of running high strength casing to pump a foam, we did not consider foamed fluids further.

Premature screenouts were also a problem with the historical treatments. The Elk Sands in this area and throughout Pennsylvania have anomalously high fracture gradients of 0.9 to 1.2 psi/ft probably due to external horizontal stresses. This is often indicative of horizontal created hydraulic fractures. Based on our experience in Elk Sand completions, we believe that, at typical formation depths in the Council Run Field, we are achieving vertical hydraulic fractures. The high fracture gradient causes surface treating pressures to be abnormally high, resulting in the premature termination of many treatments when treating pressures reach the maximum allowable surface treating pressure.

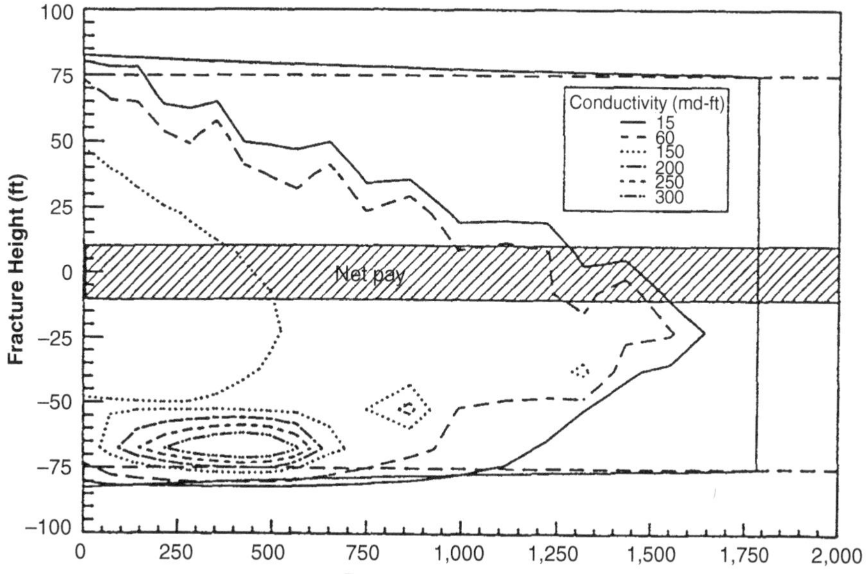

Fig. 18.2—Simulated fracture conductivity profile for typical historical fracture treatment using a low-viscosity fluid and assuming a well-contained fracture.

The use of small, low viscosity pad volumes may also have contributed to the premature termination of the treatments. Typical pad volumes were 12% or less of total treatment volume. In our simulated fracture designs, we often predicted a screenout due to depletion of the pad fluid and insufficient created fracture width. We recommended that pad volumes be increased to at least 30% of the total treatment volume in future treatments.

In summary, from our analysis of historical well data, we determined that the formation quality in the Fifth Elk Sand Reservoir varies substantially, and that with successful hydraulic fracture stimulation, large increases in production should be expected. Review of the historical stimulation treatments indicated that fracturing fluid viscosities were insufficient to provide good proppant transport and that pad volumes were too small to ensure that treatments could be pumped to completion without premature screenout. Recommended changes in the fracture treatment design included using a higher viscosity fluid (a 40 to 50 lb/1,000 gal linear gel), increasing pad volume to greater than 30%, and increasing the sand concentrations (6 lb/gal or more) to ensure adequate fracture conductivity is achieved. These changes were implemented to determine the potential benefit on post-fracture well performance.

18.4 Results of Preliminary Stimulation Design Changes

Five wells were selected for implementing the proposed fracture design changes to determine the potential benefit to the operator. Pre-fracture and post-fracture production data were collected and analyzed to determine the effect of the modified treatment designs. All five wells were substantially stimulated by the treatments; however, effective fracture half-lengths estimated from post-fracture performance data were less than we predicted with the fracture design model, indicating modifications to the assumptions used in our fracture design were needed. This section summarizes the results of the preliminary stimulation design changes on these five wells.

Perforations in the five study wells were broken down using acid and water to achieve pre-fracture flow. Achieving pre-fracture flow is essential to understanding the reservoir quality and evaluating the potential for stimulation. The wells produced for several weeks prior to stimulation, and we analyzed these data using production data analysis methods (Murtha et al. 1987; Fetkovich 1980; Fetkovich et al. 1987) for estimating formation permeability and pre-fracture skin factor. Following the pre-fracture flow period, these wells were stimulated with our modified stimulation design, cleaned up, and produced for several weeks. All five treatments were operationally successful, with no screenout problems encountered, and production increased substantially. The post-fracture production data were analyzed, assuming the pre-fracture permeability, to estimate stimulation effectiveness.

Table 18.3 summarizes the results of our pre- and post-fracture evaluation of these five wells. Permeabilities in these wells range from 0.005 to 0.04 md and the pre-fracture skin factor varied from 0.44 to –3.1 (indicating the wells were slightly damaged to slightly stimulated during drilling and pre-fracture completion operations). Prior to stimulation, these wells produced at flow rates of 22 to 50 Mscf/D. These in-line flow rates increased following stimulation to flow rates of 60 to 350 Mscf/D, and fracture half-lengths of 34 to 500 ft were estimated from analysis of the post-fracture production data (Prats 1961). These results were much better than had been achieved with previous stimulation efforts in the field; however, the apparent fracture half-lengths of 34 to 500 ft were still shorter than we expected. We designed for fracture half-lengths of 450 to 1,400 ft.

Although the post-fracture flow rates improved using the new fracture design, we did not get good agreement between the achieved and designed fracture half-lengths. To determine if our stimulation design approach was appropriate for the Council Run Field, we analyzed the detailed treatment records for the five study wells. Using our fracture design model, we calculated fracture half-lengths using two different assumptions. In one case, we assumed well-contained vertical fracture height growth (as previously discussed), and, in another case we assumed uncontained, radial fracture height growth (Geertsma and de Klerk 1969). The calculated result from our simulations are based on a dimensionless fracture conductivity (Cinco et al. 1978) cut-off of 10 for estimating effective fracture half-length.

The fracture half-lengths we calculated using our design model with the above assumptions and half-lengths from our post-fracture production evaluation are compared in **Table 18.4.** The fracture half-length predicted by the radial, uncontained fracture height growth model is

Well No.	Permeability (md)	Pre-Stimulation Skin Factor	Post-Stimulation Fracture Half-Length* (ft)	Pre-Stimulation Flow Rate (Mscf/D)	Post-Stimulation Flow Rate (Mscf/D)
CR-8	0.025	–0.9	180	24	125
CR-10	0.005	–3.1	100	22	80
CR-11	0.04	–0.7	500	50	350
CR-12	0.021	–1.7	300	38	240
CR-13	0.035	0.44	34	29	60

*Calculated using effective wellbore radius (Prats 1961).

Table 18.3—Pre- and post-fracture production data analysis results of five wells with modified fracture treatments.

	Calculated Fracture Half-Length, ft		
Well No.	Production Data Analysis	Radial Fracture Height Growth Model	Contained Fracture Height Growth Model
CR-8	180	300	625
CR-10	100	240	450
CR-11	500	320	680
CR-12	300	360	1,400
CR-13	35	260	500

Table 18.4—Comparison of fracture half-length predicted by design model and calculated from performance data.

generally closer to the calculated fracture half-length from performance data than that predicted by the contained height growth model. Rock mechanical properties measured on cores as part of this study indicated that the stress differential between the pay zone and the surrounding shales was much lower than we previously thought. The results indicate that uncontained fracture height growth might be occurring.

Treating pressure data from the five study wells were reviewed and calculated plots of net fracturing pressure (bottomhole treating pressure minus fracture closure pressure) versus pump time (Nolte and Smith 1979) were constructed. The shapes of these plots were characteristic of a radially fractured well (Economides and Nolte 1987; Smith et al. 1987). **Fig. 18.3** is a typical log-log net pressure versus pumping time plot (calculated from surface pressure data) for a Council Run well, which exhibits the classic shape of radial, uncontained fracture height growth (decreasing net pressure prior to the addition of sand).

The modified fracture design did improve stimulation results (Table 18.3). Furthermore, the radial fracture growth model appears more appropriate for fracture design than the contained, three-dimensional model which had been used previously. **Fig. 18.4** presents a fracture conductivity profile calculated using the radial fracture design model with a 50 cp fluid and a 6 lb/gal maximum sand concentration. Notice the difference in proppant placement between Figs. 18.2 and 18.4. Based on the encouraging results of this preliminary work, a final stimulation optimization study was conducted with the purpose of accurately predicting post-fracture performance; this is discussed in the following sections.

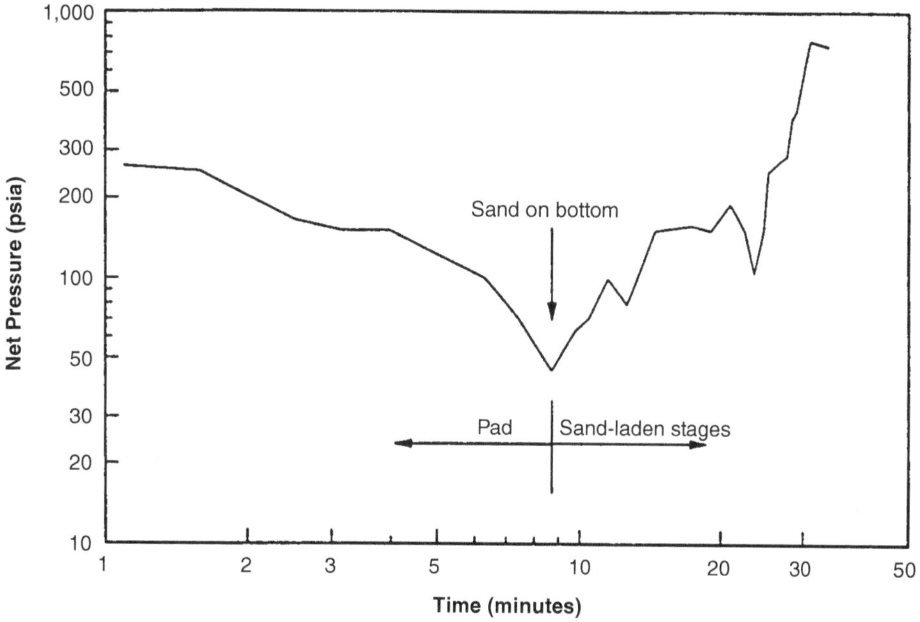

Fig. 18.3—Net pressure plot for typical well shows characteristic response for uncontained, radial fracture growth.

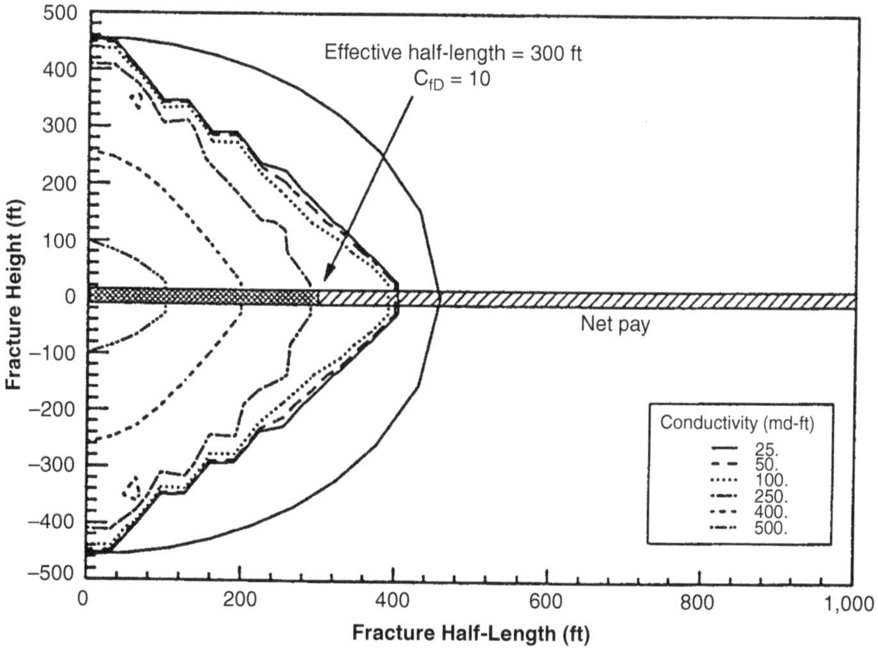

Fig. 18.4—Simulated fracture conductivity profile using a high viscosity fluid and assuming radial fracture growth.

18.5 Final Stimulation Optimization Study

Four wells were selected for a more complete stimulation optimization evaluation of the Fifth Elk Sand. Our objectives were to 1) evaluate reservoir quality from pre-fracture well test and production data, 2) design economically optimal stimulation treatments based on this reservoir quality, practical field considerations, and a radial fracture design model, and 3) evaluate the effectiveness of these treatments and the applicability of the fracture design assumptions using post-fracture well test and production data. The ultimate objective of this final stimulation optimization study was to develop a methodology which could be applied to future wells for determining the optimal stimulation treatment for individual zones based on pre-fracture permeability estimates.

18.5.1 Pre-Fracture Well Test Analysis. The four wells selected for this study were perforated and broken down using acid and water. Log analysis results for these wells are shown in **Table 18.5.** Following the breakdown treatments, the wells were flowed for three weeks and shut-in for two-week pressure buildup tests. Analysis of these data **(Table 18.6)** indicated three of the four wells exhibited dual-porosity behavior, which could be a result of natural fractures or layering effects (Murtha et al. 1987; Warren and Root 1963; Bourdet et al. 1984). For all wells, the calculated skin factors were slightly negative, indicating the breakdown treatments established good communication with the pay zone. For the purposes of our stimulation design optimization study, it is most important to determine the total system permeability and skin factor. Fracture treatment designs were modified for the potential increase in fluid leak-off that could occur if natural fracture systems do exist.

18.5.2 Fracture Design Optimization. Using the permeability calculated from the pre-fracture well test analysis, we optimized stimulation treatment designs for each well using the approach of Holditch et al. (1978) assuming limited investment funds (Holditch et al. 1978; Smith 1985; Meng and Brown 1987). The optimal fracture half-length for each well was estimated from a plot of fracture net present value (NPV) (Meng and Brown 1987) divided by investment cost versus fracture half-length. The fracture half-length and corresponding treatment size which results in the maximum value for fracture NPV to investment ratio is considered the optimal half-length. **Fig. 18.5** shows an optimization plot for a typical Council Run Field well with no-risk and risk-adjusted curves.

Well No.	Net Pay Thickness (ft)	Porosity (fraction)	Water Saturation (fraction)
CR-14	9	0.171	0.26
CR-15	17	0.112	0.30
CR-16	10	0.113	0.30
CR-17	10	0.155	0.29

Table 18.5—Log analysis results for final four study wells.

Well No.	Permeability-Thickness Product (md-ft)	Permeability (md)	Skin Factor	Initial Reservoir Pressure (psia)	ω	λ
CR-14	0.97	0.108	–3.5	1,893	0.077	1.17×10^{-4}
CR-15	0.38	0.0224	–3.6	1,950	0.048	6.88×10^{-4}
CR-16	0.064	0.00638	–3.2	—	—	—
CR-17	0.38	0.0376	–3.1	2,021	0.173	1.68×10^{-4}

Table 18.6—Pre-stimulation well test analysis results.

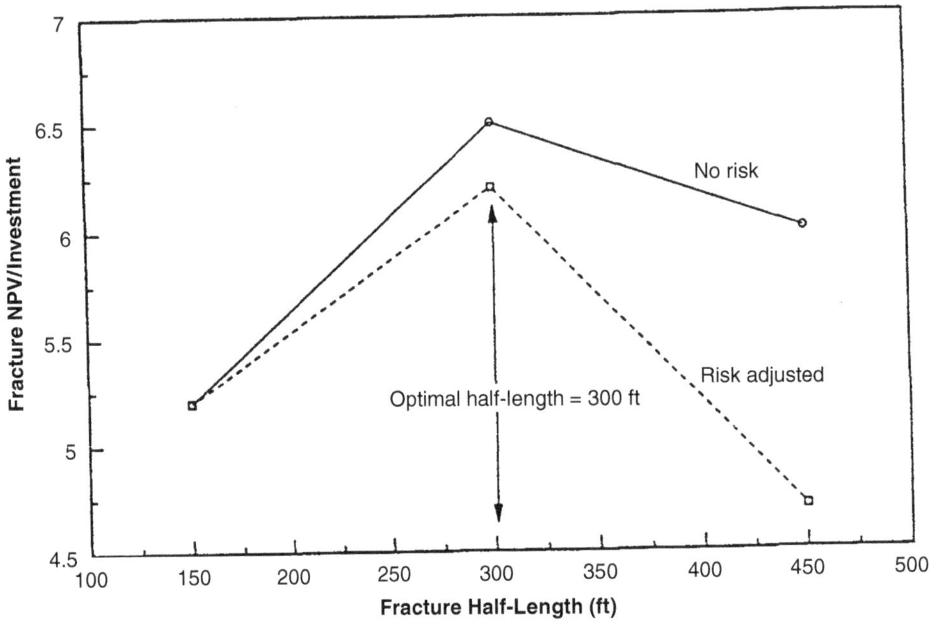

Fig. 18.5—Fracture optimization plot for typical well, with and without adjustments for risk.

Based on stimulation of numerous Council Run Field wells, a risk factor approach was developed to include technical risk in the optimization process. Essentially, the risk-adjusted curve includes the inherent technical risk associated with achieving long fracture half-lengths. Assigning technical risk will be discussed in more detail in a later section. A detailed explanation of the fracture design optimization process is provided in Appendix 18.A.

18.5.3 Post-Fracture Well Test and Production Data Analysis. The optimal fracture designs determined above were successfully pumped in each of the study wells, and the wells were flowed back using a controlled flowback procedure (Robinson et al. 1988). Two of the wells were then produced for 36 days and shut-in for 14-day pressure buildup tests. The other two wells were produced for 70 days and were not tested. We analyzed the available pressure buildup test and production data on the four wells and obtained the results summarized in **Table 18.7** and illustrated in **Fig. 18.6.**

On the basis of our analysis of these post-fracture data, we determined that post-fracture production data were sufficient for evaluating stimulation effectiveness and predicting well performance for wells in the Council Run Field. This is only true if an estimate of permeability is available from pre-fracture test data. This makes post-fracture well tests unnecessary for routine evaluation of stimulation effectiveness in the Council Run Field.

Fig. 18.6 shows the pre-fracture and post-fracture production for well CR-15. From analysis of the buildup and production data, we estimated an effective fracture half-length of 230 ft for this well. Well performance was improved substantially over pre-fracture rates. The in-line sustained flow rate (after 40 days) increased from approximately 60 to 180 Mscf/D. Fig. 18.6 shows the predicted performance (solid line) for a 230-ft achieved fracture half-length and the predicted performance (dashed line) for the 300-ft designed fracture half-length for this well. Notice that the performance projected for the well based on pre-fracture test data and the optimal fracture design half-length are very close to the performance actually achieved in the field.

This is a significant result. The methodology developed from this study can be used to describe this reservoir, design a stimulation treatment based on this reservoir description, pump the treatment successfully, and achieve the results predicted based on pre-fracture analysis. Without being able to estimate post-fracture well performance from pre-fracture data accurately, determining the economically optimal fracture treatment is not possible. This is the ultimate goal of any engineered stimulation optimization study. In three of the four wells, the achieved fracture half-lengths and well performance were close to what was predicted from pre-fracture evaluation (Table 18.7). Although the cost of the new treatments are 20 to 30% higher than for previous treatments, the success of the new treatments lead to greater profitability.

Well No.	Designed Fracture Half-Length (ft)	Achieved Fracture Half-Length (ft)	Pre-Fracture Flow Rate (Mscf/D)	Post-Fracture Flow Rate (Mscf/D)
CR-14	150	120	110	350
CR-15	300	230	100	450
CR-16	450	450	25	100
CR-17	200	34	60	100

Table 18.7—Stimulation results for final four study wells.

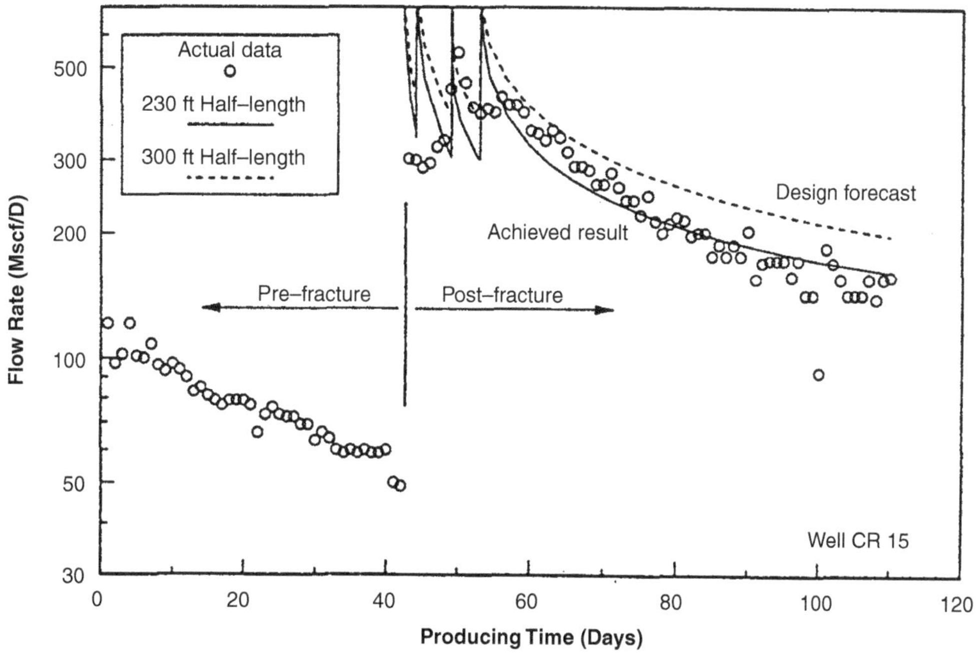

Fig. 18.6—Comparison of predicted vs. achieved post-fracture performance for Well CR 15.

18.6 Application of Results

Since performing this stimulation optimization study, over 100 Fifth Elk Sand wells have been stimulated in the Council Run Field. Most of the stimulation results have been good, some rather dramatic. "Natural" wells without stimulation have also been evaluated for stimulation; these wells would have been completed openhole without fracturing previously. One well which had an initial pre-fracture flow rate (in-line) of 850 Mscf/D, had declined to 250Mscf/D. It was stimulated and achieved a post-fracture flow rate of 3,000 Mscf/D in-line.

The stimulation optimization procedure described above has been refined to account for the technical risk of achieving a successful stimulation treatment in the field. Some screenouts and premature job terminations have still occurred, which have prevented some treatments from being pumped as designed. **Table 18.8** summarizes the percentage of designed sand-concentration stages which have been successfully pumped in the field. The treatment designs which call for maximum sand concentrations of 7 lb/gal typically place 65% of the designed sand stages. Similarly, the potential for failure is quantified for the lower sand concentrations. Using these risk-adjusted optimal fracture half-lengths, an empirical plot of optimal fracture half-length versus permeability was developed for the Council Run Field, as shown in **Fig. 18.7.** A review of Fig. 18.7 shows that (1) as permeability increases, eventually no stimulation is needed, (2) permeability can eventually become so low that fracture stimulation is uneconomical, and (3) as sand thickness increases (and thus gas-in-place), the optimal fracture half-length for a given permeability increases.

Maximum Sand Concentration (lb/gal)	Percent of Sand Placed (%)
0 to 3	100
4	91.4
5	84.5
6	74.3
7	65.0
8	0.0

Table 18.8—Risk associated with pumping higher-sand-concentration fracture treatments.

To apply the methodology developed as part of the study, new wells in the field are being flow tested after drilling (these wells are drilled on air) and before setting casing to obtain pre-fracture reservoir data. These flow test data are analyzed using the single-rate flow test analysis technique (Lee et al. 1984) to determine permeability (assuming a skin factor of zero). This permeability is then used in Fig. 18.7 to determine the optimal fracture half-length for the well.

Data collection is continuing and is being used to refine the stimulation methodology in the Council Run Field. Approaches are being evaluated for stimulating multiple zones in individual wells and for determining optimal well spacing. The results of this study have led to the development of an engineered reservoir evaluation and stimulation methodology which has yielded predictable, economically successful stimulation results in the field.

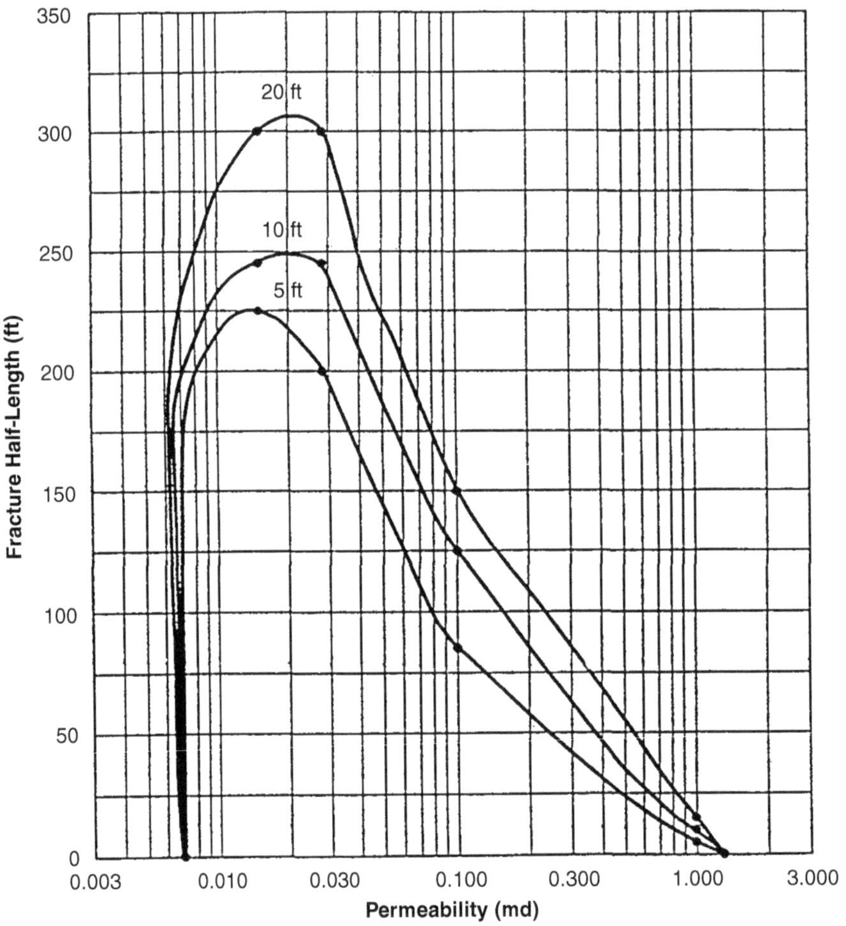

Fig. 18.7—Optimal fracture design half-length as a function of permeability for 5, 10, and 20 ft Elk Sands in the Council Run Field.

18.7 Conclusions

On the basis of the results of this study, we have drawn the following conclusions.

1. This study resulted in successful stimulations on all wells treated. In three of the four final stimulation study wells, the designed and achieved fracture half-lengths are in close agreement The methodology has proved successful in the Elk Sands of the Council Run Field and is continuing to be applied and refined in the field.
2. A pre-fracture estimate of permeability from production data (or well test data) is essential for optimizing stimulation results. In the Council Run Field, flow data after drilling are analyzed for estimating permeability and an empirical graph is used to determine optimal fracture half-lengths on a well-by-well basis.
3. Analysis of well test, production, and stimulation data for the study wells indicates that a radial, uncontained fracture height growth model is best for fracture design in the Council Run Field. This radial model provides the closest agreement between designed and achieved fracture half-lengths.
4. In the Council Run Field, analysis of post-fracture production data is adequate for estimating stimulation effectiveness and predicting future well performance when pre-fracture estimates of permeability are available.
5. Historical stimulation results in the Council Run Field were suboptimal, in part due to low fracturing fluid viscosity, low sand concentrations, small pad volumes, and optimistic assumptions for fracture geometry used in the fracture design. This study indicated that using higher viscosity fluids, sand concentrations up to and in excess of 6 lb/gal, greater pad volumes, and a radial fracture design model led to successful stimulation results.

18.8 Nomenclature

C_{fD}	=	Dimensionless fracture conductivity, $\frac{wk_f}{\pi k L_f}$, dimensionless
k	=	Formation permeability, md
k_f	=	Fracture permeability, md
k_m	=	Matrix permeability, md
L_f	=	Fracture half-length, ft
r_w	=	Wellbore radius, ft
w	=	Fracture width, ft
α	=	Interporosity flow shape factor
λ	=	Interporosity flow coefficient in dual-porosity reservoirs, $\alpha r_w^2 k_m k_f$, dimensionless
ω	=	Storativity ratio in dual-porosity reservoirs, $(\phi c_t)/\left[(\phi c_t)_f + (\phi c_t)_m\right]$, dimensionless
$(\phi c_t)_f$	=	Fracture porosity-compressibility product, psi^{-1}
$(\phi c_t)_m$	=	Matrix porosity-compressibility product, psi^{-1}

18.9 References

Bourdet, D., Ayoub, J. A., Whittle, T. M. et al. 1984. Interpreting Well Tests in Fractured Reservoirs. *World Oil*. 111–124.

Cinco, H., Samaniego, F. and Dominguez, N. 1978. Transient Pressure Behavior for a Well With a Finite-Conductivity Vertical Fracture. *SPE J.* **18** (4). SPE-6014-PA. https://doi.org/10.2118/6014-PA.

Economides, M. S. and Nolte, K. G. eds. 1987. *Reservoir Stimulation*, Chap. 7, 1–15. Houston, Texas: Schlumberger Educational Services.

Fetkovich, M. J. 1980. Decline Curve Analysis Using Type Curves. *J Pet Technol* **32** (6): 1065–1077. SPE-4629-PA. http://dx.doi.org/10.2118/4629-PA.

Fetkovich, M. J., Vienot, M. E., Bradley, M. D. et al. 1987. Decline Curve Analysis Using Type Curves: Case Histories. *SPEFE* **2** (4): 637–656. SPE-13169-PA. http://dx.doi.org/10.2118/13169-PA.

Geertsma, J. and de Klerk, F. 1969. A Rapid Method of Predicting Width and Extent of Vertical, Hydraulically Induced Fractures. *J Pet Technol* **21** (12): 1571–1581. SPE-2458-PA. https://doi.org/10.2118/2458-PA.

Holditch, S. A., Jennings, J. W., Neuse, S. H. et al.: 1978. The Optimizatic of Well Spacing and Fracture Length in Low Permeability Gas Reservoirs. Presented at the SPE Annual Technical Conference and Exhibition, Houston, Texas, 1–3 October. SPE-7496-MS. https://doi.org/10.2118/7496-MS.

Lee, W. J., Kuo, T. B., Holditch, S. A. et al. 1984. Estimating Formation Permeability from Single-Point Flow Data. Presented at the SPE Unconventional Gas Recovery Symposium, Pittsburgh, Pennsylvania, 13–15 May. SPE-12847-MS. https://doi.org/10.2118/12847-MS.

Meng, H. Z. and Brown, K. E. 1987. Coupling of Production Forecasting, Fracture Geometry Requirements and Treatment Scheduling in the Optimum Hydraulic Fracture Design. Presented at the SPE/DOE Joint Symposium on Low Permeability Reservoirs, Denver, Colorado, 18–19 May. SPE-16435-MS. https://doi.org/10.2118/16435-MS.

Murtha, J. A., Gatens, J. M. III, Lancaster, D. E. et al. 1987. Practical Analysis Methods for Well Test and Production Data. Topical Report to the Gas Research Institute, GRI Contract No. 5084-213-0980, S. A. Holditch & Associates, Inc., College Station, Texas (October 1987).

Nolte, K. G. and Smith M. B. 1979. Interpretation of Fracturing Pressures. Presented at the SPE Annual Technical Conference and Exhibition, Las Vegas, Nevada, 23–26 September. SPE-8297-MS. https://doi.org/10.2118/8297-MS.

Palmer, I. D. and Luiskutty, C. T. 1985. A Model of the Hydraulic Fracture Process and Comparison of Results With Other Models. Presented at to the SPE/DOE Symposium on Low Permeability Gas Reservoirs, Denver, Colorado, 19–22 May. SPE-13864-MS. https://doi.org/10.2118/13864-MS.

Prats, M. 1961. Effect of Vertical Fractures on Reservoir Behavior - Incompressible Fluid Case. *SPE J.* **1** (2). SPE-1575-G. https://doi.org/10.2118/1575-G.

Robinson, B. M., Holditch, S. A., Whitehead, W. S. 1988. Minimizing Damage to a Propped Fracture by Controlled Flowback Procedures. *J Pet Technol* **40** (6). SPE-15250-PA. https://doi.org/10.2118/15250-PA.

Smith, M. B., Miller, W. K., and Haga, J. 1987. Tip Screenout Fracturing: A Technique for Soft, Unstable Formations. *SPE Prod Eng* **2** (2). SPE-13273-PA. https://doi.org/10.2118/13273-PA.

Smith, M. B. 1985. Stimulation Design for Short, Precise Hydraulic Fracture. *SPE J.* **25** (3). SPE-10313-PA. https://doi.org/10.2118/10313-PA.

Warren, J. E. and Root, P. S. 1963. The Behavior of Naturally Fractured Reservoirs. *SPE J.* **3** (3): 245–255. SPE-426-PA. https://doi.org/10.2118/426-PA.

Appendix 18.A

18.A.1 Fracture Treatment Optimization

To choose the optimal fracture half-length for each study well, we predicted future well performance for different created hydraulic fracture half-lengths using a reservoir model. We subtracted the predicted pre-fracture well performance from the post-fracture predictions to determine the incremental future production due to the hydraulic fracture treatment alone. Assuming the basic well costs and operating costs as sunk costs which will be incurred whether a well is stimulated or not, we determined the net present value (NPV) of the incremental production achieved due to hydraulic fracturing. We refer to this incremental NPV as "fracture net present value" (Meng and Brown 1987).

To determine the optimal fracture treatment, we plotted the fracture net present value divided by the investment (treatment cost) required to achieve each fracture half-length versus fracture half-length, as shown in Fig. 18.5. The investment cost in this analysis was determined by simulating hydraulic fracture stimulation treatments (using the radial fracture design model) to achieve the different fracture half-lengths assumed in the performance predictions. Based on the size of the resulting design, we estimated treatment cost using data provided by stimulation service companies. The optimal treatment is determined when fracture NPV divided by investment is maximized (Fig. 18.5). We assumed a constant, square drainage area of 40 acres in this analysis.

18.A.2 About the Authors

Chris Hopkins joined S. A. Holditch & Associates Inc. in Pittsburgh, Pennsylvania, as a petroleum engineer in August 1988. His primary interests include fracture treatment design and quality control, reservoir engineering, and pressure transient analysis. Prior to joining Holditch, he was employed by Chevron USA in North Dakota and Oklahoma as a production and drilling engineer. Hopkins holds a BS degree from West Virginia University and an MS degree from Texas A&M University, both in Petroleum Engineering, and is a member of SPE. J. Michael Gatens III is Vice President of S. A. Holditch & Associates Inc. in Pittsburgh, Pennsylvania. Gatens joined Holditch in 1982 and in 1988 opened Holditch's Eastern Division office. He specializes in applied reservoir engineering and has co-authored over 30 technical articles. Previously, Gatens was with Texas A&M and Bass Enterprises. He holds BS and MS degrees in Petroleum Engineering from Texas A&M and is a member of SPE, SPWLA, and SCA. He is currently Chair of the Pittsburgh SPE Section, is a member of the SPE Reservoir Engineering Committee, and was on the SPE Gas Technology Committee during 1990–1992, serving as chair in 1992. James Caballero is Vice President, Engineering for Eastern States Exploration Company in Alexandria, Virginia. He previously worked with British Gas U.S. Inc., International Division in Houston, Texas, and Tenneco Oil Exploration and Production, International Division in Houston, Texas, and Pacific Coast Division in Bakersfield, California. Since 1980, Caballero has received experience in all aspects of petroleum reservoir engineering. He holds a BS degree in Petroleum Engineering from Texas, Tech University. Bruce Jankura is Senior Vice President, Oil and Gas Operation for Eastern States Exploration Company. He previously worked with Mitchell Energy Corporation, Appalachian Basin Area - Columbus, Ohio, and Mobil Oil Corporation, Gulf Coast - Lake Charles, Louisiana. Since 1977, Jankura has acquired experience in various aspects of petroleum engineering and corporate management. He holds a BS degree in Petroleum Engineering from Marietta College and is a registered professional petroleum engineer.

SI Metric Conversion Factors

ft	×	3.048*	E-01	=	m
psi	×	6.894 775	E+00	=	kPa
cp	×	1.0*	E-03	=	Pa·s
md	×	9.8969 233	E-04	=	μm^2
ft^3	×	2.831 685	E-02	=	m^3

*Conversion factor is exact.

Chapter 19

A Case Study of the Wilcox (Lobo) Trend in Webb and Zapata Counties, TX

B. M. Robinson, SPE, S. A. Holditch & Assocs.; **S. A. Holditch,** SPE, Texas A&M U.; and **W. J. Lee,** SPE, Texas A&M U.

SPE 11600 presents a case history of a study to determine the average properties of the Wilcox (Lobo) Formation in Webb and Zapata Counties, Texas, to be used to file for tight gas sand designation at the Federal Energy Regulation Committee. Robinson et al. summarize an integrated study of geology, completions methods, hydraulic fracture treatments, and reservoir production analyses. This study contains three examples in which reservoir simulation models were used to analyze post-fracture production and pressure data. In-depth analyses are provided to explain the mechanisms and reservoir physics behind the completions and engineering operations and to explain the discrepancies between created and effective fracture geometries. Robinson et al. conclude

- Successful development of the geopressured Lobo Formation is direct application of advanced completions, analysis, and stimulation practices
- Hydraulic fracture is required to increase gas flow rates and ultimate recovery
- Hydraulic fracture treatments must be optimized on the basis of reservoir properties
- Severe proppant crushing may be a problem in the Wilcox Lobo
- Correlations between pore pressure, fracture gradient, and temperature vs. depth for the Lobo Trend are established
- Correlations of absolute open flow rates and permeability thickness product are established
- There is need for reservoir simulation to correctly analyze complex problems and predict well performance

19.1 Summary

The Wilcox (Lobo) Trend of Webb and Zapata counties, Texas, is a series of geopressured, low-permeability sands with average depth from 5,000 to 12,000 ft [1525 to 3660 m]. More than 1,000 wells have been drilled in this prolific trend during the last 10 to 12 years. Although actively developed earlier, the trend became even more attractive after its classification by the Federal Energy Regulatory Commission (FERC) as a "tight" gas formation.

Essentially, development of the Wilcox (Lobo) has been successful because of modern technological advances. SPE 11600 presents the results of several years of study involving the geologic history, completion methods, massive-hydraulic-fracture (MHF) stimulation treatments, reservoir evaluation, and numerical analysis of hydraulically fractured wells in this trend, all of which illustrate the application of this modern technology.

19.2 Introduction

Low-permeability gas reservoirs have been developed actively in the U.S. during the past 10 to 15 years. Three of the most important criteria for the successful development of any area are (1) gas market availability, (2) gas prices adequate to justify the risks involved, and (3) proper application of modern technology. A key element is the development and transfer of the technology necessary to develop low-permeability gas sands economically.

Many authors have contributed to the transfer of technology during the past few years by presenting papers concerning fracture fluids (Conway and Harris 1982; Hsu and Conway 1981), fracture design equations (Geertsma and De Klerk 1969; Nordgren 1972; Daneshy 1973; Cleary 1980), pressure-transient analysis (Gringarten et al. 1979; Cinco-Ley et al. 1978; Lee and Holditch 1981), and other specific topics. Papers that describe a field case history (Kozik and Holditch 1981; Wyman et al. 1980; Fast et al. 1977) are normally quite useful to most petroleum engineers.

SPE 11600 presents the results of several years of work involving the geopressured Wilcox (Lobo) Formation in Webb and Zapata counties, Texas, including the geologic history, completion techniques, hydraulic-fracture stimulation techniques, and reservoir evaluation of the Lobo Formation.

In the 1960s, several wells were drilled and tested in the Wilcox (Lobo) Formation. Because of low gas prices and a lack of pipelines, however, the area was considered uneconomical and remained inactive for several years. Successful production from the Wilcox (Lobo) Sands in Mexico renewed interest in this trend in the early 1970s. Consolidated Oil and Gas Co. drilled its N. H. Clark Well 1 directly across the Rio Grande from established production in Mexico. This well produced at rates of up to 7.0 MMcf/D [198×10^3 m^3/d] and officially discovered the Wilcox (Lobo) Trend in the U.S.

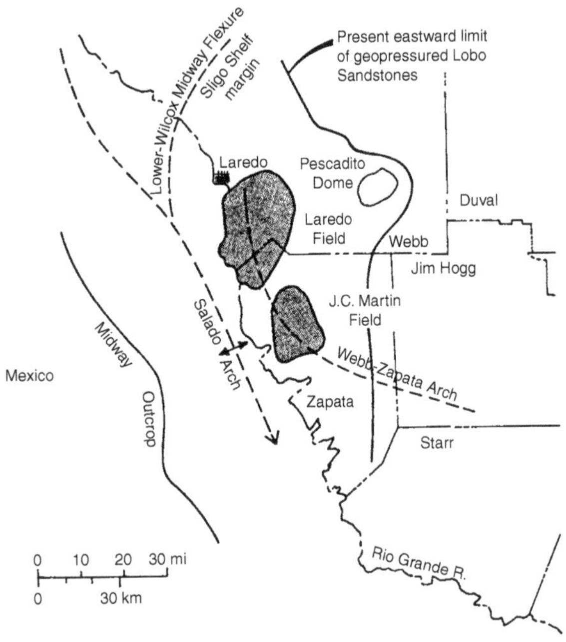

Fig. 19.1—Present area of Wilcox (Lobo) Trend (modified from Henke 1982).

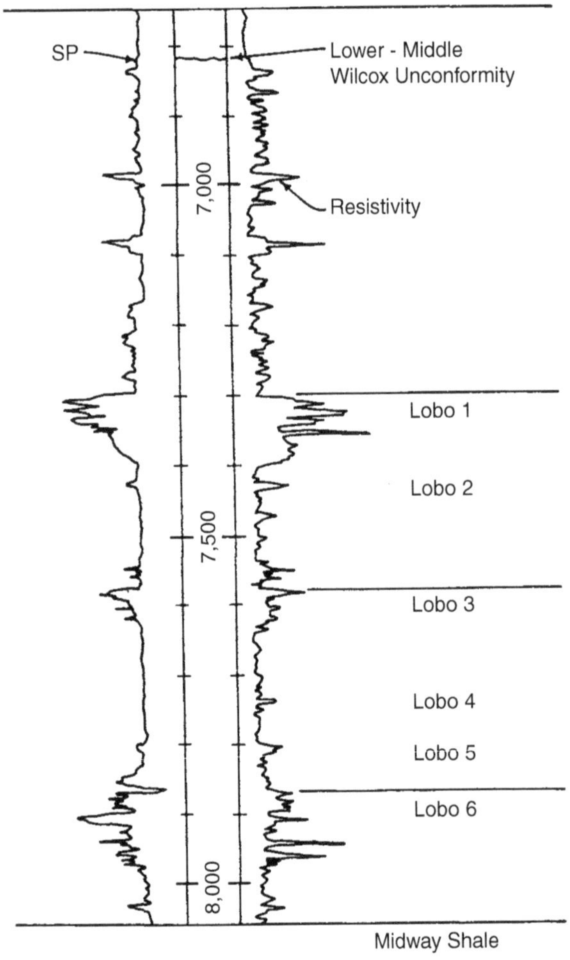

Fig. 19.2—Type log of Lobo sequence (Hilty 1980).

Early exploratory drilling, which was concentrated near the U.S./Mexico border, soon established the Wilcox (Lobo) Trend as a major gas-producing horizon. As drilling progressed farther from the border, the Lobo was encountered at greater depths and the productive extent of the Lobo was expanded to include an area of at least 1,800 sq miles [4660 km^2].

Even though the shallower sands near the border seem to produce at higher initial rates than the deeper sands, virtually every well drilled in the Lobo Trend is fracture-stimulated to increase gas flow rates and recoverable reserves. As a result, MHF technology has been used quite extensively, particularly in the deeper areas. Because fracture stimulation costs can amount to a significant fraction of the total well costs (10 to 20%), it is important to optimize this phase of the completion. State-of-the-art analysis techniques, such as numerical history-matching simulation, have been used by many operators in an effort to optimize and to ensure effective fracture stimulation treatments.

Even though more than 1,000 wells have been drilled, the Lobo Sands are far from being completely developed. The productive limits of the trend are still being explored, with wells being drilled farther to the north, east, and south of the U.S./Mexico border. In these directions, the sands dip to depths of approximately 10,000 to 12,000 ft [3050 to 3660 m]. Because of the abnormally high pore-pressure gradients associated with these depths, the drilling and completion techniques must be well integrated.

19.3 Geology

The Wilcox (Lobo) section consists of alternating sandstone, siltstone, and shale sequences occupying the lowermost unit of the Wilcox group. The Lobo conformably overlies the marine Midway Shale of Paleocene age and is unconformably overlain by the Lower Wilcox Shale of Eocene age. Generally 500 to 1,200 ft [150 to 365 m] thick, this Lower Wilcox Shale separates the Lobo from the other overlying Wilcox units.

The geo-pressured Lobo of Webb and Zapata counties is generally logged at depths between 5,000 and 12,000 ft [1525 and 3660 m]. The trend currently extends from northwestern Webb County to southwestern Zapata County and south and west into Mexico (**Fig. 19.1**). It is bounded to the north and northeast by the Pescadito salt dome, the Edwards and Sligo shelf margins, and related faults of the Wilcox fault system. It is bounded to the west by the Salado Arch in Mexico and to the south by the Rio Grande embayment (Henke 1982). Exact limits of the trend to the south and east have not been definitely established because drilling continues to expand the boundaries in these directions.

Fig. 19.2 shows the three principal correlative members that are identified in the Wilcox (Lobo) section. The lowest is the Lobo 6 Sandstone, which overlies the Midway Shale. The next higher member is the Lobo 3 Sandstone, and at the top is the Lobo 1. The Lobo 1 and Lobo 6 are the principal gas reservoirs in this series, whereas the Lobo 3 and other less extensive sandstones account for a very minor amount of production and reserves. These two principal sand members are generally separated by 100 ft [30.5 m] or more of shale. The entire Lobo section has a distinctive E-log character, and the sandstone sequences can be correlated over vast distances, indicating widespread deposition (Hilty 1980).

The Lobo sequence represents the initial cycle of sediment deposition and the regressive development of the Wilcox in the Rio Grande embayment. The Lobo Sands appear to have been deposited in various marine facies of a submarine fan. Soon after deposition, the section was subjected to intense tensional faulting, followed by severe erosion and truncation. Another period of faulting occurred, resulting in the entire section's being highly shattered and segmented. A typical cross section illustrating this intense faulting is shown in **Fig. 19.3.** Gas accumulation has been controlled largely by this faulting and, to some extent, by erosional truncation of the individual sand members (Hilty 1980).

The Lobo Sands are generally light gray to white, very fine grained, thinly bedded with a shaly matrix, and interbedded with

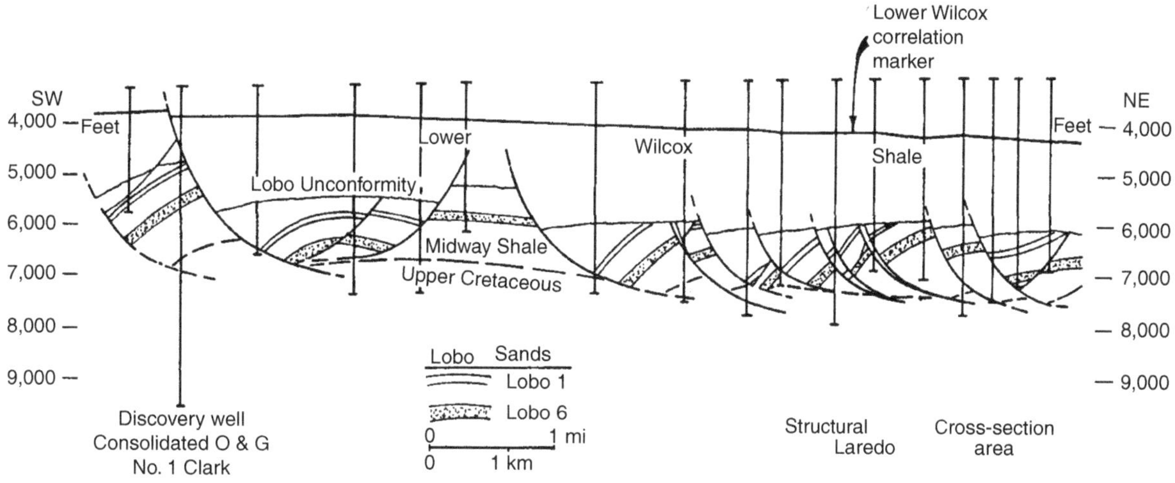

Fig. 19.3—Cross section through Laredo area (Hilty 1980).

siltstones and dark-gray-to-black shales. The original porosity and permeability of these sands have been reduced severely by compaction and diagenetic alterations, particularly quartz overgrowths (Hilty 1980).

In general, progressive development and expansion of the trend determined that the best Lobo Reservoirs occurred at the shallower depths near the Rio Grande. At greater depths where current gas exploration is being conducted, reservoir quality deteriorates, total sandstone thickness decreases, and the incidence of marginal to sub-marginal wells increases.

Generally, the mineral composition of the Lobo 1 and 6 Sands is 50 to 75% quartz, 10 to 25% feldspar, and 15 to 25% clay matrix. Scanning-electron-microscope observations reveal that the majority of the diagenetic minerals in the samples are quartz overgrowths, chlorite, illite, and calcite, all of which have combined to reduce porosity and permeability. In general, quartz overgrowths fill most of the pore space. Chlorite is the dominant clay mineral lining the pores, with calcite occurring in trace amounts (Hilty 1980).

19.4 Completion and Testing Practices

Because of the abnormally high reservoir pressures associated with this trend (**Fig. 19.4**), high-pressure control equipment and techniques must be used to drill and to complete these wells. In general, intermediate casing strings are set in the Lower Wilcox Shale for control purposes while drilling through these high-pressure sands. Although not standardized by any means, a typical completion procedure might be described by the following.

1. Set the liner from the base of the intermediate string to total depth (TD). Usually, a 5-in. [13-cm], 18-lbm/ft [27-kg/m], N-80 or P-110 casing is used (or an equivalent, depending on intermediate string size). Packoff-type liner hangers are often used for additional control and sealing at the top of the liner.
2. After cleaning out the liner and testing, circulate the well with completion fluid (2% KC1 with clay stabilizer) and run cased-hole logs.
3. Set a permanent-type packer by wireline with 8 to 15 ft [2.4 to 4.6 m] of sealbore extension, approximately 100 ft [30.5 m] above the zone(s) of interest.
4. Run the seal assembly and high-pressure seals on 2⅞-in. [7.3-cm], 7.9-lbm/ft [11.8-kg/m], N-80 tubing and sting into packer. The number of seals and length of the seal assembly will depend on the expected treating conditions.
5. Selectively perforate zone(s) of interest through tubing with 0.25-in. [0.635-cm] holes at a density of 1 shot/ft [3.3 shots/m]. (Note: Because of geopressured pore pressures, anywhere from 1,000- to 4,000-psi [6900- to 27 600-kPa] underbalance conditions will exist; therefore, pressure-control perforating equipment must be used.)

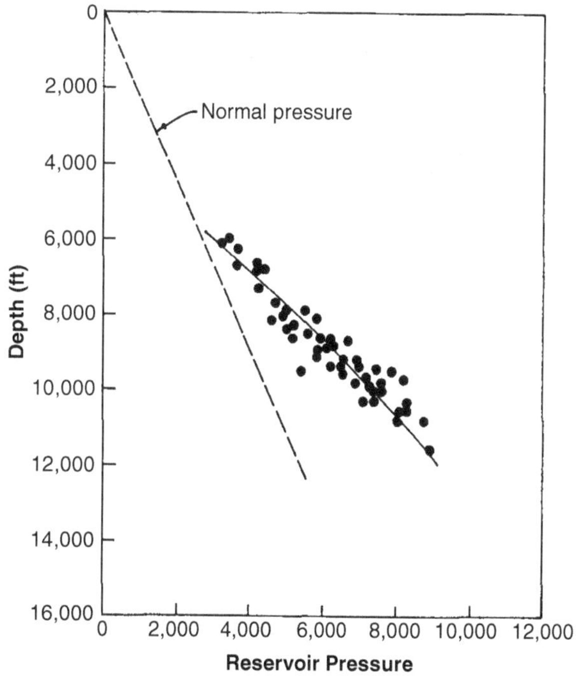

Fig. 19.4—Reservoir pressure vs. depth.

Once perforated, the Lobo should be bailed out before zone production testing. The bailout treatment serves several purposes. First, because the Lobo is a soft, geo-pressured, low-permeability rock, formation damage can occur if large pressure gradients are imposed on the formation. The bailout treatment usually will change the skin factor from a positive value (+2.0 to +5.0, caused by drilling mud and cement damage) to a slightly negative value by creating minor fractures around the wellbore. This amount of stimulation will allow production tests at moderate flow rates for short periods of time without imposing large pressure gradients around the wellbore. Second, during prefracture testing, all the net gas pay should contribute to the production tests. Because of the geo-pressures, the typical Lobo well is perforated with a through-tubing, hollow-carrier gun. As a result, all the holes will not likely be in effective communication with the formation. Therefore, the bailout treatment is necessary to open all the perforations so that a more accurate prefracture flow test and pressure-buildup test can be performed.

Acid is not usually recommended for the Wilcox because of severe damage that can result from the release of formation fines and from destruction of the delicate cementation of the matrix. The most common fluid used during a bailout treatment is 2% KC1 water with friction reducer and 5 to 10% methanol. The fluid is pumped at 5 to 10 bbl/min [0.8 to 1.6 m^3/min], and ball sealers are evenly spaced throughout the treatment for diversion and to ensure individual perforation breakdown. Commonly, a single Lobo Sand is perforated with 10 to 20 holes, and 20 to 30 ball sealers are dropped during the treatment.

A postballout, prefracture pressure-transient test is then performed to determine the productivity of the reservoir. The test can include a single-point flow test, a flow test followed by a pressure-buildup test, or a multirate test. With available analysis techniques, specific reservoir properties can be determined for the design of the MHF treatment and for projecting future well performance.

After the fracture treatment is performed, four-point absolute open-flow (AOF) tests are run to determine the productivity of the wells. In many instances, extended flow tests with postfracture pressure-buildup tests are performed to calculate the reservoir and fracture properties. Results from these types of analyses can be used to project future performance.

Once the load fluid is recovered from the stimulation treatment, water production from these reservoirs is not usually a problem. However, most wells will produce some condensate liquids. Normally, this liquid production does not cause any major production problems except when the reservoir is depleted substantially; and then only in wells with extremely low gas/liquid ratios is it necessary to use artificial lift (e.g., plunger lift, gas lift, or intermittent flow device). At this point, compressors are also installed on many wells to help extend flow life.

19.5 Hydraulic Fracture Stimulation

The MHF stimulation technique has been extremely important to the success of the Wilcox (Lobo) Trend. Virtually every well completed in the Wilcox requires fracture stimulation to increase flow rates and ultimate recovery. Since the early 1970's, the history of fracture stimulation in this trend has practically paralleled the advances in fracture stimulation technology. As exploration and development expanded into the deeper portions of the trend and as fluid and equipment technology improved, larger fracture treatments have been performed. The fracture stimulation treatments performed by one operator in this trend are summarized in **Table 19.1.**

Most current fracture stimulation treatments use crosslinked-gel systems. As illustrated in **Fig. 19.5,** reservoir temperature can exceed 300°F [149°C] in the deeper areas of the trend. Because of these extreme reservoir temperatures, it is imperative that the fluid systems be sufficiently stable to withstand high temperatures over long pumping times. In these cases, crosslinked fluids with gel concentrations of 60 to 70 lbm/1,000 gal [7 to 8 kg/m^3] are used and additional gel stabilizers are required.

In the past, almost every fracture treatment used 20/40-mesh silica sand as the proppant. Because of expansion of activity into the deeper areas of the trend, the use of higher-strength proppants has become necessary to withstand the excessive closure pressures that have been encountered. **Fig. 19.6** depicts the fracture pressure gradient as a function of depth. As can be inferred from this figure, closure pressures exerted on the proppant can exceed 8,000 to 10,000 psi [55 160 to 68 950 kPa]. Closure pressures of this magnitude will crush silica sand severely and in some instances curtail gas production and decrease the life of the well.

Note from Fig. 19.6 that fracture pressure gradients are substantially lower in wells that have been depleted somewhat by production. This observation generally supports Hubbert and Willis' (Hubbert and Willis 1957) theory, which states that the fracture gradient is a function of pore pressure.

Because of the high fracturing pressures, injection-pressure-intensifying equipment is often necessary to perform these treatments successfully. When treatments are performed down tubing, surface injection pressures can range from 6,000 to 12,000 psi [41 350 to 82 740 kPa]. Depending on the actual size of the tubing, injection rates of between 12 and 20 bbl/min [1.9 and 3.2 m^3/min] are usually achieved.

Recently, some operators have been running 4½- to 5½-in. [11.4- to 14-cm] N-80 or P-110 casing from TD to surface so that fracture treatments may be pumped down the casing. In this process, surface injection pressures can be reduced by 3,000 to 4,000 psi [20 690 to 27 600 kPa]. Before a casing-type treatment, the zone is usually perforated with a casing gun, again with pressure-control equipment. After the treatment, most of the fracturing fluid is cleaned up initially by flowing the well up the casing. Then a permanent production packer (with plug) is set by wireline, and the production tubing is run.

The Wilcox Sands are relatively soft, with typical Young's modulus values ranging from 1.0×10^6 to 3.0×10^6 psi [6.9×10^6 to 20.7×10^6 kPa]. As a result, proppant embedment has been recognized as a potential problem in these sands. To alleviate this problem, proppant concentrations in the fracture must reach about 2 lbm/ft^2 [9.8 kg/m^2]. Thus high concentrations of proppant must be pumped during the actual treatment—e.g., 10 to 12 lbm/gal [1200 to 1400 kg/m^3].

Year	Number of Wells Fracture-Treated	Average Total Fluid Volume (gal/well)	Average Total Proppant Volume (lbm/well)
1974	10	39,700	48,700
1975	42	51,100	65,700
1976	11	78,000	153,000
1977	22	86,900	156,500
1978	12	132,400	222,900
1979	25	207,600	365,500
1980	28	179,000	395,000
1981 (through July)	29	101,800	207,000

Table 19.1—Summary of fracture treatments for one operator.

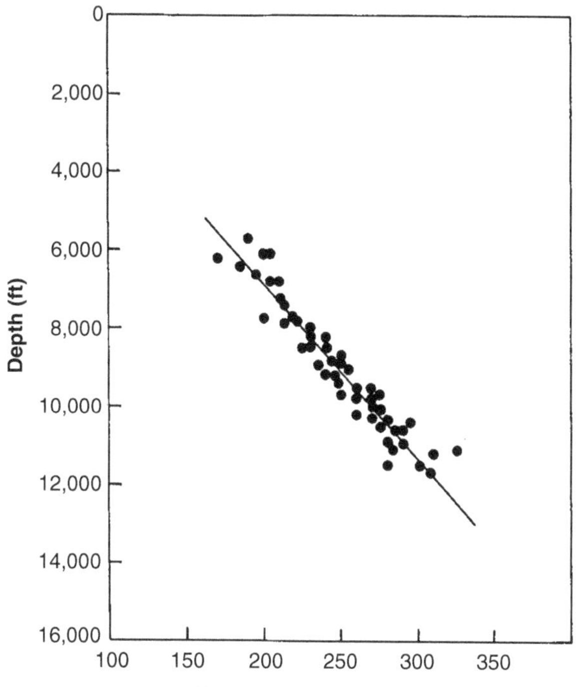

Fig. 19.5—Bottomhole temperature vs. depth.

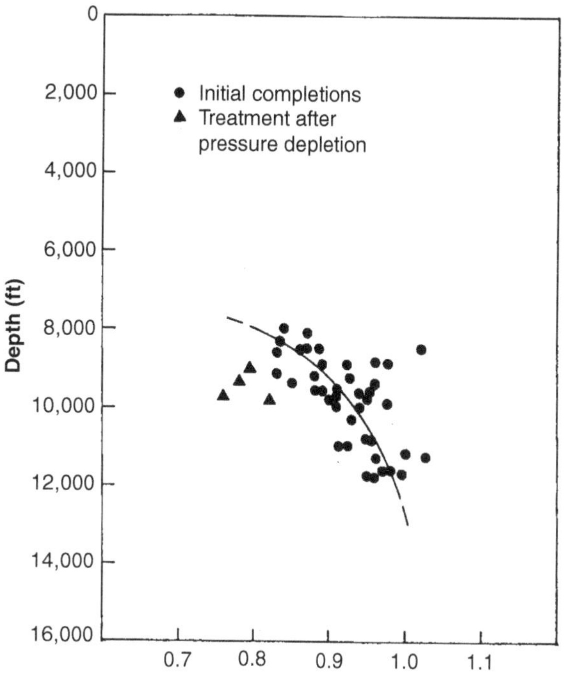

Fig. 19.6—Fracture pressure gradient vs. depth.

Generally, cleanup of the fracturing fluids is not a problem because of the abnormally high pore pressures associated with these reservoirs. In some instances, water-free production can be achieved in only a few days; however, only 25 to 50% of the total load fluid is typically recovered.

Excessive production of fracture sand has been a recurring problem experienced by many operators in this trend. In several instances, fracture sand has been crushed severely in the fracture near the wellbore and has actually been produced from the well. In such cases, overnight reductions in flow rates and flowing tubing pressures have been observed. Such problems seem to occur only in the deeper areas of the trend where excessive closure pressures exist and where the wells have been produced at very high initial flow rates. To help reduce this problem, wells are cleaned up after fracturing on very small choke sizes—e.g., 6/64 to 8/64 in. [0.24 to 0.32 cm]. After recovery of all load fluids, the choke size may be increased. To prevent shock loading of the proppant, however, such increases should be made in 1/64- to 2/64-in. [0.04- to 0.08-cm] increments.

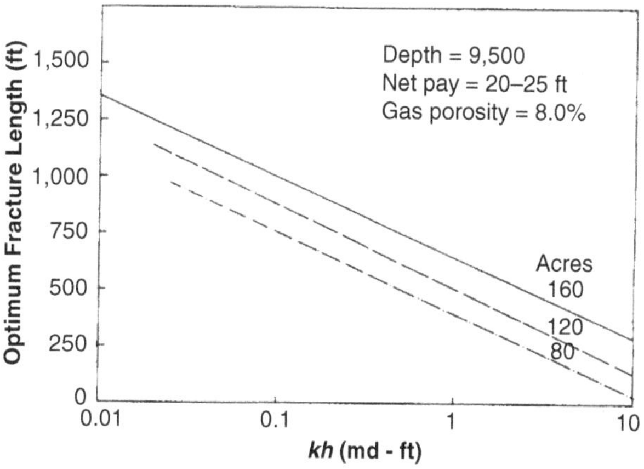

Fig. 19.7—Optimum fracture length as a function of *kh* and drainage area.

Table 19.1 shows that the average treatment size in 1981 decreased from previous years. This is not an indication of the operator's desire to reduce treatment size, but rather reflects an effort to optimize the fracture-length design on the basis of reservoir flow considerations. In late 1980, a fracture-length optimization study was performed in hopes of achieving this goal. The results of this study **(Fig. 19.7)** indicated that an optimum fracture length does exist for a given set of reservoir parameters. Accordingly, all the treatments performed in 1981 were designed to achieve the optimum fracture length, instead of a single arbitrary length. This practice resulted in substantial reductions in total job size, particularly for the higher-quality reservoirs that did not need the larger-volume treatments.

19.6 Reservoir Study

Because of the extremely complex geology associated with the Lobo Trend, it is the rule rather than the exception to offset a well directly and to find the entire Wilcox sequence altered as a result of faulting or erosion. Similarly, it is not unusual for direct-offset wells completed in the same sand member to exhibit dramatically different formation properties. In-situ gas permeabilities can range from 0.0003 to 0.5 md, and in a few isolated cases, permeabilities greater than 1.0 md have been encountered. Total porosity can typically range from 12 to 25%, while water saturations have been found to be as low as 25% and as high as 70%. Total net pay also varies considerably from well to well, with only a few feet encountered in some cases and more than 100 ft [30.5 m] penetrated in others.

Openhole log evaluation and transient test analysis can provide reliable results with respect to reservoir evaluation and projection of future reserves. One complicating factor in these types of projections, however, is the complex geology. Some wells have encountered extremely small fault-block sizes—less than 80 acres [324×10^3 m^2]—which makes the calculation of

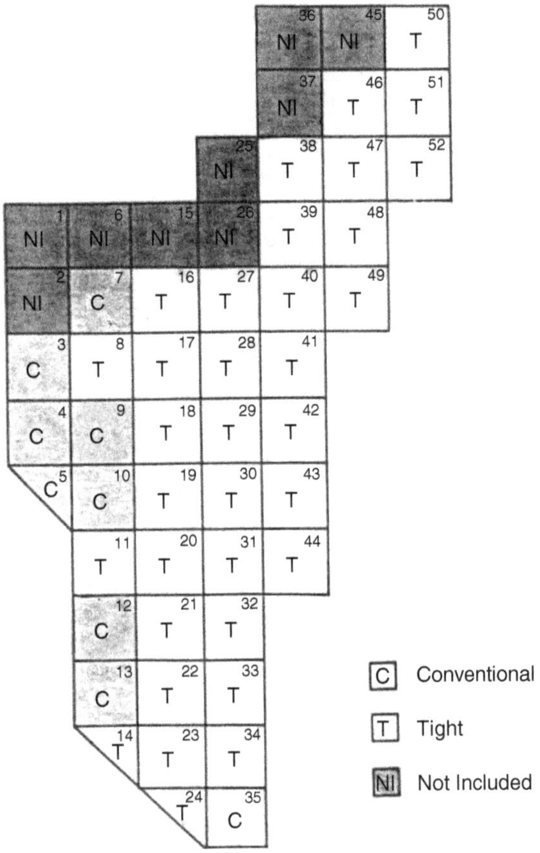

Fig. 19.8—Grid breakup for tight gas classification of Lobo Trend.

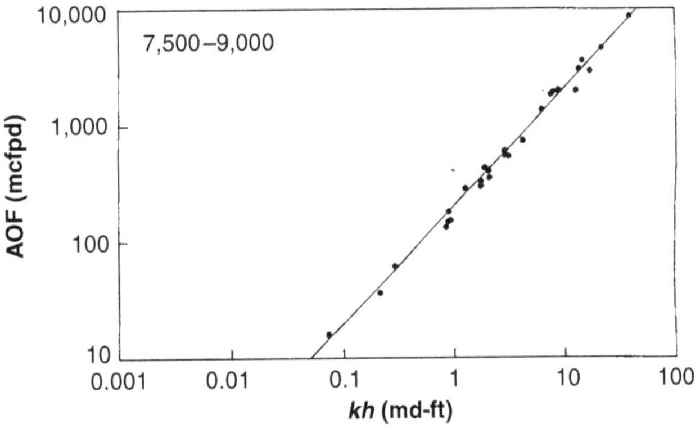

Fig. 19.9—Unstimulated AOF potential as a function of *kh* (7,500 to 9,000 ft).

reserves and projection of future recovery very difficult. Analyzing long-term production and pressure history is the only accurate method of predicting gas in place.

A fieldwide reservoir study was performed in 1980 to determine whether this trend met specific criteria outlined by the FERC for classification as a tight gas formation. For this study, information was received from the operators on nearly half of all the wells drilled in the trend. From this data base, analyses of production and pressure-transient data were performed on approximately 110 wells and were used to determine the average reservoir properties for the entire trend.

Not all the data could be analyzed because of time constraints associated with the study; however, all prefracture data were analyzed. Of the data that were not analyzed, nearly half were from wells located in areas where conclusive data concerning reservoir properties already existed or where the data were not analyzable. In the remaining wells, the data consisted mainly of postfracture multirate flow tests. The interpretation of such data usually requires complicated and time-consuming numerical history-matching techniques. Again, because of time restraints, not all the tests could be analyzed.

Data sources used for permeability determinations were, in order of preference, (1) prefracture pressure-buildup tests; (2) postfracture pressure-buildup tests; (3) prefracture single-rate flow tests; and (4) prefracture four-point deliverability tests.

Because of the complexity of the geology and the increased reservoir quality near the border, the entire trend was subdivided into arbitrarily sized gridblocks **(Fig. 19.8).** The individual blocks were classified as either tight or conventional, depending on whether the wells within that block met the specific criteria of permeability and stabilized flow rate as outlined by the FERC.

This study indicated that approximately 80% of the Wilcox (Lobo) Trend met the specific criteria for classification as a tight gas formation. As a result, the FERC subsequently ruled that the Lobo Trend (with the exception of some acreage near the border in Webb County) would be classified as tight gas. Therefore, any well drilled within this area and completed in the Wilcox (Lobo) Formation after July 15, 1979, would qualify for special price incentives according to the Section 107 pricing schedule.

The performance of these reservoirs can be correlated very well to specific formation properties. Such a correlation has been developed between well productivity (AOF) and the permeability-thickness product (*kh*, md-ft) **(Figs. 19.9 and 19.10).** These correlations were separated into two different depth categories—7,500 to 9,000 ft [2285 to 2745 m] and 9,000 to 11,000 ft [2745 to 3350 m]—mainly because of the significantly different reservoir pressures at various depths. Such variation in reservoir pressure will have a dramatic effect on well performance.

Most of these data were taken from prefracture pressure-buildup tests or multirate flow tests where the permeability and apparent skin factor could be calculated or estimated with confidence. Then the AOF was calculated with the correct permeability and a zero skin factor to represent the natural productivity of the formation. To construct and to use such correlations accurately, one must correct for any apparent skin factor because altered permeability near the wellbore will result in flow rates and pressure drops that are not representative of the true reservoir performance.

19.7 Post-Fracture Evaluation

An evaluation of post-fracture production and pressure transient data is necessary to predict reservoir and fracture parameters, as well as to project future well performance. Conventional analysis techniques (linear-flow theory, type curves, etc.) sometimes can provide reasonable estimates of permeability and fracture length; however, history-matching techniques using numerical simulation are required to analyze complex problems and to predict future performance accurately. As these relatively soft, geo-pressured reservoirs are depleted, the in-situ permeability and fracture conductivity can be reduced by an order of magnitude or more because of compaction and increasing closure stress. Because projections using conventional methods can be made assuming only constant reservoir and fracture properties, it is necessary to use numerical simulation to predict the effects of any variations in properties with pressure.

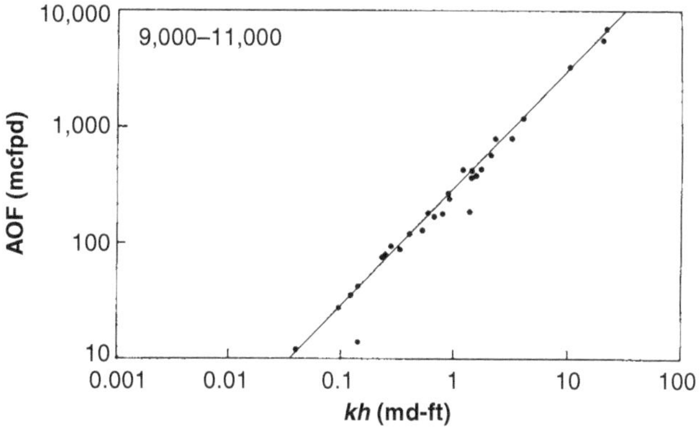

Fig. 19.10—Unstimulated AOF potential as a function of *kh* (9,000 to 11,000 ft).

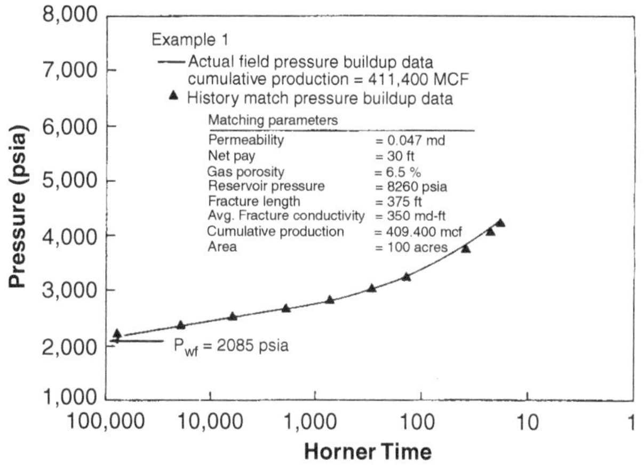

Fig. 19.11—Actual pressure-buildup data and history match (Example 22).

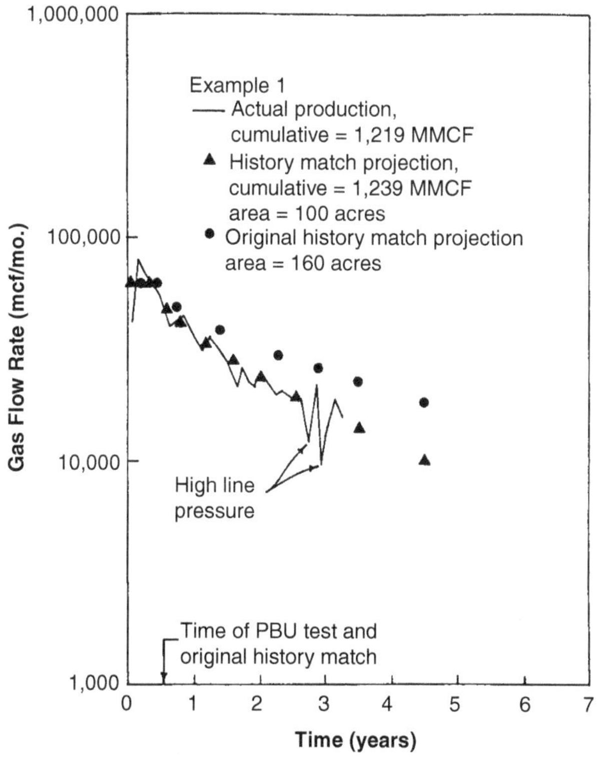

Fig. 19.12—Actual production decline and simulated forecast (Example 1).

Several field examples are included to illustrate the importance of proper evaluation.

19.7.1 Example 1. This well was completed in the Lobo 1 and 6 at an average depth of 10,360 ft [3160 m]. The Lobo was fracture-stimulated with 199,700 lbm [91 × 10³ kg] of sand; however, the treatment screened out during the latter stages. Poor sand transport and loss of pad fluid caused by insufficient gel viscosity were cited as the probable reasons for the screenout. Even though initial production was recorded at flow rates up to 2.5 MMcf/D [71 × 10³ m³/d], a pressure-buildup test was performed to determine what effect, if any, the screenout had on the created fracture properties. The original design was for a propped length of about 750 ft [230 m].

The results of a numerical history match of the post-fracture production and pressure-buildup data **(Fig. 19.11)** indicated that a relatively short propped fracture of 375 ft [114 m] was achieved. The fracture conductivity was determined to be about 350 md-ft [0.1053 md·m] at the time of the buildup, which had been decreased (because of sand crushing and embedment) from an estimated original fracture conductivity of 2,500 md-ft [0.75 md·m].

From the history-matched reservoir and fracture properties, future production was originally projected assuming a 160-acre [648 × 10³-m²] drainage area **(Fig. 19.12).** After the well had been on production for a sufficient period of time, it was apparent that the history-match projection was too optimistic. As illustrated, the production data were later matched with a drainage area of 100 acres [405 × 10³ m²], while the exact same reservoir and fracture properties determined from the pressure-buildup analysis were used.

Example 1 clearly indicates the difficulty in projecting production performance and reserves in a highly faulted area like the Wilcox (Lobo) Trend. For this reason, it is imperative that a geologist's interpretation of fault-block size be interjected into the fracture-length design and the forecasting of production and reserves.

We want to emphasize that post-fracture well test analyses with reservoir simulation history-matching techniques will result only in accurate estimates of permeability, fracture length, and fracture conductivity. These properties are combined to determine the well productivity index. Long-term production-rate forecasting, however, requires knowledge of net gas pay, average porosity, average water saturation, and drainage area. These properties are combined to determine the volume of original gas in place.

The results of post-fracture pressure-transient analysis can be used with confidence to project production performance for only a short period of time. As illustrated in Fig. 19.12, it is usually necessary to reanalyze a well after sufficient production performance is obtained to determine the correct value for original gas in place, from which the drainage area can be determined. The actual drainage area is an important input parameter when one is optimizing the size of a hydraulic fracture treatment.

19.7.2 Example 2. The second example is a well that was completed in the Lobo at a depth of approximately 9,610 ft [2930 m]. The well was fracture-treated in July 1980 with 156,000 gal [590 m³] of a 60-lbm/1,000-gal [7-kg/m³] hydroxypropyl guar fluid carrying 325,000 lbm [147 × 10³ kg] of 20/40-mesh Ottawa Sand. The final fracture gradient was 0.91 psi/ft [20.6 kPa/m].

The well was cleaned up on an 8/64-in. [0.32-cm] choke, and the flowing tubing pressure was maintained above 3,500 psi [24 133 kPa]. Sales production began Aug. 20,

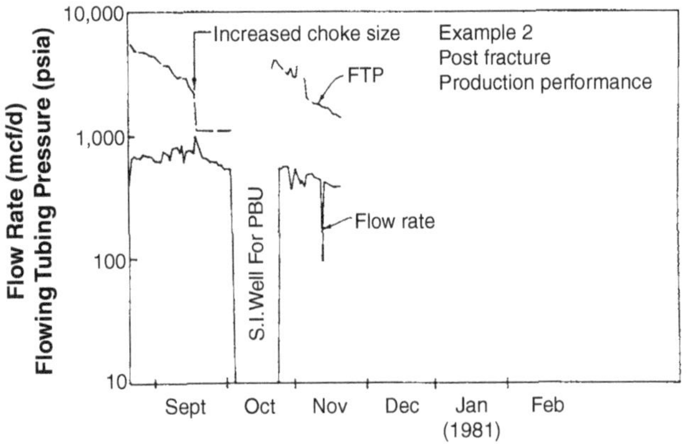

Fig. 19.13—Post-fracture production performance (Example 2).

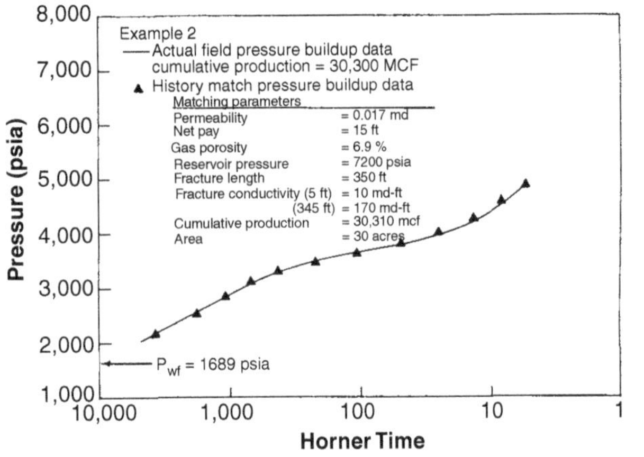

Fig. 19.14—Actual pressure-buildup data and history match (Example 2).

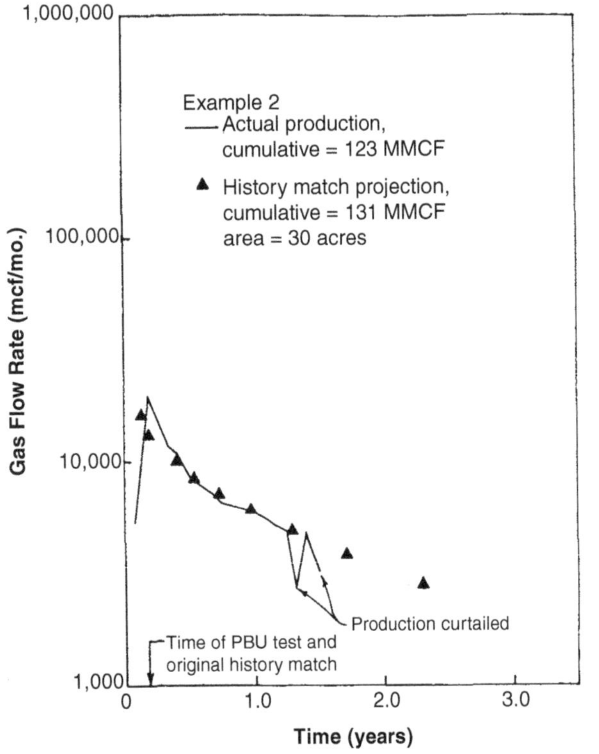

Fig. 19.15—Actual production decline and simulated forecast (Example 2).

1980, and the performance is illustrated in **Fig. 19.13.** Note that the flow rate increased from about 600 to 800 Mcf/D [17 × 10^3 to 23 × 10^3 m^3/d], while the flowing tubing pressure declined from about 5,600 to 2,200 psi [38.6 × 10^3 to 15.2 × 10^3 kPa] during Sept. 1980. At this time, a larger choke size was installed in the well.

After the larger choke was installed, the flowing tubing pressure quickly declined from 2,200 to 1,100 psi [15.2 × 10^3 to 7.6 × 10^3 kPa]. The flow rate increased from 700 to 1,000 Mcf/D [20 × 10^3 to 28 × 10^3 m^3/d] for 1 day, but then rapidly declined to 540 Mcf/D [15.3 × 10^3 m^3/d] within a matter of a few weeks. The rapid rate and pressure decline indicated a potential problem with the well; therefore, a 14-day BHP buildup survey was performed.

Fig. 19.14 presents the Homer graph of the pressure buildup data, as well as the numerical-history-match results. Notice the S-shape of the pressure-buildup curve. From a Horner time of 100 to 5.4, the shape of the pressure-buildup curve on the Horner graph is typical of a hydraulically fractured well. However, the early-time data, from Horner times of 3,600 to 400, indicate an extra pressure drop that could be attributed to damage in the fracture.

As indicated in Fig. 19.14, the post-fracture pressure buildup data were perfectly matched when 0.017 md was used for formation permeability along with a fracture length of 350 ft [106 m]. The first 5 ft [1.5 m] of the fracture, nearest the wellbore, was determined to have a fracture conductivity of only 10 md-ft [0.003 md · m], while the remaining 345 ft [105 m] of fracture was estimated to have a conductivity of 170 md-ft [0.511 md · m]. The unique shape of the Horner graph, coupled with the dramatic change in production history after the larger choke was installed, led to the conclusion that the sand in the fracture near the wellbore had been crushed severely and as a result, the productivity of the well decreased. Use of numerical-history-matching techniques was necessary to quantify the extent of damage to the fracture.

By using these properties, we projected the future production performance (**Fig. 19.15**). From the geophysical data, the geologist believed that this well was producing from a very small fault block; therefore, a 30-acre [121 × 10^3-m^2] drainage area was used in the final analysis of the pressure-buildup data and in the projection of future well performance. The excellent agreement between actual production performance and predicted performance shown in Fig. 19.15 adds a great amount of confidence to the fracture properties that were determined.

19.7.3 Example 3. Example 3 is an excellent well completed in the Lobo at an approximate depth of 8,500 ft [2590 m]. The well was fracture-stimulated with 506,000 lbm [230 × 103 kg] of sand, and no problems were encountered during the treatment. Upon recovery of the load fluid, the well produced at flow rates of between 3.0 and 4.0 MMcf/D [85 × 10^3 and 114 × 10^3 m^3/d]. After the well had produced 442.2 MMcf [12.5 × 10^6 m^3] of gas in approximately 5 months, a BHP-buildup survey was performed specifically for the purpose of calculating fracture length and fracture conductivity.

An excellent numerical history match of the post-fracture production and pressure-buildup data (**Fig. 19.16**) was achieved with an effective fracture length of 750 ft [230 m]. The original design was for a propped fracture length of approximately 850 ft [260 m]. The average fracture conductivity was determined to be 1,000 md-ft [0.3 md · m], which had been decreased from an original conductivity of approximately 3,500 md-ft [1.1 md · m].

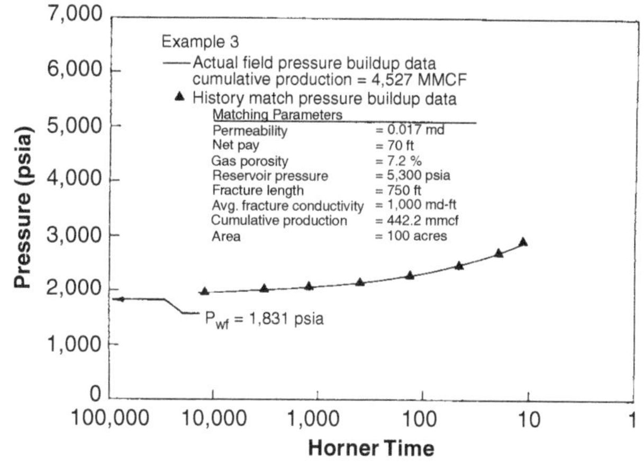

Fig. 19.16—Actual pressure-buildup data and history match (Example 3).

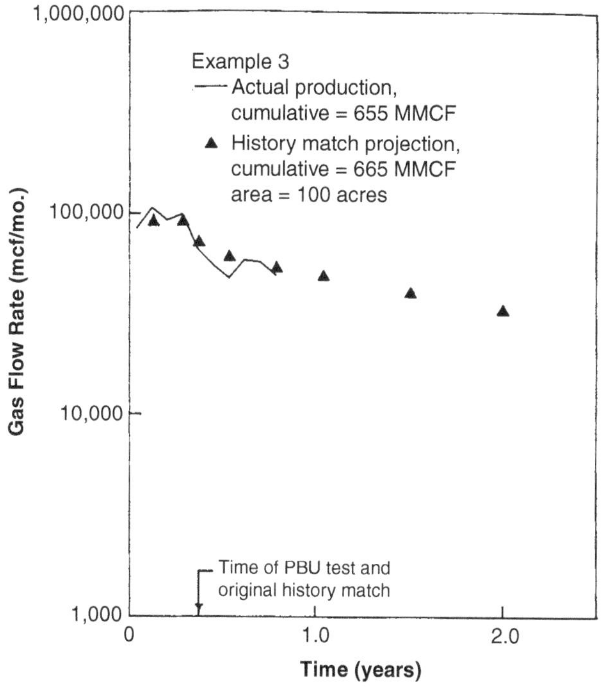

Fig. 19.17 Actual production decline and simulated forecast (Example 3).

From the reservoir and fracture properties determined from the numerical analysis, we forecasted future production **(Fig. 19.17).** The original forecast assumed 160 acres [648 × 10^3 m^2]; however, the area was later revised to 100 acres [405 × 10^3 m^2] on the basis of the actual well performance. This decrease in drainage area had no effect on the original pressure-buildup match in terms of reservoir and fracture properties.

This is an excellent example that illustrates that long, highly conductive fractures (which are essentially equivalent to the designed lengths) can be achieved in the Lobo.

One reason why MHF treatments seem to work in the Lobo is that thick, continuous shale sections separate the individual sands. These shales appear to provide excellent barriers to vertical fracture growth, which, in turn, allows reasonable estimates of fracture height in the design calculations. When a zone does contain excellent barriers to fracture growth and well-engineered stimulation treatments are performed, one can expect to achieve fracture lengths that are reasonably close to the design fracture length.

19.8 Conclusions

1. The Lobo Trend is a series of geo-pressured, low-permeability gas sands, the development of which has been technically successful because of the application of advanced completion, analysis, and stimulation practices. The high degree of technical success, adequate market conditions, and the recent classification by the FERC as a tight gas formation have all contributed to the successful development of the geo-pressured Wilcox (Lobo).
2. MHF is required to increase gas flow rates and ultimate recovery. For each reservoir, a fracture length exists that will result in maximum efficient recovery of the gas in place. This optimum fracture length is a function primarily of the permeability-thickness product and the drainage area.
3. Because of the presence of effective barriers to vertical fracture growth, designed fracture lengths can actually be achieved in the field provided that the treatments are successfully performed. Unsuccessful treatments (screenouts) can result in fracture lengths considerably shorter than the designed lengths.
4. Severe proppant crushing, caused by high closure pressures, is a major problem where excessive drawdown has occurred. Such crushing can reduce fracture conductivity to the point of affecting well performance and reducing ultimate recovery.
5. Good correlations between reservoir pore pressure, bottomhole temperature, and fracture pressure gradients vs. depth can be established for the Lobo Trend.
6. Excellent correlations between the AOF and the permeability-thickness product (md-ft) can be established, provided that a correction is made for apparent skin factor.
7. Numerical simulation and history matching are required to analyze complex problems and to predict future performance accurately.

19.9 Acknowledgments

We thank all the operators in the Wilcox (Lobo) Trend who contributed to this fieldwide study. Without the cooperative efforts of everyone involved, the study could not have been possible.

19.10 References

Cinco-Ley, H., Samaniego-V., F., and Dominguez-A., N. 1978. Transient Pressure Behavior for a Well With a Finite-Conductivity Vertical Fracture. *SPE J* **18** (04): 253–264. SPE-6014-PA. https://doi.org/10.2118/6014-PA.

Cleary, M. P. 1980. Comprehensive Design Formulae for Hydraulic Fracturing. Paper presented at the SPE Annual Technical Conference and Exhibition, Dallas, 21–24 September. SPE-9259-MS. https://doi.org/10.2118/9259-MS.

Conway, M. W. and Harris, L. E. 1982. A Laboratory and Field Evaluation of a Technique for Hydraulic Fracturing Stimulation of Deep Wells. Paper presented at the SPE Annual Technical Conference and Exhibition, New Orleans, 26–29 September. SPE-10964-MS. https://doi.org/10.2118/10964-MS.

Daneshy, A. A. 1973. On the Design of Vertical Hydraulic Fractures. *J Pet Technol* **25** (01): 83–97. SPE-3654-PA. https://doi.org/10.2118/3654-PA.

Fast, C. R., Holman, G. B., and Colvin, R. J. 1977. The Application of Massive Hydraulic Fracturing to the Tight Muddy 'J' Formation, Wattenburg Field, Colorado. *J Pet Technol* **29** (01): 10–11. SPE-5624-PA. https://doi.org/10.2118/5624-PA.

Geertsma, J. and De Klerk, F. 1969. A Rapid Method of Predicting Width and Extent of Hydraulically Induced Fractures. *J Pet Technol* **21** (12): 1571–1581. SPE-2458-PA. https://doi.org/10.2118/2458-PA.

Gringarten, A. C., Bourdet, D. P., Landel, P. A. et al. 1979. A Comparison Between Different Skin and Wellbore Storage Type-Curves for Early-Time Transient Analysis. Paper presented at the Annual Technical Conference and Exhibition, Las Vegas, Nevada, USA, 23–26 September. SPE-8205-MS. https://doi.org/10.2118/8205-MS.

Henke, K. A. 1982. *Origin of Late Paleocene-Early Eocene Wilcox Sandstones, Lobo Trend, Webb and Zapata Counties, Texas.* MS thesis, Texas A&M University, College Station, Texas, USA (December 1982).

Hilty, R. E. 1980. Geological Testimony Regarding the Application of the Wilcox (Lobo) Trend for Classification as a Tight Gas Formation. Paper presented to the Railroad Commission of Texas, Austin, Texas (May 1980).

Hsu, C. H. and Conway, M. W. 1981. Fracturing Fluids for Deep, Hot Formations. *J Pet Tecnol* **33** (11): 2,213–2,218. SPE-9707-PA. https://doi.org/10.2118/9707-PA.

Hubbert, M. K. and Willis, D. G. 1957. Mechanics of Hydraulic Fracturing. In *Transactions of the Society of Petroleum Engineers*, Vol. 210, SPE-686-G, 153–167. Richardson, Texas: Society of Petroleum Engineers.

Kozik, H. G. and Holditch, S. A. 1981. A Case History for Massive Hydraulic Fracturing the Cotton Valley Lime Matrix, Fallon and Personville Fields. *J Pet Technol* **33** (02): 229–244. SPE-7911-PA. https://doi.org/10.2118/7911-PA.

Lee, W. J. and Holditch, S. A. 1981. Fracture Evaluation With Pressure Transient Tests in Low-Permeability Gas Reservoirs. *J Pet Technol* **33** (09): 1,776–1,792. SPE-9975-PA. https://doi.org/10.2118/9975-PA.

Nordgren, R. P. 1972. Propagation of a Vertical Hydraulic Fracture. *SPE J.* **12** (04): 306–314. SPE-3009-PA. https://doi.org/10.2118/3009-PA.

Wyman, R. E., Holditch, S. A., and Randolph, P. L. 1980. Analyses of an Elmworth Hydraulic Fracture in Alberta. *J Pet Technol* **32** (09): 1,621–1,630. SPE-7935-PA. https://doi.org/10.2118/7935-PA.

SI Metric Conversion Factors

acre × 4.046 873	E+03 = m^2
ft × 3.048*	E–01 = m
ft^3 × 2.831 685	E–02 = m^3
°F (°F-32)/1.8	=°C
md-ft × 3.008 142	E–04 = md · m
mile × 1.609 344*	E+00 = km
psi × 6.894 757	E+00 = kPa

*Conversion factor is exact.

Chapter 20

Fracture Stimulation of a Horizontal Well in a Deep, Tight Gas Reservoir: A Case History From Offshore The Netherlands

W. E. Baumgartner, Nederlandse Aarodolie Mij. B. V.; **Jacob Shlyapobersky,** Shell Research B. V.; **I. S. Abou Sayed,** Mobil E&P Technical Center; and **R. C. Jacquier,** Halliburton Services B. V.

 Symposium paper SPE 26795 was presented at the Offshore Europe conference held in Aberdeen, Scotland, United Kingdom, 7–10 September 1993.

SPE 26795 presents a case history of stimulation of a horizontal well with two hydraulic fractures in the Upper Slochteren tight gas sandstone of the Permian Rotliegend Group located offshore of The Netherlands. Baumgartner et al. open the case history by way of the following:

- A concise introduction of the field and development strategy
- Discussing the economic need for improved well design based on the field production of two offset wells;
- Drilling a horizontal well with two hydraulic fractures after single-well analytical modeling, full-field simulation, and economic analysis
- Summarizing completions and stimulation design in depth, including a prefracture injection test, a lab test of fluid systems, and a mini-frac
- Summarizing the execution and onsite design using parameters derived from the analysis of mini-frac data; and at the end, summarizing gas production rates and mentioning severe proppant back-production (11%) during a 12-day cleanup operation

From their study, Baumgartner et al. conclude

- Stimulation technology is implemented successfully and may deployed to develop other poor productivity and marginally economic fields in the region
- The importance and need of proppant slugs in placing the treatment
- Extremely short perforation intervals should be used to minimize multiple fractures
- Real-time downhole pressure measurement should be used when possible

20.1 Abstract

AME-204 is a horizontal well with two hydraulic fractures draining a Rotliegendes sub-block of the Ameland East Reservoir located offshore The Netherlands. The reservoir in this part of the field is of poor quality, with a strong diagenetic impairment caused by kaolinite leading to low matrix permeability (0.2–1 mD). Economic development of this part of the field is hampered by the low well productivities.

This paper outlines the screening process which led to the decision to develop this sub-block with a multiple-fractured horizontal well and describes the well configuration chosen. Completion and stimulation designs are discussed in some depth, including selection and laboratory testing of fluid systems and fracture initiation. Additionally, the use and design of pre-stimulation injection tests and proppant pills to reduce near wellbore restrictions are shown.

Practical considerations in the execution of stimulation treatments in horizontal wells are covered. Actual and predicted production performances are compared. The actual productivity improvement achieved over a deviated unfractured well is estimated as a minimum factor of 4 at only 30% higher cost. The paper concludes with recommendations for similar projects in the future and also looks at the impact on the development of hitherto economically submarginal fields.

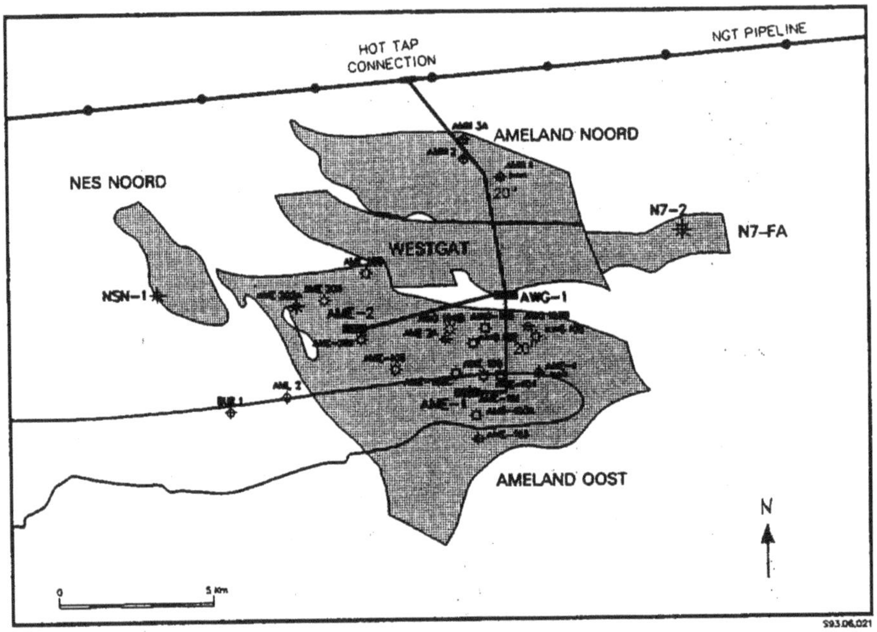

Fig. 20.1—Overview of the Ameland Fields and facilities.

20.2 Introduction

The Ameland Field, operated by NAM for its partners EBN and Mobil, is the third largest gas field in The Netherlands, with a GIIP (gas initially in place) of some 58.5×10^9 m^3. The gas is produced with wells from three separate locations, two offshore, and one onshore on the Ameland island **(Fig. 20.1).** The field is close to the Waddenzee area in an environmentally extremely sensitive location.

All Ameland Reservoirs belong to the Permian Rotliegend Group. The productive formation is the Upper Slochteren Sandstone, which has a gross thickness varying between 90 and 120 m, at a depth of 3200 to 3500 m. Original reservoir pressure was 550 bar, but the field is now depleted to 190 bar. Static bottomhole temperature is 113°C. The Rotliegendes mud/silt/sandstone sequence in the Ameland-East Field can be subdivided into six distinct reservoir units, four of which contain mainly fine-grained sandstones and two mainly tight siltstones. The field is divided laterally into six major, southward dipping, fault bounded blocks **(Fig. 20.2).** Most of the AME Field production comes from wells drilled in the highly productive central blocks E15 and Ell. Well productivity, at 50 bar drawdown, ranges from 3.5 to 5.0 × 10^{6}m^3/d in the central blocks, to 0.03×10^{6}m^3/d in flank blocks.

20.2.1 Ameland Development Strategy. In the first 5 years of production, the contractual capacity of the Ameland Field could be maintained with wells drilled in the highly productive central blocks only. It was initially assumed that the connectivity of reserves in flank blocks was adequate to ensure effective drainage towards the pressure sink in the central area. However, production data showed that material balance GIIP was substantially lower than volumetric GIIP and the flank blocks were not being drained fully.

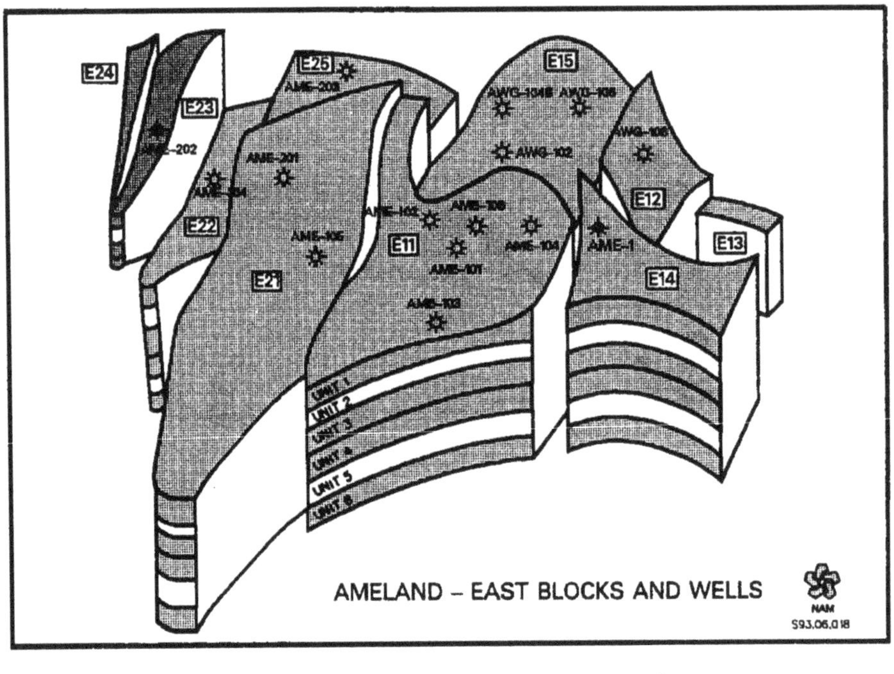

Fig. 20.2—Ameland East Field blocks and wells.

With the progressive decline in reservoir pressure and well production, a strategy had to be developed for positioning future wells that were needed to maintain production capacity. Reservoir simulation studies showed that further infill drilling in the central blocks was not economically justified and a development plan for the less productive western flank blocks was prepared. This plan comprised originally the tie-back of an existing, suspended appraisal well, AME-201, as a producer to a new mini-satellite platform (an unmanned minimum facilities installation) and the drilling of a new well, AME-203, from the same surface location as well as the option for further infill drilling. During late 1991 it became clear that an infill well, AME-204, would be required during 1992. This well was tentatively planned as a deviated fracture stimulated well drilled into the western boundary block E 22.

20.2.2 Why a Horizontal Well? Well AME-201, which had been fracture stimulated in 1978 performed as per expectation since coming onstream in 1991. A new well, AME-203, which was drilled into the northern most flank block E25, showed highly disappointing performance with actual productivity of less than 25% of the predicted value.

Results of AME-203 led to a re-evaluation of the permeability in the western flank block area. Average porosity and net thickness were found as predicted but permeability was found to be very low with average core plug air permeabilities in the main reservoir units of some 0.2 mD. This prompted a change in the plans for the infill well AME-204. The main factors which qualified AME-204 as a horizontal multiple fractured well candidate were:

1. Clear structural picture available from 3D seismic
2. High GIIP to justify a "2 for 1" concept
3. Good seal from bottom water by the underlying Ameland claystone
4. Simple well path

A detailed evaluation of well options was carried out. Two well options were screened: A deviated well with one hydraulic fracture and a horizontal well with two hydraulic fractures. Due to the low ratio of vertical to horizontal permeability, an unfractured horizontal well, which had initially been considered, was not economically viable. A single well analytical model was used to establish productivities, which were fed into a full field simulator and incremental production profiles were established for all cases. These were then used as input into an economic model. In all cases a horizontal well with two hydraulic fractures emerged as the best project.

20.3 Completion Design

The main challenge in well and completion design was to realise the well productivity which the initial study had indicated. For operational reasons, two major principles were adopted:

1. To keep the level of mechanical complexity as low as possible
2. To minimise deviation from the standard well completion for the Ameland Field

This led to the completion shown in **Fig. 20.3.** Five-inch 13 Cr production tubing was chosen to accommodate the expected production rates. To minimise drilling problems, an 8 3/8 in. hole size was selected for the 500 m horizontal section. In view of the planned stimulations, a 7 in. liner was cemented over the reservoir section. The fracture stimulations were carried out through the final production tubing rather than a dedicated workstring. This avoided open perforations in this high pressure gas well during the completion and minimised the need for tripping.

Zonal separation for the two fracture treatments was achieved with a permanent production packer/landing nipple assembly in which a plug could be set on coiled tubing. This option was preferred to the mechanically somewhat more complex dedicated horizontal well systems, as described in Damgaard et al. (1992), since only two hydraulic fractures were planned. The need for a good cement bond over the intervals which were to be fracture stimulated was realised to be a critical success factor at the planning stage.

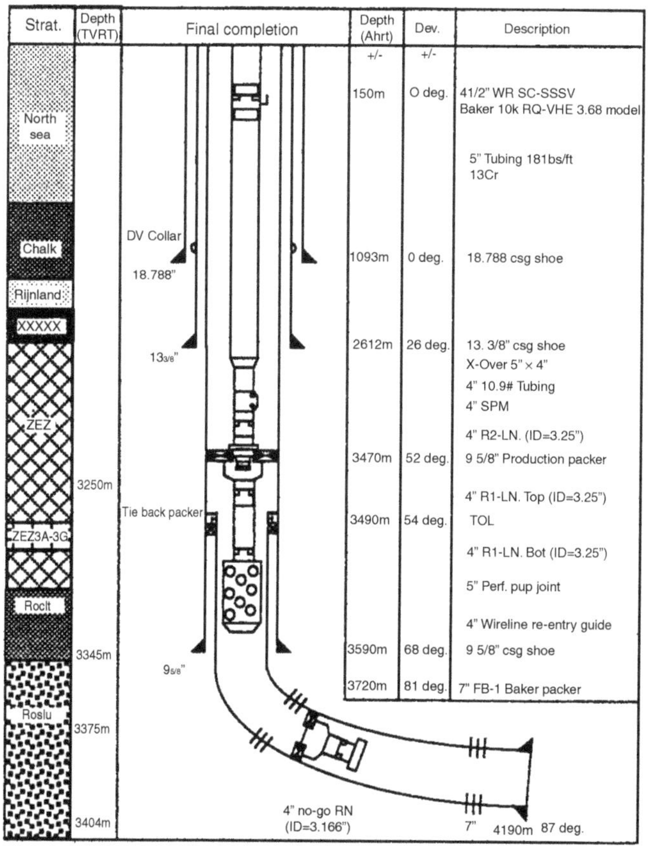

Fig. 20.3—Completion schematic of well AME-204.

20.3.1 Wellbore/Fracture Orientation. The effects of the orientation of the wellbore to the in-situ stress field have been described in many laboratory studies and theoretical papers (El Rabaa 1989; Abass et al. 1996). Experiences in conducting propped fracture stimulation treatments from horizontal wells in chalk reservoirs have been published recently (Weijers et al. 1994). Abass et al. (1996) have described the relationship of stress and wellbore orientation angle to fracture initiation pressure. They demonstrated that as the horizontal wellbore approaches the direction of the minimum principal stress the fracture initiation pressure increases. In some cases this has manifested itself in pressures which prevented the stimulation from occurring due to the pressure constraints on the surface and completion equipment. Furthermore, the subsequent reorientation of the frac may cause a severe choking effect and thus not only hamper the placement of proppant into the well but also reduce productivity considerably.

For the AME-204 well indications of drilling induced fractures on a facies log which was run in nearby well AME-203 and general data for the area both suggested a NW-SE fracture direction. Choice of target location resulted in a well trajectory at an angle of less than 15 degrees to the assumed fracture direction. Nevertheless, in view of the high uncertainty of the fracture direction the following provisions for a bigger angle between fracture and well were made:

1. Small perforated interval to minimize/exclude the formation of multiple fractures
2. Formation breakdown at the highest possible rate to get a smooth turn and a big radius of curvature in case of frac reorientation
3. Pumping of proppant slugs (small proppant stages at up to a maximum of 2 lb/gal and overdisplaced from the wellbore)

The beneficial effect of proppant slugs has been described previously (Cleary et al. 1993). This technique is believed to work due to the following three processes:

1. Erosion/cleaning up of plugged perforations
2. Plugging off multiple fractures
3. Removing tortuosity caused by fracture reorientation

In a horizontal well, the application of this technique is considered essential, particularly if the direction of the minimum horizontal stress is not fully known.

20.3.2 Perforating Considerations. The perforating strategy was designed with the following objectives:

1. Through tubing perforating to avoid running a completion in a high pressure gas well with open perforations
2. Enhance initiation of a single frac
3. Minimize pressure drop at the perforations to minimize treatment pressures and non-Darcy skin effect

Points 2 and 3 conflict with each other: To avoid multiple fractures, the perforated interval should be as short as possible; however, to avoid excessive pressure drop a certain number of perforations is required. This resulted in the selection of a 0.6-m interval perforated with 2 7/8 in. 60° phasing guns at a shot density of 18 spf. The high density was achieved by perforating the same interval three times with a 6 spf gun employing a selective switch technique. The guns were run on electric lines conveyed by coiled tubing ("stiff wireline").

20.4 Fluid System

Due to the requirement for ultra high fracture conductivity imposed by the reduced perforated interval and the anticipation of non-darcy effects, available fluid systems were reviewed to determine an optimum choice. While several crosslinked fluid types met the viscosity requirements a delayed Borate system was chosen due to its capability to:

1. Provide the highest possible fracture conductivity (Parker 1990)
2. Provide minimal interaction of crosslinker and breaker with resin coated proppant (Thesing 1993)
3. Provide a minimal friction pressure drop with the delayed crosslinker (Tan et al. 1992)

Encapsulated breaker was used in the treatments at concentrations up to 10 lb / 1000 gal. The high concentration of breaker was used in the proppant laden stages to increase the achieved conductivity of the proppant pack.

Due to the environmentally sensitive Ameland area the stimulation fluid was not allowed to contain any hydrocarbons. Water based liquid gel concentrate was used to generate the HPG polymer. Diesel fluid loss control was not an option for these treatments and silica flour was available as a contingency had the mini-frac indicated a leak-off control problem.

20.4.1 Laboratory Testing. The need for maximum conductivity generated a study of the delayed Borate fluid system's viscosity profile versus breaker loading. A computer simulator was used to calculate the temperature profile for each stage of the treatment. The temperature profile to show the heat-up rate of the fluid while pumping and the heat-up rate occurring until formation closure is shown in **Fig. 20.4.** By calculating the exact heat-up profile of each fluid stage, optimum breaker loadings were derived from static break tests and Fann Model 50 tests. A combination of these testing methods was required to determine breaker loading since proppant cannot be included in the tight clearances of the Fann Model 50 viscometer bob and sleeve.

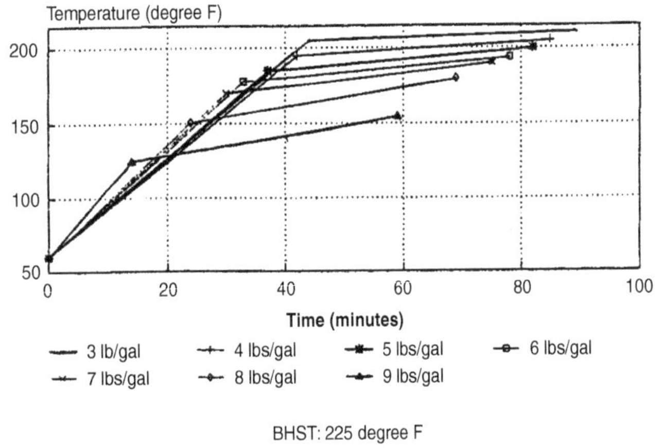

Fig. 20.4—Simulated heat-up profile for the various stages during the frac treatment.

20.4.2 Static Break Tests. An interesting observation made during the static break tests was the capability of the fluid to thermally break when exposed to bottomhole static temperature - regardless of the presence of breaker. Based on the generated temperature profiles encapsulated breaker loadings were optimized for the 5 through 9 lb/gal proppant stages.

The stability of the delayed Borate fluid system was also determined with respect to the concentration of high pH buffer, since the crosslinking is a pH controlled process. A reduction in the concentration of the pH buffer produces a fluid of lower stability. Therefore stringent pH control during the treatment was essential to the performance of the delayed Borate fluid system.

20.4.3 Fann Model 50 Tests. The rheology of Borate fluid systems is difficult to measure and modifications as presented by Gdanski et al. (1991) were used with the Fann Model 50C viscometer. These modifications include a constant shear rate due to the rehealing nature of the Borate fluid and the elimination of the void area normally located below the standard bob.

The effect of temperature on the pad fluid stage is presented in **Fig. 20.5.** The delayed Borate system exhibited a decrease in viscosity at 113°C (235°F) and stable viscosity at 80°C (175°F).

To include the effects of the resin coated proppant, base fluid was prepared by placing 9 lb/gal resin coated proppant in Ameland seawater and heating the mixture for 15 hours at 95°C (203°F). Decanted fluid from this process was used for gel preparation.

20.4.4 Magnesium Precipitation in Sea Water. The high concentration of pH buffer required to stabilize the delayed Borate fluid system caused concern regarding potential precipitation of large amounts of metal-ions from the sea water in the fracture. These ions, particularly magnesium, will fall from solution at a fluid pH of about 10.5. A test was set up which verified the reversibility of this process thereby confirming that the magnesium salts will go back into solution once the pH is lowered by production.

20.4.5 Proppant and Resin. Intermediate strength ceramic proppant was chosen on the basis of the expected closure stress. Due to the size of the perforation entrance hole, 20/40 mesh was selected. Since potential back-production of proppant gave cause for concern, mainly in view of the limited surface facilities on the unmanned platform, it was decided to use resin-coated proppant. Laboratory tests were performed on two commercially available phenolic resins to assess compatibility with frac fluid, compressive strength, and elastic properties. The tests showed significant differences with regard to unconfined compressive strength and elasticity of the proppant pack. Resin 1 was shown to form a much stronger, but far more brittle, proppant pack. Resin 2 showed superior elastic properties making it more suitable for cyclic loading on the proppant pack, which is being caused by opening and closing on wells and therefore to prevent proppant back production in the long term. This resin, however, showed some interaction with fluid breakers, rendering oxidizing breakers inactive. However, since it was planned to use encapsulated breaker and furthermore the fluid showed a high degree of thermal degradability, Resin 2 was chosen in view of its superior elastic properties.

20.5 Fracture Design

Prior experience with hydraulic fracturing in the Ameland East and the surrounding fields was used for the design of the hydraulic fractures. Fractures are believed to be contained within the Upper Slochteren Sandstone by the over- and underlying Ten Boer and Ameland Claystones, respectively. Since all the wells were deviated, no temperature logging has been attempted. Stress contrasts within the reservoir exist but were not expected to form a barrier, at least in the case of no significant differential depletion. The average Young's modulus is approximately 270,000 bar (3.9×10^6 psi). Leak-off coefficients established from prior fracturing were in the range of 0.003 - 0.005 ft/$\sqrt{}$min.

20.5.1 Fracture Length vs. Fracture Width. In a conventional well (i.e., vertical or deviated) the optimum fracture length is mainly governed by the level of matrix permeability. A dimensionless fracture conductivity (FCD) between 10 and 30 (Holditch 1979) has generally been accepted as an optimum range. In a horizontal well, due to the short perforated intervals, fracture conductivity has a far more dominant role than in a conventional well. To investigate this aspect further, a simulation study with an analytical inflow model was performed for matrix permeabilities of 0.25 mD (expectation case) and 0.9 mD, which is the average permeability of another nearby well, AME-202. For each of these permeabilities, productivities were calculated for fracture halflengths of 15, 30 and 60 m as a function of fracture width. Fracture permeability was assumed constant at 100D (equivalent to a retention factor of 27%). The results of these simulations are shown in **Fig. 20.6.** Due to the strong non-Darcy flow effects at the short perforated interval, the production rate is extremely sensitive to fracture width.

From Fig. 20.6 it becomes clear that optimization on the fracture geometry should be focused toward maximizing the propped width of the fracture, in particular near the wellbore, rather than fracture length. To achieve this, a packed fracture design ('Frac & Pack') is required. The design of such a treatment is done in two stages: (1) tip screenout (TSO) stage is calculated with conventional tools and

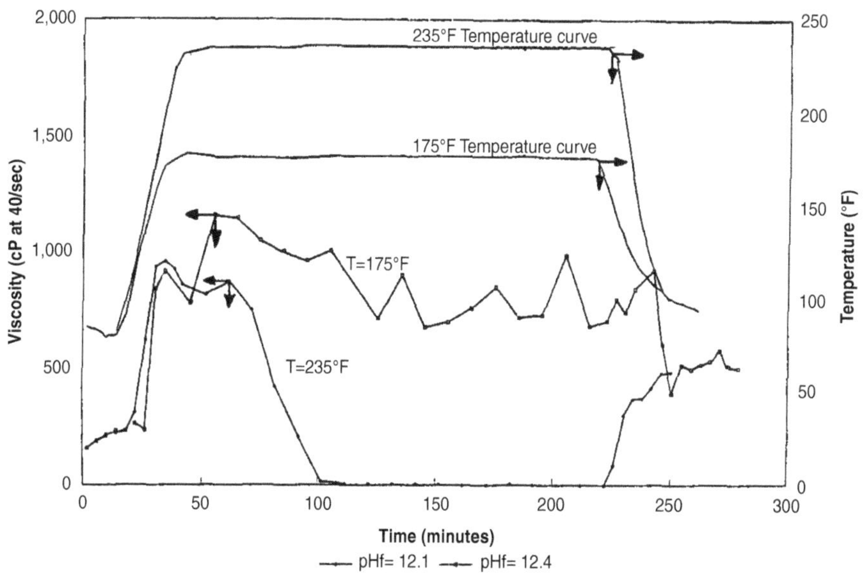

Fig. 20.5—Effect of temperature on pad fluid.

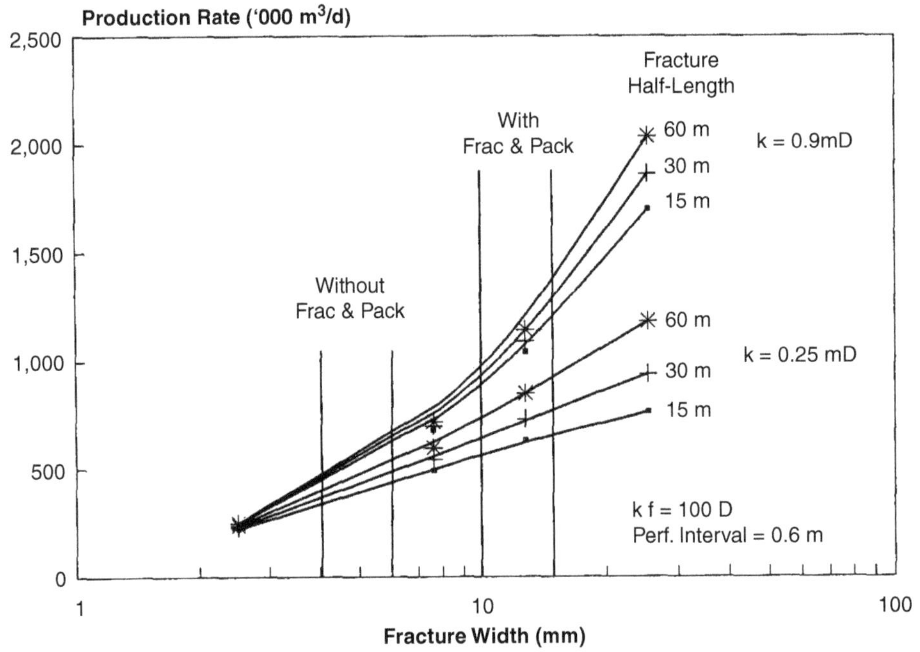

Fig. 20.6—Productivity as a function of fracture width for various fracture lengths.

(2) a fracture inflation and packing (FIP) stage, with proppant at the highest possible concentration, is added thereafter (Roodhart et al. 1993). A detailed description of this type of design, although from a slightly different angle, has been given by Martins et al. (1992).

Packing of the fracture also eliminates two other possible problems in a fractured horizontal well: proppant convection and settling. With a conventional design, if closure times and fluid break times are estimated incorrectly, the possibility of proppant settling out in the frac prior to closure exists: In a horizontal well this could mean that the propped fracture (or part of it) is effectively disconnected from the wellbore. By packing the fracture, the probability of this happening is reduced significantly.

An overpressure calibrated design model (Shlyapobersky et al. 1988a, 1988b) was chosen for the design of the TSO stage. With this model, leak-off coefficient and fracture overpressure are determined in a minifrac prior to the main treatment. Maximum proppant concentration was limited to 9 lb/gal to minimise the risk of a screenout. The FIP stage was designed such that the final slurry concentration in the fracture would not exceed 20 lb/gal.

20.5.2 Mini-Frac Programme. Based on the above considerations, a stimulation programme with the following highlights was prepared:

1. Formation breakdown with low polymer concentration linear gel pumped at highest possible rate.
2. Pump-in/shut-in tests with linear gel to establish instantaneous shut-in pressure (ISIP) and fracture closure pressure (FCP). In case this test did not show the FCP clearly, a pump-in/flow back test was programmed as a contingency.
3. Proppant slugs at 1-2 lb/gal in case of experiencing near-wellbore problems.
4. Mini-frac with cross-linked gel at 15% of main treatment volume.

For the overpressure calibrated design, leak-off coefficient and overpressure are essential design parameters. Therefore, a mini-frac for each zone was considered essential.

20.6 Execution

20.6.1 Cement Bond Evaluation and Perforation. A cement bond log was run in the 7 in. liner since a good cement bond over the perforated interval had been identified as a pre-requisite for successful fracture stimulation. It was used, together with the FEWD (Formation Evaluation While Drilling) interpretation, as the basis for selecting the perforated intervals. Following the running of the completion, the far zone was perforated with 2 1/8 in. 4 spf guns, after various attempts to feed the 2 7/8 in. gun through the separation packer had failed. To compensate for the smaller perforation diameter, the length of the perforation interval was increased to 0.9 m, which was considered the maximum to stay within the criteria for the length of the perforated interval not exceeding 4 times the diameter of the hole (El Rabaa 1989).

20.6.2 Mini-Frac Implementation and Pressure Analysis. During the breakdown stage with linear gel, unexpected high treatment pressure was encountered. The observed pressure and rate (Area 1 in **Fig. 20.7**) could have made the planned stimulation impossible to achieve. There are three possible reasons for this high pressure at the relatively low injection rates:

1. Perforation restriction (number, size and penetration)
2. Multiple fracture initiation
3. Fracture reorientation (near wellbore tortuosity)

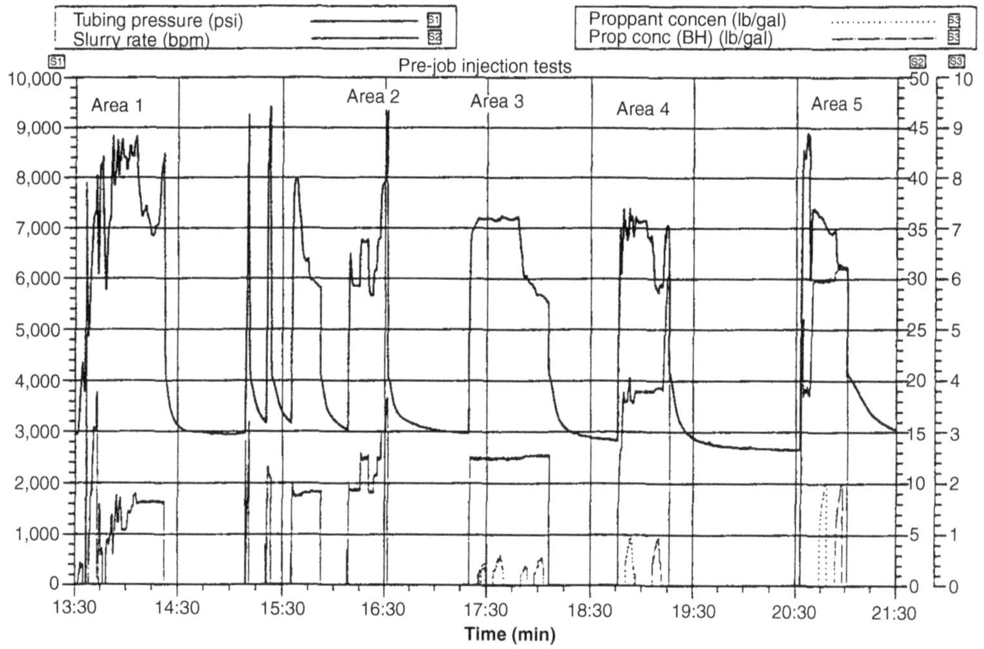

Fig. 20.7—Treatment rate and pressure record for AME-204, far zone, injection tests, proppant slugs, and mini-frac.

To identify the main reasons for this high pressure, a rate variation test was conducted **(Fig. 20.8).** A conventional step rate test, starting from a low rate, was consciously omitted due to the possible danger to create multiple fractures as seen in other areas. The pressure and rate records are shown in Fig. 20.8. According to Cleary et al. (1993) and Martinez et al. (1993) in a rate variation test, the pressure increase (Δp) is related to the injection rate increments (Δq) as follows:

$$\Delta p \,\alpha\, (\Delta q)^{\alpha}. \quad (20.1)$$

The power α depends on the restriction; for perforation dominated restriction, $\alpha = 2.0$, whereas for near wellbore curving fractures and multiple fracture initiation, $\alpha = 0.5$. From the data from Fig. 20.8, the pressure increments were proportional to square root of injection rate as shown in **Fig. 20.9.** We concluded that the observed high treatment pressure was caused by multiple fracture initiation and/or near wellbore tortuosity. To overcome this near wellbore effect and multiple fracture initiation, four different stages of proppant slugs were used. The first two stages (shown in **Fig. 20.10)** were pumped at 0.5 lb/gal and caused a surface pressure reduction of more than 110 bar (1600 psi). It is important to note that the second stage did not cause a large pressure reduction

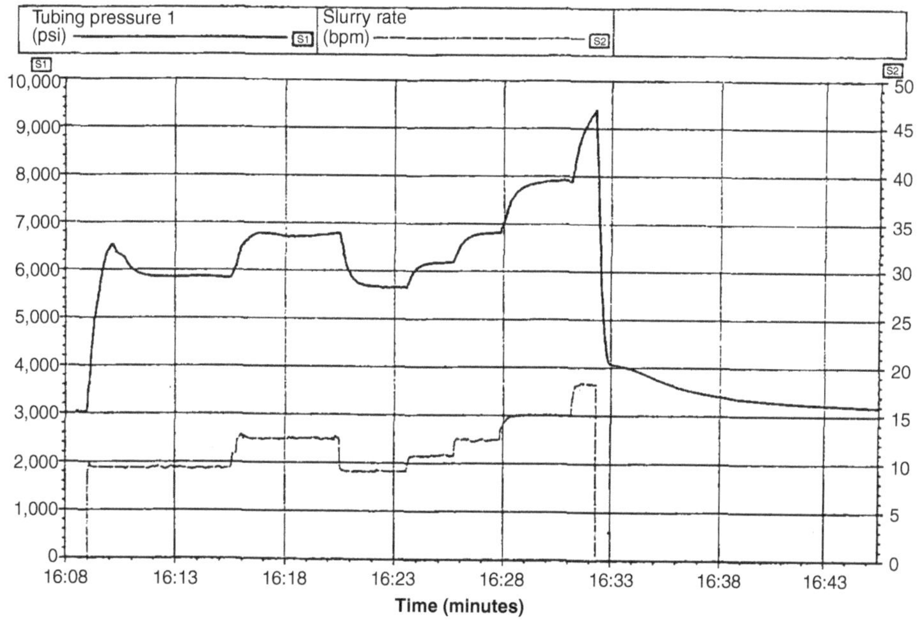

Fig. 20.8—Treatment rate and pressure record for AME-204, far zone rate variation test.

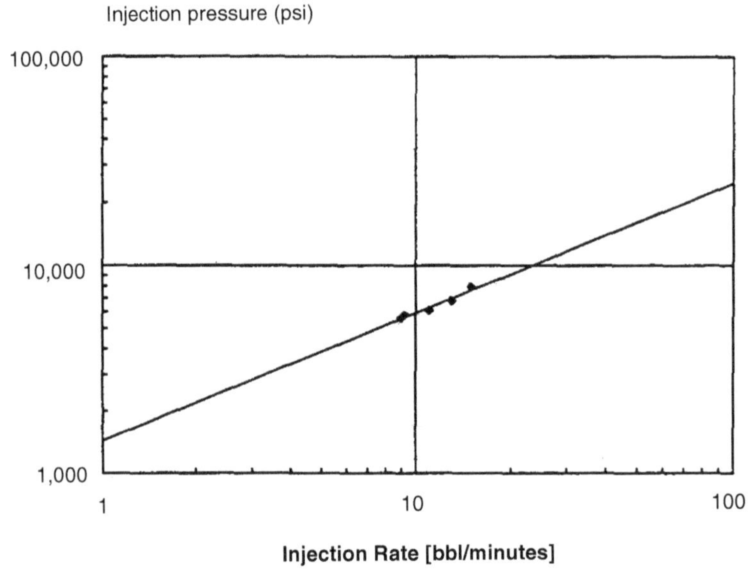

Fig. 20.9—Pressure vs. injection rate for rate variation test.

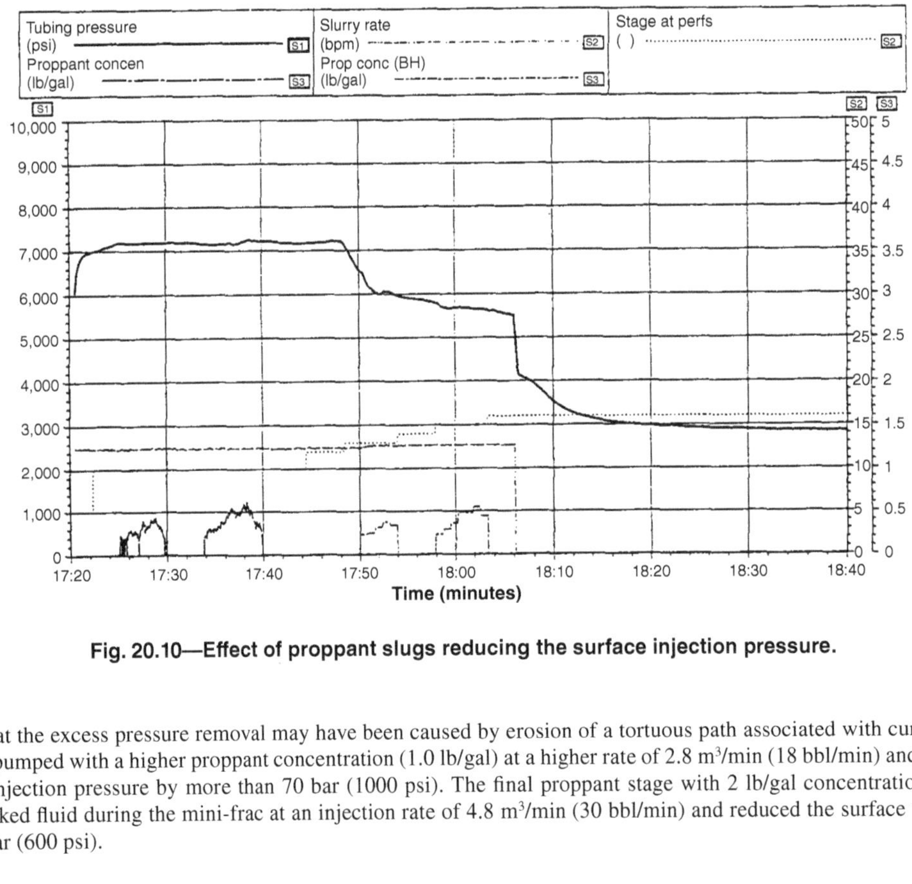

Fig. 20.10—Effect of proppant slugs reducing the surface injection pressure.

indicating that the excess pressure removal may have been caused by erosion of a tortuous path associated with curving fractures. Stage 3 was pumped with a higher proppant concentration (1.0 lb/gal) at a higher rate of 2.8 m^3/min (18 bbl/min) and again reduced the surface injection pressure by more than 70 bar (1000 psi). The final proppant stage with 2 lb/gal concentration was pumped with crosslinked fluid during the mini-frac at an injection rate of 4.8 m^3/min (30 bbl/min) and reduced the surface treatment pressure by 41 bar (600 psi).

20.6.3 On-Site Final Design. The on-site final design was carried out with the parameters derived from analysis of the mini-frac data. A pumping schedule for the TSO stage was calculated and a rather conservative FIP stage of 47.6 m^3 (300 bbl) at 9 lb/gal was added. This resulted in a total of 250,000 lb of proppant and a total slurry volume of 524 m^3 (3300 bbl). Since only surface treatment pressures were available, a more aggressive FIP stage was not considered due to the danger of screening out and leaving excessive amounts of proppant in the tubing. The far interval was treated as per programme and subsequently separated by setting a plug in the separation packer and the near interval was perforated.

Following a short flow test in which the near zone was shown unable to produce at a drawdown of about 200 bar, fracturing operations commenced. None of the high treatment pressures seen in the first treatment were noticed. Some modifications to the programme were made: the breakdown was programmed at 10 bbl/min only in order to achieve a clean breakdown in the first attempt without tripping the overpressure shut-down system. Only one proppant slug was pumped but no response on the treatment pressure was seen. Overpressure and leak-off coefficient were on the same order as in the first treatment. Design for the main treatment was very similar to the treatment on the far zone with a higher treating rate at 7.9 m^3/min (50 bbl/min) due to the lower pressures. The TSO stages of the main treatment were pumped with a slightly too low proppant concentration due to a blockage in the proppant supply. Entering the FIP stage, a rise in surface pressure was noticed and in the absence of downhole data an oncoming premature screenout was suspected and displacement was called. The surface treatment pressure dropped again showing that the signs had been misinterpreted. This incident highlighted the need for real time bottomhole pressure data for successful packed fracture treatments.

20.7 Production Performance

After a brief shut-in period to cure the resin, cleanup operations commenced. A maximum gas rate of 2.0×10^6 m^3/d (70.8 MMScf/d) was recorded during the cleanout period and proppant-free production was achieved after 12 days at 1.5×10^6 m^3/d (53 MMScf/d) at 340 bar FTHP, at a sandface drawdown of approximately 50 bar. Although still in transient flow, this exceeded the expectation of 0.9×10^6 m^3/d (32MMscf/d) at 150 bar drawdown significantly. The salinity of the produced liquid indicated that formation water was being produced. The water/gas ratio at the end of the clean-up was around 100 m/10^6 m^3 (18 bbl/MMscf). Fluid samples taken over a period of 9 to 19 days following the treatments showed a decrease in fluid pH from 7.1 to 3.7.

A total of 20.8 tons of proppant corresponding to 11 % of the total proppant pumped was produced back during the cleanout period. The high proppant back production was initially attributed to the high velocities caused by the small perforated interval in combination with the high liquid rates. However, on well AME-203, which was treated with the same proppant, resin, and fluid system as AME-204, an even more severe proppant back production problem was noticed in spite of significantly lower gas and liquid rates and a much longer perforated interval.

Proppant samples collected from the sand catchers were found to be well rounded, probably from the high abrasion occurring during the high velocity trip up the tubular string. No visible deformation existed under the light microscope or the scanning electron microscope from which any conclusions concerning the surface coating or broken bonds sites could be made. The sample was placed in a steel cell at 80°C (176°F) and 69 bar (1000 psi) for 30 minutes. The sample did not exhibit any indication of setting, indicative of the resin curing previously at downhole conditions.

A sample was tested using a Loss of Ignition procedure to determine the amount of resin still remaining on the proppant. A sample was found to have 2.1% resin by weight compared to 3.5% resin by weight for a new lab sample.

20.8 Discussion

The two most significant features in the AME-204 production performance are the higher than expected productivity and formation water production. Neither can be readily explained with the current geological model of the field. In the absence of cores, no formation permeability data are available. A higher than expected matrix permeability would be a possibility, but this would not explain the water production, since the vertical distance from the base of the reservoir to the depth of the (assumed fieldwide) gas water contact (GWC) is about 100 m. Furthermore, the GWC in this block is thought to be across the Ameland claystone, which is assumed to form a good seal. A more likely explanation would be the presence of a natural fracture system reaching down into the water zone. A remote possibility is that the water bearing zone in block E23 has been connected up. A well test in well AME-202 had previously shown this block, which is at a distance of about 250 m from the far end of the horizontal well, to be separated from the main Ameland East Field. Although the frac has almost certainly not reached 250m, it could have opened a plane of weakness connecting the both blocks.

A parameter frequently quoted in conjunction with hydraulic fracturing is the Productivity Improvement Factor (PIF), which is defined as the ratio of pseudo steady-state production at a certain drawdown before and after the frac job. Due to the low k_v/k_h ratio and the short perforated interval a meaningful comparison of pre- and post-frac production performance and thus the determination of a measured PIF is not possible. One could only make certain assumptions and calculate a theoretical improvement factor. For an assumed matrix permeability of 0.25 mD, the improvement factor compared to an unstimulated conventional well would be on the order of 6 to 9. For higher formation permeabilities the improvement factor will decrease.

A more meaningful comparison can be done with nearby wells, as presented in **Fig. 20.11,** which compares the fractured and unfractured productivity at 50 bar drawdown for all wells on the western flank of the Ameland East Field.

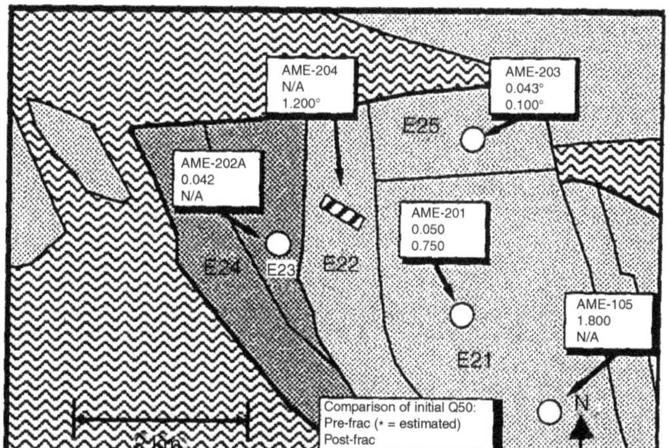

Fig. 20.11—Comparison of productivity with other wells in the AME-2 area.

20.9 Conclusions

The Ameland East Reservoir with its thick sand package and the low vertical permeabilities is not a horizontal well candidate in the classical sense. Nevertheless, once hydraulic fracturing is considered as an option the horizontal multiple fractured well will (economically) outperform a conventional well in the reservoir parts with low matrix permeability.

The successful fracture stimulation of a deep, tight, and layered sandstone reservoir with two hydraulic fractures placed through the final completion in a horizontal well has been demonstrated.

It has been shown that the placement of proppant slugs played an important role for the AME-204 stimulations and that the application of this technique can be essential for fracturing horizontal wells. It would not have been possible to place the first frac without proppant slugs.

The concept of performing the hydraulic fracture stimulation through the final completion has been applied successfully for AME-204. In case more than two fracs are required, fracture stimulation through the final completion becomes less attractive due to erosion etc., and dedicated equipment is maybe a better solution.

In horizontal well fracturing, perforated intervals are usually kept extremely short in order to minimize the formation of multiple fractures. Due to the short perforation interval fracture conductivity becomes very important because of the extremely high velocities in the proppant pack close to the wellbore. Since the possibilities of increasing fracture permeability are limited, propped fracture width becomes the dominating factor. In practice this translates into packed fracture (i.e., Frac & Pack) designs required for frac treatments in horizontal wells. This is applicable even in formations which in the classical sense qualify as tight, i.e., less than 1 mD.

To place this extensive fracture inflation and packing stage, the use of real time bottomhole data is essential. The current models calculating surface to bottomhole pressure are not considered accurate enough for this, particularly for high proppant concentrations. Only real time bottom hole pressure data during the placement of the proppant stages will allow the exploitation of this technology to its potential.

Resin 2 in combination with the high pH Borate fluid showed disappointing performance regarding prevention of proppant back-production during the cleanup. The resin was chosen due to its superior quality in terms of elasticity. This, however causes an unconfined compressive strength which is lower than for other products on the market. Nevertheless, in the long term this may still be the best choice. After more than 6 months of almost continuous production the well has not shown a severe proppant back-production problem.

There are several small fields in the Dutch sector of the North Sea, which suffer from poor productivity and show therefore marginal economics. The technology employed in AME-204 is seen as an important step in enabling economic development of these reserves located in tight reservoirs.

20.10 Acknowledgments

We would like to thank managements of NAM, Mobil, Shell and Halliburton for permission to publish this work and for making this unusual cooperation possible. We are grateful to Sophie Postif and Jan van Riezen, NAM, Sander de Kruijf and Leo Roodhart, KSEPL, Hans Kiampferer and Syd Kreuger, Halliburton Services, Peter Stevens, Mobil and many others for their respective contributions to this project.

20.11 References

Abass, H. H., Hedayati, S., and Meadows, D. L. 1996. Nonplanar Fracture Propagation From A Horizontal Wellbore: Experimental Study. *SPE Prod & Fac* **11** (3). SPE-24823-PA. https://doi.org/10.2118/24823-PA.

Cleary, M. P., Johnson, D. E., Kogsbøll, H. H. et al. 1993. Field Implementation of Proppant Slugs to Avoid Premature Screen-Out Of Hydraulic Fractures with Adequate Proppant Concentrations. Presented at the Low Permeability Reservoirs Symposium, Denver, Colorado, 26–28 April. SPE-25892-MS. https://doi.org/10.2118/25892-MS.

Damgaard, A., Bangert, D. S., Murray, D. J. et al. 1992. A Unique Method For Perforating, Fracturing, and Completing Horizontal Wells. *SPE Prod Eng* **7** (1). SPE-19282-PA. https://doi.org/10.2118/19282-PA.

El Rabaa, W. 1989. Experimental Study of Hydraulic Fracture Geometry Initiated From Horizontal Wellbores. Presented at the SPE Annual Technical Conference and Exhibition, San Antonio, Texas, 8–11. SPE-19720-MS. https://doi.org/10.2118/19720-MS.

Gdanski, R. D., Harris, P. C., Lord, D. L. et al. 1991. Analysis of Fann Model 50 for Fracturing Fluid Rheology. Presented at the SPE Production Operations Symposium, Oklahoma City, Oklahoma, 7–9 April. SPE-21647-MS. https://doi.org/10.2118/21647-MS.

Martinez, A. D., Wright, C. A., and Wright, T. B. 1993. Field Application of Real-Time Hydraulic Fracturing Analysis. Presented at the Low Permeability Reservoirs Symposium, Denver, Colorado, 12–14 April. SPE-25916-MS. https://doi.org/10.2118/25916-MS.

Martins, J. P., Leung, K. H., Jackson, M. R. et al. 1992. Tip Screen Out Fracturing Applied to the Ravenspurn South Gas Field Development. *SPE Prod Eng* **7** (3). SPE-19766-PA. https://doi.org/10.2118/19766-PA.

Parker, M. A. 1990. Effect of Gelled Fracturing Fluids on the Conductivity of Propped Fractures. Presented at the Annual Technical Meeting, Calgary, Alberta, Canada, 10–13 June. PETSOC-90-92. https://doi.org/10.2118/90-92.

Roodhart, L. P., Fokker, P. A., Davies, D. R. et al. 1993. Frac & Pack Stimulation: Application, Design, and Field Experience From The Gulf of Mexico to Borneo. Presented at the Annual Technical Conference and Exhibition, Houston, Texas, 3–6 October. SPE-26564-MS. https://doi.org/10.2118/26564-MS.

Shlyapobersky, J., Wong, G. K., and Walhaug, W. W. 1988a. Overpressure Calibrated Design of Hydraulic Fracture Stimulations. Presented at theSPE Annual Technical Conference and Exhibition, Houston, Texas, 2–5 October. SPE-18194-MS. https://doi.org/10.2118/18194-MS.

Shlyapobersky, J., Walhaug, W. W., Sheffield, R. E. et al. 1988b. Field Determination of Fracturing Parameters, for Overpressure Calibrated Design of Hydraulic Fracturing. Presented at the SPE Technical Conference and Exhibition, Houston, Texas, 2–5 October. SPE-18195-MS. https://doi.org/10.2118/18195-MS.

Tan, H. C., Wesselowski, K. S., Willingham, J. D. 1992. Delayed Borate Crosslinked Fluids Minimize Pipe Pressure. Presented at the SPE Rocky Mountain Regional Meeting, Casper, Wyoming, 18–21 May. SPE-24342-MS. https://doi.org/10.2118/24342-MS.

Thesing, A. 1993. Sea Water Based Borate Crosslinked Fluids for Use in Fracturing Applications up to 300°F. Presented at the 6th Clausthal University Symposium. February 1993.

Weijers, L., de Pater, C. J., Owens, K. A. et al. 1994. Geometry of Hydraulic Fractures Induced From Horizontal Wellbores. *SPE Prod & Fac* **9** (2). SPE-25049-PA. https://doi.org/10.2118/25049-PA.

Chapter 21

Integrated Reservoir Geomechanics Techniques in the Burgos Basin México: An Improved Gas Reservoirs Management

Sergio Berumen, SPE, Pemex E&P; **Craig Cipolla,** SPE, Pinnacle Technologies; **Thomas Finkbeiner,** SPE, Geomechanics International; **Steve Wolhart,** SPE, Gas Research Institute; and **Fernando Rodriguez,** SPE, Pemex E&P

SPE 59418 presents a case history of using hydraulic fracture mapping, fracture propagation modeling, and single-well reservoir modeling to optimize fracture design and well spacing in the Wilcox Sands in the Burgos Basin in Mexico. Berumen et al. first compares hydraulic fracture length, height, and azimuth in four wells obtained from microseismic mapping with that from 3D fracture propagation modeling. With the knowledge of fracture geometries, they investigate the effect of well spacing, types of proppants, and fracture length on ultimate gas recovery and economics. From their studies, they draw the following conclusions:

- Hydraulic fracture direction follows the regional stress trend of the NNE/SSW direction, but varies because of local stress changes
- Production causes changes in the pore pressure and in-situ stress, including re-orientation of minimal stress direction
- Fracture geometries change across the basin
- Reduction in fracture treatment size may result in better economics

21.1 Abstract

An overall geomechanical model was developed in the Wilcox Sands of the Arcabuz-Culebra Field, a tight gas reservoir in the Burgos Basin, northern Mexico. The application of multiple technologies were integrated to build a geomechanical model of the field, which was used as a primary tool to predict fracture direction and stress field distribution. Advanced fracture diagnostic technologies that measure the created fracture geometry were used to support the geomechanical modeling.

The technical approach combined geomechanical modeling, hydraulic fracture mapping, and fracture and reservoir engineering studies to optimize fracture treatment designs and improve the planning of well location and spacing. This methodology can be very important to avoid drilling unnecessary wells or leaving gas in place.

By integrating geomechanics techniques through single well reservoir modeling it was observed that ineffective well positioning could inhibit production by as much as 0.5 BCF per well. Fracture engineering analysis identified significant potential cost savings by reducing fracture treatment size and changing proppant type. In addition potential production increase of 3 BCF per well may be possible with re-fracture treatments. SPE 59418 discusses these results, and presents a case history that details the application and integration of reservoir geomechanics technologies to improve the practices in the gas reservoir management.

21.2 Introduction

Located in northern Mexico just south of the Texas border, the Burgos Basin is a gas province with a current natural gas production of 1 BCF per day. Its reserves are allocated in sandstone tight formations from Eocene through Paleocene age. Over one half of the basin's production is sustained on the Arcabuz-Culebra Field, a mature gas field producing from the units Wilcox and Mount Selman Reservoirs since 1963. These shallow marine sediments were deposited during transgressive-regressive sea level cycles with interbedded shales being deposited during transgressive sequences, and the Wilcox Sands deposited during separate regressive sequences. The Wilcox Sands in Arcabuz-Culebra Field are an extension of the same prolific Wilcox/Lobo Trend of South Texas (Berumen et al. 1998; Wolhart et al. 2000; Dickerson et al. 1995).

Interpretation of seismic and production data indicates that the Wilcox Reservoir Sands are intensively compartmentalized. These compartments are bounded by north-south trending listric growth faults. The target producing reservoirs of this field, and the primary focus for geomechanical analysis, are the Wilcox-4, Wilcox-3, and Wilcox-1 Sands. Average depths of these reservoirs range from

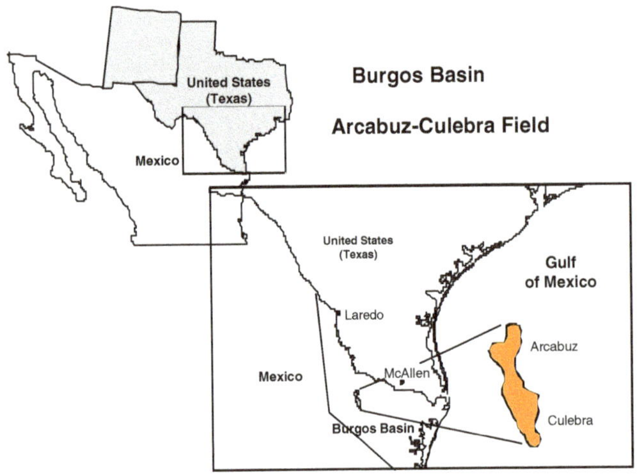

Fig. 21.1—Geographical position of the Burgos Basin and the Arcabuz-Culebra Gas Field.

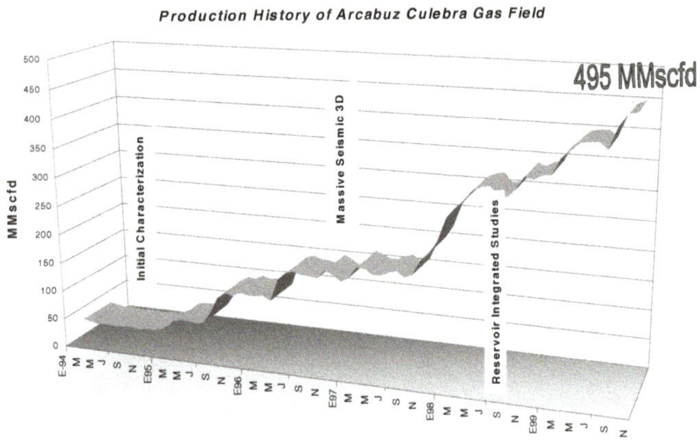

Fig. 21.2—Production performance of Arcabuz-Culebra Field.

6,500 ft in the west to 9,800 ft in the east of the field, while thickness varies between 13 ft and 131 ft. These reservoirs are clean, relatively high-porosity, low permeability gas sands that require hydraulic fracturing stimulation for commercial production.

The economic development of low permeability, overpressured reservoirs such as those of the Burgos Basin requires the efficient application of hydraulic fracturing and effective well spacing and location. Thus, the efficient management of existing gas fields in the basin is crucial. One valuable tool to improve the reservoirs development is reservoir geomechanics, which is the integration of different techniques oriented to recognize the stress field and was used in this study to improve the geological modeling and dynamic characterization of reservoirs.

Fig. 21.1 depicts the geographical position of the basin, and the location of Arcabuz-Culebra Field within the basin. The field is considered as an average prototype of the geological nature of the basin, and its gas potential is supported on the production of Wilcox Sands, and less importantly from Mount Selman Formations. **Fig. 21.2** shows the production performance experienced by this field along its entire life as well as different field management stages implemented along its development. It can readily be seen that the last stage of production has been supported by the massive application of modern technology, incorporating 3D seismic and reservoir integrated studies, as well as aggressive practices on hydraulic fracturing.

However, the integration of logs and petrophysical information as well as the fracturing data into a model that allows to forecast the fracture orientation and the stress field was a key ingredient that has improved the field management. **Fig. 21.3** depicts a flow chart showing the integration of geomechanical modeling, 3D fracture modeling and fracture mapping with reservoir modeling to optimize the Arcabuz-Culebra Field development. This methodology was followed during this study.

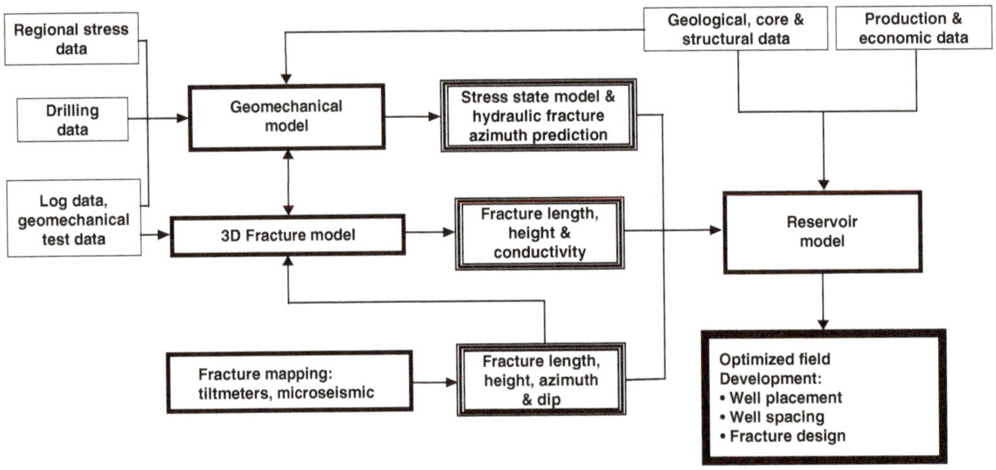

Fig. 21.3—Methodology followed for the integration of geomechanical modeling in Arcabuz-Culebra.

21.3 Stress and Physical Property Analysis in the Arcabuz-Culebra Field, México

The objective of this phase was to quantify the magnitude and orientations of all three principal stresses (vertical stress, least and maximum principal horizontal stress), as well as rock strength and pore pressure. A systematic and comprehensive program of geomechanical data collection, data handling and global analysis of the Arcabuz-Culebra Field was implemented. Based on the available information both on quality and quantity, the strategy was to pick up 29 key wells along the field. The general approach to geomechanical modeling is illustrated in **Fig. 21.4.**

The left-hand side displays the types of input data, available from the field, for preparatory analysis. The center section displays resulting parameters that are essential prerequisites for developing a geomechanical model (Gaarenstroom et al. 1993; Ibanez and Kronenberg 1993; Moos and Zoback 1990; Peska and Zoback 1995a, 1995b; Plumb and Hickman 1985; Vernik and Zoback 1992; Zoback 1989). Primary output from the model is predictions and maps of the stress field.

Regional stress data from the World Stress Map (WSM) database (Zoback 1989), and structural data were used to develop a preliminary regional model for the state of the stress in the area comprising the Arcabuz-Culebra Field. In **Fig. 21.5,** a straight-line segment at each point where an observation was made represents the orientation throughout the region of $S_{H\max}$. Stress indicators define a regional trend adjacent to the Gulf Coast that generally parallels the coastal margin and indicates active crustal extension perpendicular to the continental margin. In contrast, measurements in south central Mexico reveal roughly E-W compression, perpendicular to the Gulf Coast and not related to the Gulf of Mexico seaward extensional faulting.

Wellbore breakouts were identified using 4-arm caliper data from dipmeter logs. Strict criteria were used to distinguish stress-induced wellbore breakouts from other enlargements. The first require that the caliper tool stops rotating in the well. Secondly, the small diameter measured by the caliper must be equal to the bit size and the large diameter must be larger than the small diameter. The third criterion is to assure that the direction in which the wellbore is enlarged is not the same as the direction of wellbore deviation (Zoback 1989; Zheng et al. 1989; Vernik et al. 1993).

Sixty-three wells had 4-arm caliper-dipmeter data, but breakout intervals were identified in only 15 of them. These results led to a reliable map of in-situ stress and stress variations in the geologically complex area of Arcabuz-Culebra Field. A specific outcome of this modeling was the magnitude of the maximum horizontal principal stress ($S_{H\max}$). **Fig. 21.6** displays the resulting $S_{H\max}$ directions for the 15 wells in which breakouts were detected. There is no single overall stress orientation in the field. However, a consistent trend emerges in which $S_{H\max}$ azimuths are roughly parallel to the local strike of mapped faults.

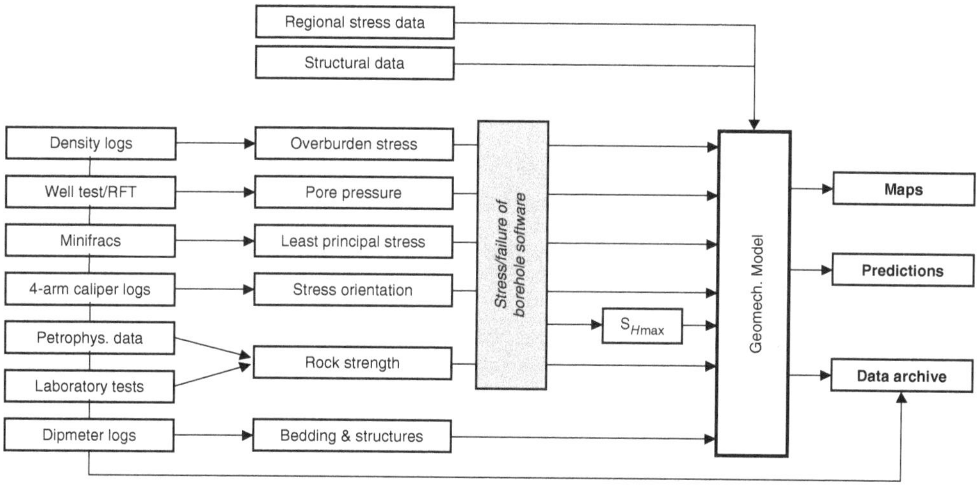

Fig. 21.4—General approach to geomechanical modeling in Arcabuz-Culebra.

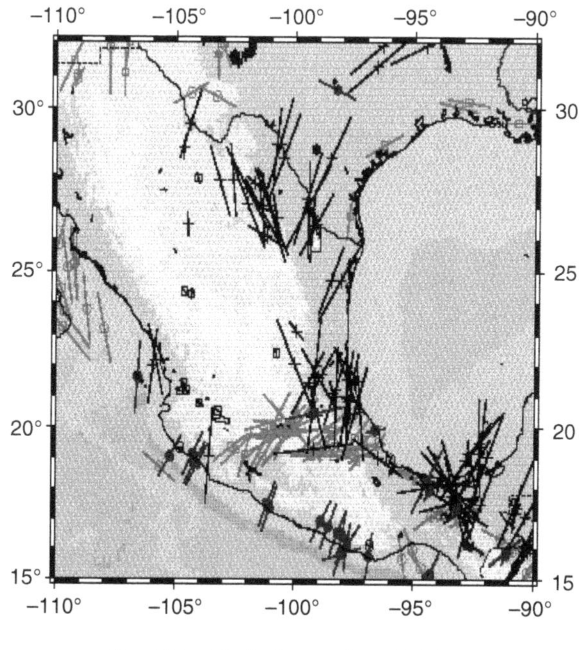

Fig. 21.5—World Stress Map: México.

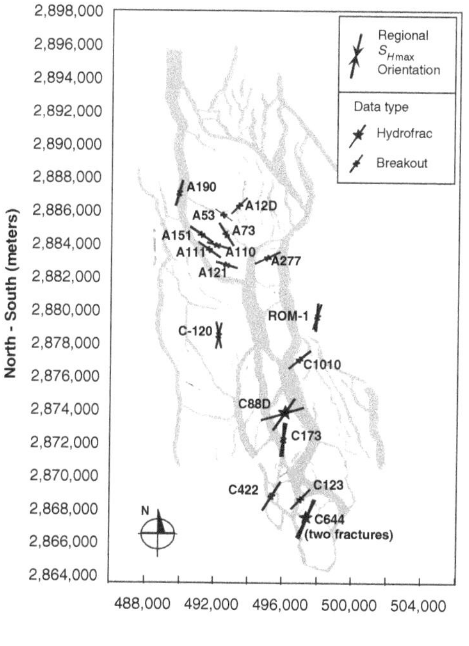

Fig. 21.6—Orientation of maximum horizontal stress in Arcabuz-Culebra.

This model was used to refine structural models from seismic data to identify sub-seismic faults and zones of drilling induced compressive wellbore failure (breakouts). Least principal stress values (S_3) were inferred from available minifrac data. The vertical stress was determined as a function of depth from density logs. Pore pressure was obtained by direct well test measurements. Using this data and information on mud weight and wellbore orientation a commercial simulator (Stanford University) was used to constrain the full stress tensor as a function of depth and location within the field. Some additional physical parameters used to carry out the modeling were coefficient of internal friction = 1, coefficient of sliding friction = 0.6, Biot's coefficient = 1, Poisson's ratio = 0.25.

Although no critically stressed features were identified in any of the Arcabuz-Culebra wells analyzed, the modelled stress in this reservoir would promote slip on favorably oriented planes. This implies that if the wells intersect fractures with optimal orientations for slip, the fractures may contribute to reservoir permeability.

Wellbore stability analysis in the field indicates that wells drilled at all orientations and inclinations are safe to drill with current drilling practices. Low stress concentrations and overbalanced drilling conditions ensure borehole stability even if the rock strength is relatively low. The analysis also indicates that, unless compressive rock strength falls below approximately 4,000 psi, even horizontal wells are stable to drill (Vernik et al. 1993; Stanford University) in Arcabuz-Culebra Field.

The observed stress changes are a consequence of poroelastic effects, which cause a decrease in horizontal total stresses due to production-related reservoir fluid pressure drop. By itself, poroelasticity could cause a change in the orientation of the horizontal stresses only if the region affected is asymmetrical. In this case the observed stress reorientation is caused by slip along older structures triggered by the poroelasticity induced change from a strike-slip to a normal faulting stress regime.

Combined with structural models of the field the geomechanical model can be used to predict the field-wide orientation of induced fractures, to evaluate the importance of mapped faults to reservoir behaviour, and to evaluate the stability of deviated wells as a function of wellbore orientation and mud weight.

21.4 Fracture Mapping and Fracture Engineering

Well placement and spacing can be optimized using a reservoir model to predict well performance combined with a geomechanical model to predict fracture direction. However, hydraulic fracture and geomechanical models must be calibrated to ensure that predictions and optimisations are reliable. In Arcabuz-Culebra Field, hydraulic fracture geometry (azimuth, dip, length and height) was measured on representative wells using a unique combination of microseismic, surface tiltmeter and downhole-tiltmeter fracture mapping technologies.

The measurements were performed in two wells, the Culebra 88D (C88D) and Culebra 644 (C644) completed in the Wilcox Sand Formations. Surface-tiltmeters and downhole-tiltmeters mapping were performed for the Culebra 88D, while microseismic event location mapping was done for the C644 completion. Fracture modeling 3D and fracture engineering were performed for these two wells, in both real time and postfrac. Features of these diagnostic techniques as well as physical principles and limitations are well documented in the modern literature.

21.4.1 Microseismic Mapping. Research on microseismic techniques among wells in mapping the growth of hydraulic fracturing was strongly developed in the late 1980s and 1990s. Microseismic imaging has the potential to produce maps of microseismic events, which define hydraulic fracture azimuth, as well as the spatial and temporal aspects of fracture length and height. Basically the technique consists in monitoring the microseismic events generated by a hydraulic fracturing, through an array of accelerometers placed on a neighbour observer well (Peterson et al. 1996; Young and Warpinski 1996; Warpinski et al. 1998a, 1998b).

Microseismic monitoring of a fracture treatment in the Wilcox-4 Sandstone was achieved by running a four-receiver accelerometer array in the C309 well and observing seismic activity induced by a fracture in the C644 well. **Fig. 21.7** is a map view of this Culebra test area showing the microseismic monitor well, C309, and the treatment well located at 1740 ft (530 m), C644. Receiver locations in the C309 were situated at depths of 8225, 8375, 8525, and 8675 ft, as shown in **Fig. 21.8.** A total of 21 microseismic events associated with the fracturing in the C644 were observed both in the field and in subsequent analysis. A least-squares fit of these data sets

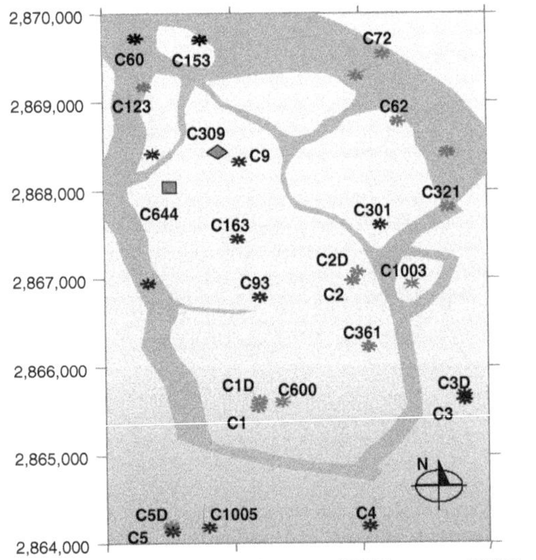

Fig. 21.7—Fracture diagnostics map view of Culebra test area.

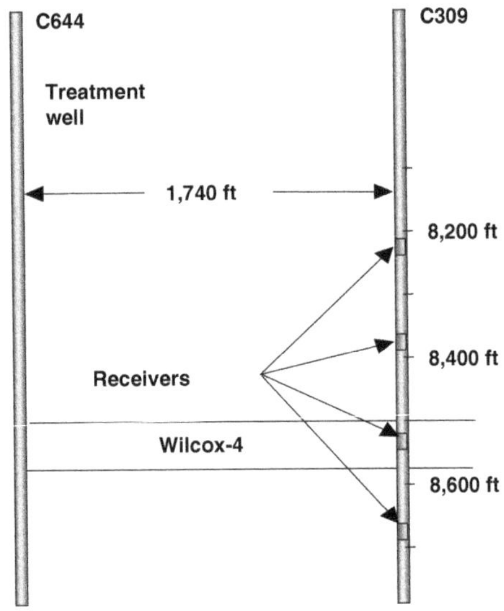

Fig. 21.8—Receiver's location in the monitor well C309, Wilcox-4.

provides a preferred fracture azimuth of N16-20°E. The N18°E azimuth is shown in **Fig. 21.9.** This represents a reasonably accurate map of the W-4 hydraulic fracture. **Fig. 21.10** shows the side or profile view of these 21 microseismic event locations on the C644 well. There are a sufficient number of events along the SSW wing of the frac to provide reasonably good information regarding wing length and frac asymmetry. From this analysis an average frac wing length of 690 ft and a total fracture height of 440 ft were detected.

By moving the four–level accelerometer array in the C309 well up to the Wilcox-2 Sand depth and observing activity induced by a fracture in the C644 well, a second microseismic monitoring was conducted. **Fig. 21.11** shows the configuration of the receivers situated in the C309 well. Monitoring of the W-2 fracturing resulted in about 25 events, from which six microseisms had sufficient energy to provide reasonable estimates of the event locations. Plan view and side view maps of the estimated event locations are shown in **Figs. 21.12 and 21.13.** The estimated fracture azimuth is N23°E based on a regression analysis; average frac wing lengths of 590 ft and total fracture height of 440 ft were estimated.

21.4.2 Surface and Downhole Tiltmeters. Tiltmeter technology as a fracture diagnostic tool has been successfully applied in the industry since 1996. As with microseismic monitoring, multiple tiltmeters are run on a wire line in an offset well and/or at surface locations around the well under a fracturing job to measure the earth's tilt due to the hydraulic fracture (Fig. 21.13). The mapping principle is based on the fact that creating a hydraulic fracture involves parting the rock and deforming the reservoir. Tiltmeter mapping involves measuring the fracture-induced deformation around the fracturing well versus time and depth. A tiltmeter is so sensitive that it can detect tilts down to one nanoradian. The data is inverted to obtain created fracture dimensions. Fracture height, length and width can be determined from analysis of this data. Downhole tiltmeters are very sensitive to fracture dimensions and less sensitive to fracture orientation–just the opposite of surface tiltmeters (Wright et al. 1998a, 1998b; Wright 1998; Wright and Conant 1995; Wright et al. 1995).

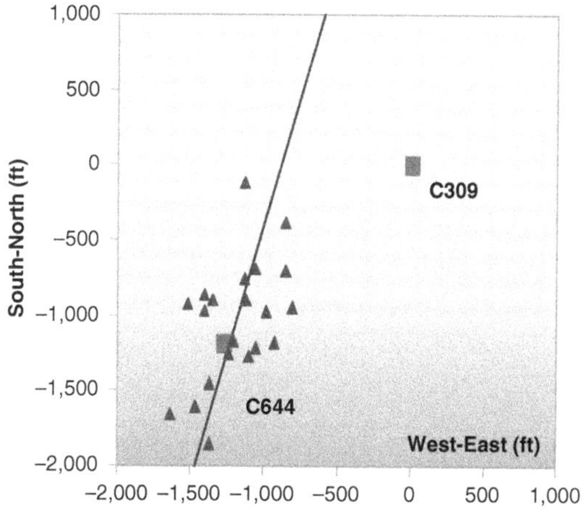

Fig. 21.9—Plan view of events in W-4 associated with the fracturing of C644.

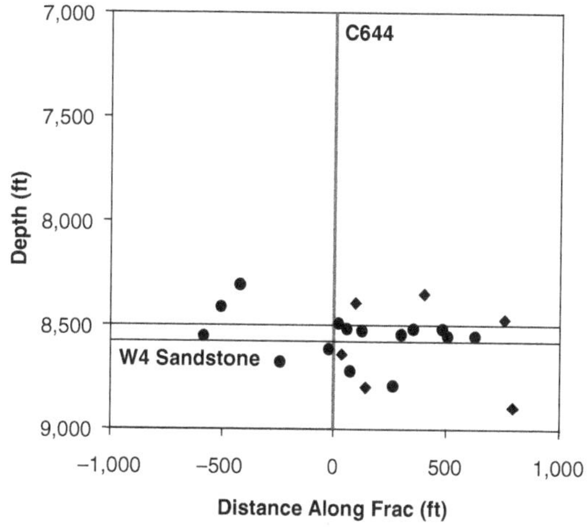

Fig. 21.10—Side view of microseismic events location on the C644 well, W-4.

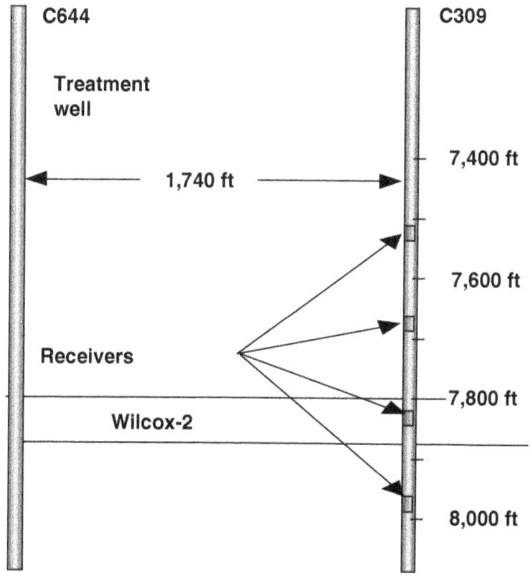

Fig. 21.11—Receiver's location in the monitor well C309, Wilcox-2.

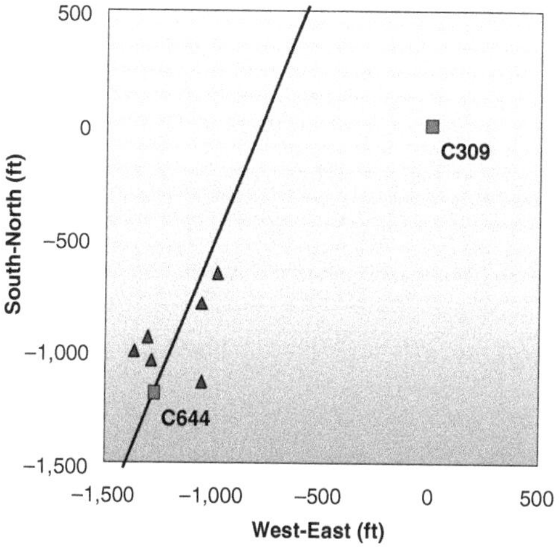

Fig. 21.12—Plan view of microseismic events in W-2 associated with the fracturing of C644.

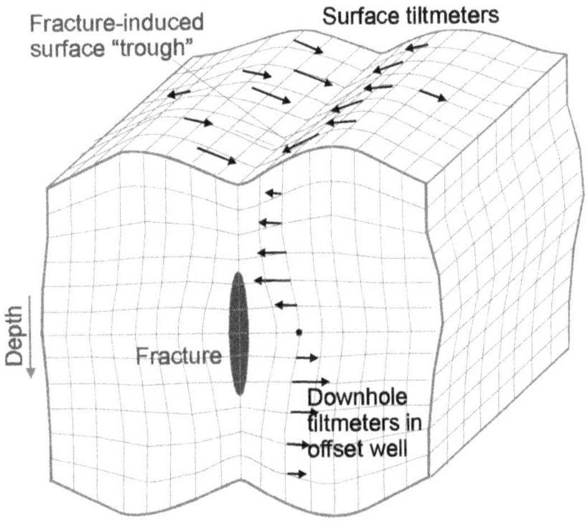

Fig. 21.13—Tilt Field due to hydraulic fracture.

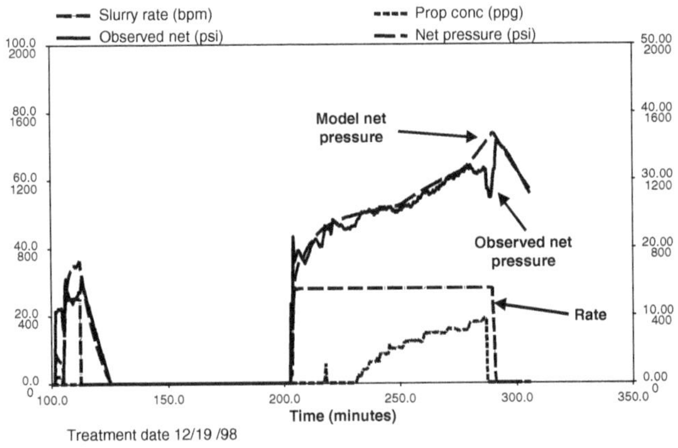

Fig. 21.14—Pressure history match for C88D, hydraulic fracturing on Wilcox-4.

An array of 23 surface tiltmeters was radially placed around the well C88D, under hydraulic fracturing treatment, and an array of 5 downhole tiltmeters was placed in an offset observation well C75, 450 m away. The tools were spread over a vertical distance of 318 ft to capture the overall motion as the approaching fracture deformed the earth. The distance between the observation well and the fracture location of the treatment well was about 1247 ft. Downhole tiltmeters electronics were severely affected by the downhole temperature (250°F to 270°F), however the hydraulic fracture dimensions were estimated acceptably. The geometry of the fracture on Wilcox-4 was characterized with $L_f = 658$ ft, $h_f = 107$ and a direction of N31°E, and dip of W82°, while the fracture geometry for the Wilcox-3 was $L_f = 735$ ft, $h_f = 433$ ft, direction N71°E, dip of S85°.

21.4.3 Fracture Engineering. Fracture modeling was performed in the Arcabuz-Culebra study using a 3D fracture propagation model to estimate fracture height, length, and conductivity. In all of the treatment wells involved in the fracture mapping the net pressure history match was performed with a 3D fracture model. Closure pressures were estimated from minifracs prior to the main jobs. **Fig. 21.14** show the history match of the treating pressure data and modelled fracture geometry for the Wilcox-4 in the well C88D. Fracture model history matching is analogous to reservoir simulation history matching, with the primary match parameter being the measured net pressure in the fracture. The measured pressure is the pressure in the fracture minus the minimum in-situ horizontal stress, also called closure pressure (Cipolla and Mayerhofer 1998; Cipolla and Kyte 1992; Cipolla et al. 1994; Cipolla et al. 1996). The minimum horizontal stress is quite low in the Wilcox-4 (0.62 psi/ft) at the C88D location, caused probably by severe reservoir depletion on that region. Similar phenomenon was observed at C644 location, with $\sigma_H = 075$ psi/ft.

Results of the fracture mapping and fracture modeling are summarized in **Table 21.1.** The results of the study showed a changing hydraulic fracture direction. Hydraulic fractures in the northern part of the field (Arcabuz) parallel the major adjacent faults. Hydraulic fractures in the southern part of the field (Culebra) generally followed the regional trend of the stress field.

21.5 Integrated Reservoir Modeling

The knowledge of fracture geometry and azimuth in these low permeability reservoirs contributed to better understand well location and spacing along the field. These were the main objectives of this study to support the reservoir management of this set of compartmentalized low permeability reservoirs. Once the geomechanical model was defined, single well reservoir modeling was used to analyze several production schemes. It was detected that sub-optimum fracture designs and improper well placement and spacing will result in production losses and/or economic inefficiency. **Fig. 21.15** shows the concept of optimum well placement, and **Fig. 21.16** illustrates the effect of improper well placement on gas recovery, where it is shown that production losses of 0.5 BCF could result if well placement is not optimized.

According to the fracture engineering analysis and production modeling, frac jobs could be optimized by reducing treatment size and changing to a lower strength proppant. It was shown that in the Wilcox-4, standard 20/40 mesh Ottawa Sand could substitute

Well	Prop (Mlbm)	Stress (psi/ft)	Fracture Geometry from Modeling		Avg. Prop. Conc. (lbm/ft^2)	Fracture Direction from Mapping	Frac Dip from STM	Fracture Geometry from Mapping	
			L_f(ft)	H_f(ft)				L_f(ft)	H_f(ft)
C88D-W4	300	0.62	658	107	2.3	N31°E	W82°	762	122
C88D-W3	230	0.89	735	433	0.7	N71°E	S85°	762	457
C644-W4	320	0.75	530	146	2.3	N18°E	-	640	408
C644-W2	250	0.87	277	607	1.0	N23°E	-	549	408

Table 21.1—Results of fracture mapping and fracture modeling.

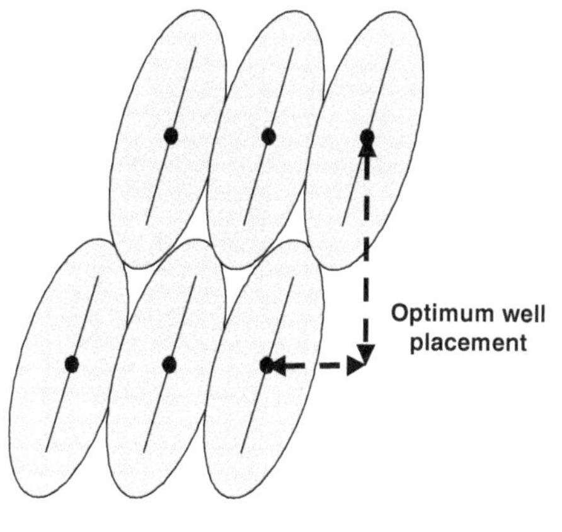

Fig. 21.15—Optimum well placement and spacing in low permeability reservoirs with a long hydraulic fracture.

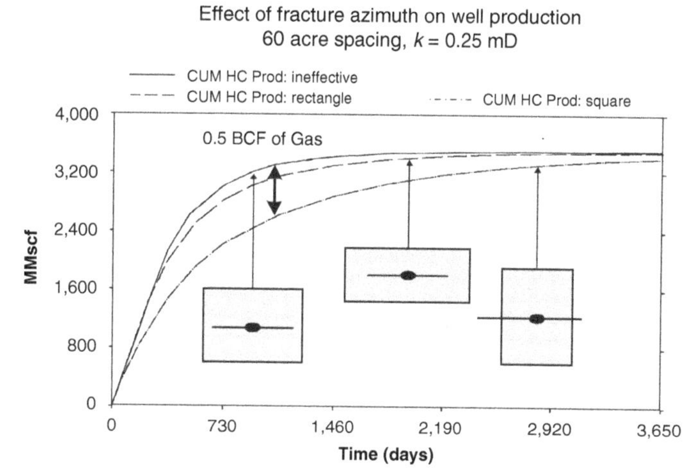

Fig. 21.16—Effect of well placement on gas recovery.

for higher strength resin-coated sand (RCS). **Fig. 21.17** depicts the expected production using sand and RCS for the C88D Wilcox-4, showing almost identical results. Same results were observed for lower permeability sands Wilcox-2 and -3 since fracture conductivity becomes less important for lower reservoir permeability. Using uncoated sands instead RCS could result in cost savings of US $45,000 per fracture treatment.

Re-fracture treatments in older Wilcox-4 wells may significantly improve production due to lower closure stress and the ability to achieve much longer fracture lengths compared to the initial treatments. **Fig. 21.18** illustrates this finding showing the corresponding production increase. A potential was observed to increase gas recovery by 3 BCF per well after 6.5 years and increase net cash flow by US $2,900,000.

Based on the fracture engineering analysis several reservoir simulation runs were carried out for different fracture lengths in the range observed in treatments for Wilcox-4 along the field (Berumen et al. 2000; Cipolla and Mayerhofer 1996). The results showed that a 40% reduction of current job sizes could optimize economic returns while minimizing costs as can be observed in **Fig. 21.19.** The optimum fracture length is 500 ft, which is 25% less than the 650-ft propped length modelled in the Culebra 88D. Such a reduction in treatment size could reduce completion costs by US $65,000 per fracture treatment, sacrificing little production.

From the integrated results of the study, it was concluded that well spacing and location are controlled by the in-situ stress state (which controls fracture geometry and direction), reservoir permeability, and the distribution and orientation of faults. These results have allowed improvement in well-placement decisions, improvement in fracture treatment designs, and optimization of production economics of Arcabuz-Culebra Gas Field. These new practices in the reservoir management of this field have provided new opportunities to redevelop the Burgos Basin in México.

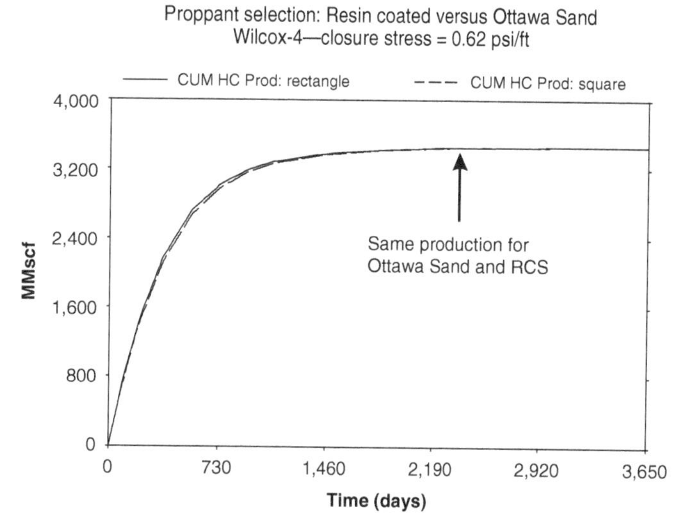

Fig. 21.17—Cumulative gas production for RCS and uncoated sand.

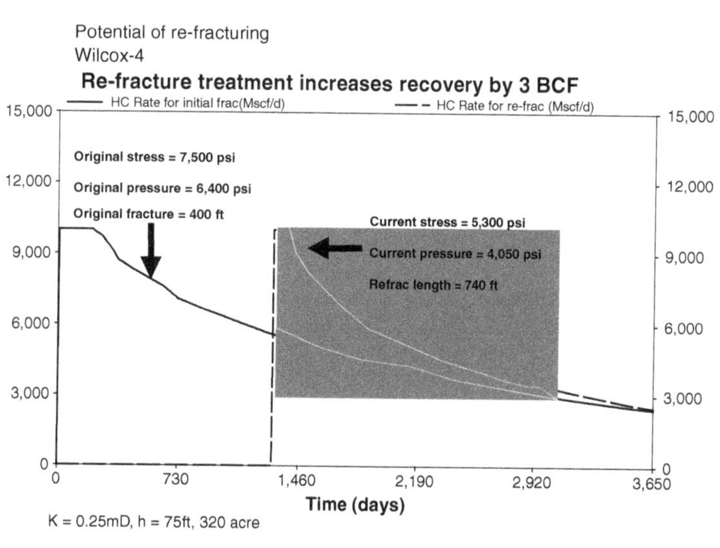

Fig. 21.18—Potential increase recovery by re-fracturing old wells in Arcabuz-Culebra Gas Field.

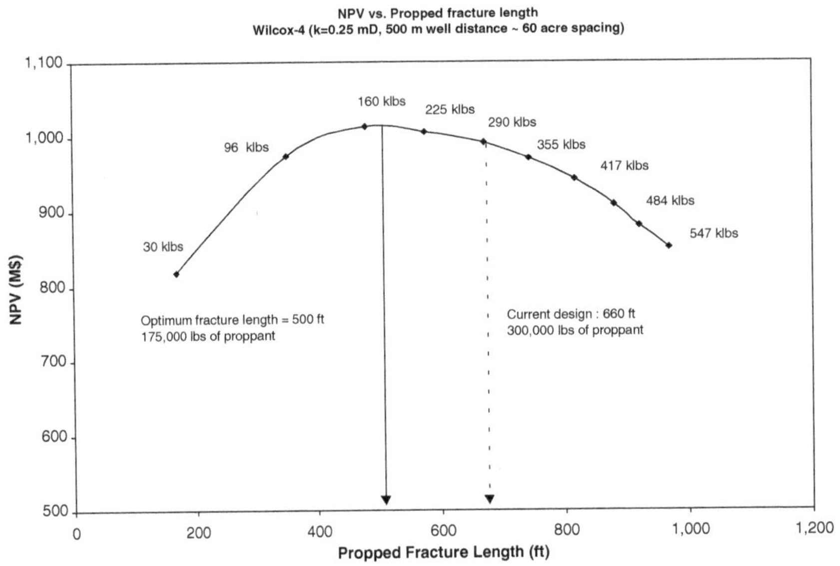

Fig. 21.19—Optimum frac length in Wilcox-4 of Arcabuz-Culebra Gas Field. Reduction in treatment size optimizes economic returns.

21.6 Conclusions

The study has determined the following major conclusions:

- The integration of geomechanical techniques, such as fracture mapping, fracture and reservoir engineering, and geomechanical studies, has proved to be a powerful methodology to improve the efficiency of the reservoir management in the Burgos Basin of México. Significant economic benefits are being obtained in the Arcabuz-Culebra Gas Field (a representative field of the basin) through the application of this methodology.
- While the regional stress trend follows a general NNE-SSW orientation, hydraulic fracture directions are expected to vary locally due to local changes in stress orientation. In the Culebra Field, fracture directions predicted from wellbore-breakout analyses are consistent with the regional stress orientation. In contrast, stress orientations indicate that in the Arcabuz Field hydraulic fractures are expected to propagate parallel to the nearest faults.
- Production-related depletion has resulted in significant pore pressure and stress changes. Stress orientation changes may accompany these stress and pore pressure changes that by themselves may cause local changes in fracture orientation, or may re-activate local normal faults resulting in a stress state that follows the strike of nearby faults.
- Fracture geometry changed from one area of the field to another. This should be considered when planning well spacing and location. Production could be reduced by as much as 0.5 BCF due to ineffective well placement.
- Re-fracturing potential exists for some wells in Arcabuz-Culebra Gas Field with the opportunity to increase ultimate recovery by 3 BCF per well after 6.5 years and increase net cash flow by US $2,900,000.
- Fracture treatments could be modified resulting in potential cost savings. The results showed that a 40% reduction of current job sizes could optimize economic returns while minimizing costs. The optimum fracture length is 500 ft, which is 25% less than the 650-ft propped length modelled in an average well in Culebra Field. Such a reduction in treatment size could reduce completion costs by US $65,000 per fracture treatment, sacrificing little production.

21.7 Acknowledgments

The authors would like to thank Pemex E&P for permission to publish the results of the project presented in this paper. We would also like to acknowledge Projeito for their assistance and logistical support along the project. Also, special thanks to Heber Cinco-Ley for his critical comments on this project and supporting the publication of this work.

21.8 References

Berumen, S., Tiab, D. and Rodriguez, F. 2000. Constant Rate Solutions for a Fractured Well with an Asymmetric Fracture, *J Petrol Sci Eng* **25** (1–2): 49–58. https://doi.org/10.1016/S0920-4105(99)00053-4.

Berumen, S., Sanchez J. M. and Rodriguez, F. 1998. A Strategy for Additional Developments in the Burgos Basin: The Arcabuz-Culebra Gas Field. Paper presented at the International Petroleum Conference and Exhibition of Mexico, Villahermosa, México, 3–5 March. SPE-39845-MS. https://doi.org/10.2118/39845-MS.

Cipolla, C. L. and Mayerhofer, M. 1998. Understanding Fracture Performance by Integrating Well Testing & Fracture Modeling. Paper presented at the SPE Annual Technical Conference and Exhibition, New Orleans, Louisiana, 27–30 September. SPE-49044-MS. https://doi.org/10.2118/49044-MS.

Cipolla, C. L. and Kyte, D. G. 1992. Infill Drilling in the Moxa Arch: A Case History of the Frontier Formation. Paper presented at the SPE Annual Technical Conference and Exhibition, Washington, D.C., 4–7 October. SPE-24909-MS. https://doi.org/10.2118/24909-MS.

Cipolla, C. L., Meehan, D. N., Dunleavy, J. R. et al. 1994. Field Application of Static Tubing Pressure Data and On-Site Fracture Modeling. Paper presented at the SPE Annual Technical Conference and Exhibition, New Orleans, Louisiana, 25–28 September. SPE-28491-MS. https://doi.org/10.2118/28491-MS.

Cipolla, C. L., Meehan, D. N., and Stevens, P. L. 1996. Hydraulic Fracture Performance in the Moxa Arch Frontier Formation. *SPE Prod & Fac* **11** (4). SPE-25918-PA. https://doi.org/10.2118/25918-PA.

Cipolla, C. L. and Mayerhofer, M. 1996. Infill Drilling & Reserve Growth Determination in Lenticular Tight Gas Sands. Paper presented at the SPE Annual Technical Conference and Exibition, Denver, Colorado, 7–9 October. SPE-36735-MS. https://doi.org/10.2118/36735-MS.

Dickerson, P. W., Hamlin, H. S., Hentz, T. F. et al. 1995. Regional Domains of the Wilcox Lobo Natural Gas Trend, South Texas. GRI Topical Report No. GRI-95/0027, Contract No.5082-211-0708, (March 1995).

Gaarenstroom, L., Tromp, R. A. J., deJong, M. C. et al. 1993. Over-pressures in the Central North Sea: Implications for trap integrity and drilling safety. Petroleum Geology Conference Series: Proceedings of the 4th Conference, *The Geological Society* **4**: 1305–1313. https://doi.org/10.1144/0041305.

Ibanez, W. D. and Kronenberg, A. K. 1993. Experimental deformation of shale: Mechanical properties and microstructural indicators of mechanisms. *Int. J. Rock Mech. Min. Sci. & Geomech. Abstr.* **30** (7): 723–734.

Moos, D. and Zoback, M. D. 1990. Utilisation of observations related to wellbore failure to constrain the orientation and magnitude of crustal stresses: Application to continental, DSDP and ODP boreholes. *J. Geophys. Res.* **95**: 9305–9325.

Peska, P. and Zoback, M. D. 1995a. Compressive and tensile failure of inclined wellbores and determination of in situ stress and rock strength. *J. Geophys. Res.* **100** (12): 791–811.

Peska, P. and Zoback, M. D. 1995b. Observations of borehole breakouts and tensile wall-fractures in deviated boreholes—A technique to constrain in situ stress and rock strength. Paper presented at the 35th U.S. Symposium on Rock Mechanics (USRMS), Reno, Nevada, 5–7 June. ARMA-95-0319.

Peterson, R. E., Wolhart, S. L., Frohne, K. H. et al. 1996. Fracture Diagnostics Research at the GRI/DOE Multi-Site Project: Overview of the Concept and Results. Paper presented at the SPE Annual Technical Conference and Exhibition, Denver, Colorado. 6–9 October. SPE-36449-MS. https://doi.org/10.2118/36449-MS.

Plumb, R. A. and Hickman, S. H. 1985. Stress-induced borehole elongation—A comparison between the four-arm dipmeter and the borehole televiewer in the Auburn geothermal well. *J. Geophys. Res.* **90**: 5513–5521.

Stanford University, "GMI•SFIB™ (Stress and Failure of Inclined Boreholes)", licensed to GeoMechanics International.

Vernik, L. and Zoback, M. D. 1992. Estimation of the maximum horizontal principal stress magnitude from stress-induced wellbore breakouts in the Cajon Pass scientific research borehole. *J. Geophys. Res.* **97**: 5109–5120.

Vernik, L., Bruno, M., and Bovburg, B. 1993. Empirical relations between compressive strength and porosity of siliciclastic rocks. *Int. J. Rock Mech. Min. Sci. & Geomech. Abstr.* **30** (7): 677–680.

Warpinski, N. R., Branagan, P. T., Peterson, R. E. et al. 1998a. Mapping Hydraulic Fracture Growth and Geometry Using Microseismic Events Detected by a Wireline Retrievable Accelerometer Array. Paper presented at the SPE Gas Technology Symposium, Calgary, Alberta, Canada, 15–18 March. SPE-40014-MS. https://doi.org/10.2118/40014-MS.

Warpinski, N. R., Branagan, P. T., Peterson, R. E. et al. 1998b. An Interpretation of M-Site Hydraulic Fracture Diagnostic Results. Paper presented at the SPE Rocky Mountain Regional/Low Permeability Reservoirs Symposium and Exhibition, Denver, Colorado, 5–8 April. SPE-39950-MS. https://doi.org/10.2118/39950-MS.

Wolhart, S., Berumen, S., Cipolla, C. et al. 2000. Use of Hydraulic Fracture Diagnostics to Optimize Fracturing Jobs in the Arcabuz-Culebra Field. Paper presented at the SPE Rocky Mountain Regional/Low-Permeability Reservoirs Symposium, Denver, Colorado, 12–15 March. SPE-60314-MS. https://doi.org/10.2118/60314-MS.

Wright, C. A., Davis, E. J., Weijers, L. et al. 1998a. Downhole Tiltmeter Fracture Mapping: A New Tool for Directly Measuring Hydraulic Fracture Dimensions. Paper presented at the SPE Annual Technical Conference, New Orleans, Louisiana, 27–30 September. SPE-49193-MS. https://doi.org/10.2118/49193-MS.

Wright, C. A. 1998. Tiltmeter Fracture Mapping: From the Surface And Now Downhole. *Petroleum Engineer International* **71** (1).

Wright, C. A., Davis, E. J., Golich, G. M. et al. 1998b. Downhole Tiltmeter Fracture Mapping: A Step Forward in Fracture Diagnostics. *GasTIPS* **4** (3):18–26.

Wright, C. A. and Conant, R. A. 1995. Hydraulic Fracture Reorientation in Primary and Secondary Recovery from Low-Permeability Reservoirs. Paper presented at the SPE Annual Technical Conference and Exhibition, Dallas, Texas, 22–25 October. SPE-30484-MS. https://doi.org/10.2118/30484-MS.

Wright, C. A., Conant, R. A., Golich, G. M. et al. 1995. Hydraulic Fracture Orientation and Production/Injection Induced Reservoir Stress Changes in Diatomite Waterfloods. Paper presented at the SPE Western Regional Meeting, Bakersfield, California, 8–10 March. SPE-29625-MS. https://doi.org/10.2118/29625-MS.

Young, C. and Warpinski, N. R. 1996. Microseism Location Analysis. GRI Topical Report No. GRI-96/0338, (September 1996).

Zheng, Z., Kemeny, J., and Cook, N. G. 1989. Analysis of borehole breakouts. *J. Geophys. Res.* **94**: 7171–7182.

Zoback, M. L. 1989. Global patterns of intraplate stresses: a status report on the world stress map project of the International Lithosphere Program: Nature. In *Reservoir Geomechnaics*, vol. 341, 291–298. United Kingdom: Cambridge University Press.

Chapter 22

Case Histories—Combining Crossed Dipole Sonic Anisotropy and Oriented Perforating to Optimize Hydraulic Fracturing in the Burgos Basin—Reynosa, Mexico

Antonio Hernández Patino and **Eduardo Rodríguez Garcia,** PEMEX, Reynosa, México; **Juan Páblo Torne** and **Jorge Lazo,** Halliburton Energy Services, Reynosa, México; and **Calvin Kessler,** Halliburton Energy Services, Houston, Texas

 Symposium paper SPE 92014 was presented at the SPE International Petroleum Conference in Mexico held in Puebla, Mexico, 7–9 November 2004.

SPE 92014 presents a case history of perforations oriented to minimize tortuosity in two wells: One in the Vicksburg Formation and the other in unidentified formation in the Burgos Basin in Mexico. Patino et al. discuss the identification of stress orientation using multiaxis caliper log, borehole imaging data, and crossed dipole sonic anisotropy analysis. The cross-dipole anisotropy analysis technique is applied to identify far-field maximum stress orientation. Then, perforating guns with a phasing angle of 180° are oriented with the help of a gyroscope. Afterward, step-down tests are conducted to quantify near-wellbore fractional pressure loss and perforation frictions, which are 50% of offset wells without using oriented perforations. The authors' final conclusion is that oriented perforations can reduce near-wellbore fraction and tortuosity, if done successfully.

22.1 Proposal

Based on local experience, a well-accepted work flow or set of best practices has been developed for optimizing stimulation treatments in the Burgos Basin. Recently, these best practices have been enhanced by combining stress field orientation and gyroscopically oriented wireline perforating. Orienting perforations in the direction of maximum horizontal stress has greatly reduced near wellbore tortuosity and friction in hydraulic fracturing treatments. The result has been improved proppant placement resulting in higher proppant conductivity and increased production.

The best practices for optimization of the stimulation design and economic modeling include many steps (Kessler et al. 1994; Voneiff and Holditch 1994; Cipolla et al. 1994; Whitehead et al. 1986; Katahara 1996; Boonen et al. 1998; Anderson and Walker 1972; Fogal et al. 2002; Kessler and Varsamis 2001; Águila et al. 2003). The candidate selection process evaluates the net-to-gross sand thickness ratio, reservoir pressure, and permeability for each zone. The net-to-gross ratio provides information on productive and non-productive fracture height, which is needed for inflow analysis. The reservoir fluid, pressure, and permeability of the formation are vital for determining the fluid leakoff/fracturing fluid rheology requirements and required fracture conductivity. Geological and geophysical data are reviewed to identify potential reservoir boundaries such as faults or pinch-outs that could impede hydraulic fracture propagation. The vertical stress profile between the zone to be treated and the underlying and overlying formations is used in 3D fracture simulators to evaluate the fracture height (and any possible unwanted out-of-zone fracture height) versus fracture half-length and proppant concentration.

To implement the oriented perforating enhancement to the best practices, crossed dipole sonic anisotropy analysis is used to determine the far-field stress orientation and magnitude combined with the six electrode dipmeter for orientation and ovality visualization. Borehole breakout analysis and anelastic strain recovery of whole cores can provide near-field stress orientation data. These factors are then used to determine optimum perforation direction. In this paper, several case histories will be presented comparing the results of stimulation treatments using the normal practices and traditional perforating methods versus stimulation treatments using the same practices but incorporating oriented perforating.

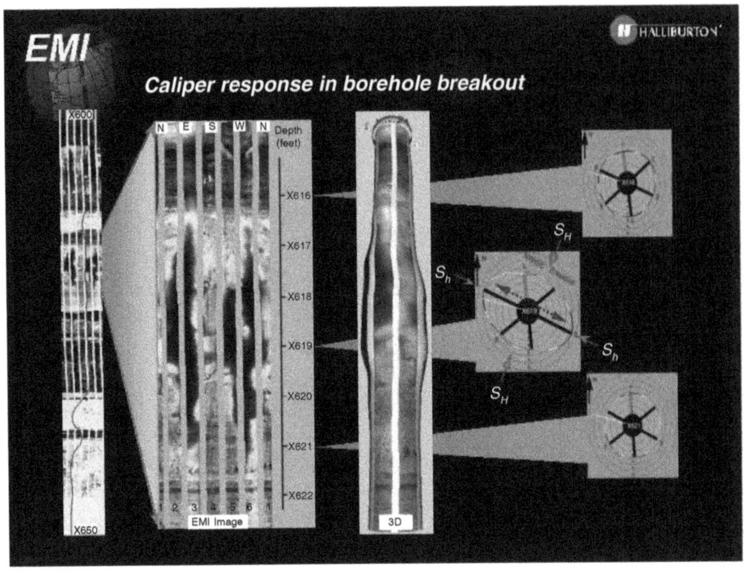

Fig. 22.1—Borehole breakout analysis from 6-arm caliper data.

22.2 Stress Orientation and Borehole Breakout

The near field stress orientation can be determined from borehole breakout analysis, which requires at a minimum multi-axis calipers and borehole orientation data. In both vertical and deviated wellbores, borehole breakout analysis from X-Y caliper data can sometimes provide suitable information. The borehole ovality can also be a result of the drill string and its rotation or sliding on the low side of the borehole in deviated well conditions. Both of these phenomena can result in a key shaped borehole. If only X-Y caliper data is used, the pipe groove/key shaped borehole can be interpreted as borehole ovality. These conditions can generally be identified with the long axis of the borehole, constantly aligned with high side/low side of the borehole. Multi-axis calipers with more than X-Y measurements **(Fig. 22.1)** provide additional data which can enhance the borehole geometry/borehole break out analysis.

Borehole imaging data **(Fig. 22.2)**, either ultrasonic or electrical, have much greater radial and vertical sampling than multi-axis caliper data. The imaging data not only provide high resolution borehole geometry data, but yield essentially a picture of the detailed bedding. Fractures can readily be identified from imaging data.

22.3 Stress Orientation and Crossed Dipole Anisotropy Analysis

The depth of investigation of sonic tools is in excess of one wave length, which varies primarily with frequency of the acoustic wave, and slowness of the formation. Using one particular crossed dipole sonic tool in the formations encountered in the Burgos Basin, the radial depth of investigation is on the order of 5 ft. Crossed dipole sonic anisotropy analysis can provide far field stress orientation and the fast and slow shear wave travel times and their corresponding orientations. In isotropic formation conditions, the fast and slow shear waves travel at the same slowness. Since the slowness values are equal there is no slowness anisotropy and the orientation of the anisotropy is undefined **(Fig. 22.3)**. By definition in isotropic formations, there is no preferred stress orientation.

In formations that are anisotropic, the fast and slow shear waves propagate at different travel times. The fast shear wave (low numerical value of travel time) is aligned with the maximum stress direction **(Fig. 22.4)**. In this example the travel time of the fast shear wave is 125 μsec/ft and the slow shear wave travel time is 132 μsec/ft. The azimuth of the fast shear is 25° and the azimuth of the slow shear, by wave theory, is 90° from the fast shear azimuth which is 335°. The magnitude of the slowness anisotropy in this example is 5½%, which is determined by the numerical difference between the slow and fast travel times divided by the mean.

Fractures, either natural or hydraulically induced, will propagate in the direction of maximum stress. In fractured formations, the fast shear wave travel time propagates longitudinal to the natural fractures. The slow shear wave travel time is transverse to the fractures. Crossed dipole anisotropy analysis, borehole breakout analysis, and borehole image data in a naturally fractured formation **(Fig. 22.5)** confirm that the fast shear azimuth, which is the direction of maximum stress (NE-SW), and the strike of the fractures from the imaging data (NE-SW) are in agreement. The borehole breakout analysis determines the azimuth of the long axis of the borehole, which is the direction of the minimum stress (NW-SE) and is orthogonal to the maximum stress direction.

22.4 Dynamic and Static Elastic Moduli

Log derived elastic moduli are referred to as dynamic properties, since they are measured in-situ. Rock laboratory measurements on core samples are referred to as static measurements. The conversion of dynamic to static measurements, like most log analysis parameters, is formation specific. There are no dynamic to static conversion constants that can be universally applied. A practical approach for the dynamic to static conversion is to normalize the calculated fracture closure pressure from the petrophysical data to the observed fracture closure pressure from the fluid efficiency test.

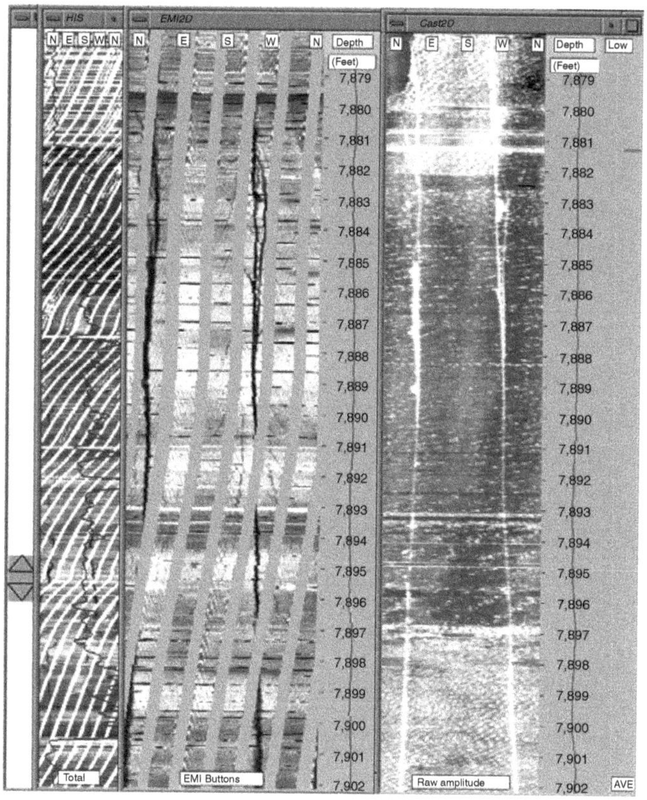

Fig. 22.2—Electrical and ultrasonic imaging data after openhole minifrac treatment.

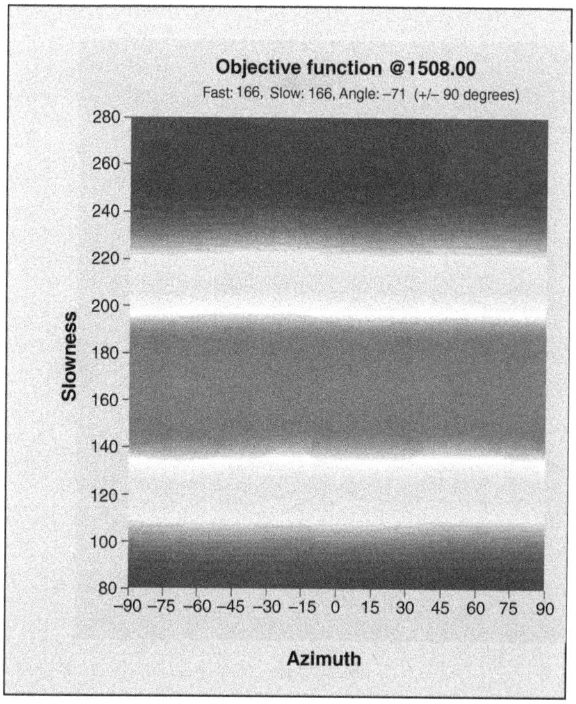

Fig. 22.3—Anisotropy analysis single depth QC plot in an isotropic formation shows that fast and slow shear wave travel times are identical. The azimuth of the fast shear is undefined.

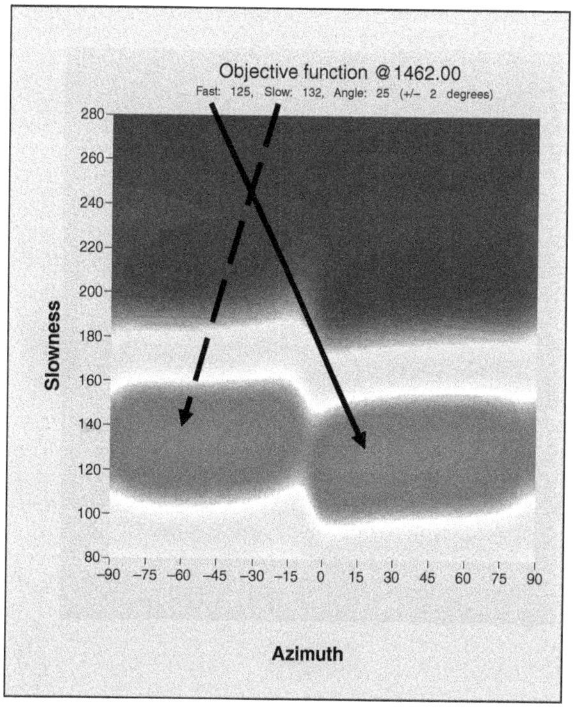

Fig. 22.4—Anisotropy analysis single depth QC plot in an anisotropic formation, fast shear wave travel time is 125 and the slow shear wave travel time is 132 μsec/ft. The azimuth of the fast shear arrival is 25° and the azimuth of the slow shear is 335°.

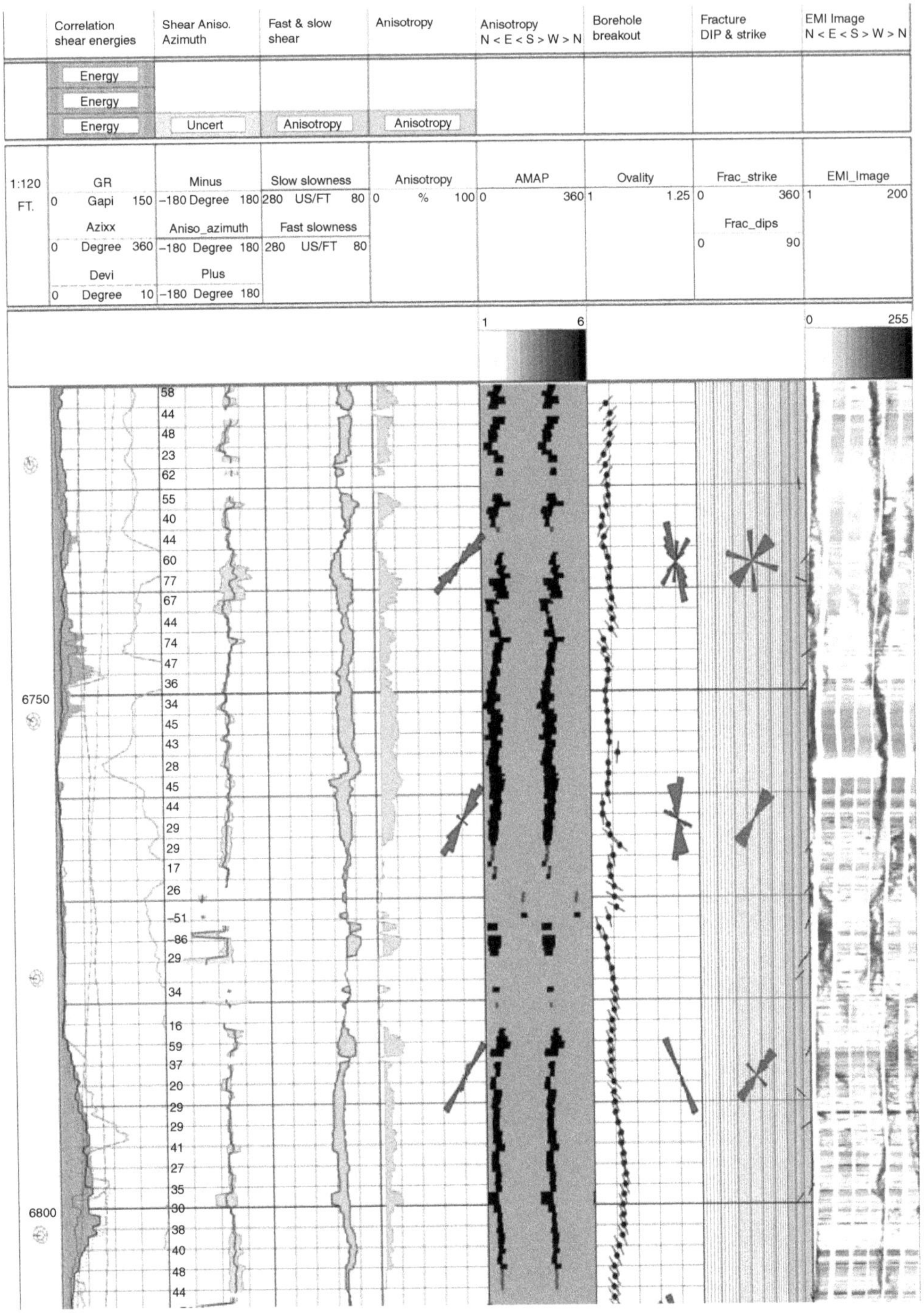

Fig. 22.5—Crossed dipole anisotropy analysis and borehole image data in a naturally fractured formation. Fast shear azimuth (maximum stress direction) in predominately NE-SW. Fracture strike from imaging data is NE-SW. From borehole breakout analysis, the long axis of the borehole minimum stress direction is NW-SE.

$$\sigma_h = P_o\left(\frac{\mu}{1-\mu}\right) + \alpha P_p\left(1-\frac{\mu}{1-\mu}\right),$$

$$\sigma_H = \sigma_h + T,$$

$$P_{bd} = 3\sigma_h - \sigma_H - P_m - \alpha P_p,$$

where

σ_h = the minimum horizontal stress

σ_H = the maximum horizontal stress

P_o = the overburden pressure

P_p = the pore pressure

P_m = the hydrostatic pressure
P_{bd} = the break down pressure
μ = Poisson's ratio (requires compressional-wave and shear-wave data)
a = Biot's constant

$$\alpha = 1 - \frac{C_{ma}}{C_b}$$

C_{ma} = the matrix compressibility
C_b = the bulk compressibility
T = tectonic stress (obtained from the anisotropy)

22.5 Perforating for Low Permeability Hydraulic Stimulation

To minimize perforation friction and near wellbore tortuosity the perforations should be oriented within 30° of the maximum horizontal stress direction (Kessler et al. 2004) **(Fig. 22.6).** By definition, such orientation can be obtained by using a perforating gun with 60° charge-to-charge phasing **(Fig. 22.7).** In isotropic formations **(Fig. 22.8)** multiple near-field fractures and possible screenouts can be a result of low charge-to-charge phasing angles.

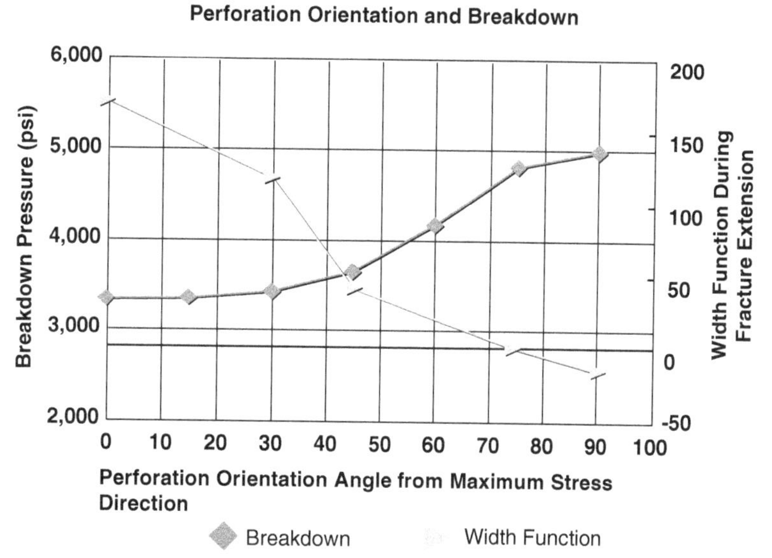

Fig. 22.6—Effects of perforation orientation from maximum stress direction on breakdown pressure and fracture width.

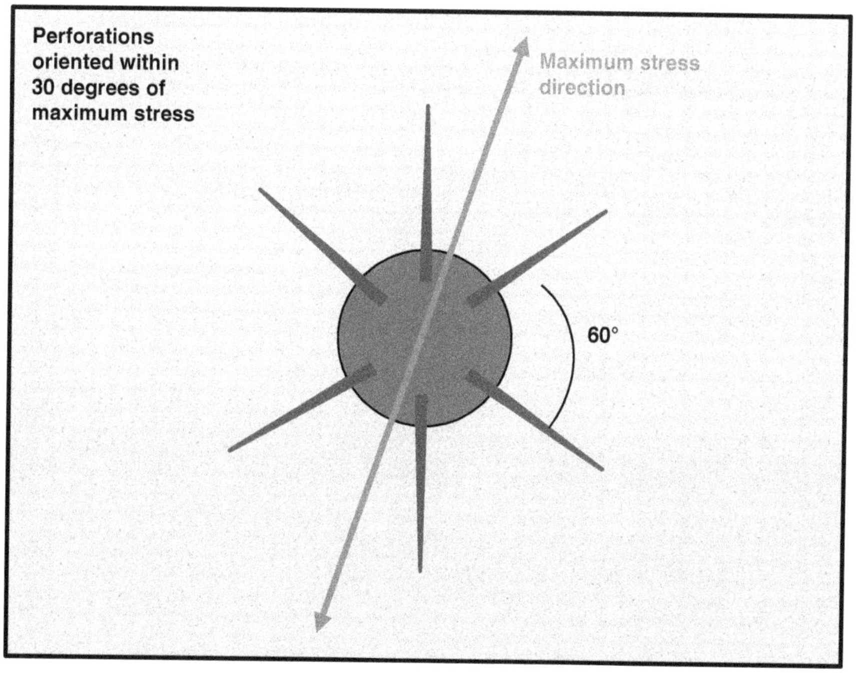

Fig. 22.7—Perforating gun with 60 degree charge-to-charge phasing.

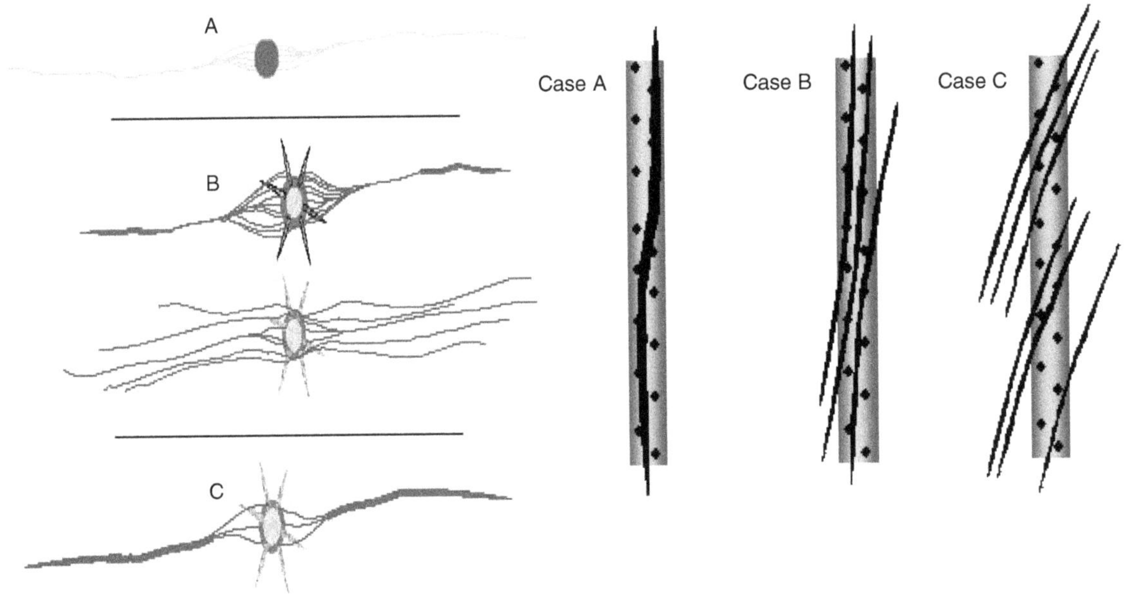

Fig. 22.8—Multiple near wellbore fractures can be a result of non-optimal perforating gun phasing in isotropic formations.

In anisotropic formation conditions using a perforating gun with 60° phasing, may result in too many perforations for effective hydraulic stimulation treatment. There are many perforations that are not aligned with the maximum stress direction. The number of perforations should be determined based on the anticipated pumping rate, the fluid rheology, proppant concentration, the permeability of the formation, and leak off. To minimize possible screenouts, the perforation diameter at downhole conditions should be at least 6 times the diameter of the proppant.

22.5.1 Case History 1—Anisotropy Analysis Combined with Oriented Perforating to Reduce Tortuosity. This well is in the Burgos Basin; the west-east seismic cross section **(Fig. 22.9)** shows that there are numerous faults in this area. The zones of interest for well completion are within the box on the large scale seismic section. The detailed seismic section of these horizons is shown on the right hand side of the figure. The magnitude and orientation of the tectonic stresses obviously vary with both depth and distance from the faults, and the strike/orientation of the faults.

Borehole breakout analysis from oriented multi-axis caliper data can be used to determine the near field stress orientation. Hydraulic fractures propagate in the direction of maximum stress, which is the short axis of the wellbore. The crossed dipole anisotropy analysis **(Fig. 22.10)** in the lower interval to be completed (2,930–2,947 m) indicates that the anisotropy is approximately 6% to 8% and the orientation of the far field maximum stress is approximately N 45 W to N 30 W.

To minimize near wellbore tortuosity oriented perforating was performed on this well. The well was perforated in the direction of maximum far field stress obtained from the cross dipole anisotropy analysis. The wireline conveyed hollow steel carrier perforating

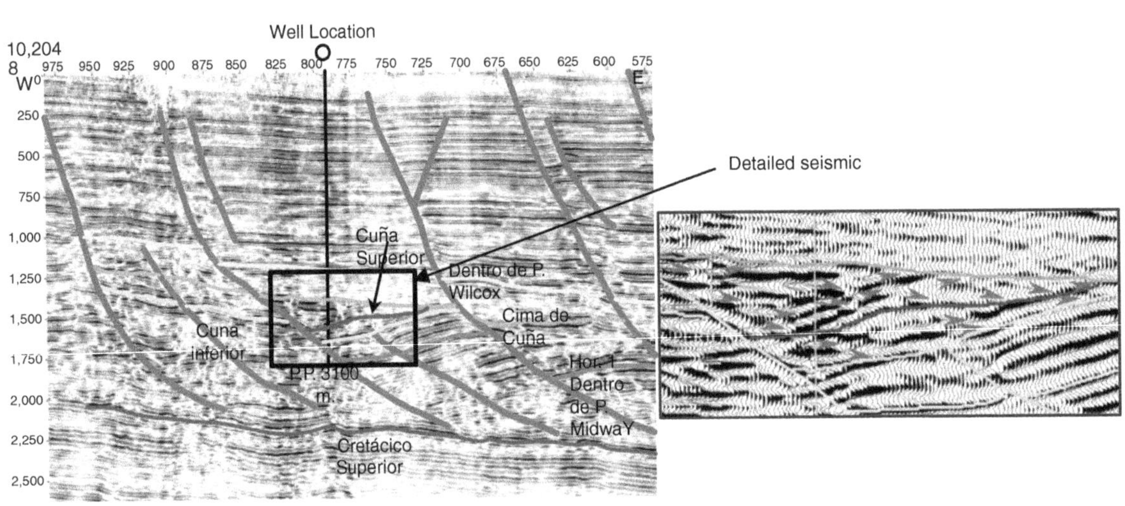

Fig. 22.9—Case History 1—East-West seismic section showing well location and multiple faults, which can cause stress anisotropy.

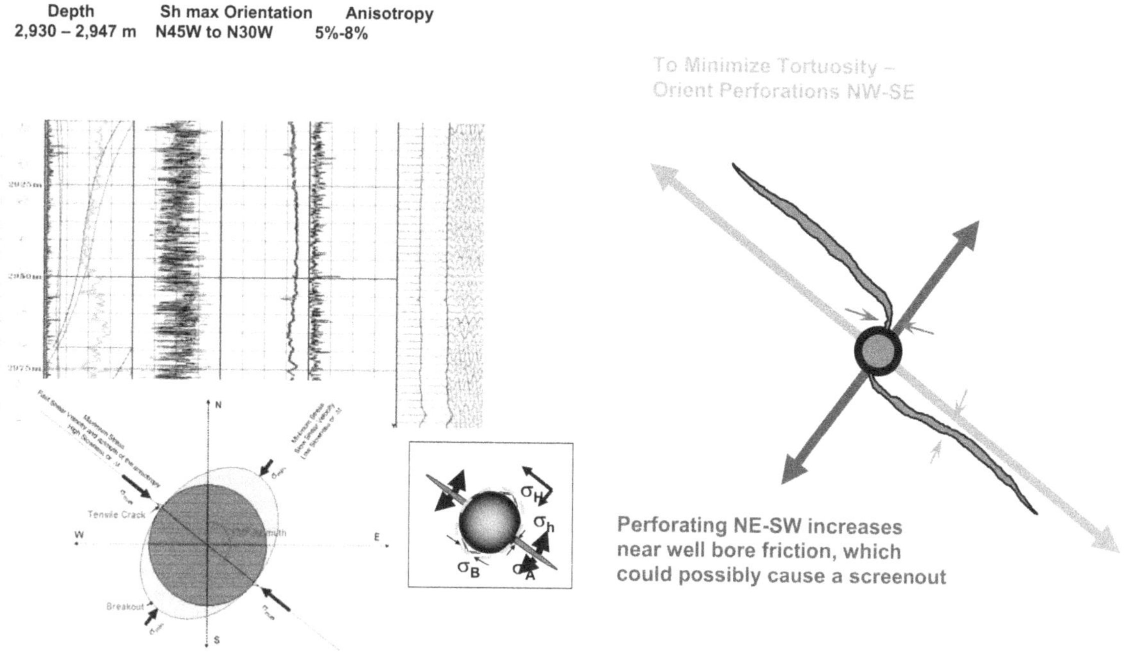

Fig. 22.10—Case History 1—Crossed dipole anisotropy analysis indicates that the maximum stress orientation is approximately N 45 W to N 30 W. Borehole breakout analysis from multiple axis caliper data indicates that the long axis of the borehole, which is the minimum near field stress orientation, is oriented NE-SW.

guns were loaded with charge-to-charge phasing angles of 180°. A gyroscope was used to verify the orientation of the guns, which utilize gravity for alignment with the low side of the borehole.

Post-oriented-perforating mini-frac analysis **(Fig. 22.11)** from a four rate step down injection fall-off test at 20, 15, 10, and 5 bbl/min (BPM) indicated that the maximum near wellbore friction at 20.5 BPM was 164 psi and the perforation friction was 124 psi. Friction values from offset wells perforated with 60 degree phasing and using the same fluids and injection rates range from 250 to 500 psi. The oriented perforation reduced the near well bore tortuosity by approximately 50%. Combining far field stress orientation and oriented perforating has been performed on several other wells with similar results.

Perf and near-wellbore friction - F8

Entry friction versus time

	Time (min)	Rate #1 (bpm)	Rate #2 (bpm)	Near-wellbore friction (psi)	Perf. Friction (psi)	Perf. Coef. Mult.	Perfs open (effective)
1	18.64	20.53	0.00	164	124	50.50	41.51
2	0.00	0.00	0.00	0	0	0.00	0.00
3	0.00	0.00	0.00	0	0	0.00	0.00
4	0.00	0.00	0.00	0	0	0.00	0.00
5	0.00	0.00	0.00	0	0	0.00	0.00
6	0.00	0.00	0.00	0	0	0.00	0.00

Perforation data

	Top MD (m)	Bot MD (m)	#Perfs	Diam (in)
1	2,925	2,940	295	0.280
2	0	0	0	0.000
3	0	0	0	0.000
4	0	0	0	0.000
5	0	0	0	0.000
6	0	0	0	0.000
7	0	0	0	0.000

Perforation pressure drop model: Default fracproPT model; Use cramer's perf erosion model

Rate step-down test analyser

Previous Test | Curr Test: 1 | Next Test

	Time (min)	Rate #1 (bpm)	Rate #2 (bpm)	Change in friction (psi)
1	17.54	20.63	15.73	51
2	18.22	15.73	9.93	74
3	18.95	9.93	4.92	70
4	19.75	4.81	0.00	81
5	0.00	0.00	0.00	0

Note: Flow rate needs to go all the way to zero.

Display step-down plot | New step-down analysis

Step-down friction analysis

	Estimated value
Total friction power B	0.85
Total friction coeff K	21.46
Perf friction coeff K	0.29
NWB Friction coeff K	36.16
Effective perfs open	41.50
Perf friction (psi)	124
NWB Friction (psi)	164
Max flow rate (bpm)	20.53

Fig. 22.11—Case History 5—Mini-frac data analysis after oriented perforating. At 20.5 BPM, the near wellbore friction is 164 psi; and perforation friction is 124 psi. These values are approximately 100% lower than values observed in offset wells which were perforated with traditional charge-to-charge phasing and not oriented.

22.5.2 Cased History 2—Anisotropy Analysis Combined with Oriented Perforating to Optimize Hydraulic Fracturing of Deep Tight Gas Formations. This well is located in the Burgos Basin also, and one of the objectives was the Vicksburg Formation which is normally a tight sandstone when depth exceeds 4000 m; hydraulic fracturing the area has shown perforating and near well bore friction on the order of 400 to 450 psi, and this high tortuosity results in partial fractures and premature sanding with excessive well head pressures.

The north-south seismic cross section **(Fig. 22.12)** shows that the well crosses a fault near the zones of interest within the oval in the large seismic section. The magnitude and orientation of the tectonic stresses obviously vary with both depth and distance from the faults, and the strike/orientation of the faults. To determine the direction and magnitude of the stress field is important in these formation conditions where downhole fracture treatment pressures exceed 10,000 psi.

Borehole break out analysis from oriented multiaxis caliper data can be used to determine the near field stress orientation. The far field stress orientation is obtained from the crossed dipole anisotropy analysis. For the deeper interval to be completed (4,352–4,359 m), the analysis indicated that the direction of the maximum stress is N 20 W **(Fig. 22.13).** The well direction in the same zone is N 80 E which is near to the minimum stress direction. The well deviation in this interval is approximately 14.5 degrees, which allows the use of wireline gravity oriented perforating systems.

Rock mechanical properties were calculated using the RockXpert™ software and the dynamic fracture pressure gradient is approximately 0.91 psi/ft. At 4,350 m. the estimated fracture pressure is 12,800 psi **(Fig. 22.14).** The critical pressure (shear failure) of the formation was calculated to be approximately 7,900 psi. If the bottom hole pressure decreases to less than the critical pressure, shear failure of the formation could possibly occur. (These values were considered only as a reference because there was no core data available for laboratory calibration of dynamic to static moduli). The critical flowing bottom hole pressure data is used in the inflow and economic modeling comparing a natural completion to a completion with production enhancement treatment.

Oriented perforating was performed in this well to minimize near wellbore tortuosity. The wireline conveyed perforating system utilized a 2.5-in. hollow steel carrier and super deep penetrating perforating charges. Perforating modeling and down hole charge performance calculations were used for charge selection. At downhole conditions the formation compressive strength is approximately 15,700 psi, the formation penetration is approximately 9 in., inches and the average entry hole diameter in the 4 ½-in. casing is 0.27-in. **(Fig. 22.15).** The entry hole diameter is greater than the minimum entry hole requirements to minimize friction losses using 20/40 proppant (Gruesbeck and Collins 1982).

The wireline conveyed 2.5-in. hollow steel carrier perforating guns were loaded with charge-to-charge phasing angles of 180° degrees and the gravity oriented perforating technique for wireline was used to orient the charges plane in the maximum stress field direction, which was N 20 W.

The mini-frac and a complete hydraulic fracture were successfully performed. Post-oriented-perforating mini-frac analysis **(Fig. 22.16)** from a four rate step down injection fall-off test (31, 27, 11, and 6 BPM) indicated that the maximum near wellbore friction at 23.34 BPM was 208 psi and the perforation friction was 243 psi. Friction values from a previous similar well perforated with 60° charge-to-charge phasing angle and using similar fluids and injection rates range from 400 to 450 psi. The 180° oriented perforation reduced the near well bore tortuosity by approximately 50%.

The designed production enhancement treatment based on the results of the mini-frac was completed with 100% success. The bottom hole treatment pressures were above 13,500 psi in the 4.5-in. monobore completion, with 2 to 3 different concentrations being in the well at the same time **(Fig. 22.17).**

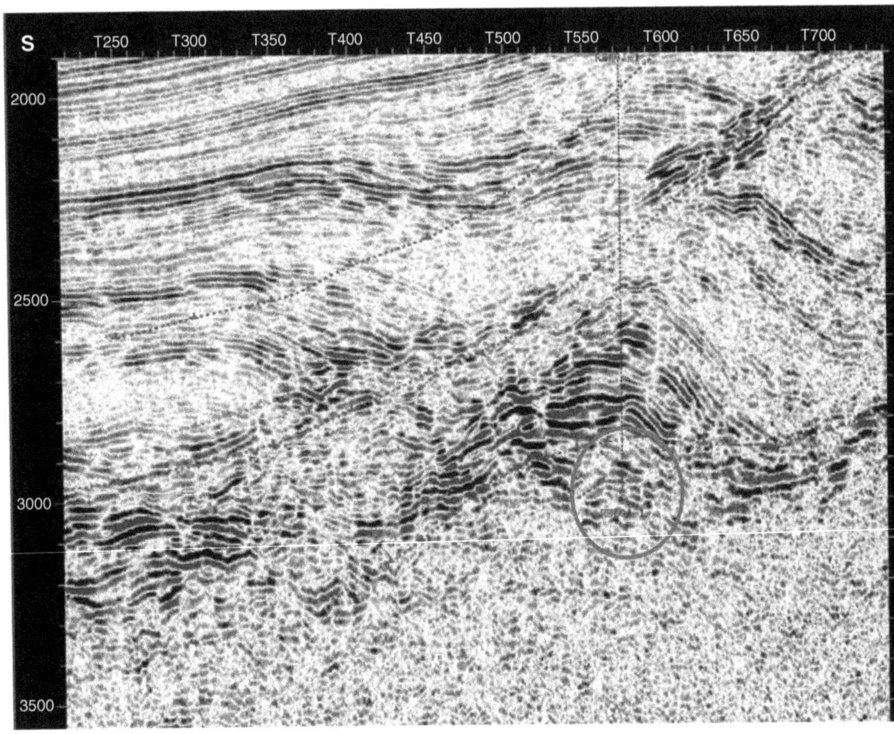

Fig. 22.12—Case History 2—North-South seismic cross section.

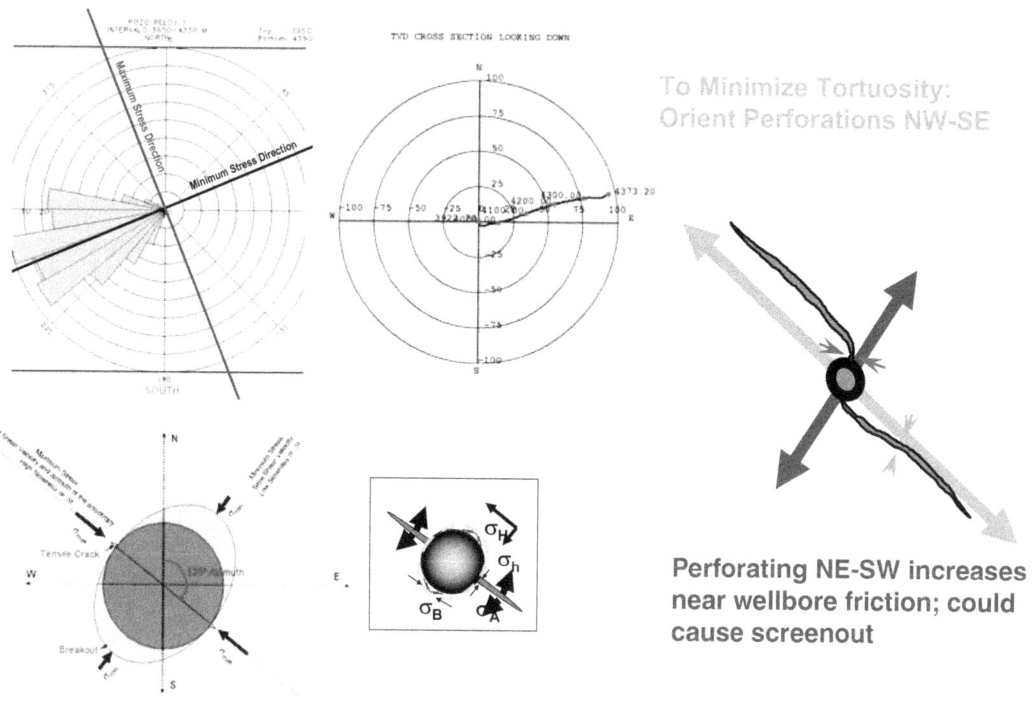

Fig. 22.13—Case History 2—Crossed dipole anisotropy analysis indicates that the maximum stress orientation is approximately N 45 W to N 30 W. Borehole breakout analysis from multiple axis caliper data indicates that the long axis of the borehole, which is the minimum near field stress orientation, is oriented NE-SW.

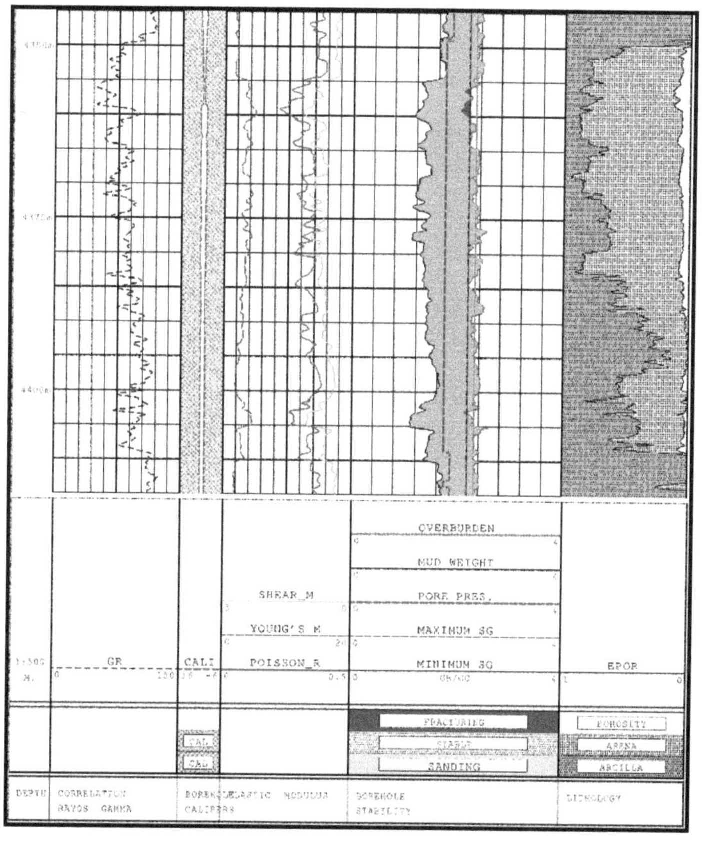

Fig. 22.14—Case History 2—Mechanical properties using the program RockXpert™.

Gun 1: 2-1/2" Millennium

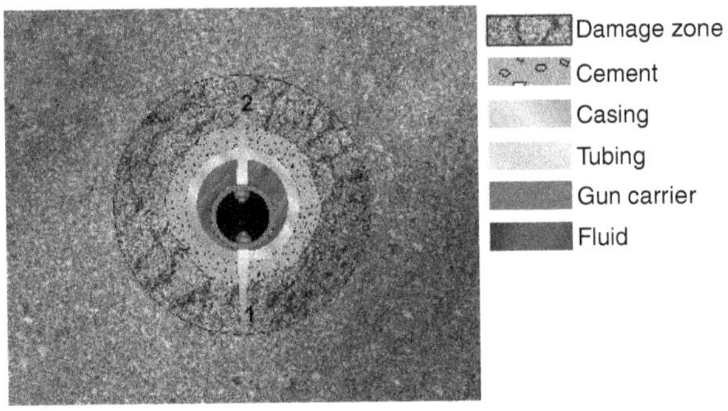

Gun/charge type	2-1/2" Millennium		Avg formation penetration	9.06 In
gun position	eccentered		avg exit hole dia	0.27 In
shot phasing	180 Deg			
Shot No.	1	2		
Orientation, deg	0.0	180.0		
Gun clearance, in	0.0	1.33		
Formation penetration, In	8.87	9.25		
Exit hole dia 1st Csg, In	0.31	0.26		

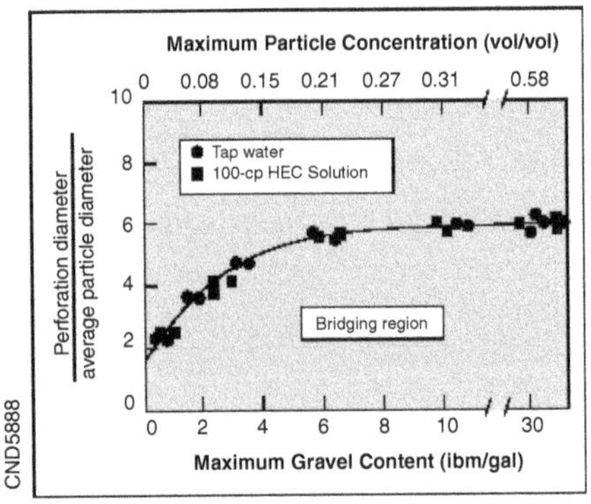

Bridging of particles in perforations

Simulation using perf-pro for 12% porosity, 360°F reservoir temperature and 15,747 psi formation compressive strength

Fig. 22.15—Case History 2—charge performance simulation using Perf-Pro to minimize perforation friction losses.

Perf and near-wellbore friction - F8

Entry friction versus time

	Time (min)	Rate #1 (bpm)	Rate #2 (bpm)	Near-wellbore friction (psi)	Perf. Friction (psi)	Perf. Coef. Mult.	Perfs open (effective)
1	25.71	23.34	0.00	208	231	10.46	43.30
2	0.00	0.00	0.00	0	0	0.00	0.00
3	0.00	0.00	0.00	0	0	0.00	0.00
4	0.00	0.00	0.00	0	0	0.00	0.00
5	0.00	0.00	0.00	0	0	0.00	0.00
6	0.00	0.00	0.00	0	0	0.00	0.00

Perforation data

	Top MD (m)	Bot MD (m)	#Perfs	Diam (in)
1	4,352	4,359	140	0.250
2	0	0	0	0.000
3	0	0	0	0.000
4	0	0	0	0.000
5	0	0	0	0.000
6	0	0	0	0.000
7	0	0	0	0.000

Perforation pressure drop model: Default fracproPT model — Use cramer's perf erosion model

Rate step-down test analyser

Previous test — Curr test 1 — Next test

	Time (min)	Rate #1 (bpm)	Rate #2 (bpm)	Change in friction (psi)
1	23.96	31.38	27.09	108
2	24.84	26.98	18.96	142
3	26.57	11.05	6.04	83
4	27.46	6.01	0.00	120
5	0.00	0.00	0.00	0

Note: Flow rate needs to go all the way to zero.

Display step-down plot — New step-down analysis

Step-down friction analyser

	Estimated value
Total friction power B	0.98
Total friction coeff K	19.95
Perf friction coeff K	0.45
NWB Friction coeff K	43.00
Effective perfs open	42.26
Perf friction (psi)	243
NWB Friction (psi)	208
Max flow rate (bpm)	23.34

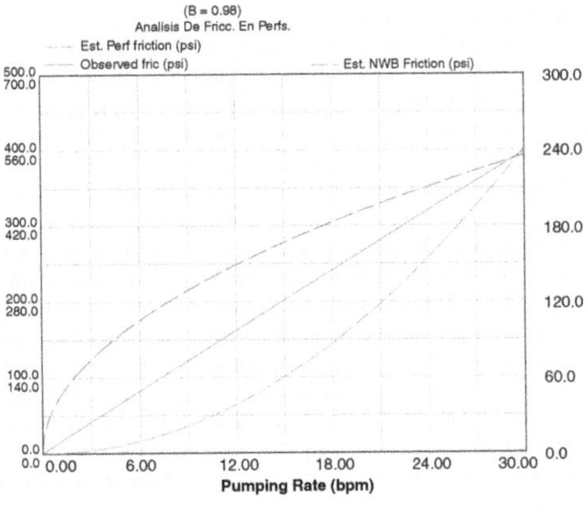

Mini-frac perf and near wellbore friction

Fig. 22.16—Case History 2—Perforation and near wellbore friction analysis from mini-frac.

22.6 Observations and Recommendations

- Hydraulic fracturing in tight gas formations and deep wells needs to be carefully planned. The effect of anisotropy and near well bore tortuosity should be an integral part of the production enhancement design.
- It is recommended to select the perforating charges using a Charge Performance Simulation program to plan for sufficient entry hole size at downhole conditions for the propellant to be used and with maximum formation penetration.
- The Gravity Oriented Perforating technique has proved to be successful with well deviations as low as 5 degrees.
- Using 180 degree charge-to-charge phasing, perforating guns in isotropic formation conditions can reduce the possibility of multiple near field fractures.
- In anisotropic formation conditions (where the anisotropy exceeds approximately 5%), orientation of 180 degree phasing perforation guns in the direction of maximum stress can reduce near well bore friction and tortuosity.
- The stress regime for the formation must be carefully determined and based on the percentage of anisotropy and the well conditions. If anisotropy values are less than 5%, it is recommended to use a 60 degree phasing gun.
- It is recommended to limit the maximum underbalance to the critical pressure during perforating to minimize possibilities of shear failure and possible fines flow back.

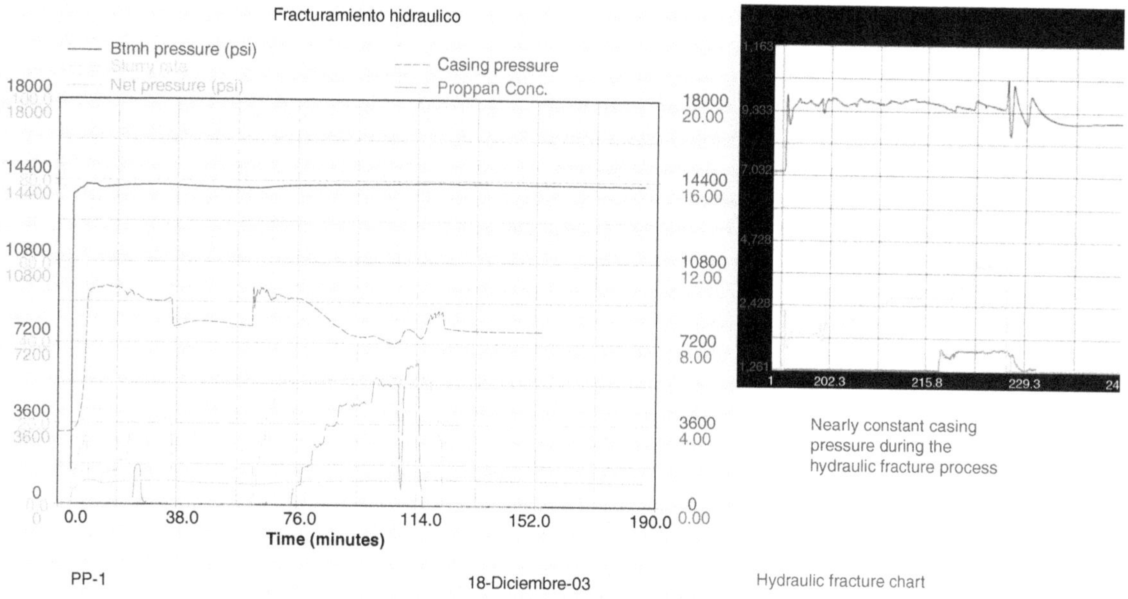

Fig. 22.17—Case History 2—hydraulic fracturing treatment data.

22.7 References

Abass, H., Brumley, J. L., and Venditto, J. J. 1994. Oriented Perforating - A Rock Mechanics View. Paper presented at the SPE Annual Technical Conference and Exhibition, New Orleans, Louisiana., 25–28 September. SPE-28555-MS. https://doi.org/10.2118/28555-MS.

Águila, F. J., Barajas, J. S., Mesa, H. et al. 2003. Using Cross Dipole Sonic Anisotropy Data to Improve Reservoir Understanding in the Southern/Marine Areas of Mexico. Paper presented at the SPE Annual Technical Conference and Exhibition, Denver, Colorado, 5–8 October. SPE-84204-MS. https://doi.org/10.2118/84204-MS.

Anderson, T. and Walker, T. 1972. Log Derived Rock Properties for Use in Stimulation Design. Paper presented at the Fall Meeting of the Society of Petroleum Engineers of AIME, San Antonio, Texas, 8–11 October. SPE-4095-MS. https://doi.org/10.2118/4095-MS.

Boonen, P., Byrd, A. C., Frisch, G. et al. 1998. Improved Selection of Candidates for Stimulation Treatment in the Wilcox Play in the South Texas/Mexico Border Area. Paper presented at the International Petroleum Conference and Exhibition of Mexico, Villahermosa, Mexico, 3–5 March. SPE-39895-MS. https://doi.org/10.2118/39895-MS.

Cipolla C. L., Diabin Liu, and Kyte, D. G. 1994. Practical Application of In-Situ Stress Profiles. Paper presented at the SPE Annual Technical Conference and Exhibition, New Orleans, Louisiana, 25–28 September. SPE-28607-MS. https://doi.org/10.2118/28607-MS.

Fogal, J., Kessler, C., and Varsamis, G. 2002. Application of Shear Anisotropy from a New Generation Crossed Dipole Acoustic Tool. Paper presented at the SPE Annual Technical Conference and Exhibition, San Antonio, Texas, 29 September–2 October 2002. SPE-77792-MS. https://doi.org/10.2118/77792-MS.

Gruesbeck, C. and Collins, R. E. 1982. Particle Transport Through Perforations. *SPE J.* **22** (6). SPE-7006-PA. https://doi.org/10.2118/7006-PA.

Katahara, K. W. 1996. Estimation of In-Situ Stress Profiles from Well Logs. Paper presented at the SPWLA 37th Annual Logging Symposium, New Orleans, Louisiana, 16–19 June. SPWLA-1996-I.

Kessler, C., Varsamis, G., Blanch, J. et al. 2004. New Developments in Monopole Acoustic Logging. Paper presented at the SPWLA 45th Annual Logging Symposium, Noordwijk, Netherlands, 6–9 June 6-9. SPWLA-2004-Z.

Kessler, C. and Varsamis, G. 2001. An New Generation Crossed Dipole Logging Tool: Design and Case Histories. Paper presented at the SPE Annual Technical Conference and Exhibition, 30 September–3 October. SPE-71740-MS. https://doi.org/10.2118/71740-MS.

Kessler, C., Venditto, J., and McMechan, D. 1994. A Synergistic Approach to Optimizing Hydraulic Fracturing. Paper presented at the SPE Asia Pacific Oil and Gas Conference, Melbourne, Australia, 7–10 November. SPE-28802-MS. https://doi.org/10.2118/28802-MS.

Voneiff, G. W. and Holditch, S. A. 1994. Economic Assessment of Applying Advances in Fracturing Technology. *J Pet Technol* **46** (1). SPE-24888-PA. https://doi.org/10.2118/24888-PA.

Whitehead, W. S., Hunt, E. R., Finley, R. J. et al. 1986. In-Situ Stresses: A Comparison Between Log-Derived Values and Actual Field-Measured Values in the Travis Peak Formation of East Texas. Paper presented at the SPE Unconventional Gas Technology Symposium, Louisville, Kentucky, 18–21 May. SPE-15209-MS. https://doi.org/10.2118/15209-MS.

Appendix 22.A

22.A.1 About the Authors

Antonio Hernández Patiño is a geologist with PEMEX in Reynosa, Mexico. Patiño is the exploration geological operations manager for the Reynosa Activo. He has over 20 years with PEMEX. Eduardo Rodriguez García is a petroleum engineer with PEMEX in Reynosa, Mexico. Garcia is in the geological operations group. He is responsible for drilling, well completions, perforating optimization, and stimulation treatment design. He has over 10 years experience with PEMEX. Juan Pablo Torne is the Technical Manager for Halliburton Energy Services Wireline Logging and Perforating product lines in the Northern Mexico Region. Torne has a degree in Electronic Engineering from Columbia. He is a member of SPE and SPWLA, and has been with Halliburton Energy Services for 21 years. Jorge Lazo is currently the Global Service Quality Coordinator for Halliburton Energy Services Production Enhancement. His prior assignment was Business Development Manager for Production Enhancement Services in Mexico. Lazo has a BS in Petroleum Engineering from Universidad Nacional Autonoma de Mexico. He is a member of SPE and has 13 years experience with Halliburton. Calvin Kessler is the Business Development and Commercialization Manager for Wireline and LWD Sonic and Borehole Seismic services at Halliburton Energy Services. He has a BS PE and MS-Mining from New Mexico Institute of Mining and Technology. He is a member of SPWLA, SPE, API, AADE, and CCSG, and has 32 years of experience with Halliburton.

Chapter 23

Economic Assessment of Applying Advances in Fracturing Technology

G. W. Voneiff, SPE, and **S. A. Holditch,** SPE, S. A. Holditch & Assocs Inc.

SPE 24888 presents a case history of economic assessments of the cost and benefit of applying three new technologies in six tight gas formations in U.S. Voneiff and Holditch first discuss the necessity of new fracture technologies through in-depth analyses of fracture fluid problems with no quality control, fracture height overestimate or underestimate without accurate in-situ stress information, and the inadequacy of 2D fracture design models when a formation has multiple pay zones or when stress contrast is not large enough (> 1,500 psi) to contain a fracture. The authors then estimate the costs of applying each technology at three different levels (i.e., none with zero added cost; standard; and intense, with the highest added cost). They compile an "average" well with representative reservoir and fluid properties for each of the six formations using the approach discussed in Case Histories Chapter 9 of this book and compute an optimal fracture length and treatment size. From there, Voneiff and Holditch determine the probability of errors that may be encountered in fracture design and implementation, with and without applying the new technologies. They first compute the gas production and revenue for each possible outcome by use of a 2D reservoir simulation model, and then compute the benefits of applying new technologies. From this research, Voneiff and Holditch conclude the following:

- One needs to use 3D fracture design models with accurate in-situ stress profiles to design optimized fracture treatments
- To pump the optimal fracture treatment as designed, onsite quality control must be applied
- New technologies help eliminate unwanted outcomes or errors from the fracture design and implementation process, which increases the probability of achieving optimal fracture lengths and expected production
- These new technologies can increase gas recovery by 10 to 15% at an added cost of 5 to 10%
- Benefits of applying the new technologies are maximized when formation permeability ranges between 0.01 to 0.1 md, but remain cost effective for permeability in the range between 0.001 to 0.3 md
- Investment in applying the new technologies are lower risk and more efficient compared with the total investment in the well

23.1 Summary

To obtain optimal results from the fracturing process, several new fracturing technologies can be applied to prevent inappropriate stimulation designs and catastrophic fracture-fluid problems. Despite industry awareness of these technologies, they are not applied routinely because of their added expense. In addition, the benefits are difficult to quantify and often are not clearly understood. In SPE 24888, we present an economic assessment of fracture-treatment quality control, in-situ stress profiling, and 3D fracture- propagation models that indicates that these technologies increase gas reserves by 10 to 20% for a 5 to 10% increase in the well cost. The application of these technologies is profitable when the formation gas permeability is between 0.001 and 0.3 md.

23.2 Introduction

In recent years, hydraulic fracturing research performed by the Gas Research Inst. (GRI) and other industry organizations has generated a wealth of new technology, including the GRI Mobile Testing and Control Facility, the GRI Rheology Unit, FRACPRO (a 3D fracture design program), the Treatment Analysis Unit, procedures for developing in-situ stress profiles, and microseismic processing techniques to determine fracture height and azimuth. Also, field experiments performed on Cooperative Research and Staged Field Experiment (SFE) wells have provided comprehensive data sets for developing and testing new ideas (Holditch et al. 1987; Robinson et al. 1991, 1992; S. A. Holditch & Assocs. Inc. 1992).

Our field research shows that the in-situ stress contrasts between rock layers and fracture-fluid viscosity are often quite different from the values assumed or estimated by the design engineer (Holditch et al. 1987; Robinson et al. 1991, 1992; S. A. Holditch & Assocs. Inc. 1992; Ely and Holditch 1988; Whitehead et al. 1986). Errors in assumed values of the in-situ stress contrast lead to inaccurate reservoir descriptions and inappropriate stimulation designs. Fracture-fluid problems compound the situation through poor proppant transport or fracture plugging with unbroken gel. The application of new fracture technology will minimize undesirable stimulation problems and consistently create the optimal length fracture.

A 1990 survey indicates that the industry does not apply new fracture technology routinely because "it costs too much." Unfortunately, the economic benefits of new fracture technology are difficult to quantify and often not clearly understood (Carter and Holditch 1990). In SPE 24888, we quantify the benefits of three new fracturing technologies: fracture-treatment quality control, in-situ stress profiling, and 3D fracture models. We discuss the types of fracture design and fracture-fluid problems that occur regularly and quantify the additional gas that can be produced when technology application consistently eliminates these problems. Results and conclusions are provided for six tight gas formations and a wide range of formation permeabilities.

23.3 Need for New Fracture Technology

23.3.1 Fracture-Treatment Quality Control. Tight gas reservoirs require the placement of a long, propped, hydraulic fracture to produce gas at economical flow rates. The most crucial aspect of successfully creating these long fractures is the ability of the fracture fluid (1) to propagate and open the fracture enough to allow proppant-laden fluid to enter the fracture, (2) to suspend and transport the proppant to the fracture extremities, and (3) to break sufficiently to leave an undamaged fracture face and a high-conductivity proppant pack.

Fluids used to conduct a hydraulic fracture treatment are complex mixtures of chemicals that must meet or exceed the specifications designed by the engineer in the office. Only by being in the field and making fracture- fluid property measurements will one ever know the magnitude of the actual fluid problems that occur during a treatment. Application of fracture-treatment quality control will prevent and/or correct most problems (Ely and Holditch 1988).

Since 1988, S. A. Holditch & Assocs. Inc. has monitored or supervised more than 200 fracture treatments. We found quality-control problems with 70% of those treatments (Ely and Holditch 1988). Chevron Geosciences Co. reported that 80% of its treatments failed to meet specifications because of fracture-fluid problems (Hall and Larkin 1989). Typical fracture-fluid problems are centered on crosslinkers, breakers, bacteria growth, proppant contamination, gel mixing procedures, and buffers.

Many times the chemicals used during a fracture treatment either become contaminated or experience self-degradation. When these problems occur, one must either adjust the concentration of the chemicals or replace them with fresh chemicals.

We do not believe that quality-control problems result from neglect or disinterest by operating or service company personnel. Rather, the fluids used to stimulate tight gas reservoirs are complex systems that require intense quality control to mix and pump correctly. Fracture fluids are very sensitive to chemical concentrations and mixing procedures.

To eliminate gels that are too thin or that will not break, the operator must perform fracture-treatment quality control in the field. We defined three levels of fracture-treatment quality control (GRI 1990, 1992a, 1992b): none, standard, and intense. Increased levels of quality control reduce the frequency of poor proppant transport and gel plugging, creating consistently superior stimulation treatments. **Table 23.1** shows the costs of performing each level of quality control. A typical operator applies technology at either the "none" or "standard" level.

Level	Frequency of Significant Problems (%)	Per-Treatment Cost ($)	Description
None	50%	0	Do not test fluids or inventory materials
Standard	32%	1,000	Inventory materials and check fluid rheology using rotational viscometer and heated cup
Intense	5%	10,000	Inventory materials; test gels before and during treatment at bottomhole temperature and shear rate for length of pump time

Table 23.1—Definition and cost of fracture-treatment quality control.

23.3.2 In-Situ Stress Profiling. An in-situ stress profile is a plot of depth vs. the minimum pressure required to create a hydraulic fracture. These plots are usually constructed by breaking the vertical section into layers and determining the minimum stress in each layer.

The in-situ stress profile clearly controls the shape of a fracture propagating through a layered reservoir (Gidley et al. 1990). The stress contrast between the pay interval and barrier rock (stress in barrier minus stress in pay zone) determines the degree of vertical confinement: a large value of in-situ stress contrast will confine the fracture height and create a longer fracture; a low stress contrast will result in significant height growth and a shorter fracture. The GRIs field research provided strong evidence that significant fracture height growth occurs in formations with a low stress contrast (Holditch et al. 1987; Robinson et al. 1991). **Fig. 23.1** illustrates the influence of in-situ stress contrast on fracture geometry.

When the design engineer does not know the in-situ stress contrast, accurately predicting the length and height of the created fracture becomes difficult. Without knowledge of the created-fracture geometry, the optimal treatment volume cannot be computed

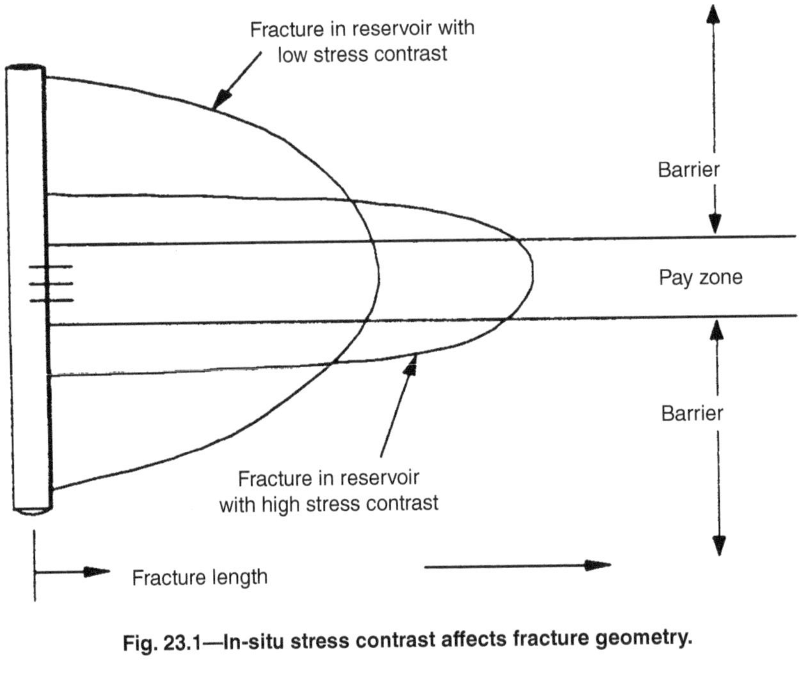

Fig. 23.1—In-situ stress contrast affects fracture geometry.

consistently, resulting in stimulation treatments that can be inappropriate for a given reservoir.

To compute an in-situ stress profile economically, one must use primarily data from logs. A long-spaced sonic log or digital sonic log measures the compressional travel time and the shear travel time through the various rock layers. These travel times determine the velocity ratio, Poisson's ratio, and minimum horizontal stress. Several publications (Holditch et al. 1987; Robinson et al. 1991, 1992; S. A. Holditch & Assocs. Inc. 1992; Gidley et al. 1990) describe the mechanics of performing these calculations.

Despite the ability to estimate the in-situ stress profile from logs, one should calibrate the logs with stress tests. Various correlations (Gidley et al. 1990) allow the engineer to calibrate the log data against the stress tests. **Fig. 23.2** is an example of a calibrated stress profile measured on the SFE No. 3 Cotton Valley interval (Robinson et al. 1992). Note how lithology provides a good indication of the in-situ stress. In this example, the stress profile has been calibrated with cased-hole stress tests. Fig. 23.2 also shows profiles of reservoir gas volume and gas flow capacity.

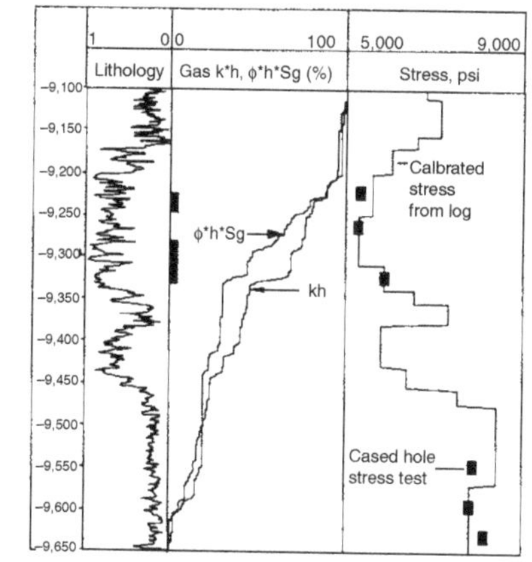

Fig. 23.2—Permeability-thickness and stress profile in SFE No. 3.

On the basis of our experience in generating and using in-situ stress profiles, we defined three levels of stress-profiling technology application: none, standard, and intense. **Table 23.2** describes these technology levels. As with fracture-treatment quality control, the typical operator applies stress technology at either the "none" or "standard" level.

Many engineers tend to overestimate the stress contrast (i.e., assume a fracture height that is too small). However, field research shows that very little fracture height containment exists in most formations, with 200- to 400-ft fracture heights common. In-situ stress profiles estimated without the use of technology may contain enough error to result in fracture lengths that are only 50% of the optimal length. Simple application of in-situ stress profiling combined with 3D fracture models will eliminate treatment volumes that are significantly under- or overdesigned.

23.3.3 3D Fracture Models. In almost every formation investigated through the GRI Tight Gas Sands Program, excessive fracture height growth is part of the engineering problem (Holditch et al. 1987; Robinson et al. 1991, 1992; S. A. Holditch & Assocs. Inc. 1992). The height growth results in fracture geometries that do not resemble the fracture shape predicted by 2D models. In most formations, accurate estimate of the fracture geometry requires a 3D fracture model. Correct modeling of fracture geometry is important because the wrong geometry will cause a very poor estimate of fracture length and can result in inappropriate stimulation designs. **Fig. 23.3** illustrates this problem.

Despite their shortcomings, 2D fracture models have been used successfully for years to design fracture treatments. Many treatments designed with 2D fracture models are pumped to completion and create a profitable well. As a result, it seems difficult to justify using a 3D fracture propagation model to design a stimulation treatment. However, our research shows that a profitable well might not contain the optimal fracture dimensions. Achieving the optimal fracture dimensions will increase gas production or maintain the same gas production with reduced stimulation costs.

Today, 3D fracture models are easy to use. Input data, such as in-situ stresses, Poisson's ratio, and Young's modulus, can be derived through lithology-based correlations if they are not measured with mechanical property logs or stress tests. Commercially available 3D models typically run on personal computers, are menu driven, and return usable results in only minutes· For a small investment of time for each fracture design, the engineer will significantly improve the fracture geometry description and consistently design better stimulation treatments.

Level	95% of Stress Contrast Errors Are Between (psi)	Per-Well Cost ($)	Description
None	± 600	0	Use nonacoustic openhole logs to estimate fracture height
Standard	± 300	10,000	Use acoustic logs to compute and define in-situ stress profile
Intense	± 100	15,000	Calibrate acoustic log stresses with stresses measured from pump-in/ falloff tests performed on one out of every six wells

Table 23.2—Definition and cost of in-situ stress profiling.

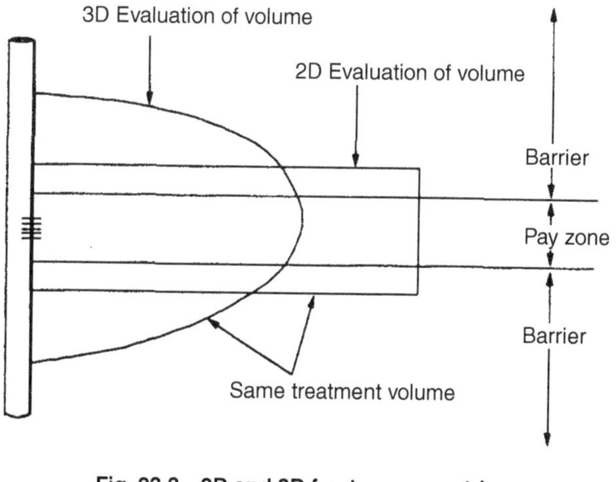

Fig. 23.3—2D and 3D fracture geometries.

23.4 Quantifying Technology-Application Benefits

Even though the application of new fracture technology will improve stimulation designs and eliminate potential fracture-fluid problems, these qualitative benefits do not guarantee that these technologies will increase profit; the benefits must be quantified. However, quantifying the benefits is difficult. New fracture-treatment technology does not result in a precise improvement for every application; rather, it improves results in the majority of applications. When fracture-treatment quality control is applied, fracture-fluid problems are not eliminated, but the chances of a problem occurring are greatly reduced. Likewise, when a stress profile is developed, the profile may contain some degree of error and the resulting fracture treatment may not be optimum, but the stimulation treatment design will be consistently better than a design where the stress profile is not considered or the fracture height is assumed.

23.4.1 Method. We view the fracture design and implementation process as a statistical distribution of possible outcomes. One such outcome might be a fracture design where the in-situ stress contrast is in error by 600 psi and, after the treatment is pumped, the fracture is plugged with unbroken gel. Another possible outcome is that the correct in-situ stress profile is used and no significant fluid problems occur (the optimal outcome). Probability distributions describe how often a particular stress error or fluid problem occurs. Applying technology changes the probability distributions by reducing the frequency of unwanted in-situ stress profile errors and fluid problems.

We use the probability distributions of in-situ stress errors and fluid problems to determine what can be expected on average in terms of incremental cost and incremental profit. Technology application is recommended when the incremental profit is larger than the incremental cost. Listed below are the steps we followed to quantify the benefits of technology application for six typical tight gas basins.

1) Define costs and measurement accuracies of new fracture-treatment technologies.
2) Compile representative reservoir properties for all six formations.
3) Construct probability distributions of in-situ stress profile errors and fracture-fluid properties. These are 2D, *x-y* plots of the magnitude of error vs. the frequency of occurrence. A new distribution is developed for each level of technology application.
4) Compute the fracture-treatment design an engineer would recommend for each incorrect stress profile defined in Step 3. An incorrect stress profile causes the design engineer to recommend a nonoptimal stimulation treatment.
5) Use the correct stress profile to compute the fracture length and revenue that would occur for each incorrectly designed fracture treatment (owing to stress contrast errors) and each fracture-fluid problem.
6) Compute the expected revenue by combining the revenue for each outcome with the probability of each outcome.
7) Compute the benefit as the change in expected revenue when various levels of technology are applied.

We repeated these steps for five permeabilities in each of the six formations. The result is a quantification of the benefits of applying these new fracture-treatment technologies over a wide range of conditions. GRI 1992b gives details of these calculations.

23.4.2 Reservoir Descriptions. We derived data for the six formations from information in our files, industry literature, and GRI technical reports (Holditch et al. 1987; Robinson et al. 1991, 1992; S. A. Holditch & Assocs. Inc. 1992). **Table 23.3** summarizes some of these data. We chose these six formations because they contain large, tight gas resources and are typical areas where the tight gas resource is being developed.

23.4.3 Optimization of Fracture-Treatment Size. In-situ stress profiles and 3D fracture models create an economic benefit by enhancing our ability to design optimal fracture treatments. To perform the optimization, a stimulation treatment volume is found

	Travis Peak, East Texas	Cotton Valley Taylor, East Texas	Canyon, West Texas	Frontier, Wyoming	Corcoran-Cozzette, Colorado	Medina-Clinton, Appalachian Basin
Porosity	0.085	0,08	0.08	0.12	0.08	0.08
Reservoir pressure, psi	4,000	4,900	2,400	3,600	4,000	1,500
Water saturation	0.4	0.4	0.4	0.5	0.5	0.3
Typical gas permeability, md	0.04	0.011	0.037	0.066	0.03	0.088
Permeability range, md						
Minimum	0.001	0.001	0.001	0.001	0.001	0.001
Maximum	0.5	0.2	0.2	0.5	0.5	0.5
Average depth, ft	8,500	9,500	6,000	8,000	7,500	5,000
Net pay, ft	50	75	50	25	50	30
Bottomhole temperature, °F	250	275	175	200	200	125
Gas specific gravity	0.6	0.6	0.6	0.6	0.6	0.6
Well spacing, acres	160	320	160	160	160	160
Bottomhole flowing pressure, psi	1,000	1,000	500	1,000	1,000	200
Typical proppant	Sand	Sand	Sand	Sand	Sand	Sand
Total fluid-loss coefficient, ft/min	0.0005	0.001	0.001	0.001	0.0005	0.001

Table 23.3—Summary of reservoir descriptions for six formations.

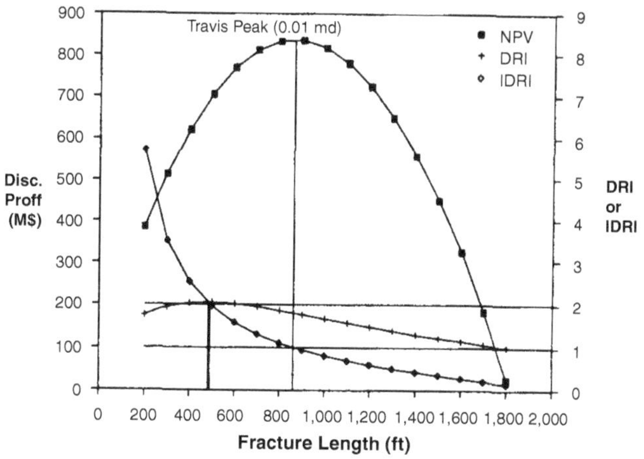

Fig. 23.4—Economic indicators as functions of fracture length for the Travis Peak Formation when the gas permeability is 0.01 md.

that maximizes or achieves some predetermined level of an economic parameter, such as profit or investment efficiency. To optimize a treatment, one must compute the stimulation cost and discounted revenue as functions of fracture length. We will briefly describe how we perform a fracture-treatment optimization for this analysis. Additional details are available elsewhere (GRI 1990, 1992a, 1992b).

We calculated stimulation cost as a function of fracture length using FRACPRO, a lumped 3D fracture-propagation model. We converted treatment sizes to stimulation costs using typical costs for proppant, fracture fluid, and pumping equipment.

We used a single-phase, 2D reservoir model to compute gas production as a function of propped fracture length. Table 23.3 indicates much of the input data. The gas production was discounted at 10% to obtain a discounted gas volume and multiplied by a constant gas price of $2.00/Mcf to compute a present-value revenue.

For each fracture length, we computed three economic indicators using the stimulation costs and the discounted revenue. Two indicators, present-value profit and total investment efficiency, were used to compare our stimulation designs with designs that maximize these indicators. We used a third indicator, incremental investment efficiency, to optimize the treatment size for our analysis. We define the three indicators as follows.

1. Net present value (NPV) = (discounted revenue) – (total well cost).
2. Discounted revenue-to-investment ratio (DRI) = (discounted revenue)/(total well cost).
3. Incremental DRI (IDRI) = (incremental discounted revenue)/(incremental cost).

Although the IDRI is the most complicated of the three indicators, it provides direct quantification of the benefit of pumping incrementally larger fracture treatments. **Fig. 23.4** plots the three economic indicators vs. fracture length for the Travis Peak Formation when the formation permeability is 0.01 md. Similar graphs were constructed for the other five formations for every value of gas permeability investigated.

IDRI = 1 marks the point of maximum profit (maximum NPV). At this point, the last dollar spent on stimulation returns $1 of incremental revenue, and incremental expenditures to create longer fractures will not be profitable. Fig. 23.4 shows this to occur at an 850-ft fracture length. The fracture length where the IDRI and DRI curves cross indicates a treatment size that maximizes the investment efficiency of the total well cost. Creating fractures longer than this length will continue to increase the total profit, but incremental expenditures will return a smaller revenue-to-cost ratio than the total well investment. In Fig. 23.4, this occurs at a fracture length of 490 ft.

We chose the optimal fracture length as the length where IDRI = 2. This fracture length usually falls between the fracture lengths that maximize NPV and DRI. This is a compromise of the various indicators typically used in the petroleum industry. When IDRI = 2, the last dollar spent on the fracture treatment will generate $2 of additional discounted revenue. In Fig. 23.4, this corresponds to a fracture length of 492 ft, which happens to correspond to the length that achieves the maximum DRI.

23.4.4 Effect of Permeability on Optimal Fracture Length. The optimal fracture length depends heavily on the value of in-situ gas permeability in the formation. As permeability decreases below 0.1 md, optimal lengths increase. In two formations, the Cotton Valley Taylor and Corcoran-Cozzette, we computed optimal fracture half-lengths greater than 1,500 ft. In these cases, we reduced the optimal fracture half-length to 1,500 ft to stay within the probable limits of proppant transport. Proppant transport within a fracture is a very complex process that is not thoroughly understood. Although some proppant-transport models may predict very long propped fractures, there is little evidence from production performance that conductive fractures longer than 1,500 ft can be created consistently in layered, complex, tight gas reservoirs.

Fig. 23.5 illustrates the influence of formation permeability on optimal fracture half-lengths. Only in high-stress-contrast formations, such as the Cotton Valley Taylor and Corcoran-Cozzette, are fracture lengths greater than 1,000 ft recommended.

23.4.5 Statistical Fracturing Description—The Expected Outcome. Even though the fracture design and implementation process results in a distribution of possible outcomes, we can compute what is expected to occur on average. Simple statistical equations multiply the fracture length and revenue of each possible outcome by the probability of each outcome to derive expected values (Milton and Arnold 1990) of fracture length and revenue. These expected values represent what should occur on average.

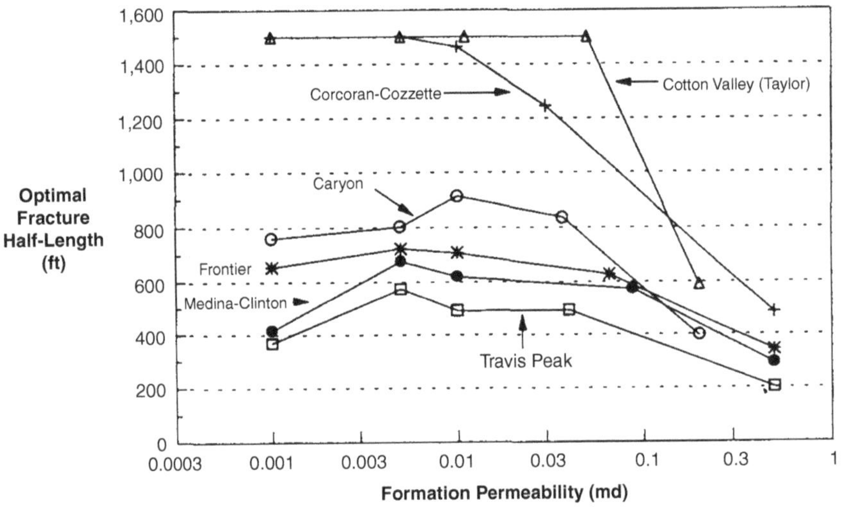

Fig. 23.5—Effect of formation permeability on optimal fracture length.

By applying technology, we can change how often certain outcomes occur (change the distribution of outcomes), causing the expected outcome to change as well.

Our database shows that unwanted outcomes (i.e., plugged fracture or no proppant transport combined with a sizable stress profile error) occur less often when fracture-fluid quality-control technology is applied. **Figs. 23.6 through 23.8** show how technology application changes the probability distribution of potential fluid viscosity and in-situ stress contrast errors, focusing the most likely outcomes toward the optimal outcome. Fig. 23.6 represents a "none" level application of technology; Fig. 23.7 represents a "standard" level of technology application; and Fig. 23.8 represents an "intense" level of technology application. GRI 1992b describes how these figures were constructed.

Note that the most likely outcome in Fig. 23.6 is "more height growth than anticipated" and no fluid problems. Also apparent from Fig. 23.6 are the significant probabilities of many unwanted outcomes, such as gel plugging and poor proppant transport. Figs. 23.7 and 23.8 show that, as more technology is applied, these unwanted events become less likely.

The probability distributions in Figs. 23.6 through 23.8 are used with statistical equations (Milton and Arnold 1990) to determine what is expected on average. Here, we represented the distributions shown in these figures with 35 specific outcomes. For each specific outcome, we calculated the type of stimulation treatment an engineer would recommend, the actual fracture length that would be created, and the amount of discounted revenue and profit that would result. The revenue and profit for each specific outcome were multiplied by the probability of that outcome, and then all 35 resultants were summed to compute the expected revenue and expected profit. GRI 1992b details these computations.

Table 23.4 shows the results of applying our method to a Travis Peak data set where the formation permeability is 0.04 md. By applying technology, we reduce the occurrence of fluid problems and stress estimation errors and increase the expected revenue toward the optimal revenue. The benefit of applying technology is the difference between the average revenue with technology and the average revenue without technology. Table 23.4 shows this benefit to be \$441, 000 (a 22% increase) for an "intense" application of technology in the Travis Peak Formation. The cost of the technology application is \$25,000 (\$10,000 controlling quality +\$15,000 in-situ stress profiling).

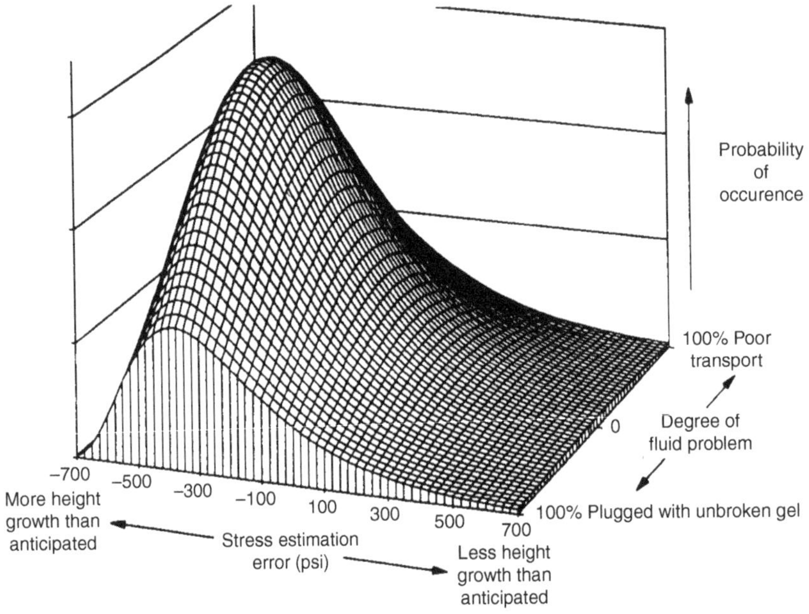

Fig. 23.6—Probability distribution of fracture outcomes (no technology application).

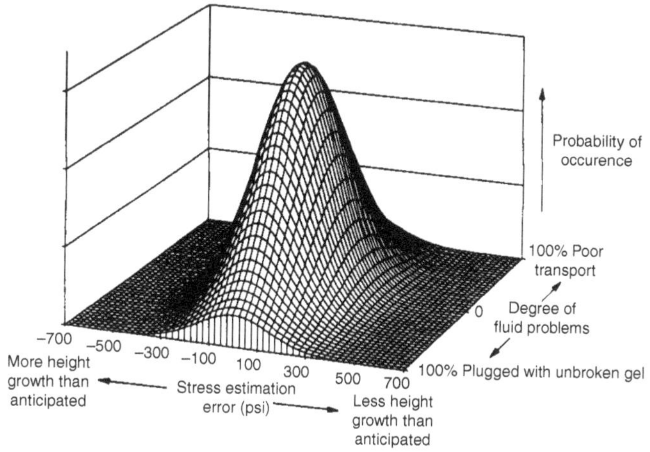

Fig. 23.7—Probability distribution of fracture outcomes (standard technology application).

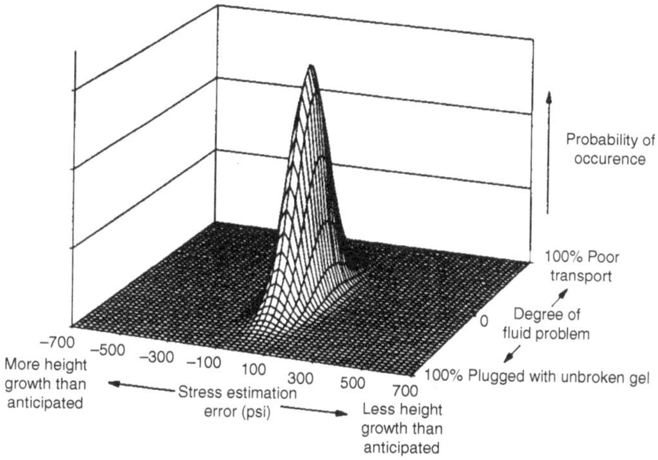

Fig. 23.8—Probability distribution of fracture outcomes (intense technology application).

Outcome of Fracture Design and Implementation Process	Discounted Revenue ($)	Technology Cost ($)
Range of outcomes		
Optimal	2,651,000	NA
Worst conditions	1,800,000	NA
Statistical evaluation of outcomes		
Average with intense technology	2,475,000	25,000
Average without technology	2,034,000	—
Technology benefit and cost	441,000	25,000

NA = not applicable.

Table 23.4—Example benefit calculation for the Travis Peak Formation when the formation permeability is 0.04 md and technology is applied at an intense level.

We also evaluated the use of 3D fracture models vs. 2D models. We computed the fracture lengths, revenues, and costs that result when either a 2D or 3D model is used to design the treatment volume (GRI 1990). When 2D models are applied, the industry commonly uses either a Geertsma-deKlerk (GDK) (Gidley et al. 1990) or a Perkins-Kerns-Nordgren (PKN) (Gidley et al. 1990) model to design a fracture treatment.

The 2D models usually overestimate optimal length and result in treatment volumes that are either larger or smaller (depending on reservoir conditions) than the true optimal volume. The differences are often so large that either significant sums of money are omitted from worthy projects or fracture-treatment costs become unnecessarily large, sometimes to the point that the treatment is not profitable.

	Travis Peak	Canyon	Frontier	Medina Clinton	Cotton Valley Taylor	Corcoran-Cozzette
	Recommended Treatment Size (lbm sand)					
FRACPRO	1,451,400	1,744,000	848,000	671,200	2,162,000	1,734,000
GDK	1,882,000	1,872,000	942,000	631,600	2,504,000	1,596,000
PKN	1,410,000	1,662,000	673,000	344,800	1,204,000	765,000
	Predicted Length (ft)					
FRACPRO	492	830	628	573	1,500	1,245
GDK	1,134	927	986	914	1,500	1,363
PKN	1,500	1,356	1,482	1,500	1,500	1,500
	Actual Length (ft)					
FRACPRO	492	830	628	573	1,500	1,245
GDK	556	870	644	554	1,651	1,185
PKN	476	805	560	438	1,030	748

Table 23.5—"Optimal" fracture length and recommended treatment volumes from various fracture design models.

Table 23.5 shows optimal length estimates and treatment volumes developed with 2D models. This table also shows the length that would result if these designs were pumped. The "actual" values were computed with FRACPRO. Even when a 2D model adequately estimates the optimal treatment volume, it tends to overpredict the created length. Optimistic fracture-length estimates can lead to incorrect well spacing and inadequate field development planning.

23.5 Summary of Benefits

Fig. 23.9 shows the results of our benefits analysis. Three bars are shown for each formation. The first and second bars are the benefit-to-cost ratios of applying fracture-treatment quality control and in-situ stress profiling; the third bar represents a combined application of fracture-treatment quality control, in-situ stress profiling, and 3D fracture models. Bars exceeding a value of one indicate a profitable application of technology. Fig. 23.9, reflecting results at each formation's average permeability, shows that technology application is profitable in a wide variety of tight gas reservoirs. Stress profiling in the Medina-Clinton Formation is marginally profitable, indicating a need for less expensive stress profiling.

The benefits in Fig. 23.9 represent increases in total gas recovery of 10 to 20% (50 to 400 MMscf, depending on the formation). These technologies do not turn bad wells into great wells, but they do consistently improve recovery from hydraulically fractured tight gas formations. The moderate increases in gas recovery come at a cost of 5 to 10% of the total well cost. This makes application of these technologies a desirable investment opportunity, considering the return one expects on total well investment.

When evaluating Fig. 23.9, note that the benefit-to-cost ratio, not the benefit, is plotted. Although fracture-treatment quality control generates the largest benefit-to-cost ratios, it does not necessarily generate the largest incremental gas reserves. The combined application of all three technologies (black bars) will recover more additional gas than fracture-treatment quality control alone, but fracture-treatment quality control costs proportionately less and therefore achieves a greater benefit-to-cost ratio.

Fig. 23.9 was constructed for a single permeability in each formation. We repeated these benefit calculations for a permeability range spanning roughly 90% of the permeabilities found in each formation (GRI 1992b). **Fig. 23.10** shows the effect of permeability on the benefit-to-cost ratio for an "intense" level of technology application in the Travis Peak Formation.

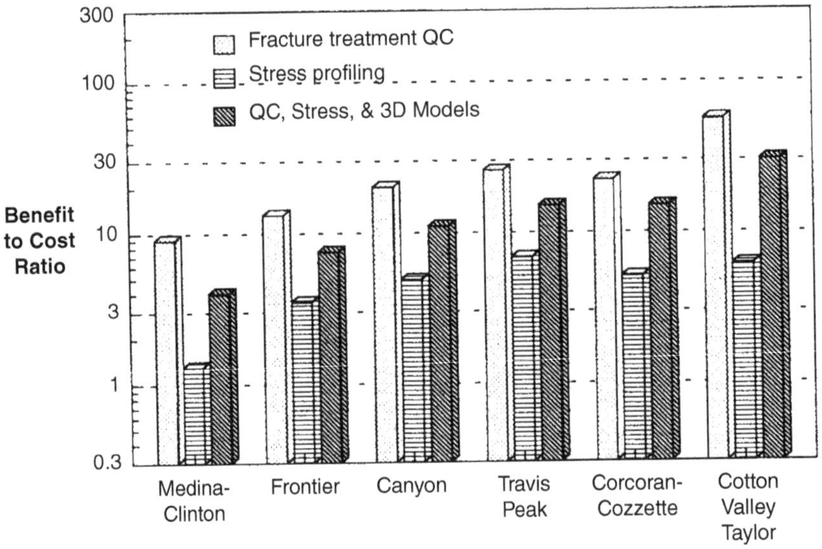

Fig. 23.9—Benefit-to-cost ratio of new fracture technology for six formations.

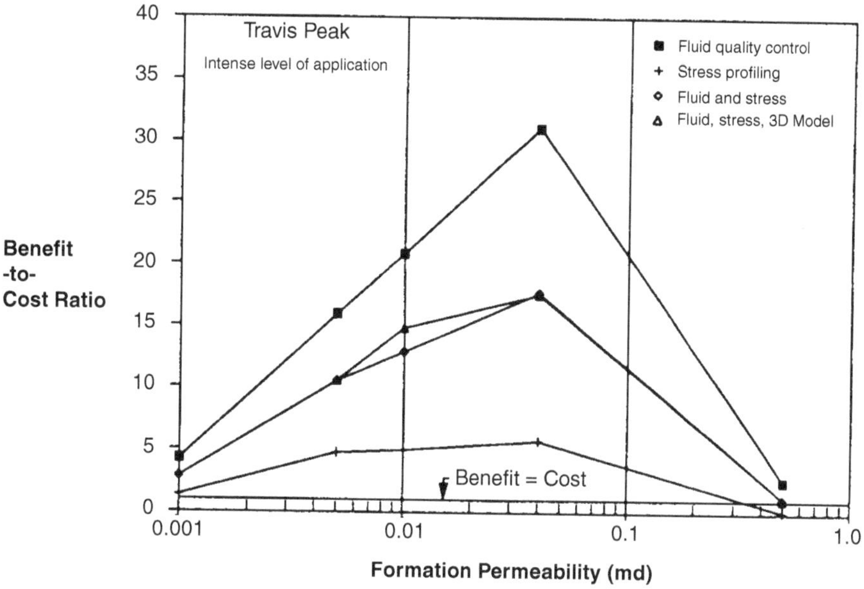

Fig. 23.10—Effect of permeability on the benefit-to-cost ratio of applying technology in the Travis Peak.

The permeability extremes where new fracture technology loses its cost-effectiveness correlate roughly with the economic limits of hydraulic fracturing itself. At permeabilities below 0.001 md, the well as a whole will not be profitable unless a very thick pay zone exists. At permeabilities above 1 md, the sensitivity of gas recovery to fracture length is reduced greatly.

Graphs similar to **Fig. 23.10** constructed for the other five formations generate the same conclusions: technology benefits are maximized when the formation permeability is between 0.01 and 0.1 md, and the range of profitable application is 0.001 to 0.3 md. In all cases, fracture-treatment quality control appears to have a slightly larger range of application than stress profiling or 3D fracture models.

23.6 Conclusions

1. Applying new fracture-treatment technologies in typical tight gas sand reservoirs can increase the total discounted gas recovery by 10 to 20%, for an increase in the total well cost of only 5 to 10%.
2. The investment efficiency of expenditures to apply new fracture technology generally is greater than the investment efficiency of the total well cost. This makes application of these technologies competitive with the well investment in tight gas formations.
3. The benefits of applying new fracture technology are maximized when the formation gas permeability is between 0.01 and 0.1 md. These technologies are profitable for gas permeabilities between 0.001 and 0.3 md.
4. To design the optimal treatment consistently, the design engineer must determine the in-situ stress profile and use a 3D fracture-propagation model.
5. To ensure that the optimal fracture treatment is pumped as designed, fracture-treatment quality control must be applied in the field.
6. Application of new fracture technology reduces the occurrence of unwanted outcomes from the fracture design and implementation process. Preventing these unwanted outcomes focuses the expected outcome toward the optimal outcome, increasing profit.

23.7 Nomenclature

h = porosity, fraction

kh = formation gas flow capacity, md-ft

S_g = gas saturation, fraction

ϕ = porosity, fraction

23.8 References

Carter, R. H. and Holditch, S. A. 1990. An Evaluation of Present and Future Fracture Treatment Design, Control and Fluid Quality Control Practices—An Industry Survey. GRI, Chicago, Illinois.

Ely, J. W. and Holditch, S. A. 1988. Field Experience with the GRI Rheology Unit. Paper presented at the SPE Gas Technology Symposium, Dallas, Texas, 13–15 June. SPE-17715-MS. https://doi.org/10.2118/17715-MS.

Gas Research Institute. 1990. Evaluating the Benefits of Tight Gas Sands Research—Part I: Parametric Study To Evaluate Benefits of Fracture Fluid Quality Control and In-Situ Stress Research. GRI, Chicago, Illinois.

Gas Research Institute. 1992a. Evaluating the Benefits of Tight Gas Sands Research—Part II: Parametric Study To Evaluate Benefits of Fracture Fluid Quality Control and In-Situ Stress Profiling in Six Tight Gas Formations. Contract No. 5091-221-2124, Report No. GRI/92-0040, GRI, Chicago, Illinois (1992).

Gas Research Institute. 1992b. Evaluating the Benefits of Tight Gas Sands Research—Part III: A Statistical Approach. Topical Report. Contract No. 5091-221-2129, GRI, Chicago, Illinois (December 1992).

Gidley, J. L., Holditch, S. A., Nierode, D. E. et al. eds. 1990. *Recent Advances in Hydraulic Fracturing,* Vol. 12, Richardson, Texas: Monograph Series, Society of Petroleum Engineers.
Hall, B. E. and Larkin, S. D. 1989. On-Site Quality Control of Fracture Treatments. *J Pet Technol* **41** (5). SPE-15014-PA. https://doi.org/10.2118/15014-PA.
Holditch, S. A., Robinson, S. M., Whitehead, W. S. et al. 1987. The GRI Staged Field Experiment. Paper presented at the SPE/DOE Joint Symposium on Low Permebility Reservoirs, Denver, Colorado, 18–19 May. SPE-16429-MS. https://doi.org/10.2118/16429-MS.
Milton, J. S. and Arnold, J. C. 1990. *Introduction to Probability and Statistics.* New York City: McGraw-Hill Book Co.
Robinson, B. M., Holditch, S. A., and Peterson, R. E. 1991. The Gas Research Institute's Second Staged Field Experiment: A Study of Hydraulic Fracturing. Paper presented SPE Gas Technology Symposium, Houston, Texas. 22–24 January. SPE-21495-MS. https://doi.org/10.2118/21495-MS.
Robinson, B. M., Holditch, S. A., Whitehead, W. S. et al. 1992. Hydraulic Fracturing Research in East Texas: Third GRI Staged Field Experiment. *J Pet Technol* **44** (1). SPE-22878-PA. https://doi.org/10.2118/22878-PA.
S. A. Holditch & Assocs. Inc. 1992. Staged Field Experiment No. 4—Application of Advanced Technologies in Tight Gas Sandstones—Frontier Formation, Wyoming. CER Corporation, prepared for the Gas Research Institute, Contract Report No. GRI-92/0394 (1992).
Whitehead, W. S. Hunt, E. R., Finley, R. J. et al. 1986. In-Situ Stresses: A Comparison Between Log-Derived Values and Actual Field-Measured Values in the Travis Peak Formation of East Texas. Paper presented at the SPE Unconventional Gas Technology Symposium, Louisville, Kentucky, 18–21 May. SPE-15209-MS. https://doi.org/10.2118/15209-MS.

Distinguished Author Series articles are general, descriptive representations that summarize the state of the art in an area of technology by describing recent developments for readers who are not specialists in the topics discussed. Written by individuals recognized as experts in the area, these articles provide key references to more definitive work and present specific details only to illustrate the technology.

SI Metric Conversion Factors

acre × 4.046 873	E–01 = ha
ft × 3.048*	E–0 = m
ft^3 × 2.831 685	E–02 = m^3
°F (°F–32)/1.8	=°C
lbm × 4.535 924	E–01 = kg
psi × 6.894 757	E+00 = kPa

*Conversion factor is exact.

Appendix 23.A

23.A.1 About the Authors

Voneiff

Holditch

George W. Voneiff is a senior petroleum engineer at S. A. Holditch & Assocs. Inc. in College Station, TX. He previously worked for Enserch Exploration Inc. as a reservoir engineer for several large tight gas fields in the East Texas basin. Current interests include Voneiff Holditch reservoir simulation of hydraulically fractured wells, statistical analysis of production and completion characteristics, and analysis of long-term production data. Voneiff holds BS and MS degrees in petroleum engineering from Texas A&M U. Stephen A. Holditch was president of S. A. Holditch & Assocs. Inc. and professor of petroleum engineering at Texas A&M U. in College Station, where he held the Shell Distinguished Chair in Petroleum Engineering during 1983–87. Holditch received the 1981 SPE Distinguished Achievement Award for Petroleum Engineering Faculty. He taught the SPE Short Course on hydraulic fracturing and was an associate editor of *Recent Advances in Hydraulic Fracturing,* Vol. 12, in the SPE Monograph Series. He was a member of the Technology Today Series Committee and was 1991–93 chair of the Reprint Series Committee. Holditch chaired the program committee for the 1988 Gas Technology Symposium and was a member of the 1988-89 Distinguished Author Series, 1978–80 Editorial Review, and 1973–74 Education and Accreditation committees. He was an SPE Distinguished Member. Holditch holds BS, MS, and PhD degrees in petroleum engineering from Texas A&M U.

Bibliography

The collection of papers in this book represents the top works we felt were right to include. However, they do not constitute the full body of work on these topics. To read more, check out the additional authors and topics below. Papers appearing in this reprint series volume are indicated with an **[R]** after the entry; those not appearing in this series of case studies do not.

Abacioglu, Y., Sebastian, H. M., and Bonney, K. E. 2013. Case History of Using Integrated Reservoir Modeling Workflow for Tight Gas Reservoirs. Paper presented at the SPE Unconventional Gas Conference and Exhibition, Muscat, Oman, 28–30 January. SPE-163989-MS. https://doi.org/10.2118/163989-MS.

Adeyeye, A., Perez, R., Boyer, M. et al. 2013. Multizone Tight Gas Completions in the Piceance Basin: Over a Decade of Learnings. Paper presented at the International Petroleum Technology Conference, Beijing, China, 26–28 March. IPTC-17129-MS. https://doi.org/10.2523/IPTC-17129-MS.

Ahmed, U., Schatz, J. F., Abou-Sayed, A. S. et al. 1982. A Step by Step Approach to Hydraulic Fracture Treatment Design, Implementation, and Analysis for Tight Gas Sands. Paper presented at the SPE Unconventional Gas Recovery Symposium, Pittsburgh, Pennsylvania, USA, 16–18 May. https://doi.org/10.2118/10829-MS.

Al-Ali, A. H., Al-Anazi, H. A., and Aziz, A. A. A. 2015. Successful Exploitation of Heterogeneous Unconsolidated Clastic Gas Reservoirs in Saudi Arabia: Case Histories. Paper presented at the SPE Middle East Oil & Gas Show and Conference, Manama, Bahrain, 8–11 March. SPE-172617-MS. https://doi.org/10.2118/172617-MS.

Arevalo-Villagran, J. A., Cinco-Ley, H., Wattenbarger, R. A. et al. 2003. Transient Analysis of Tight Gas Well Performance–More Case Histories. Paper presented at the SPE Annual Technical Conference and Exhibition, Denver, Colorado, 5–8 October. SPE-84476-MS. https://doi.org/10.2118/84476-MS.

Arevalo-Villagran, J. A., Wattenbarger, R. A., and El-Banbi, A. H. 2000. Production Analysis of Commingled Gas Reservoirs–Case Histories. Paper presented at the SPE International Petroleum Conference and Exhibition in Mexico, Villahermosa, Mexico, 1–3 February. SPE-58985-MS. https://doi.org/10.2118/58985-MS. **[R]**

Arevalo-Villagran, J. A., Wattenbarger, R. A., Samaniego-Verduzco, F. et al. 2001. Production Analysis of Long-Term Linear Flow in Tight Gas Reservoirs: Case Histories. Paper presented at the SPE Annual Technical Conference and Exhibition, New Orleans, Louisiana, 30 September–3 October. SPE-71516-MS. https://doi.org/10.2118/71516-MS.

Bassett, L. 2010. Case History Using ESP. Paper presented at the SPE Asia Pacific Oil and Gas Conference and Exhibition, Brisbane, Queensland, Australia, 18–20 October. SPE-133464-MS. https://doi.org/10.2118/133464-MS.

Baumgartner, W. E., Shlyapobersky, J., Abou Sayed, I. S. et al. 1993. Fracture Stimulation of a Horizontal Well in a Deep, Tight Gas Reservoir: A Case History From Offshore The Netherlands. Paper presented at the Offshore Europe conference, Aberdeen, Scotland, United Kingdom, 7–10 September. SPE-26795-MS. https://doi.org/10.2118/26795-MS. **[R]**

Berumen, S., Cipolla, C., Finkbeiner, T. 2000. Integrated Reservoir Geomechanics Techniques in the Burgos Basin Mexico: An Improved Gas Reservoirs Management. Paper presented at the SPE Asia Pacific Conference on Integrated Modeling for Asset Management, Yokohama, Japan, 25–26 April. SPE-59418-MS. https://doi.org/10.2118/59418-MS. **[R]**

Bezilla, M., Adewumi, M. A., Ertekin, T. et al. 1989. Production Strategies for Tight Gas Sands: A Case Study of the Upper Cozzette Blanket Sand. Paper presented at the SPE Gas Technology Symposium, Dallas, Texas, 7–9 June. SPE-19108-MS. https://doi.org/10.2118/19108-MS.

Boutaud de la Combe, J. L., Jaffrezic, V., and Loizel, H. 2013. Dynamic Characterisation of a Chinese Fluvial Tight Gas Reservoir. Paper presented at the International Petroleum Technology Conference, Beijing, China, 26–28 March. IPTC-16882-MS. https://doi.org/10.2523/IPTC-16882-MS.

Charles, J. G., Hydock, K. D., Udick, T. H. et al. 1982. Stimulation Evaluation in the Appalachian Tight Gas Sands. Paper presented at the SPE Eastern Regional Meeting, Washington, DC, 3–5 November. SPE-11257-MS. https://doi.org/10.2118/11257-MS.

Charles, J. G., Hudock, K. D., Udick, T. H. et al. 1983. Evaluation of Stimulation Treatments in Appalachian Tight Gas Sands. Paper presented at the SPE/DOE Low Permeability Gas Reservoirs Symposium, Denver, Colorado, 14–16 March. SPE-11601-MS. https://doi.org/10.2118/11601-MS.

Cipolla, C. L. and Mayerhofer, M. 1996. Infill Drilling & Reserve Growth Determination in Lenticular Tight Gas Sands. Paper presented at the SPE Annual Technical Conference and Exhibition, Denver, Colorado, 6–9 October. SPE-36735-MS. https://doi.org/10.2118/36735-MS.

Cipolla, C. L., McCarley, D. L., Peterman, F. et al. 2005. Effect of Well Placement on Production and Frac Design in a Mature Tight Gas Field. Paper presented at the SPE Annual Technical Conference and Exhibition, Dallas, Texas, 9–12 October. SPE-95337-MS. https://doi.org/10.2118/95337-MS.

Clayton, F. M. and Gordon, N. C. 1990. The Leman F and G Development: Obtaining Commercial Production Rates From a Tight Gas Reservoir. Paper presented at the European Petroleum Conference, The Hague, The Netherlands, 21–24 October. SPE-20993-MS. https://doi.org/10.2118/20993-MS.

Colwell, D. A. F., O'Brien, C. G., and Gates, T. D. 2006. Evolution of Completion Practices in the Wild River Tight Gas Field. *SPE Drill & Compl* **21** (1): 54–61. SPE-89719-PA. https://doi.org/10.2118/89719-PA.

Craig, D. P., Odegard, C. E., Pearson, W. C. et al. 2000. Case History: Observations From Diagnostic Injection Tests in Multiple Pay Sands of the Mamm Creek Field, Piceance Basin, Colorado. Paper presented at the SPE Rocky Mountain Regional/Low-Permeability Reservoirs Symposium and Exhibition, Denver, Colorado, 12–15 March. SPE-60321-MS. https://doi.org/10.2118/60321-MS.

Davis, J. S., Belien, I., Liu, X. et al. 2011. Modeling Gas Migration, Distribution, and Saturation in a Structurally and Petrologically Evolving Tight Gas Reservoir. Paper presented at the International Petroleum Technology Conference, Bangkok, Thailand, 15–17 November. IPTC-14621-MS. https://doi.org/10.2523/IPTC-14621-MS.

de Oliveira Neto, J. M. and Pritchard, T. 2016. Deploying a Fit-For-Purpose Formation Evaluation Program to De-Risk Tight Gas Sand Plays. Paper presented at the SPWLA 57th Annual Logging Symposium, Reykjavik, Iceland, 25–29 June. SPWLA-2016-N.

Duda, J. R. and Hancock, J. S. 1989. Tight-Sand Development Potential in the Southern Rulison Area, Garfield County, Colorado. *J Pet Technol* **41** (5): 551–557. SPE-16426-PA. https://doi.org/10.2118/16426-PA.

Eberhard, M. J., Mullen, M. J., Seal, C. A. et al. 2000. Integrated Field Study for Production Optimization: Jonah Field–Sublette County, Wyoming. Paper presented at the SPE/CERI Gas Technology Symposium, Calgary, Alberta, Canada, 3–5 April. SPE-59790-MS. https://doi.org/10.2118/59790-MS. **[R]**

Eller, J. G., Garner, J. J., Snider, P. et al. 2002. A Case History: Use of a Casing-Conveyed Perforating System to Improve Life of Well Economics in Tight Gas Sands. Paper presented at the SPE Western Regional/AAPG Pacific Section Joint Meeting, Anchorage, Alaska, 20–22 May. SPE-76742-MS. https://doi.org/10.2118/76742-MS.

Elshahawi, H., Hashem, M. N., Ardila, M. et al. 2005. The Power of Real-Time Monitoring and Interpretation in Wireline Formation Testing-Case Studies. Paper presented at the SPE Annual Technical Conference and Exhibition, Dallas, Texas, 9–12 October. SPE-94708-MS. https://doi.org/10.2118/94708-MS.

Engineer, R. 1985. Cal Canal Field, California: Case History of a Tight and Abnormally Pressured Gas Condensate Reservoir. Paper presented at the SPE California Regional Meeting, Bakersfield, California, 27–29 March. SPE-13650-MS. https://doi.org/10.2118/13650-MS.

Fairhurst, D. L., Indriati, S., Reynolds, B. W. et al. 2007. Advanced Technology Completion Strategies for Marginal Tight Gas Sand Reservoirs: A Production Optimization Case Study in South Texas. Paper presented at the SPE Annual Technical Conference and Exhibition, Anaheim, California, USA, 11–14 November. SPE-109863-MS. https://doi.org/10.2118/109863-MS.

Finley, R. J. and O'Shea, P. A. 1983. Geologic and Engineering Analysis of Blanket-Geometry Tight Gas Sandstones. Paper presented at the SPE/DOE Low Permeability Gas Reservoirs Symposium, Denver, Colorado, 14–16 March. SPE-11607-MS. https://doi.org/10.2118/11607-MS.

Franco, G., Davis, T. L., and Grechka, V. 2007. Seismic Anisotropy of Tight-Gas Sandstones, Rulison Field, Piceance Basin, Colorado. Paper presented at the SEG Annual Meeting, San Antonio, Texas, 23–28 September. SEG-2007-1461.

Fyten, G. C., Taylor, R. S., and Price, D. 2015. Viking Stimulation: Case History. Paper presented at th SPE/CSUR Unconventional Resources Conference, Calgary, Alberta, Canada, 20–22 October. SPE-17955-MS. https://doi.org/10.2118/17955-MS.

Greenfield, H. S. and Ahmed, U. 1983. Fracture Stimulation Treatment of a Western Tight Sand Reservoir: A Case History. Paper presented at the SPE Rocky Mountain Regional Meeting, Salt Lake City, Utah, USA, 22–25 May. SPE-11825-MS. https://doi.org/10.2118/11825-MS.

Guyatt, R. C. P. and Allen, J. P. 1996. Application of Horizontal Wells to a Tight Gas Sandstone Reservoir: A Case History. *SPE Res Eng* **11** (3): 203–209.SPE-35640-PA. https://doi.org/10.2118/35640-PA.

Han, J. and Wang, J.Y. 2014. Development and Application of A Type Curve for Gel Damage Identification in Tight Gas Wells. *Intl J of Pet Eng* **1** (1): 34–48. https://doi.org/10.1504/IJPE.2014.059419.

Harms, L. K. 2010. Wellhead Compression on Tight Gas Wells in the Long Run: A Follow-Up Case History on Seven Years of Success in Lobo. Society of Petroleum Engineers. Paper presented at the Tight Gas Completions Conference, San Antonio, Texas, 2–3 November. SPE-138488-MS. https://doi.org/10.2118/138488-MS.

Harting, T. A., Cox, D. L., Murillo, R. et al. 2004. Application of Mini-Falloff Test to Determine Reservoir Parameters and Optimize Fracture Designs in a Tight Gas Field. Paper presented at the SPE Annual Technical Conference and Exhibition, Houston, Texas, 26–29 September. SPE-90455-MS. https://doi.org/10.2118/90455-MS.

Hathaway, S. C. and Salehi, I. 1986. The Gas Research Institute's Tight Gas Sands Program in East Texas. Paper presented at the SPE East Texas Regional Meeting, Tyler, Texas, 21–22 April. SPE-14664-MS. https://doi.org/10.2118/14664-MS.

Hinn, R. L. 1983. Case History: A Multizone Tight Gas Completion Technique for the Blocker Cotton Valley Field, East Texas. Paper presented at the SPE Annual Technical Conference and Exhibition, San Francisco, California, USA, 5–8 October. SPE-11934-MS. https://doi.org/10.2118/11934-MS.

Hinn Jr., R. L., Glenn, J. M., and McNichol, K. C. 1988. Case History: Use of a Multiwell Model To Optimize Infill Development of a Tight-Gas-Sand Reservoir. *J Pet Technol* **40** (7): 881–886. SPE-14659-PA. https://doi.org/10.2118/14659-PA.

Holditch, S. A. and Ely, J. 1973. Successful Stimulation of Deep Wells Using High Proppant Concentrations. *J Pet Technol* **25** (8): 959–964. SPE-4118-PA. https://doi.org/10.2118/4118-PA. **[R]**

Holditch, S. A., Robinson, B. W., Whitehead, W.S. et al. 1988. The GRI Staged Field Experiment. *SPE Form Eval* **3** (3): 519–533. SPE-16429-PA. https://doi.org/10.2118/16429-PA. **[R]**

Hopkins, C. W., Gatens, J. M., Caballero, J. S. et al. 1991. Stimulation Optimization in a Low-Permeability Upper Devonian Sandstone Reservoir: A Case History. Paper presented at the SPE Gas Technology Symposium, Houston, Texas, 23–25 January. SPE-21499-MS. **[R]**

Kang, Y., Luo, P., Jin, X. et al. 2000. Employing Both Damage Control and Stimulation: A Way to Successful Development for Tight Gas Sandstone Reservoirs. Paper presented at the International Oil and Gas Conference and Exhibition in China, Beijing, China, 7–10 November. SPE-64707-MS. https://doi.org/10.2118/64707-MS. **[R]**

Kostenuk, N. and Thiessen, R. 2007. After Closure Analysis in Tight Gas Reservoirs: A Case History of Pre-Fracture Injection Tests Performed in the Brassey Field. Paper presented at the Canadian International Petroleum Conference, Calgary, Alberta, Canada, 12–14 June. PETSOC-2007-101. https://doi.org/10.2118/2007-101.

Kozik, H. G. and Holditch, S. A. 1981. A Case History for Massive Hydraulic Fracturing of the Cotton Valley Lime Matrix, Fallon and Personville Fields. *J Pet Technol* **33** (2): 229–244. SPE-7911-PA. https://doi.org/10.2118/7911-PA. **[R]**

Larkin, S. D., Brown, E. K., Bazan, L. W. et al. 2002. Stimulation Design and Post Fracture Production Analysis: A Tight Gas Sand Case History. Paper presented at the SPE International Petroleum Conference and Exhibition in Mexico, Villahermosa, Mexico, 10–12 February. SPE-74361-MS. https://doi.org/10.2118/74361-MS.

Li, H. and Wang, J. Y. 2014. Optimization of Breaker Schedules in the Stimulation of Tight Gas Reservoirs. *J of Natural Gas Science and Engineering* **16** (January): 31–43. https://doi.org/10.1016/j.jngse.2013.10.006.

Luo, H., Mahiya, G., Pannett, S. et al. 2014. The Use of Rate-Transient-Analysis Modeling To Quantify Uncertainties in Commingled Tight Gas Production-Forecasting and Decline-Analysis Parameters in the Alberta Deep Basin. *SPE Res Eval & Eng* **17** (2): 209–219. SPE-147529-PA. https://doi.org/10.2118/147529-PA.

McKetta, S. F., Koziar, G., and Cook, L. S. 1980. Eastern Tight Sands, A Stimulation Assessment. Paper presented at the SPE Eastern Regional Meeting, Morgantown, West Virginia, USA, 5–7 November. SPE-9536-MS. https://doi.org/10.2118/9536-MS.

McKinney, P. D., Rushing, J. A., and Sanders, L. A. 2002. Applied Reservoir Characterization for Maximizing Reserve Growth and Profitability in Tight Gas Sands: A Paradigm Shift in Development Strategies for Low-Permeability Gas Reservoirs. Paper presented at the SPE Gas Technology Symposium, Calgary, Alberta, Canada, 30 April–2 May. SPE-75708-MS. https://doi.org/10.2118/75708-MS.

Milligan, M. R. 1989. Panther River Well Stimulations: A Case History. Paper presented at the SPE Annual Technical Meeting, Calgary, Alberta, Canada, 12–16 June. PETSOC-88-39-51. https://doi.org/10.2118/88-39-51.

Modeland, N., Buller, D., Middaugh, R. L. et al. 2013. Applying Petrophysics and Systematic Completion Science to Increase Potential: A Terryville Field Tight Gas Case History. Paper presented at the SPE Unconventional Gas Conference and Exhibition, Muscat, Oman, 28–30 January. SPE-163968-MS. https://doi.org/10.2118/163968-MS.

Newhouse, C. C. 1991. Successfully Drilling Severely Depleted Sands. Paper presented at the SPE/IADC Drilling Conference, Amsterdam, The Netherlands, 11–14 March. SPE-21913-MS. https://doi.org/10.2118/21913-MS.

Parrot, D. I. and Long, M. G. 1979. A Case History of Massive Hydraulic Refracturing In The Tight Muddy "J" Formation. Paper presented at the SPE Symposium on Low Permeability Gas Reservoirs, Denver, Colorado, 20–22 May. SPE-7936-MS. https://doi.org/10.2118/7936-MS. **[R]**

Patino, A. H., Garcia, E. R., Torne, J. P. et al. 2004. Case Histories–Combining Crossed Dipole Sonic Anisotropy and Oriented Perforating to Optimize Hydraulic Fracturing In the Burgos Basin–Reynosa Mexico. Paper presented at the 2004 SPE International Petroleum Conference in Mexico, Puebla, Mexico, 7–9 November. SPE-92014-MS. https://doi.org/10.2118/92014-MS. **[R]**

Pearce, K. W., Heiken, K., Samuelson, M. L. et al. 2002. Successfully Pushing the Limits in Tight Gas Fracturing. Paper presented at the SPE Western Regional/AAPG Pacific Section Joint Meeting, Anchorage, Alaska, 20–22 May. SPE-76721-MS. https://doi.org/10.2118/76721-MS.

Pferdehirt, D. J., Brown, J. E., and Rucker, L. R. 1990. New Stimulation Techniques Improve Canyon Sand Gas Production: A Case Study. Paper presented at the Permian Basin Oil and Gas Recovery Conference, Midland, Texas, 8–9 March. SPE-20133-MS. https://doi.org /10.2118/20133-MS.

Reeves, S. R., Hill, D. G., Tiner, R. L. et al. 1999. Restimulation of Tight Gas Sand Wells in the Rocky Mountain Region. Paper presented at the SPE Rocky Mountain Regional Meeting, Gillette, Wyoming, USA, 15–18 May. SPE-55627-MS. https://doi.org/10.2118/55627-MS. **[R]**

Rivera, J. G., Carrillo, A. B., Luna, J. B. et al. 2008. Sustaining Fracture Conductivity Increases Cumulative Production in Tight-Gas Reservoir: Case History. Paper presented at the SPE North Africa Technical Conference & Exhibition, Marrakech, Morocco, 12–14 March. SPE-111992-MS. https://doi.org/10.2118/111992-MS.

Roberts, C. N. 1981. Fracture Optimization In A Tight Gas Play: Muddy "J" Formation, Wattenberg Field, Colorado. Paper presented at the SPE/DOE Low Permeability Gas Reservoir Symposium, Denver, Colorado, 27–29 May. SPE-9851-MS. https://doi.org/10.2118/9851-MS.

Robinson, B. M., Holditch, S. A., and Lee, W. J. 1986. A Case Study of the Wilcox (Lobo) Trend in Webb and Zapata Counties, TX. *J Pet Technol* **38** (12): 1355–1366. SPE-11600-PA. https://doi.org/10.2118/11600-PA. **[R]**

Robinson, B. M., Holditch, S. A., and Peterson, R. E. 1991. The Gas Research Institute's Second Staged Field Experiment: A Study of Hydraulic Fracturing. Paper presented at the SPE Gas Technology Symposium, Houston, Texas, 22–24 January. SPE-21495-MS. https://doi.org/10.2118/21495-MS. **[R]**

Robinson, B. M., Holditch, S. A., Whitehead, W. S. et al. 1992. Hydraulic Fracturing Research in East Texas: Third GRI Staged Field Experiment. *J Pet Technol* **44** (1): 78–87. SPE-22878-PA. https://doi.org/10.2118/22878-PA **[R]**

Rosenberg, J. I., O'Shea, P., Mercer, J. et al. 1983. A Sensitivity Analysis of the NPC Study of Tight Gas. Paper presented at the SPE/DOE Low Permeability Gas Reservoir Symposium, Denver, Colorado, 14–16 March. SPE-11645-MS. https://doi.org/10.2118/11645-MS.

Rushing, J. A. and Sullivan, R. B. 2003. Evaluation of a Hybrid Water-Frac Stimulation Technology in the Bossier Tight Gas Sand Play. Paper presented at the SPE Annual Technical Conference and Exhibition, Denver, Colorado, 5–8 October. SPE-84394-MS. https://doi.org/10.2118/84394-MS.

Saunders, B. F., Robinson, B. M., Holditch, S. A. et al. 1992. Hydraulic Fracturing Research in the Frontier Formation Through the Gas Research Institute's Fourth Staged Field Experiment. Paper presented at the SPE Annual Technical Conference and Exhibition, Washington, DC, 4–7 October. SPE-24854-MS. https://doi.org/10.2118/24854-MS. **[R]**

Schrooten, R. A., Boratko, E. C., Singh, H. et al. 2007. A Case Study: Using Wireline Pressure Measurements To Improve Reservoir Characterization in Tight Formation Gas—Wamsutter Field, Wyoming. Paper presented at the SPE Annual Technical Conference and Exhibition, Anaheim, California, USA, 11–14 November. SPE-109565-MS. https://doi.org/10.2118/109565-MS.

Scott, M. P., Stephens, T., Durant, R. et al. 2013. Investigating Hydraulic Fracturing in Tight Gas Sand and Shale Gas Reservoirs in the Cooper Basin. Paper presented at SPE Unconventional Resources Conference and Exhibition–Asia Pacific, Brisbane, Australia, 11–13 November. SPE-167073-MS. https://doi.org/10.2118/167073-MS. **[R]**

Sencenbaugh, R. N., Lytle, D. M., Birmingham, T. J. et al. 2001. Restimulating Tight Gas Sand: Case Study of the Codell Formation. Paper presented at the SPE Rocky Mountain Petroleum Technology Conference, Keystone, Colorado, USA, 21–23 May. SPE-71045-MS. https://doi.org/10.2118/71045-MS.

Settari, A., Sullivan, R. B., Rother, R. J. et al. 2009. Comprehensive Coupled Modeling Analysis of Stimulations and Post-Frac Productivity - Case Study of a Tight Gas Field In Wyoming. Paper presented at the SPE Hydraulic Fracturing Technology Conference, The Woodlands, Texas, USA, 19–21 January. SPE-119394-MS. https://doi.org/10.2118/119394-MS.

Spain, D. R., Naidu, R., Dawson, W. et al. 2015. Integrated Workflow for Selecting Hydraulic Fracture Initiation Points in the Khazzan Giant Tight Gas Field, Sultanate of Oman. Paper presented at the Abu Dhabi International Petroleum Exhibition and Conference, Abu Dhabi, UAE, 9–12 November. SPE-177939-MS. https://doi.org/10.2118/177939-MS.

Sunde, A., Chen, H. Y. and Teufel, L. W. 2000. Producing Characteristics and Drainage Volume of Dakota Reservoirs, San Juan Basin, New Mexico. Paper presented at the SPE Rocky Mountain Regional/Low-Permeability Reservoirs Symposium and Exhibition, Denver, Colorado, 12–15 March. SPE-60288-MS. https://doi.org/10.2118/60288-MS. **[R]**

Voneiff, G. W. and Holditch, S. A. 1994. Economic Assessment of Applying Advances in Fracturing Technology. J Pet Technol **46** (1): 281–296. SPE-24888-PA. https://doi.org/10.2118/24888-PA. **[R]**

Wanner, J. J., Chesnut, D. A., and Cox, D. O. 1979. Gas Production From Tight Mesa Verde Sands In Wyoming — A Field Case History. Paper presented at the Symposium on Low Permeability Gas Reservoirs, Denver, Colorado, 20–22 May. SPE-7934-MS. https://doi.org/10.2118/7934-MS.

Warpinski, N. R., Branagan, P. T., Settler, A. R. et al. 1987. Fracturing and Testing Case Study of Paludal, Tight, Lenticular Gas Sands. *SPE Form Eval* **2** (04): 535–545. SPE-13876-PA. https://doi.org/10.2118/13876-PA. **[R]**

Warpinski, N. R., Branagan, P. T., Sattler, A. R. et al. 1990. Case Study of a Stimulation Experiment in a Fluvial, Tight-Sandstone Gas Reservoir (includes associated papers 23475 and 23567). *SPE Prod Eng* **5** (4): 403-410. SPE-18258-PA. https://doi.org/10.2118/18258-PA.

Warpinski, N. R., Wright, T. B., Uhl, J. E. et al. 1999. Microseismic Monitoring of the B-Sand Hydraulic-Fracture Experiment at the DOE/GRI Multisite Project. *SPE J* **4** (3): 242–250. SPE-57593-PA. https://doi.org/10.2118/57593-PA. **[R]**

Weijermans, P., Daniau, G., and Westerhof, D. 2016. Developing Marginal Near-Tight Gas Fields in a Mature Area With Long-Reach Hydraulically Fractured Wells - A Case Study. Paper presented at the SPE Europec featured at 78th EAGE Conference and Exhibition, Vienna, Austria, 30 May–2 June. SPE-180169-MS. https://doi.org/10.2118/180169-MS.

Wyman, R. E., Holditch, S. A., and Randolph, P. L. 1980. Analyses of an Elmworth Hydraulic Fracture in Alberta. *J Pet Technol* **32** (9): 1621–1630. SPE-7935-PA. https://doi.org/10.2118/7935-PA. **[R]**

Xiong, H. and Holditch, S. A. 2006. Will the Blossoming of Unconventional Natural Gas Development in North America Be Repeated in China? Paper presented at the International Oil and Gas Conference and Exhibition in China, Beijing, China, 5–7 December. SPE-103775-MS. https://doi.org/10.2118/103775-MS. **[R]**

www.ingramcontent.com/pod-product-compliance
Ingram Content Group UK Ltd.
Pitfield, Milton Keynes, MK11 3LW, UK
UKHW060223061026
15625UKWH00007B/467

9 781613 9981